Instrumentation Engineering

MCQs | Multiple Choice Questions with Detailed Solutions

Other CBS Titles on GATE

1. Karna: ***GATE Electrical Engineering:*** *Multiple Choice Questions with Detailed Solutions*
2. Karna: ***GATE Electronics and Communication Engineering:*** *Multiple Choice Questions with Detailed Solutions*
3. Karna: ***A Guidebook for GATE Electrical Engineering:*** *Theory, Objective Questions with Detailed Answers*
4. Karna: ***A Guidebook for GATE Electronics and Communication Engineering:*** *Theory, Objective Questions with Detailed Answers*
5. Karna: ***A Guidebook for GATE Instrumentation Engineering:*** *Theory, Objective Questions with Detailed Answers*
6. Karna: ***GATE Electrical Engineering:*** *Theory, Objective Questions with Detailed Answers,* 2nd edn.
7. Karna: ***Instrumentation Engineering:*** *Theory, Objective Questions with Detailed Answers,* 3rd edn.
8. Karna: ***GATE Electronics and Communication Engineering:*** *Theory, Objective Questions with Detailed Answers,* 3rd edn.
9. Karna: ***GATE Computer Science and Information Technology:*** *Topicwise Previous Years Solved Papers,* 2nd edn.
10. Karna: ***GATE Instrumentation Engineering:*** *Topicwise Previous Years Solved Papers,* 3rd edn.
11. Karna: ***GATE Electrical Engineering:*** *Topicwise Previous Years Solved Papers,* 2nd edn.
12. Karna: ***GATE Electronics and Communication Engineering:*** *Topicwise Previous Years Solved Papers,* 3rd edn.
13. Karna: ***GATE Electronics and Communication Engineering:*** *Objective Type Questions*
14. Karna: ***GATE Electrical Engineering:*** *Objective Type Questions*
15. Karna: ***GATE Instrumentation Engineering:*** *Objective Type Questions*

Instrumentation Engineering

MCQs | Multiple Choice Questions with Detailed Solutions

Satish K Karna ME

Ex-Educational Consultant
Karna Institute of Engineering
Chandigarh

CBS Publishers & Distributors Pvt Ltd

New Delhi • Bengaluru • Chennai • Kochi • Mumbai • Pune
• Hyderabad • Kolkata • Nagpur • Patna • Vijayawada

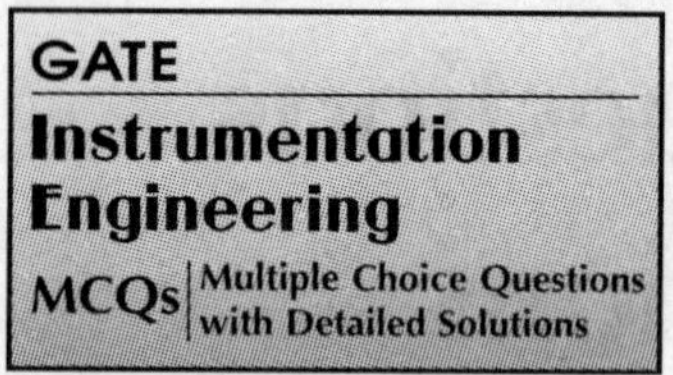

ISBN: 978-81-239-2357-4

First Edition: 2014

Published by Satish Kumar Jain for

CBS Publishers & Distributors Pvt Ltd
4819/XI Prahlad Street, 24 Ansari Road, Daryaganj, New Delhi 110 002, India.
Ph: 23289259, 23266861, 23266867 Fax: 011-23243014 Website: www.cbspd.com
e-mail: delhi@cbspd.com; cbspubs@airtelmail.in

Corporate Office: 204 FIE, Industrial Area, Patparganj, Delhi 110 092
Ph: 4934 4934 Fax: 4934 4935 e-mail: publishing@cbspd.com; publicity@cbspd.com

Branches

- **Bengaluru:** Seema House 2975, 17th Cross, K.R. Road, Banasankari 2nd Stage, Bengaluru 560 070, Karnataka
  Ph: +91-80-26771678/79 Fax: +91-80-26771680 e-mail: bangalore@cbspd.com
- **Chennai:** 20, West Park Road, Shenoy Nagar, Chennai 600 030, Tamil Nadu
  Ph: +91-44-26260666, 26208620 Fax: +91-44-42032115 e-mail: chennai@cbspd.com
- **Kochi:** 36/14 Kalluvilakam, Lissie Hospital Road, Kochi 682 018, Kerala
  Ph: +91-484-4059061-65 Fax: +91-484-4059065 e-mail: kochi@cbspd.com
- **Mumbai:** 83-C, Dr E Moses Road, Worli, Mumbai-400018, Maharashtra
  Ph: +91-9833017933 e-mail: mumbai@cbspd.com
- **Pune:** Bhuruk Prestige, Sr. No. 52/12/2+1+3/2 Narhe, Haveli (Near Katraj-Dehu Road Bypass), Pune 411 041, Maharashtra
  Ph: +91-20-64704058, 64704059, 32392277 Fax: +91-20-24300160 e-mail: pune@cbspd.com

Representatives

- **Hyderabad** 0-9885175004
- **Kolkata** 0-9831437309, 0-9051152362
- **Nagpur** 0-9021734563
- **Patna** 0-9334159340
- **Vijayawada** 0-9000660880

Printed at India Binding House, Noida (UP)

GATE
Instrumentation Engineering
MCQs | Multiple Choice Questions with Detailed Solutions

Preface

We have tried our level best to satisfy the readers. This book provides complete set of problems which appeared in competitive examination as well as fresh set of problems.

Time is a critical factor from both the points of view of preparing and solving in the examination.

This book consists of 11 chapters:

1. Analytical, Optical and Biomedical Instrumentation
2. Transducers, Mechanical Measurement and Industrial Instrumentation
3. Electrical and Electronic Measurements
4. Signals and Systems
5. Communication Systems
6. Basics of Circuit and Measurement System
7. Analog Electronics
8. Digital Electronics
9. Microprocessor
10. Control Systems and Process Control
11. Engineering Mathematics

This book is a must for GATE preparation. Theory may be followed from the standard books. But for a wide collection of problems, the different ways of solving them is the need which this book fulfills. In this book each chapter is subdivided into units.

I shall appreciate and greatly acknowledge the comments and suggestions from the readers of this book.

Satish K Karna
satish_karna@ymail.com

GATE
Instrumentation Engineering
MCQs | Multiple Choice Questions with Detailed Solutions

Acknowledgements

This book would be incomplete without expressing my gratitude to those individuals who motivated and supported me to write it. I am grateful to the following:

Dr S Chatterjee, Head, Deparment of Electrical Engineering, TTTI, Chandigarh

Mrs Manmeet Kaur ME (Electrical, Power System)

Mr Pravin Kumar ME (Instrumentation and Control)

Mrs Suman ME (Electrical, Power System)

Mr Divesh Kumar ME (Electrical, Power System)

Mrs Ritula Thakaur ME (Electrical, Power System)

Mr Gautam Khera ME (Electrical, Power Electronics and Drives)

Mr Anand Parkash ME (Electronics and Communication Engineering)

Mr Hitendar Singh Pawar ME (Microelectronics)

Mr Jitender Virmani ME (Control and Instrumentation)

Mr Anupam Anand MSc (Statistics) and MTech (Computer Sc.)

Mr Deepak Singla ME (Computer Science)

Mr Surender Singh ME (Computer Science)

Mr S. Tiwari ME (Computer Science)

Mr Ashish Saxena BE (Electronics and Communication)

And *our students:*

Miss Kritika Singh BE (Electronics and Communication)

Mrs Monika BE (Electronics and Communication)

Mr Amandeep BE (Electronics and Communication)

Mr Chirag BE (Electronics and Communication)

Mr Prateek Srivastava BE (Electronics and Instrumentation)

Mr Rohil Khanna BE (Electronics and Instrumentation)

Mr Ravi Verma BE (Electronics and Instrumentation)

Mr Rajneesh Kumar BE (Electrical Engineering)

Mr Saurabh Kamboj BE (Electronics and Communication)

Mr Gur Gaurav Singh BE (Electrical-Electronics Engineering)

Miss Isha Gupta BE (Electronics and Communication)

Miss Arzoo Khetarpal BE (Electronics and Communication)

Mr Aman BE (Electrical Engineering)

Miss Garima BE (Electrical-Electronics Engineering)

Miss Ashmi BE (Electrical-Electronics Engineering)

Miss Dupinder Kaur BE (Electrical-Electronics Engineering)

Mr Vivek BE (Electronics and Instrumentation)

Mr Arshdeep Singh BE (Electrical-Electronics Engineering)

Mr Baljinder Singh BE (Electrical-Electronics Engineering) for their guidance and support in solving the problems.

I also wish to thank Mr Y. N. Arjuna, Mr P. S. Ghuman and Ms Poonam Kapoor of CBS Publishers and Distributors, New Delhi, for their role in bringing out the book.

I owe a debt of gratitude to my wife Mrs Vibha Karna, son Mayank and daughter Nilophar for their patience and understanding.

I am thankful to Mr Mukesh Kumar Koundal, Miss Preety Rani and Mr Kamal Kumar for efficiently typing the manuscript.

Satish K Karna

GATE
Instrumentation Engineering
MCQs | Multiple Choice Questions with Detailed Solutions

Contents

Syllabus

ENGINEERING MATHEMATICS

Linear algebra: Matrix algebra, Systems of linear equations, Eigenvalues and eigenvectors.

Calculus: Mean value theorems, Theorems of integral calculus, Evaluation of definite and improper integrals, Partial Derivatives, Maxima and minima, Multiple integrals, Fourier series. Vector identities, Directional derivatives, Line, Surface and Volume integrals, Stokes, Gauss and Green's theorems.

Differential equations: First order equation (linear and nonlinear), Higher order linear differential equations with constant coefficients, Method of variation of parameters, Cauchy's and Euler's equations, Initial and boundary value problems, Partial Differential Equations and variable separable method.

Complex variables: Analytic functions, Cauchy's integral theorem and integral formula, Taylor's and Laurent' series, Residue theorem, solution integrals.

Probability and statistics: Sampling theorems, Conditional probability, Mean, median, mode and standard deviation, Random variables, Discrete and continuous distributions, Poisson, Normal and Binomial distribution, Correlation and regression analysis.

Numerical methods: Solutions of non-linear algebraic equations, single and multi-step methods for differential equations.

Transform theory: Fourier transform, Laplace transform, Z-transform.

INSTRUMENTATION ENGINEERING

Basics of Circuit and Measurement system: Kirchoff's laws, mesh and nodal analysis. Circuit theorems. One-port and two-port network functions. Static and dynamic characteristics of Measurement Systems. Error and uncertainty analysis. Statistical analysis of data and curve fitting.

Transducers, Mechanical Measurement and Industrial Instrumentation: Resistive, Capacitive, Inductive and piezoelectric transducers and their signal conditioning. Measurement of displacement, velocity and acceleration (translational and rotational), force, torque, vibration and shock. Measurement of pressure, flow, temperature and liquid level. Measurement of pH, conductivity, viscosity and humidity.

Analog Electronics: Characteristics of diode, BJT, JFET and MOSFET. Diode circuits. Transistors at low and high frequencies, Amplifiers, single and multi-stage. Feedback amplifiers. Operational amplifiers, characteristics and circuit configurations. Instrumentation amplifier. Precision rectifier. V-to-I and I-to-V converter. Op-Amp based active filters. Oscillators and signal generators.

Digital Electronics: Combinational logic circuits, minimization of Boolean functions. IC families, TTL, MOS and CMOS. Arithmetic circuits. Comparators, Schmitt trigger, timers and mono-stable multi-vibrator. Sequential circuits, flip-flops, counters, shift registers. Multiplexer, S/H circuit. Analog-to-Digital and Digital-to-Analog converters. Basics of number system. Microprocessor applications, memory and input-output interfacing. Microcontrollers.

Signals, Systems and Communications: Periodic and aperiodic signals. Impulse response, transfer function and frequency response of first- and second order systems. Convolution, correlation and characteristics of linear time invariant systems. Discrete time system, impulse and frequency response. Pulse transfer function. IIR and FIR filters. Amplitude and frequency modulation and demodulation. Sampling theorem, pulse code modulation. Frequency and time division multiplexing. Amplitude shift keying, frequency shift keying and pulse shift keying for digital modulation.

Electrical and Electronic Measurements: Bridges and potentiometers, measurement of R,L and C. Measurements of voltage, current, power, power factor and energy. AC & DC current probes. Extension of instrument ranges. Q-meter and waveform analyzer. Digital voltmeter and multi-meter. Time, phase and frequency measurements. Cathode ray oscilloscope. Serial and parallel communication. Shielding and grounding.

Control Systems and Process Control: Feedback principles. Signal flow graphs. Transient Response, steady-state-errors. Routh and Nyquist criteria. Bode plot, root loci. Time delay systems. Phase and gain margin. State space representation of systems. Mechanical, hydraulic and pneumatic system components. Synchro pair, servo and step motors. On-off, cascade, P, P-I, P-I-D, feed forward and derivative controller, Fuzzy controllers.

Analytical, Optical and Biomedical Instrumentation: Mass spectrometry. UV, visible and IR spectrometry. X-ray and nuclear radiation measurements. Optical sources and detectors, LED, laser, Photo-diode, photo-resistor and their characteristics. Interferometers, applications in metrology. Basics of fiber optics. Biomedical instruments, EEG, ECG and EMG. Clinical measurements. Ultrasonic transducers and Ultrasonography. Principles of Computer Assisted Tomography.

1 Analytical, Optical and Biomedical Instrumentation

1.1 UV, Visible and IR Spectroscopy

1. A grating that has 1200 lines/mm is irradiated with a narrow beam of white light at an incident angle of 25° to the grating normal. The first order peak observed at an angle of 15° belongs to the wavelength

   (a) 567.9 nm (b) 587.8 nm
   (c) 597.1 nm (d) 600.7 nm

2. In a spectrophotometer, the monochromator must be able to resolve two wavelengths 599.9 nm and 600.1 nm. The required resolution is

   (a) 100 (b) 1000
   (c) 3000 (d) 5000

3. Match the List I instruments with List II, the parameters they measure and select the correct answer using the codes:

List I	List II
A. Autocollimator	1. Radiation scattered
B. Hygrometer	2. Angular error
C. Nephelometer	3. Molecular weight
D. Mass spectrometer	4. Humidity
	5. Density

Codes:

	A	B	C	D
(a)	2	4	3	1
(b)	4	1	3	2
(c)	3	1	4	2
(d)	2	4	5	3

4. For alignment and testing of two surfaces at right angles, a constant deviation prism, also called an optical square, is used. It is a

   (a) Right angled prism
   (b) Square prism
   (c) Pentagonal prism
   (d) Hexagonal prism

5. The minimum number of lines that must be drawn on a grating needed to resolve the sodium having nominal wavelengths of 589.0 nm and 589.6 nm is

   (a) 600 (b) 1200
   (c) 982 (d) 300

6. IR spectroscopy

   (a) Has a use of radiation from 2.5 to 1.5 microns
   (b) Is unsuitable for analysis of mixture of metals
   (c) Is unsuitable for analysis of organic gases
   (d) Uses bolometer as one of the detectors

7. An interference filter has been made with a dielectric of magnesium fluoride (refractive index $\eta = 1.38$) for a normal wavelength of 455 nm. The required thickness of the dielectric for first order interference is

   (a) 8.2 nm (b) 16.5 nm
   (c) 82 nm (d) 165 nm

8. The transmittance of a sample solution is monitored using 599 nm wavelength. The transmittance is reduced from 100% to 10% due to addition of pollutants. As a result, the absorbance recorded by the spectrophotometer changes from

   (a) $0 \rightarrow 1$ (b) $2 \rightarrow 1$
   (c) $0 \rightarrow 2$ (d) $0 \rightarrow 10$

9. An interference filter is made of a dielectric of $M_gF_2 (\mu = 1.38)$.

   For a wavelength of 455 nm the required thickness of dielectric for first order interference is

   (a) 160 *nm* (b) 163 *nm*
   (c) 165 *nm* (d) 168 *nm*

10. A dielectric layer of R. I $\mu = 1.35$ and thickness of 185 nm is used in interference filter to produce first order, second order, third order filters. The wavelength of each is.

   (a) 166.2 *nm* (b) 166.5 *nm*
   (c) 166.7 *nm* (d) 166.9 *nm*

11. A prism spectrometer uses a flintglass prism of base 1 cm. For flint glass dispersion $d\mu / d\lambda$ is 1000 cm^{-1} in the range of $\lambda = 6000 \overset{\circ}{A}$.

The smallest difference of wavelength is

(a) 2 Å (b) 6 Å

(c) 8 Å (d) 10 Å

12. A grating with 1200 lines/mm is exposed with a narrow beam of white light at an incident anlge of 25° to grating normal.

The first order peak observed at an angle of 15° belongs to the wavelength of

(a) 567 *nm* (b) 569 *nm*
(c) 572 *nm* (d) 574 *nm*

13. A prism spectrometer uses a flint glass with glass dispersion $d\mu/d\lambda = 952\ \text{cm}^{-1}$ and change of wavelength is 6 Å centred around a wavelength of 5893 Å. The thickness is

(a) 1.026 *cm* (b) 1.037 *cm*
(c) 1.034 *cm* (d) 1.031 *cm*

14. Molar absorbability of a solution is 1200 L/molep-cm and minimum detectable absorption is 0.001. The minimum molar concentration that can be detected for 1 cm path length is

(a) 6.3×10^{-8} (b) 7.3×10^{-8}
(c) 8.3×10^{-8} (d) None of the above

15. The minimum number of lines that must be drawn on a greating needed to resolve the sodium doublet having nominal wavelengths 589.0 nm and 589.6 nm is

(a) 600 (b) 1200
(c) 982 (d) 300

16. Find the diffraction angle at which the radiation of 300nm diffracts at normal incidence of light with a grating 1180 groves/mm in first order

(a) 20.73° (b) 20.76°
(c) 20.83° (d) 20.86°

17. A glass plate is coated with thin film, the reflectance of combintion is monitored with variationof 600nm wavelength. If R.I of glass plate is greater than R.I of film coat. R.I reaches minimum when film thickness equal to

(a) 600 *nm* (b) 300 *nm*
(c) 500 *nm* (d) 150 *nm*

Common Data for Questions 18 and 19

Light is incident normally on a grating 0.5 cm wide with 2500 lines.

For the principal maximum of the two sodium line 5890 Å and 5890 Å in first order spectrum.

18. The angle of diffraction is

(a) 983.2 (b) 982.1
(c) 982.1 (d) 982.2

19. The resolution is

(a) 985.1 (b) 982.1
(c) 895.1 (d) 892.1

20. Optical communication is feasible only in IR range because
(a) it offers minimum dispersion
(b) it offers minimum attenuation
(c) both (a) and (b)
(d) none of the above

21. An equilateral prism is placed on the prism table of a spectrometer in the position of minimum deviation. If the angle of incidence is 60°, the angle of deviation of the ray is:

(a) 90° (b) 60°
(c) 45° (d) 30°

22. The refractive index of material varies with wavelength as:

(a) $\mu \propto \lambda$ (b) $\mu \propto \frac{1}{\lambda}$

(c) $\mu \propto \lambda^2$ (d) $\mu \propto \frac{1}{\lambda^2}$

23. The speed of light in air is $3\times10^8\ ms^{-1}$. What will be its speed in diamond whose refractive index is 2.4?

(a) $3\times10^8\ ms^{-1}$ (b) $322\ ms^{-1}$
(c) $1.25\times10^8\ ms^{-1}$ (d) $7.2\times10^8\ ms^{-1}$

24. A beam of monochromatic light is refracted from vacuum into a medium of refractive index 1.5. The wavelength of refracted light will be
(a) same
(b) dependent on intensity of refracted light
(c) larger
(d) smaller

25. A ray of light is travelling through a medium of refractive index $\sqrt{2}$ w.r.t. air. When it is incident on the suface at 45° which of the following will take place?
(a) Angle of refraction will be 45°
(b) Angle of refraction will be 90°
(c) The ray will be internally reflected
(d) None of the above

26. The twinking of stars is due to
(a) periodic brust of light from the star
(b) interference between light from the sun and star
(c) partial absorption of light in the atmosphere
(d) refractive index fluctuation in the atmosphere

27. The sun become visible before the actual sunrise and remains visible even after the actual sunset. It is because of

(a) scattering of light (b) diffraction of light
(c) refraction of light (d) dispersion of light

28. Huygen's principle of secondary wave is used to
(a) Obtain new position of wave front geometrically
(b) Explain the principle of superposition of waves
(c) Explain the interference phenomenon
(d) Explain the polarisation

29. Two points are situated at the same distance from a source of light but in opposite direction. The phase diference between the light waves from the two-points will be
(a) π (b) zero
(c) $\pi/2$ (d) none of the above

30. Which of the following produce a plane wave front
(a) Point source (b) Line source
(c) Extended source (d) None of the above

31. A ray of light is incident on a glass plate at 60°. Reflected and refracted rays are found to be mutually perpendicular. The refractive index of glass is
(a) $\frac{2}{\sqrt{3}}$ (b) $\sqrt{3}$
(c) 1.5 (d) 1.62

32. Incorrect statement
(a) Doppler effect in sound is symmetrical
(b) Doppler effect in light is symmetrical
(c) Red shift in doppler effect in light predicts expanding nature of universe
(d) A spectral line shift towards the blue end, if source of light move towards the observer

33. An isotropic source of 2-candela produces light flux equal to
(a) 2π lumen (b) 4π lumen
(c) 6π lumen (d) 8π lumen

34. The tangent of angle of polarisation is equal to refractive index of the material is called
(a) Malu's law (b) Brewster's law
(c) Bragg's law (d) None of the above

35. Which of the following types of waves cannot be polarised
(a) Longitudinal waves in gas
(b) Transverse waves in string
(c) Radio waves
(d) X-rays

36. In double slit experiment, the two slits are illuminated with 2-independent identical sources of light. What happens to interference pattern?
(a) No interference pattern is produced
(b) Intensity of bright fringes gets doubled
(c) Intensity of bright fringes becomes 4 times
(d) Two sets of interference patterns are produced

37. In Youngs double slit experiment, the fringe width is
(a) directly proportional to distance between the sources
(b) inversely proportional to distance between the sources
(c) inversely proportional to distance of screen from the sources
(d) inversely proportional to wavelength of light used

38. At what angle of incidence should a beam of light strike the surface of a still pond if the angle between the reflected ray and the refracted ray is to be 90°? (n = 1.33 for water).
(a) 13° (b) 23°
(c) 33° (d) 53°

39. What is the dispersive power of a prism to separate two lines of wavelengths 5997 $\overset{\circ}{A}$ and 6003 $\overset{\circ}{A}$?
(a) 0.001 (b) 0.002
(c) 0.003 (d) 0.004

40. What is the refractive index of glass (5 μm thick) introduced in the path of one of the interfering beams in a prism with λ of light used as 589 nm and central bright fringe shifts a distance equal to width of 5 fringes?
(a) 1.589 (b) 1.329
(c) 1.129 (d) 1.789

41. Refractive index of a liquid intoduced below the convex surface when diameter of 5th dark ring is redcued to (1/2) of its value is
(a) 2 (b) 4
(c) 6 (d) 8

42. Wavelength of light used in **Newton's rings experiment** when radius of curvature of plano-convex lens is 1 m and diameters of nth and (n + 14) th rings are 4.2 mm and 7.0 mm.
(a) 3600 A (b) 4600 A
(c) 5600 A (d) 6600 A

43. Find the time spent by a photon in cavity before it is absorbed/ emitted in a **GaAs optical cavity** (length 200 mm)and reflectivity 0.33. Absorption loss in cavity is $10\ cm^{-1}$.
(a) 0.31 ps (b) 0.41 ps
(c) 0.51 ps (d) 0.61 ps

44. If refractive index of water is $\eta = 1.333$ A person looks into a swimming pool at the 5-ft-deep level. How deep does it look to the person?
(a) 2.75 ft (b) 3.75 ft
(c) 4.75 ft (d) 5.75 ft

45. Light enters a glass prism having a **refracting angle** of 60°. If the angle of incidnece is 30° (incident ray is parallel to base) and the index of refraction of the glass is 1.50. What is the angle the ray leaving the prism makes with the normal?
(a) 35° (b) 44°
(c) 55° (d) 77°

46. A ray of light is incident at $30°$ on a prism with apex angle $55°$ and index of refraction 1.50. Calculate the angles the ray makes with the sides of the prism.

(a) $15.5°$ (b) $25.5°$
(c) $35.5°$ (d) $45.5°$

47. A beam of sodium light passes from air into water and then into flint glass, all with parallel surfaces. If the **angle of incidence** in the air is $45°$, what is the angle of refraction in the glass? (n = 1.333 and 1.63 for water and flint glass, respectively).

(a) $12°$ (b) $18°$
(c) $23°$ (d) $26°$

48. A layer of benzene (index of refraction = 1.50) floats on water. If the angle of incidence of the light entering the benzene from air is $60°$, what is the angle the light makes with the vertical in the benzene?

(a) $15°$ (b) $25°$
(c) $35°$ (d) $45°$

49. Superimposing $y_1 = 20 \sin \omega t$ and $y_2 = 20 \sin (\omega t + 60°)$ we get

(a) $20 \cos \omega t$
(b) $20 \sin \omega t$
(c) $20 \sin (\omega t + 60°)$
(d) $34.6 \sin (\omega t + 30°)$

50. Superimposing $x_1 = 30 \sin \omega t$ and $y_2 = 30 \cos \omega t$.

(a) $30 \sin \omega t$
(b) $30 \cos \omega t$
(c) $90 \sin \omega t \cos \omega t$
(d) $42.4 \sin (\omega t + 45°)$

51. Find the minimum change in refractive index of material if Bragg angle for a acousts-optic modulator length = 50 mm, diffraction efficieny 70%, acoustic wavelength 4.3×10^{-5} m, acoustic velocity 3500m/sec.

(a) 1.27×10^{-1}
(b) 1.27×10^{-2}
(c) 1.27×10^{-5}
(d) 1.27×10^{-4}

52. Light is **incident at an angle** θ_1 on an isosceles triangular prism of apex angle a and refractive index $n > 1$. Obtain an expression for the minimum angular deviation, δ_{min}, produced by the prism.

(a) $\dfrac{\sin \alpha + \delta_{min}}{2} = n \sin \dfrac{\alpha}{2}$

(b) $\dfrac{\sin \alpha - \delta_{min}}{2} = n \sin \dfrac{\alpha}{2}$

(c) $\dfrac{\cos \alpha + \delta_{min}}{2} = n \cos \dfrac{\alpha}{2}$

(d) $\dfrac{\cos \alpha - \delta_{min}}{2} = n \cos \dfrac{\alpha}{2}$

53. Find the wavelength in water of a blue light whose wavelength in air is 420 nm.
Given : $n = 1.33$ for water

(a) 116 nm (b) 216 nm
(c) 316 nm (d) 416 nm

54. Find the frequency of light of a certain colour has 1800 waves to the millimeter in air (h = 1.52 for glass).

(a) 5.4×10^{7} Hz
(b) 5.4×10^{9} Hz
(c) 5.4×10^{11} Hz
(d) 5.4×10^{14} Hz

55. Find the angle of refraction of light incident on a water surface at an angle of 48° to the normal. ($n = 1.333$).

(a) 23.9° (b) 33.9°
(c) 43.9° (d) 53.9°

56. The **index of refraction of** n-proply alcohol is 1.39. Find angle of refraction if light comes from air with an angle of incidence of 55°.

(a) 32° (b) 36°
(c) 40° (d) 44°

57. Light passes from air into a liquid and is deviated 19° when the angle of incidence is 52°. Find the refractive index of liquid

(a) 1.15 (b) 1.25
(c) 1.35 (d) 1.45

58. 'X' is the incident radiant energy on a sample and 'y' is the energy which is transmitted. The absorbance of the sample is

(a) $\log (x - y)$ (b) $\log (x + y)$
(c) $\log (x / y)$ (d) $\log (y / x)$

59. The transmittance of a particular solution measured is T. The concentration and thickness of solution is now doubled. Assuming Beer Lambert's law holds good for both the cases, the transmittance of second case would be

(a) T^2 (b) T^3
(c) $\sqrt{T}$ (d) $T^{1/3}$

60. The absorbance of a particular solution measured is A. The concentration increased 4 times and thickness halved. The absorbance of second case would be

(a) A^2 (b) $\sqrt{A}$
(c) $3A$ (d) $2A$

61. Instrument related error in spectrophotometer belongs to
(a) Electric Noise (b) Stray light
(c) Both (a) and (b) (d) None of the above

62. For organic compound generally used spectrophotometer is
(a) Visible (b) Ultraviolet
(c) Infrared (d) None of the above

63. Which one of the following IR detector has poor response time
(a) Golay cell (b) Bolometer
(c) Thermocouple (d) Photoconductive cell

64. The gas which cannot be used as carrier gas in Gas chromatograph is
(a) Helium (b) Nitrogen
(c) Ammonia (d) None of the above

65. The resistance of a photo - resistive cell
(a) Remain constant irrespective of the light intensity incident upon it
(b) Decrease with increase in light intensity in a exponential manner.
(c) Decrease with increase in light intensity in a linear manner.

66. A prism spectrometer uses a flint glass prism of dispersion $d\mu / d\lambda$ is 1000 cm^{-1} in the wavelength range 6000 Å The wavelength which a spectrometer can resolve is
(a) 6 Å (b) 4 Å
(c) 8 Å (d) None of the above

67. The sensitivity of photovoltaic cell
(a) is almost constant when it is short circuited
(b) decrease with increase of load.
(c) is almost negligible when the load resistance is about 10 kW
(d) All of the above.

68. Match List-I the physical principle with List - 2 the characteristic equation and select the answer using the codes given below

List - I
A. Photonic effect
B. Absorption of light
C. Total radiation pyrometry
D. Thermoelectric effect

List - II
1. $E = C_1(F_1 - F_2) + C_2(T_1^2 - T_2^2)$
2. $E = C\sigma T^4$
3. $E = hc/\lambda$
4. $A = abc$

Codes

	A	B	C	D
(a)	4	1	3	2
(b)	4	3	1	2
(c)	1	4	3	2
(d)	2	4	1	3

69. The glass prism monochromator is used in the spectrum range of
(a) Visible (b) Ultraviolet
(c) Infrared (d) X - ray

70. Which one has the poor selectivity among the following.
(a) Monochromator (b) Gratings
(c) Absorption filters (d) Interference filter

71. Higher order spectra will appear as stray light in
(a) Monochromator (b) Gratings
(c) Absorption filters (d) Interference filter

72. From the following which one does not need external power supply.
(a) Photovoltaic cell (b) Photo emissive cell
(c) Photomultiplier (d) photo - conductor

73. A light from tungsten is
(a) Neither special nor temperal coherence
(b) Not special but has temperal coherence
(c) Spacial but not temperal coherence
(d) Both special and temeperal coherence

74. A photomultiplier tube uses 10 stages. The sensitivity of each dynode of the tube is 100 mA/1m
(a) 270 mA/1m (b) 2700 A/1m
(c) 2700 mA/1m (d) 2.06 A/1m

75. The frequency of a given electromagnetic radiation is 6×10^{14} Hz. The wavelength for which radiation falls within visible spectrum is
(a) 1.5 μm (b) 0.5 μm
(c) 2.0 μm (d) None of the above

76. A photomultiplier has a current gain 3×10^6. A weak light beam produces 50 electrons per second at the photocathode. The anode to ground resistance must be used to get voltage of 3 μV is ($1e = 1.6 \times 10^{-19}$(C)
(a) 1000 Ω (b) 2 KΩ
(c) 10 KΩ (d) 125 KΩ

77. The relative efficacy of radiation of a wavelength of 0.49 μm is approximately 0.2. The international luminosity curve indicates relative efficacy of a wavelength of 0.555 μm as unity. Radiation equivalent to 20 W is produced by a wavelength of 0.49 μm. The radiation (in W) at a wavelength 0.555 μm which produces the same sensation in the human eye is
(a) 4 W
(b) 10 W
(c) 15 W
(d) None of the above

78. A photovoltaic cell produces a voltage of 0.33 V in an open circuit when illuminated by 10 W/m^3 radiant incidences. When a load of 100 Ω is connected to the cell, a current of 2.2 mA is delivered at that intensity. The open circuit voltage for a radiant incidence of 25 W/m^2 is

(a) 1.1 V (b) 5 V
(c) 0.462 (d) None of the above

79. There are two components in a sample which absorb radiation of same wavelength, The thickness of sample is 5mm and the concentration .of components is 0.25 gram per liters . If absorptivity of one component is 2 liters per gram.cm, then the absorbance of sample would be

(a) 0.75 (b) 0.075
(c) 0.625 (d) 0.0625

80. Match List - I with List - II and select the current answer using the codes given below :

List-I	List-II
A. Visible	1. Hydrogen lamp
B. Ultraviolet	2. Tungsten lamp
C. Infrared	3. Nernst filament
	4. Deuterium lamp

Code

	A	B	C
(a)	1	4	3
(b)	1	4	2
(c)	3	1	2
(d)	2	1	3

81. A reflection grating have 1600 grooves/mm and the spectrum is filter is such that only first order exist. Then selection of wave length of the spectrum is

(a) 1250 nm (b) 625 nm
(c) 1600 nm (d) 3200 nm

82. Match List 1 with List 2 and select the answer using the codes given below

List 1	List 2
A. Beer's law	1. X-ray diffraction
B. Brag's law	2. Thermo-emf
C. Peltier effect cell	3. Electro-chemical
D. Nerust equation light	4. Absorption of light

Code

	A	B	C	D
(a)	3	1	4	2
(b)	1	2	4	3
(c)	4	1	2	3
(d)	3	4	2	1

83. A quassi-monochromatic light source with average central wavelength of 630 nm has a spectral width of 10 GHz . The coherence length of the source is

(a) 3 μm
(b) 3 mm
(c) 3 cm
(d) None of the above

84. The absorptivity of a sample is 3 liters per gram -Cm and the thickness along optical path is 4 mm. If the absorbance optical path is 4 mm. If the absorbance of a sample is 0.30 then concentration in gram per liter is

(a) 0.225
(b) 0.25
(c) 0.025
(d) None of the above

85. A sample concentration of 10 mg/dl gives a spectrophotometer of transmittance 35 %. Assume Beer's law holds good. The concentrate for transmittance of 70 % is

(a) 2.5 mg/dl
(b) 3.4 mg/dl
(c) 4.4 mg/dl
(d) 5.4 mg/dl

86. A spectrophotometer is used to measure concentration of a molecule based on Beer's law using a cuvette of path length 10 mm and detectable absorption of 0.01. The molar absorptivity of solution that can be detected is

(a) 1 µ mole/lit
(b) 2 µ mole/lit
(b) 5 µ mole/lit
(d) None of the above

87. The molar absorptivity of a particular solute is 2.1×10^4. The transmittance through a cuvette with 5 cm light path for 2×10^{-6} moles solution is

(a) 0.61
(b) 0.71
(c) 0.81
(d) None of the above

1.2 Mass Spectroscopy

1. A mass spectrometer is to be used to resolve peaks corresponding to CO^+ and N_2^+. The atomic masses are $^{12}C = 12.0000$ amu, $^{16}O = 15.9949$ amu and $^{14}N = 14.0031$ amu . The resolving power of the mass spectrometer should be atleast

(a) 250
(b) 350
(c) 2500
(d) 3500

2. The time taken by an ionized atom, of mass m kg and charge e coulombs, pulsed into a field-free region with V volts, to reach a detector L meters away is

(a) $\frac{1}{L}\sqrt{\frac{m}{2eV}}$ (b) $L\sqrt{\frac{m}{2eV}}$

(c) $m\sqrt{\frac{L}{2eV}}$ (d) $\frac{2}{L}\sqrt{\frac{m}{2eV}}$

3. Two ionic species A and B of masses in the ratio 1:2 and of equal charge are simultaneously accelerated in a time of flight **mass-spectrometer**. When ion A reaches the end of a 1 m drift tube, ion B would be at a distance of

(a) $\sqrt{2}\ m$ (b) $\sqrt{2}-1\ m$

(c) $\frac{1}{\sqrt{2}}m$ (d) $\frac{1}{2}m$

4. A mass spectrometer is to be used to resolve peaks corresponding to CO_+ and z N_{2+}. The atomic masses are $^{12}C = 12.0000$, $^{16}O = 15.9949$ and $^{14}N = 14.0031$ amu. The resolving power of the mass spectrometer should be at least

(a) 250 (b) 350
(c) 2500 (d) 3500

5. Two ionic species A and B of masses in the ratio 1:2 and of equal charge are simultaneously accelerated in a time of filgth mass spectrometer when ion 'A' reaches the end of a 1 m drift tube, ion B would be at a distance of

(a) 0.709 m (b) 0.717 m
(c) 0.719 m (d) 0.707 m

6. The time taken by an ionized atom, of mass m kg and charge 'e' coulombs, pulsed into a filled free region with V volts, to reach a detector one meter away is.

(a) $m/2e$ V (b) sqrt (m/2eV)
(c) sqrt (m/2V) (d) sqrt (m/e V)

7. In a Mass sectormeter, a resolving power. Is required to distringuish oxygen of mass 31.9988 form sulphur of mass 32,0600 with equal intensities into the spectrum.

(a) 435 (b) 430
(c) 543 (d) 523

8. A time of flight mass spectrometer has a drift tube 85 cm long. If the accelerating potential is 2.0KV, calculate the difference in arrival times of ions with mass numbers 200 and 300. Take the charge of an electron to be 1.6×10^{-19} C and mass of a proton to be 1.66×10^{-27} Kg.

(a) 2.2 μs (b) 4.36 μs
(c) 36.4 μs (d) 1.2 μs

9. Absorption co-efficient μ of a substance when X-ray falls on it is (where, Z = atomic number of absorbed atom, λ = wavelength of X-ray, ρ = Lenght of absorber, C = Constants).

(a) $Z^3\lambda^2\rho$ (b) $CZ^4\lambda^3\rho$
(c) $Z^2\lambda^2\rho$ (d) $Z\lambda^2\rho^2$

10. CT number (Hounsfield unit) H of a tissue when u is attenuation co-efficient of X-ray and μ water attenuation co-efficient of water (0.205) is given by

(a) $\frac{\mu_{water}}{\mu_{water}-\mu}\times1000$ (b) $\frac{\mu}{\mu_{water}-\mu}\times1000$

(c) $\frac{\mu-\mu_{water}}{\mu_{water}}\times1000$ (d) $\frac{\mu_{water}-\mu}{\mu}\times1000$

11. In nuclear instrumentation, heaviest and flow moving particles are:

(a) Beta particles (b) Alpha particles
(c) Gamma rays (d) Protons

12. Doppler effect in sound and light are

(a) exactly identical to each other
(b) both synmmetrical
(c) both asymmertical
(d) different from each other

13. Correct statement about doppler effect

(a) Doppler effect in light merely depends upon relative velocity of source and observer
(b) Doppler effect in light is saymmetrical
(c) Doppler shift (Δv) is more when source moves with the velocity v w.r.t. stationary observer than that what it is when observer moves with velocity v w.r.t. stationary source
(d) Doppler shift (Δv) is more when source moves with the velocity w.r.t. stationary source than that what it is when source moves with velocity v w.r.t. stationary observer.

14. 6000 Å spectral line emitted by a star is found to be red shift by 20 Å. The speed with which star is recording is given by

(a) $3\times10^8\ m/s$ (b) $3\times10^5\ m/\text{sec}$
(c) $10^6\ m/\text{sec}$ (d) none of these

15. What is the separation on photographic plate in the velocity selector of **Bain bridge mass spectro-cope** with E = 200 V/cm and H = 0.2 Teasla. H in main chamber as 0.3 Teasla. Mixture of signaly charged ions of carbon isotopes C^{12} and C^{13} is analysed.

(a) 0.69 cm
(b) 0.59 cm
(c) 0.49 cm
(d) 0.39 cm

16. A wedge-shaped film of refractive index 1.4 is formed. Its

is illuminated by yellow light of wavelength 5893 *A*. When viewed normally 10 bands were observed in 9 mm. Determine the anlge of wedge.

(a) 0.0134° (b) 0.0144°
(c) 0.0154° (d) 0.0164°

17. Determine the thickness of wedge-shaped air film at fifth bright band from the edg when the film is illuminated by red light of wavelength 6500 and observed normally.

(a) 1.4×10^{-1} *cm* (b) 1.4×10^{-2} *cm*
(c) 1.4×10^{-3} *cm* (d) 1.4×10^{-4} *cm*

18. Find the Fringe separation between 2 consecutive fringes when a wedge shaped film is illuminated by light of wavelength 4650 Å angle of wedge = 40°.

(a) 0.09 cm (b) 0.12 cm
(c) 0.15 cm (d) 0.18 cm

19.40 fringes are seen with a biprism with sodium light, $\lambda = 5892$. Find the number of fringes obtained when Na lamp is replaced by mercury lamp $\lambda = 4358$ Å.

(a) 34 (b) 44
(c) 54 (d) 64

20. Find the speed of electrons with which they strike the target when potential different applied to 12.4 kV.

(a) 3.2×10^{7} *m*/sec (b) 4.5×10^{7} *m*/sec
(c) 6.6×10^{7} *m*/sec (d) 7.5×10^{7} *m*/sec

21. Wavelength of line in the target of $Z=42$ when wavelengrh of line in the target of $z=24$ is 2 Å. Taking m as 1.

(a) 0.43 Å (b) 0.53 Å
(c) 0.63 Å (d) 0.73 Å

22. Find the wavelength of scattered *X*-rays in **Compton scattering** of λ 3 Å and scattered at 30° angle w.r.t. incident beam.

(a) 2.005678 Å (b) 3.005678 Å
(c) 4.005678 Å (d) 5.005678 Å

23. Number of waves per millimeter when the light is travelling in water if frequency of light of a certain colour has 1800 waves to the millimeter in air (h = 1.52 for glass).

(a) 1399 mm^{-1} (b) 1499 mm^{-1}
(c) 1799 mm^{-1} (d) 2399 mm^{-1}

24. What is **Critical angle** if the sphere is immersed is`air ($n_2 = 1$) if index of refraction of a glass sphere, $n_1 = 1\ 76$.

(a) 14.6°
(b) 24.6°
(c) 34.6°
(d) 44.6°

25. To some extent, the brilliance of diamonds is attributable to total internal reflection. Calcuate the **critical angle** for a diamond-air surface. (For dimond $\eta = 2.42$).

(a) 12° (b) 15°
(c) 18° (d) 24°

26. At what angle must a ray of light be incident on acetone to be refracted into the liquid at 25°? ($n = 1.358$).

(a) 15° (b) 25°
(c) 35° (d) 45°

27. Which of the following is correct ?

(a) Base peak is most intense in the spectrum.
(b) The molecular ions, the fragment ions and the fragment radical ions are separated by deflection in a variable magnetic field according to their m/e ratio.
(c) The base peak is assigned the arbitrary intensity value 100, while all other peaks are given their proportional value.
(d) All are correct

28. The instruments employing the time of flight mass separators are less satisfactory than those using magnetic focusing from the stand point of

(a) Resolution
(b) Reproducibility
(c) Ease of mass identification
(d) All of the above

29. Double focusing principle is used in

(a) Magnetic deflection
(b) time of flight
(c) quadruole mass spectrometer
(d) none of the above

30. Match List - I with List - II and select the current answer using the codes given below :

List - I

A. Interferometer
B. Magnetic field
C. Column
D. Ionizationchamber

List - II

1. FTIR spectrometer
2. Radiation detector
3. Mass spectrometer
4. Gas Chromatograph

Code

	A	B	C	D
(a)	3	1	4	2
(b)	1	2	4	3
(c)	1	3	4	2
(d)	3	4	2	1

1.3 X-Rays and Nuclear Radiation Measurements

1. An X-ray source radiates two characteristic wavelength: λ_1 and λ_2, where $\lambda_1 < \lambda_2$. The shorter wavelength of the laser can be filtered out by passing the radiation through.
   (a) A narrow slit
   (b) An interference filter
   (c) A crystalline material whose absorption edge lies between λ_1 and λ_2
   (d) A polarizer that rotates the polarization state of radiation at λ_2

2. The dispersion in an X-ray differactometer, $\frac{d\theta}{d\lambda}$, is given by the expression
   (a) $\frac{m}{2d \cos \theta}$ (b) $\frac{m}{2d \sin \theta}$
   (c) $2d \sin \theta$ (d) $2d \cos \theta$

3. The Geiger Counter has high quantum efficiency in the wavelength Å range
   (a) 0.2 to 0.3 (b) 0.4 to 0.5
   (c) 1.5 to 2.1 (d) 6 to 7

4. The Duan-Hunt law represents the minimum wavelength λ. (Å) of X-rays generated for an applied voltage V (volts) and is given by:
   (a) $V = [12400 / \lambda]$ (b) $V = 12400\ \lambda$
   (c) $V = [12400 / \lambda^2]$ (d) $V = 12400\ \lambda^2$

5. If the half-life of a radiation source is 1 year, its decay constant per year is:
   (a) 1 (b) 2
   (c) 0.693 (d) 63.2

6. Attenuation of a narrow monochromic X-ray beam in a metal plate of thickness 'd' is given by the equation
   (a) $I = I_0 \exp(-\mu d)$
   (b) $I = I_0 \exp(-\mu d^2)$
   (c) $I = I_0 \exp(-\mu/d)$
   (b) $I = I_0 \exp(-\mu/d^2)$

7. Identify the correct set of matches from the following:

List I	List II
A. Beer's law	1. X-ray diffraction
B. Bragg's law	2. Thermo-emf
C. Peltier effect	3. Electro-chemical cell
D. Nernst equation	4. Absorption of light

Codes:

	A	B	C	D
(a)	4	1	3	2
(b)	3	2	1	4
(c)	4	1	2	3

   (d) None of the above

8. X-rays emitted from a molybdenum target at 35 k V have wavelength edges $K_\alpha = 70.9$ pm and $K_\beta = 63.2$ pm. These X-rays are passed through a zirconia filter of K-edge 68.9 pm. The output of the filter contains primarily X-rays of wavelengths
   (a) 63.2 pm
   (b) 70.9 pm
   (c) Both 63.2 pm and 70.9 pm
   (d) Neither 63.2 pm nor 70.9 pm

9. In an ideal X-ray tube operated at 40 kV, the short wavelength limit of X-ray emission in A is (Planck's constant $h = 6.626 \times 10^{-34}$ Js, charge of an electron $e = 1.6 \times 10^{-19}$ coulomb, velocity of light $c = 3 \times 10^8$ m/s)
   (a) 0.155
   (b) 0.31
   (c) 0.62
   (d) 0.93

10. The H-D curve of two X-ray films, F_1 and F_2, are shown below.

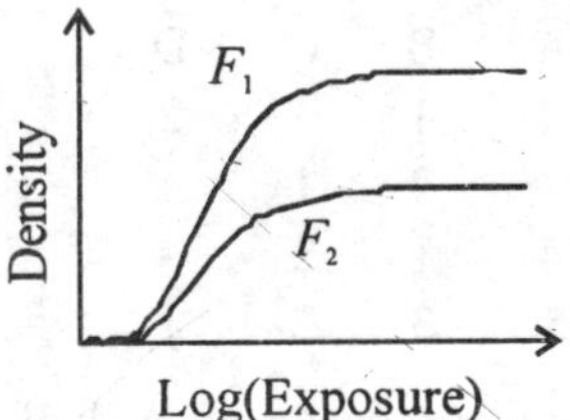

The curves show that, for the same change in exposure, F_1 has
   (a) Higher contrast than F_2 but the same fog level
   (b) Lower contrast than F_2 but the same fog level
   (c) Higher contrast and higher fog level than F_2
   (d) Lower contrast than F_2 but higher fog level

11. An X-ray tube is operated at 80 kV anode voltage. In order to filter the low intensity X-ray, a 2.5 mm thick aluminum is used. It is given that at 80 kV anode voltage the mass attenuation coefficients and densities

of aluminum are 0.02 m^2kg^{-1} and 2699 kgm^{-3} respectively and for copper these are 0.075 m^2kg^{-1} and 8960 kgm^{-3} respectively. If a copper filter is to replace the aluminum filter with the same filtering effect, the thickness of the copper filter should be.

(a) 0.2 mm
(b) 0.66 mm
(c) 1.5 mm
(d) 5 mm

12. The operating voltage of an X-ray tube is changed from 40 kV to 50 kV. The resulting change in the shortest wavelength generated is

(a) + 20% (b) – 20%
(c) + 25% (d) – 36%

13. For radioisotope imaging, an Anger camera is fitted with a parallel hole collimator. If the thickness of the collimator is increased, the camera

(a) Resolution and sensitivity will increase
(b) Resolution and sensitivity will decrease
(c) Resolution will increase and sensitivity will decrease
(d) Resolution will decrease and sensitivity will increase

14. Match the List-I name of the equation with List-II the formula of the equation and select the correct answer using the codes.

List-I	List-II
A. Bragg equation	1. A = a b c
B. Duan hunt equation	2. $N = N_o e^{-\lambda t}$
C. Radioactive decay equation	3. λ = [hc/eV]
D. Beer Lambert law	4. (μ - 1) t = nλ
	5. 2d sin θ = nλ

Codes:

	A	B	C	D
(a)	4	1	3	2
(b)	3	2	1	4
(c)	4	1	2	3
(d)	5	3	2	1

15. Find the minimum wavelength of X-rays produced by an X-tube operated @1000 KV.

(a) 1.247 þm (b) 1.237 þm
(c) 1.245 þm (d) 1.257 þm

16. If the potenital difference applied an X-ray tube is 5 KV. Find the velocity of striking electron on target.

(a) 2.1×10^7 m/s (b) 3.1×10^6 m/s
(c) 4.1×10^5 m/s (d) 4.1×10^7 m/s

Common Data for Questions 17 and 18

A certain X-ray tube operted @ 50 V.

17. The λ min is

(a) 0.48 Å (b) 24.8 Å
(c) 0.248 Å (d) 0.29 Å

18. The maximum intensity wave length is

(a) 0.172 Å (b) 0.72 Å
(c) 0.272 Å (d) 0.372 Å

19. The Daune-Hunt law represents the minimum wavelength (λ A^0) of X-rays generated for an applied voltage V volts and is given by.

(a) V = (12400 / λ) (b) V = 12400 λ
(c) V = 12400 / λ^2 (d) V = 12400 λ^2

20. A voltage applied to an X-ray tube is increased 1.5 times. The short wavelength of an X-ray has continuous spectrum shift $\Delta\lambda$ = 26pm.

Find the initial voltage applied to the tube is

(a) 0.26 (b) 0.34
(c) 0.29 (d) 0.30

21. Attenuatin of narrow monochromatic X-ray beam in metal plate of thickness 'd' is given by the equation

(a) $I = I_o e^{(-\mu d)}$ (b) $I = I_o e^{(-\mu d^2)}$
(c) $I = I_o e^{(-\mu / d)}$ (d) $I = I_o e^{(-\mu / d^2)}$

Common Data for Questions 22 and 23

For a narror beam of X-ray radiation of wavelength 62 pm, the mass absorption co-efficient and densities for Aluminimum and lead are given below

	μ	P
Aluminium	$0.02 m^2 / g$	$2699 / Kg / m^3$
Copper	$0.075\ m^2 / g$	$8960\ Kg / m^3$

22. The X-ray attenuate in aluminium screen of 2.6 cm thickness is

(a) 0.03 (b) 0.05
(c) 0.04 (d) 0.02

23. To fully attenuate the beam a thick lead screen used is

(a) 0.1 m (b) 0.2 m
(c) 0.02 m (d) 0.01 m

24. X-rays emitted from a molybdenum target at 35 KV have wavelength edges K_a = 70.9 and K_β = 63.2 pm. These X-rays are passed through a zirconia filter of K-edge 68.9 pm. The output of the filter contains primarily X-ray of wave length

(a) 63.2 pm (b) 70. pm
(c) both a and b (d) None

25. The dispersion in an X-ray diffractormeter, $d\theta / d\lambda$ is given by expression

(a) $\dfrac{m}{2d\cos\theta}$ (b) $\dfrac{m}{2d\sin\theta}$
(c) $2d\sin\theta$ (d) $2d\cos\theta$

26. An X-ray tube operated at 80 kV anode voltage. In orderto filter the low inensity X-ray, a 2.5 mm thick aluminium filter is used
Given

	μ	P
Aluminium	$0.02 m^2/g$	$2699/Kg/m^3$
Copper	$0.075\, m^2/g$	$8960\, Kg/m^3$

If copper filter is to replace the aluminium with same filtering effect, the thickness of the copper should be
(a) 0.2 *mm* (b) 0.6 *mm*
(c) 1.5 *mm* (d) 5 *mm*

27. The X-rays of wavelegth 0.36A is diffracted in a braggs psecptrometer at an alngle of $4°48'$. Find the effectie value of atomic spacing
(a) 2.153 Å (b) 2.155 Å
(c) 2.159 Å (d) 2.151 Å

28. Spacing between atomic planes of crystal is 1.82 A. If X-rays of wavelength 0.68 A is targeted towards the crystal.
The diffraction angle of first order diffraction.
(a) 12.2° (b) 12.7°
(c) 12.9° (d) 12.5°

29. If the half life radiation source is one year, its decay constant per year is ....
(a) 0.687 (b) 0.693
(c) 0.698 (d) 0.683

30. Half life of radiation is 3.8 days and its decay constant is 0.182/day.
Only 1/20th of sample left over after the days of
(a) 16.46 days (b) 16.48 days
(c) 16.40 days (d) 16.43 days

31. The activity of certain radioactive material decreases 2.5 times after 7 days.
Its half life time is
(a) 5.6 days (b) 4.6 days
(c) 4.5 days (d) 5.3 days

32. A radio nuclide decays to 95% of its value in 60 min. The mean life time is
(a) 20.11 min (b) 20.04 min
(c) 20.14 min (d) 20.08 min

33. For a nuclear radiation measurement a proportional counter compared to GM counter has.........
(a) Low dead time and high counting rate
(b) High dead time and high counting rate
(c) Higher internal amplification
(d) None

34. Match the List- I light souce with List IInature of radiations and select the correct answer by using the codes given below

List-I	List-II
A. Tungsten lamp	1. Laser
B. C0-60	2. IR
C. He-Ne	3. Gamma
D. Discharge lamp	4. UV rays

Codes:

	A	B	C	D
(a)	1	2	3	4
(b)	3	4	1	2
(c)	2	3	1	4
(d)	4	3	2	1

35. Match the List I detector light souce with List II spectrocopie regions and select the correct answer by using the codes given below

List-I	List-II
A. Bolometer	1. UV
B. Photomuliplier	2. IR
C. Gm counter	3. X-rays
D. Photographic	4. Gamma-rays

Codes:

	A	B	C	D
(a)	1	2	3	4
(b)	3	4	1	2
(c)	2	1	4	3
(d)	4	3	2	1

36. Match the List- I and List-II and select the correct answer by using the codes given below

List-I	List-II
A. Beers law	1. X-ray diffraction
B. Braggs law	2. Thermo emf
C. Peltier effect	3. Electrochemical cell
D. Nernst equatin	4. Absorption of light

Codes:

	A	B	C	D
(a)	1	2	3	4
(b)	3	4	1	2
(c)	2	3	4	1
(d)	4	1	2	3

37. Match the List-I name of equation with List-II formula of equation and select the correct answer by using the codes given below

List-I	List-II
A. Braggs equation	1. A = abc
B. Duachunt equation	2. $N = N_o e^{-\lambda T}$
C. Radio active decay equation	3. Wave length $= (he/eV)$
D. Beer lamberts law	4. $N\lambda = 2\, d \sin\theta$

Codes:

	A	B	C	D
(a)	1	3	4	2
(b)	3	4	1	2
(c)	2	3	4	1
(d)	4	3	2	1

38. Match the List-I instruments with List-II parameters they measure and select the correct answer by using the codes given below

List-I	List-II
A. Auto collimator	1. Radiation scattered
B. Hygrometer	2. Molecular weigth
C. Nephlometer	3. Humidity
D. Mass spectrometer	4. Density

Codes:

	A	B	C	D
(a)	1	3	4	2
(b)	3	4	1	2
(c)	2	3	4	1
(d)	2	3	1	4

39. Intensity of an X-ray beam of intensity I_o when passed through a substance of thickness X reduces to (μ = absorption co-efficient of substance).

(a) $I = I_0 e^{ux}$ (b) $I = I_0 e^{ux^2}$
(c) $I = I_0 e^{-ux^2}$ (d) $I = I_0 e^{-ux}$

40. Interference phenomenon deffers from diffraction in that unlike diffraction.

(a) Interference fringes are of varying intensity
(b) Interference cannot be observed with white light
(c) Interference minima may be perfectly dark.
(d) Interference fringes are of unequal width.

41. Find the wavelength of *X*-rays emitted when electrons acceletatyed through 30 kV strike a target

(a) 0.031 nm (b) 0.041 nm
(c) 0.051 nm (d) 0.061 nm

42. Find the atomic spacing when a beam of X-rays of wavelength 0.3 nm is incident on a crystal and gives a first-order maximum when the glancing angle is 9.0 degrees.

(a) 0.76 nm (b) 0.86 nm
(c) 0.96 nm (d) 1.06 nm

43. Find the wavelength of primary *X*- radiation if Comptom scattering of this radiation graphite at an angle of 60°, the wavelength of **scattered radiation** is 2.54×10^{-11}m.

(a) 0.232 Å
(b) 0.242 Å
(c) 0.252 Å
(d) 0.262 Å

44. Find the Angle at which 3rd coder reflection take place from crystal when *X*- rays from a undergo first coder reflection at a gane angle of 12° from the face of a calcuater crystal. Given grating space of calcite crying as 3.04×10^8 cm.

(a) 19.4° (b) 22.9°
(c) 25.4° (d) 27.3°

45. Find the minimum applied potential required produce *X*-rays of 1 Å wavelength.

(a) 11.7 kV (b) 12.4 kV
(c) 13.5 kV (d) 15.6 kV

46. Find the shortest λ of **X - ray produced** in a *X*-ray tube with accelerating potential as 70 kV.

(a) 0.017 (b) 0.018
(c) 0.019 (d) 0.015

47. Find the C_T number (House field unit) H of heart muscle ($\mu = 0.212$) and μ Water = 0.205 is

(a) $\frac{205}{7000}$ (b) $\frac{405}{7000}$
(c) $\frac{7000}{205}$ (d) $\frac{7000}{405}$

48. In an ***X*-ray** tube, accelerating potential 70 kV. Determine the shortest wavelength X-ray produced

(a) 0.0177 nm (b) 0.0133 nm
(c) 0.0144 nm (d) 0.0166 nm

49. In an **X-ray spectrum**, the wavelength of certain line is 2 Å for the target of $z = 24$. Calculate the wavelength of the same line the target of $z = 42$. Take *a* as unity. From Moseley's we can write.

(a) 0.33 A (b) 0.43 A
(c) 0.43 A (d) 0.63 A

50. In compton scatterijng, **X- ray** wavelength 3 Å are scattered at an angle 40° with respect to incident beam. Calculate the wavelength of the scattered X-rays.

(a) 1 Å (b) 2 Å
(c) 3 Å (d) 4 Å

51. If the half - life of a radiation source is 1 year. its decay constant per year is

(a) 1 (b) 2
(c) 0.693 (d) 63.2

52. Match List - I with List - II and select the current answer using the codes given below :

List - 1	List - 2
A. Bolometer	1. Laser
B. Photomultiplier	2. IR
C. GM Counter	3. UV
D. Discharge lamp	4. γ-ray

Code

	A	B	C	D
(a)	2	1	3	4
(b)	1	3	4	2
(c)	2	3	4	1
(d)	3	4	2	1

53. X - rays emitted from a molybdenum target at 35 kV have wavelength edges $k_\alpha = 7.09$ pm and $k_\beta = 63.2$pm . These X - rays are passed through a zircons filter of K – *edge* 68.9 pm. The output of the filter contains primarily X - rays of wave lengths of
(a) 63.2 pm
(b) 70.9 pm
(c) Both (a) and (b)
(d) Neither (a) Nor (b)

54. Match List - I with List - II and select the current answer using the codes given below :

List - 1
A. Electron capture detector
B. Scintillation counter
C. Thermopile
D. Zirconia probe

List - 2
1. IR Spectroscopy
2. Oxygen analyzer
3. Gas chromatography
4. Nuclear radiation spectroscopy

Code

	A	B	C	D
(a)	2	3	1	4
(b)	2	4	1	3
(c)	1	3	2	4
(d)	1	2	3	4

55. An ideal X - ray tube operated at 60 kV, then the cut - off wavelength in pm is
(a) 10.35 (b) 20.7
(c) 41.41 (d) None of the above

56. The voltage required for X - ray tube to produce minimum wavelength of 0.31 A° is
(a) 20 kV (b) 40 kV
(c) 60 kV (d) None of the above

Common Data for Questions 57 and 58

A certain X-ray tube at 50 kV

57. The short wave length region for the X - rays being emitted is
(a) 0.248 A° (b) 0.248 nm
(c) 0.248 mm (d) None of the above

58. The wavelength of maximum intensity is
(a) 0.248 A° (b) 0.372 A°
(c) 0.496 A° (d) None of the above

1.4 Optical Sources and Detectors

1. To a swimmer viewing a sodium lamp beside the pool, the color of the lamp seen from underneath the water appears
(a) Yellow
(b) Yellow with a reddish tinge
(c) Yellow with a greenish tinge
(d) White

2. In a photo multiplier
(a) Gain is independent of stray magnetic fields
(b) High frequency response is improved by increasing the number of dynodes (emitting surfaces)
(c) Secondary emission is used for amplification of low level photo-current
(d) The electrons are directed to the anode by applying a strong magnetic field

3. Which part of below radiation is a part of sunlight spectrum ?
(a) Near UV (b) Mid IR radiation
(c) Far IR radiation (d) all the above

4. Which of the following can be used as a filament in incandescent filament lamps ?
(a) Tantalum (b) Osmium
(c) Tungsten (d) all the above

5. Match the List - I source with List - II spectrum and select the correct answer by using the codes given below

List - I Source	List - II Spectrum
A. Sunlight	1. Continous
B. Incandescent bulb	2. Discountinuous
C. Fluorescent source	3. Continous with peaks
D Hydrogen lamp with low pressue	4. Line spectrum

	A	B	C	D
(a)	3	2	1	4
(b)	4	3	2	1
(c)	1	1	3	4
(d)	2	3	4	1

6. Black light (woods lamp) refers to
(a) UV light (b) Visible light
(c) Dark lamp (d) IR lamp

7. A vaccum type photo cell produces 10 μA of tube current for 3W of incident power.

The radiant sensitivity is

(a) 0.30 μ (b) 0.3 μ
(c) 3.33 μ (d) cannot be calculated

8. PMT's are sensitive in the Electromagnetic. The range of Spectrum is

(a) UV (b) Visible
(c) Near I.R (d) all the above

9. When light travels from one medium to another. Which are separated by a sharp boundary, the characteristic which does not change is.

(a) Velocity (b) Wavelength
(c) Frequency (d) Amplitude

10. A head on PMT has a window made of Ultraviolet glass, then the PMT can perform detection of

(a) Visible light (b) Ultra - Violet light
(c) both (a) and (b) (d) Near I.R.

11. Match the List - I optical defector with List - II sensitivity and select the correct answer by using the codes given below

	List - I Optical detector		List - II Sensitivity
A	Gas filled photo tube.	1.	20A/Im
B	PMT	2.	100 μA / Im
C	Vacuum type PEC	3.	30μA / Im

Codes:

	A	B	C
(a)	3	2	1
(b)	2	1	3
(c)	1	3	2
(d)	2	3	1

12. In a photo multiplier

(a) Gain is independent of stray magnetic fields
(b) High frequency response is improved by increasing the number of dynodes
(c) Secondary emission is used for amplification of low level photo current
(d) The electrons are directed to the anode by applying a strong magnetic field

13. The LDR a 4 lumen irradiation has a resistance of 1 kW when 10 V source is suppressed and has 5 kW resistance when not suppressed. The output voltage V for the below op-amp circuit is

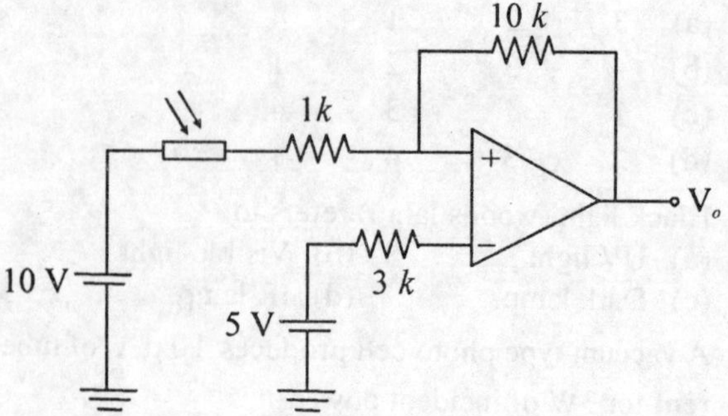

(a) 13.33 V (b) 13.30 V
(c) 13.35 V (d) 13.38 V

14. A phot diode application circt was designed for a special puupose of photo detection. The photo diodes D1 and D2 have cut-in votages 0.5 V and 0.3 V with forward resistances 20 W and 100W respectively.

The voltages across 2 kW resistors is

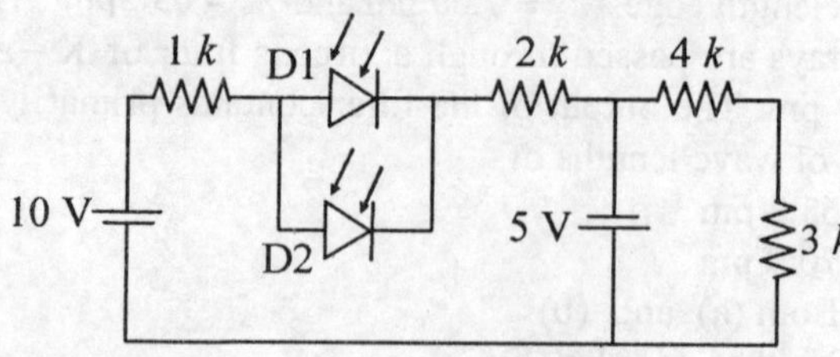

(a) 10.4839 (b) 7.4828
(c) 9.4838 (d) 6.4848

15. A photo detector circuit is given in figure. The photo-diodes has an active area of 0.01 cm² and sensitivity of 0.55 A/W When light of intensity 10 mW/cm² falls on the photo diode, the output V is

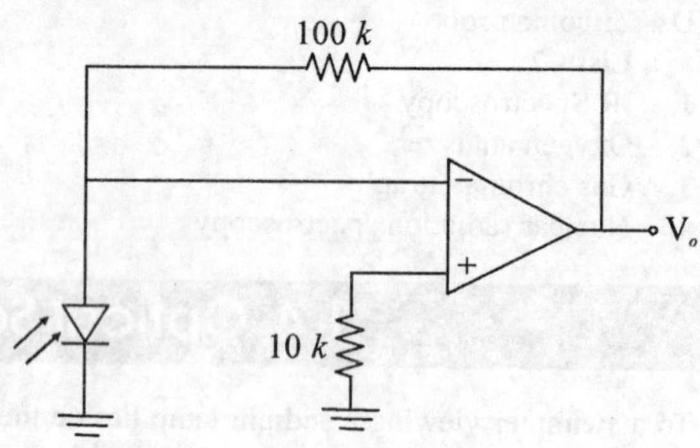

(a) −5.3 V (b) −5.7 V
(c) −5.5 V (d) −5.9 V

16. If the effective area of a photodiode is 0.4 $0.4\times10^{-6}\ m^2$ and irradiance is $300\ W/m^2$.

The incident power is

(a) 140 μW (b) 130 μW
(c) 120 μW (d) 150 μW

17. The energy of photon is 5eV then the wavelength of photon is

(a) 2.5 Å
(b) 25 Å
(c) 250 Å
(d) 2500 Å

18. When the light falls on the photodiode shown in following circuit, the reverse saturation current of the photodiode changes from 100 mA to 200 mA. Assuming the op-amp to be ideal, the output voltage V of the circuit changes from to

(a) (b)
(c) 9 to 90 V (d)

19. A PIN photodector has a junction capacitance of 16 pF, active area of 10 mm² and a responsivity of 0.5 A/W. It is illuminated by a light of intensity 1mW/cm². A volage output is obtained by connecting a 100 kΩ Load resistance resistor R_L in series with the dètector. The steady state voltage across R_L is
(a) 0.5 V (b) 1.5 V
(c) 5 V (d) 7 V

20. A PIN didoe is constructed of GaAs which has a band gap of 1.43 eV. Find the maximum wavelength that can generate current
(a) 0.859 mm (b) 0.869 mm
(c) 0.867 mm (d) 0.857 mm

21. In a 100 ns pulse, 6×10^6 photons at a wavelength of 1300nm fall on InGaAs PIN photodetector and on an average 5.4×10^6 EHPs are generated. Calculate the quantum efficiency?
(a) 40% (b) 80%
(c) 90% (d) 50%

22. Photons are incident on a PIN photodiode which has responsivity 0.65 A/W. If Optical power incident is 10 mW. Calculate photo current generated
(a) 5.7 μA (b) 5.5 μA
(c) 6.5 μA (d) 6.7 μA

23. A Si APD has quantum efficiency of 0.65 at a wavelenght of 900 nm. Suppose 0.5 mW of optical power produces a multiplied curent of 10 mA. Find multplicaion factor
(a) 45 (b) 47
(c) 43 (d) 40

24. A Si pn-diode generates heat rather than light when in forward bias.This is due to ....
(a) Radiative recombination
(b) Direct transition
(c) Two step transfer
(d) One step transfer

25. An LED is connected as shown, it should glow when V is in logic '0' [0.2 V]. Assuming minimum current through LED in active state as 1.5 mA. Calculate R
The value of R is

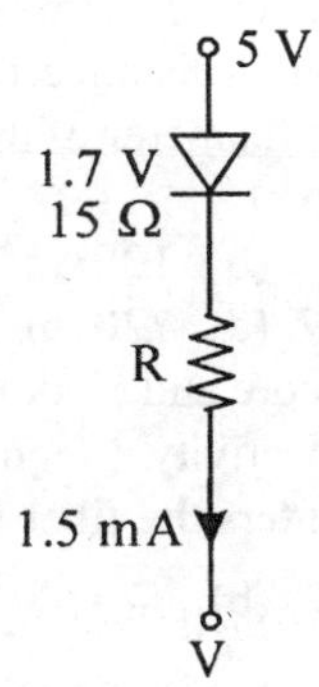

(a) 1.9 kW (b) 1.6 kW
(c) 1.2 kW (d) 1.4 kW

26. The potential V in the below LED circuit is

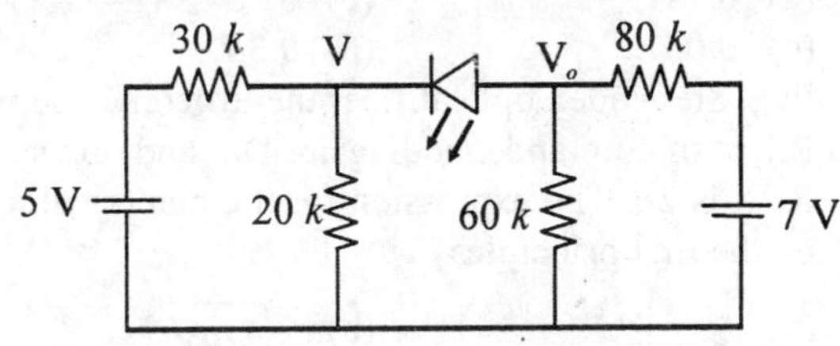

(a) 2.254 V (b) 2.249 V
(c) 2.244 V (d) 2.259 V

27. The most used He-Ne gas laser has a wavelength of
(a) 0.5 μm (b) 0.633 μm
(c) 0.7 μm (d) 0.80 μm

28. In a He-Ne laser operating at a wavelength of 632.8 nm the actual laser transistion takes place between the energy levels of
(a) He (b) Neon
(c) He and Neon (d) None

29. The coherence length of the He-Ne laser beam is 120 cm. Its coherence time in seconds is
(a) 3 *ns* (b) 5 *ns*
(c) 2 *ns* (d) 4 *ns*

30. A Laser beam ($\lambda = 632.8\ nm$) having a divergence 1 m rad is directed towards the moon at at distance of approximately
(a) 800 *km* (b) 700 *km*
(c) 500 *km* (d) 600 *km*

31. A He-Ne laser cavity length 500 nm has an oscillating bandwidth of 1500 MHz. The maximum number of longitudianal oscillating modes that are accommodated with in the B.W is
(a) 4 (b) 5
(c) 30 (d) 40

32. A frequency stabilized He-Ne laser with a wavelength of 6328A has a bandwidth of 1 MHz.

Its coherence length is

(a) 0.3 m (b) 100 m
(c) 30 m (d) 300 m

33. A He-Ne laser beam having a wavelength of 632.8 nm is expanded to a diameter of 1 m and then pointed towards the moom. If the distance of the moon from earth is 3.8 ′ 10⁵ km, the diameter of the central maxima on the surface of the moon.

(a) 1 m (b) 100 m
(c) 587 m (d) 720 m

34. In an optical fiber with core and cladding having refractive index n_1 and n_2 respectivity. ($n_2>n_1$). The numerical aperture is when light enters the fiber in air.

(a) $Sqrt(n_1^2 - n_2^2)$ (b) $Sqrt(n_2^2 - n_1^2)$
(c) $Sqrt(n_1^2 + n_1^2)$ (d) none

35. A Step index optical fiber the difference between the R.F s of core and cladding are 1.44 and 1.40 respectively is surrounded by air, its N.A is

(a) 0.12 (b) 0.75
(c) 0.06 (d) 0.34

36. In a Step index optical fiber the difference between the R.F s of core and cladding are *Dn* and the sum of the RI's is *2n.* The expression for the numerical aperature by the first principles

(a) $2n\Delta n$ (b) $\sqrt{2n\Delta n}$
(c) $\frac{\sqrt{2n\Delta n}}{3}$ (d) $\sqrt[3]{2n\Delta n}$

Common Data for Questions for 37 and 38

One end of a step index multi-mode optical fiber is connected to PIN-photo diode. The area of the photo-diode is large enough to capature all the ligh emitted out of the fiber. The data for the beber and photo diode are as follows: Optical fiber core $n_1 = 1.5$ and cladd ; length = 10 m, wavelength ≠ 800 nm, attenuation at 800 nm; Respeconsivity = 0.55 A/W

37. The numerical aperture of the fiber is

(a) 0.23 (b) 0.384
(c) 0.192 (d) 0.768

38. If light of wavelength 800 nm and power 2mW is launched into one and of the Fiber. The detector current at the end is

(a) 0.22 mA (b) 0.33 mA
(c) 0.44 mA (d) 0.11 mA

39. The relative permittivity of an optical medium is 2.5 Its refractive index is

(a) 0.4 (b) 1.25
(c) 1.48 (d) 1.58

40. A optical fiber has a R.I of 1.64 for core and 1.422 for the cladding. The critical angle above which a ray will be toally relfected is

(a) 37° (b) 41°
(c) 45° (d) 60°

41. The core/cladding index difference of a signal-mode optical fiber cables is 0.01. The R.I of the materials of the core is 1.5. The maximum angle of acceptance of fiber is paaroximately equal to

(a) 17.5 degrees. (b) 12.1 degrees.
(c) 8.6 degrees. (d) 2.0 degrees.

42. A parallel beam is incident normally on a air-glass interface. If the R.I of glass is 1.720 then loss of internsity through reflection is

(a) 7% (b) 4%
(c) 20% (d) 1%

43. A Laser light with a wave length of 633 nm is passed through 1 cm length of tissue and 2 cm length of glass. The R.Is tissue and glass are 1.33 and 1.55 resp. The velocities of laser light in the tissue and in the glass in the ratio of

(a) 1.33 : 9.757 (b) 1.33 : 3.0
(c) 1.33 : 1.5 (d) 1.5 :19.33

44. Two cross polarizers are placed in the part of a light beam. The light output is

(a) Zero (b) Plane polarized
(c) Circularly polarized (d) elliptically polarized

45. A parallel beam of light (wavelength = 515 nm) is incident on a flint glass of R.I = 1.6 the angle refraction is 15° an interference pattern is observed in reflected light. If one bright fringe is observed the thickness of plate is

(a) 0.23 (b) 0.16
(c) 0.85 (d) 0.47

46. For a change in elecvation level of 1.5 mm between an optical flat and surface, a light wave of wavelength 0.5 mm produces fringes. The number of fringes is

(a) 3 (b) 6
(c) 12 (d) 18

47. In a Michelson interferometer a shift of 100 fringes is observed at wavelength of 6328 A, when air in the tube of 20 cm length is removed. Calculate the R.I of air in tube is

(a) 0.015 (b) 1.001516
(c) 1.00015 (d) 0.1015

48. An optical flat is tested against a standared reference flat in Newtorn interferosope. If the optical flat is as good as the reference, the fringes observed are a set of

(a) Concentric rigns
(b) Elliptical rings
(c) Hyperbolic rigns
(d) Straight and parallel fringes

49. A light beam is incident normally on a glass-air interface the glass has a R.I = 1.53. Air is assumed to be R.I = 1.0. The Fresnel, intensity reflection coefficient is

(a) 0.44
(b) 4.4
(c) 44
(d) 0.044

50. The **optical spectrum** has wavelengths ranging from:
(a) 10 nm to 10^6 nm (b) 10 nm to 770 nm
(c) 370 m to 770 nm (d) 370 nm to 10^6 nm

51. The **visible spectrum** has wavelengths ranging from:
(a) 10 nm to 10^6 nm (b) 10 nm to 770 nm
(c) 370 nm to 770 nm (d) 370 nm to 10^6 nm.

52. A certain emissive material has a work function of 2.24 e V, the radiations emitted from it will have **threshold length** of approximately:
(a) 900 nm (b) 600 nm
(c) 533.6 nm (d) 527.6 nm

53. In order that **photoelectric emission** to be possible over whole of the visible spectrum of 370 nm - 770 nm, the work function of the photoemission surface should be aproximately:
(a) 1.61 to 3.35 eV (b) 1.3 to 2.25 eV
(c) 1.5 to 4.206 eV (d) 0.231 to 5.62 eV

54. When light beam is made incident on a surface which is rough as compared with the wavelength of light.
(a) Specular reflection takes place
(b) Diffused reflection takes place
(c) Absorbtion takes place
(d) Transmittance take place

55. Radiometry refers to measurement of optical rediations.
(a) from a wavelength of 370 nm - 770 nm
(b) from a wavelength of 10^{-12} nm to 10^6 *nm*
(c) from a wavelength of 10^2 nm to 10^6 nm
(d) from a wavelength of 10 nm to 10^6 nm

56. The responce of human eye under normal light conditions is called.
(a) stoppoic vision and it peaks at awavelength of 555 nm.
(b) stoppic vision an it peaks at a wvelength of 507 nm.
(c) photopic vision and it peaks at a wevalength of 507 mn
(d) photopic vision and it peaks of 555 nm

57. A **light source** may be considered as a point soutce if its diameter is.
(a) less than 10% of distance it and the detector.
(b) less than 1% of the distance between it and the dejector.
(c) more less 10% of the distance between it and the detector.
(d) more than 2% of the distance between it and the detector.

58. The units of lumin\ous lntensity are:
(a) lux (b) Cd
(c) W/sr (d) Cd/sr

59. The units of lumuance are:
(a) nit (b) lm/sr - m^2
(c) Cd/m^2 (d) all of the above.

60. The errors invovled in **photometric measurements** are of the order of the order of:
(a) 1% (b) 1–5%
(c) 10-20 % (d) 40–50%

61. The desirable properties of a **laser** beam are:
(a) it has a high radiant energy
(b) it can be polarized
(c) it is monochromatic and coherent
(d) all of the above

62. The current through a **photoemissive tube depends** upon:
(a) intensity of light incident upon it.
(b) colour and wave length of incident light
(c) voltage applied between cathode and anode of tube
(d) all of the above

63. The **luminous sensitivity** of gas filled photelectric cells varies between:
(a) 10 to 100 $\mu A/lm$
(b) 40 to 150 $\mu A/lm$
(c) 0.002 to 0.1 $\mu A/\mu W$
(d) 0.01 to 0.15 $\mu A/\mu W$

64. A **photomultiplier tube** uses 10 stages. The sensitivity of each dynnode of the tube is 100 $\mu A/lm$ the overall sensitivity of the photomultiplier tube, if the gain of each stage is 2.7, is approximately:
(a) 2.06 A/lm (b) 270 $\mu A/lm$
(c) 2700 $\mu A/lm$ (d) None of above

65. A **photoconductive cell** may be called as:
(a) photoresistive cell
(b) photo resistor
(c) light dependent resistor
(d) all of the above

66. The resistance of a **photo resistive cell**
(a) increases with increase in light intensity
(b) remains constant irrespective of the light inten sity incident upon it
(c) decreases with increase in light intensity in a lin ear manner
(d) decreases with increase in light intensity in an expenential manner.

67. Photodiodes
(a) do not have any response in darkness
(b) have a reponse even in darkness
(c) have a leakage current even in darkness which is called dark current that doubles for every $10°C$ increase in the ambient temperature that which decrease the signal to noise ratio.
(d) have a leakage current even in darkness which is called current that doubles for every increase of $5°C$ in temperture and which increase the signal to noise ratio.

68. The sensitivity of a photovaltaic cell
(a) is almost constant when it is short circuited.
(b) decrease with increase of load resistance is abouT 1000 W
(c) is most negligible when the load resistance
(d) all of the above
(e) none of the above

69. In case of measurement of Intensity of a test source is used. The photometric head used when comparing two lights to different colours, is:
(a) Bunsen photometer head
(b) Constast type Lummer Brodhun photometer head
(c) Comparion type Lummer Brodhum photometeR head.
(d) Flicker photometer head.

70. The optical sensors may be classified as:
(a) Intensity and intrinsic types
(b) Interferometric and extrinsic types
(c) Intensity or Instrinsic type and interfermetric or extrimsic types
(d) all of the above

71. Silicon photodetector can only used to detect wavelength in the
(a) 1st window of optical communication
(b) 2nd window of optical communication
(c) 3rd window of optical communication
(d) both 2nd and 3rd window.

72. Noise equivalent power of a photodetector is:
(a) inversely proportional to the dark current
(b) proportional to the square root of the dark current
(c) directly proportional to the dark current
(d) none of the above

73. To be used as photodetector PIN diode should be connected
(a) forward bias (b) reverse bias
(c) zero bias (d) either (a) or (c)

74. An optical repeater consists of
(a) series of receivers
(b) series of transmitter
(c) A receiver and a transmitter placed back to back
(d) none of the above

75. Photo detector is a
(a) traingular device
(b) square law device
(c) linear device
(d) both (a) and (b)

76. Photo-detector converts
(a) optical power into electric voltage
(b) optical power into electric current
(c) electrical currents into optical power
(d) none of the above

77. In the wavelength range (11–6) μm, the principal optical source material is
(a) In GaAsBr
(b) In GaAsF
(c) In GaAsP
(d) None of (a), (b) and (c)

78. Square-law device in optical source
(a) Linear variation in drive voltage results in a corresponding linear change in the optical output power.
(b) Linear variation in the drive current results in a corresponding linear change in the output power.
(c) Non-linear variation in the drive current result in a coreesponding linear change in the optical output power.
(d) None of the above

79. Band gap energy of a material is 1.24 eV: wavelength of peak emission is
(a) 0.75 μm (b) 1.0 μm
(c) 1.24 μm (d) 2.24 μm

80. Responsivity of a photodiode is specified as
(a) amps/watts (b) watts/amp
(c) volts/watt (d) watt/volt

81. Scattering loss in optical varies with wavelength as
(a) $\frac{1}{\lambda^2}$ (b) $\frac{1}{\lambda^3}$
(c) $\frac{1}{\lambda^4}$ (d) $\frac{1}{\lambda}$

82. Which of the following phenomenon is not explained by Huygen's construction of wavefront?
(a) Refraction (b) Reflaction
(c) Diffraction (d) Origin of spectra.

83. At sunrise or sunset, the sun looks reddish, because
(a) the sun is coldest at these times
(b) of the effects of reflection and refraction.
(c) the sun is hottest at these times
(d) of the scattrering of light.

84. The sky appears blue because.
(a) red light is absorbed
(b) blue light is scattered the most.
(c) blue light is absorbed.
(d) it is its natural colour

85. One cannot see through fog, because
(a) fog absorbs light
(b) the refractive index of fog is infinity
(c) light suffers total internal reflection at a droplet in fog.
(d) light is scattered by the droplets in fog

86. In diagnostic applications, ultrasound is both generated and detected by the piezoelectric effect in the
(a) MF and HF ranges
(b) LF range
(c) VLF range
(d) EHF range

87. Which of the following materials may be used to produce polaroids
(a) Tourmaline
(b) Calcite
(c) Quartz
(d) Crystals of quinine iodosulphate

88. Light from a sodium vapour lamp (589 nm) forms an **interference pattern** on a screen 0.8m from a pair of slits. The bright fringes is the pattern are 0.35 cm apart. Slit separatic is:
(a) 0.115 *mm* (b) 0.125 *mm*
(c) 0.135 *mm* (d) 0.145 *mm*

89. Find the maximum irradiance of four coherent and in coherent sources.
(a) $16I, 2I$ (b) $16I, 4I$
(c) $8I, 8I$ (d) $8I, 4I$

90. Find the ionization energy of hydriogen atom $(z=1)$ from $n=1$ to $n=\infty$
(a) 7.6 eV (b) 9.6 eV
(c) 11.6 eV (d) 13.6 eV

91. Find the reflectance at Ga As surface with refractive index 3.6 of Ga As.
(a) 0.12 (b) 0.22
(c) 0.32 (d) 0.42

92. Find the overall power efficiency of a low power **He-Ne laser** operating with a current of 10 mA at a dc voltage of 2500 V and given an optical output of 5 mV.
(a) 0.01 % (b) 0.02 %
(c) 0.03 % (d) 0.04 %

93. The frequency of light radiation with a wavelength of 500 nm is
(a) 6×10^{14} Hz (b) 6×10^{10} Hz
(c) 6×10^{7} Hz (d) None of the above

94. If the distance of distinct vision is 250 mm. The magnifying power of a lens with a focal length of 10 mm is
(a) 60 mm (b) 40 mm
(c) 80 mm (d) None of the above

95. Two crossed polarizer's are placed in the part of a light beam. The light output is
(a) zero
(b) plane polarized
(c) circularly plarized
(d) elliptically polarized

96. The resolving power of a telescope is proportional to
(a) λ / a (b) λa
(c) a / λ (d) None of the above

97. If a light of 0.8 μm wave length is allowed to fall on the photo conductors. If quantum efficiency 18 %. Then the responsively is
(a) 116 nm (b) 11.6 nm
(c) 1.16 nm (d) None of the above

98. If energy of photon is 5eV then the wavelength of photon is
(a) 02.5 A° (b) 25 A°
(c) 250 A° (d) None of the above

99. The effective area of a photodiode is 0.4×10^{-6} m^2 and the irradiance is 300 W / m^2. The incident power is
(a) 60 μW (b) 120 mW
(c) 100 μW (d) None of the above

1.5 LEDS and Lasers

1. In a He – Ne **laser** operating at a wavelength of 632.8 nm, the actual laser transition takes place between the energy levels of
(a) Helium (b) Neon
(c) Helium and Neon (d) None of the above

2. A light source is formed by putting a pinhole to let in sunlight. The above light source has
(a) No spatial or temporal coherence.
(b) No spatial coherence but has temporal coherence.
(c) Spatial coherence but no temporal coherence.
(d) Quasi – monochromaticity.

3. The longitudinal mode separation in a **gas laser** with a cavity length of 0.4 m (given velocity of light = 3 $\times10^{8}$ m/s) is
(a) 374 MHz (b) 375 MHz
(c) 748 MHz (d) 750 MHz

4. Collimated light beams from a He–Ne laser and a sodium vapour lamp (used for street lighting) are focused using a lens. The size of the spot at the focal point due to the laser is relatively smaller because
(a) Laser light is relatively more monochromatic.
(b) The limiting divergence of the laser light is relatively smaller.

(c) The output power of the laser is relatively larger
(d) The wavelength of the laser is relatively longer

5. A laser light with a wavelength of 6.33 nm is passed through 1 cm length of tissue and 2 cm length of glass. The refractive indices of tissue and glass are 1.33 and 1.5, respectivety .The velocities of laser light in the tissue and in the glass are in the ratio

(a) 1.33 : 0.75 (b) 1.33 : 3.0
(c) 1.33 : 1.5 (d) 1.5 :1.33

6. A mass spectrometer is to be used to resolve peaks corresponding to CO^+ and N_2^+. The atomic masses are $^{12}C = 12.0000$, $^{16}O = 15.9949$ and $^{14}N = 14.0031$ amu. The resolving power of the mass spectrometer should be atleast

(a) 250 (b) 350
(c) 2500 (d) 3500

7. A **laser** beam (λ = 632.8 nm) having a divergence of 1 milli-radian is directed towards the moon at a distance of approximately 4 × 10^5 km. The beam would have spread to a diameter of

(a) 400 km (b) 200 km
(c) 100 km (d) 4 km

8. A frequency stablised He-Ne laser with a **wavelength** of 6328 Å has a bandwidth of 1 MHz. Its coherence length is:

(a) 0.3 m (b) 3 m
(c) 30 m (d) 300 m

9. A He-Ne **laser** beam having a wavelength of 632.8 nm is expanded to a diameter of 1 meter and then pointed towards the moon. If the distance of the moon from earth is 3.8 × 10^5 km, the diameter of the central maxima on the surface of the moon is

(a) 1 m (b) 100 m
(c) 587 m (d) 720 m

10. An unpolarized He - Ne **laser** beam of wavelength 632.8 nm is reflected from the flat face of a dielectric [see figure]. A detector analyzes the reflected beam. It is found that when the angle between the incident and reflected beam is 115°34′ the light is completely plane polarized. The refractive index of the dielectric at this wavelength is

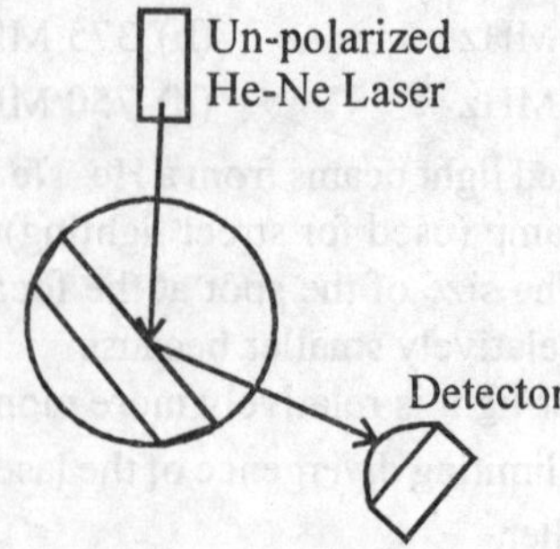

(a) 1.521 (b) 1.587
(c) 1.621 (d) 1.571

11. A quasi-monochromatic light source with average central wavelength of 630 nm has a spectral width of 10 GHz. The coherence length of the source is

(a) 3 μm (b) 3 mm
(c) 3 cm (d) 3 m

12. A He-Ne laser of cavity length 500 mm has an oscillating bandwidth of 1500 MHz. The maximum number of longitudinal oscillating modes that is accommodated within the bandwidth is

(a) 4 (b) 5
(c) 30 (d) 40

13. The output from a 633 nm He-Ne **laser** comes out from the mirror with a beam diameter of 1mm and diverges to the far field. It is brought to focus by a convex lens of focal length of 17 mm. The spot size diameter of the beam at the focal point is

(a) 20 μm (b) 26 μm
(c) 52 μm (d) 13 μm

Common Data for Questions 14 and 15

Laser light is generated by energizing helium-neon gas in a chamber. The ground and metastable states of helium are 0 eV and 20.61 eV, respectively. The ground, higher and metastable energies of neon are 0 eV, 18.70 eV and 20.66 eV, respectively. The values of speed of light, Planck constant and charge on an electron are 3 × 10^8 *m/*s, 6.625 × 10^{-34} *Js* and 1.6 × 10^{-19} C respectively.

14. In this process, helium molecules

(a) Play no role
(b) Produce laser light
(c) Give energy to neon molecules
(d) Absorb energy from neon molecules

15. Wavelength of laser light generated in this process is

(a) 61.6 nm (b) 66.4 nm
(c) 633.8 nm (d) 650.3 nm

16. The **crossed** length of the He–Ne laser beam is 120 cm. Its coherence time in seconds is

(a) 4 × 10^{-1} (b) 4 × 10^{-3}
(c) 4 × 10^{-5} (d) 4 × 10^{-9}

17. The colour of a LED can be changed by

(a) using different band gap semiconductor
(b) by changing the doping level of the semiconductor
(c) by increasing applied voltage
(d) none of the above

18. Q-Switched Laser is a

(a) continuous Laser
(b) short burst Laser

(c) Laser produced by Switching operation
(d) low powered laser

19. In comparison to LED, Laser has
(a) no tuning arrangements
(b) higher emission efficiency
(c) narrow spectral width
(d) provision for confinement

20. Which of the semiconductor can be used to fabricate a LED.
(a) Si (b) Ge
(c) GaAs (d) none of these

21. Which of the following condition should be obeyed by every Laser
(a) rate stimulated emission should be greater than the rate of spontaneous emission
(b) there should be no spontaneous emission
(c) rate of stimulated emission should be equal to the rate of spontaneous
(d) there should be no stimulated emission

22. Which of the following is not applicable for Laser
(a) spatial coherence
(b) temperal cohorence
(c) both (a) and (b)
(d) none of these

23. Which of the following is not applicable for Laser
(a) no tuning arrangement
(b) higher emission efficiency
(c) narrow spectral width
(d) provision for confinement

24. The configuration utilized by a line communication system capable of transmitting a data at the rate of 100 Mb/s is
(a) coaxial cable system
(b) open line system
(c) optical fibre system
(d) twisted wire system

25. Which of the folliwing is more sensitive
(a) PIN diode
(b) APD
(c) neither (a) or (b)
(d) either (a) or (b)

26. Spectral width of LED is typically.
(a) 20–100 nm (b) 1–10 nm
(c) 500–1000 nm (d) 100–500 nm

27. The photo diode as compared to a photo transistor has:
(a) faster switching time
(b) lower sensitivity
(c) higher size for the same value of output current
(d) all of the above

28. A photo diode function as:
(a) photo conductive device while working with re verse voltage.
(b) photo voltaic devices while working without a reverse voltage.
(c) photo voltaic devces while working with or without reverse voltage
(d) none of the above.

29. Laser light (630 mm) incident on a pair of slits produces an interference pattern in which the bright fringes are separated by 8.3 mm. A second light produces an interference pattern in which the bright fringes are separated by 7.6 mm. What is the wavelength of this second light?
(a) 377 mm (b) 477 mm
(c) 577 mm (d) 677 mm

30. Find the radiated power from a source of area $10^{-5} m^2$ at a temperature of 2000 K.
Given the emissivity of surface is 0.7 .
(a) 4.35 W (b) 5.35 W
(c) 6.35 W (d) 7.35 W

31. Calculate the total radiant power output W of an LED chip dia 0.2 nm viewed from a distance of 1m, $\lambda = 550\, nm$

$\eta_e = 0.1\,\%$ operated at 2 V and 50 mA.
(a) 1.13×10^{-1} watts (b) 1.13×10^{-2} watts
(c) 1.13×10^{-3} watts (d) 1.13×10^{-4} watts

32. Coherence length given that n^{w} at 0.8 μm 1.5019 and n^{2w} at 0.4 μm is 1.4802 in KDP
(a) $10^{-1} m$ (b) $10^{-2} m$
(c) $10^{-3} m$ (d) $10^{-4} m$

33. Responsivity R_λ with a quantum efficiency $\eta = 0.25$ at $\lambda = 0.5\ \mu m$.
(a) 0.1 A/W (b) 0.2 A/W
(c) 0.3 A/W (d) 0.4 A/W

34. Minimum detectable signal for **pohotomultipiler** with bandwith 1 Hz .
(a) $2\times10^{-16}\, W$ (b) $4\times10^{-16}\, W$
(c) $6\times10^{-16}\, W$ (d) $8\times10^{-20}\, W$

35. What material dispersion over 1 km of silica based fiber when a Ga As LEO operating at 850 nm and with a 50 nm line width is used as a source is
(a) $1.3\times10^{-9} s$ (b) $2.3\times10^{-9} s$
(c) $3.3\times10^{-9} s$ (d) $4.3\times10^{-9} s$

36. A laser beam has a wavelength of 8×10^{-7} m and an aperture of 5×10^{-3} m . The laser beam is sent to moon. If the distance of moon from the earth is 4×10^{5} m , then angular spread of the beam is
(a) 1.6×10^{5} radians (b) 1.6×10^{-3} radians
(c) 1.6×10^{-4} radians (d) None of the above

37. A laser beam has a power of 50 nm and an aperture of 5×10^{-3} m and emits a light of wavelength 7200 A° units. The beam is focused with less of focal length of 0.1 m. The intensity of images is

(a) 24.1×10^5 W / m^2 (b) 24.1×10^5 W / m^2
(c) 241×10^5 W / m^2 (d) None of the above

38. A laser beam of $\lambda = 6000$ A° is focused by a mirror of diameter 2 m on to later on the moon. If the distance of the moon is 4×10^{-8}. The size of spot on moon is

(a) $1.44\times10^{-30}\,m^2$ (b) $3.3\times10^{-30}\,m^2$
(c) $0.33\times10^{-30}\,m^2$ (d) None of the above

Common Data for Questions 39 and 40

The coherence length of a Na light is $2.945\times10^{-2}\,m$. The λ of Na light is 5890 A° units. Calculate

39. The no of oscillations corresponding to the coherence length is

(a) 5×10^4 (b) 50×10^4
(c) 5×10^4 (d) None of the above

40. The Coherence time is

(a) 987×10^7sec (b) 9.87×10^7sec
(c) 98.7×10^7sec (d) None of the above

41. An unpolarized He - Ne laser 632.25 nm is reflected from a flat surface of a dielectric. A detector analysis the deflected beam. It is found that incident and reflected beam is 115°.34' The light is completely plane polarized. The reflection coefficient μ of dielectric at this wavelength is

(a) 1 (b) 1.33
(c) 1.414 (d) 1.58

42. A He - Ne laser $\lambda = 632.8$ nm is expanded to a 10 m, and then distance of moon from earth in 3.8×10^5 km. Diameter of central maximum of the moon is

(a) 271 m (b) 487 m
(c) 587 m (d) None of the above

43. Match List 1 with List 2 and select the answer using the codes given below:

List - I		List- II
A. NdYag	1.	514.5 nm
B. CO_2	2.	632.8 nm
C. Argon ion	3.	1.06 μm
D. He Ne	4.	10.6 μm

Code

	A	B	C	D
(a)	3	2	1	4
(b)	2	4	1	3
(c)	4	1	3	2
(d)	3	2	4	1

44. Match List 1 with List 2 and select the answer using the codes given below

List-I		List-II
A. Nd YAG laser	1.	10^9 Hz
B. He Ne laser	2.	10^{13} Hz
C. Semiconductor laser	3.	10^{11} Hz
D. LED	4.	10^{12} Hz
	5.	10^5 Hz

Codes

	A	B	C	D
(a)	3	5	1	4
(b)	2	4	1	5
(c)	5	3	1	4
(d)	3	5	4	1

45. The coherence length of the He-Ne laser bemas 120 cm. Its coherence time in seconds is

(a) 4×10^{-1} (b) 4×10^{-3}
(c) 4×10^{-5} (d) None of the above

46. A frequency stabilized He - Ne laser with a wavelength of 6328 A° has a bandwidth of 1 MHz. Its coherence length is

(a) 0.3 m (b) 3 m
(c) 30 m (d) None of the above

47. A He - Ne laser of cavity length 500 m has an oscillating bandwidth of 1500 MHz. The maximum number of longitudinal oscillating modes that is accommodated within bandwidth is

(a) 4 (b) 5
(c) 30 (d) None of the above

48. The output from a 633 nm He - Ne laser comes out from the mirror with a beam diameter of 1 mm and diverges to the far field. It is brought to a focus by a convex lens of focal length of 17 mm. The spot size diameter of the beam at the focal point is

(a) 20 μm (b) 26 μm
(c) 52 μm (d) None of the above

49. A He - Ne laser beam having a wavelength of 632.8 mm. is expanded to a diameter of 1 meter and then pointed towards the moon. If the distance of the moon from earth is 3.8×10^5 *km*, the distance of the central maxima on the surface of the moon is

(a) 1 *m* (b) 100 *m*
(c) 587 *m* (d) None of the above

50. Match List-I with List -II maximum output power and select the answer using the codes given below

List-I		List-II
A. He - Ne laser	1	10 *mW*
B. Argon ion laser	2	100 *mW*
C. Carbon dioxide laser	3	10 W
D. Diode laser	4	500 W

Code

	A	B	C	D
(a)	1	2	3	4
(b)	1	2	4	3
(c)	4	1	2	3
(d)	2	4	3	1

51. A laser beam ($\lambda = 632.8\ nm$) having a divergence of 1 milliradian is directed towards the moon at a distance of approximately 4×10^5 km. The beam would have spread to a diameter of

(a) 400 *km* (b) 200 *km*
(c) 100 *km* (d) None of the above

52. Match List I with List II and select the answer using the codes given below

List - I

A. Optical pumping
B. Electrical Discharge
C. Direct conversion
D. Chemical reaction

List - II

1. CO_2 laser
2. Semiconductor laser
3. Ruby laser
4. Argon lon laser

Codes

	A	B	C	D
(a)	3	4	2	1
(b)	1	2	3	4
(c)	4	3	2	1
(d)	3	4	1	2

Common Data for Questions 53 and 54

A laser beam at wavelength 700 nm and aperature 5×10^{-3} m is sent to moon which is 4×10^5 km away from the earth. Find

53. Angular spread is

(a) 1.70×10^5 (b) 1.54×10^{-4} rad
(c) 1.54×10^7 rad/sec (d) 1.70×10^{-4} rad

54. Area spread is

(a) $1.45\times10^{10}\text{m}^2$ (b) $1.45\times10^8\text{m}^2$
(c) $0.0145\times10^8\text{m}^2$ (d) $1.45\times10^{12}\ \text{m}^2$

55. A laser emits light near visible band. If $E_2 - E_1$ is 2.36 eV. The wavelength of the radiation is

(a) 5360 A° (b) 5460 A°
(c) 5256 A° (d) 5260 A°

56. A He - Ne laser emit light of wavelength 672.8 nm has an output power of 2.3 mW/sec. The number of photons emitted per minute by the laser source are

(a) 4.6×10^{17} photons/min (b) 4.2×10^{16} photons/min
(c) 9.8×10^{16} photons/min (d) 5.6×10^{17} photons/min

57. A laser beam of 50 m Watt and aperture 5 mm and $\lambda = 720$ nm is fused with lens of local length 0.1 m. The intensity of the image is

(a) $51.23\ MW/m^2$ (b) $52.8\ MW/m^2$
(c) $51.6\ MW/m^2$ (d) $50.32\ MW/m^2$

58. A He - Ne laser cavity length 500 nm an oscillating bandwidth (BW) of 15000 MHz. The maximum number of longitudnal oscillating modes that are accomodated within bandwidth is.

(a) 6 modes (b) 5 modes
(c) 8 modes (d) 12 modes

59. The longitudinal modes of a semiconductor laser at wavelength of 0.94 μm are separated by 200 GHz. The length of optical cavity (Refractive index of semiconductor laser = 3.3) is

(a) 0.05 mm (b) 0.25 mm
(c) 0.15 mm (d) 0.35 mm

60. Monochromaticity of laser means it has

(a) narrow width (b) wide width
(c) multiple frequency (d) single frequency

61. A laser beam consists of

(a) light material
(b) highty coherent photons
(c) electrons
(d) cosmic tays

62. What is the cut - off voltage or the minimum voltage above which light emission takes place in case of an LED.

(a) 0.7 volts (b) 1.3 volts
(c) 1.0 volts (d) 0.3 volts

63. A LED has a minimum drop of 1.5 V if the supply voltage be 10 V and $R_s = 470\ \Omega$, then maximum value of LED current will be

(a) 16 mA (b) 17 mA
(c) 18.1 mA (d) 20 mA

Common Data for Questions 64 and 65

When 3×10^{11} photons each with a wvelength of 0.85 μm are incident on a photodiode, on an average 1.2×10^{11} electrons are collected at the terminals of the device.

64. The quantum efficiency is

(a) 45% (b) 40%
(c) 50% (d) 60%

65. The responsivity of the photodiode at 0.85 μm is

(a) 0.385 AW^{-1} (b) 0.52 AW^{-1}
(c) 0.62 AW^{-1} (d) 0.274 AW^{-1}

66. A photodode has a quantum efficiency of 62 % when the photons of energy 2×10^{-19} joules is allowed to incident upon it. The incmdent optical power required to obtain a photocurrent of 2.8 μA is

(a) 3.24 mW (b) 6.64 mW
(c) 1.21 mW (d) 7.84 mW

67. A photodiode mad eof GaAs has a energy gap of 1.43 eV . The wavelength at whcih it will start to operete is
(a) 0.68 *μm* (b) 0.87 *μm*
(c) 0.73 *μm* (d) 0.84 *μm*

68. With the increase in the illumination across a photoresistor, its resistance
(a) increases linearly
(b) decreases linearly
(c) increases exponantially
(d) decreases exponantially

69. A Cds cell has a dark resistance of 100 kW and a resistance of 30 kW in a day light beam. If the time constant is 72 m sec then the resistance of the cell after 10 msec of application of a beam is.
(a) 39.1 *k*Ω (b) 38.6 *k*Ω
(c) 41.2 *k*Ω (d) 28.1 *k*Ω

70. A CdS cell has heighest response for a light for wavelength of nearly
(a) 300 nm (b) 800 nm
(c) 500 nm (d) 600 nm

71. Photoconductive cell is useful for
(a) high frequency applications
(b) medium frequency applications
(c) low frequency applications
(d) all of the above

1.6 Interferometers

1. An optical flat is tested against a standard reference flat in a Newton's interferoscope. If the optical flat is as good as the reference, the fringes observed are a set of:
(a) Concentric rings
(b) Elliptical rings
(c) Hyperbolic fringes
(d) Straight and parallel fringes.

2. In a wave propagation, a path difference of x_o corresponds to a phase difference of
(a) λx_o (b) πx_o
(c) $\left(\frac{2\pi}{\lambda}\right)x_o$ (d) $\frac{x_o}{\lambda}$

3. A **laser** beam (λ = 632.8 nm) having a divergence of 1 milli-radian is directed towards the moon at a distance of approximately 4 × 10^5 km. The beam would have spread to a diameter of
(a) 400 km (b) 200 km
(c) 100 km (d) 4 km

4. Match the List I instrument to the List II characteristic and select the correct answer using the codes:

List I	List II
A. Interferometer	1. Wavelength dispersion
B. Monochromator	2. Magnifying power
	3. Flatness measurement

Codes:

	A	B
(a)	3	2
(b)	1	2
(c)	3	1
(d)	2	1

5. For a change in elevation level of 1.5 μm between an **optical** flat and a test surface, a light wave of wavelength 0.5 μm produces fringes. The number of fringes is
(a) 3
(b) 6
(c) 9
(d) 12

6. Koster's modification of Michelson **interferometer** is used to determine the thickness of a gage G. The gage G is placed on the mirror M_2 as shown in figure (a). The field of view observed through the lens L is shown in figure(b). The fractional fringe shift e as shown in figure (b), is measured to be 0.15 when the wavelength of light used is 0.63 μm. A quick check with a micrometer shows that the gage length is approximately 100 μm. Then an accurate estimate of gage length, in μm, is

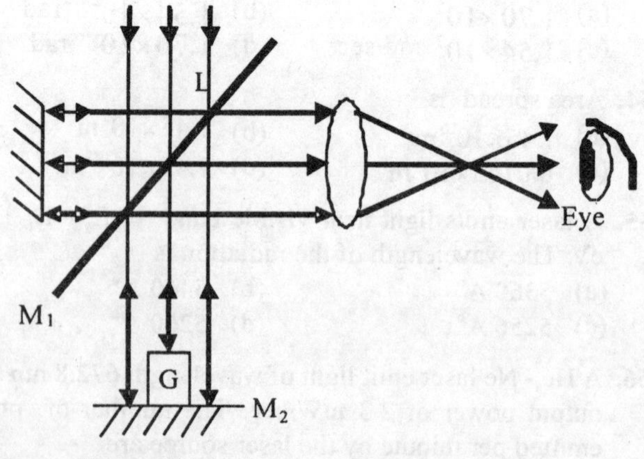

Figure (a)

Figure (b)

(a) 99.902 (b) 100.804
(c) 100.000 (d) 101.706

7. In a Michelson's interferometer illuminated by a coherent and quasi-monochromatic light source, a very small frequency shift is introduced in the light returning from one of the arms. The effect on the observed fringe pattern is

(a) A decrease in contrast
(b) An increase in contrast
(c) Appearance of two sets of the same pattern
(d) Uniform movement across the observation plane

8. A glass plate is overcoated with a thin film as shown below. The reflectance of the combination is monitored with 600 nm wavelength. If n_g and n_f are refractive indices of the glass plate and film material, respectively with $n_g > n_f$, then reflectance reaches a minimum when the geometrical film thickness equals

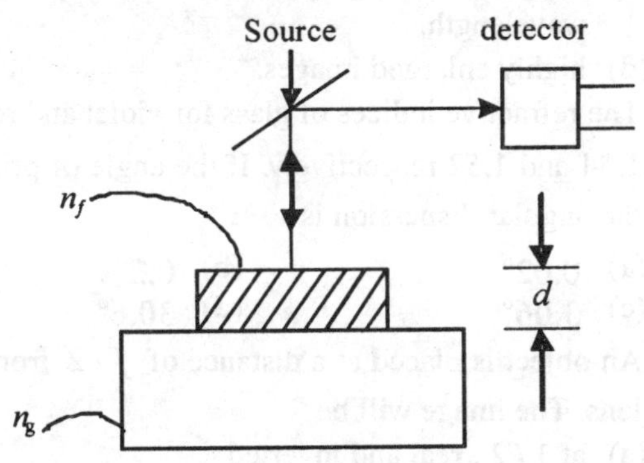

(a) 600 nm (b) 300 nm
(c) 150 nm (d) 100 nm

9. A tissue with a refractive index 1.33 is introduced in one of the light paths of a Michelson interferometer operating with a monochromatic coherent light source of wavelength 589 nm. After the introduction of a tissue sample of thickness Δt, the fringe pattern will shift by 50 fringes. If the thickness is 2 Δt, the frignge pattern is observed to shift by

(a) 25 fringes (b) 50 fringes
(c) 100 fringes (d) 200 fringes

Statement for Linked Answer Questions 10 and 11

A Michelson interferometer illuminated with a source of central wavelength λ_o and spectral width Δl is adjusted for equal path difference for the beams returning from the two mirrors. When one of the mirrors is moved by a distance of 0.1 mm from this position, 300 fringes move past the field of view. When the mirror is moved further, the fringes completely disapper when the mirror is approximately 4 cm from the initial position.

10. The central wavelength of the source is

(a) 540 nm (b) 632.8 nm
(c) 667 nm (d) 720 nm

11. The spectral width of the source $\Delta\lambda$ is approximately

(a) 0.0056 nm (b) 0.0100 nm
(c) 0.0500 nm (d) 0.1000 nm

12. Two narrow straight slits, 0.25 mm apart are illuminated by an unknown monochromatic source of light. If the fifth bright fringe is 12 mm away from the central fringe when the screen constant is 0.8, the wavelength of the unknown source is

(a) 540 nm (b) 600 nm
(c) 700 nm (d) 750 nm

13. To produce a magnified image atleast distance of distinct vision with the help of a simple microscope, the object should be placed at:

(a) focus
(b) a point far away form focus
(c) a point between focus and optical centre
(d) none of the above

14. If the least distance of distinct vision is 25 cm, then a convex lens of focal length 5 cm acts as a magnifier of magnifying power.

(a) 5 (b) less then 5
(c) 6 (d) more then 6

15. The final image formed by a microscope is:

(a) real and magnified (b) real and diminished
(c) virtual and magnified (d) virtual and diminished

16. When the length of a microscope tube is increased, its maginfying power.

(a) decreases (b) increases
(c) does not change (d) cannot be predicted

17. To increase the magnifying power of a telescope, one should increase

(a) the focal length of the objective
(b) the focal lenght of eye piece
(c) aperture of the objective
(d) aperture of the eye piece

18. A telescope in normal adjustment has a distance between the objective and eye-piece equal to 100 cm. If the

magnifying power of the telescope is 24, the focal length of the two lenses are
(a) 96 cm, 4 cm (b) 124 cm, 100 cm
(c) 100 cm, 76 cm (d) 62 cm, 38 cm

19. An observer looks at a tree of height 15 m with a telecope of magnifying power 10. To him, tree appears.
(a) 10 times taller (b) 15 times taller
(c) 10 time nearer (d) 15 times nearer

20. Magnifying power of a telescope is *m*. If the focal length of the eye-peice is doubled, then its magnifying power will become.
(a) 2 m (b) m/2
(c) $\sqrt{2}m$ (d) 3 m

21. A film projector lense magnifies 10 cm square film strip on a screen. If the linear magnification is 4, the area of the magnified film on the screen is:
(a) 1600 cm^2 (b) 400 cm^2
(c) 800 cm^2 (d) 400 cm^2

22. The resolving power of telescope depends upon.
(a) the focal length of the eye lens
(b) the focal length of the object lens
(c) the length of the telescope
(d) the diameter of the object lens.

23. A fly is sitting on the objective of a telescope pointed to the moon. What effect is expected in the photograph of the moon taken through the telescope.
(a) The entire yield of vision is blocked.
(b) There is an image of the fly on the photograph
(c) There is no effect at all
(d) the image of the moon is of lesser intensity.

24. For observing a circket match, binocular is preferred to terrestrial telescope for the reason that
(a) the binocular is very easy to handle
(b) the binocular provides three dimenison vision
(c) the binocular produces image free chromatic aberration
(d) the binocular produces erect image

25. Angular magnitication of a telescope if α = angle subtended by object at eye and β = angle subtended by image at eye.
(a) $\frac{\alpha}{\beta}$ (b) $\alpha\beta$
(c) $\frac{\beta}{\alpha}$ (d) $\frac{\alpha}{\beta^2}$

26. Resolving power of a telescope is proportional to:
(a) $\frac{\lambda}{a}$ (b) λa
(c) $\frac{a}{\lambda}$ (d) $\frac{\lambda}{a^2}$

27. When light passes form a medium of refractive index n_1 into a medium of refractivbe index n_2, reflection at boundary.
(a) $(n_2 - n_1)^2$ (b) $\left(\frac{n_2 - n_1}{n_2 + n_1}\right)^2$
(c) $\frac{1}{(n_2 - n_1)^2}$ (d) $(n_2 + n_1)^2$

28. Angle of minimum deviation for a prism of refractive index 1.5 is equal to the angle of prism. The angle of prism is:
(a) 62° (b) 41°
(c) 82° (d) 31°

29. The refracting angle of a prism is A and the retractive index of the prism is cot A/2. The angle of minimum deviation is:
(a) $180° - 3A$ (b) $180° + 2A$
(c) $90° - A$ (d) $180° - 2A$

30. A ray is incident at an angle of incidence is on one surface of a prism of small angle A and emerges normally from the opposite surface. If the refractive index of the material of the prism is μ, the angle of incidence *i* is nearly equal to:
(a) A/μ (b) $A/2\mu$
(c) μA (d) $\mu A/2$

31. An achromatic combination of lenses produces
(a) images in black and white
(b) colouled images
(c) images unaffected by variation of refractive index with wavelength.
(d) highly enlarged images.

32. The refractive indices of glass for violet and red light are 1.54 and 1.52 respectively. If the angle of prism is 10°, the angular dispersion is
(a) 0.02° (b) 0.2°
(c) 3.06° (d) 30.6°

33. An object is placed at a distance of $f/2$ from a convex lens. The image will be
(a) at 3 $f/2$, real and inverted
(b) at $2f$, virtual and erect
(c) at *f*, real and inverted
(d) at one of the foci, virtual and double its size.

34. A film projector magnifies a 100 cm^2 film strip on a screen. If the lineat magnification is 4, the area of the magnified film on the screen is:
(a) 1600 cm^2 (b) 400 cm^2
(c) 800 cm^2 (d) 200 cm^2

35. A converging lens is used to form an image on a screen. When the upper half of the lens is convered by an opaque screen.

(a) half the image will disappear
(b) complete image will be formed
(c) intensity of image will decrease
(d) intensity of image will increase

36. If in a plano convex lens radius of curvature of convex surface is 10 cm and the focal length of the lens is 30 cm, the refractive index of the material of the lens will be:
(a) 1.5 (b) 1.66
(c) 1.33 (d) 3

37. Focal length of a convex lens of refractive eindex 1.5 and 2 cm. Focal length of lens when immersed in a liquid of refractifve index of 1.25 will be
(a) 10 cm (b) 7.5 cm
(c) 5 cm (d) 2.5 cm

38. A convex lens of glass is immersed in water. Compared to its poower in air, its power in water will
(a) diminish for red light (b) remain the same
(c) increase (d) diminish

39. A lens behaves as a converging lens in air and a diverging lens in water. The refractive index of the material of the lens is:
(a) equal to unity
(b) equal to 1.33
(c) between unity and 1.33
(d) greater than 1.33

40. Two lense having powers + 6D and $-4D$ are placed in contact. The power of the combination is:
(a) $-2D$ (b) $-4D$
(c) $+4D$ (d) $+2D$

41. Two coherent sources are virtual in:
(a) Young's double slit experiment
(b) Leoyd's mirror
(c) Fresnel's prism
(d) All of the above

42. A convex mirror froms an image which is $\frac{1}{n}$ times the size of the object. I the focal length of the mirror is f, then the distance of the object is:
(a) nf (b) f/n
(c) $\frac{n-1}{f}$ (d) $(n+1)f$

43. Focal length of concave mirror produicing magnification m is:
(a) $\frac{(m-1)f}{m}$ (b) $\frac{(m+1)f}{m}$
(c) $\frac{(m-1)}{f}$ (d) $\frac{(m+1)}{f}$

44. Time taken by the sum light to pass through window of thickness 4 mm whose refractive index is 1.5 is
(a) $2\times10^{-8}\ s$ (b) $2\times10^{8}\ s$
(c) $2\times10^{-11}\ s$ (d) $2\times10^{11}\ s$

45. As a plane wave front propagates, its radius of curvature.
(a) decreases
(b) increases
(c) first increases then decteases
(d) remains infinity

46. True about a plane wave front propagating in a medium.
(a) it propagates parallel to itself
(b) it cannot propagate in medium
(c) it changes to spherical wavefront
(d) it changes to cylindrical wavefront

47. Width of diffraction fringes varies
(a) Directly as distance between slit and screen
(b) Inversely as wavelength
(c) Directly as width of slit
(d) None of the above

48. Replacing red light by blue light in a single slit diffraction pattern.
(a) No change in diffraction pattern
(b) Diffraction pattern disappears
(c) Diffraction fringes become narrower and crowded together.
(d) Diffraction fringes become broader and farther about

49. Scattering of light is a direct consequence of.
(a) Interference (b) Dispersion
(c) Diffraction (d) Reflection

50. What is the μ of air with a shift of 150 fringes of wave lenght 4000 Å in a **Michelson interferometer** when all air is removed tube length is 20 cm.
(a) 1.00000 (b) 1.00015
(c) 1.00050 (d) 1.00075

51. In **Young's interference experiment** two slits are illuminated with orange light of wavelength 6000 Å. The interferece pattern is observed on a screen very far from the slits. If the central bright fringe is numbered zero. What must be the path difference for the light from the two slits at the fouth-right fringe?
(a) 1.4 μm (b) 2.4 μm
(c) 3.4 μm (d) None of the above

52. Two wavelengths of light λ_1 and λ_2 are sent through a **Young's double-slit apparatus** simulrancously. What most be true concerning λ_1 and λ_2 if the third - order λ_1 bright fringe is to coincide with the fourth - order λ_2 fringe?

(a) $\lambda_1 = \lambda_2$ (b) $\lambda_1 = 2\lambda_2$

(c) $\lambda_1 = 3\lambda_2$ (d) $\lambda_2 = \frac{3}{4}\lambda_1$

53. Find the sine of the crictical angle of incidence for an air-water interface.

(a) 0.550 (b) 0.650

(c) 0.750 (d) 0.850

54. Two wavelengths λ_1 and λ_2 are used in the **double-slit experiment**. If one is 430 nm, what value must the other have for the fourth-order bright fringe of one to fall on the sixth-order bright fringe of the other?

(a) $2\lambda_1 = \lambda_2$ (b) $3\lambda_1 = 2\lambda_2$

c) $4\lambda_1 = 6\lambda_2$ (d) $\lambda_1 = 3\lambda_2$

55. Two identical radiators have a separartion $d = \lambda/8$, where λ is the wavelength of the waves emitted by either source. The phase diffrence $\Delta\phi$ of the sources is $\pi/4$. Find the **intensity distribution** in the radiatioin field as a function of the angle θ which specifies the direction from the radiators to the distant observation point P.

(a) $I_0 \sin^2\left[\frac{\pi}{8}(\sin\theta + 1)\right]$

(b) $4I_0 \cos^2\left[\frac{\pi}{8}(\sin\theta + 1)\right]$

(c) $I_0 \cos^2\left[\frac{\pi}{8}(\sin\theta + 1)\right]$

(d) $4I_0 \sin^2\left[\frac{\pi}{8}(\sin\theta + 1)\right]$

56. A flake of glass of index of refraction 1.5 is placed over one of the openings of a **double-slit apparatus.** There is a displacement of the interference pattern through seven successive maxima toward the side where the flake was placed. If the wavelenght of the diffracted $\lambda = 600\ nm$, how thick is the flake?

(a) 7400 *nm* (b) 8400 *nm*

(c) 9400 *nm* (d) 1040 *nm*

57. What is the wavelenght of light when inital and final reading of a **Michelson interferometer** screw are 10.7347 mm and 10.7051 mm with 100 fringes pan.

(a) 4920 Å (b) 5920 Å

(c) 6920 Å (d) 7920 Å

58. What is the wavelenths in the region 4000 Å to 6000 Å for which interface minimum exists in a **Fabry Periot interferometer** with a plate separation of 2×10^{-6} and normal incidence of light

(a) 5714 Å, 5000 Å, 4444 Å

(b) 5714 Å, 6000 Å, 4444 Å

(c) 5714 Å, 6000 Å, 5555 Å

(a) 5714 Å, 7000 Å, 5555 Å

59. A thin parallel film of thickness 0.2 mm is formed using a liquid of refractive index 1.25. It is illuminated with a light of wavelength ranging from 4000 to 5000 Å. It is observed at an angle of 35°. Determine the wavelenghs, absent in reflected light.

(a) 1442 *A* (b) 2442 *A*

(c) 3442 *A* (d) 4442 *A*

60. A film of liquid of refractive index 1.39 is formed and illuminated with white light. When it is observed at an angle of 30°, two consecutive dark bands are observed corresponding to wavelengths 5125 *A* and 5000 *A*. Determine the thickness of the film.

(a) $5.9\times10^{-4}\ cm$ (b) $6.9\times10^{-4}\ cm$

(c) $7.9\times10^{-4}\ cm$ (d) $8.9\times10^{-4}\ cm$

61. Wavelength of light with shift of 400 fringes when movable mirror of **Michelson interferometer** moves a distance of 0.1 mm.

(a) 4000 Å (b) 5000 Å

(c) 6000 Å (d) 7000 Å

62. Through what distance does the mirror in **Michelson interferometer** is to be moved between 2-positions of disapperance of frings with a source which consists of 2-wavelenghts 4882 Å and 4886 Å.

(a) 0.049 mm (b) 0.059 mm

(c) 0.069 (d) 0.079 mm

63. Thickness of a thin glass plate ($\mu = 15$) introduced in one of the arms of a **Michelson interferometer** using 5890 Å with a shift of 10 fringes.

(a) 4.89 μm (b) 5.89 μm

(c) 6.89 μm (d) 7.89 μm

64. In a certain Young's double-slit experiment for which $D = 1.00m$ and 0.10 cm. bright fringes are 0.5 mm a part. What wavelength of light is being used?

(a) 100 mm (b) 200 mm

(c) 300 mm (d) 500 mm

65. A Young's double-slit exeriment is set up using a light source of wavelength 500 *nm* and placing slits 2 m from an observer. Given that the angular resolution of the observer's eye is 1 in (0.000291rad), how far a part are the two slits. If he can just distinguish between the interference frings

(a) 1.32 mm (b) 1.42 mm

(c) 1.62 mm (d) 1.72 mm

66. In a Young's double-slit experiment, the slits are 2 mm apart and are illuminated with a mixture of two wavelengths, $\lambda = 750nm$ and $\lambda' = 900nm$. At what minimum distance from the common central bright fringe on a screen 2 m from the slits will a bright fringe from one **interference pattern** coincide with a bright fringe from the other?

(a) 1.5 mm (b) 2.5 mm
(c) 3.5 mm (d) 4.5 mm

67. Find the wave length of light incident on a convex lens of 350 cm (R) placed on a flat plate and illuminated by monochromatic light and dia of 6th bright ring as 0.68 cm.

(a) 3080 Å (b) 4080 Å
(c) 5080 Å (d) 6080 Å

68. Find the angle of wedge when light of $\lambda = 6000$ Å incident normally on a wedge shaped film ($\mu = 1.35$) and fringes are 2 mm apart.

(a) 0.0032° (b) 0.0064°
(c) 0.0048° (d) 0.0080°

69. In Fresnels Biprism experiment on putting a thin plate of glass in path of one of interfering beams, centeral bright fringe shifts into posiotion pereviously occupied by 5th bright fringe, $\lambda = 6000$ Å, refractive index of glass plate = 1.5. Thickness of plate is:

(a) 4 μm (b) 5 μm
(c) 6 μm (d) 7 μm

70. In a **Doppler - shift flow velocity meter**,. If f is the transmitted frequency,

C is velocity of sound in blood and

θ is angle of inclination of incident wave to direction of blood flow

Δf is doppler shift,

the velocity of blood cells is given as.

(a) $\dfrac{2f\,\Delta f\,C}{\cos\theta}$ (b) $\dfrac{2f\,C}{\Delta f\cos\theta}$

(c) $\dfrac{\Delta f\,C}{2f\cos\theta}$ (d) $\dfrac{2f}{\Delta f C\cos\theta}$

71. In **interference** pattern due to **young's double slit arrangement**, maximas are located by

(a) $y = \dfrac{D\lambda}{d}$ (b) $y = \dfrac{m\,D\lambda}{d}$

(c) $y = \dfrac{m\lambda}{d}$ (d) $\dfrac{md}{\lambda}$

72. The **interference pattern** of two identical slits separated by a distance $d = 0.25mm$ is observed on a screen at a distance of 1 m from the plane of the slits. The slits are illuminated by monochromatic light of wavelength 589.3 mm travelling perpendicular to the plane of the slits. Bright bands are observed on each side of the central maximum. The separation between adjacent bright bands is:

(a) 1.86 mm (b) 1.96 mm
(c) 2.16 mm (d) 2.36 mm

73. In **Newton's rings experiment** with air film the diametr of the fifth brignt ring is 0.5 cm. The wavelength of light used is 5893 A. The light enters the air film with the angle of 30°. Determine the radius of curvature of lens.

(a) 104 cm (b) 144 cm
(c) 184 cm (d) 204 cm

74. Find the Width of central band if slit width 0.6 mm when wavelength of light used is 6000 Å in a **Fraunhofer diffraction** due to single slit screen is 1 m away from slit.

(a) 1 nm (b) 2 nm
(c) 3 nm (d) 4 nm

75. A diffraction grating has 6000 lines per cm. The diffracted second oder spectral line is observed at 50°. Calculate the wavelength of radiation.

(a) 4383 Å (b) 5383 Å
(c) 6383 Å (d) 6383 Å

76. A transmission grating has 15000 LPI. What is the **maximum order of diffraction** that can be observed when a light of wavelength 6000 Å is used?

(a) $n = 1$ (b) $n = 2$
(c) $n = 3$ (d) $n = 4$

77. Calculate the ratio of **intensity** of central maxima to frist secondary maxima.

(a) 0.025 (b) 0.035
(c) 0.045 (d) 0.055

78. The light of wavelength 6400 Å is used to obtain **single-slit Fraunhofer diffraction patron.** The slit width is 0.3 mm. The screen is at a distance of 110 cm from the slit. Calculate the distance between the centre of the central maximum and the third dark fringe (third piont zer intensity).

(a) 5 mm (b) 6 mm
(c) 7 mm (d) 8 mm

79. A ray of light is incident on the left vertical face of a glass cube of **refractive index** n_2. The plane of incidence is the plane of the page, and the cube is surrounded by liquid (n_1). What is the largest angle of incidnece θ_1 for which total internal reflection occurs at the top surface?

(a) $\sqrt{\dfrac{n_1}{n_2} - 1}$ (b) $\sqrt{\dfrac{n_1}{n_2} + 1}$

(c) $\sqrt{\left(\frac{n_1}{n_2}\right)^2 - 1}$ (d) $\sqrt{\left(\frac{n_2}{n_1}\right)^2 - 1}$

80. If the speed of light in ice is $2.3\times10^8 m/s$. what is its **index of refraction**? What is the **critical angle** if incidence for light going from ice to air.
(a) 1.1 and 20° (b) 1.304 and 30°
(c) 1.3 and 40° (d) 1.304 and 50°

81. The index of refraction of a flint glass is 1.64. Find the speed of light in the glass, the **angle of refraction** in glass if light is incident from water at an angle of 44°, and the cirtical angle of incidence for a glass-water interface.
(a) $1.83\times10^8 m/s$ and 14.4°
(b) $1.38\times10^8 m/s$ and 24.4°
(c) $1.83\times10^8 m/s$ and 34.4°
(d) $1.38\times10^8 m/s$ and 44.4°

82. Find the Angle of incidence when a ray of light shines down vertically on the mirror which lies face up making an angle of 15° with horizontal.
(a) 10° (b) 15°
(c) 20° (d) 25°

83. Find Angle between reflected ray, and the horizontal in above situation.If **Find the Angle of incidence** when a ray of light shines down vertically on the mirror which lies face up making an angle of 15° with horizontal.
(a) 60° (b) 90°
(c) 120° (d) 150°

84. How many wavelengths thick is the plate is terms of beam within it ifA beam of 436-nm light is incident perpendicular to a glass plate that is 2.0 cm thick and has $n=1.66$.
(a) 53127 (b) 43717
(c) 53417 (d) 71647

85. What is the Critical angle of indicence when a ray light passes from crown glass to water A beam of 436-nm light is incident perpendicular to a glass plate that is 2.0 cm thick and has $n=1.66$.
(a) 41.3° (b) 51.3°
(c) 61.3° (d) 71.3°

86. What is the angle of refraction when angle of incident in glass is 55° .A beam of 436-nm light is incident perpendicular to a glass plate that is 2.0 cm thick and has $n=1.66$.
(a) 59° (b) 69°
(c) 79° (d) 89°

87. Interferometer is the heart of
(a) Visible - ultraviolet Spectrometer
(b) IR Spectrometer
(c) FTIR Spectrometer
(d) Mass Spectrometer

88. Green light of λ 500 Å from a narrow split is incident on a double slit. If an overall separation of 10 fringes on screen 200 cm away in 2 cm. The separation is
(a) 0.51 *cm* (b) 5.1 *cm*
(c) 0.051 *cm* (d) None of the above

89. In a young's double slit experiment, the separation of the slit is 1.99 mm and the fringe spacing is 1 mm of a distance of 1 m from the slits. The wavelength of light is
(a) 5000 Å (b) 5980 Å
(c) 5098 Å (d) None of the above

90. Two straight and narrow slits 1 mm apart are illuminated by monochromatic light, fringes formed on the screen is held at a distance of 100 cm from the slit are 0.05 apart. The wavelength of light is
(a) 500 Å (b) 500 Å
(c) 50 Å (d) None of the above

91. A light source emits two wavelength $\lambda_1 = 4300$ Å and $\lambda_2 = 5100$ Å the source is used in a young's double slit interference experiment the distance between source and screen is 1.5 m and the distance between slits in 0.025 mm. The separation between the 3rd order bright fringe due to these two wavelengths is
(a) 14.4 *cm* (b) 144 *cm*
(c) 1.44 *cm* (d) None of the above

92. In moving are mirror in Michelson interferometer through a distance of 0.1474 mm. 500 fringes cross the center of the field of the view. The wavelength of light used is
(a) 5698 Å (b) 5986 Å
(c) 6589 Å (d) None of the above

Common Data for Questions 93 and 94

A monochromatic source of radiation of wavelength 5500 Å from a narrow slit is incident on a double slit the overall separation of five fringes on a screen 200 cm away is 1 cm.

93. The fringe width is
(a) 0.055 cm (b) 0.2 cm
(c) 2 cm (d) 3 cm

94. The slit separation is
(a) 0.2 cm
(b) 0.055 cm
(c) 5.5 cm
(d) 0.55 cm

95. In Michelson interferometer 200 fringes cross the field of view when the mobable mirror is displaced through 0.589 mm. The wavelength of transmitting light is used is
(a) 50.98 Å (b) 58.90 Å
(c) 59.80 Å (d) 50.00 Å

96. Fringes of equal inclination are observed in Michelson interferometer as one of the mirror is move by 1 mm 3663

fringes move out of from center of pattern. The wavelength is
(a) 50.46 Å (b) 56.40 Å
(c) 54.60 Å (d) 45.60 Å

97. In Michelson interferometer 100 fringes cross the field of view when the movable mirror is displaced by 0.22948 mm. The wavelength of monochrome light is
(a) 458.96 Å (b) 497.65 Å
(c) 456.78 Å (d) 485.96 Å

98. When a movable mirror Michelson interferometer is moved through a distance of 0.03 mm, 100 fringes are displaced. The wavelength of light used is
(a) 6000 Å (b) 600 Å
(c) 60 Å (d) 6 Å

99. In an experiment with Michelson interferometer the distance traveled by a mirror for two successive position of maximum distance was 0.2945 *mm*. If the mean wavelength of the two components of Na beams light is 5893 A°. The difference between them is
(a) 1 Å (b) 10 Å
(c) 100 Å (d) 589 Å

100. In a Michelson Interferometer 100 fringes cross the field of view. When the movable mirror is dislaced through 0.294 mm. The wavelength of source of radiation is used in experiment is
(a) 588 Å (b) 5880 Å
(c) 58.8 Å (d) 58 Å

101. In an experiment for determining the μ of air with Michelson interferometer a shift of 150 fringes is observed when all the air was removed from the tube. If the λ of light used is 4000 A° in air and the length of the tube is 20 cm. The Refractive index μ of air is
(a) 1 (b) 1.5
(c) 1.15 (d) 1.015

102. When all the gas removed from a tube, if λ of light used is 5460 Å and length of tube is 20 cm,. The refractive index of gas is
(a) 1.02 (b) 1
(c) 1.5 (d) 1.05

103. Two narrow straight slips 0.25 nm apart are illuminated by unknown monochromatic source of light of the 5th bright fringe is 12 mm away from central from where screen list is 0.8 m. The wave length λ of unknown source is
(a) 0.75×10^{-3} (b) 75×10^{-3}
(c) 750×10^{-3} (d) 7500×10^{-3} Å

104. Straight fringes are observed in a Michelson interferometer with a light of wavelength of 4800 Å in are of the paths liquid film of gradually changing thickness is introduced and 500 fringes cross the field of view. If μ of liquid is 1.40. The change in thickness of film is
(a) 0.03 *cm* (b) 0.003 *cm*
(c) 0.3 *cm* (d) 3 *cm*

105. A turn plate of μ 1.5 displacement 10 fringes when it si introduced is on of the arms of Michael son interferometer. The thickness of the plate is
(a) 6×10^{-5} (b) 5×10^{-4}
(c) 5×10^{-5} (d) 6×10^{-4}

106. A parallel beam of light (λ – 515 nm) is incident on a flint glass of refractive index 1.6. The angle is $45°$. An interference pattern is observed in reflected light. The thickness of the plate if dark fringe observed is
(a) 1.138 nm (b) 11.38 nm
(c) 113.8 nm (d) 1138 nm

107. Match List-I instrument with List-II the characteristic and select the answer using the codes given below

List - I
A. Interferometer
B. Monochromator

List-II
1. Wavelength dispersion
2. Magnifying power
3. Flatness measurement

Codes

	A	B
(a)	2	1
(b)	3	2
(c)	3	1
(d)	2	3

108. A quassi-monochromatic light source with average central wavelength of 630 nm has a spectral width of 10 GHz. The coherence length of the source is
(a) 3 *μm* (b) 3 *mm*
(c) 3 *cm* (d) 3 *m*

109. Brewster angle is the angle of incidence of monochromatic light at the interface of two media, for which
(a) the incidence beam would be internally reflected.
(b) a polarized beam would become unppolarised after reflection.
(c) an unpolarised beam would become polarized after reflection.
(d) all the above

110. The narrow straight slits, 0.25 mm apart are illuminated by an unknown monochromatic source of light. If the fifth fringe is 12 mm away from the central fringe when the screen distance is 0.8 m. The wavelength of the unknown source is

(a) 540 *nm* (b) 600 *nm*
(c) 700 *nm* (d) 750 *nm*

111. A Microscope uses a micro - objective 10X, numberical aperture 0.25 and an eyepiece of focal length of 25 mm. The magnification of the microscope is

(a) 25 (b) 50
(c) 100 (d) 125

112. In a Michelson interferometer 200 fringes cross the field, when movable mirror is displacewd through 0.0589 mm.The wavelength of the monochromatic light is

(a) 580 nm (b) 582 nm
(c) 589 nm (d) 602 nm

113. A movable mirrror michelson interferometer is moved through a distance of 0.02603 nm. The number frnges shifted if the wavelength is 5246 A is

(a) 100 (b) 200
(c) 300 (d) 400

114. A transparent glass sheet with $\mu = 1.5$ is indroduced with wavelength of source 4800 A°. This causes 500 fringes to cross the field. The thickness of the film is

(a) 102 μm (b) 206 μm
(c) 240 μm (d) 320 μm

115. When thin glass sheet with thickness 3.8×10^{-3} *cm* is placed in the interfering beam is michelson interferometer. It is found that 5 fringes are shifted with a alight of wavelength equal to 6.4×10^{-4} cm. The refractive index of the glass sheet is

(a) 1.42 (b) 1.38
(c) 1.46 (d) 1.34

116. The distance between two successive position of the movable mirror of michelosn interferometer giving best fringes in the case of sodium source with lines of $\lambda = 5890$ A° and 5896 A° is

(a) 280 nm (b) 282 nm
(c) 308 nm (d) 289 nm

117. In a Michelson interferometer scale readings for a pair of maximum brightness are around 0.6939 mm and 0.9884 mm. If the mean wavelength of two lines is 5893 A°. The difference between the wavelength $(\lambda_1 - \lambda_2)$ is

(a) 5.896 A° (b) 5.625 A°
(c) 6.210 A° (d) 6.411 A°

118. When a thin glass sheet with thickness 4.2×10^{-3} *cm* is inserted the interfering beam tath in a Bi - prism arrangement. It is found that 5 fringes are shifted with a light of wavelength $\lambda = 6.3 \times 10^{-4}$ cm. The refractive index of the glass sheet is

(a) 2.45 (b) 3.25
(c) 1.75 (d) 1.95

119. The number of fringes over a distance 'L' is 10. The source of light is cadmium having wavelength equal to ($\lambda = 9.426\ \mu m$).

The difference in height 'd' between two blocks as shown in the figure given below is

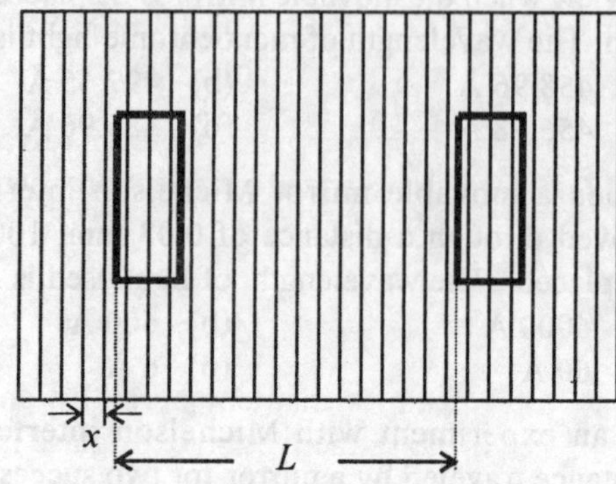

(a) 47.13 μm (b) 46.22 μm
(c) 44.87 μm (d) 45.84 μm

120. The diameters of two rollers, each having approximatedly 12 mm are to be compared by teh means of interference fringes produced by an optical flat and the monochromatic light of wavelength 0.8 μm as given in the figure.

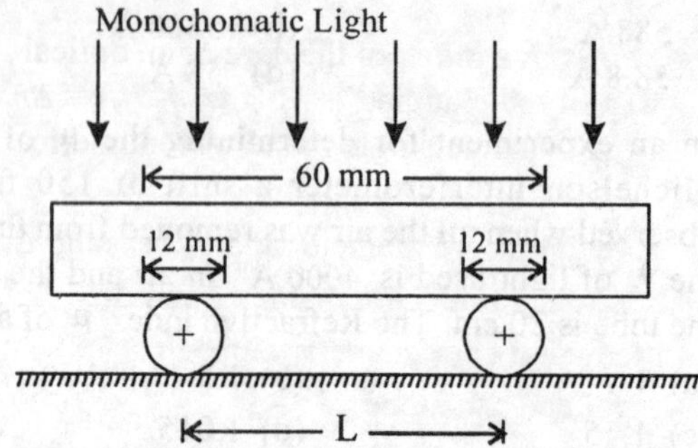

Ten fringes are seen in a length of 24 mm over the centre portion of the flat. The difference between the two diameters is

(a) 8 μm (b) 9 μm
(c) 5 μm (d) 10 μm

121. A block placed on a flat surface was observed with the help of an optical flat in the monochromatic light of wavelength 0.622 μm. It was observed that there are $3\frac{1}{2}$ fringes along the length and $6\frac{1}{2}$ fringes long the base. The change in thickness along its length is

(a) 0.853 μm
(b) 0.748 μm
(c) 0.993 μm
(d) 0.984 μm

1.7 Fiber Optics

1. The refractive indices of glass and water are 1.50 and 1.33, respectively. If the glass is immersed in water its relative **refractive** index is

(a) 0.89 (b) 1.13
(c) 1.99 (d) 2.83

2. Light is incident on a glass plane of **refractive index** 1.732. It is plane polarized on reflection when the incident angle is:

(a) 30° (b) 45°
(c) 60° (d) 75°

3. A parallel beam is incident normally on a air glass interface. If the refractive index of glass is 1.720, then the loss of intensity through reflection is:

(a) 7% (b) 4%
(b) 2% (d) 1%

4. A light beam is incident normally on a glass-air interface. The glass has a refractive index 1.53. Air is assumed to be of refractive index 1.0. The **Fresnel, intensity reflection coefficient** is

(a) 0.53 (b) 0.25
(c) 0.044 (d) 0.022

5. The **refractive index** of the core of an optical fiber is n_1 and that of the cladding is n_2 If $(n_1 - n_2) = \Delta n$, then the fiber can be made single mode with numerical aperture unchanged by

(a) Reducing core diameter and increasing Δn
(b) Reducing both core diameter and Δn
(c) Reducing core diameter alone
(d) Reducing Δn alone

6. In a fiber optic displacement sensor shown in the following figure, the ratio of the output light intensity DOES NOT depend on

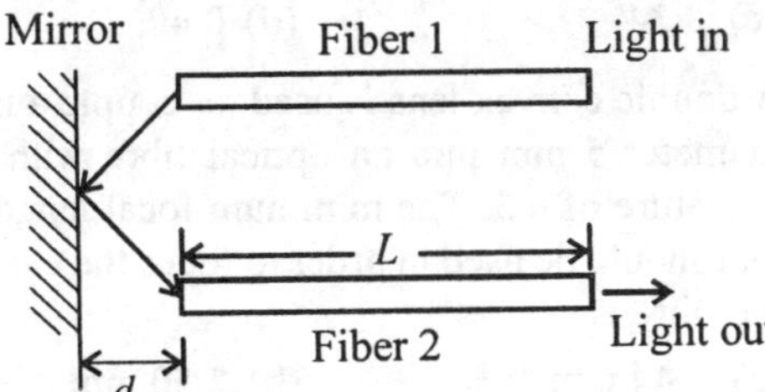

(a) Numerical aperture of the fibers
(b) Length L of the fibers
(c) Distance d
(d) Reflectivity of the mirror

7. A beam of unpolarized light is first passed through a linear polarizer and then through a quarter wave plate. The emergent beam is

(a) Unpolarized
(b) Linearly polarized
(c) Circularly polarized
(c) Elliptically polarized

8. The core/cladding index difference of a single-mode optical fiber cable is 0.01. The refractive index of material of the core is 1.5. The maximum angle of acceptance of the fiber is approximately equal to:

(a) 17.5° (b) 12.1°
(c) 8.6° (d) 2.0°

9. An LED emitting at 1 *μm* with a spectral width of 50 *nm* is used in a Michelson interferometer. To obtain a sustained interference, the maximum optical path difference between the two arms of the interferometer is

(a) 200 *μm* (b) 20 *μm*
(c) 1 *μm* (d) 50 *nm*

10. Light of wavelength 630 nm in vacuum, falling normally on a biological specimen of thickness 10 μm splits into two beams that are polarized at right angles. The refractive index of the tissue for the two polarizations are 1.32 and 1.333. When the two beams emerge, they are out of phase by

(a) 0.13° (b) 74.3°
(c) 90.0° (d) 128.6°

11. An optical fibre is characterized by

(a) Total internal relfection
(b) A core material of refractive index lower than that of cladding
(c) Scattering loss
(d) Diffraction

12. For a glass of **refractive index** 1.732, the angle of incidence of ordinary light, for which the reflected beam is completely polarized is

(a) 30 degrees (b) 60 degrees
(c) 45 degrees (d) 75 degrees

13. A **step index optical fiber**, whose refractive indices of the core and cladding are 1.44 and 1.40, respectively is surrounded by air. Its **numerical aperture** is

(a) 0.12 (b) 0.75
(c) 0.06 (d) 0.34

14. When **light** passes from a medium of refractive index n_1 into a medium of refractive index n_2, the reflection at the boundary is expressed as

(a) $(n_2 - n_1)^2$ (b) $[(n_2 - n_1)/(n_2 + n_1)]^2$
(c) $[1/(n_2 - n_1)^2]$ (d) $(n_2 + n_1)^2$

15. Two crossed polarizer's are placed in the path of a light beam. The **light** output is

(a) Zero (b) Plane polarized
(c) Circularly polarized (d) Elliptically polarized

16. An **optical fibre** has refractive index of 1.641 for the core and 1.422 for the cladding. The critical angle above which a ray will be totally internally reflected is

(a) 37° (b) 41°
(c) 45° (d) 60°

17. The **relative permittivity** (ε_r) of an optical medium is 2.5. Its refractive index is:

(a) 0.4 (b) 1.25
(c) 1.58 (d) 1.27

18. **Brewster angle** is the angle of incidence of monochromatic light at the interface of two media, for which:

(a) The incidence beam would be totally internally reflected
(b) A polarized beam would become unpolarised after refection
(c) An unpolarised beam would become polarised after reflection
(d) None of the above

19. The refractive indices of the core and cladding of an optical fibre are 1.46 and 1.455, respectively. The diameter of the core is 4 μm. When operated at a wavelength of 1.31 μm the fibre functions as a

(a) Single - mode fibre with a numerical aperture of 0.1
(b) Multimode fibre with a numerical aperture of 0.15
(c) Single - mode fibre with a numerical aperture of 0.12
(d) Ten-mode fibre with a numerical aperture of 0.12

20. A step index fibre has a core refractive index of 1.46 and a cladding refractive index of 1.45. The pulse dispersion in ns/km due to the effect of numerical aperture alone is

(a) 33.6 (b) 31.5
(c) 29.5 (d) 25.2

21. The maximum solid **angle is acceptance of light** coupled into a step index fibre having core and cladding refractive indices of 1.48 and 1.45, respectively is

(a) 0.28 steradians (b) 0.30 steradians
(c) 0.32 steradians (d) 0.34 steradians

Statement for Linked Answer Questions 22 and 23

In a step–index optical fibre, the refractive indices of the core and the cladding are equal to 1.501 and 1.499, respectively, at the wavelength of 850 nm. The fibre is surrounded by air.

22. The numerical aperature of the optical fibre is

(a) 0.077 (b) 0.501
(c) 1.499 (d) 1.501

23. An LED emitting light around wavelength of 850 nm is butt–coupled to the optical fibre. If the LED is assumed to be a Lambertain source, then efficiency of coupling LED light to the fiber is

(a) 7.70% (b) 1.02%
(c) 0.60% (d) 0.57%

Statement for Linked answer Questions 24 and 25.

The numerical aperture of a step index fibre in air (refractive index = 1) is 0.39. The diameter of the core is 200 μm.

24. The angle of acceptance when the fibre is used in water (refractive index = 1.33) is closed to

(a) 15° (b) 16°
(c) 17° (d) 18°

25. Two experiments are conducted in which light is launched into the fibre from a uniformly distributed planar source kept 5 mm away from the tip. In the first experiment (E_1) both the source and the fibre are in air. In the second experiment (E_2) both source and the fibre are in water. Neglecting absorption in the medium between the source and the tip, the ratio of the amount of light coupled into the fiber in E_1 to the amount of light coupled into the fiber in E_2 is closest to

(a) 0.5 (b) 1.0
(c) 1.4 (d) 1.9

26. Light coming out of an optical fiber is incident on a plane perpendicular to the fiber axis and 50 mm away from the end of the fibre. The light coming out creates a circular spot that can at most be of 20 mm diameter. Neglecting the diameter of the fibre, the numerical aperture of the fibre is, approximateiy.

(a) 0.14 (b) 0.20
(c) 0.34 (d) 0.40

27. A double convex lens is used to couple a laser beam of diameter 5 mm into an optical fibre with a numerical aperature of 0.5. The minimum focal length of the lens that should be used in order to focus the entire beam into the fibre is

(a) 1.44 mm (b) 2.50 mm
(c) 4.33 mm (d) 5.00 mm

Common Data for Questions 28 and 29:

The deflection profile $y(x)$ of a cantilever beam due to application of a point force F (in Newton), as a function of distance x from its base, is given by

$y(x) = 0.001F\,x^2\left(1 - \frac{x}{3}\right)m$. The angular deformation θ at the end of the cantilever is measured by reflecting a laser beam off a mirror *M* as shown in the figure.

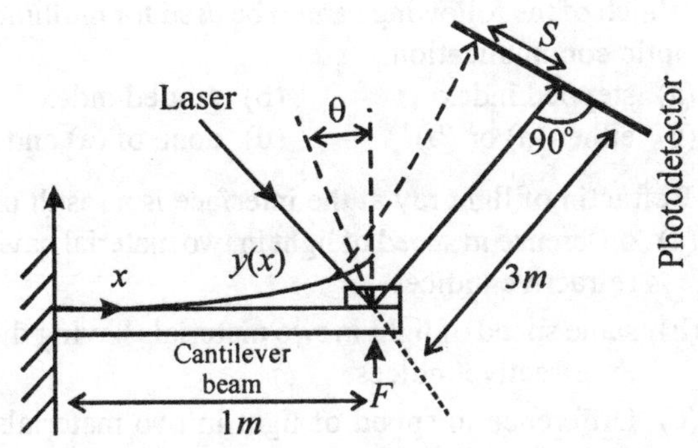

28. The translation *S* of the spot of laser light on the photodetector when a force $F = 1\ N$ is applied to the cantilever is

(a) 1 mm (b) 3 mm
(c) 6 mm (d) 12 mm

29. If linear variable differential transformers (LVDTs) are mounted at $x = \frac{1}{2}m$ and $x = \frac{1}{4}m$ on the cantilever to measure the effect of time varying forces, the ratio of their outputs is

(a) $\frac{12}{7}$ (b) $\frac{40}{11}$
(c) $\frac{176}{23}$ (d) $\frac{112}{15}$

30. Operating wavelength window of 1550 nm is not yet in use with fibre optic system because
(a) loss in fibre is higher at 1300 nm
(b) suitable Laser devices have not yet been obtained
(c) wavelength multiplexing is not possible at this wavelength
(d) both (b) and (c)

31. An optical fibre consists of a cylindrical dielectric rod of refractive index μ_1 surrounded by another dielectric of refractive index u_2, where $u_2 < \mu_1$. If a ray is incident from air at an angle *i* to the axis, then it udergoes total internal reflection at the interface AB if:

(a) $i \geq \sin^{-1}\sqrt{\mu_1^2 - \mu_2^2}$

(b) $i =< \sin^{-1}\sqrt{\mu_1 - \mu_2}$

(c) $i =< \sin^{-1}\sqrt{\mu_1^2 - \mu_2^1}$

(d) $i => \sin^{-1}\sqrt{\mu_1 - \mu_2}$

32. Total internal reflections can take place when light travels from
(a) diamond to glass (b) water to glass
(c) air to water (d) air to glass

33. Loss in fibre is due to (indicate false statement)
(a) Impurities
(b) microbending
(c) attenuation in fibre
(d) stepped index operation

34. Material dispersion of an optial fibre vansihes if:
(a) refractive index of the core varies lineraly with wavelength
(b) refractive index of core remains constant with wavelength
(c) both (a) and (b)
(d) neither (a) nor (b)

35. Invention of optical windows are:
(a) Five (b) three
(c) two (d) four

36. The cable that offers lowest dispersion is
(a) SMSI (b) MMSI
(c) MMGI (d) None of these

37. A graded index profile in optical fibre cable provides:
(a) Less waveguide dispersion than SI profile
(b) Less material dispersion than SI profile
(c) Less attenuation in SI profile
(d) Less modal dispersion than SI profile

38. Normalised frequency (v) for a single mode cable is
(a) $v \leq 2.4$ (b) $2.4 \leq n \leq 3.98$
(c) $n \leq 2.4$ (d) $0 \leq n \leq 3.8$

39. If BW is the bandwidth of an optical fibre and NA is the numerical aperture then

(a) BW μ NA (b) BW μ $\frac{1}{NA}$
(c) BW μ $\frac{1}{(NA)^2}$ (d) BW μ $\frac{1}{(NA)^3}$

40. In a dispersion shifted fibre desired wavelength of poeration is
(a) 1.33 mm (b) 1.55 mm
(c) 0.85 (d) none of the above

41. The configuration utilized by a line communication system capable of transmitting a data at the rate of 100 Mb/s is
(a) coaxial cable system (b) open line system
(c) optical fibre system (d) twisted wire system

42. Which of the following is the transmission frequency in optical fibre.
(a) 10^9Hz (b) 10^{11}Hz
(c) 10^{14}Hz (d) None

43. Optical fibre was invented in
(a) 1950 (b) 1960
(c) 1970 (d) 1975

44. Light propagates along optical fibre by
(a) total internal reflection
(b) total internal refraction
(c) both (a) and (b)
(d) either (a) or (b).

45. Optical fibre can be satisfactorily operated if
(a) RI of the core is lesser than that of cladding
(b) RI of the core is greater than that of cladding
(c) RI of the core is equal to the that of cladding
(d) None of the above

46. RI of the core of an optical fibre is higher than that of cladding because of
(a) better confinement of light
(b) maximum distance operation
(c) easy to handle
(d) higher life time of the material

47. In the fabrication of optical fibre silica is used because
(a) it is cheap and easily avialable
(b) it is obtained in abundance
(c) it is obtained in purest form
(d) none of above

48. Attenuation in optical can be measure in
(a) dB/km (b) dB/m
(c) Kdb/m (d) dBm/m.

49. Transmission of signal through optical fibre is of the form of
(a) sound (b) electricity
(c) light (d) speed

50. For extremely low transmission losses in optical fibre, operating wavelength is in the range of
(a) (0.1 to 1) μm (b) (0.01 to 2) μm
(c) (0.2 to 8) μm (d) (0.0001 to 2) μm.

51. The material used for optical fibre for least losses is
(a) SiF_4 (b) NaF_4
(c) ZrF_4 (d) $NaSiF_4$

52. In communication repeaters are used to
(a) amplify the signal (b) reshape the signal
(c) either (a) or (b) (d) both (a) and (b)

53. The ability of a receiver used in optical fibre communication is to achieve a certain performance levels depends on
(a) type of the photo-detector
(b) effect of noise in the system
(c) the characteristics of the successive amplifying stages
(d) all the three mentioned above

54. At critical angle of incidence
(a) light ray in air is perpendicular to the glass surface
(b) light ray in air is parallel to the glass surface
(c) there is no such relation
(d) none of the above

55. Which of the following cannot be used for multimode fibre optic communication
(a) stepped index (b) graded index
(c) either (a) or (b) (d) none of (a) and (b)

56. Refractin of light ray at the interface is a result of
(a) difference in speed of light in two material having same refractive indices
(b) same speed of light in two materials having difference in refractive indices
(c) Difference in spped of light in two materials having different refractive indices
(d) none of the above

57. Which of the following is having the highest refractive index
(a) diamond (b) air
(c) water (d) glass

58. Function of receiver in optical fibre is to
(a) reshape the degraded signal only
(b) only amplify of degraded signal
(c) both amiplify and reshaped the degraded signla
(d) none of the above

59. Photodetectors used in optical fibre is
(a) PIN, APDs (b) PIN, Gunn diodes
(c) APD, Gunn diodes (d) none of the above

60. Photo-detector converts
(a) optical power into electric voltage
(b) optical power into electric current
(c) electrical currents into optical power
(d) none of the above

61. Source of light for optical fibre is
(a) PIN diode (b) Photodiode
(c) LED (d) none of these

62. In the first window of optical fibre, light sources are generally.
(a) GaAIP (b) GaABr
(c) GaAIAs (d) GeAIAs

63. Dispersion in silica fibrees is minimum at λ =
(a) 850 nm (b) 1300 nm
(c) 1550 nm (d) 1800 μm

64. Graded index fibre with the following value of index profile, (α), results in minimum dispersion.
(a) 1.0 (b) 2.0
(c) 3.0 (d) 4.0

65. The V-parameter for an optical fibre is 50. The number of modes in that fibre is approximately.
(a) 50 (b) 1250
(c) 2500 (d) 1400

66. One of the combination of materials used for **optical fibers** is:
(a) Copper core and glass cladding
(b) Glass core and aluminium cladding
(c) Glass core and plastic cladding
(d) Plastic core and glass cladding

67. For **fiber optical sensors:**
(a) single mode transmission is used.
(b) multimode transmission is used.
(c) both single and multimode transmissions are commonly used
(d) none of the above

68. In **optical fibres**, if the total reflection is to takes place, the following conditions must be satisfied.
(a) $n_2 < n_1$ *and* $\phi_1 > \cos^{-1}(n_2/n_1)$
(b) $n_1 < n_2$ *and* $\phi_1 < \cos^{-1}(n_2/n_1)$
(c) $n_2 < n_1$ *and* $\phi < \cos^{-1}(n_2/n_1)$
(d) $n_2 > n_1$ *and* $\phi_1 > \cos^{-1}(n_2/n_1)$

69. In a **graded index fibre**, the total reflected light takes a
(a) straight line path (b) parabolic path
(c) cliptical pathe (d) circular path

70. The bottom of a glass vessel is a thick plane plate that has an index of **refraction** of 1.50 The vessel is filled with water. What is the largest angle of incidence in the water for which a ray can pass through the glass bottom and into the air below? How does this compare with the critical angle of incidence for a water-air interface which it 48.6° ? What is the **angle of refraction** in the glass?
(a) 38.6° (b) 48.6°
(c) 58.6° (d) 68.6°

71. Light is passing from air into a liquid and is deviated 19° when the **angle of incidence** is 52°Under what conditions will total reflection occur at this interface?
(a) 23.7° (b) 33.7°
(c) 43.7° (d) 53.7°

72. Find the Angle of refraction in water, when the angle of incidence in air is 40° and a beam of light is passing through air to oil to water.
(a) 14.9° (b) 18.9°
(c) 22.9° (d) 28.9°

73. The **optical fibre** is 2 m long and has a diameter of 20 μm. If a ray of light is incident on one end of the fiber at an angle $\theta_1 = 40°$. How many reflections does it make before emerging from the other end (The index of refraction of the fibre is 1.30)?
(a) 27000 (b) 37000
(c) 47000 (d) 57000

74. In practice, **optical fibres** have a coating of glass ($n_3 = 1.512$) to protect the optical surface of the fibre. If the fibre itself has an index of refraction $n_2 = 1.700$, what is the **critical angle for** total reflection of a ray inside the fibre?
(a) 42.7° (b) 52.7°
(c) 62.7° (d) 72.7°

75. The index of refraction (for sound) of water $(v = 1498)$ when index of refraction for sound is defined as ratio of speed in air (343 m/s) to speed of sound in medium.
(a) 0.119 (b) 0.229
(c) 0.339 (d) 0.449

76. Find the critical angle θc for total reflection of sound from water if the index of refraction (for sound) of water $(v = 1498)$ when index of refraction for sound is defined as ratio of speed in air (343 m/s) to speed of sound in medium.
(a) 7.2° (b) 9.2°
(c) 11.2° (d) 13.2°

77. A ray of light incident at an angle θ_1 on one end of an **optical fiber**. Its angle of refraction is θ_2 and it strikes the side of the fiber at an angle ϕ_2. If the index of refraction of the fibre is 1.30, what is the largest angle of incidence θ_2 that a ray can have and still be totally reflected from the side of the fibre?
(a) 36.2° (b) 46.2°
(c) 56.2° (d) 66.2°

78. Find the **Bragg angle** for a acousts-optic modulator length = 50 mm, diffraction efficieny 70% acoustic wavelenght 4.3×10^{-5} m acoustic velocity 3500m/sec.
(a) 0.22° (b) 0.32°
(c) 0.42° (d) 0.52°

79. Energy separation in a GaAs laser with a cavity length of 200 μm.
(a) 0.63 m e V (b) 0.73 m e V
(c) 0.83 m e V (d) 0.93 m e V

80. Smallerst value of internal angle θ_c in a fiber in air with $n_1 = 1.53$, $n_2 = 1.50$ and $n_0 = 1$.
(a) 58.6°
(b) 78.6°
(c) 68.6°
(d) none of above

81. Numerical aperture if fiber in air with $n_1 = 1.53$, $n_2 = 1.50$ and $n_0 = 1$.
(a) 0.101 (b) 0.202
(c) 0.303 (d) 0.404

82. Fibre acceptance angle if internal angle θ_c in a fiber in air with $n_1 = 1.53$, $n_2 = 1.50$ and $n_0 = 1$.
(a) 15.6° (b) 16.6°
(c) 17.6° (d) 18.6°

83. What maximum pulse broadening in a **step index fiber** with NA as 0.2, core refractivbe index as 1.45 over a distance 1 km is:
(a) 26 ns/km (b) 36 ns/km
(c) 40 ns/km (d) 46 ns/km

84. What is the **Rayleigh scattering** for silica fibre at $\lambda = 1\,\mu m$, $n = 1.46$, $p = 0.286$, $\beta = 7\times10^{-11}\ m^2/N$, TF = 1400 K is:
(a) $1.64\times10^{-4}/m$ (b) $1.74\times10^{-4}/m$
(c) $1.84\times10^{-4}/m$ (d) $1.94\times10^{-4}/m$

85. A plane harmonic infrared wave traveling through a transparent medium is given by:

$$E_x = (y,t) = E_{ox}\sin 2\pi\left(\frac{y}{5\times10^{-7}} - 3\times10^{14}t\right)$$

in SI as usual. Determine the **refractive index** of the medium at that frequency?
(a) 1 (b) 2
(c) 3 (d) 4

86. Light from a sodium lamp ($\lambda_0 = 589\ nm$) passes through a tank of glycerin (refractive index 1.47) 20 m long in a time t_1. If it takes a time t_2 to traverse the same tank when filled with carbon disulphide (index 1.63). Determine the difference $t_2 - t_1$.
(a) $1.07\times10^{-5}s$ (b) $1.07\times10^{-6}s$
(c) $1.07\times10^{-7}s$ (d) $1.07\times10^{-8}s$

87. A beam of 436-nm light is incident perpendicular to a glass plate that is 2.0 cm thick and has $n = 1.66$. How long does it take a point on the beam to pass through the plate?
(a) 0.57 ns (b) 0.67 ns
(c) 0.87 ns (d) 0.111 ns

88. When light passes from a medium of refractive index n_1 into a medium of refractive index n_2, the reflection at boundary is expressed as
(a) $(n_2 - n_1)^2$ (b) $\left(\frac{n_2 - n_1}{n_2 + n_1}\right)^2$
(c) $\frac{1}{(n_2 - n_1)}$ (d) $(n_2 + n_1)^2$

89. Light is incident on a glass platc of refractive index 1.732. It is polarized on reflection when the incidence angle is
(a) 30° (b) 45°
(c) 60° (d) 75°

90. In a step index optical fiber the difference between the refractive indeces of the core and cladding is Δn and the sum of refractive indices is $2n$. Then numerical aperture is
(a) $\sqrt{\Delta n/n}$ (b) $\sqrt{2n\times\Delta n}$
(c) $\sqrt{\Delta n\times 2n}$ (d) $\sqrt{n\times\Delta n}$

91. The refractive index of the core of an optical fiber is n_1 and that of cladding is n_2. If $(n_1 - n_2) = \Delta n$, then the fiber can be made single mode with numerical aperture unchanged by
(a) reducing core diameter and increasng Δn.
(b) reducing both core diameter and Δn.
(c) reducing core diameter alone.
(d) reducing Δn alone.

92. The total internal reflection principle is derived from
(a) Snell's law (b) Beer's law
(c) Brag's law (d) Nernst equation

93. When a light travel in graded index multimode optical fiber.
(a) Change in speed and frequency
(b) Change in frequency and wavelength
(c) Change in speed and wavelength
(d) No change

94. When a light travel in step index multimode optical fiber.
(a) Change in speed and frequency.
(b) Change in frequency and wavelelngth
(c) Change in speed and wavelength
(d) No change in speed, wavelength and frequency

95. For fiber optical sensors
(a) single mode transmission is used
(b) single and multimode transmissions are commonly used
(c) multimode ransmission is used
(d) None of the above

96. The optical sensors my be classified as
(a) Intensity (or) intrinsic type and interferometer(or) extrinsic type
(b) Electromagnetic and Ferromagnetic type
(c) Intensity (or) intrinsic type and interferometer (or) extrinsic type
(d) All of the above

97. One of the combinations of materials used for optical fiber is
(a) Copper core and glass caldding
(b) Glass core and aluminum cladding
(c) Glass core and plastic cladding
(d) Platstic core and glass cladding

98. The optical fiber has highest bandwidth
(a) Single mode
(b) Step-index multimode
(c) Graded index multimode
(d) Step-index ten mode

99. For a glass of refractive index 1.732, the angle of incidence of ordinary light, for which the reflected beam is cmletely polarized is

(a) 30 degress (b) 60 degress
(c) 45 degress (d) 75 degress

100. A step index fiber, whose refractive indices of the core and cladding are 1.44 and 1.40 respectively, is surrounded by air whose NA is

(a) 0.12 (b) 0.75
(c) 0.06 (d) 0.34

101. An optical fiber has refractive index of 1.64 for the core and 1.422 for the cladding. The crictical angle above which a ray will be totally internally reflected is

(a) 37° (b) 41°
(c) 45° (d) 60°

102. The refractive indices of glass and water are 1.50 and 1.33, respectively. If the glass insimmersed in Water, its refractive index is

(a) 0.89 (b) 1.13
(c) 1.99 (d) 2.83

103. A ray of light enters a glass optical fiber (refractive index fiber 1.45) surrounded by air. The angle of incidence grater than which causes total internal reflection

(a) 36.8° (b) 43.6°
(c) 53.6° (d) None of the above

104. The maximum solid angle of acceptance of the light into a step index fiber having core and cladding refractive indices of 1.48 and 1.45 respectively, i s

(a) 0.28 steradins (b) 0.30 steradians
(c) 0.32 steradians (d) None of the above

105. The acceptance angle for a single fiber with a numerical aperture of 0.0 96 is

(a) 55° (b) 5.5°
(c) 45° (d) 4.5°

106. A light beam is incident normally on a glass-air interfce. The glass has a refractive index 1.53. Air is assumed to be of refractive index 1.0. The Fresnel, intensity reflection coefficient is

(a) 0.53 (b) 0.25
(c) 0.044 (d) 0.022

107. The NA of an optical fiber is 0.3775 and the refractive index of core is 1.45 the refractive index of cladding

(a) 1.44 (b) 1.40
(c) 0.40 (d) 0.45

108. If the Acceptance angle of optical fiber is 20.48°, then the NA is

(a) 0.30 (b) 0.35
(c) 0.40 (d) 0.45

109. If the critical angle is 75° and the refractive index of core is 1.45 then the accetance angle of potical fiber is

(a) 9° (b) 11°
(c) 18° (d) 22°

110. A monohromatic light source which is physiclly broad of optical fiber is

(a) Neither special nor temperal coherence
(b) No special but has temperal coherence
(c) Spacial cohernce but no temperal coherence
(d) Both special and temperal coherence

111. A light ray travelling in a medium of refractive inex 1.4 strikes the second medium of refractive index 1.2.If the angle of incidence is 25° then the angle of refractin would be

(a) 45° (b) 60°
(c) 30° (d) 15°

112. Yellow light of wavelength 589 nm is incident on a boundary separating air and glass. The refractive index of glass for yellow light 1.5. The speed of reflected light is

(a) 3×10^8 m/s (b) 3×10^8 m/s
(c) 2×10^8 m/s (d) 2×10^8 m/s

113. For an optical fiber to gather all the light falling on it, its numerical aperature (N.A.) should be

(a) < 1 (b) > 1
(c) = 1 (d) zero

114. A optical fiber with R.I. of $\mu_{core} = 1.641$, $\mu_{cladding} = 1.422$. A angle above which a ray of light will get totally internally reflected is ?

(a) 59.20° (b) 60.05°
(c) 61.30° (d) 58.07°

115. Find the acceptance angle for a single mode fiber with a N.A. of 0.096 .

(a) 5.1° (b) 5.2°
(c) 5.5° (d) 5.8°

116. An optical filber is characterised by

(a) $\mu_{core} < \mu_{cladding}$ (b) TIR
(c) $\mu_{core} > \mu_{cladding}$ (d) both (b) and (c)

117. The refractive index (R.I.) of core and cladding is 1.46 and 1.455 respectively. Diameter of core is $4\ \mu m$. When operated at a wavelength of $1.31\ \mu m$. The fiber functions as a.

(a) single mode with N.A. = 0.10
(b) multi mode with N.A. = 0.15
(c) single mode with N.A. = 0.12
(d) multi mode with N.A. = 0.12

118. Find the core radius necessary for single mode operation at a wavelength of 820 nm for a step index fiber with $n_1 = 1.48$ and $n_2 = 1.478$.

(a) 4.08 μm (b) 4.02 μm

(c) 3.08 μm (d) 3.07 μm

119. Find the loss in optical fiber (dB/km) for 80 % transmission /km

(a) 1 dB/km (b) 2 dB/km
(c) 3 dB/km (d) 4 dB/km

120. A step index fiber with $\mu_{core} = 1.42$ and $\mu_{cleadding} = 1.40$. The pulse dispersion in n sec/km is

(a) 67.12 (b) 68.42
(c) 66.13 (d) 67.61

121. Find the distance for which the output power in a beam is attenuated by a factor of 10, When propagating in fiber having attenuation coefficient $\alpha = 20\ dB/km$

(a) 0.5 *km* (b) 1 *km*

(c) 1.5 *km* (d) 2 *km*

122. The refractive index of core is 1.46 and $\Delta = 0.01$ this optical fiber is used to transfer the information. This transfer rate of the information for the step index fiber is

(a) 1 Mbps (b) 10 Mbps
(c) 1000 Mbps (d) 100 Mbps

Linked Questions and Answers 123 and 124

A multimode step index fiber core diameter 80 μm. $\Delta = 1.5\%$ is operated at a wave length of 0.85 μm. If R.I. of core (n_1) = 1.48.

123. Normalized frequency (V) is

(a) 24 (b) 74
(c) 37 (d) 104

124. The Number of modes is

(a) 2764 (b) 37
(c) 6354 (d) 2738

Common Data for Questions 125 and 126

A manufacturer wants to make a Si step index fiber with a V number of 75. NA = 0.3, used t 820 nm. If n_1 \1.458

125. The Core size is

(a) 32 μm (b) 42 μm
(c) 52 μm (d) 62 μm

126. The cladding refractive index is

(a) 1.43 (b) 1.33
(c) 1.33 (d) 1.42

127. Find the time taken by light to travel a distance of 600 m in water.

(a) 0.26 n sec (b) 0.43 n sec
(c) 0.15 n sec (d) 0.13 n sec

128. A light ray incident on air - glass interface with an angle of 30°. If glass refractive index is 1.5. Find the refracted angle.

(a) 20.30° (b) 18.70°
(c) 19.47° (d) 21.80°

129. A ray of light strikes a glass plate at an angle of incidence 60°. If the reflected and refracted rays are perpendicular to each other, find the refractive index of the glass.

(a) 1.732 (b) 1.62
(c) 1.84 (d) 1.68

130. A light beam incident normally at a glass - air interface, Glass has refractive index of 1.53 and air has refractive idex of 1. The reflection loss coefficient is

(a) 5.2 % (b) 5.1 %
(c) 4.4 % (d) 4.1 %

1.8 ECG, EEG and CMG

1. Match the bipolar limb leads to the correct combination of electrodes used in ECG and select the correct answer using codes:

Group - I	Group - II
A. Lead I	1. Left leg and Right arm
B. Lead II	2. Left leg and Right leg
C. Lead III	3. Left arm and Right arm
	4. Left leg and Left arm

Codes:

	A	B	C
(a)	1	3	4
(b)	3	1	4
(c)	4	1	3
(d)	2	1	4

2. The P wave, QRS complex and T wave in one heart cycle of the ECG waveform correspond respectively to

(a) Polarization of atria; depolarization of ventricles and repolarization of atria.

(b) Polarization of atria; polarization of ventricles and repolarization of atria.

(c) Depolarization of atria; depolarization of ventricles and repolarization of atria.

(d) Depolarization of atria; repolarization of ventricles and depolarization of atria.

3. Choose the correct match for signal and biological activity and select the correct answer with the codes given below:

A. EEG	1. Eye movement
B. EOG	2. Gastric mobility
C. EMG	3. Muscle activity
D. EGG	4. Brain activity
	5. Heart activity
	6. Bladder activity

Codes:

	A	B	C	D
(a)	4	5	4	3
(b)	4	1	3	2
(c)	3	2	6	1
(d)	2	3	1	6

4. A cranial generator has a peak-to-peak voltage of 100 μV and a source resistance of 10 kΩ. If the electrode resistance are 20 kΩ, the voltage across the amplifier terminals with an input impedance of 10 MΩ, is

(a) 93.5 μV (b) 95.5 μV
(c) 97.5 μV (d) 99.5 μV

5. For bio–potential recording, the impedance of an *Ag*/*AgCl* electrode, as compared to that of an Ag electrode of the same size, is

(a) Lower at all frequencies.
(b) Lower for frequencies less than 10 Hz and lower for frequencies greater than 300 Hz.
(c) Higher for frequencies less than 10 Hz and higher for frequencies greater than 300Hz.
(d) Higher at all frequencies

6. During intravascular measurementof arterial blood pressure, catherers may by introduced in different configurations, as shown in the figure below

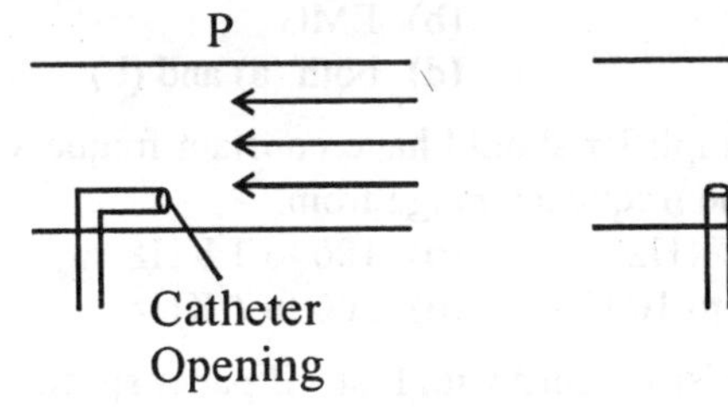

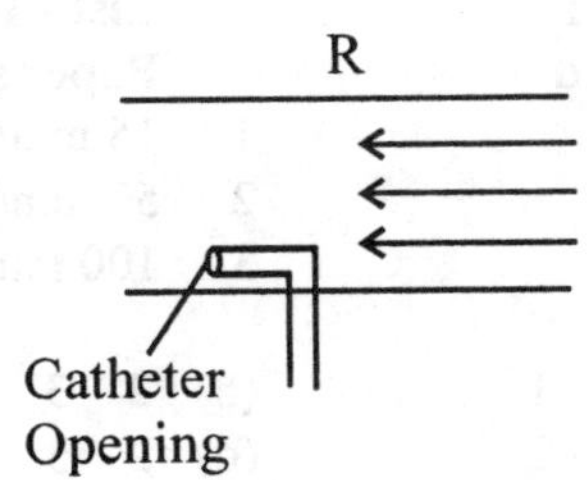

The static pressure will be measured correctly in the configuration (s)

(a) P and Q but not in R (b) R only
(c) Q only (d) P and R but not in Q

7. In the standard 12-lead ECG recording system, the minimum number of electrodes required to be attached to a human subject for recording any one of the unipolar chest lead signals is

(a) 1 (b) 2
(c) 4 (d) 5

8. The Treadmill Test is used to diagnose

(a) The balancing style during walk of the patient
(b) The auditory activity of the patient
(c) The visual activity of the patient
(d) The cardiac activity of the patient

9. The bandwidth of an electrocardiogram (ECG) amplifier is:

(a) dc to 0.01 Hz (b) 0.05 to 500 Hz
(c) 550 to 1500 Hz (d) 2000 to 10000 Hz

10. Korotkoff sounds are used

(a) As a reference for sound level measurement
(b) For studying heart muscle functioning
(c) For blood pressure measurement
(d) For study of heart value functioning

11. Which of the following gauges can measure the lowest vacuum pressure?

(a) McLeod gauge (b) Pirani gauge
(c) Ionisation gauge (d) None of the above

12. Among the electrodes which are used to record ECG signals, the one used for ground referencing is placed at,

(a) Left arm (b) Right leg
(c) Left leg (d) None of the above

13. For a ECG amplifier the input signal is 10 mV corrupted by a common mode noise of 5 m V output of the amplifier contians 50 mV of signal and 0.001 mV of noise. The CMRR of the amplifier is

(a) 40 dB (b) 60 dB
(c) 80 dB (d) 100 dB

14. In the driven–right–leg system for reducing the common mode interference in ECG recording, the common mode voltage is sensed and connected to the right leg electrode. As a result, the desirable condition for the amplifier gain is

(a) A large positive gain (b) A large negative gain
(c) A gain of 1 (d) A gain of –1

15. The ECG of a patient is being recorded using the three standard frontal plane leads. If the cardiac vector is oriented at an angle of 45 degrees to Lead I and has a magnitude of 3 mV, the voltages seen on Leads I, II and III are

(a) 1.50 , 2.27 , and 0.77 mV
(b) 2.12 , 0.77 , and 2.89 mV
(c) 2.12 , 3.15 , and 1.03 mV
(d) 2.12 , 2.89 , and 0.77 mV

16. A standard three-lead frontal plane ECG is taken of a person with a normal heart. The peak amplitude of the R-wave is

(a) Greatest in lead I
(b) Greastest in lead II
(c) Greatest in lead III
(d) Equal in all the leads

17. The band width of ECG amplifier is.
(a) Dc to 0.01 Hz (b) 0.05 to 500 Hz
(c) 550 to 1500 Hz (d) 2000 to 10,000 Hz

18. If V_w is Wilson's central terminal potential and potentials of right arm, left arm and left leg are V_1, V_2, V_3 respectively then relation between WCT and V_1, V_2, V_3 is.

(a) $V_w = 0.333(V_1 + V_2 + V_3)$
(b) $V_w = 0.5(V_1 + V_2 + V_3)$
(c) $V_w = 0.25(V_1 + V_2 + V_3)$
(d) $V_w = (V_1 + V_2 + V_3)$

19. Which of the following instrument uses an ultra sonic after transducer ?
(a) Echo-cardiograph
(b) Electro-cardiograph
(c) Electroencephalograph
(d) Electro-retinograph

20. A standard ECG machine has .... limb electrodes.
(a) 4
(b) 12
(c) 10
(d) depends on manufacturing

21. Among the Electrodes which are used to record ECG signals,
(a) Left arm (b) chest
(c) left lag (d) none

22. For an ECG amlifier the input signal is 100 mV corrupted by a common noise of 5 mV. Output of the amplifier contains 50 mV of signal and 0.001mV of noise. The CMMR of the amplifier is
(a) 40 dB (b) 60 dB
(c) 80 dB (d) 100 dB

23. Total number of Limb leads present in ECG........
(a) 3 (b) 4
(c) 6 (d) none

24. The base line voltage of ECG is knwon as........
(a) iso-electric line (b) iso-tropic line
(c) iso-mesomer line (d) iso-potential line

25. Match the bipolar limb leads to the correct combination of electrodes used in ECG

A. Lead-I	1. LL and RA
B. Lead-II	2. LL and RL
C. Lead-III	3. LA and RA
	4. LL and LA

Codes:

	A	B	C
(a)	1	2	3
(b)	2	3	4
(c)	3	1	4
(d)	4	2	1

26. Enlarged T-wave indicates ....
(a) hyper-kalemia (b) hypo-kalemia
(c) hyper-natrim (d) hypo-natrium

27. In the standard 12-Lead ECG recording system, minimum number of electrodes required to be attched to human subject for recording any one of the unipolar chest lead signals is ....
(a) 1 (b) 2
(c) 4 (d) 5

28. A cranial generator has a peak to peak votlage of 100 μV and a source resistanfe of 10 kΩ. If the electrode resistance are 20 kΩ. the voltage across the amlifier terminals withand input impedance of 10 MΩ is
(a) 93.5 μV (b) 95.5 μV
(c) 97.5 μV (d) 99.5 μV

29. Graphic record of heart sounds is known as ....
(a) electrocardio gram (b) vector cardiogram
(c) phonocardiogram (d) both (b) and (c)

30. A 15mV signal voltage can be seen in ....
(a) ECG (b) EMG
(c) EEG (d) both (a) and (b)

31. An EMG amplifier should have constant frequency response in the frequency range from.
(a) 10 to 1 KHz (b) 100 to 1 KHz
(c) 1 KHz to 10 KHz (d) 500 to 1 KHz

32. Match the List I record with List - II paper speed

	List - I Record		List - II Paper speed
A	EC G	1	15 mm/sec
B	EMG	2	50 mm/sec
C	EE G	3	100 mm/sec

Codes
(a) 3 2 1 (b) 2 3 1
(c) 1 3 2 (d) 2 3 1

33. CMRR in an ECG amplifier when input impedance is 2 MW and difference of electrode skin contant as 1 kΩ is is
(a) 1000 (b) 2000
(c) 3000 (d) 5000

34. A graphic record of heart sounds is known
(a) Electrocardiogram (b) Phonocardiogram
(c) Vectorcardiogram (d) Electroencephatogram

35. The triangle whose sides represent the line along which 3 projections of ECG are measured is known a
(a) Galvani tringle (b) Einthoven triangle
(c) Abbott triangle (d) Korotkoff triangle

36. Sequence, direction and distribution of forces generated by heart are showm by:
(a) E C G (b) Vectorcardiogram
(c) Electroencephalogram (d) Rheoencephalogram

37. Heart sounds range in the frequency form
(a) 2 kHz to 6 kHz (b) 6 kHz to 10 kHz
(c) 5 to 2 kHz (d) 5 to 100 kHz

38. Cardiac output is measured in
(a) Ltrs / sec (b) Ltrs / min
(c) ml/sec (d) ml/min

39. Natural pacemaker is
(a) SA node (b) AV node
(c) Bundle of His (d) None

40. Defiberillator with a synchronizing circuit called
(a) Pacemaker (b) Pacing defibrillator
(c) Cardiometer (d) Sync. Defibrillator

41. The pulmonary vein contains
(a) Oxygenated blood (b) deoxygenated blood
(c) carboxy blood (d) white blood

42. P^H value of Blood is
(a) 0 (b) 7
(c) 7.4 (d) 14

43. Safety can be improved by
(a) Equiptential grounding
(b) Isolation transformers
(c) Low voltage systems
(d) All of these

44. Counter shock in difibrillatin should not be delivered (gives) during the
(a) R wave (b) T wave
(c) P wave (d) None of the above

45. In QRSW complex signal QRS interval is
(a) 10 sec (b) 1 sec
(c) 0.09 sec (d) None of the above

46. The major effect(s) due to the electric current passing throuhg the body are
(a) Injury to tissues
(b) uncontrollable muscle contraction
(c) fibrillation of the heart
(d) None of the above

47. The effect of current applied directly to the heart is known as
(a) Macro Shock (b) Micro Shcok
(c) minishock (d) None of the above

48. Too Fast heart rate is called
(a) Bradycardia (b) Tachycardia
(c) bratycardia (d) None of the above

49. Intra cellular fluid contains large concentration of
(a) Potassium (b) Sodium
(c) Choride (d) None of the above

50. The value of resting potential is
(a) -20 mV (b) -70 mV
(c) -90 mV (d) None of the above

51. Each action potential in the heart originates at
(a) A V Node (b) SA Node
(c) Purkinje Fibers (d) None of the above

52. If P-R interval in the a ECG waveform is greater than 0.2 s it may suggest that
(a) Blockage of AV Node (b) Heart Block
(c) Arrhythmia (d) None of the above

53. Meterial used in medical electronds is
(a) Silver (b) Gold
(c) Copper (d) None of the above

54. Which of the following is the natrual transducer
(a) Ear (b) Skin
(c) Eye (d) None of the above

55. A typical medical instrumentation amplifier has CMRR of
(a) 10^6 (b) 1
(c) 0 (d) None of the above

56. Wandering base line results from
(a) Losse elecrodes
(b) Movement of patient cable
(c) Both (a) and (b)
(d) None of the above

57. What is the Bandwidth of an ECG amplifier
(a) 50 Hz - 10 kHz (b) 0 Hz - 100 Hz
(c) 5 Hz - 1 kHz (d) None of the above

58. To reduce the effect of variation on Electrode impedance all Biomedical Insturemnts used at input stages is
(a) Buffer amplifier
(b) Inverting amplifier
(c) Non inverting amplifier
(d) None of the above

59. Among the electrodes which ar used to record ECG signals the one used for ground referecne is
(a) Left arm (b) Right arm
(c) Left leg (d) Right Leg

60. In case of ERG. The type of electrodes are used to pick up singal is
(a) Disc elctrodes
(b) Retinal electrodes
(c) Vacuum electrodes
(d) None of the above

61. Loud speaker is also used in the recorder of
(a) EMG
(b) ECG
(c) EOG
(d) None of the above

62. Wax coated graph graph sheets are used for recording bcrasue
(a) to write easily
(b) to avoid spreading of ink
(c) to improve the constrast
(d) None of the above

63. The heart dipole field measured by lead system called
(a) Wilson lead systems
(b) Orthogonal lead system
(c) Vector lead sysem
(d) None of the above

64. Arrhythmia can be diagnosed by
(a) EEG
(b) ECG
(c) Vector cardiogram
(d) None of the above

65. The T wave is produced during
(a) atrial depolarization
(b) artrial repolarization
(c) repolarization of the ventricles
(d) None of the above

66. In ECG, the calibratin signal amplitude is
(a) 1 mV
(b) 1 V
(c) 1 mV
(d) None of the above

67. EEG waves include frequency content ranging from
(a) less than 1 Hz to 50 Hz
(b) DC to 2000 Hz
(c) 10 Hz to 1000 Hz
(d) None of the above

68. In the case of EEG with unipolar leads, the potential of each electrode is measured with respect to
(a) adjacent electrode
(b) indifferent electrode
(c) forehead electrode
(d) ground

69. If V_w is Wilson's central terminal potential and potentials of right arm, left arm and left leg are V_1, V_2, V_3 respectively then relation between WCT and V_1, V_2, V_3 is.
(a) $V_w = 0.333(V_1 + V_2 + V_3)$
(b) $V_w = 0.5(V_1 + V_2 + V_3)$
(c) $V_w = 0.25(V_1 + V_2 + V_3)$
(d) $V_w = (V_1 + V_2 + V_3)$

70. For an ECG amlifier the input signal is 100 mV corrupted by a common noise of 5 mV. Output of the amplifier contains 50 mV of signal and 0.001mV of noise. The CMMR of the amplifier is
(a) 40 dB (b) 60 dB
(c) 80 dB (d) 100 dB

71. The ECG of a patient is being recorded using the three standard frontal plane leads. If the cardiac vector is oriented at an angle of 45 degrees to Lead I and has a magnitude of 3 mV, the voltages seen on Leads I, II and III are
(a) 1.50 , 2.27 , and 0.77 mV
(b) 2.12 , 0.77 , and 2.89 mV
(c) 2.12 , 3.15 , and 1.03 mV
(d) 2.12 , 2.89 , and 0.77 mV

72. A cranial generator has a peak to peak votlage of 100 μV and a source resistanfe of 10 kΩ. If the electrode resistance are 20 kΩ. the voltage across the amplifier terminals with an input impedance of 10 MW is
(a) 93.5 μV (b) 95.5 μV
(c) 97.5 μV (d) 99.5 μV

73. Given input impedance of amplifier is 10 MΩ. While recording EEG signal, Cranial Potential of the brain is represented by a current source of 20 pA and with source resistance 5 MΩ. The electrode scalp resistance is 5 MΩ. The problem faced by the EEG amplifier if sensing out the problem facted by the EEG amplifier if sensing capability is 60 μV is
(a) Amplifier goes into saturation
(b) Amplifier does not sence input
(c) With such input amplifier will be burnt out
(d) None of the above

74. How many leads are there in ECG ?
(a) 10 (b) 1
(c) 12 (d) 13

75. Of the given lead voltage, which of the set of voltages for Lead I, II, and III is correct.
(a) 1.8 mV, 2.9 mV, 1.2 mV
(b) 2.9 mV, 1.8 mV, 1.2 mV
(c) 2.1 mV, 3.7 mV, 1.6 mV
(d) 1.1 mV, 3.2 mV, 9 mV

1.9 Clinical Measurements

1. A **gas chromatograph** is used for
   (a) Measuring flow rate of a gas
   (b) Measuring the temperature of a gas
   (c) Measuring the pressure of a gas
   (d) Analysing the composition of a gas

2. The following method is widely accepted to determine oxides of Nitrogen in an **automobile emission**.
   (a) Orsat analysis
   (b) Gas – chromatography
   (c) Chemi luminescence
   (d) Flame ionization detection

3. In a particular sample, the absorbance is 0.6. For a molar concentration of the solute of 1.0×10^{-4} M and a 2.0 cm path length, the molar absorptivity (litre per mol per cm) is:
   (a) 1.2 (b) 1200
   (c) 3000 (d) 3×10^{-4}

4. The transmittance of a coloured solution is 0.5. The **absorbance** of the solution is:
   (a) 0.3 (b) 0.69
   (c) 3.16 (d) – 1.5

5. The transmittance of a particular solution measured is T. The concentration of the solution is now doubled. Assuming that **Beer-Lambert's law** holds good for both the cases, the transmittance for the second would be
   (a) T / 2 (b) 2T
   (c) T^2 (d) $\sqrt{T}$

6. A spectrophotometer is used to measure concentration of a molecule in a solution based on Beer's law using a cuvetter of path 10 mm and detectable absorbance of 0.01. The molar absorptivity of the solution is 10^4 litres / mol/ cm. The minimum connectration of the solution that can be dectected is
   (a) 1 μ mol/ litre (b) 2 μ mil/ litre
   (c) 5 μ mol/ litre (d) 10 μ mol/ litre

7. Light of intensity I_o is equally divided and passed through 2 cuvettes P_1 and P_2 containing an analyte at concentrations c and 0.5 c, respectively. The corresponding path lengths in P_1 and P_2 are 4 cm and 1 cm. The cross sectional areas are 1 cm² and 3 cm², respectively. The ratio of the absorbance in P_1 and P_2 is
   (a) 1.5 (b) 8 / 3
   (c) 3 (d) 8

8. A solution "P" is put in a spectrophotometer cuvette of optical path length 1cm. The transmittance is found to be 10%. Another solution "Q" has a transmittance of 40% under the same circumstances. If equal volumes of P and Q are mixed together, the transmittance of the resulting solution (assuming the constituents of P and Q do not react with each other) is, approximately.
   (a) 15% (b) 20%
   (c) 25% (d) 30%

9. The time taken for one heart beat cycle are
   1. 0.7s
   2. 0.8 s
   3. 0.9 s
   (a) 85, 75, 66 (b) 75, 85, 66
   (c) 66, 75, 85 (d) none

10. Korotkoff sounds are used
    (a) As a reference for sound level mesurement
    (b) For studying heart muscle functioning
    (c) For blood pressure measurement
    (d) For study of heat vlaue functioning

11. For Bio-Potential recording, the impedacne of an Ag/ AgCl electrode as compared to that of an Ag elec trode of the same size is
    (a) Lower at all frequencies
    (b) Lower at all frequencies less than 10 Hz nad higher for freqiencies greater than 300 Hz.
    (c) Higher at all frequencies less than 19 Hz and lower for frequencies greater than 300 Hz
    (d) Higher at all frequencies

12. In a particular sample, the absorbance is 0.6. For a molar concentation of the solute of 1.0×10^{-4} M nad a 2.0 cm path length, the molar concentration absorptivity (liter/mol/cm) is
    (a) 1.2 (b) 1200
    (c) 3000 (d) 3×10^{-4}

13. A gas chromatography is used for
    (a) Measuring flow rate of a gas
    (b) Measuring temperature of the gas
    (c) Measuring pressure of the gas
    (d) Analyzing composition of the gas

14. The following method is widely accepted to determine oxides of nitrogen in an automobile emission.
    (a) ORSAT-analysis
    (b) gas chromatography
    (c) chemi-luminescence
    (d) FID

15. The stroke volume of heart, if the cardiac output is 5.2 lit and pulse rate is 72/min is ....

(a) 70 ml (b) 75 ml
(c) 72 ml (d) 77 ml

16. A cell in the resting state is said to be ....
(a) Depolarized (b) Polarized
(c) active (d) none

17. The preferred electrode in general to record bio - potentials would be.
(a) Plate electrode (b) needle electrode
(c) micro electrode (d) all the above

18. Korotkoff sound are used in:
(a) Blood flow measurements
(b) Blood pressure measurements
(c) Heart valve functioning
(d) As a reference for sound level measurement.

19. Instrument measuring volume changes in any part of the body due to pulsations of blood occuring with each heart beat is:
(a) Plethysmograph (b) Arithmetic unit.
(c) Ventilator (d) Cardiac pacemaker

20. In NMR blood flowmeter, **alarms** frequencies is given as:
(a) $\frac{2\pi}{v}$ (b) $\frac{r}{B_o}$
(c) rB_0 (d) $\frac{v}{B_0}$

21. For a patient pulse rate is 100 bpm. The time taken for 1 cardiac cycle is
(a) 0.6 min (b) 1 min
(c) 10 min (d) None of the above

22. Which of the following instruments uses an ultrasonic transducer is
(a) Echo-cardiograph
(b) Electro-cardiograph
(c) Electro-enchephalograph
(d) None of the above

23. Which of the following equipment is used to estimate erythrocytes present in the blood
(a) Electrophoresis (b) Colorimeter
(c) Cell counter (d) None of the above

24. Which type of filter is used to remove 50 Hz line interference is
(a) Band stop filter (b) Low pass filter
(c) High pass filter (d) None of the above

25. Which one of the following is not an optical system
(a) Monochromator (b) Colorimeter
(c) Flame photmeter (d) None of the above

26. Which one of the following is expressed as % of total blood volume
(a) Packed cell volume
(b) Mean cell volume
(c) Mean cell Hemoglobin
(d) Mean cell hemoglobin concentration

27. If the packed cell volume is 45% and 5 ′ 10^{12} red blood celles are present per lit, then mean cell volume is
(a) 9 plit (b) 0.9 plit
(c) 0.09 plit (d) None of the above

28. Which of the following is used to measure % transmittance or absorbance?
(a) Coulter (b) colourimeter
(c) Chloridimeter (d) None of the above

29. The volume of the packed cells is called
(a) Hemotocrit
(b) Mean cell volume
(c) Mean cell hemoglobin concentration
(d) None of the above

30. Optical density is given by
(a) $\log_{10}(1/T)$ (b) $\log_{10}(100/T)$
(c) $\log_{10}(T)$ (d) None of the above

31. The color emitted by sodium ions in the flame photometer is
(a) Yellow (b) Red
(c) Violet (d) Brown

32. The time taken for one heart beat cycle are
1. 0.7s
2. 0.8 s
3. 0.9 s
Then pulse rate for the above is –, – and –
(a) 85, 75, 66 (b) 75, 85, 66
(c) 66, 75, 85 (d) none

Linked Answer Questions
Statement for linked Answer Questions 33 and 34

The intra cellular K^+ concentration of a group of cell is 150×10^{-6} moles/cm³. The extra cellular conc. Of K^+ averages to 6×10^{-6} moles/cm³ . Find the

33. Concentration ratio is
(a) $\frac{1}{10}$ (b) $\frac{1}{25}$
(c) $\frac{1}{30}$ (d) $\frac{1}{15}$

34. The diffusion potential is
(a) ± 61.9 (b) ± 25.3
(c) ± 85.3 (d) ± 22.5

35. In a particular sample, the absorbance is 0.6. For a molar concentation of the solute of 1.0×10^{-4} M nad a 2.0 cm path length, the molar concentration absorptivity (liter/mol/cm) is
(a) 1.2 (b) 1200
(c) 3000 (d) 3×10^{-4}

Common Data for Questions 36 and 37

A particular solution sample of colored sobstance known to follow beers law shown 80% transmittance when measured in a cuvette of 1 cm optical path length

36. The percentage transmittance for the solution of twice the concentraion in the sme curve.

(a) 60% (b) 64 %
(c) 30% (d) 32 %

37. The path lenght inthe cuvette to give the same transmissin 80% for the solution of twice the original concentration is

(a) 2 cm (b) 1 cm
(c) 4 cm (d) 0.5 cm

38. The transmittance of colored solution is 0.5 the absorbance of the solution is

(a) 0.3 (b) 0.69
(c) 3.16 (d) −1.5

39. The transmittance of a particular solution measured is T. The concentration of the solution is now doubled assuming that beer law holds good for both the cases, the transmittance for the second would be

(a) T/2 (b) 2T
(c) T^2 (d) $\sqrt{T}$

40. The transmitted sampled solution is monitored using 500 mm wavelength. The transmittance is reduced from 100% to 10% due to addition of pollutants. As a result, the absorbance recorded by the spectrophotometer chnges from

(a) 0 to 1 (b) 2 to 1
(c) 0 to 2 (d) 0 to 10

41. Light of intensity I_o is eqully divided and passed throuhg 2 curvettes P1 and P2 containing an analyte at concentrations 'C' and '0.5 C' respectively. The corresponding path lengths in P1 and P2 are 4cm and 1 cm. The cross sectional areas are $1cm^2$ and $3cm^2$ respectively. The ratio of the absorbance in P1 and P2 is

(a) 1.5 (b) 8/3
(c) 3 (d) 8

42. A solution 'P' is put in a psectrophotometer cuvette of optical path length 1cm. The transmittance is found to be 10%. Another solution Q has a transmittance of 40%. under the same circumstances. If equal volumes of P and Q are mixed together, the trasmittance of the resulting solution [assuming the contitutents of P and Q do not react with each other] approx

(a) 5% (b) 6%
(c) 4% (d) none

Common Data for Questions 43, 44 and 45

If given RBC = 5 million/μl; Hb = 15 gm/dl and HCT = 45%

43. The value of MCV (μm^3) is

(a) $0.9\times10^9\ \mu m^3$ (b) $0.9\times10^{15}\ \mu m^3$
(c) $0.3\times10^9\ \mu m^3$ (d) $0.3\times10^{10}\ \mu m^3$

44. The value of MCV (πγμ) is

(a) 20 *pgm* (b) 25 *pgm*
(c) 30 *pgm* (d) 35 *pgm*

45. The value of MCV (gm/dl) is

(a) $\frac{15}{2}$ (b) $\frac{10}{3}$
(c) $\frac{15}{3}$ (d) $\frac{10}{2}$

46. The stroke volume of heart, if the cardiac output is 5.2 lit and pulse rate is 72/min is ....

(a) 70 ml (b) 75 ml
(c) 72 ml (d) 77 ml

47. The preferred electrode in general to record bio - potentials would be.

(a) Plate electrode (b) needle electrode
(c) micro electrode (d) all the above

48. The cardiac output is determined with the help of indicator dilution method. 10 mg of the indicator is injeced in the blood stream and the average concentrton of the indicator as calaculated from the dilution curve was found to be 5 mg /*l* for a curve duraton of 20 sec. Find the cardiac ouptut.

(a) $2l/min$ (b) $3l/min$
(c) $10l/min$ (d) $6l/min$

49. The caridac output is measured with the help of indicator dilution method. The quantity of indicator injected is 20 mg and the area under the die dilution curve is found o be 180 mg sec/ litre. Find the caridac output.

(a) $6l/min$ (b) $666l/min$
(c) $8l/min$ (d) $8.66l/min$

50. Cardiac output is

(a) The amount of air pumped by lungs per minute.
(b) The amount of air pumped by lungs per second.
(c) The amount of blood delivered by the heart to the aorta per minute.
(d) The amount of blood delivered by the heart to the aorta per second.

51. The transmittance of a particular solution measurement is T, The concentration of he solution is now doubled. Assuming that Beer - Lambert's law holds good for both the cases, the transmittance fr the second would be.

(a) T/2 (b) 2 T
(c) T^2 (d) $\sqrt{T}$

52. A blood sample of patient is taken and is found tha the packed cell volume of the blood sample is 0.45 and the number of red blood corpuscles are found to be 5×10^{12}, then find the mean cell volume.

(a) 90 *f*/*l* (b) 80 *f*/*l*
(c) 150 *f*/*l* (d) 45 *f*/*l*

53. For a blood sample 15 g of Hb is present per deciletre (d) of the blood and there are 5×10^{12} RBC's per

litre of the blood find the mean cell haemoglobin
(a) 29 pg (b) 300 pg
(c) 39 pg (d) 60 pg

54. On testing a blood sample it is found that it contains 15 g of Hb per declitre of blood sample and a PCV of 0.45. Find the mean cell heamoglobin concenration for he blood sample.
(a) 33.3 g/dl (b) 28.3 g/dl
(c) 33.9 g/dl (d) 31.2 g/dl

55. The counter metod used for counting of the blood cells is based in the principle of the change in.
(a) electrical conductivity
(b) volume
(c) weight
(d) number of blood cells

56. Number of cells in the blood can be estimated by
(a) Bicycle test (b) Coulter counter
(c) Treadmill test (d) Beam balance

57. An expreiment was carried out to count the number of cells present in the blood with the help of a 'Picoscale'. I was observed that the dilution factor is 63000 and the result displayed on the display was 520, the scaling factor of the counter is 60 and the measurd volume was 0.378 cm^3. Find the number of cells in a unit volume of the blood.
(a) 5.15×10^6 per mm^3
(b) 4.89×10^6 per mm^3
(c) 5.35×10^6 per mm^3
(d) 5.20×10^6 per mm^3

58. Spirometer is used to mesure;
(a) Lung volume and capacity
(b) Cardiac volume
(c) Breathing rate

1.10 Ultrasonic Transducers and Ultrasonography

1. In an **electromagnetic blood flow meter,** the induced voltage is directly proportional to the
(a) Blood flow rate
(b) Square root of the blood flow rate
(c) Square of the blood flow rate
(d) Logarithm of the blood flow rate

2. An ultransonic beam of frequency 1 MHz and intensity 0.5 W/cm^2 passes through a layer of soft tissue of thickness t with an attenuation coefficient of 1.18 cm^{-1}. The ratio of output to input power is $1/e^2$. The thickness of the tissue is
(a) 1 cm (b) 1.695 cm
(c) 2.408 cm (d) 3.721 cm

Statement for Linked Answer Questions 3 and 4:

Blood flow through a straight segment of an artery is measured using a Doppler flow meter. The probe is oriented at an angle of 45 degrees to the longitudinal axis of the artery. It is known that Doppler shift is proportional to the velocity of the blood in the direction of propagation of sound from the probe and the frequency generated from the probe. The velocity of sound in blood is 1500 m/s and the maximum blood flow velocity to be measured is 110 cm/s.

3. Signal processing limitations constrain the maximum Doppler shift to be less than 3 kHz. The maximum source frequency (probe output), to the nearest MHz, should be.
(a) 2 MHz (b) 3 MHz
(c) 5 MHz (d) 10 MHz

4. In above equation if the piezoelectric crystal of the probe is made of quartz (Young's modulus,

$$Y = 80 \times 10^9 \text{ Nm}^{-2},\ \rho = 2.65 \times 10^3 \text{ kg m}^{-3})$$

the maximum thickness of the crystal required to resonate at the frequency determined above, to the nearest mm, is
(a) 1 mm (b) 2 mm
(c) 3 mm (d) 4 mm

5. A 5MHz acoustic pulse travels from a transducer through a 2 cm thick fat tissue before it encounters an interface with a liver tissue at normal incidence. The amplitude attenuation factors of fat and liver are 0.075 $Npcm^{-1}/MHz$ and 0.1 $Npcm^{-1}/MHz$ respectively. The amplitude reflectivety coefficient of fat liver interface is 0.1. Taking both attenuation and reflection losses into account. the amplitude loss (in dB) of echo pulse when it returns to the transducer is
(a) 0.74 (b) – 2.6
(c) – 6 (d) – 33

6. In a pulsed ultrasound imaging system, a single 5 MHz crystal is used both as source and as detector. Bursts of at least 20 cycles are needed for acceptable image quality. The velocity of sound in the tissue being imaged is 1500 m/s. The minimum distance of the objects to be imaged should be
(a) 12 mm (b) 6 mm
(c) 1 mm (d) 3 mm

7. High frequency Ultra sonic transducer is used for
(a) Infant (b) peripheral muscles
(c) thin persons (d) all the above

8. The density of material A is 1.3 g/c.c and that of B is 0.88g/c.c. The velocity of sound in A is 1400m/s and that of B is 1100m/s. Find the percentage of fraction of reflected energy of ultra sound beam.
(a) 9.3 % (b) 10.3 %
(c) 11 % (d) 14 %

9. A 5 MHz acoustic pulse travels from a transducer throuhg a 2 cm thick fat tissue before it encounter an interface with the liver tissue at normal incidence. The amplitude attenuation factors of fat and liver are 0.075 N pcm^{-1}/MHz nd 0.1 Npcm^{-1}/MHz respectively. the amplitude reflectivity co-efficient of fat-liver interface i s 0.1. Taking both attenuation and reflection losses into account, the amplitude losses (dB) of Echo pulse when it returns to the transducer is
(a) 0.74 (b) –2.6
(c) –6 (d) –33

10. The ultra sound image is generally termed as
(a) Ultragram (b) mechogram
(c) sonogram (d) all the above

11. Transit time blood flow meter piezo electric crystals 2cms a part with an angle $45°$. Find the mean flow velocity of blood if trasit time and velocity of sound are 0.25μ sec and $1200 m/s$ respectively ?
(a) $5.72 m/s$ (b) $120.6 m/s$
(c) $12.72 m/s$ (d) $2.7 m/s$

12. In an electromagnetic blood flow meter, the induced voltage is directly proportional to ....
(a) cube of blood flow rate
(b) blood flow rate
(c) square of blood flow rate
(d) logarithm of blood flow rate

13. A doppler flow meter uses 2 trasmitters one at $45°$ angle adn other at $30°$ with the flow axis. Find the doppler shift of 7 MHz is flow velocity of bleed is $8 cm/s$ and speed of sound in blood is $1100 m/s$.
(a) 210 Hz (b) 1601.73 Hz
(c) 2000 Hz (d) 1000 Hz

14. An ultra sonic beam of frequency 1 MHz and intensity $0.5\ W/cm^2$ passes through a layer of soft tissues of thickness 't' with an attenuation co-efficient of $1.18\ cm^{-1}$. The ratio of output to input power is $1/e^2$. The thickness of tissue is ....
(a) 1 cm (b) 1.695 cm
(c) 2.408 cm (d) 3.721 cm

15. Signal processing limitations constration the maximu m doppler shift to be less than 3KHz. The maximum source freqency to the nearest MHz, should be.
(a) 2 MHz (b) 3 MHz
(c) 5 MHz (d) 10 MHz

16. If the piezo electric crystal required to resonate at the frequency
(a) 2 MHz (b) 3 MHz
(c) 5 MHz (d) 10 MHz

17. Attenuation of ultrasounds is maximum in ......
(a) bone (b) fat
(c) blood (d) muscle

18. The density of material A is 1.3 g/c.c and that of B is 0.88g/c.c. The velocity of sound in A is 1400m/s and that of B is 1100m/s. The percentage of fraction of reflected energy of ultra sound beam is
(a) 9.3 % (b) 10.3 %
(c) 11 % (d) 14 %

19. The energy of ultrasound for a half value layer of a soft tissue is 17 mJ. The incident enrgy on tissue is
(a) 17 mJ (b) 51 mJ
(c) 24 mJ (d) 34 mJ

20. Ultrasound waves are passed through soft tissue-compact bone interface. ultra sound echo wave received after $80\mu s$ if velocity of ultra sound in soft tissue and compact bone are $1500 m/s$ and $4000 m/s$ respectively. The thickness of the soft tissue is
(a) 6 cm (b) 8 cm
(c) 10 cm (d) 15 cm

21. A 5 MHz acoustic pulse travels from a transducer throuhg a 2 cm thick fat tissue before it encounter an interface with the liver tissue at normal incidence. The amplitude attenuation factors of fat and liver are 0.075 N pcm^{-1}/MHz nd 0.1 Npcm^{-1}/MHz respectively. the amplitude reflectivity co-efficient of fat-liver interface i s 0.1. Taking both attenuation and reflection losses into account, the amplitude losses (dB) of Echo pulse when it returns to the transducer is
(a) 0.74 (b) –2.6
(c) –6 (d) –33

22. Transit time blood flow meter piezo electric crystals 2cms a part with an angle $45°$. If trasit time and velocity of sound are 0.25μ sec and $1200 m/s$ respectively . The mean flow velocity of blood is
(a) $5.72 m/s$ (b) $120.6 m/s$
(c) $12.72 m/s$ (d) $2.7 m/s$

23. A doppler flow meter uses 2 trasmitters one at $45°$ angle adn other at $30°$ with the flow axis. If flow velocity of blood is 8 cm/s and speed of sound in blood is $1100 m/s$. The doppler shift of 7 MHz is
(a) 210 Hz (b) 1601.73 Hz
(c) 2000 Hz (d) 1000 Hz

24. An ultrasonic beam of frequency 1 MHz and intensity $0.5\ W/cm^2$ passes through a layer of soft tissues of thickness 't' with an attenuation co-efficient of $1.18\ cm^{-1}$. The ratio of output to input power is $1/e^2$. The thickness

of tissue is
(a) 1 *cm*
(b) 1.695 *cm*
(c) 2.408 *cm*
(d) 3.721 *cm*

25. Attenuation of ultrasounds is maximum in ......
(a) bone
(b) fat
(c) blood
(d) muscle

26. In an electromagnetic blood flow meter, the induced voltage is directly proportional to
(a) Blood flow rate
(b) Square root of the blood flow rare
(c) Square of the blood fow rate
(d) Logarithm of the blood flow rate

27. Which of the following instrument uses an ultrasonic transducer
(a) Echo-cardiograph
(b) Wlectro caridograph
(c) Electro encephalograph
(d) Manometer

28. The trasnducer used for direct measurement of Blood pressure is
(a) LVDT
(b) Strain Gauge
(c) Capacitive
(d) Piezo electric

29. Echoencephalograph employs the following display
(a) A mode (b) B mode
(c) M mode (d) B scan

30. With a stroboscop synchronism has been obtained at the flashing rates of 25 and 20 pulses per second the speed of rotation is
(a) 0.6 rev/min
(b) 50 rev/min
(c) 100 rev/min
(d) 6000 rev/min

31. Plethsymograph is used to measure
(a) Blood pressure
(b) Blood flow
(c) Glucose content in blood
(d) None of the above

32. A well known chartersitics of ultrasound frequently used in Biomedical instrumental in
(a) Doppler dffect
(b) Early effect
(c) Peltier effect
(d) None of the above

33. In a Electromagnetic blood flow meter the induced volt is directly proportional
(a) Square root of blood flow rate
(b) Square root flow rate
(c) Blood flow rate
(d) Log of blood flow rate

34. The velocity of ultrasould wave in blood is 1500 m/s. Angle between direction of flow and direction of incident beam is 30º, Doppler shift in frequency is 231 Hz and and incident frequency is 2 MHz then velocity blood is
(a) 0.1 m/s
(b) 0.2 m/s
(c) 0.3 m/s
(d) None of the above

1.11 Computed Tomography

1. A substance has linear attenuation coefficient of 0.190 cm^{-1} then the expected tissue is ....
(a) blood
(b) fat
(c) CSF
(d) muscle

2. The principle of obtaining X-ray images from a number of vantage points is used in the following tomography
(a) linear
(b) Simple
(c) Axial
(d) None of the above

3. The following crystal used to measure the intensity of X-rays after penstrating the body is
(a) Sodium iodide
(b) Barium iodide
(c) Calcium chloride, X-enon
(d) Both (a) and (c)

4. In which tomography a narrow pencil like X-ray beam is made to scan a single perpendicular to body axis?
(a) linear
(b) Simple
(c) Axial
(d) none

5. Which of the following is used as the most popular image reconstructing principle?
(a) FFT based method
(b) FT based method
(c) Holographic method
(d) None of the above

6. For a monochromatric X-ray beam, the tissue attenuation characteristics can be dscribed by the equation.

   (a) $I_o = I_t e^{-\mu x}$
   (b) $I_o = -I_t e^{-\mu x}$
   (c) $I_t = I_o e^{-\mu x}$
   (d) $I_t = -I_o e^{-\mu x}$

7. Find the thickness of the tissue which attenuate the incident beam of X - Ray to 1/20 of the initial value when the attenuation coefficient of the tissue is 1.62/cm.

   (a) 135 cm
   (b) 1.85 cm
   (c) 2.05 cm
   (d) 2.75 cm

8. Which of the following type of image produced by a CT scan machine

   (a) 1 - D image
   (b) 2 - D image
   (c) 3 - D image
   (d) all of the above

9. Find the CT number (Hounsfield unit) of the heart muscle ($\mu = 0.212$) if the attenuation coefficient of water is 0.205.

   (a) $\frac{205}{7000}$
   (b) $\frac{405}{7000}$
   (c) $\frac{7000}{205}$
   (d) $\frac{7000}{405}$

Answers

1.1 UV, Visible and IR Spectroscopy

1. **Ans. (a)**

Solution

$$n\lambda = d[\sin i + \sin r]$$

$$\lambda = \frac{1\,mm}{1200}[\sin 25° + \sin 15°]$$

$$\lambda = \frac{10^{-3}\,m}{1200}[\sin 25° + \sin 15°]$$

$$\lambda = \frac{10^{-3}\,m}{1200}[0.4226 + 0.2588]$$

$$\lambda = \frac{10^{-3}\,m}{1200}[0.6814]$$

$$\lambda = 567.9\,nm$$

2. **Ans. (c)**

Solution:

$$\text{Resolution} = \frac{\lambda_2 + \lambda_1}{2(\lambda_2 - \lambda_1)} = \frac{1}{2} \times \frac{(600.1 + 599.9)}{600.1 - 599.9}$$

$$= \frac{2 \times 600}{0.02 \times 2} = \frac{6000}{2} = 3000$$

3. **Ans. (d)**

4. **Ans. (a)**

5. **Ans. (c)**

Solution:

We know that for grating

$$\frac{\lambda}{d\lambda} = nL$$

For minimum number of lines take $n = 1$

$$\therefore\ L\min = \frac{589}{589.6 - 589}$$

$$= \frac{589}{0.6 - 589} \cong 982$$

6. **Ans. (d)**

7. **Ans. (d)**

Solution:

$$t = \frac{n\lambda}{2\mu}$$

$$t = \frac{1 \times 455 \times 10^{-9}}{2 \times 1.38}$$

$$t = 165\,nm$$

8. **Ans. (a)**

Solution:

Relation between absorbance and transmittance is

$$A = \log\frac{1}{T}$$

for 100% transmittance $T = 1 \Rightarrow A = 0$

for 10% transmittance $T = 0.1 \Rightarrow A = \log\frac{1}{0.1}$

$$A = 1$$

9. **Ans. (c)**

Solution:

Given, Ref index, $\mu = 1.38$

Wavelength, $\lambda = 455\,nm$

Thickness of dielectric, $t = ?$

order, $n = 1$

$$n\lambda = 2t\mu$$

$$t = \frac{n\lambda}{2\mu}$$

$$= \frac{1 \times 455 \times 10^{-9}}{2 \times .38}$$

$$= 165\,nm$$

10. **Ans. (b)**

Solution:

Given R.I. $\mu = 1.35$

$$t = 1.85\,nm$$

$$n = 1$$

$$n = 2$$

$$n = 3$$

$$n\lambda = 2t\mu$$

$$\lambda_1 = \frac{2t\,\mu}{n}$$

$$= \frac{2 \times 185 \times 10^{-9} \times 1.35}{1}$$

$$= 499.5\ nm$$

Similarly,

$$\lambda_2 = \frac{2t\,\mu}{2} = 249.75\ nm$$

$$\&\ \lambda_3 = \frac{2t\,\mu}{3} = 166.5\ nm$$

11. Ans. (b)

Solution:

Given, $t = 1\ cm,$

$$\frac{d\mu}{d\lambda} = 1000\ cm^{-1}$$

$$\lambda = 6000\,\mathring{A}$$

$$\Delta\lambda = ?$$

it is known that

$$R = \frac{\lambda}{\Delta\lambda}\ \&\ R = \frac{k\,d\mu}{d\lambda}$$

where R = resolution

$$\frac{\lambda}{\Delta\lambda} = t\frac{d\mu}{d\lambda}$$

$$\Delta\lambda = \frac{\lambda}{t}\left(\frac{d\mu}{d\lambda}\right)$$

$$= \frac{6000 \times 10^{-10}}{1} \times \frac{1}{1000} = 6\,\mathring{A}$$

12. Ans. (a)

Solution :-

Given, $1\ mm \rightarrow 1200$ Lines

$$\therefore 1\ line \rightarrow \left(\frac{1}{1200}\right) mm = d$$

$$\theta_i = 25^\circ$$

$$n = 1$$

$$\theta_d = 15^\circ$$

$$\lambda = ?$$

$$\because\ n\lambda = d(\sin\theta_i + \sin\theta_d)$$

$$\lambda = \frac{1}{1200 \times 10^{-3}} \times (\sin 15^\circ + \sin 25^\circ)$$

$$\lambda = 567\ nm$$

13. Ans. (d)

Solution:

Given, $\frac{d\mu}{d\lambda} = 952\ cm^{-1}$

$$\Delta\lambda = 6\,\mathring{A}$$

$$\lambda = 5893\,\mathring{A}$$

$$R = t\frac{d\mu}{d\lambda} = \frac{\lambda}{\Delta\lambda}$$

$$t = \frac{5893}{6} \times \frac{1}{952\ cm^{-1}}$$

$$t = 1.031\ cm$$

14. Ans. (c)

15. Ans. (c)

Solution:

Given, $\lambda_1 = 589\ nm$

$$\lambda_2 = 589.6\ nm$$

$$n = ?$$

$$R = \frac{\lambda}{\Delta\lambda} = nL$$

$$L = \frac{\lambda}{\Delta\lambda} = 982\left[\begin{array}{l}\because \lambda = \left(\frac{\lambda_2 + \lambda_1}{2}\right) \\ \&\ (\Delta\lambda = \lambda_2 - \lambda)\end{array}\right]$$

16. Ans. (a)

Solution:

Given, $\lambda = 300\ nm$

$$\theta_i = 0^\circ$$

$$1mm \rightarrow 1180\ groves$$

$$\Rightarrow 1\ groves \Rightarrow \left(\frac{1}{1180}\right) mm$$

$n = 1$

$\theta_d = ?$

$n\lambda = d(\sin\theta_i + \sin\theta_d)$

$$1\times 300\times 10^{-9} = \left(\frac{1}{1180}\right)\times 10^{-3}(0+\sin\theta_d)$$

$\Rightarrow \theta_d = 20.73°$

17. Ans. (d)

Solution:

Given R.I. *of air*, $\mu = 1$

$\lambda = 600\ nm$

$$t = \frac{\lambda}{4\mu}$$

$$= \frac{600}{4\times 1}$$

$t = 150\ nm$

18. Ans. (b)

Solution:

Given, $0.5\ cm \rightarrow 2500\ lines$

$$1\ line \rightarrow \left(\frac{0.5}{2500}\right) cm = d$$

$\lambda_1 = 5890\ \mathring{A}$

$\lambda_2 = 5896\ \mathring{A}$

$n\lambda = d\sin\theta_d$

$$\theta_{d_1} = \sin^{-1}\left(\frac{n\lambda}{d}\right)$$

$$= \sin^{-1}\left(\frac{1\times 5890\times 2500\times 10^{-10}}{0.5\times 10^{-2}}\right)$$

$= 17.12°$

$$\&\ \theta_{d_2} = \sin^{-1}\left(\frac{1\times 5896\times 2500\times 10^{-10}}{0.5\times 10^{-2}}\right)$$

$= 17.14°$

19. Ans. (b)

Solution:

$$R = \frac{\lambda}{\Delta\lambda}$$

$$= \frac{5896+5890}{2(5896-5890)}$$

$$= \frac{5893}{6}$$

$= 982.1$

20. Ans. (b)

21. Ans. (b)

22. Ans. (c)

23. Ans. (d)

24. Ans. (b)

25. Ans. (d)

26. Ans. (c)

27. Ans. (b)

28. Ans. (a)

29. Ans. (d)

30. Ans. (d)

31. Ans. (b)

32. Ans. (a)

33. Ans. (a)

34. Ans. (b)

35. Ans. (a)

36. Ans. (a)

37. Ans. (c)

38. Ans. (d)

Solution:

For angle of incidence and refraction θ, ϕ we must have $\theta + \phi = 90°$ or $\theta = \phi - 90°$.

From Snell's law, $\sin\theta = n\sin(90° - \theta)$

$= n\cos\theta$

or $\theta = \tan^{-1} n = \tan^{-1} 1.33 = 53°$

39. Ans. (a)

Solution:

$\lambda_1 = 5997\ \mathring{A}$

$\lambda_2 = 6003\ \mathring{A}$

Dispersive power = ratio of wavelength

$$= \frac{6003}{5997}\times 1000$$

$= 1\times 10^{-3}$

$= 0.001$

40. Ans. (a)

Solution:

Shift in central fringe.

$$\Delta = \frac{D}{2d}(\mu - 1)t$$

Given that $\frac{D}{2d}(\mu - 1)\,t = \frac{D}{2d}5\lambda$

or $(\mu - 1) = 5\lambda = 5 \times 589 \times 10^{-9}$

or $\mu = 1.589$

41. Ans. (b)

Solution:

Diameter of *n*th dark ring with liquid

$$D_n^2 = \frac{4n\lambda R}{\mu}$$

$$D_{na}^2 = 4n\,\lambda\,R \text{ (with air)}$$

Given that $\frac{D_{na}}{D_n} = 2$

Hence, $\mu = \frac{D_{na}^2}{D_n^2} = 4$

42. Ans. (c)

Solution:

$$\lambda = \frac{D_{n+p}^2 - D_n^2}{4\,pR}$$

Here $D_n = 4.2\ mm = 0.42\ cm$

$D_{n+p} = D_{n+14} = 7\ mm = 0.7\ cm$

$P = 14$

$R = 1\text{ m} = 100\text{ cm}$

Hence $\lambda = 5600\ \text{Å}$

43. Ans. (c)

Solution:

Loss co-effcient for photon

$$a_{\omega t} = a_{loss} + a_R = a_{loss}$$

$$= \frac{1}{2}\ell n\ R$$

$$= 65.43/cm$$

$$\tau_{ph} = \frac{1}{v\,a_{\omega t}}$$

$$= \frac{3.6}{3 \times 10^6 \times 65.43}$$

$$= 0.51\ ps$$

44. Ans. (b)

Solution:

We find that, n = actual depth/apparent depth, so $1.333 = (5\,ft)/y$ and $y = 3.75\ ft$.

45. Ans. (d)

Solution:

Analyze the refraction in sequence from the left

$$n = \frac{\sin\theta_1}{\sin\theta_2}$$

or $1.50 = \frac{\sin 30^\circ}{\sin\theta_2}$

$\sin\theta_2 = \frac{0.5}{1.50} = 0.3333$ and

$$\theta_2 = 19.5^\circ$$

$$\theta_2 + \theta_3 + 120^\circ = 180^\circ$$

or $\theta_2 + \theta_3 = 60^\circ$

The prism angle. Then $19.5^\circ + \theta_3 = 60^\circ$ and $\theta_3 = 40.5^\circ$

$n = \frac{\sin\theta_4}{\sin\theta_3}$, $1.50 = \frac{\sin\theta_4}{\sin 40.5^\circ}$

$0.6494(1.50) = \sin\theta_4$ and $\theta_4 = 77^\circ$

46. Ans. (c)

Solution:

The angles are obtained from snell's law

$$\theta_2 + \phi_2 + (180^\circ - A) = 180^\circ$$

$$\theta_2 + \phi_2 = A$$

$$\sin\theta_2 = \frac{n_1}{n_2}\sin\theta_1$$

$$= \frac{\sin 30^\circ}{1.50} = 0.333$$

$$\theta_2 = 19.5^\circ$$

$$\phi_2 = A - \theta_2 = 55^\circ - 19.5^\circ = 35.5^\circ$$

47. Ans. (d)

Solution:

Use Snell's law for the air - water inter surface.

$$n_w = \frac{\sin\theta_1}{\sin\theta_2}$$

$$1.333 = \frac{\sin 45^\circ}{\sin\theta_2} = \frac{0.7071}{\sin\theta_2}$$

$$\sin\theta_2 = \frac{0.7071}{1.333} = 0.5305$$

and $\theta_2 = 32^\circ$

Now use Snell's law for the water - glass inter surface.

$$n_{BA} = \frac{n_g}{n_w} = \frac{\sin\theta_2}{\sin\theta_3}$$

$$\frac{1.63}{1.333} = \frac{0.5305}{\sin\theta_3}$$

$$\sin\theta_3 = \frac{0.5305\,(1.333)}{1.63} = 0.4338$$

and $\theta_3 = 26^\circ$

48. Ans. (c)

Solution:

Here the light rays bend inward in the benzene and outward in the water. Use Snell's law first at the air - benzene surface and then at the benzene - water surface.

$$n_b = \frac{\sin\theta_1}{\sin\theta_2}$$

$$1.50 = \frac{\sin 60^\circ}{\sin\theta_2}$$

$$\sin\theta_2 = \frac{0.8660}{1.50} = 0.5773$$

and $\theta_2 = 35$

49. Ans. (d)

Solution:

$y = 20\sin\omega t + 20\sin(\omega t + 60^\circ)$.

We can think of y_1 and y_2 as projections on the y-axis of vectors of length 20 rotating with angular velocity ω and making angles ωt and $\omega t + 60^\circ$, respectively, with the x – axis. The projection of the vector sum of these two vectors on the y axis is then our result $y = y_1 + y_2$. $A^2 = 20^2 + 20^2 + 2(20)\cos 60^\circ = 3(20^2)$;

$A = 34.6$; and from the isosceles triangle $2\phi = 60^\circ$; $\phi = 30^\circ$.

Thus, $y = 34.6\sin(\omega t + 30^\circ)$.

50. Ans. (d)

Solution:

$\cos\omega t = \sin\left(\omega t + \frac{\pi}{2}\right)$. Thus y_2 "leads" y_1 by 90°.

$A^2 = 30^2 + 30^2 = 2(30^2)$; $A = 42.4$. For the right isosceles tringle, $\phi = 45^\circ$ and

$y = y_1 + y_2 = 42.4\sin(\omega t + 45^\circ)$.

51. Ans. (c)

Solution:

$$\phi = 2\sin^{-1}\sqrt{\eta} = 2\sin^{-1}\sqrt{0.7}$$

$$= 113.6^\circ$$

$$\Delta n = \frac{\phi\lambda\cos\theta_s}{2\pi}$$

$$= 1.27\times10^{-5}$$

52. Ans. (a)

Solution:

Because the prism is symmetrical and light paths are reversible, it is easy to see that the deviation δ will be a minimum when gives at the two inter faces.

$$\sin K = n\sin\theta_2$$

and $n\sin(a - \theta_2) = \sin K$

which imply that $\theta_2 = a/2$.

Then, $\sin K = n\sin\frac{a}{2}$

But, $\delta_{min} = \theta_1 + \theta_3 - \alpha = 2k - \alpha$

or $K = \frac{\alpha + \delta_{min}}{2}$

Hence, $\sin\frac{\alpha + \delta_{min}}{2} = n\sin\frac{a}{2}$

53. Ans. (c)

Solution:

$$\lambda' = \frac{\lambda}{n} = \frac{420\ nm}{1.33} = 316\ nm$$

54. Ans. (d)

Solution:

$$\lambda = \frac{1\,mm}{1800} = 5.556 \times 10^{-7}\,m$$
$$= 556\,nm$$

$$v = \frac{3 \times 10^{8}}{5.556 \times 10^{-7}} = 5.4 \times 10^{14}\,Hz$$

55. Ans. (b)

Solution:

From Snell's law,

$$1 \sin 48^\circ = 1.333 \sin \theta_w$$

$$\sin \theta_w = 0.5575;$$

$$\theta_w = 33.9^\circ$$

56. Ans. (b)

Solution:

$$\frac{\sin 55^\circ}{\sin \theta_{al}} = 1.39$$

$$\sin \theta_{al} = 0.58932$$

$$\theta_{al} = 36.1^\circ$$

57. Ans. (d)

Solution:

$$\theta_2 = 52^\circ - 19^\circ = 33^\circ$$

$$n = \frac{\sin 52^\circ}{\sin 33^\circ} = \frac{0.7880}{0.5446} = 1.45$$

58. Ans. (c)

59. Ans (c)

60. Ans (d)

61. Ans (c)

62. Ans (c)

63. Ans (c)

64. Ans (a)

65. Ans (b)

66. Ans (a)

67. Ans (d)

68. Ans (c)

69. Ans (a)

70. Ans (c)

71. Ans (b)

72. Ans (a)

73. Ans (a)

74. Ans (a)

75. Ans (b)

76. Ans (d)

77. Ans (a)

78. Ans (c)

79. Ans (c)

80. Ans (d)

81. Ans (a)

82. Ans (c)

83. Ans (c)

84. Ans (c)

85. Ans (b)

86. Ans (a)

87. Ans (a)

1.2 Mass Spectroscopy

1. Ans. (c)

Solution:

Resolution of a mass spectrometer measures the ability of the mass spectrometer to distinguish the two peaks of slightly different mass to charge ratio (ΔM) in a mass-spectrum.

$$\text{Resolution } (R) = \frac{M}{\Delta M}$$

* A higher resolution of a mass spectrometer indicates a better peak seperation in the mass-spectrum.

Finding m/z ratio for CO^+

m_1= mass of CO = 12 amu + 15.9949 amu

= 27.9949 amu

$$z_1 = +1$$

$$M_1 = \frac{m_1}{z_1} = \frac{27.9949}{1}$$

or $M_1 \simeq 27.9949$

Finding the m/z ratio for N_2^+

m_2= mass of N_2

= 2 × 14.0031 amu

= 28.0062 amu

$z_2 = +1$

Therefore, $M_2 = \dfrac{m_2}{z_2}$

$$= \frac{28.0062}{1}$$

or $\quad M_2 = 28.0062$

Now, Resolution (R) $= \dfrac{M}{\Delta M}$

$M =$ maximum of (M_1, M_2)

$\Delta M = 28.0062 - 27.0049$

$= 0.0113$

Therefore, $R = \dfrac{28.0062}{0.0113}$

$= 2478.42$

$\simeq 2500$

2. **Ans. (b)**

Solution:

$$\frac{1}{2}mv^2 = eV$$

or $\quad v = \sqrt{\dfrac{2\,eV}{m}}$

and $\quad vt = L\,,\ t = \dfrac{L}{V}$

or $\quad t = L\sqrt{\dfrac{m}{2\,eV}}$

3. **Ans. (c)**

Solution:

The time-to-flight type mass spectrometer depends on the principle that the ions of different m/e ratio will take different times to travel over an identical path from the source to the collector.

The essential part of the time of flight type mass spectrometer are:

- An electron gun to produce the ions of the molecules present in the sample compound
- A grid system which accelerates the ions to the uniform velocities
- An evacuated tube called the drift tube in which the ions travel
- An ion detector and an electronic circuitry which helps in translating the time-dependent arrival of the ions at the different velocities into a time base which is related to the mass number

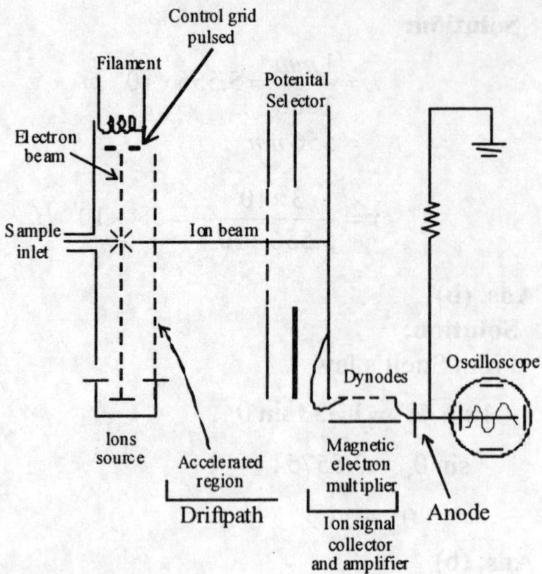

In Present Case

$$\frac{1}{2}mv^2 = eV$$

or $\quad v = \sqrt{\dfrac{2\,eV}{m}}$

and $\quad vt = L\,,\ t = \dfrac{L}{V}$

or $\quad t = L\sqrt{\dfrac{m}{2\,eV}}$

For ionic species A

$$t_A = L_A\sqrt{\frac{m_A}{2eV_A}}$$

For ionic species B

$$t_B = L_B\sqrt{\frac{m_B}{2eV_B}}$$

$$t_A = t_B$$

$$\therefore \quad L_A\sqrt{\frac{m_A}{2eV_A}} = L_B\sqrt{\frac{m_B}{2eV_B}}$$

Given $\quad L_A = 1\,\text{m},\ \dfrac{m_A}{m_B} = \dfrac{1}{2}$

$$V_A = V_B$$

$$\therefore \quad 1 \times \sqrt{\frac{m_A}{m_B}} = L_B$$

$$L_B = \frac{1}{\sqrt{2}}\,\text{m}$$

4. Ans. (c)
Solution:

Given, $m_1(CO^+) = 27.9949$

$$m_2(N_2^+) = 28.0062$$

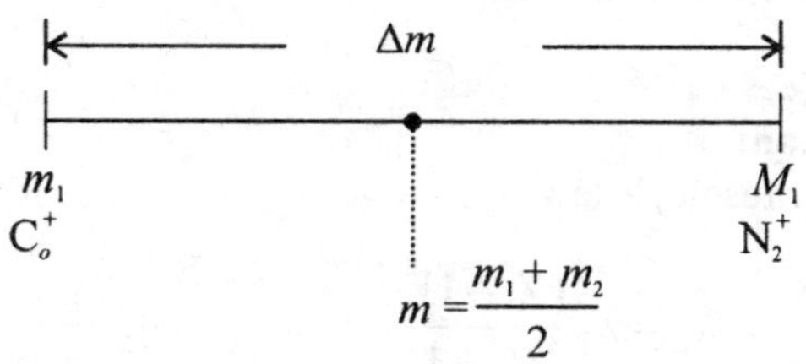

$$m = 28.00055$$
$$\Delta m = 0.0113$$
$$R = 2477.92 \ (nearest\ to\ 2500)$$

5. Ans. (d)
Solution:

Given, $\frac{m_A}{m_B} = \frac{1}{2}$

drift tube, $\ell = 1m$

$$t = \ell\sqrt{\frac{m}{2\,eV}}$$

ℓ_A = dist trauelled by A.

ℓ_B = dist tranelled by B.

$t_A = t_B$

$$\ell_A\sqrt{\frac{m_A}{2eV}} = \ell_B\sqrt{\frac{m_B}{2eV}}$$

$$\frac{\ell_A}{\ell_B} = \sqrt{\frac{m_B}{m_A}}$$

$$\frac{1}{\ell_B} = \sqrt{2}$$

$$\Rightarrow \ell_B = \frac{1}{\sqrt{2}} = 0.707\ m$$

6. Ans. (a)
Solution:

$$t = \sqrt{\frac{m}{2eV}}\ [\because \ell = 1]$$

7. Ans. (d)
Solution:

Given, $m_{oxygen} = 31.9988$

$$m_{sulphur} = 32.0600$$

$$m = \frac{31.9988 + 32.0600}{2} = 32.0294$$

$$\Delta m = 0.0612$$

$$R = \frac{m}{\Delta m}$$

$$= \frac{32.0294}{0.0612}$$

$$= 523$$

8. Ans. (c)
Solution:

Given, $l = 85cm = 0.85m$

$$v = 2\ kV = 2000V$$
$$m_A = 200$$
$$m_B = 300$$
$$e = 1.6 \times 10^{-19}$$

mass of protrom $= 1.66 \times 10^{-27}\ kg$

$$\Delta t = \left(\ell\sqrt{\frac{m_B}{2eV}} - \ell\sqrt{\frac{m_B}{2eV}}\right) \quad \left[\begin{matrix} m_2 > m_1 \\ \therefore t_B > t_A \end{matrix}\right]$$

$$= \frac{\ell}{\sqrt{2eV}}\left(\sqrt{m_B} - \sqrt{m_A}\right)$$

$$= \frac{0.85}{\sqrt{2 \times 1.6 \times 10^{-19} \times 2000}}\left(\begin{matrix}\sqrt{300 \times 166 \times 10^{-27}} \\ -\sqrt{200 \times 1.66 \times 10^{-27}}\end{matrix}\right)$$

$$= 4.36\ \mu s$$

9. Ans. (b)

10. Ans. (c)

11. Ans. (b)

12. Ans. (a)

13. Ans. (c)

14. Ans. (c)

15. Ans. (a)
Solution:

$$V = \frac{\overline{E}}{B} = \frac{2 \times 10^4}{0.2} = 10^5 m/s$$

Separation $= 2\,(R_2 - R_1) = \frac{2\,V\,(m_2 - m_1)}{qB}$

$$= \frac{(2 \times 10^5)(13 - 12)(1.67 \times 10^{-27})}{1.6 \times 10^{-19} \times 0.3}$$

= 0.69 cm.

16. Ans. (a)
Solution:

$$x = \frac{\lambda}{2\mu \tan\beta \cos(\beta + r)}$$

$$= \frac{\lambda}{2\mu \tan\beta \cos\beta} = \frac{\lambda}{2\mu \sin\beta}$$

Therefore, $\beta = \sin^{-1}\frac{\lambda}{2\mu x}$

$$= \sin^{-1}\frac{5893\times10^{-8}}{2\times1.4\times0.09} = 0.0134°$$

17. Ans. (d)
Solution:
Given that

n = 4, $\mu = 1$,

$$2\ \mu t = \left(n + \frac{1}{2}\right)\lambda$$

Therefore, $t = \frac{4.5\times6500\times10^{-8}}{2}$

$$= 1.4625\times10^{-4}\ cm.$$

18. Ans. (b)
Solution:

Fringe separation, $\beta = \frac{\lambda}{2a}$

$$a = 40 = \left(\frac{40}{3600}\right)^{0}$$

$$\beta = \frac{4650\times10^{-10}}{2\times1.9\times10^{-4}} = 0.12\ cm$$

19. Ans. (c)
Solution:`

Fringe width $= \frac{D\lambda}{2d}$

$$W_1 = \frac{D\lambda_1}{2d} = \frac{D\eta_1\lambda}{2d}$$

$$W_2 = \frac{D\lambda_2}{2d_2} = \frac{D\eta_2\lambda}{2d}$$

$$\therefore \quad \frac{W_1}{W_2} = \frac{\lambda_1}{\lambda_2} = \frac{\eta_1}{\eta_2}$$

Also $\quad \eta_2 W_1 = 40\ W_2$

$$\eta_2 = \frac{40\times5996}{4358} = 54$$

20. Ans. (c)
Solution:

$$V = 5.93\times10^5\sqrt{V}$$

$$= 5.93\times10^5\sqrt{12.4\ kV}$$

$$= 6.6\times10^7\ m/\text{sec}.$$

21. Ans. (c)
Solution:
From Moseley's law

$$\lambda_2 = \lambda_1\frac{(Z_1 - 1)^2}{(Z_2 - 1)}$$

$$= \frac{2(24-1)^2}{(42-1)}$$

$$= 0.63\ \mathring{A}$$

22. Ans. (b)
Solution:

$$\Delta\lambda = \frac{h}{mC}(1 - \cos\theta)$$

$$= \frac{(6.626\times10^{-34})(1 - \cos 40)}{9.11\times10^{31}\times3\times10^8}$$

$$= 0.005678\ \mathring{A}$$

$$\therefore \quad \lambda = \lambda_1 + \Delta\lambda$$

$$= 3 + 0.005678$$

$$= 3.005678\ \mathring{A}$$

23. Ans. (d)
Solution:
No of waves per millimeter in glass.

$$1800\times1.333 = 2399\ mm^{-1}$$

24. Ans. (c)
Solution:

$$\sin C_c = \frac{n_2}{n_1}\theta_c$$

$$= \text{arc}\sin\frac{1}{1.76}$$

$$= 34.6°$$

25. Ans. (d)

Solution:

$$\theta_c = \theta_2 = \text{arc sin}\ \frac{n_A}{n_B} = \text{arc sin}\ \frac{1}{2.42}$$

$$= \text{arc sin}\ 0.4312 = 24°$$

26. **Ans. (c)**

Solution:

We use snell's law : $n_1 \sin\theta_1 = n_2 \sin\theta_2$ with measured from normal. These $= n_2 \sin\theta_2$ with measured from normal. These

$$\sin\theta_1 = \frac{n_2 \sin\theta_2}{n_1}$$

$$= \frac{(1.358)(\sin 25°)}{1.00} = 0.574$$

27. **Ans. (d)**
28. **Ans. (d)**
29. **Ans. (a)**
30. **Ans. (c)**

1.3 X-rays and Nuclear Radiation Measurements

1. **Ans. (d)**

Solution:

A **crystal monochromatar** is used as wavelength selector depending on material. One can select for crystal required wavelengths and eliminate other wavelengths. If a crystalline material absorption edge lies between λ_1 and λ_2 then crystal will eliminate wavelength.

2. **Ans. (a)**

Solution:

Bragg's Law

$2\, d \sin\theta = m\lambda$

Take derivative w.r.t θ and λ

$2\, d \cos\theta\, d\theta = m d\lambda$

$$\frac{d\theta}{d\lambda} = \frac{m}{2d\cos\theta}$$

3. **Ans. (d)**
4. **Ans. (a)**
5. **Ans. (c)**

Solution:

$$\text{Half-life}\left(\frac{t}{2}\right) = 1 \text{ year}$$

$$\lambda\,(\text{decay constant}) = \frac{0.693}{\frac{t}{2}}$$

$$= 0.693/\text{year}$$

6. **Ans. (a)**

Solution:

$$E_y = E_0 e^{-\alpha x}$$

$$E_y = E_0 e^{-\alpha x}\, e^{-1\ \beta x}$$

(in terms of attenuations)

7. **Ans. (c)**

Solution: The correct match is given below:

Beer's law	→	Absorption of light
Bragg's law	→	X-ray diffraction
Peltier effect	→	Thermo-emf
Nernst equation	→	Electro-chemical cell

8. **Ans. (b)**

Solution:

Absorption of X-rays

X-rays can be absorbed by the matter and the degree of absorption is controlled by the nature and amount of the absorbing material.

When the thickness of an absorbing material is of primary interest, it is convenient to write **Beer's law** in the form

$$P_x = P_o\, e^{-\mu x}$$

$$ln\left(\frac{P_o}{P_x}\right) = \mu x$$

Where,

P_o = Initial power of the radiation

P_x = Power after passage through an absorbing sample (x cm in length)

μ = Linear absorption coefficient, it is the fraction of energy absorbed for centimeter of a given element.

Monochromatic X-ray Sources.

a. It is quiet simple matter to obtain a narrow band of wavelength at various discrete points in the X-ray region of the spectra, but to design a monochromator or that can be varied to any derived wavelength is difficult

A nearly monochromatic beam can be obtained by the use for filter made of an element (or of its compound) that has a critical absorption edge at just the correct wavelength to isolate the characteristic line from the source target.

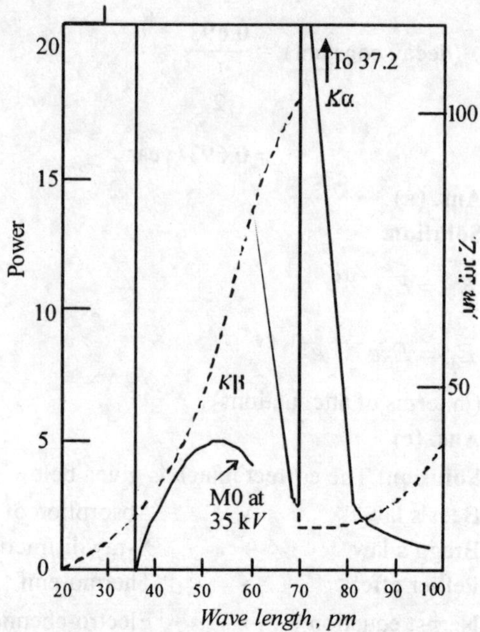

a. The figure above shows the absorption spectra (in dashed lines) of the Zirconium superimposed on the emission spectrum of the molybdenum.

The Zirconia filter with the cutting edge 68.9 pm will block all the wavelength upto 68.9 pm and pass the rest. So out of 70.9 pm and 63.2 pm, only the 70.9 pm wavelength will come as the output

9. Ans. (b)

Solution:

$$\frac{hc}{\lambda} = E = eV$$

or, $$\lambda = \frac{hc}{eV}$$

$$= \frac{6.626 \times 10^{-34} \times 3 \times 10^{8}}{1.6 \times 10^{-19} \times 40 \times 10^{3}}$$

or, $\lambda = 0.31$ ú

Second Method

$$\lambda_{min} = \frac{12400\,\mathring{A}}{V}$$

$$= \frac{12400\,\mathring{A}}{40 \times 10^{2}}$$

$$= 0.3$$

10. Ans. (c)

Solution:

X-ray Film:

a. X-ray film is a photographic print of the X-ray on a film

b. It is the print which can be seen is given by a doctor when he takes the X-ray of a particular body part in order to find fracture in the bone

c. The characteristic of the film are better visualized by the help of the HD curves. The HD curve is nothing but the plot of the density vs the exposure level.

d. Contrast is the most significant characteristic of an image reused on the film. Contrast is the variation in the film density (shades of grey) that actually forms the image. Without contrast there is no image. A film can be considered as a contrast converter. One of its function is to convert the difference in exposure (subject contrast) into film contrast difference in the density

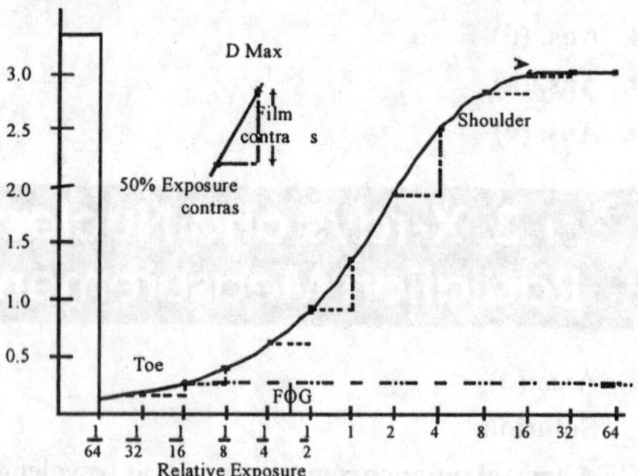

The film characteristic curve has three distinct regions with different contrast characteristic. These are:

Toe: The part of the curve associated with the low exposure is called as the toe. It also corresponds to the light or low density portions of an image. When an image is exposed so that area fall within the toe region, little or no contrast in transferred to the image. In the image shown in the above graph, the areas on the left corresponds to the toe of the characteristics curve.

Shoulder: The film also have a reduced ability to transfer contrast in areas that receive relatively high exposures. The condition corresponds to the upper portions of the characteristics curve while the slope decreases with the increasing exposure. This portion of the curve is transitionally referred to as the shoulder.

Contrast: The highest level of contrast is produced within a range of exposures falling between the toe and the shoulder. This part of the curve is characterized by the relatively straight and very steep (nearly perpendicular) slope in comparison to the toe and the shoulder regions. It is desirable in most of the applications, to expose the film with this range so as to obtain the maximum contrast.

Latitude: The exposure range over which the film produces a useful contrast is designated as the latitude.

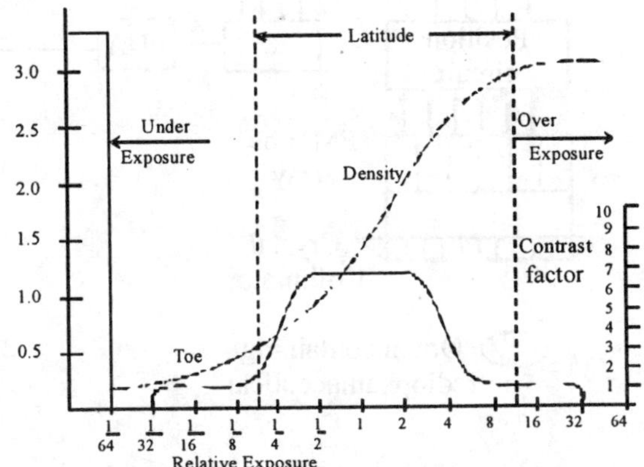

Fog: Any density in the film that is not produced as a part of the image forming exposure is reffered as fog. It is basically a defect and it must be low

The fog in the film may be due to:

1. Its inherent properties
2. Worse chemical
3. Heat and Age
4. Exposure of the radiation on the film.

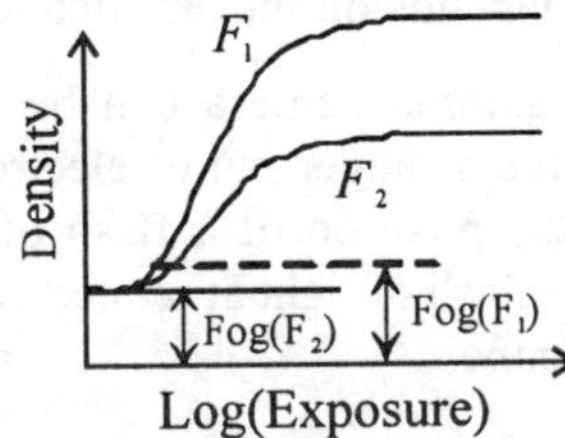

From the above figure it can be observed

1. Clearly the slope of F_1 is more steep than the slope of F_2
2. Hence, $(\text{Contrast})_{F_1} > (\text{Contrast})_{F_2}$ and

$$(\text{fog})_{F_1} > (\text{fog})_{F_3}$$

11. Ans. (a)

Solution:

CT Scan is the photograph of the intensity of the beam coming out of the tissue, when the X-ray are made to fall on the tissue from various angles.

For the case of a monochromatic X-ray the tissue attenuation characteristics can be described by

$$I_t = I_o e^{-\mu x}$$

Where,

I_o = Incident radiation intensity

I_t = Transmitter radiation intensity

X = Thickness of the tissue

μ = Characteristic attenuation coefficient of tissue.

For X-ray filters

$$\frac{I}{I_o} = e^{-\mu_i p_i}$$

Case-1; Aluminium

$$\frac{I_t}{I_0} = e^{-\mu_A x_A} \quad ...(1)$$

Case-II: Copper

$$\frac{I_t}{I_0} = e^{-\mu_C x_C} \quad ...(2)$$

From equations (1) and (2)

$$e^{-\mu_C x_C} = e^{\mu_A x_A}$$

Because $x_A \mu_A = \mu_C x_C$

therefore $$x_C = \frac{x_A \mu_A p_A}{\mu_C p_C}$$

$$= \frac{25 \times 10^{-3} \times 0.02 \times 2699}{0.075 \times 8960}$$

$$= 0.2mm$$

12. Ans (b)

Solution:

X-ray Generation

X-ray are generated when the free electrons give up some of their energy when they intersect with the orbital electrons of nucleus of an atom. The energy given up by the electrons during this interaction appears as the electromagnetic energy known as X-radiation. There are two different process for producing X-radiations.

* When the beam of energetic electrons imprings on a target material, the elecrons are then in general slowed down by the multiple interactions with the electrons of the traget. The energy lost is converted into X-radiation called the *Breamsstrahlung*
* This X-radiation thus emitted show a sharp minimum wavelength λ_{min} (maximum frequency) corresponding to the maximum energy of the electrons which cannot be exceeded

This cut-off wavelength (in nanometers) is given by

$$E = hf = \frac{hc}{\lambda}$$

Energy $(E) = V_e$

$$V_e = \frac{hc}{\lambda}$$

$$\lambda_{min} = \frac{hc}{V_e} = \frac{1240}{V}$$

Where,

$h =$ Plank's constant

$c =$ Velocity of electromagnetic radiations in vaccum (Velocity of striking electrons)

$e =$ Electronic charge

$V =$ Accelerating potential across the X-ray tube (in k Volts)

For 40 *KV*

$$\lambda_1 = \frac{1240}{V} = \frac{1240}{40\ kV}$$

$$= 31 \times 10^{-3} m = 31\ mm$$

For 50 *KV*

$$\lambda_2 = \frac{1240}{50\ KV}$$

$$= 24.80 \times 10^{-3} = 24.8\ mm$$

So resulting change in wavelength in terms of (%)

$$\%\Delta\lambda = \frac{\lambda_2 - \lambda_1}{\lambda_1} \times 100$$

$$= \frac{24.8 - 31}{31} \times 100 = \frac{-6.2}{31} \times 100$$

$$= -20\ \%$$

13. Ans. (c)

Solution:

Nuclear medicine imaging system technique uses the nuclear medicine which is a special medicine and uses in medial imaging useds the radionuclides and relies on the process of radiocative decay for the diagnosis and the treatment of diseases.

The informatin obtined using the imagin method is mainly related to the physiological functioning of an organ as opposed to the mainly anatomial obtained by the X-rray imazing techniques.

NMIS Technique:

In carrying out NMIS techniques a radio pharmaceutical is a administered to the patient and it accumulates in the organ of interest. So, the Gamma rays are emitted in all directions from the organ. There gamma rays are then observed with the help of a gamma camera also called the **Anger camera.**

The basic design of a typical anger cemra is given as under

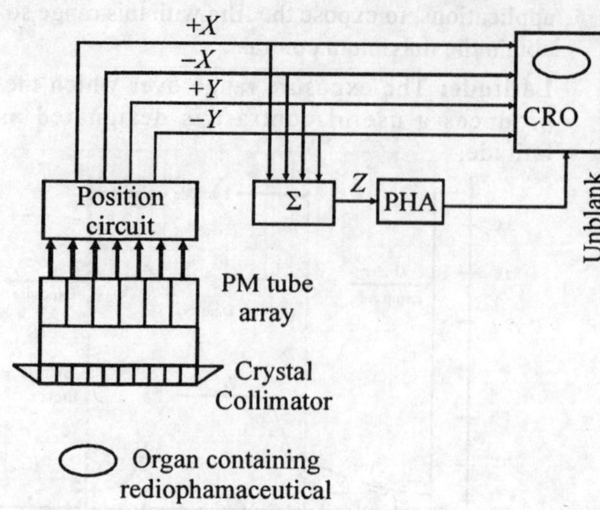

It consists of a large diameter scintillation cycstal which is viewed by a larger number of photomulitplier tubes (PM tubes).

The PM tubes are housed in cylindrical shaped housing commonly called the gamma head, the cross-sectional view of this is given above.

The diameter of the crystal can be about 25 cm 40 cm and it can be around cm in thickness. However the diameter depends on the amplications of the device.

Thus the gamma camera can be considered to be a sophisticated arrangement of electronic circuits used to translate the position of a flash of light a scintillaion crystal to a flash of light at related point on the screen of an osclloscope.

Essentials of a gamma camera.

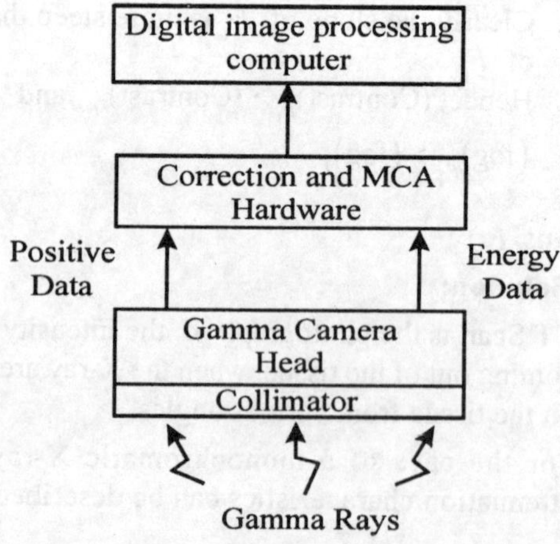

Collimation:

The collimator is a device which is atteched to the front

of a gamma camera head. The figher below shlows a parallel-hole collimated attached to a crystal

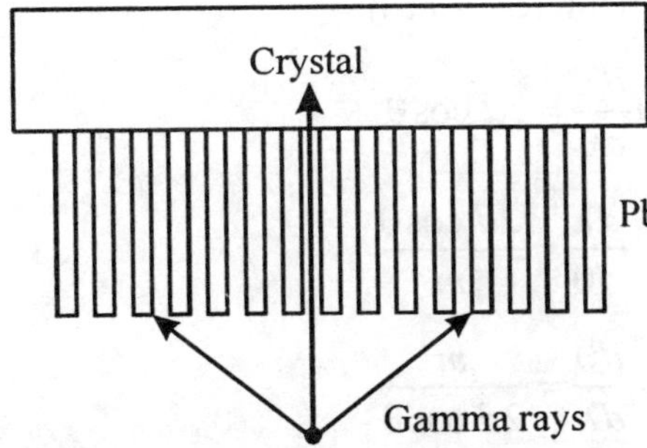

The gamma ray entering at an angle gets absorbed by the lead plates and only those along the direction of the hole get through to cause scintilations in the crystal and if the collimator was not in place these obliquely incident γ-ray would get absorbed and the image would become have the blurred, i.e. the image would not be clear.

The different collimators have the same base design, they only vary in terms of diameter of each hole. the depth of the hole and the thickness of lead between each hole (also called the *septum thickness*)

Collimator Holes:

Collimator holes can be round, square or hexagonal.

The round holes easy to make and can be kept closely together, but they leave dead spaces in between them from where no rays can pass through.

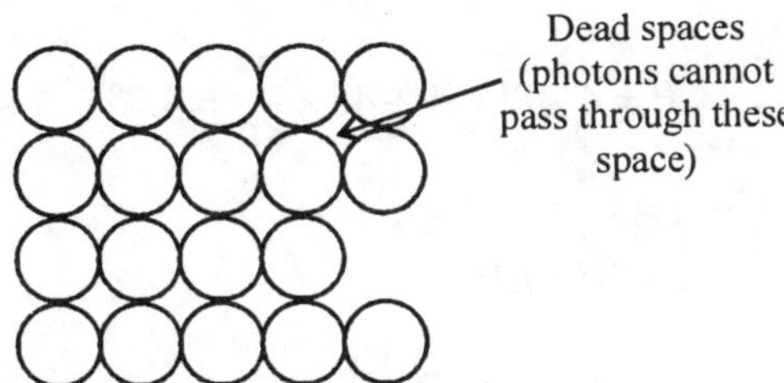

So the sensitivity of a round hole colimator is less.

* The square holes are easy to manufacture, but because the lenght of the holes diagona in larger than the length of its sides so the resolution varies withthe direciton of measurement
* The maximum sensitivity is obtained by the hexagonal closed packed hles.

Design parameter for collimators are:

1. Hole diameter
2. Hole length
3. Distance

4. Septal Thickness:

The thickness of the lead septa between the collimaor holes plays a major role in the determination of the penetration compoent. Septal peneration is the ability of a the γ-rays to cross from one hole to the another without being absorbed in the septum.

It is known from the nuclear physics that the photon absorption is a probablistic function and it is not guranged that any thickness of lead will stop the enrgy of gamma rays. So one will always have a penertration fucntion. But if one get much spatial penetration than the resolution-get-reduced, and the resuling image is less clear.

As the energy of the photon increases the septal thickness must also increase or too many of the γ-rays will pass the collimator. But the septa are made thick enough to minimize the peneration, the collimator losses the senesivity and becomes very heavy.

14. Ans. (d)

15. Ans. (a)

Solution:

Given $v = 1000\ kV$

$$x_{min} = \frac{1.242}{10^6} = 1.247\ \text{þ}m$$

16. Ans. (d)

Solution:

Given $v = 5000$ V

$$m = 9.1 \times 10^{-31}$$

$$\mu = \sqrt{\frac{2\ eV}{m}}$$

$$= \sqrt{\frac{2 \times 1.6 \times 10^{-19} \times 500}{9.1 \times 10^{-31}}}$$

$$= 4.1 \times 10^7\ m/s$$

17. Ans. (c)

Solution:

Given $v = 50\ kV$

$$\lambda_{min} = \frac{1.2420}{50000}$$

$$= 0.248\ \text{Å}$$

18. Ans. (c)

Solution:

wave length of x - ray emitted for maximum intersity is given by

$$\lambda = 1.5\ \lambda_{min}$$

$$\lambda = 1.5 \times 0.248\ \text{A}^\circ$$

$$\lambda = 0.372\ \text{A}^\circ$$

19. Ans. (c)

Solution:

$$v = (12400/\lambda)$$

20. Ans. (a)

Solution:

By daune - hunt equation $\lambda_{min} = \dfrac{12400}{v}\,\mathring{A}$

$$\lambda_1 = \frac{12400}{v}\,\mathring{A}$$

$$\lambda_2 = \frac{12400}{1.5\,v}\,\mathring{A}$$

$$\Delta\lambda = \lambda_1 - \lambda_2 = \frac{12400}{v} - \frac{12400}{1.5v} = 0.26\,\mathring{A}$$

$$\Rightarrow \frac{1}{v}\left[12400 - \frac{12400}{1.5}\right] = 0.26$$

$$\Rightarrow v = 15.897\ kV$$

21. Ans. (a)

Solution:

$$I = I_0 e^{(-\mu d)}$$

22. Ans. (d)

Solution:

$$I_t = I_i e^{-\mu e_x}$$

$$\text{Attenuation} = \frac{I_t}{I_i}$$

$$= \frac{I_i e^{-\mu\, px}}{I_i}$$

$$= \exp\,(-0.02 \times 2699 \times 10^3 \times .026)$$

23. Ans. (b)

Solution:

$$\frac{I_t}{I_i} = \exp(-\mu\ px)$$

$$= e^{(-0.075 \times 10^3 \times 8960 \times x)}$$

$$= \exp(-0.02 \times 2699 \times 10^3 \times .0026)$$

$$\Rightarrow x = 0.02$$

24. Ans. (b)

Solution:

Given, 70.9 pm

K_{edge} = upto that λ the x - rays get abscribed

25. Ans. (a)

Solution:

$$m\lambda = 2d \sin\theta$$

Different w.r.t. θ

$$m\frac{d\lambda}{d\theta} = 2d\cos\theta$$

$$\frac{d\lambda}{d\theta} = \frac{2d\cos\theta}{m}$$

$$\Rightarrow \frac{d\theta}{d\lambda} = \frac{m}{2d\cos\theta}$$

26. Ans. (a)

Solution:

$$I_t = I_i \exp(-\mu_1 \rho_1 x_1)$$

$$I_t = I_i \exp(-\mu_2\, \rho_2,\, x_2)$$

$$x_2 = \frac{\mu_1 \rho_1 x_1}{\mu_2 \rho_2}$$

$$= \frac{0.02 \times 2699 \times 2.5}{0.075 \times 8960}$$

$$= 0.2\ mm$$

27. Ans. (d)

Solution:

Given, $\lambda = 0.36\,\mathring{A}$

$$\theta = 4°48' = 4' + 48 \times \frac{1}{60} = 4.8°$$

$$n = 1$$

$$n\lambda = 2d\sin\theta$$

$$d = \frac{n\lambda}{\sin\theta} = \frac{1 \times .36\,\mathring{A}}{2\sin 4.8°}$$

$$= 2.151\,\mathring{A}$$

28. Ans. (d)

Solution:

Given, $\lambda = 0.68\,\mathring{A}$

$$d = 1.82\,\mathring{A}$$

$$n = 1$$

$$n\lambda = 2d\sin\theta$$

$$\theta = \sin^{-1}\left(\frac{n\lambda}{2d}\right)$$

$$= \sin^{-1}\left(\frac{1\times.68}{2\times1.82}\right)$$

$$= 12.5°$$

29. Ans. (b)
Solution:

$$t_{1/2} = 1 \text{ year}$$

$$t_{1/2} = \frac{0.693}{\lambda}$$

$$\lambda = 0.693 \text{ per year}$$

30. Ans. (a)
Solution:

Given, $\lambda = 0.182$ per day

$$NR = N_i\, e^{-\lambda t}$$

$$\frac{Ni}{20} = N_i\, e^{-\lambda t}$$

$$\log\frac{1}{20} = -\lambda t$$

$$t_1 = 16.46 \text{ days}$$

31. Ans. (d)
Solution:

$$NR = \frac{N_i}{2.5}$$

$$\frac{1}{2.5} = e^{-\lambda 7}$$

$$\lambda = 0.13$$

$$t_{1/2} = 5.3 \text{ days}$$

32. Ans. (b)
Solution:

$$N_R = 5\,\% \text{ of } Ni$$

$$N_R = N_i\, e^{-\lambda t}$$

$$\frac{5}{100}\times Ni = Ni\, e^{-\lambda\times 60}$$

$$\lambda = 0.0499/\text{min}$$

Mean life time $\tau_a = \frac{1}{\lambda} = 20.04$ min

33. Ans. (a)
Solution:
low dead time and high counting rate.

34. Ans. (c)
Solution:
The correct match is givne below

Tungsten lamp → IR
C0-60 → Gamma
He-Ne → Laser
Discharge lamp → UV rays

35. Ans. (c)
Solution:
The correct match is given below

Bolometer → IR
Photomuliplier → UV
Gm counter → Gamma-rays
Photographic → X-rays

36. Ans. (d)
Solution:
The correct match is given below

Beers law → Absorption of light
Braggs law → X-ray diffraction
Peltier effect → Thermo emf
Nernst equatin → Electrochemical cell

37. Ans. (d)
Solution:
The correct match is given below

Braggs equation → $N\lambda = 2\,d\sin\theta$
Duachunt equation → Wave length = (he/eV)
Radioactive decay equation → $N = N_o\, e^{-\lambda T}$
Beer lamberts law → A = abc

38. Ans. (a)
Solution:

Auto collimator → Radiation scattered
Hygrometer → Humidity
Nephlometer → Density
Mass spectrometer → Molecular weigth

39. Ans. (d)

40. Ans. (a)

41. Ans. (b)
Solution:

$$\frac{1}{2}mv^2 = eV = hf$$

Hence frequency

$$f = \frac{eV}{h} = \frac{1.6\times10^{-19}\times3\times10^4}{6.63\times10^{-34}}$$

$$\therefore\quad \lambda = \frac{C}{f} = 0.41\times10^{-10}\,m = 0.041\,nm.$$

42. Ans. (c)
Solution:
From Bragg's law,

$$2d \sin \theta = m\lambda$$

$$\therefore \quad d = \frac{\lambda}{2 \sin \theta} = \frac{0.3}{2 \times 0.156} = 0.96\ nm$$

43. Ans. (b)
Solution:

$$\Delta\lambda = 0.024\ (1 - \cos\theta)\ \mathring{A}$$

$$= 0.024(1 - \cos 60)\ A$$

$$= 0.012\ \mathring{A}$$

Now,

$$\lambda + \Delta\lambda = 0.254 \times 10^{-11} m$$

$$\therefore \quad \lambda = 0.254 - 0.012$$

$$= 0.242\ \mathring{A}$$

44. Ans. (b)
Solution:

Given : $\theta = 12°$, $n = 1$,

$$d = 3.04 \times 10^{-10} m$$

$$2d \sin \theta = n\lambda$$

$$\therefore \quad 2 \times 3.04 \times 10^{-10} \sin 12 = 1 \times \lambda$$

$$\lambda = 0.79 \times 10^{-10}$$

For 3rd order reflection,

$$2 \times 3.04 \times 10^{-10} \sin\theta_3 = 3 \times 0.79 \times 10^{-10}$$

$$\sin\theta_3 = 0.3898$$

Hence $\theta_3 = 22.9°$

45. Ans. (b)
Solution:

$$\lambda_{\min} = \frac{12400}{V}\ \mathring{A}$$

$$\therefore \quad V = \frac{12400}{\lambda_{\min}}$$

$$= 12.4\ kV.$$

46. Ans. (a)
Soution:

$$\lambda_{\min} = \frac{hC}{eV} = \frac{6.626 \times 10^{-34} \times 3 \times 10^{8}}{1.6 \times 10^{-19} \times 70 \times 10^{3}}$$

$$= 0.017\ nm$$

47. Ans. (c)
Solution:

$$H = \frac{\mu - \mu_{water}}{\mu_{water}} \times 1000$$

$$\mu = 0.212$$

$$\mu_{water} = 0.205$$

$$H = \frac{0.212 - 0.205}{0.205} \times 1000$$

$$= \frac{0.007 \times 1000}{205} = \frac{7000}{205}$$

48. Ans. (a)
Solution:

$$\lambda_{\min} = \frac{hC}{eV} = \frac{6.6}{1.6 \times 10^{-19} \times 70 \times 10^{3}}$$

$$= 0.0177$$

49. Ans. (d)
Solution:

$$\lambda_2 = \frac{\lambda_1 (Z_1 - a)^2}{(Z_2 - a)^2}$$

$$= \frac{2(24-1)^2}{(42-1)^2} = 0.63\ \mathring{A}$$

50. Ans. (c)
Solution:

$$\Delta\lambda = \frac{h}{mC}(1 - \cos\theta)$$

$$= \frac{(6.626 \times 10^{-34})(1 - \cos 40°)}{9.11 \times 10^{-31} \times 3 \times 10^{8}}$$

$$= 0.005678\ \mathring{A}$$

$$\lambda = \lambda_1 + \Delta\lambda = 3 + 0.005678$$

$$= 3.005678\ \mathring{A}$$

51. Ans. (c)
52. Ans. (a)
53. Ans. (c)
54. Ans. (b)
55. Ans. (b)
56. Ans. (b)
57. Ans. (a)
58. Ans. (b)

1.4 Optical Sources and Detectors

1. **Ans. (a)**
2. **Ans. (c)**

   Solution: In a photo multiplier secondary emission is used for amplification of low level photo current.
3. **Ans. (a)**

   Solution:

   λ sunlight – 275 *nm* to 2850 *nm*

   i.e. near UV
4. **Ans. (d)**

   Solution:

   All the above
5. **Ans. (c)**

   Solution:

   The correct match is given below

   Sunlight → Continuous
   Incandesent bulb → Continuous
   Fluorescent source .→ Continous with peaks
   Hydrogen lamp with low pressue .→ Line spectrum
7. **Ans. (c)**

   Solution:

   $I_b = 10\,\mu A$

   $\phi = 3\ watt$

   $$S_R = \frac{I_b}{\phi}$$

   $$= \frac{10}{3} = 3.33\ (\mu\ A/W)$$
8. **Ans. (d)**
9. **Ans. (c)**

   Solution:

   Frequency
10. **Ans. (c)**

    Solution:

    both visible and *uv*
11. **Ans. (b)**

    Solution :-

    $A \rightarrow 2,\ B \rightarrow 1,\ C \rightarrow 3$

    Gas filled photo tube → $100\,\mu A/Im$
    PMT → 1.20A/Im
    Vacuum type PEC → $30\mu A/Im$
12. **Ans. (c)**
13. **Ans. (a)**

    Solution:

    Given when 10 v is supprened $R = 1\,k\Omega$

    when 10 v is not supprened $R = 5\,k\Omega$

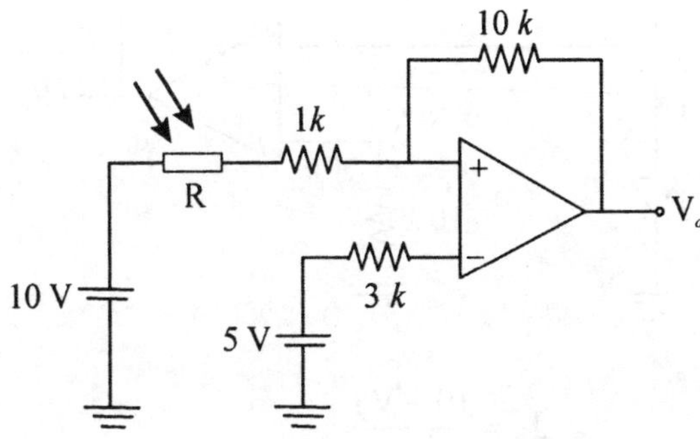

$$V_0 = V_{01}\big|_{5V} + V_{02}\big|_{10V}$$

$$= \left(1 + \frac{10k}{2k}\right)5 + \left(\frac{-10k}{6k}\right)10$$

$$= 13.33\ V$$

14. **Ans. (c)**

    Solution:

    Equivalent circuit:

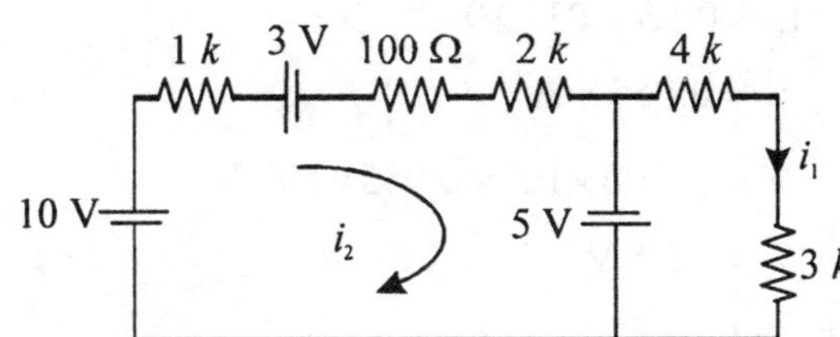

$D_1 \rightarrow OFF$

$D_2 \rightarrow ON$

[As cut in voltage for D_2 is 0.3 V i.e. lower than D_1]

$$i_1 = \frac{-5}{4k + 3k} = \frac{-5}{7k}$$

$$V_{3k} = i_1 \times 3k = .\frac{-5}{7k} \times 3k = \left(\frac{-15}{7}\right)V$$

Using KVL at loop 2

$$-10 + i_2(1k + 100\Omega + 2k) + 0.3 - 5 = 0$$

$$i_2 = \frac{14.7}{3100\,\Omega} = 4.74\,\mu A$$

$$V_{2k} = 9.4838\ V$$

15. **Ans. (c)**

    Solution:

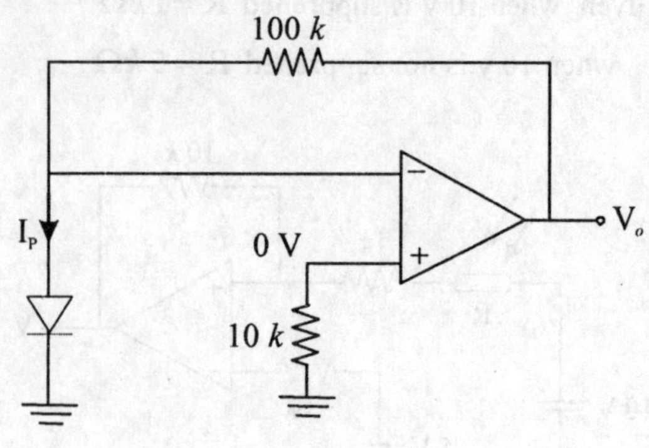

$$I_P = \frac{(0 - V)}{100k}$$

$$V = (-100\,k)I_P$$

Given, $A = 0.01\,cm^2, I = 100 mW/cm^2$

$$S = 0.55\,A/W = \frac{I_P}{A \times I}$$

$$= \frac{I_P}{0.01\,cm^2 \times \frac{100\,mW}{cm^2}} = 0.55$$

$$I_P = 0.55 \times .01 \times 10$$

$$= 0.055\,mA$$

$$V = (-100 \times 10^3 \times 0.055 \times 10^3)V$$

$$= -5.5\,V$$

16. Ans. (c)

Solution:

Given, $A = 4.0 \times 10^{-6} m^2$

$I = 300\,W/m^2$

Incident Power $= A \times I$

$= 0.4 \times 10^{-6} \times 300$

$= 120\,\mu W$

17. Ans. (d)

Solution:

$$\lambda = \frac{12400}{V} A^\circ$$

$$= \frac{12400}{5} A^\circ$$

$$= 2480\,A^\circ \approx 2500\,A^\circ$$

18. Ans. (c)

Solution:

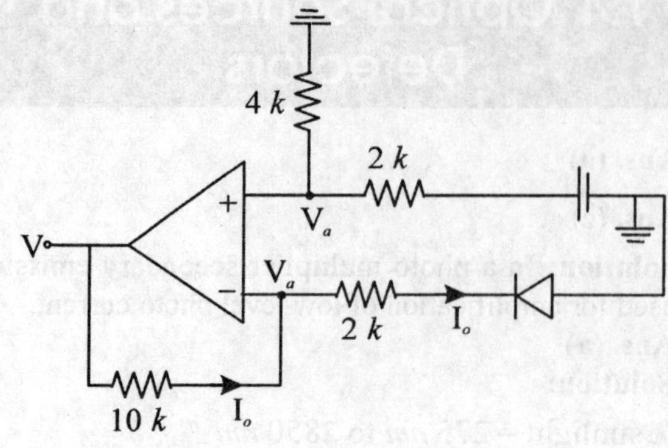

Given

$$I_{01} = 100\,\mu A$$

$$I_{02} = 200\,\mu A$$

$$V_a = \frac{4k \times 12}{4k + 2k} = 8V$$

$$I_0 = \left(\frac{V - 8}{10}\right)$$

for I_{01}

$$V = (100\,\mu \times 10k + 8)$$

$$= 9V$$

for I_{02}

$$V = (200\,\mu A \times 10\,k + 8)$$

$$= 10$$

$\therefore$ V changes from (9 to 10V)

19. Ans. (c)

Solution:

Given, $C_j = 16\,pF$

$A = 10\,mm^2$

$R = 0.5\,A/W$

$I = 1\,mW/cm^2$

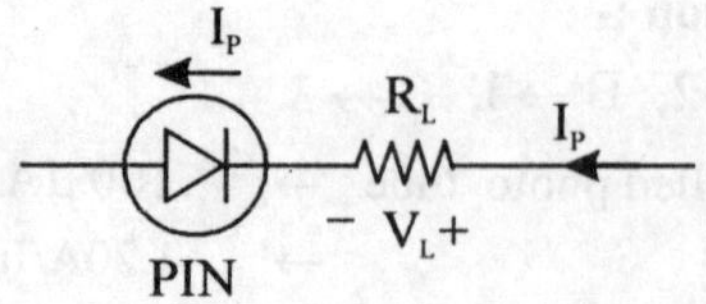

$R_L = 100\,k\Omega$

$$R = 0.5\,\frac{A}{W} = \frac{I_P}{P_0} = \frac{I_P}{A \times I}$$

$$I_P = 0.5 \times 10\, mm^2 \times \frac{1\, mW}{cm^2}$$

$$= 0.5 \times 10^{-4} \text{ A}$$

$$V_L = I_P \times R_L$$

$$= 0.5 \times 10^{-4} \times 100\, k\Omega$$

$$V = 5V$$

20. Ans. (c)

Solution:

$$\lambda_{max} \approx \frac{1.24}{V}$$

$$= \frac{1.24}{1.43}$$

$$= 0.867\, \mu m$$

21. Ans. (c)

Solution:

Given number of electron hole pairs (EHP) $= 5.4 \times 10^6$

number of photons $= 6 \times 10^6$

$$\text{Efficiency} = \frac{\text{number of EHP}}{\text{number of photons}}$$

$$= \frac{5.4 \times 10^6}{6 \times 10^6}$$

$$= 0.9 = 90\%$$

22. Ans. (c)

Solution:

Given, $R = 0.65$ A/W

$$P = 10\, \mu W$$

$$I_P = 10 \times 0.65$$

$$= 6.5\, \mu A$$

23. Ans. (c)

Solution:

Given, $n = 0.65$

$$\lambda = 900\, nm$$

$$P_0 = 0.5\, \mu W$$

$$I_m = 10\, \mu A$$

$$E = \frac{hc}{\lambda}$$

$$= \frac{6.63 \times 10^{-34} \times 3 \times 10^8}{900 \times 10^{-9}}$$

$$= 2.21 \times 10^{-19} \text{ V}$$

$$\eta = \frac{E\, I_P}{P_0 q}$$

$$\Rightarrow I_P = \frac{\eta P_o q}{E}$$

Multiplication factor $m = \frac{I_m}{I_P}$

$$= \frac{I_m}{\eta P_{0q}} E$$

$$= \frac{10 \times 10^{-6} \times 2.21 \times 10^{-19}}{0.65 \times 0.5 \times 10^{-6} \times 1.6 \times 10^{-19}}$$

$$= 42.5 \approx 43$$

24. Ans. (c)

Solution:

Two step transfer

25. Ans. (a)

Solution:

by KVL ⇒

$$-5 + 1.7 + 1.5m(15 + R) + 0.2 = 0$$

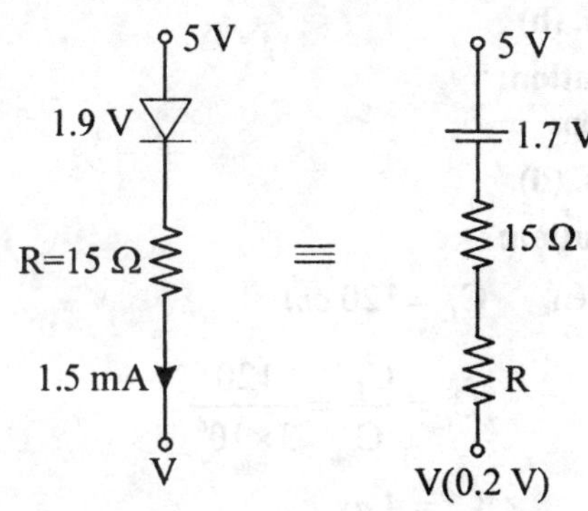

$$R \approx 1.9\, k\Omega$$

26. Ans. (d)

Solution:

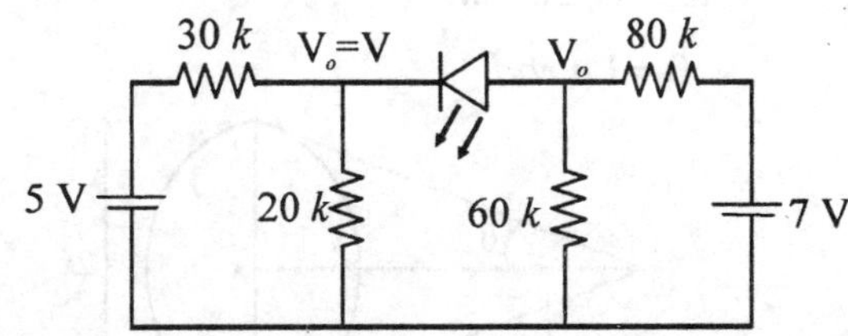

Remove diode and check whether diode is 0N or 0FF by checking the terminal voltage.

$$V_a = \frac{5 \times 20k}{20k + 30k} = 2\text{V}$$

$$V_b = \frac{7 \times 60k}{60k + 80k} = 3\text{V}$$

$$V_b > V_a$$

$\therefore$ LED is ON

$\Rightarrow$ short circuit i.e.

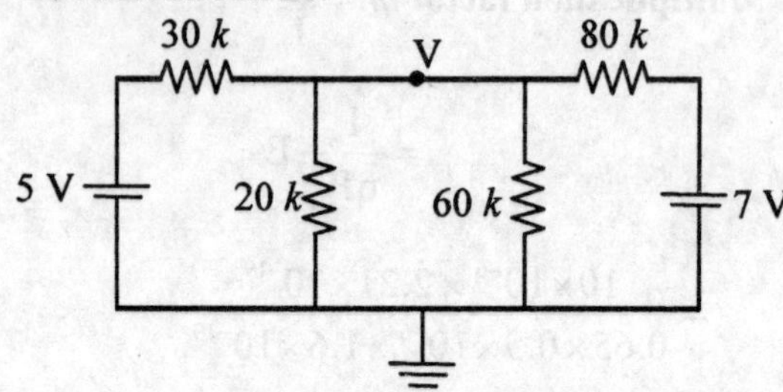

$\therefore$ by nodal,

$$\frac{V-5}{30k} + \frac{V}{20k} + \frac{V}{60k} + \frac{V-7}{80k} = 0$$

$$\Rightarrow \quad V = 2.259\text{ V}$$

27. Ans. (b)

Solution:

$0.633\ \mu m$

28. Ans. (b)

Solution:

Neon

29. Ans. (d)

Solution:

Given, $C_L = 120\ cm$

$$C_t = \frac{C_L}{C} = \frac{120}{3 \times 10^8}$$

$$\therefore C_t = 4\ ns$$

30. Ans. (a)

Solution:

Given, $\lambda = 632.8\ nm$

$$d = 4 \times 10^5\ km$$

$$\theta = 1\ n\ rad$$

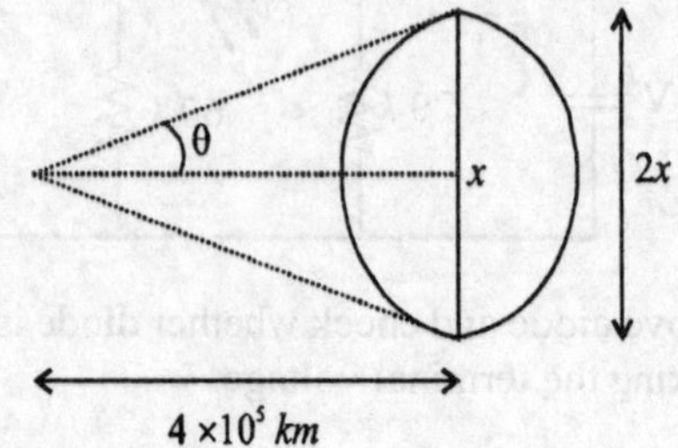

$$\tan\theta = \frac{x}{d}$$

$$\Rightarrow x = d\tan\theta$$

$$\Rightarrow 2x = 2 \times 4 \times 10^5 \times 10^{-3}$$

$$= 800\ km$$

31. Ans. (b)

Solution:

Given. $L = 500\ mm$

$BW = 1500\text{ MHz}$

$$\text{number of nodes} = \frac{\dfrac{1500}{3 \times 10^8}}{2 \times 500 \times 10^{-3}} = 5$$

32. Ans. (d)

Solution:

Given, $BW = 1\text{MHz} = f$

$$\ell = c/f$$

$$= \frac{3 \times 10^8}{1 \times 10^6}$$

$$= 300\ m$$

33. Ans. (c)

Solution:

Given, $\lambda = 632.8\ nm$

$$d = 1m$$

$$d = 3.8 \times 10^5\ km$$

$$\theta = \frac{1.22\lambda}{d}$$

$$= \frac{1.22 \times 632.8 \times 10^{-9}}{1m}$$

$$= 772 \times 10^{-7}$$

$$\tan\theta = \frac{x}{3.8 \times 10^5\ km}$$

$$x = \tan(7.72 \times 10^{-7}) \times 3.8 \times 10^8\ m$$

For diameter,

$$2x = 2 \times \tan(7.72 \times 10^{-7}) \times 3.8 \times 10^8$$

$$= 587\ m$$

34. Ans. (d)

Solution:

n_2 can't be greater than n_1.

35. Ans. (d)

Solution:

$$\text{N.A.} = \sqrt{n_1^2 - n_2^2}$$
$$= 0.34$$

36. Ans. (b)

Solution:

$$\text{N.A.} = \sqrt{n_1^2 - n_2^2}$$
$$= \sqrt{(n_1 - n_2)(n_1 + n_2)}$$
$$\text{N.A.} = \sqrt{2n\,\Delta n}$$

37. Ans. (b)

Solution:

Given, $n_1 = 1.5$

$n_2 = 1.45$

$\lambda = 800\,nm$

Attenuation $= 10\,d\text{B}$

Reponsivity $= 0.55\text{ A/W}$

$$\text{N.A.} = \sqrt{n_1^2 - n_2^2} = 0.384$$

38. Ans. (b)

Solution:

$$\text{P}_i = 2\,m\text{W}$$

$$\text{Attenuation} = 10\log\left(\frac{\text{P}_i}{\text{P}_o}\right)$$

$$\Rightarrow 10 = 10\log\frac{\text{P}_i}{\text{P}_o}$$

$$\Rightarrow \text{P}_o = \frac{\text{P}_i}{\text{P}_{10}} = 0.2\,mW$$

$$\text{R} = \frac{\text{I}_p}{\text{P}_o}$$

$$\Rightarrow \text{I}_p = 0.55 \times \text{P}_o = 0.55 \times 0.2$$
$$= 0.11\,mA$$

39. Ans. (d)

Solution :

Permittivity, $\varepsilon_r = 2.5$

$$n = \sqrt{\mu_r\,\varepsilon_r}$$
$$= \sqrt{2.5} \qquad \mu_r = 1\,(air)$$
$$= 1.58$$

40. Ans. (d)

Solution :

Given, $n_1 = 1.64$

$n_2 = 1.422$

$$\frac{\sin\theta_c}{\sin 90^\circ} = \frac{n_2}{n_1}$$

$$\theta_c \approx 60^\circ$$

41. Ans. (b)

Solution :

$$\Delta = \frac{n_1 - n_2}{n_1} = 0.01$$

$n_1 = 1.5$

$n_2 = 1.485$

$\text{NR} = n_o$ number $\sin\theta_o$

$$\sqrt{n_1^2 - n_2^2} = n_o \sin\theta_o$$

θ_o = acceptance angle

$$= \sin^{-1}\sqrt{1.5^2 - 1.485^2}$$
$$= 12.1^\circ$$

42. Ans. (a)

Solution :

$n_g = 1.72$

$$\text{R} = \left(\frac{n_g - n_{air}}{n_g + n_{air}}\right)^2 \times 100\,\%$$

$$= \left(\frac{1.72 - 1}{1.72 + 1}\right)^2 \times 100$$

$$= 7\,\%$$

43. Ans (d)

Solution :

Given, $\mu_l = 1.33$

$\mu_g = 1.55$

$$\frac{\upsilon_l}{\upsilon_g} = \frac{\mu_g}{\mu_l}$$

$$= \frac{1.55}{1.33}$$

44. Ans. (a)

Solution :

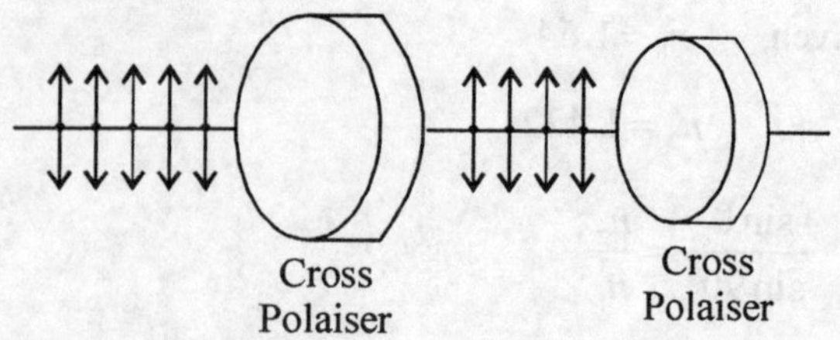

$\Rightarrow$ zero

45. Ans. (c)

Solution :

$\lambda = 515\ nm$

first glass

$\mu = 1.6$

$n = 1$

$t(\mu - 1) = n\lambda$

$$t = \frac{n\lambda}{\mu - 1} = \frac{1 \times 515 \times 10^{-9}}{(1.6 - 1)}$$

$= 0.85\ \mu m$

46. Ans. (b)

Solution :

$t = 1.5\ \mu m$

$\lambda = 0.5\ \mu m$

$$t = (2n+1)\frac{\lambda}{4}$$

$$n = \left(t - \frac{\lambda}{4}\right)\frac{2}{\lambda}$$

$= 5.5 = 6$

47. Ans. (c)

Solution :

$n = 100$

$\lambda = 6328$

$t = 20\ cm$

$\mu = ?$

$2t(\mu - 1) = n\lambda$

$$\mu - 1 = \frac{n\lambda}{2t}$$

$\mu = 1.0001582$

48. Ans. (a)

Solution :

Concentric rings

49. Ans. (d)

Solution :

$\mu_g = 1.53$

$\mu_{air} = 1.0$

$$\underset{\substack{\uparrow \\ fresnel's \\ coeff}}{R} = \left(\frac{\mu_g - \mu_{air}}{\mu_g + \mu_{air}}\right)^2$$

$$R = \left(\frac{53}{2.53}\right)^2 = 0.044$$

$R\ \% = 4.4\%$

50. Ans. (a)

51. Ans. (c)

52. Ans. (c)

53. Ans. (a)

54. Ans. (b)

55. Ans. (d)

56. Ans. (d)

57. Ans. (a)

58. Ans. (b)

59. Ans. (d)

60. Ans. (c)

61. Ans. (d)

62. Ans. (d)

63. Ans. (b)

64. Ans. (a)

65. Ans. (d)

66. Ans. (d)

67. Ans. (c)

68. Ans. (d)

69. Ans. (d)

70. Ans. (c)

71. Ans. (a)

72. Ans. (b)

73. Ans. (b)

74. Ans. (c)

75. Ans. (b)

76. Ans. (c)

77. Ans. (c)

78. Ans. (b)

79. Ans. (b)

80. Ans. (a)

81. Ans. (c)

82. Ans. (d)

83. Ans. (d)

84. Ans. (b)

85. Ans. (d)

86. Ans. (b)

87. Ans. (b)

88. Ans. (c)

Solution:

With $y = x_n$ the position of the nth maximum on the screen, measured from the central maximum. The distance between adjacent fringes is:

$$\Delta x = x_{n+1} - x_n = \frac{D\lambda}{d}(n+1) - \frac{D\lambda}{d}n = \frac{D\lambda}{d}$$

So, $$d = \frac{D\lambda}{\Delta x} = \frac{(0.8m)(589\times10^{-9}m)}{0.35\times10^{-2}m}$$

$$= 1.35\times10^{-4}m = 0.135\,mm$$

89. Ans. (b)

Solution:

For coherent sources $I\max = n^2 I = 16I$

For incoherent sources $I\max = nI = 4I$

90. Ans. (d)

Solution:

$$E_{ion} = \frac{9.1\times10^{-31}\times(1.6\times10^{-19})^4}{8\times(6.6\times10^{-34}\times8.85\times10^{-12})^2}$$

$$= 13.6\,eV$$

91. Ans. (c)

Solution:

$$R = \left(\frac{n_2 - n_1}{n_2 + n_1}\right)^2 = \left(\frac{3.6-1}{3.6+1}\right)^2 = 0.32$$

92. Ans. (b)

Solution:

Overall power efficiency

$$= \frac{5\times10^{-3}}{2500\times1\times10^{-2}} = 0.02\%$$

93. Ans. (a)

94. Ans. (b)

95. Ans. (c)

96. Ans. (a)

97. Ans. (d)

98. Ans. (d)

99. Ans. (b)

1.5 LEDS and Lasers

1. Ans. (b)

2. Ans. (c)

Solution:

Coherence

a. Laser beam has a high degree of ordering in its light field or in other words it has high degree of coherence.

b. Coherence is a measure of the degree of phase correlation that exists in the radiation field of light source at different locations and at different times.

c. Coherence is the characteristics which distinguish leaser from other source of light

d. Light from a conventional source is a jumble of tiny separate waves that cancel each other in a random order. The wavefront so produced changes from on point to other and varies from instant to instant.

Hence there are two separate concepts of coherence

1. Spatial coherence
2. Temporal coherence

Spatial Coherence

- A light source emits light in all directions it has light in is surroundings
- If two fields at two different point on a wavefront of a electromagnetic wave has constant phase difference over any time '*t*' they posses spatial coherence.

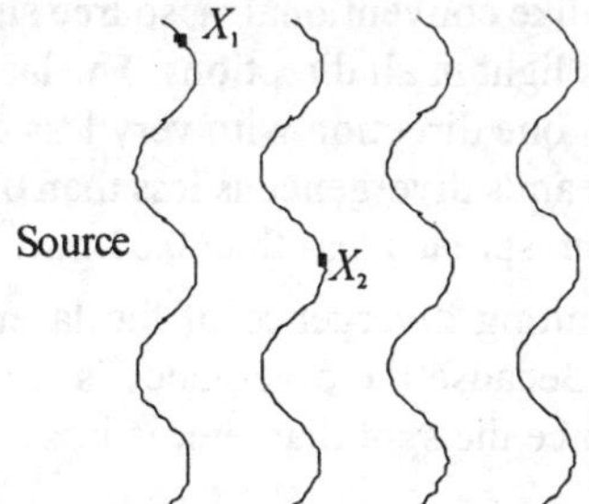

- Consider any two point in space X_1 and X_2 in the plane surrounding the source in while the light is preset The correlation between the light waves a these two point and is called as the spatial coherence.

Temporal Coherence

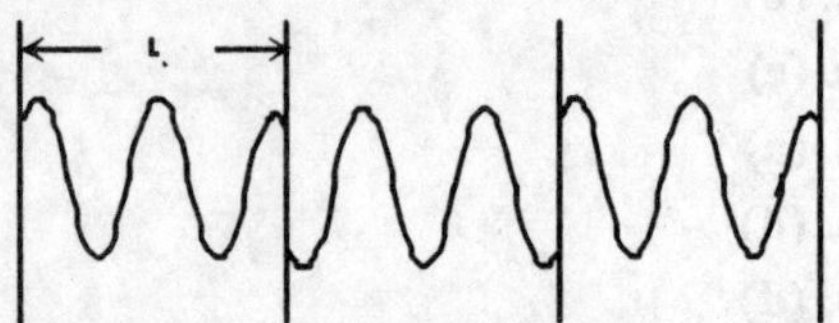

(Electromagnetic wave with coherence time τ_c)

- Temporal coherence refer to the correlation between the field at a point and the field at the same point at a later time. If the phase difference between the two fields in constant during period of time, than a wave is said to possess temporal coherence. The time for which the wave remain sinusoidal and do not break in phase is called as the coherence time (τ_c).

The distance travelled by the wave in the time interval τ_c is called as the length of coherence (l)

$$l = c\tau_c$$

Bandwidth (Δf) is related to $_c$ as

$$\Delta f = \frac{1}{\tau_c}$$

3. **Ans. (b)**

Solution:

$$\Delta f = \frac{C}{2L}$$

$C = 3\times10^8$ m/sec

$L = 0.04\ m$

$$\Delta f = \frac{3\times10^8}{2\times0.04\ m}$$

$$\Delta f = 375\ MHz$$

4. **Ans. (b)**

Solution:

Directionally: Leaser beam is highly directional, it has very less spread. The spread for an ideal laser is taken as zero unlike conventional at source such as a lamp which spreads light in all directions. The laser beam emits light in only one direction with very less deviation. A typical laser beam's divergence is less than 0.01 milliradian, i.e. the beam spreads less than 0.01 mm for every meter

The limiting divergence of the laser light is relatively small. Because the divergence is small so the spread is less hence the spot diameter is less

5. **Ans (d)**

Solution:

$$\frac{\text{Velocity in the tissue}}{\text{Velocity in the glass}} = \frac{n_g}{N_t} = \frac{1.5}{1.33}$$

[Since-velocities of light in the medium is inversely propertional to refractional to refractive index of the medium].

6. **Ans (c)**

Solution:

Resolution of a mass spectrometer measures the ability of the mass spectrometer to distinguish the two peaks of slightly different mass to charge ratio (ΔM) in a mass-spectrum.

$$\text{Resolution } (R) = \frac{M}{\Delta M}$$

* A higher resolution of a mass spectrometer indicates a better peak seperation in the mass-spectrum.

Finding m/z ratio for CO_+

m_1 = mass of CO $= 12\text{amu} + 15.9949\text{amu}$

$= 27.9949\text{amu}$

$z_1 = +1$

$$M_1 = \frac{m_1}{z_1} = \frac{27.9949}{1}$$

or $M_1 = 27.9949$

Finding the m/z ratio for N_2^+

m_2 = mass of $N_2 = 2\times14.0031\text{amu}$

$= 28.0062\text{amu}$

$z_2 = +1$

Therefore, $M_2 = \frac{m_2}{z_2} = \frac{28.0062}{1}$

or $M_2 = 28.0062$

Now, Resolution (R) $= \frac{M}{\Delta M}$

M = maximum of (M_1, M_2)

$\Delta M = 28.0062 - 27.0049$

$= 0.0113$

Therefore, $R = \frac{28.0062}{0.0113}$

$= 2478.42$

$\simeq 2500$

7. **Ans. (a)**

Solution:

Given,

Divergence $\Delta\theta = 1$ m rad

$= 10^{-3}$ rad

Distance/ radius $= 4\times10^5\ km$

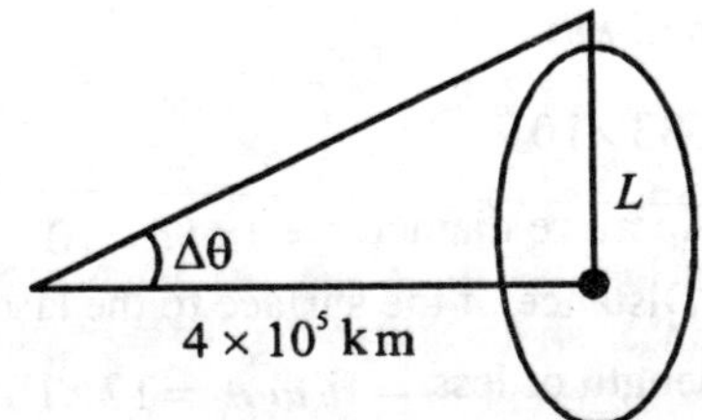

Beam diameter $L = r(\Delta\theta)$

$$= (4\times10^5\ km)(10^{-3})$$

$$= 400\ km$$

8. Ans. (d)

Solution:

$$\Delta\lambda = \frac{\lambda^2 \Delta f}{c}$$

$$= \frac{(632.8\times10^{-9})^2 \times10^6}{3\times10^8}$$

$$= 1.33\times10^{-15}$$

$$\ell_c = \frac{\lambda^2}{\Delta\lambda} = \frac{(632.8\times10^{-9})^2}{1.33\times10^{-15}} = 300\ m$$

Directly as $\lambda = \frac{C}{f}$

Coherence length $= \frac{C}{\Delta f}$

$$= \frac{3\times10^8\ m}{1\times10^6}$$

$$= 300\ m$$

9. Ans. (c)

Solution:

Angle of divergence $(\Delta\theta) = \frac{1.22\lambda}{d}$

$$= \frac{1.22\times632.8\times10^{-9}}{1\ m}$$

$$= 772\times10^{-9}$$

Diameter of central maxima = 2 × distance × Δθ

$$= 2\times3.8\times10^5\ km\times772\times10^{-9}$$

$$= 587\ m$$

10. Ans. (b)

Solution:

From the given figure

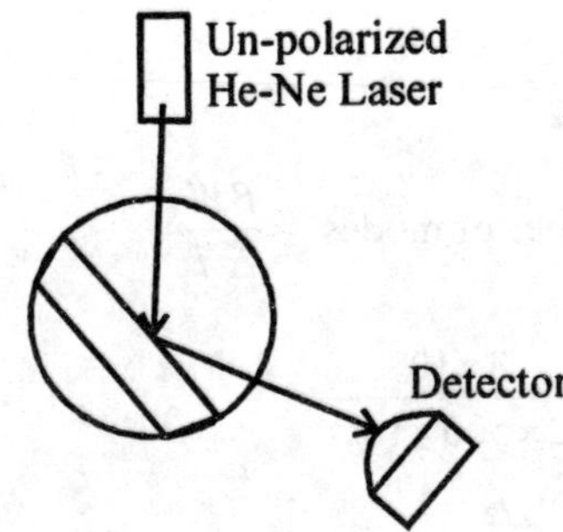

$$2i = 115°.34' = 115.566°$$

$$i = 57.783°$$

Reflective index

$$n = \tan i = \tan 57.783° = 1.587$$

11. Ans. (c)

Solution:

We know that

$$L_C = C\Delta t$$

$$= \frac{C}{\Delta V}$$

ΔV = Band width

C = velocity of light

$$L_C = \frac{3\times10^8\ m/sec}{10\ GHz}$$

$$= \frac{3\times10^8\ m/sec}{10\times10^9\ Hz}$$

$$= \frac{3\times10^8\ cm/sec}{10^{10}\ Hz}$$

$$= 3\ cm$$

12. Ans. (b)

Solution:

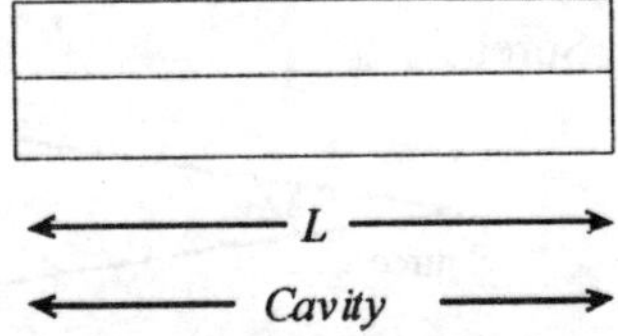

The difference in frequency between two consecutive modes are same, i.e. longitudinal mode separation

$$V = \frac{d}{t}$$

$$t = \frac{d}{V}$$

$$F = \frac{V}{d} = \frac{V}{2L}$$

$$\Delta f = \frac{C}{2L}$$

So, number of modes $= \frac{B.W}{\Delta f}$

$$\Delta f = \frac{3 \times 10^8}{2 \times 500 \times 10^{-3}}$$

$$\Rightarrow 3 \times 10^8 \ Hz$$

Number of modes $= \frac{B.W}{\Delta f}$

$$= \frac{1500 \times 10^6}{3 \times 10^8} = 5 \text{ modes}$$

13. Ans. (b)

Solution:

Directionally: Leaser beam is highly directional, it has very less spread. The spread for an ideal laser is taken as zero unlike conventional at source such as a lamp which spreads light in all directions. The laser beam emits light in only one direction with very less deviation. A typical laser beam's divergence is less than 0.01 milliradian, i.e. the beam spreads less than 0.01 mm for every meter

1. Angle of divergence or angular spread

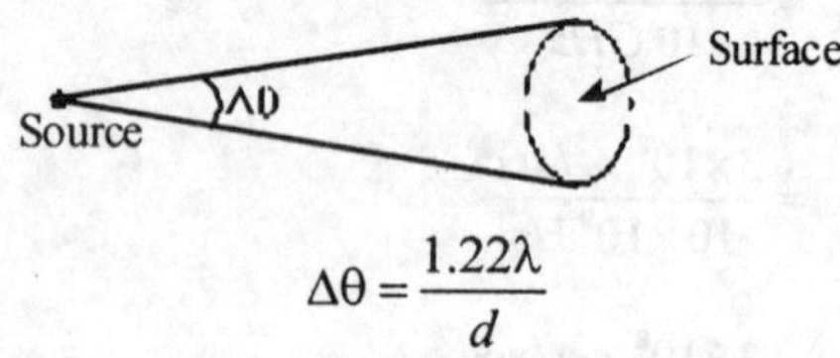

$$\Delta\theta = \frac{1.22\lambda}{d}$$

Where,

$d =$ Aperture diameter of light source from where the light is comming

$\lambda =$ Wavelength of the light coming from the laser source

$\Delta\theta =$ Angular spread of the laser beam.

2. Areal Spread

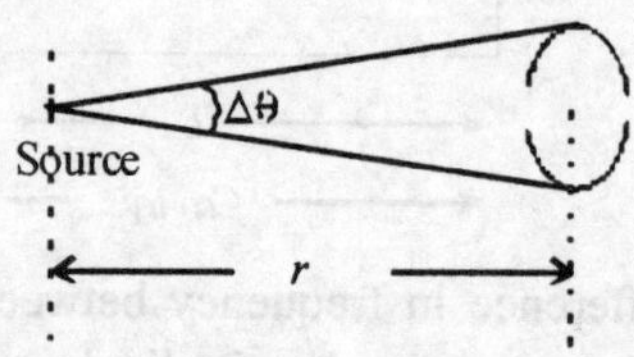

Areal spread $= \pi(r/\Delta\theta)^2$

$r =$ distance of the surface from the laser source

As given

$\lambda =$ wave length of light

$= 633 \ nm$

$= 633 \times 10^{-9}$

$d =$ Aperture diameter $= 1mm = 10^{-3}$

$r =$ Distance of the surface to the laser source, i.e. focal length of less $= 17mm = 17 \times 10^{-3}$

$$\Delta\theta = \frac{1.22\lambda}{d}$$

$$= \frac{1.22 \times 633 \times 10^{-9}}{10^{-3}}$$

$$= 772.26 \times 1^{0-6}$$

Spot size diameter $= D$

$$D = 2r\Delta\theta$$

$$D = 2 \times 17 \times 10^{-3} \times 772.26 \times 10^{-6}$$

$$= 26.25 \ \mu m$$

14. Ans (c)

Solution:

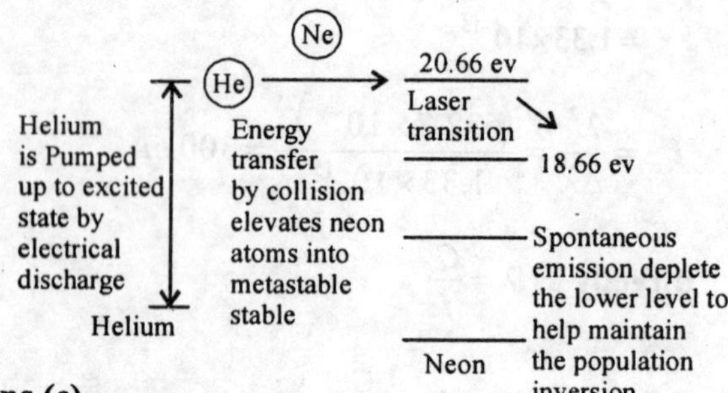

15. Ans (c)

Solution:

The transition for the major element Ne is to be considered, since He is present just to excite the Ne atoms. The transition between heigher and the atom stability in the meta-stabel state are considered.

Energy difference $= E_2 - E_1$

$$= (20.66 - 18.70)eV = 1.96 \ eV$$

It is known,

$$E = hf = \frac{hc}{\lambda}$$

$$1.96 \times 1.6 \times 10^{-19} = \frac{6.62 \times 10^{-34} \times 3 \times 10^8}{\lambda}$$

$$\lambda = \frac{6.62 \times 10^{-34} \times 3 \times 10^8}{1.96 \times 1.6 \times 10^{-9}}$$

$$= 6.3329 \times 10^{-7}$$

$$= 633.29 \times 10^{-9}$$

$$\simeq 633.8 \text{ nm}$$

16. Ans. (d)

Solution:

$$\lambda = \frac{c}{f} = cT = 3 \times 10^8 \times T$$

or $$T_o = \frac{1.2}{3 \times 10^8} = 4 \times 10^{-9} \text{ sec}$$

17. Ans. (a)

18. Ans. (b)

19. Ans. (c)

20. Ans. (c)

21. Ans. (a)

22. Ans. (c)

23. Ans. (a)

24. Ans. (c)

25. Ans. (b)

26. Ans. (a)

27. Ans. (d)

28. Ans. (c)

29. Ans. (c)

Solution:

From $\lambda = (d/D)\Delta x$, Where d and D are fixed,

$$\lambda' = \frac{\Delta x'}{\Delta x}\lambda = \frac{7.6}{8.3} \times 630 = 577 nm.$$

30. Ans. (c)

Solution:

$$W = 0.7 \times 5.67 \times 10^{-8} \times 10^{-5} \times (2000)$$

$$= 6.35\, W$$

31. Ans. (d)

Solution:

$$W = \frac{hC}{\lambda} \times \eta_e \times \text{ Photon emission rates}$$

$$= \frac{hc}{550 \times 10^{-9}} \times 0.001 \times \frac{50 \times 10^{-3}}{e}$$

$$= 1.13 \times 10^{-4} \text{ watts.}$$

32. Ans. (d)

Solution:

$$\lambda_c = \frac{\lambda_0}{4(n^w - n^{2w})}$$

$$= \frac{0.8 \times 10^{-6}}{4 \times 0.00217} = 10^{-5} m.$$

33. Ans. (a)

Solution:

$$R_\lambda = \frac{\eta e \lambda}{hc}$$

$$= \frac{0.25 \times 1.6 \times 10^{-19} \times 0.5 \times 10^{-4}}{6.6 \times 10^{-34} \times 3 \times 10^8}$$

$$= 0.1\, A/W.$$

34. Ans. (c)

Solution:

$$W_{\min} = \frac{(2 \times 2 \times 10^{-14} \times 1.6 \times 10^{-19} \times 1)^{1/2}}{0.1}$$

$$= 8.10\, W$$

35. Ans. (d)

Solution:

Material dispersion

$$\Delta C = \frac{10^3 \times 850 \times 10^{-9} \times 3 \times 10^{10} \times 50 \times 10^{-9}}{3 \times 10^8}$$

$$= 4.3 \times 10^{-9} s$$

36. Ans. (c)

37. Ans. (a)

38. Ans. (d)

39. Ans. (a)

40. Ans. (b)

41. Ans. (d)

42. Ans. (c)

43. Ans. (c)

44. Ans. (b)

45. Ans. (d)

46. Ans. (b)

47. Ans. (b)

48. Ans. (c)

49. Ans. (c)

50. Ans. (d)

51. Ans. (d)

52. Ans. (a)

53. Ans. (d)

Solution:

Angular spread $(\Delta\theta)$

$$\Delta\theta = \frac{1.22\,\lambda}{d} = \frac{1.22\times700\times10^{-9}}{5\times10^{-3}}$$

$$= 1.70\times10^{-4} \text{ radian}$$

54. Ans. (a)
Solution:
Area Spread (DA)

$$(\text{DA}) = \pi(r\theta)^2$$

$$= 3.14(4\times10^{8}\times1/70\times10^{-4})^2$$

$$= 145\,1936\times10^{8}$$

$$= 1.45\times10^{10}\,\text{m}^2$$

55. Ans. (d)
Solution:

$\therefore \quad \Delta E = hf$

$$E_2 - E_1 = hf = \frac{hc}{\lambda}$$

$$2.36\times.16\times10^{-19} = \frac{6.62\times10^{-34}\times10^{8}}{\lambda}$$

$$\lambda = 5.26\times10^{-7}$$

$$= 5260\times10^{-10}$$

$$= 5260\ \text{A}^\circ$$

56. Ans. (a)
Solution:
Energy of a single photon

$$= hf = \frac{hc}{\lambda}$$

$$= \frac{6.624\times10^{-34}\times3\times10^{8}}{672.8\times10^{-9}}$$

$$= 0.0296\times10^{-17}$$

$$= 2.96\times10^{19}\ \text{Joule}$$

Suppose there are *n* photons emitted in one minute

output power $= 23\ mW/\text{sec}$

$$n = \frac{2.3\times10^{-3}}{2.95\times10^{-19}}$$

$$= 77.968\times10^{16}\ \text{photons/sec}$$

$$= 77.968\times10^{16}\times60$$

$$= 4.6\times10^{17}\ \text{photons / minute}$$

57. Ans. (c)

Solution:

$$\text{Areal spread} = \pi(\theta r)^2$$

$$\theta = \frac{1.22\,\lambda}{d}$$

$$= \frac{1.22\times720\times10^{-9}}{5\times10^{-3}}$$

$$= 292.8\times10^{-6}$$

$\therefore$ areal spread $= 3.14(292.8\times10^{-6}\times0.1)^2$

$$= 9.69\times10^{-10}\ \text{m}^2$$

$\therefore$ intensity (I)

$$= \frac{50\times10^{-3}}{9.69\times10^{-10}} = 5.159\times10^{7}$$

$$= 51.59\times10^{6} = 51.6\ \text{MW}/m^2$$

58. Ans. (b)
Solution:
The difference in frequencies between consecutive mode is DV and length of cavity is L

$$\Delta v = \frac{c}{2\text{L}} = \frac{3\times10^{4}}{2\times500\times10^{-3}}$$

$$= 3\times10^{8}$$

$$\text{No of modes} = \frac{\text{Bandwidth}}{\Delta v}$$

$$= \frac{150\times10^{6}}{3\times10^{8}}$$

$$= 5 \text{ modes}$$

59. Ans. (c)
Solution:

For gas laser, $\quad \Delta v = \dfrac{c}{2\text{L}}$

For semiconductor laser,

$$\Delta v = \frac{c}{2\mu\text{L}}$$

$$= \frac{3\times10^{8}}{2\times3.3\times300\times10^{12}}$$

$$= 0.1515\times10^{-6}$$

$$= 0.15\ \mu m$$

60. Ans. (d)
61. Ans. (b)
62. Ans. (c)
63. Ans. (c)
Solution:

$$\text{I}_{\max} = \frac{V_s - (V_{LED})\text{min}}{Rs}$$

$$=\frac{10-15}{470}\text{A}$$

$$=18.1\text{ mA}$$

64. Ans. (b)

Solution:

Given $r_e = 1.2\times10^{11}$

$r_p = 3\times10^{11}$

$$\eta = \frac{r_e}{r_p}$$

$$=\frac{1.2\times10^{11}}{3\times10^{11}} = 0.4$$

Therefore the quanturn efficiency of the photodiode at 0.85 μm is 40 %.

65. Ans. (d)

Solution:

$$\text{Responsivity (R)} = \frac{\eta e\lambda}{hc}$$

$$=\frac{0.4\times1.602\times10^{-19}\times0.85\times10^{-6}}{6.626\times10^{-34}\times3\times0^{8}}$$

$$=0.274\text{ AW}^{-1}$$

Therefor, Hence the responsivity of the photodiode at 0.85 μm is 0.274 AW^{-1}

66. Ans. (b)

Solution:

$$\text{Responsivity (R)} = \frac{\eta e\lambda}{hf} = \frac{\eta e}{hf}$$

$$E = hf$$

$$\therefore \quad R = \frac{\eta e}{E}$$

$$=\frac{0.62\times1.602\times10^{-19}}{2\times10^{-19}}$$

$$=0.496\text{ AW}^{-1}$$

Now, $$R = \frac{I_p}{P_o}$$

or, $$P_o = \frac{I_p}{R} = \frac{2.8\times10^{-6}}{0.496}$$

$$=5.645\ \mu W$$

Therefore, the incident optical power required is 6.645 μW.

67. Ans. (b)

Solution:

$$\lambda_c = \frac{hc}{E_g} = \frac{6.628\times10^{-34}\times3\times10^{8}}{1.43\times1.6\times10^{-19}}$$

$$=0.868 \approx 0.87\ \mu m$$

68. Ans. (d)

Solution:

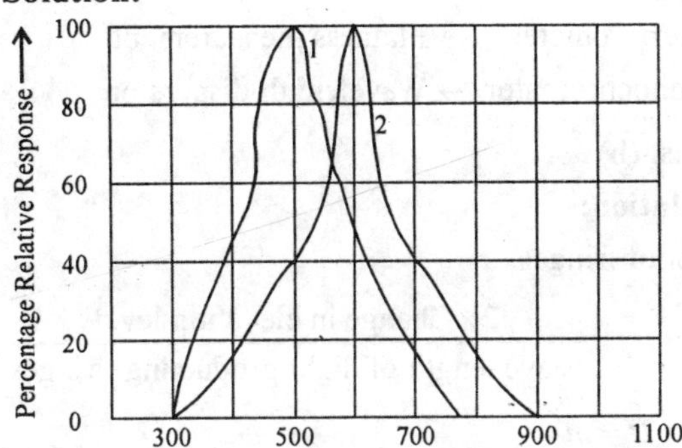

Fig: Relative response Vs wavelenght of

(1) Cadninum Sulphide Cell (Cds

(2) Cadnium Sulpho - Selanide Cell

* Materials used to make photoresistors are CdS, CdSe, PbS, PbSe

69. Ans. (a)

Solution:

$$R_e = R_1 + (R_t - R_i)[1-\exp(-t/\tau)]$$

$$=30+(100-30)[1-\exp(-10/72)]$$

$$=30+70[1-e^{-10/72}]$$

$$=39.1\ k\Omega$$

70. Ans. (c)

71. Ans. (c)

1.6 Interferometers

1. Ans. (a)

2. Ans. (c)

3. Ans. (a)

Solution:

Given,

Divergence $\Delta\theta = 1$ m rad

$= 10^{-3}$ rad

Distance/radius $= 4\times10^{5}\ km$

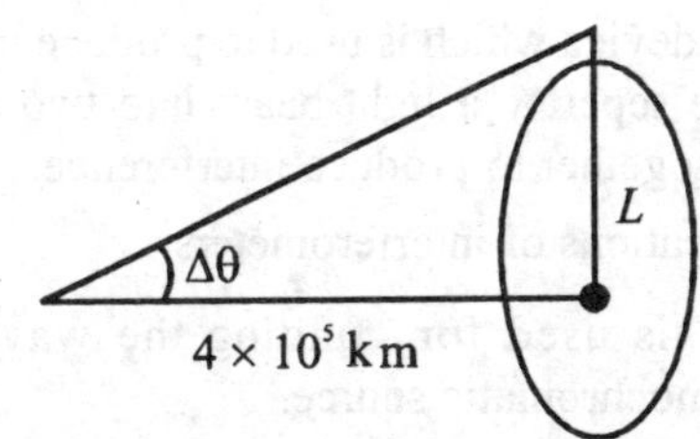

Beam diameter $L = r(\Delta\theta)$

$$= (4\times10^5\ km)(10^{-3})$$

4. Ans. (c)

Solution: The correct match is given below

Interferometer → Flatness measurement.

Monochromator → Wavelength dispersion

5. Ans. (b)

Solution:

No. of fringes

$$= \frac{2\times\text{Change in elevation level}}{\text{Wave length of light producting fringes}}$$

i.e. $2t = m\lambda$

$$m = \frac{2\times1.5\ \mu m}{0.5\ \mu m}$$

$$m = 6\ \text{fringes}$$

6. Ans. (a)

Solution:

Here, $\lambda = 0.63\ \mu m$

The error encountered due to the faction fringe shift 'e' is

Error $= e\times\lambda$

$= 0.15\times0.63\ \mu m$

$= 0.0945\ \mu m$

Approximated length of gauge = 100 μm

Therefore actual length $= 100\mu m - 0.095\mu m$

$= 99.902\ \mu m$

7. Ans. (c)

8. Ans. (b)

Solution:

Reflection reaches minimum

When wave length is $\frac{\lambda}{4}$ i.e. $\frac{600\ nm}{4}$

$= 150\ nm$

9. Ans (c)

Solution:

Interferometer

It is a device which is used to produce interference. The device seperate a light beam into two beam and bring them together to produce interference.

Applications of interferometers.

1. It is used for finding the wavelength of the nonochromatic source.
2. To calculate the thickness of the transparant sheets (e.g. glass, mica)
3. To calculate the refractive index of sheat (μ)
4. To find the resolution of the spectral line.
5. To find difference between neighbouring wavelengths.

Michelson interferometer

In michelson interferometer, the amplitude of a light beam comming from an extended source is divided into two parts of equal intensity by parital reflection and transmittance through the mirros, the two beams then obtained are sent at right angles to each other and are than brough together by reflection through mirros and focused on a telescope for observence.

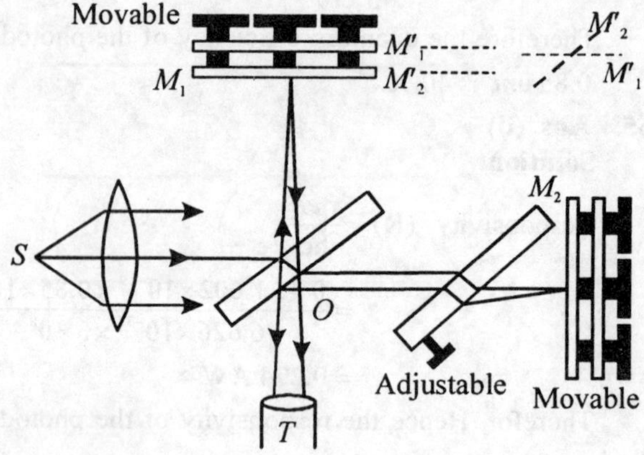

Basic parts:

1. Monochromatric light source: The source must be monochromatic, light has to be splitted into two beams
2. Beam splitter: Beam splitter (x) is plced at 45°. It is semi transparent, i.e. partly relfecting and party transmitting
3. Two Mirros: Two mirrors are present one is removable called the movable mirror. The movable mirror is mounted with the micro meter screw to move slowly.

If $OM_1 = OM_2 \Rightarrow$ Path difference = 0 (Bright band)

When we move $M_2 \Rightarrow$

If path difference $= \frac{\lambda}{2} \Rightarrow$ Dark band

If path difference $\lambda \Rightarrow$ Bright band

If path difference $= \frac{3\lambda}{2} \Rightarrow$ Dark band

If instead of using a monochromatic light we use a white (polychromatic) light, different colour fringes appears, and the firnge pattern is not useful and not readable, hence we always use a monochromatic light (colour fringes may overlap)

Micheleson interferometer is used for the standaradization of metre

Type of Fringes:

Circular fringes: For circular fringes to be formed the mirror M_1 and the virtual mirror M_2' (which is the image of M_2) must be parallel, i.e. $M_1 \parallel M_2'$

Localised fringes:

When the mirrors M_1 and M_2' are seperate by distance d, the fringes are as shown in figure (a)

When $d = 0$, the fringes become straight as shown in figure (c) and beyond this they will become convexs to the other side as is shown in figure (b)

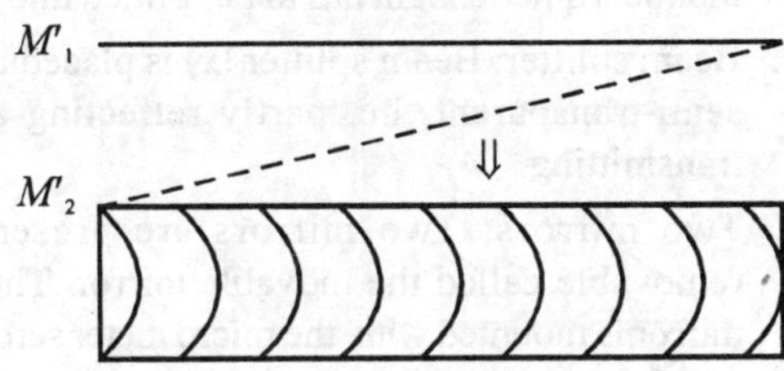

(Convex toward thin edge)
Figure (a)

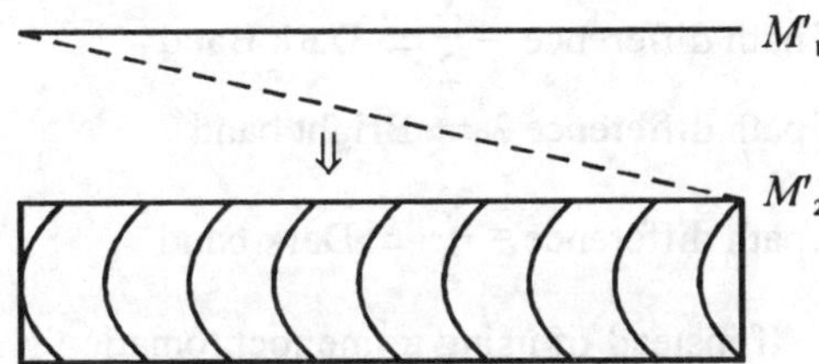

(Convex torwards thin edge) figure (b)

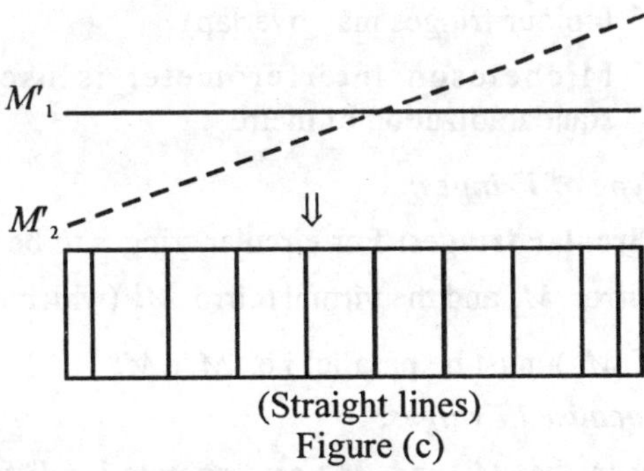

(Straight lines)
Figure (c)

The fringes produced can be viewed by a telescope

Application of Michelson interferometer

Finding the wavelength of monochromatic light

$$n\lambda = 2(\Delta x) = 2(x_2 - x_1)$$

Where,

$n =$ Number of fringes

$\lambda =$ Wavelength of monochromatic light

$\Delta x =$ Micrometer screw movement

$x_2 =$ Final position

$x_1 =$ Initial position

* Calculation of thinckness and the refractive index (μ) of any transparent sheet.

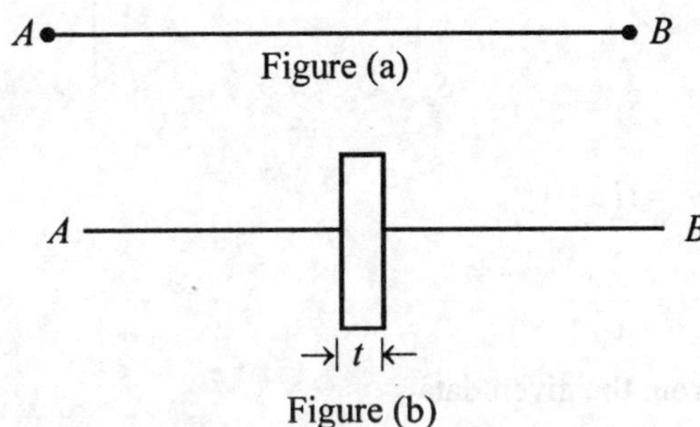

Figure (a) shows the path between A and B without any slab of refractive index (μ)

Figure (b) shows the case when the slab of Rim and the thickness 't' is inserted the slab introduces a path of 'μt' instead of 't'

So the path difference = Distance between A, B in case 2 – Distance between A, B in case 1

$$= (AB - t + \mu t) - (AB)$$

$$= AB - t + \mu t - AB$$

$$= \mu t - t$$

$$= t(\mu - 1)$$

$$\Delta x = t(\mu - 1)$$

$$n\lambda = 2(\Delta x) = 2t(\mu - 1)$$

Resoulution spectral lines: Michelson interferometer has proved very useful in resolving very close lines by studying the change in the visibility of the fringes with the path difference

Find $n\lambda = 2(\Delta x)$

For n[th] fringe, let wavelength be λ_1

$$n\lambda_1 = 2(\Delta x)$$

For $(n+1)^{th}$ finge, let the wavelength be λ_2

$$(n+1)\lambda_2 = 2(\Delta x)$$

$$\lambda_1 - \lambda_2 = \frac{\lambda_1\lambda_2}{2x} = \frac{(\lambda_{av})^2}{2x}$$

* Calculation of thinckness of sheet width by Young's Method

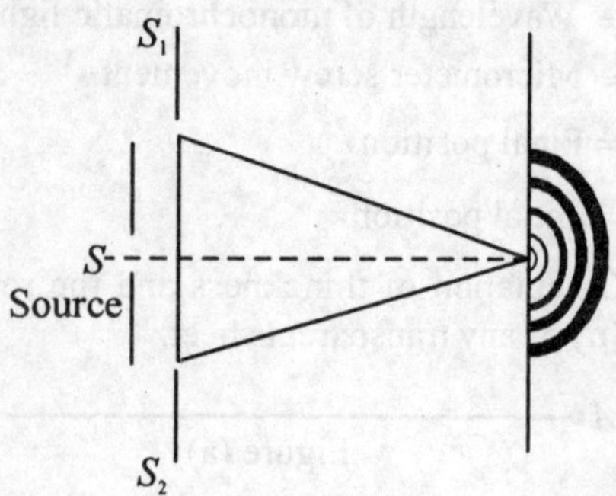

$n\lambda = t(\mu - 1)$

$t = \frac{n\lambda}{\mu - 1}$

From the given data

$n\lambda = 2\Delta x \ = 2t(\mu - 1)$

$\therefore n \propto t$

as Δt is replaced to $(2(\Delta t))$.

therefore *n* will also gets doubled.

$n' = 2n = 2 \times 50$

$= 100$ fringes

10. Ans (c)

Solution:

Interferometer

It is a device which is used to produce interference. The device seperate a light beam into two beams and bring them together to produce interference.

Applications of interferometers.

1. It is used for finding the wavelength of the monochromatic source.
2. To calculate the thickness of the transparant sheets (e.g. glass, mica)
3. To calculate the refractive index of sheet (μ)
4. To find the resolution of the spectral line.
5. To find difference between neighbouring wavelengths.

Michelson interferometor

In Michelson interferometer, the amplitude of a light beam coming from an extended source is divided into two parts of equal intensity by partial reflection and transmittance through the mirrors, the two beams then obtained are sent at right angles to each other and are then brought together by reflection through mirrors and focused on a telescope for observence.

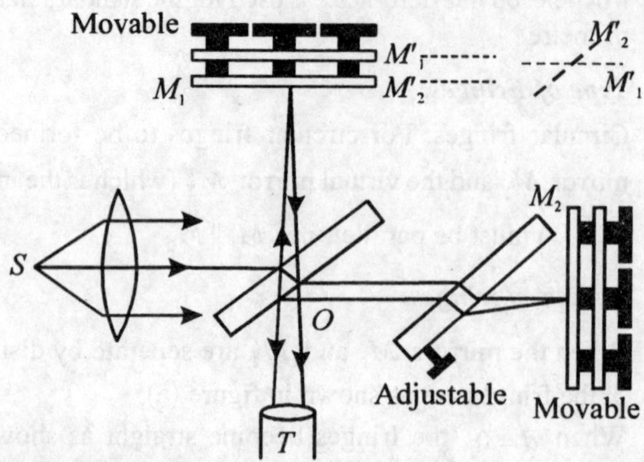

Baisc Parts:

1. Monochromatric light souce: The source must be monochromatic, light has to be splitted into two beams
2. Beam splitter: Beam splitter (x) is placed at 45°. It is semi-transparent, i.e. partly relfecting and partly transmitting
3. Two mirrors: Two mirrors are present, one is removable called the movable mirror. The movable mirror is mounted with the micro meter screw to move slowly.

If $\quad OM_1 = OM_2 \Rightarrow$ Path difference $= 0$ (Bright band)

When we move $M_2 \Rightarrow$

If path difference $= \frac{\lambda}{2} \Rightarrow$ Dark Band

If path difference $\lambda \Rightarrow$ Bright band

If path difference $= \frac{3\lambda}{2} \Rightarrow$Dark band

* If instead of using a monochromatic light we use a white (polychromatic) light, different colour fringes appears, and the firnge pattern is not useful and not readable,hence we always use a monochromatic light (colour fringes may overlap)
* Micheleson interferometer is used for the standaradization of metre

Type of Fringes:

Circular fringes: For circular fringes to be formed the mirror M_1 and the virtual mirro M_2' (which is the image of M_2) must be parallel i.e. $M_1 \parallel M_2'$

Localised Fringes:

when the M_1 and M_2' are seperate by distance d, the friges are as shown in figure (a)

when $d = 0$, the fringes become straight as shown in figure (c)

and beyond this they will become convexs to the other side as is shown in figure (b)

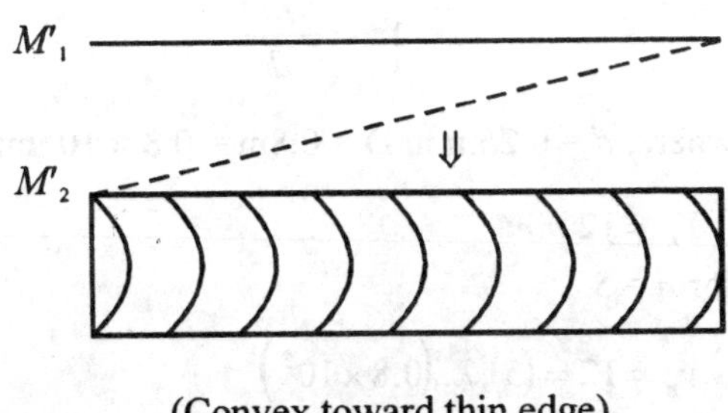

(Convex toward thin edge)
Figure (a)

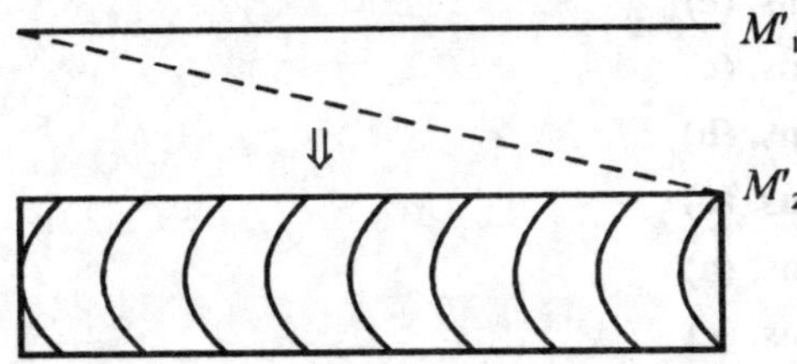

(Convex towards thin edge) figure (b)

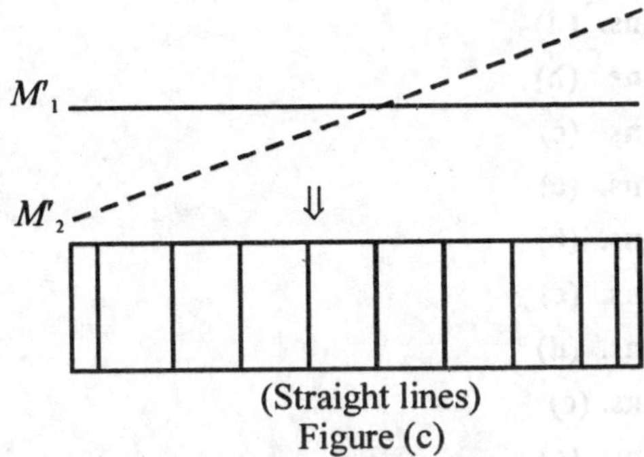

(Straight lines)
Figure (c)

The fringes produced can be viewed by a telescope

Application of Michelson interferometer

* Finding the wavelength of monochromatic light

$$n\lambda = 2(\Delta x) = 2(x_2 - x_1)$$

Where,

$n =$ Number of fringes

$\lambda =$ Wavelength of monochromatic light

$\Delta x =$ Micrometer screw movement

$x_2 =$ Final position

$x_1 =$ Initial position

* Calculation of thickness and the refractive index (μ) of any transparent sheet.

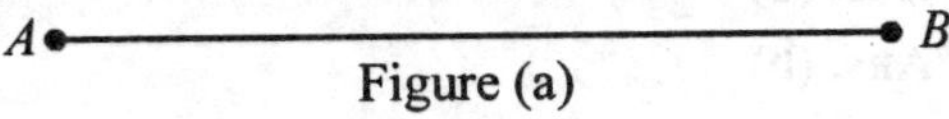

Figure (a)

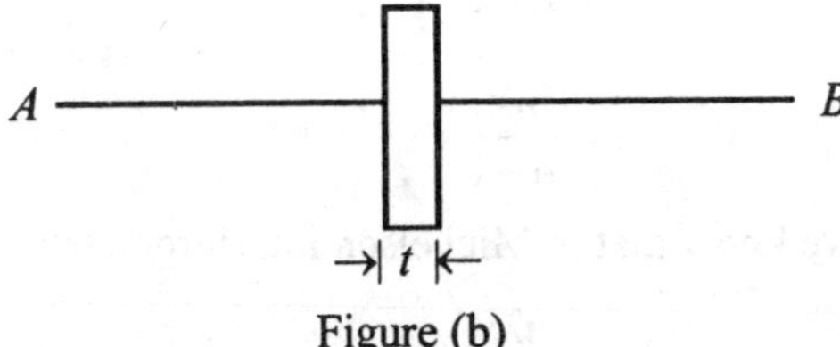

Figure (b)

Figure (a) shows the path between A and B without any slab of refractive index (μ)

Figure (b) shows the case when the slab of Rim and the thickness 't' is inserted the slab introduces a path of 'μt' instead of 't'

So the path difference = Distance between A, B in case 2 – Distance between A, B in case 1

$$= (AB - t + \mu t) - (AB)$$

$$= AB - t + \mu t - AB$$

$$= \mu t - t$$

$$= t(\mu - 1)$$

$$\Delta x = t(\mu - 1)$$

$$n\lambda = 2(\Delta x) = 2t(\mu - 1)$$

* Resoulution spectral lines: Michelson interferometer has proved very useful in resolving very close lines by studying the change in the visibility of the fringes with the path difference

Find $\quad n\lambda = 2(\Delta x)$

For n[th] fringe, let wavelength be λ_1

$$n\lambda_1 = 2(\Delta x)$$

For $(n+1)^{th}$ finge, let the wavelength be λ_2

$$(n+1)\lambda_2 = 2(\Delta x)$$

$$\lambda_1 - \lambda_2 = \frac{\lambda_1\lambda_2}{2x} = \frac{(\lambda_{av})^2}{2x}$$

* Calculation of thickness of sheet width by Young's Method

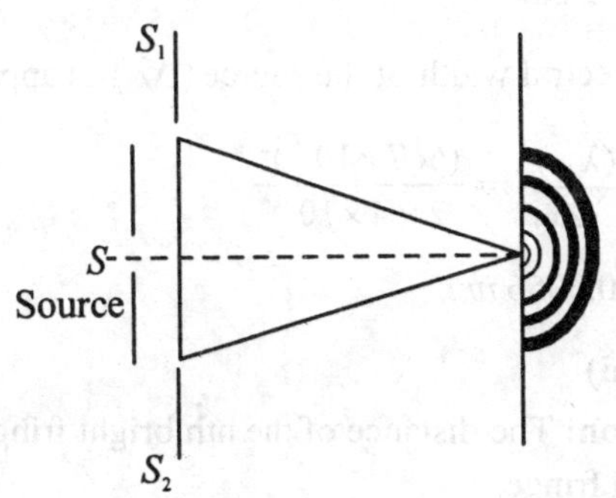

$$n\lambda = t(\mu - 1)$$

$$t = \frac{n\lambda}{\mu - 1}$$

We know that in Michelson interferometer

$$\Delta x = \frac{n\lambda}{2} \text{ where}$$

$\Delta x =$ Distance moved by mirror

$\lambda = ?$

$n =$ no of fringer

$$0.1\,mm = \frac{(300) \times \lambda}{2}$$

$$0.1 \times 10^{-3} \times 2 = 300 \times \lambda$$

$$\lambda = \frac{1 \times 10^{-4} \times 2}{3 \times 10^{2}}$$

$$\lambda = \frac{2}{3} \times 10^{-6}$$

$$\lambda = \frac{2000}{3} \times 10^{-9}$$

$$\lambda = 667\,nm$$

11. Ans. (a)

Solution:

First find $\Delta\lambda$

$$\Delta\lambda = \lambda_1 - \lambda_2$$

$\lambda_{av} = 667\,nm$ From 1 st part

$x = 4\,cm$

It is known that

$$x = \frac{\lambda_1 \times \lambda_2}{2(\lambda_1 - \lambda_2)}$$

$$= \frac{(\lambda_{av})^2}{2\,\Delta\lambda}$$

The spectral width of the source $(\Delta\lambda)$ is approximately

$$\Delta\lambda = \frac{(\lambda_{av})^2}{2x} = \frac{(667 \times 10^{-9})^2}{2 \times 4 \times 10^{-2}}$$

$$= 0.0056\,nm$$

12. Ans. (d)

Solution: The distance of the nth bright fringe from the central fringe.

$$Y_n = \frac{n\lambda D}{d}$$

where, $d = 0.25$ mm D = 0.8m = 0.8 × 10^3 mm

$Y_n = 12\,mm$

for $n = 5$,

$$y_n = 12 = (5).\lambda..(0.8 \times 10^3)$$

or $\lambda = 0.75 \times 10^{-3}$ mm

13. Ans. (c)
14. Ans. (c)
15. Ans. (c)
16. Ans. (b)
17. Ans. (a)
18. Ans. (a)
19. Ans. (c)
20. Ans. (b)
21. Ans. (a)
22. Ans. (d)
23. Ans. (d)
24. Ans. (b)
25. Ans. (c)
26. Ans. (c)
27. Ans. (b)
28. Ans. (c)
29. Ans. (d)
30. Ans. (c)
31. Ans. (c)
32. Ans. (b)
33. Ans. (d)
34. Ans. (a)
35. Ans. (c)
36. Ans. (c)
37. Ans. (d)
38. Ans. (d)
39. Ans. (c)
40. Ans. (d)
41. Ans. (d)
42. Ans. (b)
43. Ans. (c)
44. Ans. (d)
45. Ans. (a)
46. Ans. (d)
47. Ans. (b)

48. Ans. (c)

49. Ans. (b)

50. Ans. (b)

Solution:

$t = 20\ cm$

$= 0.2\ m$

$$\Delta = t(\mu_a - 1) = \frac{n\lambda}{2}$$

$$\mu_a = \frac{150 \times 4000 \times 10^{-10}}{2 \times 0.2} = 1.00015$$

51. Ans. (b)

Solution:

Path difference $= 4\lambda$

$= 4\,(6000\ \mathring{A})$

$= 24000\ \mathring{A}$

$= 2.4\ \mu m.$

52. Ans. (d)

Solution:

The distance along the screen for the third order fringe is

$$x_1 = [(3\lambda_1)D]d\ :$$

for λ_2 it is $x_2 = [(4\lambda_2)D]/d$.

Since they fall at the same location,

$\therefore \quad x_1 = x_2$

Hence $\quad \lambda_2 = \frac{3}{4}\lambda_1.$

53. Ans. (c)

Solution:

For light passing from water to air the angle of refraction is greater than the angle of incidence, so for angle of incidence greater than $i_e < 90°$, the refraction angle will be greater than $90°$, implying no refraction. The angle i_c is the critical angle and corresponds to a refraction angle of exactly $90°$. Then,

$$1 \sin 90° = 1.333 \sin i_c$$

$$\sin i_c = 0.750$$

54. Ans. (c)

Solution:

The condition for constructive interference twice is $(4\lambda_1)/d = x/D$

and $(6\lambda_2)/d = x/d,$

so that $4\lambda_1 = 6\lambda_2$.

If $\lambda_1 = 430\ nm$, then $\lambda_2 = 287\ nm.$

55. Ans. (c)

Solution:

Since $\Delta r = (\lambda/8)\ \sin\theta$,

the phase difference at P is

$$\phi = \frac{2\pi}{\lambda}\Delta r + \Delta\phi = \frac{\pi}{4}(\sin\theta + 1)$$

and $\quad I(\theta) = 4I_0 \cos^2\frac{\phi}{2}$

$$= 4I_0 \cos^2\left[\frac{\pi}{8}(\sin\theta + 1)\right]$$

56. Ans. (b)

Solution:

$$t = \frac{\Sigma n\lambda}{\eta - 1} = \frac{7(600)nm}{0.5} = 8400\ nm$$

57. Ans. (b)

Solution:

$$X_2 - X_1 = N\lambda/2$$

$$\lambda = \frac{2(X_2 - X_1)}{N}$$

$$= \frac{2\,(10.7347 \times 10^{-3} - 10.7051 \times 10^{-3})}{100}$$

$$= 5920\ \mathring{A}$$

58. Ans. (a)

Solution:

For a fabry Periot interferometer the conduct of maxima in transimted beam is given by

$$2t\cos\theta = n\lambda$$

t = Sepration between heater

For normal incidence, $\theta = 0$

$\therefore \quad 2t = n\lambda$

or, $\quad n = \frac{2t}{\lambda}$

For $4000\ \mathring{A}$, $n = \frac{2 \times 2 \times 10^{-6}}{400 \times 10^{-10}} = 10$

For $6000\ \mathring{A}$, $n = \frac{2 \times 2 \times 10^{-6}}{6000 \times 10^{-10}} = 6.67$

Hence intermediate wavelengths orders will be 7, 8, 9, Corresponding wavelengths are

$$\lambda_1 = \frac{2t}{7} = 5714\ \text{Å}$$

$$\lambda_2 = \frac{2t}{8} = 5000\ \text{Å}$$

$$\lambda_3 = \frac{2t}{9} = 4444\ \text{Å}$$

59. Ans. (d)
Solution:
The condition for the wavelength absent in the reflected light is

$$2\mu t \cos r = \lambda,\ 2\lambda\ 3\lambda, \ldots\ldots,\ etc$$

also $\mu = \dfrac{\sin i}{\sin r}$

Therefore $r = \sin^{-1}\left(\dfrac{\sin i}{\mu}\right)$

$$= \sin^{-1}\left(\frac{\sin 35°}{1.25}\right) = 27°18$$

$$\lambda_1 = 2\mu t \cos r$$

$$= 2 \times 0.2 \times 10^{-6} \times 1.25 \times \cos 27°18$$

$$= 4442\ \text{Å}$$

60. Ans. (c)
Solution:

$$r = \sin^{-1}\left(\frac{\sin 30°}{1.39}\right) = 21.082524°$$

Therefore, cos r = 0.9330633

$$2\mu t \cos r = n\lambda \quad \text{and}$$

$$2\mu t \cos r = (n+1)\lambda$$

Therefore, n (5125) = (n + 1) 5000

or $\quad n = 40$

Therefore, $\quad t = \dfrac{40 \times 512510^{-8}}{2 \times 1.39 \times 0.9330633}$

$$= 7.9031 \times 10^{-4}\ cm$$

61. Ans. (b)
Solution:
Distance moved for shift of 1 fringe $= \lambda/2$

Displacement for n-fringe, $x = n\lambda/2$

Given that $\quad x = 0.1\ mm,\ n = 400$

Hence, $\quad \lambda = \dfrac{2 \times 0.1 \times 10^{-3}}{400} = 500\ \text{Å}$

62. Ans. (b)
Solution:

$$\Delta\lambda = \lambda_1 - \lambda_2 = \frac{\lambda_a^2}{2d}$$

$$\lambda_a = \frac{4882 + 4886}{2} = 4884\ \text{Å}$$

$$\Delta\lambda = 4886 - 4882 = 4\ \text{Å}$$

$$\therefore d = \frac{(4884 \times 10^{-10})}{4 \times 10^{-10}} = 0.059\ mm$$

63. Ans. (b)
Solution:

$$2(\mu - 1)t = N\lambda$$

$$t = \frac{N\lambda}{2(\mu - 1)}$$

Given that $\quad N = 10,$

$$\lambda = 590 \times 10^{-10}\ m$$

$$\mu = 1.5$$

Hence, $\quad t = \dfrac{10 \times 5890 \times 10^{-10}}{2 \times 0.50}$

$$= 5.89\ \mu m.$$

64. Ans. (d)
Solution:

$$\lambda = \frac{d\Delta r}{D}.$$

65. Ans. (d)
Solution:
Bright fringes are located at

$$\sin\theta_m = \theta_m = \frac{m\lambda}{d}$$

the angular separation of adjacent fringes is λ/d. Thus the condition for seeing adjacent fringes as distinct is

$$\frac{\lambda}{d_{\max}} = 0.000291$$

or $\quad d_{\max} = \dfrac{500 \times 10^{-9}}{0.000291} = 1.72\ mm$

66. Ans. (d)
Solution:

The mth bright fringe of the λ pattern and the mth bright fringe of the λ' pattern are located at

$$y_m = \frac{mD\lambda}{d} \text{ and } y_m = \frac{m'D\lambda'}{d}$$

Equating these distances gives

$$\frac{m}{m'} = \frac{\lambda'}{\lambda} = \frac{900}{750} = \frac{6}{5}$$

Hence, the first position at which overlappling occurs is

$$y_6 = y_5' = \frac{(6)(2)(750\times10^{-9})}{2\times10^{-3}} = 4.5mm$$

67. Ans. (c)
Solution:

$D_n^2 = 2(2n+1)\lambda R/\mu,$

Given that $\mu = 1$ (Air)

$n = 6$, $D = 0.68\,cm$, $R = 350\,cm$

Hence, $$\lambda = \frac{D_n^2}{2(2n+1)R}$$

$$= \frac{(60)^2}{2\times(2\times6+1)\times350} = 5080\ \text{Å}$$

68. Ans. (b)
Solution:
Fringe width

$$\beta = \frac{\lambda}{2\mu \tan\alpha \cos(r+\alpha)}$$

α = angle of wedge

For normal incidence r = 0

$$\beta = \frac{\lambda}{2\mu\left(\dfrac{\sin\alpha}{\cos\alpha}\right)\cos\alpha} = \frac{\lambda}{2\mu\sin\alpha}$$

or, $$\sin\alpha = \frac{\lambda}{2\,\mu\beta}$$

Given that $\lambda = 6000\times10^{-10}\,m$

$\mu = 1.35$

$\beta = 2\,mm = 2\times10^{-3}\,m$

$\alpha = 1.11\times10^{-4}\times\dfrac{180}{\pi}$

$\alpha = 0.0064°$

69. Ans. (c)
Solution:
Displacement of fringe,

$$\frac{\beta}{\lambda}(\mu-1)t = 5\beta$$

$$\therefore \quad t = \frac{5\lambda}{\mu-1} = 6\times10^{-6} = 6\,\mu m$$

70. Ans. (c)
Solution:
When blood is moving towards transmitter, the apparent Frequency, f_1

$$f_1 = \frac{f(C - v\cos\theta)}{C}$$

If incident and scattered radiations are both inclined at θ to the direction of flow.

$$f_2 = f_1\left[\frac{C}{C + v\cos\theta}\right]$$

Resultant doppler shift,

$$\Delta f = f - f_2 = f\cdot\frac{f(C - v\cos\theta)}{C(C + C\cos\theta)}$$

Since C >> V

$$\therefore \quad v = \frac{\Delta f C}{2f\cos\theta}$$

71. Ans. (b)
Solution:
Where the two slits act as in-phase source of equal intensities and we assume $d << D$, intensity maxima are located by (for any integer m).

$$\Delta r = m\lambda$$

or $$d\sin\theta = m\lambda$$

$$\sin\theta = \frac{m\lambda}{d}$$

$$\frac{y}{D} = \frac{m\lambda}{d}$$

or $$y = \frac{mD\lambda}{d}$$

72. Ans. (d)
Solution:
For the first bright band, $n = 1$.
The path difference is one wavelength i.e.

$$n\lambda = \frac{dy}{D}$$

$$5839\times10^{-10}=\frac{0.25\times10^{-3}y}{1}$$

$$y=\frac{5839\times10^{-10}}{0.25\times10^{-3}}$$

$$=.2.36\times10^{-3}m=2.36\ mm$$

73. Ans. (d)
Solution:

$$2\mu t\cos r=\left(n+\frac{1}{2}\right)\lambda \text{ and } t=\frac{r_n^2}{2R}$$

Therefore, $R=\dfrac{d_n^2\mu\cos r}{2(2n+1)\lambda}$

$$=\frac{(0.5)^2\cos 30^\circ}{2(2\times4+1)5093\times10^{-8}}$$

= 204.1 cm.

74. Ans. (b)
Solution:

$$2x_1=\frac{2D\lambda}{W}$$

$$=\frac{2\times1\times6000\times10^{-10}}{0.6\times10^{-3}}=2mm.$$

75. Ans. (c)
Solution:

$$\lambda=\frac{d\sin\theta}{n}=\frac{1}{6000}$$

$$\frac{\sin 50^\circ}{2}=6383.7\ \mathring{A}$$

76. Ans. (b)
Solution:

$$n=\frac{d\sin\theta}{\lambda}=\frac{1}{6000\times6000\times10^{-8}}$$

$$=\frac{100}{36}=2.78<3$$

Therefore, $n=2$.

77. Ans. (c)
Solution:

$$\frac{I}{I_0}=\left(\frac{\sin B}{B}\right)^2$$

$$=\left(\frac{\sin(3\pi/2)}{3\pi/2}\right)^2=0.045$$

78. Ans. (c)
Solution:

$$Sin\,\theta=\frac{x}{D}=\frac{n\lambda}{d}$$

Therefore, $x=\dfrac{n\lambda D}{d}$

$$=\frac{3\times6400\times10^{-10}\times1.1}{0.3\times10^{-3}}$$

$$=7.04\ mm$$

79. Ans. (d)
Solution:
Appluing Snell's law at A.

$$n_1\sin\theta_1=n_2\sin\theta_2 \quad (1)$$

But $\theta_2=90^\circ-\theta_c$,

and so $\cos\theta_2=\sin\theta_c=\dfrac{\eta_1}{\eta_2}$ (2)

Elimination of θ_2 from equation (1) and (2) gives

$$\sin\theta_1=\sqrt{\left(\frac{n_2}{n1/}\right)^2-1} \quad (3)$$

80. Ans. (d)
Solution:
From the data we can determine the index of refraction of ice, $n=\dfrac{c}{v}=\dfrac{(3\times10^8)}{(2.3\times10^8)}=1.304$,

Then for the critical angle,

$$1\sin 90^\circ=1.304\sin i_c$$

and $\sin i_c=0.7667$

$$i_c=50.1^\circ$$

81. Ans. (c)
Solution:
Speed of light in glass

$$v_g=\frac{c}{n}=\frac{3\times10^8}{1.64}=1.83\times10^8\,m/s$$

Then, $\dfrac{\sin 44^\circ}{\sin\theta_g}=\dfrac{v_w}{v_g}$

or $1.333\sin 44^\circ=1.64\sin\theta_g$

$$=1.64\sin\theta_g=\frac{v_w}{v_g}$$

and $\sin\theta_g = 0.56462$

$$\theta_g = 3.44^\circ$$

82. Ans. (b)
Solution:
The angle of incident is the angle made with the normal to mirror. Thus $i = 15^\circ$.

83. Ans. (a)
Solution:
Angle of reflection is also $15(i = r)$. Hence the reflected ray makes angles of 30° with vertical and 60° with horizontal.

84. Ans. (d)
Solution:

$$\lambda' = \frac{436\,nn}{1.65}.$$

No of complete waves $= \dfrac{d}{\lambda'} = 71647$

85. Ans. (c)
Solution:

$$1.52\sin\theta_c = 1.333\sin 90$$

$$\sin\theta_c = 0.877\ \theta_c = 61.3^\circ$$

86. Ans. (b)
Solution:

$$1.52\sin 55 = 1.33\sin\theta_w$$

$$\theta_w = 69^\circ$$

$$\theta_1 = 35^\circ$$

87. Ans. (c)
88. Ans. (c)
89. Ans. (d)
90. Ans. (d)
91. Ans. (c)
92. Ans. (d)
93. Ans. (a)
94. Ans. (a)
95. Ans. (c)
96. Ans. (c)
97. Ans. (a)
98. Ans. (c)
99. Ans. (d)
100. Ans. (c)
101. Ans. (d)
102. Ans. (a)
103. Ans. (a)
104. Ans. (a)
105. Ans. (d)
106. Ans. (b)
107. Ans. (c)
108. Ans. (c)
109. Ans. (a)
110. Ans. (d)
111. Ans. (d)
112. Ans. (c)
Solution:

$$n\lambda = 2(\Delta x)$$

$$\lambda = \frac{2(\Delta x)}{n}$$

$$= \frac{2\times 0.0589}{200}$$

$$= 0.0589\times 10^5$$

$$\lambda = 589\ nm$$

113. Ans. (a)
Solution:

$$n\lambda = (2\Delta x)$$

$$n\lambda = \frac{(2\Delta x)}{\lambda}$$

$$\therefore \quad n = \frac{2\times 0.026003\times 10^{-3}}{0.5206\times 10^{-6}}$$

$$= 0.00001\times 10^7$$

$$n = 100$$

114. Ans. (c)
Solution:

$$n\lambda = 2(\mu - 1)$$

$$2t = \frac{n\lambda}{\mu - 1}$$

$$= \frac{500\times 4800\times 10^{-10}}{0.5}$$

$$t = \frac{48}{2}\times 10^{-5}$$

$$= 240\ \mu m$$

115. Ans. (a)
Solution:

$$n\lambda = 2(\mu - 1)$$

$$\frac{n\lambda}{2t} = \mu - 1$$

$$\mu = \frac{n\lambda}{2t} + 1$$

$$= \frac{5\times 6.4\times 10^{-6}}{2\times 3.8\times 10^{-3}} + 1$$

$= 4.2\times10^{-1} + 1$

$= 0.42 + 1$

$= 1.42$

116. Ans. (d)
Solution:

$$\lambda_1 - \lambda_2 = \frac{\lambda_1\lambda_2}{2x}$$

$$= \frac{5890\times5896\times10^{-20}}{(5890-5896)\times10^{-10}}$$

$= 289$ nm

117. Ans. (a)
Solution:

$$\lambda_1 - \lambda_2 = \frac{\lambda_1\lambda_2}{2x} = \frac{(\lambda_{av})^2}{2x}$$

$$= \frac{(5893\times10^{-10})^2}{2(0.984-0.6939)\times10^{-3}}$$

$$= \frac{(5893\times10^{-10})^2}{2\times0.2945\times10^{-3}}$$

$= 5.896$ A°

118. Ans. (c)
Solution:
For Bi - prism arrangement

$$n\lambda = 1(\mu - 1)$$

$$\frac{n\lambda}{t} = \mu - 1$$

$$\mu = \frac{n\lambda}{t} + 1$$

$$\therefore \quad \mu = \frac{5\times6.3\times10^{-6}}{4.2\times10^{-5}}$$

$= 7.5\times10^{-1} + 1$

$= 0.75 + 1$

$= 1.75$

$\mu = 1.75$

119. Ans. (a)
Solution:
x = Length between the adjacent fringes
L = Horizontal distance between the blocks

No of frings $= 10 = \frac{L}{x}$

$$d = L.\frac{\lambda}{2x} = \left(\frac{L}{x}\right).\frac{\lambda}{2}$$

$$= 10\times\frac{9.426\times10^{-6}}{2}$$

$= 47.13\ \mu m$

120. Ans. (a)

Solution:
The number of fringes present over a distance of 24 mm are 10

$$\therefore \quad 10 = \frac{L}{x}$$

$$10 = \frac{24\times10^{-3}}{x}$$

x = 2 mm

The distance between the centre of rollers

$= 60 - (6+6) = 48$ mm

$$d = L\frac{\lambda}{2x}$$

$$= 48\times10^{-3}\times\frac{0.8\times10^{-6}}{2\times2.4\times10^{-3}}$$

$= 8\times10^{-6} m = 8\ \mu m$

∴ The difference between the diameter of the two spheres is equal to $8\ \mu m$.

121. Ans. (c)
Solution:
Given

There are $3\frac{1}{2}$ fringes along the length and $6\frac{1}{2}$ fringes along the base.

The difference in fringes is $= 6\frac{1}{2} - 3\frac{1}{2} = 3$

A single fringe represents half of the wavelength between top and bottom surface.

∴ Change in thickness $= 3\times\frac{1}{2}\times0.622\times10^{-6}$

s = 0.993 μm

1.7 Fiber Optics

1. Ans. (a)
Solution:

$$\mu_{rg} = \frac{\mu_w}{\mu_g} - \frac{1.33}{1.7} = 0.886$$

2. Ans. (c)
Solution:

$$n_g = 1.732 = \sqrt{3}$$

$$= \frac{\sin i}{\sin 30}$$

$$\sin i = \sqrt{3}\times\sin 30° = \frac{\sqrt{3}}{2}$$

$$i = 60°$$

3. Ans. (a)
Solution:

Reflectivity, $R = \left(\frac{n_2 - n_1}{n_2 + n_1}\right)^2 = \left(\frac{1.720 - 1}{1.720 + 1}\right)^2$

$= 0.07$

Thus, loss of intensity through refraction is 7%

4. Ans. (c)

Solution:

$n_2 = 1$

$n_1 = 1.53$

So Fresnel, intensity reflection coefficient is

$\Rightarrow \left(\frac{n_2 - n_1}{n_2 + n_1}\right)^2$

$\Rightarrow \left(\frac{1.53 - 1}{1.53 + 1}\right)^2$

$\Rightarrow \left(\frac{0.53}{2.53}\right)^2$

$= 0.044$

5. Ans. (c)

6. Ans. (b)

7. Ans. (c)

8. Ans. (b)

Solution:

$N.A. = n_1\sqrt{2\Delta}$

Δ = core/cladding index difference.

$n_1 = 1.5$

$\Delta = 0.01$

N.A. = $1.5\sqrt{2(.01)} \Rightarrow 0.21$

Acceptance Angle = $\sin^{-1}$ (N.A)

$= \sin^{-1}(0.21) \Rightarrow 12.2^\circ$

9. Ans. (b)

Solution:

The optical path difference between the two arms

$$= \frac{1\ \mu m}{50\ nm} = \frac{1\times10^{-6}}{50\times10^{-9}} = \frac{1}{50\times10^{-3}} = 20\ \mu m$$

10. Ans.(b)

Solution:

Given

Wavelength of light, $\lambda = 630\ nm$

$= 63\times10^{-8} m$

Thickness of specimen, $d = 10^{-5}\ m$

Refractive Index of first beam $\mu_1 = 1.32$

Refractive Index of second beam $\mu_2 = 1.333$

Path difference between two beams

$t = (\mu_2 - \mu_1)d$

$= 0.013\times10^{-5} m$

$= 13\times10^{-8} m$

For the given light, the phase difference between these two emerging beams $= \frac{360}{\lambda}t$

$$= \frac{360}{63\times10^{-8}}\times13\times10^{-8} = 74.3^\circ$$

11. Ans. (a)

12. Ans. (a)

Solution:

For angle of incidence,

$\tan\theta = \frac{\mu_1}{\mu_2} = \frac{1}{1.752}$

or $\theta = \tan^{-1}\left(\frac{1}{1.752}\right) = \tan^{-1}\left(\frac{1}{\sqrt{3}}\right) = 30^\circ$

13. Ans. (d)

Solution: Numerical aperture = $\sqrt{\mu_1^2 - \mu_2^2}$

$= \sqrt{(\mu_1 + \mu_1)\times(\mu_1 - \mu_1)}$

$= \sqrt{2.84\ (1.44 - 1.4)}$

$= \sqrt{2.84\times0.04}$

$= \sqrt{0.1136} = 0.34$

14. Ans. (b)

15. Ans. (b) / (a)

16. Ans. (d)

Solution:

$$\cos\phi_c = \sqrt{\frac{n_1^2 - n_2^2}{n_1}}$$

$$= \sqrt{\frac{1.641^2 - 1.422^2}{1.641}} = 0.499$$

$\phi = \cos^{-1} 0.5$

$\phi = 60°$

17. Ans. (c)

Solution: $\eta = \sqrt{\varepsilon_r} = 2.5$

$=1.58$

18. Ans. (a)

19. Ans. (c)

Solution:

$n_1 = 1.46$

$n_2 = 1.455$

Numerical Aperture $= \sqrt{n_1^2 - n_2^2}$

If core diameter is in 2–10 μm then it is single mode

20. Ans. (a)

Solution:

The broadening of the light pulses at the output of the optical fiber cable is called as the **dispersion**. Its unit is (n sec/km)

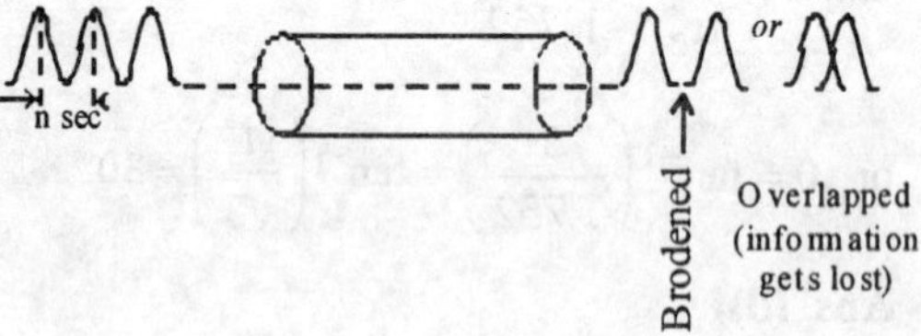

To avoid distortion keep some time gap between the pulses (denoted by ΔT)

$$\Delta T = \frac{n_1 L}{c}\Delta \quad \text{(For step index)}$$

$$\Delta T = \frac{n_2 L}{2c}\Delta^2 \quad \text{(for graded index)}$$

$$\Delta = \frac{n_1}{n_2} - 1$$

Where, $c =$ speed of light

ΔT for graded index $> \Delta T$ for step index

According to given condition

$$\Delta = \frac{n_1}{n_2} - 1 = \frac{1.46}{1.45} - 1$$

$$= \frac{1.46 - 1.45}{1.45} = \frac{0.01}{1.45}$$

$$= 6.89 \times 10^{-3}$$

$$\frac{\Delta T}{L} = \frac{n_1 \Delta}{C}$$

For step index fibers

$$= \frac{1.46}{3 \times 10^8} \times 6.89 \times 10^{-3}$$

$$= 3.33 \times 10^{-11} \text{ sec/ m}$$

$$= 33.3 \times 10^{-9} \text{ sec/ km}$$

$$= 33.3 \text{ ns/ km}$$

Another Method

For a step index fibre

$$\frac{\delta T_s}{L(m)} = \frac{(NA)^2}{2n_1 C} = \frac{n_1 \Delta}{C}$$

$$= \frac{n_1^2 - n_2^2}{2n_1 C}$$

$$= \frac{1.46^2 - 1.45^2}{2 \times 1.40 \times 3 \times 10^8}$$

$$= \frac{0.0291}{8.76 \times 10^8}$$

$$= 0.003322 \times 10^{-8}$$

$$= 0.03322 \times 10^{-9}$$

$$= 0.03322 \times 10^{-9} / km$$

$$= 33.22 \; ns/km$$

21. Ans. (a)

Solution:

In step index fibre:

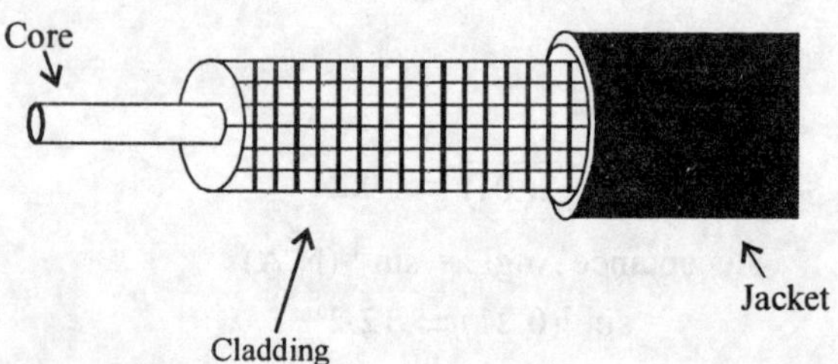

$\mu_{clerd} = 1.45$

$\mu_{core} = 1.48$

If $n_2 =$ refractive index of core

$n_1 =$ refractive index of cladding

$$NA = \sqrt{(n_2)^2 - (n_1)^2} = \sqrt{(1.48)^2 - (1.45)^2}$$

$$= \sqrt{2.1904 - 2.1025}$$

$$= 0.2964$$

Maximum solid angle $= \pi(\theta)^2$

$= 0.28$ steradians

22. Ans. (a)

Solution:

N. A (Numerical aperture of optical fibre)

$= \sqrt{(n_2)^2 - (n_1)^2}$

n_2 = refractive index of core

n_1 = refractive index of cladding

$= \sqrt{(1.501)^2 - (1.499)^2}$

$= \sqrt{2.253 - 2.247}$

$= 0.077$

23. Ans. (c)

Solution:

Numerical aperture is the light gathering ability of the optical fibre. It determines the efficiency of coupling

Coupling efficiency $= (NA)^2 \times 100\%$

$= (0.077)^2 \times 100\%$

$= 0.59 \approx 0.6\%$

24. Ans. (c)

Solution:

Numerical Aperture in Air = 0.39

$(\mu_a) = 1$

The Numerical Aperture in H_2O having

$(\mu_H) = 1.33$

$\Rightarrow \frac{0.39}{1.33} = 0.2932$

Where θ_A is acceptance angle

$\theta_A = \sin^{-1}(0.2932)$

$\theta_A = 17°$

25. Ans. (c)

Solution:

Numerical aperture is the light gathering ability of the optical fibre. Mathematically, is the sine of the acceptance angle (θ_a)

Angle of Acceptance for air medium $= \sin^{-1}(0.39)$

$= 22.9°$

Ratio $\frac{E_1}{E_2} = \frac{22.9°}{17°}$

$= 1.4$

26. Ans (b)

Solution:

From the given data figure can be drawn as

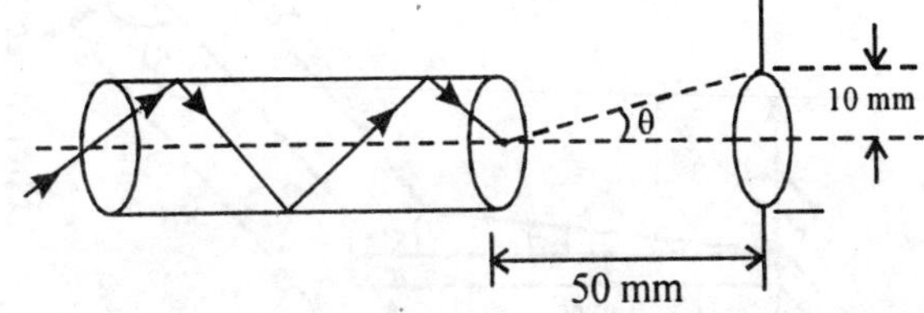

From figure

$\theta = \tan^{-1}\left(\frac{10}{50}\right)$

$= 11.45°$

$NA = \sqrt{n_1^2 - n_2^2}$

$= \sin\theta$

$= \sin(11.45°)$

$= 0.19 \simeq 0.2$

27. Ans.(c)

Solution:

Numerical aperture $= \sin\theta_a = 0.5$

$\theta_a = \sin^{-1} 0.5 = 30°$

Diameter of lens

$D = 5\ mm$

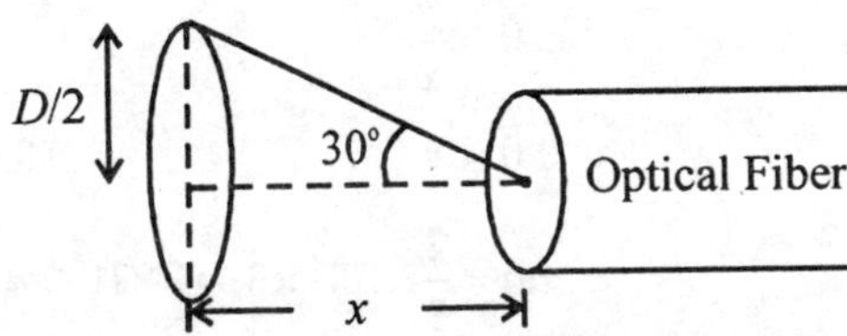

Finding this distance 'x', as it is the focal length of the lens

$\tan 30° = \frac{D/2}{x}$

$\frac{1}{\sqrt{3}} = \frac{5}{2x}$

$x = \frac{5\sqrt{3}}{2}$

$= 4.33\ mm$

28. Ans. (b)

Solution:

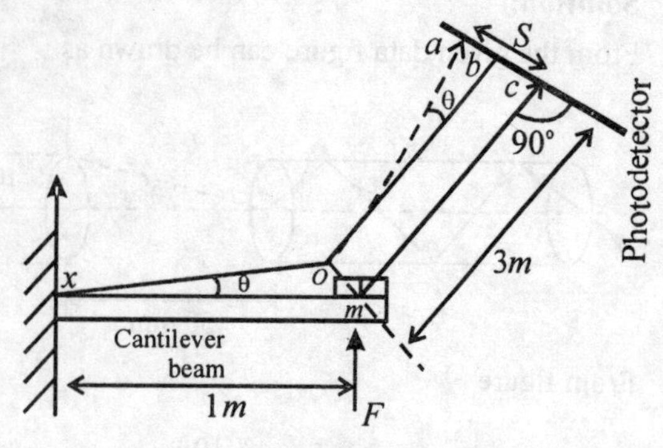

$F = 1\,N$ Given

Let, $x = 1\,m$

$$y = 0.001\,Fx^2\left(1-\frac{x}{3}\right)m$$

$$= 0.001\times1\times1\left(1-\frac{1}{3}\right)$$

$$= 0.001\times\frac{2}{3} = \frac{2}{3}\,mm$$

$$om = y = bc$$

$$bc = \frac{2}{3}\,mm$$

in triangle *oab*

$$\theta = \frac{ab}{ob}$$

$$ab = \theta(ob)$$

also $\theta = \dfrac{y}{x}$

$$\theta = y \qquad [\text{for } x = 1m]$$

$$ab = \frac{2}{3}\times10^{-3}\times3 = 2\times10^{-3}\,m$$

$$s = bc + ab$$

$$= \left(\frac{2}{3}+2\right)\times10^{-3}\,m$$

$$= 2.7\,mm$$

$$s \approx 3mm$$

29. Ans. (b)

Solution:

Output of LVDT is proportional to deflection.

$$\therefore \quad = \frac{\text{Ouput of LVDT at}\left(x=\frac{1}{2}m\right)}{\text{Ouput of LVDT at}\left(x=\frac{1}{4}m\right)}$$

$$= \frac{0.001\,F\left(\frac{1}{2}\right)^2\left[1-\frac{1}{2\times3}\right]}{0.001\,F\left(\frac{1}{4}\right)^2\left[1-\frac{1}{4\times3}\right]}$$

$$= \frac{\frac{1}{4}\times\frac{5}{6}}{\frac{1}{16}\times\frac{11}{12}} = \frac{40}{11}$$

30. Ans. (b)
31. Ans. (a)
32. Ans. (a)
33. Ans. (d)
34. Ans. (c)
35. Ans. (b)
36. Ans. (a)
37. Ans. (c)
38. Ans. (a)
39. Ans. (d)
40. Ans. (b)
41. Ans. (c)
42. Ans. (b)
43. Ans. (b)
44. Ans. (a)
45. Ans. (b)
46. Ans. (a)
47. Ans. (c)
48. Ans. (a)
49. Ans. (c)
50. Ans. (c)
51. Ans. (c)
52. Ans. (d)
53. Ans. (d)
54. Ans. (b)
55. Ans. (b)
56. Ans. (c)
57. Ans. (a)
58. Ans. (c)
59. Ans. (a)
60. Ans. (c)
61. Ans. (c)
62. Ans. (c)
63. Ans. (b)
64. Ans. (b)
65. Ans. (b)

66. Ans. (c)

67. Ans. (a)

68. Ans. (a)

69. Ans. (b)

70. Ans. (b)

Solution:

For largest angle of incident $\theta_a = 90^o$ for light from glass. Then, for angle in glass.

$$1 \sin 90^\circ = 1.50 \sin \theta_{gt}$$

$$\sin \theta_{gt} = 0.66667$$

$$\sin \theta_{gt} = 41.8^\circ$$

Then for angle in water

$$1.50 \sin \theta_{gl} = 1.333 \sin \theta_w$$

$$\sin \theta_W = 0.7500$$

$$\theta_W = 48.6^\circ$$

71. Ans. (c)

Solution:

When light travels from the liquid to the air at angles of incidence greater than the crical angle, we have total internal reflection at the interface. To find the critical angle, we need n_1, the index of refeaction in the liquid. Now for the liquid.

$$\theta_1 = 52^\circ - 19^\circ$$

$$= 33^\circ$$

and $1 \sin 52^\circ = n_1 \sin 33^\circ$

so, $0.7889 = 0.54464\, n_1$

$$n_1 = 1.447$$

Then for the ciritical angle.

$$2 \sin 90^\circ = 1.447 \sin \theta_d$$

$$\sin \theta_d = 0.69116$$

and $\theta_d = 43.7^\circ$

72. Ans. (d)

Solution:

At the air oil interface.

$$\sin 40^\circ = 1.45 \sin \theta_{oil}$$

At the oil- water interface

$$1.45 \sin \theta_{oil} = 1.33 \sin \theta_w$$

$$\theta_w = \sin^{-1}\left[\frac{\sin 40^\circ}{1.33}\right] = 28.9^\circ$$

73. Ans. (d)

Solution:

If the condition for total internal reflection is fulfilled. Then,

$$\sin \theta_2 = \frac{n_1}{n_2} \sin \theta_1$$

$$= \frac{\sin 40^\circ}{1.3} = 0.495$$

and $\theta_2 = 29.7^\circ$

The length along fiber to the first reflection is

$$l = \frac{d}{\tan \theta_2} = \frac{2 \times 10^{-5} m}{0.570} = 3.51 \times 10^{-5} m$$

By the law of reflection, l also represents the length along the fiber between successive reflections, the number of reflections is

$$\frac{L}{l} = \frac{2m}{3.51 \times 10^{-5} m} = 57000$$

74. Ans. (c)

Solution:

The critical angle is determined by the interface between fibre and coating.

Then,

$$\sin \theta_c = \frac{n_3}{n_2} = \frac{1.512}{1.700}$$

$$= 0.899$$

$$\theta_c = 62.7^o$$

75. Ans. (b)

Solution:

$$\eta_{water} = \frac{V_{air}}{V_{water}} = \frac{343\ m/s}{1498\ m/s} = 0.229$$

76. Ans. (d)

Solution:

$$\sin \theta_c = \frac{\eta_{water}}{\eta_{air}} = 0.229$$

$$\theta_c = 13.2^\circ$$

77. Ans. (c)

Solution:

$$n_1 = 1$$

$$\sin \theta_1 = \sqrt{1.30^2 - 1} = 0.831$$

$$\therefore \quad \theta_1 = 56.2^\circ$$

78. Ans. (c)

Solution:

$$\sin \theta = \frac{m\lambda}{2d}$$

$$\phi = \sin^{-1}\left(\frac{1\times 4.3\times 10^{-5}}{2\times 50\times 10^{-3}}\right)$$

$$= 0.42^{\circ}$$

79. Ans. (c)
Solution:

$$hf = \frac{6.6\times 10^{-34}\times 2\times 10^{11}}{1.6\times 10^{-19}}$$

$$= 0.83\ me\,V$$

80. Ans. (b)
Solution:

$$\sin\theta_c = \frac{1.50}{1.53}$$

$$\therefore \quad \theta_c = 78.6^{\circ}$$

81. Ans.(c)
Solution:

$$NA = 1.53\times\sqrt{2\times 0.0196}$$

$$= 0.303$$

82. Ans. (c)
Solution:

$$\alpha_{\max} = \sin^{-1}(0.303) = 17.6^{\circ}$$

83. Ans. (d)
Solution:
Maximum pulse broadening

$$= \frac{d\times NA}{2\times C\times\eta}$$

$$\Delta_C\ step = \frac{10^3(0.2)^2}{2\times 3\times 10^8\times 1.45} S\ km^{-1}$$

$$= 46\ ns/km$$

84. Ans. (c)
Solution:

$$a_R = \left(\begin{array}{l}\dfrac{8\pi^2}{3(10^{-6})^4}\times(1.46)^6\times)(0.286)^2\\ \times 7\times 10^{-11}\times 1.38\times 10^{-23}\times 1400\end{array}\right)$$

$$= 1.84\times 10^{-4}/m$$

85. Ans. (b)
Solution:
The phase is familiar in the form $k(y-vt)$; accordingly, we rewrite the above as $E_{0x}\sin\phi$ where

$$\phi = \frac{2\pi}{5\times 10^{-7}}(y - 15\times 10^7 t)$$

Clearly, $\lambda = 5\times 10^{-7} m$

and $v = 1.5\times 10^8 m/s$

Thus, $n = c/v = 2$.

86. Ans. (d)
Solution:

Since $v = \dfrac{c}{n}$

$$t_1 = \frac{20}{c/n} = \frac{20(1.7)}{e}$$

and $t_2 = \dfrac{20(1.63)}{c}$

Accordingly, $t_2 - t_1 = \dfrac{20}{c}(1.63 - 1.47)$

$$= 10.7\times 10^{-8}\ s.$$

87. Ans. (d)
Solution:

$$v \text{ within the plate} = \frac{c}{n}$$

$$= \frac{c}{1.66} = 1.807\times 10^8 m/s$$

$$t = \frac{\theta}{d} = \frac{0.020}{v} = 0.111\,ns$$

88. Ans. (b)
89. Ans. (c)
90. Ans. (b)
91. Ans. (c)
92. Ans. (a)
93. Ans. (b)
94. Ans. (d)
95. Ans. (a)
96. Ans. (c)
97. Ans. (c)
98. Ans. (c)
99. Ans. (a)
100. Ans. (d)
101. Ans. (d)
102. Ans. (a)
103. Ans. (b)
104. Ans. (b)
105. Ans. (b)
106. Ans. (d)

107. Ans. (b)

108. Ans. (c)

109. Ans. (d)

110. Ans. (a)

111. Ans. (c)

112. Ans. (d)

113. Ans. (c)

Solution:

For optical fiber N.A.=1, for all light gathering

114. Ans. (a)

Solution:

Angle above which Total internal reflection (TIR) occurs is critical angle (θ_c)

$$\therefore\ \theta_c = \sin^{-1}\left(\frac{\mu_{cladding}}{\mu_{core}}\right)$$

$$= \sin^{-1}\left(\frac{1.422}{1.641}\right)$$

$$= 60.05^\circ$$

115. Ans. (c)

Solution:

Given

N.A. = 0.096

Acceptance angle $\text{A.A.} = \sin^{-1}\left(\frac{\text{N.A.}}{\mu}\right)$

For surrounding medium to be air, $\mu = 1$

$$\text{AA} = \sin^{-1}\left(\frac{0.096}{1}\right) = 5.5^\circ$$

116. Ans. (d)

117. Ans. (c)

Solution:

Single mode with N.A. = 0.12

$$\text{N.A.} = \sqrt{n_1^2 - n_2^2}$$

$$= \sqrt{(1.46)^2 - (1.455)^2} = 0.12$$

* Since the core diameter of the fiber is given to be $= 4\ \mu\text{m}$ that lies between $(2-10\ \mu m)$.

Hence it is a single mode fiber.

118. Ans. (a)

Solution:

For the single mode operation of a step index fiber. V - number must be

$$V = \frac{2\pi r}{\lambda}\sqrt{n_1^2 - n_2^2} \le 2.4045$$

$$2.404 = \frac{2\times 3.14\times r}{820\times 10^{-9}}\sqrt{(1.48)^2 - (1.478)^2}$$

$$r = \frac{2.404\times 820\times 10^{-9}}{2\times 3.14\times 0.07691}$$

$$= 4.08\ \mu m$$

119. Ans. (a)

Solution:

$$\text{Loss} = \frac{10}{\text{L}}\log\frac{\text{P}_o}{\text{P}_i}$$

$$= \frac{10}{1}\log\frac{80}{100}$$

$$= 10\log(0.8) = -0.96$$

$$\simeq -1\ dB/km$$

Negative sign here shows losses

120. Ans. (d)

Solution:

$$\Delta = \frac{n_1}{n_2} - 1 = \frac{1.42}{1.40} - 1 = 0.014$$

for step index fiber

$$\Delta\text{T} = \frac{n_1\text{L}}{c}\Delta = \frac{1.42\times 1000}{3\times 10^8}$$

$$= 67.12\ n\ \text{sec}/\ km$$

121. Ans. (a)

Solution:

When $\alpha = 20\ dB/km$

$$\alpha = \frac{10}{\text{L}}\log\left(\frac{\text{P}_o}{\text{P}_i}\right)$$

$$\text{L} = \frac{10\log\left(\frac{1}{10}\right)}{20}$$

$$= \frac{\log 0.1}{2} = 0.5\ km$$

122. Ans. (a)

Solution:

When, $n_1 = 1.46$

$\Delta = 0.01$

For step index fiber $\Delta\text{T} = \frac{n_1 L}{c}\Delta$

$$\therefore\ \Delta\text{T} = \frac{1.46\times 1000}{3\times 10^8}\times 0.01 = 48.66\ n\ \text{sec}$$

$$\text{No of pulses} = \frac{1}{48.66\times 10^{-9}}$$

$$= 20.54\times10^6 \text{ bp sec}$$
$$= 20.54 \text{ Mbps}$$

So, though this fiber the bit rate of less than or equal to 20.54 Mbps can only be transferred, so from given options only (a) 1 Mbps can be transferred.

123. Ans. (b)

Solution:

$$N.A = n_1\sqrt{2\Delta}$$

$$N.A = 1.48\sqrt{2\times0.015} = 0.25$$

$$N.A. = \sqrt{n_1^2 - n_2^2}$$

$$\sqrt{n_1^2 - n_2^2} = 0.25$$

$$V = \frac{2\pi r}{\lambda} = \sqrt{n_1^2 - n_2^2}$$

$$= \frac{2\times3.14}{0.85\times10^{-6}}\times\frac{8\times10^{-6}}{2}\times0.25$$

$$= 73.88$$

= 74 Normalized frequency

124. Ans. (d)

Solution:

Number of modes $\frac{V^2}{2} = \frac{(74)^2}{2} = 2{,}738$ modes

125. Ans. (a)

Solution:

$$V = \frac{2\pi r}{\lambda}\sqrt{n_1^2 - n_2^2}$$

$$75 = \frac{2\pi r}{820\times10^{-9}}(0,3)$$

$$r = \frac{75\times820\times10^{-9}}{2\times3.17\times0.3}$$

$$= 32.334\times10^{-6}$$

$$= 32\ \mu m$$

126. Ans. (d)

Solution:

$$n_1 = 1.458$$

$$N.A = 0.3$$

$$N.A = \sqrt{n_1^2 - n_2^2}$$

$$(0.3)^2 = (1.458)^2 - n_2^2$$

$$n_2^2 = (1.458)^2 - (0.3)^2$$

$$n_2 = \sqrt{2.125 - 0.09}$$

$$n_2 = 1.42$$

127. Ans. (a)

Solution:

$$V_{water} = \frac{3\times10^8}{4/3}$$

$$\therefore \left[V_{water} = \frac{V_{air}}{\text{RI of water}}\right]$$

$$t = \frac{d}{V_{water}}$$

$$= \frac{600}{2.25\times10^8}$$

$$= 0.26\ n\text{ sec}$$

128. Ans. (c)

Solution:

$$n_1 \sin i = n_2 \sin r$$

$$1 \sin 30^\circ = 1.5 \sin r$$

$$\frac{1}{2} = 1.5 \sin r$$

$$\sin r = \frac{1}{3}$$

$$r = \sin^{-1}\left(\frac{1}{3}\right) = 19.47^\circ$$

129. Ans. (a)

Solution:

When the reflected and refracted rays are perpendicular, the angle of incidence is equal to the polarizing angle, thus

$$\mu = \tan j$$

$$= \tan 60^\circ$$

$$= 1.732$$

130. Ans. (c)

Solution:

$$\text{Reflection loss (RL)} = \left(\frac{\mu-1}{\mu+1}\right)^2$$

$$= \left(\frac{1.53-1}{1.53+1}\right)^2$$

$$= \left(\frac{0.53}{2.53}\right)^2$$

$$= 0.044$$

$$\text{RL} = 0.044\times100$$

$$= 4.4\ \%$$

1.8 ECG, EEG and CMG

1. **Ans. (b)**

   Solution:

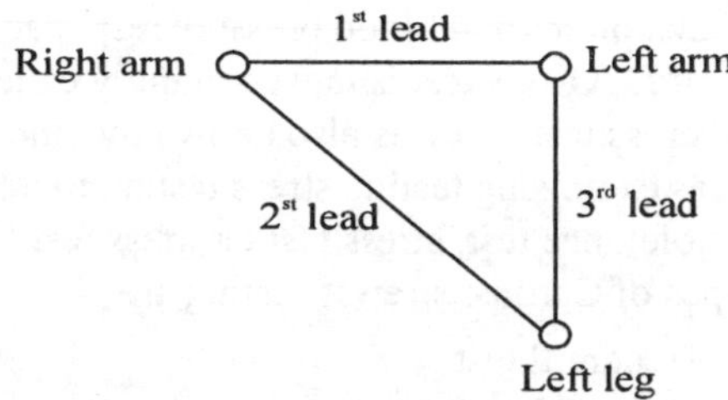

So

Lead 1 ⇒ Left Arm and Right arm

Lead 2 ⇒ Left leg and Right arm

Lead 3 ⇒ Left arm and left leg

2. **Ans. (c)**

   Solution:

The waveform generated as a result of the hearts electrical activity is known as the **electrocardiogram.**

Electro + Cardio +gram

Electrical activity | Of heart | Graphical Representation

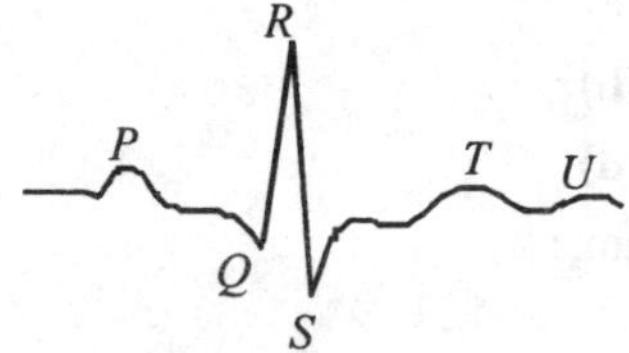

The waveform of typical electrocardiogram is given as above.

The waveform preceding the P wave is termed as the **baseline** or the isopotential line.

a. The P wave represents the depolarization of the atria muscles.
b. The QRS complex is obtained as a combined result of the repolarizaton of the atria and the depolarization of the ventricles, both of these processes occur almost simultaneously.
c. T wave is generated as result of ventricular repolarization
d. Sometimes U-wave is also present which is obtained as a result of the after potentials in the ventricular muscle.
e. The P-Q interval represents the time for which the excitation wave is delayed in the fibers near the A.V. node.

Note: The shape and the features of this waveform varies with the location at measuring electrodes are placed with respect to the heart.

3. **Ans. (b)**

   Solution:

(EEG): The bioelectric potentials generated by the neuronal activity of the brain is called as the **electroencephalogram (EEG).** The output generated by EEG is quiet complex and is difficult to read as compared to ECG.

(EMG): The bioelectric potentials generated by the muscle activity is called as the **electromyogram (EMG).** These type of potentials are measured directly by penetrating needle electrodes into the skin of the muscle.

(ERG): The record of the pattern obtained by the bioelectric potentials generated by the retina of the eye is called as the **electroretinogram (ERG).**

(EOG): It is a measure of the variations in the cornal-retina potential as is affected by the position and movement of eye is called as the **electroovulogram (EOG).**

(EGG): It is the pattern generated due to the gastric movements in the stomach with the help of EMG. It is called **Electro-gastrogram**

4. **Ans. (d)**

   Solution:

Cranial generator peak to peak voltage $= 100\ \mu V$

Source resistance $= 10\ k\Omega$

Electrode resistance $= 20\ k\Omega$

Input impedance of amplifier $= 10\ M\Omega$

$$i = \frac{\text{Input Voltage}}{\text{Total impedance}}$$

$$i = \frac{100\ \mu V}{30\ k\Omega + 10\ M\Omega}$$

$$= \frac{100\ \mu V}{10.03 \times 10^6}$$

Voltage across amplifier terminals $= i \times 10\ M\Omega$

$$= \frac{100 \times 10^{-6}}{10.03 \times 10^6} \times 10 \times 10^6$$

$$= 99.5\ \mu V$$

5. **Ans. (c)**

6. **Ans. (b)**

7. Ans (b)

Solution: Unipolar chest leads.

In the standard 12 lead ECG recording system minimum 4 leads are used as below.

Out of the four limb leads, there are used to take signals and one is set to reference, also the chest electrode participates in getting the output, *so a minimum of four electrode are used to get output in the case of unipolar chest leads.*

The unipolar chest lead arrangement is given in the figure below:

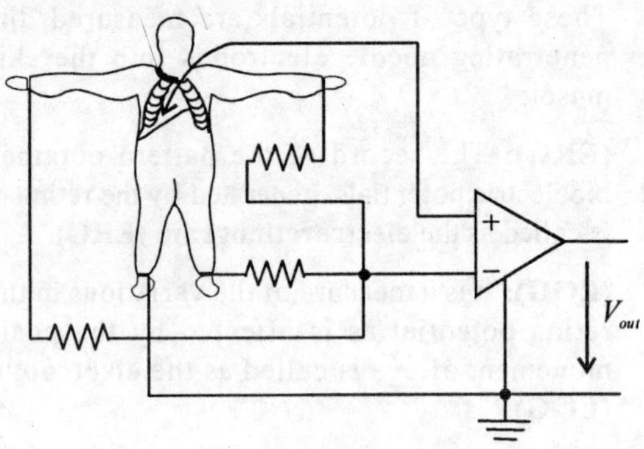

The various locations for placing the chest lead i.e. from V_1 to V_6 is given in the figure below:

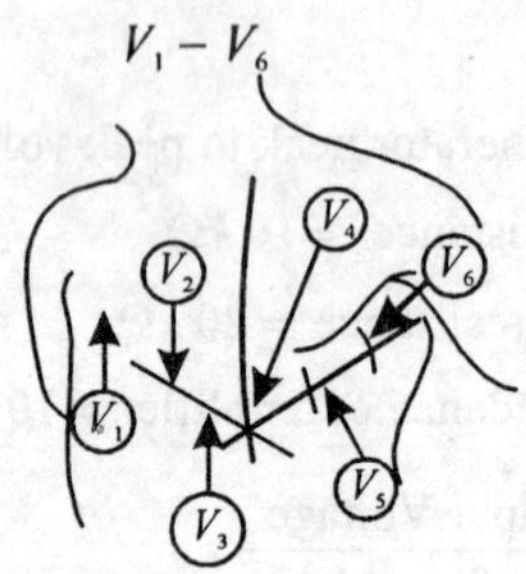

Where,

V_1 = Fourth intercostal space, at right sternal margin

V_2 = Fourth intecostal space, at left sternal margin.

V_3 = Midways between V_2 and V_4.

V_4 = Fifth intercostal space, at mid-clavivular line.

V_5 = Same level as V_4, on an terior axillary line.

V_6 = Same level as V_4, on midaxillary line.

8. Ans. (d)

Solution:

Cardic Stress Testing

* Cardiac stress testing is donw with heart simulation either by exercise on a tread mill or with a introveneous phamachological simulation, with thre patient connected to an electrocardiagram (ECG)
* The level of mechanical test in progressively increased by adjusting the diffeiculty and speed. The test administrator or attending physician examines the symptoms and blood pressure response with the help of ECG, this test is most commoly called the cardiac stress test, but it is also known by other names, such as exercising testing stress testing treadmills exercse tolerance test, stress test or stress test ECG.

Types of Cardiac strewss testing are

* Treadmill test
* Bicycle test

In Treadmill Test the dynamic exercise is performed by the patient who walks on the treadmill on which the speed and the elevation can be adjusted either physically or automatically.

There are two type of treadmill test:

The balke-wase protocol: It includes a constant speed of 5.3 km/hr with the progressive increments in the load at every 2 minutes.

The Bruce Protocol: In this treadmill speed and elevation both are increased simultaneously at the intervals of 3 mintues.

9. Ans. (b)

10. Ans. (c)

11. Ans. (c)

12. Ans. (b)

13. Ans. (d)

Solution:

$$\textbf{CMMR} = 20 \log\left(\frac{A_d}{A_{cm}}\right)$$

$$A_d = \frac{50\ m}{100\ \mu V}$$

$$= 0.5 \times 10^3$$

$$A_{cm} = \frac{0.11\ mV}{5\ mV}$$

$$= 2 \times 10^{-4}$$

$$\textbf{CMMR} = 20 \log\left(\frac{500}{2 \times 10^{-4}}\right)$$

$$= 127.95\ dB$$

$$\approx 100\ dB$$

14. Ans. (b)/(a)

15. Ans. (d)

Solution:

In case of these three basic limb (i.e. Lead I ,II, III), a scientist named **Einthoven postulated** that at any given instant of the cardiac cycle, the frontal plane representation of the electrical axis of the heart is a two dimensional vector, and the ECG produced from any one of the three basic limb leads is a time-varying single dimensional component of that vector.

The heart is the basic origin of the vector, and can be consideration as near the center of an equilateral triangle, the apexes of which are the right and right and left shoulder and the crotch.

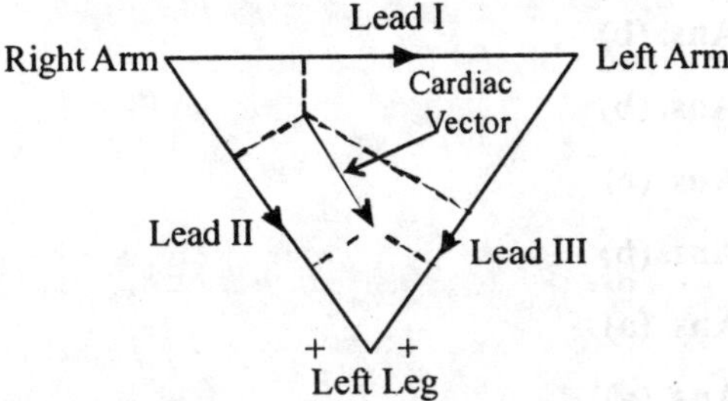

The points of the edges of the triangle represents the electrode positions of the three limb leads. This triangle is know as the Einthoven Triangle, as shown in the figure above.

If we know the magnitude of the cardiac vector (p) and the cardiac vector is oriented at an angle. $'\alpha'$ w.r.t Lead I, then we can find the voltage of lead I, II and III.

$$V_I = P\cos\alpha$$

$$V_{II} = -\frac{1}{2}P\cos\alpha - \frac{\sqrt{3}}{2}P\cos\alpha$$

$$V_{III} = \frac{1}{2}P\cos\alpha - \frac{\sqrt{3}}{2}P\cos\alpha$$

In present case

$$V_I = P\cos\alpha = 3\cos 45^0\, mV$$

$$|V_1| = 2.12mV$$

$$V_{II} = -\frac{1}{2}p\cos\alpha - \frac{\sqrt{3}}{2}p\sin\alpha$$

$$= \frac{1}{2}3\cos 45^0 - \frac{\sqrt{3}}{2}3\sin 45^0\, mV$$

$$= -1.06 - 1.86mV$$

$$= 2.89mV$$

$$|V_{II}| = 2.89mV$$

$$V_{III} = \frac{1}{2}p\cos\alpha - \frac{\sqrt{3}}{2}p\cos\alpha$$

$$= \frac{1}{2}3\cos 45^0 - \frac{\sqrt{3}}{2}3\sin 45^0\, mV$$

$$= 1.06 - 1.83mV = -077mV$$

$$|V_{III}| = 0.77mV$$

It is known

$$|V_{II}| = |V_I| + |V_{II}|$$

as 2.89 = 2.12 + 0.77

16. Ans (b)

Solution:

ECG Leads

Out of the five electrodes availabel for the recording of the electrocardiograph, four electrodes are used to record the electrocadiogram and right leg (RL) electrode is only for the ground reference. But the ECG recorder have only two terminals. So a selection has to be done among the available electrodes.

There are 12 standard leads which are used to record the electrocardiograph

These are:

(a) 3 Bipolar leads
(b) 3 Unipolar limb leads.
(c) 6 Unipolar chest leads.

In the **bipolar leads**, three electrodes are used out of these three electrodes the electrocardigram is recorded from two electrodes and the third electrodes is taken as reference as are shown in the figure below.

Since only the two electrodes are used to obtain the differential output, hence it is known as the bipolar leads. Bipolar leads are of three type:

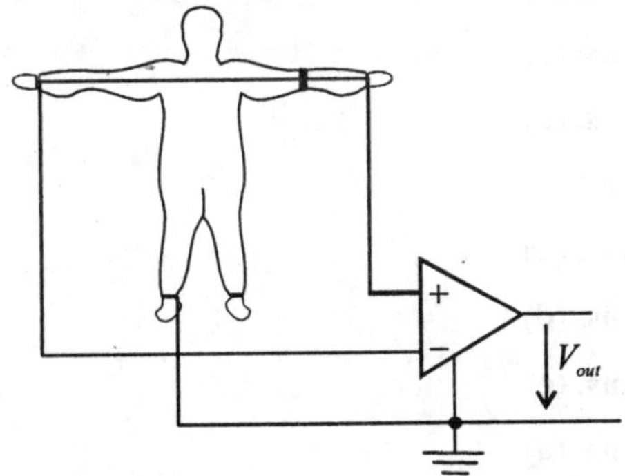

Figure: Lead-I: Left arm (LA) and Right arm (RA).

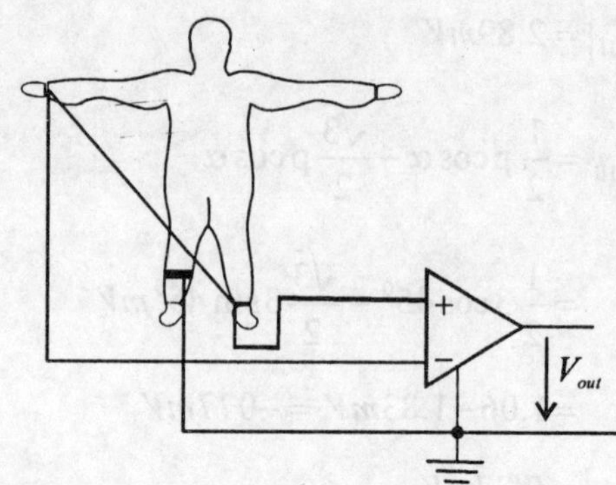

Figure: Lead-II: Left leg (LL) and Right arm (RA)

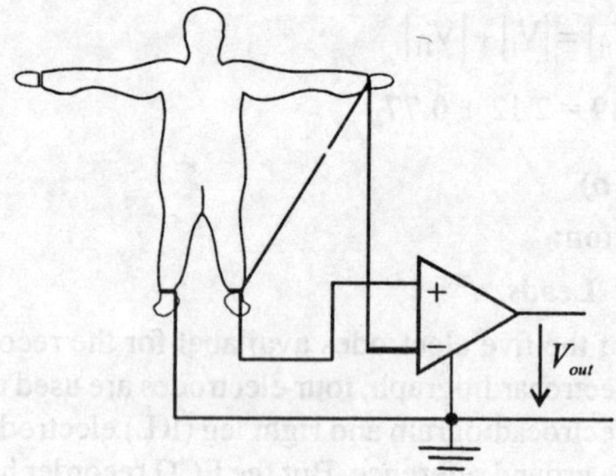

Figure: Lead III: Left leg (LL) and left arm (LA)

* Different leads products defferent magnitude of the R-waves. The magnitude of the R- wave produced is the greatest in Lead-II
* The magnitude of the voltage produced for R-wave for the Lead-II is the sum of those produced for Lead-1 and Lead-III

that is $V_{II} = V_I + V_{III}$

If ECG is taken of a person with 3-Lead frontal plane ECG. With a normal heart person

Then lead 2 produces the greatast $R-$ wavepotential. Thus when the amplitude of 3- limds leads are measured $R-$wave amplitude of lead – 2 is equal to sum of lead – 1 lead –3.

17. Ans. (b)

18. Ans. (a)

19. Ans. (a)

20. Ans. (a)

21. Ans. (d)

22. Ans. (d)

23. Ans. (c)

24. Ans. (a)

25. Ans. (c)

26. Ans. (a)

27. Ans. (d)

28. Ans. (d)

29. Ans. (c)

30. Ans. (b)

31. Ans. (a)

32. Ans. (b)

33. Ans. (a)

34. Ans. (b)

35. Ans. (b)

36. Ans. (b)

37. Ans. (c)

38. Ans. (b)

39. Ans. (a)

40. Ans. (a)

41. Ans. (b)

42. Ans. (c)

43. Ans. (d)

44. Ans. (c)

45. Ans. (c)

46. Ans. (d)

47. Ans. (a)

48. Ans. (b)

49. Ans. (b)

50. Ans. (c)

51. Ans. (b)

52. Ans. (b)

53. Ans. (a)

54. Ans. (b)

55. Ans. (a)

56. Ans. (c)

57. Ans. (d)

58. Ans. (c)

59. Ans. (d)

60. Ans. (b)

61. Ans. (a)

62. Ans. (b)

63. Ans. (b)

64. Ans. (b)

65. Ans. (c)

66. Ans. (a)

67. Ans. (a)

68. Ans. (b)

69. Ans (d)

Solution :

$$V_w = \frac{1}{3}(V_1 + V_2 + V_3)$$

70. Ans (d)

Solution :

Given $V_{id} = 100\ \mu V$

$V_{icm} = 5\ mV$

$V_{od} = 50\ mV$

$V_{oc} = 0.001\ mV$

$$CMRR = 20\log\left(\frac{A_d}{A_c}\right) = 20\log\left(\frac{V_{od}/V_{id}}{V_{oc}/V_{ic}}\right)$$

$$= 20\log\left(\frac{50}{100}\times\frac{5}{0.001}\right)$$

$$= 100\ dB$$

71. Ans (d)

Solution :

$V_I = p\cos\alpha$

$= 3\times\cos 45°$

$= 2.12\ mV$

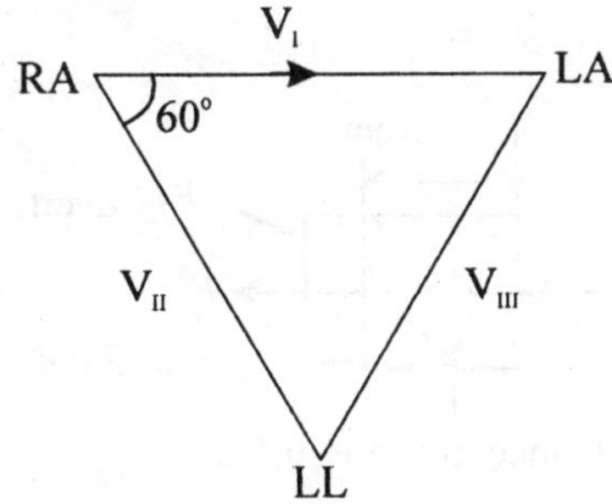

$V_{II} = p\cos(60-\alpha)$

$= 3\times\cos(60-45°)$

$= 2.89\ mV$

$V_{III} = V_{II} - V_I$

$= 0.77\ mV$

72. Ans (d)

Solution :

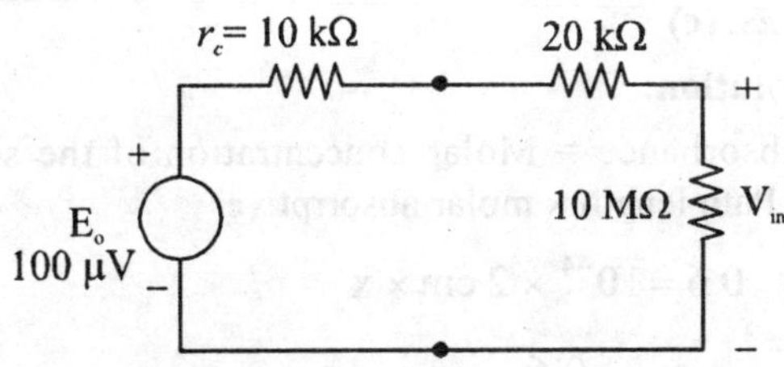

$$V_i = \frac{V\times R_1}{R_1 + R_2 + R_3}$$

$$V_{in} = \frac{(100\ \mu V)\times 10\ M\Omega}{(10\ k + 20\ k + 10\ M)}$$

$$= 99.7\ \mu V$$

73. Ans (b)

Solution :

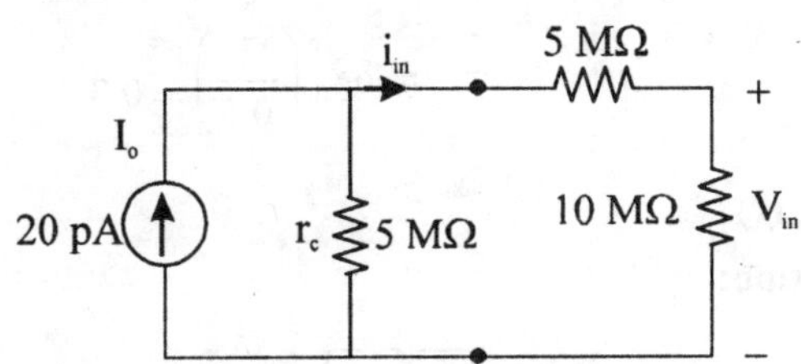

$$V_{in} = \frac{20\,pA\times 5\ M}{(5\ M + 5\ M + 10\ M)}$$

$$= 5\ pA$$

$$V_{in} = (10\ M\pi\times 5\ pA)$$

$$= 50\ \mu V$$

$\because V_{in}$ is less than $60\ \mu V$

Amplifier will not sense

74. Ans. (c)

75. Ans. (c)

Solution:

It is known that the voltage in Lead II is the maximum, and satisfy the relation :

$$V_{II} = V_I + V_{III}$$

Hence, it is clear from , the option (c)

$$3.7 = 2.1 + 1.6$$

Which satisfy the condition

1.9 Clinical Measurements

1. **Ans. (d)**
2. **Ans. (a)**
3. **Ans. (c)**

Solution:

Absorbance = Molar concentration of the solute mc × Path length × molar absorrptive.

or $0.6 = 10^{-4} \times 2\text{ cm} \times x$

$$x = \frac{0.6}{10^{-4} \times 2}$$

$$x = 3000$$

4. **Ans. (a)**

Solution:

Given transmittance of solution $T = 0.5$

Absorbance of solution $A = \log_{10}\left(\frac{1}{T}\right)$

$$= \log_{10}\left(\frac{1}{0.5}\right) = 0.3$$

5. **Ans. (c)**

Solution:

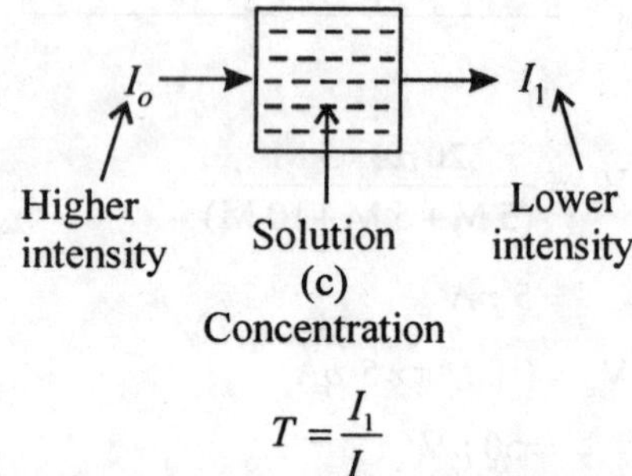

$$T = \frac{I_1}{I_o}$$

Because I_1 has lower intensity and I_o has higher intensity so T has lower value than 1. If concentration of solution is doubled then.

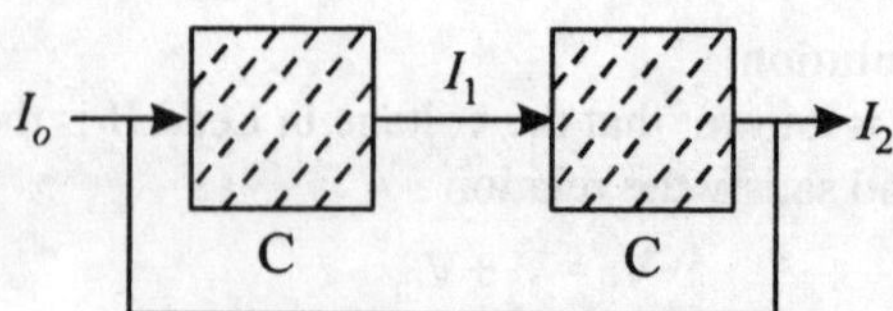

Then $I_2 = T\,I_1$ and $I_1 = T\,I_o$

$$I_2 = T\,(T\,I_o)$$

$$I_2 = T^2\,I_o$$

$$\frac{I_2}{I_o} = T^2$$

6. **Ans. (a)**

Solution:

$b = 10\ mm \qquad A = 0.01 \qquad a = 10^4$ lit/mol/cm

$= 1\ cm$

$A = abc$

$0.01 = 10^4 \times 1 \times C$

$C = 10^{-6}$

$= 1\ \mu$ mol/lit

7. **Ans. (b)**

Solution:

We know that absorbance

$$A = K\,a\,c\,L$$

Where

a = Areas of cross-section of cuvettes

L = Path length of cuvette

C = Concentration of absorbing substance

K = Proportionality constant

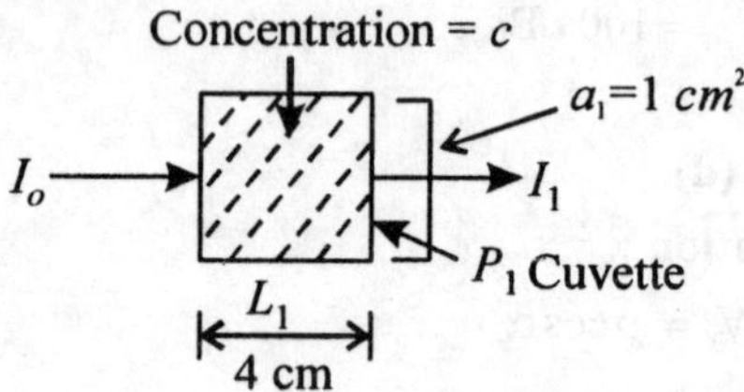

$A_1 = 1 \times c \times 4\ cm$

$A_1 = 4\ c$

Now

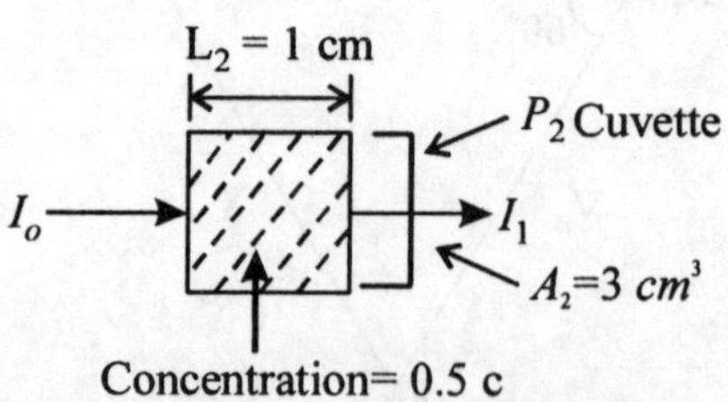

$A_2 = 3 \times 0.5 \times 1 = 1.5\ c$

The ratio of absorbance in P_1 and P_2

$$\frac{4\ c}{1.5c} = \frac{40}{15} = \frac{8}{3}$$

8. Ans. (c)

Solution:

Since volumes are equal, resulting solution is having each type solution 50% volume of total volume. Hence solution transmittance.

$$= 0.1 \times 50\% + 0.4 \times 50\%$$

$$= 25\%$$

9. Ans. (a)

10. Ans. (c)

11. Ans. (c)

12. Ans. (c)

13. Ans. (d)

14. Ans. (a)

15. Ans. (c)

16. Ans. (b)

17. Ans. (a)

18. Ans. (b)

19. Ans. (a)

20. Ans. (c)

21. Ans. (a)

22. Ans. (c)

23. Ans. (b)

24. Ans. (d)

25. Ans. (b)

26. Ans. (a)

27. Ans. (c)

28. Ans. (b)

29. Ans. (a)

30. Ans. (b)

31. Ans. (a)

32. Ans. (a)

Solution:

(1) Given $t = 0.7s$

$$\text{Pulse rate} = \frac{60}{0.7} \approx 85$$

(II) Given $t = 0.8\ s$

$$\text{Pulse rate} = \frac{60}{0.8} \approx 75$$

(III) Given $t = 0.9\ s$

$$\text{Pulse rate} = \frac{60}{0.9} \approx 66$$

33. Ans (b)

Solution :

Given

extracellular concentration of K^+,

$$C_e = 6 \times 10^{-6}\ moles/cm^3$$

and intracellular conc of, K^+

$$C_i = 150 \times 10^{-6}\ moles/cm^3 \text{ sd}$$

$$\therefore \text{ Concentration ratio} = \frac{C_e}{C_i} = \frac{6 \times 10^{-6}}{150 \times 10^{-6}}$$

$$= \frac{1}{25}$$

34. Ans. (c)

Solution:

Diffusion potential

$$= \pm 61 \log_{10}\left(\frac{C_e}{C_i}\right)$$

$$= \pm\ 85.27$$

35. Ans (c)

Solution :

Given, Absorbance $A = 0.6$

Concentration, $C = 1.0 \times 10^{-4}$ M

length, $= 2\ cm = 0.02\ m$

$$\therefore \quad A = a(\lambda)\ bc$$

$$a(\lambda) = \frac{A}{bc} = \frac{0.6}{1.0 \times 10^{-4}} \times 0.02$$

$$= 3000$$

36. Ans (b)

Solution :

Given $T = 0.08$

$b = 1\ cm = 0.01\ m$

$$A = -\log, \quad T_1 = abc \qquad (1)$$

When concentration is doubled

$$A = -\log_e T_2 = ab2c \qquad (2)$$

Equation (1) / (2)

$$\frac{+\log_e 0.08 = abc}{+\log_e T_2 = ab2c}$$

$$+\log T_2 = 2.52 \times 2$$

$$= 5.05$$

$$T_2 = 64\ \%$$

37. Ans. (d)

Solution

For first solution

$$A = ac\ 0.01 \qquad (1)$$

For 2nd solution

$$A = a\ 2\ c\ b \qquad (2)$$

Equation (1) / (2)

$$b = \frac{0.01}{2} = 0.005\ cm$$

$$= 0.5\ cm$$

38. Ans (a)

Solution :

Given, $T = 0.5$

absorbance, $A = -\log T$

$$= -\log 0.5$$

$$= 0.3010$$

$$\approx 0.3$$

39. Ans (c)

Solution :

Given, $T_1 = T$, and $T_2 = T^2$

$C_1 = C$, and $C_2 = 2C$

$A = abc$

$\Rightarrow -\log T_1 = abc$ (1)

$A = ac$

$\Rightarrow -\log T_2 = ab2c$ (2)

Equation (1) / (2)

$$\frac{-\log T_1}{-\log T_2} = \frac{abc}{ab2c}$$

$$\Rightarrow -2\log T_1 = -\log T_2$$

$$\Rightarrow (T_1)^2 = T_2$$

$$T_1^2 = T_2$$

40. Ans (a)

Solution :

Given, $T \rightarrow 100\,\%$ to 10%

$A \rightarrow 0\ to\ 1$

$A = -\log T$

$$= -\log\frac{10}{100} \approx 1\,\%$$

$\therefore$ (*a*) 0 to 1

41. Ans (d)

Solution :

Given , $C_1 = C$, $C_2 = 0.5\,C$

$b_1 = 4\ cm = 0.04\ m$

$b_2 = 1\ cm = 0.01\ m$

$$\therefore \frac{A_1}{A_2} = \frac{a \times 0.04 \times C}{a \times 0.01 \times 0.5\,C}$$

$$= \frac{4 \times 10}{5}$$

$$= 8$$

42. Ans (c)

Solution :

Given,

For solution P

$T = 0.10 \Rightarrow A_p = -\log_{10} 0.10$

For solution Q

$T = 0.4 \Rightarrow A_Q = -\log_{10} 0.4$

$A_{total} = 1.39$

$T_{total} = 10^{-A} = 10^{-1.39} \times 100 -$

$0.04 \approx 4.10$

43. Ans (b)

Solution :

Given, RBC = 5 *million* / μl

$Hb = 15\ gm/dl$

$HCT = 45\,\%$

$$MCV = \frac{PCV}{RBC\ count} = \frac{0.45 \times 10^{-6} \times 10^{15}}{5 \times 10^{6}}$$

$$= 0.9 \times 10^{15}\ \mu m^3$$

[$\because$ $1m^3 = 1000\ lt$

$1\ \mu m = (10^{-6}\ m)^3 = 1000\ l$

$= 10^{-18} \times 1000\ l$

$= 10^{-15}\ l$

$\therefore$ $1\mu = 10^{15}\ \mu m$]

44. Ans (c)

Solution :

$$MCH = \frac{[Hb]}{RBC\ count}$$

$$= \frac{15\ gm \times 10^{-5}}{10^{-1}\ lt \times 5 \times 10^{6}}$$

$$= 30\ pgm$$

45. Ans (b)

Solution :

$$MCHC = \frac{[Hb]}{PCV}$$

$$= \frac{15\ gm}{0.45} = \frac{10}{3}$$

46. Ans. (c)

Solution :

Given, $C_o = 5.2\,l$

$HBR = 72/\min$

$\therefore \quad C_o = SV \times HBR$

$$SV = \frac{C_o}{HBR}$$

$$= \frac{5.2}{72} = 0.722$$

$\approx 72\ ml$

47. Ans. (a)

Solution :

(a) plate electrode

Since surface area is more in plate type

48. Ans. (d)

Solution :

Given

Curve duration $= 20$ sec

$\text{Mass} = 10\ mg$

Average concentration of indicator recorded $= 5\ mg/l$

$$\therefore \text{ Cardiac output, } Q = \frac{10}{5 \times 20} l/\sec$$

$= 0.1\,l/\sec$

$= 0.1 \times 60\,l/\min$

$= 6l/\min$

49. Ans. (b)

Solution:

Given mass $m = 20\ mg$

$$Q = \frac{M}{\text{Area under the dilution curve}} l/\sec$$

$$= \frac{20}{180} l/\sec. = \frac{20 \times 60}{180} l/\min$$

$= 6.66\,l/\min$

50. Ans. (c)

51. Ans. (c)

Solution:

The relation between absobance A at wave length J and the ransmisttance T is given as

$$A = \log \frac{1}{T}$$

Since $A = k\,C\,b$

When C is double, $A \to 2A$

$$\therefore \quad \frac{A}{2A} = \frac{\log \frac{1}{T}}{\log \frac{1}{T'}}$$

$$\left[\log \frac{1}{T'}\right]\frac{1}{2} = \left[\log \frac{1}{T}\right]$$

$$\log\left(\frac{1}{T'}\right) = 2\log\left(\frac{1}{T}\right)$$

$$\therefore \quad T' = T^2$$

52. Ans. (a)

Solution :

Mean Cell Volume

$$MCV = \frac{PCV}{\text{No. of RBC/liter of blood sample}}$$

$$= \frac{0.45}{5 \times 10^{12}} = 90 \times 10^{-15}$$

$= 90 f/l$

53. Ans. (a)

Solution

Mean cell Heamoglobine

$$MCH = \frac{\text{Weight of Hb/litre of blood}}{\text{Number of RBC's/litre of blood}} pg$$

$15 g/dl = 150\ g/litre$

$$\therefore \quad MCH = \frac{150}{5 \times 10^{12}} = 30\,pg$$

54. Ans. (a)

Solution :

$$MCHC = \frac{15}{0.45} g/dl$$

$= 33.3\ g/dl$ s

55. Ans. (a)

56. Ans. (b)

57. Ans. (d)

Solution :

We know that,

$$N = \frac{HLE}{v}$$

Here, $H = 63000 \times 6.3 \times 10^4$

$L = 60$

$E = 520 \times 5.20 \times 10^2$

$$\therefore \quad N = \frac{(6.3\times10^4)\times(60)\times(5.20\times10^2)}{(3.78\times10^2)}$$

$$= 5.20\times10^6 \text{ per mm}^3$$

58. Ans. (a)

1.10 Ultrasonic Transducers and Ultrasonography

1. Ans. (a)

2. Ans. (a)

Solution:

Absorption mechanism converts energy of an acoustic wave to heat as the wave propagates through a medium. A plane, ultrasonic wave in an absorbing medium will loose intensity as

$$I(x) = I_o\, e^{-2\alpha x}$$

Where,

α = Amplitude absorption coefficient of the medium, (cm^{-1})

x = Represents the distance raveled in the tissue

I_o = Intensity at $x = 0$

At each point in the absorbing medium heat is generated at a rate of

$$2\alpha I(x) \;\; W/cm^2$$

As given

$\alpha = 1.18\ cm^{-1}$

$x = t$

The ratio of output to input is

$$\frac{I(x)}{I_o} = \frac{1}{e^2} \quad \ldots(1)$$

$$I(x) = I_o\, e^{-2\alpha x}$$

$$\frac{I(x)}{I_o} = e^{-2\alpha x} \quad \ldots(2)$$

From equation (1) and (2)

$$\frac{1}{e^2} = e^{-2\alpha x}$$

$$e^{-2} = e^{-2\alpha x}$$

$$-2 = -2\alpha x$$

$$-2 = -2\times1.18\times t$$

$$t = 0.847\ cm \quad [\because x = t]$$

$$\approx 1\ cm$$

3. Ans. (b)

Solution:

$$\Delta f = \frac{2FV\cos\theta}{C}$$

$\Delta f = 3\ kHz$ [doppler shift]

$f = ?$

$\theta = 45°$

$$3k = \frac{2\times f\times110\times10^{-2}\times\cos 45}{1500}$$

$$f = \frac{3\times10^3\times1500}{2\times110\times10^{-2}\times\frac{1}{\sqrt{2}}}$$

$$f = 3\ MHz$$

4. Ans. (a)

Solution:

$Y = 80\times10^9\ N/m^2$

$S = 2.65\times10^3\ kg/m^3$

$$f = \frac{1}{2t}\sqrt{\frac{80\times10^9}{2.65\times10^3}}$$

$$3\ MHz = \frac{1}{2t}\sqrt{\frac{80\times10^6}{2.65}}$$

$$t = \frac{1}{2\times3\times10^6}\sqrt{\frac{80\times10^6}{2.65}}$$

$$t = 1\ mm$$

5. Ans (d)

Solution:

Given $f = 5\ MHz$

Thick $t = 25\ m$

It is we know that 1 Neper = 8.6 dB

Fat attenuation loss in dB

= 0.075 Np/cm/MHz × 2cm × 5MHz × 8.6

= 6.45 dB

≈ 6.5 dB

Total attenuation losses = 2×.65 dB

= 13 dB of losses

i.e. -13 dB

Reflection losses in $dB = 20\log\frac{A_2}{A_1}$

$= 20\log 0.1$

$= -20$ dB

Total losses $= -(20+13)$ dB

$= -33$ dB

6. Ans (b)

Solution:

Minimum distance, $d = v \, . \, f'$

$= 1500 \times \frac{20}{5} = 6$ mm

Second Method

We know that

Distance = velocity × time

$$d = \frac{Velocity}{Frequency}$$

20 cycles having frequency of $= 5\times10^6$ Hz

1 cycles having frequency of $\frac{5\times10^6}{20} Hz$

$$d = \frac{1500\times20}{5\times10^6} = 6\times10^{-3}$$

$= 6$ mm

7. Ans. (d)

8. Ans. (a)

9. Ans. (d)

10. Ans. (c)

11. Ans. (c)

12. Ans. (b)

13. Ans. (b)

14. Ans. (b)

15. Ans. (b)

16. Ans. (a)

17. Ans. (a)

18. Ans (a)

Solution :

Given, $\rho_A = 1.3\ g/cc$

$\rho_B = 0.88\ g/cc$

$\upsilon_A = 1400\ m/s$

$\upsilon_B = 1100\ m/s$

$\therefore\ Z_A = \rho_A\ \upsilon_A = 1800$

$Z_B = \rho_B\ \upsilon_B = 900$

$$R = \left(\frac{Z_A - Z_B}{Z_A + Z_B}\right)^2 \times 100\ \%$$

$$= \left(\frac{1820-968}{1820+968}\right)^2 \times 100\ \%$$

$= 9.3\%$

19. Ans. (d)

Solution :

$\because\ E_o = \frac{1}{2}E_i$

$\Rightarrow E_I = 2\,E_o$

$= 2\times17\ mJ$

$= 34\ mJ$

20. Ans. (a)

Solution :

Given $\upsilon_{st} = 1500\ m/s$

$\upsilon_{cb} = 4000\ m/s$

Time $(t) = 90\ \mu s$

Thinkness $t_H = ?$

$$\therefore \qquad \upsilon_{st} = \frac{2t_H}{\text{Time}(t)}$$

$$t_H = \frac{1500\times80\times10^{-6}}{2}$$

$= 0.06 = 6\ cm$

21. Ans. (d)

Solution :

$$\left(\frac{A_o}{A_i}\right)_{dB} \rightarrow \left[20\ \log\left(\frac{A_o}{A_i}\right)\right]_{dB}$$

$P = 8.68\ dB$

Amplitude attenuation factor, fat =

$= 0.75\ \ \text{Npcm}^{-1}/\text{MHZ}$

Total amplifier loss in fat $= -2(0.075\times8.68\times2\times5)$

$= -13\ dB$

Reflectivity coeff $= 0.1$

$$\left(\frac{A_2}{A_1}\right)_{dB} = 20\ \log\ 0.1 = 20\ dB$$

net total loss $= -20 - 13$

$= -33\ dB$

22. Ans. (c)

Solution :

$$\Delta T = \frac{2\upsilon D}{C^2}\cos\theta$$

$$\upsilon = \frac{\Delta T \times C^2}{2\times D\cos\theta} = \frac{0.25\times 10^{-6}\times(1200)^2}{2\times 0.02\times\cos 45^\circ}$$

$$\Rightarrow \quad \upsilon = 12.72\ m/s$$

23. Ans. (c)

Solution :

$$\Delta f = 2f\frac{V_b}{V_{sb}}[\cos\theta_1 + \cos\theta_2]$$

$$= \frac{2\times 7\times 10^6\times 8\times 10^{-2}}{1100}\left[\cos 45^\circ + \cos 30^\circ\right]$$

$$= 1601.73\ \text{Hz}$$

24. Ans. (b)

Solution :

Given,

Attenuation Coefficient $\alpha = 1.18\ cm^{-1}$

$$\frac{P_o}{P_i} = \frac{1}{c^2}$$

$$\because \quad \frac{E_o}{t} = \frac{E_i}{t}e^{-\alpha t}$$

$$P_o = P_i\, e^{-\alpha t}$$

$$\frac{P_o}{P_i} = e^{-\alpha t}$$

$$e^{-2} = e^{-\alpha t}$$

$$-2 = -\alpha t \qquad t = \frac{2}{\alpha}$$

$$\Rightarrow t = \left(\frac{2}{1.18}\right) cm$$

$$= 1.695\ cm$$

25. Ans. (a)
26. Ans. (b)
27. Ans. (c)
28. Ans. (a)
29. Ans. (c)
30. Ans. (d)
31. Ans. (b)
32. Ans. (a)
33. Ans. (a)
34. Ans. (a)

1.11 Computed Tomography

1. Ans. (b)

Solution :

Given,

$$\mu_{H20} = 0.0205\ cm^{-1}$$

$$\mu_{sample} = 0.190\ cm^{-1}$$

$$\text{CT value} = \left(\frac{\mu_{sample} - \mu_{H20}}{\mu_{H20}}\right)\times 1000\ \text{H}\mu$$

$$= \frac{0.190 - 0.205}{0.205}\times 1000$$

$$\approx -73 \approx -70$$

$\therefore$ The tissue is fat

2. Ans. (c)
3. Ans. (d)
4. Ans. (c)
5. Ans. (a)
6. Ans. (a)
7. Ans (b)

Solution :

$$I = I_0\, e^{-\mu x}$$

Given $\quad \mu = 1.62$

$$\frac{I}{I_0} = \frac{1}{20}$$

$$\therefore \quad \frac{I}{I_0} = e^{-\mu x}$$

$$\frac{1}{20} = e^{-1.62x}$$

$$20 = e\,1.62x$$

$$ln\,20 = 1.62x\,ln\,e$$

$$x = \frac{ln\,20}{1.62} = 1.85\ \text{cm}$$

8. Ans. (c)
9. Ans. (c)

Solution :

It is known that house field unit

$$H = \frac{\mu - \mu_{water}}{\mu_{water}}\times 1000$$

$$= \frac{0.212 - 0.205}{0.205}\times 1000 = \frac{7000}{205}$$

2 Transducers, Mechanical Measurement and Industrial Instrumentation

2.1 Basic of Transducers

1. **Bellow's expansion** is usually against a spring. The spring is provided to
   (a) Increase sensitivity
   (b) Increase operating range
   (c) Increase linearity
   (d) Decrease hysteresis effect

2. The natural frequency of a **seismic mass type** pickup is f_n. The pick-up will give the correct motion information, if the exciting frequency f_e is
   (a) f_n (b) $[1/\sqrt{2}]\, f_n$
   (c) $>> f_n$ (d) $<< f_n$

3. A **seismic type of transducer** has a damping constant of 10. The transducer is designed to measure.
   (a) Acceleration (b) Velocity
   (c) Displacement (d) Force

4. Different types of transducers and signal conditioning circuits are given below. Match the transducer with the appropriate signal conditioning circuit commonly used with it and select the correct match using the codes.

A. Thermocouple	1. DC Bridge
B. Strain gauge	2. Phase sensitive detector
C. Piezoelectric	3. Charge amplifier sensor
D. L.V.D.T	4. Cold junction compensation
	5. Instrumentation amplifier

Codes:

	A	B	C	D
(a)	2	3	5	1
(b)	1	5	2	3
(c)	4	1	2	5
(d)	4	1	3	2

5. Match the List I device with List II its working principles and select the correct answer using the codes:

List I	List II
A. Bourdon guage	1. Thermal conductivity
B. Pirani gauge	2. Thermal co-efficient of expansion
C. Semi-conductor strain gauge	3. Piezo-resistive effect
D. Bimetallic strip	4. Phote-electric effect
	5. Deformation of element

Codes:

	A	B	C	D
(a)	2	4	3	1
(b)	4	1	3	2
(c)	5	1	3	2
(d)	4	2	1	3

6. Match the List I instruments with the List II measurements and select the correct answer using the codes:

List I	List II
A. Rotameter	1. Rotational speed
B. Tachometer	2. Roughness
C. Tracer	3. Temperature
D. Pyrometer	4. Flow rate

Codes:

	A	B	C	D
(a)	2	4	3	1
(b)	4	1	2	3
(c)	3	1	4	2
(d)	2	4	1	3

7. Match the List I **pressure measuring** devices with the List II pressure levels in torr and select the correct answer using the codes:

List I	List II
A. Bourdon tube	1. 10^3
B. Ionization gauge	2. 10^{-10}
C. Water manometer	3. 1.0
D. Pirani	4. 10^{-5}

Codes:

	A	B	C	D
(a)	2	4	3	1
(b)	3	4	1	2
(c)	3	1	4	2
(d)	1	2	3	4

8. Identify the correct set of matches from the following

List I	List II
A. LVDT	1. Reference junction compensation
B. Piezoelectric crystal	2. Charge amplifier
C. pH probe	3. Phase sensitive detector
D. Thermocouple	4. High input impedance amplifier

Codes:

	A	B	C	D
(a)	1	3	4	2
(b)	3	2	4	1
(c)	4	1	2	1
(d)	None of the above			

9. Identify the correct set of matches from the following

List I	List II
A. Pirani gauge	1. Temperature measurement
B. Venturi meter	2. Vacuum pressure measurement
C. Radiation pyrometer	3. Force measurement
D. Load cell	4. Flow measurement

Codes:

	A	B	C	D
(a)	2	1	4	3
(b)	4	3	2	1
(c)	2	4	1	3
(d)	None of the above			

10. In the context of **transducers,** identify the correct matches using the code:

A. Mean free path	1. Optical pyrometer
B. Humdity	2. Knudsen gauge
C. Heat transfer Coefficient	3. Sling psychrometer
D. Intensity of radiation	4. Hot wire anemometer

Codes:

	A	B	C	D
(a)	1	2	3	4
(b)	2	1	4	3
(c)	2	4	1	2
(d)	2	3	4	1

11. Match the following

P. Radiation pyrometer

Q. Dall tube

R. Pirani gauge

S. Gyroscope

W. Angular velocity measurement

X. Vacuum pressure measurement

Y. Flow measurement

Z. Temperature measurement

(a) P-Z, Q-W, R-X, S-Y
(b) P-Z, Q-Y, R-X, S-W
(c) P-W, Q-X, R-Y, S-Z
(d) P-Z, Q-X, R-W, S-Y

12. The **transducer** element in a measurment system server to.
(a) Codify the input signal
(b) Indicate the measurement
(c) Amplify and transmit the measurement
(d) Convert the input signal from one physical form into another

13. The intermediate stage of an instrument is used to
(a) Sense the primary signal
(b) Condition and transmit the signal after it has been transduced.
(c) Provide a record or indication of the measurement
(d) Give a signal that the desired value has been reached.

14. The output stage of a generalised measurement system may comprise of
(a) detector - transducer unit
(b) transmission element to telemeter the data for remote reading and recording.
(c) indicating or recording unit
(d) any one of the above depending upon the nature and application of measurement.

15. Which of the following **does not fit into** the functions of an instrument?
(a) acquisition of information by sensing and perception.
(b) Processing of the acquired information.
(c) Presentation of measured data to a human observer.
(d) Changing of the signal to be measured

16. The measurement of the speed of a rotating shaft by means of an electrical tachometer is a
(a) direct measurement
(b) non-contact type measurement
(c) secondary measurement
(d) tertiary measurement

17. The most appropriate clause pertaining to the development of digital instruments is
(a) fast reading
(b) no observation error
(c) better accuracy
(d) possibility of output being fed to a memory circuit

18. The **proximity or non-contact instrument** is one which measures the desired input even though it may not be in physical contact with the system being investigated. Which of the following devices constitutes a proximity instrument?
(a) strain gauge load cell (b) stroboscope
(c) clinical thermometer (d) dial gauge

19. In a **self-generated instrument**, the energy requirements are met entirely by the input signal itself. Which of the following is a self-generating type instrument?

(a) piezoelectric pickup
(b) thermistor
(c) potentiometric device
(d) linear variable differential transformer

20. Indicate the **externally-powered** type of instrument transducer.

(a) spring balance
(b) centrifugal tachometer
(c) thermocouple or thermopile
(d) photoemissive cell

21. A **null type instrument** generates an equivalent opposing effect to nullify the physical effect casued by the quantity being measured. The equivalent null causing effect then provides a measure of the quantity. Amongst the following, an instrument with null output is

(a) Bourdon gauge
(b) mercury manometer
(c) platform weighing scale
(d) millivoltmeter used to measure the emf generated in a thermocouple circuit

22. Pointout the device that refer to **self-generating** transducer.

(a) resistive (b) capacitive
(c) piezoelectric (d) thermometer

23. A transducers converts

(a) mechanical energy into electrical energy.
(b) mechanical displacement into electrical signal.
(c) one form of energy into another form of energy
(d) electrical energy into mechanical form

24. In a transducer, the experimentally obtained transfer function is different from the theoretical transfer function. The errors result from their difference are called

(a) Zero errors
(b) Sensitivity errors
(c) Non-conformity errors
(d) Dynamic errors

25. In a transducer, the observed output deviates from the correct value by a constant factor the resulting error is called

(a) Zero error (b) Sensitivity error
(c) Nonconformity error (d) Hysterisis error

26. A transducer has an output impednace of 1 *k*W and a load resistance of 1 MW, the transducer behave as

(a) Constant current source
(b) Constant voltage source
(c) Constant power source
(d) None of these

27. If a transducer has an output impedance of 1 *k*W and a load resistance of 1 W, it then it behave as

(a) Constant current source
(b) Constant voltage source
(c) Constant impedance source
(d) None of these

28. While selecting a transducer for a particular application:

(a) Only input characteristic should be considered
(b) Only output characteristic should be considered
(c) Only transfer characteristics should be considered
(d) Input, output and transfer characteristics should be considered

29. An **inverse transducer** converts

(a) Electrical energy to any other form of energy
(b) Electrical energy to light energy
(c) Mechanical displacement into electrical
(d) Electrical energy to mechanical form

30. Which one of the following act as **inverse transducer**?

(a) Electrical resistance potentiometer
(b) LVDT
(c) Capacitive transducer
(d) Piezoelectrical crystal

31. Which one of the following is an **active transducer**?

(a) Strain gauge (b) selsyn
(c) photovoltaic cell (d) photo emissive cell

32. A **strip chart** recorder is an

(a) active transducer (b) inverse transducer
(c) Output transducer (d) (b) and (c)

33. The lower limit of useful working range of transducer is determined by its

(a) Error
(b) Noise
(c) Both (a) and (b)
(d) None of these

34. While selecting a transducer for a particular application

(a) Only the input characteristic should be considered
(b) Only the output characteristics should be considered
(c) Only the transfer characteristics should be considered
(d) Input, output and transfer characteristics should be considered.

35. The part of a transducer which responds to a physical phenomenon or change in a physical phenomenon is

(a) Sensing (b) Transduction
(c) Resistive (d) Inductive

36. A Bourdon tube senses the pressure and converts the pressure into a displacement which is converted into an analogous voltage by L.V.D.T. In this the primary transducer and secondary transducer, respectively are

(a) LVDT, Bourdon tube (b) Bourdon tube, LVDT
(c) Strain gauge, LVDT (d) None of the above

2.2 Piezoelectric Seismic Transducers

1. A piezoelectric type 50 Mpa full scale pressure sensor with built in electronics has a sensitivity of 100 mV per Mpa. If this sensor is subjected to a static pressure of 10 Mpa, then its output will be

(a) 0 mV (b) 100 mV
(c) $1000/\sqrt{2}$ mV (d) 1000 mV

2. For signal conditioning of a piezoelectric type transducer, we require

(a) A charge amplifier
(b) A differential amplifier
(c) An instrumentation amplifier
(d) A transconductance amplifier

3. The seismic mass of an accelerometer oscillates sinusoidally at 100 Hz with a maximum displacement of 10 mm from its mean position. The peak acceleration of the seismic mass is

(a) 3947.84 m/s^2 (b) 3141.50 m/s^2
(c) 314.15 m/s^2 (d) 100.00 m/s^2

4. A piezoelectric transducer is directly connected through a cable to an electronic voltmeter. The minimum operating frequency of measurement is 1000 Hz. If the connecting cable length is doubled, the new minimum operating frequency is

(a) 500 Hz (b) 1000 Hz
(c) 2000 Hz (d) 4000 Hz

5. A seismic vibration sensor having natural frequency ω_n and damping ratio ζ, used for measuring amplitude of vibration with frequency ω, should have

(a) $\omega_n \ll \omega$ and ζ much greater than 1
(b) $\omega_n = \omega$ and ζ slightly less than 1
(c) $\omega_n \gg \omega$ and ζ slightly less than 1
(d) $\omega_n \ll \omega$ and ζ slightly less than 1

6. A quartz crystal (Young's modulus, $E = 9 \times 10^{10}$ N/m^2) with **piezo-electric** properties has a diameter of 10 mm and thickness of 2 mm. Its voltage sensitivity constant is 4500 V / mm. If the voltage output is 127.3 V, the applied load is approximately

(a) 100 N (b) 200 N
(c) 127.3 N (d) 6.4 N

7. A temperature measuring instrument is modeled as a first order system with a time constant of 5 s. The sensor of the instrument is placed inside an oil bath whose temperature has a sinusoidal variation with amplitude of 10°C and a period of 20 s around an average temperature of 200°C. The sinusoidal component at the output of the instrument will have amplitude

(a) 0°C
(b) 5.37°C
(c) 8.57°C
(d) 10°C

8. A piezoelectric transducer with a sensitivity of 2.0 pC/N having a capacitance of 1600 pF and a leakage resistance of 10^{12} W is connected to a charge amplifier as shown in figure. If a force of 0.1 sin 10 t N is applied to the transducer, the output amplitude of the charge amplifier is

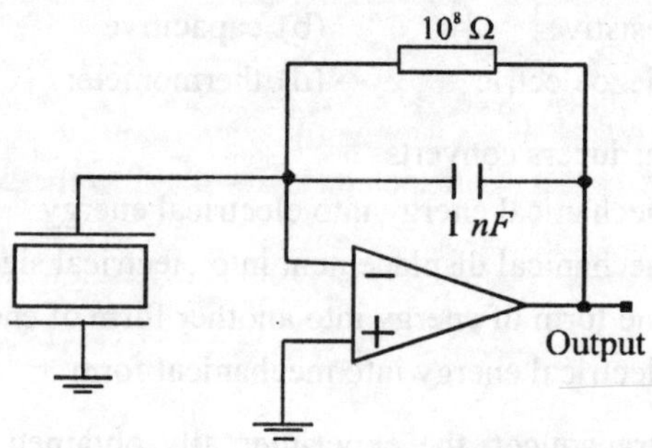

(a) 0 .141 mV (b) 0.232 mV
(c) 1.414 mV (d) 1.732 mV

9. In a **seismic acceleration** sensor with a **displacement transducer** as the secondary element, decreasing the mass while maintaining all other parameters of the sensor unchanged will

(a) Reduce both natural frequency and steady state sensitivity
(b) Increase both natural frequency and steady state sensitivity
(c) Increase natural frequency but reduce steady state sensitivity
(d) Increase natural frequency without affecting steady state sensitivity.

10. An accelerometer has a **seismic mass** of 100 μ gram and natural frequency of 1 kHz. With *g* the acceleration due to gravity, the static sensitivity of the acceleration in (nm/g) is

(a) 2.58 nm/g (b) 9.81 nm/g
(c) 7.28×10^2 nm/g (d) 6.28 nm/g

11. A piezoelectric type accelerometer has a sensitivity of 100 mV/g. The transducer is subjected to a constant acceleration of 5g. The steady state output of the transducer will be

(a) 0 V (b) 100 mV
(c) 0.5 V (d) 5 V

12. A dynamometer arm make contact with the piezoelectric load cell as shown. The g-constant of the piezoelectric material is 50×10^{-3} *Vm/N* and the surface area of the load cell is 4 cm². If a torque $\tau = 20$ *Nm* is applied to the dynamometer, the output voltage V_o of the load cell is

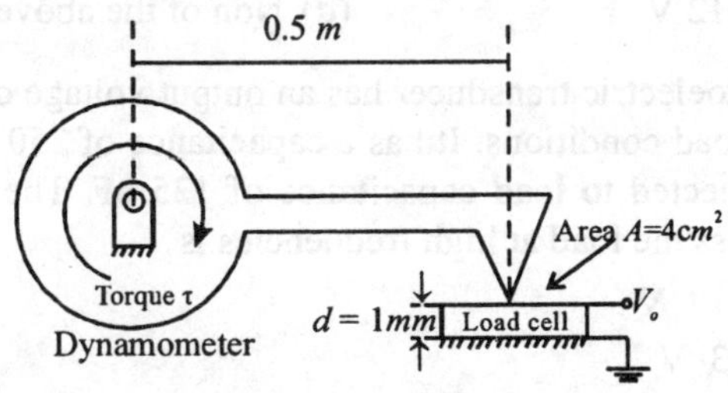

(a) 4 *V* (b) 5 *V*
(c) 10 *V* (d) 16 *V*

13. Given the voltage sensitivity constant of the crystal is 250 V/mm. A piezoelectric accelerometer has a natural frequency of 10,000 Hz and a damping ratio of 0.2. The voltage output for an input acceleration amplitude of 500 g (g being acceleration due to gravity) at a frequency of 2000 Hz is

(a) 200 V/mm (b) 300 V/m
(c) 400 V/mm (d) 250 V/m

14. A quartz crystal having a thickness of 2mm and a voltage sensitivity of 0.05 Volt-m/newton is subjected to a pressure of 15×10^5 N/m².
The voltage developed by the piezo-electric pick up is

(a) 120 V (b) 150 V
(c) 100 V (d) 200 V

Linked Answer Questions

Statement for linked Answer Questions 15, 16 and 17

A body is stated to have vibrational acceleration amplitude of 16g at 8 Hz.

15. The displacement is

(a) 3.124 m/s (b) 13.24 m/s
(c) 31.24 m/s (d) 312.4 m/s

16. Velocity amplitude is

(a) 6.22 m (b) 2.11 m
(c) 0.0622 m (d) 0.211 m

17. The velocity vibrational level is

(a) 0.69 dB (b) 6.98 dB
(c) 98.9 dB (d) 69.89 dB

18. A seismic vibrometer sensing displacement hs undamped natural frequency of 20 Hz and damping ratio of 0.7. The amplitude ratio and phase angle between input and output if input is sinusoidal displacement of 30 Hz is

(a) 0.67 and 128.8° (b) 0.92 and 120.8°
(c) 0.45 and 12.88° (d) 0.82 and 12.8°

19. Piezo-electric transducers are used for the measurement of

(a) Dynamic displacement
(b) Static displacement
(c) Both (a) and (c)
(d) None of the above

20. A piezo-electric transducer has an output voltage of 3 V at no load conditions. It has a capacitance 250 pF. It is connected to load capacitance of 125 pF. The voltage across the load at high frequency is

(a) 1 V
(b) 0.5 V
(c) 9 V
(d) Cannot be determined by the given data

21. A piezoelectric crystal transducer of 0.5 cm² area and 1 mm thickness is connected to a charge amplifier having charge sensitivity r = 2pC/N. The crystal is subjected to the sinusoidal force of 30×10^{-3} sin 150 t N.

The peak to peak voltage of the crystal

(a) 0.135 V (b) 0.27 V
(c) 0.54 V (d) 0.35 V

22. What is the order of minimum displacement that can be sensed with capacitive transducers

(a) 1cm (b) 1 mm
(c) 1 mm (d) 1 pm

23. Piezoelectric type of load cells can be used for measurement of :

(a) dynamic forces only
(b) dynamic forces and static forces providing, that the load cells have a small time constant.
(c) dynamic forces and static forces providing the load cells have a large time constant.
(d) none of the above

24. Elastic elements which are used for the measurement of force give :

(a) low sensitivity and low response if they are stiff.
(b) low sensitivity and fast response if they are stiff.
(c) low sensitivity and slow response if they are stiff.
(d) none of the above

25. Ceramic materials are used for piezoelectrical transducers becuase
(a) They are poly crystalline in nature
(b) They are basically made of barium titanate
(c) They do not have piezoelectric properties in their original state but these properties are produced by special polarizing treatment.
(d) All of the above

26. Piezoelectric transducers are
(a) passive transducers (b) active transducers
(c) inverse transducers (d) Both (b) and (c)

27. Quartz and Rochelle salt belongs to
(a) natural group of piezoelectric materials
(b) synthetic group of piezoelectric materials
(c) Both (a) and (b)
(d) None of the above

28. The output of the piezoelectric crystal has
(a) low amplitude and low impedance.
(b) high amplitude and high impedance.
(c) low amplitude and high impedance
(d) high amplitude and low impedance

29. In a piezoelectric crystal, application of a mechanical stress would produce
(a) Plastic deformation of the crystal
(b) magnetic dipoles in the crystal
(c) electrical polarization in the crystal
(d) shift in the fermi level

30. In piezoelectric crystal, voltage sensitivtiy is defined as
(a) voltage developed per unit stress
(b) field developed per unit stress
(c) voltage developed per unit force
(d) field developed per unit force

31. Piezoelectric quartz crystal resonators can be used as
(a) signal amplifier
(b) rectifier
(c) signal frequency control
(d) modulator

32. Which of the following are piezoelectric substances
1. Barium titanate
2. Lead titanate
3. Lead zirconate
4. Cadmium sulphate
(a) 1, 2 and 4 (b) 1, 3 and 4
(c) 1, 2 and 3 (b) 2, 3 and 4

33. The voltage sensitivities of barium titanate and quartz are respectively 12×10^{-3} Vm / N and 50×10^{-3} Vm / N. Their respective permittivities are 12.5×10^{-9} F / m and 40.6×10^{-12} F / m. Their charge sensitivities are
(a) 1.04×10^{6} C/N, 1.23×10^{9} C/N
(b) 150 *p* C/N, 2 *p*C/N
(c) 20 *p*C/N, 50 pC/N
(d) None of the above

34. A barium titanate crystal has a thickness of 2 mm. Its voltage sensitivity is 12×10^{-3} Vm / N. It is subjected to a pressure of 0.5 MN / m^2. The votlage generated is
(a) 3 V (b) 6 V
(c) 12 V (d) Non of the above

35. Piezoelectric transducer has an output voltage of 3 V at no load conditions. It has a capacitance of 250 pF. It is connected to load capacitance of 125 pF. The voltage across the load at high frequencies is
(a) 12 V
(b) 3 V
(c) 9 V
(d) cannot be determined with the data given

Common Data for Questions 36 and 37

A quartz peizoelectric crystal having a thickness of 2 mm and voltage sensitivity of 0.055 V-m/N is subjected to a pressure of 1.5 MN / m^2.

If the permittivity of quartz is 40.6×10^{-12} F / m

36. The output voltage
(a) 165 V (b) 170 V
(c) 175 V (d) 185 V

37. Its charge sensitivity
(a) 2.23 pC/N (b) 3.23 pC/N
(c) 4.23 pC/N (d) 2.23 pC/N

38. A piezoelectric crystal having dimensions of 5 mm×5 mm×1.5 mm and a voltage sensitivity of 0.055 V-m / N is used for force measurement. If the voltage developed is 100 V, then the force is
(a) 20 N (b) 30 N
(c) 40 N (d) None of the above

Common Data for Questions 39 and 40

A barium titanate pickup has the dimension of 5 mm×5 mm×1.25 mm. The force acting on it is 5 N.

The charge sensitivity of barium titanate is 150 pC/N and its permittivity is $12.5\times10^{-9}\,F/m$. If the modulus of elasticity of barium titanate is $12\times10^{-9}\,N/m^2$

39. The strain is
(a) 0.0167 (b) 1.67
(c) 0.167 (d) None of the above

40. The charge is
(a) 1000 pC (b) 750 pC
(c) 500 pC (d) 250 pC

41. The capacitance is
(a) 200 pF (b) 650 pF
(c) 450 pF (d) 250 pF

Statement for linked answers Q. 42 to Q. 43 is given below:

In a piezoelectric transducer a flat frequency response within 5% is required

42. The value of minimum frequency in terms of time constant for which it can be used.
(a) 0.3 (b) 3.04
(c) 2.04 (d) 4.04

43. If the time constant of the transducer is 1.5 ms, the value of minimum frequency is
(a) $2.02\times10^3\,rad/s$ (b) $3.02\times10^3\,rad/s$
(c) $4.02\times10^3\,rad/s$ (d) $1.02\times10^3\,rad/s$

44. The phase shift at this frequency
(a) 19.2 (b) 20.2
(c) 18.2 (d) 17.2

Statement for Linked answers Q. 45 to Q. 46.

A piezoelectric transducer has a capacitance of 1000 pF and a charge sensitivity of 40×10^{-3} C/m. The connecting cable has a capacitance of 300 pF, while the oscilloscope used for read out has a readout input resistance of 1 MW with a parallel capacitance of 50 pF.

45. The sensitivity of the transducer is
(a) $40\times10^6\,V/m$ (b) $30\times10^6\,V/m$
(c) $50\times10^6\,V/m$ (c) $60\times10^6\,V/m$

46. The high frequency sensitivity of the entire system is
(a) $39.63\times10^6\,V/m$ (b) $29.63\times10^6\,V/m$
(c) $59.63\times10^6\,V/m$ (d) $49.63\times10^6\,V/m$

47. The lowest frequency that can be measured with 5% amplitude error by the entire system is
(a) 358.7 Hz (b) 458.7 Hz
(c) 258.7 Hz (d) 558.7 Hz

48. The value of an external shunt capacitance that can be connected in order to extend the range of 5 % error down to 10 Hz is
(a) 47030 pF (b) 48.38 pF
(c) 46.38 pF (d) 45.38 pF

49. With external capacitance calculated in the above question connected in the circuit, the system high frequency sensitivity is
(a) 827 *kV/m* (b) 927 *kV/m*
(c) 1027 *kV/m* (d) 727 *kV/m*

50. A piezoelectric crystal has Young's modulus of 130 Gpa. The uniaxial stress to increase its polarization from 550 to 555 cm^{-2} is nearly
(a) 2.798 Gpa (b) 2.175 Gpa
(c) 1.593 Gpa (d) 1.182 Gpa

51. A piezoelectric accelerometer has a transfer function of $61\,mV/g$ and a natural frequency of 45000 Hz. In a vibration test at 110 Hz, a reading of 3.6 V peak is obtained. The vibration peak displacement is
(a) 1.21 mm (b) 3.2 mm
(c) 2.99 mm (d) none of the above

2.3 Resistive-Inductive-Capacitive Transducer

1. **The least suitable transducer for static pressure measurement is**
(a) Semi-conductor strain guage
(b) Variable capacitor for transducer
(c) Metal wire strain guage
(d) Piezoelectric transducer

2. **A semi-conductor strain guage**
(a) Has a much higher guage factor than that of a metal wire guage
(b) Employs piezoelectric property of undoped silicon
(c) Does not require temperature compensation
(d) Exhibits very little guage factor variation as compared to that of metal wire gauges

3. The emf developed by a **thermocouple** depends on
(a) The length of wires and temperature difference between the hot and cold junction
(b) Materials used, diameter of wires used and temperature difference between the hot and cold junctions

(c) Materials used, temperature of hot junction and temperature of cold junction

(d) Materials used, shape and size of materials, resistance of the wires and temperature difference between the hot and cold junctions.

4. A unity ratio **quarter bridge** strain measuring circuit produces an output of 1 mV for a strain of 500 micro strain when the **bridge excitation** is 4 volts. The gauge factor of the element is then

(a) 1 (b) 2

(c) 3 (d) 4

5. Choose the **thermocouple** that can measure a temperature in the range of 1300°C to 1400°C

(a) Iron-constantan

(b) Copper-constantan

(c) Platinum-platinum rhodium

(d) Chromel-alumel

6. The two secondary coils of an LVDT have wrongly been connected as shown in figure below. Then the input-output relationship would be

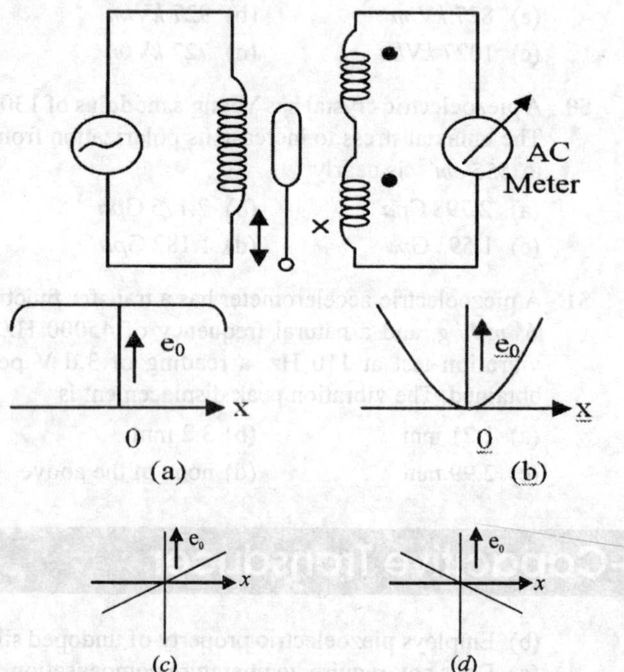

7. A strain gauge has a gauge factor G = –100 . The type of the strain gauge is

(a) Unbounded metal type

(b) Bonded metal foil type

(c) p-type semi conductor

(d) n-type semi conductor.

8. The emf-temperature data for a **thermocouple** with reference junction at 0°C is as follows.

Temperature (°C)	20	180	200
Emf (mV)	1.2	11.8	13.5

The emf developed in mV with the two junctions at 200° C is

(a) 11.8 (b) 12.3

(c) 13.0 (d) 13.5

9. The resistance **temperature** characteristic for a temperature transducer is shown in Figure. The transducer is

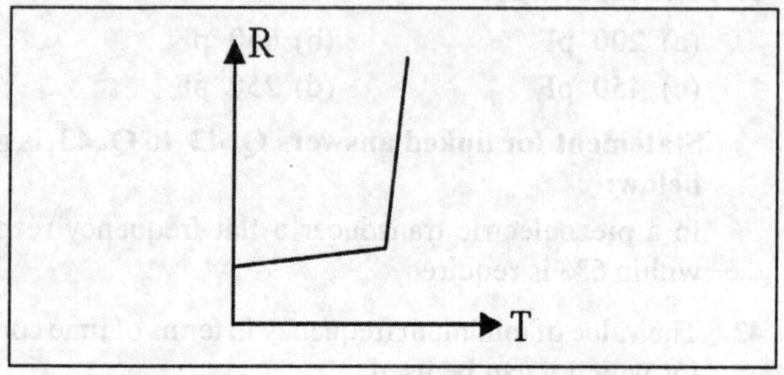

(a) Nicket RTD (b) NTC thermistor

(c) Platinum RTD (d) PTC themistor

10. **Linear variable differential transformer** has

(a) Two primary coils connected in phase and a secondary coil.

(b) Two primary coils connected in opposition and a secondary coil.

(c) One primary coil and two secondary coils connected in phase.

(d) One primary coil and two secondary coils in opposition.

11. A **thermocouple** is suddenly immersed in a medium of high temperature. The approximate time taken by the thermocouple to reach 98% of the steady state value is.

(a) Equal to the time constant of the thermocouple

(b) Equal to twice the value of the time constant of the thermocouple

(c) Equal to four times the value of the time constant of the thermocouple.

(d) Independent of the time constant

12. A **thermocouple** can be considered as a single capacity system with a time constant of 100 ms and unity gain. It is immersed in a heat bath whose temperature is varying at 1°C/s from 0°C/s. When the actual temperature of the head bath is 70°C, the reading of the thermocouple will be

(a) 70.1°C (b) 71°C

(c) 69.9°C (d) 69°C

13. To avoid the effect of fringing, the potential of the guard ring of a capacitance transducer, shown in figure is to be made equal to

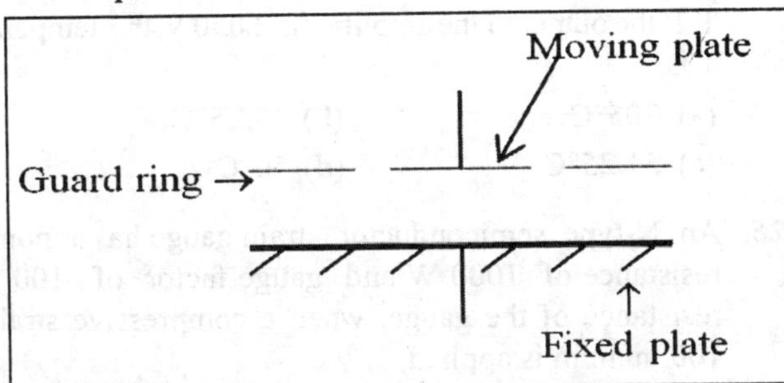

(a) The ground potential
(b) Half of the excitation voltage
(c) The voltage of the fixed plate
(d) The voltage of the moving plate

14. The excitation frequency of an LVDT is 2 kHz. The maximum frequency of displacement should be limited to

(a) 200 Hz (b) 1.5 kHz
(c) 2 kHz (d) 2.5 kHz

15. A thermometer at room temperature 30°C is dipped suddenly into a bath of boiling water at 100°C. It takes 30 seconds to reach 96.5°C. The time required to reach a temperature of 98°C is

(a) 32.5 s (b) 34.6 s
(c) 35.6 s (d) 38.6 s

16. A type J (iron-constantan) thermocouple has a voltage sensitivity of 55 m V/°C. A digital voltmeter (DVM) is used to measure the voltage under the condition shown in the following figure

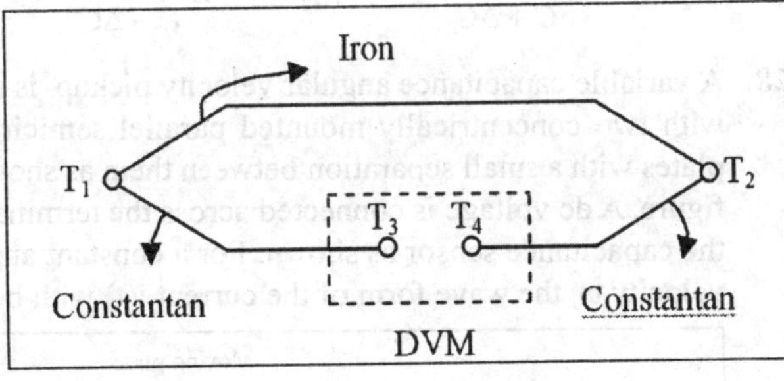

Given that T_1 = 300°C, T_2=100°C and T_3 = T_4=20°C, the meter will indicate a voltage of

(a) 11.0 mV (b) 15.4 mV
(c) 16.5 mV (d) 17.6 mV

17. The figure below shows various configuration of bonding a strain gauge to a cantilever subjected to a bending force F.

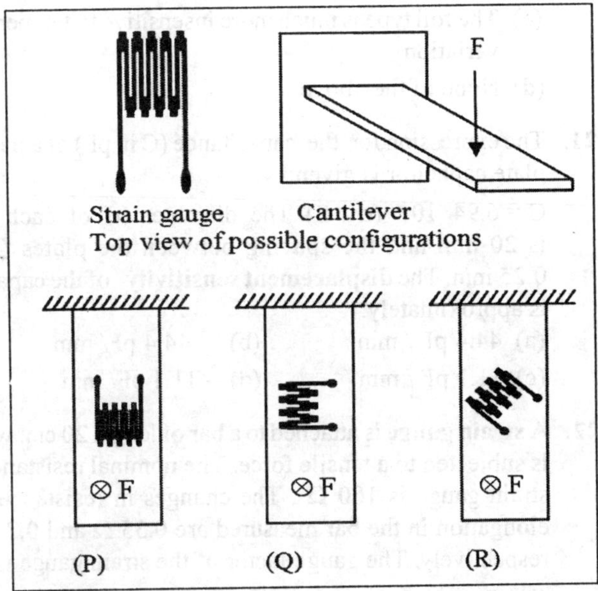

Which configuration gives the maximum change in resistance for the this force?

(a) P
(b) Q
(c) R
(d) All have equal change in resistance

15. The output of the LVDT with input mechanical motion of 10 Hz and excitation frequency 400 Hz will contain frequencies

(a) 10 Hz and 400 Hz (b) 400 Hz only
(c) 10 Hz only (d) 390 Hz and 410 Hz

19. Match the List I type of the thermocouple with List II its characteristics and select the correct answer using the codes:

List I	List II
A. Type E Chromel–Constantan	1. Low cost
B. Type J Fe – Alumel	2. Good stability
C. Type T Cu – constantan	3. High precision
D. Type S Pt – PtRh	4. High output

Codes:

	A	B	C	D
(a)	2	4	3	1
(b)	4	1	3	2
(c)	3	1	4	2
(d)	4	2	1	3

20. Among the bonded metal **strain gauges**, the foil type is more popular than the wire type because,

(a) Error due to transverse strain is much less in foil type
(b) Gauge factor is much higher in foil type

(c) The foil type is much more insensitive to temperature variation

(d) None of the above

21. The expression for the capacitance (C in pF) of a parallel plate capacitor is given by:

$C = 6.94\ 10^{-3}\ (d^2 / S)$. The diameter (d) of each plate is 20 mm and the spacing between the plates (S) is 0.25 mm. The **displacement sensitivity** of the capacitor is approximately:

(a) 44.4 pF / mm (b) – 44.4 pF/ mm

(c) 11.1 pF / mm (d) – 11.1 pF/ mm

22. A **strain gauge** is attached to a bar of length 20 cm. which is subjected to a tensile force. The nominal resistance of strain gauge is 100 Ω . The changes in resistance and elongation in the bar measured are 0.35 Ω and 0.2 mm, respectively. The gauge factor of the strain gauge is.

(a) 2 (b) 3.5

(c) 10 (d) 100

23. An RTD is assumed to be represented in the range 0°C to 100°C by a linear model as $R_T = R_0 (1 + 0.004\ T)$ Where T is the temperature in °C . R_0 = 100 Ω with a variation of ± 2Ω. The true model of the RTD is.

$$R_T = R_0 (1 + 0.004T + 6 \times 10^{-7} T^2).$$

The worst case error magnitude in the resistance value that will be introduced, if the linear model is used over a range of 0 to 100°C, is.

(a) 2 Ω (b) 2.8 Ω

(c) 3.0 Ω (d) 3.4 Ω

Data given below should be used for answering question Q.24 and Q 25

The table provides the thermo-emf sensitivity of five materials with reference to platinum around 273 K.

Material	Sensitivity (m VK^{-1})
Constantan	–35
Nickel	–25
Copper	+6
Iron	–18.5
Nichrome	–25

24. The thermocouple pair that gives the maximum sensitivity around 273 K is.

(a) Platinum – Constantan

(b) Nichrome – Constantan

(c) Nickel – Constantan

(d) Copper - Nickel

25. Two copper – constantan thermocoup les are connected such that the two constantan wires are joined together. The two copper wires are connected to the input of a low noise chopper stabilized differential amplifier having a gain of 1000. One of the thermocouple junctions is immersed in a flask containing ice and water in equal proportion. The other thermocouple is at a temperature T. If the output of the amplifier is 2.050 V, the temperature T is.

(a) 205°C (b) 102.5°C

(c) 51.25°C (d) 50°C

26. An N-type semiconductor strain gauge has a nominal resistance of 1000 W and gauge factor of –100. The resistance of the gauge, when a compressive strain of 100 m m/m is applied, is

(a) 900 W (b) 990 W

(c) 1010 W (d) 1100 W

27. A signal conditioning circuit suitable for a push - pull type capacitive transducer is given in figure. The output V_o is .

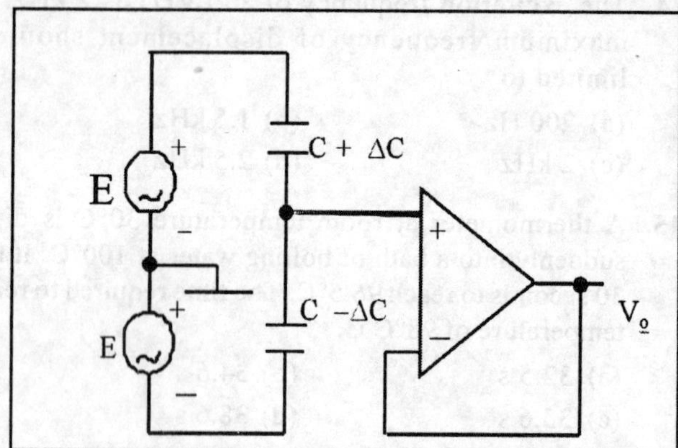

(a) $V_o = E\dfrac{C}{\Delta C}$ (b) $V_o = E\dfrac{\Delta C}{C}$

(c) $V_o = E\dfrac{C-\Delta C}{C+\Delta C}$ (d) $V_o = E\dfrac{C+\Delta C}{C-\Delta C}$

28. A variable capacitance angular velocity pickup is made with two concentrically mounted parallel semicircular plates with a small separation between them as shown in figure. A dc voltage is connected across the terminals of the capacitance sensor as shown. For a constant angular velocity ω, the wave form of the current i(t) with be

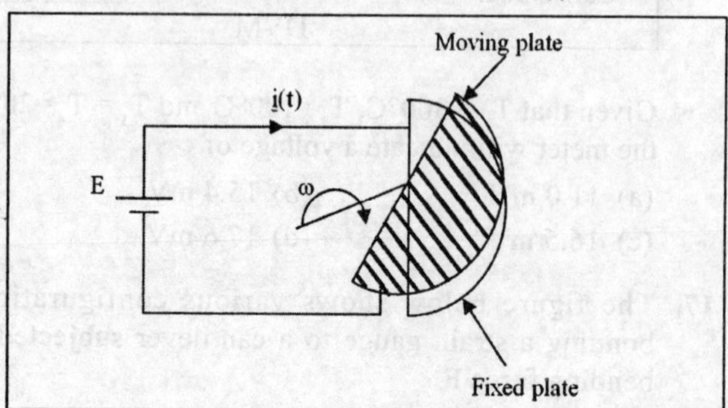

(a) dc (b) Sinusoidal
(c) Triangular (d) Square

29. A strain gauge of resistance 120 Ω and gauge factor 2.0 is at zero strain condition. A 200 kΩ fixed resistance is connected in parallel with it. Then the combination will represent an equivalent strain of.

(a) 290 μm/m (b) zero
(c) – 123.8 μm/m (d) – 300 μm/m

30. An iron-constantan thermocouple is to be used to measure temperature between 0 and 300°C. With reference junction at 0°C, the emf induced at 100°C and 300°C are

$$E_{100} = 5268\ \mu V, \quad E_{300} = 16325\ \mu V$$

Then the nonlinearity at 100 °C as a percentage of full scale is

(a) + 3.23% (b) – 1.06 %
(c) – 0.96 % (d) + 3.23 %

31. A mercury **barometer** reads h mm Hg with the temperature of the mercury at T°C. The barometer reading corrected for the standard temperature 0°C with β denoting the volumetric expansion coefficient of mercury in 0°C^{-1}, is

(a) $\frac{h}{1+\beta T}$ (b) $h(\beta + T)$
(c) $h(1+\beta T)$ (d) $h(\beta - T)$

Common Data for 32 and 33

The temperature e.m.f. characteristics of the iron constantan and copper-constantan junctions at 0°C are given by E_{IC} = 53 T μV and E_{CC} = 43 T μV, over an appropriate temperature range, where T is 0°C. It is known that *AD* 590 is an *IC* temperature sensor which can drive a current of (273 + T) mA at T°C into an appropriate load resistance.

32. An iron-constantan **thermocouple** has its measuring junction in a both whose temperature can vary between 100°C–200°C. The reference junction is maintained at ambient temperature. If the e.m.f developed is to correspond to that with the reference junction at °C, the maximum error, expressed as a percentage of the e.m.f developed, due to ambient temperature variation in the range 0°C – 45°C, is

(a) 8.182 (b) 22.5
(c) 45 (d) 81.82

33. A thermocouple is made using a copper and an iron wire. Its measuring junction is kept at a temperature of 100°C and the reference junction is kept at the ambient temperature of 30°C. The e.m.f. developed would be

(a) 700 μV (b) 1000 μV
(c) 3010 μV (d) 5300 μV

34. A voltage generated using a AD590 sensor is used to compensate for ambient temperature error for the iron-constantan thermocouple in the circuit shown in the figure such that the output voltage V_0 is proportional to the measuring junction temperature T_m. The values of R_L and E for exact compensation, assuming ideal components, are

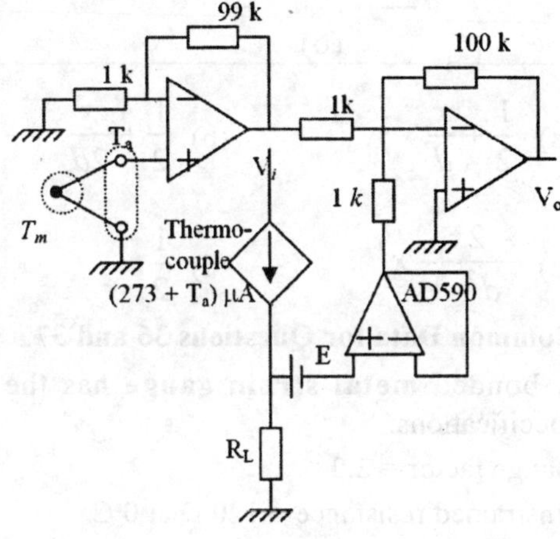

(a) 1 kΩ, 273 mV
(b) 4.3 kΩ, 1.174 V
(c) 5.3 kΩ, 1.44 V
(d) 1 kΩ, 303 mV

35. Three parallel-plate air capacitor of equal plate area 'A' are formed as shown in the figure (*a*), for measuring zero mean vibrational displacement. The capacitors have three separate fixed plates and one common moving plate, which are connected as shown figure (*b*). When the displacement, '*x*', is zero, the distance separating the fixed plates from the common moving plate is the same , equal to d. The circuit is excited by a sinusoidal voltage $V_i \sin \omega t$. Neglecting ***fringe effects*** and the possibility of saturation of the Op-amp output, the output voltage V_0 normalized by V_i for a displacement '*x*' is

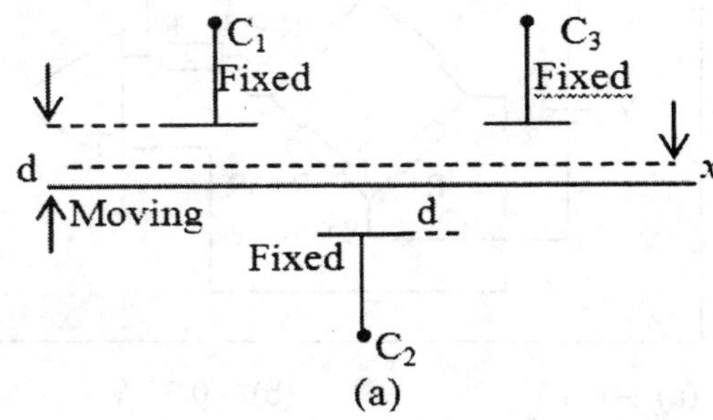

(a)

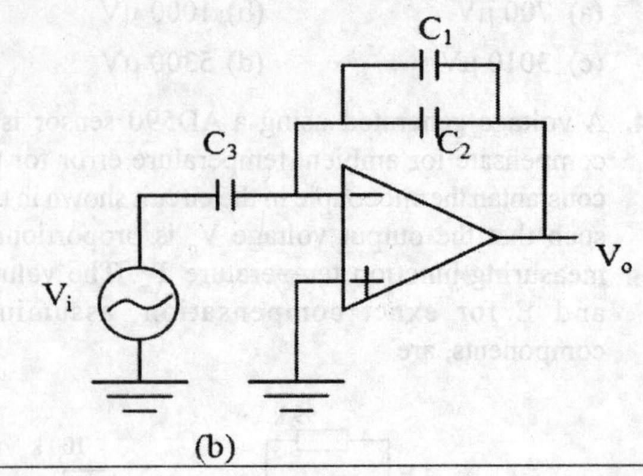

(a) $\frac{1}{2}+\frac{x}{d}$ (b) $\frac{1}{2}+\frac{x}{2d}$

(c) $\frac{-2x}{d^2}-x^2$ (d) $\frac{1}{2}$

Common Data for Questions 36 and 37

A bonded **metal strain gauge** has the following specifications:

Gauge factor = 2.0

Unstrained resistance = 120 Ω at 0°C

Maximum operating temperature = 150°C

Temperature coefficient of resistance = (2×10^{-5})/ $°C^{-1}$

Maximum tensile and compressive strain = 2×10^{-2}

36. The temperature-induced resistance error at the maximum operating temperature expressed as the percentage on resistance change due to maximum tensile strain only, assuming self-heating effects to be negligible, is

(a) 0.0625 (b) 0.625

(c) 6.26 (d) 7.5

37. Two strain gauges, with the specifications given below are mounted on a structural member in a push-pull configuration, connected in a bridge along with two standard resistance, whose output is fed to an amplifier as shown in the figure. The output voltage V_0 per unit battery voltage, under maximum loading, is

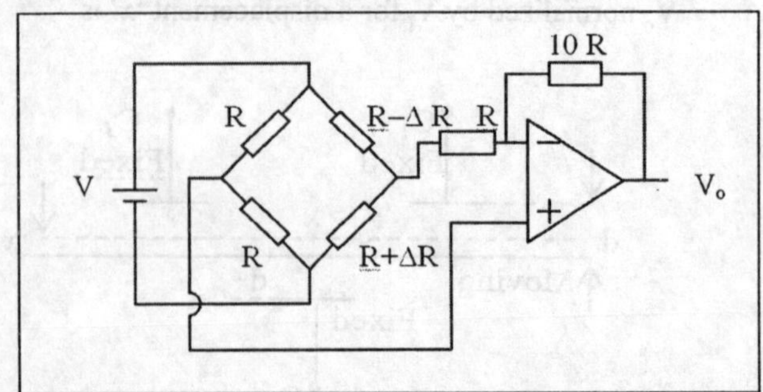

(a) – 0.4 V (b) – 0.37 V

(c) 0.37 V (d) 0.4 V

38. A metal wire has a uniform cross–section A, length ℓ, and resistance R between its two end points. It is uniformly stretched so that its length became $\alpha\ell$. The new resistance is

(a) αR (b) $\alpha^2 R$

(c) $\sqrt{\alpha}\, R$ (d) $e^{\alpha} R$

39. The secondary induced voltages of a LVDT, shown in the following figure, at null position are $v_1 = 1.0\text{ V} \angle 0^0$, $v_2 = 1.0\text{ V} \angle 10^0$. respectively. Then the null voltage of the LVDT is

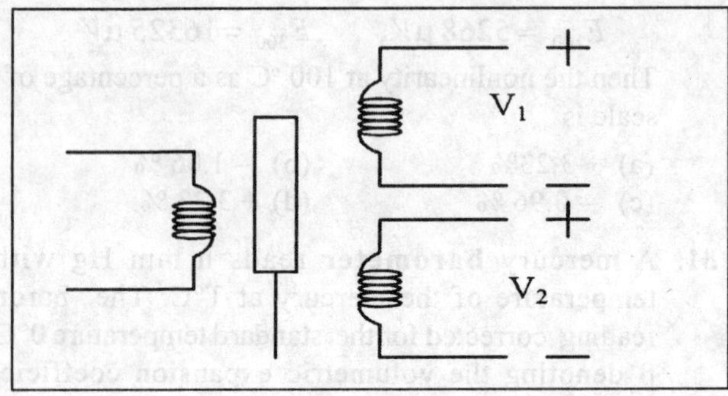

(a) 0 V (b) 0.014 V

(c) 0.174 V (d) 2 V

40. A variable air gap type capacitor consists of two parallel plates: a fixed plate and a moving plate at a distance x. If a potential V is applied across the two plates, then force of attraction between the plates is related to as

(a) $F \propto x^2$ (b) $F \propto 1/x^2$

(c) $F \propto 1/x$ (d) $F \propto x$

Statement for the Linked Answer Questions 41 and 42:

A capacitive type displacement transducer consists of two triangular plates, placed side by side with negligible gap in between them and a rectangular plate moving laterally with an uniform air gap of 1 mm between the fixed plates and the moving plate. The schematic diagram is shown below with appropriate dimension.

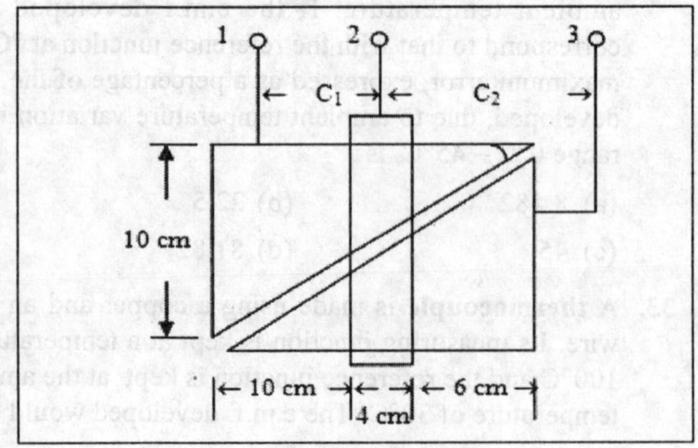

41. With the position of the moving plate shown in the figure above, the values of the capacitance C_1 and C_2 thus formed are

(a) C_1= 2.14 pF, C_2 =1.29 pF
(b) C_1 = 14.17 pF, C_2 = 21.25 pF
(c) C_1 = 339 Pf, C_2 = 212 pF
(d) C_1 =18.16 nF, C_2 =23.2 nF

42. The below sensor is incorporated in a capacitance measuring circuit as shown in the following figure. Assuming ideal op-amp, the output voltage under the conditions mentioned above is

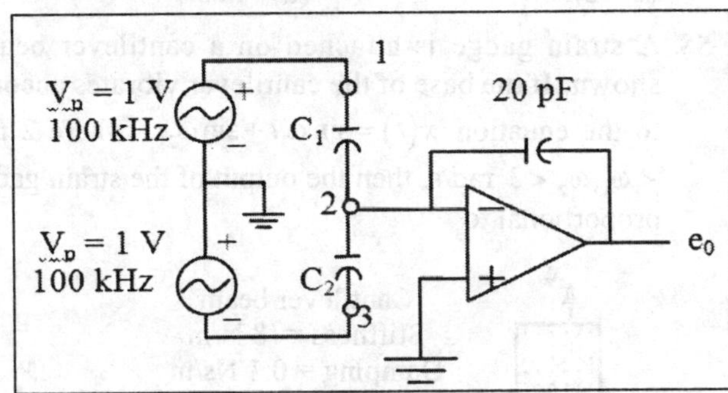

(a) 0.012 V (b) 0.354 V
(c) 1.23 V (d) 2.541 V

43. A thermometer with time constant τ, initially at the ambient temperature, is used to measure the temperature of a liquid in a bath. The excess temperature of the thermometer and the liquid over the ambient are $\theta(t)$ and $\theta_1(t)$, respectively, where t denotes the time. If $\theta_1(t) = kt$, where k is a constant, the steady state error, defined as $\lim_{t\to\infty}[\theta(t) - \theta_I(t)]$, is

(a) ∞ (b) 0
(c) $-$ k (d) $-$ k τ

44. A thermistor has a resistance of 10 $k\Omega$ at 25°C and 1 $k\Omega$ at 100°C. The range of operation is 0°C – 150°C. The excitation voltage is 5 V and a series resistor of 1 $k\Omega$ is connected to the thermistor.

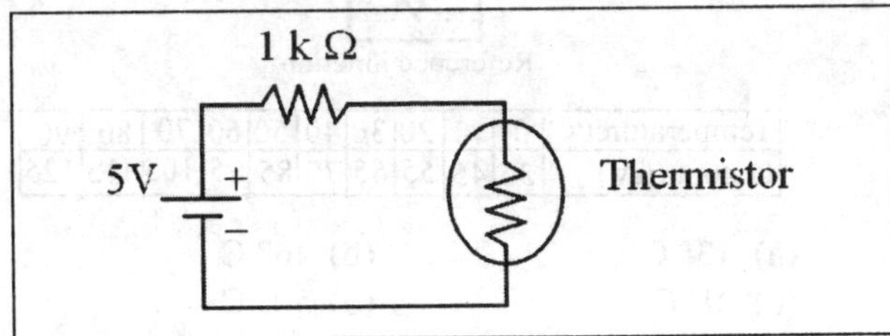

The power dissipated in the thermistor at 150° C is

(a) 4.0 mW (b) 4.7 mW
(c) 5.4 mW (d) 6.1 mW

Statement for Linked Answer Questions 45 and 46:

A push-pull capacitive displacement transducer is interfaced to a differential amplifier and an ADC shown in the figure below:

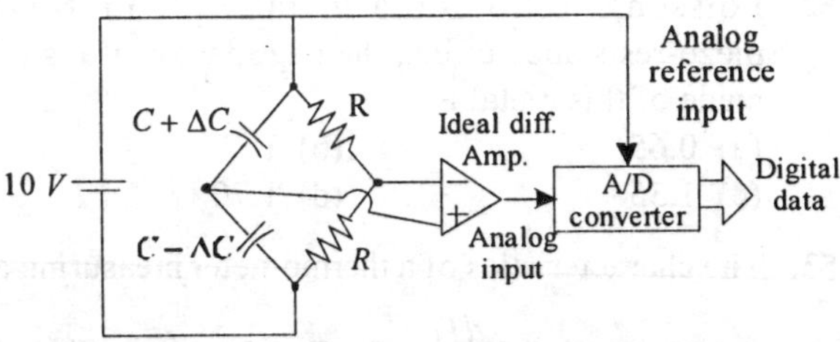

Not: that the bridge supply and the analog reference input for the ADC are derived from the same 10 V DC source.

45. The change in capacitance for full scale displacement is ± 5 % for the capacitors. The gain of the differential amplifier for utilization of the full range of the ADC (which is ± 10 V) is

(a) 10 (b) 20
(c) 30 (d) 40

46. If the supply voltage to the bridge decreases by 5%, the sensitivity of the measurement system

(a) Decrease by 5 % (b) Does not change
(c) Increases by 5 % (d) Increase by 200 %

47. All metal resistive strain gas s have a gauge factor (GF) nearly 2.5 due to the

(a) Young's modulus is the same for all metals and alloys.
(b) Poison's ratio is the same for all metals and alloys.
(c) The conductivity of the material changes with applied strain the elastic region
(d) The conductivity of the material is independent of the applied strain.

48. The temperature being sensed by a negative temperature coefficient (NTC) type thermistor is linearly increasing. Its resistance will.

(a) Linearly increase with temperature
(b) Exponentially increase with temperature
(c) Linearly decrease with temperature
(d) Exponentially decrease with temperature

49. A linear variable differential transformer (LVDT) is

(a) A displacement transducer
(b) An impedance matching transformer
(c) A differential temperature sensor
(d) An auto transformer

50. An LVDT is supplied with a sinusoidal voltage of amplitude 5 V and frequency 1 kHz. The output is connected to an ac voltmeter. The reading of the voltmeter is 1 V for a displacement of 1 mm from the null position. When the displacement is 1 mm in the opposite direction from the null position, the reading of the voltmeter is

(a) -1 V (b) -0.2 V
(c) 1 V (d) 5 V

51. The output voltage of a transducer with an output resistance of 10 $k\Omega$ is connected to an amplifier. The minimum input resistance of the amplifier so that the error in recording the transducer output does not exceed 2% is

(a) 10 $k\Omega$ (b) 49 $k\Omega$
(c) 490 $k\Omega$ (d) 1.2 $M\Omega$

52. Poisson's ratio for a metal is 0.35 Neglecting piezo-resistance effect, the gage factor of a strain gage made of this metal is

(a) 0.65 (b) 1
(c) 1.35 (d) 1.70

53. The characteristics of a thermometer measuring ambient temperature is $2\frac{dT_i}{dt}+T_i-T_a=0$, where T_i and T_a are the indicated and ambient temperatures, respectivley, both in ^{0}C, and time is in seconds. The –3dB cut - off frequency response of the thermometer is

(a) $\frac{1}{4\pi}$ Hz (b) $\frac{1}{2}$ Hz
(c) 1 Hz (d) 2 π Hz

54. The primary and secondary of an LVDT (stroke length $\pm 50mm$) are connected to a 3 kHz sinusoidal source and ideal semiconductor diode bridge based phase sensitive demodulator circuit.

The core of the LVDT remains static at 15 mm above the ideal null position. The frequency of the voltage observed at the input of the low-pass filter is

(a) 1 kHz (b) 1.5 kHz
(c) 3 kHz (d) 6 kHz

55. For a copper-constantan (Type T) thermocouple, the junction potential E (in mV) at q^{0}C is given by $E = 38.74\,q + 3.3\times10^{-2}\,q^2 + 2.07\times10^{-4}\,q^3 - 2.2\times10^{-6}\,q^4$ + higher order terms, assuming the cold junction compensation. The sensitivity of thermocouple at 100^{0}C is approximately.

(a) 45.34 μV/^{0}C
(b) 42.75 μV/^{0}C
(c) 38.74 μV/^{0}C
(d) 0.06 μV/^{0}C

56. The temperature of a furnace is monitored at a distance of 50 m away. The temperture transmitter has a range of 0 – 500^{0}C and provides 4 – 20 mA current output. The measured temperture and the output current have a straight line relationship with positive slope. The temperture is determined from the voltage measured across a resistance of 500Ω in the current loop. If the voltage measured across the resistance is 4V, the temperture of the furnace is

(a) 100°C (b) 125°C
(c) 150°C (d) 200°C

57. A capacitive motion transducer circuit is shown. The gap d between the parallel plates of the capacitor is varied as $d(t)=10^{-3}\left[1+0.1\sin(1000\,\pi t)\right]m$. If the value of the capacitance is 2 pF at t = 0 ms, the output voltage V_o at t = 2 ms is

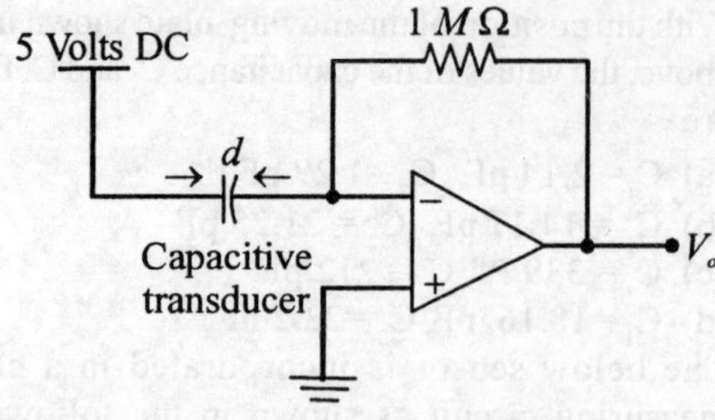

(a) $\frac{\pi}{2}\,mV$ (b) $\pi\,mV$
(c) $2\pi\,mV$ (d) $4\pi\,mV$

58. A strain gauge is attached on a cantilever beam as shown. If the base of the cantilever vibrates according to the equation $x(t)=\sin\omega_1 t+\sin\omega_2 t$, where 2 rad/ s $<\omega_1,\omega_2<$ 3 rad/s, then the output of the strain gauge is proportional to

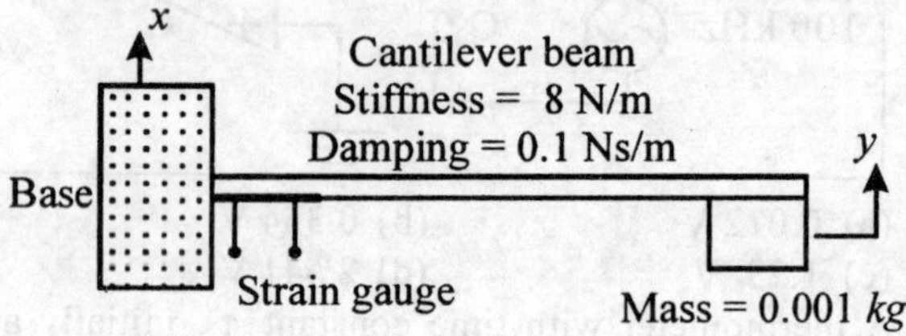

(a) x (b) $\frac{dx}{dt}$
(c) $\frac{d^2x}{dt^2}$ (d) $\frac{d(x-y)}{dt}$

59. A pair of identical thermocouples is employed for measuring the temperature of a specimen as shown below. The characteristic of the thermocouple is tabulated below. The reference junction is at 2°C. The meter reads 48 μV. The correct temperature of the specimen is.

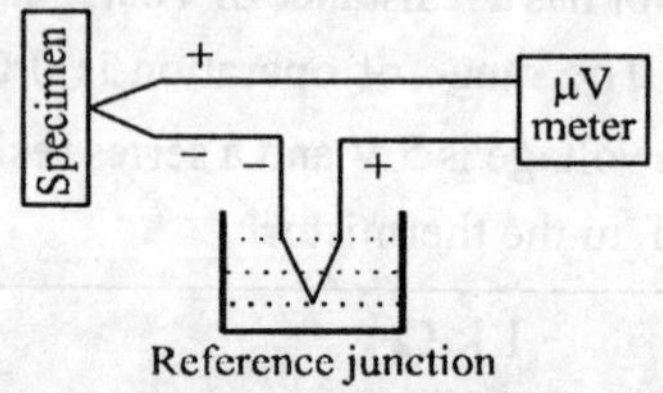

Temperature(°C)	0	10	20	30	40	50	60	70	80	90
Output (μV)	35	45	55	65	75	85	95	105	115	125

(a) 13° C (b) 46° C
(c) 48°C (d) 50° C

60. A strain gauge has a nominal resistance of 600 W and gauge factor of 2.5. The strain gauge is connected in a dc bridge with three other resistances of 600 W each. The bridge is excited by a 4 V battery. If the stain gauge is subjected to strain of 100 $\mu m / m$, the magnitude of the bridge output will be.

(a) 0 V (b) 250 μ V
(c) 500 μ V (d) 750 μ V

61. To reduce the effect of fringing in a capacitive type transducer.

(a) The transducer is shielded and the shield is kept at ground potential.

(b) A guard ring is provided and it is kept at ground potential.

(c) The transducer is shielded and the shiedl is kept at the same potential as the moving plate.

(d) A guard ring is porvided and it is kept at the same potential as the moving plate.

62. The values of the material constant β for thermistors P and Q are 4000 K and 3000 K, respectively. The resistance of each thermistor at 298 K is 2 $k\Omega$. At 373 K the ratio of the resistance of thermistor P to that of thermistor Q will be closest to

(a) 1.33 (b) 1.00
(c) 0.75 (d) 0.50

63. Four strain gauges are fixed on a cylindrical shaft to measure torque, as shown in the figure.

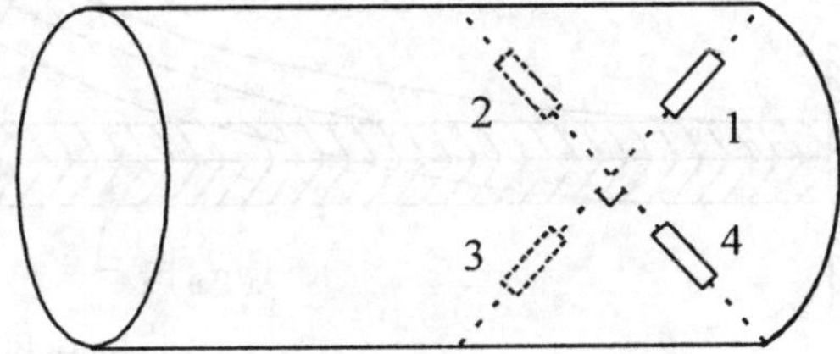

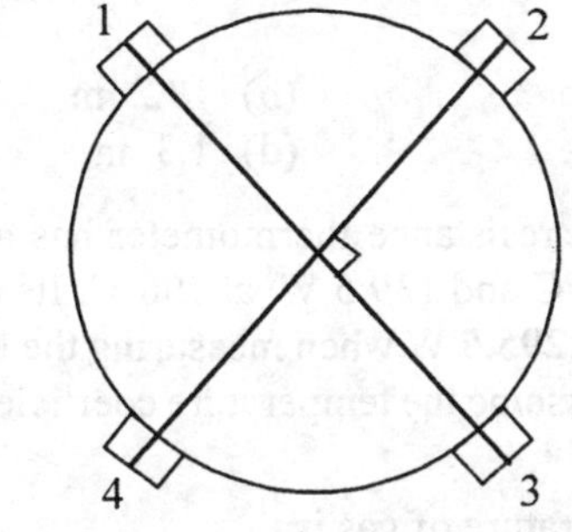

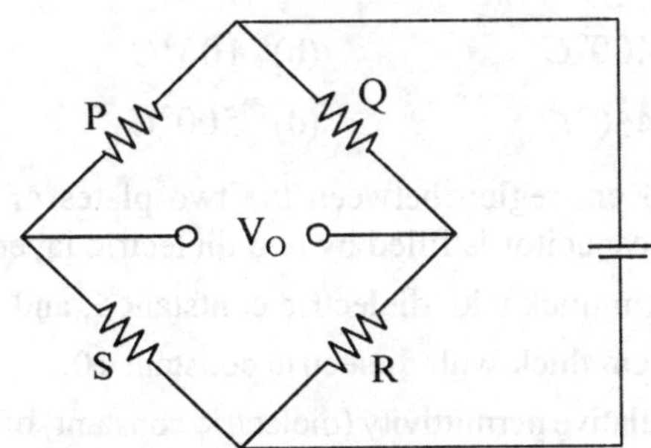

A correct way to place these gauges in the bridge is

(a) P-1, Q-2, R-3, S-4 (b) P-1, Q-3, R-2, S-4
(c) P-3, Q-1, R-2, S-4 (d) P-2, Q-1, R-3, S-4

64. A thermocouple based temperature measurement system is shown in the adjoining figure. Relevant thermocouple emf data (in mV) is given below. The cold junction is kept at 0°C. The temperature is 30°C in the other part of the system. The emf V_o is measured to be 26.74 mV. The temperature of the hot liquid is

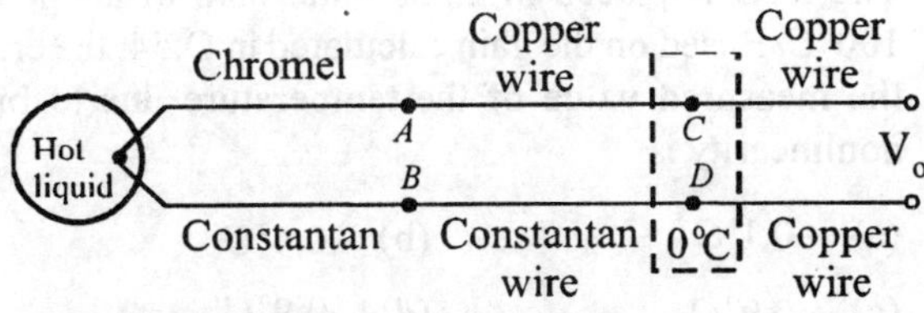

Temperature	emf of Chromel-Constantan	emf of Copper-Constantan
10 °C	0.591	0.391
20 °C	1.192	0.789
30 °C	1.801	1.196
370 °C	26.549	19.027
380 °C	27.345	19.638

(a) 370.0 °C (b) 372.4 °C
(c) 376.6 °C (d) 380.0 °C

65. A differential push-pull type capacitive displacment sensor (nomial capacitance C_0 = 0.01μF) is connected in two adjacent arms of an ac bridge in such a way that the output voltage of the bridge is independent of the freqeuncy of the supply voltage. Supply to the bridge is 1V at 1 kHz, and two equal resistance (R = 3.9 kΩ) are placed in the other arms of the bridge. The bridge sensitivity is

(a) 0.001 mV/pF (b) 0.05 mV/pF
(c) 1 mV/pF (d) 0.5 V/pF

Statement for Linked Answer Questions 66 and 67

Consider a temperature measurement scheme shown in the adjoining figure. It uses an RTD whose resistance at 0°C is 100Ω and temperature coefficient of resistance (α) is 0.00392 /° C

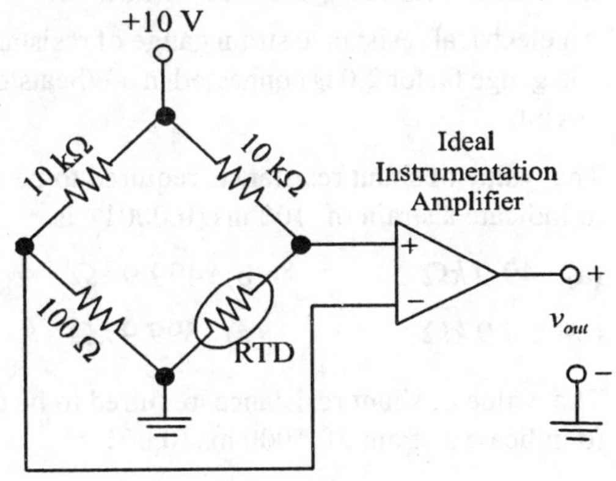

66. The differential gain of the instrumentation amplifier to achieve a voltage sensitivity of 10 mV/°C at 0 °C, should be approximately

(a) 13.41 (b) 26.02
(c) 57.53 (d) 90.14

67. The RTD is placed in a hot water bath of temperature 100°C, Based on the gain calculated in Q.54, the error in the measured value of the temperature due to bridge nonlinearity is

(a) $-0.1°C$ (b) $-0.4°C$
(c) $-0.9°C$ (d) $-0.9°C$

68. An electrical resistance strain gauge has a gauge factor of 2.0 and resistance of 120 ohms. The change in resistance of the gauge caused by a strain of 1 mm/m is

(a) 0.002% (b) 0.0002%
(c) 0.001% (d) 0.0001%

Common Data for Questions 69 and 70:

An electrical resistance strain gauge of resistance 120 Ω and gauge factor 2.0 is bonded to a specimen of steel.
For the gauge,
Temperature coefficient of resistance
$= 20 \times 10^{-6}/°C$
Thermal coefficient of linear expansion
$= 16 \times 10^{-6}/°C$
For the steel,
Modulus of elasticity E = 200 GN/m²
Thermal coefficient of linear expansion = $12 \times 10^{-6}/°C$

69. The resistance change of the gauge due to a tensile stress of 50 MN/m² in the specimen is

(a) 0.72 Ω (b) + 0.06 Ω
(c) 7.2 Ω (d) −0.72 Ω

70. The resistance change of the gauge due to an increase of 30°C in temperature

(a) 0 .72 (b) −0.18
(c) −0.028 (d) −0.0028

Common Data for Questions 71 and 72:

An electrical resistance strain gauge of resistance 100W and gauge factor 2.0 is connected in a Wheatstone bridge circuit.

71. The value of shunt resistance required to be connected to indicate a strain of 100 ms (0.0001) is

(a) 49.9 $k\Omega$ (b) 499.9 $k\Omega$
(c) 59.9 $k\Omega$ (d) 599.9 $k\Omega$

72. The value of shunt resistance required to be connected to indicate a strain of 1000 ms (0.001) is

(a) 499.9 $k\Omega$ (b) 49.9 $k\Omega$
(c) 49.9 $k\Omega$ (d) 69.9 $k\Omega$

73. An electrical resistance strain gauge of gauge factor F = 2.0 and resistance R_g = 120 Ω is placed in an Wheatstone bridge in which all resistances are equal to 120 Ω. The excitation voltage is 4.0 V. The unbalanced current for a strain of 1 mm/m indicated by a galvanometer of resistance 100W is

(a) 9.09 þA (b) 9.09 μA
(c) 90.9 þA (d) 90.9 μA

74. A bimetal strip 55 mm long is made of strips of nickel-chrome alloy and Invar bonded to ether at 30°C. Each material has a thickness of 1.5 mm. Assume following data:

For Invar → $a_1 = 1.5 \times 10^{-6}$°C^{-1}
$E_1 = 1.5 \times 10^{11}$ N/m²
For Nickel-chrome alloy → $a_2 = 12.5 \times 10^{-6}$°C^{-1},
$E_2 = 2.3 \times 10^{11}$ N/m²

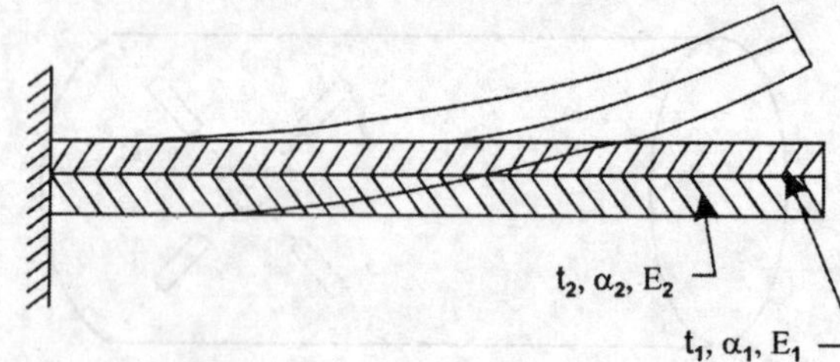

The radius of curvature of the strip subjected to 110° C is.

(a) 1.32 m (b) 13.2 m
(c) 2.3 m (d) 1.3 m

75. A platinum resistance thermometer has a resistance of 100 W at 0°C and 139.5 W at 100°C. Its resistance was found to be 295.5 W when measuring the temperature of a hot gas.Assume the temperature coefficient of platinum = 0.0039°C⁻¹.

The temperature of gas is

(a) $400°C$ (b) $100°C$
(c) $450°C$ (d) $500°C$

76. The 3 cm region between the two plates of a parallel plate capacitor is filled by two dielectric layers:

(*i*) 1 cm thick with dielectric contstant 5, and

(*ii*) 2 cm thick with dielectric constant 10.

The relative permittivity (dielectric constant) of a material which gives the same capacitance if it completely fills the region between the plates is

(a) 6.5 (b) 7.5
(c) 4.5 (d) 5.5

77. A capacitive transducer using two quartz diaphragms of area 800 mm^2 and separated by a distance of 4 mm has a capacitance of 350 μF. When a pressure of 1 MN/m^2 is applied to one of the diaphragms, a deflection of 0.75 mm is produced. The change in the capacitance of the system is

(a) 67.8 *n*F (b) 77.8 mF
(c) 80.7 mF (d) 08.7 *n*F

78. A parallel plate air spaced capacitor has an effective palte area of 6.5 × 10^{-4} m^2 and the distance between the plates is 1mm.

The displacement sensitivity of the device if air has a relative permittivity of unity is

(a) 575.5 *n*F/m (b) 75.5 *n*F/m
(c) 875.5 *n*F/m (d) 675.5 *n*F/m

79. A capacitive transducer consists of five plates arranged as shown in figure. Each plate measures 25 mm ′ 25 mm and the distance between the adjacent plates is 0.25 mm. This arrangement is used to measure the displacement x by observing the resulting change in capacitance.

The sensitivity of the transducer if air is the medium separating the plates is

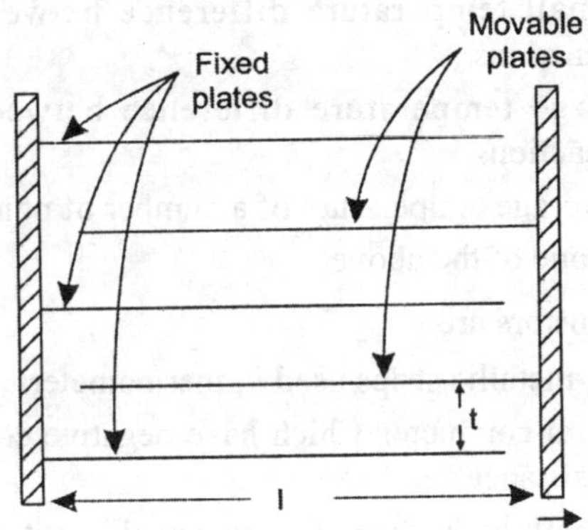

(a) 354.16 mF/*m*
(b) -3.541 mF/*m*
(c) -3541.6 mF/*m*
(d) -35.16 mF/*m*

80. The diaphragms of a capacitive transducer has a capacitance of 300 pF and is connected in an oscillator circuit having a frequency of 100 kHz. When a pressure of 500 kN/m^2 is applied, an average deflection of 0.28 mm is observed in the diaphragm of the transducer.

The change in frequency of the oscillator after the application of pressure to the transducer is

(a) 5.36 *kHz*
(b) 53.6 *kHz*
(c) 35.6 *kHz*
(d) 3.56 *kHz*

81. A 100Ω strain gauge is bonded to a low carbon steel bar which has been subjected to a tensile load. The bar has a preload uniform cross-sectional area of 0.5×10^{-4} m^2 and Young's modulus for low carbon steel is 200 GN/m^2. If a load of 50 kN produces a change of 1Ω in the gauge resistance.

The gauge factor for the strain gauge is

(a) 2 (b) 0.2
(c) 4 (d) 0.4

82. A metallic electric resistance strain gauge of 120Ω resistance with gauge factor of 2 is cemented onto an alumnium test surface. If the yield stess is 200 × 10^6 N/m^2 and Young's sodulus of elasticity is 700 × 10^9 N/m^2. The change in resistance of the gauge that would be causedby loading the material to yield point is

(a) 1.686 W
(b) 16.86 W
(c) 6.86 W
(d) 0.686 W

Linked Answer Questions

Statement for linked Answer Questions 83 and 84

A strain gauge potentiometer circuit uses a strain gauge of 120Ω resistance and a gauge factor 2. When working under maximum sensitivity conditions, the current flow through the system is limited to 20mA.

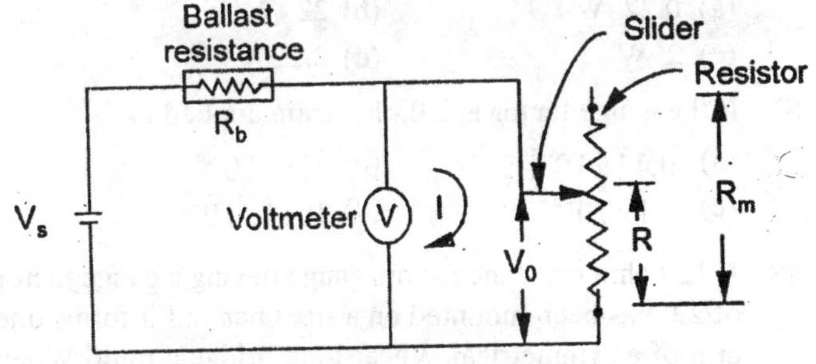

83. The supply voltage is

(a) 48 V (b) 2.4 V
(c) 4.8 V (d) 24 V

84. The change in output voltage when a strain of 500 microstrain impressed upon the gauge is

(a) 0.0012 V (b) 12 V
(c) 0.12 V (d) 1.2 V

Linked Answer Questions

Statement for linked Answer Questions 85, 86 and 87

In a strain-gauge Wheatstone bridge, figure the four limbs were:

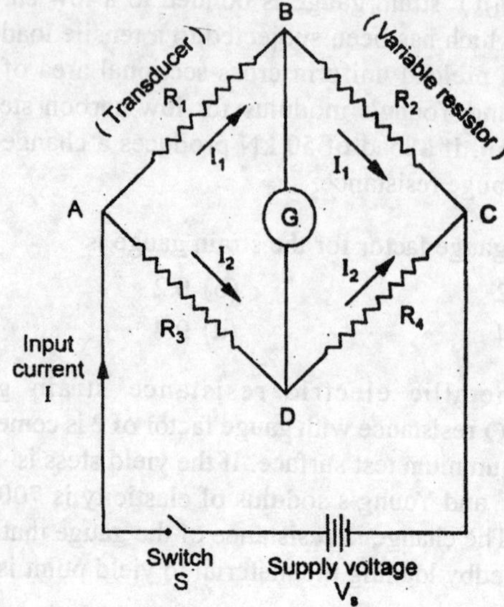

R_1 = =99.75 (strain gauge resistance);
R_2 = variable
$R_3 = 100\Omega$; $R_4 = 10000\Omega$

85. The value of variable resistance R_2 at initial balance with zero strain is

(a) 997 Ω (b) 9975 Ω
(c) 99 Ω (d) 975 Ω

86. When the bridge was rebalanced after the strain was applied, this resistance increased to 9997Ω. The change in resistance of the gauge is

(a) 0.22 W (b) 22 W
(c) 2 W (d) 2.2 W

87. If the gauge factor is 2.0, the strain applied is

(a) 0.11×10^{-3} (b) 1.1×10^{-3}
(c) 1.1×10^{-2} (d) 0.11×10^{-2}

88. A 120 ohm resistance strain gauge having a gauge factor of 2.0 has been mounted on a steel bar and it forms one arm of a symmetrical Wheatstone bridge circuit. When the steel bar is subjected to a tensile load, the galvanometer indicates an output voltage equivalent to 5mV. If the gauge specifications supplied by the manufacturer indicate an operating current of 15 mA the mechanical strain is

(a) 2.78×10^{-2} (b) 27.8×10^{-2}
(c) 27.8×10^{-3} (d) 2.78×10^{-3}

89. The LVDT is an inductive trannducer which function due to:

(a) change in the air gap
(b) change in the amount of core material
(c) mutual inductance
(d) variation in the position of the core

90. Thermocouples are generally used for temperature measurements upto

(a) 500°C
(b) 1000°C
(c) 1500°C
(d) 2000°C

91. The electrical output from a thermocouple circuit is detected by using

(a) wheatstone bridge
(b) current sensitive device
(c) voltage - balancing circuit
(d) voltage sensitive device

92. These units are employed where it is not possible to have physical contact with the system whose temperature is to be measured.

(a) thermocouples
(b) resistance thermometers
(c) radiation pyrometers
(d) thermstors

93. A parallel combination of thermocouples is used for the measurement of

(a) small temperature difference between the two junctions
(b) large temperature difference between the two junctions
(c) average temperature of a number of points
(d) None of the above

94. Thermistors are

(a) bi-metallic strips used in thermometers
(b) semi-conductor which have negative coefficient of resistance
(c) metals having small postive coefficient of resistance
(d) transducers used for measuring the thermal properties of materials

95. A resistance-potentiometer is a

(a) first order instrument
(b) zero order instrument
(c) second order instrument
(d) none of these

96. In a resistance potentiometer, the non-linearity

(a) Increase with increase of load of potentiom eter resistance (R_m / R_p)
(b) Decrease with increase of load to potentiom eter resistance (R_m / R_p)
(c) Is not dependent upon load to potentiometer resistance (R_m / R_p)
(d) None of these

97. In a resistance potentiometer high value of resistance of potentiometer leads to

(a) high value of sensitivity

(b) low value of sensitivity

(c) low value of non-linearity

(d) low value of error

98. Which of the following type of strain gauges must be used for determining the stress when the direction of principle stess is not known

(a) Unbonded metal wire strain gauge

(b) Bonded metal wire strain gauge

(c) Semiconductor strain gauge

(d) Rosettes

99. Platinum is commonly used metal for RTD's . This is because

(a) Platinum has constant value of resistance temperature coefficient of 0.004/°C for a temperature range between 0°C to 100°C

(b) The resisitivity of platinum tends to increases less rapidly at higher temperature

(c) Platinum is available in pure form for commercial applications and has a stability over higher ranges of temperature

(d) All of the above

100. An LVDT:

(a) exhibits linear character upto a displacement of ±5 mm.

(b) has a linearity of 0.05%

(c) has an infant resolution and a high sensitivity which is of the order of 40 V/mm.

(d) all of the above

101. What is the order of minimum displacement that can be sensed with capacitive transducers?

(a) 1cm (b) 1 mm

(c) 1 mm (d) 1 pm

102. In **wire wound strain gauges**, the change in resistance of POT leads to

(a) Change in length of the wire

(b) Change in diameter of wire

(c) Change in both length and diameter of wire

(d) Change in resistivity

103. Unbounded strain gauges are

(a) Exclusively used for transducer applications

(b) Exclusively used for stress analysis

(c) Commonly used for both transducer applications as well as for stress analysis

(d) None of the above

104. Metal foil gauges use fat end turns in order to

(a) Reduce transverse sensitivity

(b) Reduce longitudinal sensitivity

(c) Reduce cross sensitivity

(d) All of the above

105. In semiconductor strain gauges, when a tensile strain is applied

(a) Resistance increases in n-type of materials

(b) Resistance increases in p-type of materiels

(c) Resistance increases in both p and n type materials

(d) Resistance decreases in both p and n type

106. The sensitivity of **thermistors** as compared with sensitivity of platinum resistance temperature detector over a temperature range of -100 °C to 400 °C to changes in temperature is

(a) 100 times

(b) 10^6 times

(c) 10^7 times

(d) 10^3 times

107. A **thermistor** exhibits

(a) Only a negative change of resistance with increase in temperature

(b) Only a positive change of resistance with increase in temperature

(c) Can exhibit either a negative or positive change of resistance with increase of temperature depending upon the type of material used

(d) None of the above

108. Three types of **temperature transducer** are compared with regards to their sensitivity. The order in which they exhibit their sensitivities (highest to lowest) is

(a) Thermistors, RTDs thermocouples

(b) Thermocouples, RTDs thermistors

(c) RTDs thermistors, thermocouples

(d) RTDs thermocouples , thermistors

109. The **semiconductor strain gauge** has a gauge factor of

(a) 2 (b) 10

(c) 1000 (d) 100

110. A **semiconductor strain gauge** has an advantage over the normal strain gauge that

(a) It is more sensitive

(b) It is more linear

(c) It is less temperature dependent

(d) Its cost is low

111. Which of the following is not an advantage of semiconductor strain gauges as compared to conventional strain gauges?

(a) Excellent hysteresis characteristics
(b) Least sensitive to temperature changes
(c) High fatigue life
(d) Smaller size

112. The resistance of a 125 Ω **strain gauge** change by 1 Ω for 4000 micro strain. The gauge factor of the strain gauge is

(a) 1.5 (b) 2
(c) 2.5 (d) 3

113. Acoustic strain gauges

(a) Have good linearity of measurement
(b) Do not have long-term stability
(c) Have a large range of measurement
(d) Have an accuracy of 0.1–1% depending on environment

114. The principle of **strain gauge** is based on

(a) Combustibility of the medium
(b) Thermal conductivity of the medium
(c) Humidity of the medium
(d) Electrical conductivity

115. The **strain gauge** having the highest gauge factor is

(a) Platinum-tungsten (b) Semi-conductor
(c) Constantan (d) Isoelastic

116. A constantan wire of small diameter has a gauge factor of 2. Determine the **Poisson ratio** of the wire.

(a) 1 (b) 2
(c) 1/2 (d) 1/4

117. Two strain gauges are used to measure strain in a cantilever. One gauge is mounted on top of the cantilever and the other is placed at the bottom. The two strain gauges form two arms of a voltage sensitive Wheatstone bridge. This bridge configuration is called

(a) Quarter bridge (b) Half bridge
(c) Full bridge (d) Null bridge

118. Dummy strain gauges are used for

(a) Compensation of temperature changes
(b) Increasing the sensitivity of bridge in which they are included
(c) Compensation for different expansion
(d) Calibration of strain gauges

119. The output of a strain bridge using two active strain gauges used in Poisson's configuration is

(a) $\frac{vG_f\varepsilon e_i}{4}$ (b) $\frac{vG_f\varepsilon e_i}{2}$
(c) $\frac{(1+v)G_f\varepsilon e_i}{2}$ (d) $\frac{(1+v)G_f\varepsilon e_i}{4}$

120. A 100 Ω resistance of potentiometer is used with an input supply voltage of 10 V. If the thermal resistance 30 °C/w and the ambient temperature is 40 °C. Then the temperature of the POT is

(a) 60 °C (b) 80 °C
(c) 70 °C (d) None of the above

121. A helipot is used for measurement of linear displacement of 0.1 m. It uses a 10 turn potentiometer (3000° of movement full scale) with total of 5000 turns of wire. Its resolution is

(a) 0.1 mm (b) 0.2 mm
(c) 0.01 mm (d) 0.02 mm

122. Metal foil type of strain gauges are superior to wire wound metal strain gauges because

(a) They have a higher heat dissipation capacity
(b) They have a large surface area which makes the bonding easier.
(c) They can be fabricated economically using techniques which are similar to those used for fabrication of micro-electronic circuitry
(d) All of the above

123. In **semi-conductor strain gauges**, the change in resistance on application of strain

(a) Is much higher than that in the case of metal strain gauges and this is mainly due to change in dimensions
(b) Is much higher than that in the case of metal strain gauges and this is mainly due to change in resistivity
(c) Is lower than that in the case of metal strain gauges
(d) None of the above

124. The resistance of a thermometer is 5 Ω at 30° C and 6.5 W at 60° C. Using linear approximation, the value of resistance temperature coefficient at 40° C is

(a) 0.0009/ °C (b) 0.0087/ °C
(c) 0.0085/ °C (d) None of the above

125. The resistance value of a thermistor is 5000 Ω at 25° C and its resistance temperature coefficient is 0.04° C. A measurement with a lead resistance of 10 Ω will cause an error of

(a) 0.02 °C (b) 0.05 °C
(c) 0.1 °C (d) None of the above

126. The temperature transducer exhibit non-linear behaviour.

The order in which they exhibit non-linearity (highest to lowest) is

(a) Thermocouple, RTD, thermistors

(b) Thermistors, thermocouple, RTDs

(c) RTDs thermocouples, thermistors

(d) Thermistors, RTDs thermocouples

127. The gauge factor of strain gauge is 2, stress is 1050 kg/cm^2, Y = 2.1 × 106 kg. cm^2 and R is 1000 Ω. The value of DR will be

(a) 2 Ω (b) 3 Ω
(c) 4 Ω (d) 1 Ω

128. A single element strain gauge bridge has two fixed resistance R_1 and R_2 of 120 Ω each and a variable resistance which give full deflection at 120 Ω at zero strain and at 120.6 Ω with a strain e. If the gauge factor of the given strain gauge is 2, then the value of strain e in the beam will be

(a) 0.00025 (b) 0.0035
(c) 0.005 (d) None of the above

129. A wire strain gauge element subjected to a strain has gauge factor of 2, a resistance of 125 Ω and is of 1 m length. For a change in length of the wire by 0.005 m, the change in the inductance will be

(a) 0.25 Ω (b) 0.5 Ω
(c) 1.25 Ω (d) None of the above

130. A 350 Ω Ni-Cr strain gauge with G = 2 is working at 2000 microstomia. The output level for 1/4 bridge excited at 20 V is

(a) 19.96 V (b) 19.96 mV
(c) 16.96 mV (d) None of the above

131. A metal strain gauge has a gauge factor 2. Its normal resistance is 120 Ω. If it undergoes a strain of 10^{-5}, the resistance change in response to the strain is

(a) 240 Ω (b) 2×10^{-5} W
(c) 2.4×10^{-3} W (d) None of the above

132. A unity ratio quarter being strain measuring circuit produces an output of 1 mV for a strain of 500 micro strain when the bridge excitation is 4 V. The gauge factor of the element is

(a) 2 (b) 4
(c) 2.25 (d) None of the above

133. A strain having a resistance 100 Ω and gauge factor of 2 is connected in series with a ballast resistance of 100 Ω across a 12 V supply. The modulus of elasticity is 200 GN/m^2. The difference between the output voltage with no stress applied and a stress of 140 MN/m^2 is

(a) 0.4 mV (b) 0.42 mV
(c) 0.5 mV (d) 0.52 mV

134. A single strain gauge is mounted to measure the axial strain in a simple tensile member. The recorded strain is 380 microstrain. The modulus of elasticity of steel is 200 G N/m^2. If the member is of steel then the axial stress is

(a) 67 MN/m^2 (b) 76 MN/m^2
(c) 86 MN/m^2 (d) None of the above

135. A helipot is provided with 400 turns/mm. The gearing arrangement is such that the motion of the main shaft cause 5 revolution of potentiometer shaft. The resolution of the potentiometer will be

(a) 5 mm (b) 5 mm
(c) 7 mm (d) None of the above

136. A resistance wire strain gauge uses a soft iron wire of small diameter. The gauge factor is +4.2. Neglecting the piezoresistive effects, the Poisson's ratio is

(a) 0.6 (b) 1.6
(c) 2.6 (d) None of the above

137. A resistance wire strain gauge with gauge factor of 2 is bounded to a steel structural member subjected to a stress of 100 MN/m^2. The modulus of elasticity of steel is 200 N/m^2.

The percentage change in the value of gauge resistance due to the applied stress is

(a) 0.1 % (b) 0.2%
(c) 0.3% (d) None of the above

138. Air cored inductive transducers are suitable for use

(a) At lower frequencies

(b) At higher frequencies

(c) At equal frequencies

(d) None of the above

139. The size of air cored transducers as compared with their iron cored transducer parts is

(a) Smaller

(b) Bigger

(c) Same

(d) All of the above

140. Inductive transducers are used in differential configuration because the output is influenced by

(a) External magnetic fields

(b) Temperature changes

(c) Variations of supply voltage and frequency

(d) All of the above

141. In an LVDT, the two secondary windings are connected in differential to obtain

(a) Higher output voltage
(b) An output voltage which is phase sensitive
(c) In order to establish the null or the reference point for the displacement of the core
(d) Both (b) and (c)

142. An LVDT has

(a) Linear characteristics upto a displacement of $\pm 5\ mm$
(b) Linearity of 0.05%
(c) Infinite resolution and a high sensitivity which is of the order of 40 V/mm
(d) All of the above

143. An iron cored transducer is excited with a frequency of 20 kHz. The maximum value of frequency of displacement to be measured with this transducer should not exceed

(a) 0.5 kHz (b) 1 kHz
(c) 1.5 kHz (d) 2 kHz

144. Two inductive transducers working on the principle of change of self inductance L are connected in a push-pull arrangement. If the change of inductance of transducers is DL, the change of inductance exhibited at the output terminals is

(a) DL (b) 2 DL
(c) $\pm$ 2 DL (d) None of the above

145. In an LVDT, the core is made up of a

(a) Non-magnetic material
(b) A solid ferro-electric material
(c) High permeability, nickel-iron hydrogen annealed material in order to produce low harmonics, low null voltage and high sensitivity. The core is slotted to reduce eddy current losses
(d) All of the abvoe

146. A LVDT has linearity of 0.05%. It is linear upto a displacement of $\pm$ 5 mm. it has

(a) An infinite resolution and high sensitivity of the order of 40 V/mm
(b) good frequency response
(c) no effect due to stray magnetic fields
(d) A zero voltage at null point even if the supply voltage has harmonics and unsymmetrical configuration of the secondary windings

147. The output voltage of a LVDT is 1.5 V at maximum displacement. At a load of 0.5 mW, the deviation from linearity is maximum and it is $\pm$ 0.003 V from straight line through origin. The linearity at the given load is

(a) $\pm$ 0.3% (b) $\pm$ 0.2%
(c) $\pm$ 0.4% (d) None of the above

148. Capacitive transducers are normally used for

(a) Static measurements
(b) Dynamic measurements
(c) Both (a) and (b)
(d) Transient measurements

149. The dynamic characteristic of capacitive transducers are similar to those of

(a) Low pass filter (b) High pass filter
(c) Notch filter (d) None of the above

150. A capacitive transducer working on the principle of change of capacitance with change of displacement exhibits a non-linear characteristic. The response of these transducers can be made linear by using

(a) Differential arrangement
(b) Use of an operational amplifier
(c) Working them over a small displacement range
(d) All of the above

151. What is the order of minimum displacement that can be measured with capacitive transducer?

(a) 1 cm (b) 1mm
(c) 1 mm (d) None of the above

152. Capacitive transducers can be used for measurement of liquid level. The principle of operation used in the case

(a) Change of capacitance with change of distance between plates
(b) Change in area of plates
(c) Change in dielectric medium
(d) None of the above

153. Capacitive transducer is superior to inductive type for the measurement of displacement because of

(a) Absence of non-linearity
(b) High frequency response
(c) Small size
(d) None of the above

154. A capacitor microphone transducer is used for

(a) Thickness measurements
(b) Displacement measurements
(c) Speech, music and noise measurements
(d) None of the above

155. A capacitive potential divider is generally used for extending the range of

(a) Electrodynamometer instrument

(b) Electrostatic instrument
(c) Moving iron instrument
(d) None of the above

156. A capacitor losses energy due to dielectric distortion. This type of loss is known as
(a) Dielectric absorption
(b) Leakage loss
(c) Dielectric distortion loss
(d) None of the above

157. The capacitive transducers have a high output impedance and therefore
(a) To reduce loading effects long lengths of cable should be used
(b) To increase the value of resonant frequency long lengths of cables should be used
(c) Long cable lengths should be used in conjunction with capacitive transducers in order to improve their frequency response
(d) None of the above

158. A diaphragm has a natural frequency of 30 kHz. If both diameter and thickness are halved, then the natural frequency will be
(a) 15 kHz (b) 60 kHz
(c) 120 kHz (d) 240 kHz

159. A capacitive transducer uses two quartz diaphragms of area 750 mm^2 separated by a distance of 3.5 mm. A pressure of 900 kN/m^2 when applied to the top diaphragm produces deflection of 0.6 mm. The capacitance is 370 pF when no pressure is applied to diaphragms. The value of capacitance after the application of pressure of 900 kN/^{2} is
(a) 546.5 pF (b) 446.5 pF
(c) 346.5 pF (d) None of the above

Statement for linked answers Q.160 and Q.161 is given below

A parallel plate capacitive transducer uses plates of area 500 mm^2 which are separated by a distance 0.2 m.

160. The value of capacitance when the dielectric is air having a permittivity of 8.8 ´ 10^{-12} F/m is
(a) 32.125 pF (b) 22.125 pF
(c) 42.125 pF (d) None of the above

161. The change in capacitance if a linear displacement reduces the distance between the plates to 0.18 mm is
(a) 3.458 pF (b) 2.458 pF
(c) 4.458 pF (d) None of the above

2.4 Optical Flow Pressure pH Transducers

1. The operation of a **Pirani guage** is based on
(a) Ionization of gas at low pressures
(b) Variation of volume with pressure
(c) Variation of viscosity with pressure
(d) Variation of thermal conductivity of gas with pressure

2. In a **seismic pick-up for getting** an output proportional to acceleration it is desirable to have
(a) A natural frequency very small in comparison to frequency of input and a damping ratio around 0.7
(b) A natural frequency very large in comparison to frequency of input and a damping ratio around 0.7
(c) A natural frequency equal to the frequency of input signal and a damping ratio around 0.7
(d) A natural frequency very small in comparison to frequency of input and a damping ratio higher that unity.

3. Assuming uniform **pressure** over the supported area, the friction moment in an axial bearing of a shaft of diameter d for an axial load of F (With co-efficient of friction μ) is given by
(a) [4 / 3] Fμd (b) [1 / 3] Fμd
(c) [2 /3] Fμd (d) Fμd

4. Measurement of **viscosity** involves measuring
(a) Frictional force (b) Crioll's force
(c) Centrifugal force (d) Buoyant force

5. An example of variable area device for **measuring flow** is
(a) Flow nozzle (b) Corioli's meter
(c) Venturimeter (d) Rotameter

6. The primary standard for calibrating vacuum is
(a) McLeod gauge
(b) Deadweight tester
(c) Thermocouple gauge
(d) Kundsen guage

7. A **pressure gauge** used to measure vacuum indicates a guage pressure of 5 kPa, If the atmospheric pressure is 100 kPa, the absolute pressure is
(a) 105 kPa (b) 0.05 kPa
(c) 95 kPa (d) 20 kPa

8. A **cantilever** is used as a primary sensing element for the measurement of force. The measured deflection is 1 mm. Suppose each of the dimensions, length width and thickness of the cantilever are doubled for the same material and force, the deflection will be

(a) 0.125 mm (b) 0.25 mm
(c) 0.5 mm (d) 2 mm

9. **Light** from a mains powered fluorescent lamp is falling on a gear rotating clockwise. The gear wheel has one tooth painted white. The white tooth appears to be rotating in the anticlockwise direction with a speed of 60 r.p.m. if the mains frequency is 50 Hz, the true speed of the gear wheel is:
(a) 2940 r.p.m. (b) 3060 r.p.m.
(d) 5940 r.p.m. (d) 5960 r. p.m.

10. A **pitol static tube** is used to measure the airflow rate in a square tube. If p_d is the differential pressure, the flow rate is proportional to
(a) $\sqrt{(P_d)}$ (b) P_d
(c) $[1/\sqrt{(p_d)}]$ (d) P_d^2

11. The **sound pressure level** (SPL) measured in open space (free field), at a distance of 6 m from a noise source is 80 dB. At a distance of 60 m the SPL is.
(a) 80 dB (b) 60 dB
(c) 8 db (d) 1.34 dB

12. Value of **pH** of a solution is 4. It indicates that concentration of hydrogen ions is
(a) 10^{-4} g/ litre and the solution is acidic
(b) 10^{-4} g/ litre and the solution is alkaline
(c) 10^{-4} mg/ litre and the solution is acidic
(d) 10^{-4} mg/ litre and the solution is alkaline.

13. For **flow measurement**, a rotameter can be installed in a pipe line.
(a) Horizontally with flow inlet in a specific direction
(b) Horizontally with flow inlet in any direction
(c) Vertically with flow inlet at the bottom and the outlet at the top.
(d) Vertically with flow inlet at the top and outlet at the bottom.

14. An example of a **positive displacement flow meter** is.
(a) Orifice meter
(b) Rotary vane type meter
(c) Turbine type meter
(d) Ultrasonic flow meter.

15. A Pirani gauge sensor is used to measure pressure of the order of .
(a) 10 MPa (b) 1 MPa
(c) 100 Pa (d) 1 Pa

16. A shaft encoder attached to adc motor has sensitivity of 500 pulses per revolution. A frequency meter connected to the output of the encoder indicates the frequency to be 5500 Hz. The speed of the motor in RPM is.
(a) 110 (b) 220
(c) 550 (d) 660

17. Which of the following **flow meters** has the lowest pressure drop for a given range of flow?
(a) Orifice meter (b) Venturimeter
(c) Flow Nozzle (d) Rotameter

18. The diameter of an aluminium sphere is found to be 50 mm at an atmospheric pressure of 10^5 Pa. The bulk modulus of aluminium is 68.9 GPa. The change in the diameter of the sphere in microns, when placed in vacuum is
(a) 0.018 (b) 0.024
(c) 0.032 (d) 0.038

19. In a **fly ball-type *angular speed magnitude measuring device***, the centrifugal force, given $10^{-6}\,\omega^2$, where ω is the angular speed in rad/s, is transmitted to a spring with the non-linear characteristic $F = 16\pi^2 x^2$, where F is the applied force and x is the spring compression in meters. The sensitivity of the device in mm.s is
(a) $1/\pi$ (b) $1/2\pi$
(c) $1/3\pi$ (d) $1/4\pi$

20. In a **variable-area flowmeter** (rotameter), the inlet tube diameter is the same as that of the effective float diameter and the tube taper is small. When the volume flow rate of a fluid is Q_1, the float stands at a height of 70 mm from the inlet and when the volume flow rate of the fluid is Q_2, the float moves to a height of 20 mm form the inlet. The ratio of the flow rates Q_1/Q_2 is
(a) 3.5 (b) 0.29
(c) 0.082 (d) 12.3

21. "Vena Contracta" is the cross section where the flow area is a minimum for a ***restriction type flowmeter.*** For an *orifice meter*, if d is the diameter of the orifice opening, then area of the vena contracta is approximately
(a) $\frac{\pi d^2}{4}$ (b) $\frac{0.99\pi d^2}{4}$
(c) $\frac{0.8\pi d^2}{4}$ (d) $\frac{0.6\pi d^2}{4}$

22. A Pirani gauge measures vacuum pressure and works on the principle of
(a) Change in ionizing potential
(b) Change in thermal conductivity
(c) Deformation of elastic body
(d) Change in self-inductance

23. A weight of 4.9 N rests on a rough horizontal floor that has a co-efficient of friction of 0.5. If g = 9.8 m/s², the smallest horizontal force that gives the weight an acceleration of 4m/s² is approximately
(a) 2.0 N (b) 2.5 N
(c) 4.5 N (d) 19.6 N

24. With the **stroboscope**, synchronism has been obtained at the flashing rates of 2.5 and 20 pulses per second. The speed of rotation is
    (a) 0.6 rev/min. (b) 50 rev/min.
    (c) 100 rev/min. (d) 6000 rev/min.

25. An **elastic transducer** is used to measure pressure in a vessel and it indicates a pressure of 3.2 bar. Atmospheric pressure is 1.01 bar. The absolute pressure in the vessel in bar is
    (a) 1.01 (b) 2.19
    (c) 3.20 (d) 4.21

26. A photo conductive **transducer** works on the principle that when a light beam strikes
    (a) The material, its resistance decreases, which is sensed by an external circuit
    (b) The barrier between transparent metal layer and a semi-conductor material, a voltage is generated
    (c) The barrier between transparent metal layer and a semi-conductor material, a current is generated in the external circuit
    (d) The cathode, it releases electrons, which are attracted towards anode, thereby producing electric current in the eternal circuit.

27. Which of the following gauges can measure the lowest vacuum pressure?
    (a) McLead gauge (b) Pirani gauge
    (c) Ionisation gauge (d) None of the above

28. Which of the following restrictors has the **highest discharge co-efficient**?
    (a) Orifice plate (b) Flow nozzle
    (c) Venturi bube (d) None of these

29. In a rotameter, used for measuring flow rate of a fluid P_a: Pressure above the float, P_b: pressure below the float, A: area of float, V : volume of the float, d_1: density of float material, d_2: density of the fluid g: acceleration due to gravity. The following equation describes the equilibrium of the float.
    (a) $(P_b - P_a) A = V_g (d_2 - d_1)$
    (b) $(P_a + P_b) A = V_g (d_2 + d_1)$
    (c) $(P_a - P_b) A = V_g (d_2 - d_1)$
    (d) $(P_b - P_a) A = V_g (d_2 + d_1)$

30. For a steady **flow of liquid,** the float of the **rotameter** will remain in a particular position when.
    (a) Drag force is balanced by the weight of the float and buoyancy force.
    (b) Weight of the float is balanced by the drag force and buoyancy force.
    (c) Buoyancy force is balanced by the drag force and weight of the float.
    (d) Drag force and buoyancy together is slightly greater than the weight of the float.

31. A photo detector circuit is given in Figure. The photo – diode has an active area of 10^{-2} cm² and sensitivity of 0.55 A/W. When light of intensity 10 m W/cm² falls on the photo diode, the output V_o is

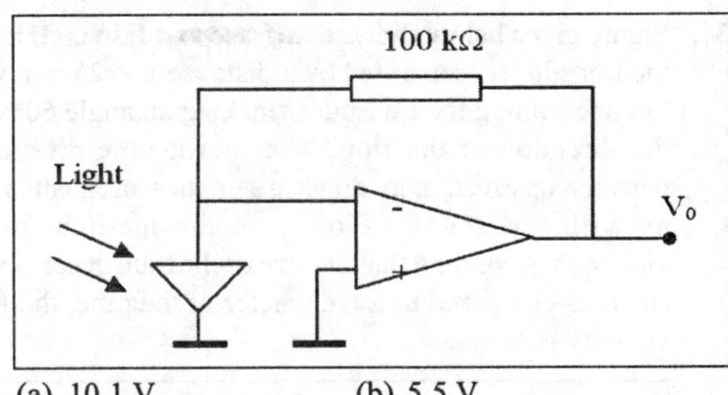

    (a) 10.1 V (b) 5.5 V
    (c) –5.5 V (d) –10.1 V

32. A cantilever-type micro accelerometer is designed by scaling down each dimension of a macro-accelerometer by a factor of 100. If the natural frequency of the macro-accelerometer is w, then the natural frequency of the micro-accelerometer will be
    (a) 100 ω (b) 10 ω
    (c) 0.1 ω (d) 0.01 ω

33. A pressure sensor has the following specifications Sensitivity at the design temperature = 10 V/MPa, Zero drift = 0.01 V/°C, Sensitivity drift = 0.01 (V/MPa)/°C
    When the sensor is used in an ambient 20 °C above the design temperature, the output from the device is 7.4 V. The true value of the pressure will be
    (a) 0.71 MPa (b) 0.68 MPa
    (c) 0.65 MPa (d) 0.61 MPa

34. A quartz clock employs a ceramic crystal with nominal resonance frequency of 32.768 kHz. The clock losses 30.32 s every 23 days. The actual resonance frequency of the crystal is
    (a) 32.7685 kHz (b) 32.768 kHz
    (c) 32.7675 kHz (d) 32.7670 kHz

35. A single-sensor, constant-type ultrasonic **flaw detector** uses a frequency of 330 kHz. When testing a specimen, an echo from a flaw is recorded 0.05 mass after the transmitted pulse. If the velocity of sound in the test object is 6.0 km/s then the flaw is at a depth of
    (a) 120 cm (b) 60 cm
    (c) 30 cm (d) 15 cm

36. An **electromagnetic flowmeter** is used to measure the average flow rate of a liquid in a pipe of diameter 50

mm. The flux density in the liquid has A peak values of 0.1 T. The output from the flow meter is taken to ana amplifier of gain 1000 and input impedance of 1 MW. If the impedance of the liquid between the electrodes is 200 kW, then the peak value of the amplifier output voltage for an average flow velocity of 20 mm/s is

(a) 1.365 V (b) 1 V

(c) 0.1 V (d) 0.083 V

37. Figure given below shows as **ultrasound flowmeter** with the transducers separated by a distance L = 25 mm with the line joining the transducer making an angle 60° with the direction of the flow. The transit time difference between upstream and downstream measurements is 10 ns with the sound velocity in the medium being 1000 m/s. Assuming that the size of the transducer is very small as compared to the diameter of the pipe, the flow velocity is

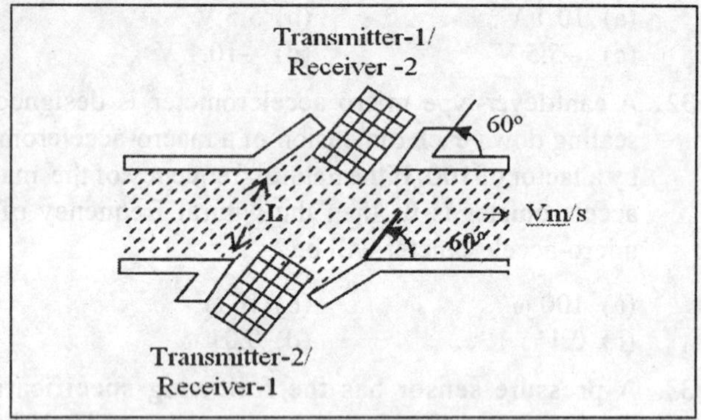

(a) 0.2 m/s (b) 2 m/s

(c) 0.4 m/s (d) 40 m/s

38. A well of cross-sectional area is connected to an inclined tube of cross-sectional area to form a differential pressure gauge as shown in the figure below. When the common liquid level is denoted by A. When , the liquid level in the well is depressed to B, and the level in the tube rises by along its length such that the difference in the tube and well level is.

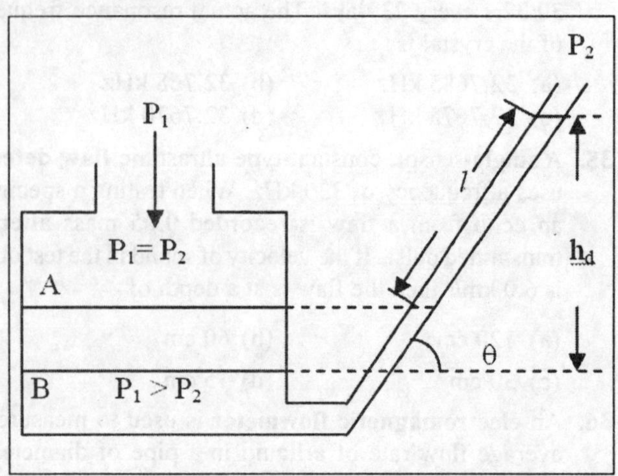

The angle of inclination θ of the tube with the horizontal is

(a) $\sin^{-1}\left[\frac{1}{h_d} - \frac{a_w}{a_t}\right]$ (b) $\sin^{-1}\left[\frac{h_d}{\ell} + \frac{a_t}{a_w}\right]$

(c) $\sin^{-1}\left[\frac{h_d}{\ell} - \frac{a_1}{a_W}\right]$ (d) $\sin^{-1}\left[\frac{h_d}{\ell} - \frac{a_w}{a_t}\right]$

39. The accompanying figure shows a VERTICAL venturimeter with upward water flow. When the measured static pressure difference, , between the inlet and throat is 30 kPa, the flow rate is found to be 50 liters per second. Assume that the coefficient of discharge remains the same

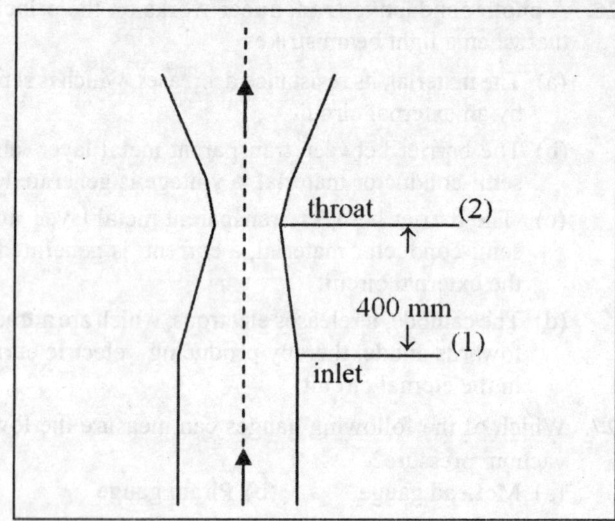

When kPa, the flow rate, in liters per second, is

(a) 33.3 (b) 39.3

(c) 40.8 (d) 54.2

40. In a laminar flow experiment. Fluid A is pumped through a straight tube and the volumetric flow rate and pressure drop per unit length are recorded. In a second straight tube having twice the internal diameter of the first one. Fluid B records the same pressure drop per unit length at the same volumetric flow rate. Assuming fully developed flow conditions in the tubes, the ratio of the dynamic viscosity of Fluid B to that of Fluid A is

(a) 16 (b) 32

(c) 64 (d) 128

41. A measurement system for the rotational speed of a motor is shown in the figure below. The system consists of an opaque disk attached to the motor shaft with a hole as shown. A light source and photo detector are placed on two sides of the disk so that whenever the hole crosses the light path, the photo detector receiver light through the hole. The photo detector circuit is also shown below. Assume sufficient light intensity and TTL logic levels for the inverter.

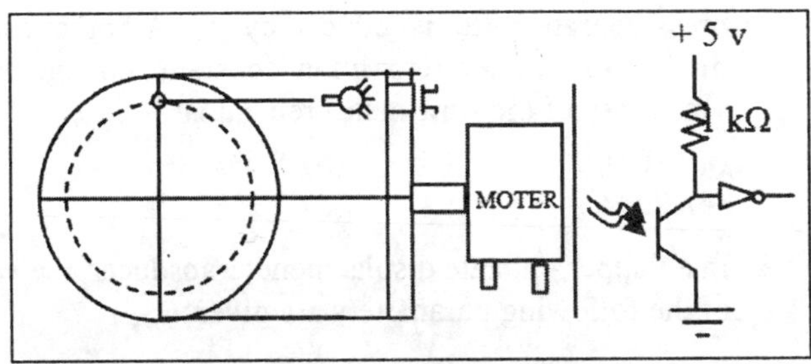

The output of the photodetector is a

(a) Triangular wave
(b) Square wave with 50 % duty cycle
(c) Rectangular wave with duty cycle close to unity
(d) Rectangular wave with duty cycle close to zero

42. A pH ***electrode***, being used at 25°C, has a source resistance of $10^{10}\Omega$. The electrode obeys the Nernst equation perfectly. The electronic voltmeter, with which the potential is being measured, has an input impedance of $10^{11}\ \Omega$ and a gain of 100. If the pH of the analyte changes from 6.5 to 7.8, the change in voltage observed on the voltmeter is

(a) Less than 6.8 V
(b) Between 6.8 V and 7.19 V
(c) Between 7. and 7.49 V
(d) Greater than 7.49 V

43. Assuming complete dissociation, the pH of a 1 ml solution of H_2SO_4 is closest to

(a) 0 (b) 2.7
(c) 2.4 (d) 2.1

44. The conventional way of expressing vibration is in terms of:

(a) Richter scale
(b) Acceleration due to gravity
(c) Speed of sound
(d) Atmospheric pressure

45. A psychrometric chart is used to determine

(a) pH
(b) Sound velocity in gleasses
(c) CO_2 concentration
(d) Relative humidity

46. The responsivity of the PIN photodiode shown is 0.9 A/W. To obtain V_{out} of $-1V$ for an incident optical power of 1 mW, the value of R to be used is

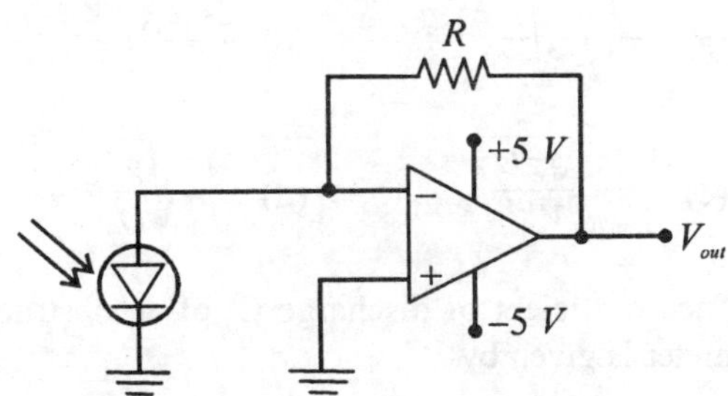

(a) 0.9 Ω (b) 1.1 Ω
(c) 0.9 kΩ (d) 1.1 kΩ

47. The speed of a gear having 60 teech is measured using a proximity sensor. The output of the proximity sensor is fed to a counter with a gating time of is. The counter indicates a value of 120. The speed at which the gear is rotating is.

(a) 60 rpm (b) 120 rpm
(c) 600 rpm (d) 1200 rpm

48. The torque in a rotating shaft is measured using strain gauges. The strain gauges must be positioned on the shaft such that the axes of the strain gauges are at.

(a) 0° with respect to the axis of the shaft
(b) 30° with respect to the axis of the shaft
(c) 45° with respect to the axis of the shaft
(d) 90° with respect to the axis of the shaft

49. A stroboscopic system is used for measuring the speed of a rotating shaft. The shaft has one target mark on it.The maximum strobe rate at which synchronism is achieved is r_1 flashes per minute.The next lower flash rate at which synchronism is achieved is r_2 flashes per minute. The speed of the shaft in rpm is

(a) $\frac{r_1 r_2}{r_1 - r_2}$ (b) $\frac{r_1 r_2}{r_1 + r_2}$

(c) $\frac{r_2^2}{r_1 + r_2}$ (d) $\frac{r_1^2}{r_1 + r_2}$

50. The figure shows the cross-sectional diagram of an orifice flow meter with an orifice radius R point 'a' is 30R upstream while points 'b' and 'c' are 0.8R and 30R downstream from the orifice respectively. The pressures at points a,b and c are $P_a P_b$ and P_c respectively. Then

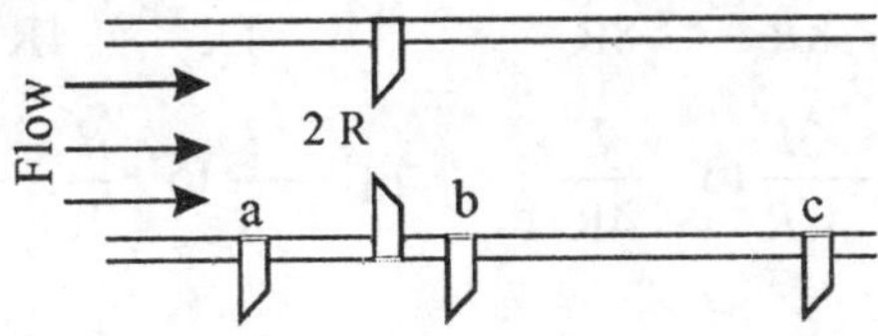

(a) $P_c > P_b > P_a$ (b) $P_b > P_c > P_a$
(c) $P_a > P_b > P_c$ (d) $P_a > P_c > P_b$

Common Data for Questions 51 and 52:

The figuer shows an arrangement for measuring small angular displacement in a vertical plane. A non-conducting tube of length $2l$ and rectangular cross section (width w, height d) is bent along an arc of a circle with radius $R >> d$, centered at P. Four electrode plates of length l and width w are placed to form two curved parallel plate capacitors C_1 and C_2 with a negligible gap between them. The tube contains water with an air bubble of rectangular cross section (with *w*, height *d*) and length $l/4$. The capacitors are connected in a bridge circuit as shown in the figure, where the bridge circuit as shown in the figure, where the bridge has ac excitation v_i. Angular displacement $\Delta\theta$ occurs about the point P.

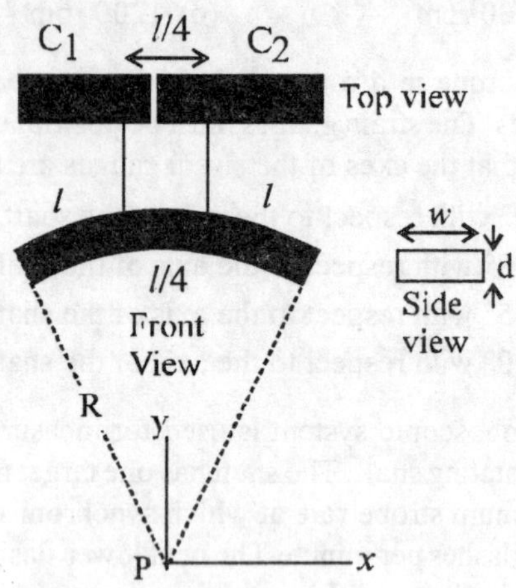

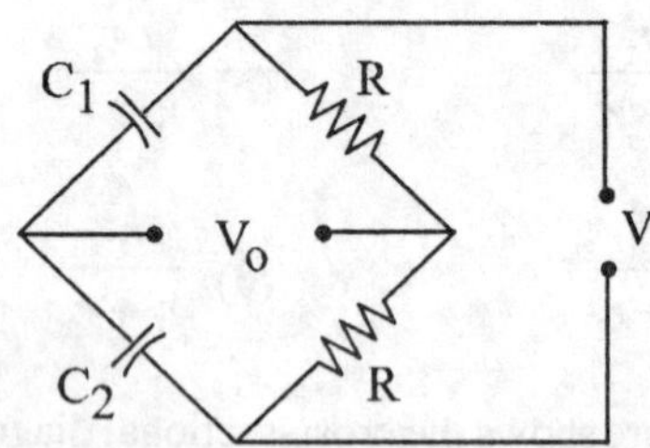

51. The range of angular displacement (in radians) this system can measure is

(a) $-\frac{l}{8R}$ to $+\frac{l}{8R}$ (b) $-\frac{l}{4R}$ to $+\frac{l}{4R}$

(c) $-\frac{3l}{4R}$ to $+\frac{l}{4R}$ (d) $-\frac{l}{R}$ to $+\frac{l}{R}$

52. The sensitivity $\frac{v_o/v_i}{\Delta\theta}$ is

(a) Inversely proportional to *R* and *l*
(b) Inversely proportional to *R* and directly propor tional to *l*
(c) Directly proportional to *R* and *l*
(d) Directly proportional to *R* and inversely propor tional to *l*

53. A differential pressure transmitter is used to measure the flow rate in a pipe. Due to aging, the sensitivity of the pressure transmitter is reduced by 5%. All other aspects of the flow meter remaining constant, change in the sensitivity of the flow measurement is

(a) 10.0% (b) 5.0%
(c) 2.5% (d) 2.2%

54. In a flapper - nozzle displacmenet transducer, the values of the following parameters are given:

Diameter of the orifice = 0.2 mm,
Diameter of the nozzle = 0.8 m,
Supply pressure = 1.4×10^2 kpa(gauge),
Ambient pressure = 0 (gauge).
The maximum value of the sensitivity is

(a) 4.0 MP a/mm (b) 5.6 MPa/mm
(c) 6.4 MPa/mmI (d) 7.3 MPa/mm

55. A turbine flowmeter is rotating at 72 rpm. The flux ψ linked to the nearby magnet and coil assembly is given by $\psi(\theta)=3+\cos(4\theta)$ mWb, where θ is the anuglar position (in radian). The amplitude and freqeuncy of the output voltage signal, respectivley, are

(a) 4 mV and 45.8 Hz
(b) 30.2 mV and 4.8 Hz
(c) 30.2 mV and 30.2 Hz
(d) 288 mV and 45.8 Hzb

56. Water (density 1000 kgm^{-3}) stored in a cylindrical durm of diameter 1 m is emptied through a horizontal pipe of diameter 0.05 m. A pitot-static tube is placed inside the pipe facing the flow. At the time when the difference between the stagnation and static pressures measured by the pitotstatic tube is 10 *kPa*, the rate of reduction in water level in the drum is.

(a) $\frac{1}{200\sqrt{5}} ms^{-1}$ (b) $\frac{1}{75\sqrt{10}} ms^{-1}$

(c) $\frac{1}{50\sqrt{10}} ms^{-1}$ (d) $\frac{1}{40\sqrt{5}} ms^{-1}$

57. A U-tube manometer of tube diameter *D* is filled with a liquid of zero viscosity. If the volume of the liquid filled is *V*, the natural frequency of oscillation in the liquid level about its mean position, due to small perturbations, is

(a) $\frac{D}{2\sqrt{2\pi}}\sqrt{\frac{g}{V}}$ (b) $\frac{2\sqrt{2}}{\sqrt{\pi}}\frac{\sqrt{gV}}{D^2}$

(c) $\frac{1}{2\sqrt{\pi}}\frac{\sqrt{gD}}{V^{1/3}}$ (d) $\frac{1}{\sqrt{\pi}}\sqrt{\frac{g}{D}}$

58. The coefficient of discharge C_d of an obstruction flow meter is given by

$$C_d = \frac{m}{t\rho A\sqrt{2gh}}$$

m = mass of water collected = (400 ± 0.2) kg
t = time = (550 ± 2)s
r = density = (1000 ± 0.1%) kg/m³

$A = \text{Area} = \frac{\pi}{4} d^2$, d = (13 ± 0.02) mm
g = (9.81 ± 0.1%) m/s²
h = pressure head = (3.5 ± 0.005) m

For above data, the value of C_d is

(a) 0.66 (b) 0.55
(c) 0.44 (d) 0.33

59. A U-tube mercury manometer is used to measure a differential pressure of 50 k Pa of oil density 800 kg/m². The difference in the mercury column in two arms is observed as

(a) 37.5 mm (b) 37.4 cm
(c) 37.1 cm (d) 37.2 mm

Common Data for Questions 60 and 61

A diaphragm pressure gauge is constructed of steel of diameter 5 cm. It is designed to measure a maximum pressure of 1.5 M *Pa*.

Assume following data:
E = 200 GN/m², v = 0.3
s \ 7.9 × 10³ kg/m³

60. The thickness of the diaphragm to limit its maximum delfection to one-third its thickness is

(a) 1.01 *mm* (b) 1.1 *mm*
(c) 0.11 *mm* (d) 11.01 *mm*

61. The natural frequency of the diaphragm is.

(a) 435.3 *kHz* (b) 4.353 *kHz*
(c) 4.28 *kHz* (d) 43.52 *kHz*

62. A McLeod gauge has bulb and capillary volume of (till cut off point) 100cm² and a capillary of 1 mm diameter. The pressure indicated by a reading of h = 2.5 cm of the gauge is

(a) 0.049 torr (b) 49 torr
(c) 0.0049 torr (d) 4 torr

63. The sound pressure level corresponding to sound pressures of 1 N/m²

(a) 0.93 dB (b) 0.938 dB
(c) 93.98 dB (d) 9.398 dB

64. Find the sound pressure level corresponding to sound pressures of 10 N/m²

(a) 1139 dB (b) 113.9 dB
(c) 93.98 dB (d) 9.398 dB

65. A pitot-static tube is placed in front of a submarine moving horizontally in the sea under water. A U-tube mercury manometer shows a deflection of 15 cm between stagnation pressure and static pressure.

The speed of the submarine if the density of sea water is 1026 kg/m³ is

(a) 16.6 m/s (b) 6.0 m/s
(c) 1.66 m/s (d) 6.6 m/s

66. A pitot static tube is fitted at the nose of an airplane. A U-tube water manometer connected to it shows a deflection of 10 cm of water.

The air velocity assuming density of air 1.20 kg/m³ is

(a) 140. 43 m/s (b) 4.43 m/s
(c) 14.43 m/s (d) 40.43 m/s

67. A 30 cm × 15 cm venturimeter is used to measure the flow of water in a pipeline. A U-tube mercury manometer connected at the entrance and throat shows a deflection of 25 cm. Take the coefficient of discharge of venturimeter C_d = 0.97.

The rate of flow of water is

(a) 0.246 m³/s (b) 0.0246 m³/s
(c) 2.46 m³/s (d) 24.6 m³/s

68. A pitot tube is inserted in a pipe of 30 cm diameter carrying water. The difference between stagnation and static pressure head is found to be 2.36 m of water at the centre of the pipe.Assume C_v = 0.8.

The rate of flow of water, if the mean velocity of low is 0.85 times the central velocity is

(a) 4.006 m³/s (b) 44.06 m³/s
(c) 0.40066 m³/s (d) 14.06 m³/s

69. Given the voltage sensitivity constant of the crystal is 250 V/mm. A piezoelectric accelerometer has a natural frequency of 10,000 Hz and a damping ratio of 0.2.

The voltage output for an input acceleration amplitude of 500 g (g being acceleration due to gravity) at a frequency of 2000 Hz is

(a) 200 V /mm (b) 300 V/m
(c) 400 V/mm (d) 250 V/m

70. Assume the thermostal to be a first order instrument. A thermostal for on-off control of room temperature produces a cycle with 30 minute perod and 3°C amplitude. If a thermometer with a time lag of 4 minutes is placed beside it, the cycle it indicates is

(a) 34.72 minute period
(b) 3.472 minute period
(c) 0.34272 minute period
(d) 43.72 minute period

Linked Answer Questions

Statement for linked Answer Questions 71, 72 and 73

A thermocouple system with a time constant of 8 seconds has been used to measure the temperature of a furnace whose temperature fluctuates sinusoidally between 400°C and 450°C with a periodic time of 60 seconds.

71. The maximum and minimum values that will be indicated by the thermocouple are

(a) $400.2°C,\ 350.2°C$
(b) $444.2°C,\ 408.8°C$
(c) $333.7\ °C,\ 303.7°C$
(d) $466.2\ °C,\ 444.2°C$

72. The phase angle is

(a) 0.0069 rad (b) 69.2 rad
(c) 6.92 rad (d) 0.693 rad

73. The corresponding time lag between the temperature signals and the thermocouple output signals is

(a) 6.669 sec (b) 5.432 sec
(c) 66.69 sec (d) 54.32 sec

74. Consider the pressure transducer to behave as a second-order system. During the act of force measurement by a pressure transducer of the diaphragm type, the system is stated to have a natural frequency of 1000 cps and a damping ratio of 60%.

The frequency range over which the amplitude ratio corresponding to sinusoidal input deviates by a maximum amount of 10% is

(a) 0 - 91 cps (b) 0 - 51 cps
(c) 0 - 81 cps (d) 0 - 916 cps

75. The discharge over a triangular weir varies as $H^{2.5}$ where H is the head above the weir. If uncertainty in the head reading is 2% of the absolute value, no error results in the measurement of other quantities. The percentage error in discharge measurement is

(a) 1 % (b) 3 %
(c) 5 % (d) 2 %

76. If atmospheric pressure $p_a = 10\times10^4$ N/m². One end of a U-tube manometer is connected to a piezometer opening in a air duct and the other end is left open to atmosphere. The manometer has mercury as the balancing liquid. If the mercury level in the manometer on the duct side stads 0.2m higher than that on the open end. The absolute pressure of air inside the duct is .

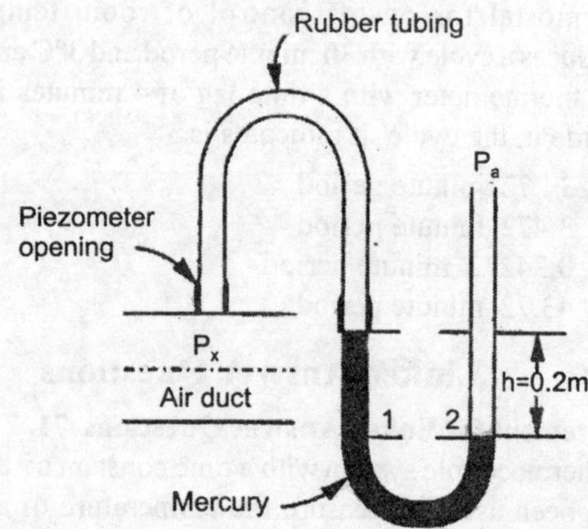

(a) $73.316\ kN/m^3$ (b) $733.16\ kN/m^3$
(c) $633.16\ kN/m^3$ (d) $63.16\ kN/m^3$

77. The right limb of a U-tube manometer constaining mercury is open to the atmosphere while the left limb is connected to a pipe through which flows a fluid of specific gravity 0.85. The center of the pipe lies 15cm below the level of mercury in the right limb. If the difference of mercury level in the two limb is 25 cm, the pressure of fluid in the pipe is

(a) $3.252\ kN/m^3$ (b) $32.52\ kN/m^3$
(c) $3.52\ kN/m^3$ (d) $3.252\ kN/m^3$

78. The right limb of a simple U-tube manometer containing mercury is open to atmosphere and the left limb is connected to a pipe through which flows a fluid of specific gravity 0.8. If the difference of mercury level in the limbs is 30cm and the level of fluid in the limb is 10cm below the centre of pipe, the vacuum pressure in the pipe is

(a) $-20.8\ kN/m^3$ (b) $+40.8\ kN/m^3$
(c) $+20.8\ kN/m^3$ (d) $-40.809\ kN/m^3$

Common data for Questions 79 and 80

Given that density of water is 1000 kg/m³ and that of mercury is 13560 kg/m³. The differential pressure across an obstruction in a water-filled pipe has been measured by a U-tube mercury manometer. If the difference in the mercury level in the two limbs is 50mm,

79. The differential pressure being measured.

(a) 6160.98 (b) 6200.23
(c) 6660.98 (d) 6000.45

80. The error introduced by neglecting the water density in the limb is

(a) 5.2 % (b) 6.44%
(c) 7.64% (d) 7.96%

81. A single-column manometer is using mercury of specific gravity 13.6 as the manometric liquid.

The wider and the narrow limb diameters are 160 mm and 6 mm, respectively. If a differential pressure of 75 kN/m² is applied the mercurry height rise in narrow limb is

(a) 42.2 cm (b) 5.62 cm
(c) 4.22 cm (d) 56.2 cm

82. A well type manometer using mercury (specific gravity 13.56) as manometric liquid has well and tube diameters as 20mm and 4 mm, respectively. This manometer has a scale graduated accurately in mm to measure the pressure differenc, i.e. 1 mm scale division indicates 1mm head of pressure change.

The angle at which the tube is inclined to the vertical to accomplish this task is

(a) 24°　　(b) 20°
(c) 16°　　(d) 32°

83. With walled inverted bell-type differential pressure gauge instrument, the bell diameter is 150mm and it moves vertically 20 mm when a full load pressure of 250 mm water gauge is applied.

The spring rate of the opposing spring
(a) 21.66 N/m　　(b) 1.133 N/m
(c) 11.33 N/m　　(d) 2.166 N/m

84. It is required to design a diaphragm pressure gauge to measure a maximum pressure of 1.5 MPa gauge. The diaphragm is to be constructed of spring steel 5 cm in diameter; modulus of elasticity E = 200 GPa and poission's ratio $\mu = 0.3$

If the maximum deflection is limited to one-third of thickness, the diaphragm thickness is
(a) 0.11 cm　　(b) 0.01 cm
(c) 11 cm　　(d) 0.22 cm

85. A Mcleod gauge of $V_0 = 200$ cm³ and capillary cross-sectional area $a = 0.1$ cm² indicates 1 cm of mercury.

The pressure in microns is
(a) 10 microns　　(b) 20 microns
(c) 5 microns　　(d) 15 microns

86. The piston and the platform of a dead-weight tester weight 20 N and the piston diameter is 1.25 cm.

The weight must be added to the platform to produce a pressure of 175 kPa is
(a) 14. 64 N　　(b) 1.464 N
(c) 28.32 N　　(d) 2.8 32 N

Linked Answer Questions

Statement for linked Answer Questions 87 and 88

A 30 cm × 15 cm venturimeter is provided in a vertical pipe line carrying oil of specific gravity 0.9; the flow being upwards. The difference in elevations of the throat section and entrance section of the venturimeter is 30 cm. The differential U-tube mercury manometer shows a gauge deflection of 25 cm. Take the discharge coefficient as 0.98 and the specific gravity of mercury as 13.6.

87. The discharge of the oil is
(a) $14.87\ m^3/s$　　(b) $25.74\ m^3/s$
(c) $2.574\ m^3/s$　　(d) $0.1487\ m^3/s$

88. The pressure difference between the entrance and throat section is
(a) $3379\ N/m^3$　　(b) $33.79\ KN/m^2$
(c) $379\ N/m^2$　　(d) $3.79\ KN/m^3$

89. Air flows through a duct and its velocity is being measured by a pitot-static tube placed properly along the duct axis. The leads from the tubes have been attached to a differential manometer which shows a difference of hed of 0 cm of water Take density of air 1.22 kg/m³ and coefficient of velocity for the pitot-tube is 0.98.

The air velocity is
(a) 3.92 m/s　　(b) 39.2 m/s
(c) 0.392 m/s　　(d) 392 m/

90. A pivot-static probe with a thermometer attached is used to measure the velocity of air in a duct and a differential pressure of 10 cm of water is recorded. The thermocouple shows an air temperature of 300 K in the duct, and a separate static tapping shows that the air pressure in the ducts 10 kN/m². The local barometer reads 760 mm of mercury. If the coefficient for the pitot-static probe is 0.98, take the gas constant R as 287 J/kgK for air and the relative density of mercury as 13.6.

The air velocity is
(a) 3822 m/s　　(b) 3. 822 m/s
(c) 38.22 m/s　　(d) 382.2 m/s

91. A bourdon tube temperature indicator has a range of 0°C to 250°C and produces a corresponding rotation of 0 to 270°.

If the mechanical levers and gears have an amplification of 30, then the sensitivity of the bourdon tube in radians per degree celcius is
(a) 6.28×10^{-2} rad/°C
(b) 62.8×10^{-4} rad/°C
(c) 628×10^{-4} rad/°C
(d) 6.28×10^{-4} rad/°C

92. A platinum resistance thermometer has a resistance of 150Ω at 100°C. When exposed to high temperature gas, the resistance is 300Ω. It may be presumed that at 0°C, the resistance is 100Ω and that the temperature coefficient of resistance for platinum is 0.0039°C⁻¹.

The temperature of hot gas is
(a) $4.846°C$　　(b) $484.6°C$
(c) $4846°C$　　(d) $48.46°C$

93. A thermistor is characterised by the relation: $R_1 = R_0\, a e^{b/T}$
where R_0 is resistance at ice point, R_1 is the resistance at the absolute temperature T K, and a and b are constants.
If $R_1 = 1050\Omega, R_0 = 4000\Omega, \ a = 300\times10^{-6}$ and $b = 2850$
The temperature indicated by such a thermistor is
(a) 110 K　　(b) 220 K
(c) 320 K　　(d) 420 K

Common Data for Questions 94 and 95

An electrical resistance strain gauge of resistance 100Ω and gauge factor 2.0 is bounded to steel having an elastic-limit stress of 400 MN/m^2 and modulus of elasticity 200 GN/m^2.

The gauge is made of advance alloy for which temperature coefficient of resistance

$$\alpha_g = 20 \times 10^{-6} \text{ per } °C$$

temperatue coefficient of linear expansion $\lambda_g = 16 \times 10^{-6}$ per °C

For steel, the temperature coefficient of linear expansion is $\lambda_s = 12 \times 10^{-6}$ per °C

94. The resistance change of the gauge due to change of stress equal to 10% of elastic range is

(a) 40.4 Ω (b) 404 Ω
(c) 202 Ω (d) 4.04 Ω

95. The resistance change of the gauge due to an increase of 25°C in temperature is

(a) 0.02 Ω (b) – 0.02 Ω
(c) 0.01 Ω (d) –0.01 Ω

Linked Answer Questions

Statement for linked Answer Questions 96 and 97

A load of 26500 N was applied to a tension member and strains of 1600 and 450 microstrain were measured in the axial and transverse directions, respectively. Presuming that the tension member has an area of 125 mm^2,

96. The modulus of elsticity is

(a) 13.25 GN/m^2 (b) 1.325 GN/m^2
(c) 0.32 GN/m^2 (d) 132.5 GN/m^2

97. Poissions's ratio for the material of the tension member is

(a) 281 (b) 0.282
(c) 14 (d) 0.14

98. If measured vibration velocity amplitude is 3.125 m/s and the reference level is 10^{-3} m/s.

The following vibration amplitudes in decibels of a certain mechanism is

(a) 6.89 dB (b) 68.9 dB
(c) 69.89 dB (d) 6.89 dB

99. If measured vibration acceleration amplitude is 1.5g and the reference level is 10^{-5} m/s^2.

The following vibration amplitudes in decibels of a certain mechanism is

(a) 12.35 dB (b) 32.53 dB
(c) 32.153 dB (d) 123.35 dB

100. Specify the photo-electric device which converts the light information to resistance information :

(a) photo-emissive cell
(b) photo-conductive cell
(c) photo-voltatic cell
(d) All of the above

101. Identify the pressure sensing element commonly used in **ameroid barometer**

(a) flat diaphragm (b) bellows
(c) capsule (d) sprial tube

102. The **stagnation (total) pressure** at a point is measured by

(a) static tube
(b) piezometer tube
(c) impact tube
(d) pitot-static probe

103. A **pitot-static** tube measures

(a) undisturbed fluid pressure
(b) dynamic pressure of a moving stream
(c) pressure difference between two fluids
(d) difference between the dynamic and static pressure

104. Small pressure changes is low velocity flows can be measured accurately by

(a) U-tube manometer
(b) single column manometer
(c) compound U-tube manometer
(d) draft gauge

105. A U-tube differential manometer is

(a) always used upright
(b) used upright when pressure difference is large
(c) used inverted when pressure difference is small
(d) used inverted when pressure difference is large

106. **A well-type manometer** is used in preference to a simple U-tube manometer to obtain

(a) better accuracy (b) better precision
(c) high sensitivity (d) high resolution

107. Which of the following type of manometer is likely to have the highest sensitivity and accuracy?

(a) U-tube manometer
(b) inverted U-tube manometer
(c) single column manometers
(d) U-tube double reservoir manometer

108. Mercury is used in barometers because it

(a) does not stick to the walls of the tubing
(b) is shining and can be read only
(c) has practically zero vapour pressure
(d) has high density

109. Which of the following type of Bourdon tube shape has a small tip travel and necessitates amplification?

(a) C-type (b) spiral
(c) helical (d) twisted tube

110. In a Bourdon tube pressure gauge, incorrect readings may be encountered due to

(a) hysteresis (b) fatigue
(c) creep (d) drift

111. All of the following devices are used for measuring vacuum, except

(a) kenometer
(b) hydrometer
(c) pirani gauge
(d) thermal conductivity gauge

112. High vacuum pressure is most commonly expressed in

(a) cm of water (b) pascal
(c) torr (d) micron

113. The **hot wire anemometer** is used to measure:

(a) pressure in gases
(b) liquid discharges
(c) gas velocities
(d) wind velocities at airports

114. In a venturimeter, pressure at the throat is

(a) less than in the entrance pipe
(b) greater than that in the entrance pipe
(c) equal to that in the entrance pipe
(d) independent of the rate of flow

115. The range of coefficient discharge for a **venturimeter** is

(a) 0.6 – 0.7 (b) 0.74 – 0.8
(c) 0.8 – 0.9 (d) 0.95 – 0.97

116. For a verturimeter, the coefficient of discharge

(a) does not depend on the Reynolds number
(b) increase with increase in Reynolds number
(c) decrease with increase in Reynolds number
(d) increase upto a certain value of Reynolds number

117. Positive displacement flow meter are

(a) the quantity flow meters
(b) the constant area-variable head meters
(c) the constant head-variable area meters
(d) the variable area-variable head meters

118. Which of the following instruments is not a **rate meter**?

(a) weirs and flumes (b) orifice plate
(c) nutating disk (d) rotameter

119. Cylinder and piston type flow meters are

(a) variable area flow meters
(b) variable head flow meters
(c) vortex flow meters
(d) positive displacement flow meters

120. Which amongst the follwing is a constant head variable area flow meter?

(a) Venturimeter (b) weir
(c) vane anemometer (d) rotameter

121. Which amongst the following gives the flow rate as well as totalised flow?

(a) flow nozzle
(b) nutating disk flow meter
(c) electro-magnetic flow meter
(d) weirs and flumes

122. Consider the following statements in the context of **flow meters:**

(1) Hot wire anemometer is preferred to measure highly fluctuating velocities.
(2) In an electromagnetic flow meter the induced voltage is proportional to flow rate
(3) The obstruction type flow meters have a linear relation ship between the pressure difference and the rate of fluid flow
(4) Rotameter is a variable area flow meter and it should always be installed in a vertical pipeline

Select the correct answer using the codes given blow:

(a) 1 and 2 (b) 1, 2 and 4
(c) 2 and 4 (d) 1, 2, 3 and 4

123. Orifice plates have:

(a) Coefficient of discharge which is typically about 0.6 and loss of head of about 60 - 70%.
(b) Coefficient of discharge which is about 0.95 and loss of head of about 10%
(c) Coefficient of discharge which is typically about 0.6 and loss of head of about 10%
(d) None of the above

124. Nozzles are useful for the measurement of:

(a) fluids which are clean.
(b) fluids which are clean, but may contain sol-ids.
(c) fluid which are clean, but may contain solids and also for high pressure/temperature steam.
(d) fluid which are clean, but may contain solids and also corrosive liquids.

125. Which of the following device are used for the measurement of liquid flow through a channel.

(a) Venturimeter (b) Flotameters
(c) Wiers and flumes (d) Pitot tube

126. A flow meter that measures flow which are independent of density is

(a) rotameter
(b) Electromagnetic flowmeter
(c) Venturimeter
(d) Orifice meter

127. A laser doppler anemometer can function properly if the fluid
(a) Contains small tracer particles
(b) Contains large concentration of tracer particles
(c) Contains no tracer particles
(d) None of above

128. The meter which is suitable for flow totalization is
(a) turbine meter
(b) venturimeter
(c) Ultrasonic flow meters
(d) orifice plate

129. Manometers measures unknown pressure by
(a) measuring liquid levels
(b) measuring height of liquid columns
(c) balacing the unknown force produced by pressure against known force.
(d) notify the deflection of a pointer

130. The desirable property of a manometric fluid is:
(a) high viscosity
(b) high - coefficient of thermal expansion.
(c) low vapour pressure
(d) corrosiveness and strictness

131. Least suitabel tranducer for static pressure measurement is :
(a) Semiconductor strain gauge
(b) Variable capacitor for transducer
(c) Metal wire strain gauge
(d) Piezoelectric transducer

132. A Mc Leod gauge having volume of bulbs and measuring capillary, $V = 100 \times 10^{-6} m^3$ and the diameter of measuring capillary is 1 mm. For the reading of 60 mm, the pressure
(a) 14 μm (b) 28 μm
(c) 42 μm (d) 56 μm

133. A thermopile is :
(a) a combination of a number of thermocouples connected in series.
(b) a combination of a number of thermocouples connected in parallel.
(c) a combination of a number of thermocouples some of which are connected in series and some in parallel.
(d) none of the above

134. The thermocouples are :
(a) Passive transducers (b) Active transducers
(c) Both (a) and (b) (d) Output transducers

135. The thermocouples:
(a) are most commonly used temperature transducers.
(b) requires reference junctions compensation
(c) have low output voltage level
(d) all of the above

136. The device which are used for the measurement of very high temperature are :
(a) thermistors (b) thermocouples
(c) pyrometers (d) RTD's

137. Dunmore hygrometer has
(a) linear resistance vs relative humidity characteristics
(b) non-linear resistances vs relative humidity characteristics
(c) linear inductance vs relative humidity characteristics.
(d) non-linear inductance vs relative humidity characteristics

138. Which of the following instrument is used for the measurement of humidity.
(a) Autocollimeter (b) Hygrometer
(c) Nephelometer (d) Mass spectrometer

139. In a laminar flow experiment, fluid A is pumped through a straight tube and the volumetric flow rate and pressure drop per unit length are recorded in a second straight tube having twice the internal diameter of the first one, fluid B records the same pressure drop per unit length at the same volumetric flow rate. Assuming fully developed flow condition in the tubes, the ratio of the dynamic viscosity of fluid B to that A is
(a) 16 (b) 32
(c) 64 (d) 128

140. In using capillary tube viscometer corrections have to be made for
(a) Change in density
(b) Change in temperature
(c) Change in gravitational constant
(d) Loss in head

2.5 Miscellaneous

1. A torsion bar of radius 15 mm is used to measure a torque of 12 Nm. If the shear modulus of mild steel is 80 GN/m^2, the angle of twist will then be :
(a) 0.2832 degree (b) 0.2832 rad
(c) 0.02832 rad (d) $0.2832 \times 10^{-3} rad$

2. A variable reactance type tachometer has number of teeth on rotor as 60. The counter records 3600 counts per second. Determine the speed in rpm.
(a) 3500 rpm
(b) 600 rpm

(c) 800 rpm
(d) 3600 rppm

3. A resistance R_T of a **resistive transducer** is modeled as $R_T = R(1 - Kx)$, where K is the constant of transformation and x the input quantity being sensed. The resistor is connected to an ideal, op-amp signal conditioning circuit of figure. The value of R is 100 W, K is 0.004 and x is 75. The output V_o is

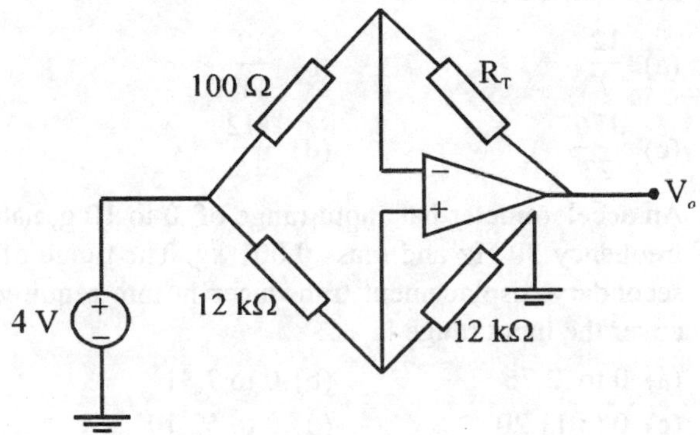

(a) – 400 mV
(b) – 4 mV
(c) – 300 mV
(d) + 600 mV

Statement for Linked Answers Questions 4 and 5:

A differential amplifier with signal terminals X, Y, Z is connected as shown in Fig. (a) below for CMRR measurement where the differential amplifier has an additional constant offset voltage in the output. The observations obtained are: When $V_i = 2V$ $V_o = 3$ mV and when $V_i = 3$ V, $V_o = 4$ mV

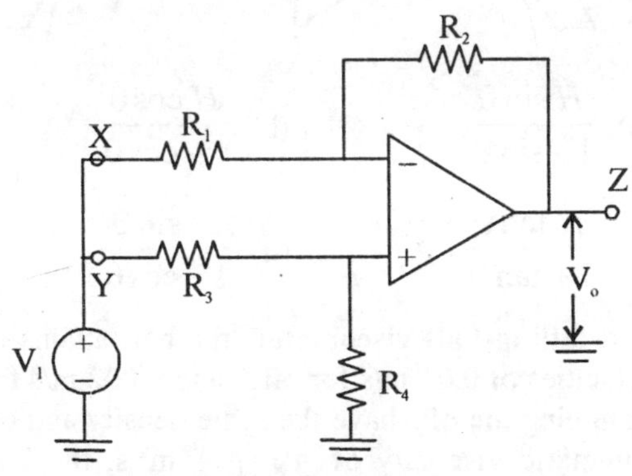

Figure: (a)

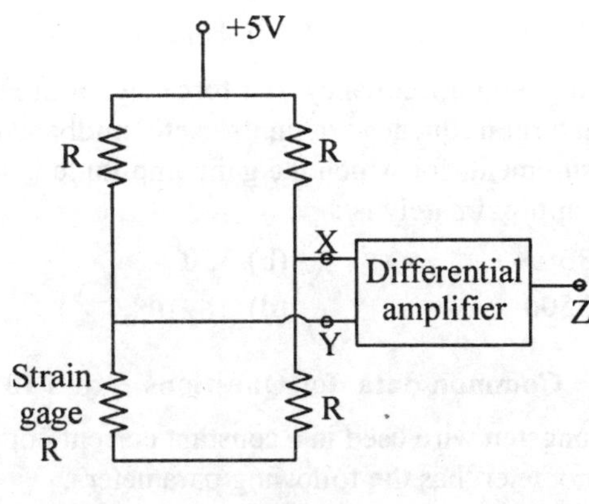

Figure (b)

4. Assuming its differential gain to be 10 and the op-amp to be otherwise ideal, the CMRR is

(a) 10^2 (b) 10^3
(c) 10^4 (d) 10^5

5. The differential amplifier is connected as shown in Figure (b) above to a single strain gage bridge. Let the strain gage resistance vary around its no load resistance R by $\pm 1\%$. Assume the input impedance of the amplifier to be high compared to the equivalent source resistance of the bridge, and the common mode characteristic to be as obtained above. The output voltage in mV varies approximately from

(a) +128 to –128 (b) +128 to –122
(c) +122 to –122 (d) +99 to –101

6. A quartz peizoelectric type pressure sensor has a built in charge amplifier. The sensor has a sensitivity of 1 m V/Pa. It is subjected to a constant pressure of 120 kPa. The output of the transducer at steady state is

(a) 0 mV (b) 100 mV
(c) 120 mV (d) 120 mV

Common data for Questions 7 and 8

A piezoelectric force sensor, connected by a cable to a voltage simplifier has the following parameter:
Crystal properties: stiffness 10^9 N/m, damping ratio 0.01, Natural frequency 10^5 rad/s, force-to-charge sensitivity 10^{-9} C/N, capacitance 10^{-9} F with its loss angle assumed negligible
Cable properties: Capacitance $2 \times 10^{-9} F$ with its resistance assumed negligible
Amplifier properties: input impedance 1 MΩ, bandwidth 1 MHz, gain 3

7. The minimum frequency of a force signal in Hz with in its useful mid-band range of measurement for which the gain amplitude is more than 0.95 approximately is

(a) 16 (b) 160
(c) 1600 (d) 16×10^3

8. The maximum frequency of a force signal in Hz below the natural frequency within its useful midband range of measurement, for which the gain amplitude is less than 1.05 approximately is :

(a) 35 (b) 350
(c) 3500 (d) 16×10^3

Common data for Questions 9 and 10

A tungsten wire used in a constant current hot wire anemometer has the following parameters:
Resistance at 0° C is 10 Ω, Surface area is 10^{-4} m². Linear temperature coefficient of resistance of the tungsten wire is 4.8×10^{-3} /° C, Convective heat transfer coefficient is $25.2\, W/m^2/°C$, flowing air temperature is 30 °C wire current is 100 mA, mass-specific heat product is $2.5\times10^{-5} J/°C$

9. At steady state, the resistance of the wire in Ω is

(a) 10.000 (b) 10.144
(c) 12.152 (d) 14.128

10. The thermal time constant of the hot wire under flowing air condition in ms is

(a) 24.5 (b) 12.25
(c) 6.125 (d) 3.0625

11. A differential pressure transmitter of a flow meter using venturi tube reads 2.5×10^5 Pa for a flow rate of $0.5\, m^3/s$. The approximate flow rate in m^3/s for a differential pressure of 0.9×10^5 Pa is

(a) 0.30 (b) 0.18
(c) 0.83 (d) 0.60

Common Data for Questions 12 and 13:

The deflection profile $y(x)$ of a cantilever beam due to application of a point force F (in Newton), as a function of distance x from its base, is given by $y(x) = 0.001F\, x^2\left(1-\frac{x}{3}\right) m$. The angular deformation θ at the end of the cantilever is measured by reflecting a laser beam off a mirror M as shown in the figure.

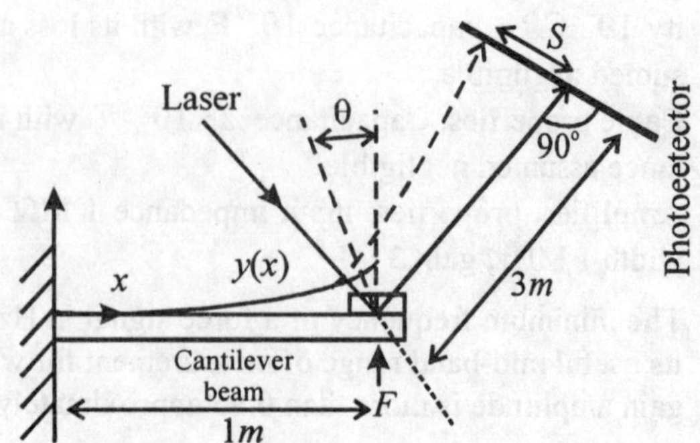

12. The translation S of the spot of laser light on the photodector when a force F =1 N is applied to the cantilever is

(a) 1 mm (b) 3 mm
(c) 6 mm (d) 12 mm

13. If linear variable differential transformers (LVDTs) are mounted at $x=\frac{1}{2}m$ and $x=\frac{1}{4}m$ on the cantilever to measure the effect of time varying forces, the ratio of their outputs is

(a) $\frac{12}{7}$ (b) $\frac{40}{11}$
(c) $\frac{176}{23}$ (d) $\frac{112}{15}$

14. An accelerometer has input range of 0 to 10 g, natural frequency 30 Hz and mass 0.001 kg. The range of the secondary displacement transducer in mm required to cover the input range is

(a) 0 to 2.76 (b) 0 to 9.81
(c) 0 to 11.20 (d) 0 to 52.10

15. Measurement of large external radius is possible with the help of an angle piece and a screw gage as shown in figure. The radius R can be expressed from the measurement of H and q as

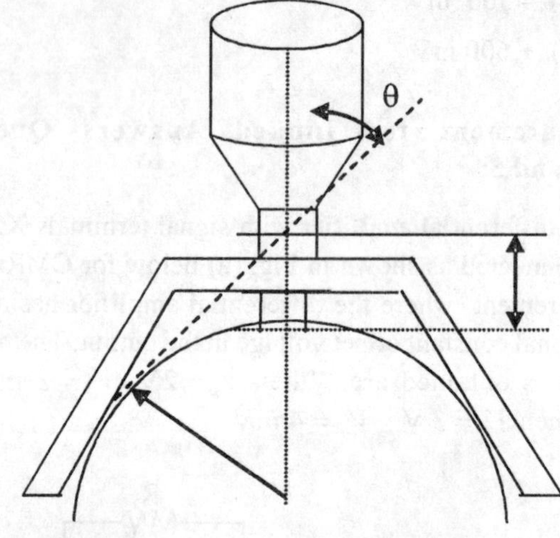

(a) $\frac{H\sin\theta}{1-\sin\theta}$ (b) $\frac{H\cos\theta}{1-\cos\theta}$
(c) $\frac{H\tan\theta}{1+\tan\theta}$ (d) $\frac{H\sin\theta}{1-\cos\theta}$

16. In a **falling-ball viscometer**, the ball attains terminal velocities of 0.01 m/s for oil A and 0.002 m/s for oil B. Assuming the oils have the same density and oil A has kinematic viscosity of 5×10^{-3} m²/s, the kinematic viscosity of oil B in m²/s is

(a) 15×10^{-3} (b) 20×10^{-3}
(c) 25×10^{-3} (d) 30×10^{-3}

Answers

2.1 Basic of Transducers

1. **Ans. (d)**

Solution:

Bellow's expansion is usually against a spring. The spring is provided to decrease hysteresis effect.

2. **Ans. (d)**

Solution:

$f_c << f_n$

Where f_e is excitation frequency

f_n is natural frequency of seismic mass type pick up

3. **Ans. (a)**

Solution:

A seismic type of transducer has damping constant of 10. The transducer is designed to measure acceleration.

4. **Ans. (d)**

Solution: The correct match is given below

Thermocouple	Cold junction compensation
Strain gage	DC bridge
Piezoelectric	Charge amplifier sensor
L.V.D.T	Phase sensitive detector

A **thermocouple** is suddenly immersed in a medium of high temperature. The approximate time taken by the thermocouple to reach 98% of the steady state value is equal to the time constant of the thermocouple.

5. **Ans. (c)**

Solution: The correct match is given below

Bourdon guage	Deformation of element
Pirani gauge	Thermal conductivity
Semi-conductor strain guage	Piezo-resistive effect
Bimetallic strip	Thermal co-efficient of expansion

Thermal-Conductivity Gauge

Thermal-conductivity gauges measure pressure by measuring the changes in the ability of a gas to conduct heat. The ability of a material to carry heat by conduction change when the pressure changes, until the pressure drops below about one torr. As the pressure continues to drop, the conductivity of the gas decreases and the gas loses its ability to conduct heat. Thus, at low pressure, the conductivity of a gas has a direct relationship to its pressure the relationship between changes in conductivity and changes in pressure work over a pressure range form about 10^{-4} torr up to about 10^{-2} torr. It is used for the absolute pressure (very low pressure) measurement.

Pirani Gauge

It consists of two wire filament only. Filament serves as a reference and is sealed in an evacuated glass, while the other filament is kept in a container connected to the source of pressure. These two filaments are connected in a bridge circuit.

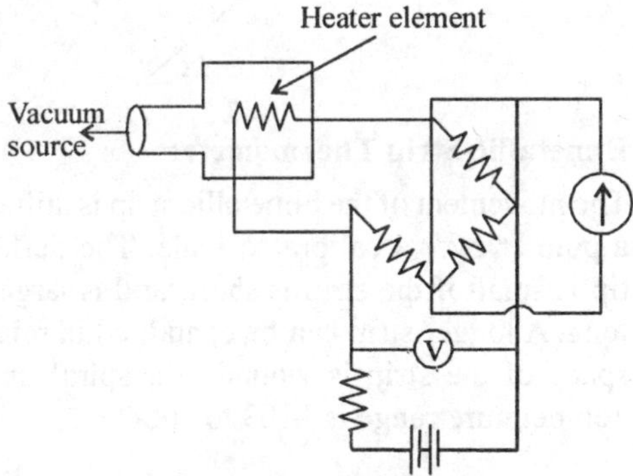

Figure: Pirani Gauge

The operation of pirani gauge depends on variation of the **thermal conductivity** of a gas with pressure. At low pressure there is a relationship between pressure and the thermal conductivity of the gas, the heat conductivity decreases with decrease of pressure. Pressure range of the pirani gauge is 10^{-5} to 1 torr.

Bimetallic thermometers

The commercial solid expansion type is a bimetallic thermometers. Two strips A and B of dissimilar metals are bonded firmly together at temperature T_o The strips have thickness t_1 and t_2 and coefficients of expansions are α_1 and α_2. One sides of the bonded member is held rigidly. If now the temperature changes from that of the bonding temperature to 'T' because of unequal expansions in the two strips will form a curve of radius 'r'.

$$r = \frac{t[3(1+m)^2 + (1+m)(m^2 + 1/mn)]}{6(\alpha_2 - \alpha_1)(T - T_0)(1+m^2)}$$

n→ ratio of young's modulus of low to high

expansion materials $\left(n = \frac{E_1}{E_2}\right)$

m → ratio of the thickness of low $\left(m = \frac{t_1}{t_2}\right)$ to high expansion materials.

$\alpha_1 \rightarrow$ higher coefficient of expansion

$T \rightarrow$ Operating temperature

$T_o \rightarrow$ initial bonding temperature

The deflection of the free tip is thus directly proportional to temperature.

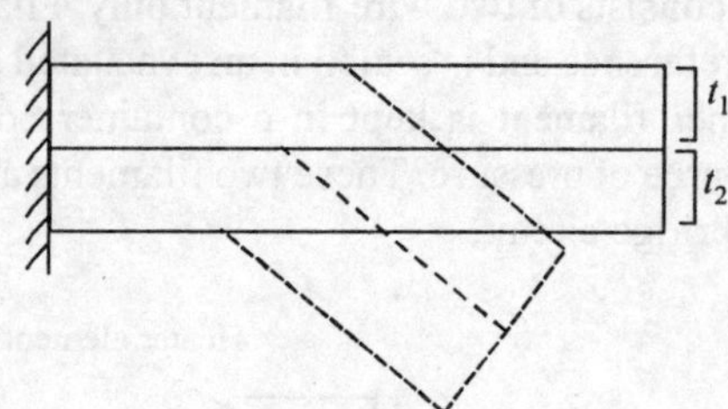

Bimetallic Strip Thermometers

The movement of the bimetallic strip is utilized to deflect a pointer over a calibrated scale. The deflection of the tip is small if the strip is short, and is large if the tip is long. A longer strip can be contained in relatively small space of the strip is wound in a spiral or helix form. Temperature range is –103 to 1004°F .

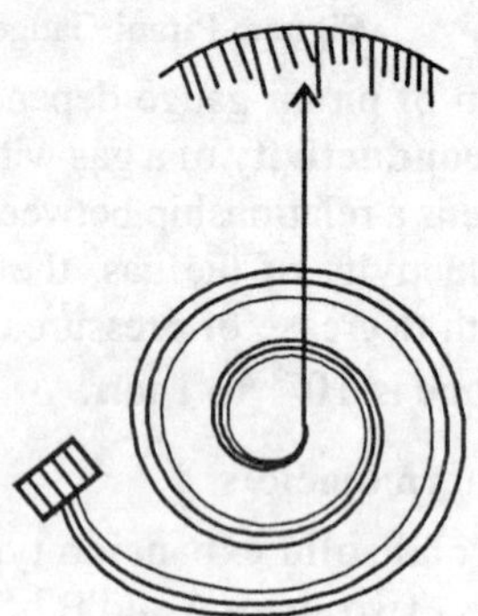

Figure: Bimetallic strip thermometers

Expansion of liquids

(a) Liquid-in-glass thermometer
(b) Liquids-in-metal thermometer.

6. **Ans. (b)**

Solution: The correct match is given below

Rotameter	Flow rate
Tachometer	Rotational speed
Tracer	Roughness
Pyrometer	Temperature

7. **Ans. (b)**

Solution:

Bourdon tube	1.0
Ionization gauge	10^{-5}
Water manometer	10^3
Pirani	10^{-10}

8. **Ans. (b)**

Solution:

The correct match is given below

LVDT	Phase sensitive detector
Piezoelectric crystal	Charge amplifier
pH probe	High input impedance amplifier
Thermocouple	Reference junction compensation

pH-Meter

Measurement of pH requires a special hydrogen ion detector associated circuit. The most common method of pH determination is by a special glass electrode.

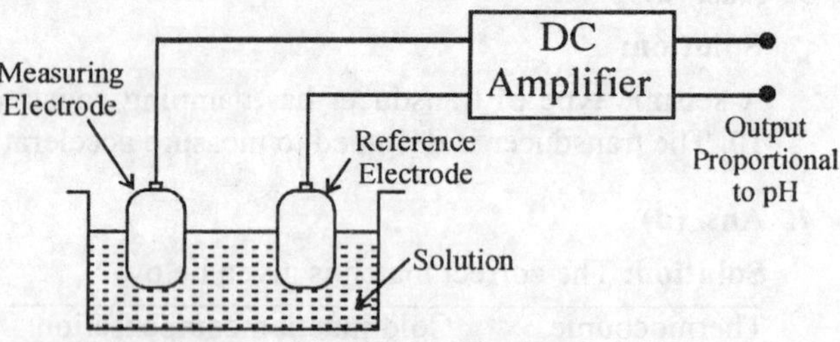

Figure: Schematic of glass-electrode pH-meter

The arrangement consists of two electrodes which are to be immersed in the solution whose pH is to determined and measuring the voltage developed across the two electrodes. One the electrodes called measuring electrode is a **glass electrode** which is permeable only to hydrogen ions. The very thin bottom acts as a membrane. The bulb contains chemical solution. The other electrode called **reference electrode** is also glass bulb which contains solution of calomel.

Both electrodes are mounted to form an assembly that is dipped into the solution. The difference of potential between two electrodes, is a function of the pH of the solution, can be measured by a voltmeter calibrated in the units of pH value. Application of the voltage signal is preferred. The pH sensitive glass electrode produces a voltage change of about 9 mV for 1.0 pH solution at 25°C.

The glass electrode method of pH measurement has some problem associated with it.

1. Measuring probe has a very high resistance of the order of $10^8\,\Omega$ and very low output. The probe is to be properly electrically screened to prevent spurious signal pick-up and a DC amplifier of very high input impedance is required.

2. Deterioration in accuracy may occur as the glass membrane becomes coated with substance that may be present in the measured solution.
3. The response time is high with a time constant of 60–120 seconds. The probe therefore, is to be kept for long time is solution before reading is taken.

9. Ans. (c)

Solution: The correct match is given below:

Pirani gauge	Vacuum pressure measurement
Venturi meter	Flow measurement
Radiation pyrometer	Temperature measurement
Load cell	Force measurement

Thermocouples

The working principle of a thermocouple depends on the thermo-electric effect. If two dissimilar metals are joined to gather so as to form a closed circuit, there will be two junctions where they meet each other. If one of these junctions is heated, then, a current flows in the circuit which can be detected by a galvanometer. The amount of the current produced depends on the difference in temperature between the two junctions and on the characteristics of the two metals. This was first observed by **Seebeck** in 1821 and is known as Seebeck effect.

Instruments which record the variations in current flow are calibrated in terms of temperature and are known as Thermocouple Pyrometers.

Strain Gauge Load Cell

It is an electromechanical transducer which translates changes in force into change in voltage. The change in voltage can be calibrated directly in terms of the force (or load) applied to the cell.

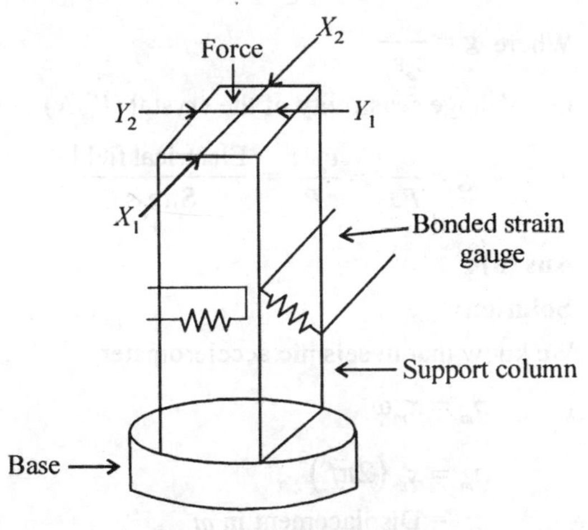

Figure: Strain gauge load cell

Advantage:

1. Strain gauge load cells are small and compact in size.
2. They are well suited for measurements where an electrical output signal is desired.
3. They respond rapidly to load variations. But they have to be protected from angular or non-axial loads as they have no way to discriminate between bending and axial loads.

10. Ans. (d)

Solution: The correct match is given below

Mean free path	Knudsen gauge
Humidity	Sling psychrometer
Heat transfer coefficient	Hot wire anemometer
Intensity of radiation	Optical pyrometer

11. Ans (d)

Solution:

In this we have to match the lists, so we know that

Radiation pyrometer: It is used for measuring the temperature of sun.

Dall Tube: It is used for flow measurement

Pirani Gauge: It is used for vacuum pressure measurement

Gyroscope: It is used in aircraft to measure angular velocity

12. Ans (d)

13. Ans (b)

14. Ans (c)

15. Ans (d)

16. Ans (d)

17. Ans (d)

18. Ans (b)

19. Ans (b)

20. Ans (d)

21. Ans (c)

22. Ans (c)

23. Ans (c)

24. Ans (c)

25. Ans (b)

26. Ans (b)

27. Ans (a)

28. Ans (d)

29. Ans (a)

30. Ans (d)

31. Ans (c)

32. Ans (d)

33. Ans (c)

34. Ans (d)

35. Ans (a)

36. Ans (b)

2.2 Piezoelectric Seismic Transducers

1. Ans. (a)

Solution:

Piezoelectric type of pressure sensor is a dynamic type of instrument, i.e. output voltage is generated if the input pressure is dynamic that is changes. Hence for static (fixed) pressure input output voltage is 0 mV.

2. Ans (a)

Solution:

The mechanical force (F) applied to the ***piezo-electric crystal*** has to be converted into the electrical output.

The piezoelectric material may be thought as a charge generator and a capacitor. The mechanical deformation produced by the force (F) generates a charge and this charge appears as a voltage across the electrodes. The voltage is

$$e_o = \frac{Q}{C}$$

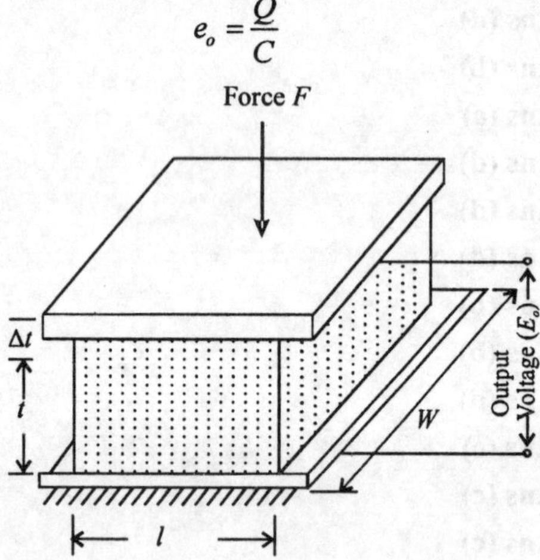

Main point:

1. The piezoelectric effect is direction sensitive, a tensil force produces a voltage of one polarity while a compressive force produce a voltage of opposite polarity.
2. The magnitude and polarity of the induced surface charge are proportional to the magnitude and direction of the applied force (F)

Charge $Q = d \times F$ coulomb

where

d = charge sensitivity of the crystal (C/N) (it is constant for a given crystal)

F = applied force (N)

Young's Modulus,

$$E = \frac{Stress}{Strain} = \frac{(F/A)}{(\Delta t/t)}$$

$$= \frac{Ft}{A(\Delta t)} N/m^2$$

A = Area of the crystal (m^2)

$A = W \times l$

W = Width of the crystal (m)

l = Length of the crystal (m)

$$Q = \frac{d \times AE(\Delta t)}{t}$$

This charge at the electrodes gives rise to an output voltage (e_o)

$$e_o = \frac{Q}{C_p}$$

Where,

C_p = Capacitance between electrodes

$$C_p = \frac{A\varepsilon_o\varepsilon_r}{t}$$

$$E_0 = \frac{Q}{C_p} = \frac{dF}{A\varepsilon_1\varepsilon_r/t}$$

$$= \frac{dt}{\varepsilon_o\varepsilon_r} \times \frac{F}{A}$$

$$E_0 = \frac{d}{\varepsilon_o\varepsilon_r} \cdot t \cdot p$$

$\frac{F}{A} = P$ i.e. Pressure or stress (N/m²)

Where $g = \frac{d}{\varepsilon_o\varepsilon_r}$

i.e. Voltage sensitivity of the crystal (V_m/N)

$$g = \frac{e_o}{P.t} = \frac{e_o/t}{P} = \frac{\text{Electrical field}}{\text{Stress}}$$

3. Ans (a)

Solution:

We know that in seismic accelerometer

$$a_m = x_m\omega^2$$

$$a_m = x_m(2\pi f)^2$$

x_m = Displacement in m

$$x_m = 100mm = 0.01m$$

$$f = 100 \text{ Hz}$$

$$a_m = (0.01) \times 4 \times \left(\frac{22}{7}\right)^2 \times (100)^2$$

$$a_m = 3943.84\ m/s^2$$

4. **Ans. (a)**

Solution: If the length of cable is increased then the capacitance increases therefore capacitance becomes double if the length becomes double.

$$Frequencies \propto \frac{1}{\text{Capacitance}} = \frac{1}{2C}$$

$$\frac{f_{new}}{f_{old}} = \frac{1/2C}{1/C} = \frac{1}{2}$$

$$f_{new} = \frac{1}{2} f_{old} = \frac{1000}{2} = 500\ Hz$$

5. **Ans. (d)**

Solution:

Seismic Instrument: Acceleration Response

Impressed harmonic motion is,

$$x = x_o \cos \omega t$$

Acceleration of impressed motion

$$x'' = -\omega^2 x_o \cos \omega t$$

Amplitude of acceleration

$$A_o = \omega^2 x_o \qquad \text{...(7)}$$

Expression of equation (6)

$$\frac{z_o}{x_o} = \frac{(\omega/\omega_n)^2}{\sqrt{\left(1-\left(\frac{\omega}{\omega_n}\right)^2\right)^2 + 2\delta\left(\frac{\omega}{\omega_n}\right)^2}}$$

Can be written as

$$\frac{z_o}{\omega^2 x_o} = \frac{1}{\omega_n^2\sqrt{\left(1-\left(\frac{\omega}{\omega_n}\right)^2\right)^2 + 2\delta\left(\frac{\omega}{\omega_n}\right)^2}}$$

$$\frac{z_o}{A_o} = \frac{1}{\omega_n^2}\frac{1}{\sqrt{(1-r^2)^2 + (2\delta r)^2}}$$

Where $r = \frac{\omega}{\omega_n}$

The relative displacement z_o between the seismic mass and the frame, therefore, can be used as measured of the amplitude of acceleration of the frame. The term ω_n^2 is fixed for a instrument. The term under the root sign depends on the frequency of excitation w, damping remaining constant for a instrument.

Seismic Accelerometer

Let the input frequency ω be much smaller that the natural frequency ω_n, then

$$(x_2 - x_1) = \frac{\omega^2 A}{\omega_n^2} = \frac{1}{\omega_n^2} \times \text{maximum acceleration}$$

And this relation remains valid for $\frac{\omega}{\omega_n} \le 0.4$.

This if the pick up is to be used for acceleration measurement, ω_n should be large, i.e. the system should be have a stiff spring and small mass. That would enable to operate the system over a wide range of frequencies and still keep the response linear.

6. **Ans. (a)**

Solution:

$E = 9 \times 10^{10}\ N/m^2$

g = voltage sensitivity = 4500 V/μm

$$= \frac{4500 \times 10^6}{E}\left(\frac{V/m}{N/m^2}\right)$$

V = voltage output = 127.3 V

We know that,

$$V = g.t.\frac{F}{A}$$

Where t = thickness
F = applied load
A = Area

$$\therefore \quad F = \frac{VA}{g.t}$$

$$= \frac{127.3 \times \pi \times (0.01)^2 \times 9 \times 10^{10}}{4 \times 4500 \times 0.002 \times 10^6}$$

$$= 99.98\ n \approx 100\ N$$

7. **Ans. (d)**

Solution:

First order temperature measuring system equation

$$T_o = T_i\left(1 - e^{\frac{-t}{\tau}}\right)$$

As the temperature of oil bath is sinusoidal, the sinusoidal component of output also has same amplitude, i.e. 10°C

8. **Ans. (a)**

Solution:

Charge Q = dF = 2 × 10^{-12} (C/N) × 0.1 sin 10t
= 0.2 sin 10t (pC)

Voltage developed,

$$V = \frac{Q}{C_P} = \frac{0.2}{1600}\sin(10t)\,volts$$

Output voltage, $$V_o = \left[\frac{\omega C_P R_L}{\sqrt{1+\omega^2 (C_P + C_L)^2 R_L^2}}\right] \times \frac{dF}{C_p}$$

$$= \left[\frac{10\times1600\times10^{-12}\times10^{8}}{\sqrt{1+100(2600)^2\times10^{-24}\times10^{+16}}}\right]\frac{0.2}{1600}$$

= 0.071 mV

For negative cycle V_o = 0.071 mV

Then amplitude = 0.141 mV

Another Method

Charge Q = Sensitivity × Force

$= 2\times10^{-12}\times0.1\sin10t$

$= 0.2\times10^{-12}\sin10t$

$= 0.2\sin10t(pC)$

$\omega = 10$ rad/sec

Voltage developed (V) $= \frac{Q}{C_p}$

$$= \frac{0.2}{1600}\sin10t\,V$$

$$= 0.125\sin10t(mV)$$

$R_F \Rightarrow$ Resistance of feedback circuit.

$$R_F = 10^8 \parallel X_C$$

$$X_C = \frac{1}{\omega C} = \frac{1}{10\times10^{-9}}$$

$$= 10^8\ \Omega$$

$$\therefore \quad R_F = 10^8 \parallel 10^8$$

$$= 0.5\times10^8\Omega$$

$R_1 \Rightarrow$ resistance of the inverting terminal of the op-am

$$C_p = 1600\,pF$$

$$= 1600\times10^{-12}F$$

Piezo transducer reactance

$$X_{Cp} = \frac{1}{\omega C_p}$$

$$= \frac{1}{10\times1600\times10^{-12}}$$

$$= \frac{10^{12}}{16000}$$

Leakage resistance $R_L = 10^{12}$

$$R_1 = R_L \parallel X_{Cp}$$

$$= \frac{10^{12}\times\frac{10^{12}}{16000}}{10^{12}+\frac{10^{12}}{16000}} \approx \frac{10^{12}}{16000}$$

$$V_{out} = \frac{-R_F}{R_1}$$

$$|V_{out}| = \frac{R_F}{R_1}(V)$$

$$V_{out} = \frac{0.5\times10^8}{10^{12}}\times16000\times0.125\times10^{-3}$$

$$= 0.1\times10^{-3}$$

$$= 0.1mV$$

V_{out} is nearest to 0.141 mV

9. Ans. (d)

Solution:

In Seismic accelerometers, decreasing mass alone will increase nature frequency without affecting the steady state sensitivity.

10. Ans. (a)

Solution:

Given mass m = 100 μ gm

Frequency f = 1 kHz

Sensitivity of accelerometer $= \frac{O/P}{I/P}$

$\omega_n = \sqrt{\frac{K}{m}}$ where K is spring constant.

$$\omega_n^2 = \frac{K}{m},\ [\because \omega = 2\pi f]$$

$$(2\pi\times1k)^2 = \frac{K}{100\,\mu gm}$$

$$\therefore K = 39.47$$

$$\text{Sensitivity} = \frac{O/P}{I/P}$$

$$= \frac{\Delta x}{a} = \frac{m}{K} = \frac{100\,\mu\,gm}{39.47} = 2.58\ nm/s^2$$

Sensitivity in nm/g $= \frac{2.58}{g}\,nm/g$

$$= 2.58\ nm/g$$

11. Ans (c)

Solution:

Sensitity of piezoelectric type accelerometer

$$= 100 \frac{mv}{g}$$

$$= \frac{100 \times 10^{-3} v}{gm}$$

Means

For 1 gm of input , output changes by $= 100 \times 10^{-3}$

For 1 gm of input, output changes by

$$= 5 \times 100 \times 10^{-3} V$$

$$= 5 \times 10^{-1} V = 0.5 V$$

12. Ans. (b)

Solution:

Given piezoelectric material constant

$$g = 50 \times 10^{-3} \text{ Vm/N}$$

Surface area of load cell $A = 4\ cm^2$

Torque $\tau = 20\ Nm$

Torque = force × Arm distance

$$T = Fr \Rightarrow F = \frac{T}{r} = \frac{20}{0.5} = 40\ N$$

Polarization $P = \frac{F}{A}$

Thickness of load circuit

Output voltage of the piezoelectric

$$= g\,t\,P$$

$$= g\,t\left(\frac{F}{A}\right)$$

$$= 50 \times 10^{-3} \times 1 \times 10^{-3} \times \frac{40}{4 \times 10^{-4}}$$

$$= 5 \text{ Volts}$$

13. Ans.(d)

Solution:

$$A_0 = 500g = 500 \times 9.81 = 4905 \text{ m/s}^2$$

$$\xi = 0.2$$

$$\omega_n = 10000 \text{ Hz} = 2\pi \times 10000$$

$$= 6.28 \times 10^4 \text{ rad/s}$$

$$\omega = 2000 \text{ Hz} = 2\pi \times 2000$$

$$= 1.256 \times 10^4 \text{ rad/s}$$

$$r = \frac{\omega}{\omega_n} = 0.2$$

$$\frac{Z_0}{A_0} = \frac{1}{\omega_n^2 \sqrt{(1-r^2)2 + (2\xi r)^2}}$$

$$Z_0 = \frac{A_0}{\omega_n^2 \sqrt{(1-r^2)2 + (2\xi r)^2}}$$

$$Z_0 = \frac{4905}{(6.28 \times 10^4)^2} \times \frac{1}{\sqrt{(1-(0.2)^2)^2 + (2 \times 0.2 \times 0.2)^2}}$$

$Z_0 = 1.295 \times 10^{-6}$m

(Amplitude of output displacement)

$= 1.295 \times 10^{-3}$m

Sensitivity of piezoelectric crystal = 250 V/mm.

14. Ans. (b)

Solution:

The output voltage for a piezo-electric pick up is given by:

$$V_0 = g\,t\,P$$

$$= 0.05 \times 0.002 \times (15 \times 10^5) = 150 \text{V}$$

15. Ans. (a)

Solution:

$$\omega = 2\pi f = 2\pi \times 850.24 \text{ rad/s}$$

acceleration amplitude

$$= 16g = 16 \times 9.81 = 156.96 \text{ m/s}^2$$

$$\text{velocity amplitude} = \frac{\text{acceleration amplitude}}{\omega}$$

$$= \frac{156.96}{50.24} = 3.124 \text{ m/s}$$

16. Ans. (c)

Solution:

$$\text{displacement amplitude} = \frac{\text{velocity amplitude}}{\omega}$$

$$= \frac{3.124}{50.24} = 0.0622 \text{ m}$$

17. Ans. (d)

Solution:

$$\text{velocity vibration level} = 20 \log_{10} \frac{A_1}{A_0}$$

$$= 20\log_{10}\frac{3.124}{10^{-3}} = 69.89 \text{ dB}$$

The above calculation is with respect to velocity reference level of 10^{-3} m/s.

18. Ans. (b)

Solution: $\xi = 0.7$ ω = 2p × 20 rad/s

ω = 2π × 30 rad/s

$$r = \frac{\omega}{\omega_n} = \frac{30}{20} = 1.5$$

$$\text{Amplitude ratio} = \frac{z}{x} = \frac{r^2}{\sqrt{(1-r^2)+(2\xi r)^2}}$$

$$= \frac{(1.5)^2}{\sqrt{(1-(1.5)^2+(2\times 0.7\times 1.5)^2}}$$

$$\frac{z}{x} = 0.92$$

At this frequency (30Hz) output error of −8% occurs

$$\phi = \tan^{-1}\frac{2\xi r}{1-r^2}$$

$$= \tan^{-1}\frac{-20.7\times 1.5}{1-(1.5)^2}$$

$$= \tan^{-1}(-1.68)$$

$$\phi = 120.8°$$

19. Ans (a)

20. Ans (b)

Solution:

$$\text{Sensitivity} = \frac{100\ mV}{g} = \frac{output\ voltage}{5\ g}$$

$$\text{Output voltage} = 500\times 10^{-3}\ V$$

$$= 0.5\ V$$

21. Ans (b)

Solution:

Given

Area $A = 0.5\times 10^{-4}\ \text{m}^2$

Density $\rho = 2$ pC/N

Force $F = 30\times 10^{-3}\sin 150t$

Thickness $t = 1\times 10^{-3}$ m

$$\text{Voltage sensitivity } g = \frac{\rho}{\varepsilon_o \varepsilon_r}\ Vm/N$$

$$= \frac{2\times 10^{-12}}{8.5\times 10^{-12}\times 1} = \frac{2}{8.85}$$

Output voltage across X-tal $V = gtP$

$$= \frac{2}{8.85}\times 10^{-3}\times 1\times \frac{30\times 10^{-3}}{0.5\times 10^{-4}}\sin 150\,t$$

$$= \frac{60\times 2}{8.85}\times 10^{-2}\sin 150\,t$$

$$V_{pp} = 2\,V$$

$$= 2\times 0.135\sin 150\,t$$

$$= 0.27 \text{ volts}$$

22. Ans (c)

23. Ans. (c)

Solution:

Piezoelectric transducers are generally used for the measurement of dynamic forces. It can be used to measure if the time constant ($\tau = RC$) is high. The static forces. Magnitude of output (M) for piezo–electric transducers is given as :

$$M = \frac{1}{\sqrt{1+(\omega\tau)^2}}$$

For low time constant (τ) and low frequency

$$M = \frac{1}{\sqrt{1+(\text{low})^2}} = \frac{1}{low} = \infty$$

Hence, the output will become infinite which cannot be observed

If time constants τ is high and for static system, i.e. ω = low

$$M = \frac{1}{\sqrt{1+(low\times high)^2}} = \text{finite}$$

Hence, one can measure static output for high time constants with piezo-electric load cells.

24. Ans (b)

25. Ans (d)

26. Ans (d)

27. Ans (a)

28. Ans (c)

29. Ans (c)

30. Ans (b)

31. Ans (c)

32. Ans (a)

33. Ans (b)

34. Ans (c)
35. Ans (d)
36. Ans (a)
37. Ans (a)
38. Ans (b)
39. Ans (a)
40. Ans (b)
41. Ans (d)
42. Ans (b)
43. Ans (a)
44. Ans (c)
45. Ans (a)
46. Ans (b)
47. Ans (a)
48. Ans (a)
49. Ans (a)
50. Ans (d)
51. Ans (a)

3.3 Resistive-Inductive-Capacitive Transducer

1. Ans. (c)

Solution:

Metal wire strain guage is having lowest gauge factor least suitable for static pressure measurement

Metal wire strain gauge frequently used for dynamic measurement.

2. Ans. (a)

Solution:

Semi conductor strain gauge has higher gauge factor as act $\pm$ 130 which is more than the metal wire strain gauge.

3. Ans. (c)

Solution:

Emf developed by the thermocouple mainly depends on materials used, hot and cold junction temperature.

4. Ans. (b)

Solution: Gauge factor = strain × excitation

$$= 500\times10^{-3}\times4$$

$$= 2000\times10^{-3}$$

$$= 2$$

Second Method

For quarter bridge output $V_o = \frac{V_s}{4}\times G.F\times\varepsilon$

$$1\,mV = \frac{4}{4}\times G.F\times500\times10^{-6}$$

or, $$G.F = \frac{10^{-3}}{500\times10^{-6}}$$

$$= 2$$

5. Ans. (c)

Solution:

Thermocouple that can measure temperature in the range of 1300°C to 1400°C is R type of thermocouple which has range of $-50°\ C$ to $1700°\ C$. Its composition is 87% platinum, 13% (platinum + Rhodium).

6. Ans. (a)

Solution:

LVDT output winding is connected wrongly so its output is constant.

7. Ans. (d)

Solution:

Strain gauge which has negative Gauge factor and high gauge factor is N-type semiconductor.

Strain Gauge

Strain gauge consists of a fine resistance wire which is pasted on the member subjected to strain. The resistance of the gauge change with strain and this principle is used to measure stress, force, pressure, torque etc.

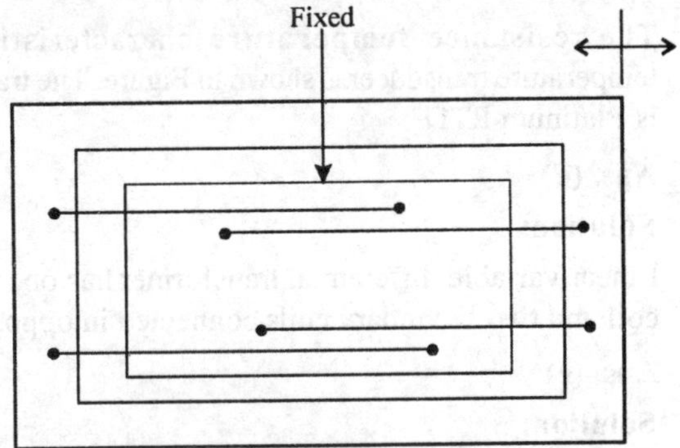

Figure: Unbounded gauge

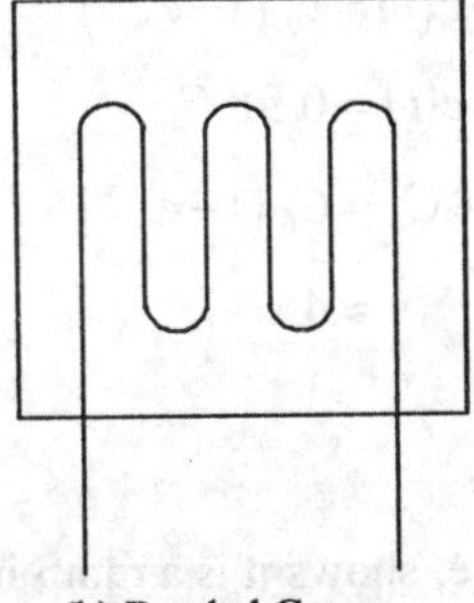

(b) Bonded Gauge

Figure: Bonded gauge

Deposited thin metal film gauges and sputter-deposited thin metal film gauges are termed as thin film gauges that are recently developed. The sensitivity of the strain gauge is expressed as gauge factor which is given by $\frac{\Delta R/R}{\Delta l/l}$. Where R is the resistance value of the strain gauge, ΔR is the change in resistance due to a strain of $\Delta l/l$. Some of the materials used for the construction of metallic strain gauge are Advance (GF = 0.21), Isoelastic (GF = 3.6), Nickel (GF = −12), Nichrome $V(GF = 2.5)$ and Manganise (GF = 0.47). Semiconductor gauges have a large value of gauge factor. The typical values of gauge factor are given below:

Silicon (p-type) → 175;

Silicon (n-type) → −133

Germanium (p-type) → 102;

Germanium (n-type) → 157

when compared with metallic strain gauges semiconductor gauges have sensitivity but also have higher non-linearly and temperature instability

8. Ans. (b)

Solution:

If reference junction is at $20°C$ and the other junction is at $200°C$ then the measured voltage $= 13.5 - 1.2 = 12.3\ mV$

9. Ans. (c)

Solution:

The resistance **temperature** characteristic for a temperature transducer is shown in Figure. The transducer is Platinum RTD

10. Ans. (d)

Solution:

Linear variable differential transformer has one primary coil and two secondary coils connected in opposition.

11. Ans. (c)

Solution:

The output at line t

$$c(t) = c_0\left(1 - e^{-t/\tau}\right)$$

For, $c(t) = 0.98\, C_0$

We have $0.98\, C_0 = C_0\left(1 - e^{-t/\tau}\right)$

Or $t \approx 4\,\tau$

12. Ans. (c)

Solution:

1°C/sec change, shows it is a ramp input

$$q_o = K\left(t - T\left\{1 - e^{-t/T}\right\}\right)$$

$$q_o = 1\left(70 - 0.1\left\{1 - e^{\frac{-70}{0.1}}\right\}\right)$$

$$q_o = 69.9°C$$

Second Method

Error is measurement when a ramp of magnitude 'K' is applied is $e_{ss} = K\,\tau$

Where τ = time constant = 100 m sec.

$$e_{ss} = \frac{1°C}{\text{sec}} \times 100 \times 10^{-3}\ \text{sec}$$

$$= 0.1°C$$

The reading of thermocouple is

$$= 70° - 0.1\,°C = 69.9\,°C$$

13. Ans. (a)

Solution:

To avoid the effect of fringing, the potential of guard ring of a **capacitance** transducer is to be made equal to the ground potential.

14. Ans. (a).

Solution:

For accurate filtering and demodulation of the signal, a frequency ratio of 10:1 is followed:

∴ Maximum frequency of displacement is limited to

$$= \frac{2kHz}{10} = \frac{2000}{10} = 200Hz$$

15. Ans. (c)

Solution:

$$\theta_i = 30°\,C;\ \theta_F = 100°C\ ;\ t = 30$$

$T =$ Time constants

$$\theta_t = 96.5°C$$

$$96.5 = 100\left[1 - e^{\frac{30}{T}}\right] + [30]\,e^{\frac{-30}{T}}$$

$$96.5 = 100 - 100e^{\frac{30}{T}} + 30\,e^{\frac{-30}{T}}$$

$$-3.5 = -70\,e^{\frac{30}{T}}$$

$$\ln\left(\frac{3.5}{70}\right) = -\frac{30}{T}$$

$$T = 10$$

Time required to each $98°C$

$$98 = 100\left[1 - e^{-t/10}\right] + [30]e^{-t/10}$$

$$98 = 100 - 70\,e^{-t/10}$$

$$\ln\left(\frac{1}{35}\right) = e^{-t/10}$$

$$t = 35.6 \text{ sec}$$

16. Ans. (d)

Solution:

Themocouple is a temperature measuring device, which is based on the principle of ***seeback effect i.e*** when the two dissimilar metals are joined together, an emf will exist between the two point A and B which is function of the function temperature.

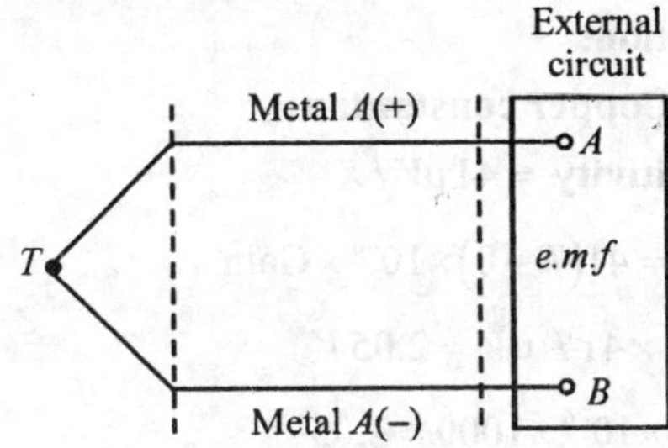

Figure (a)

Figure: Phenomenon of seeback effect is shown diagrammatically

The e.m.f. so produced between the terminals A and B is given by

$$E = a(\Delta\theta) + b(\Delta\theta)^2 + c(\Delta\theta)^3$$

Where,

$E =$ emf developed in mv

$\Delta\theta =$ temperature different between the measuring (hot) junction and the reference junction (°C)

$a, b, c =$ constant

The curve between the emf generated an the change in temperature is parabolic

$$a >> b >> c$$

$$E = a\,\Delta\theta$$

Which tries to improve the non-linearity of the thermocouple.

Measurement of temperature with thermocouple

The figure below shows the correct arrangement for the functioning of the thermocouple. The two junctions, junction (A) which is the hot junction (The junction whose temperature is to be measured) and other is the junction (B) which is the cold (the reference junction) are shown. Since, according to seeback effect, the emf depends on the difference between the temperature of the hot junction and the reference junction. The reference junction should be held constant so that there are no error on account of the change in the ambient temperature. The temperature of the reference junction is usually kept at 0°C

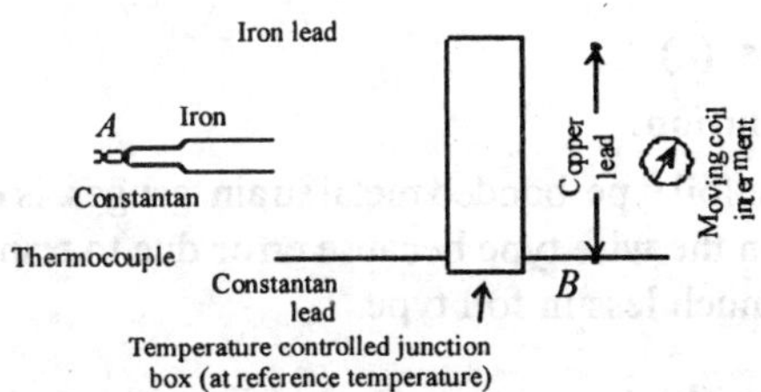

Figure (b)

Figure: Practical thermocouple

In present case

Difference in the temperature between the hot and the cold junction is

$$\Delta T = 300\,°C - 100\,°C$$

The temperatures T_3 and T_4 do not have any effect since the metal is same (i.e. constantan)

According to first law, the application of heat on a single homogenous metal do not produce any electric current.

Sensitivity $= 55\,\mu V/°C$

$\therefore$ for $1\,°C$ output $= 55\,\mu V$

for $200\,°C$ output $= 55 \times 200\,\mu V$

$= 110 \times 100\,\mu V$

$= 11\,mV$

17. Ans (a)

Solution:

The correct answer is option (a) because here we will fined maximum change in resistance as

$$R = \rho\frac{l}{a}$$

So $R \propto l$

So maximum length will change in figure P

Second Method

The expansion the strain gauge is in the same direction of the expansion of the canilever beam. Hence maximum change in resistance occur 'P' configuration.

18. Ans. (d)

Solution: $f = f \pm f_m$

$= 400 \pm 10\,Hz$ $= 390\,Hz$ and $410\,Hz$

19. Ans. (d)

Solution: The correct match is given below

Type E Chromel – Constantan	High output
Type J Fe – Alumel	Good stability
Type T Cu – constantan	Low cost
Type S Pt – PtRh	High precision

20. Ans. (a)

Solution:

The foil type bonded metal strain gauges, is more popular than the wire type because error due to transverse strain is much less in foil type

21. Ans. (b)

Solution:

$$C = 6.94 \times 10^{-3}\left(\frac{d^2}{s}\right)$$

Displacement sensitivity,

$$\frac{dC}{dS} = -\left(6.94 \times 10^{3}\right) d^2/S^2$$

$$= -\left(6.94 \times 10^{-3}\right)\frac{20^2}{(0.25)^2} = -44.4\ pF/mm$$

Second Method

We know that

$$C = \frac{\varepsilon A}{d}$$

Displacement sensitivity $\frac{\partial C}{\partial d} = -\frac{\varepsilon A}{d^2}$

$$S = \frac{-6.94 \times 10^3 \times \left(20 \times 10^{-3}\right)^2}{\left(0.25 \times 10^{-3}\right)^2}$$

$$S = -44.4\ pF/mm$$

22. Ans. (b)

Solution:

$$C_{gauge}\ factor = \frac{\frac{\Delta R}{R}}{\frac{\Delta l}{l}} = \frac{\frac{35}{100}}{\frac{0.02}{20}} = \frac{7}{2} = 3.5$$

23. Ans. (b)

Solution:

If the linear model of RTD is used in the range 0-100ºC the governing equation is given by

$$R_T = R_0\left(1 + 0.004\,T\right)$$

Let $R_0 = 100\ \Omega$

$$R_T = 100\left(1 + (0.004 \times 100)\right)$$

$$= 140\ \Omega$$

If $R_0 = 98\ \Omega$

$$R_{T1} = 98(1 + 0.004 \times 100)$$

$$R_{T1} = 137.2\ \Omega$$

Worst case error $= R_T - R_{T1}$

$$= 140 - 137.2$$

$$= 2.8\ \Omega$$

24. Ans. (b)

Solution:

Among the four, Nichrome constantan has highest sensitivity

$$25 - (-35) = +60\ \mu V/K$$

25. Ans. (d)

Solution:

For Copper constantan

Sensitivity $= 41\ \mu V/K$

emf $= 41(T - 0) \times 10^{-6} \times \text{Gain}$

$$\text{Gain} \times 41\,T\ \mu V = 2.05\ V$$

$$41\,T \times 10^{-6} \times 1000 = 2.05$$

$$T = 50^\circ C$$

26. Ans. (b)

Solution:

$G_f = -100$, $R = 1000$ and Strain $= 100 \times 10^{-6}$

Gauge factor, $G_f = \frac{\Delta R/R}{\Delta l/l}$

Or $\frac{\Delta R}{R} = -100 \times 100 \times 10^{-6} = -10^{-2}$

Or $\Delta R = 1000 \times \left(\frac{1}{-100}\right) = -10\ \Omega$

Then, resistance of the gauge = 1000 – 10 = 990 Ω

27. Ans. (a)

Solution:

$$V_- = V_+ = v_o$$

equal charge on both capacitors $Q = VC$

Then $(v_o - E)(C + \Delta C) = |(C - \Delta C)(v_o + E)$

or $\frac{v_o - E}{v_o + E} = \frac{C - \Delta C}{C + \Delta C}$

or $\frac{v_o}{E} = \frac{C}{\Delta C}$

or $v_o = \frac{EC}{\Delta C}$

Second Method

From the figure writing the nodal equation at '+' terminal

Let voltage at '+' terminal of Op-amp $= V_i$

$$= \frac{V_i - E}{1/S(C+\Delta C)} + \frac{V_i - (-E)}{1/S(C-\Delta C)} = 0$$

$$(V_i - E)(C+\Delta C) = (V_i + E)(C - \Delta C)$$

$$V_i(C+\Delta C - C + \Delta C) = E(C - \Delta C + C + \Delta C)$$

$$2V_i \Delta C = 2EC$$

$$V_i = E\frac{C}{\Delta C}$$

Here $V_o = V_i$

28. Ans. (a)

Solution:

Capacitance between the plates at angle θ,

$$C_\theta = \frac{\varepsilon_o \pi r^2 (\pi - \theta)}{2d}$$

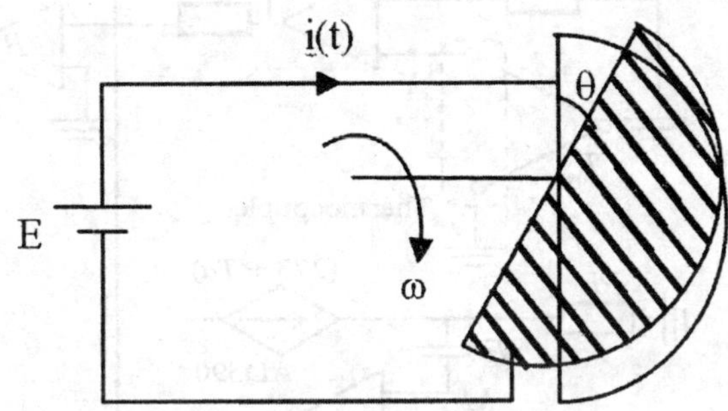

Where, d = plate separation

Then,

$$q_\theta = C_\theta E$$

$$q_\theta = \frac{E\varepsilon_o \pi r^2}{2d} \cdot (\pi - \theta)$$

$$i(t) = \frac{dq_\theta}{dt} = -\frac{E\varepsilon_o \pi r^2}{2d} \cdot \frac{d\theta}{dt}$$

$$= -\frac{E\varepsilon_o \pi r^2}{2d} \cdot \omega$$

$i(t) = A$ constant for dc voltage

Second Method

It is observed from that the figure the current $i(t)$ is a function of 'ω' which is constant and hence $i(t)$ is also constant.

29. Ans. (d)

Solution:

When 200 kΩ resistance is connected in parallel to 120 Ω, then the net resistance is given by 119.928 Ω.

Here, $\Delta R = (119.928 - 120) = -0.0719$

$$G.F = \frac{\frac{\Delta R}{R}}{\varepsilon}$$

$$\varepsilon = \frac{\frac{\Delta R}{R}}{G.F} = \frac{\frac{0.0719}{120}}{2}$$

$$= -299.82\ \mu\ m/m$$

$$\approx -300\ \mu\ m/m$$

–sign represents compressive strain.

30. Ans. (b)

Solution:

$$E_{100} = 5268\ \mu V$$

$$E_{300} = 16325\ \mu V$$

It there is linear relationship then E_{100} must be

$$= \frac{E_{300}}{3}$$

$$\therefore E_{100} = \frac{16325}{3}$$

$$= 5441.66\ \mu V$$

$$\%\text{ age error} = \frac{5268 - 5441.66}{16325} \times 100$$

$$= -1.06\%$$

Second Method

Given, that the voltages developed by iron-constantan thermocouple with reference at 0°C are

$$E_{100} = 5268\ \mu V$$

$$E_{200} = 16325\ \mu V$$

Sensitivity of thermocouple according to linearity relation

$$= \frac{16325 - 0}{300 - 0}$$

$$= 54.41\ \mu V / ^\circ C$$

According to linearity, voltage at 100 °C

$$= V_{o100} = 54.41\ \mu V / ^\circ C \times 100\ ^\circ C$$

$$= 5441\ \mu V$$

Error due to Non linearity at 100 °C as percentage of full scale

$$= \frac{5268-5441}{16325} \times 100$$

$$= -1.06\%$$

31. Ans. (a)

Solution: $V(T) = V(1+\beta T)$

Where, V(T) is function of volume as a function of temperature.

Since cross-section area is constant, therefore

$$ah = V(1+\beta T)$$

$$\Rightarrow \quad V = \frac{ah}{1+\beta T}$$

$$\therefore \quad a\,h' = a\frac{h}{1+\beta T}$$

So, $$h' = \frac{h}{1+\beta T}$$

32. Ans. (c)

Solution:

Change in measured junction temperature.

100 to 200

Change in reference junction temperature

$= 0^\circ$ to $45^\circ C$

% Error $= \frac{M-T}{T} \times 100$

E_{IC} (Measured value) $= 53 \times (100-45)$

$= 2915$

E_{IC} (True value) = 53 × (100 – 0) = 5300

% Error $= \frac{2915-5300}{5300} \times 100 = 45\%$

Second Method

At 100 °C

$E_{IC} = 53\,T\,\mu V = 53(100-45)\,\mu V = 2915\,\mu V$

At $200^\circ C$

$E_{IC} = 53(200-45) = 53 \times 155 = 8215\,\mu V$

From the analysis

Measured value	True value	Error
At 100°C	2915 μV 5300 μV	45%
At 200°C	8215 μV 10,600 μV	22.5%

Maximum error = 45%

33. Ans. (a)

Solution:

As given

$$E_{IC(iron\text{-}constantan)} = 53 \times T\,\mu V$$

$$E_{CC(constantan\text{-}copper)} = 43 \times T\,\mu V$$

$$\therefore E_{(ICO)(iron\text{-}copper)} = (53-43)\,T\mu V$$

$$= (10T\mu V)$$

When the measuring junction is at 100°C and the reference junction is at 30°C then result change in temperature

$$T = 100-30$$

$$= 70\ ^\circ C$$

Emf developed $= 10 \times 70\,\mu V$

$= 700\,\mu V$

34. Ans. (c)*

Solution:

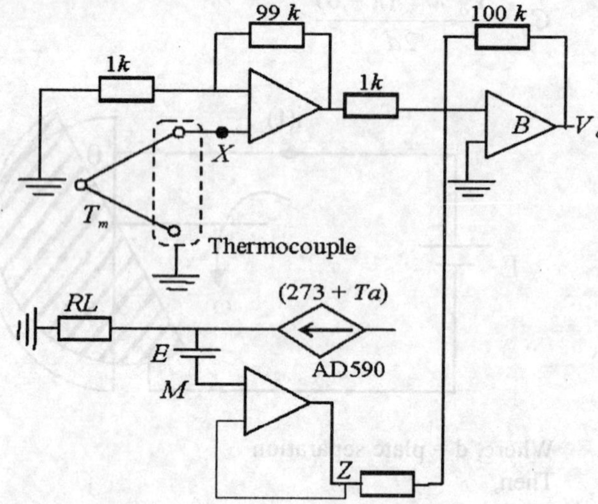

It can be observed from the above circuit diagram, that the op-amp 'B' is behaving as a summer circuit, with two arms as the input (i.e. am A and am C).

The thermocouple is placed in the circuit A, measuring temperature TM and the reference junction at the ambinent temperature (T_a), for output to be error free, T_a must be kept 0°C but here it has some finite temperature. So we have to compensate for this error by using a compensation circuit, i.e. (current C) here.

Let the T_m be 100°C, and T_a be at 30°C

∴ The emf produced by the thermocouple at point Y

$= (100-30)53 \times 10^{-3}\,\mu V = 3710\,\mu V$ (for T_a = 30°C)

The op-amp 'A' is configured in the non-inverting configuration

$$E_x = \left[1+\frac{99}{1}\right] E_y = 100\times 3710\,\mu V = 371\,mV$$

But to have the zero error. T_a must be at 0°C

So Ey for T_a = 0°C is

$$E_y = (100-0)53\times10^{-3}\,\mu V = 5300\,\mu V$$

And $E_x = \left(1+\frac{99}{1}\right)5300\,\mu V$

$$= 530\,\mu V$$

Therefore, error $= (E_x)T_a = 0^o - (E_x)_{T_a} = 30^o$

$$= 530 - 372\,mV = 159\,mV$$

This error must be compensated by the output of the op-amp 'C'. For op-amp 'C' all the current coming from AD 590 sensor will pas through the resistor R, (since no current goes inside the op-amp)

$\therefore$ E_m = Voltage across $R_L - E$

$$= (273+30)\times10^{-6}\times R_L - E$$

$$= 330\times10^{-6}\times R_L - E \quad \text{...(1)}$$

Op-amp 'C' is used in the voltage follower configuration

$\therefore$ $E_2 = E_M$

and inorder to provide temperature compensation

(E_z = error voltage i.e. = 159 mV) ...(2)

Putting equation (2) in equation (1) we get

$$330\times10^{-6}\times R_L - E = 1598\times10^{-3}$$

By hit and trial clearly option (c) satisfies the above equation

Hence $R_L = 5.3\,k\Omega$

$E = 1.447\,V$

35. Ans. (b)

Solution:

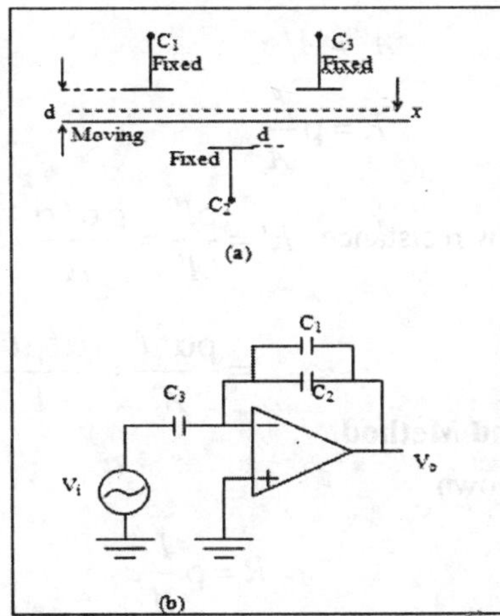

Let movable plate move to up direction, then

$$C_1 = \frac{A\varepsilon_0}{d-x},$$

$$C_2 = \frac{A\varepsilon_0}{d+x}$$

$$C_3 = \frac{A\varepsilon_0}{d-x}$$

From figure (b), op-amp is used in non-inverting mode

$$C_F = C_1 \,||\, C_2 = \frac{A\varepsilon_0}{d-x} + \frac{A\varepsilon_0}{d+x}$$

$$= A\varepsilon_0 \frac{(d+x+d-x)}{(d+x)(d-x)}$$

$$= \frac{A\varepsilon_0(2d)}{(d+x)(d-x)}$$

$$X_F = \frac{1}{\omega C_F}$$

$$C_3 = \frac{A\varepsilon_0}{(d-x)}$$

$$X_1 = \frac{1}{\omega C_3}$$

$$\left|\frac{V_0}{V_i}\right| = \left|\frac{X_F}{X_1}\right|$$

or

$$\frac{V_0}{V_i} = \frac{X_F}{X_1}$$

$$= \frac{\dfrac{(d+x)(d-x)}{2d\,A\varepsilon_0}}{\dfrac{(d-x)}{A\varepsilon_0}}$$

$$= \frac{d+x}{2d} = \frac{d}{2d} + \frac{x}{2d}$$

$$= \frac{1}{2} + \frac{x}{2d}$$

36. Ans. (d)

Solution:

The strain gauge is considered at maximum operating temperature

Hence due to maximum temperature, the resistance of the gauge change to $R_t = R_0(1+\alpha\Delta t)$

$$= 120(1+2\times10^{-5}\times150)$$

$$= 120(1+0.03)$$

$$= 120.36$$

Change in resistance due to temperature variation is

$$(\Delta R) = R_t - R_0$$
$$= 120.36 - 120$$
$$= 0.36$$

or, $\frac{(\Delta R)_t}{R} = \frac{0.36}{120}$ ... (1)

Resistance error due to the tensile strain only is $\frac{\Delta R}{R}$

$$\frac{\Delta R}{R} = G_f \frac{\Delta L}{L} = G_f \varepsilon \quad \text{... (2)}$$

Where, $\varepsilon = \frac{\Delta L}{L} \rightarrow$ Strain $= 2\times10^{-2}$

$G_f \rightarrow$ gauge factor = 2

$$\% \text{ resistance change} = \frac{\Delta R / R}{G_f \varepsilon} \times 100$$

$$= \frac{\frac{0.36}{120}}{G_f \in} .\times 100$$

$$= \frac{0.36}{2\times2\times10^{-2}\times120} .\times 100$$

$$= 0.075\times100$$
$$= 7.5$$

37. Ans. (d)

Solution:

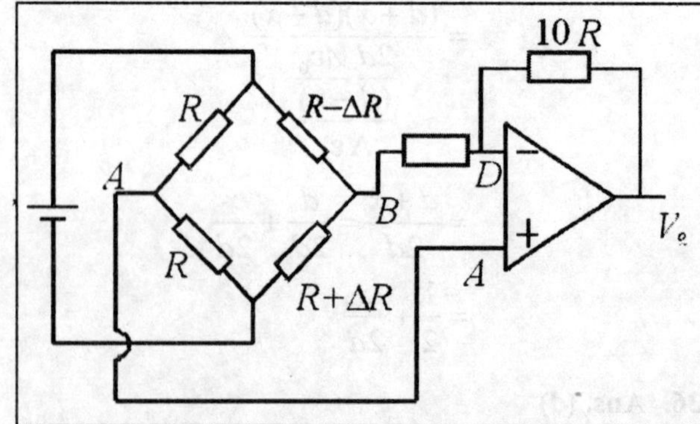

Voltage at point A

$$V_A = \frac{R}{R+R}V = \frac{V}{2} \quad \text{... (1)}$$

Voltage at point B, $V_B = \frac{R+\Delta R}{R+\Delta R+R-\Delta R}V$

$$= \frac{R+\Delta R}{2R}V = \left(\frac{1}{2} + \frac{\Delta R}{R}\right)V \quad \text{... (2)}$$

By the concept of virtual ground in op-amps. $V_D = V_A$

$$\therefore \quad V_D = \frac{V}{2} \quad \text{... (3)}$$

Apply KCL at point D

$$\frac{V_B - V_D}{R} = \frac{V_D - V_O}{10R}$$

$$\frac{V_B - V_D}{1} = \frac{V_D - V_O}{10}$$

$$10V_B + V_O = 11V_D \quad \text{... (4)}$$

From equation (2) and (4)

$$10\left(\frac{1}{2} + \frac{\Delta R}{R}\right)V + V_0 = 11V_D$$

From equation (2) and (5)

$$V\left(5 + 10\frac{\Delta R}{R}\right) + V_0 = 11\frac{V}{2}$$

$$V\left(5 + 10\frac{0.36}{120}\right) + V_0 = 11\frac{V}{2}$$

$$V(5+0.03) + V_0 = 5.5V$$

$$V_0 = (5.5 - 5.03)V$$

$$\frac{V_0}{V} = 0.47$$

$$\frac{V_0}{V} \simeq 0.4$$

38. Ans. (b)

Solution:

Since volume is constant

$$lA = l'A'$$
$$lA = \alpha \ell . A'$$
$$\Rightarrow \quad A' = A/\alpha$$
$$R = \rho\frac{\ell}{A}$$

$\therefore$ New resistance, $R' = \frac{\rho l'}{A'} = \frac{\rho.\alpha l.\alpha}{A}$

$$= \frac{\rho\alpha^2\ell}{A} = \frac{\alpha^2\rho\ell}{A} = \alpha^2 R$$

Second Method

As known

$$R = \rho\frac{l}{A}$$

Where ρ is the property of the material, so it will be constant for same material.
For constant volume

$$R = kI^2$$

$$\therefore \quad R' = k(\alpha I)^2$$

$$= \alpha^2 R$$

39. Ans. (c)

Solution:

As given

$\overline{V_1} = 1.0V\angle 0^0,$

$\overline{V_2} = 1.0V\angle 10^0$

The null voltage of the above shown LVDT is given by

Null voltage $= \overline{V_1} - \overline{V_2}$

$$= 1.0\angle 0^\circ - 1.0\angle 10^\circ$$

$$= 1(\cos 0^\circ + i\sin 0^\circ) - 10(\cos 10^\circ + i\sin 10^\circ)$$

$$= 1(1 + i0) - 10(0.9848 + i\,0.1736)$$

$$= 0.0152 - i\,0.1736$$

Null voltage, $\Delta V_o = |V_1 - V_2|$

Magnitude of null voltage

$$= \sqrt{(0.0152)^2 + (0.1736)^2}$$

$$= \sqrt{0.03036}$$

$$= 0.1742$$

40. Ans. (b)

Solution:

Characteristic of parallel plate capacitor transducer

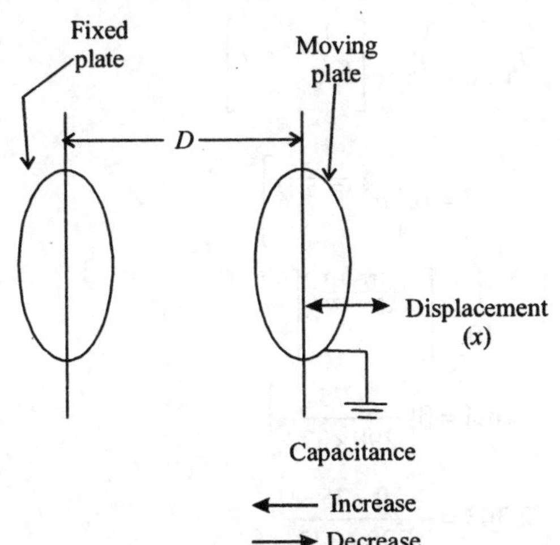

The capacitance can be varied by the change of distance 'd' between the plates

One of the two plates is kept fixed and the other plate is made moveable, the displacement to be sensed is applied to the movable plate.

$$C = \frac{A\varepsilon}{d}$$

$$C \propto \frac{1}{d}$$

The capacitance increase as the distance decreases and vice-versa

There is a inverse relationship between the capacitance and the distance 'd', hence the response become completely non-linear, i.e. hyperbolic

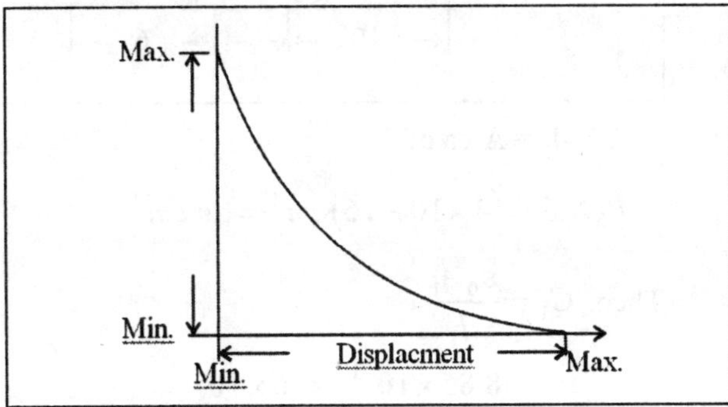

Sensitivity $S = \frac{\partial C}{\partial x} = \frac{\varepsilon A}{x^2}$

The sensitivity of this transducer is not constant but varies over the range of transducer

Main Point

1. Transducer using the change in distance between plates is used only for the measurement of small displacement
2. The sensitivity of this transducer can be increased to any desirable value by making the distance the plates extremely smaller.

Energy stored in capacitor

$$We = \frac{1}{2}CV^2 = \frac{1}{2}\cdot\frac{\varepsilon_o A}{x}V^2$$

$We =$ force ×distance

$$= F \cdot x$$

$$F = -\frac{\partial We}{\partial x} = \frac{1}{2}\frac{\varepsilon_o A}{x^2}V^2 \propto \frac{1}{x^2}$$

41. Ans. (b)

Solution:

$$\tan\theta = \frac{10}{20} = \frac{1}{2}$$

$$\tan\theta = \frac{EC}{DE} = \frac{1}{2}$$

⇒ BC = 2, BD, = AC – EC = 5 – 3 = 2

$$A_1 = \text{Area of } ABCD = 4.3 + \frac{1}{2}.4.2 = 16\,cm^2$$

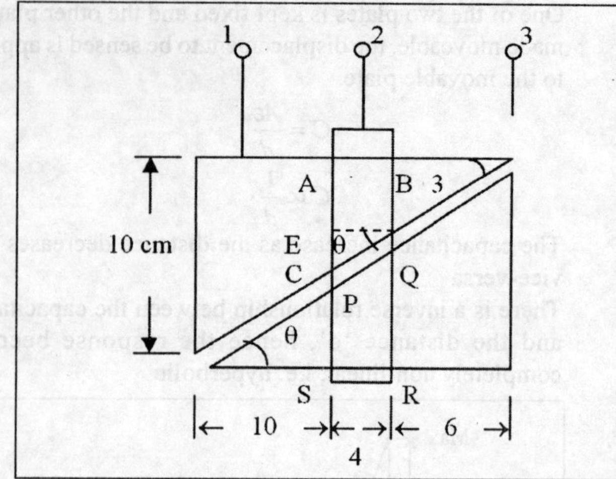

A_2 = Area of

$$PQRS = (4 \times 10 - 16)\,cm^2 = 24\,cm^2$$

Then, $C_1 = \dfrac{\varepsilon_0 A_1}{d_1}$

$$= \frac{8.85 \times 10^{-12} \times 16 \times 10^{-4}}{10^{-3}} = 14.17\ pF$$

$$C_2 = \frac{\varepsilon_0 A_1}{d_2}$$

$$= \frac{8.85 \times 10^{-12} \times 24 \times 10^{-4}}{10^{-3}} = 21.25\ pF$$

42. Ans. (b)

Solution:

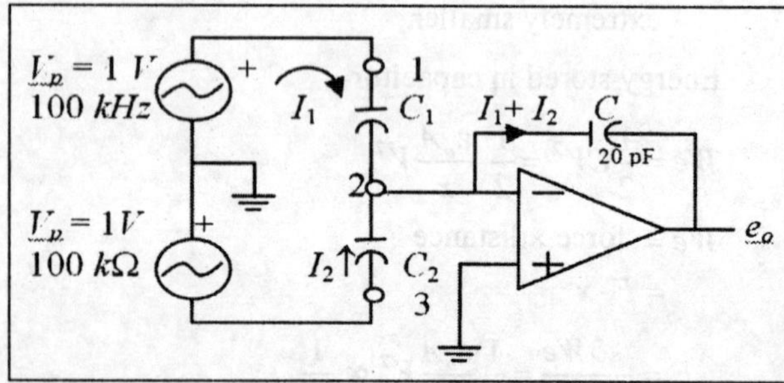

$$I_1 = V_p\, j\omega C_1$$

$$I_2 = V_p j\omega C_2$$

$$e_o = (I_1 + I_2)\frac{1}{j\omega C} = \left(\frac{C_1 + C_2}{C}\right)V_p$$

$$= 0.354 \text{ volts}$$

Second Method

Apply KCL at node 2

$$\frac{0-1}{1/sC_1} + \frac{0-(-1)}{1/sC_2} + \frac{0-V_o}{1/sC_1} = 0 \quad \text{...(1)}$$

Substituting the value of

$C_1 = 14.17\ pF$ and

$C_2 = 21.25\ pF$ in equation (1)

$\Rightarrow V_o = 0.345\ V$

43. Ans (d)

Solution:

$$E_{ss} = \lim_{t \to \infty}[\theta(t) - \theta_L(t)]$$

We known that E_{ss} = input – output. But in this case output – input is given. So we are getting negative answer. We know that $\theta_L(t)$ = Ramp. Signal, so by derivation we are getting $= -k\tau$

44. Ans (b)

Solution:

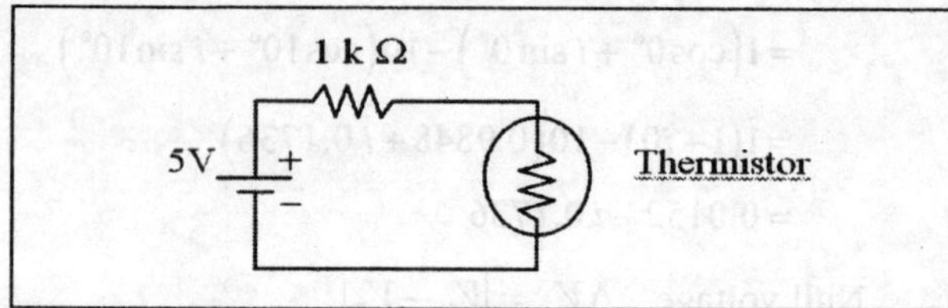

$$P = I^2 R$$

$$R_{25°C} = 10\ k\Omega \quad \text{... (1)}$$

$$R_{100°C} = 1\ k\Omega \quad \text{... (2)}$$

$$R_{150°C} = ?$$

$$R_{150} = R_{25}\left[e^{\beta\left(\frac{1}{423} - \frac{1}{298}\right)}\right] \quad \text{... (3)}$$

For calculation of β, solve equations (1) and (2)

$$R_{100°C} = R_{25°C}\left[e^{\beta\left(\frac{1}{373} - \frac{1}{298}\right)}\right]$$

$$1 = 10\left[e^{\beta\left(\frac{1}{373} - \frac{1}{298}\right)}\right]$$

$$\frac{1}{10} = \left[e^{\beta\left(\frac{298-373}{298 \times 373}\right)}\right]$$

$$\ell n.1 = \beta\left[\frac{-75}{298 \times 373}\right]$$

$$-2.303 = -\frac{\beta \times 75}{298 \times 373}$$

$$\beta = \frac{2.303 \times 298 \times 373}{75}$$

$\beta = 3412.55$

Put β in equation (3)

$$R_{150} = 10\,k\Omega\left[e^{3412.55\left(\frac{1}{423}-\frac{1}{298}\right)}\right]$$

$$R_{150} = 10\,k\Omega\left[e^{\left\{\frac{3412.55}{423\times298}\times(298-423)\right\}}\right]$$

$$R_{150} = 10\,k\Omega\left[e^{\left\{\frac{3412.55}{423\times298}\times(-125)\right\}}\right]$$

$$R_{150} = 10\,k\Omega\left[e^{(-3.384)}\right]$$

$$R_{150} = 10\times10^3\,[0.03391]$$

$$R_{150} = 339.11\,\Omega$$

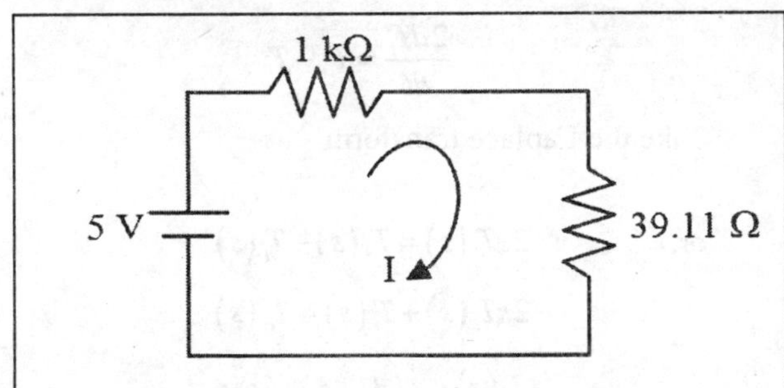

$$I = \frac{5}{1339.11}$$

$$I = 3.73\,mA$$

$$P = I^2 R_{Themistor} = (3.73)^2 \times 339.11$$

$$= 4.7\,mW$$

45. Ans (d)

Solution:

$$X_C = \frac{1}{j\omega C}$$

Given $\frac{\Delta C}{C} = 5\%$

$$= \frac{5}{100}$$

The input voltage to be differential amplifier is determined as (V_{in})

$$V_{in} = 10\left[\frac{R}{R+R} - \frac{\frac{1}{(C-\Delta C)}}{\frac{1}{C+\Delta C}+\frac{1}{C-\Delta C}}\right]$$

$$= 10\left[\frac{1}{2} - \frac{C+\Delta C}{C-\Delta C+C+\Delta C}\right]$$

$$= 10\left[\frac{1}{2} - \frac{C+\Delta C}{2C}\right]$$

$$= 10\left[\frac{1}{2} - \frac{C}{2C} - \frac{\Delta C}{2C}\right]$$

$$= -10\left[\frac{\Delta C}{2C}\right] = -\frac{10\Delta C}{2C}$$

$$\therefore \quad V_{output} = V_{in} \times \text{gain}(A)$$

$$\therefore \quad A = \frac{V_{out}}{V_{in}}$$

For the full utilization $V_o = 10$

$$A = \frac{2C}{\Delta C} = 2\times\frac{C}{\Delta C}$$

$$= 2\times\frac{100}{5} = 40$$

46. Ans (b)

Solution:

The sensitivity of the above system is independent of supply voltage.

47. Ans (b)

Solution:

All metal gauges have low guage factor of the factor of the order 2.5 because the conductivity of the material is almost independent of the applied strain, i.e. piezoresistive effect is almost indelible.

48. Ans (d)

Solution:

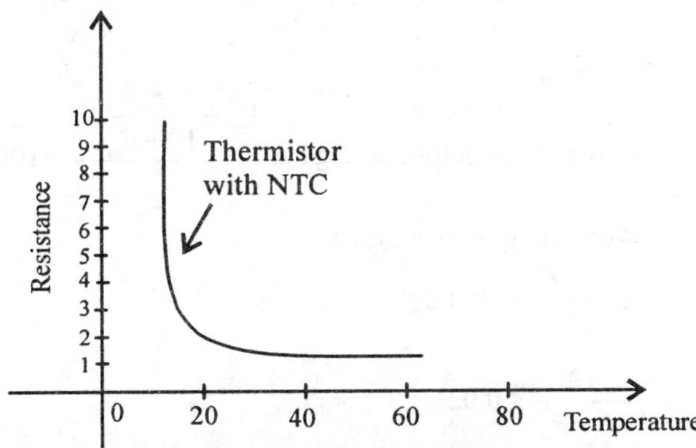

N.T.C. = It means-Negative temperatame resistance

$$R_T = a\,R_o\,e^{(b/T)}$$

here a,b are constant

R_o = Resistance at $0^\circ\ C$ i.e., constant for any material

$R_T \propto e^{1/T}$

So as Temperature increases So resistance decreases. Resistance decreases exponentially

49. Ans (a)

Solution:

LVDT :- It means linear variable differential Transformer.

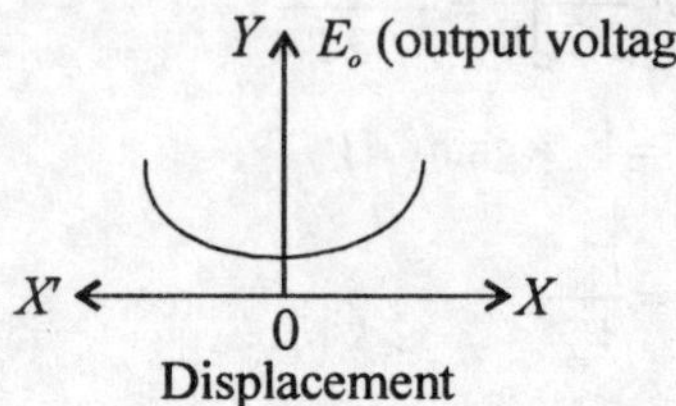

Because output voltage depends upon displacement So it is a displacement Transducer.

50. Ans. (c)

Solution:

When the displacment is same in the opposite direction then the output voltage will have the same magnitude but the output is 180° out of phase.

Hence the reading of voltmeter = 1 V

51. Ans (c)

Solution:

The outpt resistance of transducer $= 10\ k\Omega$

Let the input resisance of Amplifier = R_{in}

Transducer and amplifier connectin is given below

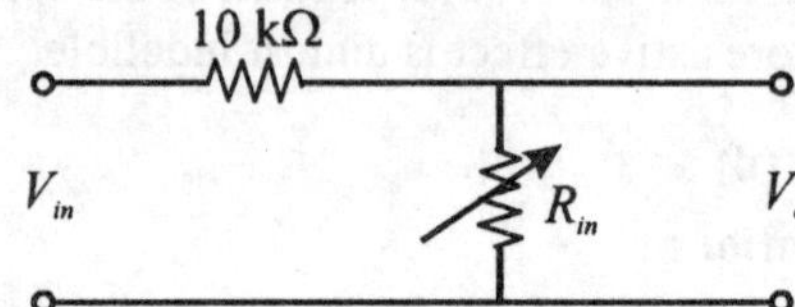

Error in the output voltage $= \dfrac{V_{in} - \dfrac{R_{in}}{10 + R_{in}} V_{in}}{V_{in}} \times 100\%$

Accroding to given data

$$1 - \frac{R_{in}}{10 + R_{in}} \le 0.02$$

$$\frac{10}{10 + R_{in}} \le 0.02$$

$$10 \le 0.2 + 0.02\ R_{in}$$

$$0.02\ R_{in} \ge 9.8$$

$$R_{in} \ge 490$$

$$R_{min} = 490\ k\Omega$$

52. Ans (d)

Solution:

Poisson Ratio $= 35$

Gauge Factor $\Rightarrow$?

We know that

$GF \Rightarrow 1 + 2V$

where $GF \Rightarrow$ Gauge Factor.

$V \Rightarrow$ Poisson Ratio.

$$GF = 1 + 2(0.35)$$

$$\Rightarrow 1 + 0.70 = 1.7$$

53. Ans. (a)

Solution: Characteristic of thermometer

$$\frac{2dT_i}{dt} = T_a - T_i$$

Take the Laplace transform

or, $\quad 2sT_i(s) = T_a(s) - T_i(s)$

$\quad 2sT_i(s) + T_i(s) = T_a(s)$

or, $\quad \dfrac{T_i(s)}{T_o(s)} = \dfrac{1}{2s+1} = \dfrac{0.5}{s+0.5}$

$\Rightarrow \quad 0.5\, e^{-0.5t}$

Hence, $\quad \omega = 0.5$

$$f = \frac{\omega}{2\pi} = \frac{1}{4\pi} Hz$$

54. Ans. (c)

Solution:

Given the inpput frequency of sinusoidal source given to LVDT = 3 kHz

For LVDT the input freqency is same as the output frequency.

Hence the freqency of voltage observed at the input of the low-pass filter is also 3 kHz.

55. Ans. (b)

Solution: $E = 38.740\theta + 3.3 \times 10^{-2}\theta^2 + 2.07 \times 10^{-4}\theta^3 - 2.2 \times 10^{-6}\theta^4$

$$\frac{dE}{d\theta} = 38.74 + 3.3 \times 2 \times 10^{-2}\theta + 2.07 \times 10^{-4} \times 3 \times \theta^2 - 2.2 \times 10^{-6} \times 4 \times \theta^3$$

Put $\theta = 100^\circ C$

$$\frac{dE}{d\theta} = 38.74 + 6.6\times10^{-2}\times10^{2} + 2.07\times3\times10^{-4}\times10^{4}$$

$$-2.2\times10^{-6}\times4\times10^{6}$$

$$\frac{dE}{d\theta} = 38.74 + 6.6 + 6.21 - 8.8$$

$$\frac{dE}{d\theta} = 42.75\mu V/{}^{0}C$$

56. Ans. (b)

Solution: Range = 0 – 500°C

Current ⇒ 4 – 20 mA

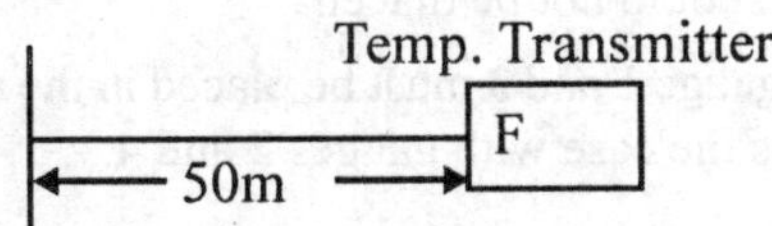

R = 500 W

v = 4 V

$$I = \frac{v}{R} = \frac{4}{500} = 8\text{mA}$$

$$\text{Temp. measured} \Rightarrow \frac{\text{Measured} - \text{Min}}{\text{Max} - \text{Min}} \times \text{temp}$$

$$= \frac{8-4}{20-4}\times500 \Rightarrow \frac{4}{16}\times500 = 125^{0}C$$

57. Ans. (b)

Solution:

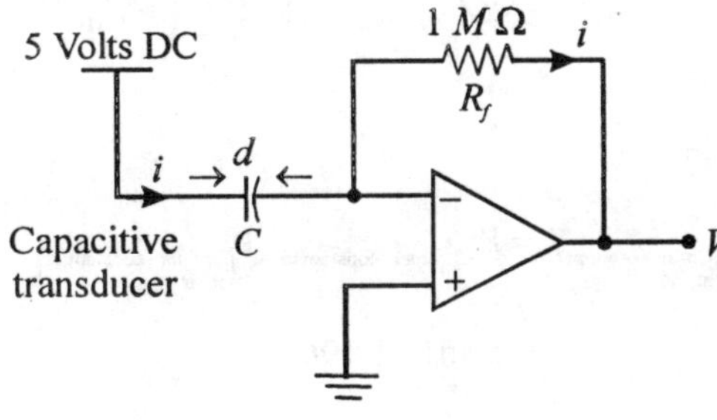

$$\because\ i = \frac{d\theta}{dt}$$

$$i = \frac{d}{dt}(CV)$$

Here V is constant i.e. 5 volts

$$i = V\frac{d}{dt}C$$

$$C = \frac{\varepsilon_o A}{d}$$

$$i = V\frac{d}{dt}\left(\frac{A\varepsilon_o}{d}\right)$$

$$= V A\varepsilon_o\left(\frac{-1}{d^2}\right)\cdot\frac{d}{dt}\left[10^{-3}(1+0.1)\sin(1000\,\pi t)\right]$$

$$= \left[\frac{-VA\varepsilon_o\left[10^{-3}\times0.1\times1000\,\pi\cos(100\pi t)\right]}{d^2}\right]$$

$$= \left[\frac{-VA\varepsilon_o\left[10^{-3}\times0.1\times1000\,\pi\cos(100\pi t)\right]}{\left[10^{-3}\{1+0.1\times1000\,\pi\cos(100\pi t)\}\right]^2}\right]$$

at $t = 0$ msec. C = 2 pF

$$\therefore\ 2\times10^{-12} = \frac{A\varepsilon_o}{10^{-3}[1+0.1\sin0]}$$

$$= \frac{A\varepsilon_o}{10^{-3}}$$

$$A\varepsilon_o = 2\times10^{-15}$$

at $t = 2$ msec

$$i = \frac{-5\times2\times10^{-15}\left[10^{-3}\times0.1\times1000\,\pi\cos\left(1000\,\pi\times2\times10^{-3}\right)\right]}{\left[10^{-3}\left(1+0.1\sin\{1000\,\pi\times2\times10^{-3}\}\right)\right]^2}$$

$$= \frac{-10\times10^{-15}\times0.1\,\pi}{10^{-6}}$$

$$= -10^{-9}\,\pi$$

$$V_o = -i\,R_F$$

$$= 10^{-9}\times10^{6}\pi = \pi\times10^{-3}\text{ volts}$$

$$= \pi\,mV$$

58. Ans. (c)

Solution:

Here cantilever beam is used for measurement of acceleration. Hence the output of the strain gauge is proportional to $\frac{d^2x}{dt^2}$

59. Ans (d)

Solution

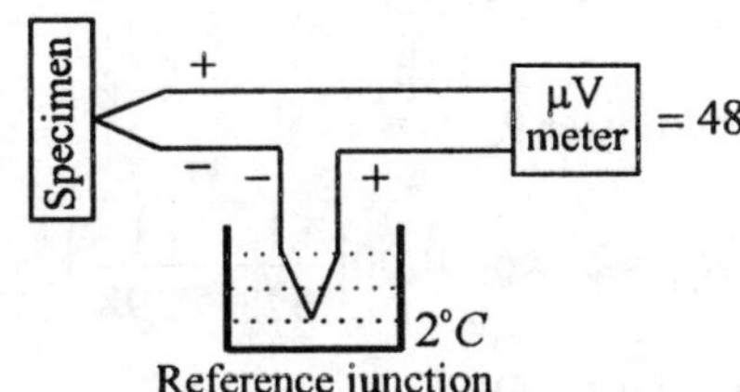

For general tempeature analysis initial value is not consider due to Hystreris loss.
In general central value to be considered
Thermocouple Sensitivity (S)

$$=\frac{\Delta\,\text{output}}{\Delta\,\text{input}}=\frac{\Delta\,\text{output}(\mu\,\text{V})}{\Delta\,\text{temperature}(^\circ\text{C})}$$

For any two readings at mid place,

Let 40^oC and 50^oC

$$S=\frac{(85-75)}{(50-40)}=\frac{10}{10}=1\,\mu V/^oC$$

$$=1\times10^{-6}V/C$$

Hence

$$S\left[\text{T}_{\text{specimen}}-\text{T}_{\text{reference}}\right]=\text{meter reading (output)}$$

therefore $1\times10^{-6}\ (T_{specimen}-2)\ =48\times10^{-6}$

$T_{specimen}-2=48$

Therefore, $T_{specimen}=50^oC$

60. Ans (b)

Solution :

It seem to be **the case of a quarter bridge**

Gauge factor $G_f=2.5$

strain $\varepsilon=100\,\mu m/m=100\times10^{-6}m/m$

$$E=4V$$

$$V_0=\frac{E}{4}G_f.\varepsilon$$

$$=\frac{4}{4}\times2.5\times100\times10^{-6}=250\mu V$$

61. Ans (b)

Solution:

- To reduce/avoid the effect of fringing in capacitive type tranducer a guard ring is provided which is kept at ground potential

62. Ans (d)

Solution:

$$R_{T1}=R_{T2}\,.\exp\left[\beta\left(\frac{1}{T_1}-\frac{1}{T_2}\right)\right]$$

For thermistor P,

$$R_{T1}=2\exp\left[4000\left(\frac{1}{373}-\frac{1}{298}\right)\right]=0.1345$$

For thermistor Q,

$$R_{T1}=2\exp\left[3000\left(\frac{1}{373}-\frac{1}{298}\right)\right]=0.264$$

Then, ratio of resistance of thermistor P to thermistor Q

$$=\frac{0.1345}{0.264}=0.50$$

63. Ans (a)

Solution:

The compresive and tensile strain guages should be adjacent in bridge for measurement of output
1 and 3 should not be adjacent
2 and 4 should not be djacent
i.e. the gauge 1 and 3 must be placed in the oppsite arms and so is the case with gauges 2 and 4.

64. Ans (d)

Solution:

The e.m.f produced at the point (B) and (C) will be zero. Since it is a combination of the similar metals (in order to generate e.m.f the junction must be made up of dissimilar metals). Point (A) is at 30°C, from the given data and is a combination of chromel and copper.

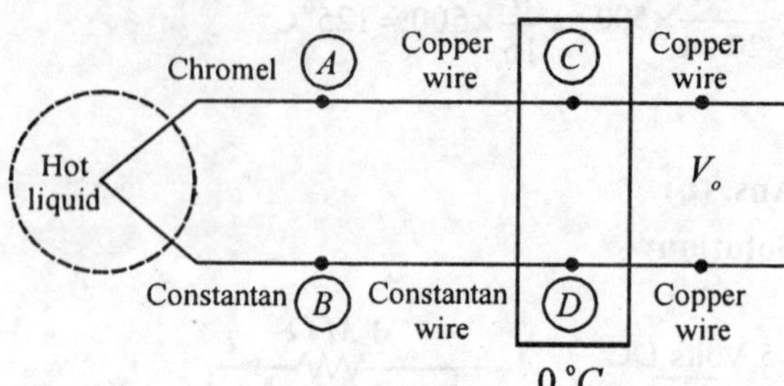

at 30 °C

$$V_{\binom{\text{Chromel-copper}}{\text{at }A}}=V_{(\text{Chromel-constantan})}-V_{\binom{\text{Copper-constantan}}{\text{at D}}}$$

$$=1.801-1.196$$

$$=0.650\,mV$$

Hence $\quad V_o=V_{\text{hot junction}}-0.605$

$$V_{\text{hot junction}}=26.74+0.605$$

$$=27.345\,mV$$

It is clearly seen from the figure A that the hot junction is made up of chromel-constantan.

From the table for chromel-constantan junction the voltage output of 27.345 μV corresponds to 380°C.

65. Ans. (b)

Solution:

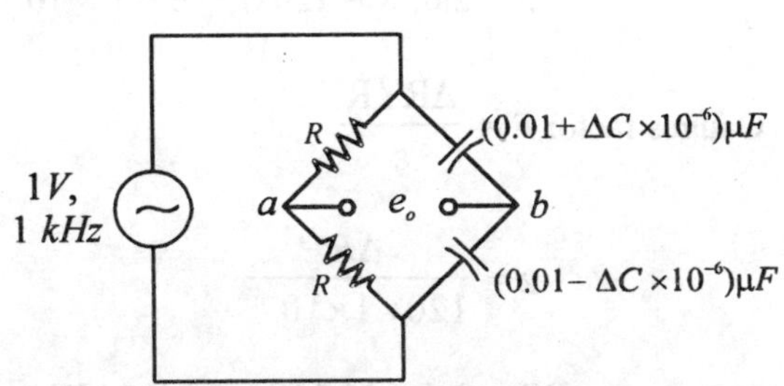

Taking ΔC of the order of pF($10^{-12}F$)

$$V_o = \frac{R}{R+R}\times 1 = \frac{1}{2} = 0.5\,V$$

$$V_o = \frac{0.01+\Delta C+10^{-6}}{0.01+\Delta C\times 10^{-6}+0.01-\Delta C\times 10^{-6}}\times 1$$

$$= \frac{0.01+\Delta C+10^{-6}}{0.02}$$

$$V_o = V_b - V_a$$

$$= \frac{0.01+\Delta C\times 10^{-6}-0.5\times 0.02}{0.02}$$

$$= \frac{\Delta C\times 10^{-6}}{0.02}$$

$$= 50\times 10^{-6}\Delta C$$

$$\frac{V_o}{\Delta C} = 50\times 10^{-6}\,V/pF$$

$$= 0.05\,mV/pF$$

66. Ans (b)

Solution:

Given that $\alpha_o = 0.00392/^oC$

$$R_o = 100\,\Omega$$

As we know $\alpha_o = \frac{1}{R_o}\cdot\frac{R_{Q2}-R_{Q1}}{Q_2-Q_1}$

$$\therefore\quad 0.00392 = \frac{1}{100}\left[\frac{\Delta R}{\Delta Q}\right]$$

$$\Rightarrow\quad \frac{\Delta R}{\Delta Q} = 0.392$$

$$\Rightarrow \Delta R = 0.392\,\Delta Q$$

Now in the given bridge

$$v_0 = v_i\left[\frac{\Delta R_1}{R_1}-\frac{\Delta R_2}{R_2}+\frac{\Delta R_3}{R_3}-\frac{\Delta R_4}{R_4}\right]\frac{R_1\cdot R_2}{(R_1+R_2)^2}$$

Since only one resistance is variable, therefore and $v_i = 10\,V$

$$\therefore\quad v_0 = 10\left[\frac{\Delta R_1}{R_2}\right]\frac{R_1\cdot R_2}{(R_1+R_2)^2}$$

$$v_0 = 10\frac{\Delta R_1}{100}\times\frac{(100\times 10{,}000)}{(100+10000)^2}$$

$$= 10\times\frac{0.392\,\Delta Q}{100}\frac{(100\times 10{,}000)}{(10{,}000+100)^2}$$

$$= \frac{3.92}{10201}\Delta Q$$

$$\therefore\quad \frac{v_0}{\Delta Q} = \frac{3.92}{10201}V/^oC$$

$$= 0.384\times 10^{-3}V/^oC$$

Now sensitivity of $10\,mV/^oC$ for the system is required. If the amplifier gain is K.

Then $0.384\times 10^{-3}\times K = 10\times 10^{-3}$

or $K = 26.02$

67. Ans (b)

Solution:

Characteristic of Resistance Thermometer (RTD)

1. RTD a resistance transducer, whose resistance changes with the change in the temperature.
2. RTD's have positive tempeatue coefficient, i.e. their resistance increase with the increses in temperature.
3. The variation of the resistance with temperature is given as

$$R_T = R_o\left(1+\alpha_1\Delta T+\alpha_2\Delta T^2+\alpha_3\Delta T^3+\ldots+\alpha_n\Delta T^n+\ldots\right)$$

Where

R = Resistance at temperature (0 °C)

ΔT = Change in temperature

R_T = Resistance at temperature (T °C)

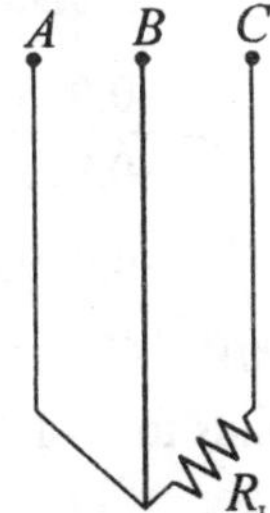

4. An RTD have three leads, a resistance (R_T) as shown in the figure is put in the two arms of the RTD and the third arm in used as reference lead to find the

resistance of the two other leads as it is same as its own since all three leads are made up of same meterial.

$$R_{lead} = R_A = R_B = R_C$$

the resistance returned by arms BC is $R_{BC} = R_T + R_B + R_C$

but we are supposed to find R_T only.

We can find R_T as

$$R_T = R_{BC} - R_B - R_C$$
$$= R_{BC} - R_A - R_A$$
$$= R_{BC} - 2R_A$$

In present case

At 100 °C the resistance of RTD becme

$$R_{RTD} = R_o(1+0.00392\times 100)$$

$$=100(1+0.392)$$

$$=100\times 1.392$$

$$=139.2\,\Omega$$

actual gain $=25$

Dut ot bridge non-linearity gain is 26.02

$$V_A = 0.099V$$

$$V_B = \frac{139.2}{10000+139.2}\times 10\,V$$

$$= \frac{139.2}{10139.2} = 0.1370\,V$$

$$V_{BC} = 0.137 - 0.099$$

$$= 0.0396$$

$$(V_{out})_{26.02} = 0.0382\times 26.02$$

$$= 0.00396$$

$$\%\text{ error in voltage} = \frac{(V_{out})_{25} - (V_{out})_{26.02}}{(V_{out})_{25}}$$

$$= \frac{0.95-0.993}{0.955}$$

$$= \frac{-0.038}{-0.955}$$

$$= -0.0397$$

Sensitivity $=10\; mV/^{o}\,C$

10 mV corresponds to $\Rightarrow$ 1 °C

$$-0.0397 \text{ correspond to } \Rightarrow \frac{1}{10\times 10^{-3}}\times(-0.0397)$$

$$= 0.397$$

$$= -0.4$$

68. Ans. (b)

Solution:

F = 2.0, R = 120W, e = 1 × 10^{-6}

$$\text{Gauges factor F} = \frac{\Delta R/R}{\varepsilon}$$

$$2 = \frac{\Delta R}{120\times 1\times 10^{-6}}$$

or DR = 2.4 × 10^{-4} W = 0.00024 W

Percentage change in resistance

$$= \frac{0.00024}{120}\times 100$$

$$= 0.0002\%$$

To measure changes in resistance as small as of this magnitude, required sensitive instrumentation or circuits.

69. Ans. (b)

Solution:

(a) Due to stress, s = 50 MN/m² = 50 × 10^6 N/M²

$$\text{Strain induced, e} = \frac{\sigma}{E} = \frac{50\times 10^6}{200\times 10^9}$$

$$= 2.5\times 10^{-4}\text{ ms}$$

$$F = \frac{\Delta R/R}{\varepsilon}$$

DR = F e R = 2.0 × 2.5 × 10^{-4} × 120 = + 0.06W

70. Ans. (c)

Solution:

1. Due to temperature: It consists of two components, Due to temperature coefficient of resistance,

$(DR)_{(i)}$ = R g DT = 120 × 20 × 10^{-6} × 30

= (+) 0.072 Ω

2. Due to differential expansion of gauge material and steel specimen

$(DR)_{(ii)}$ = FR $(a_s - a_g)$ DT

= 2 × 120 (12 – 16) 10^{-6} × 30

= (–) 0.028 Ω $(a_g > a_s)$

Note here the relative magnitude of resistance changes. Positive DR is an increase and negative one is decrease in resistance. Further, the two components of resistance change due to temperature have opposite signs.

71. Ans. (b)

Solution:

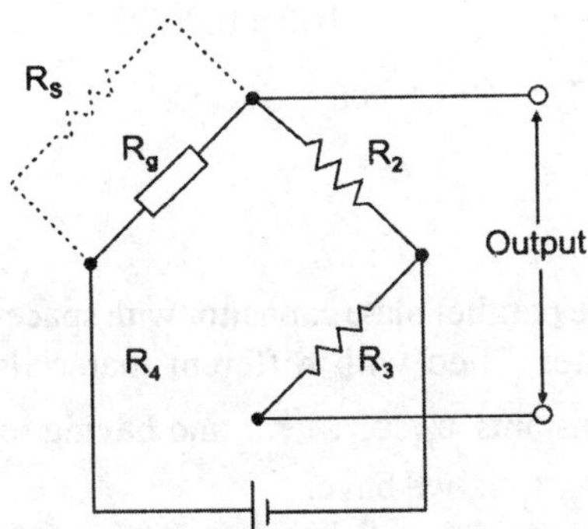

Figure shows a WB circuit shunted by a resistance R_s, strain induced is,

$$\varepsilon = \frac{R_g}{F(R_g + R_s)}$$

$F = 2.0, \quad R_g = 100W$

For, $\varepsilon = 100$ ms $= 100 \times 10^{-6}$

$$100 \times 10^{-6} = \frac{100}{2(100 + R_s)}$$

$$100 + R_s = \frac{100}{2} \times \frac{1}{100 \times 10^{-6}}$$

$$= 50 \times 10^4$$

$R_s = 500000 - 100$

$= 499900\ \Omega$

$R_s = 499.9\ k\Omega$

72. Ans. (c)

Solution:

For, e = 1000 ms = 1000×10^{-6}

$$1000 \times 10^{-6} = \frac{100}{2(100 + R_c)}$$

$100 + R_s = 50 \times 10^3 = 50000$

$R_s = 49900\Omega$ or $49.9\ k\Omega$

73. Ans. (a)

Solution :

F = 2.0

$R_1 = R_2 = R_3$

$= R_4 = 120\Omega$

V = 4.0 V

e = 1mm/m

$R_G = 100W$ (resistance of galvanometer)

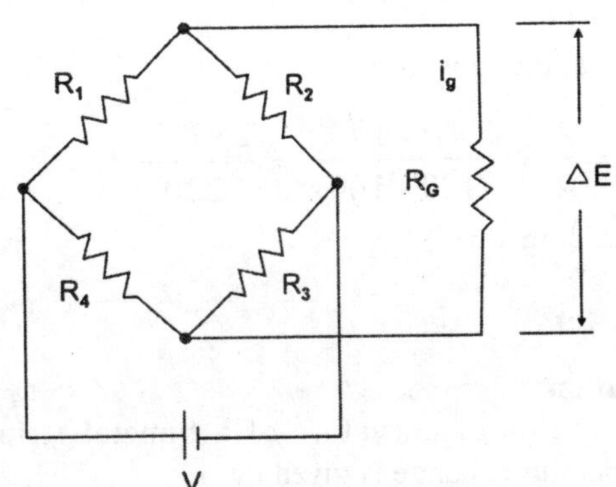

Output voltage (when R_G is very large or open circuit voltage)

$$\Delta E = \frac{V R_1 R_2}{(R_1 + R_2)^2} \times \left(\frac{\Delta R_1}{R_1} - \frac{\Delta R_2}{R_2} + \frac{\Delta R_3}{R_3} - \frac{\Delta R_4}{R_4}\right)$$

$$\Delta E = V \frac{(R_1)^2}{(2R_1)^2}\left[\frac{\Delta R_1}{R_1}\right] = \frac{V}{4}\frac{\Delta R_1}{R_1}$$

$$F = \frac{\Delta R_1 / R_1}{\varepsilon}$$

$$\frac{\Delta R_1}{R_1} = F\,\varepsilon$$

$$\Delta E = \frac{V}{4} = F\varepsilon$$

Substituting value,

$$\Delta E = \frac{4}{4} \times 2 \times 1 \times 10^{-6}\ V = 2\mu V$$

Unbalanced current, considering galvanometer resistance

$$i_g = \frac{V}{4(R_1 + R_G)} \times \left(\frac{\Delta R_1}{R_1} - \frac{\Delta R_2}{R_2} + \frac{\Delta R_3}{R_3} - \frac{\Delta R_4}{R_4}\right)$$

when only one arm has a strain gauge.

$$i_g = \frac{V}{4(R_1 + R_G)}\frac{\Delta R_1}{R_1}$$

$$F = \frac{\Delta R_1 / R_1}{\varepsilon_1}$$

$$\frac{\Delta R_1}{R_1} = F\,\varepsilon$$

$$i_g = \frac{V}{4}\frac{1}{(R_1 + R_G)} = F\,\varepsilon$$

$\varepsilon = 1 \times 10^{-6}$ µm/m

F = 2.0

Substituting values,

$$i_g = \frac{4}{4} \times \frac{2\times(1\times10^{-6})}{(120+100)} = \frac{2\times10^{-6}}{220}$$

$i_g = 9.09 \times 10^{-3}$ µA.

74. Ans. (c)

Solution:

The radius of curvature of a bimetal strip due to temperature change is given by

$$r = \frac{t\left[3(1+m)^2 + (1+mn)\left[m^2 + 1/mn\right]\right]}{6(\alpha_2 - \alpha_1)(T - T_0)(1+m)^2}$$

where,

t = combined thickness of the strip = $t_1 + t_2$

$m = \dfrac{t_2}{t_1}$ = ratio of thickness of low to high expansion materials

$n = \dfrac{E_2}{E_1}$ = ratio of modulii of elasticity of low to high expansion materials

a_1 = lower coefficient of thermal expansion

a_2 = higher coefficient of thermal expansion

T = operating temperature

T_0 = initial bonding temperature

$t = 1.5 + 1.5 = 3 \times 10^{-3}$ m

$m = 1.5/1.5 = 1.0$

T = 110°C

$$n = \frac{1.5\times10^{11}}{2.3\times10^{11}} = 0.652$$

T_0 = 30°C

$$r = \frac{3\times10^{-3}\left[3(1+1)^2 + (1+1\times0.652)\left(1^2 + \frac{1}{1\times0.625}\right)\right]}{6(12.5-1.5)10^{-6}(110-30)(1+1)^2}$$

$$r = \frac{1\times[3\times4 + 1.652\times2.6]10^3}{2(11\times80\times4)}$$

$r = 2.3$ m.

75. Ans. (d)

Solution:

The temperature resistance relation for platinum is,

$R_2 = R_1 + R_0\, a\,(T_2 - T_1)$

$$T_2 = T_1 + \frac{(R_2 - R_1)}{R_0\,\alpha}$$

$$T_2 = 100 + \frac{(295 - 5 - 139.5)}{100\times0.0039}$$

$T_2 = 100 + 400 = 500$°C.

76. Ans. (b)

Solution:

For a multiple parallel plate capacitor with space between adjacent plates filled with different materials having dielectric constants $\varepsilon_{r1}, \varepsilon_{r2}, \varepsilon_{r3}$... and having respective distances t_1, t_2, t_3 ..., we have:

$$C = \frac{\varepsilon_0 A}{\frac{t_1}{\varepsilon_{r1}} + \frac{t_2}{\varepsilon_{r2}} + \frac{t_3}{\varepsilon_{r3}}}$$

Substituting the values from the given data,

$$C = \frac{\varepsilon_0 A}{\frac{0.01}{5} + \frac{0.02}{10}} = \frac{\varepsilon_0 A}{0.004} \quad \text{... (i)}$$

Let e_r be the dielectric constant of the single medium which completely fills the gap of 3 cm between the two parallel plates of area A and gives the same capacitance. Then

$$C = \varepsilon_0\varepsilon_r \frac{A}{t} = \varepsilon_0\varepsilon_r \frac{A}{0.03} \quad \text{... (ii)}$$

Equating the capacitance values given by expressions (*i*) and (*ii*).

$$\frac{\varepsilon_0 A}{0.004} = \varepsilon_0\varepsilon_r \frac{A}{0.03}$$

$$\therefore \quad \varepsilon_r = \frac{0.03}{0.004} = 7.5$$

77. Ans. (c)

Solution:

Before application of pressure:

$$C_1 = \varepsilon_0\varepsilon_r \frac{A}{t_1}$$

After application of pressure: $C_2 = \varepsilon_0\varepsilon_r \dfrac{A}{t_2}$

Given: C_2 = 350µF; t_1 = 4 mm and $t_2 = 4 - 0.75 = 3.25$ mm

$$\therefore\ C_2 = 350\times\frac{4}{3.25} = 430.77\ \mu\text{F}$$

$\therefore$ Change in capacitance $\Delta C = C_2 - C_1$

$= 430.77 - 350 = 80.77$ µF

78. Ans. (a)

Solution:

For a parallel plate capacitor,

$$C = \varepsilon_0 \varepsilon_r \frac{A}{t}$$

Upon differentiation, we get the sensitivity as

$$\frac{dC}{dt} = -\varepsilon_0 \varepsilon_r \frac{A}{t^2}$$

The minus sign indicates a reduction in the capacitance value for increasing t; the the distance between the plates. Apparently, the sensitivirty is high for small values of t.

$$\text{Sensitivity } \frac{dC}{dt} = \frac{8.854 \times 10^{-12} \times 1.00 \times 6.5 \times 10^{-4}}{(0.001)^2}$$

$$= 57.55 \times 10^{-10} \text{ F/m}$$

79. Ans. (c)

Solution:

For a multiple capacitor, the capacitance C is given by

$$C = \varepsilon_0 \varepsilon_r \frac{A}{t}(N-1) \text{ farads}$$

where A = overlapping or effective area between plates
T = distance between plates (m)
N = number of capacitor plates
e_0 = permittivity of free space = 8.95 10^{-12} F/m
e_r = relative permittivity or dielectric constant of the medium between the places. For air, $\varepsilon_r = 1$.

With the given dimension l, width of each plate w and displacement x given to the movable plate on the right side.

$$C = \varepsilon_0 \varepsilon_r \frac{(l-x)w}{t}(N-1)$$

and sensitivity of the transducer is

$$\frac{\partial C}{\partial x} = -\varepsilon_0 \varepsilon_r \frac{w}{t}(N-1)$$

$$= -8.854 \times 10^{-12} \times 1 \times \frac{25 \times 10^{-3}}{0.25 \times 10^{-3}} \times (5 \times 1)$$

$$= -3541.6 \times 10^{-12} \text{ F/m}$$

* Disregarding the sign, the sensitivity of the transducer is 3541.6 pF/m.

80. Ans. (d)

Solution:

Let subscripts 1 and 2 refer to the conditions before and after the application pressure. Then

t_1 = 4 mm, t_2 = 4 – 0.28 = 0.28 = 3.72mm, C_1 = 300 pF and fr_1 = 100 kHz

For an oscillator circuit having inductance L and capacitance C, the resonant frequency is prescribed by the relation.

$$f = \frac{1}{2\pi\sqrt{LC}}$$

For the given setup, inductance is constant and the capacitance is variable. Accordingly

$$f_1 = \frac{1}{2\pi\sqrt{LC_1}} \text{ and } f_2 = \frac{1}{2\pi\sqrt{LC_2}}$$

$$\text{Hence } \frac{fr_2}{fr_1} = \sqrt{\frac{C_1}{C_2}}$$

Since $C_1 = \varepsilon_0 \varepsilon_r \frac{A}{t_1}$ and $C_2 = \varepsilon_0 \varepsilon_r \frac{A}{t_2}$, we get

$$\frac{f_2}{f_1} = \sqrt{\frac{\varepsilon_0 \varepsilon_r A/t_1}{\varepsilon_0 \varepsilon_r A/t_2}} = \sqrt{\frac{t_2}{t_1}}$$

or

$$f_2 = f_1\sqrt{\frac{t_2}{t_1}} = 100\sqrt{\frac{3.72}{4}}$$

∴ Change in frequency of oscillator
= 100 – 96.44 = 3.56 kHz

81. Ans. (a)

Solution:

Within elastic limits

$$\text{Young's modulus} = \frac{\text{stress}}{\text{strain}}$$

$$200 \times 10^9 = \frac{50 \times 10^3/(0.5 \times 10^{-4})}{dl/l}$$

$$\therefore \quad \frac{dl}{l} = \frac{50 \times 10^3}{0.5 \times 10^{-4} \times 200 \times 10^9} = 0.005$$

The gauge factor of a resistance strain gauge is defined as the fractional change in gauge resistance divided by the strain applied to it, i.e.

$$F = \frac{dR/R}{dl/l} = \frac{1/100}{0.005} = 2$$

82. Ans.(d)

Solution: The stain at the yield point would be:

$$\text{Strain} = \frac{\text{yield point stress}}{\text{Young's modulus}}$$

$$\frac{dl}{l} = \frac{200\times10^6}{700\times10^6}0.286\times10^{-2}$$

This is the maximum possible strain that can be applied to the test-specimen.

Now, gauge factor $F = \frac{dR/R}{dl/l}$

∴ Change in the strain gauge resistance,

$$dR = F\frac{dl}{l}R$$

$$= 2 \times (0.286 \times 10^{-2}) \times 120 = 0.686\ \Omega$$

Evidently the change in resistance is very small; only about 0.57% over the whole possible region of strain. Such small changes must be accurately determined by employing a wheatstone bridge circuit incorporating very sensitive galvanometer.

83. Ans. (c)

Solution:

Refer figure. The current flowing through the system is given by:

$$i = \frac{V}{R + R_b}$$

For maximum sensitivity $R = R_b$ and therefore

$$i = \frac{V_s}{2R}$$

and supply voltage

$$V_s = 2iR = 2\times(20\times10^{-3})\times120 = 4.8\text{ V}$$

84. Ans. (d)

Solution:

When strained, the change in output voltage is:

$$dV_0 = \frac{V_s}{4}F\,\varepsilon$$

$$= \frac{4.8}{4}\times2\times(500\times10^{-6}) = 1.2\times10^{-3}$$

$$dV_0 = 1.2\text{ mV}$$

85. Ans. (b)

Solution:

Initially when the bridge is balanced with zero strain, the variable resistor R_2 can be worked out from the equality:

$$\frac{R_1}{R_2} = \frac{R_3}{R_4}$$

$$R_2 = R_1\times\frac{R_3}{R_4} = 99.75\times\frac{10000}{100} = 9975\Omega$$

86. Ans. (a)

Solution:

When strain is applied, let dR_2 be the change in variable resistance R_2 to rebalance the bridge. Then

$$dR_2 = 9997 - 9975 = 22\Omega$$

Change in resistance dR_1 of the strain gauge R_1 is then given by:

$$dR_1 = dR_2\times\frac{R_3}{R_4} = 22\times\frac{100}{10000} = 0.22\Omega$$

87. Ans. (b)

Solution:

By definition: Gauge factor $F = \frac{dR_1/R_1}{dl/l}$

$$\therefore\ \text{Strain}\ \frac{dl}{l} = \frac{1}{F}\frac{dR_1}{R_1} = \frac{1}{2.0}\times\frac{0.22}{99.75} = 1.103\times10^{-3}$$

It would be a tensile strain since R_2 and hence R_1 have increased.

88. Ans. (d)

Solution:

Voltage across one resistor

$$= IR = (15\times10^{-3})\times120 = 1.8\text{ V}$$

Since there are two resistors in each parallel path, the total exciting volatage will be twice the voltage across each resistor.

Exitation (supply) voltage = 1.8 × 2 = 3.6 V

For the quarter bridge (*i.e.*, one active strain gauge), the output voltage is given by

$$V_0 = \frac{V_s}{4}\frac{dR}{R} = \frac{V_s}{4}F\frac{dl}{l}$$

$$\therefore\qquad \text{Strain}\ \frac{dl}{l} = \frac{4V_0}{V_s F}$$

$$= \frac{4\times(5\times10^{-3})}{3.6\times2} = 2.78\times10^{-3}$$

89. Ans. (d)

90. Ans. (c)

91. Ans. (c)

92. Ans. (c)

93. Ans. (c)

94. Ans. (b)

95. Ans. (b)

96. Ans. (b)

97. Ans. (a)

98. Ans. (d)

99. Ans. (d)

100. Ans. (d)

101. Ans. (c)

102. Ans. (d)

103. Ans. (a)

104. Ans. (a)

105. Ans. (b)

106. Ans. (b)

107. Ans. (c)

108. Ans. (a)

109. Ans. (d)

110. Ans. (a)

111. Ans. (b)

112. Ans. (b)

113. Ans. (d)

114. Ans. (b)

115. Ans. (b)

116. Ans. (c)

117. Ans. (b)

118. Ans. (a)

119. Ans. (d)

120. Ans. (c)

121. Ans. (d)

122. Ans. (d)

123. Ans. (b)

124. Ans. (b)

125. Ans. (c)

126. Ans. (b)

127. Ans. (d)

128. Ans. (a)

129. Ans. (c)

130. Ans. (b)

131. Ans. (c)

132. Ans. (a)

133. Ans. (b)

134. Ans. (b)

135. Ans. (a)

136. Ans. (b)

137. Ans. (a)

138. Ans. (b)

139. Ans. (b)

140. Ans. (b)

141. Ans. (c)

142. Ans. (d)

143. Ans. (d)

144. Ans. (c)

145. Ans. (c)

146. Ans. (a)

147. Ans. (b)

148. Ans. (b)

149. Ans. (b)

150. Ans. (d)

151. Ans. (c)

152. Ans. (c)

153. Ans. (b)

154. Ans. (c)

155. Ans. (b)

156. Ans. (d)

157. Ans. (d)

158. Ans. (d)

159. Ans. (b)

160. Ans. (b)

161. Ans. (b)

3.4 Optical, Flow, Pressure, pH Transducers

1. Ans. (d)

Solution:

Pirani gauge operation is based on variation of thermal conductivity of gas with pressure.

2. Ans. (b)

Solution:

In a **seismic pick-up for getting** an output proportional to acceleration it is desirable to have a natural frequency very large in comparison to frequency of input and a damping ratio around 0.7.

3. Ans. (d)

Solution:

Given

Coefficient of friction = μ

Axial load / Force = F

Shaft diameter = d

If the pressure is uniform then the friction moment

$$= F \times \mu \times d$$

$$= F\mu d$$

4. Ans. (a)

Solution:

Measurement of **viscosity** involves measuring Frictional force

Viscosity is a property of a fluid which affect its behavior. If a shearing stress is applied to a fluid it will deform.

Viscosity of a fluid is the property which **resists shearing** and is a **comparative measure of resistance to flow within a fluid**. For Newtonian fluids, the shearing stress experienced by two layers is directly proportional to the velocity gradient between them. Newton's law of viscosity can be expressed as,

Shearing stress $\propto$ velocity gradient

$$\tau = \text{constant}\ \frac{dv}{dy} = \mu \frac{dv}{dy}, \quad \tau = \frac{F}{A}$$

μ is termed as dynamic /absolute viscosity

Figure illustrates an infinitely thin film of fluid of thickness dy. The two faces are having velocity of V and $V + dV$, caused by a shearing force F acting on area A.

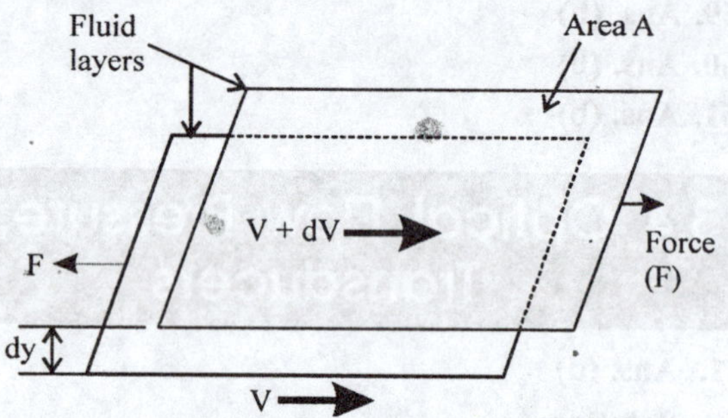

S.I. unit of dynamic viscosity (μ) is Ns/m^2

cgs unit is poise $\left(\frac{\text{dyne-second}}{\text{centrimetre}}\right)$

Another term that occurs frequency is called kinematics viscosity (V)

$$V = \frac{\mu}{\rho} \quad \text{Where, } \rho \text{ is mass density}$$

S. I. unit is m^2/s

Cgs unit is cm^2/s

Measurement of viscosity is important in lubrication and wear studies. Also, in the atomisation of fuel oils such as in fuel injection and for paint spraying, viscosity is an important factor.

5. Ans. (d)

Solution:

Example of variable area flowmeter is rotameter which is designed in special way for which direct reading is there.

Second Method

Flow Nozzles:

A flow nozzle is similar to a venturimeter but without a diverging core Unlike an orifice plate, the jet contraction in case of flow nozzle is less because of its curved shaped on the upstream side.

Note: Since Nozzle have a wise open area, they can be used for the *measurement of flow of solids.*

Rotameter:

- Rotameter is a constant pressure drop type **variable area flowmeter,** •
- It has a vertical tapered tube with a flaot which is free to move within the tube.
- The fluid flow from bottom to top in a rotameter. The float is made of such a diameter such that when no fluid is flowing across the rotameter, the float completely blocks are inlet of the meter.
- The density of float is greater than the density of the fluid flowing through it, so that the buoyant force alone is not sufficient to lift the float
- Rota meters are the constant pressure drop, and variable area type flow meters, all other meters like orifice meter, venture meter and flow nozzle are differential pressure meters.

6. Ans. (a)

Solution:

The primary standard for calibrating vacuum is McLeod gauge

McLeod Gauge

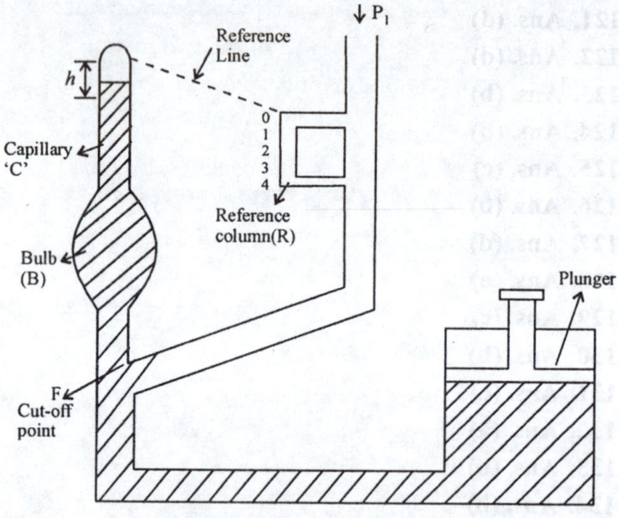

Figure: McLeod Gauge

It is based on the principle of compressing sample of known volume of low pressure gas to a known volume and high pressure which can be read buy a simple manometer.

The McLeod gauge is used for measuring very low pressures down to one hundred thousandth of an inch of mercury (or 10 m bar down to 10^{-5} m bar) The McLeod gauge amplifies pressure by compressing a gas into a small volume. The pressure of the compressed gas is than measured with a mercury manometer.

Construction and Working

The gas enters the gauge through B and fills the tubes down to the level of the mercury reservoir .The pressure is equal throughout the tubes and the bulb mercury is pumped up from the reservoir G as the mercury rises above the cut off, it traps gas inside the bulb The mercury is then pumped higher in the gauge until all the gas in the bulb is compressed into the closed capillary tube The operator allows the mercury to rise until it reference line on the closed capillary tube .The mercury rises faster in the open capillary tube A.

The compression of gas in the closed capillary tube makes the pressure of the trapped gas higher than the measured pressure, This pressure difference causes a difference mercury levels in the two tubes. The difference in height is used to calculate the pressure.

The volume of the bulb can be made quite large and the zero reference line on the closed capillary tube can be placed near the top of the tube. Thus, a large volume of gas can be compressed in to very small volume. This compression multiples the pressure by time. The pressure can be calculated by using the following equations. $P = KH\ H_0(1 - KH)$

Where

P = measured pressure

K = a constant, determined by the geometry of the gauge

H = difference in heights of the two mercury Columns

H_0 = height of the top of the closed Capillary tube above the line marked one the tube

The McLeod gauge is a very accurate pressure measuring device and often serves as *standard for calibrating other low pressure gauges, it can be designed to measure pressures as low as 0.05 microns (00005 torr).*

7. **Ans. (a)**

Solution: McLeod Gauge

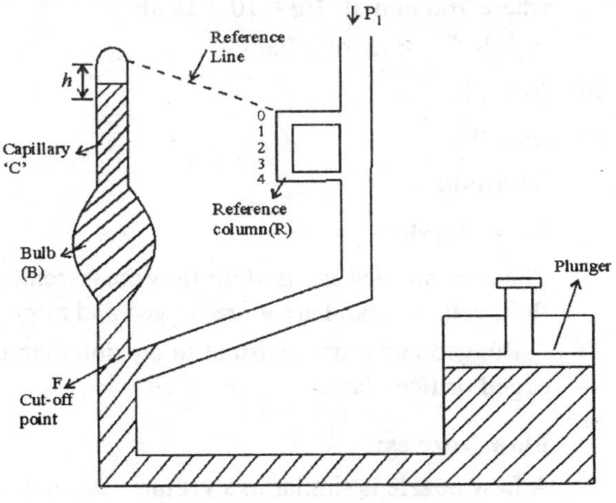

Figure: McLeod Gauge

It is based on the principle of compressing sample of known volume of low pressure gas to a known volume and high pressure which can be read buy a simple manometer.

The McLeod gauge is used for measuring very low pressures down to one hundred thousandth of an inch of mercury (or 10 m bar down to m bar) The McLeod gauge amplifies pressure by compressing a gas into a small volume. The pressure of the compressed gas is than measured with a mercury manometer.

Construction and Working

The gas enters the gauge through B and fills the tubes down to the level of the mercury reservoir .The pressure is equal throughout the tubes and the bulb mercury is pumped up from the reservoir G as the mercury rises above the cut off, it traps gas inside the bulb The mercury is then pumped higher in the gauge until all the gas in the bulb is compressed into the closed capillary tube The operator allows the mercury to rise until it reference line on the closed capillary tube .The mercury rises faster in the open capillary tube A.

The compression of gas in the closed capillary tube makes the pressure of the trapped gas higher than the measured pressure, This pressure difference causes a difference mercury levels in the two tubes. The difference in height is used to calculate the pressure.

The volume of the bulb can be made quite large and the zero reference line on the closed capillary tube can be placed near the top of the tube. Thus, a large volume of gas can be compressed in to very small volume. This compression multiples the pressure by time. The pressure can be calculated by using the following equations

$$P = KH\ H_0(1 - KH).$$

Where

P = measured pressure

K = a constant, determined by the geometry of the gauge

H = difference in heights of the two mercury columns

H_0 = height of the top of the closed Capillary tube above the line marked one the tube

The McLeod gauge is a very accurate pressure measuring device and often serves as standard for calibrating other low pressure gauges, it can be designed to measure pressures as low as 0.05 microns (00005 torr).

Absolute pressure = atmospheric presser + guagepressure

= 100 + 5

= 105 kPa

8. **Ans. (c)**

Solution:

Cantilever Beam: In this a force F is applied at a distance L from the fixed end. The force causes a bending moment and the deflection of the free end. Strains ε_1 and ε_2 induced are equal in magnitude but opposite in sign.

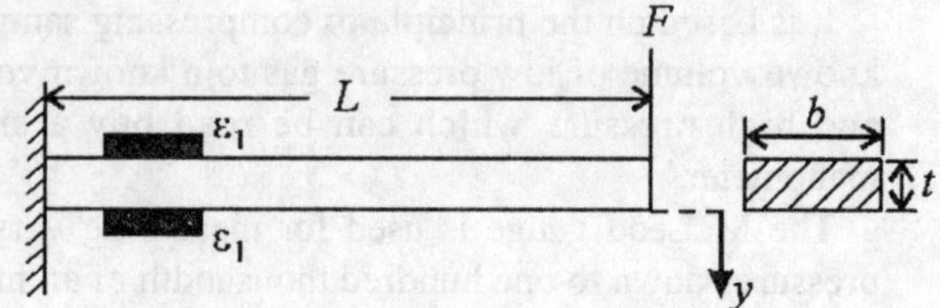

Deflection of the free end $y = \frac{1}{3}\frac{FL^3}{EI} = \frac{4FL^3}{E\,b\,t^3}$

Stiffness $k = \frac{3EI}{L^3} = \frac{E\,b\,t^3}{4\,L^3}$

Where, $I = \frac{b\,t^3}{12}$ → the moment of inertia of the section about the centroidal axis

Strain, $\varepsilon_1 = \frac{6\,FL}{E\,b\,t^2}$ (tension)

$\varepsilon_2 = \frac{6\,FL}{E\,b\,t^2}$ (compression)

The opposite sign of strains ε_1 and ε_2 (in above case) is of advantage while using strain gauges employing a bridge circuit.

In Present Case

For Cantilever

Deflection (y) $= \frac{Fl^3}{3EI}$ and $I = \frac{bt^3}{12}$

If length, width and thickness are doubled, then

$$y_{new} = \frac{x}{2} = \frac{1}{2} = 0.5\ mm$$

9. **Ans. (a)**

Solution:

Speed = 50 × 60 rpm = 3000 rpm

One tooth moving in anti-clock wise direction then the true speed of gear wheel

= 3000 – 60

= 2940 rpm

10. **Ans. (a)**

Solution:

We know that

$$P_d = \frac{1}{2}\int V^2$$

or $V^2 \propto P_d$

Therefore $V \propto \sqrt{P_d}$

11. **Ans. (b)**

Solution:

$$SPL_2 - SPL_1 = 20\log\frac{d_1}{d_2}$$

$= -20dB\ SPL_2 = 80 - 20 = 60\ dB$

$\therefore\ 80 - 20 = 60\ dB$

Second Method

Solution pressure level $\propto \frac{1}{\text{distance}}$

Sound pressure level = 80 $dB = 10^4$

If it is at 60 m then pressure level $=10^3 = 60\ dB$

12. **Ans. (a)**

Solution:

Given

PH of solution = 4

$-\log\left[H^+\right] = 4$

$\log\left[H^+\right] = -4$

Acidic with concentration 10^{-4} gm/litre

13. **Ans. (c)**

14. **Ans. (b)**

Solution:

Positive displacement flow meters are

Rotary vane type meter,

Sliding vane type meter.

Notating Disc type meter

15. **Ans. (d)**

Solution:

Pirani gauge is used for the low pressure measurement in the range of 10^{-5} to 1 torr, where 1 torr

* Measure pressure of order of 10^{-1} to 10^{-3} mm of Hg where 760 mm of $Hg = 10^5$ Pascals

∴ It is the order of 1 Pa

16. **Ans. (d)**

17. **Ans. (b)**

Solution:

Flow Nozzles

The flow nozzles are used for flow measurement at high fluid velocities and are more rugged and more.

Rugged and more resistant to erosion than the sharp edged orifice plate.

Flow Nozzles:

A flow nozzle is similar to a venturimeter but without a diverging core Unlike an orifice plate, the jet contraction in case of flow nozzle is less because of its curved shaped on the upstream side.

Note: Since Nozzle have a wise open area, they can be used for the *measurement of flow of solids.*

Advantages of Nozzle are:

1. They are cheaper and easier to install as compared to the Veturimeters.
2. They have high coefficient of discharge, higher than the orifice meters
3. They are useful for measurement of flow of fluids containing solids that settle and also widely used for high pressure/temperature steam flow
4. They can be used for measurement of high flow velocity

Variable Area Flowmetrs

In the variable area flow meters, the size of the restriction is adjusted by an amount necessary to keep the pressure differential constant when the flow rate changes and the amount of adjustment required proportional to the flow rate.

Rotameters

The rotameter is the most extensively used form of the variable area flowmeter It consists of a vertical tapered tube with a float which is free to move up or down within the tube.

Rotameter:

- Rotameter is a constant pressure drop type variable area flowmeter,
- It has a vertical tapered tube with a flaot which is free to move within the tube.
- The fluid flow from bottom to top in a rotameter. The float is made of such a diameter such that when no fluid is flowing across the rotameter , the float completely blocks are inlet of the meter.
- The density of float is greater than the density of the fluid flowing through it, so that the buoyant force alone is not sufficient to lift the float

Advantages of Rotameter:

1. The scale of the Rotameters is linear unlike the venturimeter which have the non-linear scale.
2. In case of venturimeter $Q \propto \sqrt{P}$, hence the scale becomes non-linear
3. In case of roameters $Q \propto X$, hence the scale is linear.
4. They can be used for the measurement of flow of liquid, gases and vapours
5. They are useful for measurement of low flow rates
6. They are independent form the effect of viscosity upto a certain range.
7. The accuracy of the rotameter is higher than the accuracy of the obstruction type metes.

- Rota meters are the constant pressure drop, and variable area type flow meters, all other meters like orifice meter, venture meter and flow nozzle are differential pressure meters.
- As the obstruction is sloppy the venturimetre has the lowest pressure drop when compared to others.

18. Ans. (b)

Solution:

We Know Bulk Modulus $= -\dfrac{p}{(\Delta V/V)}$

$$\Rightarrow \qquad 68.6\times10^9 = \frac{10^5}{\dfrac{d'^3 - d^3}{d^3}}$$

$$\Rightarrow \qquad 1-\left(\frac{d'}{d}\right)^3 = 1.45\times10^{-6}$$

$$\Rightarrow \qquad d' - d = 0.024\ \mu$$

19. Ans. (d)

Solution:

$$F_c = 10^{-6}\ \omega^2, \text{ and}$$

$$F_s = 16\,\pi^2 x^2$$

At equilibrium, $10^{-6}\omega^2 = 16\pi^2 x^2$

$$\Rightarrow \qquad x = \frac{10^{-3}}{4\pi}$$

$$\Rightarrow \qquad \frac{dx}{d\omega} = \frac{1}{4\pi}\ mm.s.$$

20. Ans. (a)

Solution:

Variable area Type Flowmeters

These are those flow meters in which the area through which the fluid is allowed to pass is made variable. The most variable area type folometer are rotameters.

Rotameter:

- Rotameter is a constant pressure drop type variable area flowmeter,
- It has a vertical tapered tube with a flaot which is free to move within the tube.
- The fluid flow from bottom to top in a rotameter. The float is made of such a diameter such that when no fluid is flowing across the rotameter , the float completely blocks are inlet of the meter.
- The density of float is greater than the density of the

fluid flowing through it, so that the buoyant force alone is not sufficient to lift the float

- The passage remains closed until the pressure of the flowing material, plus the fluid buoyancy effect, exceeds the downward pressure due to the weight of the float. The makings on the rotameter are calibrated in terms of the fluid flow.
- The float when rises up an annular passage keeps an increasing for the fluid flow. The floats attains the dynamic equilibrium when the upward and the downward forces balances each other.
- This increase in area of the passage make the rotameter a variable area flowmeter.
- The increase in the fluid flow at the inlet, rises the float up in the rotameter tube and an decrease in flow brings down the position of the float, hence the float position depends on the inlet fluid flow.

Note: He discharge (volume flow rate) for the rotameter is directly proportional to the displacement of the float (X)

$$Q \propto X$$

Hence, the relationship betweens the flow the (Q) and float position (X) is linear.

In order to make the volume flow rate independent of the fluid, the density of the float is made the density to the following fluid

$$\rho_f = 2\rho$$

Advantages of Rotameter:

1. The scale of the Rotameters is linear unlike the venturimeter which have the non-linear scale.
2. In case of venturimeter $Q \propto \sqrt{P}$, hence the scale becomes non-linear
3. In case of roameters $Q \propto X$, hence the scale is linear.
4. They can be used for the measurement of flow of liquid, gases and vapours
5. They are useful for measurement of low flow rates
6. They are independent form the effect of viscosity upto a certain range.
7. The accuracy of the rotameter is higher than the accuracy of the obstruction type metes.

In present case

Since the scale of a Rota meter is linear, so the rising of the float in the tube depends directly on the applied flow i.e. $Q \propto x$

$$\therefore \frac{Q_1}{Q_2} = \frac{70}{20}$$

$$= 3.5$$

21. Ans. (d)

Solution:

Vena contracta is the cross section where the flow area is minimum for a restrictions type flow meter

Vena contracta $\Rightarrow$ minimum area point

$\Rightarrow$ Velocity at maximum at this point.

Area of Jet at venacontracta = $C_c \times$ area of orifice

Contracta coefficient are considered as

$$C_C = 0.65$$

$$= 0.65 \times \frac{\pi d^2}{4}$$

22. Ans (b)

Solution:

Pirani Gauge

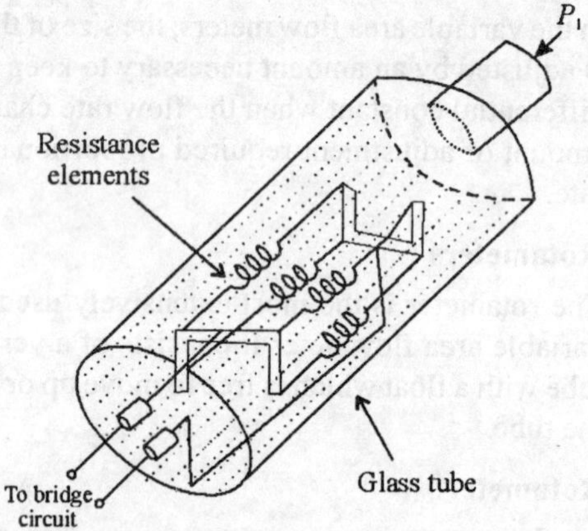

It consists of four coils connected to each other and forming one of the arms of the whetstone bridge. Pirani gauge is based on the principle of the change in conductivity of the gas with pressure which changes the temperature of the heated element

23. Ans. (c)

Solution:

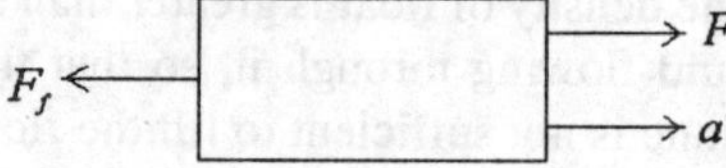

Coefficient of friction $\mu = 0.5$

Weight $m_g = 4.9\ N$

$$m = \frac{4.9}{9.8} = 0.5\ g$$

Acceleration $a = 4\ m/\sec^2$

Force due to friction $F_f = \mu mg$

$$= 0.5 \times 4.9 = 2.45\ N$$

$$F - F_f = ma$$

$$F = ma + F_f$$

$$= (0.5)(4) + 2.45$$

$$= 2 + 2.45$$

$$F = 4.44N \approx 4.5\,N$$

24. Ans. (d)

Solution:

Flashing rates are

$$f_1 = 25 \text{ pulses/sec}$$

$$f_2 = 20 \text{ pulses/sec}$$

$$\text{Speed of rotation} = \frac{f_1 f_2}{f_1 - f_2} = \frac{25 \times 20}{25 - 20}$$

$$= \frac{25 \times 20}{5} = 100 \text{ rev/sec}$$

$$= 100 \times 60 \text{ rev/min}$$

$$\approx 6000 \text{ rev/min}$$

25. Ans. (d)

Solution:

Gauge Pressure

A gauge that indicates zero at atmospheric pressure measures the difference between actual and atmospheric pressure ,This difference is called "gauge pressure", It is abbreviated as psig (pounds per square inch gauge.)

Absolute Pressure

Absolute pressure is actual total pressure (including atmospheric pressure) acting on a surface. It is abbreviated as Pisa (pounds per square inch absolute).

Absolute pressure

$$= \text{Guage pressure} + \text{Atmosphereic pressure}$$

$$= 3.2 \text{ bar} + 1.01 \text{ bar}$$

$$= 4.21 \text{ bar}$$

26. Ans. (d)

Solution:

A photo-conductive transducer works on the principle that when light beam strikes the cathode it releases electrons which are attracted toward anode producing electric current in the external circuit

27. Ans. (c)

Solution:

The order of pressure measurement is
Pirani guage > McLeod guage > Ionisation guage

28. Ans. (c)

Solution:

Venturi bube has the **highest discharge co-efficient**

29. Ans. (a)

Solution:

By Bernoulie is equation

(Difference in pressure) × A = weight of float – Buoyancy force

or $$(p_b - p_a)A = V_g(d_2 - d_1)$$

30. Ans. (b)

Solution:

Rotameter:

- Rotameter is a constant pressure drop type variable area flowmeter,
- It has a vertical tapered tube with a flaot which is free to move within the tube.
- The fluid flow from bottom to top in a rotameter. The float is made of such a diameter such that when no fluid is flowing across the rotameter, the float completely blocks are inlet of the meter.
- The density of float is greater than the density of the fluid flowing through it, so that the buoyant force alone is not sufficient to lift the float

Advantages of Rotameter:

1. The scale of the Rotameters is linear unlike the venturimeter which have the non-linear scale.
2. In case of venturimeter $Q \propto \sqrt{P}$, hence the scale becomes non-linear
3. In case of roameters $Q \propto X$, hence the scale is linear.
4. They can be used for the measurement of flow of liquid, gases and vapours
5. They are useful for measurement of low flow rates
6. They are independent form the effect of viscosity upto a certain range.
7. The accuracy of the rotameter is higher than the accuracy of the obstruction type metes.

- Rota meters are the constant pressure drop, and variable area type flow meters, all other meters like orifice meter, venture meter and flow nozzle are differential pressure meters.
- As the obstruction is sloppy the venturimetre has the lowest pressure drop when compared to others.

For steady state flow of liquid, the float of rotameter will remain in particular position, when weight of float is balanced by the drag force and buoyancy force.

31. Ans. (c)

Solution:

Photodiode current = Sensitivity × Intensity × Area

$$= 10^{-2} \times 10 \times 10^{-3} \times 0.55$$

$$= 0.055\,mA$$

$$\text{Output } V_o = -(100\,k\Omega) \times 0.055 \times 10^{-3}$$

$$= -5.5\,V$$

32. Ans. (d)

Solution:

Cantilever Beam: In this a force F is applied at a distance L from the fixed end. The force causes a bending moment and the deflection of the free end. Strains and induced are equal in magnitude but opposite in sign.

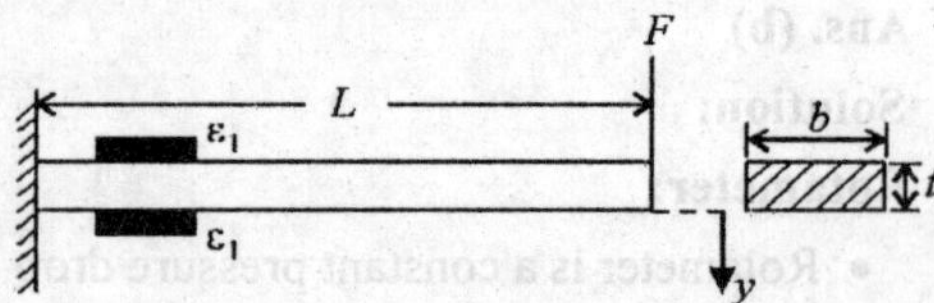

Deflection of the free end $y = \frac{1}{3}\frac{FL^3}{EI} = \frac{4FL^3}{E\,b\,t^3}$

Stiffness $k = \frac{3EI}{L^3} = \frac{E\,b\,t^3}{4\,L^3}$

Where,

$I = \frac{b\,t^3}{12} \rightarrow$ the moment of inertia of the section about the centroidal axis

Strain, $\varepsilon_1 = \frac{6\,FL}{E\,b\,t^2}$ (tension)

$\varepsilon_2 = \frac{6\,FL}{E\,b\,t^2}$ (compression)

The opposite sign of strains ε_1 and ε_2 (in above case) is of advantage while using strain gauges employing a bridge circuit.

For Cantilever the deviation $= \frac{Fl^3}{3\,EI}$

$I = \frac{bt^3}{12}$

If each dimension of Cantilever is scaled by 100, then the $x_{new} = \frac{x_{old}}{100}$

Where deviation is directly proportional to frequency,

Hence ω_n of micro accelerometer $= \frac{\omega}{100} = 0.01\,\omega$

33. Ans. (a)

Solution:

Output voltage

$$V_o = \left(0.01\,\frac{V}{i}\right) \times 20 + \left[0.01\left(\frac{V}{MPa}\right).20\;P_{tture} + 10P_{true}\right]$$

or $7.4 = 0.2 + 0.01 \times 20\,P + 10\,P$

$P = 0.71\,MPa$

34. Ans. (a)

Solution:

Since the frequency generated by the crystal is less, as clock is we are loosing 30.32 sec, every 23 days. So the actual frequency should be more than 32.768 kHz i.e. 32.7685.

35. Ans. (d)

Solution:

Depth $d = v \times \frac{T}{2}$

$= 6\;km/s \times 0.025 \times 10^{-3} = 15\;cm$

36. Ans. (d)

Solution:

Electromagnetic flowmeters do not put any obstruction in the flow of the fluid through them, hence these are called as the *non-instruction type flow meters*

- These are used for the flow measurement of slurries, and electrically conducting liquid.

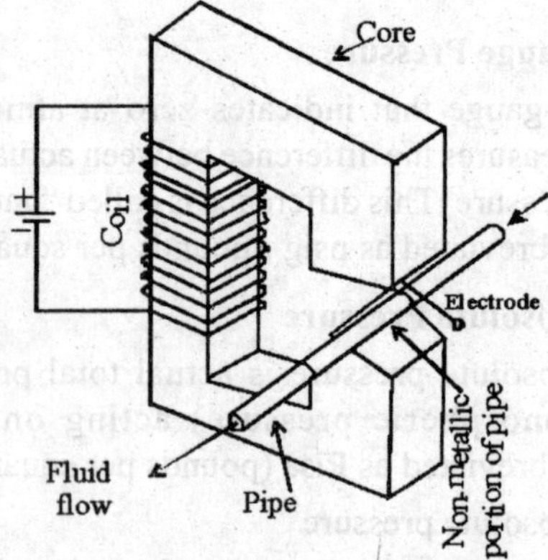

- The electromagnetic flow water consists of an non-conducting non-magnetic pipe carrying the liquid whose flow is to be measured. The pipe is surrounded by an electromagnet which produces a magnetic field. The fluid flowing through the pipe is conducing in nature. This type of arrangement is analogus to a conductor moving in a magnetic field and thus generates an e.m.f. across the electrodes. This voltage generated is than given as $e = B\,d\,v$

  Where, B = Flux density

  d = length of conductor-diameter of pipe

  v = average velocity of the conductor

  At constant magnetic field

  Voltage across the electrodes $\propto$ (average velocity × diameter)

- Electromagnetic flowmeter is a volume flow meter.

Main Point

1. Non-conducting pipe has to be used, because if a conducting pipe, the output will get short circuited
2. At low fluid flow, the output voltage genera rated

will be low, hence an amplifier has to be used to bring the output voltage in the suitable usable form.

3. The output voltage of electromagnetic flow meter is directly calibrated in terms of the fluid flow

As given

Flux density $B = 0.1T$

Pipe diameter $d = 50\times10^{-3} m$

Flow velocity $v = 20\ mm/\text{sec}$

$= 20\times10^{-3} m/\text{sec}.$

Voltage induced

$e = Bdv$

Pipe diameter

$e = 0.1\times50\times10^{-3}\times20\times10^{-3}$

$= 100\times10^{-6} = 100\ \mu V$

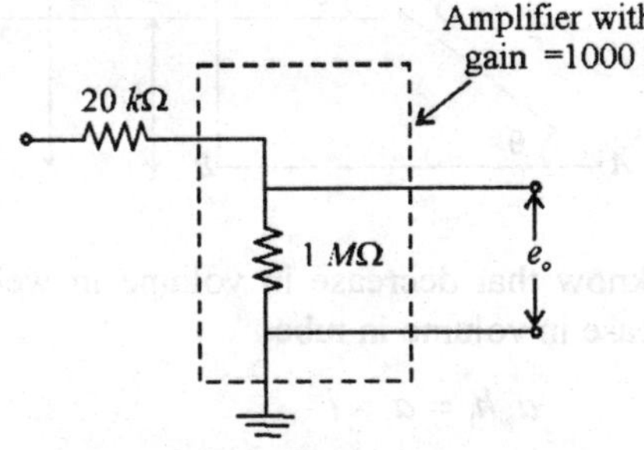

Clearly by voltage divider rule output

$$e_0 = \frac{1M\Omega}{1M\Omega + 20k\Omega} e \times gain$$

$$= \frac{1M\Omega}{1.2M\Omega}\times e\times gain$$

$$= \frac{1}{1.2}\times(100\times10^{-6})\times1000$$

= 0.083 Volts

37. Ans. (c)

Solution:

Ultrasonic Flowmeters: An ultrasonic flowmeter that is used to measure the velocity of liquid or a gas by using the principle of ultrasound. Using the ultrasonic trasducer (like piezo-electric crystal) the flowmeter can measure the average velocity along the path of an emitted beam of ultrasound, by averaging the difference in measured transient time between the pulses of ultrasound propagating into and against the direction of flow.

Characteristic:

1. Ultrasonic flow meters are affected by the temperature, density and viscosity of the flowing medium.
2. They are inexpressive to use and maintain because they do not use moving parts, unlike the mechanical flow meters.
3. They are called as ultrasonic because their working frequency is about several MHz which is above the range of audible frequency
4. There are various arrangement in which the crystal are placed to measure the flow

Case-1: The crystal are placed inside the tube:

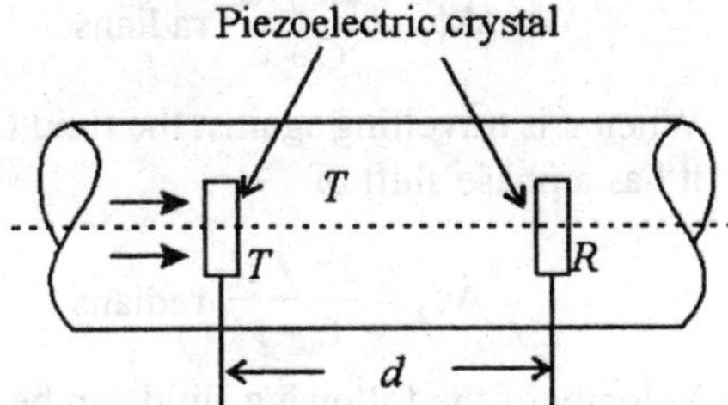

In the case the transmitting crystal (*T*) and the receiving crystal (*R*) are placed inside the tube and hence they are in contact with the fluid so this is called as the *intrusive case.*

Working:

To the transmitter the electrical energy in the form of short burst, a high frequency voltage is applied to the piezoelectric crystal causing it to vibrate. If the crystal is in contact with the fluid, the vibration will be communicated to the fluid and propagated through it. The receiving piezoelectric crystal exposed to these pressure fluctuations and responds by vibrating. The vibratory motion produces an electrical signal in proportional to the flowing fluid

The transmitter emits an ultrasonic pulse along the direction of flow, (i.e. from T-1 to R-2), the pulse is received at the receiver a time Δt_1, later. The transient time in the direction of flow is

$$\Delta t_1 = \frac{d}{C+v} \text{ (i.e. } T_1 \text{ to } R_2)$$

Where

d = distance between T and R (in meter)

C = velocity of sound propagation in the medium (m/sec)

v = linear velocity of flow (m/sec)

When the signal is travelling in the opposite direction (i.e. against the direction of fluid flow or when the position of T and R interchanged.)

$$\Delta t_2 = \frac{d}{C-v} \text{ (i.e. from } T_2 \text{ to } R_1)$$

The difference in the transient time is given by

$$\Delta t = \Delta t_2 - \Delta t_1$$

$$= \frac{2dv}{C^2 - v^2}$$

Usually (C>>v)

$$\Delta = \frac{2dv}{C^2}$$

Main Points:

1. Δt is linearly proportional to flow velocity (v)
2. When the sinusoidal signal of frequency f Hz travelling along the fluid flow direction, it has a phase shift of

$$\Delta\phi_1 = \frac{2\pi\ f\ d}{C+v} \text{ radians}$$

3. When it is travelling against the fluid flow direction, it has a phase shift of

$$\Delta\phi_2 = \frac{2\pi\ f\ d}{C-v} \text{ radians}$$

Velocity of the following fluid can be calculated by either measuring the transient time of the phase shift

Case-II: The crystals (T and R) are placed outside the tube

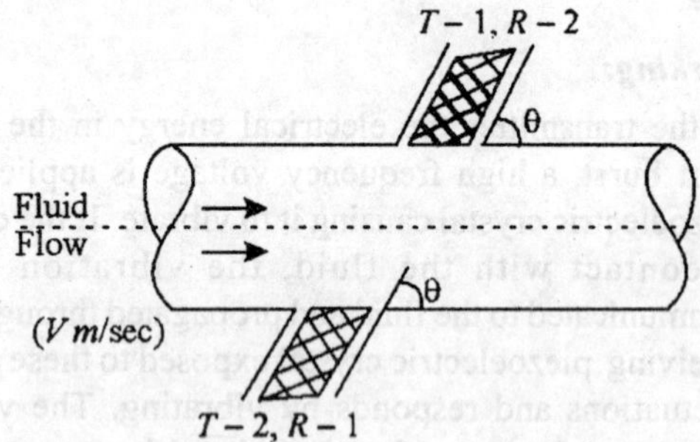

Hence the T and R are placed at some inclination 'θ' and function as transmitter and receiver alternately Δt in this case is given as

$$\Delta t = \frac{2\ d\cos\theta}{C^2}v$$

And velocity can then be gives as

$$v = \frac{\Delta t\ C^2}{2\ d\cos\theta}$$

In present case

$L = 25\ mm = \frac{25}{1000}\ m = 0.025\ m$

Sound velocity $= 1000\ m/s$

Δt transit time $= 10\ ns = 10\times10^{-9}$ sec

θ (angle of path w.r.t. pipe axis) = 60°

V(flow velocity) = ?

We know $V = \dfrac{\Delta t}{\dfrac{2L\cos\theta}{C^2}}$

$$V = \frac{C^2\Delta t}{2L\cos\theta}$$

$$= \frac{(1000)^2\times(10\times10^{-9})}{2\times0.025\times\cos 60^\circ}$$

$$V = 0.4\ m/s$$

38. Ans (c)

Solution:

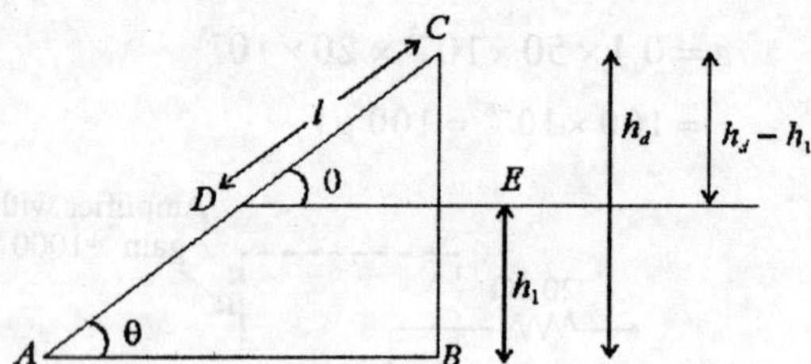

We know that decrease in volume in well is equal to increase in volume in tube

So, $a_w h_1 = a_t \times l$

$$\frac{h_1}{l} = \frac{a_t}{a_w} \quad (1)$$

In Δ DEC

$$\sin\theta = \left[\frac{h_d - h_1}{l}\right]$$

$$\sin\theta = \frac{h_d}{l} - \frac{h_1}{l}$$

$$\theta = \sin^{-1}\left(\frac{h_d}{l} - \frac{h_1}{l}\right)$$

$$\frac{h_1}{l} = \frac{a_t}{a_w}$$

$$\theta = \sin^{-1}\left(\frac{h_d}{l} - \frac{a_t}{a_w}\right)$$

39. Ans (c)

Solution:

The venturimeter is a device which is used for the measurement of discharge through a pipe. It is a short pipe that first tapes to a small cross-section known as throat. The diameter is then again gradually increased to its original size, as shown in the figure

- The veturimeter consists of a inlet main (inlet pipe), convergent cone Cylindrical throat, Diverging cone and an exit main (outlet).
- The ventuimete may be used in any position, horizontal, vertical or inclined
- The size of the venturimete is expressed in terms of the inlet and the throat diameter

For example a 300 × 150 mm Venturimete, first in a pipe of diameter 300 mm and the throat diameter of 150 mm.

In present case

Since $Q \propto \sqrt{\Delta P}$ (in venturimeter)

Hence, $\frac{50}{Q} = \sqrt{\frac{30}{20}}$

Q = liters/sec

40. Ans (a)

Solution:

$$\eta = \frac{\pi Ptr^4}{8 lV}$$

$$\eta \propto r^4$$

$$\left(\frac{P}{l}\right) = \text{constant}$$

$$\left(\frac{V}{t}\right) = \text{constant}$$

$$\frac{n_b}{n_a} = \frac{(2r)^4}{r^4} = 16$$

so, $n_b = 16\, n_a$

So dynamic viscosity of fluid B to that of Fluid A is 16

41. Ans (d)

Solution:

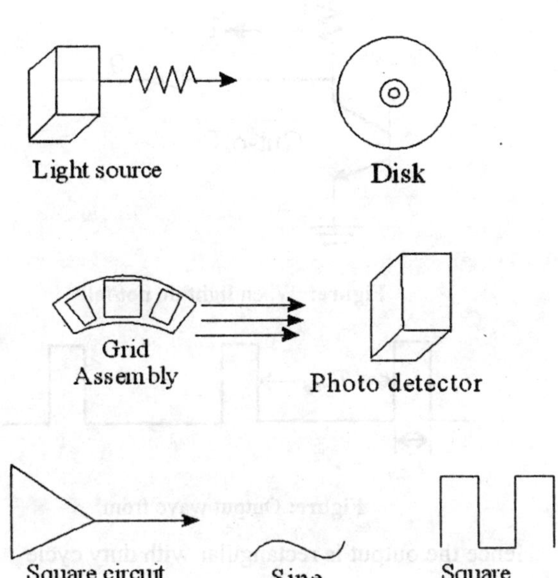

An encoder is an electrical mechanical device that can monitor motion or position . An encoder uses optical sensors to provide an series of pulses that are translated into motion, position or direction. Above figure shows rotary encoder . Above figure shows a thin disk and stationary LED. The light will be continually focused through the glass disk. A light activated transistor is mounted on other side which detects the light from LED. When shaft runs the disk runs and when the disk lines up so that light from LED is focused on the photodector it will go into saturation and an electrical square wave pulse is produced

Since the disk has only a single hole on its periphery. So the light falls on the detector (i.e) the base of the transistor for a very small duration in one revolution. The transistor is the npn transistor. The npn transistor comes in saturation when a high is applied on his base and is cut off when a when light fails:

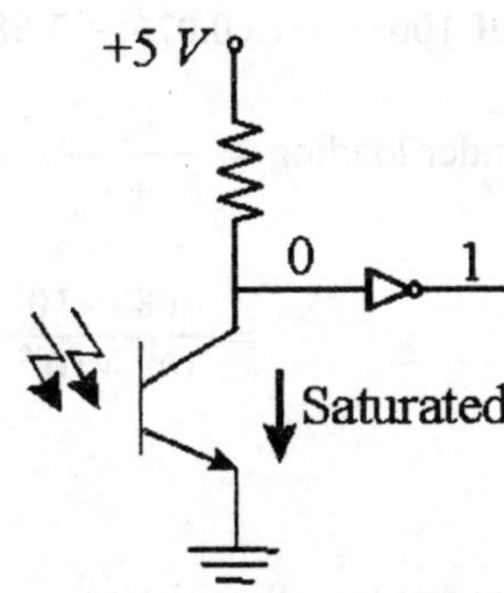

Figure: When light fall

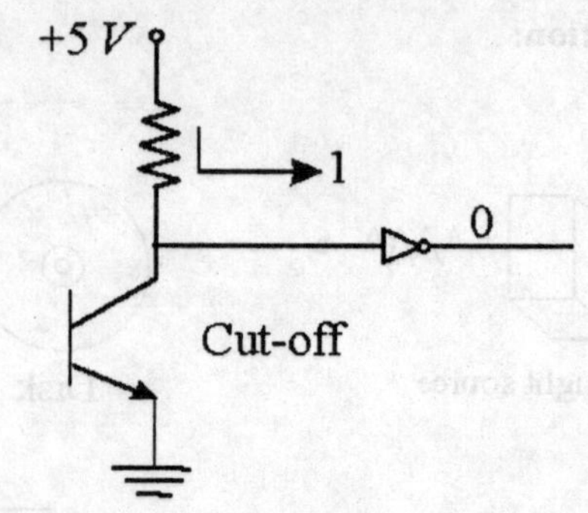

Figure: When light do not fall

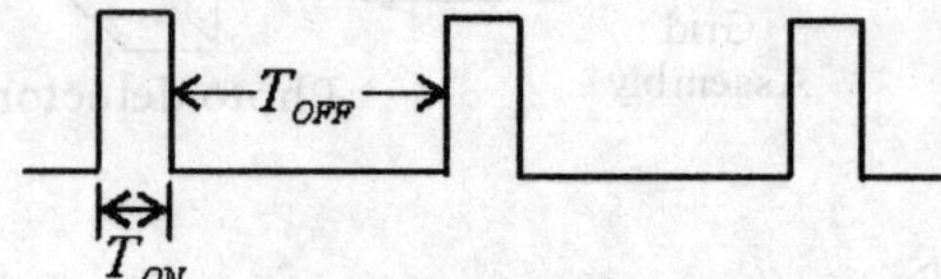

Figure: Output wave from

Hence the output is rectangular with duty cycle

$$\frac{T_{On}}{T_{ON}+T_{OFF}}$$

$$=\frac{1}{1+\frac{T_{OFF}}{T_{On}}}=\frac{1}{1+\infty}$$

$$\cong 0 \text{ as } T_{ON} << T_{OFF}$$

42. Ans (b)

Solution:

We know that slope factor value changes $54\ mV|_{1\ PH\ unit}$ at °C and $59.16\ mV|_{1\ PH\ unit}$ at 25°C. These two are Ideal case

So for 1 PH change voltage changes

For change voltage changes

$= 1.3 \times 59.16 = 0.07683\ V$

For gain of $100 = 100 \times 0.076 = 7.683$

Voltage under loading $= \frac{V_o}{R_o + R_L} \times R_L$

$$= \frac{7.683 \times 10^{11}}{10^{11} + 10^{10}} \Rightarrow 6.99\ V$$

43. Ans (b)

Solution:

Assuming complete dissociation

PH of 1 mM is

$$PH = -\log\left[H^+\right]$$

$$= -\log\left[1\times10^{-3}\right]$$

$$= 2.7$$

44. Ans (b)

Solution

The conventional way of expressing vibration is in terms of:) acceleration due to gravity

45. Ans.(d)

Solution:

Psychrometric chart is used to determine the Relative Humidity.

46. Ans.(d)

Solution:

Responsivity - 0.9 A/W

Incident Power = 1 mW

The current throuhg diode and 'R' from input to output

$$= I = 0.9\times10^{-3}\ A = 9\times10^{-4}\ A$$

$$V_{out} = -IR$$

$$V_{out} = -1\ V$$

$$R = \frac{10^4}{9}\Omega$$

$$= 1.1\ k\Omega$$

47. Ans (b)

Solution:

Gear Teeth = 60

Gating Time = 1 sec.

GATING TIME: It is the time taken by the counter to count the total pulse.

Because a gear has 60 which means 60 teeth is equal to 1 revolution and in 1 sec. Sensor read 120 teeth which means.

In 1 sec it is making 2 revolutione. In 60 sec it is making 2 × 60 revolutions = 120 *rpm.*

48. Ans (c)

Solution:

* **Magneto-Strictive Transducer** are based on the principle of the chance in the permeability of mangetic materials when they are subjected to stain.

  The permeability of a material decreases when a positive strain is applied and increases when negative srain is applied

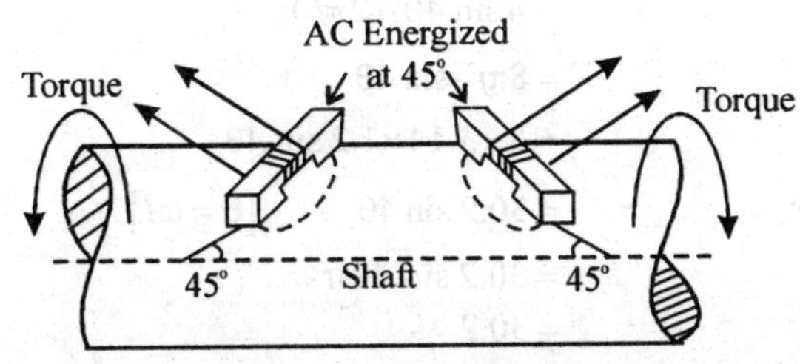

* **When a shaft is subjected to torsion, the heightest strain (positive and negative) are experienced at 45° to the shaft axis.** Thus, if we establish a flux paths in theses areas, it gets modified due to the strain. For the purpose of measurement of the torque by this principle, two a.c., energized coils are placed on the shaft as shown in the figure.

When no torque is applied the bridge is balanced (these two coil are made a part of an ac bridge) and when a torque applied the inductance of one of the coil increases while that of the other coil decreases, thus making the bridge unbalanced and the amount of current flowing through the galvanometer gives the measure of the applied torque.

49. Ans (a)

Solution:

Speed of the shaft,

$$\eta = \frac{f_m f_1 (m-1)}{f_m - f_1}$$

where f_m = maximum flashing rate $= r_1$

f_1 = lowest flashing rate $= r_2$

m = number of flashing frequency $= 2$

then, $$\eta = \frac{r_1 r_2}{r_1 - r_2}$$

50. Ans (d)

51. Ans (a)

Solution:

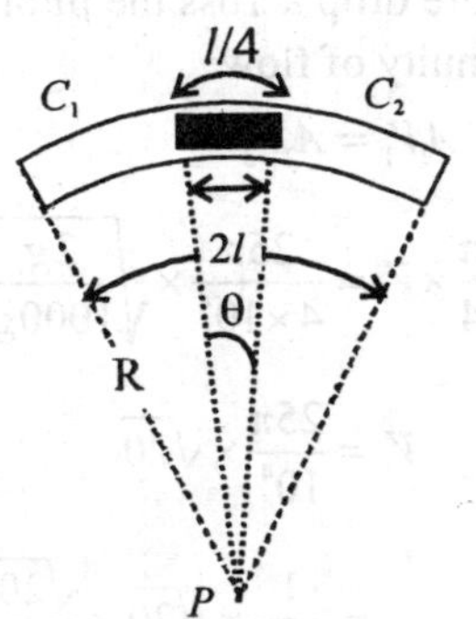

From the given figure

$$\frac{l}{8} = R\theta$$

Angular displacement $\theta = \dfrac{l}{8R}$

It varies from $\dfrac{-l}{8R}$ to $\dfrac{l}{8R}$

52. Ans (d)

Solution:

$$\left[v_i \frac{R}{R+R} - \frac{\frac{1}{j\omega C_1}}{\frac{1}{j\omega C_1} + \frac{1}{j\omega C_2}} \right] = v_o$$

$$\Rightarrow \frac{v_o}{v_i}\left[\frac{1}{2} - \frac{C_2}{C_1 + C_2}\right] = 1$$

$$\Rightarrow \frac{v_o}{v_i}\left[\frac{1}{2} - \frac{C_2}{C_1 + C_2}\right] = \left(\frac{C_1 - C_2}{C_1 + C_2}\right)$$

and $$\frac{v_o v_i}{\Delta\theta}\left(\frac{C_1 + C_2}{C_1 - C_2}\right)\frac{4R}{l}$$

$$\Rightarrow \frac{v_o / v_i}{\Delta\theta} \propto \frac{R}{l}$$

53. Ans (b)

Solution:

As given all other aspects, i.e. parameter are constant. Hence sensitivity of flow meter is constant

Sensitivity of flow measurement

= sensitivity of flow meter × sensitivity of pressure transmitter

Hence, change in sensitivity of flow measurement = 5%

54. Ans. (b)

Solution:

Radius of nozzle $r_n = \dfrac{0.8}{2} = 0.4\,mm$

Radius of ofice $r_o = \dfrac{0.2}{2} = 0.1\,mm$

Sensitivity $S = \dfrac{P \times r_n}{r_o^2}$

$$= 1.4\times10^2 \times \frac{0.4\times10^3}{(0.1)^2}$$

$$= 5.6\times10^6 \; Pa/mm$$

$$= 5.6 \; M \; Pa/mm$$

55. Ans. (b)

Solution:

Turbine Flowmeter

Turbine flowmeter are the devices, in which a small wheel is placed in the pipe along the path of the fluid flow. The flowing fluid rotates the blades of the trubine and hence the output voltage is generated in the turbine. The generated output is a function of the fluid flow.

* Turbine flow meters are the volumetric flow meters.
* The output of these meter is in the form of digital electric signal whose frequency is directly proportional to the flow rate quantity where each pulse represents a discrete volume.
* It has a permanent magnet scaled inside the rotor body and is polarized at 90° to the axis of rotation.
* As the rotor rotates, so does the magnet, hence a magnetic field (ϕ) is produced. The field (ϕ) produces a (E) in the pick-up coil located external to the meter housing

$$E = -\frac{d\phi}{dt}$$

* The frequency of this generated voltage is directly proportional to the speed of the turbine and hence the flow rate.
* The pulses are totalized, the total number of pulses during a particular interval of time, and the total flow in that interval can be known, hence this meter is used to find the totalized flow.

In present case

Flux linkage

$$\psi(\theta) = 3 + \cos 4\theta$$

rotation $= 72 \; rpm$

$$f = \frac{72}{60} rev/\sec$$

$$= 1.2 \; rev/\sec$$

Voltage induced

$$V = \frac{-d\psi(\theta)}{dt} = \frac{-d\psi}{d\theta}\times\frac{d\theta}{dt}$$

$$= \frac{d}{d\theta}(3+\cos 4\theta)\times\frac{d\theta}{dt}$$

$$= 4\sin 4\theta\cdot\frac{d\theta}{dt}; \quad \left[\text{since } \omega = \frac{d\theta}{dt}\right]$$

$$= 4\sin 4\theta\cdot(2\pi f)$$

$$V = 8\pi f\cdot\sin 4\theta$$

$$= 8\times3.14\times1.2\sin 4\theta$$

$$= 30.2\sin 4\theta \qquad [\theta = \omega t]$$

$$= 30.2 \sin 4\,\omega t$$

$$= 30.2$$

Amplitude $= 30.2 \; mV$

Frequency of output voltage

$$= 4\times f = 4.8 \; Hz$$

56. Ans. (d)

Solution:

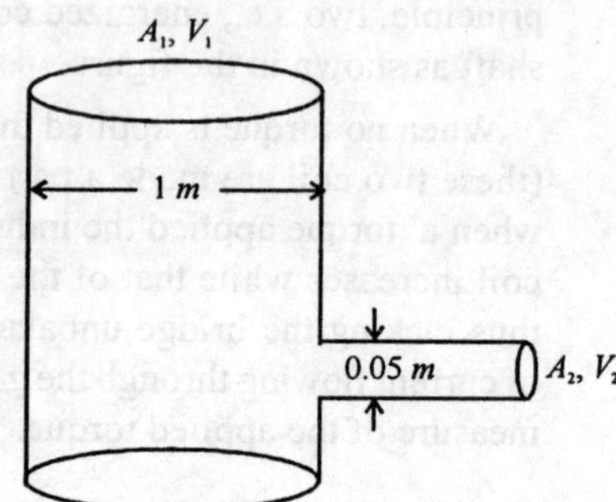

Base area of dram $A_1 = \pi r_1^2$

$$= \pi\times\left(\frac{1}{2}\right)^2 = \frac{\pi}{4}m^2$$

Pipe flow area $A_2 = \pi r_2^2$

$$= \pi\times\left(\frac{0.05}{2}\right)^2$$

$$= \pi\times\left(\frac{5}{200}\right)^2 = \frac{\pi\times25}{4\times10^4}m^2$$

Velocity of water out flow $V_2 = \sqrt{\frac{2g}{w}(\Delta P)}$

Where,

$w =$ Specific weight of water

$= \rho g$

$\Delta P =$ Pressure drop across the pitot tube

As per continuity of flow

$$A_1V_1 = A_2V_2$$

$$\therefore \quad \frac{\pi}{4}\times V_1 = \frac{25\pi}{4\times10^4}\times\sqrt{\frac{2g}{1000g}\times10\times10^3}$$

$$V_1 = \frac{25\pi}{10^4}\times\sqrt{20}$$

$$= \frac{1}{400}\times\sqrt{20}\times\frac{\sqrt{20}}{\sqrt{20}}$$

$$= \frac{1}{400} \times \frac{20}{\sqrt{20}}$$

$$= \frac{1}{20\sqrt{20}} = \frac{1}{20 \times 2\sqrt{5}}$$

$$= \frac{1}{40\sqrt{5}} \text{msec}^{-1}$$

57. Ans. (a)

Solution:

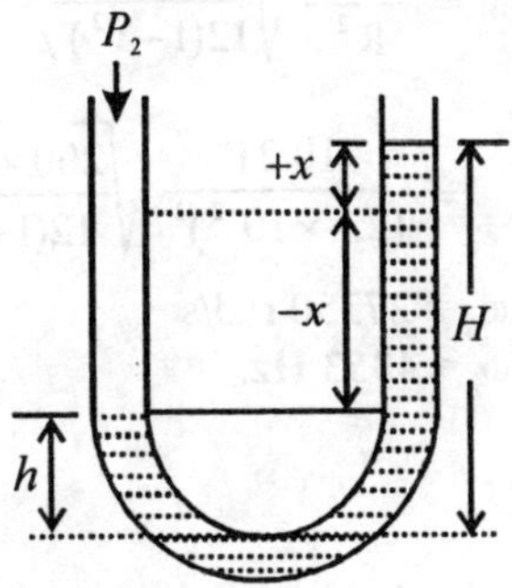

Consider the case of V-tube monometer as per above figure.

Total force exerted on the liquid

$$= \rho\, a\, g\,(H\text{-}h)$$

$$= 2\, \rho\, a\, g\, x$$

Force acting due to friction of the liquid

$$= B\frac{dx}{dt} = 0 \text{ (given)}$$

Force acting due to inertia $= \rho V \dfrac{d^2x}{dt^2}$

Where,

$a =$ area of the tube (m^2)

$x =$ displacement of the liquid (m)

$p =$ applied pressure

$V =$ total volume of manomeer fluid (m^3)

Total force due to pressure P = total force on the liquid + force due to inertia + force due to viscous friction

$$Pa = \rho V = \frac{dx^2}{dt^2} + \frac{B}{a}\frac{dx}{dt} + 2\rho\, g\, x$$

$$P = \frac{PV}{a}\frac{dx^2}{dt^2} + \frac{B}{a}\frac{dx}{dt} + 2\rho\, g\, x$$

Taking Laplace transform both side

$$P(s) = \frac{\rho V}{a} s^2 X(s) + \frac{B}{a} s\, X(s) + 2\, \rho\, g\, X(s)$$

$$G(s) = \frac{\text{LT of output}}{\text{LT of input}} \frac{X(s)}{P(s)}$$

normalized transfer function is

$$G(s) = 2\, \rho\, a\, g \frac{X(s)}{P(s)}$$

$$= \frac{2\, a\, g/V}{s^2 + \dfrac{Bs}{\rho V} + \dfrac{2ag}{V}}$$

Now since $Bs = 0$ (Given)

Hence $G(s) = \dfrac{2\, a\, g/V}{s^2 + \dfrac{2a\, g}{V}}$

$$s^2 + \frac{2a\, g}{V} = 0$$

Replacing s by ($j\omega$) we get

$$(j\omega)^2 + \frac{2\, ag}{V} = 0$$

$$-\omega = -\frac{2\, ag}{V}$$

$$\omega = \sqrt{\frac{2\, ag}{V}}$$

Area of tube

$$a = \pi\left(\frac{D}{2}\right)^2 = \frac{\pi D^2}{4}$$

$$\omega = \sqrt{2\frac{\pi D^2}{4}\frac{g}{V}}$$

$$= \sqrt{\frac{\pi D^2 g}{2V}} = D\sqrt{\frac{\pi\, g}{2V}}$$

$$f = \frac{1}{2\pi} D\sqrt{\frac{\pi\, g}{2V}}$$

$$f = \frac{1}{2\sqrt{2\pi}}\sqrt{\frac{g}{V}}$$

58. Ans. (a)

Solution :

Nominal value of C_d

$$C_d = \frac{m}{t\, \rho\, A\sqrt{2gh}} \quad (1)$$

Substituting values, (dimensions of all the variables are to be consistent)

$$A = \frac{\pi}{4} d^2 = \frac{\pi}{4}\left(\frac{13}{1000}\right)^2 = 1.33 \times 10^{-4}\ \text{m}^2$$

$$C_d = \frac{400}{(550)(1000)(1.33\times10^{-4})\sqrt{2\times9.81\times3.5}}$$

$$= 0.66$$

59. Ans. (b)

Solution:

$$p_1 > p_2$$

$$p_1 - p_2 = 50\ \text{k Pa}$$

$$= 50000\ \text{N/m}^2$$

density of oil $r_0 = 800\ \text{kg/m}^3$

density of mercury $= r_m = 13600\ \text{kg/m}^3$

$$p_1 - p_2 = r_m g\,h - r_0\,g\,h$$

$$50000 = gh\,(r_m - r_0)$$

$$50000 = 9.81 \times h\,(13600 - 800)$$

$$h = \frac{50000}{9.81\times12600} = 0.40\ \text{m} = 40\ \text{cm}$$

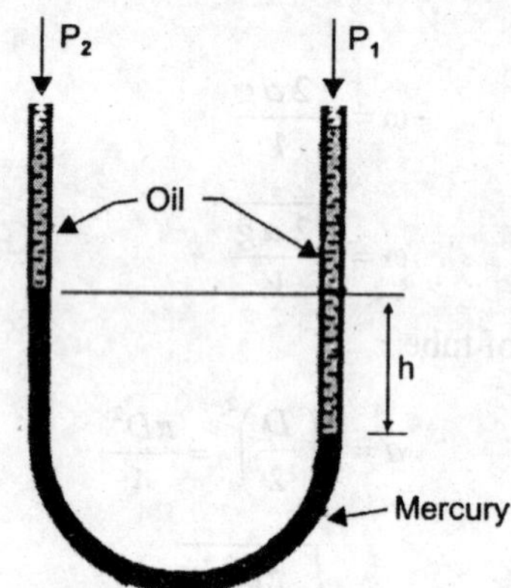

If the same manometer were to measure a differential pressure of 50 k Pa of air, the second term in the formula is to be neglected as density of air compared to that of mercury is negligibily small. Corresponding deflection

$$h = \frac{50000}{9.81\times13600} = 0.375\text{m} = 37.4\ \text{cm}$$

60. Ans. (a)

Solution:

Max. deflection is when $r = 0$

$$y_{max} = \frac{3}{16}\frac{p}{Et^3} R^4 (1 - v_2)$$

$p = 1.5\ \text{M Pa} = 1.5 \times 10^6\ \text{N/m}^2$

$E = 200\ \text{GN/m}^2 = 200 \times 10^9\ \text{N/m}^2$

$R = 2.5\ \text{cm} = 2.5 \times 10^{-2}\text{m}$

$$y_{max} = \frac{t}{3}$$

$$\frac{t}{3} = \frac{3}{16} \times \frac{1.5\times10^6(2.5\times10^{-2})^4[1-(0.3)^2]}{200\times10^9\,t^3}$$

$$t^4 = 3 \times \frac{3}{16} \times \frac{1.5\times10^6\times39.06\times10^{-8}\times0.91}{200\times10^9}$$

$$t = 1.1\ \text{mm} = 1.1 \times 10^{-3}\ \text{m}.$$

61. Ans. (b)

Solution:

$$\omega_n = \frac{10.21}{R^2}\sqrt{\frac{Et^2}{12(1-v^2)\,\rho}}$$

$$= \frac{10.21}{(2.5\times10^{-2})^2}\sqrt{\frac{200\times10^9\times(1.1\times10^{-3})^2}{12(1-0.3^2)\times7.9\times10^3}}$$

$$\omega_n = 27350\ \text{rad/s}$$

$$\omega_n = 4353\ \text{Hz.}$$

62. Ans (c)

Solution:

Pressure indicated in terms of mercury column.

$$\frac{p_i}{\rho g} = \frac{Ah^2}{V - Ah}$$

$$h = 2.5\ \text{cm} = 26\ \text{mm},\ d = 1\ \text{mm}$$

$$A = \frac{\pi}{4} d^2 = \frac{\pi}{4}\,(1)^2 = 0.705\ \text{mm}^2$$

$$V = 100\ \text{cm}^3 = 10^5\ \text{mm}^3$$

Pressure in mm of mercury

$$\frac{p_i}{\rho g} = \frac{0.785\times(25)^2}{1\times10^5 - 0.785\times25} = \frac{490.625}{10^5 - 19.625}$$

$$= 490.6 \times 10^{-5}$$

In the above equation, since the unit used is millimetre, the pressure head calculated shall be in mm of Hg. Also, as V >>> Ah, ($10^5 > 19.265$) so using the approximation,

$$\frac{p_i}{\rho g} \approx \frac{Ah^2}{V}$$

will not introduce much of an error.

Pressure = 490.6×10^{-5} mm of mercury

= 0.0049 mm of mercury

= 0.0049 torr

63. Ans. (c)

Solution:

$$\text{SPL} = 20 \log \frac{1}{0.00002}$$

$$= 93.98\ \text{dB}$$

64. Ans. (b)

Solution:

$$\text{SPL} = 20 \log \frac{10}{0.00002}$$

$$= 113.99 \text{ dB}$$

65. Ans. (b)

Solution:

Density of mercury, $r_m = 13.6 \times 1000$ kg/m³

Density of sea water, $r_m = 1026$ kg/m³

Manometer deflection = 15 cm of mercury

Pressure difference in terms of sea water column = h_m

$$\left(\frac{\rho_m}{\rho_m} - 1\right)$$

$$h_w' = 15\left(\frac{13.6 \times 1000}{1026} - 1\right) = 183.8 \text{ cm}$$

Velocity of fluid relative to submarine

$$V = \sqrt{2\, g\, h_w} = \sqrt{2 \times 9.81 \times 1.838} = 6.0 \text{ m/s}$$

66. Ans. (d)

Solution:

Pressure difference $h_w' = 10$ cm of water

Pressure difference in terms of air column

$$h_a = h_w\left(\frac{\rho_w}{\rho_a} - 1\right)$$

$$\approx h_w \frac{\rho_w}{\rho_a}$$

$$= \frac{10}{100}\left(\frac{1000}{1.2}\right)$$

Flow velocity = $\sqrt{2 g h_a} = \sqrt{2 \times 9.81 \times 83.3}$

= 40.43 m/s

In the above solution, air has been treated as incompressible.

67. Ans. (b)

Solution:

Using, $$Q = C_d \frac{A_1 A_2}{\sqrt{A_1^2 - A_2^2}} \sqrt{2g}\sqrt{h}$$

where h is differential pressure in terms of water column and is,

$$h = h_m\left(\frac{\rho_m}{\rho_w} - 1\right)$$

where, r_w = density of mercury

r_w = density of water

$$= \frac{25}{100}\left(\frac{13.6}{1} - 1\right) m = 3.15 \text{ m}$$

h = 3.15 m

$$A_1 = \frac{\pi}{4}(0.3)^2 = 0.0707 \text{ m}^2$$

$$A_2 = \frac{\pi}{4}(0.15)^2 = 0.0176 \text{ m}^2$$

$$Q = 0.97 \times \frac{0.0707 \times 0.017}{\sqrt{(0.0707)^2 - (0.0176)^2}} \sqrt{2 \times 9.81 \times 3.15}$$

$$Q = 0.0246 \text{ m}^3/\text{s}$$

As the value of D_d was given, there is no need to find flow coefficient.

*Flow coefficients are tabulated for a meter size and varying range of Reynold's numbers.

68. Ans. (c)

Soluton:

Velocity at the centre = $C_v \sqrt{2gh}$

$$= 0.98\sqrt{2 \times 9.81 \times 2.36}$$

$$= 6.668 \text{ m/s}$$

Mean velocity = 5.6678 m/s

$$\text{Area of flow} = \frac{\pi}{4} d^2$$

$$= \frac{\pi}{4}(0.3)^2 = 0.7606 m^2$$

Rate of flow of water = Mean velocity × Area

= 5.6678 × 0.7068

Rate of flow = 0.4006 m³/s

69. Ans. (d)

Solution:

A_0 = 500g = 500 × 9.81 = 4905 m/s²

x = 0.2

w_n = 10000 Hz = 2p × 10000 = 6.28 × 10⁴ rad/s

w = 2000 Hz = 2p × 2000 = 1.256 × 10⁴ rad/s

$$r = \frac{\omega}{\omega_n} = 0.2$$

$$\frac{Z_0}{A_0} = \frac{1}{\omega_n^2 \sqrt{\left(1 - r^2\right)2 + \left(2\xi r\right)^2}}$$

$$Z_0 = \frac{A_0}{\omega_n^2 \sqrt{\left(1 - r^2\right)2 + \left(2\xi r\right)^2}}$$

$$Z_0 = \frac{4905}{\left(6.28 \times 10^4\right)^2} \times \frac{1}{\sqrt{\left(1 - (0.2)^2\right)^2 + \left(2 \times 0.2 \times 0.2\right)^2}}$$

$Z_0 = 1.295 \times 10^{-6}$m

(Amplitude of output displacement)

$= 1.295 \times 10^{-3}$m

Sensitivity of piezoelectric crystal = 250 V/mm.

70. Ans. (a)

Solution:

$$\text{Frequency } w = 2p f = \frac{2\pi}{t} = \frac{2\pi}{30}$$

$$= 0.209 \text{ rad/min}$$

$$\text{Time lag} = \frac{1}{\omega} \times \text{phase lag in radians}$$

$$= \frac{1}{\omega}(\tan^{-1} wt) = 4 \text{ minutes (given)}$$

The output has a frequency response given by

$$= \sin\omega\left(t - \frac{1}{\omega}\tan^{-1}\omega\tau\right)$$

$$= \sin 0.209\,(30 - 4) = \sin 5.434$$

If w_c is changed angular frequency, then $\sin w_c t = \tan$ 5.434.

$$\therefore \quad \omega_c = \frac{5.434}{t} = \frac{5.434}{30} = 0.1811 \text{ rad/min}$$

$$f_c = \frac{\omega_c}{2\pi} = \frac{0.1811}{2\pi} = 0.0288 \text{ cpm}$$

Therefore, the thermostat has a cycle of

$$\frac{1}{0.0288} \approx 34.72 \text{ minute period.}$$

71. Ans. (b)

Solution:

The temperature would oscillate abut the mean value of 425°C with an amplitude of 25°C.

Frequency of oscillation,

$$\omega = 2\pi f = \frac{2\pi}{t} = \frac{2\pi}{60} = 0.104 \text{ rad/s}$$

$$\omega\tau = 0.104 \times 8 = 0.832$$

After the transient effects die away, the amplitude ratio is

$$\frac{\theta_0}{\theta_i} = \frac{1}{\sqrt{1+(\omega\tau)^2}} = \frac{1}{\sqrt{1+(0.832)^2}} = 0.768$$

$$q_0 = 0.768 \times 25 = 192°\text{C}$$

Therefore:

Maximum temperature = 425 + 19.2 = 444.2°C

Minimum temperature = 425 – 19.2 = 408.8°C

72. Ans. (d)

Solution:

Phase sift f = $\tan^{-1}$ (wt)

= $\tan^{-1}$ (0.832)

= 39.76°C = 0.6935 radians.

73. Ans. (a)

Solution:

$$\text{Time lag} = \frac{1}{\omega} \times \text{phase shift in adians}$$

$$= \frac{0.6935}{0.104} = 6.669 \text{ seconds}$$

74. Ans. (d)

Solution:

Corresponding to a maximum deviation of 10%, the amplitude ratio is either 0.9 or 1.1. Further, the amplitude ratio is prescribed by the relation.

$$M = \frac{1}{\sqrt{(1-r^2)+(2\zeta r)^2}}$$

where r = normalised frequency = $\frac{\omega}{\omega_n}$

(*i*) When M = 1.1

$$1.1 = \frac{1}{\sqrt{(1-r^2)+(2\times 2.6\times r)^2}}$$

or $(1-r^2)^2 + (2\times 0.6\times r)^2 = (1/1.1)^2 = 0.826$

This results into the following quadratic equation,

$$r^4 - 0.56r^2 + 0.174 = 0$$

The solution consist of the imaginary roots and this implies that with $\zeta = 0.6$ the output would never be 1.1 times the input for any frequency.

(*ii*) When $M = 0.9$

$$0.9 = \frac{1}{\sqrt{(1-r^2)^2+(2\times 0.6\times r)^2}}$$

This would give,

$$r^4 - 0.56r^2 - 2.34 = 0$$

$$r^2 = \frac{0.56 \pm \sqrt{(0.56)^2 + 4\times 0.234}}{2} = 0.84$$

$$r = \sqrt{0.84} = 0.916$$

$$\therefore \ \omega = \omega_n \times r = 1000 \times 0.916 = 916 \text{ cps}$$

Apparently, the deviation remains within 10% of output for the frequency range 0–916 cps.

75. Ans. (c)

Solution:

The expression relating discharge Q to heat H above the weir is

$$Q = K\,H^{2.5}$$

where K is a constant.

The error specified pertains to statistical bonunds, therefore the error or uncertainty in discharge measurement is

$$e_Q = \sqrt{\left(\frac{\partial Q}{\partial H} e_H\right)^2}$$

$$\frac{\partial Q}{\partial H} = 2.5\,K\,H^{1.5}$$

e_H = 2% of absolute value = 0.02 H

$$\therefore\; e_Q = \left[(2.5\,K\,H^{1.5} \times 0.02\,H)^2\right]^{1/2} = 0.05\,K\,H^{2.5}$$

The percentage error in discharge measurement becomes,

$$e_Q \frac{0.05\,K\,H^{2.5}}{K\,H^{2.5}} \times 100 = 5\%$$

76. Ans. (a)

Solution:

Refer figure. Let p_x be the gauge pressure in the air duct. Neglecting the pressure exerted by the air column above mercury in the left limb of the manometer, the governing manometric equation for the points 1 and 2 is:

$$p_x = w_m h = 0\;;\; p_x = -w_m h$$

$$p_x = -(9810 \times 13.6) \times 0.2$$

$$= -26683.2 \text{ N/m}^2 \text{ (gauge)}$$

$$= 26683.2 + 10 \times 10^4 = 73316.8 \text{ N/m}^2 \text{ (absolute)}$$

77. Ans. (b)

Solution:

Refer figure. Let p_x be gauge pressure of the fluid in the pipeline. Consider pressure balance in the horizontal plane 0–0; the pressure in the left and right limbs at this plane are equal. That is:

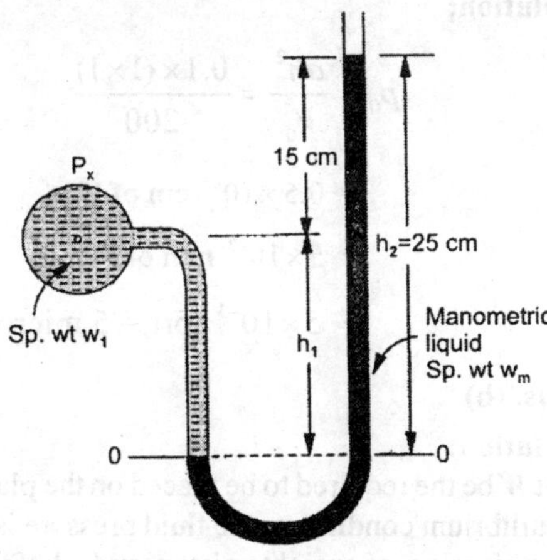

$$p_x + w_1 h_1 = w_m h_2$$

or $p_x = w_m h_2 - w_1 h_1$

$$= (9810 \times 13.6) \times 0.25 - (9810 \times 0.85) \times (0.25 - 0.15)$$

$$= 33354 - 833.85 \simeq 32520 \text{ N/m}^2$$

78. Ans. (d)

Solution:

Refer figure

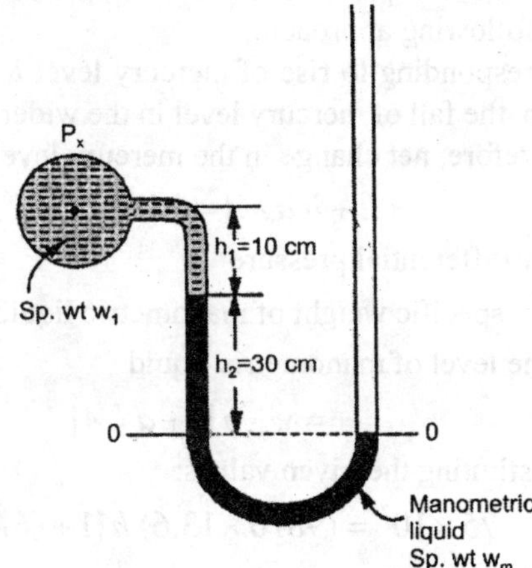

Let p_x be gauge pressure of fluid in the pipeline. Consider pressure balance in the horizontal plane 0–0; the pressure in the left and right limb at this level are equal. That is:

$$p_x + w_1 h_1 + w_m h_2 = 0$$

$$px = -(w_1 h_1 + w_m h_2)$$

$$= -(9810 \times 0.8 \times 0.1 + 9810 \times 13.6 \times 0.3)$$

$$+9810 \times 13.6 \times 0.3)$$

$$= -(784.8 + 40024.8) = -40809.6 \text{ N/m}^2$$

79. Ans. (a)

Solution:

The pressure differential is worked out from the relation.

$$p_x - p_y = (\rho_m - \rho_w) g\,h$$

$$= (13560 - 1000) \times 9.81 \times \frac{50}{1000}$$

$$= 6160.98 \text{ N/m}^2$$

80. Ans. (d)

Solution: Neglecting the density of water,

$$p_x - p_y = \rho_m\,g\,h$$

$$\text{Percentage error} = \frac{\rho_w\,g\,h}{(\rho_m - p_w)\,g\,h} \times 100 = \frac{\rho_w}{\rho_m - \rho_w}$$

$$= \frac{1000}{13560-1000} \times 100 = 7.96\%$$

81. Ans. (d)

Solution:

For a single column manometer, the differential pressure is given by:

$$\Delta p = w_m h\,[1 + a/A]$$

The above expression can be set up easily by adopting the following approach:

Corresponding to rise of mercury level h in the narrow limb, the fall of mercury level in the wider limb is $h\,a/A$. Therefore, net change in the mercury level becomes

$$h + h\,a/A = h\,[1 + a/A]$$

Now differential pressure

Δp = specific weight of manometric liquid × net change in the level of manometric liquid

$$= w_m\, h\,[1 + a/A]$$

Substituting the given values:

$$75 \times 10^2 = (9810 \times 13.6)\, h\,[1 + (6/160)^2]$$

$[\because a/A = (d/D)^2]$

Manometric height h = 0.562 m = 56.2 cm

82. Ans. (c)

Solution:

For a single column manometer, the differential pressure is given by:

$$\Delta p = \rho g h(1 + a/A) = \rho g l \sin\theta\,(1 + a/A)$$

Substituting: Δp = 1mm of mercury column

$$= \frac{1}{1000} \times (1000 \times 13.56 \times 9.81) = 133 \text{ N/m}^2$$

$$(\because p = \rho h g)$$

$$\frac{\alpha}{A} = \left(\frac{d}{D}\right)^2 = \left(\frac{4}{20}\right)^2 = \frac{1}{25}$$

we get:

$$133 = (1000 \times 13.56 \times 9.81) \times \frac{1}{1000} \times \sin\theta$$

$$\left[(1 + d/D)^2\right]$$

$$\sin q = \frac{133 \times 1000}{1000 \times 13.56 \times 9.81} \times \frac{25}{26} = 0.9614$$

$$q = \sin^{-1}(0.9614) = 74°$$

Hence the angle at which the tube is inclined to vertical = 90° – 74° = 16°.

83. Ans. (d)

Solution:

For a thin walled inverted bell-type differential gauge, the equilibrium equation is

$$\Delta P = \frac{Fh}{a}$$

where F is the spring constant, h is the bell movement and a is cross sectional area of the bell. Substituting: DP = 250 mm water gauge.

$$= 1000 \times 9.81 \times \frac{250}{1000} = 2452.5 \text{ N/m}^2$$

$(\because P = \rho g h)$

$$= 2452.5 \times 10^{-6} \text{ N/mm}^2$$

$$h = 20 \text{ mm and } a = \frac{\pi}{4}(150)^2 17662.5 \text{ mm}^2$$

wet get : $2452.6 \times 10^{-6} = \dfrac{F \times 20}{17662.5}$;

$$F = 2.166 \text{ N/mm}$$

84. Ans.(a)

Solution:

Using the relation:

$$y_{max} = \frac{t}{3} = \frac{3\Delta P}{16Et^3} r^4 (1 - \mu^2)$$

$$\therefore \quad t^4 = \frac{9\Delta P r^4 (1 - \mu^2)}{16E}$$

$$= \frac{9 \times 1.5 \times 10^6 \times .25^4\,[1 - (0.3)^2]}{16 \times 200 \times 10^9}$$

$$= 149.96 \times 10^{-6} \text{ cm}^4$$

$$t = 0.1106 \text{ cm}$$

85. Ans. (c)

Solution:

$$p_0 = \frac{ah^2}{V_0} = \frac{0.1 \times (1 \times 1)}{200}$$

$$= 0.5 \times 10^{-3} \text{ cm of Hg}$$

$$= 5 \times 10^{-3} \text{ mm of Hg}$$

$$= 5 \times 10^{-3} \text{ torr} = 5 \text{ microns}$$

86. Ans. (b)

Solution:

Let W be the required to be placed on the platfrom. Under equilibrium conditions, the fluid pressure is given by the combination of weight, piston and platform assembly divided by the effective piston-cylinder area. That is:

$$\text{Pressure} = \frac{\text{total weight}}{\text{effective area}}$$

or $$175 \times 10^3 = \frac{W+20}{\frac{\pi}{4}\left(1.25/100\right)^2}$$

$$W + 20 = 175 \times 10^3 \times \frac{\pi}{4}\left(1.25/100\right)^2 = 21.464$$

$$\therefore \quad W = 21.464 - 20 = 1.464 \text{ N}$$

87. Ans. (d)

Solution:

The actual discharge through a venturimeter is given by the relation:

$$Q = C_d\, M\, A_2\, E\sqrt{2\left\{\frac{p_1 - p_2}{\rho} + g(z_1 - z_2)\right\}}$$

Presuming thermal expansion factor $E = 1$ and recalling that velocity of approach factor.

$$M = \frac{1}{\sqrt{1-(A_2/A_1)^2}} = \frac{A_1}{\sqrt{A_1^2 - A_2^2}}$$

we may write:

$$Q = C_d \frac{A_1 A_2}{\sqrt{A_1^2 - A_2^2}}\sqrt{2g\left\{\frac{p_1 - p_2}{\rho g} + (z_1 - z_2)\right\}}$$

Now, $A_1 = \frac{\pi}{4}(0.3)^2 = 0.07065 \text{ m}^2$

and $A_2 = \frac{\pi}{4}(0.15)^2 = 0.01766$

$$\frac{p_1 - p_2}{\rho g} + (z_1 - z_2) = h\left(\frac{\rho_m}{\rho} - 1\right)$$

$$= 0.25\left(\frac{13.6}{0.9} - 1\right) = 3.528 \text{m of oil}$$

Substituting the given data in the discharge equation,

$$Q = 0.98 \times \frac{0.07065 \times 0.01766}{\sqrt{0.07065^2 - 0.01766^2}} \times \sqrt{2 \times 9.813.528}$$

$$= 0.1487 \text{ m}^3/\text{s}$$

88. Ans. (b)

Solution:

$$\frac{p_1 - p_2}{\rho g} + (z_1 - z_2) = 3.528$$

$$\therefore \quad \frac{p_1 - p_2}{\rho g} = 3.528 - (z_1 - z_2)$$

$$= 3.528 - (-0.3) = 2.828 \text{ m of oil}$$

Hence the pressure difference between the entrance and throat section,

$$p_1 - p_2 = \rho g \times 3.528$$

$$= (1000 \times 0.9) \times 9.81 \times 3.828$$

$$= 33797.4 \text{ N/m}^2$$

89. Ans.(b)

Solution:

Manometric head h_m = 10 cm of water

Equivalent head of air column,

$$h = h_m\left(\frac{\rho_m}{\rho} - 1\right)$$

$$= \frac{10}{100}\left(\frac{1000}{1.22} - 1\right) = 91.87 \text{ m of air}$$

$\therefore$ Flow velocity, $V = C_v\sqrt{2gh}$

$$= 0.9\sqrt{2 \times 9.81 \times 81.87} = 39.2 \text{ m/s}$$

90. Ans.(c)

Solution:

The atmospheric pressure is,

$$p_{at} = \rho gh\,(13.6 \times 10^3) \times 9.81 \times \frac{760}{1000}$$

$$= 101.39 \times 10^3 \text{ N/m}^2 = 101.39 \text{ kN/m}^2$$

$\therefore$ Static pressure in the duct = 101.39 + 10 = 111.39 kN/m²

From the characteristic gas equation, $pV = mRT$.

$\therefore$ Density of air, $\rho = \frac{m}{V} = \frac{p}{RT} = \frac{111.39 \times 10^3}{287 \times 300}$

$$= 1.29 \text{ kg/m}^3$$

Manometric head h_m = 10 cm of water

Equivalent head of air column,

$$h = h_m \frac{\rho_m}{\rho} = \frac{10}{100} \times \frac{1000}{1.29} = 77.52 \text{ m of air}$$

Flow velocity $V = C_v\sqrt{2gh}$

$$= 0.98\sqrt{2 \times 9.81 \times 77.52}$$

$$= 38.22 \text{ m/s}$$

91. Ans. (d)

Solution:
(*b*) Mechanical ampification

$$= \frac{\text{movement of pointer}}{\text{angular displacement of the bourdon tube}}$$

$$30 = \frac{270}{\text{angular displacement of the bourdon tube}}$$

$\therefore$ Angular displacement of bourdon tube

$$= \frac{270}{30} = 9° \text{ or } 0.157 \text{ radian}$$

Sensitivity of Bourdon tube

$$= \frac{\text{angular displacement}}{\text{temperature}}$$

$$= \frac{0.157}{250} = 6.28 \times 10^{-4} \text{ rad/°C}$$

92. Ans. (b)

Solution:
Over a limited temperature range around 0°C;

$$R = R_0(1+\alpha\theta);\ R_1 = R_0(1+\alpha\theta_1)$$

and $R_2 = R_0(1+\alpha\theta_2)$

Rearrangement gives: $\theta_2 = \theta_1 + \frac{R_2 - R_1}{\alpha R_0}$

Inserting the appropriate values,

$$\theta_2 = 100 + \frac{300-150}{0.0039 \times 100} = 484.61\text{°C}$$

The temperature of the gas is therefore, 484.61°C.

93. Ans. (d)

Solution:

$$R_1 = R_0\, a e^{b/T}$$

$$\frac{b}{T} = \log_e R_1 - \log_e (R_0 \times a)$$

$$T = \frac{b}{\log_e R_1 - \log_e (R_0 \times a)}$$

$$= \frac{2850}{\log_e 1050 - \log_e (4000 \times 300 \times 10^{-6})}$$

$$= \frac{2850}{6.956 - 0.1823} = 420 \text{ K}$$

94. Ans. (d)

Solution:
Change of stress

$$\alpha = \frac{10}{100} \times (400) = 40 \text{ MN/m}^2$$

Change of strain $\varepsilon = \frac{\sigma}{E} = \frac{40 \times 10^6}{200 \times 10^9} = 2 \times 10^{-4}$

$$\therefore dR_1 = F \varepsilon R$$

$$= 2 \times (2 \times 10^{-4}) \times 100 = 4.04\Omega$$

Change in resistance due to temperature coefficient of resistance of the gauge is

$$dR_2 = R\alpha_g\, dt$$

$$= 10 \times (20 \times 10^{-6}) \times 25 = 0.05\Omega$$

95. Ans. (b)

Solution: Change in resistance due to differential expansion of the gauge metal and the steel test-specimen is

$$dR_3 = FR(\lambda_s - \lambda_g)\, dt$$

$$= 2 \times 100\,(12-16) \times 10^{-6} \times 25 = -0.02\Omega$$

It is to be noted that changes of resistance due to measured strains and due to temperature changes are of the same order. Evidently, the strain measurements under varying temperature conditions would be hopelessly inaccurate unless means for temperature compensation are provided.

96. Ans. (d)

Solution:
Axial stress

$$\sigma_x = \frac{26500}{125 \times 10^{-6}} = 212 \times 10^6 \text{ N/m}^2$$

Transverse stress $\sigma_y = 0$

Axial strain $\varepsilon_x = \frac{1}{E}(\sigma_x - \mu\sigma_y) = \frac{\sigma_x}{E}$

$\therefore$ Modulus of elasticity $E = \frac{\sigma_x}{\varepsilon_x}$

$$= \frac{212 \times 10^6}{1600 \times 10^{-6}} = 132.5 \times 10^6 \text{ N/m}^2$$

$$= 132.5 \text{ GN/m}^2$$

97. Ans. (b)

Solution:
Strain in the transverse direction,

$$\varepsilon_y = \frac{1}{E}(\sigma_y - \mu\sigma_x) = -\frac{\mu\sigma_x}{E}$$

$$-450 \times 10^{-6} = -\mu \times \frac{212 \times 10^6}{132.5 \times 10^9}$$

Upon solution, we get
Poission's ratio $\mu = 0.281$.

98. Ans. (c)

Solution:
The vibration level expressed in decibels is given by

$$\text{vibration level} = 120\ \log_{10} \frac{A_1}{A_0}\text{dB}$$

where A_1 is the measured amplitude and A_0 is the reference amplitude.

$\therefore$ velocity vibration level

$$= 20\ \log_{10} = \frac{3.125}{10^{-3}}$$

$$= 69.897\text{ dB}$$

99. Ans. (d)

Solution:

acceleration vibration level

$$= 20\ \log_{10} = \frac{1.5\times 9.81}{10^{-5}}$$

$$= 123.355\text{ dB}$$

100. Ans. (b)

110. Ans. (c)

111. Ans. (b)

112. Ans. (c)

113. Ans. (b)

114. Ans. (a)

115. Ans. (d)

116. Ans. (d)

117. Ans. (a)

118. Ans. (c)

119. Ans. (d)

120. Ans. (d)

121. Ans. (b)

122. Ans. (b)

123. Ans. (a)

124. Ans. (c)

125. Ans. (c)

126. Ans. (b)

127. Ans. (b)

128. Ans. (a)

129. Ans. (c)

130. Ans. (c)

131. Ans. (c)

132. Ans. (b)

133. Ans. (a)

134. Ans. (b)

135. Ans. (d)

136. Ans. (c)

137. Ans. (b)

138. Ans. (b)

139. Ans. (a)

Solution :

Volumetric flow/discharge is given as

$$Q = \frac{\pi r^4(P_1 - P_2)}{8\mu L}$$

Since (P_1 - P_2) is constant and all other variables are also made constant in the second experiment. So

$$\frac{\pi(P_1 - P_2)}{8\mu L} = K(\text{say})$$

$$Q_A = Kr_A^4 \qquad (1)$$

and also $Q_B = Kr_B^4$

Put, $r_B = 2\, r_A$

Hence $Q_B = K(2\, r_A)^4$

$$= 16\, k_A^4 \qquad (2)$$

Dividing equation (2) by (1)

$$\frac{Q_B}{Q_A} = \frac{16\ Kr_a^4}{K\ r_a^4} = 16$$

140. Ans. (d)

3.5 Miscellaneous

1. Ans. (d)

Solution:

It is known

Torque for torsion bar

$$\text{T} = \frac{\pi}{16} Gd^3\theta$$

$$\therefore \qquad \theta = \frac{16T}{\pi Gd^3}$$

$$= \frac{16\times 120}{3.14\times 80\times 10^8\times(2\times 15\times 10^{-3})3}$$

$$= 0.2830\times 10^{-3}\, rad$$

$$= 0.2832\times 10^{-3}\ rad$$

2. Ans. (d)

Soultion:

Speed of the shaft (n)

$$= \frac{\text{Count recorded per second}}{\text{Number of teeths on rotor}}\text{rps}$$

$$= \frac{3600}{60} rps$$

$$= 60\times 60\ rpm$$

$$= 3600\ rpm$$

3. Ans. (d)

Solution:

By voltage division voltage at positive

$$= V_+ = \frac{12}{24} \times 14 = 2\ volts$$

$$V_+ = V_-$$

Using KCL at inverting terminal, we have

$$\frac{4-2}{100} = \frac{2-V_o}{R_T}$$

or $$2 - V_o = \frac{2}{100} \cdot R_T = \frac{2}{10} \times R(1 - Kx)$$

$$= \frac{2}{100} \cdot 100(1 - kx)$$

$$= 2(1 - 0.004 \times 75)$$

$$= 2[1 - 0.3]$$

or $$2 - V_o = 1.4$$

or $$V_o = 2 - 1.4 = 0.6\,V = 600\ mV$$

4. Ans. (c)

Solution:

From the given data make a graph for $V_i - V_o$ relation

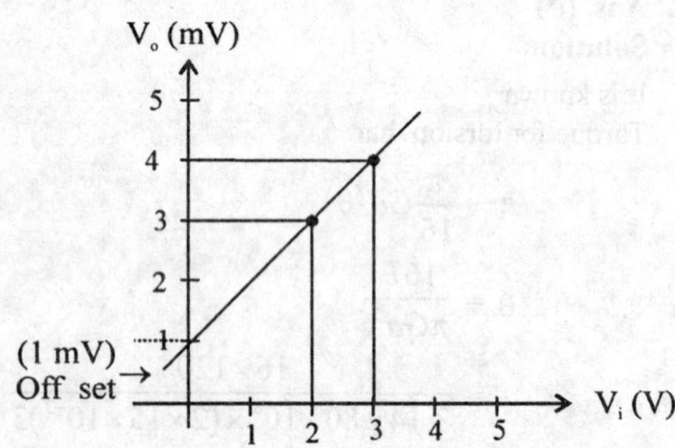

Given, $A_d = 10$

From the graph

offset voltage $= 1\,\text{mV}$

i.e. $A_c = 10^{-3}$

$$\therefore\ \text{CMMR} = \frac{A_d}{A_c} = \frac{10}{10^{-3}}$$

$$= 10^4$$

5. Ans. (b)

Solution:

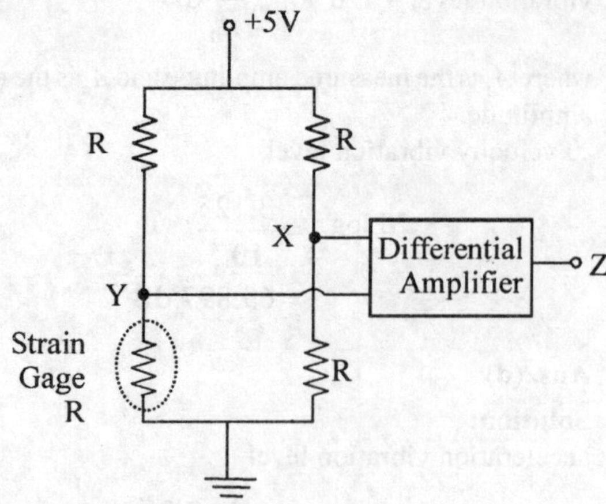

For Variation in R by +1%

$$V_x = \frac{R}{R+R} \times 5$$

$$= \frac{5}{2}$$

$$V_y = \frac{1.01\,R}{R + 1.01\,R} \times 5$$

$$= \left(\frac{1.01}{2.01}\right) \times 5$$

$V_z = A_d V_d + A_c V_c +$ offset voltge

$$= 10\,(V_y - V_x) + 10^{-3}\frac{(V_y + V_x)}{2} + 1\,\text{mV}$$

$$= +128\,\text{mV}$$

For variation in R by −1%

$$V_x = \frac{R}{R+R} \times 5$$

$$= \left(\frac{1}{2}\right)5$$

$$V_y = \frac{0.99R}{R + 0.99R}$$

$$= \left(\frac{0.99}{1.99}\right)5$$

$V_z = A_d V_d A_c V_c +$ Off set voltage

$= -122\ \text{mV}$

6. Ans. (d)

Solution: Output voltage = Sensitivity ′ Pressure

$$= 1\,\mu v/Pa \times 120\,kPa$$

$$= 120\,mV$$

7. Ans. (c)

8. Ans (b)

Solution:

Equivalent circuit of a piezoelectric transdcer with cable capacitance and amplifier, is

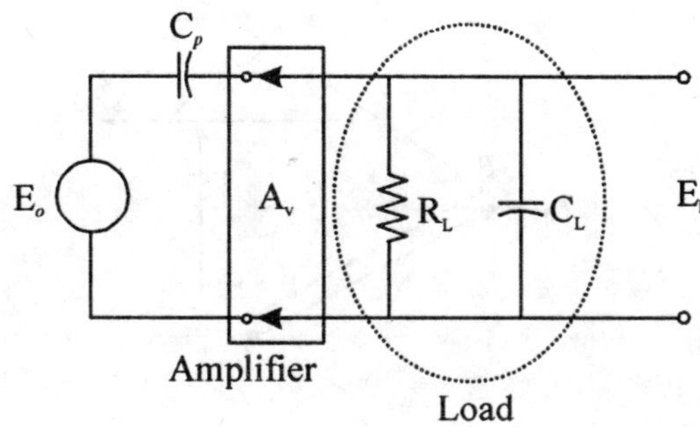

$$E_L = \left[\frac{\omega C_p R_L}{\sqrt{1+\omega^2(C_p+C_L)^2 R_L^2}}\right] \times A_v \times E_o$$

$$\frac{E_L}{E_o} = \frac{\omega C_p R_L}{\sqrt{1+\omega^2(C_p+C_L)^2 R_L^2}} \times Av$$

$$\frac{E_L}{E_o} = 0.95$$

Given

$A_v = 3$

$C_p = 1\times10^{-9} F$

$C_L = 2\times10^{-9} F$

$R_L = 1 M\Omega$

From the given condition

$$\frac{\omega\times10^{-9}\times10^{6}}{\sqrt{1+\omega^2(3\times10^{-9})\times(10^6)^2}}\times3 = 0.95$$

$$\frac{(\omega\times10^{-3})^2}{1+\omega^2\times9\times10^{-6}}\times9 = (0.95)^2$$

$$9\omega^2\times10^{-6} = (1+9\omega^2\times10^{-6})(0.95)^2$$

$$[1-(0.95)^2](9\times\omega^2\times10^{-6}) = (0.95)^2$$

$$9\times\omega^2\times10^{-6} = 9.25641$$

$$\omega^2 = 1028490.02$$

$$f = 161.4\ Hz$$

9. Ans (d)

10. Ans (b)

11. Ans (a)

Solution:

Here, $Q \rightarrow$ Flow rate

$R \rightarrow$ Pressure

$$Q \propto \sqrt{P}$$

$$Q_1 = \sqrt{P_1}$$

$$Q_2 = \sqrt{P_2}$$

$$\frac{Q_2}{Q_1} = \sqrt{\frac{P_2}{P_1}}$$

$$Q_2 = Q_1\sqrt{\frac{P_2}{P_1}}$$

$$= \frac{5\ m^3}{\text{second}}\sqrt{\frac{0.9\times10^5}{2.5\times10^5}}$$

$$= 0.5\times\frac{3}{5} = 0.30$$

12. Ans. (b)

Solution:

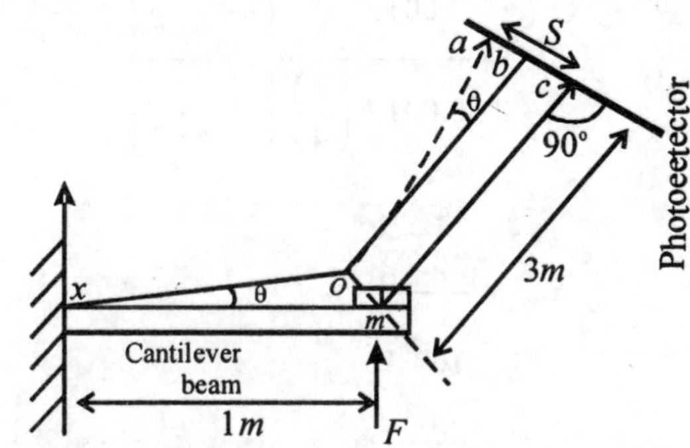

$F = 1\ N$ Given

Let, $x = 1\ m$

$$y = 0.001\ Fx^2\left(1-\frac{x}{3}\right)m$$

$$= 0.001\times1\times1\left(1-\frac{1}{3}\right)$$

$$= 0.001\times\frac{2}{3} = \frac{2}{3}mm$$

$$om = y = bc$$

$$bc = \frac{2}{3}mm$$

in triangle oab

$$\theta = \frac{ab}{ob}$$

$$ab = \theta(ob)$$

also $\theta = \dfrac{y}{x}$

$\theta = y \qquad [\text{for } x = 1m]$

$ab = \frac{2}{3} \times 10^{-3} \times 3 = 2 \times 10^{-3} m$

$s = bc + ab$

$= \left(\frac{2}{3} + 2\right) \times 10^{-3} m = 2.7\ mm$

$s \approx 3mm$

13. Ans. (b)
Solution:

Output of LVDT is proportional to deflection.

$$\therefore \quad = \frac{\text{ouput of LVDT at} \left(x = \frac{1}{2} m\right)}{\text{ouput of LVDT at} \left(x = \frac{1}{4} m\right)}$$

$$= \frac{0.001\,F\left(\frac{1}{2}\right)^2\left[1 - \frac{1}{2\times 3}\right]}{0.001\,F\left(\frac{1}{4}\right)^2\left[1 - \frac{1}{4\times 3}\right]}$$

$$= \frac{\frac{1}{4}\times\frac{5}{6}}{\frac{1}{16}\times\frac{11}{12}}$$

$$= \frac{40}{11}$$

14. Ans (a)

Solution:

Given $f = 30\ Hz$, $m = 0.001\ kg$

$a = \omega^2 x$

$$\Rightarrow \qquad x = \frac{a}{(2\pi f)^2}$$

for range of $a = 0 - 10\ g$

$x = 10 \times 0.001\ kg \times 1000$

$$x = \frac{10 \times 0.001\ kg \times 1000}{(2\pi \times 30)^2}$$

$x \Rightarrow 0\ to\ 2.76$

15. Ans. (a)

Solution:

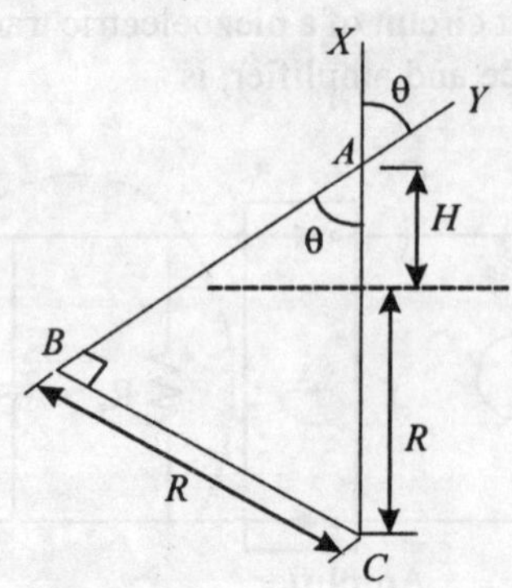

R is radius of the circle.

In $\Delta\ ABC$, $\angle B = 90°$

Since, $\angle XAY = \theta$

Therefore, $\angle BAC$ is also θ

i.e. $(\angle XAY = \angle BAC)$ (vertically opposite angles)

In $\Delta\ ABC$

$$\sin\theta = \frac{\text{Perpendicular}}{\text{hypotenuse}}$$

$$= \frac{R}{R+H}$$

$(R+H)\sin\theta = R$

$R\sin\theta + H\sin\theta = R$

$R - R\sin\theta = H\sin\theta$

$R(1-\sin\theta) = H\sin\theta$

Therefore, $R = \frac{H\sin\theta}{(1-\sin\theta)}$

16. Ans. (c)
Solution:

$$\eta = \frac{2\,r^2 g(\rho_b - \rho_L)}{9\,V}$$

$$\frac{\eta_1}{\eta_2} \Rightarrow \frac{V_2}{V_1}$$

$x \times 0.002 = 5 \times 10^{-3} \times 0.01$

3 Electrical and Electronic Measurements

3.1 Fundamental and Error Analysis

1. A metal strain gauge has factor of two. Its normal resistance is 120 ohms. If it undergoes a strain of 10^{-5}, the value of change of resistance in response to the strain is

(a) 240 ohms (b) 2×10^{-5} ohms
(c) 2.5×10^{-5} ohms (d) 1.2×10^{-3} ohms

2. Two 100 mA full scale PMMC meters are empoyed to construct a 10 V and a 100 V full scale voltmeter. These meters will have figure of merit (sensitivities) as

(a) 10 kΩ /V and 10 kΩ / V.
(b) 100 kΩ/V and 10 kΩ/V
(c) 10 kΩ/V and 100 kΩ/V.
(d) 10 kΩ/V. and 1 kΩ/V.

3. A variable w is related to three other variables x, y, z as $w = xy/z$. The variables are measured with meters of accuracy ± 0.5% reading, ± 1 % of full scale value and ± 1.5% reading. The actual readings of the three meters are 80, 20 and 50 with 100 being the full scale value for all three. The maximum uncertainty in the measurement of w will be

(a) ± 5% rdg (b) ± 6.5% rdg
(c) ± 6.7% rdg (d) ± 7.0 rdg

4. The term 'precision' used in instrumentation means

(a) Gradual departure of the measured value from the calibrated value
(b) Smallest increment in the measurand that can be detected by the instrument
(c) Maximum distance or angle through which any part of a mechanical system may be moved in one direction without causing motion of the next part
(d) The ability of the instrument to give output readings close to each other, when the input is constant.

5. Threshold of a measurement system is

(a) The smallest change in input which can be detected
(b) A measure of linearity of the system.
(c) The smallest input which can be detected
(d) A measure of precision of the system.

6. A large number of 230 Ω resistors are obtained by combining 120 Ω resistors with a standard deviation of 4.0 Ω and 110 Ω resistors with a standard deviation of 3.0. The standard deviation of the 230 Ω resistors thus formed will be

(a) 3.5 Ω (b) 5.0 Ω
(c) 7.0 Ω (d) 12.0 Ω

7. The following terms used in the context of an instrument are numbered as shown:

1. Accuracy 2. Sensitivity
3. Precision 4. Resolution

Match these with their possible definitions listed below

P. Repeatability of readings on successive observations
Q. Smallest perceptible change in the output
R. Deviation of the output from the true value
S. Minimum value of the input from the true value
T. Ratio of the change in the instrument reading to the change in the measured variable

(a) 1-P, 2-Q, 3-R, 4-S (b) 1-S, 2-Q, 3-P, 4-T
(c) 1-R, 2-T, 3-P, 4-Q (d) 1-T, 2-Q, 3-P, 4-R

8. For a proving ring used in a press, the value of force and its uncertainty are given as 12366 ± 2354 N without any rounding off. If the error should be rounded off at the maximum to two significant figures only, the value of the reading is rounded off as

(a) 12300 ± 2300 N (b) 12360 ± 2350 N
(c) 12400 ± 2300 N (d) 12400 ± 2400 N

9. The measurement of a source voltage are 5.9 V, 5.7 V and 6.1 V. The sample standard deviation of the reading is

(a) 0.013 (b) 0.04
(c) 0.115 (d) 0.2

10. The **threshold** of an instrument is normally defined

(a) As the smallest measurable input change (non-zero value) which can be detected.
(b) As the smallest measurable input which can be detected.

(c) In terms of linearity of scale.

(d) As a function of drift.

11. Match the terms and their definitions

List I	*List II*
A. Accuracy	1. Smallest perceptible change in the output
B. Sensitivity	2. Error between the measured value and the absolute value.
C. Repeatability	3. Closeness of the measurement to the mean value
D. Resolution	4. Ratio of change in the instrument reading to the change in the measured quantity
	5. The ability to display the same reading while measuring a quantity

Codes:

	A	B	C	D
(a)	1	2	3	4
(b)	3	5	2	1
(c)	3	4	5	1
(d)	3	2	4	5

12. Three DC current I_1, I_2 and I_3 meet at a node with I_1 entering and I_2 and I_3 leaving the node. I_1 and I_2 are measured as 100 mA and 99 mA with a ± 1% accuracy. Then the value of I_3 and the accuracy of I_3 are:

(a) 1 m A ± 2%

(b) 199 m A ± 2%

(c) 1 m A ± 2%

(d) 1m A ± 199%

3.2 Indicating Instruments

1. A galvanometer with a full scale current of 10 mA has a resistance of 1000 Ω. The multiplying power (the ratio of measured current to galvanometer current) of a 100 Ω shunt with this galvanometer is

(a) 110

(b) 100

(c) 11

(d) 10

2. An ammeter has a current range of $0-5\,A$, and its internal resistance is $0.2\,\Omega$. In order to change the range to $0-25\,A$, we need to add a resistance of

(a) $0.8\,\Omega$ in series with the meter.

(b) $1.0\,\Omega$ in series with the meter.

(c) $0.04\,\Omega$ in parallel with the meter.

(d) $0.05\,\Omega$ in parallel with the meter.

3. The instrument that does not have any restoring torque is:

(a) D'Arsonval galvanometer

(b) Flux meter

(c) Ballistic galvanometer

(d) Moving iron instrument

4. A dc voltage of magnitude 11 kV is to be measured without loading the source under steady condition. The best-suited instrument is

(a) A rectifier type voltmeter

(b) An electrodynamics type voltmeter

(c) A megger

(d) An electrostatic voltmeter

5. A voltmeter has range [4 V, 20 V] and a resolution of 1mV. The dynamic range is

(a) 24 dB (b) 60 dB

(c) 72 dB (d) 84 dB

6. The PMMC ammeter A in the adjoining figure has a range of 0 to 3 mA. When switch S_1 is opened, the pointer of the ammeter swings to the 1 mA mark, returns and settle at 0.9 mA. The meter is

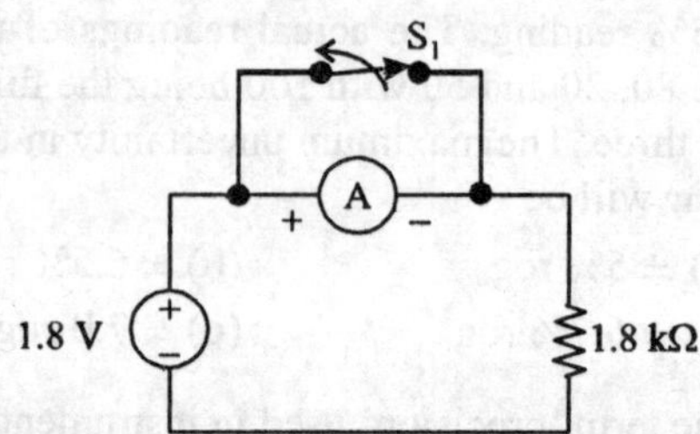

(a) Critically damped and has a coil resistance of 100 Ω

(b) Critically damped and has a coil resistance of 200 Ω

(c) Under damped and has a coil resistance of 100 Ω

(d) Under damped and has a coil resistance of 200 Ω

7. The deflection angle of the pointer of an ideal moving iron ammeter is 20° for 1.0 ampere dc current. If a current of 3sin (314 *t*) amperes is passed through the ammeter then the deflection angle is

(a) 0° (b) 42°

(c) 60° (d) 90°

8. In the context of a PMMC instrument, identify the correct matches.

A. A pair of springs	1. To provide controlling Torque
B. Aluminiumformer	2. To provide damping Torque
	3. To act as a base for the coil
	4. To provide current into and a out of the moving coil

(a) A – 1 and 2; B – 3 and 4
(b) A – 3 and 4; B – 1 and 2
(c) A – 1 and 4; B – 3 and 2
(d) A – 3 and 2; B – 1 and 4

3.3 Measurement of Current and Voltage

1. An unshielded moving iron voltmeter is used to measure the voltage in an a. c. circuit. If a stray d. c. magnetic field having a component along the axis of the meter coil appears. The meter reading would be
(a) Unaffected
(b) Decreased
(c) Increased
(d) Either decreased or increased depending on the direction of the d. c. field

2. A 0-10 mA PMMC ammeter reads 4 mA in a circuit. Its bottom control spring snaps suddenly. The meter will now read nearly
(a) 10 mA (b) 8 mA
(c) 2 mA (d) Zero

3. Four ammeters M1, M2, M3 and M4 with the following specifications are available.

Instrument	Type	Full scale Value (A)	Accuracy % of FS
M1	$3^{1/2}$ digit dual slope	20	± 0.10
M2	PMMC	10	± 0.20
M3	Electrodynamic	5	± 0.50
M4	Moving iron	1	±1.00

A current 1 A is to be measured. To obtain minimum error in the reading, one should select meter
(a) M1 (b) M2
(c) M3 (d) M4

4. An advantage of a permanent magnet moving coil instrument is that it is
(a) Free form friction error
(b) Has high (torque /weight of the moving parts) ratio
(c) Has low (torque/weight of the moving parts) ratio
(d) Can be used on both a.c and d.c

5. The scale of a voltmeter is uniform. Its type is
(a) Moving iron
(b) Induction
(c) Moving coil permanent magnet
(d) Moving coil dynamometer

6. The DC voltmeter has a sensitivity of 1000 Ω/volt. When it measures half full scale in 100 V range, the current through the voltmeter is
(a) 100 mA (b) 1 mA
(c) 0.5 mA (d) 50 mA

7. A Manganin swamping resistance is connected in series with a moving coil ammeter consisting of a milli-ammeter and a suitable shunt in order to
(a) Minimise the effect of temperature variation
(b) Obtain large deflection torque
(c) Reduce the size of the meter
(d) Minimize the effect of stray magnetic fields

8. The effect of stray magnetic fields on the actuating torque of a portable instrument is maximum when the operating field of the instrument and the stray fields are
(a) Perpendicular
(b) Parallel
(c) Inclined at 60°
(d) Inclined at 30°

9. A PMMC voltmeter connected across a series combination of a DC voltage source $V_1 = 2$V and an AC voltage source $V_2(t) = 3 \sin(4t)$ V. The meter reads

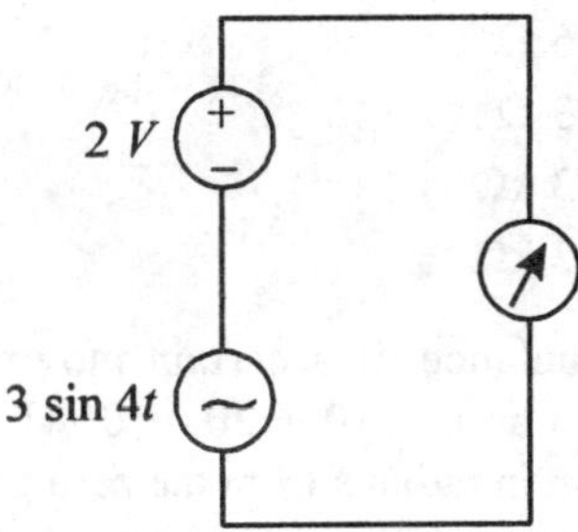

(a) 2 V
(b) 5 V
(c) $(2 + \sqrt{3}/2)$ V
(d) $(\sqrt{17}/2)$ V

10. A periodic voltage waveform observed on an oscilloscope across a load is shown.

A permanent magnet moving coil (PMMC) meter connected across the same load reads

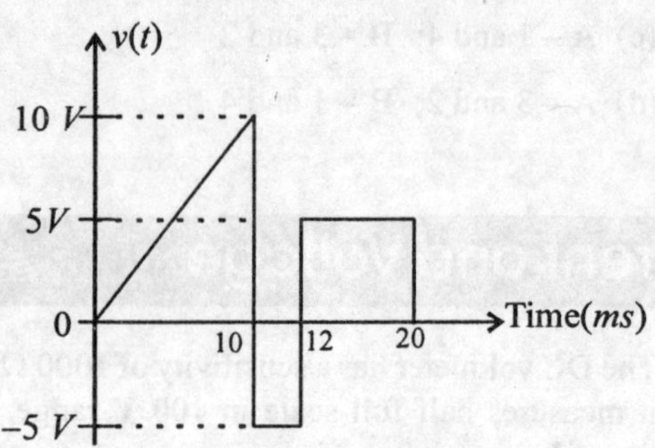

(a) 4 V
(b) 5 V
(c) 8 V
(d) 10 V

11. A rectifier type ac voltmeter consists of a series resistance R_s, and ideal full-wave rectifier bridge and PMMC instrument as shown in figure. The internal resistance of the instrument is 100 Ω a full scale deflection is produced by a dc current of 1 mA. The value of Rs required to obtain full scale deflection with an ac voltage of 100 V (rms) applied to the input terminal is:

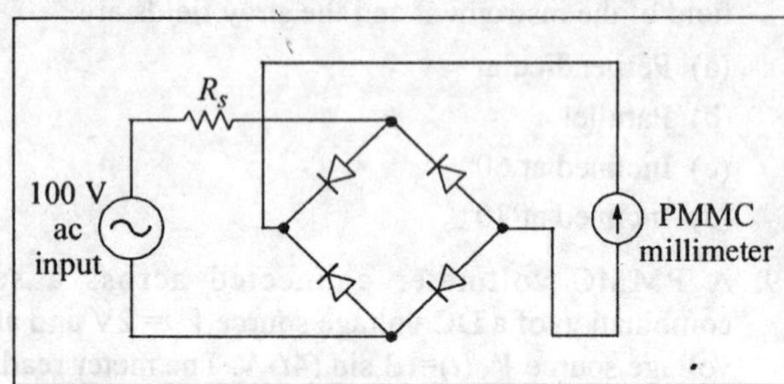

(a) 63.56 Ω
(b) 69.93 Ω
(c) 89.93 kΩ
(d) 141.3 kΩ

12. The inductance of a certain moving-iron ammeter is expressed as $L = 10 + 3\theta - [\theta^2/4]$ μH, where θ is the deflection in radians from the zero position. The control spring torque in 25×10^{-6} Nm/radian. The deflection of the pointer in radian when the meter carries a current of 5A, is

(a) 2.4
(b) 2.0
(c) 1.2
(d) 1.0

13. A moving coil of a meter has 100 turns, and a length and depth of 10 mm and 20 mm respectively. It is positioned in a uniform radial flux density of 200 mT. The coil carrier a current of 50 mA. The torque on the coil is

(a) 200 μNm
(b) 100 μNm
(c) 2 μNm
(d) 1 μNm

14. A 1000 V DC supply has two 1-core cables as its positive and negative leads: their insulation resistance to earth are 4 MΩ and 6 MΩ, respectively, as shown in the figure. A voltmeter with resistance 50 kΩ is used to measure the insulation of the cable. When connected between the positive core and earth, then voltmeter reads

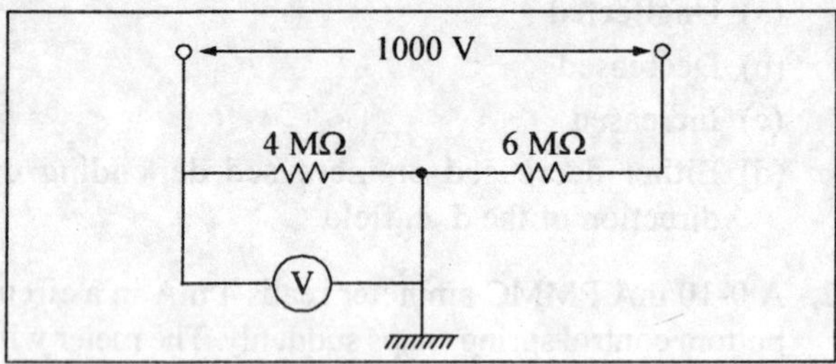

(a) 8 V
(b) 16 V
(c) 24 V
(d) 40 V

15. A current of $-8 + 6\sqrt{2}$ $(\sin \omega t + 30°)$ A is passed through three meters. They are a center zero PMMC meter, a true rms meter and a moving iron instrument. The respective readings (in A) will be

(a) 8, 6, 10
(b) 8, 6, 8
(c) – 8, 10, 10
(d) – 8, 2, 2

16. An average reading digital multimeter reads 10V when fed with a triangular wave, symmetric about the time axis. For the same input an rms reading meter will read.

(a) $20/\sqrt{3}$
(b) $10/\sqrt{3}$
(c) $20\sqrt{3}$
(d) $10\sqrt{3}$

17. An analog voltmeter uses external multiplier settings. With a multiplier setting of 20 kΩ, it reads 440 V and with a multiplier setting of 80 kΩ it reads 352 V. For a multiplier setting of 40 kΩ, the voltmeter reads

(a) 371 V
(b) 383 V
(c) 394 V
(d) 406 V

3.4 Measurement of Resistance

1. A Kelvin double bridge is best suited for the measurement of
   (a) Inductance
   (b) Capacitance
   (c) Low resistance
   (d) High resistance

2. Resistance R1 and R2 have, respectively nominal value of 10 Ω and 5 Ω, and tolerance of ± 5 % and 10 %. The range of values for the parallel combination of R_1 and R_2 is
   (a) 3.077 Ω to 3.636 Ω
   (b) 2.805 Ω to 3.371 Ω
   (c) 3.237 Ω to 3.636 Ω
   (d) 3.192 Ω to 3.435 Ω

3. The set-up in the figure is used to measure resistance R. The ammeter and voltmeter resistances are 0.01 Ω and 2000 Ω, respectively. Their readings are 2 A and 180 V, respectively, giving a measured resistance of 90 Ω. The percentage error in the measurement is

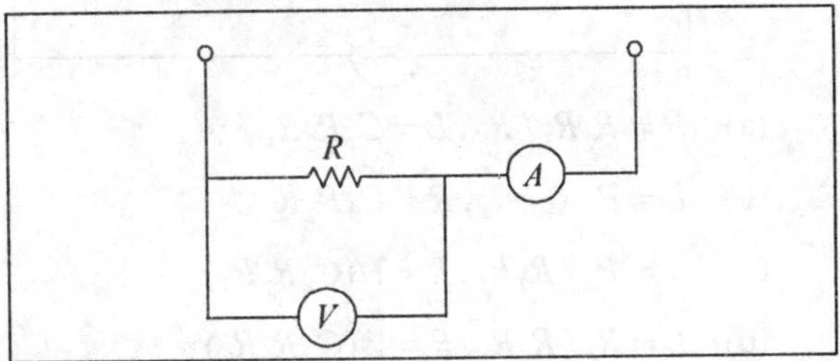

   (a) 2.25 %
   (b) 2.35 %
   (c) 4.5 %
   (d) 4.71 %

3.5 Bridge, Measurement of Inductance and Capacitance

1. The bridge circuit shown in the figure below is used for the measurement of an unknown element Z_x. The bridge circuit is best suited when Z_x is

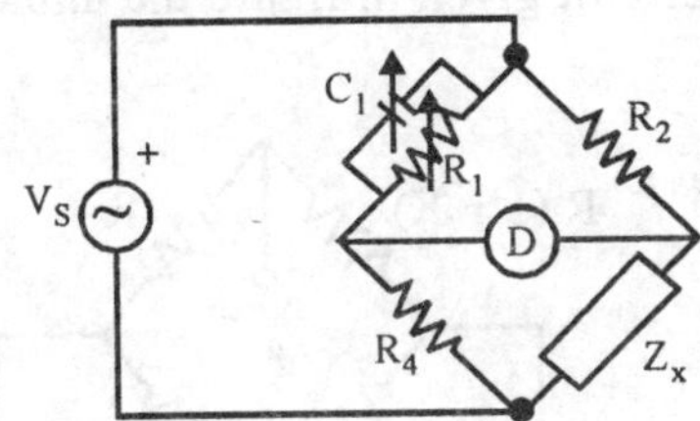

   (a) Low resistance
   (b) High resistance
   (c) Low Q inductor
   (d) Lossy capacitor

2. The items in group I represent the various types of measurements to be made with a reasonable accuracy using a suitable bridge. The items in Group-II represent the various bridges available for this purpose. Select the correct choice of the item in Group II for the corresponding item in Group I form the following

Group I

P. Resistance in the milli-Ohm range
Q. Low values of Capacitance
R. Comparison of resistance which are nearly equal
S. Inductance of a coil with a large time-constant

Group II

1. Wheat stone Bridge
2. Kelvin Double Bridge
3. Schering Bridge
4. Wien's Bridge
5. Hay's Bridge
6. Carey-Foster Bridge

Codes:

	P	Q	R	S
(a)	2	3	6	5
(b)	2	6	4	5
(c)	2	3	5	4
(d)	1	3	2	6

3. A moving iron ammeter produces a full scale torque of 240 μNm with a deflection of 120° at a current of 10 A. The rate of change of self inductance (μH/radian) of the instrument at full scale is
   (a) 2.0 μH/radian
   (b) 4.8 μH/radian
   (c) 12.0 μH/radian
   (d) 114.6 μH/radian

4. R_1 and R_4 are the opposite arms of a Wheatstone bridge as are R_3 and R_2. The source voltage is applied across R_1 and R_3. Under balanced conditions which one of the following is true
   (a) $R_1 = R_3 R_4/R_2$
   (b) $R_1 = R_2 R_3/R_4$
   (c) $R_1 = R_2 R_4/R_3$
   (d) $R_1 = R_2 + R_3 + R_4$

5. The Maxwell's bridge shown in the figure is at balance. The parameters of the inductive coil are

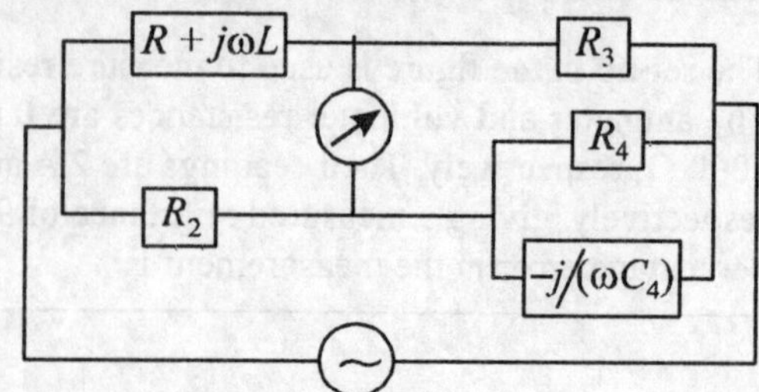

(a) $R = R_2R_3/R_4, L = C_4R_2R_3$
(b) $L = R_2R_3/R_4, R = C_4R_2R_3$
(c) $R = R_4/R_2R_3, L = 1/(C_4R_2R_3)$
(d) $L = R_4/R_2R_3, R = 1/(C_4R_2R_3)$

6. In the a.c. bridge circuit shown in figure at balance the parameters of the coil are calculated to the 50 mH and 20 Ω. Another coil is connected in parallel to the previous coil, and balance is restored by changing R_2 to 2.5 kΩ. The parameters of the new coil are than.

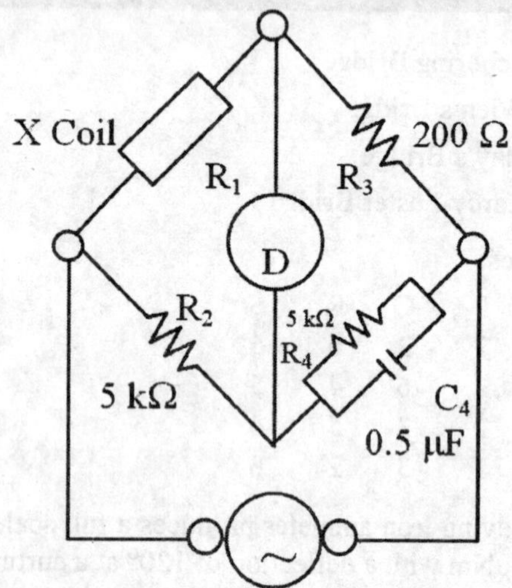

(a) 25 mH, 20 Ω
(b) 50 mH, 10Ω
(c) 50 mH, 20 Ω
(d) 100 mH, 40

7. The equations under balance conditions for a bridge are:
$R_1 = [(R_2R_3)/R_4]$ and $L_1 = R_2 R_3 C_4$, where R_1 and L_1 respectively unknown resistance and inductance.
In order to achieve faster balance which of the following variables are chosen
(a) R_2 and R_3 (b) R_2 and C_4
(c) R_4 and C_4 (d) R_3 and C_4

8. In the bridge circuit shown in Figure, when $[X_C/R] = 1$, the voltmeter reads.

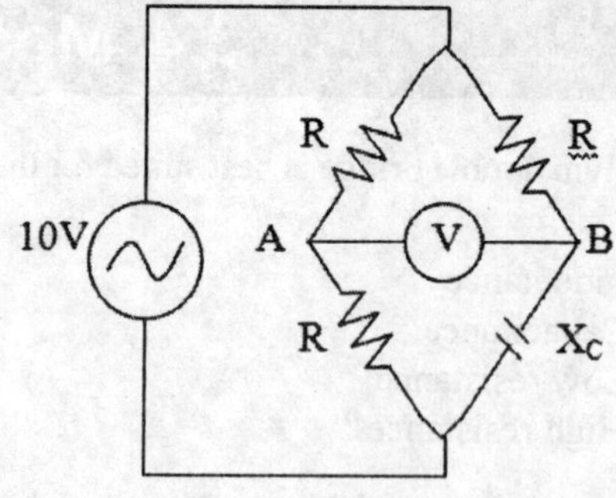

(a) 5 V (b) 0.0 V
(c) 2.5 V (d) 10 V

9. In figure the stored energy of the capacitor is

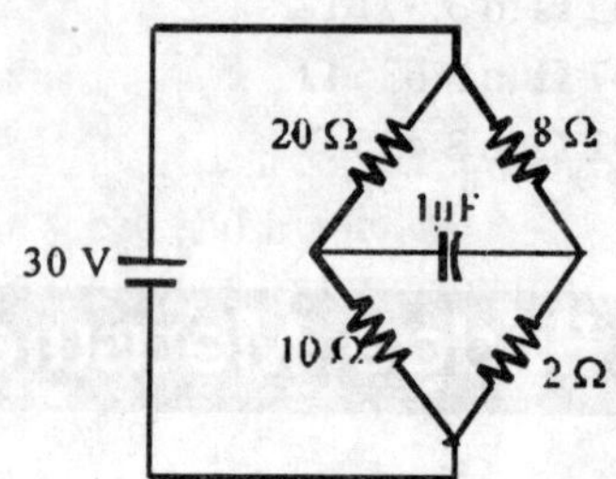

(a) 8 μJ (b) 30μJ
(c) 450 μJ (d) 900 μJ

10. For the circuit given in figure the unbalance voltage e_o is:

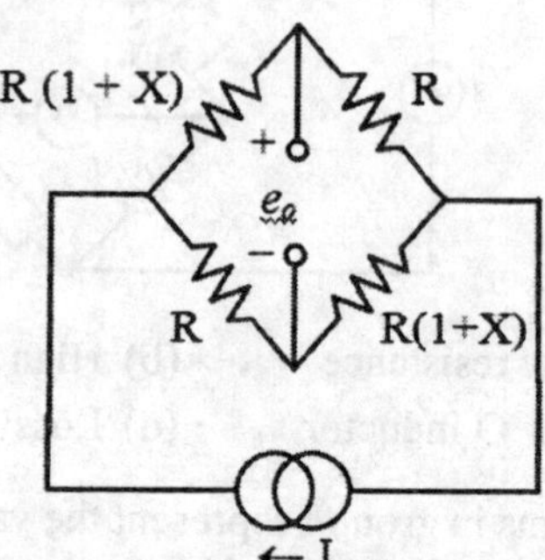

(a) e_o = IRX (b) e_o = [IRX / 2]
(c) e_o = 2 IRX (d) e_o = IR (1 + X)

11. The a.c bridge in figure remains balanced if Z comprises of:

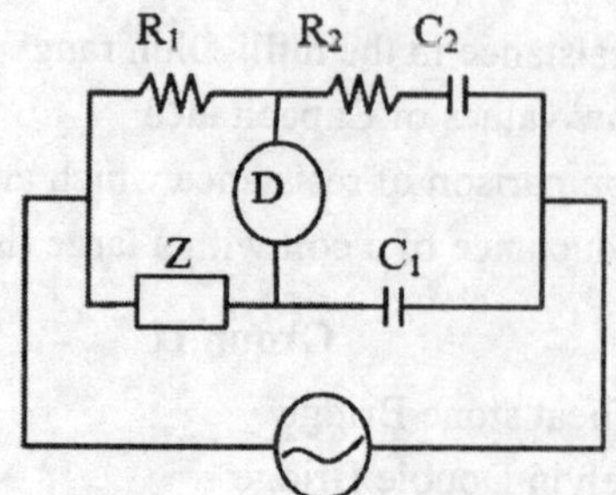

(a) Resistance only
(b) Capacitance only
(c) Resistance and capacitance in parallel
(d) Resistance and inductor in series.

12. The bridge circuit in figure is balanced. The magnitude of current I is:

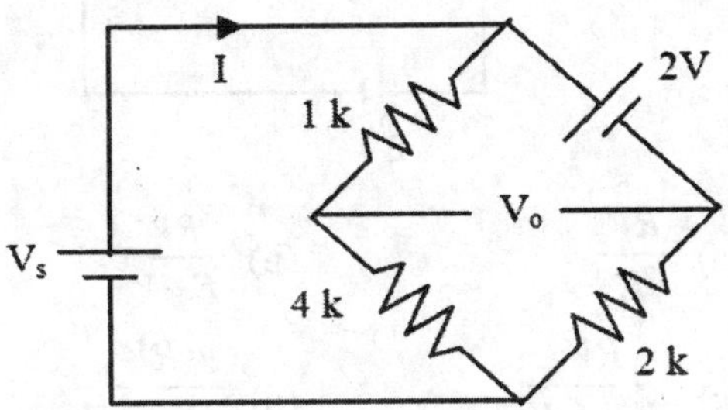

(a) 2 mA (b) 4 mA
(c) 5 mA (d) 6 mA

13. **Kelvin double bridge** is best suited for the measurement of:
(a) Resistances of very low value
(b) Low value capacitance
(c) Resistance of very high value
(d) High value capacitance

14. **Wien bridge** is best suited for the measurement of
(a) Frequency (b) Capacitor
(c) Inductor (d) Resistor

15. If the ac bridge circuit shown below is balanced, the element Z can be a

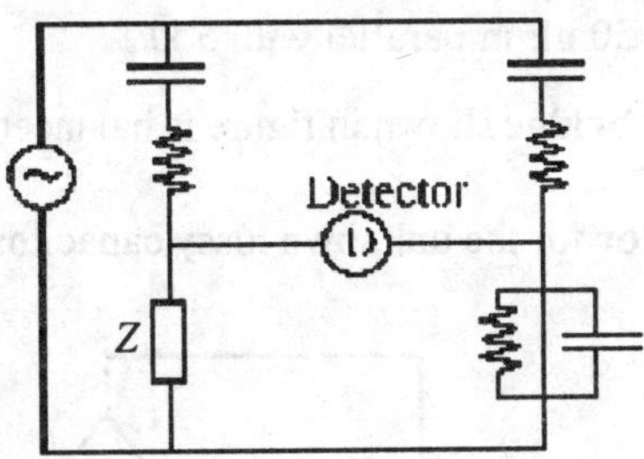

(a) Pure capacitor
(b) Pure inductor
(c) $R-L$ Series combination
(d) $R-L$ Parallel combination

16. The bridge method commonly used for finding mutual inductance is
(a) Heaviside Campbell bridge
(b) Schering bridge
(c) De sauty bridge
(d) Wien bridge

17. Below figure shows a balanced ac bridge excited by a voltage source of fixed frequency.

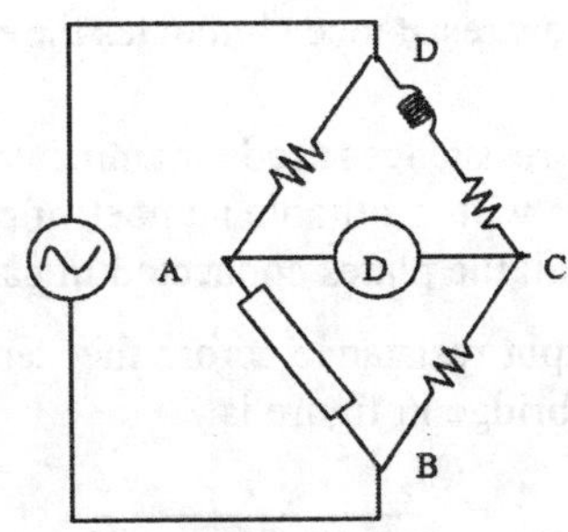

(a) An R and L in series
(b) An R and C in series
(c) An R and C in parallel
(d) An R and L in parallel

18. Match the List I parameter with the List II type of bridge used for measurement and select the correct answer using the codes:

List I	List II
A. Low value of R	1. Schering bridge
B. High value of Q	2. Maxwell-Wien bridge
C. Inductance L	3. Kelvin double bridge
D. Capacitance C	4. Hay bridge

Codes:

	A	B	C	D
(a)	2	4	3	1
(b)	4	1	3	2
(c)	3	4	1	2
(d)	3	4	2	1

19. Match the measured variables with the bridges and select the correct answer using the codes:

List I	List II
A. Self inductance	1. Anderson
B. Capacitance	2. Wien–Robinson
C. Mutual inductance	3. Campbell
D. Frequency	4. Schering

Codes:

	A	B	C	D
(a)	2	4	3	1
(b)	1	4	3	2
(c)	3	1	4	2
(d)	2	4	1	3

20. In the context of low resistances, identify the INCORRECT statement.
(a) Standard low resistances have four terminal construction.

(b) In Kelvin bridge for measuring low resistance, precision depends on the detector sensitivity.

(c) A Pair of ratio arms in kevin bridge for measuring the low resistance eliminates the error due to thermo emf.

(d) Low resistances used for ammeter shunts are usually made with a suitable number of plates of large area and all the plates connected in parallel

21. The output resistance across the terminals 1 and 2 of the DC bridge in figure is

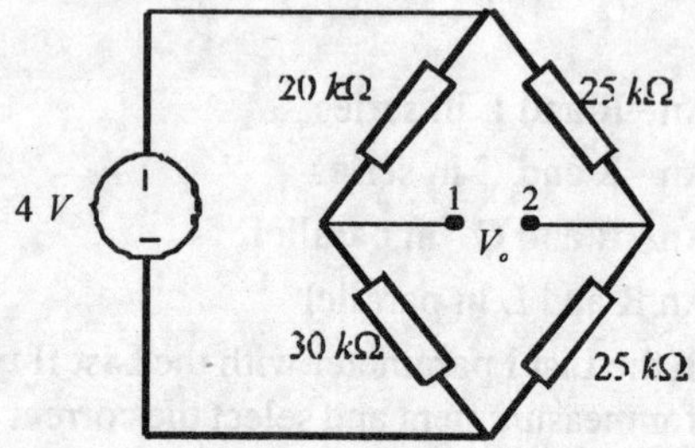

(a) 12.5Ω (b) 24.5 kΩ

(c) 25.0Ω (d) 100 kΩ

22. For the **bridge circuit** in figure the voltages are

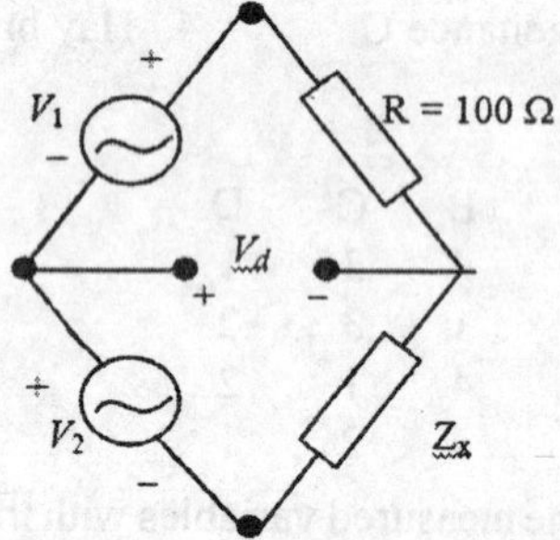

$V_1 = \sqrt{2}\cos 1000\,t\,V,$

$V_2 = 2\cos\left(1000\ t + 45^\circ\right)V,$ and

$V_d = 0$

The value of R = 100 Ω. Then Z_x will be

(a) 100 Ω resistor in series with 100 mH inductor.

(b) 100 Ω resistor in series with 100 μF inductor.

(c) 100 Ω resistance in parallel with 100 mH inductor.

(d) 100 Ω resistance in parallel with 100 μF capacitor.

23. The resistance values of the bridge circuit shown in figure are $R_1 = R_2 = R_3 = R$ and $R_4 = R + \Delta R$. The bridge is balanced by introducing a small voltage V. The value of ΔR is

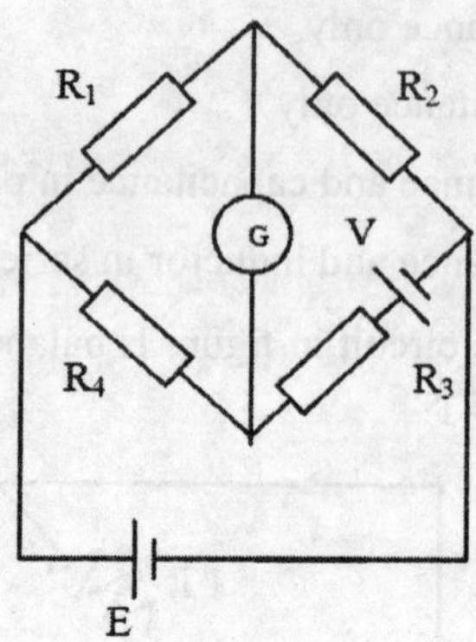

(a) $\dfrac{RV}{E}$ (b) $\dfrac{RV}{E-V}$

(c) $\dfrac{RV}{E-2V}$ (d) $\dfrac{RV}{E+2V}$

24. In the ac bridge shown in figure the detector D reads zero. Then Z is made of

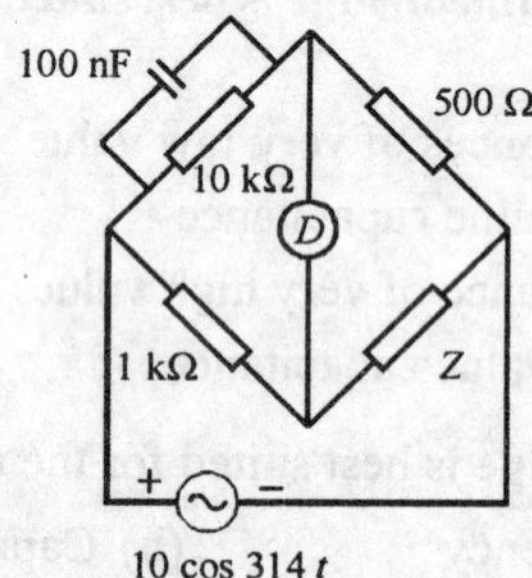

(a) 50 mH in parallel with 50 Ω

(b) 50 mH in series with 50 Ω

(c) 50 nF in series with 10 kΩ

(d) 50 nF in parallel with 5 kΩ

25. The **bridge** shown in figure is balanced. The **dissipation factor** for the unknown lossy capacitor $R_x + \dfrac{1}{j\,\omega C_x}$ is

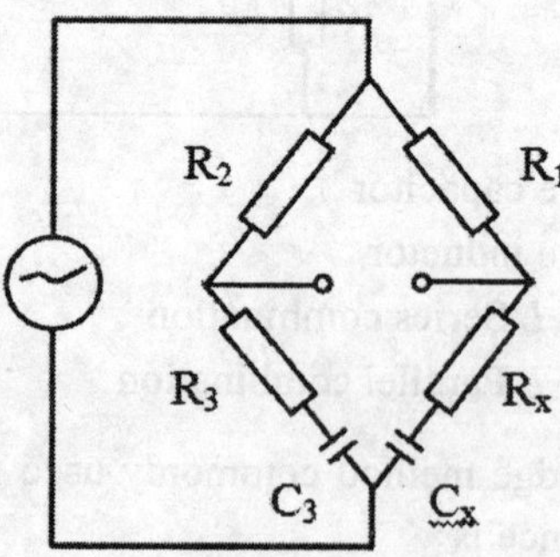

(a) $\omega R_2 C_3$

(b) $\dfrac{\omega R_1^2 C_3}{R_2}$

(c) $\omega R_3 C_3$

(d) $\frac{R_1}{R_2}$

26. Consider the AC bridge shown below. If $\omega RC = 1$ and $\frac{\Delta C}{C} < 0.01$, then ratio $\left|\frac{V_0}{V_s}\right|$ is approximately equal to.

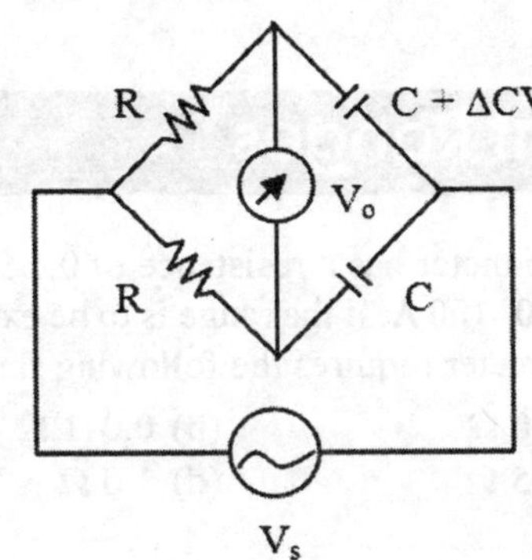

(a) $\sqrt{1+\frac{\Delta C}{C}}$ (b) $\frac{2\Delta C}{C}$

(c) $\frac{\Delta C}{C}$ (d) $\frac{\Delta C}{2C}$

Statement for the linked Answer Questions 27 and 28:

In the circuit shown below, the current through the PMMC meter is assumed to be zero. The ideal switch toggles between position 1 and position 2. For each position, it is connected for time $\frac{T}{2}$. Assume $R_4\ C_X$ << and $R_2\ C_X >> \frac{T}{2}$

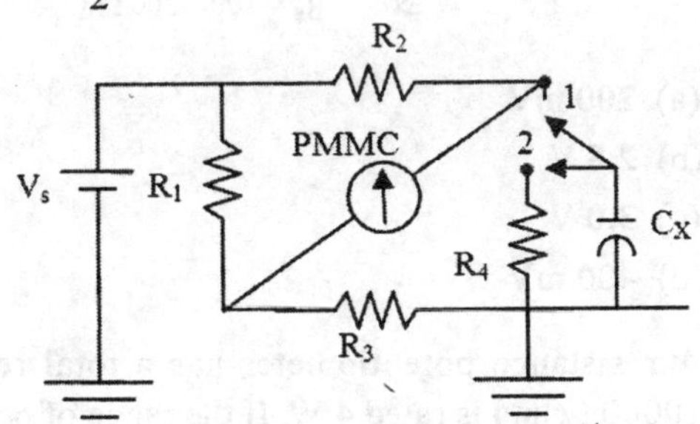

27. The capacitor C_x can be charged up to the maximum voltage.

(a) $\frac{V_s T}{2R_2 C_X}$ (b) $\frac{V_s T}{2R_4 C_X}$

(c) $\frac{2V_s R_2 C_X}{T}$ (d) $\frac{2V_s R_4 C_X}{T}$

28. In the bridge circuit shown in the figure above, if the PMMC meter shown null deflection, then the value of C_x is

(a) $\frac{T}{R_4}$ (b) $\frac{T(R_1+R_3)}{2R_2R_3}$

(c) $\frac{T(R_3+R_1)}{4R_2(R_3-R_1)}$ (d) $\frac{TR_3}{2R_2(R_1-R_3)}$

29. Consider the AC bridge shown in the figure below, with A, L and C having positive finite values.Then

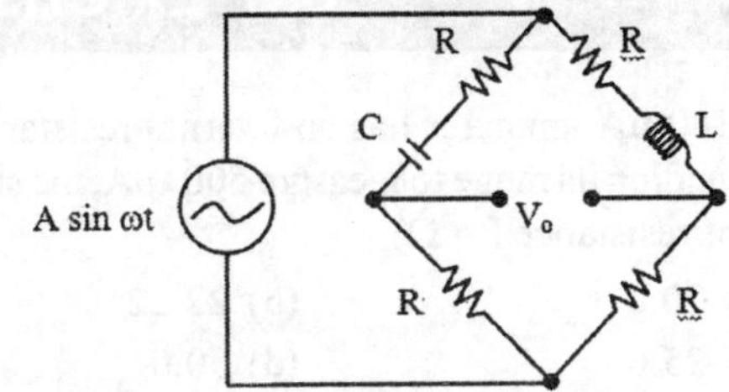

(a) $V_o = 0$ if $\omega L = \frac{1}{\omega C}$

(b) $V_o = 0$ if $L = C$

(a) $V_o = 0$ if $R = \frac{1}{\omega\sqrt{LC}}$

(b) V_o cannot be made zero

30. If the supply voltage to the bridge decreases by 5 %, the sensitivity of the measurement system

(a) Decrease by 5%
(b) Does not change
(c) Increases by 5%
(d) Increase by 200 %

Statement for Linked Answer Questions 31 and 32

In the Wheatstone bridge shown below the galvanometer G has a current sensitivity of 1 μA/mm, a resistance of 2.5 kΩ and a scale resolution of 1 mm. Let ΔR be the minimum increase in R from its nominal value of 2 kΩ that can be detected by this bridge.

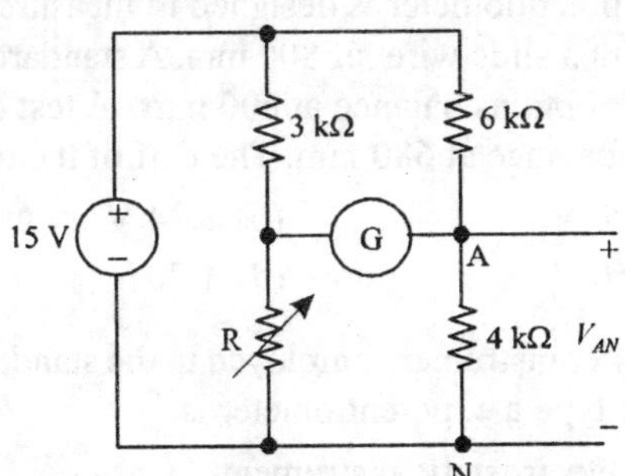

31. When R is 2 kΩ + ΔR, V_{AN} is.

(a) 6 V (b) 6.0024 V

(c) 6.0038 V (d) 6.005 V

32. The value of ΔR is approximately.
(a) 2.8 Ω
(b) 3.4 Ω
(c) 5.2 Ω
(d) 12 Ω

33. In the wheatstone bridge shown below, when the resistance R_1 increases by 1Ω, the current through the galvanometer is

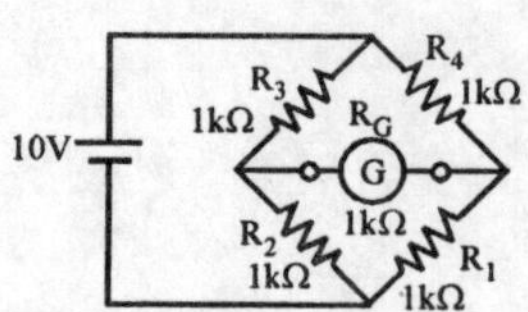

(consider the Thevnin equivalent resistance of the bridge in the calculations)
(a) 1.25 μA (b) 2.5 μA
(c) 12.5 μA (d) 25 μA

3.6 Extension of Instrument Ranges

1. A 100 μA ammeter has an internal resistance of 100 Ω extending its range to measure 500 μA, the shunt required is of resistance (in Ω)
(a) 20.0 (b) 22.22
(c) 25.0 (d) 50.0

2. A DC ammeter has a resistance of 0.1 Ω and its current range is 0–100 A. if the range is to be extended to 0–500 A, then meter requires the following shunt resistance
(a) 0.010 Ω (b) 0.011 Ω
(c) 0.025 Ω (d) 1.0 Ω

3.7 Potentiometers

1. In d.c. potentiometer measurements, a second reading is often taken after reversing the polarities of the DC supply and the unknown voltage, and the average of the two reading is taken. This is with a view to eliminate the effects of
(a) Ripples in the DC supply
(b) Stray magnetic fields
(c) Stray thermal emf's
(d) Erroneous standardization

2. A potentiometer is basically a
(a) Deflection type instrument
(b) Null type instrument
(c) Deflection as well as null type instrument
(d) A digital instrument

3. A dc potentiometer is designed to measure up to about 2 V with a slide wire of 800 mm. A standard cell of emf 1.18 V obtains balance at 600 mm. A test cell is sen to obtain balance at 680 mm. The emf of the test cell is
(a) 1.00 V (b) 1.34 V
(c) 1.50 V (d) 1.70 V

4. A transfer instrument employed in the standardization of a polar type a.c. potentiometer is
(a) An electrostatic instrument
(b) A thermal instrument
(c) A dynamometer instrument
(d) A moving coil instrument

5. In the potentiometric circuit shown in figure balance is obtained. The unknown E_x is then

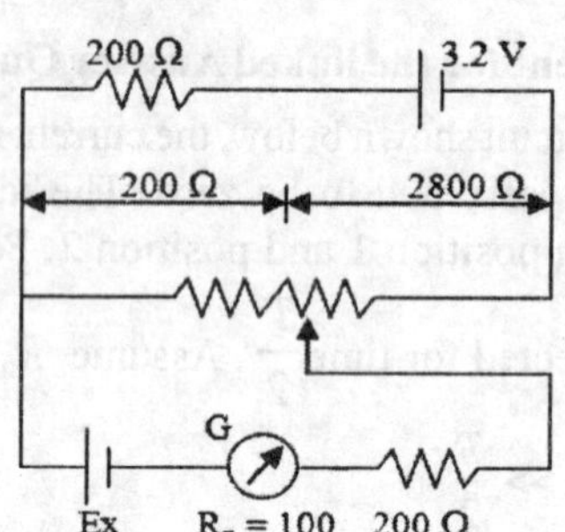

(a) 200 mV
(b) 2.8 V
(c) 3.0 V
(d) 400 mV

6. A resistance potentiometer has a total resistance of 10000 Ω and is rated 4 W. If the range of potentiometer is 0 to 100 mm, then its sensitivity in V/mm is
(a) 1.0
(b) 2.0
(c) 2.5
(d) 25

7. A 4 kΩ, 0.02W potentiometer is used in the circuit shown below. The minimum value of the resistance R_s in order to protect the potentiometer is

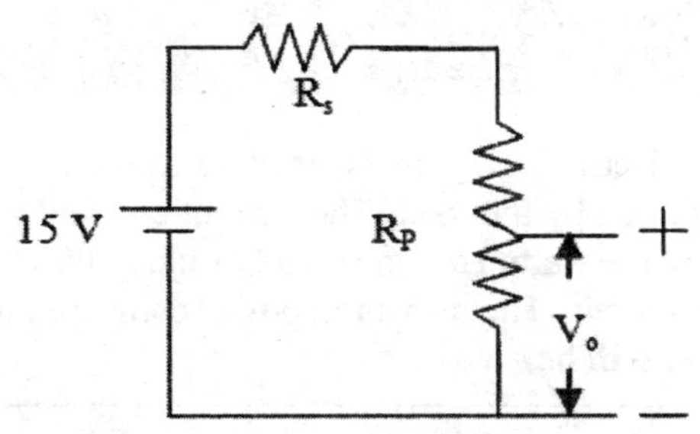

(a) 2.23 kΩ (b) 2.71 kΩ
(c) 3.82 kΩ (d) 8.92 kΩ

8. A **potentiometer** circuit is shown in figure. The jockey of the potentiometer is kept exactly at the mid point and the ammeter reads the current, I, as 0.2 mA When V_x is reversed the ammeter reads 3.8 mA. The internal resistance of the ammeter and the unknown potential V_x are respectively.

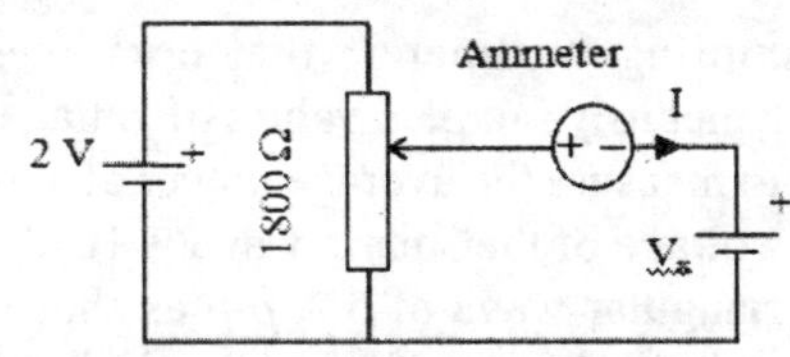

(a) 500 Ω, 1.1 V (b) 50 Ω, 1.1 V
(c) 500 Ω, 0.9 V (d) 50 Ω, 0.9 V

9. A wire potentiometer length 11 m and resistance 1 Ω/m balance a standard cell voltage of 1.018 V at a length of 10 m 18 cm. If the voltage of the battery supply the current through the potentiometer is 2.0 V, then the value of the series resistance connected to the potentiometer is

(a) 9 (b) 90 Ω
(c) 900 Ω (d) 80 Ω

10. An unknown voltage source E_u (with negligible internal resistance) is connected to a potentiometer circuit as shown in the following figure. If the galvanometer current is 10 μ A with the direction as indicated, then value of E_u is

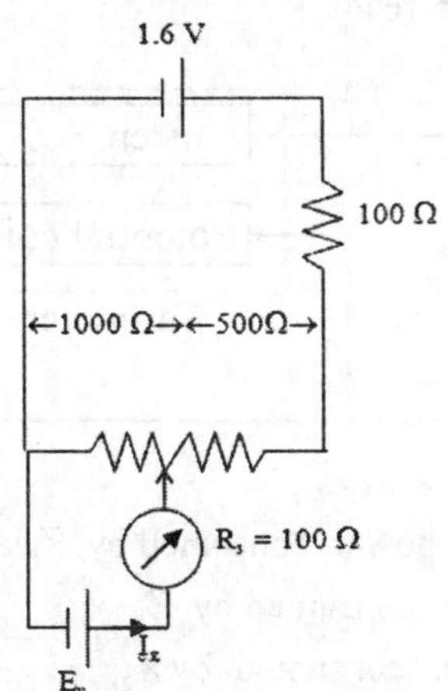

(a) 0.9910 V (b) 0.9991 V
(c) 1.0012 V (d) 1.0048 V

11. The figure shows a potentiometer of total resistance R_T with a Sliding contact:

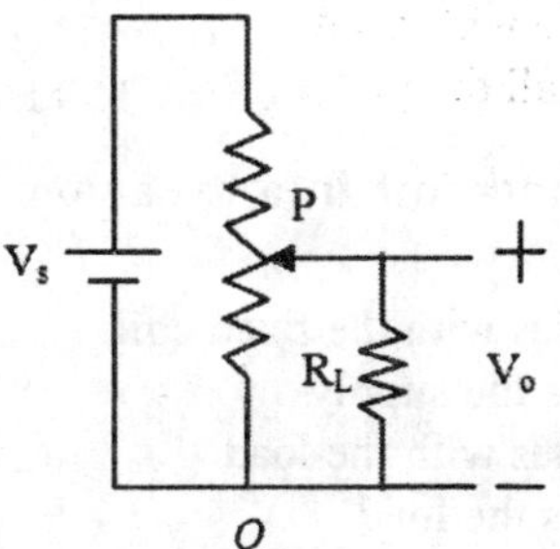

The resistance between the points P and Q of the potentiometer at the position of the contact shown is R_C and the voltage ratio $\frac{V_o}{V_s}$ at this point is 0.5. If the ratio $\frac{R_L}{R_T} = 1$, the ratio $\frac{R_C}{R_T}$ is

(a) $\frac{-1+\sqrt{5}}{2}$

(b) $\frac{1+\sqrt{5}}{2}$

(c) $-1+\sqrt{5}$

(d) $1+\sqrt{5}$

12. The dc potentiometer shown in the figure has a working current of 10 mA with switch S open. Let $R_g + R_1 = 100\,\Omega$. The galvanometer G can only detect currents greater then 10 μA. The maximum percentage error in the measurement of the unknown emf as calculated from the slider position shown in closest to

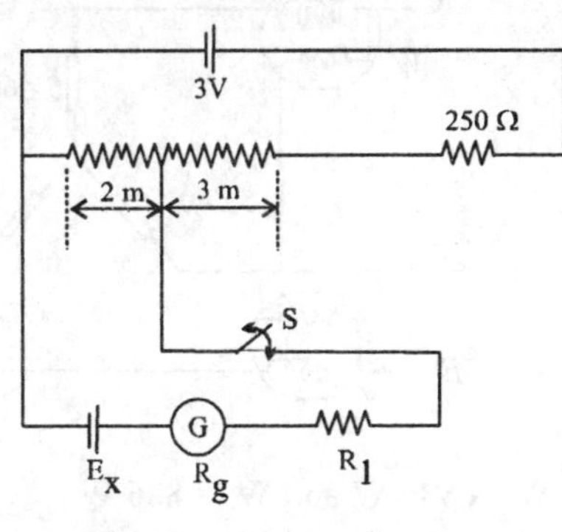

(a) 0.3 (b) 0.5
(c) 0.6 (d) 1.0

3.8 Measurement of Power and Energy

1. A water boiler at home is switched on to the a.c. mains supplying power at 230 V/ 50 Hz. The frequency of instantaneous power consumed by the boiler is

(a) 0 Hz (b) 50 Hz
(c) 100 Hz (d) 150 Hz

2. The moving coil in a dynamometer wattmeter is connected

(a) In series with the fixed coil
(b) Across the supply
(c) In series with the load
(d) Across the load

3. A dynamometer type wattmeter responds to the

(a) Average value of active power
(b) Average value of reactive power
(c) Peak value of active power
(d) Peak value of reactive power

4. Electro-dynamic type watt-meters have large errors while measuring power is ac circuits at low power factor conditions, since the voltage across and the current through the

(a) Current coil are not in phase
(b) Current coil are not in Quadrature
(c) Pressure coil are not in phase.
(d) Pressure coil not in Quadrature

5. The minimum number of wattmeter(s) required to measure 3-phase, 3-wire balanced or unbalanced power is

(a) 1 (b) 2
(c) 3 (d) 4

6. The line-to-line input voltage to the 3 phase, 50Hz, ac circuit shown in Figure is 100 V rms. Assuming that the phase sequence is RYB, the watt-meters would read

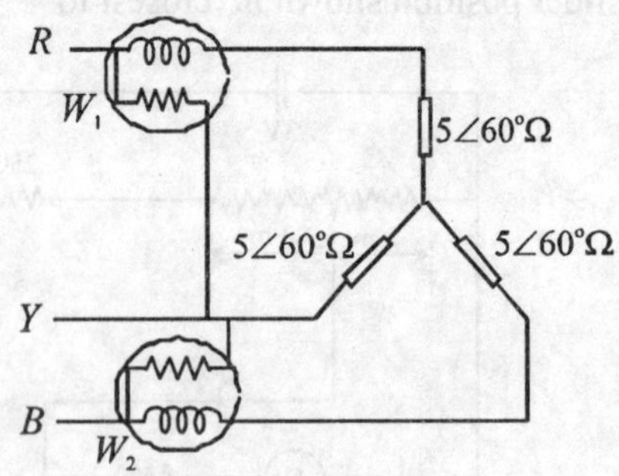

(a) W_1 = 886 W and W_2 = 886 W
(b) W_1 = 500 W and W_2 = 500 W
(c) W_1 = 0 W and W_2 = 1000 W
(d) W_1 = 250 W and W_2 = 750 W

7. The circuit in figure is used to measure the power consumed by the load. The current coil and the voltage coil of the wattmeter have 0.02 Ω and 1000 Ω resistance respectively. The measured power compared to the load power will be

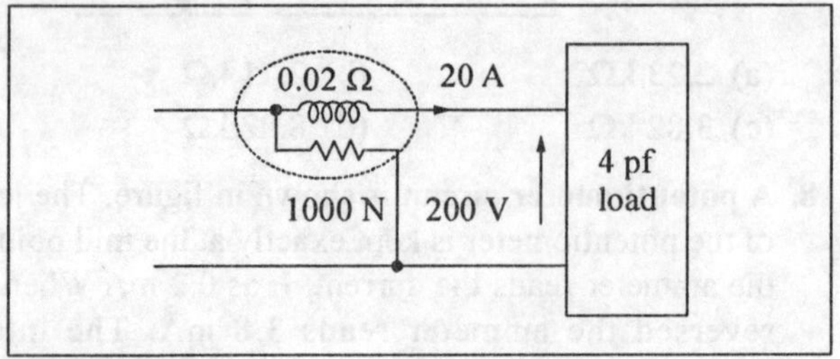

(a) 0.4 % less (b) 0.2 % less
(c) 0.2 % more (d) 0.4 % more

8. A sampling wattmeter (that computes power from simultaneously sampled values of voltage and current) is used to measure the average power of a load. the peak to peak voltage of the square wave is 10 V and the current is a triangular wave of 5 A *p-p* as shown in the figure. The period is 20 ms. The reading in *W* will be

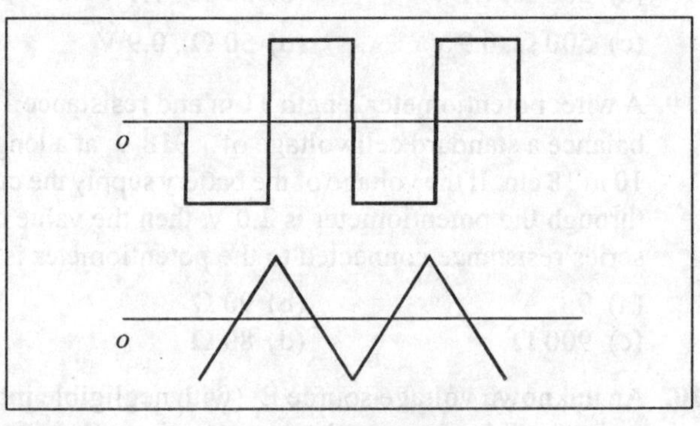

(a) 0 W (b) 25 W
(c) 50 W (d) 100 W

9. A wattmeter is connected as shown in the figure. The wattmeter reads

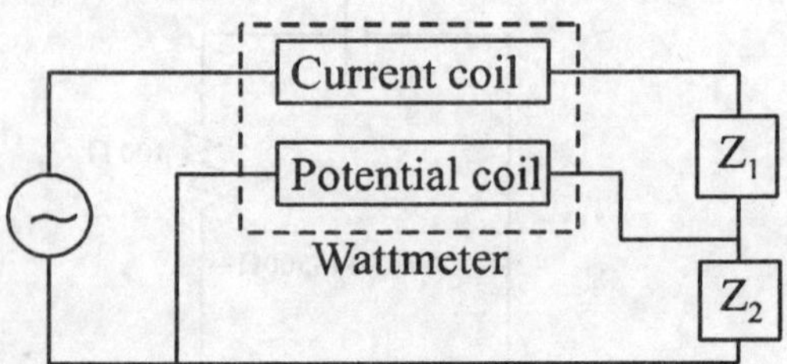

(a) Zero always
(b) Total power consumed by Z_1 and Z_2
(c) Power consumed by Z_1
(c) Power consumed by Z_2

10. For the circuit shown in the figure, the voltage and current expressions are $v(t) = E_1 \sin(\omega t) + E_3 \sin(3\omega t)$ and $i(t) = I_1 \sin(\omega t - \phi_1) + I_3 \sin(2\omega t - \phi_3) + I_5 (5\omega t)$. The average power measured by the Wattmeter is

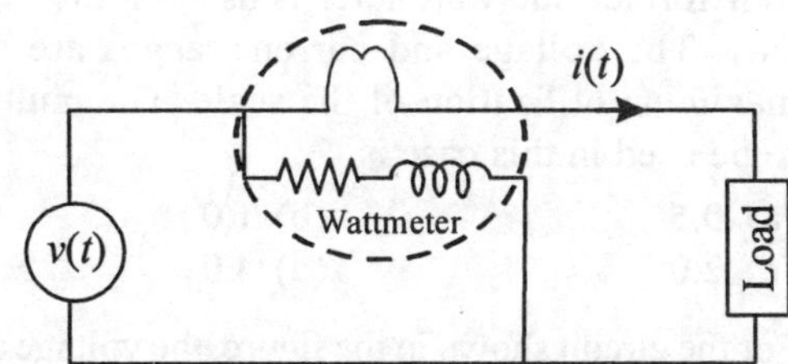

(a) $\frac{1}{2} E_1 I_1 \cos\phi_1$

(b) $\frac{1}{2}\begin{bmatrix} E_1 I_1 \cos\phi_1 + \\ E_1 I_3 \cos\phi_3 + E_1 I_5 \end{bmatrix}$

(c) $\frac{1}{2}\begin{bmatrix} E_1 I_1 \cos\phi_1 + \\ E_3 I_3 \cos\phi_3 \end{bmatrix}$

(d) $\frac{1}{2}[E_1 I_1 \cos\phi_1 + E_3 I_1 \cos\phi_1]$

11. A wattmeter 400 W when its current coil is connected in the R phase and its pressure coil is connected between this phase and the neural of a symmetrical 3-phase system supplying a balanced star connected 0.8 p.f inductions load. The phase sequence is RYB. What will be the reading of this wattmeter if its pressure coil alone is reconnected the B and Y phases, all other connections remaining as before?

(a) 400.0 (b) 519.6
(c) 300.0 (d) 692.8

12. The voltage-flux adjustment of a certain 1-phase 220 V induction watt-hour meter is altered so that the phase angle between the applied voltage and the flux due to it is 85° (instead of 90°). The errors introduced in the reading of this meter when the current is 5 A at power factors of unity and 0.5 lagging are respectively

(a) 3.8 mW, 77.4 mW
(b) – 3.8 mW, – 77.4 mW
(c) – 4.2 W, – 85.1 W
(d) 4.2 W, 85.1 W

13. A dc A-h meter is rated for 15 A, 250 V. the meter constant is 14.4 A-sec/rev. The meter constant at rated voltage may be expressed as

(a) 3750 rev/kWh
(b) 3600 rev/kWh
(c) 1000 rev/kWh
(d) 960 rev/kWh

14. A single-phase load is connected between *R* and *Y* terminals of a 415 V, symmetrical, 3-phase, 4 wire system with phase sequence *RYB*. A wattmeter is connected in the system as shown in figure. The power factor of the load is 0.8 lagging. The wattmeter will read

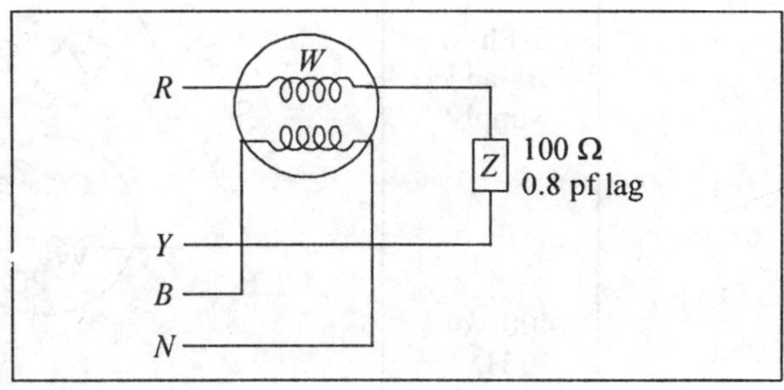

(a) – 795 W (b) – 597 W
(c) + 597 W (d) + 795 W

15. Two wattmeters, which are connected to measure the total power on a three-phase system supplying a balanced load, read 10.5 kW and – 2.5 kW, respectively. The total power and the power factor, respectively, are

(a) 13.0 kW, 0.334
(b) 13.0 kW, 0.684
(c) 8.0 kW, 0.52
(d) 8.0 kW, 0.334

16. An energy meter connected to an immersion heater (resistive) operating on an AC 230 V, 50 Hz, AC single phase source reads 2.3 units (kWh) in 1 hour. The heater is removed from the supply and now connected to a 400 V peak to peak square wave source of 150 Hz. The power in kW dissipated by the heater will be

(a) 3.478
(b) 1.739
(c) 1.540
(d) 0.870

17. A 3 V dc supply with an internal resistance of 2 Ω supplie a passive non-linear resistance characterized by the relation $V_{NL} = I_{NL}^2$. The power dissipated in the non-linear resistane is

(a) 1.0 W
(b) 1. 5 W
(c) 2..5 W
(d) 3.0 W

18. The figure shows a three phase delta connected load supplied from a 400V, 50 Hz, 3 phase balanced source. The pressure coil (PC) and current coil (CC) of a wattmeter are connected to the load as shown , with the coil polarities suitably selected to ensure a positive deflection. The wattmeter reading will be

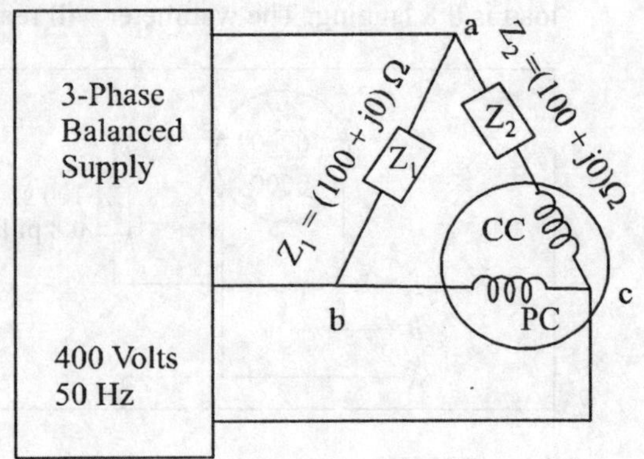

(a) 0 (b) 1600 Watt
(c) 800 Watt (d) 400 Watt

19. The two wattmeter method is used to measure active power on a three phase, three wire system. If the phase voltage is unbalanced, then the power reading is

(a) Affected by both negative sequence and zero sequence voltage.
(b) Affected by negative sequence voltage but not by zero sequence voltages.
(c) Affected by zero sequence voltages but not by negative sequence voltages.
(d) Not affected by negative sequence voltage.

20. A voltage drop across a non-linear element is given as $v = i + 2i^2$. The power loss across it for a current of sin t is

(a) 1.0 W (b) 2.0W
(c) 3.0 W (d) 0.5 W

21. The current coil of a 200 V, 5A, **electrodynamometer** type LPF **wattmeter carries** a current of $\sqrt{2}\cos(100\pi t)$ A. The voltage across the pressure coil is $\sqrt{2}\, 200 \sin(100\pi t)$ V. The meter will indicate

(a) 0 W (b) 100 W
(c) 200 W (d) 400 W

22. The **brake magnet** of an error free electromechani-cal, induction type **energy meter** is shifted from its position and moved a small distance towards the edge of the disc. Then the meter will.

(a) Continue to work correctly without any charge in its characteristics
(b) Continue to work, but registers, a lower kWhr
(c) Continue to work, but registers a higher kWhr
(d) Stop working

23. A dynamometer type wattmeter with a single scale marked for the smallest power range, has two current ranges, namely, 0-5 A and 0-10 A as well as two voltage ranges, namely, 0-150 V and 0–300 V. To carry out a load list on a 230 V / 115 V , 1kVA, Single phase transformer, the wattmeter is used on the high voltage side. The voltage and current ranges are chosen for maximum utilization of the scale. The multiply factor to be used in this case is

(a) 0.5 (b) 1.0
(c) 2.0 (d) 4.0

24. For the circuit shown in the figure, the voltage and current expressions are $v(t) = E_1 \sin(\omega t) + E_3 \sin(3\omega t)$ and $i(t) = I_1 \sin(\omega t - \phi_1) + I_3 \sin(3\omega t - \phi_3) + I_5 \sin(5\omega t)$. The average power measured by the Wattmeter is

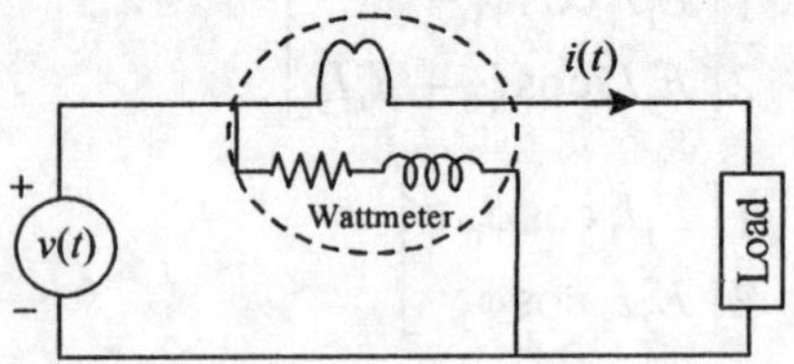

(a) $\frac{1}{2} E_1 I_1 \cos\phi_1$

(b) $\frac{1}{2}[E_1 I_1 \cos\phi_1 + E_1 I_3 \cos\phi_3 + E_1 I_5]$

(c) $\frac{1}{2}[E_1 I_1 \cos\phi_1 + E_3 I_3 \cos\phi_3]$

(d) $\frac{1}{2}[E_1 I_1 \cos\phi_1 + E_3 I_1 \cos\phi_1]$

25. Disc in an **induction type energy-meter** rotates in the opposite direction when.

(a) A large capacitor is connected on the load side
(b) A large inductor is connected on the load side
(c) Current coil and voltage coil are wrongly connected
(d) Either current coil or voltage coil is wrongly connected

26. A 5 A, 110 V **electrodynamics** type **wattmeter** has a scale having 110 divisions. Its pressure coil is fed a voltage of [110 $\sqrt{2}$ cos (314 t) + $\sqrt{2}$ sin (942t)] V and its current coil carries a current of [5 $\sqrt{2}$ cos (314 t + 60) + 2 $\sqrt{2}$ sin (628t + 90) + $\sqrt{2}$cos (942t + 90)] A. The needle of the watt meter will move to

(a) 110 divisions
(b) 50 divisions
(c) 54 divisions
(d) 55 divisions

27. A 3-phase balanced R-L load is connected to a 3-phase balanced symmetrical 400 V, 50 Hz, RYB sequence mains as shown in figure. The wattmeter W reads 1.732 kW. Then.

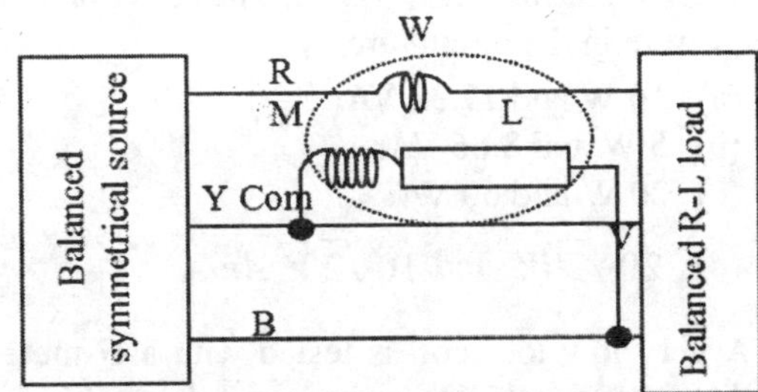

(a) The active power in the load is 1.732 kW
(b) The reactive power of the load is 1.732 kVAr
(c) The active power in the load is 3 kW
(d) The reactive power of the load is 3 kVAr

28. An unknown impedance Z is shunted across an ac current source of value 2 cos (10 t) A. The current source developed a voltage of $\sqrt{2}\,6\cos\left(10t+45^{o}\right)V$. The values of the impedance is.
(a) 3 Ω resistance in series with 0.3 H inductor
(b) 3 Ω resistance in parallel with 0.3 H inductor
(c) 3 Ω resistance in series with 0.3 F capacitor
(d) 3 Ω resistance in parallel with 0.3 F capacitor

29. The **two-wattmeter method** is used to measure power in a balanced 3-phase circuit, drawing lagging current. The power factor if one of the wattmeters reads zero is.
(a) 0.5 (b) 0.8
(c) 1.0 (d) 0.0

30. The two–wattmeter method is used to measure power in a 3–phase, 3wire balanced inductive circuit The line voltage and line current are 400 V and 10 A respectively. If the load power factor is 0.866 lagging, then readings of the two wattmeters are.
(a) 6000 W and 0 W
(b) 5000 W and 1000 W
(c) 4500 W and 1500 W
(d) 4000W and 2000 W

31. A 230 *V, 5 A, 50 Hz* single phase house service energy meter has a meter constant of 360 rev/k Whr. The meter takes 50 s for making 51 revolutions of the disc when connected to a 10 kW, unity power factor load. The error in the reading of the meter is.
(a) 0%
(b) + 0.5%
(c) – 2.0%
(d) + 2.0 %

3.9 Measurement of Phase and Frequency

1. For a given frequency, the deflecting torque of an induction ammeter is directly proportional to
(a) Current ^{2}
(b) Current ^{3}
(c) √current
(d) Current

2. A two phase load draws the following phase currents:

$i_1(t) = I_m \sin(\omega t - \phi_1)$,

$i_2(t) = I_m \cos(\omega t - \phi_2)$,

These currents are balanced if ϕ_2 is equal to

(a) $-\phi_2$
(b) ϕ_2
(c) $(\pi/2 - \phi_2)$
(d) $(\pi/2 + \phi_2)$

3. The ac bridge shown in the figure is used to measure the impedance Z.

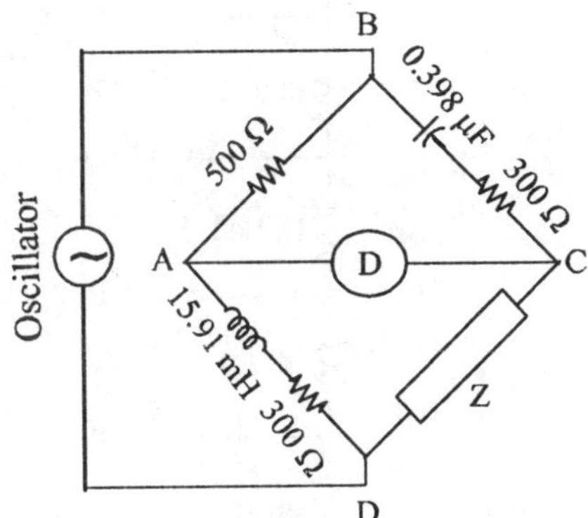

If the bridge is balanced for oscillator frequency f = 2 kHz. then the impedance Z will be
(a) $(260 + j0)\,\Omega$ (b) $(0 + j200)\,\Omega$
(c) $(260 - j200)\,\Omega$ (d) $(260 + j200)\,\Omega$

3.10 Q-meter and Potentiometric Recorders

1. A reading of 120 is obtained when a standard inductor was connected in the circuit of a Q meter and the variable capacitor is adjusted to a value of 300 pF. A loss-less capacitor of unknown value C_x is then connected in parallel with the variable capacitor and the same reading was obtained when the variable was obtained when the variable capacitor is readjusted to a value of 200 pF. The value of C_x in pF is
   (a) 100
   (b) 200
   (c) 300
   (d) 500

2. The Q-meter works on the principle of
   (a) Mutual inductance
   (b) Self inductance
   (c) Series resonance
   (d) Parallel resonance

3. The figure shows input attenuator of a multimeter. The meter full-scale with 12 V at M with the range swtich at position B obtain full-scale deflection with the range switch position at D?

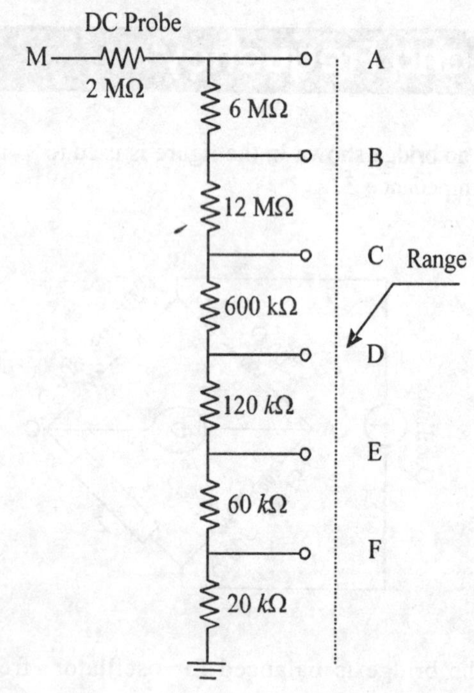

   (a) 1 V
   (b) 150 V
   (c) 120 V
   (d) 147 V

4. The voltage phasor of a circuit is 10 $\angle 15°$ V and the current phasor is 2 $\angle -45°$ A. The active and the reactive powers in the circuit are
   (a) 10 W and 17.32 VAr.
   (b) 5 W and 8.66 VAr
   (c) 20 W and 60 VAr
   (d) $20\sqrt{2}\ W$ and $10\sqrt{2}\ V\ Ar$

5. A very low loss coil is tested with a **Q-meter** and the distributed (self) capacitance of the coil is found to be 820 pF. Resonance occurred at an angular frequency of 10^6 rad/s with a capacitance of 9.18 nF. The inductance of the coil is
   (a) 100 pH
   (b) 100 μH
   (c) 100 nH
   (d) 100 mH

6. A **Q-meter** having an insertion resistance of 0.02 Ω is used to measure the inductance of a coil. Resonance occurs at an angular frequency of 10^6 rad/s with a capacitance of 40 pF. The inductance of the coil is
   (a) 250 μH (b) 2.5 mH
   (c) 25 mH (d) 0.25 H

7. Figure Shows a circuit which has a coil of resistance 'R' and inductance 'L'. At resonace, the **Q-factor** of the coil is given by

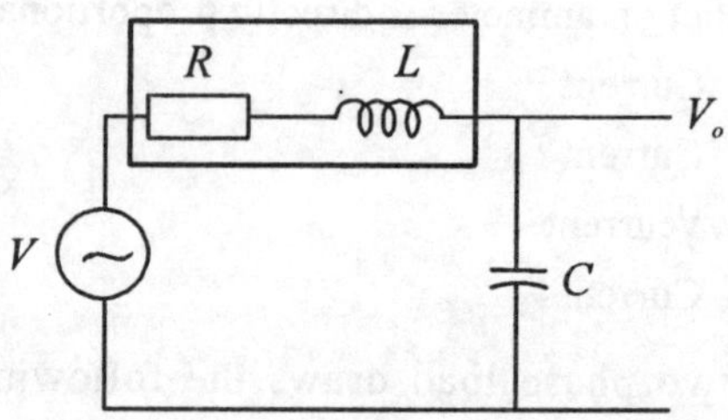

   (a) $\frac{V - V_o}{V}$

   (b) $\frac{V_o}{V}$

   (c) $\frac{V_o - V}{V}$

   (d) $\frac{V}{V_o}$

3.11 Oscilloscope

1. A Lissajous pattern, as shown in Figure below, is observed on the screen of a CRO when voltages of frequencies f_x and f_y are applied to the x and y plates respectively. $f_x : f_y$ is then equal to

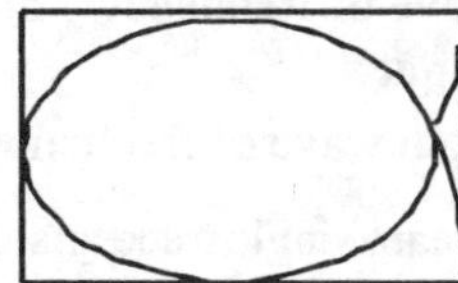

(a) 3 : 2 (b) 1 : 2
(c) 2 : 3 (d) 2 : 1

2. A certain oscilloscope with 4 cm by 4 cm screen has its own sweep output fed to its input. If the x and y sensitivities are same, the oscilloscope will display a
(a) Triangular wave (b) Diagonal line
(c) Sine wave (d) Circle

3. Two in-phase, 50 Hz sinusoidal waveforms of unit amplitude are fed into channel 1 and channel 2 respectively of and oscilloscope. Assuming that the voltage scale, time scale and other setting are exactly the same for both the channels, what would be observed if the oscilloscope is operated in X-Y mode ?
(a) A circle of unit radius
(b) An ellipse
(c) A parabola
(d) A straight line inclined at 45° with respect to the x-axis

4. The time/div and voltage/div axes of an oscilloscope have been erased. A student connects a 1 kHz, 5 V p-p square wave calibration pulse to channel 1 of the scope and observe the screen to be as shown is the upper trace of the figure. An unknown signal is connected to channel 2 (lower trace) of the scope. It the time/div and V/div on both channel are the same, the amplitude (*p-p*) and period of the unknown signal are respectively

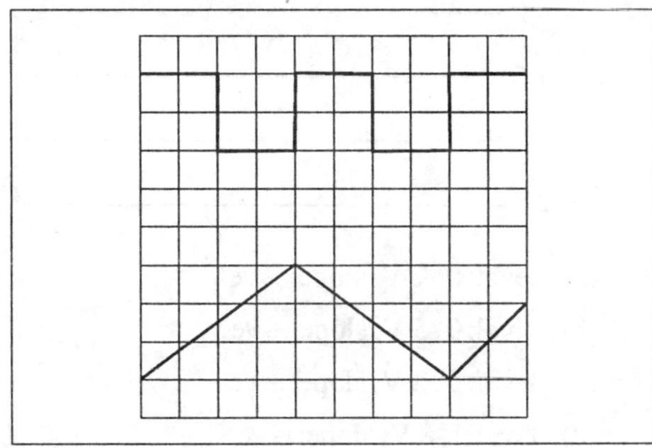

(a) 5 V, 1 ms (b) 5 V, 2 ms
(c) 7.5 V, 2 ms (d) 10 V, 1 ms

5. The probes of a non-isolated, two-channel oscilloscope are clipped to points *A*, *B* and *C* in the circuit of the adjacent figure. V_{in} is a square wave of a suitable low frequency. The display on Ch_1 and Ch_2 are as shown on the right. Then the "Signal" and "Ground" probes S_1, G_1 and S_2, G_2 of Ch_1 and Ch_2 respectively are connected to points:

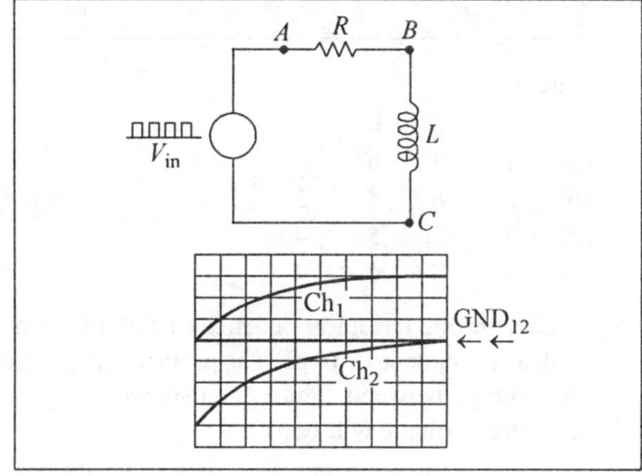

(a) *A*, *B*, *C*, *A* (b) *A*, *B*, *C*, *B*
(c) *C*, *B*, *A*, *B* (d) *B*, *A*, *B*, *C*

6. The two inputs of a CRO are fed with two stationary periodic signals. In the X-Y mode, the screen shows a figure which changes from ellipse to circle and back to ellipse with its major axis changing orientation slowly and repeatedly. The following inference can be made from this.
(a) The signals are not sinusoidal
(b) The amplitude of the signals are very close but not equal
(c) The signals are sinusoidal with their frequen cies very close but not equal
(d) There is a constant but small phase difference between the signals

7. Group I represents the figure obtained on a CRO screen when the voltage signals $V_x = V_{xm} \sin \omega t$ and $V_y = V_{ym} \sin (\omega t + \Phi)$ are given to its *X* and *Y* plates respectively and Φ is changed. Choose the correct value of Φ from Group I to match with the corresponding figure of Group II

Group I
P. $\Phi = 0$
Q. $\Phi = \pi/2$
R. $\pi < \Phi < 3\pi/2$
S. $\Phi = 3\pi/2$

Group II

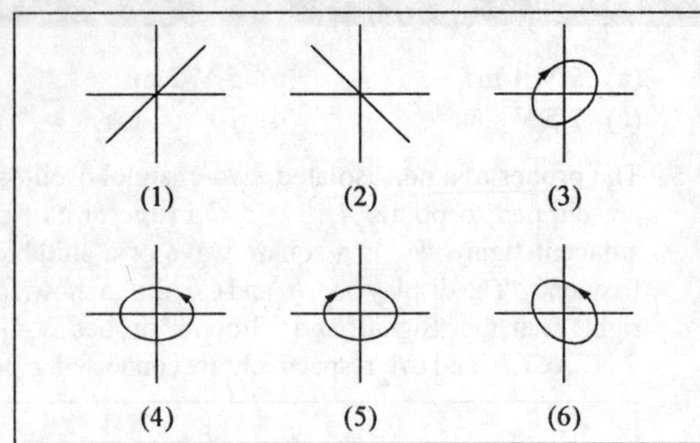

Codes:

	P	Q	R	S
(a)	1	3	6	5
(b)	2	6	4	5
(c)	2	3	5	4
(d)	1	5	6	4

8. A CRO probe has an impedence of 500 kΩ in parallel with a capacitance of 10 pF. The probe is used to measure the voltage between P and Q as shown in figure. The measured voltage will be

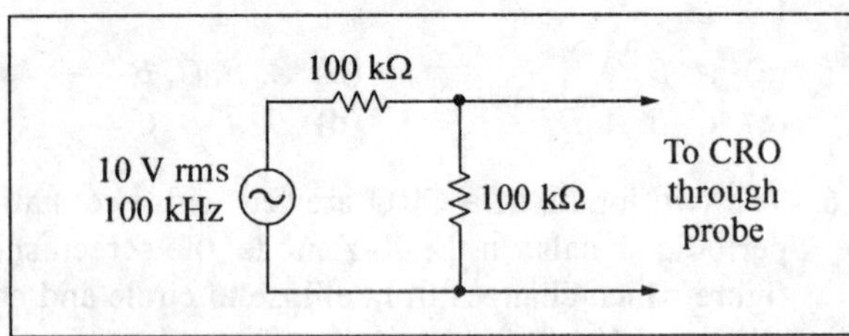

(a) 3.53 V (b) 4.37 V
(c) 4.54 V (d) 5.00 V

9. The simultaneous application of signals $x(t)$ and $y(t)$ to the horizontal and vertical plates, respectively, of an oscilloscope, produces a vertical figure-of-8 display. If P and Q are constants, and $x(t) = P \sin(4t + 30)$, then $y(t)$ is equal to

(a) $Q \sin(4t - 30)$
(b) $Q \sin(2t + 15)$
(c) $Q \sin(8t + 60)$
(d) $Q \sin(4t + 30)$

10. An oscilloscope is powered from 230 V mains having a nominal frequency of nearly 50 Hz. The triggering mode of the oscilloscopa is set to LINE. A square wave from a function generator is fed to the Y input and a stable display (properly triggered) is obtained when the frequency of the function generator is 297 Hz. The actual frequency of the power supply is.

(a) 50 Hz (b) 49.5 Hz
(c) 49.0 Hz (d) 48.5 Hz

11. An **oscilloscope** screen displays a line inclined at 45°. Its Y-input is a sine wave of frequency f then the X-input should be.

(a) Sine wave of frequency f and 0° phase shift with the Y-input.
(b) Sine wave of frequency f and 45° phase shift with the Y-input
(c) Sine wave of frequency f and 90° phase shift with the Y-input.
(d) Saw tooth wave of frequency f

12. In a single-beam single trace analog oscilloscope, a delay line is inserted between

(a) X-input and X-plants
(b) Y-input and Y-plants
(c) X-input and the trigger circuit
(d) Y-input and the trigger circuit

13. A voltage signal $v(t) = A + B \sin \omega t$ is fed an osciloscope with the coupling mode set to DC, other setting being appropriate. The displayed wave can be expressed as

(a) $A + B \sin \omega t$ (b) $B \sin \omega t$
(c) A (d) $| A + B \sin \omega t |$

14. The figure shows a periodic waveform to be displayed on a CRO. A tigger setting which ensures a stigger setting which ensures a stationary display is

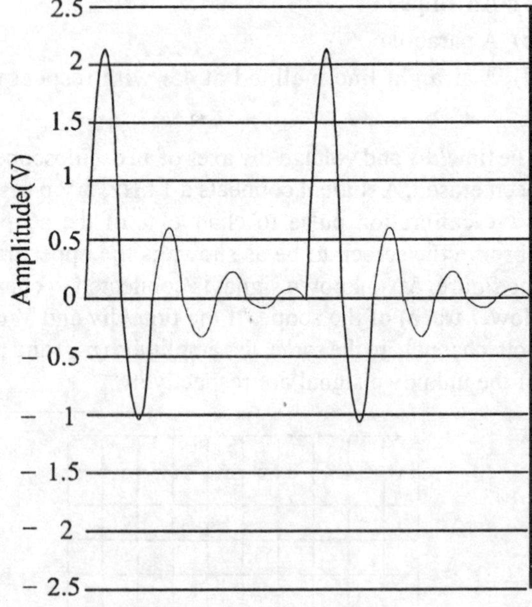

(a) level: 0.2 V, slope: – ve
(b) level: 0.5 V, slope: – ve
(c) level: 0.2 V, slope: + ve
(d) level: 1.8 V, slope: – ve

15. The input impedance of a CRO is equivalent to a 1 MΩ resistance in parallel with a 45 pF capacitance. It is used with a compensated 10-to-1 attenuation probe. The effective input capacitance at the probe tip is

(a) 4.5 *pF* (b) 5 *pF*
(c) 45 *pF* (d) 450 *pF*

16. A **symmetrical square wave** of frequency 25 kHz and a peak to peak amplitude of 10 V is fed to *Y*-input of a single channel **oscilloscope.** The screen appears as shown in figure. Then the *X* and *Y* sensitivities and the trigger settings respectively are

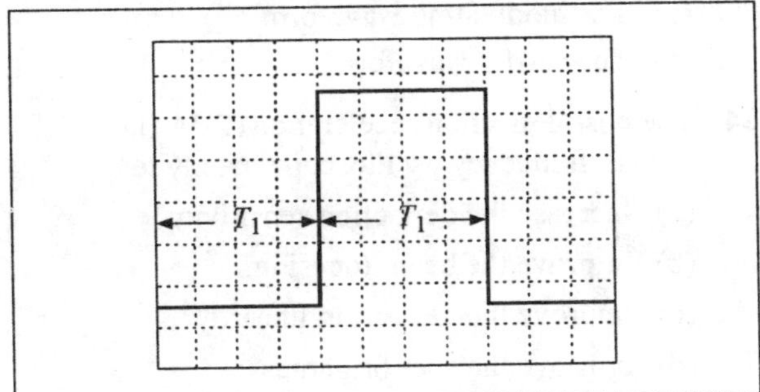

(a) 5 μs/cm, 2 V/cm and + ve slope
(b) 5 μs/cm, 2 V/cm and – ve slope
(c) 10 μs/cm, 1 V/cm and + ve slope
(d) 20 μs/cm, 2 V/cm and – ve slope

17. The linear sweep for the time base in an **oscilloscope** has deviation from its nominal waveform. The nominal (dashed line) and actual (solid line) sweep waveform are shown in the following figure.

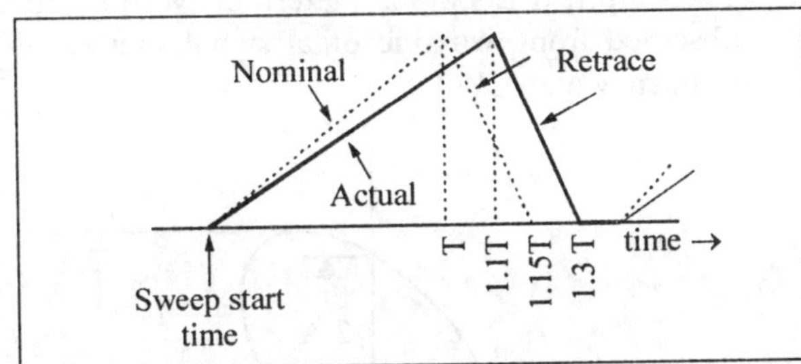

A 5 V p-p sine wave with a frequency of 1 kHz will be measured on the oscilloscope as a sine wave with

(a) 4.45 V p-p and 1 kHz frequency
(b) 5 V p-p and 1 kHz frequency
(c) 5 V p-p and 1.1 kHz frequency
(d) 5 V p-p and 1.15 kHz frequency

18. Two signals of peak to peak voltage 5 V and 2 V are being fed to Channel 1 and Channel 2, respectively, of an **oscilloscope** with a single time base. The vertical sensitively of both channels is 1 V/division. The two sine wave have identical frequency and phase. The trigger is on manual mode an triggers at a level of + 1.25 V on Channel 1, as shown in the figure below:

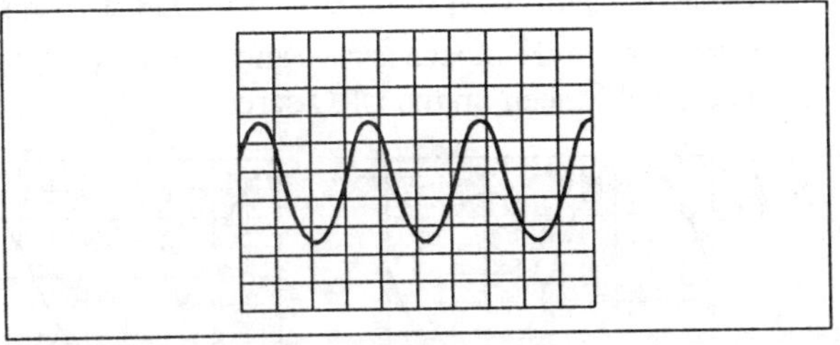

Which of the following figure correctly depicts the trace seen in Channel 2?

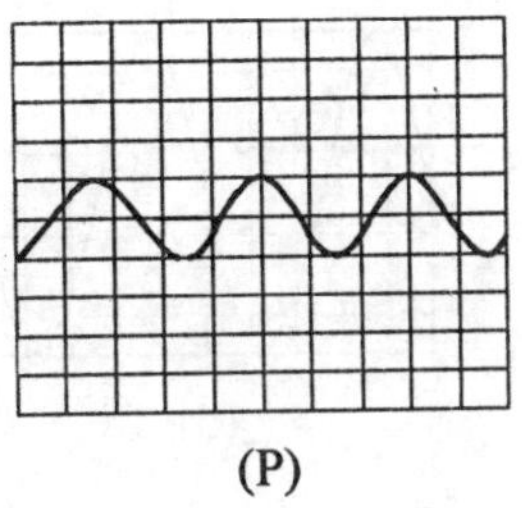
(P)

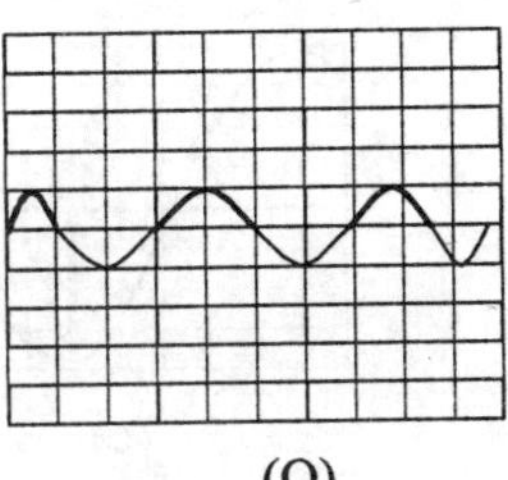
(Q)

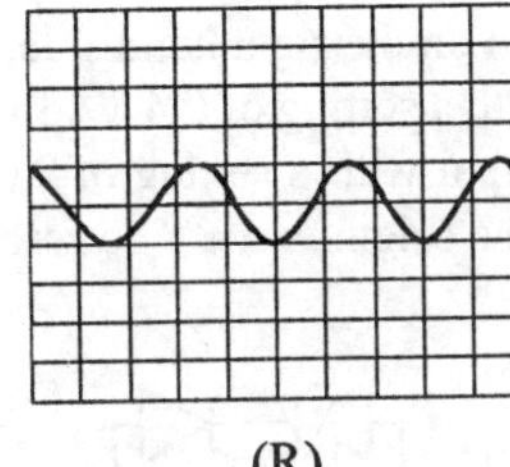
(R)

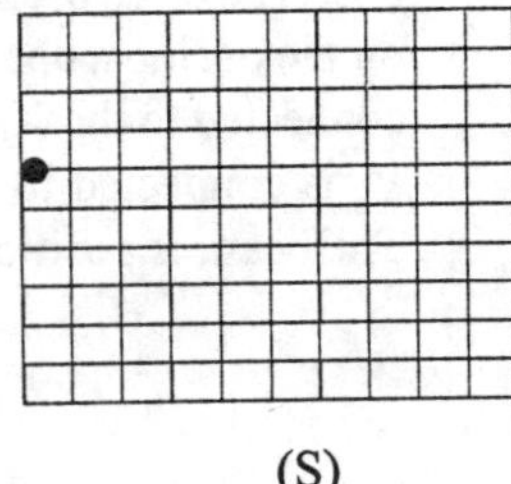
(S)

(a) *P* (b) *Q*
(c) *R* (d) *S*

19. The *x* and *y* sensitivities of an analog oscilloscope are set as 2 ms/cm and 1V/cm respectively. The trigger is set at 0V with negative slope. An input of 2 cos (100 pt + 30°) V is fed to the *y* input of the oscilloscope. The waveform seen on the oscilloscope will be.

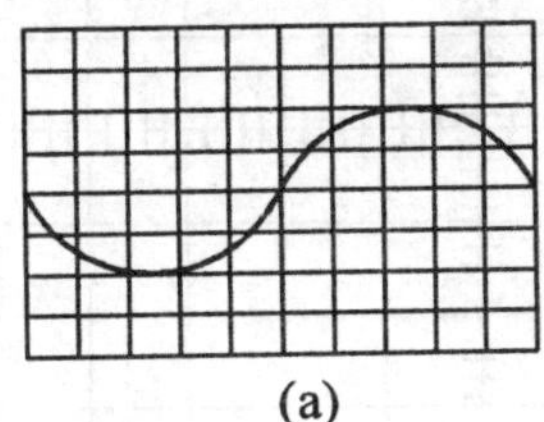
(a)

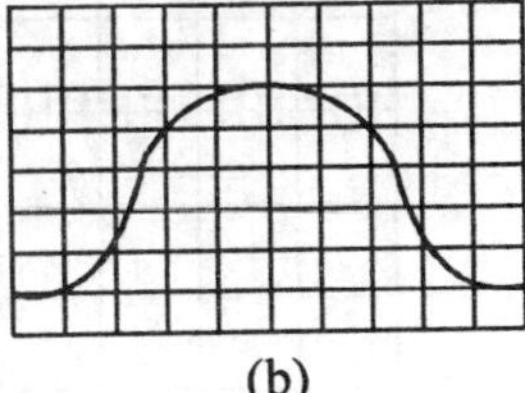
(b)

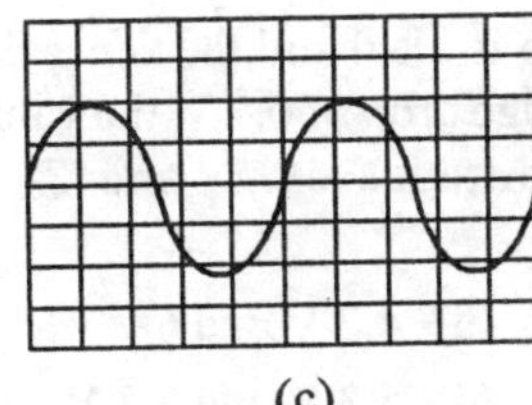
(c)

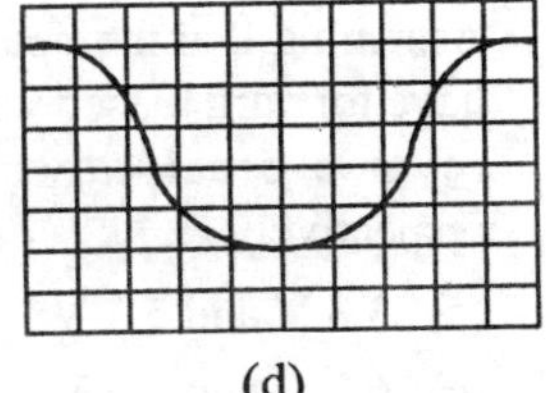
(d)

20. In an analog single channel cathode ray oscilloscope (CRO), the x and y sensitivities are set as 1 ms/div. and 1V/div. respectively. The y-input is connected to a voltage signal $4\cos(200\pi t - 45°)$ V. The trigger source is internal, level chosen is zero and the slope is positive. The display seen on the CRO screen is

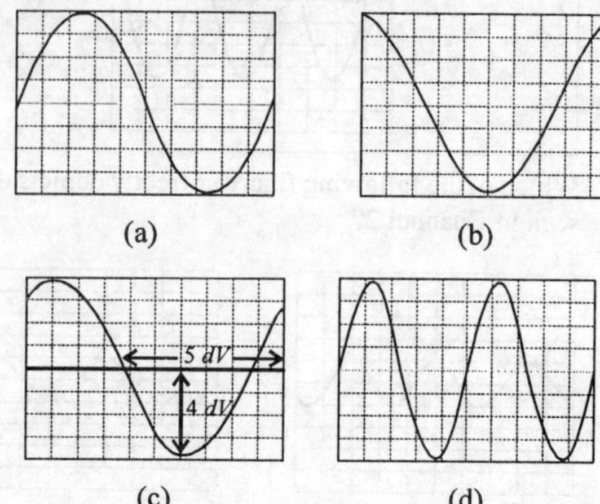

21. The transfer characteristics of the circuit drawn below is observed on an oscilloscope used in XY mode. The display on the oscilloscope is shown on the right hand side. V_i is connected to the X input with a setting of 0.5 V/ div, and V_0 is connected to the Y inptu with a setting of 2V/div. The beam is positioned at the origin when V_i is zero.

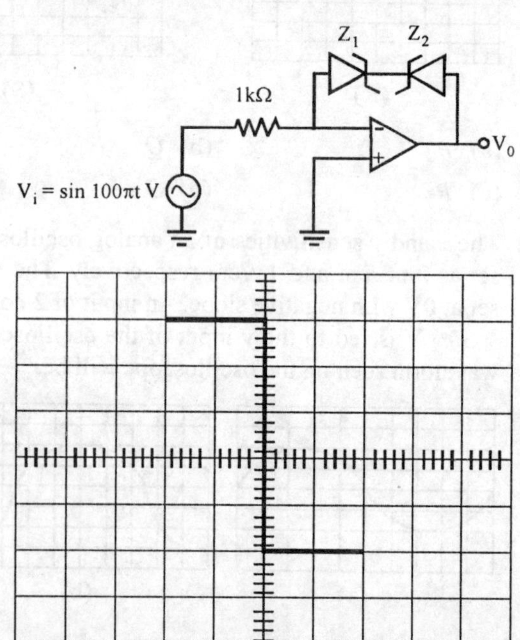

Assuming that the opamp is ideal and the zener diodes have forward biased voltage drop of 0.7 V, the values of reverse break-down voltages of Z_1 and Z_2 are, respectively,

(a) 3.3 V and 5.3 V (b) 4.7 V and 6.7 V
(c) 6.7 V and 4.7 V (d) 5.3 V and 3.3 V

22. The **deflection sensitivity** of a Cathode ray tube is

(a) Directly proportional to the deflection voltage
(b) Directly proportional to the square of the deflection voltage
(c) Inversely proportional to the deflection voltage
(d) Independent of the deflection voltage.

23. An amplitude modulated signal can be observed on a CRO by applying the following waveform to the "external trigger" input:

(a) The modulated waveform itself
(b) Derivative of the modulated waveform
(c) The modulating waveform
(d) The carrier waveform

24. The post-deflection acceleration (PDA) in a CRT is used in high frequency oscilloscope mainly to

(a) Increase the deflection sensitivity
(b) Improve the beam focusing.
(c) Improve the deflection linearity
(d) Improve the spot brightness.

25. An oscilloscope in put impedance consists of 1 MΩ in parallel with 100pF. A compensated 20:1 attenuator is obtained by connecting a parallel combination of:

(a) 19 MΩ and [100 / 19] pF
(b) 20 MΩ and [00 / 20] pF
(c) 19 MΩ and 1900 pF
(d) 20 MΩ and 2000 pF.

26. The elliptical Lissajous pattern shown in figure was observed from two sinusoidal signal sources of same frequency on a CRO

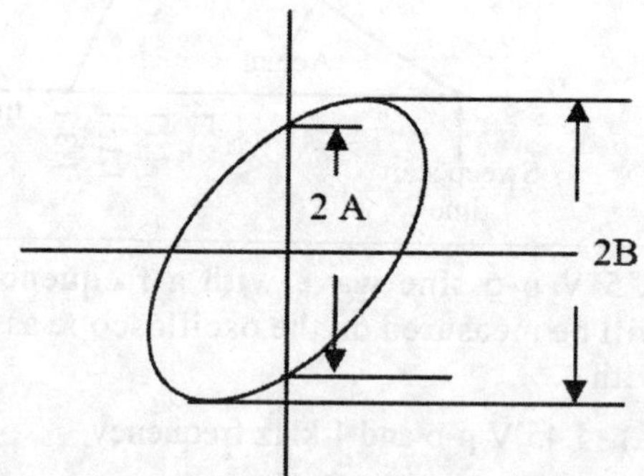

The phase angle between the signals is.

(a) $\cos^{-1}\left(\dfrac{A}{B}\right)$

(b) $\tan^{-1}\left(\dfrac{A}{B}\right)$

(c) $\sin^{-1}\left(\frac{A}{B}\right)$

(d) $\left(\frac{A}{B}\right)$

27. A double beam **CRO** has a bandwidth of 10 MHz. Two signals.

$V_1(t) = 10 \sin (2 \pi \times 20 \times 10^3 t)$ and

$V_2(t) = 10 \sin (2 \pi \times 20 \times 10^6 t)$

are applied to the two channels. If P_1 and P_2 are the peak to peak values of the signals displayed, then.

(a) $P_1 = 20, P_2 = 20$

(b) $P_1 < 20, P_2 > 20$

(c) $P_1 = 20, P_2 < 20$

(d) $P_1 > 20, P_2 < 20$

3.12 Electronic Multimeters, Analog and Digital Voltmeters

1. A digital-to-analog converter with a full-scale output voltage 3.5 V has a resolution close to 14 mV. Its bit size is

(a) 4 (b) 8

(c) 16 (d) 32

2. Two 8-bit ADCs, one of single slope integrating type and other of successive approximation type, take T_A and T_B times to convert 5 V analog input signal to equivalent digital output.If the input analog signal is reduced to 2.5 V, the approximate time take by the two ADCs will respectively be

(a) T_A, T_B (b) $T_A/2, T_B$

(c) $T_A, T_B/2$ (d) $T_A/2, T_B/2$

3. Two sinusoidal signals $p(\omega, t) = A$ sin and $q(\omega_2\ t)$ are applied to X and Y inputs of a dual channel CRO .The Lissajous figure displayed on the screen is shown below:

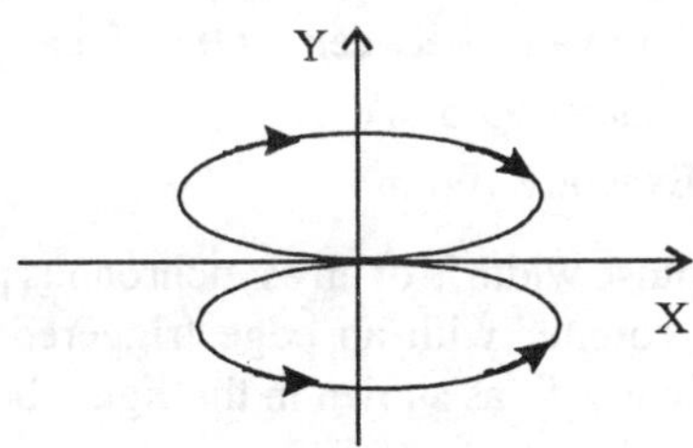

The signal $q(\omega_2 t)$ will be represented as

(a) $q(\omega_2 t) = A \sin \omega_2 t,\ \omega_2 = 2\omega_1$

(b) $q(\omega_2 t) = A \sin \omega_2 t,\ \omega_2 = 2\omega_1$

(c) $q(\omega_2 t) = A \cos \omega_2 t,\ \omega_2 = \omega_1/2$

(d) $q(\omega_2 t) = A \cos \omega_2 t,\ \omega_2 = \omega_1/2$

4. The simplified block diagram of a 10-bit A/D converter of dual slope integrator type is shown in Figure. The 10-bit counter at the output is clocked by a 1 MHz clock. Assuming negligible timing overhead for the control logic, the maximum frequency of the analog signal that can be converted using this A/D converter is approximately

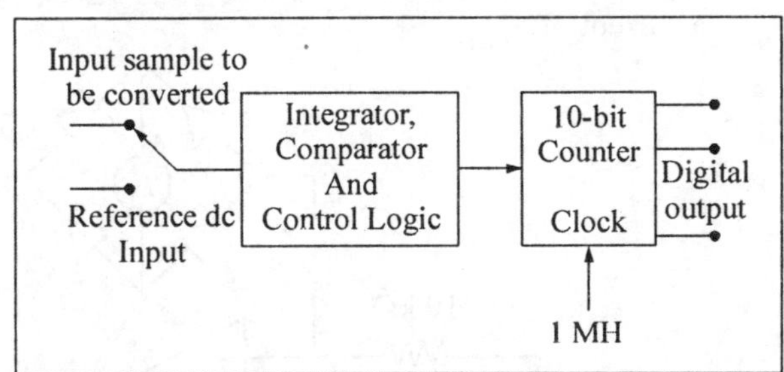

(a) 2 kHz (b) 1 kHz

(c) 500 Hz (d) 250 Hz

5. A 4½ digit DMM has the error specification as: 0.2% of reading +10 counts. If a dc voltage of 100 V is read on its 200 V full scale, the maximum error that can be expected in the reading is

(a) ±0.1% (b) ±0.2%

(c) ±0.3% (d) ±0.4%

6. The op-amp based circuit of a half-wave rectifier electronic voltmeter shown below uses a PMMC ammeter with a full scale deflection (FSD) current of 1mA and a coil resistance of 1 $k\Omega$. The value of R that gives FSD for a sinusoidal input voltage of 100 mV *(RMS)* is.

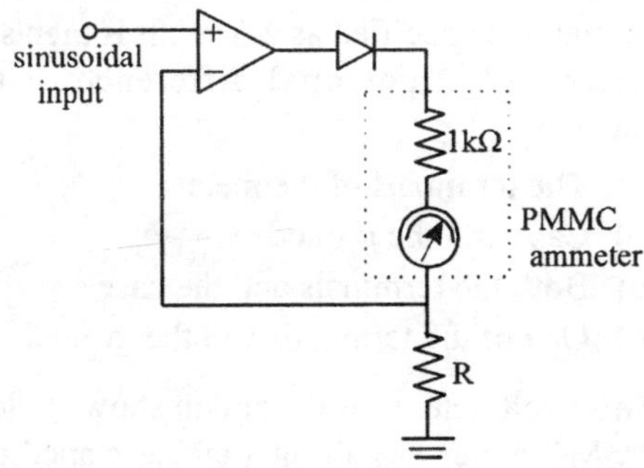

(a) 45 Ω (b) 67.5 Ω
(c) 100 Ω (d) 144.4 Ω

7. A PMMC type ammeter has full scale current of 100 μA and a coil resistance of 100 Ω.

The resistance required to convert the 100 μA ammeter into a 1A full scale dc ammeter is

(a) 10 mΩ in series with the meter
(b) 10 mΩ in parallel with the meter
(c) 1 mΩ in series with the meter
(d) 1 mΩ in parallel with the meter

8. A PMMC type ammeter has full scale current of 100 μA and a coil resistance of 100 Ω.

The above PMMC meter is connected in the circuit shown in the adjoining figure. The op-amp is ideal. The voltage $v_i(t) = 1.0 \sin 314t\ V$. Assuming the source impedance of $v_i(t)$ to be zero, the ammeter will indicate a current of

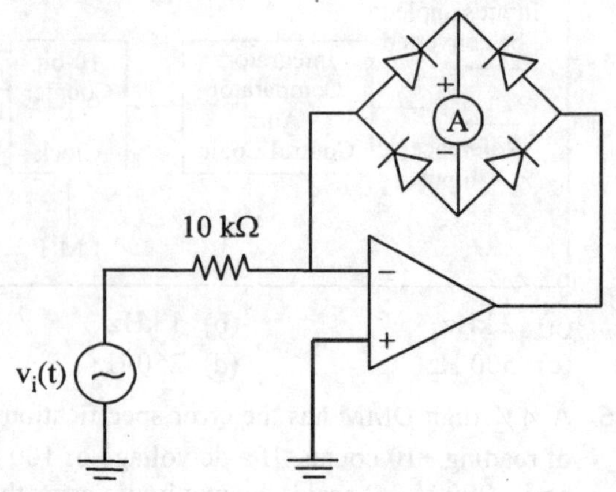

(a) 100 μA (b) 70.7 μA
(c) 63.7 μA (d) 31.8 μA

9. An analog voltmeter uses external multiplier settings. With a multiplier setting of 20 $k\Omega$, it reads 440 V and with a multiplier setting of 80 $k\Omega$, it reads 352 V. For a multiplier setting of 40 $k\Omega$, the voltmeter reads

(a) 371 V (b) 383 V
(c) 394 V (d) 406 V

10. The insulation level for a 5 A full scale rectifier type ammeter is specified as 2.5 kV. It is then safe to use the meter with a potential difference of upto 2.5 kV between.

(a) The terminals of the meter
(b) Case and the ground
(c) Both the terminals and the case
(d) One of the terminals and the ground

11. An ac voltmeter uses the circuit shown below, where the PMMC meter has an internal resistance of 100 Ω and requires a dc current of 1 mA for full scale deflection. Assuming the diodes to be ideal, the value of R_s to obtain full scale deflection with 100 V(ac rms) applied to the input terminal would be

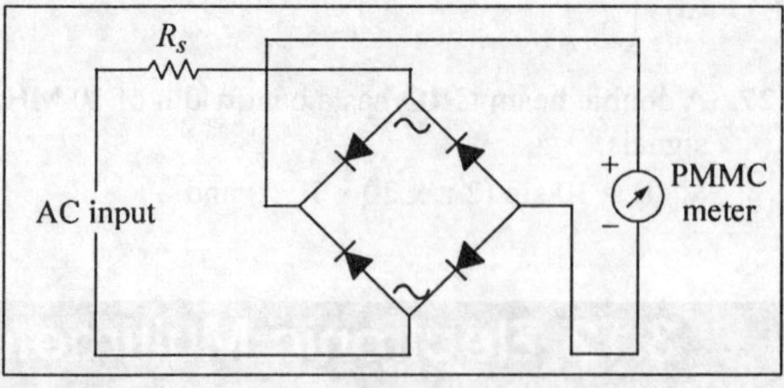

(a) 80 kΩ (b) 89 kΩ
(c) 89.9 kΩ (d) 90 kΩ

12. The clock frequency of a **timer counter** is 10 MHz. The timer counter is used in the period mode and the input to the timer counter is a square wave of frequency 2 kHz. The display of the timer counter will show a value

(a) 200 (b) 2000
(c) 5000 (d) 50000

13. A 3½ digit multimeter has an accuracy specification of (± 0.5 % of reading ± 5 counts) If the meter reads 2.00 mA, on a full scale of 20 mA, the worst case error in the reading is

(a) 0.5 % (b) 2.5%
(c) 3.0% (d) 5.5%

14. A 3½ digit, 200 mV full scale DVM has an accuarcy specification of ± 0.5% of reading plus 5 counts. When the meter reads 100 mV the voltage being measured is

(a) Any vlaue between 99.5 mV and 100.5 mV
(b) Any value between 99.0 mV and 101.0 mV
(c) Exactly 99.5 mV
(d) Exactaly 100 mV

15. The pulse width T of an asynchronous pulse is measured by a counter with an edge triggered clock of known frequency f_c as shown in the figure below:

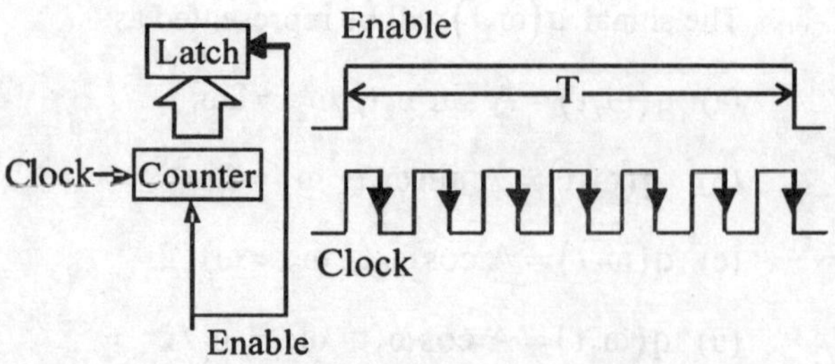

The pulse, whose width is to be measured, is applied to the Enable pin of the counter. The counter counts while the Enable is high and is held reset to zero otherwise. The counter output is latched by the negative edge of the Enable signal. The measured pulse width is taken to be N times the clock period, where N is the count reached at the end of a count cycle.

Assuming no overflow, the measurement error will be limited to x % of T if.

(a) $T > \frac{100}{x f_c}$ (b) $T < \frac{100}{x f_c}$

(c) $T > \frac{200}{x f_c}$ (d) $T < \frac{200}{x f_c}$

16. An electronic voltmeter consisting of a full-wave precision rectifier and an averaging circuit gives correct RMS value of square wave inputs. Its reading for a 2V peak- to-peak sinusoidal input will be

(a) [2/π] V (b) [1/π] V

(c) [2/√(2π)] V (d) [√(2) / π] V

17. The scale of a full wave rectifier type average sensing meter is calibrated to read the r.m.s. for ideal sine wave signal (form factor 1.11). Determine the multiplying factor of the meter if it is used to measure the r.m.s. value of square waves

(a) 1.11

(b) 1/1.11

(c) 1.0

(d) Depends on the frequency of the square wave.

18. In the dc millivoltmeter circuit shown in figure the input for full scale deflection is:

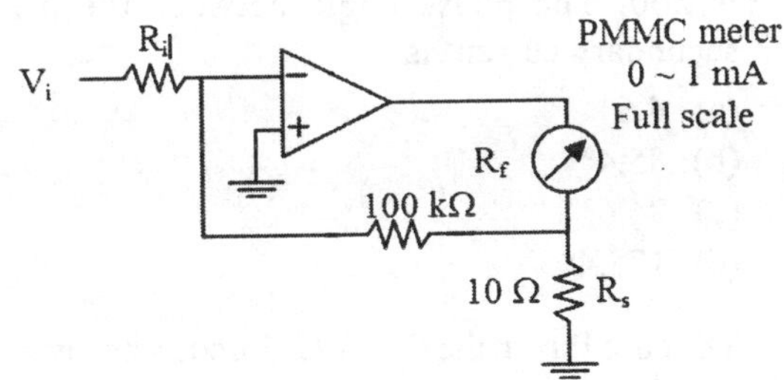

(a) 10 V

(b) 1 V

(c) 100 mV

(d) 10 mV.

19. **A rectifier type PMMC voltmeter** has a sensitivity of 20 kΩ/V. A reading of 4.5 V is obtained when measuring a voltage source with an internal resistance on its 5 V scale. When the scale is changed to 10 V full scale, a reading of 6 V is obtained. The value of the voltage and its internal resistance R are:

(a) 10 V and 100 kΩ (b) 9 V and 100 kΩ

(c) 10 V and 200 kΩ (d) 9 V and 200 kΩ

20. A 3 ½ digit **DVM** has a lowest measuring range of 200 mV. Hence it can be concluded that

(a) Its best resolution is 0.1 mV

(b) Its poorest resolution is 0.2 mV

(c) Its accuracy is atleast 0.05 %

(d) The maximum voltage that can be measured in this range is 199.9 mV.

21. The type of A/D converter normally used in a 3 ½ digit multimeter is

(a) Dual-slope integrating type

(b) Voltage - to - frequency converter type

(c) Flash (or parallel) type

(d) Successive approximation type

22. A 3½-digit digital multimeter can display

(a) 999.5 (b) 5.999

(c) 5555 (d) 1.999

23. Integrated output waveform for the dual slope ADC is shown in figure. The time T for an 8-bit counter with 4 MHz clock will be

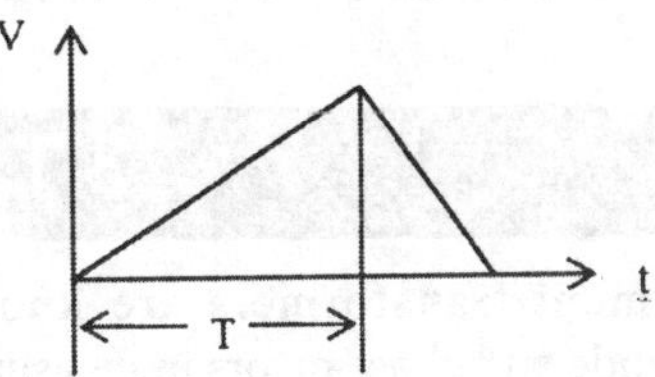

(a) 0.032 ms (b) 0. 0.64 ms

(c) 0. 64 ms (d) 0.024 ms

24. The advantage of a dual slope converter over successive approximation converter is that the dual slope converter

(a) Is faster

(b) Eliminates error due to drift

(c) Can reduce the errors due to power supply

(d) Does not require a stable voltage reference

25. Majority of digital voltmeters are built with a dual-slope ADC because

(a) Dual slop ADCs are less complex than other types of ADCs.

(b) Dual slope ADCs are faster than other types of ADCs.

(c) Dual slope ADCs can be designed to be insensitive to noise and interference.

(d) Dual slop ADCs provide BCD outputs.

26. A **counter timer** has a basic clock of 16 MHz. The count value displayed is in error by 1 count. The frequency at which the error in the displayed value is the same whether the counter timer is used in the frequency mode of operation or period mode of operation is:

(a) 16 MHz

(b) 10 MHz

(c) 8 MHz

(d) 4 MHz.

27. In a dual slope type digital voltmeter, an unknown signal voltage is integrated over 100 cycles of the clock. If the signal has a 50 Hz pick up, the maximum clock frequency can be.

(a) 50 Hz
(b) 5 kHz.
(c) 10 kHz
(d) 50 kHz.

28. A 3½ **digit voltmeter** has an internal reference of 200 mV. Its full scale count and resolution are:

(a) 2000 and 1 mV

(b) 4000 and 0.1 mV

(c) 2000 and 0.1 mV

(d) 4000 and 1 m V

29. A 3 ½ **digital multimeter** has an accuracy specification of 0.5% of reading plus 5 counts. The value of an unknown resistance is read as 50 .0Ω the 200 Ω scale of the meter. The value of the resistance is:

(a) 50.0 ± 0.25 Ω

(b) 50.0 ± 0.5 Ω

(c) 50.0 ± 0.75 Ω

(d) 50.0 ± 1.0 Ω

30. In a dual slope ADC, the reference voltage is 100 mV and the first integration period is set as 50 ms. The input resistor of the integrator is 100 kΩ and the integrating capacitor .047 μF.

For an input voltage of 120 mV., the second integration (de – integration) period will be.

(a) 50 ms
(b) 60 ms
(c) 100 ms
(d) 120 ms

31. In a dual slope ADC, the reference voltage is 100 mV and the first integration period is set as 50 ms. The input resistor of the integrator is 100 kΩ and the integrating capacitor .047 μF.

If the input of 120 mV is corrupted by power supply interference at 50 Hz having peak amplitude of 3π mV, the worst case error introduced by the interference in the reading is .

(a) 0% (b) 1%
(c) 3% (d) π%

3.13 Instrument Transformer

1. Instrument transformers are known to introduce magnitude and phase errors in measurements. These are primarily due to
(a) Improper connections on the primary side
(b) Measurement error inherent in the meter connected to the transformer secondary
(c) Open and short circuit parameters of the instrument transformers
(d) None of the above.

2. A 500 A/5 A, 50 Hz current transformer has a bar primary. The secondary burden is a pure resistance of 1 Ω and it draws a current of 5A. If the magnetic core requires 250 AT for magnetization, the percentage ratio error is

(a) 10.56
(b) – 10.56
(c) 11.80
(d) – 11.80

3. A 50 Hz, bar primary CT has a secondary with 500 turns. The secondary supplies 5 A current into a purely resistive burden of 1Ω. The magnetizing ampere-turns is 200. The phase angle between the primary and secondary current is

(a) 4.6°
(b) 85.4°
(c) 94.6°
(d) 175.4°

4. The core flux in the CT of Q. 3 under the given operating conditions

(a) 0

(b) 45.0 μWb

(c) 22.5 mWb

(d) 100.0 mWb

5. A 200/1 Current transformer (CT) is wound with 200 turns on the secondary on a toroidal core. When it carries a current of 160 A on the primary, the ratio and phase errors of the CT are found to be – 0.5% and 30 minutes

respectively. If the number of secondary turns is reduced by 1, the new ratio error (%) and phase error (min) will be respectively

(a) 0.0, 30
(b) – 0.5, 35
(c) –1.0, 30
(d) – 1.0, 25

6. The primary of a 200/1 A, 5 VA, 0.8 PF current transformer (CT) carriers 100 A. The secondary current is 0.495 mA. The ratio error of the CT is

(a) – 1.0%
(b) – 0.5%
(c) + 0.5%
(d) + 1.0%

7. The **ratio error** of a **current transformer** depends mainly on

1. Type of the ammeter used on the secondary
2. Eddy current loss of the core
3. Magnetising current of the core
4. Power factor of the load

From the above statements following is correct

(a) 1 and 2
(b) 2 and 3
(c) 3 and 4
(d) 1 and 4

8. Care should be taken not to open-circuit

(a) The secondary of a **voltage transformer** when the primary is energized at rated voltage
(b) The secondary of a **current transformer** when the primary is energized at rated current
(c) The primary of a **current transformer** when the secondary is connected to rated burden
(d) The primary of a **voltage transformer** when the secondary is connected to rated burden

Answers

3.1 Fundamental and Error Analysis

1. **Ans. (c)**

Solution:

Given

Gauge Factor $G_f = 2$

Normal resistance $R = 120\ \Omega$

$\varepsilon = 10^{-5}$ $\Delta R = ?$

Strain $\varepsilon = \frac{\Delta L}{L} = 10^{-5}$

$$G_f = \frac{\Delta \frac{R}{R}}{\Delta \frac{L}{L}}$$

$$\Delta R = R \times \varepsilon \times G_f$$

$$= 120 \times 10^{-5} \times 2$$

$$\Rightarrow 2.4 \times 10^{-3}\ \Omega$$

2. **Ans. (a)**

Solution:

For 10 V total input resistance

$$R_v = \frac{V_{fsd}}{I_{m\ fsd}}$$

$$= 10/100\ \mu A = 10^5$$

Sensitivity $\frac{R_v}{I_{fsd}} = \frac{10^5}{10} = 10\ k\Omega/V$

For 100 V input resistance $R_v = \frac{10}{100\mu A} = 10^6\Omega$

Sensitivity $= \frac{R_v}{V_{fsd}} = \frac{10^6}{100}$

$$= 10\ k\Omega / \text{V}$$

3. **Ans. (a)**

Solution: Given , ω = [*xy*/*z*]

Taking log, we get, log ω = log *x* + log *y* – log *z*

Differentiating with respect to ω, we get

$$\frac{d\omega}{\omega} = \left[\frac{dx}{x} + \frac{dy}{y} - \frac{dz}{z}\right] \times 100$$

$$dx = \frac{0.5 \times 80}{100} = 4$$

$$dy = \frac{1 \times 20}{100} = 0.2$$

$$dz = \frac{1.5 \times 50}{100} = 0.75$$

$$\frac{d\omega}{\omega} = \left[\frac{4}{80} + \frac{0.2}{20} - \frac{0.75}{50}\right] \times 100$$

$$= \frac{20 + 4 - 6}{400} \times 100$$

$$= \frac{18}{4} = 4.5 = 5\%$$

4. **Ans. (d)**

Solution:

The most repeatable value or reproductable value out of set of recorders is called precision.

5. **Ans. (a)**

6. **Ans. (b)**

Solution:

When the number of observation is less than 20, then the standard deviation of a finite number of of data having (Standard deviation) as $d_1, d_2, d_3, \ldots, d_n$ is expressed as

$$SD = \sqrt{\frac{d_1^2 + d_2^2 + \ldots + d_n^2}{n-1}}$$

Here, n = 2 and $d_1 = 4\ \Omega$, $d_2 = 8\ \Omega$

$$\therefore \quad SD = \sqrt{\frac{4^2 + 3^2}{1}} = 5\ \Omega$$

7. **Ans. (c)**

Solution:

Accuracy

It is the closeness with which an instrument reading approaches the true value of the quantity being measured. It can be specified in term of inaccuracy or limit of error.

Precision

It refers to repeatability or to how closely the individual measured values of a quantity agree with each other when the measurements are carried out under identical conditions at a short interval of time. It can be defined as the ability of the instrument to reproduce a group of measurements of the same measured quantity under the same conditions. Imprecision is caused due to random errors and can be quantified by statistical analysis.

Accuracy Vs Precision

It may be noted that a precise measurement may not necessarily be accurate and viceversa. This statement is illustrated by an analogy from shooting practice on a target as shown in figure.

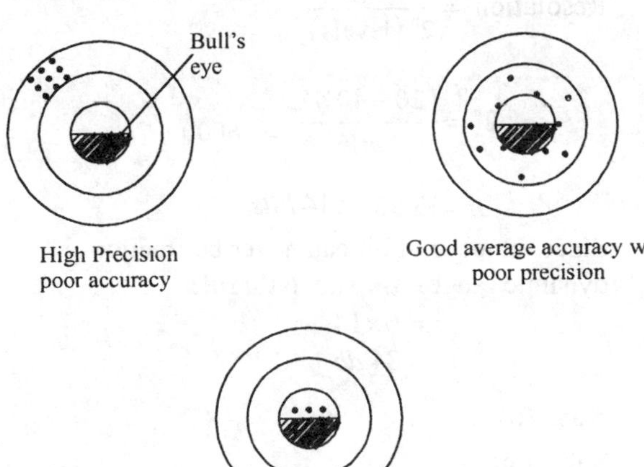

Figure: Accuracy and precision in a typical target shooting

Sensitivity

It is defined as the ratio of the magnitude of output signal to the magnitude of input signal.

$$\text{Sensitivity} = \frac{\text{Change of output signal}}{\text{Change in input signal}}$$

A 1 mV recorder might have a 10 cm scale. Assuming a linear scale, its sensitivity would be 10 cm/mV.

The sensitivity is constant in a linear instrument and usually it is required to be high.

Resolution

This is the smallest change in input or measured value which can be detected by the instrument. The least count of an instrument can be taken as the resolution of an instrument.

8. **Ans. (d)**

Solution:

Consider 12355 and rounded off two significant figure, then 12400 will be obtain. Similarly 2354 rounded off to 2400

9. **Ans. (d)**

Solution:

$$\text{Arithmetic mean } \bar{x} = \frac{5.9+5.7+6.1}{3}$$

$$= \frac{17.7}{3} = 5.9$$

$$\text{Deviation } d_1 = x_1 - \bar{x}$$

$$= 5.9 - 5.9 = 0$$

$$\text{Deviation } d_2 = x_2 - \bar{x} = 5.7 - 5.9$$

$$= -0.2$$

$$\text{Deviation } d_3 = x_3 - \bar{x}$$

$$= 6.1 - 5.9$$

$$= 0.2$$

Standard deviation $\sigma = \sqrt{\frac{\Sigma d^2}{n-1}}$

When n = number o f experiment

$$\sigma = \sqrt{\frac{(0)^2 + (-0.2)^2 + (0.2)^2}{(3-1)}}$$

$$= 0.2$$

10. **Ans. (a)**

Solution: The threshold of an instrument is normally defined as the smallest measurable input charge (non zero value) which can be detectable.

If the instrument input is increased very gradually from zero there will be some minimum value below which no output change can be detected. This minimum value defines the threshold of the instrument.

11. **Ans. (c)**

Solution:

Accuracy: It is the closeness with which an instrument reading approaches the true value of the quantity being measured. It may be specified in terms of limits of error.

$$\textbf{Sensitivity} = \frac{\text{change in output}}{\text{change in input}}$$

Repeatability: Reproducibility and repeatability are a measure of closeness with which a given input may be measured over and over again.

Resolution: The smallest increment in input which can be detected with certainty by an instrument is the resolution.

12. **Ans. (c)**

Solution:

$$-I_1 + I_2 + I_3 = 0 \text{ (Using KCL at the node)}$$

$$I_3 = I_1 - I_2$$

$$= 100\ mA - 99\ mA = 1\ mA$$

$$\frac{dI_3}{I_3} = \pm\frac{dI_1}{I_1} \pm \frac{dI_2}{I_2}$$

$$= \pm 1 \pm 1 = \pm 2\,\%$$

3.2 Indicating Instruments

1. Ans. (c)

Solution: For galvanometer,

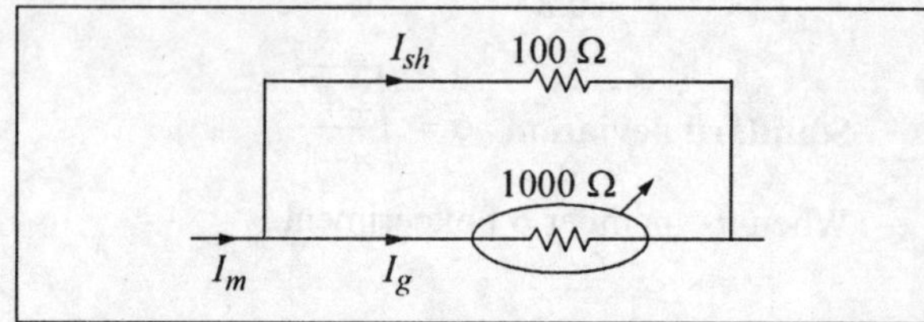

$$\text{Multiplying Power} = \frac{I_m}{I_g}$$

$$\frac{I_g}{I_{sh}} = \frac{100}{1000}$$

$$I_{sh} = 10\,I_g$$

$$I_m = I_g + I_{sh} = I_g + 10\,I_g = 11\,I_g;$$

$$\frac{I_m}{I_g} = 11$$

2. Ans (d)

Solution:

$I_m = 5A$

$I = 25A$

$R_m = 0.2\,\Omega$

$R_{shunt} = ?$

Multiplying factor

$$m = \frac{I}{I_m} = 1 + \frac{R_m}{R_{shunt}}$$

$$R_{shunt} = \frac{R_m}{(m-1)} = \frac{0.2}{\left(\frac{25}{5}\right) - 1}$$

$$= \frac{0.2}{5-1} = \frac{0.2}{4} = 0.05\,\Omega$$

$$m = \frac{25}{5}$$

$$R_{sh} = \frac{0.2}{m-1} = \frac{0.2}{4} = 0.05\,\Omega$$

3. Ans. (d)

4. Ans. (d)

Solution:

Only electrostatic voltmeter is suitable to measure high dc voltage without loading

5. Ans. (d)

Solution:

$$\text{Resolution} = \frac{\text{range}}{2^n\,(\text{levels})}$$

$$2^n = \frac{(20-40)V}{1\,mV} = 16000$$

$$n = 13.96 \cong 14\ bits$$

Since number of bit can never be fraction dynamic range = $6n$, i.e. 6 db rule

$$= 6 \times 14$$

$$= 84\ db$$

6. Ans. (d)

Solution:

When Switch s is open, then

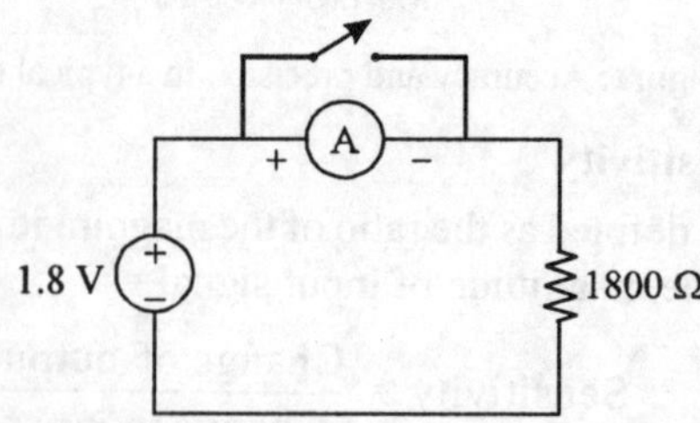

Let the resistance of coil $x = -2$ and I_1 is the current flowing in the circuit.

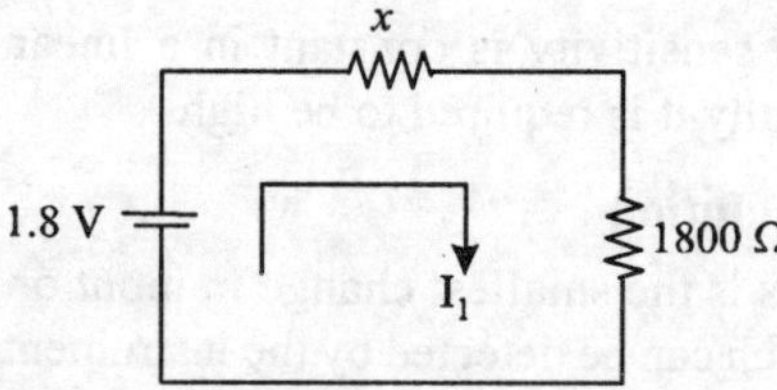

It is given that $I_1 = 0.9\ mA$

$$I_1 = 0.9 \times 10^{-3}\ A$$

$$1.8 - I_1(x) - I_1(1800) = 0$$

$$(9\times10^{-4})x - (9\times10^{-4})(1800) = 1.8$$

$$(9\times10^{-4})x = 0.18$$

$$x \Rightarrow \frac{18}{100\times9} \times 10^4$$

$$x = 200\,\Omega$$

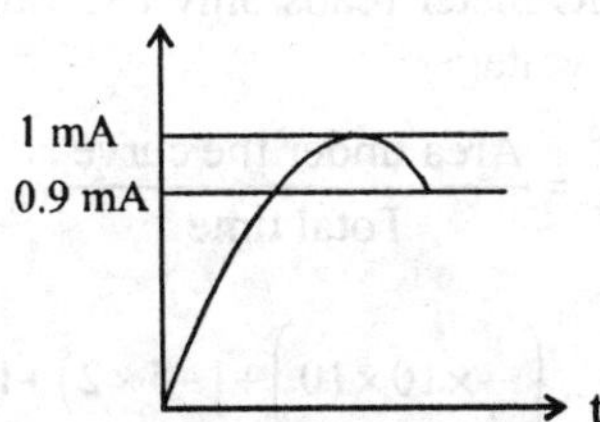

In this the system is having oscillation so it is known as under damped system.
So coil is under damped with resistance of 200 Ω

7. **Ans (d)**

Solution:

$I_1 = 1$ Amp (d.c.)

$\theta_1 = 20°$

$I_2 = \frac{3}{\sqrt{2}}$ [r.m.s.] [a.c. current]

$\theta_2 = ?$

$$\frac{\theta_1}{\theta_2} = \frac{I_1^2}{I_2^2} \Rightarrow \left[\frac{20°}{\theta_2} \Rightarrow \frac{1}{\left(\frac{3}{\sqrt{2}}\right)^2}\right]$$

$$\frac{20°}{\theta_2} = \frac{2}{9}$$

$$\theta_2 = \frac{180}{2} = 90°$$

8. **Ans. (c)**

Solution:

- Spring of PMMC meter are in series with the coil and provide controlling torque and provide current in and out of the MC.
- Aluminium former holds the coil and provide damping torque.

3.3 Measurement of Current and Voltage

1. **Ans. (d)**

Solution: Either decreased or increased depending on the direction of flux

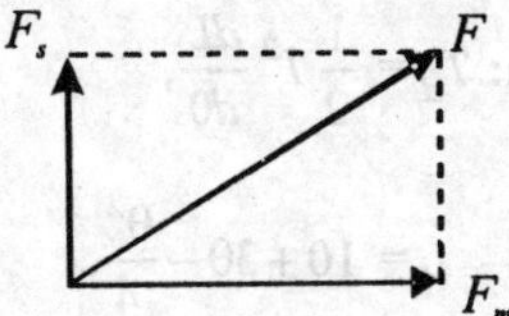

Let F_m be the flux due to meter coil and F_s is due to stray magnetic field, then resultant will be F.
Similarly when F_s is in opposite direction.

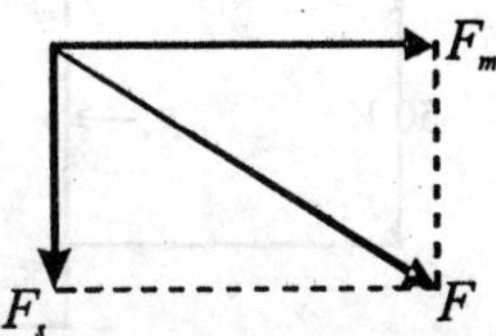

So the reading would be either increased or decreased

2. **Ans. (d)**

Solution: The spring gives the controlling torque and is connected in series with the coil. If the spring is snapped, there will be no deflection.

3. **Ans. (d)**

Solution: M1 : 20 × 0.001 = 0.02,

Relative error at 1A = [0.02 / 1] = 2 %

M2: 10 × 0.002 = 0.02,

error = [0.02 / 1] = 2 %

M3: 5 × 0.005 = 0.025,

error = [0.025 / 1] = 2.5 %

M4: 1 × 0.01 = 0.01,

error = [0.01 / 1] = 1 %

So, M4 has least error on full scale

4. **Ans. (b)**

Solution:

The friction error of the PMMC instrument is less, this error is decided by torque/ weight ratio. The friction error is less only when torque/weight ratio is high

5. **Ans. (c)**

6. **Ans. (c)**

Solution:
Sensitivity $S_{dc} = 1000\ \Omega/\text{volt}$

$$S_{dc} = \frac{1}{I_{mfsd}}$$

$$\Rightarrow \quad I_{mfsd} = \frac{1}{1000} = 1\ mA$$

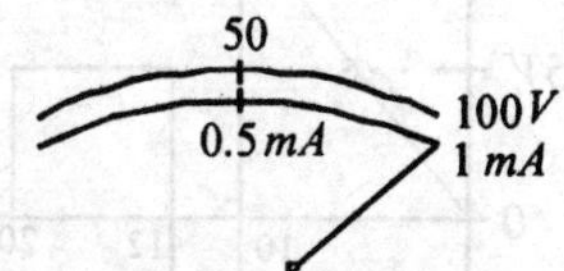

∴ At half full scale, the reading will be 0.5 *mA*

Second Method

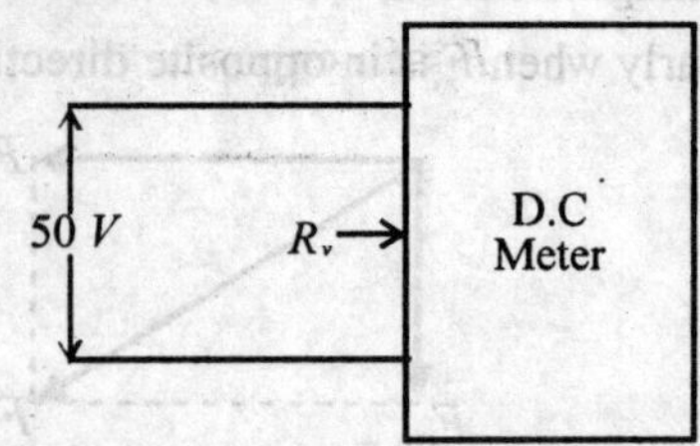

$$S = \frac{R_v}{V_{fsd}}$$

$$R_v = 1000 \times 100$$

$\Rightarrow$ 100 $k\Omega$

$$I_m = \frac{50}{100 \times 10^3} = 0.5\ mA$$

7. **Ans. (a)**

Solution:

In PMCC instruments, as temperature increases the coil resistance increases. Swamp resistors are connected in series with the moving coil to provide temperature compensation. Swamping resistor is made of magnin, which has a zero-temperature coefficient.

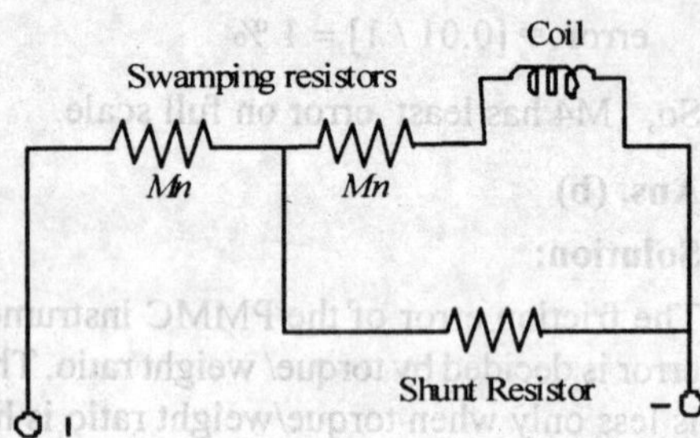

8. **Ans. (b)**

9. **Ans. (a)**

Solution: PMMC instruments measure DC value only.

10. **Ans. (a)**

Solution:

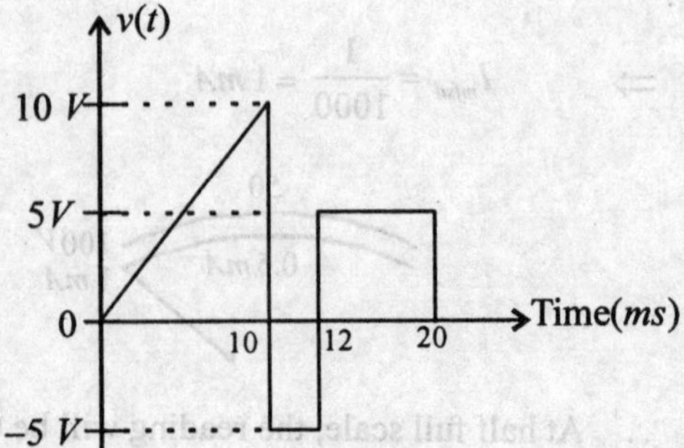

A PMMC meter reads only DC value or average value average voltage

$$V_{avg} = \frac{\text{Area under the curve}}{\text{Total time}}$$

$$V_{avg} = \frac{\left(\frac{1}{2} \times 10 \times 10\right) + (-5 \times 2) + (5 \times 8)}{20}$$

$$= \frac{50 - 10 + 40}{20}$$

$$= \frac{80}{20} = 4\ V$$

11. **Ans. (c)**

Solution:

Given $R = 100\ \Omega$

$$V_{rms} = 100\ V$$

Output of full wave rectifier.

$$V_{dc} = \frac{2V_m}{\pi} = \frac{2\sqrt{2}}{\pi} V_{rms} = \frac{2\sqrt{2} \times 100}{\pi} = 90.03$$

For full deflection

$$R_{Total} = \frac{V_{dc}}{I_{fsd}}$$

$$R_m + R_s = \frac{90.03}{1 \times 10^{-3}}$$

$$R_s = 90.03 \times 10^3 - 0.1 \times 10^3 = 89.93\ k\Omega$$

Second Method

d.c. sensitivity is $S_{dc} = \frac{1}{I_{fs}} = \frac{1}{1\ mA} = 10^3\ \Omega/V$

For full-wave rectifier a.c. sensitivity,

$$S_{ac} = 0.9\ S_{dc} = 0.9 \times 10^3 = 900\ \Omega/V$$

Resistance of multiplier,

$$R_s = S_{ac}\ V - R_m - 2\ R_d$$

Since diodes are ideal, $R_d = 0$

The $R_s = 900 \times 100 - 100 = 89.90\ k\Omega$

12. **Ans. (b)**

Solution: $T_d = \frac{1}{2} I^2 \frac{dL}{d\theta}$,

$$L = 10 + 3\theta - \frac{\theta^2}{4}$$

$$\frac{dL}{d\theta} = \left(3 - \frac{2\theta}{4}\right)\mu\text{H}$$

$$25 \times 10^{-6} = \frac{1}{2} \times 5^2 \times \left(3 - \frac{\theta}{2}\right) \times 10^{-6}$$

$$1 = \frac{1}{2}\left(3 - \frac{2\theta}{4}\right)$$

$$2 = 3 - \frac{\theta}{2}$$

$$\frac{\theta}{2} = 3 - 2$$

$$\theta = 2$$

13. Ans. (a)

Solution:

Given $N = 100$, $B = 200\ m$, $I = 50\ mA$, $L = 10\ mm$, $D = 20\ mm$

$$A = 10 \times 20 \times \mu m$$

Deflecting torque

$$T_d = NBIA$$

$$= 100\left(200 \times 10^{-3}\right)\left(50 \times 10^{-3}\right)\left(20 \times 10^{-3}\right)\left(10 \times 10^{-3}\right)$$

$$= 20 \times 10^{-3} \times 10 \times 10^{-3}$$

$$= 200\ \mu\ Nm$$

14. Ans. (a)

Solution:

$$R_{eft} = 4\ M\Omega \parallel 50\ k\Omega$$

$$R_T = R_{eft} + 6\ M\Omega$$

Since $R_T = \dfrac{4 \times 0.05}{4.05} + 6 = 6.05\ M\Omega$

$$\therefore \quad I_T = \frac{1000}{6.05\ \text{M}\Omega}\ \text{A}$$

$\therefore$ Voltage reading $V_T = \dfrac{1000}{6.05} \times 0.05 = 8\ \text{V}$

15. Ans. (c)

Solution: PMMC instrument reads only DC value and since it is centre zero type so it will give – 8 value

RMS meter will read rms value of the current, so

$$\text{RMS} = \sqrt{8^2 + 6^2} = 10\ \text{A}$$

Moving iron also reads rms value, so its reading will also be 10 A.

Hence ultimately it will be – 8, 10, 10.

16. Ans (a)

Solution :

For triangular wave,

Avg value $\rightarrow \dfrac{V_m}{2}$

rms value $\rightarrow \dfrac{V_m}{\sqrt{3}}$

$$\therefore \frac{V_m}{2} = 10V \Rightarrow V_m = 20\ V$$

rms value $= \dfrac{20}{\sqrt{3}}$

17. Ans. (d)

Solution:

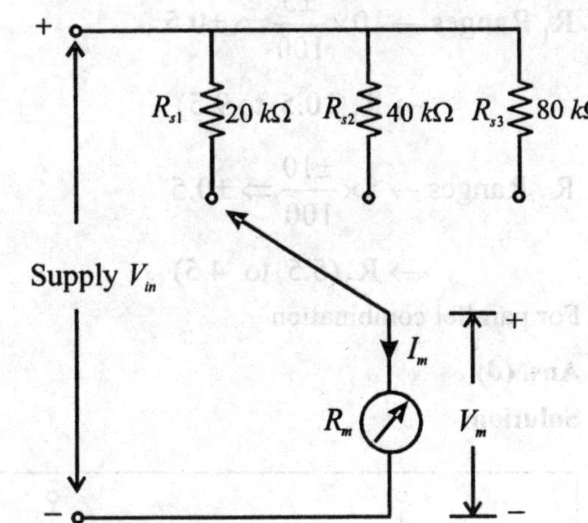

When $R_{s1} = 20\ k\Omega$,

$$V_m = 440\ V$$

$$I_m = \frac{V_m}{R_m}$$

$$V = V_{Rs1} + V_{m1}$$

$$V_{in} = \frac{440}{R_m} \times 20 + 440 \qquad \text{...(1)}$$

When $R_{s3} = 80\ k\Omega$, $V_{m3} = 352\ V$

$$V_{in} = \frac{352}{R_m} \times 80 + 352 \qquad \text{...(2)}$$

From equation (1) and (2)

or, $\dfrac{352 \times 80 - 440 \times 20}{88} = R_m$

$$R_m = 220\ k\Omega,\ V_{in} = 480\ V$$

When $R_{s2} = 40\ k\Omega$ and $V_{in} = 480\ V$

$$480 = \frac{V_{m3}}{R_m} \times 40 + V_{m3}$$

$$480 = \frac{13}{11} V_{m3}$$

or $V_{m3} = 406.15$ Volt

3.4 Measurement of Resistance

1. **Ans. (c)**
2. **Ans. (a)**

Solution:

Given, $R_1 = 10\Omega \pm 5\%$ and $R_2 = 5\Omega \pm 10\%$

$$R_1 \text{ Ranges} \rightarrow 10 \times \frac{\pm 5}{100} \Rightarrow \pm 0.5$$

$$\rightarrow R_1 (10.5 \text{ to } 9.5)$$

$$R_2 \text{ Ranges} \rightarrow 5 \times \frac{\pm 10}{100} \Rightarrow \pm 0.5$$

$$\rightarrow R_2 (5.5 \text{ to } 4.5)$$

For parallel combination

$$\frac{10.5 \times 5.5}{10.5 + 5.5} \text{ and } \frac{9.5 \times 4.5}{9.5 + 4.5} = 3.60\ \Omega \text{ and}$$

$$3.05\ \Omega\ \frac{10.5 \times 5.5}{10.5 + 5.5} \text{ and } \frac{9.5 \times 4.5}{9.5 + 4.5}$$

$$= 3.60\ \Omega \text{ and } 3.05\ \Omega$$

3. **Ans. (d)**

Solution:

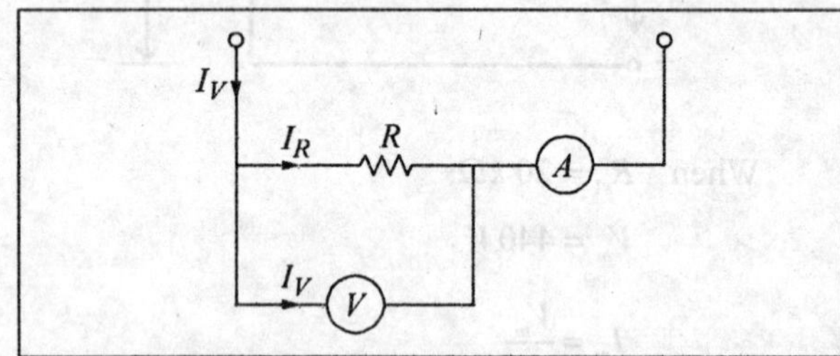

$R_A = 0.01\Omega$; $R_V = 2000\ \Omega$

$I_R = 2$ A ; $V_R = V_V = 180$ V

$I_A = I_R + I_V$; $I_R = I_A - I_V$

Since, $I_V = \dfrac{180}{2000}$

$$\therefore \quad I_R = 2 - \frac{9}{100} = 1.91\ \text{A}$$

Since, $1.91\ R = 180$

$$\therefore \quad R = 94.24\ \Omega$$

$$\text{Error} = \frac{\text{True value} - \text{measured value}}{\text{measured value}}$$

$$\therefore \quad \text{Error} = \frac{94.24 - 90}{90} \times 100 = 4.71\%$$

3.5 Bridge, Measurement of Inductance and Capacitance

1. **Ans. (c)**

Solution:

The circuit is of Maxwell's bridge and used for measurement of low Q coils. The impedance Z_x is Maxwell inductance bridge consists of resistance in series with inductance. The expression for Q is $Q = \dfrac{\omega L}{R}$ where L and R is independant Z_x.

or $Z_x = R + j\omega L$

$$Q = \frac{\omega L}{R} = \omega C_1 R_1 \quad \text{... (1)}$$

Also $L = R_4 R_4 C_1$ .... (2)

$$\text{and} \quad R = \frac{R_2 R_4}{R_1} \quad \text{... (3)}$$

R_1 and C_1 are the two variation for following the equations (2) and (3).

From eqn (1) we see the measurement of high Q coil demands a large resistance R_1. This may be in the range 10^5 or $10^6\ \Omega$. The resistance box of such high value will be very expensive. Thus for $Q > 10$, the Maxwell bridge is unsuitable.

* If Q is high, high capacitor C_1 is required for balancing, so it is not suitable for measurement of high Q-coils, i.e. $Q > 10$.
* It is also not suitable for measurement of very low Q-coils, i.e. ($Q < 1$). Mainly Maxwell's bridge is suitable for medium Q-coils $(1 < Q < 10)$

2. **Ans. (a)**
3. **Ans. (b)**

Solution: At full scale position, $\dfrac{1}{2} I^2 \dfrac{dL}{d\theta} = T_d$

or $\dfrac{1}{2}(10)^2 \dfrac{dL}{d\theta} = 240 \times 10^{-6}$

or $\frac{dL}{d\theta} = 4.8\ \mu H/\text{radian}$

4. Ans. (b)

Solution:

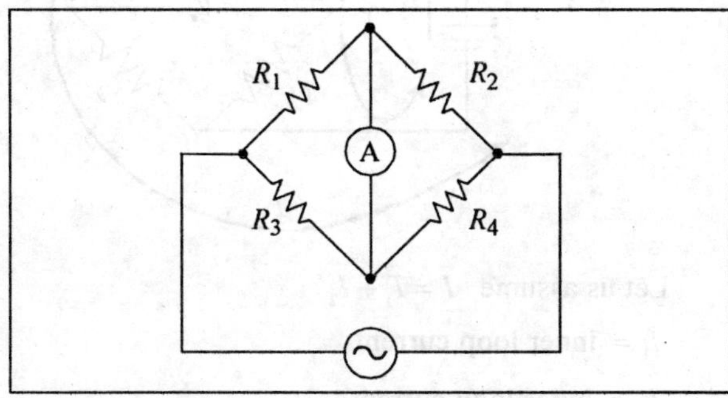

When the bridge is balanced,

$$\frac{R_1}{R_2} = \frac{R_3}{R_4} \Rightarrow R_1 = \frac{R_2R_3}{R_4}$$

5. Ans (a)

Solution:

The product of cross arm impedance are equal at the null deflection

$$(R + j\omega L)\left(\frac{R_4 \times \{-j/\omega C_4\}}{R_4 - j/\omega C_4}\right) = R_2R_3$$

$$(R + j\omega L)\left(\frac{-jR_4}{\omega R_4C_4 - j}\right) = R_2R_3$$

$$(R + j\omega L)\left(\frac{R_4}{1 + j\omega R_4C_4}\right) = R_2R_3$$

$$RR_4 + j\omega LR_4 = R_2R_3 + j\omega R_2R_3R_4C_4$$

Comparing real and imaginary parts

$$R R_4 = R_2R_3$$

$$R = \frac{R_2R_3}{R_4} \text{ and}$$

$$LR_4 = R_2R_3R_4C_4$$

$$L = R_2R_3C_4$$

6. Ans. (c)

Solution:

In the given ac bridge

When new coil is connected parallel to the X_{coil} then balance to the circuit is obtained by reducing the resistance of R_2 to half of its value. Hence the impedance value of the equivalent X_{coil} has to be half. This can be obtained only when the unknown parameter of the coil are equal to X_{coil} parameters.

7. Ans. (c)

8. Ans. (b)

Solution:

Since $\frac{X_c}{R} = 1 \Rightarrow \frac{1}{j\omega RC} = 1$

$$V_A = \frac{R}{2R} \times 10 = 5V$$

$$V_B = \frac{\frac{1}{j\omega C}}{R + \frac{1}{j\omega C}} \times 10$$

$$= \frac{1}{j\omega RC + 1} \times 10 = \frac{10}{1 + j}$$

$$V = V_A - V_B = 5 - \frac{10}{1+j} = \frac{5 + 5j - 10}{1 + j}$$

$$= \frac{5(j-1)}{1+j} = 5\ V$$

$$V_{AB} = V_A - V_B = 5 - 5 = 0\ V$$

9. Ans. (a)

Solution:

$$V_{ab} = V_a - V_b = \frac{30 \times 10}{30} - \frac{30 \times 2}{10} = 4$$

Energy stored in capacitor $= \frac{1}{2}CV_{ab}^2$

$$= \frac{1}{2} \times 1 \times 10^{-6} \times 4^2 = 8\ \mu J$$

10. Ans. (b)

Solution:

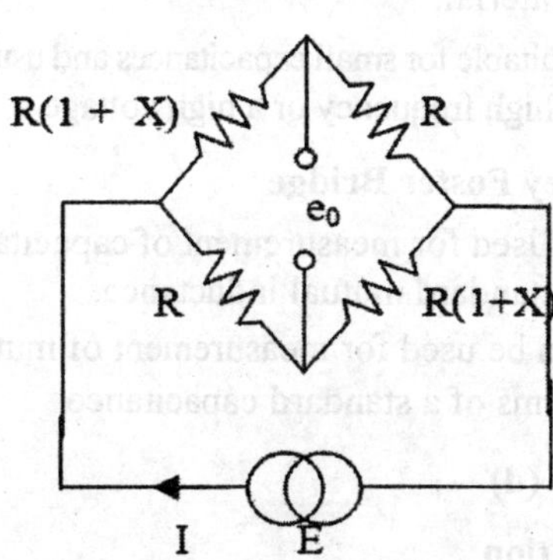

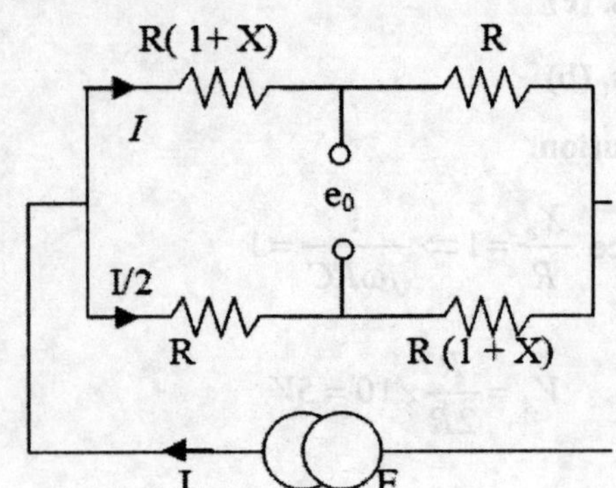

I divides equally as resistance in two parallel branches are same

$$e_0 = \frac{I}{2} \times R(1+X) - \frac{I}{2}R$$

$$e_0 = \frac{I}{2}[R + RX - R] = I\frac{RX}{2}$$

11. Ans. (c)

Solution:

Shering Bridge

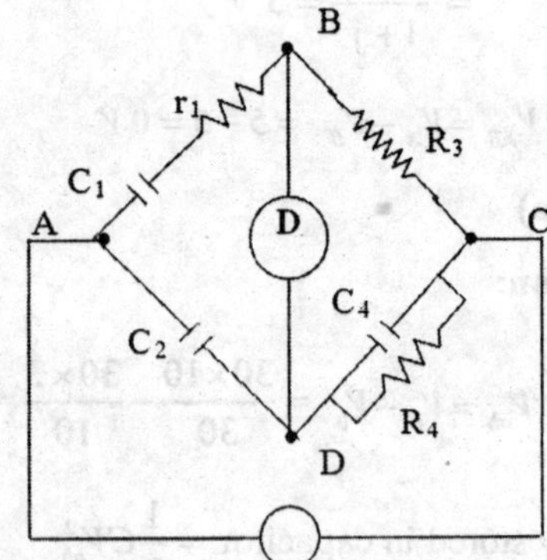

$$r_1 = \frac{R_3 C_4}{C_2} \text{ and } C_1 = C_2\left(\frac{R_4}{R_3}\right).$$

- Used for measurement of the properties of insulators, capacitor bushings insulating oil and other insulating material.
- Suitable for small capacitances and usually supplied from a high frequency or a high voltage source.

Carey Foster Bridge

(a) Used for measurement of capacitance in terms of a standard mutual inductance.

It can be used for measurement of mutual inductance in terms of a standard capacitance

12. Ans. (d)

Solution:

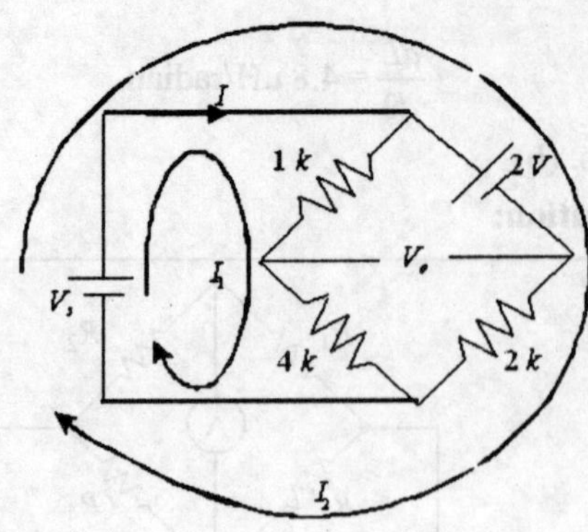

Let us assume $I = I_1 + I_2$

I_1 = inner loop current

I_2 = outer loop current

Apply KVL in these loops

$$-V_s + 5\times10^3 I_1 = 0 \quad \text{...(1)}$$

$$-V_s + 2 + 2\times10^3 I_2 = 0 \quad \text{...(2)}$$

Under blacked condition $V_o = 0$

Voltage across 4 kΩ = voltage across 2 kΩ.

$$4\times10^3 I_1 = 2\times10^3 I_2$$

or $I_2 = 2I_1$ ...(3)

on solving (1), (2) and (3)

$I_1 = 2\,mA$

$I_2 = 4\,mA$

$\therefore\ I = I_1 + I_2 = 2 + 4 = 6\,mA$

13. Ans. (a)

Solution:

Kelvin double bridge is best suited for the measurement of resistances of very low value

14. Ans. (a)

Solution:

Wien bridge is best suited for the measurement of frequency

15. Ans (a)

Solution:

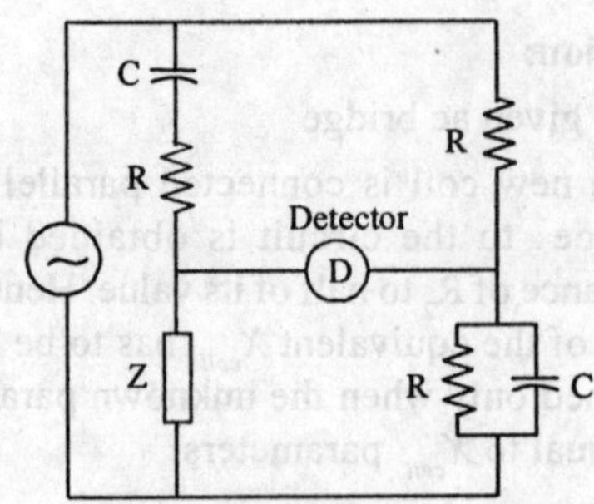

This can be modified as

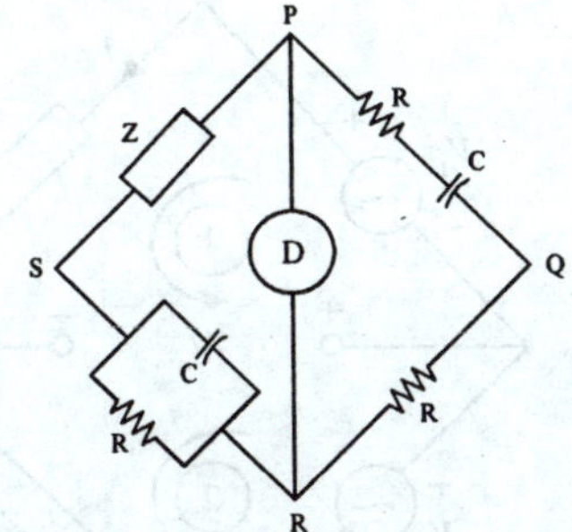

So if bridge is balanced then

$$Z \times R = \left[R + \frac{1}{Cs} \right] \left[\frac{R \times \frac{1}{Cs}}{R + \frac{1}{Cs}} \right]$$

$$Z \times R = \left[\frac{RCs+1}{Cs} \right] \left[\frac{R}{RCs+1} \right]$$

$$Z = \frac{1}{Cs}$$

Means Z is only capacitor

16. Ans. (a)

Solution:

Heaviside Mutual Inductance Bridge:

This bridge measures mutual inductance in terms of a known self-inductance. The same bridge, slightly modified, by Campbell known as Heaviside Campbell's bridge is used to measure a self-inductance in terms of a known mutual inductance.

(a)

(b)

Figure: Positive and negative couplings for mutual inductors

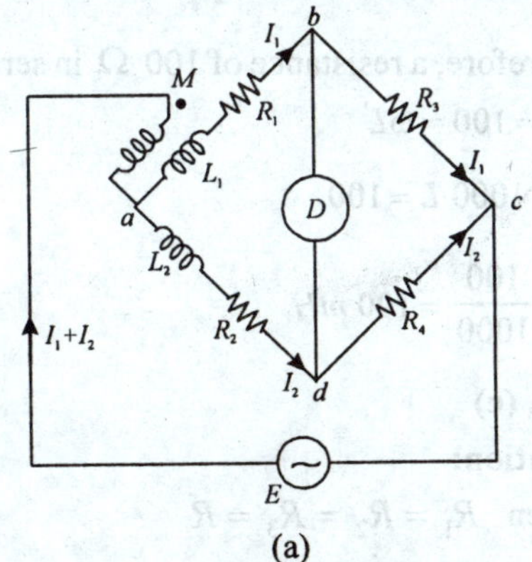

Figure: Heaviside mutual inductance bridge

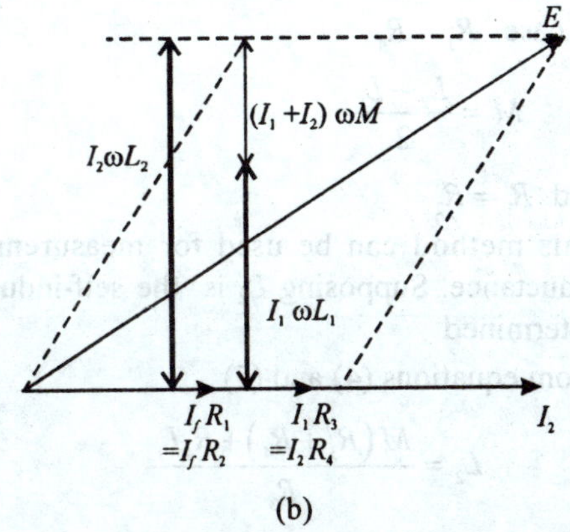

(b)

Let,

M = unknown mutual inductance.

L_1 = self-inductance of secondary of mutual inductance,

L_2 = known self-inductance

and R_1, R_2, R_3, R_4 = non-inductive resistors.

At balance, voltage drop between b and c must be equal to the voltage drop between d and c. Also the voltage drop across a-b-c must equal the voltage drop across a-d-c. Thus following equations are obtained at balance.

$$I_1 R_3 = I_2 R_4 \quad \text{...(1)}$$

and

$$(I_1 + I_2)(j\omega M) + I_1(R_1 + R_3 + j\omega L_1) = I_2(R_2 + R_4 + j\omega L_2) \quad \text{...(2)}$$

from equations (1) and (2)

$$I_2\left(\frac{R_4}{R_3}+1\right) j\omega M + I_2 \frac{R_4}{R_3}(R_1 + R_3 + j\omega L_1) = I_2(R_2 + R_4 + j\omega L_2) \quad \text{...(3)}$$

or

$$j\omega M\left(\frac{R_4}{R_3}+1\right) + \frac{R_4}{R_3}R_1 + R_4 + j\omega L_1 \frac{R_4}{R_3} = R_2 + R_4 + j\omega L_2$$

Thus, $$R_1 = \frac{R_2 R_3}{R_4} \quad \text{...(4)}$$

and $$M = \frac{L_2 - \frac{L_1 R_4}{R_3}}{\frac{R_4}{R_3}+1}$$

$$= \frac{R_3 L_2 - R_4 L_1}{R_3 + R_4} \quad \text{...(5)}$$

It is clear from equation (5) that L_1, the self-inductance of the secondary of the mutual inductor must be known in order that M be measured by this method.

In case $R_3 = R_4$

$$\Rightarrow \quad M = \frac{L_2 - L_1}{2} \quad \text{...(6)}$$

and $R_1 = R_2$ ...(7)

This method can be used for measurement of self-inductance. Supposing L_2 is the self-inductance to be determined

From equations (4) and (5)

$$L_2 = \frac{M(R_3 + R_4) + R_4 L_1}{R_3}$$

$$= M\left(1 + \frac{R_4}{R_3}\right) + \frac{R_4}{R_3} L_1$$

and $\quad R_2 = \dfrac{R_1 R_4}{R_3}$

In case $R_4 = R_3$, we have

$$L_2 = 2M + L_1$$

and $\quad R_2 = R_1$

17. Ans. (c)

Solution:

The given bridge is Maxwell inductance capacitance bridge. Therefore, the unknown element is R and C in parallel

18. Ans. (d)

Solution: The correct match is given below

Low value of R	Kelvin double bridge
High value of Q	Hay bridge
Inductance L	Maxwell-Wien bridge
Capacitance C	Schering bridge

19. Ans. (b)

Solution: Correct match is given below

Self inductance	Anderson
Capacitance	Schering
Mutual inductance	Campbell
Frequency	Wien – Robinson

20. Ans. (c)

21. Ans. (b)

Solution:

Shorting the voltage source, we have

$$R_{12} = [20 \parallel 30] + [25 \parallel 25] k\Omega$$

$$= 24.5\ k\Omega$$

22. Ans. (a)

Solution:

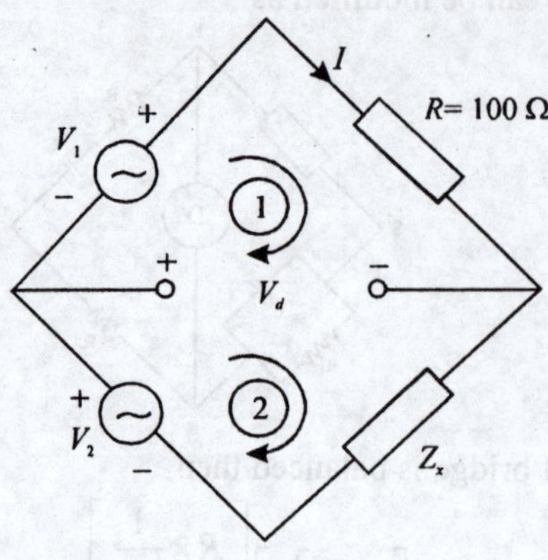

Given

$$V_1 = \sqrt{2}\cos 1000t$$

$$V_2 = 2\cos(1000t + 45^o)$$

Here, $\omega = 1000$

Apply KVL in loop 1

$$-V_1 + IR - V_d = 0$$

or, $V_1 = IR$

or, $I = \dfrac{V_1}{R}$

$$I = \frac{V_1}{R} = \frac{\sqrt{2}}{100}\cos 1000t \quad \text{...(1)}$$

Apply KVL in loop 2,

$$IZ_x - V_2 + V_d = 0$$

Or, $V_2 = I\, Z_x$

$$2\cos\ (1000t + 45^o) = \frac{\sqrt{2}}{100}(\cos 1000t) \times Z_x$$

LHS = RHS, If Z_x provide 45° of phase and magnitude of $100\sqrt{2}$

That is $Z_x = 100 + 100j$

Therefore, a resistance of 100 Ω in series with inductor

$$X_L = 100 = \omega L$$

or $\quad 1000\ L = 100$

$$L = \frac{100}{1000} = 100\ mH$$

23. Ans. (c)

Solution:

Given $R_1 = R_2 = R_3 = R$

$$R_4 = R + \Delta R$$

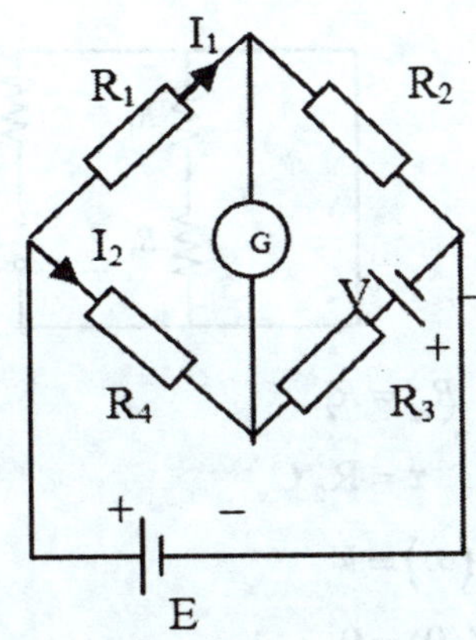

When balanced, current through galvanometer G would be zero

and $$I_1R_1 = I_2R_4 \quad ...(1)$$

$$I_1R_2 = I_2R_3 + V \quad ...(2)$$

$$E = (R_1 + R_2)I_1 \quad ...(3)$$

From equations (1) and (2)

$$I_2 = I_1\frac{R_1}{R_4}$$

$$I_1R_2 = \frac{I_1R_1}{R_4}\cdot R_3 + V$$

$$I_1R_2 - I_1\frac{R_1R_3}{R_4} = V$$

$$I_1\left(R - \frac{R^2}{R+\Delta R}\right) = V$$

$$I_1 = \frac{V(R+\Delta R)}{R\,\Delta\,R} \quad ...(4)$$

From equations (3) and (4)

$$E = \frac{(R_1+R_2)(R+\Delta\,R)}{R\,\Delta\,R}$$

$$= \frac{2RV\,(R+\Delta\,R)}{R\,\Delta\,R}$$

$$E\Delta R = 2VR + 2V\Delta R$$

$$E\Delta R - 2V\Delta R = 2VR$$

$$\Delta R = \frac{2VR}{E-2V}$$

24. Ans. (b)

Solution: When bridge is balance then,

$$Z_1Z_4 = Z_2Z_3$$

$$Z_4 = \frac{Z_2Z_3}{Z_1}$$

$$\text{When } Z_1 = \frac{\frac{1}{C_1s}\times R_1}{\frac{1}{C_1s+R_1}} = \frac{R_1}{R_1C_1s+1}$$

$$\text{Then, } Z_4 = Z = \frac{R_2R_3}{R_1}(R_1C_1\,s+1)$$

$$= \frac{1000\times 500}{10\times 10^3}\left(10^4\times 100\times 10^{-9}\times j314+1\right)$$

$$= \left(50 + j\,314\times 50\times 10^{-3}\right)$$

$$= 50\,\Omega \text{ series with 50 mH}$$

25. Ans. (d)

Solution:

Given circuit is:

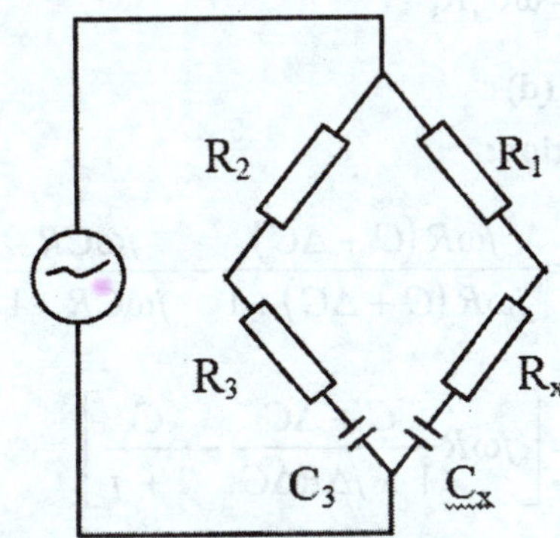

For bridge balancing condition

$$R_2\left[R_x + \frac{1}{j\omega C_x}\right] = R_1\left[R_3 + \frac{1}{j\omega C_3}\right]$$

$$R_2R_x + \frac{R_2}{j\omega C_x} = R_1R_3 + \frac{R_1}{j\omega C_3}$$

$$\frac{R_2R_xj\omega C_x + R_2}{j\omega C_x} = \frac{R_1R_3j\omega C_3 + R_1}{j\omega C_3}$$

$$j\omega R_2R_xC_xC_3 + R_2C_3 = j\omega R_1R_3C_3C_x + R_1C_x$$

Equating the real and the imaginary parts

$$R_2C_3 = R_1C_x \quad ...(1)$$

$$R_2R_xC_xC_3 = R_1R_3C_3C_x \quad ...(2)$$

From equation (1)

$$C_x = \frac{R_2 C_3}{R_1} \quad \text{... (3)}$$

From equations (2) and (3)

$$R_x = \frac{R_1 R_3}{R_2}$$

Dissipation factor

$$D = \frac{1}{Q}$$

$$= \frac{1}{\frac{1}{\omega C_x R_x}}$$

$$= \omega C_x R_x$$

$$= \omega \frac{C_3 R_2}{R_1} \times \frac{R_1 R_3}{R_2}$$

$$= \omega C_3 R_3$$

26. Ans. (d)

Solution:

$$\frac{V_o}{V_s} = \frac{j\omega R\left(C+\Delta C\right)}{j\omega R\left(C+\Delta C\right)+1} - \frac{j\omega CR}{j\omega CR+1}$$

$$\frac{V_o}{V_s} = \left[j\omega R\frac{C+\Delta C}{1+j\Delta R\Delta C} - \frac{C}{1+j}\right]$$

$$= j\omega R\left[\frac{C+jC+\Delta C+\Delta Cj-C-Cj-j\omega R\Delta C.C}{(1+j)\,(1+j+j\omega R\Delta C)}\right]$$

$$= \frac{\Delta C}{2C}$$

27. Ans. (a)

Solution:

PMMC is zero then modified circuit is as below

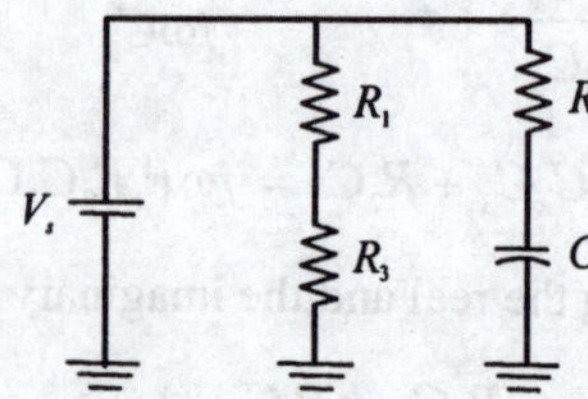

Time constant $\tau = R_{eq}\,C$

Net resistance across C is obtained as

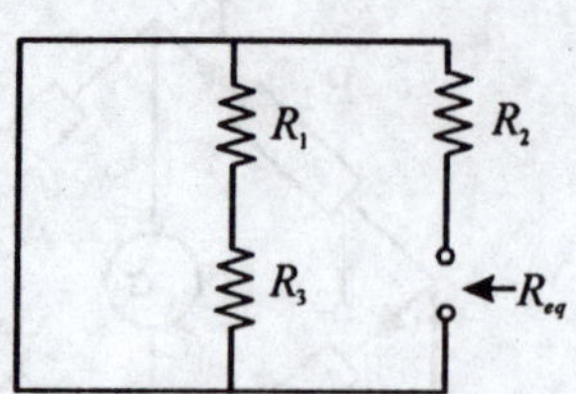

$$R_{eq} = R_2$$

$$\therefore \quad \tau = R_2\,C_x$$

Now, $V(\infty) = V_s$

$$V(0) = 0$$

So, charging equation for capacitor will be

$$V_c(t) = V(\infty) - \left[V(\infty) - V(0)\right]e^{-t/\tau}$$

$$= V_s - \left[V_s - 0\right]e^{-tR_2C_x}$$

$$= V_s\left[1 - e^{-tR_2C_x}\right]$$

As time increases, voltage across capacitor increases. So at $t = T/2$, the maximum voltage during charging is

$$V_c = V_s\left[1 - e^{-T/2R_2C_x}\right]$$

Because $T/2 << R_2C_x$

So $\frac{T}{2R_2C_x} << 1$

Hence $e^{-x} = 1 - x$ to take linear approximation

Where, $x = \frac{T}{2R_2C_x}$

Now, $V_c = V_s\left[1 - \left(1 - \frac{T}{2R_2C_x}\right)\right]$

$$= \frac{V_s T}{2R_2C_x}$$

During discharging, $R_4C_x << T/2$,

Therefore, it discharges rapidly at position 2.

28. Ans. (b)

Solution:

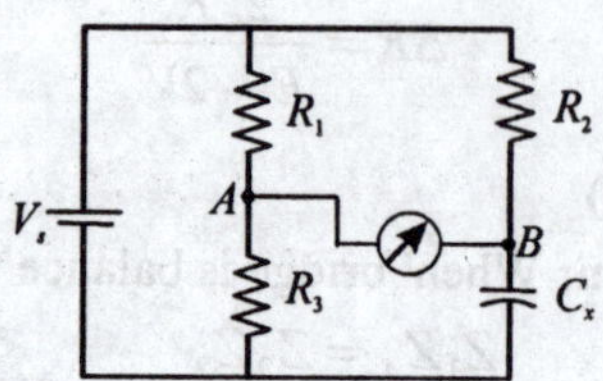

Figure: Bridge circuit when switch will be as (at position 1) voltage at B at $t = T/2$ is

$$V_B = V_C = \frac{V_s T}{2R_2 C_x} \qquad \text{...(1)}$$

And voltage at A is $V_A = \frac{R_3}{R_1 R_3} \times V_s$ ...(2)

For null deflection both V_A and V_B must be equal

$$\therefore \frac{R_3}{R_1 + R_3} V_s$$

$$= \frac{V_s T}{2R_2 C_x}$$

Solving above equation

$$C_x = \frac{T(R_1 + R_3)}{2R_2 R_3}$$

29. Ans (d)

Solution:

$$V_0 = \frac{(A\sin\omega t)R}{2R + \frac{1}{j\omega C}} - \frac{(A\sin\omega t)R}{2R + j\omega L}$$

$$= \frac{(A\sin\omega t)j\omega CR}{2Rj\omega C + 1} - \frac{(A\sin\omega t)R}{2R + j\omega L}$$

$$= \frac{\left[2R^2 j\omega C - R\omega^2 LC - 2R^2 j\omega C - R\right]A\sin\omega t}{(2R + j\omega L)(2Rj\omega C + 1)}$$

$$V_o = \frac{-(R + R\omega^2 LC)A\sin\omega t}{(2R + j\omega L)(2R\omega C + 1)}$$

As required $V_0 = 0$

$$R + R\omega^2 LC = 0$$

or $R(1 + \omega^2 LC) = 0$

$$\omega = \frac{-1}{LC}$$

$$\omega = \frac{\sqrt{-1}}{LC} = \frac{j}{LC}$$

Frequency cannot be imaginary. Hence $V_o \neq 0$ r

Second Method

When bridge is balanced then $V_o = 0$

$$R\left(R + \frac{1}{j\omega C}\right) = R(R + j\omega L)$$

$$R\left(R - \frac{j}{\omega C}\right) = R(R + j\omega L)$$

$$R^2 - \frac{jR}{\omega C} R^2 + j\omega RL$$

Comparing the real and imaginary parts, $-\frac{R}{\omega C} = \omega RL$

$$\omega^2 = -\frac{1}{LC} = \frac{\sqrt{-1}}{LC} = \frac{j}{\sqrt{LC}}$$

Therefore, frequency is imaginary which is impossible.

Hence, $V_o \neq 0$

30. Ans. (b)

Solution:

Sensitivity is independent of supply voltage

31. Ans (b)

Solution:

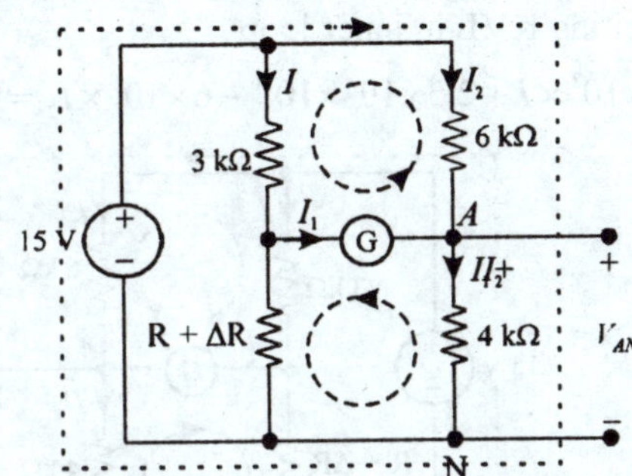

As given galvanometer current senstivity and scale resolution = 1*mm*

Hence bridge will only detect when current through galvanometer is more than $1\mu A$

Hence $I_1 = 1\mu A = 10^{-6} A.$

Now KVL in outer Loop

$$-15 + 6\times10^3 I^2 + 4\times10^3(I_1 + I_2) = 0$$

or, $15 - 4\times10^3 I_1 = 10^4 I_2$

or, $15 - 0.004 = 10^4 I_2$

or, $14.996 = 10^4 I_2$

$$I_2 = \frac{1.499}{10^4}$$

$$= 14.99\times10^{-3}$$

Hence $V_{AN} = 4\times10^3(I_1 + I_2)$

$$= 4\times10^3(10^{-6} + 1.4996\times10^{-3})$$

$$0.004 + 5.9984 = 6.0024V$$

32. Ans (b)

Solution:

KVL in outer loop

$$-15 + 6\times10^3 I^2 + 4\times10^3(I_1 + I_2) = 0$$

or, $15 - 4\times10^3 I_1 = 10^4 I_2$

or, $15 - 0.004 = 10^4 I_2$

or, $14.996 = 10^4 I_2$

$$I_2 = \frac{1.499}{10^4}$$

$$= 14.99\times10^{-3}$$

Hence $V_{AN} = 4\times10^3(I_1 + I_2)$

$$= 4\times10^3(10^{-6} + 1.4996\times10^{-3})$$

$$0.004 + 5.9984 = 6.0024V$$

$$I_1 = 10^{-6} A$$

$$I_2 = 1.499\times10^{-3} A$$

Writing KVL in inner loop

$$3\times10^3\times I + 2.5\times10^3\times10^{-6} - 6\times10^3\times I_2 = 0$$

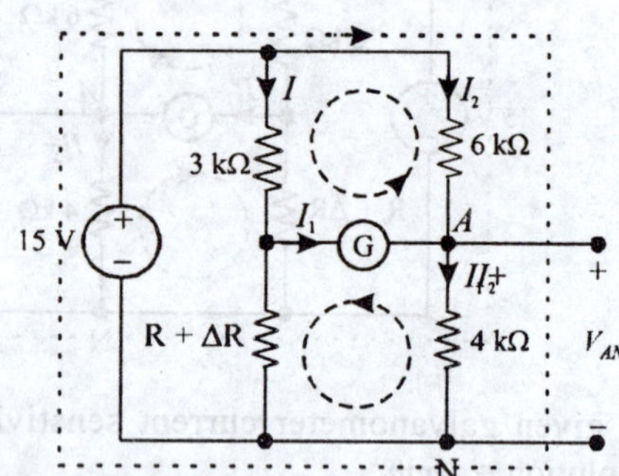

or $\left[3\times10^3 I + 0.0025 - 6\times10^3\times1.499\times10^{-3}\right]$

or $3\times10^3 I = 0.025 + 8.9976$

Writing KVL in another inner loop we get,

$$(R+\Delta R)\times(I - I_1) - 0.0025 - 6.0024 = 0$$

$$(R+\Delta R) = \frac{6.0049}{(2.9983 - 0.001)\times10^3}$$

$$R + \Delta R = 2003.436$$

$$\Delta R = 3.436\Omega$$

33. Ans. (a)

Solution:

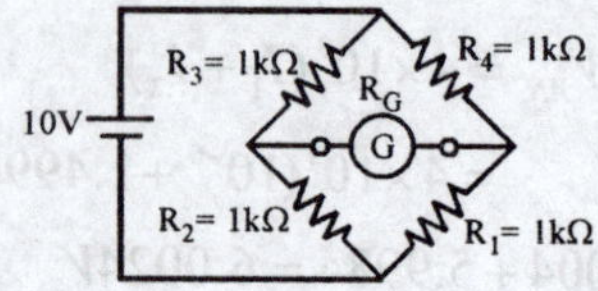

R_1 increases by 1 ohm.

$R_1 = 1000 + 1 = 1001$ ohms.

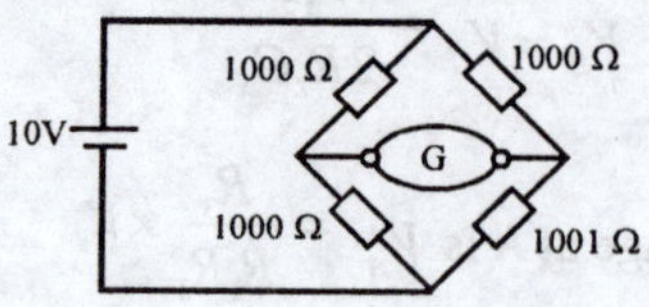

Find V_{TH}

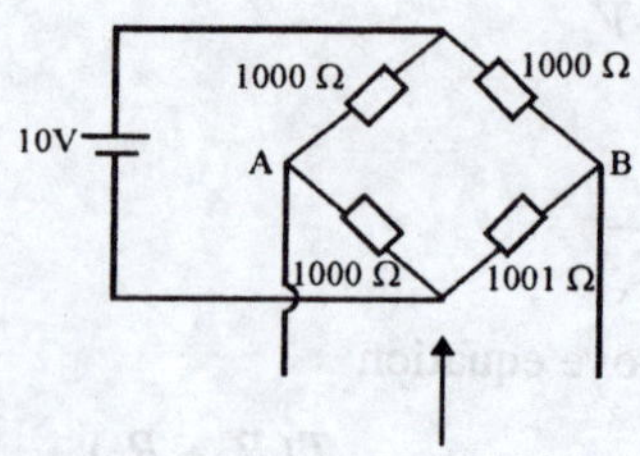

$$V_A = 5V$$

$$V_B = 5.002498$$

$$V_{TH} = V_B - V_A$$

$$V_{TH} = 0.0025$$

$R_{TH} \Rightarrow$

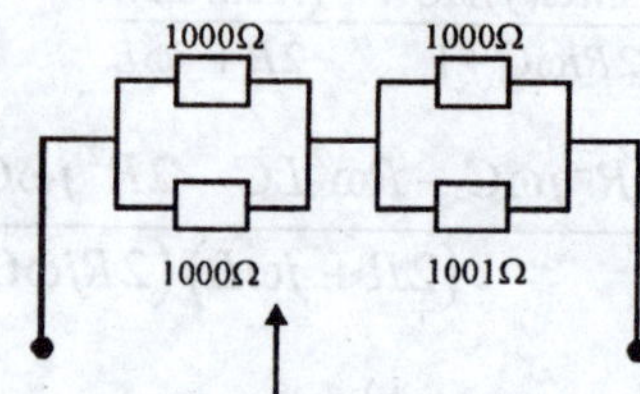

$R_{TH} = R_{TH} = 500 + 500.249$

$R_{TH} = R_{TH} = 1000.2498$

$$I = \frac{.002498}{1000 + 1000.2498} \Rightarrow 1.25\times10^{-6}$$

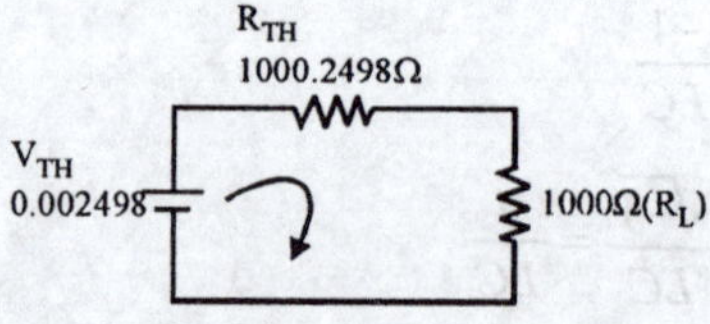

$$I = \frac{0.002498}{1000 + 1000.2498}$$

$$= 1.25\times10^{-6}$$

$$I = 1.2\ \mu A$$

3.6 Extension of Instrument Ranges

1. **Ans. (c)**

Solution

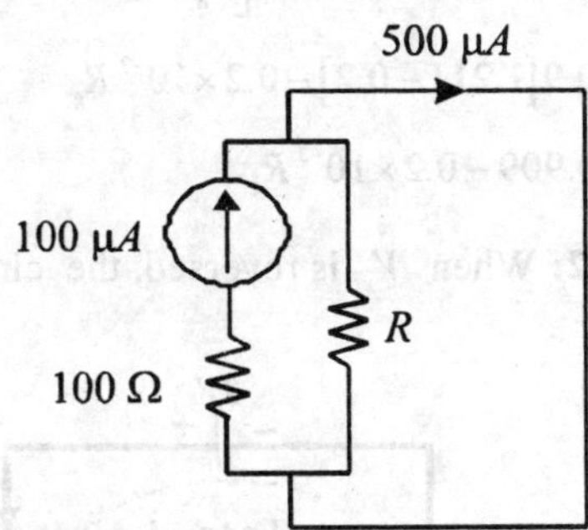

Let required shunt resistance be R.
Then by current division

$$100 = \left(\frac{R}{R+100}\right) \times 500$$

$$R = 25\,\Omega$$

Second Method

$R_m = 100\Omega \quad R_{sh} = ?$

Multiplying factor (m)

$$m = \frac{I_{fsd}}{I_m} = \frac{500}{100} \Rightarrow 5$$

$$\therefore R_{sh} = \frac{R_m}{m-1} \Rightarrow \frac{100}{4} \Rightarrow 25\Omega$$

2. **Ans. (c)**

Solution: $I_m = 100$ A, $R_m = 0.1\Omega$

$$I_m = \frac{V_m}{R_m} = \frac{V_m}{0.1} = 100, \Rightarrow V_m = 10\text{ V}$$

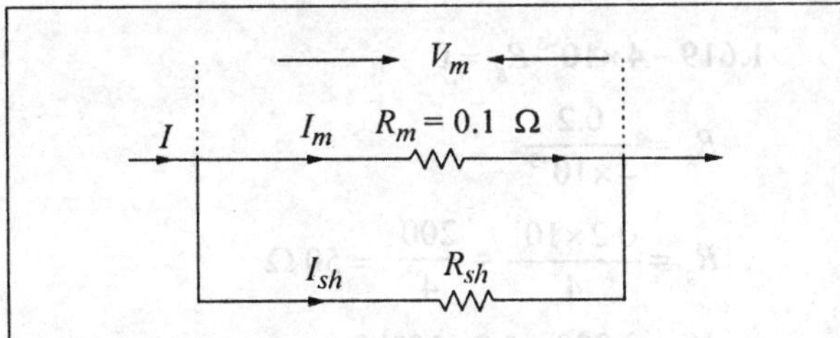

Meter current remain same, $V_m = \left(\frac{R_m . R_{sh}}{R_m + R_s}\right) I$

Now, $10\text{ V} = 500 \times \frac{0.1R_{sh}}{0.1 + R_{sh}}$

$\Rightarrow \quad R_{sh} = 0.025\ \Omega$

Second Method

Multiplying factor

$$m = \frac{I_{fsd}}{I_m} = \frac{500}{100} = 5$$

$$R_{sh} = \frac{R_m}{m-1}$$

$$= \frac{0.1}{5-1} = \frac{0.1}{4}$$

$$= 0.025\ \Omega$$

3.7 Potentiometers

1. **Ans. (c)**

Solution: Error in the potentiometer is due to thermo electric emf' s set up at junctions of dissimilar metals and also by the heat from the operator's hand during adjustment of the working parts of the potentiometer. Reversing of polarity and taking reading will give error in the negative direction. Suppose the reading with error due to one direction is D + DR and in the other direction is D – DR, then the means value.

$$= \frac{\text{D} + \text{DR} + \text{D} - \text{DR}}{2}$$

= D, which is the actual value

2. **Ans. (b)**

3. **Ans. (b)**

Solution:

Standard cell e.m.f = 1.18 V

And balanced at 600 mm

Working current $I_w = \frac{1.16\,V}{600 \times 10^{-3}}$

$= 1.967 \times 10^{-3}\ A$

Test cell balanced at 680 mm

Test cell voltage (V) = $680 \times I_w$

$= 680 \times 1.96 \times 10^{-3}$

$= 1.34$ V

4. **Ans. (b)**

5. **Ans. (a)**

Solution:

In balanced condition, current through G (Galvanometer) is zero, therefore voltage across 200 Ω resistance

$$V = \frac{200}{2800 + 200 + 200} \times 3.2$$

$$= \frac{200 \times 3.2}{3200} = 200 \text{ mV}$$

6. Ans. (b)

Solution:

Given total resistance $R = 10000\ \Omega$

Power Rating $I^2R = 4$ Watt

$$I^2 = \frac{4}{10000}$$

Or, $I = 20\ mA$

$$\text{Sensitivity} = \frac{V}{mm} = \frac{IR}{mm}$$

$$= \frac{20 \times 10^{-3} \times 1000}{100} = 2\ V/mm$$

7. Ans. (b)

Solution: $P = I^2R$

$$0.02 = I^2 \times 4\ k$$

$$I^2 = \frac{0.02}{4 \times 10^3} = \frac{2}{4 \times 10^5} = \frac{1}{20} \times 10^{-4}$$

or $I = \frac{10^{-2}}{\sqrt{20}}$

and $I = \frac{15}{R_s + R_p} = \frac{10^{-2}}{\sqrt{20}} = \frac{10^{-2}}{4.47}$

$\Rightarrow R_s + R_p = 6.71\ k\Omega$

$R_s = 6.71 - 4$

$\Rightarrow R_s = 2.71\ k\Omega$

8. Ans. (d)

Solution:

Case 1:

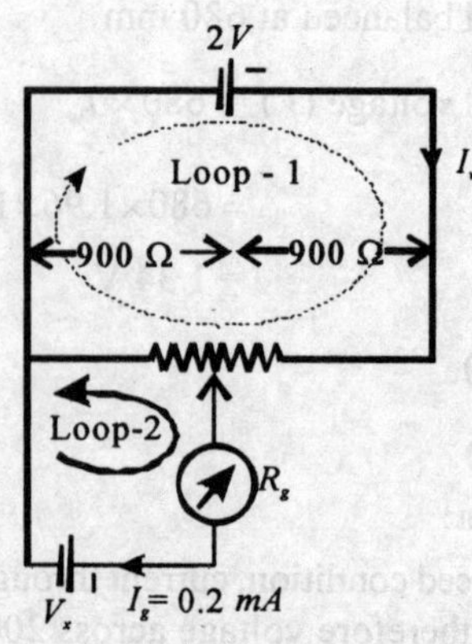

Apply KVL in loop 1:

$$-2V + 900I_w + 900(I_w - 0.2 \times 10^{-3}) = 0$$

or $I_w = 1.211\ mA$

Apply KVL in loop 2:

$$-V_x - 0.2 \times 10^{-3} R_g + 900[I_w - 0.2 \times 10^{-3}] = 0$$

$$V_x = 0.9[1.211 - 0.2] - 0.2 \times 10^{-3} R_g$$

$$V_x = 0.909 - 0.2 \times 10^{-3} R_g \qquad ...(1)$$

Case 2: When V_x is reversed, the circuit is

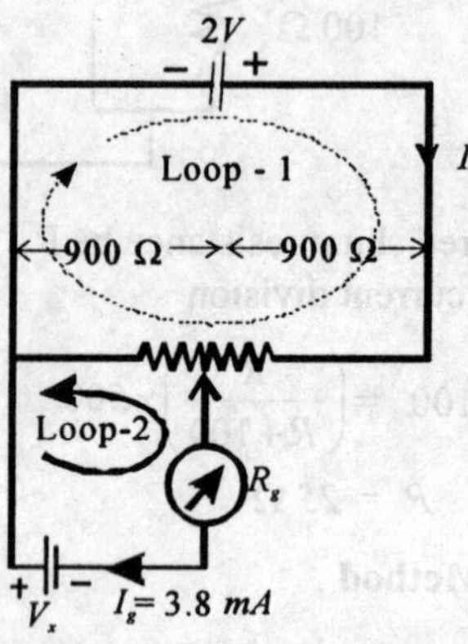

Apply KVL in loop 1:

$$-2 + 900\ I_w + 900[I_w - 3.8 \times 10^{-3}] = 0$$

or, $I_w = 3.011\ mA$

Apply KVL in loop 2:

$$+V_x + 900[3.011 \times 10^{-3} - 3.8 \times 10^{-3}] - 3.8 \times 10^{-3} R_g = 0$$

$$-V_x = -0.710 - 3.8 \times 10^{-3} R_g \qquad ...(2)$$

On solving (1) and (2), one gets

$$V_x = 0.909 - 0.2 \times 10^{-3}\ R_g$$

$$-V_x = -0.7101 - 3.8 \times 10^{-3}\ R_g$$

$$1.619 - 4 \times 10^{-3} R_g = 0$$

$$R_g = \frac{0.2}{4 \times 10^{-3}}$$

$$R_g = \frac{0.2 \times 10^3}{4} = \frac{200}{4} = 50\ \Omega$$

$$V_x = 0.909 - 0.2 \times 10^{-3} R_g$$

$$= 0.909 - 0.2 \times 10^{-3} \times 50$$

$$= 0.909$$

$$= 0.909 - 10 \times 10^{-3}$$

$$= 0.909 - 0.001$$

$$= 0.908\ V$$

9. Ans. (a)

Solution:

Given resistance = 1 Ω/m

For 11 m → 11 Ω

For 10 m + 18 cm→ 10.18 Ω

From circuit

$1.018 = I_w\,(10.18)$

$\Rightarrow \quad I_w = 0.1\text{ A}$

and $I_w = \dfrac{2}{11 + R_h}$

$$R_h = \frac{2}{0.1} - 1 = 9\,\Omega$$

10. Ans. (d)

Solution:

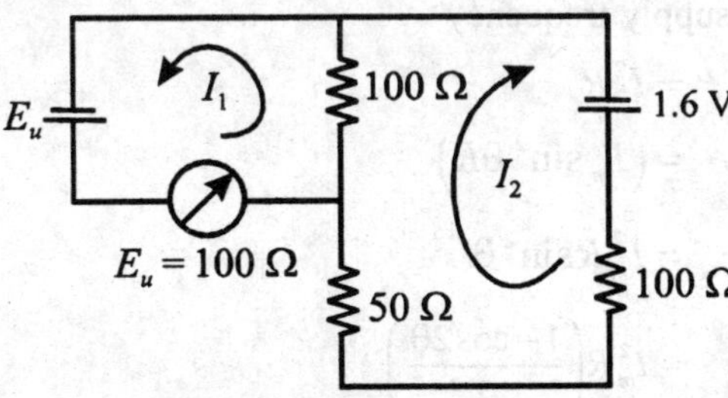

$$E_u = 100\,I_1 + 100(I_1 + I_2)$$

and $1.6 = (100 + 500)I_2 + 1000(I_2 + I_1)$

Putting value of $I_1 = 10\times10^{-6}\,A$ equations (1) and (2)

$$E_u = 100\times10\times10^{-6} + 100(10\times10^{-6} + I_2)$$

$$E_u = 1\times10^{-3} + 1000I_2 + 0.01$$

$$= 0.001 + 1000I_2 + 0.01$$

$E_u = 0.011 + 1000I_2$ and

$$1.6 = 1600I_2 + (1000\times10\times10^{-6})$$

$$1.6 = 1600I_2 + 0.01$$

Solving equations (3) and (4)

$$E_u = 0.011 + 100I_2 \times 1.6$$

$$1.6 = 0.011 + 1600I_2$$

$$1.6\,Eu = 0.0176 + 1600I_2$$

$$\underline{1.6\,Eu = 0.0176 + 1600I_2}$$

$$1.6\,Eu - 1.6 = 0.0176 - 0.01$$

$$1.6\,Eu = 0.0176 + 1.6$$

$$\frac{Eu = 0.0176 - 0.01 + 1.6 = 1.00475}{1.6} \qquad say = 1.0048$$

11. Ans. (a)

Solution

Since R_L and R_C are in parallel

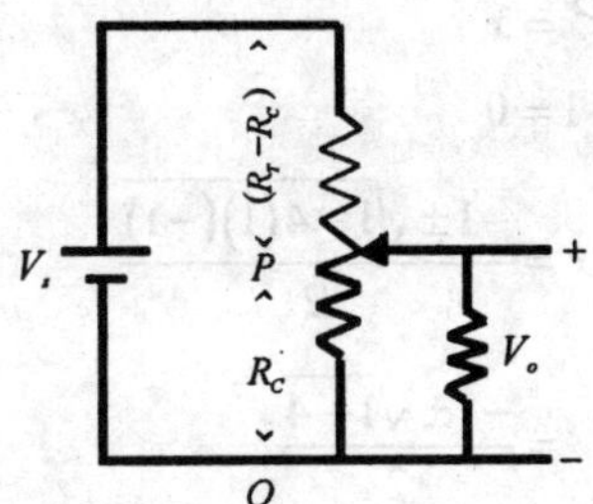

Apply potential divider rule

$$V_o = \frac{\dfrac{R_C R_L}{R_C + R_L}}{\dfrac{R_C R_L}{R_C + R_L} + (R_T - R_C)} V_s$$

$$\frac{V_o}{V_s} = \frac{1}{1 + \dfrac{(R_C + R_L)(R_T - R_C)}{R_C R_L}}$$

As given at this point $\dfrac{V_o}{V_s} = 0.5 = \dfrac{1}{2}$

Hence

$$\frac{1}{2} = \frac{1}{1 + \dfrac{(R_C + R_L)(R_T - R_C)}{R_C R_L}}$$

$$1 + \frac{(R_C + R_L)(R_T - R_C)}{R_C R_L} = 2$$

$$\frac{(R_C + R_L)(R_T - R_C)}{R_C R_L} = 1$$

As given

$\therefore\ R_T = R_L$

Given

Therefore, $(R_T + R_C)(R_T - R_C) = R_C R_T$

$$\frac{R_L}{R_T} = 1$$

$$R_T^2 - R_C^2 = R_C R_T$$

$$\frac{R_T^2}{R_T R_C} - \frac{R_C^2}{R_T R_C} = 1$$

$$\frac{R_T}{R_C}-\frac{R_C}{R_T}=1$$

Let $\frac{R_C}{R_T}=x$, then

$$\frac{1}{x}-x=1$$

$$1-x^2=x$$

$$x^2+x-1=0$$

$$x=\frac{-1\pm\sqrt{1-4(1)(-1)}}{2}$$

$$=\frac{-1\pm\sqrt{1+4}}{2}$$

$$=\frac{-1\pm\sqrt{5}}{2}$$

As resistance value cannot be negative
Therefore, neglecting the negative value

$$\frac{R_C}{R_T}=\frac{-1+\sqrt{5}}{2}$$

12. Ans (b)

Solution:

When the switch 's' is open,

$$I_W=10mA$$

$$R_g+R_1=100\,\Omega$$

$$I_W=\frac{3}{R_{pot}+250}$$

$$(10\times10^{-3})(R_{pot}+250)=3$$

$$R_{pot}+250=300$$

Therefore, $R_{pot}=50$

Length of pot = 5m

$$5\,m=50\Omega$$

therefore $1m=\frac{50}{5}\Omega$

$$2m=\frac{50}{5}\times2$$

$$=20\Omega$$

$$V_{2m}=(20\Omega)\times(10mA)$$

$$=200mV$$

But the errors in the reading are encountered due to the presence of extra resistance due to the galvanometer.

But the galvanometer can only detect current greater than 10 μA.

Now the error is nothing but voltage drop across R_g+R_1

$$I_g\left(R_g+R_1\right)=10\times10^{-6}(100)=1\,mV$$

$$\text{Percentage error}=\frac{1\times10^{-2}\times100}{10.2}=0.5$$

3.8 Measurement of Power and Energy

1. Ans. (c)

Solution:

The instantaneous power in a.c. circuit is double the supply frequency

$$P=I^2R$$

$$=\left(I_m\sin^2\theta R\right)$$

$$=I_m^2R\sin^2\theta$$

$$=I_m^2R\left(\frac{1-\cos2\theta}{2}\right)$$

$$\text{P}\Rightarrow(1-\cos2\omega t)/2$$

i.e. $50\times2=100$ Hz

2. Ans. (b)

3. Ans. (a)

4. Ans. (c)

5. Ans. (b)

6. Ans. (c)

Solution: In two watt meter method

$W_1=\sqrt{3}\,V_P\,I_P\cos(30+\theta)$

$W_2=\sqrt{3}\,V_P\,I_P\cos(30+\theta)$

$\theta=60°$

Hence

$W_1=\sqrt{3}\,V_P\,I_P\cos30=0$

$W_2=\sqrt{3}\,[V_P^2/R]\cos(30-60)$

$=1000\,W$

Second Method

Given

$$V_L=100V$$

$$V_{ph}=\frac{100}{\sqrt{3}}$$

In star connected load

$$I_{ph} = I_L = \frac{V_{ph} \angle 0}{Z \angle \phi}$$

$$= \frac{100/\sqrt{3}}{5\angle -60^\circ} = 11.547\angle -60^\circ$$

Wattmeter W_1 Reading is $V_L I_L \cos(30^\circ - \theta)$

$$= 100 \times 11.547 \cos 90^\circ$$
$$= 0\ W$$

Wattmeter W_2 reading is $V_L I_L \cos(30^\circ + \theta)$

$$= 100 \times 11.547 \cos 30^\circ$$
$$= 1000\ W$$

7. Ans. (c)

Solution:

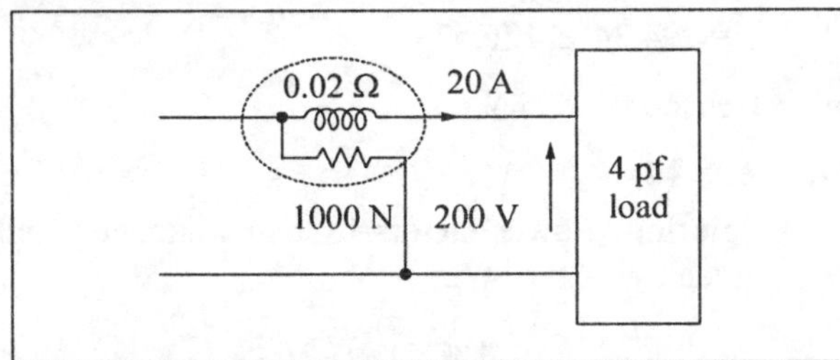

Power indicated by wattmeter,

= Power consumed by load + Power loss in current coil

$= VI \cos\phi + I^2 R$

$= 200 \times 20 + 20^2 \times 0.02 = 4008$ W

Load power = 4000 W

Then % increase or error $= \frac{4008 - 4000}{4000} \times 100$

$= 0.2\%$ more

8. Ans. (a)

Solution:

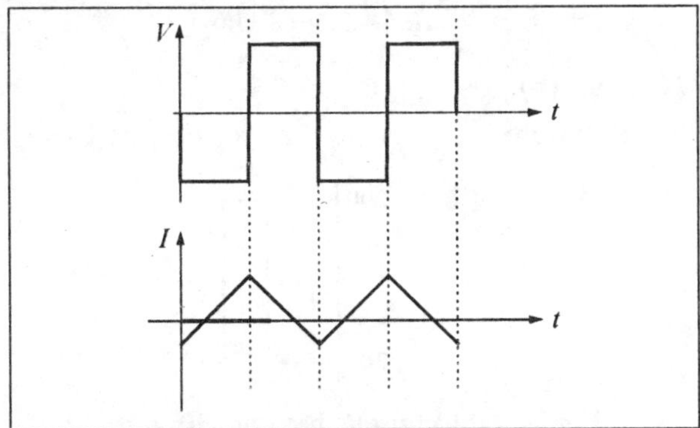

If we consider both the waves, we can see positive power and negative power in each cycle is equal which gets cancelled

Therefore, Net power = 0

9. Ans (d)

Solution:

Circuit can be redrawn as

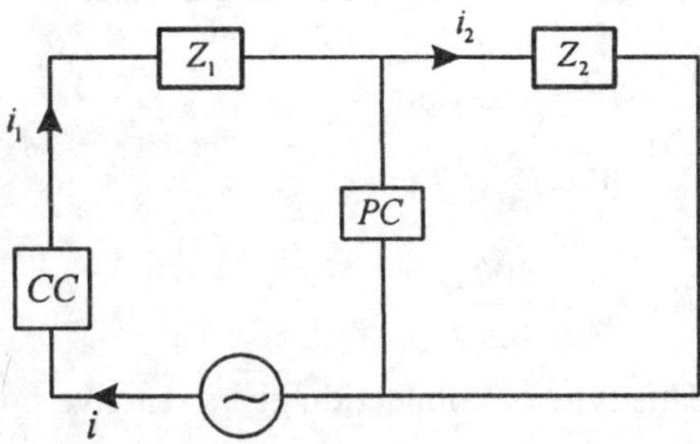

As we can see from the diagram

Current coil measures the total current i.

and $i = i_1 = i_2$ as both Z_1 and Z_2 are in series with each other and pressure coil measures the potential across only Z_2 so $PC = V_2$ (voltage drop across Z_2)

So watt meter reading $= i_1 V_2$

= Power consumed by Z_2

10. Ans. (c)

Solution:

Given

$$\upsilon(t) = E_1 \sin \omega t + E_3 \sin 3\omega t$$

$$i(t) = I_1 \sin(\omega t - \phi_1) + I_3 \sin(3\omega t - \phi_3) + I_5 \sin 5\omega t$$

Average power

$$P_{avg} = \frac{1}{2\pi} \int_0^{2\pi} \upsilon(t)\, i(t)\, d(\omega t)$$

$$= \frac{1}{2\pi} \int_0^{2\pi} \left[\left\{ \begin{array}{l} (E_1 \sin \omega t + E_3 \sin 3\omega t)\{I_1 \sin(\omega t - \phi_1)\} \\ + I_3 \sin(3\omega t - \phi_3) + I_5 \sin 5\omega t \end{array} \right\} d(\omega t) \right]$$

$$= \frac{1}{2} E_1 I_1 \cos\phi_1 + \frac{1}{2} E_3 I_3 \cos\phi_3$$

As the product of different frequency terms have zero average vaue.

11. Ans. (b)

Solution:

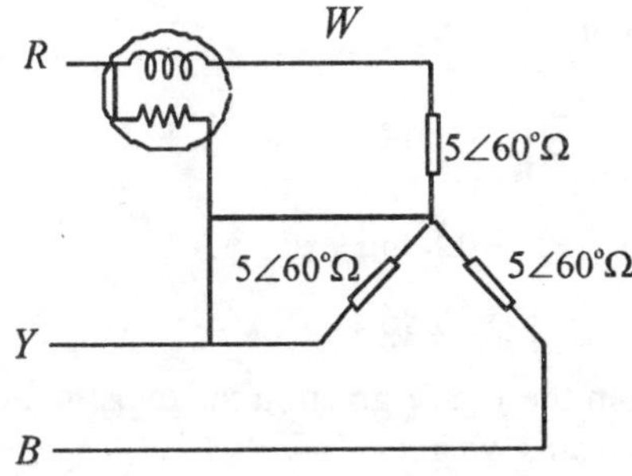

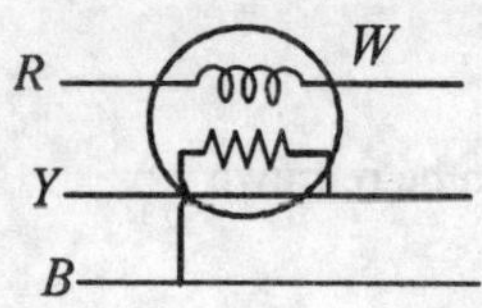

$W = 400$ watt

$W = V_{ph}\, I_{ph} \cos\phi$

$V_{ph}\, I_{ph} = \dfrac{400}{0.8}$

This type of connection gives reactive power

$W = \sqrt{3}\, V_p\, I_p \sin\phi$

$= \sqrt{3} \times \dfrac{400}{0.8} \times 0.6$

$= 519.6$

12. Ans. (c)

Solution:

Given V = 220 V, I = 5 A

At $\Delta = 85°$, $\phi = 60°$ (pf = 0.5)

Energy registered = $VI \sin(\Delta - \phi)$

Then error = $VI \sin(\Delta - \phi) - VI\cos\phi$

$= 200 \times 5\,(\sin 85° - \cos 60°)$

$= -85.1$ W

At $\Delta = 85°, \phi = 0°$ (pf = 0°)

error = $VI \sin(\Delta - \phi) - VI\cos\phi$

$= VI(\sin 85° - \cos 0°)$

$= -4.2$ W

13. Ans. (c)

Solution: Meter constant is 14.4 A sec per rev.

i.e., $\dfrac{14.4}{3600}\text{Ahr} = \dfrac{14.4 \times 250}{3600}$

Where = 1 Whr per rev. Hence 1rev/Whr = 1000 rev/kWhr.

14. Ans. (b)

Solution:

Since PF = 0.8 log = cos ϕ

$\Rightarrow \quad \phi = 36.86$

From the figure given in the question, draw the phaser diagram. We get

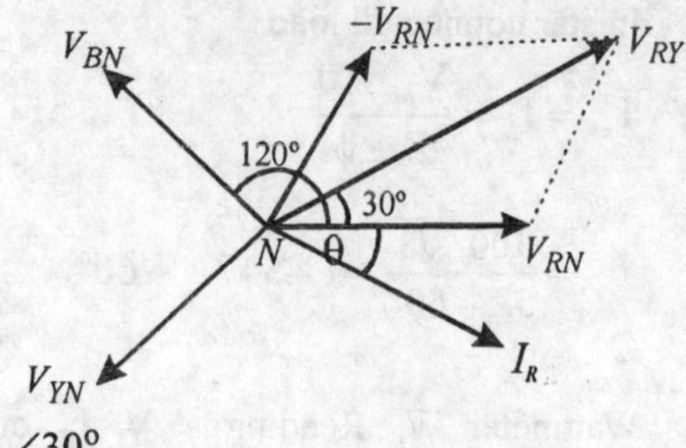

$V_{RY} = 415\,\angle 30°$

$V_{BN} = \dfrac{415}{\sqrt{3}}\angle 120°$

Now, current through current coil is,

$I_c = \dfrac{V_{RY}}{Z} = \dfrac{415\angle 30°}{100\angle 36.87°} = 4.15\angle -6.87°$

$S = VI^* = 415/\sqrt{3}\ \angle\ 120 \times 4.15\ \angle 6.87$

$= 994.34\ \angle\ 126.57$

Hence $P = -597$ W.

15. Ans. (d)

Solution: Power factors by two wattmeter method is given by

$$\tan\phi = \frac{\sqrt{3}(w_1 - w_2)}{w_1 + w_2}$$

Where w_1 and w_2 are the readings of two wattmeter.

$$\therefore \quad \phi = \tan^{-1}\left(\frac{\sqrt{3}(w_1 - w_2)}{w_1 + w_2}\right)$$

Given $w_1 = 10.5$ w,

$w_2 = -2.5$ w

$\therefore \quad \cos\phi = 0.334$

Total power $= w_1 + w_2$

$= 10.5 - 2.5 = 8\ kW$

16. Ans. (b)

Solution:

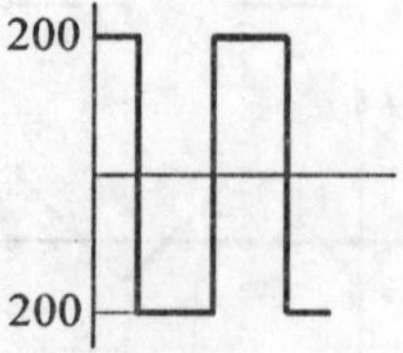

2.3 units (khh) for one hour at 230 V.

Energy consumed = kWh

$2.3 = \dfrac{V^2}{R} \times t$

$$2.3 = \frac{230^2}{R} \times 1\, hr$$

$\Rightarrow \quad R = 23\,\Omega$

Now heater connected to 400 V peak to peak square wave

$$\therefore \quad P = \frac{V^2}{R}$$

$$= \frac{200^2}{23}$$

$$= 1.736\, kW$$

17. Ans. (a)

Solution:

For the circuit diagram

$$3 = 2I + I_{NL}^2$$

To satisfy the above equation, the current must be one amphere.

So $\quad I = 1\,A$

$$V_{NL} = 1\,V$$

Power dissipated ithat resistance is $I \times V_{NL}$

$\Rightarrow 1 \times 1 = 1\,W$

18. Ans. (c)

Solution:

$$\text{Wattmeter reading} = \begin{bmatrix} \text{Current through the coil(I)} \\ \times \text{Voltage across Pressure coil(V)} \\ \times \ \cos(\ \text{phase angle between V\&I}) \end{bmatrix}$$

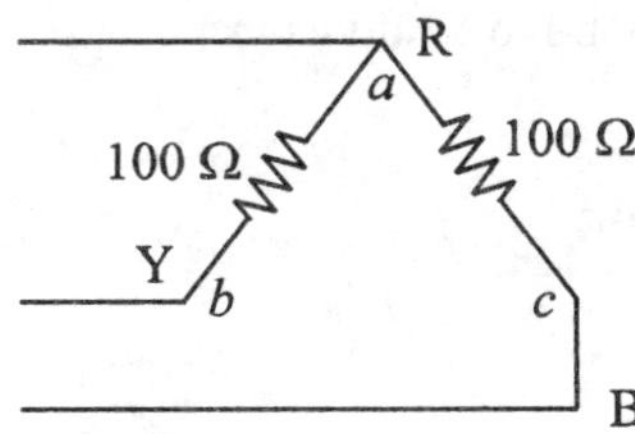

$$I_{BR} = I_{CC} = \frac{400\angle 120°}{100} = 4\angle 120°$$

$$V_{YB} = V_{PC} = 400\angle -120$$

$$W = 4\angle -120 \times 400\angle -120$$

$$= 4 \times \cos 240$$

$$= 4 \times 400 \ \cos 240$$

$$= 1600\left(\frac{-1}{2}\right)$$

$$= -800 \ \text{Watt}$$

19. Ans. (d)

Solution:

Power measured by two wattmeters is independent of nature of waveform.

20. Ans. (d)

Solution:

Given $I = \sin t$

$$V = i + 2_i^2 = \sin t + 2\sin^2 t \ = \sin t + 1 - \cos 2t$$

$$\text{Power loss } P = \frac{1}{2}\int_0^{2\pi} V.I\, dt$$

$$= \frac{1}{2} \times \cos(0)$$

$$= 0.5\, W$$

21. Ans. (a)

Solution:

Given $\quad V = \sqrt{2}\ 200 \sin(100\ \pi t)$

$$I = \sqrt{2}\cos(100\ \pi t)$$

Power $\quad P = VI$

$$= \sqrt{2}\ 200 \sin(100\ \pi t) \cdot \sqrt{2}\cos(100\, t)$$

$$= \frac{1}{2} \times \sqrt{2} \times \sqrt{2} \times 200 \times \sin(0°)$$

$$= 0 \ \text{watt}$$

[as $2 \sin A \cos B = \sin(A - B)$]

Second Method

$$P = \frac{1}{2\pi}\int_0^{2\pi} vi\, d\theta$$

$$P = \frac{1}{2\pi}\int_0^{2\pi} \sqrt{2}\ 200 \sin(100\pi t) \times \sqrt{2}\cos(100\pi t)\, dt$$

or $\quad P = \frac{1}{2\pi}\int_0^{2\pi} \sqrt{2}\ 200 \sin(100\pi t) \times \sqrt{2}\cos\theta\, d\theta$

or $\quad P = \frac{400}{2\pi}\int_0^{2\pi} \frac{2 \sin\theta\, d\theta}{2}$

$$P = \frac{400}{2\pi}\int_0^{2\pi} \frac{1}{2}\sin 2\theta\, d\theta$$

$$P = \frac{400}{2\pi}\int_0^{2\pi} \sin 2\theta \, d\theta$$

$$P = \frac{100}{2\pi}\int_0^{2\pi} \sin 2\theta \, d\theta = 0$$

Since power consumed over one full cycle is zero for sinusoidal waveform

22. Ans. (b)

Solution:

In Induction type energy meter

Breaking torque $(T_B) = K_2 \phi \, iR$

$$T_B \propto R$$

Where,

$R =$ effective radius from the axis of the disc

So, if break magnetics moved a small distance towards the edge, then the R will increase, hence T_B will increase

Therefore if breaking torque $T_B \uparrow$ meter will move slowly and hence meter read a lower reading.

23. Ans. (c)

Solution:

The wattmeter is connected towards high voltage side

The rated current $I = \frac{1\,kVA}{230} = 4.34\,A$

The suitable range is 5 A and 300 V wattmeter.

The load is connected across low voltage side

$\therefore$ The rated current $= \frac{1\,kVA}{115} = 8.69\,A$

The multiplying factor $= \frac{8.69}{4.34} = 2$

24. Ans. (c)

Solution:

Given

$$v(t) = E_1 \sin \omega t + E_3 \sin 3\omega t$$

$$i(t) = I_1 \sin(\omega t - \phi_1) + I_3 \sin(3\omega t - \phi_3) + I_5 \sin 5\omega t$$

Average power

$$P_{avg} = \frac{1}{2\pi}\int_0^{2\pi} v(t)\, i(t)\, d(\omega t)$$

$$= \frac{1}{2\pi}\int_0^{2\pi}\left[\begin{Bmatrix} E_1 \sin \omega t + E_3 \sin 3\omega t + \\ (I_1 \sin \omega t - \phi_1) + I_3 \sin(3\omega t - \phi_3) \\ + I_\pi \sin \pi\omega t \end{Bmatrix} d(\omega t)\right]$$

$$= \frac{1}{2}E_1 I_1 \cos \phi_1 + \frac{1}{2}E_3 I_3 \cos \phi_3$$

As the product of different frequency terms have zero average value.

25. Ans. (d)

Solution:

Induction energy meter rotate in opposite direction by reversing either the potential coil terminals or current coil terminals

26. Ans. (d)

Solution:

Watt meter reading, $P = \frac{1}{2\pi}\int_0^{2\pi} vi\,d\theta$

or

$$P = \frac{1}{2\pi}\int_0^{2\pi}\left[110\sqrt{2}\cos(314\,t) + \sqrt{2}\cdot\sin(942\,t)\right]$$

$$\begin{bmatrix} 5\sqrt{2}\cos(314\,t + 60^o) + 2\sqrt{2}\sin(628t + 90^o) \\ +\sqrt{2}\cos(942t + 90^o)\cdot \end{bmatrix} d\theta$$

$$p = \frac{1}{2\pi}\int_0^{2\pi} 500\left[(\cos 628t + 60^o) + \cos 60^o\right]$$

$$+220\left[\sin(942\,t) + \sin(314\,t + 90^o)\cdot\right]$$

$$+110\left[\cos(1156\,t + 90^o) + \cos(628t + 90^o)\right]$$

$$+5\left[\sin(1256\,t + 60^o) + \sin(628\,t - 60^o)\cdot\right]$$

$$\begin{bmatrix} +220\left[\sin(1256\,t + 90^o) + \sin(314\,t + 90^o)\right] \\ +110\left[\cos(1256\,t + 90^o) + \cos(628\,t - 60^o)\right] \end{bmatrix} .d\theta$$

$$= \frac{1}{2\pi} \times 550\,(\cos\,60^o)\int_0^{2\pi} d\theta = \frac{550}{2} = 275$$

110 divisions compared to 550 watt, then 275 watt will correspond to 55 divisions

27. Ans. (c)

Solution:

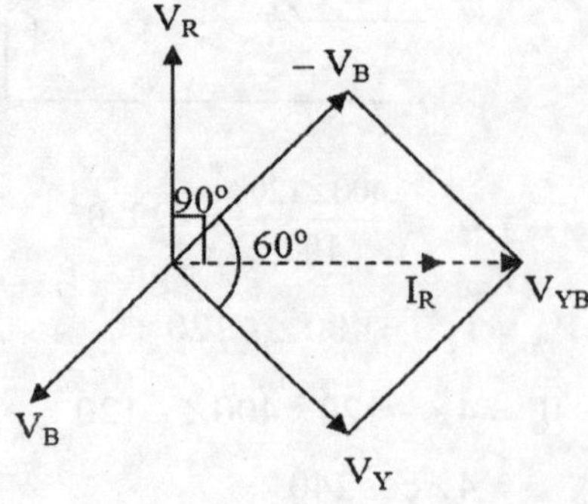

Wattmeter reading, $W = V_{YB} \cdot I_R \cos 0^o$

$= 1.732\, kW$

and, power in the load $= \sqrt{3}\, V_{YB} \cdot I_R \cos 0^o$

$= \sqrt{3} \times \sqrt{3} = 3\, kW$

[Active power because phase between $V_{YB} \cdot I_R$ is zero]

28. Ans. (a)

Solution:

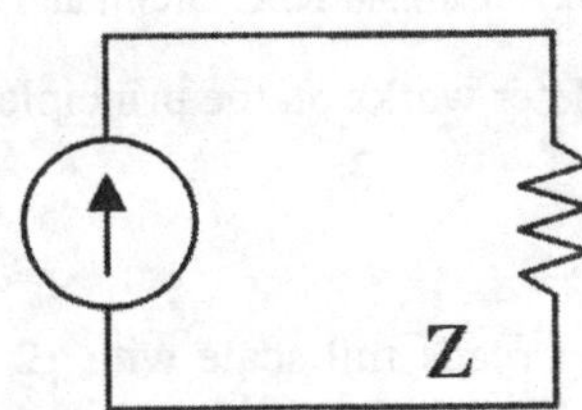

$$Z = \frac{V}{I} = \frac{6\sqrt{2}\cos(10t + 45^o)}{2\cos(10t)} = 3\sqrt{2}\,.\angle 45^o$$

$$= 3\sqrt{2}(\cos 45^o + j \sin 45^o)$$

$$3\sqrt{2}\left(\frac{1}{\sqrt{2}} + \frac{j}{\sqrt{2}}\right) = 3 + 3j = R + j\omega L$$

$$= R + j\,10\,L$$

Then, R = 3 Ω, L = 0.3 H

29. Ans. (a)

Solution:

In two watt meter method,

$$\phi = \tan^{-1}\left[\frac{\sqrt{3}\,(W_2 - W_1)}{(W_2 - W_1)}\right]$$

Since, $W_2 = 0$, therefore $\phi = 60^o$

Power factor cos $\phi = 1/2 = 0.5$

Second Method

$W_1 = V_L I_L \cos(30° - \phi)$

$W_1 = V_L I_L \cos(30° + \phi)$

If $W_2 = 0$

Then

$\Rightarrow V_L I_L \cos(30° + \phi) = 0$

$\Rightarrow \cos(30° + \phi) = 0$

$\Rightarrow \cos 90° = 0$

$\phi = 90° - 30°$

$= 60°$

Hence power factor $\cos\phi = \cos 60° = 0.5$

30. Ans. (d)

Solution:

Given $\cos\phi = 0.866$

$$\cos \tan^{-1} \sqrt{3}\left(\frac{W_1 - W_2}{W_1 + W_2}\right) = 0.866$$

$$\Rightarrow \quad W_1 - W_2 = \left(\frac{W_1 + W_2}{3}\right) \quad \text{...(1)}$$

$$W_1 + W_2 = \sqrt{3}\,.400.10\frac{\sqrt{3}}{2} = 600\,W \quad \text{...(2)}$$

By equation (1) and (2)

$W_1 = 4000\, W$

$W_2 = 2000\, W$

Second Method

Power factor $= 0.866$ log

$\phi = 30°$

$W_1 = V_L I_L \cos(30 - \phi)$

$= 4000 \times 10 \cos(0)$

$= 4000$ watt

$W_2 = V_L I_L \cos(30 + \phi)$

$= 4000 \times 10 \times \cos 60°$

$= 2000$ watt

31. Ans (d)

Solution:

Meter reading $10\, kW \times \frac{50}{60 \times 60} = \frac{5}{36} kW / hV$

Total revolution $= \frac{5}{36} \times 360 = 50$

Percentage error $= \frac{51 - 50}{50} \times (100) = 2\%$

3.9 Measurement of Phase and Frequency

1. **Ans. (a)**
2. **Ans. (b)**

Solution:

$$i_1(t) = I_m \sin(\omega t - \phi_1)$$

$$= I_m \cos[90 - (\omega t - \phi_1)]$$

$$= I_m \cos(\omega t - \phi_1 - 90°)$$

$$i_2(t) = I_m \cos(\omega t - \phi_2)$$

Angle difference between two currents should be −180° or 180° for banalced

$$-\phi_1 + \phi_2 - 90 = -180°$$

$$\Rightarrow \quad \phi_1 = 90 + \phi_2$$

3. **Ans. (a)**

Solution:

Bridge is balanced

$$Z_1 Z_4 = Z_2 Z_3$$

$$Z_2 = 300 + 15.91 \times 10^{-3} \times 2000j$$

$$Z_3 = 300 + 0.398 \times 10^{-6} \times 2000j$$

$$500 \times Z = (300 + 200j)(300 - 200j)$$

$$500Z = 130000$$

$$Z = \frac{130000}{500} = 260$$

$$Z = 260\,\Omega$$

and $q(\omega_2 t)$ will lead $p(\omega_1 t)$ by 90°

$$\therefore q(\omega_2 t) = A \sin q(\omega_2 t + 90°)$$

$$\Rightarrow q(\omega_2 t) = A \cos(\omega_2 t)$$

3.10 Q-meter and Potentiometric Recorders

1. **Ans. (a)**

Solution: Let C_V = 200 pF be the readjusted value and C = 300 pF.

Quality factor is given by $(Q) = \frac{1}{RX_C}$

According to given condition

$$\frac{1}{\omega RC} = \frac{1}{\omega R(C_V + C_X)}$$

or $\quad C = C_V + C_x$

or $\quad C_x = C - C_V = 300 - 200 = 100$ pF

2. **Ans. (c)**

Solution: Q-Meter

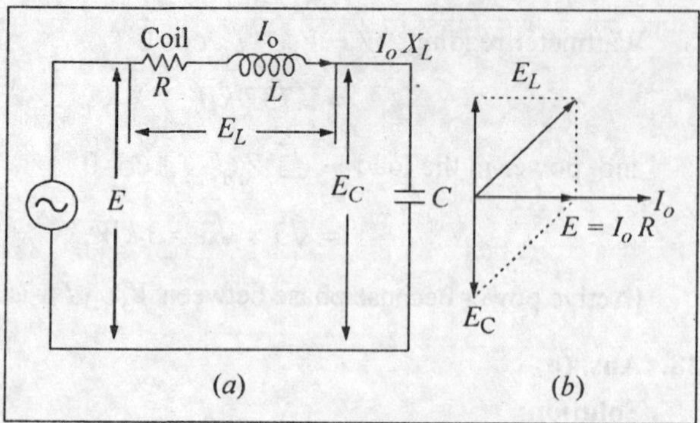

Figure: A resonant RLC circuit and its phasor diagram.

The Q-Meter works on the principle of series resonance

3. **Ans. (a)**

Solution:

The meter reads full scale with 12 V at M and range swtich at B.

Now range switch is a D and meter reads full scale value. Voltage at M is

At position B:

$$V_o = V_s\left(\frac{1.2 + 0.6 + 0.12 + 0.06 + 0.02}{2 + 6 + 1.2 + 0.6 + 0.12 + 0.06 + 0.02}\right)$$

$$= 12 \times 0.2 = 2.4$$

At position D:

$$V_o = V_s\left(\frac{0.12 + 0.06 + 0.02}{2 + 6 + 1.2 + 0.6 + 0.12 + 0.06 + 0.02}\right)$$

$$2.4 = V_s \times 0.02$$

$$V_s = \frac{2.4}{0.02} = 120 \text{ V}$$

4. **Ans. (a)**

Solution:

$V = 10\angle 15°$ and $I = 2\angle -45°$

$$\phi = 45 + 15°$$

$$= 60°$$

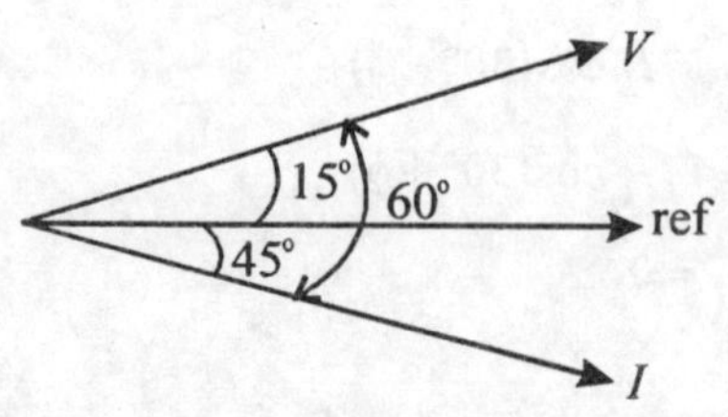

Active power $= VI \cos\phi$

$$= 10 \times 2 \times \sin 60°$$

$$= 10\,W$$

Reactive power $= VI \sin \phi$

$$= 10 \times 2 \sin 60^\circ$$

$$= 17.32 \text{ VAR}$$

5. Ans. (b)

Solution:

$$\omega = \frac{1}{\sqrt{L(C + C_d)}}$$

or $$L = \frac{1}{\omega 2(C + C_d)}$$

$$= \frac{1}{10^{12}(0.820 + 9.18) \times 10^{-9}}$$

$$= 100\, \mu H$$

6. Ans. (c)

Solution: $\omega_o = 10^6$ rad/s; C = 10 pF

$$Q = \frac{\omega_0 L}{R} = \frac{1}{\omega_0\, RC}$$

$$L = \frac{R}{\omega_0^2\, RC} = \frac{1}{\omega_0^2 C}$$

$$= \frac{1}{10^{12} \times 40 \times 10^{-12}} = 25\, mH$$

7. Ans. (b)

Soution:

A series RLC circuit exhibits voltage magnification property at resonance, i.e. the votage across the capacitor will be equal to Q-times of the applied voltage.

Given, V = applied voltage

V_o = Voltage across capacitor

$\therefore$ $$Q = \frac{V_{c\max}}{V_{in}}$$

$\Rightarrow$ $$Q = \frac{V_o}{V}$$

3.11 Oscilloscope

1. Ans. (c)

Solution

$$\frac{f_x}{f_y} = \frac{\text{Points of intersection to vertical line}}{\text{Points of intersection to horizontal line}} = \frac{2}{3}$$

Second Method

$$\frac{f_y}{f_x} = \frac{n_x}{n_y}$$

where n_x = number of horizontal touches

n_y = number or vertical touches

$$\frac{f_x}{f_y} = \frac{n_y}{n_x} = \frac{1}{1 + \frac{1}{2}}$$

$$= \frac{2}{3} = 2:3$$

2. Ans. (b)

Solution:

x and y sensitivities are same

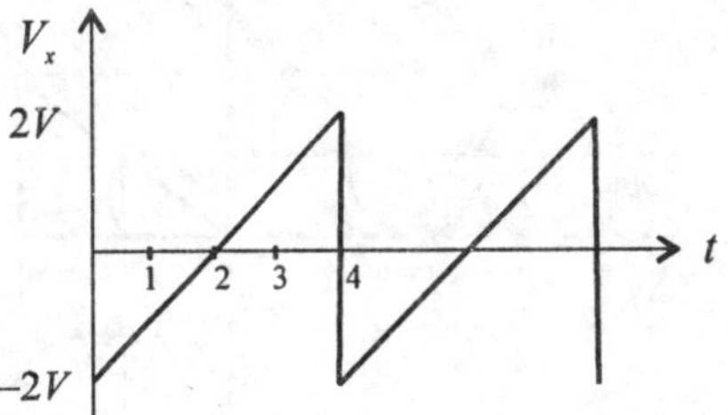

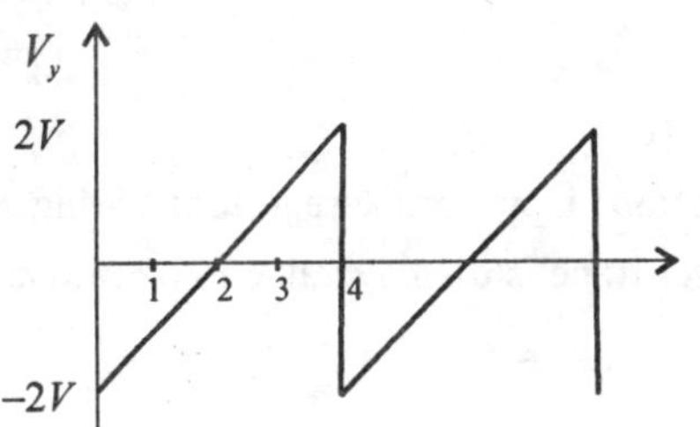

Where K_y and K_x are sensitivities and are given equal.

Let $K_y = K_x = 2\ volt/div$

t	V_y	V_x	$d_y = k_y V_y$	$d_x = k_x V_x$	Points
1	−2	−2	−4	−4	(−4,−4)
2	−1	−1	−2	−2	(−2,−2)
3	0	0	0	0	(0,0)
4	1	1	2	2	(2,2)(4,4)
5	2	2	4	4	

By using points $(-4, 4), (-2, -2)$, $(0, 0), (2, 2)$ and $(4, 4)$ make the points on screen and join them, then the diagonal line is obtained on the screen as shown below:

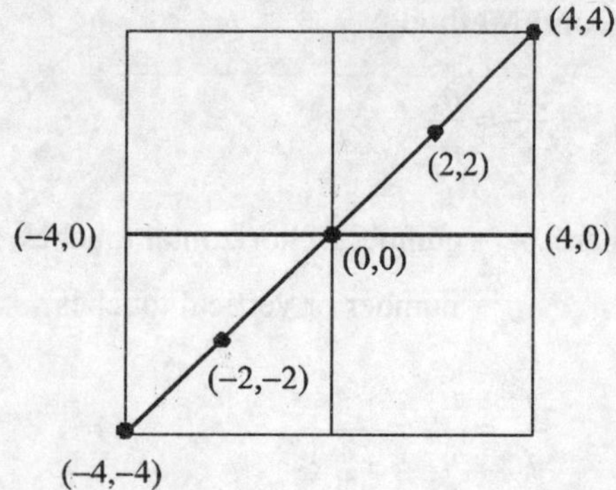

3. **Ans. (d)**

Solution:

Since both the signal are of same frequency, same phase and of same amplitude hence these are applied at the X-Y input of the CRO, hence on the CRO's screen, these will be produced a straight line with 45° with respect to x-axis. This can be understood from the following figure

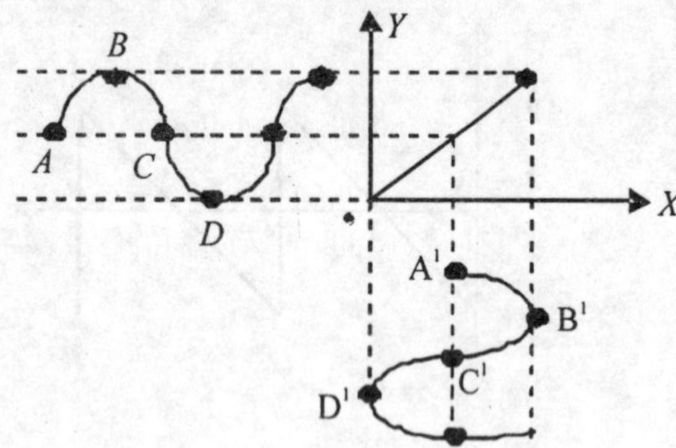

4. **Ans. (c)**

Solution: Upper trace represent setting of CRO

Lower trace is display value to be measured

2 divisions → 5 V P – P

3 divisions → $\frac{3 \times 5}{2} = 7.5$ VP – P

$$\left.\begin{array}{r}\text{Frequency} = 1\text{ kHz} \Rightarrow T = 1\text{ ms} \\ \text{Division on } x\text{-axis} = 4\end{array}\right\}\text{represent setting}$$

Result, 8 division on x-axis = 2 ms

5. **Ans. (b)**

Solution:

Since, Ch_2 shows a maximum step ramp curve so it is used against inductor.

As current through L is = $\frac{1}{L}\int V\,dt$

Integral of square is ramp.

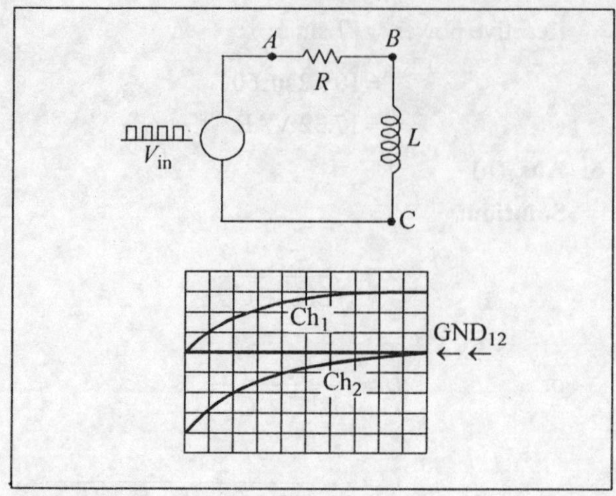

6. **Ans (d)**·

Solution :.

* When two equal voltages of equal frequency , but with 90° phase displacement are applied to a CRO, the trace on the screen is a circle.
* When two equal voltages of equal frequency but with a phase shift ϕ (not equal to 0° or 90°) are applied to a CRO , we obtain an ellipse .

7. **Ans. (d)**

Solution:

We can obtain the Lissaju pattern (in X-Y mode) by following method.

For $\phi = 0°$,

$V_X = V_{xm} \sin \omega t$

$V_y = V_{ym} \sin(\omega t + 0°) = \sin \omega t$

Draw V_X and V_y as shown below

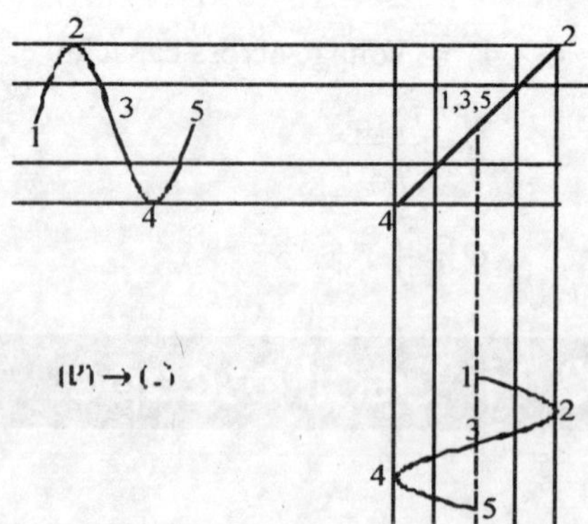

Divide both V_y and V_X into equal parts and match the corresponding points on the screen.

Similarly for $\phi = 90^\circ$

$$V_X = V_{xm} \sin \omega t$$

$$V_y = V_{ym} \sin (\omega t + 90^\circ)$$

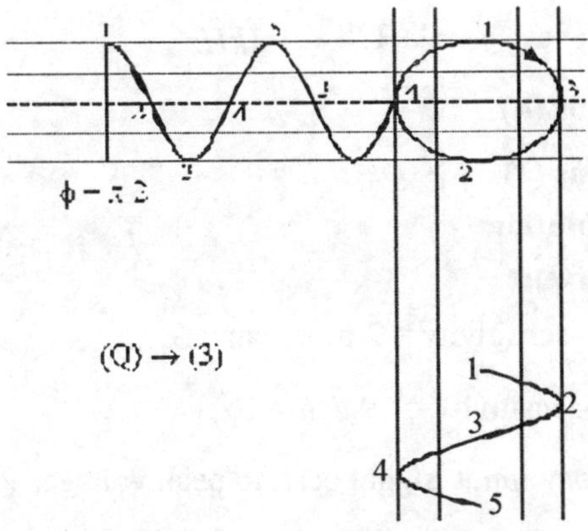

Similarly for $\phi = \dfrac{3\pi}{2}$

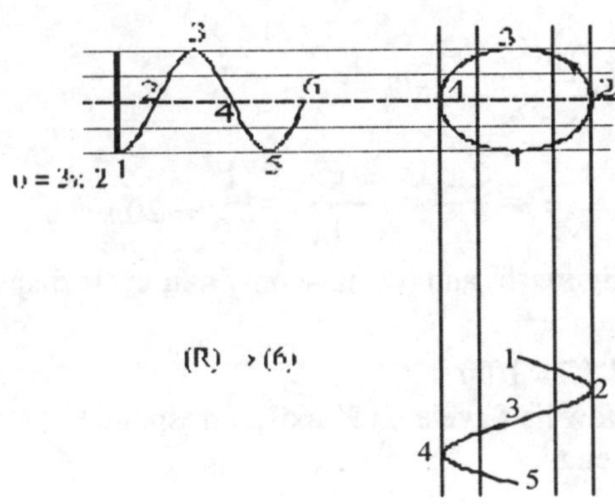

We can also obtain for $0 < \phi < \dfrac{3\pi}{2}$

8. **Ans. (a)**

Solution: $f = 100$; $\omega = 2\pi f$; $C = 10$ pF

Here, $X_c = \dfrac{1}{\omega C} = 159$ k

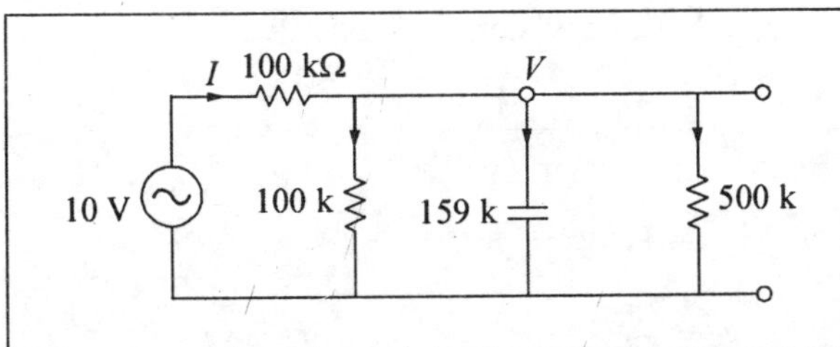

Apply the KCL

$$\therefore \quad \frac{10 - V}{100} = V\left(\frac{1}{100} + \frac{1}{159} + \frac{1}{500}\right) = 0.018 \text{ V}$$

$$\Rightarrow \quad 10 - V = 1.82 \text{ V}$$

$$V = \frac{10}{2.82} = 3.53 \text{ V.}$$

9. **Ans. (b)**

Solution: $\dfrac{f_y}{f_x} = \dfrac{x\text{-peak}}{y\text{-peak}}$

Here, x-peak = 1, and y-peak = 2

$$\therefore \quad f_y = \frac{f_x}{2}$$

Since $x(t) = P \sin (4t + 30)$

$$\therefore \quad y(t) = Q \sin\left(\frac{4t + 30}{2}\right) = Q \sin(2t + 15)$$

10. **Ans. (b)**

Solution:

Frequency of square wave $= 297 Hz$

So fundamental frequencies of square wave

$$= 297 Hz = f_1$$

So if X input has sine wave with no phase difference thenonly we get a line.
This is possible if Y sine wave frequency is integral multiple of X sine wave.

$\therefore$ of X wave $= 49.5 Hz$

Then, $\dfrac{f_1}{f} = \dfrac{297}{49.5} = 6$

11. **Ans. (a)**

Solution:

Let ϕ_1 and ϕ_2 are the phases of X and Y-inputs, then

$$\tan 45^o = \frac{\phi_1}{\phi_2}$$

or $\quad \phi_1 = \phi_2$

Since the output is a line, then there must be same frequency of both the X and Y inputs

12. **Ans. (b)**

Solution:

Delay line is introduced between Y- inputs and Y- plate to synchronize the (X) horizontal and (Y) deflection system.

13. Ans. (a)

Solution:

* When a voltage signal is fed to an oscilloscope with DC coupling mode then, entire waveform will be displayed including the DC component.
* Whereas, AC coupling mode removes DC portion of the signal and displayed waveform will be contered a zero volts.

14. Ans (d)

Solution:

A trigger setting that ensures a stationary display is with tigger voage level as 1.8 V and trigger slope as –ve.

15. Ans (b)

Solution:

Input capacitance,

$$C_{in} = \frac{C}{k-1} = \frac{45}{10-1} = 5\ pF$$

16. Ans. (b)

Solution:

Here, Y- direction-amplitude axis
X - direction time axis
5 cm on Y-axis correspond to 10 V

$$\text{Y-Sensitivity} = \frac{\text{peak – to – peak voltage}}{\text{length of vertical cells}}$$

$$= \frac{10\,V}{5\,cm} = 2\,V/cm$$

Here, display only for half-cycle is shown for which cells can be counted

$$\frac{T}{2} = \frac{1}{2f} = \frac{1}{2\times 25\times 10^3} = 0.02 \text{ msec} = 20\ \mu\text{ sec X}$$

$$\text{– sensitivity} = \frac{\text{Time period}}{\text{Length of horizontal cell}}$$

$$= \frac{T/2}{4cm}$$

$$= \frac{20\mu\text{sec}}{4cm} = 5\mu\text{sec}/cm$$

Trigger circuit triggers the UJT oscillator which is a –Ve resistance device so slope will be – Ve.

17. Ans (c)

Solution:

It is clear from the figure that there is no change in amplitude.

So it display 5V peak.

Amplitude is function of Y plates so no change.

Here slope during sweep will change and that is responsible for change in frequency.

Slope is degraded by a factor 1.1, so frequency is to be multiplied by 1.1 (as $f \alpha 1/T$).

Hence $f = 1\times 1.1 = 1.1kHz$

18. Ans (d)

19. Ans (a)

Solution:

Given

X- sensitivity = 2 msec/cm = S_x

Y-sensitivity = 1 V/cm = S_y

from input signal peak to peak voltage, $V_{P-P} = 4V$

Angalor frequency $\omega = 100\,\pi = 2\pi f$

$$\frac{2\pi}{T} = 100\,\pi$$

so length on Y axis from peak to peak,

$$L_y = \frac{V_{PP}}{S_y} = 4cm$$

$$\rightarrow\ T = \frac{2\pi}{\omega} = \frac{2\pi}{100\pi} = \frac{1}{50} = 20m\sec$$

Figure (b) and (d) have only half cycle display
Hence,

$T/2 = 10m\sec$

Now half cycle on X-axis correspond to Length,

$$L_x = \frac{T/2}{S_x} = \frac{10m\sec}{2m\sec/cm} = 5cm$$

This is shown in figure (a) only.

20. Ans (c)

Solution:

$$Y_{input} = 4\cos(200\pi t - 45°)$$

$$Y_P = 4\,V$$

$$\omega = 200\pi$$

Number of devision on Y-axis,

$$d_Y = \frac{4}{S_Y} = \frac{4}{1V/div}$$

$$= 4\,div$$

$$\omega = \frac{2\pi}{T}$$

$$T = \frac{2\pi}{\omega}$$

or $T = \frac{2\pi}{200\pi} = 10\,m\text{sec}$

or $\frac{T}{2} = 5\,m\text{sec}$

Always cheek for half cycle as full cycle may not be displyed as CRO number of devision on x-axis is

$$d_x = \frac{T/2}{S_x} = \frac{5\,m\text{sec}}{1\,m\text{sec}/\,div}$$

$$= 5\,div$$

The display is cosine wave shifted by $\pi/4$. Hence it may display as

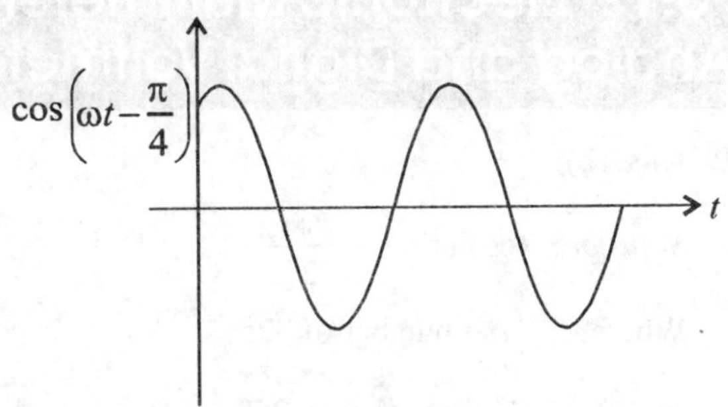

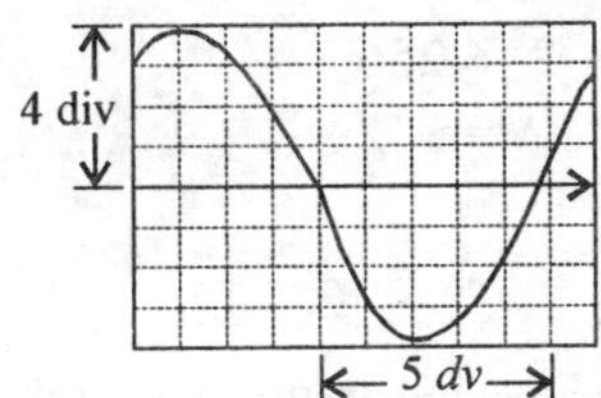

21. Ans (d)

Solution:

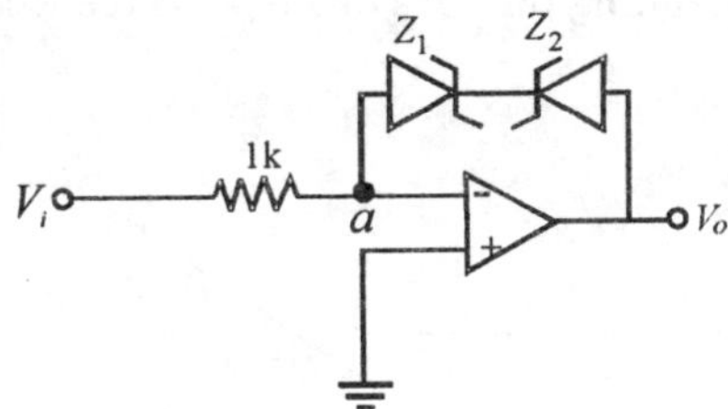

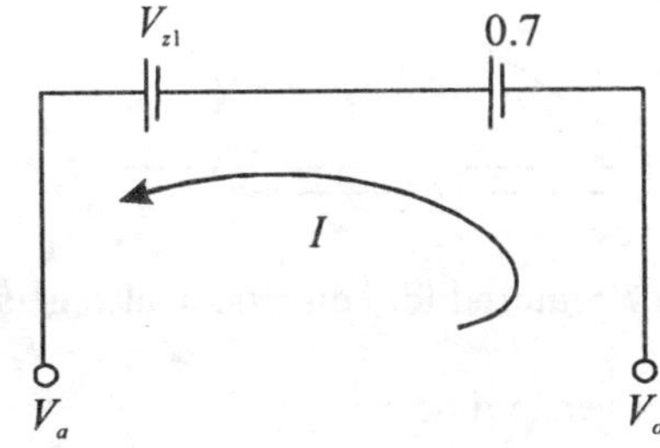

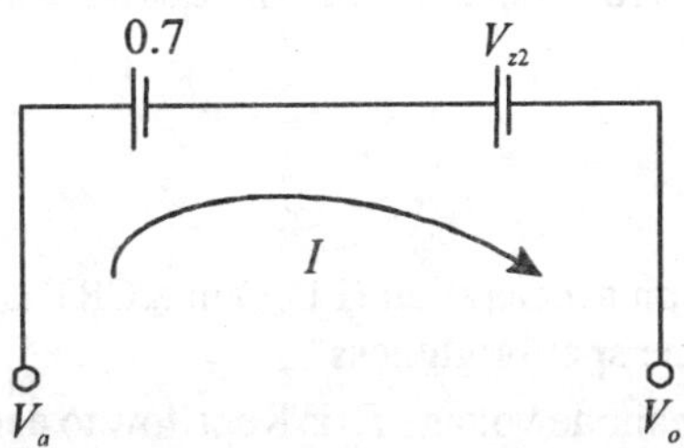

When input voltage $V_i > 0$ (During +ve cycle)

$Z_1 \rightarrow$ Forward biased

$Z_2 \rightarrow$ Reverse biased

$V_a - 0.7 - V_{Z2} - V_o = 0$

$V_o = -0.7 - V_{Z2}$ or $V_{Z2} = -0.7 - V_o$

$= -0.7 - (-4) = 3.3\,V$

When input voltage $V_i < 0$ (During –ve cycle)

$\left.\begin{array}{l} Z_1 \rightarrow \text{Reverse biased} \\ Z_2 \rightarrow \text{Forward biased} \end{array}\right\} V_o = 6\,V$

$V_{Z1} - 0.7 - V_{Z1} = 0$

$V_{Z1} = 6 - 0.7$

$V_{Z1} = 5.3\,V$

22. Ans. (d)

Solution: Deflection of CRT screen is

$$D = \frac{L\,\ell_d\,E_d}{2d\,E_a}$$

Deflection sensitivity of CRT is defined as deflection of screen per unit deflection voltage

$$S = \frac{D}{E_d} = \frac{L\ell_d E_d}{2d\,E_a \times E_d} = \frac{L \cdot \ell_d}{2d\,E_a}$$

From above equation it can be explained as

- Deflection sensitivity of CRO is independent of the deflection voltage.
- Deflection sensitivity is directly proportional to Length of Vertical Deflecting Plates, (VDP) Distance between screen and centre of VDPs.
- Defection sensitivity is inversely proportional to separation distance between VDPs accelerating anode voltage .

23. Ans. (c)

Solution:

An amplitude modulated signal (trapezoidal pattern) can be observed on a CRO by applying the modulating signal to the external Horizontal input and carrier wave to vertical input.

24. Ans. (a)

Solution:

- Post deflection acceleration (PDA) in a CRT is used to improve the spot brightness.
- Accelerating anode voltage E_a is Kept low to achieve high sensitivity $\left(\text{as } S_v \alpha \frac{1}{E_a}\right)$.
- But, in turn the velocity of e^- beam will be low $\left(\text{as } V_x \alpha \sqrt{E_a}\right)$.
- Speed of e^- beam is one of the deciding factors of brightness of images displayed on screen.
- Post to deflection, e^- beam once again is accelerated with high velocity due to very high positive voltage (PDA) attached to the screen. This is for improving the brightness.

25. Ans. (a)

Solution:

$$\frac{R_i}{R_i + R_p} = \frac{1}{20}$$

$$R_1 + R_2 = 20$$

$$R_p = 20 - 1 = 19\ M\Omega$$

$$\frac{X_{ci}}{X_{cp}} = \frac{R_i}{R_p}$$

$$\frac{C_p}{C_i} = \frac{R_i}{R_p}$$

$$C_p = 100 \times \frac{1}{19} = \frac{100}{19}\ pF$$

26. Ans. (c)

Solution:

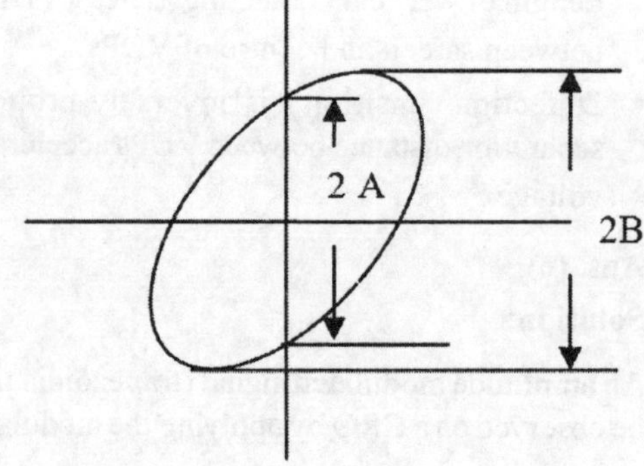

The phase angle between the signals is.

The given ellipse is having major Axis in 1st and 3rd Quadrants

Therefore, the phase angle between the signals is

$$\phi = \sin^{-1}\left(\frac{y_{\text{intercept}}}{y_{\max}}\right)$$

$$= \sin^{-1}\left(\frac{2A/2}{2B/2}\right)$$

$$= \sin^{-1}\left(\frac{A}{B}\right)$$

27. Ans. (a)

3.12 Electronic Multimeters, Analog and Digital Voltmeters

1. Ans. (b)

Solution: Resolution = $\frac{FSV}{2^n}$

Where, n = number of bits.

Then $2^n = \frac{3.5}{14 \times 10^{-3}} = 250$

$2^8 = 256$

$N = 8.$

2. Ans. (b)

Solution:

In single slope integrating type, reference voltage $-V_R$ is integraed and this ramp is compared to input voltage V_i when output of integrator V_o reaches V_i the counter is taken during time T. Now if V_i is reduced to half.

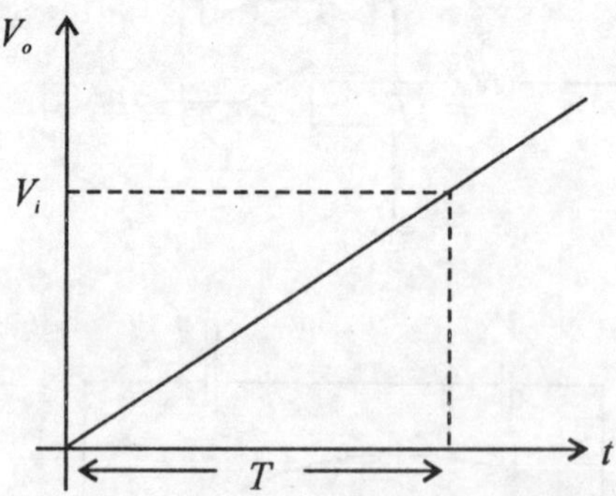

Time T required for conversion also gets halved

$\therefore$ conversin time $\frac{T_A}{2}$.

In successive Approximation type

Conversiontime $= n\tau$,

where, $n =$ Number of bits

$\tau =$ Clock time period

The conversion time remain same for any input voltage, i.e. same time, T_B.

3. **Ans. (d)**

Solution:

The signal aplied to X-plates

$x(t) = p(\omega_1, t) = A \sin \omega_1 t$

And signal applied to the Y-plates

$y(t) = q(\omega_2 t)$

Lissajous figure given as

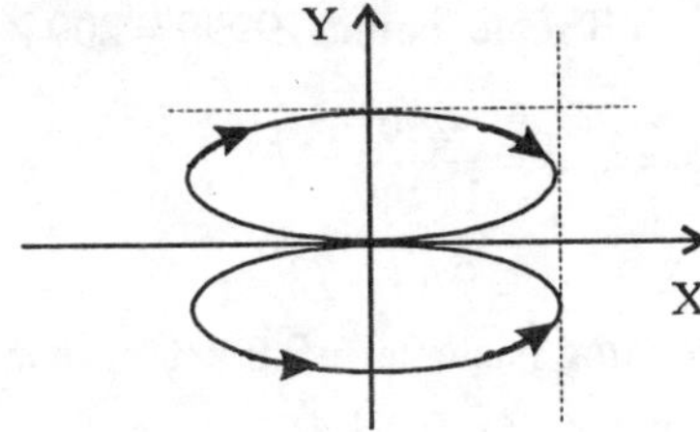

$$\frac{\omega_2}{\omega_1} = \frac{\text{number of vertical maxima}}{\text{number of horizontal maxima}}$$

$$\frac{\omega_2}{\omega_1} = \frac{1}{2}$$

$$\omega_2 = \frac{\omega_1}{2}$$

Time period $T_2 = 2T_1$

Now To determine $y(t)$ sine or cosine function

if $y(t) = q(\omega_2, t) = A \sin \omega_2 t$

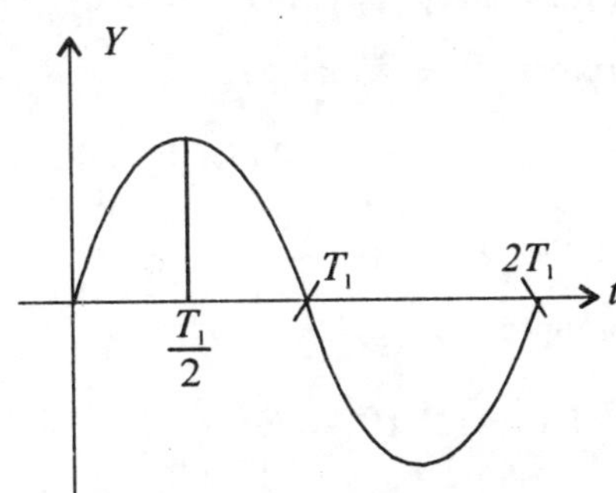

Figure: Y input signal waveform

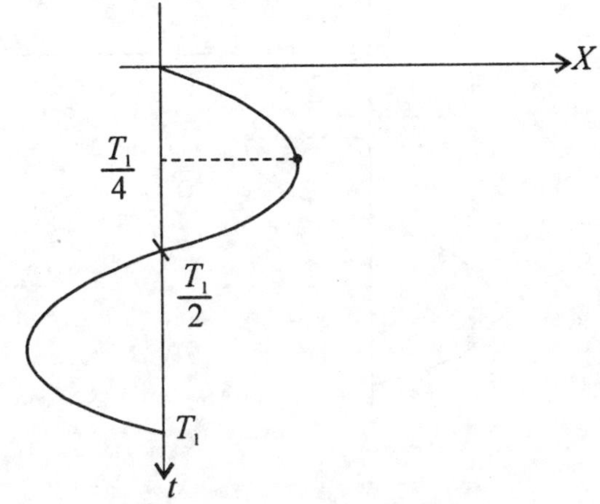

Figure: X input signal wave form.

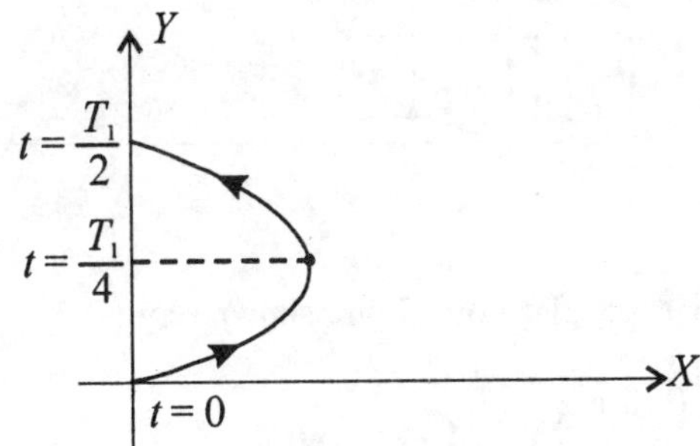

Figure: Lissajous patern

A part of Lissajous figure shown represent

$$t = 0 \text{ for } \begin{cases} x = 0 \\ t = 0 \end{cases}$$

$$t = \frac{T_1}{4} \text{ for } \begin{cases} x = +A \text{ maxima} \\ y < 0 \end{cases}$$

$$t = \frac{T_1}{2} \text{ for } \begin{cases} x = 0 \\ y = +A \end{cases}$$

Now if $y(t) = A \cos \omega_2 t$

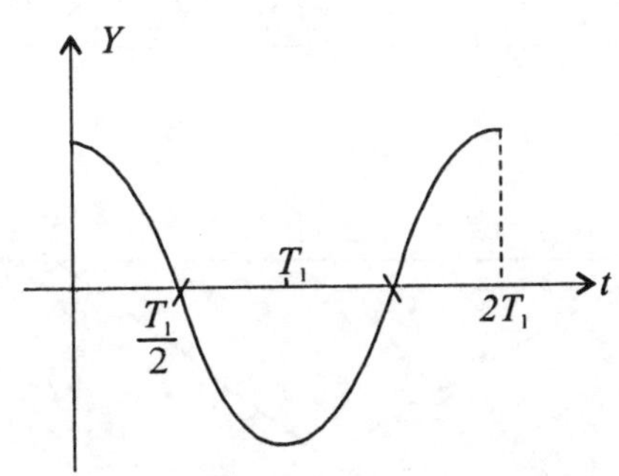

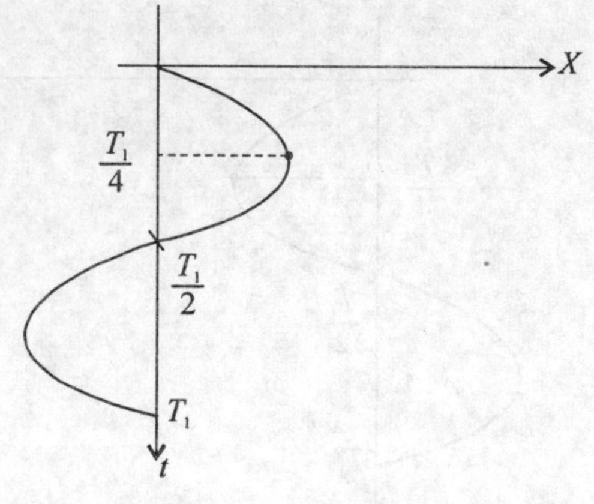

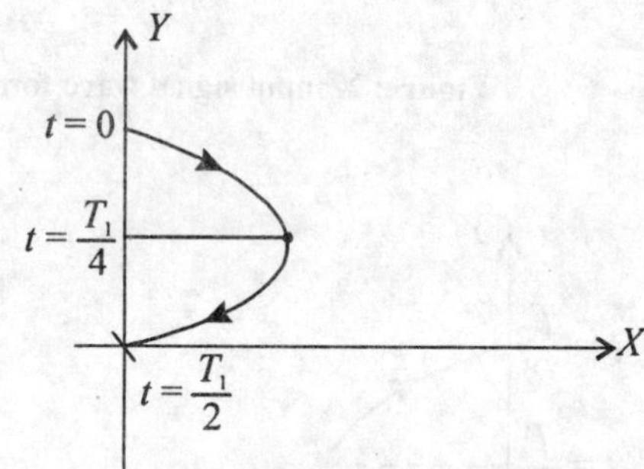

A part of Lissajous figure shown represent.

$t = 0$ for $\begin{cases} x = 0 \\ y = A \rightarrow \text{maxima} \end{cases}$

$t = \frac{T_1}{4}$ for $\begin{cases} x = A \rightarrow \text{maxima} \\ y < A \end{cases}$

$t = \frac{T_1}{2}$ for $\begin{cases} x = 0 \\ y = 0 \end{cases}$

The lissajous figure sketched for $0 < t < \frac{T}{2}$ shown is same as given.

Hence $q(\omega_2, t) = A \cos \omega_2 t$

$$\omega_2 = \frac{\omega_1}{2}$$

4. **Ans. (b)**

Solution:

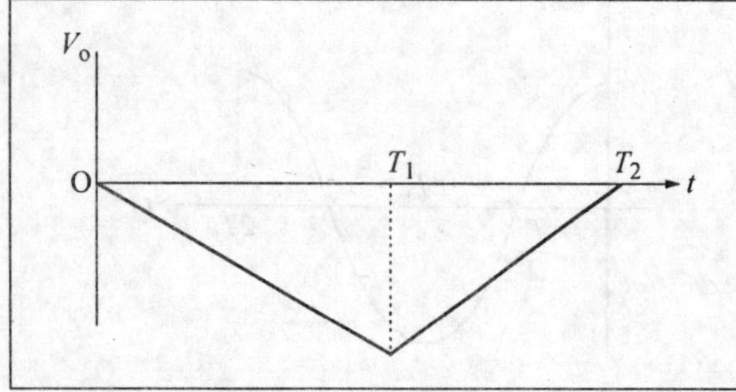

The waveform for dual slope integrater is shown. The maximum frequency can be attained when $T_2 - T_1 = 0$ and, as

$$T_1 = 2^n T_C$$

where T_C = clock period and

n = number of bits.

$$\therefore \quad f_{max} = \frac{1}{2^n . T_C} = \frac{f_c}{2^n} = \frac{10^6}{1024} \approx 1 \text{ kHz}$$

5. **Ans. (c)**

Solution:

$4\frac{1}{2}$ digit DMV has full scale voltage $V_{FS} = 200\ V$

Maximum count on $4\frac{1}{2}$ digit display $= 19999$

Hence full scale count $19999 = 200\ V$

or, 10 counts $= \frac{200}{19999} \times 10$

$\cong 0.1\ V$

The reading has error of 0.2% of reading +10 counts

0.2% of 100 $V = 0.2\ V$

To measure 100 V dc

Total error $= 0.2\ V + 0.1\ V = 0.3\ V$

Reading $= 100\ V \pm 0.3\ V$

Error $= \pm 0.3\ V$

$= \pm 0.3\ \%$

6. **Ans (a)**

Solution:

$$V_B = V_{avg} = \frac{100\sqrt{2}}{\pi} = 45mV$$

$$V_A = V_B$$

(By the concept of virtual ground)

therefore $V_A = 45mV$

also, it can be observed that $V_C = V_B$

therefore $V_C = 45mV$

at point 'c', $V_C = I\ R$

$45 \times 10^{-3} = 1 \times 10^{-3} \times R$

therefore $R = \frac{45 \times 10^{-3}}{1 \times 10^{-3}} = 45\Omega$

7. **Ans (b)**

Solution:

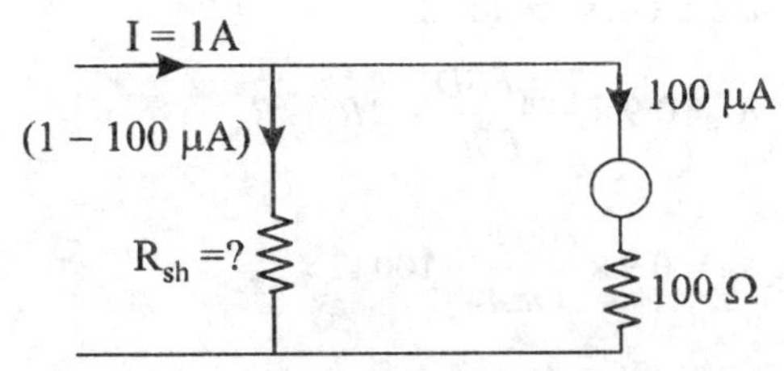

$$(1-100\ \mu A)\ R_{sh}=100\ \mu A\times 10^2$$

$$\left(1-10^{-4}\right)\ R_{sh}\Rightarrow 10^4\ \mu A$$

$$\left(\frac{10^4-1}{10^4}\right)\ R_{sh}\Rightarrow 10^4\ \mu A$$

$$R_{sh}\Rightarrow\frac{10^4\times 10^4\ \mu A}{\left(10^4-1\right)}$$

$$\left(10^4-1\right)\cong 10^4$$

$$R_{sh}\Rightarrow\frac{10^8\ \mu A}{10^4}\Rightarrow 10^4\ \mu\Omega$$

$$\Rightarrow 10^4\times 10^{-3}\times 10^{-3}\Omega$$

$$\Rightarrow 10\times 10^{-3}\Omega$$

$$\Rightarrow 10\ m\Omega$$

∴ 10 mΩ in parallel with meter

8. **Ans (c)**

Solution:

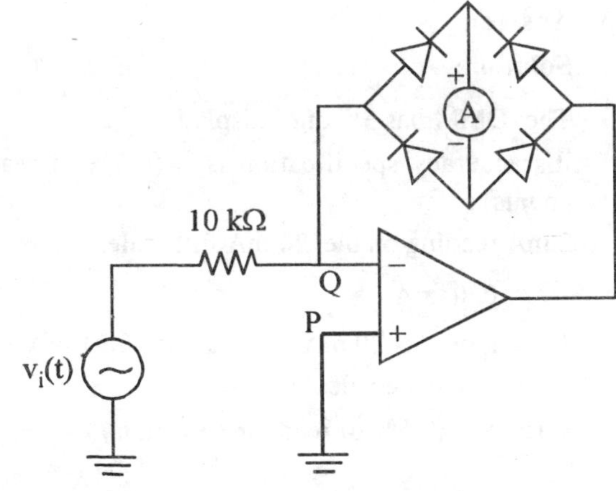

$$v_i(t)=V_m\sin\omega t\ V$$

$$=1\sin 314\ V$$

$$V_m=1$$

By virtual ground concept voltage at $Q=0$

$$I_m=\frac{1-0}{10\ K\Omega}=100\ \mu A$$

Because it is a PMMC meter so it will measure

$$I_{dc}=2\frac{\text{Im}}{\pi}$$

$$=\frac{2\times 100}{3.14}\times 10^{-6}=63.7\ \mu A$$

9. **Ans.(d)**

Solution:

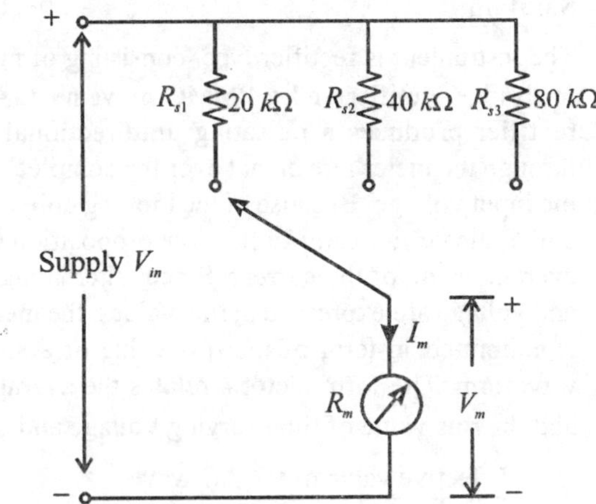

When $R_{s1}=20\ k\Omega$,

$$V_m=440\ V$$

$$I_m=\frac{V_m}{R_m}$$

$$V=V_{Rs1}+V_{m1}$$

$$V_{in}=\frac{440}{R_m}\times 20+440 \qquad \text{...(1)}$$

When $R_{s3}=80\ k\Omega$, $V_{m3}=352\ V$

$$V_{in}=\frac{352}{R_m}\times 80+352 \qquad \text{...(2)}$$

From equations (1) and (2)

$$\text{or,}\quad \frac{352\times 80-440\times 20}{88}=R_m$$

$$R_m=220\ k\Omega,\ V_{in}=480\ V$$

When $R_{s2}=40\ k\Omega$ and $V_{in}=480\ V$

$$480 = \frac{V_{m3}}{R_m} \times 40 + V_{m3}$$

$$480 = \frac{13}{11} V_{m3}$$

or $V_{m3} = 406.15$ Volt

10. Ans. (d)

Solution:

It is safe to use the meter with a potential difference of upto 2.5 kV between one of the terminals and the ground.

11. Ans. (d)

Solution:

The instrument is rectifier type, consisting of multiplier R_s, bridge rectifier and a PMMC movement. A bridge rectifier produces a pulsating unidirectional current through the meter movement over the complete cycle of the input voltage. Because of the moving coil, the meter coil indicate a steady deflection propositional to the average value of the current. Since, alternating current and voltage are expressed in rms values, the meter scale is calibrated in term of the rms value of a sinusoidal waveform. The form factor k_f relates the average value and the rms value of time varying voltage and current.

$$k_f = \frac{\text{Effective value of the AC wave}}{\text{Average value of the AC wave}}$$

$$= \frac{\frac{V_m}{\sqrt{2}}}{\frac{V_m \times \pi}{4}} = 1.11 \text{ for sinusoidal waveform}$$

Applied voltage $= 100\,V$ (rms)

Internal resistance (r) = $= 100\ \Omega$

Since current through the meter for full scale deflection is mA

$$\therefore \quad I = \frac{V}{Rs + r}$$

$$1 \times 10^{-3} = \frac{100\,V}{Rs + 100\ \Omega}$$

$$Rs = \frac{100}{1 \times 10^{-3}} - 100 = 99.9\ \Omega$$

$$Rs = \frac{99.9\ \Omega}{1.11} = \frac{9990}{111} = 90\ \Omega$$

Second Method

Given data; Full wave Bride Rectifier AC voltmeter's AC volt range is 0–100 V.

The PMMC ammeter used in the design has full scale current rating of 2 mA and internal resistance of 100 Ω and diodes are ideal

$$R_s = 0.9 \times \frac{V_{rms} FSD}{I_{dc} FSD} - 2R_d - R_m$$

$$= 0.9 \times \frac{100\,V}{1\,mA} - 100\ \Omega$$

$$= 90\Omega - 100\ \Omega$$

$$= 89.9\ k\Omega$$

12. Ans. (c)

Solution:

The timer-counter is used in period mode. Clock frequency, i.e. oscillator frequency is 10 MHz.

The incoming or applied signal is square wave has frequency of 2 kHz.

We know $n = fT$

Where,

n = Number of count display

f = Clock frequency

T = Time period of applied signal

i.e. Counter Display $= \dfrac{\text{clock frequency}}{\text{input frequency}}$

$$= 10\ MHz \times \frac{1}{2\ kHz}$$

$$= 5 \times 10^3 \text{ cycle}$$

$$= 5000 \text{ cycle}$$

∴ The display will show a value of 5000

13. Ans. (c)

Solution:

The DMM has $3^{1/2}$ digit display

it's accuracy specification is ±(0.5% of reading +5 counts)

2 mA reading on the 20 mA full scale.

0 2 . 0 0 mA

1 count on this 20 mA full scale is 00.01 mA

∴ error can be calculated as

Error = ±(0.5% of reading +5 counts)

$$= \pm\left(\frac{0.5}{100} \times 02.00mA + 5 \times 00.01\ mA\right)$$

$$= \pm\left(\frac{0.5}{100} \times 2mA + 5 \times 0.01\ mA\right)$$

$$= \pm(0.01mA + 0.05\,mA)$$
$$= \pm(0.06\,mA)$$

% error in reading of 2 mA is

$$\%\text{ error} = \pm\frac{0.06\,mA}{2\,mA}\times 100$$
$$= \pm 3\%$$

14. Ans (b)

Solution:

The DVM has 3½ digit display

It's accuracy specification is ± (0.5% of reading + 5 counts)

100 mV reading on the 200 mV full scale.

1 0 0 . 0 mV

1 count on this 200 mV full scale is 00.01 mV

∴ error can be calculated as

Error = ± (0.5% of reading +5 counts)

$$= \pm\left(\frac{0.5}{100}\times 100.0\,mV + 5\times 000.1\,mV\right)$$
$$= \pm\left(\frac{0.5}{100}\times 100\,mV + 5\times 0.1\,mV\right)$$
$$= \pm(0.5\,mV + 0.5\,mV)$$
$$= \pm 1mV$$

The voltage being mesured in $100\,mV \pm 1\,mV$, i.e. any value between 99.0 mV and 101.0 mV

15. Ans (a)

Solution:

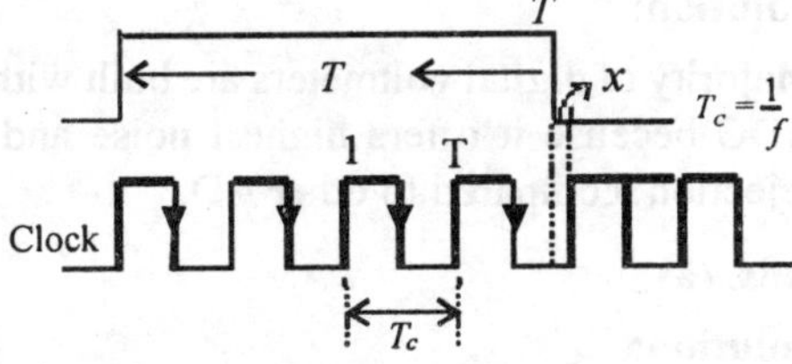

Maximum error possible is one clock period, if this occurs when negative edge of enable occurs just before negative edge of clock so that count is 1 less than actual count.

Since error should be limited to x% of T.

So maximum error should not be greater than x% of T.

$$T_c < x\%\,T$$
$$T_c < \frac{x\times T}{100}$$
$$\frac{1}{f_c} < \frac{xT}{100}$$
$$T > \frac{100}{xf_c}$$

16. Ans. (a)

Solution:

Here $V_m = 1$ V or $V_{pp} = 2$ V

Voltmeter reading = rms value of square wave

= average value of sine wave

$$= \frac{2V_m}{\pi} = \frac{2}{\pi}$$

17. Ans. (b)

Solution:

The scale of a full wave rectifier type voltmeter is calibrated to read r.m.s for ideal sine wave, i.e. reading $= 1.11\,V_{dc}$.

Where V_{dc} is average voltage of output of FWR for given input

This voltmeter is used for square wave input

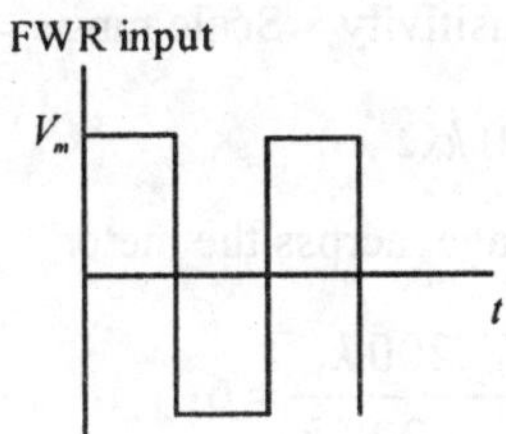

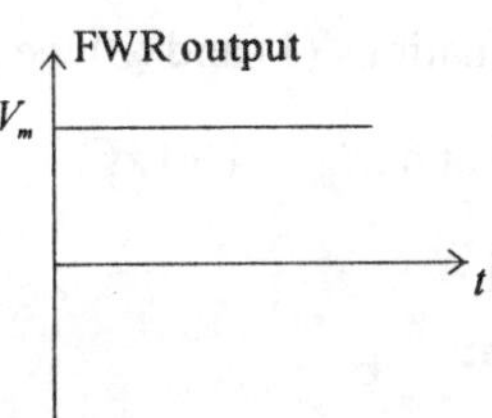

- DC voltmeter measures V_{dc} of output of FWR

$$V_{dc} = V_m$$

- Reading is indicated rms $= 1.11\,V_{dc} = 1.11\,V_m$

But true rms of square wave is V_m.

Therefore, the multiplying factor of the meter for correction is $\frac{1}{1.11}\left(\text{as } 1.11\,V_m \times \frac{1}{1.11} \Rightarrow V_m\right)$

18. Ans. (d)

Solution: $A_d = 1$

Full scale deflection = 1 mA

$V_0 = 1\,mA \times 10 = 10\,mV$

$V_2 = V_0 = 10\,mV$

19. Ans. (b)

Solution:

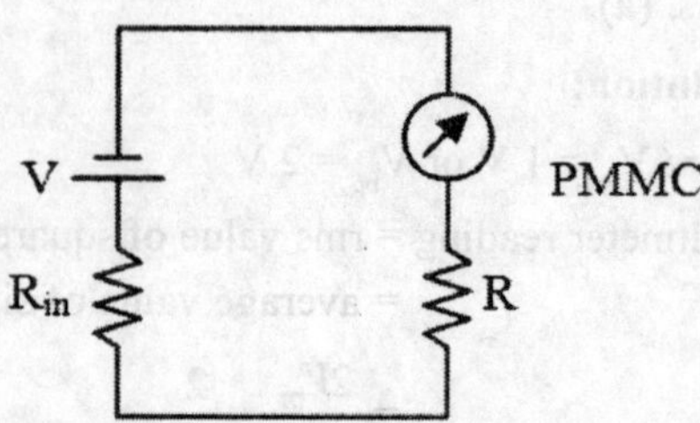

Let R_{in} be the internal resistance of the unknown voltage source V.

On 5 V scale

R_n = Sensitivity × Scale range $= (20)5 = 100\ k\Omega$

then Voltage across meter $= \dfrac{V_{in}100\,k}{R_{in} + 100\,k} = 4.5$ ...(1)

On 10 V scale,

R_m = Sensitivity × Scale range $= (20\ k\Omega) = 10\ V$

$= 200\ k\Omega$

then voltage across the meter

$= \dfrac{V_{in}.200\,k}{R_{in} + 200\,k} = 6$ ...(2)

From equations (1) and (2) we get

$V = 9\,V$ and $R_{in} = 100\ k\Omega$

20. Ans. (d)

Solution:

- The DVM has $3\frac{1}{2}$ digit display

Therefore, its scale resolution is 0.001

- Its resolution in 200 mV range is 100 mV
- The maximum voltage that can be measured in this 200 mV lowers Range: 1999 mV

21. Ans. (a)

Solution:

The type of A/D converter normally used in 3½ digital multimeter is dual-slope integrating type. It offers highest accuracy, highest noise rejection and highest stability than other any A/D converters.

22. Ans. (d)

Solution:

1. The DVM has 3½ digit display
2. The ½ digit at most significant side of digital readout can display either 0 or 1. It cannot display 5 or 9
3. Therefore, only 1.999 can be displayed 3½ digit DMM

23. Ans. (b)

Solution:

In a dual slope ADC, the counter counts upto its maximum count and overflows during the 1st integration time -period

Therefore, $T_1 = 2^n \times T_{clk}$

$= 2^n \times \dfrac{1}{f_{clk}}$

Given that the counter is 8 bit counter and clock frequency is 4 MHz

$\therefore \quad T_1 = 2^8 \times \dfrac{1}{4\ MHz}$

$= 2^6\ \mu s$

$= 64\ \mu s$

$= 0.064\ ms$

24. Ans. (c)

Solution:

The dual slope converter has very high noise rejection. Normal mode noise refers to the type of noise that appears superimposed on the input signal in the form of ripples or spikes. In dual slope integrating type converter, such noise is rejected by averaging due to integration.

25. Ans. (c)

Solution:

Majority of digital voltmeters are built with a dual slope ADC because it offers highest noise and interference rejection compared to other AD.

26. Ans. (a)

Solution:

We know $n = f.t = \dfrac{t}{T}$

Accumulated count $= \dfrac{\text{Control pulse period}}{\text{Clock period}}$

Accumulated count $= \dfrac{\text{Clock frequency}}{\text{Control pulse frequency}}$

Therefore, the frequency is 16 MHz .

27. Ans. (b)

Solution:

In a dual slope type DVM, the analog dc voltage (to be measured) is integrated for a fixed period T_1.

To eliminate or reject power-line noise $T_1 = n\,T_s$

Where

T_1 = 1st integration period

T_S = Time period of sinusoidal component riding on dc voltage to be measured, having a frequency of f_s.

And $n \geq 1$

Given data $f_s = 50\ Hz$ and $T_1 = 100\ T_{clk}$

$$T_1 = n \times T_s$$

$$\Rightarrow \quad 100\ T_{clk} = n \times T_s$$

$$\Rightarrow \quad 100 \times \frac{1}{f_{clk}} = n \times \frac{1}{f_s}$$

$$\Rightarrow \quad f_{clk} = \frac{100}{f_{clk}} \times f_s$$

n should be minimum (i.e. 1) for maximum clock frequency

$$f_{clk}(max) = \frac{100}{1} \times 50\ Hz$$

$$= 5000\ Hz$$

$$= 5\ kHz$$

28. Ans. (a)

Solution:

$$\text{Resolution} = \frac{1}{10^n} = \frac{1}{10^3} = 0.001\ V = 1\ mV$$

There are three full digits in this DVM, and thus meter will be able to count from zero to 999 the fourth digit placed to the left of active digits indicates either 0 or 1. Thus count goes from above from 999 to 1999, thereby making total 2000 counts

29. Ans. (c)

Solution:

Error in reading 0.5 % of 50 Ω

Then, reading error $= \frac{0.5}{100} \times 50 = 0.25\ \Omega$

5 count will contribute error of $5 \times 0.1\ \Omega = 0.5\ \Omega$

Thus, total error = 0.75 Ω

and the value of resistance = 50 ± 0.75 Ω

30. Ans. (b)

Solution:

Given: For dual slope ADC

$$V_{ref} = 100\ mV,\ T_1 = 50\ ms$$

$$R = 100\ k\Omega,\ C = 0.047\ \mu F$$

$$\bar{V}_m = 120\ mV$$

We known $V_m T_1 = V_{ref} T_2$

$$\Rightarrow T_2 = \frac{V_m}{V_{ref}} \times T_1$$

$$\Rightarrow T_2 = \frac{120\ mV}{100\ mV} \times 50\ ms$$

$$T_2 = 60\ ms$$

31. Ans. (a)

Solution:

The input of 120 mV is corrupted by power supply interference at 50 Hz having peak amplitude of 3π mV. But, 1st integration (where input signal is integrated) is carried output for 50 ms which is 2.5 periods of supply frequency 50 Hz. In dual slop ADC in dual slop ADC power supply noise is completely rejected by averaging due to integration. Hence error is 0%

3.13 Instrument Transformer

1. Ans. (c)

2. Ans. (b)

Solution: K_T, Nominal ratio $= \frac{500}{5} = 100$

$$\text{Magnetizing current} = \frac{\text{AT}}{\text{primary turns}} = \frac{250}{1} = 250$$

$$\text{Primary current } I_p = \sqrt{(K_T I_s)^2 + (I_m)^2}$$

$$= \sqrt{(100 \times 5)^2 + (250)^2} = 559$$

$$\text{Actual Turns ratio} = \frac{I_P}{I_S} = \frac{559}{5} = 111.8$$

$$\text{Ratio error} = \frac{\text{nominal ratio} - \text{actual ratio}}{\text{actual ratio}}$$

$$= \frac{100 - 111.8}{111.8} = -\frac{11.8}{111.8} \times 100 = -10.56\%$$

3. Ans. (a)

Solution: Phase angle between the I_s and I_p is,

$$\theta = \frac{180}{\pi}\left(\frac{I_o \cos\{\alpha + \delta\}}{\eta I_s}\right)$$

$$\theta \approx \text{degree} \approx \left(\frac{180}{\pi}\right) \cdot \left(\frac{200 \times 1}{500 \times 5}\right) \approx 4.6^\circ$$

4. Ans. (b)

Solution: Voltage induced in the secondary,

$$E_s = I_s \times Z_s = 5 \text{ Volts; Now } E_s = 4.44 f \phi_m N_s.$$

$$\therefore \quad \phi_m = \frac{E_s}{4.44 f N_s} = \frac{5}{4.44 \times 50 \times 500} = 45\ \mu\text{Wb}.$$

5. Ans. (b)

Solution: Ratio error $= \frac{1}{200} \times 100 = 0.5\,\%$ and will be negative because one turn is reduced. Reduction in turn has no effect on phase error. So, net

Ratio error = – (0.5% + 0.5%) = – 1%

Phase error = 30 min (no charge)

6. Ans. (a)

Solution:

Actual transformation ratio

$$R = +\frac{I_p}{I_s} = +\frac{1}{0.495} = 202.02,$$

$$\text{Nominal ratio } K_n = \frac{200}{1} = 200$$

Ratio error

$$= \frac{K - R}{R} \times 100$$

$$= \frac{200 - 202.02}{202.02} \times 100$$

$$= -1.0\%$$

7. Ans. (b)

Solution: Ratio error of current transformer

$$= \frac{\text{nominal ratio – actual ratio}}{\text{actual ratio}} = \frac{R_n - R_a}{R_a}$$

$$K = n + \frac{I_m \sin\delta + I_c \cos\delta}{I_s}$$

So ratio error of current transformer depends upon the I_m magnetize current and I_e eddy current

8. Ans. (b)

Solution:

If secondary winding of current transformer gets open circuited, then there is no opposing mmf (load mmf $N_2 I_2$) to the primary. Hence the resultant mmf in the C.T is to be high, due to this high mmf the induced voltage in secondary winding is very high and which may cause severe insulation stress in C.T and risk to operating persons.

4 Signals and Systems

4.1 Signals

1. The rms value of the resultant current in a wire which carries a dc current of 10 A and a sinusoidal alternating current of peak value 20 A is
(a) 14.1A (b) 17.3 A
(c) 22.4 A (d) 30.0 A

2. The Fourier series for a periodic signal is given as:
$x(t) = \cos(1.2\pi t) + \cos(2\pi t) + \cos(2.8\pi t)$
The fundamental frequency of the signal is
(a) 0.2 Hz (b) 0.6 Hz
(c) 1.0 Hz (d) 1.4 Hz

3. The fundamental period of the discrete time signal $x[n] = e^{j\left(\frac{5\pi}{6}\right)n}$ is
(a) $\frac{6}{5\pi}$ (b) $\frac{12}{5}$
(c) 6 (d) 12

4. The fundamental period of $x(t) = 2\sin 2\pi t + 3\sin 3\pi t$, with t expressed in seconds, is
(a) 1s (b) 0.67s
(c) 2s (d) 3s

5. The root mean squared value of
$x(t) = 3 + 2\sin(t)\cos(2t)$ is
(a) $\sqrt{3}$
(b) $\sqrt{8}$
(c) $\sqrt{10}$
(d) $\sqrt{11}$

6. Which of the following signals is non-periodic ?
(a) $s(t) = \cos 2t + \cos 3t + \cos 5t$
(b) $s(t) = \exp(j\,8\pi t)$
(c) $s(t) = \exp(-7t)\sin 10\pi t$
(d) $s(t) = \cos 2t \cos 4t$

7. The fundamental period of the sequence $x[n] = 2\sin(1.3\pi n + 0.5\pi) + 5\sin(1.2\pi n)$ is
(a) 20 (b) $\frac{2\pi}{1.3\pi}$
(c) $\frac{2\pi}{1.2\pi}$ (d) 10

8. Consider the periodic signal $x(t) = (1 + 0.5\cos 40\pi t)\cos 200\,2\pi$, where t is in seconds. Its fundamental frequency, in Hz is
(a) 20 (b) 40
(c) 100 (d) 200

9. Consider the discrete-time signal $x(n) = \left(\frac{1}{3}\right)^n u(n)$,
where $u(n) = \begin{cases} 1, n \geq 0 \\ 0, n < 0 \end{cases}$.
Define the signal $y(n)$ as $y(n) = x(-n)$, $-\infty < n < \infty$.
Then $\sum_{n=-\infty}^{\infty} y(n)$ equals
(a) $-\frac{2}{3}$
(b) $\frac{2}{3}$
(c) $\frac{3}{2}$
(d) 3

10. A point Z has been plotted in the complex plane, as shown in figure below.

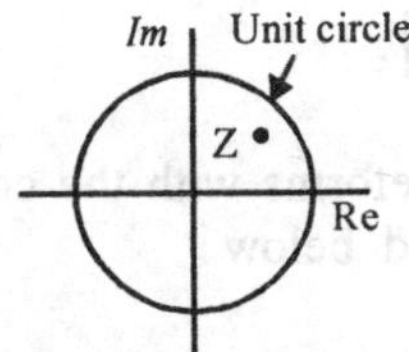

The plot of the complex number $y = \frac{1}{z}$ is

(a)

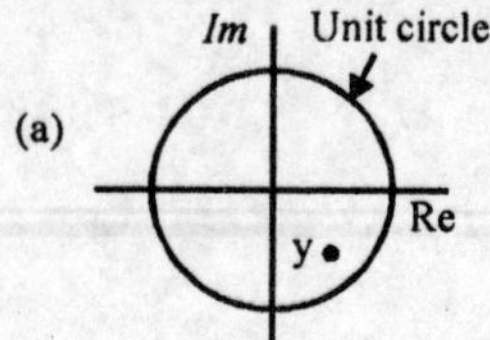

(b)

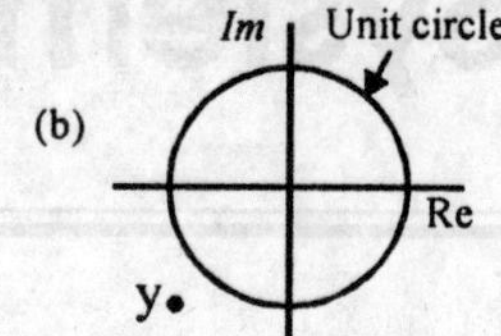

(c)

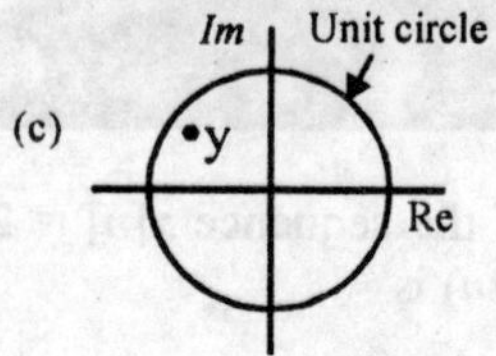

(d)

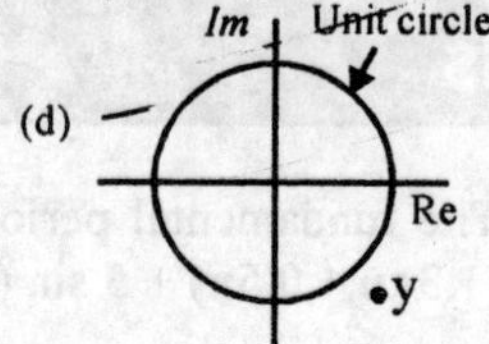

11. Given a sequence $x[n]$, to generate the sequence $y[n] = x[3-4n]$, which one of the following procedures would be correct ?

(a) First delay $x[n]$, by 3 samples to generate $z_1[n]$, then pick every 4th sample of $z_1[n]$ to generate $z_2[n]$, and then finally time reverse $z_2[n]$ to obtain $y[n]$

(b) First advance $x[n]$ by 3 samples ot generate $z_1[n]$, then pick every 4th sample of $z_1[n]$ to generate $z_2[n]$, and then finally time reverse $z_2[n]$ to obtain $y[n]$

(c) First pick every fourth sample of $x[n]$ to generate $v_1[n]$, time reverse $v_1[n]$ to obtain $v_2[n]$, and finally advance $v_2[n]$ by 3 samples to obtain $y[n]$.

(d) First pick every fourth sample of $x[n]$ to generate $v_1[n]$, time reverse $v_1[n]$ to obtain $v_2[n]$, and finally delay $v_2[n]$ by 3 samples to obtain $y[n]$.

12. Match the waveforms with the correct mathematical description listed below

Wave from $f(t)$

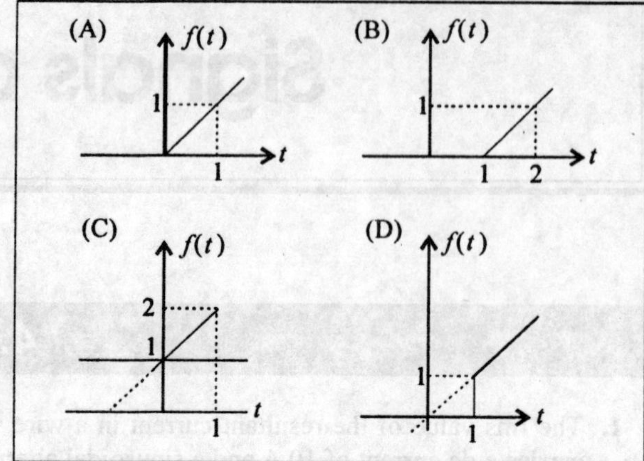

1. $t \cdot u(t-1)$
2. $(t+1) \cdot u(t-1)$
3. $t \cdot u(t)$
4. $(t+1) \cdot u(t)$
5. $(t-1) \cdot u(t)$
6. $(t-1) \cdot u(t-1)$

Codes:

	A	B	C	D
(a)	1	4	3	6
(b)	3	6	1	6
(c)	1	2	6	1
(d)	3	6	4	1

13. A current impulse, $5d(t)$, is forced through a capacitor C. The voltage, $V_c(t)$, across the capacitance is given by:

(a) $5t$ (b) $5u(t) - C$
(c) $5/Ct$ (d) $5u(t)/C$

14. If a current of $\left[-6\sqrt{2}\sin(100\pi t) + 6\sqrt{2}\cos\left(300\pi t + \frac{\pi}{4}\right) + 6\sqrt{2}\right]$ A is passed through a true RMS ammeter, the meter reading will be.

(a) $6\sqrt{2}\,A$ (b) $\sqrt{126}A$
(c) $12\,A$ (d) $\sqrt{216}A$

15. If the bandwidth of a low pass signal $g(t)$ is 3 kHz, The bandwidth of $g^2(t)$ will be.

(a) $\frac{3}{2}kHz$ (b) $3\,MHz$
(c) $6\,kHz$ (d) $9\,kHz$

16. The integral $\int_{-\infty}^{\infty} \delta(t-\pi/6)6\sin(t)\,dt$ evaluates to

(a) 6 (b) 3
(c) 1.5 (d) 0

17. The continuous-time signal $x(t) = \sin \omega_0 t$ is a periodic signal. However, for its discrete-time counterpart $x[n] = \sin \omega_0 n$ to be periodic, the necessary condition is:

(a) $0 \le \omega_0 < 2\pi$

(b) $2\pi/\omega_0$ to be an integer

(c) $2\pi/\omega_0$ to be a ratio of integers

$cos(100\pi t) - sin(100\pi t)$

(d) None

18. The integral

$$\frac{1}{\sqrt{2\pi}}\int_{-\infty}^{\infty} t^2 e^{-t^2/2}\,\delta(1-2t)dt \text{ is equal to}$$

(a) $\frac{1}{8\sqrt{2\pi}}e^{-1/8}$

(b) $\frac{1}{4\sqrt{2\pi}}e^{-1/8}$

(c) $\frac{1}{\sqrt{2\pi}}e^{-1/2}$

(d) 1

19. Given $x[n] = \frac{\sin \omega_c n}{\pi n}$, the energy of the signal given by $\sum_{n=-\infty}^{\infty} x|[n]|^2$ is

(a) ω_c/π (b) $\pi\omega_c$

(c) Infinite (d) $2\,\pi\omega_c$

20. The plot of a function $f(x)$ is shown in the following figure. A possible expression for the function $f(x)$ is

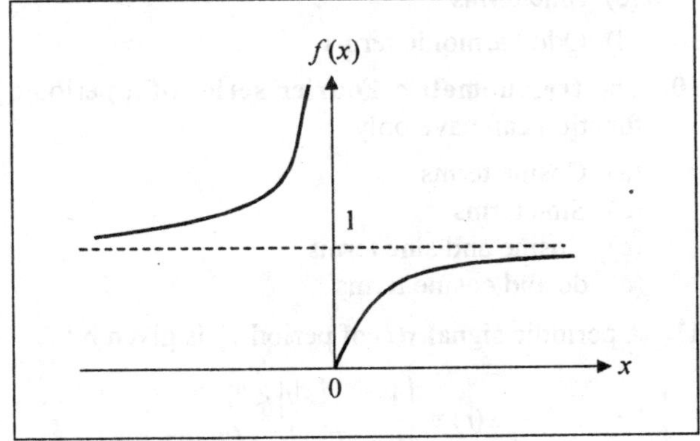

(a) $e^{(|x|)}$ (b) $e^{\left(\frac{-1}{x}\right)}$

(c) $e^{(-x)}$ (d) $e^{\left(\frac{1}{x}\right)}$

21. The period of the signal

$$x(t) = 8\sin\left(0.8\pi t + \frac{\pi}{4}\right) \text{ is}$$

(a) $0.4\pi\ s$

(b) $0.8\pi\ s$

(c) $1.25\ s$

(d) $2.5\ s$

22. The RMS value of the voltage v(t) = 3+4cos (3t) is

(a) $\sqrt{17}$ V

(b) 5 V

(c) 7 V

(d) $\left(3+2\sqrt{2}\right)V$

23. Which of the following is true

(a) A finite signal is always bounded

(b) A bounded signal is always possesses infinite energy

(c) A bounded signal is always zero outside the interval $[-t_0, t_0]$ for some t_0

(d) A bounded signal always finite

24. Given a sequence $x[n]$, to generate the sequence $y[n] = x[3-4n]$, which one of the following pocedures would be correct ?

(a) First delay $x[n]$, by 3 samples to generate $z_1[n]$, then pick every 4th sample of $z_1[n]$ to generate $z_2[n]$, and then finally time reverse $z_2[n]$ to obtain $y[n]$

(b) First advance $x[n]$ by 3 samples ot generate$z_1[n]$, then pick every 4th sample of $z_1[n]$ to generate $z_2[n]$, and then finally time reverse $z_2[n]$ to obtain $y[n]$

(c) First pick every fourth sample of $x[n]$ to generate $v_1[n]$, time reverse $v_1[n]$ to obtain $v_2[n]$, and finally advance $v_2[n]$by 3 samples to obtain $y[n]$.

(d) First pick every fourth sample of $x[n]$ to generate$v_1[n]$, time reverse $v_1[n]$ to obtain $v_2[n]$, and finally delay $v_2[n]$ by 3 samples to obtain $y[n]$.

4.2 Fourier Series

1. The rms value of the periodic waveform given in figure is

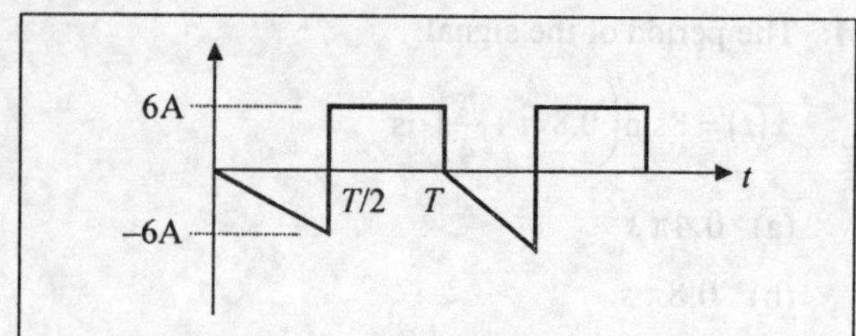

(a) $2\sqrt{6}$ A (b) $6\sqrt{2}$ A

(c) $\sqrt{\frac{4}{3}}$ A (d) 1.5 A

2. For the triangular wave form shown in the figure, the RMS value of the voltage is equal to

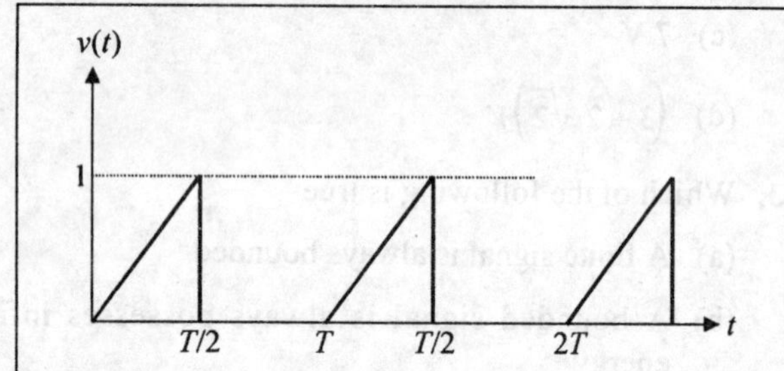

(a) $\sqrt{\frac{1}{6}}$ (b) $\sqrt{\frac{1}{3}}$

(c) $\frac{1}{3}$ (d) $\sqrt{\frac{2}{3}}$

3. The Fourier series for the function $f(x) = \sin^2 x$ is

(a) $\sin x + \sin 2x$ (b) $1 - \cos 2x$

(c) $\sin 2x + \cos 2x$ (d) $0.5 - 0.5 \cos 2x$

4. $x(t)$ is a real valued function of a real variable with period T. Its trigonometric Fourier Series expansion contains no terms of frequency $\omega = 2\pi(2K)/T$; $K = 1, 2, ...$ Also, no sine terms are present. Then $x(t)$ satisfies the equation

(a) $x(t) = -x(t-T)$

(b) $x(t) = x(T-t) = -x(-t)$

(c) $x(t) = x(T-t) = -x(t-T/2)$

(d) $x(t) = x(t-T) = x(t-T/2)$

5. If a current of $\left[-6\sqrt{2}\sin(100\pi t) + 6\sqrt{2}\cos\left(300\pi t + \frac{\pi}{4}\right) + 6\sqrt{2}\right]$ A is passed through a true RMS ammeter, the meter reading will be

(a) $6\sqrt{2}$ A (b) $\sqrt{126}$A

(c) 12 A (d) $\sqrt{216}$A

6. The Fourier series expansion of a real periodic signal with fundamental frequency f_o is given by

$$g_p(t) = \sum_{n=-\infty}^{\infty} C_n e^{j2\pi n f_o t}$$

It is given than $C_3 = 3 + j5$. Then C_{-3} is

(a) $5 + j3$ (b) $-3 - j5$

(c) $-5 + j3$ (d) $3 - j5$

7. A signal $x(t)$ is given by

$$x(t) = \begin{cases} 1, & \frac{-T}{4} < t \le \frac{3T}{4} \\ -1, & \frac{3T}{4} < t \le \frac{7T}{4} \\ -x(t+T) \end{cases}$$

Which among the following gives the fundamental Fourier term of $x(t)$?

(a) $\frac{4}{\pi}\cos\left(\frac{\pi t}{T} - \frac{\pi}{4}\right)$ (b) $\frac{\pi}{4}\cos\left(\frac{\pi t}{2T} + \frac{\pi}{4}\right)$

(c) $\frac{4}{\pi}\sin\left(\frac{\pi t}{T} - \frac{\pi}{4}\right)$ (d) $\frac{4}{\pi}\sin\left(\frac{\pi t}{2T} + \frac{\pi}{4}\right)$

8. The Fourier Series of an odd periodic function, contains only

(a) Odd harmonics (b) Even harmonics

(c) Cosine terms (d) Sine terms

9. The trigonometric Fourier series of an **even function** of time does not have

(a) dc term

(b) Cosine terms

(c) Sine terms

(d) Odd harmonic terms

10. The **trigonometric Fourier series** of a periodic time function can have only

(a) Cosine terms

(b) Sine terms

(c) Cosine and sine terms

(d) dc and cosine terms

11. A periodic signal $x(t)$ of period T_o is given by

$$x(t) = \begin{cases} 1 & |t| < T_1 \\ 0 & T_1 < |t| < (T_o/2) \end{cases}$$

The dc component of $x(t)$ is

(a) $\frac{T_1}{T_o}$ (b) $\frac{T_1}{2T_o}$

(c) $\frac{2T_1}{T_o}$ (d) $\frac{T_o}{T_1}$

12. The Fourier series representation of an impulse train denoted by

$$s(t) = \sum_{n=-\infty}^{\infty} \delta(t - nT_o) \text{ is given by}$$

(a) $\frac{1}{T_o}\sum_{n=-\infty}^{\infty} e^{\left(\frac{-j2\pi nt}{T_o}\right)}$ (b) $\frac{1}{T_o}\sum_{n=-\infty}^{\infty} e^{\left(\frac{-j\pi nt}{T_o}\right)}$

(c) $\frac{1}{T_o}\sum_{n=-\infty}^{\infty} e^{\left(\frac{j\pi nt}{T_o}\right)}$ (d) $\frac{1}{T_o}\sum_{n=-\infty}^{\infty} e^{\left(\frac{j2\pi nt}{T_o}\right)}$

13. One period (0, T) each of two periodic waveforms W_1 and W_2 are shown in figure. The magnitudes of the nth Fourier series coefficients of W_1 and W_2 for $n \geq 1$, n odd, are respectively proportional to

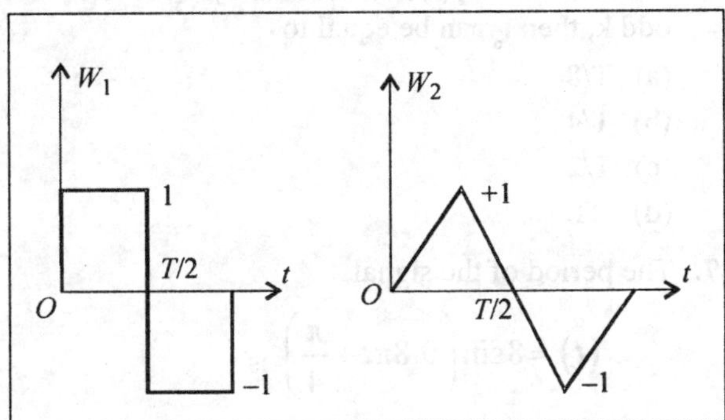

(a) $|n^{-3}|$ and $|n^{-2}|$ (b) $|n^{-2}|$ and $|n^{-3}|$

(c) $|n^{-1}|$ and $|n^{-2}|$ (d) $|n^{-4}|$ and $|n^{-2}|$

14. Which of the following cannot be Fourier series expansion of a periodic signal?

(a) $x(t) = 2\cos t + 3\cos 3t$

(b) $x(t) = 2\cos \pi t + 7\cos t$

(c) $x(t) = \cos t + 0.5$

(d) $x(t) = 2\cos 1.5\pi t + \sin 3.5\pi t$

15. Which of the following signals is non periodic ?

(a) $s(t) = \cos 2t + \cos 3t + \cos 5t$

(b) $s(t) = \exp(j\,8\pi t)$

(c) $s(t) = \exp(-7t)\sin 10\pi t$

(d) $s(t) = \cos 2t \cos 4t$

16. The Fourier series of a real periodic function has only

P. cosine terms if it is even Q. sine terms if it is even

R. cosine terms if it is odd S. sine terms if it odd

Which of the above statements are correct?

(a) P and S

(b) P and R

(c) Q and S

(d) Q and R

17. A function is given by $f(t) = \sin^2 t + \cos 2t$. Which of the following is true ?

(a) f has frequency components at 0 and $1/2\pi$ Hz

(b) f has frequency components at 0 and $1/\pi$ Hz

(c) f has frequency components at $1/2\pi$ and $1/\pi$ Hz

(d) f has frequency components at 0, $1/2\pi$ and $1/\pi$ Hz

18. The trigonometric Fourier series for the waveform $f(t)$ shown below contains

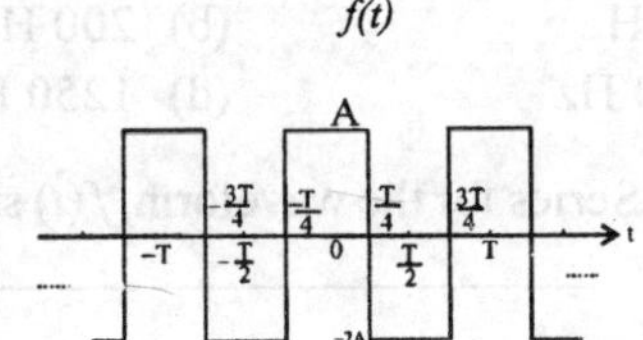

(a) Only cosine terms and zero value for the dc component

(b) Only cosine terms and a positive value for the dc component

(c) Only cosine terms and a negative value for the dc component

(d) Only sine terms and a negative value for the dc component

19. The trignometric Fourier series of an even function does not have the

(a) dc term (b) cosine terms

(c) sine terms (d) odd harmonic terms

20. Choose the function $f(t)$; $-\infty < t < \infty$, for which a Fourier series cannot be defined.

(a) 3 sin (25 t)

(b) 4 cos (20t + 3) + 2 sin (10 t)

(c) exp (– | t |) sin (25 t)

(d) 1

21. The rms value of the periodic waveform $e(t)$, shown in figure below is

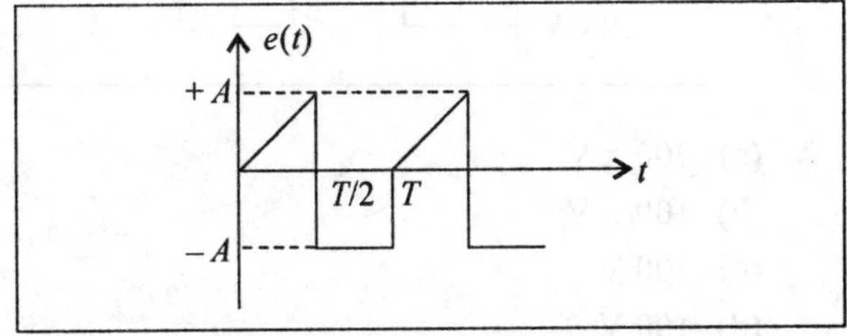

(a) $\sqrt{3/2}$ A (b) $\sqrt{2/3}$ A

(c) $\sqrt{1/3}$ A (d) $\sqrt{2}$ A

22. A periodic rectangular signal, $x(t)$ has the wave form shown in figure Frequency of the fifth harmonic of its spectrum is

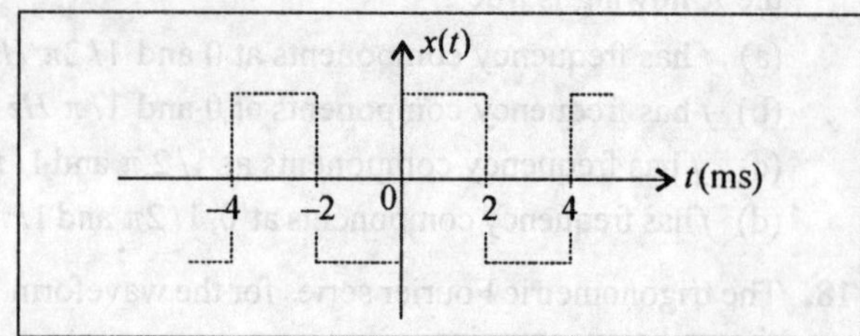

(a) 40 Hz (b) 200 Hz

(c) 250 Hz (d) 1250 Hz

23. Fourier Series for the waveform, $f(t)$ shown in Figure is:

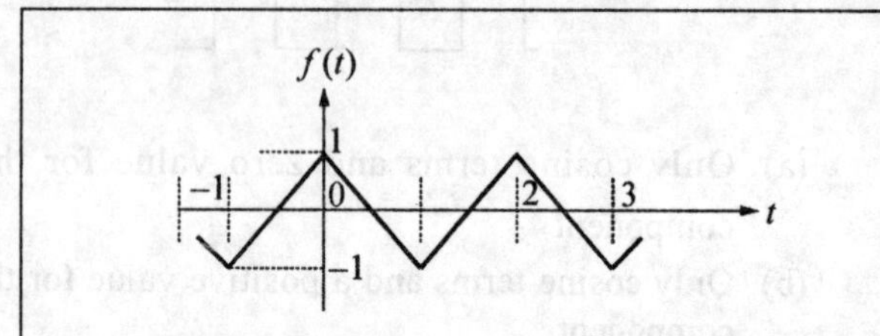

(a) $\frac{8}{\pi^2}\left(\sin(\pi t)+\frac{1}{9}\sin(3\pi t)+\frac{1}{25}\sin(5\pi t)+...\right)$

(b) $\frac{8}{\pi^2}\left(\sin(\pi t)-\frac{1}{9}\sin(3\pi t)+\frac{1}{25}\sin(5\pi t)+...\right)$

(c) $\frac{8}{\pi^2}\left(\cos(\pi t)+\frac{1}{9}\cos(3\pi t)+\frac{1}{25}\cos(5\pi t)+...\right)$

(d) $\frac{8}{\pi^2}\left(\cos(\pi t)-\frac{1}{9}\cos(3\pi t)+\frac{1}{25}\cos(5\pi t)+...\right)$

24. What is the rms value of the voltage waveform shown in Figure?

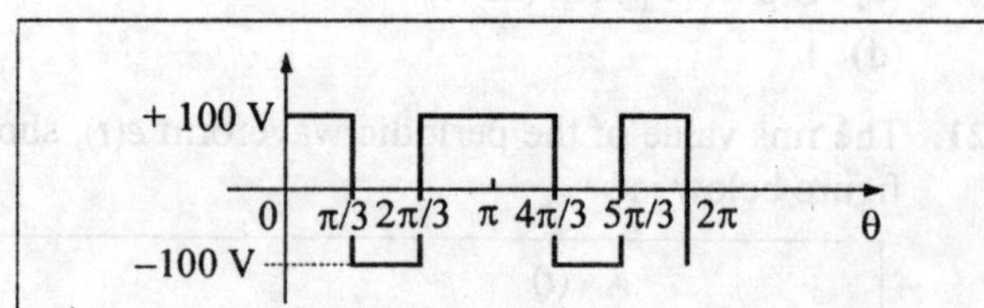

(a) 200/π V

(b) 100/π V

(c) 200 V

(d) 100 V

25. The Fourier Series coefficients, of a periodic signal $x(t)=\sum_{k=-\infty}^{\infty} a_k e^{j2\pi kt/T}$ are given by

$a_{-2}=2-j1;\ a_{-1}=0\cdot5+j0\cdot2;$

$a_0=j2;$

$a_1=0\cdot5-j0\cdot2;$

$a_2=2+j1;$ and

$a_k=0;$ for $|k|>2.$

Which of the following is true?

(a) $x(t)$ has infinite energy because only finitely many coefficients are non zero .

(b) $x(t)$ has zero average value because it is periodic

(c) The imaginary part of $x(t)$ is constant

(d) The real part of $x(t)$ is even

26. Let x(t) be a periodic signal with time period T. Let y(t) = x (t - t_o) + x (t + t_o) for some t_o. The Fourier series coefficients of y(t) are denoted by b_k. If b_k = 0 for all odd k, then t_o can be equal to

(a) T/8

(b) T/4

(c) T/2

(d) 2T

27. The period of the signal

$$x(t)=8\sin\left(0.8\pi t+\frac{\pi}{4}\right) \text{ is}$$

(a) 0.4π *s*

(b) 0.8π *s*

(c) 1.25 *s*

(d) 2.5 *s*

28. The second harmonic component of the periodic waveform given in the figure has an amplitude of

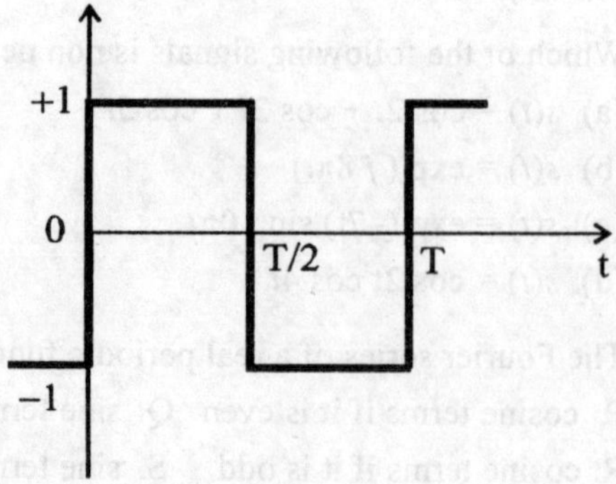

(a) 0
(b) 1
(c) $2/\pi$
(d) $\sqrt{5}$

29. The Fourier series expansion

$$f(t) = a_0 + \sum_{n=1}^{\infty} a_n \cos n\omega t + b_n \sin n\omega t$$

of the periodic signal shown below will contain the following non-zero terms

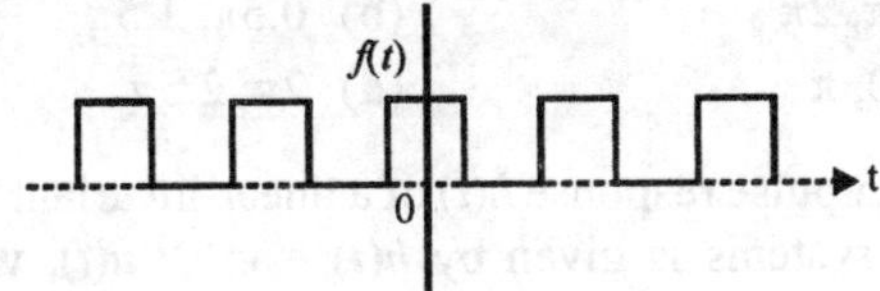

(a) a_0 and $b_n, n = 1,3,5,...\infty$
(b) a_0 and $a_n, n = 1,2,3,...\infty$
(c) a_0, a_n and $b_n, n = 1,2,3,...\infty$
(d) a_0 and $a_n, n = 1,3,5,...\infty$

30. The RMS value of a rectangular wave of period T, having a value of + V for a duration, T_1 (<T) and – V for the duration, $T - T_1 = T_2$, equals
(a) V
(b) $\frac{T_1 - T_2}{T}V$
(c) $\frac{V}{\sqrt{2}}$
(d) $\frac{T_1}{T_2}V$

31. $f(x)$, shown in the adjoining figure is represented by

$$f(x) = a_0 + \sum_{n=1}^{\infty}\{a_n \cos(nx) + b_n \sin(nx)\}.$$

The value of a_0 is.

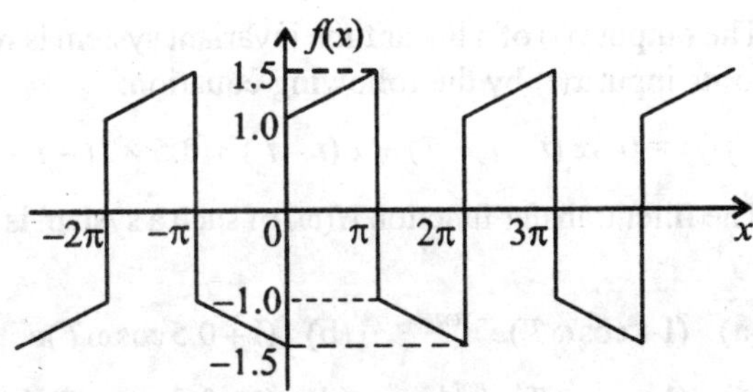

(a) 0
(b) $\pi/2$
(c) π
(d) 2π

32. Consider a periodic signal $x(t)$ as shown below

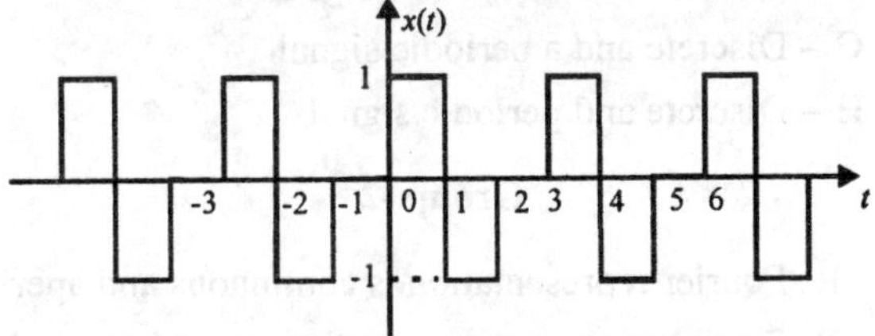

It has a Fourier series representation

$$x(t) = \sum_{k=-\infty}^{\infty} a_k e^{j(2\pi/T)kt}$$

Which one of the following statements is **TRUE**?
(a) $a_k = 0$, for k odd integer and T = 3
(b) $a_k = 0$, for k even integer and T = 3
(c) $a_k = 0$, for k even integer and T = 6
(d) $a_k = 0$, for k odd integer and T = 6

4.3 Fourier Transform

1. The Fourier transform of a conjugate symmetric function is always
(a) Imaginary
(b) Conjugate anti-symmetric
(b) Real
(d) Conjugate symmetric

2. Let $x(t)$ and $y(t)$ (With Fourier transforms $X(f)$ and $Y(f)$ respectively) be related as shown in Figure

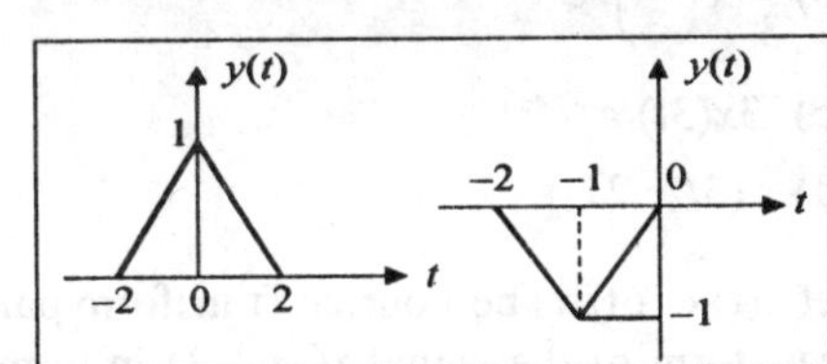

Then $Y(f)$ is

(a) $-\frac{1}{2}X\left(\frac{f}{2}\right)e^{-j2\pi f}$ (b) $-\frac{1}{2}X\left(\frac{f}{2}\right)e^{j2\pi f}$

(c) $-X\left(\frac{f_1}{2}\right)e^{j2\pi f}$ (d) $-X\left(\frac{f}{2}\right)e^{-j2\pi f}$

3. The output $y(t)$ of a linear time invariant system is related to its input $x(t)$ by the following equation:

$$y(t) = 0.5x(t - t_d + T) + x(t - t_d) + 0.5 \times (t - t_d - T).$$

The filter transfer function $H(\omega)$ of such a system is given by

(a) $(1+\cos\omega T)e^{-j\omega t_d}$ (b) $(1+0.5\cos\omega T)e^{-j\omega t_d}$

(c) $(1+\cos\omega T)e^{j\omega t_d}$ (d) $(1-0.5\cos\omega T)e^{-j\omega t_d}$

4. Match the following and choose the correct combination.

Group-1

E – continuous and a periodic signal
F – continuous and periodic signal
G – Discrete and a periodic signal
H – Discrete and periodic signal

Group -2

1. Fourier representation is continuous and aperiodic
2. Fourier representation is discrete and aperiodic
3. Fourier representation is continuous and periodic
4. Fourier representation is discrete and periodic

	E	F	G	H
(a)	3	2	4	1
(b)	1	3	2	4
(c)	1	2	3	4
(d)	2	1	4	3

5. For a signal $x(t)$ the Fourier transform is $X(f)$. Then the inverse Fourier transform of $X(3f + 2)$ is given by

(a) $\frac{1}{2}x\left(\frac{t}{2}\right)e^{j3\pi t}$

(b) $\frac{1}{3}x\left(\frac{t}{3}\right)e^{-j4\pi t/3}$

(c) $3x(3t)\,e^{-j4\pi t}$

(d) $x(3t + 2)$

6. Let $x(t) \leftrightarrow X(jw)$ be Fourier Transform pair. The Fourier Transform of the signal $x(5t - 3)$ in terms of $X(jw)$ is given as.

(a) $\frac{1}{5}\exp\left(\frac{-j3\omega}{5}\right)X\left(\frac{j\omega}{5}\right)$

(b) $\frac{1}{5}\exp\left(\frac{j3\omega}{5}\right)X\left(\frac{j\omega}{5}\right)$

(c) $\frac{1}{5}\exp(-j3\omega)X\left(\frac{j\omega}{5}\right)$

(d) $\frac{1}{5}\exp\left(\frac{-j3\omega}{5}\right)X\left(\frac{j\omega}{5}\right)$

7. The signal $x(t)$ is described by

$$x(t) = \begin{cases} 1 & \text{for } -1 \le t \le +1 \\ 0 & \text{otherwise} \end{cases}$$

Two of the angular frequencies at which its Fourier transform becomes zero are

(a) π, 2π (b) 0.5π, 1.5π

(c) 0, π (d) 2π, 2.5π

8. The impulse response $h(t)$ of a linear-invariant continous time systems is given by $h(t) = e^{(-2t)}u(t)$, where $u(t)$ denotes the unit step function.
The frequency response $H(\omega)$ of this systems in terms of angular ω, is given by $H(\omega) =$

(a) $\frac{1}{1+j2\omega}$ (b) $\frac{\sin(\omega)}{\omega}$

(c) $\frac{1}{2+j\omega}$ (d) $\frac{j\omega}{2+j\omega}$

9. The impulse response $h(t)$ of a linear-invariant continous time systems is given by $h(t) = e^{(-2t)}u(t)$, where $u(t)$ denotes the unit step function.
The output of this system, to the sinusoidal input $x(t) = 2\cos(2t)$ for all time t, is

(a) 0
(b) $2^{-0.25}\cos(2t - 0.125\pi)$
(c) $2^{-0.5}\cos(2t - 0.125\pi)$
(d) $2^{-0.5}\cos(2t - 0.25\pi)$

10. If $G(f)$ represents the Fourier transform of a signal $g(t)$ which is real and odd symmetric in time, then

(a) $G(f)$ is complex
(b) $G(f)$ is imaginary
(c) $G(f)$ is real
(d) $G(f)$ is real and non-negative

11. The Fourier transform of a real valued time signal has

(a) Odd symmetry
(b) Even symmetry
(c) Conjugate symmetry
(d) No Symmetry

12. The function $f(t)$ has the Fourier Transform $g(\omega)$. The Fourier Transform of

$$g(t) = \left(\int_{-\infty}^{\infty} g(t)\,e^{-j\omega t}\,dt\right) \text{ is}$$

(a) $\frac{1}{2\pi}F(\omega)$ (b) $\frac{1}{2\pi}F(-\omega)$

(c) $2\pi\, F(-\omega)$ (d) None of the above

13. The Fourier transform of a function $x(t)$ is $X(f)$. The Fourier transform of $[dx(t)/dt]$ will be

(a) $\frac{d\,X(f)}{df}$ (b) $j\,2\pi f\, X(f)$

(c) $jfX(f)$ (d) $\frac{X(f)}{jf}$

14. A signal $x(t)$ has a Fourier transform $X(\omega)$. If $x(t)$ is a real and odd function of t, then $X(\omega)$ is

(a) A real and even function of ω
(b) An imaginary and odd function of ω
(c) An imaginary and even function of ω
(d) A real and odd function of ω

15. The Fourier Transform of the signal $x(t) = e^{(-3t^2)}$ is of the following form, where A and B are constants

(a) $Ae^{-B|f|}$ (b) Ae^{-Bf}

(c) $A + B|f|^2$ (d) $Ae^{(-Bf^2)}$

16. The Fourier transform $F\{e^{-t}\, u(t)\}$ is equal to $\frac{1}{1+j2\pi f}$

Therefore, $F\left\{\frac{1}{1+j2\pi f}\right\}$ is

(a) $e^{f}\,u(f)$
(b) $e^{-f}\,u(f)$
(c) $e^{f}\,u(-f)$
(d) $e^{-f}\,u(-f)$

17. A real function $f(t)$ has a Fourier transform $F(\omega)$. The Fourier transform of $[f(t) - f(-t)]$ is .

(a) Zero (b) Real

(c) Real and odd (d) Imaginary

18. The continuous-time signal $x(t) = \frac{1}{a^2+t^2}$ has the Fourier transform $\frac{\pi}{a}e^{(-a|\omega|)}$. The signal $x(t)\cos bt$ has the Fourier transform

(a) $\frac{\pi}{2a}[e^{(-a|\omega-b|)} + e^{(-a|\omega+b|)}]$

(b) $\frac{\pi}{2a}[e^{(-a|\omega|)} + e^{(-a|\omega|)}]$

(c) $\frac{\pi}{2a}[e^{(-a|\omega|)}\cos b\omega]$

(d) $\frac{\pi}{2a}[e^{(-a|\omega-b|)} - e^{(-a|\omega+b|)}]$

19. If the waveform shown in the following figure, corresponds to the second derivative of a given function $f(t)$, then the Fourier transform of $f(t)$ is

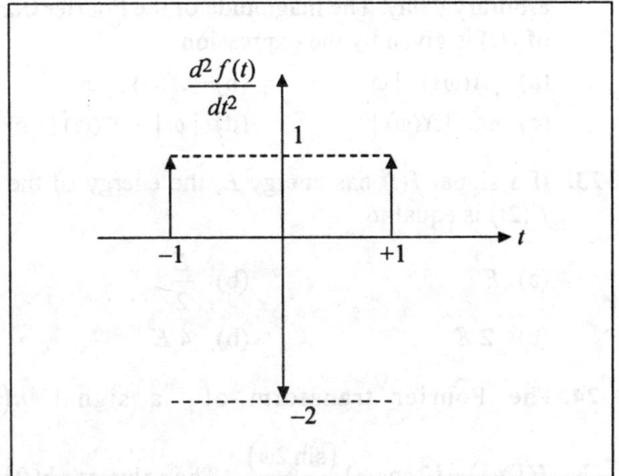

(a) $1 + \sin\omega$ (b) $1 + \cos\omega$

(c) $\frac{2(1-\cos\omega)}{\omega^2}$ (d) $\frac{2(1+\cos\omega)}{\omega^2}$

20. The Fourier transform of a function $g(t)$ is given as.

$$G(\omega) = \frac{\omega^2+21}{\omega^2+9}$$

Then the function $g(t)$ is given as

(a) $\delta(t) + 2\exp(-3|t|)$ (b) $\cos 3\omega t + 21\exp(-3t)$

(c) $\sin 3\omega t + 7\cos\omega t$ (d) $\sin 3\omega t + 21\exp(3t)$

21. The magnitude of Fourier transform $X(\omega)$ of a function $x(t)$ is shown below in figure (a). The magnitude of Fourier transform $Y(\omega)$ of another function $y(t)$ is shown below in Fig. (b).The phases of $X(\omega)$ and $Y(\omega)$ are zero for all. The magnitude and frequency units are identical in both the figures. The function $y(t)$ can be expressed in terms of $x(t)$ as

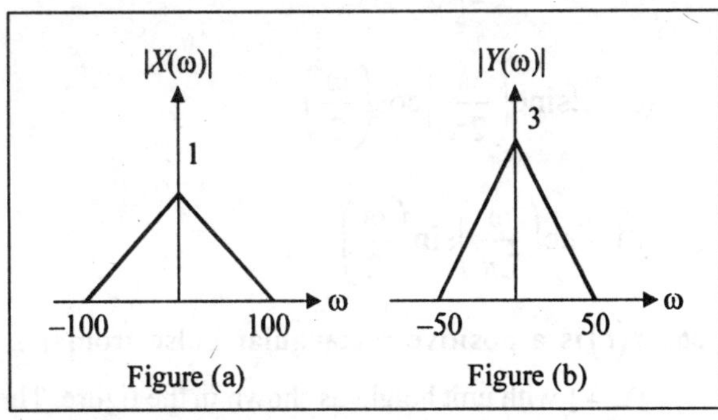

(a) $\frac{2}{3}x\left(\frac{t}{2}\right)$ (b) $\frac{3}{2}x(2t)$

(c) $\frac{2}{3}x(2t)$ (d) $\frac{3}{2}x\left(\frac{t}{2}\right)$

22. Let the signal $x(t)$ have the Fourier transform $X(\omega)$. Consider the signal $y(t) = \frac{d}{dt}[x(t-t_d)]$ where t_d is an arbitrary delay. The magnitude of the Fourier transform of $y(t)$ is given by the expression

(a) $|X(\omega)| \cdot |\omega|$
(b) $|X(\omega)| \cdot \omega$
(c) $\omega^2 \cdot |X(\omega)|$
(d) $|\omega| \times X(\omega) | \cdot e^{-j\omega t_d}$

23. If a signal $f(t)$ has energy E, the energy of the signal $f(2t)$ is equal to

(a) E
(b) $\frac{E}{2}$
(c) $2E$
(b) $4E$

24. The Fourier transform of a signal $h(t)$ is $H(j\omega) = (2\cos\omega)\frac{(\sin 2\omega)}{\omega}$. The value of $h(0)$ is

(a) 1/4
(b) 1/2
(c) 1
(d) 2

25. Let $x(t) = \text{rect}\left(t - \frac{1}{2}\right)$ (where $\text{rect}(x) = 1$ for $-\frac{1}{2} \le x \le \frac{1}{2}$ and zero otherwise). Then if $\text{sinc}(x) = \frac{\sin(\pi x)}{\pi x}$, the Fourier transformer of $x(t) + x(-t)$ will be given by

(a) $\text{sinc}\left(\frac{\omega}{2\pi}\right)$

(b) $2\text{sinc}\left(\frac{\omega}{2\pi}\right)$

(c) $2\text{sinc}\left(\frac{\omega}{2\pi}\right)\cos\left(\frac{\omega}{2}\right)$

(d) $\text{sinc}\left(\frac{\omega}{2\pi}\right)\sin\left(\frac{\omega}{2}\right)$

26. $x(t)$ is a positive rectangular pulse from $t = -1$ to $t = +1$ with unit height as shown in the figure. The value of $\int_{-\infty}^{\infty} |X(\omega)|^2 d\omega$ {where $X(\omega)$ is the Fourier transform of $x(t)$ }is

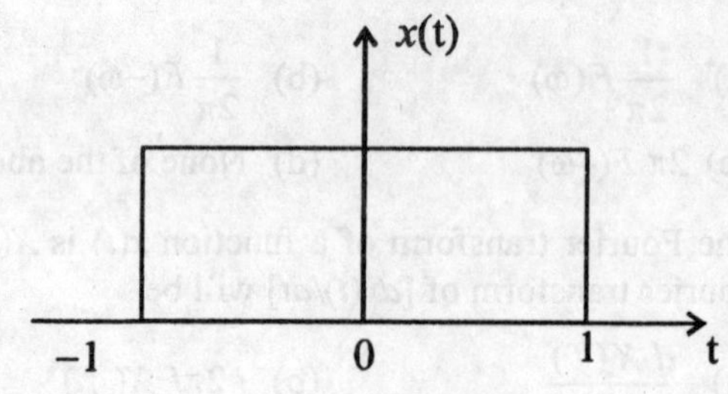

(a) 2
(b) 2π
(c) 4
(d) 4π

27. The Fourier transform of a signal $h(t)$ is $H(j\omega) = (2\cos\omega)\frac{(\sin 2\omega)}{\omega}$. The value of $h(0)$ is

(a) 1/4
(b) 1/2
(c) 1
(d) 2

28. The amplitude spectrum of a Gaussian pulse is

(a) Uniform
(b) A sine function
(c) Gaussian
(d) An impulse function

29. The Fourier transform of a voltage signal $x(t)$ is $X(f)$. The unit $|X(f)|$ is

(a) Volt
(b) Volt-sec
(c) Volt /sec
(d) Volt²

30. Consider the signal $x(t) = \begin{cases} e^{-t} & t \ge 0 \\ 0, & t < 0 \end{cases}$.

Let $X(\omega)$ denote the Fourier transform of this signal.

The integral $\frac{1}{2\pi}\int_{-\infty}^{\infty} X(\omega)d\omega$ is:

(a) 0
(b) 1/2
(c) 1
(d) ∞

31. A real function $f(t)$ has a Fourier transform $F(\omega)$. The Fourier transform of $[f(t) - f(-t)]$ is.

(a) Zero
(b) Real
(c) Real and odd
(d) Imaginary

32. If the Fourier transform of x[n] is $X(e^{jw})$, then the Fourier transform of $(-1)^n x[n]$

(a) $(-j)^{\omega} X(e^{j\omega})$
(b) $(-1)^{\omega} X(e^{j\omega})$
(c) $X(e^{j(\omega-\pi)})$
(d) $\frac{d}{d\omega}(X(e^{j\omega}))$

33. The continuous-time signal $x(t) = \frac{1}{a^2 + t^2}$ has the Fourier transform $\frac{\pi}{a}e^{(-a|\omega|)}$. The signal $x(t)$ cos bt has the Fourier transform

(a) $\frac{\pi}{2a}[e^{(-a|\omega-b|)}+e^{(-a|\omega+b|)}]$

(b) $\frac{\pi}{2a}[e^{(-a|\omega|)}+e^{(-a|\omega|)}]$

(c) $\frac{\pi}{2a}[e^{(-a|\omega|)}\cos b\omega]$

(d) $\frac{\pi}{2a}[e^{(-a|\omega-b|)}-e^{(-a|\omega+b|)}]$

34. If the waveform shown in the following figure, corresponds to the second derivative of a given function $f(t)$, then the Fourier transform of $f(t)$ is

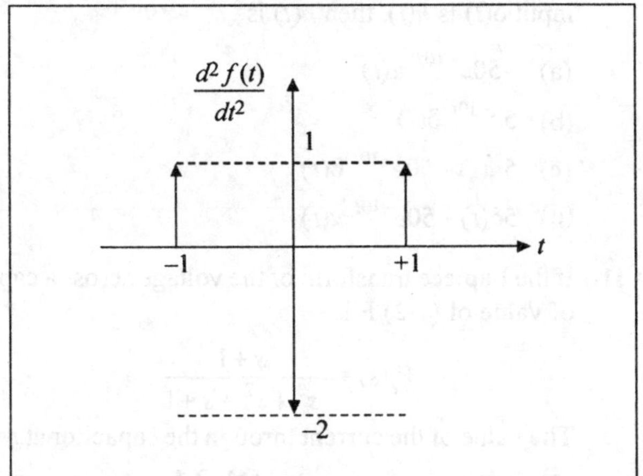

(a) $1 + \sin \omega$ (b) $1 + \cos \omega$

(c) $\frac{2(1-\cos\omega)}{\omega^2}$ (d) $\frac{2(1+\cos\omega)}{\omega^2}$

35. The Fourier transform of a function $g(t)$ is given as.

$$G(\omega) = \frac{\omega^2+21}{\omega^2+9}$$

Then the function $g(t)$ is given as

(a) $\delta(t) + 2\exp(-3|t|)$

(b) $\cos 3\omega t + 21 \exp(-3t)$

(c) $\sin 3\omega t + 7\cos \omega t$

(d) $\sin 3\omega t + 21 \exp(3t)$

36. The magnitude of Fourier transform $X(\omega)$ of a function $x(t)$ is shown below in figure (a). The magnitude of Fourier transform $Y(\omega)$ of another function $y(t)$ is shown below in Fig. (b). The phases of $X(\omega)$ and $Y(\omega)$ are zero for all. The magnitude and frequency units are identical in both the figures. The function $y(t)$ can be expressed in terms of $x(t)$ as

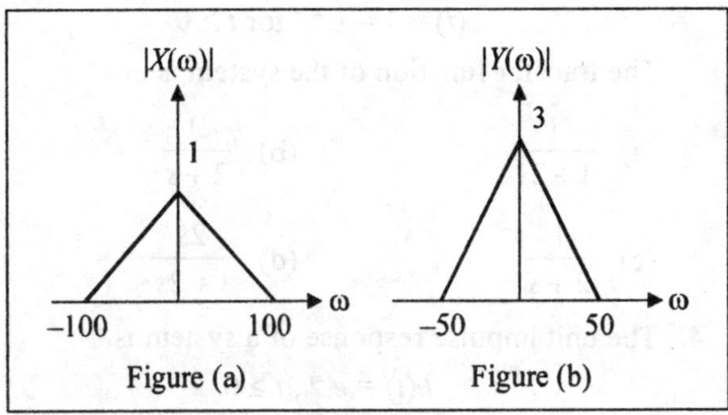

(a) $\frac{2}{3}x\left(\frac{t}{2}\right)$

(b) $\frac{3}{2}x(2t)$

(c) $\frac{2}{3}x(2t)$

(d) $\frac{3}{2}x\left(\frac{t}{2}\right)$

37. Let the signal $x(t)$ have the Fourier transform $X(\omega)$. Consider the signal $y(t) = \frac{d}{dt}[x(t-t_d)]$ where t_d is an arbitrary delay. The magnitude of the Fourier transform of $y(t)$ is given by the expression

(a) $|X(\omega)|\cdot|\omega|$ (b) $|X(\omega)|\cdot\omega$

(c) $\omega^2\cdot|X(\omega)|$ (d) $|\omega|\times X(\omega)|\cdot e^{-j\omega t_d}$

4.4 Laplace Transform

1. In the system shown below, $x(t) = (\sin t)\, u(t)$. In steady state, the response $y(t)$ will be.

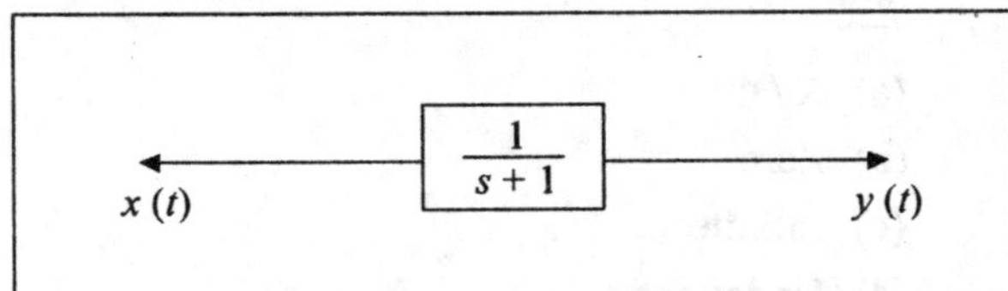

(a) $\frac{1}{\sqrt{2}}\sin[t-(\pi/4)]$

(b) $\frac{1}{\sqrt{2}}\sin[t+(\pi/4)]$

(c) $\frac{1}{\sqrt{2}}e^{-1}\sin t$

(d) $\sin t - \cos t$

2. Consider the function $f(t)$ having Laplace transform

$$F(s) = \frac{\omega_0}{(s^2 + \omega_0^2)} \text{ Re}[s] > 0$$

The final value of $f(t)$ would be:

(a) 0 (b) 1
(c) $-1 \le f(\infty) \le 1$ (d) ∞

3. The unit–step response of a system starting from rest is given by.

$$c(t) = 1 - e^{-2t} \text{ for } t \ge 0$$

The transfer function of the system is .

(a) $\frac{1}{1+2s}$ (b) $\frac{2}{2+s}$
(c) $\frac{1}{2+s}$ (d) $\frac{2s}{1+2s}$

4. The unit impulse response of a system is.

$$h(t) = e^{-t}, t \ge 0$$

For this system, the steady-state value of the output for unit step input is equal to.

(a) – 1 (b) 0
(c) 1 (d) ∞

5. If the Laplace of a signal $y(t)$ is $Y(s) = \frac{1}{s(s-1)}$, then its final value is

(a) – 1 (b) 0
(c) 1 (d) Unbounded

6. The 3-dB bandwidth of the low-pass signal $e^{-t}u(t)$, where $u(t)$ is the unit step function , is given by

(a) $\frac{1}{2\pi}$ Hz (b) $\frac{1}{2\pi}\sqrt{\sqrt{2}-1}$ Hz
(c) ∞ (d) 1 Hz

7. The Laplace transform of a function $f(t)$ if

$$F(s) = \frac{5s^2 + 23s + 6}{s(s^2 + 2s + 2)}$$. As $t \to \infty$, $f(t)$ approaches

(a) 3 (b) 5
(c) $\frac{17}{2}$ (d) ∞

8. A continuous time LTI system is described by

$$\frac{d^2y(t)}{dt^2} + 4\frac{dy(t)}{dt} + 3y(t) = 2\frac{dx(t)}{dt} + 4x(t)$$

Assuming zero initial conditiones, the response $y(t)$ of the above system for the input $x(t) = e^{-2t}u(t)$ is given by

(a) $(e^{t} - e^{3t})u(t)$
(b) $(e^{-t} - e^{-3t})u(t)$
(c) $(e^{-t} + e^{-3t})u(t)$
(d) $(e^{t} + e^{3t})u(t)$

9. A Linear Time Invariant system with an impulse response $h(t)$ produces output $y(t)$ when input $x(t)$ is applied. When the input $x(t - \tau)$ is applied to a system with impulse response $h(t - \tau)$, the output will be

(a) $y(t)$ (b) $y(2(t - \tau))$
(c) $y(t - \tau)$ (d) $y(t - 2\tau)$

10. The step response of a linear time imvariant system is $y(t) = 5e^{-10t}\,u(t)$, where $u(t)$ is the unit step function. If the output of the system corresponding to an impulse input $\delta(t)$ is $h(t)$, then $h(t)$ is

(a) $-50e^{-10t}\,u(t)$
(b) $5e^{-10t}\,\delta(t)$
(c) $5u(t) - 50e^{-10t}\,\delta(t)$
(d) $5\delta(t) - 50e^{-10t}\,u(t)$

11. If the Laplace transform of the voltage across a capacitor of value of (1/2) F is

$$V_c(s) = \frac{s+1}{s^3 + s^2 + s + 1}$$

The value of the current through the capacitor at $t = 0^+$ is

(a) 0 A (b) 2 A
(c) $\frac{1}{2}$ A (d) 1 A

12. The voltage across an impedance in a network is $V(s) = Z(s)\,I(s)$, where $V(s)$, $Z(s)$ and $I(s)$ are the Laplace Transforms of the corresponding time functions $v(t)$, $z(t)$ and $i(t)$. The voltage $v(t)$ is

(a) $v(t) = z(t)\; i(t)$
(b) $v(t) = \int_0^t i(\tau)\cdot z(t-\tau)\,d\tau$
(c) $v(t) = \int_0^t i(\tau)\cdot z(t+\tau)\,d\tau$
(d) $v(t) = z(t) + i(\tau)$

13. If $F(s)$ = FT $[f(t)]$ and $s\,F(s) = \frac{K}{(s+1)(s^2+4)}$ the $\lim_{t\to\infty} f(t)$.

(a) $K/4$
(b) Zero
(c) Infinite
(d) Undefined

14. The Laplace transform of a unit ramp function starting at $t = a$ is

(a) $\frac{1}{(s+a)^2}$ (b) $\frac{e^{-as}}{(s+a)^2}$

(c) $\frac{e^{-as}}{s^2}$ (d) $\frac{a}{s^2}$

15. The inverse Laplace transform of the function $\frac{s+5}{(s+1)(s+3)}$ is

(a) $2e^{-t} - e^{-3t}$ (b) $2e^{-t} + e^{-3t}$
(c) $e^{-t} - 2e^{-3t}$ (d) $e^{-t} + 2e^{-3t}$

16. Given that $F(s)$ is the one-sided Laplace transform of $f(t)$, the Laplace transform of $\int_0^t f(\tau)\,d\tau$ is

(a) $sF(s) - f(0)$ (b) $\frac{1}{s}F(s)$

(c) $\int_0^s F(\tau)\,d\tau$ (d) $\frac{1}{s}[F(s) - f(0)]$

17. Given $f(t) = LT^{-1}\left[\frac{3s+1}{s^3+4s^2+(K-3)s}\right]$.

If $\lim_{t\to\infty} f(t) = 1$, then the value of K is
(a) 1 (b) 2
(c) 3 (d) 4

18. If $F(s) = L[f(t)] = \frac{2(s+1)}{s^2+4s+7}$ then the initial and final values of $f(t)$ are respectively
(a)0.2 (b) 2, 0
(c)0, 2/7 (d) 2/7, 0

19. If $[f(t)] = F(s)$, then $[f(t-T)]$ is equal to
(a) $e^{sT}F(s)$ (b) $e^{-sT}F(s)$

(c) $\frac{F(s)}{1+e^{sT}}$ (d) $\frac{F(s)}{1-e^{-sT}}$

20. The Laplace transformation of $f(t)$ is $F(s)$. Given $F(s) = [\omega/(s^2+\omega^2)]$, the final value of $f(t)$ is
(a) Infinity (b) Zero
(c) One (d) None of the above

21. The Laplace transform of $(t^2 - 2t)\,u(t-1)$ is

(a) $\frac{2}{s^2}e^{-s} - \frac{2}{s^2}e^{-s}$ (b) $\frac{2}{s^3}e^{-2s} - \frac{2}{s^2}e^{-s}$

(c) $\frac{2}{s^3}e^{-s} - \frac{2}{s^2}e^{-s}$ (d) None of the above

22. A rectangular current pulse of duration T and magnitude I has the Laplace transform
(a) I/s (b) $(I/s)\exp(Ts)$
(c) $(I/s)\exp(Ts)$ (d) $(I/s)[1-\exp(Ts)]$.

23. Let $Y(s)$ be the Laplace transformation of the function $y(t)$, then the final value of the function is
(a) $\lim_{s\to 0} Y(s)$ (b) $\lim_{s\to 0} Y(s)$
(c) $\lim_{s\to 0} sY(s)$ (d) $\lim_{s\to 0} sY(s)$

24. The unilateral Laplace transform of $f(t)$ is $\frac{1}{s^2+s+1}$. The unilateral Laplace transform of $t\,f(t)$ is

(a) $-\frac{s}{(s^2+s+1)^2}$ (b) $-\frac{2s+1}{(s^2+s+1)^2}$

(c) $\frac{s}{(s^2+s+1)^2}$ (d) $\frac{2s+1}{(s^2+s+1)^2}$

25. Let the Laplace transform of a function $f(t)$ which exists for $t > 0$ be $F_1(s)$ and the Laplace transform of its delayed version $f(t-\tau)$ be $F_2(s)$. Let $F_1^*(s)$ be the complex conjugate of $F_1(s)$ with the Laplace variable set as $s = \sigma + j\omega$. if $G(s) = \frac{F_2(s).F_1^*(s)}{|F_1(s)|^2}$,

then the inverse Laplace transform of $G(s)$ is
(a) An ideal impulse $\delta(t)$
(b) An ideal delayed impulse $\delta(t-\tau)$
(c) An ideal step function $u(t)$
(d) An ideal delayed step function $u(t-\tau)$

26. The Laplace transform of a function $f(t)$ if

$F(s) = \frac{5s^2+23s+6}{s(s^2+2s+2)}$. As $t \to \infty$, $f(t)$ approaches
(a) 3 (b) 5

(c) $\frac{17}{2}$ (d) ∞

27. If $L[f(t)] = \frac{2(s+1)}{s^2+2s+5}$ then $f(0^+)$ and $f(\infty)$ are given by
(a) 0, 2 respectively (b) 2, 0 respectively
(c) 0, 1 respectively (d) 2/5, 0 respectively

28. The final value theorem is used to find the
(a) Steady state value of the system output.
(b) Initial value of the system output

(c) Transient behavior of the system output
(d) None of these

29. The Laplace Transform of $e^{\alpha t} \cos(\alpha t)$ is equal to

(a) $\dfrac{(s-\alpha)}{(s-\alpha)^2+\alpha^2}$ (b) $\dfrac{(s+\alpha)}{(s-\alpha)^2+\alpha^2}$

(c) $\dfrac{1}{(s-\alpha)^2}$ (d) None of the above

30. If $L[f(t)] = \dfrac{\omega}{(s^2+\omega^2)}$, then the value of $\lim_{t\to\infty} f(t)$

(a) Cannot be determined
(b) Is zero
(c) Is unity
(d) Is infinite

31. Given that

$$L[f(t)] = \frac{s+2}{s^2+1}, \; L[g(t)] = \frac{s^2+1}{(s+3)(s+2)},$$

$$h(t) = \int_0^t f(\tau)\, g(t-\tau)\, d\tau$$

$L[h(t)]$ is

(a) $\dfrac{s^2+1}{s^2+3}$ (b) $\dfrac{1}{s+3}$

(c) $\dfrac{s^2+1}{(s+3)(s+2)} + \dfrac{s+2}{s^2+1}$ (d) None of the above

32. The Laplace transform of $i(t)$ is given by

$$I(s) = \frac{2}{s(1+s)}$$

As $t \to \infty$, the value of $i(t)$ tends to

(a) 0 (b) 1
(c) 2 (d) ∞

33. The laplace transform of a continuous-time signal $x(t)$ is $X(s) = \dfrac{5-s}{s^2-s-2}$. If the Fourier transform of this signal exists, then $x(t)$ is

(a) $e^{2t}u(t) - 2e^{-t}u(t)$ (b) $-e^{2t}u(-t) + 2e^{-t}u(t)$

(c) $-e^{2t}u(-t) - 2e^{-t}u(t)$ (d) $e^{2t}u(-t) - 2e^{-t}u(t)$

34. In what range should Re(*s*) remain so that the Laplace transform of the function $e^{(a+2)t+5}$ exits ?

(a) $\text{Re}(s) > a+2$ (b) $\text{Re}(s) > a+7$
(c) $\text{Re}(s) < 2$ (d) $\text{Re}(s) > a+5$

35. For the equation $\ddot{x}(t) + 3\dot{x}(t) + 2x(t) = 5$, the solution $x(t)$ approaches which of the following values as $t \to \infty$?

(a) 0 (b) 5/2
(c) 5 (d) 10

36. $u(t)$ represents the unit step function. The Laplace transform of $u(t-\tau)$ is

(a) $\dfrac{1}{s\tau}$ (b) $\dfrac{1}{s-\tau}$

(c) $\dfrac{e^{-s\tau}}{s}$ (d) $e^{-s\tau}$

37. The unilateral Laplace transform of $f(t)$ is $\dfrac{1}{s^2+s+1}$. The unilateral Laplace transform of $t\,f(t)$ is

(a) $-\dfrac{s}{(s^2+s+1)^2}$ (b) $-\dfrac{2s+1}{(s^2+s+1)^2}$

(c) $\dfrac{s}{(s^2+s+1)^2}$ (d) $\dfrac{2s+1}{(s^2+s+1)^2}$

4.5 Z-Transform

1. A sequence $x(n)$ with the *z*-transform $X(z) = z^4 + z^3 - 2z + 2 - 3z^{-4}$ is applied as an input to a linear, time-invariant system with the impulse response $h(n) = 2\delta(n-3)$ where

$\delta(n) = 1; \quad n = 0$

$\qquad 0; \quad$ otherwise

The output at $n = 4$ is

(a) -6
(b) zero
(c) 2
(d) -4

2. The *z*-transform of a system is $H(z) = \dfrac{z}{z-0.2}$ If the ROC is $|z| < 0.2$, then the impulse response of the system is

(a) $(0.2)^n u[n]$ (b) $(0.2)^n u[-n-1]$
(c) $-(0.2)^n u[n]$ (d) $-(0.2)^n u[-n-1]$

3. The region of con convergence of Z-transform of sequence $\frac{5}{6}\, u(n) - [6/5]^n\, u(-n-1)$ must be

(a) $|z| < \frac{5}{6}$ (b) $|z| > \frac{6}{5}$

(c) $\frac{5}{6} < |z| < [6/5]$ (d) $\frac{6}{5} < |z| < \infty$

4. If the region of convergence of $x_1[n] + x_2[n]$ is $[1/3] < |z| < [2/3]$, then the region of convergence of $x_1[n] - x_2[n]$ includes .
(a) $[1/3] < |z| < 3$ (b) $[2/3] < |z| < 3$
(c) $[3/2] < |z| < 3$ (d) $[1/3] < |z| < [2/3]$

5. If $u(t)$ is the unit step and $\delta(t)$ is the unit impulse function, the inverse z-transform of $F(z) = \frac{1}{z+1}$ for $k > 0$ is
(a) $(-1)^k \delta(k)$ (b) $\delta(k) - (-1)^k$
(c) $(-1)^k u(k)$ (d) $u(k) - (-1)_k$

6. The discrete-time signal

$$x[n] \leftrightarrow X(z) = \sum_{n=0}^{\infty} \frac{3^n}{2+n} z^{2n}$$

where $\leftrightarrow$ denotes a transform-pair relationship, is orthogonal to the signal

(a) $y_1[n] \leftrightarrow Y_1(z) = \sum_{n=0}^{\infty} \left(\frac{2}{3}\right)^n z^{-n}$

(b) $y_2[n] \leftrightarrow Y_2(z) = \sum_{n=0}^{\infty} (5^n - n) z^{-(2n+1)}$

(c) $y_3[n] \leftrightarrow Y_3(z) = \sum_{n=-\infty}^{\infty} 2^{-|n|} z^{-n}$

(d) $y_4[n] \leftrightarrow Y_4(z) = 2z^{-4} + 3z^{-2} + 1$

7. Consider the z-transform $X(z) = 5z^2 + 4z^{-1} + 3;\ 0 < |z| < \infty$. The inverse z-transform $x[n]$ is
(a) $5\delta[n+2] + 3\delta[n] + 4\delta[n-1]$
(b) $5\delta[n-2] + 3\delta[n] + 4\delta[n+1]$
(c) $5\delta[n+2] + 3\delta[n] + 4u[n-1]$
(d) $5\delta[n-2] + 3\delta[n] + 4u[n+1]$

8. $H(z)$ is a transfer function of a real system. When a signal $x[n] = (1+j)^n$ is the input to such a system, the output is zero. Further , the Region of Convergence (ROC) of $\left(1 - \frac{1}{2}z^{-1}\right)H(z)$ is the entire Z-phase (except $Z = 0$), It can then be inferred that $H(z)$ can have a minimum of
(a) One pole and one zero
(b) One pole and one zeros
(c) Two poles and two zero
(d) Two poles and two zeros

9. Given $X(z) = \frac{z}{(z-a)^2}$ with $|z| > a$ the residue of $X(z)z^{n-1}$ at $z = a$ for $n \geq 0$ will be
(a) a^{n-1} (b) a^n
(c) na^n (d) na^{n-1}

10. Given the finite length input $x[n]$ and the corresponding finite length output $y[n]$ of an LTI system as shown below, the impulse response $h[n]$ of the system is

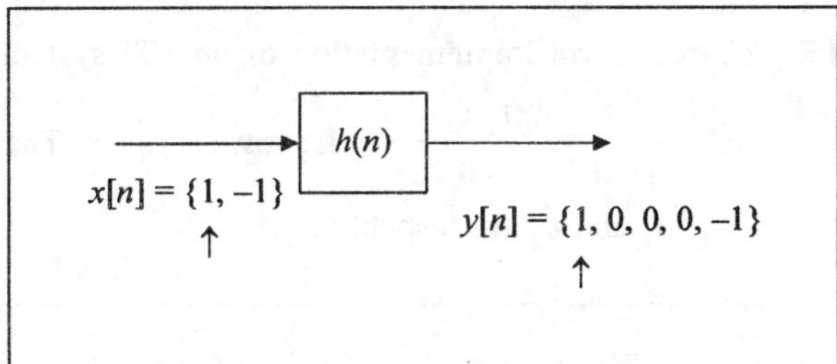

(a) $h[n] = \{1, 0, 0, 1\}$ (↑ under first element) (b) $h[n] = \{1, 0, 1\}$ (↑ under first element)

(c) $h[n] = \{1, 1, 1, 1\}$ (↑ under first element) (d) $h[n] = \{1, 1, 1\}$ (↑ under first element)

11. Consider a discrete time system for which the input $x[n]$ and the output $y[n]$ are related as $y[n] = x[n] - \frac{1}{3}y[n-1]$. If $y[n] = 0$ for $n < 0$ and $x[n] = \delta[n]$, they $y[n]$ can be expressed in terms of the unit step $u[n]$ as
(a) $\left(\frac{-1}{3}\right)^n u[n]$ (b) $\left(\frac{1}{3}\right)^n u[n]$
(c) $(3)^n u[n]$ (d) $(-3)^n u[n]$

12. The region of convergence of the z-transform of the discrete-time signal $x[n] = 2^n u[n]$ will be
(a) $|z| > 2$
(b) $|z| < 2$
(c) $|z| > \frac{1}{2}$
(d) $|z| < \frac{1}{2}$

13. $H(z)$ is a discrete rational transfer function. To ensure that both $H(z)$ and its inverse are stable its
(a) Poles must be inside the unit circle and zeros must be outside the unit circle.
(b) Poles and zeros must be inside the unit circle.
(c) Poles and zeros must be outside the unit circle.
(d) Poles must be outside the unit circle and the zeros should be inside the unit circle.

14. Given $X(z) = \dfrac{\frac{1}{2}}{1-az^{-1}} + \dfrac{\frac{1}{3}}{1-bz^{-1}}$, $|a|$ and $|b| < 1$ with the ROC specified as $|a| < |z| < |b|$, $x[0]$ of the corresponding sequence is given by

(a) $\frac{1}{3}$ (b) $\frac{5}{6}$

(c) $\frac{1}{2}$ (d) $\frac{1}{6}$

15. A direct form implementation of an LTI system with $H(z) = \dfrac{1}{1-0.7z^{-1}+0.13z^{-2}}$ is shown in figure. The value of a_0, a_1 and a_2 are respectively

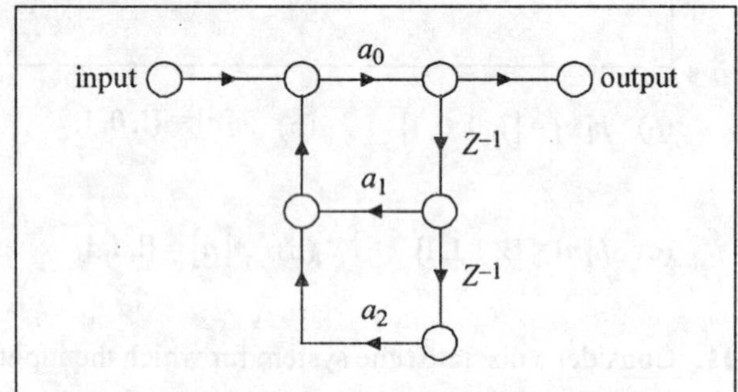

(a) 1.0, 0.7 and – 0.13 (b) – 0.13, 0.7 and 1.0
(c) 1.0, – 0.7 and 0.13 (d) 0.13, – 0.7 and 1.0

16. A causal, analog system has a transfer function $H(s) = \dfrac{a}{s^2+a^2}$. Assuming a sampling time of T seconds, the poles of the transfer function $H(z)$ for an equivalent digital system obtained using impulse invariance method are at

(a) (e^{aT}, e^{-aT}) (b) $\left(j\frac{a}{T}, -j\frac{a}{T}\right)$

(c) (e^{jaT}, e^{-jaT}) (d) $(e^{aT/2}, e^{-aT/2})$

17. A discrete-time signal $x[n]$, suffered a distortion modeled by an LTI system with $H(z) = (1 - az^{-1})$, a is real and $|a| > 1$. The impulse response of a stable system that exactly compensates the magnitude of the distortion is

(a) $\left(\frac{1}{a}\right)^n u[n]$ (b) $-\left(\frac{1}{a}\right)^n u[-n-1]$

(c) $a^n u[n]$ (d) $a^n u[-n-1]$

18. A digital notch filter with a notch frequency of 60 Hz is to be transformed into one operating at a new notch frequency of 120 Hz. The sampling frequency is 400 Hz. A lowpass to lowpass transformation $z^{-1} = \dfrac{1-\alpha\bar{z}}{\bar{z}-\alpha}$ is used on the first to obtain the new notch filter. Then the value of should be

(a) – 0.95
(b) 0.95
(c) – 0.46
(d) 0.46

19. The Z-transform of the time function

$$\sum_{k=0}^{\infty} \delta(n-k)$$

(a) $\dfrac{Z-1}{Z}$ (b) $\dfrac{Z}{(Z-1)^2}$

(c) $\dfrac{Z}{Z-1}$ (d) $\dfrac{(Z-1)^2}{Z}$

20. The z-transform $F(z)$ of the function of $f(nT) = a^{nT}$ is

(a) $\dfrac{z}{z-a^T}$ (b) $\dfrac{z}{z+a^T}$

(c) $\dfrac{z}{z-a^{-T}}$ (d) $\dfrac{z}{z+a^{-T}}$

21. The ROC of Z-transform of the discrete time sequence $x(n) = \left(\frac{1}{3}\right)^n u(n) - \left(\frac{1}{2}\right)^n u(-n-1)$ is

(a) $\frac{1}{3} < |z| < \frac{1}{2}$ (b) $|z| > \frac{1}{2}$

(c) $|z| < \frac{1}{3}$ (d) $2 < |z| < 3$

22. If $x[n] = \left(\frac{1}{3}\right)^{|n|} - \left(\frac{1}{2}\right)^n u[n]$, then the region of convergence (ROC) of its Z-transform in the Z-plane will be

(a) $\frac{1}{3} < |z| < 3$

(b) $\frac{1}{3} < |z| < \frac{1}{2}$

(c) $\frac{1}{2} < |z| < 3$

(d) $\frac{1}{3} < |z|$

Statement for Linked Answer Questions 23 and 24:

In the following network, the switch is closed at $t = 0^-$ and the sampling starts from $t = 0$. The sampling frequency is 10 Hz.

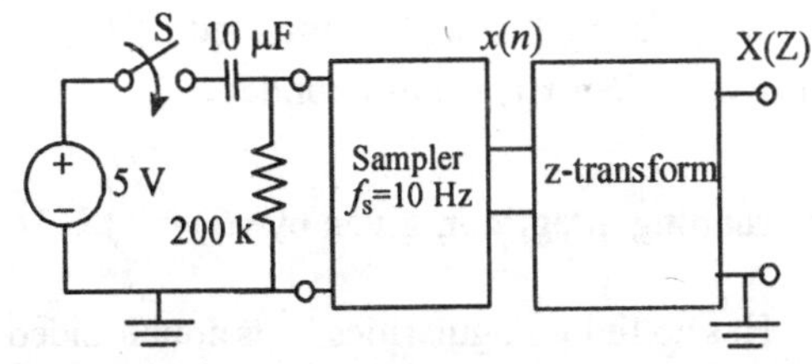

23. The samples x(n) (n = 0, 1, 2, ...) are given by
(a) $5\left(1-e^{-0.05}\right)$ (b) $5\,e^{-0.05n}$
(c) $5\left(1-e^{-5n}\right)$ (d) $5\,e^{-5n}$

24. The expression and the region of convergence of the z-transform of the sampled signal are
(a) $\dfrac{5z}{z-e^{5}},\ |z|<e^{-5}$ (b) $\dfrac{5z}{z-e^{-0.05}},\ |z|<e^{-0.05}$
(c) $\dfrac{5z}{z-e^{-0.05}},\ |z|>e^{-0.05}$ (d) $\dfrac{5z}{z-e^{-5}},\ |z|>e^{-5}$

25. Given $X(z)=\dfrac{z}{(z-a)^2}$ with $|z|>a$. the residue of $X(z)z^{n-1}$ at $z=a$ for $n\ge 0$ will be
(a) a^{n-1} (b) a^{n}
(c) na^{n} (d) na^{n-1}

26. The region of convergence of he Z-transform of a unit step function is
(a) $|Z|>1$ (b) $|Z|<1$
(c) (Real part of z) > 0 (d) (Real part of z) < 0

27. The z-transform of the following real exponential sequence:
$$x(nT)=a^n,\quad nT\ge 0$$
$$=0,\quad nT<0,\ a>0$$
is given by
(a) $\dfrac{1}{1-z^{-1}};|z|>1$ (b) $\dfrac{1}{1-az^{-1}};|z|>a$
(c) 1 for all z (d) $\dfrac{1}{1-az^{-1}};|z|<1$

28. The z-transform of a signal is given by $C(z)=\dfrac{1}{4}\dfrac{z^{-1}(1-z^{-4})}{(1-z^{-1})^2}$. Its final value is
(a) 1/4 (b) Zero
(c) 1.0 (d) Infinity

29. The z-transform X[z] of a sequence x[n] is given by $X[z]=\dfrac{0.5}{1-2z^{-1}}$. It is given that the region of convergence of $X[z]$ includes the unit circle. The value of x[0] is
(a) – 0.5 (b) 0
(c) 0.25 (d) 0.5

30. The sequence x [n] whose z–transform is $X(z)=e^{1/z}$ is.
(a) $\dfrac{1}{n\,!}u[n]$ (b) $\dfrac{1}{-n\,!}u[-n]$
(c) $(-1)^n\dfrac{1}{n\,!}u[n]$ (d) $\dfrac{1}{-(n+1)\,!}u[n-1]$

4.6 LTI Systems

1. Let x(t) be the input to a linear , time-invariant system. The required output is 4 × (t – 2). The transfer function of the system should be
(a) $4\,e^{j4\pi f}$
(b) $2\,e^{-j8\pi f}$
(c) $4\,e^{-j4\pi f}$
(d) $2\,e^{j8\pi f}$

2. Let P be linearity, Q be time-invariance, R be causality and S be stability. A discrete-time system has the input-output relationship,
$$y(n)=\begin{cases}x(n) & n\ge 1\\ 0 & n=0\\ x(n+1) & n\le -1\end{cases}$$
where x(n) is the input and y(n) is the output. The above system has the properties
(a) P, S but not Q, R (b) P, Q, S but not R
(c) P, Q, R, S (d) Q, R, S but not P

3. The impulse response h[n] of a linear time-invariant system is given by
$$h[n]=u[n+3]+u[n-2]-2u[n-7]$$
where u[n] is the unit step sequence. The above system is
(a) Stable but not causal
(b) Stable and causal
(c) Causal but unstable
(d) Unstable and not causal

4. A causal LTI system is described by the difference equation

$$2y[n] = \alpha y[n-2] - 2x[n] + \beta x[n-1]$$

The system is stable only if

(a) $|\alpha| = 2, |\beta| < 2$

(b) $|\alpha| > 2, |\beta| > 2$

(c) $|\alpha| < 2$, any value of β

(d) $|\beta| < 2$, any value of α

5. Which of the following can be impulse response of a causal system?

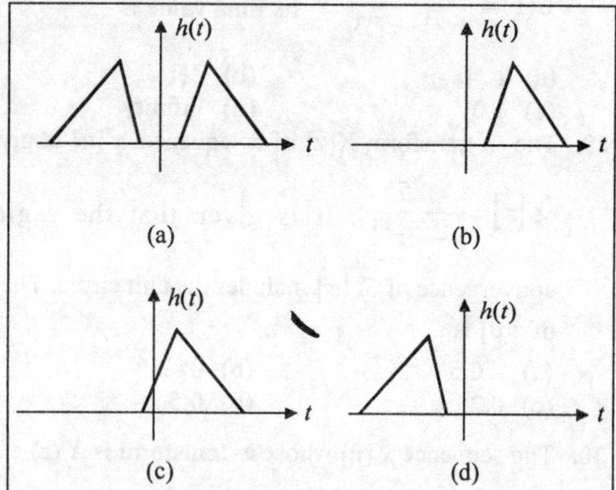

6. A system with input $x[n]$ and output $y[n]$ is given as

$$y[n] = \left(\sin\left(\frac{5}{6}\right)\pi n\right)x(n).$$

The system is.

(a) Linear, stable and invertible

(b) Non – linear, stable and non – invertible

(c) Linear, stable and non – invertible

(d) Linear, unstable and invertible

7. A discrete real all pass system has a pole at $z = 2\angle 30°$: it, therefore

(a) Also has a pole at $\frac{1}{2}\angle 30°$

(b) Has a constant phase response over the z-plane: arg $|H(z)|$ = constant

(c) Is stable only if it is anticausal

(d) Has a constant phase response over the unit circle: arg $|H(e^{j\Omega})|$ = const.

8. A continuous-time system is described by $y(t) = e^{-|x(t)|}$, where $y(t)$ is the output and $x(t)$ is the input. $y(t)$ is bounded

(a) Only when $x(t)$ is bounded

(b) Only when $x(t)$ is non-negative

(c) Only for $t \geq 0$ if $x(t)$ is bounded $t \geq 0$

(d) Even when $x(t)$ is not bounded

9. The running integrator, given by $y(t) = \int_{-\infty}^{t} x(\tau)\, d\tau$

(a) Has no finite singularities in its double sided Laplace Transform $Y(s)$

(b) Produces a bounded output for every causal bounded input

(c) Produces a bounded output for every anticausal bounded input

(d) Has no finite zeroes in its double sided Laplace Transform $Y(s)$

10. A signal is processed by a causal filter with transfer function $G(s)$. For a distortion free output signal waveform $G(s)$ must.

(a) Provide zero phase shift for all frequency

(b) Provide constant phase shift for all frequency

(c) Provide linear shift that is proportional to frequency

(d) Provide a phase shift that is inversely properly proportional to frequency

11. Consider the **discrete-time system** shown in the figure where the impulse response of $G(z)$ is $g(0) = 0$, $g(1) = g(2) = 1$, $g(3) = g(4) = ... = 0$

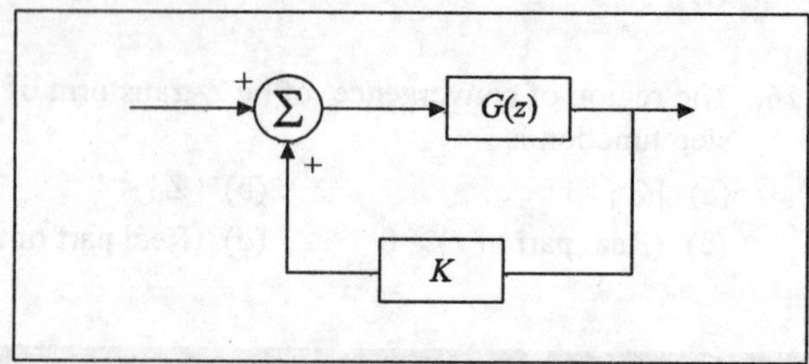

The system is stable for range of values of K

(a) [–1, 1 /2]

(b) [–1, 1]

(c) [–1/2, 1]

(d) [–1/2, 2]

12. $X(z) = 1 - 3z^{-1}$, $Y(z) = 1 + 2z^{-2}$ are Z-transforms of two signals $x[n]$, $y[n]$ respectively. A **linear time invariant system** has the impulse response $h[n]$ defined by these two signals as

$$h[n] = x[n-1] * y[n]$$

where * denotes discrete time convolution. Then the output of the system for the input $\delta[n-1]$

(a) Has Z-transform $z^{-1} X(z)\, Y(z)$

(b) Equals $\delta[n-2] - 3\delta[n-3] + 2\delta[n-4] - 6\delta[n-5]$

(c) Has Z-transform $1 - 3z^{-1} + 2z^{-2} - 6z^{-3}$

(d) Does not satisfy any of the above three

13. The input and output of a continuous time system are respectively denoted by $x(t)$ and $y(t)$. Which of the following descriptions corresponds to a causal system?

(a) $y(t) = x(t-2) + x(t+4)$

(b) $y(t) = (t-4)\, x(t+1)$

(c) $y(t) = (t+4)\, x(t-1)$

(d) $y(t) = (t+5)\, x(t+5)$

14. The impulse response $h(t)$ of a linear time invariant continuous time system is described by

$$h(t) = \exp(\alpha t)\, u(t) + \exp(\beta t)\, u(-t)$$

where $u(t)$ denotes the unit step function, and α and β are real constants. This system is stable if

(a) α is positive and β is positive

(b) α is negative and β is negative

(c) α is positive and β is negative

(d) α is negative and β is positive

15. A linear, time-invariant, causal continuous time system has a rational transfer function with simple poles at $s = -2$ and $s = -4$ and one simple zero at $s = -1$. A unit step $u(t)$ is applied at the input of the system. At steady state, the output has constant value of 1. The impuse responsee of this system is

(a) $[e^{(-2t)} + e^{(-4t)}]\, u(t)$

(b) $[-4e^{(-2t)} + 12e^{(-4t)} - e^{(-t)}]\, u(t)$

(c) $[-4e^{(-2t)} + 12e^{(-4t)}]\, u(t)$

(d) $[-0.5e^{(-2t)} + 1.5e^{(-4t)}]\, u(t)$

16. A discrete time linear shift-invariant system has an impulse response $h[n]$ with $h[0] = 1, h[1] = -1, h[2] = 2$ and zero otherwise. The system is given an input sequence $x[n]$ with $x[0] = x[2] = 1$, and zero othrewise. The number of nonzero samples in the output sequence $y[n]$, and the value of $y[2]$ are, respecively

(a) 5, 2 (b) 6, 2

(c) 6, 1 (d) 5, 3

17. Let $x(t)$ be the input and $y(t)$ be the output of a continous time system. Match the system properties P_1, P_2 and P_3 with system relations R_1, R_2, R_3, R_4

Properties

P_1: Linear but NOT time-invariant

P_2: Time-invariant but NOT linear

P_3: Linear 'and time-invariant

Relations

R_1: $y(t) = t^2 x(t)$ R_2: $y(t) = t|x(t)|$

R_3: $y(t) = |x(t)|$ R_4: $y(t) = x(t-5)$

(a) $(P_1, R_1), (P_2, R_3), (P_3, R_4)$

(b) $(P_1, R_2), (P_2, R_3), (P_3, R_4)$

(c) $(P_1, R_3), (P_2, R_1), (P_3, R_2)$

(d) $(P_1, R_1), (P_2, R_2), (P_3, R_3)$

18. Consider a system whose input x and output y are related by the equation $y(t) = \int_{-\infty}^{\infty} x(t-\tau)\, h(2\tau)\, d\tau$ where $h(t)$ is shown in the graph.

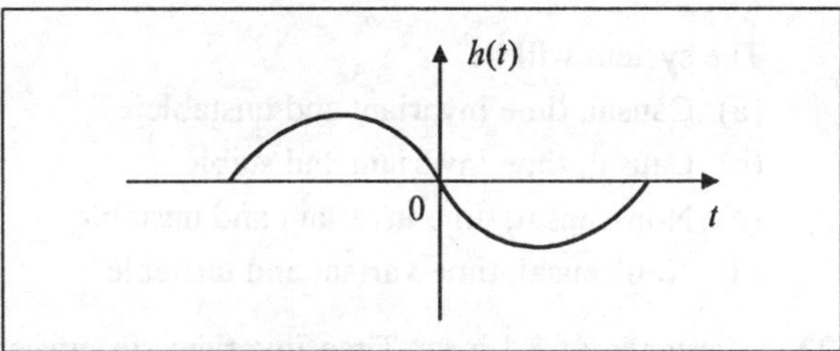

Which of the following four propperties are possessed by the system ?

BIBO : Bounded input gives a bounded output.

Causal : The system is causal.

LP : The system is low pass.

LTI : The system is linear and time-invariant

(a) Causal, LP (b) BIBO, LTI

(c) BIBO, Causal, LTI (d) LP, LTI

19. The transfer function of a discrete time LTI system is given by

$$H(z) = \frac{2 - \frac{3}{4}z^{-1}}{1 - \frac{3}{4}z^{-1} + \frac{1}{8}z^{-2}}$$

Consider the following statements:

S_1: The system is stable and causal for ROC: $|z| > \frac{1}{2}$

S_2: The system is stable but not causal for ROC: $|z| < \frac{1}{4}$

S_3: The system is neither stable nor casual for ROC: $\frac{1}{4} < |z| < \frac{1}{2}$

Which one of the following statements is valid?

(a) Both S_1 and S_2 are true

(b) Both S_2 and S_3 are true

(c) Both S_1 and S_3 are true

(d) S_1, S_2 and S_3 are all true

20. The impulse response of a causal linear time invariant system is given as $h(t)$.Now consider the following two statements:

Statement (I) : Principle of superposition holds

Statement (II) : $h(t) = 0$ for $t < 0$

Which one of the following statements is correct?

(a) Statement (I) is correct and Statement (II) wrong

(b) Statement (II) is correct and statement (I) is worng

(c) Both Statement (I) and Statement (II) are worng

(d) Both Statement (I) and Statement (II) are correct

21. A system with input $x(t)$ and output $y(t)$ is defined by the input-output relation:

$$y(t) = \int_{-\infty}^{-2t} x(\tau)d\tau$$

The system will be

(a) Causal, time invariant and unstable

(b) Causal, time invariant and stable

(c) Non-causal, time invariant and unstable

(d) Non causal, time variant and unstable

22. A cascade of 3 Linear Time Invariant system is causal and unstable . From this, we conclude that

(a) Each system in the cascade is individually causal and unstable

(b) At least one system is unstable and at least one system is causal

(c) At least one system is causal and all systems are nstable

(d) The majority are unstable and the majority are causal.

23. A system is described by the following state and out put equations .

$$\frac{dx_1(t)}{dt} = -3x_1(t) + x_2(t) + 2u(t)$$

$$\frac{dx_2(t)}{dt} = -2x_2(t) + u(t)$$

$$y(t) = x_1(t)$$

where $u(t)$ is the input and $y(t)$ is the output

The system transfer function is

(a) $\frac{s+2}{s^2+5s-6}$ (b) $\frac{s+3}{s^2+5s+6}$

(c) $\frac{2s+5}{s^2+5s+6}$ (d) $\frac{2s-5}{s^2+5s-6}$

24. Which one of the following discrete time systems is time invariant ?

(a) $y[n] = nx[n]$

(b) $y[n] = x[3n]$

(c) $y[n] = x[-n]$

(d) $y[n] = x[n-1]$

25. For input $x(t)$, an ideal impulse sampling system produces the output

$$y(t) = \sum_{k=\infty}^{c} x(kT)\delta(t-kT)$$

Where $\delta(t)$ is the Dirac delat function.

The system is

(a) Nonlinear and time invariant

(b) Nonlinear and time varying

(c) Linear and time invariant

(d) Linear and time varying

26. An excitation is applied to a system at $t = T$ and its response is zero for $-\infty < t < T$. Such a system is

(a) Non-causal system

(b) Stable system

(c) Causal system

(d) Unstable system

27. The impulse response functions of four linear systems S_1, S_2, S_3, S_4 are given respectively by

$$h_1(t) = 1$$

$$h_2(t) = u(t)$$

$$h_3(t) = \frac{u(t)}{t+1}$$

$$h_4(t) = e^{-3t}u(t)$$

Where $U(t)$ is the unit step function. Which of these system is time invariant causal, and stable?

(a) S_1 (b) S_2

(c) S_3 (d) S_4

28. Given $h[n]$ = [1, 2, 2,], $f[n]$ is obtained by convolving $h[n]$ with itself and $g[n]$ by correlating $g[n]$ with itself. Which one of the following statements is true ?

(a) $f[n]$ is causal and its maximum value is 9.

(b) $f[n]$ is non causal and its maximum value is 8

(c) $g[n]$ is causal and its maximum value is 9

(d) $g[n]$ is non - causal and its maximum value is.

29. A discrete-time system is shown in figure. The system function, $H(z)$, of the network is given by

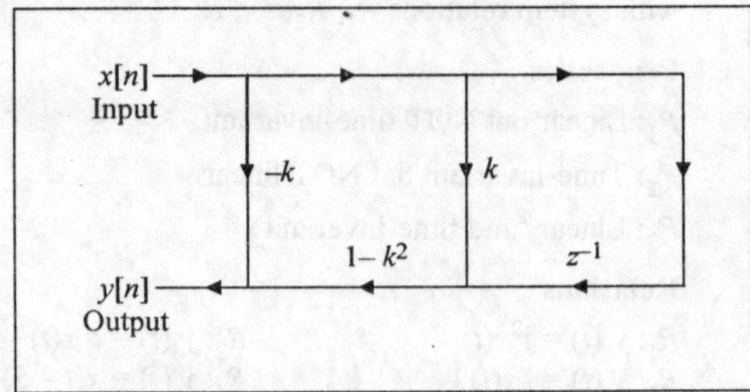

(a) $\frac{-k+z^{-1}}{1-(1-k^2)z^{-1}}$

(b) $\frac{(1-k^2)+z^{-1}}{1-k\,z^{-1}}$

(c) $\frac{1-(1-k^2)z^{-1}}{1-kz^{-1}}$

(d) $\frac{-k+z^{-1}}{1-kz^{-1}}$

30. The voltage across an impedance in a network is $V(s) = Z(s)\, I(s)$, where $V(s)$, $Z(s)$ and $I(s)$ are the Laplace Transforms of the corresponding time functions $v(t)$, $z(t)$ and $i(t)$. The voltage $v(t)$ is

(a) $v(t) = z(t)\, i(t)$

(b) $v(t) = \int_0^t i(\tau)\cdot z(t-\tau)\, d\tau$

(c) $v(t) = \int_0^t i(\tau)\cdot z(t+\tau)\, d\tau$

(d) $v(t) = z(t) + i(\tau)$

31. Let $u(t)$ be the unit step function. Which of the waveform in figure $(A) - (D)$ corresponds to the convolution of $u(t) - u(t-1)$ with $u(t) - u(t-2)$?

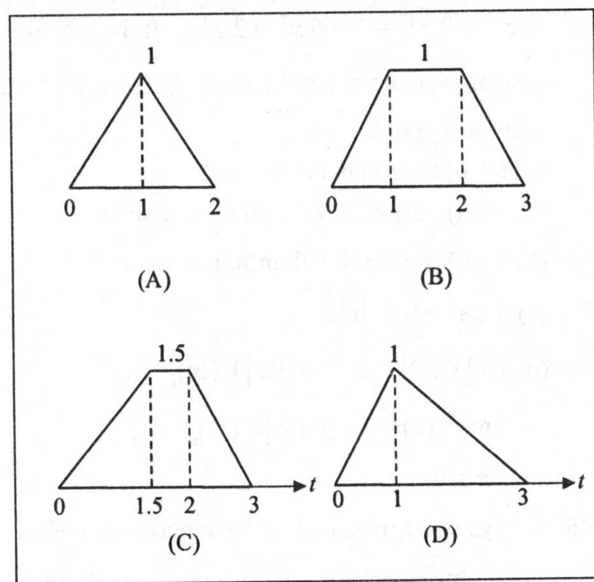

(a) Figure-A (b) Figure-B

(c) Figure-C (d) Figure-D

32. The transfer function of a linear system is the

(a) Ratio of the output, $V_o(t)$, and input, $V_i(t)$

(b) Ratio of the derivative of the output and the input

(c) Ratio of the Laplace transform of the output and that of the input with all initial conditions zeros

(d) None of these

33. Non-minimum phase transfer function is defined as the transfer function:

(a) Which has zeros in the right-half S-plane

(b) Which has zeros only in the left-half S-plane

(c) Which has poles in the right-half S-plane

(d) Which has poles in the left-half S-plane.

34. The transfer function of a zero-order-hold system is:

(a) $\frac{1}{s}(1+e^{-sT})$

(b) $\frac{1}{s}(1-e^{-sT})$

(c) $1-\frac{1}{s}e^{-sT}$

(d) $1+\frac{1}{s}e^{-sT}$

35. The differential equation $10\frac{d^2y}{dt^2} - 20\frac{dy}{dt} + y = x(t)$ describes a system with an input $x(t)$ and an output $y(t)$. The system, whcih is initially relaxed, is exicted by a unit step input. The output $y(t)$ cna be represented by the waveform.

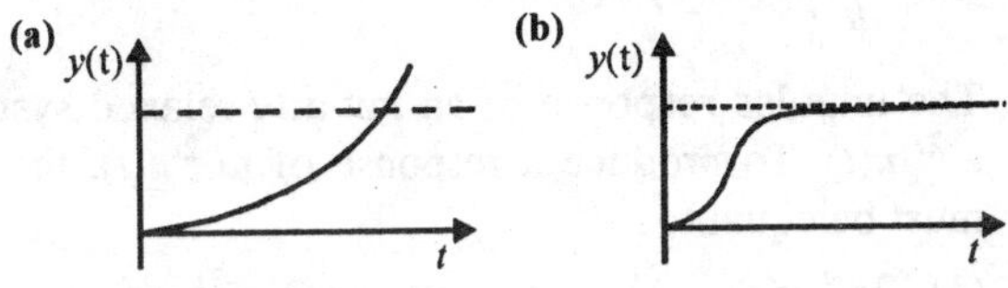

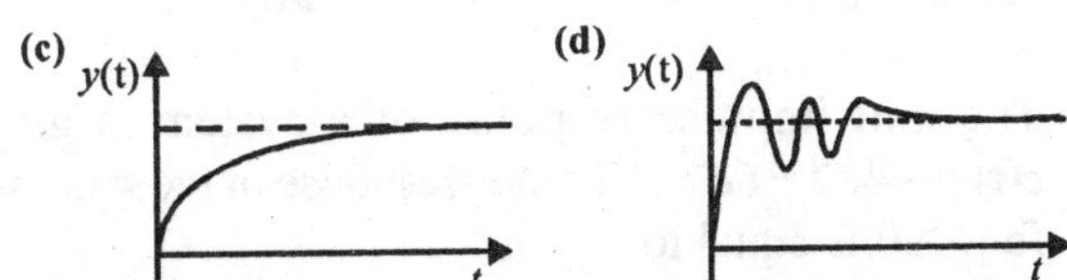

36. An input $x(t) = \exp(-2t)u(t) + \delta(t-6)$ is applied to an LTI system with impulse response $h(t) = u(t)$. THe output is

(a) $[1-\exp(-2t)]\, u(t) + u(t+6)$

(b) $[1-\exp(-2t)]\, u(t) + u(t-6)$

(c) $0.5[1-\exp(-2t)]\, u(t) + u(t+6)$

(d) $0.5[1-\exp(-2t)]\, u(t) + u(t-6)$

37. A continuous time LTI system is described by

$$\frac{d^2y(t)}{dt^2}+4\frac{dy(t)}{dt}+3y(t)=2\frac{dx(t)}{dt}+4x(t)$$

Assuming zero initial conditions, the response $y(t)$ of

the above system for the input $x(t)=e^{-2t}u(t)$ is given by

(a) $\left(e^{t}-e^{3t}\right)u(t)$ (b) $\left(e^{-t}-e^{-3t}\right)u(t)$

(c) $\left(e^{-t}+e^{-3t}\right)u(t)$ (d) $\left(e^{t}+e^{3t}\right)u(t)$

38. The input $x(t)$ and output $y(t)$ of a system are related as $y(t)=\int_{-\infty}^{t}x(\tau)\cos(3\tau)d\tau$. The system is

(a) Time-invariant and stable

(b) Stable and not time-invairant

(c) Time-invariant and not stable

(d) Not-time-invariant and not stable

39. $s(t)$ is step response and $h(t)$ is impulse response of a system. Its response $y(t)$ for any input $u(t)$ is given by

(a) $\frac{d}{dt}\int_{0}^{t}s(t-\tau)u(\tau)\,d\tau$

(b) $\int_{0}^{t}s(t-\tau)\ u(\tau)\ d\tau$

(c) $\int_{0}^{t}\int_{0}^{t}s(t-\tau_1)\ u(\tau_1)\ d\tau_1\ d\tau$

(d) $\frac{d}{dt}\int_{0}^{t}h(t-\tau)u(\tau)\,d\tau$

40. The impulse response of an initially relaxed system is $e^{-2t}\,u(t)$. To produce a response of $te^{-2t}u(t)$, the input must be equal to

(a) $2e^{-t}\,u(t)$ (b) $(1/2)e^{-2t}\,u(t)$

(c) $e^{-2t}\,u(t)$ (d) $e^{-t}\,u(t)$

41. The unit impulse response of a system is given as $c(t)=-4e^{-t}+6e^{-2t}$. The step response of the same system for $t\geq 0$ is equal to

(a) $-3e^{-2t}-4e^{-t}+1$ (b) $-3e^{-2t}+4e^{-t}-1$

(c) $3^{-2t}-4e^{-t}-1$ (d) $3e^{-2t}+4e^{-t}-1$

42. A signal $e^{-at}\sin(\omega t)$ is the input to a real Linear Time Invariant system. Given K and ϕ are constants. the output of the system will be of form $Ke^{-\beta t}\sin(vt+\phi)$ where

(a) β need not be equal to α but v equal to ω

(b) V need not be equal to ω but β equal to α

(c) β equal to α and v equal to ω

(d) β need not be equal to α and v need be equal to ω

43. A signal $x(t)=\text{sinc}(\alpha t)$ where α is a real constant $\left(\text{sinc}(x)=\frac{\sin(\pi x)}{\pi x}\right)$ is the input to a linear Time invariant system whose impulse response $h(t)=\text{sinc}(\beta t)$ where β is a real constant. If min (α,β) denotes the minimum of α and β, and similarly max, (α,β) denotes the maximum of α and β and K is a constant, which one of the following statements is true about the output of the system?

(a) It will be of the form K sinc (γt) where γ = min (a, b)

(b) It will be of the form K sinc (γt) where γ = max (α, β)

(c) It will be of the form K sinc (αt)

(d) Cannot be a sinc type of signal

44. A Linear Time Invariant system with an impulse response $h(t)$ produces output $y(t)$ when input $x(t)$ is applied. When the input $x(t-\tau)$ is applied to a system with impulse response $h(t-\tau)$, the output will be

(a) $y(t)$ (b) $y(2(t-\tau))$

(c) $y(t-\tau)$ (d) $y(t-2\tau)$

45. The z transform of a signal $x[n]$ is given by $4z^{-3}+3z^{-1}+2-6z^{2}+2z^{3}$. It is applied to a system, with a transfer function $H(z)=3z^{-1}-2$

Let the output be $y(n)$.

Which of the following is true ?

(a) $y(n)$ is not causal with finite support

(b) $y(n)$ is causal with infinite support

(c) $y(n)=0;\ |n|>3$

(d) $\text{Re}[Y(z)]_{z=e^{j\theta}}=-\text{Re}[Y(z)]_{z=e^{-j\theta}}$;

$\text{Im}[Y(z)]_{z=e^{j\theta}}=\text{Im}[Y(z)]_{z=e^{-j\theta}}$;

$-\pi\leq\theta<\pi$

46. The system represented by the input-output relationship

$y(t)=\int_{-\infty}^{5t}x(\tau)d\tau$, $t>0$ is

(a) Linear and causal

(b) Linear but not causal

(c) Causal but not linear

(d) Neither linear nor causal

47. For the system $2/(s+1)$, the approximate time taken for a step response to reach 98% of its final value is

(a) $1\,s$ (b) $2\,s$

(c) $4\,s$ (d) $8\,s$

48. Given the finite length input $x[n]$ and the corresponding finite length output $y[n]$ of an LTI system as shown below, the impulse response $h[n]$ of the system is

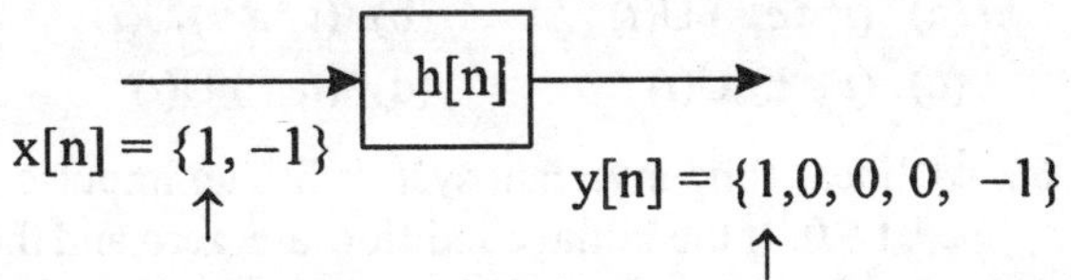

(a) $h[n]=\{1, 0, 0, 1\}$ $\uparrow$

(b) $h[n]=\{1, 0, 1\}$ $\uparrow$

(c) $h[n]=\{1, 1, 1, 1\}$ $\uparrow$

(d) $h[n]=\{1, 1, 1\}$ $\uparrow$

49. Given two continuous time signals $x(t)=e^{-t}$ and $y(t)=e^{-2t}$ which exist for $t>0$, the convolution $z(t)=x(t)*y(t)$ is

(a) $e^{-t}-e^{-2t}$ (b) e^{-3t}

(c) e^{+t} (d) $e^{-t}+e^{-2t}$

50. A low-pass filter with a cut-off frequency of 30 Hz is cascaded with a high-pass filter with a cut-off frequency of 20 Hz. The resultant system of filters will funtion as

(a) An all-pass filter
(b) An all-stop filter
(c) A band stop (band-reject) filter
(d) A band-pass filter

51. The response $h(t)$ of a linear time invariant system to an impulse $\delta(t)$, under initially relaxed condition is $h(t)=e^{-t}+e^{-2t}$. The response of this system for a unit step input $u(t)$ is

(a) $u(t)+e^{-t}+e^{-2t}$ (b) $(e^{-t}+e^{-2t})u(t)$

(c) $(1.5-e^{-t}-0.5e^{-2t})u(t)$ (d) $e^{-t}\delta(t)+e^{-2t}u(t)$

52. The input $x(t)$ and output $y(t)$ of a system are related as

$y(t)=\int_{-\infty}^{t} x(\tau)\cos(3\tau)\,d\tau$. The system is

(a) Time-invariant and stable
(b) Stable and not time-invariant
(c) Time-invariant and not stable
(d) Not-time-invariant and not stable

53. $x[n]=0;\ n<-1,\ n>0,\ x[-1]=-1,\ x[0]=2$ is the input and $y[n]=0;\ n<-1,\ n>2,\ y[-1]=-1=y[1],\ y[0]=3,\ y[2]=-2$ is the output of a discrete-time LTI system. The system impulse response $h[n]$ will be

(a) $h[n]=0;\ n<0,\ n>2,\ h[0]=1,\ h[1]=h[2]=-1$
(b) $h[n]=0;\ n<-1,\ n>1,\ h[-1]=1,\ h[0]=h[1]=2$
(c) $h[n]=0;\ n<0,\ n>3,\ h[0]=-1,\ h[1]=2,\ h[2]=1$
(d) $h[n]=0;\ n<-2,\ n>1,\ h[-2]=h[1]=h[-1]=H(0)=3$

54. $G(z)=\alpha z^{-1}+\beta z^{-3}$ is a low-pass digital filter with a phase characteristics same as that of the above question if.

(a) $\alpha=\beta$ (b) $\alpha=-\beta$

(c) $\alpha=\beta^{1/3}$ (d) $\alpha=\beta^{-(1/3)}$

55. A **differentiator** has transfer function whose

(a) Phase increases linearly with frequency
(b) Amplitude remains constant
(c) Amplitude increases linearly with frequency
(d) Amplitude decreases linearly with frequency

56. The output of a **linear time invariant control system** is $c(t)$ for a certain input $r(t)$. If $r(t)$ is modified by passing it through a block whose transfer function is e^{-s} and then applied to the system, the modified output of the system would be

(a) $\dfrac{c(t)}{1+e^{t}}$ (b) $\dfrac{c(t)}{1+e^{-t}}$

(c) $c(t-1)\,u(t-1)$ (d) $c(t)\,u(t-1)$

57. A **linear time-invariant system initially** at rest, when subjected to a unit-step input, givens a response $y(t)=te^{-t},\ t>0$. The transfer function of the system is.

(a) $\dfrac{1}{(s+1)^2}$ (b) $\dfrac{1}{s(s+1)^2}$

(c) $\dfrac{s}{(s+1)^2}$ (d) $\dfrac{1}{s(s+1)}$

58. The **transfer function** of the system described by $\dfrac{d^2y}{dt^2}+\dfrac{dx}{dt}=\dfrac{du}{dt}+2u$ with u as input and y as output is

(a) $\frac{(s+2)}{(s^2+s)}$ (b) $\frac{(s+1)}{s^2+s}$

(c) $\frac{2}{s^2+s}$ (d) $\frac{2s}{(s^2+s)}$

59. A control system is defined by the following mathematical relationship

$$\frac{d^2x}{dt^2}+6\frac{dx}{dt}+5x=12(1-e^{-2t})$$

The response of the system as $t \to \infty$ is

(a) $x = 6$ (b) $x = 2$
(c) $x = 2.4$ (d) $x = -2$

60. A control system with certain excitation is governed by the following mathematical equation

$$\frac{d^2x}{dt^2}+\frac{1}{2}\frac{dx}{dt}+\frac{1}{18}x=10+5e^{-4t}+2e^{-5t}$$

The natural time constants of the response of the system are

(a) 2s and 5s (b) 3s and 6s
(c) 4s and 5s (d) $\frac{1}{3}$ s and $\frac{1}{6}$ s

61. The unit impulse response of a second order under-damped system starting from rest is given by

$$C(t) = 12.5\, e^{-6t} \sin 8\, t,\ t \geq 0$$

The steady –state value of the unit step response of the system is equal to

(a) 0 (b) 0.25
(c) 0.5 (d) 1.0

62. In the system shown in figure the input $x(t) = \sin t$. In the steady-state, the response $y(t)$ will be

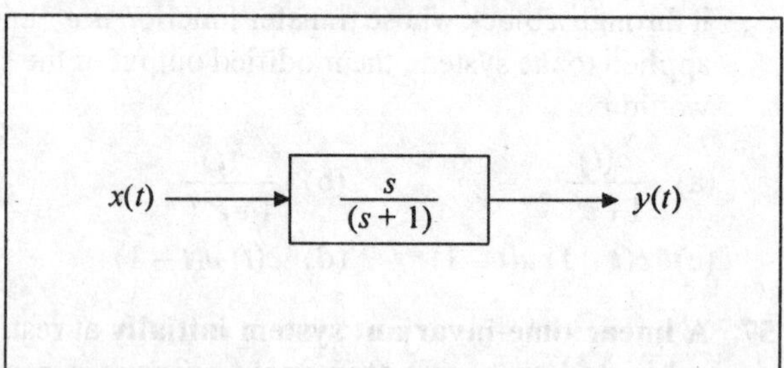

(a) $\frac{1}{\sqrt{2}}\sin(t-45°)$ (b) $\frac{1}{\sqrt{2}}\sin(t+45°)$

(c) $\sin(t-45°)$ (d) $\sin(t-45°)$

63. The unit impulse response of a linear time invariant system is the unit step function $u(t)$. For $t > 0$, the response of the system to an excitation $e^{-at}\, u(t)$, $a > 0$ will be

(a) ae^{-at} (b) $(1/a)(1-e^{-at})$
(c) $a(1-e^{-at})$ (d) $1 - e^{-at}$

64. A system with an input $x(t)$ and an output $y(t)$ is described by the relation: $y(t) = t\, x(t)$. This system is

(a) Linear and time-invariant
(b) Linear and time varying
(c) Non-linear and time-invariant
(d) Non-linear and time-varying

65. The transfer function of a system is a given by $H(s)=\frac{1}{s^2(s-2)}$ The impulse response of the system is

(a) $(t^2 * e^{-2t})\,U(t)$ (b) $(t * e^{-2t})\,U(t)$
(c) $(t^2e^{-2t})\,U(t)$ (d) $(te^{-2t})\,U(t)$

66. A linear time invariant system has an impulse response e^{2t}, $t > 0$. If the initial condition are zero and the input is e^{3t}, the output for $t > 0$ is

(a) $e^{3t} - e^{2t}$ (b) e^{5t}
(c) $e^{3t} + e^{2t}$ (d) None of the above

67. A system described by the following differential equation

$$\frac{d^2y}{dt^2}+3\frac{dy}{dt}+2y=x(t)$$

is initially at rest. For input $x(t) = 2\,u(t)$, the output y(t) is

(a) $(1 - 2\,e^{-t} + e^{-t})\,u(t)$
(b) $(1 - 2\,e^{-t}\ 2\,e^{-t})\,u(t)$
(c) $(0.5 + e^{-t} + 1.5\,e^{-t})\,u(t)$
(d) $(0.5 - 2\,e^{-t} + 2\,e^{-t})\,u(t)$

68. A causal system having the transfer function $H(s)=\frac{1}{s+2}$ is excited with 10u(t). The time at which the output reaches 99% of its steady value is

(a) 2.7 sec (b) 2.5 sec
(c) 2.3 sec (c) 2.1 sec

69. The output y(t) of a linear time invariant system is related to its input x(t) by the following equation:

$y(t) = 0.5\,x(t - t_d + T) + x(t - t_d) + 0.5\,x(t - t_d - T)$.

The filter transfer function $H(\omega)$ of such a system is given by

(a) $(1+\cos\omega T)e^{-j\omega t_d}$ (b) $(1+0.5\cos\omega T)e^{-j\omega t_d}$

(c) $(1-\cos\omega T)e^{-j\omega t_d}$ (d) $(1-0.5\cos\omega T)e^{-j\omega t_d}$

70. The frequency response of a **linear, time-invariant** system is given by $H(f)=\frac{5}{1+j10\pi f}$. The step response of the system is

(a) $5\left(1-e^{-5t}\right)u(t)$ (b) $5\left(1-e^{-t/5}\right)u(t)$

(c) $\frac{1}{5}\left(1-e^{-5t}\right)u(t)$ (d) $\frac{1}{5}\left(1-e^{-\frac{t}{5}}\right)u(t)$

71. Let a signal $a_1 \sin(\omega_1 t + \phi_1)$ be applied to a stable linear time-invariant system. Let the corresponding steady state output be represented as $a_2 F(\omega_2 t + \phi_2)$. Then which of the following statements is true?
   (a) F is not necessarily a "sine" or "cosine" function but must be periodic with $\omega_1 = \omega_2$
   (b) F must be a "sine" or "cosine" function with $a_1 = a_2$
   (c) F must be a "sine" function with $\omega_1 = \omega_2$ and $\phi_1 = \phi_2$
   (d) Must be a "sine" or "cosine" function with $\omega_1 = \omega_2$

72. Given the relationship between the input u(t) and the output y(t) to be

$$y(t) = \int_0^t (2 + t - \tau)e^{-3(t-\tau)}u(\tau)d\tau$$

The transfer fucntion $\frac{Y(s)}{U(s)}$ is

(a) $\frac{2e^{-2s}}{s+3}$ (b) $\frac{s+2}{(s+3)^2}$

(c) $\frac{2s+5}{s+3}$ (c) $\frac{2s+7}{(s+3)^2}$

73. Consider a discrete time system for which the input $x[n]$ and the output $y[n]$ are related as $y[n] = x[n] - \frac{1}{3}y[n-1]$. If $y[n] = 0$ for $n < 0$ and $x[n] = \delta[n]$, then $y[n]$ can be expressed in terms of the unit step $u[n]$ as.

(a) $\left(\frac{-1}{3}\right)^n u[n]$ (b) $\left(\frac{1}{3}\right)^n u[n]$

(c) $(3)^n u[n]$ (d) $(-3)^n u[n]$

74. $H(z)$ is a discrete rational transfer function. To ensure that both $H(z)$ and its inverse are stable its.
   (a) Poles must be inside the unit circle and zeros must be outside the unit circle.
   (b) Poles and zeros must be inside the unit circle.
   (c) Poles and zeros must be outside the unit circle.
   (d) Poles must be outside the unit circle and the zeros should be inside the unit circle

75. The input $x(t)$ and the corresponding output $y(t)$ of a system are related by $y(t) = \int_{-\infty}^{5t} x(\tau)d\tau$. The system is
   (a) Time invariant and causal
   (b) Time invariant and non-causal
   (c) Time variant and non-causal
   (d) Time variant and causal

76. Consider a system with input x(t) and output y(t) related as follows

$$y(t) = \frac{d}{dt}\{e^{-t}x(t)\}$$

Which one of the following statement is **TRUE**?
   (a) The system is nonlinear
   (b) The system is time - invariant
   (c) The system is stable
   (d) The system has memory

77. The characteristics of a thermometer measuring ambient temperature is $2\frac{dTi}{dt} + T_i - T_a = 0$, where T_i and T_a are the indicated and ambient temperatures, respectivley, both in °C, and time is in seconds. The –3dB cut - off frequency response of the thermometer is

(a) $\frac{1}{4\pi}$ Hz (b) $\frac{1}{2}$ Hz

(c) 1 Hz (d) 2p Hz

78. Consider the difference eqation $y[n] - \frac{1}{3}y[n-1] = x[n]$ and suppose that $x[n] = \left(\frac{1}{2}\right)^n u[n]$. Assuming the condition of initial rest, the solution for $y[n], n \geq 0$ is

(a) $3\left(\frac{1}{3}\right)^n - 2\left(\frac{1}{2}\right)^n$ (b) $-2\left(\frac{1}{3}\right)^n + 3\left(\frac{1}{2}\right)^n$

(c) $\frac{2}{3}\left(\frac{1}{3}\right)^n + \frac{1}{3}\left(\frac{1}{2}\right)^n$ (d) $\frac{1}{3}\left(\frac{1}{3}\right)^n + \frac{2}{3}\left(\frac{1}{2}\right)^n$

79. A discrete – time system with input x[n]-output y[n] relation y [n] = x [n] – 2 x [n – 1] + x [n – 2] is a good approximation to a
   (a) High – pass filter
   (b) Band stop filter blocking $(\pi / 8) \leq |\omega| \leq \pi / 4$
   (c) Low pass filter
   (d) Band pass filter passing $(\pi / 8) \leq |\omega| < \pi / 4$.

80. Bilinear transformation avoid the problem of aliasing encountered with the use of impulse invariance through.
   (a) Mapping the entire imaginary axis of the *s*-plane on to the unit circle in the *z*-plane.

(b) Pre-filtering the input signal to impose bank limitedness.

(c) Mapping zeros of the left half of the s- plane inside the unit circle in the z-plane.

(d) Up sampling so that the bandwidth is reduced.

81. Given $x = [a, b, c, d]$ as the input, a linear time invariant system produces an output $y = [x, x, x, x, \ldots$ <repeated N times>]. The impulse response of the system is .

(a) $\sum_{i=0}^{N-1} \delta[n-4i]$ (b) $u[n] - u[n-N]$

(c) $u[n] - u[n-N-1]$ (d) $\sum_{i=0}^{N-1} \delta[n-i]$

82. A discrete-time signal $x[n]$, suffered a distortion modeled by an LTI system with $H(z) = (1 - az^{-1})$, a is real and $|a| > 1$. The impulse response of a stable system that exactly compensates the magnitude of the distortion is

(a) $\left(\frac{1}{a}\right)^n u[n]$ (b) $-\left(\frac{1}{a}\right)^n u[-n-1]$

(c) $a^n u[n]$ (d) $a^n u[-n-1]$

4.7 Convolution

1. If $u(t)$, $r(t)$ denote the unit step and unit ramp function respectively and $u(t) * r(t)$ their convolution, then the function $u(t+1) * r(t-2)$ is given by

(a) $(1/2)(t-1)(t-2)$ (b) $(1/2)(t-1)(t-2)$

(c) $(1/2)(t-1)^2 u(t-1)$ (d) None of the above

2. Let $u(t)$ be the unit step function. Which of the waveform in figure (A) – (D) corresponds to the convolution of $u(t) - u(t-1)$ with $u(t) - u(t-2)$?

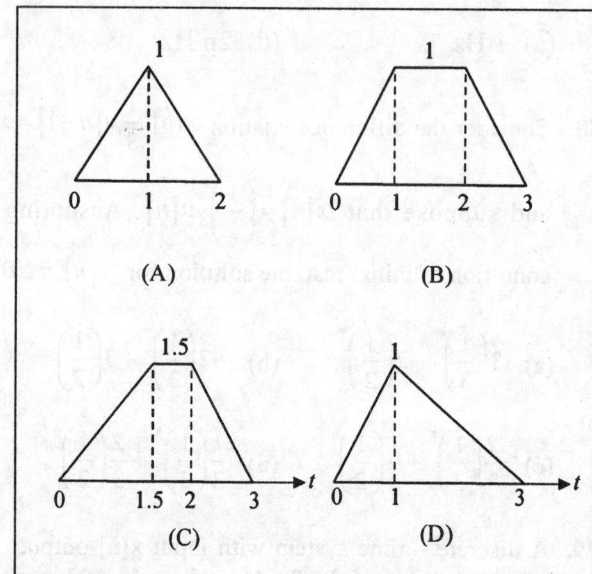

(a) Figure-A (b) Figure-B

(c) Figure-C (d) Figure-D

3. Convolution of $x(t+5)$ with impulse function $\delta(t-7)$ is equal to

(a) $x(t-12)$ (b) $x(t+12)$

(c) $x(t-2)$ (d) $x(t+2)$

4. Let $y[n]$ denote the convolution of $h[n]$ and $g[n]$, where $h[n] = \left(\frac{1}{2}\right)^n u[n]$ and $g[n]$ is a causal sequence. If $y[0] = 1$ and $y[1] = \frac{1}{2}$ then $g[1]$ equals

(a) 0 (b) 1/2

(c) 1 (d) 3/2

5. Given $h[n] = [1, 2, 2,]$, $f[n]$ is obtained by convolving $h[n]$ with itself and $g[n]$ by correlating $g[n]$ with itself. Which one of the following statements is true?

(a) $f[n]$ is causal and its maximum value is 9.

(b) $f[n]$ is non causal and its maximum value is 8

(c) $g[n]$ is causal and its maximum value is 9

(d) $g[n]$ is non - causal and its maximum value is 9.0.

6. Given

$$x(t) * x(t) = t\, e^{(-2t)}\, u(t)$$

the function $x(t)$ is

(a) $e^{(-2t)} u(t)$ (b) $e^{(-t)} u(t)$

(c) $t\, e^{(-t)} u(t)$ (d) $0.5\, t\, e^{(-t)} u(t)$

7. The signal $x(t)$ and $h(t)$ shown in the figure are convolved to yield $y(t)$

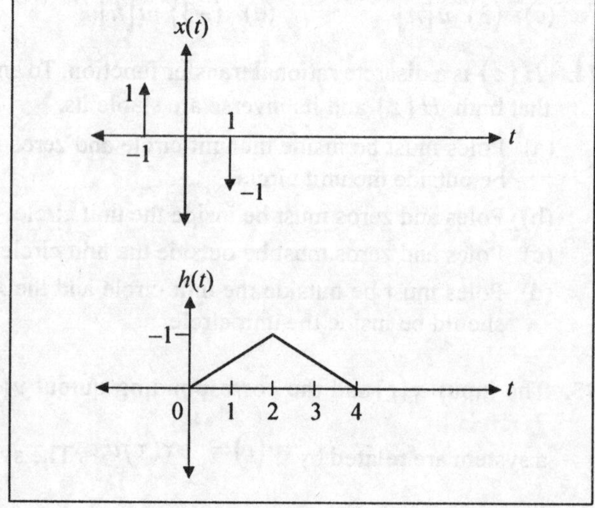

Which one of the following figures represents the output $y(t)$?

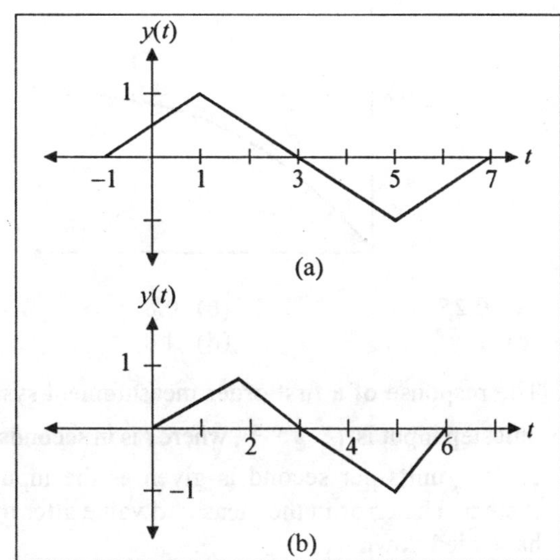

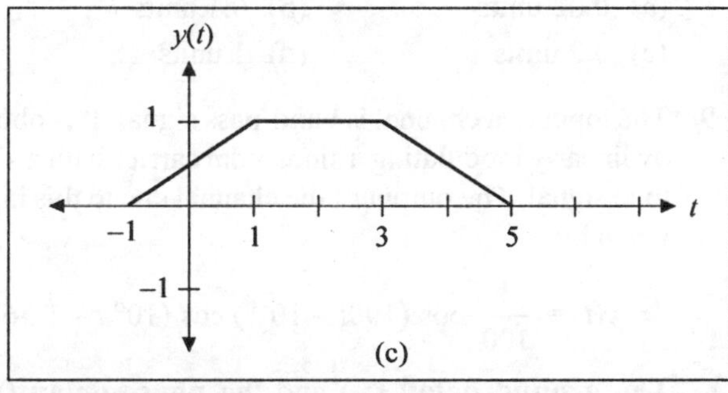

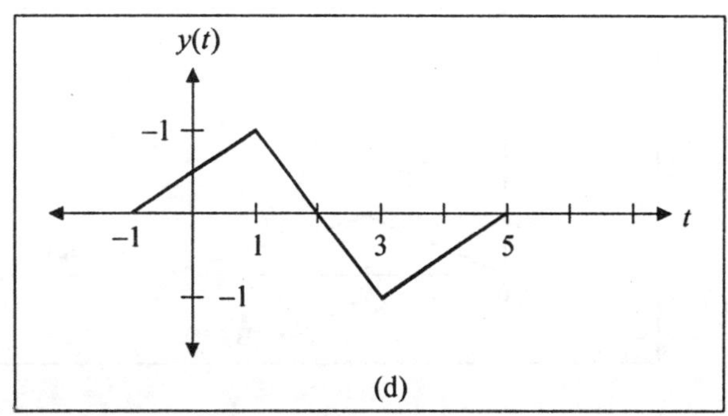

8. Consider a discrete time LTI system with input $x[n]=\delta[n]+\delta[n-1]$ and impulse response $h[n]=\delta[n]-\delta[n-1]$. The output of the system will be given by.

(a) $\delta[n]-\delta[n-2]$

(b) $\delta[n]-\delta[n-1]$

(c) $\delta[n-1]+\delta[n-2]$

(d) $\delta[n]+\delta[n-1]+\delta[n-2]$

4.8 Frequency Response

1. Two discrete time systems with impulse responses $h_1[n] = \delta[n-1]$ and $h_2[n] = \delta[n-2]$ are connected in cascade. The overall impulse response of the cascaded system is

(a) $\delta[n-1]+\delta[n-2]$ (b) $\delta[n-4]$
(c) $\delta[n-3]$ (d) $\delta[n-1]\delta[n-2]$

2. A signal $x(t)$ = sinc (α t) where α is a real constant $\left(\text{sin}\,c\,(x)=\dfrac{\sin(\pi x)}{\pi x}\right)$ is the input to a linear Time invariant system whose impulse response $h(t) = \text{sin}\,c\,(\beta t)$ where β is a real constant. If min (α, β) denotes the minimum of α and β, and similarly max, (α, β) denotes the maximum of α and β and K is a constant, which one of the following statements is true about the output of the system?

(a) It will be of the form K sinc (γt) where γ = min (α, b)

(b) It will be of the form K sinc (γt) where γ = max (α, β)

(c) It will be of the form K sinc (αt)

(d) Cannot be a sinc type of signal

3. The response of an initially relaxed linear constant parameter network to a unit impulse applied at $t = 0$ is $4e^{-2t}\,u(t)$. The response of this network to a unit step function will be:

(a) $2[1-e^{-2t}]\,u(t)$ (b) $4[e^{-t}-e^{-2t}]\,u(t)$
(c) $\sin 2t$ (d) $[1-4e^{-4t}]\,u(t)$

4. The impulse response and the excitation function of a linear time invariant causal system are shown in Figure. A and B respectively. The output of the system at $t = 2$ sec, is equal to

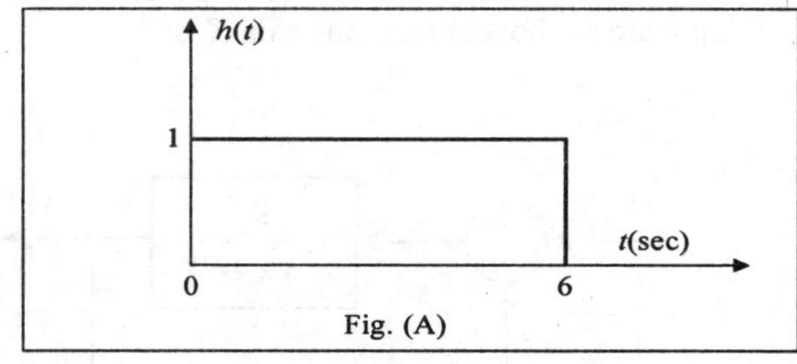

Fig. (A)

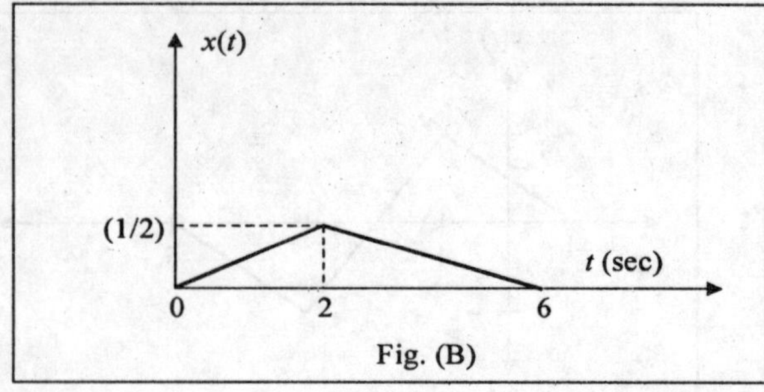

Fig. (B)

(a) 0 (b) $\frac{1}{2}$

(c) $\frac{3}{2}$ (d) 2

5. Let $h(t)$ be the impulse response of a linear time invariant system. Then the response of the system for any input $u(t)$ is

(a) $\int_0^t h(\tau)\, u(t-\tau)\, d\tau$

(b) $\frac{d}{dt}\int_0^t h(\tau)\, u(t-\tau)\, d\tau$

(c) $\int_0^t \left(\int_0^t (h(\tau)\, u(\tau)\, d\tau)\right) dt$

(d) $\int_0^t h^2(\tau)\, u(t-\tau)\, dt$

6. A linear time-invariant causal system has a frequency response given in polar form as $\frac{1}{\sqrt{1+\omega^2}} \angle \tan^{-1}\omega$.

For input $x(t) = \sin(t)$, the output is

(a) $\frac{1}{\sqrt{2}}\cos t$ (b) $\frac{1}{\sqrt{2}}\cos\left(t-\frac{\pi}{4}\right)$

(c) $\frac{1}{\sqrt{2}}\sin t$ (d) $\frac{1}{\sqrt{2}}\sin t\left(t+\frac{\pi}{4}\right)$

7. If a first order system and its time response to a unit step input are as shown below, the gain K is.

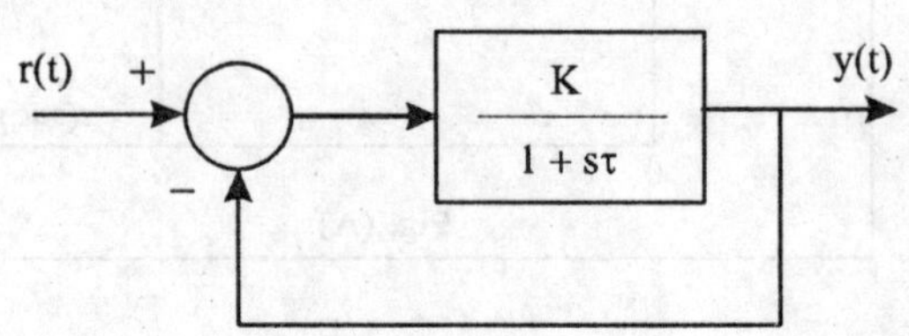

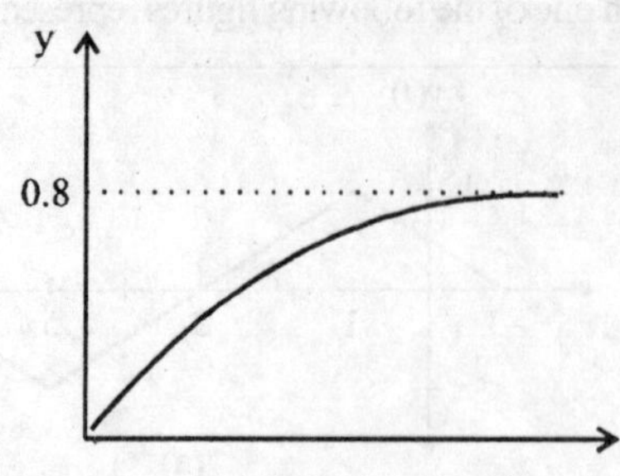

(a) 0.25 (b) 0.8

(c) 1 (d) 4

8. The response of a first order measurement system to a unit step input is $1 - e^{-0.5t}$, where t is in seconds. A ramp of 0.1 units per second is given as the input to this system. The error in the measured value after transients have died down is

(a) 0.02 units (b) 0.1 units

(c) 0.2 units (d) 1 units

9. The input to a channel is band-pass signal. It is obtained by linearly modulating a sinusoidal carrier with a single-tone signal. The output of the channel due to this input is given by

$$y(t) = \frac{1}{100}\cos(100t - 10^{-6})\cos(10^6 t - 1.56).$$

The ground delay (t_g) and the **phase delay** (t_p) in second, of the channel are

(a) $t_g = 10^{-6}$, $t_p = 1.56$

(b) $t_g = 1.56$, $t_p = 10^{-6}$

(c) $t_g = 10^{-8}$, $t_p = 15.6 \times 10^{-6}$

(d) $t_g = 10^{-8}$, $t_p = 1.56$

10. A linear phase channel with **phase delay T_P and ground delay** T_g must have

(a) $T_P = T_g =$ constant

(b) $T_P \propto f$ and $T_g \propto f$

(c) $T_g \propto f$ and $T_p =$ constant

(d) $T_P \propto f$ and $T_g =$ constant

11. A system with the transfer function $\frac{Y(s)}{X(s)} = \frac{s}{s+p}$ has an output $y(t) = \cos\left(2t - \frac{\pi}{2}\right)$ for the input signal $x(t) = p\cos\left(2t - \frac{\pi}{2}\right)$. Then, the system parameter 'p' is

(a) $\sqrt{3}$ (b) $\frac{2}{\sqrt{3}}$

(c) 1 (d) $\frac{\sqrt{3}}{2}$

12. The magnitude and phase transfer functions for a distortion-less filter should respectively be:

	(Magnitude)	*(Phase)*
(a)	Linear	Constant
(b)	Constant	Constant
(c)	Constant	Linear
(d)	Linear	Linear

13. A system has a phase response given by $\phi(\omega)$ where w is the angular frequency. The phase delay and group delay at $\omega = \omega_o$ are respectively given by

(a) $\frac{-\phi(\omega_o)}{\omega_o}, \left.\frac{-d\phi(\omega)}{d\omega}\right|_{\omega=\omega_o}$

(b) $\phi(\omega_o), \left.\frac{-d^2\phi(\omega)}{d\omega^2}\right|_{\omega=\omega_o}$

(c) $\frac{(\omega_o)}{\phi(\omega_o)}, \left.\frac{-d\phi(\omega)}{d\omega}\right|_{\omega=\omega_o}$

(d) $\omega_o\phi(\omega_o), \int_{-\infty}^{\omega_o}\phi(\lambda)\, d\lambda$

Statement for Linked Answer Questions 14 and 15:

The impulse response $h(t)$ of a linear-invariant continous time systems is given by $h(t) = e^{(-2t)}\, u(t)$, where $u(t)$ denotes the unit step function.

14. The frequency response $H(\omega)$ of this systems in terms of angular ω, is given by $H(\omega) =$

(a) $\frac{1}{1+j2\omega}$ (b) $\frac{\sin(\omega)}{\omega}$

(c) $\frac{1}{2+j\omega}$ (d) $\frac{j\omega}{2+j\omega}$

15. The output of this system, to the sinusoidal input $x(t) = 2\cos(2t)$ for all time t, is

(a) 0

(b) $2^{-0.25}\cos(2t - 0.125\,\pi)$

(c) $2^{-0.5}\cos(2t - 0.125\,\pi)$

(d) $2^{-0.5}\cos(2t - 0.25\,\pi)$

Data for Q. 16 and 17 are given below. Solve the problems and choose the correct answers

The system under consideration is are RC Low-Pass filter (RC-LPF) with $R = 1.0$ kΩ and $C = 0.01$ μF.

16. Let $H(f)$ denote the frequency response of the *RC-LPF*. Let f_1 be the highest frequency such that

$0 \le |f| \le f_1, \frac{|Hf_1|}{H(0)} \ge 0.95$. Then f_1 (in Hz) is

(a) 327.8 (b) 163.9

(c) 52.2 (d) 104.4

17. Let $t_g(f)$ be the group delay function of the given RC-LPF and $f_2 = 100$ Hz. Then $t_g(f_2)$ in ms, is

(a) 0.717 (b) 7.17

(c) 71.7 (d) 4.505

18. A system has poles at 0.01 Hz, 1 Hz and 80 Hz; zeros at 5Hz. 100 Hz and 200 Hz. The approximate phase of the system response at 20 Hz is:

(a) –90°

(b) 0°

(c) 90°

(d) –180°

19. A system with transfer function $H(z)$ has impulse response $h(.)$ defined as $h(2) = 1, h(3) = -1$ and $h(k) = 0$ otherwise. Consider the follwing statements.

$S1: H(z)$ is a low pass filter

$S2: H(z)$ is an FIR filter.

Which of the following is correct

(a) Only S2 is true

(b) Both S1 and S2 are flase

(c) Both S1 and S2 are true, and S2 is a reason for S1

(d) Both S1 and S2 are true, but S2 is not a reason for S1

20. A signal $x(n) = \sin(\omega_0 n + \phi)$ is the input to a linear time-invariant system having a frequency response $H(e^{j\omega})$. If the output of the system $Ax(n - n_0)$, then the most general form $\angle H(e^{j\omega})$ will be

(a) $-n_0\omega_0 + \beta$ for any arbitrary real β.

(b) $-n_0\omega_0 + 2\pi k$ for any arbitrary integer k.

(c) $n_0\omega_0 + 2\pi k$ for any arbitrary integer k.

(d) $-n_0\omega_0\,\phi$

21. $G(z) = \alpha z^{-1} + \beta z^{-3}$ is a low-pass digital filter with a phase characteristics same as that of the above question if.

(a) $\alpha = \beta$ (b) $\alpha = -\beta$

(c) $\alpha = \beta^{1/3}$ (d) $\alpha = \beta^{-(1/3)}$

22. The pole-zero plot given below corresponds to a

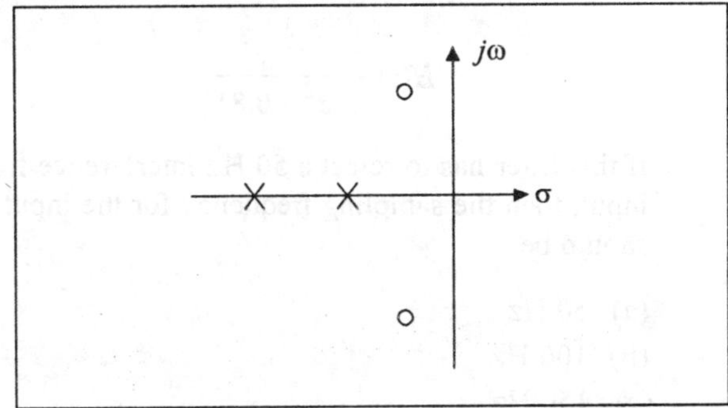

(a) Low pass filter
(b) High pass filter
(c) Band pass filter
(d) Notch filter

23. Identify the transfer function corresponding to an all-pass filter form the following

(a) $\frac{1-s\tau}{1+s\tau}$ (b) $\frac{1+s\tau_1}{1+s\tau_2}$

(c) $\frac{1}{1+s\tau}$ (d) $\frac{s\tau}{1+s\tau}$

24. An analog low-pass filter is needed with the following specifications:

Pass-band: $0.9 \le |H(j\omega)| \le 1, \quad 0 \le \omega\ 2\pi \times 1000;$

Stop-band: $|H(j\omega)| \le 0.2, \quad |\omega| \ge 2\pi \times 1500;$

The minimum order N of the Butterworth filter to approximate the above is

(a) 6 (b) 11
(c) 12 (d) 15

25. In the IIR filter below, a is a variable gain. For which of the following cases, the system will transit form stable to unstable condition ?

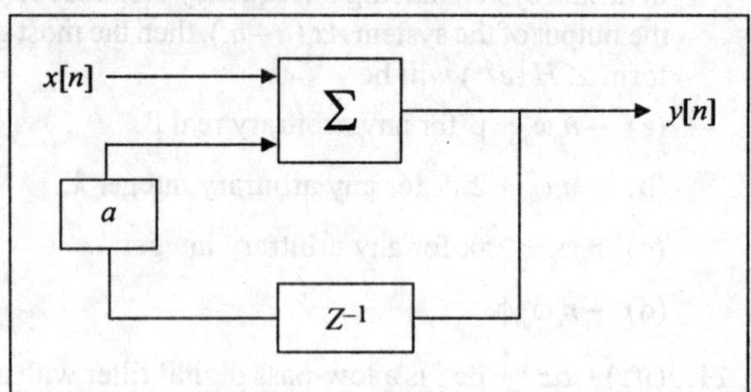

(a) $0.1 < a < 0.5$ (b) $0.5 < a > 1.5$
(c) $1.5 < a < 2.5$ (d) $2 < a < \infty$

26. A digital filter has the transfer function.

$$H(z) = \frac{z^2+1}{z^2+0.81}$$

If this filter has to reject a 50 Hz interference from the input, then the sampling frequency for the input signal should be

(a) 50 Hz
(b) 100 Hz
(c) 150 Hz
(d) 200 Hz

27. A low-pass filter has sampling frequency $f_T = 10$ kHz. The passband (f_p) and stop band (f_s) edge frequencies are 2 kHz and 3 kHz respectively. The frequency axis is pre-warped before applying bilinear transformation to get new angular frequencies, $\tilde{\Omega}_p$ and $\tilde{\Omega}_S$. The ratio of the warped edge frequencies $(\tilde{\Omega}_s/\tilde{\Omega}_p)$ is

(a) – 1.213
(b) 1.213
(c) 1.575
(d) 1.894

28. The circuit in figure is a

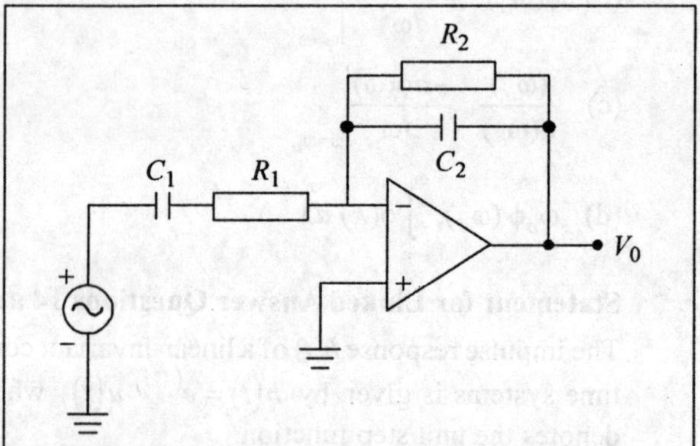

(a) Band-pass filter with lower cut-off $\omega_L = \frac{1}{R_1C_1}$ and higher cut off $\omega_H = \frac{1}{R_2C_2}$

(b) Band-reject filter with lower cut-off $\omega_L = \frac{1}{R_1C_1}$ and higher cut off $\omega_H = \frac{1}{R_2C_2}$

(c) Band-pass filter with lower cut-off $\omega_L = \frac{1}{R_2C_2}$ and higher cut off $\omega_H = \frac{1}{R_1C_1}$

(d) Band-pass filter with lower cut-off $\omega_L = \frac{1}{R_2C_2}$ and higher cut off $\omega_H = \frac{1}{R_1C_1}$

29. Identify the transfer function corresponding to an all-pass filter form the following

(a) $\frac{1-s\tau}{1+s\tau}$ (b) $\frac{1+s\tau_1}{1+s\tau_2}$

(c) $\frac{1}{1+s\tau}$ (d) $\frac{s\tau}{1+s\tau}$

30. Shown below is the pole-zero plot of a digital filter

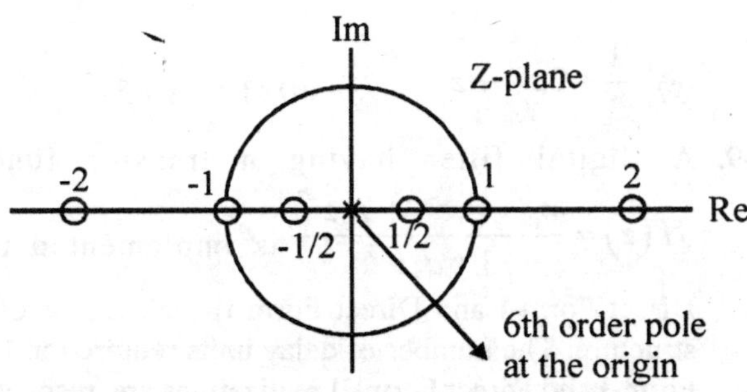

Which one of the following statements is **TRUE**?

(a) This is a low pass filter

(b) This is a high pass filter

(c) This is an IIR filter

(d) This is an FIR filter

31. A linear phase FIR system with impulse response real has a zero at z $z = \frac{1}{2}e^{j\frac{\pi}{4}}$. The largest set of remaining zeros that can be obtained from the above information is.

(a) $\frac{1}{2}e^{j\frac{\pi}{2}}, \frac{1}{2}e^{j\frac{3\pi}{4}}, \frac{1}{2}e^{-j\frac{\pi}{2}}$

(b) $\frac{1}{2}e^{-j\frac{\pi}{2}}, \frac{1}{2}e^{-j\frac{3\pi}{4}}, \frac{1}{2}e^{-j\frac{\pi}{2}}$

(c) $2e^{-j\frac{\pi}{4}}, 2e^{j\frac{\pi}{4}}, 2e^{-j\frac{\pi}{4}}$

(d) $2e^{j\frac{\pi}{4}}. \frac{1}{2}e^{-j\frac{\pi}{4}}$

32. A direct form implementation of an LTI system with $H(z) = \frac{1}{1 - 0.7z^{-1} + 0.13z^{-2}}$ is shown in figure. The value of a_0, a_1 and a_2 are respectively

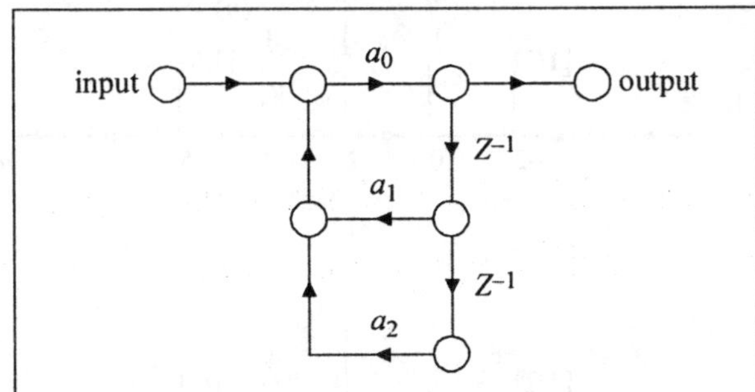

(a) 1.0, 0.7 and – 0.13 (b) – 0.13, 0.7 and 1.0

(c) 1.0, – 0.7 and 0.13 (d) 0.13, – 0.7 and 1.0

33. A low-pass filter has sampling frequency f_T = 10 kHz. The passband (f_p) and stop band (f_s) edge frequencies are 2 kHz and 3 kHz respectively. The frequency axis is pre-warped before applying bilinear transformation to get new angular frequencies, $\tilde{\Omega}_p$ and $\tilde{\Omega}_S$. The ratio of the warped edge frequencies ($\tilde{\Omega}_s/\tilde{\Omega}_p$) is

(a) – 1.213

(b) 1.213

(c) 1.575

(d) 1.894

34. An analog low-pass filter is needed with the following specifications:

Pass-band: $0.9 \le |H(j\omega)| \le 1, \quad 0 \le \omega\ 2\pi \times 1000;$

Stop-band: $|H(j\omega)| \le 0.2, \quad |\omega| \ge 2\pi \times 1500;$

The minimum order N of the Butterworth filter to approximate the above is

(a) 6 (b) 11

(c) 12 (d) 15

35. A digital notch filter with a notch frequency of 60 Hz is to be transformed into one operating at a new notch frequency of 120 Hz. The sampling frequency is 400 Hz. A lowpass to lowpass transformation $z^{-1} = \frac{1 - \alpha\bar{z}}{\bar{z} - \alpha}$ is used on the first to obtain the new notch filter. Then the value of should be

(a) – 0.95

(b) 0.95

(c) – 0.46

(d) 0.46

36. In the IIR filter below, a is a variable gain. For which of the following cases, the system will transit form stable to unstable condition ?

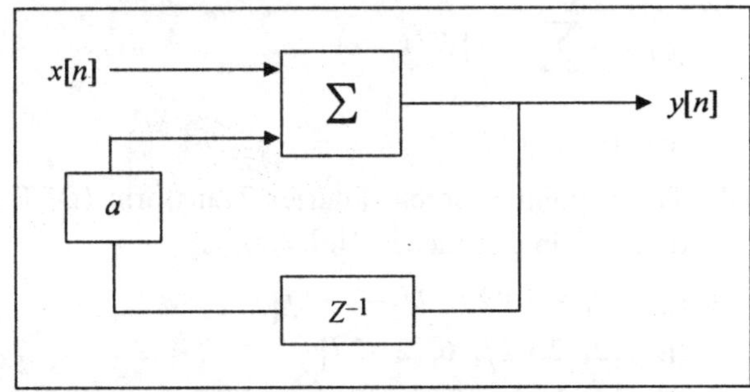

(a) $0.1 < a < 0.5$ (b) $0.5 < a > 1.5$

(c) $1.5 < a < 2.5$ (d) $2 < a < \infty$

37. An analog signal is sampled at 9 kHz. The sequence so obtained is filtered by an FIR filter with transfer function $H[z]=1-z^{-6}$. One of the analog frequencies for which the magnitude response of the filter is zero is

(a) 0.75 kHz (b) 1 kHz
(c) 1.5 kHz (d) 2 kHz

38. The transfer function $H(z)$ of a fourth-order linear phase FIR system is given by

$H(z)=(1+2z^{-1}+3z^{-2}),G(z)$

Then $G(z)$ is

(a) $3+2z^{-1}+z^{-2}$ (b) $1+\frac{1}{2}z^{-1}+\frac{1}{3}z^{-2}$
(c) $\frac{1}{3}+2z^{-1}+z^{-2}$ (d) $1+2z+3z^2$

39. A digital filter having a transfer function $H(z)=\dfrac{p_0+p_1z^{-1}+p_3z^{-3}}{1+d_3z^{-3}}$ is implemented using Direct Form-I and Direct Form-II realizations of IIR structure. The number of delay units required in Direct Form-I and Direct Form-II realizations are, respectively

(a) 6 and 6 (b) 6 and 3
(c) 3 and 3 (d) 3 and 2

4.9 Discrete Fourier Transform

1. For an N-point FFT algorithm with $N=2^m$, which one of the following statements is TRUE?

(a) It is not possible to construct a signal flow graph with both input and output in normal order
(b) The number of butterflies in the mth stage is N/m
(c) In-place computation requires storage of only 2N node data
(d) Computation of a butterfly requires only one complex multiplication

2. $\{x(n)\}$ is a real-valued periodic sequence with a period N. $x(n)$ and $X(k)$ form N-point Discrete Fourier Transform (DFT) pairs. The DFT $Y(k)$ of the sequence

$y(n)=\frac{1}{n}\sum_{r=0}^{N-1}x(r)x(n+r)$ is

(a) $|X(k)|^2$
(b) $\frac{1}{N}\sum_{r=0}^{N-1}X(r)X^*(k+r)$
(c) $\frac{1}{N}\sum_{r=0}^{N-1}X(r)X(k+r)$
(d) 0

3. The 4-point Discrete Fourier Transform (DFT) of a discrete time sequence $\{1,0,2,3\}$ is

(a) $[0,\ -2+2j,\ 2,\ -2-2j]$
(b) $[2,\ 2+2j,\ 6,\ 2-2j]$
(c) $[6,\ 1-3j,\ 2,\ 1+3j]$
(d) $[6,\ -1+3j,\ 0,\ -1-3j]$

4. The first six points of the 8 - point DFT of a real valued seqeunce are 5, 1– j3, 0, 3 – j4, 0 and 3 + j4. The last two points of the DFT are respectivley.

(a) 0, 1– j3 (b) 0, 1 + j3
(c) 1 + j3, 5 (d) 1 – j3, 5]

Statement for linked Answer Questions 5 and 6

A sequence $x(n)$ has non-zero values as shown in the figure

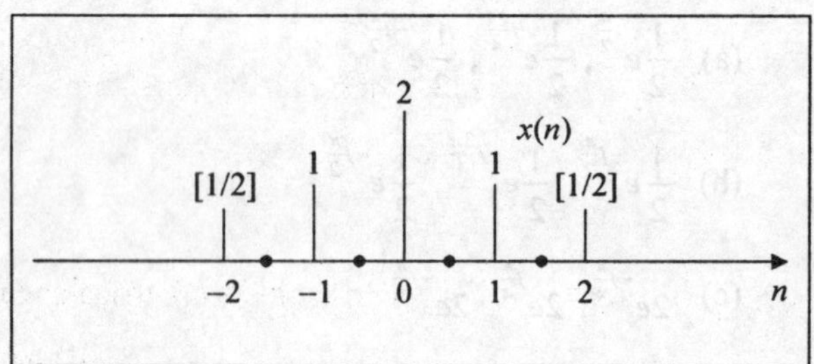

5. The sequence

$$y(n)=\begin{cases}x\left(\frac{n}{2}-1\right) & \text{for } n \text{ even}\\ 0 & \text{for } n \text{ odd will be}\end{cases}$$

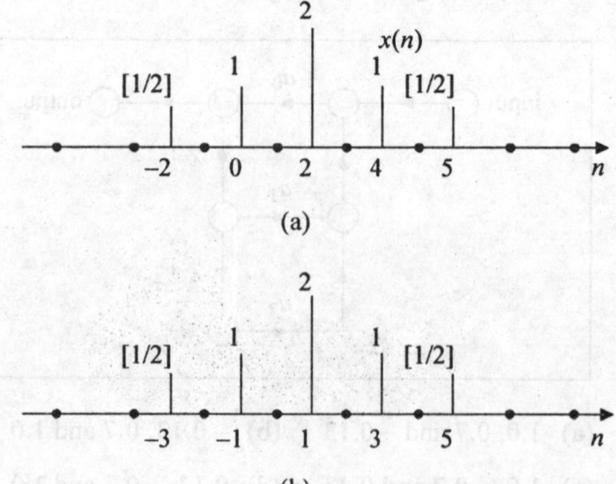

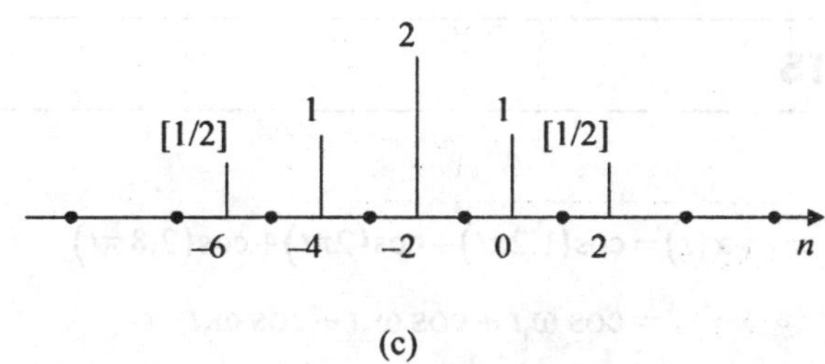

(c)

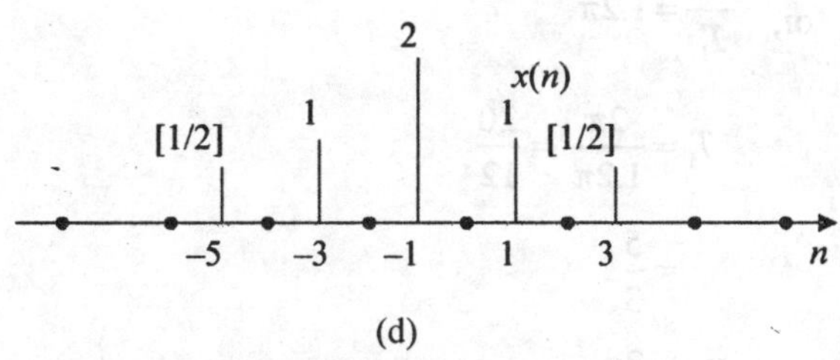

(d)

6. The Fourier transform of $y(2n)$ will be

(a) $e^{-j2\omega}[\cos 4\omega + 2\cos 2\omega + 2]$

(b) $\cos 2\omega + 2\cos \omega + 2]$

(c) $e^{-j\omega}[\cos 2\omega + 2\cos \omega + 2]$

(d) $e^{-j2\omega}[\cos 2\omega + 2\cos 2\omega + 2]$

7. A 5-pointy sequence $x[n]$ is given as $x[-3]=1$, $x[-2]=1$, $x[-1]=0$, $x[0]=5$, and $x[1]=1$

Let $X(e^{j\omega})$ denote that discrete-time Fourier transform of [n]. The value of $\int_{-\pi}^{\pi} X(e^{j\omega})$ is

(a) 5

(b) 10π

(c) 16π

(d) $5 + j10\pi$

8. Let $x(n) = \left(\frac{1}{2}\right)^n u(n)$, $y(n) = x^2(n)$ and $Y(e^{j\omega})$ be the Fourier transform of $y(n)$. Then $Y(e^{j0})$ is.

(a) $\frac{1}{4}$

(b) 2

(c) 4

(d) $\frac{4}{3}$

4.10 Miscellaneous

1. A full charged mobile phone with a 12 V battery is good for a 10 minute talk-time Assume that, during the talk-time, the battery delivers a constant current of 2 A and its voltage drops linearly from 12 V to 10 V as shown in the figure. How much energy does the battery deliver during this talk time?

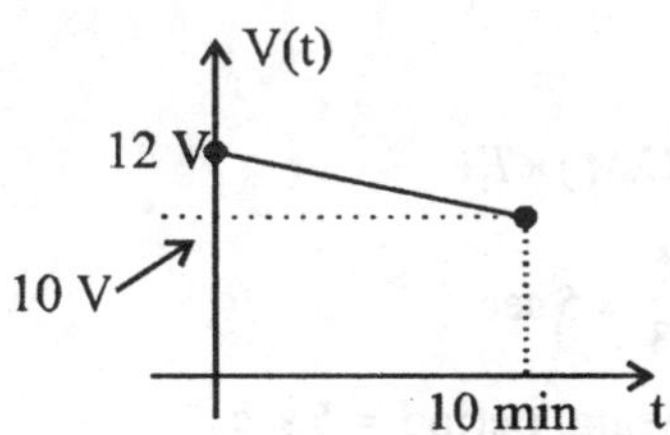

(a) 220 J

(b) 12 kJ

(c) 13.2 kJ

(d) 14.4 kJ

2. An LTI system having transfer function $\frac{s^2+1}{s^2+2s+1}$ and input $x(t)=\sin(t+1)$ is in steady state. The output is sampled at a rate ω_s rad/s to obtain the final output $\{y(k)\}$. Which of the following is true?

(a) $y(.)$ is zero for all sampling frequencies ω_s

(b) $y(.)$ is nonzero for all sampling frequencies ω_s

(c) $y(.)$ is nonzero for $\omega_s > 2$ but zero for $\omega_s < 2$

(d) $y(.)$ is zero for $\omega_s > 2$ but nonzero for $\omega_s < 2$

3. Consider an angle modulated signal

$$x(t) = 6\cos\begin{bmatrix} 2\pi\times10^6 t + 2\sin(8000\pi t) \\ +4\cos(8000\pi t) \end{bmatrix} V$$

The average power of $x(t)$ is

(a) $10W$ (b) $18W$

(c) $20W$ (d) $28W$

Answers

4.1 Signals

1. Ans. (b)

Solution:

$i = A \sin \omega t$; $\omega = 2\pi/T$

i(t) = 10 +20 sin t

$$I_{rms} = \left[\frac{1}{2\pi}\int_o^{2\pi} (10+20\sin t)^2 .dt\right]^{1/2}$$

$$= \left[\frac{1}{2\pi}\int_o^{2\pi} (100+400\sin^2 t+400\sin t).dt\right]^{1/2}$$

$$= \left[\frac{1}{2\pi}\int_o^{2\pi} (100+200(1-\cos 2t)+400\sin t).dt\right]^{1/2}$$

$$= \left[\frac{1}{2\pi}\int_o^{2\pi} (300-200\cos 2t+400\sin t).dt\right]^{1/2}$$

$$= \left[\frac{1}{2\pi}\int_o^{2\pi} 300.dt\right]^{1/2} = \sqrt{300} = 17.3\ A.$$

Second Method

Given

dc current = 10 A

$$V_{rms} = \sqrt{10^2 + \left(\frac{20}{\sqrt{2}}\right)^2}$$

$$= \sqrt{100 + \frac{400}{2}}$$

$$= \sqrt{100+200}$$

$$= \sqrt{300} = 10\sqrt{3} = 17.3\text{A}$$

sinusoid ac current peak value = 20 A

2. Ans. (a)

Solution:

$$\omega = 2\pi f \ : \ f = \frac{\omega}{2\pi f} = \frac{1.2\pi}{2\pi} = 0.6\ Hz$$

Fourier series represented by sinusoidal function and its harmonic component. Therefore, all terms frequency must be multiple of fundamental frequency. In present case that is only possible with 0.2 Hz.

Another Method

Given

$$x(t) = \cos(1.2\pi t) + \cos(2\pi t) + \cos(2.8\,\pi t)$$

$$= \cos \omega_1 t + \cos \omega_2 t + \cos \omega_3 t$$

$$\omega_1 = 1.2\pi$$

or, $\dfrac{2\pi}{T_1} = 1.2\pi$

$$T_1 = \frac{2\pi}{1.2\pi} = \frac{20}{12}$$

$$= \frac{5}{3}$$

$$\omega_2 = 2\pi$$

or, $\dfrac{2\pi}{T_2} = 2\pi$

or, $T_2 = \dfrac{2\pi}{2\pi} = 1$

$$\omega_3 = 2.8\pi$$

or, $\dfrac{2\pi}{T_3} = 2.8\pi$

or, $T_3 = \dfrac{2\pi}{2.8\pi} = \dfrac{20}{28} = \dfrac{5}{7}$

Step-2:

$$\frac{T_1}{T_2} = \frac{5}{3},\ \frac{T_1}{T_3} = \frac{7}{3}$$

Step-3:

L.C.M of denominators of step 2 is L.C.M = 3

Step-4:

$$T = (L.C.M) \times T_1$$

$$= 3 \times \frac{5}{3} = 5 \text{ sec}$$

Fundamental period = 5 sec

Fundamental frequency of the signal $= \dfrac{1}{T} = \dfrac{1}{5}$

$$= 0.2\ Hz$$

3. Ans (d)

Solution:

A discrete signal $x\,[n]$ is said to be peridic if $x[n+N] = x[n]$ for all n.

When N is period of the signal

A discret time sinusoida is periodic for all vale of n only if its frequency is rational, i.e

$$f = \frac{m}{N}$$

Where N is fundamental perid and m is some integer.

Given

$$x[N] = e^{j\left(\frac{5\pi}{6}\right)n}$$

$$e^{jx} = \cos x + j\sin x$$

$$e^{j\left(\frac{5\pi}{6}\right)n} = \cos\left(\frac{5\pi}{6}n\right) + j\sin\left(\frac{5\pi}{6}n\right)$$

$$\frac{\omega_o}{2\pi} = \frac{m}{N} = \text{rational number}$$

$$\omega_o = 2\pi f$$

$$f = \frac{m}{N}$$

or, $\frac{m}{N} = \frac{5\pi/6}{2\pi} = \frac{5}{12}$

Therefore, Fundamental period $N = 12$

4. Ans. (c)

Solution:

Given

$$x(t) = 2\sin 2\pi t + 3\sin 3\pi t$$

$$\omega = 2\pi f = \frac{2\pi}{T}$$

Here, $\omega_1 = 2\pi$

or $\frac{2\pi}{T_1} = 2\pi$

or $T_1 = 1$

Here, $\omega_2 = 3\pi$

or $\frac{2\pi}{T_2} = 3\pi$

or $T_2 = \frac{2}{3}$

$\frac{T_1}{T_2} = \frac{1}{2/3} = \frac{3}{2}$ is a rational number

$\therefore$ Fundamental period

$$T_0 = 3T_2 = 2T_1$$

$$= 3 \times \frac{2}{3} = 2 \text{ sec}$$

5. Ans. (d)

Solution:

$$x(t) = 3 + 2\sin t\cos 2t$$

$$x(t) = 3 + \sin 3t - \sin t$$

$\therefore$ Root mean square value

$$= \sqrt{3^2 + 1^2 + 1^2} = \sqrt{11}$$

6. Ans. (c)

Solution: $s(t) = \cos 2t + \cos 3t + \cos 5t$

Period = 2π

fundamental frequency = 1 rad/sec.

The ratio of any two freqencies = m/n

where m and n are integers.

$\therefore s(t) = \cos(2t) + \cos(3t) + \cos(5t)$ is periodic

$$s(t) = e^{j8\pi t}$$

$$s(t+T) = e^{j8\pi(t+T)} = e^{j8\pi t}\, e^{j8\pi T}$$

$T = \frac{2\pi}{8\pi} = \frac{1}{4}$ period and $\omega = 8\pi$

$\therefore\ s(t) = e^{j8\pi t}$ is periodic

$$s(t) = e^{-7t}\sin 10\pi t$$

$$s(t+T) = e^{-7(t+T)}\sin 10\pi(t+T)$$

$$= e^{-7t}e^{7T}\sin 10\pi t$$

* Exponentially decreasing function value becomes zero after some time, hence its is non periodic

e^{-7t} is not periodic as it is not complex exponential of the form $e^{\pm j\omega t}$

$\sin(10\pi t)$ is periodic with $\omega = 10\pi$

Hence the product $s(t) = e^{-7t}\sin(10\pi t)$ is not periodic

$$s(t) = \cos 2t \times \cos 4t$$

$$= \frac{1}{2}[\cos 6t + \cos 2t]$$

$$s(t+T) = \cos(2t+2T)\cos(4t+4T)$$

$T = \pi$ Periodic

$\omega = 2$ rad/sec

$\therefore\ s(t) = \cos(2t)\cos(4t) = \frac{1}{2}[\cos(6t) + \cos(2t)]$ is periodic

7. Ans. (a)

Solution: Given sequence is

$$x[n]=3\sin[1.3\pi n+0.5\pi]+5\sin[1.2\pi n]$$

$$x[n]=3\sin[\omega_1 n+\phi]+5\sin[\omega_2 n]$$

Apply formula, $\frac{\omega_o}{2\pi}=\frac{M}{N}$

or, $\frac{M_1}{N_1}=\frac{\omega_1}{2\pi}=\frac{1.3\pi}{2\pi}$

$$=\frac{13}{20}\Rightarrow N_1=20$$

and $\frac{M_2}{N_2}=\frac{\omega_2}{2\pi}=\frac{1.2\pi}{2\pi}$

$$=\frac{12}{20}=\frac{3}{5}\Rightarrow N_2=5$$

Fundamental period

$$N=L.C.M(N_1,N_2)=20$$

8. Ans (a)

Solution:

Given signal is

$$x(t)=(1+0.5\cos 40\pi t)\cos 200\,\pi t$$

$$x(t)=\cos 200\pi t+\frac{1}{2}(\cos 40\pi t\times\cos 200\pi t)$$

$$=\cos 200\pi t+\frac{1}{4}\cos 240\pi t+\frac{1}{4}\cos 160\pi t$$

Fundamental of frequency (HCF of frequency component)

$$\omega=40\pi$$

$$2\pi f=40\pi$$

Therefore $f=20Hz$

9. Ans (c)

Solution

Given, $x(n)=\left(\frac{1}{3}\right)^n u(n);\ n\geq 0$

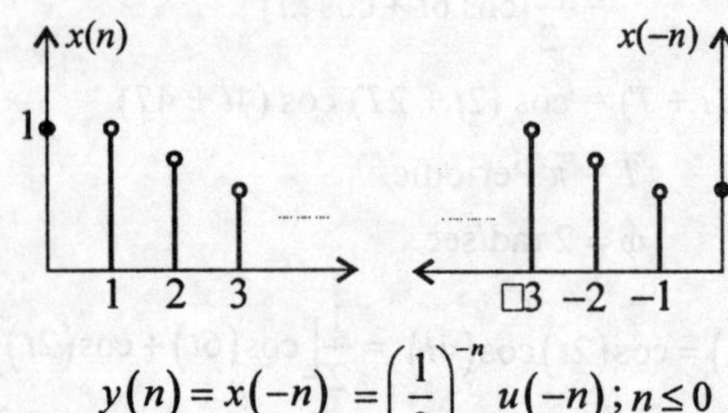

$$y(n)=x(-n)=\left(\frac{1}{3}\right)^{-n}u(-n);\ n\leq 0$$

Then $S=\sum_{n=-\infty}^{\infty}y(n)$

$$=\sum_{n=-\infty}^{0}\left(\frac{1}{3}\right)^{-n}u(-n)$$

$$=\sum_{n=-\infty}^{0}\left(\frac{1}{3}\right)^{-n}$$

$$=1+\frac{1}{3}+\frac{1}{9}+......$$

$$=\frac{1}{1-\frac{1}{3}}$$

$$=\frac{3}{2}$$

10. Ans. (d)

Solution:

$|Z|<1,\ so\ |Y|>1$

Z is having positive real part and positive imaginary part (∴ from the characteristics)

So Y should have positive real part and negative imaginary part.

11. Ans.(c)

Solution:

(1) $x(4n)\rightarrow$ scalling $\rightarrow$ Pick every fourth sample
(2) $x(-4n)\rightarrow$ folding $\rightarrow$ Time reverse sample
(3) $x(3-4n)\rightarrow$ Shifting $\rightarrow$ Advance

12. Ans.(d)

13 Ans.(d)

Solution:

$$v_c(t)=\frac{1}{C}\int_{-\infty}^{t}i(t)dt$$

$$=\frac{1}{C}\int_{-\infty}^{t}5\,\delta(t)dt$$

$$=\frac{5}{C}u(t)$$

14. Ans (c)

Solution:

$$I=\left[\begin{array}{l}-6\sqrt{2}\sin(100\pi t)+6\sqrt{2}\cos\\ \qquad\left(300\pi+\frac{\pi}{4}\right)+6\sqrt{2}\end{array}\right]A$$

Find Reading if it passes through true R.M.S. Ammeter R.M.S. Value = It is that value of current or voltage which produce the Same heat as it produce by the peak voltage or current.

In a.c. current $I_{rms} = \dfrac{I_{peak}}{\sqrt{2}}$

In D.C. current $I_{rms} = I_{peak}$

So $I = \sqrt{\left(\dfrac{-6\sqrt{2}}{\sqrt{2}}\right)^2 + \left(\dfrac{6\sqrt{2}}{\sqrt{2}}\right)^2 + \left(6\sqrt{2}\right)^2}$

$\sqrt{36+36+72}$

$= \sqrt{14\,A} = 12\,A$

15. Ans (c)

Solution:

$g(t) = a_m \cos\omega_m t$

$g^2(t) = a_m^{\;2}[1+\cos 2\omega_m t]$

$\therefore \text{BW} = f_{m'} = \dfrac{2\omega_m}{2\pi} = 2f_m = 6\ kHz$

16. Ans (b)

Solution:

$$\int_{-\infty}^{\infty} \delta\left(t - \frac{-\pi}{6}\right) 6\sin(t)\, dt$$

$$= 6\int_{-\infty}^{\infty} \delta\left(t - \frac{-\pi}{6}\right) 6\sin t\, dt$$

[Using Property $6\int_{-\infty}^{\infty} x(t)\ \delta(t-t_0)dt = x(t_0)$]

$$= 6\sin\left(\frac{\pi}{6}\right)$$

$$= 6\left(\frac{1}{2}\right) \Rightarrow 3$$

17. Ans. (c)

Solution: $x(t) = \sin w_0 t$ **is periodic**

So For $x(n) = \sin w_0 n$ to be periodic $\left[\dfrac{2\pi}{\omega_0}\right]$ to be a ratio of integer

18. Ans. (a)

Solution:

$$\frac{1}{\sqrt{2\pi}} \int_{-\infty}^{\infty} t^2 e^{-t^2/2} \delta(1-2t)dt$$

$$\left[\begin{array}{l} \delta(1-2t) \Rightarrow \delta(-2(t-\frac{1}{2})) \\ \delta(-at) = \frac{1}{a}\delta(t) \Rightarrow \frac{1}{2}\delta(t-\frac{1}{2}) \end{array}\right]$$

$$= \frac{1}{\sqrt{2\pi}} \int_{-\infty}^{\infty} t^2 e^{-t^2/2} \frac{1}{2}\delta(t-\frac{1}{2})dt$$

$$= \frac{1}{2\sqrt{2\pi}} \int_{-\infty}^{\infty} \left(\frac{1}{2}\right)^2 e^{-t^2/2} \times \delta\left(t-\frac{1}{2}\right)dt$$

$$= \frac{1}{8\sqrt{2\pi}} e^{-t^2/2} \int_{-\infty}^{\infty} \delta(t-\frac{1}{2})dt$$

$$\Rightarrow \frac{1}{8\sqrt{2\pi}} e^{\frac{-1}{8}}.$$

19. Ans. (a)

Solution:

The given signal is sampling signal and it Fourier Transform is gate function

The Fourier transform of $x[n]$ is shown below.

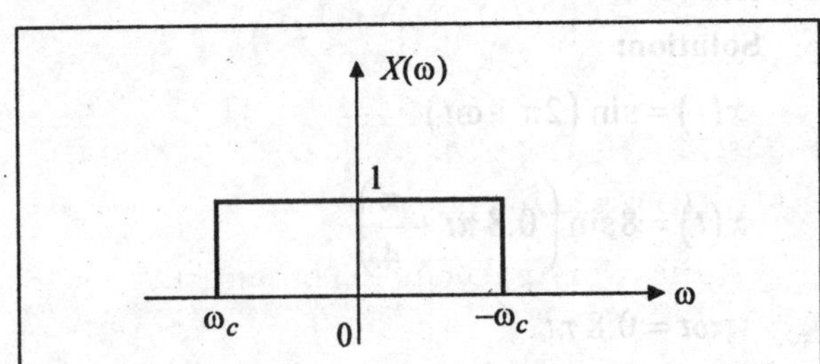

$$\text{Energy} = \frac{1}{2\pi}\int_{-\infty}^{\infty} |X(\omega)|^2\, d\omega$$

$$= \frac{1}{2\pi}\int_{-\omega_C}^{\omega_C} 1 \cdot d\omega$$

$$= \frac{1}{2\pi}\cdot[\omega]_{-\omega_c}^{\omega_c}$$

$$= \frac{1}{2\pi}\cdot[\omega_c - (-\omega_c)]$$

$$= \frac{2\omega_c}{2\pi} = \frac{\omega_c}{\pi}$$

$$\text{Energy} = \frac{\omega_c}{\pi}$$

20. Ans. (b)

Solution:

$$f(x) = \exp\left(-\frac{1}{x}\right)$$

when $x \to -\infty$

$$f(x) = \exp\left(\frac{-1}{-\infty}\right) \to 1$$

when $x \to 0^-$

$$f(x) = \exp\left(\frac{-1}{0^-}\right) \to \infty$$

when $x \to 0^+$

$$f(x) = \exp\left(\frac{-1}{0^+}\right) \to 0$$

and when $x \to 0 + \infty$

$$f(x) = \exp\left(\frac{-1}{+\infty}\right) \to 1$$

$\therefore$ The possible expression for $f(x)$ is $\exp\left(-\frac{1}{x}\right)$

21. Ans (c)

Solution:

$$x(t) = \sin(2\pi + \omega t)$$

$$x(t) = 8\sin\left(0.8\,\pi t + \frac{\pi}{4}\right)$$

$$\omega t = 0.8\,\pi t$$

$$\omega = \frac{2\pi}{T}$$

$$0.8\,\pi = \frac{2\pi}{T}$$

$$T = \frac{2\pi}{0.8\,\pi} = 2.5\ s$$

22. Ans. (a)

Solution.

Given that u (t) = 3 + 4 cos 3t

Which mean, $\omega = 3$; $\omega = 2\pi f = \frac{2\pi}{T}$

$$T = \frac{2\pi}{\omega} = \frac{2\pi}{3}$$

$$V_{rms} = \sqrt{\frac{1}{T}\int_0^T [v(t)]^2\, dt}$$

$$= \sqrt{\frac{3}{2\pi}\int_0^{2\pi/3} (3 + 4\cos 3t)^2\, dt} = \sqrt{17}$$

Second Method

Given $v(t) = 3 + 4\cos(3t)$

Assuming that the peak value of cos term is 4 V.

$$V_{rms} = \sqrt{3^2 + \left(\frac{4}{\sqrt{2}}\right)^2}$$

$$= \sqrt{9 + 8}$$

$$V_{rms} = \sqrt{17}$$

23. Ans. (c)

24. Ans. (c)

Solution:

1. $x(4n) \to$ scalling $\to$ Pick every fourth smaple
2. $x(-4n) \to$ folding $\to$ Time reverse sample
3. $x(3 - 4n) \to$ Shifting $\to$ Advance

4.2 Fourier Series

1. Ans. (a)

Solution: Straight line equation, $y = mt + C$

When m = slop of line = tan θ.

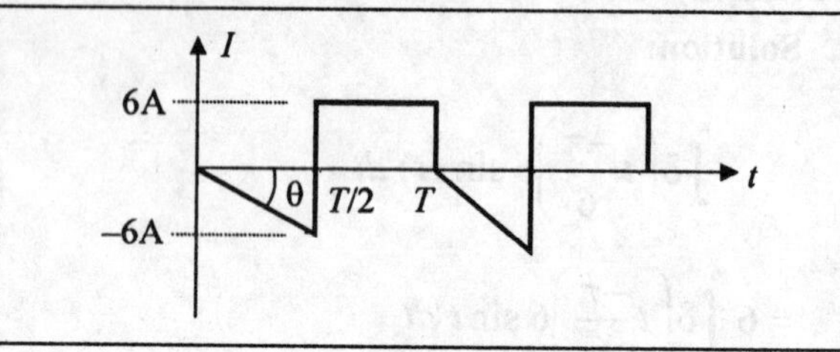

then, $m = \frac{-6}{T/2} = \frac{-12}{T}$ and C = 0

$$f(t) = mt\ ; \quad 0 \le t \le T/2$$
$$= 6\ ; \quad T/2 \le t \le T$$

rms value for any waveform $= \sqrt{\frac{1}{T}\int_o^T \{i(t)\}^2}$

$$= \left[\frac{1}{T}\left(\int_o^{T/2} (mt)^2 \cdot dt + \int_{T/2}^{T} 6^2 \cdot dt\right)\right]^{1/2}$$

$$= \left[\frac{1}{T}\left(\frac{144}{T^2} \times \frac{T^3}{3 \times 8} + 36 \times \frac{T}{2}\right)\right]^{1/2}$$

$$= [6 + 18]^{1/2} = \sqrt{24} = 2\sqrt{6}\ A$$

2. Ans. (a)

Solution: $V(t) = \frac{2t}{T}, 0 \le t \le \frac{T}{2}$

$$= 0; \frac{T}{2} \le t \le T$$

$$V(t)_{rms} = \sqrt{\frac{1}{T}\int_o^T |V(t)|^2\, dt}$$

$$= \sqrt{\frac{1}{T}\int_o^{\frac{T}{2}} \frac{4}{T^2} t^2\, dt}$$

$$= \sqrt{\frac{1}{T}\cdot\frac{4}{T^2}\left.\frac{t^3}{3}\right|_o^{\frac{t}{2}}} = \sqrt{\frac{4}{T^3}\cdot\frac{T^3}{8\times 3}} = \sqrt{\frac{1}{6}}$$

3. **Ans. (d)**

Solution: $f(x) = \sin^2 x$

Since, $f(x)$ is an even function, hence it has no sine terms.

$$a_o = \frac{2}{T}\int_o^T f(x)dx$$

$$a_o = \frac{2}{2\pi}\int_o^{2T} f(x)dx = 1$$

$$\therefore \quad f(x) = \frac{a_o}{2} + \sum_{n=1}^{\infty} a_n \cos nx$$

$f(x) = 0.5 +$ terms containing cosine

Second Method

$$f(x) = \sin^2 x = \left(\frac{1-\cos 2x}{2}\right)$$

$$f(x) = 0.5 - 0.5\cos 2x$$

4. **Ans. (c)**

Solution: This is care of

Half wave and even symmetry

$a_o =$ Find out.

$a_n = 0$

$$a_n = \frac{8}{T}\int_0^{\frac{T}{4}} f(t)\cos(n\,\omega_o t) \quad n \text{ odd}$$

$b_n = 0$ all n

* Since there are no sine terms, the function is even (i.e. symmetric along y-axis)

$$x(t) = x(-t) = x(T-t) \quad \text{...(1)}$$

* Since the even harmonics are absent, the function exhibits half wave smmetry, i.e.

$$x(t) = -x\left(t - \frac{T}{2}\right) \quad \text{...(2)}$$

∴ From equation (1) an (2)

$$x(t) = x(T-t) = -x\left(t - \frac{T}{2}\right)$$

5. **Ans. (c)**

Solution:

$$I = \left[\begin{array}{l}-6\sqrt{2}\sin(100\pi t) + 6\sqrt{2}\cos \\ \left(300\pi + \frac{\pi}{4}\right) + 6\sqrt{2}\end{array}\right] A$$

Find Reading if it passes through true R.M.S. ammeter

R.M.S. Value = It is that value of current or voltage which produce the same heat as it produce by the peak voltage or current

In a.c. current $I_{rms} = \frac{I_{peak}}{\sqrt{2}}$

In D.C. current $I_{rms} = I_{peak}$

So $\quad I = \sqrt{\left(\frac{-6\sqrt{2}}{\sqrt{2}}\right)^2 + \left(\frac{6\sqrt{2}}{\sqrt{2}}\right)^2 + (6\sqrt{2})^2}$

$\sqrt{36+36+72} = \sqrt{14}\text{ A} = 12\text{ A}$

6. **Ans. (d)**

Solution: For real Periodic signal, Fourier Series coefficient are complex conjugate of each other

$$C_{-K} = \bar{C}_K$$

$$\therefore \quad C_{-3} = \bar{C}_3 = \overline{3+5j} = 3-5j$$

According to property of Fourier series

If a periodic signal g(t) is real its exponential Fourier

Series coefficient for the nth harmonic satisfies the relation

$$C_n = C_{-n}^*$$

or $C_{-n} = C_n^*$

or C_n is conjugate even Symmetry function (where * denote complex conjugaion)

Given $C_3 = 3 + j5$

Therefore, $C_{-n} = C_n^*$

$= 3 - j5$

7. **Ans. (a)**

Solution:

Given signal has the following wave-form

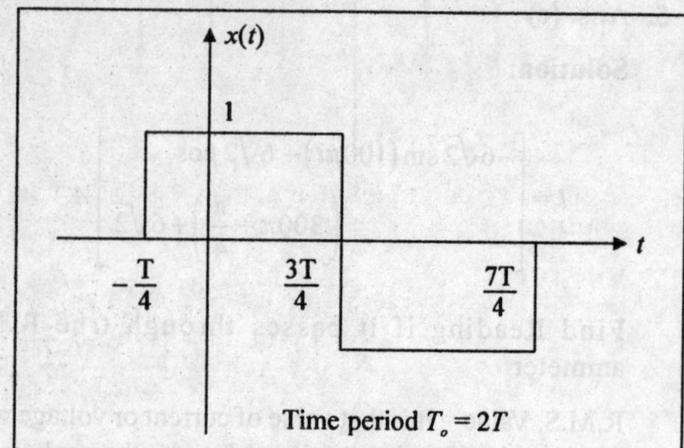

Function $x(t)$ is periodic with period $2T$ and given that $x(t) = -x(t + T)$ (Half-wave symmetric)

The given waveform in figure is

For fundamental term phase shift $= \frac{\pi}{4}$

Hence, fundamental term

$$= \frac{4}{\pi}\cos\left[\frac{\pi t}{\pi} - \frac{\pi}{4}\right]$$

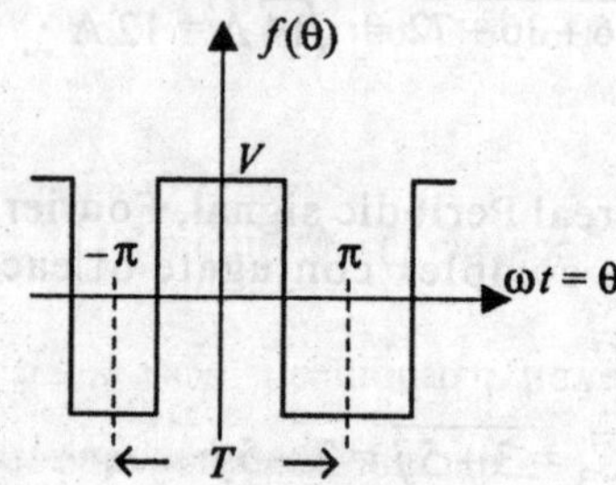

$$f(t) = +V;\ 0 > \theta < \pi/2$$

$$= -V;\ \pi/2 < \theta < \pi$$

$T \rightarrow -\pi$ to π i.e. $T = 2\pi$

$T/2 \rightarrow 0$ to π

1. d.c. value = 0 ; $\quad a_0 = 0$
2. even symmetry ; $\quad b_n = 0$
3. $a_n = \dfrac{4}{T}\displaystyle\int_0^{T/2} f(\theta)\cos n\theta\, d\theta$

$$= \frac{4}{T}\int_0^{\pi} f(\theta)\cos n\theta\, d\theta$$

$$= \frac{4}{2\pi}\int_0^{\pi/2} V\cos n\theta\, d\theta$$

$$= +\frac{4}{2\pi}\int_{\pi/2}^{\pi} -v\cos n\theta\, d\theta$$

$$= \frac{4v}{2\pi}\left(\frac{\sin n\theta}{n}\Bigg|_0^{\pi/2} - \frac{\sin n\theta}{n}\Bigg|_{\pi/2}^{\pi}\right)$$

$$= \frac{2V}{n\pi}\left(\left\{\sin\frac{n\pi}{2} - 0\right\} - \left\{\sin n\pi - \sin\frac{n\pi}{2}\right\}\right)$$

$$= \frac{2V}{n\pi}.2\sin\frac{n\pi}{2} \quad [\sin n\pi = 0]$$

$$a_n = \frac{4V}{n\pi}\sin\frac{n\pi}{2}$$

$n_1 = 1\ ;\ a_1 = 4\ V/\pi$

$n_2 = 2\ ;\ a_2 = 0$

$n_3 = 3\ ;\ a_3 = -4\ V/3\pi$

$n_4 = 4\ ;\ a_4 = 0$

$$f(\theta) = \frac{a_0}{2} + \sum_{n=1}^{\infty}(a_n\cos n\theta + b_n\sin n\theta)$$

$$= \sum_{n=1}^{\infty} a_n\cos n\theta$$

$$= a_1\cos\theta + a_2\cos 2\theta + a_3\cos 3\theta + \ldots$$

$$f(\theta) = \frac{4V}{\pi}\cos\theta + 0 - \frac{4V}{3\pi}\cos 3\theta + 0 + \ldots$$

$$f(t) = \frac{4V}{\pi}\left[\cos\omega t - \frac{1}{3}\cos 3\omega t + \ldots\right]$$

In Present Case

$$f(t) = x(t)$$

$$\omega_o = \frac{2\pi}{T} = \frac{2\pi}{2T} = \frac{\pi}{T}$$

$$V = 1$$

$$\text{Hence } f(t) = \frac{4}{\pi}\left[\cos\frac{\pi t}{T} - \frac{1}{3}\cos\frac{3\pi t}{T} + \ldots..\right]$$

Second Method

Let us consider a signal $x_1(t)$ given as:

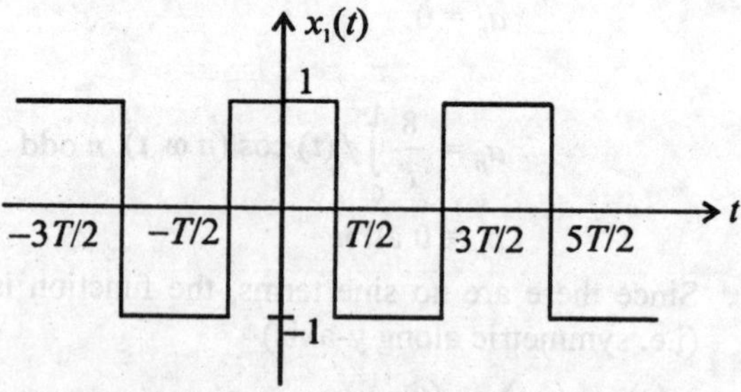

This signal has a time period = $2T$

Its Fourier series can be given as:

$$x_1(t) = \frac{4}{\pi}\sum_{n=1}^{\infty}\frac{(-1)^{n-1}}{(2n-1)}\cos\left[\frac{2\pi}{2T}(2n-1)t\right]$$

Hence its fundamntal frequency terms is

$$= \frac{4}{\pi}\cos\left(\frac{\pi t}{T}\right)$$

Now, the signal $x(t)$ in the question is as follows:

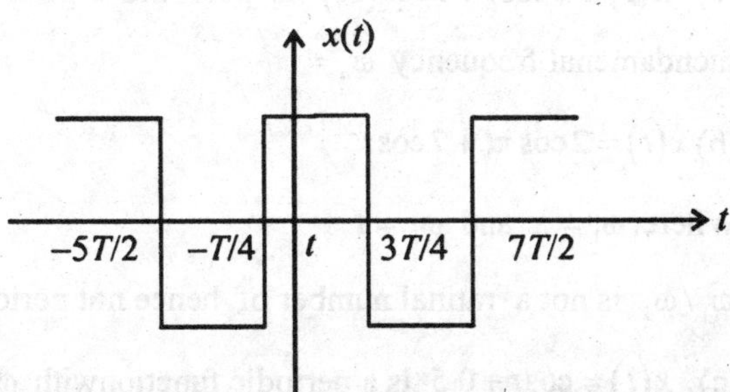

Hence the given signal is shifted version of let function

i.e. $x(t) = x_1\left(t - \frac{T}{4}\right)$

Hence, Fourier series term of $x(t)$ will have a phase shift equal to $\left[-2\pi n f_o \frac{T}{4}\right]$ where $f_o = \frac{1}{2T}$

$\Rightarrow$ For fundamental frequency terms, phase shift $= -\frac{\pi}{4}$

$\Rightarrow$ The fundamental frequency terms is $\frac{4}{\pi}\cos\left[\frac{\pi t}{T} - \frac{\pi}{4}\right]$

8. Ans. (d)

Solution: $x(t) = a_0 + \sum_{n=1}^{\infty} a_n \cos n\omega_s t + \sum_{n=1}^{\infty} b_n \sin n\omega_s t$

If $x(t)$ is odd function then it will contain sine term only, i.e. sin $\omega_s t$ is odd function.

9. Ans. (c)

Solution: $x(t) = a_0 + \sum_{n=1}^{\infty} a_n \cos n\omega_s t + \sum_{n=1}^{\infty} b_n \sin n\omega_s t$

- **If $x(t)$ is even function then it will not contain sine term sin $\omega_s t$ is odd function.**
- **It can have dc term and consine terms of all harmonics.**

It may be observe that dc term may be/not be present and it does not contain evern harmonic if it satisifies half-wave or odd harmonic symmetry.

10. Ans. (c)

Solution:

The Trigonometric Fourier series of a periodic function $x(t)$ is given by

$$x(t) = \sum_{n=0}^{\infty} a_n \cos(n\omega_o t) + \sum_{n=1}^{\infty} b_n \cos(n\omega_o t)$$

∴ The sereis has cosine terms of all harmoics:

$n\omega_o$, $n = 0, 1, 2, \ldots\ldots$

where 0th harmonic = dc term

= average or mean value = a_o

and sine terms of all harmonics;l: $n\omega_o$, $n = 1, 2, \ldots\ldots$

$$x(t) = a_o + \sum_{n=0}^{\infty}\left[a_n \cos(n\omega_o t) + b_n \sin(n\omega_o t)\right]$$

11. Ans. (c)

Solution: The periodic signal $X(t)$ of period T_o

$X(t) = 1, \quad |t| < T_1$

$X(t) = 0, \quad T_1 < |t| < T_o/2$

$x(t)$

1

$-\frac{T_o}{2}$ $\quad -T_1 \quad 0 \quad T_1 \quad \frac{T_o}{2}$ $\quad t$

Dc component $A_o = \frac{1}{T_o}\int_{-T_o/2}^{T_o/2} x(t)\,dt$

$$= \frac{1}{T_o}\int_{-T_1}^{T_1} 1\cdot dt = \frac{2}{T_o}\int_0^{T_1} 1\cdot dt = \frac{2T_1}{T_o}$$

12. Ans. (d)

Solution:

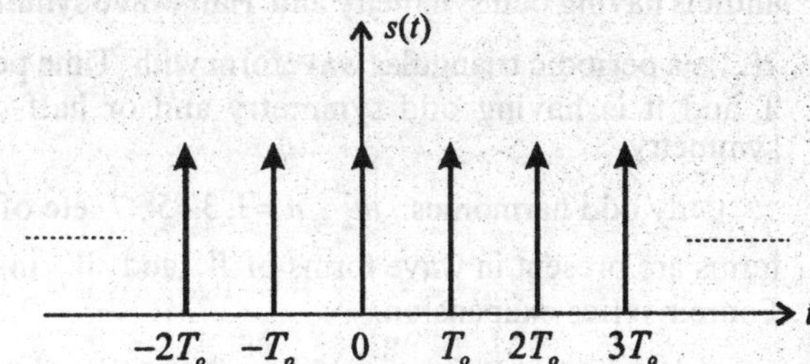

The given impulse tran, $s(t)$ with strenght of each impulse as 1, is periodic function with periodic, T_o as shwon in figure

$$s(t) = \sum_{n=-\infty}^{\infty} F_n e^{jn\omega_o t} \text{ where } \omega_o = \frac{2\pi}{T_o}$$

and F_o is given by

$$F_n = \frac{1}{T_o}\int_{-T_o/2}^{T_o/2} \delta(t)\, e^{-jn\omega_o t}\,dt$$

$$= \frac{1}{T_o},$$

$$\therefore \quad s(t) = \frac{1}{T_o}\sum_{n=-\infty}^{\infty} e^{j\frac{2\pi}{T_o}nt}$$

13. Ans. (c)

Solution: $W_1 = 1$ for 0 to $\frac{T}{2}$

$= -1$ for T / 2 to T

W_1 will have integral form $\frac{1}{T_P}\int_{-\frac{T_o}{2}}^{\frac{T_o}{2}} 1\, e^{-jn\omega_o t}\, dt$

i.e. $W_1 \propto \frac{1}{n} = |n^{-1}|$

$W_2 \rightarrow$ is linear for 0 to $\frac{t}{4}$

$\rightarrow$ is linear for $\frac{t}{4}$ to $\frac{3t}{4}$

$\rightarrow$ is linear for $\frac{3t}{4}$ to T

W_2 will have integral form

$$\frac{1}{T_P}\int_{-\frac{T_P}{2}}^{\frac{T_P}{2}} n\cdot e^{-jn\omega_o t}\, dt$$

i.e. $W_2 \propto \frac{1}{n^2} = |n^{-2}|$

Second Method

W_1 is a periodic square waveform with Time period T andit is having odd symmetry and Half-wave symmetry.

W_2 is a periodic triangular waveform with Time period T and it is having odd symmetry and or half-wave symmetry.

$\therefore$ Only odd harmonics: nf_o, $n = 1, 3, 5, 7$ etc of sine terms are present in wave forms of W_1 and W_2 in their Foureir series expansion.

It can be observed that waveform, W_2 can be obtained by integating the waveform, W_1.

If F_n is exponential Fourier series coefficient of the nth harmonic component $F_n e^{jn\omega_o t}$

$|F_n| \propto \left|\frac{1}{n}\right| = |n^{-1}|$ for wave form W_1

$|F_n| \propto \left|\frac{1}{n^2}\right| = |n^{-2}|$ for waveform W_2

14. Ans. (b)

Solution: $x(t) = 2\cos \pi t + 7\cos t$

Where, $\omega_1 = \pi$ and $\omega_2 = 1$

The ratio of $\frac{T_1}{T_1} \neq$ integeral

Hence not a peirodic signal

Second Method

(a) $x(t) = 2\cos t + 3\cos(3t)$ is periodic signal with fucndamenal frequency $\omega_o = 1$

(b) $x(t) = 2\cos \pi t + 7\cos t$

Where, $\omega_1 = \pi$ and $\omega_2 = 1$

ω_1/ω_2 is not a ratinal number of hence not periodic

(c) $x(t) = \cos t + 0.5$ is a periodic functionwith $\omega_o = 1$

(d) $x(t) = 2\cos(1.5\pi)t + \sin(3.5\pi)t$

Where $\omega_1 = 1.5\,\pi$ and $\omega_2 = 3.5\,\pi$

$$\frac{\omega_1}{\omega_2} = \frac{1.5\,\pi}{3.5\,\pi} = \frac{1.5}{3.5}$$

$$= \frac{3\times 0.5}{7\times 0.5} = \frac{3}{7} \Rightarrow \text{is a rational number.}$$

$x(t)$ is a periodic signal with fundamental frequency $\omega_o = 0.5\,\pi$

* As $x(t)$ in option (b) is not periodic Henc it can be the Fourier sereis expansion, althought it has consine terms.

15. Ans. (c)

Solution: $s(t) = \cos 2t + \cos 3t + \cos 5t$

Period $= 2\pi$

fundamental frequency = 1 rad/sec.

The ratio of any two freqencies = m/n

where m and n are integers.

$\therefore$ $s(t) = \cos(2t) + \cos(3t) + \cos(5t)$ is periodic

$$s(t) = e^{j8\pi t}$$

$$s(t+T) = e^{j8\pi(t+T)} = e^{j8\pi t}\, e^{j8\pi T}$$

$$T = \frac{2\pi}{8\pi} = \frac{1}{4} \text{ period and } \omega = 8\pi$$

$\therefore$ $s(t) = e^{j8\pi t}$ is periodic

$$s(t) = e^{-7t}\sin 10\pi t$$

$$s(t+T) = e^{-7(t+T)}\sin 10\pi\,(t+T)$$

$$= e^{-7t}e^{7T}\sin 10\pi t$$

* Exponentially decreasing function value becomes zero after some time, hence its is non periodic

e^{-7t} is not periodic as it is not complex exponential of the form $e^{\pm j\omega t}$

$\sin(10\pi t)$ is periodic with $\omega = 10\pi$

Hence the product $s(t) = e^{-7t}\sin(10\pi t)$ is not periodic

$$s(t) = \cos 2t \times \cos 4t$$

$$= \frac{1}{2}[\cos 6t + \cos 2t]$$

$$s(t + T) = \cos(2t + 2T)\cos(4t + 4T)$$

$$T = \pi \text{ Periodic}$$

$$\omega = 2 \text{ rad/sec}$$

$\therefore\ s(t) = \cos(2t)\cos(4t) = \frac{1}{2}[\cos(6t) + \cos(2t)]$ is periodic

16. Ans (a)

Solution:

When a periodic function is an odd function, then

$$a_0 = 0$$

$$a_n = 0$$

$$b_n = \frac{2}{\pi}\int_0^{\pi} f(x)\sin nx\, dx$$

when a periodic function is an even function, then

$$a_o = \frac{2}{\pi}\int_0^{\pi} f(x)\, dx$$

$$a_n = \frac{2}{\pi}\int_0^{\pi} f(x)\cos nx\, dx$$

$$b_n = 0$$

Thus, if function is odd, then its Fourier expansion contains only sine terms and if function is even, then its Fourier expansion contains only cosine terms.

17. Ans (b)

Solution:

Given that

$$f(t) = \sin^2 t + \cos 2t$$

$$= \frac{1 - \cos 2t}{2} + \cos 2t$$

$$= \frac{1 - \cos 2t + 2\cos 2t}{2} = \frac{1 + \cos 2t}{2}$$

$$f(t) = \frac{1}{2} + \frac{1}{2}\cos 2t$$

$$f(t) = f_1(t) + f_2(t)$$

where $f_1(t) = \frac{1}{2}$

$$\omega_1 = 0$$

$$f_1 = 0 \text{ Hz}$$

And, $f_2(t) = \frac{1}{2}\cos 2t$

$$\omega_2 = 2$$

$$2\pi f_2 = 2$$

$$f_2 = \frac{1}{\pi} \text{ Hz}$$

Therefore, it has 0 and $\frac{1}{\pi}$ Hz frequency component.

18. Ans. (c)

Solution:

Since given waveform is even, thus it will have only cosine terms. Function f(t) has more negative value then positive, so it will have negative DC Component.

Second Method

Trigonometric Fourier series is expressed as

$$f(t) \Rightarrow \frac{a_o}{2} + \left[\sum_{n=1}^{\infty}(a_n \cos n\omega_o t) + b_n \sin n\,\omega_o t\right]$$

$$\omega_0 \rightarrow \frac{2\pi}{T}$$

$a_0 \rightarrow$ Reprcnts the d.c. value

i.e. the area of the waveform for one period. T above and below t axis

Area from $-\frac{T}{4}$ to $\frac{T}{4}$

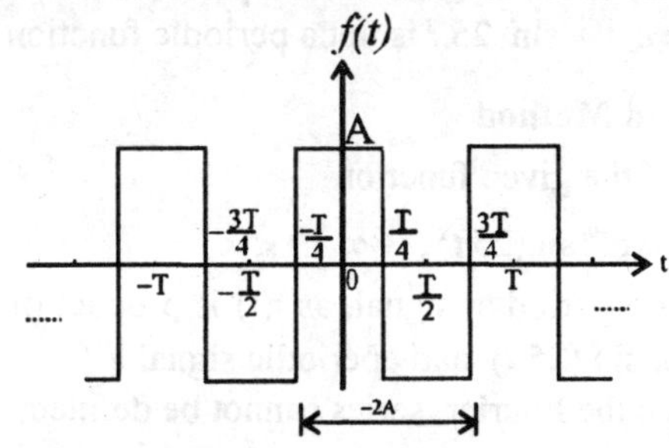

$$\Rightarrow \frac{T}{2} \times A \rightarrow \frac{AT}{2}$$

Area from $\frac{T}{4}$ to $\frac{3T}{4}$

$$\Rightarrow \frac{T}{2} \times -2A = -AT$$

$\Rightarrow$ Net area is in negative

$\therefore$ Negative value of d.c. component

More over the waveform has even symmetry

$\therefore$ Only a_n coefficient are present i.e. cosine terms are present

19. Ans. (c)

Solution: f (t) is even function,

Hence $b_n = 0$

Where 'b_n' is the coefficient of sine terms.

If $f(t)$ is an even function, i.e.

$f(t) = f(-t)$. For example cos t, t^2, t sin t.

$$a_n = \frac{4}{T}\int_0^{T/2} f(t)\cos(n\omega_o t)\,dt$$

(a) $b_n = 0$. The graph of an even function will be symmetrical with respect the to y-axis and only cosine terms are present (dc term optional)

(b) The sum or product of two or more even function is an even function.

20. Ans. (c)

Solution

Fourier series are defined for a function which satisfy dirichlet condition, i.e.

* f (t) has only a finite number of maxima and minima in any one period.
* f (t) is discontinuous and has a finite number of discontinuities in any one period.
* f(t) discontinuities are bonded, i.e. it has a finite average values over the period T

Fourier series is defined for a periodic function. Here $e^{-|t|}$ sin 25 t is not a periodic function

Second Method

Out of the given function

$f(t) = e^{-|t|}\sin(25\,t)$, $-\infty < t < \infty$

is not a periodic signal, as $f(t)$ is product of a periodic signal sin (25 t) and aperiodic signal $e^{-|t|}$.

Hence the Fourier series cannot be defined.

21. Ans. (b)

Solution:

$$s(t) = \frac{At}{(T/2)} = \frac{2At}{T};\ 0 < t < T/2$$

$$= -A;\ [T/2] < t < T$$

$$e_{rms} = \sqrt{\frac{1}{T}\int_0^T e^2(t)\,d\tau}$$

$$= \sqrt{\frac{1}{T}\int_0^{T/2}\left(\frac{2At}{T}\right)^2 dt + \int_{T/2}^{T} A^2\,dt}$$

$$= \sqrt{\frac{1}{T}\left(\frac{4A^2}{T^2}\frac{1}{3}\frac{T^3}{8} + A\frac{T}{2}\right)}$$

$$= \sqrt{\frac{A^2}{6} + \frac{A^2}{2}} = \sqrt{\frac{2A^2}{3}}$$

$$= \sqrt{\frac{2}{3}}A = A\sqrt{\frac{2}{3}}$$

22. Ans. (d)

Solution: The periodic time (T) = 4ms = 4×10^{-3} sec

$\therefore$ Fundamental frequency (f) = (1/T) = [10^3/4]

= 250 Hz

$\therefore$ Frequency of the 5th harmonic = 250 × 5

= 1250 Hz

23. Ans. (c)

Solution:

* The function $f(t)$ exhibits half wave symmetry. Hence, there will be only odd harmonics and there will be no dc term.

* $f(t)$ is an even function. Hence there will no be sine terms.

Hence Fourier series of $f(t)$ will be of the form

$$f(t) = a_1\cos\omega t + a_3\cos 3\omega t + a_5\cos 5\omega t + \ldots$$

Second Method

For the given periodic wave form $f(t)$, peak to peak value = 2

$$T_o = 2,\ \omega_o = \frac{2\pi}{T_o} = \pi$$

* It has dc value = 0 as positive and negative peak are equal
* It has no sine terms in it FS as $f(t) = f(-t)$

3. It is half wave symmetric as $f(t) = -f\left(t - \frac{T}{2}\right)$

In this case even harmonics are absent, i.e. only odd harmonics are present

Hence FS contains only consine terms of odd harmonics.

Note that this restriction is satisfied only for the FS is option (c)

* The FS for standard even triangular wave is given by

$$f(t)=\frac{k}{2}\left[\frac{1}{1^2}\cos(\pi t)+\frac{1}{3^2}\cos(3\pi t)+\frac{1}{5^2}\cos(5\pi t)+\ldots\right]$$

Where, $k=\left(\frac{2\times \text{Peak}-\text{Peak value}}{\pi}\right)^2=\frac{16}{\pi^2}$

24. Ans. (d)

Solution:

$$V_{rms}=\sqrt{\frac{1}{T}\int_0^T x(t)^2\,dt}$$

$$=\sqrt{\frac{1}{2(\pi/3)}\int_0^{2(\pi/3)} 100^2\,dt}$$

$$=\sqrt{\frac{3}{2\pi}\times 100^2 (t)\Big|_0^{2\pi/3}}$$

$$=\sqrt{\frac{3}{2\pi}\times 100^2\,\frac{2\pi}{3}}$$

$$V_{rms}=100\text{ V}$$

Second Method

$$V_{rms}=\sqrt{\frac{A}{2\pi}\left[\int_0^{\pi/3} d\theta+\int_{\pi/3}^{2\pi/3} d\theta+\int_{2\pi/3}^{4\pi/3} d\theta+\int_{4\pi/3}^{5\pi/3} d\theta+\int_{5\pi/3}^{2\pi} d\theta\right]}$$

$$=A\sqrt{\frac{1}{2\pi}\left[\frac{\pi}{3}-0+\frac{2\pi}{3}-\frac{\pi}{3}+\frac{4\pi}{3}-\frac{2\pi}{\pi}+\frac{5\pi}{3}-\frac{4\pi}{3}+2\pi-\frac{5\pi}{3}\right]}$$

$=A$ = 100 Volt

25. Ans. (c)

Solution:

Given

$$x(t)=\sum_{k=-\infty}^{\infty} a_k\, e^{\frac{j2\pi kt}{t}}$$

$a_{-2}=2-j$

$a_{-1}=0.5+j0.2$

$a_0=j\,2$

$a_1=0.5-j\,0.2$

$a_2=2+j$

$a_k=0$ **for** $|k|>2$

Using given Fourier series coefficent fucntion $x(t)$ is as

$$x(t)=\begin{bmatrix}\frac{0.5}{2}\cos\left(\frac{2\pi t}{T}\right)-\frac{0.2}{2}\sin\left(\frac{2\pi t}{T}\right)\\ +\frac{2}{2}\cos\left(\frac{2\pi 2t}{T}\right)+\frac{1}{2}\sin\left(\frac{2\pi 2t}{T}\right)+j\,2\end{bmatrix}$$

From above fucntion it indicate that real part of $x(t)$ is a function of time while the imaginary part is a contant

26. Ans.(b)

Solution:

Let $x(t)\rightarrow$ Periodic signal

$a_k\rightarrow$ Fourier sereis coefficent of signal $x(t)$

Now, Given

$y(t)=x(t-t_o)+x(t+t_o)\rightarrow$ Periodic signal

The Fourier series coefficent of $y(t)$

$$b_k=e^{-jk\omega t_o}a_k+e^{jk\omega t_o}a_k$$

$$=a_k\left(e^{jk\omega t_o}+e^{-jk\omega t_o}\right)$$

$$=2a_k\cos k\omega t_o$$

Now $b_k=0$ for k odd

or, $k\omega t_o=\frac{\pi}{2}$ for $k\rightarrow$ odd

or, $k\frac{2\pi}{T}t_o=\frac{\pi}{2}$

For $k=1$

$$t_o=\frac{\pi}{2}\left(\frac{T}{2\pi}\right)$$

$$t_o=\frac{T}{4}$$

27. Ans (d)

Solution:

$$x(t)=\sin(2\pi+\omega t)$$

$$x(t)=8\sin\left(0.8\,\pi t+\frac{\pi}{4}\right)$$

$$\omega t=0.8\,\pi t$$

$$\omega=\frac{2\,\pi}{T}$$

$$0.8\,\pi = \frac{2\,\pi}{T}$$

$$T = \frac{2\,\pi}{0.8\,\pi} = 2.5\,s$$

28. Ans. (a)

Solution:

The given periodic waveform is haivng half and odd symmertry.

29. Ans. (d)

Solution:

⇒ It satisfies the even symmetry. So $b_n = 0$

⇒ it satisfies thr half wave symmetry,

so that it contains only odd harmonics.

30. Ans. (a)

Solution

The rectangular wave of period T for $0 < t < T$ is shown in figuer where $T = T_1 + T_2$

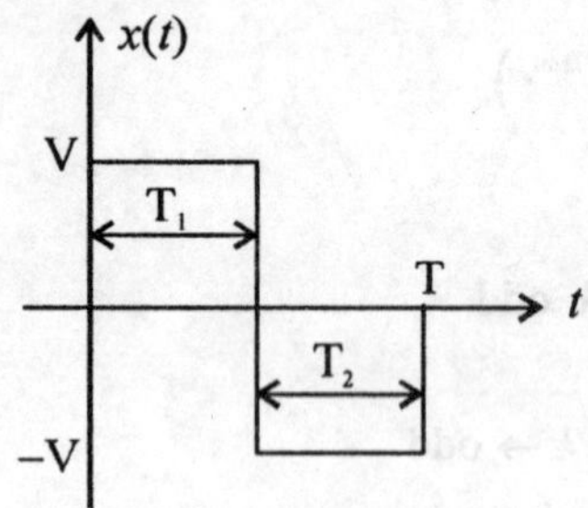

$$\text{RMS value} = \sqrt{\frac{1}{T}\int_0^T x^2(t)\,dt}$$

$$= \sqrt{\frac{1}{T}[V^2 T_1 + V^2 T_2]}$$

$$= \sqrt{\frac{1}{T} V^2 (T_1 + T_2)}$$

$$= V\sqrt{\left(\frac{T_1 + T_2}{T}\right)}$$

$$= V\sqrt{\frac{T}{T}}$$

$$= V$$

31. Ans. (a)

Solution:

It this we have to find out the a_0 which is known as d,c, component. The given diagram is symmetrical about Real axis and it is having the same positive and negative area.

So according to formula.

$$a_0 = \frac{2}{T}\int_0^T f(t)\,dt$$

$$a_0 = \frac{2}{T}\int_0^T f(t)\,dt = 0$$

So $a_0 = 0$

32. Ans. (a)

Solution:

$$n(t) = a_o + \sum_1^{\infty} a_n \cos n\omega_o t + \sum_1^{\infty} b_n \sin n\,\omega_o t$$

$$a_o = \frac{1}{T}\int_T n(t)\,dt$$

For $T = 3$ and $\omega_o = \frac{2\pi}{3}$

$$a_n = \frac{2}{T}\int_T n(t)\cos n\omega_o t\,dt$$

$$= \frac{2}{3}\left[\int_0^1 \cos n\omega_o t\,dt + \int_1^2 (-1)\cos n\,\omega_o t\,dt + \int_2^3 0\right]$$

$$= \frac{2}{3}\left[\left[\frac{\sin n\,\omega_o t}{n\,\omega_o}\right]_0^1 - \left[\frac{\sin n\,\omega_o t}{n\,\omega_o}\right]_1^2\right]$$

$$= \frac{2}{3n\omega_o}[\sin n\,\omega_o - \sin 2n\omega_o + \sin n\omega_o]$$

$$= \frac{2}{3n\omega_o}[2\sin n\,\omega_o - 2\sin n\omega_o \cos n\,\omega_o]$$

$$= \frac{4}{3}\frac{\sin n\,\omega_o}{n\,\omega_o}[1 - \cos n\,\omega_o] = \frac{4}{3}$$

$$= \frac{4}{3}\frac{\sin n\,\omega_o}{n\,\omega_o}\left[1 - \cos n \cdot \frac{2\pi}{3}\right]$$

For $n = 3$ i.e odd integer

Second Method

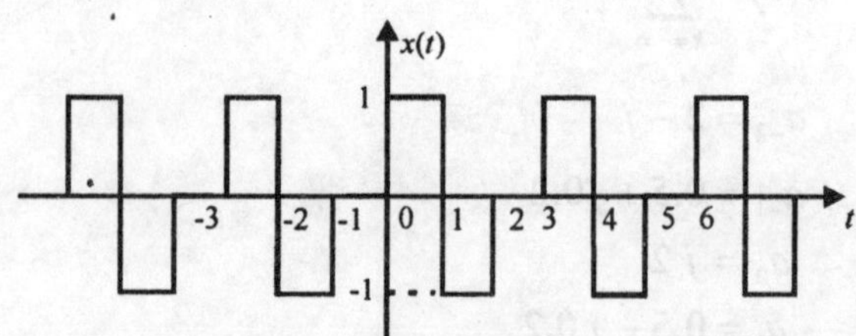

If one shift the signal by $t = 1$ then it becomes a odd symmetry signal and half wave symmetry, therefore a_k will be zero for even integer value of K.

Time period of given waveform is 3.

4.3 Fourier Transform

1. **Ans. (c)**

Solution:

A function $g(t)$ is conjugate symmetric if

$$g(t) = g^*(-t)$$

if $F\{g(t)\} = G(f)$

Then $F\{g^*(-t)\} = G^*(f)$

i.e. $G(f)$ is real

Hence Fourier transform of a conjugate symmetric function is always real

2. **Ans. (b)**

Solution:

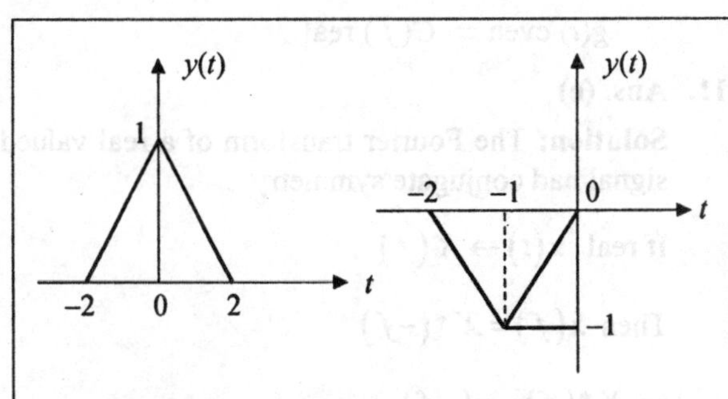

$x(t - t_o) \rightarrow X(f)\, e^{-j2\pi f t - o}$

$X(a\, t) \rightarrow [1 / |a|]\, X(f / a)$

$y(t) = -x[2(t+2)] \rightarrow -[1/2]\, X(f/2)\, e^{-j2\pi f \cdot 2/2}$

$$\rightarrow -\frac{1}{2} \times \left(\frac{f}{2}\right) e^{j2\pi f}$$

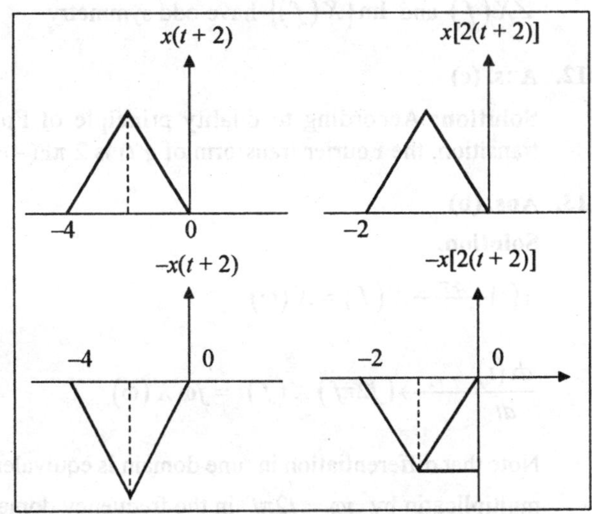

3. **Ans. (a)**

Solution: Fourier transform relation for shifted function

$$x(t - t_o) \leftrightarrow e^{-j\omega t_o} X(\omega)$$

$$y(t) = 0.5x\,(t - t_d + T) + x(t - t_d) + 0.5x\,(t - t_d - T).$$

$$Y(\omega) = 0.5e^{-j\omega(t_d - T)}\, X(\omega) + e^{-j\omega t_d} X(\omega)$$

$$+0.5e^{-j\omega(t_d + T)} X(\omega)$$

$$= 0.5e^{-j\omega(\,td - T)}\, X(\omega) + 0.5e^{-j\omega c(td + T)}\, X(\omega)$$

$$+ e^{-j\omega td}\, X(\omega)$$

$$= 0.5\, X(\omega)\, [e^{-j\omega(\,td - T)} + e^{-j\omega c(td + T)}]$$

$$+ e^{-j\omega td}\, X(\omega)$$

$$= e^{-j\omega td}\, X(\omega)\, [1 + 0.5\,(2 \cos \{\omega T\})]$$

$$\frac{Y(\omega)}{X(\omega)} = e^{-j\omega t_d}\; [1 + \cos \omega T] = H(\omega)$$

4. **Ans. (c)**

5. **And. (b)**

Solution: $f(t) \rightarrow F(\omega)$

$f(at) \rightarrow [1/|a|]\, F[\omega/a]$

$f(t - T) \rightarrow F(\omega)\, e^{-j\omega T}$

$f(at - T) \rightarrow [1/|a|]\, F(\omega/a)\, e^{-j\omega T}.$

If $\quad x(t) \Leftrightarrow X(f)$

Then $\quad x(at) \Leftrightarrow X\,[1/|a|]\,[f/(|a|)]$

Now, X has $3f$

$\therefore \quad |a| = [1/3]$

Inverse Fourier transform of

$X(af + b)\, [1/a]\, e^{-j2\pi\,[b\backslash a]t}\, x[t/a]$

$a = 3,\, b = 2 \leftrightarrow [1/3]\, x\,[t/3] \exp - j(4\pi t/3)$

By scaling property

Second Method

$$X(f) \xrightarrow{F^{-1}} x(t)$$

$$\therefore X(f+2) \xrightarrow{F^{-1}} x(t) e^{-j2\pi.2t}$$

$$\therefore X(3f+2) \xrightarrow{F^{-1}} \frac{1}{3} x\left(\frac{t}{3}\right) e^{-j2\pi.2\frac{t}{3}}$$

$$\frac{1}{3} x\left(\frac{t}{3}\right) e^{-j\frac{4\pi t}{3}}$$

6. Ans. (a)

Solution: By scaling Property

$$f(t) \to F(\omega)$$

$$f(at) \to \frac{1}{|a|} F(\omega/a)$$

By shifting property

$$f(t - T) \to F(\omega)\, e^{-j\omega T}$$

By shifting and scaling

$$f(at - T) \to \frac{1}{|a|} F(\omega/a) e^{\frac{-j\omega T}{a}}$$

7. Ans. (a)

Solution:

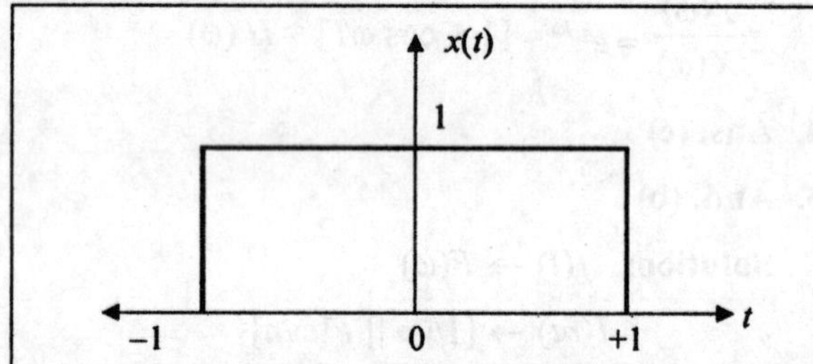

$$X(j\omega) = \int_{-\infty}^{\infty} x(t) e^{-j\omega t} dt$$

$$X(j\omega) = \int_{-1}^{1} 1.e^{-j\omega t} = \frac{2\sin\omega}{\omega} \Rightarrow \text{Sampling function}$$

At $\omega = 0$, $X(0) = 2$

At $\omega = \pi$, $X(\pi) = 0$

$\omega = 2\pi$, $X(2\pi) = 0$

8. Ans. (c)

Solution: $H(\omega) = \int_0^{\infty} e^{-2t} e^{-j\omega t} dt = \int_0^{\infty} e^{-(2+j\omega)} dt$

$$= \left.\frac{e^{-(2+j\omega)t}}{-(2+j\omega)}\right|_0^{\infty} = \frac{-1}{2+j\omega}[0-1]$$

$$= \frac{1}{2+j\omega}$$

9. Ans. (d)

Solution: When input is sinusoidal then output is also sinusodal with same frequency but amplitude and phase and phase changes according to transfer function

Thus amplitude is

$$2 \times \left|\frac{1}{j\omega + 2}\right|_{\omega=2} = 2 \times \left|\frac{1}{j2+2}\right| = 2\left|\frac{1}{\sqrt{4+4}}\right|$$

$$= \frac{2}{2\sqrt{2}} = 2^{\frac{-1}{2}} = 2^{-0.5}$$

Phase is $\tan^{-1}\frac{\text{imaginary}}{\text{Real}} = \tan^{-1}\left.-\left(\frac{\omega}{2}\right)\right|_{\omega=2}$

$$= -45° = -0.25\,\pi$$

Thus $Y = 2^{-0.5}\cos(2t - 0.25\pi)$

10. Ans. (b)

Solution: It $g(t)$ is real and odd then $G(t)$ will be imaginary and odd.

Here, $g(t)$ is odd symmetric

$g(t)$ odd $\Rightarrow G(f)$ imaginary

$g(t)$ even $\Rightarrow G(f)$ real

11. Ans. (c)

Solution: The Fourier transform of a real valued time signal had conjugate symmetry

If real $x(t) \to X(f)$

Then $x(f) = X^*(-f)$

or $X^*(f) = x(-f)$

If $X(f)$ satisfies the above condition, then $X(f)$ is said to posses conjugate even symmetry.
From this condition it can be observed that

i.e. $|X(f)|$ and $\text{Re}\{X(f)\}$ have even symmetry

$\angle X(f)$ and $\text{Im}\{X(f)\}$ have odd symmetry

12. Ans. (c)

Solution: According to duality principle of Fourier transition, the Fourier transform of $g(t)$ is $2\,\pi F(-\omega)$.

13. Ans. (b)

Solution:

$$x(t) \xrightarrow{FT} X(f) = X(\omega)$$

$$\frac{dx(t)}{dt} \xrightarrow{FT} (j2\pi f)\, X(f) = j\omega\, X(\omega)$$

Note that differentiation in time domain is equivalent to multiplicatin by $j\omega = j2\pi f$ in the frequency domain

14. Ans. (b)

Solution: It $x(t)$ is real and odd function of and then $X(\omega)$ is imaginary and odd function of ω.

15. Ans. (d)

Solution: Fourier Transform of Gaussian pulse is also Gaussian, i.e.

$$\frac{1}{\sqrt{2\pi}} e^{-\frac{t^2}{2}} \xrightarrow{FT} e^{\frac{-\omega^2}{2}}$$

The given signal $x(t) = e^{-3t^2}$ is of Gaussian form, it Fourier transform will also be Gaussian form.

$$e^{-3t^2} \leftrightarrow Ae^{-3f^2} \Rightarrow Ae^{-Bf^2}$$

Second Method

The Fourier Transform of a normalized Gaussian pulse is also a normalized Gaussian pulse in the form of

$$g(t) = e^{-\pi t^2} \to G(f) = e^{-\pi f^2}$$

$$\therefore\ x(t) = e^{-3t^2} \to X(f) = Ae^{-Bf^2}$$

Here, the constants A and B can be found if necessary by using time scaling prperty:

If $g(t) \to G(f)$

$$g(a\ t) \to \frac{1}{|a|} G\left(\frac{f}{a}\right)$$

For $g(t) = e^{-\pi t^2}$

$$g(at) = e^{-\pi a^2 t^2}$$

If $x(t) = e^{-3t^2} = g(at) = e^{-3t^2}$

$$e^{-3t^2} = e^{-\pi a^2 t^2}$$

Then $\pi a^2 = 3$

or, $a = \sqrt{\frac{3}{\pi}}$

$$e^{-3t^2} \to \sqrt{\frac{\pi}{3}} e^{-\pi\frac{\pi}{3}f^2} = \sqrt{\frac{\pi}{3}} e^{-\pi\frac{\pi}{3}f^2}$$

$$A = \sqrt{\frac{\pi}{3}},\ B = \frac{\pi^2}{3}$$

General formula can be made like below function.

$$Ke^{-at^2} \to K\ A\ e^{-Bf^2}$$

Where $A = \sqrt{\frac{\pi}{a}}$, $B = \pi A^2$

16. Ans. (c)

Solution: $FT\,[e^{-t}\,u(t)] \leftrightarrow \frac{1}{1+1j2\pi f}$

$$FT\frac{1}{1+1j2\pi f} \leftrightarrow e^{f}\,u(-f)$$

By applying duality theory $g(t) \to G(f)$

$$G(t) \to g(-f)$$

17. Ans. (d)

Solution:

$$f(t) \to F(\omega),\ f(-t) \to F^*(\omega)$$

F^* is conjugate of $F(\omega)$

The Fourier transform pair

$$f(t) - f(-t) = F(\omega) - F^*(\omega)$$

it must be imaginary

Then, $F^*(\omega)$ as difference of $F(\omega)$ and $F^*(\omega)$

18. Ans. (a)

Solution:

As $\cos bt = \frac{e^{jbt} + e^{-jbt}}{2}$

$$y(t) = x(t)\cos bt = \frac{x(t)}{2}e^{jbt} + \frac{x(t)}{2}e^{-jbt}$$

or, $x(t)\cos bt = \frac{1}{2}\left[x(t)e^{jbt} + x(t)e^{-jbt}\right]$

By frequency shift property

Using, $x(t)e^{j\omega_o t} \leftrightarrow X(\omega - \omega_o)$

$\therefore\ FT[x(t)\cos bt]$

$$= \frac{1}{2}\left[\exp(-a|\omega - b| + \exp(-a(\omega + b)\right]$$

$$\therefore\quad Y(\omega) = \frac{X(\omega - b)}{2} + \frac{X(\omega + b)}{2}$$

Given $X(\omega) = \frac{\pi}{a} e^{(-a[\omega])}$

$$\therefore\ Y(\omega) = \frac{\pi}{2a}\left[e^{(-a|\omega - b|)} + e^{(-a|\omega + b|)}\right]$$

19. Ans. (c)

Solution:

$$\frac{d^2(t)}{dt^2} = \delta(t-1) + \delta(t+1) - 2\,\delta(t)$$

Taking Fourier transform both sides

$$\Rightarrow (j\omega^2)F(j\omega) = e^{-j\omega} + e^{j\omega} - 2$$

$$\Rightarrow F(j\omega) = \frac{2(1 - \cos\omega)}{\omega^2}$$

20. Ans. (a)

Solution:

$$g(t) = \delta(t) + 2\,e^{(-3|t|)}$$

Taking Lapace transform both sides

$$G(\omega) = 1 + 2\int_{-\infty}^{0} e^{(3t)}\, e^{(-j\omega t)}.\,dt$$

$$+ 2\int_{0}^{\infty} e^{(-3t)}\, e^{(-j\omega t)}.\,dt$$

$$= 1 + 2\int_{-\infty}^{0} e^{(3-j\omega)t}\, dt + 2\int_{0}^{\infty} e^{(-3-j\omega t)t}.\,dt$$

$$= 1 + \frac{2}{3 - j\omega} + \frac{2}{3 + j\omega}$$

$$\frac{\omega^2 + 9 + 6 + 2j + 6 \cdot 2j}{\omega^2 + 9}$$

$$G(\omega) = \frac{\omega^2 + 21}{\omega^2 + 9}$$

21. Ans. (d)

Solution:

As given

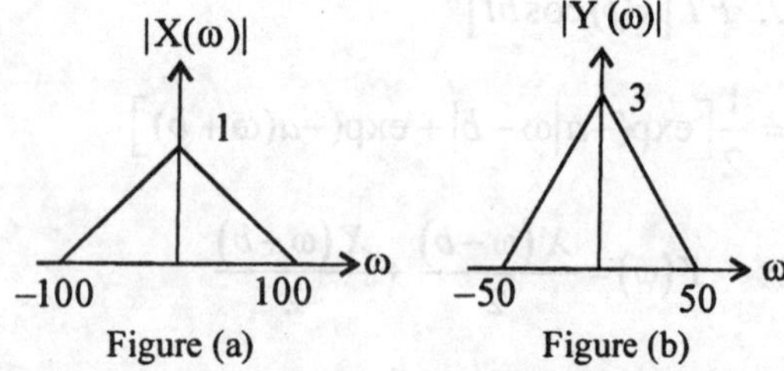

Figure (a) Figure (b)

It is known that expansion in frequency domain result in compression in time domain and viceversa.

As given compression is done in frequency domain by factor of 2 which result expansion in time domain by same amount.

$$x(t) \Leftrightarrow x(\omega)$$

$$x(at) \Leftrightarrow \frac{1}{|a|} \times \left(\frac{\omega}{a}\right)$$

So for $Ax\left(\frac{t}{2}\right) \leftrightarrow 2A\,X(2\,j\omega)$

$$2A = 3$$

$$A = \frac{3}{2}$$

$$y(t) = \frac{3}{2}x\left(\frac{t}{2}\right)$$

So, $y(t) = \frac{3}{2}x\left(\frac{t}{2}\right)$

22. Ans (d)

Solution:

Using

$$x'(t) = \frac{d}{dt}x(t) \leftrightarrow j\omega X(\omega)$$

$$x(t) \leftrightarrow X(\omega)$$

$$x(t - t_d) \leftrightarrow e^{-j\omega t_d} X(\omega)$$

$$\frac{d[x(t - t_d)]}{dt} \leftrightarrow (j\omega)\,X(\omega)\,e^{-j\omega d}$$

$$\text{Magnitude} = |j\omega||X(\omega)||e^{-j\omega t_d}|$$

$$= |X(\omega)||\omega||1|$$

$$= |X(\omega)||\omega|$$

23. Ans. (b)

Solution:

$$E = \int_{-\infty}^{\infty} (f(t))^2 dt$$

$$E' = \int_{-\infty}^{\infty} (f(2t))^2 dt\,,$$

Put $2t = m \Rightarrow dt = \frac{dm}{2}$

$$E' = \frac{1}{2}\int_{-\infty}^{\infty} (f(m))^2 dm = \frac{E}{2}$$

* It can be observed that $f(2t)$ is obtained by compressing $f(t)$ in time by a factor of 1/2 and the energy is also reduced by a factor of 1/2

24. Ans.(c)

Solution:

Fourier transform of Gate function.

$f(t) = A$; $-T/2 < t < T/2$

$= 0$; otherwise

is derived as

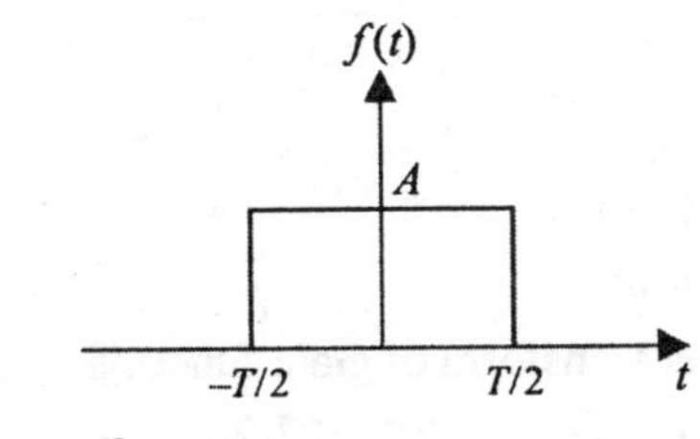

$$F(\omega) = \int_{-\infty}^{\infty} f(t)e^{-j\omega t}dt$$

$$= \int_{-T/2}^{T/2} A\, e^{-j\omega t}dt$$

$$= A\left|\frac{e^{-j\omega t}}{-j\omega}\right|_{-T/2}^{T/2} = \frac{-A}{j\omega}[e^{-j\omega T/2} - e^{+j\omega T/2}]$$

$$= \frac{-A}{j\omega}\left(\frac{e^{j\omega T/2} - e^{-j\omega T/2}}{2j}\right) \times [2j]$$

$$= \frac{2A}{\omega}\sin\frac{\omega(T)}{2}$$

$$= \frac{2A}{\omega}\frac{\sin(\omega T/2)}{(\omega T/2)}[\omega T/2]$$

$F(\omega) = AT(\sin x / x)$; where, $x = \omega T/2$
= Sampling function

Hence

$$F(\omega) = AT \sin\frac{\sin\omega T/2}{\omega T/2}$$

In peresent case

$$\frac{\sin 2\omega}{\omega} = \frac{2\sin 2\omega}{2\omega}$$

Here, $AT = 2$

and $\frac{\omega T}{2} = 2\omega$

or $T = 4$

$\Rightarrow$ $\frac{T}{2} = 2$

and $A.4 = 2$

$$A = \frac{2}{4} = \frac{1}{2}$$

Multiplication with $e^{j\omega}$ in frequency domain result shift in time domain, i.e.

$$F(j\omega)e^{j\omega T} \to f(t-T)$$

$\frac{\sin 2\omega}{\omega} \longleftrightarrow$

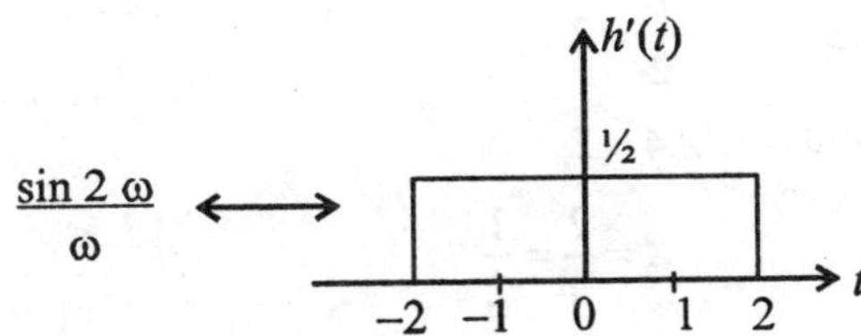

$$2\cos\omega\left(\frac{\sin 2\omega}{\omega}\right)$$

$$= \left[e^{j\omega} + e^{-j\omega}\right]\left(\frac{\sin 2\omega}{\omega}\right) \longleftrightarrow h(t) = h'(t-1) + h'(t+1)$$

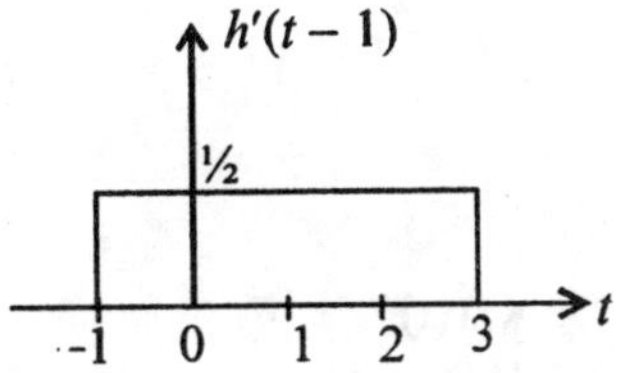

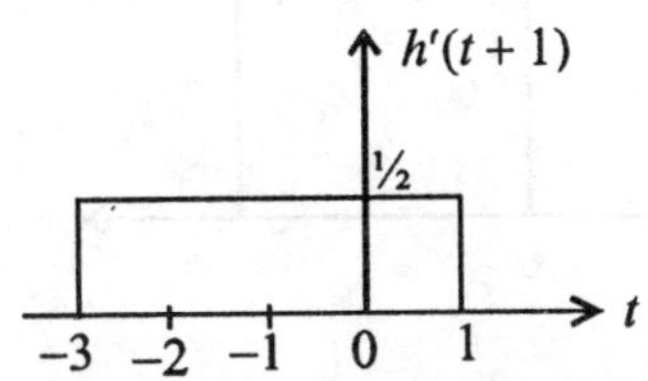

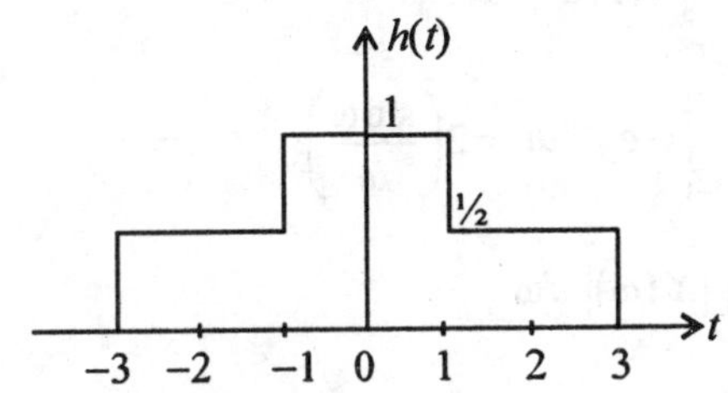

25. Ans.(b)

Solution:

$$r_{ect}(x) = 1 \text{ for } \frac{-1}{2} \le x < \frac{1}{2}$$

$= 0$ otherwise

$$x(t) = r_{ect}\left(t - \frac{1}{2}\right)$$

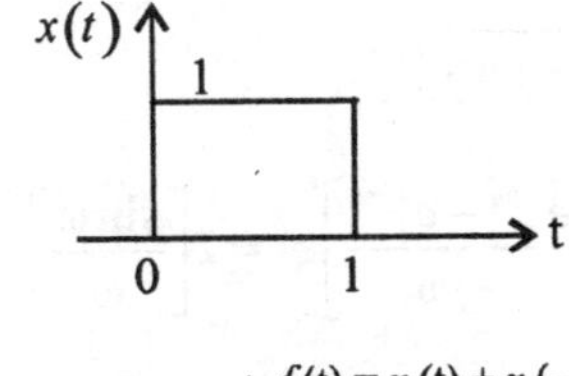

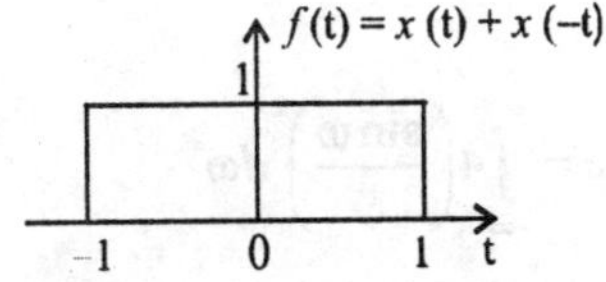

$f(t)$ is Gate function and its FT is sampling function

$$F(\omega)=\int_{-\infty}^{\infty} f(t)e^{-j\omega t}dt=\int_{-1}^{+1} e^{j\omega t}dt$$

$$=2\text{sinc}\frac{\omega}{2\pi}$$

26 . Ans (d)

Solution:

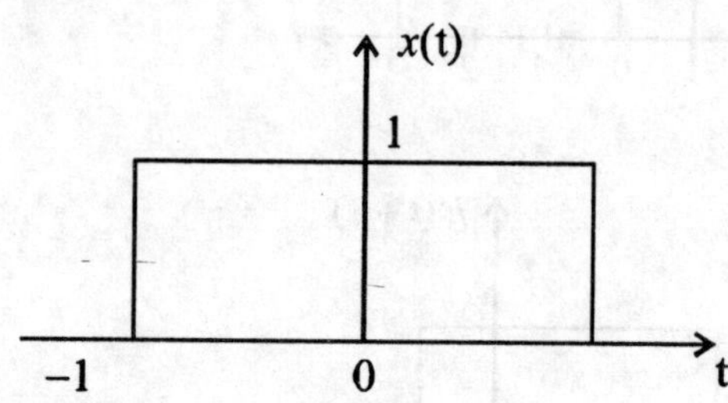

$$X(\omega)=\int_{-\infty}^{\infty} x(t)\,e^{-j\omega t}dt$$

$$=\int_{-1}^{+1} 1\cdot e^{-j\omega t}dt=2\left(\frac{\sin\omega}{\omega}\right)$$

find $\int_{-\infty}^{\infty} |X(\omega)|^2 d\omega$

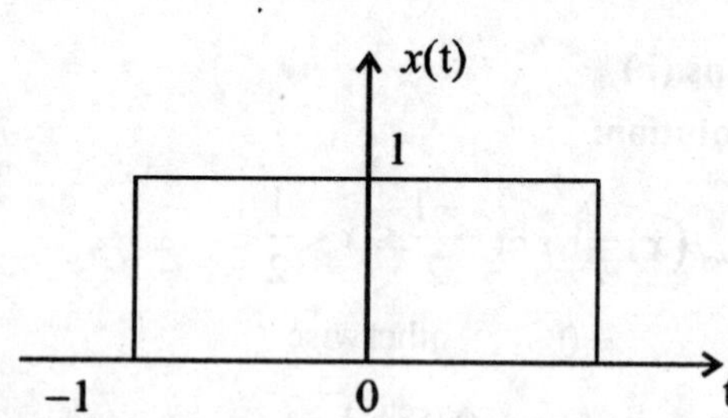

$$X(\omega)=\int_{-\infty}^{\infty} x(t)\,e^{-j\omega t}dt$$

$$=\int_{-1}^{+1} 1\cdot e^{-j\omega t}dt$$

$$=\left[\frac{e^{-j\omega t}}{-j\omega}\right]_{-1}^{1}$$

$$=\left[\frac{e^{-j\omega}-e^{+j\omega}}{-j\omega}\right]_{-1}^{1}=2\left[\frac{\sin\omega}{\omega}\right]$$

$$\int_{-\infty}^{\infty} |X(\omega)|^2 d\omega=\int_{-\infty}^{\infty} 4\left(\frac{\sin\omega}{\omega}\right)^2 d\omega$$

$$=4\int_{-\infty}^{\infty}\left(\frac{\sin\omega}{\omega}\right)^2 d\omega$$

$$=4\pi$$

27. Ans. (c)

Solution:

Fourier transform of Gate function.

$f(t) = A$; $-T/2 < t < T/2$

$= 0$; otherwise

is derived as

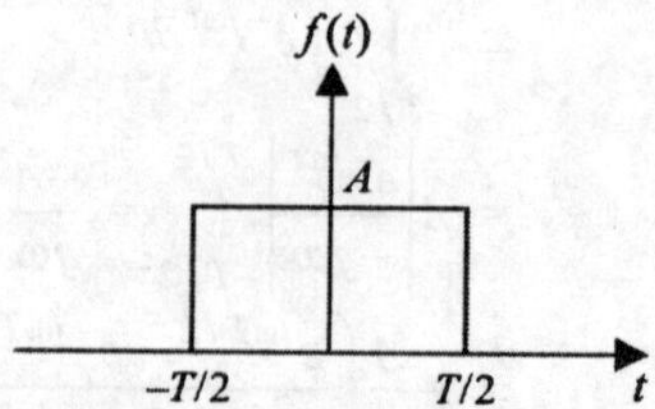

$$F(\omega)=\int_{-\infty}^{\infty} f(t)e^{-j\omega t}dt$$

$$=\int_{-T/2}^{T/2} A\,e^{-j\omega t}dt$$

$$=A\left|\frac{e^{-j\omega t}}{-j\omega}\right|_{-T/2}^{T/2}=\frac{-A}{j\omega}\,[e^{-j\omega T/2}-e^{+j\omega T/2}]$$

$$=\frac{-A}{j\omega}\left(\frac{e^{j\omega T/2}-e^{-j\omega T/2}}{2j}\right)\times[2j]$$

$$=\frac{2A}{\omega}\sin\frac{\omega(T)}{2}$$

$$=\frac{2A}{\omega}\frac{\sin(\omega T/2)}{(\omega T/2)}\;[\omega\,T/2]$$

$F(\omega) = AT(\sin x/x)$; where, $x = \omega T/2$

$=$ Sampling function

Hence

$$F(\omega)=AT\sin\frac{\sin\omega T/2}{\omega T/2}$$

In peresent case

$$\frac{\sin 2\omega}{\omega}=\frac{2\sin 2\omega}{2\omega}$$ Here, $AT = 2$

and $\frac{\omega T}{2}=2\omega$

or $T = 4$

$\Rightarrow$ $\frac{T}{2}=2$

and $A.4 = 2$

$$A=\frac{2}{4}=\frac{1}{2}$$

Multiplication with $e^{j\omega}$ in frequency domain result shift in time domain i.e.

$$F(j\omega)e^{j\omega T} \rightarrow f(t-T)$$

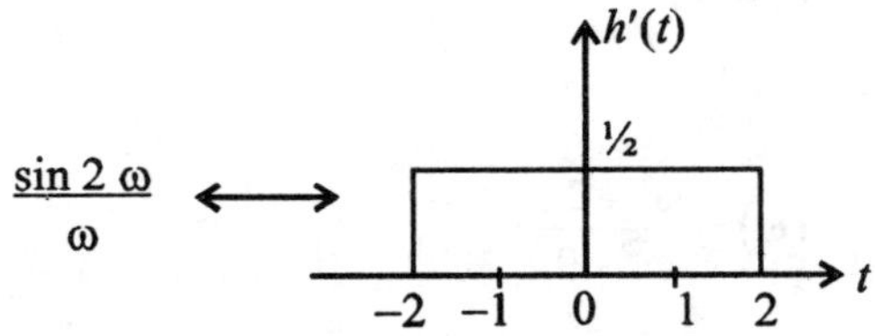

$$2\cos\omega\left(\frac{\sin 2\omega}{\omega}\right)$$

$$= \left[e^{j\omega} + e^{-j\omega}\right]\left(\frac{\sin 2\omega}{\omega}\right) \longleftrightarrow h(t) = h'(t-1) + h'(t+1)$$

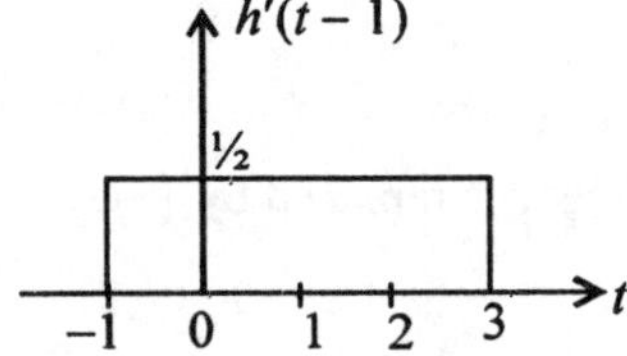

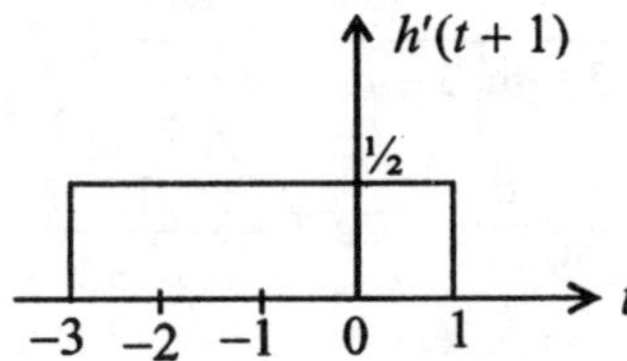

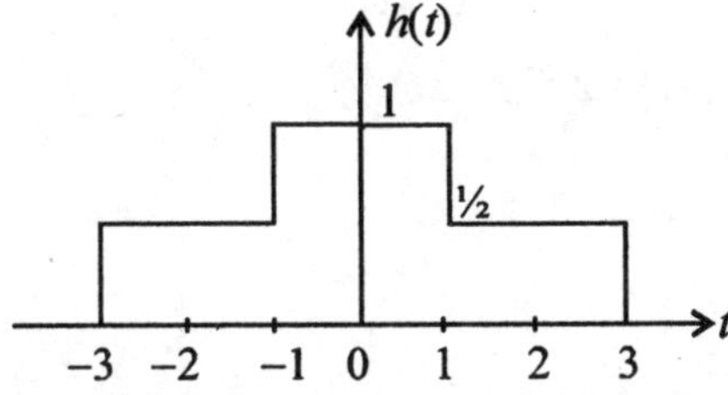

28. Ans. (c)

Solution:

A Ganssian pulse is defined as $x(t) = e^{-\pi t^2}$ for all t with

$x(0) = 1$

If A = Ara under $x(t)$

$$= \int_{-\infty}^{\infty} x(t)\, dt = 1$$

Then

$x(t) \rightarrow X(f) = e^{-\pi f^2}$.

* $X(f)$ is also a Gaussian pulse in frequency domain having the same expression $x(t)$ in time-domain.

* Even if the Gauusian pulse is not normalized, i.e. $x(0) \neq 1$ and $A \neq 1$, its Fourier Transform is also a Gaussian of the form $Be^{-C\pi f^2}$ where B and C are constants.

29. Ans. (b)

Solution:

Definition of Fourier transform

$$X(f) = \int_{t=-\infty}^{\infty} x(t)e^{-j2\pi ft}dt$$

As $x(t)$ is in volts, $\int x(t)\ e^{-j2\pi ft}dt$ is in voltage

$\therefore\ X(f)$ is in volt-sec.

30. Ans. (c)

Solution:

$$x(t) = \frac{1}{2\pi}\int_{-\infty}^{\infty} X(\omega)\, e^{j\omega t}\, d\omega$$

At, $t = 0$

$$x(0) = \frac{1}{2\pi}\int_{-\infty}^{\infty} X(\omega)\, d\omega$$

$$e^{-o} = 1$$

31. Ans. (d)

Solution:

$$f(t) \rightarrow F(\omega), f(-t) \rightarrow F^*(\omega)$$

F^* is conjugate of $F(\omega)$

The Fourier transform pair

$$f(t) - f(-t) = F(\omega) - F^*(\omega)$$

it must be imaginary

Then, $F^*(\omega)$ as difference of $F(\omega)$ and $F^*(\omega)$

32. Ans. (c)

Solution:

$$x[n] \leftrightarrow X\left(e^{j\omega}\right)$$

$$e^{j\omega_o n}x[n] \leftrightarrow X\left(e^{j(\omega-\omega_o)}\right)$$

$$(-1)^n\ x[n] = \left(e^{j\pi}\ x[n]\right) \qquad (\because\ e^{j\pi} = -1)$$

$$e^{j\pi n}x[n] \leftrightarrow X[e^{j(\omega-\pi)}]$$

* Multiplication in time domain result shifting in frequency domain

33. Ans. (d)

Solution:

As $\cos bt = \dfrac{e^{jbt} + e^{-jbt}}{2}$

$$y(t) = x(t)\cos bt = \frac{x(t)}{2}e^{jbt} + \frac{x(t)}{2}e^{-jbt}$$

or, $x(t)\cos bt = \frac{1}{2}\left[x(t)e^{jbt} + x(t)e^{-jbt}\right]$

By frequency shift property

Using, $x(t)e^{j\omega_o t} \leftrightarrow X(\omega - \omega_o)$

$\therefore FT\left[x(t)\cos bt\right]$

$= \frac{1}{2}\left[\exp(-a|\omega - b| + \exp(-a(\omega + b)\right]$

$\therefore \quad Y(\omega) = \frac{X(\omega - b)}{2} + \frac{X(\omega + b)}{2}$

Given $X(\omega) = \frac{\pi}{a}e^{(-a[\omega])}$

$\therefore Y(\omega) = \frac{\pi}{2a}\left[e^{(-a|\omega - b|)} + e^{(-a|\omega + b|)}\right]$

34. Ans. (c)

Solution: $\frac{d^2(t)}{dt^2} = \delta(t-1) + \delta(t+1) - 2\,\delta(t)$

Taking Fourier transform both sides

$\Rightarrow (j\omega^2)F(j\omega) = e^{-j\omega} + e^{j\omega} - 2$

$\Rightarrow F(j\omega) = \frac{2(1 - \cos\omega)}{\omega^2}$

35. Ans. (a)

Solution: $g(t) = \delta(t) + 2\,e^{(-3|t|)}$

Taking Laplace transform on both sides

$$G(\omega) = 1 + 2\int_{-\infty}^{0} e^{(3t)}\,e^{(-j\omega t)}.\,dt$$

$$+ 2\int_{0}^{\infty} e^{(-3t)}\,e^{(-j\omega t)}.\,dt$$

$$= 1 + 2\int_{-\infty}^{0} e^{(3-j\omega)t}\,dt + 2\int_{0}^{\infty} e^{(-3-j\omega t)t}.\,dt$$

$$= 1 + \frac{2}{3 - j\omega} + \frac{2}{3 + j\omega}$$

$$\frac{\omega^2 + 9 + 6 + 2j + 6\cdot 2j}{\omega^2 + 9}$$

$$G(\omega) = \frac{\omega^2 + 21}{\omega^2 + 9}$$

Second method

$g(t) \rightarrow G(\omega)$

Given $G(\omega) = \frac{\omega^2 + 21}{\omega^2 + 9}$

$$G(\omega) = \frac{\omega^2 + 9 + 21}{\omega^2 + 9}$$

$$= 1 + \frac{12}{\omega^2 + 9}$$

Use the F.T. pairs:

$\delta(t) \rightarrow 1$

$e^{-3t}u(t) \rightarrow \frac{1}{3 + j\omega}$

If t is replaced by $(-t)$, w is replaced by $(-\omega)$

$e^{+3t}u(-t) \rightarrow \frac{1}{3 - j\omega}$

$e^{-3t}u(t) + e^{3t}u(-t) \rightarrow \frac{1}{3 + j\omega}\frac{1}{3 - j\omega} = \frac{6}{9 + \omega^2}$

$e^{-3|t|} \rightarrow \frac{1}{3 + j\omega}\frac{1}{3 - j\omega} = \frac{6}{9 + \omega^2},\ 2e^{-3|t|} \rightarrow \frac{12}{\omega^2 + 9}$

In general, $K\,e^{-\alpha|t|} \rightarrow K\frac{2\alpha}{\alpha^2 + \omega^2}$

$\therefore \quad g(t) = 1\,\delta(t) + 2^{-3|t|}$

36. Ans. (d)

Solution:

As given

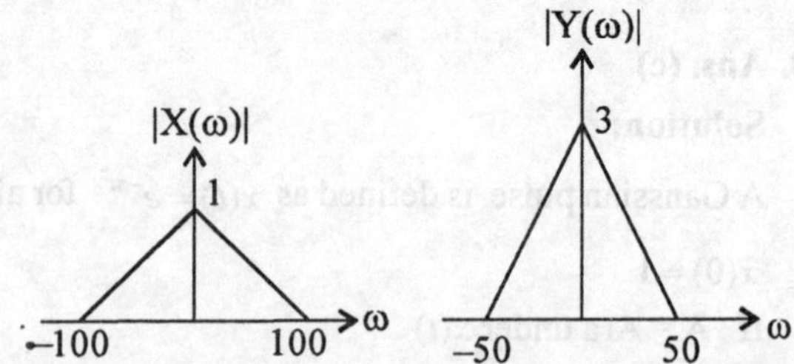

It is known that expansion in frequency domain result in compression in time domain and vice-versa.

As given compression is done in frequency domain by factor of 2 which result expansion in time domain by same amount.

$$x(t) \Leftrightarrow X(\omega)$$

$$x(at) \Leftrightarrow \frac{1}{|a|} \times \left(\frac{\omega}{a}\right)$$

So for $Ax\left(\frac{t}{2}\right) \leftrightarrow 2A\,X(2\,j\omega)$

$$2A = 3$$

$$A = \frac{3}{2}$$

$$y(t) = \frac{3}{2} x\left(\frac{t}{2}\right)$$

So, $y(t) = \frac{3}{2} x\left(\frac{t}{2}\right)$

37. Ans (a)

Solution:

Using

$$x'(t) = \frac{d}{dt} x(t) \leftrightarrow j\omega X(\omega)$$

$$x(t) \leftrightarrow X(\omega)$$

$$x(t - t_d) \leftrightarrow e^{-j\omega t_d} X(\omega)$$

$$\frac{d\left[x(t - t_d)\right]}{dt} \leftrightarrow (j\omega)\,X(\omega)\,e^{-j\omega d}$$

$$\text{Magnitude} = |j\omega||X(\omega)||e^{-j\omega t_d}|$$

$$= |X(\omega)||\omega||1|$$

$$= |X(\omega)||\omega|$$

4.4 Laplace Transform

1. Ans. (a)

Solution: System $[1/(s + 1)]$ gives change in output as

$$\frac{1}{\sqrt{(1+1)}} \angle - \tan^{-1}\frac{1}{1} = \frac{1}{\sqrt{2}} \angle -45°$$

Therefore, output = $[1/\sqrt{2}\,]\sin[t - (\pi/4)]$

2. Ans. (c)

Solution: $F(s) = \dfrac{\omega_0}{s^2 + \omega_0}$; $\text{Re}(s) > 0$

$$f(t) = \sin \omega_0 t$$

Final value lie between – 1 and 1,

i.e. at infinity the signal will be oscillating between –1 and +1.

* Since $f(t)$ is oscillating between two boundaries, i.e. –1 and +1, final value theorem is not applcable to it.

3. Ans. (b)

Solution: $c(t) = 1 - e^{-2t}$; $r(t) = u(t)$

$$R(s) = [1/s]$$

$$C(s) = \frac{1}{s} - \frac{1}{s+2} = \frac{2}{s(s+2)}$$

$$H(s) = \frac{C(s)}{R(s)} = \frac{2}{(s+2)}$$

4. Ans. (c)

Solution: $h(t) = e^{-t}$; $H(s) = \dfrac{1}{s+1}$

$$Y(s) = H(s)\,X(s) = \frac{1}{s+1} \times \frac{1}{s}$$

$$y(t) = 1 - e^{-t}$$

In steady state, $t \to \infty$; $y(\infty) = 1$

5. Ans. (d)

Solution: $Y(s) = \dfrac{1}{s(s-1)}$

$$y(t) = LT^{-1}\left(\frac{-1}{s} + \frac{1}{s-1}\right) = (-1 + e^t)\,u(t)$$

$$y(\infty) = \infty$$

This result is unbounded

* Final value theorem $\lim_{t\to\infty} y(t) = \lim_{s\to 0} sY(s)$ is not applicable if poles are lying in positive half of s-plane.

6. Ans. (a)

Solution:

$$h(t)e^{-t}\,u(t) \to \frac{1}{s+1} = \frac{1}{1+j\omega}$$

$$H(f) = \frac{1}{1 + j2\pi f} = \frac{1}{1 + j\frac{f}{f_o}}$$

Where f_o is known as the 3dB BW

Hence $f_o = \dfrac{1}{2\pi}$

7. Ans. (a)

Solution: Apply final value theorem

$$x(\infty) = \lim_{t\to\infty} f(t) = \lim_{s\to 0} sF(s)$$

$$= \lim_{s\to 0} \frac{s(5s^2+23s+6)}{(s^2+2s+2)s} = 3$$

8. Ans. (b)

Solution: $s^2Y(s) + 4s\,Y(s) + 3Y(s)$

$$= 2s\,X(s) + 4\,X(s)$$

$$Y(s)\,[s^2 + 4s + 3] = X(s)\,[2s + 4]$$

$$\frac{Y(s)}{X(s)} = \frac{2s+4}{s^2+4s+3} = \frac{2(s+2)}{(s+1)(s+3)}$$

if $\quad x(t) = e^{-2t}\,u(t)$

$$X(s) = \frac{1}{s+2}$$

Then $Y(s) = \dfrac{2(s+2)}{(s+1)(s+3)} \times \dfrac{1}{s+2}$

$$= \frac{2}{(s+1)(s+3)}$$

$$= \frac{\left.\frac{2}{(s+3)}\right|_{s=-1}}{s+1} + \frac{\left.\frac{2}{(s+1)}\right|_{s=-3}}{s+2}$$

$$= \frac{1}{s+1} - \frac{1}{s+2}$$

$$= [e^{-t} - e^{-2t}]\,u(t)$$

9. Ans. (d)

Solution:

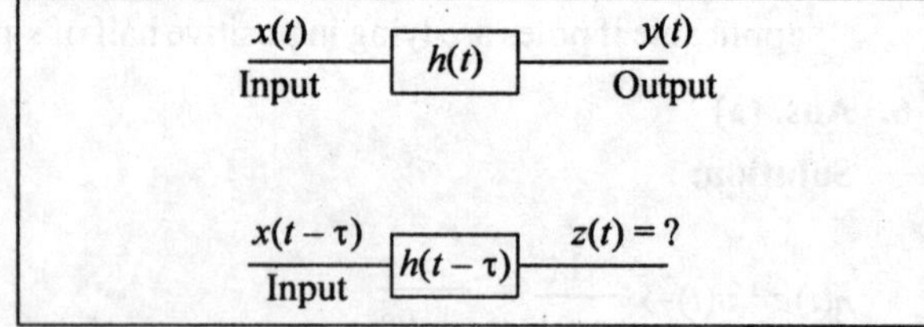

From the given data

$$y(t) = x(t)\,h(t) \qquad \text{...(1)}$$

$$x(t-\tau)\,h(t-\tau) = ?$$

Suppose output = $z(t)$

$$x(t-\tau) \times h(t-\tau) = z(t)$$

Taking laplace transform

$$X(s)\,(e^{-\tau s}).\,H(s)\,(e^{-\tau s}) = Z(s)$$

$$X(s).\,H(s)\,(e^{-2\tau s}) = Z(s)$$

From equation (1)

$$H(s)\;X(s) = Y(s)$$

$$Y(s)\,e^{-2\tau s} = Z(s)$$

Taking inverse Laplace

$$y(t-2\tau) = z(t)$$

$$\therefore\; x(t-\tau)\,h(t-\tau) = y(t-2\tau)$$

10. Ans. (d)

Solution: $y(t) = 5e^{-10t}u(t)$

We know that $y(t) = x(t) * h(t)$

Here $x(t) = \delta(t)$

$$y(t) = \delta(t) * h(t)$$

Convolution in time domain is equal to multiplication in frequency domain

$$Y(s) = H(s)$$

$$H(s) = Y(s) \;\Rightarrow\; \frac{5s}{s+10}$$

$$h(t) = ?$$

Because degree of numerator and denominator is same so will firstly divide N by D

$$H(s) = s+10\,\overline{)\,5s\,}\,(5$$
$$\underline{5s+50}$$
$$\underline{-50}$$

$$H(s) \Rightarrow 5 - \frac{50}{s+10}$$

$$h(t) = 5\delta(t) - 50e^{-10t}u(t)$$

11. Ans. (c)

Solution: $C = \dfrac{1}{2}F$

$$I_c(s) = \frac{V(s)}{Z(s)} = \frac{V_c(s)}{1/sC}$$

$$= \frac{(s+1)}{\frac{1}{sC}(s^3+s^2+s+1)} = \frac{sC(s+1)}{(s^3+s^2+s+1)}$$

$$= \frac{s}{(s^2+1)\cdot 2}$$

Apply initial value theorem

$$I_c(0^+) = \mathop{\text{Lt}}_{s\to\infty} sI_c(s)$$

$$= \underset{s\to\infty}{\text{Lt}}\, sI_c(s)\times\frac{s}{(s^2+1)\,2}$$

$$= \underset{s\to 0}{\text{Lt}}\frac{s^2}{(s^2+1)2} = \underset{s\to\infty}{\text{Lt}}\frac{1}{2+\frac{2}{s^2}} = \frac{1}{2}$$

12. Ans. (b)

Solution: Multiplication in time domain corresponds to convolution in time domain

$V(s) = Z(s) \times I(s)$

$$V(t) = \int_0^t i(\tau)\, z(t-\tau)\, d\tau$$

13. Ans. (d)

Solution: $LT\,[f(t)] = F(s)$

Final value theorem, $\lim_{t\to\infty} f(t) = \lim_{s\to 0} sF(s)$ is only aplicable when all poles are in LH of s-plane

Here $F(s)$ has poles at $s = -1$, $s = \pm 2$

As all the poles are not in the left half of the s-plane Hence, final value theorem cannot be applied and $\lim_{t\to\infty} f(t)$ is undefined

14. Ans. (c)

Solution: $r(t) = (t - a)\, u(t - a)$

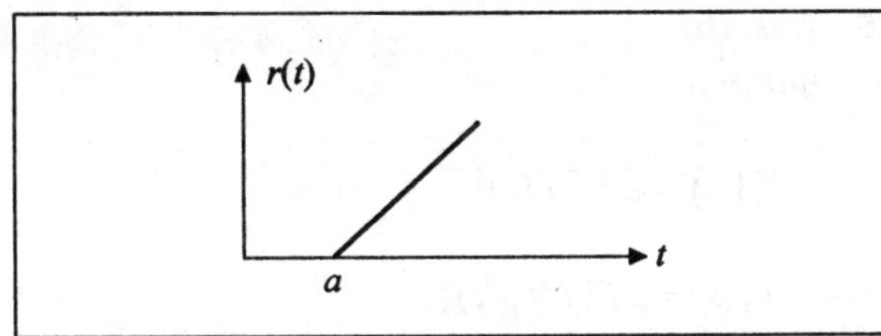

$LT[r(t)] = \dfrac{e^{-as}}{s^2}$ by the use of time shifting property.

15. Ans. (a)

Solution:

$$F(s) = \frac{s+5}{(s+1)(s+3)} = \frac{A}{s+1} + \frac{B}{s+3}$$

$$A = \left.\frac{s+5}{s+3}\right|_{s=-1} = 2,$$

$$B = \left.\frac{s+5}{s+1}\right|_{s=-3} = \frac{2}{-2} = -1$$

$$F(s) = \frac{2}{s+1} - \frac{1}{s+3}$$

$\therefore\ L^{-1} F(s) = [2\,e^{-1} - e^{-3t}\,]\, u(t)$

16. Ans (b)

Solution: For both sided

The laplace transform of $\int_{-\infty}^{t} f(\tau)\,dz$

$$= \frac{1}{s}\left[F(s) - f^{-1}(0)\right]$$

Since it is one sided, take initial conditions equal to zero

$$\int_0^t f(\tau)\, d\tau = \frac{1}{s}F(s)$$

17. Ans. (d)

Solution:

Apply final value theorem $\underset{t\to\infty}{Lim} f(t) = \underset{s\to 0}{Lim}\, s\cdot F(s)$

$$1 = \left.\frac{s(3s+1)}{s^3+4s^2+(k-3)s}\right|_{s\to 0} = \left.\frac{3s+1}{s^2+4s+(k-3)}\right|_{s\to 0}$$

or $\quad 1 = \dfrac{1}{K-3}$

or $\quad K - 3 = 1$

$K = 4$

18. Ans. (b)

Solution: $\underset{t\to 0}{Lt}\, f(t) = \underset{s\to\infty}{Lt}\, s\,F(s)$

$$= \underset{s\to\infty}{Lt}\frac{s(2s+1)}{s^2+4s+7} = 2$$

$$\underset{t\to\infty}{Lt}\, f(t) = \underset{s\to 0}{Lt}\, s\,F(s)$$

$$= \underset{s\to 0}{Lt}\frac{s(2s+1)}{s^2+4s+7} = 0$$

Second Method

Initial Value

$\lim_{s\to\infty} sF(0)$

$$F(s) = \frac{2(s+1)}{s^2+4s+7} = \frac{2s\left(1+\frac{1}{s}\right)}{s^2\left(1+\frac{4}{s}+\frac{7}{s^2}\right)}$$

$$\lim_{s\to\infty} sF(s) = \frac{2\left(1+\frac{1}{s}\right)}{\left(1+\frac{4}{s}+\frac{7}{s^2}\right)}$$

$= 2$

Final value

$\lim_{s\to 0} sF(s) = 0$

19. Ans (b)

Solution: $L[f(t)] = F(s)$ then

$L[f(t-T)]$ is equal to $e^{-sT}F(s)$.

Apply time shifting property of Laplace transform.

20. Ans. (d)

Solution :

Given

$$F(s) = \frac{\omega}{s^2+\omega^2},$$

$\Rightarrow f(t) = \sin(\omega t)\,u(t)$

The final value of $f(t)$, i.e. $\underset{t\to\infty}{Lt} f(t)$ is indeterminate (not defined) as $f(t)$ may have any value between $-1 \,\&\, +1$.

* The final value theorem of LT is not applicable to function have has poles on the imaginary axis in the s-plane and not strictly in the left half of the s-plane.

21. Ans. (c)

Solution:

Let $x(t) = (t^2 - 2t)\,u(t-1)$

$= t^2u(t-1) - 2t(t-1)$

Using the properties of LT:

$$u(t) \to \frac{1}{s},\ 2u(t-1) \to \frac{2}{s}e^{-s}$$

$$t^2u(t-1) \to -\frac{2}{s^3}e^{-1s}$$

$$X(s) = \frac{2e^{-1s}}{s^3} - \frac{2e^{-s}}{s}$$

22. Ans. (d)

Solution:

Given the rectangular pulse of current of duration, T and magnitude, 1:

$p(t) = 1\,u(t) - 1\,u(t-T)$

$$P(s) = \frac{1}{s} - \frac{1}{s}e^{-Ts}$$

$$= \frac{1}{s}\left(1 - e^{-Ts}\right)$$

23. Ans. (c)

Solution:

According to the final value theorem :

Final value of $y(t) = \lim_{s\to\infty} y(t) = \lim_{s\to 0}[s\,Y(s)]$

* Provided the function is $s\,Y(s)$ has poles strictly in the left half of the s-plane.

24. Ans.(d)

Solution:

If $f(t) \leftrightarrow F(s)$,

$LT[f(t)] = F(s)$

Then $tf(t) \leftrightarrow -\frac{d}{ds}F(s)$

$$LT[tf(t)] = (-1)\frac{dF(s)}{ds}$$

Thus if $F(s) = \dfrac{1}{s^2+s+1}$

$$tf(t) \to -\frac{d}{ds}\left(\frac{1}{s^2+s+1}\right)$$

$$= (-1)\left[\frac{-(2s+1)}{(s^2+s+1)^2}\right] = \frac{2s+1}{(s^2+s+1)^2}$$

25. Ans. (b)

Solution:-

$F_1(s) = LT[f(t)]$

$F_2(s) = LT\{f(t-\tau)\}$

$= e^{-ST}F_1(S)$

$$G(s) = \frac{F_2(s)\,F_1^*(s)}{|F_1(s)|^2}$$

$$G(s) = \frac{e^{-s\tau}F_1(s)F_1^*(s)}{|F_1(s)|^2} = e^{-s\tau}$$

$g(t) = LT^{-1}\{G(s)\} = \delta(t-\tau)$

26. Ans. (a)

Solution: Apply final value theorem

$$x(\infty) = \underset{t\to\infty}{\text{Lim}} f(t) = \underset{s\to 0}{\text{Lim}}\, sF(s)$$

$$= \underset{s\to 0}{\text{Lim}} \frac{s(5s^2+23s+6)}{(s^2+2s+2)s} = 3$$

27. Ans.(b)

Solution:

$$F(s) = \frac{2(s+1)}{s^2+2s+5}$$

$$f(0^+) = \lim_{s\to\infty} sF(s) = s\frac{2(s+1)}{(s^2+2s+5)}$$

$$= \frac{2s^2+s}{s^2+2s+5}$$

For $\lim_{s\to\infty}$ consider the highest order term.

The highest order term,

$$f(0^+) = \frac{2s^2}{s^2} = 2$$

$f(\infty) = \lim_{s\to 0} sF(s)$, provided poles of $sF(s)$ lie only in the open left half of the s-plane.

Here, the poles are $p_1 = -1+j2$, $p_2 = -1-j2$

$\therefore \quad f(\infty) = 0$

28. Ans. (a)

Solution:

The final value theorem is used to find the final value or the steady state value of the system output.

$\text{Lt}_{t\to\infty} f(t) = \text{Lt}_{s\to 0} sF(s)$,

provided poles of $sF(s)$ lie only in the left half of the s-plane.

29. Ans.(a)

Solution:

We known that

$$\cos(\alpha t)\, u(t) \to \frac{s}{s^2+\alpha^2}$$

$$e^{\alpha t} \to \frac{1}{s-\alpha}$$

According to shift in 's' property

$$e^{\alpha t}\cos(\alpha t)\, u(t) \to \frac{(s-\alpha)}{(s-\alpha)^2+\alpha^2}$$

30. Ans.(a)

Solution:

$$L[f(t)] = F(s) = \frac{\omega}{s^2+\omega^2}$$

According to Final value theorem

$$\lim_{t\to\infty} f(t) = \lim_{s\to 0} [sF(s)]$$

Provided all the poles of $sF(s)$ lie in the left half of s-plane.

Poles of $sF(s)$ in this question are $s = \pm j\omega$ lying on the imaginary axis.

Hence, the theorem cannot be applied

In fact $f(t)$= sin (ωt) $u(t)$, for the given F(s)

The value $f(t)$ as $t\to\infty$ cannot be determined

31. Ans.(b)

Solution:

Observe that $h(t)$ is given by

$$h(t) = f(t) * g(t)$$

Where * denotes convolution

* Convolution in time domain is equal to multiplication in frequency domain.

$\therefore \quad H(s) = F(s)\times G(s)$

$$H(s) = \frac{(s+2)}{(s^2+1)}\times\frac{(s^2+1)}{(s+3)(s+2)} = \frac{1}{s+3}$$

32. Ans.(c)

Solution:

$$i(t) \to I(s) = \frac{2}{s(1+s)}$$

According to Final value theorem

$\lim_{t\to\infty} i(t) = \lim_{s\to 0} [sF(s)] = 2$ as $sF(s)$ has poles omy in the left half of s-plane

33. Ans. (c)

Solution:

$$x(t) \to X(s) = \frac{5-s}{s^2-s-2}$$

$$= \frac{5-s}{(s+1)(s-2)}$$

Taking partial fractions, $X(s) = -\frac{2}{s+1}+\frac{1}{s-2}$

If the FT of $x(t)$ exists then the ROC of $x(t)\to X(s)$ pair includes the jω-axis in the s-plane

Here Bidirectional Laplace transform is to be applied using ROC concept.

$\therefore$ ROC of $X(s)$ is to be understood as $-1<\sigma<2$

$$X(s) = -\frac{2}{s+1}+\frac{1}{s-2}, \quad -1<\sigma<2$$

Use the standard pairs for Bidirectional Laplace transform.

$$e^{\alpha t}u(t) \to \frac{1}{s-\alpha}, \quad \sigma>\alpha$$

$$-e^{\alpha t}u(-t) \to \frac{1}{s-\alpha}, \quad \sigma < \alpha$$

$$\therefore \; x(t) = -2\, e^{-t}u(t) - 1e^{2t}u(-t)$$

34. Ans. (a)

Solution:

As it is known the LT pair:

$$e^{at}u(t) \to \frac{1}{s-a} \text{ for } \operatorname{Re}(s) > a, \text{ a real}$$

$$\therefore \; e^{(a+2)t+5} = e^5\, e^{(a+2)t} \to e^5 \frac{1}{s-(a+2)}$$

$$\text{ROC}: \operatorname{Re}(s) > (a+2)$$

35. Ans. (b)

Solution:

For the given differential equation

$$\ddot{x}(t) + 3\,\dot{x}(t) + 2x(t) = 5$$

Taking LT

$$(s^2 + 3s + 2)\, X(s) = \frac{5}{s}$$

$$X(s) = \frac{5/s}{s^2 + 3s + 2}$$

Apply final value theorem.

$$x(\infty) \lim_{t\to\infty} x(t) = \lim_{s\to 0} s\, X(s)$$

$$= \lim_{s\to 0} \frac{s \times 5/s}{s^2 + 3s + 2}$$

$$= \frac{5}{0+0+2}$$

$$= \frac{5}{2}$$

36. Ans (c)

Solution:

$$f(t) = u(t-\tau)$$

$$F(s) = ?$$

If $f(t-\tau) = F(s)e^{-\tau s}$

If $f(t) = u(t)$, then $F(s) = \dfrac{1}{s}$

$$FT\left[u(t-\tau)\right] = \frac{1}{s}e^{-\tau s}$$

37. Ans. (d)

Solution:

If $f(t) \leftrightarrow F(s)$,

$$LT\left[f(t)\right] = F(s)$$

Then $tf(t) \leftrightarrow -\dfrac{d}{ds}F(s)$

$$LT\left[tf(t)\right] = (-1)\frac{dF(s)}{ds}$$

Thus, if $F(s) = \dfrac{1}{s^2+s+1}$

$$tf(t) \to -\frac{d}{ds}\left(\frac{1}{s^2+s+1}\right)$$

$$= (-1)\left[\frac{-(2s+1)}{\left(s^2+s+1\right)^2}\right]$$

$$= \frac{2s+1}{\left(s^2+s+1\right)^2}$$

4.5 Z-Transform

1. Ans. (b)

Solution: $h[n] = 2\delta[n-3]$

$$H(z) = 2\, Z^{-3}$$

The output, $Y[z] = H(z) \cdot X[z]$

$$= 2z^{-3}[z^4 + z^2 - 2z + 2 - 3z^4]$$

$$= 2[z + z^{-1} - 2\, z^{-1} + 2z^{-3} - 3z^{-7}]$$

Taking inverse Z-transform, we have

$$y[n] = 2\left[\begin{array}{r}\delta[n+1] + \delta[n-1] - 2\delta[n-2] + 2\delta[n-3] \\ -3\,\delta[n-7]\end{array}\right]$$

At $n = 4;\; y[n] = 0$

2. Ans. (d)

Solution: Using the following Z-transform pair,

$$-A_k(dk)^n\, u[-n-1] \xrightarrow{z} \frac{A_k}{1 - dk\, z^{-1}}$$

with ROC $1\,|\,Z\,| < dk$

Comparing with above equation

$$\frac{z}{z-0.2} \xleftrightarrow{z} -(0.2)^n\, u[-n-1]$$

$$H(z) = \frac{z}{z-0.2}$$

$$= \frac{1}{1-0.2\, z^{-1}}; \quad ROC: |z| < 0.2$$

3. Ans. (c)

Solution: The given sequence is a 2 sided sequence so its ROC is a circle

ROC for $\left(\frac{5}{6}\right)^n u(n)$ is $|z| > \left(\frac{5}{6}\right)$

ROC for $\left(\frac{6}{5}\right)^n u(-n-1)$ is

$$\left(\frac{5}{6}\right) < |z| < \left(\frac{6}{5}\right)$$

4. Ans. (d)

Solution: ROC remains same in both cases, i.e. for $x_1(n)+x_2(n)$ and $x_1(n)-x_2(n)$

5. Ans. (b)

Solution: $F(z) = \frac{1}{z+1} = \frac{z+1-z}{z+1}$

$$= 1-\frac{z}{z-(-1)}$$

$\therefore \quad ZT^{-1}[F(Z)] = \delta(t) - (-1)^n$

$$\left[\therefore ZT^{-1}\left(\frac{z}{z-a}\right) = a^n\right]$$

6. Ans. (b)

Solution:

$$X(z) = \sum_{n=0}^{\infty} \frac{3^n}{2+n} z^{2n}$$

This means $x(n)$ has non-zero values only for even value of n, i.e. of the form:

$$x(n) = \{a_0, 0, a_2, 0, a_4, 0, a_6, 0, \ldots\}$$

* Two sequences are orthogonal if their inner [dot] product is equal to zero,

i.e. if $x(n)$ and $y(n)$ are orthogonal, then

$\langle x(n), y(n)\rangle = 0$

$\Rightarrow x_0y_0 + x_1y_1 + x_2y_2 + \ldots = 0$

For any even numbered sequence, all the odd numbered sequences are always orthogonal.

Among all the four options given, option (b) corresponds to $y(n)$ with non-zero values only for odd values of n.

7. Ans. (a)

Solution: $X(z) = 5z^2 + 4z^{-1} + 3$

$$= 5z^2 + 3 + 3z^{-1}$$

$$X(n) = \left\{\begin{matrix}5, 0, 3, 4 \\ \uparrow\end{matrix}\right\}$$

$$x(n) = 5\,\delta[n+2] + 3\,\delta[n] + 4\,\delta[n-1]$$

8. Ans. (a)

Solution: $x[n] = (1+j)^n \rightarrow \boxed{H(z)} \rightarrow$

$$X[z] = \frac{z}{z-(1+j)}$$

$\because$ O/P is zero for $x(n) = (1+j)^n$. The ROC of $\left(1+\frac{1}{2}z^{-1}\right)$H(z) implies that H(z) has 1 pole and 1 zero

9. Ans. (c)

Solution: Reside of

$$z^p X(z) = \frac{1}{(n-1)!}\frac{d^n}{dz^{n-1}} z^p X(z)(z-a)^n\Big|_{z=a}$$

where $(z-a)^n$ at $(z=a)$ $n =$ order of pole $n = 2$

$$\text{Resides } (z^{n-1}X(z)) = \frac{1}{(2-1)!}\frac{d}{dz}(z^{n-1}X(z))\Big|_{z=a}$$

$$= \frac{1}{1!}\frac{d}{dz}(z^{n-1})\frac{z}{(z-a)^2}(2-a)^2$$

$$= \frac{d}{dz}(z^n)\Big|_{z=a} = nz^{n-1}\Big|_{z=a} = na^{n-1}$$

10. Ans. (c)

Solution: $X(z) = 1 - z^{-1}$

$$Y(z) = 1 - z^{-4}$$

$$H(z) = \frac{Y(z)}{X(z)} = \frac{1-z^{-4}}{1-z^{-1}}$$

$$1-z^{-1}\big)\ 1-z^{-4}\ \big(z^{-3}+z^{-2}+z^{-1}+1$$

$$\underline{z^{-3}-z^{-4}}$$

$$1^{-2}-z^{-3}$$

$$\underline{z^{-2}-z^{-3}}$$

$$1-z^{-2}$$

$$\underline{z^{-1}-z^{-2}}$$

$$1-z^{-1}$$

$$\underline{1-z^{-1}}$$

$$X$$

$$H(s) = 1 + z^{-1} + z^{-2} + z^{-3}$$

$$h[n] = [\underset{\uparrow}{1}, 1, 1, 1]$$

11. Ans. (a)

Solution: $y(n) = x(n) - \frac{1}{3}y(n-1)$

when $y(n) = 0$ for $n < 0$

and $X[n] = \delta[n]$

$$Y(n) = x(n) * h(n)$$

$$Y(z) = X(z)\, H(z)$$

$$X[z] = 1$$

$$Y(z) = H(z)$$

$$y(n) = h(n)$$

$$y(n) = x(n) - \frac{1}{3}y(n-1)$$

$$Y(z) = X(z) - \frac{1}{3}Y(z)z^{-1}$$

$$Y(z) + \frac{1}{3}Y(z)z^{-1} \Rightarrow 1$$

$$Y(n) = \frac{1}{1+\frac{1}{3}z^{1}}$$

$$Y(n) = \frac{1}{1-\left(\frac{-1}{3}\right)z^{-1}}$$

$$y(n) = \left(\frac{-1}{3}\right)^n u(n)$$

12. Ans. (a)

Solution:

$$x(n) = 2^n u(n)$$

$$Z(z) = \frac{1}{1-2z^{-1}}$$

$$X(z) = \frac{z}{z-2}$$

For function to be stable, the $X(z)$ should not be negative or ∞ $|Z-2| > 0$

$$|Z| > 2$$

13. Ans. (b)

Solution: If $H(z)$ is discrete rational transfer function then,

where $\frac{Y[z]}{X[z]} = \frac{\text{zero}}{\text{pole}}$

For $H(z)$ to be stable, the poles must be in unit circle

$$[H(z)]^{-1} = \frac{X(z)}{Y(z)}$$

The zeroes must be inside the circle.

So poles and zeroes must be inside the unit circle

14. Ans. (d)

Solution:

$$a^n u[n] \leftrightarrow \frac{1}{1-a^{z-1}}; \quad |a| < z$$

$$b^n u[-n] \leftrightarrow \frac{-1}{1-b^{z-1}}; \quad |b| > z$$

$$x[n] = \frac{1}{2}a^n - \frac{1}{3}a^n$$

$$x[0] = \frac{1}{2} - \frac{1}{3} = \frac{1}{6}$$

Second Method

Given $X(z) = \frac{1/2}{1-az^{-1}} + \frac{1/3}{1-bz^{-1}}$, $|a| < 1$, $|b| < 1$

$$\text{ROC}: |a| < |z| < |b|$$

Remember the standard Z.T pairs:

$$a^n u(n) \xrightarrow{ZT} \frac{1}{1-az^{-1}} = \frac{z}{z-a}, \ |z| > |a|$$

$$-b^n u(-n-1) \xrightarrow{ZT} \frac{1}{1-bz^{-1}} = \frac{z}{z-b}, \ |z| < |b|$$

$\therefore$ For the given ROC

$$X(n) = \frac{1}{2}a^n u(n) - \frac{1}{3}b^n u(n-1)$$

$$x(0) = \frac{1}{2}$$

15. Ans. (a)

Solution:

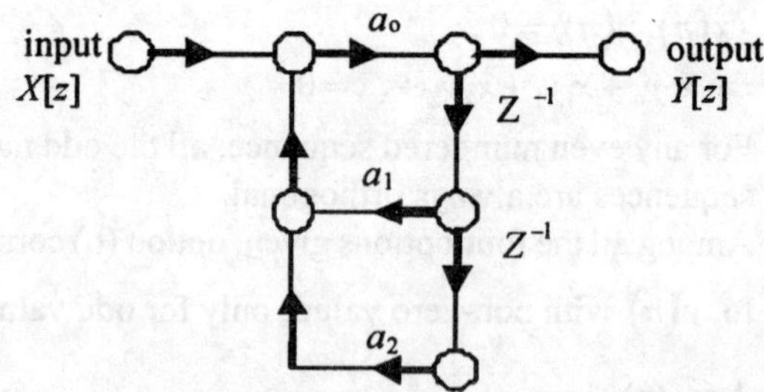

$$Y(z) = \left[X(z) + Y(z)\cdot z^{-2}a_2 + Y(z)a_1 \cdot z^{-1}\right] \times a_0$$

$$Y(z) - Y(z)a_0 a_2 z^{-2} - Y(z)a_0 a_1 z^{-1} = X(z)\, a_0$$

or $$\frac{Y(z)}{X(x)} = H(z) = \frac{a_0}{1 - a_0 a_1 z^{-1} - a_0 a_2 z^{-2}}$$

$$= \frac{1}{\frac{1}{a_0} - \frac{a_1}{a_0} z^{-1} - \frac{a_2}{a_0} z^{-2}}$$

Given $H(z) = \frac{1}{1 - 0.7z^{-1} + 0.13z^{-2}}$

Comparing with the given H(z), we get

$$\frac{1}{a_0} = 1, \; -\frac{a_1}{a_0} = 0.7, \; -\frac{a_2}{a_0} = 0.13$$

$$a_0 = 1, \; a_1 = 0.7, \; a_2 = -0.13$$

16. Ans. (c)

Solution:

Given

$$H(s) = \frac{a}{(s^2 + a^2)}$$

$h(t) = \text{sinat}$

$h(nt) = \sin \text{a nT, c}$

(Where T is sampling period)

$h(nT) = \sin \text{aTn}$

$$H(s) = \frac{aT}{s^2 + (aT)^2}$$

For analog to digital transformation z

$$z = e^s$$

$$\Rightarrow \log_e z = s$$

$$ln\, z = s$$

So, $H(z) = \frac{aT}{(lnz)^2 + (aT)^2}$

To get poles $(lnz)^2 + (aT)^2 = 0$

$(lnz)^2 = -(aT)^2$

$lnz = \pm jaT$

$z = e^{jaT}, e^{-jaT}$

17. Ans. (b)

Solution:

For compensation, a element is cascaded which make the overall gain unity and no delay (no distortion).

$$H'(z)H(z) = 1$$

Where $H'(z) \rightarrow$ compensator

$$\therefore H'(z) = \frac{1}{H(z)} = \frac{1}{(1 - az^{-1})}$$

$h'(n) = -(a)^n u[-n-1],$

$([z] < [a])$

Because system $H'(z)$ has to be stable

Therefore, unit circle must be inside the ROC.

$[a] > 1,$

If ROC is $[z] < [a]$, then it will include unit circle.

Neglecting the –ve sign,

$h'(n) = a^n u[n-1]$

18. Ans. (a)

19. Ans. (c)

Solution: Given time function as $x[n] = \sum_{K=0}^{\infty} \delta(n - K)$

$$X(z) = \sum_{K=0}^{\infty} z^{-k} = \frac{1}{1 - Z^{-1}} = \frac{Z}{Z - 1}, \; |z| > 1$$

20. Ans. (a)

Solution:

Standard relation

$x[n] = a^n u[n]$

$$x(z) = \frac{1}{1 - az^{-1}} = \frac{z}{z - a} \; ; \; |z| > a$$

- Using the same concept

$f(nT) = a^{nT}$

$$F(z) = \sum_{n=0}^{\infty} (aT) z^{-n} = \left(\frac{z}{z - a^T}\right) \; ; \; |z| > a^T$$

21. Ans. (a)

Solution: Given that

$$x(n) = \left(\frac{1}{3}\right)^n u(n) - \left(\frac{1}{2}\right)^n u(-n-1)$$

$$X(z) = \frac{1}{1 - \frac{1}{3} z^{-1}} + \frac{1}{1 - \frac{1}{2} z^{-1}}$$

$X(z) = X_1(z) + X_2(z)$

Where $X_1(z) = \frac{1}{1 - \frac{1}{3} z^{-1}}$

ROC $\Rightarrow |z| > \frac{1}{3}$

And $X_2(z) = \dfrac{1}{1-\frac{1}{2}z^{-1}}$

$ROC \Rightarrow |z| < \frac{1}{2}$

$\therefore$ $X(z)$ has ROC

$\frac{1}{3} < |z| < \frac{1}{2}$

22. Ans.(c)

Solution:

$$x(n) = \left[\left(\frac{1}{3}\right)^{|n|} - \left(\frac{1}{2}\right)^{n}\right]u(n)$$

$$= \left(\frac{1}{3}\right)^{n} u(n) + \left(\frac{1}{3}\right)^{-n} u(-n-1) - \left(\frac{1}{2}\right)^{n} u(n)$$

$$x(n) = \left(\frac{1}{3}\right)^{n} u(n) + (3)^{n} u(-n-1) - \left(\frac{1}{2}\right)^{n} u(n)$$

As known $a^{|n|}\, u[n]$

$$= a^{n}\, u[n] + a^{-n} u[-n-1]$$

$$= a^{n}\, u[n] \rightarrow \frac{1}{1-a\,z^{-1}},\ ROC|z| > |a|$$

and $a^{n}u[-n-1] \rightarrow \dfrac{a^{-1}z}{1-a^{-1}z}$

$$\rightarrow \frac{z}{a-z}\ ROC,\ |z| < a$$

$\left(\frac{1}{3}\right)^{n} u(n) \rightarrow ROC: |z| > \left|\frac{1}{3}\right|$

$3^{n}u(-n-1)\ ROC: |z| < |3|$

$\left(\frac{1}{2}\right)^{n} u(n)\ ROC: |z| > \left|\frac{1}{2}\right|$

Common ROC: $\frac{1}{2} < |z| < 3$

23. Ans. (b)

Solution: Initially capacitor is short circuit then voltage across $R = 5\,V$

Finally capacitor is open circuit

Then voltage across $R = 0\,V$

Voltage across resistor at any time t is

$$V_R = V_{in} e^{-\frac{t}{\tau}}$$

When τ is the time constant

$\tau = RC = 10\,\mu F \times 200\,k = 2$

and $V_{in} = 5\,V$

In present case $V_R = 5e^{-0.5t}$

Sampling is done at $f_s = 10\,Hz$

$$t = T_s n = \frac{1}{f_s} n = \frac{1}{10} n$$

$$\therefore\ x[n] = 5e^{\frac{-0.5}{10}n} = 5e^{-0.05n}$$

24. Ans. (c)

Solution: $X(z) = \sum_{n=0}^{\infty} 5.e^{-0.05n} \cdot Z^{-n}$

$$= \frac{5}{1-e^{-0.05}Z^{-1}} |Z| > e^{-0.05}$$

$$= \frac{5Z}{Z-e^{-0.05}} \qquad |Z| > e^{-0.05}$$

25. Ans. (d)

Solution: Residue of

$$z^{p}\, X(z) = \frac{1}{(n-1)!}\frac{d^{n}}{dz^{n-1}} z^{p} X(z)(z-a)^{n}\Big|_{z=a}$$

where $(z-a)^{n}$ at $(z=a)$ $\quad n=$ order of pole

given problem $n = 2$

$$\text{Resides } \left(z^{n-1}X(z)\right) = \frac{1}{(2-1)!}\frac{d}{dz}\left(z^{n-1}X(z)\right)\Big|_{z=a}$$

$$= \frac{1}{1!}\frac{d}{dz}\left(z^{n-1}\right)\frac{z}{(z-a)^{2}}(2-a)^{2}$$

$$= \frac{d}{dz}\left(z^{n}\right)\Big|_{z=a} = nz^{n-1}\Big|_{z=a} = na^{n-1}$$

26. Ans. (a)

Solution:

$u(n)$ = unit step function or sequence

$u(n) \xrightarrow{ZT} \frac{z}{z-1}$ ROC $|z| > |a|$

ROC is outside the unity circle with its center at the origin in the z-plane.

27. Ans. (b)

Solution:

The given sequence $x(n) = a^{n}\ u(n)$ where $a > 0$

$$x(n) \rightarrow X(z) = \frac{z}{z-a}\ |z| > |a|$$

Divide N(z) and D(z) by z and getting

$$X(z)=\frac{1}{1-az^{-1}}\ ;\ |z|>|a|$$

28. Ans. (c)

Solution:

$$c(n)\rightarrow C(z)=\frac{1z^{-1}(1-z^{-4})}{4(1-z^{-1})^2}$$

$$c(\infty)=\lim_{z\to 1}(1-z^{-1}).\,C(z)$$

$$=\lim_{z\to 1}\frac{(1-z^{-1})\,z^{-1}(1+z^{-2})\ (1-z^{-1})\,(1+z^{-1})}{4(1-z^{-1})^2}$$

$$=\lim_{z\to 1}\frac{z^{-1}(1+z^{-2})\ (1+z^{-1})}{4}$$

$$=\lim_{z\to 1}\frac{z^{-1}(1+z^{-2})\ (1+z^{-1})}{4}$$

$$c(\infty)=1$$

29. Ans. (b)

Solution:

$$X(z)=\frac{0.5}{1-2z^{-1}}$$

Now there are two cases.

If $x(n)$ is a causal sequence, i.e. for right sided sequence

ROC $\Rightarrow |2z^{-1}|<1$

$\Rightarrow\ |z|>2$

In this case, R.O.C does not include unit circle

If $x(n)$ is a non-causal sequence, then

ROC $\Rightarrow\ |2z^{-1}|>1$

$\Rightarrow |z|<2$

In this case, R.O.C includes unit circle.

Hence, $x(n)$ must be a non-causal sequence.

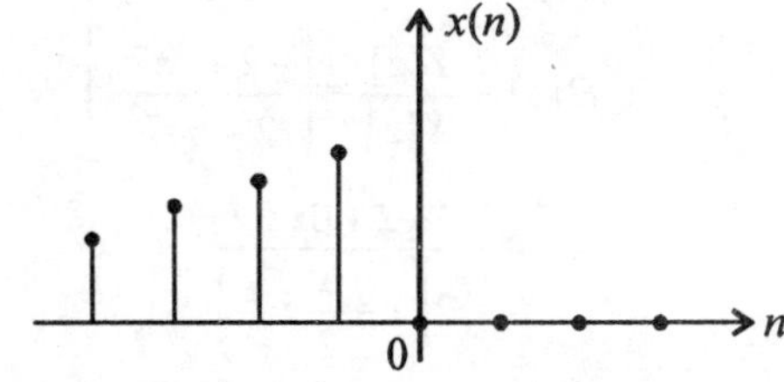

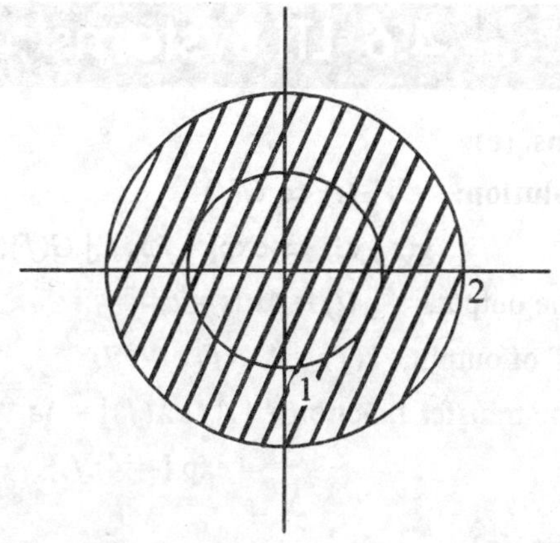

In present case

$x(n)=-K.2^n\,u(-n-1)$, where K is a constant

$$x[n]\ =\ -\frac{1}{2}2^n\ U[-n-1]$$

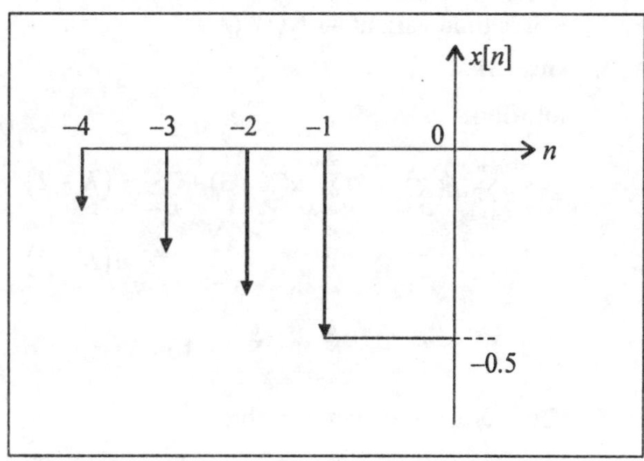

$\Rightarrow x(0)=0$

30. Ans. (c)

Solution:

Given $X(z)=2^{\frac{1}{z}}$

To find $x(n)$ expand the exponential in power series:

$$X(z)=1+\frac{1}{z}+\frac{1}{\angle 2}\frac{1}{z}+\frac{1}{\angle 3}\frac{1}{z^3}+\ldots+\frac{1}{\angle n}\frac{1}{z^n}+\ldots$$

$$=\sum_{n=0}^{\infty}\frac{1}{\angle n}z^{-n}=\sum_{n=0}^{\infty}x(n)z^{-n}$$

$$\therefore\ x(n)=\frac{1}{n!},\ n\geq 0\ \text{ or }\ x(n)=\frac{1}{n!}u(n)$$

4.6 LTI Systems

1. **Ans. (c)**

Solution: $V_g(t) \Leftrightarrow G(f)$

$g(t-t_o) \Leftrightarrow \exp[-j\omega_o t_o]\, G(f);\quad \omega = 2\pi f$

The output, $y(t) = 4x(t-2)$

FT of output, $Y(f) = 4X(f)\, e^{-j4\pi f}$

The transfer function $= [Y(f)/X(f)] = 4e^{-j4\pi f}$

$= 4\exp[-j4\pi f]$

2. **Ans. (a)**

Solution:

- $y(n) = x(n) \Rightarrow$ Thus, it is linear $\rightarrow P$
- $y(n) = x(n+1) \Rightarrow$ Thus it is not causal $\rightarrow$ not R
- If $x(n)$ is bounded, output also going to be bounded $\Rightarrow$ Thus it is stable system $\rightarrow S$
- $y(n)$ has different value for the same time shift in input, i.e. $y(n)$ is not shifted by same amount as $x(n) \Rightarrow$ Thus it is time variant $\rightarrow$ NOT Q

3. **Ans. (a)**

Solution:

$$\sum_{L=-\infty}^{+\infty} h(K) = \sum_{K=-3}^{\infty} u(K+3) + \sum_{K=-2}^{\infty} u(K-2) - 2\sum_{K=7}^{\infty} u(K-7)$$

$$= \sum_{K=-3}^{6} 1 + \sum_{K=2}^{6} 1 = 10+5 = 15 < \infty$$

Therefore, the system is stable

Also, the three system described by $h[n]$ is non-causal, because the response depends upon the future value of the input signal, namely $u[n+3]$.

Second Method

$$h(n) = u(n+3) + u(n-2) - 2u(n-7)$$

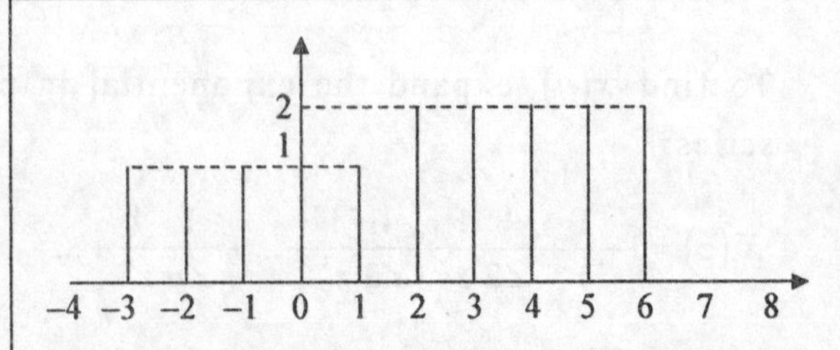

$$\sum_{-\infty}^{\infty} h(n) = 1+1+1+1+1+1+2+2+2+2+2 = 15$$

We can see that $\sum_{-\infty}^{\infty} h(n)$ is finite hence system is stable.

Since $h(n)$ depends on $u(n+3)$, i.e. future values, so it is not a causal system.

4. **Ans. (c)**

Solution: $H(z) = \dfrac{Y(z)}{X(z)} = \dfrac{b_0 + b_1 Z^{-1} + b_2 Z^{-2}}{1 + a_1 Z^{-1} + a_2 Z^{-2}}$ ... (1)

$$q[n+1] = \begin{vmatrix} -a_1 & 1 \\ -a_2 & 0 \end{vmatrix} q[n] + \begin{vmatrix} b_1 & -a_1 b_0 \\ b_2 & -a_2 b_0 \end{vmatrix} x[n] \quad \text{... (2)}$$

$$y[n] = [1\ 0]\, q[n] + b_0\, x[n]$$

Given $2Y[n] = \alpha Y[n-2] - 2x[n] + \beta x[n-1]$

$$y[n] = \frac{\alpha}{2} y[n-2] - x[n] + \frac{\beta}{2} \times [n-1]$$

z transforming.

$$Y[z]\left(1 - \left(\frac{\alpha}{2}\right) z^{-2}\right) = X[z]\left[-1 + \left(\frac{\beta}{2}\right) z^{-1}\right]$$

$$H(z) = \frac{Y(z)}{X(z)} = \frac{-1 + \left(\frac{\beta}{2}\right) z^{-1}}{1 - \left(\frac{\alpha}{2}\right) z^{-2}}$$

Comparing with (1)

$$b_0 = -1,\ b_1 = \left(\frac{\beta}{2}\right),\ a_1 = 0,\ a_2 = -\left(\frac{\alpha}{2}\right)$$

From (2)

$$= q[n+1] = \begin{pmatrix} 0 & 1 \\ \alpha/2 & 0 \end{pmatrix} q[n] + \begin{pmatrix} (\beta/2) \\ -(\alpha/2) \end{pmatrix} x[n]$$

$$c(\lambda) = [\lambda_1 - A] = \begin{vmatrix} \lambda & -1 \\ (\alpha/2) & \lambda \end{vmatrix}$$

Eigen value $= \lambda^2 - \left(\frac{\alpha}{2}\right) = 0 \Rightarrow \lambda^2 = \left(\frac{\alpha}{2}\right)$

$$\Rightarrow \quad \lambda = \pm\sqrt{\frac{\alpha}{2}}$$

For stability $\lambda < \pm 1$, $\therefore\ \alpha < 2$

β does not figure in stability

Method 2

Given system differential equation is

$$2y(n) = \alpha y(n-2) - 2x(n) + \beta x(n-1)$$

Taking Z-transform.

$$2Y[z] = \alpha z^{-2} Y[z] - 2X[z] + \beta z^{-1} X[z]$$

$$Y[z][2 = \alpha z^{-2}] = X[z][-2 + \beta z^{-1}]$$

$$H[z] = \frac{Y[z]}{X[z]} = \left[\frac{-2 + \beta z^{-1}}{2 - \alpha z^{-2}}\right]$$

$$= \frac{-2 + \beta z^{-1}}{2\left(1 - \frac{\alpha}{2} z^{-2}\right)}$$

For stable system ROC should include unit circle $H[z]$ i.e. its pole must lie in the unity circle.

$$1-\frac{\alpha}{2}z^{-2} > 0$$

or $$1 > \frac{\alpha}{2z^{-2}}$$

or $$z^2 > \frac{\alpha}{2}$$

or $$1 > \frac{\alpha}{2}$$

Thus, $\alpha < 2$

β does not appear in the analysis.

5. Ans. (b)

Solution: For causal system,

$$h(t) = 0, \text{ for } t < 0$$

6. Ans. (c)

Solution:

Given output function is

$$y(n) = \left(\sin\frac{5}{6}\pi n\right)x(n)$$

The system is linear, since the terms $y(n)$ and $x(n)$ have equal degree, i.e. 1.

The system is stable, since input $\sin\left(\frac{5}{6}\pi n\right)$ is limited between -1 and 1. So if $x(n)$ is bounded, $y(n)$ is also bounded.

The system is not invertible since the mapping is not one into one.

7. Ans. (c)

Solution:

Z-transform of a discrete all pass system is given as

$$H(z) = \frac{z^{-1} - z_0^*}{1 - z_0 z^{-1}}$$

It has a pole at z_0 and a zero at $1/z_0^*$.

Given system has a pole at

$$z_0 = 2\angle 30$$

$$= 2\left[\cos 30° + j\sin 30°\right]$$

$$= 2\frac{(\sqrt{3}+j)}{2} = (\sqrt{3}+j)$$

$|z_0| = 2$, i.e. $|z_0| > 1$

The system is stable only if the unit circle lies inside the R.O.C. For this R.O.C. must lie interior to $|z| = |z_0|$ circle.

This is only possible when $h(n)$ is defined for $n < 0$, i.e. the system is stable only if it is anti-causal.

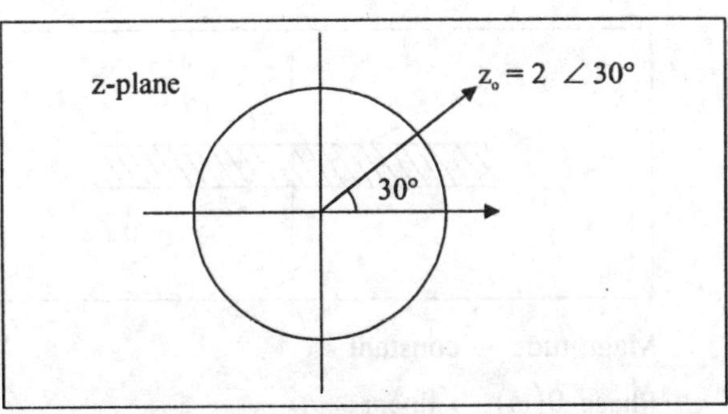

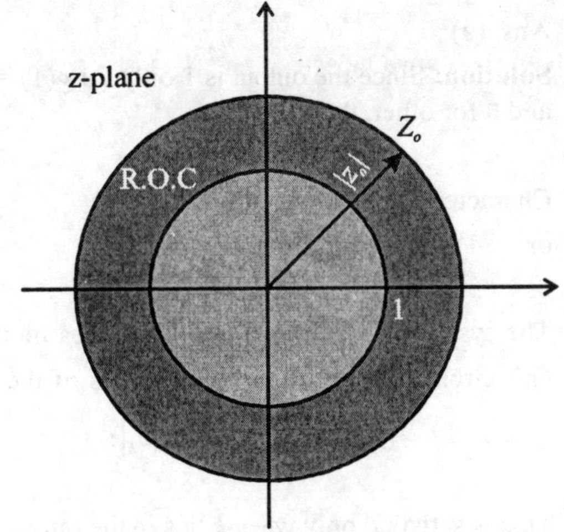

System is stable if $|z| < 1$ and for this it is anti-causal.

8. Ans. (d)

Solution:

$$y(t) = e^{-|x(t)|}$$

When $|x(t)| \to \infty$; then $y(t) \to 0$

when $|x(t)| \to 0$; then $y(t) \to 1$

i.e. $y(t)$ is bounded irrespective of $x(t)$

* Therefore, e^{-x} is always convergent even when x is not bounded

9. Ans. (b)

Solution: The running integrator $\int_{-\infty}^{t} x(\tau)\,d\tau = 0$ for every causal system.

As causal system have no memory and initial values, is zero the output is followed by input so, $y(t)$ will always be bounded if this function is causal bounded system

10 Ans. (c)

Solution:

For Distortion less filter

$$Y(j\omega) = A\,(j\omega)\,.\,X(j\omega)$$

$$H(j\omega) = A\,e^{-j\omega}$$

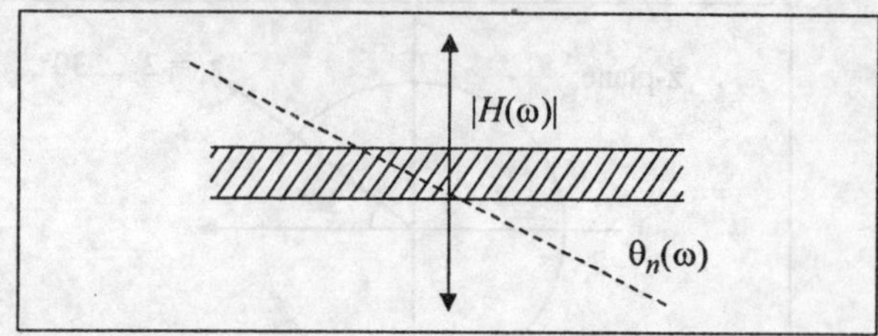

Magnitude → constant

Phase $\theta(\omega)$ → linear

11. Ans. (a)

Solution: Since the output is 1 only for $g(1) = g(2) = 1$ and 0 for other, then

$$G(z) = z^{-1} + z^{-2}$$

Characteristic equation, $1 - KG(z) = 0$

or $\quad 1 - K(z^{-1} + z^{-2}) = 0$

or $\quad z^2 - Kz - K = 0$

The system is stable only when its poles lie inside the unit circle, i.e. if z_1 and z_2 are roots of the equation $z^2 - kz - k = 0$, then $|z_1| \le 1$ and $|z_2| \le 1$

This is satisfied only when k lies in the range $\left(-1, \frac{1}{2}\right)$

If $k = -1$; then the roots are $z_{1,2} = -\frac{1}{2} + j\frac{\sqrt{3}}{2}$.

If $k = \frac{1}{2}$; then the roots are $z_{1,2} = 1, -\frac{1}{2}$

If $k = 1$; then the roots are $z_{1,2} = \frac{1 \pm \sqrt{5}}{2}$; one of the roots lie outside the unit circle.

12. Ans. (b)

Solution: $\quad X(z) = 1 - 3z^{-1}$; $Y(z) = 1 + 2z^{-2}$

$$h[n] = x(n-1) * y(n)$$

$$ZT[x(n-1)] = z^{-1} \quad X(z) = z^{-1}(1 - 3z^{-1})$$

$$= z^{-1} - 3z^{-2}$$

$\therefore$ $$H(z) = (z^{-1} - 3z^{-2})(1 + 2z^{-2})$$

$$= z^{-1} - 3z^{-2} + 2z^{-3} - 6z^{-4}$$

For $\quad u[n] = \delta(n-1)$

$$U(z) = z^{-1}$$

$$Y(z) = H(z) \times U(z)$$

$$= (z^{-1} - 3z^{-2} + 2z^{-3} - 6z^{-4})(z^{-1})$$

$$= z^{-2} - 3z^{-3} + 2z^{-4} - 6z^{-5}$$

$$Y[n] = (\underset{\uparrow}{0}, 0, 1, -3, 2, -6)$$

$$h[n] = (0, 1, -3, 2, -6)$$

$$u[n] = \delta(n-1)$$

$$Y[n] = h[n] * u[n]$$

$$= h[n-1]$$

$$Y[n] = (\underset{\uparrow}{0}, 0, 1, -3, 2, -6)$$

13 Ans. (c)

Solution: Causal system are those in which changes in the output are dependent only on the present and the past values of the input and/or previous output values, and are not dependent on future input values. For a linear time invariant system, the causality condition is given by $h[n] = 0$ for $n < 0$

Only option (c) satisfies the condition of causality.

14. Ans. (d)

Solution: $\quad h(t) = e^{+\alpha t}u(t) + e^{\beta t}u(-t)$

For $h(t)$ to be stable $\int |h(t)| dt < \infty$.

$\therefore$ $$H(s) = \frac{1}{s-\alpha} - \frac{1}{s+\beta}$$

$$= \frac{(\beta + \alpha)}{(s-\alpha)(s+\beta)}$$

For the system to be stable, poles must be –ve

⇒ α is negative and β is positive

It will happen only when α is negative and β is positive.

15. Ans. (c)

Solution: $\quad H(s) = \frac{K(s+1)}{(s+2)(s+4)}$

input → $u(t)$ → $\frac{1}{s}$

at steaty state, i.e. $t = \infty$, output = 1

$$V_o(s) = \frac{K(s+1)}{s(s+2)(s+4)}$$

$$V(\infty) = \underset{s \to 0}{\text{Lt}}\ s \cdot V_o(s)$$

$$= \underset{s \to 0}{\text{Lt}} \frac{K(s+1)s}{s(s+2)(s+4)}$$

$$= \frac{K(1)}{2 \times 4} = \frac{K}{8} = 1 \quad \text{(given)}$$

$$K = 8$$

$$h(t) = \text{LT}^{-1}\left[\frac{8(s+1)}{(s+2)(s+4)}\right]$$

$$= LT^{-1}\left[\frac{8(-4+1)}{(-4+2)(s+4)}+\frac{8(-2+1)}{(-2+4)(s+2)}\right]$$

$$= LT^{-1}\left[\frac{8(-3)}{(-2)(s+4)}+\frac{8(-1)}{(+2)(s+2)}\right]$$

$$= LT^{-1}\left[\frac{12}{(s+4)}-\frac{4}{(s+4)}\right]$$

$$= [12\, e^{-4t} - 4\, e^{-2t}]\, u(t)$$

16. Ans. (d)

Solution: Use convolution result to get the total number of terms

$$h[n] = [1, -1, 2]$$

$$x[n] = [1, 0, 1]$$

$$y[n] = x[n] * h[n] = \sum_{k=-\infty}^{\infty} x(K)\, h[n-K]$$

$$y(2) = \sum_{k=-\infty}^{\infty} x(K)\, h(2-K)$$

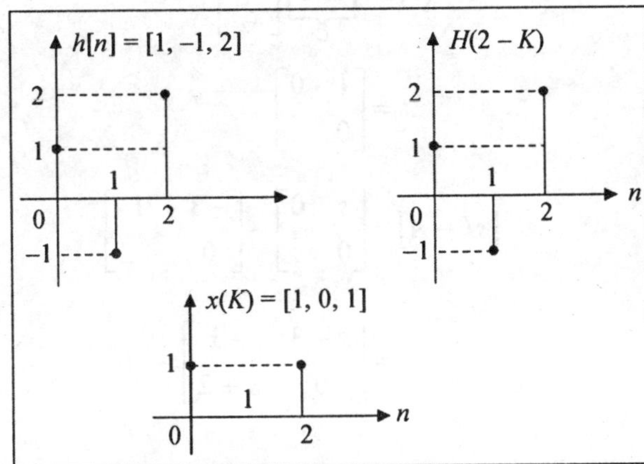

$$y(2) = 2 \times 1 + (-1) \times 0 + 1 \times 1 = 3$$

$$y(-1) = 0, y(0) = 1, y(1) = -1, y(2) = 3,$$

$$y(3) = -1, y(4) = 2, y(5) = 0$$

Hence, five non-zero samples.

17. Ans. (a)

Solution: R_1: $y(t) = t^2 x(t)$

Linear time variant

$$R_2 : y(t) = t|\, x(t)\, |$$

Non, linear non-time invariant

$$R_3 : y(t) = |\, x(t)\, |$$

Non linear time invariant

$$R_4 : y(t) = x(t-5)$$

Liner time invariant

18. Ans. (b)

Solution:

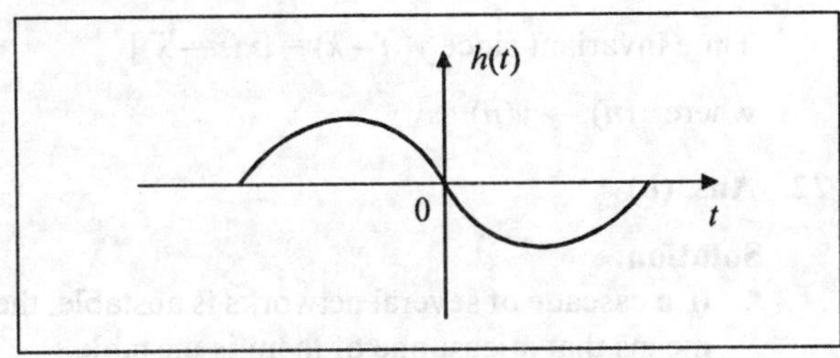

$$y(t) = \int_{-\infty}^{\infty} x(t-\tau)\, h(2\tau) d\tau$$

As, $\lim_{t\to\infty} h(t)$ or $\int_{-\infty}^{\infty} h(t)dt$ is finite for the given $h(t)$

So system is Bounded

So answer can be (b) or (c)

Now system is not causal $h(t)$ is defined for $t < 0$ so correct answer answer is (b)

19. Ans. (c)

Solution: Given that:

$$H(z) = \frac{2-\frac{3}{4}z^{-1}}{1-\frac{3}{4}z^{-1}+\frac{1}{8}z^{-2}}$$

$$H(z) = \frac{\left(1-\frac{1}{4}z^{-1}\right)+\left(1-\frac{1}{2}z^{-1}\right)}{\left(1-\frac{1}{2}z^{-1}\right)\left(1-\frac{1}{4}z^{-1}\right)}$$

By partial fraction

$$H(z) = \frac{1}{\left(1-\frac{1}{2}z^{-1}\right)}+\frac{1}{\left(1-\frac{1}{4}z^{-1}\right)}$$

S_1: System will be causal for $|z| > \frac{1}{2}$; $|z| = \infty$

S_1: System will be stable for $|z| > \frac{1}{4}; |z| = 1$

System will be neither causal nor stable for ROC

$$\frac{1}{4} < |z| < \frac{1}{2}$$

S_1 true, S_2 false (System is stable nor causal), S_3 true.

20. Ans. (d)

Solution: LTI system holds superposition theorem causal system means $h(t) = 0$ for $t < 0$.

21. Ans. (c)

Solution: Non-causal because the signal is defined for $t < 0$

Time invariant since $y(n-k) = [x(n-k)]$

where $x(n) \rightarrow y(n)$

22. Ans. (b)

Solution:

* If a cascade of several networks is unstable, then this means that atleast one of them is unstable.
* If the overall cascaded network is causal, then at least one of them is causal.

For example:

Let the impulse response of three individual systems be

$h_1(t) = \delta(t+1)$; non-causal but stable

$h_2(t) = \delta(t+2)$; non-causal but stable

$h_3(t) = u(t-4)$; causal but unstable

Overall impulse response

$h(t) = h_1(t) * h_2(t) * h_3(t)$

or, $h(t) = u(t-1)$; causal but unstable

23. Ans. (c)

Solution:

$$\frac{dx_1(t)}{dt} = -3X_1(t) + x_2(t) + 2u(t)$$

$$\Rightarrow \quad sX_1(s) = -3X_1(s) + X_2(s) + 2U(s) \qquad ...(1)$$

$$\frac{dx_2(t)}{dt} = -2x_2(t) + u(t)$$

$$\Rightarrow \quad sX_1(s) = -2X_2(s) + U(s)$$

$$(s+2)X_2(s) = U(s)$$

$$X_2(s) = \frac{U(s)}{(s+2)}$$

From [1], $(s+3)X_1(s) = 2U(s) + \dfrac{U(s)}{s(s+2)}$

$$X_1(s) = \frac{2s+5}{(s+2)(s+3)}$$

$$y(t) = x_1(t)$$

$$\Rightarrow H(s) = \frac{Y(s)}{X(s)}$$

$$= \frac{X_1(s)}{U(s)} = \frac{2s+5}{(s+2)(s+3)}$$

$$= \frac{2s+5}{s^2+5s+6}$$

Second Method

$$x_i'(t) = -3x_1(t) + x_2(t) + 2u(t)$$

$$x_2'(t) = -2x_2(t) + u(t)$$

$$y(t) = x_1(t)$$

$$\begin{bmatrix} x_1' \\ x_2' \end{bmatrix} = \underbrace{\begin{bmatrix} -3 & 1 \\ 0 & -2 \end{bmatrix}}_{A} \begin{bmatrix} x_1 \\ x_2 \end{bmatrix} + \underbrace{\begin{bmatrix} 2 \\ 1 \end{bmatrix}}_{B} u(t)$$

$$y(t) = \underbrace{\begin{bmatrix} 1 & 0 \end{bmatrix}}_{C} \begin{bmatrix} x_1 \\ x_2 \end{bmatrix}$$

$$I = \begin{bmatrix} 1 & 0 \\ 0 & 1 \end{bmatrix}$$

$$[sI - A] = \begin{bmatrix} s & 0 \\ 0 & s \end{bmatrix} - \begin{bmatrix} -3 & 1 \\ 0 & -2 \end{bmatrix}$$

$$= \begin{bmatrix} s+3 & -1 \\ 0 & s+2 \end{bmatrix}$$

$$[sI - A]^{-1} = \frac{1}{(s+3)(s+2)} \begin{bmatrix} s+2 & 1 \\ 0 & s+3 \end{bmatrix}$$

$$= \begin{bmatrix} \frac{1}{s+3} & \frac{1}{(s+3)(s+2)} \\ 0 & \frac{1}{s+2} \end{bmatrix}$$

$$C = [1 \quad 0]$$

$\therefore$ The required transfer function

$$H(s) = C[sI - A]^{-1} B$$

$$= \begin{bmatrix} 1 & 0 \end{bmatrix} \begin{bmatrix} \frac{1}{s+3} & \frac{1}{(s+3)(s+2)} \\ 0 & \frac{1}{(s+2)} \end{bmatrix} \begin{bmatrix} 2 \\ 1 \end{bmatrix}$$

$$= \begin{bmatrix} \frac{1}{s+3} & \frac{1}{(s+3)(s+2)} \end{bmatrix} \begin{bmatrix} 2 \\ 1 \end{bmatrix}$$

$$= \begin{bmatrix} \frac{2}{s+3} + \frac{2}{(s+3)(s+2)} \end{bmatrix}$$

$$= \frac{2(s+2)+1}{(s+3)(s+2)}$$

$$= \frac{2s+4+1}{s^2+5s+6}$$

$$= \frac{2s+5}{s^2+5s+6}$$

24. Ans. (d)

Solution: $y[n] = F\,x(n)$

$\Rightarrow y(n-K) = F(x(n-K))$

Then it is time invariant

Apply this property in options

(a) $y[n] = n\,x[n]$

$y[n-x] = [n-x]x[n-x]$

$Fx[n] = n\,x[n]$

$F\,x(n-K) = n\,x(n-K)$

$y(n-K) \neq F\,x(n-K)$

So it is time variant

(b) $y[n] = x[3n] = F\,x(n)$

$y(n-K) = x[3(n-K)]$

$= x[3n-3K]$

$Fx(n-K) = x[3n-K]$

$y(n-K) \neq F\,x(n-K)$

So it is time variant

(c) $y[n] = x[-n] = Fx(n)$

$y(n-K) = x[-(n-K)] = x(-n+K)$

$Fx(n-K) = x[-n-K)]$

$y(n-K) \neq Fx(n-K)$

So it is time variant.

(d) $y[n] = x[n-3] = Fx(n)$

$y(n-K) = x[n-K-3]$

$Fx(n-K) = x[n-K-3]$

$y(n-K) = Fx(n-K)$

So it is time invariant.

25. Ans. (d)

Solution:

Given signal is

$$y(t) = \sum_{k=-\infty}^{c} x(kT)\delta(t-kT)$$

This is the representation of a signal $x(t)$ in weighted and sum form and it obeys the principle of superposition and homogenity. So this is linear as for $x(t-t_0)$,

$y'(t) = \Sigma x(kT-t_0)\delta(t-kT)$

and $y(t-t_0) = \Sigma x(kT)\delta(t-t_0-kT)$,

therefore $y(t-t_0) \neq y'(t)$,

so time varying.

26. Ans. (c)

Solution:

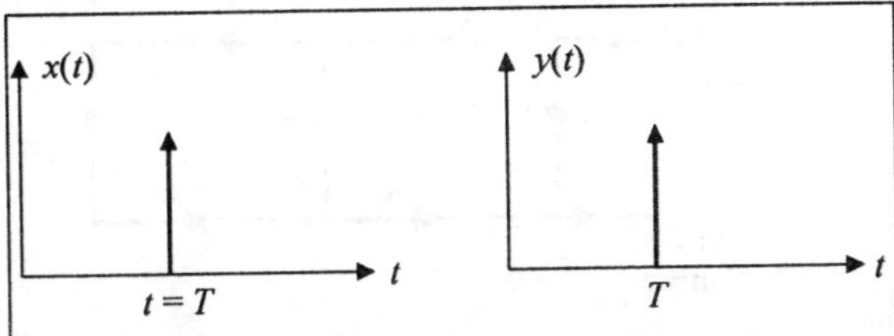

Causality sequence is defined as

$x(t) = 0$ for $t < 0$

$h(n) = 0$ for $n < 0$

Since value of the signal start at T and its response is zero for all interval of time below T. Hence it is a causal system.

27. Ans. (d)

Solution: For time invariance system input and output should not be related by a function of time.

Hence $h_1(t)$, $h_2(t)$ and $h_4(t)$ are time invariant.

For causal system, output depends only on present and past value of input

Hence $h_1(t)$, $h_2(t)$, $h_3(t)$, and $h_4(t)$ are causal.

For stability, impulse response must be finite

$$|h_1| = \int_{-\infty}^{\infty} 1 \cdot dt = \infty$$

$$|h_1| = \int_{0}^{\infty} e^{-3t} dt = \frac{1}{3}$$

Thus, $h_4(t)$ is time invariant, causal and stable.

By Experience Method

From the observation of the impulse responses given, $h_4(t) = e^{-3t}u(t)$, is the familiar expression for the voltage across the C of a series RC circuit when the input applied is an impulse of voltage.

RC circuit system is linear and time-invariant, causal and stable.

28. Ans. (d)

Solution:

Given, $h[n] = (1, 2, 2)$

Result of auto-correlation function

$$f(n) = h[n] * h(n) = \{\underset{\uparrow}{1}, 4, 8, 8, 4\}$$

Result of auto-correlation function

$$g(n) = h[n] * h[-n] = \{2, 6, \underset{\uparrow}{9}, 6, 2\}$$

$\therefore g[n]$ is non-causal and having maximum value 9.0

29. Ans. (d)

Solution: Given figure is

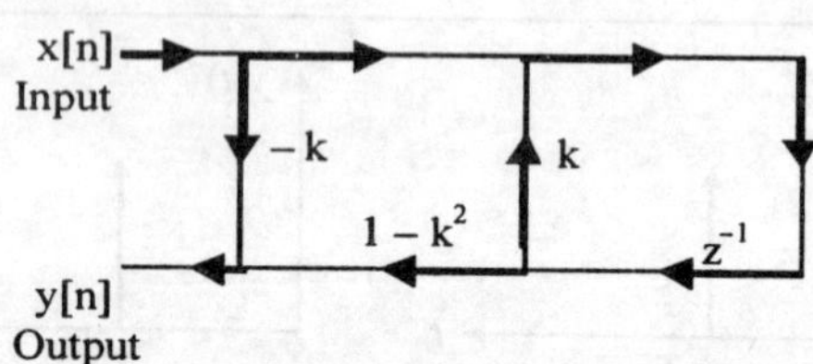

Using Mason's gain formula, or by block diagram algebra, we can calculate

$$\frac{Y(z)}{X(z)} = -k + \frac{(1-k^2)\, z^{-1}}{1-kz^{-1}} = \frac{-k + z^{-1}}{1-kz^{-1}}$$

30. Ans. (b)

Solution: Multiplication in time domain corresponds to convolution in time domain

$V(s) = Z(s) \times I(s)$

$$V(t) = \int_0^t i(\tau)\, z(t-\tau)\, d\tau$$

31. Ans. (b)

Solution: $x_1(t) = u(t) - u(t-1)$

$$X_1(s) = \frac{1}{s}[1 - e^{-s}]$$

$$x_2(t) = u(t) - u(t-2)$$

$$X_2(s) = \frac{1}{s}[1 - e^{-2s}]$$

$$x(t) = x_1(t) * x_2(t) \rightarrow X_1(s)\, X_2(s)$$

$$= \frac{1}{s}[1 - e^{-s}] \times \frac{1}{s}[1 - e^{-2s}]$$

$$= \frac{1}{s^2}[1 - e^{-s} - e^{-2s} + e^{-3s}]$$

$$x(t) = r(t) - r(t-1) - r(t-2) + r(t-3)$$

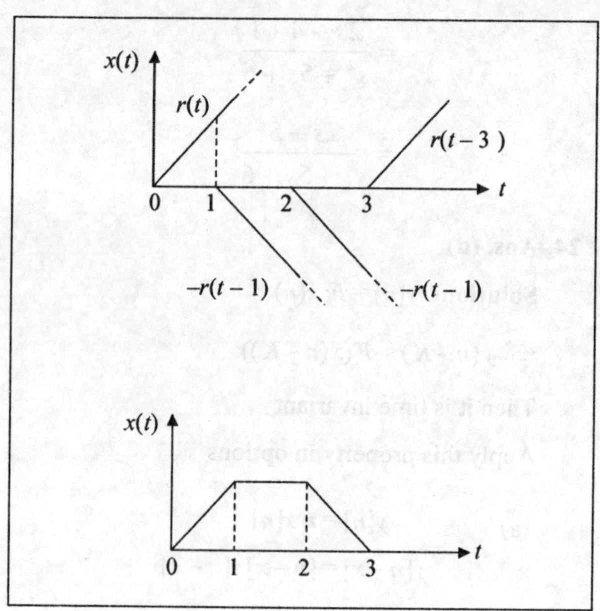

Second Method

Let $x(t) = u(t) - u(t-1)$, $x_2(t) = u(t) - u(t-2)$ then $y(t) = x_1(t) * x_2(t)$

$x(t)$ and $g(t)$ and $\frac{dx(t)}{dt}$ are shown in figures 1, 2 and 3.

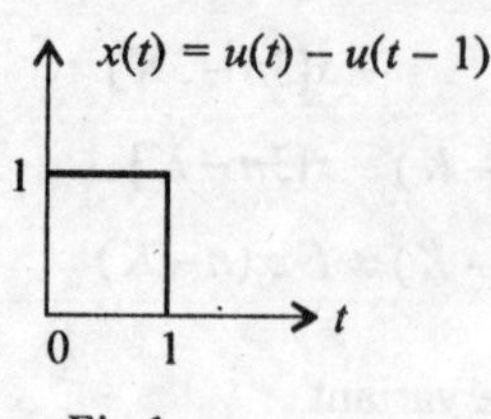

Fig.1

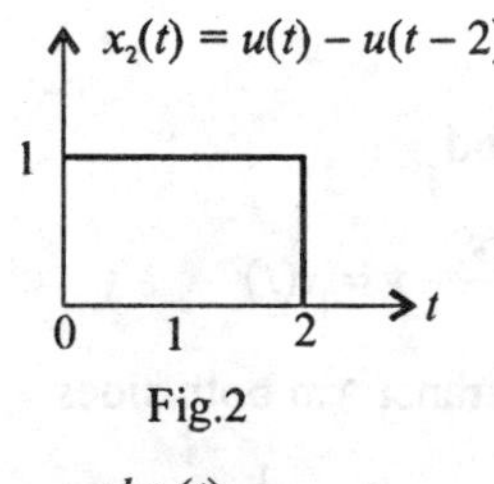

Fig.2

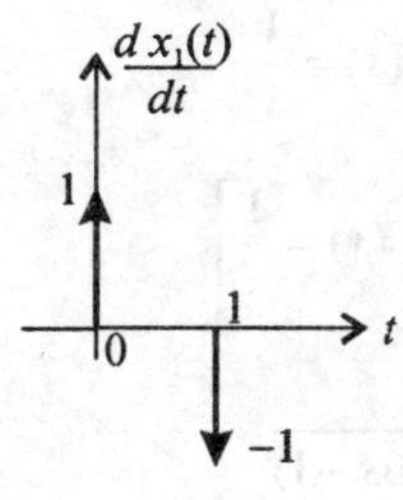

Fig.3

Let $y_1(t) = \dfrac{d\,x_1(t)}{dt} * x_2(t)$,

Then $y_1(t)$ is shown in Figs. 4 and 5

$$y(t) = x_1(t) * x_2(t) = \int_{t=-\infty}^{t} y_1(t)\,dt$$

$y(t)$ is shown in Fig. 6

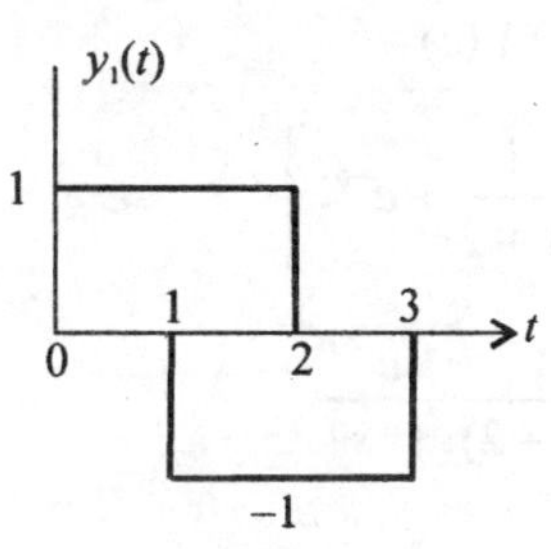

Fig.4

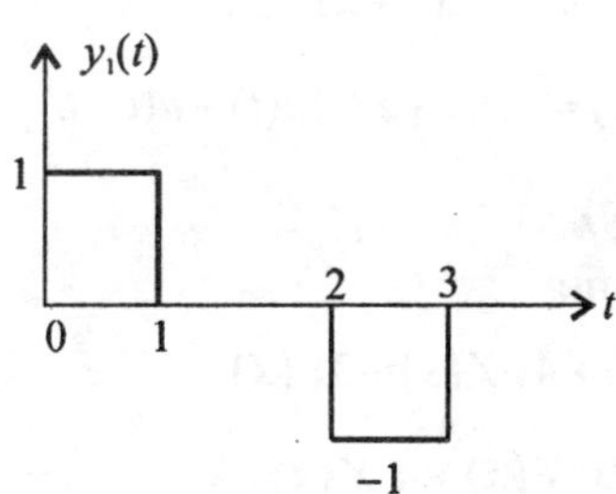

Fig.5

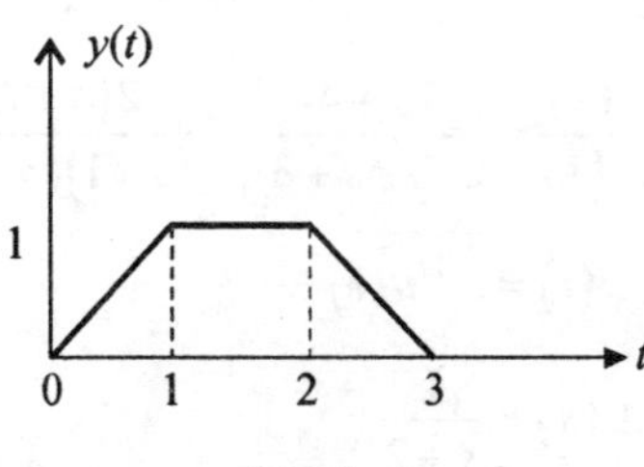

Fig.6

32. Ans. (c)

Solution: $H(s) = \dfrac{F_o(s)}{F_i(s)}$

33. Ans. (a)

Solution: A transfer function which has one or more zero in the right half S-plane is known as non minimum-phase transfer function.

34. Ans. (b)

Solution: Zero order system transfer function is represented by $TF = u(t) - u(t-T)$

$$H(s) = [\frac{1}{s} - \frac{1}{s}\,e^{-sT} = \frac{1}{s}[1 - e^{-sT}]$$

Second Method

Analysis of Sampler and Zero-Order Hold

Figure below shows a sampler with Zero-Order Hold (ZOH). As ZOH holds the input signal value for a period T, it means that for a short duration (D) input pulse, it produces an output pulse of duration T, the sampling period. This is illustrated in figure.

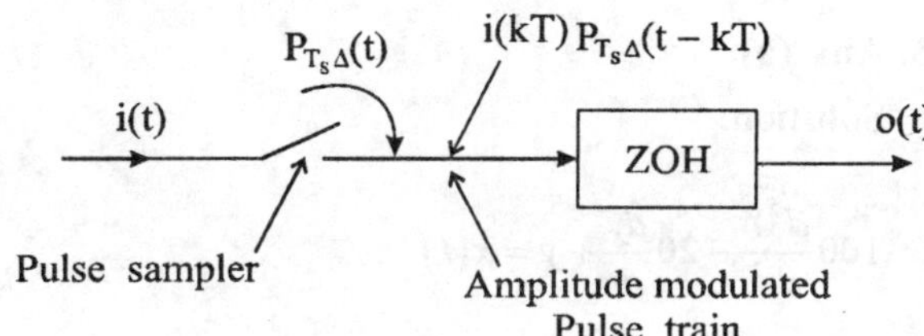

Fig. Pulse sampler with ZOH.

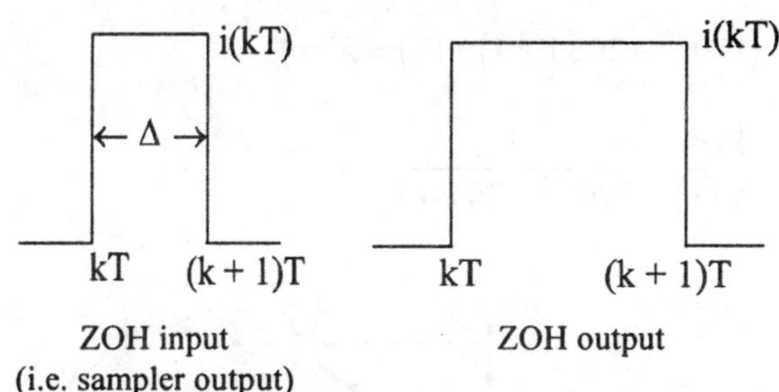

Fig. ZOH input and output pulses.

The ZOH output pulse appearing at kT instant can be expressed as

$$i(kT)\left[u(t-kT) - u(t-\overline{k+1T})\right]$$

We can therefore write the ZOH output as

$$o(t) = \sum_{k=0}^{\infty} i(kT)\left[u(t-kT) - u(t-\overline{k+1T})\right] \quad \text{...(1)}$$

Taking the Laplace transform, we have

$$O(s) = \sum_{k=0}^{\infty} i(kT)\left[\frac{e^{-skT} - e^{-s(k+1)T}}{s}\right]$$

$$= \left(\frac{1-e^{-sT}}{s}\right)\sum_{k=0}^{\infty} i(kT)e^{-skT}$$

$$L^{-1}\left[\sum_{k=0}^{\infty} i(kT)e^{-skT}\right]$$

$$= \sum_{k=0}^{\infty} i(kT)\,\delta(t-kT) = i(t)\,\delta_T(t)\ ..(2)$$

We can give a new input-output interpretation to above equation. Taking the inverse Laplace transform of the infinite sum.

$$L^{-1}\left[\sum_{k=0}^{\infty} i(kT)e^{-skT}\right]$$

$$= \sum_{k=0}^{\infty} i(kT)\,\delta(t-kT) = i(t)\,\delta_T(t)\ ...(3)$$

Thus the output $o(t)$ of pulse sampler and ZOH can be produced by impulse sampled $i(t)$ when pass through a transfer function.

$$G_o(s) = \frac{1-e^{-sT}}{s} \quad ...(4)$$

35. Ans. (a)

Solution:

$$100\frac{d^2Y}{dt^2} - 20\frac{dy}{dt} + y = x(t)$$

$$100s^2Y(s) - 20\,sY(s) + Y(s) = X(s)$$

$$\left(100s^2 - 20\,sY + 1\right)Y(s) = X(s)$$

$$\frac{Y(s)}{X(s)} = \frac{1}{100\,s^2 - 20\,s + 1}$$

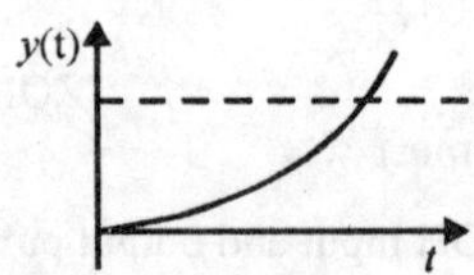

$$= \frac{1}{100\left(s^2 - \frac{1}{5}s + \frac{1}{100}\right)}$$

$$\omega_n = \frac{1}{10}$$

$$\delta_1 = -1;\ 2\delta_1\omega_n = \frac{-1}{5}$$

Initially relaxed

$$Y'(0) = 0$$

$$Y''(0) = 0$$

Second Method

$$\frac{100d^2y}{dt^2} - \frac{20dy}{dt} + y = x(t)$$

Take Laplace transform both sides

$$(100s^2 - 20s + 1)Y(s) = \frac{1}{s}$$

$$\left[\because x(t) = u(t) \Rightarrow X(s) = \frac{1}{s}\right]$$

$$Y(s) = \frac{1}{s(100s^2 - 20s + 1)}$$

So we have poles with positive real part $\Rightarrow$ system is unstable.

36. Ans. (d)

Solution:

$$x(t) = e^{-2t} + \delta(t-6)$$

$$h(t) = u(t)$$

$$X(s) = \frac{1}{s+2} + e^{-6s} \text{ and } H(s) = \frac{1}{s}$$

$$Y(s) = H(s)X(s)$$

$$= \frac{1}{s}\left[\frac{1}{s+2} + e^{-6s}\right]$$

$$= \frac{1}{s(s+2)} + \frac{e^{-6s}}{s}$$

$$= \frac{1}{2}\frac{1}{s} - \frac{1}{2(s+2)} + \frac{e^{-6s}}{s}$$

or, $y(t) = 0.5(1 - e^{-2t})u(t) + u(t-6)$

37. Ans. (b)

Solution:

$$s^2Y(s) + 4s\,Y(s) + 3Y(s)$$

$$= 2s\,X(s) + 4\,X(s)$$

$$Y(s)\left[s^2 + 4s + 3\right] = X(s)[2s + 4]$$

$$\frac{Y(s)}{X(s)} = \frac{2s+4}{s^2+4s+3} = \frac{2(s+2)}{(s+1)(s+3)}$$

if $\quad x(t) = e^{-2t}u(t)$

$$X(s) = \frac{1}{s+2}$$

Then $Y(s)=\frac{2(s+2)}{(s+1)(s+3)}\times\frac{1}{s+2}$

$$=\frac{2}{(s+1)(s+3)}$$

$$=\frac{\left.\frac{2}{(s+3)}\right|_{s=-1}}{s+1}+\frac{\left.\frac{2}{(s+1)}\right|_{s=-3}}{s+3}$$

$$=\frac{1}{s+1}-\frac{1}{s+3}$$

$$=\left[e^{-t}-e^{-3t}\right]u(t)$$

38. Ans. (d)

Solution:

The input $x(t)$ and output $y(t)$ related as

$$y=\int_{-\infty}^{t}x(\tau)\cos(3\tau)d\tau$$

Delayed output

$$y(t-t_o)=\int_{-\infty}^{t-t_o}x(\tau)\cos(3\tau)d\tau$$

Output for delayed input:

$y'(t)$ for input $x(t-t_o)$ is

$$y'(t)=\int_{-\infty}^{t}x(\tau-t_o)\cos 3\tau\, d\tau$$

$$y'(t)=\int_{-\infty}^{(t-t_o)}x(\tau)\cos 3(\tau+t_o)\, d\tau$$

Since

$y'(t)\neq y(t-t_o)$ so system is not time invariant for input $x(\tau)=\cos(3\tau)$, i.e. bounded input

$$y(t)=\int_{-\infty}^{t}\cos^2(3\tau)\to\infty \text{ as } t\to\infty$$

So for bounded input, output is not bounded therefore system in not stable.

39. Ans. (a)

Solution: Response for any input $u(t)$ is,

$$y(t)=u(t)*h(t)$$

$$y(t)=\int_{0}^{t}h(t-\tau)\,u(\tau)\,d\tau$$

where impulse response, $h(t)=[d/dt]\,[s(t)]$

and $s(t)$ is the step response

Hence $y(t)=\frac{d}{dt}\int_{0}^{t}s(t-\tau)\,y(\tau)\,d\tau$

* In LTI system: If the input is differentiaed , output also is differentiated
* The property of Convolution

$$\frac{d}{dt}[x(t)]*g(t)=\frac{dx(t)}{dt}*g(t)$$

$$=x(t)*\frac{d}{dt}g(t)$$

40. Ans. (c)

Solution:

$$H(s)=\frac{1}{s+2} \text{ and } Y(s)=\frac{1}{(s+2)^2}$$

$$X(s)=\frac{Y(s)}{H(s)}=\frac{1}{s+2} \text{ and } x(t)=e^{-2t}\,|\,u(t)$$

41. Ans. (b)

Solution: Step response is integral of unit impulse response

$$u(t)=\int_{0}^{t}(-4e^{-t}+6e^{-2t})\,dt$$

$$=\left[\frac{-4}{-1}e^{-t}+\frac{6}{-2}e^{-2t}\right]_0^t=[4e^{-t}-3e^{-2t}]_0^t$$

$$=4e^{-t}-3e^{-2t}-4+3$$

$$=4e^{-t}-3e^{-2t}-1$$

42. Ans. (c)

Solution:

When we apply any signal to LTI system the output amplitude and phase changes that depends on transfer function.

When we apply $e^{-\alpha t}\sin\omega t$ to LTI system output of form $Ke^{-\alpha t}\sin(\omega t+\phi)$, so compare this equation with given equation $Ke^{-\beta t}\sin(vt+\phi)$.

So $\beta=\alpha$, $v=\omega$.

43. Ans. (b)

Solution:

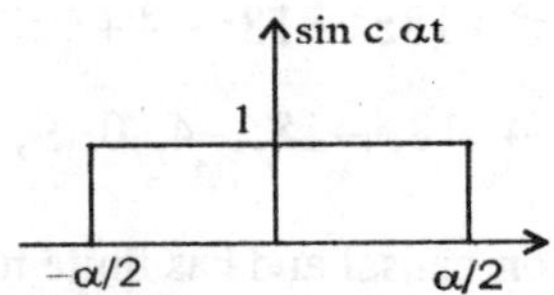

Convolution in time domain is equivalent to multiplication in frequency domain.

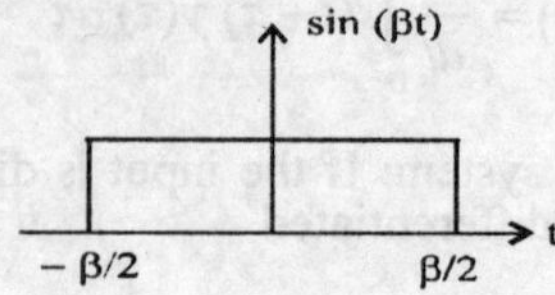

so it will be max (α,β)

44. Ans. (d)

Solution :

x(t) Input → h(t) → y(t) Output

x(t − τ) Input → h(t−τ) → z(t) = ?

From the given data

$$y(t) = x(t).h(t) \quad \text{...(1)}$$

$$x(t-\tau).h(t-\tau) = ?$$

Suppose output = $z(t)$

$$x(t-\tau).h(t-\tau) = z(t)$$

Taking Laplace transform

$$X(s)\overline{e}^{\tau s}.H(s)\overline{e}^{\tau s} = Z(s)$$

$$X(s).H(s)\overline{e}^{2\tau s} = Z(s)$$

From equation (1)

$$H(s)\,X(s) = Y(s)$$

$$Y(s)\,e^{-2\tau s} = Z(s)$$

Taking inverse laplace

$$y(t-2\tau) = z(t)$$

$$\therefore\ x(t-\tau).h(t-\tau) = y(t-2\tau)$$

45. Ans. (a)

Solution:

Given

Signal, $X(z) = 4z^{-3} + 3z^{-1} + 2 - 6z^{-2} + 2z^{3}$

System $H(z) = 3z^{-2} - 2$

Output $Y(z) = X(z)\cdot H(z)$

$$= \left\{4z^{-3} + 3z^{-1} + 2 - 6z^{2} + 2z^{3}\right\}\left\{3z^{-3} - 2\right\}$$

$$= 4z^{3} + 18z^{2} + 18z - 4 + 9z^{-2} + 12z^{-4}$$

$$y(n) = \left\{-4,\ 18,\ -18,\ \underset{\uparrow}{-4},\ 0,\ 9,\ 0,\ 12\right\}$$

$y(n)$ is non-causal and has finite number of elements.

46. Ans. (b)

Solution:

Given input and output

$$y(t) = \int_{-\infty}^{5t} x(\tau)d\tau,\quad t > 0$$

For first input $x_1(\tau)$

$$y_1(t) = \int_{-\infty}^{5t} x_1(\tau)\,d\tau$$

For second input $x_2(\tau)$

$$y_2(t) = \int_{-\infty}^{5t} x_2(\tau)d\tau$$

For input $x_1(\tau) + x_2(\tau)$

$$y_3(t) = \int_{-\infty}^{5t} \left[x_1(\tau) + x_2(\tau)\right]d\tau$$

$$y_3(t) = y_1(\tau) + y_2(\tau)$$

So system is linear.

Causal: Since system depends on the value for $\tau < 0$, so that the system is non-causal.

47. Ans. (c)

Solution:

$$H(s) = \frac{2}{s+1}$$

Input step $X(s) = \frac{1}{s}$

Step response, $Y(s) = \frac{2}{s(s+1)}$

$$\Rightarrow y(t) = 2\left(1 - e^{t}\right)u(t)$$

$$\Rightarrow 2\left(1 - e^{t}\right) = 0.98 \times 2$$

$$\Rightarrow t = 3.9\ \text{sec}$$

48. Ans. (c)

Solution:

$$X(z) = 1 - z^{-1}$$

$$Y(z) = 1 - z^{-4}$$

$$H(z) = \frac{Y(z)}{X(z)}$$

$$= \frac{1 - z^{-4}}{1 - z^{-1}}$$

$$1-z^{-1}\big)\ 1-z^{-4}\left(z^{-3}+z^{-2}+z^{-1}+1\right.$$

$$\frac{z^{-3}-z^{-4}}{1^{-2}-z^{-3}}$$

$$\frac{z^{-2}-z^{-3}}{1-z^{-2}}$$

$$\frac{z^{-1}-z^{-2}}{1-z^{-1}}$$

$$\frac{1-z^{-1}}{X}$$

$$H(s)=1+z^{-1}+z^{-2}+z^{-3}$$

$$h[n]=[1,\ 1,\ 1,\ 1]$$
$\uparrow$

49. Ans. (a)

Solution:

$$z(t)=x(t)*y(t)$$

$$Z(s)=X(s).Y(s)$$

$$=\frac{1}{(s+1)}.\frac{1}{(s+2)}=\frac{1}{(s+1)}-\frac{1}{(s+2)}$$

$$LT^{-1}\{z(s)\}=z(t)\ =e^{-t}-e^{-2t}$$

50. Ans. (d)

Solution:

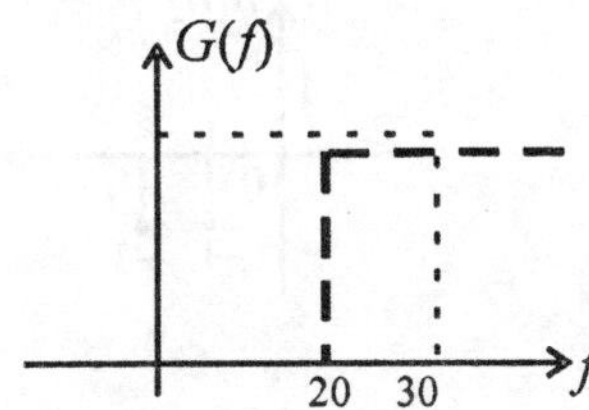

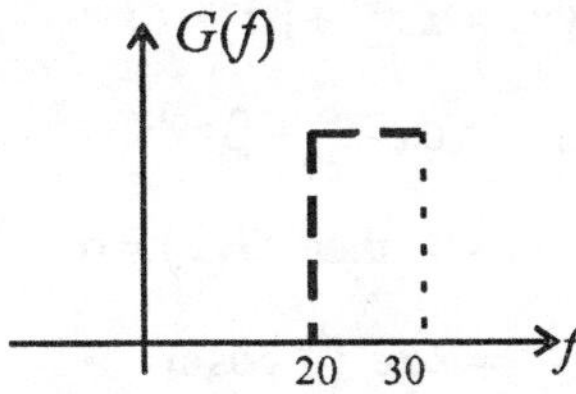

So it is a band pass filter

51. Ans. (c)

Solution:

LT (Impulse response) $=$ Transfer Function

$$=\frac{1}{(s+1)}+\frac{1}{(s+2)}$$

Step response $=LT^{-1}\left[\frac{1}{s(s+1)}+\frac{1}{s(s+2)}\right]$

$$=LT^{-1}\left[\left(\frac{1}{s}-\frac{1}{s+1}\right)+\left(\frac{0.5}{s}-\frac{0.5}{(s+2)}\right)\right]$$

$$=(1-e^{-t}+0.5-0.5e^{-2t})u(t)$$

$$=(1.5-e^{-t}-0.5e^{-2t})u(t)$$

52. Ans.(d)

Solution:

The input $x(t)$ and ouput $y(t)$ related as

$$y=\int_{-\infty}^{t}x(\tau)\cos(3\tau)d\tau$$

Delayed output

$$y(t-t_o)=\int_{-\infty}^{t-t_o}x(\tau)\cos(3\tau)d\tau$$

Output for delayed input:

$y'(t)$ for input $x(t-t_o)$ is

$$y'(t)=\int_{-\infty}^{t}x(\tau-t_o)\cos 3\tau\ d\tau$$

$$y'(t)=\int_{-\infty}^{(t-t_o)}x(\tau)\cos 3(\tau+t_o)\ d\tau$$

Since

$y'(t)\neq y(t-t_o)$ so system is not time invariant for input $x(\tau)=\cos(3\tau)$, i.e. bounded input

$$y(t)=\int_{-\infty}^{t}\cos^2(3\tau)\rightarrow\infty \text{ as } t\rightarrow\infty$$

So for bounded input, output is not bounded

53. Ans. (a)

Solution:

If $x(n)$ has N_1 terms and

$h(n)$ has N_2 terms, then

$y(n)$ has N_1+N_2-1 terms.

Hence in this case, $N_1=2$ and

$N_y=N_1+N_2-1=4$

$\Rightarrow N_2=3$ (i.e. $h(n)$ has three terms).

Moreover, if $x(n)$ spreads from $-M_1$ to $+M_2$ and $h(n)$ spread from $-L_1$ and $+L_2$, then $y(n)$ spreads from $-(M_1+L_1)$ to $+(M_2+L_2)$.

Hence, in the given case $h(n)$ must lie between $n=0$ to $n=2$, which is satisfied only by option (a) and (c)

The given $y(n)$ can be obtained by convolving $h(n)$ in option (a) with $x(n)$.

Given data input and output sequences are

$$x[n]=\{-1,\underset{\uparrow}{2}\},-1\le n\le 0$$

$$y[n]=\{-1,\underset{\uparrow}{3},-1,-2\},-1\le n\le 2$$

If impulse response of system is $h[n]$ then output

$$y[n]=h[n]*x[n]$$

Since length of convolution $(y[n])$ is – 1 to 2, $x[n]$ is of length – to 0

So length of $h[n]$ is 0 to 2.

Let $h[n]=\{a,b,c,\}$

Convolution

↓	a	b	c
-1	$-a$	$-b$	$-c$
→ 2	$2a$	$2b$	$2c$

$$y[n]=\{-a,2a-b,2b-c,2c\}$$

$$y[n]=\{-1,\underset{\uparrow}{3},-1,-2\}$$

So,

$$a=1$$

$$2a-b=3\Rightarrow b-1$$

$$2a-c=-1\Rightarrow c=-1$$

Impulse response $h[n]=\{\underset{\uparrow}{1},-1,-1\}$

Second Method

Given $x(n)=\{-1,\ \underset{\uparrow}{2}\}$ and

$$y(n)=\{-1,\ \underset{\uparrow}{3},\ -1,\ -2\}$$

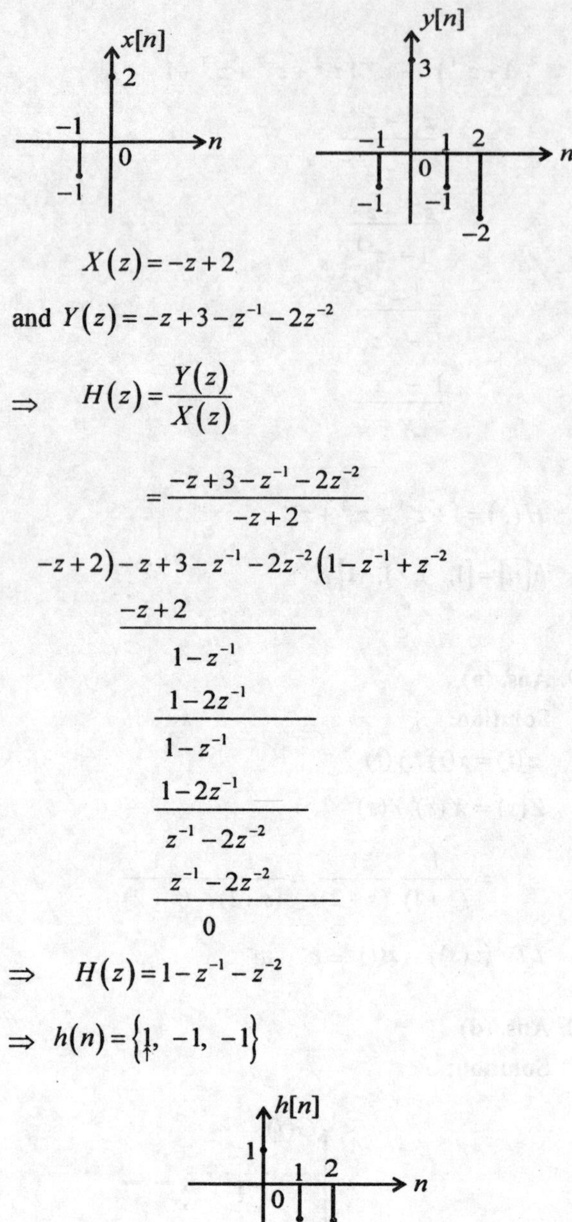

$$X(z)=-z+2$$

and $Y(z)=-z+3-z^{-1}-2z^{-2}$

$$\Rightarrow\quad H(z)=\frac{Y(z)}{X(z)}$$

$$=\frac{-z+3-z^{-1}-2z^{-2}}{-z+2}$$

$$-z+2)\,-z+3-z^{-1}-2z^{-2}\,(1-z^{-1}+z^{-2}$$

$$-z+2$$

$$1-z^{-1}$$

$$1-2z^{-1}$$

$$1-z^{-1}$$

$$1-2z^{-1}$$

$$z^{-1}-2z^{-2}$$

$$z^{-1}-2z^{-2}$$

$$0$$

$$\Rightarrow\quad H(z)=1-z^{-1}-z^{-2}$$

$$\Rightarrow\quad h(n)=\{\underset{\uparrow}{1},\ -1,\ -1\}$$

54. Ans. (a)

Solution:

$$G(z)=\alpha z^{-1}+\beta z^{-3}$$

$$G(z)\Big|_{z=e^{j\omega}}=\alpha e^{-j\omega}+\beta e^{-3j\omega}$$

Let $\alpha=\beta$, then $G(z)=\alpha z^{-j2\omega}\left(e^{j\omega}+e^{-j\omega}\right)$

$$=2\alpha e^{-j2\omega}\cos\omega$$

$$\left|G\left(e^{j\omega}\right)\right|=|2\alpha\cos\omega|\text{ and }\theta(\omega)=-2\omega$$

$$\Rightarrow\frac{\theta(\omega)}{\omega}=2=\text{constant}$$

$\Rightarrow\theta(\omega)$ is proportional to ω.

55. Ans. (c)

Solution: Let $f(t) = A \sin \omega t$

$$\frac{df(t)}{dt} = A\,\omega \cos \omega t$$

A differentiator has transfer function whose amplitude increases linearly with frequency.

56. Ans. (c)

Solution: Linear time invariant control system in shown as below:

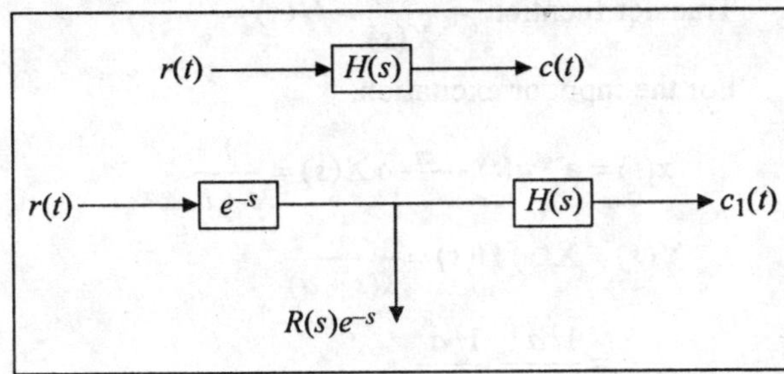

We know, $C(s) = H(s)\,R(s)$

$\therefore$ output is $H(s)\,R(s)\,e^{-s} = C_1(s)$

or $C_1(s) = e^{-S}\,C(s)$

$\Rightarrow$ $c_1(t) = c(t-1)\,u(t-1)$

57. Ans. (c)

Solution: $x(t) = u(t) \Rightarrow X(s) = \frac{1}{s}$

$$y(t) = t\,e^{-t} \Rightarrow Y(s) = \frac{1}{(s+1)^2}$$

$$G(s) = \frac{Y(s)}{H(s)} = \frac{s}{(s+1)^2}$$

58. Ans. (a)

Solution: $\frac{d^2y}{dt^2} + \frac{dy}{dt} = \frac{du}{dt} + 2u$

Taking Laplace transform and initial value zero,

$$s^2\,Y(s) + sY(s) = sU(s) + 2U(s)$$

$$(s^2 + s)\,Y(s) = U(s)\,(s+2)$$

$$\frac{Y(s)}{U(s)} = \frac{(s+2)}{s^2+s}$$

59. Ans. (c)

Solution: By applying Laplace transform on both sides, we get

$$\left[s^2\,X(s) - sx(0) - x'(0)\right] + 6\left[s\,X(s) - x(0)\,\right] + 5X(s)$$

$$= 12\left(\frac{1}{s} - \frac{1}{s+2}\right)$$

It ignition condition not given assume zero

$$[s^2 + 6s + 5]\,X(s) = 12\left(\frac{1}{s} - \frac{1}{s+2}\right) = \frac{24}{s(s+2)}$$

or $X(s) = \dfrac{24}{s(s^2+6s+5)\,(s+2)}$

then, $\lim\limits_{t\to\infty} x(t) = \lim\limits_{s\to\infty}\left[sX(s)\right] = \lim\limits_{s\to 0}\dfrac{24}{(s^2+6s+5)\,(s+2)}$

$$= \frac{24}{5\times 2} = 2.4$$

60. Ans. (b)

Solution: Natural response is when differential equation is zero, i.e. forcing function is zero

$$\left(s^2 + \frac{1}{2}s + \frac{1}{18}\right) = 0.$$

$$s_{1,2} = \frac{-\frac{1}{2} \pm \sqrt{\frac{1}{4} - \frac{4}{18}}}{12} = -\frac{1}{4} \pm \frac{1}{12} = \frac{1}{3}, \frac{1}{6}$$

The time constant $T_1 = 3$ sec

$T_2 = 6$ sec

Second Method

$$T(s) = \frac{C(s)}{R(s)} = \frac{1}{s^2 + \frac{s}{2} + \frac{1}{18}} = \frac{18}{18s^2 + 9s + 1}$$

$$= \frac{18}{(3s+1)(6s+1)}$$

As, $T(s)$ have the form, $\dfrac{A}{(T_1 s+1)(T_2 s+1)}$

The time constant are $T_1 = 3$ sec,

$T_2 = 6$ sec.

Third Method

For natural response

$$\frac{dx^2}{dt^2} + \frac{1}{2}\frac{dx}{dt} + \frac{x}{18} = \delta(t)$$

$$X(s) = \frac{1}{s^2 + \frac{s}{2} + \frac{1}{18}} = \frac{1}{\left(s+\frac{1}{3}\right)\left(s+\frac{1}{6}\right)}$$

Thus $s = -\frac{1}{3}$ and $s = -\frac{1}{6}$

Natural time constant is 3 seconds and 6 seconds.

61. Ans. (d)

Solution: $H(s) = C(s) = \dfrac{12.5 \times 8}{(s+6)^2 + 8}$

$$= \frac{100}{s^2 + 12s + 100}$$

Then, output for unit step response

$$C_u(s) = \frac{1}{s} \cdot \frac{100}{s^2 + 12s + 100}$$

and steady-state value $= \lim_{s \to 0} sC_u(s)$

$$= \lim_{s \to 0} \frac{100}{s^2 + 12s + 100} = 1$$

62. Ans. (b)

Solution: $H(s) = \dfrac{s}{(s+1)}$

$$\phi = 90 - 45 = 45°$$

$$|H(s)| = \frac{1}{\sqrt{1+1}} = \frac{1}{\sqrt{2}}$$

Therefore this transfer function circuit will give amplitude change of $\dfrac{1}{\sqrt{2}}$ and phase 45°

Hence, in steady state output is $\dfrac{1}{\sqrt{2}} \sin(t + 45°)$

Second Method

$x(t) = \sin t$; $x(s) = \dfrac{1}{s^2 + 1}$; $Y(s) = X(s)\dfrac{s}{s+1}$

$$Y(s) = \frac{1}{(s^2+1)} \cdot \frac{s}{s+1} = \frac{A}{(s+j)} + \frac{B}{(s-j)} + \frac{C}{(s+1)}$$

$$= \frac{1}{2(+j)}\left(\frac{1}{s-j}\right) + \frac{1}{2(1-j)(s+j)} - \frac{1}{2(s+1)}$$

Taking inverse Laplace transformation, we have

$$y(t) = \frac{1}{2(1+j)} e^{jt} + \frac{1}{2(1-j)} e^{-jt} - \frac{1}{2} e^{-t}$$

$$= \frac{1}{2}(\sin t + \cos t) - \frac{1}{2} e^{-t}$$

$$= \frac{1}{\sqrt{2}} \sin(t + 45^o) - \frac{1}{2} e^{-t}$$

Then, steady state value $= \lim_{t \to \infty} y(t)$

$$\frac{1}{\sqrt{2}} \sin(t + 45°)$$

63. Ans. (b)

Solution:

$x(t) \longrightarrow$ [LTI System] $\longrightarrow y(t)$

$X(s)$ $\qquad$ $Y(s)$

For the LTI system unit impulse response, $h(t)$ is given by

$$h(t) = 1\,u(t) \xrightarrow{LT} H(s) = \frac{1}{s}$$

$$\text{Transfer fucntion } = \frac{Y(s)}{X(s)} = H(s) = \frac{1}{s}$$

For the input or excitation,

$$x(t) = e^{-at}u(t) \xrightarrow{LT} X(s) = \frac{1}{s+a}$$

$$Y(s) = X(s)\,H(s) = \frac{1}{s(s+a)}$$

$$= \frac{1/a}{s} - \frac{1/a}{s+a}$$

$$y(t) = \frac{1}{a}[1 - e^{-at}]\,u(t)$$

$$= \frac{1}{a}(1 - e^{-at}),\ t \geq 0$$

$$= 0,\ t < 0$$

Second Method

The output $y(t)$ can also be obtained by integrating the input $x(t)$ as the system is an integrator

$$y(t) = \int_{-\infty}^{t} e^{-at}u(t)dt$$

$$= \frac{1}{a}(1 - e^{-at})$$

64. Ans. (b)

Solution:

* As $y(t) \propto x(t)$ the system is linear
* As $x(t)$ is multiplied by function of time the system is time varying

Therefore, the system is linear and time varying

65. Ans. (b)

Solution:

$$H(s) = \frac{1}{s^2(s-2)} = \frac{1}{s^2} \times \frac{1}{(s-2)}$$

Impulse response $h(t) = L^{-1}\left[\dfrac{1}{s^2} \times \dfrac{1}{s-2}\right]$

* Convolution in time domain is equivalent to multiplication in s-domain,

Use the pairs: $tU(t) \rightarrow \frac{1}{s^2}$

$$e^{2t}U(t) \rightarrow \frac{1}{s-2}$$

$$h(t) = [t\,U(t)] * [e^{2t}\,U(t)] = (t * e^{2t})\,U(t)$$

66. Ans. (a)

Solution:

Impulse resposne, $h(t) = e^{2t}\,u(t) \xrightarrow{L.T} H(s) = \frac{1}{s-2}$

Input, $x(t) = e^{3t}u(t) \xrightarrow{L.T} X(s) = \frac{1}{s-3}$

If $y(t)$ is the output, then

$$Y(s) = X(s)\,H(s)$$

$$Y(s) = \frac{1}{(s-2)\,(s-3)}$$

$$= \frac{-1}{s-2} + \frac{1}{s-3}$$

$$y(t) = -1.e^{2t}u(t) + e^{3t}u(t)$$

$$y(t) = [e^{3t} - e^{2t}] \text{ for } t > 0$$

67. Ans. (a)

Solution:

$y(t)$ can be calculated by using Laplce transform.

Let $x(t) \xrightarrow{LT} X(s)$

$y(t) \xrightarrow{LT} Y(s)$

From the given differential equation, using the properties of LT, the transfer fucntion H(s) is obtained as

$$\frac{d^2y}{dt^2} + 3\frac{dy}{dt} + 2y = x(t)$$

$$x^2Y(s) + 3X(s) + 2\,Y(s) = X(s)$$

$$Y(s)[s^2 + 3s + 2] = X(s)$$

$$H(s) = \frac{Y(s)}{X(s)}$$

$$= \frac{1}{s^2 + 3s + 2} = \frac{1}{(s+1)\,(s+2)}$$

For the given $x(t) = 2\,u(t) \xrightarrow{LT} X(s) = \frac{2}{s}$

$$Y(s) = H(s)\,X(s) = \frac{2}{s(s+1)\,(s+2)}$$

Using the partial fractions:

$$Y(s) = \frac{1}{s} - \frac{2}{s+1} + \frac{1}{s+2}$$

Apply the standard LT pair

$$u(t) \longrightarrow \frac{1}{s}$$

$$e^{-\alpha t}u(t) \longrightarrow \frac{1}{s+\alpha}$$

$$y(t) = [1 - 2e^{-t} + e^{-2t}]u(t)$$

68. Ans. (c)

Solution:

$$H(s) = \frac{Y(s)}{X(s)} = \frac{1}{s+2}$$

Where, the input $x(t) \xrightarrow{LT} X(s)$

and the output $y(t) \xrightarrow{LT} Y(s)$

For $x(t) = 10\,u(t) \xrightarrow{LT} X(s) = \frac{10}{s}$

$$Y(s) = X(s)\,H(s) = \frac{10}{s(s+2)} = \frac{5}{s} - \frac{5}{s+2}$$

$$= 5\left[\frac{1}{s} - \frac{1}{s+2}\right]$$

$$y(t) = 5[1 - e^{-2t}]u(t)$$

Steady state value of $y(t) = y(\infty) = 5$

Let at $t = T$ the response $y(t)$ reaches 99%

$$Y_{ss} = 0.99 \times 0.5 = 4.95$$

So, $\quad 4.95 = 5(1 - e^{-2T})$

$\Rightarrow \quad T = 2.3$ seconds

69. Ans. (a)

Solution: Fourier transform relation for shifted function

$$x(t - t_o) \leftrightarrow e^{-j\omega to}\,x(\omega)$$

$$y(t) = 0.5x\,(t - t_d + T) + x(t - t_d) + 0.5x\,(t - t_d - T).$$

$$Y(\omega) = 0.5e^{-j\omega(t\,d - T)}\,X(\omega) + e^{-j\omega td}\,X(\omega) + 0.5e^{-j\omega c(td+T)}\,X(\omega)$$

$$= 0.5e^{-j\omega(\,td-T)}\,X(\omega) + 0.5e^{-j\omega c(td+T)}\,X(\omega) + e^{-j\omega td}\,X(\omega)$$

$$= 0.5\,X(\omega)\,[e^{-j\omega(\,td-T)} + e^{-j\omega c(td+T)}] + e^{-j\omega td}\,X(\omega)$$

$$= e^{-j\omega td}\,X(\omega)\,[1 + 0.5\,(2\cos\{\omega T\})]$$

$$\frac{Y(\omega)}{X(\omega)} = e^{-j\omega t_d}\,[1 + \cos\omega T] = H(\omega)$$

70. Ans. (b)

Solution:

For step response $Y(f)$, $x(t)$ is $u(t)$

$$X(f) = \frac{1}{1 + j10\pi f}$$

$$Y(f) = X(f) \times H(f)$$

$$= \left(\frac{1}{j2\pi f}\right)\left(\frac{5}{1 + j2\pi f(5)}\right)$$

$$= 5\left(\frac{1}{j2\pi f} - \frac{1}{\frac{1}{5} + j2\pi f}\right)$$

$$y(t) = 5(1 - e^{-t/5})\, u(t)$$

$$\therefore \quad j2\pi f = S$$

Second Method

$$H(f) = \frac{5}{1 + 10\pi f} = \frac{1}{0.2 + j2\pi f}$$

Using the FT pair $e^{-\alpha t}u(t) \to \dfrac{1}{\alpha + j2\pi f}$

Therefore, the impulse resposne $h(t) = e^{-0.2t}u(t)$

For LTI system, if the input is integrated, the output is also integrated.

As $u(t) = \int_{t=-\infty}^{t} \delta(t)dt$, the step response is given by

$$s(t) = \int_{t=-\infty}^{t} h(t)dt$$

$$= \int_0^t e^{-0.2t}dt = 5[1 - e^{-0.2t}] \text{ for } t > 0$$

71. Ans. (d)

Solution:

Let $H(\omega)$ be the frequency response of stable LTI system.

For input, $x(t) = a_1 \sin(\omega_1 t + \phi_1)$,

the steady state output,

$$y(t) = a_1 |H(\omega_1)| \sin((\omega_1 + \phi_1 + \angle H(\omega_1)) \quad \text{... (1)}$$

Given output equation $y(t) = a_2 F(\omega_2 t + \phi_2)$ ... (2)

Comparing equations (1) and (2)

F is a sine function with $\omega_2 = \omega_1$,

$$y(t) = a_2 F(\omega_2 t + \phi_2)$$

$$y(t) = a_2 F(\omega_2 t + \phi_2)$$

F may be a cosine function if $\frac{\pi}{2} < [\phi_1 + \angle H(\omega_1)] < \pi$, but

$$\omega_2 = \omega_1$$

$a_2 \neq a_1$ unless $|H(\omega_1)| = 1$

$\phi_2 \neq \phi_1$ unless $\angle H(\omega_1) = 0$

72. Ans. (d)

Solution:

Given

Relation between input u(t) and output y(t) is as below

$$y(t) = \int_0^t (2 + t - \tau)e^{-3(t-\tau)}u(\tau)d\tau$$

It is observed that the above integral is a convolution operation between the input u(t) and the impulse response of the system, i.e.

$$y(t) = [(2 + t)e^{-3t}] * u(t)$$

$$\therefore \quad h(t) = (2 + t)e^{-3t}\, u(t) = 2e^{-3t} + 2e^{-3t}$$

Use the LT pairs

$$e^{-3t}\, u(t) \to \frac{1}{s+3}$$

$$t\, u(t) \to \frac{1}{s^2}$$

$$t \,.\, e^{-3t}\, u(t) \to \frac{1}{(s+3)^2}$$

$$H(s) = \frac{2}{s+3} + \frac{1}{(s+3)^2}$$

$$= \frac{2(s+3)+1}{(s+3)^2} = \frac{2s+6+1}{(s+3)^2}$$

$$= \frac{(2s+7)}{(s+3)^2}$$

73. Ans (a)

Solution:

$$y(n) = x(n) - \frac{1}{3}y(n-1)$$

when $y(n) = 0$ for $n < 0$

and $X[n] = \delta[n]$

$$Y(n) = x(n) * h(n)$$

$$Y(z) = X(z).\, H(z)$$

$$X[z] = 1$$

$$Y(z) = H(z)$$

$$y(n) = h(n)$$

$$y(n) = x(n) - \frac{1}{3}y(n-1)$$

$$Y(z) = X(z) - \frac{1}{3}Y(z)z^{-1}$$

$$Y(z) + \frac{1}{3}Y(z)z^{-1} \Rightarrow 1$$

$$Y(z) = \frac{1}{1 + \frac{1}{3}z^{1}}$$

$$Y(z) = \frac{1}{1 - \left(\frac{-1}{3}\right)z^{-1}}$$

$$y(n) = \left(\frac{-1}{3}\right)^n u(n)$$

74. Ans. (b)

Solution:

If $H(z)$ is discrete rational transfer function then,

where $\frac{Y[z]}{X[z]} = \frac{\text{zero}}{\text{pole}}$

For $H(z)$ to be stable the poles most be in unit circle

$$[H(z)]^{-1} = \frac{X(z)}{Y(z)}$$

The zeroes must be inside the circle.

So poles and zeroes must be inside the unit circle

75. Ans. (c)

Solution:

$$y(t) = \int_{-\infty}^{5t} x(\tau)d\tau$$

for $t = 1$

$$y(1) = \int_{-\infty}^{5} x(\tau)d\tau$$

Here output depends on future values of $x(t)$ as even though $t = 1$ output takes upto 5 times unit so system is non-causal.

Now if shift in input $x(t-t_o)$, then output is

$$y'(t) = \int_{-\infty}^{5t} x(\tau - t_o)d\tau$$

$$= \int_{-\infty}^{5t} x(\tau - t_o)d\tau$$

$$\tau - t_o = \tau'$$

or $$y'(t) = \int_{-\infty}^{(5t-t_o)} x(\tau')d\tau'$$

and $$y(t-t_o) = \int_{-\infty}^{(5t-t_o)} x(\tau)d\tau \neq y'(t)$$

So system is time variant.

76. Ans. (c)

Solution: $y(t) = \frac{d}{dt}\left[e^{-t}x(t)\right]$

$$y(t) = e^{-t}x'(t) + x(t)e^{-t}(-1)$$

$$y(t) = e^{-t}x'(t) - e^{-t}x(t)$$

$$y(t) = e^{-t}\left[x'(t) - x(t)\right]$$

At $t \to \infty$; $y(t) = 0$

So it's o/p is stable, because it is bounded even for $t \to \infty$.

77. Ans. (a)

Solution:

Characteristic of thermometer

$$\frac{2dT_i}{dt} = T_a - T_i$$

Take the Laplace transform

or, $$2sT_i(s) = T_a(s) - T_i(s)$$

$$2sT_i(s) + T_i(s) = T_a(s)$$

or, $$\frac{T_i(s)}{T_o(s)} = \frac{1}{2s+1} = \frac{0.5}{s+0.5}$$

$\Rightarrow$ $$0.5\, e^{-0.5t}$$

Hence, $$\omega = 0.5$$

$$f = \frac{\omega}{2\pi} = \frac{1}{4\pi} Hz$$

78. Ans. (b)

Solution:

$$x[n] = \left[\frac{1}{2}\right]^n u[n]$$

$$X[z] = \frac{z}{(z - \frac{1}{2})}$$

$$y[n]-\frac{1}{3}Y[n-1]=x[n]$$

$$Y(z)-\frac{1}{3}z^{-1}Y(z)=X(z)$$

$$Y(z)\left[1-\frac{1}{3}z^{-1}\right]=\frac{z}{\left(z-\frac{1}{2}\right)}$$

$$Y(z)=\frac{z}{\left(z-\frac{1}{2}\right)(1-\frac{1}{3}z^{-1})}$$

$$Y(z)=\frac{z^2}{(z-\frac{1}{2})(z-\frac{1}{3})}$$

$$\frac{Y(z)}{z}=\frac{z}{(z-\frac{1}{2})(z-\frac{1}{3})}$$

$$\frac{z}{\left(z-\frac{1}{2}\right)\left(z-\frac{1}{3}\right)}=\frac{A}{(z-\frac{1}{2})}+\frac{B}{(z-\frac{1}{3})}$$

$$z=A\left(z-\frac{1}{3}\right)+B\left(z-\frac{1}{2}\right)$$

Put $z=\frac{1}{3}$

$$\frac{1}{3}=B\left(\frac{1}{3}-\frac{1}{2}\right)$$

$$\frac{1}{3}=B\left(\frac{2-3}{6}\right)$$

$$B=-2$$

Put $z=\frac{1}{2}$

$$\frac{1}{2}=A\left(\frac{1}{2}-\frac{1}{3}\right)$$

$$\frac{1}{2}=A\left(\frac{3-2}{6}\right)$$

$$A=3$$

$$\frac{Y(z)}{z}=\frac{3}{(z-\frac{1}{2})}-\frac{2}{(z-\frac{1}{3})}$$

$$Y(z)=\frac{3z}{(z-\frac{1}{2})}-\frac{2z}{\left(z-\frac{1}{3}\right)}$$

$$Y(z)=\frac{3}{\left(1-\frac{1}{2}z^{-1}\right)}-\frac{2}{\left(1-\frac{1}{3}z^{-1}\right)}$$

$$y(n)=3\left(\frac{1}{2}\right)^n-2\left(\frac{1}{3}\right)^n$$

$$y(n)=-2\left(\frac{1}{3}\right)^n+3\left(\frac{1}{2}\right)^n$$

79. Ans. (d)

Solution:

$$y[n]=x[n]-2x[n-1]+x[n-2]$$

Take z- transform,

$$Y(z)=X(z)-2z^{-1}X[z]+z^{-2}X[z]$$

$$\frac{Y(z)}{X(z)}=H(z)=1-2z^{-1}+z^{-2}$$

or $H(j\omega)=1-2e^{-j\omega}+e^{-2j\omega}$

or $H(j\omega)=(1-e^{-j\omega})^2$

$$=\{1-[\cos\omega-j\sin\omega]\}^2$$

Second Method

Given $y(n)=x(n)-2x(n-1)+x(n-2)$ ...(1)

$$H\left(e^{j\omega}\right)=\frac{Y\left(e^{j\omega}\right)}{X\left(e^{j\omega}\right)}$$

$$=1-2\left(e^{j\omega}\right)+\left(e^{j\omega}\right)$$

$$=\left(1-e^{j\omega}\right)^2$$

$$H\left(e^{jo}\right)=0$$

$$H\left(e^{j\frac{\pi}{2}}\right)=(1+j)^2=2\angle 90^\circ$$

$$H\left(e^{j\pi}\right)=4$$

$\therefore$ **The given discrete time system is a good approximation of high pass filter**

80. Ans. (a)

Solution:

In the design of IIR digital filters by the method of impulse invariance, the impulse response $h_d(n)$ of the digital filter is obtained by sampling the impulse response, $h_a(t)$ $h_d(n) = h_a(t)\big|_{t=nT_s}$

As no practical analog filter is band limited, problem of aliasing (overlapping of the adjacent analog spectra is bound to occur) occur. Hence frequency response of digital filter is not a scaled version of the frequency response of the analog filter. In the design of IIR digital filters by the method of Bilinear Transformation, $s = \frac{z-1}{z+1}$ aliasing is avoided, because the mapping from s-plane to z-plane is one to one and the entire imaginary axis of a s-plane is mapped into the unit circle in the z-plane

81. Ans. (a)

Solution:

Given,

Input $x = [a, b, c, d]$

Output $y = [x, x, x, x \ldots \text{N times}]$

This is clearly observed for different inputs outputs are same and periodic with fundamental period N = 4, i.e.

$$y = [\underbrace{a, b, c, d}, \underbrace{a, b, c, d}, \underbrace{a, b, c, d} \ldots \text{N-time}]$$

$$= [\delta(n) \quad + \delta(n-4) \quad + \delta(n-8) + \cdots\cdots +]$$

This can be written as

$$\sum_{i=0}^{n-1} \delta(n-4i)$$

82. Ans. (d)

Solution:

For compensation, a element is cascaded which make the overall gain unity and no delay (no distortion).

$H'(z)H(z) = 1$

Where $H'(z) \rightarrow$ compensator

$$\therefore H'(z) = \frac{1}{H(z)}$$

$$= \frac{1}{(1 - az^{-1})}$$

$$h'(n) = -(a)^n u[-n-1], \ ([z] < [a])$$

Because system $H'(z)$ has to be stable

Therefore unit circle must be inside the ROC.

$[a] > 1,$

If ROC is $[z] < [a]$, then it will include unit circle.

Neglecting the –ve sign,

$h'(n) = a^n \ u[n-1]$

4.7 Convolution

1. Ans. (c)

Solution: $LT\,[u(t+1)] = \frac{1}{s} e^{s}$

$$LT[r(t-2)] = \frac{1}{s} \cdot e^{-2s}$$

$$u(t) * r(t) \rightarrow U(s) \cdot R(s)$$

$$u(t-1) * r(t-2) \rightarrow \frac{e^{s}}{s} \cdot \frac{e^{-2s}}{s^2}$$

$$\rightarrow \frac{e^{-s}}{s^3}$$

$$\text{Then function output} = LT^{-1}\left(\frac{e^{-s}}{s^3}\right)$$

$$= \frac{1}{2}(t-1)^2 \, u(t-1)$$

2. Ans. (b)

Solution : $x_1(t) = u(t) - u(t-1)$

$$X_1(s) = \frac{1}{s}[1 - e^{-s}]$$

$$x_2(t) = u(t) - u(t-2)$$

$$X_2(s) = \frac{1}{s}[1 - e^{-2s}]$$

$$x(t) = x_1(t) * x_2(t) \rightarrow X_1(s)\, X_2(s)$$

$$= \frac{1}{s}[1 - e^{-s}] \times \frac{1}{s}[1 - e^{-2s}]$$

$$= \frac{1}{s^2}[1 - e^{-s} - e^{-2s} + e^{-3s}]$$

$$x(t) = r(t) - r(t-1) - r(t-2) + r(t-3)$$

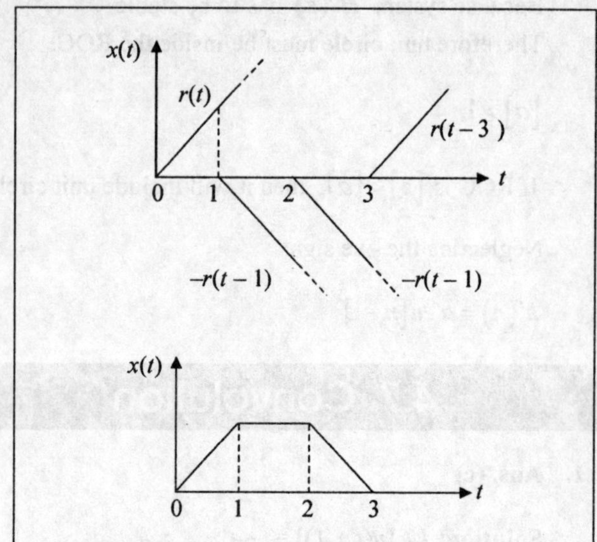

Second Method

Let $x(t)=u(t)-u(t-1)$, $x_2(t)=u(t)-u(t-2)$ then

$y(t)=x_1(t)*x_2(t)$

$x(t)$, $g(t)$ and $\dfrac{dx(t)}{dt}$ are shown in figures 1, 2 and 3.

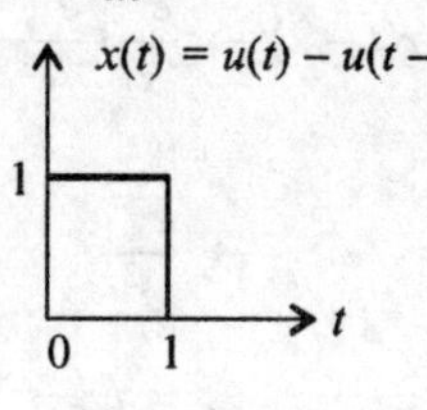

Fig.1

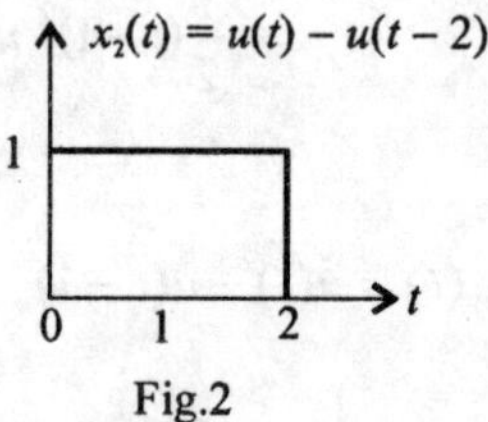

Fig.2

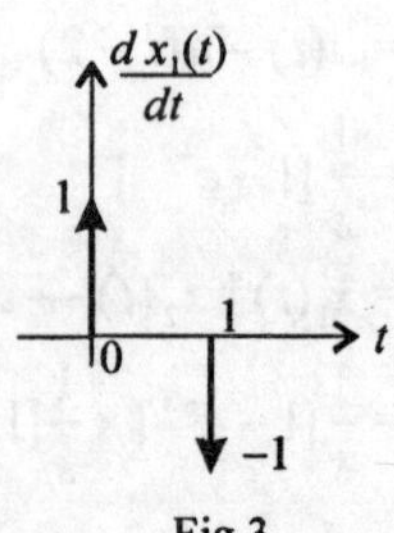

Fig.3

Let $y_1(t)=\dfrac{d\,x_1(t)}{dt}*x_2(t)$,

Then $y_1(t)$ is shown in Figs. 4 and 5

$$y(t)=x_1(t)*x_2(t)=\int_{t=-\infty}^{t} y_1(t)\,dt$$

$y(t)$ is shown in Fig. 6

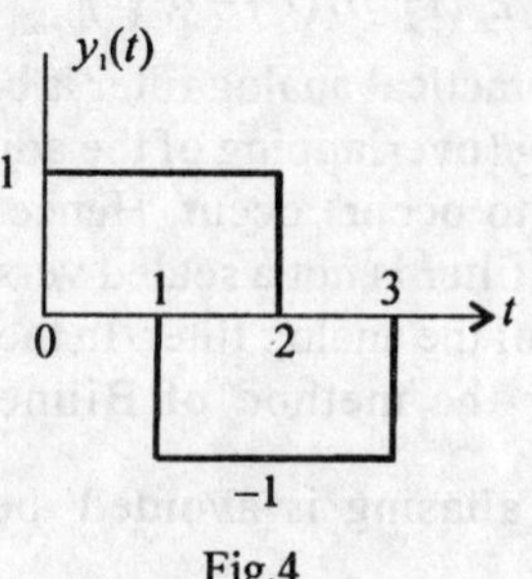

Fig.4

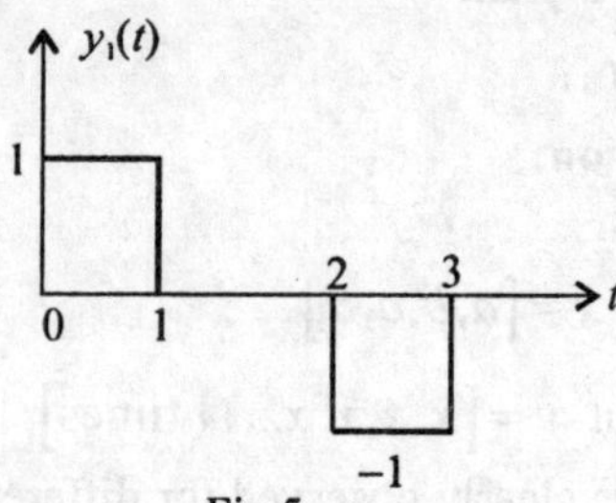

Fig.5

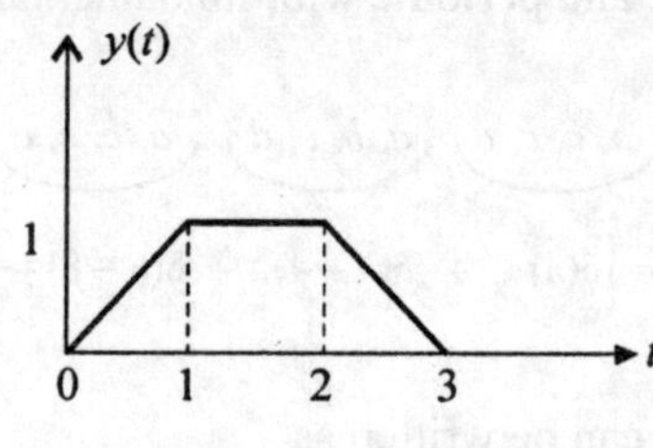

Fig.6

3. **Ans. (c)**

Solution: $\delta(t) * \delta(t-T) = g(t)|_{t=T}$

$$x(t+5) * \delta(t-7) = x(t+5)|_{t=7}$$

$$= x(t-7+5) = x(t-2)$$

4. **Ans. (a)**

Solution:

Given

$$h[n]=\left(\frac{1}{2}\right)^n u(n)$$

$y(0)=1$ and

$$y(1)=\frac{1}{2}$$

Since $y(n) = g(n) * h(n)$

$$= \sum_{K=-\infty}^{\infty} g(K)\, h(n-K)$$

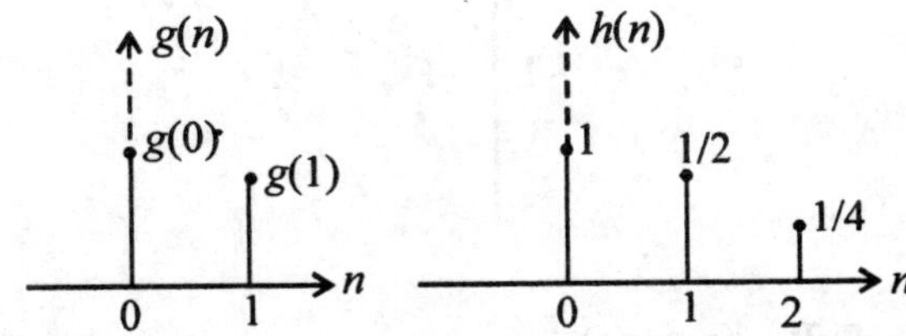

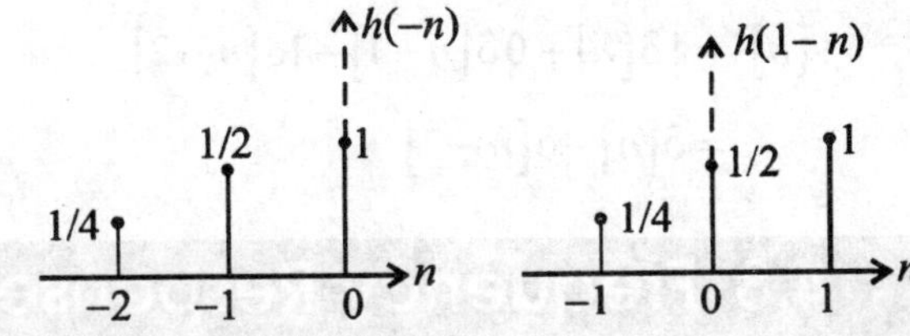

multiplying $g(n)$ and $h(-n)$

$$y(0) = \frac{1}{4} \times 0 + \frac{1}{2} \times 0 + 1 \times g(0)$$

$$y(1) = \frac{1}{2} = \frac{1}{2} g(0) + g(1)$$

$$g(1) = 0$$

Second Method

$$Y(n) = \sum_{K=0}^{\infty} \left(\frac{1}{2}\right)^K g(n-K)$$

$$Y(0) = \sum_{K=0}^{\infty} \left(\frac{1}{2}\right)^K g(-K) = 1$$

or, $\left(\frac{1}{2}\right)^0 g(0) = 1$

or, $g(0) = 1$

Since $g(n)$ is causal sequence

$g(-1), g(-2), \ldots = 0$

$$Y(1) = \sum_{K=0}^{\infty} \left(\frac{1}{2}\right)^K g(1-K)$$

$$\frac{1}{2} = \left(\frac{1}{2}\right)^0 g(1) + \left(\frac{1}{2}\right) g(0)$$

$$= g(1) + \frac{1}{2} \times 1$$

or $= g(1) = \frac{1}{2} - \frac{1}{2} = 0$

5. **Ans. (d)**

Solution:

Given, $h[n] = (1, 2, 2)$

Result of auto-correlation function

$$f(n) = h[n] * h(n) = \{\underset{\uparrow}{1}, 4, 8, 8, 4\}$$

Result of auto-correlation function

$$g(n) = h[n] * h[-n] = \{2, 6, \underset{\uparrow}{9}, 6, 2\}$$

$\therefore$ $g[n]$ **is non-causal and having maximum value 9.0**

6. **Ans. (a)**

Solution: $x(t) * x(t) = t\, e^{(-2t)} \cdot u(t)$

Taking Laplace transform both sides,

$$X(s).X(s) = \frac{1}{(s+2)^2}$$

$$\Rightarrow \quad X(s) = \frac{1}{s+2}$$

$$x(t) = e^{(-2t)}.u(t)$$

7. **Ans (d)**

Solution:

$$x(t) = \delta(t+1) - \delta(t-1)$$

$$h(t) = \frac{t}{2}[U(t) - U(t-2)] + \frac{(4+2)}{2}\,[U(t-2) - U(t-4)]$$

$$y(t) = h(t) * x(t)$$

$$[u \sin g\; x(t) - \delta(t - t_0) = x(t_0)]$$

We have

$$y(t) = \frac{(t+2)}{2}[U(t+1) - U(t-1)] - \frac{(t-1)}{2}$$

$$[U(t-1) - U(t-3)]$$

$$+ \frac{(3-t)}{2}[U(t-1) - U(t-3)] + \frac{(5-t)}{2}$$

$$[U(t-3) - U(t-5)]$$

and the graph would be as shown below

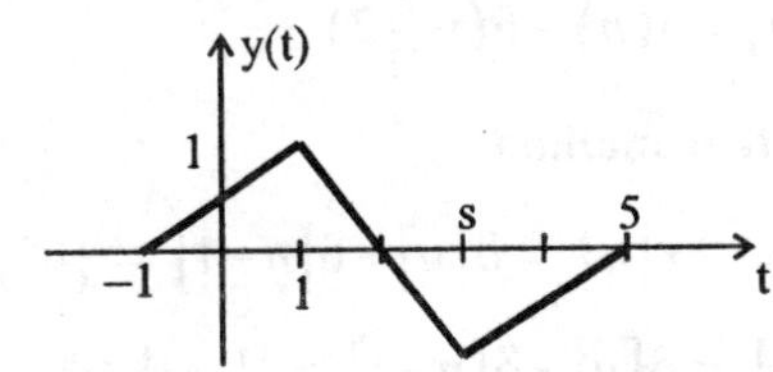

Graphical convolution method

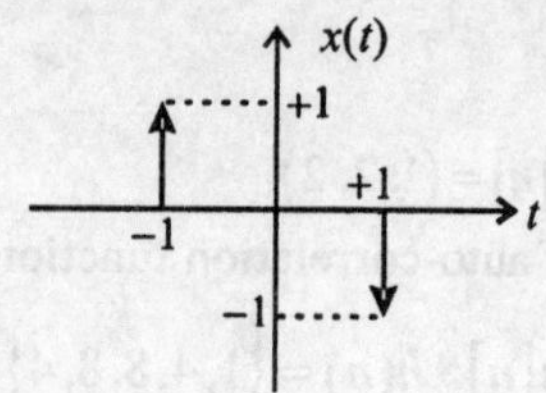

$x(t) = \delta(t+1) - \delta(-t-1)$

Now $y(t) = x(t) * h(t)$

$= [\delta(t+1) - \delta(t-1)] * h(t)$

$y(t) = h(t+1) - h(t-1)$

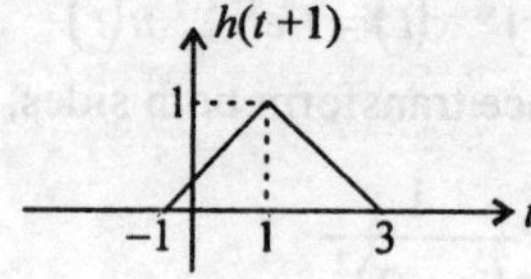

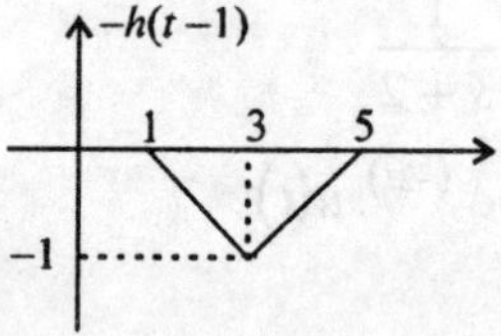

Hence,

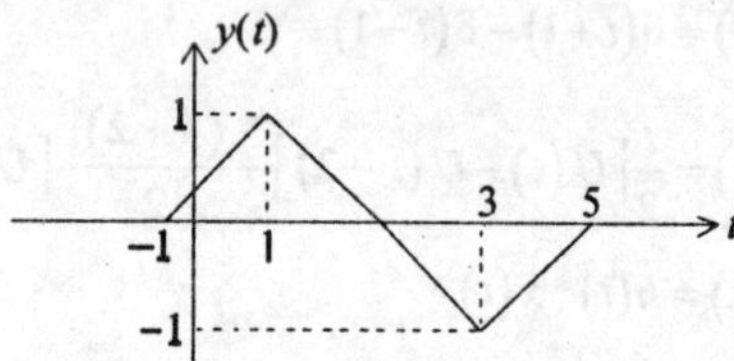

8. Ans (a)

Solution

$x[n] = \delta[n] + \delta[n-1]$

$h[n] = \delta[n] - \delta[n-1]$

$y(n) \Rightarrow x(n) * h(n)$

$[\delta(n) + \delta(n-1)] * [\delta(n) - \delta(n-1)]$

$\delta(n) + \delta(n-1) - \delta(n-1) - \delta(n-1-1)$

$y(n) = \delta(n) - \delta(n-2)$

Another method

Given, $x[n] = \delta[n] + \delta[n-1] = \{1,1\}$

$h[n] = \delta[n] - \delta[n-1] = \{1,-1\}$

Use table method of convolution

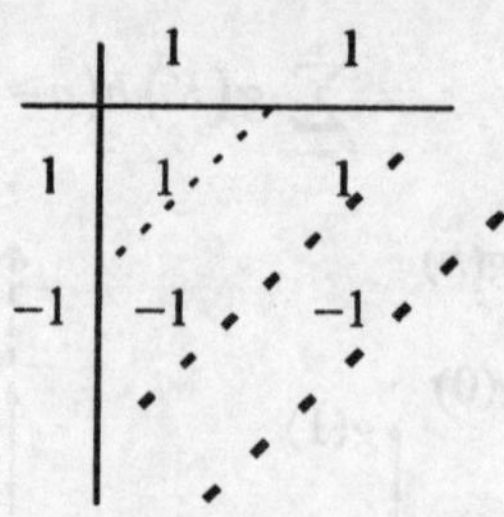

$y[n] = \{1, 0, -1\}$

$y[n] = 1\delta[n] + 0\delta[n-1] - 1\delta[n-2]$

$= \delta[n] - \delta[n-2]$

4.8 Frequency Response

1. Ans. (c)

Solution: $h_1(n) \rightarrow$ delay by 1

$h_2(n) \rightarrow$ delay by 2

Multiplication in frequency domain is equal to convolution in time domain

$h_1(n) * h_2(n) = \delta(n-1) * \delta(n-2)$

$= \delta(n-3)$ delay by 3

2. Ans. (b)

Solution:

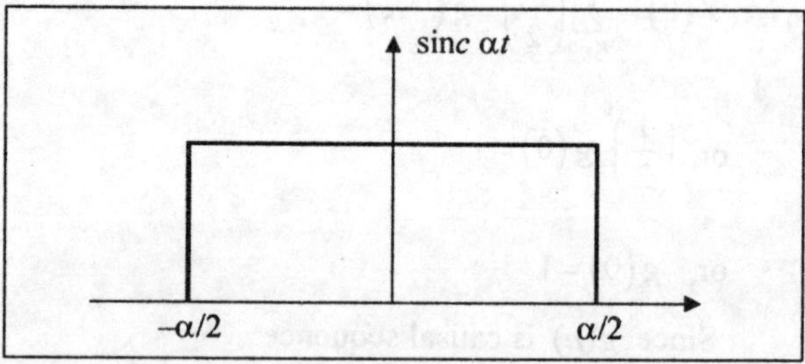

Convolution in time domain is equivalent to multiplication in frequency domain.

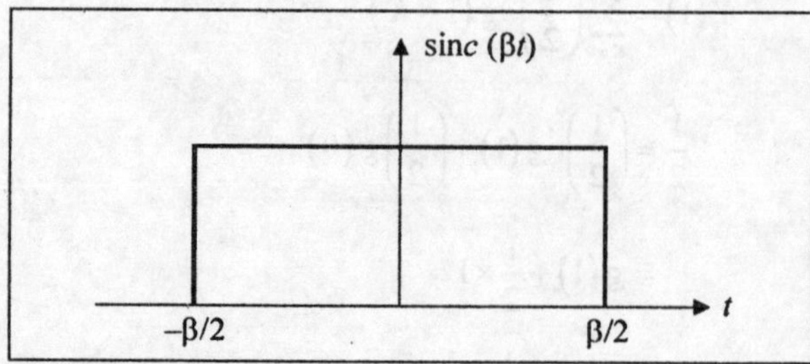

So it will be max (α, β)

3. Ans. (a)

Solution: $h(t) = 4\,e^{-2t}$; $\quad H(s) = \dfrac{4}{s+2}$

$Y(s) = H(s)\,X(s)$;

$x(t)$ = unit step, $X(s) = \dfrac{1}{s}$

$$Y(s) = \frac{4}{s(s+2)} = \frac{A}{s} + \frac{B}{s+2}$$

$$4 = A\,(s+2) + B(s) = (A+B)s + 2\,A$$

Compare coefficient

$$2\,A = 4; \quad A = \frac{4}{2} = 2$$

$$A + B = 0; \quad B = -A = -2$$

$$Y(s) = \frac{2}{s} - \frac{2}{s+2}$$

$$y(t) = 2(1 - e^{-2t})\,u(t)$$

Second Method

$$h(t) = 4e^{-2t}\,u(t)$$

Impulse response

Step response = $\int$ impulse response

$$= \int_0^t 4\,e^{-2t} dt = \left.\frac{4\,e^{-2t}}{-2}\right|_0^t$$

$$= -2[e^{-2t} - 1] = 2[1 - e^{-2t}]\,u(t)$$

4. Ans. (b)

Solution:

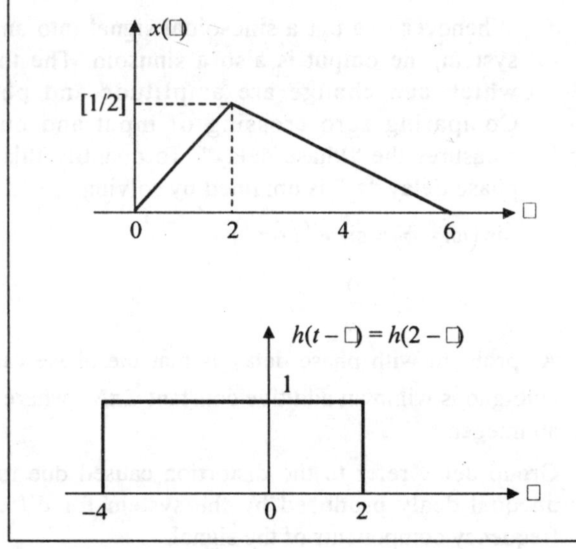

The above figure demonstrate the process of convolution. The output $y(t)$ of the system is given by

$$y(t) = \int x(\tau)\; h(t-\tau)\; d\tau$$

$$y(2) = \int x(\tau)\; h(2-\tau)\; d\tau$$

$$= \int_0^2 \left[\left(\frac{1}{4}\right)\cdot\tau\right]\cdot 1\cdot d\tau$$

$$= \frac{1}{4}\times\left.\frac{\tau^2}{2}\right|_0^2 = \frac{1}{4}[2-0] = \frac{1}{2}$$

5. Ans. (a)

Solution: Response of the system is convolution of impulse response $h(t)$ and input $u(t)$

i.e. $\quad \int_0^t h(\tau)\,u(t-\tau)\,d\tau$

Second Method

$$y(t) = \int_{\tau=-\infty}^{\infty} h(\tau)u(t-\tau)d\tau$$

or $\quad = \displaystyle\int_{\tau=-\infty}^{\infty} h(t-\tau)u(\tau)d\tau$

$= \displaystyle\int_{\tau=-0}^{\infty} h(\tau)h(t-\tau)d\tau$, if the system is causal, i.e. $h(t) = 0$ for $t < 0$.

$= \displaystyle\int_{\tau=0}^{t} h(\tau)h(t-\tau)d\tau$, if the system is causal and also the input $u(t)$ is causal, i.e. $u(t) = 0$ for $t < 0$

6. Ans (b)

Solution:

Frequency response

$$H(s) = \frac{1}{1+s}$$

$$|H(\omega)| = \frac{1}{\sqrt{1+\omega^2}}$$

$$C(s) = H(s)\,R(s)$$

$$= \frac{\omega}{s^2+\omega^2}\cdot\frac{1}{s+1}$$

As $x(t) = \sin t$

Hence $\omega = 1$

For $\omega = 1$,

$$|H(s)| = \frac{1}{\sqrt{2}}(1)\ \angle -\tan^{-1} 1$$

$$= \frac{1}{\sqrt{2}} \angle -45^\circ$$

$$C(t) = \frac{1}{\sqrt{2}} \sin\left(t - \frac{\pi}{4}\right)$$

7. **Ans. (d)**

Solution:

$$\frac{y(s)}{U(s)} = \frac{\dfrac{K}{(1+s\tau)}}{1+\left(\dfrac{K}{1+s\tau}\right)} = \frac{K}{1+s\tau+K}$$

$$\therefore y(s) = \frac{K}{s+\dfrac{K+1}{\tau}} \cdot \frac{1}{s} = \left[\frac{1}{s} - \frac{1}{s+\dfrac{K+1}{\tau}}\right] \cdot \frac{K}{K+1}$$

$$y(t) = \frac{K}{K+1}\left[1 - e^{\frac{-K+1}{\tau}t}\right]$$

At steady-state $y(\infty) = 0.8 = \dfrac{K}{K+1}$ or $K = 4$

8. **Ans (c)**

Solution:

Unit step response $y(t) = 1 - e^{-0.5t}$

therefore, impulse response is $= \dfrac{dy(t)}{dt}$

$$h(t) = 0.5\, e^{-0.5t}$$

or $$H(s) = \frac{0.5}{(s+0.5)}$$

Let us assume unity feedback system

or $$H(s) = \frac{G(s)}{1+G(s)} = \frac{0.5}{s+0.5}$$

$$= \frac{1/2}{s+\frac{1}{2}} = \frac{1}{2s+1}$$

or $$G(s)(2s+1) = 1 + G(s)$$

or $$G(s)2s + G(s)$$

$$= 1 + G(s)$$

or $$G(s) = \frac{1}{2s}$$

So velocity error constant

$$K_V = \lim_{s \to 0} sG(s) = \frac{1}{2}$$

$$e_{ss} = \frac{A}{K_V}$$

So for 0.1 $r(t)$

$A = 0.1$

$$e_{ss} = \frac{0.1}{1/2} = 0.2$$

9. **Ans. (c)**

Solution:

$$y(t) = \left(\frac{1}{100}\right)\cos\left(100t - 10^{-6}\right)\cos\left(10^6 t - 1.56\right)$$

$$= \left(\frac{1}{100}\right)\cos\left\{100\left(t - 10^{-8}\right)\right\}\cos\left\{10^6\left(t - 1.56\times10^{-6}\right)\right\}$$

Comparing it with $y(t) = K\, f(t - \tau_g)\cos\{\omega(t - \tau_p)\}$

we get $\tau_g = 10^{-8}$

$\tau_p = 1.56\times10^{-6}$

and function $f(t - \tau_g) = \cos\{100(t - 10^{-8})\}$

* Whenever we put a sinusoidal signal into an LTI system, the output is also a sinusoid. The things which can change are amplitude and phase. Comparing zero crossing of input and output measures the "Phase delay". To quantify this, the phase delay "τ_p" is obtained by solving

$$\sin(\omega t - \phi) = \sin\omega(t - \tau_p)$$

$$\Rightarrow \quad \tau_p = \frac{\phi}{\omega}$$

A problem with phase delay is that the phase can be ambiguous wihin an additive constant $2\pi N$, where N is an integer.

Group delay refer to the distortion caused due to the unequal dealy produced by the system for different frequency components of the signal.

To understand this, let us consider that the input signals contains two frequency components close to each other,

i.e. $x(t)=\cos(\omega_1 t)+\cos(\omega_2 t)$

$$\Rightarrow \; x(t)=2\cos\left\{\left(\frac{\omega_1-\omega_2}{2}\right)t\right\}+\cos\left\{\left(\frac{\omega_1+\omega_2}{2}\right)t\right\}$$

Since the two frequency components are delayed by different amounts, the output of the LTI system can be represented as:

$$y(t)=\cos(\omega_1 t-\phi)\cdot\cos(\omega_2 t-\phi_2)$$

$$y(t)=2\cos\left\{\begin{matrix}\left\{\left(\frac{\omega_1-\omega_2}{2}\right)t-\left(\frac{\phi_1-\phi_2}{2}\right)\right\} \\ \cos\left\{\left(\frac{\omega_1+\omega_2}{22}\right)t-\left(\frac{\phi_1+\phi_2}{2}\right)\right\}\end{matrix}\right\}$$

Sicne $\omega_1 \approx \omega_2$

or $\frac{\omega_1+\omega_2}{2}\approx\omega_1(\text{or }\omega_2)=\omega_C$

and $\frac{\omega_1-\omega_2}{2}=\frac{\Delta\omega}{2}$

or,

$$y(t)=2\cos\left\{\frac{\Delta\omega t}{2}-\left(\frac{\phi_1-\phi_2}{2}\right)\right\}\cos\left\{\omega_C t-\left(\frac{\phi_1-\phi_2}{2}\right)\right\}$$

The above equation can be rewritten in the form:

$$y(t)=2\cos\left\{\frac{\Delta\omega t}{2}-(t-\tau_g)\right\}\cos\left\{\omega_C(t-\tau_p)\right\}$$

where, $\tau_g=\frac{\phi_1-\phi_2}{\omega_1-\omega_2}=\frac{d\phi}{d\omega}$

or, $y(t)=K\ f(t-\tau_g)\cos\{\omega(t-\tau_p)\}$

10. Ans. (b)

Solution:

$$\tau_p=\frac{-\beta(f_c)}{2\pi f_c}$$

$$\tau_g=+\left.\frac{d(\beta f_c)}{df}\right|_{t=f_c}$$

When phase response β, i.e. linear with frequency and $\beta(0)=0$, τ_p take a common value, while it is stated linear, $\beta(0)=0$ is not stated

11. Ans. (b)

Solution:

$$|y(t)|=|M(\omega)||x(t)|$$

$$M(\omega)=\frac{y(\omega)}{X(s)}=\frac{1}{P}$$

$$|M(\omega)|=\frac{\omega}{\sqrt{\omega^2+P^2}}$$

$$1=\frac{\omega p}{\sqrt{\omega^2+p^2}}$$

$$\omega^2+p^2=\omega^2p^2$$

$$4+p^2=4p^2$$

$$3p^2=4,\ p=\frac{2}{\sqrt{3}}$$

Second Method

Given $\omega=2\,\text{rad/s}$

Phase difference between input and output,

$$\phi=-\frac{\pi}{3}-\left(-\frac{\pi}{2}\right)=\frac{\pi}{6}=30^\circ$$

Phase angle from transfer function,

$$\phi=90^\circ-\tan^{-1}\frac{\omega}{p}$$

$$90^\circ-\tan^{-1}\frac{2}{p}=30^\circ$$

$$\tan^{-1}\frac{2}{p}=60^\circ$$

$$\frac{2}{p}=\tan 60^\circ=\sqrt{3}$$

$$p=\frac{2}{\sqrt{3}}$$

12. Ans. (c)

Solution:

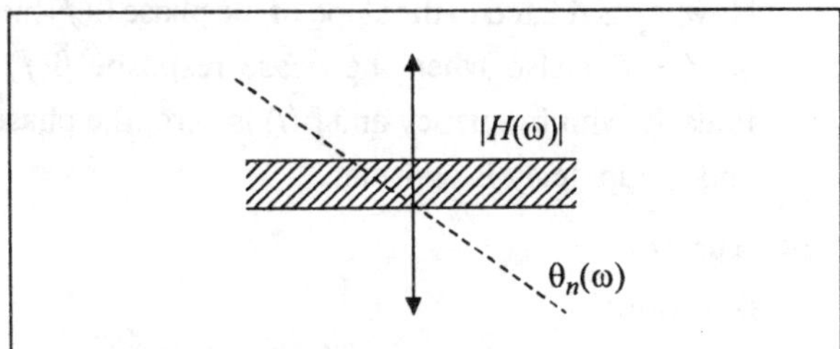

For Distortion less filter

$$Y(j\omega)=A(j\omega)\ X(j\omega)$$

$$H(j\omega)=Ae^{-j\omega}$$

Magnitude → constant

Phase → linear

13. Ans. (a)

Solution:

Phase delay and *Group delay* of LTI system.

Phase delay: Let a sinusoidal signal at frequency f_c is transmission through a dispersive channel which has a total phase-shift of $\beta(f_c)$ radians at that frequency. The time taken by the received signal phase to sweep out this phase lag is simply equal to $[\beta(f_c)/(2\pi f_c)]$ second. this time is called phase delay of the channel

Group delay: Let the dispersive channel is described by the transfer function

$$H(f) = K\, e^{j\beta(f)} \qquad \text{... (1)}$$

Where the phase $\beta(f)$ is a non-linear function of frequency.

The input signal $x(t)$ consists of a narrow-band signal defined by

$$x(t) = x_c(t)\,.\,\cos(2\pi f_c t) \qquad \text{... (2)}$$

Expanding $\beta(f)$ by Taylor's series at the point $f = f_c$ only two terms,

$$\beta(f) = \beta(f_c) + (f - f_c)\left.\frac{\partial\beta(f_c)}{\partial f}\right|_{f=f_c} \qquad \text{... (3)}$$

Define $$\tau_p = -\frac{\beta(f_c)}{2\pi f_c} \qquad \text{... (4)}$$

and $$\tau_g = \frac{1}{2\pi}\left.\frac{\partial\beta(f)}{\partial f_c}\right|_{f=f_c} \qquad \text{... (5)}$$

Then finally, the received signal is given by

$$y(t) = Kf(t-\tau_g)\cos\left[2\pi f_c(t-\tau_p)\right] \qquad \text{... (6)}$$

From the received output signal of above equation, we conclude that,

1. The sinusoidal carrier wave $\cos(2\pi f_c t)$ is delayed by τ_p second, hence τ_p represent the **phase delay**
2. The envelope $f(t)$ is delayed by τ_p seconds, hence τ_g represents the envelope or **group delay**

Now τ_g is related to the slope of the phase $\beta(f)$ measured at $f = f_c$. Also when the phase resposne $\beta(f)$ varies linearly with frequency and $\beta(f)$ is zero, the phase delay and group delay is constant.

14. Ans. (c)

Solution:

$$H(\omega) = \int_0^\infty e^{-2t}e^{-j\omega t}\,dt = \int_0^\infty e^{-(2+j\omega)}\,dt$$

$$= \left.\frac{e^{-(2+j\omega)t}}{-(2+j\omega)}\right|_0^\infty = \frac{-1}{2+j\omega}[0\ \ -1]$$

$$= \frac{1}{2+j\omega}$$

15. Ans. (d)

Solution: When input is sinusoidal then output is also sinusoidal with same frequency but amplitude and phase changes according to transfer function

Thus amplitude is

$$2\times\left|\frac{1}{j\omega+2}\right|_{\omega=2} = 2\times\left|\frac{1}{j2+2}\right| = 2\left|\frac{1}{\sqrt{4+4}}\right|$$

$$= \frac{2}{2\sqrt{2}} = 2^{\frac{-1}{2}} = 2^{-0.5}$$

$$\text{Phase is } \tan^{-1}\frac{\text{imaginary}}{\text{Real}} = \tan^{-1} - \left.\left(\frac{\omega}{2}\right)\right|_{\omega=2}$$

$$= -45^\circ = -0.25\pi$$

Thus $Y = 2^{-0.5}\cos(2t - 0.25\pi)$

Second Method

$$x(t) = 2\cos 2t$$

$$X(s) = 2\frac{s}{s^2+4}$$

$$h(t) = e^{-2t}u(t)$$

$$H(s) = \frac{1}{s+2}$$

$$Y(s) = X(s)H(s)$$

$$= 2\frac{s}{(s^2+4)}\cdot\frac{1}{(s+2)}$$

$$= \frac{As+B}{s^2+4} + \frac{C}{s+2}$$

or $2s = As^2 + Bs + 2As + 2B + Cs^2 + C4$

$$= (A + +C)s^2 + (2A+B) + 2B + 4C$$

Comparing the coefficients on both sides

$$A + C = 0$$

$$2A + B = 2$$

$$4C + 2B = 0$$

On solving

$$A = \frac{1}{2};\ B = 1;\ C = \frac{-1}{2}$$

$$Y(s) = \frac{1}{2}\left[\frac{s+2}{s^2+4} - \frac{1}{s+2}\right]$$

$$= \frac{1}{2}\left[\frac{s}{(s^2+4)} + \frac{2}{s+4} - \frac{1}{s-1}\right]$$

$$y(t) = \frac{1}{2}\left[\cos 2t + \sin 2t - e^{-2t}\right]$$

$$= \frac{1}{2}\left[\sqrt{2}\cos\left(2t - \frac{\pi}{4}\right) - e^{-2t}\right]$$

Hence for steady state response,

$$\lim_{t\to\infty} y(t) = \frac{1}{\sqrt{2}}\cos\left(2t - \frac{\pi}{4}\right)$$

$$= 2^{-0.5}\cos(2t - 0.25\,\pi)$$

16. Ans. (c)

Solution:

$$H(f) = \frac{1}{j2\pi R\,f\,C + 1},\ H(0) = 1$$

Now, as $\dfrac{|H(f_1)|}{H(0)} \geq 0.95$

Then $\dfrac{1}{\sqrt{4\pi^2 R^2 f_1^2 C^2 + 1}} \geq 0.95$

or $4\pi^2 R^2 f_1^2 C^2 \leq 0.108$

or $(f_1)_{\max} = \dfrac{\sqrt{0.018}}{2\pi\,RC} = \dfrac{\sqrt{0.018}}{2\times 3.14\times 10^{-3}} = 52.2$ Hz

17. Ans. (a)

Solution:

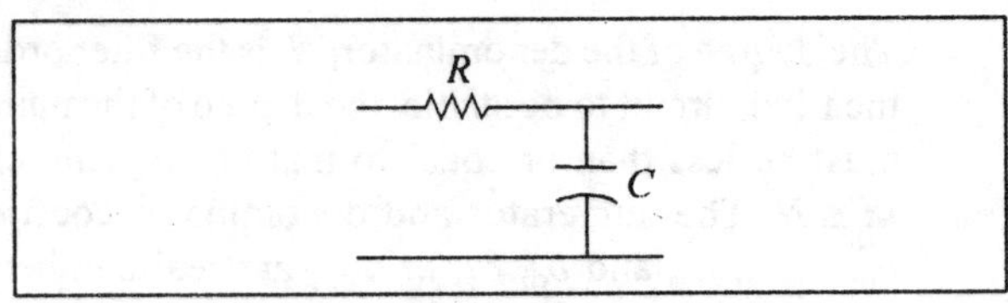

$$H(\omega) = \frac{1}{1 + j\omega RC}$$

$$\theta(\omega) = -\tan^{-1}\omega RC$$

Group delay $\tau_g = \dfrac{-d\theta(\omega)}{-d\omega} = \dfrac{RC}{1+\omega^2 R^2 C^2}$

$$= \frac{10^{-3}}{1 + 4\pi^2 \times 10^4 \times 10^{-6}}$$

$$= 0.717 \text{ ms}$$

18. Ans. (a)

Solution: Transfer function

$$= G(s) = \frac{\left(\frac{s}{5}+1\right)\left(\frac{s}{100}+1\right)\left(\frac{s}{200}+1\right)}{\left(\frac{s}{.01}+1\right)\left(\frac{s}{1}+1\right)\left(\frac{s}{80}+1\right)}$$

$$\angle G(j\omega) = \tan^{-1}\frac{\omega}{5} + \tan^{-1}\frac{\omega}{100} + \tan^{-1}\frac{\omega}{200}$$

$$- \tan^{-1}\frac{\omega}{.01} - \tan^{-1} - \tan^{-1}\frac{\omega}{80}$$

At $\omega = 20$, transfer function. $\angle G(j\omega)$

$$= \tan^{-1}(4) + \tan^{-1}\frac{1}{5} + \tan^{-1}\frac{1}{10}$$

$$- \tan^{-1}(2000) - \tan^{-1} 20 - \tan^{-1}\frac{1}{4} = -90°$$

19. Ans (a)

Solution:

The diagram of response is

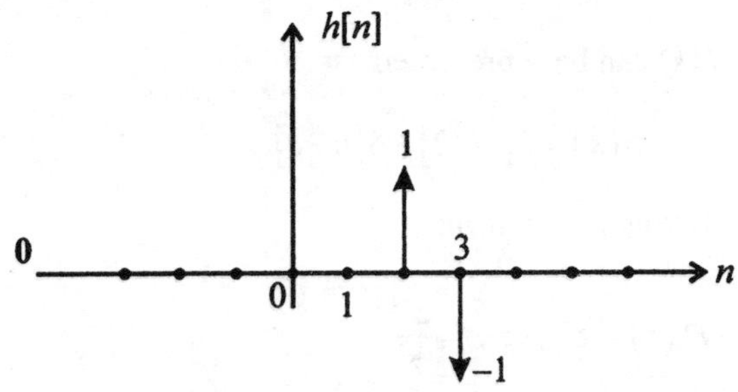

$$H(z) = z^{-2} - z^{-3}$$

or $H(e^{j\omega}) = e^{-j2\omega} - e^{-j3\omega}$

$$= e^{-j\frac{5}{2}\omega}\left[e^{-j\frac{\omega}{2}} - e^{-j\frac{\omega}{2}}\right]$$

$$= e^{-j\frac{5}{2}\omega}\left[j2\sin\frac{\omega}{2}\right]$$

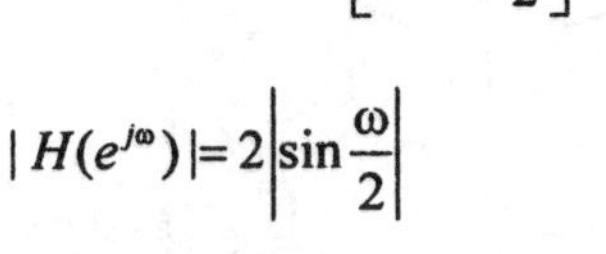

$$|H(e^{j\omega})| = 2\left|\sin\frac{\omega}{2}\right|$$

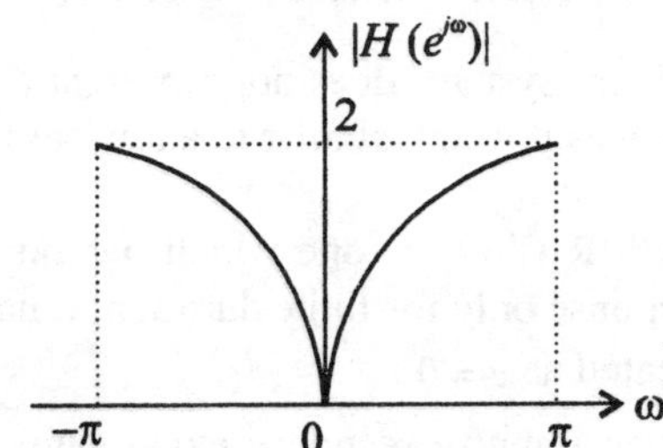

This is the impulse response of a HPF

Also, $h(n)$ contains finite lenght and finite magnitude value, so, it is FIR high pass filter

Hence it represent a FIR filter.

Second method

* In a low pass filter poles are located in unit circle at points corresponding to $\omega = 0$ and zero are located near or on unit circle a points corresponding to high frequencies (near $\omega = \pi$). Such locations are as shown below

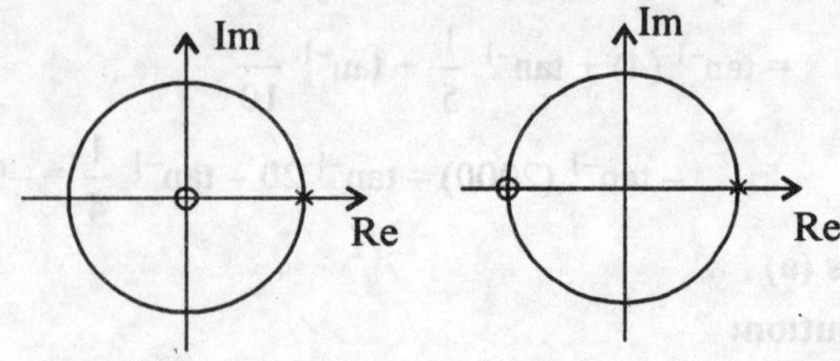

Given $h[2]=1$, $h[3]=-1$ and $h[k]=0$, otherwise

$h(k)$ can be represented as

$$h(k)=\delta[n-2]-\delta[n-3]$$

Taking z-transform,

$$H(z)=z^{-2}-z^{-3}=\frac{z-1}{z^3}$$

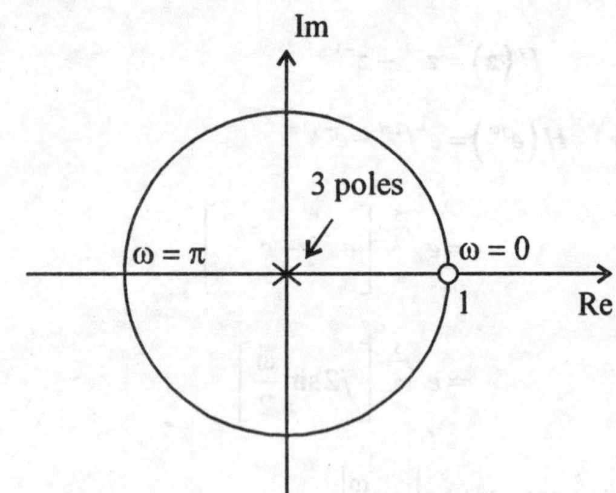

Figure: Pole-zero diagram of $\frac{(z-1)}{z}$

The given system does not represent a low pass filter because as it is not similar to a low pass filter.

* An FIR filter is one which has non-zero impulse response only for finite duration. It has all the poles located at $z=0$.

As given impulse response exists only for $k=2$ and $k=3$, therefore, given response represents a FIR filter.

20. Ans. (b)

Solution: $x(n)=\sin(\omega_o n+\phi)$

$$X(_n-n_o)=\sin[\omega_o(n-n_o)+\phi]$$

$$=\sin[\omega_o n-\omega_o n_o+\phi]$$

$$y(n)=Ax(n-n_o)$$

$$Y(e^{j\omega})=A\exp[-j\omega n_o]X(e^{j\omega})$$

$$H(e^{j\omega})=[Y(e^{j\omega})/X(e^{j\omega})]$$

$$=A\exp[-j\omega n_o]=|H(e^{j\omega})|\angle H(e^{j\omega})$$

$$\angle H(e^{j\omega})=-\omega n_o$$

$$\angle H(e^{j\omega})=-\omega_o n_o+2\pi k$$

General solution; k is a arbitrary integer

21. Ans. (a)

Solution:

$$G(z)=\alpha z^{-1}+\beta z^{-3}$$

$$G(z)\big|_{z=e^{j\omega}}=\alpha e^{-j\omega}+\beta e^{-3j\omega}$$

Let $\alpha=\beta$, then $G(z)=\alpha z^{-j2\omega}\left(e^{j\omega}+e^{-j\omega}\right)$

$$=2\alpha e^{-j2\omega}\cos\omega$$

$\left|G(e^{j\omega})\right|=|2\alpha\cos\omega|$ and $\theta(\omega)=-2\omega$

$$\Rightarrow\ \frac{\theta(\omega)}{\omega}=2=\text{constant}$$

$\Rightarrow\theta(\omega)$ is proportional to ω.

22. Ans. (d)

Solution: The filter transfer function $T(s)$ can be written as the ratio of two polynomials as

$$T(s)=\frac{a_M s^M+a_{M-1}s^M+\cdots+a_0}{s^N+a_{N-1}s^N+\cdots+b_0}$$

The degree of the denominator, N, is the filter order. For the filter circuit to be stable, the degree of the numerator must be less than or equal to that of the denominator: $M\le N$. The numerator and denominator coefficients, $a_0, a_1, \ldots a_M$ and $b_0, b_1, \ldots a_{N-1}$ are real numbers. The polynomials in the numerator and denominator can be factored, and $T(s)$ can be expressed in the form

$$T(s)=\frac{a_M(s-z_1)((s-z_2))\cdots(s-z_M)}{(s-p_1)(s-p_2)\cdots(s-p_N)}$$

The numerator roots, $z_1, z_2, \ldots, z_M$, are the transfer-function zeros, or **transmission zeros**; and the denominator roots, $p_1, p_2, \ldots, p_N$, are the transfer function poles, or the natural poles. Each transmission zero or pole can be either a real or a complex number. Complex zeros and poles, however, must occur in conjugate pairs. Thus, if $-1+j2$ happens to be a zero, then $-1+j2$ also must be a zero

Write equation of all filter

$$T(s)=\frac{a_m s^m+a_{m-1}s^{m-1}+\ldots a_o}{b_n s^n+a_{n-1}s^{n-1}+\ldots b_o}$$

The degree of denominator N, is the filter order for the circuit to be stable.

If the transmission zero are located on the $j\omega$ axis, at the complex-conjugate locations $\pm j\omega_n$, then the mag nitude reponse exhibits zero transmission at $\omega = \omega_n$ and ω_n is known as the **notch frequency**. Three cases of the second-order notch filter are possible: the regular notch,obtained when $\omega_n = \omega_0$, the low-pass notch obtained when $\omega_n > \omega_0$, and the the high-pass notch, obtained when $\omega_n < \omega_0$ in all notch cases, the transmission at dc and at $s = \infty$ is finite. This is so because there are no transmission zeros at either $s = 0$ or $s = \infty$.

Since in the filter stopband the transmission requried to be zero or small, the filter transmission zero are usually placed onthe $(j\omega)$ axis at stopband frequencies. There is complex conjugate zero near the origine which give attenuation. Then indicate notch filter given attenuation on perperticular frequency only

23. Ans. (a)

Solution: Since, pole zero plot of all pass filter is symmetrical about $j\omega$ axis, hence only possible answers is.

24. Ans. (a)

Solution: Magnitude Function for Nth order Butterworth filter with passband edge ω_p is given by

$$|T(j\omega)| = \frac{1}{\sqrt{1+\varepsilon^2\left(\frac{\omega}{\omega_p}\right)^{2N}}}$$

At $\omega = \omega_p$, $|T(j\omega_p)| = \frac{1}{\sqrt{1+\varepsilon^2}}$

Here, $\omega_p = 2\pi \times 1000$

$$\frac{1}{\sqrt{1+\varepsilon^2}} = 0.9$$

$$\Rightarrow \quad 1+\varepsilon^2 = \frac{1}{0.81}$$

$$\Rightarrow \quad \varepsilon = 0.4843$$

$\omega_s = 2\pi \times 1500\, rad\, s^{-1}$

$$|T(j\omega)| = \frac{1}{\sqrt{1+0.23\left(\frac{3}{2}\right)^{2N}}}$$

$$\Rightarrow \quad 1+0.23\left(\frac{3}{2}\right)^{2N} = \frac{1}{|T(j\omega)|^2}$$

At stop band edge $|T(j\omega)| = 0.2$

$$\therefore \quad 1+0.23\,(1.5)^{2N} = \frac{1}{0.04}$$

Calculating it, we get N = 6

Second Method

Given,

Pass band

$0.9 \le |H(j\omega)| \le 1$

$0 \le 2\pi \times 1000$

Stop Band: $|H(j\omega)| \le 2$

$|1(\omega)| \ge 2\pi \times 1500$

Then

$\delta_1 = 0.9$

$\delta_2 = 0.2$

$\omega_2 = 2\pi \times 1000$

$\omega_2 = 2\pi \times 1500$

$$\frac{\omega_2}{\omega_1} = \frac{1500}{1000} = 1.5$$

order of the filter must be,

$$N \ge \frac{\frac{1}{2}\log\left\{\left(\frac{1}{\delta_2^2}-1\right)/\left(\frac{1}{\delta_1^2}-1\right)\right\}}{\log\left(\frac{\omega_2}{\omega_1}\right)}$$

$N \ge 5.70$

So, $N_{min} = 6$

25. Ans. (a)

26. Ans. (a)

Solution:

In digital filter to reject the 50 Hz interference, sampling frequency must be as low as the noise frequency so that they are separated in time domain and has less interference.

Therefore, $f_s = 50Hz$

27. Ans. (c)

Solution:

Given, sampling frequency

$f_T = 10\,\text{kHz}$

$$T = \frac{1}{f_T} = 10^4 \text{ sec}$$

Pass band edge frequency

$f_p = 2kHz$

Stop Band edge frequency

$f_s = 3kHz$

Angular frequencies are

$\omega_p = 4\pi \times 10^3$ rad / sec

$\omega_s = 6\pi \times 10^3$ rad / sec

Warped edge frequencies are

$$\Omega_s = 2\tan^{-1}\frac{\Omega T}{2}$$

$$= 2\tan^{-1}(3\pi \times 0.1)$$

$$\Omega_p = 1.511$$

$$\Omega_p 2\tan^{-1}(2\pi \times 0.1) = 1.12$$

$$\therefore \quad \frac{\Omega_s}{\Omega_p} = \frac{1.511}{1.12} = 1.349$$

28. Ans. (a)

Solution:

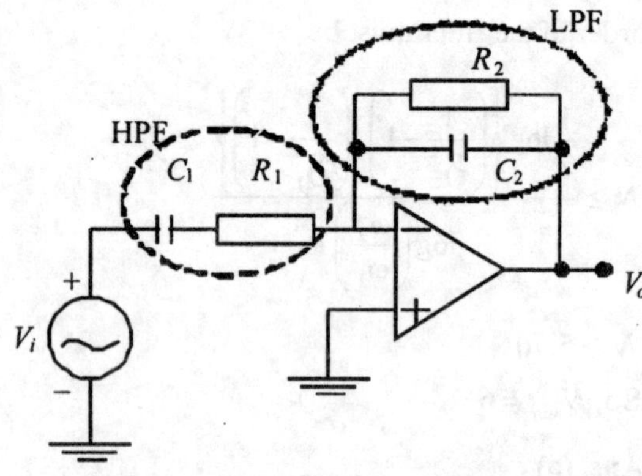

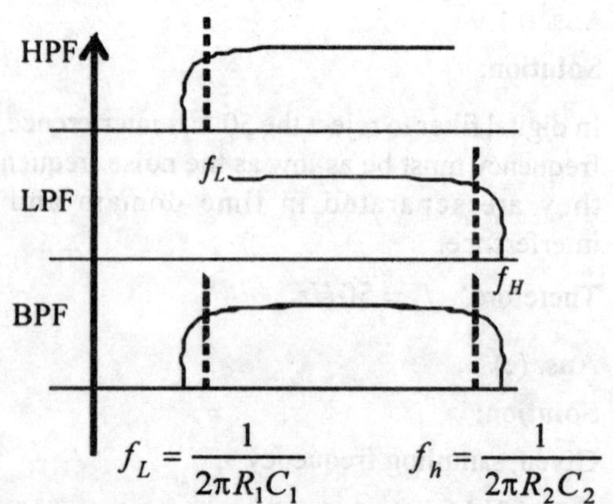

$$f_L = \frac{1}{2\pi R_1 C_1} \qquad f_h = \frac{1}{2\pi R_2 C_2}$$

Second Method

KCL at the inverting terminal

$$\frac{V_i - 0}{R_1 + \frac{1}{C_1 s}} = \frac{0 - V_0}{\frac{R_2}{R_2 C_2 s + 1}}$$

$$\text{or, } A = \frac{V_0}{V_i} = \frac{-R_2 C_1 s}{(R_2 C_2 s + 1)(R_1 C_1 s + 1)}$$

$$\text{or, } A_v(f) = \frac{-R_2/R_1}{1 - j(f_L/f)} \cdot 1 + j(f/f_H)$$

$$= [-A_L(jf) \cdot A_H(jf)]$$

$$\text{Where, } f_L = \frac{1}{2\pi R_1 C_1}, f_H = \frac{1}{2\pi R_2 C_2}$$

29. Ans. (a)

Solution: Since, pole zero plot of all pass filter is symmetrical about jw axis, hence only possible answers is

$$\frac{1 - s\tau}{1 + s\tau}$$

Second Method

For an all-pass analog filter

$|H\omega| = \text{constant}$

$$\text{If } |Hs| = \frac{1 - s\tau}{1 + s\tau},$$

$$H(\omega) = \frac{1 - j\omega\tau}{1 + j\omega\tau}$$

$$|H(\omega)| = \frac{\sqrt{1 + \omega^2\tau^2}}{1 + \omega^2\tau^2} = 1$$

30. Ans. (d)

Solution:

From pole zero plot Transfer function of digital filter

$$H(z) = \frac{(z-1)(z-2)(z+1/2)(z-1/2)(z+1)(z+2)}{z^6}$$

$$= \frac{(z^2 - 4)(z^2 - 1)(z^2 - 1/4)}{z^6}$$

$$= \frac{z^6 - \frac{21}{4}z^4 + \frac{21}{4}z^2 - 1}{z^6}$$

$$= 1 - \frac{21}{4}z^{-2} + \frac{21}{4}z^{-4} - z^{-6}$$

$$h[n] = \delta[n] - \frac{21}{4}\delta[n-2] + \frac{21}{4}\delta[n-4] - \delta[n-6]$$

In FIR system, the impulse response consist of finite number of samples hence it is FIR filter.

31. Ans. (c)

Solution:

FIR system with impulse response has a zero at $z=\frac{1}{2}e^{+j\pi/4}$, Since impulse response has to be real therefore corresponding to this zero will be another zero $z=\frac{1}{2}e^{-j\pi/4}$, The largest set will have zero in conjugate pairs which available.0 only in option (c)

Second Method

Remember the property of linear phase FIR filters with a real impulse response regarding the location of its zero.

If $z_1 = r\,e^{j\phi}$ is a complex zero.

Then 3 more complex zeros occur as given below

$$z_2 = \frac{1}{z_1} = \frac{1}{r}e^{j\phi}$$

$$z_3 = z_1 = r\,e^{j\phi}$$

and $z_1 = z_1^{*} = \frac{1}{r}e^{j\phi}$

are also zeros

However the number of zeros at $z=\pm 1$ depends upon the Type of Filter.

(Type 1 or 2 or 3 and 4)

For a zero at $z_1=\frac{1}{2}e^{j\frac{\pi}{4}}$, the largest set of remaining zeros that can be obtained from this information are

$$\frac{1}{2}e^{-j\frac{\pi}{4}},\ 2e^{-j\frac{\pi}{4}},\ 2e^{j\frac{\pi}{4}}$$

32. Ans. (a)

Solution:

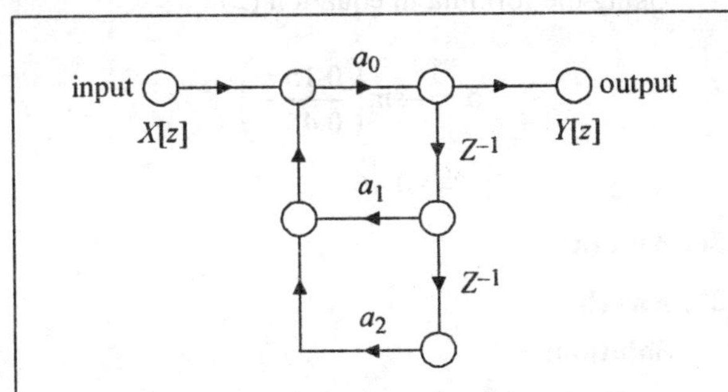

$$Y(z)=\left[X(z)+Y(z)\cdot z^{-2}a_2+Y(z)a_1\cdot z^{-1}\right]\times a_0$$

$$Y(z)-Y(z)a_0a_2\,z^{-2}-Y(z)a_0a_1z^{-1}=X(z)\,a_0$$

or $$\frac{Y(z)}{X(x)}=H(z)=\frac{a_0}{1-a_0a_1\,z^{-1}-a_0a_2\,z^{-2}}$$

$$=\frac{1}{\frac{1}{a_0}-\frac{a_1}{a_0}z^{-1}-\frac{a_2}{a_0}z^{-2}}$$

Given $H(z)=\dfrac{1}{1-0.7z^{-1}+0.13z^{-2}}$

Comparing with the given H(z), we get

$$\frac{1}{a_0}=1,\ -\frac{a_1}{a_0}=0.7,\ -\frac{a_2}{a_0}=0.13$$

$$a_0=1,\ a_1=0.7,\ a_2=-0.13$$

33. Ans. (c)

Solution:

Given, Sampling frequency

$$f_T = 10\,\text{kHz}$$

$$T=\frac{1}{f_T}=10^4\ \text{sec}$$

Pass band edge frequency

$$f_p = 2kHz$$

Stop Band edge. Frequency

$$f_s = 3kHz$$

Angular frequencies are

$$\omega_p = 4\pi\times10^3\ \text{rad/sec}$$

$$\omega_s = 6\pi\times10^3\ \text{rad/sec}$$

Warped edge frequencies are

$$\Omega_s = 2\tan^{-1}\frac{\Omega T}{2}$$

$$=2\tan^{-1}(3\pi\times0.1)$$

$$\Omega_s = 1.511$$

$$\Omega_p = 2\tan^{-1}(2\pi\times0.1)$$

$$=1.12$$

$$\therefore\ \frac{\Omega_s}{\Omega_p}=\frac{1.511}{1.12}$$

$$=1.349$$

34. Ans. (a)

Solution: Magnitude function for *n*th order Butterworth filter with passband edge ω_p is given by

$$|T(j\omega)| = \frac{1}{\sqrt{1+\varepsilon^2\left(\frac{\omega}{\omega_p}\right)^{2N}}}$$

At $\omega = \omega_p$, $|T(j\omega_p)| = \frac{1}{\sqrt{1+\varepsilon^2}}$

Here, $\omega_p = 2\pi \times 1000$

$$\frac{1}{\sqrt{1+\varepsilon^2}} = 0.9$$

$$\Rightarrow \quad 1+\varepsilon^2 = \frac{1}{0.81}$$

$$\Rightarrow \quad \varepsilon = 0.4843$$

$\omega_s = 2\pi \times 1500\, rad\, s^{-1}$

$$|T(j\omega)| = \frac{1}{\sqrt{1+0.23\left(\frac{3}{2}\right)^{2N}}}$$

$$\Rightarrow \quad 1+0.23\left(\frac{3}{2}\right)^{2N} = \frac{1}{|T(j\omega)|^2}$$

At stop band edge $|T(j\omega)| = 0.2$

$$\therefore \quad 1+0.23\,(1.5)^{2N} = \frac{1}{0.04}$$

Calculating it, we get N = 6

Second Method

Given,

Pass band

$0.9 \le |H(j\omega)| \le 1$

$0 \le 2\pi \times 1000$

Stop Band: $|H(j\omega)| \le 2$

$|1(\omega)| \ge 2\pi \times 1500$

Then $\delta_1 = 0.9$

$\delta_2 = 0.2$

$\omega_2 = 2\pi \times 1000$

$\omega_2 = 2\pi \times 1500$

$$\frac{\omega_2}{\omega_1} = \frac{1500}{1000} = 1.5$$

order of the filter must be,

$$N \ge \frac{\frac{1}{2}\log\left\{\left(\frac{1}{\delta_2^2}-1\right)\right\}\Big/\left(\frac{1}{\delta_1^2}-1\right)}{\log\left(\frac{\omega_2}{\omega_1}\right)}$$

$N \ge 5.70$

So, $N_{\min} = 6$

35. Ans. (a)

Solution:

Lowpass-to-Lowpass transformation

$$z^{-1} = \frac{1-\alpha\,\hat{z}}{\hat{z}-\alpha} \quad \text{... (1)}$$

$$\text{or } e^{-j\omega} = \frac{1-\alpha\, e^{j\hat{\omega}}}{e^{j\hat{\omega}}-\alpha}$$

Where ω is the frequency of the prototype, $\hat{\omega}$ is the frequency after transformation

If can be shown that $\alpha = \dfrac{\sin\left(\frac{\omega-\hat{\omega}}{2}\right)}{\sin\left(\frac{\omega-\hat{\omega}}{2}\right)}$ ... (2)

Given: Notch frequency of prototype digital filter, $f_N = 60\ Hz$

With $f_s = 400\ Hz$, $\omega_N = 2\pi\frac{60}{400} = 0.3\,\pi$ rad

Notch frequency after transformation $\hat{f}_N = 120\ Hz$

$$\hat{\omega}_N = \frac{2\pi \times 120}{400} = 0.6\,\pi \text{ rad}$$

$$\frac{\omega_N - \hat{\omega}_N}{2} = -0.15\,\pi$$

$$\frac{\omega_N + \hat{\omega}_N}{2} = -0.45\,\pi$$

Using the formula in equation (2)

$$\alpha = -\sin\left(\frac{0.15\,\pi}{0.45\,\pi}\right)$$

$$= -0.46$$

36. Ans. (a)

37. Ans (b)

Solution:

$$H(z) = 1 - z^{-6}$$

$$H(e^{j\omega}) = 1 - e^{-6j\omega}$$

$$= 1-(\cos 6\omega 0\, j \sin 6\omega)$$

$= (1-\cos 6\omega) + j\sin 6\omega$

$\left|H\left(e^{j\omega}\right)\right| = \sqrt{(1-\cos^2 6\omega)^2 + \sin^2 6\omega}$

$= \sqrt{1+(\cos^2 6\omega + \sin^2 6\omega) - 2\cos 6\omega}$

$= \sqrt{2-2\cos 6\omega}$

$= \sqrt{2-2(1-2\sin^2 3\omega)}$

$= \sqrt{2\times 2\sin^2 3\omega}$

$= \left|H\left(e^{j\omega}\right)\right| = 2\sin 3\omega$ **for** $0 \le \omega \le 2\pi$

so, $\left|H(e^{j\omega})\right| = 0$, if $3\omega = n\pi$

(where $n_{max} = 6$)

$\omega = \frac{n\pi}{3}$

$= \frac{6\pi}{3}$

$= 2\pi$

$\omega = 2\pi f$

$\therefore \quad f = 1kHz$

38. Ans (b)

39. Ans (b)

Solution: Given, digital filter transfer function as

$$H(z) = \frac{P_0 + P_1 z^{-1} + P_3 z^{-3}}{1 + d_3 z^{-3}}$$

$$= H_1(z)\, H_2(z)$$

here, $H_1(z) = P_0 + P_1 z^{-1} + P_3 z^{-3}$

$$H_2(z) = \frac{1}{1 + d_3 z^{-3}}$$

For direct form-1,

realise it as

$X(z) \longrightarrow \boxed{H_1} \xrightarrow{W} \boxed{H_2} \longrightarrow Y(z)$

Hence it is realised as

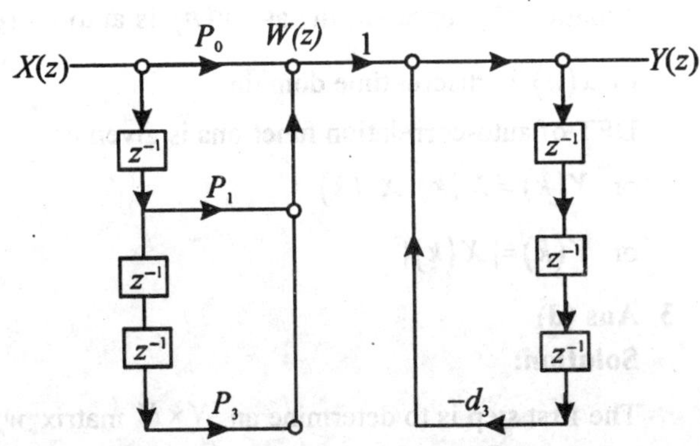

Hence delay elements required are 6

In direct form 2

$H_1(z)$ and $H_2(z)$ are interchanged so realization is as below

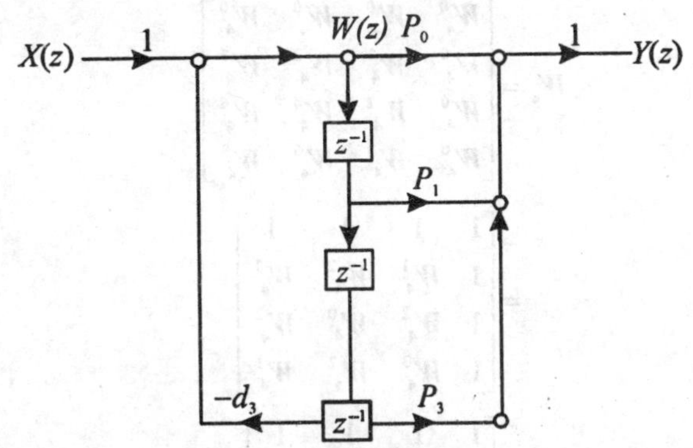

Hence delay elements required are 3

4.9 Discrete Fourier Transform

1. **Ans. (d)**

   Solution:

   Computation of butterfly requires one complex multiplication and two complex addition

2. **Ans. (a)**

   Solution: Direct property

   Convolution in time domain $\longleftrightarrow$ Multiplication in frequency domain

   Second Method

   Given

   $x(n) \xleftrightarrow{DFT} X(k)$ ...(1)

   $y(n) = \sum_{r=0}^{N-1} x(r)\, x(n+r)$ ...(2)

equation (2) represent the as $y(n)$ is auto-correlation of $x(n)$ in discret time domain

DFT of auto-correlation functions is given as

or $Y(k) = X(k) \cdot X^*(k)$

or $Y(k) = |X(k)|^2$

3. Ans (d)

Solution:

The first step is to determine an $N \times N$ matrix w_N

$$W_N = \begin{bmatrix} W_N^0 & W_N^0 & --- & W_N^0 \\ W_N^0 & W_N^1 & W_N^2 --- & W_N^{N-1} \\ W_N^0 & W_N^2 & W_N^4 --- & W_N^{2(N-1)} \\ W_N & W_N^{N-1} & W_N^{2(N-1)} & W_N^{(N-1)(N-1)} \end{bmatrix}$$

Here $N = 4$

$$W_4 = \begin{bmatrix} W_4^0 & W_4^0 & W_4^0 & W_4^0 \\ W_4^0 & W_4^1 & W_4^2 & W_4^3 \\ W_4^0 & W_4^2 & W_4^4 & W_4^6 \\ W_4^0 & W_4^3 & W_4^6 & W_4^9 \end{bmatrix}$$

$$= \begin{bmatrix} 1 & 1 & 1 & 1 \\ 1 & W_4^1 & W_4^2 & W_4^3 \\ 1 & W_4^2 & W_4^0 & W_4^2 \\ 1 & W_4^3 & W_4^2 & W_4^1 \end{bmatrix}$$

$$= \begin{bmatrix} 1 & 1 & 1 & 1 \\ 1 & -j & -1 & j \\ 1 & -1 & 1 & -1 \\ 1 & j & -1 & -j \end{bmatrix}$$

Then, $x_4 = W_4 x_4 = \begin{bmatrix} 1 \\ 0 \\ 2 \\ 3 \end{bmatrix}$

$$= \begin{bmatrix} 1 & 1 & 1 & 1 \\ 1 & -j & -1 & j \\ 1 & -1 & 1 & -1 \\ 1 & j & -1 & -j \end{bmatrix} \begin{bmatrix} 1 \\ 0 \\ 2 \\ 3 \end{bmatrix}$$

$$= \begin{bmatrix} 6 \\ -1+3j \\ 0 \\ -1-3j \end{bmatrix}$$

4. Ans. (c)

Solution:

DFT points are complex conjugates of each other and they are symmetric to the middle point.

$x(0) = x^*(7)$

$x(1) = x^*(6)$

$x(2) = x^*(5)$

$x(3) = x^*(4)$

$\Rightarrow$ Last two points will be and $x^*(1)$ $x^*(0)$

$= 1 + j3$ and 5

5. Ans. (a)

Solution:

y(*n*) is time shifted and time scaled form of *x*(*n*)

One of the simplest method of finding *y*(*n*) is to first plot $x(n-1)$.

Then time expand $x(n-1)$ by a factor of 2 to obtain $x\left(\frac{n}{2}-1\right)$

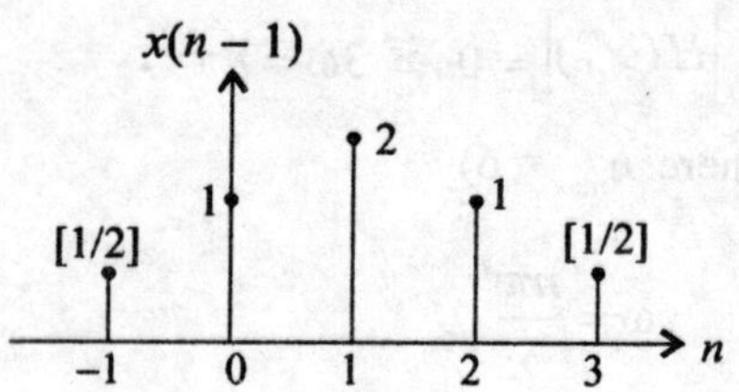

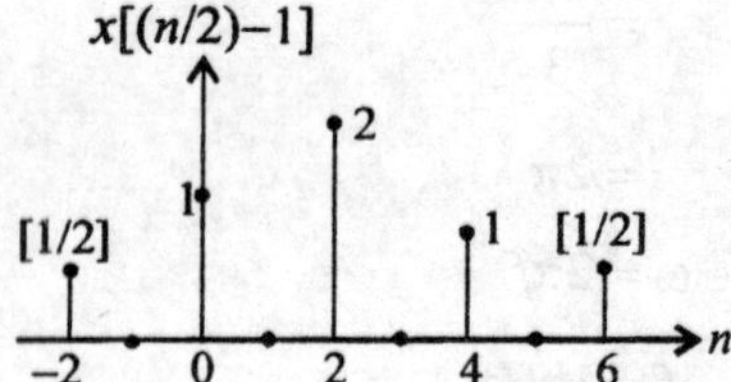

$y(2) = x(0) = 2$

$y(4) = x(1) = 1$

6. Ans. (c)

Solution:

$$y(2n) = \left\{ \frac{1}{2}, 1, 2, 1, \frac{1}{2} \right\}, \ldots \quad \underset{n=0}{\uparrow}$$

Fourier transform of $y(n) = \sum_{n=-\infty}^{\infty} y(n) e^{-j\omega n}$

$= [1/2]\, e^{j\omega} + 1 + 2e^{-j\omega} + e^{j2\omega} + [1/2]\, e^{-j3\omega}$

$= e^{j\omega}\, [\cos 2\omega + 2\cos\omega + 2]$

7. Ans. (b)

Solution: $X(e^{j\omega}) = \sum_{-\infty}^{\infty} x[n]\, e^{-j\omega n}$

$= x(-3)\, e^{3j\omega} + x(-2)\, e^{2j\omega} + x(-1)\, e^{j\omega} + x(0)\, e^{-j0} + x(1)\, e^{-j\omega}$

$= e^{3j\omega} + e^{2j\omega} + 5 + e^{-j\omega}$

$\therefore \int_{-\pi}^{\pi} X(e^{j\omega})\, d\omega$

$= \int_{-\pi}^{\pi} (e^{3j\omega} + e^{2j\omega} + 5 + e^{-j\omega})\, d\omega$

$= \left| \frac{e^{3j\omega}}{3j} + \frac{e^{2j\omega}}{2j} + 5\omega + \frac{e^{-j\omega}}{-j} \right|_{-\pi}^{\pi}$

$= \left[\left(\frac{-1}{3j} + \frac{1}{2j} + 5\pi + \frac{1}{j} \right) - \left(\frac{-1}{3j} + \frac{1}{2j} - 5\pi + \frac{1}{j} \right) \right]$

$= 2\,(5\pi) = 10\,\pi$

8. Ans. (d)

Solution: $x(n) = [1/2]^n\, u(n)$; $x^2(n) = [1/2]^{2n}\, u(n)$

$y(n) = [1/4]^n\, u(n)$

$Y(e^{j\omega}) = \frac{1}{1 - [1/4]e^{-j\omega}} \Rightarrow Y(e^{j\omega}) = \frac{1}{1 - [1/4]}$

$Y(e^{jo}) = [4/3]$

4.10 Miscellaneous

1. Ans (c)

Solution: $E = pt = E_1 + E_2$

$E_1 = 10\,V \times 2A \times 10 \text{ min}$

$= 20\,VA \times 60 \times 10 = 12000\,J$

$E_2 = \frac{1}{2} \times 2 \times 2 \times 60 \times 10 = 1200\,J$

$E = 12000\,J + 1200\,J$

$= 13200\,J = 13.2\,kJ$

2. Ans (a)

Solution:

$x(t) = \sin(t+1)$

Then $\omega = 1$

$H(s) = \frac{s^2+1}{s^2+2s+1}$

$H(j\omega) = \left. \frac{-\omega^2+1}{-\omega^2+2j\omega+1} \right|_{\omega=1} = 0$

Thus output is zero for all sampling frequencies

Second Method

LTI system, $H(s) = \frac{s^2+1}{s^2+2s+1}$

Input, $x(t) = \sin(t+1) = \sin \omega(t+1)$

$\rightarrow \omega = 1$

or $X(s) = \frac{e^s}{s^2+1}$

or $Y(s) = X(s) \cdot H(s)$

$Y(s) = \left(\frac{e^s}{s^2+1} \right) \cdot \left(\frac{s^2+1}{s^2+2s+1} \right)$

or $Y(s) = \frac{e^s}{(s+1)^2}$

$y(\infty) = \lim_{s \to 0} sY(s)$

$= \lim s \cdot \frac{e^s}{(s+1)^2} = 0$

The output of system becomes zero at steady state, so $y(1)$ remain zero for all sampling frequency.

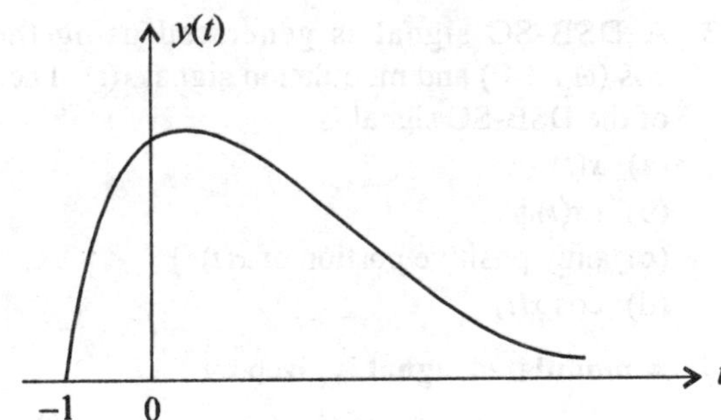

Or in otherway, taking inverse Laplace transform

$y(t) = (t+1)\, e^{-(t+1)}$

Hence when $t \rightarrow \infty$ (i.e. in steady state)

$y(t) \rightarrow 0$

3. Ans. (b)

Solution:

$I_{rms} = \frac{A_C}{\sqrt{2}}$

Average Power $= \frac{A_C^2}{2} = \frac{6^2}{2} = \frac{36}{2} = 18$ Watt

5 Communication Systems

5.1 Amplitude Modulation

1. The maximum power-efficiency of an AM modulation is
(a) 25 %
(b) 50 %
(c) 75 %
(d) 100 %

2. Which of the following demodulator(s) can be used for demodulating the signal $x(t) = 5\,(1 + 2\cos 2000\,\pi t)\cos 2000\,\pi t$.
(a) Envelope demodulator
(b) Square-law demodulator
(c) Synchronous demodulator
(d) None of the above

3. A DSB-SC signal is generated using the carrier $\cos(\omega_c t + \theta)$ and modulation signal $x(t)$. The envelope of the DSB-SC signal is
(a) $x(t)$
(b) $|x(t)|$
(c) only positive portion of $x(t)$
(d) $\cos x(t)$

4. **A modulated signal** is given by

$$s(t) = m_t(t)\cos(2\pi f_c t) + m_2(t)\sin(2\pi f_c t)$$

where the base-band signal $m_1(t)$ and $m_2(t)$ have bandwidth of 10 kHz and 15 kHz, respectively. The bandwidth of the modulated signal, in kHz is
(a) 10
(b) 15
(c) 25
(d) 30

5. The amplitude modulated waveform

$$s(t) = A_c[1 + K_a m(t)]\cos \omega_c t$$

is fed to an ideal envelope detector. The maximum magnitude of $K_a m(t)$ is greater than 1. Which of the following could be the detector output?
(a) $A_c m(t)$
(b) $A_c^2[1 + K_a m(t)]^2$
(c) $|A_c[1 + K_a m(t)]|$
(d) $A_c[1 + K_a m(t)]^2$

6. A message $m(t)$ band-limited to the frequency f_m has a power of P_m. The power of the output signal in the figure is

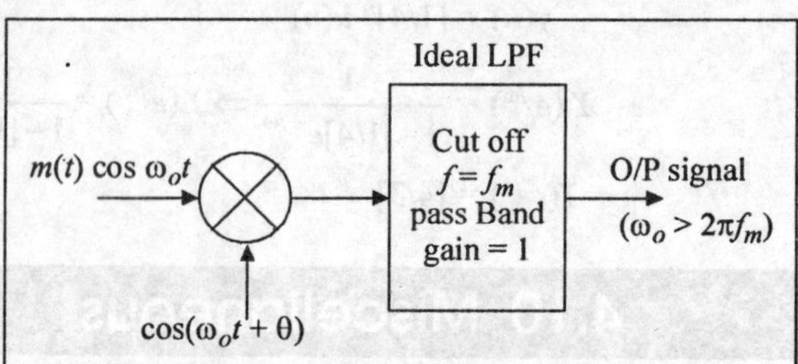

(a) $\dfrac{P_m \cos\theta}{2}$
(b) $\dfrac{P_m}{4}$
(c) $\dfrac{P_m \sin^2\theta}{4}$
(d) $\dfrac{P_m \cos^2\theta}{4}$

7. A 1 MHz sinusoidal carrier is amplitude modulated by a symmetrical square wave of period 100 sec. Which of the following frequency will NOT be present in the modulated signal?
(a) 990 kHz
(b) 1010 kHz
(c) 1020 kHz
(d) 1030 kHz

8. The input to a coherent detector is DSB-SC signal plus noise. The noise at the detector output is
(a) The in-phase component
(b) The quadrature-component
(c) Zero
(d) The envelope

9. A DSB-SC signal is to be generated with a carrier frequency $f_c = 1$ MHz using a non-linear device with the input-output characteristic

$$v_0 = a_0 v_i + a_1 v_i^2$$

where a_0 and a_1 are constants. The output of the non-linear device can be filtered by an appropriate band pass filter.

Let $\quad v_i = A_{ci}\cos(2\pi f_{ci} t) + m(t)$

Where $m(t)$ is the message signal.

Then the value of f_{ci} in MHz is

(a) 1.0 (b) 0.333
(c) 0.5 (d) 3.0

Common Data for Questions 10 and 11

Let $m(t) = \cos[(4\pi \times 10^3)\, t]$ be the message signal and $c(t) = 5\cos[(2\pi \times 10^6)\, t]$ be the carrier

10. $c(t)$ and $m(t)$ are used to generate an AM signal. The modulation index of the generated AM signal is 0.5. Then the quantity [Total sideband power / Carrier power] is

(a) $\frac{1}{2}$ (b) $\frac{1}{4}$
(c) $\frac{1}{3}$ (d) $\frac{1}{8}$

11. $c(t)$ and $m(t)$ are used to generate an FM signal. If the peak frequency deviation of the generated FM signal is three times the transmission bandwidth of the AM signal, then the coefficient of the term $\cos[2\pi\ (1008 \times 10^3 t)]$ in the FM signal (in terms is the Bessel coefficients) is

(a) $5\, J_4(3)$ (b) $\frac{5}{2}\, J_8(3)$
(c) $\frac{5}{2}\, J_8(4)$ (d) $5\, J_4(6)$

12. An AM signal is detected using an envelope detector. The carrier frequency and modulating signal frequency are 1 MHz and 2 kHz respectively. An appropriate value for the time constant of the envelope detector is

(a) 500 μsec (b) 20 μsec
(c) 0.2 μsec (d) 1 μsec

13. A 100 MHz carrier of 1 V amplitude and a 1 MHz modulating signal of 1 V amplitude are fed to a balance modulator. The output of the modulator is passed through and ideal high-pass filter with cut-off frequency of 100 MHz. The output of the filter is added with 100 MHz signal of 1 V amplitude and 90° phase shift as shown in figure. The envelope of the resultant signal is

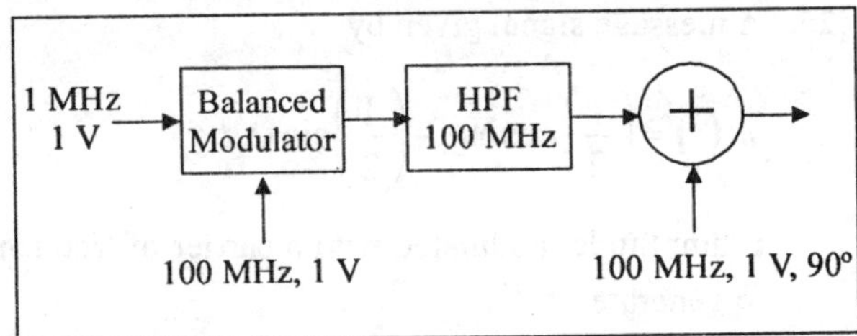

(a) Constant

(b) $\sqrt{1 + \sin(2\pi \times 10^6\, t)}$

(c) $\sqrt{\frac{5}{4} - \sin(2\pi \times 10^6\, t)}$

(d) $\sqrt{\frac{5}{4} + \cos(2\pi \times 10^6\, t)}$

14. Which of the following analog modulation scheme requires the minimum transmitted power and minimum channel bandwidth?

(a) VSB (b) DSB -SC
(c) SSB (d) AM

15. The diagonal clipping in amplitude demodulation (using envelope detector) can be avoided if RC time-constant of the envelope satisfies the following condition, (here W is message band width and ω_c is carrier frequency both in rad/sec)

(a) $RC < \frac{1}{W}$ (b) $RC > \frac{1}{W}$
(c) $RC < \frac{1}{\omega_c}$ (d) $RC > \frac{1}{\omega_c}$

Common data for Questions 16 and 17

Let $g(t) = p(t) * p(t)$, where $*$ denotes convolution and $p(t) = u(t) - u(t-1)$ with $u(t)$ being the unit step function.

16. The impulse response of filter matched to the signal $s(t) = \delta(t) - \delta(t-2) * g(t)$ is given as :

(a) $s(1-t)$ (b) $-s(1-t)$
(c) $-s(t)$ (d) $s(t)$

17. An amplitude modulated signal is given as

$$x_{AM}(t) = 100\,[p(t) + 0.5\, g(t)] \cos \omega_c t$$

in the interval $0 \le t \le 1$. One set of possible values of the modulating signal and modulation index would be.

(a) t, 0.5 (b) t, 1.0
(c) t, 2.0 (d) t^2, 0.5

Statement for linked Answer Questions 18 and 19

Consider the following Amplitude Modulated (AM) signal,

$$x_{AM}(t) = 10\,(1 + 0.5 \sin 2\pi f_m t) \cos 2\pi f_c t.$$

where $f_m < B$.

18. The average side band power for the AM signal given above is

(a) 25
(b) 12.5
(c) 6.25
(d) 3.125

19. The AM signal gets added to a noise with power spectral density $S_n(f)$ given in the figure below. The ratio of average sideband power to mean noise would be

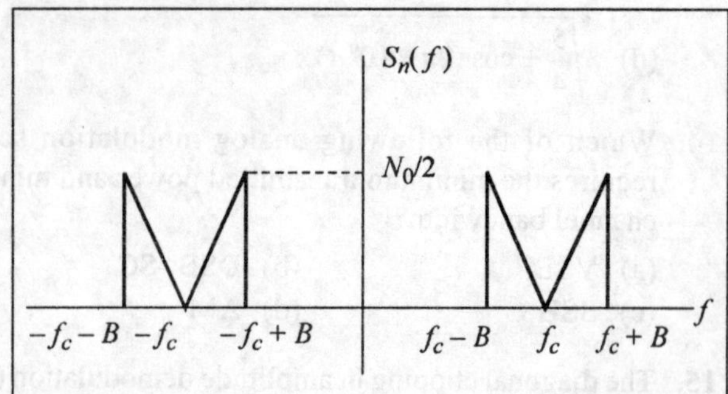

(a) $\dfrac{25}{8N_0B}$ (b) $\dfrac{25}{4N_0B}$

(c) $\dfrac{25}{2N_0B}$ (d) $\dfrac{25}{N_0B}$

20. Consider as system shown in Figure. Let X(*f*) and Y(*f*) denote the Fourier transforms of x(t) respectively. The ideal HPF has the cut-off frequency 10 KHz.

The positive frequency where Y(*f*) has spectral peaks are

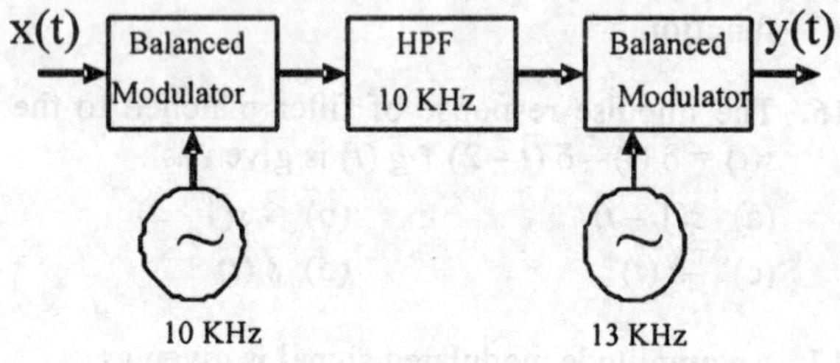

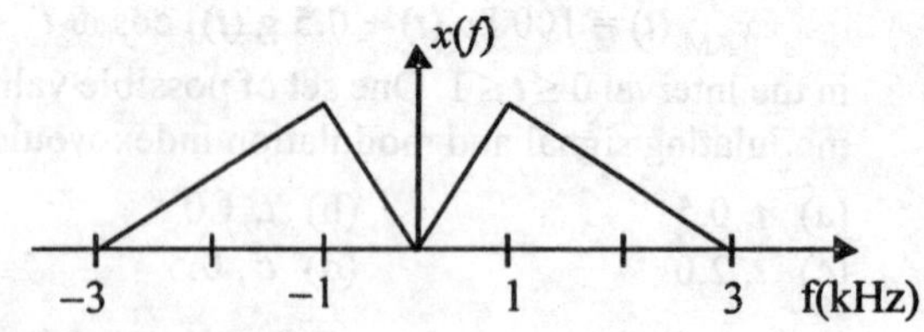

The positive frequency were Y(f) has spectral peaks are

(a) 1 kHz and 24 kHz

(b) 2 kHz and 24 kHz

(c) 1 kHz and 14 kHz

(d) 2 kHz and 14 kHz

21. The column-I lists the attributes and the column-2 lists the modulation systems. Match the attribute to the modulation system that best meets it.

Column – 1

P. Power efficient transmission of signals

Q. Most bandwidth efficient transmission of voice signals

R. Simplest receiver structure

S. Bandwidth efficient transmission of signals with significant dc compoent

Column - II

I. Conventional AM

II. FM

III. VSB

IV. SSB-SC

(a) P - IV, Q - II, R - I, S - III

(b) P - II, Q - IV, R- I, S - III

(c) P - III, Q - II, R - I, S- IV

(d) P- II, Q- IV, R - III, S - I

22. A message signal $m(t) = \cos 2000\,\pi\tau + 4\cos 4000\,\pi\tau$ modulates the carrier $c(t) = \cos 2\pi f_c t$ where $f_c = 1\text{MHz}$ to produce an AM signal. For demodulating the generated AM signal using an envelope detector, the time constant RC of the detector circuit should satisfy

(a) 0.5 ms < RC < 1ms (b) 1μs << RC < 0.5 ms

(c) RC << 1μS (d) RC >> 0.5 ms

23. Suppose that the modulating signal is $m(t) = 2\cos(2\pi f_m t)$ and the carrier signal is $x_c(t) = A_c\cos(2\pi f_c t)$. Which one of the following is a conventional AM signal **without** over-modulation ?

(a) $x(t) = A_c m(t)\cos(2\pi f_c t)$

(b) $x(t) = A_c[1+m(t)]\cos(2\pi f_c t)$

(c) $x(t) = A_c\cos(2\pi f_c t) + \dfrac{A_c}{4}m(t)\cos(2\pi f_c t)$

(d) $x(t) = A_c\cos(2\pi f_m t)\cos(2\pi f_c t) + A_c\sin(2\pi f_m t)\sin(2\pi f_c t)$

24. For a message signal $m(t) = \cos(2\pi f_m t)$ and carrier of frequency f_c which of the following represents a single-side -band (SSB) signal ?

(a) $\cos(2\pi f_m t)\cos(2\pi f_c t)$

(b) $\cos(2\pi f_c t)$

(c) $\cos[2\pi(f_c + f_m)t]$

(d) $[1+\cos(2\pi f_m t)]\cos(2\pi f_c t)$

25. A message signal given by

$$m(t) = \left(\frac{1}{2}\right)\cos\omega_1 t - \left(\frac{1}{2}\right)\sin\omega_2 t$$

is amplitude modulated with a carrier of frequency ω_c to generate

$s(t) = [1+m(t)]\cos\omega_c t$.

What is the power efficiency achieved by this modulation scheme?

(a) 8.33% (b) 11.11%

(c) 20 % (d) 25 %

26. Consider the amplitude modulated (AM) signal $A_c \cos\omega_c t + 2\cos\omega_m t \cos\omega_c t$. For demodulating the signal using envelope detector, the minimum value of A_c should be

(a) 2 (b) 1
(c) 0.5 (d) 0

27. For signal **amplitude modulated** to a depth of 100 by a sinusoidal signal, power is

(a) Same as the power of un modulated carrier
(b) Twice as the power of un modulated carrier
(c) 3/2 times the power of un modulated carrier
(d) 2/3 times the power of un modulated carrier

28. A carrier is **simultaneously modulated** by two sine waves with modulation indices of 0.3 and 0.4. If the modulated power is 10 kW, what is the total modulated power?

(a) 11.25 kW (b) 10 kW
(c) 10.125 kW (d) 10.5 kW

29. The **envelope detector** is a/an

(a) Synchronous detector
(b) Asynchronous detector
(c) Product demodulator
(d) Coherent detector

30. An image uses 512 × 512 picture elements. Each of the picture elements can take any of the 8 distinguishable intensity levels. The **maximum entropy** in the above image will be

(a) 2097152 bits
(b) 786432 bits
(c) 648 bits
(d) 144 bits

31. A RF carrier of 12 kV at 1 MHz is **amplitude modulated** by a 1 KHz signal of 6 kV peak. The modulation pattern is observed on a CRO calibrated suitably. The voltage indicated will be

(a) 18 kV (b) 1.001 kV
(c) 18 V (d) 36 kV

32. Which one of the following statements regarding the following signal $x(t) = 5 \sin(2\pi \times 10^3 t) \sin(2\pi \times 10^6 t)$ is corrected

(a) The upper side band frequency is 1001000
(b) The lower side band frequency is 999000
(c) The carrier amplitude is 5
(d) $x(t)$ is a DSB-SC signal

33. An **AM broadcast radio transmitter** radiates 10 kW power, when the modulation percentage is 60. The power of the carrier is

(a) 5.00 kW (b) 7.69 kW
(c) 8.47 kW (d) 9.17 kW

34. In a **low level modulation AM system,** the amplifiers following the modulated stage can be

(a) Linear amplifiers
(b) Harmonic generators
(c) Class C power amp
(d) Class B untuned amp

35. The choice of the product **RC is an envelope detector** using a diode and RC circuit is governed by

(a) Both the lowest and the highest modulation frequencies
(b) Only the depth of modulation
(c) The depth of modulation and the lowest modulation frequency
(d) The depth of modulation and the highest modulation frequency

36. In an **amplitude modulated system,** if the total power is 600 W and power in carrier is 400 W, then the modulation index is

(a) 0.5 (b) 0.75
(c) 0.9 (d) 1

37. For an AM wave, the maximum voltage was found to be 10 V and the minimum voltage was found to 5 V. The **modulation index** of the wave would be

(a) 0.33 (c) 0.52
(d) 0.40 (d) 0.1

38. An arbitrary signal $m(t)$ has zero average value and it is band-limited to 3.2 0Hz. It is sampled at the rate of 8 K samples/s. The samples are passed through an ideal band-pass filter with center frequency of 32 kHz and bandwidth of 6.4 kHz. The output of the band-pass filter is

(a) AM-DSB signal with suppressed carrier
(b) AM-DSB signal with carrier
(c) AM-SSB signal with carrier
(d) A sequence of exponentially decaying sine waves

39. If the radiated power of AM transmitted is 10 KW, the power in the carrier for modulation index of 0.6 is nearly

(a) 8.24 kW (b) 8.47 kW
(c) 9.26 kW (d) 9.6 kW

40. A balanced modulator is used in the generation of which of the following?

(a) DSB - SC signal (b) FM signal
(c) PM signal (d) PAM signal

41. An amplitude modulated signal occupies a frequency range from 395 kHz to 405 kHz. It can be demodulated by which of the following:

(a) Using an envelope detector and filter .
(b) Multiplying with a 395 kHz local signal
(c) Multiplying with a 405 kHz local signal
(d) Low pass filtering with cut off at 400 kHz.

42. An amplitude modulated signal can be observed on a CRO by applying the following waveform to the "external trigger" input:

(a) The modulated waveform itself
(b) Derivative of the modulated waveform
(c) The modulating waveform.
(d) The carrier waveform

43. For an amplitude modulated signal with carrier power 500 W and modulation index 0.75, the total power is

(a) 540.5 W (b) 640.5 W
(c) 440.5 W (d) 740.5 W

44. For a suppressed carrier **amplitude modulator** (AM - SC) system, the carrier and the modulating inputs are $x_c(t) = \cos\omega_c t$ and $x_m(t) = 0.5 \sin\omega_m t$, respectively. The output of the system is proportional to.

(a) $\sin(\omega_c + \omega_m)t - \sin(\omega_c - \omega_m)t$
(b) $\sin(\omega_c + \omega_m)t + \cos(\omega_c - \omega_m)t$
(c) $(1 + 0.5 \sin\omega_m t)\cos\omega_c t$
(d) $(1 - 0.5 \sin\omega_m t)\cos\omega_c t$

45. Due to an amplitude modulation by a sine wave, if the total current in the antenna increase 4A to 4.8A the depth of modulation in percentage is

(a) 93.8 (b) 80.1
(c) 40.4 (d) 20.2

46. Consider the AM signal $s(t) = [1 + m(t)]\cos(2\pi f_c t)$. It is given that the bandwidth of the real, low pass message signal $m(t)$ is 2 kHz.If $f_c = 2\ MHz.$, the bandwidth of the band pass signal $s(t)$ will be.

(a) 2.004 MHz (b) 2 MHz
(c) 4 kHz (d) 2 kHz

5.2 Frequency Modulation

1. $V(t) = 5[\cos(10^6 \pi t) - \sin(10^3 \pi t) \times \sin(10^6 \pi t)]$ represent

(a) DSB suppressed carrier signal
(b) AM signal
(c) SSB upper sideband signal
(d) Narrow Band FM signal

2. A modulated signal is given by

$$s(t) = e^{-at} \cos[(\omega_c + \Delta\omega)t]\, u(t)$$

where a, ω_c and $\Delta\omega$ are positive constants, and $\omega_c >> \Delta\omega$. The complex envelope of $s(t)$ is given by

(a) $\exp(-at)\exp[j(\omega_c + \Delta\omega)t]\, u(t)$
(b) $\exp(-at)\exp(j\Delta\omega t)\, u(t)$
(c) $\exp(j\Delta\omega t) \times u(t)$
(d) $\exp[j(\omega_c + \Delta\omega)t]$

3. In an FM system a carrier of 100 MHz is modulated by a sinusoidal signal of 5 kHz. The bandwidth by Carson's approximation is 1 MHz.

If $y(t)$ = (modulated waveform)^3

than by using Carson's approximation the bandwidth of $y(t)$ around 300 MHz and the spacing of spectral components are, respectively.

(a) 3 MHz, 5 kHz (b) 1 MHz, 15 kHz
(c) 3 MHz, 15 kHz (d) 1 MHz, 5 kHz

4. An angle-modulated signal is given by

$$S(t) = \cos 2\pi(2 \times 10^6 t + 30 \sin 150t + 40 \cos 150t)$$

The maximum frequency and phase deviations of $s(t)$ are

(a) 10.5 kHz, 140 πrad
(b) 6 kHz, 80 πrad
(c) 10.5 kHz, 100 πrad
(d) 7.5 kHz 100 πrad

5. In figure, $m(T) = \dfrac{2\sin 2\pi t}{t}$, $s(t) = \cos 200\pi t$ and $n(t) = \dfrac{\sin 199\pi t}{t}$. The output $y(t)$ will be

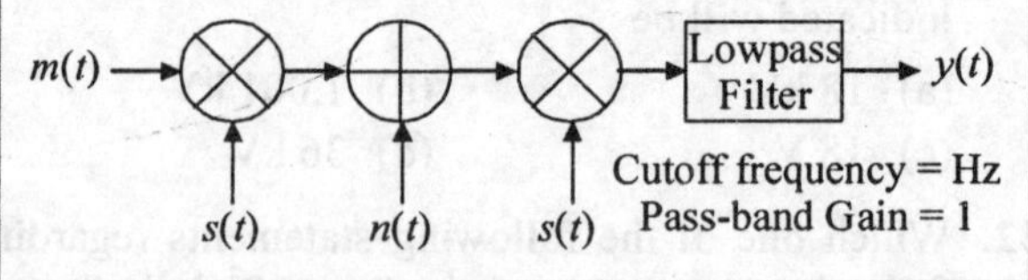

(a) $\dfrac{\sin 2\pi t}{t}$

(b) $\dfrac{\sin 2\pi t}{t} + \dfrac{\sin \pi t}{t}\cos 3\pi t$

(c) $\dfrac{\sin 2\pi t}{t} + \dfrac{\sin 0.5t}{t}\cos 1.5\pi t$

(d) $\dfrac{\sin 2\pi t}{t} + \dfrac{\sin \pi t}{t}\cos 0.75\pi t$

6. An AM signal and a narrow-band FM signal with identical carriers, modulating signals and modulation indices of 0.1 are added together. The resultant signal can be closely approximated by
(a) Broadband FM (b) SSB with carrier
(c) DSB-SC (d) SSB without carrier

7. Two sinusoidal signal of same amplitude and frequencies 10 kHz and 10.1 kHz are added together. The combined signal is given to an ideal frequency detector. The output of the detector is
(a) 0.1 kHz sinusoid
(b) 20.1 KHz sinusoid
(c) A linear function of time
(d) A constant

8. Find the correct match between group 1 and group 2.

Group-I

$P-\ [1+km(t)]A\sin(\omega_c t)$

$Q-\ km(t)A\sin(\omega_c t)$

$R-\ A\sin[\omega_c^t + k]_{-\infty}^{t} m(t)dt$

$S-\ A\sin\left[\omega_c t + k\int_{-\infty}^{t} m(t)dt\right]$

Group-II

W – Phase modulation
X – Frequency modulation
Y – Amplitude modulation
Z – DSB-SC modulation

	P	Q	R	S
(a)	Z	Y	X	W
(b)	W	X	Y	Z
(c)	X	W	Z	Y
(d)	Y	Z	W	X

9. A carrier is phase modulated (PM) with frequency deviation of 10 kHz by a single tone frequency of 1 kHz. If the single tone frequency is increased to 2 kHz, assuming that **phase deviation remains unchanged**, the bandwidth of the PM signal is
(a) 21 kHz (b) 22 kHz
(c) 42 kHz (d) 44 kHz

10. A message signal with bandwidth 10 kHz is Lower-Side Band SSB modulated with carrier frequency $f_{c1} = 10^6$ Hz. The resulting signal is then passed through a narrow-Band Frequency Modulator with carrier frequency $f_{c2} = 10^9$ Hz.

The bandwidth of the output would be
(a) 4×10^4 Hz (b) 2×10^6 Hz.
(c) 2×10^9 Hz. (d) 2×10^{10} Hz

11. The signal *m(t)* as shown is applied both to a phase modulator (with k_p as the phase constant) and a frequency modulator (with k_f as the frequency constant) having the same carrier frequency

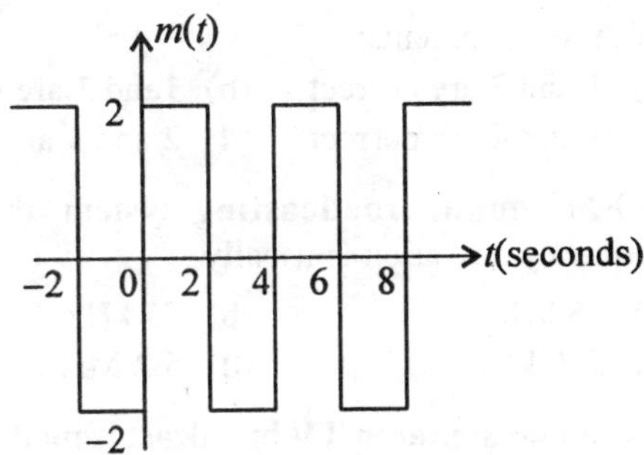

The ratio $\frac{k_p}{k_f}$ (in rad/*Hz*) for the same maximum phase deviation is
(a) 8π
(b) 4π
(c) 2π
(d) π

12. Consider an angle modulated signal

$$x(t) = 6\cos\left[\begin{array}{l}2\pi\times10^6 t + 2\sin(8000\pi t)\\ +4\cos(8000\pi t)\end{array}\right]V$$

The average power of $x(t)$ is
(a) 10 W (b) 18 W
(c) 20 W (d) 28 W

13. Consider the frequency modulated signal $10\cos\left[\begin{array}{l}2\pi\times10^5 t + 5\sin(2\pi\times1500t)\\ +7.5\sin(2\pi\times1000t)\end{array}\right]$ with carrier frequency of 10^5 Hz. The modulation index is
(a) 12.5 (b) 10
(c) 7.5 (d) 5

14. The signal $\cos\omega_c t + 0.5\cos\omega_m t\sin\omega_c t$ is
(a) FM only (b) AM only
(c) Both AM and FM (d) Neither AM nor FM

15. An **FM signal** with a deviation δ is passed through a mixer and has its frequency reduced fivefold. The deviation in the output of the mixer is
(a) 5δ (b) Intermediate
(c) $\frac{\delta}{5}$ (d) δ

16. Consider the following statements:
    1. The **bandwidth of the A.M** wave depends on the bandwidth of the modulating signal.
    2. The bandwidth of the A.M. wave depends on the modulation index
    3. The **bandwidth of F.M.** wave, for all practical purposes, depends on the amplitude of the carrier

    Of these statements:
    (a) 1 and 2 are correct (b) 1and 3 are correct
    (c) 1 and 4 are correct (d) 2 and 4 are correct

17. In **FM sound broadcasting** system, the maximum frequency deviation is usually
    (a) 15 kHz (b) 75 kHz
    (c) 200 kHz (d) 5.2 MHz

18. The **sound signal in TV** broadcast is modulated in
    (a) SSB (b) VSB
    (c) FM (d) DSB-SC

19. In the **generation of modulated** signal, a **varactors diode** can be used for
    (a) FM generation only
    (b) AM generation only
    (c) PM generation only
    (d) Both AM and PM generation

20. **Pre-emphasis in FM systems** involves.
    (a) Compression of the modulating signal
    (b) Expansion of the modulating signal
    (c) Amplification of lower frequency components of the modulating signal
    (d) Amplification of higher frequency components of the modulating signal

21. As the **modulation index of a FM** signal with sinusoidal modulation is increased from zero to three, the power in the sideband component will
    (a) Increase continuously
    (b) Decrease continuously
    (c) First increase, attain a maximum and then decrease
    (d) First decreases, become zero and then increase

22. Two carriers 40 MHz and 80 MHz respectively are frequency modulated by a signal of frequency 4 KHz, such that the bandwidth of the FM signal in the two cases are same. The **peak deviation** in the two cases are in the ratio of
    (a) 1 : 4 (b) 1 : 2
    (c) 1 : 1 (d) 2 : 1

23. In **phase modulation,** the frequency deviation is
    (a) Independent of the modulating signal frequency
    (b) Inversely proportional to the modulating signal frequency
    (c) Directly proportional to the modulating signal frequency
    (d) Inversely proportional to the square root of the modulating frequency

24. A signal $m(t) = 5 \cos 2\pi 100t$ frequency modulates a carrier. The resulting FM signal is

    $$10 \cos [(2\pi 10^5 t)] + 15 \sin (2\pi 100 t)$$

    The approximate bandwidth of the FM would be
    (a) 0.1 kHz (b) 1 kHz
    (c) 3.2 kHz (d) 100 kHz

25. An **angle-modulated** signal is expressed by $f_a(t) = \cos (2 \times 108 \pi t + 75 \sin 2 \times 10^3 \pi t)$. **The peak frequency deviation** of the carrier is then
    (a) 1 kHz (b) 7.5 kHz
    (c) 75 kHz (d) 100 MHz

26. An FM signal with a modulation index 9 is applied to a frequency Tripler. The modulation index in the output signal will be
    (a) 0 (b) 3
    (c) 9 (d) 27

27. Which one of the following is an indirect way of generating FM ?
    (a) Reactance FET Modulator
    (b) Varacter diode Modulator
    (c) Armstrong Moudlator
    (d) Reactance tube Modulator

28. Why does and FM radio station perform better than an AM station radiating the same total power?
    (a) FM is immune to noise
    (b) AM has only two sidebands while FM has more
    (c) FM uses larger bandwidth for large modulation depth
    (d) Capture effect appears in FM

29. A carrier signal at frequency 100 MHz is frequency modulation with modulation index 2 by a signal at 2 kHz. At what frequencies are the sidebands produced?
    (a) 102 MHz and 98 MHz
    (b) 100 MHz + 2 kHz , and 100 MHz – 2 kHz
    (c) 100 MHz $\pm \Delta f_d$, where f_d is frequency deviation
    (d) 100 MHz $\pm 2n$ kHz ,where n is 1, 2, ..., 5

30. In an FM broadcast, the maximum frequency deviation allowed is 75 kHz and the maximum modulation frequency is 15 kHz. The bandwidth is closed to
    (a) 180 kHz (b) 60 kHz
    (c) 105 kHz (d) 120 kHz

31. A sinusoidal signal of frequency 1 kHz is used to produce an FM signal with a modulation index $\beta = 5$. The bandwidth (where 98% of power is contained) of the FM signal is.

(a) 2 kHz (b) 3 kHz
(c) 6 kHz (d) 12 kHz

32. In a **frequency modulated system,** the carrier and the output signals are $x_c(t) = \cos \omega_c(t)$ and $y(t) = \cos(\omega_c t + \sin 2t - \cos t)$, respectively. The modulation input $x_m(t)$ is proportional to

(a) 2 cos 2t + sin t (b) sin 2t – cos t
(c) cos 2t + sin t (d) 2 cos 2t – sin t

5.3 Sampling

1. A deterministic signal has the power spectrum given in Figure. The minimum sampling rate needed to completely represent this signal is

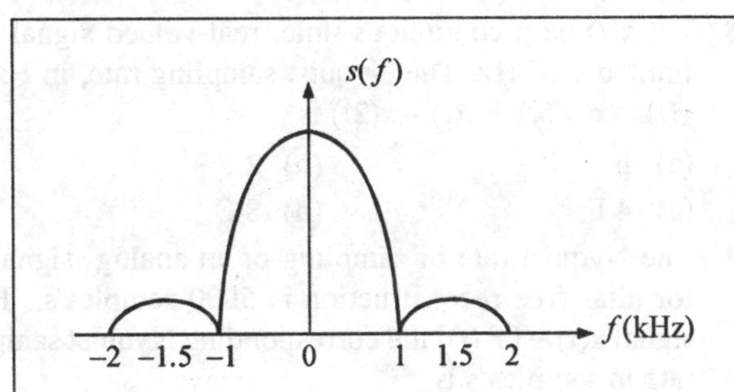

(a) 1 kHz (b) 2 kHz
(c) 3 kHz (d) None of the above

2. A signal $x(t) = 100 \cos(24\pi \times 10^3)t$ is ideally sampled with a sampling period of 50 sec and then passed through an ideal low-pass filter with cut-off frequency of 15 KHz. Which of the following frequency is/are present at the filter output?

(a) 12 kHz only
(b) 8 kHz
(c) 12 kHz and 9 kHz
(d) 12 kHz and 8 kHz

3. The Nyquist sampling frequency (in Hz) of a signal given by $6 \times 10^4 \sin c^2 (400\, t) * 10^6 \sin c^3 (100\, t)$ is

(a) 200 (b) 300
(c) 500 (d) 1000

4. The minimum samling frequency (in samples/sec) required to reconstruct the following signal from its samples without distortion.

$$x(t) = 5\left(\frac{\sin 2\pi 1000t}{\pi t}\right)^3 + 7\left(\frac{\sin 2\pi 1000t}{\pi t}\right)^2$$

Would be

(a) 2×10^3 (b) 4×10^3
(c) 6×10^3 (d) 8×10^3

5. The Nyquist sampling rate for the signal $S(t) = \dfrac{\sin(500\pi t)}{\pi t} \times \dfrac{\sin(700\pi t)}{\pi t}$ is given by

(a) 400 Hz (b) 600 Hz
(c) 1200 Hz (d) 1400 Hz

6. An audio signal is band limited to 4 kHz. It is sampled at 8 kHz. What will be the spectrum of the sampled signal?

(a) – 4 kHz to 4 kHz
(b) – 8 kHz to 8 kHz
(c) Every 4n kHz and repeating
(d) Every ± 8 kHz and repeating as well as at zero (k integer)

7. A signal occupies a band 5 kHz to 10 kHz .For proper error free reconstruction at what rate it should be sampled?

(a) 10 kHz (b) 20 kHz
(c) 5 kHz (d) $-(10+5)\times 2$ kHz

8. The spectral range of a band pass signal extends from 10 MHz to 10.4 MHz. What is the minimum sampling frequnency required for reconstruction?

(a) 20 MHz (b) 20.8 MHz
(c) 20.4 MHz (d) 0.8 MHz

9. Falt top sampling of low pass signals

(a) Gives rise to aperture erect
(b) Implies over-sampling
(c) Leads to aliasing
(d) Introduces delay distortion

10. The spectrum of a band pass signal spans from 20 kHz to 30 kHz. The signal can be recovered ideally from the sample values when the sampling rate is at least:

(a) 20 kHz (b) 60 kHz
(c) 50 kHz (d) 40 kHz

11. The Nyquist rate for the signal $x(t) = 2\cos(2000\pi t)\cos(5000\pi t)$ is:

(a) 7 kHz (b) 5 kHz
(c) 14 kHz (d) 10 kHz

12. A signal occupies a band 5 kHz to 10 kHz .For proper error free reconstruction at what rate it should be sampled?

(a) 10 kHz (b) 20 kHz
(c) 5 kHz (d) $-(10+5)\times 2$ kHz

13. The spectral range of a band pass signal extends from 10 MHz to 10.4 MHz. What is the minimum sampling frequency required for reconstruction ?

(a) 20 MHz (b) 20.8 MHz
(c) 20.4 MHz (d) 0.8 MHz

14. When $x(t) = \sin 2\pi 10^3 t + 2 \sin 2\pi 660t$, at what sampling frequency should this signal be sampled to avoid aliasing?

(a) 2×660 Hz (b) 2×1000 Hz
(c) $2[1000 + 660]$ Hz (d) $2[1000 - 660]$ Hz.

15. An analog signal has significant spectral components from 1 kHz to 5 kHz. What is the Nyquist sampling rate for this signal?

(a) 5 k samples/s (b) 4 k samples/s
(c) 8 k samples/s (d) 10 k samples/s

16. The frequency spectrum of a signal is shown in the figure. If this signal is ideally sampled at intervals of 1 ms, then the frequency spectrum of the sampled signal will be

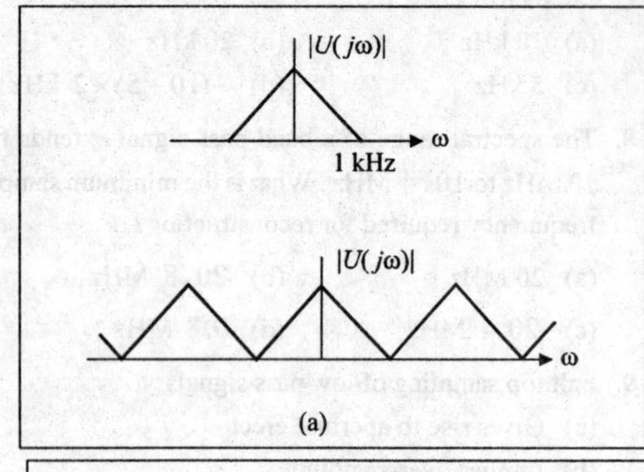

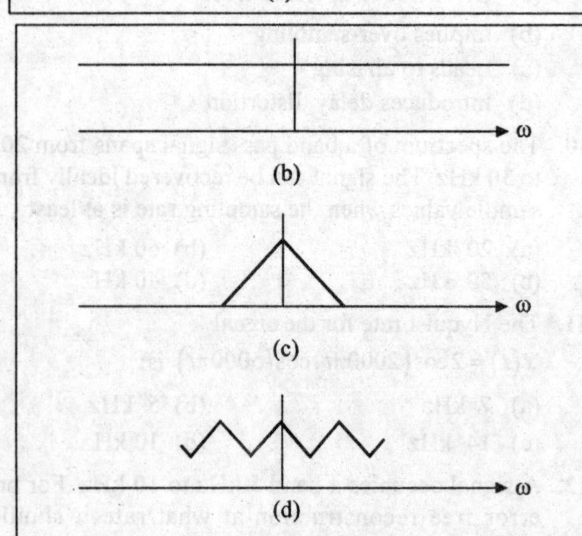

17. The spectrum of a band limited singal after sampling is shown below. The value of sampling interval is

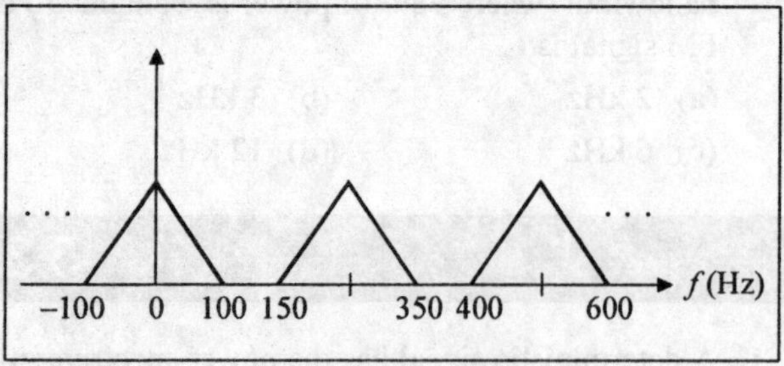

(a) 1 ms (b) 2 ms
(c) 4 ms (d) 8 ms

18. Let $x(t)$ be a continuous-time, real-valued signal band limited to F Hz. The Nyquist sampling rate, in Hz, for $y(t) = x(0.5t) + x(t) - x(2t)$ is

(a) F (b) 2 F
(c) 4 F (d) 8 F

19. The Nyquist rate of sampling of an analog signal s(t) for alias free reconstruction is 5000 samples/s. For a signal $x(t) = [s^2(t)]$ the corresponding Nyquist sampling rate in samples/s is

(a) 2500 (b) 5000
(c) 10000 (d) 25000

20. A signal $x(t) = 5 \cos(150\, pt - 60)$ is sampled at 200 Hz. The fundamental period of the sampled sequence x[n] is

(a) 1/200 (b) 2/200
(c) 4 (d) 8

21. A signal with frequency components 50 Hz, 100 Hz and 200 Hz only is sampled at 150 samples/s. The ideally reconstructed signal will have frequency component (s) of

(a) 50 Hz only
(b) 75 Hz only
(c) 50 Hz and75 Hz
(d) 50 Hz,75 Hz and100 Hz

22. The continuous time signal

$x(t) = \cos(100\pi t) + \sin(300\pi t)$ is sampled at the rate 100 Hz to get the signal $x_s(t) = \sum_{n=-\infty}^{\infty} x(nT_s)\delta(t - nT_s)$,

T_s = sampling period

The signal $x_s(t)$ is passed through an ideal low pass filter with cutoff frequency 100 Hz. The output of the filter is proportional to

(a) $\cos(100\pi t)$
(b) $\cos(100\pi t) + \sin(100\pi t)$
(c) $\cos(100\pi t) - \sin(100\pi t)$
(d) $\sin(100\pi t)$

23. The most commonly used amplifier in sample and hold circuit is
(a) A unity gain inverting amplifier
(b) A unity gain non-inverting amplifier
(c) An inverting amplifier with a gain of 10
(d) An inverting amplifier with a gain of 100

24. Consider a sampled signal

$$y(t) = 5 \times 10^{-6}\, x(t) \sum_{n=-\infty}^{+\infty} \delta(t - nT_s)$$

Where $x(t) = 10 \cos (8\pi \times 10^3)t$ and $T_s = 100$ μsec. When $y(t)$ is passed through an ideal low-pass filter with a cutoff frequency of 5 kHz, the output of the filter is
(a) $5 \times 10^{-6} \cos (8\pi \times 10^3)\, t$
(b) $5 \times 10^{-5} \cos (8\pi \times 10^3)\, t$
(c) $5 \times 10^{-1} \cos (8\pi \times 10^3)\, t$
(d) $10 \cos (8\pi \times 10^3)\, t$

25. Let x(t) = 2 cos (800 π t) + cos (1400 π t) is sampled with the rectangular pulse train shown in Figure. The only spectral components (in kHz) present in the sampled signal in the frequency range 2.5 kHz are to 3.5 kHz

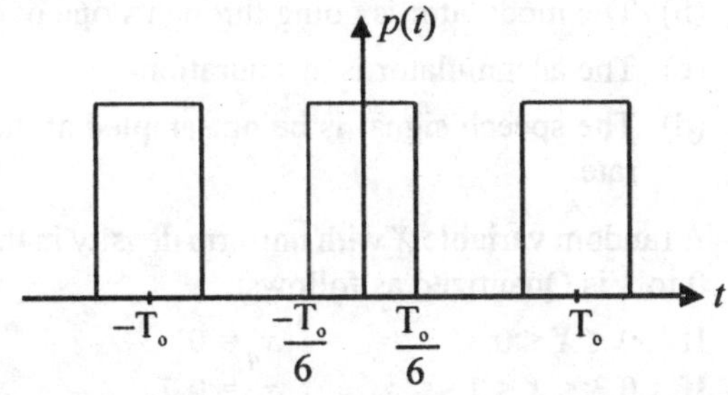

(a) 2.7 , 3.4 (b) 3.3 , 3.6
(c) 2.6, 2.7. 3.3. 3.4, 3.6 (d) 2.7 , 3.3

26. A 1 kHz sinusoidal signal is ideally sampled at 1500/sec and the sampled signal is passed through an ideal low-pass filter with cut-off frequency 800 Hz. The output signal has the frequency
(a) Zero Hz (b) 0.75 kHz
(c) 0.5 kHz (d) 0.25 kHz

27. A rectangular pulse train s(t) as shown in figure is convolved with the signal $\cos^2 (4\pi \times\times 10^3$ t). The convolved signal will be a

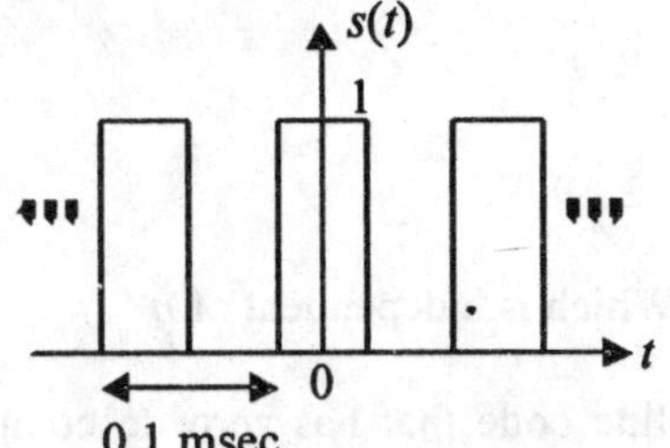

(a) DC (b) 12 kHz sinusoid
(c) 8 kHz sinusoid (d) 14 kHz sinusoid

28. Which one of the following is the code that is very close to trellis coded modulation?
(a) Combines analog and digital modulations
(b) Combines modulation are encoding
(c) Encodes following trellis diagram
(d) Combines amplitude and frequency modulation

29. A band limited signal with highest frequency content of 1000 Hz is undergoing sampling at uniform intervals. For recovery of the signals in an unambiguous way, the sampling frequency should be necessarily greater than
(a) 500 Hz (b) 100Hz
(c) 1500Hz (d) 2000Hz

5.4 Pulse Code Modulation

1. A 4 GHz carrier is DSB/SC modulated by a low-pass massage signal with maximum frequency of 2 MHz. The resultant signal is to be ideally sampled. The minimum, frequency of the sampling in train should be:
(a) 4 MHz (b) 8 MHz
(c) 8 GHz (d) 8.004 GHz

2. Increased pulse-width in the flat-top sampling, leads to
(a) Attenuation of high frequency in reproduction
(b) Attenuation of low frequency in reproduction
(c) Greater Aliasing errors in reproduction
(d) No harmful erects in reproduction

3. If the number of bits per sample in a PCM system is increased from 5 to 6 bit/sample, the improvement in signal-to quantization noise ratio will be
(a) 3 dB (b) 6 dB
(c) $2n$ dB (d) 0 dB

4. A 1.0 kHz signal is flat-top sampled at the rate of 1800 samples/sec and the samples are applied to an ideal rectangular LPF with the cut-off frequency of 1100 Hz, then the output of the filter contains.
(a) Only 800 Hz component
(b) 800 Hz and 900 Hz components
(c) 800 Hz and 1000 Hz components
(d) 800 Hz, 900 Hz and 1000 Hz components

5. The signal to quantization noise ratio in an n-bit PCM system
   (a) Depends upon the sampling frequency employed
   (b) Is independent of the value of 'n'
   (c) Increases with increasing value of 'n'
   (d) Decreases with the increasing value of 'n

6. The number of bits in a binary PCM system is increased from n to $n + 1$. As a result, the signal to quantization noise ratio will improve by a factor
   (a) $\frac{n+1}{n}$
   (b) $2^{\frac{(n+1)}{n}}$
   (c) $2^{\frac{2(n+1)}{n}}$
   (d) Which is independent of n

7. The line code that has zero dc component for pulse transmission of random binary data is
   (a) Non-return to zero (NRZ)
   (b) Return to zero (RZ)
   (c) Alternate Mark inversion (AMI)
   (d) None of the above

8. In a PCM system with uniform quantization, increasing the number of bits from 8 to 9 will reduce noise power by a factor of
   (a) 9 (b) 8
   (c) 4 (d) 2

9. Compression in PCM refers to relative compression of
   (a) Higher signal amplitudes
   (b) Lower signal amplitudes
   (c) Lower signal frequencies
   (d) Higher signal frequencies

10. The peak-to-peak input to an 8-bit PCM coder is 2 volts. The signal power-to-quantization noise power ratio (in dB) for an input of 0.5 cos ($\omega_m t$) is
    (a) 47.8 (b) 43.9
    (c) 95.6 (d) 99.6

11. A sinusoidal signal with peak-to-peak amplitude of 1.536 V is quantized into 128 levels using a mid-rise uniform quantizer. The quantization-noise power is
    (a) 0.768 V (b) 48×10^{-6} V^2
    (c) 12×10^{-6} V^2 (d) 3.072 V

12. The input to a linear delta modulator having a step-size $\Delta = 0.628$ is sine wave with frequency f_m and peak amplitude E_m. If the sampling frequency $f_s = 40$ kHz, the combination of the sine-wave frequency and the peak amplitude, where slopes overload will take place is

	E_m	f_m
(a)	0.3 V	8 kHz
(b)	1.5 V	4 kHz
(c)	1.5 V	2 kHz
(d)	3.0 V	1 kHz

13. A signal is sampled at 8 kHz and is quantized using 8-bit uniform quantizer. Assuming SNR$_q$ for a sinusoidal signal, the correct statement for PCM signal with a bit rate of R is
    (a) $R = 32$ kbps, SNR$_q$ = 25.8 dB
    (b) $R = 64$ kbps, SNR$_q$ = 49.8 dB
    (c) $R = 64$ kbps, SNR$_q$ = 55.8 dB
    (d) $R = 32$ kbps, SNR$_q$ = 49.8 dB

14. In PCM system, if the code word length is increased from 6 to 8 bits, the signal to quantization noise ratio improves by the factor
    (a) $\frac{8}{6}$ (b) 12
    (c) 16 (d) 8

15. In the output of a DM speech encoder, the consecutive pulses are of opposite polarity during time interval $t_1 \le t \le t_2$. This indicates that during this interval
    (a) The input to the modulator is essentially constant
    (b) The modulator is going through slope overload
    (c) The accumulator is in saturation
    (d) The speech signal is being sampled at the Nyquist rate

16. A random variable X with uniform density in the interval 0 to 1 is Quantized as follows:
    If $0 \le X \le 0.3$, $x_q = 0$
    If $0.3 \le X \le 1$, $x_q = 0.7$
    Where X_q is the quatized value of x
    The root mean square value of the quantization noise is
    (a) 0.573 (b) 0.198
    (c) 2.205 (d) 0.266

17. Three analog signals, having bandwidth 1200 Hz, 600 Hz and 600 Hz, are sampled at their respective Nyquist rates, encoded with 12 bit words, and time division multiplexed. The bit rate for the multiplexed signal is
    (a) 115.2 kbps
    (b) 28.8 kbps
    (c) 57.6 kbps
    (d) 38.4 kbps

18. In the following figure value of the constant "C" which is to be added to $y_1(t)$ such that $y_1(t)$ and $y_2(t)$ are different, is

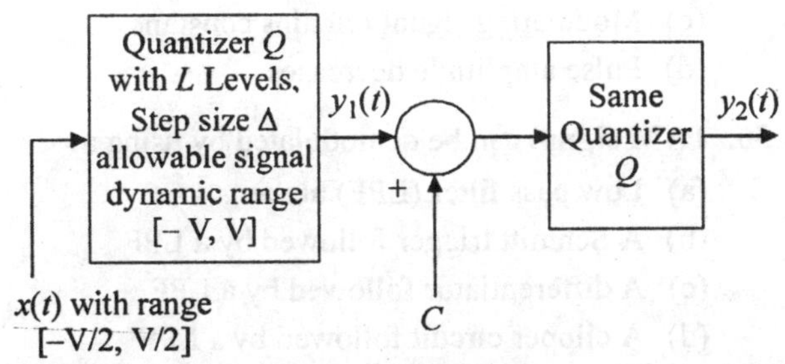

(a) Δ (b) $\frac{\Delta}{2}$

(c) $\frac{\Delta^2}{12}$ (d) $\frac{\Delta}{L}$

19. In delta modulation, the slope overload distortion can be reduced by

(a) Decreasing the step size
(b) Decreasing the granular noise
(c) Decreasing the sampling rate
(d) Increasing the step size

Common data for Question 21 and 22

Asymmetric three-level midtread quantizer is to be designed assuming equiprobable occurence of all quantization level.

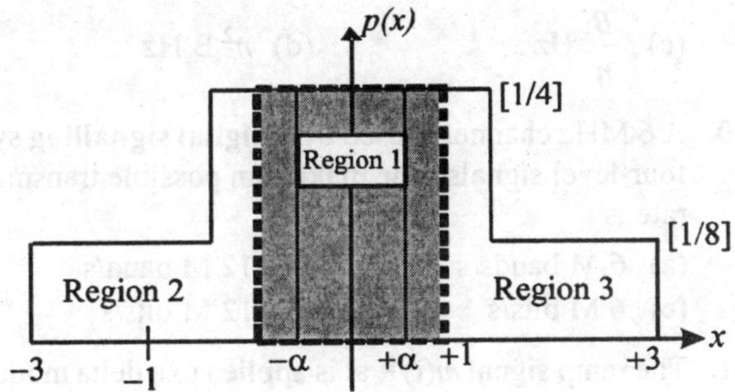

20. If the probability density function is divided in the regions as shown in the figure, the value of a in the figure is

(a) [1 / 3] (b) [2 / 3]

(c) [1 / 2] (d) [1 / 4]

21. The quantization noise power for the quantization region between – a and + a in the figure is

(a) [4 / 81] (b) [1 / 9]

(c) [5 / 81] (d) [2 / 81]

Statement for linked Answer Questions 22 and 23

An input to a 6-level quantizer has the probability density function $f(x)$ as shown in the figure. Decision boundaries of the quantizer are chosen so as to maximize the entropy of the quantizer output. It is given that 3 consecutive decision boundaries are '–1' '0' and '1'.

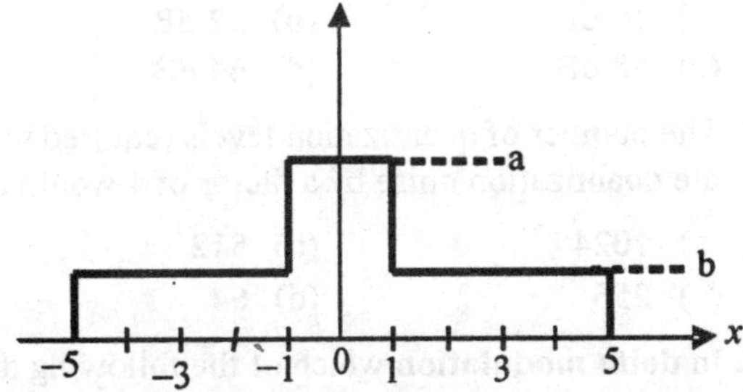

22. The value of a and b are

(a) $a=\frac{1}{6}$ and $b=\frac{1}{12}$ (b) $a=\frac{1}{5}$ and $b=\frac{3}{40}$

(c) $a=\frac{1}{4}$ and $b=\frac{1}{16}$ (d) $a=\frac{1}{3}$ and $b=\frac{1}{24}$

23. Assuming that the reconsturtion levels of the quantizer are the mid-points of the decision boundaries, the ratio of signal power of quantization noise power is

(a) 152/9 (b) 64/3

(c) 76/3 (d) 28

Common Data for Questions 24 and 25

The amptitude of a random signal is uniformly distributed between –5 V and 5 V.

24. If the signal to quantization noise ratio required in uniformly quantizing the signal is 43.5 dB , the step size of the quantization is approximately

(a) 0.0333 V (b) 0.05 V

(c) 0.0667 V (d) 0.10 V

25. If the positive values of the signal are uniformly quantized with a step size of 0.05 V, and the negative values are uniformly quantized with a step size of 0.1 V, the resulting signal to quantization noise ratio is approximately

(a) 46 dB (b) 43.8 dB

(c) 42 dB (d) 40 dB

Common Data Questions

Common Data for Questions 26 and 27.

A speech signal, band limited to 4 kHz and peak voltage varying between + 5 V and –5 V, is sampled at the Nyquist rate. Each sample is quamnitzed and represented by 8 bits

26. If the bits 0 and 1 are transmitted using bipolar pulses, the minimum bandwidth required for distrotion free transmission is

(a) 64 kHz (b) 32 kHz

(c) 8 kHz (d) 4 kHz

27. Assuming the singal to be uniformly distributed between its peak to peak value, the singal to noise ratio at the quantizer output is

(a) 16 dB (b) 32 dB
(c) 48 dB (d) 64 dB

28. The number of quantization levels required to reduce the quantization noise by a factor of 4 would be

(a) 1024 (b) 512
(c) 256 (d) 64

29. In **delta modulation** which of the following drawbacks are existing?

1. Slope overload
2. Serration noise
3. Granular noise

(a) 1 and 2 only (b) 2 and 3 only
(c) 1 and 3 only (d) 1, 2 and 3

30. In a **PCM** the amplitude levels are transmitted in a 7 unit code. The sampling is done at the rate of 10 Hz. The bandwidth should be

(a) 5 kHz (b) 35 kHz
(c) 70 kHz (d) 5 MHz

31. Pulse code modulation **(PCM)** employing 4-bit code is used to transmit a data signal having frequency components from dc to 2 kHz. The **minimum bandwidth of the carrier** channel should be

(a) 2 kHz (b) 4 kHz
(c) 8 kHz (d) 16 kHz

32. In a **PCM** system the number of quantization levels are 16 and the maximum signal frequency is 4 KHz, the bit transmission rate is

(a) 64 K bits/sec (b) 16 K bits/sec
(c) 32 K bits/sec (d) 32 bits/sec

33. PCM system require regenerative repeaters over long distance. The correct sequence of the operations which such a repeater performs is

(a) Timing, equalization and decision making
(b) Equalization, timing and decision making
(c) Timing, thresholding (decision) and equalization
(d) Thresholding, timing and equalization

34. A message signal band limited to 5 KHz is **sampled** at the minimum rate as dictated by the sampling theorem. The number of **quantization levels is 64.** If the samples are encoded in binary form, the transmission rate is

(a) 60 kbps (b) 50 kbps
(c) 32 kbps (d) 10 kbps

35. In a **DM (delta modulation)** system, the granular (idling) noise occurs when the

(a) Modulation signal increases rapidly
(b) Pulse rate decreases
(c) Modulating signal remains constant
(d) Pulse amplitude decreases

36. PAM signal can be demodulated by using a

(a) Low pass filter (LPF) alone
(b) A Schmitt trigger followed by a LPF
(c) A differentiator followed by a LPF
(d) A clipper circuit followed by a LPF

37. The use of **non-uniform quantization** leads to

(a) Reduction in transmission bandwidth
(b) Increase in maximum SNR
(c) Increase in SNR for low level signals
(d) Simplication of quantization process

38. The bandwidth of a 'N' bit binary coded PCM signal for modulating a signal having bandwidth of 'f' Hz is

(a) $\frac{f}{N}$ Hz (b) $\frac{f}{N^2}$
(c) Nf Hz (d) N^2 F Hz

39. The bandwidth of an *n*-bit binary coded PCM signal for an original signal bandwidth of B Hz is

(a) B Hz (b) n B Hz
(c) $\frac{B}{n}$ Hz (d) n^2 B Hz

40. A 6 MHz channel is used by a **digital signalling** system four-level signals. The maximum possible transmission rate is

(a) 6 M bauds/s (b) 12 M baud/s
(c) 6 M bits/s (d) 12 M bits/s

41. The ramp signal $m(t)$ = at is applied to a delta modulator with sampling period T_s step size δ Slope overload distortion would occur if

(a) $\delta < \alpha$ (b) $\delta > \alpha$
(c) $\delta < \alpha T_s$ (d) $\delta > \alpha T_s$

42. In a **PCM system,** each quantization level is encoded into 8 bits. The signal to quantization noise is equal to

(a) $\left(\frac{1}{12}\right)\cdot\left(\frac{1}{256}\right)^2$ (b) 48 dB
(c) 64 dB (d) 256 dB

43. Four signals each band-limited to 5 kHz are sampled at twice the Nyquist Rate. The resulting PAM samples are transmitted over a single channel after time division multiplexing. The theoretical minimum transmissions bandwidth of the channel should be equal to

(a) 5 kHz (b) 20 kHz
(c) 40 kHz (d) 80 kHz

44. For **10-bit PCM system,** the signal to quantization noise ratio is 62 dB. If the number of bits is increased by 2, then the signal to quantization noise ratio will

(a) Increase by 6 dB
(b) Increase by 12 dB
(c) Decrease by 6 dB
(d) Decrease by 12 dB

45. The minimum step size required for a Delta-Modulator operating at 32 K samples/sec to track the signal (here $u(t)$ is the unit-step function)

$x(t) = 125t\,[u(t) - u(t-1)] + (250 - 125t)\,[u(t-1) - u(t-2)]$

So that slope-overload is avoided, would be.

(a) 2^{-10} (b) 2^{-8}
(c) 2^{-6} (d) 2^{-4}

46. A 13 bit PCM system performance is better than an 8 bit PCM system because:

(a) Noise is lower and is proportional to reciprocal of bandwidth
(b) Bandwidth is larger and detection is easier
(c) Quantization noise is lesser other things being equal
(d) Quantization noise is nearly zero since 2^{13} levels are available

47. Which of the following pulse modulations are digital?

1. PCM
2. Differential PCM
3. PWM

Select the correct answer using the code given below:

(a) 1 and 2 only (b) 2 and 3 only
(c) 1 and 3 only (d) 1, 2 and 3

48. In a PCM system, if we increase the quantization levels from 2 to 8, how do the relative bandwidth requirements vary?

(a) Get doubled
(b) Get tripled
(c) Remain same
(d) Become eight times

49. How can the quantizing noise be reduced?

(a) By using de-emphasis circuit
(b) By using RF amplifier in the receiver
(c) By increasing the number of quantization levels
(d) By sending sloping pulses

50. How much is contributed by each bit in the code word of a PCM system under uniform and symmetric quantization to the signal to noise ratio?

(a) 3 dB (b) 4 dB
(c) 5 dB (d) 6 dB.

51. In delta modulation, when is the slope over loading noise absent? (h is the pulse height and sampled every T_s seconds $f(t)$ is the input signal, h and T_s are step-size of integration and sampling period, respectively)

(a) $\dfrac{df(t)}{dt} \le (h/T_s)$ (b) $\dfrac{df(t)}{dt} > (h/T_s)$

(c) $\dfrac{df(t)}{dt} \ge (h.T_s)$ (d) $\dfrac{df(t)}{dt} > (h.T_s)$.

52. The ramp signal m (t) = at is applied to a delta modulator with sampling period T_s step size δ. Slope overload distortion would occur if

(a) $\delta < a$
(b) $\delta > a$
(c) $\delta < aT_s$
(d) $\delta > aT_s$.

53. In a PCM system, each quantization level is encoded into 8 bits. The signal to quantization noise is equal to

(a) $[1/12]\,.\,(1/256)^2$
(b) 48 dB
(c) 64 dB
(d) 256 dB.

54. In a pulse code modulated (PCM) signal sampled at f_s and encoded into an n-bit code, the minimum bandwidth required for faithful reconstruction is.

(a) $2nf_s$
(b) nf_s
(c) $nf_s/2$
(d) f_s

5.5 Multiplexing and Satellite

1. Quadrature multiplexing is

(a) The same as FDM
(b) The same as TDM
(c) A combination of FDM and TDM
(d) Quite different from FDM and TDM

2. Four independent message have bandwidths of 100 Hz, 100 Hz, 200 Hz and 400 Hz, respectively. Each is sampled at the Nyquist rate, and the samples are time division multiplexed (TDM) and transmitted. The transmitted sample rate (in Hz) is

(a) 1600 (b) 800
(c) 400 (d) 200

3. In a **GSM system,** 8 channels can co-exist in 200 kHz bandwidth using TDMA. A GSM based cellular operator is allocated 5 MHz bandwidth. Assuming a frequency reuse factor of 1/5, i.e. a five-cell repeat pattern, the maximum number of simultaneous channels that can exist in one cell is

(a) 200 (b) 40
(c) 25 (d) 5

4. In a **Direct Sequence CDMA system** the chip rate is 1.2288×10^6 chips per second. If the processing gain is desired to be LEAST 100, the data rate

(a) Must be less than or equal to 12.288 ′ 10^3 bits per sec
(b) Must be grater than 12.288×10^3 bits per sec
(c) Must be exactly equal to 12.288×10^3 bits per sec
(d) Can take any value less than 122.88×10^3 bits per sec

5. Four messages band limited to W, W, 2 W and 3W respectively are to be multiplexed using Time Division Multiplexing (TDM). The minimum bandwidth required for transmission of this TDM signal is

(a) W (b) 3 W
(c) 6 W (d) 7 W

6. The frequency range for satellite communication is

(a) 1 KHz to 100 KHz
(b) 100 KHz to 10 MHz
(c) 10 MHz to 100 MHz
(d) 1 GHz to 30 GHz

7. A comparison of **frequency division and time division multiplexing** systems shows that

(a) FDM requires lower bandwidth, but TDM has greater noise immunity
(b) FDM has greater noise immunity and requires lower bandwidth than TDM
(c) FDM requires channel synchronization, while TDM has greater noise immunity
(d) FDM requires more while TDM requires band pass filter.

8. A **TDM link** has 20 signal channels and each channel is sampled 8000 times/sec. Each sample is represented by seven binary bits and contains an additional bit for synchronization. The total bit rate for the TDM link is

(a) 1180 K bits/sec (b) 1280 K bits/sec
(c) 1180 M bits/sec (d) 1280 M bits/sec

9. Time division multiplexing requires

(a) Constant data transmission
(b) Transmission of data samples
(c) Transmission of data at random
(d) Transmission of data of only one measured

10. The output stage of a transponder onboard a satellite has a maximum power output of 10 watts. However, it is not operated at the maximum power output in order to

(a) Conserve the available limited battery power
(b) Reduce noise due to devices
(c) Avoid intermodulation distortion
(d) Avoid heating up of the satellite beyond a pre set value

11. In a typical satellite communication system, which of the following could be the up-link and down-link frequency respectively?

(a) 40 GHz and 60 GHz
(b) 60 GHz and 40 GHz
(c) 6 GHz and 4 GHz
(d) 4 GHz and 6 GHz

12. Multiplexing is possible if signals are sampled. Two signals have bandwidths A = 0 to 4 kHz and B = 0 to 8 kHz respectively. The sampling frequency chosen is 12 kHz.

Which one of the followig is correct ?
This choice of the sampling frequency

(a) Is correct since A and B have an integral relationship of 2
(b) Will not lead to aliasing
(c) Does not obey sampling theorem
(d) Can never lead to multiplexing

13. Which is most important sub-system for recovering and reconstructing signals in a TDM system?

(a) Envelop detector followed by a low-pass filter
(b) Synchronization circuit for proper timing
(c) Band pass filters to segregate channels
(d) Coherent detector to ensure frequency and phase connection.

14. A 12 channel TDM system where each channel 4 kHz is sampled at 8 kHz. What is the bandwidth requirement?

(a) 12 kHz (b) 12×4 kHz
(c) 12×8 kHz (d) $12 \times 8 \times 4$ kHz.

15. Four signals g_1 (t), g_2 (t), g_3 (t) and g_4 (t) are to be multiplexed and transmitted. g_1 (t) and g_4 (t) have a bandwidth of 4 kHz, and the remaining two signals have bandwidth of 8 kHz. Each sample requires 8 bit for encoding. What is the minimum transmission bit rate of the system?

(a) 512 kbps (b) 16 kbps
(c) 192 kbps (d) 384 kbps.

16. Three analog signals, having bandwidth 1200 Hz,

600 Hz and 600 Hz, are sampled at their respective Nyquist rates, encoded with 12 bit words, and time division multiplexed. The bit rate for the multiplexed signal is

(a) 115.2 kbps (b) 28.8 kbps
(c) 57.6 kbps (d) 38.4 kbps

17. Ten real, band pass message signals, each of bandwidth 3 kHz, are to be frequency division multiplexed over a band pass channel with bandwidth B kHz. If the guard band in between any two adjacent signals should be of 500 Hz width and there is no need to provide any guard band at the edges of the band pass channel, the value of B should be at least.

(a) 30 (b) 34.5
(c) 35 (d) 35.5

5.6 Miscellaneous

1. The Hilbert transform of $\cos \omega_1 t + \sin \omega_2 t$ is

(a) $\sin \omega_1 t - \cos \omega_2 t$ (b) $\sin \omega_1 t + \cos \omega_2 t$
(c) $\cos \omega_1 t - \sin \omega_2 t$ (d) $\sin \omega_1 t + \sin \omega_2 t$

2. Match List-I with List-II and select the correct answer using the codes given below the Lists:

List-I	*List-II*
A. Ring modulator	1. Clock recovery
B. VCO	2. Demodulation of FM
C. Foster-Seely	3. Frequency conversion discriminator
D. Mixer	4. Summing the two inputs
	5. Generation of FM
	6. Generation of DSB-SC

Codes:

	A	B	C	D
(a)	1	3	2	4
(b)	6	5	2	3
(c)	6	1	3	2
(d)	5	6	1	3

3. In the following scheme, if the spectrum $M(f)$ of $m(t)$ is as shown, then the spectrum $Y(f)$ of $y(t)$ will be

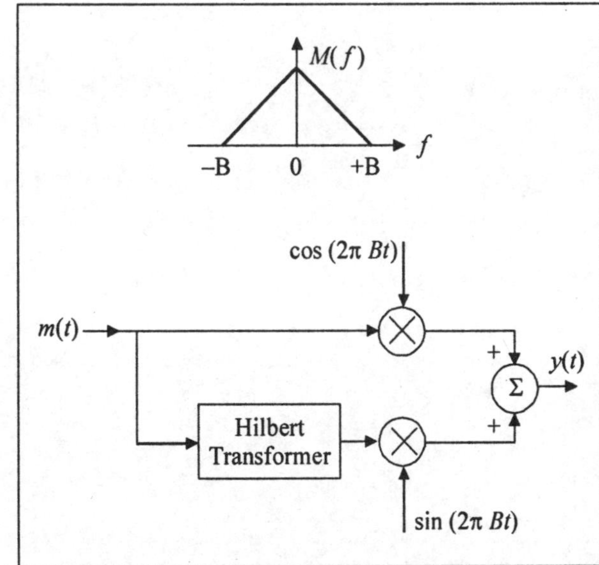

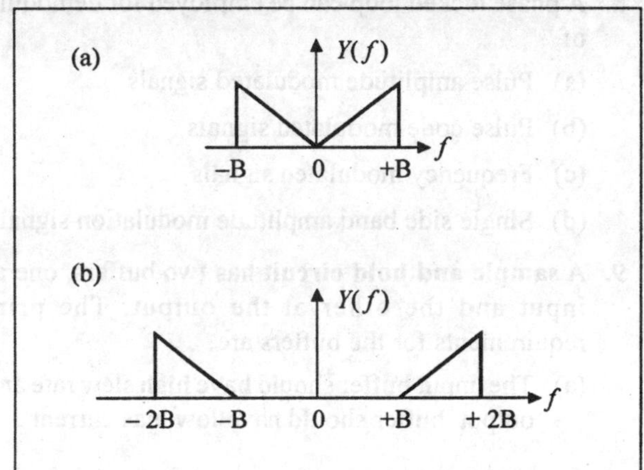

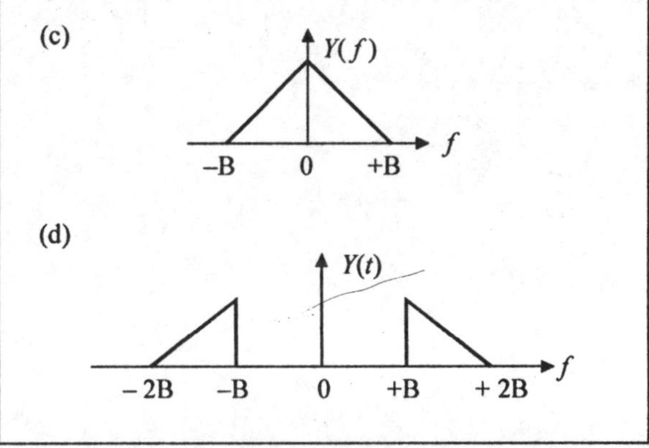

4. An amplifier operating over the frequency range of 18 to 20 MHz has a 10 KΩ input resistance. The **RMS noise voltage** at the input to the amplifier at ambient temperature of 27°C is (Assume the Boltzman constant value as 1.38×10^{-23} J/K)

(a) 18.2 micro-volts (b) 17.2 micro-volts
(c) 16.2 micro-volts (d) 14.2 micro-volts

5. The **noise figure of a receiver** is 1.6. Its **equivalent noise temperature is**

(a) 464.00 K (b) 174.00 K
(c) 108.75 K (d) 181.25 K

6. A system has a **receiver noise** resistance of 50Ω. It is connected to an antenna with an input resistance of 50Ω. The **noise figure** of the system is

(a) 1 (b) 2
(c) 50 (d) 101

7. A person with a receiver is 5 Km away form the transmitter. What is the distance that this person must move further to detect a –dB decrease in signal strength?
(a) 942 m
(b) 2070 m
(c) 4978 m
(d) 5320 m

8. A phase locked loop can be employed for demodulation of
(a) Pulse amplitude modulated signals
(b) Pulse code modulated signals
(c) Frequency modulated signals
(d) Single side band amplitude modulation signals.

9. **A sample and hold circuit** has two buffers, one at the input and the other at the output. The primary requirements for the buffers are.
(a) The input buffer should have high slew rate and the output buffer should have low bias current .
(b) The input buffer should have low bias current and the output buffer should have high slew rate
(c) Both the buffers should have low bias currents
(d) Both the buffers should have high slew rate.

10. A linear phase low pass filter with a cut-off frequency 1500 Hz shifts a 50 Hz signal by 3 m – sec. It will shift a signal of 150 Hz by
(a) 3 m-sec (b) 9 m-sec
(c) 1 m-sec (d) 30 m-sec

11. Match the List I modulation with the List II type of demodulators and select the correct answer using the codes:

List I	*List II*
A. AM	1. Envelope detector
B. FM	2. Slop detector
	3. Peak detector

Codes:

	A	B
(a)	3	2
(b)	3	1
(c)	1	2
(d)	2	3

Answers

5.1 Amplitude Modulation

1. **Ans. (b)**

Solution: $P_t = P_c\left[1+\frac{m_a^2}{2}\right]$

Use full power, i.e. *SB* power $P_{SB} = P_c\frac{m_a^2}{2}$

Carrier power $P_c = P_c$

Efficiency $\eta = \frac{P_{SB}}{P_c} = \frac{m_a^2}{2}$

But $m_a \le 1$, Hence $\eta_{max} = \frac{1}{2}\times 100 = 50\,\%$

2. **Ans. (a)**

Solution: 5 cos 200 πt + 5.2 cos 200 t . cos 200 πt

↑ Carrier ↑ DSB

The signal $x(y)$ = 5 (1 + 2 cos 200 πt) cos 2000 πt is essentially an AM–DSB/FC signal which can be demodulated by any of the 3 demodulator. But envelope detector is the most suitable one.

3. **Ans. (b)**

Solution:

The output of a DSB/SC generator,

$y(t) = x(t)\cos(\omega_c t+\theta)$

The envelope of DSB /SC takes the shape of $|x(t)|$

4. **Ans. (d)**

Solution:

The highest frequency is 15 kHz = f_m

For AM, $BW = 2f_m$ = 2 × 15 = 30 kHz

Second metod

$s(t) = m_1(t)\cos(2\pi f_c t) + m_2(t)\sin(2\pi f_c t)$

$s(t)$ is a ***Quadrature carrier multiplexed*** (QCM) also called QAM- Quadrature amplitude multiplexing signal with two carrier signals $\cos 2\pi f_c t$ and $\sin 2\pi f_c t$ having same frequency but are orthogonal to each other.

Consider only the component $m_1(t)\cos(2\pi f_c t)$; this is a DSB-SC signal.

Hence transmission BW,

$B_{T_1} = 2\times W_1$

$= 2\times 10\ kHz = 20\ kHz$

Similarly the component $m_2(t)\sin 2\pi f_c t$ is also DSB-SC signal.

Hence transmission BW,

$B_{T_2} = 2\times W_2 = 2\times 15\ kHz = 30\ kHz$

Hence overall transmission bandwidth

$B_T = \max\left(B_{T_1}, B_{T_2}\right) = 30\ kHz$

5. **Ans. (c)**

Solution:

Signal fed to the envelop detector is:

$s(t) = A_C\left[1+K\ m(t)\right]\cos\omega_C t$ ... (1)

The $s(t)$ is simply a ***bandpass signal*** which can be written in its canonical form as:

$s(t) = S_I(t)\cos(\omega_C t) - S_Q(t)\sin(\omega_C t)$ ... (2)

Where $S_I(t) \rightarrow$ inphase component and

$S_Q(t) \rightarrow$ quadrature phase components.

Comparing equation (1) and (2)

$S_I(t) = A_C\left[1+K_a m(t)\right]$

and $S_Q(t) = 0$

The output of an envelop detector is the natural envelope of the singal.

Natural envelope, $a(t) = \sqrt{S_I^2(t)+S_Q^2(t)}$

or, $a(t) = |S_I(t)|$

or, $a(t) = \left|A_C\left[1+K_a m(t)\right]\right|$

Coments:

If $|k_a m(t)| > 1$; even then the output of the envelope detector will be the natural envelope. The only difference is that, when the carrier signal is overmodulated, the natural envelope no more contains the information carrying signal. There is phase reversal at zero crossings which results in envelope distortion.

6. **Ans. (d)**

Solution: Power in $m(t) = P_m$

$y(t)$ = $[m(t)/2]\,[2\cos\omega_o t\cos(\omega_o t+\theta)]$

= $[m(t)/2]\,[\cos(2\omega_o t+\theta)+\cos\theta]$ after passing LPF

$y(t)$ = $[m(t)/2]\cos\theta$

$$\text{Power} = \frac{1}{T}\int_0^T \left|\frac{m(t)}{2}\cos\theta\right|^2 dt$$

$$= \int_0^T \frac{m^2(t)}{4}\cos^2\theta\, dt = \frac{\cos^2\theta}{4}.P_m$$

7. **Ans. (c)**

Solution: The sinusoidal carrier modulated with symmetrical square wave results the signal which contain the odd multiple of f_m of square wave

i.e. $f_s = [f_c \pm (2n-1) f_m]$; Where $n = 0, 1, 2, 3$

where,

f_s = Frequency of modulated signal

f_c = Carrier frequency (Sin) = 1 MHz

f_m = Modulating signal frequency (symmetrical square wave)

= 10 kHz

$$f_m = \frac{1}{T_m} = \frac{1}{100\ \mu s} = \frac{1}{10\times10^{-6}} = \frac{1000}{100}\text{ kHz}$$

= 10 kHz

$f_s = [f_s \pm (2n-1) f_m]$

= [1000 kHz (2n − 1) 10 kHz]

= 990 kHz, 1010 kHz, 1030 kHz

∴ The frequency 1020 kHz is not present in modulated output.

Second Method

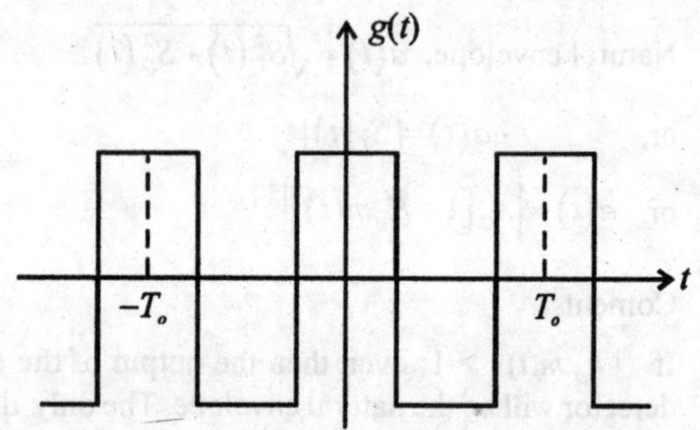

Since $g(t)$ exhibits half wave symmetry, there won't be any even harmonic present in it.

If $f(t)$ is having half wave symmetry $-f(t \pm T/2) = f(t)$

$$a_0 = 0$$

$$a_n = \frac{4}{T}\int_0^{T/2} f(t)\cos(n\omega_0 t)\, dt$$

$$b_n = \frac{4}{T}\int_0^{T/2} f(t)\sin(n\omega_0 t)\, dt$$

(a) Only odd harmonic terms are present.

(b) If $f(t \pm T/2) = f(t)$, then only even harmonic are present.

The sinusoidal carrier modulated with symmetrical square wave results the signal which contain the odd multiple of f_m of square wave.

i.e. $f_s = [f_c \pm (2n-1) f_m]$; Where $n = 0, 1, 2, 3$

where, f_s = Frequency of modulated signal

f_c = Carrier frequency (sin) = 1 MHz

f_m = Modulating signal frequency symmetrical squarewave)

= 10 kHz

$$f_m = \frac{1}{T_m} = \frac{1}{100\,\mu s} = \frac{1}{10\times10^{-6}} = \frac{1000}{100}\text{ kHz}$$

= 10 kHz

$f_s = [f_s \pm (2n-1) f_m]$

= 1000 kHz ± (2n − 1) 10 kHz

= 990 kHz, 1010 kHz, 1030 kHz

∴ The frequency 1020 kHz is not present in modulated output.

8. **Ans. (a)**

Solution: The noise at (DSB-SC) detector output is only **in phase component.**

Second Method

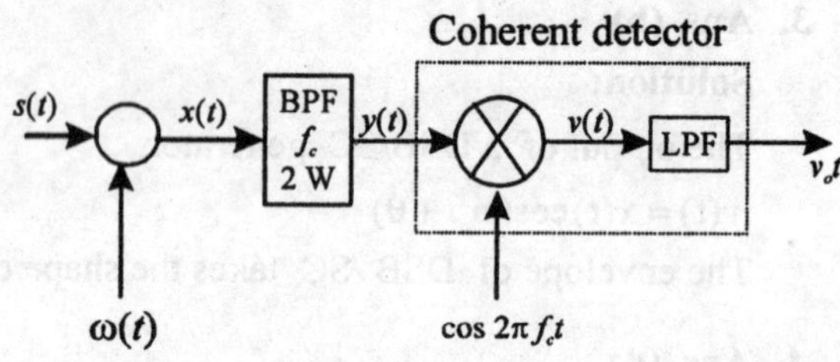

Let $s(t) = A_c m(t)\cos 2\pi f_c t$ (i.e. DSB-SC signal)

Where $m(t)$ is the meassage signal with

$$BW = 2W$$

And $\omega(t)$ be AWGN with $PSD = \frac{N_o}{2}$

or, $x(t) = s(t) + \omega(t)$

$= A_c m(t)\cos 2\pi f_c t + +\omega(t)$

or, $y(t) = A_c m(t)\cos 2\pi f_c t + n(t)$

Where $n(t)$ is the filtered noise

$n(t)$ can be written in its canonical form as:

$$n(t) = n_I(t)\cos 2\pi f_c t + n_Q(t)\sin 2\pi f_c t$$

$$v(t) = y(t)\cdot\cos 2\pi f_c t$$

$$= \begin{bmatrix} \{A_c m(t) + n_I(t)\}\cdot\cos 2\pi f_c t \\ +\{A_c m(t) + n_Q(t)\}\cdot\sin 2\pi f_c t\cdot\cos 2\pi f_c t \end{bmatrix}$$

or output of the low pass filter,

i.e. $v_o(t) = \frac{1}{2}A_c m(t) + \frac{1}{2}n_I(t)$

i.e. Noise at the detector output is the inphase component

9. Ans. (c)

Solution:

$$f_c = 1\ MHz$$

$$v_o = a_0 v_i + a_1 v_i^3$$

$$v_i = A_{ci}\cos(2\pi f_c t) + m(t)$$

or, $v_o = \begin{Bmatrix} a_0\left[A_{ci}\cos(2\pi f_c t) + m(t)\right] \\ + a_1\left[A_{ci}\cos(2\pi f_c t) + m(t)\right]^3 \end{Bmatrix}$

or $\quad v_o = \ldots + 3a_1 A_{ci}^2\cos^2(2\pi f_c t)m(t) + \ldots$

or $\quad v_o = \ldots k\{1 + \cos(2\pi 2 f_c t)m(t)\} + \ldots$

or $\quad v_o = \ldots k_m(t)\cos(2\pi 2 f_c t)m(t) + \ldots$

Hence for getting carrier frequency, $f_c = 1\ MHz$

or $\quad f_{ci} = 0.5\ MHz$

10. Ans. (d)

Solution: Modulation index = 0.5 = $\frac{1}{2}$ = m_a

$$P_t = P_c\left(1 + \frac{m_a^2}{2}\right) = P_c + P_{SB}$$

$$\frac{P_t}{P_c} = 1 + \frac{m_a^2}{2} = 1 + \frac{0.5}{2}$$

$$= 1 + \frac{0.25}{2}$$

Side Band power = $P_t - P_c$

Then, $\frac{P_t - P_c}{P_c} = \frac{1.125}{1} = \frac{0.125}{1} = \frac{1}{8}$

Second Method

$$P_t = P_c\left(1 + \frac{m_a^2}{2}\right) = P_c + P_{SB}$$

$$\frac{P_{SB}}{P_c} = \frac{\frac{m_a^2}{2}P_c}{P_c} = \frac{m_a^2}{2} = \frac{(0.5)^2}{2}$$

$$= \frac{0.25}{2} = \frac{1}{8}$$

11. Ans. (d)

Solution: Audio Frequency = ω_m

AM BW = $2\omega_m$

For FM, $\Delta\omega = \delta = 3 \times 2\omega_m = 6\omega_m$

$$\omega_m = 4\pi \times 10^3$$

$$\omega_c = 2\pi \times 10^6$$

$$m_f = \beta = \frac{\Delta\omega}{\omega_m} = 6$$

$$X_{FM}(t) = A\sum_{n=-\infty}^{\infty} J_n(\beta)\cos(\omega_c + n\omega_m)t$$

$$\omega_c + n\,\omega_m = 2\pi\,(1008 \times 10^3)$$

or, $\quad 2\pi \times 10^6 + n\,.\,4\pi \times 10^3 = 2\pi\,(1008 \times 10^3)$

or, $\quad 2\pi \times 1000 \times 10^3 + n\,.\,4\pi \times 10^3 = 2\pi\,(1008 \times 10^3)$

$n\,4\pi \times 10^3 = 8 \times 2\pi\,10^3$

$$n = \frac{8 \times 2\pi}{4\pi} \quad \text{or, } n = 4$$

$$A\,J_n(m_f) = 5\,J_4(6)$$

12. Ans. (b)

Solution: RC = Time constant,

For No Clipping $\frac{1}{RC} \geq 2\pi f_m m_a$

Typically $m_a = 0.4$ then,

$$\frac{1}{2\pi f_m m_a} \geq RC$$

or $\quad \frac{1}{2\pi \times 2 \times 10^3 \times 0.4} \geq RC$

or $\quad 10^{-3} \geq RC$

or, $\quad 4 \times 10^{-4} \geq RC$

$$RC \leq 10^{-4}$$

$$= \frac{4}{10^4} \times \frac{100}{100} = \frac{400}{10^6} = 400\ \mu s$$

The time constant of Envelop detector should be less than time period of modulating signal but grater than time period of carrier, i.e. $\frac{1}{\omega_c} < RC < \frac{1}{\omega_m}$

Where $\frac{1}{f_c} \Rightarrow$ to avoid the fluctuation at recoved output

$\frac{1}{f_m} \Rightarrow$ to avoid diagonal clipping

Hence $\frac{1}{2\pi\times 10^6} << RC << \frac{1}{2\pi\times 2\times 10^{-3}}$

0.16 μs < RC < 80 μs

The best choice is 20 μs

13. Ans. (c)

Solution:

$$y(t) = \frac{1}{2}\cos 2\pi(101\times 10^6 t) + \sin 2\pi(100\times 10^6 t)$$

$$= \frac{1}{2}\cos 2\pi(10^8 t)\cos 2\pi(10^6 t)$$

$$-\frac{1}{2}\sin 2\pi(100\times 10^6 t)\sin 2\pi 10^6 t + \sin 2\pi(100\times 10^6 t)$$

$$= \frac{1}{2}\sin 2\pi(10^8 t)\cos 2\pi(10^6 t) - \left(1 - \frac{1}{2}\sin 2\pi(10^8 t)\right)$$

So envelope detector output

$$y(t) = \left[\frac{1}{4}\cos^2(\omega_m t) + \left(1 - \frac{1}{2}\sin\omega_m t\right)^2\right]^{1/2}$$

$$= \frac{1}{4}\cos^2\left(2\pi\times 10^6 t\right) + 1 + \frac{1}{4}\sin^2\left(2\pi\times 10^6 t\right) - \sin\left(2\pi\times 10^6 t\right)$$

$$= \sqrt{\frac{5}{4} - \sin(2\pi\times 10^6)t}$$

Second Method

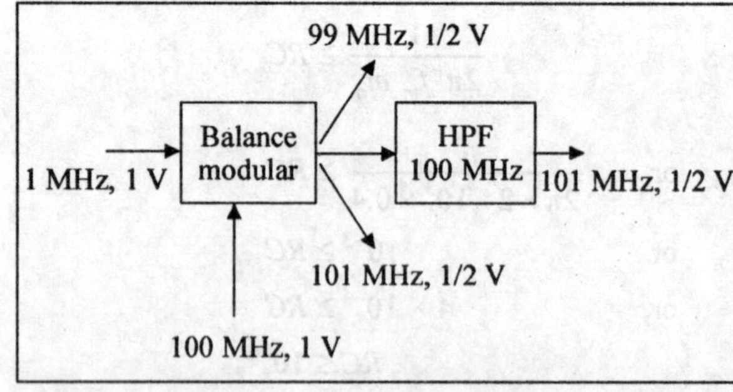

$$y(t) = \frac{1}{2}\cos\{2\pi\times 101\times 10^6\}t + \sin 2\pi\times 100\times 10^6 t$$

$$= \frac{1}{2}[\cos\{2\pi\times 100\times 10^6\}t\cos 2\pi 10^6 t.$$

$$-\sin(2\pi\times 100\times 10^6)t\sin 2\pi 10^6 t]$$

$$+\sin 2\pi\times 100\times 10^6 t$$

$$= \frac{1}{2}\cos 2\pi\, 100\times 10^6 t[\cos 2\pi\times 10^6 t]$$

$$+\frac{1}{2}\sin 2\pi\times 100\times 10^6 t[\sin 2\pi\times 10^6 t - 1]$$

$$= A\cos 2\pi\times 10^6 t - B\sin\cos 2\pi\times 10^6 t$$

$$= Y_I(t)\cos\omega_m t - Y_Q(t)\sin\omega_m t$$

$$\text{Envelope} = \sqrt{A^2 + B^2}$$

$$y(t)\sqrt{=\frac{1}{4}\cos^2 2\pi\times 10^6 t + \frac{1}{4}[\sin 2\pi\times 10^6 t - 1]^2}$$

$$\text{Envelope} = \sqrt{\frac{5}{4} - \sin 2\pi\times 10^6 t}$$

14. And. (c)

Solution: VSB $\to f_m + f_c$

DBS-SC $\to 2f_m$

SSB $\to f_m$

AM $\to 2f_m$

Thus SSB has minimum bandwidth and it required minimum power, i.e. 17% as compared to AM.

15. Ans. (a)

Solution: Discharge time Constant *RC* should be such that the rate of capacitor discharge = Slop of envelope of modulated signal. Then the output Capacitor voltage follow the envelope and *f*(*t*) is detected with high *S*/*N*

$$\frac{1}{RC} = \frac{\omega_m m_a}{\sqrt{1 - m_a^2}} \text{ for no clipping}$$

Typically $m_a = 0.4$, $m_a^2 = 0.16$

$$\frac{1}{RC} \geq m_a\,\omega_m$$

$$RC \leq \frac{1}{m_a\omega_m}$$

$$\frac{1}{\omega_C} << RC << \frac{1}{\omega_m}$$

16. Ans. (c)

Solution: Given

$$p(t) = u(t) - u(t-1)$$

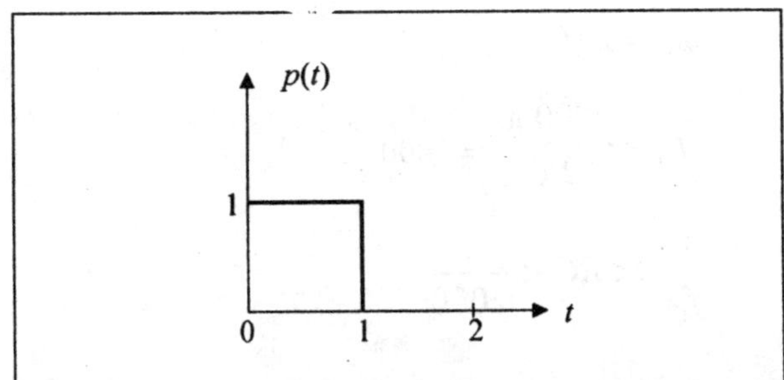

$g(t) = p(t) * p(t)$

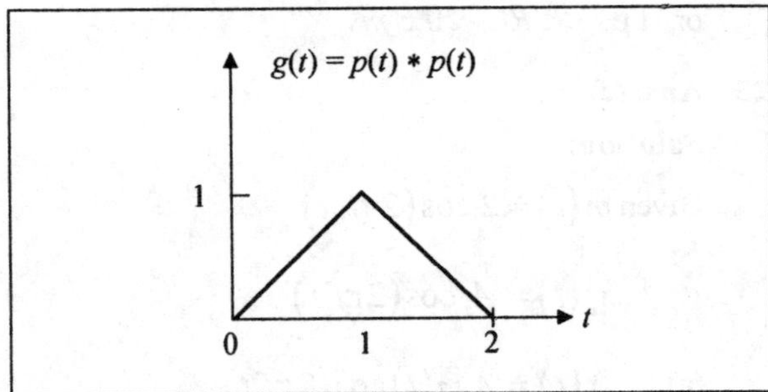

$s(t) = g(t) - g(t-2) * g(t)$

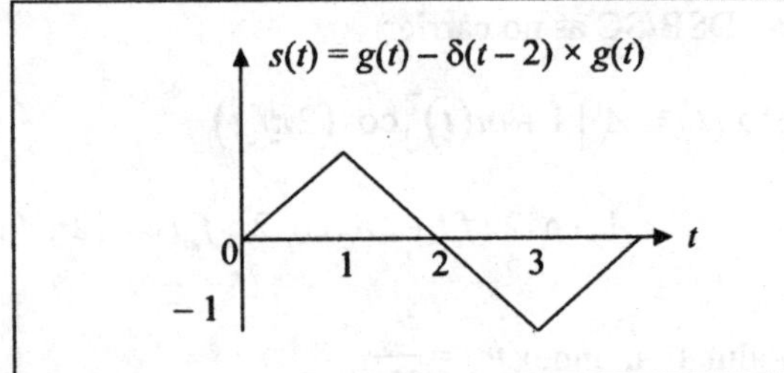

The impulse response of matched filter is

$h(t) = s(T-t)$ where T = 4

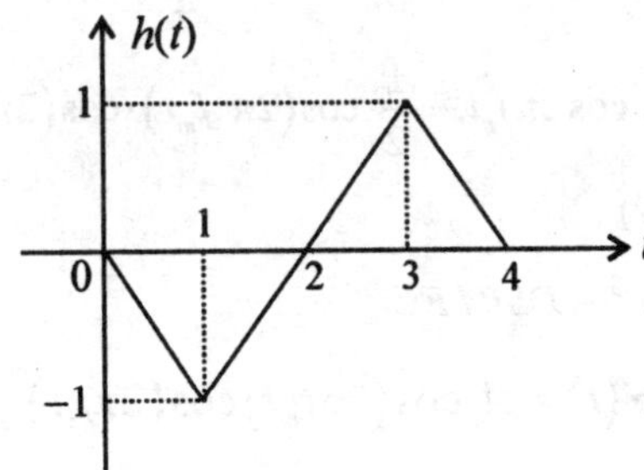

or $h(t) = -s(t)$

17. Ans. (a)

Solution: Given modulated signal

$$x_{AM}(t) = 100\,[p(t) + 0.5\,g(t)]\cos\omega_c t$$

where $p(t) = u(t) - u(t-1)$

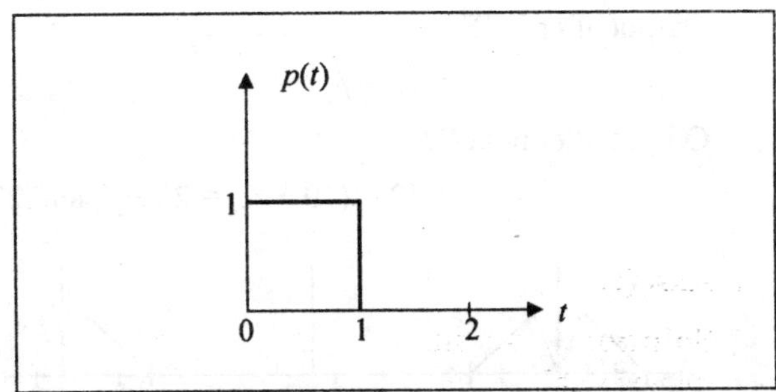

and $g(t) = r(t) - 2r(t-1) + r(t-2)$

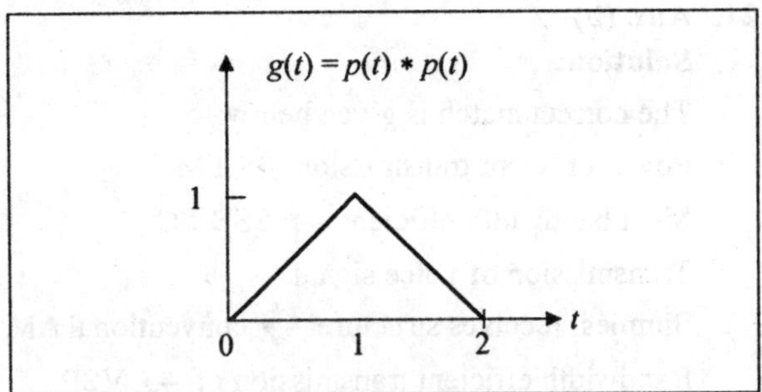

For desired interval $0 \le t \le 1$

$p(t) = 1$ and $g(t) = t$. Thus we have,

$$x_{AM}(t) = 100\,(1 - 0.5\,t)\cos\omega_c t$$

$$m(t) = t$$

Hence modulation index is 0.5.

18. Ans. (c)

Solution:

We have $x_{AM}(t) = 10(1 + 0.5\sin 2\pi f_m t)\cos 2\pi f_c t$

$m_a = 0.5$

$V_c = 10$

Carrier power $P_c = \dfrac{(10)^2}{2} = 50$

Side band power $P_c = \dfrac{m_a^2 P_c}{2} = \dfrac{(0.5)^2(50)}{2} = 6.25$ W

19. Ans. (b)

Solution: Mean noise power = Area under the PSD curve

$$= 4\left[\frac{1}{2} \times B \times \frac{N_o}{2}\right] = BN_o$$

The ratio of average sideband power to mean noise power is

$$\frac{\text{Side Band Power}}{\text{Noise Power}} = \frac{6.25}{N_o P} = \frac{25}{4N_o B}$$

20. Ans (a)

Solution:

The output after BM = $10 \pm f_x$

Output after HPF

$$= 10 + f_x$$

Output after next BM

$$= 13 \pm (10 \pm f_x) = 23 + f_x \text{ and } 23 - f_x$$

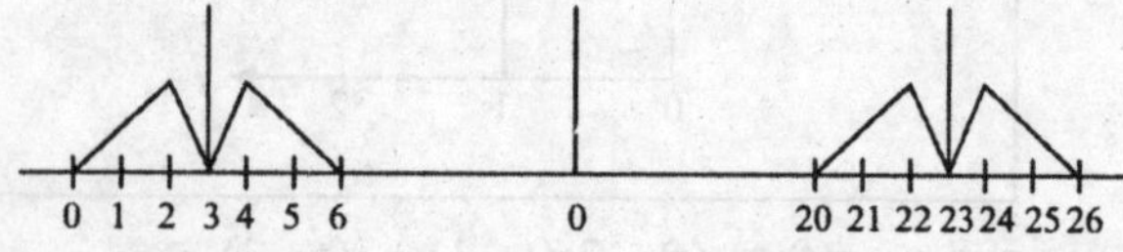

21. Ans. (b)

Solution:

The correct match is given below:

Power efficient transmission → FM

Most bandwidth efficient → SSB-SC

Transmission of voice signal

Simplest receives structure → conventional AM

Bandwidth efficient transmission of → VSB

Signals with significant DC component

Second Method

1. In conventional AM, Envelope detector is used.
2. In FM transmited power = A_C^2 = Carrier power
3. In VSB, Transmission BW is reduced by partially suppressing one of the side bands which radius the BW by 1/3 compare to AM. The vestige of the other sideband is allowed to retain the DC component of the signal.
4. In SSB-SC, transmissin BW, $B_T = W$, i.e. equal to message BW.

22. Ans. (b)

Solution: For demodulating using invelop detection with out distortion i.e. diogernal or negaive peak cliping the condition must be satisfied that is

$$\frac{1}{RC} \geq \omega_m m_a^2$$

if modulation index is less than 0.4

$$\text{Then } \frac{1}{RC} \geq 2\pi f_m \ \ m_a$$

$$\Rightarrow \ \frac{1}{RC} \geq f_m$$

in other word, circuit time consant must be inthe range of

$$\frac{1}{f_C} << RC << \frac{1}{f_m}$$

$$\omega_m = 2\pi f_m$$

$$f_m = \frac{4000\,\pi}{2\pi} = 2000$$

$$\frac{1}{f_C} << RC < \frac{1}{2000}$$

$$\frac{1}{1\,MHz} << RC < \frac{1}{2000\,Hz}$$

or, $1\,\mu s << RC < 0.5\,ms$

23. Ans. (c)

Solution:

Given $m(t) = 2\cos(2\pi f_m t)$

$$x_c(t) = A_c \cos(2\pi f_c t)$$

(a) $x(t) = A_c m(t)\cos(2\pi f_c t)$

$$= 2\,A_c \cos(2\pi f_m t)\cdot\cos(2\pi\, f_c t)$$

⇒ DSB/SC as no carrier

(b) $x(t) = A_c[1 + m(t)]\cos(2\pi f_c t)$

$$= A_C \cos 2\pi f_c t + 2A_C \cos 2\pi\, f_m t \cos(2\pi\, f_c t)$$

modulation index $m_a = \dfrac{V_m}{V_C}$

$= 2$ (over modulation)

(c) $x(t) = A_c \cos(2\pi f_c t) + \dfrac{A_c}{4} m(t)\cos(2\pi f_c t)$

$$= A_C \cos 2\pi f_c t + \frac{A_C}{2}\cos(2\pi\, f_m t)\cdot\cos(2\pi\, f_c t)$$

$m_a = 1$

$\Rightarrow AM - DSB/FC$

(d) $x(t) = A_c \cos(2\pi f_m t)\cos(2\pi f_c t)$

$$+ A_c \sin(2\pi f_m t)\sin(2\pi f_c t)$$

$\Rightarrow AM - SSB/SC$

$\Rightarrow AM - LSB/SC$

24. Ans (c)

Solution: Amplitude Modulation

Here the amplitude of the carrier is varied in accordance with the instan- taneous value of the amplitude of the modulating signal, keeping frequency and phase of the carrier to be fixed.

$f(t) = V_m \sin w_m(t)$ ... modulating signal

$= V_c \sin w_c(t)$ ... carrier

Modulated Signal (AM Signal)

$V(t) = [V_c + K_a V_m(t)] \sin w_c t$

$= V_c \sin w_c t + K_a \frac{V_m}{2} 2 \sin w_c t \, . \sin w_m t$

$= V_c \sin w_c t + K_a \frac{V_m}{2} [\cos (w_c - w_m)t - \cos (w_c + w_m)t]$

Spectrum of AM Signal

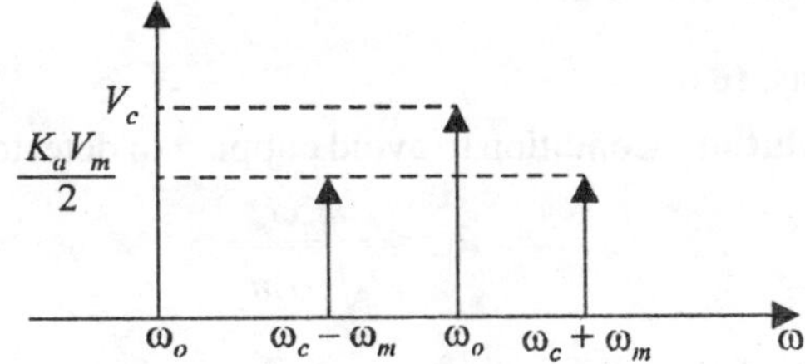

Frequency component of AM signal

1. w_c **2.** $w_c + w_m$ **3.** $w_c - w_m$

Modulation Index (m_a = 0.4–0.6)

The extent of amplitude variation in AM about an unmodulated carrier amplitude is measured in terms of a factor called modulation index. It is define as.

$$m_a = \frac{(V_m)_{\max}}{V_c} = \frac{K_a V_m}{V_c} = \frac{V_m}{V_c} \text{ (if } K_a = 1)$$

$$V_m = \frac{m_a V_c}{K_a} \text{ or } K_a V_m = m_a V_c$$

This factor m_a is also known as depth of modulation or degree of modulation or modulation factor.

Power Contained in the AM Signal

$$V(t) = V_c \sin w_c t + \frac{m_a V_c}{2} [\cos (w_c - w_m)t - \cos (w_c + w_m)t]$$

Total $T \times R$ power = Power in the carrier + Power USB + Power LSB

= Power (Carrier) + 2 Power (SB)

The option $\cos\left[2\pi\left(f_c + f_m\right)t\right]$ contains only upper side band frequency. Thus, it represents single side band signal.

25. Ans (c)

Solution: $m_a = \frac{1}{\sqrt{2}}$,

$$\text{Efficiency} = \frac{m_a^2}{2 + m_a^2}$$

$$= \frac{1/2}{2 + \frac{1}{2}} = 20\%$$

Second Method

$$m(t) = \left(\frac{1}{2}\right)\cos\omega_1 t - \left(\frac{1}{2}\right)\sin\omega_2 t$$

$$s(t) = \left[1 + m(t)\cos\omega_c t\right]$$

$$= \left[1 + \left(\frac{1}{2}\right)\cos\omega_1 t - \left(\frac{1}{2}\right)\sin\omega_2 t\right]\cos\omega_1 t$$

$$m_{a1} = \frac{1}{2} \text{ and } m_{a2} = \frac{1}{2}$$

$$m_a^2 = m_{a1}^2 + m_{a2}^2 = \left(\frac{1}{2}\right)^2 + \left(\frac{1}{2}\right)^2$$

$$= \frac{1}{2}$$

$$\text{Power effeciency } \eta = \frac{m_a^2}{m_a^2 + 2} = \frac{1/2}{\frac{1}{2} + 2}$$

$$= \frac{1}{5}$$

$$\% \, \eta = \frac{1}{5} \times 100\%$$

$$= 20\%$$

26. Ans. (a)

Solution: Given AM signal =

$A_c \cos\omega_c t + 2\cos\omega_m t \cos\omega_c t$

$= A_c \cos\omega_c t\left[1 + \frac{2}{A_c}\cos\omega_m t\right]$

For envelop detection modulation index $m_a < 1 \Rightarrow \frac{2}{A_c} < 1$

$\Rightarrow A_c$ should be at least 2

27. Ans. (c)

Solution: $P_T = P_c\left(1 + \frac{m_a^2}{2}\right)$

In case of 100 % depth of modulation

$\therefore \quad P_T = \left(\frac{3}{2}\right)P_c$

28. Ans. (a)

Solution: $m_a = m_{ef1} = \sqrt{m_1^2 + m_2^2}$

$$= \sqrt{(0.3)^2 + (0.4)^2} = 0.5$$

$$P_T = P_c\left(1 + \frac{m_a^2}{2}\right)$$

$$P_T = 10\text{ kW}\left(\frac{2 + 0.25}{2}\right)$$

$$= 10 \times 1.125 = 11.25\text{ kW.}$$

29. Ans. (b)

Solution: An envelope detector is a diode half wave rectifier followed by and RC-low pass detector. The output of the detector represents the envelope of the incoming high frequency signal. It is a asynchronous detection.

30. Ans. (b)

Solution: Picture elements (Pixel)

$$= 512 \times 512 = 262144$$

Information per picture element $n = \log_2 M$

or $M = 2^n$

or $8 = 2^n$

or $n = 3$

Information for 262144 pixels $= 262144 \times 3$

$= 784332$ bits.

Entropy is same as average information per message.

31. Ans. (d)

Solution: $V_c = 12$

$V_m = 6$

Peak $RF = 2\,(12 + 6) = 2 \times 18 = 36$ kV

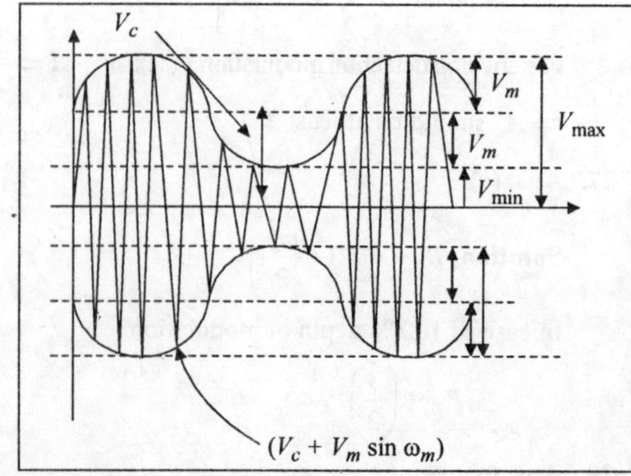

32. Ans. (c)

Solution: Carrier amplitude decreases at higher frequency $(10^6 \pm 10^3)$ Hz

33. Ans. (c)

Solution: $P_T = P_c\left(1 + \dfrac{m_a^2}{2}\right)$

$$P_c = \frac{10\text{ kW}}{\left(1 + \dfrac{0.36}{2}\right)} = \frac{10\text{ kW}}{1.18}$$

$$= 8.47\text{ kW}$$

34. Ans. (d)

Solution: In a low level modulation AM system, the amplifiers following the modulated stage can be Class B untuned amp.

35. Ans. (d)

Solution: Condition to avoid clipping of detector output

$$RC \le \frac{m_a \omega_m}{\sqrt{1 - m_a^2}}$$

36. Ans. (d)

Solution: $P_T = P_C\left(1 + \dfrac{m_a^2}{2}\right)$

$$\Rightarrow \frac{600}{400} = 1 + \frac{m_a^2}{2} \Rightarrow \frac{3}{2} = 1 + m_a^2$$

$$\Rightarrow \frac{1}{2} = \frac{m_a^2}{2}$$

$$m_a = 1$$

37. Ans. (a)

Solution: $m_a = \dfrac{V_{max} - V_{min}}{V_{max} + V_{min}} = \dfrac{10 - 5}{10 + 6} = \dfrac{1}{3}$

$= 0.33$

38. Ans. (a)

39. Ans. (b)

Solution: $P_t = 10$ kW $m_a = 0.6$

Transmitted power $P_t = P_c + P_s$

or $P_t = P_c\left(\dfrac{1 + m_a^2}{2}\right)$

or $P_c = \dfrac{10}{\left(1 + \dfrac{m_a^2}{2}\right)} = \dfrac{10}{\left(1 + \dfrac{0.6^2}{2}\right)}$

$= 8.47$ kW.

40. Ans. (a)

Solution:

A balanced modulation is used in generation of DSB - SC signal.

41. Ans. (d)

Solution:

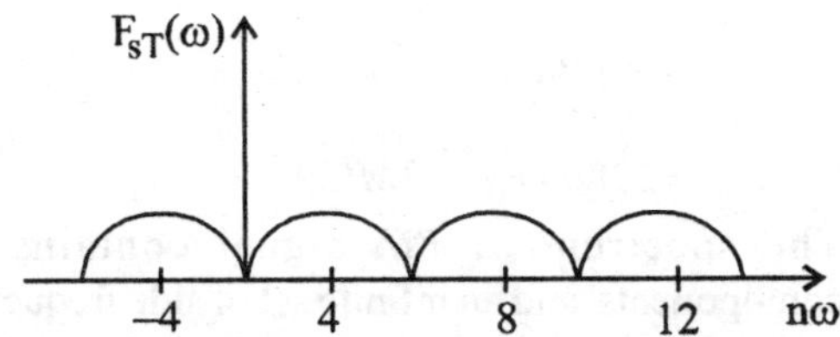

42. Ans. (a)

Solution:

The amplitude modulated signal can be observed on a CRO by applying the following waveform to the "external trigger" input the modulated waveform itself

43. Ans. (b)

Solution:

$$P_t = P_c\left(1+\frac{m_a^2}{2}\right)$$

$$= 500\left(1+\frac{(0.75)^2}{2}\right) = 500\left(1+\frac{0.5625}{2}\right)$$

$= 640.5$ Watt

44. Ans. (a)

Solution:

For AM-SC system having average two sideband these are $X_c(t)\cdot X_m(t) = (\cos\omega_c t)(0.5\sin\omega_c t)$

$= \sin(\omega_c+\omega_m)t - \sin(\omega_c-\omega_m)t$

45 Ans. (a)

Solution:

$$P_t = P_c\sqrt{1+\frac{m_a^2}{2}}$$

$$P = I^2 R$$

Where, R is same under all condition then

$I_t = I_c\sqrt{1+\frac{m_a^2}{2}}$ or, $4.8 = 4\sqrt{1+\frac{m_a^2}{2}}$

or, $1.2 = \sqrt{1+\frac{m_a^2}{2}}$ or, $1.44 = \sqrt{1+\frac{m_a^2}{2}}$

or, $0.44 = \frac{m_a^2}{2}$ or, $0.88 = m_a^2$

or, $m = 0.938$ or, $m = 93.8\ \%$

46. Ans (c)

Solution:

Given $f_m = 2\ kHz$, $f_c = 2\ Mhz$

$s(t) = [1+m(t)]\cos(2\pi f_c t)$

For AM bandwidth of band pass signal $= 2 f_m$

$= 2\times 2\ kHz$

$= 4\ kHz$

5.2 Frequency Modulation

1. Ans. (d)

Solution:

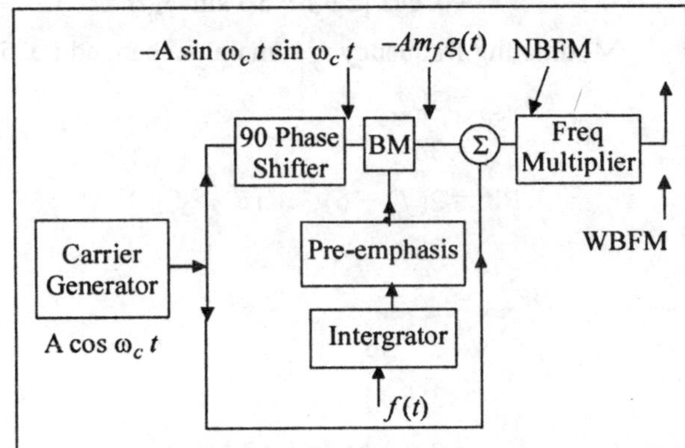

$\varphi_{FM}(t) = A\cos\omega_c t - A\, m_f\, g(t)\sin\omega_c t$

$\varphi_{PM}(t) = A\cos\omega_c t - A\, m_p\, f(t)\sin\omega_c t$

Let $f_c(t) = \cos\omega_c t$ and $f_m(t) = \cos\omega_m t$

$\int\cos\omega t\, dt = \sin\omega_m t$

Following the diagram and result is as NBFM.

2. Ans. (a)

Solution:

$$\cos x = \frac{e^{jx}+e^{-jx}}{2}$$

$s(t) = e^{-at}\cos[(\omega_C+\Delta\omega)t]\,u(t)$ ... (1)

$s(t)$ is modulated, hence a band pass signal.

Let $s_+(t)$ be the pre-envelop of $s(t)$, and

$\hat{s}(t)$ be the complex envelope of $s(t)$.

or $s(t) = \text{Re}\{s_+(t)\}$ ... (2)

$s(t)$ can be written as:

$s(t) = \text{Re}\left[e^{-at}\, e^{j(\omega_C+\Delta\omega)t}u(t)\right]$ ... (3)

Comparing equation (2) and (3), we get

$s_+(t) = e^{-at}\, e^{j(\omega_C+\Delta\omega)t}u(t)$

$= e^{-at}e^{j\Delta\omega t}\cdot e^{j\omega_C t}u(t)$

Complex envelop

$$\hat{s}(t) = s_+(t) \cdot e^{-j\omega_c t}$$

$$= e^{-at} e^{j\Delta\omega t} \cdot e^{j\Delta\omega_c t} \cdot e^{-j\omega_c t} u(t)$$

$$= e^{-at} e^{j\Delta\omega t} u(t)$$

3. Ans. (a)

Solution:

$$x(t) = A_c \cos [2\omega_c t + \beta \sin \omega_m t]$$
$$y(t) = [x(t)]^3$$
$$= k \cos [2\omega_c t + 3\beta \sin \omega_m t]$$

Modulating frequency f_m remain unchanged i.e. 5 kHz

$$\beta_{New} = 3\beta$$
$$\delta_{New} = 3\delta$$
$$BW = 2(f_m + \delta) \qquad (\delta >> f_m)$$
$$= 2\delta$$
$$BW_{new} = 2\delta_{new}$$
$$= 2 \times 3\delta$$
$$= 3[BW]$$
$$= 3 \times 1 \text{ MHz}] = 3 \text{ MHz}$$

Second Method

$$f_c = 100\ MHz$$
$$f_m = 5\ kHz$$
$$BW|_{FM} = 2(\Delta f + f_m) = 1\ MHz$$
$$\Delta f + f_m = 500\ kHz$$
$$\Delta f = 495\ kHz$$

The FM modulated signal can be represented as

$$S(t) = A_c \cos\left[2\pi f_c t + \frac{\Delta f}{f_m} \sin 2\pi f_m t\right]$$

Around 300 MHz (i.e. $3f_c$

$$y(t) = [s(t)]^3$$

The multipler multiplies both carrier and frequency duration i.e. modulation index

$$S_3(t) = K \cos\left[2\pi \times 3 f_c t + \frac{3\Delta f}{f_m} \sin 2\pi f_m t\right]$$

Frequency deviation will be increased by 3 times.

$$\Delta f_3 = 3 \times 495\ kHz$$
$$= 1485\ kHz$$
$$BW|_{FM} = 2(\Delta f_3 + f_m)$$
$$= 2(1485 + 5)$$
$$= 2(1490)$$
$$= 2980\ kHz \approx 3 MHz$$

The spectrum of FM signal contains a carrier comoponents and an infinite set of side frequency located at symetrically either side of carrier at frequencies seprate of f_m,

$2f_m$, $3f_m$..........

The spacing of the spectral components are governed by f_m, which remains unchanged

Hence spacing = 5 *kHz*.

4. Ans. (d)

Solution: $S(t) = \cos 2\pi (2 \times 10^6\, t + 30 \sin 150\, t + 40 \cos 150\, t)$

$$\theta'(t) = 2\pi\, [2 \times 10^6\, t + 30 \sin 150\, t + 40 \cos 150\, t]$$

Instantiations frequency

$$\omega' = \omega + m_f \,.\, f(t)$$
$$= \omega \pm [m_f \,.\, f(t)]$$
$$= \omega \pm \Delta\omega$$
$$\omega' = [d\theta'(t)/dt]$$
$$= 2\pi\, [2 \times 10^6 + 30 \times 150 \cos 150\, t - 40 \times 150 \sin 150\, t]$$
$$= 2\pi\left\{2 \times 10^6 + 10 \times 150 \times 5\left[\frac{3}{5}\cos 150\, t - \frac{4}{5}\sin 150\, t\right]\right\}$$
$$\Delta\omega = \omega' - \omega$$
$$\Delta f = 10 \times 150 \times 5 = 7500 \text{ Hz} = 7.5 \text{ kHz}$$
$$= 2\pi \,.\, 1500\, [3 \cos 150\, t - 4 \sin 150\, t]$$
$$= 3000\pi \times \sqrt{(3^2 + 4^2)}\, [\cos 150\, t + \alpha]$$
$$= 15000\pi\, [\cos (150\, t + \alpha)]$$
$$\Delta f = \frac{|\Delta\omega|}{2\pi} = \frac{15000\pi}{2\pi} = 7.5 \text{ kHz}$$
$$\phi(t) = 2\pi\, [30 \sin 150\, t + 40 \cos 150\, t]$$
$$= 2\pi \times 10 \times \sqrt{(3^2 + 4^2)} \sin (150\, t + \alpha^1)$$
$$= 2\pi \times 10 \times 5 \sin (150\, t + \alpha^2)$$
$$= 100\pi \sin (t + \alpha')$$

Where $\alpha' = \tan^{-1} [4/3]$

Second Method

$$s(t) = \cos 2\pi(2 \times 10^6 t + 30 \sin 150t + 40 \cos 150t)$$

or, $\theta(t) = 2\pi(2 \times 10^6 t + 30 \sin 150t + 40 \cos 150t)$

or, $\theta(t) = 2\pi \cdot 2\times10^6 t + 2\pi \cdot 30\sin150t + 2\pi \cdot 40\cos150t$

$= \omega t + \phi$

or, Maximum phase deviation

$= Max.[2\pi \cdot 30\sin150t + 2\pi \cdot 40\cos150t]$

Now, maximum value of $A\sin x + B\cos x = \sqrt{A^2 + B^2}$

Maximum phase deviatin

$= 2\pi \times \sqrt{30^2 + 40^2}$

$= 2\pi \times 50 = 100\pi\ rad$

Instantaneous frequency

$$f_i = \frac{1}{2\pi}\frac{d\theta(t)}{dt} = f_c + \Delta f$$

$= 2\times10^6 + 30 + 150\cos150t - 40\times150\sin150t$

Maximum frequency deviation Δf

$= 150 \times \sqrt{30^2 + 40^2}$

$= 150 \times 50 = 7500 = 7.5\ kHz$

5. Ans. (c)

Solution

$$y_1(t) = m(t)\cdot s(t) = \frac{2\sin 2\pi t \cdot \cos 200t}{t}$$

$$= \frac{\sin(202\pi t) - \sin(198\pi t)}{t}$$

$$y_2(t) = y_1 + n(t)$$

$$= \frac{[\sin(202\pi t) - \sin(198\pi t) + \sin 199\pi t]}{t}$$

$$y_2(t) = y_1 + n(t)$$

$$= \frac{[\sin(202\pi t) - \cos(200\pi t) - \sin(199\pi t).\cos 200\pi t]}{t} + \frac{\sin(199\pi t)\cdot\cos(200\pi t)}{t}$$

$$\frac{1}{2t}[\sin(402\pi t) + \sin 2\pi t]$$

$$-\frac{1}{2t}[\sin 398\pi t - \sin 2\pi t]$$

$$+\frac{1}{2t} + [\sin 399\pi t - \sin\pi t]$$

$$= \frac{1}{2t}[\sin 402 t + \sin 399 t - \sin 398\pi t + 2\sin 2\pi t - \sin\pi t]$$

Cut off frequency = 1 KHz

$\omega_1 = 2\pi$

$\omega_2 = 1\pi$

$$\text{After LPF} = \frac{1}{2t}[\sin 2\pi t + \sin 2\pi t - \sin\pi t]$$

$$= \frac{1}{2t}\left[\sin 2\pi t + 2\cos\left(3\pi\frac{t}{2}\right).\sin\left(\pi\frac{t}{2}\right)\right]$$

$$= \frac{1}{2t}[\sin 2\pi t + 2\cos 1.5\pi t.\sin 0.5 t]$$

$$= \left(\frac{1}{2}\right)\frac{\sin 2\pi t}{t} + \frac{\sin 0.5\pi t}{t}\cdot\cos 1.5\pi t$$

$$[\sin A - \sin B] = 2\cos\frac{A+B}{2}\cdot\sin\frac{A-B}{2}$$

6. Ans. (b)

Solution:

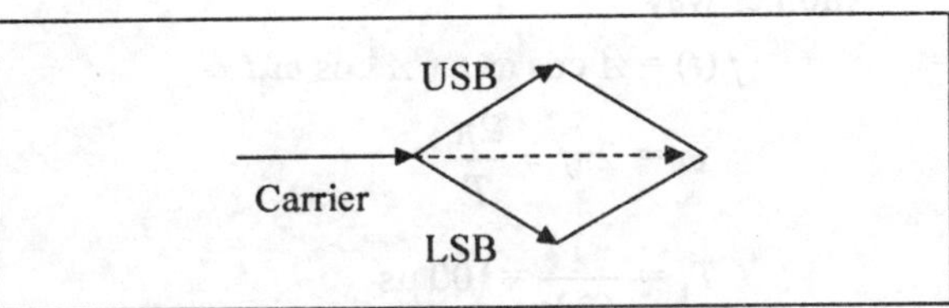

Figure (a) AM vector diagram

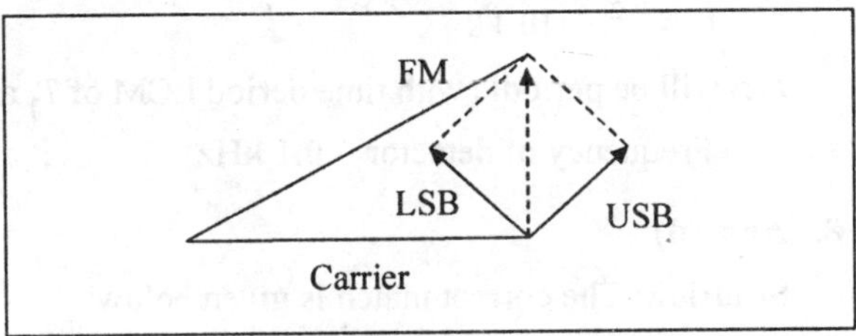

Figure (b) FM vector diagram

When we add both the vector diagram. Then one SB is cancelled out remaining with carrier and one side band i.e. AB-SSB with full carrier

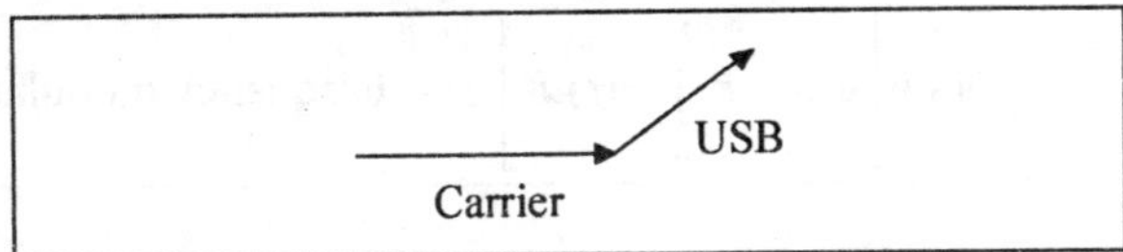

Here $\quad m_f = m_a = 0.1$

Modulating signal $f(t) = \cos\omega_m f$

$$s(t)_{FM} = A\cos\omega_c t - A\,m_f\,g(t)\sin\omega_c t$$

$$= A\cos\omega_c t + \frac{A\,m_f}{2}\cos(\omega_c + \omega_m)t - \frac{Am_f}{2}\cos(\omega_c - \omega_m)t$$

$$s(t)_{AM} = A\cos\omega_c t + Am_a\,f(t)\cos\omega_c t$$

$$= A\cos\omega_c t + \frac{A\,m_a}{2}\cos(\omega_c + \omega_m)t + \frac{Am_a}{2}\cos(\omega_c - \omega_m)t$$

$$s(t)_{FM} + s(t)_{AM} = 2A\cos\omega_c t + \left(\frac{Am_f}{2} + \frac{Am_a}{2}\right)\cos(\omega_c + \omega_m)$$

$$= \text{carrier} + \text{USB}$$

$$\Rightarrow \text{SSB with carrier}$$

7. Ans. (a)

Solution: Ideal frequency detector work as mixer whose output is $(f_2 + f_1), f_2, f_1, (f_2 - f_1)$

Detector output is connected with LPF. Hence only $(f_2 - f_1)$ signal through it i.e. (10.1 – 10) = 0.1 kHz sinusoidal

Another way

$$f(t) = A\cos\omega_1 t + A\cos\omega_2 t$$

$$\omega = 2\pi f = \frac{2\pi}{T}$$

$$T_1 = \frac{1}{10\text{ k}} = 100\ \mu\text{s}$$

$$T_2 = \frac{1}{10.1\text{ k}} = 99\ \mu\text{s}$$

$f(t)$ will be periodic with time period LCM of T_1 and T_2

∴ Frequency of detector = 0.1 kHz

8. Ans. (d)

Solution: The correct match is given below:

$[1 + km(t)]A\sin(\omega_c t)$	Amplitude modulation
$km(t)A\sin\omega_c t)$	DB-SC modulation
$A\sin[\omega_c' + k]_{-\infty}^{t} m(t)dt$	Phase modulation
$A\sin\left[\omega_c t + k\int_{-\infty}^{t} m(t)dt\right]$	Frequency modulation

9. Ans. (d)

Solution: $\Delta f = 10$ kHz $\quad f_{m\,(new)} = 2$ kHz

$f_m = 1$ kHz

By carson's Rule

$$BW = 2(\Delta f + f_m) = 2(10 + 1) = 22\text{ kHz}$$

$$\Delta f \propto f_m$$

$$\Delta f_{(new)} = 2 \times 10 = 20$$

$$BW_{(new)} = 2(20 + 2) = \mathbf{44\ kHz}$$

10. Ans. (b)

Solution: After the LSB-SSB modulation the frequency of signal is, i.e. $f_c - f_m$ i.e. 1000 – 10 kHz ≈ 1000 kHz = 10^6 Hz

The bandwidth of FM is

$$BW = 2(\beta + 1)\Delta f$$

For NBFM $\beta << 1$, thus

$$BW_{NBFM} \approx 2\Delta f = 2 \times 10^6\text{ Hz}$$

11. Ans.(b)

Solution:

FM: Here the frequency of the carrier is varied in accordance with the instantaneous value of the amplitude of the modulating signal.

PM: Here the phase of the carrier is varied in accordance with the instantaneous value of the amplitude of the modulating signal.

Let, Modulating signal = $f(t)$

Carrier = $A\cos w_c t$

$= A\cos q(t)$

For angle modulation, the modulated carrier is represented by

$$x_c(t) = A\cos[w_c t + f(t)]$$

where A and w_c are constant and the phase angle f(t) is function of the message signal $f(t)$.

If we rewrite above equation

$$x_c(t) = A\cos(t)$$

when $\theta(t) = [\omega_c t + \phi(t)]$

then we can define the instantaneous radian frequency of $x_c(t)$ defined by ω_i as

$$\omega_i = \frac{d\theta(t)}{dt} = \omega_c + \frac{d\phi(t)}{dt}$$

$$f_i = f_c + \frac{1}{2\pi}\frac{d\phi(t)}{dt}$$

The function $\phi(t)$ and $[d\phi(t)/dt]$ are known as the instantaneous **phase deviation** and instantaneous frequency deviation of $x_c(t)$. The quantity $\Delta\omega = \delta$ defined by

$$\Delta\omega = |\omega_i - \omega_c|_{max}$$

is called the maximum (or peak) radian frequency deviation of the angle modulated signal.

Where, $\theta(t) = [\omega_c t + \phi(t)]$

$$\frac{d\theta(t)}{dt} = \omega_c + \frac{d\phi(t)}{dt}$$

$$\theta(t) = \int\left[\omega_c + \frac{d\phi(t)}{dt}\right]dt$$

FM: Frequency is varied with $f(t)$

$$f(t) = E_m\cos\omega_m t$$

Instantaneous frequency

$\omega_c' = \omega_c + K_f\, f(t)$

$= \omega_C + K_f \cos \omega_m t$

$= \omega_C + \Delta\omega$

Where, $\Delta\omega = |K_f\, f(t)|_{max}$

$$\omega_c' = \int\left[\omega_c + \frac{d\phi(t)}{dt}\right]$$

$$K_f\, f(t) = \frac{d\phi(t)}{dt}$$

$$\phi(t) = \int K_f\, f(t)\, dt$$

$$\phi(t)|_{max} = 2\pi\, K_f\, f(t)|_{max}\, dt \quad \text{...(1)}$$

where K_f = the frequency deviation constant expressed in radian/sec for FM

$$\theta'(t) = \int \omega_c'\, dt = \int [\omega_c + K_f f(t)]\, dt$$

$$= \int[\omega_C + \Delta\omega \cos \omega_m t]$$

$$= \int \omega_C t + \frac{\Delta\omega}{\omega_m} \sin \omega_m t$$

$$= \int\left[\omega_c + \frac{d\phi(t)}{dt}\right]$$

FM signal $f(t) = A \cos \theta'(t)$

$= A \cos \int [\omega_c + K_f f(t)]\, dt$

$f_{FM}(t) = A \cos [\omega_c t + K_f \int f(t)\, dt]$

PM: The instantaneous **phase deviation** of the carrier is proportional to the message signal

$$\phi(t) = K_p f(t)$$

$$\phi(t)|_{max} = K_p |f(t)|_{max} \quad \text{...(2)}$$

$$\theta(t) = [\omega_c\, t + \phi(t)]$$

$$\theta'(t) = [\omega_c\, t + K_p f(t)]$$

Instantaneous frequency

$$\omega_c' = \omega_c + K_p \frac{df(t)}{dt}$$

where, K_p is the phase derived constant expressed in radian/unit of $f(t)$ for PM.

PM signal $= A \cos \theta'(t)$

$f_{PM}(t) = A \cos [\omega_c t + K_p f(t)]$

$f_{FM}(t) = A \cos [\omega_c t + K_f \int f(t)\, dt]$

In present case as derived in equation (1) and (2)

In phase modulation,

Maximum phase deviation $= K_p\,|\,m(t)\,|_{max} = K_p.2$

In Frequency modulation.

Maximum phase deviation

$$= 2\pi\, K_f \int_0^2 2\, dt$$

$$= 2\pi\, K_f \times 4$$

Now $K_p.2 = 2\pi\, k_f \times 4$

or, $\dfrac{K_p}{K_f} = 4\pi$

12. Ans. (b)

Solution:

$$I_{rms} = \frac{A_c}{\sqrt{2}}$$

$$\text{Average Power} = \frac{A_C^2}{2} = \frac{6^2}{2} = \frac{36}{2} = 18 \text{ Walt}$$

13. Ans. (b)

Solution:

Total frequency deviation

$= 5 \times 1500 + 7.5 \times 1000 = 15000$

Thus modulation index $= \dfrac{\Delta f}{f_{m_{max}}} = \dfrac{15000}{1500} = 10$

Second Method

Modulation index

$$\beta = \frac{\Delta f}{f_{max}}$$

Where ,

$\Delta f \rightarrow$ Maximum deviation of instantaneous frequency

$f_m \rightarrow$ Highest frequency component present in the modulating signal

$$f_i = \frac{1}{2\pi}\frac{d\theta_i(t)}{dt}$$

$$f_i = \begin{bmatrix} 10^5 \times 5 \times 1500 \cos(2\pi \times 1500t) \\ +7.5 \times 1000 \cos(2\pi \times 1500t) \end{bmatrix}$$

$$\Delta f = \begin{bmatrix} 5 \times 1500 \cos(2\pi \times 1500t) \\ +7.5 \times 1000 \cos(2\pi \times 1500t) \end{bmatrix}$$

$$\Delta f_{max} = 5 \times 1500 + 7.5 \times 1000$$

$$= 15000\ Hz$$

$$f_m = \max(1500,\ 1000)$$

$= 1500\ Hz$

$$\beta = \frac{15000}{1500}$$

$= 10$

14. Ans. (c)

Solution:

$$x(t) = \cos \omega_c t + 0.5 \cos \omega_m t \cdot \sin \omega_c t$$

$$= \cos \omega_c t + \frac{1}{4}\sin(\omega_c + \omega_m)t + \frac{1}{4}\sin(\omega_c - \omega_m)t$$

Frequency components presents in $x(t)$ are:

$$\omega_c,\ \omega_c + \omega_m \text{ and } \omega_c - \omega_m$$

Hence, the signal can be either AM or NBFM

15. Ans. (d)

Solution: A mixer, since it yields sum and difference frequency, will translate the spectrum of an FM signal, but will have no effect on its frequency deviation.

16. Ans. (b)

Solution: On amplitude deviation

$(\delta = k_f E_m) \rightarrow$ B.W.

$= 2(\delta + f_m)$

17. Ans. (b)

18. Ans. (c)

Solution: The sound signal in TV broadcast is modulated in FM.

19. Ans. (a)

Solution: In the generation of modulated signal, a varactors diode can be used for FM generation only.

20. Ans. (d)

Solution: Pre-emphasis in FM systems involves (d) Amplification of higher frequency components of the modulating signal.

21. Ans. (b)

Solution: As the modulation index of a FM signal with sinusoidal modulation is increased from zero to three, the power in the sideband component will Decrease continuously.

22. Ans. (c)

Solution: $BW_{FM} = 2(\delta + f_m)$

The deviation does not depend on carrier frequency.

23. Ans. (a)

Solution: In phase modulation, the frequency deviation is: (a) Independent of the modulating signal frequency.

24. Ans. (c)

Solution: $f_m = 100$ Hz

$m_f = 15$

$$m_f = \frac{\delta}{f_m} \Rightarrow \delta = m_f \cdot f_m = 1500 \text{ Hz}$$

$\text{B.W} = 2(\delta + \delta_m)$

$= 2(1500 + 100)$

$= 3.2$ kHz

25. Ans. (c)

Solution: From the given signal expression,

$$\theta(t) = 2\times10^8 \pi t + 75 \sin 2\times10^3 \pi t$$

and $\phi(t) = 75 \sin 2\times10^3 \pi t$

$$\frac{d\phi}{dt} = 75\times2\times10^3 \pi \cdot \cos$$

Then, peak frequency deviation

$$|\phi'(t)|_{\max} = \frac{7.5\times2\times10^3\pi}{2} = 75 \text{ kHz}$$

26. Ans. (d)

Solution: The frequency multiplier multiplies both carrier and frequ- ency deviation i.e. modulation index. i.e. $\delta = m_f . f_m$; where f_m = modulating frequency.

27. Ans. (c)

Solution:

Indirect Method of FM Generation

Armstrong method:

Basic principle

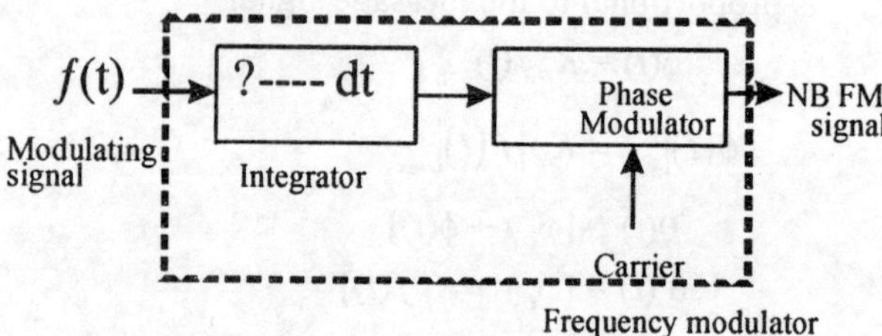

28. Ans. (a)

Solution:

The FM radio station perform better than an AM station radiating the same total power as capture effect appears in FM.

29. Ans. (d)

Solution:

Frequency Spectrum of FM Wave

Let, $f(t) = V_m \cos \omega_m t$

carrier $= A \sin \omega_c t$

$$f_{FM}(t) = A\cos\left(\omega_c t + m_f . \int f(t)\,dt\right)$$

$$f_{FM}(t) = A\sin\left(\omega_c t + m_f \sin \omega_m t\right)$$

→ It is sin of a sin function

→ The solution for this function involve the use of Bessel function.

$$f_{FM}(t) = A\sin\left(\omega_c t + m_f \sin \omega_m t\right)$$

$$= A J_o(m_f)\sin \omega_c t$$

$$+J_1(m_f)\left(\sin[\omega_c + \omega_m]t - \sin(\omega_c - \omega_m)t\right)$$

$$+J_2(m_f)\left(\sin[\omega_c + 2\omega_m]t - \sin(\omega_c - \omega_m)t\right)...$$

It is seen that the output consist of a carrier and apparently infinite sidebands each preceded by J coefficient. These are Bessel function.

30. Ans. (a)

Solution:

Bandwidth of FM is given by carson's rule, i.e. $BW|_{FM}$

$= 2(\delta + f_m)$

$= 2(75 + 15) = 2 \times 90 = 180\ kHz$

31. Ans. (d)

Solution:

Given, $f_m = 1\ kHz$

$\beta = 5$

For angle modulation index $\beta = m_f = \dfrac{\delta}{f_m}$

By carson's Rule

$$BW|_{Fm} = 2(\delta + f_m)$$

$$= 2(\beta + 1)\cdot f_m$$

$$= 2(5+1)\cdot 1\ kHz$$

$$= 12\ kHz$$

32. Ans. (a)

Solution:

Carrier signal $X_c(t) = \cos \omega_c(t)$

If modulating signal $x_m(t) = f(t)$

output signal $y(t) = \cos(\omega_c t + \sin 2t - \cos t)$ ... (1)

Modulated signal $y(t) = \cos[\omega_c t + k_f \int f(t)dt]$ ... (2)

Comparting equation (1) and (2)

$$\int f(t)dt = \sin 2t - \cos t$$

$$\therefore \quad f(t) = d\left[\frac{\sin 2t - \cos t}{dt}\right]$$

$$= 2\cos t + \sin t$$

5.3 Sampling

1. Ans. (b)

Solution: Power spectrum $G(\omega) = |F(\omega)|^2$

Since $\sin^2 at = \frac{1}{2}(1 + \cos 2at)$

In power spectrum term the frequency becomes double. In power spectrum figure the frequency is 2 kHz but originally this will have the frequency of 1 kHz. Hence minimum sampling rate of this deterministic signal is 2 × 1 kHz = 2 kHz.

2. Ans. (d)

Solution: $f_s = \dfrac{1}{T_s} = \dfrac{1}{50 \times 10^{-6}} = 20$ kHz

$$f_m = \frac{24\,\pi\,10^3}{2\,\pi} = 12 \text{ kHz}$$

Filter cut off frequency = 15 kHz

Sampled output may be $f_c, f_c \pm f_m, f_m$

i.e. 20 kHz , 20 + 12 = 32 kHz,

20 – 12 = 8 kHz, 12 kHz

As the cut off frequency of filter 15 kHz

Output contain 12 kHz and 8 kHz

3. Ans. (b)

Solution:

Let the given signal be $x(t) = x_1(t) * x_2(t)$

whre $x_1(t) = 16\times 10^4 \text{ sinc}^2(400\,t)$ and

$x_2(t) = 16^6 \text{sinc}^3(100\,t)$

Convoulution in time domain is equal to mutiplication in frequency domian

$$X(f) = X_1(f)\,X_1(f)...... \quad ...(1)$$

The spectrum $X_1(f)$ is shown in figure

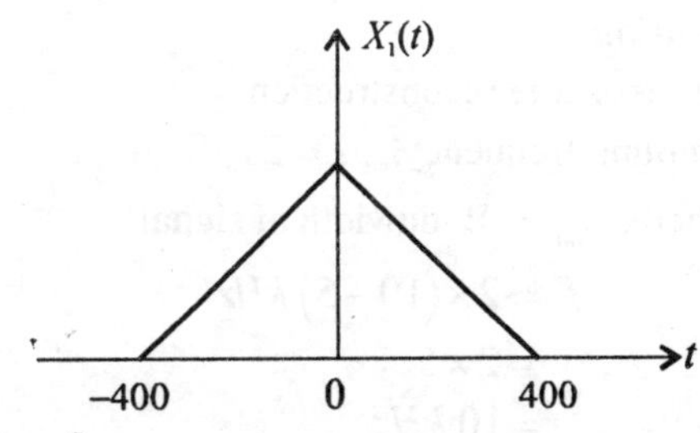

The maixmum frequency in $X_1(f)$ is 400 Hz (or) of $X_1(f)=0$ for $|f|>400\ Hz$

4. Ans. (c)

Solution: Maximum frequency of signal f_m

$= 3000$ Hz

So, sampling rate = $2f_m$ = 6000 Hz

5. Ans. (c)

Solution: $f_s \geq 2f_m$

The given signal is product of two signal

$$= 2\left[\frac{500}{2}+\frac{700}{2}\right] = 1200 \text{ Hz}$$

Second Method

Given that:

$$s(t) = \frac{\sin(500\pi t)}{\pi t} \times \frac{\sin(700\pi t)}{\pi t}$$

As we know that

$$2\sin A \sin B = \cos(A-B) - \cos(A+B)$$

$$\therefore \quad s(t) = \frac{1}{2\pi^2 t^2}[\cos(200\pi t) - \cos(1200\pi t)]$$

$$\omega_m(\max) = 1200\,\pi$$

6. Ans. (d)

Solution:

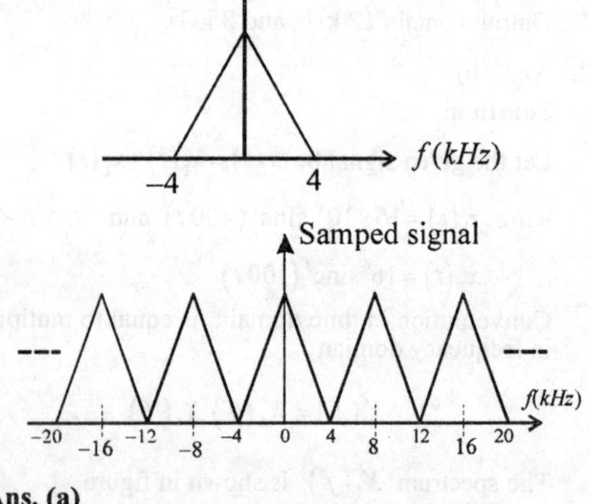

7. Ans. (a)

Solution:

For error free reconstruction,

sampling frequency, $f_s = 2f_m$

Where, f_m = Bandwidth of signal

$$f_s = 2\times(10-5)\ kHz$$
$$= 2\times 5$$
$$= 10\ kHz$$

8. Ans. (d)

Solution:

Sampling frequency, $f_s = 2f_m$

Where f_m = Bandwidth of signals

$$= 2\times(10.4-10)\ MHz$$
$$= 2\times 0.4\ MHz = 0.8\ MHz$$

9. Ans. (a)

Solution: Flate top sampling gives rise to aperture effect.

10. Ans. (a)

Solution:

$$\text{Sampling rate } = \frac{2f_H}{K}$$

$$\text{Where, } K = \text{int}\left[\frac{f_H}{f_H - f_L}\right]$$

$$K = \text{int}\left[\frac{30}{10}\right] = 3$$

$$\text{Sampling rate } = \frac{2\times 30\ kHz}{3}$$

$$= 20\ kHz$$

11. Ans. (a)

Solution:

The Nquist rate for the signal

$$x(t) = 2\cos(2000\pi t)\cdot\cos(500\pi t)$$

here $\quad f_1 = 1000$

$f_2 = 2500$

Then Nquist rate in given by

$$f_s = 2(f_1+f_2)_{\max}$$

$$f_s = 2(1000+2500) = 7\ kHz$$

12. Ans. (a)

Solution:

For error free reconstruction,

sampling frequency, $f_s = 2f_m$

Where, f_m = Bandwidth of signal

$$f_s = 2\times(10-5)\ kHz$$
$$= 2\times 5 = 10\ kHz$$

13. Ans. (d)

Solution:

Sampling frequency, $f_s = 2f_m$

Where f_m = Bandwidth of signals

$$= 2\times(10.4-10)\ MHz$$

$$= 2\times 0.4\ MHz$$

$$= 0.8\ MHz$$

14. Ans. (b)

Solution: To avoid aliasing, effect, sampling frequency

$f_s \geq 2f_m$

where f_m = Maximum signal frequency

15. Ans. (d)

Solution:

$$f_s = \frac{2f_u}{K};\ K = \left[\frac{f_u}{f_b}\right] = \left[\frac{5}{4}\right] = 1.25 \cong 1$$

where, f_a = upper cut off frequency

f_b = Bandwidth of signal

16. Ans. (b)

Solution:

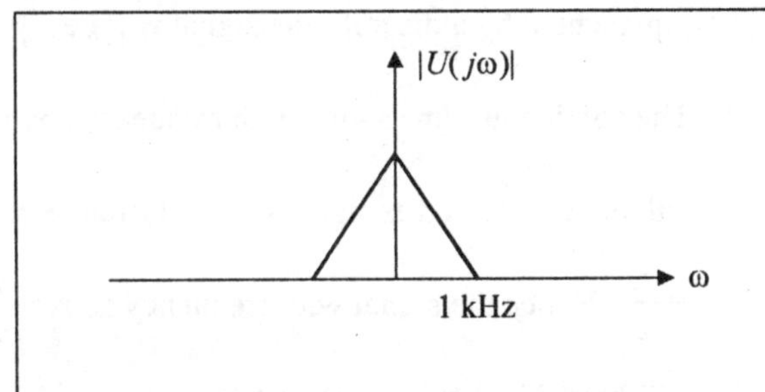

Given f_s = 1 kHz

For the given signal, Nyquist rate = 2 kHz

Because or $f_s = f_m$ or $f_s < 2f_m$

Therefore, Aliasing effect occurs and signal over loop as

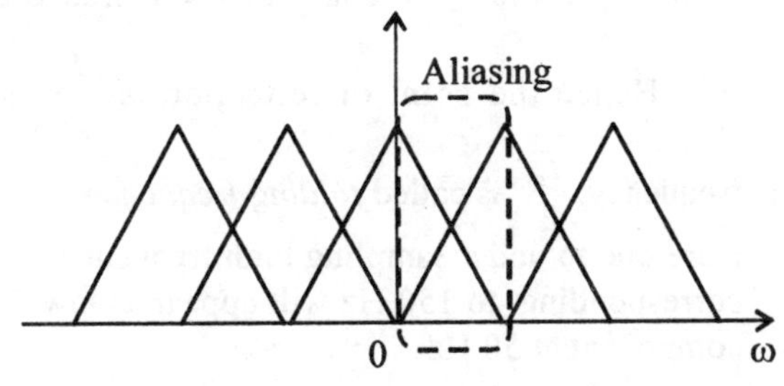

⇓

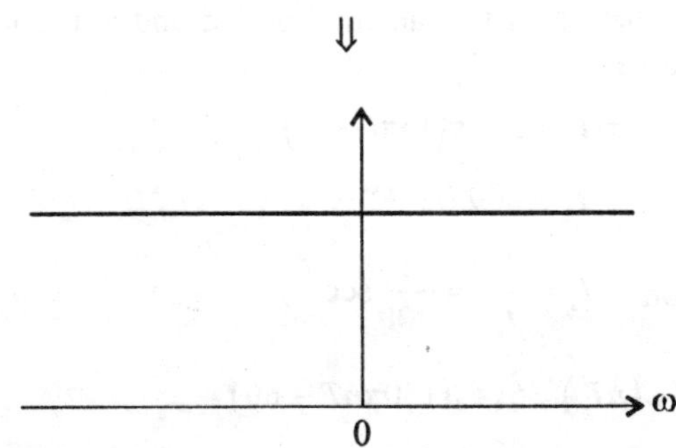

17. Ans. (c)

Solution: $f_s \geq 2f_m$

In present case

$f_s \geq 200\ Hz\quad f_s = 250\ Hz$ u

or, $T_s = \frac{1}{250} = 4ms$

18. Ans. (c)

Solution:

Expansion in time domain result as compression in frequency domain and vice –versa.

So, the maximum frequency component in given signal is $2F\ Hz$. And according to sampling theorem

Nyquist rate $= 2f_m = 2\times 2F = 4F\ Hz$

19. Ans. (c)

Solution:

$s^2(t)$ has frequency component double of that of $s(t)$

$$\sin^2\theta = \left(\frac{1+\cos 2\theta}{2}\right)$$

So sampling rate also become 2 times the sampling rate of $s(t)$

Sampling rate $= 2\times 5000$

$= 10000$ samples/sec

20. Ans. (d)

Solution:

A discrete signal $x\ [n]$ is said to be periodic if $x[n+N] = x[n]$ for all n.

When N is period of the signal

A discreet time sinusoidal is periodic for all vale of n only if its frequency is rational, i.e.

$$f = \frac{m}{N}$$

Where N is fundamental period and m is some integer.
given

$$x(t)=5\cos(15\pi t-60)$$

$$f_s=200\ Hz$$

or, $T_s=\frac{1}{f_s}=\frac{1}{200}$ sec

$$x(nT_s)=5\cos(150\pi nT_s-60)$$

$$=5\cos\left(150\pi\times\frac{1}{200}n-60\right)$$

$$=5\cos\left(\frac{3\pi}{4}n-60\right)$$

Here, $\omega_o=\frac{3\pi}{4}$

$$\frac{\omega_o}{2\pi}=\frac{m}{N}$$

or, $\frac{m}{N}=\frac{3\pi/4}{2\pi}=\frac{3\pi}{8\pi}=\frac{3}{8}$

Hence fundamental period $N=8$

21. Ans (c)

Solution:

We have given the frequency component of $50Hz$, $100Hz$, $200Hz$ and is sampled at 150 samples's To reconstruct the original signal

$$f_s>2f_m$$

So the signal having frequency less than or equal to

$$\left(\frac{150}{2}\right)=75Hz$$

Will we construct

22. Ans. (b)

Solution

Given contineous signal

$$x(t)=\cos[100\pi t]+\sin(300\pi t)$$

$$\omega=2\pi f$$

$$f=\frac{\omega}{2\pi}$$

$$f_1=\frac{100\pi}{2\pi}=50\ Hz$$

$$f_2=\frac{300\ \pi}{2\pi}=150\ Hz$$

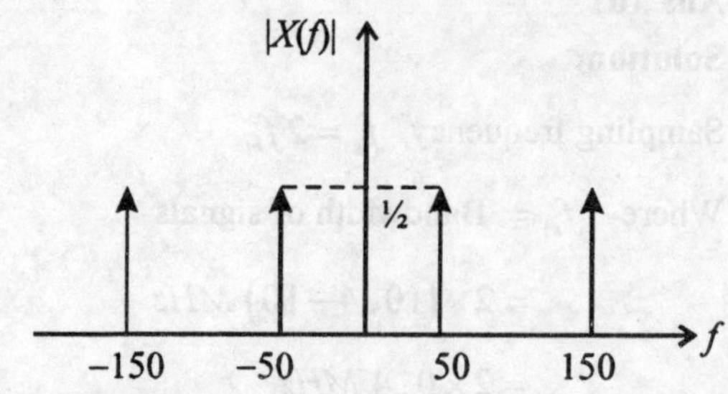

$$x_s(t)=\sum_{n=-\infty}^{\infty}x(nT_s)\delta[t-nT_s]$$

$$X_s(f)=\frac{1}{T_s}\sum_{n=-\infty}^{\infty}X(f-nf_s)$$

$$f_s=100\ Hz$$

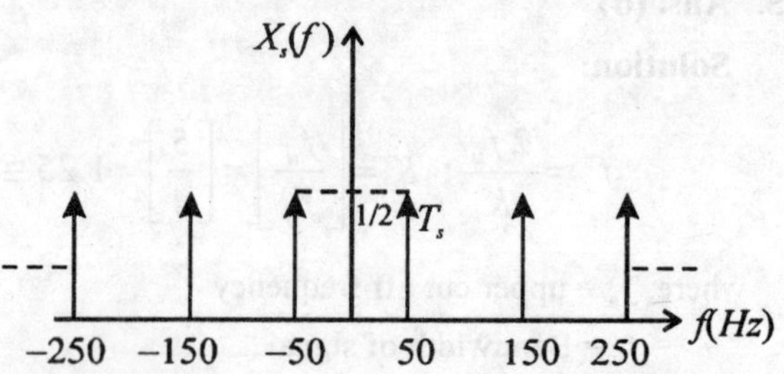

* When an analog signal is smapled at a frequency f_s the highest anglog frequency that can be uniquely represented by a discret time signal will be $\frac{f_s}{2}$.
* The contineous time signal with frequency abvoe $\frac{f_s}{2}$ will be represented as signal within the range $+\frac{f_s}{2}$ to $-\frac{f_s}{2}$. Hence the signal with frequency above $\frac{f_s}{2}$ will have identical signal with frequency belwo $\frac{f_s}{2}$ in the discret form.

Hence infinite number of high frequency contineous time signals will be represented by a single discret time signal. such signals are called alias. While sampling at f_s, the frequency abvoe $\frac{f_s}{2}$ will have alias with frequency below $\frac{f_s}{2}$. Hence the point of relfection is $\frac{f_s}{2}$ and the frequency $\frac{f_s}{2}$ is called *folding frequency.*

Here due to under sampling high frequency component corresponding to 150 Hz will appear as low frequency component at 50 Hz

So when signal filiter it by LPF having cut-off frequency of 100 Hz, we get

$$y(t) = \cos 100\pi t + \sin 100\pi t$$

Here, sin 100 πt is high frequency component which appear as low frequency component after sampling

23. Ans. (b)

Solution:

In the circuit of figure, amplifier A_1 proves a high input impedance for the analog signal and a low output impedance to allow rapid charging and discharging f the storage capacitor C. Amplifier A_2 provides a buffer between the storage capacitor and the output. As in the previous circuit, the drift rate, i.e., that rate at which the storage capacitor voltage change during the hold period, will be determined by the size of the storage capacitor, the input bias circuit current of the amplifier A_2, and the leakage through the FET switching device. We observe in the terminals may range over the full excursions of the analog signal and that a larger-amplitude gate-control voltage will therefore be required than when the switch was located at a virtual ground.

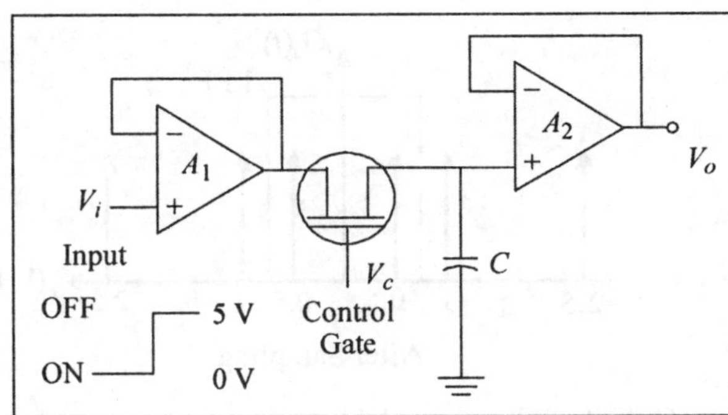

Figure: A sample and hold circuit using two ops amps connected for high input impedance having unity gain.

Similarly in the amplifier of figure, whose individual gains should be +1 with great precision, where an offset has developed the output of the individual amplifier will never quite equal the input, and so also for the operation of the entire circuit.

24. Ans. (c)

Solution: $f_s(t) = \Sigma\, f(t)\, \delta(t - nT_s)$

$$T_s = 100\ \mu s\ ;\ \omega_o = 10^4 \text{ Hz}$$

$$\text{FT}[\delta(t - nT_s)]$$

$$= \omega_0 \sum_{-\infty}^{\infty} \delta(\omega - n\omega_0)$$

$$F_s(\omega) = F(\omega) * \omega_0 \sum_{-\infty}^{\infty} \delta(\omega - n\omega_0)$$

$$= \omega_0 \sum_{-\infty}^{\infty} \delta(\omega - n\omega_0)$$

$$y(t) = 5 \times 10^{-6} \times 10 \cos(8\pi \times 10^3)t \times 10^4$$
$$= 5 \times 10^6 \times 10 \times 10^4 \cos(8\pi \times 10^3)t$$
$$5 \times 10^{-1} \cos(8\pi \times 10^3)t$$

Second Method

$$y(t) = 5 \times 10^{-6} x(t) \cdot \sum_{-\infty}^{\infty} \delta(t - nTs)$$

$$x(t) = 10 \cos(8\pi \times 10^3 t)$$

$y(t)$ = is sampled signal of $x(t)$

$$\sum_{-\infty}^{\infty} \delta(t - nT_s) = \frac{1}{T_s} + \frac{2}{T_r} \sum_{-\infty}^{\infty} \cos(n\omega_s t)$$

$$y(t) = \frac{5 \times 10^{-6} x(t)}{Ts} + \frac{2}{T_s} \sum_{-\infty}^{\infty} 5 \times 10^{-6} x(t) \cos(n\omega_s t)$$

When $y(t)$ is passed through low pass filter of 5 kHz cut off frequency only 4 kHz will pass. Thus output of low pass filter is

$$\frac{5 \times 10^{-6} \cdot 10 \cdot \cos(8\pi \times 10^3)t}{100 \times 10^{-6}}$$
$$= 5 \times 10^{-1} \cos(8\pi \times 10^3)t$$

25. Ans (d)

Solution:

Frequency compoments present in $x(t)$ are 400 Hz and 700 Hz. Sampling interval, $T_0 = 10^{-3}$ sec

or sampling rate $f_s = \frac{1}{10^{-3}} = 1\ kHz$

Pulse duration $= \frac{T_0}{6} \times 2 = \frac{T_0}{3}$

or Duty cycle $= \frac{1}{3}$

Hence $p(t)$ will not contain frequency components which are integer multiples of 3.

or Frequency component present in $p(t)$ are

(in kHz) – 1, 2, 4, 5, 7, 8, 10, 11, 13....... kHz

or Frequency components present in sampled signal are: 400 Hz, 700 Hz

1000 ± 400 Hz, 1000 ± 700 Hz

2000 ± 400 Hz, 2000 ± 700 Hz

4000 ± 400 Hz, 4000 ± 400 Hz,

which result as

400 Hz,	700 Hz
600 Hz, 300 Hz ;	1400 Hz, 1700 Hz
1600 Hz, 1300 Hz ;	2400 Hz, 2700 Hz
3600 Hz, 3300 Hz ;	4400 Hz, 4700 Hz

Hence frequency components present in the range 2.5 kHz to 3.5 kHz are 2.7 kHz and 3.3 kHz.

Method 2

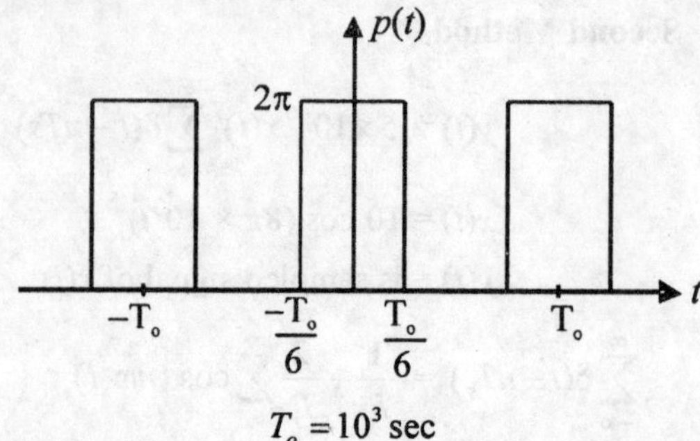

$T_o = 10^3$ sec

Fundamental frequency of pulse train $f_o = 1$ kHz.

Fourier Series representation of signal $x(t)$ is ,

$$x(t) = \sum_{n=\infty}^{\infty} C_n e^{jn\omega_o t}$$

where,

$$C_n = \frac{1}{T_o} \int_{-T_o/2}^{T_o/2} x(t) e^{-jn\omega_o t} dt$$

$$\omega_o = \frac{2\pi}{T_o}$$

Fourier series coefficient of sampling pulse train,

$$C_n = \frac{1}{T_o} \int_{-T_o/6}^{T_o/6} A e^{-j\pi\omega_o t} dt = \frac{A}{\pi n} \sin\left(\frac{n\pi}{3}\right)$$

$$\Rightarrow x(t) = \sum_{n=-\infty}^{+\infty} \left(\frac{A}{\pi n} \sin\left(\frac{n\pi}{3}\right) \right) e^{jn\omega_o t}$$

It clear from fourier series that triple harmonics such as 0, 3, 6, 9..... are absent and 1, 2, 4, 5, 7...... are present. The sampling frequencies corresponding to 1, 2, 4, 5, 7, 8, 10, 11, 13....... harmonics are 1 kHz, 2 kHz, 4 kHz, 5 kHz, 7 kHz

Given, $\quad x(t) = 2\cos(800\,\pi t) + \cos(1400\,\pi t)$

frequency of signal to be sampled, $f_{ml} = 0.4$ kHz, $f_{ml} = 0.7$ kHz

After sampling our signal becomes $x(t)p(t)$.

The spectrum of sampld signal repeats periodically with period equal to sampling frequency.

The given sampler has multiple sampling frequency of 1 kHz, 1 ± 0.4 kHz, 1 ± 0.7 kHz, 2 ± 0.4 kHz, 2 ± 0.7 kHz, 4 ± 0.4 kHz, 4 ± 0.7 kHz, 5 ± 0.4 kHz, 5 ± 0.7 kHz, and so on.

Thus, frequency components in the range of 2.5 kHz to 3.5 kHz are : 2.7 kHz, 3.3 kHz.

26. Ans (c)

Solution:

Let g(t) be 1 kHz sinusoidal signal i.e.

$f_m = 1000$ Hz

sampling frequency $f_s = 1500$ Hz

Then spectrum of $g(t)$ before and after sampling can given as:

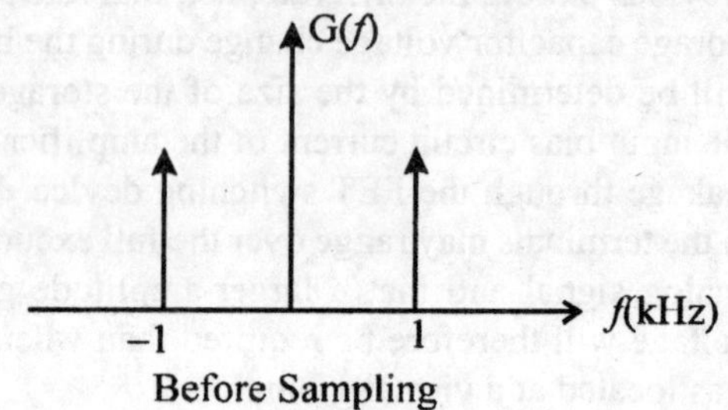

Before Sampling

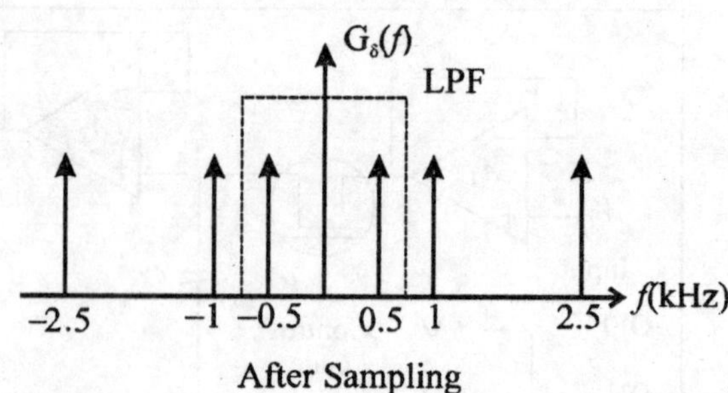

After Sampling

Output of given sampler have frequency components i.e.

f_m and $(nf_s \pm f_m)$

$n \rightarrow$ Integer

Applying a LPF with cut off frequency 800 Hz will given only one frequency component equal to 500 Hz

27. Ans. (a)

Solution:

$$g(t) = \cos^2(4\pi \times 10^3 t)$$

$$= \frac{1}{2} + \frac{1}{2}\cos(8\pi \times 10^3 t)$$

$s(t)$ is a rectangular pulse train with time period, T = 0.1 msec, or $f_0 = \frac{1}{T} = \frac{1}{0.1\times10^{-3}} = 10 \text{ kHz}$

Hence, frequency components in $s(t)$ will contain harmonics of 10 kHz.

Hence 4 kHz is not present in $S(f)$.

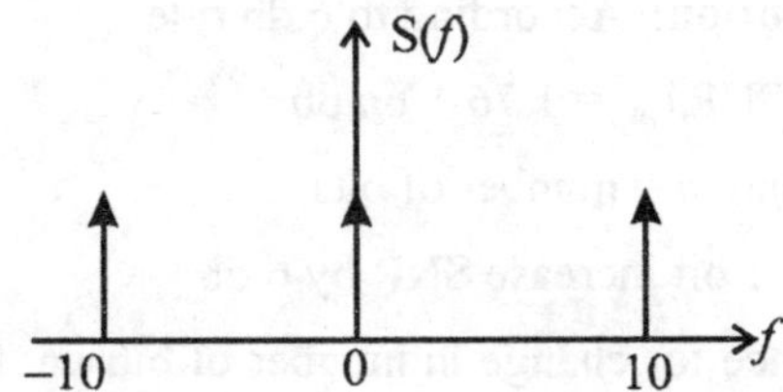

or the output will contain only DC term.

* Convolution in time domain is analogous to multiplication in frequency domain.

28. Ans. (c)

29. Ans. (d)

Solution:

Nyquist rate for band limited signal undergoing sampling frequency

= 2 times f_{max}

= 2 × 1000 Hz

= 2000 Hz

Second Method

Sampling frequency f_s is given as $f_s = 2f_{max}$, where f_{max} is the maximum frequency of modality signal

$f_s = 2\times1000\ Hz = 2000$ Hz

5.4 Pulse Code Modulation

1. Ans. (d)

Solution: 4 GHz carrier DSB/SC modulated by message signal of 2MHz.

Total bandwidth of message = 4 GHz + 2 MHz

= 4.002 GHz

Nyquist rate $f_s \geq 2f_m$ = 2 × 4.002 GHz = 8.004 GHz

2. Ans. (a)

Solution:

Increase in pulse width lead to increase in rise time and fall time. This increase lead to poor frequency response, i.e. loss of high frequency contained of signal.

2. Ans. (b)

Solution: $[S/N]_{PCM-db} = (1.76 + 6.n)$

Where n = number of bits

4. Ans. (a)

Solution: Sampling rate = 1800 samples/sec

Highest frequency rate = 1000 samples/sec

In sampling process these frequency are generated.

$$f_o = (f_{sa} - f_{si}), f_{sa}, (f_{sa} + f_{si})$$

= 800 Hz, 1.8 kHz, 2.8 kHz

Where $f_{si} \rightarrow$ signal frequency

$f_{sa} \rightarrow$ sampling frequency

Since sampled signal is passed through a LPF for recovery of signal only 800 Hz will pass, other will be attenuated.

5. Ans. (c)

Solution: $\frac{S}{N} = \frac{2^{2\times9}}{2^{2\times8}} = 2^2 = 4$

Or one bit contribute 6 db which is equal to factor of 4

6. Ans. (d)

Solution: $\frac{S}{N_{q1}} = 2^{2n}$

$$\frac{S}{N_{q2}} = 2^{2(n+1)}$$

$$\frac{\frac{S}{N_{q1}}}{\frac{S}{N_{q1}}} = \frac{2^{2(n+1)}}{2^{2n}} = 2^{(2n+2)}, 2^{-2n}$$

$$= 2^{(2n+2)-2n} = 2^2$$

7. Ans. (c)

Solution:

Alternate Mark inversion (AMI)

8. Ans. (c)

Solution: $\frac{S}{N} = \frac{2^{2\times9}}{2^{2\times8}} = 2^2 = 4$

Or one bit contribute 6 db which is equal to factor of 4.

9. Ans. (a)

Solution:

Higher signal amplitudes.

10. Ans. (b)

Solution:

Peak to peak input = 2 volts.

$m(t) = 0.5\cos(\omega_m t)$

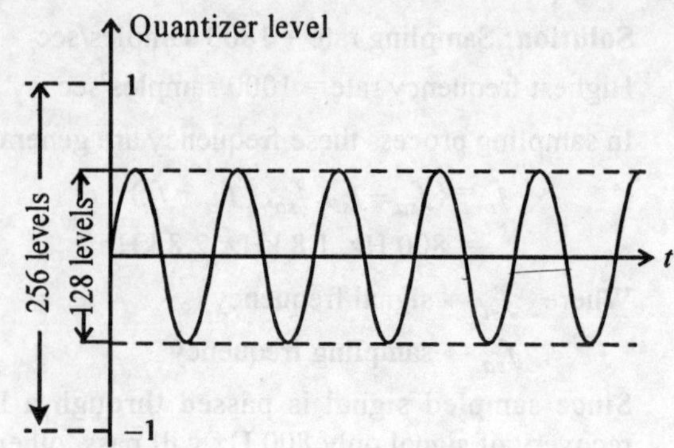

For 8-bit PCM, there are 256 quantizer levels. But, here only 128 quantizer levels will be utilized only which requires only 7-bits.

$[\text{S/N}]_{dB} = 1.76 + 6.02\, n\, dB$

Where n = number of bits, then

$$[\text{S/N}]_{dB} = 1.76 + 6.02n$$
$$= 1.76 + 6.02 \times 7$$
$$= 1.76 \times 42.14$$
$$= 43.9$$

11. Ans. (c)

Solution: Step size $S = \dfrac{V_{PP}}{M} = \dfrac{1.536}{128} = 0.012$

$$\text{Quantization Noise Power} = \frac{S^2}{12} = 0.012 \times \frac{0.012}{12}$$
$$= \frac{0.00014}{12}$$
$$= (12 \times 10^{-3})^2 = 12 \times 10^{-6}\ \text{V}^2$$

12. Ans. (b)

Solution: For slope over load noise to take place

$$\left[\frac{dx(t)}{dt}\right]_{\max} > f_s \cdot S$$

$$f_s \cdot S = 40 \times 10^3 \times 0.628$$
$$= 25.12 \text{ kHz}$$

If $x(t) = A \sin 2\pi f_m t$

$$|x'(t)|_{\max} = A \cdot 2\pi f_m$$
$$= 2 \times 3.14\, \text{A} \cdot f_m$$
$$= 6.28\, \text{A} \cdot f_m$$
$$6.28 \times 1.5 \times 4 = 37.68 \rightarrow 25.12$$

13. Ans. (b)

Solution: $f_s = 8$ kHz, $n = 8$ bits

Bit rate $R = n f_s$

$= (n \text{ bits/sample}) \times (f_s \text{ sample/sec})$

$$R = 8 \times 8$$
$$R = 64 \text{ kbps}$$
$$\text{SNR}|_{db} = 1.76 + 6 \times n = 1.76 + 6 \times 8$$
$$= 1.76 + 48 = 49.8$$

14. Ans. (c)

Solution: According to 6 db rule

$$\text{SNR}|_{db} = 1.76 + 6n \text{ db}$$

Where n = number of bits.

For 1 bit increase SNR by 6 db

Hence for change in number of bit, i.e. from 6 to 8 bit increase in bit is by 2 then SNR improve by 12 db.

$$12\ dB = 10 \log x$$
$$x = 10^{1.2}$$
$$\cong 16$$

15. Ans. (a)

Solution:

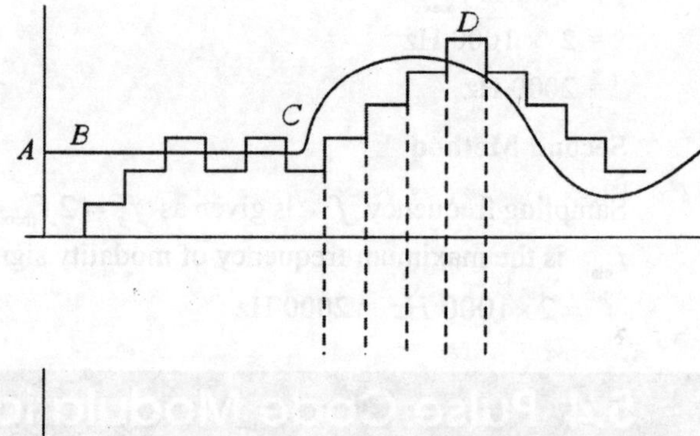

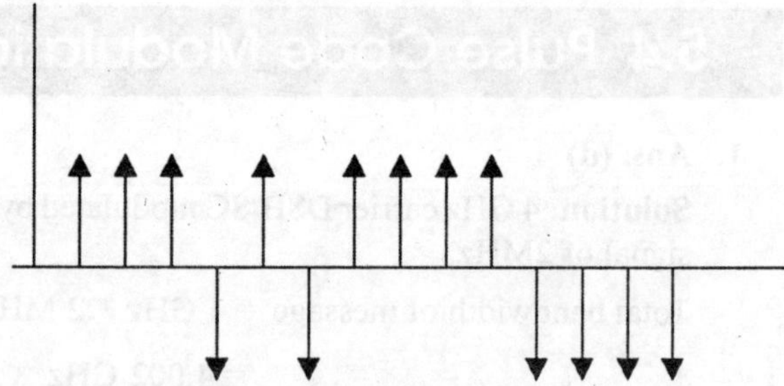

For BC constant section output pulses are opposite polarity.

Slow Varying Signal

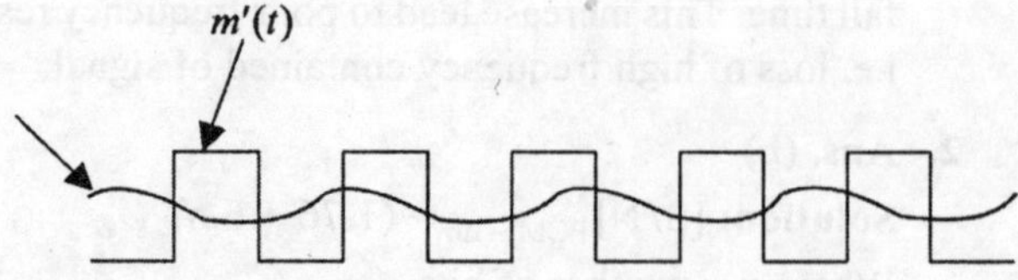

In delta modulation, if alternate pulses are of opposite polarities (i.e. 101010...); this corresponds to a very slowly varying input signal.

This phenomenon is also called granular noise.

16. Ans. (b)

Solution: Since it is uniform as

$x_q = 0$ in the range $0 \le x \le 0.3$

$x_q = 0.7$ in the range $0.3 \le x \le 1$

The square mean value is

$$\sigma^2 = \int_{-\infty}^{\infty} (x - x_q)^2 f(x)\,dx$$

$$= \int_0^1 (x - x_q)^2 f(x)\,dx$$

$$= \int_0^{0.3} (x-0)^2 f(x)\,dx + \int_{0.3}^{0.1} (x-0.7)^2 f(x)\,dx$$

$$= \left[\frac{x^3}{3}\right]_0^{0.3} + \left[\frac{x^3}{3} + 0.49\,x - 1.4\right]_{0.3}^{1}$$

or $\sigma^2 = 0.039$

The root mean square quantization noise

$$\text{RMS} = \sqrt{\sigma^2}$$

$$= \sqrt{0.039}$$

$$= 0.198$$

17. Ans. (c)

Solution: The three analog signals having BW: 1200 Hz, 600 Hz and 600 Hz are sampled at their respective Nyquist rate, i.e. at 2400, 1200, 1200 sample/sec, respectively.

The total of (2400 + 1200 + 1200) = 4800 sample/sec

The Bit rate $= n f_s$ = (4800 sample/sec) × 12

= 57. 6 Kbps

Where n = number of bits in a symbol

18. Ans. (c)

Solution:

The value of C should be equal to quantization error $= \dfrac{\Delta^2}{12}$

Second Method

The constant C will produce a dc shift to $y_1(t)$.

If $|C| \ge \dfrac{\Delta}{2}$, then $y_1(t)$ and $y_2(t)$ will be different.

19. Ans. (d)

Solution:

In delta modulation, the slope overload distortion can be reduced by increasing the step size.

20. Ans (b)

Solution:

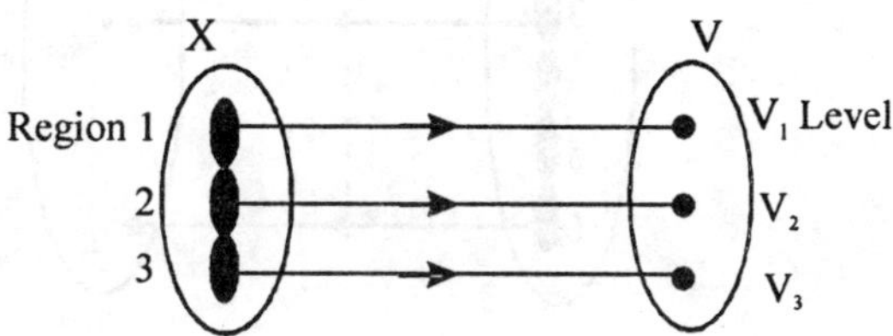

Since all the quantization levels are equiprobable, hence, area under each region will be equal $= \dfrac{1}{3}$

or $\dfrac{1}{4} \times 2a = \dfrac{1}{3}$

or $a = \dfrac{2}{3}$

21. Ans (a)

Solution :

$Q = X - V$

Since the quantizer is midtread

or $x = 0$ is a representation level

In region 1 :

$Q = X - 0$

or $E[Q^2] = E[X^2]$

As $a = \dfrac{2}{3}$

obtained in first part

or $E[Q^2] = \int_{-a}^{a} x^2\, p_x(x)\,dx$

$$= \int_{-a}^{a} x^2 \frac{1}{4}\,dx$$

$$= 2 \times \frac{1}{4} \int_0^a x^2 dx$$

$$= 2 \times \frac{1}{4} \left|\frac{x^3}{3}\right|_0^{2/3}$$

$$= 2 \times \frac{1}{4} \times \frac{1}{3} \times \left(\frac{2}{3}\right)^3$$

$$= \frac{4}{81}$$

22. Ans. (a)

Solution:

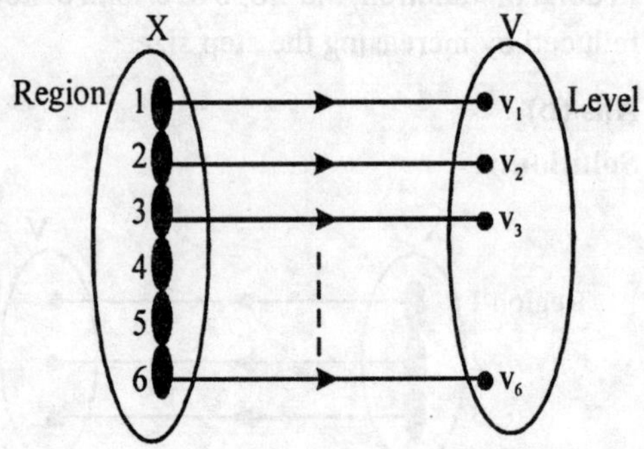

Given that the entropy is maximum.
Hence all the regions are equiprobable and area enclosed by each region is equal $= 1/6$
Now, area enclosed between decision boundaries

0 and $1 = a \times 1 = \frac{1}{6}$

or $a = \frac{1}{6}$

Total area enclosed by $f_x(x)$ curve $= 1$

or $2a + 4b + 4b = 1$

or $8b = 1 - 2 \times \frac{1}{6} = \frac{2}{3}$

or $b = \frac{2}{3 \times 8} = \frac{1}{12}$

23. Ans (d)

Solution:

The decision boundaries should be at end point, i.e −5, −3, −1, 0, 1, 3 and 5

Representation levels must lie at mid-points, i.e. −4, −2, −0.5, 0.5, 2, 4

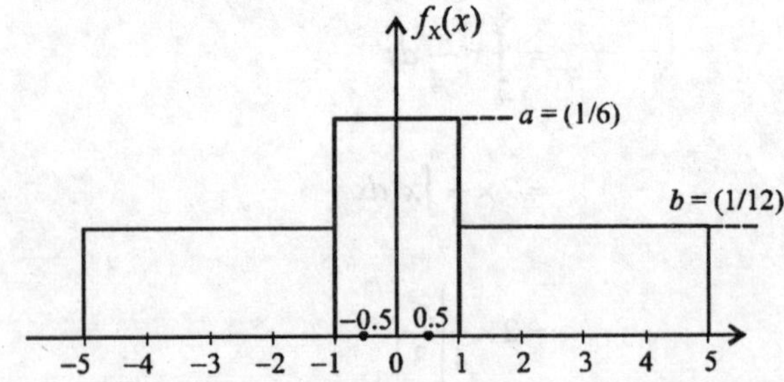

Quantization noise, $Q = X - V$

$$(SNR)_Q = \frac{E[X^2]}{E[Q^2]}$$

$$= 2\int_0^5 x^2 f_x(x)dx$$

$$= 2\int_0^5 x^2 f_x(x)dx$$

$f_x(x)$ is symmetric

or $$= 2\left[\int_0^1 x^2 \frac{1}{6} dx + \int_1^5 x^2 . \frac{1}{12} dx\right]$$

$$= 2\left[\frac{1}{6}\left|\frac{x^3}{3}\right|_0^1 + \frac{1}{12}\left|\frac{x^3}{3}\right|_1^5\right]$$

$$= 2\left[\frac{1}{6}\left(\frac{1-0}{3}\right) + \frac{1}{12}\left(\frac{5^3 - 1^3}{3}\right)\right]$$

$\Rightarrow$ $$= 7$$

$$E[Q^2] = \int_{-5}^{5} q^2 f_Q(q) dq$$

Where $q = x - v$

$$\therefore E[Q^2] = \int_{-5}^{5} (x - v)^2 f_x(x) dx$$

$$= 2\left[\int_0^1 (x - 0.5)^2 . \frac{1}{6} dx + \int_1^3 (x-2)^2 . \frac{1}{12} dx + \int_3^5 (x-4)^2 . \frac{1}{12} dx\right]$$

$$= 2\left[\frac{1}{6}\left|\frac{(x-0.5)^3}{3}\right|_0^1 + \frac{1}{12}\left|\frac{(x-2)^3}{3}\right|_1^3 + \frac{1}{12}\left|\frac{(x-4)^3}{3}\right|_3^5\right]$$

$$= \frac{1}{4}$$

$$\Rightarrow (SNR)_Q \frac{E[x^2]}{E[Q^2]} = \frac{7}{1/4} = 28$$

24. Ans (c)

Solution:

Range of quantisation $= 2m_p$

$2m_p = 10$ V

$m_p = 5$ V

Given that

SNR = 43.5 dB $\cong 22.4 \times 10^3$

SNR $\cong L^2$

$L^2 \cong 22.4 \times 10^3$

$L = 149.66$

Since step size

$$\Delta = \frac{2m_p}{L} = \frac{2\times 5}{149.66} = \frac{10}{150}$$

$$\Delta = 0.0667$$

25. Ans. (d)

Solution:

For positive values of signal

$$2m_p = 5$$

$$\Delta_1 = 0.05\,V$$

$$\therefore\ L_1 = \frac{2m_p}{\Delta} = \frac{5}{0.05} = 100$$

$$SNR \cong L_1^2 = 10000 = 40\ db$$

For negative values of signal,

$$2m_p = 5$$

$$\Delta_2 = 0.1$$

$$\therefore\ L_2 = \frac{2_{mp}}{\Delta} = \frac{5}{0.1} = 50$$

$$SNR \cong L_2^2 = 2500$$

$$SNR \cong L^2 \cong L_1^2 + L_2^2$$

$$SNR = 12500$$

$$(SNR)_{dB} = 10\log 12500 \cong 40.9\ dB$$

Second Method

Step size for positive values signal, $\Delta_p = 0.05\,V$

For uniform quantization to quantization noise for positive values can be given by,

$$N_q^+ = \frac{\Delta_p^2}{12}$$

or, $$N_q^+ = \frac{\Delta_p^2}{12} = \frac{(0.05)^2}{12} = 0.208\,mW$$

Step size for negative values signal, $\Delta_n = 0.1\,V$

For uniform quatizatin the quantization noise for negative values can be given by,

$$N_q^- = \frac{\Delta_n^2}{12}$$

or $$N_q^- = \frac{\Delta_n^2}{12} = \frac{(0.1)^2}{12} = 0.833\,mW$$

Total quantization noise power

$$N_q = N_q^+ + N_q^- = 0.208 + 0.833 = 1.401\,mW$$

The signal power, $S = \frac{V^2}{3} = \frac{5^2}{3} = \frac{25}{3}W$

The resulting signal to quantization noise ratio,

$$\left.\frac{S}{N_q}\right|_{dB} = 10\log_{10}\frac{S}{N_q} = 10\log_{10}\frac{25/3}{1.037\times 10^{-3}}$$

$$= 39.03\ dB = 40\ dB$$

26. Ans. (b)

Solution:

According to Nyquist Criteria,minimum sampling frequency $f_s = 2f_m = 2\times 4\,k = 8\,kHz$

Each sample has 8 bits.

Hence bit rate = Number of bits / sample × frequency

$= 8\times 8 = 64\ kHz$

Minimum transmission

$$BW = \frac{1}{2}\ \text{bit rate}$$

$$= \frac{1}{2}\times 64 = 32\ kHz$$

27. Ans. (c)

Solution:

$$SNR|_{PCM} = 6n + 1.7\ \text{db}$$

Where n = number of bits in a sample

$$\left.\frac{S}{N}\right|_{PCM} = 6\times 8 + 1.7 = 48 + 1.7. = 49.7$$

$$\cong 48\ \text{db}$$

28. Ans. (a)

Solution:

$$\frac{S}{N} \cong L^2 = \left(2^n\right)^2$$

$$4\left(\frac{S}{N}\right) = 4L^2 = (2L)^2$$

$$= \left(2\times 2^n\right)^2 = \left(2^{n+1}\right)^2$$

For reducing the quantization noise by a factor of 4, we need to increase the number of bits/sample by 1.

Total number of bits/ sample required $= 8 + 1 = 9$

Total number of representation levels $= 2^9 = 512$

29. Ans. (c)

30. Ans. (b)

Solution: PCM Bandwidth = $\left(\frac{1}{2}\right)nf_s$

$= \left(\frac{1}{2}\right) \times$ *Sampling rate* $\times$ *Number of unit code*

$= \frac{1}{2} \times 10 \times 7 = 35$ kHz

= 35 kHz

31. Ans. (c)

Solution: $f_s > 2f_m$; $f_s = 2 \times 2 = 4$ kHz

PCM bandwidth = $\left(\frac{1}{2}\right)n \cdot f_s$

$= \left(\frac{1}{2}\right) \times 4 \times 4 = 8$ kHz

32. Ans. (c)

Solution: $16 = 2^n \Rightarrow n = 4$

$f_m = 4$ kHz

$f_s = 2f_m = 2 \times 4 = 8$ kHz

Transmission bit rate = $n \times f_s$

$= 4 \times 8$ kHz $= 32$ kHz

33. Ans. (b)

Solution:

Equalization → Filtering operation, i.e. to reshape for the detection purpose.

Retiming → Clock recovery, i.e. this circuit provide a signal to sample the pulse at a optimum point.

Decision making → Reshaping, i.e. to correct time and is of correct width.

34. Ans. (a)

Solution: $2^n = 64 \Rightarrow n = 6$

$f_s = (2 \times 5) = 10$ kHz

Transmission rate $r = nf_s = 6 \times 10 = 60$ kHz

35. Ans. (c)

36. Ans. (d)

Solution: PAM signal can be demodulated by using a clipper circuit followed by a LPF.

37. Ans. (c)

Solution: The use of non-uniform quantization leads to increase in SNR for low level signals.

38. Ans. (c)

Solution: PCM BW = $\left(\frac{1}{2}\right)n \cdot f_s$

$= \left(\frac{1}{2}\right) \cdot n \cdot f_m = nf_m = n_f$

39. Ans. (b)

40. Ans. (d)

Solution: Transmission rate = nf_s

where, n = number of bits

f_s = sampling frequency

$= 2f_s$

$m = 2^n$

$4 = 2^n$; $n = 2$

Transmission rate = nf_s

$= 2 \times 6$ MHz

= 12 MHz

41. Ans. (d)

Solution: Slope of signal, $m_1 = \left(\frac{dm(t)}{dt}\right) = a$

Slope of signal, $m_2 = \left(\frac{\delta}{T_s}\right)$

To avoid slope over load distortion

$|m_1| \le |m_2|$

$a \le \left(\frac{\delta}{T_s}\right)$

$\delta \le aT_s$

42. Ans. (b)

Solution: $\left(\frac{S}{N}\right)_{db} = 1.76 + 6.02 \cdot n$

$= 1.76 + 6.02 \times 8 \cong 48$ dB

43. Ans. (d)

Solution: f_s (Nyquist) = $2f_m$

f_s (Present) = $2f_s$ (Nyquist)

Transmission BW = nf_s (Present)

$= 4 \times 2\ (2 \times 5$ kHz$)$

= 80 kHz

44. Ans. (b)

Solution: $\left(\frac{S}{N_q}\right)_{0dB} = 1.76 + 6.02n$

$$\left(\frac{S}{N_q}\right)_{0dB} = 1.76 + 6.02 \times 2$$

Increase in $\left(\frac{S}{N}\right)$ is 6 db for each bit.

45. Ans. (b)

Solution: $\left.\frac{dx(t)}{dt}\right|_{max} = 125$

For slope overload avoidance

$$\Delta f_s = \left.\frac{dA}{dt}\right|_{max}$$

$$\Delta_{min} = \frac{125}{32\text{ kHz}} = \frac{125}{32 \times 1000} = \frac{1}{256} = 2^{-8}$$

46. Ans. (d)

Solution:

Mean square quantization error/Quantization noise

$$= \frac{\Delta^2}{12} - \frac{V_P^2}{3M}$$

Quantization level $= M = 2^n$

Where n is number of bits.

$M = 2^n = 2^{13}$

Noise $\cong 0$

47. Ans. (a)

Solution:

In analog modulation, the modulated parameter varies continuously and take any value corresponding to the range of message. When modulated signal is mixed with noise, there is no way for the receiver to discerned the exact transmitted value.

If only two values are allowed for the modulated parameter and if separation between this parameters is large compared to noise disturbances. It is simple matter to decide which specific value are intended. Thus the effect of Random noise can be virtually eliminated. Which is the whole idea of Pulse code modulation. (Digital modulation)

48. Ans. (b)

Solution:

In PCM,

The B.W require $= nf_m$

Where,n = no. of bits

Levels are increases from 2 to 8 means.

M = 2^n So, n = 1 to 3

Initially, when

M = 2 then n = 1,

B.W = 1.$f_m = f_m$

M = 8, then n = 3,

B.W. = 3.f_m

B.W. get tripled.

49. Ans. (c)

50. Ans. (d)

Solution: The output signal-to-quantizing noise ratio of PCM system for a full-scale test tone is therefore

$$(SNR)_0 = \left(\frac{S}{N_q}\right)_0 = \frac{(A^2/2)}{A^2/(3L^2)} = \frac{3}{2}L^2$$

Expressing this in decibels, we have

$$\left(\frac{S}{N_q}\right)_0 dB = 10\log\left(\frac{S}{N_q}\right)_0$$

$= 1.76 + 20 \log L$

$= 1.76 + 20 \log 2^n$ $[L = 2^n]$

$= 1.76 + 6.02 \times n$ dB

where L = M is the number of Quantizing levels. Above equation indicates that each bit in the code word of a binary PCM system contributes 6 dB to the output signal-to-Quantizing ratio. This is called the 6 dB rule.

51. Ans. (a)

Solution: In delta modulation for no slope overload, noise, slope of the signal should be less than step size of quantizer, i.e. $\frac{df(t)}{dt} \le (h/T_s)$

52. Ans. (b).

Solution: Expressing SNR in decibels, we have

$$\left(\frac{S}{N_q}\right)_0 dB = 10\log\left(\frac{S}{N_q}\right)_0$$

$= 1.76 + 20 \log L$

$= 1.76 + 20 \log 2^n$ $[L = 2^n]$

$= 1.76 + 6.02 \times n$ dB

where L = M is the number of Quantizing levels. Above equation indicates that each bit in the code word of a binary PCM system contributes 6 dB to the output signal-to-Quantizing ratio. This is called the 6 dB rule.

SNR = (1.72 + 6n) dB

Therefore, if the number of bits is increased by 2, then SNR will increase by 6 × 2 = 12 dB.

53. Ans. (c).

Solution: Slope overload distortion would occur if, signal slop is more than the modulator output signal slope, i.e.

$$a > \frac{\delta}{T_s} \Rightarrow \delta < aT_s.$$

54. Ans. (b).

Solution: In PCM system

$$\left.\frac{S}{N}\right|_{db} = 1.76 + 6n = 1.76 + 6 \times 8 = 49.76 \text{ dB}$$

where n = number of bits in encoded system.

55. Ans (b)

Solution:

In P.C.M signal sampled at

f_S = Sampling signal

$f_S > 2f_m$

f_m = Frequency of Audio/information signal.

For n bit PCM the min bandwidth required for faithful reconstruction of signal is

$\Rightarrow nfs$

5.5 Multiplexing and Satellite

1. Ans. (d)

Solution: In Quadrature carrier multiplexing two signal modulate the same carrier but with a phase difference of 90° so that the total bandwidth is 2ω.

Quite different from FDM and TDM

2. Ans. (a)

Solution:

If a signal is combination of different frequency then maximum frequency is taken for the sampling,

i.e. $f_s \geq 2f_{m(\max)}$

In present case four independent messages are sampled then TDM. Hence transmitted sample rate is

$$f_s \geq [\text{Sum of individual sampling rate}]$$

$$\geq [2\times100+2\times100+2\times200+2\times400] \text{ Hz}$$

$$\geq [200+2000+4000+8000] \text{ Hz}$$

$$\geq 1600 \text{ Hz}$$

***Note:* In TDM, if n message are sampled with f_s sampling rate then transmitted sample rate or $BW = nf_s$**

Second Method:

$$BW|_{TDM} = nfs$$

$$= n.2fs$$

When all the signals have same bandwidth

In present case

Since the four signals are sampled at their Nyquist rates and then Time Division multiplexed.

$W_1 = 100\ Hz \Rightarrow 2W_1 = 200\ Hz$

$W_2 = 100\ Hz \Rightarrow 2W_2 = 200\ Hz$

$W_3 = 200\ Hz \Rightarrow 2W_3 = 400\ Hz$

$W_4 = 400\ Hz \Rightarrow 2W_4 = 800\ Hz$

Total number of samples present in one second, i.e. Transmitted sample rate

$$= 200+200+400+800$$

$$= 1600$$

$$= 1600 \text{ Hz}$$

3. Ans. (b)

Solution: Bandwidth allocated for 1 channel = 5 MHz

Average bandwidth for 1 channel $\frac{5}{5} = 1$ **MHz**

Total Number of simultaneous channel

$$= \frac{1\text{ M} \times 8}{200\text{ K}} = 40 \text{ channel.}$$

4. Ans. (b)

Solution: Chip Rate $R_C = 1.2288 \times 10^6$ chips/sec

Data Rate $R_b = \frac{R_C}{G}$

Since the processing gain G must be at least 100, thus for $G_{\min}$ we get

$$R_{b\max} \geq \frac{R_C}{G_{\min}} = \frac{1.2288\times10^6}{100}$$

$$\geq 1.2288 \times 10^3$$

5. Ans. (d)

Solution:

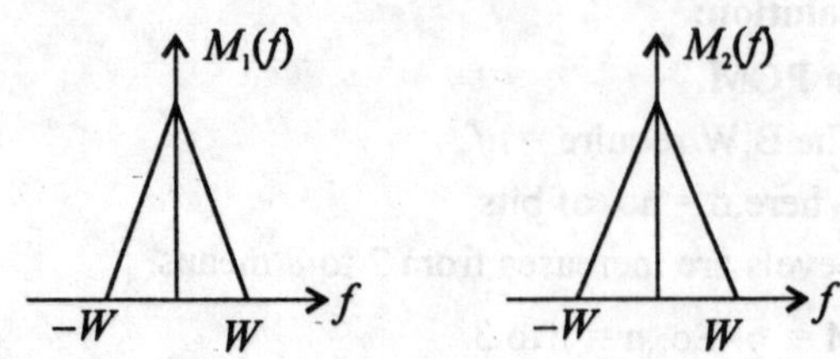

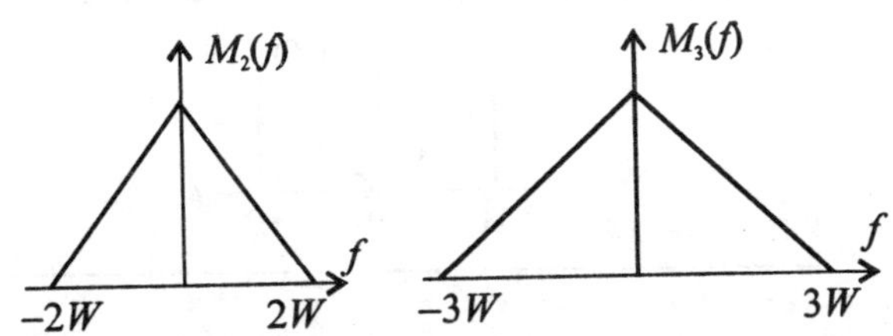

The transmission bandwidth = Half the total number of samples transmitted.

The transmissin BW is minimum for asynchronous TDM.

(a) Different sampling rate for different siganls.

* Sampling rate of a signal = twice the maximum frequency of that signal.
* Total number of samples transmitted per second is equal to sum of samples of all the signals.
* Transmission BW = Half the total number of samples transmitted .
* Bandwidth is less for asynchronous TDM
* Design of commuttor/decommutator is difficult.

Hence, for minimum transmission BW, all the signals must be sampled at their respective Nyquist rates.

Or, total number of samples transmitted per sec.

$$= 2(W + W + 2W + 3W) = 14\,W$$

or, transmissin $BW = \frac{1}{2} \times 14\,W = 7\,W$

6. Ans (d)

Solution:

The frequency band used in satellite communication are 6/4 GHz, (i.e. C-band) 14/12 GHz, (i.e. Ku-band) and 30/20 GHz (i.e. Ka-band)

Thus, the frequency range for satellite communication is 1 GHz to 30 GHz.

7. Ans. (a)

8. Ans. (b)

Solution: Number of bit

= (Number of bit + synchronization bit)

× (sampling frequency) × (number of channel)

= (7 + 1) × 8000 × 20 = 64000 × 20 = 1280000

= 1280 k bits/sec

9. Ans. (a)

Solution: Time division multiplexing requires: (a) constant data transmission.

10. Ans. (a)

Solution:

It is not operated at the maximum power output in order to conserve the available limited battery power

11. Ans. (c)

Solution:

For satellite communication system we use up-link down link frequency 6 GHz and 4 GHz respectively.

12. Ans. (c)

Solution:

Here $f_m = 8\ kHz$

Sampling frequency $f_s = 12\ kHz$

Since $f_s < 2f_m$, therefore, the requirement of sampling frequency does not obey sampling theorem

13. Ans. (b)

Solution:

The most important and most necessary circuit in TDM system is synchronization with proper timing

14. Ans. (b)

Solution:

For n channel PCM Bandwidth $= \frac{1}{2} n f_s = \frac{12 \times 8}{2}$

= 12 × 4 kHz.

15. Ans. (a)

Solution: Sampling frequency $f_s = 2\,f_m$

where f_m = max signal frequency in all the given signal = BW of signal

Transmission bit rate = n g $2f_m$

where n = number of bits

g = number of signal

Transmission bit rate = 8 ×4 × 2 × 8 = 512 Kbps.

16. Ans. (c)

Solution: The three analog signals having BW: 1200 Hz, 600 Hz and 600 Hz are sampled at their respective Nyquist rate, i.e. at 2400, 1200, 1200 sample/sec respectively.

The total of (2400 + 1200 + 1200) = 4800 sample/sec

The Bit rate = $n \cdot f_s$ = (4800 sample/sec) × 12

= 57. 6 Kbps

Where n = number of bit in a symbol

17. Ans (b)

Solution:

$B = 10B_M + 9 \times 500$

$= 10 \times 3\ kHz + 4.5\ kHz$

$= 34.5\ kHz$

5.6 Miscellaneous

1. Ans. (a)

Solution:

Hilbert transform of a function is equal to multiplying its frequency component by $-j\ \text{sgn}(f)$.

For sinusoidal function this is equivalent to giving a phase shift of $-90°$.

Hence, Hilbert transform of $\cos \omega_1 t + \sin \omega_2 t$ is equal to

$= \cos(\omega_1 t - 90°) + \sin(\omega t - 90°)$

$= \sin \omega_1 t - \cos \omega_2 t.$

Second Method

Hilbert transfrom produces a phase shift of $-90°$

$$g(t) = \cos\omega_1 t + \sin\omega_2 t$$

$$H\{\cos\omega_1 t\} = (\cos\omega_1 t - 90)$$

$$= \sin\omega_1 t$$

$$H\{\sin\omega_2 t\} = (\sin\omega_2 t - 90)$$

$$= -\cos\omega_2 t$$

$$H\{g(t)\} = \sin\omega_1 t - \cos\omega_2 t$$

2. Ans. (b)

Solution:

Ring modulation → Generation of DSB/SC

VCO → Generation of FM

Foster seely Disc → Detection of FM

Mixer → Frequency conversion.

3. Ans. (a)

Solution: This diagram represent SSB generation by phase shift method

For USB, $r(t) = m(t) \cos \pi Bt - m'(t) \sin 2\pi Bt$

For LSB, $r(t) = m(t) \cos \pi Bt + m'(t) \sin 2\pi Bt$

Since $y(t)$ is sum of both

∴ It is LSB generator and its Fourier transform is given as:

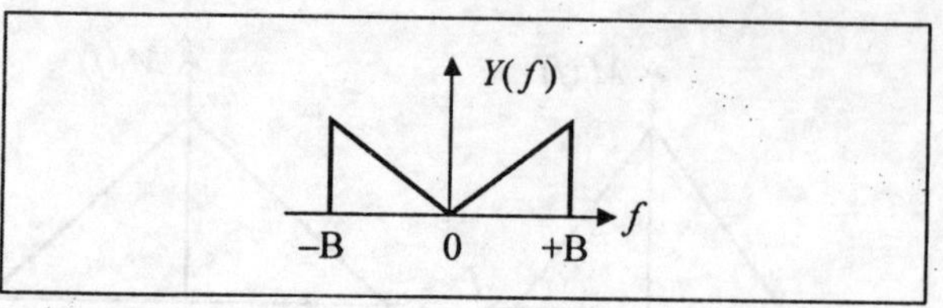

Second Method

Spectrum of signal m(t),

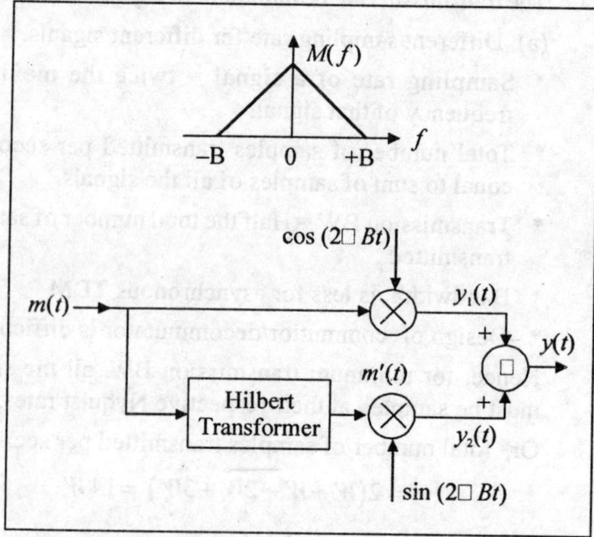

When $m(t)$ is multiplied by $\cos(2\pi Bt)$, it becomes

$$y_1(t) = m(t)\cos(2\pi Bt)$$

Fourier transform of $y_1(t)$

$$Y_1(f) = \frac{1}{2}[M(f-B) + M(f+B)]$$

The spectrum of $Y_1(f)$ can be drawn below

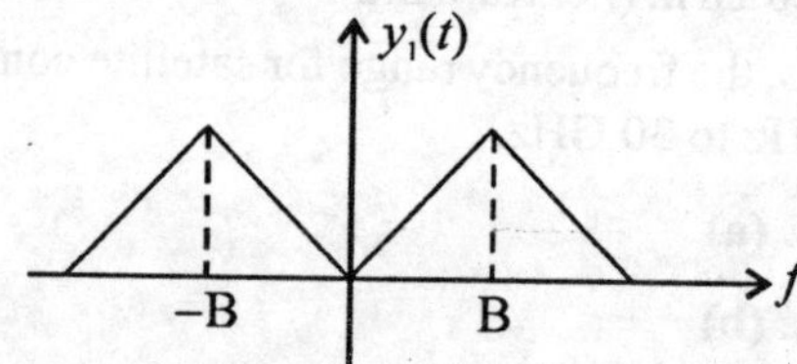

When $m(t)$ is passed through Hilbert transform block, output Hilbert transformer

$$m'(t) = m(t) * h(t)$$

$$M'(f) = M(f)\, H_Q(f)$$

Where, $H_Q(f)$ is transfer function of Hilbert transformer.

$$H_Q(f) = H_{TQ}\left\{\frac{1}{\pi t}\right\}$$

or $H_Q(f) = -j\,\text{sgn}(f)$

Here sgn (f) is called Sigmum function which is defined by,

$$\text{Sgn}(f) = +1 \quad : \quad f > 0$$
$$= -1 \quad : \quad f < 0$$
and $$H_Q(f) = -j \quad : \quad f > 0$$
$$= +j \quad : \quad f < 0$$
Then, $$M'(f) = -jM(f) \quad : \quad f > 0$$
$$= +M(f) \quad : \quad f < 0$$

The spectrum of signal M'(f) becomes as shown below,

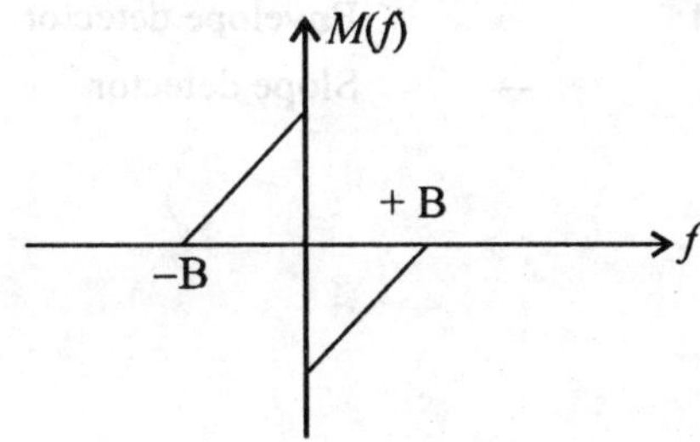

When m'(t) is multiplied by $\sin(2\pi Bt)$, it becomes

$$y_2(t) = m'(t)\sin(2\pi Bt)$$

Fourier transform of $y_2(t)$

$$Y_2(f) = \frac{1}{j2}[M(f-B) - M(f+B)]$$

The spectrum of $Y_2(f)$ is as below

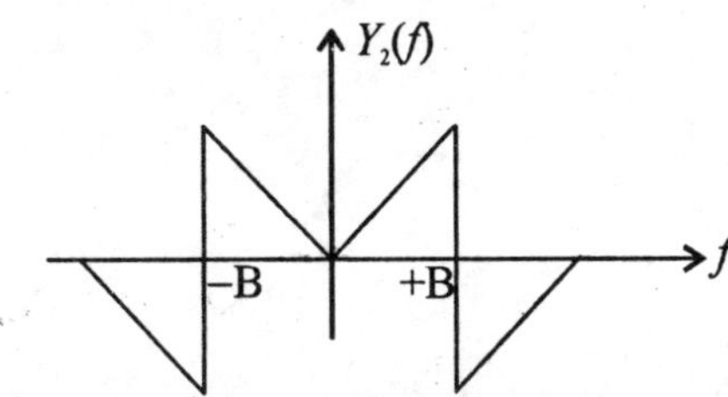

The spectrum of output summing block is obtained by adding the spectrum of $Y_1(f)$ and $Y_2(f)$, as shown under,

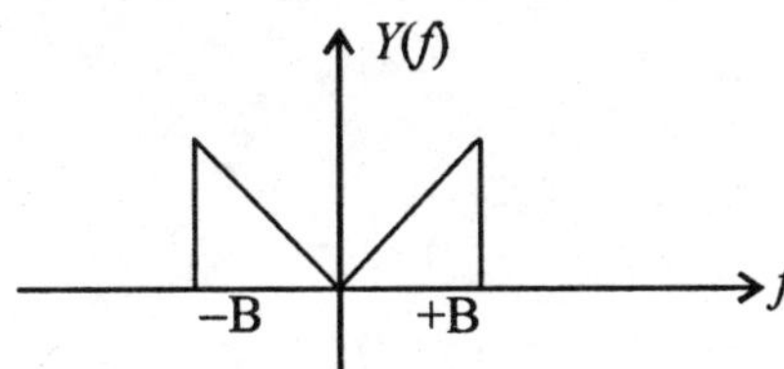

4. **Ans. (a)**

Solution:

$$\Delta f = BW = (20 - 18\text{ MHz}) = 2\text{ MHz} = 2\times 10^6$$
$$T°K = 273 + 27 = 300°\text{K}$$
$$R = 10\text{ k}\Omega = 10 \times 10^3$$

RMS Noise voltage

$$V_{\text{rms}} = \sqrt{4KTR\,\Delta f}$$
$$V_{(N)\text{rms}} = \sqrt{2\times 10^6 \times 10\times 10^3 \times 300\times 1.38\times 10^{-23}}$$

5. **Ans. (b)**

Solution: $T_e = (F-1)\cdot T_o$

Where, T_e = Temperature equivalent of noise

T_o = Room temperature in Kelvin

F = Noise figure

$$T_e = (1.6 - 2)\cdot 290$$
$$= 174.99\text{ K}$$

6. **Ans. (b)**

Solution: $F = 1 + \dfrac{R_{ep}}{R_a} = 1 + 1 = 2$

7. **Ans. (b)**

Solution:

The field strength at a distance r from a transmitting antenna is given by,

$$E = \frac{\sqrt{30P_t}}{r}$$

Where P_t is transmitted power,

$$\therefore \quad E \propto \frac{1}{r}$$

or, $$\frac{E_1}{E_2} = \frac{r_2}{r_1}$$

in dB,

$$20\log_{10}\frac{E_1}{E_2} = 20\log_{10}\frac{r_2}{r_1}$$

Given

$$20\log_{10}E_1 - 20\log_{10}E_2 = 3$$

or, $$3 = 20\log_{10}\frac{r_2}{r_1}$$

or, $$r_2 = r_1 \times 10^{0.3/2}$$

or, $$r_2 = 5000 \times 10^{0.3/2}\text{ m}$$

or, $$r_2 = 7062.69\text{ m}$$

Thus the distance required to move further so that signal strength reduces by 2 dB is,

$$d = r_2 - r_1 = 7062.69 - 5000$$
$$= 2062.69\ m \approx 2070\ m$$

8. Ans. (c)

Solution:

Common applications of a PPL include:

1. Frequency synthesizers that provide multiples of a reference signal frequency generated using a PLL.
2. FM demodulation network for FM operation with excellent linearity between the input signal frequency and the PLL output voltage.
3. Demodulation of the two data transmission or carrier frequencies in digital-data transmission used in frequency-shift keying (FSK) operation
4. A wide variety areas including modems, telemetry receivers and transmitters, tone decoders AM detectors, and tracking filters.

9. Ans. (c)

Solution:

The primary requirement of both the buffer is to keep low bias current

10. Ans. (a)

Solution:

The linear phase LPF filter is having a property that it will shift all the signals by the same amount in time domain. 50 Hz signal shifts by 3 m sec, 150 Hz signal also shifts by 3 m sec

11. Ans. (c)

Solution: The correct match is given below:

AM	→	Envelope detector
FM	→	Slope detector

6 Basics of Circuit and Measurement System

6.1 Network Elements

1. All resistances in the circuit in figure are of R ohms. The switch is initially open. What happens to the lamp's intensity when the switch is closed?

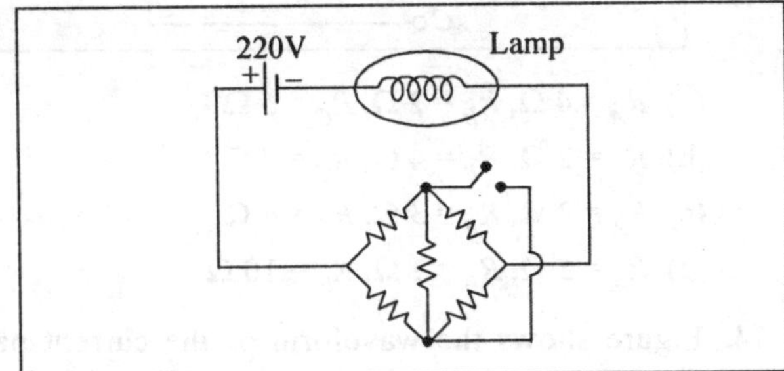

(a) Increases
(b) Decreases
(c) Remains the same
(d) Answer depends on the value of R

2. All the resistances in figure are 1 Ω each. The value of current 'I' is

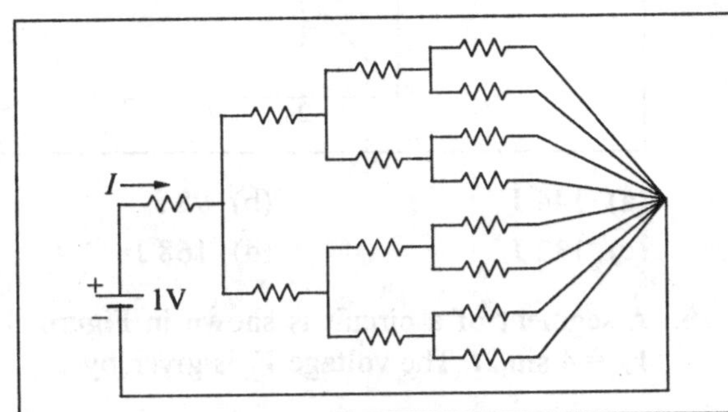

(a) 1/15 A (b) 2/15 A
(c) 4/15 A (d) 8/15 A

3. As shown in the figure, a 1 Ω resistance is connected across a source that has a load line $v + i = 100$. The current through the resistance is

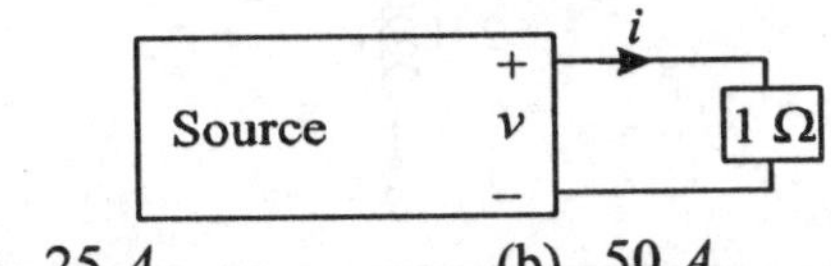

(a) 25 A (b) 50 A
(c) 100 A (d) 200 A

4. An **ideal voltage source** will charge the ideal capacitor
(a) Infinite times
(b) Exponentially
(c) Instantaneously
(d) None of the above

5. A **practical current source** is usually represented by
(a) A resistance in series with an ideal current source
(b) A resistance in parallel with ideal current source
(c) A resistance in parallel with an ideal voltage source
(d) none of the above

6. **Energy stored in a capacitor** over a cycle, when excited by an a.c. source is
(a) The same as that due to a d.c. source of equivalent magnitude
(b) Half of that due to a d.c. source or equivalent magnitude
(c) Zero
(d) None of the above

7. In the circuit shown in figure below, it is desired to have a **constant direct current** $i(t)$ through the ideal inductor L. The nature of the voltage source $v(t)$ must be

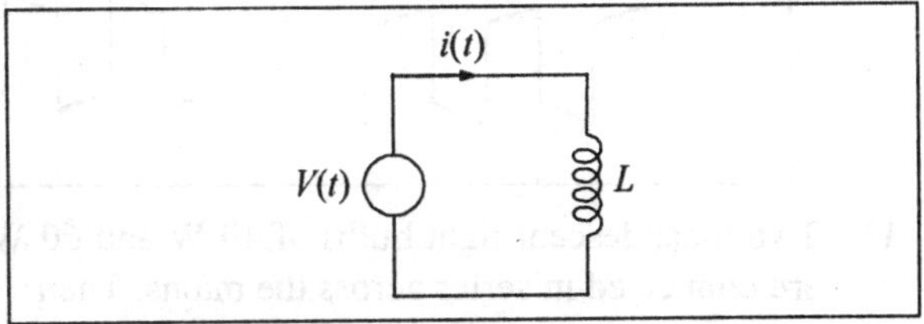

(a) Constant voltage
(b) Linearly increasing voltage
(c) An ideal impulse
(d) Exponentially increasing voltage

8. The effective inductance of the circuit across *A*, *B* in the figure shown below is

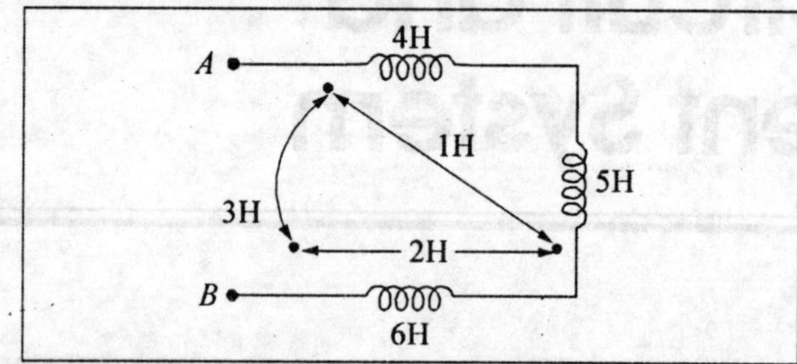

(a) 9 H (b) 21 H
(c) 11 H (d) 6 H

9. For the circuit shown in figure, the capacitance measured between terminals *B* and *Y* will be

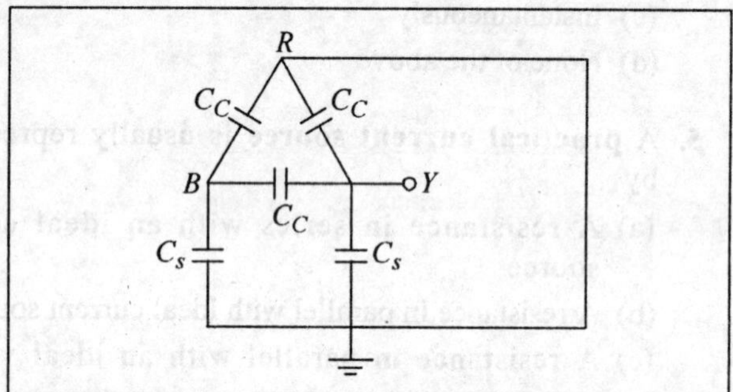

(a) $C_c + (C_s/2)$ (b) $C_s + (C_c/2)$
(c) $[(C_s + 3C_c)/2]$ (d) $3C_c + 2C_c$

10. When a *periodic triangular* voltage of peak amplitude 1V and frequency 0.5 Hz is applied to a parallel combination of 1 Ω resistance and 1 F capacitance, the current through the voltage source has wave-form

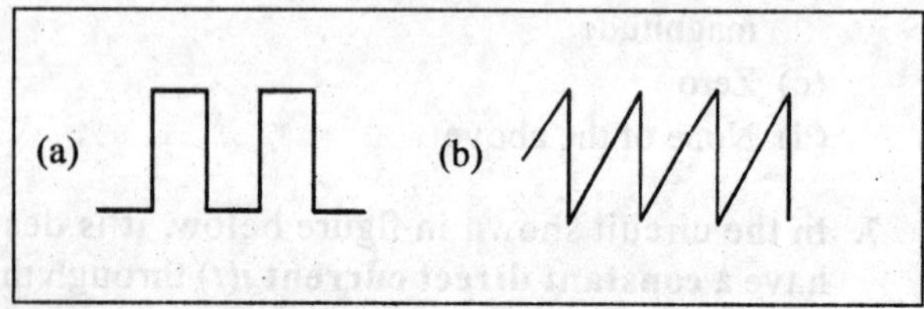

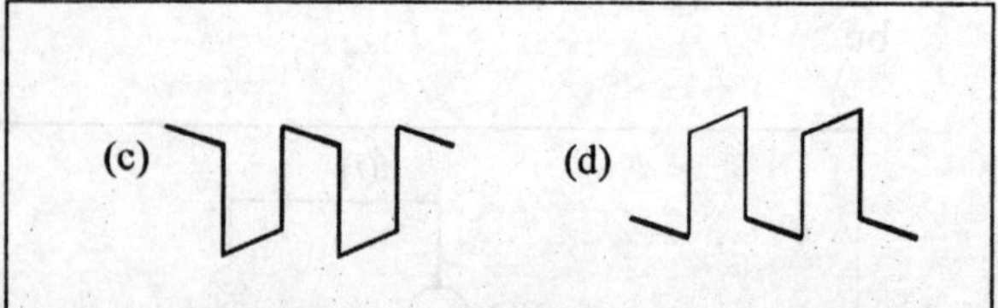

11. Two incandescent light bulbs of 40 W and 60 W rating are connected in series across the mains. Then

(a) The bulbs together consume 100 W
(b) The bulbs together consume 50 W
(c) The 60 W bulb glows brighter
(d) The 40 W bulb glows brighter

12. Given two coupled inductors L_1 and L_2, their mutual inductance *M* satisfies

(a) $M = \sqrt{(L_1^2 + L_2^2)}$ (b) $M > (L_1 + L_2)/2$
(c) $M > \sqrt{L_1 L_2}$ (d) $M \leq \sqrt{L_1 L_2}$

13. Consider the star network shown in figure. The resistance between terminals *A* and *B* with *C* open is 6 Ω, between terminals *B* and *C* with *A* open is 11Ω, and between terminals *C* and *A* with *B* open is 9Ω. Then

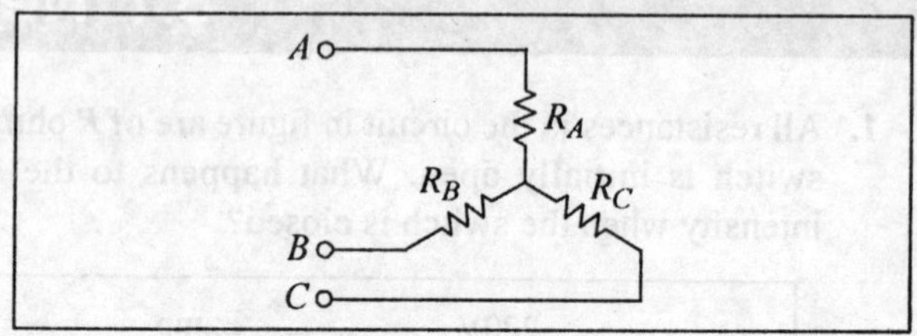

(a) $R_A = 4\ \Omega, R_B = 2\ \Omega, R_C = 5\ \Omega$
(b) $R_A = 2\ \Omega, R_B = 4\ \Omega, R_C = 7\ \Omega$
(c) $R_A = 3\ \Omega, R_B = 3\ \Omega, R_C = 4\ \Omega$
(d) $R_A = 5\ \Omega, R_B = 1\ \Omega, R_C = 10\ \Omega$

14. Figure shows the waveform of the current passing through an inductor of resistance 1Ω and inductance 2 H. The energy absorbed by the inductor in the first four seconds is

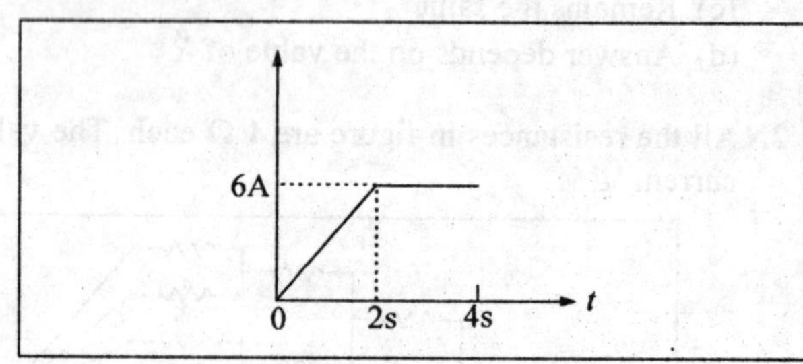

(a) 144 J (b) 98 J
(c) 132 J (d) 168 J

15. A segment of a circuit is shown in Figure. $V_R = 5$ V, $V_C = 4 \sin 2t$. The voltage V_L is given by

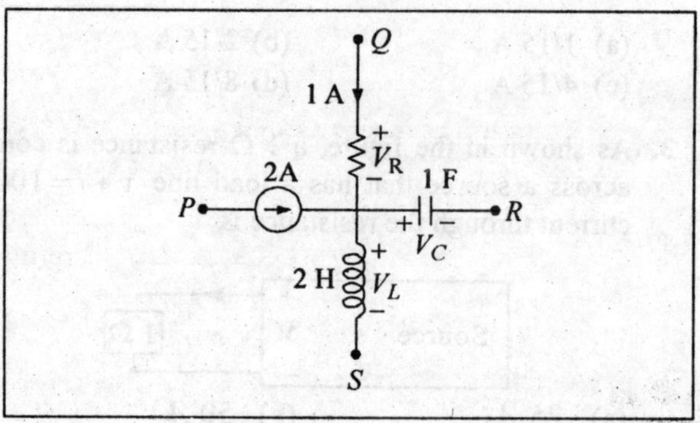

(a) $3 - 8 \cos 2t$ (b) $32 \sin 2t$
(c) $16 \sin 2t$ (d) $16 \cos 2t$

16. In figure the value of R is

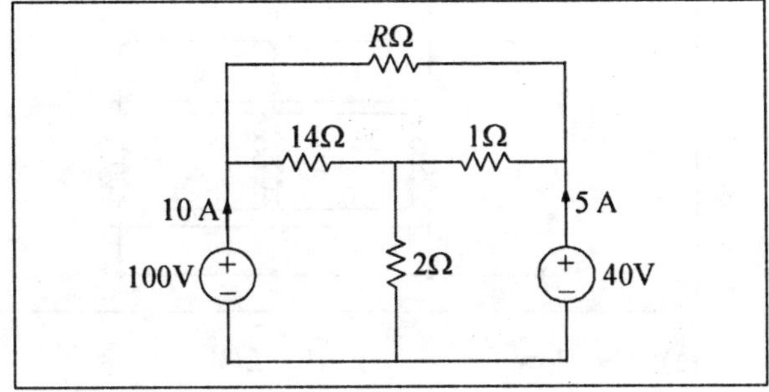

(a) 10 Ω (b) 18 Ω
(c) 24 Ω (d) 12 Ω

17. In figure the value of the source voltage is

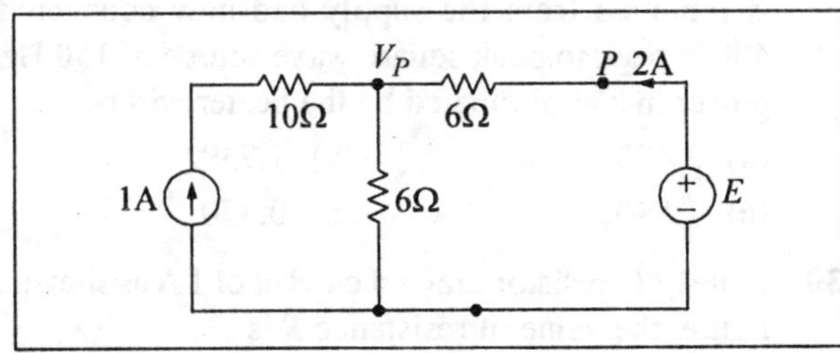

(a) 12 V (b) 24 V
(c) 30 V (d) 44 V

18. In figure R_a, R_b and R_c are 20 Ω, 10 Ω and 10 Ω respectively. The resistance R_1, R_2 and R_3 in Ω of an equivalent star-connection are

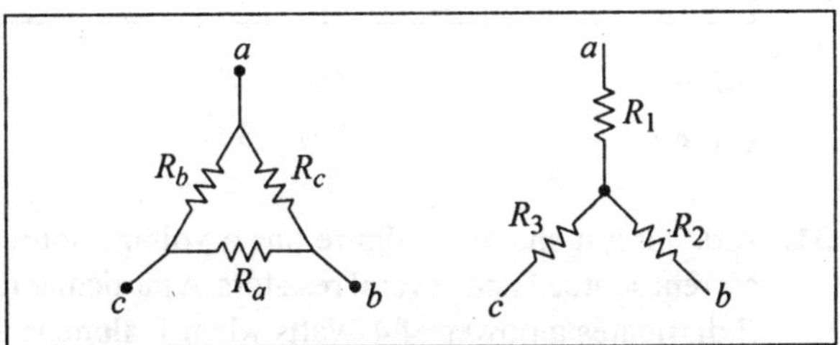

(a) 2.5, 5, 5 (b) 5, 2.5, 5
(c) 5, 5, 2.5 (d) 2.5, 5, 2.5

19. In figure the value of resistance R in Ω is

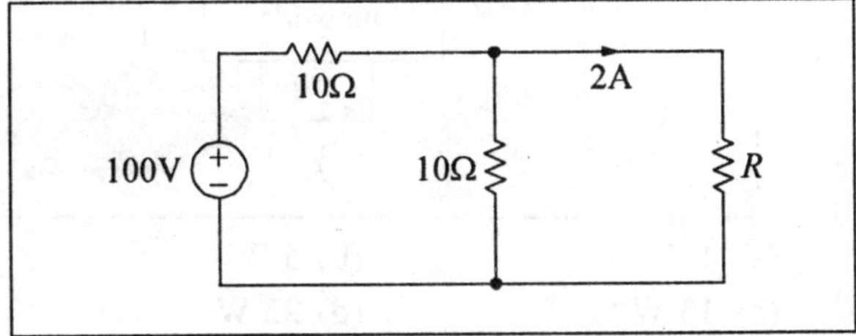

(a) 10 (b) 20
(c) 30 (d) 40

20. In the figure given below, the value of R is

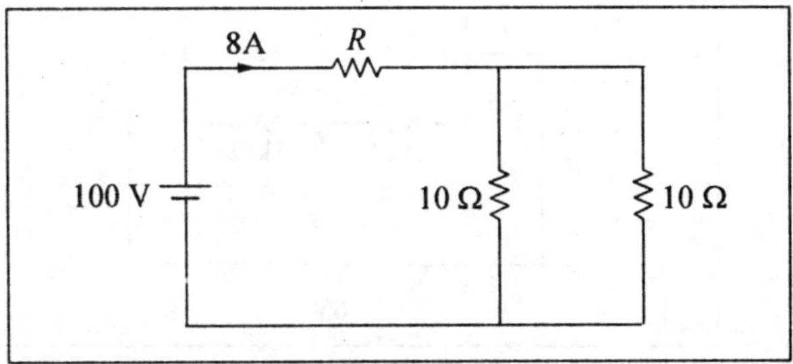

(a) 2.5 Ω
(b) 5.0 Ω
(c) 7.5 Ω
(d) 10.0 Ω

21. A 3 V dc supply with an internal resistance of 2 Ω supplies a passive non-linear resistance characterized by the relation $V_{NL} = I_{NL}^2$. The power dissipated in the non-linear resistance is

(a) 1.0 W (b) 1.5 W
(c) 2.5 W (d) 3.0 W

22. Assuming ideal elements in the circuit shown below, the voltage V_{ab} will be

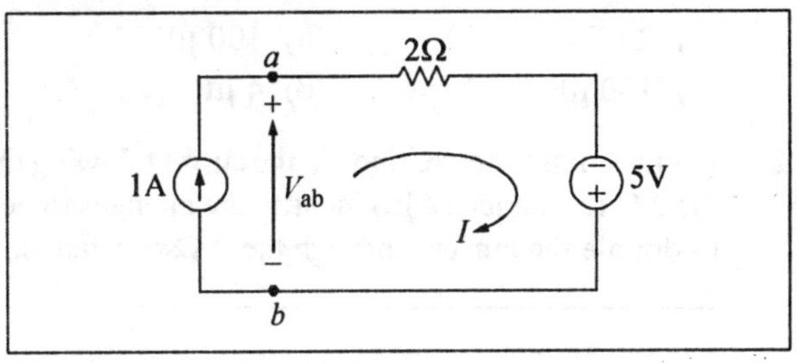

(a) –3V (b) 0 V
(c) 3V (c) 5 V

23. In the circuit shown in the figure, the value of the current i will be given by

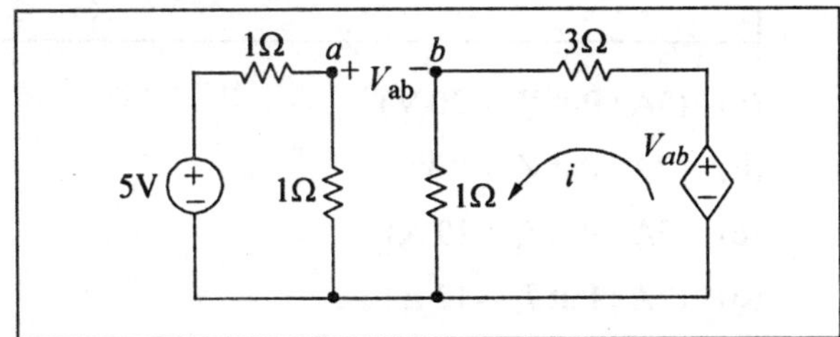

(a) 0.31 A
(b) 1.25A
(c) 1.75 A
(d) 2.5 A

24. The current through the 2 kΩ resistance in the circuit shown is

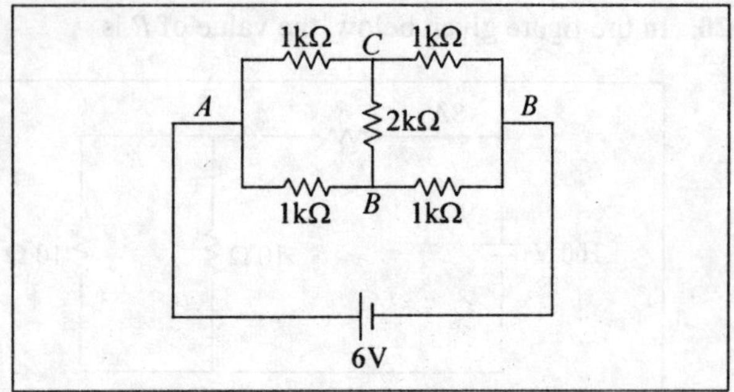

(a) 0 mA (b) 1 mA
(c) 2 mA (d) 6 mA

25. The equivalent capacitance of the input loop of the circuit shown is

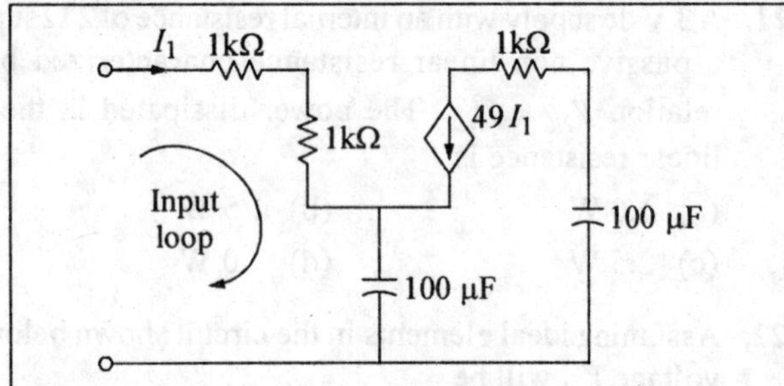

(a) 2 μF (b) 100 μF
(c) 100 μF (d) 4 μF

26. For the circuit shown, find out the current flowing through the 2 Ω resistance. Also identify the changes to be made to double the current through the 2 Ω resistance.

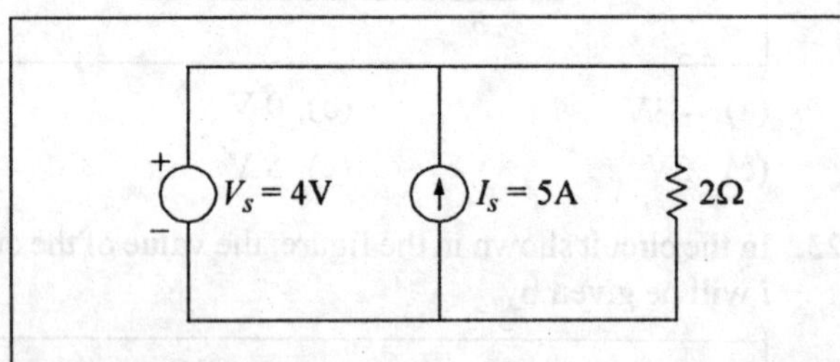

(a) (5A ; Put $V_s = 20$ V)
(b) (2A ; Put $V_s = 8$ V)
(c) (5A ; Put $I_s = 10$ A)
(d) (7A ; Put $I_s = 12$ A)

27. The color code of a 1 kΩ resistance is
(a) Black, brown, red.
(b) Red, brown, brown.
(c) Brown, black, red.
(d) Black, black, red

28. The three limbed non-ideal core shown in the figure has three windings with nominal inductances L each when measured individually with a single phase AC source. The inductance of the winding as connected will be

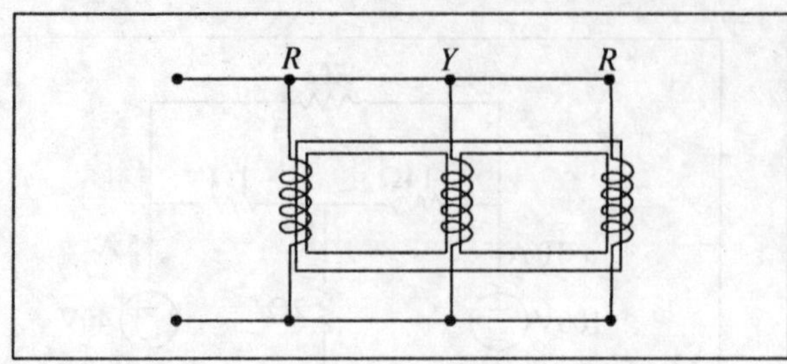

(a) Very low (b) $L/3$
(c) $3L$ (d) very high

29. An energy meter connected to an immersion heater (resistive) operating on an AC 230 V, 50 Hz, AC single phase source reads 2.3 units (kWh) in 1 hour. The heater is removed from the supply and now connected to a 400 V peak to peak square wave source of 150 Hz. The power in kW dissipated by the heater will be
(a) 3.478 (b) 1.739
(c) 1.540 (d) 0.870

30. If the 12 Ω resistor draws a current of 1 A as shown in the figure, the value of resistance R is

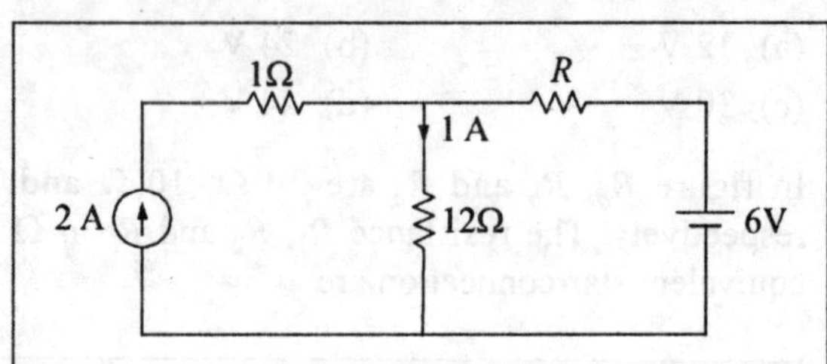

(a) 4 Ω (b) 6 Ω
(c) 8 Ω (d) 18 Ω

31. A dc circuit shown in figure has a voltage source V, a current source I and several resistors. A particular resistor R dissipates a power of 4 Watts when V alone is active. The same resistor R dissipates a power of 9 Watts when I alone is active. The power dissipated by R when both sources are active will be

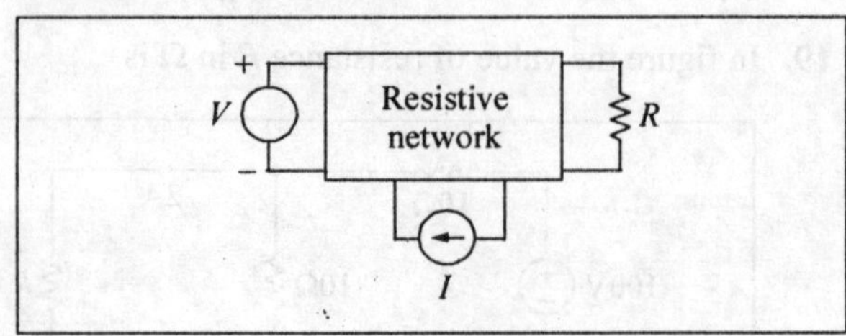

(a) 1 W (b) 5 W
(c) 13 W (d) 25 W

32. Two 2 H inductance coils are connected in series and are also magnetically coupled to each other, the coefficient of coupling being 0.1. The total inductance of the combination can be

(a) 0.4 H (b) 3.2 H
(c) 4.0 H (d) 4.4 H

33. The current i_4 in the circuit of figure is equal to

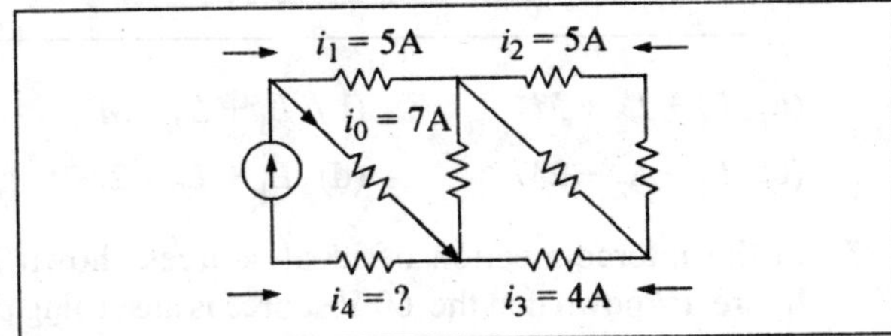

(a) 12 A (b) – 12 A
(c) 4 A (d) none of the above

34. The voltage V in figure is equal to

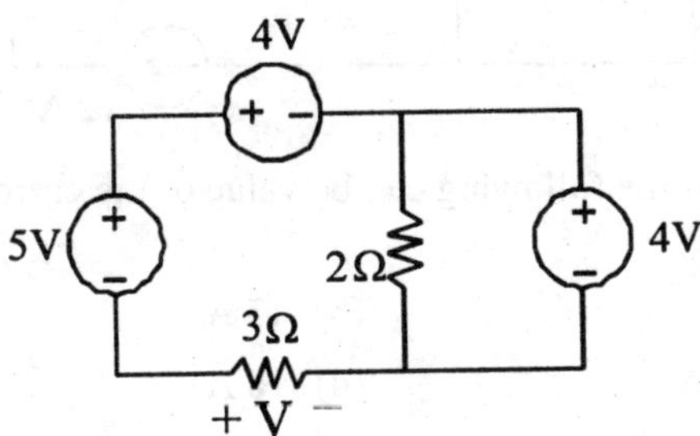

(a) 3 V
(b) – 3 V
(c) 5 V
(d) None of the above

35. The voltage V in figure is always equal to

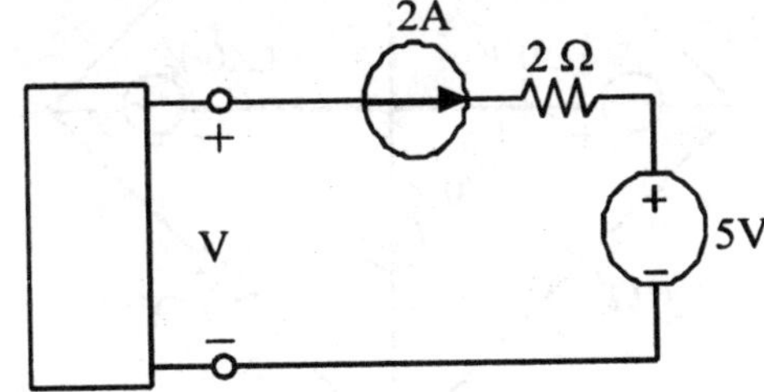

(a) 9 V
(b) 5 V
(c) 1 V
(d) None of the above

36. The voltage v in Figure is

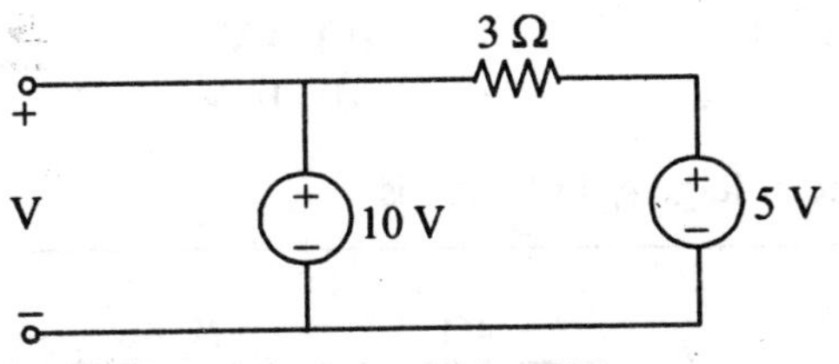

(a) 10 V (b) 15 V
(c) 5 V (d) None of the above

37. The nodal method of circuit analysis is based on

(a) KVL and Ohm's law
(b) KCL and Ohm's law
(c) KCL and KVL
(d) KCL, KVL and Ohm's law

38. The voltage across the terminals a and b in figure is

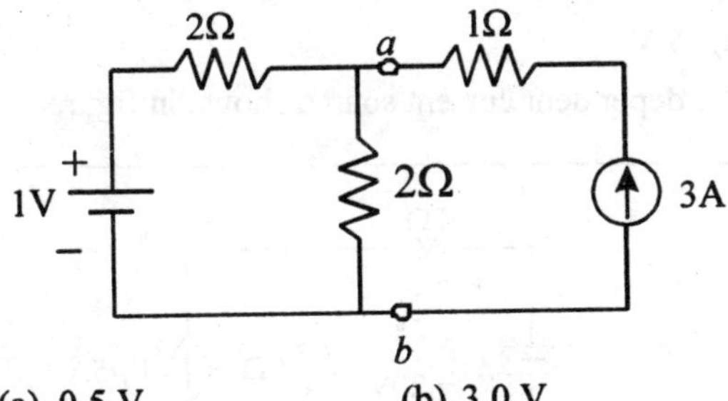

(a) 0.5 V (b) 3.0 V
(c) 3.5 V (d) 4.0 V

39. In the circuit of figure, the voltage $V(t)$ is

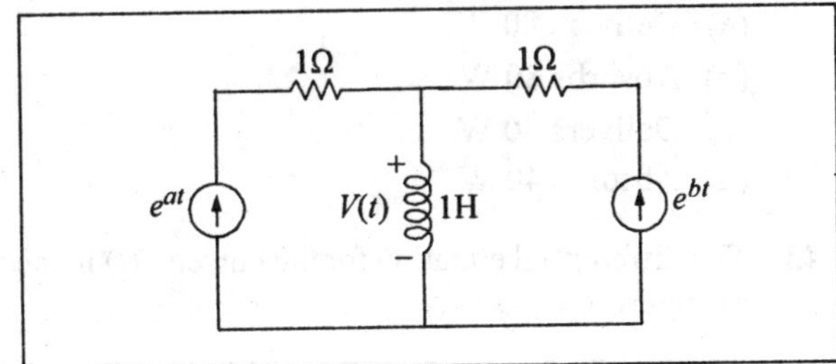

(a) $e^{at} - e^{bt}$ (b) $e^{at} + e^{bt}$
(c) $ae^{at} - be^{bt}$ (d) $ae^{at} + be^{bt}$

40. In the circuit of figure, the value of the voltage source E is

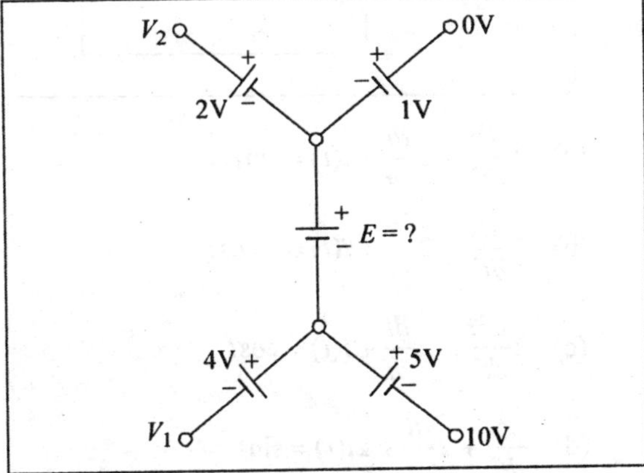

(a) –16 V (b) 4 V
(c) – 6 V (d) 16 V

41. The voltage e_0 in Figure is

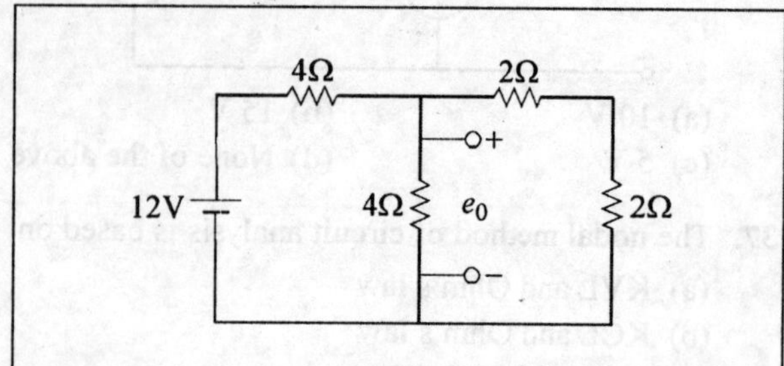

(a) 2 V
(b) [4/3] V
(c) 4 V
(d) 8 V

42. The dependent current source shown in figure

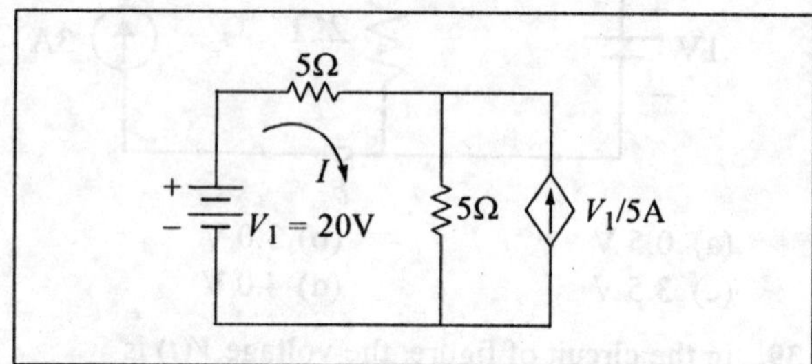

(a) Delivers 80 W
(b) Absorbs 80 W
(c) Delivers 40 W
(d) Absorbs 40 W

43. The differential equation for the current $i(t)$ in the circuit of figure is

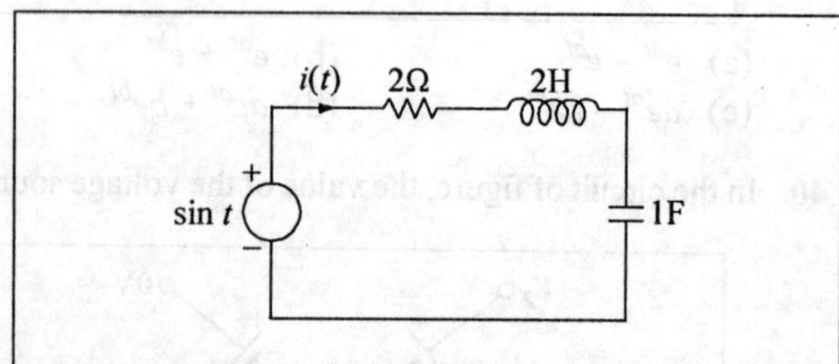

(a) $2\frac{d^2i}{dt^2}+2\frac{di}{dt}+i(t)=\sin t$

(b) $2\frac{d^2i}{dt^2}+2\frac{di}{dt}+i(t)=-\cos t$

(c) $2\frac{d^2i}{dt^2}+2\frac{di}{dt}+i(t)=\cos t$

(d) $\frac{d^2i}{dt^2}+2\frac{di}{dt}+2i(t)=\sin t$

44. The equivalent inductance measured between the terminals 1 and 2 for the circuit shown in figure is

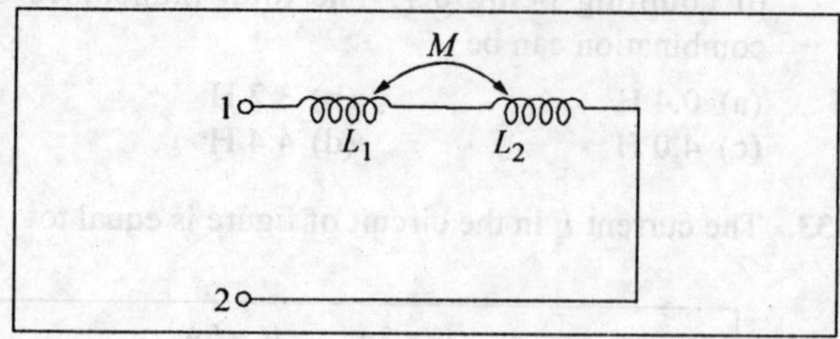

(a) $L_1 + L_2 + M$ (b) $L_1 + L_2 - M$
(c) $L_1 + L_2 + 2M$ (d) $L_1 + L_2 - 2M$

45. In the interconnection of ideal sources shown in the figure, it known that the 60 V source is absorbing power.

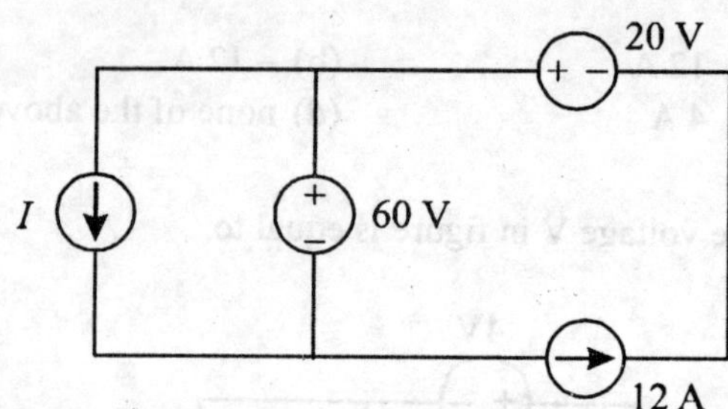

Which of the following can be value of the current source I?

(a) 10 A (b) 13 A
(c) 15 A (d) 18 A

46. In the circuit shown below, the current through the inductor is

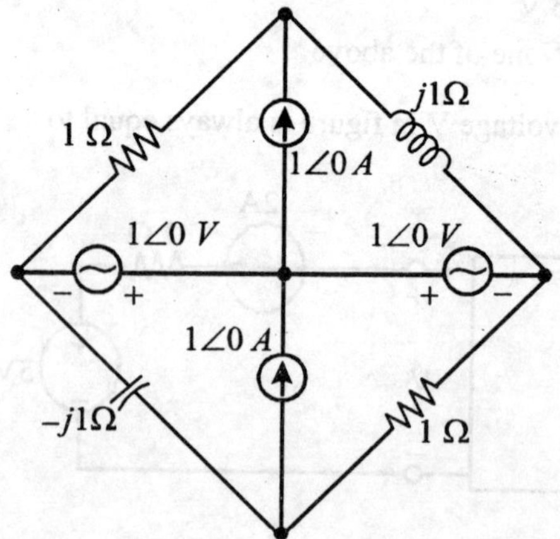

(a) $\frac{2}{1+j}A$ (b) $\frac{-1}{1+j}A$

(c) $\frac{1}{1+j}A$ (d) 0 A

47. Of the four networks, N_1, N_2, N_3 and N_4 of Figure the network having identical driving point function are

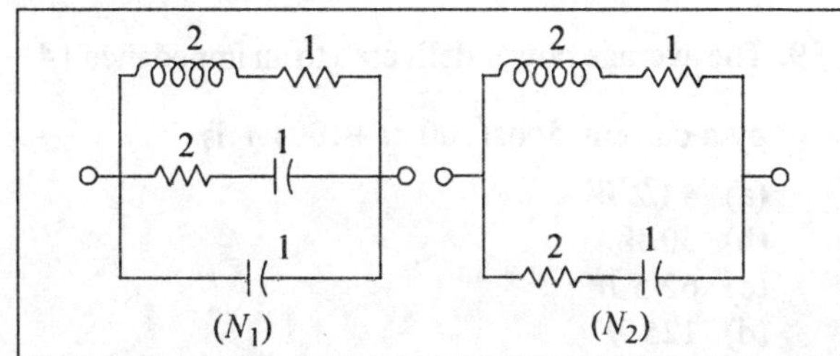

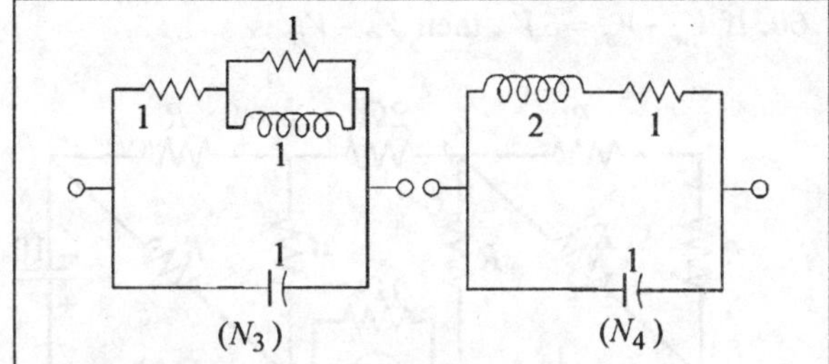

(a) N_1 and N_2 (b) N_2 and N_4
(c) N_1 and N_3 (d) N_1 and N_4

48. A network contains linear resistors and ideal voltage sources. If values of all the resistors are doubled, then the voltage across each resistor is
(a) Halved
(b) Doubled
(c) Increased by four times
(d) Not changed

49. The two electrical sub network N_1 and N_2 are connected through three resistor as shown in Figure. The voltage across 5 ohm resistor and 1 ohm resistor are given to be 10 V and 5 V, respectively. Then voltage across 15 ohm resistor is

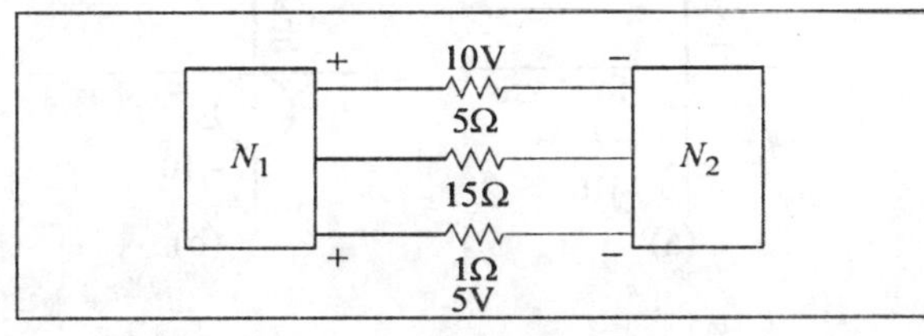

(a) − 105 V (b) + 105 V
(c) − 15 V (d) + 15 V

50. For the circuit in Figure the voltage v_o is

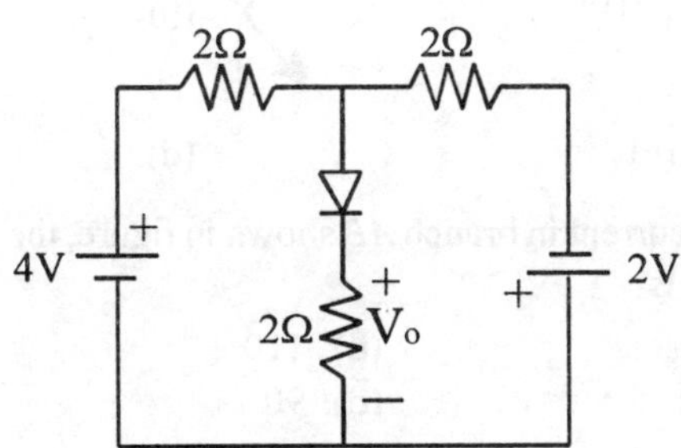

(a) 2 V (b) 1 V
(c) − 1 V (d) None of the above

51 Twelve 1Ω resistance are used as edges to form a cube. The resistance between two diagonally opposite corners of the cube is:
(a) (5/6) Ω (b) 1 Ω
(c) (6/5) Ω (d) (3/2) Ω

52. Impedance Z as shown in the given figure is

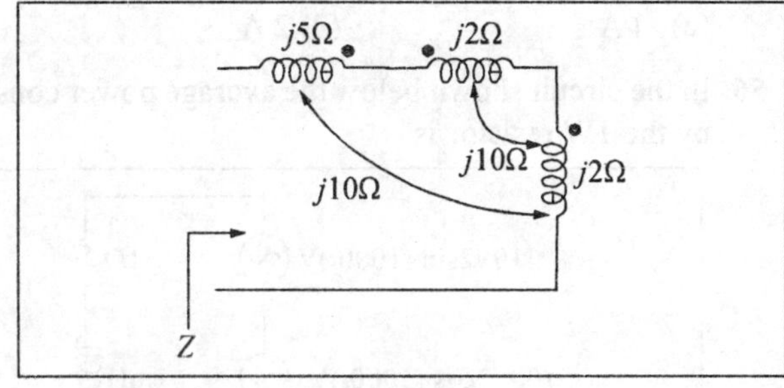

(a) j29 Ω (b) j9 Ω
(c) j19 Ω (d) j39 Ω

53. In the circuit shown the power supplied by the voltage source is

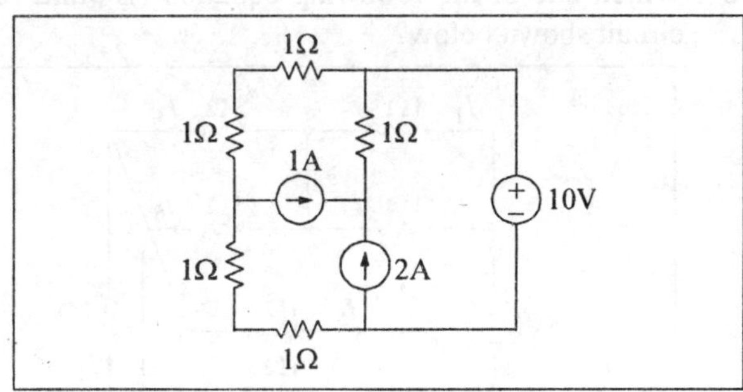

(a) 0 W
(b) 5 W
(c) 10 W
(d) 100 W

54. If $V_A - V_B = 6\,V$, then $V_C - V_D$ is

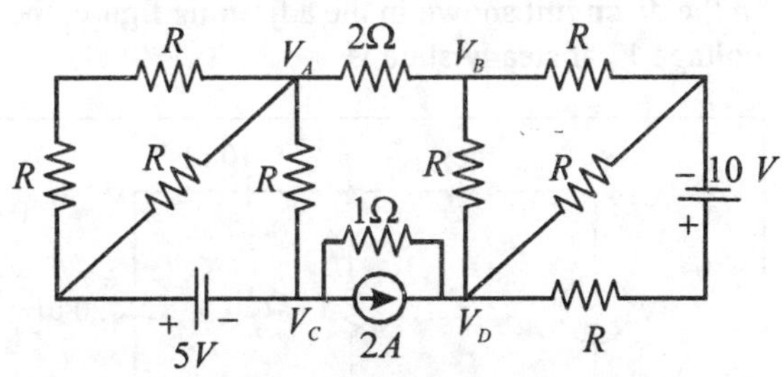

(a) −5 V (b) 2 V
(c) 3 V (d) 6V

55. The current I supplied by the dc voltage source in the circuit shown below is

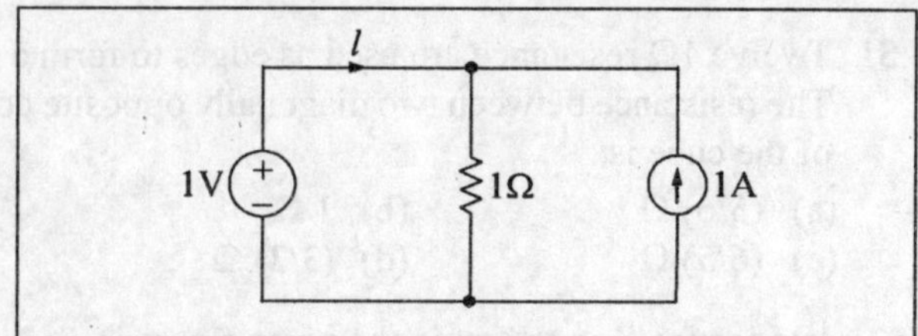

(a) 0 A (b) 0.5 A
(c) 1 A (d) 2 A

56. In the circuit shown below the average power consumed by the 1W resistor is

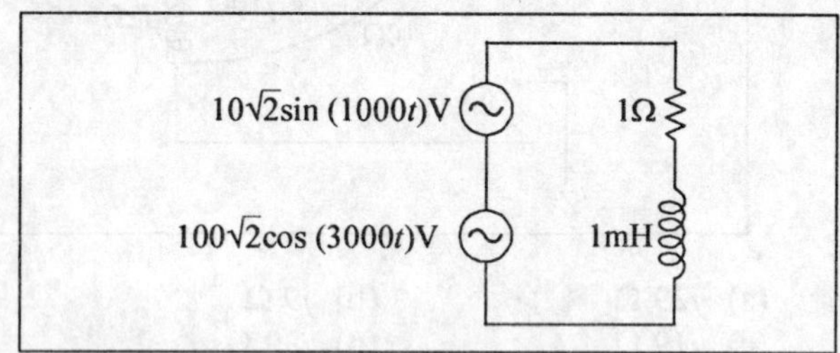

(a) 50 W (b) 1050 W
(c) 5000 W (d) 10100 W

57. Which one of the following equations is valid for the circuit shown below?

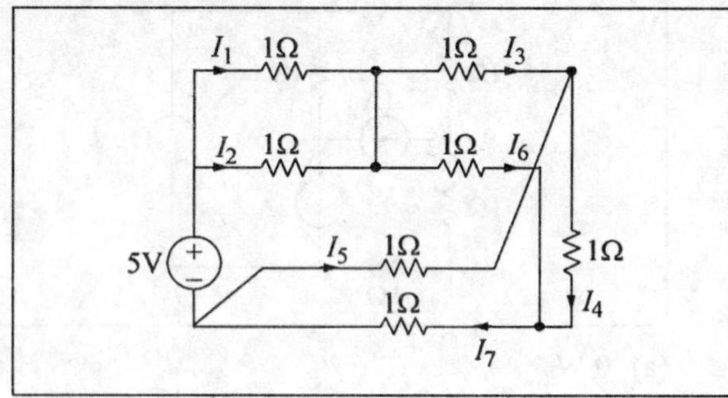

(a) $I_3 + I_5 - I_6 + I_7 = 0$
(b) $I_3 - I_5 + I_6 + I_7 = 0$
(c) $I_3 + I_5 + I_6 + I_7 = 0$
(d) $I_3 + I_5 + I_6 - I_7 = 0$

58. In the dc circuit shown in the adjoining figure, the node voltage V_2 at steady state is

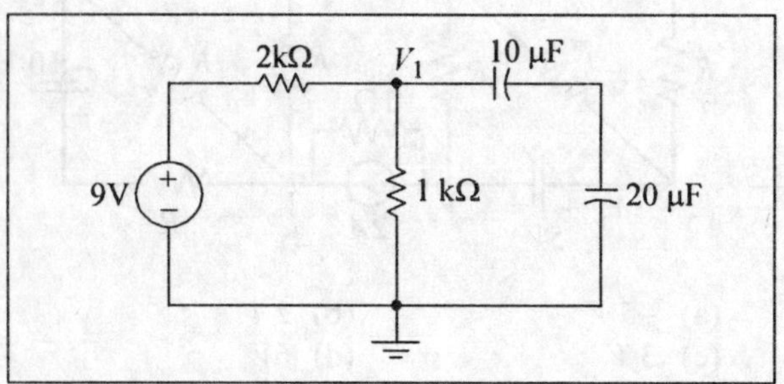

(a) 0 V (b) 1 V
(c) 2 V (d) 3 V

59. The average power delivered to an impedance $(4 - j3)\Omega$ by a current $5\cos(100\,\pi t + 100)\,A$ is

(a) 44.2 W
(b) 50 W
(c) 62.5 W
(d) 125 W

60. If $V_A - V_B = 6\,V$, then $V_C - V_D$ is

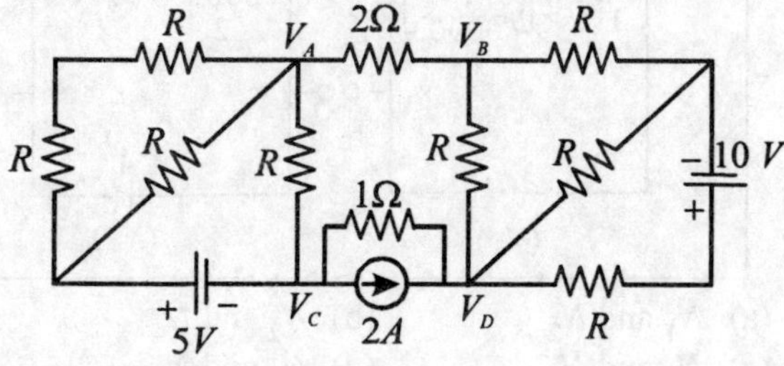

(a) $-5\ V$
(b) $2\ V$
(c) $3\ V$
(d) $6V$

61. A complex variable $Z = x + j0.1$ has its real part x varying in the range $-\infty$ to $+\infty$. Which one of the following is the locus (shown in thick lines) of $\frac{1}{Z}$ in the complex plane?

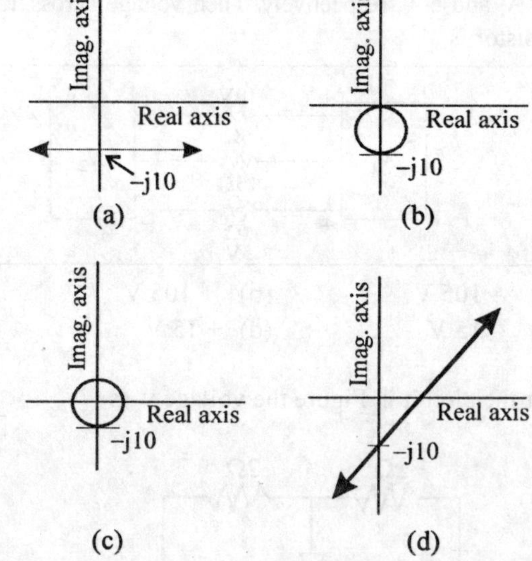

62. For the current in branch AB shown in figure, the voltage V_{in} volt is

(a) 55 (b) 110
(c) 56 (d) 90

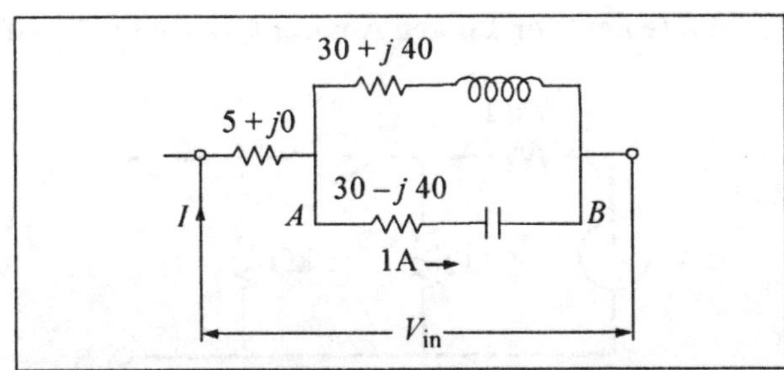

63. In figure the stored energy of the capacitor is

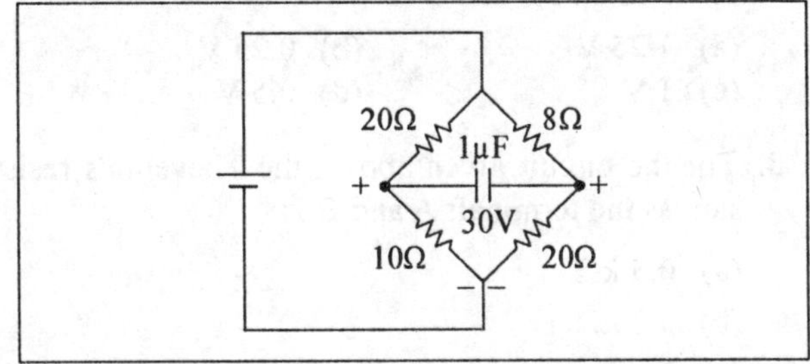

(a) 8 μJ
(b) 30μJ
(c) 450 μJ
(d) 900 μJ

64. For the waveform $V(t) = 2 + \cos[\omega t + (\pi / 6)]$ the ratio $[V_{rms}/V_{average}]$ is

(a) $[3/2(2\sqrt{2})]$

(b) $\sqrt{3/2}$

(c) π

(d) $\pi/2$

65. A voltmeter connected across the 10 kΩ resistor in the figure reads 5 V. The voltmeter is rated at 1000 ohms/volt and has a full scale reading of 10 V. The supply voltage V_s in volt is

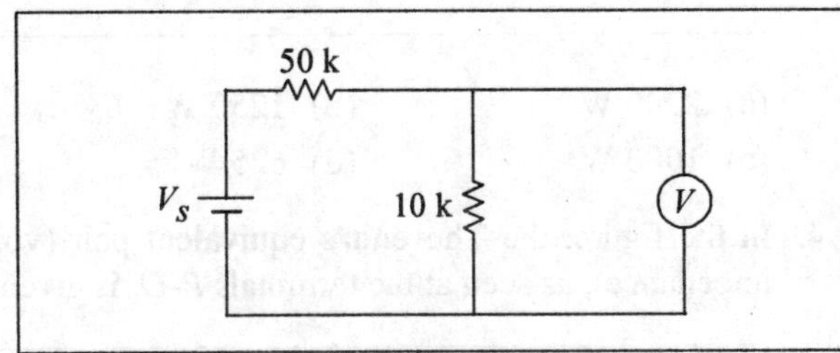

(a) 30 (b) 50
(c) 55 (d) 80

66. In the circuit shown in figure current through the 5 Ω resistor is

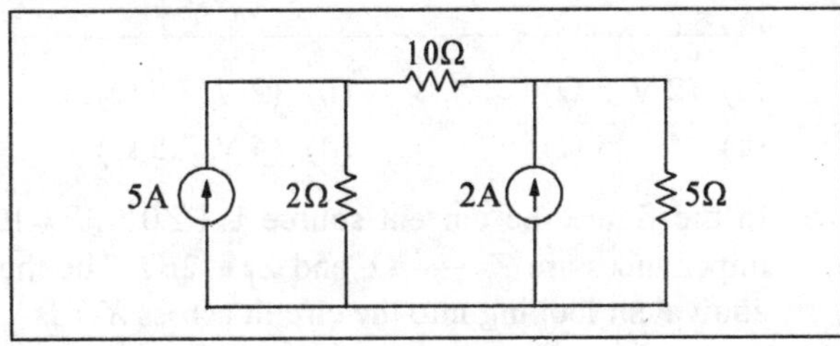

(a) 0 A (b) 2 A
(c) 3 A (d) 7 A

6.2 Network Theorems

1. In the Figure $Z_1 = 10 \angle{-60°}$, $Z_2 = 10 \angle 60°$, $Z_3 = 50 \angle 53.13°$. Thevenin impedance seen form X–Y is

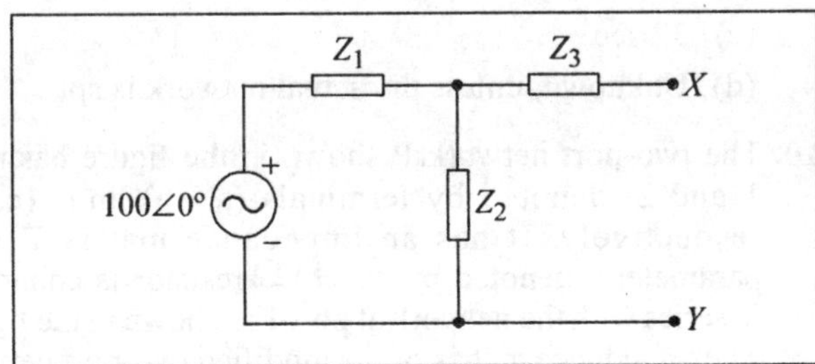

(a) 56.67 ∠45°

(b) 60 ∠30°

(c) 70 ∠30°

(d) 34.4 ∠65°

2. In Figure, the potential difference between points P and Q is

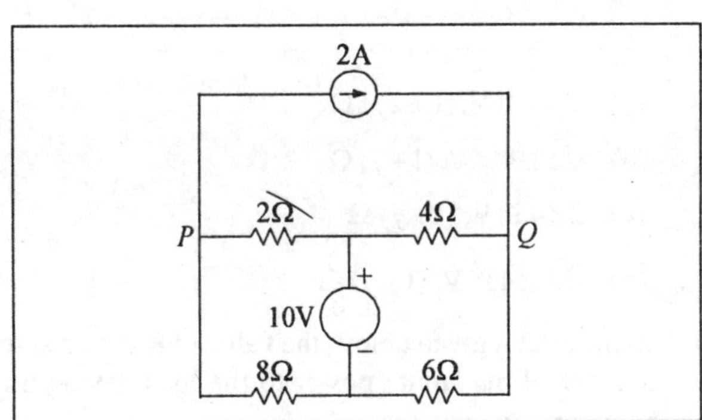

(a) 12 V (b) 10 V
(c) – 6 V (d) 8 V

3. Two ac source feed a common variable resistive load as shown in figure. Under the maximum power transfer condition, the power absorbed by the load resistance R_L is

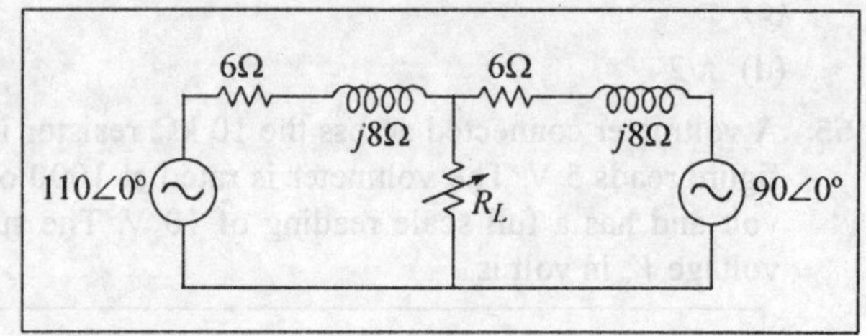

(a) 2200 W (b) 1250 W
(c) 1000 W (d) 625 W

4. In the figure, the Thevenin's equivalent pair (voltage, impedance), as seen at the terminals *P-Q*, is given by

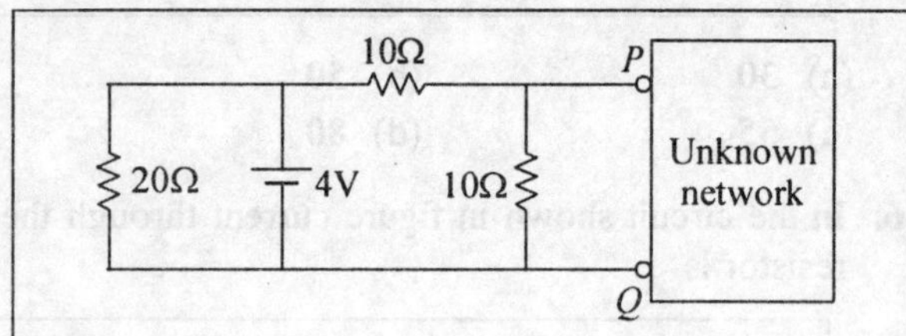

(a) (2 V, 5 Ω) (b) (2 V, 7.5 Ω)
(c) (4 V, 5 Ω) (d) (4 V, 7.5 Ω)

5. In the figure the current source is 1 ∠0 A, $R = 1\Omega$, the impedances are $Z_C = -j\Omega$, and $Z_1 = 2j\Omega$. The thevenin equivalent looking into the circuit across *X-Y* is

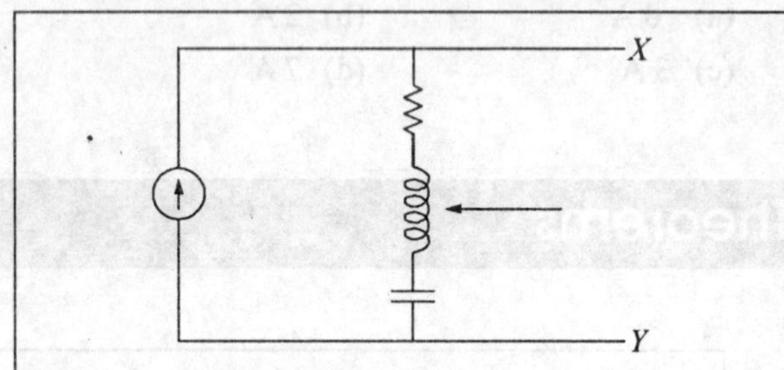

(a) $\sqrt{2}\angle 0$ V, $(1+2j)\,\Omega$

(b) $\sqrt{2}\angle 45°$ V, $(1+j)\,\Omega$

(c) $2\angle 45°$ V, $(1+j)\,\Omega$

(d) $\sqrt{2}\angle 45°$ V, $(1+j)\,\Omega$

6. In the circuit given below, the value of R required for the transfer of maximum power to the load having a resistance of 3Ω is

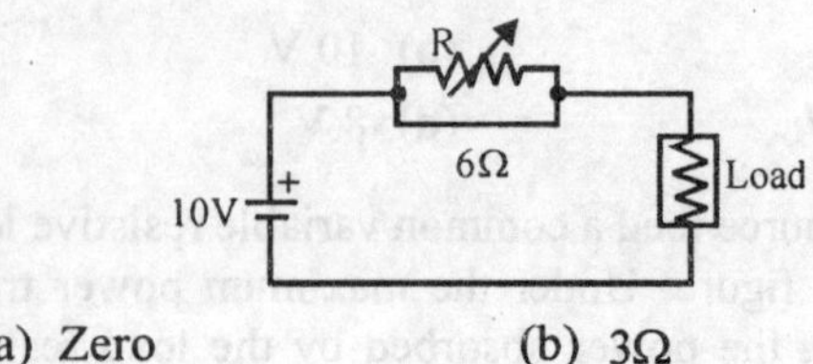

(a) Zero (b) 3Ω
(c) 6Ω (d) Infinity

Statement for Linked Answer Questions 7 and 8:

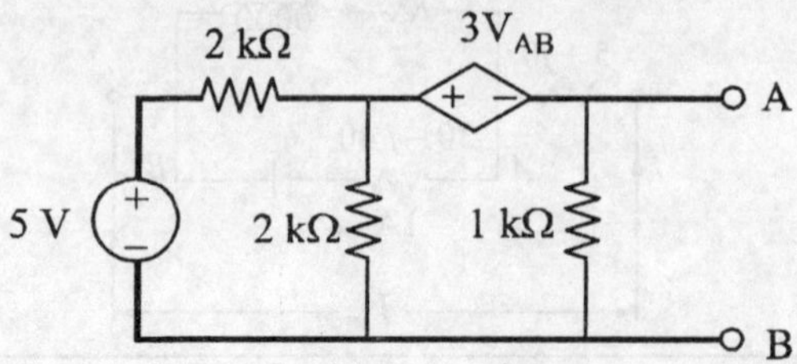

7. For the circuit given above, the Thevenin's voltage across the terminals A and B is

(a) 1.25 V (b) 0.25 V
(c) 1 V (d) 0.5 V

8. For the circuit given above, the Thevenin's resistance across the terminals A and B is

(a) 0.5 kΩ
(b) 0.2 kΩ
(c) 1 kΩ
(d) 0.11 kΩ

9. The v-i characteristic as seen from the terminal pair (*A, B*) of the network of Figure below is shown in Figure. If an inductance of value 6 mH is connected across the terminal-pair (*A, B*), the time constant of the system will be

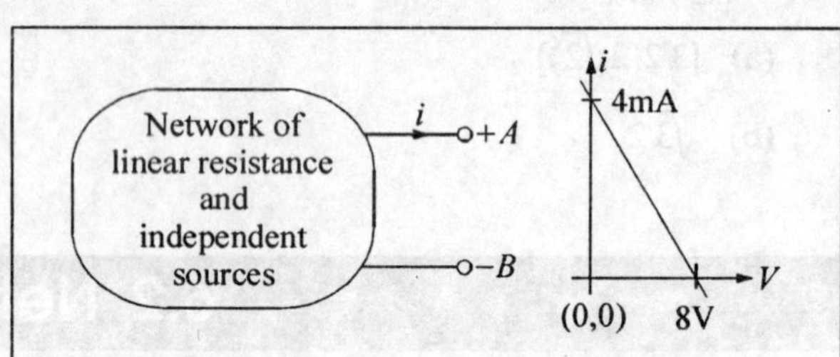

(a) 3 μ sec
(b) 12 sec
(c) 32 sec
(d) Unknown, unless the actual network is sp

10. The two-port network P shown in the figure has ports 1 and 2, denoted by terminals (a, b) and (c, d) respectively. It has an impedance matrix Z with parameters denoted by z_{ij}. A 1 Ω resistor is connected in series with the network at port 1 as shown in the figure. The impedance matrix of the modified two-port network (shown as a dashed box) is

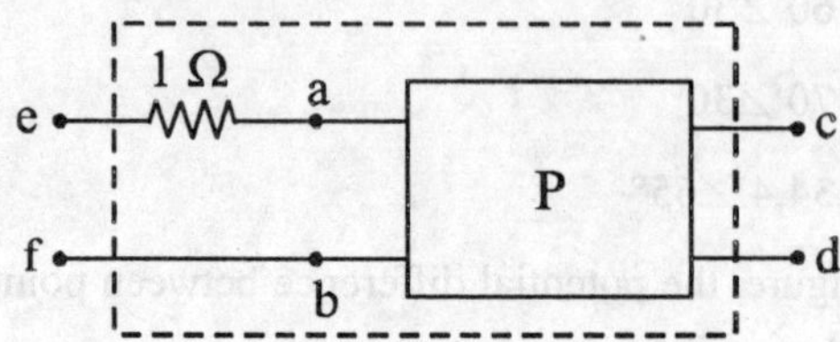

(a) $\begin{pmatrix} z_{11}+1 & z_{12}+1 \\ z_{21} & z_{22}+1 \end{pmatrix}$

(b) $\begin{pmatrix} z_{11}+1 & z_{12} \\ z_{21} & z_{22}+1 \end{pmatrix}$

(c) $\begin{pmatrix} z_{11}+1 & z_{12} \\ z_{21} & z_{22} \end{pmatrix}$

(d) $\begin{pmatrix} z_{11}+1 & z_{12} \\ z_{21}+1 & z_{22} \end{pmatrix}$

11. The *V-I* characteristic as seen from the terminal-pair (*A, B*) of the network of figure is shown in another figure. If a variable resistance R_L is connected across the terminal-pair (*A, B*) ,the maximum power that can be supplied to R_L would be

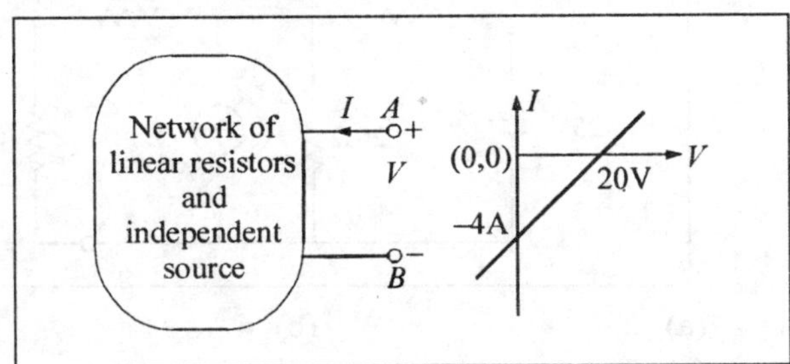

(a) 80 W
(b) 40 W
(c) 20 W
(d) Indeterminate unless the actual network is given

12. Viewed from the terminals *A, B* the following circuit shown in Figure below can be reduced to an equivalent circuit of a single voltage source in series with a single resistance with the following parameters:

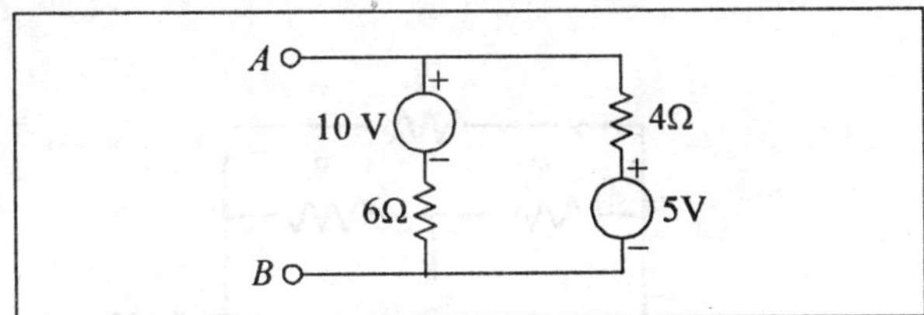

(a) 5 volt source in series with 10 Ω resistor
(b) 1 volt source in series with 2.4 Ω resistance
(c) 7 volt source in series with 2.4 Ω resistance
(d) 1 volt in series in with 10 Ω resistor

13. The circuit shown in figure is equivalent to a load of

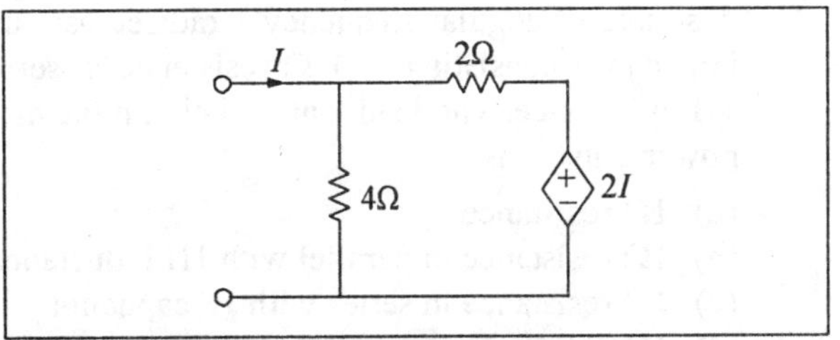

(a) [4/8] ohms (b) [8/3] ohm
(c) 4 ohm (d) 2 ohms

14. If the secondary winding of the ideal transformer shown in the circuit of figure has 40 turns, the number of turns in the primary winding for maximum power transfer to the 2 Ω resistor will be

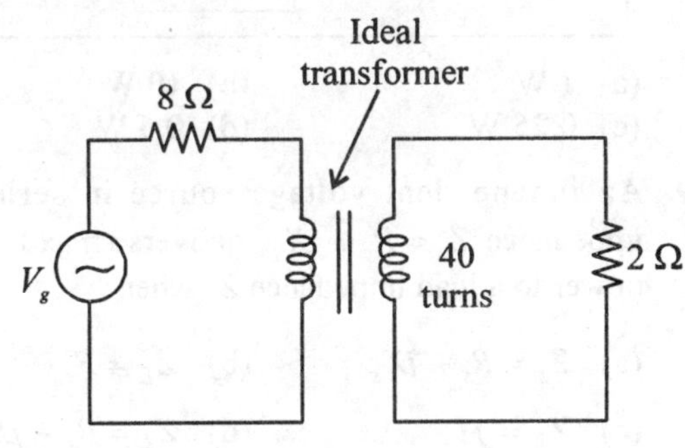

(a) 20 (b) 40
(c) 80 (d) 160

15. The value of the resistance R, connected across the terminals, A and B, (ref. Figure), which will absorb with maximum power , is

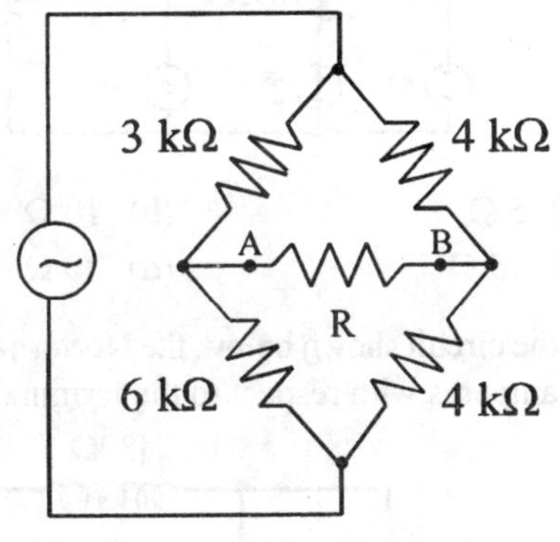

(a) 4.00 kΩ (b) 4.11 k Ω
(c) 8.00 kΩ (d) 9.00 kΩ

16. Superposition theorem is NOT applicable to networks containing
(a) Nonlinear elements
(b) Dependent voltage sources
(c) Dependent current source
(d) Transformers

17. A source of angular frequency 1 rad/sec has a source impedance consisting of 1 Ω resistance in series with 1H inductance. The load that will obtain the maximum power transfer is

(a) 1Ω resistance
(b) 1Ω resistance in parallel with 1H inductance
(c) 1Ω resistance in series with 1F capacitor
(d) 1Ω resistance in parallel with 1F capacitor

18. The maximum power that can be transferred to the load resistor *RL* from the voltage source in the figure is

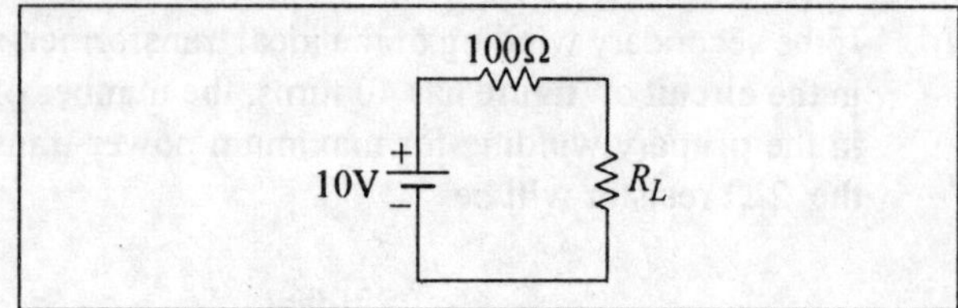

(a) 1 W (b) 10 W
(c) 0.25 W (d) 0.5 W

19. An independent voltage source in series with an impedance $Z_s = R_s + jX_s$, delivers a maximum average power to a load impedance Z_L when

(a) $Z_s = R_s + jX_s$ (b) $Z_L = R_s$
(c) $Z_L = jZ_s$ (d) $Z_L = R_s - jX_s$

20. In the circuit shown below, the value of R_L such that the power transfererd to R_L is maximum is

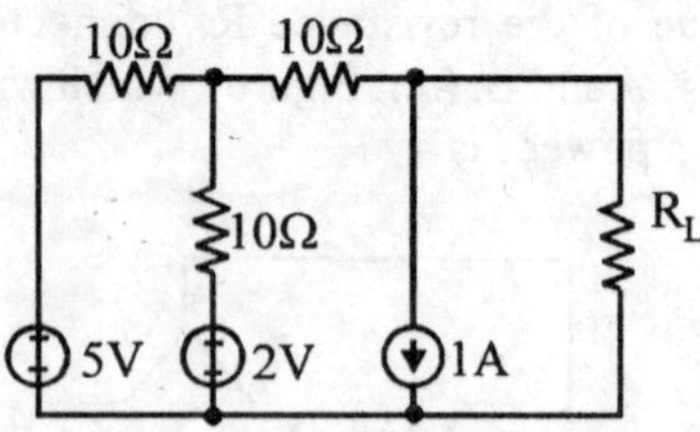

(a) 5 Ω (b) 10 Ω
(c) 15 Ω (d) 20 Ω

21. In the circuit shown below, the Nortorn equivalent current in amperes with respect to the terminals P and Q is

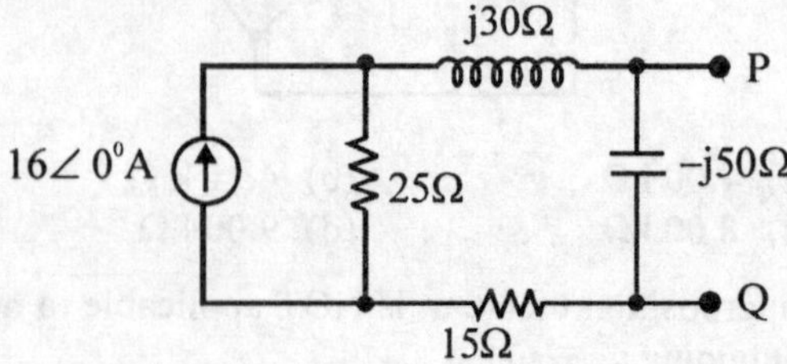

(a) 6.‹‹ – j4.8 (b) 6.56 – j7.87
(c) 10 + j0 (d) 16 + j0

22. The Thevenin equivalent voltage V_{TH} appearing between the terminals *A* and *B* of the network shown in figure is given by

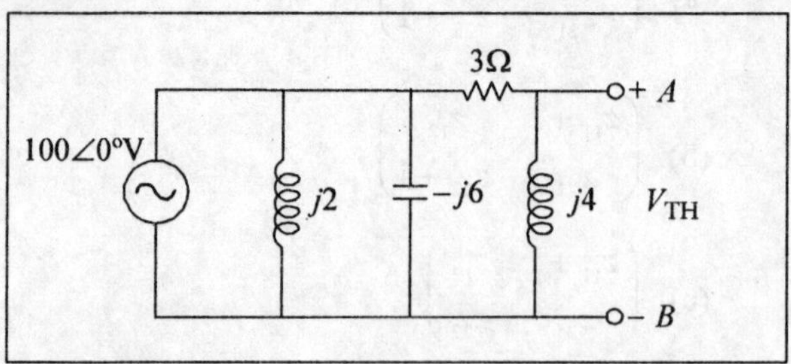

(a) *j*16 (3 – *j*4)
(b) *j*16 (3 + *j*4)
(c) 16 (3 + *j*4)
(d) 16 (3 – *j*4)

23. The value of *R*(in ohms) required for maximum power transfer in the network shown in Figure is

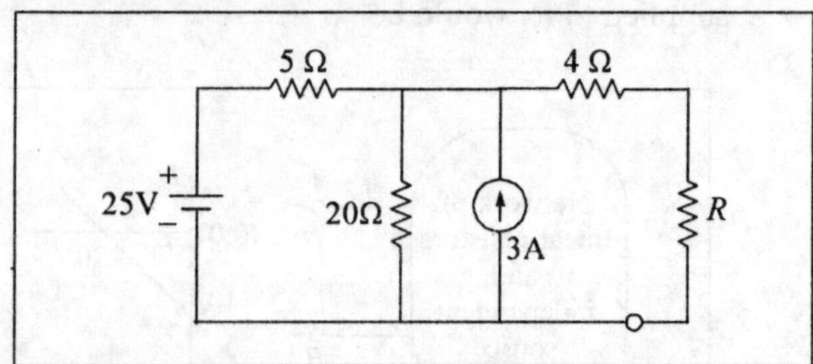

(a) 2 (b) 4
(c) 8 (d) 16

24. Use the data of Figure. The current i in the circuit of Figure is

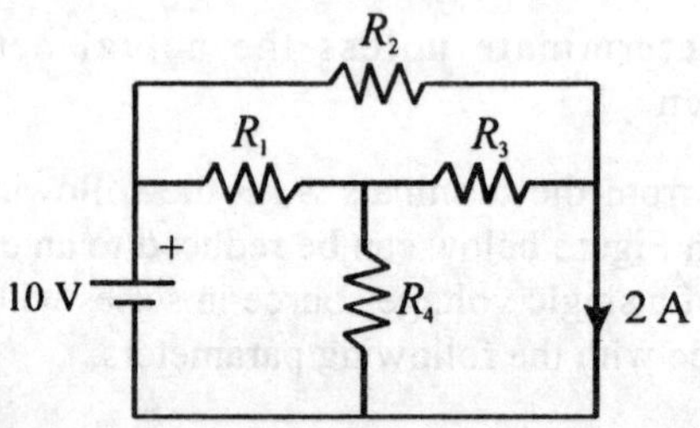

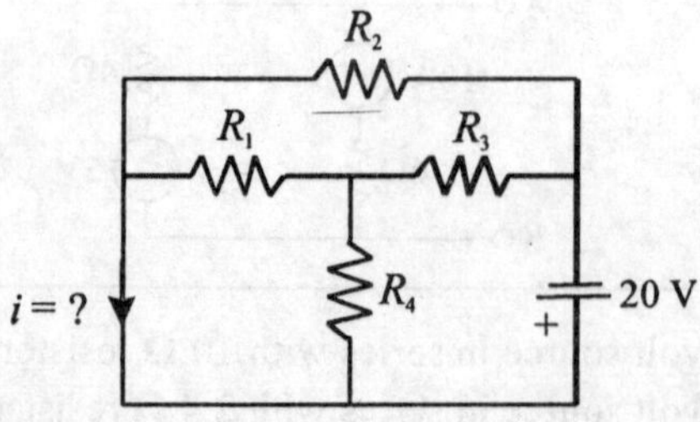

(a) –2 A
(b) 2 A
(c) –4 A
(d) + 4 A

25. The voltage e in Figure

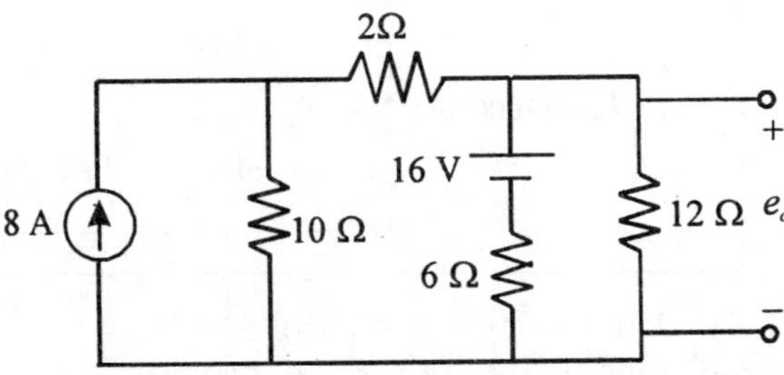

(a) 48 V (b) 24 V
(c) 36 V (d) 28 V

26. In figure the value of the load resistor R which maximizes the power delivered to it is

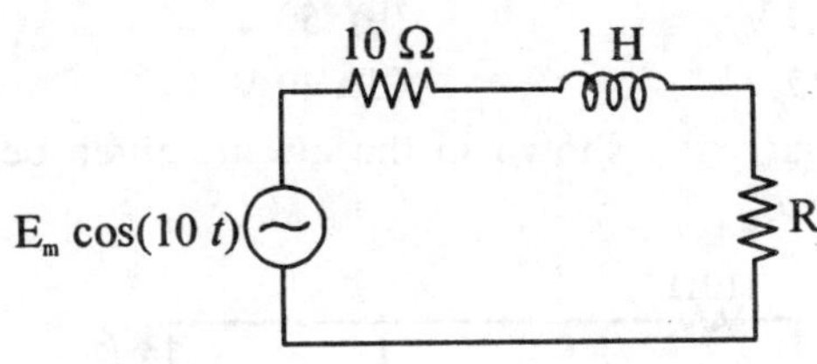

(a) 14.14 Ω (b) 10 Ω
(c) 200 Ω (d) 28.28 Ω

27. In the network of figure, the maximum power is delivered to R_L if its value is

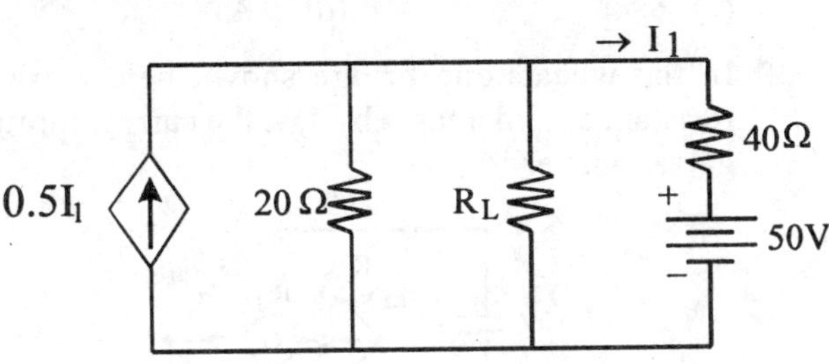

(a) 16 Ω (b) 40 / 3 Ω
(c) 60 Ω (d) 20 Ω

28. For the circuit shown in the figure, Thevenin's voltage and Thevenin's equivalent resistance at terminals *a-b* is

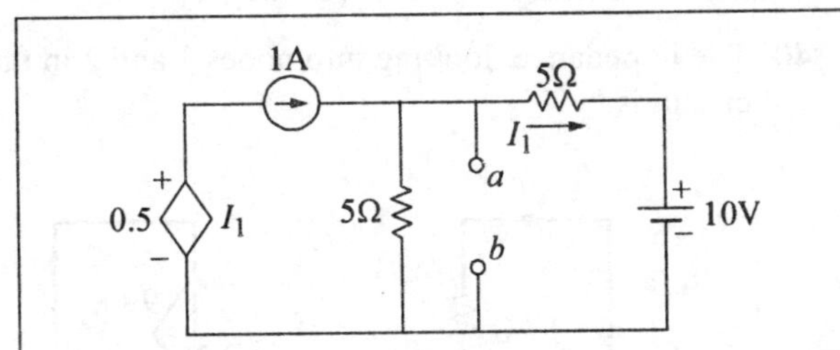

(a) 5 V and 2 Ω
(b) 7.5 V and 2.5 Ω
(c) 4 V and 2 Ω
(d) 3 V and 2.5 Ω.

29. For the circuit shown in the figure, the Thevenin voltage and resistance looking into *X-Y* are

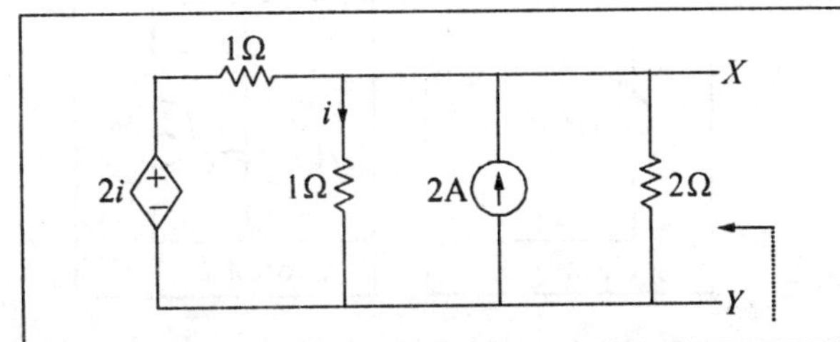

(a) $\frac{4}{3}$V, 2Ω (b) 4V, $\frac{2}{3}$Ω
(c) $\frac{4}{3}$V, $\frac{2}{3}$Ω (d) 4V, 2Ω

30. The Thevenin equivalent impedance Z_{th} between the nodes *P* and *Q* in he following circuit is

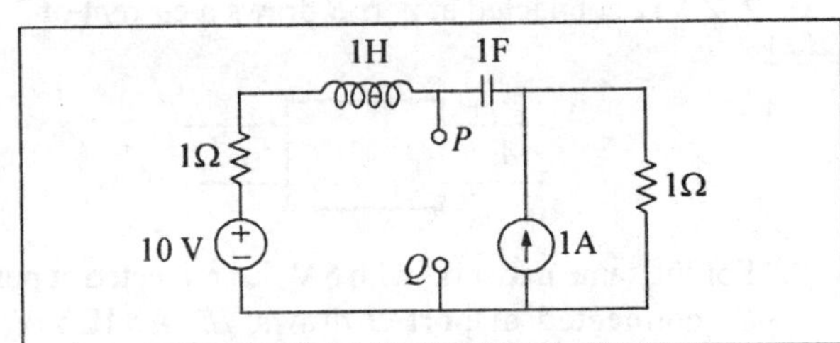

(a) 1 (b) $1+s+\frac{1}{s}$
(c) $2+s+\frac{1}{s}$ (d) $\frac{s^2+s+1}{s^2+2s+1}$

31. In the circuit shown, what value of R_L maximizes the power delivered to R_L?

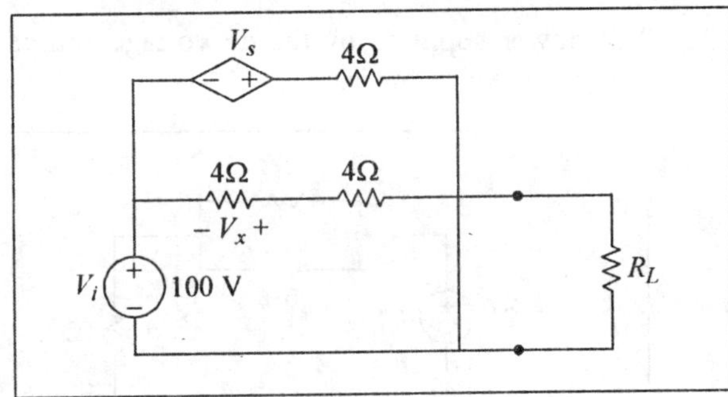

(a) 2.4 W
(b) $\frac{8}{3}$Ω
(c) 4 Ω
(d) 6 Ω

32. Assuming both the voltage sources are in phase, the value of R for which maximum power is transferred from circuit A to circuit B is

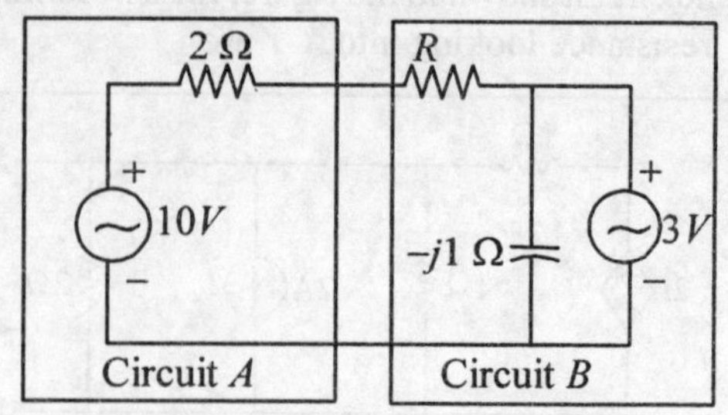

(a) 0.8 Ω (b) 1.4 Ω
(c) 2 Ω (d) 2.8 Ω

Common Data for Questions 33 and 34:

With 10 V dc connected at port A in the linear nonreciprocal two-port network shown below, the following were observed:

1. 1 Ω connected at port B draws a current of 3 A
2. 2.5 Ω connected at port B draws a current of 2 A

33. For the same network, with 6 V dc connected at port A, 1 W connected at port B draws 7/3 A. If 8 V dc is connected to port A, the open circuit voltage at port B is

(a) 6 V (b) 7 V
(c) 8 V (d) 9 V

34. With 10 V dc connected at port A, the current drawn by 7 W connected at port B is

(a) 3/7 A (b) 5/7 A
(c) 1 A (d) 9/7 A

35. The power supplied by the dc voltage source in the circuit shown below is.

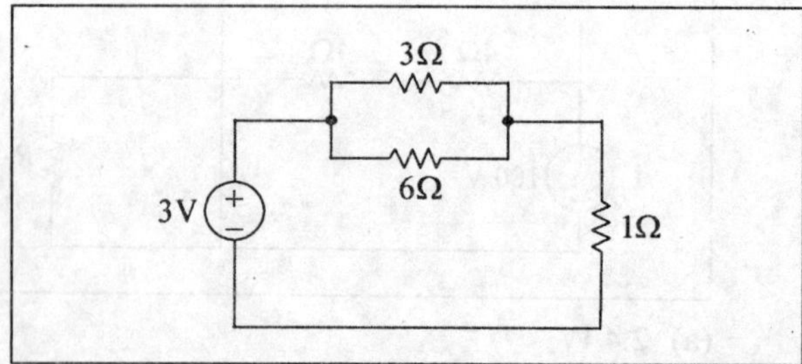

(a) 0 W
(b) 1.0 W
(c) 2.5 W
(d) 3.0 W

36. In the circuit shown below the maximum power that can be transferred to the load Z_L is

$i(t) = 10\sqrt{2}\sin(1000t)$ A

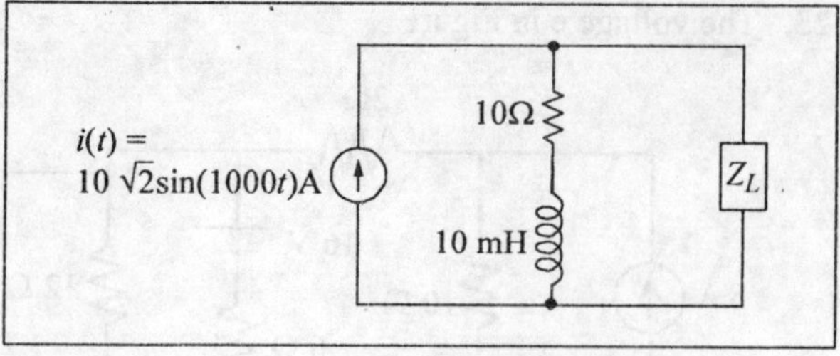

(a) 250 W (b) 500 W
(c) 1000 W (d) 2000 W

37. A 100 Ω, 1 W resistor and a 800 Ω, 2 W resistor are connected in series. The maximum dc voltage that can be applied continuously to the series circuit without exceeding the power limit of any of the resistors is

(a) 90 V (b) 50 V
(c) 45 V (d) 40 V

38. The current I shown in the circuit given below is equal to

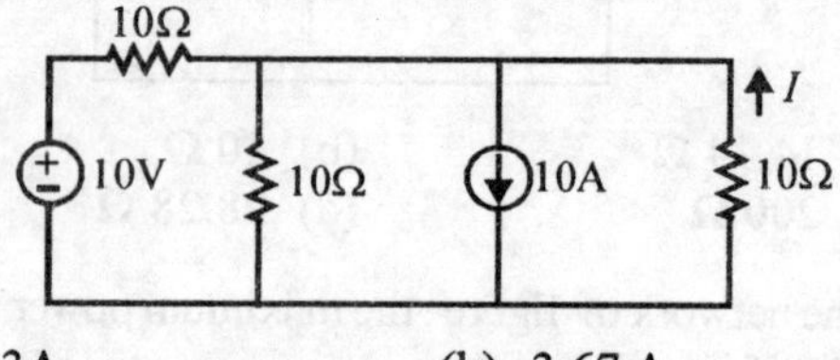

(a) 3A (b) 3.67 A
(c) 6A (d) 9 A

39. In the wheatstone bridge shown below, when the resistance R_1 increases by 1W, the current through the galvanometer is

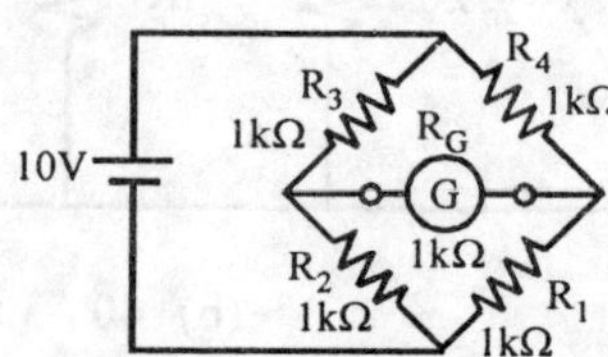

(consider the Thevenin equivalent resistance of the bridge in the calculations)

(a) 1.25 μA (b) 2.5 μA
(c) 12.5 μA (d) 25 μA

40. The impedance looking into nodes 1 and 2 in the given circuit is

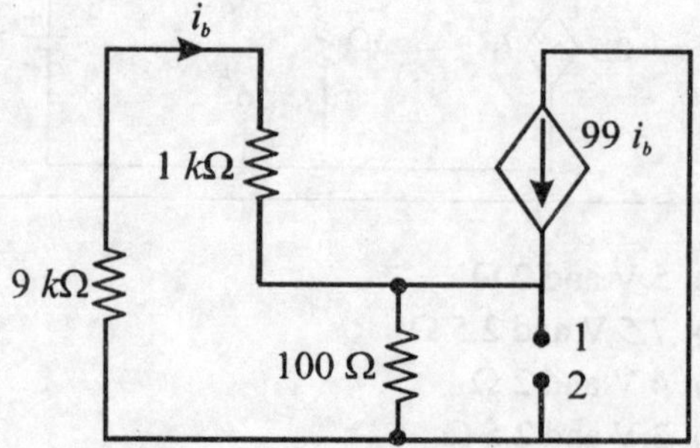

(a) 50 Ω (b) 100 Ω
(c) 5 kΩ (d) 10.1 kΩ

41. For the circuit shown in figure. $V_R = 20$ V when $R = 10\ \Omega$ and $V_R = 30$ V when $R = 20\ \Omega$. For $R = 80\ \Omega$, V_R will read as

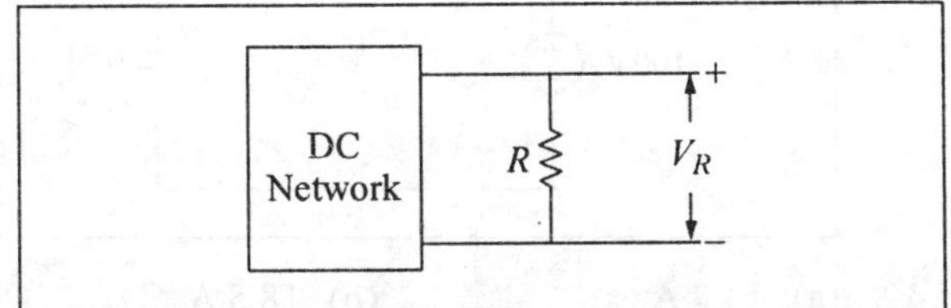

(a) 48 V (b) 60 V
(c) 120 V (d) 160 V

42. For the circuit shown in figure, R is adjusted to have maximum power transfer to it. The maximum power transferred is

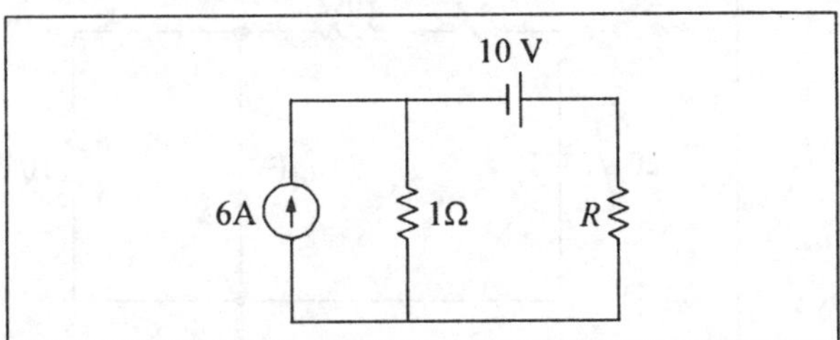

(a) 16 W
(b) 32 W
(c) 64 W
(d) 100 W

43. The value of the impedance $Z_d(s)$ between 1- 1 of the circuit shown in figure as $s \to \infty$ is

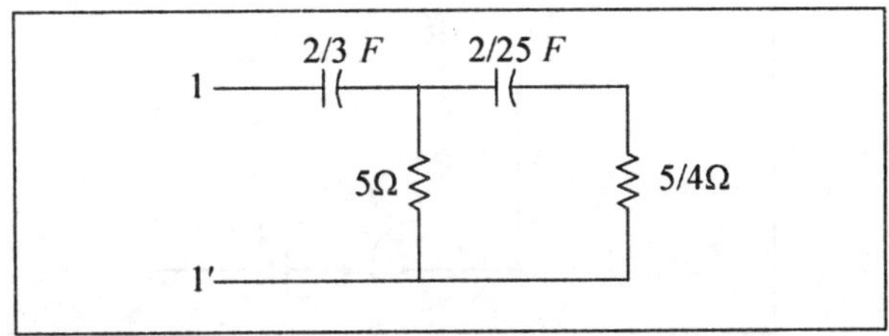

(a) 0 (b) 1
(c) 5/4 (d) 5

44. Superposition theorem does not hold for
(a) R-L-C networks
(b) Networks containing diodes and R-C elements
(c) Linear op-amp circuit
(d) None of the above

45. The output resistance across the terminals 1 and 2 of the DC bridge in figure is
(a) 12.5 kΩ
(b) 24.5 kΩ
(c) 25.0 kΩ
(d) 100 kΩ

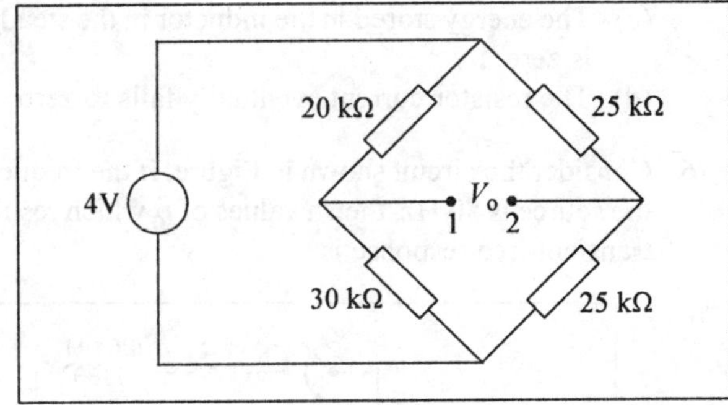

6.3 Transient Response

1. The time constant of the network shown in figure is

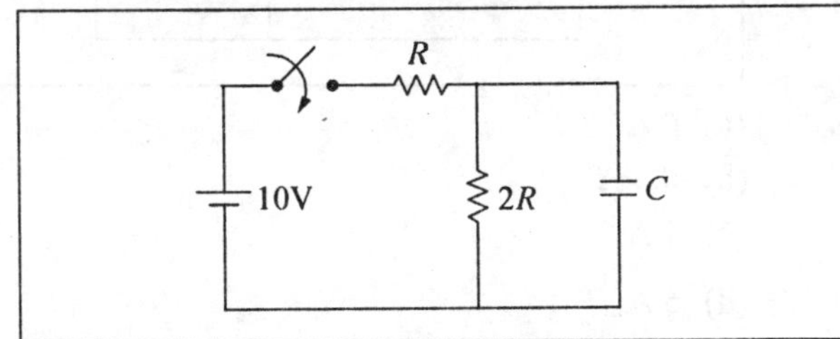

(a) 2RC (b) 3RC
(c) RC/2 (d) 2RC/3

2. In the circuit of Figure below the switch 'S' is closed at $t = 0$ with $i_L(0) = 0$ and $V_c(0) = 0$. In the **steady state** V_c equals.

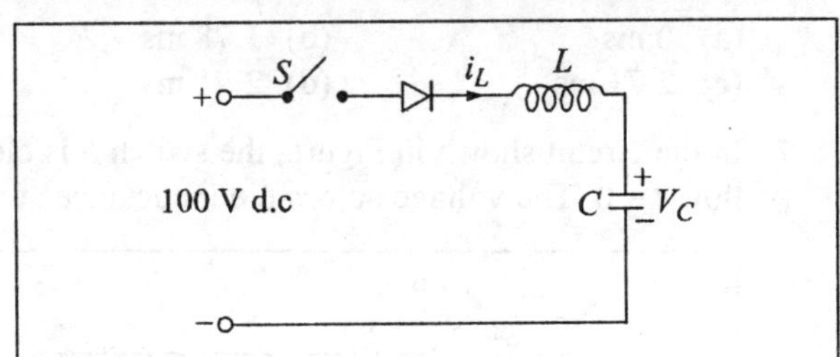

(a) 200 V (b) 100 V
(c) Zero (d) – 100

3. In the series RC circuit shown in Figure, the voltage across C starts increasing when the d.c. source is switched on. The rate of increase of voltage across C at the instant just after the switch is closed (i.e., at $t = 0^+$), will be

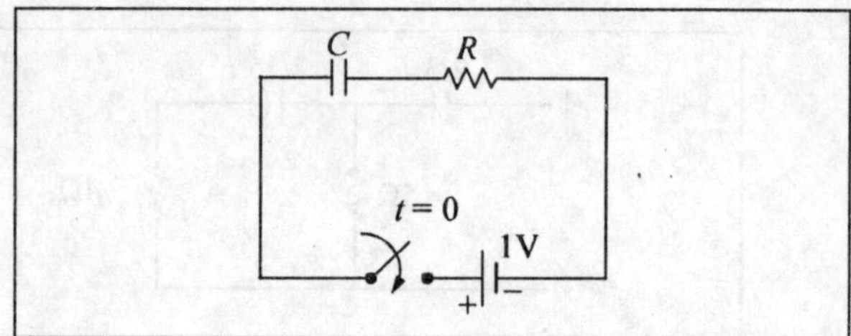

(a) Zero (b) Infinity
(c) RC (d) $1/RC$

4. A rectangular voltage pulse of magnitude V and duration T is applied to a series combination of resistance R and capacitance C. The maximum voltage developed across the capacitor is
(a) $V[1 - \exp(-T/RC)]$.
(b) VT/RC.
(c) V.
(d) $V \exp(-T/RC)$

5. A unit step voltage is applied at $t = 0$ to a series RL circuit with zero initial conditions.
(a) It is possible for the current to be oscillatory.
(b) The voltage across the resistor at $t = 0^+$ is zero.
(c) The energy stored in the inductor in the steady state is zero.
(d) The resistor current eventually falls to zero

6. Consider the circuit shown in Figure. If the frequency of the source is 50 Hz, then a values of t_0 which results in a transient free response is

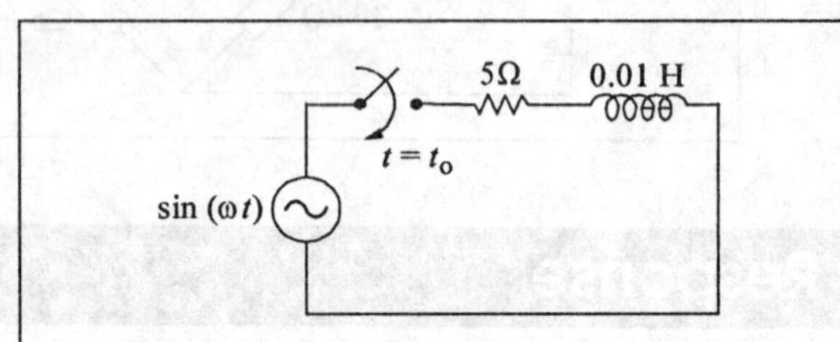

(a) 0 ms (b) 1.78 ms
(c) 2.71 ms (d) 2.91 ms

7. In the circuit shown in Figure, the switch S is closed at time $t = 0$. The voltage across the inductance at $= 0^+$, is

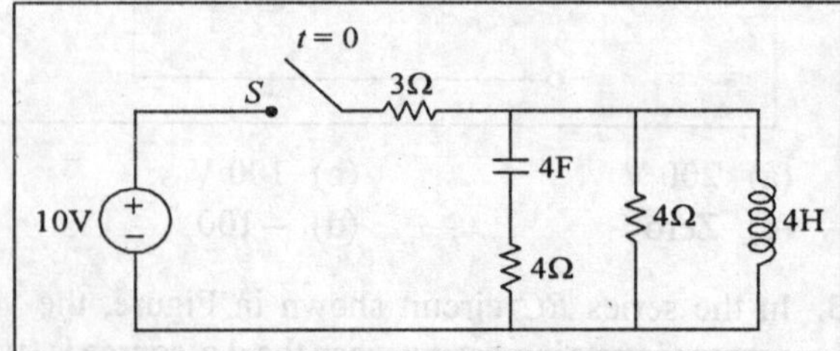

(a) 2 V (b) 4 V
(c) – 6 V (d) 8 V

8. In figure the capacitor initially has a charge of 10 Coulomb. The current in the circuit one second after the switch S is closed will be

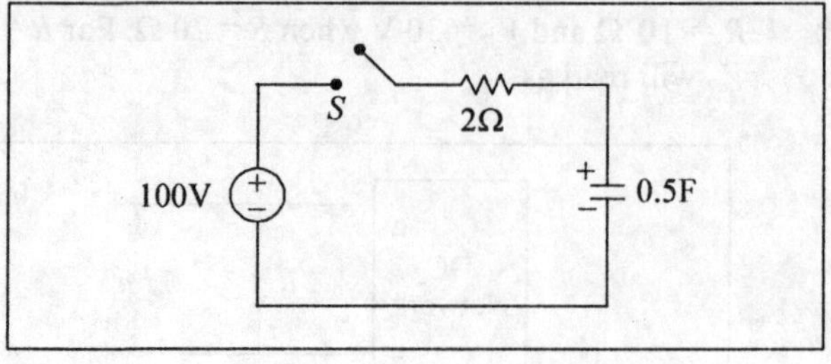

(a) 14.7 A (b) 18.5 A
(c) 40.0 A (d) 50.0 A

9. In the figure given for, the initial capacitor voltage is zero. The switch is closed at $t = 0$. The final steady-state voltage across the capacitor is

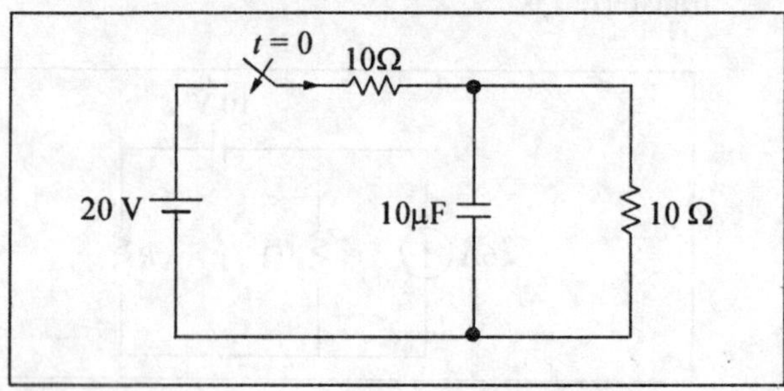

(a) 20 V (b) 10 V
(c) 5 V (d) 0 V

10. The circuit shown in the figure is in steady state, when the switch is closed at $t = 0$. Assuming that the inductance is ideal, the current through the inductor at $t = 0^+$ equals

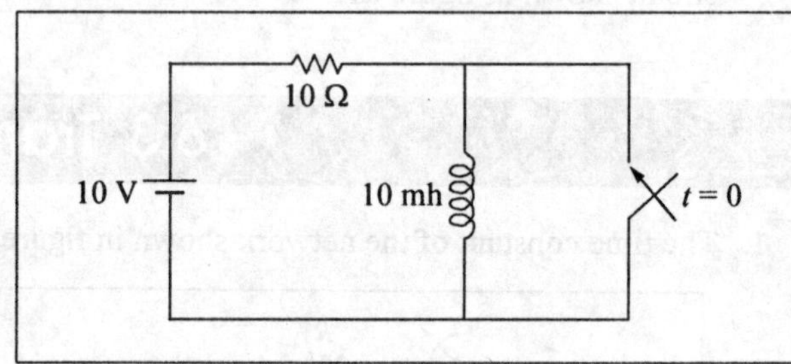

(a) 0 A
(b) 0.5 A
(c) 1 A
(d) 5 A

11. A coil of inductance 10 H and resistance 40 Ω is connected as shown in the figure. After the switch S has been in-contact with point 1 for a very long time, it is moved to point 2 at $t = 0$.
If, at $t = 0^+$, the voltage across the coil is 120 V, the value of resistance R is

(a) 0 Ω (b) 20 Ω
(c) 40 Ω (d) 60 Ω

12. A coil of inductance 10 H and resistance 40 Ω is connected as shown in the figure. After the switch S has been in-contact with point 1 for a very long time, it is moved to point 2 at $t = 0$.

For the value of obtained in (a), the time taken for 95% of the stored energy to be dissipated is close to

(a) 0.10 sec (b) 0.15 sec
(c) 0.50 sec (d) 1.0 sec

13. An ideal capacitor is charged to a voltage V_0 and connected at $t = 0$ across an ideal inductor L. (The circuit now consists of a capacitor and inductor alone). If we let $\omega_0 = \frac{1}{\sqrt{LC}}$, the voltage across the capacitor at time $t > 0$ is given by

(a) V_0
(b) $V_0 \cos(\omega_0 t)$
(c) $V_0 \sin(\omega_0 t)$
(d) $V_0 e^{-\omega_0 t} \cos(\omega_0 t)$

14. In the circuit shown in figure switch SWI is initially CLOSED and SW_2 is OPEN. The inductor L carries a current 10 A and the capacitor is charged to 10 V with polarities as indicated. SW_2 is initially CLOSED at $t = 0^-$ and SW_1 is OPENTED is $t = 0$. The current through C and the voltage across L at $t = 0^+$ is

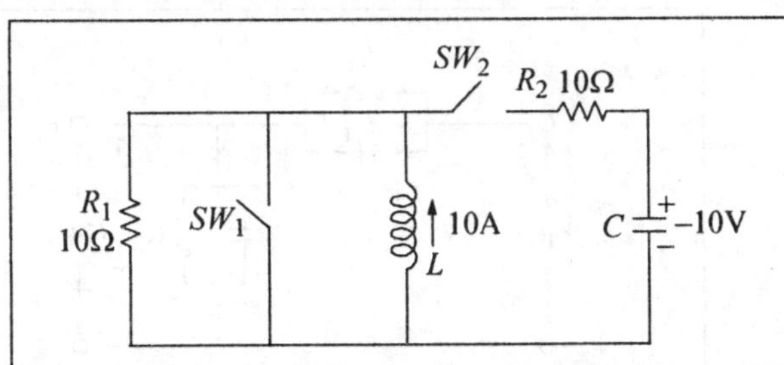

(a) 55 A, 4.5 V
(b) 5.5 A, 45 V
(c) 45 A, 5.5 V
(d) 4.5 A, 55 V

15. The time constant for the given circuit will be

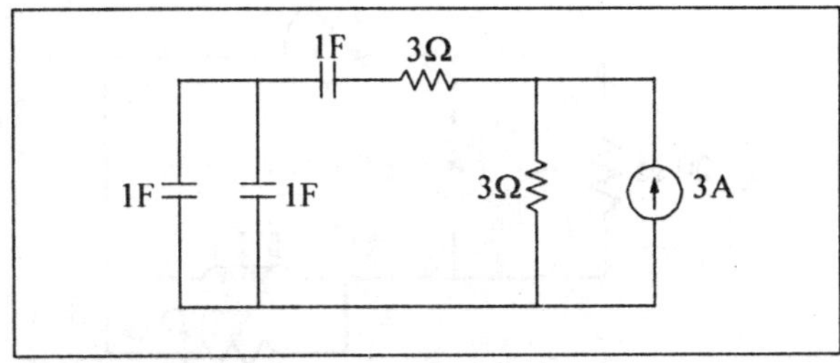

(a) 1/9 s (b) 1/4 s
(c) 4 s (d) 9 s

16. The switch in the circuit has been closed for a long time. It is opened at $t = 0$, At $t = 0^+$, the current through the 1 μF capacitor is

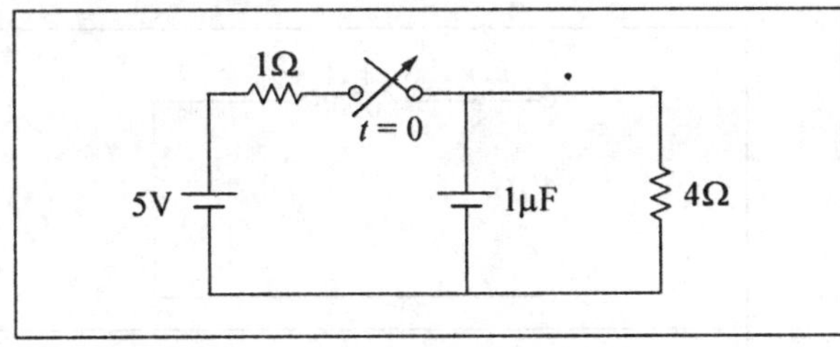

(a) 0 A (b) 1 A
(c) 1.25 A (d) 5 A

17. In the figure shown , all elements used are ideal. For time $t < 0$, S_1 remained closed and S_2 open . At $t = 0$, S_1 is opened and S_2 is closed . If the volt age V_{c2} across the capacitor C_2 at $t = 0$ is zero, the voltage across the capacitor combination at $t = 0^+$ will be .

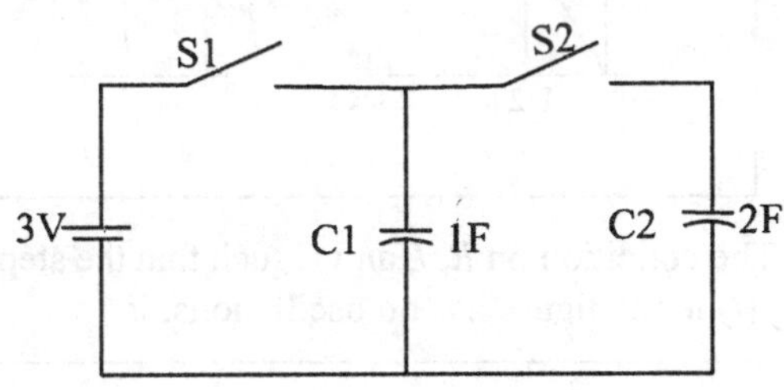

(a) 1 V (b) 2 V
(c) 1·5 V (d) 3 V

18. A ramp voltage, $v(t) = 100\,t$ volts , is applied to an RC differentiating circuit with R = 5 k W and C = 4 μF. The maximum output voltage is

(a) 0.2 volt
(b) 2.0 volts
(c) 10.0 volt
(d) 50.0 volts

19. In Figure, the switch was closed for a long time before opening at t = 0. The voltage V_x at $t = 0^+$ is

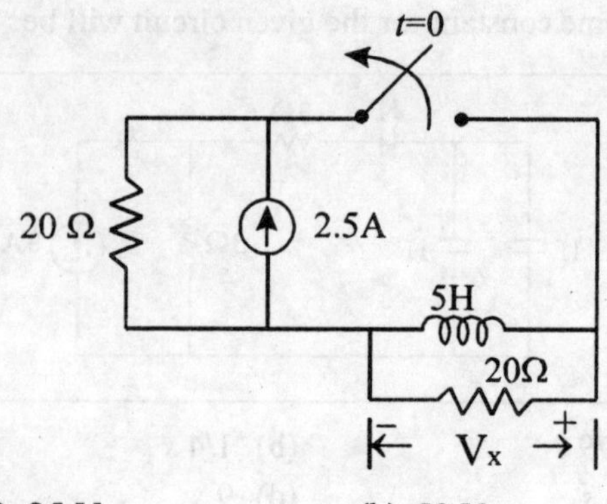

(a) 25 V (b) 50 V
(c) –50 V (c) 0 V

20. For the R-L circuit shown in Figure the input voltage $V_i(t) = u(t)$. The current $i(t)$ is

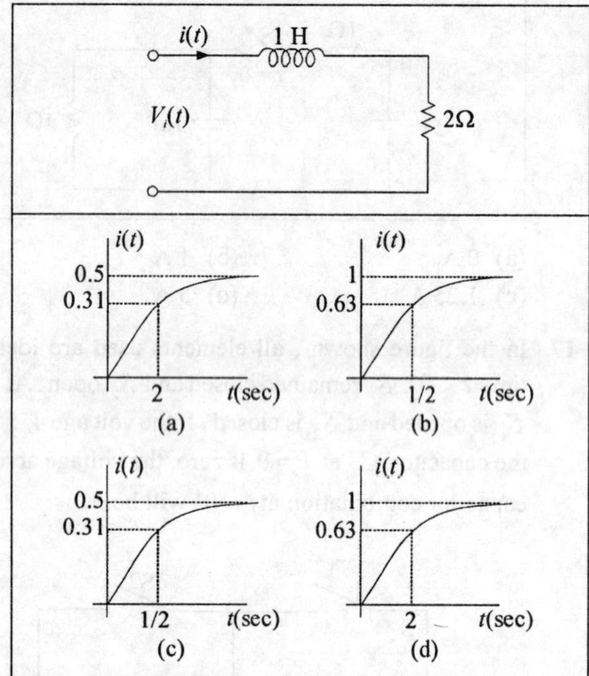

21. The condition on R, L and C such that the step response $y(t)$ in the figure has no oscillations, is

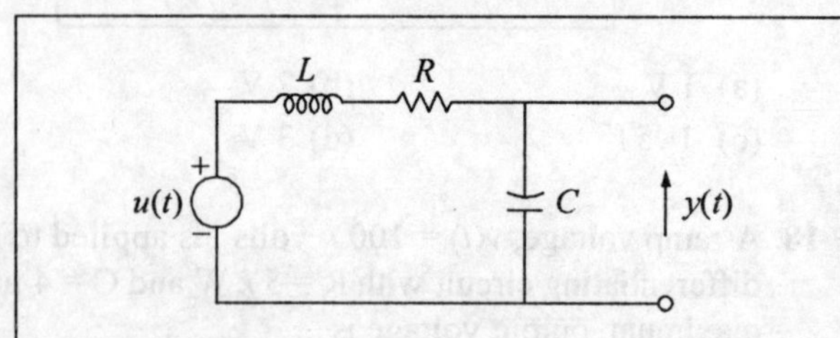

(a) $R \geq [1/2]\sqrt{(L/C)}$
(b) $R \geq \sqrt{(L/C)}$
(c) $R \geq 2\sqrt{(L/C)}$
(d) $R = [1/\sqrt{(LC)}]$

22. In the following circuit, the switch S is closed at $t = 0$. The rate of change of current $\frac{di}{dt}(0^+)$ is given by

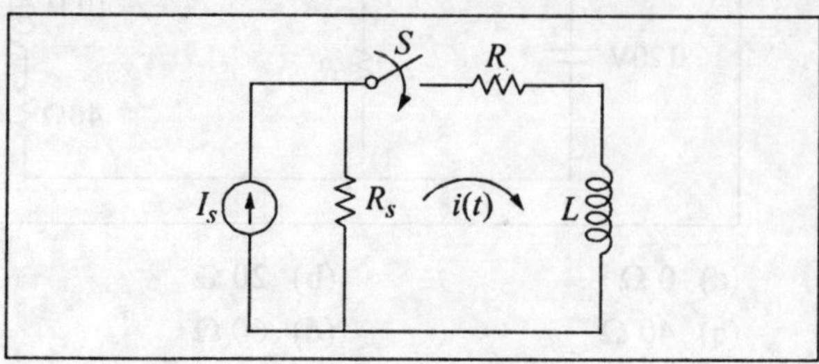

(a) 0 (b) $\frac{R_S I_S}{L}$
(d) $\frac{(R+R_S)I_S}{L}$ (d) ∞

23. If the unit step response of a network is $(1 - e^{-\alpha t})$, then tits unit impulse response is
(a) $\alpha e^{-\alpha t}$ (b) $\alpha^{-1}e^{-\alpha t}$
(c) $(1 - \alpha^{-1})e^{-\alpha t}$ (d) $(1-\alpha)e^{-\alpha t}$

24. In the following figure, C_1 and C_2 are ideal capacitors. C_1 has been charged to $12V$ before the ideal switch S is closed at $t = 0$. The current $i(t)$ for all t is

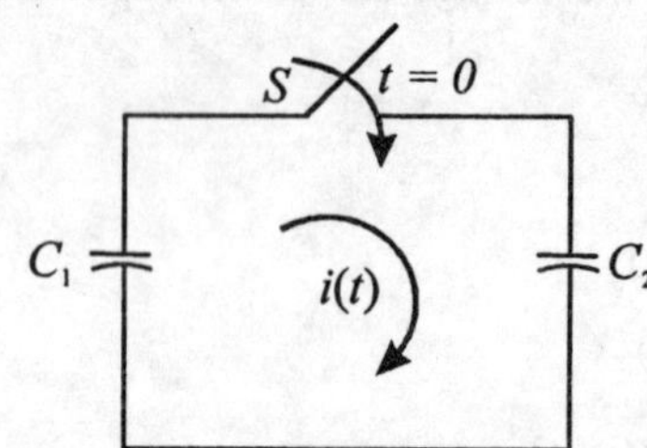

(a) Zero
(b) A step function
(c) An exponentially decaying function
(d) An impulse function

25. For the compensated attenuator of Figure The impulse response under the condition $R_1C_1 = R_2C_2$ is

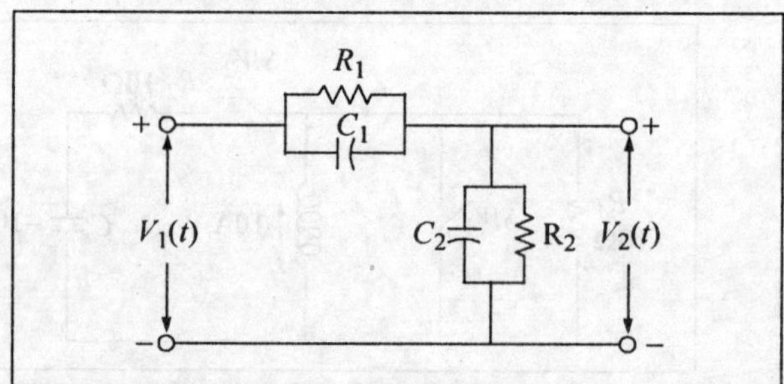

(a) $\frac{R_2}{R_1 + R_2}[1 - e^{(-1/R_1C_1)}]u(t)$

(b) $\frac{R_2}{R_1 + R_2}\delta(t)$

(c) $\frac{R_2}{R_1+R_2}u(t)$

(d) $\frac{R_2}{R_1+R_2}[1-e^{(1/R_1C_1)}]\cdot u(t)$

26. The voltage V_{C1}, V_{C2}, and V_{C3} across the capacitors in the circuit in Figure under steady state, are respectively

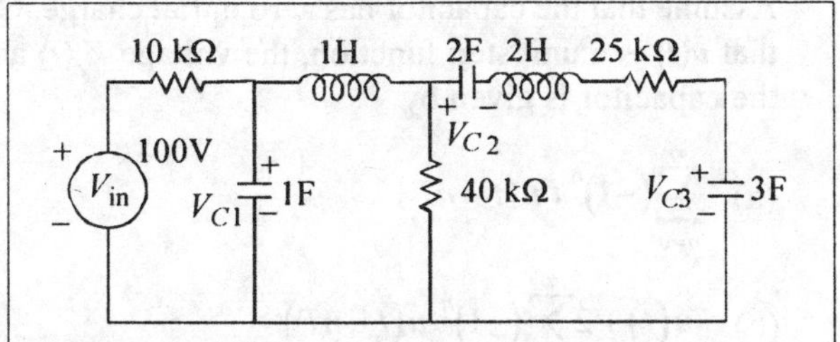

(a) 80 V, 32 V, 48 V (b) 80 V, 48 V, 32 V
(c) 20 V, 8 V, 12 V (d) 20 V, 12 V, 8 V

Common Data Questions 27 and 28

For the circuit is given below assume that the switch S is in position 1 for a long time and thrown to position 2 at $t = 0$.

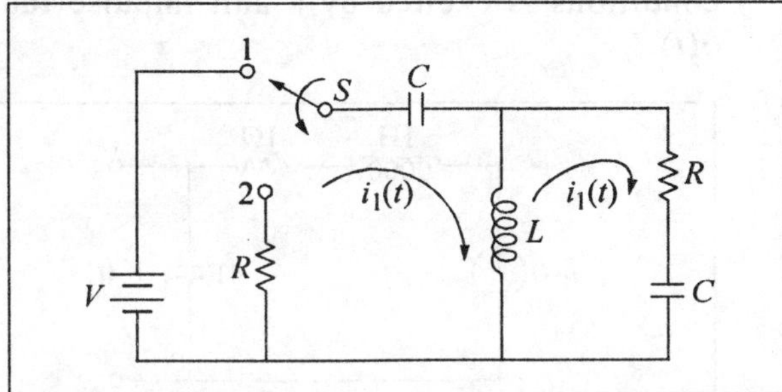

27. At $t = 0^+$, current i_1 is

(a) $\frac{-V}{2R}$ (b) $\frac{-V}{R}$

(c) $\frac{-V}{4R}$ (d) Zero

28. $I_1(s)$ and $I_2(s)$ are the Laplace transforms of $i_1(t)$ and $i_2(t)$ respectively. The equations for the loop currents $I_1(s)$ and $I_2(s)$ for the circuit shown in figure after the switch is brought from position 1 to position 2 at = 0, are

(a) $$\begin{pmatrix} R+Ls+\frac{1}{Cs} & -Ls \\ -Ls & R+\frac{1}{Cs} \end{pmatrix}\begin{pmatrix} I_1(s) \\ I_2(s) \end{pmatrix}=\begin{pmatrix} V/s \\ 0 \end{pmatrix}$$

(b) $$\begin{pmatrix} R+Ls+\frac{1}{Cs} & -Ls \\ -Ls & R+Ls+\frac{1}{Cs} \end{pmatrix}\begin{pmatrix} I_1(s) \\ I_2(s) \end{pmatrix}=\begin{pmatrix} -V/s \\ 0 \end{pmatrix}$$

(c) $$\begin{pmatrix} R+Ls+\frac{1}{Cs} & -Ls \\ -Ls & R+\frac{1}{Cs} \end{pmatrix}\begin{pmatrix} I_1(s) \\ I_2(s) \end{pmatrix}=\begin{pmatrix} -V/s \\ 0 \end{pmatrix}$$

(d) $$\begin{pmatrix} R+Ls+\frac{1}{Cs} & -Ls \\ -Ls & R+Ls+\frac{1}{Cs} \end{pmatrix}\begin{pmatrix} I_1(s) \\ I_2(s) \end{pmatrix}=\begin{pmatrix} -V/s \\ 0 \end{pmatrix}$$

29. The circuit shown in Figure has initial current $i_L(0^-) = 1$ A through the inductor and an initial voltage $v_c(0^-) = -1$V across the capacitor. For input $v(t) = u(t)$, the Laplace transform of the current $i(t)$ for $t \geq 0$ is

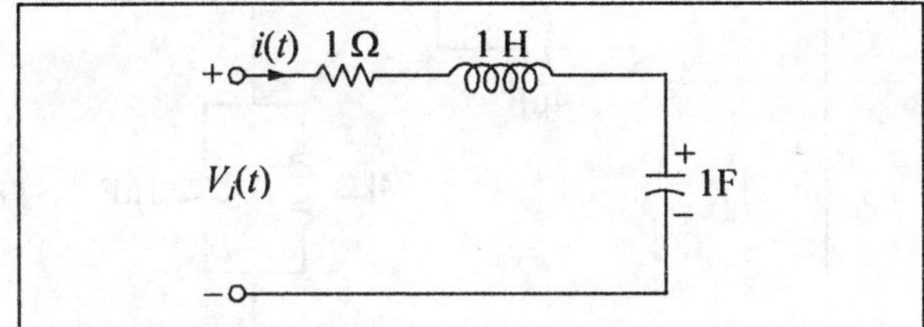

(a) $\frac{s}{s^2+s+1}$ (b) $\frac{s+2}{s^2+s+1}$

(c) $\frac{s-2}{s^2+s+1}$ (d) $\frac{s-2}{s^2+s+1}$

30. A square of 3 volts amplitude is applied to C-R circuit shown in the figure. The capacitor is initially uncharged. The output voltage V_2 at time $t = 2$ sec is

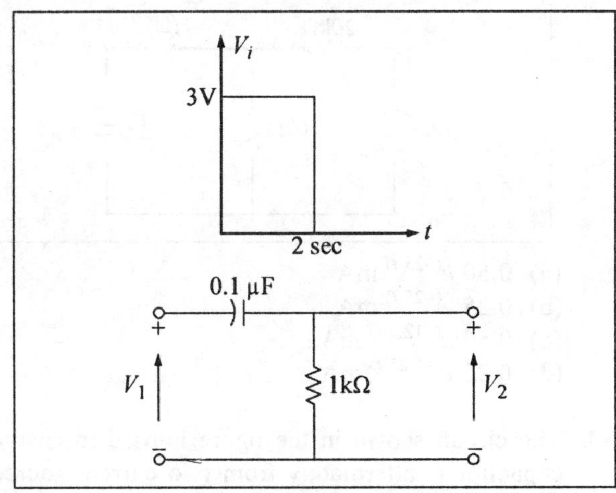

(a) 3 V (b) – 3 V
(c) 4 V (d) – 4 V

31. A 2mH inductor with some initial current can be represented as shown below, where s is the Laplace Transform variable. The value of initial current is

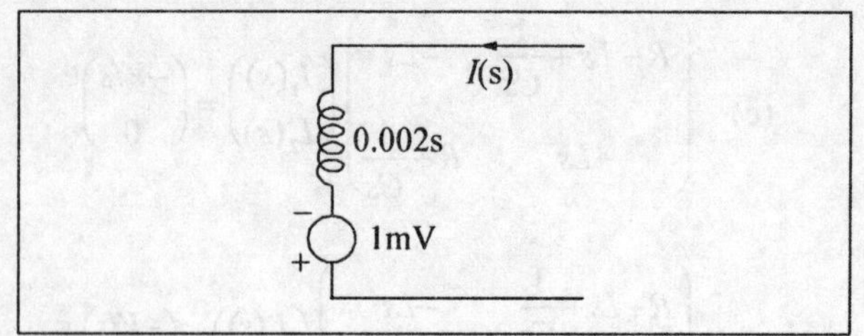

(a) 0.5 A (b) 2.0 A
(c) 1.0 A (d) 0.0 A

32. In the figure shown below, assume that all the capacitors are initially uncharged. If $v_1(t) = 10u(t)$ Volts, $V_o(t)$ is given by

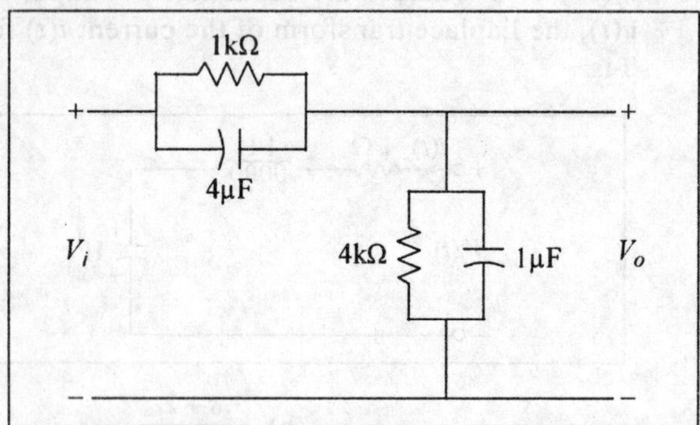

(a) $8e^{-0.004t}$ Volts (b) $8(1 - e^{+0.004t})$ Volts
(c) $8u(t)$ Volts (d) 8 Volts

33. In the circuit shown, V_C is 0 volts at $t = 0$ sec. For $t > 0$, the capacitor current $i_c(t)$, where t is in seconds, is given by

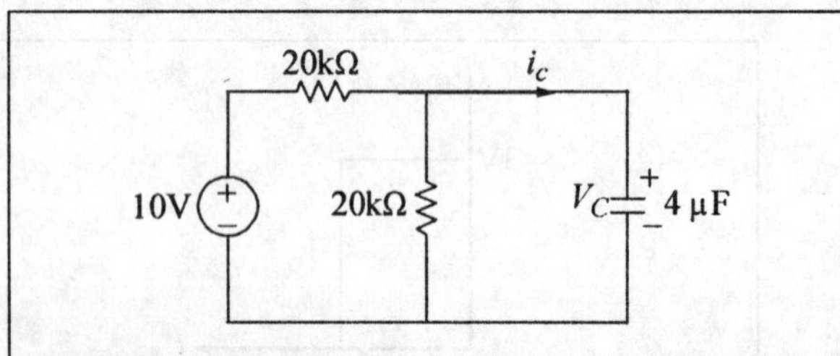

(a) $0.50\, e^{(-25\,t)}$ mA
(b) $0.25\, e^{(-25\,t)}$ mA
(c) $0.50\, e^{(-12.5\,t)}$ mA
(d) $0.25\, e^{(-6.25\,t)}$ mA

34. The circuit shown in the figure is used to charge the capacitor C altermately from two current sources as indicated. The switches S_1 and S_2 are mechanically coupled and connected as follows:

For $2nT \le t < (2n+1)T, (n = 0, 1, 2, ...)$

S_1 to P_1 and S_2 to P_2

For $(2n+1)T \le t < (2n+2)T, (n = 0, 1, 2, ...)$

S_1 to Q_1 and S_2 to Q_2

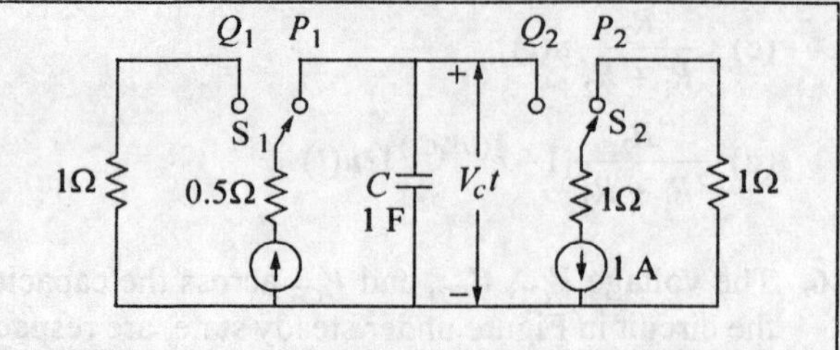

Assume that the capacitor has zero initial charge. Given that $u(t)$ is a unit step function, the voltage $V_c(t)$ across the capacitor is given by

(a) $\sum_{n=0}^{\infty} (-1)^n\, t\, u(t - nT)$

(b) $u(t) + 2\sum_{n=1}^{\infty} (-1)^n\, u(t - nT)$

(c) $t\, u(t) + 2\sum_{n=1}^{\infty} (-1)^n (t - nT) u(t - nT)$

(d) $\sum_{n=1}^{\infty} \left[0.5 - e^{-(t-2nT)} + 0.5 e^{-(t-2nT-T)} \right]$

35. The folloiwng series RLC cricuit with zero initial conditions is excited by a unit impulse function $\delta(t)$

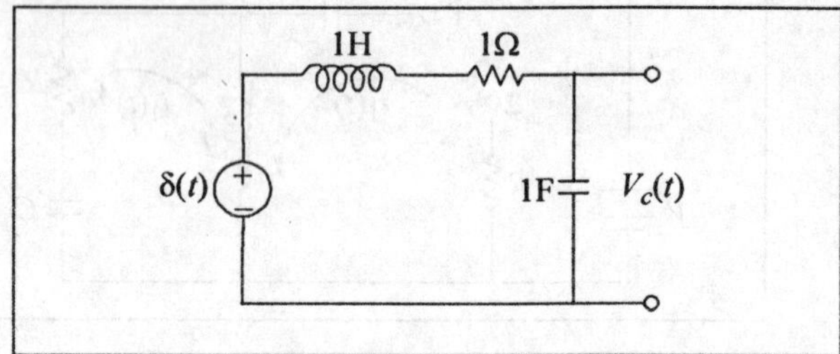

For $t > 0$, the output voltage $V_c(t)$ is

(a) $\frac{2}{\sqrt{3}}\left(e^{-\frac{1}{2}t} - e^{\frac{\sqrt{3}}{2}t}\right)$ (b) $\frac{2}{\sqrt{3}} t\, e^{-\frac{1}{2}t}$

(c) $\frac{2}{\sqrt{3}} e^{-\frac{1}{2}t} \cos\left(\frac{\sqrt{3}}{2}t\right)$ (d) $\frac{2}{\sqrt{3}} e^{-\frac{1}{2}t} \sin\left(\frac{\sqrt{3}}{2}t\right)$

36. The folloiwng series RLC cricuit with zero initial conditions is excited by a unit impulse function $\delta(t)$

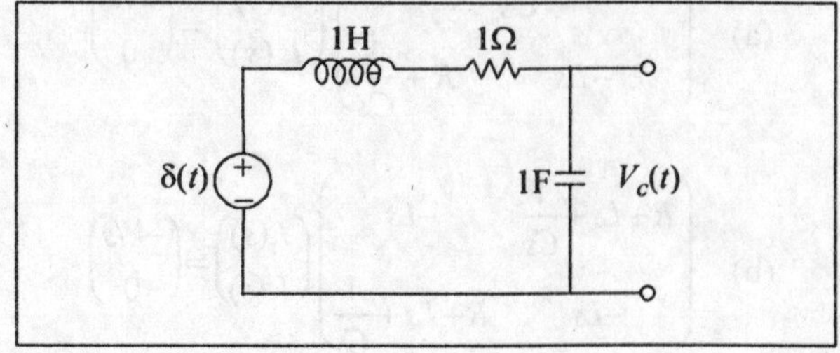

For $t > 0$, the voltage across the resistor is

(a) $\frac{1}{\sqrt{3}}\left(e^{-\frac{\sqrt{3}}{2}t} - e^{-\frac{1}{2}t}\right)$

(b) $e^{-\frac{1}{2}t}\left[\cos\left(e^{-\frac{\sqrt{3}}{2}t}\right) - \frac{1}{\sqrt{3}}\sin\left(\frac{\sqrt{3}\,t}{2}\right)\right]$

(c) $\frac{2}{\sqrt{3}}e^{-\frac{1}{2}t}\sin\left(\frac{\sqrt{3}\,t}{2}\right)$

(d) $\frac{2}{\sqrt{3}}e^{\frac{1}{2}t}\cos\left(\frac{\sqrt{3}\,t}{2}\right)$

37. The switch in the circuit shown was on position ***a*** for a long time, and is moved position ***b*** at time $t = 0$. The current $i(t)$ for $t > 0$ is given by

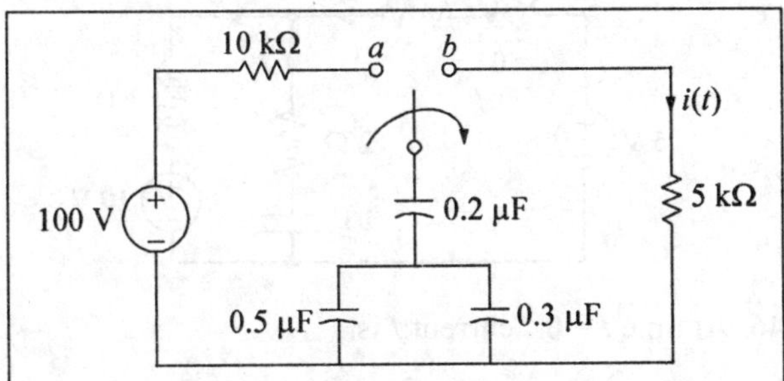

(a) $0.2e^{-125t}u(t)$ mA (b) $20e^{-1250t}u(t)$ mA

(c) $0.2e^{-1250t}u(t)$ mA (d) $20e^{-1000t}u(t)$ mA

38. The time domain behavior of an RL circuit is represented by

$$L\frac{di}{dt} + Ri = V_0\left(1 + Be^{-Rt/L}\sin t\right)u(t)$$

For an initial current of $i(0) = \frac{V_0}{R}$, the steady state value of the currents is given by

(a) $i(t) \to \frac{V_0}{R}$ (b) $i(t) \to \frac{2V_0}{R}$

(a) 1 V (b) 2 m $i(t) \to \frac{2V_0}{R}(1 + B)$

39. In the circuit shown, the switch S is open for a long time and is closed at $t = 0$.The current $i(t)$ for $t \geq 0^+$ is

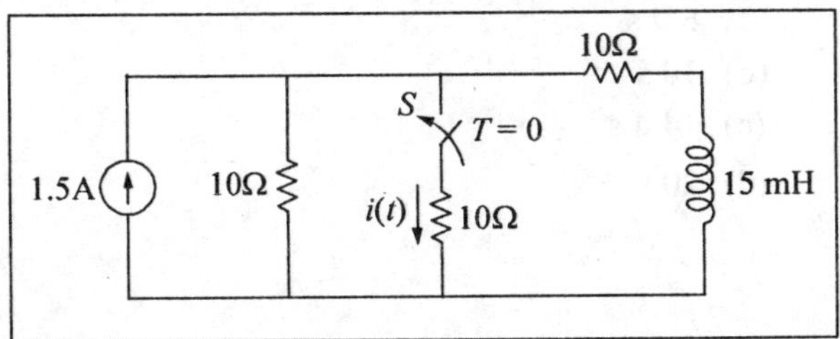

(a) $i(t) = 0.5 - 0.125e^{-1000t}$ A

(b) $i(t) = 1.5 - 0.125e^{-1000t}$ A

(c) $i(t) = 0.5 - 0.5e^{-1000t}$ A

(d) $i(t) = 0.375e^{-1000t}$ A

40. In the circuit shown below, the initial charge on the capacitor is 2.5mC, with the voltage polarity as indicated. The switch is closed at time t = 0. The current i(t) at a time t after the switch is closed is

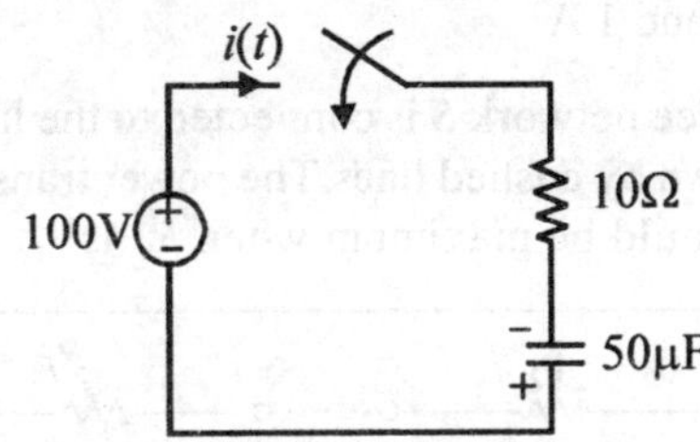

(a) $i(t) = 15\exp(-2 \times 10^3\,t)$A

(b) $i(t) = 5\exp(-2 \times 10^3\,t)$A

(c) $i(t) = 10\exp(-2 \times 10^3\,t)$A

(d) $i(t) = -5\exp(-2 \times 10^3\,t)$A

41. In the circuit shown in the figure, the switch S is closed at time $t = 0$. The supply current at $t = 0^+$ and the capacitor voltage at $t \to \infty$ are, respectively

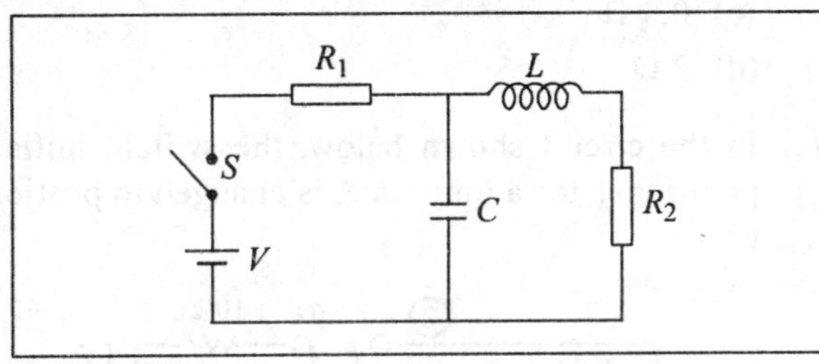

(a) 0 and V

(b) $\frac{V}{R_1}$ and $\frac{VR_2}{R_1 + R_2}$

(c) $\frac{V}{R_1 + R_2}$ and $\frac{VR_2}{R_1 + R_2}$

(d) $\frac{V}{R_1}$ and V

42. In the circuit shown in the following figure, the switch is kept closed for a long time and then opened at $t = 0$.

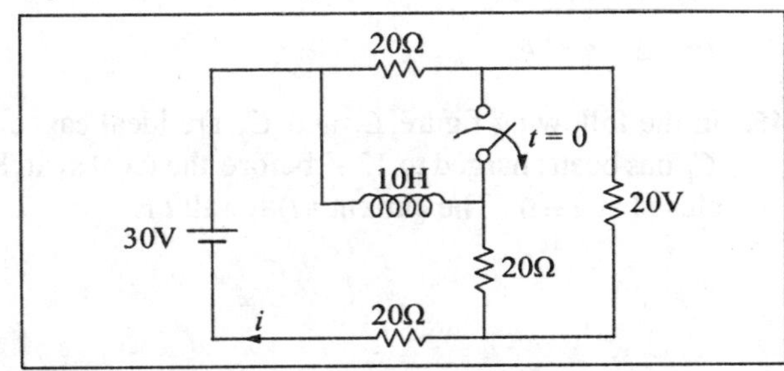

The value of the current i just before opening the switch ($t = 0^-$) and just a after opening the switch ($t = 0^+$) are, respectively

(a) $\frac{3}{4}$ A and 1 A

(b) and $\frac{5}{2}$ A

(c) 1 A and $\frac{7}{6}$ A

(d) 1 A and 1 A

43. The source network S is connected to the load network L as shown by dashed lines. The power transferred from S to L would be maximum when R_L is

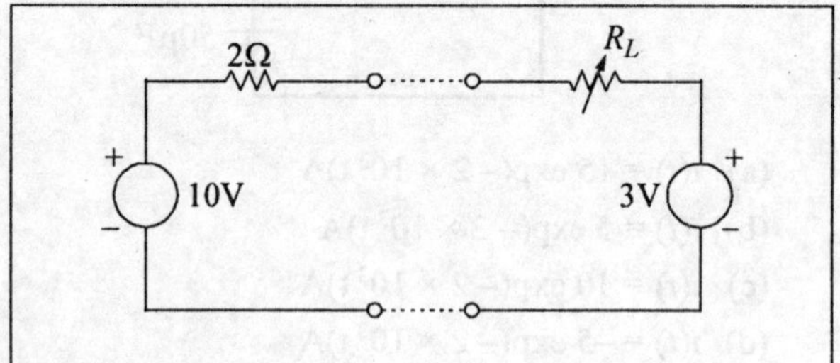

(a) 0 Ω
(b) 0.6 Ω
(c) 0.8 Ω
(d) 2 Ω

44. In the circuit shown below, the switch, initially at position 1 for a long time, is changed to postion 2 at t = 0.

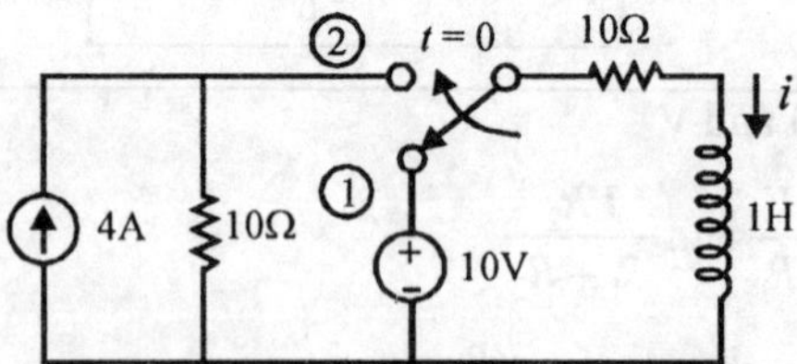

The current i through the inductor for $t \geq 0$ is

(a) $1 - e^{-20t}$ A

(b) $1 + e^{-20t}$ A

(c) $1 + 2e^{-20t}$ A

(d) $2 - e^{-20t}$ A

45. In the following figure, C_1 and C_2 are ideal capacitors. C_1 has been charged to 12 V before the ideal switch S is closed at $t = 0$. The current $i(t)$ for all t is

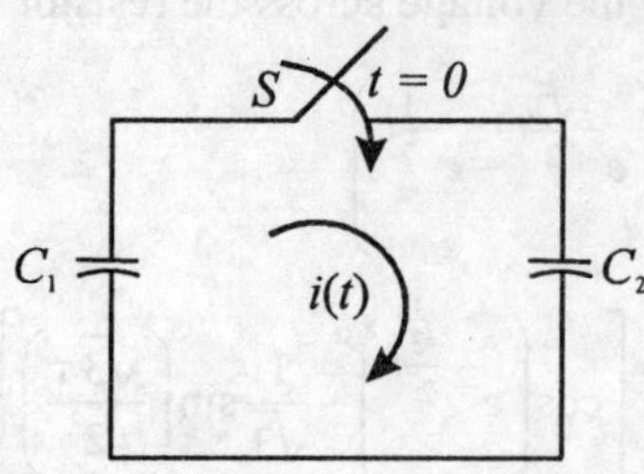

(a) Zero
(b) A step function
(c) An exponentially decaying function
(d) An impulse function

Statements for Linked Answer Questions 46 and 47

In the circuit shown below the steady-state is reached with the switch K open. Subsequently the switch is closed at time $t = 0$.

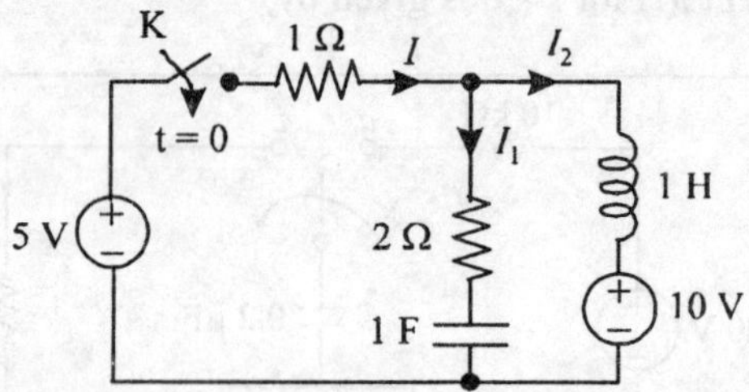

46. At time $t = 0^+$, current I is.

(a) $-\frac{5}{3}A$

(b) 0 A

(c) $\frac{5}{3}A$

(d) ∞ A

47. At $t = 0^+$, $\frac{dI_2}{dt}$ is.

(a) $-5\ A/s$
(b) $-\frac{10}{3}\ A/s$
(c) 0 A/s
(d) 5 A/s

48. A system with transfer function $[1/(s\tau + 1)]$ subjected to a step input takes 10 seconds to reach 50% of the step height. The value of τ is

(a) 6.9 s
(b) 10 s
(c) 14.4 s
(d) 20 s

6.4 Sinusoidal Steady State

1. In the following circuit Figure $i(t)$ under steady state is

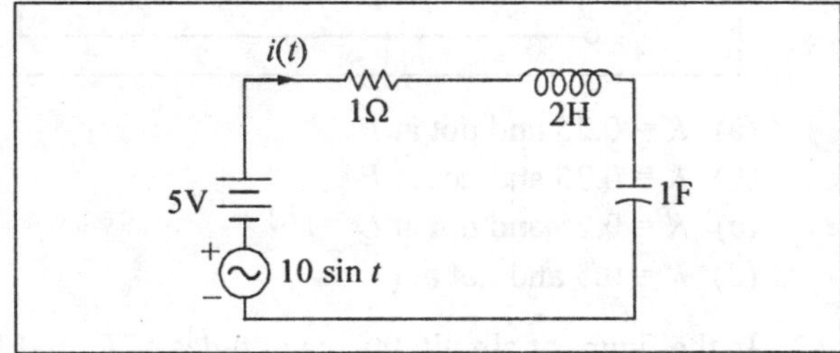

(a) Zero
(b) 5
(c) 7.07 sin t
(d) 7.07 sin (t – 45°)

2. The following circuit resonates at

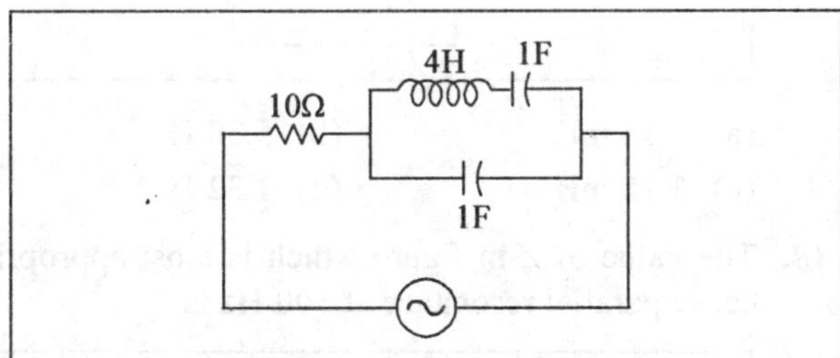

(a) All frequencies (b) 0.5 rad/sec
(c) 5 rad/sec (d) 1 rad/sec

3. At **resonance,** the **parallel circuit** of Figure below constituted by an iron-cored coil and a capacitor behaves like

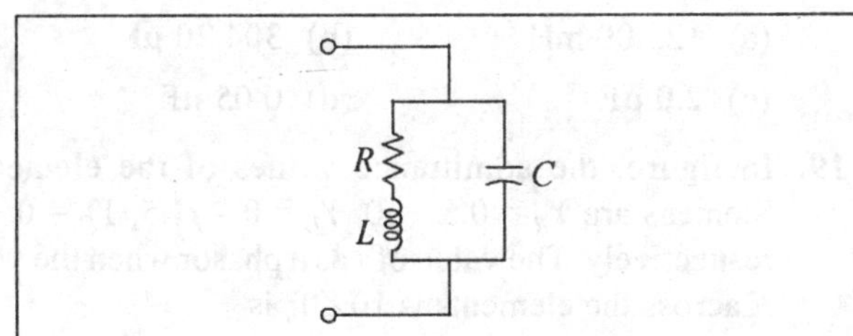

(a) An open-circuit
(b) A short-circuit
(c) A pure resistor of value R
(d) A pure resistor of value much higher than R

4. A sinusoidal source of voltage V and frequency f is connected to a series circuit of variable resistance R and a fixed reactance X. The locus of the tip of the **current-phasor** I, as R is varied from 0 to ∞ is

(a) A semicircle with diameter of V/X.
(b) A straight line with slope of R/X.
(c) An ellipse with V/R as major axis.
(d) A circle of radius R/X and origin at (0, $V/2$)

5. A circuit with a resistor, inductor and capacitor in **series** is **resonant** of f_0 Hz. If all the component values are now doubled, the new **resonant** frequency is

(a) $2f_0$ (b) Still f_0
(c) $f_0/4$ (d) $f_0/2$

6. A series *R-L-C* circuit when excited by a 10 V sinusoidal voltage source of variable frequency, exhibits resonance at 100 Hz and a 3 dB bandwidth of 5 Hz. The voltage across the inductor L at resonance is

(a) 10 V. (b) $10\sqrt{2}$ V.
(c) $[10\sqrt{2}]$ V (d) 200 V

7. The RMS value of a half-wave rectified symmetrical square wave current of 2 A is

(a) $\sqrt{2}$ A (b) 1 A.
(c) $1\sqrt{2}$ A (d) $\sqrt{3}$A.

8. The current in the circuit shown Figure is

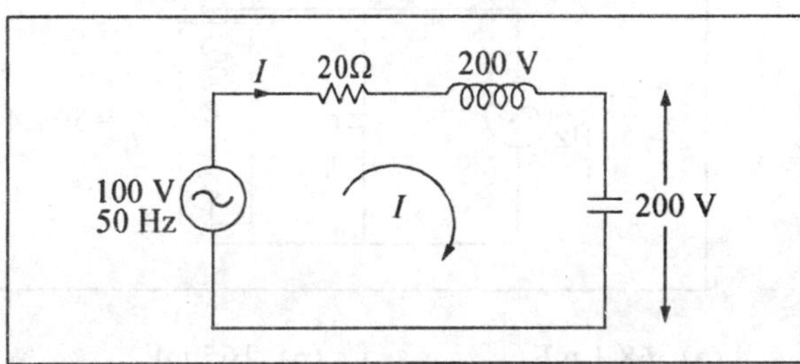

(a) 5 A (b) 10 A
(c) 15 A (d) 25 A

9. A fixed capacitor of reactance $-j\,0.02\ \Omega$ is connected in parallel across a series combination of a fixed inductor of reactance $j\,0.01\ \Omega$ and a variable resistance R. As R varied from zero to infinity, the locus diagram of the admittance of this *L-C-R* circuit will be

(a) A semi-circle of diameter j 100 and center at zero
(b) A semi circle of diameter j 50 and center at zero
(c) A straight line inclined at an angle.
(d) A straight line parallel to the x-axis.

10. The voltage phasor of a circuit is 10 ∠15° V and the current phasor is 2 ∠–45° A. The active and the reactive powers in the circuit are

(a) 10 W and 17.32 VAR
(b) 5 W and 8.66 VAr
(c) 20 W and 60 VAR
(d) $20\sqrt{2}$ W and $10\sqrt{2}$ VaR

11. The impedance seen by the source in the circuit in figure is given by

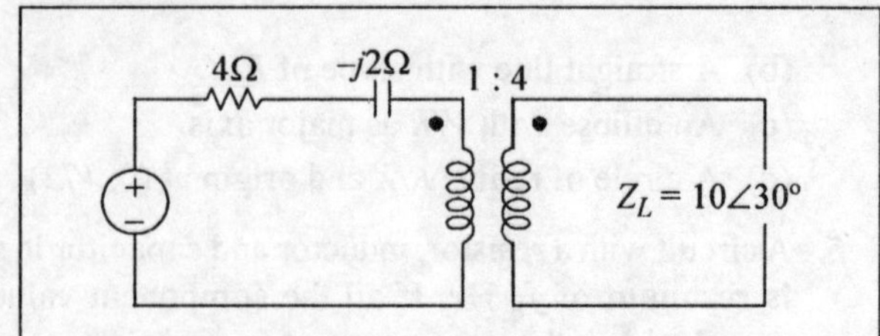

(a) $0.54 + j0.313$ ohms (b) $4 - j2$ ohms
(c) $4.54 - j1.69$ ohms (d) $4 + j2$ ohms

12. In a series *RLC* circuit at resonance, the magnitude of the voltage developed across the capacitor

(a) Is always zero
(b) Can never be greater than the input voltage
(c) Can be greater than the input voltage however, it is 90° out of phase with input voltage
(d) Can be greater than the input voltage, and is in phase with the input voltage.

13. In the circuit shown in Figure, what value of *C* will cause a unity power factor at the ac source?

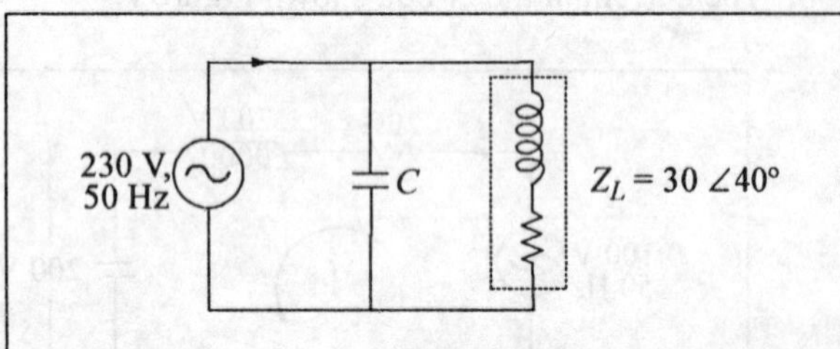

(a) 68.1 mF (b) 165 mF
(c) 0.681 mF (d) 6.81 mF

14. A first order, low pass filter is given with $R = 50\ \Omega$ and $C = 5\ \mu F$. What is the frequency at which the gain of the voltage transfer of the filter is 0.25?

(a) 4.92 kHz (b) 2.46 kHz
(c) 0.49 kHz (d) 24.6 kHz

15. A series *R-L-C* circuit has $R = 50\ \Omega$; $L = 100\ \mu H$ and $C = 1\ \mu F$. The lower half-power frequency of the circuit is

(a) 30.55 kHz (b) 3.055 kHz
(c) 51.92 kHz (d) 1.92 kHz

16. In the circuit shown in Figure, it is found that the input a.c. voltage (v_i) and current *i* are in phase. The coupling coefficient is $K = \left[M/\sqrt{L_1 L_2}\right]$, where *M* is mutual inductance between the two coils. The value of *K* and the dot polarity of the coil *P-Q* are

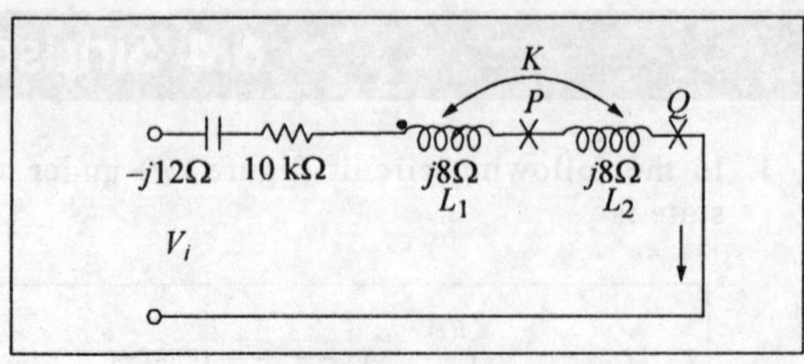

(a) $K = 0.25$ and dot at *P*
(b) $K = 0.25$ and dot at *P*
(c) $K = 0.25$ and dot at *Q*
(d) $K = 0.5$ and dot at *Q*

17. In the figure of circuit, the magnitudes of V_L and V_C are twice that of V_R. The inductance of the coil is

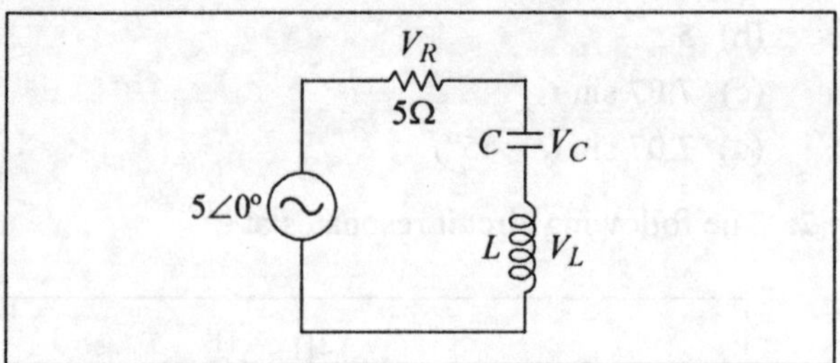

(a) 2.14 mH (b) 5.30 H
(c) 3.18 mH (d) 1.32 H

18. The value of *Z* in figure which is most appropriate to cause parallel resonance at 500 Hz is

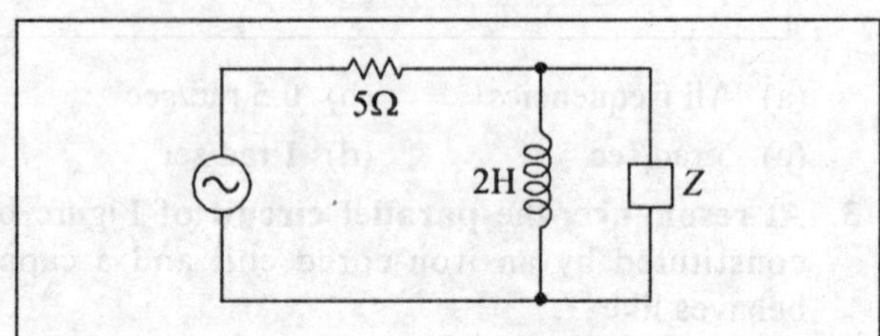

(a) 125.00 mH (b) 304.20 μF
(c) 2.0 μF (d) 0.05 μF

19. In figure, the admittance values of the elements in Siemens are $Y_R = 0.5 + j0$, $Y_L = 0 - j1.5$, $Y_C = 0 + j0.3$ respectively. The value of *I* as a phasor when the voltage *E* across the elements is $10\angle 0°$ is

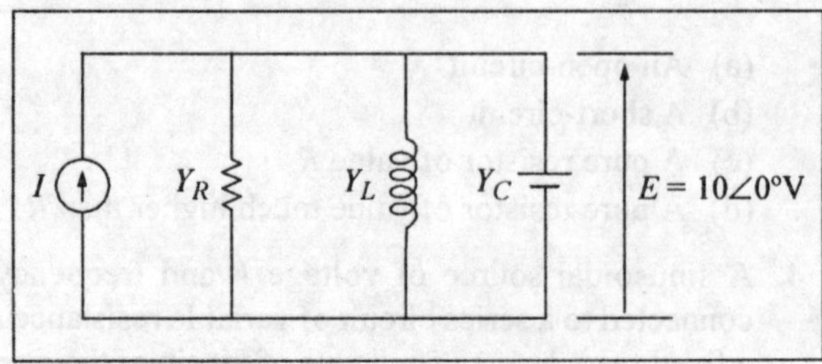

(a) $1.5 + j0.5$ (b) $5 - j18$
(c) $0.5 + j1.8$ (d) $5 - j12$

20. The *RL* circuit of the figure is fed from a constant magnitude, variable frequency sinusoidal voltage source V_{in}. At 100 Hz, the *R* and *L* elements each have a voltage drop V_{rms}. If the frequency of the source is changed to 50 Hz, then new voltage drop across *R* is

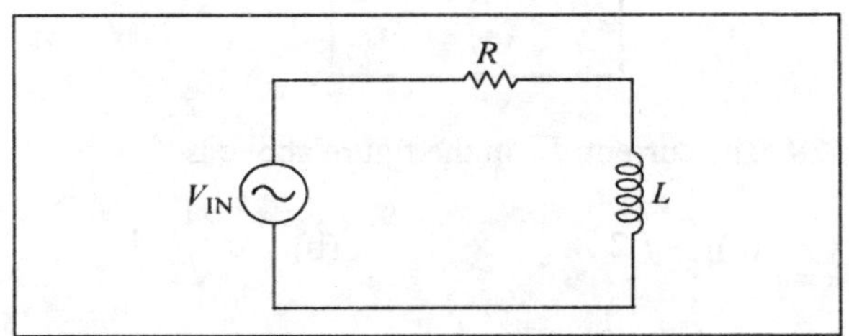

(a) $\sqrt{\frac{5}{8}}\, u_{rms}$ (b) $\sqrt{\frac{2}{3}}\, u_{rms}$

(c) $\sqrt{\frac{8}{5}}\, u_{rms}$ (d) $\sqrt{\frac{3}{2}}\, u_{rms}$

21. The circuit shown in the figure is energized by a sinusoidal voltage source V_1 at a frequency which causes resonance with a current of *I*.

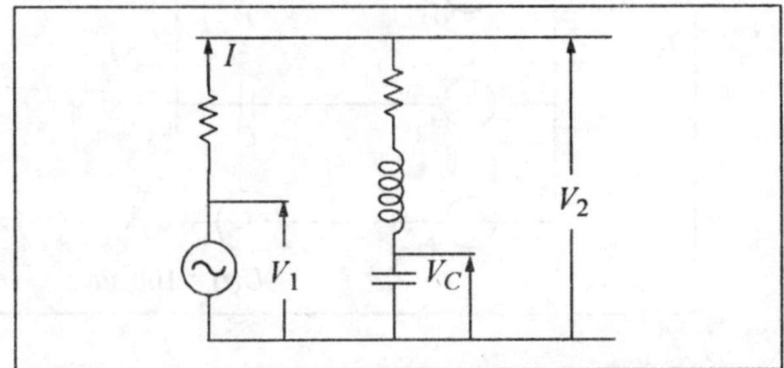

The phasor diagram which is applicable to this circuit is

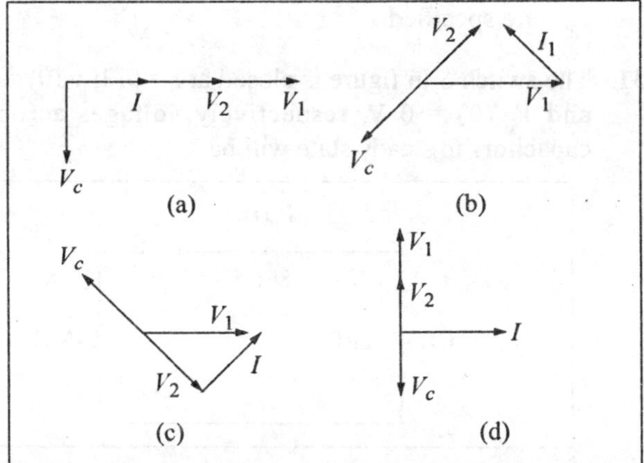

22. The R-L-C series circuit shown is supplied from a variable frequency voltage source. The admittance-locus of the R-L-C network at terminals *AB* for increasing frequency is

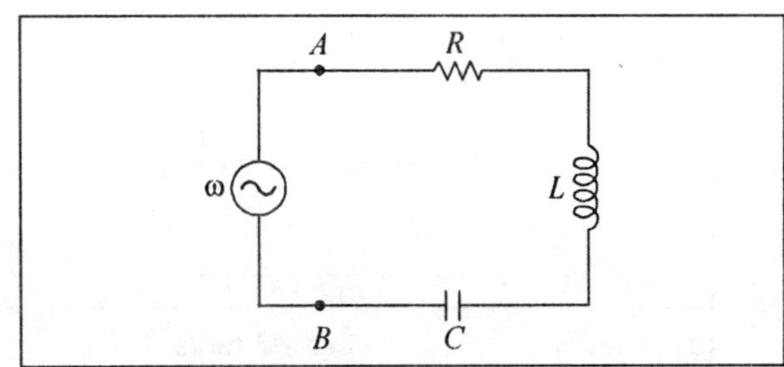

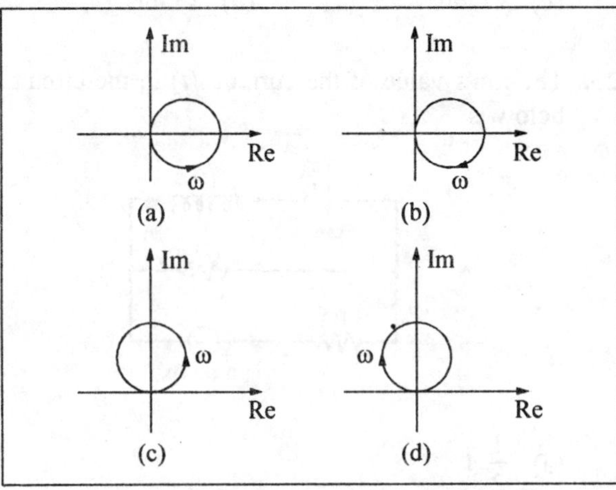

23. In the figure given below all phasors are with reference to the potential at point "*O*". The locus of voltage phasor V_{YX} as *R* is varied from zero to infinity is shown by

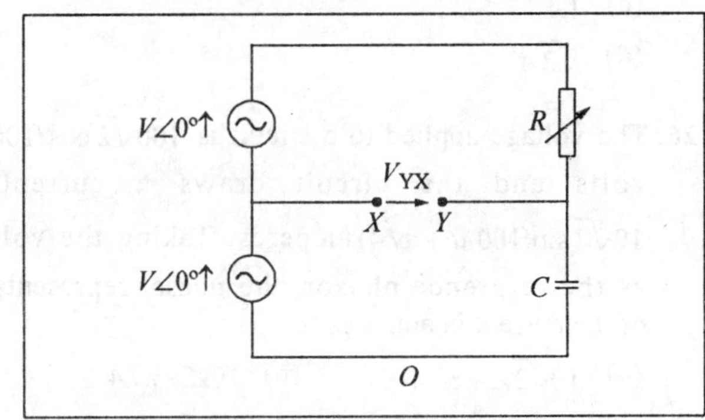

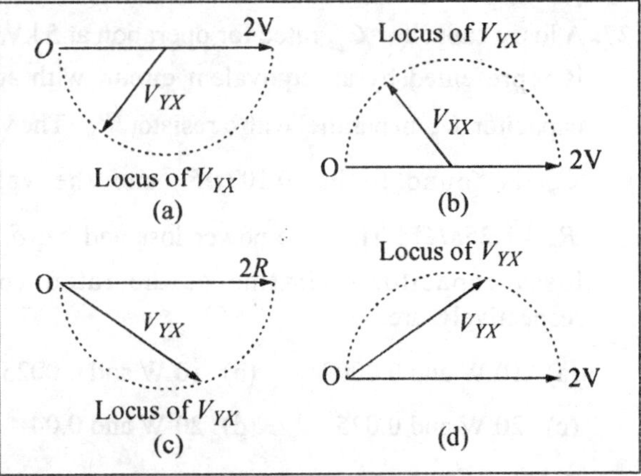

24. The resonant frequency for the given circuit will be

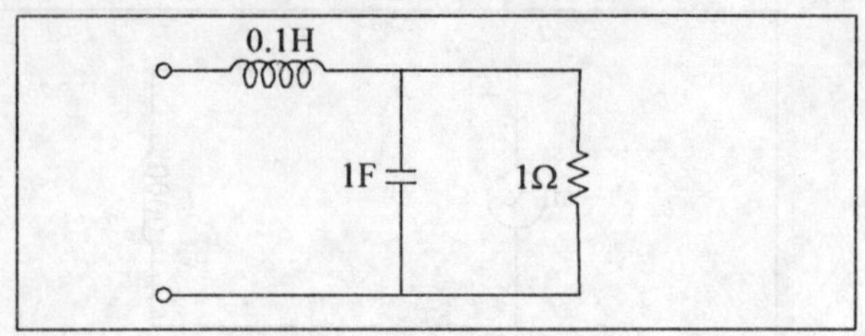

(a) 1 rad/s (b) 2 rad/s

(c) 3 rad/s (d) 4 rad/s

25. The r.m.s value of the current $i(t)$ in the circuit shown below is

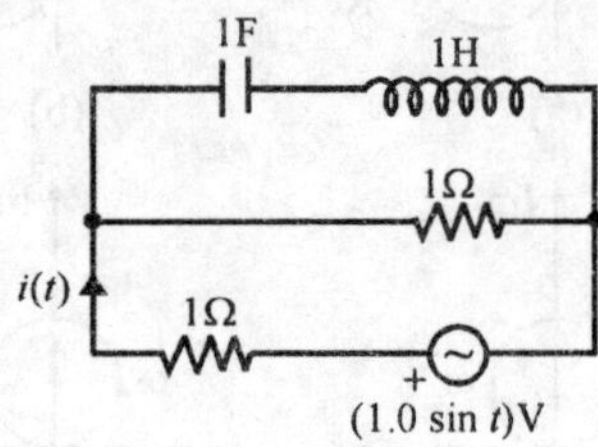

(a) $\frac{1}{2}A$

(b) $\frac{1}{\sqrt{2}}A$

(c) $1A$

(d) $\sqrt{2}A$

26. The voltage applied to a circuit is $100\sqrt{2}\cos(100\pi t)$ volts and the circuit draws a current of $10\sqrt{2}\sin(100\pi t+\pi/4)$ amperes. Taking the voltage as the reference phasor, the phasor representation of the current in amperes is

(a) $10\sqrt{2}\angle-\pi/4$ (b) $10\angle-\pi/4$

(c) $10\angle+\pi/4$ (d) $10\sqrt{2}\angle+\pi/4$

27. A lossy capacitor C_x, rated for operation at 5 kV, 50 Hz is represented by an equivalent circuit with an ideal capacitor C_p in parallel with a resistor R_p. The value of C_p is found to be $0.102\mu F$ and the value of $R_p = 1.25M\Omega$. Then the power loss and $\tan\delta$ of the lossy capacitor operating at the rated voltage, respectively, are

(a) 10 W and 0.0002 (b) 10 W and 0.0025

(c) 20 W and 0.025 (d) 20 W and 0.04

Common Data for Questions 28 and 29.

An RLC circuit with relevant data is given below.

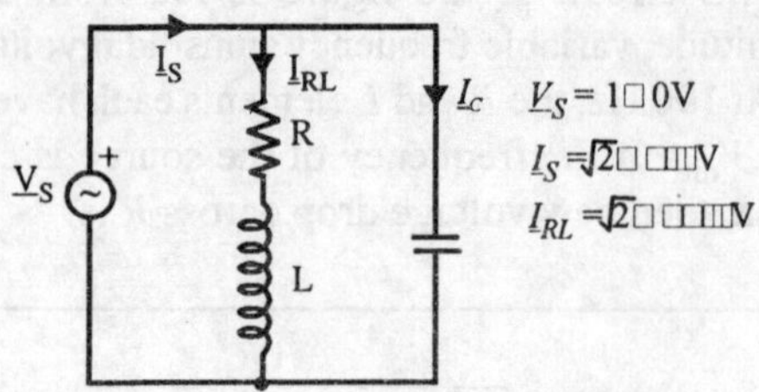

28. The current $\underline{I}_C$ in the figure above is

(a) $-j\,2\,A$ (b) $-j\frac{1}{\sqrt{2}}A$

(c) $+j\frac{1}{\sqrt{2}}A$ (d) $+j\,2\,A$

29. The power dissipated in the resistor R is

(a) 0.5 W (b) 1 W

(c) $\sqrt{2}W$ (d) 2 W

30. In the fig shown, A_1, A_2 and A_3 are ideal ammeters. If A_1 and A_3 read 5A and 13 A, respectively, reading of A_2 will be

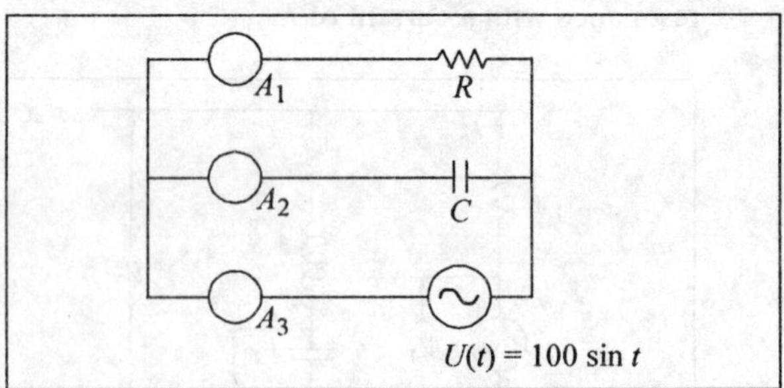

(a) 8 A
(b) 12 A
(c) 18 A
(d) Indeterminate unless the actual values of R, C and ω are specified

31. The switch S in figure is closed at $t = 0$. If $v_2(0) = 10$ V and $V_g(0) = 0$ V, respectively voltages across the capacitors in steady state will be

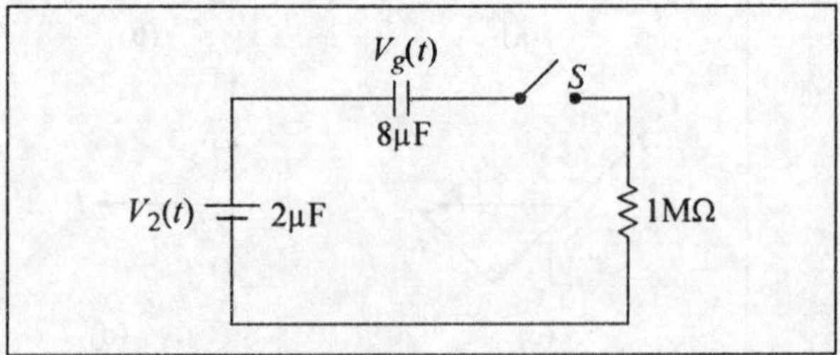

(a) $v_2(\infty) = v_g(\infty) = 0$
(b) $v_g(\infty) = 2$ V, $v_2(\infty) = 8$ V
(c) $v_2(\infty) = V_g(\infty) = 8$ V
(d) $v_2(\infty) = v_g(\infty) = 2$ V

32. A passive 2-port network is in a steady-state. Compared to its input, the steady state output can never offer

(a) Higher voltage (b) Lower impedance
(c) Greater power (d) Better regulation

33. A series RLC circuit has resonance frequency of 1 kHz and a quality factor $Q = 100$. If each of R, L and C is doubled from its original value, the new Q of the circuit is

(a) 25 (b) 50
(c) 100 (d) 200

34. The circuit shown in figure with $R = (1/3)\ \Omega$, $L = (1/4)\ H$, $C = 3F$ has input voltage $v(t) = \sin 2t$. The resulting current $i(t)$ is

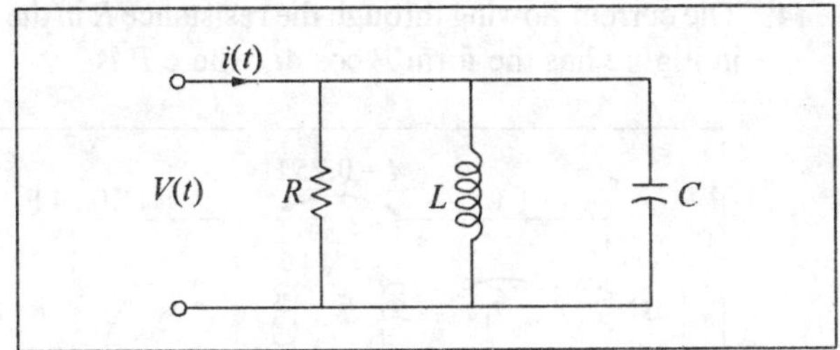

(a) $5 \sin (2t + 53.1°)$ (b) $5 \sin (2t - 53.1°)$
(c) $25 \sin (2t + 53.1°)$ (d) $25 \sin (2t - 53.1°)$

35. For the circuit shown in figure (a) the time constant $RC = 1$ ms. The input voltage is

$$Vi(t) = \sqrt{2} \sin 10^3 t.$$

The output voltage $V_o(t)$ is equal to

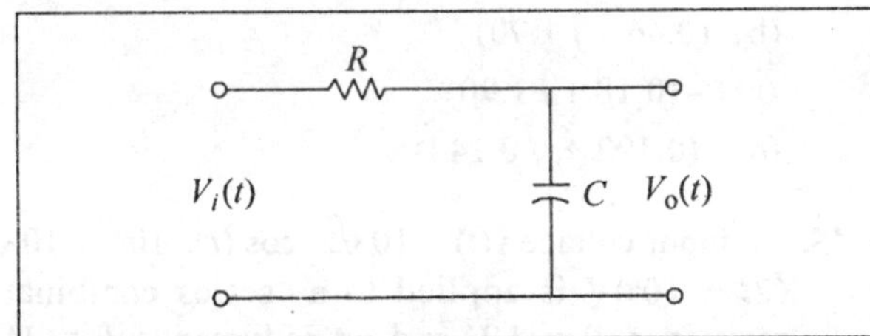

(a) $\sin (10^3 t - 45°)$ (b) $\sin (10^3 t + 45°)$
(c) $\sin (10^3 - 53°)$ (d) $\sin (10^3 t - 53°)$

36. In a series RLC circuit, R = 2kW, L = 1H, and C = [1/400] μF. The resonant frequency is

(a) 2×10^4 Hz (b) $[1/\pi] \times 10^4$ Hz
(c) 10^4 Hz (d) $2\pi \times 10^4$ Hz

37. The RC circuit shown in the figure is

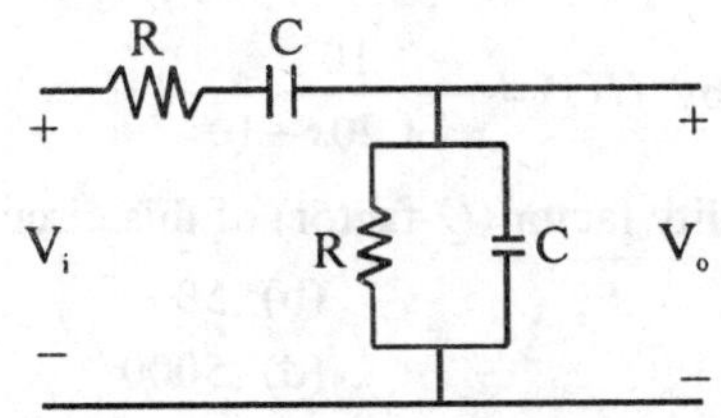

(a) A low-pas filter (b) A high-pass filter
(c) A band-pass filter (d) A band-reject filter

38. For a parallel RLC circuit, which one of the following statements is NOT correct?

(a) The bandwidth of the circuit decreases if R is increased
(b) The bandwidth of the circuit remains same if L is increased
(c) At resonance, input impedance is a real quantity
(d) At resonance, the magnitude of input impedance attains its minimum value

39. The circuit below implements a filter between the input current i_i and the output voltage v_0, Assume that the opamp is ideal. The filter implemented is a

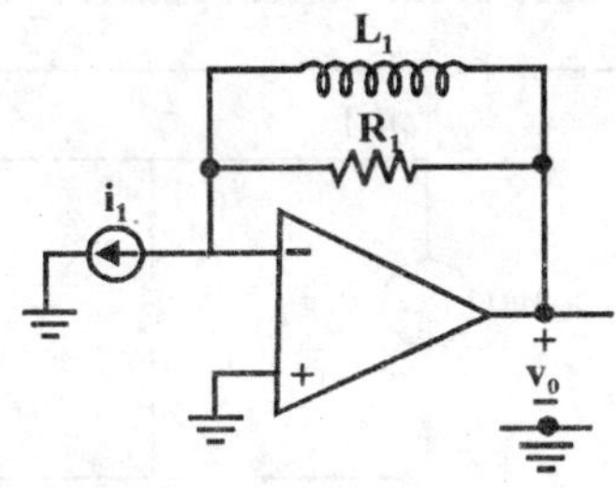

(a) Low pass filter
(b) Band pass filter
(c) Band stop filter
(d) High pass filter

40. The circuit shown below is driven by a sinusoidal input $v_i = V_P \cos(t / RC)$. The steady state output v_o is

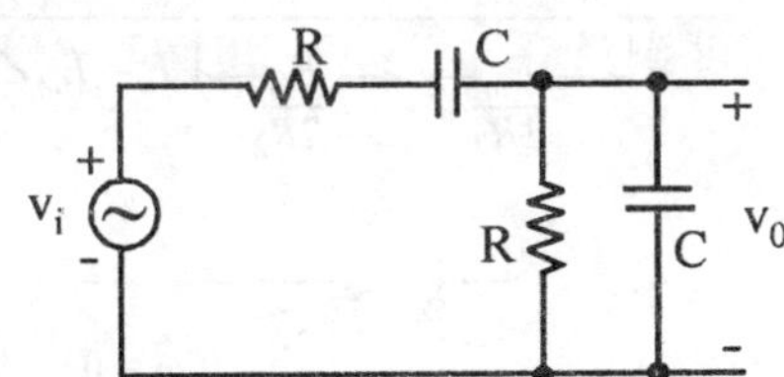

(a) $(V_p/3) \cos (t/RC)$

(b) $(V_p/3) \sin (t/RC)$

(c) $(V_p/2) \cos (t/RC)$

(d) $(V_p/2) \sin (t/RC)$

41. The average power delivered to an impedance $(4 - j3)\Omega$ by a current $5\cos(100\ \pi t + 100) A$ is

(a) 44.2 W
(b) 50 W
(c) 62.5 W
(d) 125 W

42. In Figure, the steady state output voltage corres-ponding to the input voltage $3 + 4 \sin 100t$ V is

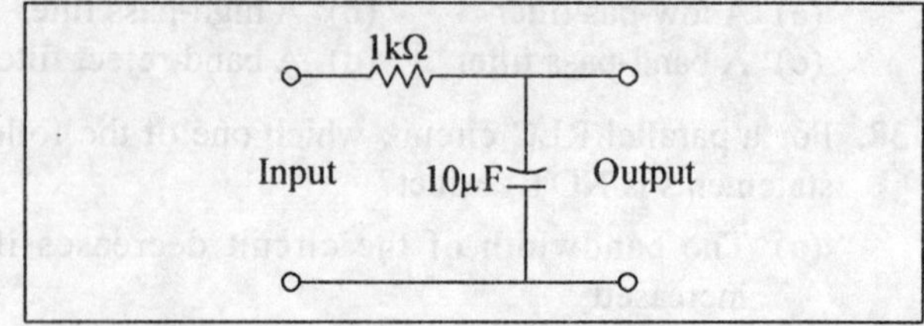

(a) $3 + (4/\sqrt{2})\ \sin(100\,t - (\pi/4)\text{V}$

(b) $3 + 4/\sqrt{2}\ \sin(100\,t - (\pi/4)\text{V}$

(c) $(3/2) + (4/\sqrt{2})\ \sin(100\,t + (\pi/4)\text{V}$

(d) $3 + 4\sin(100\,t + (\pi/4)\text{Vb}$

43. **When the angular frequency ω in Figure is varied from 0 to ∞, the locus of the current phasor I_2 is given by**

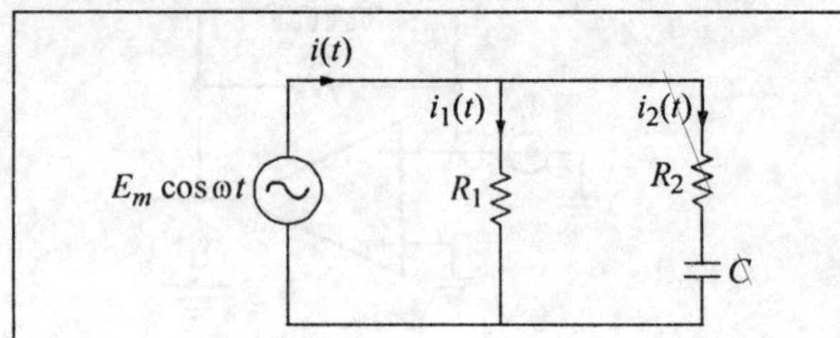

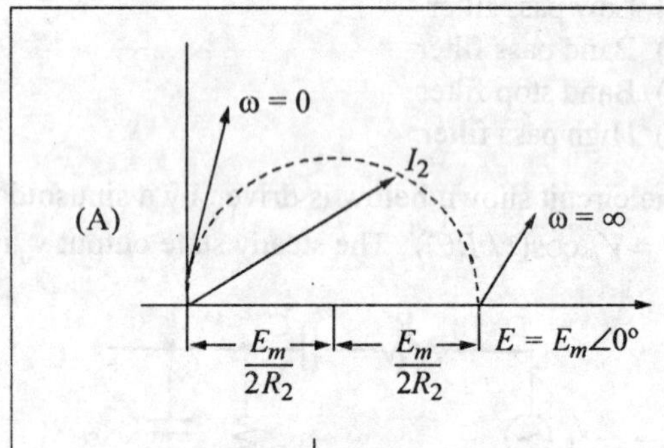

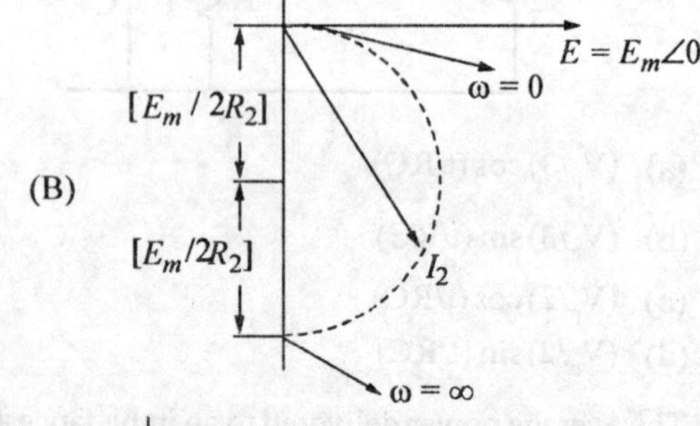

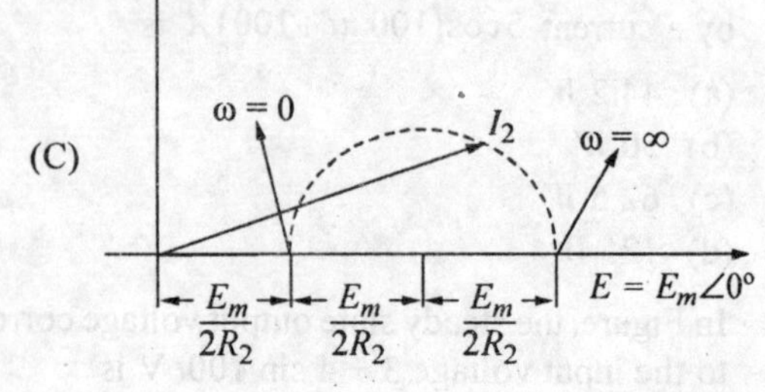

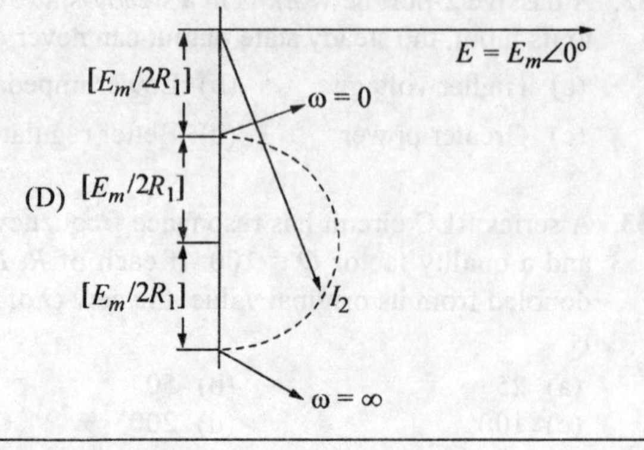

(a) Figure (A) (b) Figure (B)

(c) Figure (C) (d) Figure (D)

44. **The current flowing through the resistance R in the circuit in Figure has the form $P\cos 4t$, where P is**

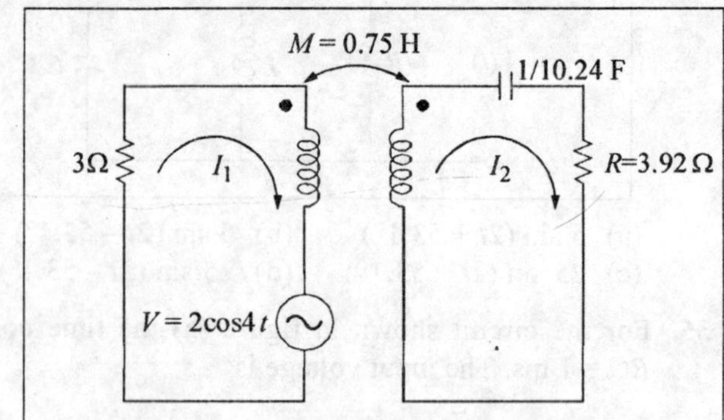

(a) $(0.18 + j\,0.72)$

(b) $(0.46 + j\,1.90)$

(c) $-(0.18 + j\,1.90)$

(d) $(0.192 + j\,0.144)$

45. An input voltage $v(t) = 10\sqrt{2}\cos(t + 10°) + 10\sqrt{5}\cos(2t + 10°)$V is applied to a series combination of resistance R = 1W and an inductance L = 1H. The resulting steady-state current $i(t)$ in ampere is

(a) $10\cos(t + 55°) + 10\cos(2t + 10° + \tan^{-1} 2)$

(b) $10\cos(t + 55°) + 10\,[\sqrt{3}/2]\cos(2t + 55°)$

(c) $10\cos(t - 35°) + 10\cos(2t + 10° - \tan^{-1} 2)$

(d) $10\cos(t - 35°) + 10\,[\sqrt{3}/2]\cos(2t + 35°)$

46. The transfer function $H(s) = \dfrac{V_o(s)}{V_i(s)}$ of an R-L-C circuit is given by $H(s) = \dfrac{10^6}{s^2 + 20s + 10^6}$

The Quality factor (Q-factor) of this circuit is

(a) 25 (b) 50

(c) 100 (d) 5000

47. Consider the following statement S_1 and S_2

S_1 : At the resonant frequency the impedance of a series R-L-C circuit is zero.

S_2 : In a parallel G-L-C circuit increasing the conductance G results is increase in its Q factor.

Which one of the following is correct?

(a) S_1 is FALSE and S_2 in TRUE
(b) Both S_1 and S_2 are TRUE
(c) S_1 is TURE and S_2 in FALSE
(d) Both S_1 and S_2 are FALSE

48. For the circuit shown in the figure, the instantaneous current $i_i(t)$ is

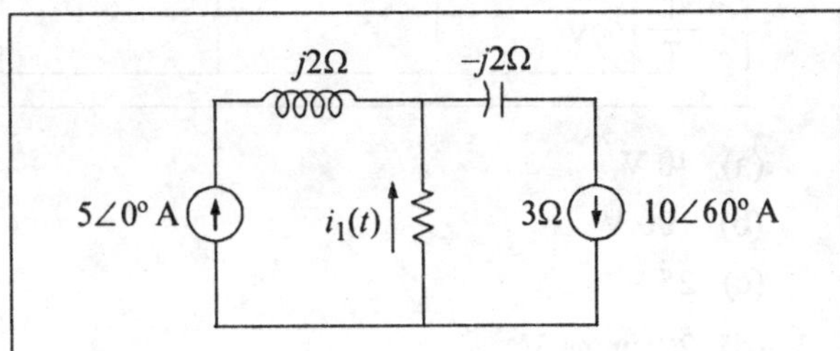

(a) $[(10\sqrt{3})/2] \angle 90°$ Amps
(b) $[(10\sqrt{3})/2] \angle -90°$ Amps
(c) $5\angle 60°$ Amps
(d) $5\angle -60°$ Amps

49. Two series resonant filter are as shown in the figure. Let the 3-dB bandwidth of Filter 1 be B_1 and that of Filter 2 be B_2. The value of B_1/B_2 is

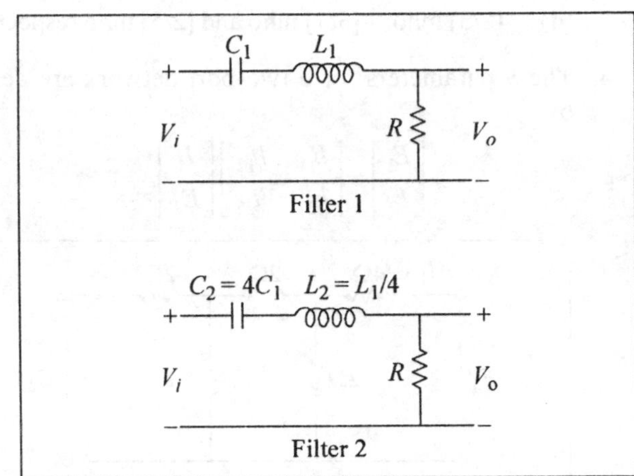

(a) 4
(b) 1
(c) 1/2
(d) 1/4

50. In the AC network shown in the figure, the phasor voltage V_{AB} (in Volts) is

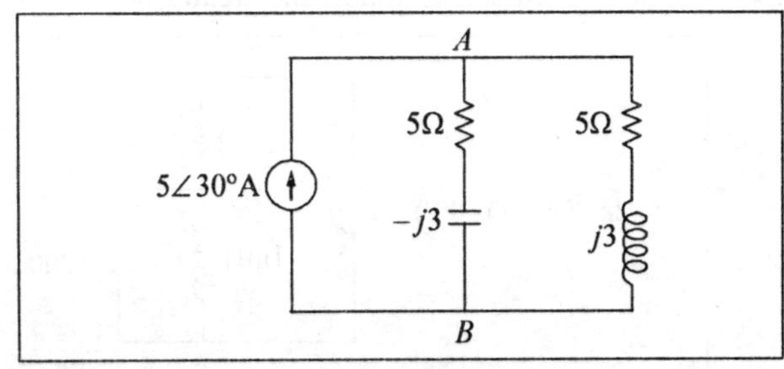

(a) 0
(b) $5\angle 30°$
(c) $12.5\angle 30°$
(d) $17\angle 30°$

51. An AC source of RMS voltage 20 V with internal impedance $Z_S = (1+2j)\,\Omega$ feeds a load of impedance $Z_L = (7+4j)\,\Omega$ in the figure below. The reactive power consumed by the load is

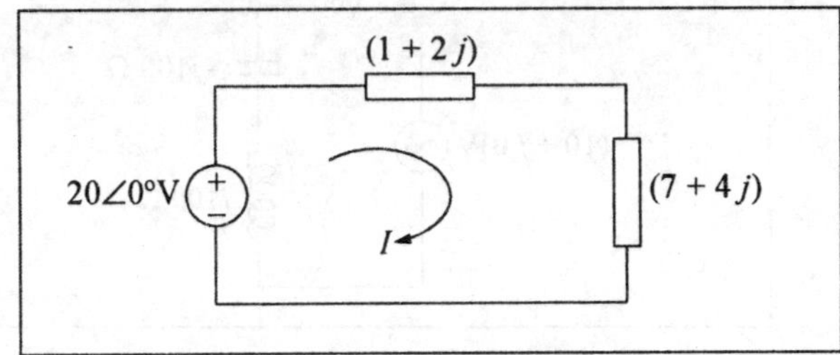

(a) 8 VAR
(b) 16 VAR
(c) 28 VAR
(d) 32 VAR

52. The current I in the circuit shown is

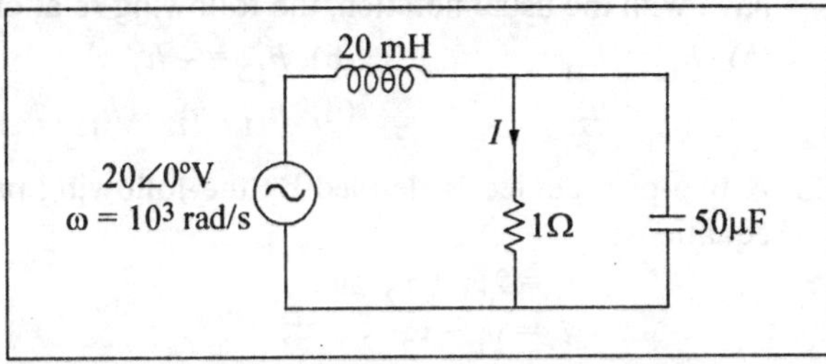

(a) $-j$ 1A
(b) j 1A
(c) 0 A
(d) 20 A

53. In the circuit shown below, the current I is equal to

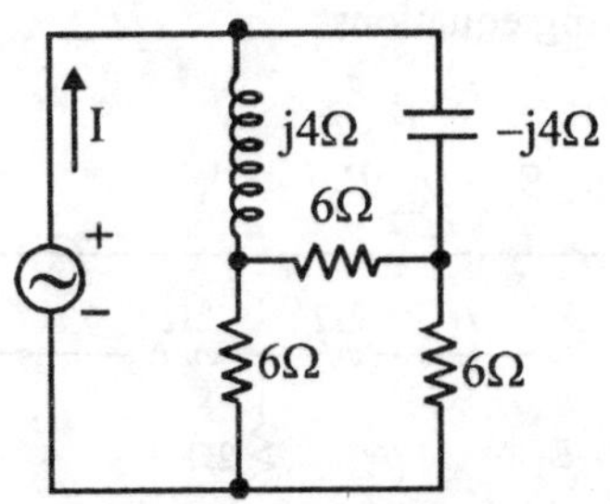

(a) $1.4\angle 0°$A
(b) $2.0\angle 0°$A
(c) $2.8\angle 0°$A
(d) $3.2\angle 0°$A

54. For the circuit shown below the steady state current I is

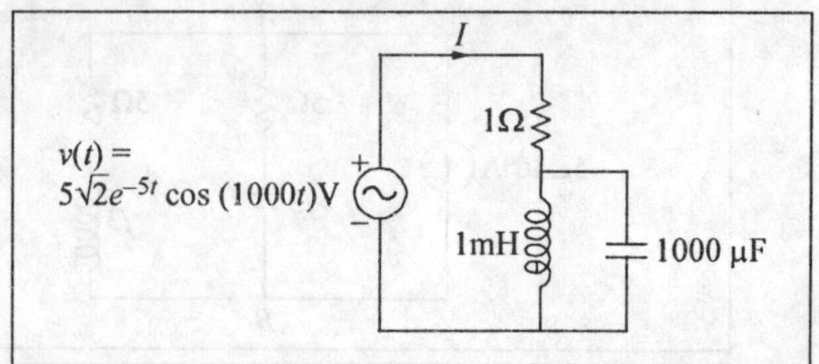

(a) 0 A (b) $5\sqrt{2}\cos(1000t)$ A

(c) $5\sqrt{2}\cos\left(1000t - \frac{\pi}{4}\right)$ A (d) $5\sqrt{2}$ A

55. For the circuit shown below the voltage across the capacitor is

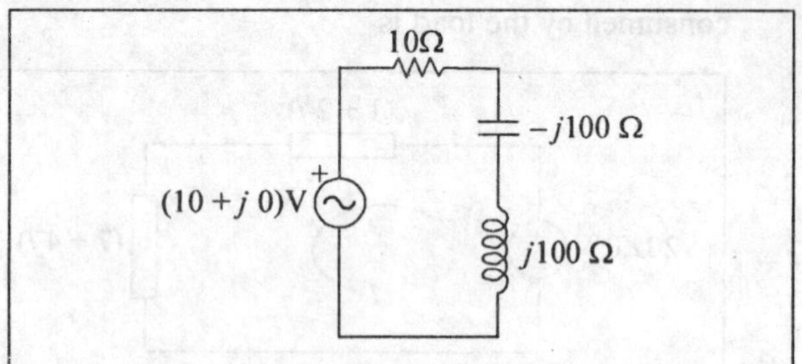

(a) $(10 + j\,0)$ V

(b) $(100 + j0)$ V

(c) $(0 + j\,100)$ V

(d) $(0 - j\,100)$ V

56. The circuit shown in figure is supplied by a dc source. The switch (S) in the circuit is closed at $t = 0$ and remains closed for a long time. The steady-state output Voltage (V_o) is.

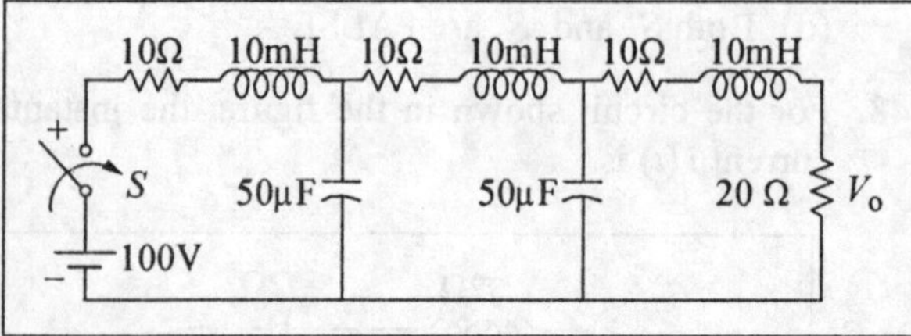

(a) 40 V

(b) 100 V

(c) 25 V

(d) 20 sin ωt V

6.5 Two Port Network

1. If a **two-port network** is passive symmetrical, then we have with the usual notation, the following relationship

(a) $h_{12} = h_{21}$ (b) $h_{12} = -h_{21}$

(c) $h_{11} = h_{22}$ (d) $h_{11} \cdot h_{22} - h_{12} \cdot h_{21} = 1$

2. A two-port device is defined by the following pair of equations:

$$i_1 = 2v_1 + v_2 \text{ and}$$
$$i_2 = v_1 + v_2,$$

Its impedance parameters $(z_{11}, z_{12}, z_{21}, z_{22})$ are given by

(a) 2, 1, 1, 1 (b) 1, –1, –1, 2

(c) 1, 1, 1, 2 (d) 2, –1, –1, 1

3. A two port network, shown in Figure, is described by the following equations

$$I_1 = y_{11}E_1 + y_{12}E_2$$
$$I_2 = y_{21}E_1 + y_{22}E_2$$

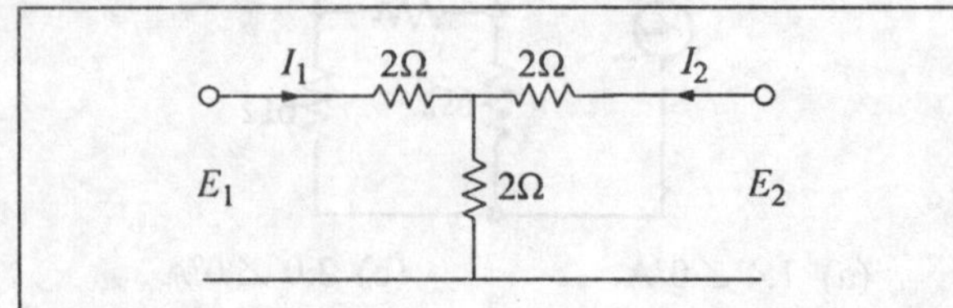

The admittance parameters, y_{11}, y_{12}, y_{21} and y_{22} for the network shown are

(a) 0.5 mho, 1 mho, 2 mho and 1 mho respectively

(b) [1/3] mho, – [1/6] mho, and [1/3] mho respectively

(c) 0.5 mho, 0.5 mho, 1.5 mho and 2 mho respectively

(d) – [2/5] mho, – [3/7] mho and [2/5] mho respectively

4. The h parameters for a two-port network are defined by

$$\begin{bmatrix} E_1 \\ I_2 \end{bmatrix} = \begin{bmatrix} h_{11} & h_{12} \\ h_{21} & h_{22} \end{bmatrix} \begin{bmatrix} I_1 \\ E_2 \end{bmatrix}$$

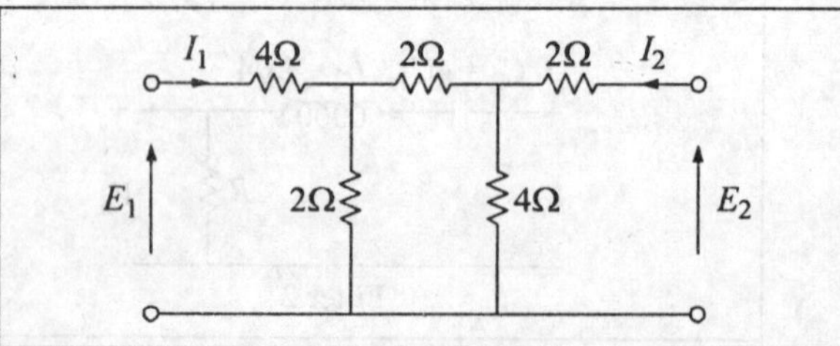

For the two-port network shown in figure the value of h_{12} is given by

(a) 0.125 (b) 0.167

(c) 0.625 (d) 0.25

5. The Z matrix of a 2-port network as given by $\begin{bmatrix} 0.9 & 0.2 \\ 0.2 & 0.6 \end{bmatrix}$

The element Y_{22} of the corresponding Y matrix of the same network is given by

(a) 1.2 (b) 0.4
(c) – 0.4 (d) 1.8

6. The two-port network P shown in the figure has ports 1 and 2, denoted by terminals (a, b) and (c, d) respectively. It has an impedance matrix Z with parameters denoted by z_{ij}. A 1 Ω resistor is connected in series with the network at port 1 as shown in the figure. The impedance matrix of the modified two-port network (shown as a dashed box) is

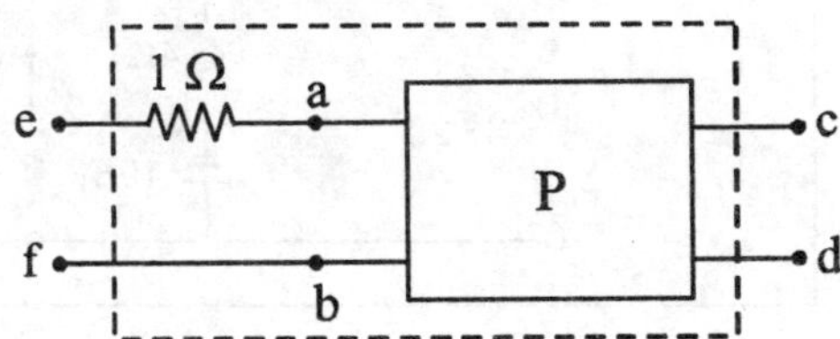

(a) $\begin{pmatrix} z_{11}+1 & z_{12}+1 \\ z_{21} & z_{22}+1 \end{pmatrix}$ (b) $\begin{pmatrix} z_{11}+1 & z_{12} \\ z_{21} & z_{22}+1 \end{pmatrix}$

(c) $\begin{pmatrix} z_{11}+1 & z_{12} \\ z_{21} & z_{22} \end{pmatrix}$ (d) $\begin{pmatrix} z_{11}+1 & z_{12} \\ z_{21}+1 & z_{22} \end{pmatrix}$

7. The terminal voltage and currents of a two-port network are indicated on the fig If the **two-port is reciprocal,** then

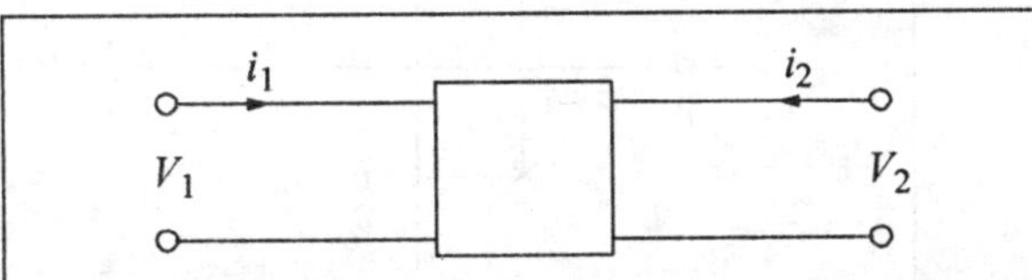

(a) $[z_{12}/y_{12}] = z_{12}^2 - z_{11} \cdot z_{22}$
(b) $z_{12} = [1/y_{22}]$
(c) $h_{12} = -h_{21}$
(d) $AD - BC = 0$

8. The parameter type and the matrix representation of the relevant two part parameter that describe the circuit shown are

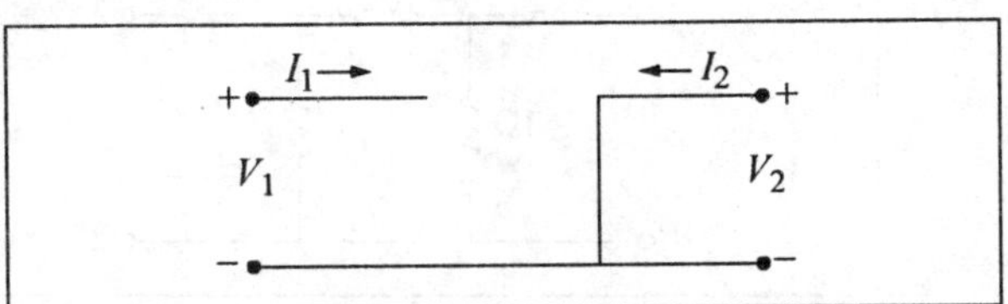

(a) z parameters, $\begin{bmatrix} 0 & 0 \\ 0 & 0 \end{bmatrix}$

(b) h parameters, $\begin{bmatrix} 1 & 0 \\ 0 & 1 \end{bmatrix}$

(c) g parameters, $\begin{bmatrix} 0 & 0 \\ 0 & 0 \end{bmatrix}$

(d) z parameters, $\begin{bmatrix} 1 & 0 \\ 0 & 1 \end{bmatrix}$

9. A 2-port network is shown in Figure. The h_{21} for the this network can be given by

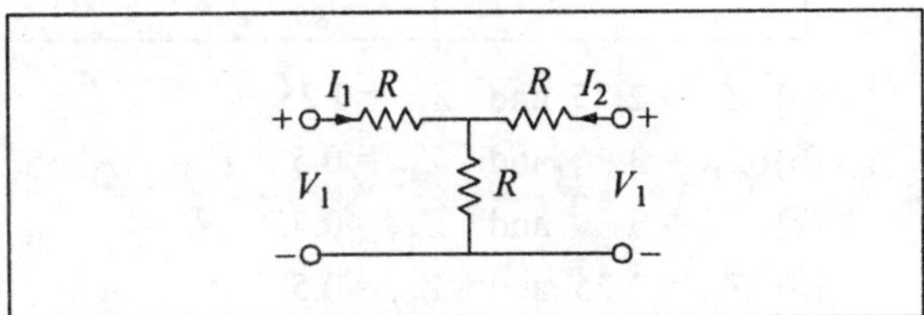

(a) – 1/2 (b) + 1/2
(c) – 3/2 (d) + 3/2

10. If each branch of a Delta circuit has impedance $\sqrt{3}Z$ then each branch of the equivalent Wye circuit has impedance

(a) $[Z/\sqrt{3}]$ (b) $3Z$
(c) $3\sqrt{3Z}$ (d) $Z/3$

11. The admittance parameter Y_{12} in the 2-port network in Figure is

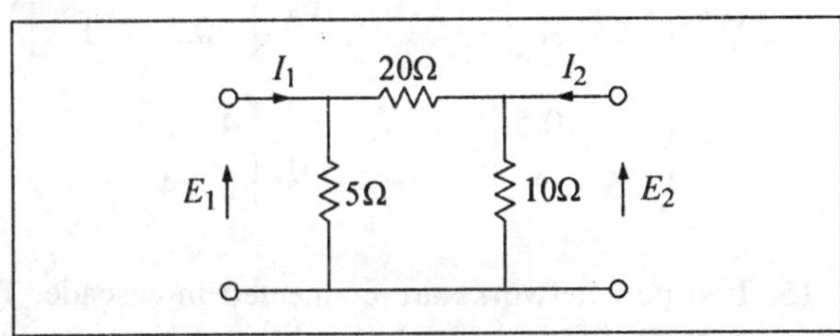

(a) – 0.2 mho (b) 0.1 mho
(c) – 0.05 mho (d) 0.05 mho

12. The $ABCD$ parameters of an ideal $n : 1$ transformer shown in the figure are $\begin{pmatrix} n & 0 \\ 0 & X \end{pmatrix}$,

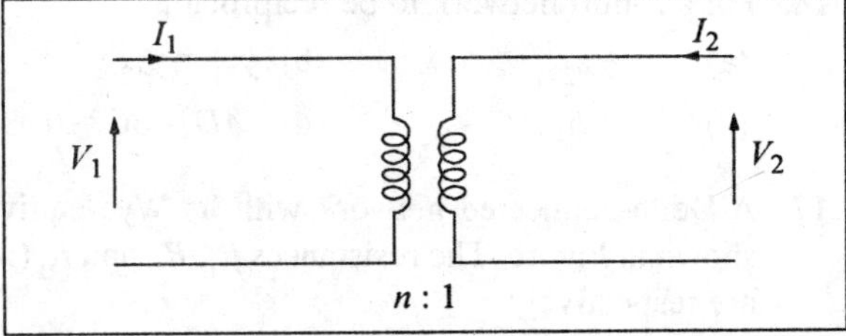

The value of X will be

(a) n
(b) $[1/n]$
(c) n^2
(d) $[1/n^2]$

13. The impedance parameters Z_{11} and Z_{12} of the two-port network in Figure are

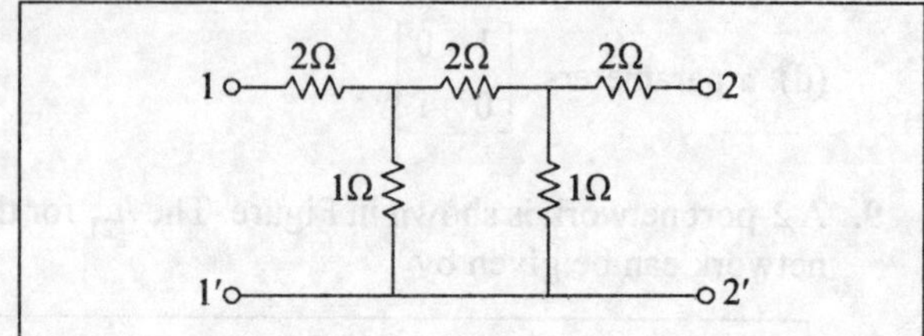

(a) $Z_{11} = 2.75$ and $Z_{12} = 0.25$
(b) $Z_{11} = 3$ and $Z_{12} = 0.5$
(c) $Z_{11} = 3$ and $Z_{12} = 0.25$
(d) $Z_{11} = 2.25$ and $Z_{12} = 0.5$

14. For the two-port network shown below, the short circuit admittance parameter matrix is

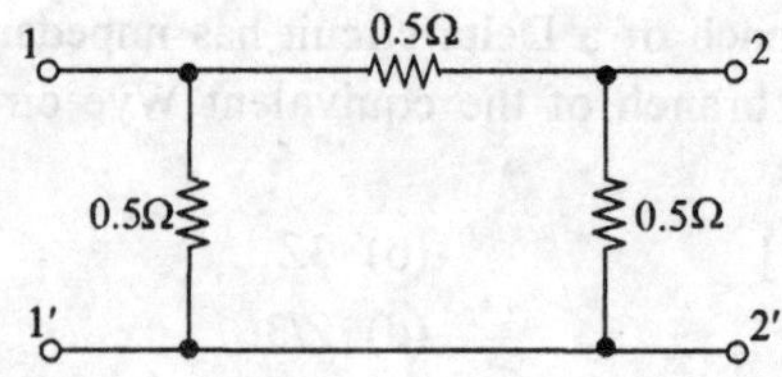

(a) $\begin{bmatrix} 4 & -2 \\ -2 & 4 \end{bmatrix} S$ (b) $\begin{bmatrix} 1 & -0.5 \\ -0.5 & 1 \end{bmatrix} S$

(c) $\begin{bmatrix} 1 & 0.5 \\ 0.5 & 1 \end{bmatrix} S$ (d) $\begin{bmatrix} 4 & 2 \\ 2 & 4 \end{bmatrix} S$

15. Two port networks are connected in cascade. The parameters of the network are obtained by multiplying the individual.

(a) z-parameter matrix
(b) h-parameter matrix
(c) y-parameter matrix
(d) $ABCD$ parameter matrix

16. For a 2-port network to be reciprocal,

(a) $z_{11} = z_{22}$ (b) $y_{21} = y_{12}$
(c) $h_{21} = h_{12}$ (d) $AD - BC = 0$

17. A Delta-connected network with its Wye-equivalent is shown in Figure. The resistances R_1, R_2, and R_3 (in ohms) are respectively

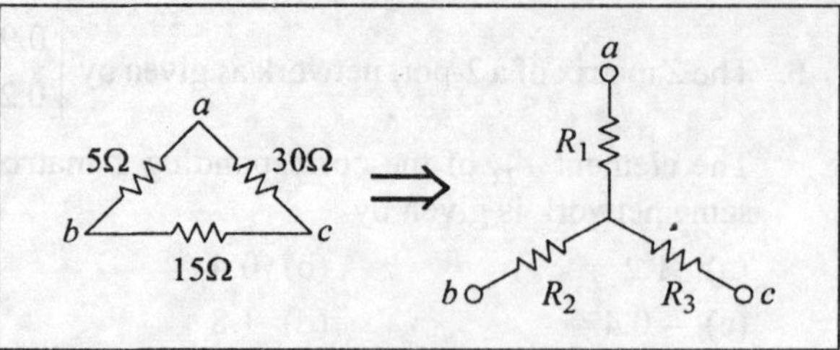

(a) 1.5, 3 and 9 (b) 3, 9 and 1.5
(c) 9, 3 and 1.5 (d) 3, 1.5 and 9

18. The Z parameters Z_{11} and Z_{21} for the 2-port network in figure are

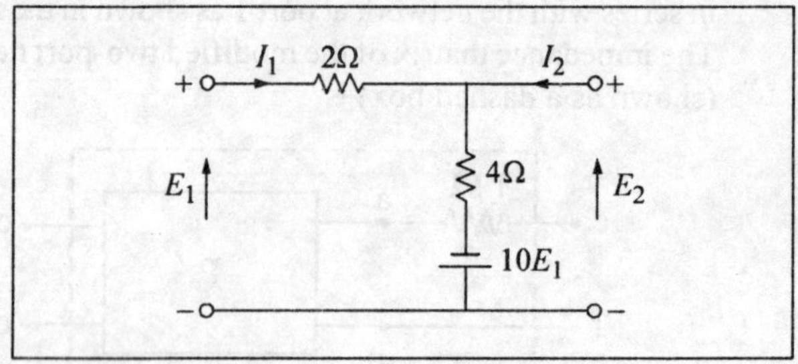

(a) $Z_{11} = -[6/11]\ \Omega$; $Z_{21} = [16/11]\ \Omega$
(b) $Z_{11} = [6/11]\ \Omega$; $Z_{21} = [4/11]\ \Omega$
(c) $Z_{11} = [6/11]\ \Omega$; $Z_{21} = -[16/11]\ \Omega$
(d) $Z_{11} = -[4/11]\ \Omega$; $Z_{21} = [4/11]\ \Omega$

19. The driving-point impedance $Z(s)$ of a network has the pole-zero locations as shown in Figure. If $Z(0) = 3$, then $Z(s)$ is

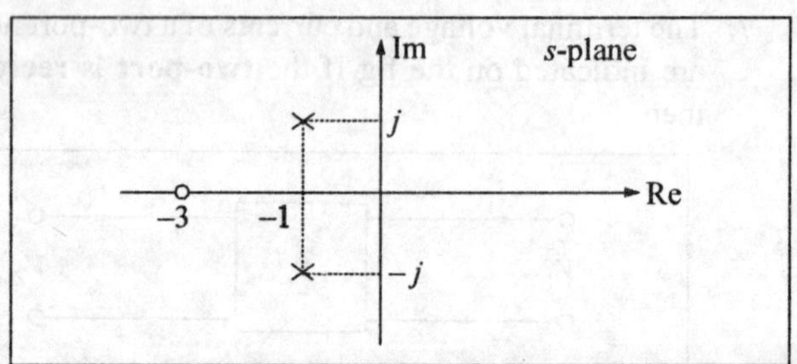

(a) $\dfrac{3(s+3)}{s^2+2s+3}$ (b) $\dfrac{2(s+3)}{s^2+2s+2}$

(c) $\dfrac{3(s+3)}{s^2-2s-3}$ (d) $\dfrac{2(s+3)}{s^2-2s-2}$

20. The impedance parameters Z_{11} and Z_{12} of the two-port network in Figure are

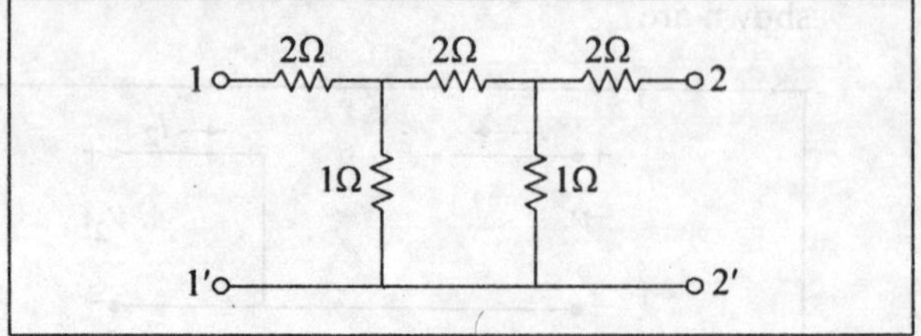

(a) $Z_{11} = 2.75$ and $Z_{12} = 0.25$
(b) $Z_{11} = 3$ and $Z_{12} = 0.5$
(c) $Z_{11} = 3$ and $Z_{12} = 0.25$
(d) $Z_{11} = 2.25$ and $Z_{12} = 0.5$

21. For the lattice circuit shown in figure $Z_a = j2\Omega$ and $Z_b = 2\Omega$. The values of the open circuit impedance parameters $Z = \begin{pmatrix} Z_{11} & Z_{12} \\ Z_{21} & Z_{22} \end{pmatrix}$ are

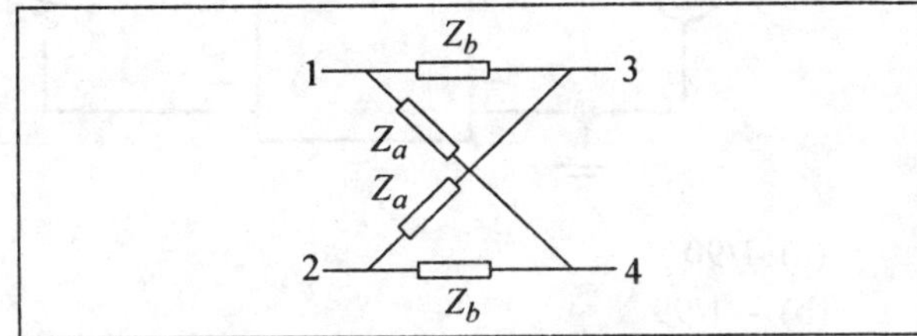

(a) $\begin{pmatrix} 1-j & 1+j \\ 1+j & 1+j \end{pmatrix}$ (b) $\begin{pmatrix} 1-j & 1+j \\ -1+j & 1-j \end{pmatrix}$
(c) $\begin{pmatrix} 1+j & 1+j \\ 1-j & 1-j \end{pmatrix}$ (d) $\begin{pmatrix} 1+j & -1+j \\ -1+j & 1+j \end{pmatrix}$

22. For the circuit shown in Figure the initial conditions are zero. Its transfer function $H(s) = \dfrac{V_o(s)}{V_i(s)}$ is

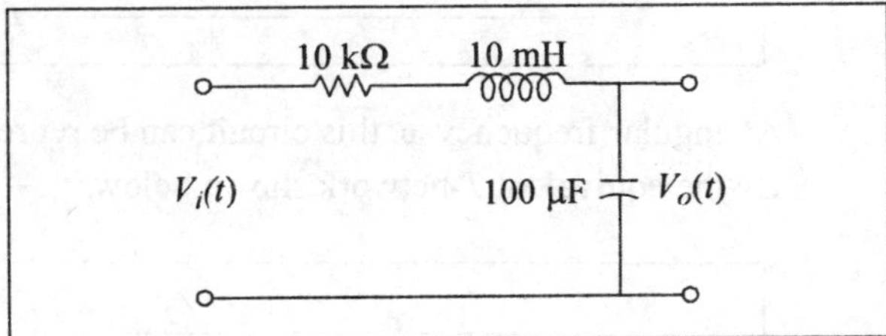

(a) $\dfrac{1}{s^2 + 10^6 s + 10^6}$ (b) $\dfrac{10^6}{s^2 + 10^3 s + 10^6}$
(c) $\dfrac{10^3}{s^2 + 10^3 s + 10^6}$ (d) $\dfrac{10^6}{s^2 + 10^6 s + 10^6}$

23. If $R_1 = R_2 = R_4 = R$ and $R_3 = 1.1R$ in the bridge circuit shown in the figure, then the reading in the ideal voltmeter connected between a and b is

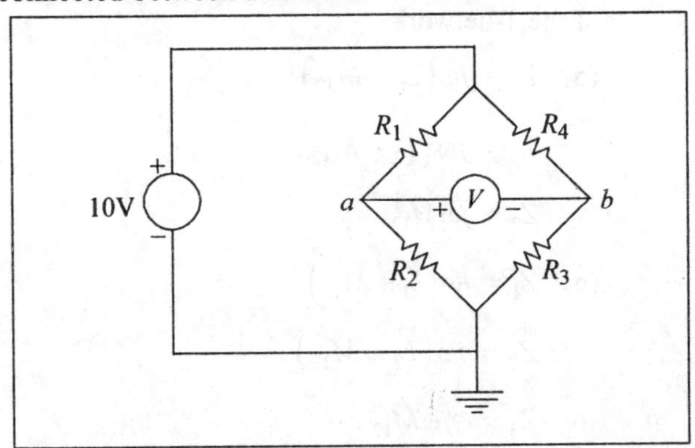

(a) 0.238 V (b) 0.138 V
(c) – 0.238 V (d) 1 V

24. The h parameters of the circuit shown in the figure are

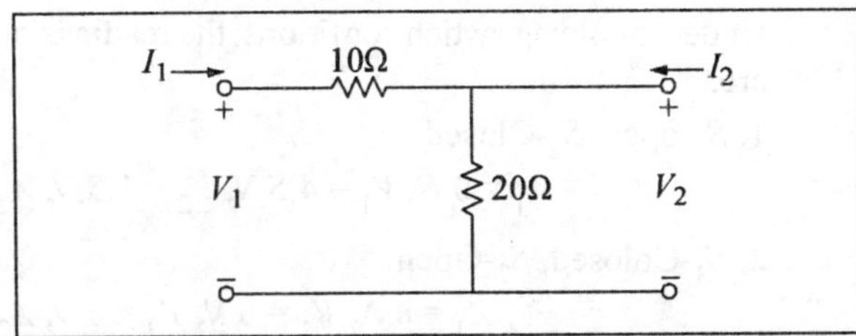

(a) $\begin{pmatrix} 0.1 & 0.1 \\ -0.1 & 0.3 \end{pmatrix}$ (b) $\begin{pmatrix} 10 & -1 \\ 1 & 0.05 \end{pmatrix}$
(c) $\begin{pmatrix} 30 & 20 \\ 20 & 20 \end{pmatrix}$ (d) $\begin{pmatrix} 10 & 1 \\ -1 & 0.05 \end{pmatrix}$

25. Two-port network is represented by $ABCD$ parameters given by.

$$\begin{pmatrix} V_1 \\ I_1 \end{pmatrix} = \begin{pmatrix} A & B \\ C & D \end{pmatrix} \begin{pmatrix} V_2 \\ -I_2 \end{pmatrix}$$

If port-2 is terminated by R_L, the input impedance seen at port-1 is given by
(a) $(A + BR_L)/(C + DR_L)$ (b) $(AR_L + C)/(BR_L + D)$
(c) $(DR_L + A)/(BR_L + C)$ (d) $(B + AR_L)/(D + CR_L)$

26. In the two port network shown in the figure below, z_{12} and z_{21} are, respectively

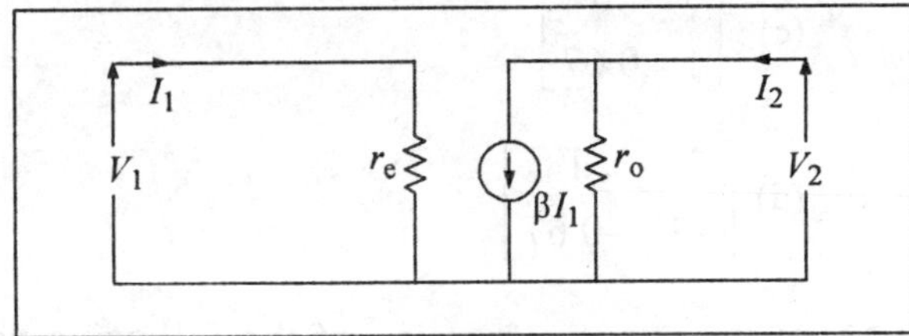

(a) r_e and βr_o (b) 0 and $-\beta r_o$
(c) 0 and βr_o (d) r_e and $-\beta r_o$

27. The driving point impedance of the following network

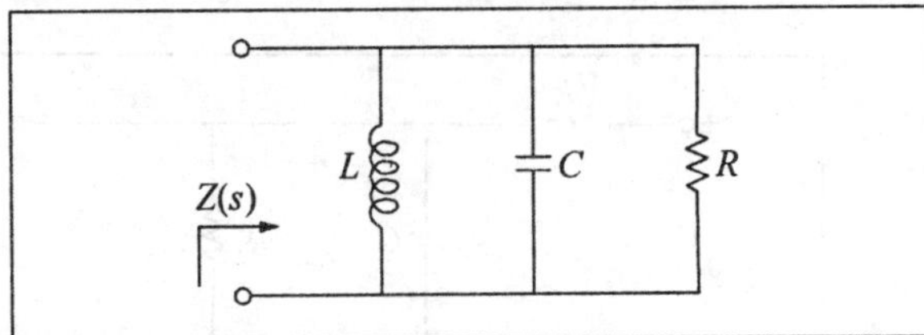

is given by $Z(s) = \dfrac{0.2s}{s^2 + 0.1s + 2}$. The component values are

(a) $L = 5$ H, $R = 0.5\ \Omega$, $C = 0.1$ F
(b) $L = 0.1$ H, $R = 0.5\ \Omega$, $C = 5$ F
(c) $L = 5$ H, $R = 2\ \Omega$, $C = 0.1$ F
(d) $L = 0.1$ H, $R = 2\ \Omega$, $C = 5$ F

28. A two-port network shown below is excited by dc sources. The voltages and the currrents are measured with voltmeters V_1, V_2 and ammeters A_1, A_2 (all assumed to be ideal), as indicated.

Under folloiwng swtich confitions, the readings obtianed are:

1. S_1-open, S_2-Closed

$$A_1 = 0\text{ A},\ V_1 = 4.5\text{ V},\ V_2 = 1.5\text{ V},\ A_2 = 1\text{ A}$$

2. S_1-Colosed, S_2-Open

$$A_1 = 4\text{ A},\ V_1 = 6\text{ V},\ V_2 = 6\text{ V},\ A_2 = 0\text{ A}$$

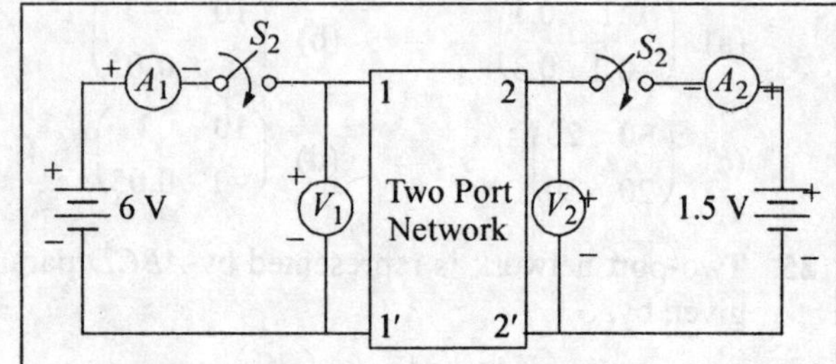

The h-parameter matrix for the network is

(a) $\begin{bmatrix} -3 & 3 \\ -1 & 0.67 \end{bmatrix}$

(b) $\begin{bmatrix} -3 & -1 \\ 3 & 0.67 \end{bmatrix}$

(c) $\begin{bmatrix} 3 & 3 \\ 1 & 0.67 \end{bmatrix}$

(d) $\begin{bmatrix} 3 & 1 \\ -3 & -0.67 \end{bmatrix}$

29. If the transfer function of the following netwrok is $\dfrac{V_o(s)}{V_i(s)} = \dfrac{1}{2+sCR}$ the value of the load resistance R_L is

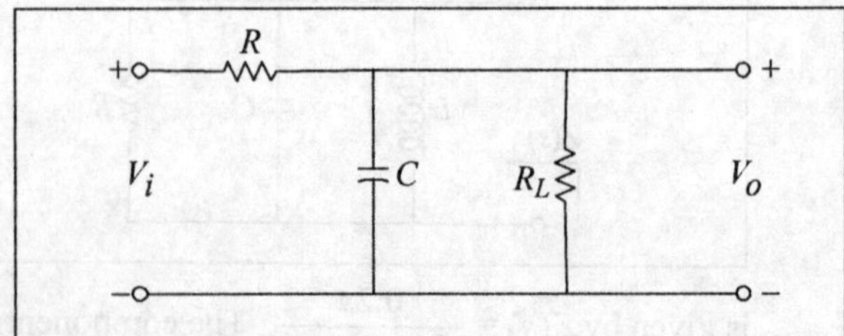

(a) $R/4$

(b) $R/2$

(c) R

(d) $2R$

30. In the circuit shown below, the network N is described by the following Y matrix:

$$Y = \begin{bmatrix} 0.1S & -0.01S \\ 0.01S & 0.1S \end{bmatrix}.$$ The voltage gain $\dfrac{V_2}{V_1}$ is

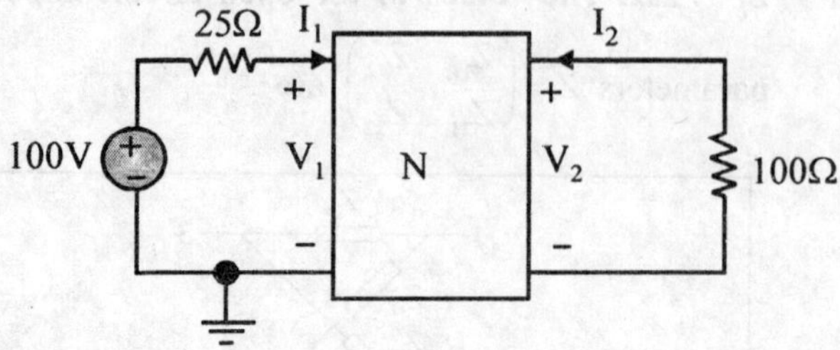

(a) 1/90

(b) – 1/99

(c) 1/99

(d) – 1/11

31. Consider the coupled circuit shown below

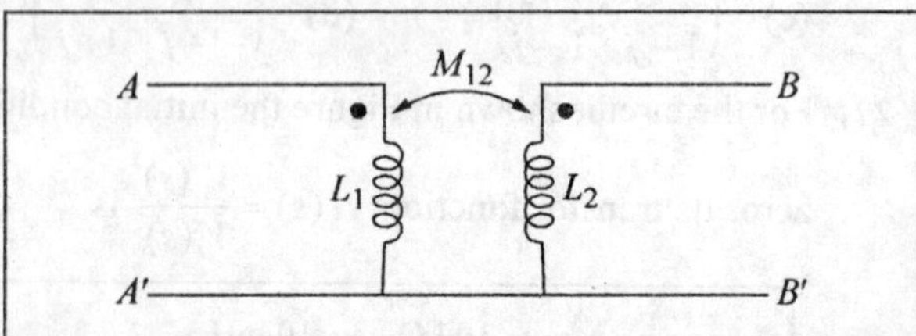

At angular frequency ω, this circuit can be represented By the equivalent *T*-network shown below.

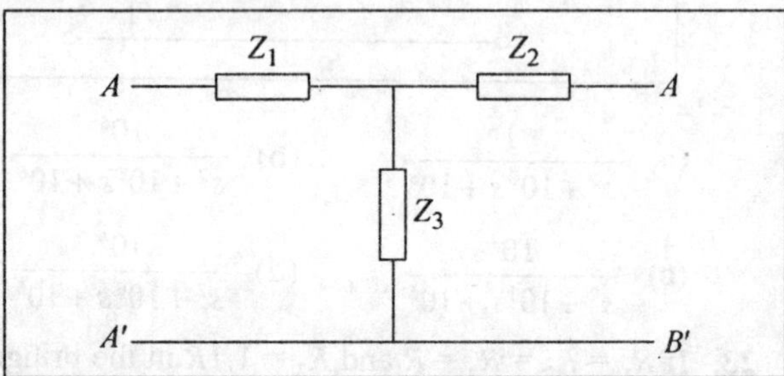

Indicate the correct set of expressions for the impedances of the T-network

(a) $Z_1 = j\omega(L_1 - M_{12})$

$Z_2 = j\omega(L_2 - M_{12})$

$Z_3 = j\omega M_{12}$

(b) $Z_1 = j\omega(L_1 + M_{12})$

$Z_2 = j\omega(L_2 + M_{12})$

$Z_3 = -j\omega M_{12}$

(c) $Z_1 = j\omega L_1$
$Z_2 = j\omega L_2$
$Z_3 = -j\omega M_{12}$

(d) $Z_1 = j\omega(L_1 - M_{12})$
$Z_2 = j\omega(L_2 - M_{12})$
$Z_3 = j\omega(L_1 + L_2 + M_{12})$

32. For the circuit shown below the input resistance

$R_{11} = \left.\frac{V_1}{I_1}\right|_{I_2=0}$ is

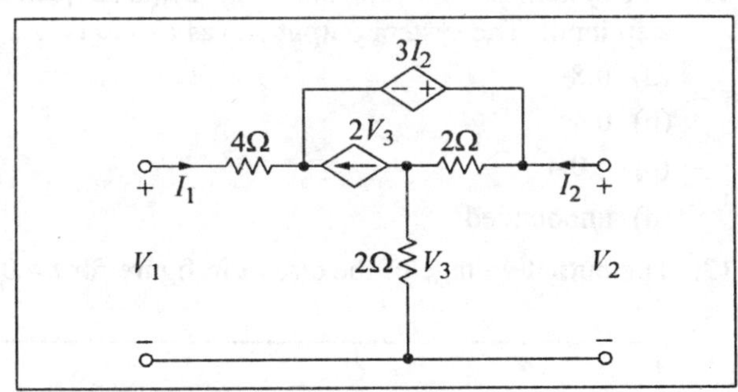

(a) $-3\ \Omega$ (b) $2\ \Omega$
(c) $3\ \Omega$ (d) $13\ \Omega$

Common Data for Questions 33 and 34:

With 10 V dc connected at port A in the linear non-reciprocal two-port network below, the following were observed:

1. 1Ω connected at port B draws a current of 3A
2. 2.5Ω connected at port B draws a current of 2A

33. With 10 V dc connected at port A, the current drawn by 7 W connected at port B is
(a) 3/7 A
(b) 5/7 A
(c) 1 A
(d) 9/7 A

34. For the same network, with 6 V dc connected at port A, 1 Ω connected at port B draws 7/3 A. If 8 V dc is connected to port A, the open circuit voltage at port B is
(a) 6 V
(b) 7 V
(c) 8 V
(d) 9 V

6.6 Laplace Transform

1. Match the following transfer functions and impulse responses:

Transfer functions

(A) $\frac{s}{s+1}$ (C) $\frac{s}{(s+1)^2}$

(B) $\frac{s}{s(s+1)+1}$ (D) $\frac{s}{s^2+1}$

Impulse Responses

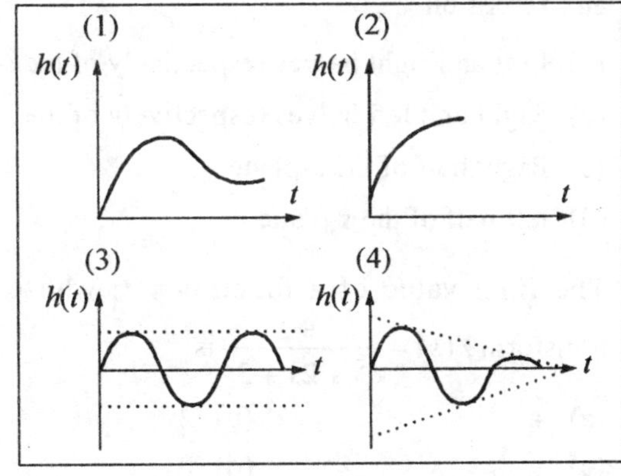

Codes:

	A	B	C	D
(a)	1	2	3	4
(b)	1	2	4	3
(c)	2	1	3	4
(d)	3	4	1	2

2. $s(t)$ is step response and $h(t)$ is impulse response of a system. Its response $y(t)$ for any input $u(t)$ is given by

(a) $\frac{d}{dt}\int_0^t s(t-\tau)\,u(\tau)\,d\tau$

(b) $\int_0^t s(t-\tau)\,u(\tau)\,d\tau$

(c) $\int_0^t \int_0^t s(t-\tau_1)\,u(\tau_1)\,d\tau_1\,d\tau$

(d) $\frac{d}{dt}\int_0^t h(t-\tau)\,u(\tau)\,d\tau$

3. The impulse response of an initially relaxed system is $e^{-2t}\, u(t)$. To produce a response of $te^{-2t}u(t)$, the input must be equal to
   (a) $2e^{-t}\, u(t)$ (b) $(1/2)e^{-2t}\, u(t)$
   (c) $e^{-2t}\, u(t)$ (d) $e^{-t}\, u(t)$
4. The Laplace transformation of $f(t)$ is $F(s)$. Given $F(s) = [\omega/(s^2 + \omega)]$, the final value of $f(t)$ is
   (a) Infinity (b) Zero
   (c) One (d) None of the above
5. The unit impulse response of a system is given as $c(t) = -4e^{-t} + 6e^{-2t}$. The step response of the same system for $t \geq 0$ is equal to
   (a) $-3e^{-2t} - 4e^{-t} + 1$ (b) $-3e^{-2t} + 4e^{-t} - 1$
   (c) $3^{-2t} - 4e^{-t} - 1$ (d) $3e^{-2t} + 4e^{-t} - 1$
6. The Laplace transform of $(t^2 - 2t)\, u(t-1)$ is
   (a) $\frac{2}{s^2}e^{-s} - \frac{2}{s^2}e^{-s}$ (b) $\frac{2}{s^3}e^{-2s} - \frac{2}{s^2}e^{-s}$
   (c) $\frac{2}{s^3}e^{-s} - \frac{2}{s^2}e^{-s}$ (d) None of the above
7. A rectangular current pulse of duration T and magnitude I has the Laplace transform
   (a) I/s
   (b) $(I/s)\exp(Ts)$
   (c) $(I/s)\exp(Ts)$
   (d) $(I/s)[1-\exp(Ts)]$.
8. Let $s(t)$ be the step response of a linear system with zero initial conditions; then the response of this system to an input $u(t)$ is
   (a) $\int_0^t s(t-\tau)\,u(\tau)\,d\tau$
   (b) $\frac{d}{dt}\left(\int_0^t s(t-\tau)\,u(\tau)\,d\tau\right)$
   (c) $\int_0^t s(t-\tau)\left(\int_0^t u(\tau_1)\,d\tau_1\right)d\tau$
   (d) $\int_0^t \{s(t-\tau)\}^2\,u(\tau)\,d\tau$
9. Let $Y(s)$ be the Laplace transformation of the function $y(t)$, then the final value of the function is
   (a) $\lim_{s\to 0} Y(s)$ (b) $\lim_{s\to 0} Y(s)$
   (c) $\lim_{s\to 0} sY(s)$ (d) $\lim_{s\to 0} sY(s)$
10. A function $y(t)$ satisfies the following differential equation:
$$\frac{dy(t)}{dt} + y(t) = \delta(t)$$
   where $\delta(t)$ is the delta function. Assuming zero initial condition and denoting the unit step function by $u(t) \cdot y(t)$ can be of the form
   (a) e^t
   (b) e^{-t}
   (c) $e^t\, u(t)$
   (d) $e^{-t}\, u(t)$
11. The system $G(s) = [0.8/\, s(s^2 + s - 2)]$ is subjected to a step input. The system output $y(t)$ as $t \to \infty$ is
   (a) 0.8
   (b) 0.4
   (c) -0.4
   (d) unbounded
12. The output voltage of the circuit in figure for $t > 0$ is

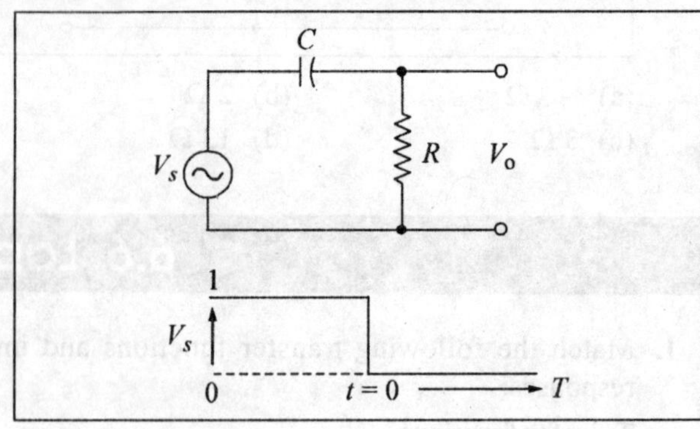

   (a) $e^{-t/RC}$
   (b) $-e^{-t/RC}$
   (c) $1 - e^{-t/RC}$
   (d) $e^{-t/RC} - 1$
13. The transfer function of a passive circuit has its poles and zeroes on
   (a) Left and right halves respectively of the s-plane
   (b) Right and left halves respectively of the s-plane
   (c) Right half of the s-plane
   (d) left half of the s-plane
14. The final value of a function $y(t)$ whose Laplace transform $Y(s) = \frac{4}{s^2 + 2s + 2}$ is
   (a) 4 (b) 2
   (c) 1 (d) 0

6.7 Miscellaneous

1. A periodic rectangular signal, $x(t)$ has the wave form shown in figure. Frequency of the fifth harmonic of its spectrum is

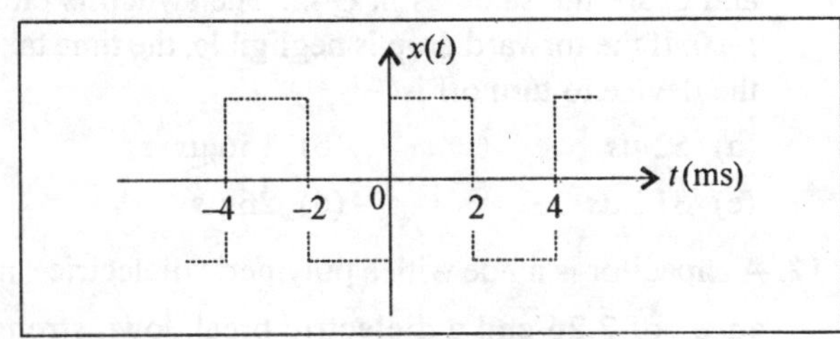

(a) 40 Hz (b) 200 Hz
(c) 250 Hz (d) 1250 Hz

2. Given the relationship between the input $u(t)$ and the output $y(t)$ to be

$$y(t)=\int_0^t (2+t-\tau)\,e^{-3(t-\tau)}\,u(\tau)\,d\tau$$

The transfer function $Y(s)/U(s)$ is

(a) $\dfrac{2e^{-2s}}{s+3}$ (b) $\dfrac{s+2}{(s+3)^2}$

(c) $\dfrac{2s+5}{s+3}$ (d) $\dfrac{2s+7}{(s+3)^2}$

3. What is the rms value of the voltage waveform shown in figure?

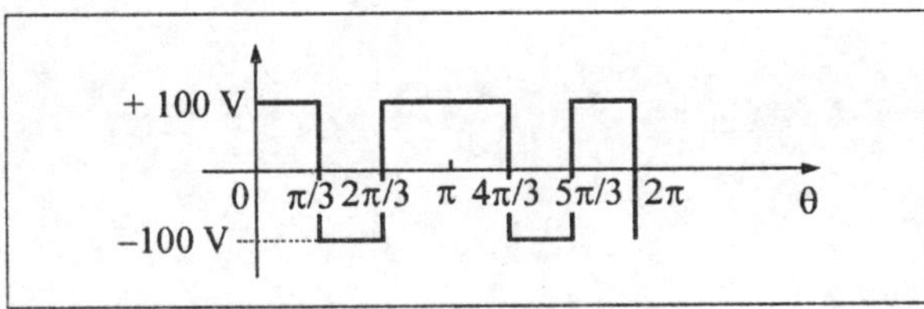

(a) 200/π V (b) 100/π V
(c) 200 V (d) 100 V

4. The rms value of the resultant current in a wire which carries a dc current of 10 A and a sinusoidal alternating current of peak value 20 A is

(a) 14.1A (b) 17.3 A
(c) 22.4 A (d) 30.0 A

5. The RMS value of the voltage $u(t) = 3 + 4\cos(3t)$ is

(a) $\sqrt{17}$ V
(b) 5 V
(c) 7 V
(d) $(3+2\sqrt{2})$ V

6. For the three-phase circuit shown in the figure, the ratio of the currents $I_R : I_Y : I_B$ is given by

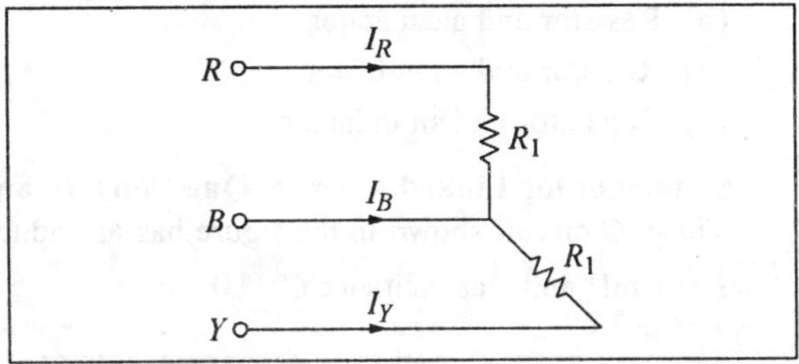

(a) $1:1:\sqrt{3}$
(b) 1: 1 : 2
(c) 1 : 1 : 0
(d) $1:1:\sqrt{3/2}$

7. Let a signal $a_1 \sin(\omega_1 t + \phi_1)$ be applied to a stable linear time-invariant system. Let the corresponding steady state output be represented as $a_2\, F(\omega_2 t + \phi_2)$. Then which of the following statements is true?

(a) F is not necessarily a "sine" or "cosine" function but must be periodic with $\omega_1 = \omega_2$
(b) F must be a "sine" or "cosine" function with $a_1 = a_2$
(c) F must be a "sine" function with and $\phi_1 = \phi_2$
(d) F Must be a "sine" or "cosine" function with $\omega_1 = \omega_2$

8. The state equation for the current I_1 shown in the network shown below in terms of the voltage V_x and the independent source V, is given by

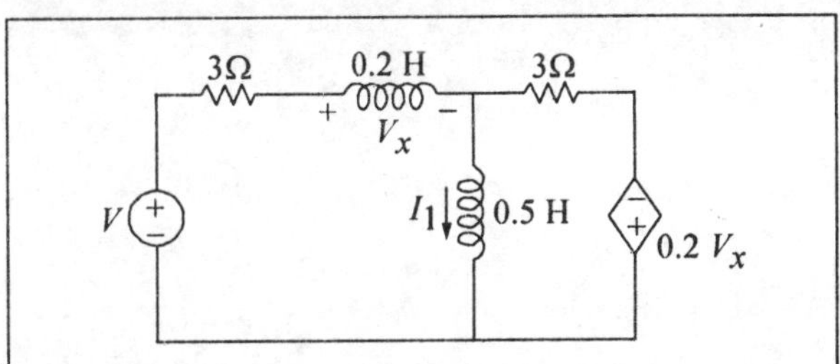

(a) $\dfrac{dI_1}{dt} = -1.4\,V_x - 3.75\,I_1 + \dfrac{5}{4}V$

(b) $\dfrac{dI_1}{dt} = -1.4\,V_x - 3.75\,I_1 - \dfrac{5}{4}V$

(c) $\dfrac{dI_1}{dt} = -1.4\,V_x + 3.75\,I_1 + \dfrac{5}{4}V$

(d) $\dfrac{dI_1}{dt} = -1.4\,V_x + 3.75\,I_1 - \dfrac{5}{4}V$

9. The Thevenin's equivalent of a circuit operating at $\omega = 5$ rad/s, has $V_{oc} = 3.71\angle -15.9°$V and $Z_0 = 2.38 - j$

0.667 Ω. At this frequency, the minimal realization of the Thevenin's impedance will have a

(a) Resistor and a capacitor and an inductor
(b) Resistor and a capacitor
(c) Resistor and an inductor
(d) Capacitor and an inductor

Statement for Linked Answers Questions 10 and 11: The *L-C* circuit shown in the figure has an inductance $L = 1$ mH and a capacitance $C = 10\ \mu F$.

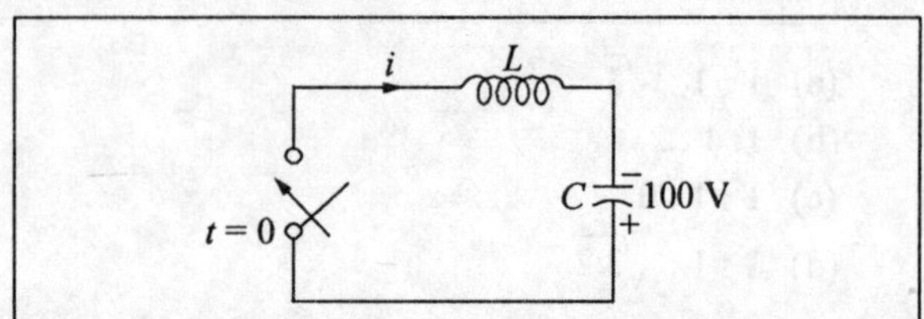

10. The initial current through the inductor is zero, while the initial capacitor voltage is 100 V. The switch is closed at $t = 0$. The current i through the circuit is:

(a) $5 \cos (5 \times 10^3\, t)$ A (b) $5 \cos (10^4\, t)$ A
(c) $10 \cos (5 \times 10^3\, t)$ A (d) $10 \cos (10^4\, t)$ A

11. The *L-C* circuit of *Q*42 is used to commutate a thyristor, which is inintially carrying a current of 5 A as shown in the figure below. The values and initial conditions of *L* and *C* are the same as in *Q*42. The switch is closed at $t = 0$. If the forward drop is negligible, the time taken for the device to turn off is

(a) 52 μs (b) 156 μs
(c) 312 μs (d) 26 μs

12. A capacitor is made with a polymeric dielectric having an ε_r of 2.26 and a dielectric breakdown strength of 50 kV/cm. The permittivity of free space is 8.85 pF/m. If the rectangular plates of the capacitor have a width of 20 cm and a length of 40 cm, then the maximum electric charge in the capacitor is

(a) 2μC (b) 4μC
(c) 8μC (d) 10μC

Answers

6.1 Network Elements

1. **Ans. (c)**

Solution:

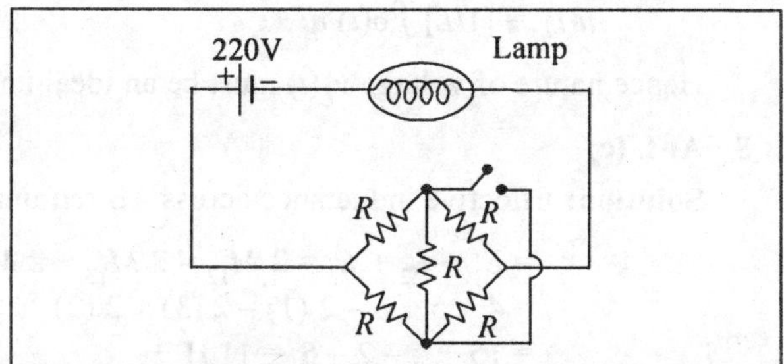

Voltage across the diagonal resistance = 20 × [R/2R] = 10 V from both directions, thus the net voltage is 0. This is equivalent to a short. When the switch is closed, then also diagonal resistance is shorted out. Hence the lamp intensity remains same. A mathematical analysis is given below:

When the switch is open

The circuit is redrawn as follows by converting delta to star.

$$R_{eq} = \frac{R \times R}{3R} = \frac{R}{3}$$

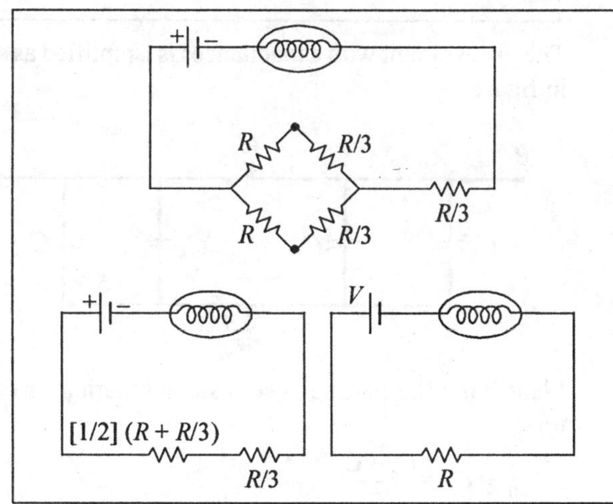

When the switch is closed

Thus the current through the lamp is same when the switch is both open and closed and also the intensity of lamp will be also same

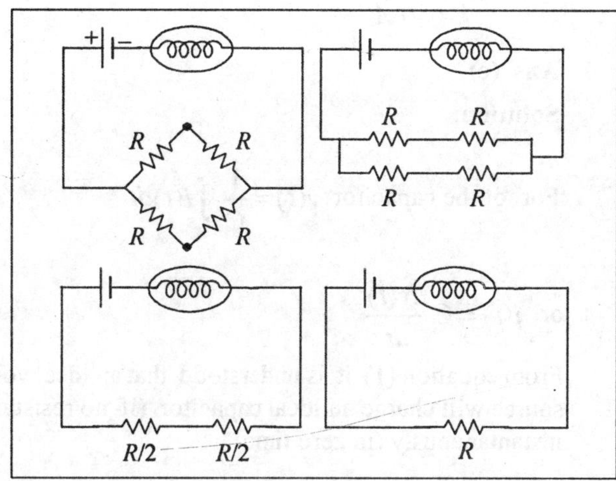

Hence choice (c) is correct.

2. **Ans. (d)**

Solution:

$$I = \frac{1}{1.875} = \frac{1000}{1875} = \frac{8}{15}\text{A}$$

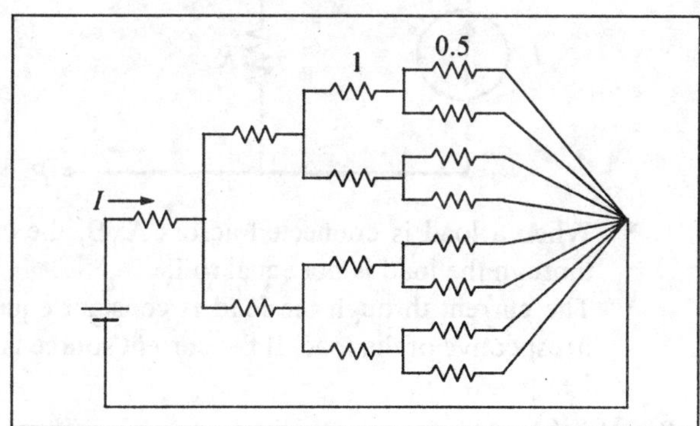

3. **Ans (b)**

Solution:

$V + i = 100$ and $\quad V = i \times 1 = i$

$i = 50\ A$

Second Method

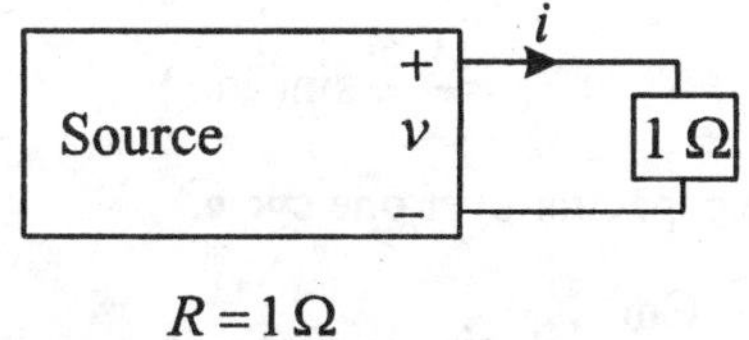

$$R = 1\,\Omega$$

$$v + i = 100$$

$$v = 100 - i$$

$$iR = 100 - i$$

Putting the value of R

$$i = 100 - i$$

$$2i = 100$$

$$i = 50\ A$$

4. Ans. (c)

Solution:

For of the capacitor $v(t) = \frac{1}{C}\int_{-\infty}^{t} i(t)dt$

or $i(t) = C\frac{dv(t)}{dt}$ (1)

From equation (1) it is understodd that an ideal voltage source will charge an ideal capacitor (if no resistance) instantaneously (in zero time).

5. Ans (b)

Solution :

A practical current source is usual;ly represented by a resistance in parallel with an ideal current source as shown in figure.

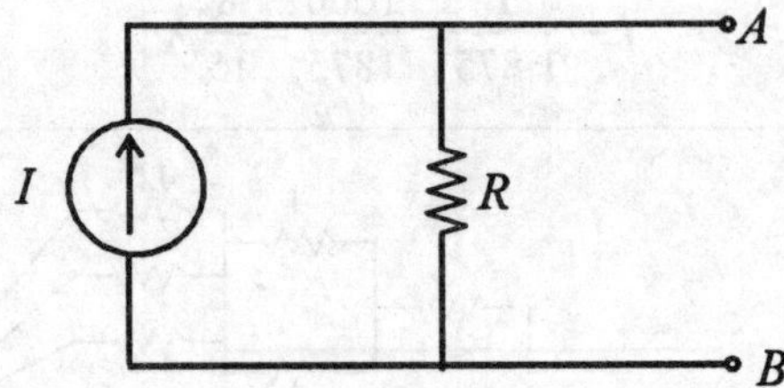

- When a load is connected across A, B, the current thorugh the load is not equal to I.
- The current through the load is constant equal to I irrespective of the load, If the current source is ideal, i.e. $R = \infty$.

6. Ans (c)

Solution :

Let $v(t) = \sin \omega t$,

$$\therefore \quad i(t) = C\frac{dv(t)}{dt} = C\,\omega \cos \omega t$$

Instantaneous power $p(t) = v(t)\ i(t)$

$$= \frac{C\omega}{2}\sin(2\omega t)$$

$\therefore$ Energy in the capacitor over one cycle,

$$\int_0^{2\pi/\omega} p(t)\,dt = \frac{C\omega}{2}\int_0^{2\pi/\omega} \sin(2\omega t)dt = 0$$

7. Ans. (c)

Solution: In the circuit given

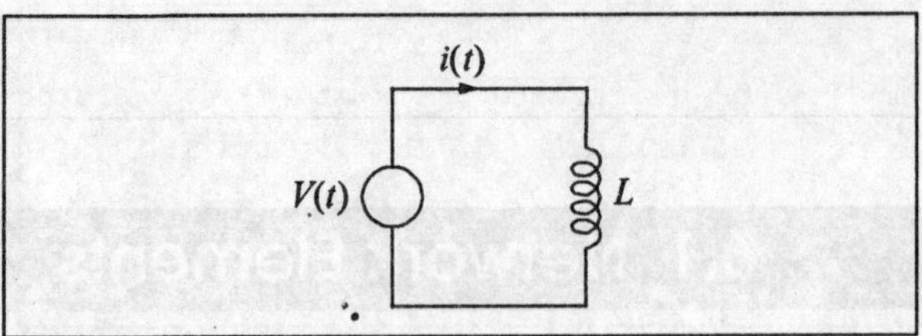

$\therefore$

Constant dc = $u(t)$

$$i_L(t) = [1/L] \int v_L\, dt$$

$$u(t) = [1/L] \int \delta(t)\, dt$$

Hence nature of voltage $v_L(t)$ must be an ideal impulse

8. Ans. (c)

Solution: Effective inductance across AB terminals.

$$= L_1 + L_2 + L_3 - 2\,M_{12} - 2\,M_{13} + 2\,M_{23}$$

$$= 4 + 5 + 6 - 2\,(1) - 2\,(3) + 2\,(2)$$

$$= 15 + 4 - 2 - 6 = 11\text{ H}$$

9. Ans. (c)

Solution:

Use Y–Δ conversion between RBY and then use condition of a balanced Bridge

Second Method

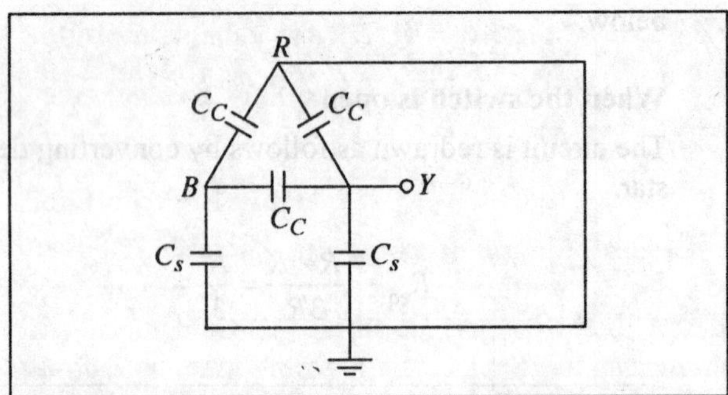

The given circuit with capacitances is ismplified as shown in figure

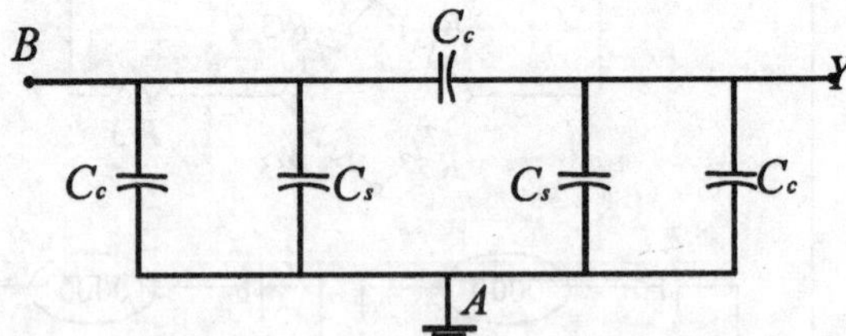

Identifying the parallel - series combination of capacitors.

$$C_{BY} = C_c + \frac{C_c + C_s}{2}$$

$$= \frac{C_s + 3C_c}{2}$$

When point A is removed, then $(C_s + C_c)$ is in series with $(C_s + C_c)$, i.e. parallel with C_c.

10. Ans. (d)

Solution:

The relevant circuit and the input are shown figures below

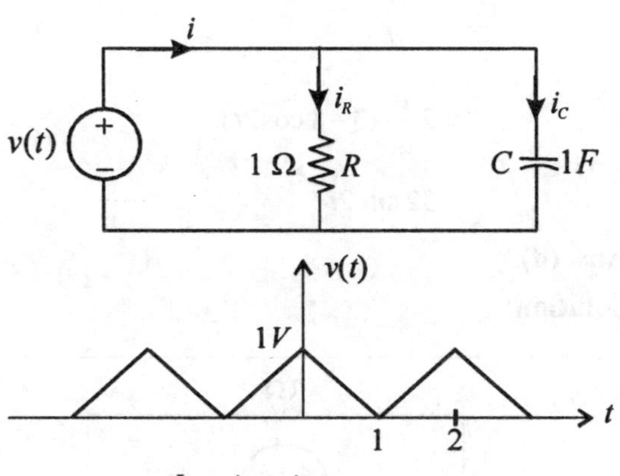

$$I_T = i_R + i_c$$

Where $i_R = [V/R]$ and $i_c = C\,[dV/dt]$

$$i = i_R + i_C$$

$$= \frac{v(t)}{1} + 1 \cdot \frac{dv}{dt}$$

$\frac{dv}{dt}$ is shown in figuer below

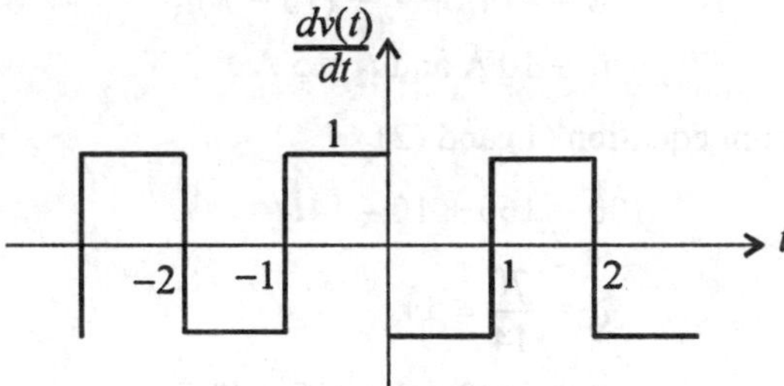

The current through the source $i(t)$ is shown in figuer below

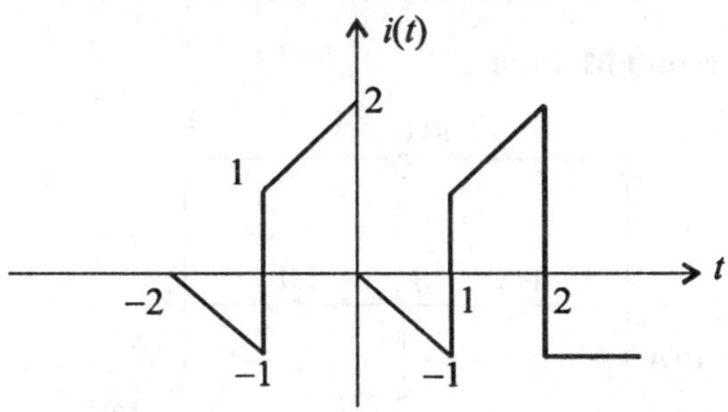

11. Ans. (d)

Solution:

The Bulbs and B_1 and B_2 are rated at $BP_1 = 40\,W$ and $BP_2 = 60\,W$ at same voltage V.

If resistance are R_1 and R_2

$$BP_1 = \frac{V^2}{R_1}$$

$$\therefore \quad R_1 = \frac{V^2}{BP_1}$$

$$\therefore \quad R_2 = \frac{V^2}{BP_2}$$

As $BP_1 < BP_2$

$$R_1 < R_2$$

$$R_1 = \frac{V^2}{40}$$

$$R_2 = \frac{V^2}{60}$$

When they are connected in series

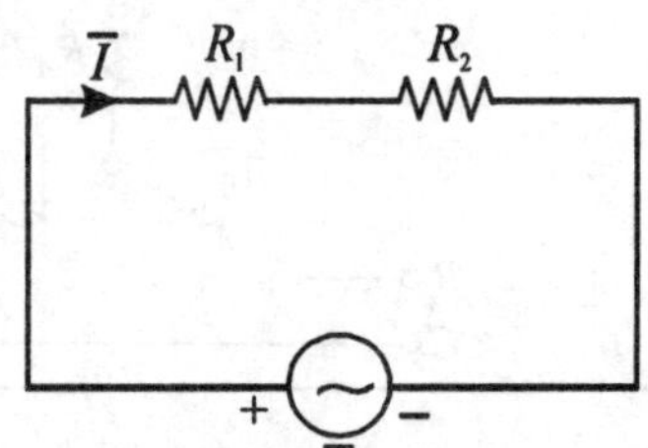

Current $\quad I = \dfrac{V}{R_1 + R_2}$

The power consumed in series combination

$$BP_{S1} = I^2 R_1$$

and $\quad BP_{S2} = I^2 R_2$

As $\quad R_1 > R_2$

$$BP_{S1} > BP_{S2}$$

So 40 bulb glows brighter.

Second Method

Power consumed by incandescent bulb: $P = \dfrac{V^2}{R}$

$$\therefore \quad P \alpha \frac{1}{R} \Rightarrow R \alpha \frac{1}{P}$$

Let R_1 be the resistance of 40 W bulb

R_2 be the resistance of 60 W bulb

Therefore, $R_1 > R_2$

When these two resistors are connected in series across a source then

In a series network, current is same.

$\therefore$ V_1 = voltage across $R_1 > V_2$ = voltage across R_2

Power consumed = VI

Power consumed by 40 W bulb $= V_1 I$

Power consumed by 60 W bulb $= V_2 I$

Therefore, $V_1 I > V_2 I$ $\quad$ [as $V_1 > V_2$]

The 40 W bulb glows brighter

12. Ans. (d)

Solution: Coefficient of coupling

$$K = \frac{M}{\sqrt{L_1 L_2}} < 1$$

$$= 0 < K < 1$$

13. Ans. (b)

Solution:

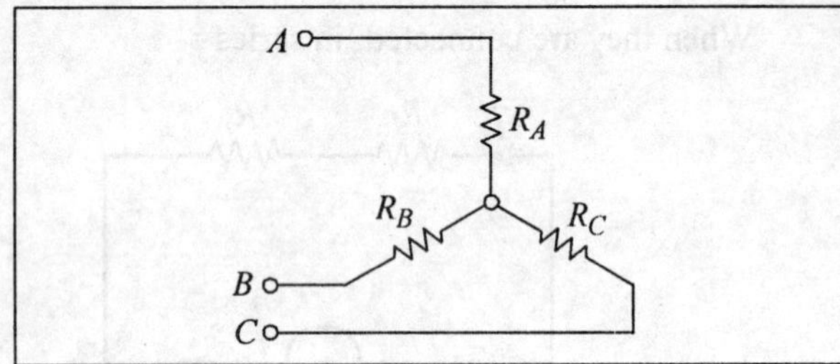

Given, $R_A + R_B = 6$ With C open

$R_B + R_C = 11$

$R_C + R_A = 9$

Then, $R_A + R_B + R_C = [26/2] = 13$

$$R_A = 2\,\Omega,\ R_B = 4\,\Omega,\ R_C = 7\,\Omega$$

14. Ans. (a)

Solution: For 0 – 2 sec, current, I changes with t i.e.

$$\frac{di}{dt} = \frac{6}{2}(\text{amp})/\text{s} = 3A/s$$

$$V_L = L\frac{di}{dt} = 2\times\frac{6}{2} = 6\text{V},$$

$$W = \text{vit} = 6 \times 6 \times 2 = 72 \text{ J}$$

For 2–4 sec; for steady current,

$L \rightarrow$ Short circuit

$$V = I \times R = 6 \times 1$$

$$= 6 \text{ V}$$

$$W = 6 \times 6 \times 2 = 72 \text{ J}$$

Total energy absorbed $= 72J + 72J = 144J$

15. Ans. (b)

Solution: Current through capacitor

$$i_c = C\frac{dv_c}{dt} = 1 \times \frac{d}{dt}(4\sin 2t) = 8\cos 2t$$

KCL at node

$$-2 - 1 + 8\cos 2t + i_L = 0$$

$$i_L = 3 - 8\cos 2t$$

$$V_L = L\frac{di_L}{dt}$$

$$= 2\frac{d}{dt}(3 - 8\cos 2t)$$

$$= 32\sin 2t$$

16. Ans. (d)

Solution:

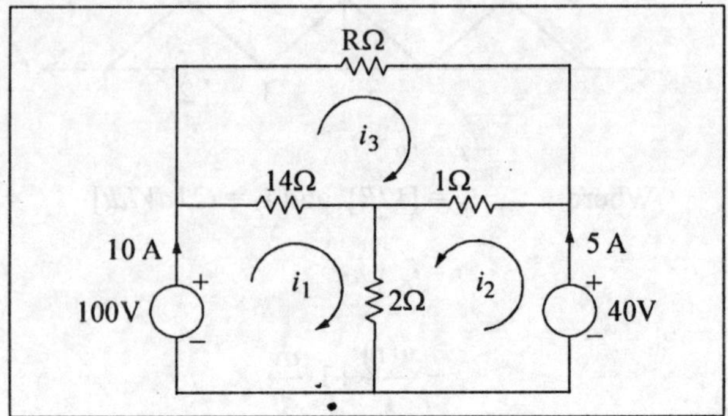

Apply the KVL in all three loops

$$100 = 16i_1 + 2i_2 - 14i_3 \quad \text{... (1)}$$

$$0 = -14i_1 + i_2 + (15 + R)i_3 \quad \text{... (2)}$$

$$i_1 = 10 \text{ A and } i_2 = 5 \text{ A}$$

From equation (1) and (2)

$$100 = 160 + 10 - 14i_3$$

$$i_3 = \frac{70}{14} = 5$$

$$0 = -140 + 5 + (15 + R)\,5$$

$$-(15 + R)\,5 = -135$$

$$R = 27 - 15 = 12\,\Omega$$

Second Method

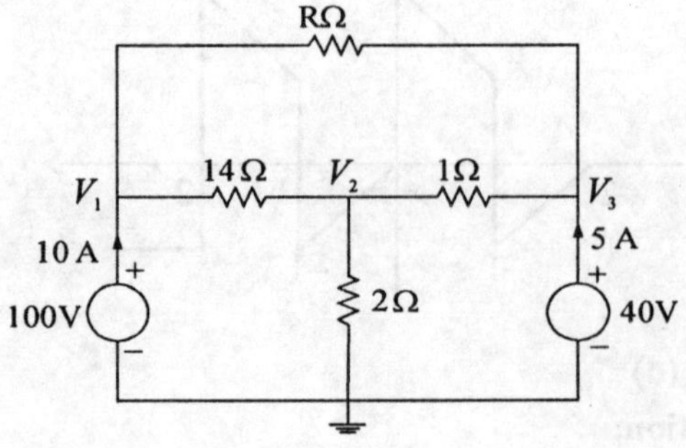

As given

$V_1 = 100\,V$ and

$V_3 = 40\ V$

Apply KCL at V_2 Node

$$\frac{V_2 - V_1}{14} + \frac{V_2}{2} + \frac{V_2 - V_3}{1} = 0$$

By solving $V_2 = 30\ V$

Now apply KCL at V_1 Node

$$-10 + \frac{V_1 - V_2}{14} + \frac{V_1 - V_3}{R} = 0$$

Hence $R = 12\ \Omega$

17. Ans. (c)

Solution:

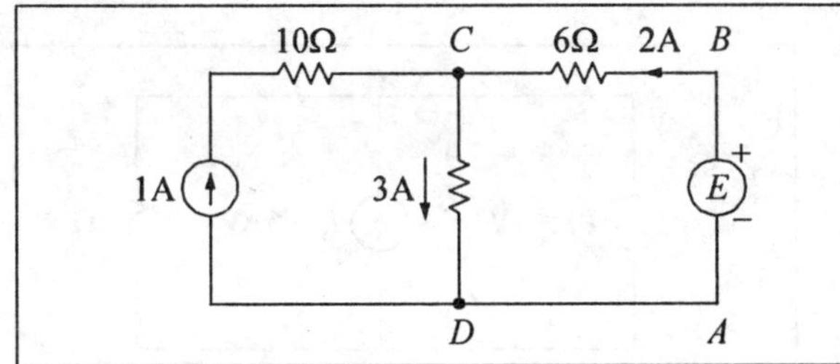

Apply KVL in loop *ABCD*

$$-E + 6 \times 2 + 6 \times 3 = 0$$
$$E = 2 \times 6 + 3 \times 6$$
$$= 12 + 18 = 30\text{V}$$

18. Ans. (a)

Solution: Star-delta conversion is carried out by applying the formula w.r.t. diagram.

$$R_1 = \frac{R_b R_c}{R_a + R_b + R_c} = \frac{100}{40} = 2.5\ \Omega$$

$$R_2 = \frac{R_a R_c}{R_a + R_b + R_c} = \frac{200}{40} = 5\ \Omega$$

$$R_3 = \frac{R_a R_b}{R_a + R_b + R_c} = \frac{200}{40} = 5\ \Omega$$

19. Ans. (b)

Solution: Applying KVL, in loop *ABCF*

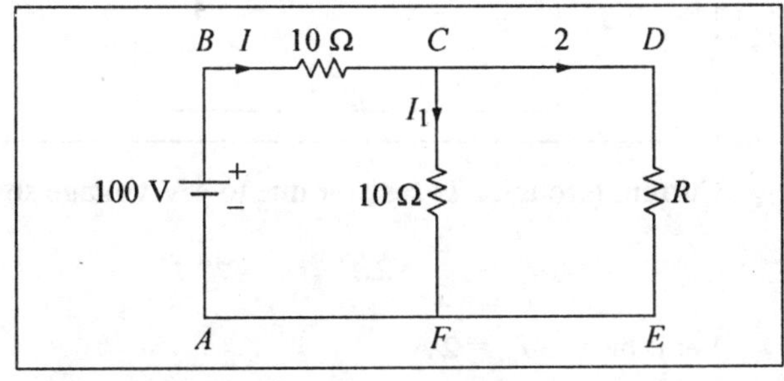

$$-100 + 10I + 10(I - 2) = 0$$

$$\Rightarrow \quad 20I = 100 + 20 = 120$$
$$\Rightarrow \quad I = 60\text{ A.}$$
$$I_1 = I - 2 = 6 - 2 = 4\text{ A}$$

Hence voltage drop across resistance R,

$$V_R = I_1.10 = 4 \times 10 = 40\text{ V} = 2R$$

$$\therefore \quad R = \frac{40}{2} = 20\Omega$$

20. Ans. (c)

Solution: $10\ \Omega \parallel 10\ \Omega = 5\ \Omega$

KVL in loop, $-100 + (R + 5)\ 8 = 0$

or $\quad -100 + 8R + 40 = 0$

or $\quad R = \frac{60}{8} = 7{\cdot}5$

21. Ans. (a)

Solution: Total voltage drop

$$V = IR + I^2$$
$$3 = 2I + I^2$$
$$I^2 + 2I - 3 = 0$$

Neglect the negative value of I

$\therefore \quad I = 1$ amp

$\therefore \quad V_{NL} = 1\text{V}$

$(\because V_{NL} = I^2)$

Power = 1

22. Ans. (a)

Solution:

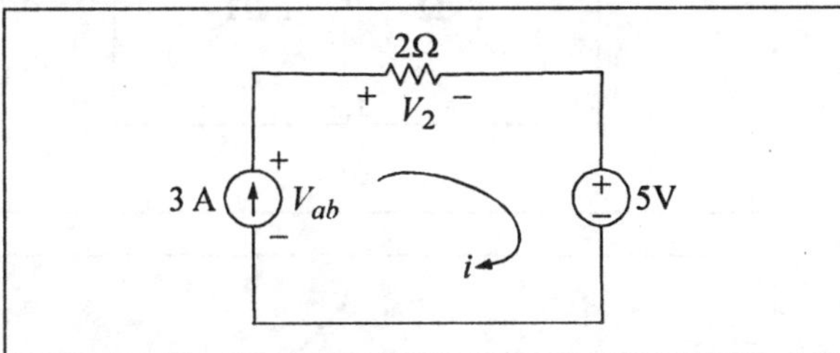

In series current is same

$$\therefore \quad i = 1\text{ A}$$
$$V_2 = 1 \times 2 = 2\text{ V}$$
$$-V_{ab} + V_2 - 5 = 0$$
$$V_{ab} = 2 - 5 = -3\text{ V}$$

23. Ans. (b)

Solution: KVL in 5 V source loop

$$-5 + 2i_1 = 0$$
$$i_1 = 2.5$$
$$V = 2.5\ V$$

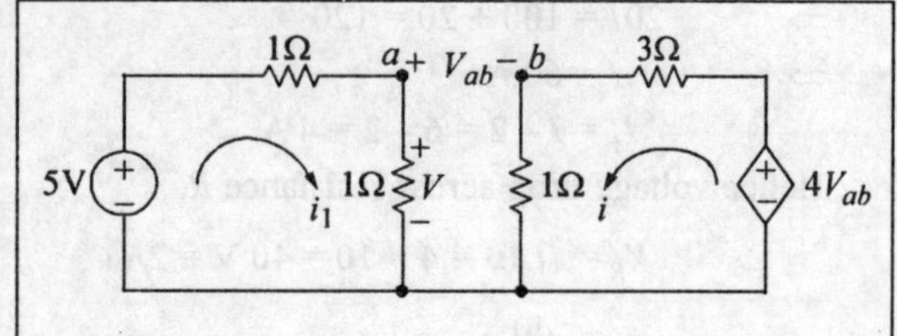

KVL in V_{ab} loop

$$-2.55 + V_{ab} + i = 0$$

$$V_{ab} + i = 2.5 \quad (1)$$

KVL in dependent source loop

$$-4V_{ab} + 3i + i = 0$$

$$i = V_{ab}$$

Put this value in equation (1)

$$i + i = 2.5$$

$$i = \frac{2.5}{2} = 1.25\ A$$

24. Ans. (a)

Solution:

It is a balanced Wheatstone bridge

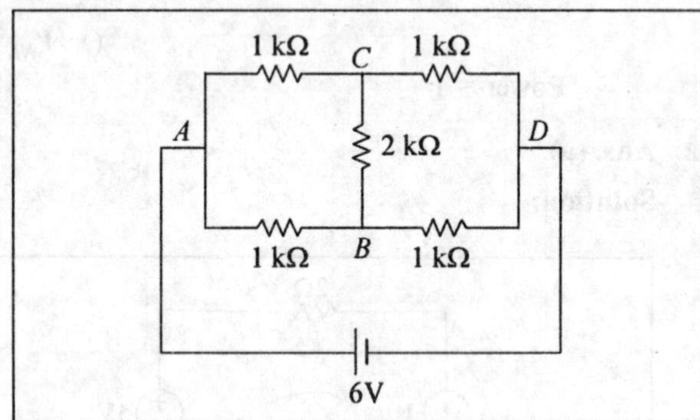

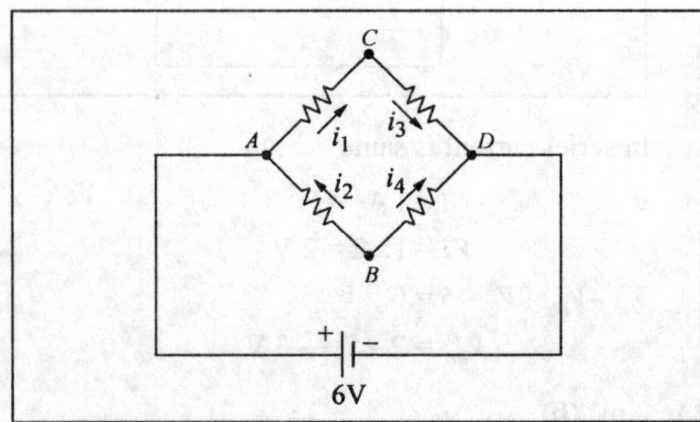

This circuit can be reduced to balance wheat stone bridge because opposite arm impedance are equal.

As it is a wheatstone bridge, so whole current will flow through four resistive arms. No current will flow through CB.

25. Ans. (a)

Solution: Appy KVL for input loop

$$V_{in} = (2\text{k}\Omega)I_1 + (49I_1 + I_1)/(100\mu\text{F})s$$

$$= (2kI_1) + \frac{50I_1}{(100\mu\text{F})s} = 2\ kI_1 + \frac{I_1}{(2\ \mu\text{F})s}$$

$$C_{in} = 2\ \mu\text{F}$$

26. Ans. (b)

Solution: By superposition theorem,

When 4V is shorted, $i^1 = 0$, as all the current will flow through short. When 5A is open, $i^{11} = 2$ A

Therefore $i^2\Omega = 2$A

Second Method

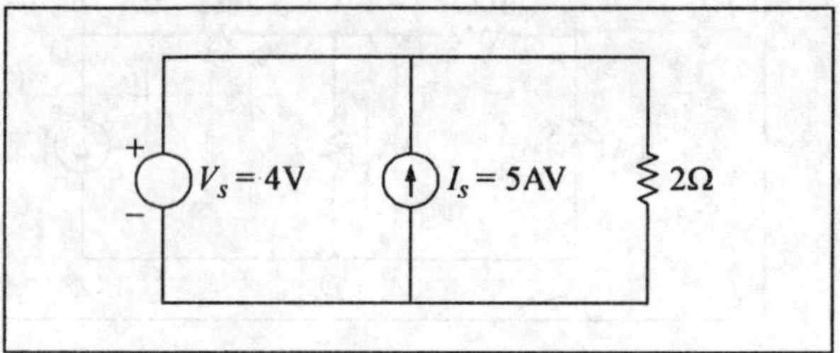

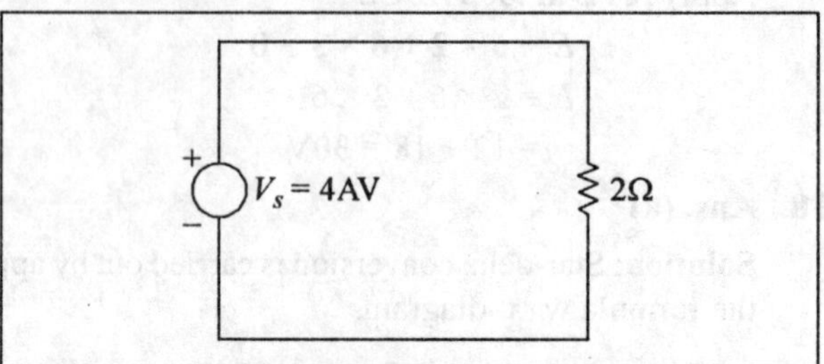

Let current flowing through 2Ω resistance = I

Apply superposition theorem to this circuit.

Let's first take voltage source alone. Open circuit the current source. Hence, the circuit look like.

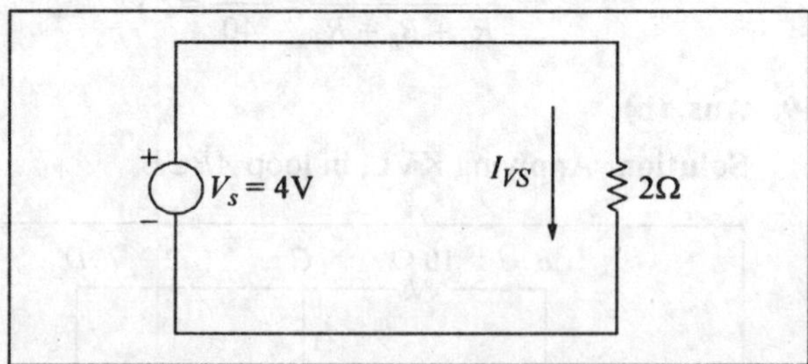

Current through 2 Ω resistor due to 4 V voltage source

$$= \frac{4}{2} = 2A$$

Let it be $\quad I_{vs} = 2$ A

Now current source alone

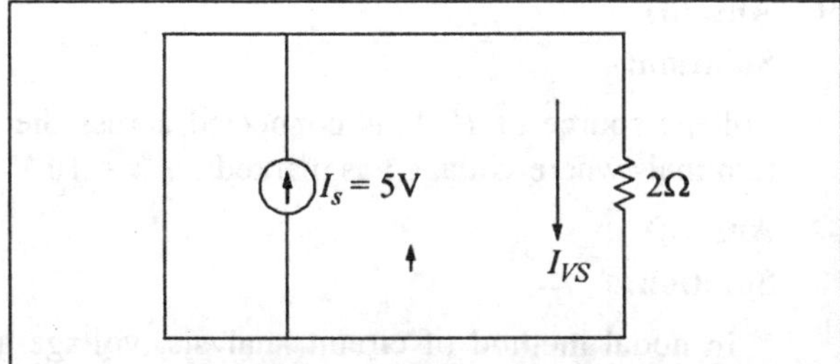

Now short circuit the voltage source and take the current source current through 2 Ω resistor due to this current source I_s = 5A is 0, because current follow through easy path. So whole of the current flows through short circuited path only.

Now Total current = $I_{vs} + I_{is}$

$= 2 + 0 = 2$ A

If we make the V_s = 8V , then current through 2 Ω resistor becomes double.

Third Method

Apply superposition theorem consider one source at a time

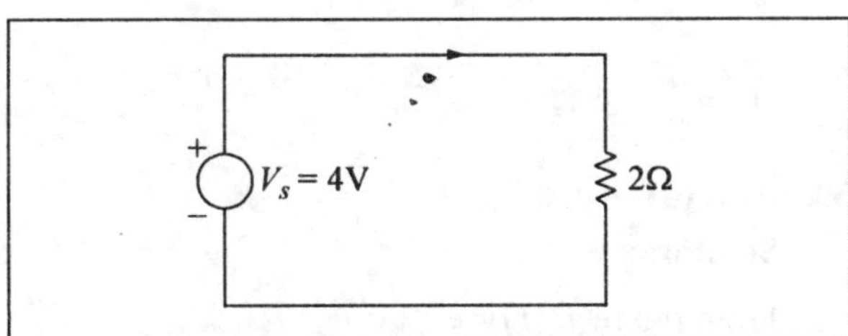

$$I = \frac{4}{2} = 2\text{A}$$

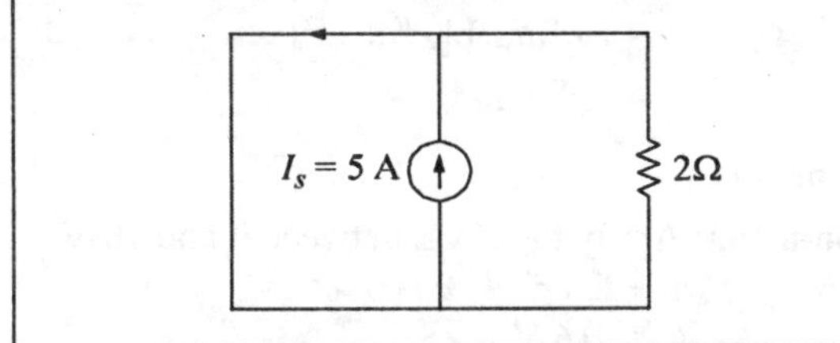

In this case no current will flow through 2 Ω resistance since the other parallel path represents a short circuit path.

Hence, the resultant current in 2 Ω resistance is 2 Amp. Also, to double the current through 2 Ω the voltage source should be doubled i.e. 8 V.

27. Ans. (c)

28. Ans. (a)

Solution:

For the same orientation of Turns like

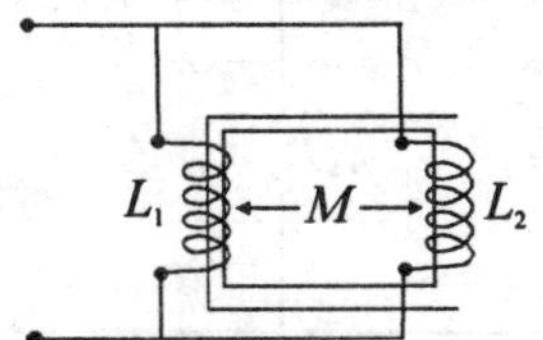

Equivalent inductance of paralled combination

$$L_{eq} = \frac{L_1 L_2 - M^2}{L_1 + L_2 + 2M}$$

Note: $M = 0$

or, $\frac{1}{L_{eq}} = \frac{1}{L_1} + \frac{1}{L_2}$

So L_{eq} decreases in parallel combination. It further decreases due to M becasue the flux of L_1 and L_2 cancels each other hence net flux decreases drastically .

So in the circuit, given in equation, L_{eq} decreases drastically.

It is much less than $\frac{L}{3}$

29. Ans. (b)

Solution: Energy = Power × time

$$\text{Power} = \frac{(\text{Voltage})^2}{\text{Resistance}}$$

$$E = [(V^2/R)]\ t$$

$$2300 = [(230)^2/R] \times 1$$

$\Rightarrow$ $R = 23\ \Omega$

$$\text{rms value} = \sqrt{\frac{200^2 \times (T/2) + (-200)^2 \times (T/2)}{T}}$$

$$= 200$$

$$p = [V^2/R] = [(200)^2/23] = 1739.13 \text{ watt}$$

$$= 1.73913 \text{ kW}$$

30. Ans. (b)

Solution: Apply the KCL at Node *A*

$$-2 + 1 + \frac{12-6}{R} = 0$$

or $\frac{6}{R} = 1$

or $R = 6\ \Omega$

31. Ans. (d)

Solution: Current and voltage follow the superposition theorem. Since power is not a linear function it does not follow superposition.

V_1 alone $P = I^2 R$

$$4 = I_1^2 R\ ;\ I_1 = [\sqrt{(4/R)}] = [2/\sqrt{R}]$$

I_2 alone $P = I^2 R$

$$9 = I_2^2 R\ ;\ I_2 = [\sqrt{(9/R)}] = [3/\sqrt{R}]$$

For, V_1 and I_2 ;

When both I_1 and I_2 are in same direction

$$I = I_1 + I_2 = \frac{2}{\sqrt{R}} + \frac{3}{\sqrt{R}} = \frac{5}{\sqrt{R}}$$

$$P = I^2 R = [25/R] \times R = 25 \text{ Watt}$$

32. **Ans. (d)**

Solution:

Given $L_1 = 2H$, $L_2 = 2H$, $k = 0.1$

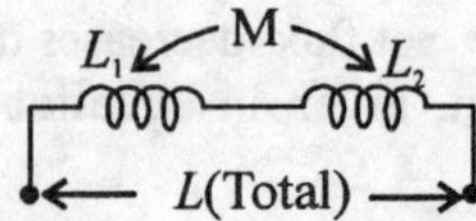

Mutual Inductance $M = k\sqrt{L_1 L_{12}}$

$= 0.1\sqrt{2 \times 2} = 0.2H$

Total Inductance of magnetically coupled series inductors

$$L = L_1 + L_2 \pm 2M$$
$$L = 2 + 2 \pm 0.4$$
$$= 4.4 \text{ H or } 3.6 \text{ H}$$

* '+' sign for series aiding and
* '–' sign for series opposing

33. **Ans. (b)**

Solution: KCL at Node

$I_0 + i_1 + i_4 = 0$

$5 + 7 + i_4 = 0$; $i_4 = -12$

34. **Ans. (a)**

Solution:

Apply KVL in the outer loop $V = 4 + 4 - 5 = 3$ Volts.

35. **Ans. (d)**

Solution:

Circuit is redrawn in figure below

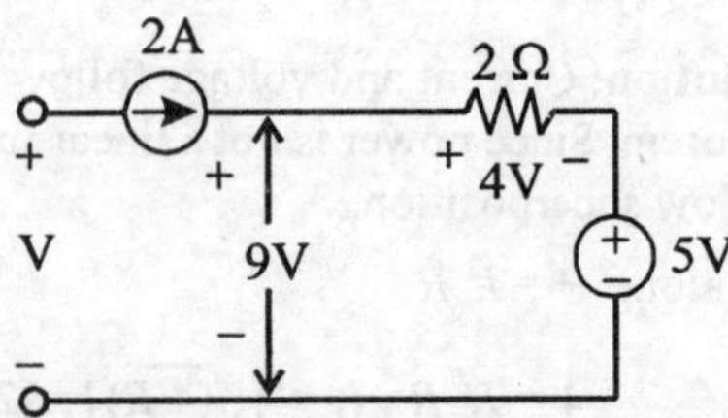

Voltage V cannot be determined as the voltage across the current source of 2 A is not known and the data is incomplete.

36. **Ans. (a)**

Solution:

Voltage source of 10 V is connected across the input terminals where voltage v is marked. ∴ $v = 10$ V.

37. **Ans. (b)**

Solution:

* In nodal method of circuit analysis, voltage at the nodes are the variables.
* KCL is applied at each node
* Ohm's law is used to find the currents in the branches meeting at the node.

38. **Ans. (c)**

Solution:

KCL at Node A

$$\frac{V_a - 1}{2} + \frac{V_a}{2} - 3 = 0$$

$$V_a - 1 + V_a - 6 = 0$$

or $2V_a = 7$

$$V_a = \frac{7}{2} = 3.5\,V$$

39. **Ans. (d)**

Solution:

From figure $i_L(t) = e^{at} + e^{bt}$

Voltage across inductor

$$V_L = L\,[di/dt]$$
$$= 1 \cdot [d/dt]\,[e^{at} + e^{bt}]$$
$$= ae^{at} + be^{bt}$$

40. **Ans. (a)**

Solution: Apply the KVL between 0 and 10 V

$1 + E + 5 + 10 = 0$

$E = -16$ V

41. **Ans. (c)**

Solution: $Z_{in} = 4 + [4 \parallel 2 + 2]$

$= 4 + [4 \parallel 4] = 4 + 2 = 6$

$I = [12/6] = 2$ A

$I_{4\Omega} = 1$ A;

$e_o = 4 \times 1 = 4$ V

Second Method

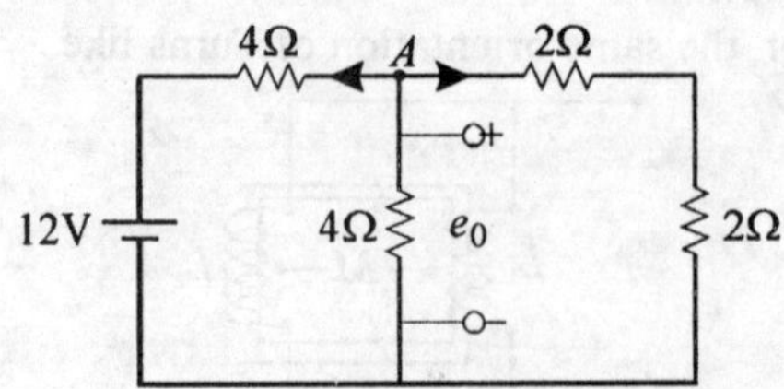

Apply KCL at node 'A'

$$\frac{e_o-12}{4}+\frac{e_o}{4}+\frac{e_o}{4}=0$$

$e_o = 4$ Volts

42. Ans. (a)

Solution:

Given $V_1 = 20\ V$

$$20 = 5I + [(V_1/5) + I]\,5$$
$$= 5I + (4+1)\,5$$
$$20 = 5I + 20 + 5I; \quad I = 0$$

Power delivered = $[(20/5)^2] \times 5 = 4^2 \times 5$
= 80 Watt

Second Method

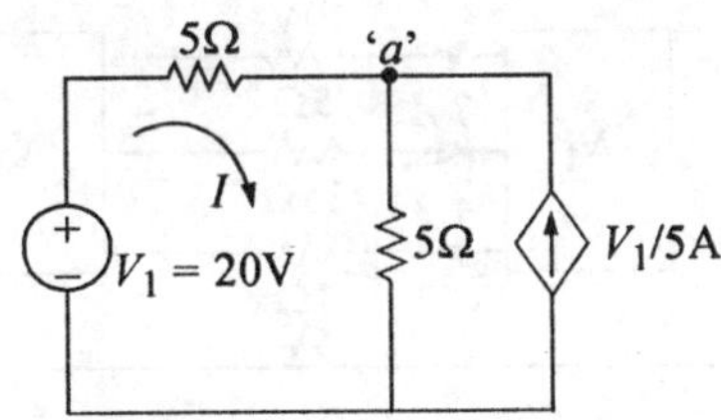

Apply KCL at node 'a'

$$\frac{V_a-V_1}{5}+\frac{V_a}{5}-\frac{V_1}{5}=0$$

or, $$\frac{V_a-20}{5}+\frac{V_a}{5}-\frac{20}{5}=0$$

$$\frac{2}{5}V_a=8$$

$$V_a = 20\ V$$

Here V_a is a positive voltage.
The power is delivered by a curent source when current leaves the positive terminal of a voltage drop across source. The curent of current source in the given circuit leaves the positive terminal so the source is delivering the power.
The power delivered by the source can be given as,

$$P = V_a \times \frac{V_1}{5}$$

$$= 20 \times \frac{20}{5} = 80\ W$$

43. Ans. (c)

Solution:

Using KVL equation the loop

$V(t) = Ri(t) + L[di(t)/dt] + [1/c] \int i(t)\,dt$

$\sin t = 2i(t) + 2[di(t)/dt] + \int i(t)\,dt$

Differentiate the above equation
$\cos t = 2\ [di(t)/dt] + 2[d^2i(t)/dt^2] + i(t)$

44. Ans. (d)

Solution:

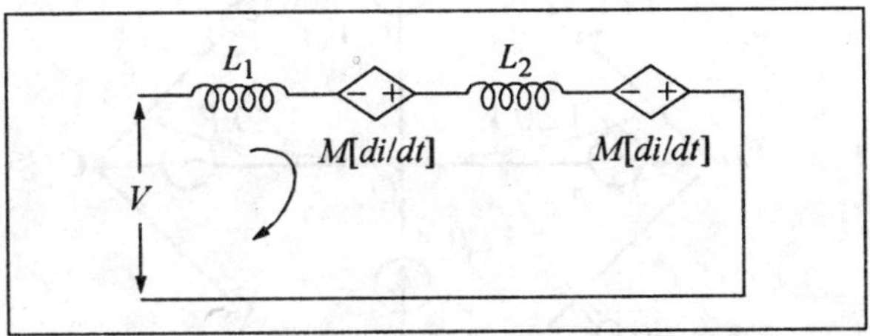

$$V = L_1\frac{di}{dt} - M\frac{di}{dt} + (L_2 - M)\frac{di}{dt}$$

$$= (L_1 + L_2 - 2M)\ [di/dt] = L[di/dt]$$

$$L = L_1 + L_2 - 2M$$

45. Ans (a)

Solution:

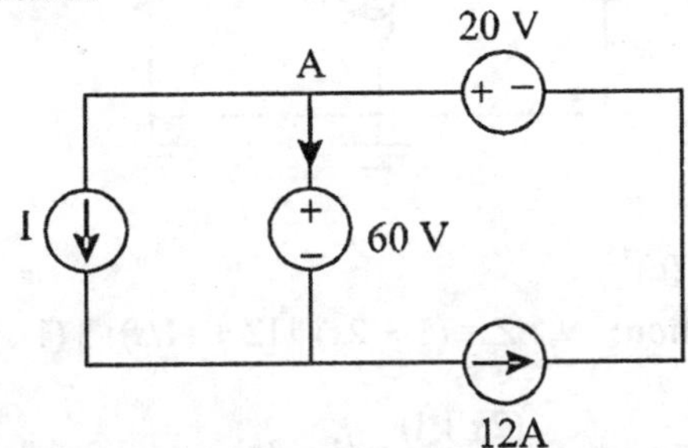

Power absorbing element is 60 V source as the current entering to positive terminal of 60 V source . At point A current division takes place the sum of I current and current through the 60 V source is 12 A.
Less than 10 A is no one option.
Hence 10 A is correct choice.
Since 12 A current is coming from one source and it is also known that 60 V source is absorbing power i.e. current is flowing inside 60 V source $12 = x + 1 \Rightarrow I = 12 - x$, thus possible option is (a)

46. Ans.(c)

Solution:

Considering OABCO loop, both voltage sources are in opposite polarity with equal magnitude then net current is zero.
Now apply the current division rule at Node B as

$$I_L = 1\angle 0A \times \frac{1}{1+j1}$$

$$= \frac{1}{1+j}A$$

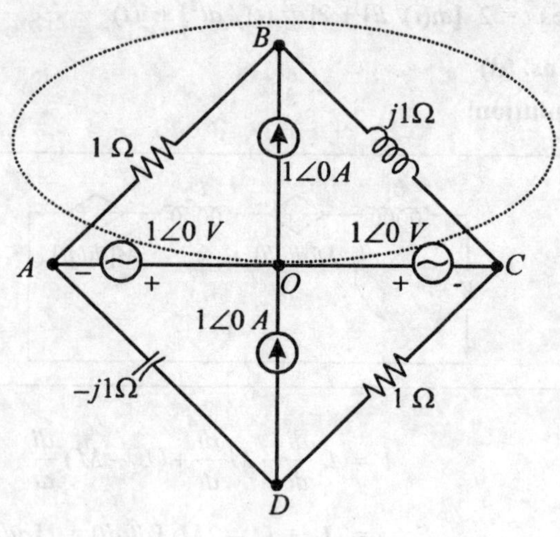

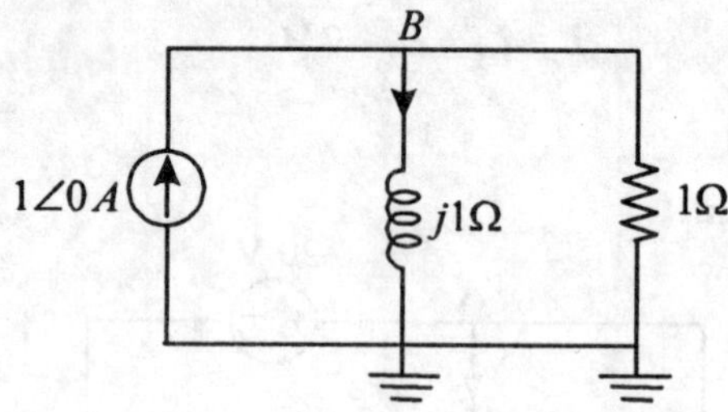

47. Ans. (c)

Solution: $N_i : Z = (1 + 2s) \parallel [2 + (1/s)] \parallel (1/s)$

$$= \frac{\frac{(2s+1)}{s} \times (1+2s)}{\frac{(2s+1)}{s} \times 1 + 2s} \parallel [1/s]$$

$$= \frac{(2s+1) \times (1+2s)}{(2s+1) + (1+2s) \cdot s} \parallel [1/s]$$

$$= \frac{(2s+1)^2}{|(2s+1) + s(1+2s)} \parallel [1/s]$$

$$= \frac{(2s+1)^2}{(2s+1) \times (s+1)} \parallel [1/s]$$

$$= \frac{(2s+1)}{(s+1)} \parallel [1/s]$$

$N_2 : Z = [2s + 1] \parallel [2 + (1/s)] = [(2s + 1)/(s + 1)]$

$N_3 : Z = [1/s] \parallel [1 + s(s + 1)]$

$= [1/s] \parallel [(2s + 1)/(s + 1)]$

$N_4 : Z = [1/s] \parallel [2s + 1]$

N_1 & N_3 are identical

48. Ans. (d)

Solution:

$$V_{R1} = \frac{V}{R_1 + R_2} \times R_1; \quad V_{R2} = \frac{V}{R_1 + R_2} \times R_2$$

If R_1 and R_2 Double

$$V_{2R1} = \frac{V}{2(R_1 + R_2)} \times 2R_1 = V_{R_1}$$

$$V_{2R2} = \frac{V}{2(R_1 + R_2)} \times 2R_1 = V_{R_2}$$

Hence voltage across resistor does not change.

49. Ans. (a)

Solution: KCL for a cutest state that algebraic sum of the current entering or leaving a cut set is equal to zero

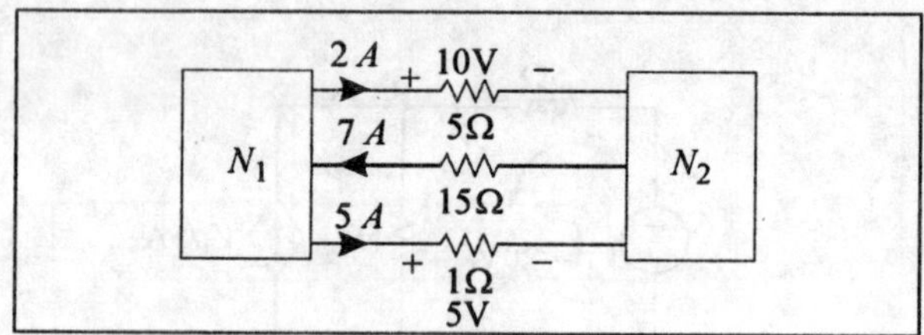

$$I_5 + I_{15} + I_1 = 0$$

or, $\quad [10/5] + I_{15} + [5/1] = 0$

$$I_{15} = -7\text{A}$$

$$V_{15} = -7 \times 15$$

$$= -105 \text{ V}$$

50. Ans. (d)

Solution:

The didoe is forward bised as anode voltage is more positive. Assuming that the didoe is ideal, the network is redrawn with node A marked as in figuer

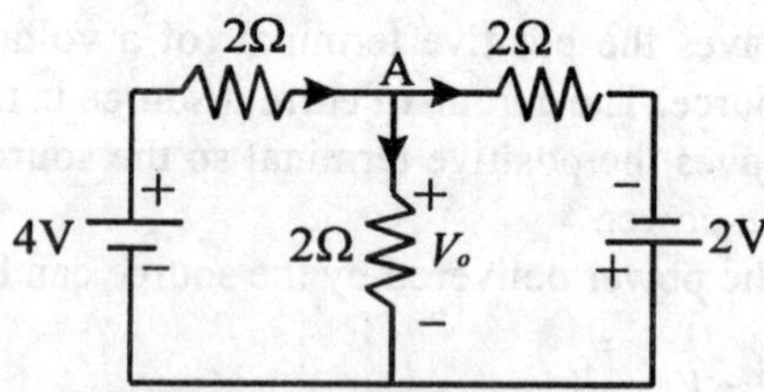

Apply KCL at node A

$$\frac{4 - V_o}{2} = \frac{V_o}{2} + \frac{V_o + 2}{2}$$

$$\frac{3V_o}{2} = 1$$

$$V_o = \frac{3}{2} V$$

51. Ans. (a)

Solution:

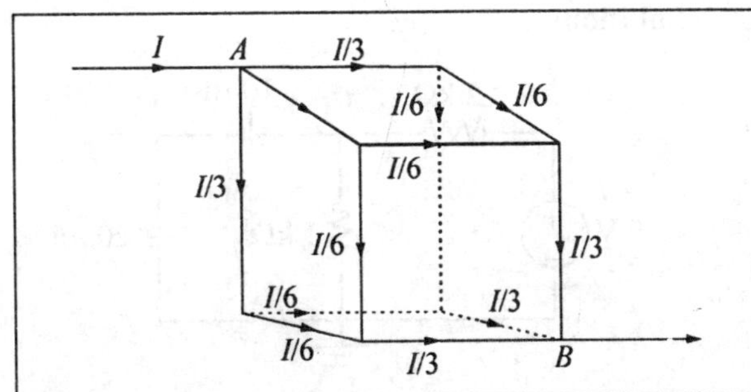

$R = 1$; $R_{AB} = [V_{AB}/1]$

$$V_{AB} = \left(\frac{1}{3} + \frac{1}{6} + \frac{1}{3}\right) R$$

$$= \frac{2I + 1I + 2I}{6}.R = \frac{5I}{6}.R = [5/6].1$$

$$V_{AB} = [5I/6]$$

$$R_{AB} = [V_{AB}/I] = [5/6]\ \Omega$$

52. Ans. (b)

Solution:

$$Z = j5\,\Omega + j2\Omega + j2\Omega + 2(j10) - 2(j10)$$

$$= j\,9\Omega$$

53. Ans. (a)

Solution:

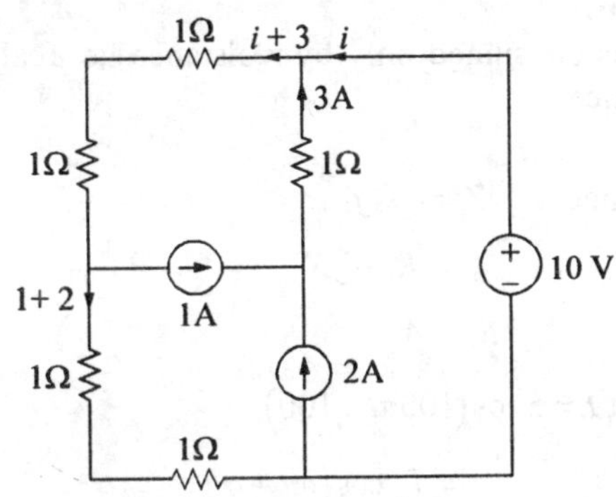

Apply KVL in loop

$$10 = 2(i+3) + 2(i+2)$$

$$10 = 10 + 4i$$

$\Rightarrow$ $i = 0$

$\therefore$ Power = 0

54. Ans. (a)

Solution:

$$I = \frac{V_A - V_B}{2} = \frac{6}{2} = 3A$$

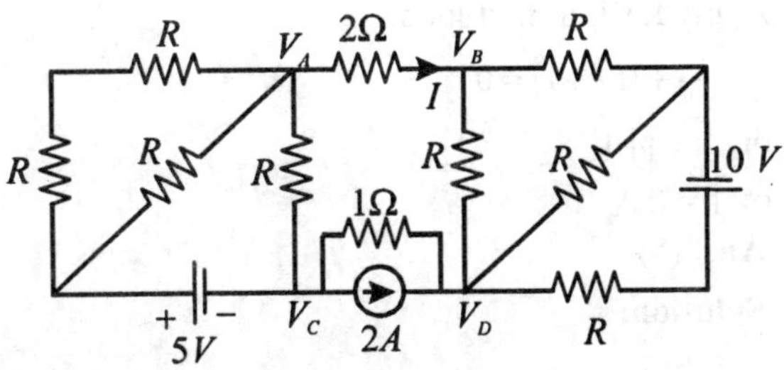

Represent the LHS and RHS circuit by network N_1 and network N_2 as below

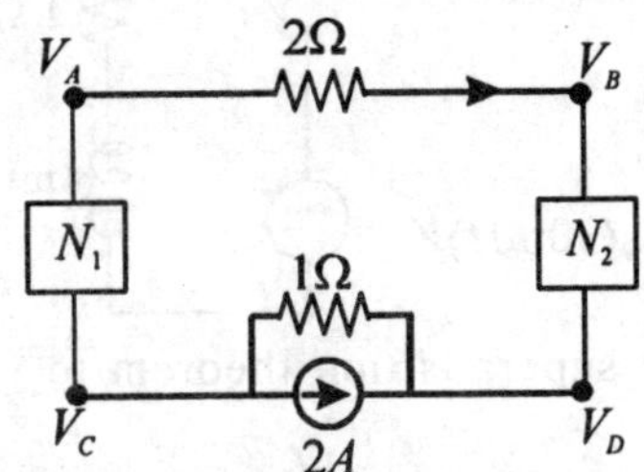

Since current entering any network is same as leaving i.e. in series circuit current is same in all element

In $V_C - V_D$ branch also it is $I = 3A$

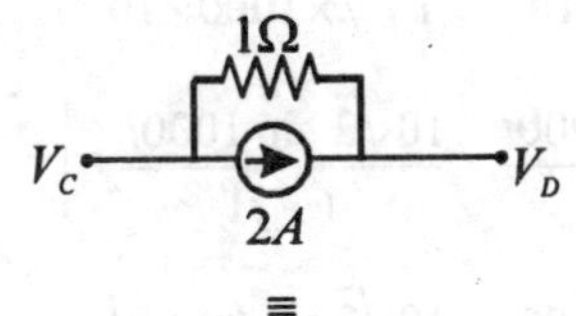

$\equiv$

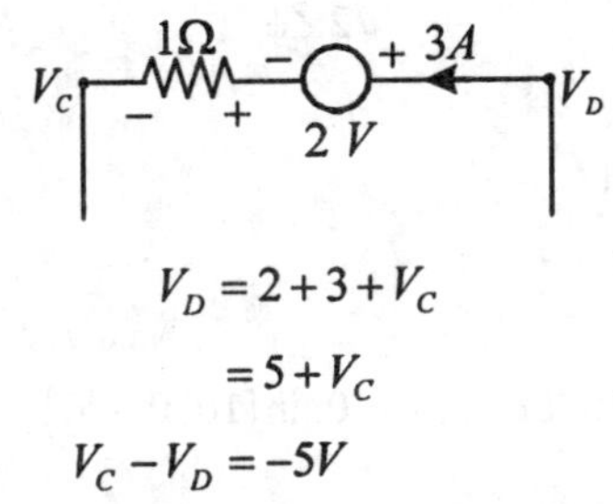

$$V_D = 2 + 3 + V_C$$

$$= 5 + V_C$$

$$V_C - V_D = -5V$$

55. Ans. (a)

Solution:

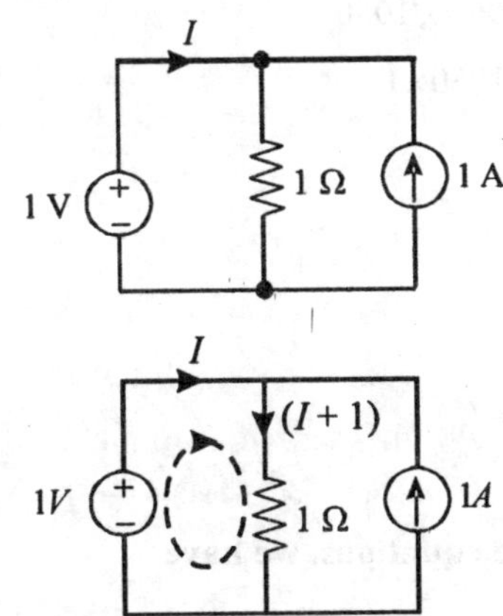

Apply KVL in first loop

$$-1+1(I+1)=0$$

or, $1 = I+1$

or $I = 0\,A$

56. Ans. (b)

Solution:

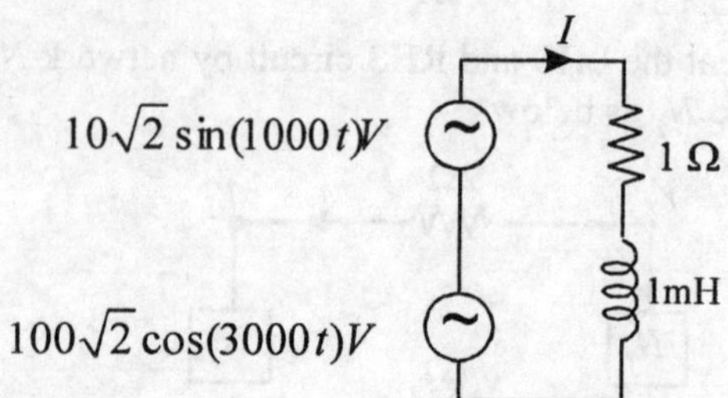

Apply the superposition theorem to get the result current I.

$$I=\frac{100\sqrt{2}\cos 3000t}{1+j\omega_1 L}+\frac{10\sqrt{2}\sin 1000t}{1+j\omega_2 L}$$

$$I=\frac{100\sqrt{2}\cos 3000t}{1+j\times 3000\times 10^{-3}}+\frac{10\sqrt{2}\sin 1000t}{1+j\times 1000\times 10^{-3}}$$

$$=\frac{100\sqrt{2}\cos 3000t}{1+j3}+\frac{10\sqrt{2}\sin 1000t}{1+j1}$$

$$=\frac{100\sqrt{2}\cos 3000t}{\sqrt{10}\angle\phi_1}+\frac{10\sqrt{2}\sin(1000)}{\sqrt{2}\angle\phi_1}$$

Where,

$\phi_1=\tan^{-1}(3)$ and

$\phi_2=\tan^{-1}(1)$

$$I=\frac{100\sqrt{2}}{\sqrt{10}}\cos(3000t-\phi_1)+10\sin(1000-\phi_2)$$

$$I_{rms}=\sqrt{\frac{1}{2}\left[\frac{10000\times 2}{10}+100\right]}$$

$$=\sqrt{1000+50}=\sqrt{1050}$$

$$P=I_{rms}^2\times R=1050\times 1$$

$$=1050\text{ Watt}$$

57. Ans (d)

Solution:

$$I_1+I_2=I_3+I_6$$

$$I_3+I_6+I_5=I_4+I_7$$

$$I_7=I_5+I_1+I_2$$

Conbining these equations, we have

$$I_7=I_3+I_5+I_6$$

58. Ans. (c)

Solution:

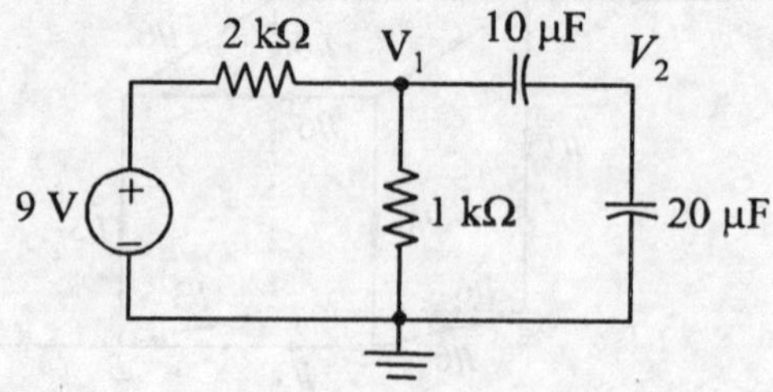

Voltage at V_1 according to potentional divider method is

$$V_1=\frac{9\times 1\,k\Omega}{(2+1)\,k\Omega}\Rightarrow 3V$$

Voltage at $V_1 = 3\,V$

Voltage at V_2 =

$$\frac{3\times\frac{1}{20\times 10^{-1}j}}{\frac{1}{j\times 10\times 10^{-6}}+\frac{1}{20\times j\times 10^{-6}}}$$

$$=\frac{3}{20}\times\frac{200}{30}=1\,V$$

$$V_2=1\,V$$

59. Ans.(b)

Solution:

Power is consumed only by resistance i.e. real part of impedance

Given

Impedance $Z=4-j3$

$$=R_L-jX_c$$

$\therefore$ $R_L=4$

Current $I=5\cos(100\pi t+100)$

$$=I_m\cos(\omega t+\alpha)$$

Average power is same as RMS power

Average power $P=\frac{1}{2}I_m^2R_L$

$$=\frac{1}{2}\times 5^2\times 4$$

$$=50\,W$$

Second Method

Impedance $=(4-3j)\Omega$

Current $=5\cos(100\,\pi t+100)\,A$

It is known that

$$P_{avg} = \frac{1}{2}\text{Re}\{VI^*\} = \frac{1}{2}\text{Re}\left(|IZ\,I^*|\right)$$

$$= \frac{1}{2}\text{Re}\left(\| I |^2 Z|\right)$$

$$= \frac{1}{2}| I |^2 \text{Re}\left(|Z|\right)$$

$$P_{avg} = \frac{1}{2} \times 5^2 \times 4 \qquad [\text{as Re}(Z) = 4]$$

$$= 50\,W$$

60. Ans. (a)

Solution:

$$I = \frac{V_A - V_B}{2} = \frac{6}{2} = 3A$$

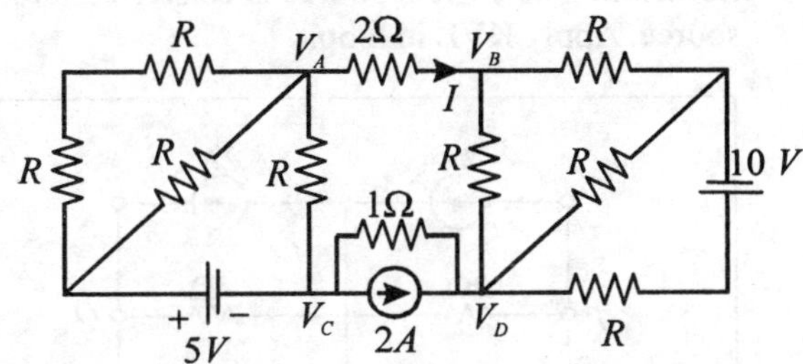

Represent the LHS and RHS circuit by network N_1 and network N_2 as below

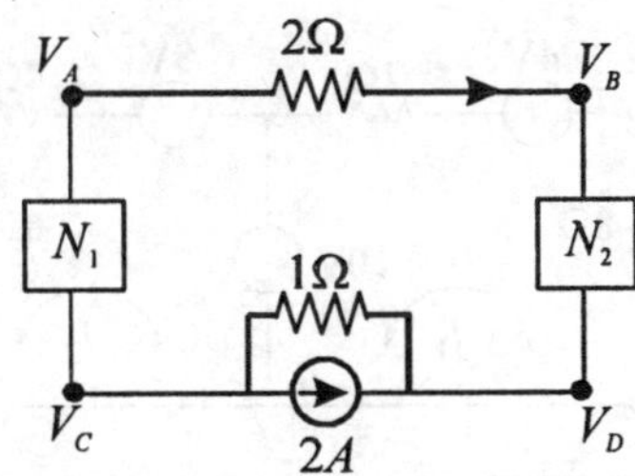

Since current entering any network is same as leaving i.e. in series circuit current is same in all element

In $V_C - V_D$ brance also it is $I = 3A$

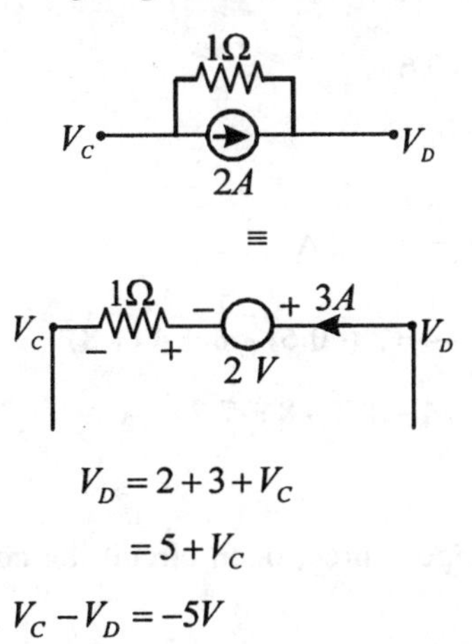

$$V_D = 2 + 3 + V_C$$

$$= 5 + V_C$$

$$V_C - V_D = -5V$$

61. Ans (b)

Solution :

Given, $\quad Z = x - j(0.1)$

Therefore $\quad \frac{1}{Z} = \frac{1}{x + j(0.1)} = \frac{x - j(0.1)}{x^2 + (0.1)^2}$

$$= \frac{x}{x^2 + (0.1)^2} - \frac{j0.1}{x^2 + (0.1)^2}$$

$$\left.\frac{1}{Z}\right|_{x=0} = 0 - \frac{j0.1}{0.01} = -j(10)$$

Real part = 0

Imaginary part = – 10

$$\left.\frac{1}{Z}\right|_{x\to\infty} = \lim_{x\to\infty}\left\{\frac{1/x}{1+\frac{0.01}{x^2}} - j\frac{0.1/x^2}{1+\frac{0.01}{x^2}}\right\}$$

$$= 0 - j0$$

Hence this represent a circle

62. Ans. (c)

Solution:

$$1A = \left(\frac{30 + j40}{30 + j40 + 30 - j40}\right) I$$

$$V_{in} = \frac{60}{30 + j40} \times 5 + (30 - j40) \times 1$$

$$= \frac{300}{30 + j40} + (30 - j40)$$

$$= \frac{300 + 900 + 1600}{30 + j40}$$

$$= \frac{2800}{50} = 56\text{ V}$$

63. Ans. (c)

Solution:

Energy stored in a capacitor $= \frac{1}{2}CV^2$

$$= \frac{1}{2} \times 1 \times 10^{-6} \times 30^2 = 450 \times 10^{-6}\text{ J}$$

64. Ans. (a)

Solution: $\quad V_{rms} = \sqrt{2^2 + \left(\frac{1}{\sqrt{2}}\right)^2} = \sqrt{\frac{9}{2}} = \frac{3}{\sqrt{2}}$

$$V_{average} = 2$$

$$\therefore \quad \frac{V_{\text{rms}}}{V_{\text{average}}} = \frac{3}{2\sqrt{2}}$$

65. Ans. (d)

Solution: Meter resistance $R_m = 5 \times 1000$

$$= 5000\ \Omega$$

$$= 5\text{ k}$$

$$\frac{5\,k \times 10\,k}{15\,k} = \frac{50}{15}k$$

$$\frac{V_S \times \frac{50}{15}k}{50\,k + \frac{50}{15}k} = 5$$

$$V_S = \frac{5 \times 80}{50} = 80\text{ V}$$

66. Ans. (b)

Solution:

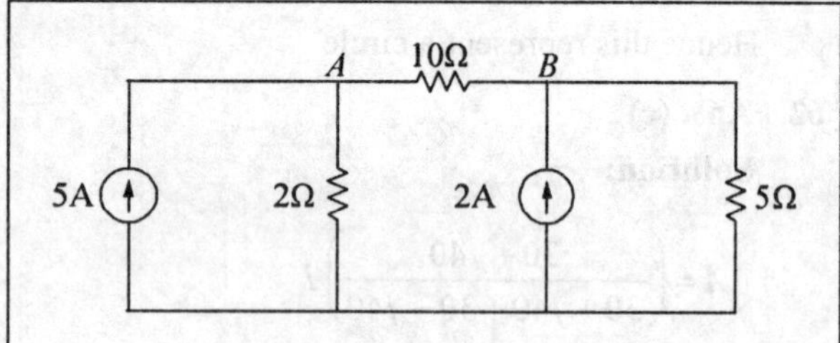

$$-5 + \frac{V_A}{2} + \frac{V_A - V_B}{10} = 0$$

$$-50 + 5V + V_A - V_B = 0$$

$$\Rightarrow \quad 6V_A - V_B = 50 \quad \ldots(1)$$

$$-2 + \frac{V_B - V_A}{10} + \frac{V_B}{5} = 0 \quad \ldots(2)$$

$$-20 + V_B - V_A + 2V_B = 0$$

$$\Rightarrow \quad -V_A + 3V_B = 20$$

From equation (1) and (2)

$$6V_A - V_B = 50$$
$$-6V_A + 18V_B = 120$$
$$-17V_B = 170$$
$$V_B = \frac{-170}{17} = 10$$
$$I_{5\Omega} = \frac{10}{5} = 2\text{ A}$$

6.2 Network Theorems

1. Ans. (a)

Solution: $Z_1 = 10\angle -60° = 10\cos(-60) + j\sin(-60)]$

$$Z_2 = 10\angle 60° = [10\cos 60 + \sin 60];$$

$$Z_3 = 50\angle 53.13° = 50[\cos 53.13 + j\sin 53.13]$$

$$Z_1 \| Z_2 = \frac{(5 - j8.66)(5 + j8.66)}{5 - j8.66 + 5 + j8.66} = \frac{25 + 75}{10} = 10$$

$$Z_{\text{Th}} = [Z_1 \| Z_2] + Z_3 = 10 + 30 + j40 = 40 + j40$$

$$= \sqrt{40^2 + 40^2}\ \angle 45° = 56.6\angle 45°$$

2. Ans. (c)

Solution: The current source is converted into voltage source. Apply KVL in loop.

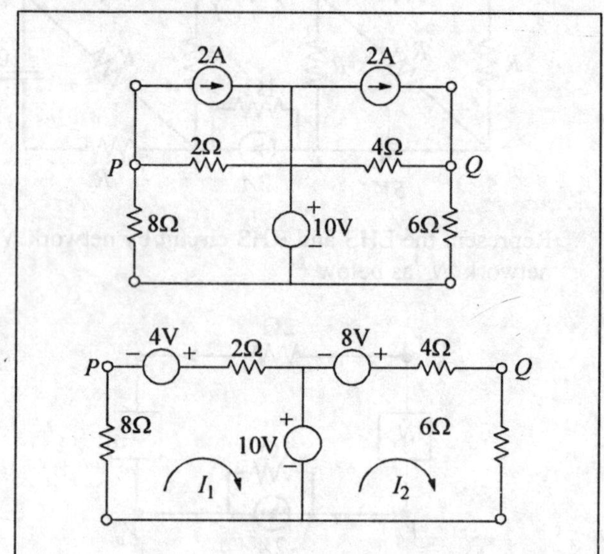

$$-4 + 10 + 10i_1 = 0$$

$$i_1 = -\frac{6}{10}$$

$$= -0.6\text{ A}$$

$$-10 - 8 + 10\,i_2 = 0,$$

$$i_2 = \frac{18}{10} = 1.8\text{ A}$$

$$V_{PQ} = -4 + 2(-0.6) - 8 + 4(1.8)$$

$$V_{PQ} = -4 - 1.2 - 8 + 7.2 = -6\text{ V}$$

Second Method

Consider only voltage source, open circuit the current source

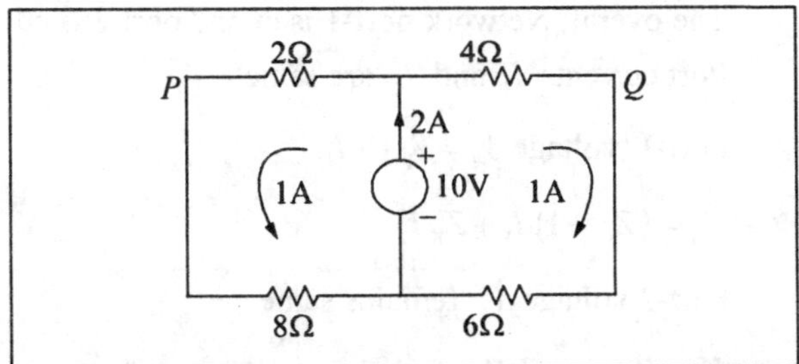

$V_P = 8 \times 1 = 8$ V
$V_Q = 6 \times 1 = 6$ V
$V_{PQ1} = 8 - 6 = 2$ V

Consider only current source, Short circuit the voltage source

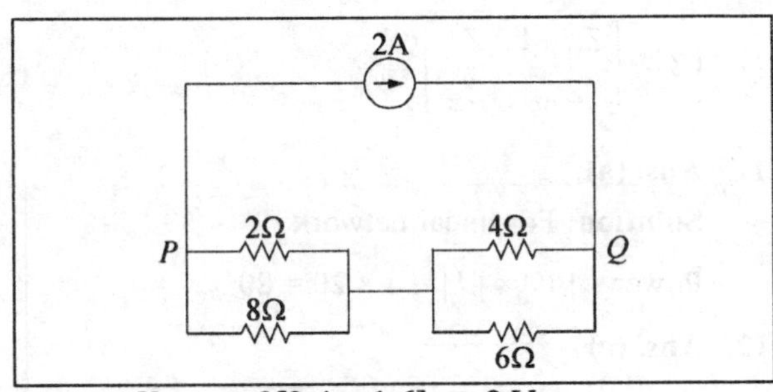

$V_{PQ2} = -2[2.4 + 1.6] = -8$ V

$V_{PQ} = -8 + 2 = -6$ V

3. **Ans. (d)**

Solution:

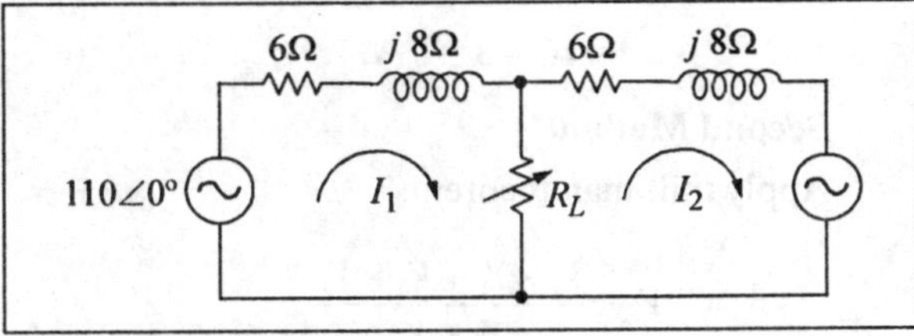

Thevenin resistance looking from R_L after shorting source is

$(6 + j8) \| (6 + j8) \quad = 3 + j4$

$R_L = \sqrt{3^2 + 4^2} = 5\Omega$

Now, $\quad 110 = (6 + j8 + 5)I_1 + 5I_2 \quad$ ...(1)

and $\quad 90 = (6 + j8 + 5)I_2 + 5I_1 \quad$ ...(2)

Solving these two equations, we get

$I_1 = 5.5 - 2.75j$

$I_2 = 4.5 - 2.2j$

$\Rightarrow$ Total current in $R_L = I_1 + I_2 = (10 - 4.55j)$ A = 11.15 A

Power absorbed by $R_L = I^2R = (11.15)^2 \times 5 = 625$ W

4. **Ans. (a)**

Solution:

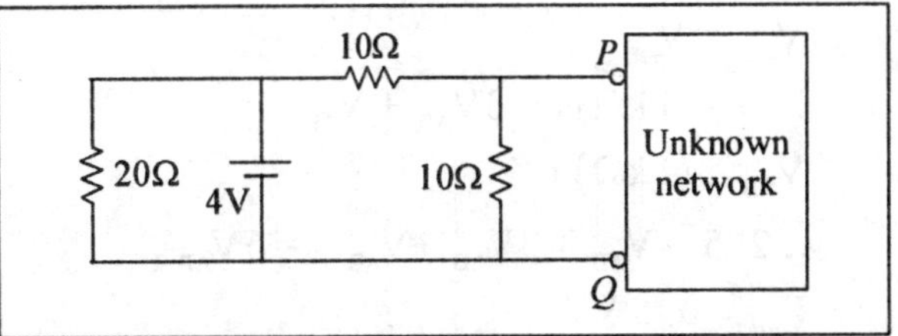

For Z_{eq}, short circuit the voltage source and open circuit the current source, in present case 20 Ω is also short circuit, therefore net impedance,

$$Z_{th} = 10 \,||\, 10 = 5\ \Omega.$$

For V_{oc}, in Parallel circuit

Current through 4 V source

$$I = \frac{4}{20 \| (10 + 10)} = \frac{4}{10} = 0.4$$

$I_{PQ} = 0.2$ A

$V_{oc} = V_{PQ} = 0.2 \times 10 = 2$V.

5. **Ans. (d)**

Solution: Open circuit the current source for Z_{th}

$Z_{th} = 1 - j + j2 = (1 + j)\ \Omega = \sqrt{2}\ \angle 45°$

$V_{th} = \sqrt{2}\ \angle 45° \times 1\angle 0° = \sqrt{2}\ \angle 45°$ V

6. **Ans. (a)**

Solution:

According to maximum power transfer theorem when the load resistance because equal to the internal resistance of the voltage source it delivers maximum power.

The internal resistance should be equla to 3 Ω (which is load resistance) But in the present case 6 Ω || R is not the source impedacne. It is the circuit element which can be made zero.

$$R = 0 : P_{max} = \frac{10^2}{3} \qquad (\therefore R_L = \text{constant})$$

7. **Ans. (d)**

Solution :

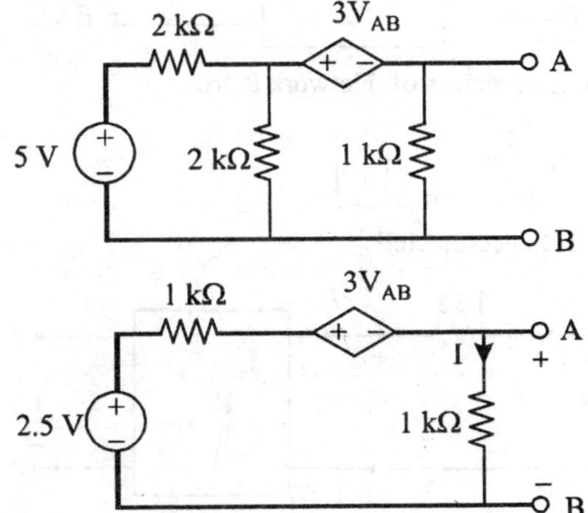

$V_{OC} = V_{AB}$

$2.5 = (1\,k\Omega)\,I + 3V_{AB} + V_{AB}$

$V_{AB} = (1\,k\Omega)\,I$

$\therefore 2.5 = V_{AB} + 3V_{AB} + V_{AB} \quad = 5V_{AB}$

$V_{AB} = 0.5\,V$

Apply short across $AB : V_{AB} = 0$

$3\,V_{AB} = 0$

$$I_{SC} = \frac{2.5\,V}{1\,k\Omega} \quad = 2.5\,mA$$

8. Ans. (b)

Solution:

$$R_{th} = \frac{V_{OC}}{I_{SC}} = \frac{0.5\,V}{2.5\,mA}$$

$$= \frac{1000}{5} = 200\,\Omega \quad = 0.2\,k\Omega$$

9. Ans. (a)

Solution: $\frac{x}{a} = \frac{y}{b} = 1$

$$\Rightarrow \quad \frac{1}{4 \times 10^{-3}} + V = 1; \quad V = I.R$$

$$R = [V/I] = [8/(4 \times 10^{-3})] = 2\,K$$

$$\text{Time constant of inductor} = \frac{L}{R} = \frac{6 \times 10^{-3}}{2 \times 10^{3}} = 3\,\mu\,sec$$

10. Ans (c)

Solution:

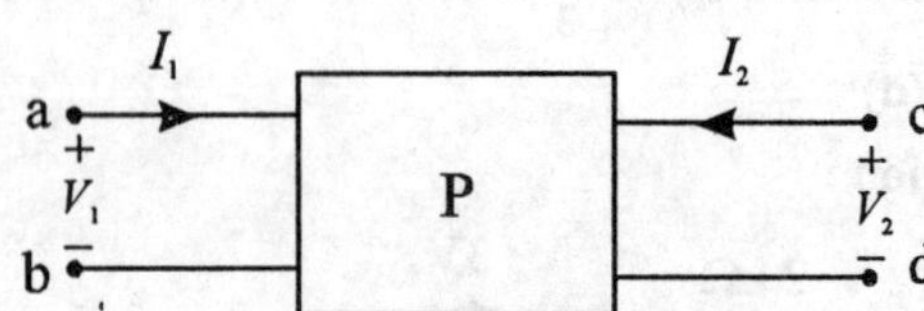

The Z-paraeters of Network P are

$$\begin{bmatrix} V_1 \\ V_2 \end{bmatrix} = \begin{bmatrix} Z_{11} & Z_{12} \\ Z_{21} & Z_{22} \end{bmatrix} \begin{bmatrix} I_1 \\ I_2 \end{bmatrix}$$

If 1Ω is connected

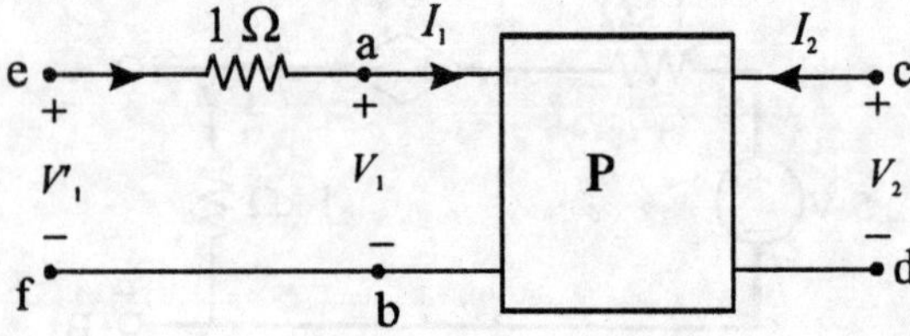

The overall Network port-1 is ef and port-2 is cd

Port currents I_1 and I_2 are same

Port-1 voltage $V_1' = V_1 + 1\,I_1$

$V_1' = (Z_{11} + 1)\,I_1 + Z_{12} I_2$

Port-2 voltage V_2 remains same

$V_2 = Z_{22}\,I_1 + Z_{12} I_2$

$$\begin{bmatrix} V_1' \\ V_2 \end{bmatrix} = \begin{bmatrix} Z_{11}+1 & Z_{12} \\ Z_{21} & Z_{22} \end{bmatrix} \begin{bmatrix} I_1 \\ I_2 \end{bmatrix}$$

Overall Z-parameter

$$[Z] = \begin{bmatrix} Z_{11}+1 & Z_{12} \\ Z_{21} & Z_{22} \end{bmatrix}$$

11. Ans. (a)

Solution: For linear network

Power = $|V| \times |I| = 4 \times 20 = 80$

12. Ans. (c)

Solution: Applying Thevenin's theorem

$$R_{eq} = 6\Omega \parallel 4\Omega = \frac{6 \times 4}{6+4} = \frac{24}{10} = 2.4\,\Omega$$

$$V_{ab} = 10 - 6 \times (5/10) = 10 - 6 \times 0.5$$

$$= 10 - 3 = \mathbf{7\,V.}$$

Second Method

Apply milliman theorem

$$V = \frac{V_1 \times Y_1 + V_2 \times Y_2}{Y_1 + Y_2}$$

$$V_{ab} = \frac{10 + [1/6] + 5 \times [1/4]}{[1/6] + [1/4]}$$

$$= \frac{[10/6] + [5/4]}{[1/6] + [1/4]}$$

$$= \frac{40 + 30}{10} = \frac{70}{10} = 7V$$

$$Y = Y_1 + Y_2\,;\; Z = Z_1 \parallel Z_2$$

$$R_{eq} = 6\,\Omega \parallel 4\,\Omega = \frac{6 \times 4}{6+4} = \frac{24}{10} = 2.4\,\Omega$$

13. Ans. (b)

Solution: Applying KCL, $\frac{V}{4} + \frac{V - 2I}{2} = I$

$\Rightarrow \quad [V/I] = [8/3]\,\Omega$

14. Ans (c)

Solution :

The impedance and number of turn relation is

$$\left|\frac{Z_L}{Z_S}\right| = \left(\frac{n_2}{n_1}\right)^2$$

where,

Z_L = Load impedance on the secondary side = 2Ω

Z_S = Source or Generator impedance on the primary side
= 8Ω

n_2 = Number of turns of the secondary winding
= 40

n_1 = Nrmber of turns of the primary winding

$$\frac{Z_L}{Z_S} = \frac{2}{8} = \frac{1}{4},$$

$$\frac{n_2}{n_1} = \sqrt{\frac{1}{4}} = \frac{1}{2},$$

$$n_1 = 2\,n_2 = 2 \times 40 = 80$$

15. Ans. (a)

Solution:

Maximum power will be absorbed by R. When $R = R_{TH}$ i.e. Thevenin's equivalent resistance across the terminals A and B. Open circuit the R and short circuit the voltage source.

Hence we get

$$R_{TH} = (3\,k\Omega \| 6\,k\Omega) + (4\,k\Omega \| 4\,k\Omega)$$
$$= 2\,k\Omega + 2\,k\Omega$$
$$= 4\,k\Omega$$

16. Ans. (a)

Solution:

Superposition theorem is applicable to

* Linear networks
* Time invariant or time- varying networ co sisting of independent sources,
* Linear dependent sources (voltage or current),
* Linear passive elements R, L, C and linear transformer.
* Superposition theorem is not applicable to nonlinear networks, containing nonlinear elements

17. Ans. (c)

Solution:

Source impedance $R_s = R + j\omega L = 1 + j$

For max power transfer $R_L = R_s^* = 1 - j$

The above equation represent 1 Ω resistance with 1 F capacitor, i.e.

1Ω —/\/\/\—||— 1F ; ω = 1

$[R + (1/j\omega C)] = [1 + (1/j)] = 1 - j$

i.e., for maximum power transfer the load impedance should be complex conjugate of source impedance.

18. Ans. (c)

Solution:

For max power delined to R_L

$$R_S = R_L = 100\,\Omega$$

$$I = \frac{V_S}{R_S + R_L} = \frac{V_S}{2R_L}$$

$$P_L = I^2 R_L$$

$$= \left(\frac{V_s}{2R_2}\right)^2 \times R_L$$

$$[V^2/4\,R_L] = [102/(4 \times 100)]$$
$$= [1/4] = 0.25\text{ W}$$

19. Ans. (d)

Solution:

$$Z_s = R_s + jX_s$$

For maximum power transfer the load impedance should be equal to complex conjugate of source impedance i.e.

$$Z_L = Z_s^* = R_s - jX_s$$

20. Ans. (c)

Solution: For maximum power transmission

$Z_L = Z_s^*$

i.e if $Z_s = A + jB$

Then $Z_L = A - jB$

In the present case

$R_L = R_{TH}^*$

For the calculation of R_{TH}

Short circuit the voltage source and open circuit the current source and find Rth as

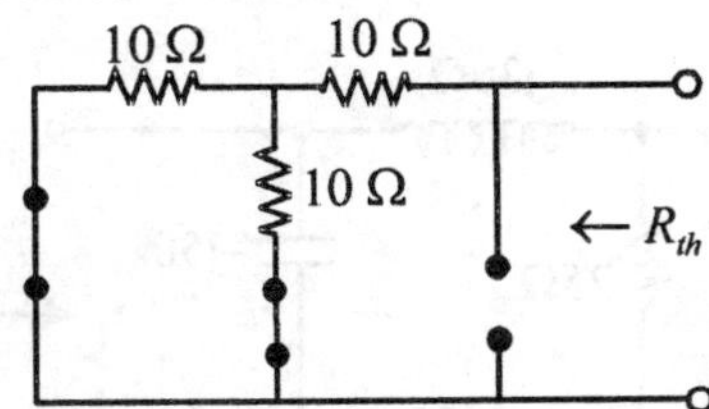

$$R_{TH} = (10 \,\|\, 10) + 10 = 15\Omega$$

Second Method

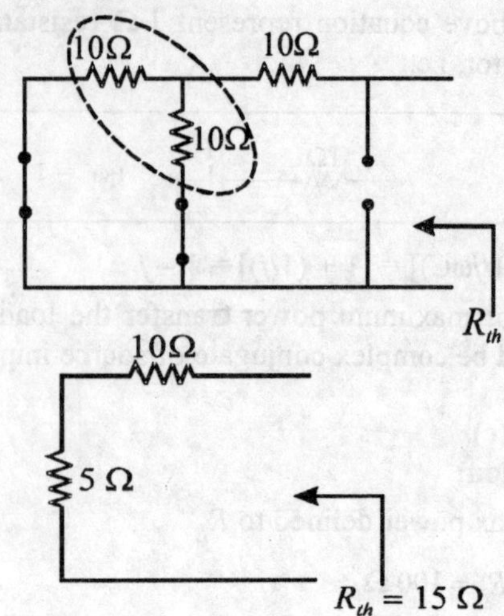

21. Ans. (a)

Solution: When terminals P and Q are short circuited Then the circuit becomes

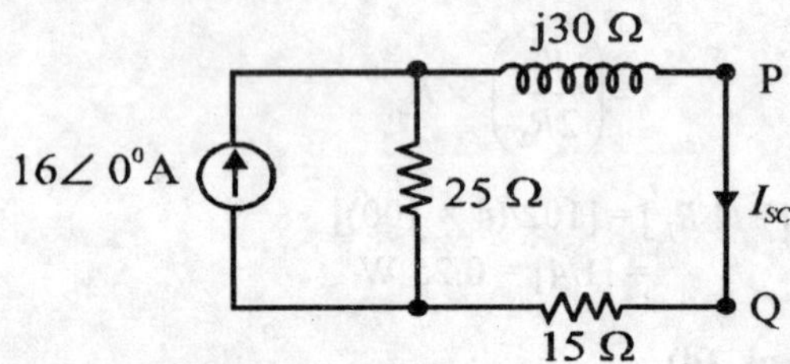

From current Division rules

$$I_{SC} = \frac{16(25)}{25+15+j30} = \frac{(16)(25)}{40+j30} = \frac{(16)(25)}{10(4+j3)}$$

$$= \frac{400}{10(4+j3)} = \frac{40\times(4-j3)}{16+9}$$

$$= \frac{40}{25}\times(4-j3) = 1.6(4-j3) = 6.4-j4.8$$

Second Method

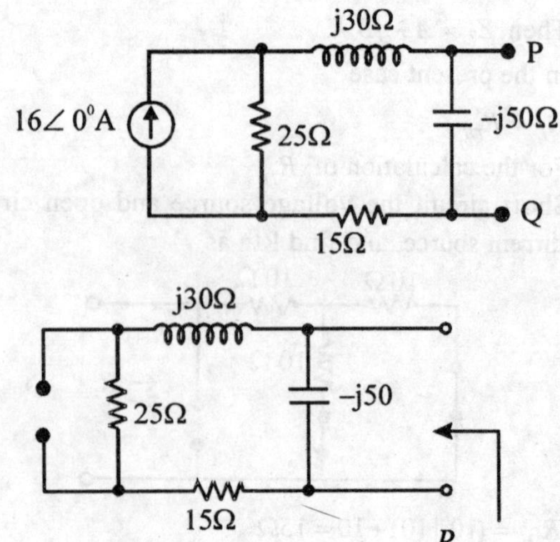

$$R_N = (-j50)\,\|\,(j30+40)$$

$$= \frac{-j50\times(j30+40)}{40+j30-j50} = \frac{50\angle-90^\circ\times 50\angle 36.86}{44.72\angle-26^\circ}$$

$$R_N = \frac{2500\angle-53.14}{44.72\angle-26} = 56\angle-26.57$$

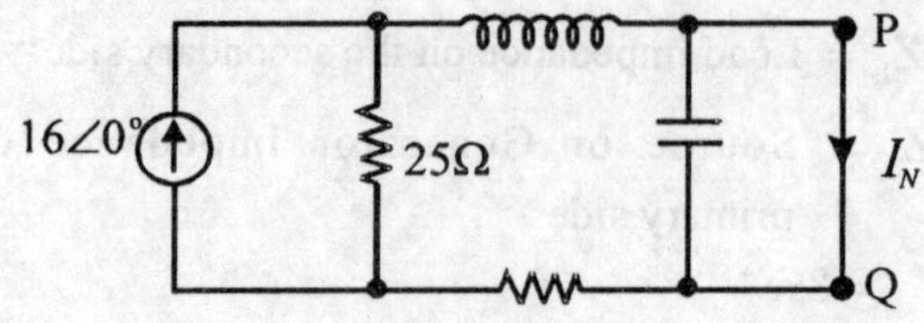

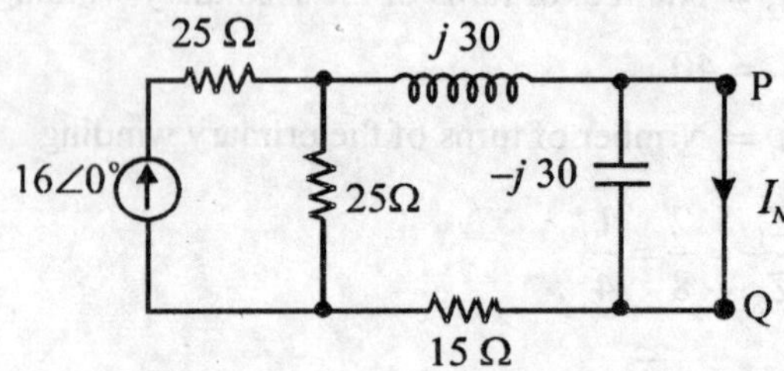

$$I_N = \frac{400}{25+15+J30} = \frac{400}{40+j30}$$

$$= \frac{400}{50\angle 36.86} = 3\angle 36.86$$

$$I_N = 6.4-4.8j$$

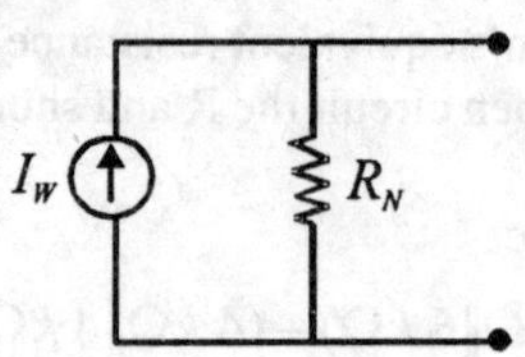

$$I_N = 6.4-j4.8$$

$$R_N = 54.6+12.46j$$

22. Ans. (a)

Solution:

$$Z = \frac{(j2\times -j6)(3+j4)}{(j2-j6)+(3+j4)} = \frac{36+j48}{3}$$

$$I = \frac{100\times 3}{36+j48}$$

$$I_1 \text{ for } j4 = \left(\frac{100\times 3}{36+j48}\right)\times\frac{(j2\times(-j6))}{(j2-j6)+(3+j4)}$$

$$= \frac{100\times 3}{36+j48}\times\frac{12}{3} = \frac{100\times 12}{12(3+j4)}$$

$$V_L = \frac{100(j4)}{3+j4} \times \frac{3-j4}{3-j4} = \frac{100 \times j4(3-j4)}{25}$$

$$V_L = j\,16\,(3-j4\,)$$

Second Method

$$\therefore \qquad j2 \parallel -j6 = j3$$

From loop (1)

$$j3(I - I_1) = 100$$

$$j3I = 100 + j3\,I_1 \qquad \ldots(1)$$

From loop (2)

$$3I_1 + j4\,I_1 - j3(I - I_1) = 0$$

From (1),

$$(3 + j4 + j3)I_1 - (100 + j3I_1) = 0$$

$$I_1 = \frac{100}{3+j4}$$

$$V_{th} = j4 \times \frac{100}{3+j4}$$

$$V_{th} = \frac{100(j4)\,(3-j4)}{(3+j4)\,(3-j4)}$$

$$V_{th} = j16(3-j4)$$

Third Method

V_{TH} is the voltage across j4 inductance

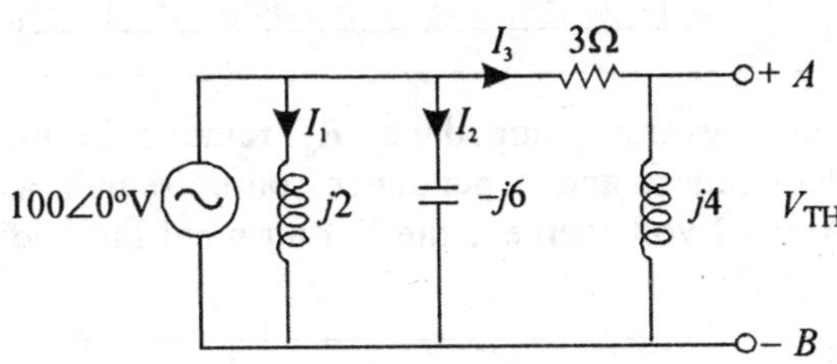

Apply voltage divider rule, to get V_{TH},

$$V_{TH} = \frac{j4}{3+j4} \times 100\angle 0^\circ$$

$$= (j16)(3-j4)\,V$$

23. Ans. (c)

Solution: For maximum power transfer $Z_L = Z_S^*$

$$R_L = R_S$$

$$Z_S = (5 \parallel 20) + 4 = \frac{5 \times 20}{5+20} + 4$$

$$= \frac{100}{25} + 4 = 4 + 4 = 8$$

24. Ans. (c)

Solution:

According to reciprocity theorem, the excitation10 V souce and 2A response are interhanged as shown as figuer (a)

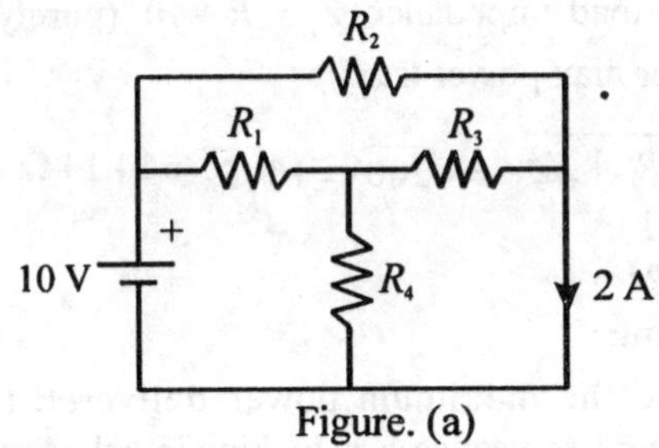

Figure. (a)

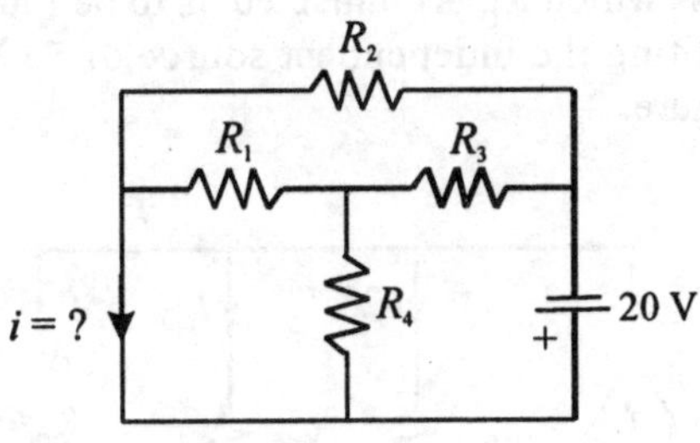

Figure. (b)

Then the current i due to 10 v in figure = 2 A

If 10 V is replaced by –20 V as shown in figure (b) the current $i = -2i_1 = -4$A

25. Ans. (d)

Solution:

Convert 8 A, 10 Ω current source to 80 V, 10 Ω voltage source.

The resulting network is shown in figuer below

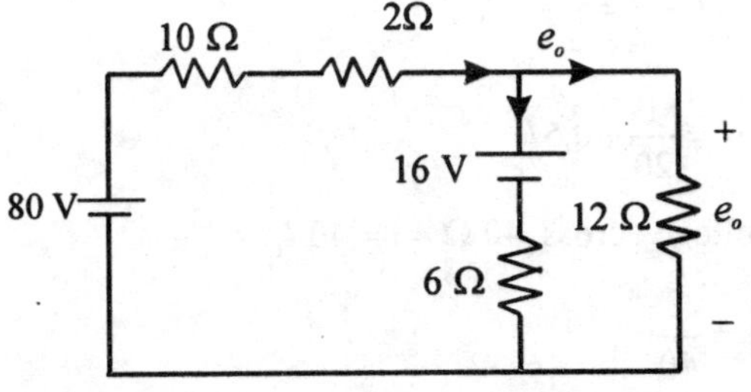

Now Apply KCL at output node.

$$\frac{e_o - 80}{12} + \frac{e_o}{12} + \frac{e_o - 16}{6} = 0$$

$$4e_o = 112$$

$$e_o = 28\ V$$

26. Ans. (a)

Solution:

Given $R_s = 10\ \Omega$, $\omega = 10\ r/s$, $L = 1\ H$

The source impedance $Z_s = Rs + j\omega Ls$

$$Z_s = 10 + j\,10$$

If the load impedance $Z_L = R + j0$ (purely resistive) then for max power transfer

$$R = \sqrt{R_s^2 + X_s^2} = \sqrt{200} = 10\sqrt{2} = 14.14\ \Omega$$

27. Ans. (a)

Solution:

To find the maximum power delivered to R_L, the Thevenin's resistance seen looking into the terminals A,B across which R_L is connected is to be found by short circuiting the independent source of 50 V as shown in figure.

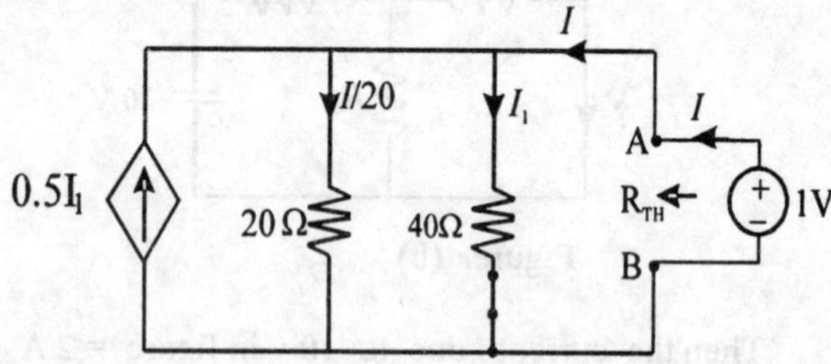

When ever there is dependent current or voltage source is present in the circuit

Connect a 1 V voltage source, across the terminals A,B and find he curent I_x

Then $R_{TH} = \frac{1}{I}$

KCL equation at A

$$0.5\,I_1 + I = \frac{1}{20} + I_1$$

$$\text{or } I = \frac{1}{20} + 0.5 I_1 \qquad \text{... (1)}$$

Voltage across $40\ \Omega = 1 = 40\ I_1$

$$I_1 = \frac{1}{40}$$

From equation (1), $I_1 = \frac{1}{20} + 0.5\frac{1}{40}$

$$= \frac{2.5}{40} = \frac{1}{16}\text{V}$$

$$R_{TH} = \frac{1}{I} = 16\ \Omega$$

According to maximum power transfer theorem

$$R_L = R_{TH} = 16\ \Omega$$

28. Ans. (b)

Solution:

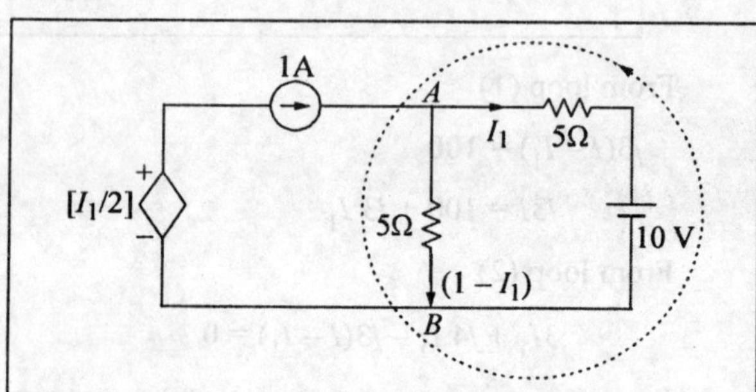

By KVL, in loop

$$10 = -5I_1 + 5\,(1 - I_1) = 5 - 10\,I_1$$

$$I_1 = -[1/2]$$

$$V_{ab} = 5\,(1 - I_1) = 5\,[1 + (1/2)]$$

$$V_{ab} = 7.5\ \text{V}$$

29. Ans. (d)

Solution:

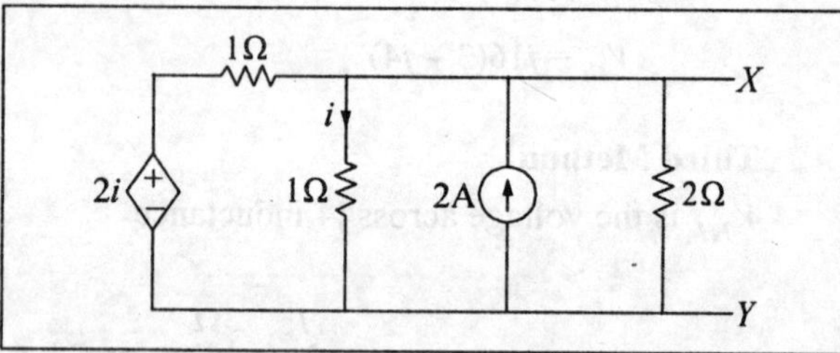

For Thevenin's impedance R_{th}, remove the independent source and dependence source remain as it is. Apply 1 volt source to the X, Y terminal and finding R_{th} = [1/I]

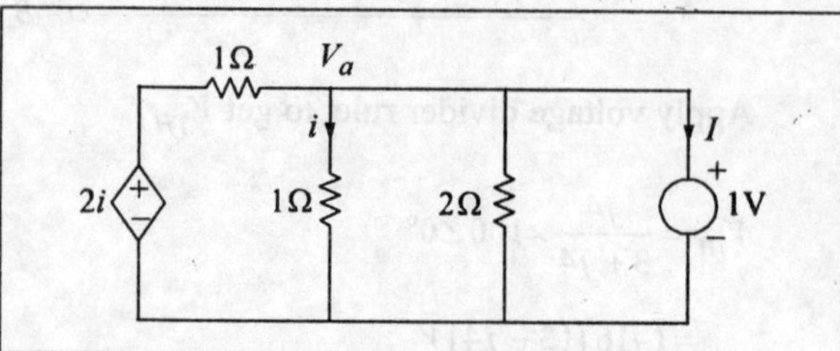

KCL at Node

$$\frac{2i - V_a}{1} = i + \frac{V_a}{2} - I$$

$$V_a = 1\text{V}, \quad i = \frac{IV}{1\Omega} = 1A, \text{ then}$$

$$\therefore \quad 1 = 1 + \frac{1}{2} - I \; : \; I = \frac{1}{2}$$

$$\therefore \quad R_{\text{th}} = \frac{1}{I} = 2\,\Omega$$

For Thevenin's voltage V_{th}

$$\frac{2i - V_a}{1} + 2 = i + \frac{V_a}{2}$$

or, $\quad 2i - i + 2 = V_a + \frac{V_a}{2}$ or, $\; i + 2 = 3\frac{V_a}{2}$

Also, $\quad Va = i \times 1 = i$

$$\therefore \quad i + 2 = 3\frac{i}{2} \Rightarrow 2 = \frac{i}{2} \Rightarrow i = 4$$

or $\quad 2 = \frac{i}{2}$

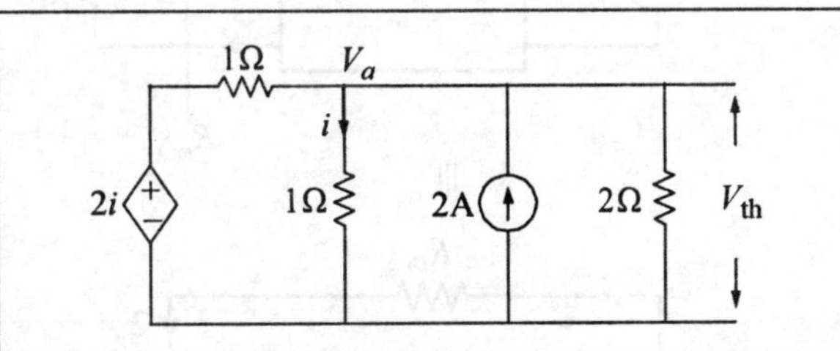

or $\quad i = 4$

$$\therefore \quad V_{\text{th}} = V_a = i = 4V$$

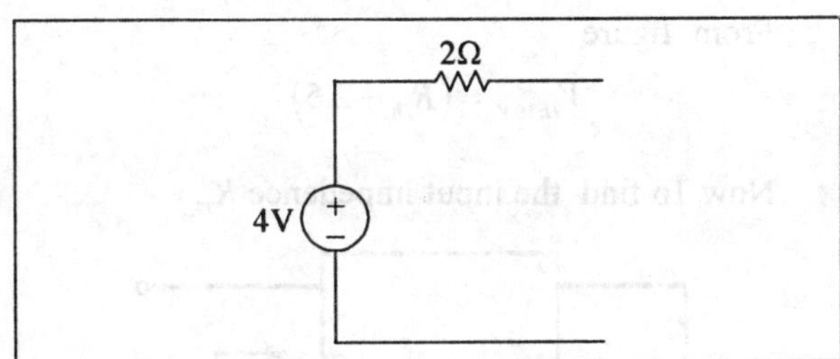

30. Ans. (a)

Solution: For finding Thevenin equivalent impedance.

Short circuit voltage source and Open circuit current source.

Thevenin equvalent imedance Z_{Th}

$$= \left(1 + \frac{1}{s}\right) \Big\| (1 + s) = \frac{\left(1 + \frac{1}{s}\right)(s+1)}{1 + \frac{1}{s} + 1 + s}$$

$$= \frac{(s+1)(s+1)/s}{\left(s^2 + 2s + 1\right)/s} = \frac{\left(s^2 + 2s + 1\right)/s}{\left(s^2 + 2s + 1\right)/s}$$

$$= 1$$

31. Ans. (c)

Solution: For maximum power transfer to load, the source impedance must be complex conjugate to the load impedance i.e. $R_S = R_L$ and $X_S = -X_L$

Hence, find the Thevenin equivalent circuit across to the load

For Z_{th}, apply 1 volt at the removed terminal of circuit.

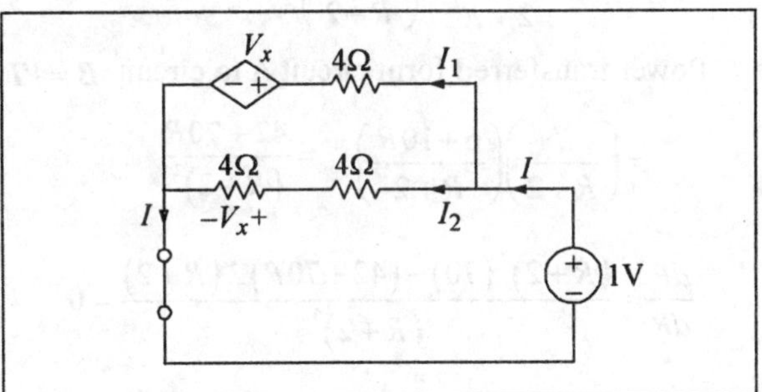

Total current, $I = I_1 + I_2$

KVL in loops

$$-1 + 4I_1 + V_x = 0 \quad (1)$$

$$-1 + 4I_2 + V_x = 0 \quad (2)$$

From equation (1) and (2)

$$4I_1 = 1 - V_x$$

$$4I_2 = 1 - V_x$$

$$\therefore \quad I_1 = I_2$$

from the circuit

$$V_x = 4I_2$$

Put the value of V_x in equation (2)

$$-1 + 4I_2 + 4I_2 = 0$$

$$8I_2 = +1$$

$$I_2 = \frac{1}{8}$$

$$I = I_1 + I_2 = \frac{1}{8} + \frac{1}{8} = \frac{1}{4}$$

$$Z_{eq} = \frac{1\text{V}}{I} = \frac{1}{(1/4)} = 4\,\Omega$$

32. Ans. (a)

Solution:

For dc, capacitor → open circuit

From the given figure power trasnsored form A to B as resultant voltage source will have voltage of

$(10-3)=7\,V$

and current through $R=\dfrac{7}{R+20}=I$

$I=\dfrac{10-3}{2+R}=\dfrac{7}{2+R}$

$V=3+IR$

$=3+\dfrac{7R}{2+R}=\left(\dfrac{6+10}{R+2}\right)$

Power transferred form circuit A to circuit $B=VI$

$=\left(\dfrac{7}{R+2}\right)\left(\dfrac{6+10R}{R+2}\right)=\dfrac{42+70R}{(R+2)^2}$

$\dfrac{dP}{dR}=\dfrac{(R+2)^2(70)-(42+70R)2(R+2)}{(R+2)^4}=0$

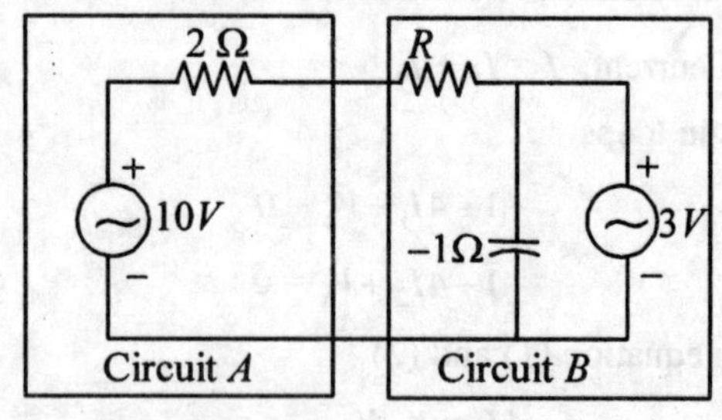

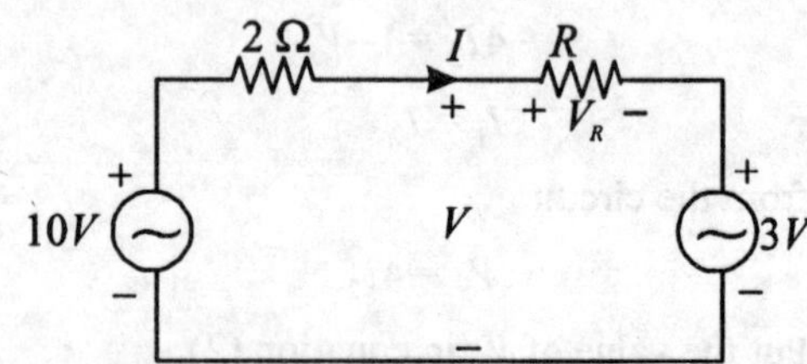

or, $70(R+2)^2=(42+70R)2(R+2)$

$=14(3+5R)2(R+2)$

or, $5(R+2)=2(3+5R)$

or, $5R+10=6+10R$

or, $10\,R-5\,R=10-6$

or, $5\,R=4$

$R=\dfrac{4}{5}=0.8\,\Omega$

33. Ans. (c)

Solution:

For given data case 1:

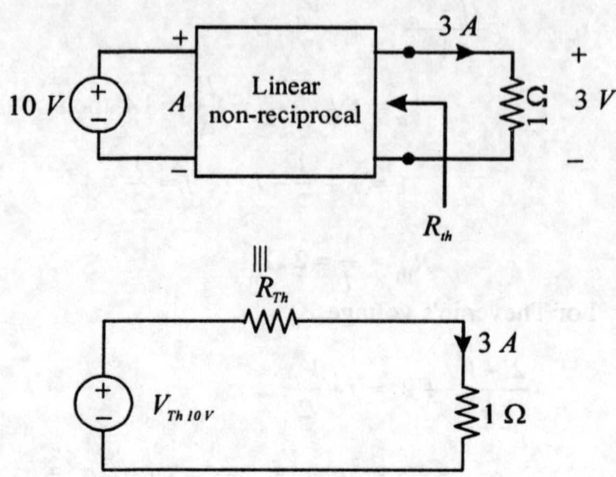

From figure,

$$V_{Th10V}=(R_{Th}+1)3 \quad \text{...(1)}$$

For given data case 2:

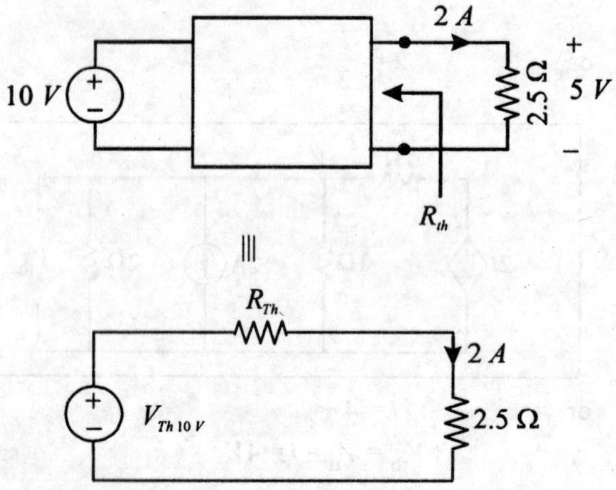

From figure

$$V_{Th10V}=(R_{Th}+2.5)2 \quad \text{...(2)}$$

Now To find the input impedance R_{TH}

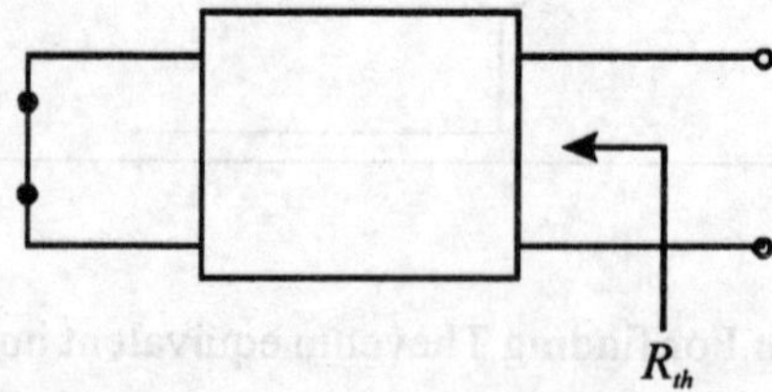

To find $V_{Th}=V_{OC}$

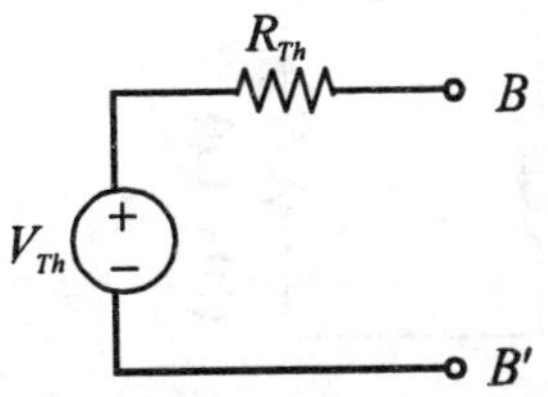

From equations (1) and (2)

$V_{Th10V} = 9\ V$

$R_{Th} = 2\ \Omega$

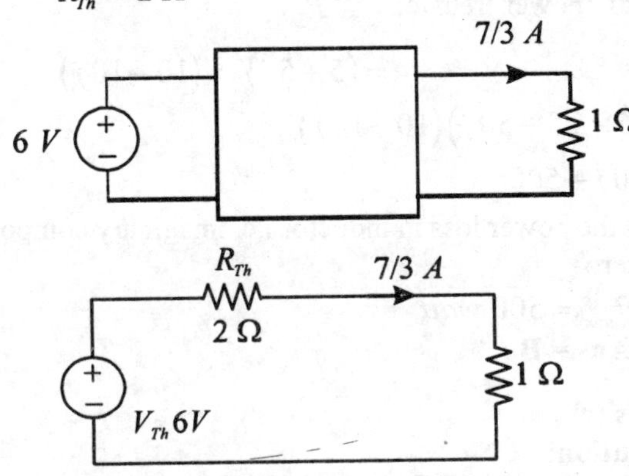

Therefore, $V_{Th\,6V} = (R_{Th}+1)\frac{7}{3} = 7\ V$

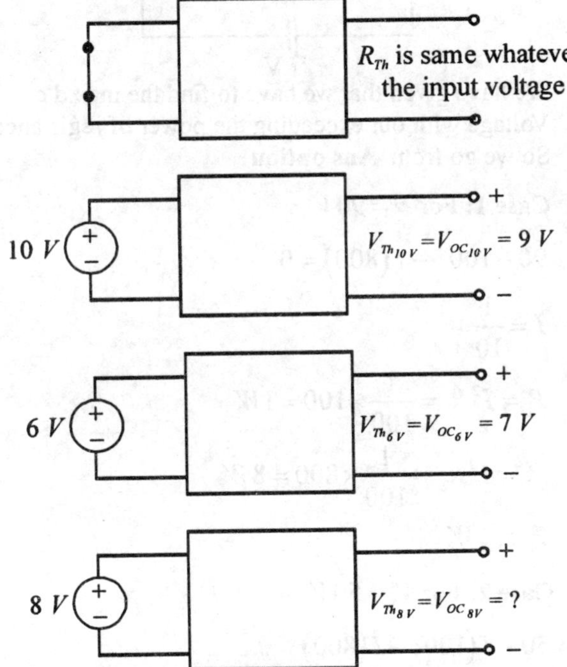

As the given two port network is linear non-reciprocal we can express output at port 2 as a linear function of output.

i.e. $V_{OC} = a\,V_{in} + b$

$9 = 10\,a + b$ ...(3)

$7 = 6\,a + b$ ...(4)

From equation (3) and (4)

$a = 0.5,\ b = 4$

Therefore, $V_{OC\,8V} = 0.5(8) + 4\ = 8\,V$

34. Ans. (c)

Solution:

Use the data obtained in part 50 Answer as $V_{Th10V} = 9\ V$ and $R_{Th} = 2\ \Omega$

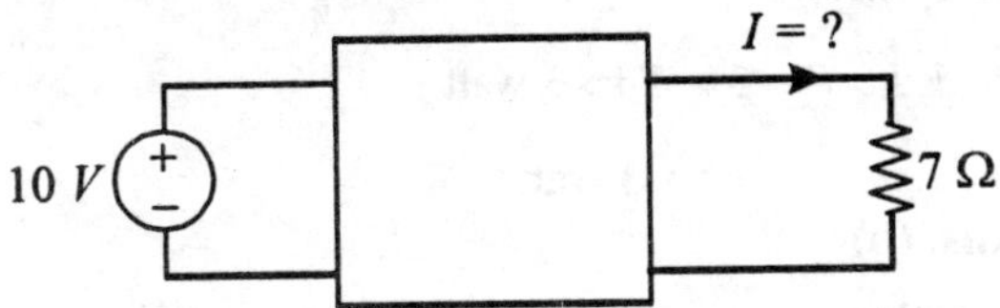

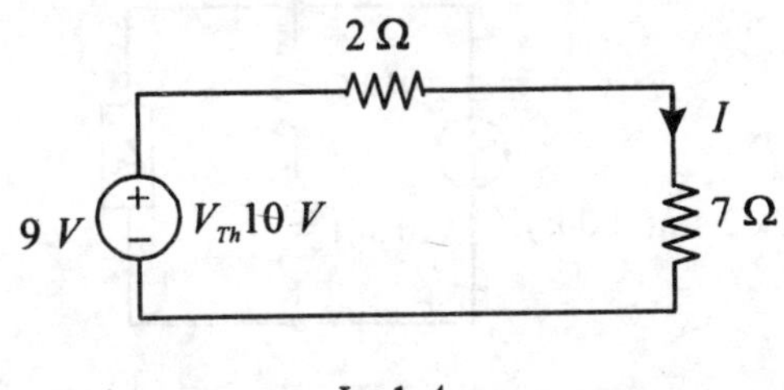

$I = 1\,A$

Second Method

The given network can be replaced by a thevenin equivalent with V_{th} and R_{th} as Thevenin voltage and thevenin resistance.

Now we can write two equations for this

$V_{th} = 3\,R_{th} + 3$

$V_{th} = R_{th} + 5$

Solving these two equations we get $R_{th} = 2$ and $V_{th} = 9$

Now using the same equation with current unkonw,

$9 = I\times 2 + 7\times I$

$= 1\,A$

35. Ans. (d)

Solution:

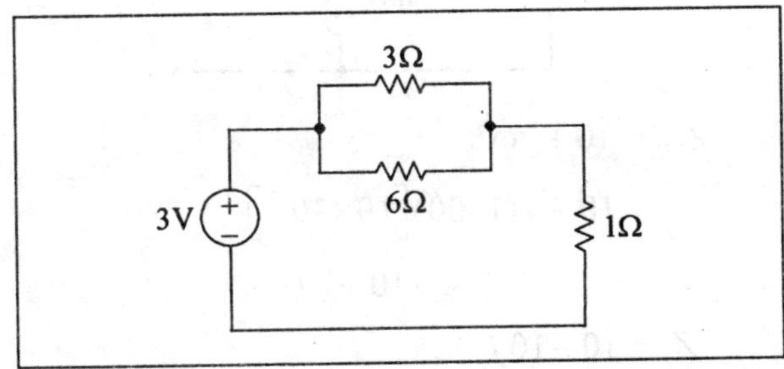

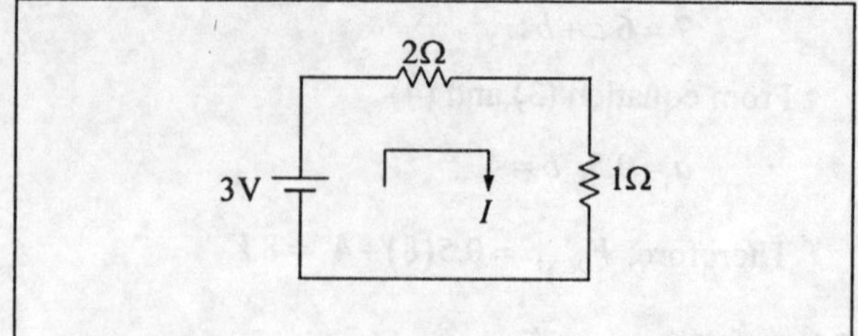

$3 - 2I - I = 0, I = 1$

We know that $P = I_2 R$ or VI

So by using

$P = VI \quad P = 3 \times 1 \Rightarrow 3$ watt

$P = 3$ watt

36. Ans. (b)

Solution:

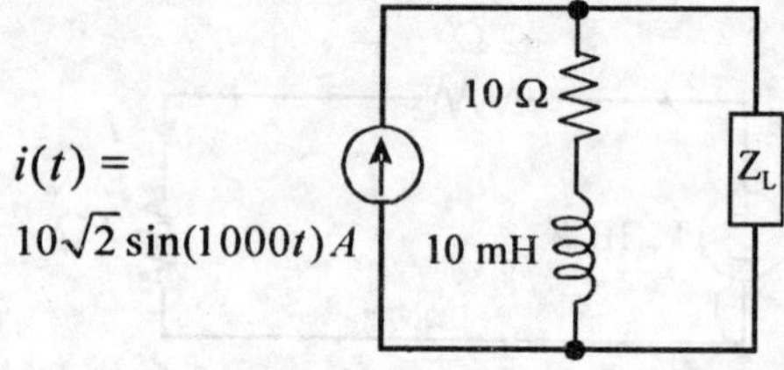

$i(t) = 10\sqrt{2}\sin(1000t)A$

$$I = I_m \sin(\omega t)$$

$$= 10\sqrt{2}\sin(1000t)$$

$$I_m = 10\sqrt{2}$$

$$I_{rms} = \frac{10\sqrt{2}}{\sqrt{2}} = 10\,A$$

$$\omega = 1000$$

For Max Power transfer

$$Z_L = Z_{TH}$$

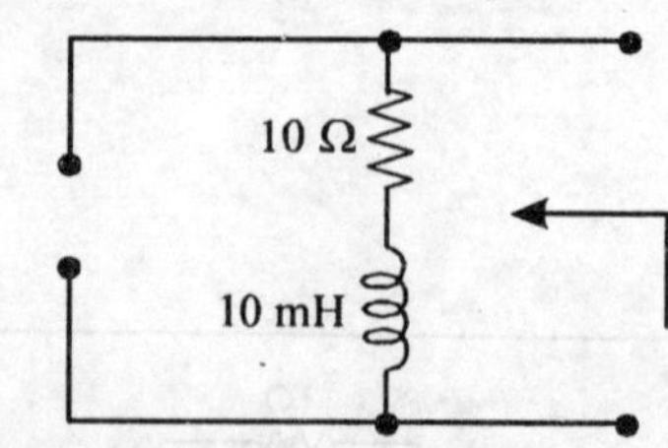

$$Z_{TH} = 10 + j\omega L$$

$$= 10 + j(1000)\left[10 \times 10^{-3}\right]$$

$$10 + 10j$$

$$Z_L = 10 - 10j$$

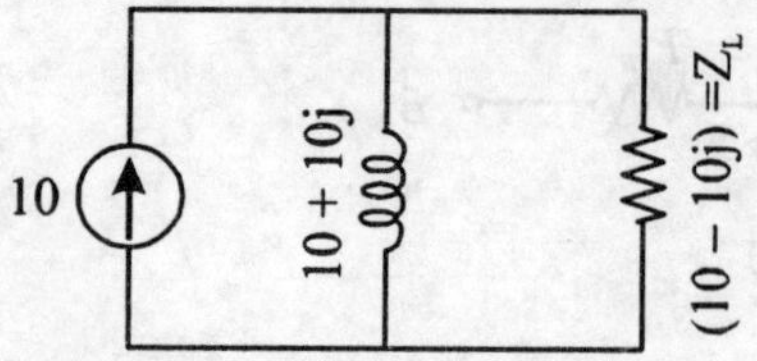

Current through Z_L

$$\frac{10 \times (10 + 10j)}{10 + 10j + 10 - 10j} = \frac{10(10 + 10j)}{20}$$

Max. power transfer $= I^2 R_L$

$$= (5 + 5j)^2 \times (10 - 10j)$$

$$(25 - 25 + 50j)(10 - 10j)$$

$$500j + 500$$

But the power loss in inductor i.e. imaginary component is zero

so $P_{max} = 500\,watt$

So Ans = B

37. Ans (c)

Solution:

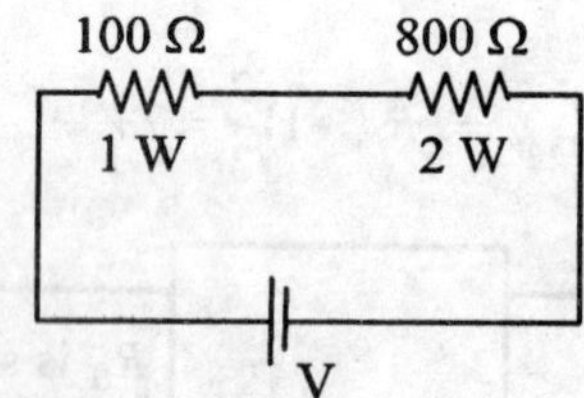

We have given that we have to find the max.d.c. Voltage with out exceeding the power of registance So we go from **Ans option**

Case 1: For $v = 90\,V$

$$90 - 100I - I(800) = 0$$

$$I = \frac{1}{10}$$

$$P_1 = I^2 R_1 = \frac{1}{100} \times 100 = 1\,W$$

$$P_2 = I^2 R_2 = \frac{1}{100} \times 800 = 8\,W$$

$$P_2 > 2\,W$$

Case 2: For $V = 50\,V$

$$50 - I(100) - I(800) = 0$$

$$I = \frac{5}{90} = \frac{1}{18}$$

$$P_1 = I^2 R_1 = \left(\frac{1}{18}\right)^2 \times 100 = 0.324\,W$$

$$P_2 = I^2R_2 > \left(\frac{1}{18}\right)^2 \times 800 = 2.469$$

$2.469 > 2W$ so rejected

Case 3: For $45\,V$

$$45 - 100I - 800I = 0$$

$$I = \frac{1}{20}$$

$$P_1 = I^2R_1 = \frac{1}{400} \times 100 = 0.25\,W$$

$$P_2 = I^2R_2 = \frac{1}{400} \times 800 = 2\,W$$

So it is ok

Case 4: For $40\,V$

$$40 - 100I - 800I = 0$$

$$I = \frac{40}{900} = \frac{2}{45}$$

$$P_1 = I^2R_1 = \left(\frac{2}{45}\right)^2 \times 100 = 0.197$$

$$P_2 = I^2R_2 = \left(\frac{2}{45}\right)^2 \times 800 = 1.58$$

Because $45\,V > 40\,V$

so maximum D.C. Voltage $= 45V$

38. Ans (a)

Solution:

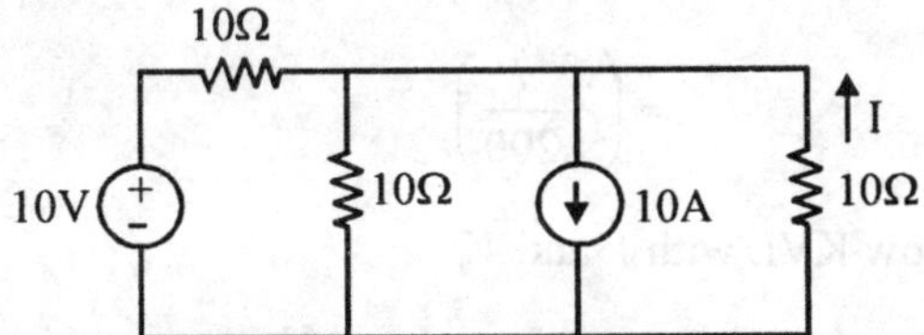

Apply super position Theorem, consider 10V source' only, open circuit the current source.

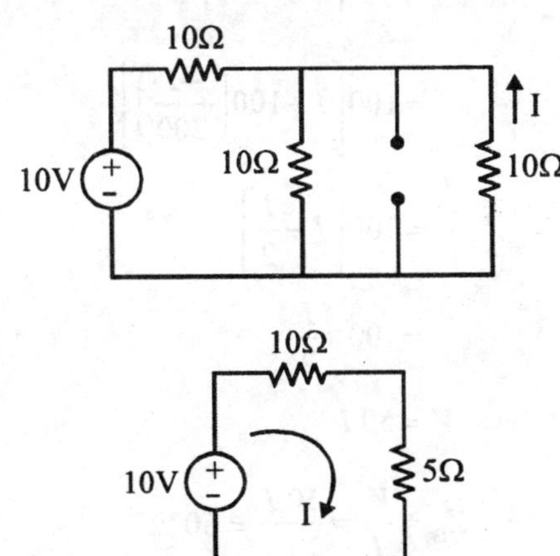

$$I = \frac{10}{15} = \frac{2}{3}A$$

$$I_1 = \frac{-2}{3 \times 2} = \frac{-2}{6} = \frac{-1}{3}A$$

Now consider the 10A current source 10A current source only short circuit the voltage source.

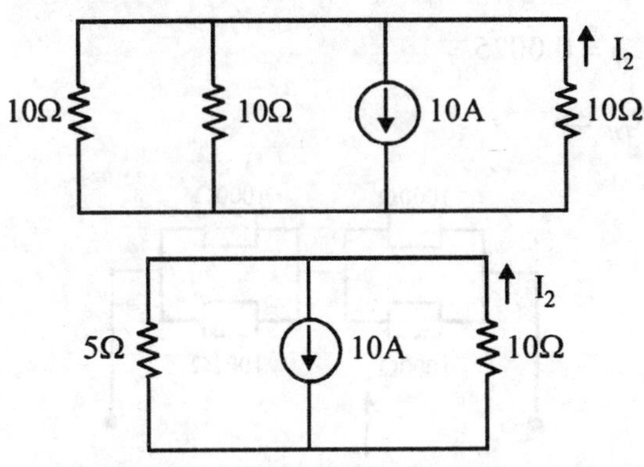

$$I_2 = \frac{10 \times 5}{10 + 5} = \frac{50}{15} = \frac{10}{3}$$

$$I_2 = \frac{10}{3}$$

$$I = I_1 + I_2$$

$$\frac{-1}{3} + \frac{10}{3} \Rightarrow \frac{9}{3} = 3A$$

$$I = 3A$$

39. Ans. (a)

Solution:

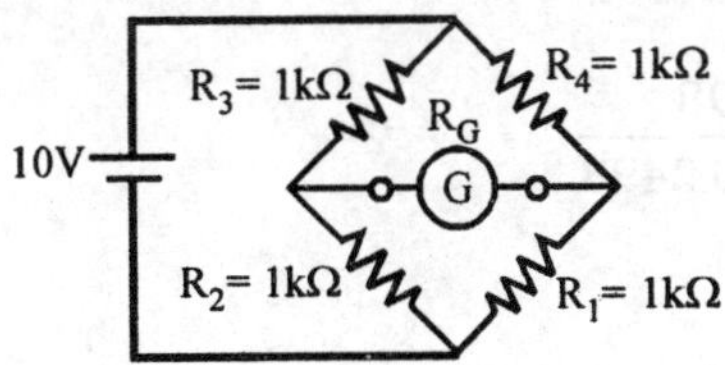

R_1 increses by 1 ohm.

$R_1 = 1000 + 1 = 1001$ ohms.

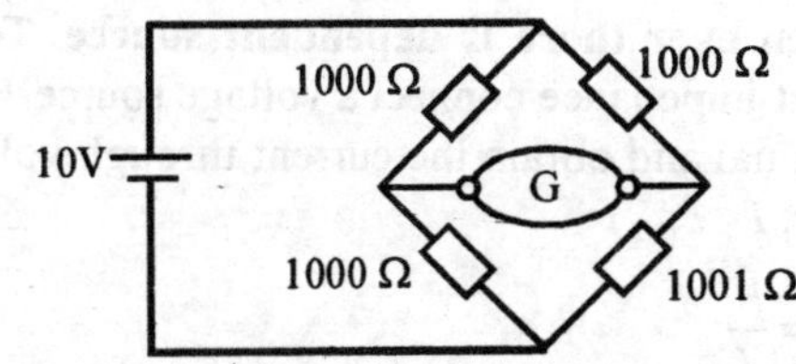

Find V_{TH} ⇒

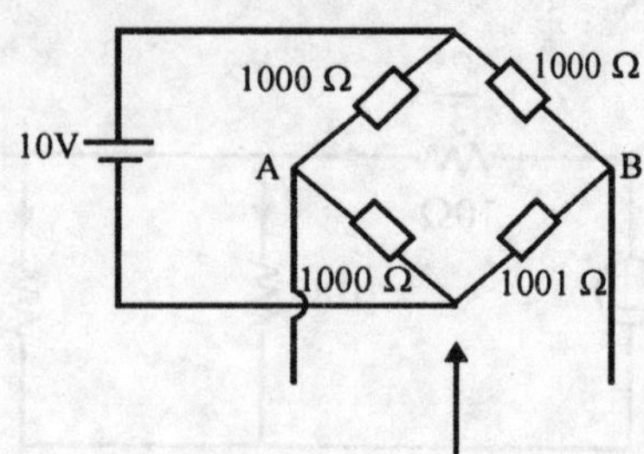

$V_A = 5\,V$

$V_B = 5.002498$

$V_{TH} = V_B - V_A$

$V_{TH} = 0.0025$

R_{TH} ⇒

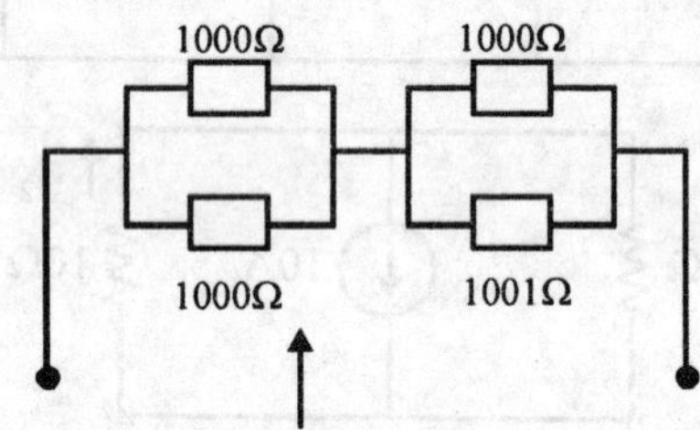

$R_{TH} = R_{TH} = 500 + 500.249$

$R_{TH} = R_{TH} = 1000.2498$

$$I = \frac{.002498}{1000 + 1000.2498} \Rightarrow 1.25 \times 10^{-6}$$

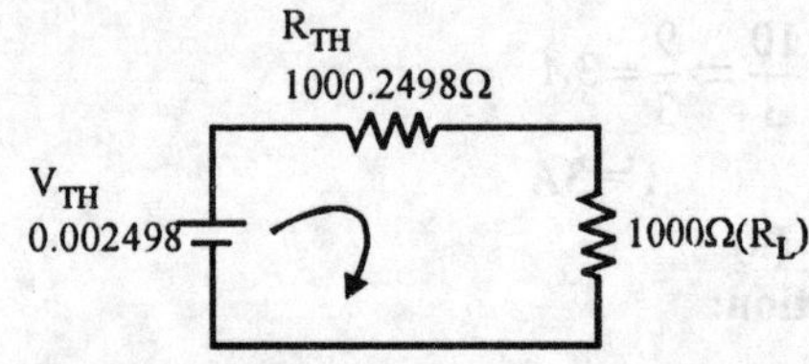

$$I = \frac{0.002498}{1000 + 1000.2498}$$

$= 1.25 \times 10^{-6}$

$I = 1.25\ \mu A$

40. Ans. (a)

Solution:

When ever there is dependent source, To fined the input impedance connect a voltage source V across the terminal and obtain the current through voltage source V i.e. I

$$Z_{th} = \frac{V}{I}$$

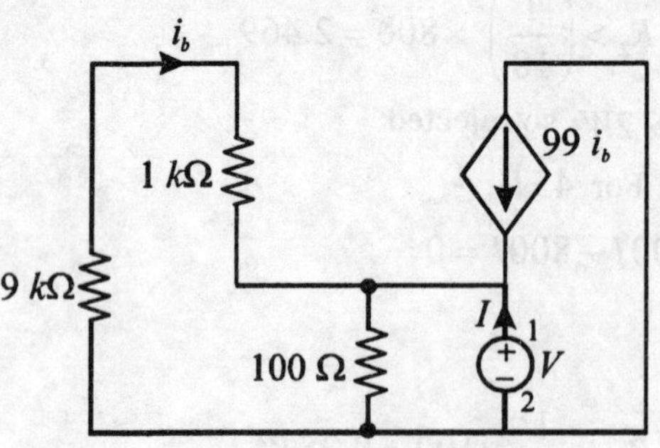

Now arrange the circuit for analysis

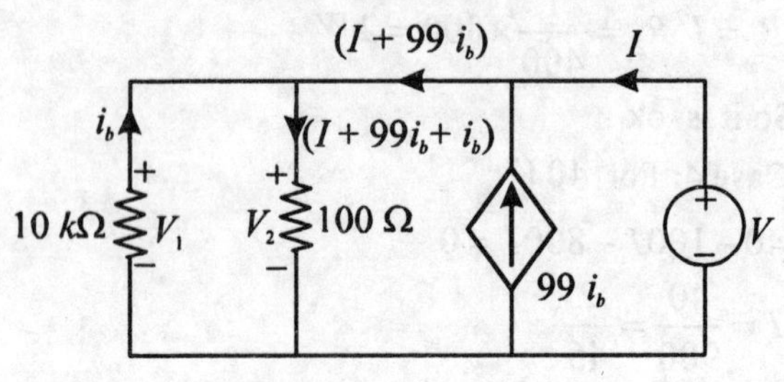

$$V_1 = V_2 \qquad (1)$$

or, $(10k)(-i_b) = 100(I + 99i_b + i_b)$

or, $-10000\, i_b = 100\, I + 100 \times 100 i_b$

$= 100\, I + 10000 i_b$

or, $-20000 i_b = 100\, I$

or, $i_b = -\left(\frac{100}{20000}\right) I$

$$= \left(-\frac{I}{200}\right) \qquad (2)$$

Now KVL with V and V_2

$V = 100[I + 99 i_b + i_b]$

$= 100[I + 100 i_b]$

$= 100\left[I + 100\left(\frac{-I}{200}\right)\right]$

$= 100\left[I - \frac{I}{2}\right]$

$= 100 \times \frac{I}{2}$

$V = 50\, I$

$$Z_{th} = \frac{V}{I} = \frac{50\, I}{I} = 50\,\Omega$$

Note: Since there is no independent source available i the circuit, no current will flow. Hence for given circuit V_{OC} and I_{SC} can not be found out and method $Z_{eq} = \frac{V_{OC}}{I_{SC}}$ is not application in this case

41. Ans. (a)

Solution: Let the thevenins's equivalent network represent the DC network

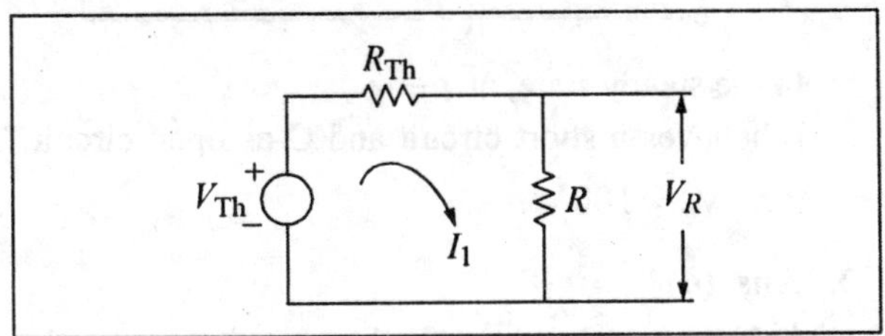

$$I_1 = \frac{V_{R1}}{R} = \frac{20}{10} = 2A \quad (1)$$

$$V_{Th} = 2R_{Th} + 20$$

Similarly, $I_2 = \frac{V_{R2}}{R} = \frac{30}{20} = 1.5\ A$

$$V_{Th} = 1.5\ R_{Th} + 30 \quad (2)$$

From (1) and (2)

$$2R_{Th} + 20 = 1.5\ R_{Th} + 30$$

$$0.5\ R_{Th} = 10$$

i.e. $R_{Th} = 20\ \Omega$

From (1)

$$V_{Th} = 2\ R_{Th} + 30$$

$$V_{Th} = 2 \times 20 + 20 = 60\ V$$

Now, when $R = 80\ \Omega$ as given then

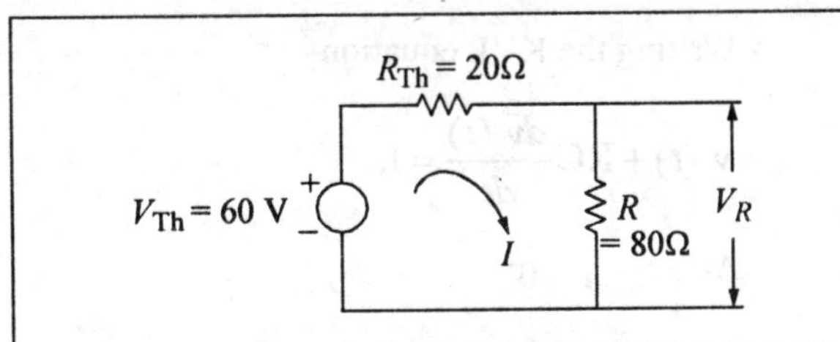

Current $I = \frac{60}{20 + 80} = 0.6\ A$

$\therefore$ $V_R = 0.6 \times 80 = 48\ V$

42. Ans. (c)

Solution: Remove the R and solve for thevenin's equivalent circuit

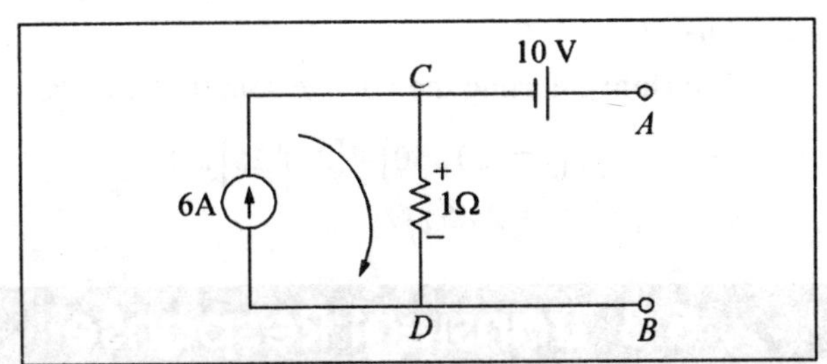

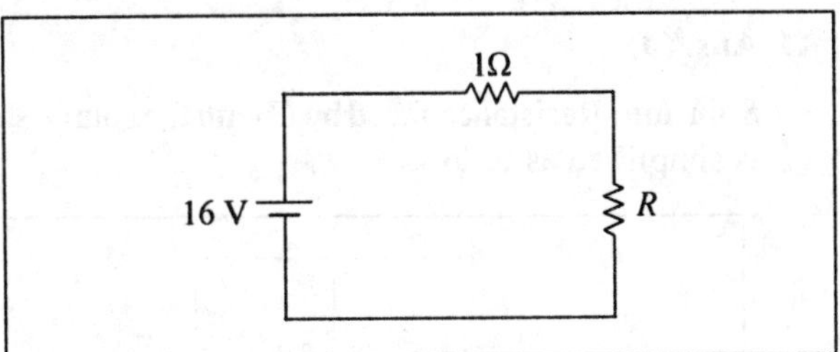

$$R_s = 1\ \Omega;\ V_{CD} = 6\ V$$

$$V_{AB} = V_{OC} = 10 + 6 = 16\ V$$

For maximum power transfer

$$Z_L = Z_S$$

$\therefore$ $R = 1\ \Omega$

$$P = I^2 R = \left(\frac{16}{2}\right)^2 \times 1 = 64\ \text{Watt}$$

43. Ans. (b)

Solution: $Z_c \to \frac{1}{sC}$

$$Z_{in} = \frac{3}{2s} + \left[5 \,||\left(\frac{25}{2s} + \frac{5}{4}\right)\right]$$

$$Z_{in} = \frac{\left(\frac{25}{2s} + \frac{5}{4}\right)5}{\left(\frac{5}{4} \times \frac{25}{2s}\right) + 5} + \frac{3}{2s}$$

$$\underset{s\to\infty}{\text{Lt}}\ Z_{in} = \frac{\left(0 + \frac{5}{4}\right)5}{\frac{5}{4} + 5} + 0 = \frac{\frac{25}{4}}{\frac{25}{4}} = 1$$

Second Method

As $t \to 0$

$C \to$ short circuit

Hence $Z_{in} = 5 \,|| \frac{5}{4}$

$= 1\ \Omega$

44. Ans. (b)

Solution: It is used for only linear network

45. Ans. (b)

Solution: Shorting the voltage source, we have

$$R_{12} = [20 \| 30] + [25 \| 25]\ \text{k}\Omega$$
$$= 24.5\ \text{k}\Omega$$

6.3 Transient Response

1. Ans. (d)

Solution: Resistance faced by *C* with the source shorted, is simplified as follows:

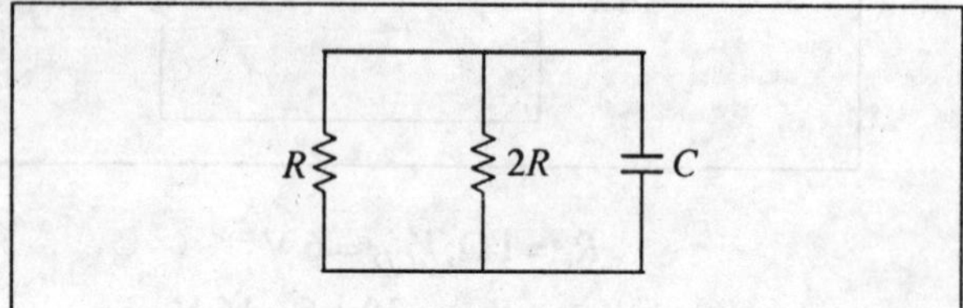

$$R_{eq} = \frac{R \times 2R}{3R} = \frac{2R}{3}$$

Time constant of the circuit,

$$t = R_{eq} \times C = \frac{2R}{3} \times C = \frac{2}{3}RC$$

2. Ans. (b)

Solution:

$$Z(s) = Ls + [1/Cs]$$

$$I_L(s) = \frac{100/s}{Ls + [1/Cs]} = \frac{100\ C}{LCs^2 + 1} = \frac{100}{L(s^2 + [1/LC])}$$

$$V_c(s) = \frac{I_L(s)}{Cs}, \text{ Since initial voltage } V_c(0^-) = 0$$

$$= \frac{100}{LCs(s^2 + [1/LC])}$$

The steady state voltage across the capacitor is

$$\lim_{t\to\infty} V_c(t) = \lim_{t\to\infty} s\,V_c(s)$$

$$= \lim_{s\to\infty} \frac{100\ \text{Volts}}{LC(s^2 + [1/LC]} = 100\ \text{V}$$

Second Method

After the switch is closed diode conducts, and offers ideally zero resistance.

For $t > 0$ the circult is shown is figure.

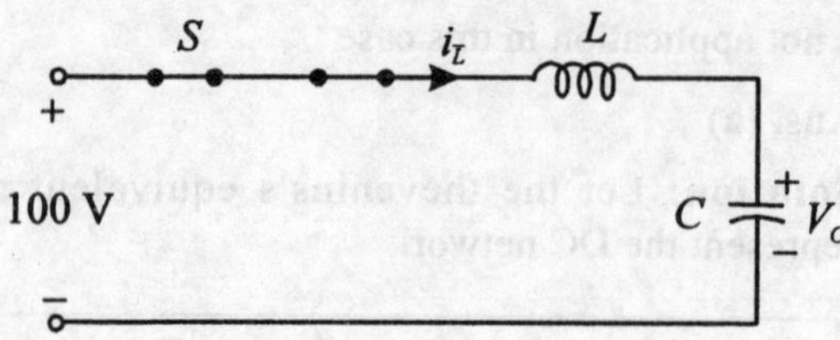

In the steady state, at $t = \infty$,

L behaves a short circuit and C as open circuit. There fore $V_C = 100\text{V}$.

3. Ans. (d)

Solution: Voltage across the capacitor at any time *t*

$$V_c = V(1 - e^{-t/RC}) = 1 - e^{-t/RC}$$

since $\quad V = 1$ Volt

$$\therefore \quad \frac{dv_c}{dt} = \frac{1}{RC}e^{-t/RC}$$

$$\text{At } t = 0^+, \quad \frac{dV_c}{dt} = \frac{1}{RC}$$

Second Method

For $t > 0$ the circuit is shown in figure

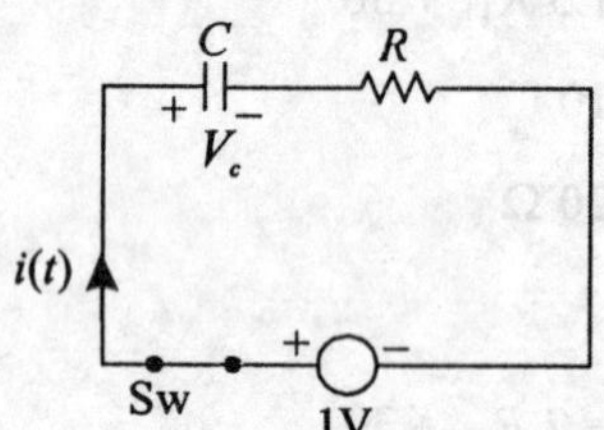

$$i(t) = C\frac{dv_c(t)}{dt} \qquad \text{... (1)}$$

Writing the KVL equation

$$v_c(t) + \text{RC}\frac{dv_c(t)}{dt} = 1, \qquad \text{... (2)}$$

At $\quad t = 0^+$

$$v_c(0^+) + \text{RC}\left.\frac{dv_c(t)}{dt}\right|_{t=0_+} = 1$$

As $v_c(0^+) = 0$,

$$\left.\frac{d\,v_c(t)}{dt}\right|_{t=0^+} = \frac{1}{\text{RC}}$$

4. Ans. (a)

Solution: $V_c = V[1 - \exp(-t/RC)]$

Where $t = T$

Second Method

According to given question figuer can be drawn as below

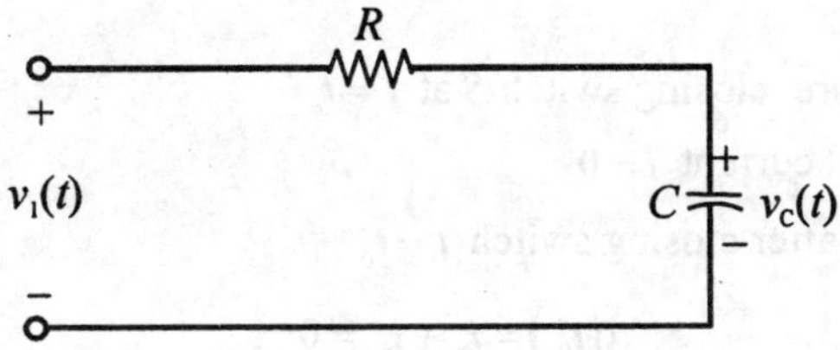

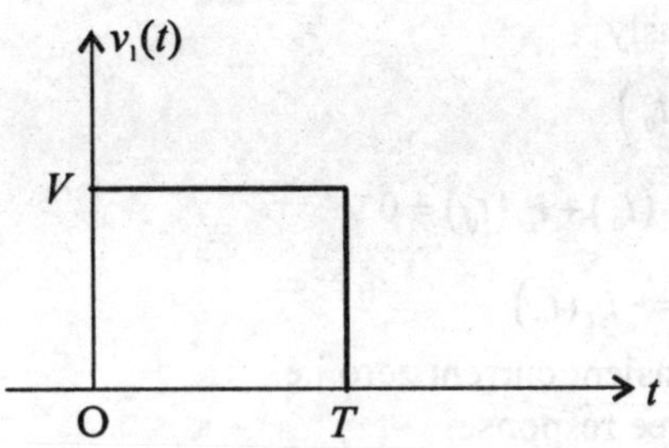

$$v_1(t) = Vu(t) - V\,u(t-T)$$

When $v_1(t) = Vu(t)$

$$v_C(t) = V\left(1 - e^{\frac{-t}{\tau}}\right),$$

Where $\tau = RC$

$$v_{C(\max)} = v_C(T) = \left(1 - e^{-\frac{T}{RC}}\right)$$

* Complete resposne is given by

$$v_C(t) = V\left(1 - e^{\frac{-t}{\tau}}\right)u(t) - V\left(1 - e^{\frac{(t-T)}{\tau}}\right)u(t-T)$$

as shown in figure below

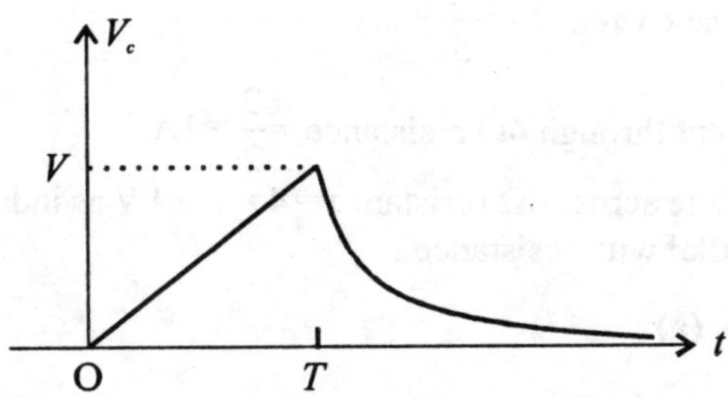

5. Ans. (b)

Solution:

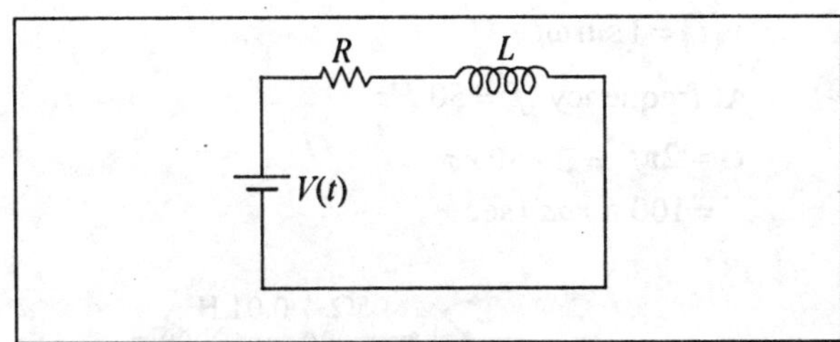

At any time t, current is given by

$$i(t) = [V(t)/R]\,[1 - e^{-(R/L)t}]$$

In present case $v(t) = u(t)$

At $t = 0^+$

$i(t) = 0$

So, $V_R = 0$

Second Method

For RL series circuit with zero initial conditions and step input.

$$i(t) = \left(1 - e^{-\frac{R}{L}t}\right)u(t)$$

* $i(t)$ grows from zero at $t = 0$ to 1 at $t = \infty$ exponentially as shown in figure below

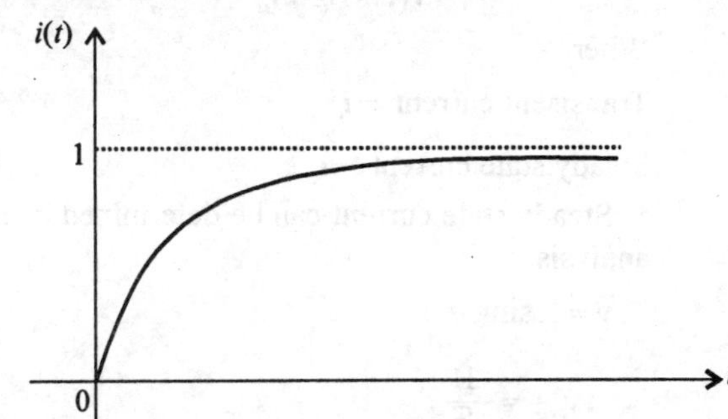

* The voltage across the resistor at $t = 0^+$ $= Ri(0^+) = 0$

* The steady state current in the inductance $i(\infty) = 1\,A$

Therefore, Energy stored in the inductor $= \frac{1}{2}Li^2$

$$= \frac{1}{2}L$$

Hence correct option is (b)

6. Ans. (b)

Solution:

The applied voltage

$v(t) = 1\sin\omega t$

At frequency $f = 50\ Hz$

$\omega = 2\pi f = 2\times 50\times\pi$

$= 100\ \pi\ rad/\sec$

5Ω 0.01 H

$t = t_o$ R L

sin (ω)t

$R = 5\ \Omega$

$X_L = \omega L$

$= 100\pi\times 0.01$

$= 3.14\ \Omega$

impedance $\bar{Z} = R + jX = 5 + j3.14$

$\bar{Z} = Z\angle\phi = 5.91\angle 32.14°$

$Z = 5.91$ and $\phi = 32.14°$

When switch S is closed at $t = t_o$, there is transient and steady state response in current

$$i(t) = i_{tr} + i_{ss}$$

Where

Transisent current $= i_t$

Steady state current $= i_{ss}$

* Steady state current can be determined using phasor analysis

$v = 1.\sin\omega t$

$$v_{rms} = \frac{1}{\sqrt{2}}$$

So phasor $\bar{V} = \frac{1}{\sqrt{2}}\angle 0°$

$$\bar{I} = \frac{\bar{V}}{\bar{Z}} = \frac{1}{2}\times\frac{1\angle 0°}{5.91\angle\phi}$$

$= 0.12\angle -\phi\ A$

rms current $I_{rms} = 0.12$

So steady state current

$i_{ss}(t) = \sqrt{2}\times 0.12\sin(\omega t - \phi)$

$= 0.17\sin(\omega t - \phi)$

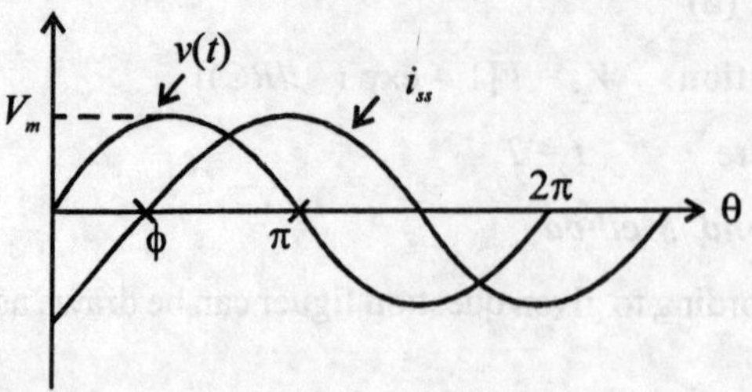

Before closing switch S at $t = t_o$

Total current $i = 0$

Just after closing switch $t = t_0^+$

$$i(t_0^+) = i_{tr} + i_{ss} = 0$$

* Current through the inductor can't change instanteniously

$i(t_0^+) = i(t_0^-)$

So $i(t_0^+) = i_{tr}(t_0) + i_{ss}(t_0) = 0$

$i_{tr}(t_0) = -i_{ss}(t_0)$

To have transient current zero i.e., Transient free response

$i_{tr}(t_0) = 0$

$i_{ss}(t_0) = 0$

the steady state current is zero at

$\theta = \omega t_0 = \phi$

$$t_o = \frac{\phi}{\omega}$$

$\phi = 32.14°$

$= 0.561\ rad$

So $t_o = \frac{0.561}{100\ \pi} = 1.786\ ms$

7. Ans. (b)

Solution: When switch is ON at $t = 0$ i.e. under initial condition capacitor is shot circuit and inductor is open circuit.

∴ $4 \parallel 4 = 2\ \Omega$,

circuit total resistance, $3 + 2 = 5\Omega$

source current $I = \frac{10}{5} = 2A$

current through 4Ω resistance $= \frac{2}{2} = 1A$

Voltage across 4Ω resistance $= 4\times 1 = 4$ V as inductor is parallel with resistance.

8. Ans. (a)

Solution:

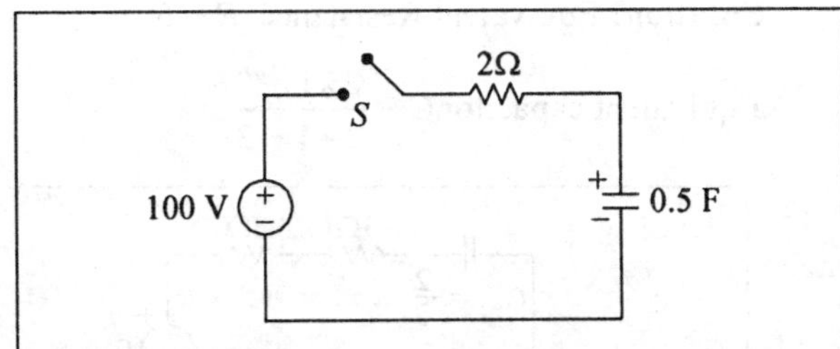

$$V = IR + \frac{1}{C}\int I dt$$

or $$R.\frac{dq}{dt} + \frac{1}{C}q = 100$$

or $$(100 - q) = RC\frac{dq}{dt}$$

$$\int_{q_o}^{q} \frac{dq}{(100\,C - q)} = \frac{1}{RC}\int_{o}^{t} dt$$

$$\Rightarrow \quad \ell n[100C - q]_{q_o}^{q} = \frac{t}{RC}$$

$$\Rightarrow \ell n\left[\frac{100C - q}{100C - q_o}\right]^{-1} = \frac{t}{RC}$$

$$\Rightarrow \quad \ell n\left[\frac{100C - q}{100C - q_o}\right] = \frac{-t}{RC}$$

$$\Rightarrow \quad 100C - q = (100C - q_o)e^{-t/RC}$$

$$\Rightarrow \quad -\frac{dq}{dt} = (100C - q_o)e^{-t/RC}\left(t\frac{1}{RC}\right)$$

$$i = \frac{(100C - q_o)e^{-t/RC}}{RC} = 40e^{-1} = 14.7$$

9. Ans. (b)

Solution: At $t = 0$, the capacitor voltage $V_c = 0$

At steady state, capacitor is open circuit i.e. at $t \to \infty$,

$$V_c = \frac{10}{10+10}\times 20 = 10\text{V}$$

10. Ans. (c)

Solution: At, $t = 0^-$ the circuit is in steady state inductor behave as short circuit, the current through the inductor $i(0^-) = 10/10 = 1$A. The current through the inductor does not change instantaneously. Therefore $i(o^-) = i(o^+) = 1$A.

11. Ans. (a)

Solution: At, $t = 0^-$, inductor is short circuit

$$i(0^-) = \frac{120}{60} = 2\text{A}$$

$$i(0^-) = i(0^+) = 2\text{A}$$

i.e. $$2 = \frac{120}{R + 40 + 20}$$

$$R = 0\ \Omega$$

12. Ans. (c)

Solution:

Let stored energy and dissipated energy are E_1 and E_2 respectively.

Then, Current

$$\frac{i_2^2}{i_1^2} = \frac{E_2}{E_1} = 0.95$$

$$i_2 = \sqrt{0.95}\,i_1 = 0.97i_1$$

Current at any time t, when the switch is in position2 is given by

$$i(t) = i_1 e^{-\frac{R}{L}t} = 2e^{-\frac{60}{10}t} = 2e^{-6t}$$

After 95% of energy dissipated current remaining in the circuit is

$$i = 2 - 2\times 0.97$$

$$= 0.05\text{ A}$$

There fore $0.05 = 2e^{-6t}$

$$t \approx 0.50\text{ sec}$$

13. Ans. (b)

Solution: By KVL,

$$V_L + V_C = 0$$

$$\frac{Ldi}{dt} + \frac{Q}{C} = 0; \quad \left\{i = \frac{dQ}{dt}\right\}$$

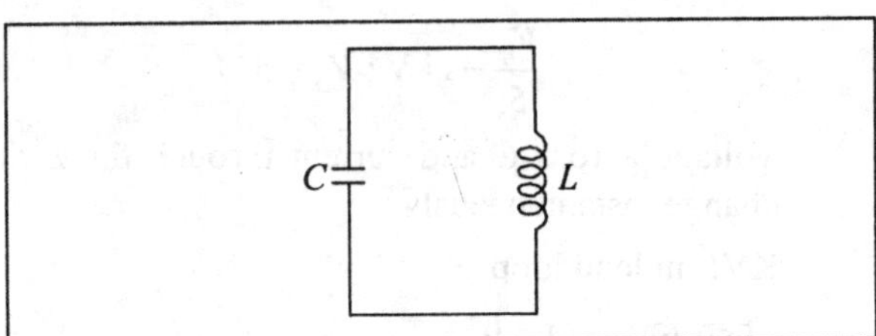

$$\frac{Ld^2Q}{dt^2} + \frac{Q}{C} = 0$$

$$Q(t) = Q_o \cos(\omega t + \phi)$$

$$Q(t) = Q_o \cos(\omega_o t)$$

$$V(t) = \frac{Q(t)}{C} = \frac{Q_o}{C}\cos\omega_o t = V_o \cos\omega(\omega_o t)$$

$$V(t) = V_o \cos(\omega_o t)$$

In other way,

It can be explained as if both capacitor and inductor an ideal then Electrical energy stored in the capacitor changes its form by storing energy in inductor in the form of magnetic energy. The change of the energy is continuous. The energy is in the from of voltage oscillate and voltage across capacitor $V_c = V_o \cos(\omega_o t)$

14. Ans. (d)

Solution:

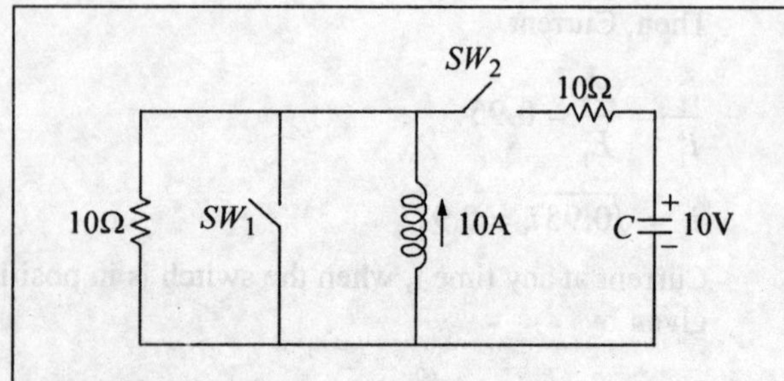

At $t = 0^-$, SW_2 is closed and SW_1 is opened

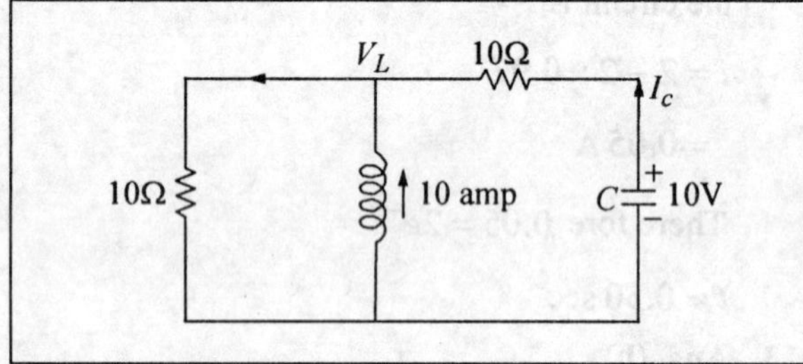

Voltage across $L = V_L$

Apply KCL at node

$$\frac{V_L}{10} = 10 + \frac{10 - V_L}{10}$$

$$\frac{V_L}{5} = 11\,\text{V};\ V_L = 55V$$

Voltage across C and current through the L does not change instantaneously.

KVL in load loop

Therefore, at $t = 0^+$

$$-10 + 10 \times I_c + V_L$$

$$V_c = 10 \times I_c + V_L$$

$$10 = 10\,I_c + 55$$

$$I_c = -4.5.$$

Another Way

Current through C will be

$$\frac{V_L - 10}{10} = \frac{55 - 10}{10} = 4.5\text{ A}$$

15. Ans. (c)

Solution: Equivalent Resistance $R = 6$

Equivalent capacitor $C = \frac{2 \times 1}{2 + 1} = \frac{2}{3}$,

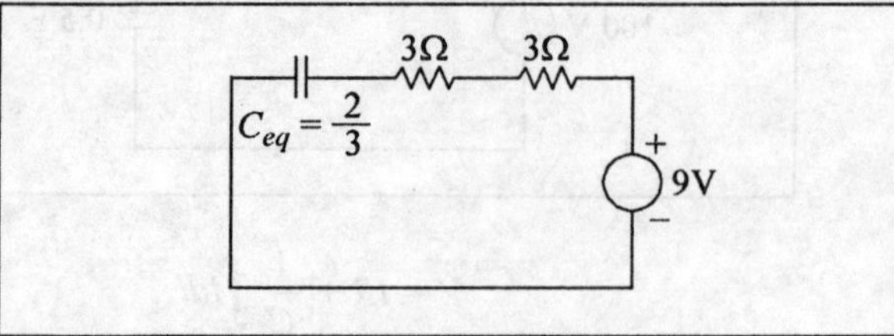

Time constant $\tau = RC = \frac{6 \times 2}{3} = 4\ s$

16. Ans. (b)

Solution:

At $t = 0^+$

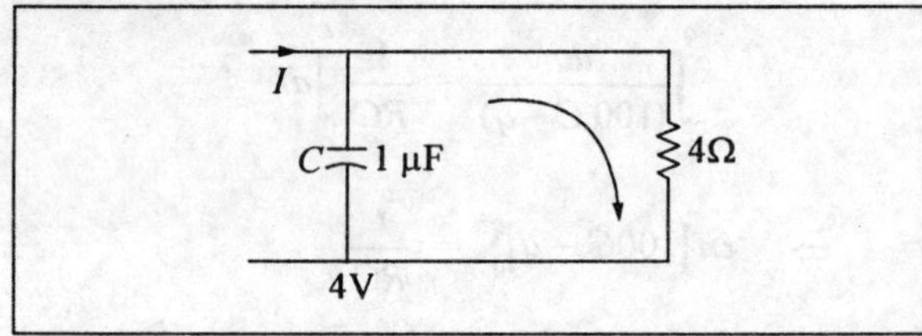

$$i(t) = \frac{V_C}{R}\left(1 - e^{-t/RC}\right)\Big|_{t=0} = \frac{V_0}{R} = \frac{4}{4} = 1\text{ A}$$

At $t = 0^-$ switch is closed for a long time, i.e. in steady state condition

$$C \to \text{Open circuit}$$

Current in the circuit from 5 V source

$$= \frac{5}{1+5} = 1\text{ A}$$

Second Method

At $t = 0^-$

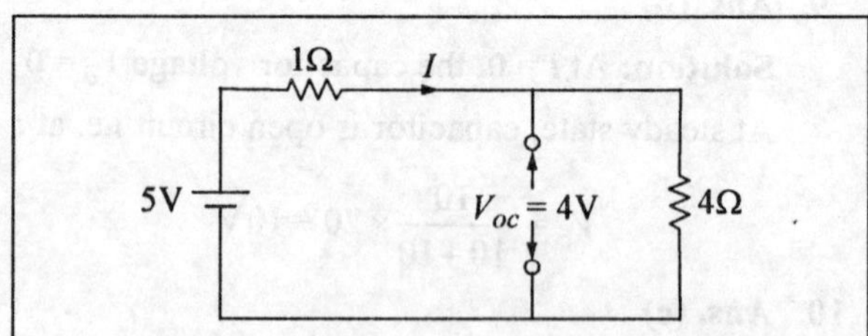

$$I = \frac{5}{5} = 1\text{ A}$$

$$V_{oc} = 5 - I \times R$$

$$= 5 - 1 \times 1$$

$$= 4\text{ V}$$

At $t = 0^+$

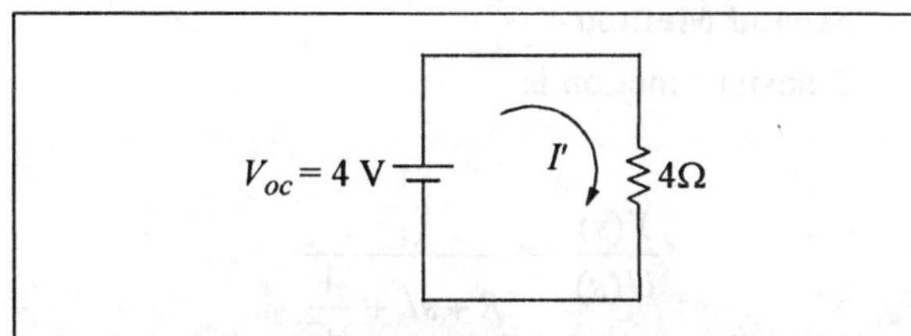

$$I' = \frac{V_{oc}}{R} = \frac{4}{4} = 1\text{A}$$

17. Ans (d)

Solution:

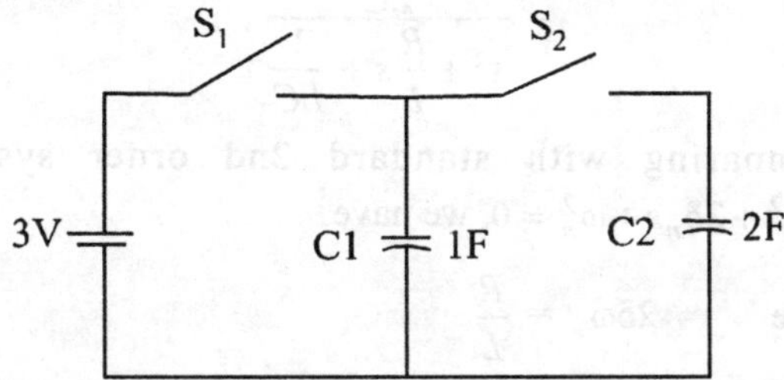

For $t < 0$, S_1 remain closed and S_2 open.

At $t = 0$, S_1 is opened and S_2 is closed , $V_{c2} = 0$

At $t < 0$, $V_{c1} = 3\,V$, $V_{c2} = 0\,V$

$t = 0^+$ means $t = 0$

At $t = 0$, S_1 is opened and capacitor acts as short circuited.

$V_{c1} = 3\,V$, $V_{c2} = 0\,V$ (given)

Total voltage across the capacitor

Combination at $t = 0$ $= V_{c1} + V_{c2}$

$= 3 + 0 = 3\,V$

Note:

This $V_{c1} = 3\,V$ at $t = 0$, because voltage at capacitor does and current in inductor does not change instanta neously.

18. Ans. (b)

Solution:

RC differentiating circuit is shown in figure below

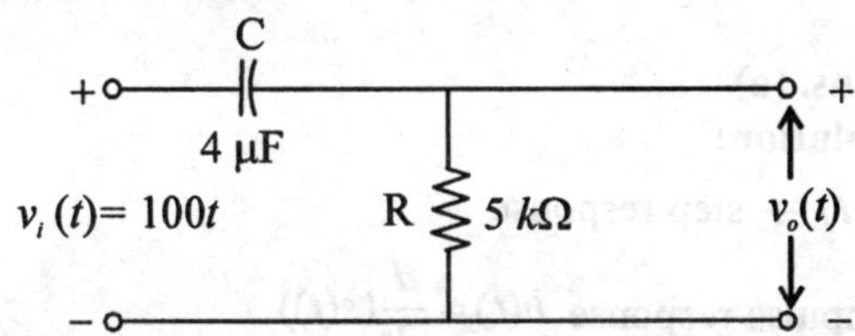

$v_i(t) = 100\,t\,u(t)$

Time constant,

$\tau = RC = 5\times10^3 \times 4\times10^{-6}$

$= 0.02$ sec

Formula : $v_0(t)\big|_{\max} = v_0(\infty)$

Final value

$V_i\tau = V_i\,RC$

$= 100\times0.02 = 2$V

Another approach

Transfer function,

$$\frac{V_o(s)}{V_i(s)} = \frac{R}{R + \frac{1}{Cs}}$$

$$= \frac{s}{s + \frac{1}{RC}} = \frac{s}{s+50}$$

$v_i(t) = 100\,t\,u(t)$

$$V_i(s) = \frac{100}{s^2}$$

$$V_o(s) = \frac{100}{s(s+50)} = \frac{2}{s} - \frac{2}{s+50}$$

$v_0(t) = 2(1 - e^{-50t})u(t)$

$v_0(t)$ is sketched in figure below

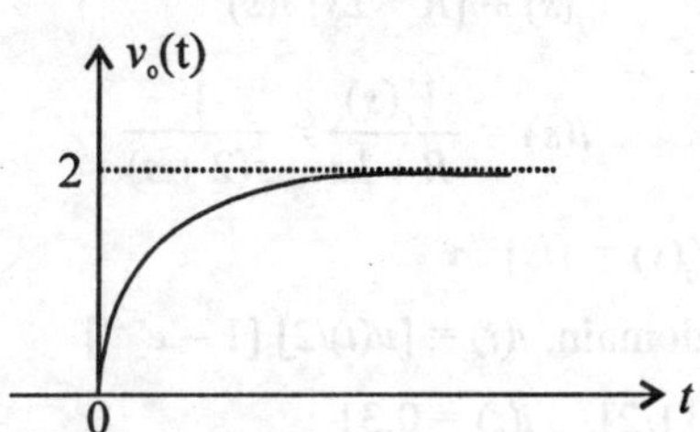

Maximum $v_0(t)$ occurs as $t \to \infty$

Maximum $v_0(t) =$ Final value = 2 V

Final value can also be obtained from

Final value theorem :

$$\underset{t\to\infty}{Lt}\ v_0(t) = \underset{s\to0}{Lt}\left[s\,V_0(s)\right] = 2$$

19. Ans (c)

Solution:

At $t = 0^-$, L behaves as a short circuit as shown in figure below.

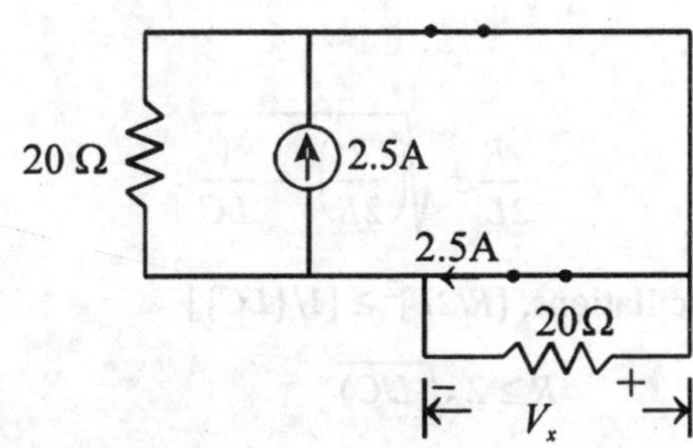

The inductor current $i_L(0^-) = 2.5$ A, $V_x(0^+) = 0$

As the current through the inductor cannot change instantaneously.

$i_L(0^+) = i_L(0^-) = 2.5$ A

At $t = 0^+$, switch is open, the status of the circuit is shown in figure below.

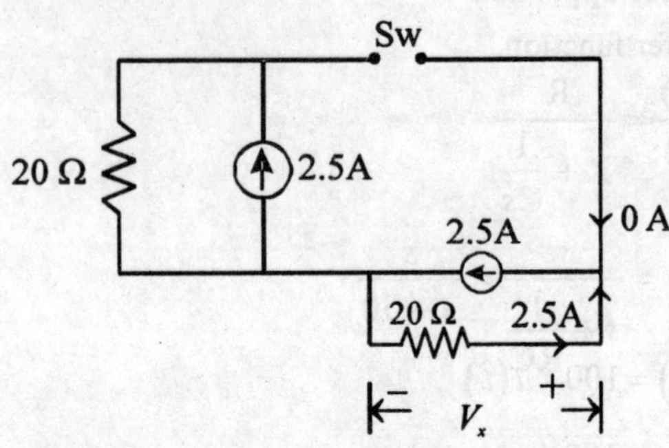

Hence, $V_x = -2.5 \times 20 = -50$ V

20. Ans. (c)

Solution:

$$V_i(t) = Ri(t) + L[di(t)/dt]$$

With Laplace transformation, we have

$$V_i(s) = [R + Ls]\ I(s)$$

or $$I(s) = \frac{V_i(s)}{R + Ls} = \frac{1}{s(2+s)}$$

[Since $V_i(s) = 1/s$]

In time-domain, $i(t) = [u(t)/2]\ [1 - e^{-2t}]$

At $t = [1/2]$, $i(t) = 0.31$

And at $t = \infty$, $i(t) = 0.5$

21. Ans. (c)

Solution:

$$u(t) = Ri + [L\ di/dt] + [1/C] \int i\ dt$$

By differentiating ,

$$0 = [Rdi/dt] + [L\ d^2 i/(dt)^2] + [1/C]\ i$$

$$\frac{d^2 i}{dt^2} + \left(\frac{R}{L}\right)\frac{di}{dt} + \left(\frac{1}{LC}\right) i = 0$$

Roots are,

$$\frac{-R}{2L} \pm \sqrt{\left(\frac{R}{2L}\right)^2 - \frac{1}{LC}}$$

For no oscillations, $[R/2L]^2 \geq [1/\{LC\}]$

$$R \geq 2\sqrt{(L/C)}$$

Second Method

Transfer function is

$$\frac{Y(s)}{U(s)} = \frac{\frac{1}{sC}}{R + sL + \frac{1}{sC}}$$

$$= \frac{1}{s^2 LC + sCR + 1}$$

$$= \frac{\frac{1}{LC}}{s^2 + \frac{R}{L}s + \frac{1}{LC}}$$

Comparing with standard 2nd order system $s^2 + 2\delta_n s + \omega_n^2 = 0$, we have

Here $$2\delta\omega_n = \frac{R}{L}$$

and $$\omega_n = \frac{1}{\sqrt{LC}}$$

Thus, $$\delta = \frac{R}{2L}\sqrt{LC} = \frac{R}{2}\sqrt{\frac{C}{L}}$$

For no oscillation $\delta \geq 1$

Thus, $$\frac{R}{2}\sqrt{\frac{C}{L}} \geq 1$$

or $$R \geq 2\sqrt{\frac{L}{C}}$$

22. Ans. (b)

Solution: At $t = 0$, i.e. under inital condition inductor acts like a open cirucit, we know $L\frac{di}{dt} = V_L$

At $t = 0^+$, inductor path is open circuit current flow through R_s then

$$V_L = I_s . R_s (\text{at } t = 0^+)$$

Thus, $$\frac{di}{dt} = \left(\frac{I_s \cdot R_s}{L}\right)$$

23. Ans. (a)

Solution:

$s(t) \rightarrow$ step response

Impulse response $$h(t) = \frac{d}{dt}(s(t))$$

$$= \frac{d}{dt}(1 - e^{-\alpha t})$$

$$= \alpha e^{-\alpha t}$$

24. Ans. (d)

Solution:

At $t = 0$, initially C is short circuited when switch is on a high current, i.e. impulse current flow

Another Method

Time constant = RC
In the given circuit, R = 0
Rise time = 0,
Hence, capacitor charges instantaneously
and the current can be represented as impulse function.

25. Ans. (b)

Solution:

$$H(s) = \frac{R_2 \parallel [1/sC_2]}{R_1 \parallel [1/sC_1] + R_2 \parallel [1/sC_2]}$$

$$= \frac{[R_2/1 + sC_2 R_2]}{\dfrac{R_1}{1 + sC_1 R_1} + \dfrac{R_2}{1 + sC_2 R_2}}$$

But, $R_1 C_1 = R_2 C_2$

$$H(s) = [R_2/(R_1 + R_2)]$$

$$h(t) = LT^{-1} [H(s)]$$

$$= [R_2/(R_1 + R_2)]\ \delta(t)$$

26. Ans. (b)

Solution: In steady state, capacitor are open and inductor, are short circuit and no current flows through the circuit.

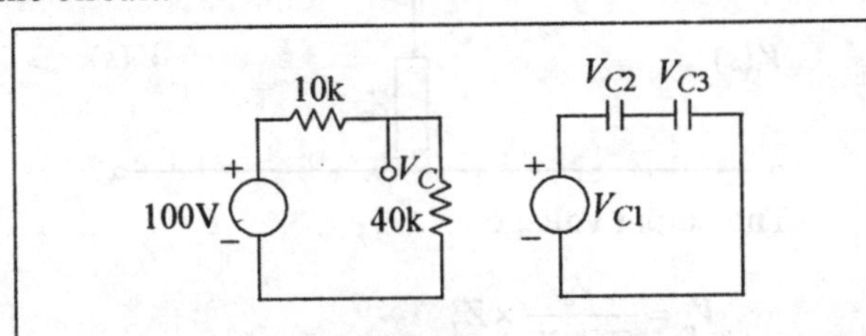

$$V_{C1} = 100 \times \frac{40}{50} = 80 \text{ V}$$

$$V_{C2} = \frac{C_3}{C_2 + C_3} \times V_{C1}$$

$$V_{C2} = 80 \times \frac{C_3}{C_2 + C_3} = 80 \times \frac{3}{5} = 48 \text{ V}$$

$$V_{C3} = 80 \times \frac{C_2}{C_2 + C_3} = 80 \times \frac{2}{5} = 32 \text{ V}$$

27. Ans. (d)

Solution:

At $t = 0^-$ i.e. final condition, it means capacitor is open circuit and $L \rightarrow$ Short circuit

At $t = 0^+$ i.e. initial condition, it means the voltage across capacitors is $-V$ i.e. $V_C(0^-) = V$

At $t = 0^+$ i.e. initial condition

$C \rightarrow$ Short and $L \rightarrow$ Open circuit

Capacitor in series with R is shorted, inductor open

Since loop is open circuit, $e_t = 0$

28. Ans. (d)

Solution:

Transformed circuit representation is as shown below

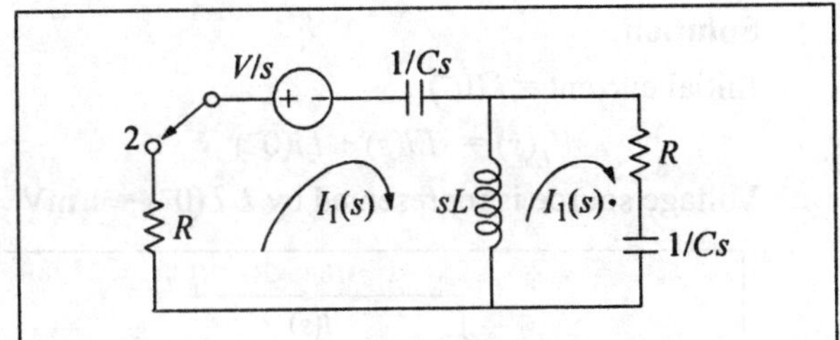

Using KVL in both the loops, we get

$$I_1(s)[1/Cs + sL + R] - I_2\, sL + [V/s] = 0$$

and $$I_2(s)\,[R + (1/Cs) + Ls] - I_1\, Ls = 0$$

Writing in matrix form,

$$\begin{pmatrix} R + sL + \dfrac{1}{Cs} & -Ls \\ -Ls & R + Ls + \dfrac{1}{Cs} \end{pmatrix} \begin{pmatrix} I_1(s) \\ I_2(s) \end{pmatrix} = \begin{pmatrix} -V/s \\ 0 \end{pmatrix}$$

29. Ans. (b)

Solution:

$$V(t) = u(t)$$

Using KVL around the loop, we have

$$V(t) = Ri(t) + \frac{Ld\,i(t)}{dt} + \frac{1}{C}\int_0^\infty i(t).dt$$

Taking Laplace transformation on both sides, we have

$$V(s) = RI(s) + Ls\,I(s) - LI\,(0^+) + \frac{I(s)}{sC} + \frac{V_c(0^+)}{s}$$

$$\frac{1}{s} = I(s) + sI(s) - 1 + \frac{I(s)}{s} - \frac{1}{s}$$

$$\frac{2}{s} + 1 = \frac{I(s)}{s}[s^2 + s + 1]$$

$$I(s) = \frac{s+2}{s^2 + s + 1}$$

30. Ans. (b)

Solution:

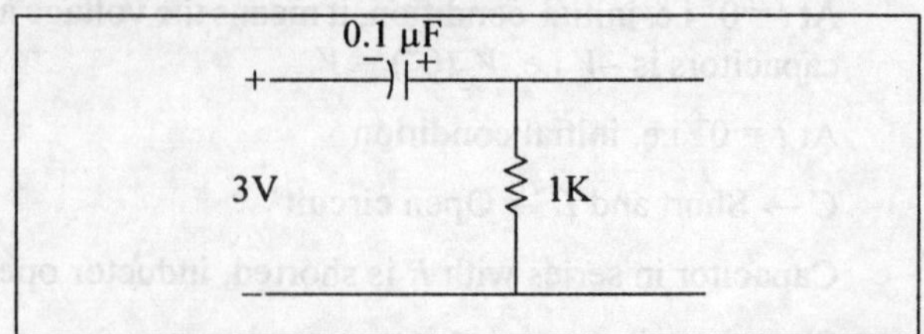

Time constant = RC = 100 μsec.

T = 2 sec >> 100 μsec.

So, C is fully charged (opposite polarity)

$$V_{out} = -3$$

31. Ans. (a)

Solution:

Initial current = $i(0^-)$

$$V_L(s) = sLI(s) + Li(0^-)$$

Voltage source is represented by $L\,i(0^-)$ = 1 mV

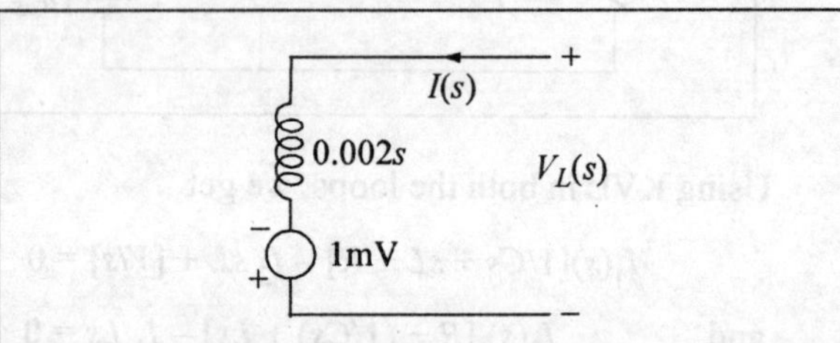

$$L \times i(0^-) = 1 \text{ mV}$$

$$\Rightarrow \quad 2 \times 10^{-3} \times i(0^-) = 1 \times 10^{-3}$$

$$\Rightarrow \quad i(0^-) = [1/2]$$

$$= 0.5 \text{ A}$$

32. Ans. (c)

Solution:

For a particular capacitor

$$i = C\frac{dv}{dt}$$

Laplace transform

$$I(s) = C\left[sV(s) - V(0)\right]$$

Initialy, voltage is zero

$$I(s) = sC\,V(s)$$

So, C can be represented by an impedance $\dfrac{1}{sC}$

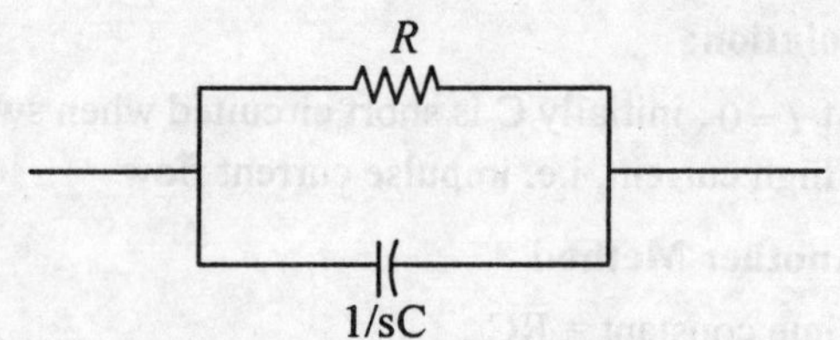

The equivalent impedance of parallel RC network

$$Z = \frac{R \times \frac{1}{sC}}{R + \frac{1}{sC}}$$

$$= \frac{R}{1 + sCR}$$

For $R_1 = 1\,k = 10^{+3}$,

$C_1 = 4\,\mu F = 4 \times 10^{-6}$

For $R_2 = 4\,k\Omega = 4 \times 10^{3}$

$C_2 = 1\,\mu F = 10^{-6}$

$R_2C_2 = 0.004$

Hence, impedance

$$Z_1 = \frac{1000}{1 + 0.004\,s}$$

$$Z_2 = \frac{4000}{1 + 0.004\,s}$$

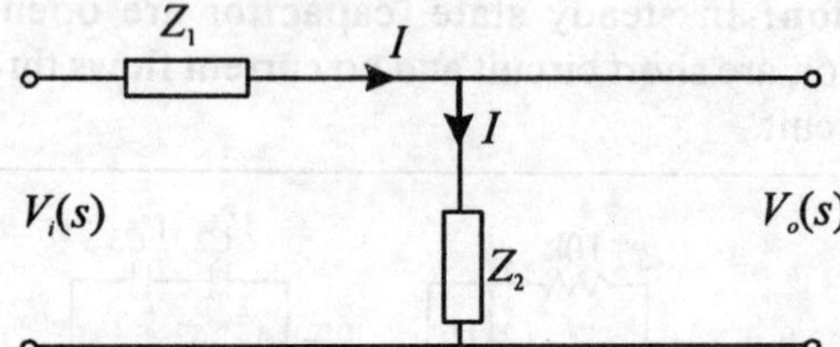

The output voltage

$$V_o = \frac{V_i}{Z_1 + Z_2} \times Z_2$$

$$= \frac{V_i}{1 + \frac{Z_1}{Z_2}}$$

$$V_o(s) = \frac{V_i(s)}{1 + \frac{1}{4}} = 0.8\,V_i(s)$$

$$v_o(t) = 0.8\,v_i(t)$$

$$= 0.8 \times 10\,u(t)$$

$$= 8\,u(t)$$

33. Ans. (a)

Solution:

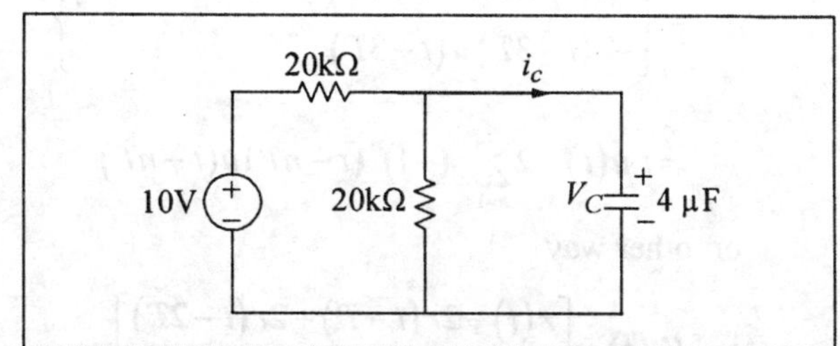

$$\frac{10}{s} = 20{,}000\, I(s) + \frac{20{,}000\, I(s)}{20{,}000 \times 4 \times 10^{-6} s + 1}$$

$$\frac{10}{s} = 20{,}000\, I(s)\left[1 + \frac{100/8}{s + \frac{100}{8}}\right]$$

$$I(s) = \frac{10}{s} \times 20{,}000 \times \left[\frac{s + \frac{100}{8}}{s + \frac{200}{8}}\right]$$

$$I_C(s) = \frac{20{,}000\, I(s)}{20{,}000 + \dfrac{1}{4 \times 10^{-6} s}}$$

$$= \left[\frac{\frac{8s}{100}}{\frac{8s}{100} + 1}\right] \times \frac{10}{s} \times \frac{1}{20{,}000}\left[\frac{s + \frac{100}{8}}{s + \frac{200}{8}}\right]$$

$$= \left[\frac{(s)}{s + \frac{100}{s}}\right] \times \frac{10}{(s)} \times \frac{1}{20{,}000}\left[\frac{s + \frac{100}{8}}{s + 25}\right]$$

$\therefore\ i_c = 0.0005\, e^{-25t}$

$i_c = 0.5\, e^{-25t}$ mA

Second Method

$$i_c(t) = i_{c\,(\text{initial})}\, e^{-\frac{t}{RC}} + i_{c\,(\text{initial})}\left(1 - e^{-\frac{t}{RC}}\right)$$

$$i_c(t) = i_{c\,(\text{final})} + \left(i_{c\,(\text{initial})} - i_{c\,(\text{final})}\right) e^{-\frac{t}{RC}}$$

$R' = 20 \parallel 20 = 10$ kohm

$i_{c\,(\text{initial})} = 0$; $i_{c\,(\text{final})} = \frac{10}{20} = 0.5$

Thus, $i_c(t) = 0.5\, e^{-25t}$

34. Ans. (a)

Solution: When S_1 is connected to P_1 and S_2 is connected to P_2, then only I_{S_1} is applied to the capacitor and circuit becomes as

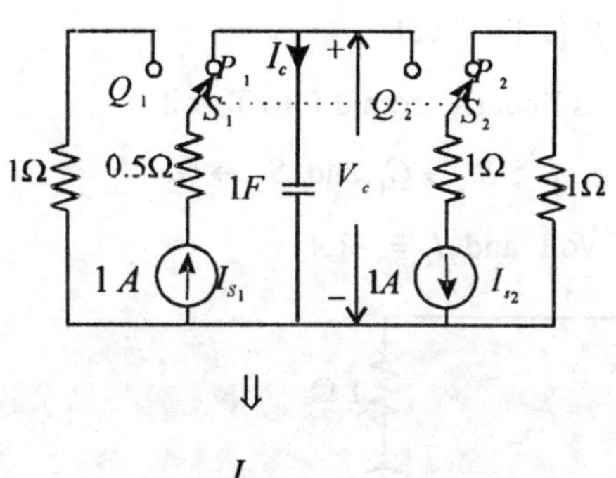

⇓

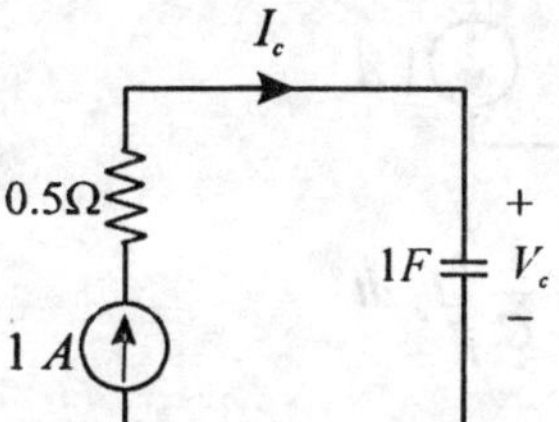

When S_1 is connected to Q_1 and S_2 is connected to Q_2, then only I_{S_2} is applied to the capacitor and circuit becomes as

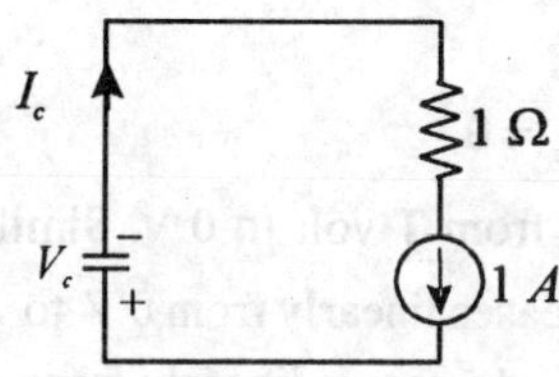

For $2nT \le t < (2n+1)T$, $n = 0, 1, 2, \ldots$

$S_1 \to P_1$ and $S_2 \to P_2$

i.e. from $0 \to T$, $2T \to 3T$, $4T \to 5T, \ldots$

For $(2n+1)T \le t < (2n+2)T$, $n = 0, 1, 2, \ldots$

$S_1 \to Q_1$ and $S_2 \to Q_2$ from

$T \to 2T$, $3T \to 4T$, $6T \to 7T, \ldots$

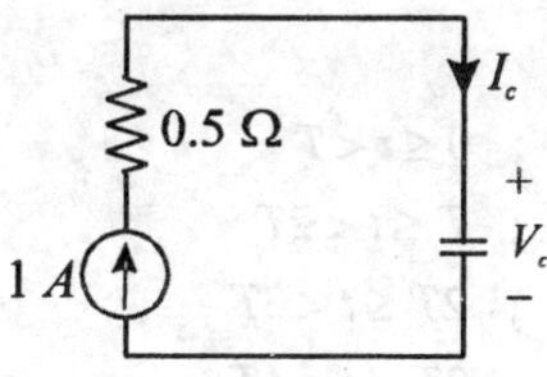

For $0 < t < T$

$I_c = 1$

$$v_c(t) = V_c(0) + \frac{1}{C}\int_0^t I_c\, dt$$

Initial voltage $V_c(0) = 0$

$$v_c(t) = 0 + \frac{1}{C}\int_0^t 1\, dt = t$$

At $t = T$, $V_c(T) = T$ volt

V_c increases linearly from 0 V to T volt

For $T < t < 2T$; $S_1 \rightarrow Q_1$ and $S_2 \rightarrow Q_2$

$V_c(T) = T$ volt and $I_c = -1A$

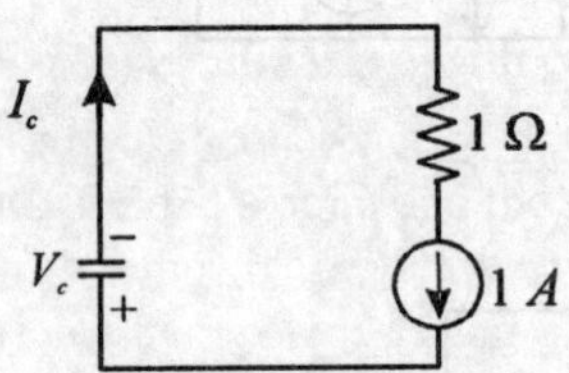

$$v_c(t) = V_c(T) + \frac{1}{C}\int_T^t I_c\, dt$$

$$v_c(t) = T + \frac{1}{1}\int_T^t -1\, dt$$

$$= T + [-t]_T^t = T - t + T$$

$$v_c(t) = 2T - t$$

At $t = 2T$ finally $V_c = 0$

V_c decreases linearly from T volt to 0 V. Similarly from $2T < t < 3T$, V_c increases linearly from 0 V to T volt and from $2T < t < 4T$, V_c decreases linearly from T volt to 0 V and so on

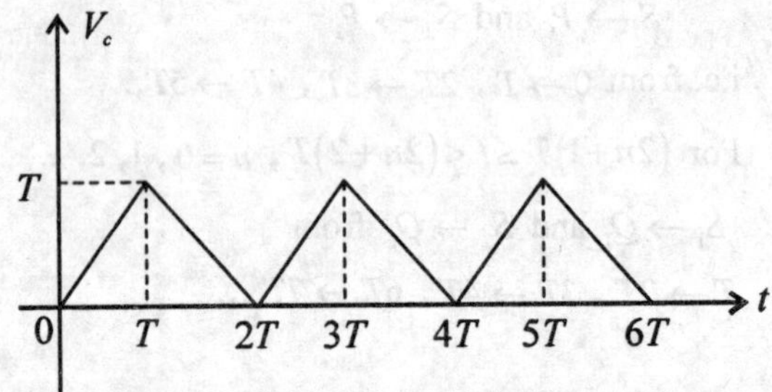

$$v_c(t) = \begin{cases} t & ; \ 0 \le t < T \\ 2T - t & ; \ T \le t < 2T \\ t - 2T & ; \ 2T \le t < 3T \\ 4T - t & ; \ 3T \le t < 4T \\ \vdots & \quad \vdots \end{cases}$$

Using unit step functions

$$v_c(t) = \left\{\begin{matrix} t[u(t) - u(t-T)]u(t-T) \\ +(2T-t)[u(t-T) - u(t-2T)] \\ +(t-2T)[u(t-2T) - u(t-3T)] + \ldots \end{matrix}\right\}$$

$$= \left\{\begin{matrix} t\,u(t) - 2(t-T) + 2(t-2T)u(t-2T) \\ -2(t-3T)u(t-3T) \end{matrix}\right\}$$

$$= t\,u(t) + 2\sum_{n=1}^{\infty} (-1)^n (t - nT)u(t - nT)$$

or other way

$$V_c(t) = \begin{bmatrix} r(t) - 2r(t-T) + 2r(t-2T) \\ -2r(t-3T) + 2r(t-4T) + \ldots \end{bmatrix}$$

$$= r(t) + 2\sum_{n=1}^{\infty}(-1)^n\, r(t - nT)$$

35. Ans. (d)

Solution: $V_C(t) = \dfrac{\frac{1}{sC}}{\frac{1}{sC} + sL + R}$

$$= \frac{\frac{1}{s}}{\frac{1}{s} + s + 1} = \frac{1}{s^2 + s + 1}$$

$$= \frac{2}{\sqrt{3}}\,\frac{\frac{\sqrt{3}}{2}}{\left(s + \frac{1}{2}\right)^2 + \left(\frac{\sqrt{3}}{2}\right)^2} = \frac{2}{\sqrt{3}} e^{-\frac{1}{2}t} \sin\frac{\sqrt{3}}{2}t$$

36. Ans. (b)

Solution: $I(s) = \dfrac{V(s)}{sL + R + \frac{1}{sC}}$

$$= \frac{sC}{s^2LC + sRC + 1} = \frac{s}{s^2 + s + 1}$$

$$= \frac{s}{\left(s + \frac{1}{2}\right)^2 + \left(\frac{\sqrt{3}}{2}\right)^2}$$

$$= \left[\begin{matrix} \dfrac{\left(s + \frac{1}{2}\right)}{\left(s + \frac{1}{2}\right)^2 + \left(\frac{\sqrt{3}}{2}\right)^2} \\ -\left[\dfrac{1}{2}\,\dfrac{\frac{\sqrt{3}}{2}}{\left(s + \frac{1}{2}\right)^2 + \left(\frac{\sqrt{3}}{2}\right)^2}\right] \times \dfrac{2}{\sqrt{3}} \end{matrix}\right]$$

$$= \left[\frac{\left(s+\frac{1}{2}\right)}{\left(s+\frac{1}{2}\right)+\left(\frac{\sqrt{3}}{2}\right)} - \frac{1}{\sqrt{3}}\left[\frac{\frac{\sqrt{3}}{2}}{\left[\left(s+\frac{1}{2}\right)^2+\left(\frac{\sqrt{3}}{2}\right)^2\right]}\right] \right]$$

$$= e^{-\frac{1}{2}t}\cos\frac{\sqrt{3}}{2}t - \frac{1}{\sqrt{3}}e^{-\frac{1}{2}t}\sin\frac{\sqrt{3}}{2}t$$

$$= e^{-\frac{1}{2}t}\left[\cos\frac{\sqrt{3}}{2}t - \frac{1}{\sqrt{3}}\sin\frac{\sqrt{3}}{2}t\right]$$

Note:

e^{-at}	$\rightarrow$	$\frac{1}{s+a}$
$\sin bt$	$\rightarrow$	$\frac{1}{s^2+b^2}$
$e^{-at}\sin bt$	$\rightarrow$	$\frac{b}{(s+a)^2+b^2}$
$e^{-at}\cos bt$	$\rightarrow$	$\frac{(s+a)}{(s+a)^2+b^2}$

37. **Ans. (b)**

Solution: Initially switch is at ***a*** for long time and circuit has achieved steady state.

At $t = 0^-$,

Capacitor has been charge to the source voltage and capacitor are open circuit.

Hence, voltage across capacitor was 100 V.

Hence, $V(0^-) = V(0^+) = 100\ V$

At $t = 0^+$,

$$I(0^+) = \frac{100}{5\ K} = 20\ \text{mA}.$$

At $t = \infty$,

the circuit will reach to steady state and the voltage across capacitor discharge to zero volt.

Hence, $I(\infty) = 0$ mA

$$i(t) = I(0^+)e^{-t/RC}u(t)$$

For the series parallel capacitor, the equivalent capacitance is given by

$$C_{eq} = \frac{0.2(0.5+0.3)}{0.2+0.5+0.3}$$

$$= \frac{0.16}{1} = 0.16\ \mu\text{F}$$

$$\frac{1}{RC_{eq}} = \frac{1}{5\times10^3\times0.16\times10^{-6}} = 1250$$

$$i(t) = 20\,e^{-1250t}\,u(t)\ \text{mA}$$

38. **Ans. (a)**

Solution:

$$L\frac{di}{dt} + Ri = V_o\,(1 + B\,e^{-\frac{R}{L}t}\sin t)\,u(t)$$

Taking laplace transform on both side we get

$$L[s\,I(s) - I(0)] + RI(s)$$

$$= V_o\left[\frac{1}{s} + \frac{B}{\left(s+\frac{R}{L}\right)^2+1}\right]$$

$$I(s)\{R+sL\} - LI(0)$$

$$= V_o\left[\frac{1}{s} + \frac{B}{\left(s+\frac{R}{L}\right)^2+1}\right]$$

$$I(s)\{R+sL\}$$

$$= L\,I(0) + V_o\left[\frac{1}{s} + \frac{B}{\left(s+\frac{R}{L}\right)^2+1}\right]$$

$$\because\quad I(0) = \frac{V_o}{R}$$

$$I(s)\{R+sL\}$$

$$= \frac{V_o L}{R} + V_o\left[\frac{1}{s} + \frac{B}{\left(s+\frac{R}{L}\right)^2+1}\right]$$

$$I(s) = \frac{V_o L}{R(R+sL)} + \frac{V_o}{R+sL}\left[\frac{1}{s} + \frac{B}{\left(s+\frac{R}{L}\right)^2+1}\right]$$

The steady state value of current is given by

$$i(t) = \lim_{s\to 0} [sI(s)]$$

$$i(t) = \lim_{s\to 0} \left[\frac{\frac{V_o Ls}{R(R+sL)} + \frac{V_o s}{R+sL}}{\left\{\frac{1}{s} + \frac{B}{\left(s+\frac{R}{L}\right)^2 + 1}\right\}}\right]$$

$$= \lim_{s\to 0} \left[\frac{\frac{V_o Ls}{R(R+sL)} + \frac{V_o}{R+sL}}{\left\{1 + \frac{B}{\left(s+\frac{R}{L}\right)^2 + 1}\right\}}\right]$$

$$\therefore \quad i(t) = \frac{V_o}{R}$$

39. Ans. (a)

Solution: $t \to \infty$ i.e find condition, $L \to$ short circuit

$$i_L(f^-) = \frac{10}{20} \times 1.5 = -0.75 \text{ A}$$

At $t = 0$, $i_L = 0^-$, $i_L(0^+) = 0.75$ i.e. initial current.

When switch is closed at $t = 0$ inductor is open circuit

$$i(t) = \frac{1.5}{2} = 0.75$$

$$i(f) = 0.5$$

$$i(t) = 0.75$$

$$i(t) = V_r + (i_i + i_f)e^{-\frac{t}{\tau}} = 0.5 - 0.125e^{-1000t}$$

Second Method

Step 1. With all independent sources Kipped

Find $\tau = \frac{L_{eq}}{R_{eq}}$ [After switching]

Circuit reduces as

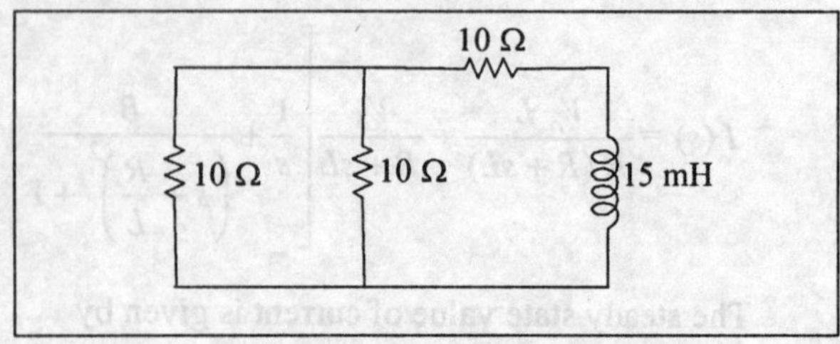

$$= \frac{15 \text{ mH}}{15} = \frac{1}{1000}$$

Step 2. Find $i_L(0^-) \to$ Use dc analysis

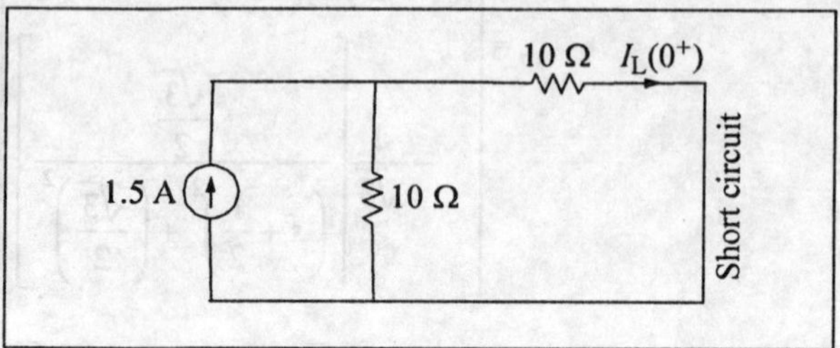

$$I_L(0^-) = 0.75 \text{ A}$$

Step. 3. Find forced response, i.e. current through in ductor after switching

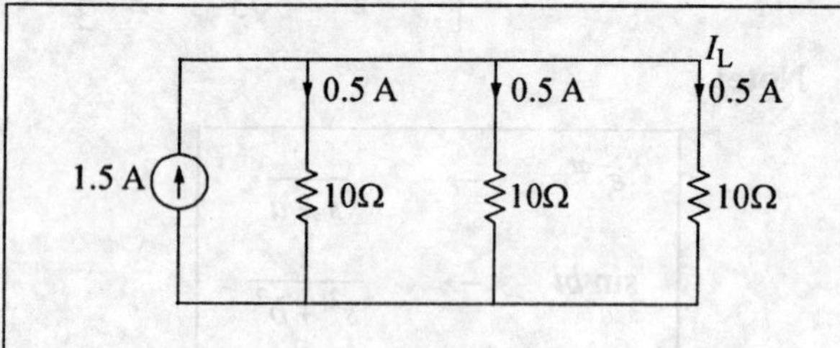

$$I_L(\infty) = 0.5 \text{ A}$$

Step 4. Total Response

$$i(t) = i(\infty) + Ae^{-\frac{t}{\tau}}$$

$i(\infty)$ through 10 Ω resistance = 0.5 A response

$$i(t) = i(\infty) + Ae^{-1000t}$$

To find A one has to use initial conditions property $i(0^+)$ i.e. current due to inductor.

$$\frac{0.75 \times 10}{20} = 0.375$$

due to 1.5 A source

$$\frac{1.5 \times 10}{20} = 0.75$$

Net current = 0.375

According to question

$$i(0^+) = i(\infty) + A$$

$$0.357 = 0.5 + A$$

$$A = -1.25$$

Now

$$i(t) = 0.5 - 1.25e^{-1000t}$$

40. Ans. (a)

Solution: $Q = 2.5$ mC

$$V_{initial} = \frac{Q}{C} = \frac{2.5\times10^{-3}C}{50\times10^{-6}F} = 50\ V$$

When the switch is closed then

The net voltage = 100 + 50 = 150 V

Initially capacitor is short circuited $i(o) = \frac{150}{10}A$

$$i(t) = i(o)e^{\frac{-t}{\tau}}$$

$$\tau = RC$$

$$= 500\ \mu s$$

$$\frac{1}{\tau} = \frac{1}{500\times10^{-6}} = 2\times10^3 \text{ sec}$$

$$i(t) = \frac{150}{10} \exp(-2\times10^3 t)A$$

$$= 15 \exp(-2\times10^3 t)A$$

Second Method

Initial charge $= 2.5\ mC$

$$q = cV$$

$$V = \frac{2.5\times10^{-3}}{50\times10^{-6}} = 50\ V$$

at t closed

Totoal voltage $= 50 + 100 = 150$

$$v(t) = \frac{150}{10}e^{-t/RC} = 15e^{-j\times10^3 t}$$

41. Ans. (b)

Solution: Given circuit is

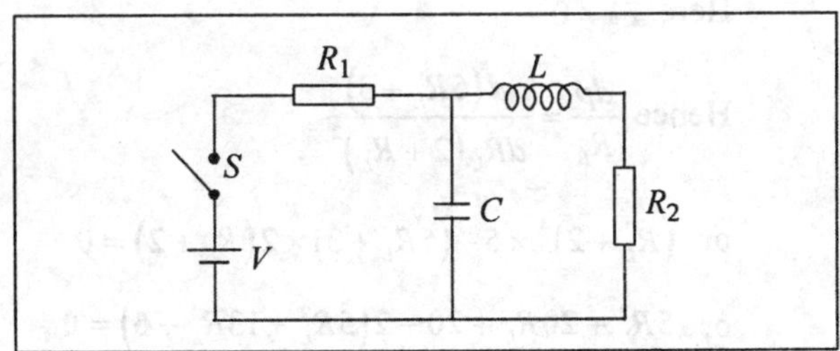

At $t = 0^+$, inductor will act as open circuited and capacitor as short circuited.

So supply current at $t = 0^+$ is $\frac{V}{R_1}$

At $t = 0^+$, inductor will act as short circuited and capacitor as short circuited

$$V_c = \frac{VR_2}{R_1 + R_2}$$

Second Method

At $t = 0$, switch is closed then current I flow.

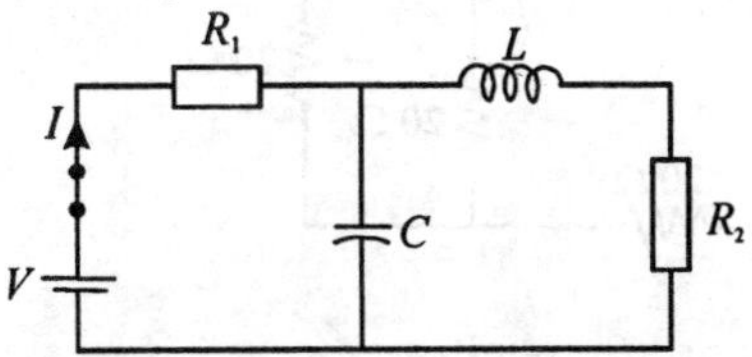

At $t = 0^+$, capacitor → short circuited, hence the supply current $I = \frac{V}{R_1}$

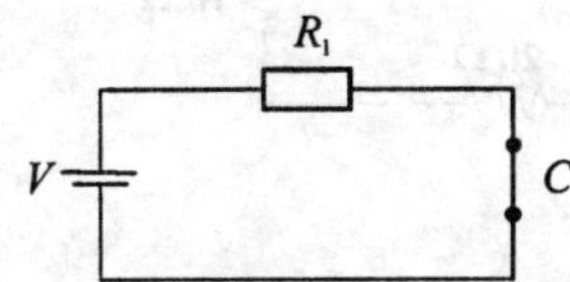

as $t \to \infty$, capacitor → open circuit and inductor → short circuit, hence voltage across capacitor $V_c = V_{R_2}$

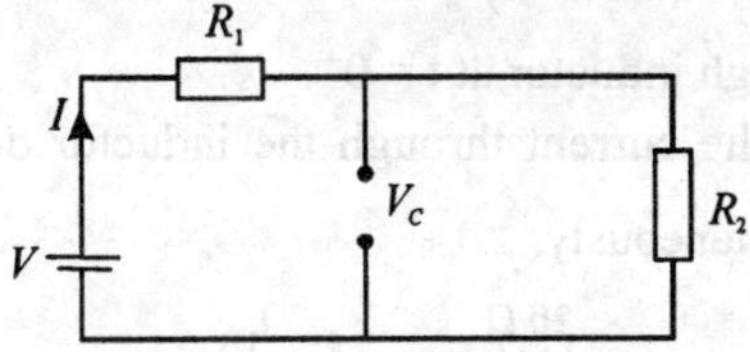

Therefore, $I = \frac{V}{R_1 + R_2}$

and $V_C = R_2 = I\ R_2$

or $V_C = \frac{R_2}{R_1 + R_2} \cdot V$

42. Ans. (c)

Solution:

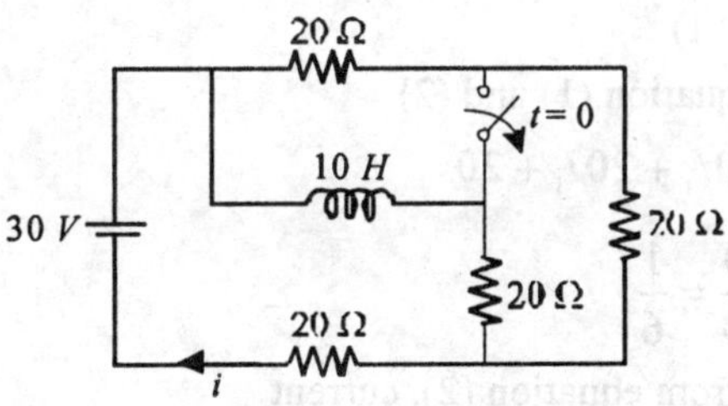

At $t = 0^-$,

Under steady state condition

$L \to$ Short circuit

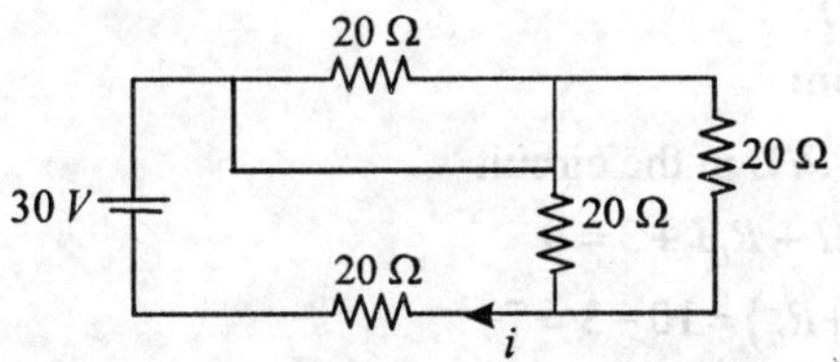

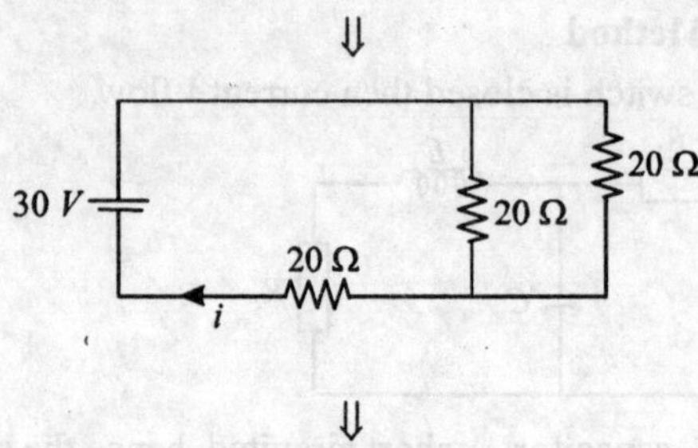

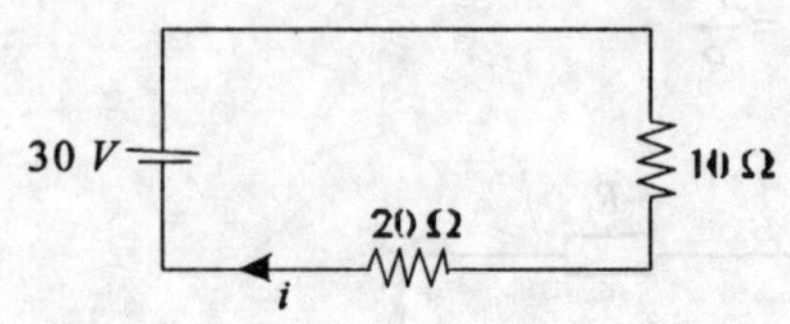

At $t = 0^-$

$$i = \frac{30}{(20 \| 20) + 20} = \frac{30}{10 + 20} = 1 \text{ A}$$

Current through inductor at $t = 0^+$

$I_L = 1A$ as the current through the inductor does not change instantaneously.

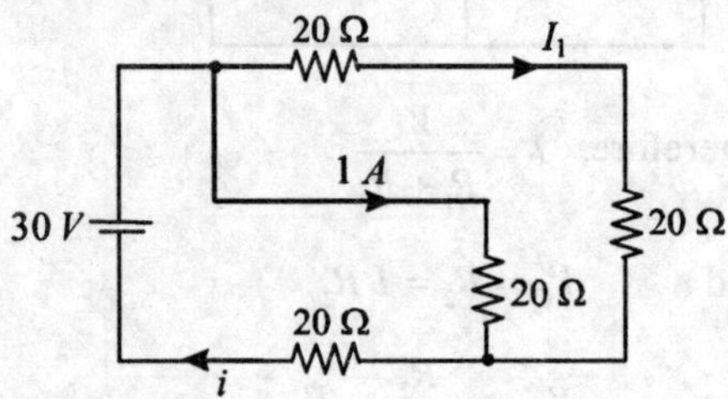

From the figure,

Applying KVL in outer loop.

$$30 = (20 + 20)I_1 + 20i \qquad (1)$$

$$i = (I_1 + 1) \qquad (2)$$

From equation (1) and (2)

$$30 = 40I_1 + 20I_1 + 20$$

$$I_1 = \frac{10}{60} = \frac{1}{6}$$

Hence from equation (2), current

$$i = I_1 + 1 = \frac{1}{6} + 1$$

$$= \frac{7}{6} A$$

43. Ans. (c)

Solution:

Apply KVL in the circuit

$$-10 + 2I + R_L I + 3 = 0$$

$$I(2 + + R_L) = 10 - 3 = 7$$

Current $I = \dfrac{7}{2 + R_L}$

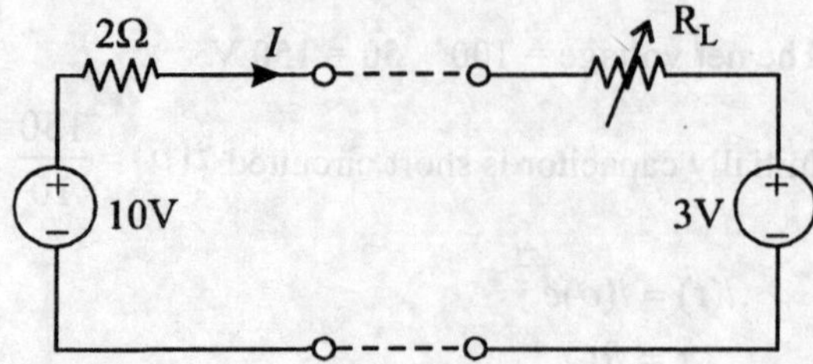

Here, R_L and 3 V source both are working as load.

Power transferred to load, $P = I^2 R_L + I \times 3$

$$= \left(\frac{7}{2 + R_L}\right)^2 R_L + 3 \times \frac{7}{2 + R_L}$$

$$= \frac{49R_L}{(2 + R_L)^2} + \frac{21}{(2 + R_L)}$$

$$= \frac{49R_L + 21R_L + 42}{(2 + R_L)^2}$$

$$= \frac{70R_L + 42}{(2 + R_L)^2}$$

$$= \frac{14(5R_L + 3)}{(2 + R_L)^2}$$

For maximum power transfer to load

$$\frac{dP}{dR_L} = 0$$

Here $14 \neq 0$

Hence $\dfrac{dp}{R_L} = \dfrac{d(5R_L + 3)}{dR_L (2 + R_L)^2}$

or $(R_L + 2)^2 \times 5 - (5R_L + 3) \times 2(R_L + 2) = 0$

or $5R_L^2 + 20R_L + 20 - 2(5R_L^2 + 13R_L + 6) = 0$

or $-5R_L^2 - 6R_L + 8 = 0$

or $5R_L^2 + 6R_L - 8 = 0$

or $R_L = 0.8$ or -2

Neglect the negative value of R_L

Hence, $R_L = 0.8\ \Omega$

44. Ans. (d)

Solution:

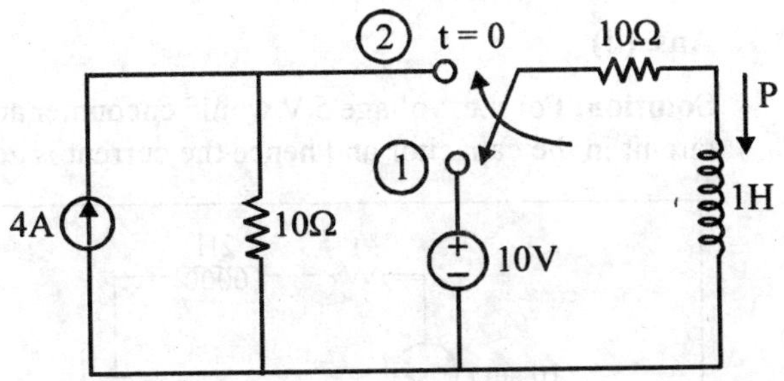

$i_L(0^-) = i_L(0^+)$

At steady state, inductor act as short circuit.

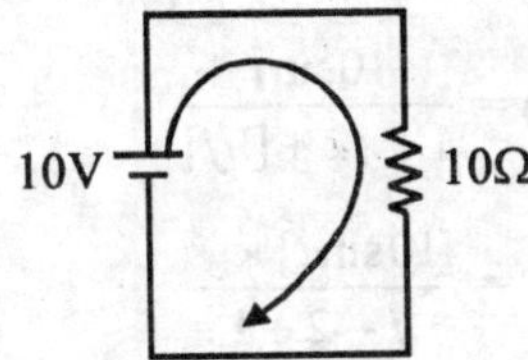

$$i_L(0^-) = \frac{10}{10} = 1A$$

$$i_L(0^+) = 1A$$

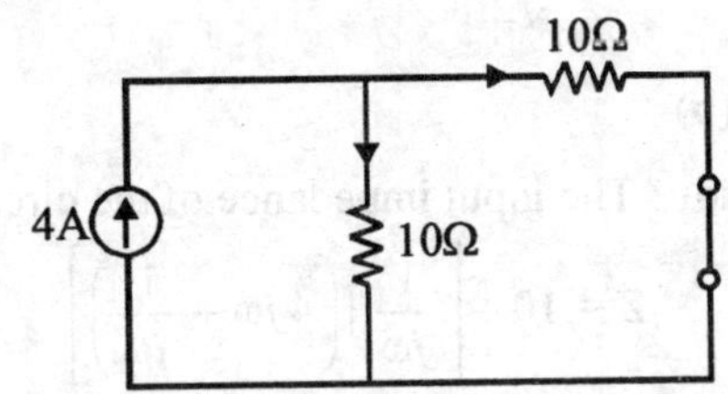

$i_L(\infty) = 2A$

At steady state inductor act as short circuit

$i(t) = (i_L(0^+) - i_L(\infty))Ae^{-t/T} + i_L(\infty)$

$= (1 - 2)e^{-t/T} + 2$

To find T open circuit the current source

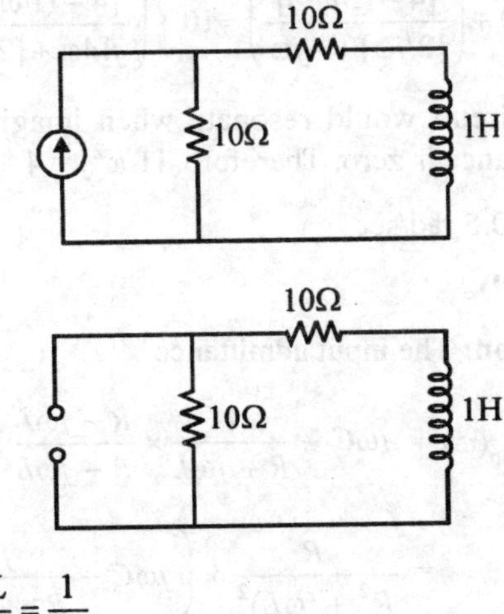

$$T = \frac{L}{R} = \frac{1}{20}$$

At $t \geq 0$

$i(t) = (1-2)e^{-t/1/20} + 2$

$i(t) = 2 - e^{-20t}$

45. Ans. (d)

Solution:

At $t = 0$, initially C is short circuited. When switch is on a high current, i.e. impulse current flow.

Another Method

Time constant = RC

In the given circuit, R = 0

Rise time = 0

Hence, capacitor charges instantaneously and the current can be represented as impulse function.

46. Ans (a)

Solution:

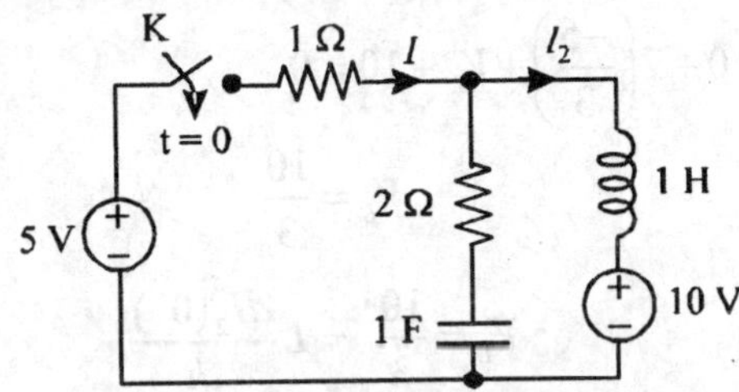

Under steady state condition, steady-state is reached at $t = 0^-$.

The capacitor is to be open, inductor is to be short. Current through inductor and voltage across capacitor will not change instantaneously i.e hence

$V_c(0^-) = V_c(0^+)$

$I_L(0^-) = I_L(0^+)$ after switch is closed.

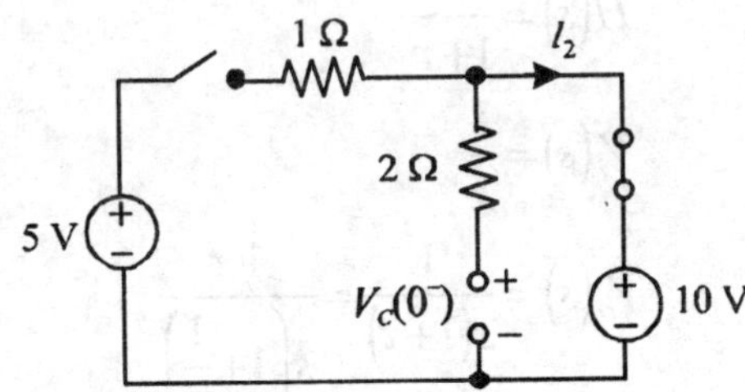

at $t = 0^-$

$I_2(0^-) = 0A$

$I_2(0^+) = 0A$

and $V_c(0^-) = 10V$

at $t = 0^+$

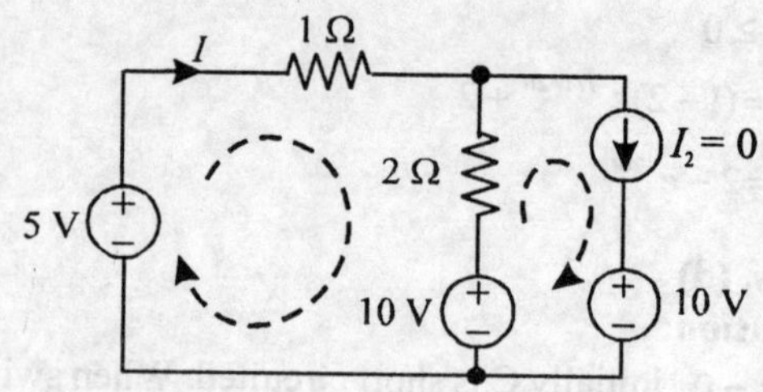

at $t = 0^+$

apply KVL in first loop

$$-5 + 1I + 2I + 10 = 0$$

or $$3I = -5$$

$$\therefore \quad I = -\frac{5}{3}A$$

47. Ans (b)

Solution:

Applying KVL in second loop

$$-10 + 2\left(\frac{-5}{3}\right) + V_L + 10 = 0$$

$$V_L = \frac{10}{3}$$

$$V_L = \frac{10}{3} = L\frac{dI_2(0^+)}{dt}$$

$$\frac{dI_2(0^+)}{dt} = \frac{V_L}{H}$$

or, $$\frac{dI_2(0^+)}{dt} = \frac{10}{3}A/s$$

48. Ans. (c)

Solution:

$$H(s) = \frac{1}{1+\tau}$$

$$V_i(s) = \frac{1}{s}$$

$$V_o(s) = \frac{1}{s(1+\tau)} = \frac{1/\tau}{s\left(1+\frac{1}{\tau}\right)}$$

$$V_o(t) = 1 - e^{-t/\tau}\sqrt{a^2+b^2}$$

$$0.5 = 1 - e^{-10/\tau}$$

$$e^{-10/\tau} = 0.5$$

$$-\frac{10}{\tau} = \log e^{0.5} = -0.69$$

$$\tau = \frac{10}{0.69} = 14.4 \text{ sec}$$

6.4 Sinusoidal Steady State

1. Ans. (d)

Solution: For d.c. voltage 5 V would encounter an open circuit in the capacitor and hence the current is zero.

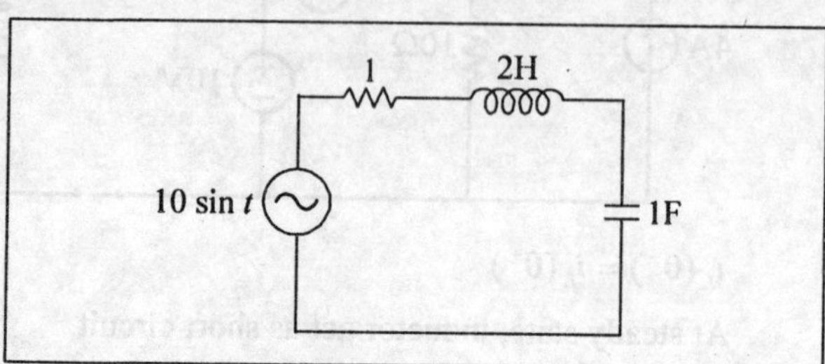

For a.c, $i(t) = \dfrac{10\sin t}{1 + j2 + [1/j]}$ $\quad [\because \ \omega = 1]$

$$= \frac{[10\sin t] \times j}{j - 2 + 1}$$

$$= \frac{j\,10\sin t}{1+j} = \frac{10}{\sqrt{2}}\sin(t)\angle(90° - 45°)$$

$$= \frac{10}{\sqrt{2}}\sin(t - 45°) = 7.07\sin(t - 45°)$$

2. Ans. (b)

Solution: The input impedance of the circuit,

$$Z = 10 + \left[\frac{1}{j\omega} \,\|\, \left(4j\omega + \frac{1}{j\omega}\right)\right]$$

$$= 10 + \left[\frac{\frac{1}{j\omega}\left(4j\omega + \frac{1}{j\omega}\right)}{\frac{1}{j\omega} + 4j\omega + \frac{1}{j\omega}}\right]$$

$$= 10 + \left(\frac{[4 - (1/\omega^2)]}{[2/j\omega] + 4j\omega}\right) = 10 + \left(\frac{[4 - (1/\omega^2)]}{j[4\omega + [2/\omega]}\right)$$

The circuit would resonate when imaginary part of impedance is zero. Therefore, $[1/\omega^2] = 4$

or ω= 0.5 rad/sec

3. Ans. (d)

Solution: The input admittance

$$Y_{in}(\omega) = j\omega C + \frac{1}{R + j\omega L} \times \frac{R - j\omega L}{R - j\omega L}$$

$$= \frac{R}{R^2 + (\omega L)^2} \times \left(j\omega C - \frac{j\omega L}{R^2 + (\omega L)^2}\right)$$

For resonance I_{in} [$Y_{in}(\omega)$] is zero

$$\therefore \quad Y_{in}(\omega) = \frac{R}{R^2 + (\omega L)^2}$$

$$\therefore \quad Z_{in}(\omega) = \frac{R^2 + (\omega L)^2}{R} = R + \frac{(\omega L)^2}{R} > R$$

Hence, the circuit behaves like a pure resistor of value higher than R.

4. Ans. (a)

Solution: A sinusoidal source of voltage V and frequency f is connected to a variable resistance R and fixed reactance X as shown

From the circuit, we have

$$I = \frac{V}{R + jX} = \frac{V(R - jX)}{R^2 + X^2}$$

Hence, $I_R = \dfrac{VR}{R^2 + X^2}$

$$I_X = \frac{-jVX}{R^2 + X^2}$$

$$\therefore \quad I_R^2 + I_X^2 = \frac{V^2}{R^2 + X^2}$$

$$= V\left(\frac{VR}{R^2 + X^2}\right) = \frac{-VI_X}{X}$$

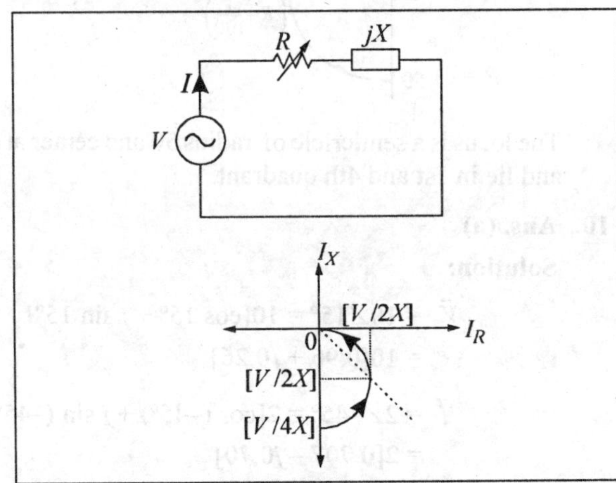

or $\quad I_R^2 = \left(I_X + \dfrac{V}{2X}\right)^2 = \dfrac{V^2}{4X^2}$

$$\Rightarrow \quad I_R = \frac{V}{2X}$$

Hence, the locus of the tip of current phasor as R is varied from 0 to ∞ is a semi-circle with diameter V/X.

Second Method

The relevant circuit is shown in figure below

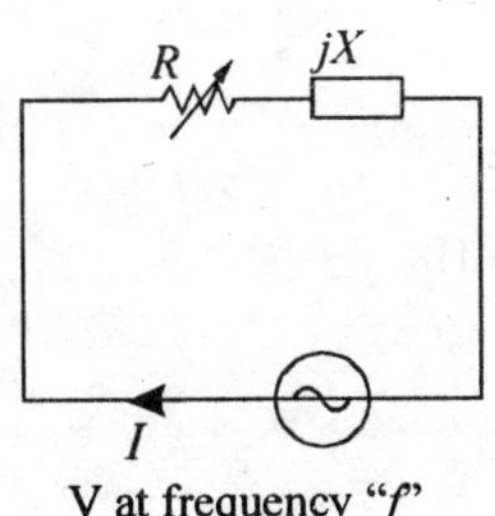

V at frequency "f"

Phasor current

$$\vec{I} = \frac{\vec{V}}{R + jX} = \frac{\vec{V}}{\sqrt{R^2 + X^2}} e^{-j\tan^{-1}\left(\frac{X}{R}\right)}$$

$$\text{for } R = 0, \quad \vec{I} = \frac{\vec{V}}{X} e^{-j90^\circ} = \frac{-V}{X}$$

$$R = X, \quad \vec{I} = \frac{\vec{V}}{\sqrt{2}X} e^{-j45^\circ}$$

$$R = \infty, \quad \vec{I} = 0\, e^{j0} = 0$$

The three points corresponding to $R = 0$, $R = X$, $R = \infty$ lie on a semicircle of diameter $\dfrac{V}{X}$ as shown in figure below.

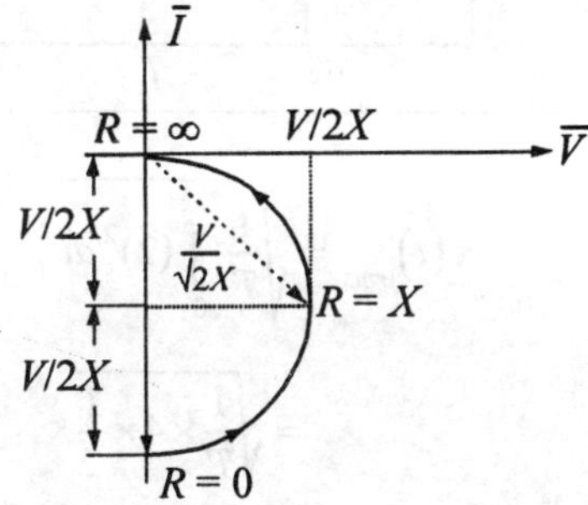

Hence, the locus of the tip of current vector is a semicircle of diameter $\dfrac{V}{X}$.

5. Ans. (d)

Solution: Resonant frequency, $f_0 = \dfrac{1}{2\pi\sqrt{(LC)}}$.

When R, L and C are doubled, resonant frequency

$$f_1 = \frac{1}{2\pi\sqrt{2L \cdot 2C}} = \frac{1}{2\pi\, 2\sqrt{LC}}$$

Hence, $f_0 = \dfrac{[1/(2\pi\sqrt{LC}]}{[1/(2\pi 2\sqrt{LC}]} = 2$

$f_1 = [f_0/2]$

6. **Ans. (d)**

Solution:

$f_0 = 100$ Hz

$\Delta f = BW = 5$ Hz

$V = 10$ V

$Q = \dfrac{f_o}{\Delta f}$

$= \dfrac{100}{5}$

$= 20$

$Q = \dfrac{V_L}{V_R}$

$V_L = Q \times V_R$

$= 20 \times 10 = 200$ V

7. **Ans. (a)**

Solution: Half-wave rectified output is given below

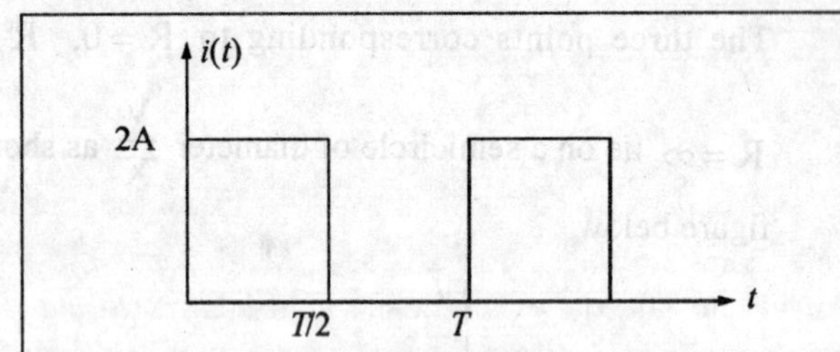

$$i(t)\Big|_{rms} = \sqrt{\frac{1}{T}\int_0^{T/2}(2)^2\,dt}$$

$$= \sqrt{\frac{1}{T}\times 4\times\frac{T}{2}}$$

$$= \sqrt{2}A$$

8. **Ans. (a)**

Solution: The circuit is in resonance as voltage across L and C is double and equal to the input

$$i = \frac{100}{20} = 5A$$

9. **Ans. (a)**

Solution:

The relevant circuit is shown in figure below

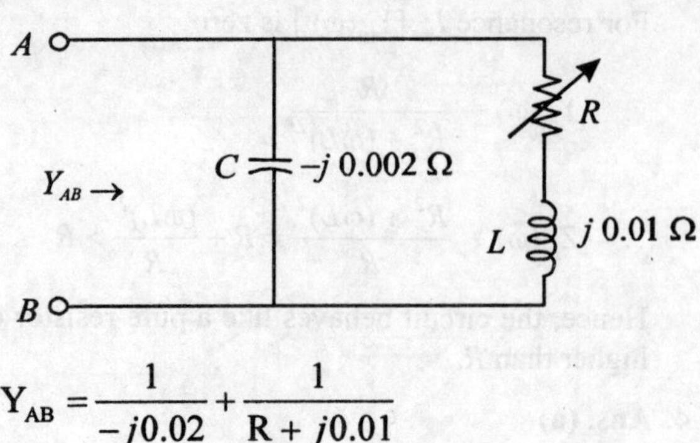

$$Y_{AB} = \frac{1}{-j0.02} + \frac{1}{R + j0.01}$$

$$= j50 + \frac{100}{100R + j1} = \frac{j5000R - 50 + 100}{100R + j1}$$

$$|Y_{AB}| = \frac{\sqrt{(50)^2 + (5000R)^2}}{\sqrt{(100R)^2 + (1)^2}}$$

$$\angle Y_{AB} = \tan^{-1}(100R) - \tan^{-1}\left(\frac{1}{100R}\right)$$

Value of R	$\lvert Y_{AB}\rvert$	$\angle Y_{AB}$
$R = 0$	50	$-90°$
$R = \frac{1}{100}$	50	$0°$
$R = \infty$	50	$90°$

The locus diagram of Y_{AB} is shown in figure below

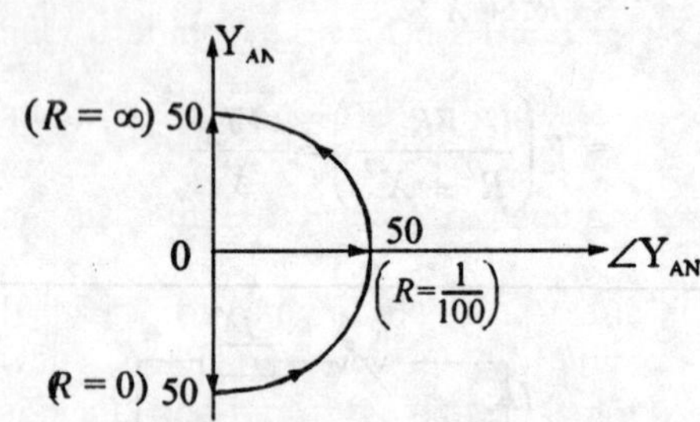

The locus is a semicricle of radius 50 and center at origin and lie in 1st and 4th quadrant.

10. **Ans. (a)**

Solution:

$\vec{V} = 10\angle 15° = 10[\cos 15° + j\sin 15°]$

$= 10[0.96 + j0.26]$

$\vec{I} = 2\angle{-45°} = 2[\cos(-45°) + j\sin(-45°)]$

$= 2[0.707 - j0.70]$

$P = V \times I^* \cong 10\,W + 17.32\,VAR$

Second Method

Given, for $\vec{V} = 10\angle 15° = 10\,e^{j15°}$ and

$\vec{I} = 2\angle -45° = 2\,e^{-j45°}$

Complex power, S is given by $\vec{S} = \bar{V}\,\bar{I}^* = 10e^{j15} \times 2e^{j45}$

$= 20\, e^{j60^\circ}$

Active power, $P = \text{Re}(S) = 20 \cos 60^\circ = 10\, W$

Reactive power, $Q = Im(S) = 20 \sin 60^\circ$

$= 10\sqrt{3}\ VAR$

$= 17.32\ VAR$

11. Ans. (c)

Solution:

Effective impedace of Z_L towards primary side

$$Z_L' = \left(\frac{N_1}{N_2}\right)^2 Z_L$$

$Z_L' = 10\angle 30^\circ\ [1/4]^2 = (0.54 + j\,0.31)\ \Omega$

Total impedance $= (4 - j2) + (0.54 + j0.3)$

$= (4.54 - j\,1.69)\ \Omega$

Second method

Given $Z_s = (4 - j2)\Omega,\ Z_L = 10\angle 30^\circ \Omega$

True ratio $\dfrac{N_1}{N_2} = \dfrac{1}{4}$

Impedance seen by the source $Z = Z_s + \left(\dfrac{N_1}{N_2}\right)^2 Z_L$

$= 4 - j2 + \dfrac{1}{4}\cos 30^\circ + j\dfrac{1}{4}\sin 30^\circ$

$= (4.54 - j\,1.69)\ \Omega$

12. Ans. (c)

Solution:

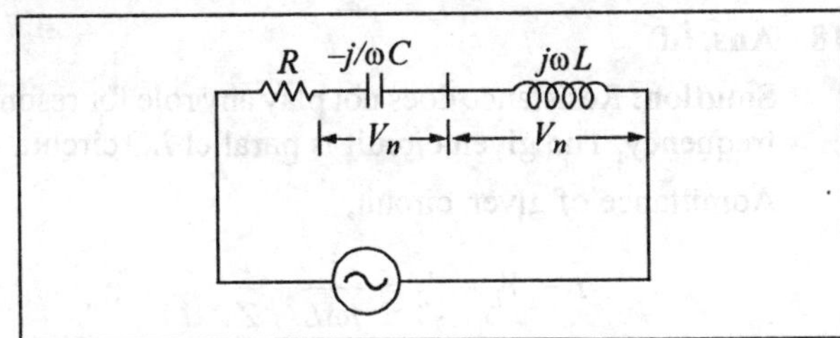

At resonance, $I = [V/R]$

$$V_c = \frac{V}{R}\frac{j}{\omega C} = \frac{V}{R\omega C}\angle -90^\circ$$

It may be $R\omega C < 1$.

Then $\quad |V_c| > V$.

13. Ans. (a)

Solution:

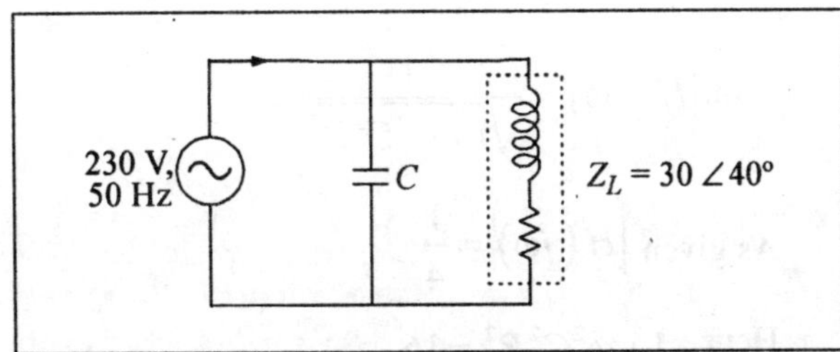

$Y = j\omega C + [1/(30\angle 40)]$

$= j\omega C + \dfrac{1}{30(0.76 + 0.64j)}$

At unity, power factor admittance of both branch is resitive. Equalizing imaginary part equal to zero, $C =$ 68.1 mF

Second Method

For unity PF at the source, Im $[Y_i] = 0$

$Y_i = Y_C + Y_L$

For $Z_L = 30\angle 40^\circ$

$Y_L = \dfrac{1}{30} e^{-j40^\circ}$

$Y_i = j\omega C + \dfrac{1}{30} e^{j40^\circ}$

$= j\omega C + \dfrac{1}{30}\cos(40^\circ) - j\dfrac{1}{30}\sin(40^\circ)$

$= \dfrac{1}{30}\cos(40^\circ) + j\left[\omega C - \dfrac{1}{30}\sin(40^\circ)\right]$

$Im[Y_i] = 0$

$\omega C = \dfrac{1}{30}\sin 40^\circ$

For $f = 50\ Hz$, $C = \dfrac{1}{30}\sin(40^\circ)$

$= \dfrac{1}{3000\,\pi} \times 0.643 \times 10^6 \mu F = 68.2\ \mu F$

14. Ans. (b)

Solution

Given $R = 50\ \Omega,\ C = 5\ \mu F$

The first order RC-LPF in figure is shown below

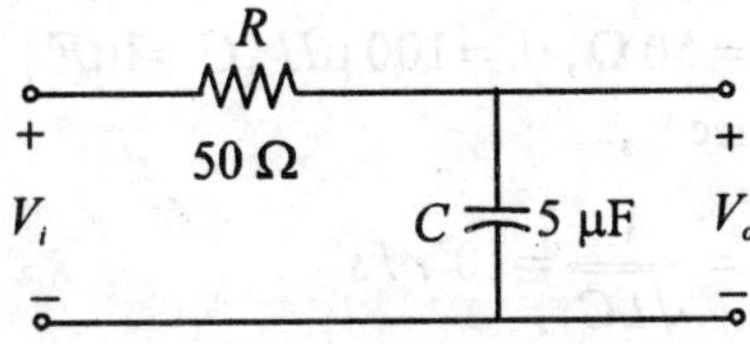

$$H(j\omega)=\frac{1}{1+j\omega CR}$$

$$\text{Gain}\,|H(j\omega)|=\frac{1}{\sqrt{1+\omega^2C^2R^2}}$$

As given $|H(j\omega)|=\frac{1}{4}$

Hence $1+\omega^2C^2R^2=16$

$$\omega^2=\frac{15}{(RC)^2}$$

$$\omega=\frac{\sqrt{15}}{RC}$$

$$f=\frac{\sqrt{15}}{2\pi\,RC}=\frac{\sqrt{15}}{25\times5\times50\times10^{-6}}$$

$$=\frac{\sqrt{15}}{25\times25\times10^{-5}}=\frac{\sqrt{15}\times10^4}{5\pi}\,Hz$$

$$f=2.46\,kHz$$

15. Ans. (b)

Solution: For series *RLC* circuit, at lower half power frequency

$$\omega_1L-[1/(\omega_1C)]=-R$$

$$\Rightarrow\quad 2\pi f_1\times100\times10^{-6}-\frac{1}{2\pi f_1\times1\times10^{-6}}=-50$$

$$-\frac{1}{2\pi f_1}10^6\triangleq-50$$

$$f_1\triangleq\frac{10^6}{2\pi\times50}$$

$$f_1=3.183\text{ kHz}$$

Second Method

The lower half-power (or 3 dB) cut off frequency of the circuit in series RLC circuit is given by

$$f_1=f_0\left(\sqrt{1+\delta^2}-\delta\right)\text{ and }\quad\delta=\frac{1}{2Q}$$

Given

$$R=50\,\Omega,\ L=100\,\mu H,\ C=1\mu F,\ LC=10^{-10}$$

Hence

$$\omega_o=\frac{1}{\sqrt{LC}}=10^5\,r/s$$

$$Q=\frac{\omega_oL}{R}=\frac{10^5\times10^{-4}}{50}$$

$$=0.2$$

$$\delta=2.5$$

$$\omega_1=0.19\times10^5\,r/s$$

$$f_1=3\,kHz$$

16. Ans. (c)

Solution: Here *V* and *I* are in phase, it means circuit is under resonance then *j* terms must be zero.

$$-j12+j8+j8+j2x=0$$

where *x* is mutual inductance

or, $\quad -j12+j16+j2x=0$

or, $\quad 2x=-4$

or, $\quad x=-2$

It means coupling should be opposing

$$K=\frac{M}{\sqrt{L_1L_2}}=\frac{2}{\sqrt{8\times8}}=\frac{2}{8}=0.25$$

Hence, dot at *Q*.

17. Ans. (c)

Solution: Given that $V_L=V_C=2V_R$,

therefore $V_L=V_C$

or $X_L=X_C$, represent a resonance condition and

$\therefore\quad$ current $I=\frac{5}{R}=\frac{5}{5}=1$ A

Now, V_L (Voltage across *L*) $=2V_R=10$

or $\quad 1\omega L=10$

$$L=\frac{10}{\omega}=\frac{10}{2\pi f}=\frac{10}{2\pi50}=31.8\text{ mH}$$

18. Ans. (d)

Solution: Resistance does not play any role for resonance frequency. The given circuit is parallel *LC* circuit. Admittance of given circuit,

$$Y=Y_L+Y_Z=\frac{1}{j\omega L}+\frac{1}{Z}$$

At resonance, *j* term must be equal to zero i.e.

$$\frac{1}{j\omega L}+\frac{1}{Z}=0$$

or $$\frac{-1}{j\omega L}=\frac{1}{Z}$$

or $\frac{j}{\omega L} = \frac{1}{Z}$

or $Z = \frac{\omega L}{j}$ i.e. Capacitive

or $\frac{1}{j\omega L} = \frac{\omega L}{j}$

$$C = \frac{1}{\omega^2 L} = 0.05\,\mu\text{F}.$$

19. Ans. (d)

Solution: From the diagram given in the question paper,

$$I_R = Y_R.E = 0.5\angle 0° * 10\angle 0° = 5\angle 0°\text{A}$$

$$I_L = Y_L.E = 1.5\angle -90° * 10\angle 0° = 15\angle -90°\text{A}$$

$$I_C = Y_C.E = 0.3\angle 90° * 10\angle 0° = 3\angle 90°\text{A}$$

$$\bar{I} = \bar{I}_R + \bar{I}_L + \bar{I}_C = 5 - j15 + j3 = (5 - j12).$$

20. Ans. (c)

Solution: At $f = 100$ Hz; $V_R = V_L$; $R = X_L = \omega L$

$$V_{\text{rms}} = \frac{V_{in}}{(\sqrt{2})}; V_{in} = (\sqrt{2})V_{rms}.$$

At $f = 50$ Hz, $X_L = \frac{\omega L}{2}$; $Z = \sqrt{R^2 + \frac{\omega^2 L^2}{4}}$

$$I_{\text{in}} = \frac{V_{\text{in}}}{\sqrt{R^2 + \frac{\omega^2 L^2}{4}}}$$

$$= \frac{V_{\text{in}}}{\sqrt{\omega^2 L^2 + \frac{\omega^2 L^2}{4}}} = \frac{2V_{\text{in}}}{\omega L\sqrt{5}}$$

$$V_{new} = \frac{2V_{in}}{\omega L\sqrt{5}} \times R = \frac{2V_{in}}{\omega L\sqrt{5}}.\omega L = \frac{2V_{in}}{\sqrt{5}}$$

$$V_{new} = \frac{2.\sqrt{2}\,V_{rms}}{\sqrt{5}} = \sqrt{\frac{8}{5}}\,V_{rms}.$$

21 Ans. (a)

Solution: In Resonance, circuit voltage V_1 and current I are in phase. The voltage across the capacitor lags the current by 90°

Hence, the answer is (a).

22. Ans. (a)

Solution:

From circuit of the given network, impedance

$$Z = R + j\left(\omega L - \frac{1}{\omega C}\right)$$

Admittance

$$Y = \frac{1}{Z} = \frac{1}{R + j\left(\omega L - \frac{1}{\omega C}\right)}$$

$$= \frac{1}{R + j\left(\omega L - \frac{1}{\omega C}\right)} \times \frac{R - j\left(\omega L - \frac{1}{\omega C}\right)}{R - j\left(\omega L - \frac{1}{\omega C}\right)}$$

$$= \frac{R - j\left(\omega L - \frac{1}{\omega C}\right)}{R^2 + \left(\omega L - \frac{1}{\omega C}\right)^2}$$

$$= \frac{R}{R^2 + \left(\omega L - \frac{1}{\omega C}\right)^2} - \frac{j\left(\omega L - \frac{1}{\omega C}\right)}{R^2 + \left(\omega L - \frac{1}{\omega C}\right)^2}$$

$$= \text{Re}(Y) + \text{Im}(Y)$$

By varying frequency for Re(Y) and Im(Y) one can obtain the admittance-locus as given below.

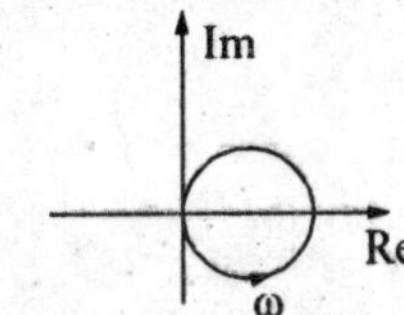

23. Ans. (b)

Solution:

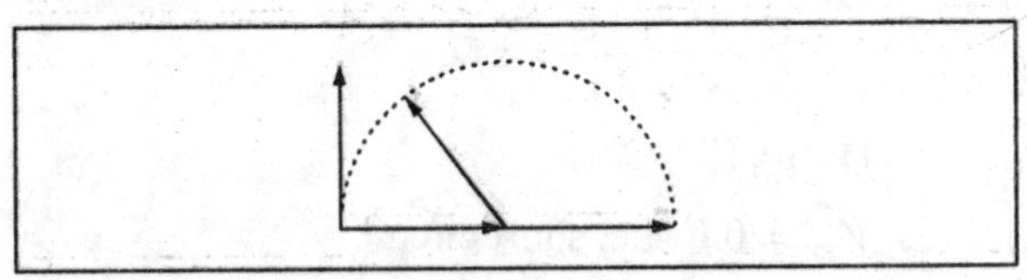

$$V_{YX} = V_X - V_Y V$$
$$= V - 0 \quad \text{for } R = \infty$$
$$= V - 2V \quad \text{for } R = 0$$

24. Ans. (c)

Solution:

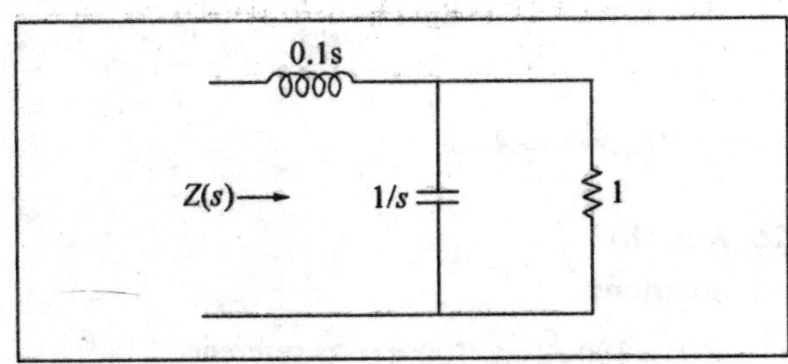

$$Z(s) = 0.1s + \frac{\left(1\times\frac{1}{s}\right)}{\left(1+\frac{1}{s}\right)} = 0.1\,s + \frac{1}{s+1}$$

$$= \frac{0.1s^2 + 0.1\,s + 1}{s+1}$$

Put $s = j\omega$

$$Z(j\omega) = \frac{-0.1\,\omega^2 + 0.1j\omega + 1}{j\,\omega + 1}$$

$$= \frac{(-0.1\,\omega^2 + 0.1j\omega + 1)(1-j\omega)}{(1+j\,\omega)(1-j\omega)}$$

$$= \left[\frac{-0.1\omega^2 + 0.1j\omega + 1 + 0.1j\omega^3 + 0.1\omega^2 - j\omega}{1+\omega^2}\right]$$

For resonance frequency, equating imaginary part equal to zero.

$$\omega(0.1\omega^2 + 0.1 - 1) = 0$$

$$0.1\omega^2 - 0.9 = 0\ ;\ \omega^2 = 9$$

$$\omega = 3$$

25. Ans: (b)

Solution:

Input $(1.0 \sin t)V$, i.e. $A \sin \omega t$

Hence, $\omega = 1\,rad/\sec$

$$X_L = \omega L = 1\times 1 = 1\,\Omega$$

$$X_C = \frac{1}{\omega C} = \frac{1}{1\times 1} = 1\,\Omega$$

$$|V_L| = |V_C|$$

$V_{XY} = 0\,V$ i.e. short circuit

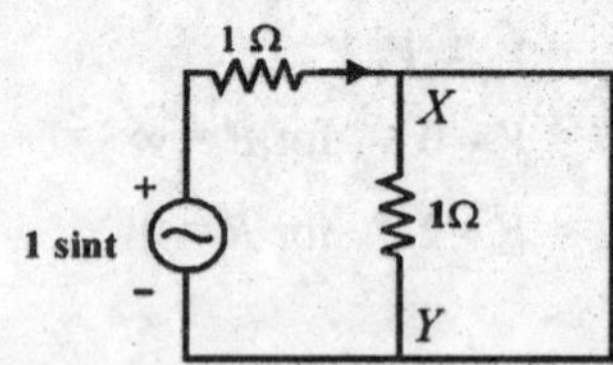

$$i(t) = \frac{\sin\, t}{1\Omega} = \sin\, t$$

$$I_{rms} = \frac{1}{\sqrt{2}}A$$

26. Ans. (b)

Solution:

$V(t) = 100\sqrt{2}\cos(100\,\pi t)$ as reference

As $\sin\alpha$

$$= \cos(90° - \alpha)$$

$$= \cos(\alpha - 90°)$$

$$i(t) = 10\sqrt{2}\sin\left(100\,\pi t + \frac{\pi}{4} + \frac{\pi}{2} - \frac{\pi}{2}\right)$$

$$= 10\sqrt{2}\cos\left(100\,\pi t - \frac{\pi}{4}\right)$$

* In phase RMS value is represented

$$So\ I = \frac{10\sqrt{2}}{\sqrt{2}}\angle -\frac{\pi}{4} = 10\angle -\frac{\pi}{4}$$

27. Ans. (c)

Solution:

The lossy capacitor C_x is represented as

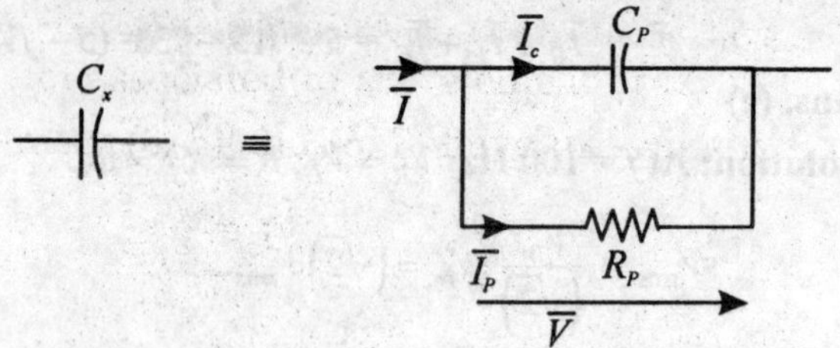

Given

$C_p = 0.102\,\mu F$, $R_p = 1.25\,M\Omega$

Current $\overline{I}_c = j\omega C\overline{V}_p$

or $\overline{I}_p$ leads $\overline{V}$ by 90°

which makes the phasor diagram as below

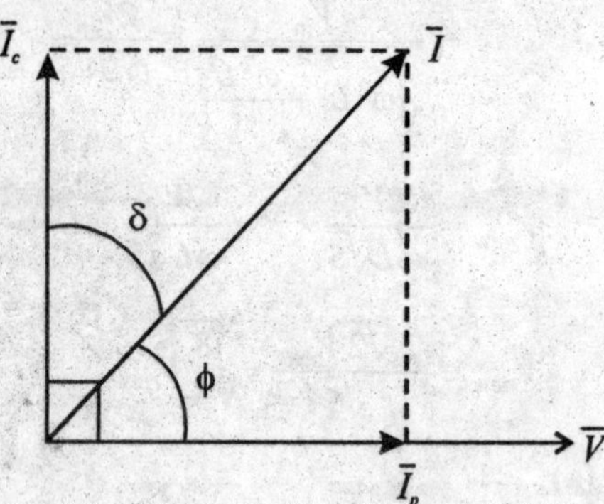

Loss tangent/dissipation factor

$$D = \tan\delta = \frac{I_p}{I_c} = \frac{1}{\omega C_p R_p}$$

$$D = \tan\delta = \frac{1}{2\pi\times 50\times 0.102\times 10^{-6}\times 1.25\times 10^{6}}$$

$$= 0.025$$

Power factor angle $\phi = 90° - \delta$

The Di-electric loss

$P_e = V\cdot I_P = VI\cos\phi = VI\sin\delta$

$$\therefore \quad P_c = V \cdot I_p = \frac{V^2}{R_p} = \frac{\left(5\times10^3\right)^2}{1.25\times10^6}$$

$$= 20 \text{ watt}$$

28. Ans. (d)

Solution:

$$I_C = I_S - I_{RL} = \sqrt{2}\angle\frac{\pi}{4} - \sqrt{2}\angle-\frac{\pi}{4}$$

$$= \sqrt{2}\left(\cos\frac{\pi}{4} + j\sin\frac{\pi}{4}\right) - \sqrt{2}\left(\cos\frac{\pi}{4} - j\sin\frac{\pi}{4}\right)$$

$$= 2\sqrt{2}j\sin\frac{\pi}{4} = j\,2\sqrt{2}\times\frac{1}{\sqrt{2}}$$

$$= j2\,A$$

29. Ans. (b)

Solution:

Total power delivered by the source

= power dissipated in 'R'

$$P = VI\cos\theta = 1\times\sqrt{2}\angle\pi/4$$

$$= 1\times\sqrt{2}\cos\left(\frac{\pi}{4}\right) = 1\,W$$

30. Ans. (b)

Solution: Mains current A_3 = 13 A is vector sum of A_1 and A_2 ; $A_3^2 = A_1^2 + A_2^2$

$$A_2^2 = A_3^2 - A_1^2 = 13^2 - 5^2 = 12 \text{ A}$$

31. Ans. (b)

Solution: In steady state font entire voltage drop across the capacitor in inverse proportion of their capacitance as

$$V_2(t) = \frac{C_1}{C_1 + C_2}\cdot V$$

where $C_1 = 8\ \mu\text{F}$

$C_2 = 2\ \mu\text{F}$

$$V_2(t) = [8/10]\times 10 = 8 \text{ V}$$

$$V_g(t) = \frac{C_2}{C_1 + C_2}\cdot V = \frac{2}{10}\cdot 10 = 2 \text{ V}$$

32. Ans. (c)

Solution:

For a passive 2-port network, there is no source (voltage and current) present. So whatever be the power input, due to the losses, the power output will be always less than its power input.

33. Ans. (b)

Solution:

Resonance frequency $\omega_o = 2\pi f_o$

$= [1/\sqrt{LC}\,]$

Quality factor $Q = [1/(\omega_o RC)] = [\omega_o L/R]$

If R, L, C are doubled then $\omega'_o = [\omega_o/2]$

$Q' = [1/\omega_o'\ R'C'] = [\omega_o'\ L'/R']$

$$= \frac{2}{\omega_o\times 2R\times 2C} = \frac{1}{2\omega_o RC} = \frac{Q_o}{2}$$

$= [100/2] = 50$

34. Ans. (a)

Solution:

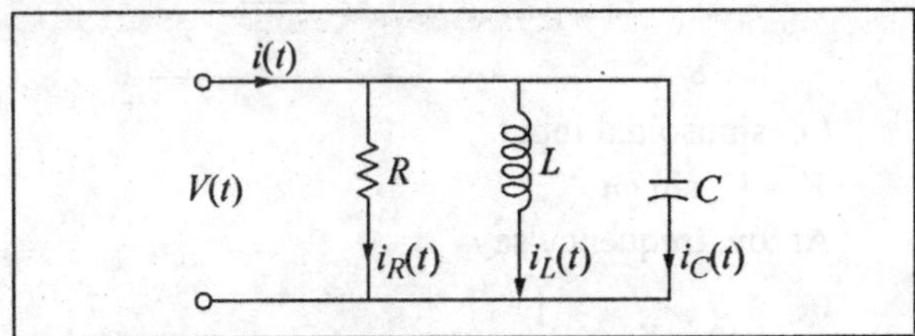

$\omega = 2$

$i(t) = i_R(t) + i_L(t) + i_C(t)$

$$i(t) = \frac{v(t)}{R} + \frac{v(t)}{j\omega L} + v(t)\times(j\omega C)$$

$\Rightarrow \quad i(t) = \sin 2t\,[3 - 2j + 6j] = (3 + 4j)\sin 2t$

$= \sqrt{(3^2 + 4^2)}\,\sin 2t\,\angle\tan^{-1}[4/3]$

$i(t) = 5\sin(2t + 53.1°)$

35. Ans. (a)

Solution:

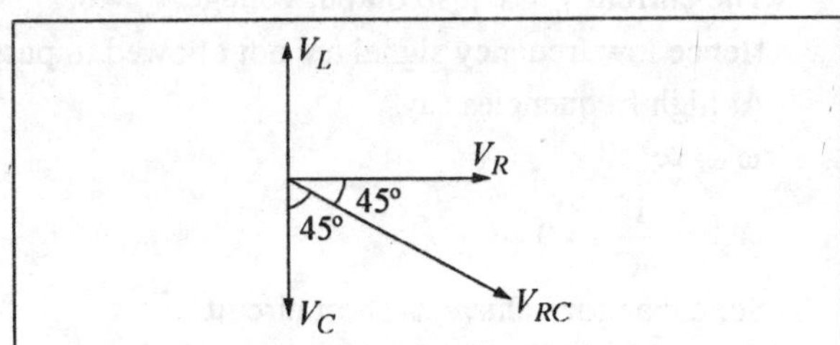

$$V_o(t) = \frac{V_i(t)}{R + \dfrac{1}{j\omega C}}\times\frac{1}{j\omega C} = \frac{\sqrt{2}\sin 10^3 t}{j+1}$$

$$= \frac{\sqrt{2}\sin 10^3 t}{\sqrt{2}\angle+45}$$

$$= \sin(10^3 t - 45°)$$

36. Ans. (b)

Solution:

Resonance frequency

$$f = [1/2\pi]\sqrt{(1/1LC)}$$

$$= [1/2\pi].1\sqrt{(1\times[1/400])\times 10^{-6}}$$

$$= \frac{20\times 10^3}{2\pi} = [10^4/\pi]$$

37. Ans. (c)

Solution:

The RC circuit is given below

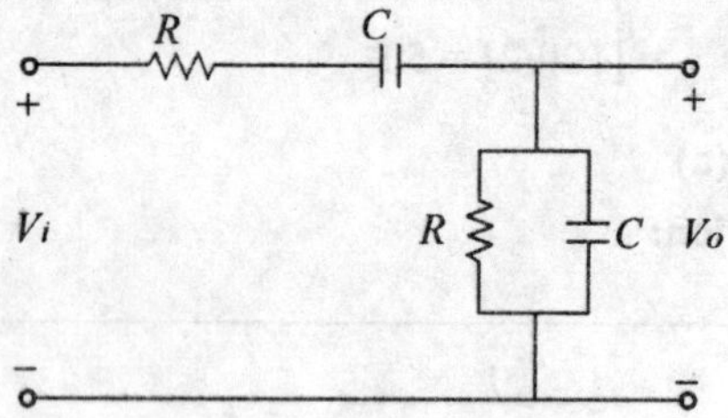

Let sinusoidal input

$V_i = V_{im}\sin\omega t$

At low frequency say

$\omega \to 0,\ X_C = \dfrac{1}{\omega C} \to \infty$

So capacitor behaves as open circuit.

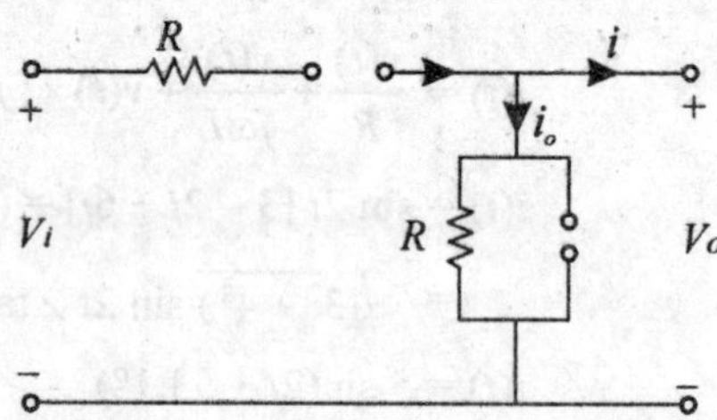

Figure: The equivalent circuit as low frequency.

The current $i_o = 0$, so output voltage $V_o = 0$

Hence low freqency signal are not allowed to pass.

At high frequencies say

$\omega \to \infty$

$X_C = \dfrac{1}{\omega C} \to 0$

So, capacitor behave as short circuit

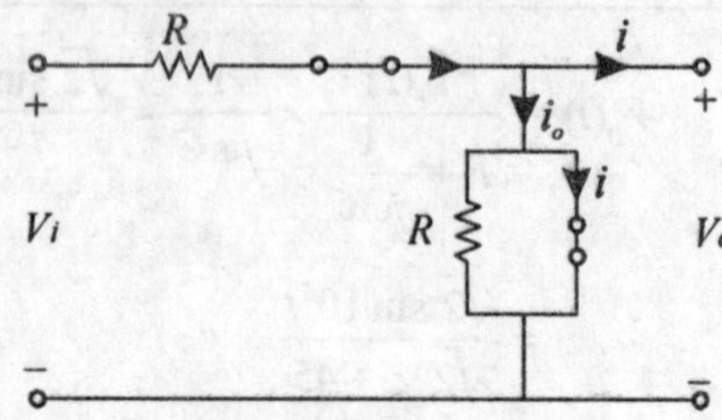

due to short circuit across output

The output voltage $V_o = 0$

Hence high frequency signal are also not allowed.

Only intermediate frequency signal are allowed to pass.

So it work as BPF

38. Ans. (d)

Solution: Parallel RLC circuit.

Bandwidth → At resonance, Z is peal so, Δ is incorrect option

A to resonance, Magnitude of input impedance attains its minimum value:

39. Ans. (d)

Solution: When $\omega = 0$; inductor acts as a short circuit

$\Rightarrow V_o = o$

And when $\omega = \infty$, inductor acts as a open circuit

$\Rightarrow V_0 = -i_1 R_1$

So it acts as a high pass filter.

40. Ans. (a)

Solution:

Given $V_i = V_p \cos\left(\dfrac{t}{RC}\right)$

$$\frac{V_o}{V_i} = \frac{Z_2}{Z_1 + Z_2}$$

$$V_o = \left|\frac{Z_2}{Z_1 + Z_2}\right| V_i$$

Where $\quad Z_2 = R \| \dfrac{1}{j\omega C}$

$$= \frac{R\times\frac{1}{j\omega C}}{R + \frac{1}{j\omega C}}$$

$$= \frac{\frac{R}{j\omega C}}{\frac{j\omega RC + 1}{j\omega C}}$$

$$= \frac{R}{j\omega RC + 1} = \frac{R}{j+1}$$

and $\quad Z_1 = R + \dfrac{1}{j\omega C}$

$$= R + \frac{1}{j\times\frac{1}{RC}\times C}$$

$$= R + \frac{R}{j} = R - jR$$

$$= R(1-j)$$

$$\frac{Z_2}{Z_1+Z_2} = \frac{\frac{R}{1+j}}{\frac{R}{1+j} + R(1-j)}$$

$$= \frac{\frac{R}{1+j}}{\frac{R+R(1^2-j^2)}{1+j}}$$

$$= \frac{R}{R[1+2]} = \frac{1}{3}$$

$$V_o = \frac{Z_2}{Z_1+Z_2} V_i$$

$$= \frac{1}{3} V_P \cos\frac{t}{RC}$$

$$= \frac{V_P}{3}\cos\left(\frac{t}{RC}\right)$$

41. Ans. (b)

Solution:

Power is consumed only by resistance i.e. real part of impedance

Given

Impedance $Z = 4 - j3$

$$= R_L - jX_c$$

$$\therefore \quad R_L = 4$$

Current $I = 5\cos(100\pi t + 100)$

$$= I_m \cos(\omega t + \alpha)$$

Average power is same as RMS power

Average power $P = \frac{1}{2} I_m^2 R_L$

$$= \frac{1}{2} \times 5^2 \times 4$$

$$= 50\,W$$

Second Method

Impedance $= (4 - 3j)\,\Omega$

Current $= 5\cos(100\,\pi t + 100)\,A$

It is known that

$$P_{avg} = \frac{1}{2}\text{Re}\{VI^*\} = \frac{1}{2}\text{Re}\left(|IZ\,I^*|\right)$$

$$= \frac{1}{2}\text{Re}\left(\| I \|^2 Z|\right)$$

$$= \frac{1}{2}| I |^2\,\text{Re}\left(|Z|\right)$$

$$P_{avg} = \frac{1}{2} \times 5^2 \times 4 \qquad [\text{as Re}(Z) = 4]$$

$$= 50\,W$$

42. Ans. (a)

Solution: Capacitor get charged with Rms voltage and it lag the main current, therefore voltage across capacitor is $3 + [4/\sqrt{2}] \sin[100\,t - (\pi/4)]$ V

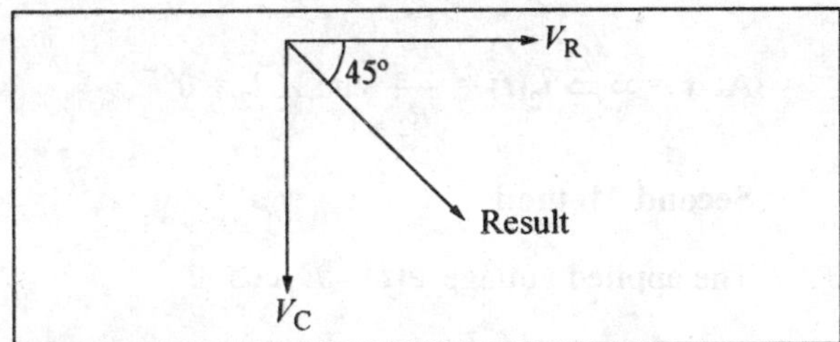

Second Method

The capacitor is open for DC as shown in figure 1.

Hence due to input of 3 V DC, the output $v_1 = 3V$,

The transfer function of the system is given by

$$H(j\omega) = \frac{1}{1 + j\omega\,RC}$$

For the input of 4 sin $(\omega_0 t)$,

Frequency, $\omega = 100$ $R = 1\,k\Omega$, and $C = 10\,\mu F$

$$\omega\,RC = 10^2 \times 10^3 \times 10 \times 10^{-6} = 1\,rad$$

$$H(j\omega) = \frac{1}{1 + j1},$$

$$|H(j1)| = \frac{1}{\sqrt{2}},$$

$$\angle H(j1) = -\frac{\pi}{4}$$

Due to the input AC voltage 4 sin $(100\,t)$, output

$$v_2 = 4 \times |H(j1)|[\sin(100t) + \angle H(j1)]$$

$$= 4 \times \frac{1}{\sqrt{2}}\left[\sin 100t - \frac{\pi}{4}\right]$$

Apply superposition theorem to get output due to $3 + 4\sin(100\,t)$

$$= v_1(t) + v_2(t)$$

$$= 3 + \frac{4}{\sqrt{2}}\left[\sin 100t - \frac{\pi}{4}\right]$$

43. Ans. (a)

Solution:

$$i_2(t) = \frac{E_m \cos \omega t}{R_2 + \dfrac{1}{j\omega C_2}}$$

$$= E_m \angle 0 \frac{j\omega C}{1 + j\omega R_2 C}$$

$$\frac{E_m \omega C}{\sqrt{1 + \omega^2 R_2^2 C^2}} \angle 90° - \tan^{-1} \omega R_2 C$$

At $\omega = 0 \Rightarrow i_2(t) = 0$ and $\angle I_2 = 90°$

At $\omega = \infty \Rightarrow i_2(t) = \dfrac{E_m}{R_2}$ and $\angle I_2 = 0°$

Second Method

The applied votlage $e(t) = E_m \cos \omega t$

rms voltage $E = \dfrac{E_m}{\sqrt{2}}$

So supply voltage phasor $\overline{E} = E\angle 0°$

$\overline{E}$ is applied across R_2 and C

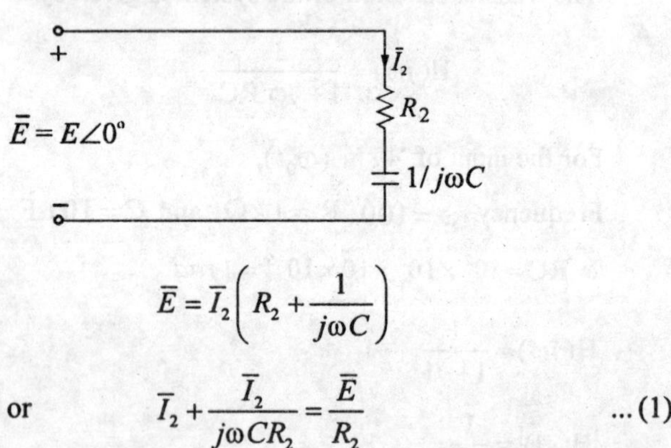

$$\overline{E} = \overline{I}_2\left(R_2 + \frac{1}{j\omega C}\right)$$

or $$\overline{I}_2 + \frac{\overline{I}_2}{j\omega CR_2} = \frac{\overline{E}}{R_2} \quad \text{...(1)}$$

Supply voltage E is constant as ω varies from 0 to ∞, the current phasor $\overline{I}_2$ (both magnitude and phase) varies. In capacitive impednace, $\overline{I}_2$ leads the supply voltage $\overline{E}$.

Hence $\dfrac{1}{j} = 0 - j1$

$$= 1\angle -90°$$

So, $$= \frac{\overline{I}_2}{j\omega CR_2} \text{ lags } \overline{I}_2 90°$$

As ω varies, $\overline{I}_2$ varies but resultant of $\overline{I}_2$ and $\dfrac{\overline{I}_2}{j\omega CR_2}$ is always constant i.e. $= \dfrac{\overline{E}}{R_2}$

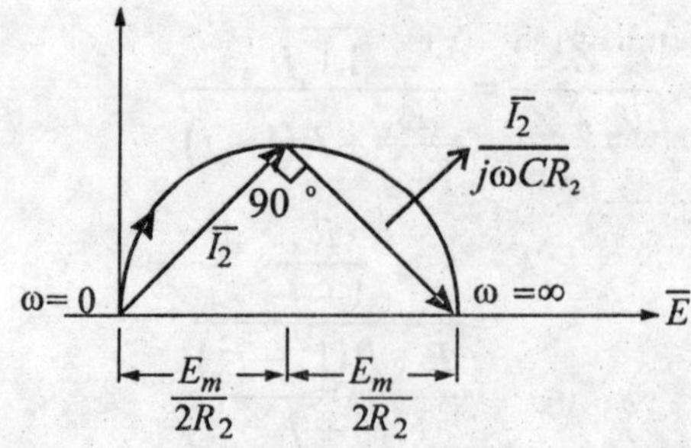

The angle between $\overline{I}_2$ and $\dfrac{\overline{I}_2}{j\omega CR_2}$ is always 90° and resultant $\dfrac{\overline{E}}{R_2}$, so locus of $\overline{I}_2$ is a semicicle of diameter $\dfrac{\overline{E}_2}{R_2}$ and centre is at $\dfrac{\overline{E}_2}{2R_2}$

From the equation (1)

$$\text{At } \omega = \infty, \quad \overline{I}_2 = \frac{\overline{E}}{R_2} \quad \text{...(2)}$$

i.e. $\overline{I}_2$ is along $\overline{E}$

$$j\omega CR_2\overline{I}_2 + \overline{I}_2 - = j\omega C\overline{E} \quad \text{...(3)}$$

$$\text{At } \omega = 0, \quad \overline{I}_2 = 0 \quad \text{...(4)}$$

Third Method

$$I_2 = \frac{E_m}{R_2 - \dfrac{j}{\omega C}},$$

$$|I_2| = \frac{E_m}{\left(R_2^{\,2} - \dfrac{I}{\omega^2 C^2}\right)^{\frac{1}{2}}},$$

$\dfrac{\overline{I}_2}{j\omega CR_2}$ 90° $\overline{I}_2$ ω= 0 ω =∞ $\overline{E} = E_m \angle 0$ $\dfrac{E_m}{2R_2}$ $\dfrac{E_m}{2R_2}$

$$\angle I_2 = +\tan^{-1}\left(\frac{1}{\omega CR_2}\right)$$

ω	0	$1/(CR_2)$	∞
$\lvert I_2\rvert$	0	$\frac{E_m}{\sqrt{2}R_2}$	$\frac{E_m}{R_2}$
$\angle I_2$	$\pi/2$	$\pi/4$	0

with the above data, the locus of I_2 is the semicircle shown in fig. (i)

44. **Ans. (d)**

Solution:

Here the value of inductance is not given we can ignore it

$$I_2 = \frac{2\cos 4t}{3} = 0.67\cos 4t \qquad \text{i.e. } \omega = 4$$

$$I_2 = \frac{-j\omega M}{R-[j/\omega C]} = \frac{-j4\times 0.75I_1}{3.92-2.56j}$$

$$= \frac{-j3\times 0.67\,\text{Cos}\,4t}{3.92-2.56j}\times\frac{3.92+2.56j}{3.92+2.5j}$$

$$= (0.23 - 0.35j)\cos 4t$$

Second Method

Here no value of L_2 and L_2 are given it means it is $1H$ each.

Let $v = 2\cos 4t$ is taken as reference.

$\overline{V} = 2\angle 0°\ V$

Let peak value is taken in phasor representation.

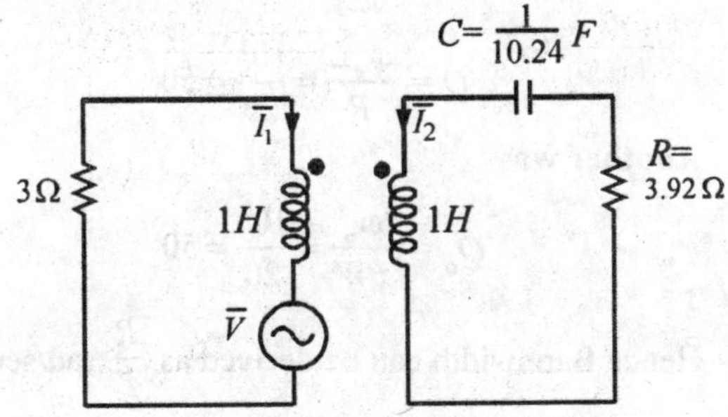

The reactance of L_1 and L_2 are $X_{L_1} = X_{L_2} = \omega L = 4\ \Omega$

Capacitor reactance $X_C = \frac{1}{\omega C} = 2.56\ \Omega$

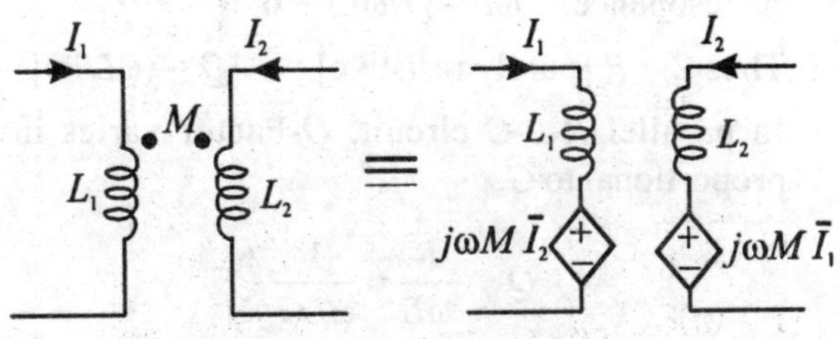

$X_M = \omega M = 3\ \Omega$

Inductor equivalent circuit diagram with mutural inductance

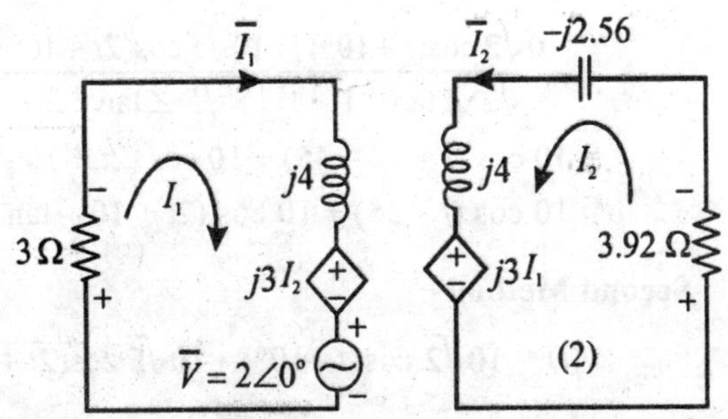

KVL in loop (1)

$$3\overline{I_1} + 4j\overline{I_1} + j3\overline{I_2} + 2\angle 0° = 0$$

$$(3+j4)I_1 + j3\overline{I_2} = -2 \qquad \text{...(1)}$$

KVL in loop (2)

$$3.92\,\overline{I_2} + (-j2.56)\overline{I_2} + j4\overline{I_2} + j3\overline{I_1} = 0$$

$$\overline{I_1} = \frac{-3.92 + j1.44}{j3}\overline{I_2} \qquad \text{...(2)}$$

Using equation (1) and (2)

$$-(3+j4)\frac{(3.92+j1.44)}{j3}\overline{I_2} + j3\overline{I_2} = -2$$

On solving

$\overline{I_2} = -(0.192 + j0.144)$

$\overline{I_2} = 0.24\angle -143.13°$

$\overline{V} = 2\angle 0°$

$v = 2\cos 4t$ reference

$\overline{I_2} = 0.24\angle -143.13°\ A$

$i_2(t) = 0.24\cos(4t - 143.13°)$

The current i_2 is of the form $P\cos 4t$ means $\cos 4t$ as reference So phasor

$\overline{P} = 0.24\angle -143.13°$

$= -(0.192 + j0.144)$

45. **Ans. (c)**

Solution:

$I(t) = i_1(t) + i_2(t)$

$$= \frac{10\sqrt{2}\cos(t+10°)}{R + j\omega_1 L} + \frac{10\sqrt{5}\cos(2t+10°)}{R + j\omega_2 L}$$

Given

$R = 1$ and $L = 1$ H

$\omega_1 = 1$ and $\omega_2 = 2$

$$I(t) = \frac{10\sqrt{2}\cos(t+10°)}{1+j} + \frac{10\sqrt{5}\cos(2t+10°)}{1+2j}$$

$$= \frac{10\sqrt{2}\cos(t+10°)}{\sqrt{2}\angle\tan^{-1}1} + \frac{10\sqrt{5}\cos(2t+10°)}{\sqrt{5}\angle\tan^{-1}2}$$

$$= 10\cos(t+10-45) + 10\cos(2t+10-\tan^{-1}2)$$

$$= 10\cos(t-35) + 10\cos(2t+10-\tan^{-1}2)$$

Second Method

$$v(t) = 10\sqrt{2}\cos(t+10°) + 10\sqrt{5}\cos(2t+10°)$$

Let, $v_1(t) = 10\sqrt{2}\cos(t+10°)$

$$v_2(t) = 10\sqrt{5}\cos(2t+10°)$$

$$\omega_1 = 1 \text{ and } \omega_2 = 2$$

$$Z_1 + R + j\omega_1 L = 1 + j1$$

$$Z_2 = R + j\omega_2 L = 1 + j2$$

$$i(t) = \frac{v_1(t)}{Z_1} + \frac{v_2(t)}{Z_2}$$

$$= \frac{10\sqrt{2}(t+10°)}{1+j} + \frac{10\sqrt{5}\cos(2t+10°)}{1+j2}$$

$$= \frac{10\sqrt{2}\cos(t+10°)}{\sqrt{1^2+1^2}\angle\tan^{-1}1} + \frac{10\sqrt{5}\cos(2t+10°)}{\sqrt{1^2+1^2}\angle\tan^{-1}2}$$

$$= \frac{10\sqrt{2}\cos(t+10°)}{\sqrt{2}\angle 45°} + \frac{10\sqrt{5}\cos(2t+10°)}{\sqrt{5}\angle 2}$$

$$i(t) = 10\cos(t-35°) + 10\cos(2t+10-\tan^{-1}2)$$

46. Ans. (b)

Solution:

R L V_i(s) I(s) C V_o(s)

$$V_o(s) = [1/sC] \times I(s)$$

$$\left(R + sL + \frac{1}{Cs}\right) I(s) = V_i(s)$$

$$\frac{I(s)}{V_i(s)} = \left(\frac{s}{L\left(s^2 + \frac{Rs}{L} + \frac{1}{LC}\right)}\right)$$

$$\frac{V_o(s)}{V_i(s)} = \frac{I(s)}{sCV_i(s)}\left(\frac{1/LC}{L\left(s^2 + \frac{Rs}{L} + \frac{1}{LC}\right)}\right)$$

$$= \frac{\omega_n^2}{s^2 + 2\delta\omega_n s + \omega_n^2}$$

$$\omega_n = [1/\sqrt{(LC)}]$$

$$\delta = \frac{R}{2L}\sqrt{(LC)} = \frac{R}{2}\sqrt{(C/L)}$$

$$\text{Given transfer function} = \frac{10^6}{s^2 + 20s + 10^6} = \frac{V_o(s)}{V_i(s)}$$

$$\omega = 10^3 = [R/Q]\ [\sqrt{(C/L)}]$$

$$Q = [\omega L/R] = [1/R\omega C]$$

$$Q^2 = [L/R^2 C]$$

$$Q = \frac{1}{R}[\sqrt{(L/C)}] = \frac{1}{2\delta} = \frac{1}{2\times 10^{-2}}$$

$$= 50$$

Second Method

$$H(s) = \frac{10^6}{s^2 + 20s + 10^6} \cong \frac{\omega_n^2}{s^2 + 2\delta\omega_n s + \omega_n^2}$$

For series RLC circuit

$$H(s) = \frac{1/LC}{s^2 + \frac{R}{L}s + \frac{1}{LC}}$$

Comparing above equations

$$\frac{1}{LC} = 10^6 \text{ and } \frac{R}{L} = 20$$

$$\omega_o = \frac{1}{\sqrt{LC}} = 10^3$$

$$Q = \frac{\omega_o L}{R} = \frac{10^3}{20} = 50$$

Another way

$$Q_o = \frac{\omega_o}{BW} = \frac{10^3}{20} = 50$$

Hence Bandwidth can be derived as $\frac{R}{L}$ rad/sec

47. Ans. (d)

Solution: In series

The impedance, $Z = R + j[\omega L - (1/\omega C)]$

At resonance, $\omega L - [1/\omega L] = 0$

Thus $Z = R$ [purely resistive] ; $[Q = (\omega L/R)]$

In parallel R-L-C circuit, Q-Factor varies inversely proportional to G.

$$Q = \frac{R}{\omega L} = \frac{1}{\omega LG}$$

48. Ans. (a)

Solution:

Given,

Source currents $\overline{I}_{S_1} = 5\angle 0° \, A$

$$\overline{I}_{S_2} = 10\angle 60° \, A$$

Modify the figure as

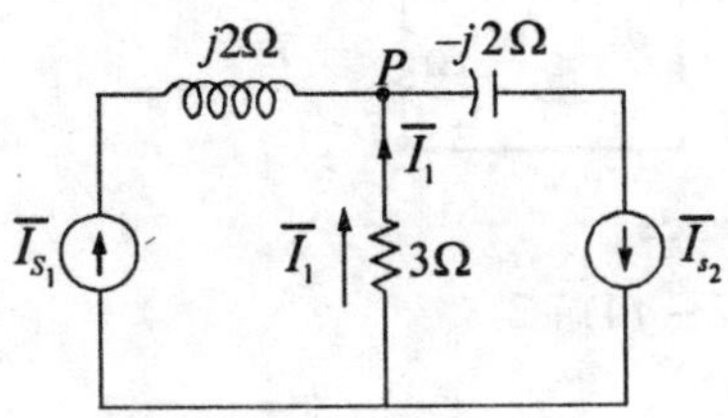

Using KCL at node P

$$-I_{S_1} - I_1 + I_{S_2} = 0$$

or
$$I_1 = Is_2 + Is_1$$

$$= -5\angle 0° + 10\angle 60°$$

$$= j5\sqrt{3}$$

$$= 5\sqrt{3}\angle 90°$$

49. Ans. (d)

Solution:

$$\frac{B_1}{B_2} = ?$$

For series RLC, Bandwidth = R/L

$$\therefore \quad B_1 = \frac{R}{L_1}; \; B_2 = \frac{R}{L_1/4}$$

$$\therefore \quad \frac{B_1}{B_2} = \frac{1}{4}$$

C_1 L_1 + + V_i R V_o − −

Bandwidth = $B_1 = [R/L_1]$

$C_2 = 4C_1$ $L_2 = L_1/4$ + + V_i R V_o − −

Bandwidth = $B_2 = \dfrac{R}{L_1/4}$

50. Ans. (d)

Solution:

$$R_{AB} = \frac{(5-3j)(5+3j)}{(5-3j)+(5+3j)} = \frac{25+9}{10} = 3.4\,\Omega$$

$$V_{AB} = R_{AB} \times I = 3.4 \times 5\angle 30 = 17\angle 30$$

51. Ans. (b)

Solution:

$$I = \frac{V}{Z_S + Z_L} = \frac{20\angle 0°}{(1+2j)+(7+4j)}$$

$$= \frac{20\angle 0°}{8+6j}$$

$$= \frac{20\angle 0°}{8+6j} \times \frac{(8-6j)}{(8-6j)} = \frac{20\angle 0°(8-6j)}{64+36}$$

$$= \frac{1}{5}(8-6j)$$

$$= \frac{10\tan^{-1}(6/8)}{5} = 2\tan^{-1}\frac{3}{4}$$

The voltage across load $V_L = I_L Z_L$

The reactive power consumed by load P_r

$$= V_L I_L = I_L Z_L \cdot I_L = Z_L \,|\, I_L \,|^2$$

$$= (7+4j)\frac{20\angle 0°}{(8+6j)} = (7+4j) \times 4$$

$$= 28 + 16j$$

Hence average power = 28

and reactive power = 16

52. Ans. (a)

Solution:

$$V = \frac{\dfrac{20}{Ls}}{\left[\dfrac{1}{Ls} + \dfrac{1}{R} + Cs\right]}$$

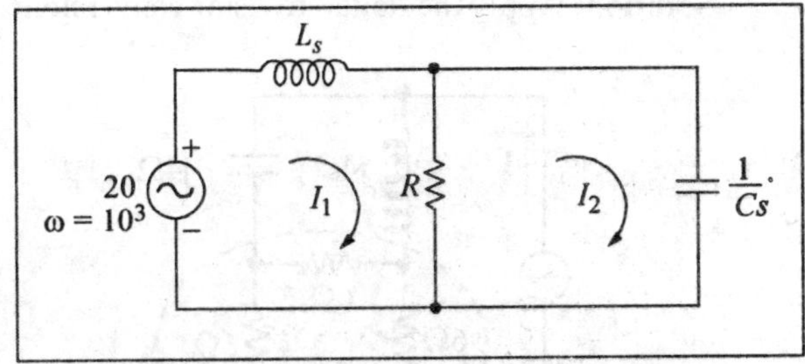

$$= \frac{20R}{R + Ls + LCRs^2} = \frac{20}{2+20j} = -j1\text{ A}$$

Second Method

Assuming two branch currents

I_1 and I_2

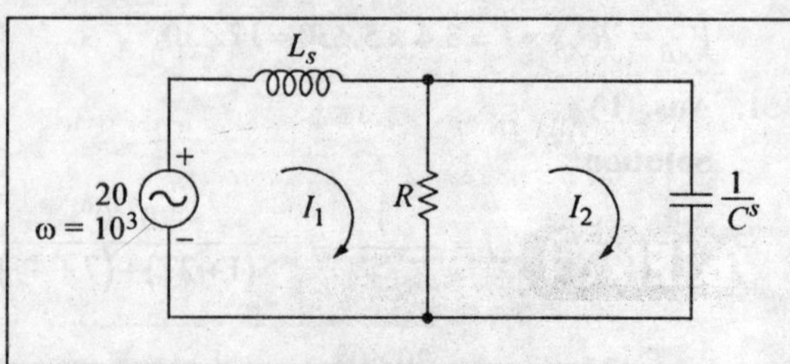

Writing KVL equations using Matrix form

$$\begin{bmatrix} 20 \\ 0 \end{bmatrix} = \begin{bmatrix} 1+j20 & -1 \\ -1 & 1-j20 \end{bmatrix} \begin{bmatrix} I_1 \\ I_2 \end{bmatrix}$$

Using Cramer's rule for finding I_1 and I_2

$$I_1 = \frac{\begin{bmatrix} 20 & -1 \\ 0 & 1-j20 \end{bmatrix}}{\begin{bmatrix} 1+j20 & -1 \\ -1 & 1-j20 \end{bmatrix}}$$

$$= \frac{20(1-j20)}{1+400-1} = \frac{1-j20}{20}$$

$$I_2 = \frac{\begin{bmatrix} 1-j20 & 20 \\ -1 & 0 \end{bmatrix}}{\begin{bmatrix} 1+j20 & -1 \\ -1 & 1-j20 \end{bmatrix}}$$

$$= \frac{+20}{1+400-1} = \frac{1}{20}$$

Net Current throuhg AB: section

$$I_1 - I_2 = \frac{1-j20}{20} - \frac{1}{20}$$

$$= \frac{-j20}{20} = -j1$$

53. Ans. (b)

Solution: Apply the delta - to - star conversion

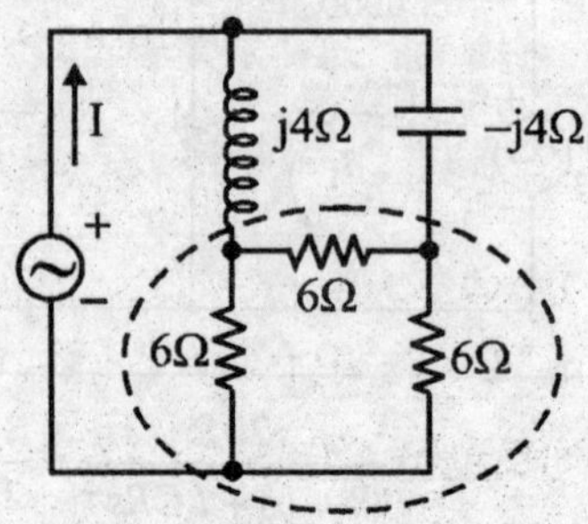

The circuit becomes

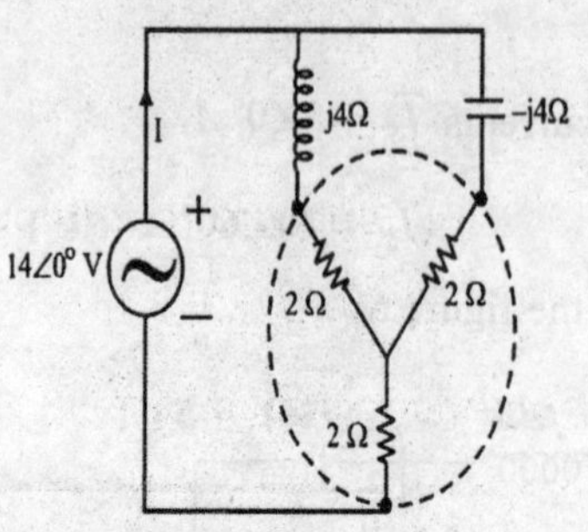

The net Impedance

$$= (2+j4) \| (2-j4) + 2$$

$$= \frac{4+16}{4} + 2 = 7\Omega$$

$$I = \frac{140\angle 0^o}{7} = 2\angle 0^o\ A$$

54. Ans. (a)

Solution

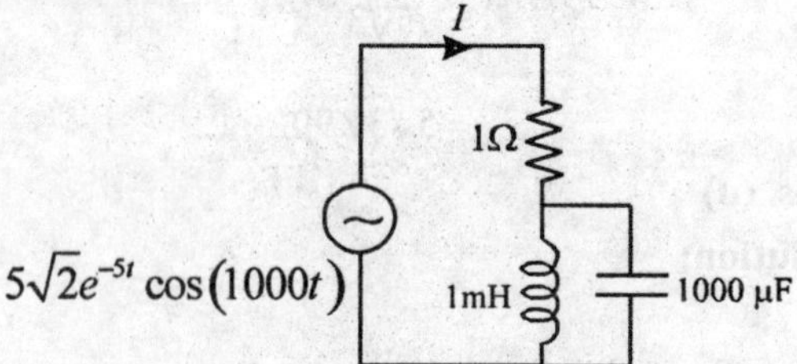

At steady state means at $t = \infty$ inductor behaves as short crercuit and capacitor behaves as open circuit

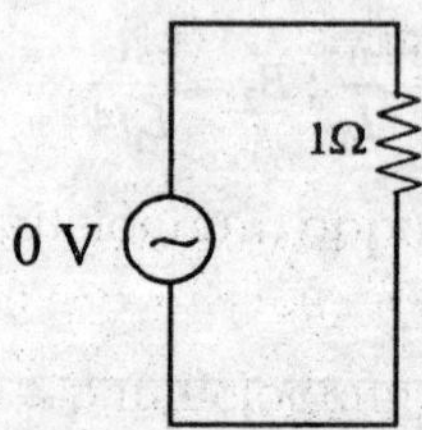

$$e^{-5t}\Big|_{t\to\infty} = 0$$

$$I = V/R = 0$$

$$I = 0$$

Another Method

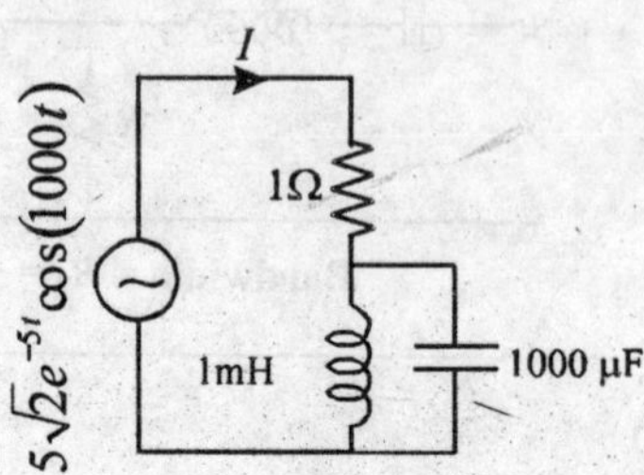

Given from source

$\omega = 1000$ rad/sec

$$Z_c = \frac{1}{j\omega C} = \frac{1}{j \times 1000 \times 10^{-3}} = -j$$

$$Z_L = j\omega L = j \times 1000 \times 10^{-3} = j$$

therefore $Z_{eq} = 1 + \frac{-j \times j}{-j + j} = 1 + \frac{1}{0}$

$Z_{eq} \to \infty$

The current in the given circuit will be zero as the input impedance is infinite

55. Ans (d)

Solution:

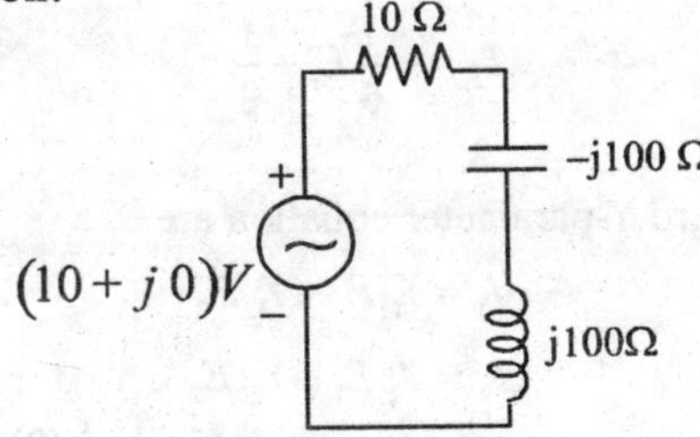

R_L = impedance of the circmit

$= 10 - 100j + 100j = 10$

Voltage across capacitor

$$V_C = \frac{10 \times (-j)(100)}{10}$$

$$V_C = 0 - 100j$$

56. Ans. (a)

Solution: Since at steady state there is no DC voltage drop across the inductor and no current flows through the capacitor as C → open circuit, so the circuit reduces to

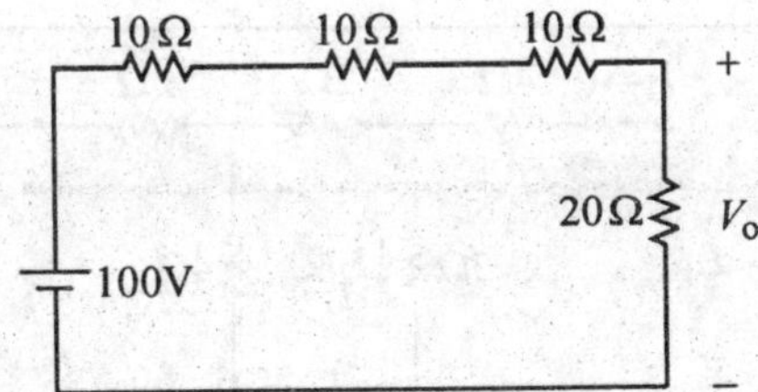

$$V_0 = \frac{20}{50} \times 100 = 40 \text{ V}$$

6.5 Two Port Network

1. Ans. (d)

Soltion:

For a two port passive network the symmetrical condition in h-parameter model is $h_{11}.h_{22} - h_{12}.h_{21} = 1$

2. Ans. (b)

Solution: y-Parameter equation

$$i_1 = 2V_1 + V_2$$

$$i_2 = V_1 + V_2$$

$$\begin{pmatrix} y_{11} & y_{12} \\ y_{21} & y_2 \end{pmatrix} = \begin{pmatrix} 2 & 1 \\ 1 & 1 \end{pmatrix},$$

$$(Z) = (Y)^{-1}$$

$$= \begin{pmatrix} 2 & 1 \\ 1 & 1 \end{pmatrix}^{-1}$$

$$Z = \begin{pmatrix} 1 & -1 \\ -1 & 2 \end{pmatrix}$$

Second Method

$$I_1 = 2V_1 + V_2 \quad \text{...(1)}$$

$$I_2 = V_1 + V_2 \quad \text{...(2)}$$

From (2)

$$V_2 = I_2 - V_1$$

Put in (1) equation

$$I_1 = 2V_1 + I_2 - V_1$$

$$V_1 = I_1 - I_2 \quad \text{...(3)}$$

$$\Rightarrow \quad V_1 = Z_{11}I_1 + Z_{12}I_2$$

$$\therefore \quad Z_{11} = 1\ \Omega, \quad Z_{12} = -1\ \Omega$$

Put (3) equation in (2) equation

$$I_2 = I_1 - I_2 + 2I_2$$

$$V_{-2} = -I_{-1} + 2I_{-2} \quad \text{...(2)}$$

$$V_2 = Z_{21}I_1 + Z_{22}I_2$$

$$\therefore \quad Z_{21} = 1\ \Omega, Z_{21} = 2\ \Omega$$

3. Ans. (b)

Solution: $I_1 = Y_{11}\,E_1 + Y_{12}\,E_2$

$$I_2 = Y_{21}\,E_1 + Y_{22}\,E_2$$

Short circuit E_2, i.e. $E_2 = 0$

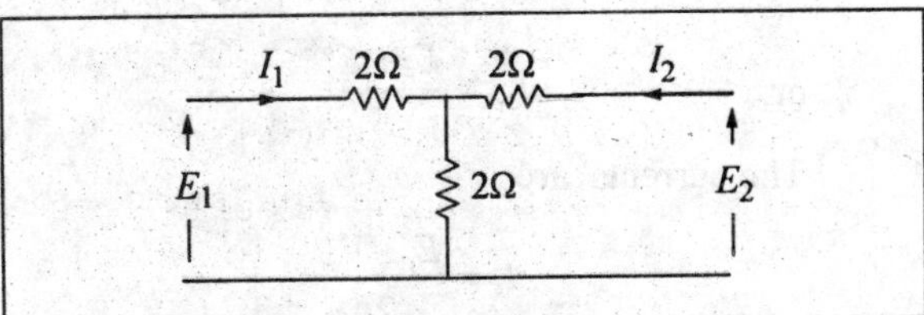

$$Y_{11} = \frac{I_1}{E_1}\bigg|_{E_2=0} = [1/3];\ Y_{21} = \frac{I_1}{E_1}\bigg|_{E_2=0} = -[1/6]$$

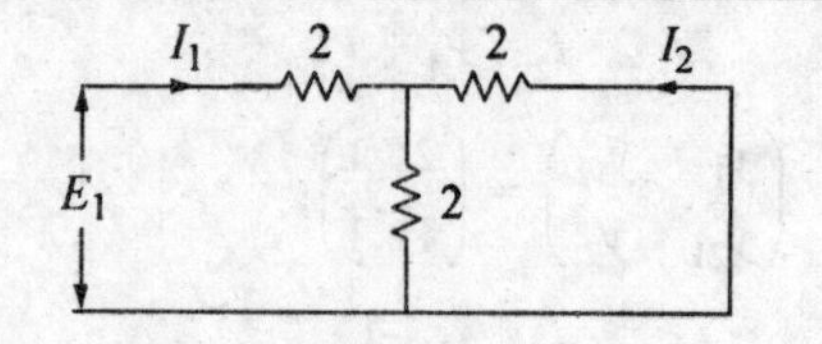

Short circuit E_1 i.e. $E_1 = 0$

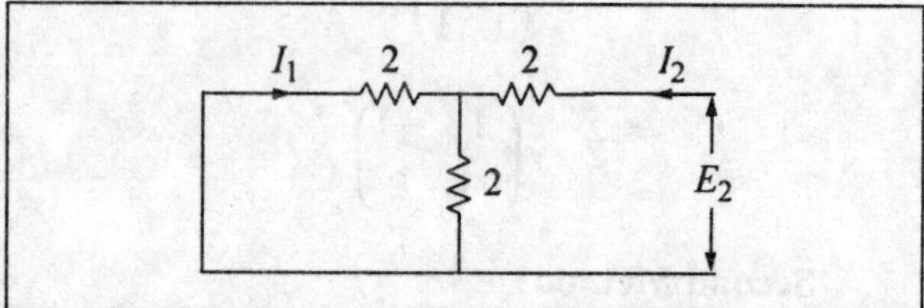

$$Y_{12} = \frac{I_1}{E_1}\bigg|_{E_1=0} = -[1/6];\ Y_{22} = \frac{I_1}{E_1}\bigg|_{E_2=0} = -[1/3]$$

Second Method

From the figure

$$Z = \begin{bmatrix} 4 & 2 \\ 2 & 4 \end{bmatrix}$$

$$|Z| = 16 - 4 = 12$$

$$Y = Z^{-1} = \frac{1}{12}\begin{bmatrix} 4 & -2 \\ -2 & 4 \end{bmatrix} = \begin{bmatrix} 1/3 & -1/6 \\ -1/6 & 1/3 \end{bmatrix}$$

Third Method

Y-parameters of the network the circuit is

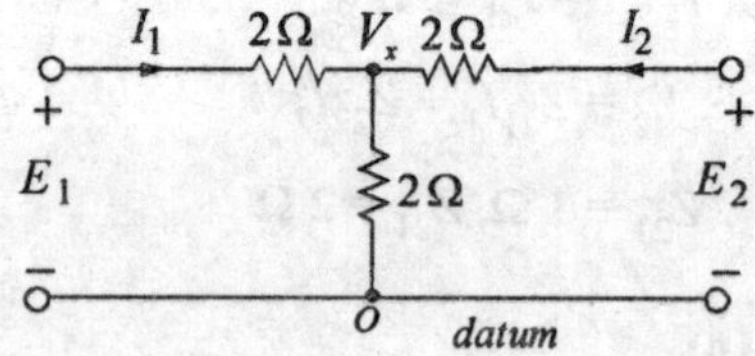

Apply KCL Using Nodal Analysis on V_x node.

Assume outgoing current from node are positive

$$\frac{V_x - E_1}{2} + \frac{V_x}{2} + \frac{V_x - E_2}{2} = 0$$

or, $$V_x = \frac{E_1 + E_2}{3}$$

The currents are

$$I_1 = \frac{E_1 - V_x}{2}$$

$$= \frac{1}{2}\left[E_1 - \left(\frac{E_1 + E_2}{3}\right)\right]$$

$$= \frac{1}{2}\left[\frac{2}{3}E_1 - \frac{E_2}{3}\right]$$

$$I_1 = \frac{1}{3}E_1 - \frac{1}{6}E_2 \quad \text{...(1)}$$

$$I_2 = \frac{E_2 - V_x}{2}$$

$$= \frac{1}{2}\left[E_2 - \left(\frac{E_1 + E_2}{3}\right)\right]$$

or, $$I_2 = -\frac{1}{6}E_1 + \frac{1}{3}E_2 \quad \text{...(2)}$$

Standard Y-parameter equation are

$$I_1 = Y_{11}E_1 + Y_{12}E_2 \quad \text{...(3)}$$

$$I_2 = Y_{21}E_1 + Y_{22}E_2 \quad \text{...(4)}$$

Comparing equation (1) with (3) and (2) with (4) the result is as

$$Y_{11} = \frac{1}{3}\mho$$

$$Y_{12} = -\frac{1}{6}$$

$$Y_{21} = -\frac{1}{6}$$

$$Y_{22} = \frac{1}{3}\mho$$

4. Ans. (d)

Solution:

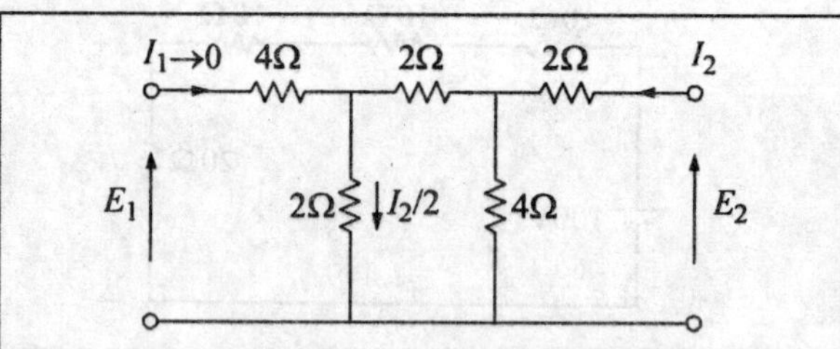

$$h_{12} = \frac{E_1}{E_2}\bigg|_{I_1=0}$$

$$E_2 = 2I_2 + \frac{I_2}{2} \times 4 = 4I_2$$

$$E_1 = \frac{I_2}{2} \times 2 = I_2$$

$$\frac{E_1}{E_2} = \frac{I_2}{4I_2} = 0.25$$

Second Method

As given $E_1 = h_{11}I_1 + h_{12}E_2$

$I_2 = h_{21}I_1 + h_{22}E_2$

Using $\Delta - Y$ conversion

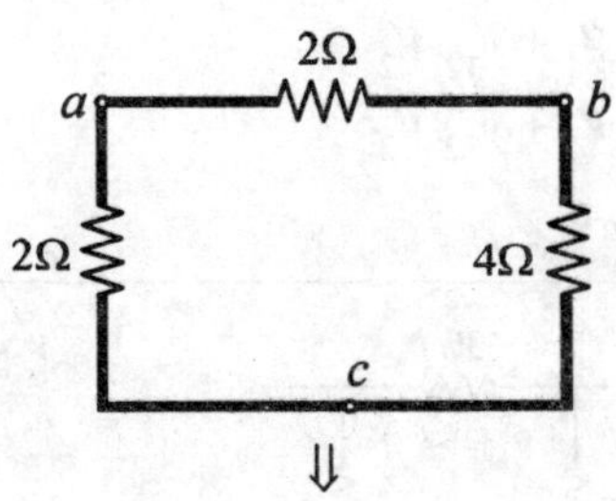

⇓

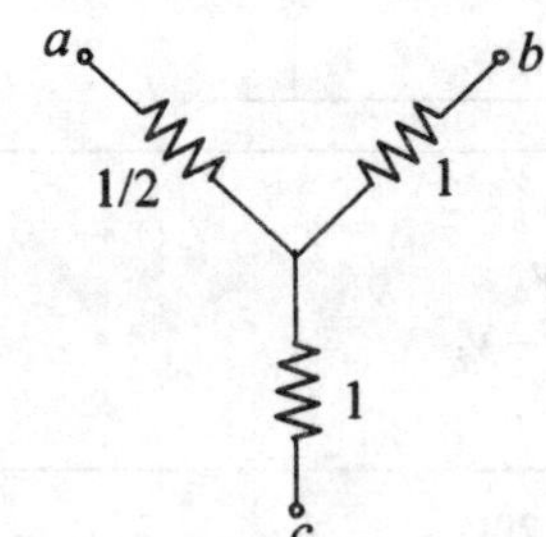

Now add with the main circuit

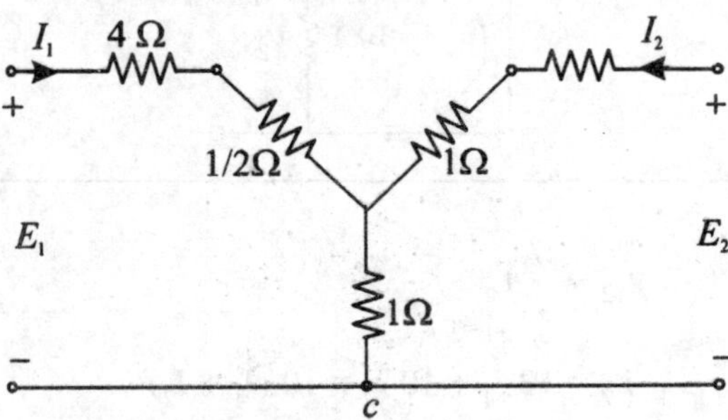

Finally circuits becomes

I_1 4.5Ω 3 Ω I_2, 1 Ω, E_1, E_2

$$E_1 = 4.5I_1 + 1(I_1 + I_2) = 5.5I_1 + I_2 \quad ...(1)$$

$$E_2 = 3I_1 + 1(I_1 + I_2) = I_1 + 4I_2 \quad ...(2)$$

From equation (2)

$$I_2 = \frac{E_2 - I_1}{4} \quad ...(3)$$

Now from equation (1) and (3)

$$E_1 = 5.5I_1 + \frac{E_2 - I_1}{4}$$

or, $E_1 = 5.25I_1 + 0.25E_2$

$h_{12} = 0.25$

5. **Ans. (d)**

Solution: $Z = \begin{bmatrix} 0.9 & 0.2 \\ 0.2 & 0.6 \end{bmatrix} = \begin{bmatrix} z_{11} & z_{12} \\ z_{21} & z_{22} \end{bmatrix}$

$|Z| = 0.9 \times 0.6 - 0.04 = 0.5$

$$Y = \frac{1}{0.5}\begin{bmatrix} 0.6 & -0.2 \\ -0.2 & 0.9 \end{bmatrix} = 2\begin{bmatrix} 0.6 & -0.2 \\ -0.2 & 0.9 \end{bmatrix}$$

$$= \begin{bmatrix} 1.2 & -0.4 \\ -0.4 & 1.8 \end{bmatrix} = \begin{bmatrix} y_{11} & y_{12} \\ y_{21} & y_{22} \end{bmatrix} = \frac{1}{|Z|}\begin{bmatrix} z_{22} & z_{21} \\ z_{12} & z_{12} \end{bmatrix}$$

Hence $y_{22} = 1.8$

6. **Ans (c)**

Solution:

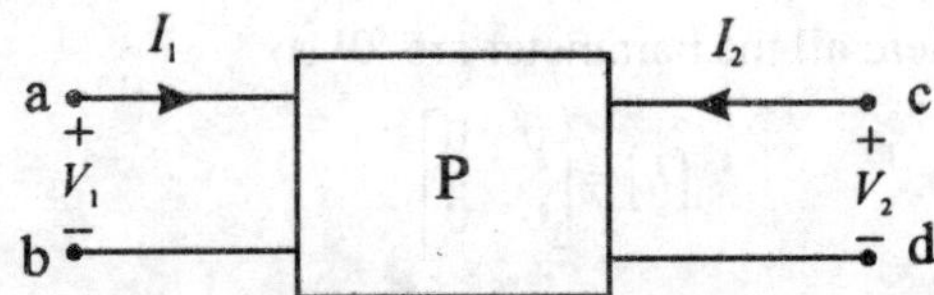

The Z-paraeters of Network P are

$$\begin{bmatrix} V_1 \\ V_2 \end{bmatrix} = \begin{bmatrix} Z_{11} & Z_{12} \\ Z_{21} & Z_{22} \end{bmatrix}\begin{bmatrix} I_1 \\ I_2 \end{bmatrix}$$

If 1Ω is connected

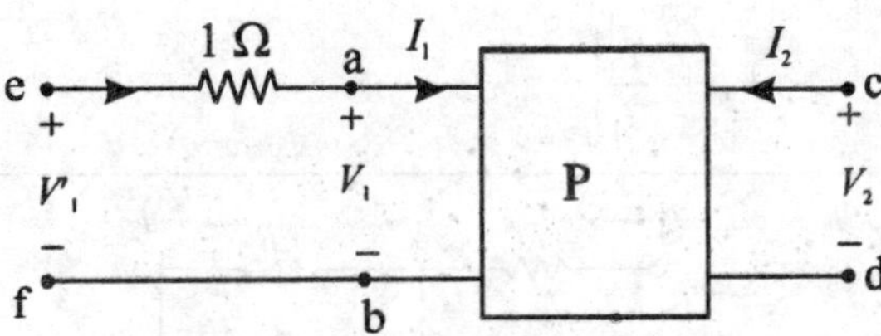

The overall Network port-1 is ef and port-2 is cd

Port currents I_1 and I_2 are same

Port-1 voltage $\quad V_1' = V_1 + 1\,I_1$

$V_1' = (Z_{11} + 1)\,I_1 + Z_{12}I_2$

Port-2 voltage $\quad V_2$ remains same

$V_2 = Z_{22}\,I_1 + Z_{12}I_2$

$$\begin{bmatrix} V_1' \\ V_2 \end{bmatrix} = \begin{bmatrix} Z_{11}+1 & Z_{12} \\ Z_{21} & Z_{22} \end{bmatrix}\begin{bmatrix} I_1 \\ I_2 \end{bmatrix}$$

Overall Z-parameter

$$[Z] = \begin{bmatrix} Z_{11}+1 & Z_{12} \\ Z_{21} & Z_{22} \end{bmatrix}$$

7. **Ans. (c)**

Solution: For two port reciprocal network;

$$h_{12} = -h_{21}$$

8. Ans. (c)

Solution:

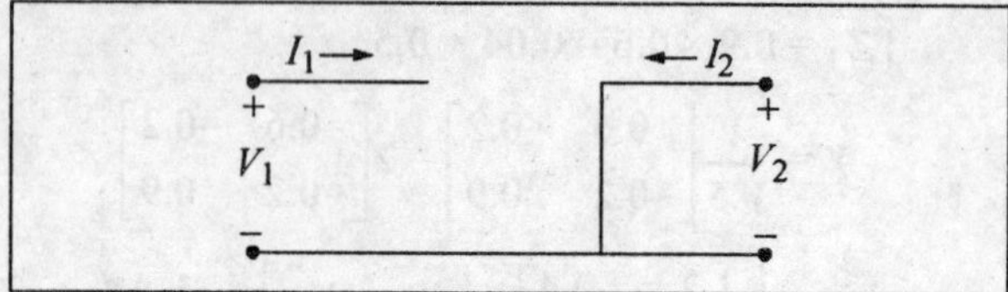

Parameter represent the component

H parameter are given by relations

$$V_1 = h_{11}I_1 + h_{12}\,V_2$$

$$I_1 = h_{11}I_1' + h_{22}\,V_2$$

here all the parameter are '0' so

$$[h] = \begin{bmatrix} 0 & 0 \\ 0 & 0 \end{bmatrix}$$

9. Ans. (a)

Solution: For *h*-parameter

$$V_1 = h_{11}\,I_1 + h_{12}\,V_2$$

$$I_2 = h_{21}\,I_1 + h_{22}\,V_2$$

$$h_{21} = \left.\frac{I_2}{I_1}\right|_{V_2=0}$$

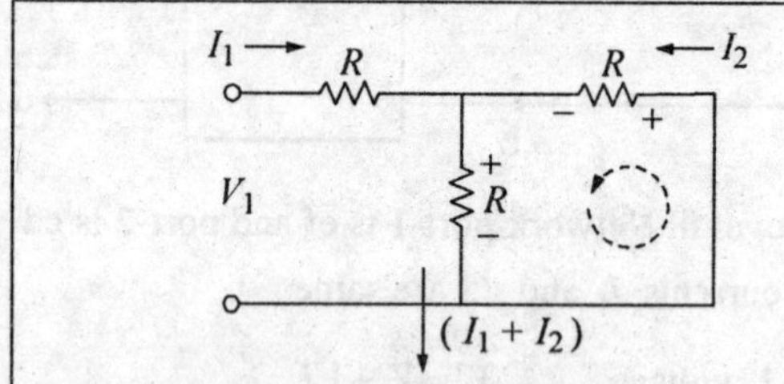

For $V_2 = 0$, output is short circuited

$R(I_1 + I_2) + RI_2 = 0$ Or, $RI_1 + 2\,RI_2 = 0$

$$\frac{I_2}{I_1} = -\,[1/2] = h_{21}$$

10. Ans. (a)

Solution: $Z_{AB} = Z_{AC} = Z_{BC} = \sqrt{3Z}$

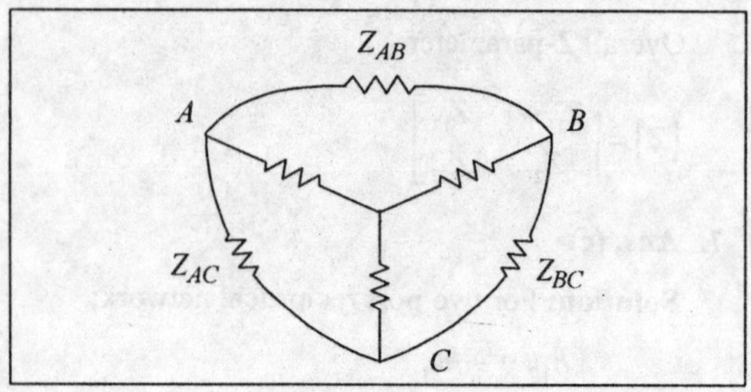

If the impedance are balance, then

$$Z_\Delta = 3\,Z_y,\; Z_y = [Z_\Delta/3]$$

$$= [(\sqrt{3Z}\,)/3] = [\,Z\sqrt{3}\,]$$

11. Ans. (c)

Solution: $I_1 = y_{11}\,V_1 + y_{12}\,V_2$

$$I_2 = y_{21}\,V_1 + y_{22}\,V_2$$

NOTE: π type network

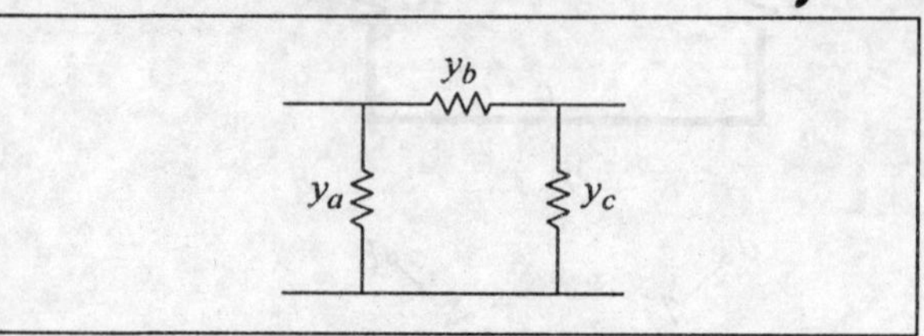

$$y_{11} = y_a + y_b$$

$$y_{22} = y_b + y_c$$

$$y_{12} = y_{21} = -\,y_b$$

Network is reciprocal

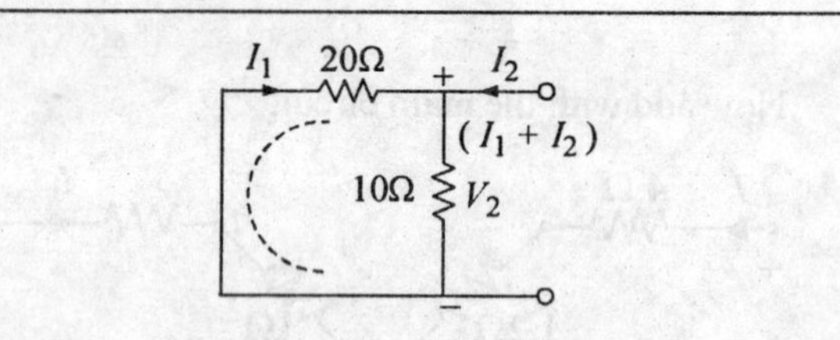

$$y_{12} = \left.\frac{I_1}{V_1}\right|_{V_1=0}$$

$$V_2 = 10\,I_1 + 10\,I_2 = 10\,(I_1 + I_2)$$

$$V_2 = -\,20\,I_1$$

$$y_{12} = [I_1/V_2] = -\,[1/20]$$

$$= -\,0.05 \text{ moh}$$

Second Method

The two-port network

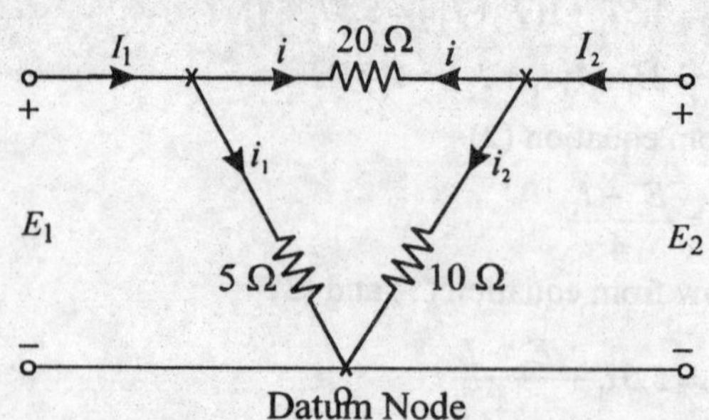

Use KCL Analysis

$$i = \frac{E_1 - E_2}{20},\quad i_1 = \frac{E_1}{50} \text{ and } i_2 = \frac{E_2}{10}$$

and $I_1 = i + i_1$

$$= \frac{E_1}{5} + \frac{E_1 - E_2}{20} \quad \ldots (1)$$

$$I_1 = \frac{1}{4}E_1 - \frac{1}{20}E_2 \quad \ldots (2)$$

Compare with equation (1) and (2)

$$I_1 = Y_{11}E_1 + Y_{12}E_2$$

$$Y_{12} = \frac{1}{20} = -0.05\ \mho$$

12. Ans. (b)

Solution:

$$n_1 = n\ ;\ n_2 = 1$$

For and ideal transformer.

$$V_1 = n_1 V_2 \quad ; \quad V_2 = n_2 V_1$$

$$I_1 = -[1/n]\, I_2$$

$$I_2 = -nI_1$$

$$A = [V_1/V_2] = n$$

$$B = -[V_1/I_2] = 0$$

$$C = 0$$

$$D = [-I_1/I_2] = [-1/n] = [1/n]$$

$$D = [1/n]$$

13. Ans. (c)

Solution:

$$Z_C = R_L \Bigg\| \frac{1}{sC} \quad Z_C = \frac{R_L \cdot \frac{1}{sC}}{R_L + \frac{1}{sC}}$$

$$= \frac{R_L}{1 + sCR_L}$$

From figure, we have

$$V_i(s) = (R + Z_C)\, I(s)$$

$$V_o(s) = Z_C\, I(s)$$

$$\frac{V_o(s)}{V_i(s)} = \frac{Z_C}{R + Z_C}$$

$$= \frac{\frac{R_L}{1 + sCR_L}}{R + \frac{R_L}{1 + sCR_L}}$$

$$= \frac{R_L}{R + sCR\,R_L + R_L}$$

Given that, $\frac{V_o(s)}{V_i(s)} = \frac{1}{2 + sCR}$

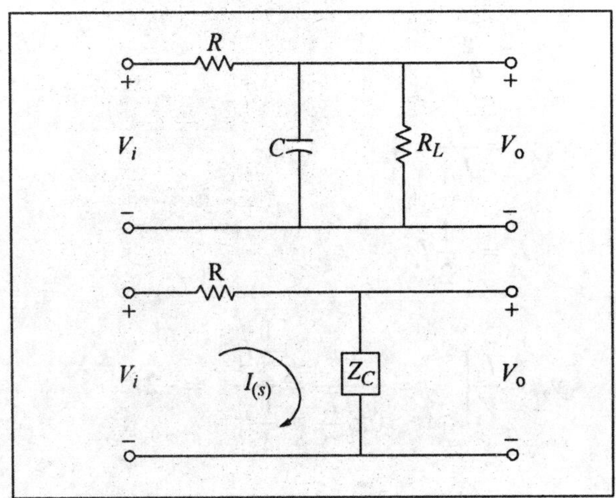

Thus if , $R_L = \mathbf{R}$

Then from (1) , we have

$$\frac{V_o(s)}{V_i(s)} = \frac{R}{R + sCR.R + R}$$

$$= \frac{R}{2R + sCR.R} = \frac{1}{2 + sCR}$$

14. Ans. (a)

Solution:

For the two porte network short circuit admittance parameter matrix is

$$I_1 = y_{11}V_1 + y_{12}V_2$$

$$I_2 = y_{21}V_1 + y_{22}V_2$$

When $V_2 = 0$, the circuit becomes.

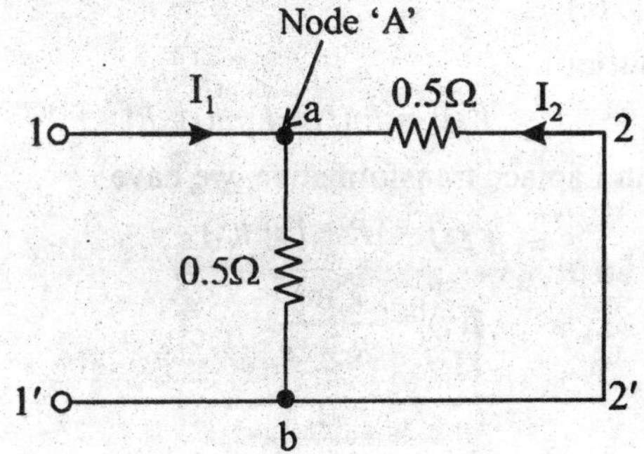

$$y_{11} = \frac{I_1}{V_1}\bigg|_{V_2=0}$$

$$y_{21} = \frac{I_2}{V_1}\bigg|_{V_2=0}$$

$$R_{eq} = \frac{0.5 \times 0.5}{0.5 \times 0.5} = \frac{1}{4}$$

$\therefore V_1 = \frac{I_1}{4}$

$\therefore y_{11} = \frac{I_1}{I_{1/4}} = 4$

$I_2 = -\frac{1}{2} I_1$

$$y_{21} = \frac{I_2}{V_1}\bigg|_{V_2=0} = \frac{I_2}{0.5\frac{I_1}{2}} = \frac{\frac{1}{2}I_1}{\frac{1}{4}I_1} = -2$$

$$[y] = \begin{bmatrix} 4 & -2 \\ -2 & 4 \end{bmatrix}$$

15. Ans. (d)

Solution:

$$\begin{vmatrix} A & B \\ C & D \end{vmatrix} = \begin{vmatrix} A' & B' \\ C' & D' \end{vmatrix}\begin{vmatrix} A'' & B'' \\ C' & D'' \end{vmatrix}$$

16. Ans. (b)

Solution: For reciprocal Network

Z-Parameter : $Z_{12} = Z_{21}$

Y-Parameter : $Y_{12} = Y_{21}$

h-parameter : $h_{12} = -h_{21}$

g-parameter : $g_{12} = -g_{21}$

T- Parameter : $AD - BC = 1$

T-Parameter : $ad - bc = 1$

17. Ans. (c)

Solution:

$V_i(t) = Ri(t) + L[di(t)/dt]$

With Laplace transformation, we have

$V_i(s) = [R + Ls]\, I(s)$

or, $I(s) = \frac{V_i(s)}{R + Ls}$

$= \frac{1}{s(2+s)}$

[Since, $V_i(s) = [1/s]$]

In time-domain, $i(t) = [u(t)/2]\,[1 - e^{-2t}]$

At $t = [1/2]$, $i(t) = 0.31$

And at $t = \infty$, $i(t) = 0.5$

18. Ans. (c)

Solution: $V_1 = Z_{11} I_1 + Z_{12} I_2$

$\boldsymbol{V_2 = Z_{21} I_1 + Z_{22} I_2}$

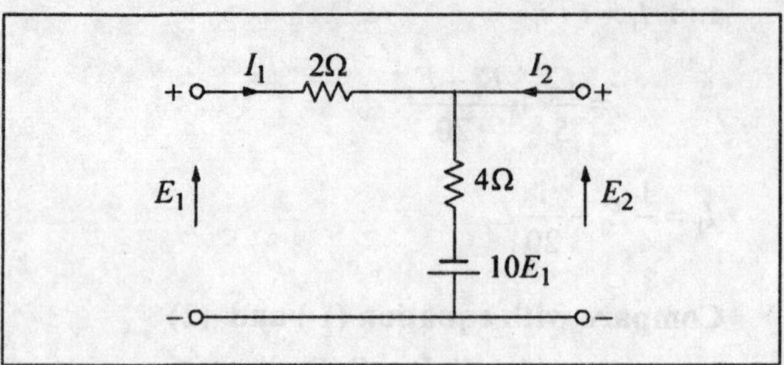

When $I_2 = 0$

$E_1 = 6\, I_1 - 10E_1$

Or, $11\, E_1 = 6\, I_1$

$Z_{11} = [V_1/I_1]$

$Z_{11} = [E_1/I_1] = [6/11]\ \Omega$

$E_2 = 4\, I_1 - 10\, E_1$

$= 4\, I_1 - 10\, [(6/11)\, I_1]$

$E_2 = \frac{44I_1 - 60I_1}{11} = \frac{-16I_1}{11}$

$Z_{21} = [E_2/I_1] = [-16/11]\ \Omega.$

19. Ans. (b)

Solution:

$$Z(s) = \frac{K(s+3)}{(s+1+j)\,(s+1-j)}$$

$$= \frac{k(s+3)}{(s+1)^2+1}$$

$Z(o) = 3 = [3K/2];\quad K = 2$

$$Z(s) = \frac{2(s+3)}{(s+1)^2+1} = \frac{2(s+3)}{s^2+2s+2}$$

20. Ans. (a)

Solution:

Convert the π into *T* configuration, i.e. Δ into *y*

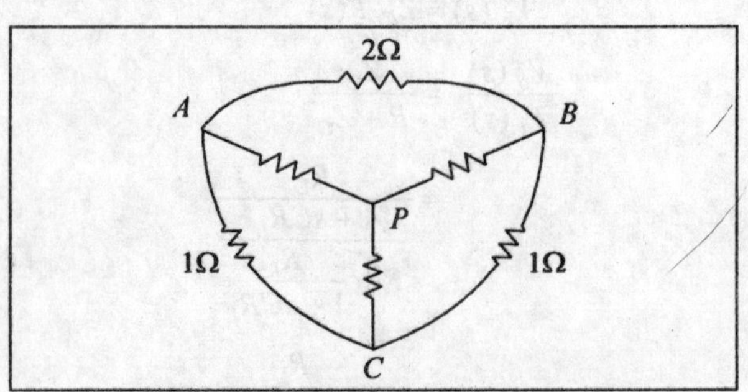

$$Z_{AP} = \frac{Z_{AB} \cdot Z_{AC}}{Z_{AB} + Z_{AC} + Z_{BC}} = \frac{2}{4} = 0.5$$

$$Z_{PB} = \frac{Z_{AB} \cdot Z_{BC}}{Z_{AB} + Z_{BC} + Z_{AC}} = \frac{2}{4} = 0.5$$

$$Z_{PC} = \frac{Z_{AC} \cdot Z_{BC}}{Z_{AB} + Z_{BC} + Z_{AC}} = \frac{1}{4} = 0.25$$

Modified Network

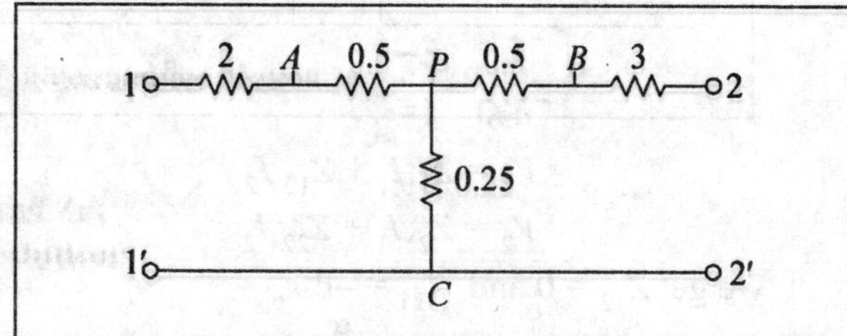

$$V_1 = 2.75\, I_1 + 0.25\, I_2$$

$$V_1 = Z_{11}\, I_1 + Z_{12}\, I_2$$

$$Z_{11} = 2.75 \text{ and } Z_{12} = 0.25$$

21. Ans. (d)

Solution:

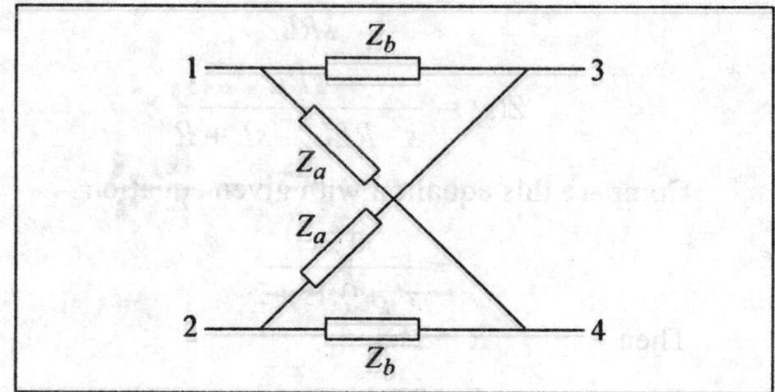

$$V_1 = I_1\, Z_{11} + I_2\, Z_{12}$$

$$V_2 = I_1\, Z_{21} + I_2\, Z_{22}$$

$$Z_{11} = \left.\frac{V_1}{I_1}\right|_{I_2=0} = 0$$

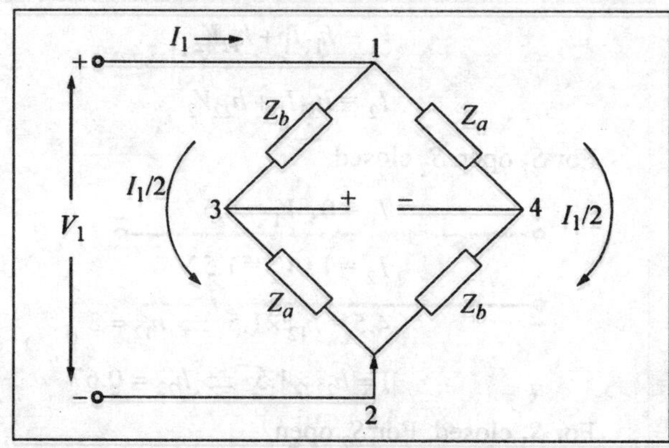

KVL

$$V_1 = [I_1/2]\,\{Z_a + Z_b\}$$

$$0 = -V_2 - [I_1/2]\, Z_b + [I_1/2]\, Z_a$$

$$V_2 = [I_1/2]\,\{Z_a - Z_b\}$$

Where, $Z_1 = Z_b = 2$

$Z_2 = Z_a = 2j$

$$Z_{11} = [V_1/I_1] = [1/2]\,\{Z_a + Z_b\} = j + 1$$

$$Z_{21} = [V_2/I_1] = [1/2]\,\{Z_a - Z_b\} = j - 1 = -1 + j$$

If series impedance in a circuit is equal then network is symmetrical. i.e. $Z_{11} = Z_{22}$

If cross arm impedance is equal then network is reciprocal i.e. $Z_{12} = Z_{21}$

$$\begin{pmatrix} Z_{11} & Z_{12} \\ Z_{21} & Z_{22} \end{pmatrix} = \begin{pmatrix} 1+j & -1+j \\ -1+j & 1+j \end{pmatrix}$$

22. Ans. (d)

Solution:

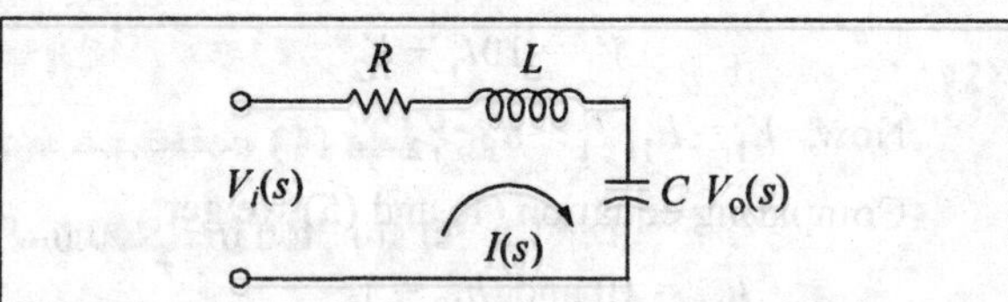

$R = 10\ \text{k}\Omega,\ L = 10\ \text{mH},\ C = 100\ \mu\text{F}$

$$V_i(s) = R_I(s) + Ls \,.\, I(s) + \frac{1}{Cs}\, I(s)$$

And $\quad V_c(s) = \dfrac{1}{Cs}\, I(s)$

Then, $\quad \dfrac{V_c(s)}{V_i(s)} = \dfrac{1}{LCs^2 + RCs + 1}$

Or, $\quad \dfrac{V_c(s)}{V_i(s)} = \dfrac{10^6}{s^2 + 10^6 s + 10^6}$

23. Ans. (c)

Solution:

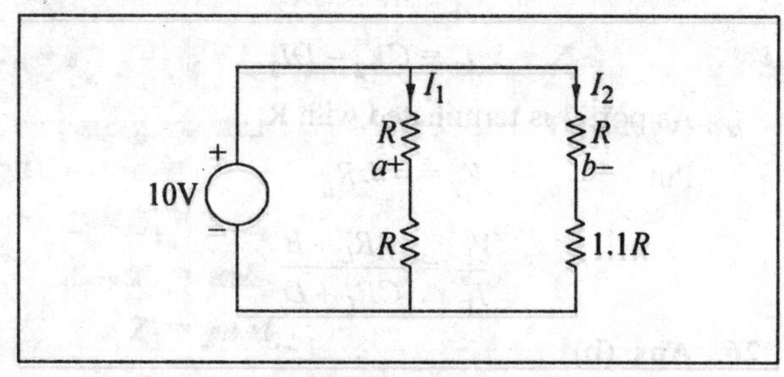

$$I_1 = [10/2R] = [5/R]$$

$$I_2 = [10/2.1R]$$

$$= 4.76/R$$

$$V_{ab} = -R\, I_1 + I_2\, R$$

$$= R\,[I_2 - I_1]$$
$$= [R/R]\,(4.76 - 5)$$
$$V_{ab} = -\,0.238$$

24. Ans. (d)

Solution:

h-parameter standard equation

$$V_1 = h_{11}\,I_1 + h_{12}\,V_2 \qquad \ldots(1)$$
$$I_2 = h_{21}\,I_1 + h_{22}\,V_2 \qquad \ldots(2)$$
$$V_1 - 10I_1 - 20(I_1 + I_2) = 0\ ; \quad V_2 - 20\,(I_1 + I_2) = 0$$
$$\Rightarrow V_1 - 30I_1 - 20I_2 = 0\ ;\ V_2 - 20I_1 - 20I_2) = 0$$
$$V_1 = 30I_1 + 20I_2 \qquad \ldots(3)$$
$$V_2 = 20I_1 + 20I_2 \qquad \ldots(4)$$
$$\Rightarrow 20I_2 = V_1 - 30I_1 \qquad \Rightarrow 20I_1 = V_2 - 20I_1$$
$$V_1 - V_2 = 10I_1 \Rightarrow V_1 = V_2 + 10I_2$$
$$V_1 = 10I_1 + V_2 \qquad \ldots(5)$$

Now, $V_1 = h_{11}\,I_1 + h_{12}\,V_2$ ... (1)

Comparing equation (1) and (5) we get

$$h_{11} = 10 \text{ and } h_{12} = 1$$

From equation (3) and (5)

$$20I_2 = V_1 - 30I_1 = V_2 + 10I_1 - 30I_1$$
$$= V_2 - 20I_1$$
$$I_2 = [1/20]\,V_2 - I_1 \qquad \ldots(6)$$
$$I_2 = h_{21}I_1 + h_{22}\,V_2 \qquad \ldots(2)$$

Comparing equation (6) and (2) we get

$$h_{21} = -1 \text{ and } h_{22} = 0.05$$

Hence, *h* parameter are $\begin{pmatrix} 10 & 1 \\ -1 & 0.05 \end{pmatrix}$

25. Ans. (d)

Solution:

$$V_1 = AV_2 - BI_2$$
$$I_1 = CV_2 - DI_2$$

As port 2 is terminated with R_L.

Put $\quad V_2 = -\,I_2R_L$

$$\frac{V_1}{I_1} = \frac{AR_L + B}{CR_L + D}$$

26. Ans. (b)

Solution:

$$V_1 = I_1\,r_e$$
$$V_2 = -r_o\,(I_2 - \beta I_1)$$
$$V_2 = -I_1\,r_o + I_2\,r_o$$

Compare with below equation

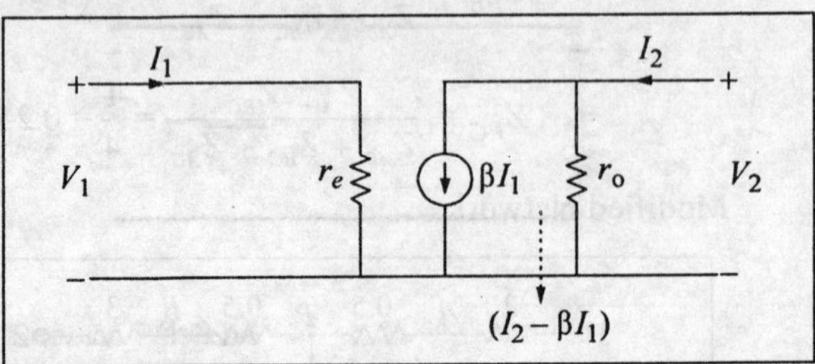

$$V_1 = Z_{11}I_1 + Z_{12}\,I_2$$
$$V_2 = Z_{21}I_1 + Z_{22}\,I_2$$

We get $Z_{12} = 0$ and $Z_{21} = -\beta\,r_o$

27. Ans.(d)

Solution: $Y(s) = \dfrac{1}{R} + Cs + \dfrac{1}{sL}$

$$= \frac{sL + s^2 R\,LC + R}{sRL}$$
$$= \frac{s^2\,RLC + sL + R}{sRL}$$
$$Z(s) = \frac{sRL}{s^2\,RLC + sL + R}$$

Compare this equaiton with given equation

$$= \frac{0.2s}{s^2 + 0.1s + 2}$$

Then $\quad R = 2\Omega$

$$L = 0.1\text{ H}$$
$$RLC = 1$$
$$\Rightarrow \quad C = \frac{1}{RL} = \frac{1}{0.2} = 5\text{F}$$

28. Ans. (a)

Solution: *h*-pararmeter equation

$$V_1 = h_{11}I_1 + h_{12}V_2$$
$$I_2 = h_{21}I_1 + h_{22}V_2$$

For S_1 open S_2 closed

$$I_1 = 0,\ V_1 = 4.5$$
$$I_2 = 1,\ V_2 = 1.5\text{ V}$$
$$4.5 = h_{12} \times 1.5 \Rightarrow h_{12} = 3$$
$$1 = h_{22} \times 1.5 \Rightarrow h_{22} = 0.67$$

For S_1 closed, For S_2 open

$$I_1 = 4A,\ V_1 = 6\text{ V}$$
$$I_2 = 0,\ V_2 = 6\text{ V}$$

$$6 = h_{11} \times 4 + 3 \times 6$$

$$h_{11} = \frac{6.18}{4} = -\frac{12}{4} = -3$$

$$0 = h_{21} \times 4 + 0.67 \times 6$$

$$h_{21} = \frac{-4.02}{4} = -1$$

$$\therefore h\text{-parameter matrix} = \begin{bmatrix} -3 & 3 \\ -1 & 0.67 \end{bmatrix},$$

29. Ans. (c)

Solution:

$$Z_C = R_L \parallel \frac{1}{sC} \quad Z_C = \frac{R_L \cdot \frac{1}{sC}}{R_L + \frac{1}{sC}}$$

$$= \frac{R_L}{1 + sCR_L}$$

From figure, we have

$$V_i(s) = (R + Z_C)\, I(s)$$

$$V_o(s) = Z_C\, I(s)$$

$$\frac{V_o(s)}{V_i(s)} = \frac{Z_C}{R + Z_C}$$

$$= \frac{\frac{R_L}{1 + sCR_L}}{R + \frac{R_L}{1 + sCR_L}}$$

$$= \frac{R_L}{R + sCR\, R_L + R_L}$$

Given that, $\frac{V_o(s)}{V_i(s)} = \frac{1}{2 + sCR}$

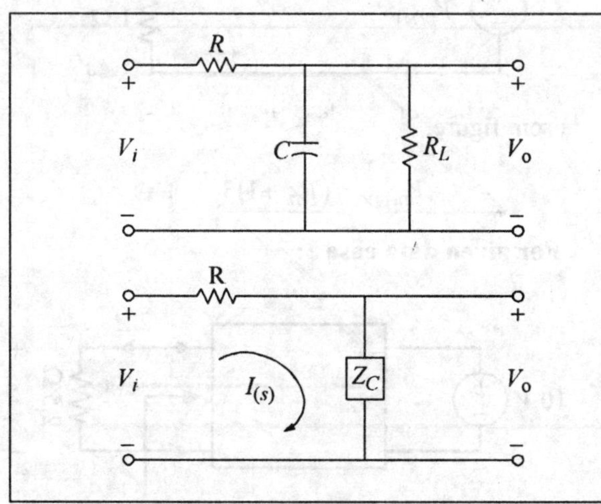

Thus if, $R_L = \mathbf{R}$

Then from (1), we have

$$\frac{V_o(s)}{V_i(s)} = \frac{R}{R + sCR \cdot R + R}$$

$$= \frac{R}{2R + sCR \cdot R}$$

$$= \frac{1}{2 + sCR}$$

30. Ans. (d)

Solution:

$$V_1 = 100V + 25I_1; \quad \text{...(1)}$$

$$V_2 = -100I_2$$

$$\Rightarrow I_2 = -0.01\,V_2$$

$$I_2 = 0.01\,V_1 + 0.1\,V_2 \quad \text{...(2)}$$

From equation (1) and (2)

or, $-0.01V_2 = 0.01V_1 + 0.1V_2$

or, $-0.01\,V_1 = 0.01\,V_1$

or, $\frac{V_2}{V_1} = \frac{-1}{11}$

31. Ans. (a)

Solution:

From given coupled circuit

$$V_1 = j\omega L_1 I_1 + j\omega M_{12} I_2 \quad \text{...(1)}$$

$$V_2 = j\omega L_2 I_2 + j\omega M_{12} I_1 \quad \text{...(2)}$$

From given network

$$V_1 = I_1 Z_1 + Z_3(I_1 + I_2) = (Z_1 + Z_3) I_1 + Z_3 I_2 \quad \text{...(3)}$$

$$V_2 = I_2 Z_2 + (I_1 + I_2) Z_3 = (Z_2 + Z_3) I_1 + Z_3 I_1 \quad \text{...(4)}$$

Comparing, equation (1) with (3) and (2) with (4) we get

$$Z_1 + Z_3 = j\omega L_1$$

$$Z_2 + Z_3 = j\omega L_2$$

$$Z_3 = j\omega M_{12}$$

$$Z_1 = j\omega(L_1 - M_{12})$$

$$Z_2 = j\omega(L_2 - M_{12})$$

$$Z_3 = j\omega M_{12}$$

32. Ans. (c)

Solution:

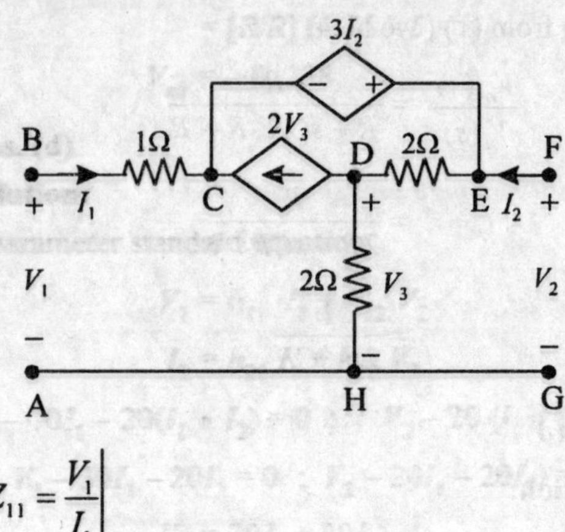

$$Z_{11} = \left.\frac{V_1}{I_1}\right|_{I_2=0}$$

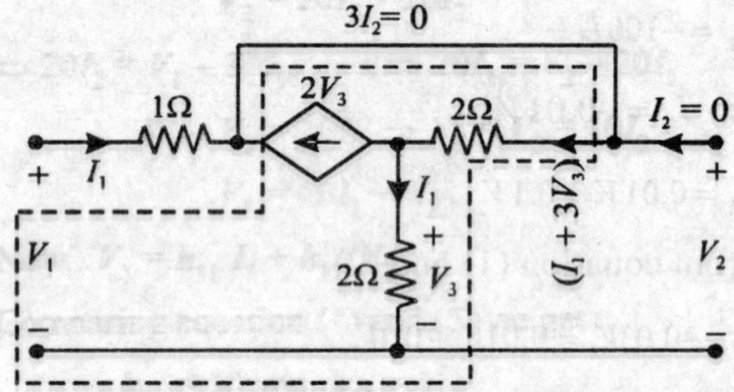

Apply KVL in input port alonge doted line as there is current source which is to be avoided.

$$-V_1 + I_1 + 2(I_1 + 2V_3) + V_3 = 0$$

or, $-V_1 + 3I_1 + 5V_3 = 0$ ...(1)

$$V_3 = 2I_1 \quad \text{...(2)}$$

From equtaion (1) and (2)

$$-V_1 + 3I_1 + 5 \times 2I_1 = 0$$

$$V_1 = 3I_1 + 10I_1$$

$$= 13I_1$$

$$\frac{V_1}{I_1} = 13$$

$$Z_{11} = 13$$

33. Ans. (c)

Solution:

Use the data obtained in part 50 Answer as $V_{Th10V} = 9\ V$ and $R_{Th} = 2\ \Omega$

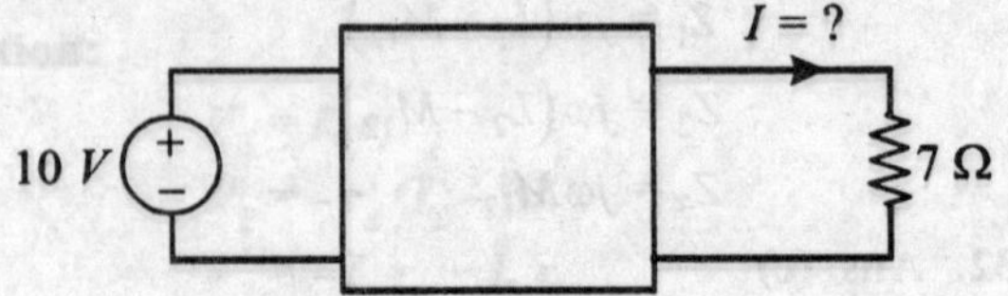

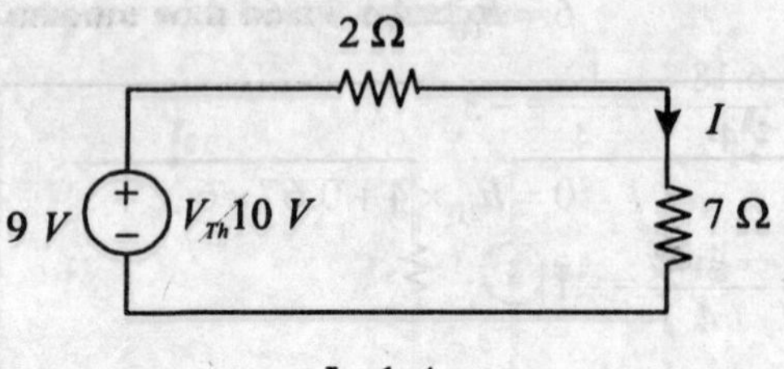

$$I = 1\ A$$

Second Method

The given network can be replaced by a thevenin equivalent with V_{th} and R_{th} as thevenin voltage and thevenin resistance.

Now, we can write two equations for this

$V_{th} = 3\ R_{th} + 3$

$V_{th} = R_{th} + 5$

Solving these two equations we get $R_{th} = 2$ and $V_{th} = 9$

Now using the same equation with current unknown,

$9 = I \times 2 + 7 \times I = 1\ A$

34. Ans. (c)

Solution:

For given data case 1:

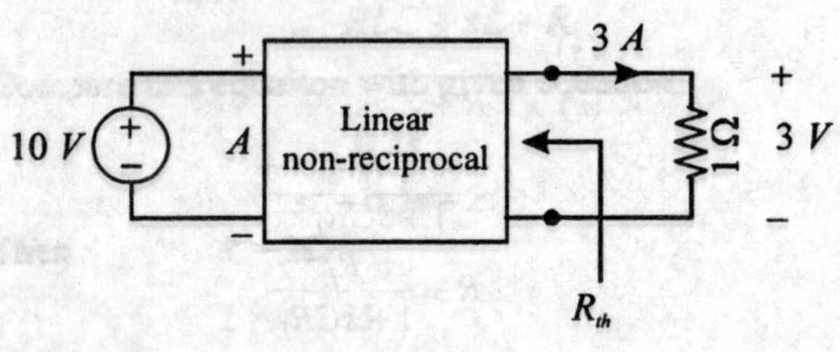

From figure,

$$V_{Th10V} = (R_{Th} + 1)3 \quad \text{...(1)}$$

For given data case 2:

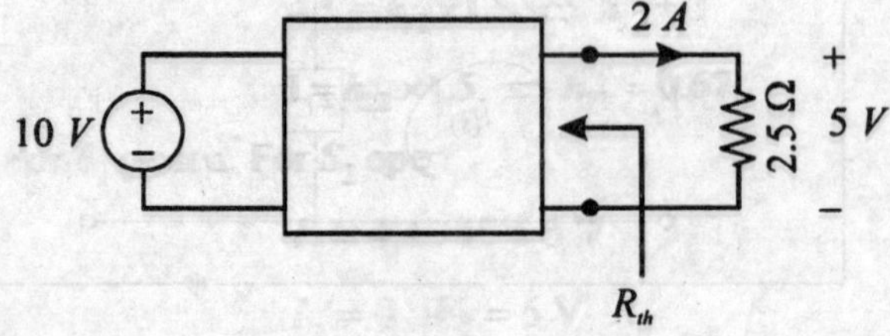

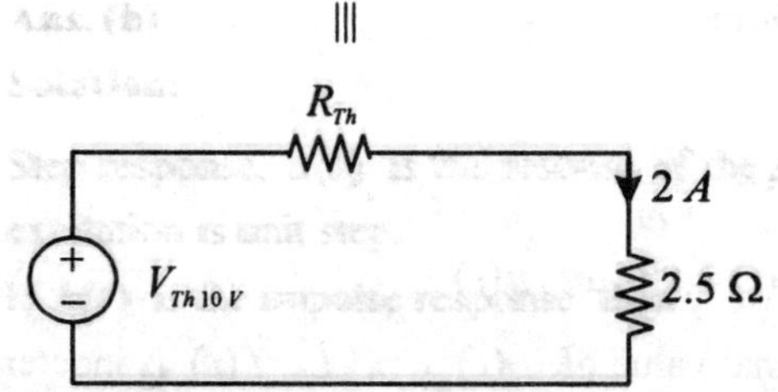

From figure

$$V_{Th\,10\,V} = (R_{Th} + 2.5)2 \qquad \text{...(2)}$$

Now to find the input impedence R_{TH}

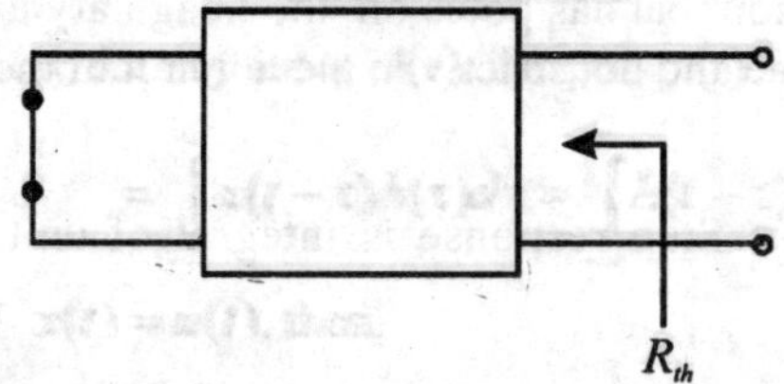

To find $V_{Th} = V_{OC}$

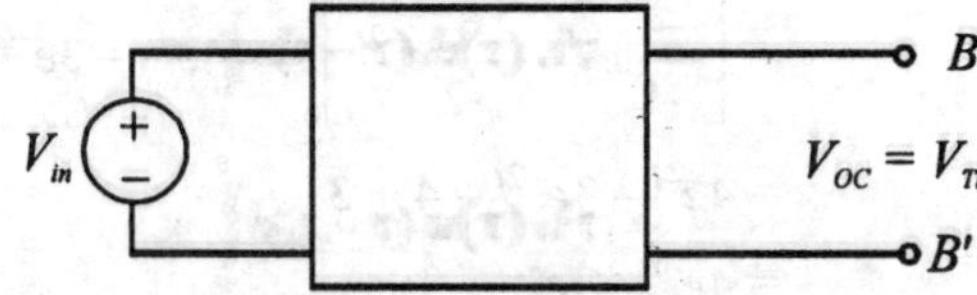

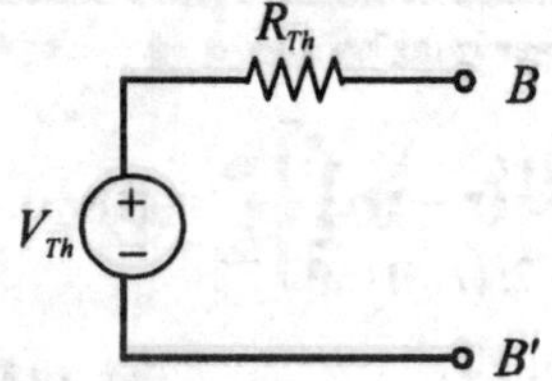

From equations (1) and (2)

$V_{Th\,10\,V} = 9\ V$

$R_{Th} = 2\ \Omega$

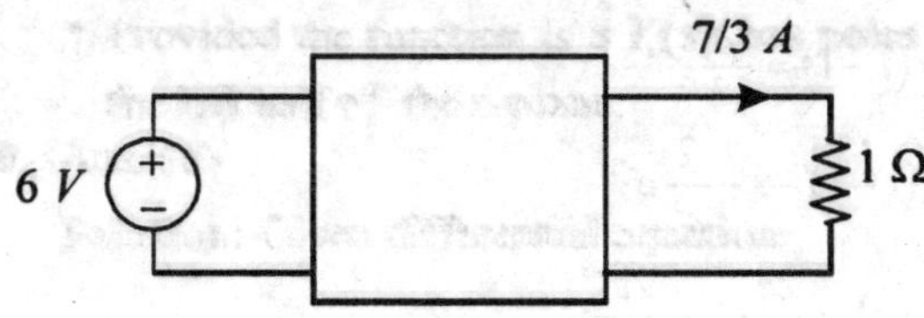

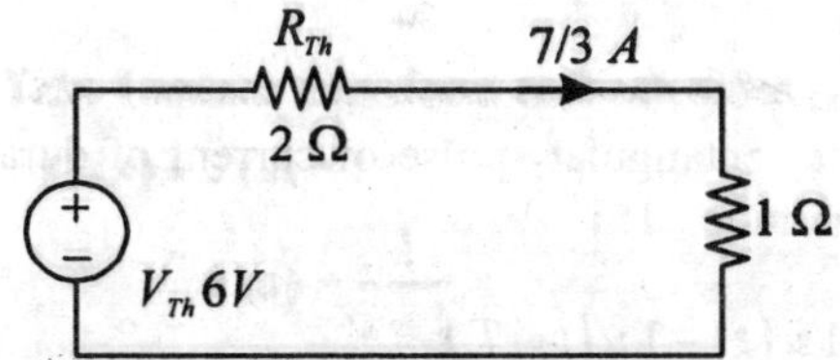

Therefore, $V_{Th\,6V} = (R_{Th} + 1)\dfrac{7}{3} = 7\ V$

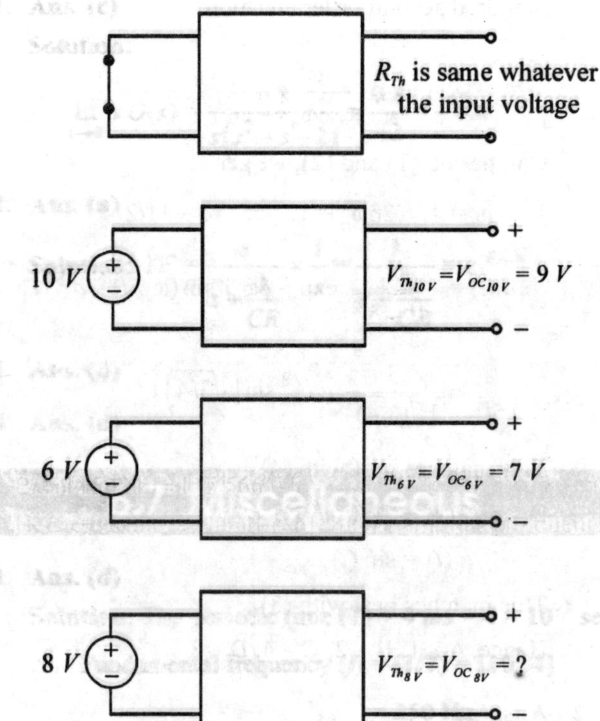

As the given two port network is linear non-reciprocal, we can express output at port 2 as a linear function of output.

i.e. $V_{OC} = a\,V_{in} + b$

$$9 = 10\,a + b \qquad \text{...(3)}$$

$$7 = 6\,a + b \qquad \text{...(4)}$$

From equation (3) and (4)

$$a = 0.5,\ b = 4$$

Therefore, $V_{OC\,8\,V} = 0.5(8) + 4 = 8\ V$

6.6 Laplace Transform

1. **Ans. (b)**

Solution:

A. $H(s) = \dfrac{s}{s+1} = \dfrac{s+1-1}{s+1} = 1 - \dfrac{1}{s+1}$

$h(t) = (t) - e^{-t}\,u(t)$

This matches best with (1)

B. $H(s) = [s/(s+1)^2]$

$h(t) = \delta(t) - e^{-t}\,\mu(t)$

This matches best with (2)

C. $H(s) = \dfrac{1}{s^2 + s + 1}$ ...(1)

Standard second order equation

$$H(s) = \frac{\omega_n^2}{s^2 + 2\delta\omega_n s + \omega_n^2} \quad ...(2)$$

Comparing (1) and (2), we get

$$\omega_n = 1, \quad 2\delta\omega_n = 1, \quad \Rightarrow \quad \delta = [1/2]$$

$$h(t) = \frac{\omega_n}{\sqrt{1-\delta^2}} \exp[-\delta\omega_n] \sin(\omega_n\sqrt{1-d^2})\, t$$

$$= \frac{1}{[\sqrt{(3/4)}]} e^{-t/2} \sin\left(\sqrt{3/4}\right) t$$

This matches best with (4)

D. $H(s) = [1/(s^2 + 1)]$

$h(t) = \sin(t)$

This matches best with (3)

Hence, A – 1, B – 2, C – 4, D – 3

2. **Ans. (a)**

Solution: Response for any input $u(t)$ is

$$y(t) = u(t) * h(t)$$

$$y(t) = \int_0^t h(t-\tau)\, u(\tau)\, d\tau$$

where impulse response $h(t) = [d/dt]\,[s(t)]$

and $s(t)$ is the step response.

Hence $y(t) = \frac{d}{dt}\int_0^t s(t-\tau)\, u(\tau)\, d\tau$

* In LTI system: If the input is differentiated, output also is differentiated
* The property of convolution

$$\frac{d}{dt}[x(t)] * g(t) = \frac{dx(t)}{dt} * g(t)$$

$$= x(t) * \frac{d}{dt} g(t)$$

3. **Ans. (c)**

Solution:

$$H(s) = \frac{1}{s+2} \quad \text{and} \quad Y(s) = \frac{1}{(s+2)^2}$$

$$X(s) = \frac{Y(s)}{H(s)} = \frac{1}{s+2} \quad \text{and} \quad x(t) = e^{-2t}\, u(t)$$

4. **Ans. (d)**

Solution:

Given

$$F(s) = \frac{\omega}{s^2 + \omega^2}$$

$$\Rightarrow f(t) = \sin(\omega t)\, u(t)$$

The final value of $f(t)$ i.e., $\underset{t\to\infty}{Lt} f(t)$ is indeterminate (not defined) as $f(t)$ may have any value between -1 & $+1$.

* The final value theorem of LT is not applicable to function that has poles on the imaginary axis in the s-plane and not strictly in the left half of the s-plane.

5. **Ans. (b)**

Solution: Step response is integral of unit impulse response

$$u(t) = \int_0^t (-4e^{-t} + 6e^{-2t})\, dt$$

$$= \left[\frac{-4}{-1}e^{-t} + \frac{6}{-2}e^{-2t}\right]_0^t = [4e^{-t} - 3e^{-2t}]_0^t$$

$$= 4e^{-t} - 3e^{-2t} - 4 + 3$$

$$= 4e^{-t} - 3e^{-2t} - 1$$

6. **Ans. (c)**

Solution:

Let $x(t) = (t^2 - 2t)\, u(t-1)$

$$= t^2 u(t-1) - 2t(t-1)$$

Using the properties of LT:

$$u(t) \to \frac{1}{s},\; 2u(t-1) \to \frac{2}{s}e^{-s}$$

$$t \to \frac{1}{s^2}$$

$$2tu(t-1) = \frac{2}{s^2}e^{-s^2}$$

$$t^2u(t-1) \to -\frac{2}{s^3}e^{-1s}$$

$$X(s) = \frac{2e^{-1s}}{s^3} - \frac{2e^{-s}}{s^2}$$

7. **Ans. (d)**

Solution:

Given the rectangular pulse of current of duration, T and magnitude, 1: I

$$p(t) = 1\,u(t) - 1\,u(t-T)$$

$$P(s) = \frac{1}{s} - \frac{1}{s}e^{-Ts} = \frac{1}{s}\left(1 - e^{-Ts}\right)$$

8. Ans. (b)

Solution:

Step response, $s(t)$ is the resonse of the system when excitation is unit step.

If $h(t)$ is the impulse response, then

$$h(t) = \frac{d}{dt}s(t)$$

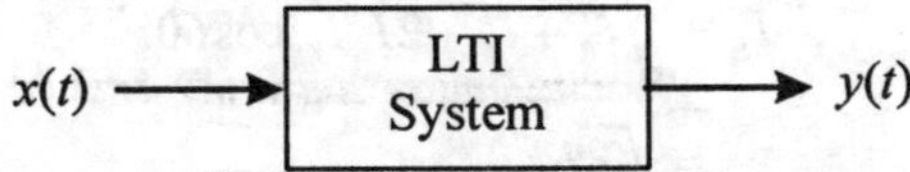

Now, for any input $x(t)$ output $y(t) = x(t) * h(t)$

$$= \int_{-\infty}^{\infty} x(t-\tau)h(\tau)d\tau = \int_{-\infty}^{\infty} h(t-\tau)x(\tau)\,d\tau$$

If $x(t) = u(t)$, then,

$$y(t) = \int_{-\infty}^{\infty} u(t-\tau)\,h(\tau)\,d\tau$$

$$= \int_{0}^{t} u(t-\tau)\,h(\tau)\,d\tau$$

$$= \int_{0}^{t} h(t-\tau)\,u(\tau)\,d\tau$$

Since convolution is commutative process hence, in terms of step response :

$$y(t) = \frac{d}{dt}\left[\int_{0}^{t} s(t-\tau)\,u(\tau)\,d\tau\right]$$

9. Ans. (c)

Solution:

According to the final value theorem :

Final value of $y(t) = \lim_{s\to\infty} y(t) = \lim_{s\to 0}\left[s\,Y(s)\right]$

* Provided the function is $s\,Y(s)$ has poles strictly in the left half of the s-plane.

10. Ans. (d)

Solution: Given differential equation

$$\frac{dy(t)}{dt} + y(t) = \delta(t)$$

Take Laplace transform on both sides

$$sY(s) + Y(s) = 1$$

$$Y(s) = \frac{1}{s+1}$$

Take inverse LT

$$y(t) = e^{-t}u(t)$$

11. Ans. (c)

Solution:

$$\underset{s\to 0}{\text{Lt}}\; s\,G(s) = \frac{0.8}{s\left(s^2+s-2\right)} = \frac{0.8}{-2} = -0.4$$

12. Ans. (a)

Solution: $TF = \dfrac{s}{s+\dfrac{1}{CR}} \times \dfrac{1}{s} = \dfrac{1}{s+\dfrac{1}{CR}} = e^{-t/CR}$

13. Ans. (d)

14. Ans. (d)

6.7 Miscellaneous

1. Ans. (d)

Solution: The periodic time (T) = 4 ms = 4×10^{-3} sec

∴ Fundamental frequency (f) = $(1/T) = [10^3/4]$

= 250 Hz

∴ Frequency of the 5th harmonic = 250 × 5

= 1250 Hz

2. Ans. (d)

Solution: From the given relationship

$$y(t) = \left(\frac{2}{3}[1-e^{-3t}] - \frac{t}{3}e^{-3t} + \frac{1}{9}[1-e^{-3t}]\right)U(t)$$

Taking Laplace both sides,

$$Y(s) = \frac{2}{3}\left(\frac{1}{s} - \frac{1}{s+3}\right) - \frac{1}{3(s+3)^2} + \frac{1}{9}\left(\frac{1}{s} - \frac{1}{s+3}\right)U(s)$$

Second Method

Given:

$$y(t) = \int_0^t (2+t-\tau)\,e^{-3(t-\tau)}u(\tau)\,d\tau$$

or $h(t) = (2+t)e^{-3t}u(t)$

or $H(s) = L\left[h(t)\right]$

$= L\left[2e^{-3t}\right] + \left[te^{-3t}\right]$

$$= \frac{2}{s+3} + \frac{1}{(s+3)^2}$$

or $$H(s) = \frac{2s+7}{(s+3)^2}$$

3. Ans. (d)

Solution:

$$V_{rms} = \sqrt{\frac{1}{T}\int_0^T x(t)^2\, dt}$$

$$= \sqrt{\frac{1}{2(\pi/3)}\int_0^{2(\pi/3)} 100^2\, dt}$$

$$= \sqrt{\frac{3}{2\pi}\times 100^2 (t)\Big|_0^{2\pi/3}}$$

$$= \sqrt{\frac{3}{2\pi}\times 100^2\,\frac{2\pi}{3}}$$

$$V_{rms} = 100 \text{ V}$$

Second Method

$$V_{rms} = \sqrt{\frac{1}{2\pi}\left[\int_0^{\pi/3} d\theta + \int_{\pi/3}^{2\pi/3} d\theta + \int_{2\pi/3}^{4\pi/3} d\theta + \int_{4\pi/3}^{5\pi/3} d\theta + \int_{5\pi/3}^{2\pi} d\theta\right]}$$

$$= \sqrt{\frac{1}{2\pi}\left[\frac{\pi}{3} - 0 + \frac{2\pi}{3} - \frac{\pi}{3} + \frac{4\pi}{3} - \frac{2\pi}{\pi} + \frac{5\pi}{3} - \frac{4\pi}{3} + 2\pi - \frac{5\pi}{3}\right]}$$

= 100 Volt

4. **Ans. (b)**

Solution: $i = A\sin\omega t \Rightarrow \omega = 2\pi/T$

$i(t) = 10 + 20\sin\omega t$

$$I_{rms} = \left[\frac{1}{2\pi}\int_0^{2\pi}(10 + 20\sin t)^2 . dt\right]^{1/2}$$

$$= \left[\frac{1}{2\pi}\int_0^{2\pi}(100 + 400\sin^2 t + 400\sin t). dt\right]^{1/2}$$

$$= \left[\frac{1}{2\pi}\int_0^{2\pi}(100 + 200(1-\cos 2t) + 400\sin t). dt\right]^{1/2}$$

$$= \left[\frac{1}{2\pi}\int_0^{2\pi} 300\, dt\right]^{1/2} = \sqrt{300} = 17 \ A$$

5. **Ans. (a)**

Solution: Given that $u(t) = 3 + 4\cos 3t$

Which mean, $\omega = 3; \omega = 2\pi f = \frac{2\pi}{T}$

$$T = \frac{2\pi}{\omega} = \frac{2\pi}{3}$$

$$U_{RMS} = \sqrt{\frac{1}{T}\int_0^T [u(t)]^2}$$

$$= \sqrt{\frac{3}{2\pi}\int_0^{2\pi/3}(3 + 4\cos 3t)^2}$$

$$= \sqrt{17}$$

6. **Ans. (a)**

Solution: $I_R : I_Y : I_B = 1 : 1 : \sqrt{3}$

$\because$ **I_B is phasor sum of I_R and I_Y at angle of 60°**

$$I_B = \sqrt{I_R^2 + I_Y^2 + 2I_R I_Y \cos 60°}$$

$$= \sqrt{3}\, I_R$$

7. **Ans. (d)**

Solution:

$$x(t) = a_1 \sin(\omega_1 t + \varphi_1)$$

$$y(t) = a_2\ F(\omega_2 t + \varphi_2)$$

System is LTI

For linear

$$y(t) = 0 \text{ for } x(t) = 0 \qquad \dots (1)$$

or $$a\, y_1(t) + b\, y_2(t) = a\, x_1(t) + b\, x_2(t) \qquad \dots (2)$$

From (1), if $x(t) = 0$

Then $\omega.t = -\phi_1$

From (2), for time invariant

$$x(t - t_o) = y(t - t_o)$$

For this $\omega_1 = \omega_2$

8. **Ans. (a)**

Solution:

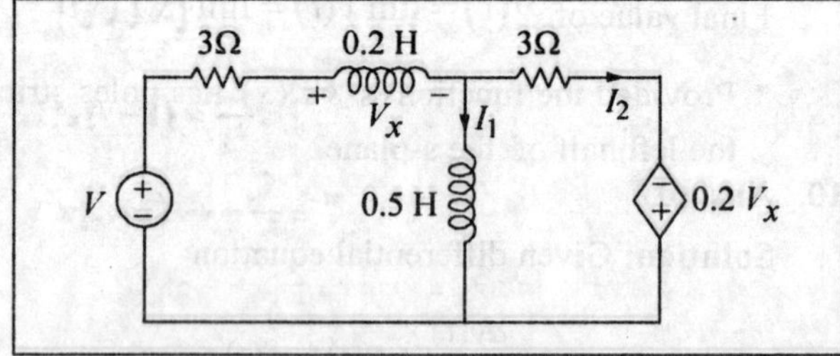

$$-V + 3(I_1 + I_2) + V_X + 0.5\frac{dI_1}{dt} = 0 \qquad \dots (1)$$

$$-0.5\frac{dI_1}{dt} + 5I_2 - 2.V_X = 0 \qquad \dots (2)$$

Solving equation (1) and (2) by eliminating I_2

$$\frac{dI_1}{dt} = \frac{5}{4}V - \frac{15}{4}I_1 - 1.4\, V_x$$

9. Ans. (b)

Solution: Given Thevenin's impedance

$$Z_0 = 2.38 - j0.667\ \Omega$$

$$Z_0 = \sqrt{(2.38)^2 + (0.607)^2}\angle \tan^{-1}\frac{-0.667}{2.38}$$

$$= 2.4\angle -8.0$$

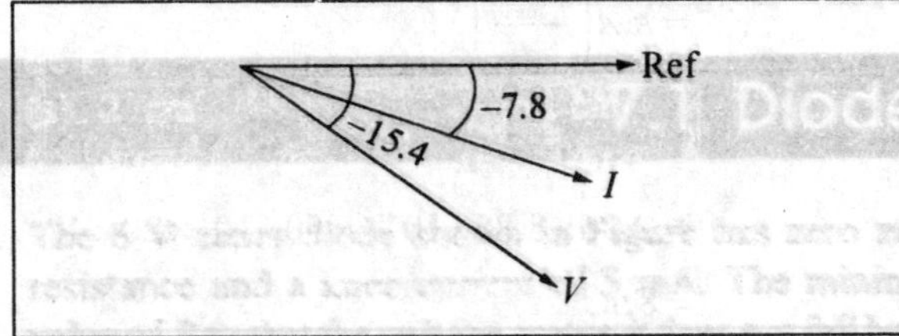

$$I_{th} = \frac{V_{th}}{Z_{th}} = \frac{3.71\angle -15.9}{2.4\angle -8.0}$$

$$= 1.54\angle -7.8$$

I lead V so it is a combination of resistance and capacitance.

Another way

Given Thevenin's impedance $Z_o = 2.38 - j0.667\ \Omega$

For minimim realization,

$$Z_o = R + \frac{1}{Cs} = R + \frac{1}{j\omega C} = R - \frac{j}{\omega C}$$

Z_o is a combination of resistor and capacitor in series.

10. Ans. (d)

Solution:

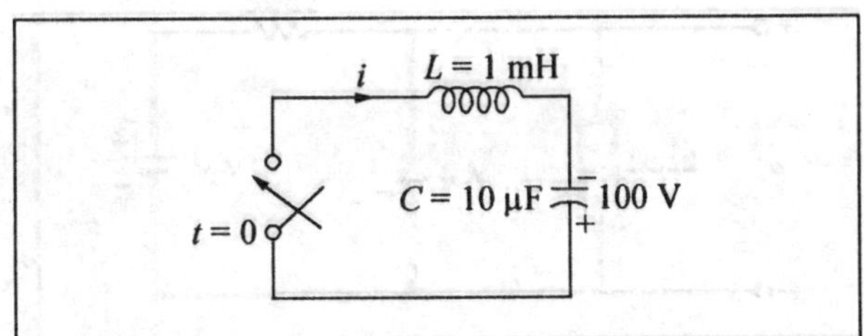

Initial $i_L(0^-) = 0$

When switch is closed current $i_L(0^+) = 0$ but current will increase

$$v_L + v_C = 0$$

$$\Rightarrow \quad v_C = C\int idt - v_{LO}$$

$$i(0^-) = 0,\ v_C(0^-) = 100\text{ V}$$

$$\therefore \quad t = 0$$

$$L\frac{di}{dt} + \frac{1}{C}\int_{-V_{C_o}} idt = 0$$

$$\Rightarrow \quad L^2\frac{di}{dt} + \frac{1}{C}i = V_{Co}$$

$$\Rightarrow \quad \frac{d^2i}{dt^2} + \frac{i}{C} = \frac{V_{Co}}{L}$$

Solution of differential equation is

$$i = V_{co}\sqrt{\frac{C}{L}}\sin \omega_o t$$

where $\omega_o = \frac{1}{\sqrt{LC}}$

$$= \frac{1}{\sqrt{10^{-3}\times 10\times 10^{-6}}} = 10^4$$

$$e = 100\times\sqrt{\frac{10\times 10^{-6}}{10^{-3}}}\sin\left(\frac{1}{\sqrt{LC}}t\right)$$

$$i = 100\times\sqrt{\frac{10\times 10^{-6}}{1\times 10^{-3}}}\sin 10^4 t$$

$$= 10\sin(10^4 t)$$

11. Ans. (a)

Solution:

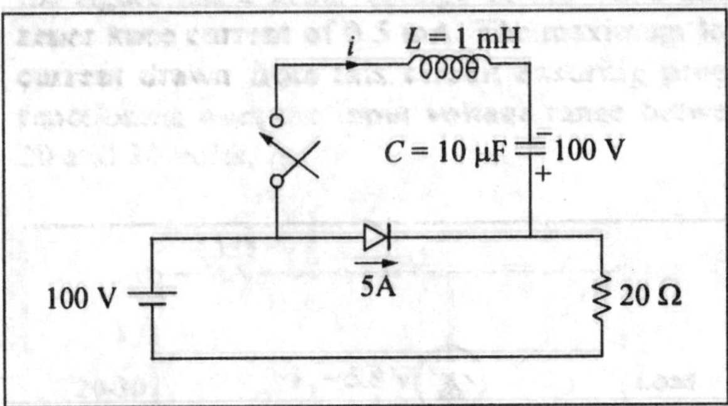

At $t = 0$ (switch closed)

So device SCR will be off, when this total 5 A current will decrese to zero, or will pass to commutation circuit using the result of Q 54

$$\therefore \quad 10\sin 10^4 t = 5$$

$$\Rightarrow \quad \sin 10^4 t = \frac{1}{2} = \sin 30^\circ$$

$$\Rightarrow \quad 0.10^4 t = 30^\circ = \frac{\pi}{6}$$

$$t = \frac{\pi}{6} \times 10^{-4}$$

$$\Rightarrow \quad = \frac{\pi \times 100}{6} \mu s$$

$$= 52\ \mu s$$

12. Ans. (c)

Solution:

Given $A = 20 \times 40\ cm^2$

$$E_{max} = 50\ kV/cm$$

$$C = \frac{\varepsilon A}{d}$$

$$= \frac{\varepsilon_o \varepsilon_r A}{d}$$

$$E_{max} = \frac{V_{max}}{d}$$

$$q_{max} = CV_{max} = \frac{\varepsilon.A}{d} \times V_{max}$$

$$= \varepsilon.A\left(\frac{V_{max}}{d}\right)$$

$$= \varepsilon_r \varepsilon_0 A \times E_{max}$$

$$= 2.26 \times 8.85 \times 10^{-14} \times 50 \times 10^3 \times 20 \times 40$$

$$= 8\mu C$$

7 Analog Electronics

7.1 Diodes Applications

1. The 6 V zener diode shown in Figure has zero zener resistance and a knee current of 5 mA. The minimum value of R so that the voltage across it does not fall below 6 V is

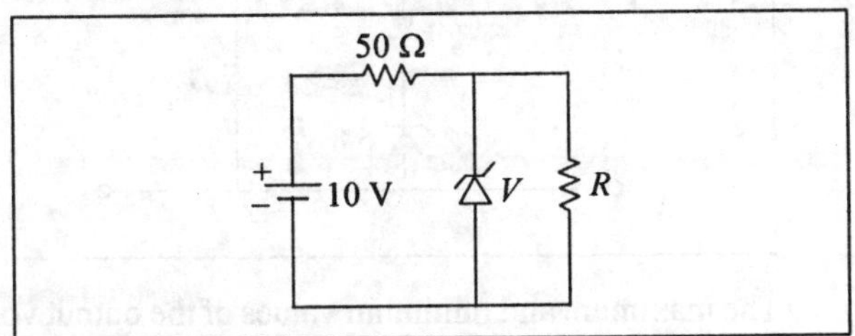

(a) 1.2 kΩ (b) 80 Ω
(c) 50 Ω (d) 0 Ω

2. A zener diode in the circuit shown in figure has a knee current of 5 mA and a maximum allowed power dissipation of 300 mW. What is the minimum and maximum load currents that can be drawn safely from the circuit, keeping the output voltage V_o constant at 6 V?

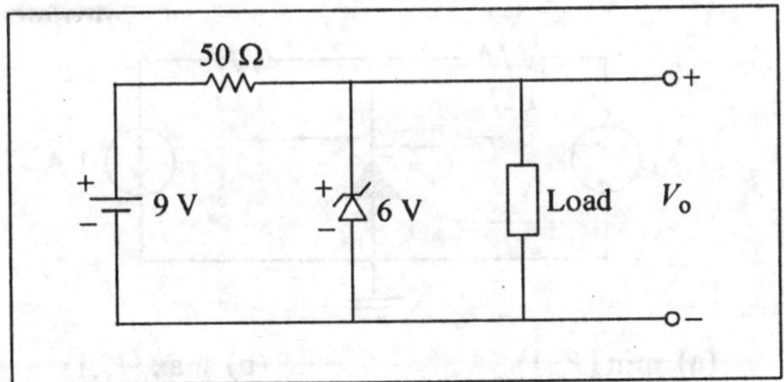

(a) 0 mA, 180 mA (b) 5 mA, 110 mA
(c) 10 mA, 55 mA (d) 60 mA, 180 mA

3. A dc power supply has a no-load voltage of 30V, and a full-load voltage of 25V at a full-load current of 1A. Its output resistance and load regulation, respectively are:

(a) 5 Ω and 20% (b) 25 Ω and 20 %
(c) 5 Ω and 16.7 % (d) 25 Ω and 16.7 %

4. A zener diode regulator in figure is to be designed to meet the specifications: I_L = 10 mA, V_0 = 10 V and V_{in} varies from 30 V to 50 V. The zener diode has V_z = 10 V and I_{zk} (knee current) = 1 mA. For satisfactory operation:

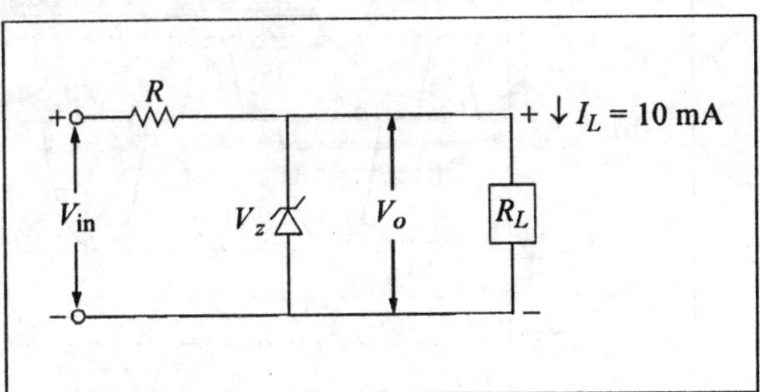

(a) $R \leq 1800\ \Omega$

(b) $2000\ \Omega \leq R \leq 2200\ \Omega$

(c) $3700\ \Omega \leq R \leq 4000\ \Omega$

(c) $R > 4000\ \Omega$

5. The zener diode in the regulator circuit shown in the figure has a zener voltage of 5.8 Volts and a zener knee current of 0.5 mA. The maximum load current drawn from this circuit ensuring proper functioning over the input voltage range between 20 and 30 volts, is:

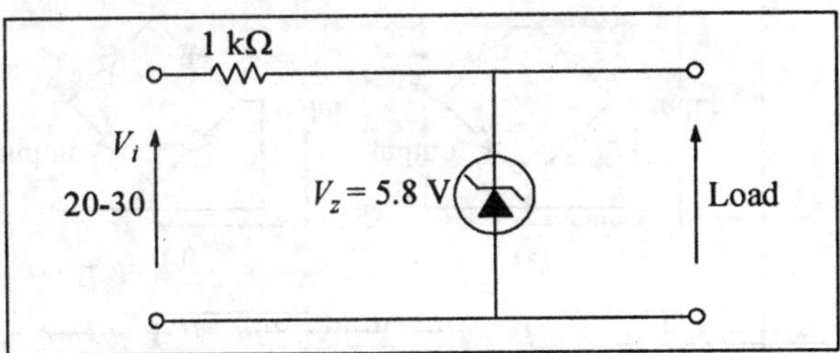

(a) 23.7 mA

(b) 14.2 mA.

(c) 13.7 mA

(d) 24.2 mA.

6. For the circuit shown below, assume that the zener diode is ideal with a breakdown voltage of 6 Volts. The waveform observed across R is:

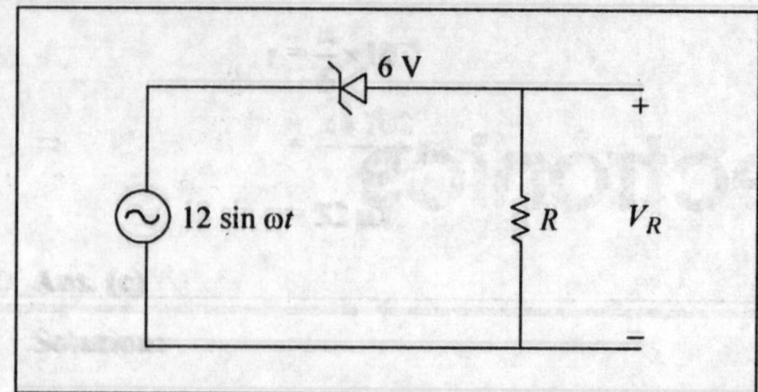

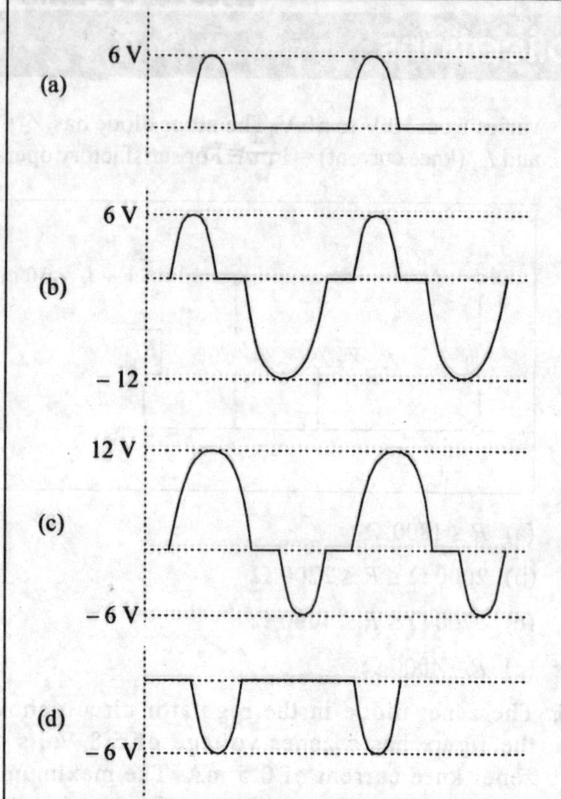

7. The correct full wave rectifier circuit is

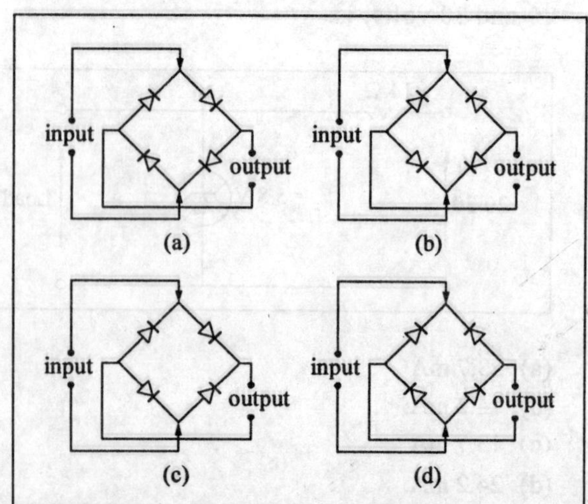

8. For the Zener diode shown in the figure, the Zener voltage at knee is 7 V, the knee current is negligible and the Zener dynamic resistance is 10 Ω. If the input voltage (V_i) range is from 10 to 16 V, the output voltage (V_o) ranges from

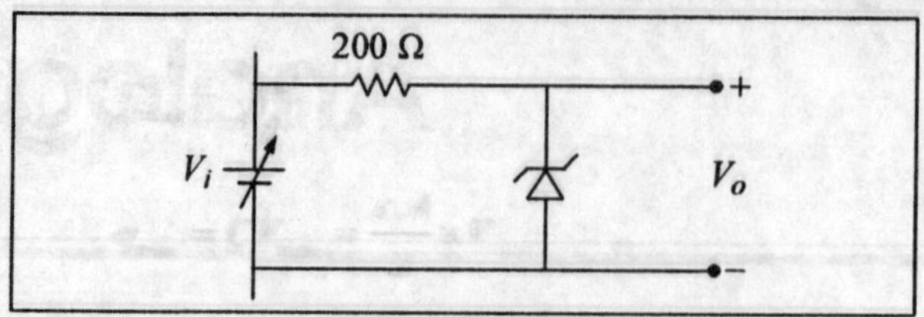

(a) 7.00 to 7.29 V (b) 7.14 to 7.29 V
(c) 7.14 to 7.43 V (d) 7.29 to 8.23 V

9. In the following limiter circuit, an input voltage $V_i = 10 \sin 100 \pi t$ is applied. Assume that the diode drop is 0.7 V when it is forward biased. The Zener breakdown voltage is 6.8 V.

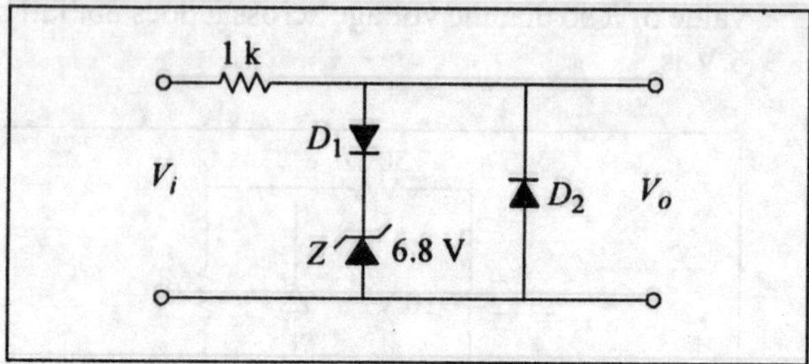

The maximum and minimum values of the output voltage respectively are

(a) 6.1 V, −0.7 V
(b) 0.7 V, −7.5 V
(c) 7.5 V, −0.7 V
(d) 7.5 V, −7.5 V

10. In the circuit below, the diode is ideal. The voltage V is given by

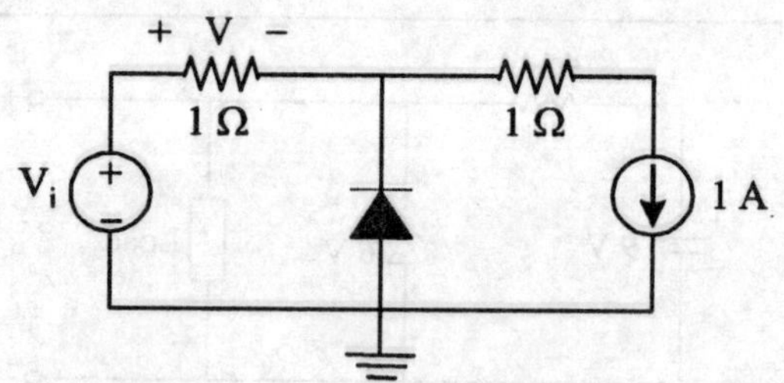

(a) min (V_i,1) (b) max (V_i,1)
(c) min ($-V_i$,1) (d) max ($-V_i$,1)

Linked Answer Questions

Statement for linked Answer Questions 11 and 12:

In the circuit shown below, assume that the voltatge drop across a forward biased diode is 0.7V. The thermal voltage $V_t = kT/q = 25$ mV. The small signal input $v_i = V_p \cos(\omega t)$ where $V_p = 10$ mV.

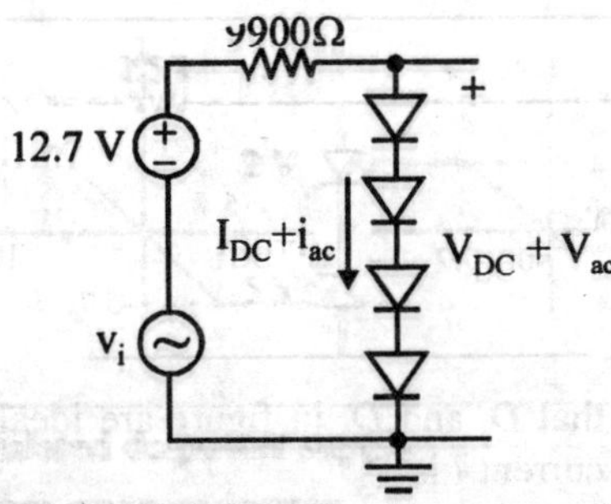

11. The bias current I_{DC} through the diodes is

(a) 1mA (b) 1.28 mA
(c) 1.5 mA (d) 2 mA

12. The ac output voltage v_{ac} is

(a) 0.25 cos (ωt) mV
(b) 1 cos (ωt) mV
(c) 22 cos (ωt) mV
(d) 2 cos (ωt) mV

13. The diode and capacitors in the circuit shown are ideal. The voltage $v(t)$ across the diode D_1 is

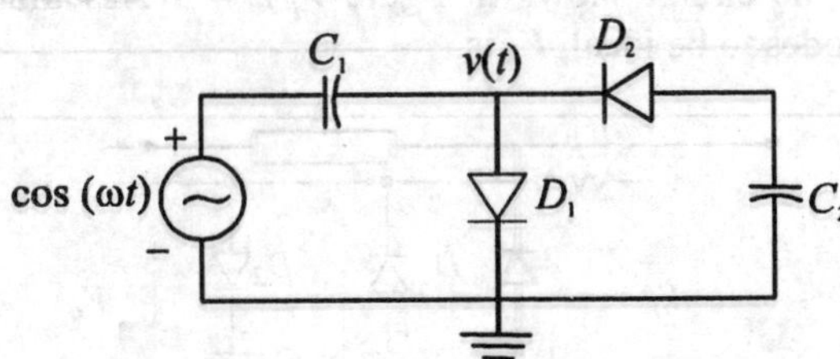

(a) $\cos(\omega t)-1$ (b) $\sin(\omega t)$
(c) $1-\cos(\omega t)$ (d) $1-\sin(\omega t)$

14. The *i-v* characteristics of the diode in the circuit given below are

$$i = \begin{cases} \frac{v-0.7}{500} A, & v \geq 0.7\,V \\ 0\,A, & v < 0.7\,V \end{cases}$$

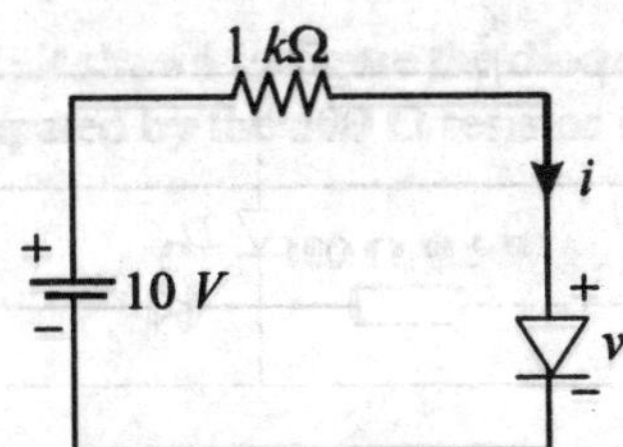

The current in the circuit is

(a) 10 *mA* (b) 9.3 *mA*
(c) 6.67 *mA* (d) 6.2 *mA*

15. A voltage signal 10 sin *wt* is applied to the circuit with ideal diodes, as shown in figure. The maximum, and minimum values of the output waveform V_{out} of the circuit are respectively.

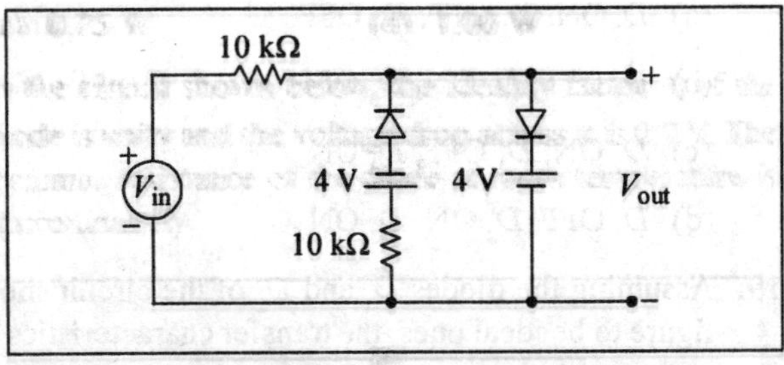

(a) + 10 V and – 10 V (b) + 4 V and – 4 V
(c) + 7 V and – 4 V (d) + 4 V and –7 V

16. The current through the Zener diode in figure is

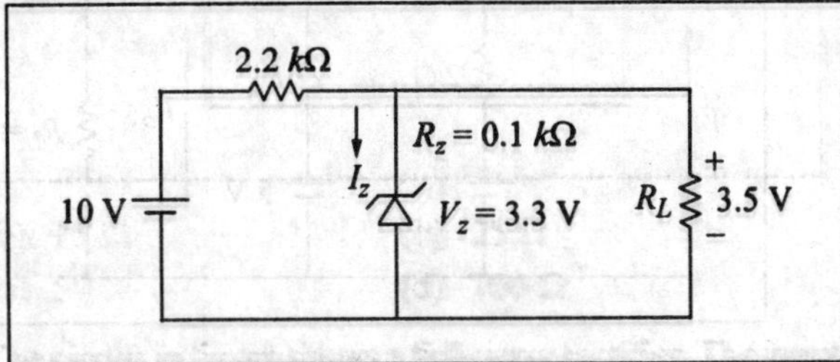

(a) 33 mA (b) 3.3 mA
(c) 2 mA (d) 0 mA

17. Assuming that the diodes are ideal in figure. The current in diode D_1 is

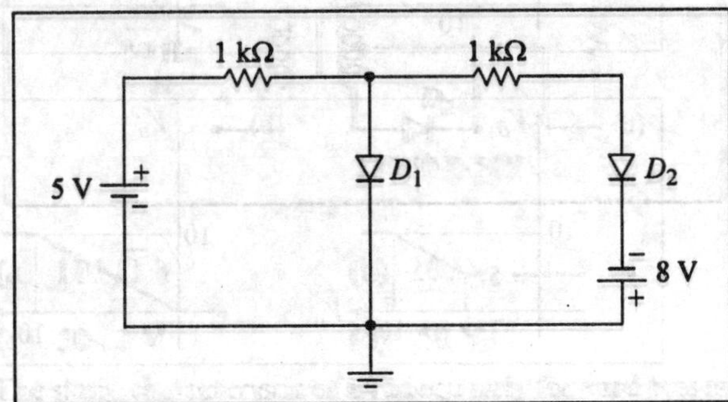

(a) 8 mA
(b) 5 mA
(c) 0 mA
(d) – 3 mA

18. What are the states of the three ideal diodes of the circuit shown in figure?

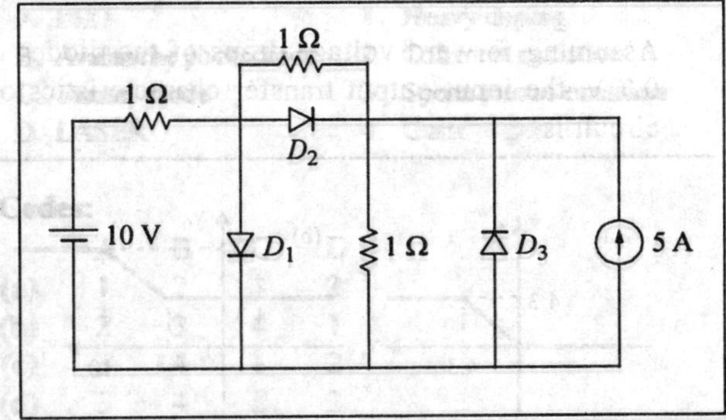

(a) D_1 ON, D_2 OFF, D_3 OFF

(b) D_1 OFF, D_2 ON, D_3 OFF

(c) D_1 ON, D_2 OFF, D_3 ON

(d) D_1 OFF, D_2 ON, D_3 ON

19. Assuming the diodes D_1 and D_2 of the circuit shown in figure to be ideal ones, the transfer characteristics of the circuit will be

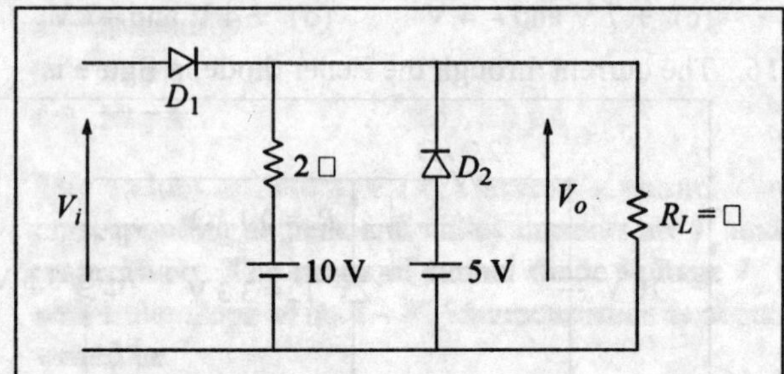

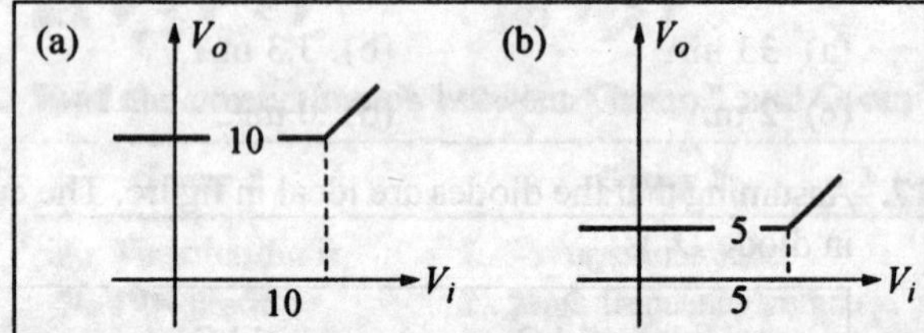

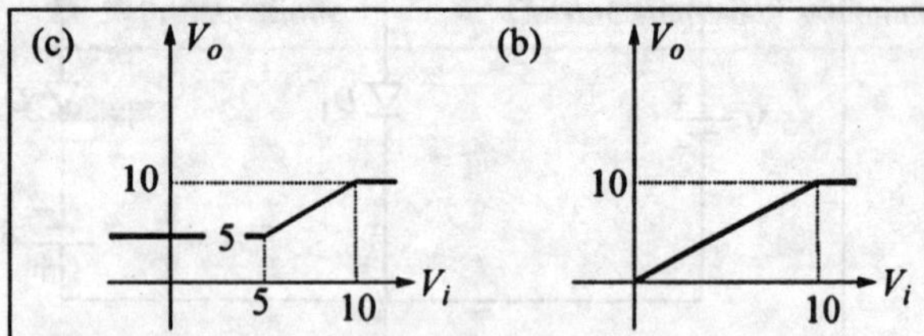

20. A clipper circuit is shown below.

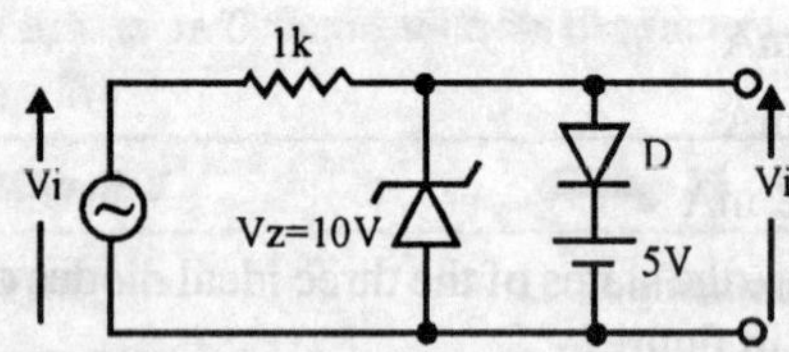

Assuming forward voltage drops of the diodes to be 0.7 V, the input-output transfer characteristics of the circuit is

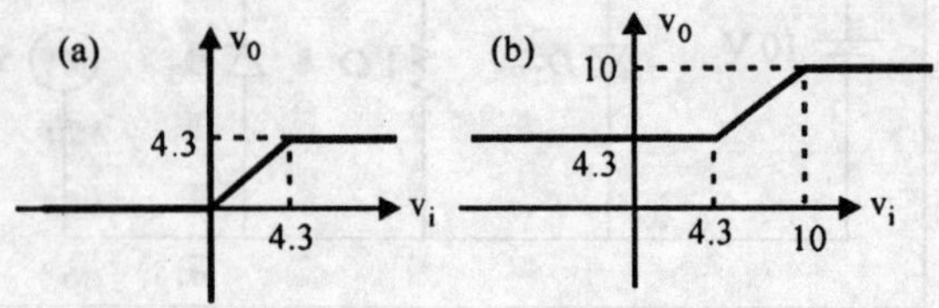

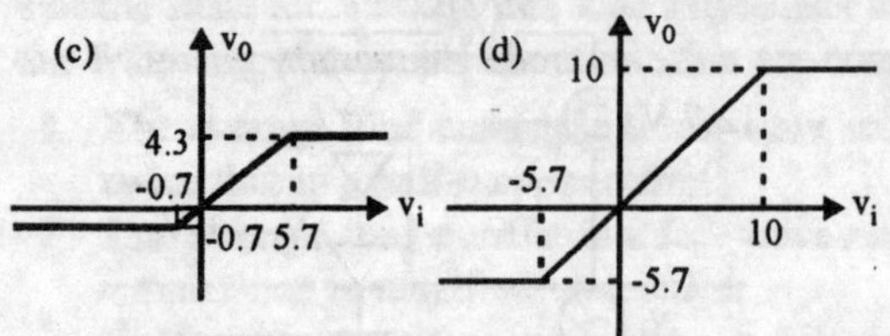

21. Assume that D_1 and D_2 in figure are ideal diodes. The value of current I is

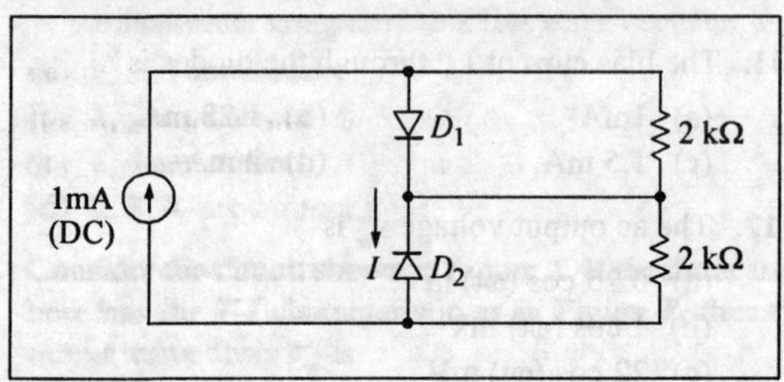

(a) 0 mA (b) 0.5 mA

(c) 1 mA (d) 2 mA

22. In the circuit shown in figure V_i is 4 V Assuming the diodes to be ideal, V_o is.

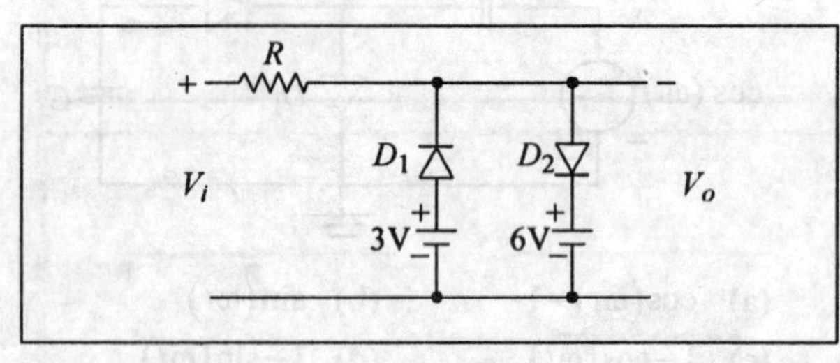

(a) 3 V (b) 4 V

(c) 4.5 V (d) 6 V

23. The input voltage, V_i is 4 + 3 sin ωt. assuming all elements to be ideal, the average of the output voltage V_o in figure is.

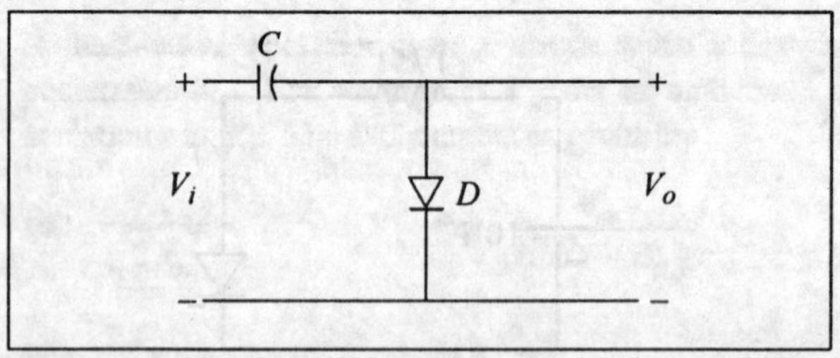

(a) − 3 V

(b) + 3 V

(c) − 7 V

(d) + 7 V

24. For a sinusoidal input 50 V amplitude, the circuit shown in figure can be used as

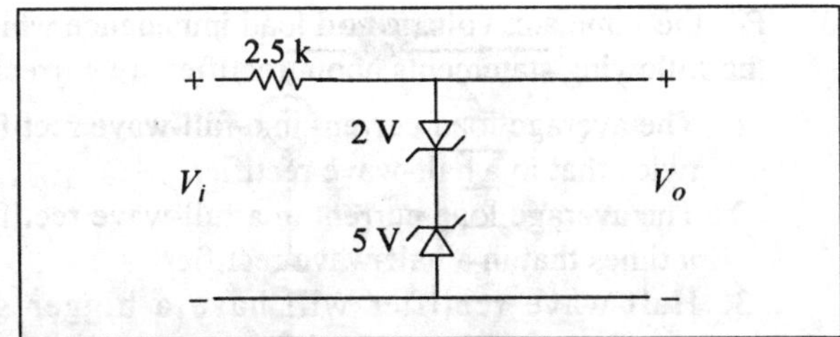

(a) Regulated dc power supply
(b) Square wave generator
(c) Half wave rectifier
(d) Full wave rectifier

25. A forward biased **silicion diode** when carrying negligible current, has a voltage drop of 0.64 V. When the current is 1 A it dissipates 1 W. The ON resistance of the diode is

(a) 0.36 Ω (b) 0.64 Ω
(c) 0.72 Ω (d) 1.0 Ω

26. Assuming ideal diode characteristics, the input/output voltage relationship for the circuit shown in figure is

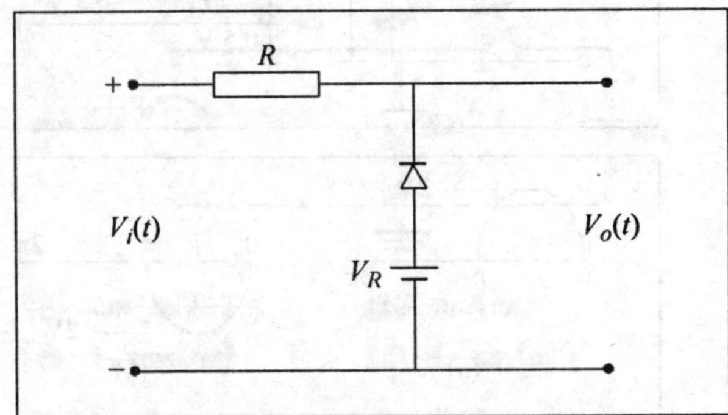

(a) $V_o(t) = V_i(t)$, for all $V_i(t)$
(b) $V_o(t) = V_i(t)$, for all $V_i(t) > V_R = 0$, other wise
(c) $V_o(t) = V_i(t)$, for $V_i(t) = V_R$, other wise
(d) $V_o(t) = V_i(t)$, for $V_i(t) > V_R = V_R$, other wise

27. For the circuit shown in figure the diode D is ideal. The power dissipated by the 300 Ω resistor is

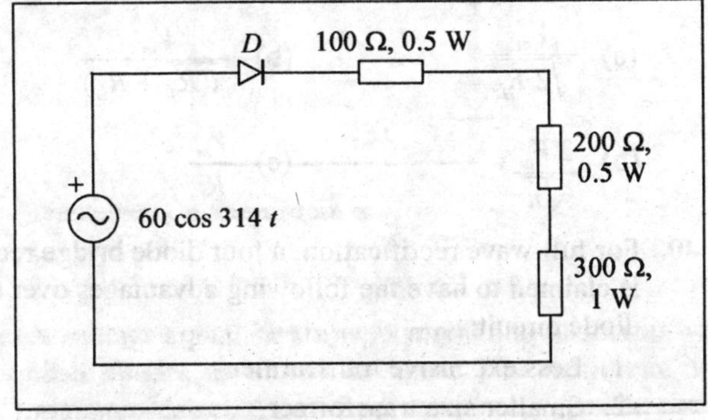

(a) 0.25 W (b) 0.50 W
(c) 0.75 W (d) 1.00 W

28. In the circuit shown below, the ideality factor η of the diode is unity and the voltage drop across it is 0.7 V. The dynamic resistance of the diode at room temperature is approximately

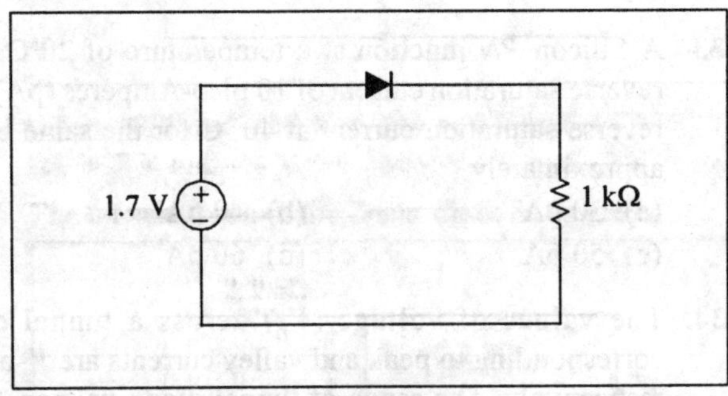

(a) 15 Ω (b) 25 Ω
(c) 50 (d) 700 Ω

29. The circuit in figure shows a full-wave rectifier. The input voltage is 230 V (rms) single-phase ac. The peak reverse voltage across the diodes D_1 and D_2 is

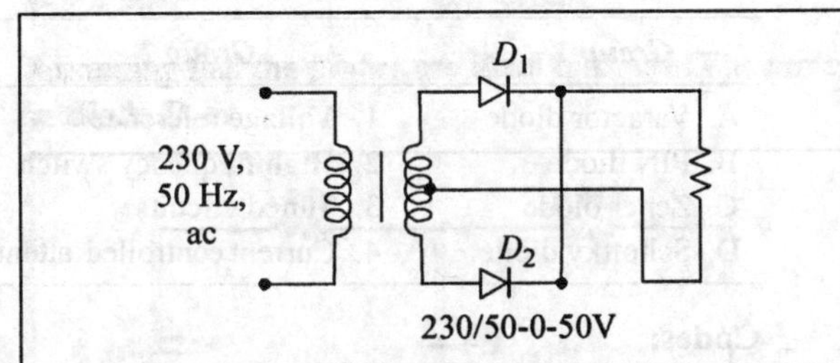

(a) $100\sqrt{2}$ V (b) 100 V
(c) $50\sqrt{2}$ V (d) 50 V

30. The static characteristic of an adequately forward biases *p-n* junction is a straight line, if the plot is of

(a) log I vs log V (b) log I vs V
(c) I vs log V (d) I vs V

31. Match items in Group 1 with items in Group 2, most suitably.

Group 1	*Group 2*
A. LED	1. Heavy doping
B. Avalanche photodiode	2. Coherent radiation
C. Tunnel diode	3. Spontaneous emission
D. LASER	4. Current gain

Codes:

	A	B	C	D
(a)	1	2	3	2
(b)	2	3	4	1
(c)	3	4	1	2
(d)	3	4	2	3

32. A particular green LED emits light of wavelength 5490 A. The energy band-gap of the semiconductor material used there is
(Planck's constant = 6.626×10^{-34} J-s)

(a) 2.26 eV (b) 1.98 eV
(c) 1.17 eV (d) 0.74 eV

33. A Silicon *PN* junction at a temperature of 20°C has a reverse saturation current of 10 pico-Amperes (pA). The reverse saturation current at 40° C for the same bias is approximately

(a) 30 pA (b) 40 pA
(c) 50 pA (d) 60 pA

34. The values of voltage (V_D) across a tunnel diode corresponding to peak and valley currents are V_p and V_v respectively. The range of tunnel diode voltage V_D for which the slope of its I – V_D characteristics is negative would be

(a) $V_D < 0$ (b) $0 \le V_D < V_p$
(c) $V_p \le V_D < V_v$ (d) $V_D \ge V_v$

35. Find the correct match between Group 1 and Group 2.

Group 1	*Group 2*
A. Varactor diode	1. Voltage reference
B. PIN diode	2. High frequency switch
C. Zener diode	3. Tuned circuits
D. Schottky diode	4. Current controlled attenuator

Codes:

	A	B	C	D
(a)	4	2	1	3
(b)	2	4	1	3
(c)	3	4	1	2
(d)	1	3	2	4

36. Group I List four different semiconductor devices. Match each device in Group I with its characteristic property in Group II

Group I	*Group II*
A. BJT	1. Population inversion
B. MOS capacitor	2. Pinch-off voltage
C. LASER diode	3. Early effect
D. JFET	4. Flat-band voltage

Codes:

	A	B	C	D
(a)	3	1	4	2
(b)	1	4	3	2
(c)	3	4	1	2
(d)	3	2	1	4

37. For the same a.c. voltage and load impedance which of the following statements about rectifier are correct?

1. The average load current in a full-wave rectifier is twice that in a half-wave rectifier.
2. The average load current in a full-wave rectifier is *n* times that in a half-wave rectifier.
3. Half-wave rectifier will have a bigger sized transforms compared to full wave-rectifier.
4. Half-wave rectifier will have a small sized transformer compared to a full wave-rectifier.

(a) 1, 2, 3 are correct
(b) 1, 3 are correct
(c) 1, 4 are correct
(d) 2, 3, 4, are correct

38. Consider the circuit shown in Figure *X*. If the diode used here has the *V-I* characteristic as in Figure *Y*, then the output wave form V_o is:

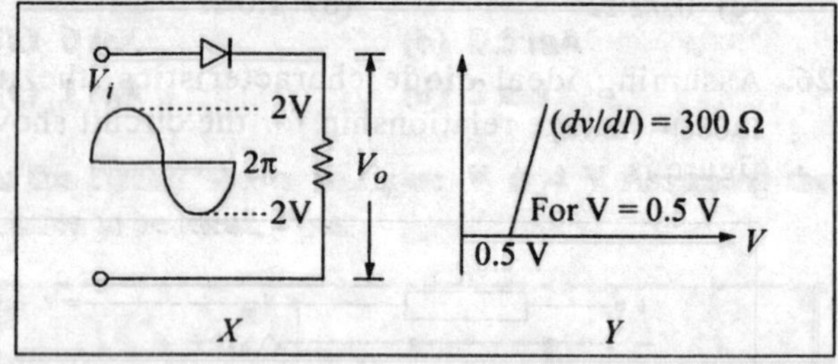

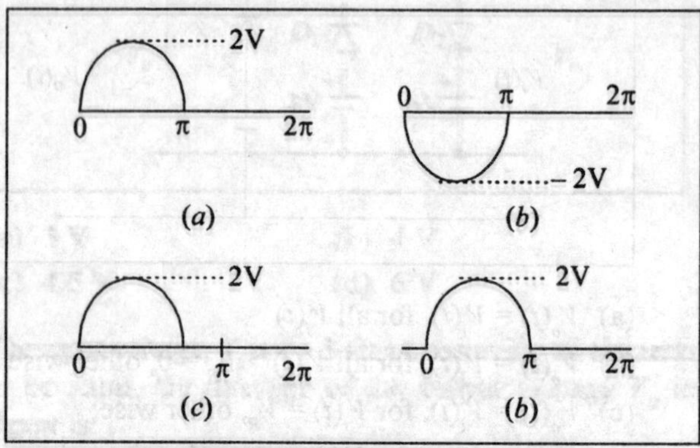

39. A half-wave rectifier uses a diode with a forward resistance R_f. The voltage is V_m sin ωt and the load resistance is R_L. The DC current is given by

(a) $\dfrac{V_m}{\sqrt{2}\, R_L}$ (b) $\dfrac{V_m}{\pi(R_f + R_L)}$

(b) $\dfrac{2\, V_m}{\sqrt{\pi}}$ (c) $\dfrac{V_m}{R_L}$

40. For full-wave rectification, a four diode bridge rectifier is claimed to have the following advantages over a two diode circuit:

1. Less expensive transformer
2. Smaller size transformer
3. Suitability for higher voltage application

Of these:

(a) Only (1) and (2) are true
(b) Only (1) and (3) are true
(c) Only (2) and (3) are true
(d) (1), (2) and (3) are true

41. For the circuit in figure the voltage V_o is:

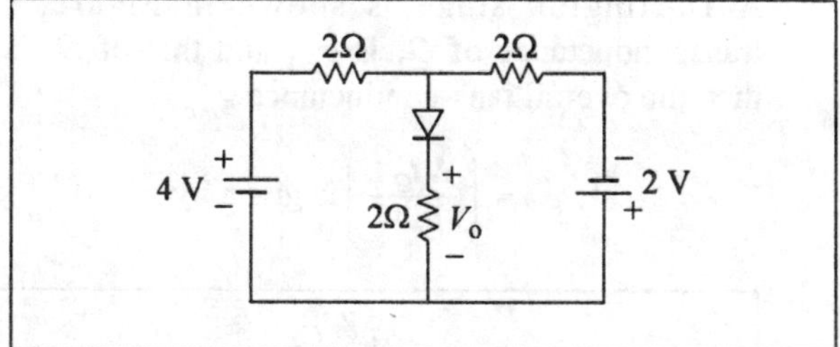

(a) 2 V
(b) 1 V
(c) −1 V
(d) None of the above

42. In a full-wave rectifier using two ideal diodes, V_{dc} and V_m are the dc and peak values of the voltage, respectively across a resistive load. If PIV is the peak inverse voltage of the diode, then the appropriate relationship for this rectifier are

(a) $V_{dc} = \left[\frac{V_m}{\pi}\right]$, PIV $= 2V_m$

(b) $V_{dc} = 2\left[\frac{V_m}{\pi}\right]$, PIV $= 2V_m$

(c) $V_{dc} = 2\left[\frac{V_m}{\pi}\right]$, PIV $= V_m$

(d) $V_{dc} = \left[\frac{V_m}{\pi}\right]$, PIV $= V_m$

43. The wave-shape of V_o in Figure will be

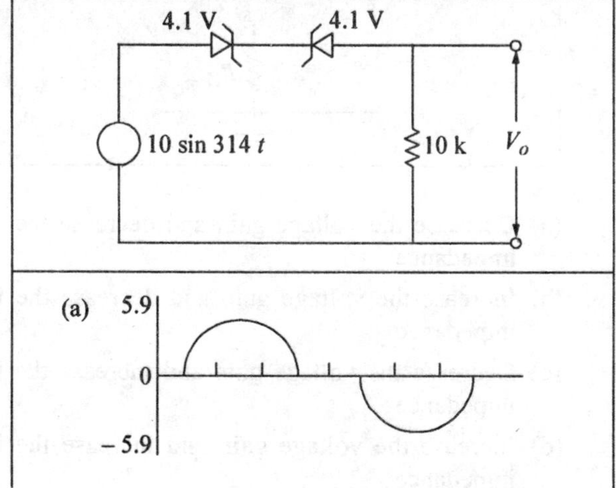

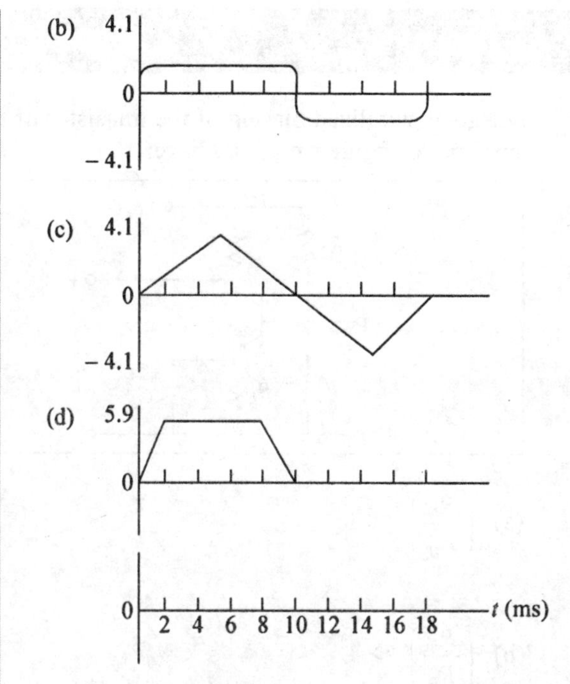

44. For the circuit shown below, assume that the zener diode is ideal with a breakdown voltage of 6 Volts. The waveform observed across R is:

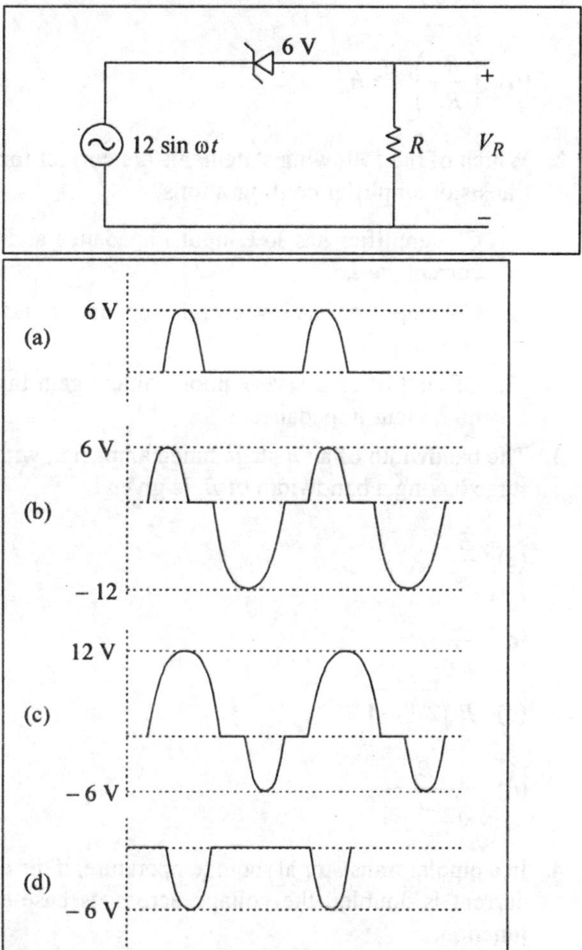

7.2 Bipolar Junction Transistor

1. For good stabilized biasing of the transistor of the CE amplifier of figure we should have:

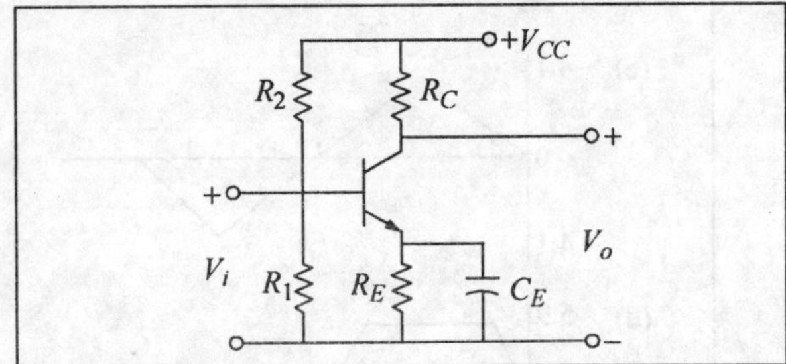

(a) $\left(\frac{R_E}{R_B}\right) << 1$

(b) $\left(\frac{R_E}{R_B}\right) >> 1$

(c) $\left(\frac{R_B}{R_E}\right) << h_{ie}$

(d) $\left(\frac{R_B}{R_E}\right) >> h_{ie}$

2. Which of the following statements are correct for basic transistor amplifier configurations?
   1. CB amplifier has low input impedance and a low current gain.
   2. CC amplifier has low output impedance and a low current gain.
   3. CE amplifier has very poor voltage gain but very high input impedance

3. The bandwidth of an n-stage tuned amplifier, with each stage having a bandwidth of B, is given by

(a) $\frac{B}{n}$

(b) $\frac{B}{\sqrt{n}}$

(c) $B\sqrt{2^{1/n}-1}$

(d) $\frac{B}{\sqrt{2^{1/n}-1}}$

4. In a bipolar transistor at room temperature, if the emitter current is doubled the voltage across its base-emitter junction.

(a) Doubles

(b) Halves

(c) Increases by about 20 mV

(d) Decreases by about 20 mV

5. **A Darlington stage is shown in Figure. If the transconductance of Q_1 is g_{m1} and that of Q_2 is G_{m2}, then the overall tans-conductance g_{mc}**

$$= \left(\frac{\Delta i_C^C}{V_{be}^C}\right) \text{ is given by}$$

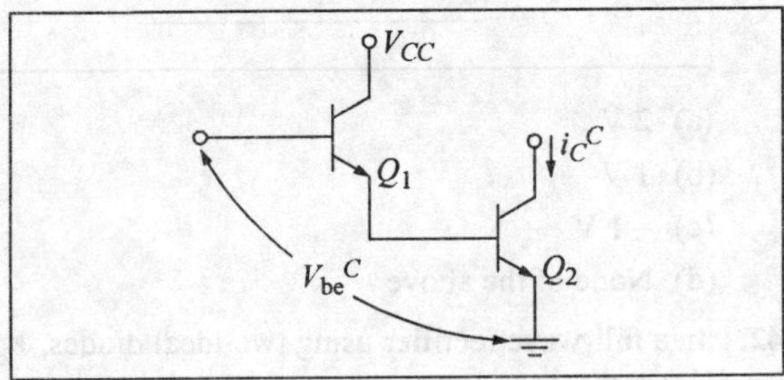

(a) g_{m1}

(b) 0.5 g_{m1}

(c) g_{m2}

(d) 0.5 g_{m2}

6. **In the BJT amplifier in Figure, the transistor is biased in the forward active region. Putting a capacitor across R_E will.**

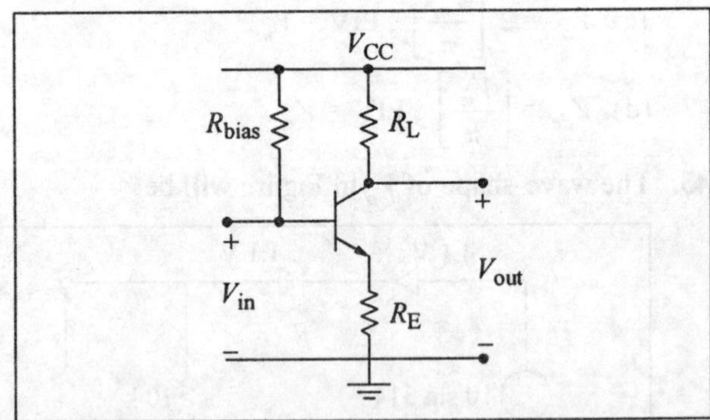

(a) Decrease the voltage gain and decrease the input impedance

(b) Increase the voltage gain and decrease the input impedance

(c) Decrease the voltage gain and increase the input impedance

(d) Increase the voltage gain and increase the input impedance

7. A cascode amplifier stage is equivalent to
   (a) A common emitter stage followed by a common base stage
   (b) A common base followed by an emitter follower
   (c) An emitter follower stage followed by a common base stage
   (d) A common base stage followed by a common emitter stage

8. The circuit of Figure is an example of feedback of the following type

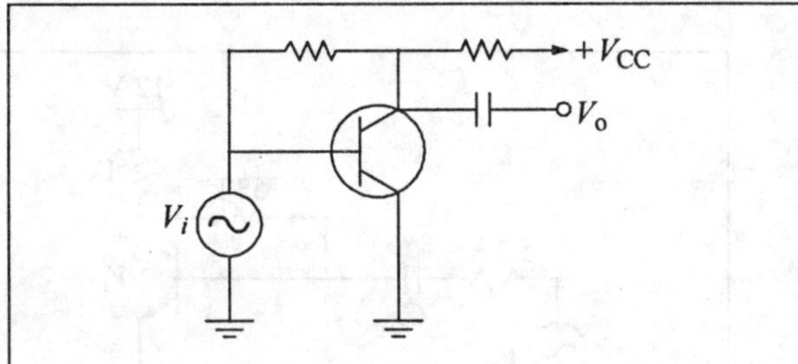

   (a) Current series
   (b) Current shunt
   (c) Voltage series
   (d) Voltage shunt

9. Introducing a resistor in the emitter of a common emitter amplifier stabilizes the dc operating point against variations in
   (a) Only the temperature
   (b) Only the β of the transistor
   (c) Both temperature and β
   (d) None of the above

10. In the circuit of figure assume that the transistor is in active region. It has a large β and its base-emitter voltage is 0.7 V. The value of I_C is:

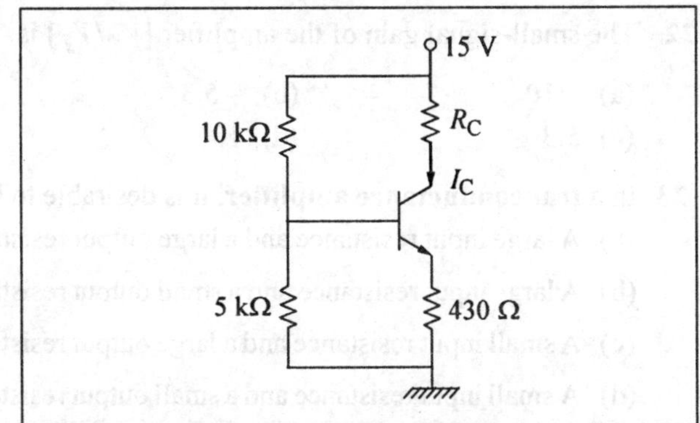

   (a) Indeterminate since R_C is not given
   (b) 1 mA
   (c) 5 mA
   (d) 10 mA

11. Choose the correct match for input resistance of various amplifier configurations shown below:

Configuration	Input resistance
CB: Common Base	*LO*: Low
CB: Common Collector	*MO*: Moderate
CE : Common Emitter	*HI*: High

   (a) *CB-LO, CC-MO, CE-HI*
   (b) *CB-LO, CC-HI, CE-MO*
   (c) *CB-MO, CC-HI, CE-LO*
   (d) *CB-HI, CC-LO, CE-MO*

12. In the amplifier circuit shown in figure the values of R_1 and R_2 are such that the transistor is operating at V_{CE} = 3 V and I_C= 1.5 mA when its β is 150. For a transistor with β of 200, the operating point (V_{CE} I_C) is:

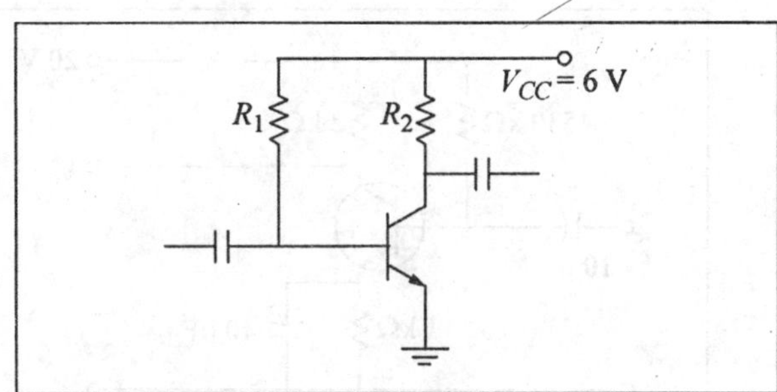

   (a) 2V, 2 mA
   (b) 3V, 2 mA
   (c) 4V, 2 mA
   (d) 4V, 1 mA

13. Three identical amplifiers with each one having a voltage gain of 50, input resistance of 1 kΩ and output resistance of 250 Ω, are cascaded. The open circuit voltage gain of the combined amplifier is
   (a) 49 dB (b) 51 dB
   (c) 98 dB (d) 102 dB

14. A bipolar transistor is operating in the active region with a collector current of 1mA. Assuming that the β of the transistor is 100 and thermal voltage (V_T) is 25 mV, the trans-conductance (g_m) and the input resistance (r_π) of the transistor in the common emitter configuration, are
   (a) g_m = 25 mA/V and r_π = 15.625 kΩ
   (b) g_m = 40 mA/V and r_π = 4.0 kΩ
   (c) g_m = 25 mA/V and r_π = 2.5 kΩ
   (d) g_m = 40 mA/V and r_π = 2.5 kΩ

15. Assuming $V_{CE\,sat} = 0.2$ V and $\beta = 50$, the minimum base current (I_B) required to drive the transistor in Figure to a saturation is:

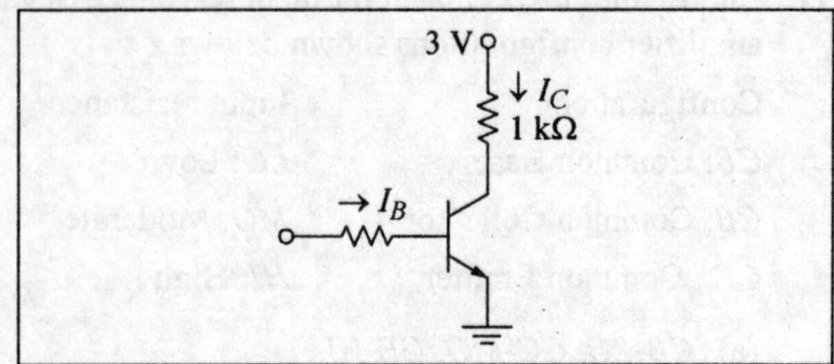

(a) 56 μA (b) 140 mA
(c) 60 μA (d) 3 mA

16. The cascode amplifier is a multistage configuration of
(a) *CC-CB* (b) *CE-CB*
(c) *CB-CC* (d) *CE-CC*

17. The circuit using a BJT with = 50 and $V_{BE} = 0.7$ V is shown in the figure The base current I_B and collector voltage by V_C are respectively

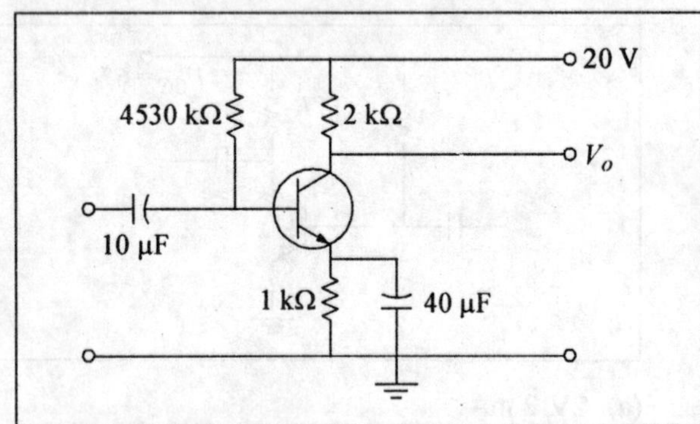

(a) 43 μA and 11.4 Volts
(b) 40 μA and 16 Volts
(c) 45 μA and 11 Volts
(d) 50 μA and 10 Volts

18. For an npn transistor connected as shown in the figure,

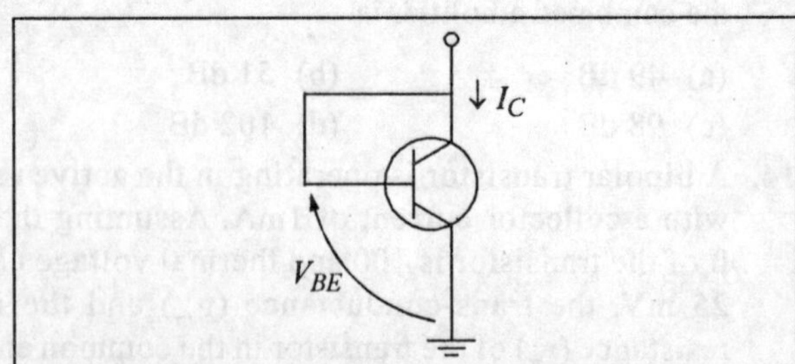

$V_{BE} = 0.7$ volts. Given that reverse saturation current of the junction at room temperature 300° K is 10^{-13} A, the emitter current is

(a) 30 mA (b) 39 mA
(c) 49 mA (d) 20 mA.

19. The input impedance (Z_i) and the output impedance (Z_o) of an ideal trans conductance (voltage controlled current source) amplifier are:

(a) $Z_i = 0, Z_o = 0$ (b) $Z_i = 0, Z_o = \infty$
(c) $Z_i = \infty, Z_o = 0$ (d) $Z_i = \infty, Z_o = \infty$

Common Data for Questions 20, 21, 22

In the transistor amplifier circuit shown in the figure below, the transistor has following parameters: $\beta_{DC} = 60$, $V_{BE} = 0.7$ V, $h_{ie} \to \infty$, $h_{fe} \to \infty$. The capacitance C_C can be assumed to be infinite.

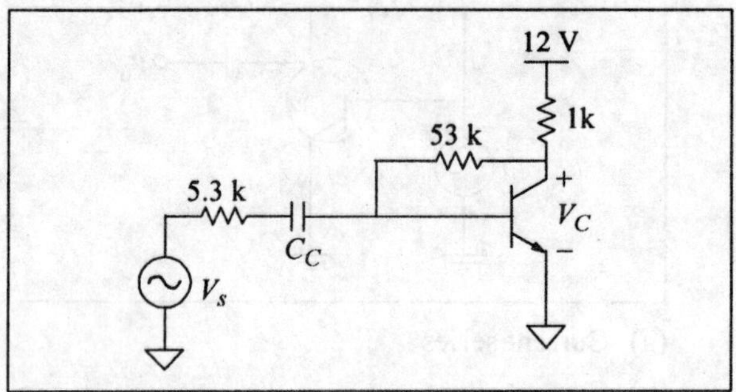

In the figure above, the ground has been shown by the symbol ∇

20. Under the DC conditions, the collector-to-emitter voltage drop is

(a) 4.8 Volts (b) 5.3 Volts
(c) 6.0 Volts (d) 6.6 Volts

21. If β_{DC} is increased by 10%, the collector-to-emitter voltage drop

(a) Increases by less than or equal to 10%
(b) Decreases by less than or equal to 10%
(c) Increases by more than 10%
(d) Decreases by more than 10%

22. The small-signal gain of the amplifier $[V_C/V_s]$ is

(a) – 10 (b) – 5.3
(c) 5.3 (d) 10

23 In a **tranconductance amplifier**, it is desirable to have
(a) A large input resistance and a large output resistance
(b) A large input resistance and a small output resistance
(c) A small input resistance and a large output resistance
(d) A small input resistance and a small output resistance

24. For the BJT circuit shown, assume that the β of the transistor is very large and $V_{BE} = 0.7$ V. The mode of operation of the BJT is:

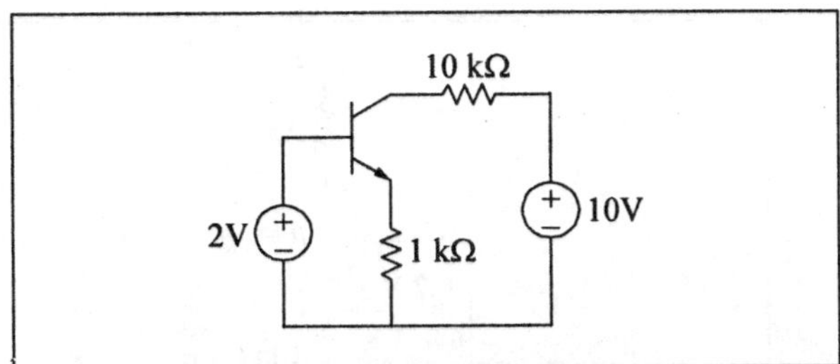

(a) Cut-off (b) Saturation
(c) Normal active (d) Reverse active

25. In the folloiwng transistor circuit $V_{BE} = 0.7\text{V}$, $r_e = 25\text{mV}/I_E$, and b and all the capacitance are very large

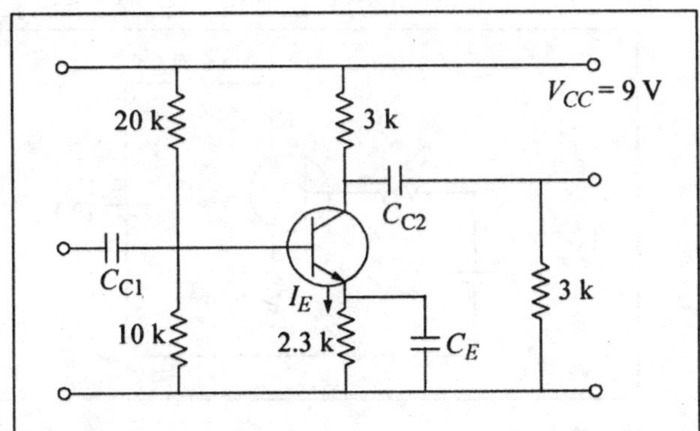

The value of DC current I_E is

(a) 1 mA
(b) 2 mA
(c) 5 mA
(d) 10 mA

26. In the folloiwng transistor circuit $V_{BE} = 0.7\text{V}$, $r_e = 25\text{mV}/I_E$, and b and all the capacitance are very large

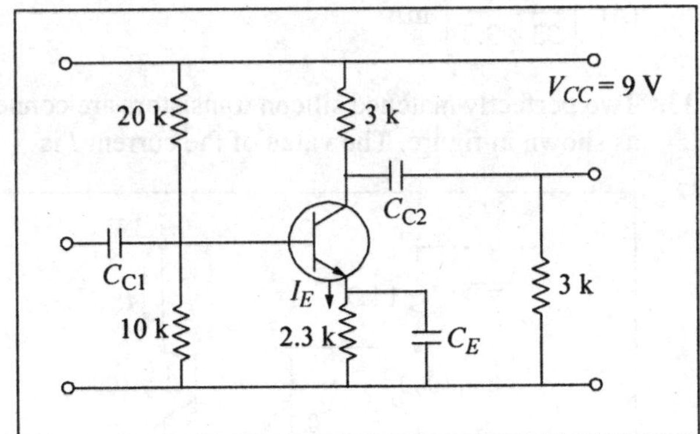

The mid-band voltage gain of the amplifier is approximately

(a) – 180
(b) – 120
(c) – 90
(d) – 60

27. A small signal source $V_i(t) = A\cos 20t + B\sin 10^6 t$ is applied to a transistor amplifier as shown below. The transistor has $\beta = 150$ and $h_{ie} = 3\ k\Omega$. Which expression best approximates $V_o(t)$?

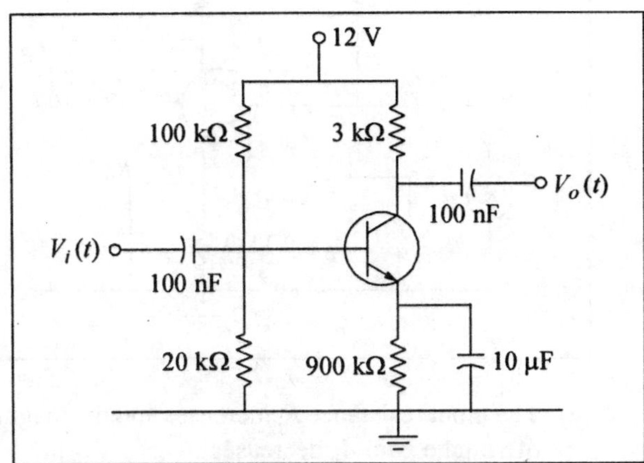

(a) $V_o(t) = -1500\,(A\cos 20t + B\sin 10^6 t)$
(b) $V_o(t) = -150\,(A\cos 20t + B\sin 10^6 t)$
(c) $V_o(t) = -1500\,B\sin 10^6 t$
(d) $V_o(t) = -150\,B\sin 10^6 t$

28. In the silicon BJT circuit shown below. assume that the emitter area of transistor Q_1 is half that of transistor Q_2.

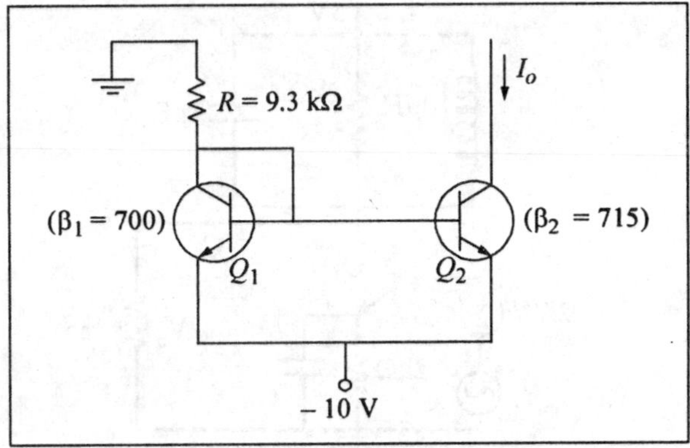

The value of current I_o is approximately

(a) 0.5 mA
(b) 2 mA
(c) 9.3 mA
(d) 15 mA

29. The amplifier circuit shown below uses a silicon transistor. The capacitors C_C and C_E can be assumed to be short at signal frequency and the effect of output resistance r_o can be ignored. If C_E is disconnected from the circuit, which one of the following statements is TRUE?

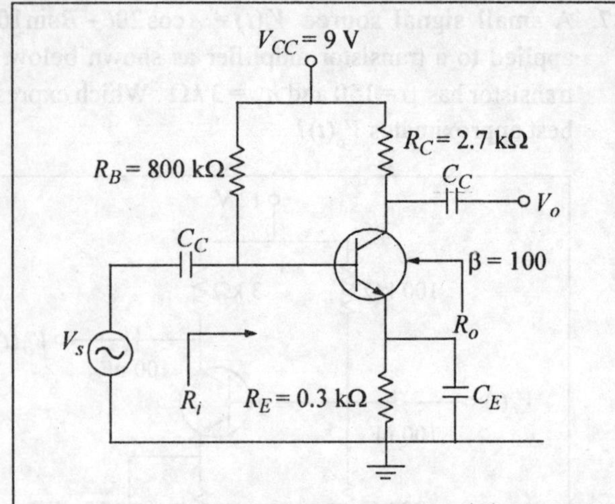

(a) The input resistance R_i increases and the magnitude of voltage gain A_v decreases
(b) The input resistance R_i decreases and the magnitude of voltage gain A_v increases
(c) Both input resistance R_i and the magnitude of voltage gain A_v decrease
(d) Both input resistance R_i and the magnitude of voltage gain A_v increase

30. In the circuit shown below, capacitors C_1 and C_2 are very large and are short at the input frequncy. v_i is a small signal input. The gain magniutde $|v_o/v_i|$ at 10 Mrad/s is

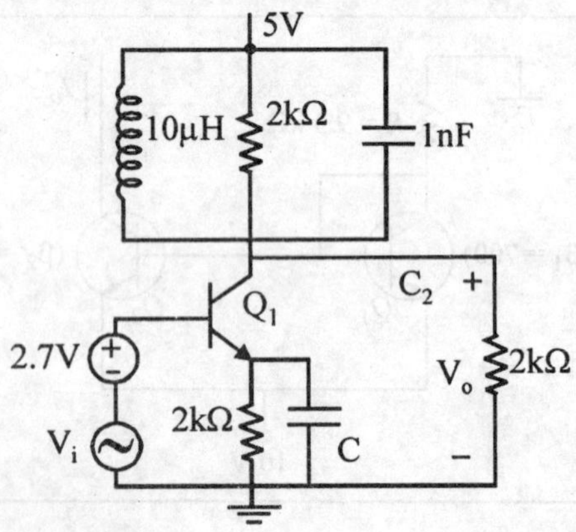

(a) Maximum
(b) Minimum
(c) Unity
(d) Zero

31. In the circuit of Figure, the value of the base current I_B will be

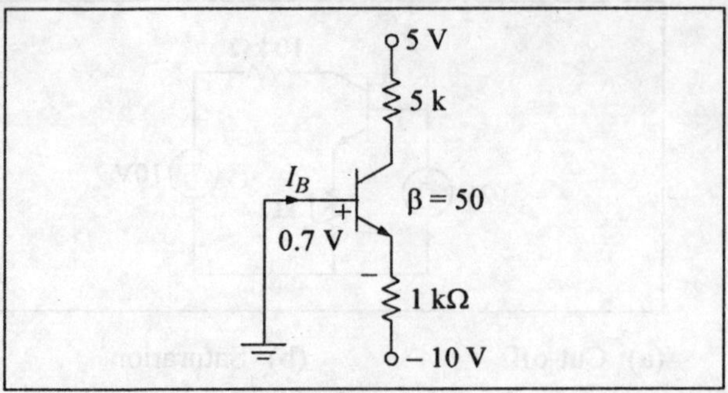

(a) 0.0 micro amperes
(b) 18.2 micro amperes
(c) 26.7 micro amperes
(d) 40.0 micro amperes

32. In the circuit of figure assume that the transistor has $h_{FE} = 99$ and $V_{BE} = 0.7$ V. The value of collector current I_C of the transistor is approximately

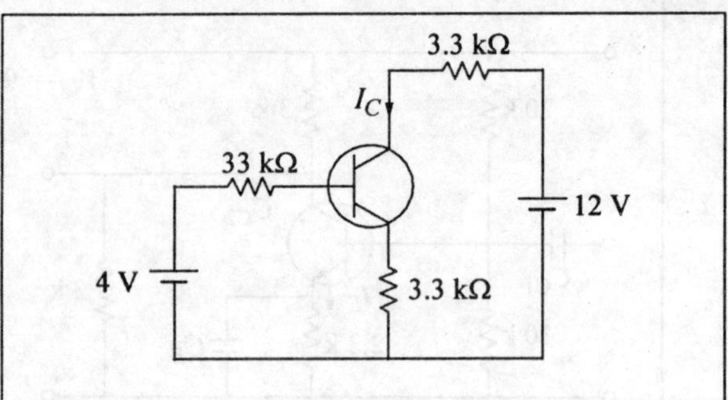

(a) $\left[\frac{33}{3.3}\right]$ mA

(b) $\left[\frac{3.3}{3.3+3.3}\right]$ mA

(c) $\left[\frac{3.3}{33}\right]$ mA

(d) $\left[\frac{3.3}{33+3.3}\right]$ mA

33. Two perfectly matched silicon transistors are connected as shown in figure. The value of the current I is

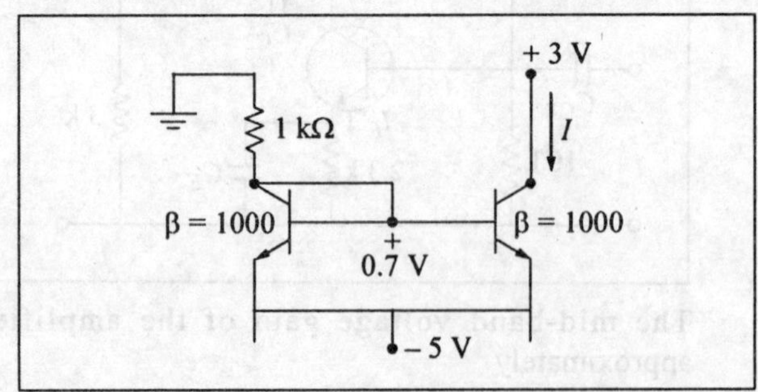

(a) 0 mA
(b) 2.3 mA
(c) 4.3 mA
(d) 7.3 mA

34. The common emitter amplifier shown in the figure is biased using a 1 mA ideal current source. The approximate base current value is

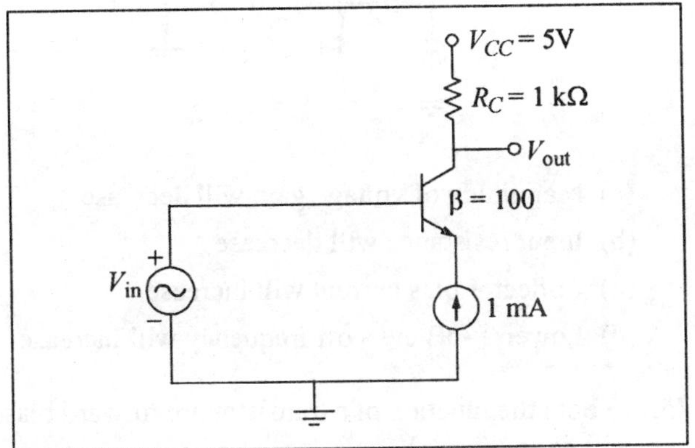

(a) 0 μA (b) 10 μA
(c) 100 μA (d) 1000 μA

35. Consider the circuit shown in figure. If the β of the transistor is 30 and I_{CBO} is 20 nA and the input voltage is + 5 V, the transistor would be operating in

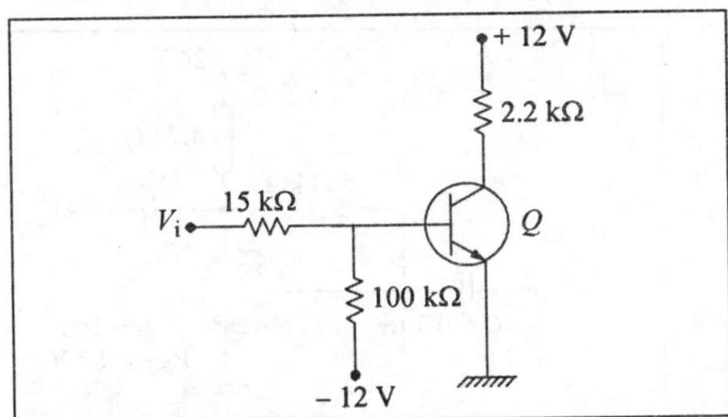

(a) Saturation region (b) Active region
(c) Breakdown region (d) Cut-of region

36. The common emitter forward current gain of the transistor shown is β_F = 100

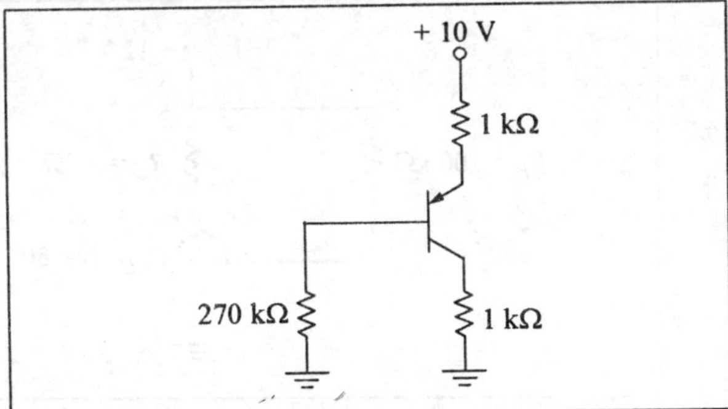

The transistor is operating in

(a) Saturation region
(b) Cutoff region
(c) Reverse active region
(d) Forward active region

37. The three-terminal linear voltage regulator is connected to a 10 Ω load resistor as shown in the figure. If V_{in} is 10 V, what is the power dissipated in the transistor?

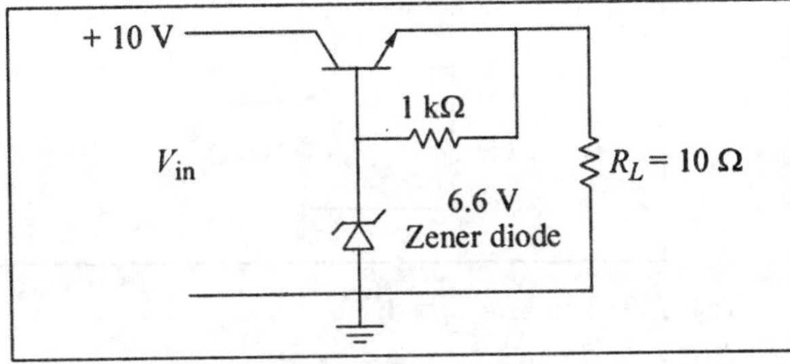

(a) 0.6 W
(b) 2.4 W
(c) 4.2 W
(d) 5.4 W

38. In the transistor circuit shown in Figure below, collector-to-ground voltage is + 20 V. Which of the following is the probable cause of error?

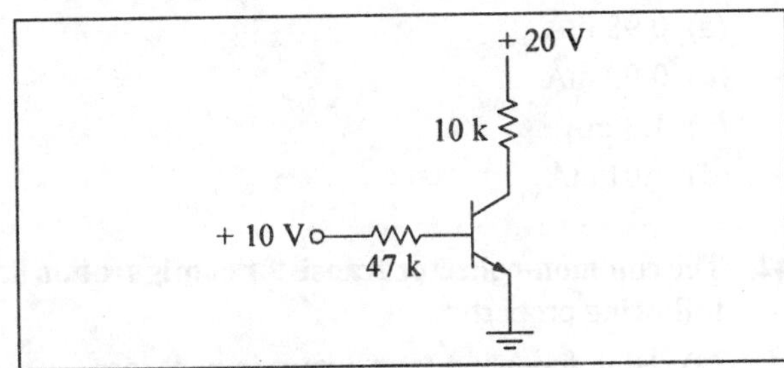

(a) Collector-emitter terminals shorted
(b) Emitted to ground connection open
(c) 10 KΩ resistor open
(d) collector-base terminals shorted

39. If the transistor in Figure is in saturation, then

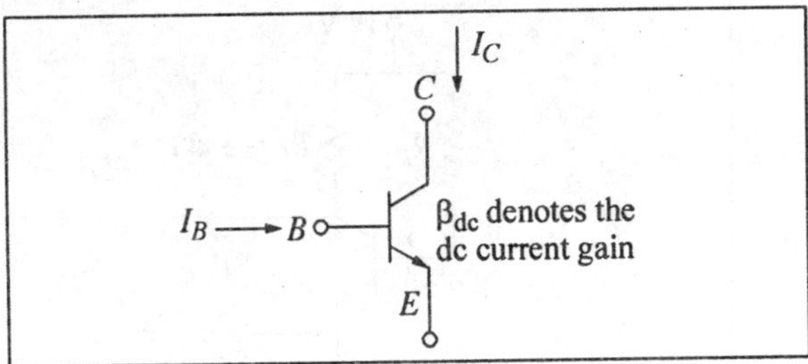

(a) I_C is always equal to $\beta_{dc} I_B$
(b) I_C is always equal to $-\beta_{dc} I_B$
(c) I_C is greater than or equal to $\beta_{dc} I_B$
(d) I_C is less than or equal to $\beta_{dc} I_{B1}$

40. Assuming that the β of the transistor is extremely large and V_{BE}=0.7 V, I_C and V_{CE} in the circuit shown in figure are:

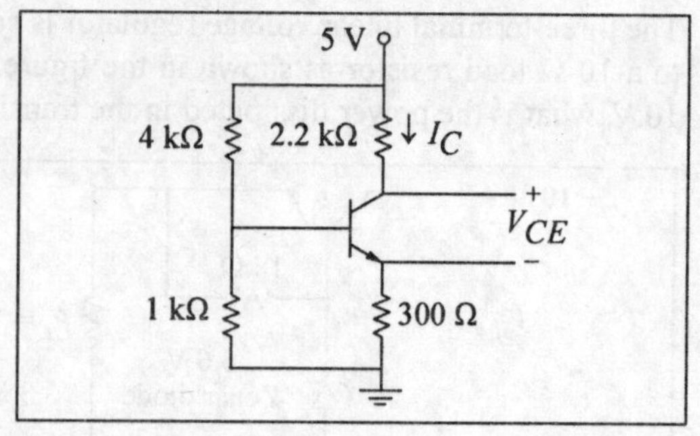

(a) $I_C = 1$ mA, $V_{CE} = 4.7$ V
(b) $I_C = 0.5$ mA, $V_{CE} = 3.75$ V
(c) $I_C = 1$ mA, $V_{CE} = 2.5$ V
(d) $I_C = 0.5$ mA, $V_{CE} = 3.9$ V

41. For a BJT, the common - base current gain a = 0.98 and the collector base junction reverse bias satruation current I_{co} = 0.6μA. This BJT is connected in the common emitter mode and operated in the active region with a base drive current I_B = 20μA. The collector current I_c for this mode of operation is

(a) 0.98 mA
(b) 0.99 mA
(c) 1.0 mA
(d) 1.01 mA

42. The **common-collector transistor configuration** has the following property:

(a) High input and Low output impedances
(b) High input and High output impedances
(c) Low input and Low output impedances
(d) Low input and High output impedances

43. For a silicon BJT shown in figure find R_B to establish V_{CE} = 2V, Assume V_{BE} = 0.7 V.

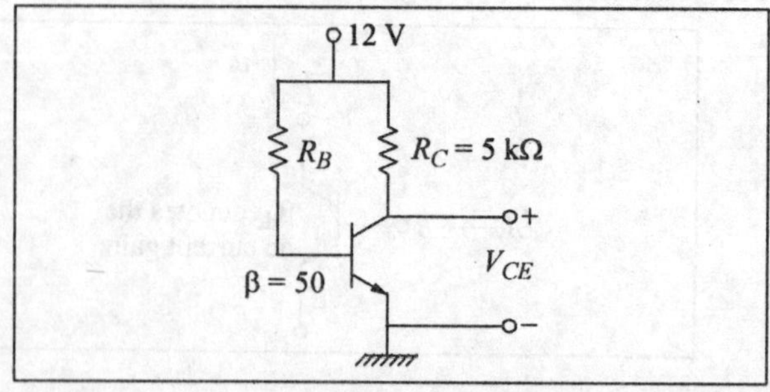

(a) 283 kΩ (b) 107 kΩ
(c) 200 kΩ (c) 242 kΩ

44. The amplifier shown below has a voltage gain of –2.5, an input resistance of 10 kΩ and a lower 3-dB cut - off frequency of 20 Hz. Whcih one of the following statements is **TRUE** when the emitter resistance R_E is doubled?

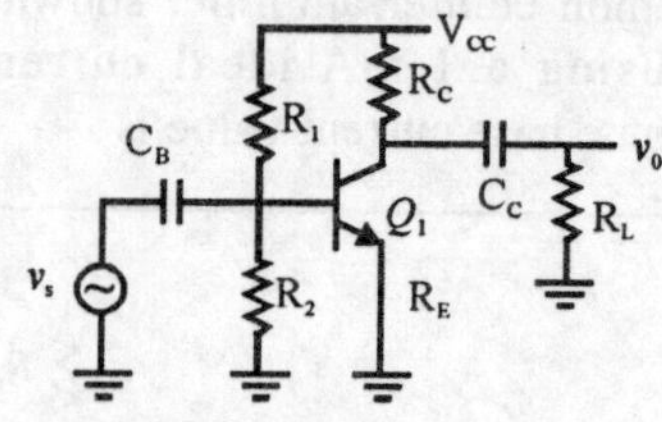

(a) Magnitude of voltage gain will decrease
(b) Input resistance will decrease
(c) Collector bias current will increase
(d) Lower 3 -dB cut - off frequency will increase

45. IF both the junction of a transistor are forward biased, it will be in

(a) Saturation mode (b) Active mode
(c) Cut off mode (d) ????

46. A transistor amplifier circuit is shown in figure. The quiescent collector current, rounded off to first decimal, is

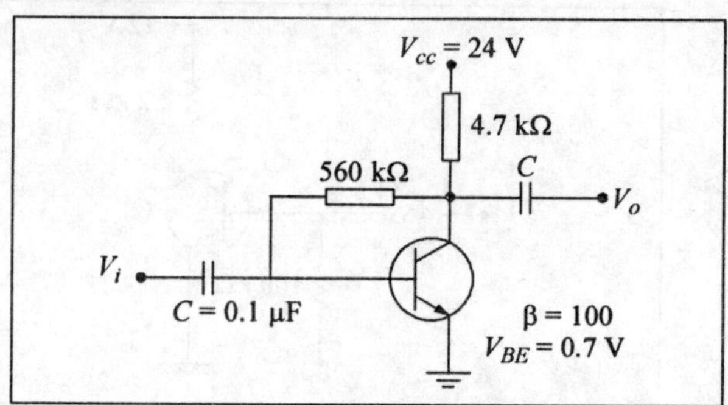

(a) 2.6 m A (b) 2.3 mA
(c) 2.1 m A (d) 2.0 m A

47. The biasing circuit of a silicon transistor is shown below. If β = 80, then what is V_{CE} for the transistor?

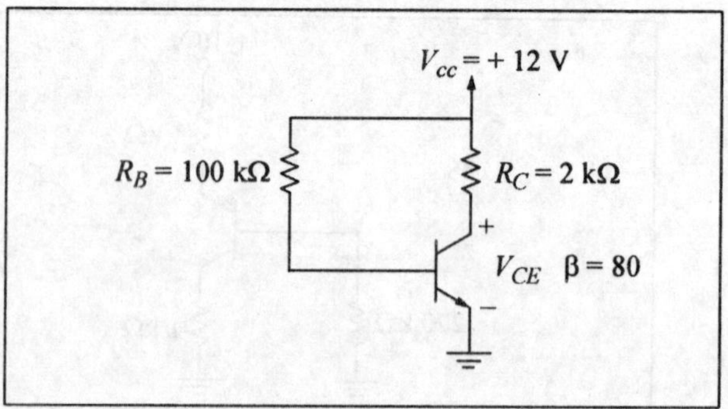

(a) – 6.08 V
(b) 0.2 V
(c) 1.2 V
(d) 6.08 V

48. In the circuit shown below, $V_{BE} = 0.7$ V.

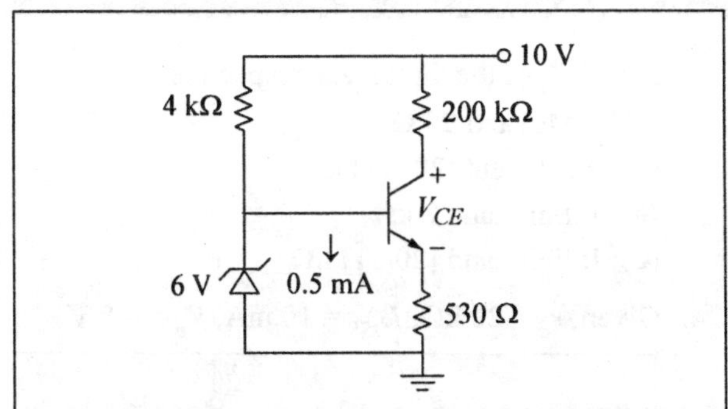

The β of the transistor and V_{CE} are respectively

(a) 19 and 2.8 V

(b) 19 and 4.7 V

(c) 38 and 2.8 V

(d) 38 and 4.7 V

49. The three transistors in the circuits shown below are identical, with $V_{BE} = 0.7$ V and β = 100

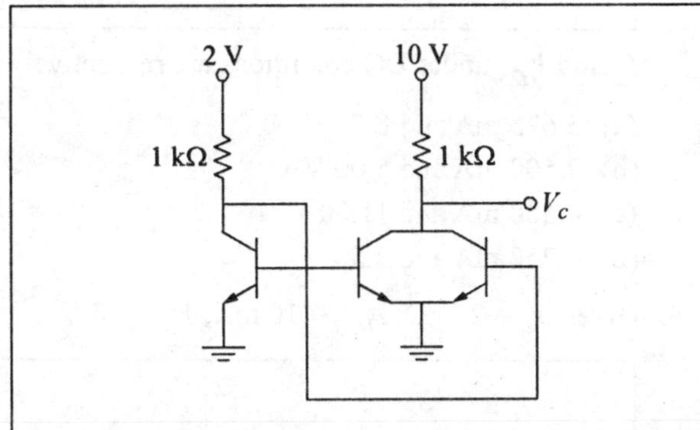

The voltage V_c is

(a) 0.2 V (b) 2 V

(c) 7.4 V (d) 10 V

50. For a single stage BJT common base amplifier.

(a) Current gain as well as voltage gain can be greater than unity

(b) Current gain can be greater than unity but voltage gain is always less than unity

(c) Voltage gain can be greater than unity but current gain is always less than unity

(d) Current gain as well as voltage gain is always less than unity

51. The matched transistors Q_1 and Q_2 shown in the adjoining figure have β = 100. Assuming the base-emitter voltages to be 0.7 V, the collector-emitter voltage V_2 of the transistor Q_2 is

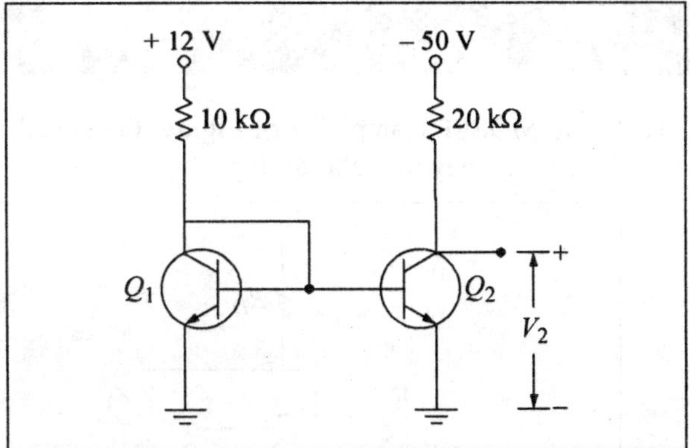

(a) 33.9 V

(b) 27.8 V

(c) 16.2 V

(d) 0.7 V

52. The neutral base width of a bipolar transistor, biased in the active region, is o.5 mm. The maximum electron concentration and the diffusion constant in the base are 10^{14}/cm^3 and $D_n = 25$ cm^2/ sec respectively. Assuming negligible recombination in the base, the collector current density is (the electron charge is 1.6×10^{-19} coulomb)

(a) 800 A/cm^2

(b) 8 A/cm^2

(c) 200 A/cm^2

(d) 2 A/cm^2

53. The phenomenon known as "Early Effect" in a bipolar transistor refers to a reduction of the effective base – width caused by.

(a) Electron–hole recombination at the base.

(b) The reverse biasing of the base–collector junction.

(c) The forward biasing of emitter-base junction.

(d) The early removal of stored base charge during saturation to cutoff switching

54. The DC current gain (b) of a BJT is 50. Assuming that the **emitter injection efficiency** is 0.995, the **base transport factor is**

(a) 0.980

(b) 0.985

(c) 0.990

(d) 0.995

55. If for a silicon *n-p-n* transistor, the base-to-emitter voltage (V_{BE}) is 0.7V and the collector-to-base voltage (V_{CB}) is 0.2 V, then the transistor is operating in the

(a) Normal active mode

(b) Saturation mode

(c) Inverse active mode

(d) Cutoff mode.

7.3 Field Effect Transistor

1. In the MOSFET amplifier of Figure, the signal output V_1 and V_2 obey the relationship

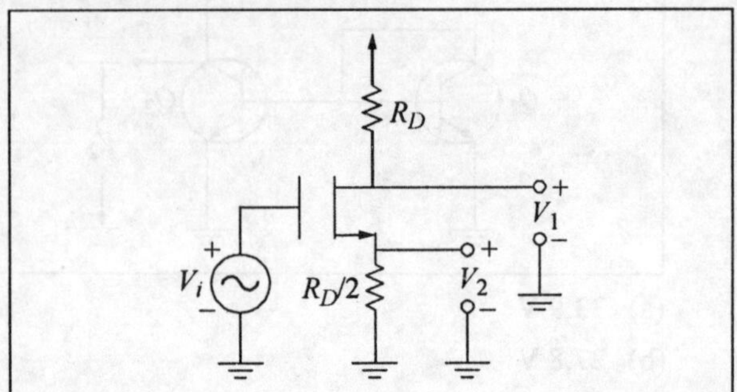

(a) $V_1 = \frac{V_2}{2}$ (b) $V_1 = -\frac{V_2}{2}$

(c) $V_1 = +2V_2$ (d) $V_1 = -2V_2$

2. The voltage gain $A_v = (V_o / V_i)$ of the JFET amplifier shown in Figure is:

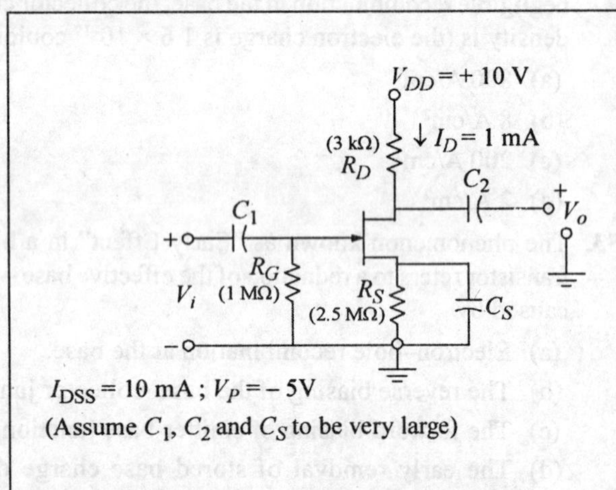

I_{DSS} = 10 mA ; V_P = − 5V
(Assume C_1, C_2 and C_S to be very large)

(a) + 18 (b) − 18

(c) + 6 (d) − 6

3. Given, r_d = 20 kΩ, I_{DSS} = 10 mA, V_p = − 8 V

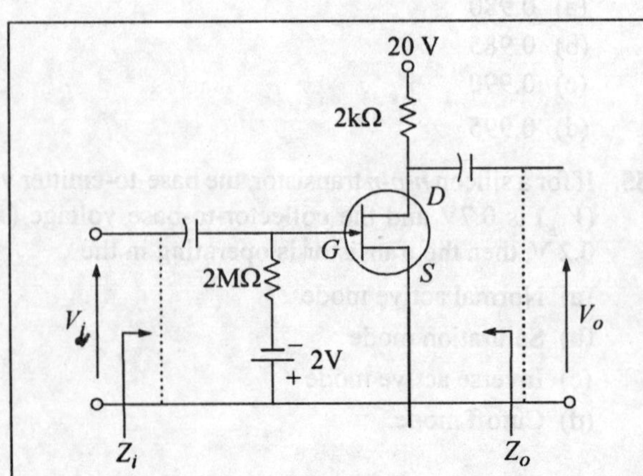

Z_i and Z_o of the circuit are respectively

(a) 2 MΩ and 2 kΩ

(b) 2 MΩ and [20 /11] kΩ

(c) Infinity and 2 kΩ

(d) Infinity and [20/11] kΩ

4. Given, r_d = 20 kΩ, I_{DSS} = 10 mA, V_p = − 8 V

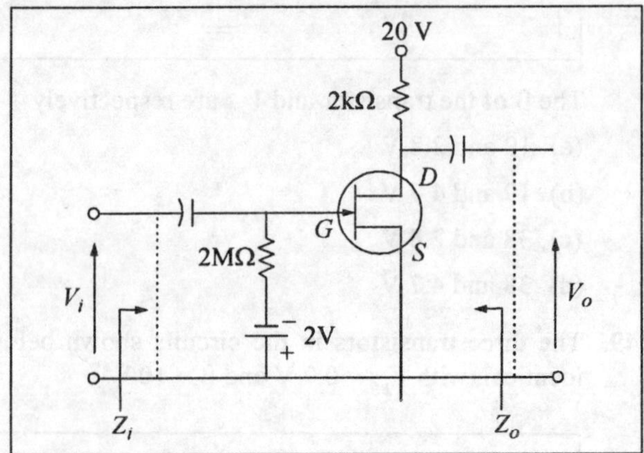

I_D and V_{DS} under DC conditions are respectively

(a) 5.625 mA and 8.75 V

(b) 7.500 mA and 5.00 V

(c) 4.500 mA and 11.00 V

(d) 6.250 mA and 7.50 V

5. Given, r_d = 20 kΩ, I_{DSS} = 10 mA, V_p = − 8 V

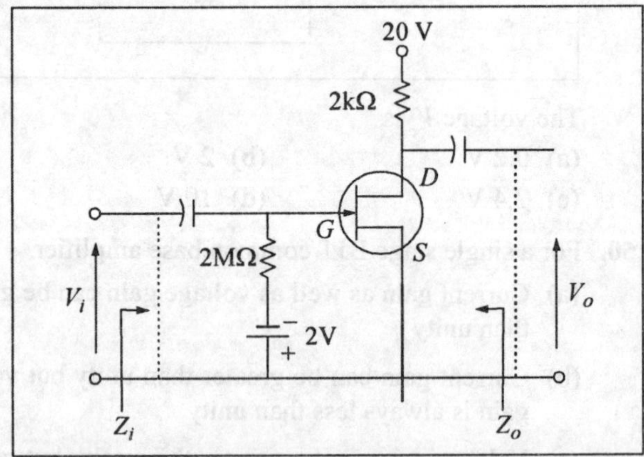

Trans-conductance in milli-Siemens (mS) and voltage gain of the amplifier are respectively

(a) 1.875 mS and 3.41 (b) 1.875 mS and − 3.41

(c) 3.3 mS and − 6 (d) 3.3 mS and 6

6. An *n*-channel depletion MOSFET has following two points on its I_D-V_{GS} curve

- V_{GS} = 0 at I_D = 12 mA and
- V_{GS} = − 6 Volts at I_D = 0 mA

Which of the following Q-point will give the highest trans-conductance gain for small signals?

(a) $V_{GS} = -6$ Volts
(b) $V_{GS} = -3$ Volts
(c) $V_{GS} = 0$ Volts
(d) $V_{GS} = 3$ Volts

7. Two identical NMOS transister M_1 and M_2 are connected as shown below. V_{bias} is chosen so that both transistors are in saturation. The equivalent g_m of the pair is defined to be $\frac{\partial I_{out}}{\partial V_i}$ at constant V_{out}.

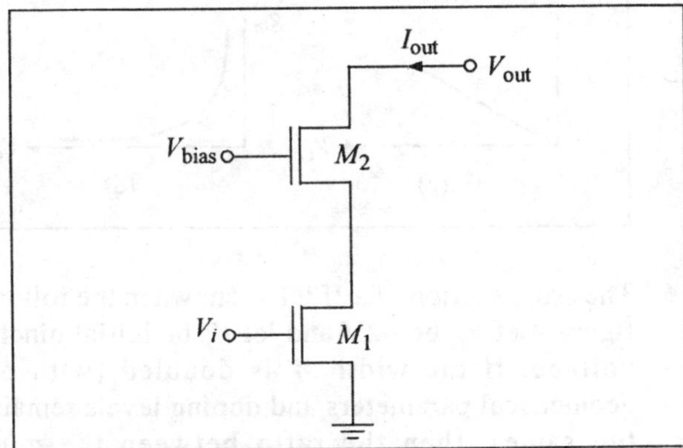

The equivalent g_m of the pair is

(a) The sum of individual g_m's of the transistors
(b) The product of individual g_m's of the transistors
(c) Nearly equal to the g_m of M_1
(d) Nearly equal to g_m of M_2

8. For the n-channel enhancement MOSFET shown in figure, the threshold voltage $V_{th} = 2$V. The drain current I_D of the MOSFET is 4 mA when the drain resistance R_D is 1 kΩ. If the value of R_D is increased to 4 kΩ, drain current I_D will become

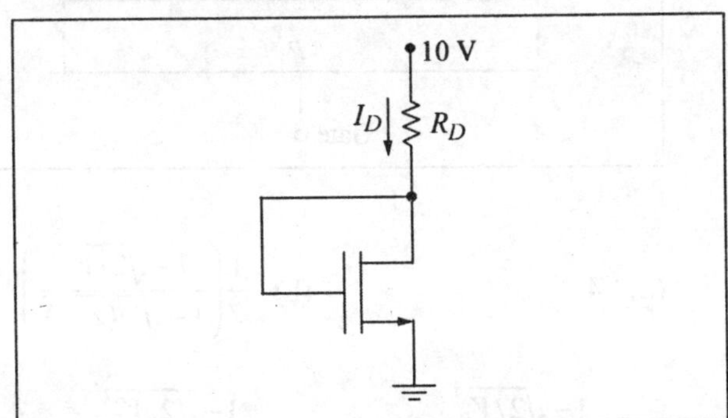

(a) 2.8 mA (b) 2.0 mA
(c) 1.4 mA (d) 1.0 mA

9. Assume that the N-channel MOSFET shown in the figure is ideal, and that its threshold voltage is +1.0V. the voltage V_{ab} between nodes a and b is

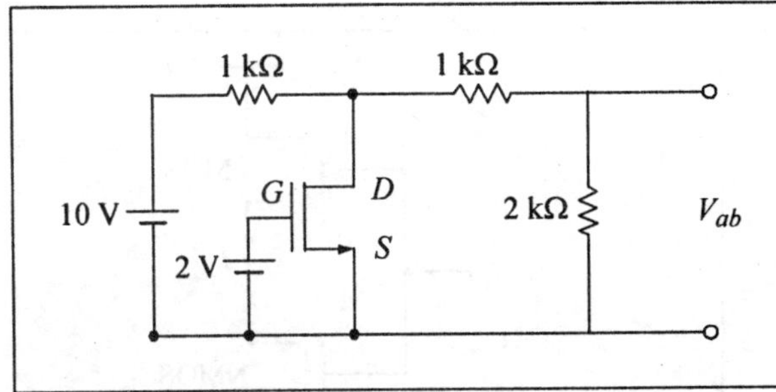

(a) 5 V
(b) 2 V
(c) 1 V
(d) 0 V

Statement for Linked Answer Questions 10 and 11

Assume that the threshold voltage of the N-channel MOSEFT shown in Fig. is + 0.75 V. The output characteristics of the MOSFET are also shown

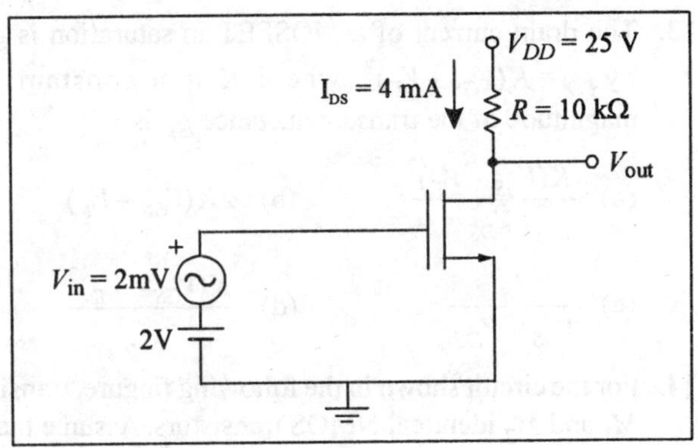

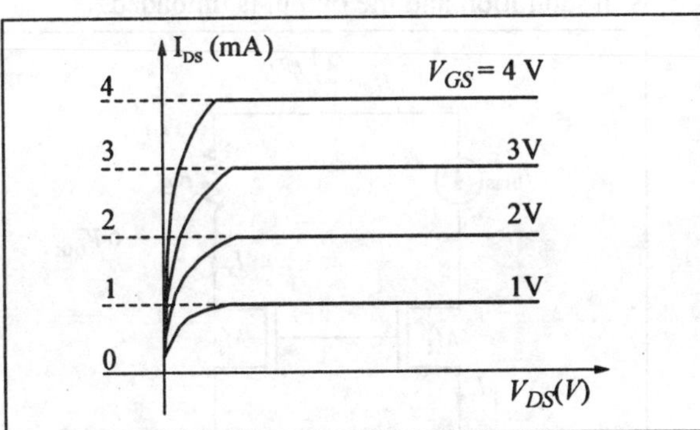

10. The trans-conductance of the MOSFET is

(a) 0.75 ms (b) 1 ms
(c) 2 ms (d) 10 ms

11. The voltage gain of the amplifier is

(a) +5 (b) – 7.5
(c) +10 (d) – 10

12. In the CMOS inverter circuit shown, if the trans-conductance parameters of the NMOS and PMOS transistors are

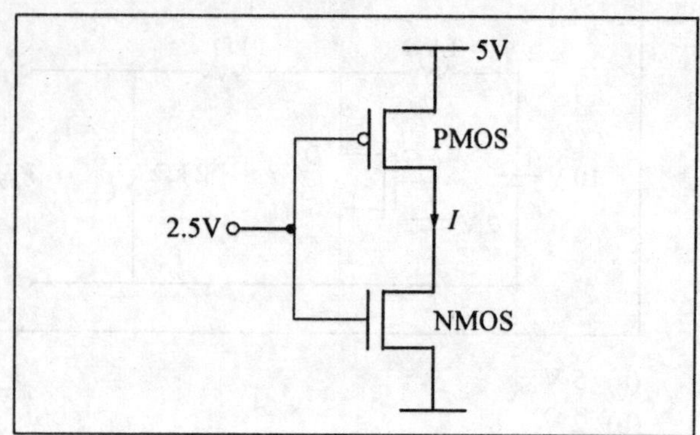

$k_n = k_p = \mu_n C_{ox} \frac{W_n}{L_n} = \mu_p C_{ox} \frac{W_p}{L_p} = 40\ \mu A/V^2$

and their threshold voltage are $V_{THn} = |V_{THp}| = 1$ V, the current I is

(a) 0 A (b) 25 μA
(c) 45 μA (d) 90 μA

13. The drain current of a MOSFET in saturation is given by $I_D = K(V_{GS} - V_T)^2$ where K is a constant. The magnitude of the transconductance g_m is

(a) $\frac{K(V_{GS} - V_T)^2}{V_{DS}}$ (b) $2K(V_{GS} - V_T)$

(c) $\frac{I_d}{V_{GS} - V_{DS}}$ (d) $\frac{K(V_{GS} - V_T)^2}{V_{GS}}$

14. For the circuit shown in the following fingure, transistors M_1 and M_2 identical NMOS transistors. Assume that M_2 is in saturation and the output is unloaded.

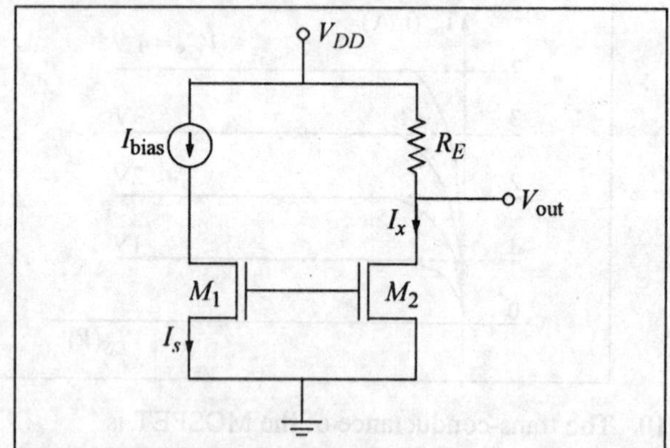

The current I_x is related to I_{bias} as

(a) $I_x = I_{bias} + I_s$
(b) $I_x = I_{bias}$
(c) $I_x = I_{bias} - I_s$
(d) $I_x = I_{bias} - \left(V_{DD} - \frac{V_{out}}{R_E}\right)$

15. The measured trasconductance g_m of an NMOS transistor oprating in the linear region is plotted against the gate voltage V_G at a constant drain voltage V_D. Which of the figures represents the expected dependence of g_m on V_G?

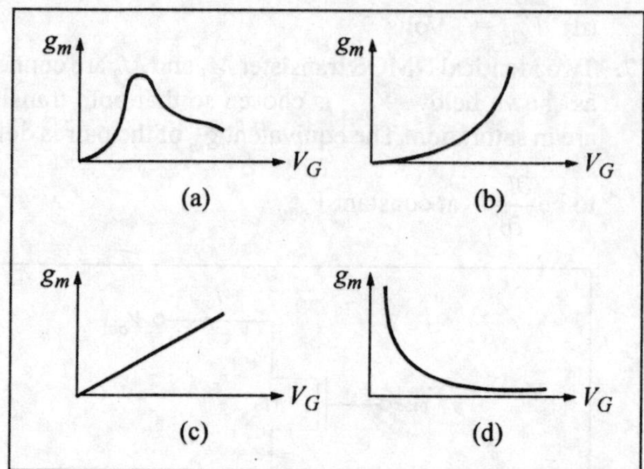

16. The cross section of a JFET is shown in the following figure. Let V_G be −2V and let V_P be initial pinch-off voltage. If the width W is doubled (with other geometrical parameters and doping levels remaining the same), then the ratio between the mutual transconductances of the initial and the modified JFET is

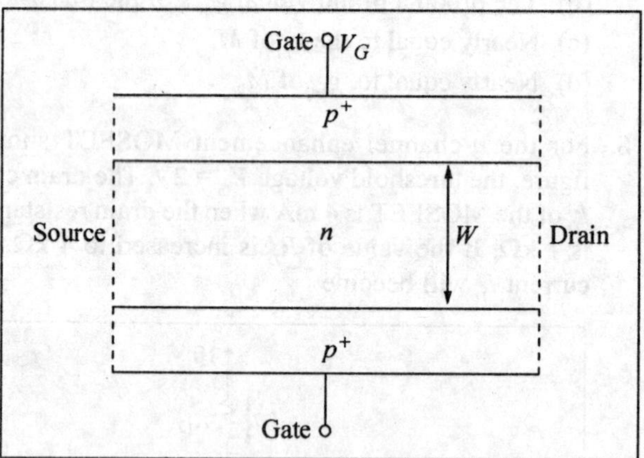

(a) 4 (b) $\frac{1}{2}\left(\frac{1-\sqrt{2/V_P}}{1-\sqrt{1/(2V_P)}}\right)$

(c) $\frac{1-\sqrt{2/V_P}}{1-\sqrt{1/(2V_P)}}$ (d) $\frac{1-\sqrt{2/V_P}}{1-\sqrt{2/(V_P)}}$

17. Match the List I figure with the List II device name and select the correct answer using the codes:

List I	*List II*
A. D G S	1. Depletion mode MOSFET
B. D G S	2. *p*-channel JFET
C. D G S	3. *n*-channel JFET
D. D G S	4. Enhancement mode MOSFET

Codes:

	A	B	C	D
(a)	2	4	3	1
(b)	4	3	2	1
(c)	2	3	4	1
(d)	2	3	1	4

18. An enhancement type *n*-channel MOSFET is represented by the symbol

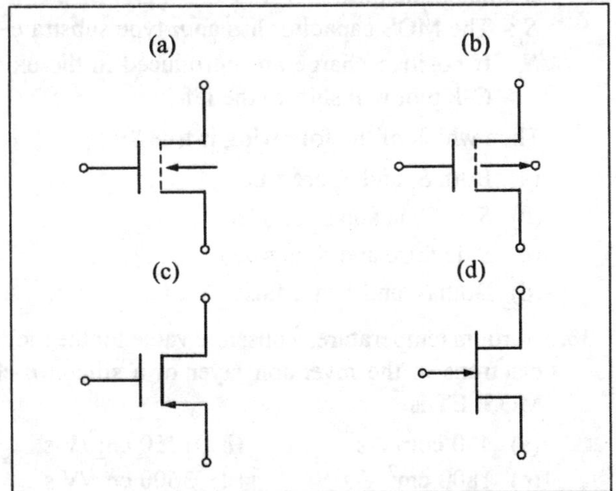

19. An *N*-channel JFET, having a pinch-off voltage (V_p) of –5 V, shows a trans-conductance (g_m) of 1 mA/V when the applied gate-to-source voltage (V_{GS}) is – 3 V. Its maximum trans-conductance (in mA/V) is

(a) 1.5 (b) 2.0
(c) 2.5 (d) 3.0

20. A *n*-channel silicon (E_g = 1.1 eV) MOSFET was fabricated using n^+ poly-silicon gate and the threshold voltage was found to be 1 V. Now, if the gate is changed to P^+ poly-silicon, other things remaining the same, the new threshold voltage should be

(a) – 0.1 V (b) 0 V
(c) 1.0 V (d) 2.1 V

21. For a MOS capacitor fabricated on a *p* type semiconductor, strong inversion occurs when

(a) Surface potential is equal to Fermi potential
(b) Surface potential is zero
(c) Surface potential is negative and equal to Fermi potential in magnitude
(d) Surface potential is positive and equal to twice the Fermi potential

22. If *P* is Passivation, *Q* is *n*-well implant, *R* is metallization and *S* is source/drain diffusion, then the order in which they are carried out in a standard *n*-well CMOS fabrication process, is

(a) P-Q-R-S
(b) Q-S-R-P
(c) R-P-S-Q
(d) S-R-Q-P

23. Consider the following statement S_1 and S_2 ·

S_1: The threshold voltage (V_T) of a MOS capacitor decreases with increase in gate oxide thickness

S_1: The threshold voltage (V_T) of a MOS capacitor decreases with increase in substrate doping concentration

Which one of the following is correct?

(a) S_1 is FALSE and S_2 is TRUE
(b) Both S_1 and S_2 are TRUE
(c) Both S_1 and S_2 are FALSE
(d) S_1 is TRUE and S_2 is FALSE

24. For an *n*-channel MOSFET and its transfer curve shown in the figure, the threshold voltage is

(a) 1 V and the device is in active region
(b) – 1 V and the device is in saturation region
(c) 1 V and the device is in saturation region.
(d) – 1 V and the device is in active region.

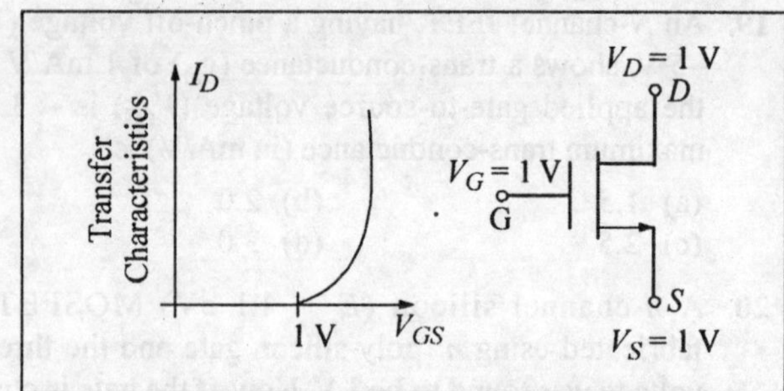

25. A MOS capacitor made using p type substrate is in the accumulation mode. The dominant charge in the channel is due to the presence of
(a) Holes
(b) Electrons
(c) Positively charged ions
(d) Negatively charged ions

26. For an n-channel enhancement type MOSFET, if the source is connected at a higher potential than that of the bulk (i.e $V_{SB} > 0$), the threshold voltage V_T of the MOSFET will
(a) Remain unchanged
(b) Decrease
(c) Change polarity
(d) Increase

27. When the gate-to-source voltage (V_{GS}) of a MOSFET with threshold voltage of 400mV, working in saturation is 900mV, the drain current is observed to be 1 mA. Neglecting the channel width modulation effect and assuming that the MOSFET is operating at saturation, the drain current for an applied V_{GS} of 1400 mV is
(a) 0.5 mA (b) 2.0 mA
(c) 3.5 mA (d) 4.0 mA

28. The action of a JFET in its equivalent circuit can best be represented as a
(a) Current Controlled Current Source
(b) Current Controlled Voltage Source
(c) Voltage Controlled Voltage Source
(d) Voltage Controlled Current Source

29. The drain of an n-channel MOSFET is shorted to the gate so that $V_{GS} = V_{DS}$. The threshold voltage (V_T) of MOSFET is 1 V. If the drain current (I_D) is 1 mA for $V_{GS} = 2$V, then for $V_{GS} = 3$ V, I_D is
(a) 2 mA (b) 3 mA
(c) 9 mA (d) 4 mA

30. The figure shows the high-frequency capacitance-voltage (C-V) characteristic of a Metal/SiO_2 /silicon (MOS) capacitor having an area of 1×10^{-4} cm². Assume that the permittivities ($\varepsilon_0 \varepsilon_r$) of silicon and SiO_2 are 1×10^{-12} F/cm and 3.5×10^{-13} F/cm respectively.

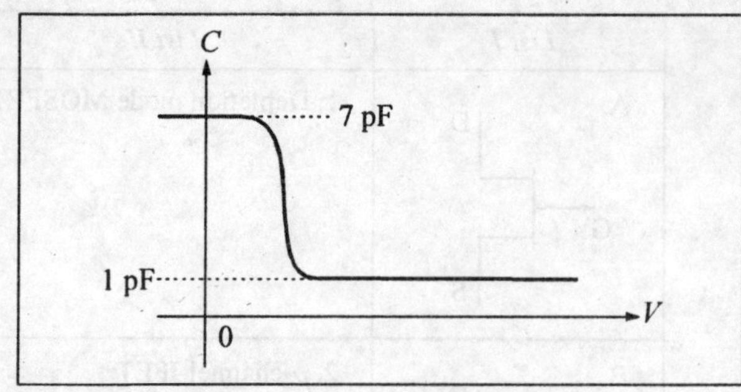

The gate oxide thickness in the MOS capacitor is
(a) 50 nm
(b) 143 nm
(c) 350 nm
(d) 1 μm

Common Data for Questions 31 and 32

The figure shows the high-frequency capacitance-voltage (C-V) characteristic of a Metal/SiO_2 /silicon (MOS) capacitor having an area of 1×10^{-4} cm². Assume that the permittivities ($\varepsilon_0 \varepsilon_r$) of silicon and SiO_2 are 1×10^{-12} F/cm and 3.5×10^{-13} F/cm respectively.

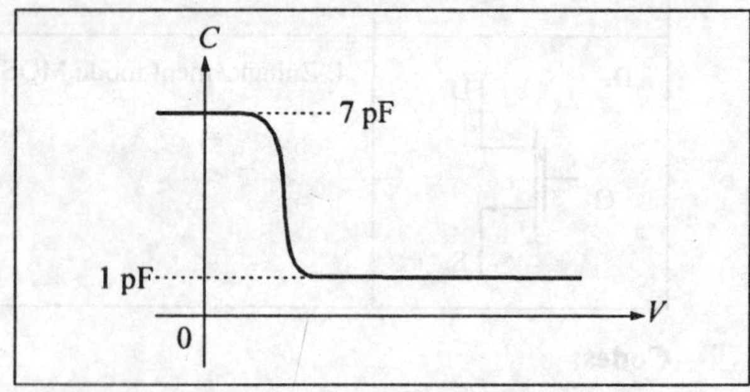

31. The maximum depletion layer width in silicon is
(a) 0.143 μm
(b) 0.857 μm
(c) 1 μm
(d) 1.143 μm

32. Consider the following statements about the C-V characteristic plot:
S_1: The MOS capacitor has an n-type substrate
S_2: If positive charge are introduced in the oxide, the C-V plot will shift to the left
Then which of the following is true ?
(a) Both S_1 and S_2 are true
(b) S_1 is True and S_2 is false
(c) S_1 is false and S_2 is true
(d) Both S_1 and S_2 are false

33. At room temperature, a possible value for the mobility of electrons in the inversion layer of a silicon n-channel MOSFET is
(a) 450 cm²/V-s (b) 1350 cm²/V-s
(c) 1800 cm²/V-s (d) 3600 cm²/V-s

7.4 Frequency Response

1. In a multi-stage R-C coupled amplifier the coupling capacitor
   (a) Limits the low frequency response
   (b) Limits the high frequency response
   (c) Does not affect the frequency response
   (d) Blocks the d.c. component without affecting the frequency response

2. In the circuit shown in figure N is a finite gain amplifier with a gain of k, a very large input impedance, and a very low output impedance. The input impedance of the feedback amplifier with the feedback impedance Z connected as shown will be

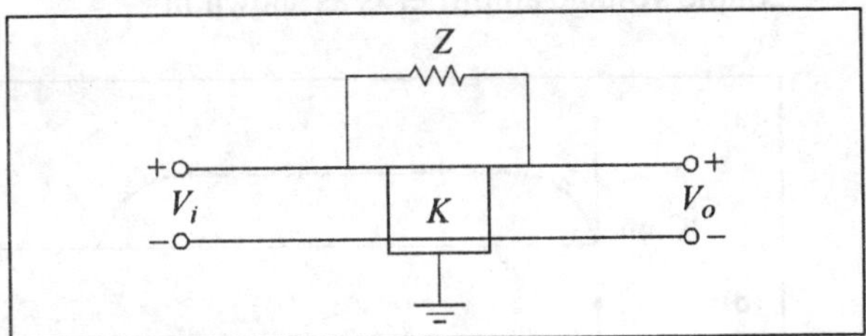

   (a) $Z\left(1-\frac{1}{k}\right)$ (b) $Z(1-k)$
   (c) $\frac{Z}{k-1}$ (d) $\frac{Z}{1-k}$

3. An npn transistor has a beta cutoff frequency f_β of 1 MHz, and common emitter short circuit low frequency current gain β_o of 200. It unity gain frequency f_T and the alpha cutoff frequency f_α respectively are
   (a) 200 MHz, 201 MHz (b) 200 MHz, 199 MHz
   (c) 199 MHz, 200 MHz (d) 201 MHz, 200 MHz

4. An npn transistor (with $C_\pi = 0.3$ pF) has a unity –gain cut off frequency f_T of 400 MHz at a dc bias current $I_C = I$ mA. The value of its C_μ (in pF) is approximately ($V_T = 26$ mV)
   (a) 15 (b) 30
   (c) 50 (d) 96

5. An amplifier is assumed to have a single-pole high-frequency transfer function. The rise time of its output response to a steps function input is 35 nsec. The upper-3dB frequency (in MHz) for the amplifier to a sinusoidal input is approximately at
   (a) 4.55 (b) 10
   (c) 20 (d) 28.6

6. The current gain of a bipolar transistor drops at high frequencies because of
   (a) Transistor capacitance
   (b) High current effects in the base
   (c) Parasitic inductive elements
   (d) The Early effect

7. An npn BJT has g_m =38 mA/V, $C_\mu = 10^{-14}$ F, $C_\pi = 4 \times 10^{-13}$ F, and DC current gain $\beta_0 = 90$. For this transistor f_T and f_β are
   (a) $f_T = 1.64 \times 10^8$ Hz and $f_\beta = 1.47 \times 10^{10}$ Hz
   (b) $f_T = 1.47 \times 10^{10}$ Hz and $f_\beta = 1.64 \times 10^{10}$ Hz
   (c) $f_T = 1.33 \times 10^{12}$ Hz and $f_\beta = 1.47 \times 10^{10}$ Hz
   (d) $f_T = 1.47 \times 10^{10}$ Hz and $f_\beta = 1.33 \times 10^{12}$ Hz

8. Three identical RC-coupled transistor amplifiers are cascaded. If each of the amplifier has frequency response as shown in Figure, the overall frequency response in as given in

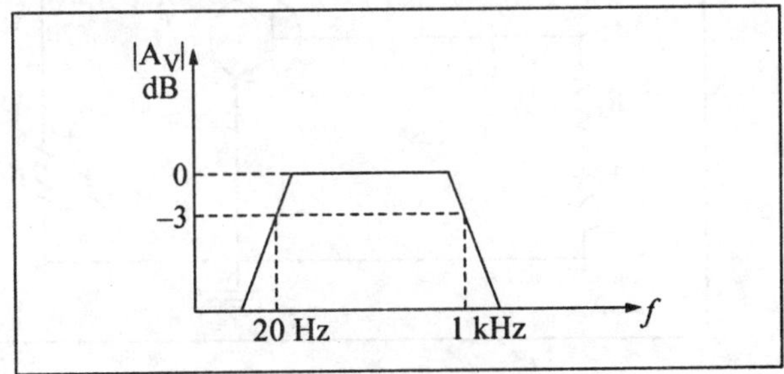

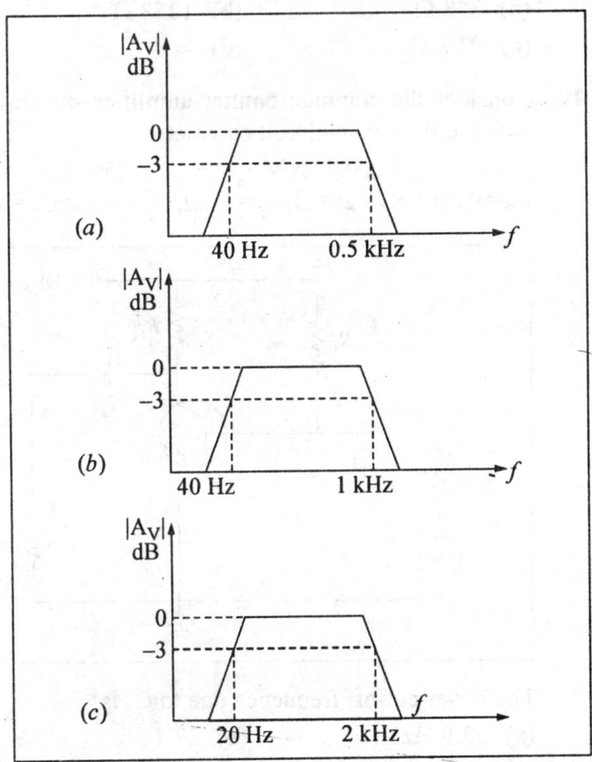

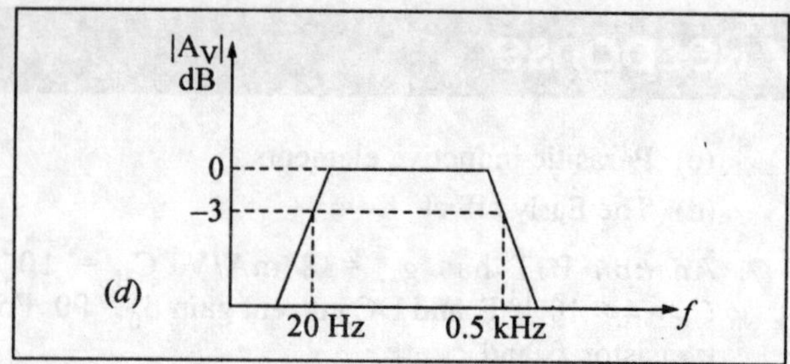

9. Consider the common emitter amplifier shown below with the following circuit parameters:
$\beta = 100$, $g_m = 0.3861$ A/V, $r_o = \infty$, $r_\pi = 259\ \Omega$, $R_S = 1$ kΩ, $R_B = 93$ kΩ, $R_C = 250\ \Omega$, $R_L = 1$ kΩ, $C_1 = \infty$ and $C_2 = 4.7$ μF.

The resistance seen by the source V_s is

(a) 258 Ω (b) 1258 Ω
(c) 93 k Ω (d) ∞

10. Consider the common emitter amplifier shown below with the following circuit parameters:
$\beta = 100$, $g_m = 0.3861$ A/V, $r_o = \infty$, $r_\pi = 259\ \Omega$, $R_S = 1$ kΩ, $R_B = 93$ kΩ, $R_C = 250\ \Omega$, $R_L = 1$ kΩ, $C_1 = \infty$ and $C_2 = 4.7$ μF.

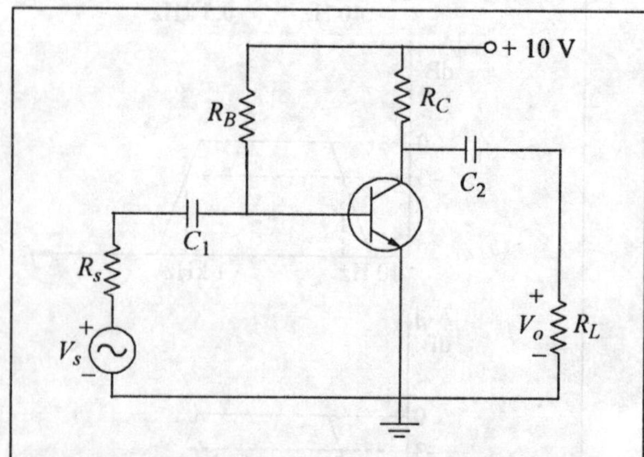

The lower cut-off frequency due to C_2 is

(a) 33.9 Hz (b) 27.1 Hz
(c) 13.6 Hz (d) 16.9 Hz

11. In an **RC-coupled common emitter amplifier**, which of the following is true?

(a) Coupling capacitance affects the High frequency response and bypass capacitance affects the Low frequency response
(b) Both coupling and bypass capacitance affect the Low frequency response only
(c) Both coupling and bypass capacitance affect High frequency response only
(d) Coupling capacitance affects the Low frequency response and the bypass capacitance affects the High frequency response.

12. The typical frequency response of a two-stage direct couple voltage amplifier is as shown in

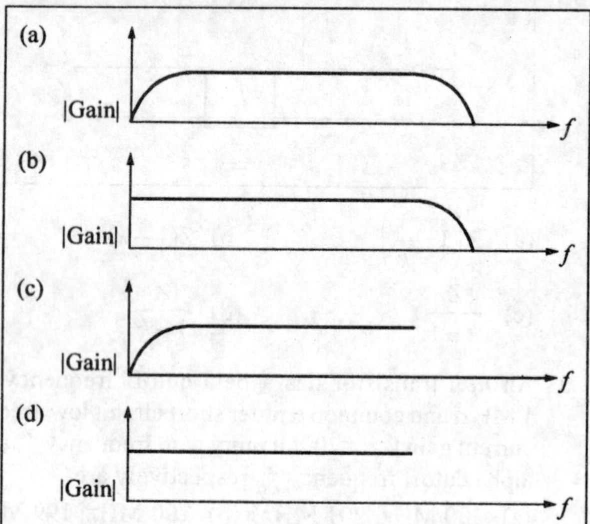

13. A unity gain buffer amplifier has a bandwidth of 1 MHz. The output voltage of the amplifier for an input of 2 V sinusoid of frequency 1 MHZ will be.

(a) 2 V
(b) $2\sqrt{2}$ V
(c) $\frac{2}{\sqrt{2}}$ V
(d) $\frac{4}{\sqrt{2}}$ V

14. An amplifier circuit is shown below. Assume that the transistor works in active region. The low frequency small-signal parameters for the transistor are $g_m = 20$ mS, $\beta_o = 50$, $r_o = \infty$, $r_b = 0$. What is the voltage gain, $A_V = \left(\frac{V_o}{V_i}\right)$, of the amplifier

7.5 Operational-Amplifier

1. The CMRR of the differential amplifier of Figure is equal to

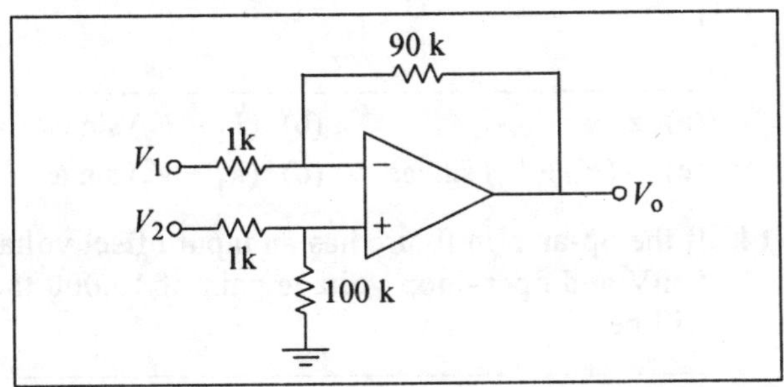

(a) ∞ (b) 0
(c) 900 (d) 1800

2. If the input to the circuit of figure is sine wave, the output will be

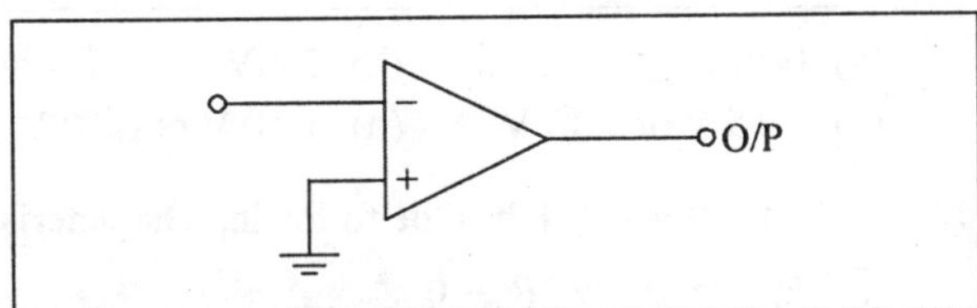

(a) A half-wave rectified sine wave
(b) A full-wave rectified sine wave
(c) A triangular wave
(d) A square wave

3. An ops-amp has an offset voltage of 1 mV and is ideal in all other respects. If this op. amp is used in the circuit shown in figure, the output voltage will be (select the nearest value)

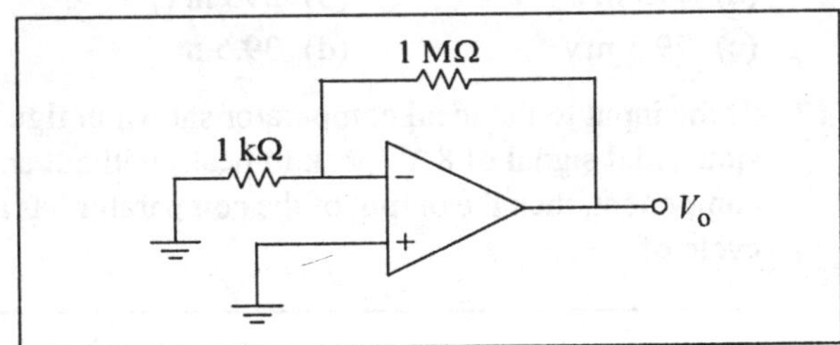

(a) 1 mV
(b) 1 V
(c) ± 1 V
(d) 0 V

4. The circuit of figure uses an ideal op amp. For small positive values of V_{in}, the circuit works as:

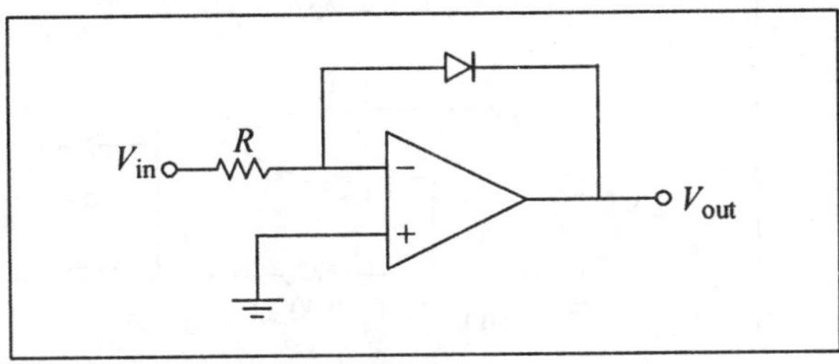

(a) A half-wave rectifier
(b) A differentiator
(c) A logarithmic amplifier
(d) An exponential amplifier

5. A change in the value of the emitter resistance, R_e in a difference amplifier
(a) Affects the difference mode gain A_d
(b) Affects the common mode gain A_c
(c) Affects both A_d and A_c
(d) does not affect either A_d or A_c

6. The circuit shown in figure, is that of

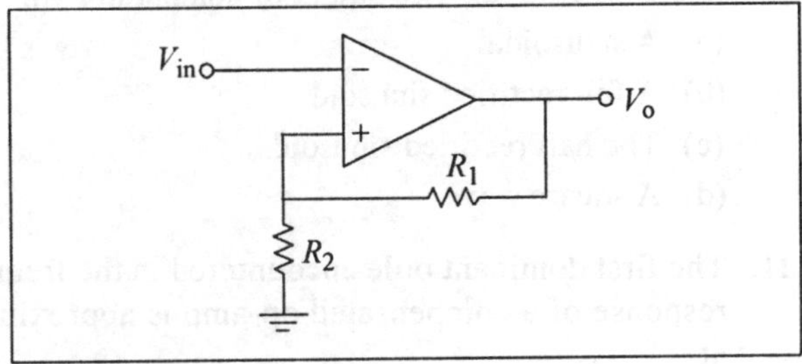

(a) A non-inverting amplifier
(b) An inverting amplifier
(c) An oscillator
(d) A Schmitt trigger

7. An amplifier A has 6 dB gain and 50 Ω input and output impedances. The noise figure of this amplifier as in figure is 3dB. Cascade of two such amplifiers as in figure will have a noise figure of

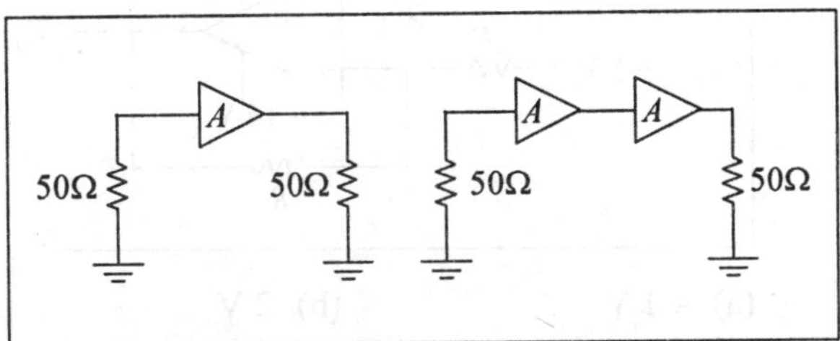

(a) 6 dB (b) 8 dB
(c) 12 dB (d) None of the above

8. The output voltage V_o of the circuit shown is

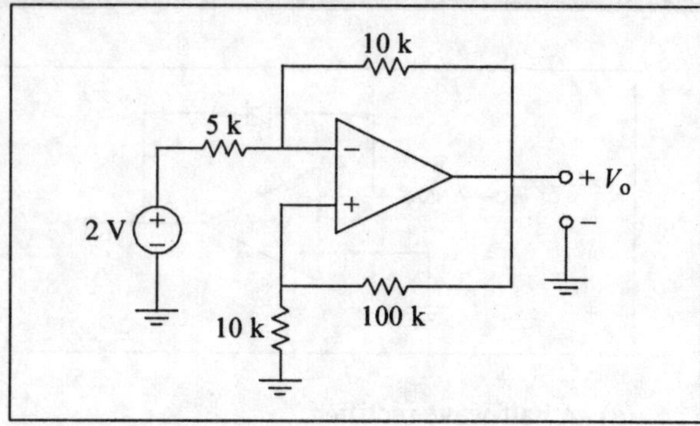

(a) – 4 V (b) 6 V
(c) 5 V (d) – 5.5 V

9. In a differential amplifier, CMRR can be improved by using and increased

(a) Emitter resistance
(b) Collector resistance
(c) Power supply voltage
(d) Source resistance

10. One input terminal of high gain comparator circuit is connected to ground and a sinusoidal voltage is applied to the other input. The output of comparator will be

(a) A sinusoidal
(b) A full rectified sinusoid
(c) The half rectified sinusoid
(d) A square wave

11. The first dominant pole encountered in the frequency response of a compensated op-amp is approximately at

(a) 5 Hz
(b) 10 Hz
(c) 1 MHz
(d) 100 MHz

12. In the circuit of figure, V_o is

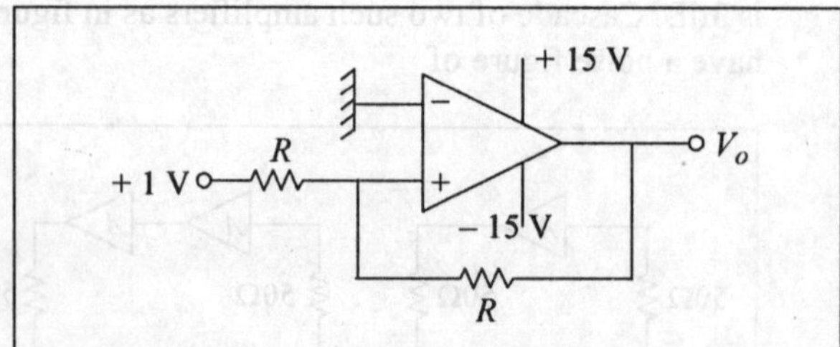

(a) – 1 V (b) 2 V
(c) + 1 V (d) + 15 V

13. If the op-amp in figure is ideal, then V_o is

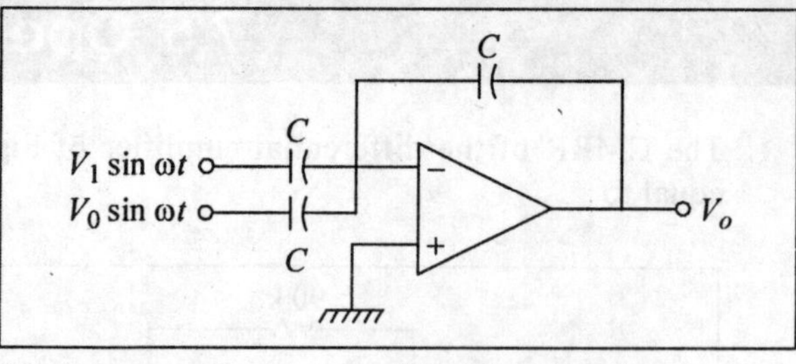

(a) zero (b) $(V_1 - V_2) \sin \omega t$
(c) $-(V_1 + V_2) \sin \omega t$ (d) $(V_1 + V_2) \sin \omega t$

14. If the op-amp in figure has an input offset voltage of 5 mV and open-loop voltage gain of 10,000 then V_o will be:

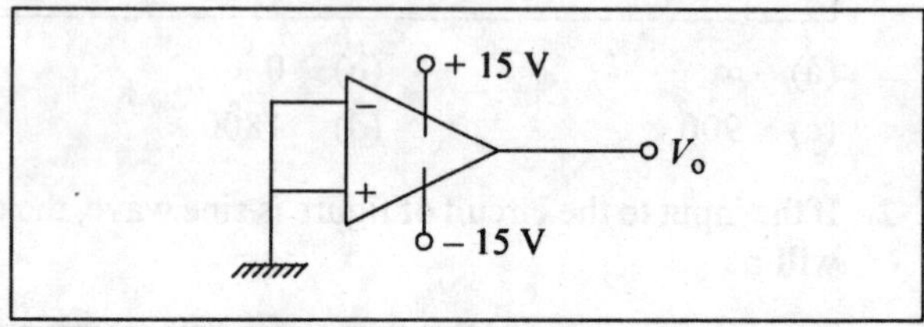

(a) 0 V (b) 5 mV
(c) + 15 V or – 15 V (d) + 50 V or – 50 V

15. The ideal OP-AMP has the following characteristics

(a) $R_i = \infty, A = \infty, R_o = 0$
(b) $R_i = 0, A = \infty, R_o = 0$
(c) $R_i = \infty, A = \infty, R_o = \infty$
(d) $R_i = 0, A = \infty, R_o = \infty$

16. An amplifier using an op-amp with a slew-rate SR = 1 V/ μ sec has a gain of 40 dB. If this amplifier has to faithfully amplify sinusoidal from dc to 20 kHz without introducing any slew-rate inducted distortion, then the input signal level must not exceed

(a) 795 mV (b) 395 mV
(c) 79.5 mV (d) 39.5 mV

17. If the input to the ideal comparator shown in figure is a sinusoidal signal of 8 V (peak to peak) without any DC component, then the output of the comparator has a duty cycle of

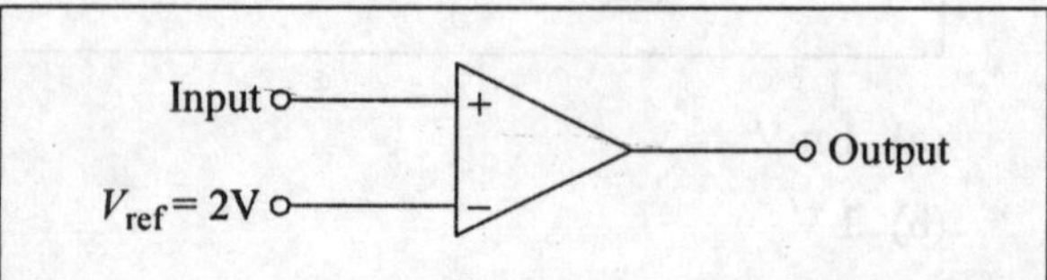

(a) $\frac{1}{2}$ (b) $\frac{1}{3}$
(c) $\frac{1}{6}$ (d) $\frac{1}{12}$

18. If the differential voltage gain and the common mode voltage gain of a differential amplifier are 48 dB and 2 dB, respectively, then its common mode rejection ratio is:

(a) 23 dB (b) 25 dB
(c) 46 dB (d) 50 dB

19. The output voltage of the regulated power supply shown in figure is:

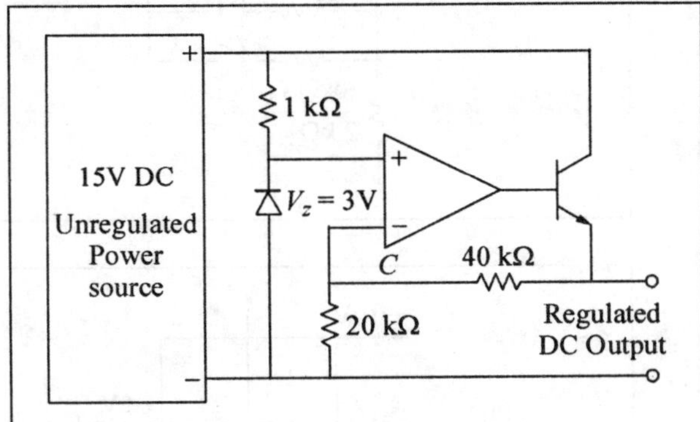

(a) 3 V (b) 6 V
(c) 9 V (d) 12 V

20. If the op-amp in figure is ideal, the output voltage V_{out} will be equal to

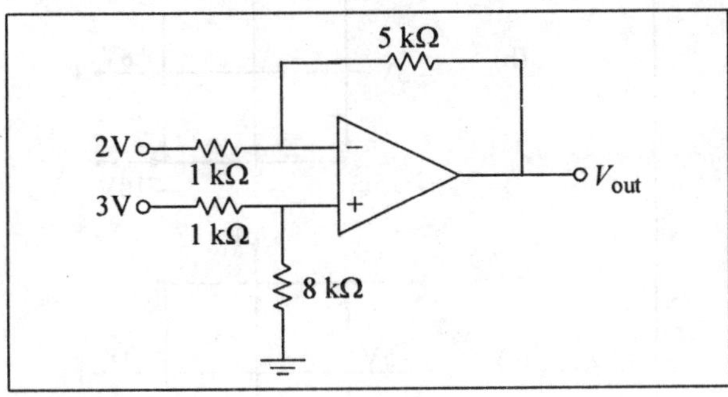

(a) 1 V (b) 6 V
(c) 14 V (d) 17 V

21. An ideal op-amp is an ideal

(a) Voltage controlled current source
(b) Voltage controlled voltage source
(c) Current controlled current source
(d) Current controlled voltage source

22. The circuit in figure is a

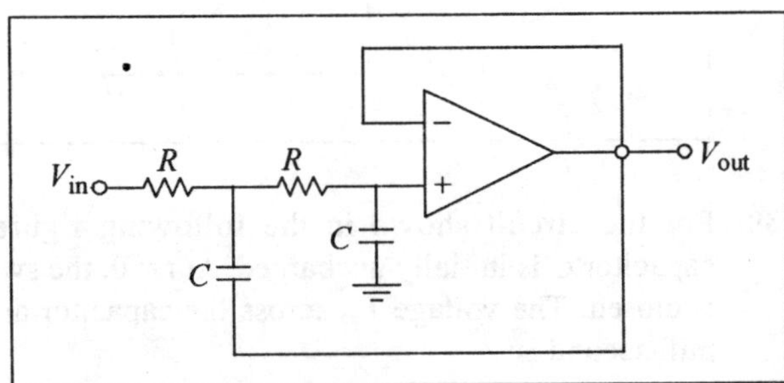

(a) Low-pass filter
(b) High-pass filter
(c) Band-pass filter
(d) Band-reject filter

23. The OP-AMP of figure has a very poor open loop voltage gain of 45 but is otherwise ideal. The gain of the amplifier equals

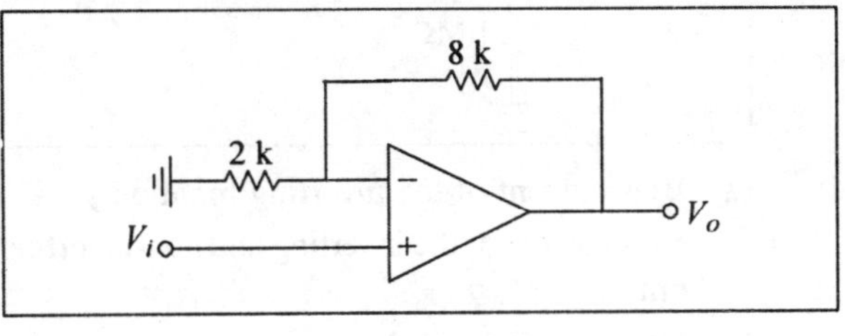

(a) 5 (b) 20
(c) 4 (d) 4.5

24. The circuit below implements a filter between the input current I_i and the output voltage V_0, Assume that the op-amp is ideal. The filter implemented is a

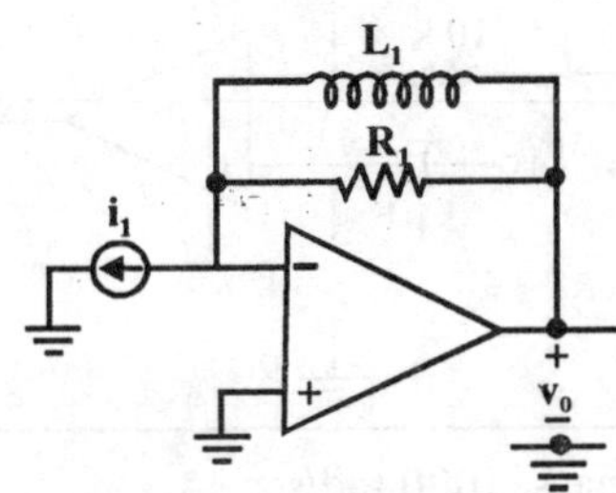

(a) Low pass filter
(b) Band pass filter
(c) Band stop filter
(d) High pass filter

25. The input resistance R_i of the amplifier shown in the figure is

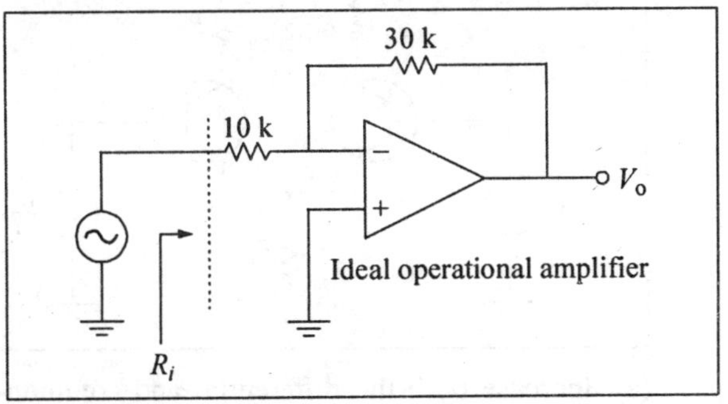

(a) $\left[\frac{30}{4}\right]$ kΩ (b) 10 kΩ
(c) 40 kΩ (d) infinite

26. The voltage e_o indication in the figure has been measured by an ideal voltmeter. Which of the following can be calculated?

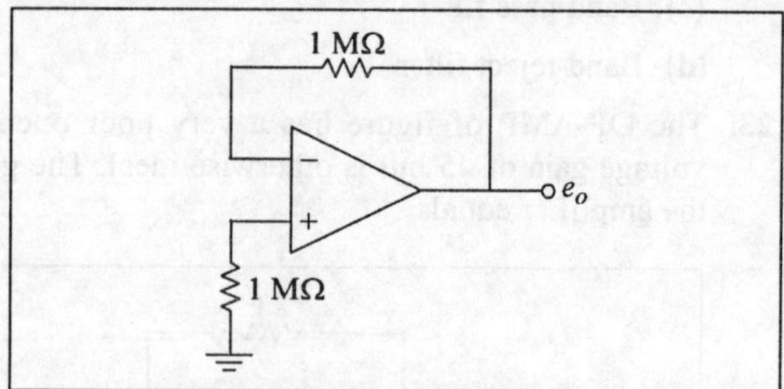

(a) Bias current of the inverting input only
(b) Bias current of the inverting and non-inverting inputs only
(c) Input offset current only
(d) Both the bias currents and the input offset current.

27. The op-amp circuit shown in the figure is a filter. The type of filter and its cut-off frequency are respectively

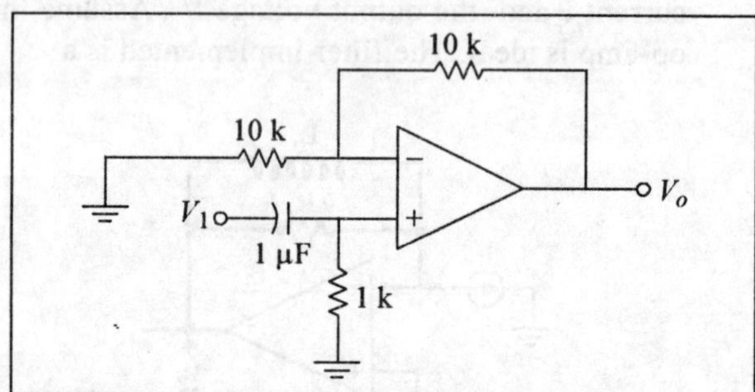

(a) High pass, 1000 rad/sec
(b) Low pass, 1000 rad/sec
(c) High pass, 10000 rad/sec
(d) Low pass, 10000 rad/sec

28. In an ideal differential amplifier shown in the figure, a large value of (R_E)

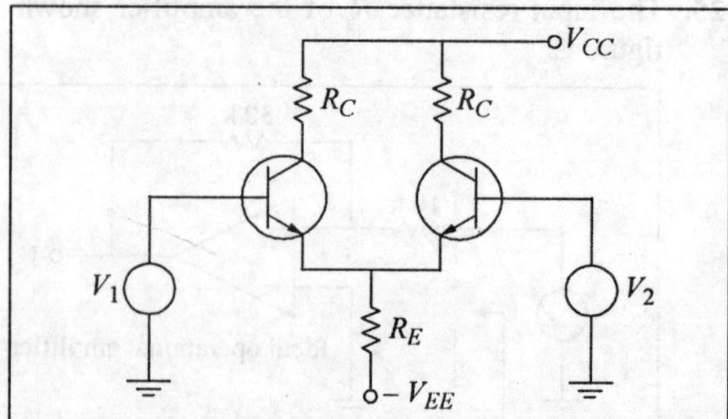

(a) Increase both the differential and common-mode gains
(b) Increase the common-mode gain only
(c) Decrease the differential-mode gain only
(d) Decrease the common-mode gain only

29. Given the ideal operational amplifier **circuit** shown in the figure, indicate the correct transfer characteristic assuming ideal diodes with zero cut-in voltage.

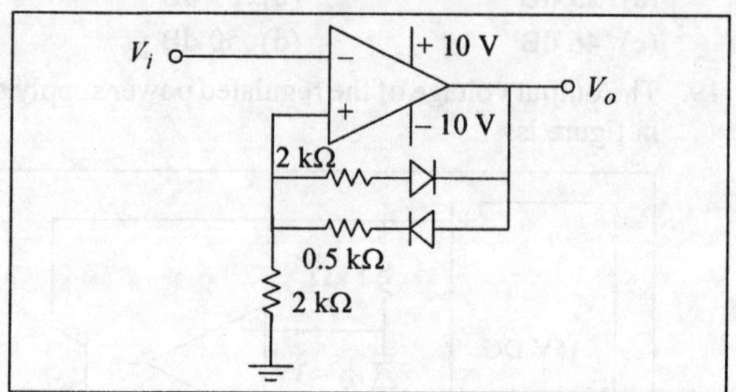

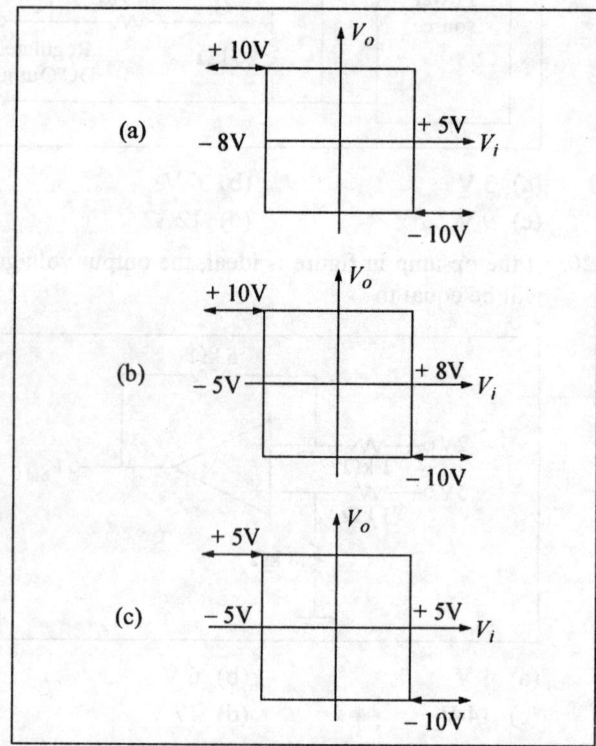

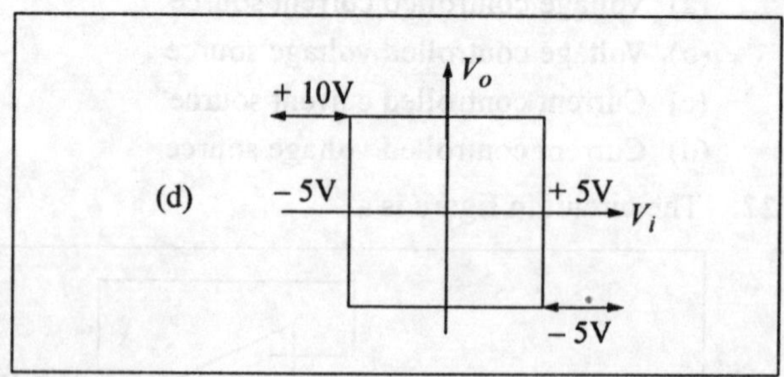

30. For the circuit shown in the following figure, the capacitor C is initially uncharged. At $t = 0$, the switch S is closed. The voltage V_C across the capacitor at $t = 1$ millisecond is:

In the figure shown above, the OP-AMP is supplied with ± 15V

(a) 0 Volt

(b) 6.3 Volts

(c) 9.45 Volts

(d) 10 Volts

31. In the op-amp circuit shown, assume that the diode current follows the equation $I = I_s e^{(V/VT)}$. For $V_i = 2$ V, $V_o = V_{o1}$, and $V_i = 4$ V, $V_o = V_{o2}$, the relationship between V_{o1} and V_{o2} is:

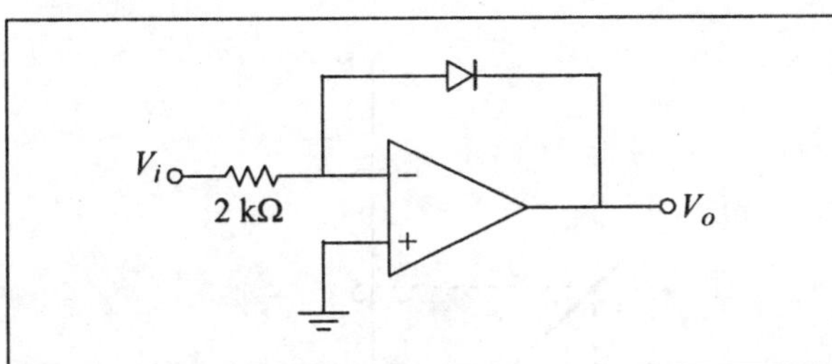

(a) $V_{o2} = \sqrt{2}\, V_{o1}$

(b) $V_{o2} = e^2 V_{o1}$

(c) $V_{o2} = V_{o1} \ln 2$

(d) $V_{o1} - V_{o2} = V_T \ln 2$

Statement for linked Questions 32 and 33

Consider the op-amp circuit shown in the figure.

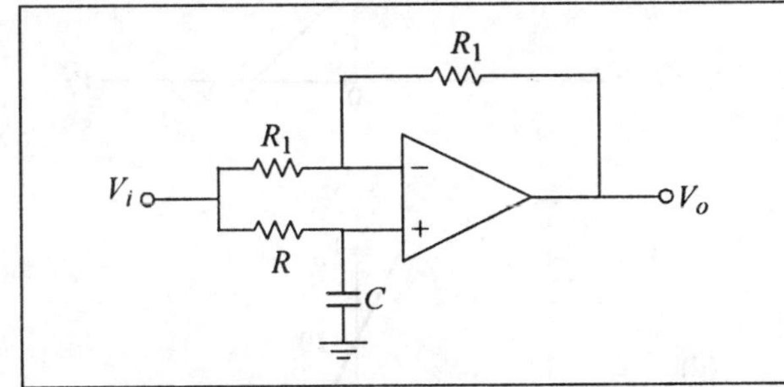

32. The transfer function $\dfrac{V_o(s)}{V_i(s)}$ is

(a) $\dfrac{1 - sRC}{1 + sRC}$

(b) $\dfrac{1 + sRC}{1 - sRC}$

(c) $\dfrac{1}{1 - sRC}$

(d) $\dfrac{1}{1 + sRC}$

33. If $V_i = V_1 \sin(\omega t)$ and $V_o = V_2 \sin(\omega t + \phi)$, then the minimum and maximum values of ϕ (in radians) are respectively

(a) $-\dfrac{\pi}{2}$ and $\dfrac{\pi}{2}$ (b) 0 and $\dfrac{\pi}{2}$

(c) $-\pi$ and 0 (d) $-\dfrac{\pi}{2}$ and 0

34 Consider the following circuit using an ideal OP-AMP. The I-V characteristics of the diode is described by the relation

$$I = I_o\left(e^{\frac{V}{V_T}} - 1\right)$$

where $V_T = 25\text{mV}$, $I_o = 1\,\mu\text{A}$ and V is the voltage across the diode (taken as positive for forward bias).

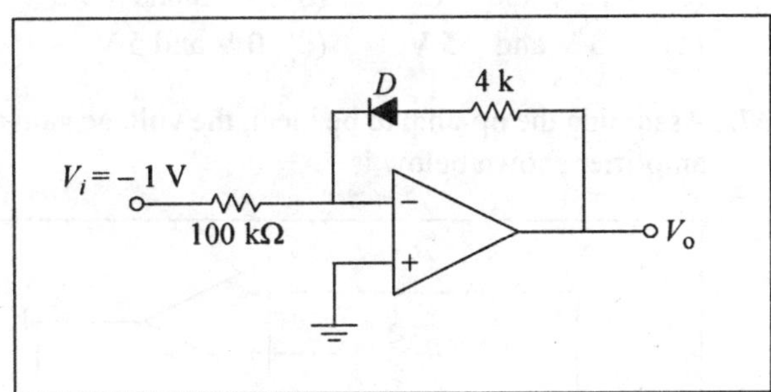

For an input voltage $V_i = -1$ V, the output voltage V_o is

(a) 0 V (b) 0.1 V

(c) 0.7 V (d) 1.1 V

35. The OP-AMP circuit shown below represents a

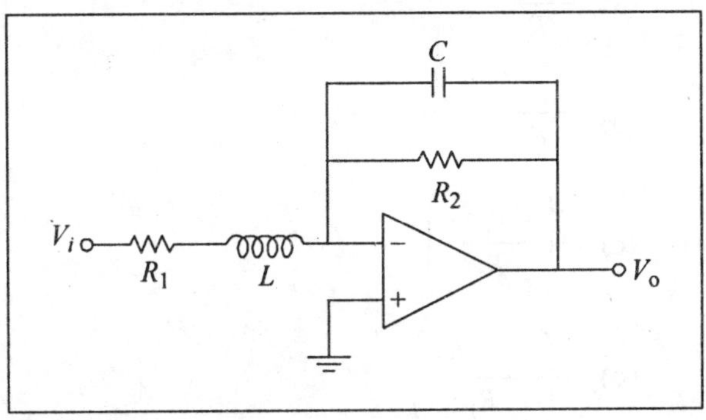

(a) High pass filter

(b) Low pass filter
(c) Band pass filter
(d) Band reject filter

36. Consider the Schmitt trigger circuit shown below.

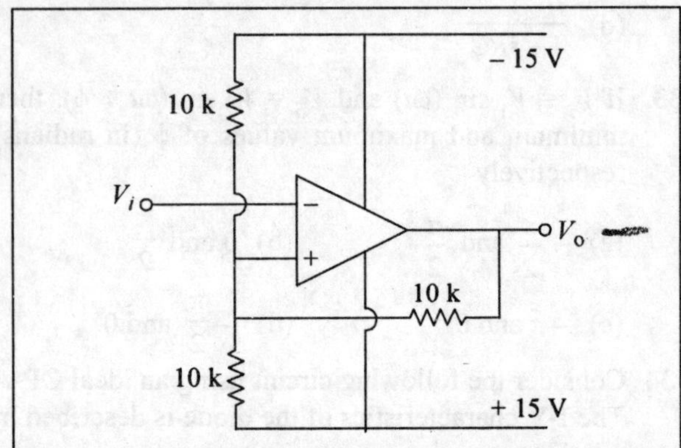

A triangular wave which goes from –12 V to 12 V is applied to the inverting input of the op-amp. Assume that the output of the op-amp swings from + 15 V to – 15 V. The voltage at the non-inverting input switches between

(a) – 12 V and + 12 V (b) – 7.5 and + 7.5 V
(c) – 5 V and + 5 V (d) 0 V and 5 V

37. Assuming the op-amp to be ideal, the voltage gain of the amplifier shown below is

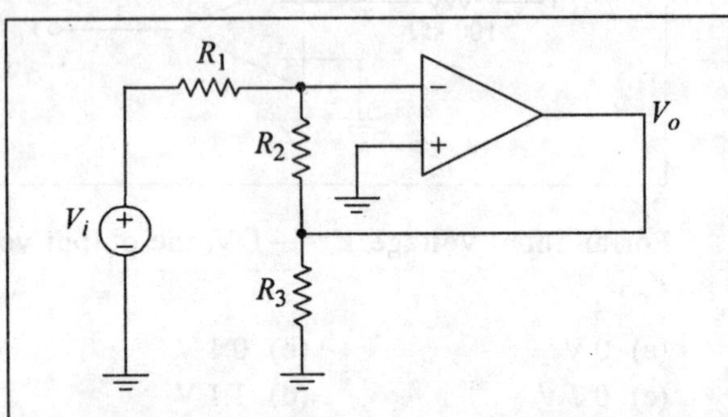

(a) $-\frac{R_2}{R_1}$

(b) $-\frac{R_3}{R_1}$

(c) $-\left(\frac{R_2 \| R_3}{R_1}\right)$

(d) $-\left(\frac{R_2 + R_3}{R_1}\right)$

38. The transfer characteristic for the precision rectifier circuit shown below is (assume ideal op-amp and practical diodes)

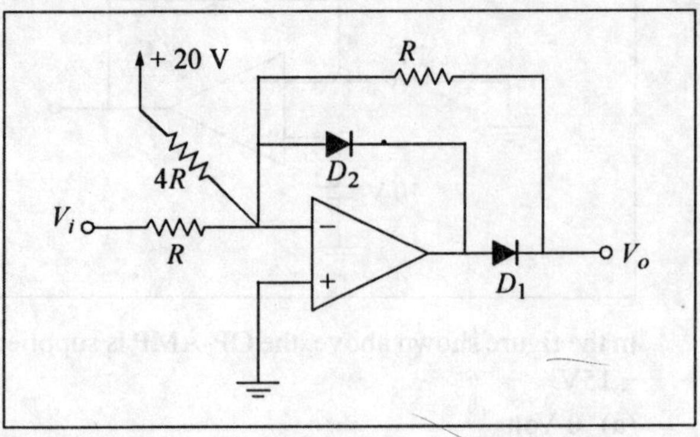

(a)

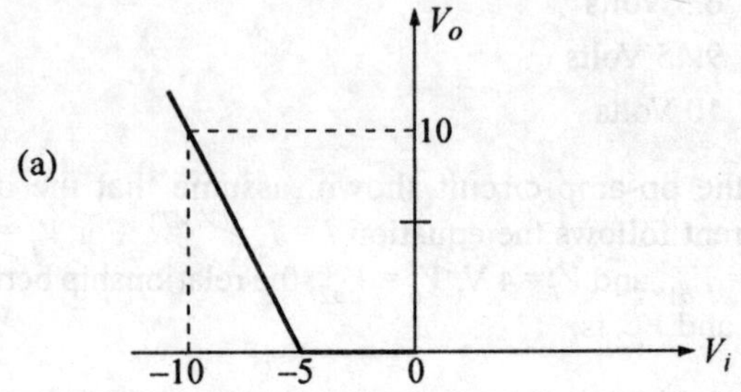

(b)

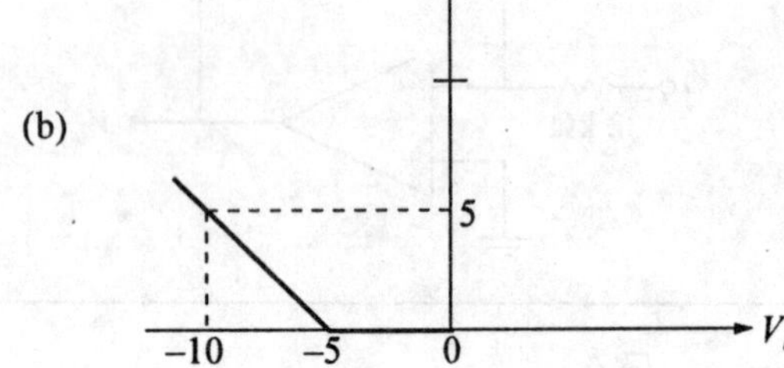

(c)

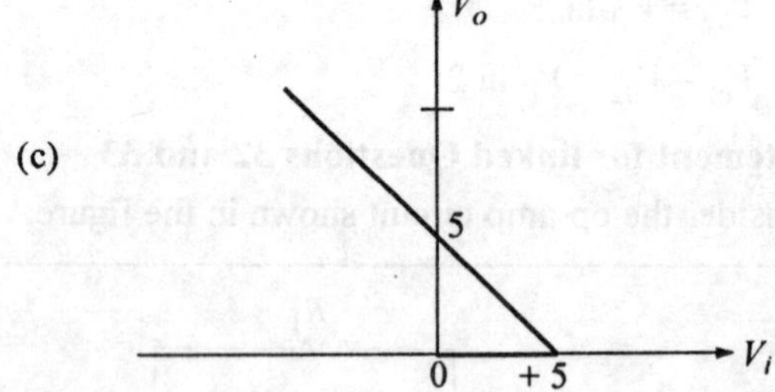

(d)

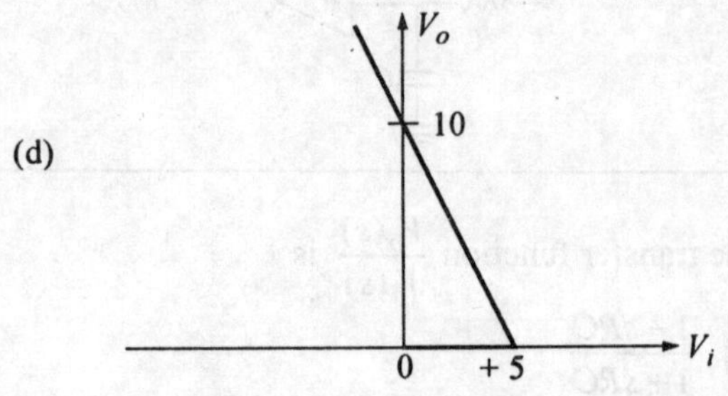

39. An ideal op-amp is used to make an inverting amplifier. The two input terminals of the op-amp are at the same potential because
   (a) The two input terminals are directly shorted internally
   (b) The input impedance of the op-amp is infinity
   (c) The open loop gain of the op-amp is infinity
   (d) CMRR is infinity

40. A non-inverting op-amp is shown in figure below. The output voltage V_o is

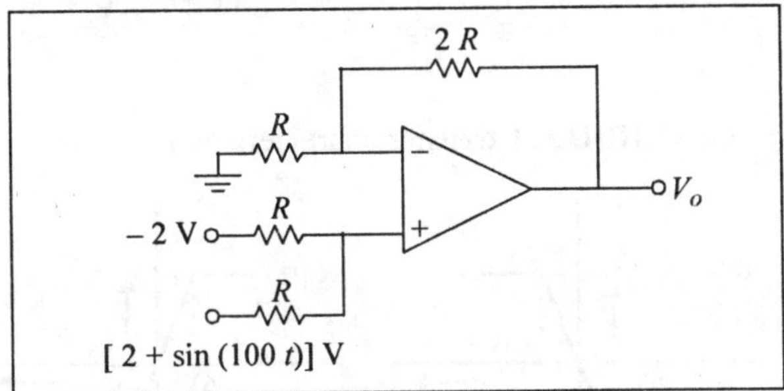

(a) $\frac{3}{2}$ sin 100 t (b) 3 sin 100 t
(c) 2 sin 100 t (d) None of the above

41. Let the magnitude of the gain in the inverting op-amp amplifier circuit shown in figure below be x with switch S_1 open. When the switch S_1 is closed, the magnitude of gain becomes

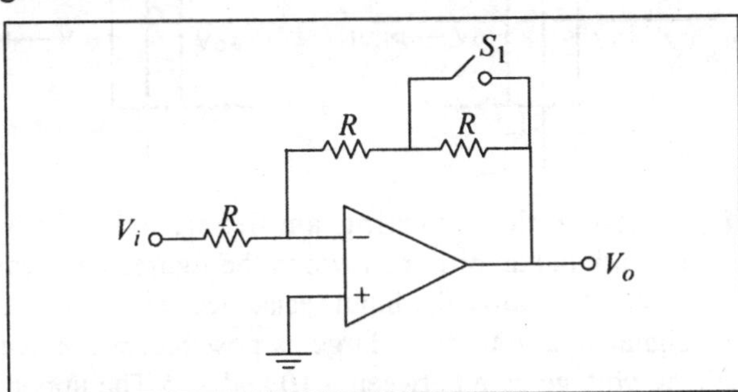

(a) $\frac{x}{2}$ (b) $-x$
(c) $2x$ (d) $-2x$

42. The circuit shown in figure uses an ideal op amp working with + 5 V and – 5 V power supplies. The output voltage V_o is equal to

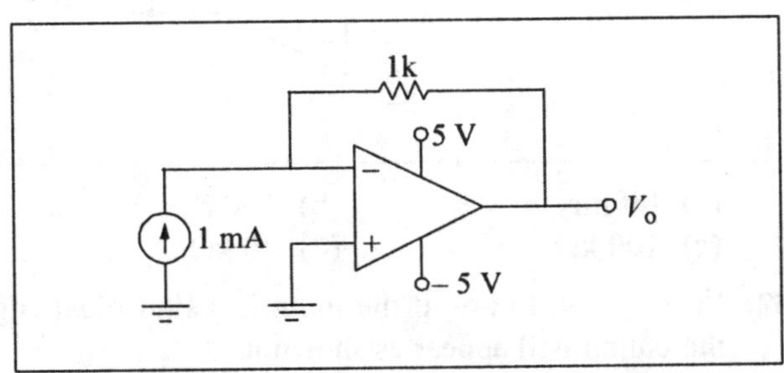

(a) + 5 V (b) – 5 V
(c) + 1 V (d) – 1 V

43. The **feedback factor** for the circuit shown in Figure is

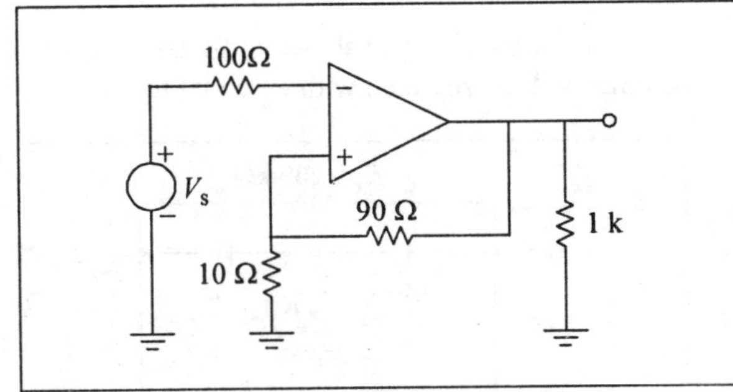

(a) 9/100 (b) 9/10
(c) 1/9 (d) 1/10

44. An **Op-amp,** having a **slew rate** of 62.8 V/μsec, is connected in a **voltage follower** configuration. If the maximum amplitude of the input sinusoid is 10 V, then the minimum frequency at which the slew rate limited distortion would set in at the output is
   (a) 1.0 MHz (b) 6.28 MHz
   (c) 10.0 MHz (d) 62.8 MHz

45. For the circuit of figure with an ideal operational amplifier, the maximum phase shift of the output V_{out} with reference to the input V_{in} is

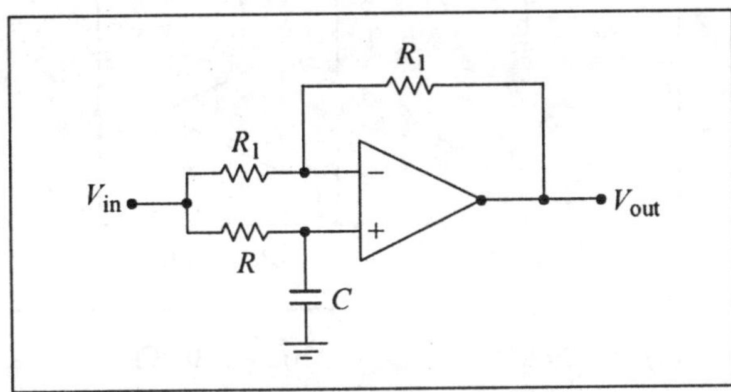

(a) 0° (b) – 90°
(c) + 90° (d) ± 180°

46. Assuming the operational amplifier to be ideal, the gain V_{out} / V_{in} for the circuit shown in figure is

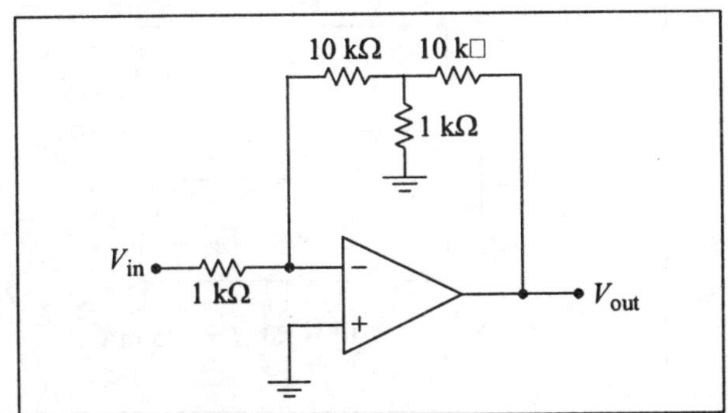

(a) – 1
(b) – 20
(c) – 100
(d) – 120

47. In the active filter circuit shown in figure, if $Q = 1$, a pair of poles will be realized with ω_o equal to

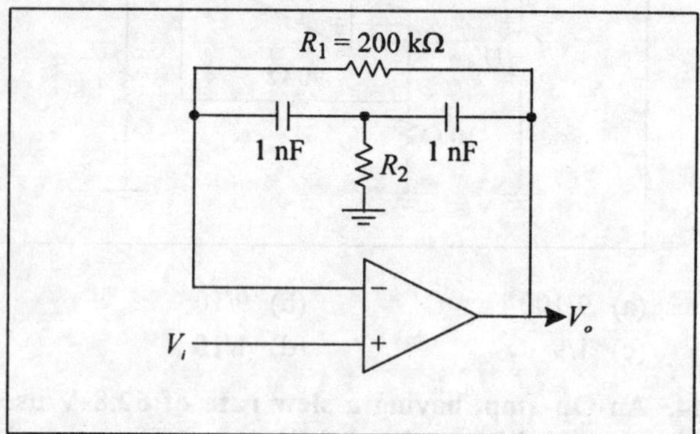

(a) 1000 rad/s
(b) 100 rad/s
(c) 10 rad/s
(d) 1 rad/s

48. The input resistance R_{IN} ($= v_x/i_x$) of the circuit in figure is

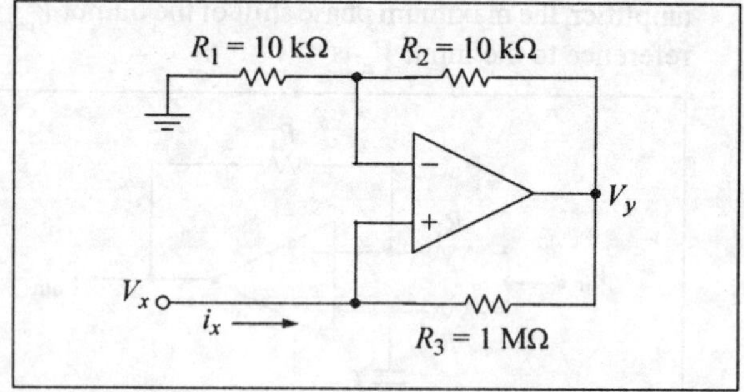

(a) + 100 kΩ (b) – 100 kΩ
(c) + 1 MΩ (d) – 1 MΩ

49. In the Schmitt trigger circuit shown in figure, if $V_{CE(sat)} = 0.1$ V, the output logic low level (V_{OL}) is

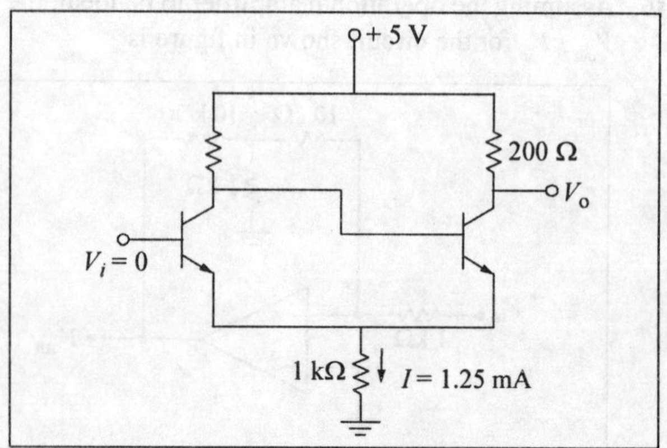

(a) 1.25 V (b) 1.35 V
(c) 2.50 V (d) 5.00 V

50. For the circuit shown below,

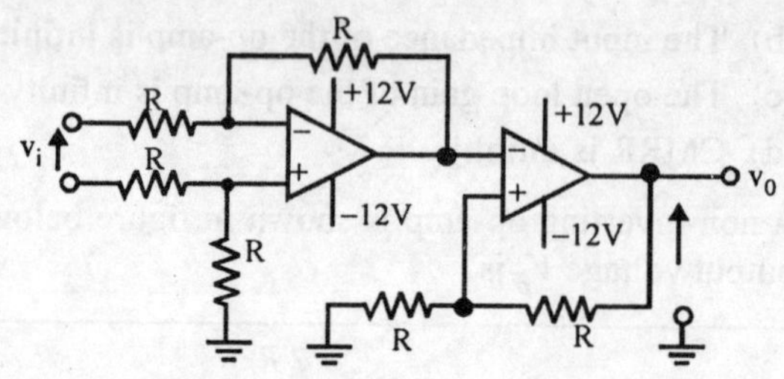

the CORRECT transfer characteristic is

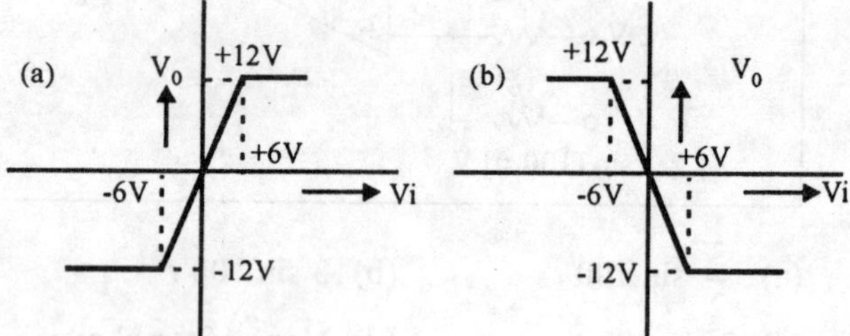

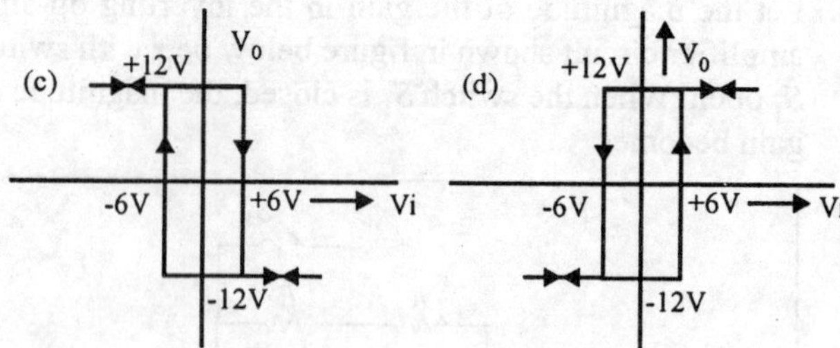

51. Consider the inverting amplifier, using an ideal operational amplifier shown in the figure. The designer wishes to realize the input resistance seen by the small-signal source to be as large as possible, while keeping the voltage gain between – 10 and –25 The upper limit of R_F is 1 MΩ. The value of R_1 should be

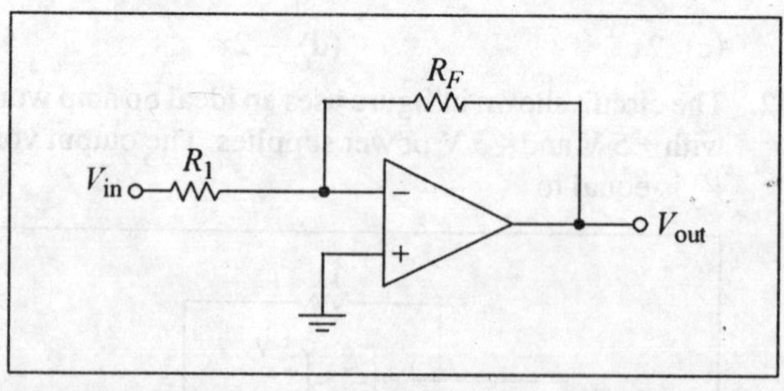

(a) Infinity (b) 1 MΩ
(c) 100 kΩ (d) 40 kΩ

52. In the given figure, if the input is a sinusoidal signal, the output will appear as shown in

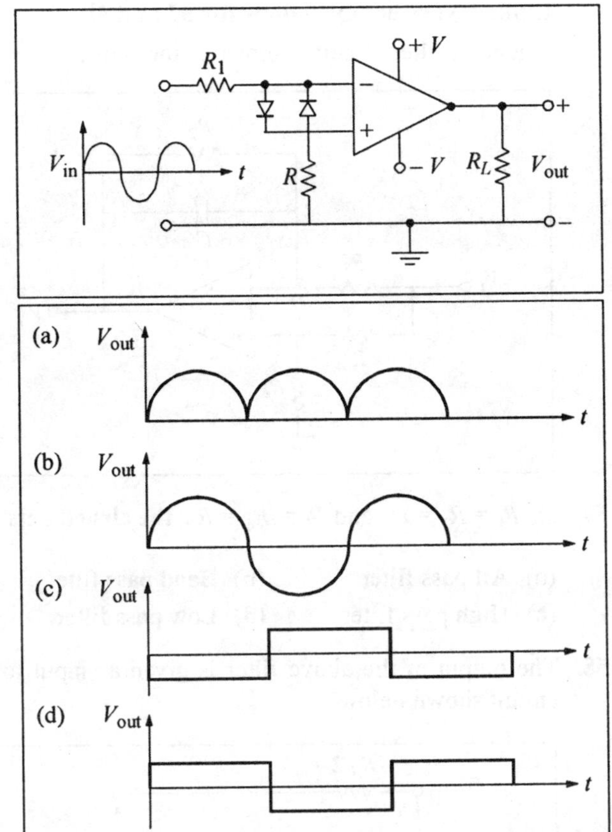

53. A relaxation oscillator is made using op-amp as shown in figure. The supply voltages of the op-amp are ± 12V. The voltage waveform at point P will be

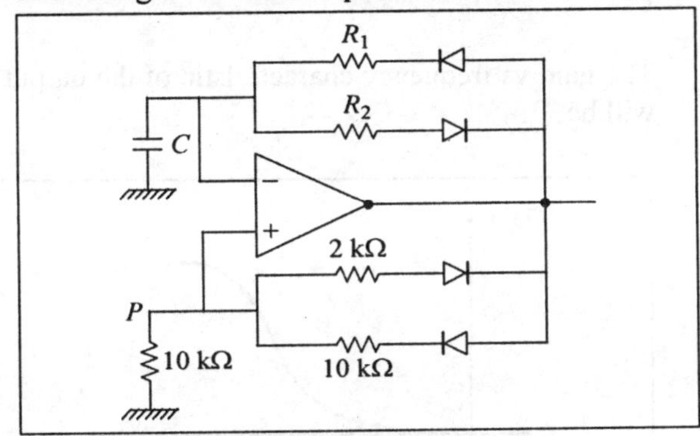

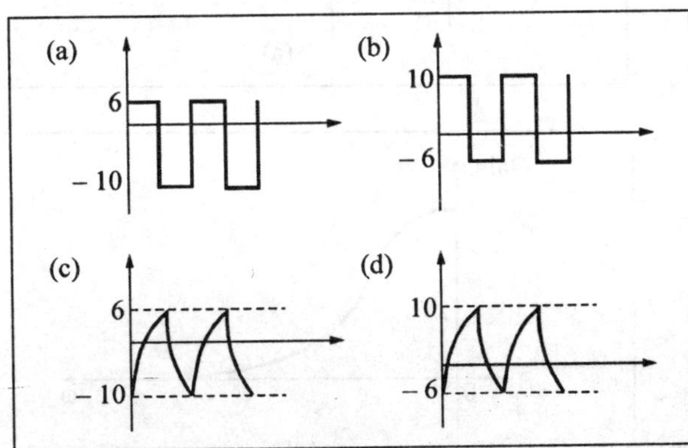

54. The circuit shown in the figure is

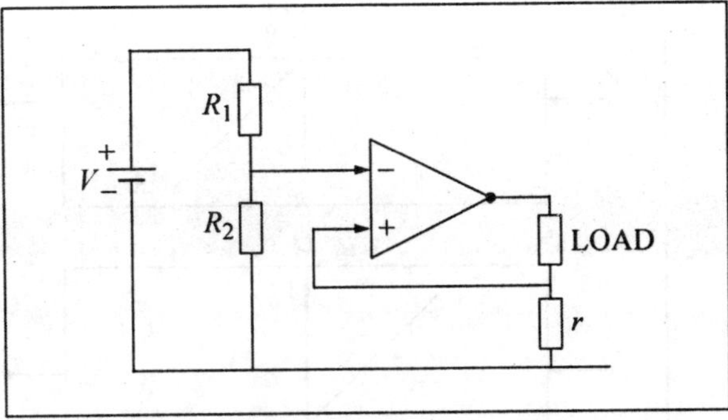

(a) A voltage source with voltage $\dfrac{rV}{R_1 \parallel R_2}$

(b) A voltage source with voltage $\dfrac{r \parallel R_2}{R_1}$V

(c) A current source with current $\dfrac{r \parallel R_2}{R_1 + R_2} \cdot \dfrac{V}{r}$

(d) A current source with current $\dfrac{R_2}{R_1 + R_2} \cdot \dfrac{V}{r}$

55. The switch S in the circuit of the figure is initially closed. It is opened at time $t = 0$. You may neglect the Zener diode forward voltage drops. What is the behaviour of V_{OUT} for $t > 0$?

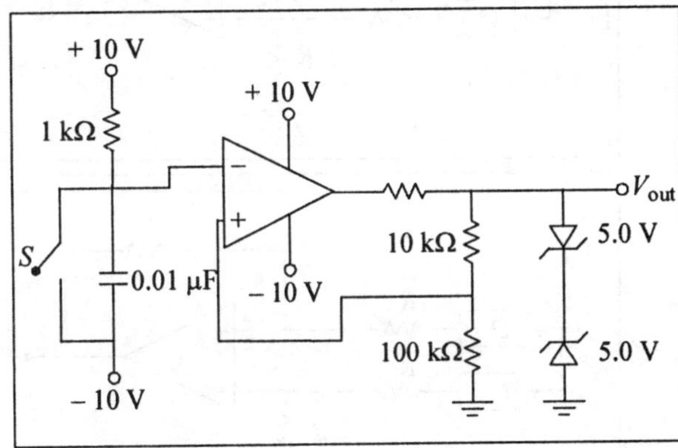

(a) It makes a transition from – 5 V to +5 V at t = 12.98 μs

(b) It makes a transition from – 5 V to +5 V at t = 2.57 μs

(c) It makes a transition from + 5 V to –5 V at t = 12.98 μs

(d) It makes a transition from + 5 V to –5 V at t = 2.57 μs

56. The block diagrams of two types of half-wave rectifier are shown in the figure. The transfer characteristic of the rectifiers are also shown within the block.

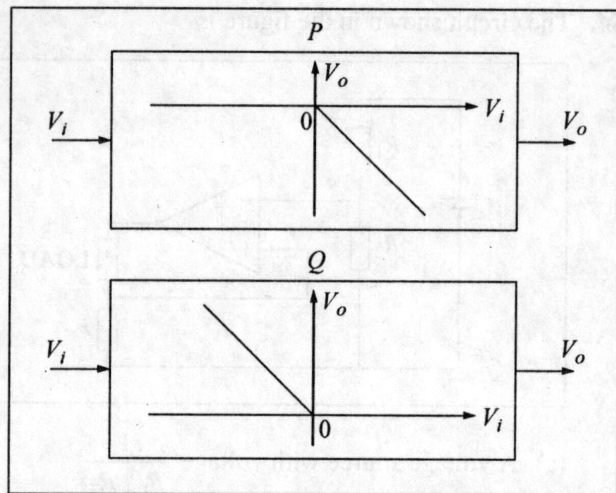

It is desired to make full-wave rectifier using above two half-wave rectifiers. The resultant circuit will be

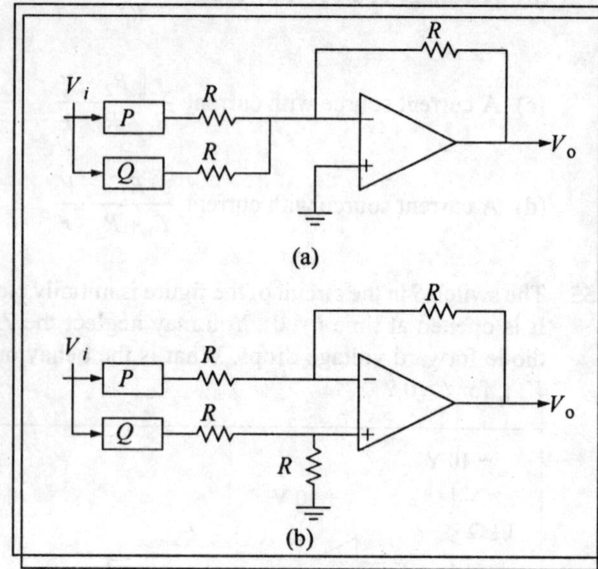

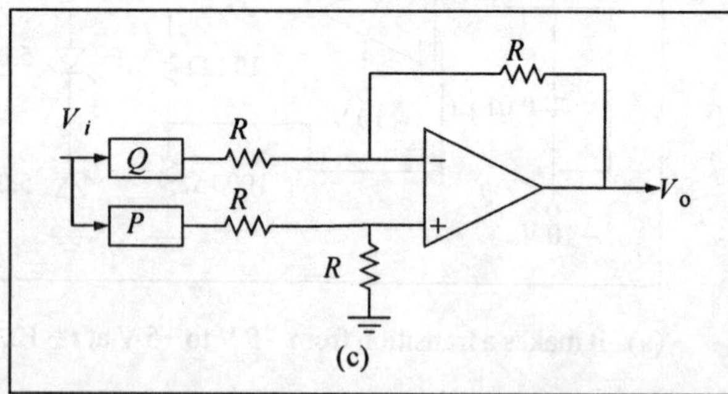

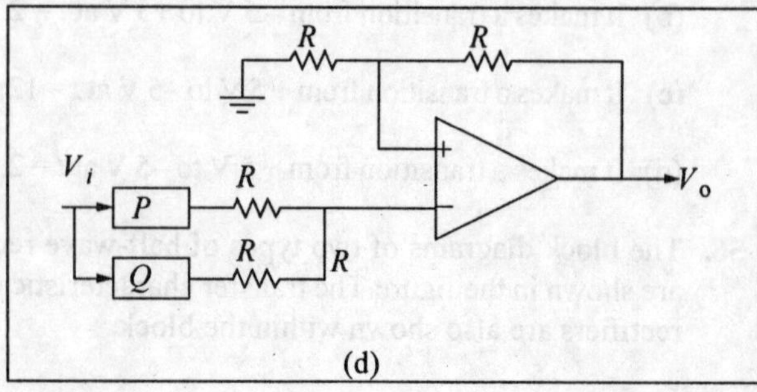

Linked Answers Questions for 57 and 58

A general filter circuit is shown in the figure.

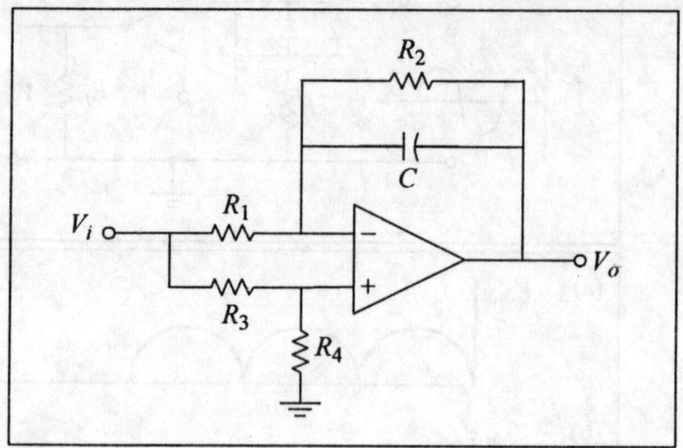

57. If, $R_1 = R_2 = R_A$ and $R_3 = R_4 = R_B$ the cirucit acts as a

(a) All pass filter
(b) Band pass filter
(c) High pass filter
(d) Low pass filter

58. The output of the above filter is given as input to the circuit shown below

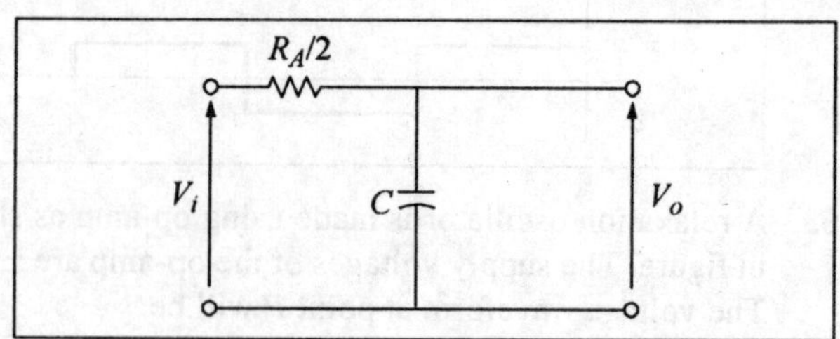

The gain vs frequency characteristic of the output (V_o) will be

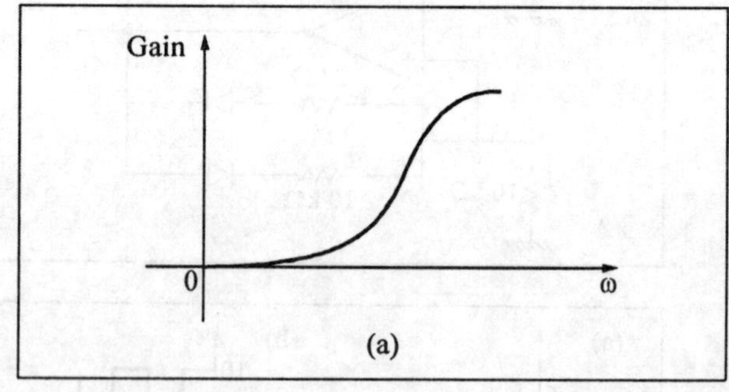

(a)

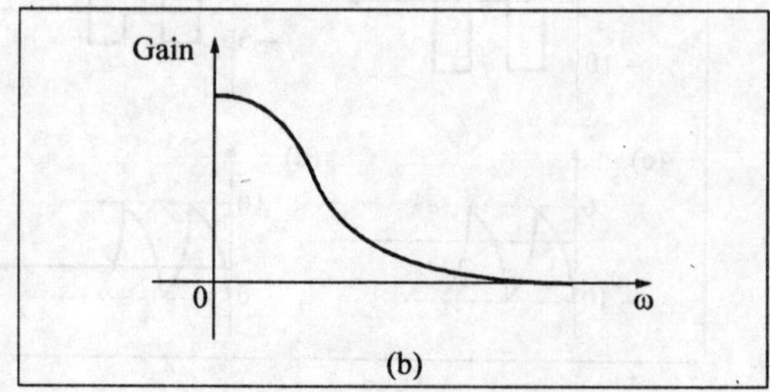

(b)

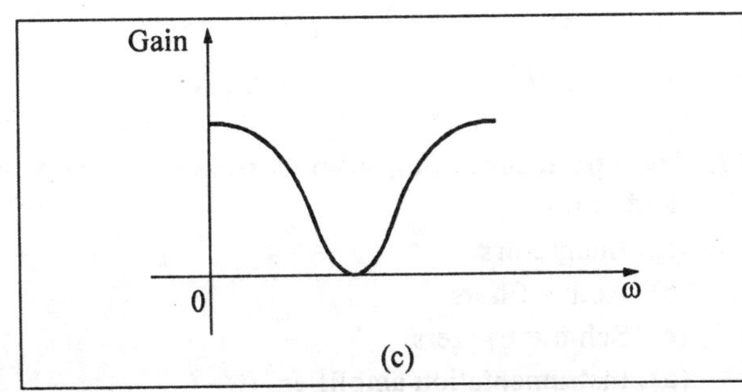

(c)

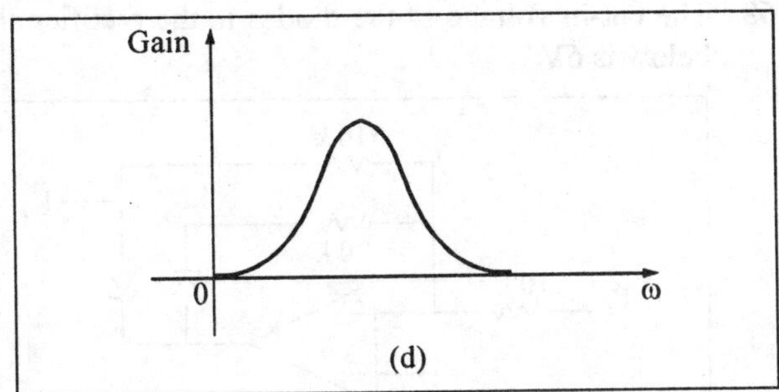

(d)

59. The nature of feedback in the opamp circuit shown is

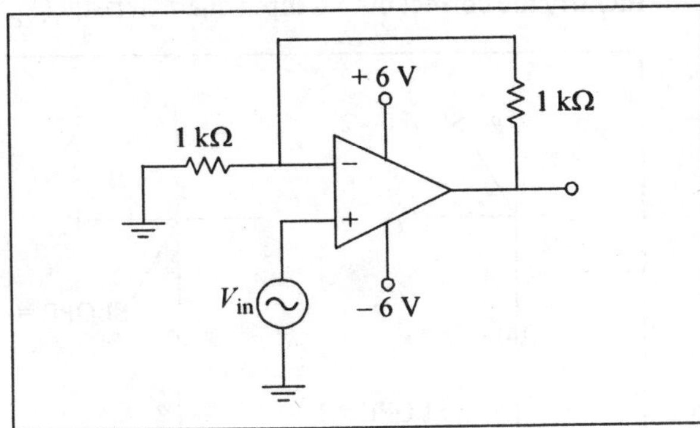

(a) Current- Current feedback
(b) Voltage-Voltage feedback
(c) Current-Voltage feedback
(d) Voltage-Current feedback

60. The following circuit has R = 10 kΩ, C = 10 μF. The input voltage is a sinusoid at 50 Hz with an rms value of 10V. Under ideal conditions, the current I_s from the source is

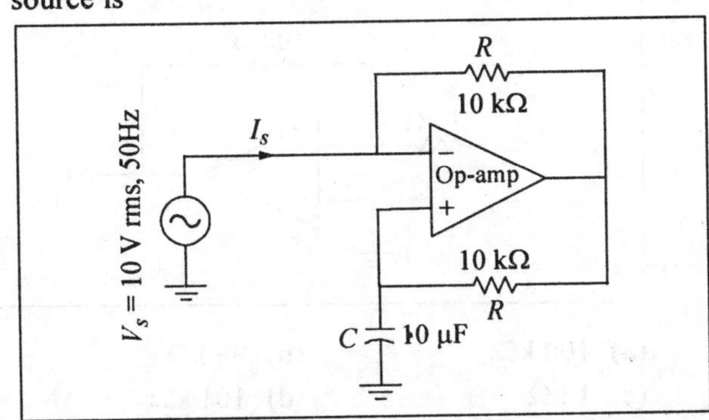

(a) 10 π mA leading by 90° (b) 20 π mA leading by 90°
(c) 10 mA leading by 90° (d) 10 π mA lagging by 90°

61. An ideal op-amp circuit and its input waveform are shown in the figure. The output waveform of this circuit will be

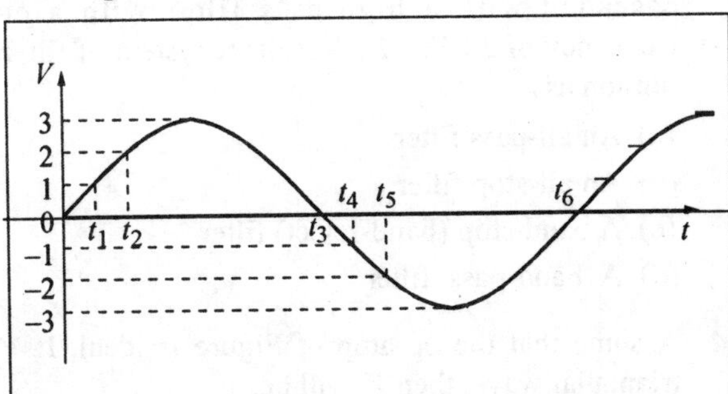

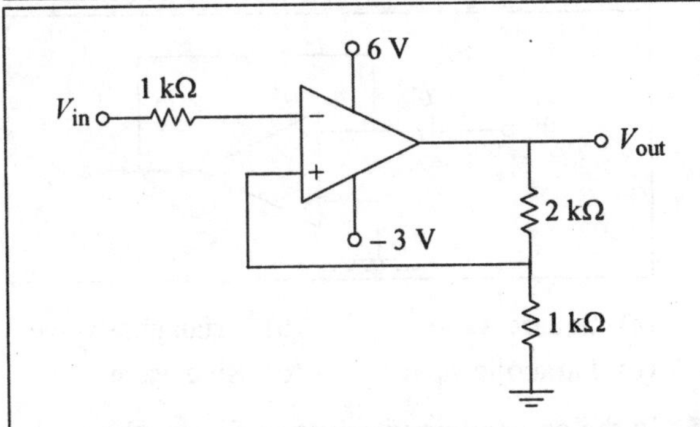

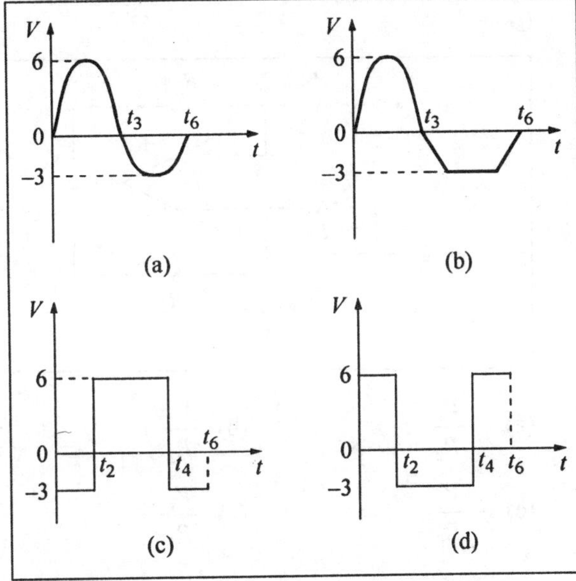

62. Given that the op-amp is ideal, the output voltage V_0 is

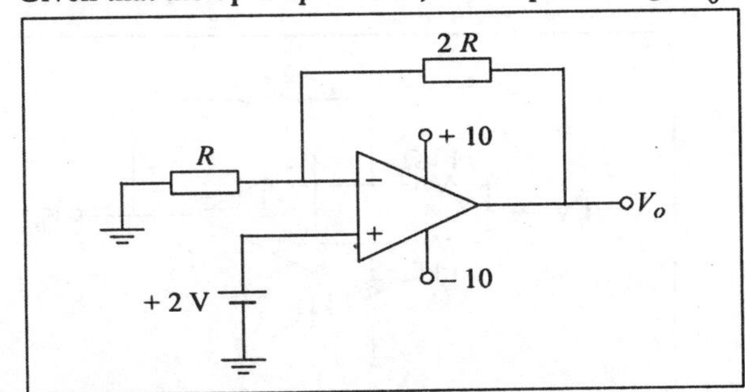

(a) 4 V (b) 6. V
(c) 7.5 V (d) 12.12 V

63. A low-pass filter with a cut-off frequency of 30 Hz is cascaded with a high-pass filter with a cut-off frequency of 20 Hz. The resultant system of filters will funtion as

(a) An all-pass filter
(b) An all-stop filter
(c) A band-stop (band-reject) filter
(d) A band-pass filter

64. Assume that the op-amp of Figure is ideal. If V_i is a triangular wave, then V_o will be:

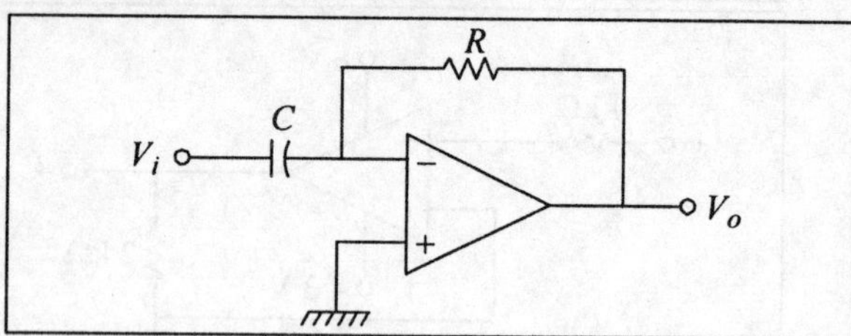

(a) Square wave (b) Triangular wave
(c) Parabolic wave (d) Sine wave

65. In the op-amp circuit given in Figure, the load current i_L, is:

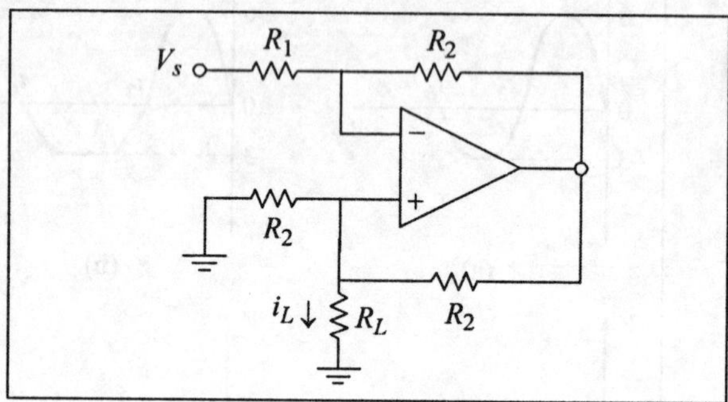

(a) $-\frac{V_s}{R_2}$ (b) $\frac{V_s}{R_2}$

(c) $-\frac{V_s}{R_L}$ (c) $\frac{V_s}{R_L}$

66. For the OP-Amp circuit shown in the figure, V_o is

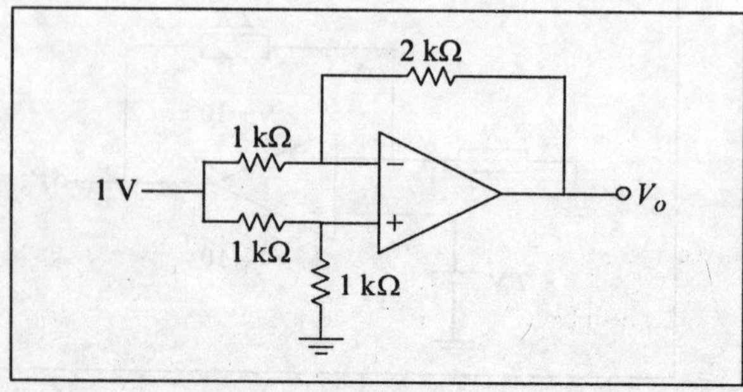

(a) – 2 V (b) – 1 V
(c) – 0.5 V (d) 0.5 V

67. The **operational amplifier** is used in the **non-linear mode** in

(a) Integrators
(b) Active filters
(c) Schmitt triggers
(d) Instrumentation amplifier

68. The cut-in voltage of the diodes in the rectifier shown below is 6V

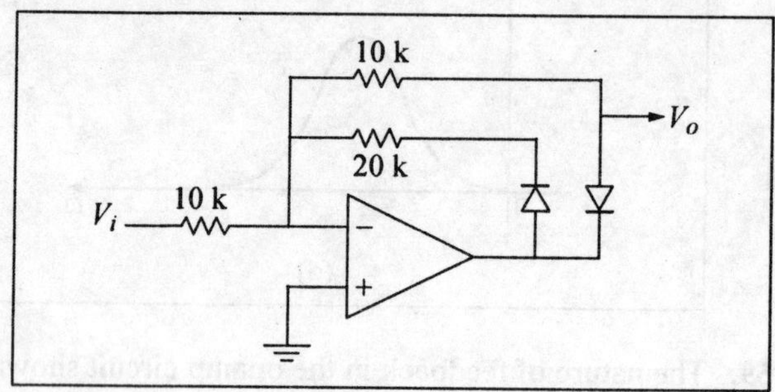

Identify the correct output input characteristic (V_o vs V_i)

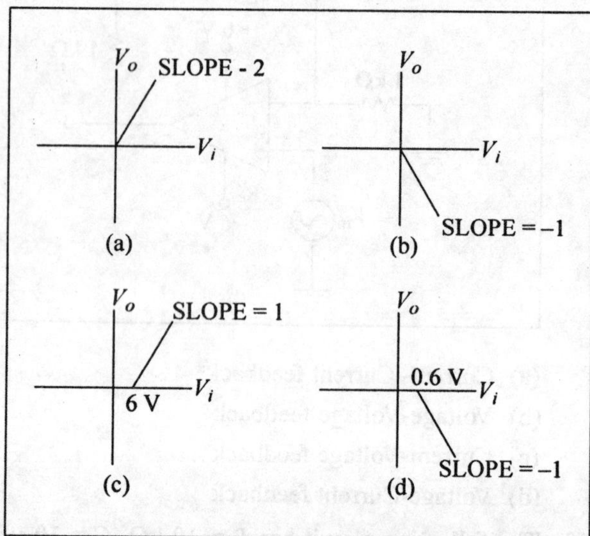

69. The input resistance of the op-amp circuit shown in figure.

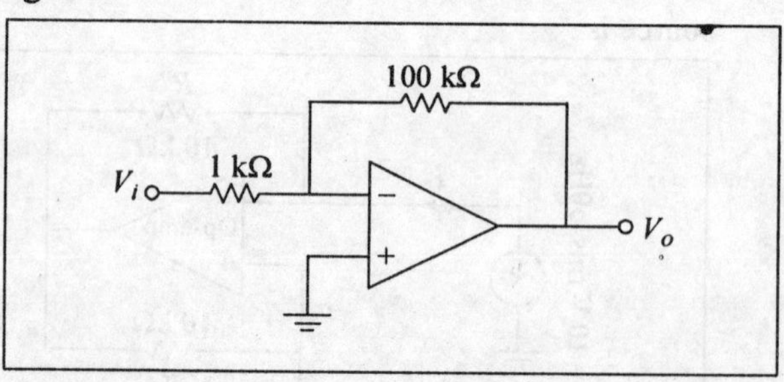

(a) 100 kΩ (b) 99 kΩ
(c) 1 kΩ (d) 101 kΩ

70. The output voltage (V_o) of the circuit shown in the figure is.

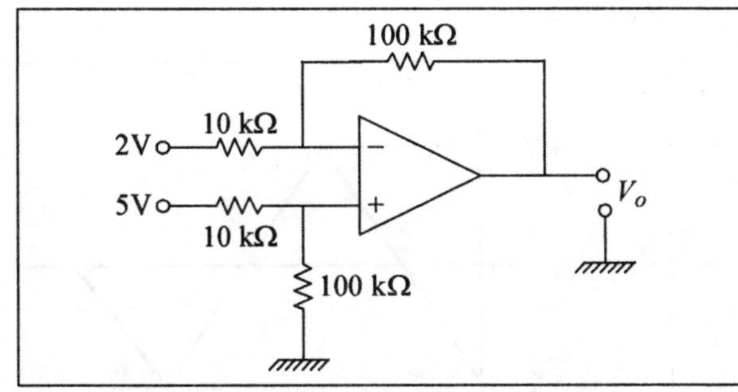

(a) – 20 V (b) 20 V
(c) – 30 V (d) 30 V

71. In the circuit shown in figure, if e_i = sin t, the voltage e_0 is

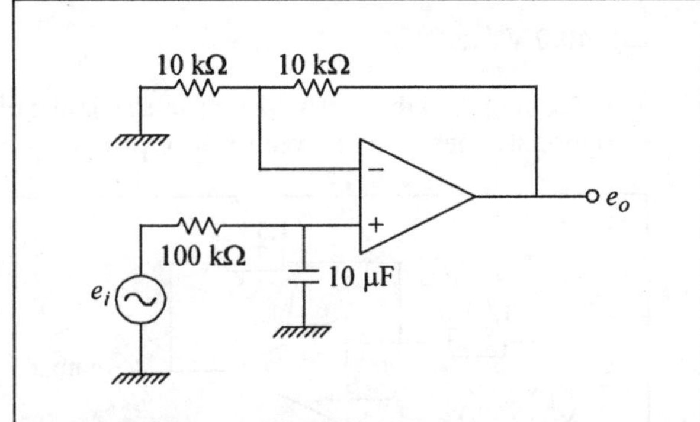

(a) $\sqrt{2}\sin\left[t-\left(\frac{\pi}{4}\right)\right]$ (b) $\frac{1}{\sqrt{2}}\sin\left[t+\left(\frac{\pi}{4}\right)\right]$

(c) $\frac{1}{\sqrt{2}}\sin\left[t-\left(\frac{\pi}{4}\right)\right]$ (d) $\sqrt{2}\sin\left[t+\left(\frac{\pi}{4}\right)\right]$

72. The differential input resistance of the circuit shown in figure is

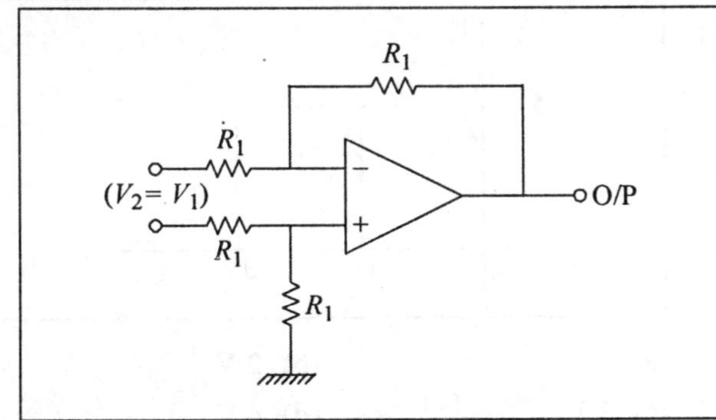

(a) R_1 (b) $\frac{R_1}{2}$
(c) $2R_1$ (d) R_f

73. The operational amplifiers use a differential input stage with a constant current source, mainly to obtain

(a) Very low common mode gain
(b) Very high differential gain
(c) Very low input noise
(d) Very high input resistance.

74. In the op-amp circuit shown in figure the output voltage V_o is equal to

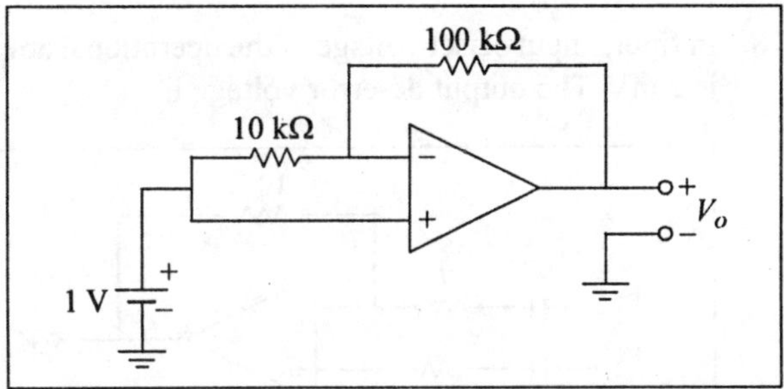

(a) 11 V (b) – 10 V
(c) 9 V (d) 1 V

75. Pick up the correct relationship for the circuit shown in figure.

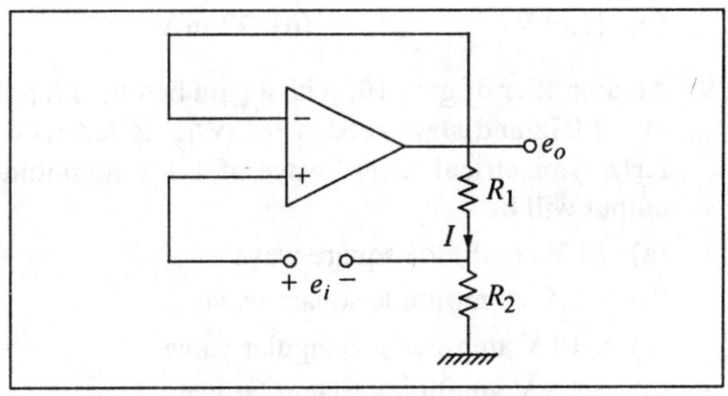

(a) $e_o = e_i$ (b) $e_o = \frac{R_2}{R_1} e_i$

(c) $1= \frac{e_i}{R_2}$ (d) $1 = \frac{e_i}{R_1}$

76. The op-amp in the amplifier circuit shown in figure has an offset voltage of 10 mV and it is ideal otherwise. If V_i is zero, the output voltage V_o is.

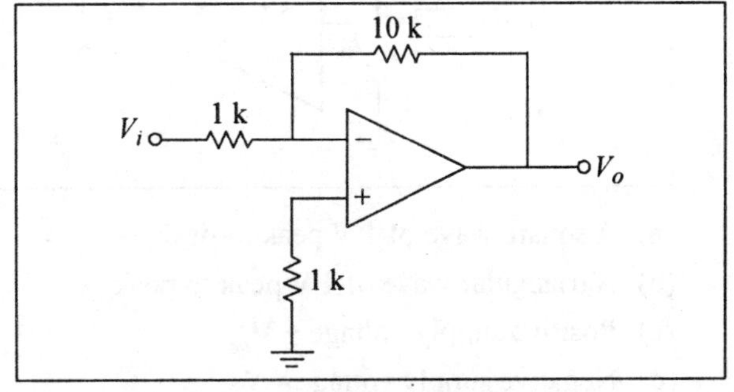

(a) 0 (b) 10 mV
(c) 100 mV (d) 110 mV

77. An op-amp with a **slew rate** of 1 V/μ sec has used to build an amplifier of gain + 10. If the input to the amplifier is a sinusoidal with peak amplitude of 1 V, the maximum allowable frequency of the input signal for undistorted output is
(a) 830 Hz (b) 15.92 Hz
(c) 31.84 kHz (d) 1.0 MHz

78. In figure input offset voltage of the operational amplifier is 2 mV. The output dc-error voltage is

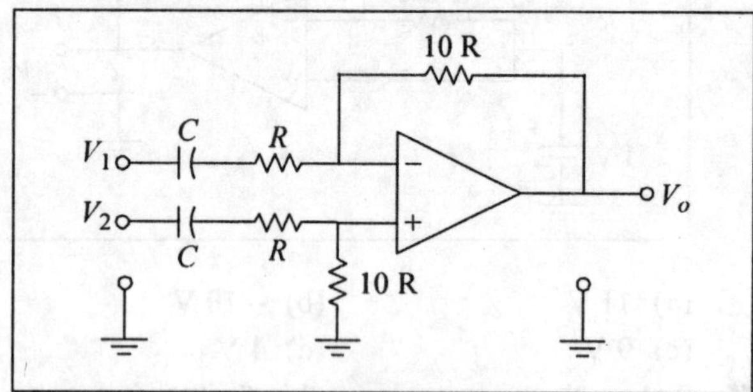

(a) 0 (b) 2 mV
(c) 11 m V (d) 22 m V

79. An amplifier of gain 10, with a gain bandwidth product of 1 MHz and **slew rate** of 0.1 V/μs is fed with a 10 KHz symmetrical square wave of ± 1 V amplitude. Its output will be
(a) 10 V amplitude square wave
(b) ± 2.5 V amplitude square wave
(c) ± 10 V amplitude triangular wave
(d) ± 2.5 V amplitude triangular wave.

80. An integrator circuit is shown in figure. The Ops-amp is of type 741 and has an **input offset current** i_{os} of 1 μA. C is 1μF and R is 1 MΩ. If the input V_i is a 1 kHz. Square wave of 1V peak to peak, the output V_o, under steady state condition, will be

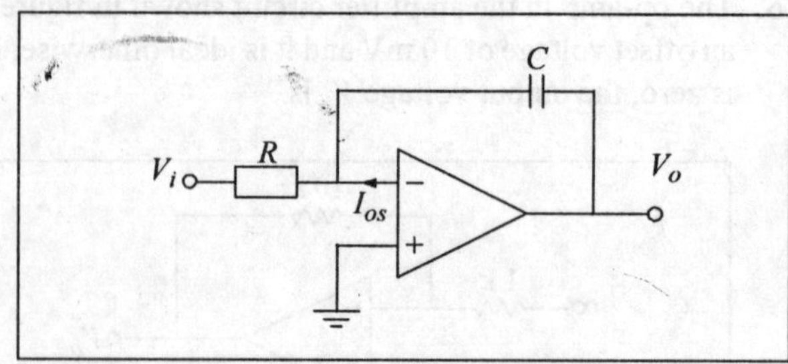

(a) A square wave of 1 V peak to peak.
(b) A triangular wave of 1 V peak to peak
(c) Positive supply voltage + V_{cc}
(d) Negative supply voltage – V_{cc}

81. The output of an **Ops-amp** whose input is a 2.5 MHz square wave is shown in figure. The **slew rate** of the Ops- amp is

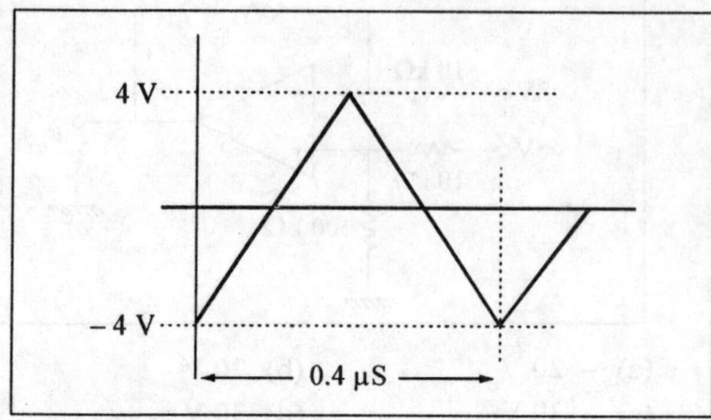

(a) 0.8 V / μs
(b) 8.0 V / μs
(c) 20.0 V / μs
(d) 40.0 V / μs

82. The Ops-amp and the 1 mA current source in the circuit of figure are ideal. The output of the Ops-amp is

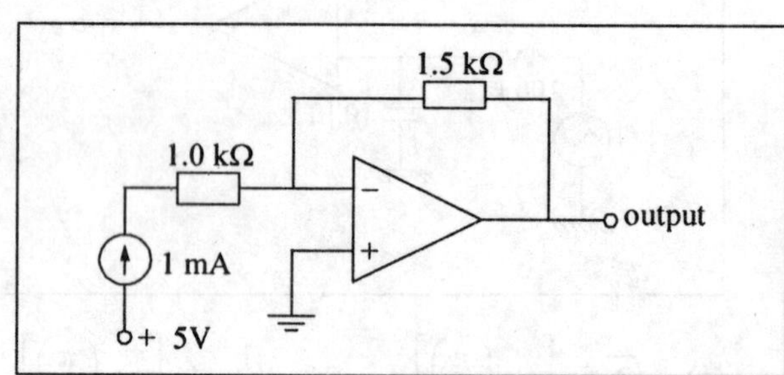

(a) –1.5mA (b) – 1.5 V
(c) –7.5 V (d) + 1.5 V

83. The input output characteristics of a **Schmitt trigger** is given in figure. The **noise margin of the Schmitt trigger** is

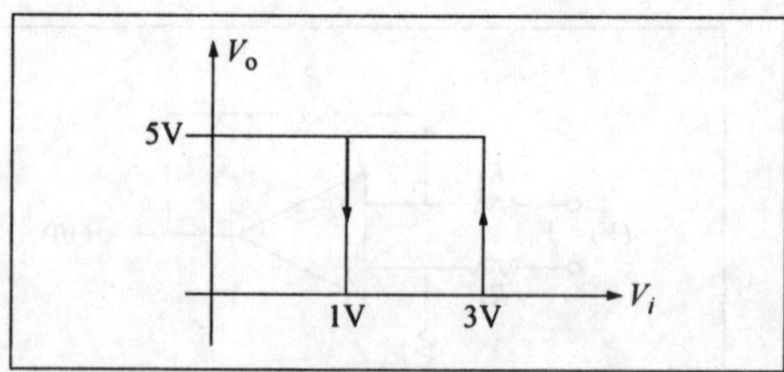

(a) 1 V (b) 2 V
(c) 3 V (d) 4 V

84. The Ops-amp used in the inverting amplifier shown in figure has an equivalent input offset voltage V_{ios} of 5 mV. The output offset voltage is.

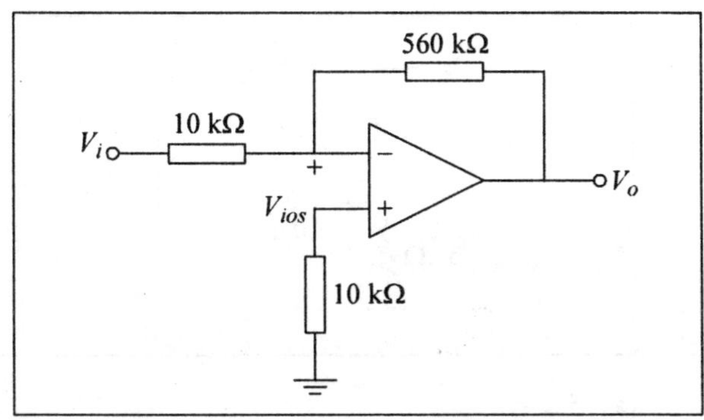

(a) 5 mV (b) 280 mV
(c) 285 mV (d) 560 mV

85. In the circuit shown in figure the Ops-amps used are ideal. The output V_o is.

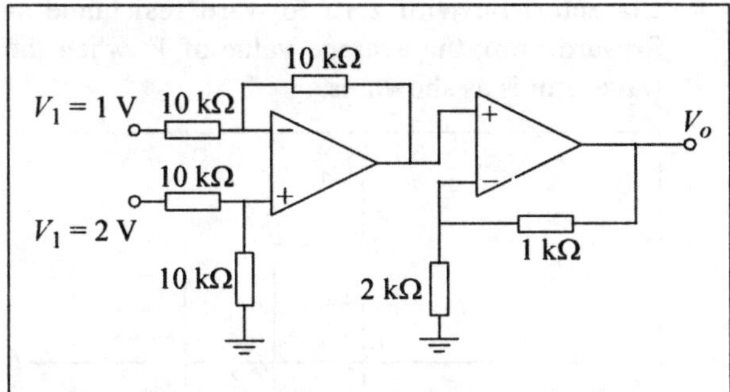

(a) 3.0 V
(b) 1.5 V
(c) 1.0 V
(d) 0.5 V

86. The circuit shown in figure is that of a waveform generator. Assuming ideal devices, and ± 12 V supply the output V_o is a

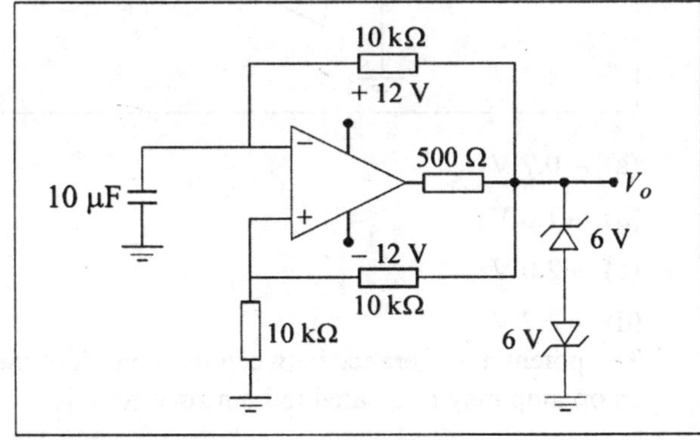

(a) Triangular wave of period 120 ms and amplitude ± 6V.

(b) Square wave of period 60 ms and amplitude ± 6V.

(c) Square wave of period 120 ms and amplitude ± 6V.

(d) Square wave of period 60 ms and amplitude ± 12V

87. The output of the **Op-amp in the circuit** of figure is

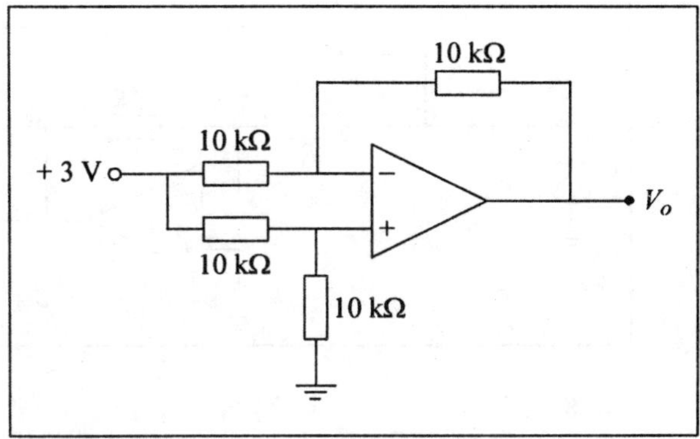

(a) 0 V (b) – 3 V
(c) + 1.5 (d) + 3 V

88. The wiper of the 20 kΩ potentiometer in figure is positioned half way. Then the voltage V_o of the circuit is

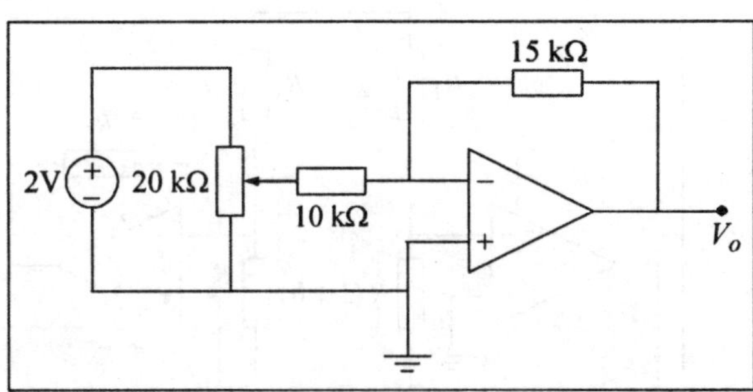

(a) – 1.5 V
(b) – 1.0 V
(c) – 0.75 V
(d) + 2.5 V

89. The value of V_o in the circuit, shown in figure is

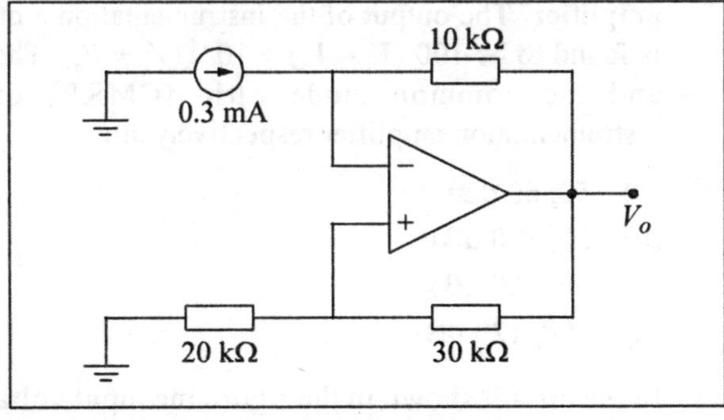

(a) – 5 V (b) – 3 V
(c) + 3V (d) + 5V

90. The gain (V_o / V_i) of the amplifier circuit shown in figure is

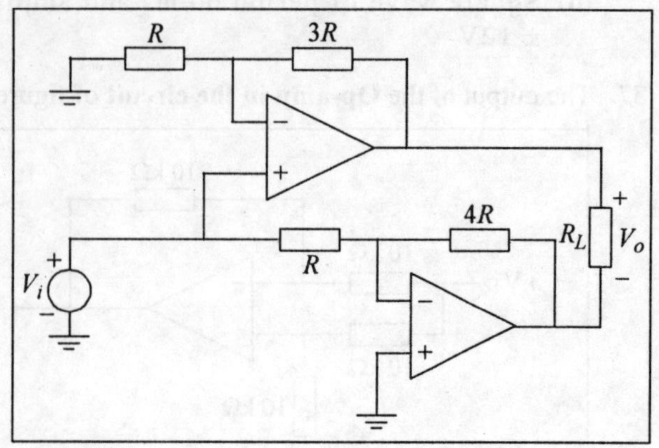

(a) 8 (b) 4

(c) – 4 (d) $\frac{3R_L}{R}$

91. The output voltage V_o in the circuit in figure is

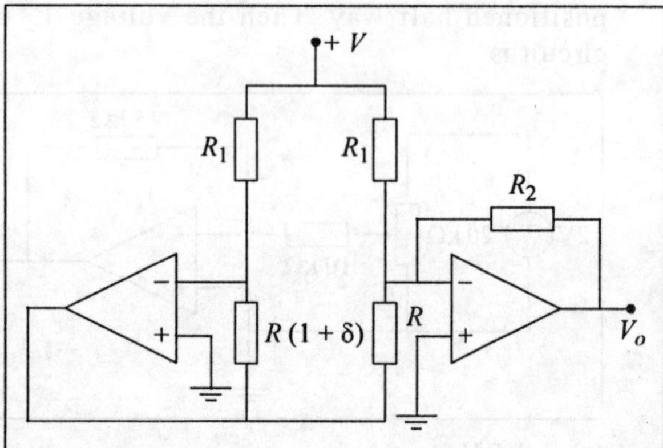

(a) $\frac{R_2}{R} V\,\delta$ (b) $\frac{R_2}{R_1} V\,\delta$

(c) $\frac{R_2}{R_1\delta} V$ (d) $\frac{R_2}{R(1+\delta)} V$

92. V_1 and V_2 are the input voltages of an **instrumentation amplifier**. The output of the instrumentation amplifier is found to be $100\,(V_1 - V_2) + 10^{-4}\,(V_1 + V_2)$. The gain and the common mode ratio (CMRR) of the instrumentation amplifier respectively are

(a) (50, 60 dB)
(b) (50, 120 dB)
(c) (100, 60 dB)
(d) (100, 120 dB)

93. In the circuit shown in the figure the input voltage V_i is a symmetrical saw-tooth wave of average value zero positive slope and peak to peak value 20 V. The average value of the output, assuming an ideal operational amplifier with peak to peak symmetrical swing of 30 V, is

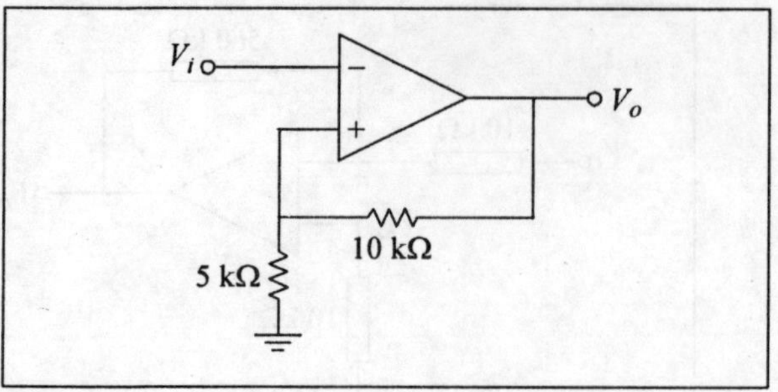

(a) 5 V
(b) 10 V
(c) – 5 V
(d) 7.5 V

94. In the circuit shown in the figure, assuming ideal diode characteristic with zero forward resistance and 0.7 forward drop, the average value of V_o when the input waveform is as shown, is

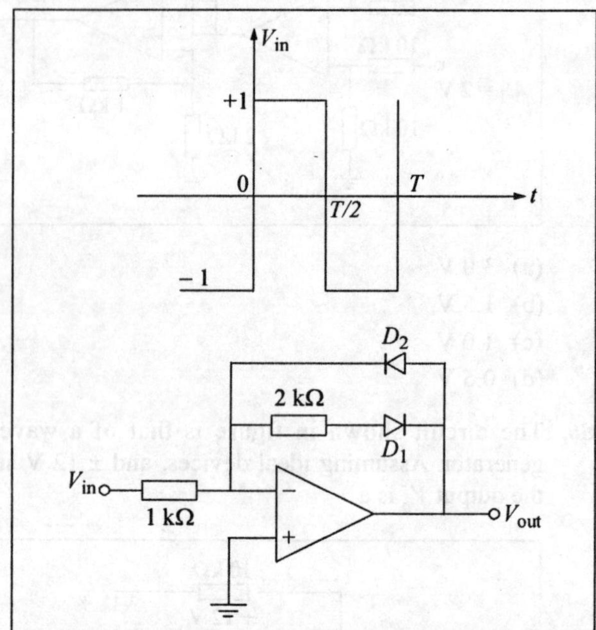

(a) – 0.7 V
(b) – 1.0 V
(c) – 2.0 V
(d) – 2.7 V

95. The potential difference between the input terminals of an op amp may be treated to be nearly zero, if

(a) The two supply voltages are balanced
(b) The output voltage is not saturated
(c) The op amp is used in a circuit having negative feedback
(d) There is a dc bias path between each of the input terminals and the circuit ground.

96. The figure shows a single-Op-amp differential amplifier circuit

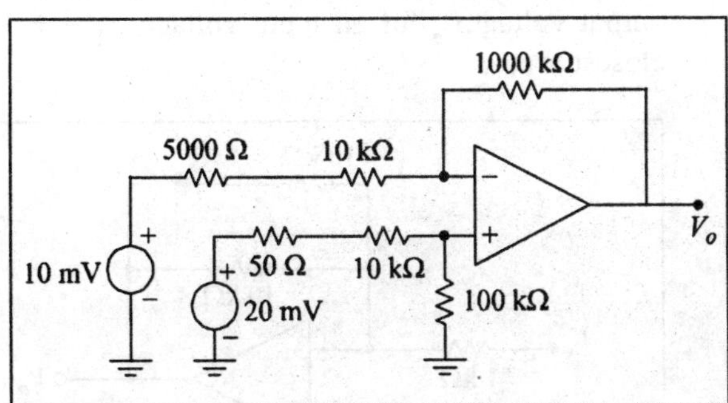

Which one of the following statements about the output is correct ?

(a) $V_o \le 95$ mV

(b) 95 mV $< V_o \le 98$ mV

(c) 98 mV $< V_o \le 101$ mV

(d) $V_o > 101$ mV

97. Consider the circuit shown below:

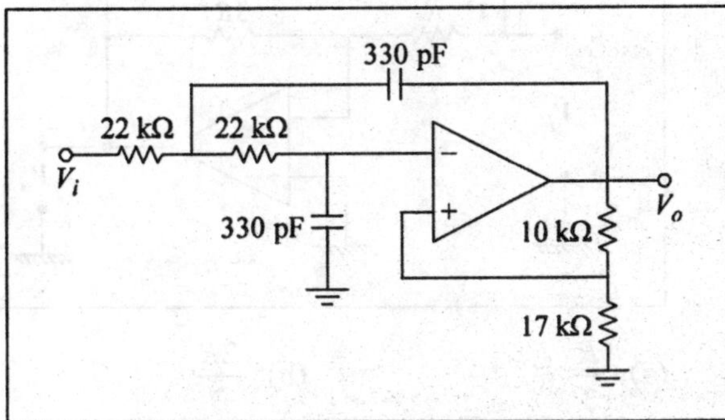

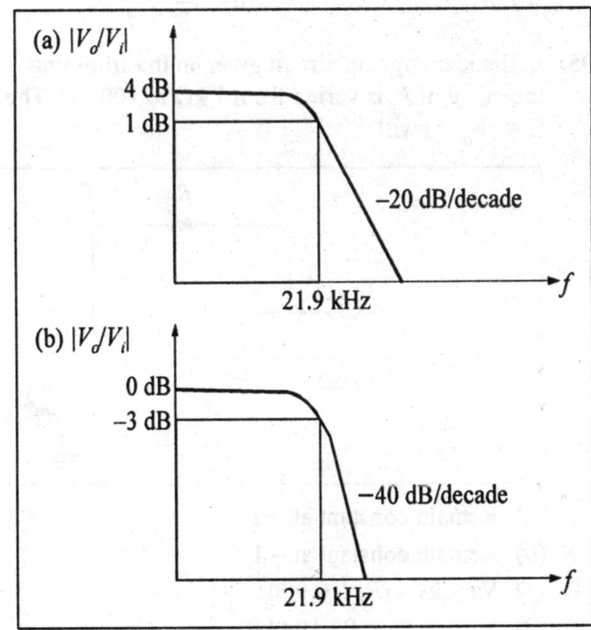

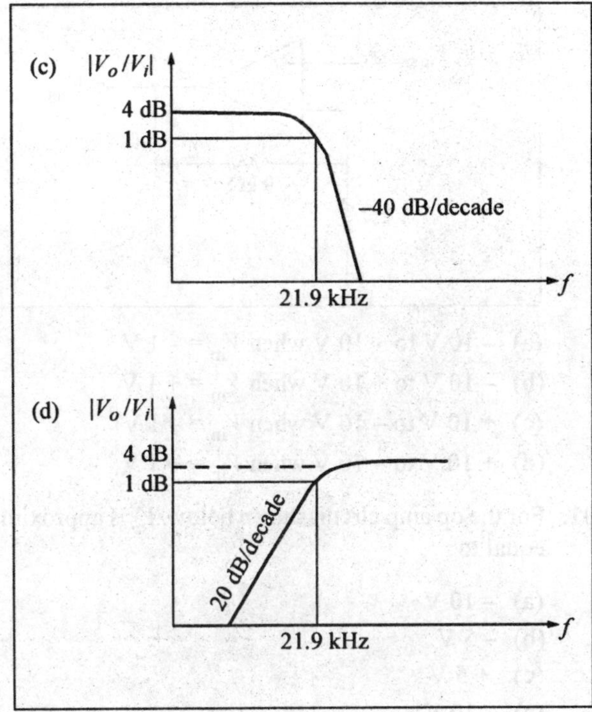

98. A differential amplifier shown below has a diffrential mode gain of 100 and a CMRR of 40 dB. If $V_1 = 0.55$ V and $V_2 = 0.45$ V the output V_o is

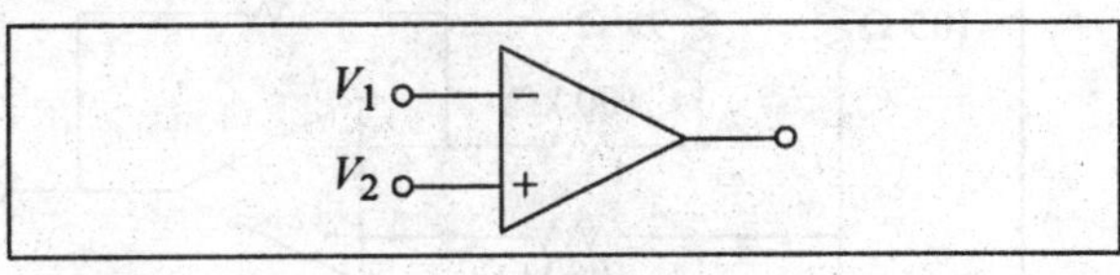

(a) 10 V (b) 10.5 V

(c) 11 V (d) 15 V

99. The op amp circuit shown below is that of a

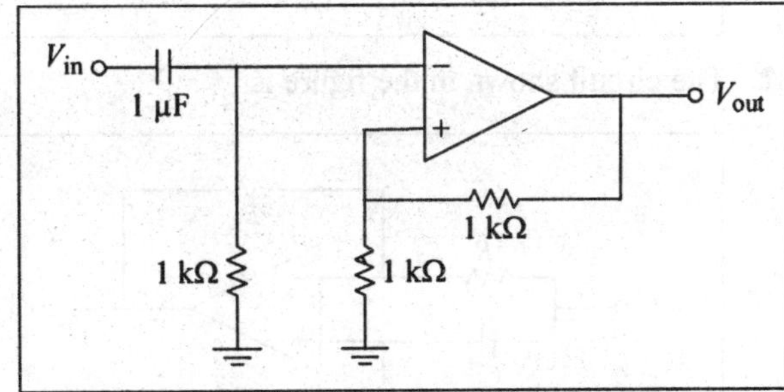

(a) Low pass filter with a maximum gain of 1

(b) Low pass filter with a maximum gain of 2

(c) High pass filter with a mixımum gain of 1

(d) High pass filter with a maximum gain of 2

100. In the op -amp circuit shown below, the input voltage V_{in} is gradually increased from – 10 V to + 10 V. Assuming that the output voltage V_{out} saturates at – 10 V and + 10 V, V_{out} will change from

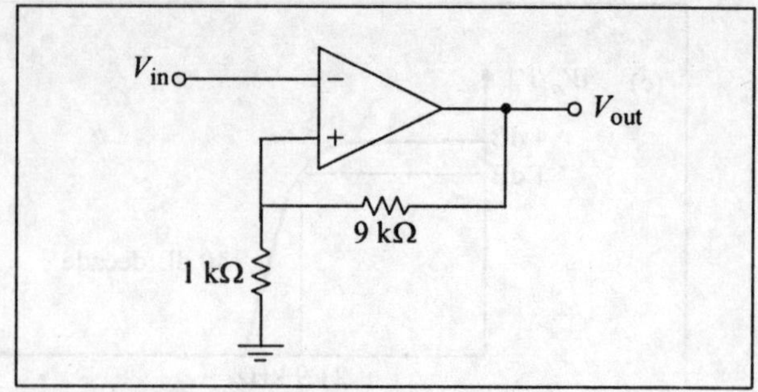

(a) – 10 V to + 10 V when V_{in} = – 1 V
(b) – 10 V to + 10 V when V_{in} = + 1 V
(c) + 10 V to – 10 V when V_{in} = – 1 V
(d) + 10 V to – 10 V when V_{in} = + 1 V

101. For the op amp circuit shown below, V_o is approximately equal to

(a) – 10 V
(b) – 5 V
(c) + 5 V
(a) – 10 V

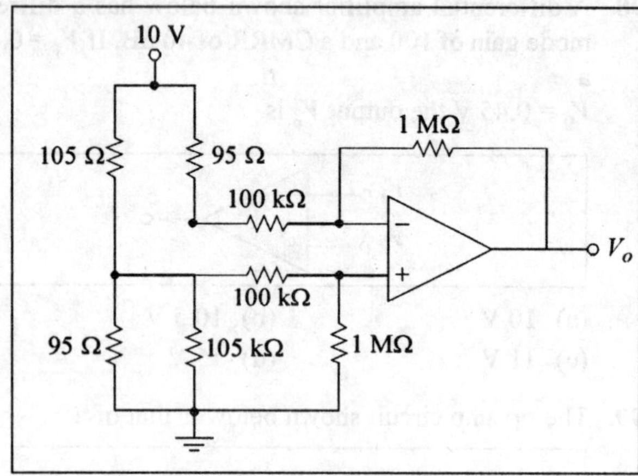

102. The circuit shown in the figure is

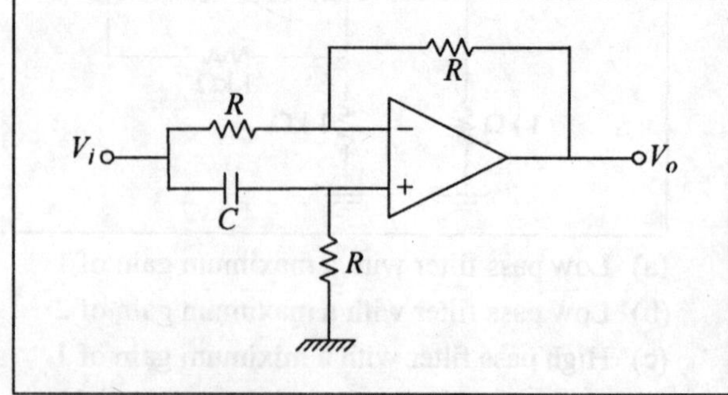

(a) An all-pass filter
(b) A bandpass filter
(c) A highpass filter
(d) A lowpass filter

103. In the circuit shown, the Zener diode has ideal characteristics and a breakdown voltage of 3.2 V. The output voltage V_o for an input voltage V_1 = + 1 V is closest to

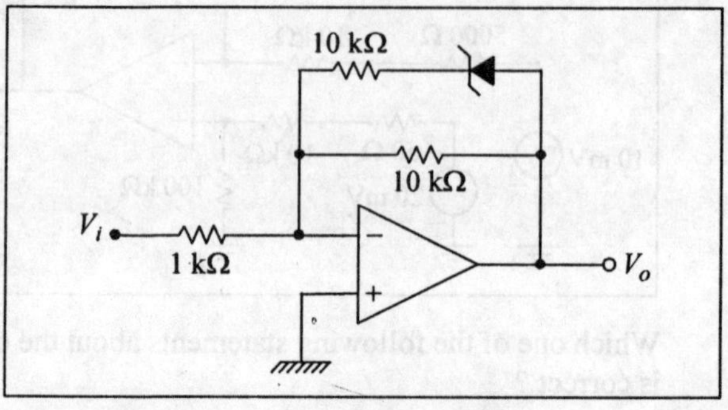

(a) – 10 V
(b) – 6.6 V
(c) – 5 V
(d) – 3.2 V

104. The input resistance of the circuit shown in the figure, assuming an ideal op amp, is

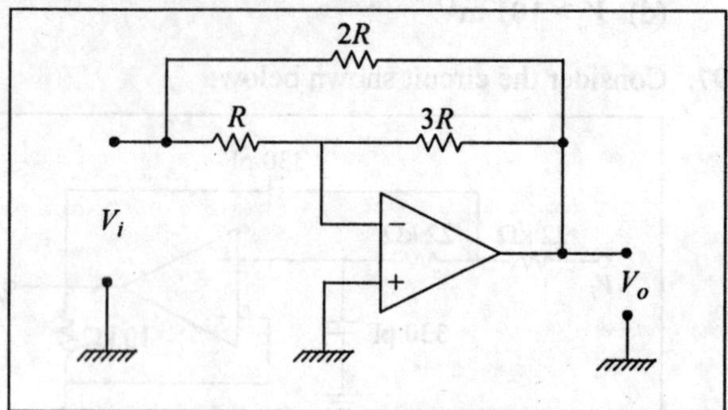

(a) $\frac{R}{3}$
(b) $\frac{2R}{3}$
(c) R
(d) $\frac{4R}{3}$

105. In the ideal opamp circuit given in the adjoining figure, the value of R_f is varied from 1 kΩ to 100 kΩ. The gain G = (V_o/V_i) will

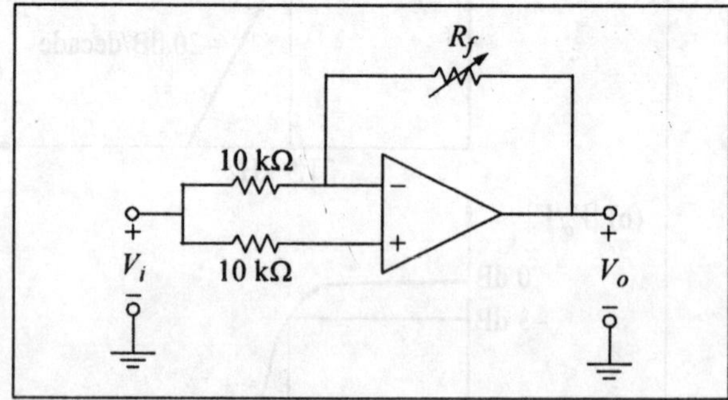

(a) Remain constant at + 1
(b) Remain constant at – 1
(c) Vary as – $(R_f/10,000)$
(d) Vary as $(1 + R_f/10,000)$

106. An active filter is shown in the adjoining figure. The dc gain and the 3 dB cut-off frequency of the filter respectively, are, nearly

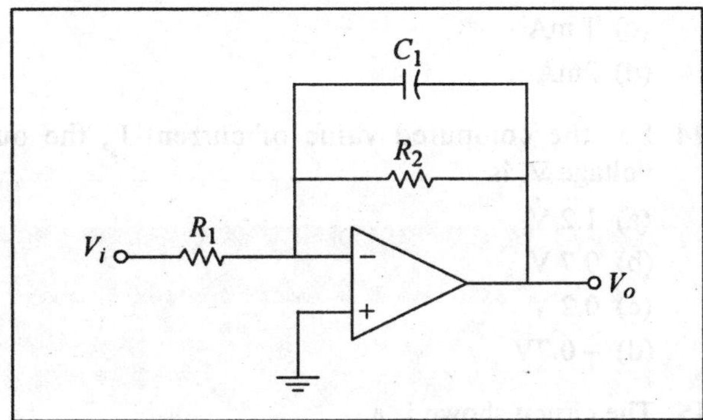

$R_1 = 15.9\ k\Omega, R_2 = 159\ k\Omega, C_1 = 1.0\ \text{nF}$

(a) 40 dB, 3.14 kHz
(b) 40 dB, 1.00 kHz
(c) 20 dB, 6.28 kHz
(d) 20 dB, 1.00 kHz

107. A differential amplifier is constructed using an ideal opamp as shown in the adjoining figure. The values of R_1 and R_2 and 47 kΩ and 470 kΩ respectively.

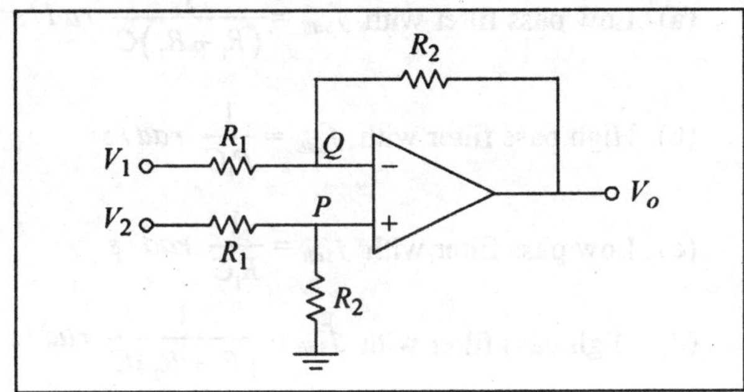

The input impedances seen looking into the terminals V_1 and V_2, with respect to ground, respectively are

(a) 47 kΩ and 43 kΩ
(b) 47 kΩ and 47 kΩ
(c) 47 kΩ and 517 kΩ
(d) 517 kΩ and 517 kΩ

108. n the circuit shown in the figure, the switch S has been in position 1 for a long time. It is then moved to position 2. Assume the Zener diodes to be ideal. The time delay between the switch moving to position 2 and the transition in the output voltage V_o is

(a) 5.00 ms
(b) 8.75 ms
(c) 10.00 ms
(d) 13.75 ms

109. Assuming base-emittr of 0.7 V and β = 99 of transistor Q_1, the output voltage V_0 in the ideal opamp circuit shown below is

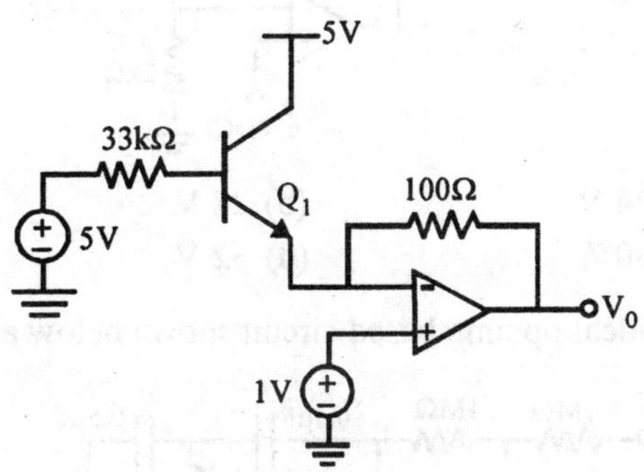

(a) –1V
(b) –1/3.3 V
(c) 0V
(d) 2V

110. The transfer characteristics of the circuit drawn below is observed on an oscilloscope used in XY mode. The display on the oscilloscope is shown on the right hand side. V_i is connected to the X input with a setting of 0.5 V/ div, and V_0 is connected to the Y inptu with a setting of 2V/div. The beam is positioned at the origin when V_i is zero.

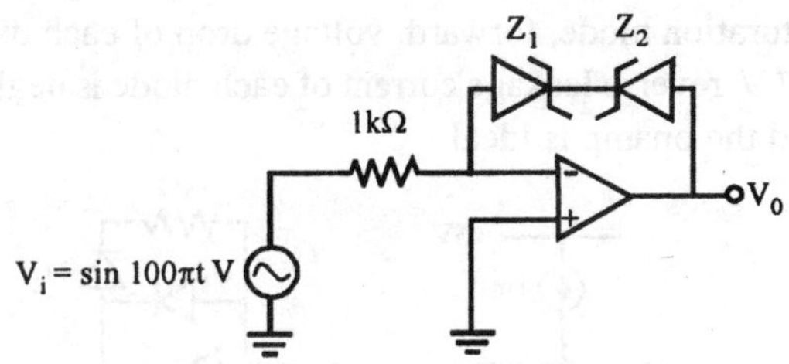

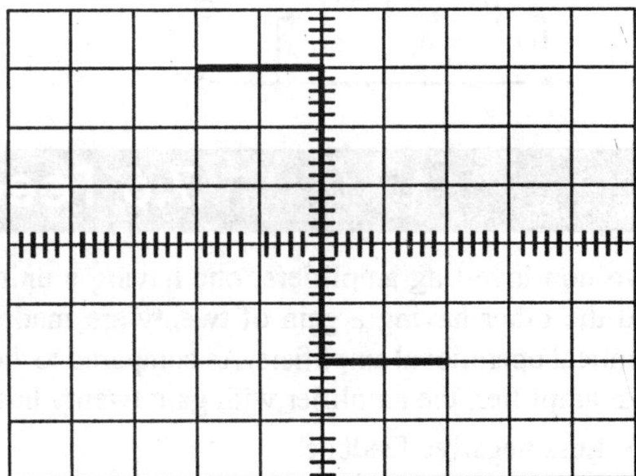

Assuming that the opamp is ideal and the zener diodes have forward biased voltage drop of 0.7 V, the values of reverse break - down voltages of Z_1 and Z_2 are, respectively,

(a) 3.3 V and 5.3 V
(b) 4.7 V and 6.7 V
(c) 6.7 V and 4.7 V
(d) 5.3 V and 3.3 V

111. The value of V_0 of the series regulator shown below is

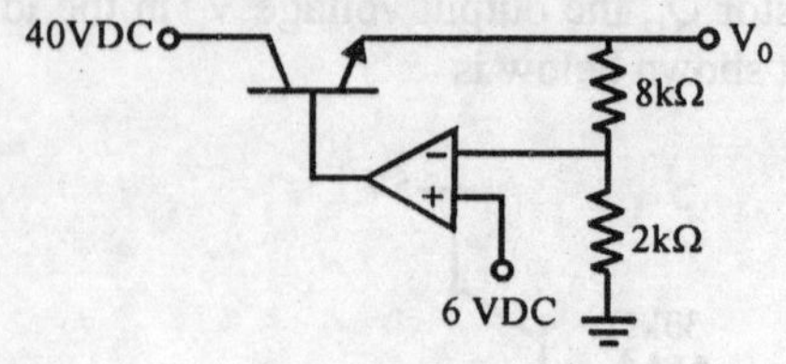

(a) 24 V (b) 28 V
(c) 30V (d) 32 V

112. The ideal op-amp based circuit shown below acts as a

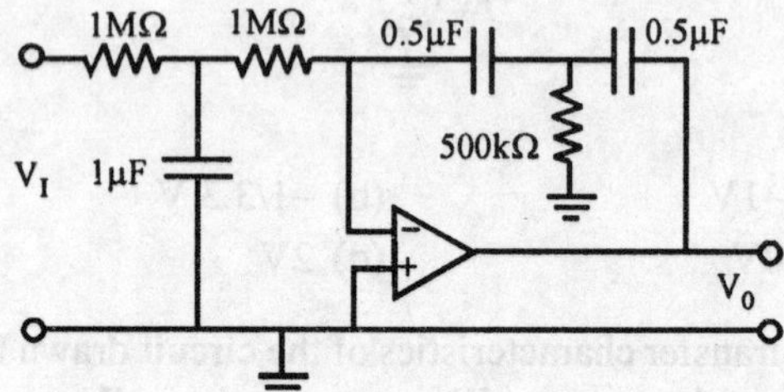

(a) Low - pass filter (b) High - pass filter
(c) Band - pass filter (d) Band - reject filter

Statement for Linked Answer Questions 113 and 114

M1, M2 and M3 in the circuit shown below are matched N - channel enhancement mode MOSFETs operating in saturation mode, forward voltage drop of each diode is 0.7 V, reverse leakage current of each diode is negligible and the opamp is ideal.

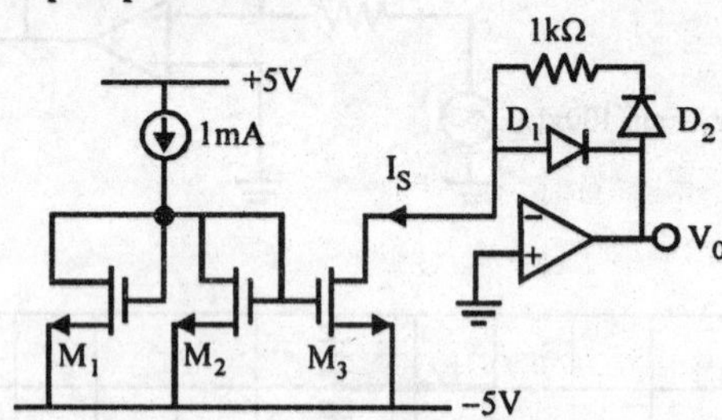

113. The current I_s in the circuit is

(a) – 1mA
(b) 0.5 mA
(c) 1 mA
(d) 2mA

114. For the computed value of current I_s, the output voltage V_0 is

(a) 1.2 V
(b) 0.7 V
(c) 0.2 V
(d) – 0.7V

115. The ciruict shown is a

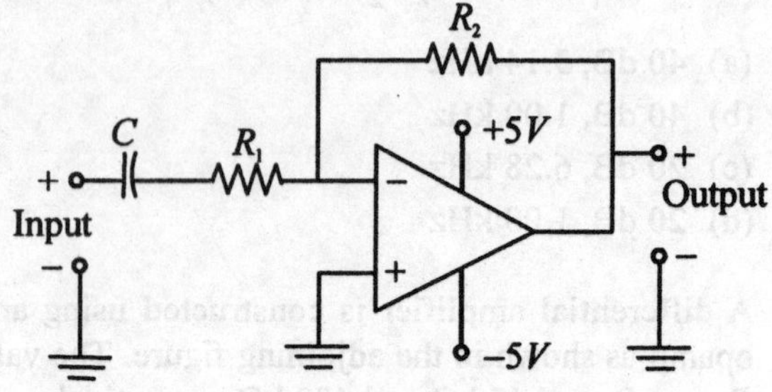

(a) Low pass filter with $f_{3dB} = \dfrac{1}{(R_1 + R_2)C}$ *rad/s*

(b) High pass filter with $f_{3dB} = \dfrac{1}{R_1C}$ *rad/s*

(c) Low pass filter with $f_{3dB} = \dfrac{1}{R_1C}$ *rad/s*

(d) High pass filter with $f_{3dB} = \dfrac{1}{(R_1 + R_2)C}$ *rad/s*

7.6 Feedback Amplifier

1. Two non-inverting amplifiers, one having a unity gain and the other having a gain of twenty are made using identical operational amplifiers. As compared to the unity gain amplifier, the amplifier with gain twenty has

(a) Less negative feedback
(b) Greater input impedance
(c) Less bandwidth
(d) None of the above

2. Negative feedback in amplifiers

(a). Improve the signal to noise ratio at the input
(b). Improve the signal to noise ratio at the Output
(c). Does not affect the signal to noise ratio at the output
(d). Reduces distortion

3. In a shunt-shunt negative amplifier, as compared to the basic amplifier

(a) Both input and output impedance decreases
(b) Input impedance decrease but output impedance increase
(c) Input impedance increase but output impedance decrease
(d) Both input and output impedance increases

4. Negative feedback in an amplifier
   (a) Reduce gain
   (b) Increases frequency and phase distortions
   (c) Reduce bandwidth
   (d) Increases noise

5. An amplifier has an open-loop gain of 100, an input impedance of 1 kΩ, and an output impedance of 100Ω. A feedback network with a feedback factor of 0.99 is connected to the amplifier is a voltage series feedback mode. The new input and output impedance, respectively are:
   (a) 10 Ω and 1Ω
   (b) 10 Ω and 10 kΩ
   (c) 100 kΩ and 1 Ω
   (d) 100 kΩ and 10 kΩ

6. In a negative feedback amplifier using voltage-series (i.e. voltage-sampling, series mixing) feedback
   Where R_i and R_o denote the input and output resistance respectively
   (a) R_i decreases and R_o decreases
   (b) R_i decreases and R_o increases
   (c) R_i increases and R_o decreases
   (d) R_i increases and R_o increases

7. The effect of current shunt feedback in an amplifier is to
   (a) Increase the input resistance and decrease the output resistance
   (b) Increase both input and output resistance.
   (c) Decrease both input and output resistance.
   (d) Decrease the input resistance and increase the output resistance

8. Voltage series feedback (also called series-shunt feedback) results in
   (a) Increase in both input and output impedances
   (b) Decrease in both input and output impedances
   (c) Increase in input impedance and decrease in output impedance
   (d) Decrease in input impedance and increase in output impedance.

9. An amplifier without feedback has a voltage gain 50, input resistance of 1 KW and output resistance of 2.5 KW. The input resistance of the current –shunt negative feedback amplifier using the above amplifier with a feedback factor of 0.2 is
   (a) 1/11 kΩ (b) 1/5 kΩ
   (c) 5 kΩ (d) 11 kΩ

10. In a **Common Emitter Amplifier**, the **unbypassed** emitter resistance provides
   (a) Voltage-shunt feedback
   (b) Current-series feedback
   (c) Negative-voltage feedback
   (d) Positive-current feedback

11. Figure below shows a **two-stage small signal transistor feedback amplifier.** Match the List-I (defective component) with List II (its probable effect on the circuit)

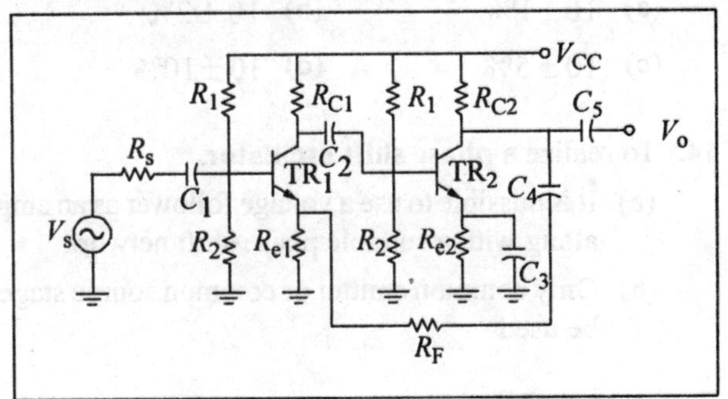

List-I	*List-II*
A. Capacitor C_1 is open	1. All dc voltage normal, V_o increases marginally
B. Capacitor C_3 is open	2. Collector or TR_2 is at V_{CC}, $V_o = 0$
C. Capacitor C_4 is open	3. All dc voltage normal, gain of 2nd stage increases, V_o decreases
D. Rc_2 in shorted	4. All dc voltage normal, $V_o = 0$
	5. All dc voltage normal, overall gain of the amplifier increases, V_o increase
	6. No change

Codes:

	A	B	C	D
(a)	1	3	5	6
(b)	2	3	6	5
(c)	4	3	5	2
(d)	2	3	4	5

12. An **Op-amp** has an **open-loop** gain of 10^5 and an open-loop upper cutoff frequency of 10 Hz. If this op-amp is connected as an amplifier with a closed-loop gain of 100, then the new **upper cutoff frequency** is
   (a) 10 Hz (b) 100 Hz
   (c) 10 kHz (d) 100 kHz

13. As shown in the figure, a negative feedback system has an amplifier of gain 100 with ±10% tolerance in the forward path, and an attenuator of value 9/100 in the feedback path. The overall system gain is approximately:

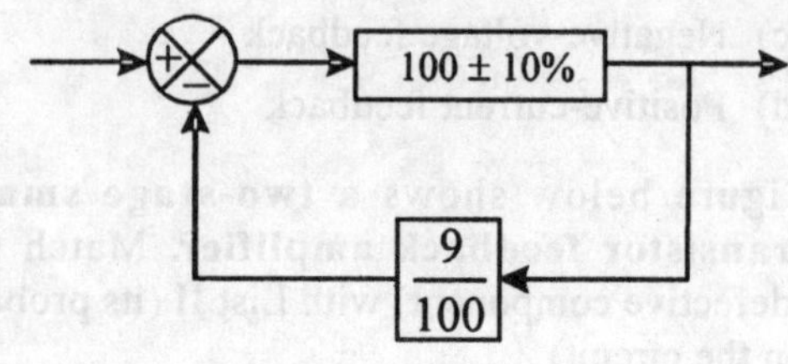

(a) 10±1%　(b) 10±2%

(c) 10±5%　(d) 10±10%

14. To realize a **phase shift oscillator.**

(a) It is possible to use a voltage follower as an amplifier along with a suitable phase shift network

(b) Only common emitter or common source stages can be used

(c) Given any phase shifting network, it is always possible to find the emitter follower

(d) None of the above

15. A FET source follower is shown in the figure below: The nature of feedback in this circuit is

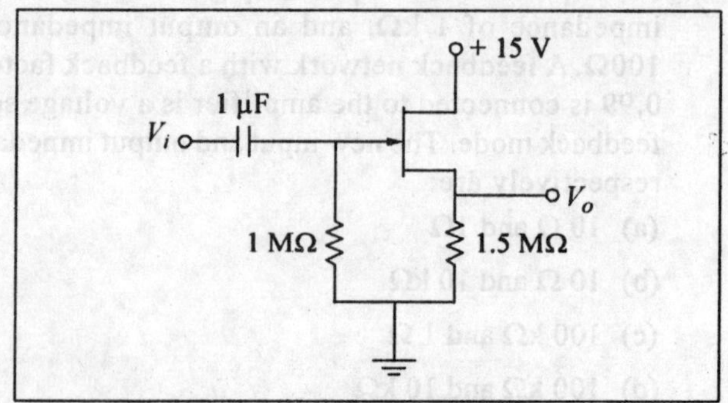

(a) Positive current　(b) Negative current

(c) positive voltage　(d) Negative voltage

7.7 Oscillators

1. Value of R in the oscillator circuit shown in figure is so chosen that it just oscillates at an angular frequency of ω. The value of ω and the required value of R will respectively be

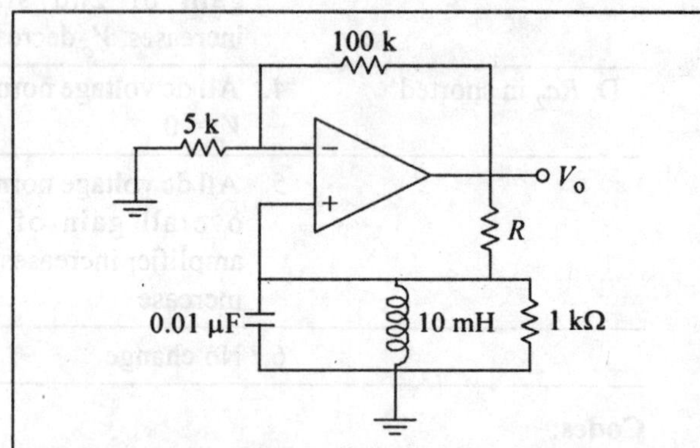

(a) 10^5 rad/sec, 2×10^4 Ω

(b) 2×10^4 rad/sec, 2×10^4Ω

(c) 210^4 rad /sec, 10^5 Ω

(d) 10^5 rad/sec, 10^5 Ω

2. The circuit in figure employs positive feedback and is intended to generate sinusoidal oscillation. If at a frequency f_0, $B(f) = [V_f(f)/V_o(f)] = (1/6) \angle 0°$, then to sustain oscillation at this frequency

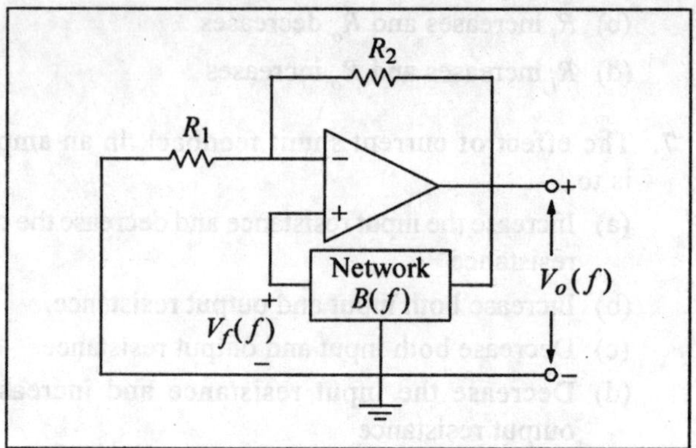

(a) $R_2 = 5R_1$　(b) $R_2 = 6\,R_1$

(c) $R_2 = \frac{R_1}{6}$　(d) $R_2 = \frac{R_1}{5}$

3. The oscillator circuit shown in figure has an ideal inverting amplifier. Its frequency of oscillation (in Hz) is:

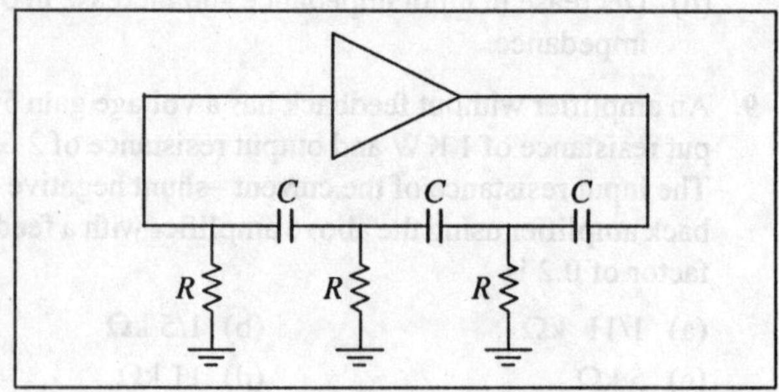

(a) $\frac{1}{(2\pi\sqrt{6}\,RC)}$

(b) $\frac{1}{(2\pi\,RC)}$

(c) $\frac{1}{(\sqrt{6}\,RC)}$

(d) $\frac{\sqrt{6}}{(2\pi\,RC)}$

4. The value of C required for sinusoidal oscillations of frequency 1 kHz in the circuit of Figure is:

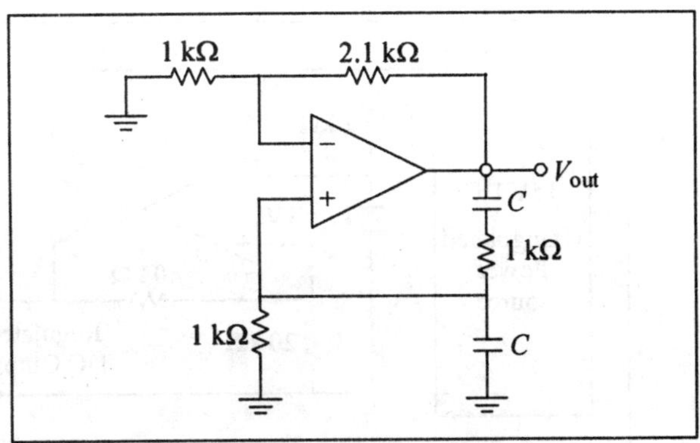

(a) $\frac{1}{2\pi}$ μF (b) 2π μF

(c) $\frac{1}{2\pi\sqrt{6}}$ μF (d) $2\pi\sqrt{6}$ μF

7.8 Power Amplifier

1. In a transistor push-pull amplifier

1. There is no d.c present in the output
2. There is no distortion in the output
3. There are no even harmonics in the output
4. There are no odd harmonics in the output

(a) 1, 2 , 3 are correct
(b) 1, 3 are correct
(c) 1, 2, 4 are correct
(d) 2, 3, 4, are correct

2. A class A transformer coupled, transistor power amplifier is required to deliver a power output of 10 watts. The maximum power rating of the transistor should not be less than

(a) 5 W (b) 10 W
(c) 20 W (d) 40 W

3. A class B push pull complementary symmetry amplifier uses.

(a) Two *n-p-n* transistor
(b) One *p-n-p* and one npn transistor
(c) One pre-amplifier of n transistor followed
(d) Amplifier of two *p-n-p* transistors.

7.9 Power Supplies

Statement for Linked Answer Questions 1 and 2

A regulated power supply, shown in figure below, has an unregulated input (UR) of 15 Volts and generates a regulated output V_{out} Use the component values shown in the figure.

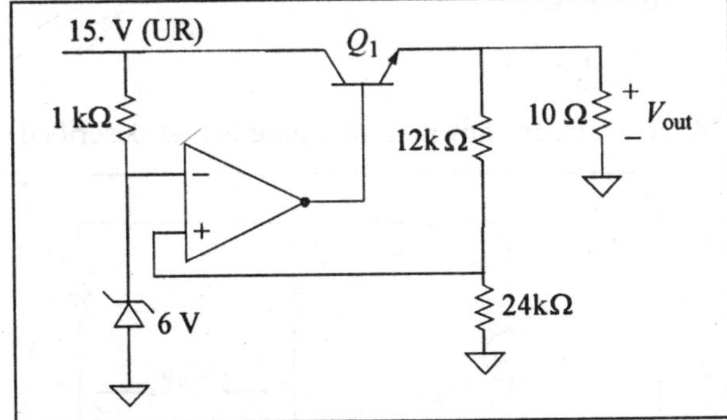

In the figure above, the ground has been shown by the symbol ∇.

1. The power dissipation across the transistor Q_1 shown in the figure is

(a) 4.8 Watts
(b) 5.0 Watts
(c) 5.4 Watts
(d) 6.0 Watts

2. If the unregulated voltage increases by 20%, the power dissipation across the transistor Q_1.

(a) Increases by 20%
(b) Increases by 50%
(c) Remains unchanged
(d) Decreases by 20%

3. The output voltage of the regulated power supply shown in figure is:

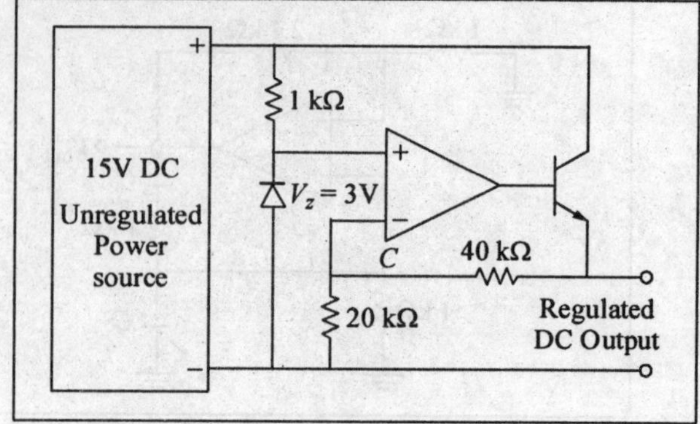

(a) 3 V (b) 6 V
(c) 9 V (d) 12 V

4. A dc power supply has a no-load voltage of 30V, and a full-load voltage of 25V at a full-load current of 1A. Its output resistance and load regulation, respectively are:
(a) 5 W and 20%
(b) 25 W and 20 %
(c) 5 W and 16.7 %
(d) 25 W and 16.7 %

5. In a series regulated power supply circuit, the voltage gain A_v of the "pass" transistor satisfies the condition:
(a) $A_v \to \infty$ (b) $1 << A_v < \infty$
(c) $A_v \cong 1$ (d) $A_v << 1$.

6. The three-terminal linear voltage regulator is connected to a 10 Ω load resistor as shown in the figure. If V_{in} is 10 V, what is the power dissipated in the transistor?

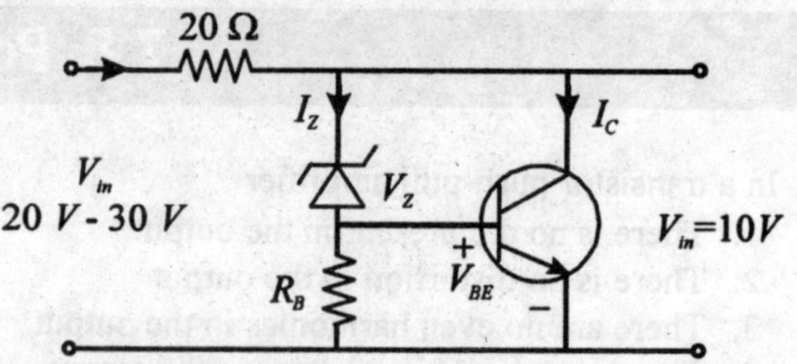

(a) 0.6 W (b) 2.4 W
(c) 4.2 W (d) 5.4 W

7. The transistor shunt regulator shown in Figure has a regulated output voltage of 10 V, when the input varies from 20 V to 30 VC . The relevant parameters for the zener diode and the transistor are : V_Z 9.5 V, V_{BE}= 0.5 V, β =99. Neglect the current through R_B. Then the

20 Ω
I_Z
I_C
V_{in}
20 V - 30 V
V_Z
V_{in}=10V
R_B
V_{BE}

maximum power dissipated in the zener diode (P_Z) and the transistor (P_T) are
(a) P_z = 75 mW, P_T = 7.9 W
(b) P_z = 85 mW, P_T = 8.9 W
(c) P_z = 95 mW, P_T = 9.9 W
(d) P_z = 115 mW, P_T = 11.9 W

7.10 Miscellaneous

1. The emitter coupled pair of BJT's gives a linear transfer relation between the differential output voltage V_{od} and the **differential** input voltage V_{id} only when the magnitude of V_{id} is less α times the thermal voltage, where α is
(a) 4 (b) 2
(c) 2 (d) 1

2. A **multistage amplifier** has a low-pass response with three real poles at $s = -\omega_1, -\omega_2$ and $-\omega_3$. The approximate overall bandwidth B of the amplifier will be given by
(a) $B = \omega_1 + \omega_2 + \omega_3$
(b) $\frac{1}{B} = \frac{1}{\omega_1} + \frac{1}{\omega_2} + \frac{1}{\omega_3}$
(c) $B = (\omega_1\, \omega_2\, \omega_3)^{1/3}$
(d) $B = \sqrt{(\omega_1^2 + \omega_2^2 + \omega_3^2)}$

3. In a series regulated power supply circuit, the voltage gain A_v of the "pass" transistor satisfies the condition:
(a) $A_v \to \infty$
(b) $1 << A_v < \infty$
(c) $A_v \cong 1$
(d) $A_v << 1$

4. The circuit shown in the figure is best described as a

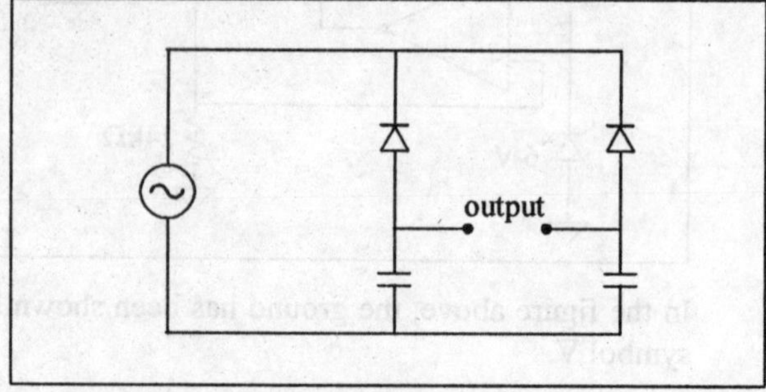

(a) Bridge rectifier

(b) Ring modulator

(c) Frequency discriminatory

(d) Voltage doubler.

5. α-cut off frequency of a bipolar junction transistor

(a) Increases with the increase in base width

(b) Increases with the increases in emitter width

(c) Increases with increases in the collector width

(d) Increases with decrease in the base width

6. The common emitter short-circuit current gain β of a transistor

(a) Is a monotonically increasing function of the collector current I_C

(b) Is an monotonically decreasing function of I_C

(c) Increases with I_C, for low I_C, reaches a maximum and then decreases with further increases

(d) Is not a function of I_C

7. A half wave rectifier uses a diode with a forward resistance R_f. The voltage is $V_m \sin \omega t$ and the load resistance is R_L. The DC current is given by

(a) $\dfrac{V_m}{\sqrt{2}\,R_L}$

(b) $\dfrac{V_m}{\pi(R_f + R_L)}$

(b) $\dfrac{2V_m}{\sqrt{\pi}}$

(c) $\dfrac{V_m}{R_L}$

8. A high-Q quartz crystal exhibits series resonance at the frequency ω_s and parallel resonance at the frequency ω_p. Then

(a) ω_s is very close to, but less than ω_p

(b) $\omega_s << \omega_p$

(c) ω_s is very close to but greater than ω_p

(d) $\omega_s >> \omega_p$

9. An amplifier with resistive negative feedback has two left half plane poles in its open-loop transfer function. The amplifier

(a) Will always be unstable at high frequencies

(b) Will be stable for all frequencies

(c) May be unstable, depending on the feedback factor

(d) Will oscillate at low frequencies

10. The configuration of figure is

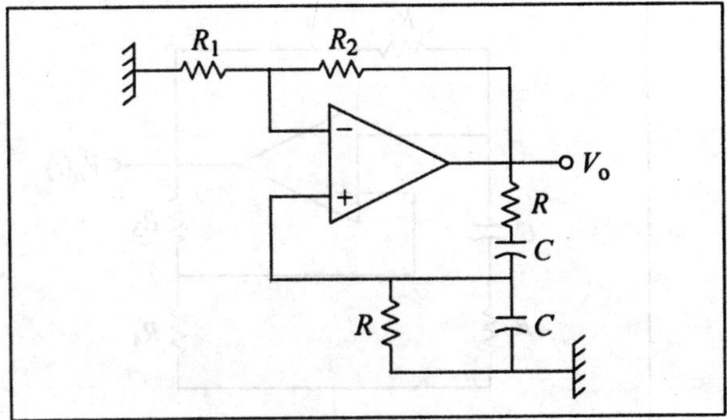

(a) Precision integrator

(b) Hartley oscillator

(c) Butter-worth high pass filter

(d) Wien-bridge oscillator

11. A ramp input applied to an unity feedback system results in 5% steady state error. The type number and zero frequency gain of the system are respectively.

(a) 1 and 20

(b) 0 and 20

(c) 0 and $\dfrac{1}{20}$

(d) 1 and $\dfrac{1}{20}$

12. Consider two transfer function

$$G_1(s) = \frac{1}{(s^2 + as + b)}$$

and $G_2(s) = \dfrac{s}{(s^2 + as + b)}$

The 3 dB bandwidths of their frequency responses are, respectively.

(a) $\sqrt{a^2 - 4b}, \sqrt{a^2 + 4b}$

(b) $\sqrt{a^2 + 4b}, \sqrt{a^2 - 4b}$

(c) $\sqrt{a^2 - 4b}, \sqrt{a^2 - 4b}$

(d) $\sqrt{a^2 + 4b}, \sqrt{a^2 + 4b}$

13. In the following astable multivibrator circuit, which properties of $v_o(t)$ depend on R_2

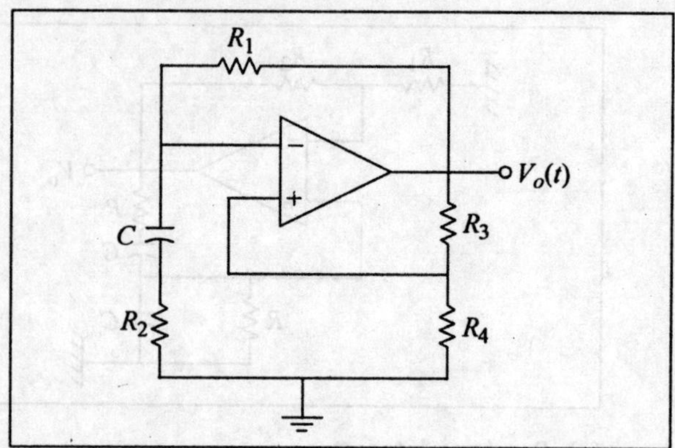

(a) Only the frequency
(b) Only the amplitude
(c) Both the amplitude and the frequency
(c) Neither the amplitude nor the frequency

14. In the circuit shown below, the op-amp is ideal, the transistor has $V_{BE} = 0.6$ V and $\beta = 150$. Decide whether the feedback in the circuit is positive or negative and determine the voltage V at the output of the op-amp.

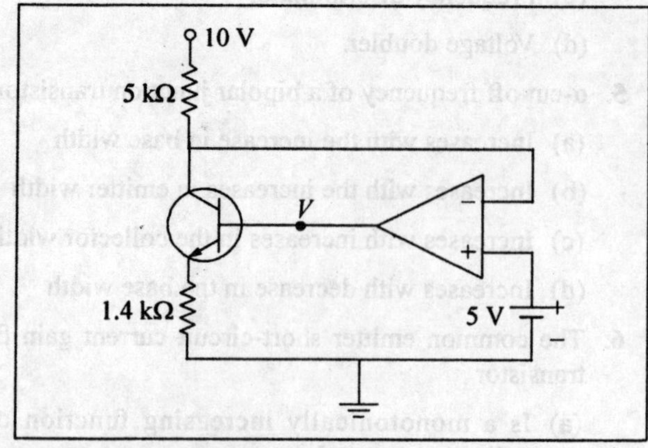

(a) Positive feedback, $V = 10$ V
(b) Positive feedback, $V = 0$ V
(c) Negative feedback, $V = 5$ V
(d) Negative feedback, $V = 2$ V

Answers

7.1 Diodes Applications

1. **Ans. (b)**

Solution:

$$I = \frac{10-6}{50} = \frac{4}{50} = 0.08\,A$$

$$= 80\,mA$$

Knee current is Zener diode minimum current

Hence load current I_L will be maximum.

I_L max = 80 – 5 = 75 mA

$$R_{(min)} = \frac{V}{I_{L(max)}}$$

$$R = \frac{6}{75 \times 10^{-3}} = 80 \text{ ohm.}$$

2. **Ans. (c)**

Solution: $I_m = \dfrac{9-6}{50} = 0.06\text{ A} = 60\text{ mA}$

Maximum current through zener diode

$$= \frac{300 \times 10^{-3}}{6} = 50\text{ mA}$$

∴ Minimum current through load = 60 – 50 = 10 mA

And Maximum current through load = 60 – 5 = 55 mA

3. **Ans. (b)**

Solution:

$$\text{Load regulation} = \frac{V_{(\text{no load})} - V_{(\text{Full load})}}{V_{(\text{Full load})}} \times 100$$

$$= \frac{30-25}{25} \times 100$$

$$= 20\%$$

$$\text{Output resistance} = \frac{V_{(\text{Full load})}}{I_{(\text{Full load})}} = \frac{25}{1} = 25\,\Omega$$

4. **Ans. (a)**

Solution: $V_{R\max} = V_{IN} - V_Z = 30V - 10V = 20V$

$$I_{R\min} = I_L + I_{zk} = (10 + 1)\text{ mA} = 11\text{ mA}$$

$$R_{\max} = \frac{20\text{ V}}{11\text{ mA}} = 1800\,\Omega$$

$$R \le 1800\,\Omega.$$

Second Method

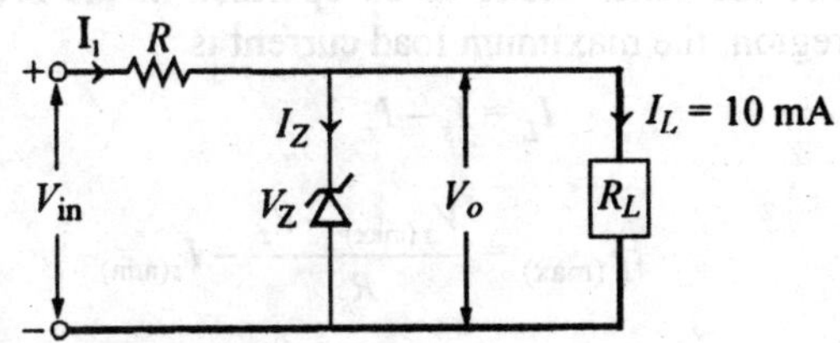

Given $I_Z = 1\,mA$, $I_L = 10\,mA$

$$I_1 = \frac{V_{in} - V_o}{R} \ge I_Z + I_L$$

if $V_{in} = 30V$

$$I_1 = \frac{30-10}{R} \ge (10+1)mA$$

$$= \frac{20}{R} \ge 11 \times 10^{-3}$$

$$R \le 1818\,\Omega \quad \text{...(1)}$$

If $V_{in} = 50V$

$$I_1 = \frac{50-10}{R} \ge (10+1)mA$$

$$= \frac{40}{R} \ge 11 \times 10^{-3}$$

$$R \le 3636 \quad \text{...(2)}$$

From equation (1) and (2) it is clear that

$$\therefore R \le 1818\Omega$$

5. **Ans. (a)**

Solution:

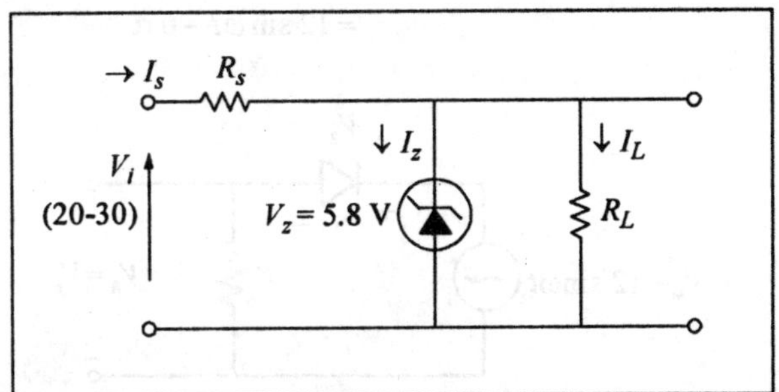

$$I_s = I_L + I_z$$

$$I_L = I_s - I_z$$

For design consideration

Load current is maximum when diode is just operate in break down region, As V_{in} increases, the load current remain but zener current increase.

For the zener diode to be operated in the breakdown region, the maximum load current is

$$I_L = I_s - I_z$$

$$i_{L\,(max)} = \frac{V_{s\,(max)} - V_z}{R_s} - I_{z(min)}$$

$$= \frac{30 - 5.8}{1\text{ k}} - \left[\frac{1}{2}\right]\text{mA}$$

$$i_{L\,(max)} = 23.7\text{ mA}$$

6. Ans. (b)

Solution:

* Transfer characteristic of ideal zener diode can be drawn as:

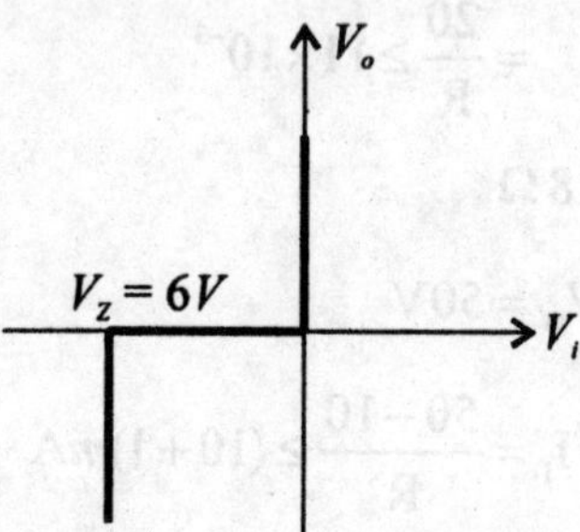

* **During positive half-cycle** the didoe is reverse biased till $V_{in} < V_z$ and operate for $V_{in} > V_z$. Where V_z is breakdown voltage of diode. Therefore

For $V_{in} < V_z$; $V_R = 0$

For $V_{in} > V_z$

During angle α; $V_R = V_{in} - V_z$

$$= 12 \sin \omega t - 6\,\alpha$$

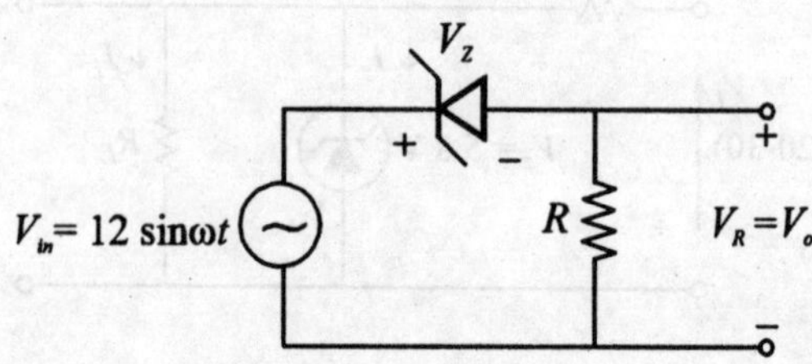

* **During negative half cycle** the didoe is forward biased and **Therefore,**

$$V_R = V_{in} = -ve = 12 \sin \omega t$$

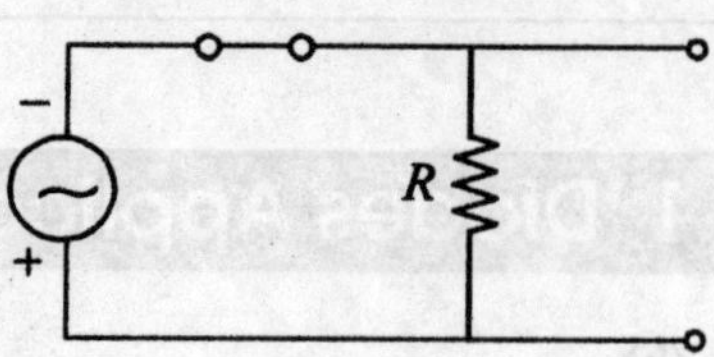

Waveform of V_R :

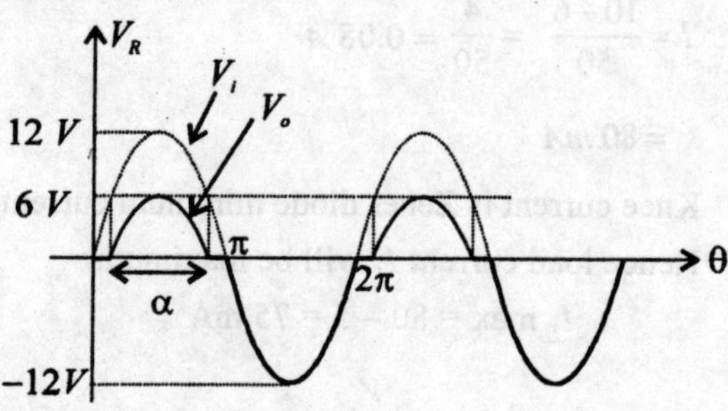

7. Ans. (c)

Solution: Full wave rectifier: In all case of supply only two diode conduct and gives terminal A + Ve and terminal B –Ve

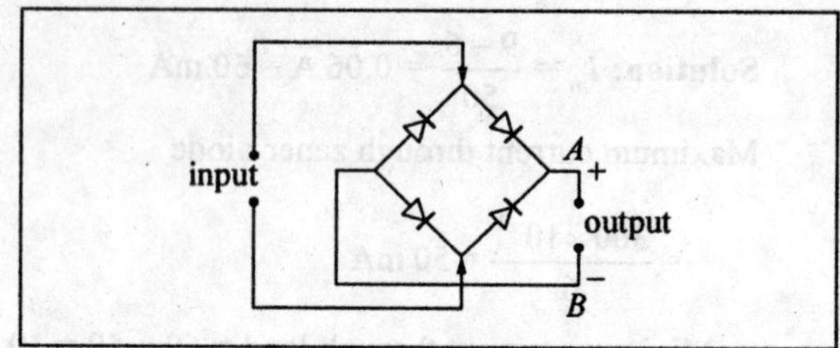

8. Ans. (c)

Solution:

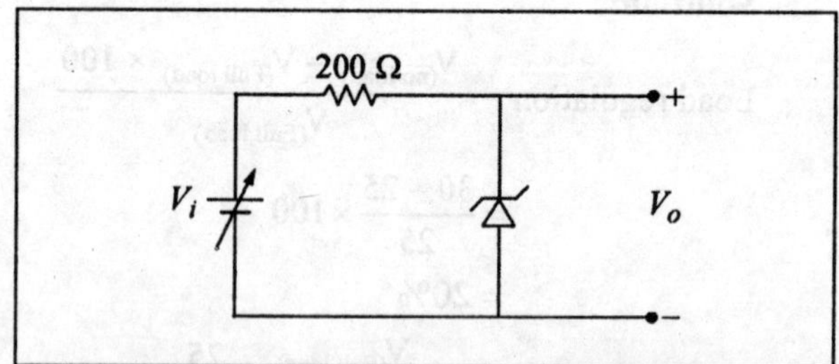

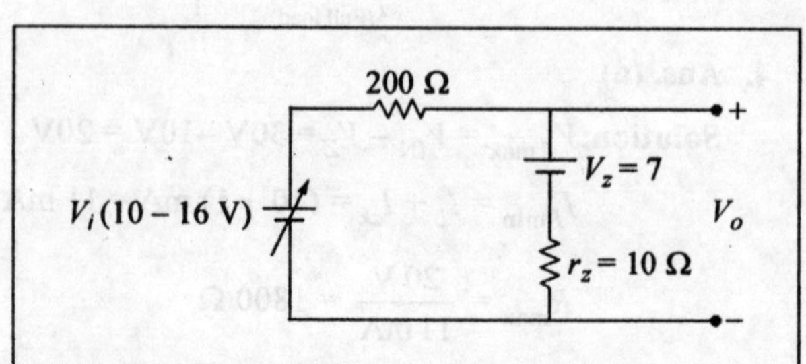

$$V_{r1} = \frac{10-7}{200} \times 10 = 0.15$$

$$V_{r2} = \frac{16-7}{200} \times 10 = 0.45$$

Zener regulator output
= Zener breakdown voltage + Voltage drop due to Zener dynamic resistance.
Thus $V_o = (7 + V_{r1})$ to $7 + V_{r2} = 7.15$ to 7.45

9. Ans. (c)

Solution: **During positive cycle of V_i**

D_2 is reversed biased.

D_2 is forward biased and Zener diode is reverse biased.
Then net output votlage $6.8 + 0.7 = 7.5\,V$

During negative cycle of V_i

D_2 is forward biased and D_1 is reverse biased.

Then net output voltage = – 0.7 V

10. Ans (d)

Solution:

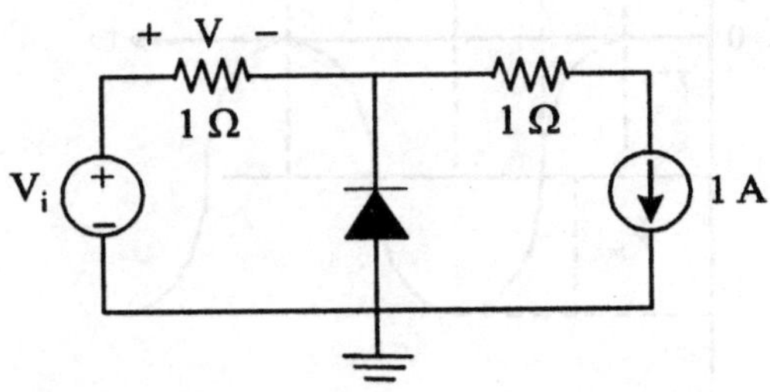

Case 1: **When V_i is positive voltage, then diode is reverse biased and acts as open switch. So current reduce to**

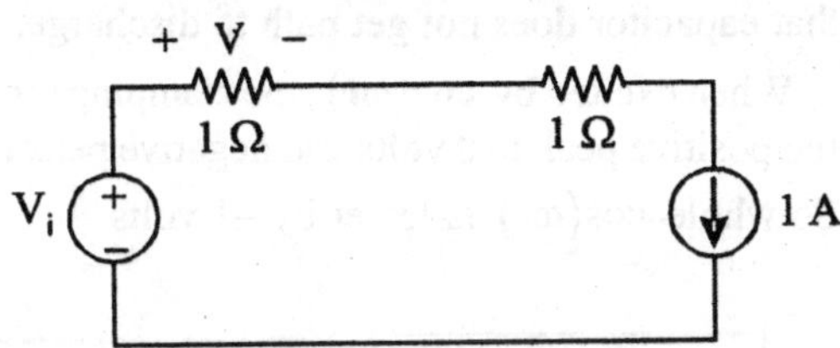

Hence $V = 1 \times 1 = 1\,V$ **and**

$V_i = 2 \times 1 = 2\,V$

Case 2: **When V_i is negative voltage, then diode is forward biased and atcs as close switch. Hence circuit reduce as**

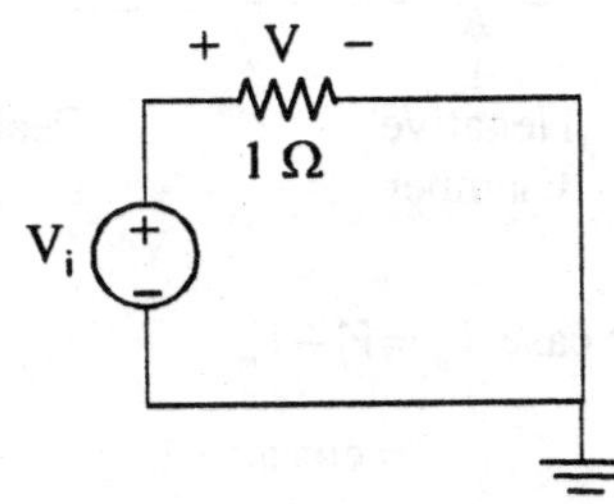

As current will follow the shortest resistance path.

Hence $V = -V_i$ as whole of voltage applied will drop across. In resistor as diode gives an ideal diode

Hence V can have maximum voltage either

as $-V_i$ **or 1 V**

or $V = \max(-V_i, 1)$

11. Ans. (a)

Solution:

Voltage drop across each forward biased diode $= 0.7\,V$

Biasing voltage = 12.7,

$$I_{DC} = \frac{12.7 - (0.7 + 0.7 + 0.7 + 0.7)}{9900} = 1mA$$

12. Ans. (d)

Solution:

As forward biased voltage drop = 0.7

That-indicate that it is *Si* diode

Hence $\eta = 2$

AC dynamic resistance,

$$r_d = \frac{\eta V_T}{I_D} = \frac{2 \times 25mV}{1mA} = 50\Omega$$

The ac dynamic resistance offered by each diode $= 50\Omega$

There are 4 diode

Hence, total dynamic resistance $= 4 \times 50$

$$= 200\,\Omega$$

$$\therefore V_{ac} = V_i(ac)\left[\frac{200\,\Omega}{9900 + 200}\right]$$

$$\frac{100 \times 200}{10100} = \frac{200}{101}\,mV$$

$$V_{ac} = 2\cos(\omega t)\,mV$$

Second Method (b)

Diode Forward Resistance,

$$R_f = \frac{V_T}{I} = \frac{25\,mV}{1\,mA} = 25\,\Omega$$

Replacing diode by the AC model, i.e. replacing them by their forward resistance R_f

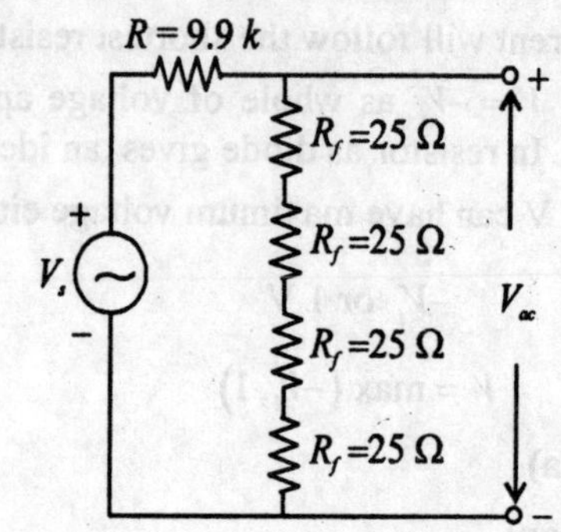

$$V_{ac} = \frac{4R_f}{R+4R_f} \times V_s$$

$$= \frac{4\times 25}{9000+4\times 25} \times V_s = \frac{100}{10,00} V_s$$

or $\quad V_{ac} = 100 \cos \omega t \; mV$

or $\quad V_{ac} = \cos \omega t \; mV$

13. Ans.(a)

Solution:

Clamping

The clamping circuit establisher a recurrent positive or negative extremity at some constant reference voltage level VR. *The clamping circuit introduces a d.c. component to a waveform that has lost its* dc component after passing through a capacitive coupling network. For the reason, clamping circuit is often referred to as dc restorer or dc inerter.

Wave Form

Consider the given circuit which consist of signal source V_i of neglecting output impedance a capacitor and a diode D.

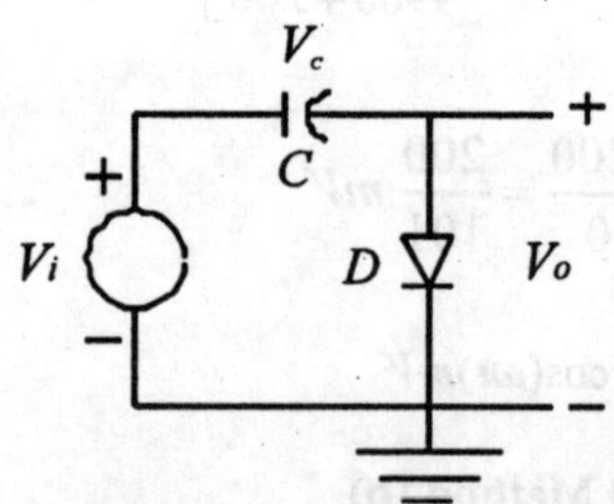

Assuming the ideal diode i.e. $V_y = 0$ and $R_i = 0$. The input is sinusoidal. During first quarter cycle of input signal the capacitor get charged as the diode is forward biased, no output voltage.

$\therefore \quad V_o = 0.$

At the end of first quarter cycle $V_c = V_m$. After the diode comes reverse bias. The input signal reduces but the capacitor is not able to discharge due to diode

$$V_o = V_i - V_m$$

During succeeding cycles cycle the positive execution of the signal just barely reaches zero. The diode need never again conduct and the positive extremity of the signal has been clamped or restore to zero.

If the amplitude of the input signal increases, the diode will conduct for first quarter cycle. The capacitor will again get charged and positive extremity will get clamped to zero voltage.

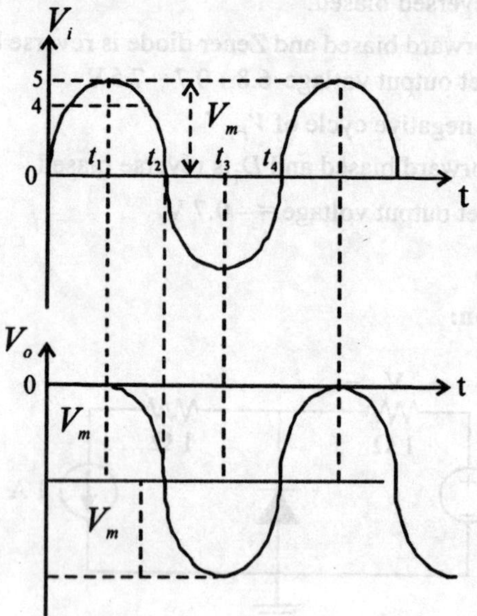

However, if the amplitude of input signal decrease. The signal does not remain clamped to zero. The reason being that capacitor does not get path to discharge.

When excited by $\cos(\omega t)$, the clamping section clamp the positive peak to 0 volts and negative peak to –2 volts. So whole $\cos(\omega t)$ is lower by –1 volts

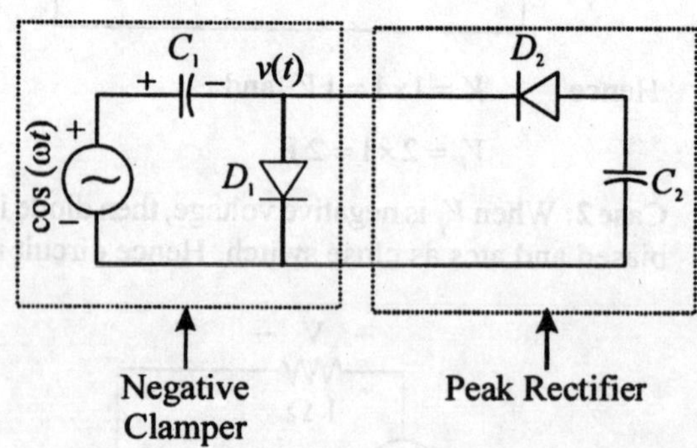

In present case $V_o = V_i - V_m$

$$= \cos \omega t - 1$$

Second Method

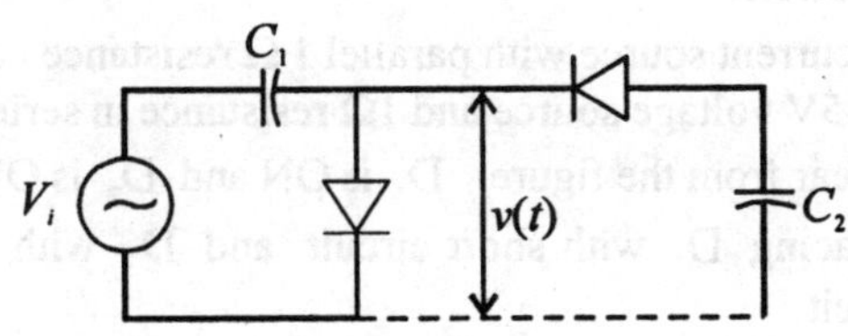

For sine as input all positive peaks clamped to zero and all negative peaks clamped to $-2V_m$ from $-V_m$

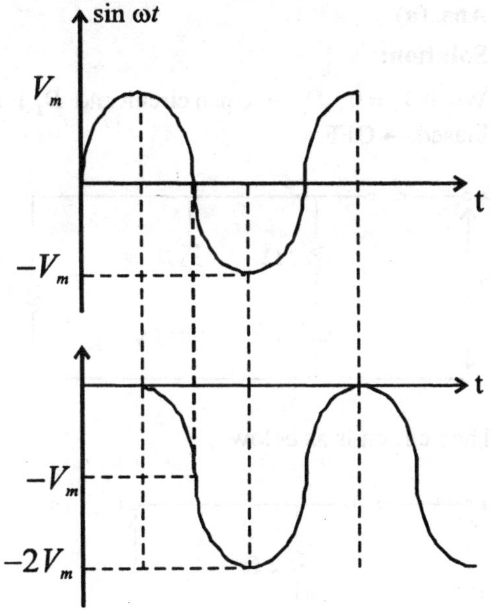

All positive peaks clamped to zero and all negative peaks clamped to $-2V_m$ from V_m for sine as input

At $\omega t = 0$

$$\cos(\omega t) - 1 = 0$$

At $\omega t = \frac{\pi}{3}$

$$\cos\left(\frac{\pi}{3}\right) - 1 = \frac{1}{2}$$

At $\omega = \frac{1}{2}$

$$\cos(\omega t) - 1 = -1$$

14. Ans. (d)

Solution:

Apply KVL in the given circuit

$$10 = (1000)i + v$$

$$= \frac{1000(v-0.7)}{500} + v$$

$$= 2(v-0.7) + v$$

$$= 2v - 1.4 + v$$

$$10 = 3v - 1.4$$

$$3v = 10 + 1.4$$

$$= 11.4$$

$$V = \frac{11.4}{3} = 3.8$$

$$i = \frac{v-0.7}{500}$$

$$= \frac{3.8-0.7}{500}$$

$$= \frac{3.1}{500}$$

$$= 6.2\ mA$$

Second Method

$$i = \frac{V-0.7}{500}$$

$$V = 500i + 0.7$$

Dynamic resistance of diode

$$r_d = \frac{dV}{di}$$

$$= 500$$

$$\frac{di}{dV} = \frac{1}{500}$$

or, $r_d = 500\ \Omega$

Since applied voltage is more than cut in voltage of diode is be forward biased and voltage across diode is 0.7 V and it exhibits a dynamic resistance

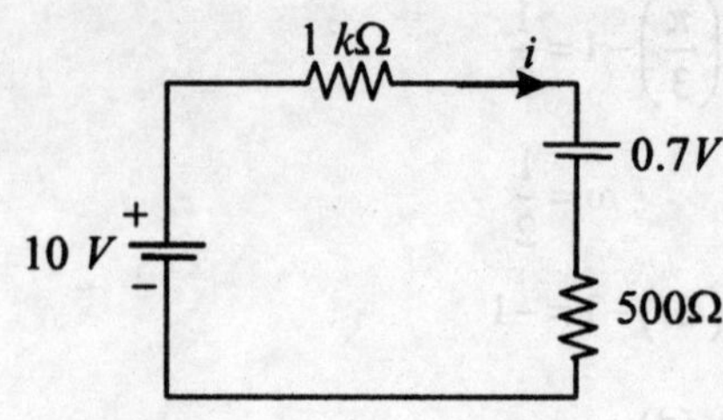

$$i = \frac{10-0.7}{1000+500}$$

$$= \frac{9.3}{15000}$$

$$= 6.2\ mA$$

15. Ans. (d)

Solution: Positive swing + 4 V

Negative swing is $-4+-6\times\frac{10}{20} = -7$ V

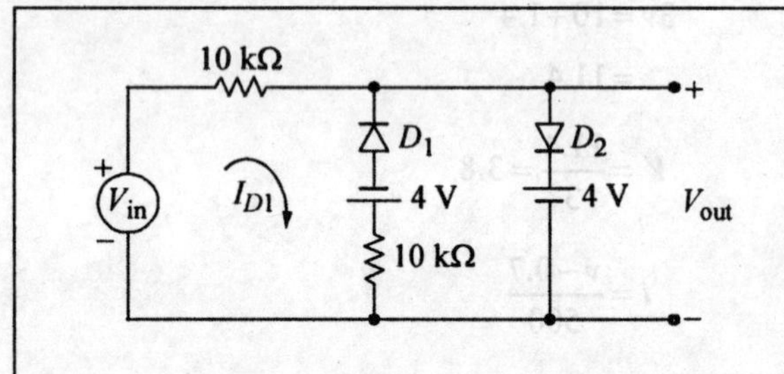

For positive cycle D_2 ON, D_1 OFF then $V_{out} = 4$ V

For negative cycle D_1 ON then D_2 OFF,

$$I_{D1} = \frac{V_{in}+4}{20} = \frac{-10+4}{20} = \frac{-3}{10}\text{mA}$$

Then $V_o = I_{D1} R\,(10\text{ k}) - 4 = -7$ volts.

16. Ans. (c)

Solution: For zener diode, Dynamic resistance

$$R_z = \frac{\Delta V_z}{\Delta I_z} = \frac{V_L - V_z}{\Delta I_z} = \frac{3.5-3.3}{\Delta I_z}$$

$$\Delta I_z = \frac{0.2}{0.1k} = 2\text{ mA}$$

17. Ans. (c)

Solution:

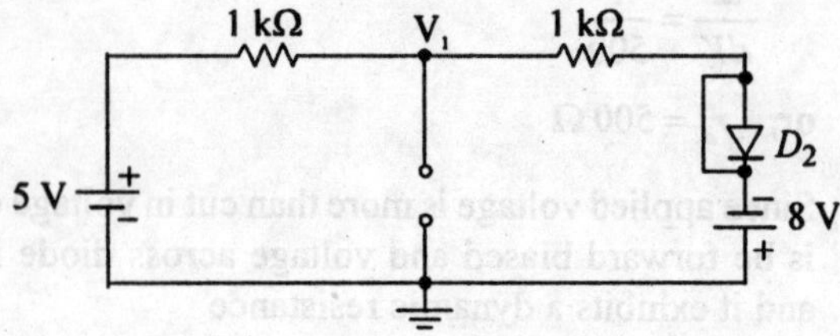

For the circuit D_2 must be in forward bias where D_1 first replace by O.C.

Apply KCL at V_1

$$\frac{V_1-5}{10^3} + \frac{V_1+8}{10^3} = 0$$

$$V_1 - 5 + V_1 + 8 = 0$$

$$2V_1 = -3$$

$$V_1 = -1.5\text{V}$$

Therefore D_1 is Reverse biased

$$I_{D_1} = 0$$

18. Ans. (a)

Solution:

5 A current source with parallel 1 Ω resistance converted into 5V voltage source and 1Ω resistance in series then it is clear from the figure. D_1 is ON and D_3 is OFF. Now replacing D_1 with short circuit and D_3 with an open cirucit

Now, Anode voltage of D_2 < cathod voltae of D_2,

$V_a = 0 \Rightarrow V_a < V_b$ or D_2 is reversed bised

Hence ; D_1 is ON ; D_2 and D_3 are OFF

19. Ans. (a)

Solution:

When $V_i = 0$, $D_1 \rightarrow$ Open circuit and D_2 is also reverse biased $\rightarrow$ OFF

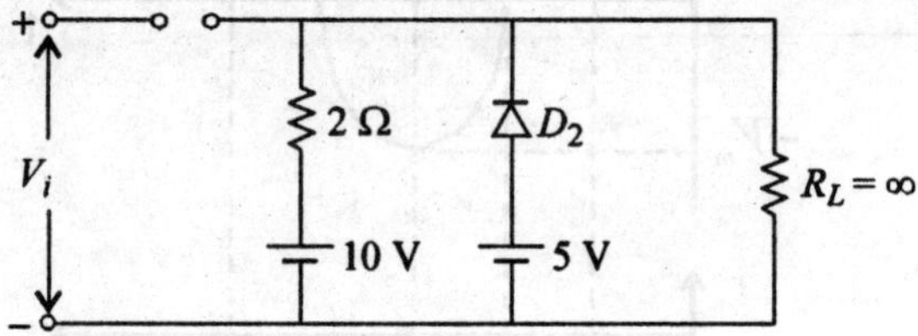

Then cirucit is as below

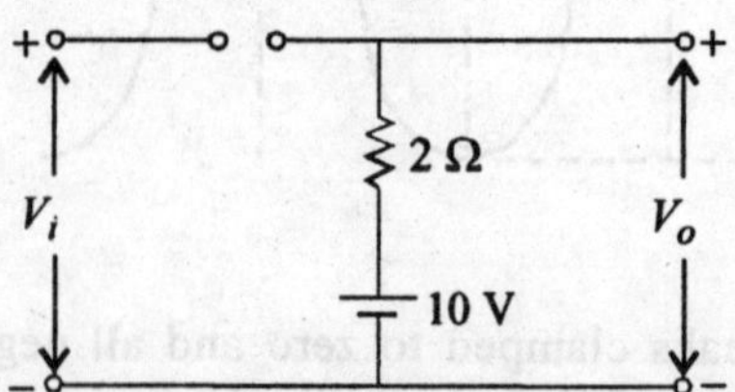

Current passing throuhg 2 Ω resisance is 0.

$V_o = 10\ V$

For $V_i \le 10\ V$ D_2 is always reverse biased

$V_o = 10$ Volts

For $V_1 > 10$ V D_1- ON and D_2- OFF

$V_o = V_i$

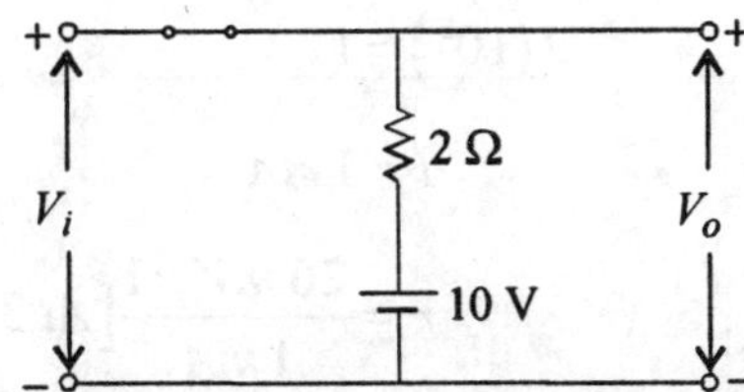

Finally transfer characteristic is as below

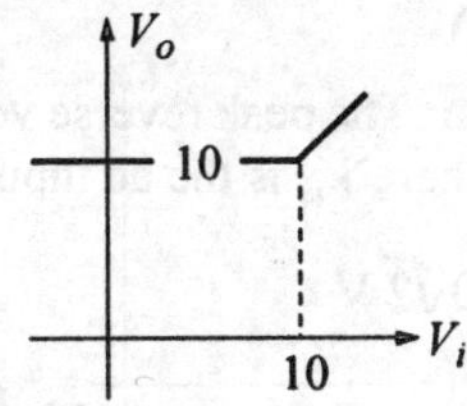

20. Ans. (c)

Solution:

When input in the range $-0.7V < V_i < 5.7\,V$ no diode conduct

Hence output will follow input,

When input $V_i \le -0.7V$ Zener diode forward bias and D is reverse bias

Hence, $V_0 = -0.7V$

When input $V_i \ge 5.7V$ diode D is forward bias

21. Ans. (a)

Solution: In present case diode D_1 is forward biased and D_2 is reversed biased. Hence D_2 current is zero.

22. Ans. (b)

Solution: For 4 V input both diodes are open circuit $V_o = V_i$

23. Ans. (c)

Solution:

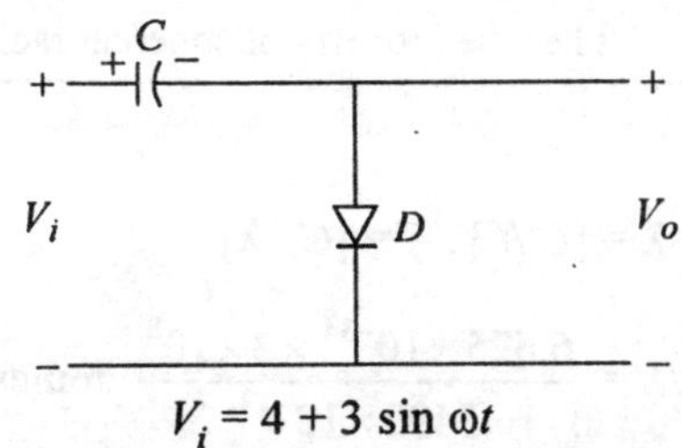

Capacitor in this circuit acts as a DC restorer, It gets charged posture half cycle to + 4 V i.e. negative clamper.

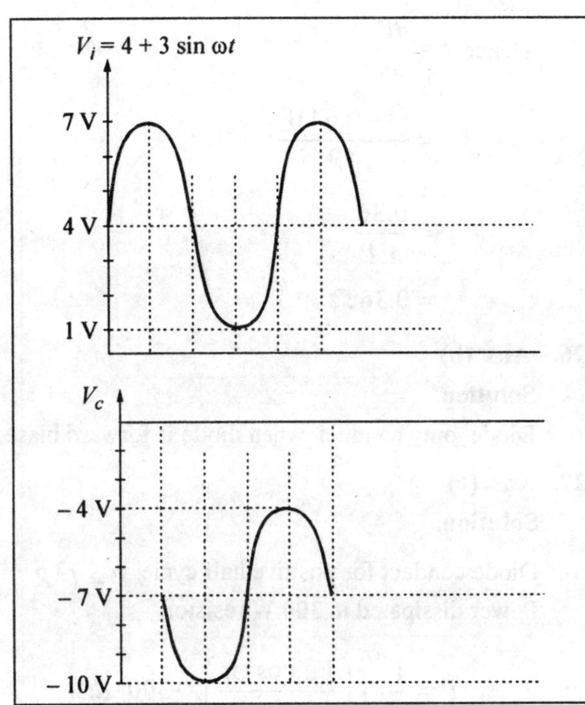

AC signal is using on – 7V. Average of AC over are cycle = 0

∴ Resultant output = – 7 V

24. Ans. (b)

Solution:

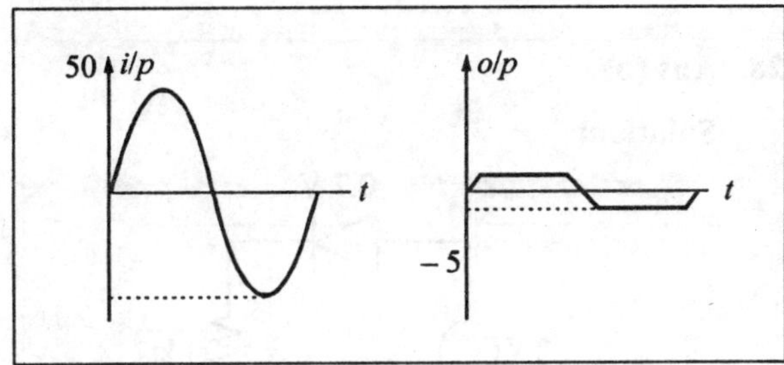

25. Ans. (a)

Solution:

$P = 1W$; $I = 1A$

$P = VI \Rightarrow V = 1V$

- At negligible current i.e. zero, the voltage drop is 0.64 V.
- At $1A$ power dissipated is 1 watt it means voltage drop of $1V$

$$dI = 0 \text{ to } 1A$$

$$= 1A$$

$$dV = 0.64 \text{ to } 1V$$

$$= 0.36\,V$$

Hence $r = \frac{dV}{dI}$

$$= \frac{(1-0.64)V}{1A}$$

$$= \frac{0.36}{1A}$$

$$= 0.36\,\Omega$$

26. Ans. (b)

Solution:

Diode only conduct when diode is forward biased.

27. Ans. (c)

Solution:

Diode conduct for positive half cycle. $P = I^2 R$

Power dissipated in 300 W resistor,

$$P = \frac{1}{T}\int_0^{T/2}\left(\frac{V_m \cos \omega t}{600}\right)^2 \cdot 300 \cdot d\omega t$$

or $$P = \frac{1}{T}\int_0^{T/2}\frac{1}{100}\cos^2 \omega t \cdot 300 \cdot d\omega t$$

$= 0.75$ Watts

28. Ans (b)

Solution:

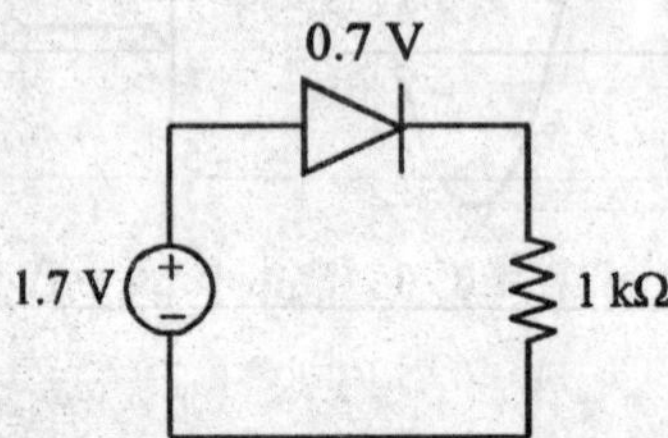

Given $\eta = 1$

By solving the equation of diode we find the find relation

Dynamic resistance $(r) = \frac{\eta V_T}{I_{(mA)}}$

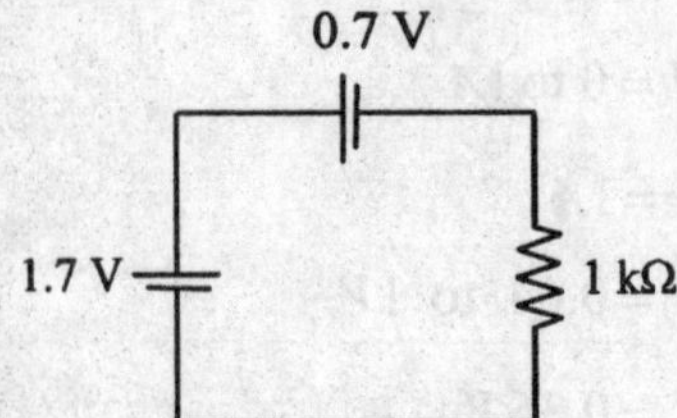

$$1.7 - 0.7 - I(10^3) = 0$$

$$I(10^3) = 1$$

$$I = 1\,mA$$

$$r = \frac{26\,mV \times 1}{1\,mA}[At\ 300\ k\Omega]$$

$$r = 26\,\Omega \cong 25\,\Omega$$

29. Ans. (a)

Solution: The peak reverse voltage in full-wave rectifier is $2V_m$, here V_m is the ac input peak value.

$V_m = 50\sqrt{2}$ V

Peak reverse voltage = $100\sqrt{2}$.

30. Ans. (b)

Solution: Static characteristic of forward biased *p-n* junction,

$$I = I_o\,[\exp\,[(V/V_T) - 1]$$

$$\log\,[(I/I_o) + I] = (V/V_T)$$

$$V = V_T\,\ell n\left[\frac{I}{I_s} + 1\right]$$

$$= V_T\,\ell n\,I - V_T\,\ell n\,I_s \qquad \left[\text{as } \frac{I}{I_s} >> 1\right]$$

$y = mx + c$ [as V_T and I_s both are cost]

i.e. Straight line is between log I vs V.

31. Ans. (c)

Solution: The correct match is given below

LED	Works on the principal of spontaneous emission
Avalanche photodiode	Due to action it have large Current gain
Tunnel Diode	Has very large doping of the ratio of 1:10³
LASER	It has the property of coherent radiation

32. Ans. (a)

Solution: $\lambda = [C/f]$; $f = [C/\lambda]$

$$E = hf = hC/\lambda = \frac{6.625 \times 10^{-34} \times 3 \times 10^8}{5490 \times 10^{-10}} \text{ Joules}$$

$= 3.62 \times 10^{-19}$ Joules.

In eV, the energy B and Gap Energy = $[E/e]$

$$= \frac{3.62 \times 10^{-19}}{1.6 \times 10^{-19}}$$

$$= 2.26 \text{ eV}$$

33. Ans. (b)

Solution: Reverse saturation current becomes doubles for every 10 degree, rise in temperature i.e.

$$I_o(T_1) = I_{oT} \times 2^{\frac{(T-T_1)}{10}}$$

$$I_s = 10 \Rightarrow \text{pA at } 20^\circ \text{C}$$
$$I_s = 20 \Rightarrow \text{pA at } 30^\circ \text{C}$$
$$I_s = 40 \Rightarrow \text{pA at } 40^\circ \text{C}$$

34. Ans. (c)

Solution: The tunnel diode has characteristic

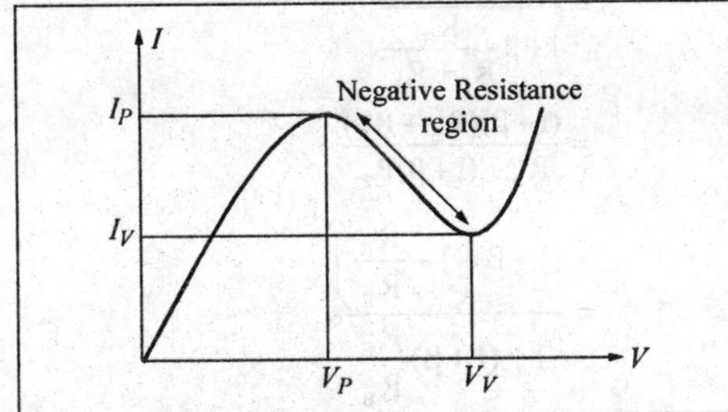

35. Ans. (c)

Solution: The correct match is given below

Varactor diode	It is used in tunned circuit as it work as capacitor specially in *VCO*
PIN diode	Current controlled attenuator
Zener diode	Used in regulated power supply as reference voltage source
Schottky diode	It is a metal-semiconductor fabri-cation used for fast switching action

36. Ans. (c)

Solution: BJT → Early effect (base width modulation)

MOS Capacitor → Flat-band voltage

Laser diode → Population inversion

JFET → Pinch off voltage

37. Ans. (c)

Solution: Half wave and Full wave Average load current are $\left[\frac{I_m}{\pi}\right], \left[\frac{2I_m}{\pi}\right]$. There Transformer ratio are 0.287, 0.693

38. Ans. (c)

Solution: The output voltage, $V_0 = 0$ for $V_1 < 0.5$ V For $V_1 \geq 0.5$, As the diode starts conducting, therefore

$$V_0 = \frac{V_i - 0.5}{900} \times 600$$

$$\text{At } V_i = 2 \text{ volts, } V_0 = \frac{2 - 0.5}{900} \times 600 = 1\text{V}$$

39. Ans. (b)

40. Ans. (d)

Solution: In four diode bridge rectifier the transformer is used with no center tape and small in size and low cost.

Since four diode are used higher current can be drawn form the rectifier.

41. Ans. (b)

Solution:

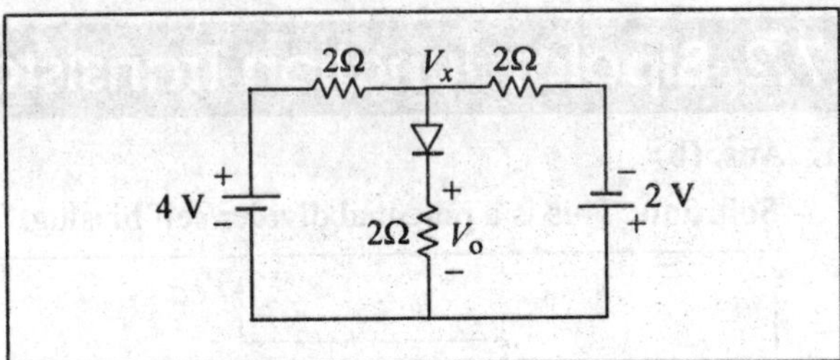

Apply KCL at Node V_x

$$\frac{V_x - 4}{2} + \frac{V_x + 2}{2} = 0$$

$$V_x = 1$$

42. Ans. (b)

43. Ans. (a)

Solution:

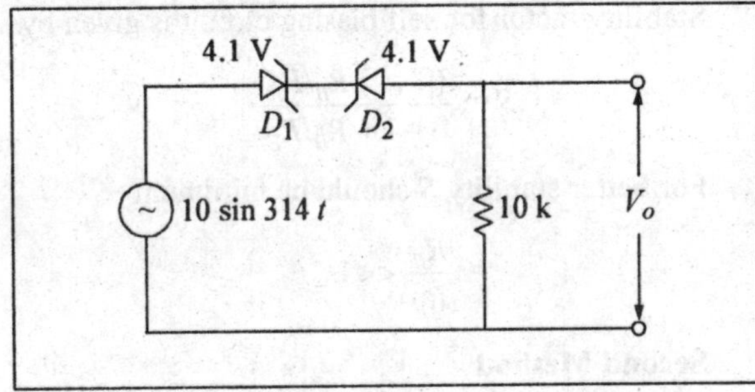

During positive half cycle for input less than 4.1 V

$D_1 \rightarrow$ Forward biased

$D_2 \rightarrow$ Reverse biased

Current through load resistance is zero. Thus $V_o = 0$

During positive half cycle for more than 4.1 V

$D_1 \rightarrow$ Forward biased

$D_2 \rightarrow$ Reverse biased but Zener break down take place and voltage across is 4.1 Volt.

Thus maximum voltage drop across load resistance is

$= 10 - 4.1 = 5.9$ V

Similarly, During the negative half cycle output is maximum to – 5.9 V

44. Ans. (d)

Solution: When $V_{in} < 0$, Zener diode works as a normal diode in forward biased. Hence output $V_R = V_{in}$

When $0 < V_{in} < 6$, Diode OFF and $V_R = 0$

When $V_{in} < 0$, Diode is having voltage of 6 V. Hence output $V_R = V_{in} - 6$

$= 12 \sin \omega t - 6$

Maximum it can go up to 6 V.

7.2 Bipolar Junction Transistor

1. Ans. (b)

Solution: This is a potential divider/self biasing.

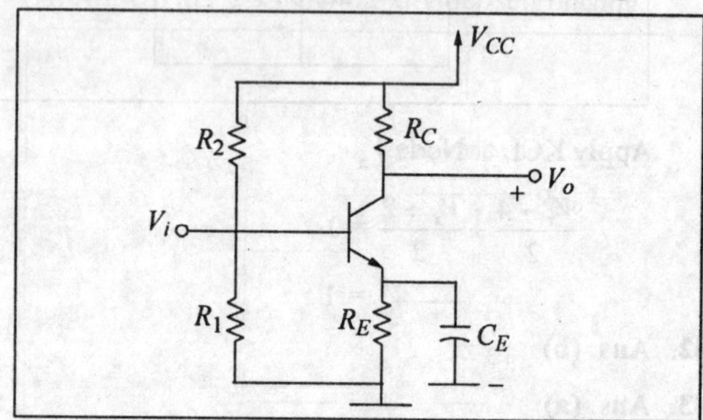

$R_1 || R_2 = R_B$

Stability factor for self biasing circuit is given by.

$$S = \frac{(\beta + 1) R_B/R_E}{\beta + 1 + R_B/R_E}$$

For better stability *S*-should be minimum

$$= \frac{R_B}{R_E} << 1$$

Second Method

For 'DC' analysis all capacitors are considered as open circuit

$$V_{th} = \frac{V_{CC}.R_1}{R_1 + R_2}$$

$$R_{th} = R_1 \| R_2 = \frac{R_1 \times R_2}{R_1 + R_2}$$

$$I_{E(DC)} = \frac{V_{th} - V_{BE}}{R_E + \left(\frac{R_{th}}{\beta + 1}\right)}$$

Since $R_E >> \frac{R_B}{\beta+1}$ $(\because R_{th} \cong R_B)$

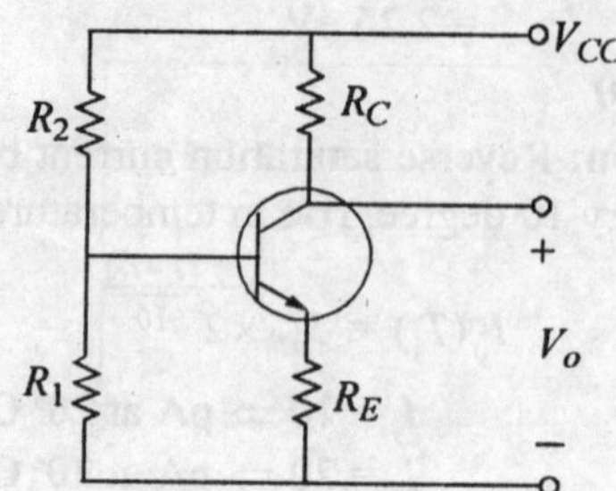

$$\frac{R_E}{R_B} >> 1$$

$$S = \frac{1+\beta}{1+\beta\frac{R_E}{R_E + R_B}}$$

$$= \frac{(1+\beta)(R_E + R_B)}{R_B + (1+\beta)R_E}$$

$$= \frac{(1+\beta)\left(1 + \frac{R_E}{R_B}\right)}{1 + (1+\beta)\frac{R_E}{R_B}}$$

The ratio $\frac{R_E}{R_B}$ controls the value of stability factor

If $\frac{R_E}{R_B} >> 1$

$$\Rightarrow S = \frac{1+\beta}{1+\beta} = 1$$

So, it gives good stability

2. Ans. (b)

3. Ans. (c)

Solution: As tuned amplifier is specified, high frequency response of *n*-stage in $B' = B\sqrt{2^{1/n} - 1}$ i.e. *B* decrease.

4 Ans (c)

Solution :

$$\frac{I_{E2}}{I_{E1}} = \frac{I_{ES}\left[e^{V_{BE2}/V_T} - 1\right]}{I_{ES}\left[e^{V_{BE1}/V_T} - 1\right]}$$

as given $I_{E2} = 2I_{E1}$

or, $$\frac{2I_{E1}}{I_{E1}} = \frac{e^{V_{BE2}/V_T}}{e^{V_{BE1}/V_T}} = e^{(V_{BE2} - V_{BE1})/V_T}$$

or, $V_{BE2} - V_{BE1} = V_T \ln(2) = 18mV$

$\therefore \Delta V_{BE} = 18mV$ i.e. increase in V_{BE}

5. Ans. (d)

Solution:

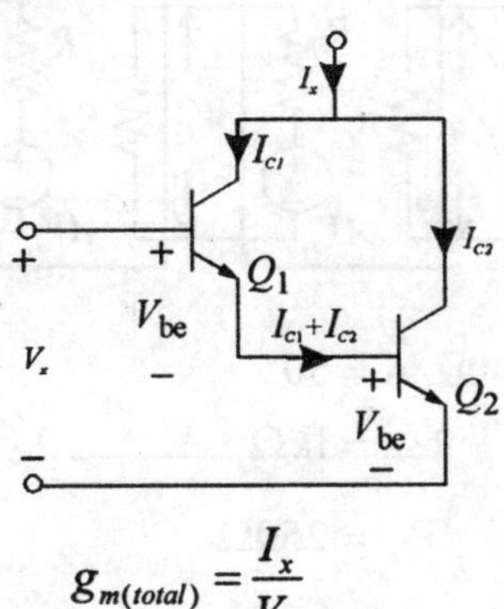

$$g_{m(total)} = \frac{I_x}{V_x}$$

Since $\quad I_x = I_{C1} + I_{C2} \cong I_{C2}$

as $\quad I_{C_1} = I_{B_2} << I_{C_2}$

$$g_{m(total)} = \frac{I_x}{V_x} = \frac{I_{C2}}{2V_{be}} = \frac{g_{m2}}{2} = 0.5\, g_{m2}$$

6. Ans. (b)

Solution: Without the capacitor there is a negative feedback and gain is reduced. By connecting a high value capacitor the ac signals are by passed and gain is increased.

Before connecting capacitor, $Z_{in} = h_{ie} + (1 + B) R_E \cdot A$ capacitor in shunt of high value reduces the value of R_E (almost shorts) and so decreases the input impedance.

7. Ans. (a)

Solution:

Cascode connection

This CE-CB connection

* Common emitter stage drives a common base amplifier.
- Gain can be improved by using large value of R_L over a wide range of frequencies.
- Overall current gain $A_I - \frac{I_{C2}}{I_{B2}} = \beta_0$.
- Voltage gain $= A_v = \frac{-\beta_0 R_L}{(R_s + r_{be})}$.
- Output resistance of a CB stage is high, so we can connect a large load resistance R_L.
- This input resistance for the CB stage is low and this is the load resistance for the CE stage. The low input resistance of CB stage makes Q_1 operate into a short-circuit load, so voltage gain of common emitters stage is low. This cause reduces the input capacitances so its Bandwidth gets increased.
- It is used in *wide Band amplifiers* due to its large bandwidth.

8. Ans. (d)

Solution: Voltage shunt

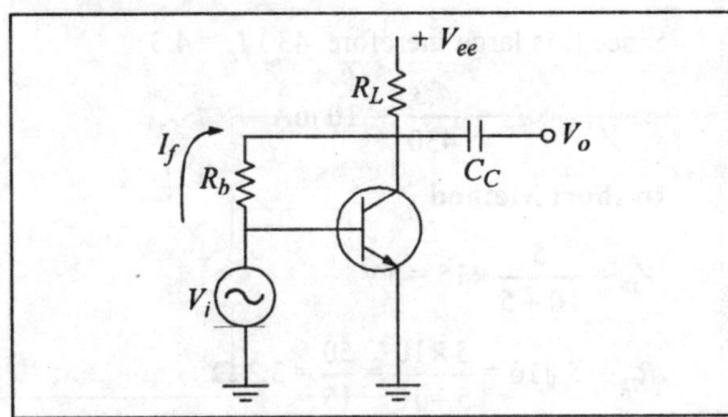

$V_o >> V_i$

$$I = \frac{V_i - V_o}{R_b} = \frac{-V_o}{R_b} = KV_o$$

Where $K = \frac{-1}{R_b}$

Since feedback current is proportional to output voltage, i.e. voltage shunt feedback.

9. Ans. (c)

Solution:

* R_E stabilizes I_C by changing I_B according
* Variations in temperature or β ultimately leads to destablization of I_C

Hence, R_E stabilizes the DC opeation point against variation in both.

10. Ans. (d)

Solution:

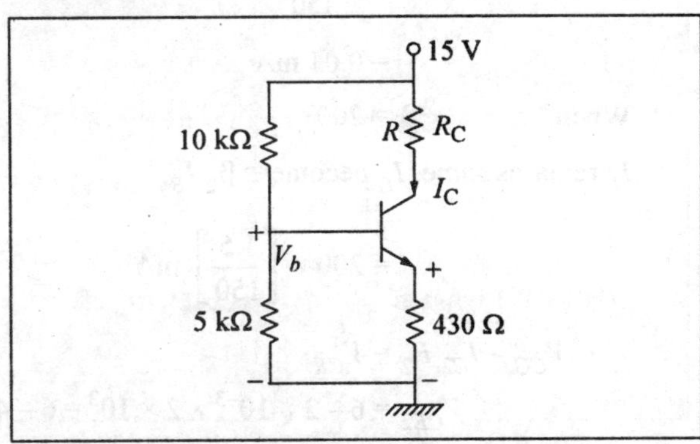

$$V_b = \left(\frac{5}{5+10}\right) 15 = 5\text{V}$$

$$I_c = \beta\, I_b; \quad I_o = \frac{I_C}{\beta}$$

$$I_e = I_C + I_b = I_C + \frac{I_C}{\beta}$$

KVL in base loop

$$-5 + 0.7 + 430\left(\frac{I_C}{\beta} + I_C\right) = 0$$

Since β is large therefore, 430 I_c = 4.3

$$\Rightarrow \quad I_c = \frac{4.3}{430} = 10 \text{ mA}$$

In short Method

$$V_{Th} = \frac{5}{10+5} \times 15 = 5\text{V}$$

$$R_{Th} = 5 \parallel 10 = \frac{5\times 10}{5+10} = \frac{50}{15} = 3.3\ \Omega$$

Since β is large $I_\beta = 0$.

$$I_C = \frac{V_{Th} - V_{BE}}{R_E}$$

$$= \frac{5-0.7}{430} = \frac{4.3}{430} = 10 \text{ mA}$$

11. Ans. (b)

12. Ans. (a)

Solution: $V_{CC} - I_C R_2 = V_{CE}$

$$R_2 = \frac{(6-3)}{1.5\text{ mA}} \text{V} = 2\text{k}\Omega$$

$$I_C = I_b \beta + (1+\beta) I_{CO}$$

$$I_{B1} = \frac{I_{C1}}{\beta}$$

$$= \frac{1.5\text{ mA}}{150}$$

$$= 0.01 \text{ mA}$$

When, β = 200

I_B remains same, I_C become = $\beta_2 I_B$

$$= 200 \times \left[\frac{1.5}{150}\right] \text{mA}$$

$$V_{CC} - I_{C2} R_2 = V_{CE}$$

$$V_{CE2} = 6 - 2\times 10^{-3} \times 2 \times 10^3 = 6 - 4 = 2\text{V}$$

This operating point is (2V, 2 mA)

13. Ans. (c)

Solution:

When three identical voltage amplifier are cascaded the block diagram of overall amplifier can be drawn as below

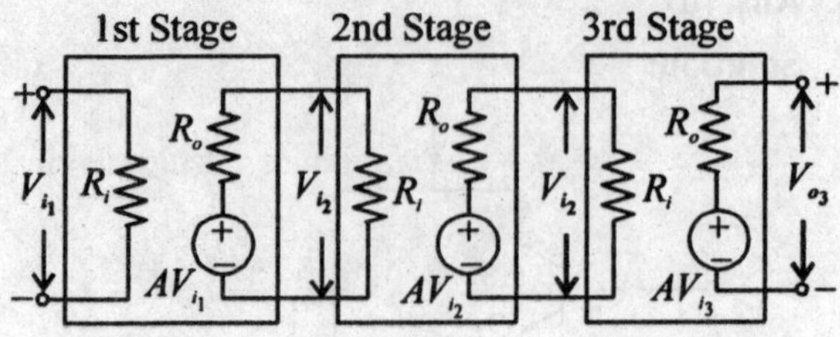

For each state ; gain, A = 50

Input resistance $R_i - = 1k\Omega$

Output resistance $R_o = 250\Omega$

Overall voltage gain $A_v = \dfrac{V_{o3}}{V_{i1}}$

or, $$A_V = \frac{V_{o_3}}{V_{i_3}} \cdot \frac{V_{i_3}}{V_{i_2}} \cdot \frac{V_{i_2}}{V_{i_1}}$$

Since $$V_{o_3} = AV_{i_3}$$

$$\therefore \quad \frac{V_{o_3}}{V_{i_3}} = A$$

Since $$V_{i3} = \frac{AV_{i2}}{R_i + R_o}$$

$$\frac{V_{i_3}}{V_{i_2}} = \frac{AV_{i_2}}{R_i + R_o}$$

Since $V_{i_2} = AV_{i_1}\dfrac{R_i}{R_i + R_o}$

$$\frac{V_{i_2}}{V_{i_1}} = \frac{AR_i}{R_i = R_o}$$

Over all gain $A_V = \dfrac{V_{o_3}}{V_{i_1}}$

$$= A\frac{AR_i}{R_i + R_o}\frac{AR_i}{R_i + R_o}$$

$$= A^3\left(\frac{R_i}{R_i + R_o}\right)^2$$

$$= 50^3\left(\frac{1000}{1000 + 250}\right)^2$$

$$\therefore \quad A_V = 80000$$

in dB, $A_{VdB} = 20\log_{10}(80000)$

$= 98$ dB

14. Ans. (d)

Solution:

$$g_m = \frac{|I_C|}{V_T} = \frac{1 \text{ mA}}{25\times(1000)^{-1}} = 40 \text{ mA/V}$$

Using relation between low frequency and high frequency parameter

$$h_{fe} = g_m r_\pi$$

But, $h_{fe} = \beta$

$$\therefore \quad r_\pi = \frac{\beta}{g_m} = \frac{100}{40 \text{ mV}} = 2.5 \text{ k}\Omega$$

15. Ans. (a)

Solution: $\beta = h_{fe} = 50$

$$V_{CC} = I_C \cdot R_C + V_{CE\text{ sat}}$$

$$I_C = h_{fe} \cdot I_B$$

$$\Rightarrow \quad I_C = \frac{3-0.2}{1\text{k}} = 2.8 \text{ mA}$$

$$I_B = \frac{2.8 \text{ mA}}{50} = 0.056 \text{ mA} = 56\ \mu\text{A}.$$

16. Ans. (b)

Solution: Cascode connection

This CE-CB connection

- Common emitter stage drives a common base amplifier.
- Gain can be improved by using large value of R_L over a wide range of frequencies.

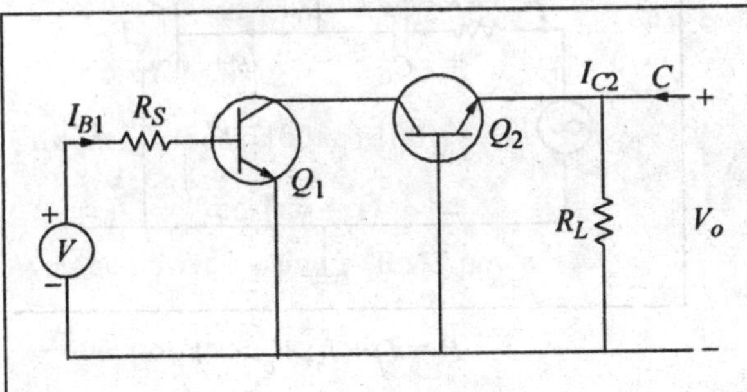

- Overall current gain $A_I - \frac{I_{C2}}{I_{B2}} = \beta_0$
- Voltage gain $= A_v = \frac{-\beta_0 R_L}{(R_s + r_{be})}$
- Output resistance of a *CB* stage is high, so we can connect a large load resistance R_L.
- This input resistance for the *CB* stage is low and this is the load resistance for the *CE* stage. The low input resistance of *CB* stage makes Q_1 operate into a short-circuit load, so voltage gain of common emitters stage is low. This cause reduces the input capacitances so its Bandwidth gets increased.

It is used in wide band amplifiers due to its large bandwidth.

17. Ans. (b)

Solution:

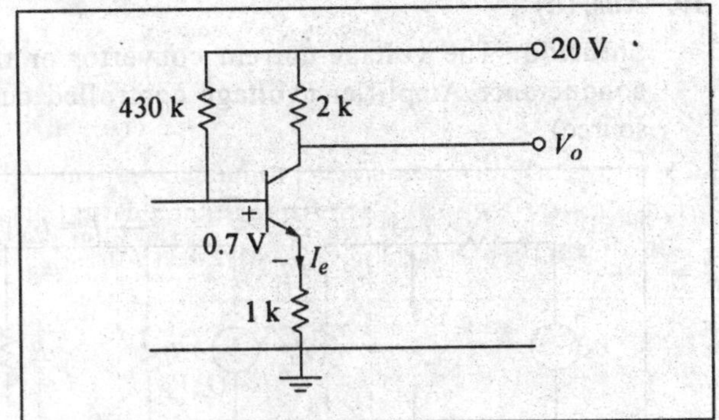

$$I_e = I_b + I_c$$

$$I_c = \beta I_b + I_{co}$$

$$\beta = 50,\ \alpha = \frac{50}{51} = 0.98039 = \frac{I_c}{I_e}$$

$$V_{BE} = 0.7 \text{ V}$$

$$20\text{V} = (430 \text{ k}) I_b + 0.7 + I_e (1 \text{ k})$$

$$20 = (430 \text{ k}) I_b + 0.7 + (51 I_b) I \text{ k}$$

$$I_b = 40\ \mu\text{A}$$

$$I_c = \beta I_b = 2000\ \mu\text{A}$$

$$V_o = 20 - 2 \text{ k} \times 2000\ \mu\text{A} = 16 \text{ V}$$

18. Ans (c)

Solution:

Since the collector and base terminals are shorted

$$\therefore \quad V_{CE} = V_{BE} = 0.7 \text{ V}.$$

Due to this the transistor will act as a forward-biased pn junction diode.

$$I = I_o\left[\exp\left\{\frac{V}{nV_T}\right\} - 1\right]$$

or $\quad I_E = I_0(e^{V_{BE}/V_T} - 1)$

Given $\quad I_0 = 10^{-13}\text{ A} = 10^{-10}\ mA$

$$V_T = \frac{T}{11{,}600}$$

$$= \frac{300}{11,600}$$

$$= 0.026 \text{ V}$$

$$\therefore \quad I_E = 10^{-10}\left(e^{0.7/0.026} - 1\right)$$

$$\therefore \quad I_E = 49.3\, mA$$

19. Ans. (d)

Solution: The voltage-current convertor or trans-conductance Amplifier (Voltage controlled current source)

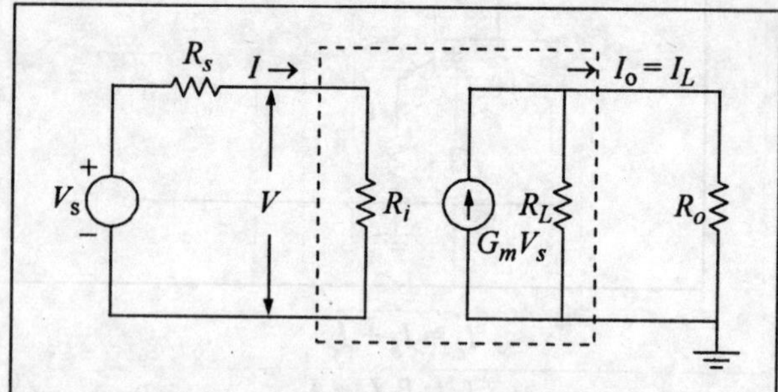

This is ideal voltage-controlled current source.

For a voltage controlled current source input is voltage so input impedance R_i is kept as high as possible i.e. $Z_i = \infty$. Output is current so output impedance R_o is kept high as possible so that max a full current can flow to load i.e. $Z_o = \infty$.

$$|R_i| >> R_s \text{ and } R_L << |R_o|$$

It mean, $Z_i = \infty$, $Z_o = \infty$

$$I_o = G_m V_s$$

20. Ans. (c)

Solution: $\beta_{DC} = 60$, $V_{BE} = 0.7$ V, $h_{ie} \rightarrow \infty$, $h_{fe} \rightarrow \infty$

The capacitance C_c can be assumed to be infinite.

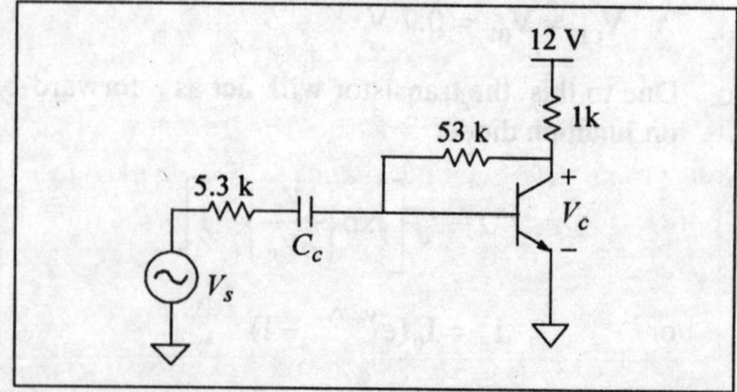

This is collector to Base biasing circuit diagram

$$\beta_{dc} = \frac{I_C}{I_b}$$

$$I_b \times \beta_{dc} = I_C$$

$$\left(\frac{V_C - 0.7}{53 \text{ k}}\right) \times 60 = \frac{12 - V_C}{1\text{k}}$$

$$60\, V_C - 42 = 53 \times 12 - 53\, V_C$$

$$V_c = 6.0 \text{ V}$$

$$V_{CC} = R_L\, [I_C + I_b] + R_b\, I_b + V_{BE}$$

$$12 = [I_C + I_b] \times 1 + 53 \cdot I_b + V_{BE.}$$

$$\Rightarrow \quad 12 = 61\, I_b + 53\, I_b + V_{BE} \Rightarrow I_b = .099 \text{ mA}$$

Now $\quad V_{CE} = 53\, I_b + V_{CE} = 5.95$ Volt

21. Ans. (b)

Solution: $\quad V_{CE} = V_{CC} - R_L\,(I_c + I_b)$

$$= V_{CC} - R_L\,(\beta + 1)\, I_b$$

From above expression if β is increased then V_{CE} is decrease by less than equal to same percentage.

22. Ans. (a)

Solution: $\quad \beta_{DC} = 60$, $V_{BE} = 0.7$ V,

$$h_{ie} \rightarrow \infty,\ h_{fe} \rightarrow \infty$$

The capacitance C_C can be assumed to be infinite.

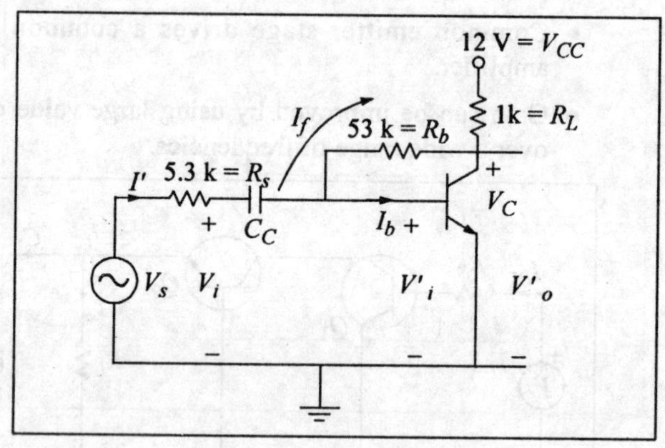

$$I' = I_f + I_b;\ V_o >> V_i$$

$$I_f = \frac{V_i - V_o}{R_b}$$

$$= \frac{-V_o}{R_b} = KV_o$$

Where, $K = \dfrac{-1}{R_b}$

Since feed back current is proportional to output voltage. It represent voltage shunt feed back

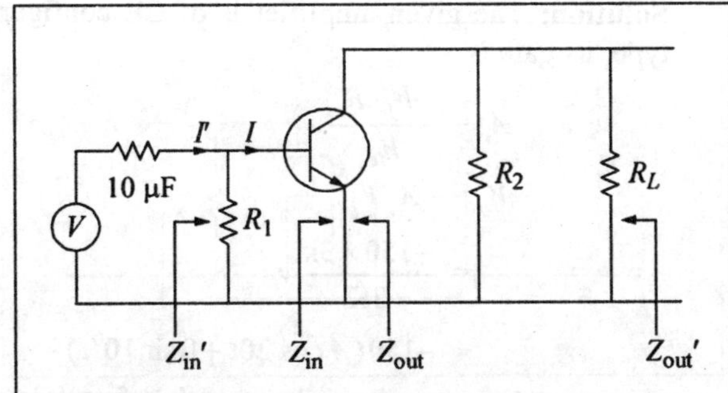

By Millar Theorem,

$$R_1 = \frac{R_b}{1-A_v};\ R_2 = \frac{R_b}{[1/1(1/A_v)]} = \frac{R_b\ A_V}{A_V - 1}$$

In *CE*, $A_v >> 1$

$$R_1 = \frac{R_b}{-A_v};\ R_2 \cong R_b$$

For *CE*: $A_v = \dfrac{-h_{fe}\ R_L}{h_{ie}};\ R_1 = \dfrac{R_B \cdot h_{ie}}{h_{fe} \cdot R_L}$

$$Z_{in} = h_{ie}\ ;\quad Z_{in}' = hi_e \,||\, R_1\ ;\quad A_I = -h_{fe}$$

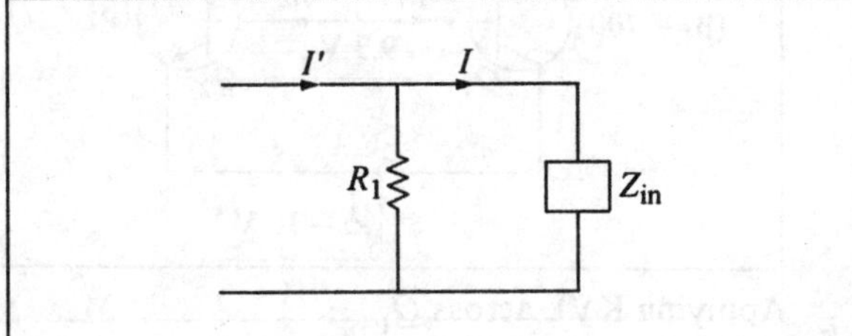

$$\frac{I}{I'} = \frac{R_1}{R_1 + Z_{in}}$$

$$A_{If} = A_I' = \frac{I_L}{I'} = \frac{I_L}{I} \times \frac{I}{I'}$$

$$= AI = \frac{R_1}{R_1 + Z_{in}} = \frac{-h_{fe}\ R_1}{R_1 + Z_{in}} = A_{If}$$

$$A_v = A_I \cdot [Z_L / Z_{in}]$$

$$= \frac{-h_{fe}(R_2 \,||\, R_1)}{h_{ie}}$$

$$Z_{out} = \infty$$

$$Z_{in} = \infty \,||\, R_2 \,||\, R_L = R_2 \,||\, R_L$$

$$= \frac{R_b \times R_L}{R_b + R_L}$$

$$A_{vf} = \frac{-h_{fe} \cdot R_b \cdot R_L}{h_{ie}(R_b + R_L)}$$

$$r_L = 1\text{ k}$$

$$A_i = h_{fe}$$

$$R_i = h_{ie}$$

$$A_v = \frac{-h_{fe}}{h_{ie}}\ r_L = -r_L$$

$$R_{B1} = \frac{53 \times 100}{1-(-1000)} = 53\ \Omega$$

$$A_{vs} = \frac{V_c}{V_s} = 1000 \times \frac{53}{5.3 \times 1000} = -10$$

23 Ans. (a)

Solution: The voltage-current convertor or transconductance Amplifier is a Voltage controlled current source

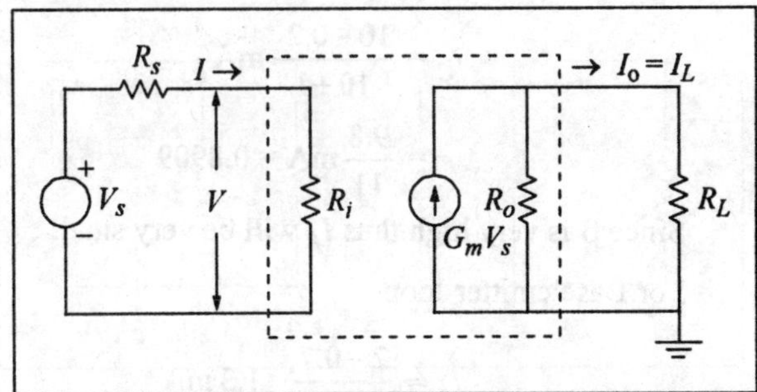

Transductions amplifier has voltage source at input end and current source at output end. Thus input resistance must be high and output resistance must be low.

$$|R_i| >> R_s \text{ and } R_L << |R_o|$$

$$I_o = G_m\ V_s$$

24. Ans. (b)

Solution:

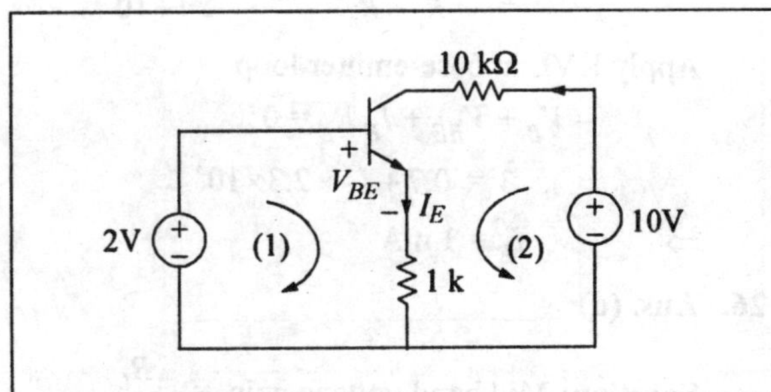

First Method

$V_{BE} = 0.7$ V; If we assume β is very large

KVL in base emitter loop

$$2 - V_{BE} = I_E\ R_E$$

$$1.3 = 1000\ I_E$$

$$I_C \cong I_E\ (\beta \text{ very large})$$

$\therefore \quad I_C \cong 1.3$ mA

KVL in collector emitter loop

$$V_{CC} = +I_C(R_C + R_E) + V_{CE}$$

$$V_{CE} = V_{CC} - I_C(R_C + R_E)$$

$$V_{CB} + V_{BE} = 10 - 1.3\times10^{-3}\times11\times10^3$$

$$V_{CB} = -4.6\text{ V} - V_{BE} = -4.6 - 0.7 = -5.3$$

Collector junction is reverse bias and emitter junction is already forward bias as $V_{BE} = 2$ V applied.

Therefore, BJT is in Saturation

Second Method

Assuming transistor in saturation

$$i_C = \frac{10-0.2}{10+1}\text{ mA}$$

$$= \frac{9.8}{11}\text{ mA} = 0.8909$$

Since β is very high thus I_B will be very small

For Base emitter loop

$$I_E = \frac{2-0.7}{1\text{ K}} = 1.3\text{ mA}$$

$$I_E = I'_B + I_C$$

$$I'_B = 1.3 - 0.89 = 0.41$$

Thus $I'_B > I_B$ which shows transistor in saturation.

25. Ans. (a)

Solution: As β very large then, $I_B \approx 0$

$$V_B = \frac{R_2}{R_1+R_2}\times V_{CC} = \frac{9\times10}{20+10} = 3\text{ V}$$

Apply KVL in base-emitter loop

$$-V_B + V_{BE} + I_E R_E = 0$$

$$3 = 0.7 + I_E\times2.3\times10^3$$

$$\Rightarrow \quad I_E = 1\text{ mA}$$

26. Ans. (d)

Solution: Mid band voltage gain $= -\dfrac{R_L}{r_e}$

From solutin of previous problem.

$$r_e = \frac{25}{I_E} = 25\Omega$$

$$R'_L = R_c \| R_L = 3\text{k}\|3\text{k} = 1.5\text{k}$$

Thus $\quad A_V = -\dfrac{1500}{25} = -60$

27. Ans. (d)

Solution: The given amplifier is of CE configuration type, its gain is

$$A_v = \frac{-h_{fe}R_c}{h_{ie}}$$

$$V_o = -A_v V_i$$

$$= \frac{-150\times3\text{k}}{3\text{K}}v_i$$

$$= -150(A\cos 20t + \beta\sin10^6 t)$$

Since coupling capacitor is large, so low frquency effect is to be considered that filter out the low frequency component,

Hence best approximation is

$$V_o = -150\,B\sin10^6 t$$

28. Ans. (b)

Solution:

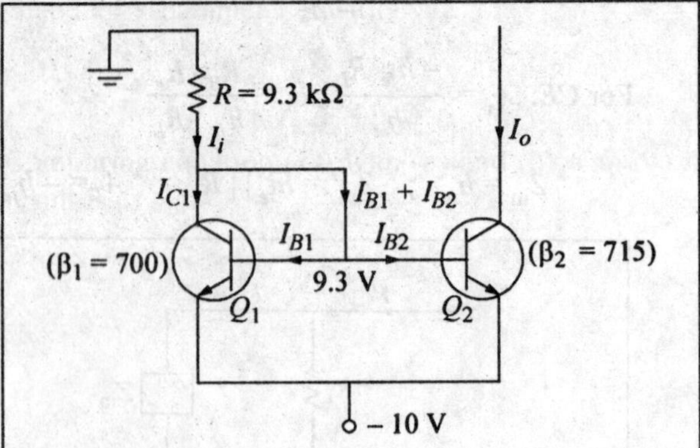

Applying KVL across Q_1

$$-10 = -0.7 - I_1 R_1$$

$$-10 + 0.7 = -I_1 R_1$$

$$9.3 = I_1 R_1$$

$$I_1 = \frac{9.3}{I_1} = \frac{9.3}{9.3\times10^3} = 1\text{ mA}$$

$$I_1 = 1\text{ mA}$$

Again emitter Area $E_1 = \dfrac{1}{2}$ EmitterArea E_2

$$E_1 = \frac{1}{2}E_2$$

$$\Rightarrow \quad I_{E1} = \frac{1}{2}I_{E2}$$

Now $\quad I_1 = I_{C1} + (I_{B1} + I_{B2})$

$$1 = \beta I_B + I_{\beta1} + I_{\beta2}$$

$$1 = (1+\beta)I_{B1} + I_{B2}$$

$1 = 701\,I_{B1} + I_{B2}$ and $701\,I_{B1} = I_{E1}$

$$1 = \frac{1}{2}I_{B2} + I_{B2}$$

$$1 = \frac{1}{2}[I_{B2} + I_o] + I_{B2}$$

$$1 = \frac{1}{2}I_{B2} + \frac{1}{2}I_o + I_{B2}$$

$$1 = \frac{3}{2}I_{B2} + \frac{1}{2}I_o$$

$$2 = 3I_{B2} + I_o$$

$$2 = \frac{3 \times I_o}{715} + I_o$$

$$2 = \left(\frac{3}{715} + 1\right)I_o$$

$$2 = \left(\frac{718}{715}\right)I_o$$

$$70 = \frac{2 \times 715}{718} = 1.99 \text{ mA}$$

$$\approx 2 \text{ mA}$$

29. Ans. (a)

Solution: Given circuit after removing C_E will behave as current-series feedback. Overall voltage gain will decrease as feedback comes into picture and since it is current-series feedback, input impedance increases.

Second Method

The effect of by pass capacitor C_E is to avoid a degenerative effect due to negative feedback i.e. to provide a separate path for a.c. in the input.

By removing C_E, a degenerative effect is produced in the feedback circuit, which results in the decrease of voltage given A_V

Again, since the circuit is current series feedback, the input impedance increases.

30. Ans. (a)

Solution: In the given parallel RLC circuit

$L = 10\mu H$ and $C = 1nF$

$$\omega_o = \frac{1}{\sqrt{LC}} = \frac{1}{\sqrt{10 \times 10^{-6} \times 10^{-9}}}$$

$$= 10^7 \, rad/s = 10 Mrad/s$$

The operating frequency is 10 M rad.

Hence it is a tuned amplifier which gain is maximum at resonant frequency

31. Ans. (b)

Solution: $R_E = 1$ kΩ,

KVL in base loop

$$V_{BE} + R_e I_e - 10 = 0$$

$$I_e = \frac{-0.7 - (-10)}{1 \times 10^3} = 9.3 \text{ mA}$$

$$I_e = I_b + I_c$$

$$= I_b + \beta I_b$$

$$I_c = (\beta + 1) I_b$$

$$\therefore \quad 9.3 = 51\, I_b$$

$$I_b = 18.2 \; \mu A$$

32. Ans. (b)

Solution: Apply KVL in B-E loop, we get

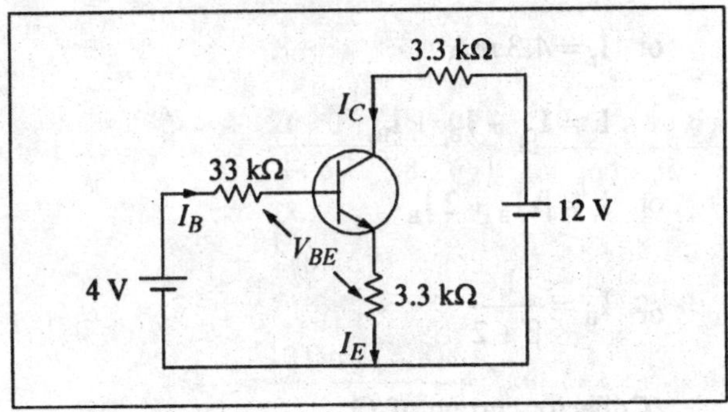

$$-4 + 33\, I_B + V_{BE} + 3.3\, I_E = 0 \quad [\therefore I_E = (1 + \beta)\, I_B]$$

or $$I_B = \frac{3.3}{33 + 3.3\,(\beta + 1)}$$

or $$I_B = \frac{3.3}{33 + 330} \text{ mA}$$

and $$I_C = \beta\, I_B$$

$$= \frac{99\,(3.3)}{33 + 330} \text{ mA}$$

or $$I_C \approx \frac{3.3}{3.3 + 0.33} \text{ mA}$$

33. Ans. (c)

Solution: The given diagram represent the current Mirror case in which both the collector current are same. Therefore,

$$I = \frac{5 - 0.7}{1}$$

$$= 4.3 \text{ A}$$

Second Method

Since the two transistors are matched ; $I_{B_1} = I_{B_2}$

Apply KVL across Base - Emitter Loop of Q_1

$$I_1 \times R_{ref} + V_{BE_1} + V_{EE} = 0$$

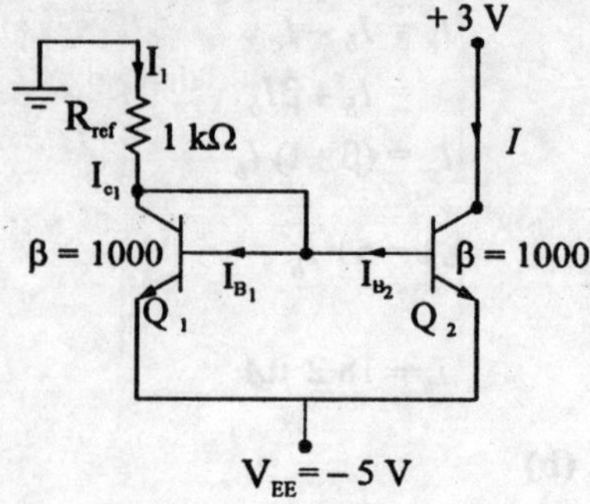

or $I_1 \times 1k + 0.7 = 5$

or $I_1 = 4.3\ mA$

$I_1 = I_{C_1} + I_{B_1} + I_{B_2}$

or $I_1 = \beta I_B + 2 I_B$

or $I_B = \dfrac{I_1}{\beta + 2}$

Collector current of Q_2

$I = I_{C_2}$

or $I = \beta I_B$

or $I = \dfrac{\beta I_1}{\beta + 2} \approx I_1$ $[\because \beta \text{ is very large}]$

$\Rightarrow I = 4.3\ mA$

34. Ans (a)

Solution:

Since the transistor is biased such that the Base - Emitter junction is reverse biased therefore the transistor will operate in cut off region

∴ Base current = zero.

35. Ans. (b)

Solution:

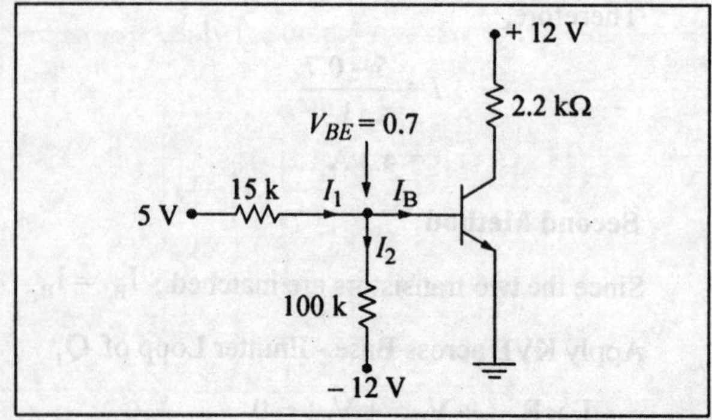

Thevenin equivalent voltage and resistance at Base of transistor are $V_{th} = 2.8\,V$, $R_{th} = 13\,k$, which indicate $B-E$ junction is Forward biased and transistor in cut off region

$$V_{BE} = 0.7 \text{ V}$$

$$I_1 = \frac{5 - 0.7}{15 \text{ k}}$$

$$= \frac{4.3}{15 \text{ k}}$$

$$= 0.286 \text{ mA}$$

$$I_1 = 286.67\ \mu\text{A}$$

$$I_2 = \frac{0.7 + 12}{100 \text{ k}}$$

$$= \frac{12.7}{100 \text{ k}}$$

$$= 0.127 \text{ mA}$$

$$I_2 = 127\mu\text{A}$$

$$I_B = I_1 - I_2 = 0.286 - 0.127$$

$$= 0.159 \text{ mA}$$

$$I_B = 159.67\ \mu\text{A}$$

Let us assume transistor in active region then

$$I_c = \beta\, I_B + (\beta + 1)\, I_{CBO}$$

$$= 30 \times 159.7\ \mu\text{A} + 31 \times 20 \times 10^{-9}$$

$$I_c = 4.79 \text{ mA}$$

$$V_{CE} = 12 - 2.2 \times 4.79 = 1.462 \text{ V}$$

$$V_{CB} = V_{CE} - V_{BE}$$

$$= 1.462 - 0.7 = 0.762$$

V_{CB} is positive so transistor is operating in the active region.

* $V_{CE} = 1.46 > V_{CE(sat)} = 0.2\,V$

Hence Transisor in active region

36. Ans. (d)

Solution:

* Since B-E junction forward biased, the transistor is not operating in cut off region.

Let us assume that the transistor is operating in active region.

Apply KVL in base emitter loop

$10 - I_E.R_E - 0.7\ I_B\ R_B = 0$

or $I_B(R_B + (1+\beta)R_E) = 10 - 0.7$

or $I_B = \dfrac{9.3}{270+101\times1} = 0.025\text{ mA}$

$I_C = \beta I_B = 2.5\text{ mA}$

$I_E = (1+\beta)I_B \approx 2.5\text{ mA}$

Apply KVL in collector loop

$10 - I_E R_E - V_{EC} - I_C R_C = 0$

or, $V_{EC} \approx 10 - 2.5(1+1)$

or, $V_{EC} \approx 5\,V > V_{CE,sat}$

Hence the transistor is operating in active region.

37. Ans. (b)

Solution:

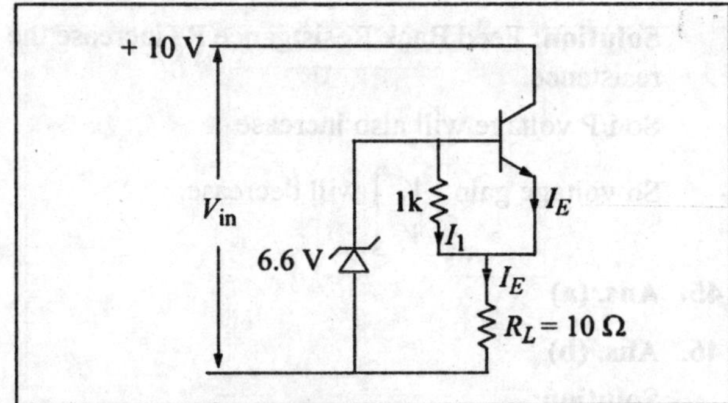

Power dissipated in transistor

$= V_{CE} \times I_C$

$V_{BE} = 0.6\text{ V}$

$I_1 = \dfrac{0.6}{1k}$

$= 0.5\ mA$

$\therefore$ Voltage, across $R_L = 6.6 - 0.6 = 6\text{ V}$

$\therefore \quad I_E = \dfrac{6}{10} = 0.6\text{ A}$

$I_E \cong I_C$ as I_B is nearly 0.

$\therefore \quad V_{CE} = 10 - 6 = 4\text{ V}$

$\therefore \quad P = 4 \times 0.6 = 2.4\text{ W}$

38. Ans. (b)

Solution:

If the NPN transistor base is connected to positive 10 V then transistor operate in active or saturation region $V_{CEsat} \cong 0$. If base register is open circuit the NPN transistor is in OFF condition whole of the V_{CC} is available to collector then collector to ground is + 20 V.

Second Method

Given that collector to ground voltage is 20 V, i.e. collector to ground is nothing but collector to emitter voltage = 20 V. So no current passing through 10 $k\Omega$. So from their we conclude that $I_E = I_C = 0$, $I_B = 0$.

or, Emitter to ground is open circited.

39. Ans. (d)

Solution:

* In saturation region $I_C < \beta I_B$
* In active region $I_C = \beta I_B$

40. Ans. (c)

Solution:

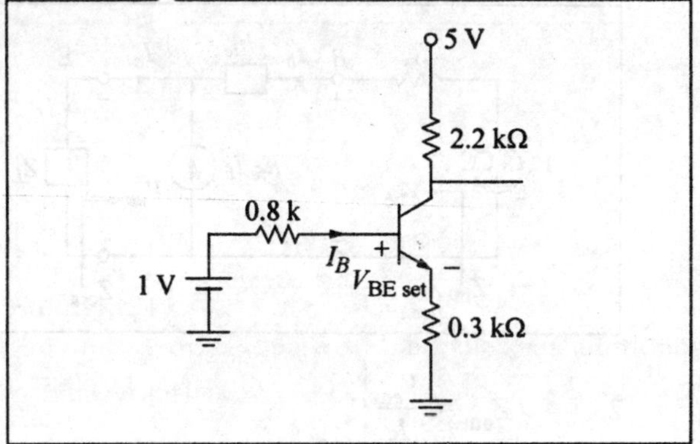

$R_{th} = \dfrac{1\times4}{1+4} = 0.8\text{ k}\Omega$

$V_{th} = \dfrac{1}{5} \times 5 = 1\text{ V}$

KVL in base loop

$-1 + 0.8 \times I_b + V_{be\ sat} + 0.3 \times I_C = 0$

$I_b \rightarrow = 0$ as β is very large

$0.3\ I_C = 1 - 0.7 = 0.3$

$I_C = 1\text{ mA}$

$V_{CE\ sat} = 5 - 2.2 \times 1 - 0.3 \times 1 = 2.5\text{ V}$

41. Ans. (d)

Solution:

$I_c = \beta I_B + (1+\beta)I_{CBO}$ (1)

$\beta = \dfrac{\alpha}{1-\alpha} = \dfrac{0.98}{1-.098} = 49$

$I_B = 20\mu A$

$I_{CBO} = 0.6\mu A$

Putting the value in equation (1)

$\therefore \quad I_c = 1.01mA$

Second Method

$$I_C = \beta I_B + (1+\beta) I_{CO}$$

$$\beta = \frac{\alpha}{1-\alpha} = 49$$

$$= 49 \times 20 \times 10^{-3} + 50 \times 0.6 \times 10^{-3} = 1.01\, mA$$

42. Ans. (a)

Solution: *CC*/ Emitter follower:

1. $$Z_{in} = \frac{V_{bc}}{I_b} = \frac{h_{ie} \cdot I_b + I_L Z_L}{I_b}$$

$$= \frac{h_{ie} I_b + Z_L (+h_{fe}) I_b}{I_b}$$

$$Z_{in} = h_{ie} + (1 + h_{fe}) Z_L \text{ ... highest}$$

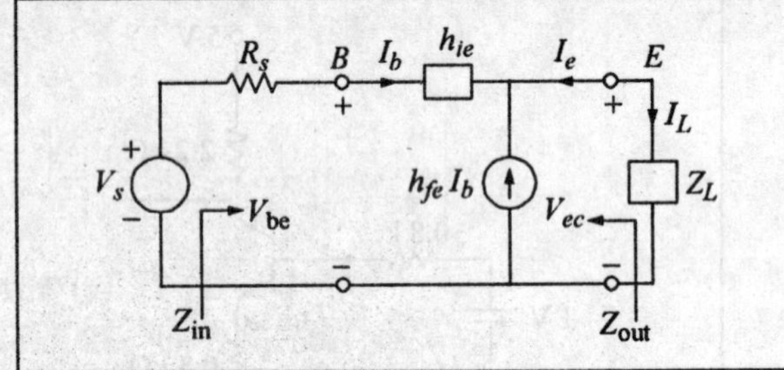

2. $$Z_{out} = \left.\frac{V_{ec}}{I_e}\right|_{V_s=0}$$

$$V_{ec} = -h_{ie} I_b - R_s I_b + V_s$$

$$= -(h_{ie} + R_s) I_b$$

$$I_e = -(1 + h_{fe}) I_b.$$

$$Z_{out} = \frac{h_{ie} + R_s}{(1 + h_{fe})} \Rightarrow \text{ lowest}$$

43. Ans. (a)

Solution:

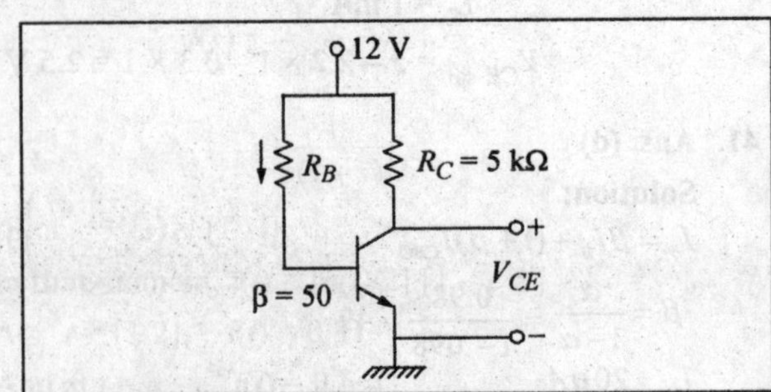

This diagram represent fixed biased

Apply KVL in CE loop

$$V_{CC} = I_C R_C + V_{CE}$$

$$V_{CE} = V_{CC} - I_C R_C$$

$$2 = 12 - I_C \times 5000$$

$$\therefore \quad I_C = 2 \text{ mA}$$

KVL in V_{CC} - *B* - Emittar - loop

$$V_{CC} = I_B R_B + V_{BE}$$

$$12 = I_B R_B + 0.7$$

$$I_B = \frac{11.3}{R_B} = \frac{I_C}{\beta} = \frac{2 \text{ mA}}{50}$$

$$R_B = \frac{11.3 \times 50}{2 \times 10^{-3}} = \frac{11.3 \times 50}{2} \text{ k}\Omega$$

$$R_B = 283 \text{ k}\Omega$$

44. Ans. (a)

Solution: **Feed Back Resistgance R_E increase the i/p resistance.**

So i/P voltage will also increase

So voltage gain $\left[\frac{V_o}{V_i}\right]$ will decrease.

45. Ans. (a)

46. Ans. (b)

Solution:

For condition DC analysis is done

Apply KVL in CBE loop

$$24 - (I_C + I_B)\, 4.7 \text{ K} - I_B\, 560 \text{ K} - 0.7 = 0$$

as $I_C = \beta I_B$

$$\text{or } I_B = \frac{23.3}{560\,K + 101 \times 4.7K} = 0.0225$$

then, $I_C = 2.3\, mA$

47. Ans. (b)

Solution:

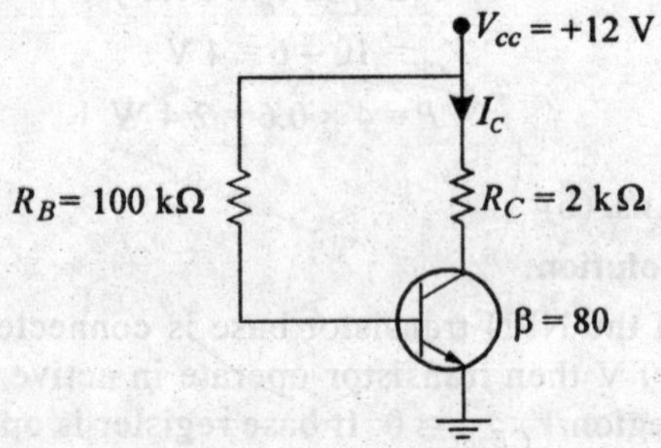

V_{OC} across base –emitter junction $= 12\text{V}$ since $V_{OC} > V_{BEactive}$

Therefore transistor is on, it may be in active region or saturation region.

Let us assume that this is in active region

Apply KVL in V_{CC} – base circuit

$$V_{CC} - R_B \cdot I_B - V_{BE} = 0$$

$$\Rightarrow \quad I_B = \frac{11.3}{100\ k\Omega} = 0.113\ mA$$

Apply KVL in Collector – Emitter circuit

$$V_{CC} - I_C R_C - V_{CE} = 0$$

$$\therefore \quad V_{CE} = V_{CC} - I_C R_C$$

$$= 12 - \beta I_B \times R_C$$

$$= 12 - 80.(0.113)2$$

$$= -6.08 \text{ volts}$$

Since V_{CE} must be greater than 0.2 V for active Region, so our assumption is wrong and transistor is in saturation region.

$$\therefore V_{CE\,sat} = 0.2\,V$$

48. Ans. (a)

Solution:

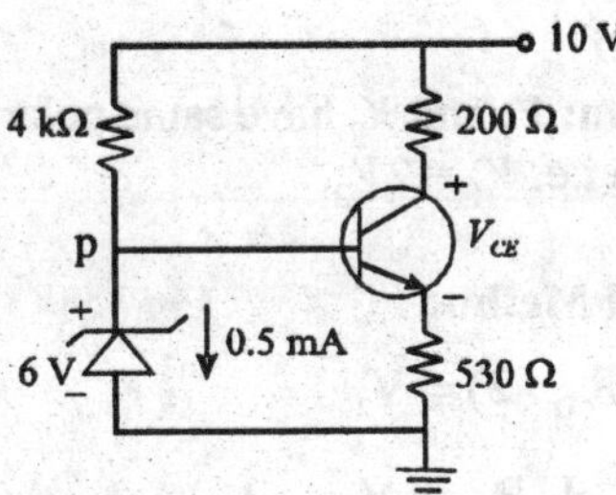

The voltage at P = 6 *V*

$$I_b = \frac{10-6}{4\ k\Omega} = 1\ mA$$

The current through zener diode is 0.5 *mA* is $I_b = 0.5$

$$6 - 0.7 - I_E R_E = 0$$

$$5.3 = (I_b + I_c).R_E$$

$$5.3 = (I_b + \beta I_b)$$

$$(1+\beta) = \frac{5.3}{5.30 \times 0.5} \quad 1+\beta = 20$$

$$\beta = 19$$

So $\quad I_c = \beta I_b = 19 \times 0.5 = 9.5\ mA$

$$V_{CE} = 10 - I_C.200 - 530 I_E$$

$$= 10 - \beta.I_B 200 - (\beta+1)530 I_B$$

$$= 10 - 19 \times 0.5 \times 200 \times 10^{-3}$$

$$-20 \times 530 \times 0.5 \times 10^{-3} = \mathbf{2.8}$$

49. Ans. (c)

Solution

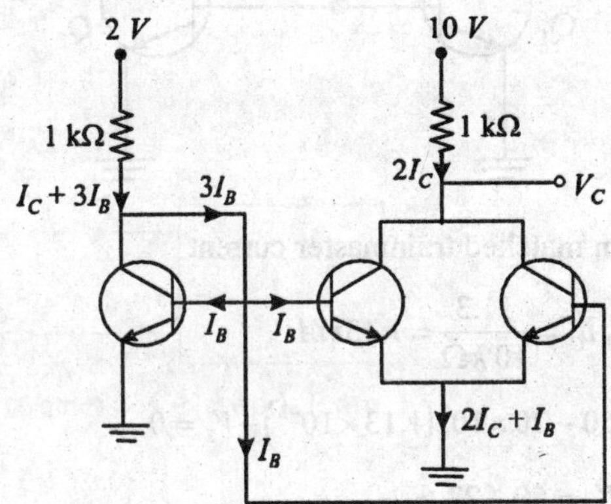

Using KVL in the left-hand single transistor,

$$2 - (I_C + 3I_B) - V_{BE} = 0$$

$$2 - 0.2 = 3I_B + \beta I_B = 103 I_B$$

$$I_B = \frac{1.3}{103}$$

$$V_C = 10 - 2I_C.1 = 10 - \frac{2 \times 1.3}{103} = 7.4$$

50. Ans. (c)

Solution:

In B.J.T with common Base

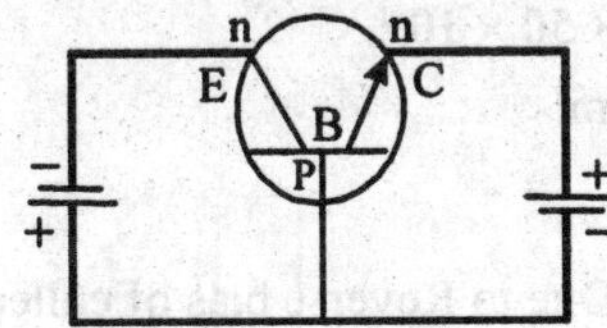

So we know in C.B. current gain

$$\alpha = \frac{I_C}{I_E} \Rightarrow \frac{I_C}{I_C + I_B} \Rightarrow < 1$$

$$A_V = \frac{V_{out}}{V_{in}} > 1$$

51. Ans. (b)

Solution:

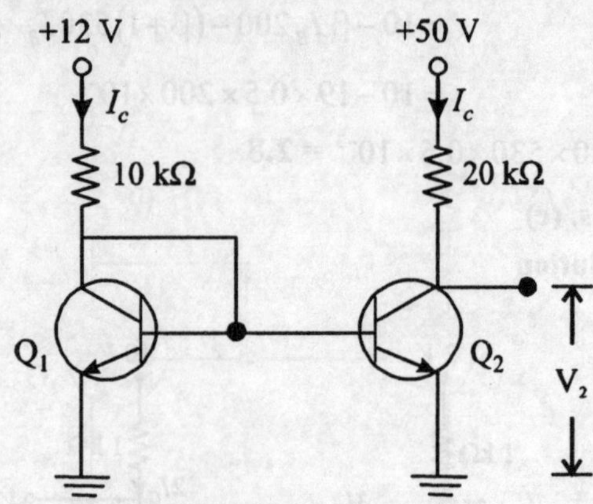

In matched trainmaster current

h $I_C = \dfrac{11.3}{10\,k\Omega} = 1.13mA$

$50 - 20 \times 10^3 (1.13 \times 10^{-3}) - V_2 = 0$

$V_2 = 50 - 22.6$

$V_2 = 27.4$

52. Ans. (b)

Solution: Diffusion current density

$J_n = e \cdot D_n \cdot [dn/dx]$

where, $dx = 0.5$ μm $= 0.5 \times 100$ μcm

$dn = 10^{14}$ / cm³

$D_n = 25$ cm²/s

$e = 1.6 \times 10^{-19}$ C

$J_n = e \cdot D_n \cdot [dn/dx]$

$= 1.6 \times 10^{-19} \times 25 \times \dfrac{10^{14}}{(0.5 \times 10^{-6}) \times 10^2}$

$= 1.6 \times 50 \times 10^{-1}$

$= 8$ A/m²

53. Ans. (b)

Solution: Due to Reverse bias of collector junction the depletion layer in base region increase and base width decreases.

54. Ans. (b)

Solution: DC current gain β = 50

Emitter injection efficiency = 0.995

base transportation factor = ?

⇒ Emitter efficiency = $\dfrac{I_{PE}}{I_{ETOT}}$

Base transportation factor, $B_T = \dfrac{I_{PC}}{I_{PE}}$

α = Emitter efficiency × Base transportation factor

$\therefore\ B_T = \dfrac{\alpha}{0.995} = \dfrac{\beta}{1+\beta} \times \dfrac{1}{0.995}$

$= \dfrac{50}{51} \times \dfrac{1}{0.995} = 0.985$

55. Ans. (a)

Solution:

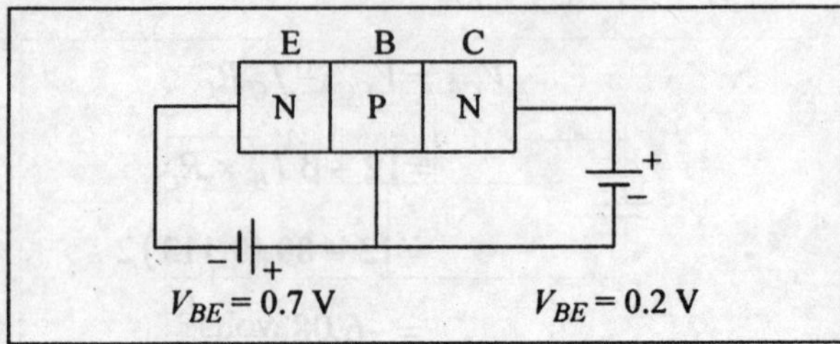

If, $V_{BE} = 0.7$ V --- Forward biased

$V_{CB} = 0.2$ V --- Reverse biased

Transistor operate in normal active mode.

7.3 Field Effect Transistor

1. Ans. (c)

Solution: V_1 and V_2 have same polarity, there is drop in voltage i.e. $V_1 = 2V_2$.

Second Method

$I_D(R_D/2) = V_2$

$I_D R_D = V_1$

Hence, $2V_2 = V_1$

2. Ans. (d)

Solution: $A_V = -g_m (r_d \| R_D)$

$g_m = \pm \left[\dfrac{2}{V_P}\right] \sqrt{I_D \cdot I_{DSS}}$

Assuming $r_d \geq 10\,R_D$ then, $A_V = -g_m R_D$

$g_m = g_{mo} \left[1 - \left(\dfrac{V_{gs}}{V_P}\right)\right]$

$= \left[\dfrac{2\,I_{DSS}}{|V_P|}\right] \sqrt{\dfrac{I_D}{I_{DSS}}}$

$V_{gs} = I_D \cdot R_S$

$$= -2.5\text{ k} \times 1\text{mA} = -2.5$$

$$g_m = \left[\frac{2\,I_{DSS}}{|V_P|}\right]\left[1-\left(\frac{V_{GS}}{V_P}\right)\right]$$

$$\frac{2\times 10\text{ mA}}{5}\left[1-\left(\frac{-2.5}{5}\right)\right] = 4\text{ mA}\times\frac{2.5}{5}$$

$$= 2\text{ mA/V}$$

$$A_V = -g_m R_D$$

$$= -2\text{ mA/V} \times 3.0\text{ k}\Omega$$

$$= -6\text{ V}$$

Second Method

$$A_v = -g_m(R_D \| r_d)$$

$$g_m = \frac{2I_{DSS}}{|V_P|}\left(1-\frac{V_{GS}}{V_P}\right)$$

$$V_s = I_D R_s = 1\text{ mA}\times 2.5\text{ k} = -2.5\text{ V}$$

$$V_G = 0$$

$$V_{GS} = V_G - V_S = -2.5\text{ V}$$

$$g_m = \frac{2\times 10\text{ mA}}{5}\left[1-\left(\frac{-2.5}{-5}\right)\right] = 2\text{ ms}$$

$A_v = -2\text{ ms} \times 2\text{ k}$ (as r_d is not given assume as ∞)

$$= -6\,V$$

3. Ans. (b)

Solution: The equivalent circuit of JFET is

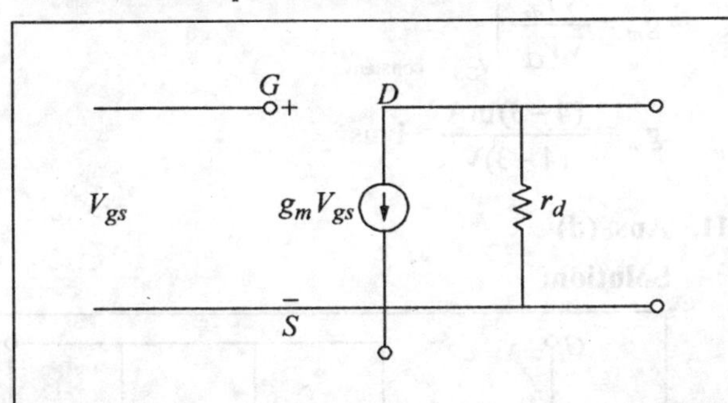

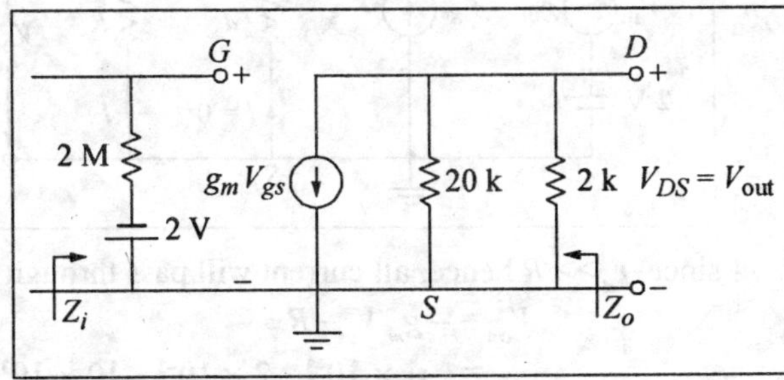

$$Z_i = 2M$$

$$Z_{out} = 20\text{ k} \| 2\text{ k} = \frac{20\times 2}{20+2} = \frac{20}{11}\text{ k}\Omega$$

4. Ans. (a)

Solution:

$$I_D = I_{DSS}\left(1-\frac{V_{GS}}{V_p}\right)^2 = 10\left(1-\frac{2}{8}\right)^2 = 5.625$$

$$I_D = 5.625\text{ mA}$$

$$V_{DS} = 20 - 2\text{ K}\times 5.625 = 8.75\text{ V}$$

$$V_{DS} = 8.75\text{ V}$$

5. Ans. (b)

Solution:

$$g_m = -\frac{2I_{DSS}}{V_p}\left(1-\frac{V_{GS}}{V_p}\right) = \frac{-2\times 10\text{ mA}}{-8}\left(1-\frac{(-2)}{-8}\right)$$

$g_m = 1.875$ msiemens

For CS amplifier

$$A_v = -g_m(r_d \| R_L) = -1.875\times\frac{20}{11}$$

$$A_v = \frac{V_o}{V_{gs}} = -\left[\frac{20}{11}\right]1.875 = -3.41$$

$$A_V = -3.41$$

6. Ans. (c)

Solution: $g_m = \dfrac{-2I_{dss}}{V_p}\left(1-\dfrac{V_{gs}}{V_p}\right)$

$$g_m = \left|\frac{2\text{I}_{\text{DSS}}}{\text{V}_p}\right|\left(1-\frac{\text{V}_{\text{GS}}}{\text{V}_p}\right)$$

⇒ g_m is maximum, when $V_{GS} = o$

The I_D-V_{GS} characteristic of *n*-channel Deplation MOSFET is given below which indicate that current is maximum for positive value of V_{GS}

$$g_m = \frac{\Delta I_D}{\Delta V_{GS}}$$

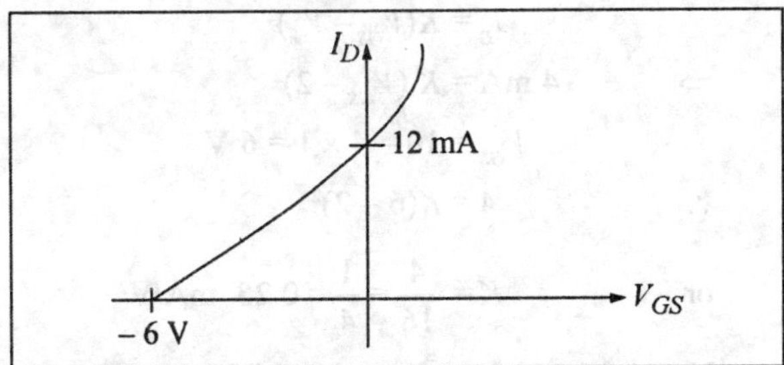

$Z_0 = \dfrac{V_o}{I_D} = \infty$ **correspondes to $I_D = o$**

7. Ans. (c)

Solution:

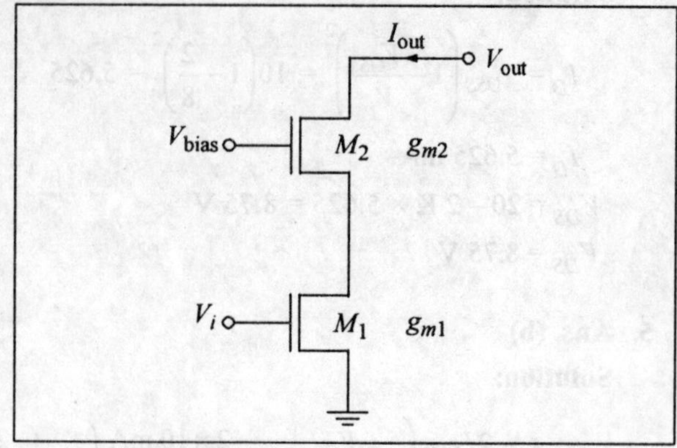

$$\frac{1}{g_m} = \frac{1}{g_{m1}} + \frac{1}{g_{m2}}$$

$\therefore$ M_2 is always saturated due to bias but g_{m1} changes according to V_i

$$g_m = \frac{g_{m1}\, g_{m2}}{g_{m1} + g_{m2}} = \frac{g_{m1}\, g_{m2}}{g_{m2}\left(1 + \frac{g_{m1}}{g_{m2}}\right)}$$

But $\quad g_{m2} >> g_{m1}$

$\therefore \quad g_m = g_{m1}$

8. Ans. (c)

Solution:

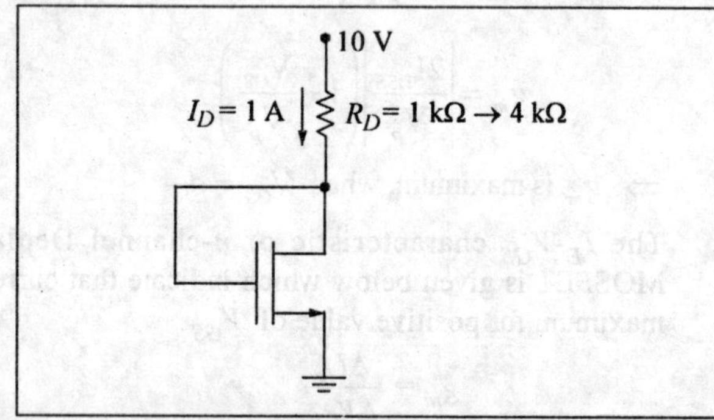

$$I_D = K(V_{GS} - V_{th})^2$$

$\Rightarrow \quad 4 \text{ mA} = K(V_{GS} - 2)^2$

$$V_{GS} = 10 - 4 \times 1 = 6 \text{ V}$$

$\therefore \quad 4 = K(6 - 2)^2$

or $\quad K = \frac{4}{16} = \frac{1}{4} = 0.25 \text{ mA/V}^2$

Now when R_D is increased to 4 kΩ.

$$V_{GS} = 10 - 4\, I_D$$

Now $\quad I_D = \frac{1}{4}(10 - 4I_D - 2)^2$

or $\quad 4\, I_D = (8 - 4\, I_D)^2 = 64 + 16\, I_D^2 - 64\, I_D$

or $\quad 16\, I_D^2 - 68\, I_D + 64 = 0$

or $\quad 4\, I_D^2 - 17\, I_D + 16 = 0$

$$I_D^2 - 4.25 I_D + 4 = 0$$

$$I_D = 2.8 \text{ mA or.4 mA}$$

if $\; I_D = 2.84\, m\text{A}, \;$ then

$$V_{GS} = V_{DS} = 10 - 4 \times 2.84$$

or $\; V_{GS} = -1.36 \text{ V} < V_{Th}$

Hence I_D cannot be 2.84 mA .

If $\; I_D = 1.4\, m\text{A}, \;$ then

$$V_{GS} = V_{DS} = 10 - 1.4 \times 4 = 4.6 \text{ V} > V_{Th}$$

Hence, $I_D = 1.4\, m\text{A}$ is the correct answer.

9. Ans. (d)

Solution: MOSFET is n-channel. Since $Vgs > V_{th}$ and Gate voltage is so connected that MOSFET will be in Enhance mode and so conductivity of the channel will be increase very much, and effectively *D* terminals acts as short circuited.

10. Ans. (b)

Solution:

From figure

$$g_m = \left.\frac{\nabla I_{ds}}{\nabla V_{Gs}}\right|_{V_{DS} = \text{constant}}$$

$$g_m = \frac{(4-3)\text{mA}}{(4-3)\text{V}} = 1 \text{ ms}$$

11. Ans. (d)

Solution:

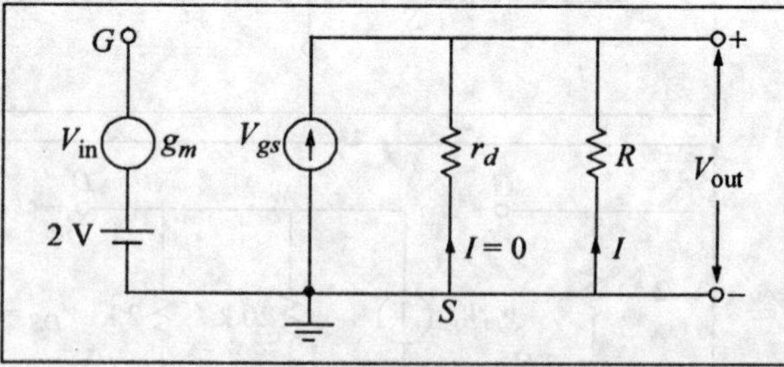

since, $r_d >> R$ hence, all current will pass through R

$\therefore \quad V_{out} = -g_m V_{gs} \cdot R$

$$= -1 \times 10^{-3} \times 2 \times 10^{-3} \times 10 \times 10^3$$

$$= -20 \text{ mV}$$

$$\text{Voltage gain} = \frac{V_{out}}{V_{in}} = -\frac{20 \text{ mV}}{2 \text{ mV}} = -10$$

Second Method

Voltage gain, $A_v = -g_m R_L$

$$= -1\times10^{-3}\times10\times10^{3}$$
$$= -10$$

12. Ans. (c)

Solution: For high Gate input (2.5 V) NMOS and PMOS in saturation region

$$I_d = \frac{1}{2}k_n(|V_{gs}| - V_{THn})^2 = \frac{1}{2}40\times10^{-6}(2.5-1)^2$$

$$\therefore \quad = \frac{1}{2}40\times10^{-6}\times2.25 = 45\ \mu A$$

Second Method

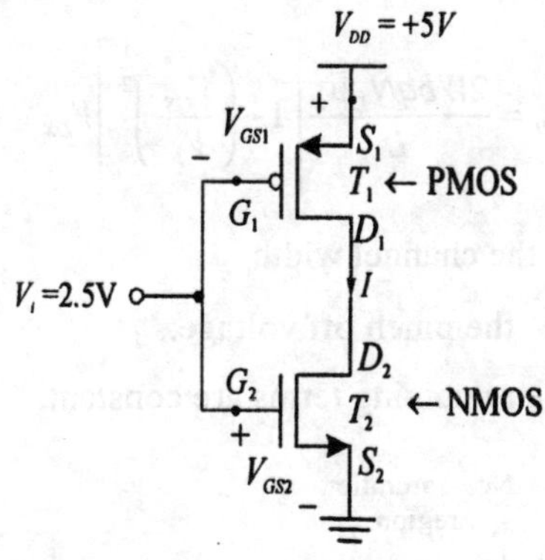

Given

$$K_n = K_p = \mu_n C_{ox}\frac{W_n}{L_n} = \mu_p C_{ox}\frac{W_p}{L_p} = 40\ \mu A/V^2$$

$$V_{t_n} = |V_{t_p}| = 1V$$

When $V_i = \frac{V_{DD}}{2}$ both T_1 PMOS and T_2 NMOS are in saturation.

Also, I is the drain current for both transistors

$$I = \frac{1}{2}K_p\left(V_{SG1} - |V_{tp}|\right)^2$$

$$= \frac{1}{2}K_n(V_{GS2} - V_{t_n})^2$$

$$V_{SG_1} = 5 - 2.5 = 2.5V = V_{GS_2}$$

$$\Rightarrow \quad I = \frac{1}{2}\times40\times(2.5\times1)^2$$

$$= \frac{1}{2}\times40\times2.25$$

$$\Rightarrow \quad I = 45\ \mu A$$

13. Ans. (b)

Solution: Given

$$I_D = K(V_{GS} - V_T)^2$$

Then $\quad \dfrac{\delta I_O}{\delta V_{GS}} = 2K(V_{GS} - V_T) = g_m$

14. Ans. (b)

Solution: Given circuit is current mirror

$$I_G = 0$$

If $\quad \left(\dfrac{W}{L}\right)_1 = \left(\dfrac{W}{L}\right)_2$

Then $\quad I_x = I_{bias}$

15. Ans. (a)

Solution: We know that drain current of MOSFET is

$$I_D = K\left(V_{GS} - V_T\right)^2$$

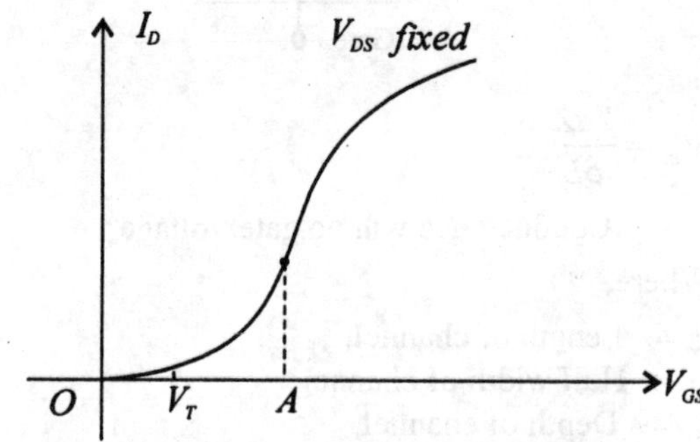

Fig: (a) Input characteristic of NMOS

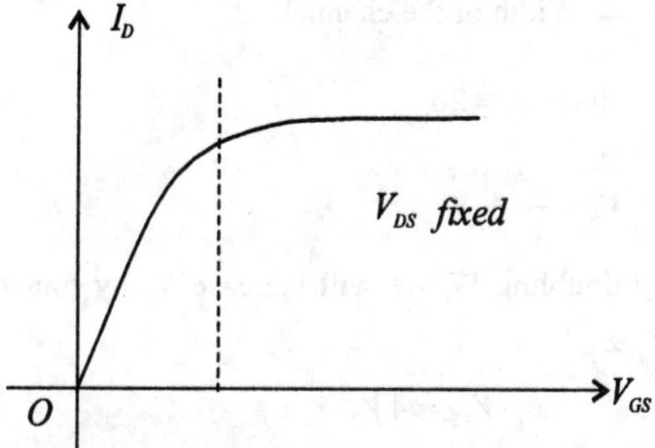

Fig (b): Output characteristic of NMS

$$g_m = \left.\frac{\partial I_D}{\partial V_{GS}}\right|_{V_{DS}=\text{constant}}$$

Hence g_m can be obtained by taking the derivative of the curve in figuer (a)

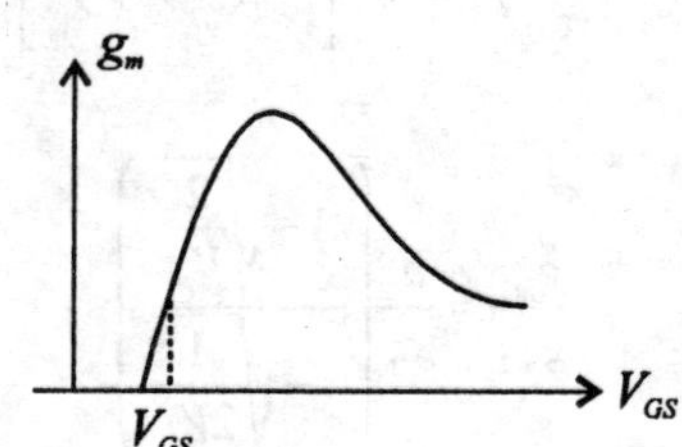

16. Ans. (b)

Solution: Transconductance,

$$g_m = g_{mo}\left[1-\left(\frac{V_{GS}}{V_P}\right)^{\frac{1}{2}}\right]$$

$$g_{m1} = g_{mo1}\left[1-\left(\frac{2}{V_P}\right)^{\frac{1}{2}}\right]$$

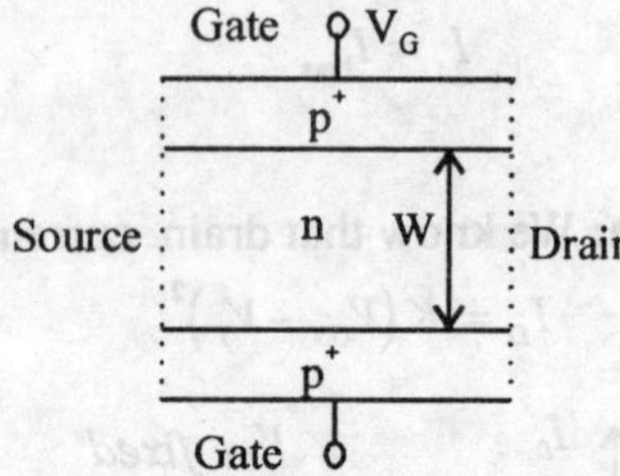

$$g_{mo} = \frac{2aZ}{\rho L}$$

= Conductance wih no gate voltage

Where,

$L \rightarrow$ Length of channel,

$a \rightarrow$ Half width of channel

$Z \rightarrow$ Depth of channel,

$\rho \rightarrow$ Resitivity of channel

$W \rightarrow$ Width of the channel.

As given $a_2 = 2a_1$

$$\because\ V_p = \frac{qN_d W^2}{2\varepsilon}$$

By doubling W, we will increase V_P by four times as $V_p \propto a^2$

$$\Rightarrow \quad V_{P2} = 4\,V_P$$

$$g_{m2} = g_{mo2}\left[1-\left(-\frac{V_{GS}}{V_{P2}}\right)^{\frac{1}{2}}\right]$$

$$= g_{mo2}\left[1-\left(\frac{V_{GS}}{2V_p}\right)^{\frac{1}{2}}\right]$$

$$\therefore \quad \frac{g_{m1}}{g_{m2}} = \frac{a_2}{a_1}\left(\frac{1-\sqrt{\frac{2}{V_P}}}{1-\sqrt{\frac{1}{2V_P}}}\right)$$

$$= \frac{1}{2}\left(\frac{1-\sqrt{\frac{2}{V_P}}}{1-\sqrt{\frac{1}{2V_P}}}\right)$$

Second Method (b)

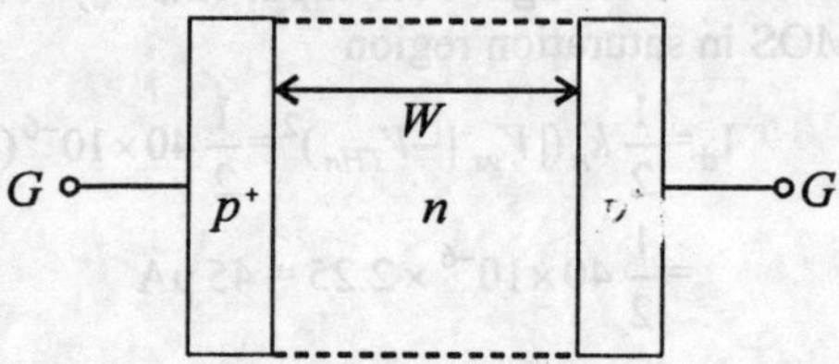

In linear region i.e. non-saturation region

$$I_D = \frac{2WbqN_D\mu_n}{L}\left[1-\left(\frac{V_{GS}}{V_P}\right)^{\frac{1}{2}}\right]V_{DS} \qquad \dots(1)$$

W $\rightarrow$ the channel width

V_P $\rightarrow$ the pinch off voltage.

All the remaining terms are constant.

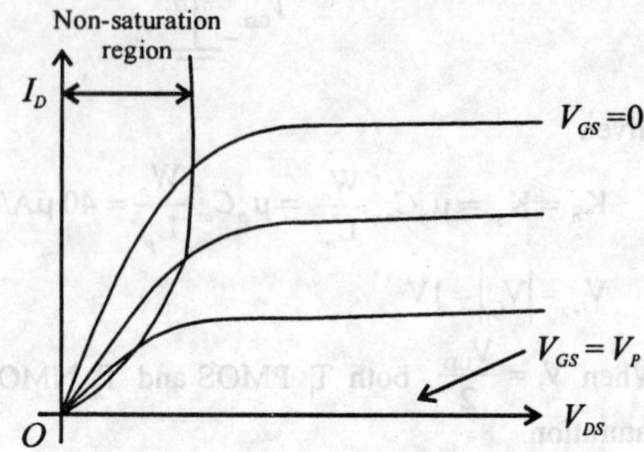

For small currents, $I_D \approx KV_{DS}$

Where K is a constant

I_D becomes a function of V_{DS}

Now, $g_m = \dfrac{\partial I_D}{\partial V_{GS}} \approx \dfrac{I_D}{V_{GS}}$ for small values of I_D

$$V_P = \frac{qN_D}{2\,\varepsilon}W^2$$

Hence, when W is doubled V_P becomes four times.

For intital JFET

$$V_{P_1} = V_P$$

$$V_{GS} = -2\,V$$

$$W_1 = W$$

or $$I_D = W \cdot K_1 \left[1 - \left(\frac{V_{GS}}{V_P}\right)\right] V_{DS}$$

Where $K_1 = \frac{2WbqN_D\mu_n}{L}$

For Modified JFET

$$V_{P_2} = 4\,V_P$$

$$W_2 = 2\,W$$

$$V_{GS} = -2\,V$$

$$I_D = 2W \cdot K_1 \left[1 - \left(\frac{V_{GS}}{4\,V_P}\right)^{\frac{1}{2}}\right] V_{DS}$$

$$\frac{g_{m_1}}{g_{m_2}} \approx \frac{I_{D_1}/V_{GS}}{I_{D_2}/V_{GS}}$$

$V_{GS} = -2\,V$ for both the cases

or, $$\frac{g_{m_1}}{g_{m_2}} \approx \frac{I_{D_1}}{I_{D_2}} = \frac{1}{2}\left[\frac{1-\left(\frac{2}{V_P}\right)^{\frac{1}{2}}}{1-\left(\frac{2}{4\,V_P}\right)^{\frac{1}{2}}}\right]$$

V_{GS} and V_P have same signs, we can consider only the magnitude of V_{GS}

or, $$\frac{g_{m_1}}{g_{m_2}} = \frac{1}{2}\left[\frac{1-\sqrt{\frac{2}{V_P}}}{1-\sqrt{\frac{2}{2V_P}}}\right]$$

17. Ans. (c)

18. Ans. (a)

19. Ans. (c)

Solution:

As, $$g_m = g_{mo}\left(1 - \frac{V_{GS}}{V_P}\right)$$

Given, $g_m = 1$ mA/V.

Then, $$g_{mo} = \frac{g_m}{\left(1 - \frac{V_{GS}}{V_P}\right)} = \frac{1}{1-\frac{3}{5}} = 2.5 \text{ mA/V}$$

Second Method:

$$I_D = \frac{I_{DSS}}{-V_P}\left(1 - \frac{V_{GS}}{V_P}\right)^2$$

$$\frac{dI_D}{d\,V_{GS}} = \frac{2\,I_{DSS}}{-V_P}$$

$$= \left(\frac{2\,I_{DSS}}{|V_P|}\right)$$

$$g_m = g_{mo}\left[1 - \frac{V_{gs}}{V_P}\right]$$

$$1 \text{ mA/V} = g_{mo}\left[1 - \frac{-3}{-5}\right]$$

$$g_{mo} = g_m \text{ max}$$

$$g_{mo} = 2.5 = g_m \text{ max}$$

20. Ans. (c)

21. Ans. (d)

22. Ans. (b)

Solution: For *n*-well CMOS fabrication, the following steps are followed:

1. *n*-well implant
2. Source drain diffusion
3. Metallization
4. Passivation

23. Ans. (c)

Solution: For *N*-Channel MOSFET

V_T = Threshold voltage

N_D = Doner concentration

C = Capacitance

t = thickness = d

$$V_T \propto \sqrt{N_D}$$

$$V_T \propto \frac{1}{C} = \frac{1}{(\varepsilon_o A)/d} \propto t$$

$$V_T \uparrow \propto V_D \uparrow \text{ and}$$

$$V_T \propto t \downarrow$$

V_T of a MOS capacitor increase with increase substrate doping concentration

24. Ans. (c)

Solution:

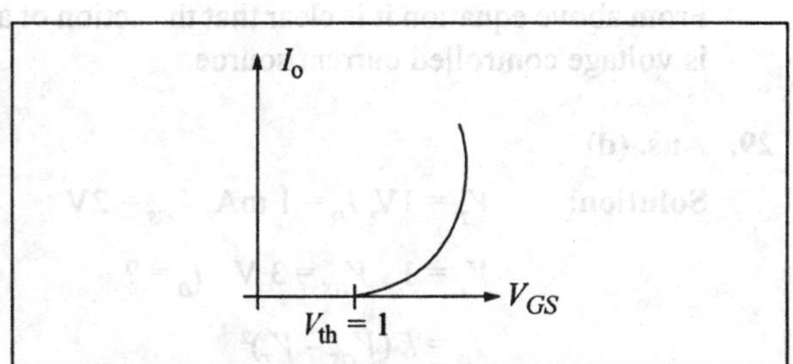

$V_{DS} = 5 - 1 = 4$

$V_{GS} = 3 - 1 = 2$

$V_{GS} - V_{th} = 2 - 1 = 1$

In present case: $V_{DS} >, V_{GS}$ i.e. 4 > 1

So device is in saturation

25. Ans. (a)

Solution: In accumulation mode for N-MOS having P-substrate V_G is Negative. When negative V_G is applied to the gate, the hole in the type substrate are attracted to the semiconductor oxide interface. This condition is called carrier accumulation on the surface.

26. Ans. (a)

Solution: V_T is independent of circuit parameter, it depends on fabrication process and material used.

27. Ans. (d)

Solution:

Given that,

For $V_T = 400$ mV

$I_D = 1$ mA

$V_{gs} = 900$ mV

For $V_{gs} = 1400$ mV

$I_D = ?$

Using, $I_D = K(V_{gs} - V_T)^2$

$$\frac{1\times10^{-3}}{I_D} = \left(\frac{900-400}{1400-400}\right)^2$$

$$\frac{1\times10^{-3}}{I_D} = \frac{1}{4}$$

$I_D = 4$ mA

28. Ans. (d)

Solution:

For a JEET in active region we have

$$I_{DS} = I_{DSS}\left(1-\frac{V_{GS}}{V_P}\right)^2$$

From above equation it is clear that the action of a JFET is voltage controlled current source.

29. Ans. (d)

Solution: $V_T = 1\text{V}, I_D = 1 \text{ mA}, V_{GS} = 2\text{V}$

$V_T = 1, V_{GS} = 3\text{ V}, I_D = ?$

$I_D = k\,(V_{GS} - V_T)^2$

$$\frac{I_{D2}}{I_{D1}} = \frac{(V_{GS2}-V_T)^2}{(V_{GS_1}-V_T)^2} = \frac{(3-1)^2}{(2-1)^2}$$

$I_{D2} = [4 / 1] \times 1 = 4$mA

30. Ans. (a)

Solution:

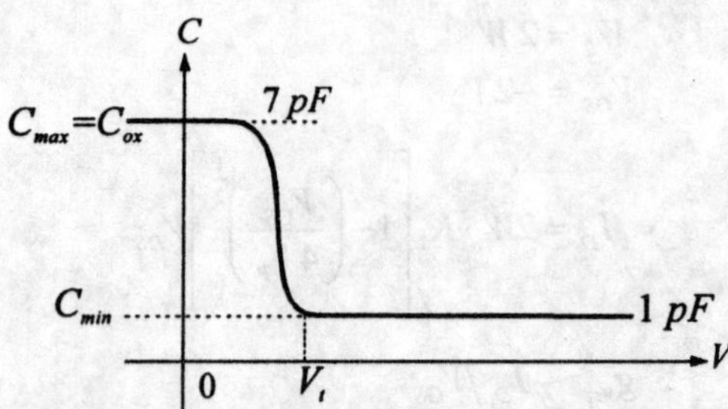

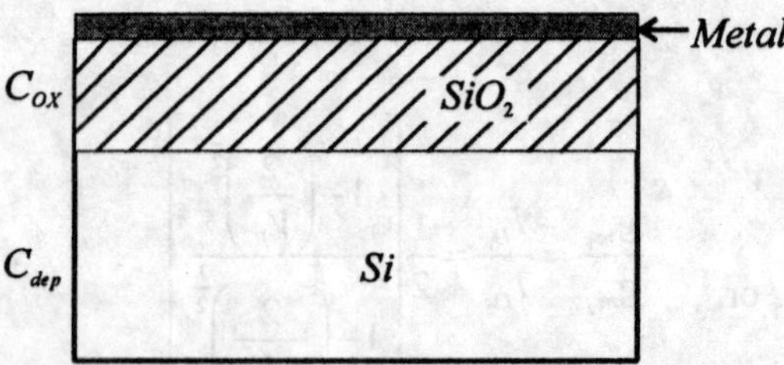

* C_{MOS} is maximum when MOSFET in an Accumulation mode.

In Accumulation mode, deplection region is not formed and *Si* layer beneath oxide layer contains plenty of majority charge carriers.

$$C_{MOS} = C_{OX} = \frac{\varepsilon_o \varepsilon_{r\,SiO_2} A}{t_{OX}}$$

Given $C_{Max} = 7\ pF = C_{OX}$

$A = 10^{-4}\ cm^2$

$$t_{OX} = \frac{3.5\times10^{-13}\times10^{-4}}{7\times10^{-12}}$$

$t_{OX} = 0.5\times10^{-4}\ cm$

$= 0.05\ \mu m$

$= 50\ nm$

31. Ans. (b)

Solution:

* C_{MOS} is minimum when MOS operates in depletion mode. Under such condition the Si layer just below the oxide layer is depleted of charge carrier.

Then $\frac{1}{C_{MOS}} = \frac{1}{C_{OX}} + \frac{1}{C_{dep}}$

Given $C_{min} = 1\ pF$

and $C_{OX} = 7\ pF$

In present case

$$C_{MOS} = C_{min}$$

or, $$C_{MOS} = \frac{C_{OX} \cdot C_{dep}}{C_{OX} + C_{dep}}$$

or, $$1 = \frac{7 \cdot C_{dep}}{7 + C_{dep}}$$

or, $7 + C_{dep} = 7 \cdot C_{dep}$

or, $$C_{dep} = \frac{7}{6}\ pF$$

Also $$C_{dep} = \frac{\varepsilon_o \varepsilon_{r,Si} \cdot A}{t_{dep}}$$

Where t_{dep} is the thickness of deplection region

Given $$\varepsilon_o \varepsilon_{r,Si} = \frac{10^{-12} \times 10^{-4}}{t_{dep}}$$

or, $$\frac{7}{6} \times 10^{-12} = \frac{10^{-12} \times 10^{-4}}{t_{dep}}$$

or, $$t_{dep} = \frac{10^{12} \times 10^{-4}}{10^{12}} \times \frac{6}{7}$$

or, $t_{dep} = 0.857 \times 10^{-4}\ cm$

or, $t_{dep} = 0.857\ \mu m$

32. **Ans. (c)**

Solution:

Since V_t is positive

∴ The MOSFET is N-channel and

the substrate of the MOSFET is p-type

Hence S_1 is false.

If positive charges are introduced in the oxide, the accumulation mode will be obtained for lesser values of V_G i.e., the threshold occur for lower vlaues of V_G.

Hence, the high frequency capacitance voltge plot is shfit toward the left

$\Rightarrow$ S_2 is true

33. **Ans. (b)**

Solution:

The mobility of Si at room temperature is1350 cm^2/V-s

7.4 Frequency Response

1. **Ans. (a)**

Solution: Coupling capacitance comes in series with the circuit and hence affects low frequency response as at low frequency its impedance becomes high. Its also blocks dc.

2 **Ans. (d)**

Solution:

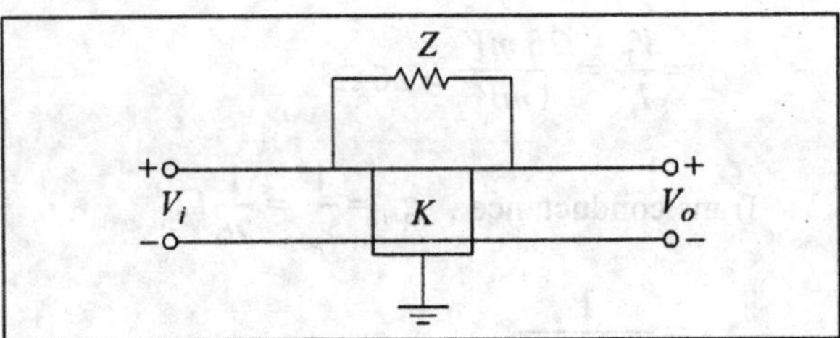

Using Miller theorem

$$Z_1 = \frac{Z}{1-K} \text{ and } Z_2 = \frac{ZK}{K-1}$$

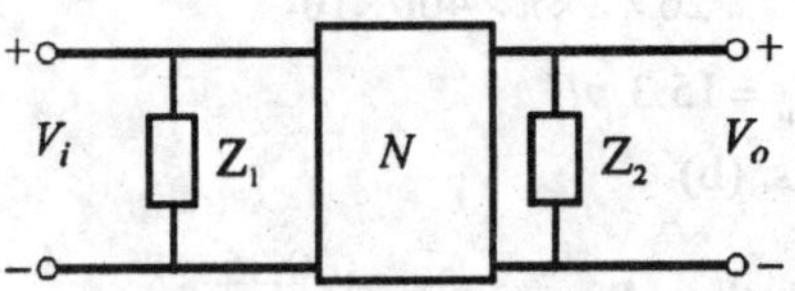

3. **Ans. (a)**

Solution: $f_T = \beta_o f_\beta = 200 \times 1\text{ MHz} = 200\text{ MHz}$.

$$f_\alpha = \frac{f_\beta}{1-\alpha} = \frac{f_\beta}{1 - \frac{\beta}{1+\beta}} = [f_\beta/1](\beta + 1)$$

$$= 1 \times 10^6 \times 201 = 201\text{ MHz}.$$

4. **Ans. (a)**

Solution: $C_\mu = 0.3$ **pF**

$$g_m = \left|\frac{I_{CQ}}{V_T}\right| = \frac{1\text{mA}}{26\text{mA}} = \left[\frac{1}{26}\right]\text{A/V}$$

$$f_T = \frac{g_m}{2\pi(C_\pi + C_\mu)}$$

or $$400 \times 10^6 = \frac{\left(\frac{1}{26}\right)}{2\pi(C_\pi + C_\mu)}$$

$$C_\mu + C_\pi = \frac{g_m}{2\pi f_T} = \frac{1/26}{2\pi \times 400 \times 10^6}$$

$$= \frac{1\times10^{-8}}{26\times 8\pi}\times10^{-8} = \frac{10^{-8}}{653.45}$$

$$= 1.53 \times 10^{-3} \times 10^{-8}$$

$$C_\mu + C_\pi = 15.3 \text{ pF}$$

$$C_\pi = 15.3 - 0.3 = 15 \text{ pF}$$

Second Method

Given, $f_T = 400\ MHz$, $C = 0.3\ pF$

$$I_C = 1\ mA\ , V_T = 26\ mV$$

$$r_E = \frac{V_T}{I_C} = \frac{26\ mV}{1\ mV} = 26\ \Omega$$

Trans conductance , $g_m = \frac{1}{r_E} = \frac{1}{26}\Omega^{-1}$

$$= \frac{1}{2\pi C_u f_T}$$

$$C_\mu = \frac{g_m}{2\pi\ f_T}$$

$$= \frac{1}{26\times2\times\pi\times400\times10^6}$$

$$C_\mu = 15.3\ pF$$

5. Ans. (b)

Solution: Rise time $t_r = \frac{0.35}{f_h}$

$$f_h = \frac{0.35}{35\ ns} = 1 \times 10^7$$

$$= 10 \text{ MHz}$$

Another way

$$\text{BW} = \frac{0.35}{\text{Rise time}}$$

$$= \frac{0.35}{35\times10^{-9}} = 10 \text{ MHz}$$

6. Ans. (a)

Solution:

The current gain of a BJT drops at high frequency because of transister parasitic capacitances i.e. internal (junction) capacitances.

7. Ans. (b)

Solution: In the high frequency, Hybrid π model

$$\beta_o = g_m\, r_\pi,\quad r_\pi = \frac{\beta_o}{g_m},\ \text{ or }\ \frac{1}{r_\pi} = \frac{g_m}{\beta_o}$$

$$f_B = \frac{1}{r_\pi \cdot 2\pi\,(C_\pi + C_\mu)}$$

$$= \frac{g_m}{\beta_o \times 2\pi\,(C_\pi + C_\mu)} = 1.64\times10^8 \text{ Hz}$$

$$f_T = \frac{B_0}{2\pi\, r_\pi (C_\pi + C_\mu)}$$

$$= \frac{g_m}{2\pi(C_\pi + C_\mu)} = 1.47\times10^{10} \text{ Hz}$$

$$f_B\, h_{fe} = f_T,$$

or $f_B\, \beta_o = f_T$

8. Ans. (a)

Solution: RC couple Amplifier

$n = 3$ stage are couple

$$f'_L = \frac{f_L}{\sqrt{2^{1/n} - 1}}$$

$$= \frac{20\ Hz}{\sqrt{2^{\frac{1}{3}} - 1}} = 40\ Hz$$

$$f'_H = f'_H \sqrt{2^{1/n} - 1}$$

$$= 1\ kHz \times \sqrt{2^{\frac{1}{3}} - 1}$$

$$= 0.5\ kHz$$

9. Ans. (b)

Solution: hfe = β

$$\text{hie} = \frac{hfe}{gm}$$

$$= \frac{100}{0.3861}$$

$$= 259$$

$$Zs = R_S + (\text{Rb} \parallel \text{hfe} \cdot \text{hie})$$

$$= 1000 + (259 \parallel 93000) = 1258\ \Omega.$$

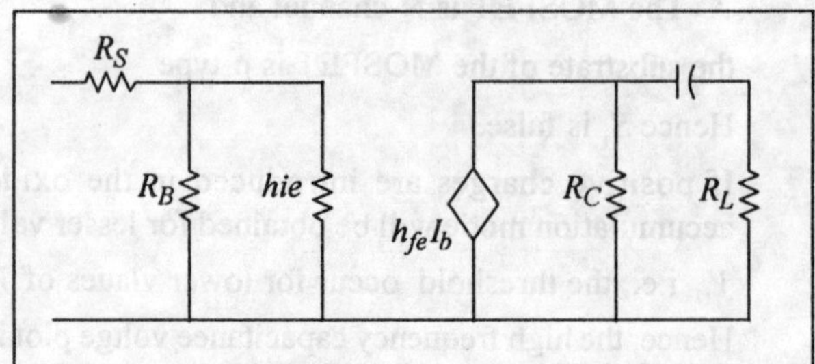

$$= 1.258\ k\Omega$$

10. Ans. (b)

Solution: Lower cut off frequency

$$f_L = \frac{1}{2\pi(R_L + R_C)C_2}$$

$$= \frac{1}{2\pi \times 1250 \times 4.7 \times 10^{-6}}$$

$$= \frac{1}{2\pi \times 1250 \times 4.7 \times 10^{-6}}$$

$$= 27.1 \text{ Hz}$$

11. Ans. (d)

12. Ans. (b)

Solution:

* For a direct coupled amplifier, there is no lower cut off frequency, since there is no coupling capacitor.
* The higher cut off frequency in direct coupled amplifier is decided by the parasitic capacitance of the transistor.

13. Ans. (a)

Solution: $V_o = V_i = V_m \sin \omega t$

Then, output voltage = V_m = 2V

14. Ans. (d)

Solution:

The given circuit diagram represent CC amplifier it simplified circuit is

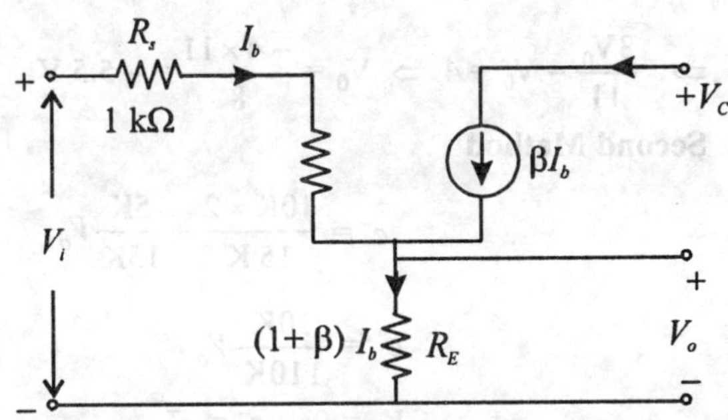

$$V_o = R_E(1+\beta)I_b$$

$$V_i = R_s I_b + R_E(1+\beta)I_b$$

$$\frac{V_o}{V_i} = \frac{R_E(1+\beta)}{R_s + R_E(1+\beta)} \frac{2(51)}{1+51\times 2} = 0.998$$

7.5 Operational-Amplifier

1. Ans. (c)

Solution: Node 1

$$\frac{V_a - V_1}{1} + \frac{V_a - V_o}{90} = 0$$

or, $$91\, V_a = V_0 + 90\, V_1 \quad \dots(1)$$

$$V_o = -90\, V_1 + 91\, V_a$$

Node 2

$$\frac{V_a - V_2}{1} + \frac{V_a - 0}{100} = 0$$

$$\frac{100\, V_a - 100\, V_2 + V_a}{100} = 0$$

or, $$V_a = \left[\frac{100}{101}\right] V_2$$

$$101\, V_a = 100\, V_2$$

From equation number (1)

or, $$V_0 = -90\, V_1 + \frac{100}{101} V_2 \times 91$$

∴ $$V_1 = V_C + \left[\frac{V_d}{2}\right]$$

and $$V_2 = V_C - \left[\frac{V_d}{2}\right]$$

Hence

$$V_0 = -90\left[V_c + \left(\frac{V_d}{2}\right)\right] + \left[\frac{100}{101}\right] \times 91 \left[V_c - \left(\frac{V_d}{2}\right)\right]$$

$$= V_c\left[\left(\frac{9100}{101}\right) - 90\right] - \left[\frac{V_d}{2}\right]\left[90 + \left(\frac{9100}{101}\right)\right]$$

$$= V_c\left[\frac{10}{101}\right] - V_d\left[\frac{9095}{101}\right]$$

$$V_o = V_c A_c + V_d \cdot A_d$$

Therefore,

$$A_c = \left[\frac{10}{101}\right],\; A_d = \left[\frac{9095}{101}\right]$$

CMRR, $\left[\frac{A_d}{A_c}\right] = 909.5 = 900.$

2. Ans. (d)

Solution:

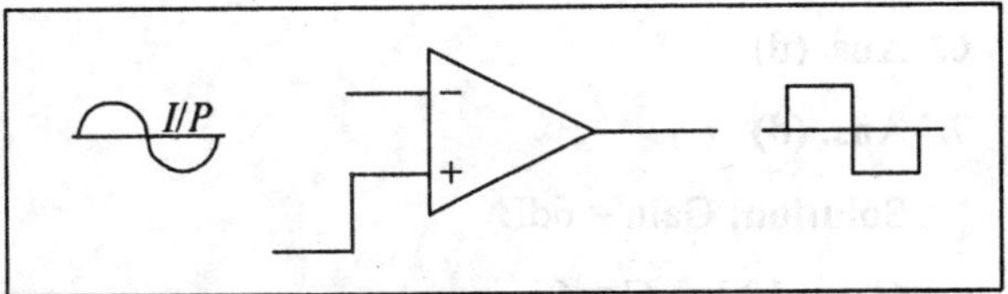

Since op amp is in open loop mode, it will give $+V_{sat}$ or $-V_{sat}$ i.e. square wave But if its slew rate is high, then triangular.

3. Ans. (c)

Solution: Output voltage = offset Voltage × Gain

$$= V_{os} \times \frac{R_2}{R_1}$$

$$= 1 \times 10^{-3} \times \frac{10^6}{10^3} = 1\text{V}$$

Since offset voltage can be ± 1mV

Therefore output voltage is ± 1V

4. Ans. (c)

Solution:

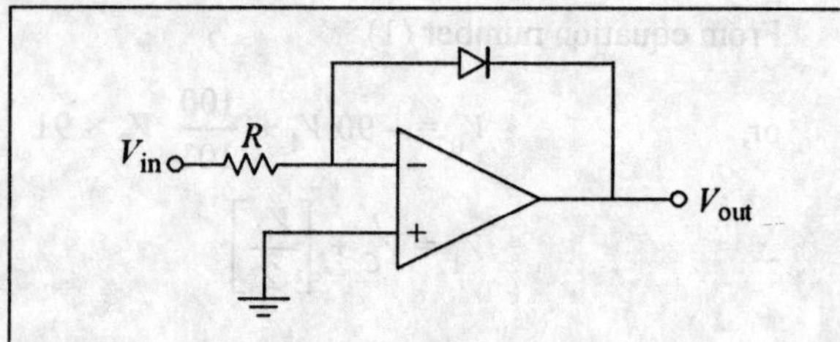

$$i = \frac{V_{in}}{R}$$

$$= I_s\left[\exp\left(\frac{V_{out}}{V_T}\right) - 1\right]$$

$$V_{out} = V_T \log\left(\frac{V_{in}}{R \cdot I_s} + 1\right)$$

Or, $V_T = \left[\frac{KT}{e}\right]$

$$V_{out} = V_T \ln\left[\frac{V_{in}}{R \cdot I_s}\right]$$

$$= 2.3\, V_T \log_{10}\left[\frac{V_{in}}{RI_S}\right]$$

Thus the circuit work as logarithmic amplifier.

5. Ans. (b)

Solution: $A_{CM} = -\frac{R_L}{2\,R_E}$

$$A_{DM} = g_m R_L$$

6. Ans. (d)

7. Ans. (d)

Solution: Gain = 6dB

$G_{dB} = 10 \log G = 6$

$G = (10)^{.6} = 3.98$

$10 \log F = 3 \Rightarrow = (10)^{[3/10]} = 2$

Overall Noise figure $F = F_1 + \frac{F_2 - 1}{G}$

$$= 2 + \frac{2-1}{3.98} = 2.25$$

$F = 10 \log 2.5 = 3.5$ dB.

8. Ans. (d)

Solution:

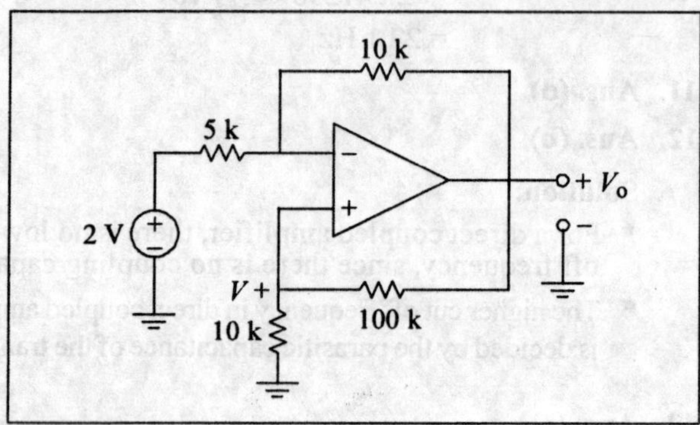

$$V+ = \left[\frac{10}{110}\right] \times V_0 = \left[\frac{V_0}{11}\right]$$

$$\Rightarrow \frac{[(V_0/11) - 2]}{5} + \frac{[(V_0/11) - V_0]}{10} = 0$$

$$\frac{2V_0}{11} - 4 + \frac{V_0}{11} - V_0 = 0$$

$$\Rightarrow \frac{3V_0}{11} - V_0 = 4 \Rightarrow V_0 = \frac{-4 \times 11}{8} = -5.5\text{ V}$$

Second Method

$$e^- = \frac{10\text{K} \times 2}{15\text{ K}} + \frac{5\text{K}}{15\text{K}} V_o$$

$$e^+ = \frac{10\text{K}}{110\text{K}} V_o$$

$$V_d = 0; \quad e^- = e^+$$

or, $\frac{4}{3} + \frac{V_o}{3} = \frac{V_o}{11}$

or, $V_o\left[\frac{1}{3} - \frac{1}{11}\right] = -\frac{4}{3}$

or, $V_o\left[\frac{11-3}{33}\right] = V_o\left[\frac{8}{33}\right] = -\frac{4}{3}$

or, $V_o = \frac{-4}{3} \times \frac{33}{8} = -5.5$ V.

9. Ans. (a)

Solution: The differential Amplifier is design to amplify differential signal. Hence we required

$$A_{DM} >> A_{CM}$$

$$C_{MRR} = \frac{A_{DM}}{A_{CM}} = \frac{-g_m R_C}{-R_C / 2R_E} \cong 2g_m R_E$$

Where R_E is emitter resistance.

10. Ans. (d)

Solution:

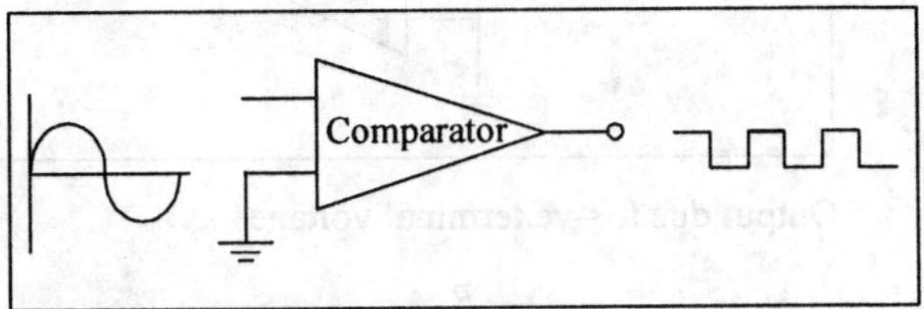

The output of comparator is either 0 or 1. When input signal is more than zero, its output is 1, when input signal is less than zero its output is 0.

11. Ans. (a)

Solution: In general dominant poles are near to the origins. Hence it may be 5 Hz.

12. Ans. (d)

Solution: The present diagram represent Non-inverting positive feed back amplifier. In positive feed back, Ops amplifier works in saturation region $\pm V_{sat}$.

Here applied voltage is positive

Hence, $V_o = +V_{sat} = +15$ V

13. Ans. (c)

Solution:

$$V_o = -\frac{R_f}{R_1} V_{in} = -\left(\frac{\omega_c}{\omega_c}\right) V_1 \sin \omega t - \left(\frac{\omega_c}{\omega_c}\right) V_2 \sin \omega t$$

$$V_o = -V_1 \sin \omega t - V_2 \sin \omega t$$

$$= -(V_1 + V_2) \sin \omega t$$

14. Ans. (c)

Solution: The DC input offset voltage V_{io} appears across the amplifier input and this voltage will be amplified / integrated and will appear at the output.

$$V_o \text{(off set)} = V_{io}\left[1 + \left(\frac{R_f}{R}\right)\right] \rightarrow \text{Feedback}$$

$$= V_{io} \times A_v - \text{Open loop.}$$

$V_o = 5 \times 10^{-3} \times 10{,}000 = \pm 50$ V i.e. $>> V_{CC}$

But since output voltage can not exceed supply voltage i.e. ± 15 V thus $V_o = \pm 15$ V

15. Ans. (a)

Solution: The ideal Ops-amplifier has the following characteristics

$$R_i \rightarrow \infty, R_o \rightarrow 0, A \rightarrow V$$

16. Ans. (c)

Solution: Slew rate is the maximum time rate of change of the closed loop amplifier output voltage under large signal condition.

$$\left|\frac{dV_o}{dt}\right|_{max} = \text{slew rate}$$

$$A_{db} = 40 = 20 \log A_V$$

$$A_V = 100$$

$$\frac{IV}{\mu\text{ second}} = [10 \times 10^6 \text{ V/s}]$$

$$\frac{IV}{\mu\text{ second}}\left[\frac{I_V}{\mu_s}\right] = \frac{dV_o}{dt} = A_V \cdot V_m \cdot \omega$$

$$= A_V \cdot V_m 2\pi f$$

$$V_m = \frac{SR}{A_V \cdot 2\pi f}$$

$$= \frac{1 \times 10^6}{2\pi \times 20 \times 10^3 \times 100} = 79.5 \text{ mV}$$

17. Ans. (b)

Solution:

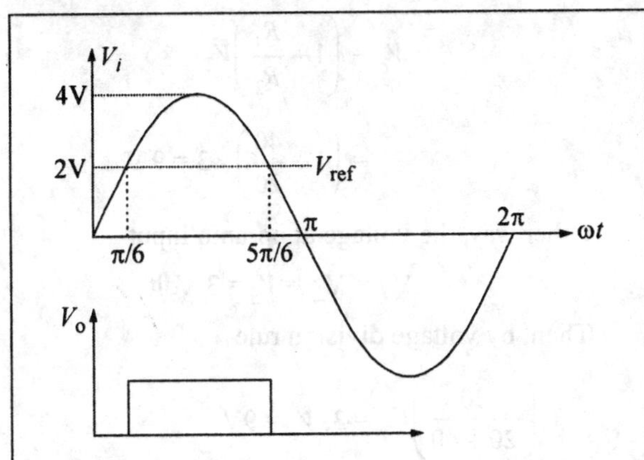

The output is positive only when the reference input is greater than 2 V.

$V_i = 8\ V_{PP}$ sinusoidal, i.e.

$= 4 \sin \omega t$

At $V_i = 2 \Rightarrow \sin \omega t = \frac{2}{4} = \frac{1}{2}$

$\Rightarrow \omega t = \sin^{-1}\frac{1}{2} = \frac{\pi}{6}$

Another cross over at $\pi - \frac{\pi}{6} = \frac{5\pi}{6}$

$$\text{Duty cycle} = \frac{T_{ON}}{T} = \frac{\frac{5\pi}{6} - \frac{\pi}{6}}{2\pi} = \frac{\frac{4\pi}{6}}{2\pi} = \frac{1}{3}$$

18. Ans. (c)

Solution: CMMR = $\frac{A_d}{A_c}$, take log on both the sides

$$\text{Log (CMMR)} = \log\left[\frac{A_d}{A_c}\right]$$

or $20 \log (\text{CMMR}) = 20 \log A_d - 20 \log A_c$
$= 48 - 2 = 46$ db

19. Ans. (c)

Solution:

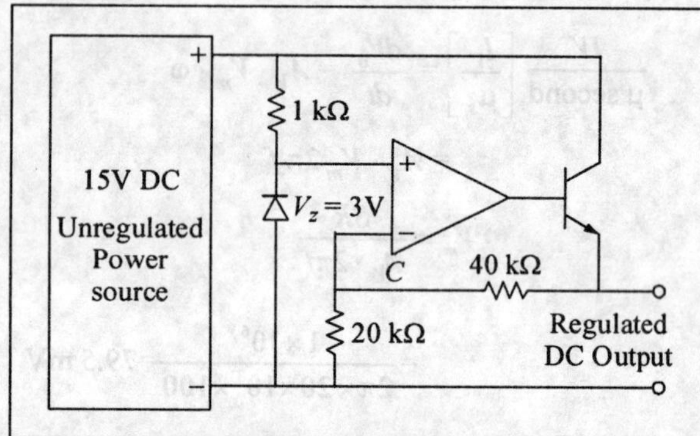

In an amp series regulator the *o/p* is

$$V_o = \left(1 + \frac{R_1}{R_2}\right) V_Z$$

$$= \left(1 + \frac{40}{20}\right) \times 3 = 9 \text{ V}$$

Other way, the voltage at op-amp input

$$V_+ = V_- = V_z = 3 \text{ Volt}$$

Then, by voltage division rule

$$\left(\frac{20}{20+40}\right) V_o = 3, \; V_o = 9\text{V}$$

Another Approach

As voltage at non-inverting terminal is 3 V due to zener diode, voltage at inverting terminal will be same as 3 V because of virtual ground concept.

∴ Current through 20 kΩ resistor is

$$\frac{3}{20 \text{ k}} = \frac{2}{20} \text{ mA}$$

$$\therefore \quad V_o = \frac{3}{20 \text{ k}} \times 60 \text{ k} = 9 \text{ V}$$

20. Ans. (b)

Solution:

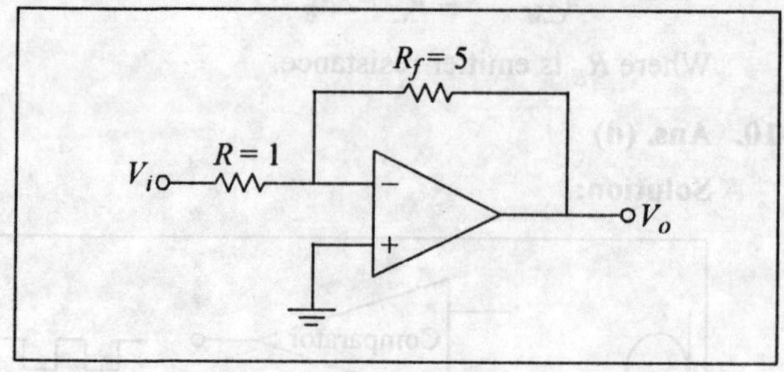

Output due to –ve terminal voltage

$$V_{o_1} = \frac{R_f}{R_1} \times \text{V} = \frac{-5}{1} \times 2 = -10 \text{ V}$$

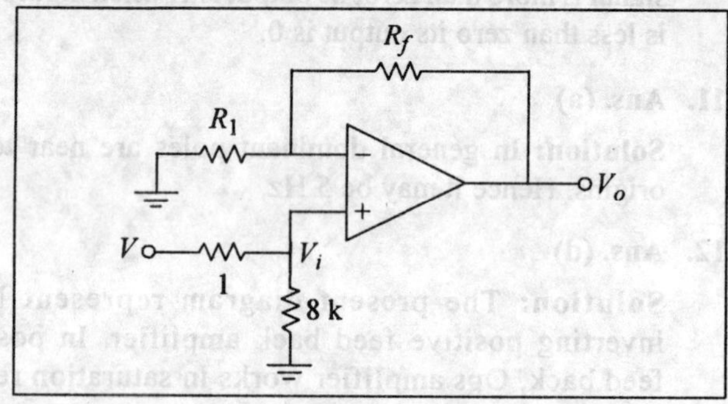

Output due to + ve terminal voltage

$$V_{o2} = \left(1 + \frac{R_f}{R}\right) V_i$$

in present case,

$$V_i = \left[\frac{8}{8+1}\right] \times 3 = \left[\frac{8}{3}\right] \text{V}$$

$$V_{o_2} = \left(1 + \frac{R_f}{R}\right) V_i = \left[1 + \frac{1}{5}\right] \times \left[\frac{8}{3}\right]$$

$$= 6 \times \left[\frac{8}{3}\right] = 16\text{V}$$

V_o (Net *o/p*) $= V_{o2} + V_{o1} = 16 - 10 = 6$ V.

21. Ans. (b)

Solution:

Voltage controlled voltage source

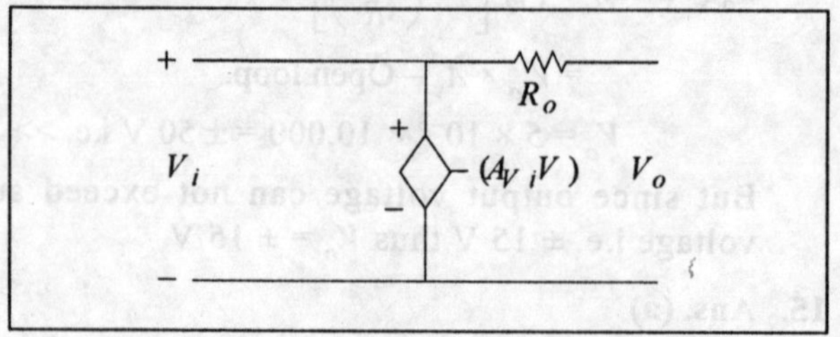

22. Ans. (a)

Solution: The combined effect of two resistor and two capacitor give a fall off rate of 40 db/decode. If is a second order LPF.

At $\omega = \infty$, $C \to$ short circuit, $\frac{V_o}{V_{in}} = 0$

At $\omega = 0$, $C \to$ open circuit, $\frac{V_o}{V_{in}} = 1$

23. Ans. (d)

Solution: If an amplifier has very high gain, i.e. $A_V \to \infty$ then

$$A_{vf} = -\frac{R_2}{R_1}, \text{ otherwise as below}$$

Close loop gain $A_{vf} = \frac{V_o}{V_i} = \frac{A_V}{1 + A_V K}$

Where, $K = \frac{R_1}{R_1 + R_2} = \frac{2}{2+8} = \frac{2}{10} = 0.2$

$$A_{Vf} = \frac{45}{1 + 45(0.2)} = \frac{45}{10} = 4.5$$

24. Ans. (d)

Solution: When $\omega = 0$; inductor acts as a short-circuit

$\Rightarrow V_o = 0$

And when $\omega = \infty$ inductor acts as a open circuit

$\Rightarrow V_o = -i_1 R_1$

So it acts as a high pass filter

25. Ans. (b)

Solution: The input resistance R_i is effective value of parallel combination of 10 k and input impedance of amplifier, i.e.

$$R_i = 10 \text{ k} \parallel \infty = 10 \text{ k}\Omega$$

26. Ans. (c)

Solution: Since no input is applied at the input side and still an output e_o is available, thus one can measure input offset current only.

27. Ans. (a)

Solution: At 1000 Hz frequency, $Z_c \to \infty$

Hence, high pass filter

Since input is at (+) terminal, therefore

$$\omega_c = \frac{1}{RC} = \frac{1}{1 \times 10^3 \times 1 \times 10^6} = 1000 \text{ rad/sec}$$

28. Ans. (d)

Solution: Equivalent circuit of BJT is

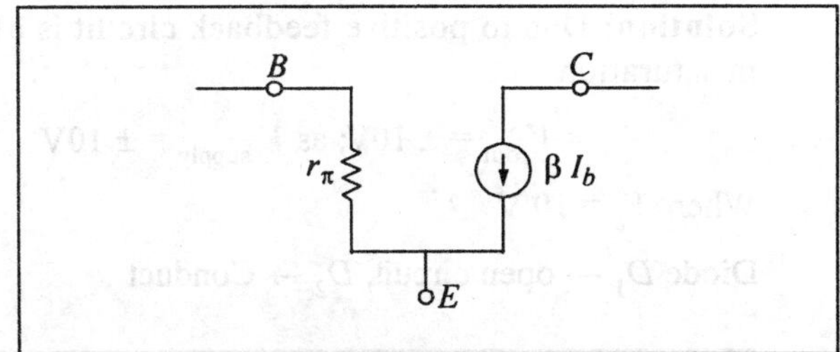

Differential mode gain (A_d):

Small signal model for differential mode

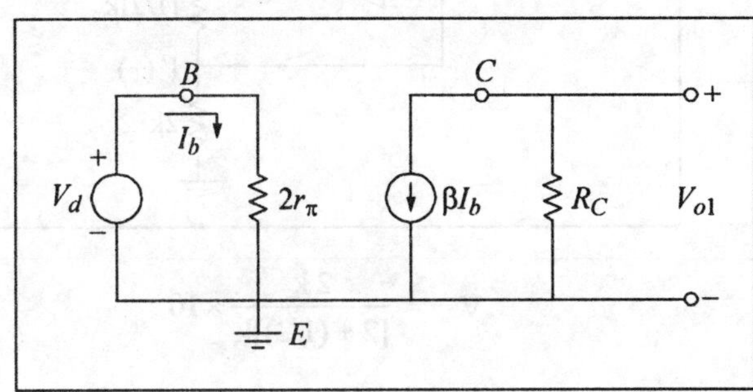

$$A_d = \frac{V_{o1}}{V_d} = \frac{-\beta R_C}{2r_\pi} = \frac{g_m R_C}{2}$$

$$A_d = [-g_m R_C / 2]$$

Common mode gain (A_c):

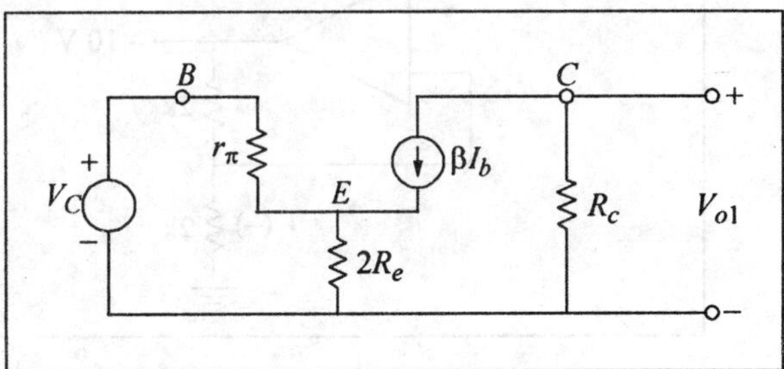

KVL in input loop

$$V_c = I_b r_\pi + 2 R_e [1 + \beta] I_b$$

Output voltage $V_o = -\beta I_b R_c$

$$A_c = \frac{V_{o1}}{V_c} = \frac{-\beta R_c I_b}{[2(1+\beta)R_e + r_\pi] I_b}$$

$$A_c = \left[-\frac{R_c}{2R_e}\right] \text{ for } \beta >> 1; r_\pi << 2R_e$$

Decreases A_c with $R_e \uparrow$

$$\text{CMRR} = \left[\frac{A_d}{A_c}\right]$$

$$\text{CMRR} = g_m R_e$$

29. Ans. (b)

Solution: Due to positive feedback circuit is always in saturation

$$V_{out} = \pm 10\text{V; as } V_{supply} = \pm 10\text{V}$$

When $V_o = 10$ V

Diode $D_1 \rightarrow$ open circuit, $D_2 \rightarrow$ Conduct

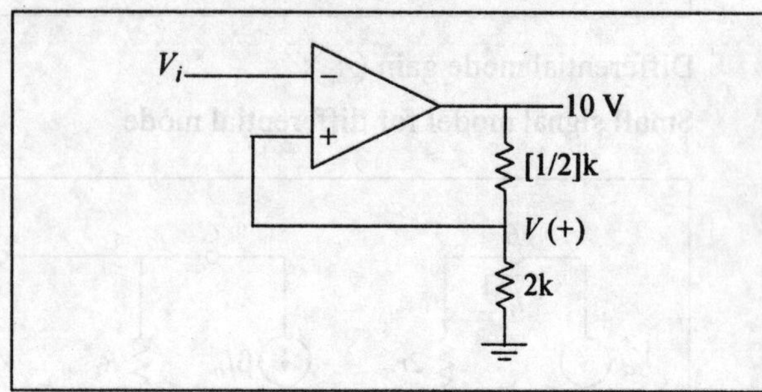

$$V_+ = \frac{2\text{ k}}{[2+(1/2)]\text{k}} \times 10$$

$$V_+ = 8\text{ V} = V_i$$

When $V_o = -10$ V,

Diode $D_1 \rightarrow$ Conduct, $D_2 \rightarrow$ open circuit

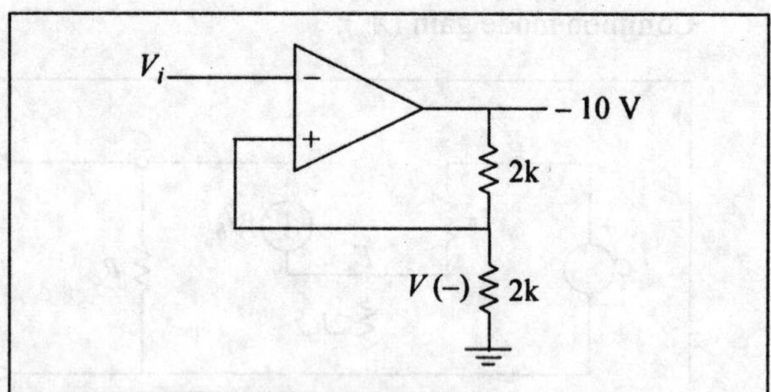

$$V_- = \frac{2\text{k}}{4\text{k}} \times (-10)$$

$$V_- = -5\text{ V} = V_i$$

Hysteresis loop is

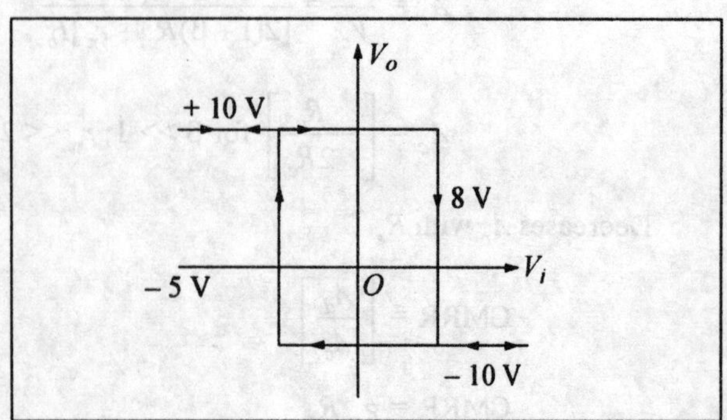

30. Ans. (d)

Solution: $V_c^+ = V_c^- = 10$ V

$$i_c = \frac{10}{1\text{ k}\Omega} = 10^{-2}\text{ A}$$

$$V_c = \frac{1}{c}\int_0^t i_c\, dt = \frac{1}{10^{-6}} \times i_c \times t$$

$$\Rightarrow \quad V_c = 10^4\, t$$

For $t = 1$ ms, $\quad V_c = 10$ V

Another way

Apply KCL at node A

$$\frac{10}{1\text{ k}\Omega} = c\frac{dV_c}{dt}$$

$$\frac{dV_c}{dt} = 10^4$$

$$\Rightarrow \quad V_c = 10^4 \times t$$

at $t = 1$ ms

$$V_c = 10\text{ V}$$

31. Ans. (d)

Solution:

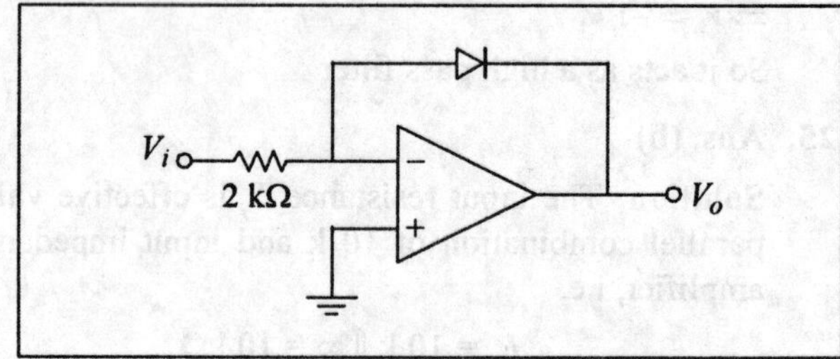

$V_o = V_{o1}$ for $V_1 = 2$ V; $\quad V_o = V_{o2}$ for $V_i = 4$V

Relation between V_{o1} and V_{o2} is as follows

For algorithmic amplifier

$$V_o = -V_T \ln\left(\frac{V_i}{R_o I_o}\right)$$

$$\therefore \quad V_{o1} = -V_T \ln\left(\frac{2}{R_o I_o}\right) = -V_T(\ln 2 - \ln R_o I_o)$$

$$V_{o2} = -V_T \ln\left(\frac{4}{R_o I_o}\right)$$

$$= -V_T(\ln 4 - \ln R_o I_o)$$

$$= -V_T(2\ln 2 - \ln R_o I_o)$$

$$\therefore V_{o1} - V_{o2} = V_T \ln 2$$

Second Method

$$V_{o1} - V_{o2} = V_T \ln\left(\frac{V_{2i}}{V_{1i}}\right) = V_T \ln\left(\frac{4}{2}\right) = V_T \ln(2)$$

32. Ans. (a)

Solution: For –ve terminal analysis

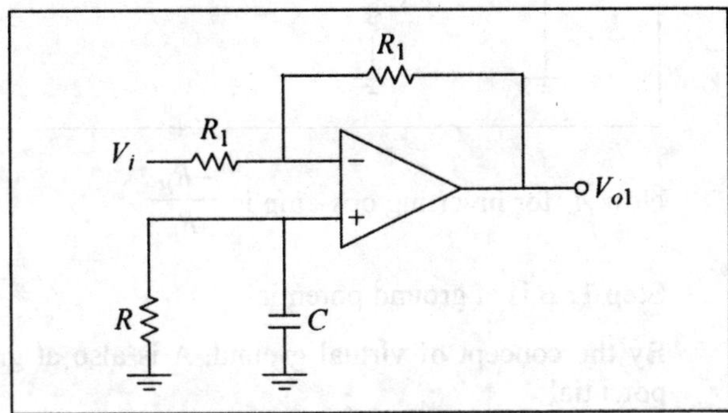

$$V_{o1} = -V_i$$

For +ve terminal analysis

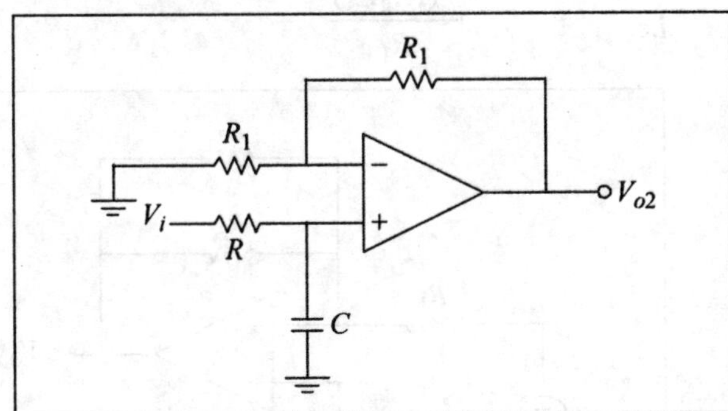

$$V_{o2} = \left(1 + \frac{R_1}{R_1}\right)V_1 = 2V_1$$

$$\text{But } V_1 = \frac{\frac{1}{sC}}{R + \frac{1}{sC}}V_i = \frac{V_i}{1 + RCs} = \frac{2V_i}{1 + RCs}$$

$$V_o = \frac{2V_i}{1 + RCs} - V_i$$

$$\frac{V_o}{V_i} = \frac{1 - RCs}{1 + RCs}$$

33. Ans. (c)

Solution:

$$T(s) = \frac{V_o(s)}{V_i(s)}$$

Argument of

$$T(s) = -\tan^{-1}\omega RC - \tan^{-1}\omega RC$$
$$= -2\tan^{-1}\omega RC = -\pi \text{ to } 0$$

34. Ans. (b)

Solution:

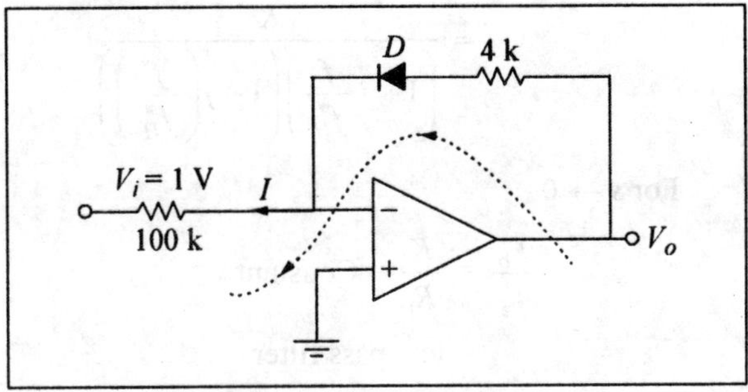

$$I = \frac{1\text{ V}}{100\text{ k}\Omega} = 10\ \mu\text{A}$$

$$I = I_0\left(e^{\frac{V}{V_T}} - 1\right)$$

Given, $I_0 = 1\ \mu\text{A}$

Then, $10\ \mu\text{A} = 1\ \mu\text{A}\left[e^{\frac{V}{25}} - 1\right]$

or $V = 60$ mV

Voltage drop across 4 kΩ resistance

$= 4\text{ k}\Omega \times 10\ \mu\text{A} = 40$ mV

Total voltage output

$= 60\text{ mV} + 40\text{ mV} = 100\text{ mV} = 0.1$ V

35. Ans. (b)

Solution: Given circuit is low pass filter as $X_L = \omega L$ is in sereis to input.

Mathematically:

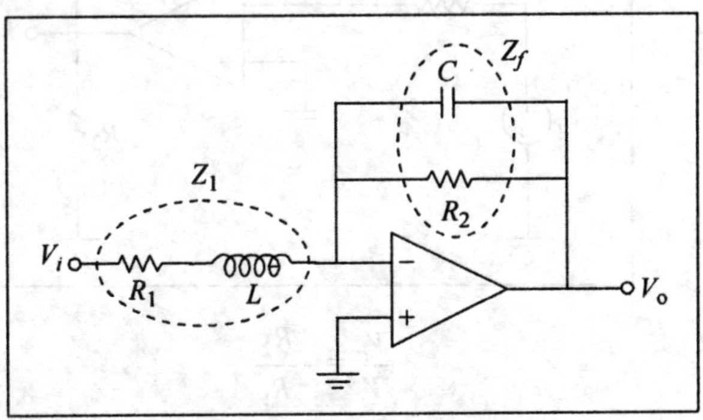

$$\frac{V_o}{V_i} = -\frac{Z_f}{Z_1}$$

$$Z_f = R_2 \parallel \frac{1}{sC} = \frac{R_2}{R_2Cs+1}$$

$$Z_1 = R_1 + sL$$

$$\frac{V_o}{V_i} = \frac{K}{(R_1+sL)(R_2Cs+1)}$$

$$= \frac{K}{\left(1+j\dfrac{f}{f'_H}\right)\left(1+j\left(\dfrac{f}{f''_H}\right)\right)}$$

For $s \to 0$

$$\frac{V_o}{V_i} = \frac{K}{R_1} = \text{Constant}$$

= low pass filter

36. Ans. (c)

Solution: When output is + 15 V,

let voltage at non, inverting terminal is V_N, then

KCL at node (assume all out going current)

$$\therefore \quad \left(\frac{V_N-15}{10}\right)2+\left(\frac{V_N+15}{10}\right)=0$$

$$\Rightarrow \quad V_N = +5 \text{ volts}$$

When output is –15 V, then KCL at node.

$$\left(\frac{V_N-15}{10}\right)+2\left(\frac{V_N+15}{10}\right)=0$$

$$\Rightarrow \quad V_N = -5 \text{ volts}$$

37. Ans. (a)

Solution: By rearranging the amplifier circuit becomes

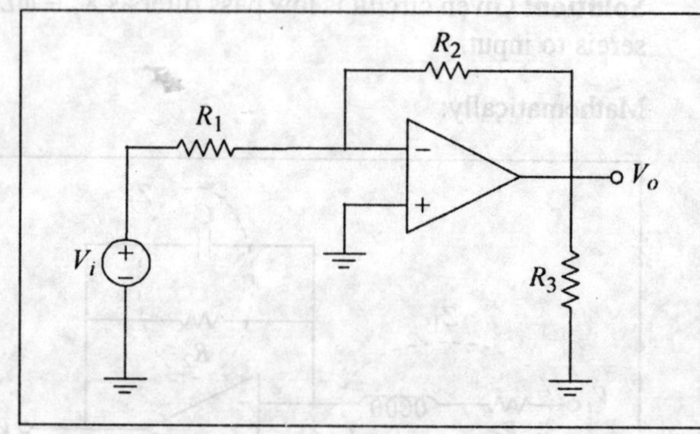

$$\frac{V_o}{V_i} = -\frac{R_2}{R_1}$$

Note: Load resistor is never considered as input side impedance only, feedback element have effect on input and output impedance.

Second Method

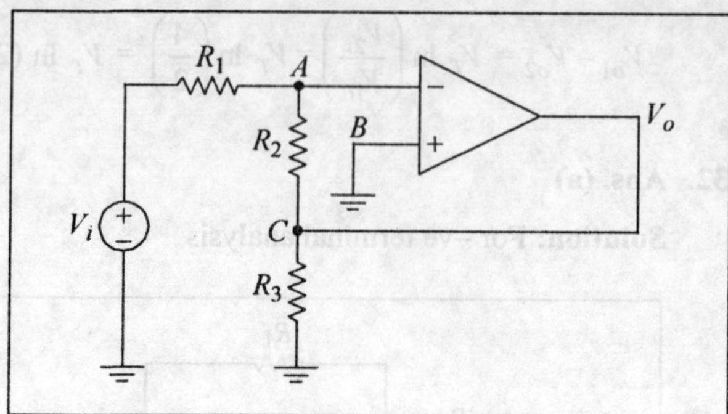

Now A_v for inverting ops-amp is $\dfrac{-R_F}{R_i}$

Step 1: *B* is at ground potential

By the concept of virtual ground, A is also at ground potential

Resistance $R_2 \neq R_3$ comes is parallel circuit modifies

$(R_2 \parallel R_3)$ is the *RF* resistor and R_1 is the input resistor

i.e. $\dfrac{-(R_2 \parallel R_3)}{R_1}$

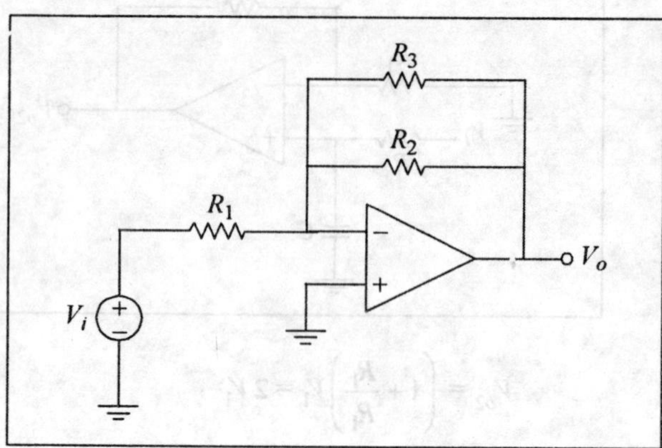

38. Ans. (b)

Solution:

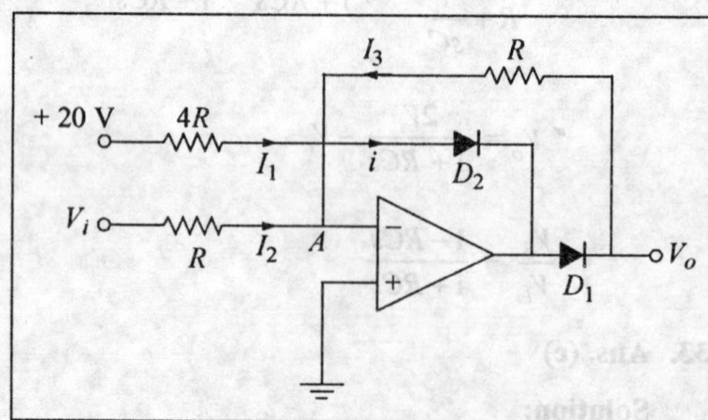

Since input is connected to negative terminal,

Operational amplifier is working as inverting amplifier.

Hence output always positive.

Since $V_A = 0$ (virtually grounded)

KCL at node A

Case 1: $i_1 + i_2 + i_3 = i$

$$\frac{20}{4R} + \frac{v_i}{R} + \frac{v_o}{R} = i$$

If V_o = + ve, diode do not coundct, so $i = 0$

$$\therefore \quad \frac{20}{4R} + \frac{V_i}{R} + \frac{V_o}{R} = 0$$

or $\quad 5 + V_i + V_o = 0$

This represent a straight line with slope = –1

If $V_i \geq -5$, both the diode conduct and $V_o = 0$

Hence correct option is (b)

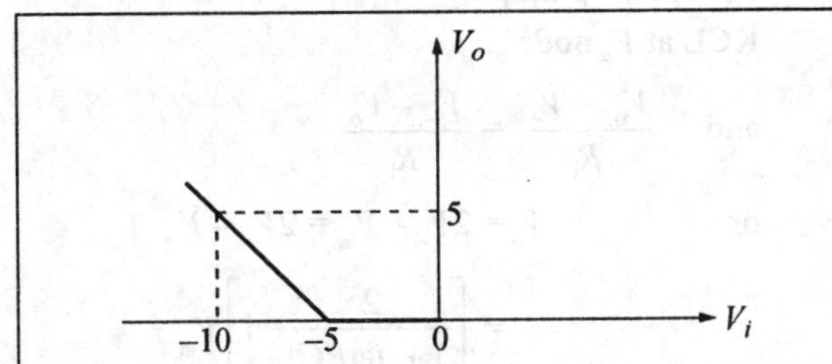

$$V_{ac} = \frac{4R_f}{R + 4R_f} \times V_s$$

$$= \frac{4 \times 25}{9000 + 4 \times 25} \times V_s$$

$$= \frac{100}{10{,}00} V_s$$

or $V_{ac} = 100 \cos \omega t \; mV$

or $V_{ac} = \cos \omega t \; mV$.

39. Ans. (c)

Solution: Since the gain of the op-amp is infinity

$$\therefore \quad V_i = \frac{V_o}{\infty} = 0$$

Hence the two input terminals are considered at the same potential

$$\text{CMRR} = \frac{A_{DM}}{A_{CM}}$$

For CMRR to be infinity,

$$A_{CM} = 0, \; A_{CM} = \left.\frac{V_o}{V_c}\right|_{V_d = 0}$$

$V_d = 0$, when $V_1 = V_2$ and $V_o = (A_{OL} \times V_d)$

40. Ans. (a)

Solution:

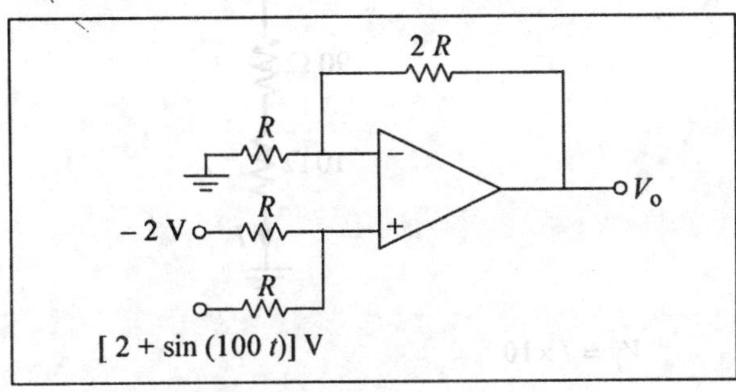

$$V_o = -2\left[1 + \frac{2R}{R}\right]\left[\frac{R}{2R}\right]$$

$$+ (2 + \sin 100\,t)\left(1 + \frac{2R}{R}\right)\left(\frac{R}{2R}\right)$$

$$= -3 + 3 + \frac{3}{2}\sin 100t$$

$$= \frac{3}{2}\sin 100t$$

41. Ans. (a)

Solution: When switch S_1 is open

$$\frac{V_o}{V_i} = A_V = -\frac{2R}{R} = -2 = x$$

When switch S_1 is closed, $A'_V = -\frac{R}{R} = -1 = \frac{x}{2}$.

42. Ans. (d)

Solution: Since it is an ideal operational amplifier, it has infinite input resistance

$$V_o = -10^{-3} \times 10^3 = -1\text{V}$$

Second Method

Apply KCL at inverting terminal

$$10^{-3} = \frac{V_a - V_o}{10^3}$$

or $V_a - V_o = 1$

$V_a \to 0$ as virtual ground

$V_o = -1$ Volt

43. Ans. (d)

Solution: Feedback factor $= \dfrac{10}{10 + 90}$

$$= \frac{10}{100}$$

$$= \frac{1}{10}$$

Second Method

The feedback network is

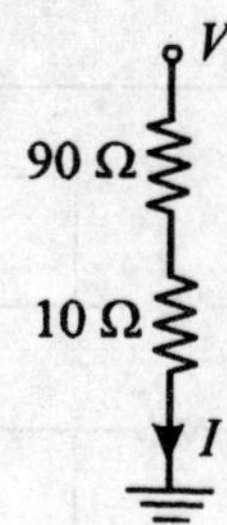

$$V_f = I \times 10$$

$$= \frac{V_o}{100} \times 10$$

$$\frac{V_f}{V_o} = \beta = \frac{1}{10}$$

44. Ans. (a)

Solution:

Slew rate $[SR]$ = maximum rate at which amplifier output can change (in V/μs).

$$SR = \left[\frac{dV_o}{dt}\right] \text{V/μs}$$

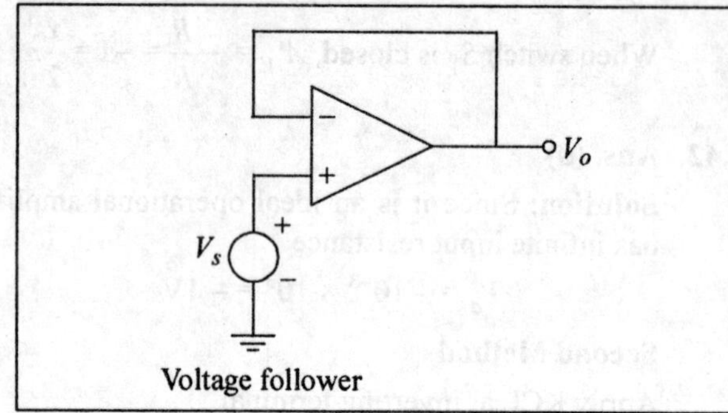

Input $V_s = V_m \sin \omega t$

$V_m = 10$ V

In a voltage follower,

Output voltage $V_o = V_s$ (input voltage)

Then, $V_o = V_m \sin \omega t$

$$\text{Slew rate} = \frac{dV_o}{dt} = \omega\, V_m \cos \omega t$$

$$\text{maximum } \frac{dV_o}{dt} = \omega V_m \geq S \text{ [Slew Rate]}$$

$$\Rightarrow \quad 2\pi f V_m \geq S$$

$$\Rightarrow \quad f \geq \frac{S}{2\pi V_m}$$

$$\geq \frac{62.8}{10^{-6} \times 6.28 \times 10}$$

$$\geq 10^6 \text{ Hz}$$

$$f_{min} = 1 \text{ MHz.}$$

45. Ans. (d)

Solution: From the given circuit, we get

$$V_+ = \frac{\frac{1}{j\omega C}}{R + \frac{1}{j\omega C}} \times V_{in} = \frac{V_{in}}{1 + j\omega RC}$$

For ideal op-amp $V_- = V_+$

KCL at V_- node

and $$\frac{V_{in} - V_-}{R_1} = \frac{V_- - V_o}{R_1}$$

or $$V_o = 2V_- - V_{in} = 2V_+ - V_{in}$$

$$= \left[\frac{2}{1 + j\omega RC} - 1\right] V_{in}$$

$$= \frac{1 - j\omega RC}{1 + j\omega RC} V_{in}$$

Then, $\angle (V_o/V_i) = -2\tan^{-1} \omega RC$

For $-90° \leq \theta \leq 90°$,

Phase-shift $\angle (V_o/V_i) = \pm 180°$

46. Ans. (d)

Solution:

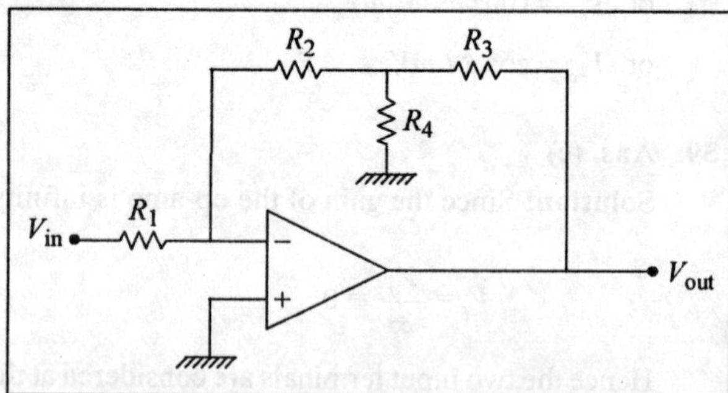

$$\frac{V_{out}}{V_{in}} = -\left[\frac{R_2 + R_2 + \frac{R_2 R_3}{R_4}}{R_1}\right]$$

$$= -\frac{10 \times 10^3 + 10 \times 10^3 + \frac{10 \times 10^3 \times 10 \times 10^3}{10^3}}{10^3}$$

$$= -\frac{20\times10^3+100\times10^3}{10^3} = -120$$

Second Method

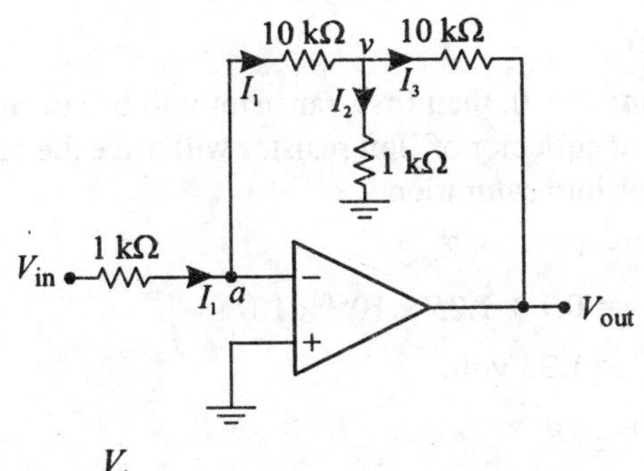

$$I_1 = \frac{V_{in}}{1\,k\Omega} \qquad \text{...(1)}$$

$$\text{Also, } I_1 = \frac{0-V}{10\,k\Omega} = \frac{-V}{10\,k\Omega} \qquad \text{...(2)}$$

From equations (1) and (2) $\frac{-V}{10} = V_{in}$

$$V = -10\,V_{in} \qquad \text{...(3)}$$

Applying KCL at node 'b'

$$I_1 = I_2 + I_3$$

$$\text{or } \frac{V_{in}}{1} = \frac{V}{1} + \frac{V-V_o}{10}$$

$$\text{or } V_{in} = -10\,V_{in} + \frac{-10V_{in} - V_o}{10}$$

$$\text{or } 11\,V_{in} = -V_{in} - \frac{V_o}{10}$$

$$\text{or } \frac{V_o}{10} = -12\,V_{in}$$

$$\text{or } \frac{V_o}{V_{in}} = -120$$

Since $V_o >> V_i$

$$I_f = \frac{V_i - V_o}{R_F} = \frac{-V_o}{R_F} = KV_o$$

Where $K = -\frac{1}{R_b}$

Since feedback current is proportional to output voltage, therefore this is voltage shunt feedback. If voltage is used as feedback then series case and current is used as feedback then shunt case. Hence present case is shunt –Shunt feedback.

* As one terminal of feedback element is connected to output voltage, so it is voltage sampling or shunt sampling and another terminal of feedback element is connected to base of the transistor, so it is shunt mixing. Hence it is shunt-shunt feedback amplifier.

47. Ans. (a)

Solution:

The given circuit is bridge T network whose transfer function is as below

$$\frac{V_o(s)}{V_o(s)} = \frac{s/R_2C_1}{s^2 + \frac{C_1+C_2}{R_1C_1C_2} + \frac{1}{R_1R_2C_1C_2}}$$

$$A_v(s) = \frac{(\omega_o/Q)A_o s}{s^2 + (\omega_o/Q)s + \omega_o{}^2}$$

$$\text{comparing we have, } \frac{\omega_o}{Q} = \frac{C_1+C_2}{R_1C_1C_2}$$

$$\text{or} \qquad \omega_o = Q\left(\frac{C_1+C_2}{R_1C_1C_2}\right) = 1\,\frac{10^{-9}\times2}{200\times10^3\times10^{-8}}$$

$$= 1000 \text{ rad/sec.}$$

Second Method

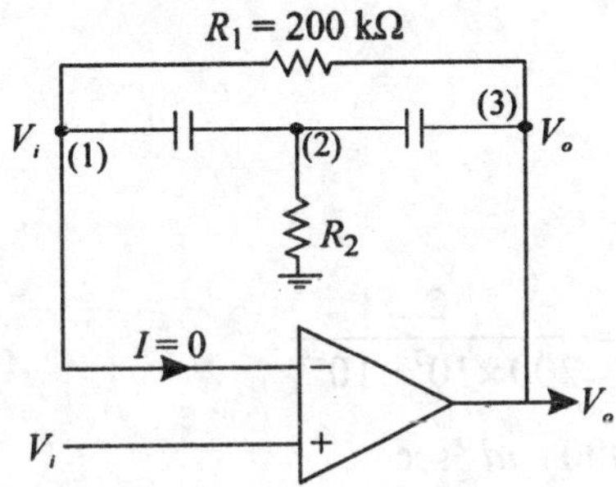

Let the voltage at node 2 = V

Apply KCL at node 1,

$$(V_i - V)sC + \frac{(V_i - V_o)}{R_1} = 0$$

$$\text{or } VsC = V_i\left(sC + \frac{1}{R}\right) - \frac{V_0}{R_1}$$

$$\text{or } V = \frac{V_1(1+sR_1C)}{sR_1C} - \frac{V_0}{sR_1C} \qquad \text{...(1)}$$

Apply KCL at node 2

$$(V - V_i)sC + \frac{V}{R_2} + (V - V_o)sC = 0$$

$$\text{or } V\left[sC + sC + \frac{1}{R_2}\right] = V_i sC + V_o sC$$

$$\text{or } \left(2sC + \frac{1}{R_2}\right)\left[\frac{V_1(1+sR_1C)}{sR_1C} - \frac{V_0}{sR_1C}\right]$$

$$= V_i sC + V_o sC$$

or $V_o\left[\frac{1+2s\,R_2C}{sR_2R_1C}+sC\right]$

$$=V_i\left[\frac{(1+sR_1C)(1+2sR_2C)}{sR_1R_2C}-sC\right]$$

or $\frac{V_o}{V_i}=\frac{(1+sR_1C)(1+2sR_2C)-s^2R_1R_2C^2}{1+2sR_2C+s^2R_1\,R_2\,C^2}$

Let us consider the characteristic equation

$$s^2R_1R_2C^2+2sR_2C+1=0$$

or $s^2+\frac{2s}{R_1C}+\frac{1}{R_1R_2C^2}=0$

Compairing with the standard second order equation

$$s^2+\left(\frac{\omega_o}{Q}\right)s+\omega_0^2=0$$

or $\frac{\omega_0}{Q}=\frac{2}{R_1C}$

Since $Q=1$

or $\omega_0=\frac{2}{R_1C}=\frac{2}{200\times10^3\times10^{-9}}$

$$=10{,}000\ rad/\sec$$

48. Ans. (b)

Solution:

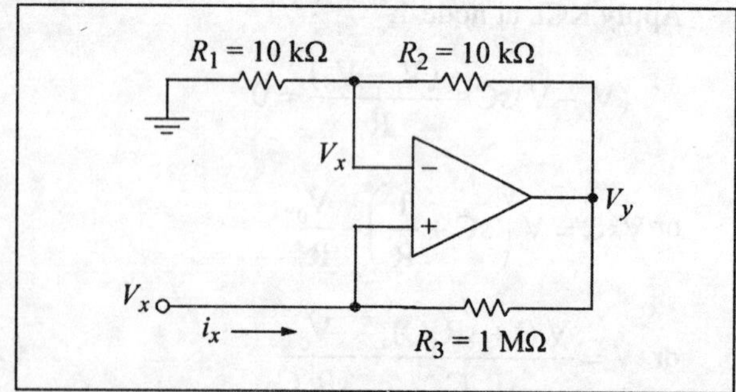

Using KCL at inverting terminal, we have

$$\frac{v_x-v_y}{R_2}+\frac{v_x-0}{R_1}=0$$

or $\frac{v_x}{v_y}=\frac{R_1}{R_1+R_2}$ ... (1)

Using same at non-inverting terminal, we get

$$i_x=\frac{v_x-v_y}{R_3}=\frac{v_x-v_x\left(\frac{R_1+R_2}{R_1}\right)}{R_3}$$

or $i_x=-\frac{v_xR_2}{R_1R_3}$

or $R_{IN}=\frac{v_x}{i_x}=-\frac{R_1R_3}{R_2}-\frac{10^4\times10^6}{100\times10^3}=-100\ \text{k}\Omega$

49. Ans. (b)

Solution: $V_i=0$, then first transistor will be cut-off and voltage at collector of left resistor will drive the second transistor into saturation.

Then, $V_o=V_{CE,\,sat}+V_R$

$=0.1+1.25\times10^{-3}\times10^3$

$=1.35$ volt.

50. Ans. (d)

Solution:

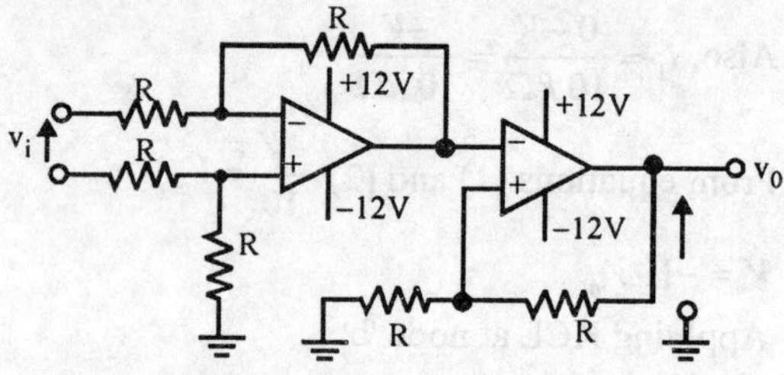

Output of op-amp 1

$$V_{o1}=-V_1+V_2=V_2-V_1=-V_{in}$$

The first state of the given circuit represent subractor

As $V_i\uparrow^{+}_{-}$ than is output is $V_{o1}=-V_i$ then

Let $V_{in}=V_m\sin\omega t$

Op-amp 2 has positive feedback so its operate under saturation and its output is either +12 V or –12 V.

The circuit of op-amp 2 is schmitt trigger with

$$V_f=\frac{R}{R+R}\times V_o$$

When $V_o=12\,V$

When $V_f=V_{UT}$

$=\frac{R}{R+R}\times12$

$=+6\,V=$ **Upper threshold voltage**

When $V_o=-12$

$V_f=V_{LT}=\frac{R}{R+R}\times(-12\,V)$

$=-6V=$ **Lower threshold**

Waveforms

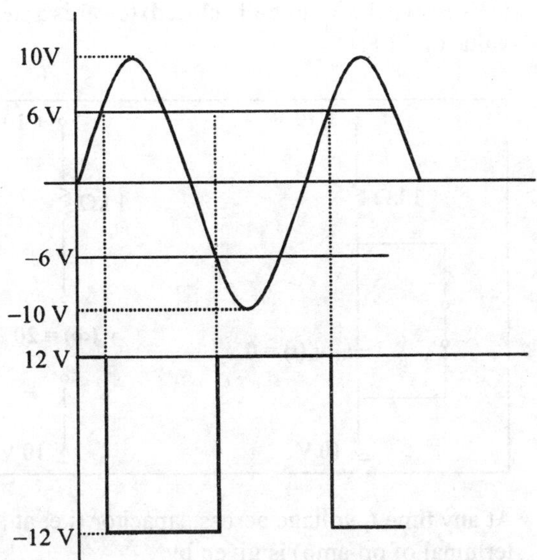

Transfer characteristic of overall circuit,

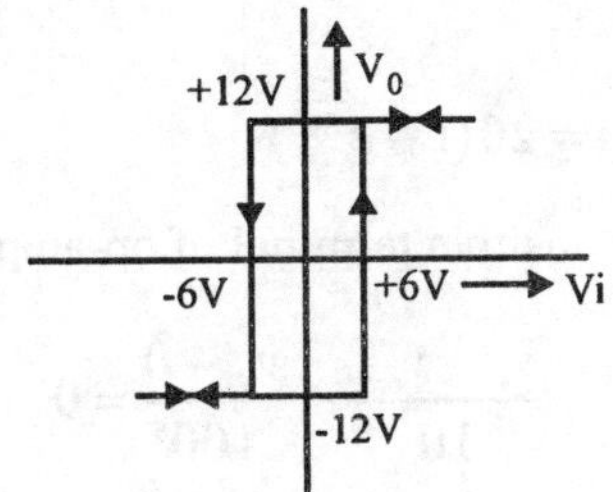

51. Ans. (c)

Solution: For inverting amplifier, the gain

$$A_v = \frac{-Z_2}{Z_1} = \frac{R_f}{R_i}$$

Case 1:

$$\text{gain} = -25 = -\frac{1\times 10^6}{R_i}$$

$$\Rightarrow \quad R_i = \frac{10^6}{25} = 40\,\text{k}\Omega$$

Case 2 :

If gain = −10, then $R_i = \frac{10^6}{10}$, $R_i = 100$ kΩ

Therefore, highest impedance $R_i = 100$ kΩ.

52. Ans. (c)

Solution:

For the +ve half cycle D_1 is ON.

or $\quad v^- - v^+ = 0.7$ V

Here $v^- > v^+$

Hence $V_{out} = -V_{cc}$

For −ve half-cycle

D_2 is ON.

or $\quad v^+ - v^- = 0.7$

Here $\quad v^+ > v^-$

Hence $V_{out} = +V_{CC}$

Output will appear 180° out of phase to the input with clipped shape.

53. Ans. (c)

Solution:

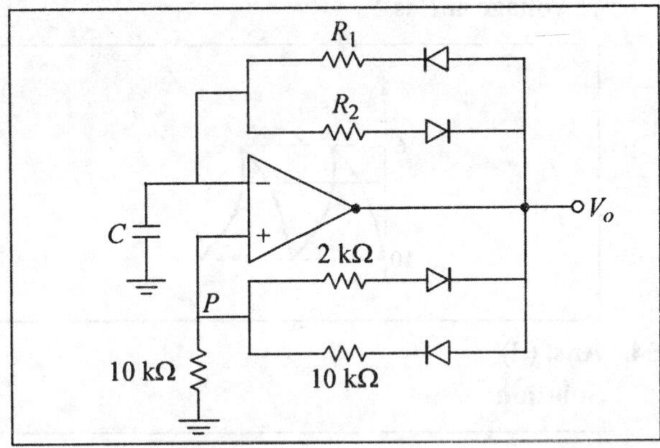

Given circuit is astable multivibrator whose output is square wave and waveform across cpacitor is exponential waveform. Therefore potential across 'C' and 'P' are same

When output voltage is +12 V, the 'C' changes toward maximum value through R_1.

Case 1: When V_o is positive

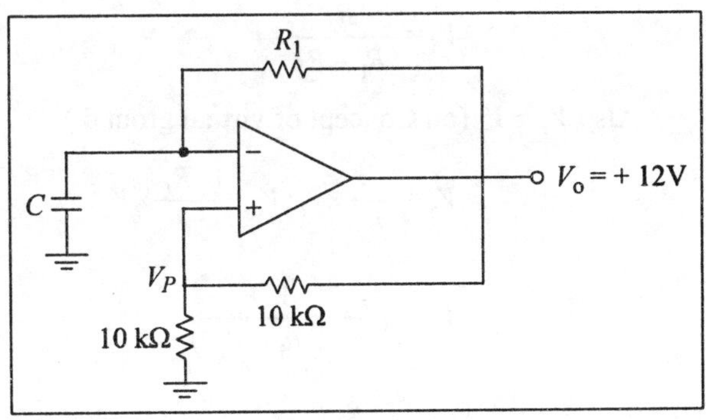

$$V_p = \frac{12}{20}\times 10$$

$$= +6\text{ V}$$

$V_{c\,max}$ is + 6 V

Case 2: When V_o is negative

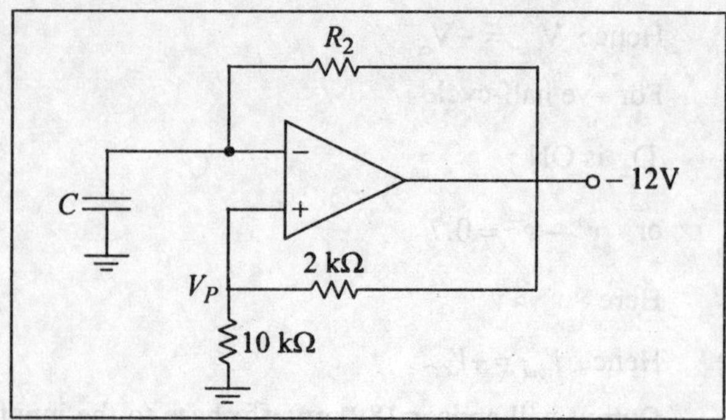

$$V_p = \frac{10}{12} \times -12$$

$$= -10 \text{ V}$$

∴ Voltage at P is

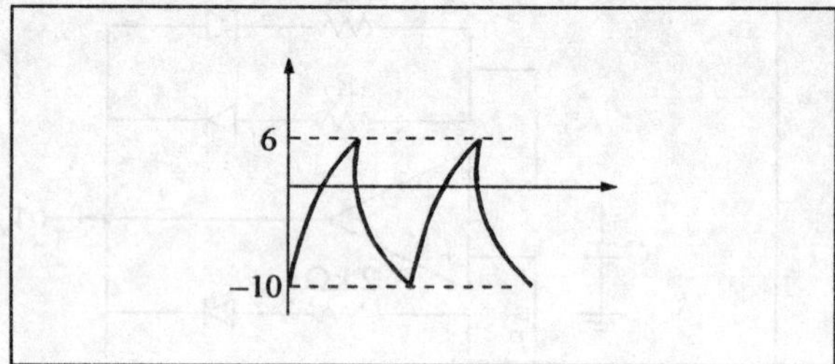

54. Ans. (d)

Solution:

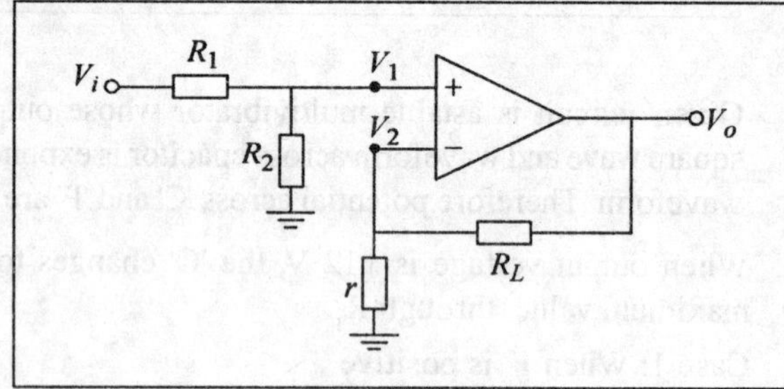

$$V_1 = \frac{R_2}{R_1 + R_2} \times V$$

Also $V_1 \cong V_2$ (on Concept of virtual ground)

$$V_2 = \frac{-r}{R_L} \times V_o; V_o = \frac{-R_L}{r} \times V_2$$

$$\therefore \quad V_o = \frac{-R_L}{r}\left(\frac{V\,R_L}{R_1 + R_L}\right)$$

or $$-\frac{V_o}{R_L} = \frac{R_2}{R_1 + R_2} \cdot \frac{V}{r}$$

This shows volt to current conversion which results current source with current = $\frac{R_2}{R_1 + R_2} \times \frac{V}{r}$.

55. Ans. (d)

In the given circuit, the capacitor starts charging from 0 V (as switch was initially closed) towards a steady state value of 20 V.

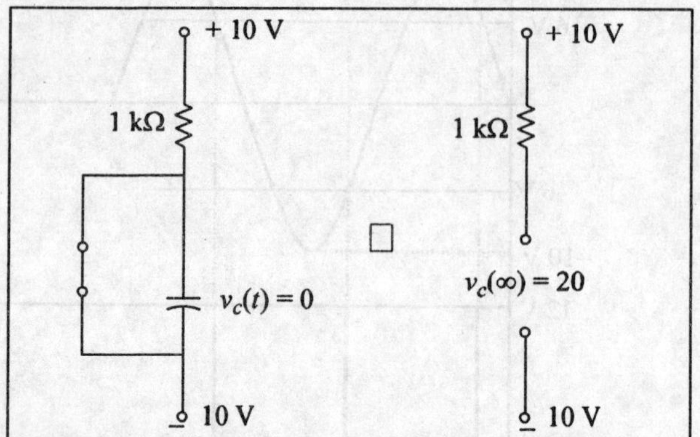

At any time t, voltage across capacitor (i.e. at inverting terminal of op-amp) is given by

$$v_c(t) = v_c(\infty) + [v_c(0) - v_c(\infty)]e^{\frac{-t}{RC}}$$

$$v_c(t) = 20(1 - e^{\frac{-t}{RC}})$$

Voltage at positive terminal of op-amp

$$\frac{v_+ - v_{out}}{10} + \frac{v_+ - 0}{100} = 0$$

$$v_+ = \frac{10}{11} v_{out}$$

Due to zener diodes, $-5 \le V_{out} \le +5$

Hence, $$v_+ = \frac{10}{11} \times 5\,V$$

Transition from +5 V to − 5 V occurs when capacitor charge upto v_+ as op-amp operated with inverting mode.

Hence, $$20(1 - e^{\frac{-t}{RC}}) = \frac{10 \times 5}{11}$$

$$1 - e^{\frac{-t}{RC}} = \frac{5}{22}$$

$$\frac{17}{22} = e^{\frac{-t}{RC}}$$

$$t = RC \text{ in}\left(\frac{22}{17}\right) = 1 \times 10^3 \times .01 \times 10^{-6} \times 0.257$$

$$= 2.57 \text{ μ sec.}$$

56. Ans. (b)

Solution:

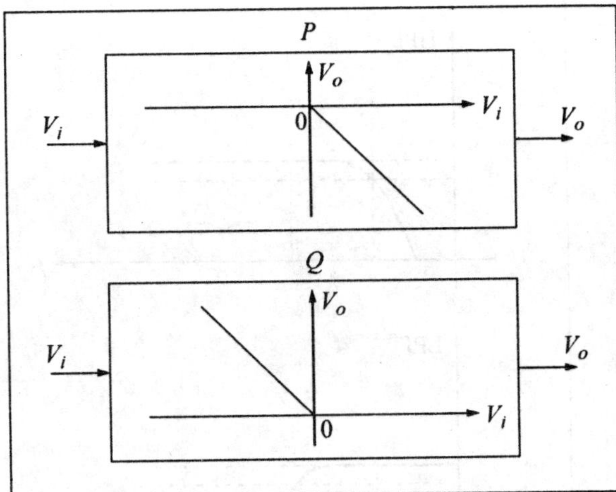

For $V_{in} > 0 \Rightarrow V_o(P) = -$ ve and $V_o(Q) = 0$

For $V_{in} < 0 \Rightarrow V_o(P) = 0$ and $V_o(Q) = +$ ve

Now check each option.

If $V_i > 0$

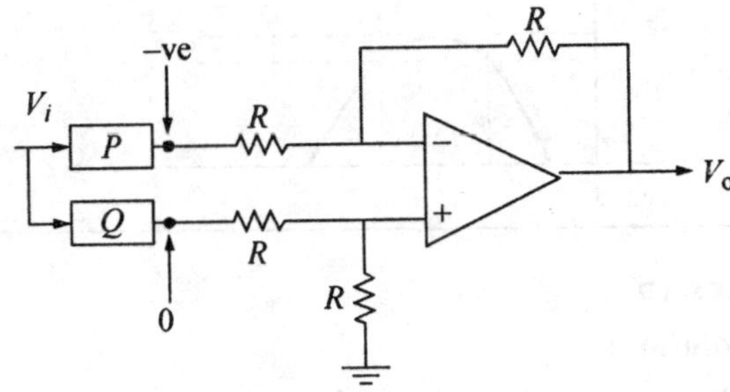

Output will be positive due to inverting action

If $V_i < 0$

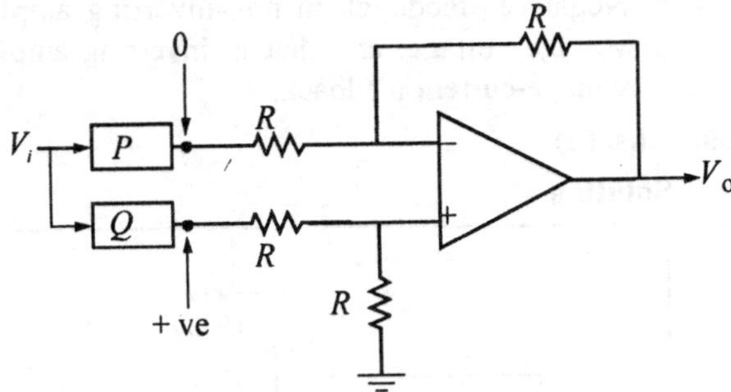

Output will be postive due to non-inverting action. So in all case the output is always rectified.

Second Method

From the transfer characteristic of the rectifier P is

$V_o = -V_i$ for $V_i > 0$

$V_o = 0$ for $V_i < 0$

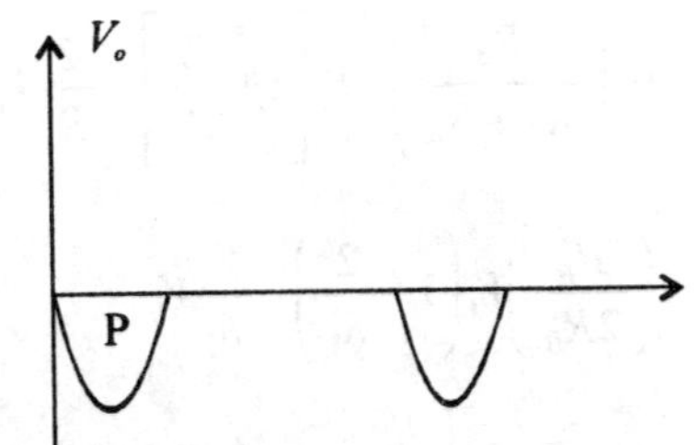

From the transfer characteristic of the rectifier Q

$V_o = 0$ for $V_i > 0$

$V_o = V_i$ for $V_i < 0$

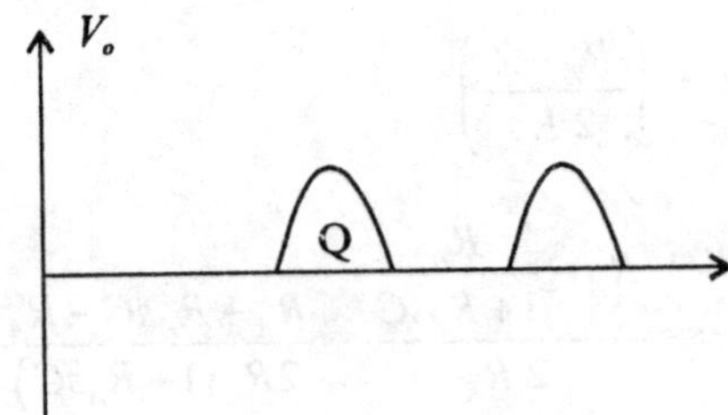

The full-wave rectifier output is

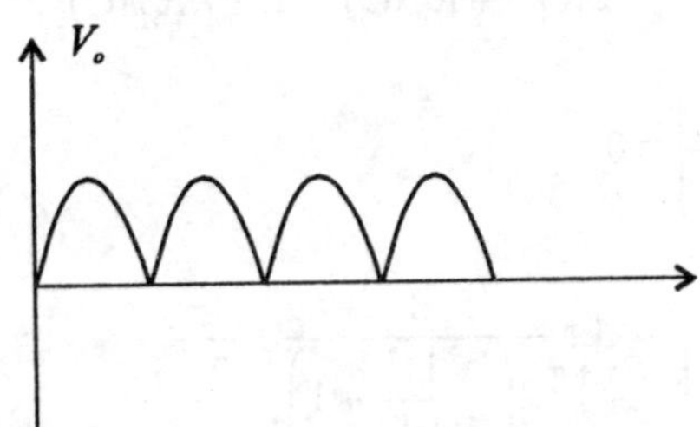

So the *HWR* 'P' must be connected to inverting and 'Q' must be connected to non-inverting terminal of the op-amp.

57. **Ans. (c)**
Solution:

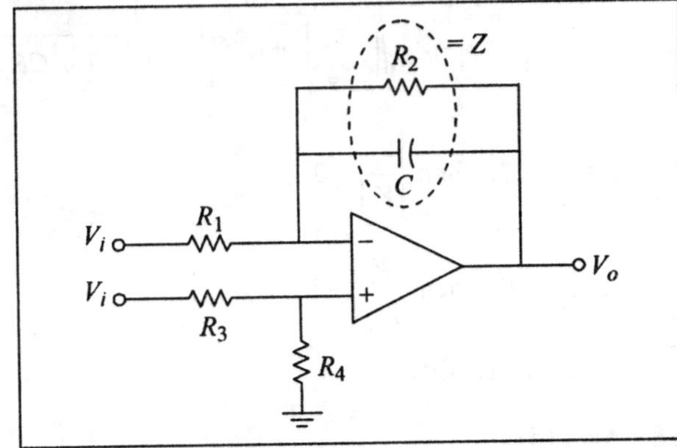

$$Z = \frac{R_2 \times \frac{1}{sC}}{R_2 + \frac{1}{sC}} = \frac{R_2}{1 + R_2 sC}$$

$$V_o = \left(\frac{R_4}{R_3 + R_4}\right) \times V_i\left[1 + \frac{Z}{R_1}\right] - \frac{Z}{R_1}V_i$$

$$\left(\frac{R_B}{2R_B}\right)V_i\left(1+\frac{2}{R_1}\right) - \frac{Z}{R_1}V_i$$

$$\frac{1}{2}V_i + \frac{Z}{2R_A}V_i - \frac{Z}{R_A}V_i = \frac{1}{2}V_i + V_i\left(\frac{Z-2Z}{2R_A}\right)$$

$$= \frac{1}{2}V_i - \frac{Z}{2R_A}V_i = \left(\frac{R_A - Z}{2R_A}\right)V_i$$

$$\frac{V_o}{V_i} = \left(\frac{R_A - Z}{2\,R_A}\right)$$

$$= \frac{R_A - \dfrac{R_A}{1 + R_A sC}}{2\,R_A} = \frac{R_A + R_A^2 sC - R_A}{2R_A(1 + R_A sC)}$$

$$= \frac{R_A^2 sC}{2R_A(1 + R_A sC)} = \frac{R_A sC}{2(1 + R_A sC)}$$

$$s \to 0 \quad 1\left|\frac{V_o}{V_i}\right| = 0$$

$$s \to \infty \quad \left|\frac{V_o}{V_i}\right| = \underset{s\to\infty}{\text{Lt}} \frac{1}{2\left(\dfrac{1}{RAsC} + 1\right)} = \frac{1}{2}$$

⇒ High pass filter

Another way

$$\left|\frac{V_o}{V_i}\right| = \frac{\omega C}{\sqrt{\left(\dfrac{1}{R_A}\right)^2 + \omega^2 C^2}} = \frac{1}{\sqrt{1 + \dfrac{1}{(\omega^2 C R_A)}}}$$

As $\omega \to 0$, Then $\left|\frac{V_o}{V_i}\right| = 0$,

As $\omega \to 0$, Then $\left|\frac{V_o}{V_i}\right| = 1$

So high pass filter.

58. Ans. (d)

Solution:

The operational amplifier is HPF and RC is LPF

HPF followed by LPF result as BPF.

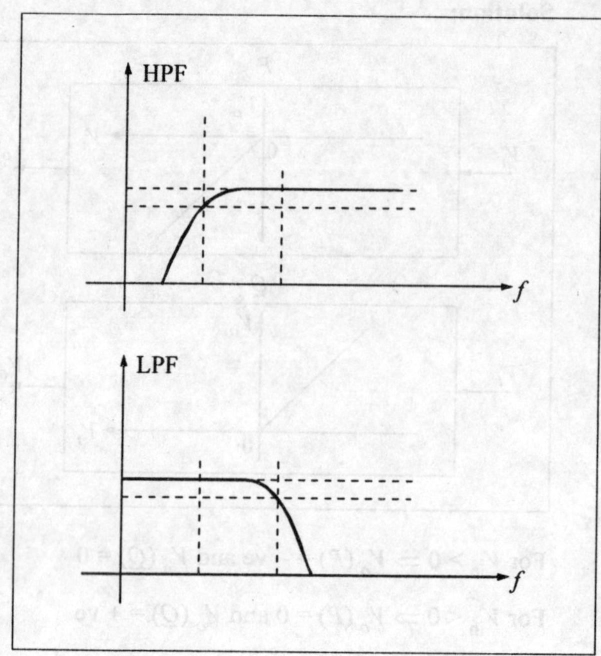

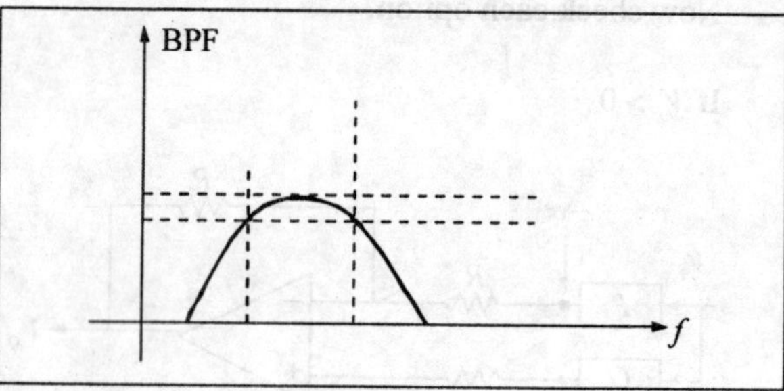

59. Ans. (b)

Solution:

The output parameter sampled is voltage and it is mixed in series with the input. Hence it is a voltage-voltage feedback arrangement also called voltage-series or shunt-series feedback.

* **Negative feedback in non-inverting amplifier is voltage-voltage; and that in inverting amplifier is voltage-current feedback.**

60. Ans. (d)

Solution:

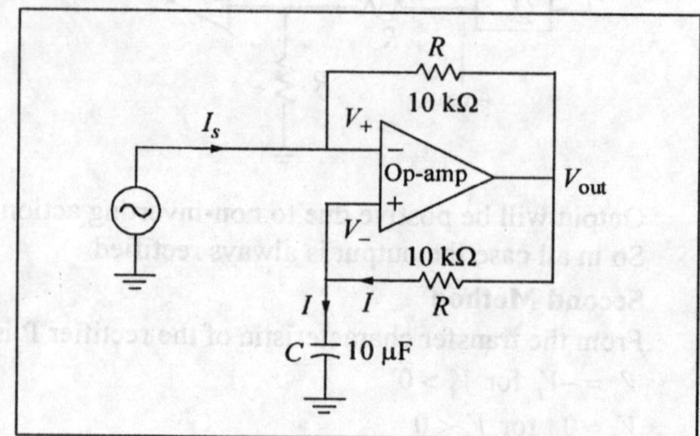

$$X_c = \frac{1}{j\omega C} = \frac{1}{j2\pi \times 50 \times 10 \times 10^{-6}}$$
$$= \frac{-j1000}{\pi}$$
$$I = \frac{V_-}{X_c} = \frac{10}{X_c} = \frac{j\pi}{100}$$
$$V_{out} = 10 + \frac{(10,000)\, j\pi}{100}$$
$$= 10 + j100\pi$$
$$i_s = \frac{V_+ - V_{out}}{R}$$
$$= \frac{10 - 10 - j100\pi}{10000}$$
$$= -j10\,\pi \text{ mA}$$
$$= 10\pi \angle -90^\circ \text{mA}$$

61. Ans. (d)

Solution:

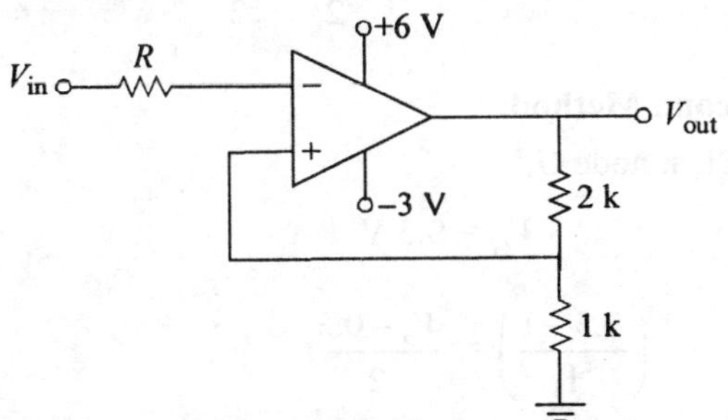

This figure is known as positive feedback Schmitt trigger or squaring circuit. The input voltage V_{in} trigger the output V_o every time it exceeds certain voltage level called the upper threshold voltage V_{out} and lower threshold voltage V_{th}. For above circuit, Hysteresis diagram is as below

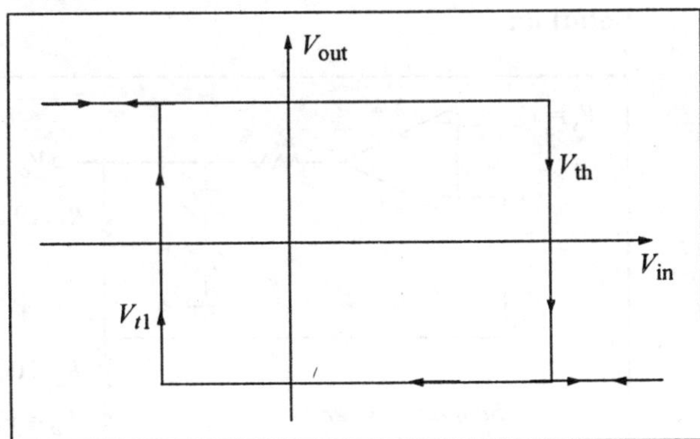

$$V_{th} = \left(\frac{1k}{3k}\right) 6 = 2\text{ V} = t_2$$

After $t_2, V_{out} = -3\text{ V}$

$$V_{th} = \left(\frac{1k}{3k}\right)(-3) = -1\text{V} = t_4$$

After $t_4, V_{out} = 6\text{ V}$

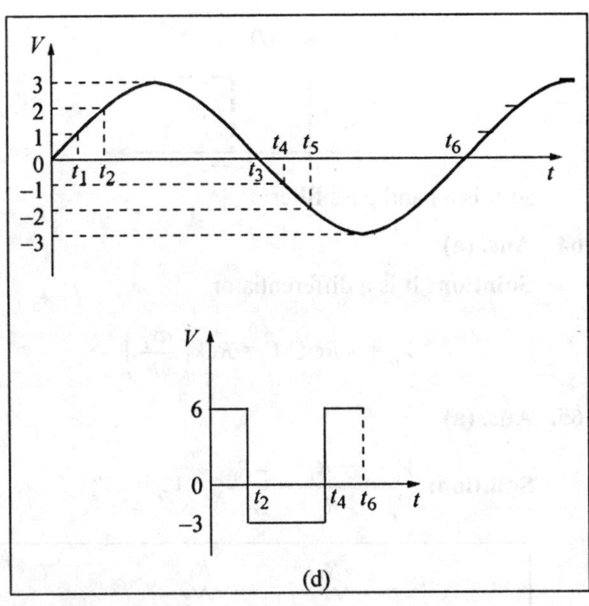

(d)

Second Method

The given circuit represents a positive feedback (Schmitt trigger)

When $v^+ > v^- \Rightarrow V_o = +6\,V$

$v^+ < v^- \Rightarrow V_o = -3\,V$

$$v^+ = \frac{v_o + 1}{1+2} = \frac{v_o}{3}$$

or, When $V_o = +6\,V$

or $v^+ = 2\,V$

and when $V_o = -3\,V$

or, $v^+ = -1\,V$

62. Ans. (b)

Solution: For non-inverting amplifier

$$A_{vf} = \frac{V_o}{V_o} = \left(1 + \frac{R_f}{R}\right)$$

$$V_o = 2\left(1 + \frac{2R}{R}\right) = 6\text{V}$$

63. Ans. (d)

Solution:

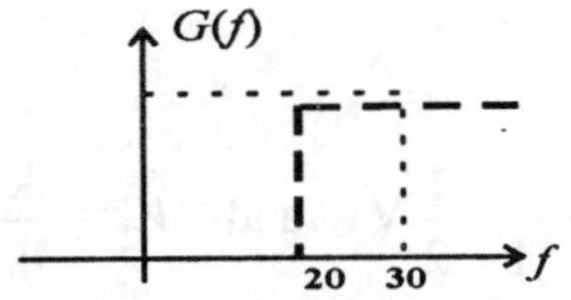

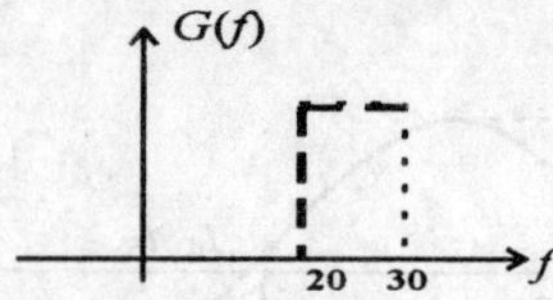

So it is a band pass filter

64. Ans. (a)

Solution: It is a differentiator

$$V_o = -R\omega C\, V_i = RC\left(\frac{dv_i}{dt}\right)$$

65. Ans. (a)

Solution: $\dfrac{V_0}{V_s} = \dfrac{-R_f}{R_s} = \dfrac{-R_1}{R_1};\ V_0 = -V_s$

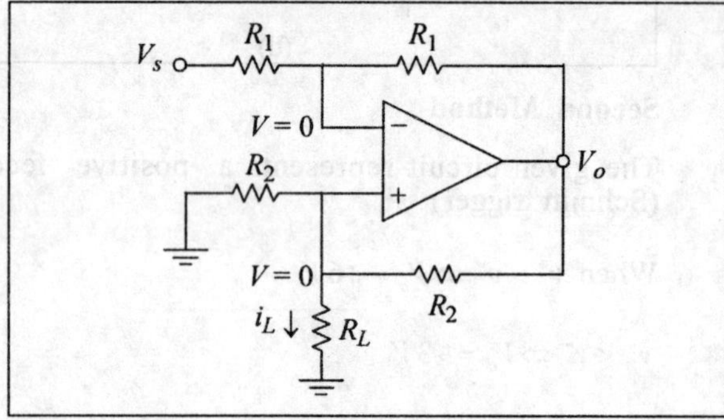

KVL, $\dfrac{0 + V_s}{R_2} + i_L = 0$

$$i_L = \frac{-V_s}{R_2}$$

66. Ans. (c)

Solution: For –ve terminal

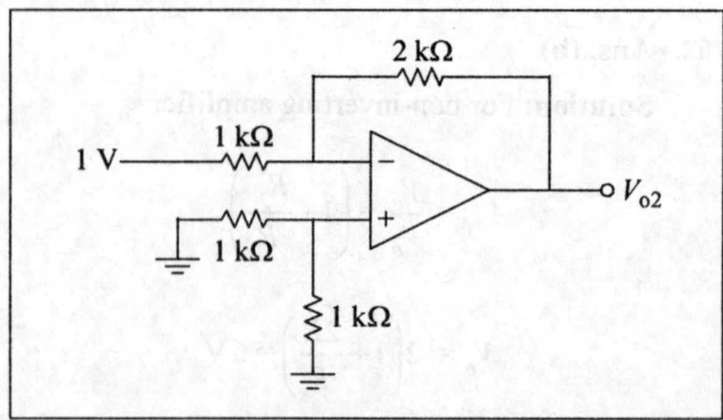

$$V_{o1} = -\frac{2}{1}(1) = -2\text{V} \quad \text{... (1)}$$

For + ve terminal

$$V_2 = \frac{1}{1+1} = \frac{1}{2}\text{ V} \quad \text{and} \quad V_2 = \frac{1\text{k}(V_{o2})}{1\text{k}+2\text{k}}$$

$$\frac{1}{2} = \frac{1}{3}V_{o2};\ V_{o2} = \frac{3}{2}\text{ V}$$

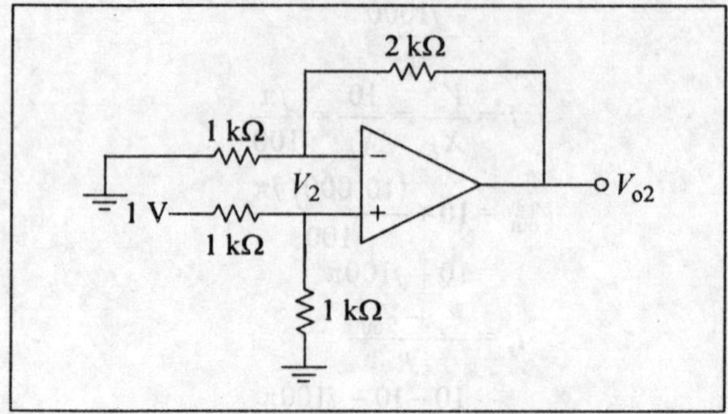

or $$V_{o2} = \left(1 + \frac{2}{1}\right)V_2 = 3 \times \frac{1}{2} = \frac{3}{2}\text{ V}$$

$$V_o = V_{o1} + V_{o2}$$

$$= -2 + \frac{3}{2} = -\frac{1}{2} = -0.5\text{ V}$$

Second Method

KCL at node G,

$$V_G = 0.5\text{ V}$$

$$\left(\frac{0.5-1}{1}\right) = \frac{V_o - 0.5}{2}$$

$$-1 = V_o - 0.5$$

$$V_o = -0.5\text{ V}$$

67. Ans. (c)

Solution:

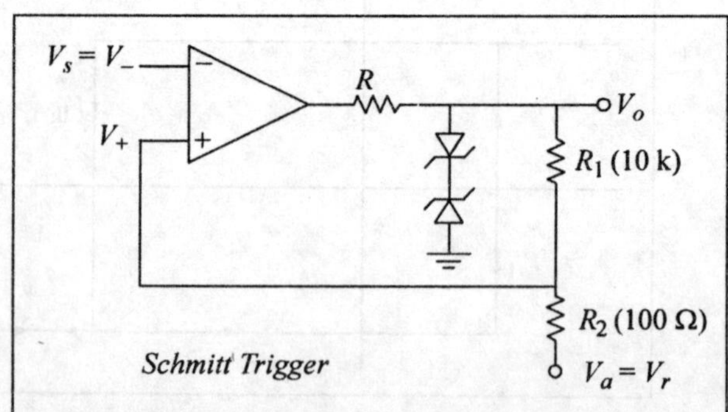

Schmitt Trigger

Schmitt trigger is an electronic circuit that is used to detect whether the voltage is crossed over a given reference level. Output of the Schmitt trigger is rectangular pulse. So the Schmitt trigger operate in non-linear mode.

68. Ans. (c)

69. Ans. (c)

Solution: Since it is inverting amplifier, therefore input impedance $R_i \cong R_1 \cong 1\ \text{k}\Omega$

70. Ans. (d)

Solution:

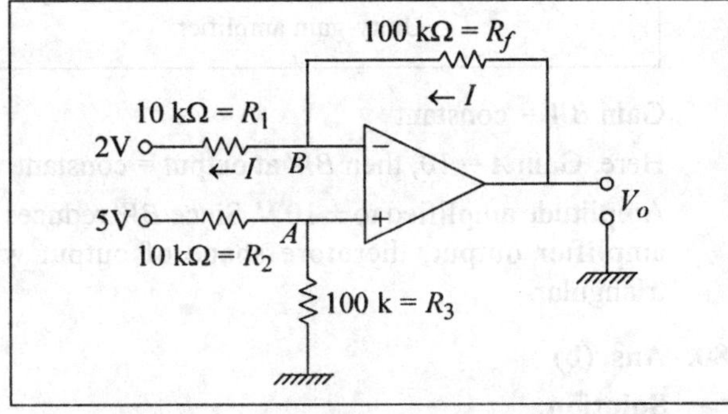

$$V_A = \frac{100\text{ k}}{(100+10)\text{ k}} \times 5 = \frac{500}{110} = \frac{50}{11}\text{ V} = V_B$$

$$I = \frac{\frac{50}{11} - 2}{10\text{ k}} = \frac{50-22}{110\text{ k}} = \frac{28}{110\text{ k}}$$

$$I = \frac{V_O - V_B}{100\text{ k}} \Rightarrow V_o - V_B = I \times 100\text{ k}$$

$$V_o = V_B + I \times 100\text{ k}$$

$$= \frac{50}{11} + \frac{28}{110\text{ k}} \times 100\text{ k} = \frac{50}{11} + \frac{2800}{110}$$

$$= \frac{50+280}{11} = \frac{330}{11} = 30\text{ V}$$

Another Method

Present case is of differential amplifier

Where $\frac{R_f}{R_1} = \frac{R_3}{R_2}$

Here $R_f = R_3 = 100\ \text{k}\Omega$

$R_1 = R_2 = 10\ \text{k}\Omega$

$V_1 = 2\text{V},\ V_2 = 5\text{V}$

Then output $= \frac{-R_f}{R_1}(V_1 - V_2)$

$$= \frac{-100}{10}(2-5) = -10 \times -3 = 30\text{ V}$$

71. Ans. (a)

Solution:

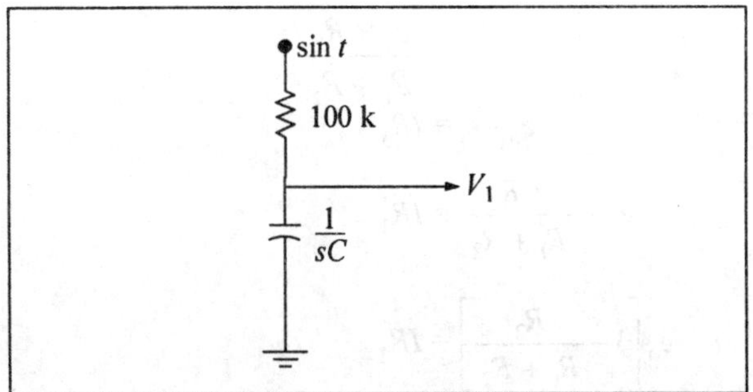

$$V_1 = \frac{1/sC}{\left(100 + \frac{1}{sC}\right)} \times \sin t$$

$$V_o = \left(1 + \frac{R_f}{R_1}\right) V_i$$

$$= (1+1)\, V_f = 2\, V_1$$

$$= \frac{2\sin t}{(100\, sC + 1)}$$

$$= \frac{2\sin t}{100\, k \times 10 \times 10^{-6} s + 1}$$

$$= \frac{2\sin t}{s+1} = \frac{2\sin t}{\sqrt{2}\angle 45^\circ}$$

$$= \sqrt{2}\sin\left(t - \frac{\pi}{4}\right)$$

72. Ans. (c)

Solution: We know that gain of the op-amp is very high

$V_1 = V_2$

So that V_i is resulting as short-circuited

$Z_{in} : R_1 + R_1 = 2R_1$

73. Ans. (a)

74. Ans. (d)

Solution: $V_{o1} = 11$ V

$V_{o2} = -10$ V

$\text{V} = V_{o1} + V_{o2} = 11\text{ V} + (-10\text{ V}) = 1\text{ V}$

75. Ans. (c)

Solution:

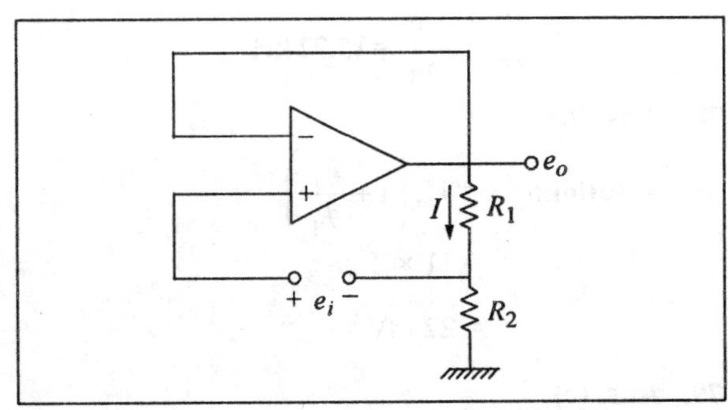

$$e_i = \frac{e_o \times R_2}{R_1 + R_2} \quad ...(1)$$

$$e_o - e_i = IR_1 \quad ...(2)$$

$$e_o - \frac{e_o R_2}{R_1 + R_2} = IR_1$$

$$e_o\left[1 - \frac{R_2}{R_1 + R_2}\right] = IR_1$$

$$e_o\left[\frac{R_1}{R_1 + R_2}\right] = IR_1$$

$$e_o = I(R_1 + R_2) \quad ...(3)$$

From (2) and (3)

$$I(R_1 + R_2) - e_i = IR_1$$

$$e_i = IR_1 + IR_2 - I R_1$$

$$e_i = IR_2$$

$$I = \frac{e_i}{R_2}$$

76. Ans. (d)

Solution: $V_o = V_i \times \left(1 + \frac{R_f}{R_i}\right)$

$$= 10 \text{ mV} \times \left(1 + \frac{10}{1}\right) = 110 \text{ mV}$$

77. Ans. (b)

Solution:

$$\text{Slew rate} = \frac{d}{dt} V_m \sin \omega_m t\Big|_{\max}$$

$$= \frac{1V}{\mu s} = V_m \cdot \omega_m$$

$$= \frac{1V}{\mu s} = V_m \cdot 2\pi f_m$$

$$= 1 \times 2\pi f_m$$

$$1 \times 10^6 = 2\pi f_m$$

$$f_m = \frac{10^6}{2\pi} = 15.92 \text{ kHz}$$

78. Ans. (d)

Solution: $= V_{io}\left(1 + \frac{R_f}{R_1}\right)$

$$= 11 \times 2$$

$$= 22 \text{ mV}$$

79. Ans. (c)

Solution:

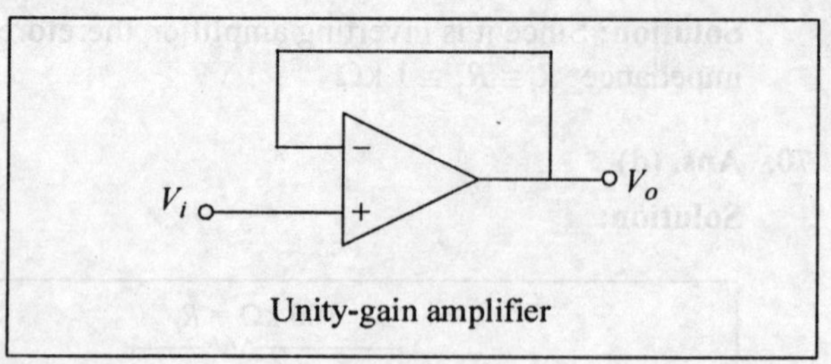

Unity-gain amplifier

Gain BW = constant

Here, Gain A = 10, then BW at output = constant

Amplitude amplified to ± 10 V. Since BW reduces at the amplifier output, therefore shape of output will be triangular.

80. Ans. (b)

Solution:

Using KVL at the inverting terminal, we have

$$\frac{V_i - 0}{R} = (0 - V_o) Cs$$

$$\text{or, } V_o = -\frac{1}{RC}\int_0^t V_i \, dt$$

The given circuit behaves as an integrator, therefore integration of square wave result in triangles.

Second Method

We know output of an integrator

$$v_o(t) = -\frac{1}{RC}\int v_i(t)\, dt$$

Given $v_i(t)$ = square wave with 1 V peak to peak

Given frequency = 1 kHz

From frequency $2a = T$

(Time period of square wave)

$$\text{or } \quad 2a = \frac{1}{f}$$

$$2a = \frac{1}{1}$$

$$a = \frac{1}{2} ms$$

Representing square wave in the form of step function.

$$v_i(t) = s(t)$$

$$= u(t) - u(t-a) + u(t-2a) - u(t-3a)$$

$$R = 1\,M\Omega,\; C = 1\,\mu F = 1 \times 10^{-6} F$$

$$V_o = \frac{1}{RC}\int V_i(t)\, dt$$

$$= \frac{-1}{1 \times 10^6 \times 1 \times 10^{-6}} \left[\int u(t) - u\left(t - \frac{1}{2}\right) + u\left(t - 2 \times \frac{1}{2}\right) \ldots \right]$$

$\because \; a = \frac{1}{2} ms$

Integrate individual term

$$= -\left[+u(t) - \left(t - \frac{1}{2}\right) u\left(t - \frac{1}{2}\right) + (t-1)\, u(t-1) \right]$$

which is the equation of triangular wave

From equation, $v_{p-p} = 1\,V$

81. Ans. (d)

Solution:

$$\text{Slew rate} = \frac{dv}{dt} = \frac{8}{T/2} = \frac{8}{0.2 \times 10^{-6}} = 40\,V/\mu s$$

82. Ans. (b)

Solution:

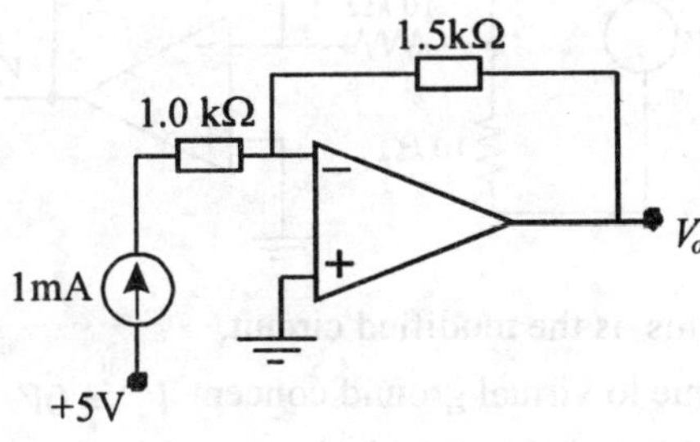

$$V_- = V_+ = 0 \text{ volt}$$

This is due to concept of virtually grounded

Ideal ops-amp $Z_{in} = \infty$, it as negligible current. Whole of the current flow through feedback register.

Then, $V_o = -1.5k \times 1\,mA = -1.5$ volts

83. Ans. (b)

Solution:

Noise margin = 3 V – 1 V = 2 V

84. Ans. (c)

Solution:

$$V_{oo} = V_{io}\left(1 + \frac{R_f}{R_i}\right)$$

$$= 5mV\left(1 + \frac{560}{10}\right)$$

$$= 57 \times 5mV = 285mV$$

85. Ans. (b)

Solution:

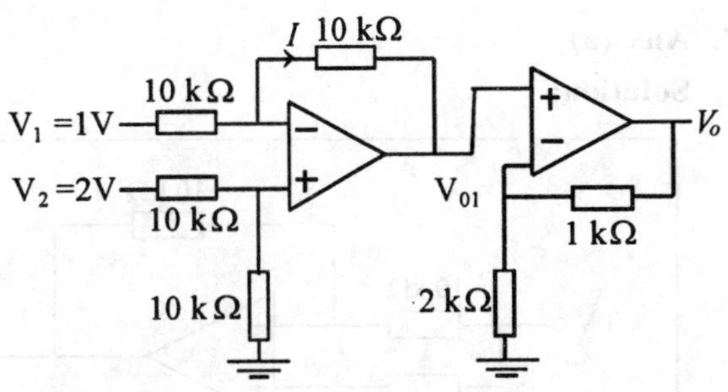

At ops-amp

Then $I = \frac{1-1}{10} = 0$

Using voltage division

$V = V_- = V_+ = 1\,volt$

$\therefore \quad V_{01} = 1\,V$

For ops-amp, by voltage division

$V_{o1} = 1$

$= \frac{2}{3} V_{o2}$

$$V_{02} = \frac{3}{2} = 1.5 \; Volt$$

or $V_0 = 1.5\,volts$

Second Method

The given first amplifier is difference amplifier as $R_1 = R_2$ and $R_f = R_3$ then $V_o = V_2 - V_1$

The non-inverting amplifier with gain $\left(1 + \frac{R_f}{R_1}\right) = \left(1 + \frac{1}{2}\right)$

Hence $V_o = 1.5\,(2-1)$

$= 1.5$ Volt

86. Ans. (c)

Solution:

The given circuit shown in the figure is a stable multi-vibrator so output will be a square wave.

There are two zener diodes due to which output will not be allowed to go beyond 6 V. So output will be ± 6 V.

$\therefore \; V_{ut} = V_{it} = 3V.$

So it will be a square wave of 50% duty cycle and

$T = 2RC\; In2$

$= 2 \times 10 \times 10^{+3} \times 10 \times 10^{-6} ln2$

$$= 2\times10^{-1}\times0.693 \text{ sec}$$

$$= 138.6\, m \triangleq 120\, ns$$

87. Ans. (a)

Solution:

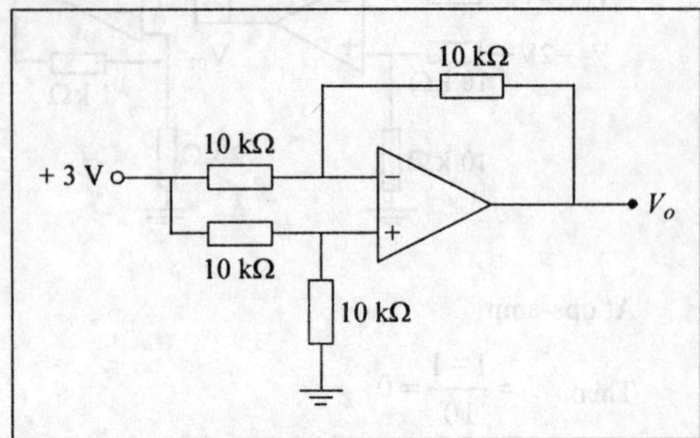

Voltage at non-inverting terminal,

$$V_+ = \frac{3\times10}{10+10} = 1.5 \text{ volts}$$

In ops-amplifier $V_+ = V_-$ due to virtual ground concept

Using KCL at inverting node, we have

$$\frac{3-V_-}{10\,K} = \frac{V_- - V_o}{10\,K}$$

or $\quad V_o = 2V_- - 3$

$$= 2\times1.5 - 3 = 0 \text{ volts}$$

88. Ans. (b)

Solution:

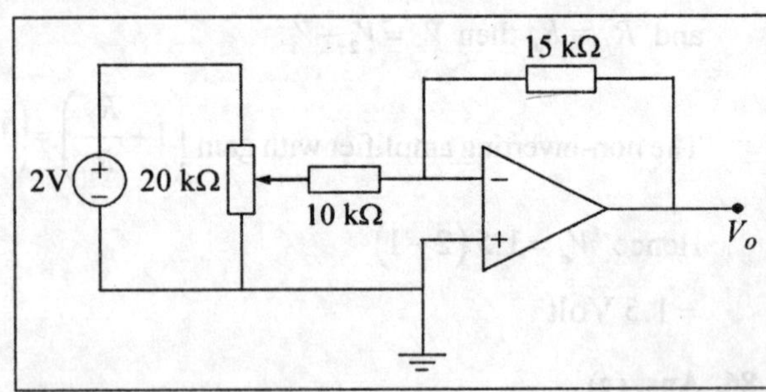

Basically point is to be converted into Thevenin's equivalent network, then op-amplifier application to be applied.

Voltage at the half-way of wiper,

$$V_A = \frac{2\,V}{20\,k\Omega}\times10\,k\Omega \qquad = 1\,\text{V}\Omega$$

$$Z_{eq} = 10 \,||\, 10$$

$$= 5\,k\Omega$$

The output of inverting amplifier

Hence

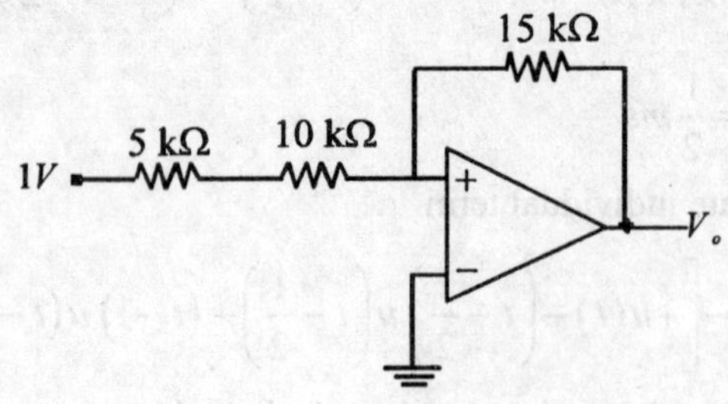

$$V_o = \frac{-R_f}{R}V_i$$

$$V_o = \frac{-15}{5+10}1\,V$$

$$= -1\,V$$

Second Method

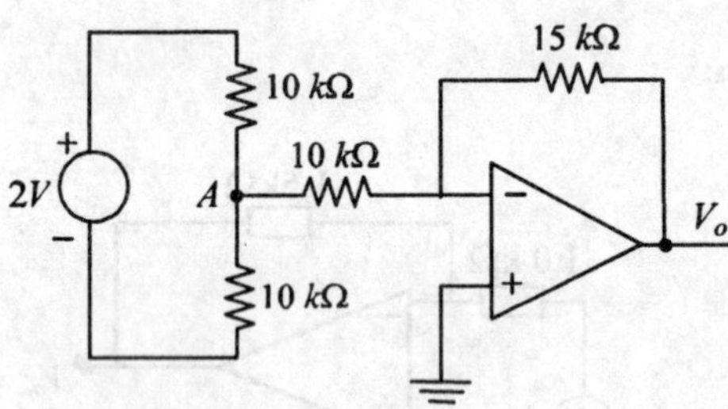

This is the modified circuit.

Due to virtual ground concept $V^- = 0V$

Writing KCL at node A, we get

$$\frac{V_A}{10} + \frac{V_A}{10} + \frac{V_A - 2}{10} = 0$$

$$3V_A - 2 = 0$$

$$V_A = \frac{2}{3}\,Volts$$

O/P from amplifier

$$V_o = \frac{-R_f}{R}V_i$$

$$V_o = \frac{-2}{3}\times\frac{15}{10}$$

$$V_o = -1 \text{ volt}$$

89. Ans. (a)

Solution:

Using Potential Divider

$$V_+ = \frac{20}{50}\times V_o = \frac{2\,V_o}{5}$$

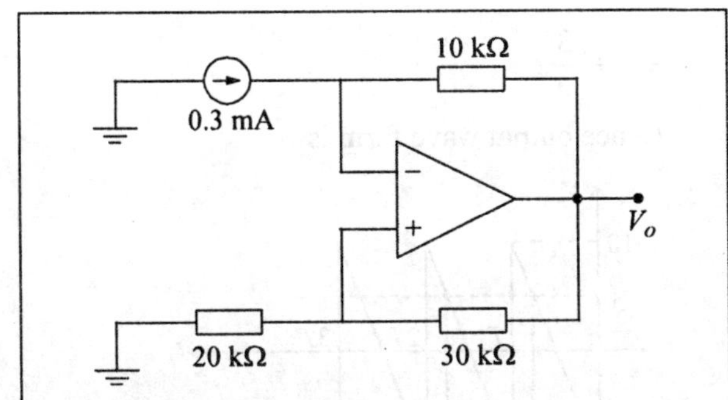

Using virtual ground concept

$V_+ = V_-$

KCL at inverting input

$$\frac{\frac{2V_o}{5} - V_o}{10\,k} = 0.3 \text{ mA}$$

or $$\frac{2V_o - 5V_o}{5} = 3V$$

or $$-3V_o = 15V$$

$$V_o = -5V$$

90. Ans. (a)

Solution:

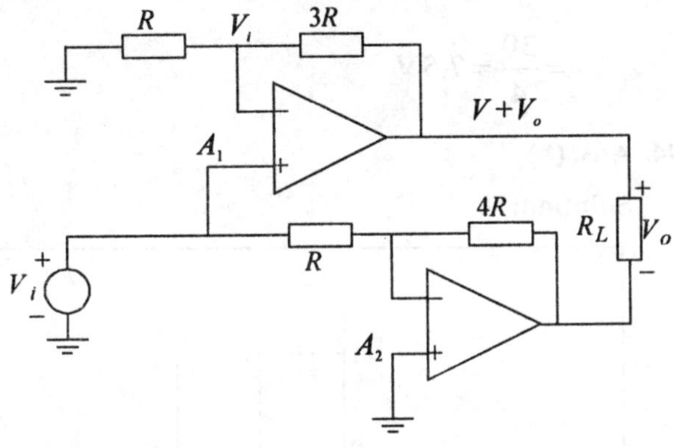

Using KCL at inverting node of A_1

$$\frac{V_i - 0}{R} + \frac{V_i - (V + V_0)}{3R} = 0 \quad \text{...(1)}$$

The same at op-amp 2 gives

$$\frac{0 - V_i}{R} + \frac{0 - V}{4R} = 0 \quad \text{...(2)}$$

From equations (1) and (2)

$$\frac{V_0}{V_i} = 8$$

91. Ans. (b)

Solution:

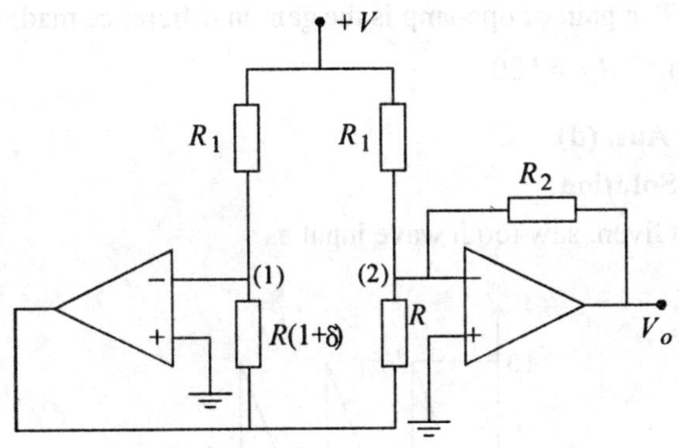

$V_+ = V_- = 0$; due to virtual ground concept

KCL at node (1)

$$\frac{0 - V}{R_1} + \frac{0 - V_1}{R(1+\delta)} = 0 \quad \text{...(1)}$$

$$\frac{V_1}{R} = \frac{-V}{R_1}(1+\delta)$$

KCL at node (2)

$$\frac{0 - V}{R_1} + \frac{0 - V_1}{R} + \frac{0 - V_0}{R_2} = 0 \quad \text{...(2)}$$

From equation (1) and equation (2)

$$V_0 = -R_2\left[\frac{V}{R_1} + \frac{V_1}{R}\right]$$

$$= -R_2\left[\frac{V}{R_1} - \frac{V}{R_1}(1+\delta)\right] = \frac{R_2}{R_1}V\delta$$

92. Ans. (d)

Solution:

For an op-amp the output

$$V_0 = A_d V_d + A_c V_c \quad \text{...(1)}$$

Where, $V_d = V_1 - V_2$

and $V_c = \frac{V_1 + V_2}{2}$

Given output equation

$$V_0 = 100(V_1 - V_2) + 10^{-4}(V_1 + V_2) \quad \text{...(2)}$$

Comparing equations (1) and (2)

$A_d = 100$; $A_c = 2 \times 10^{-4}$

Then $\text{CMRR} = \frac{A_d}{A_c} = \frac{100 \times 10^4}{2} = 50 \times 10^4$

In decibels, $\text{CMRR} = 20 \log 50\times10^4$

$= 0.114\ dB$

The gain of ops-amp is the gain in difference made only i.e. $A_d = 100$

93. Ans. (d)

Solution:

Given, saw tooth wave input as

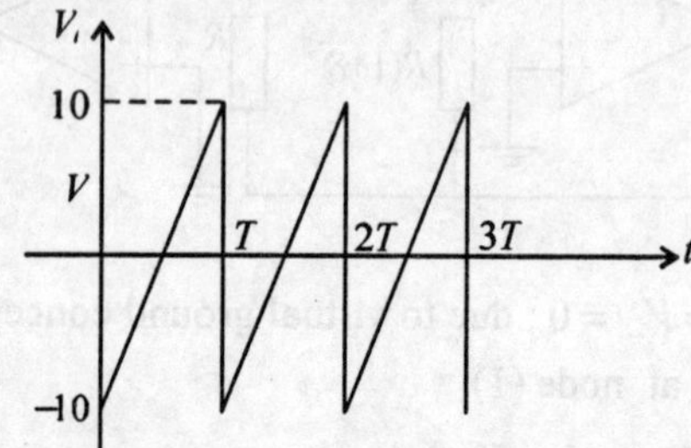

So equation of input waveform for one time-period is

$$y = \frac{20}{T}t - 10$$

The circuit given is as below

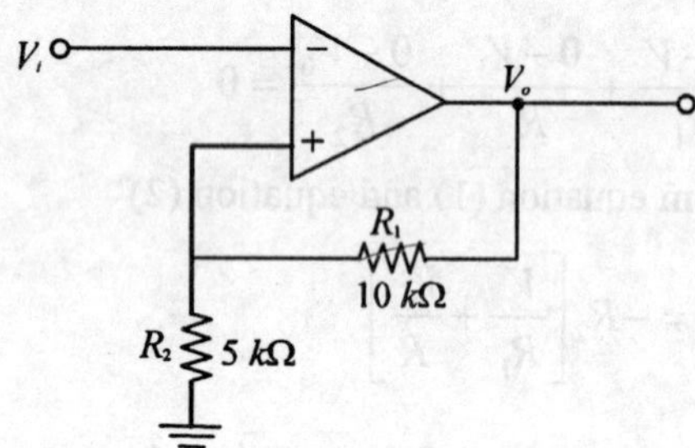

Since output is symmetrical and peak to peak = 30 V

Hence, $+V_{sat} = 15\text{V}$

$-V_{sat} = -15V$

Upper threshold input is $V_{ut} = \frac{R_2}{R_1+R_2}\times V_{sat}$

$$= \frac{5}{15}\times 15 = 5\text{V}$$

Lower threshold input is

$$V_{lt} = \frac{R_2}{R_2+R_1}\times(-V_{sat})$$

$$= -5\text{V}$$

So time at which input is $+5\text{V}$

$$5 = \frac{20}{T}t - 10$$

or $t = \frac{3}{4}T$

Hence output wave form is

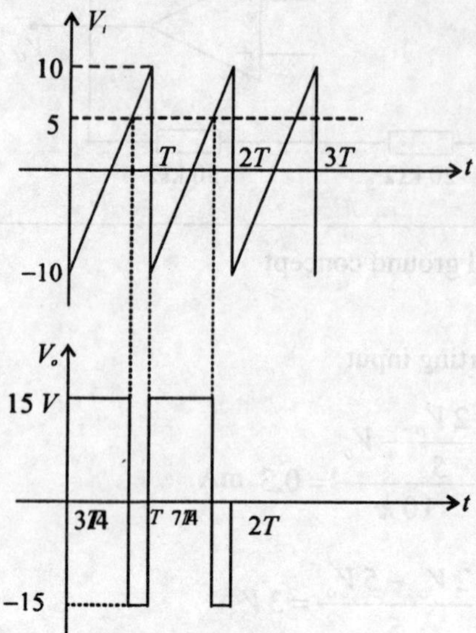

$\therefore$ average value of output

$$V_{avg} = \frac{15\times 3T/4 + (-15)\times T/4}{T}$$

$$= \frac{45}{4} - \frac{15}{4}$$

$$= \frac{30}{4} = 7.5\text{V}$$

94. Ans. (b)

Solution:

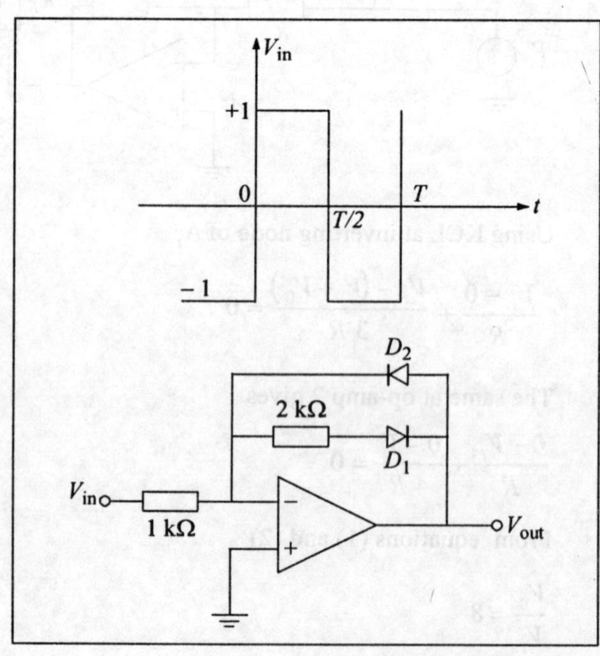

When V_{in} is $+Ve$

Then current $1-1\,k\Omega(I)=0$

$I=\frac{1}{10^3}=1\,mA$

Apply KVL

The D_2 will be open circuit

The circuit becomes

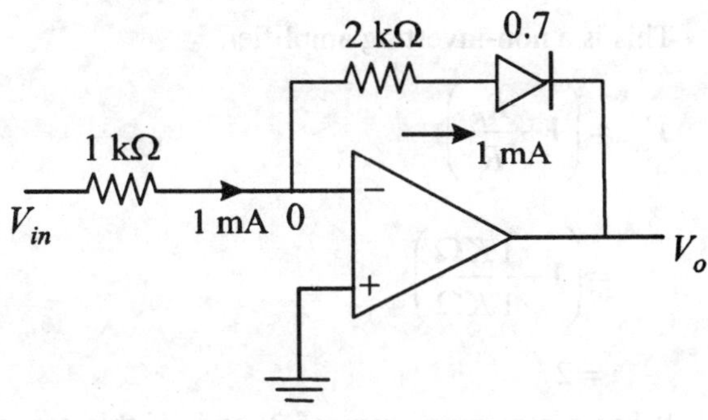

$0-2\times10^3\times10^{-3}-0.7=V_o$

$V_o=-2.7$

When $V_1=-1$

Then D_1 will be open circuit and D_2 will be short circuit then

$0+0.7=V_o$

$V_o=+0.7$

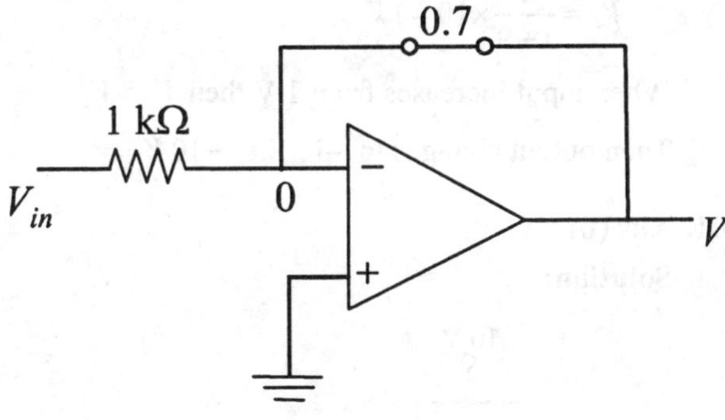

$$V_{o(avg)}=\frac{0.7+(-2.7)}{2}$$

$$=-1\,V$$

95. Ans. (c)

Solution:

When op-amp is used in negative feedback then voltage difference between the input terminals is treated as zero.

96. Ans. (b)

Solution:

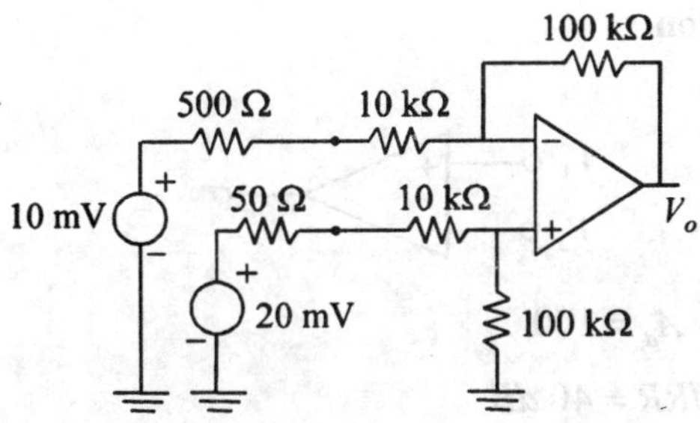

KCL at V_+ node

$$\frac{V_+}{100}+\frac{V_+-20}{10.05}=0$$

or $$\frac{V_+}{100}+\frac{V_+}{10.05}-\frac{20}{10.05}=0$$

or, $$\frac{V_+}{100}+\frac{V_+}{10.05}=1.99$$

or, $(0.01+0.0995)V_+=1.99$ ...(1)

KCL at V_- node

and $$\frac{V_--V_0}{100}+\frac{V_--10}{10.5}=0$$ ...(2)

For ideal op-amp $V_+=V_-$ ...(3)

From equations (1), (2) and (3)

$\Rightarrow$ $0.109V_+=1.99$

$V_+=18.25$

or $$\frac{18.25-V_0}{100}+\frac{18.25-10}{10.5}=0$$

or, $V_o=100(0.1825+0.7857)$

$\Rightarrow$ $=96.82\,mA$

97. Ans. (b)

Solution:

Because it is a second order low pass filter, therefore it has a – 40 dB decade. Also it has 3 dD cut-off frequency

of $$f=\frac{1}{2\pi\sqrt{R_1C_1R_2C_2}}$$

$$f=\frac{1}{2\pi\times22\times10^3\times330\times10^{-12}}$$

$=21.9\,kHz$

It has decade from 0 dB to –3dB in mid frequency range, so answer is b

98. Ans. (b)

Solution:

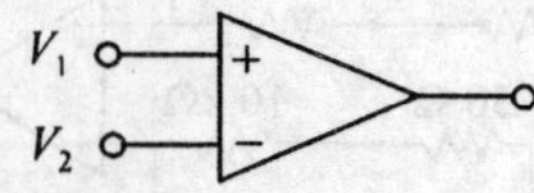

$$A_d = 100$$

$$CMRR = 40\ dB$$

$$20 \log K = 40$$

$$\log K = 2$$

$$k = 10^2 = 100$$

$$\text{CMRR} = 100$$

$$\text{CMRR} = \frac{A_d}{A_c}$$

$$100 = \frac{100}{A_c}$$

$$A_d = 100$$

$$A_c = 1$$

$$V_1 = 0.55$$

$$V_2 = 0.45$$

$$V_{ic} = \frac{0.55 + 0.45}{2} = 0.5$$

$$V_{id} = 55 - 45 = 0.1$$

$$V_o = A_d\ V_{id} + A_c\ V_{ic}$$

$$= [100][0.1] + [1][0.5]$$

$$= 10 + 0.5 = 10.5$$

99. Ans (d)

Solution:

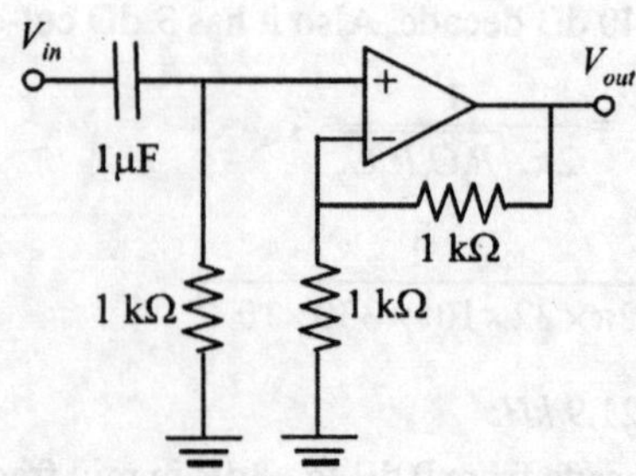

Arranging the circuit,

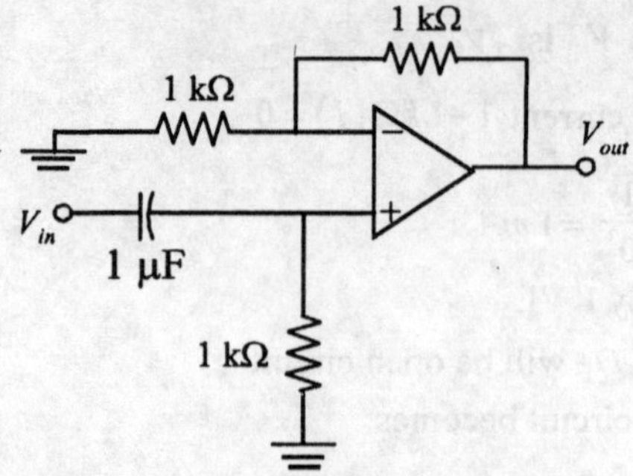

This is a non-inverting amplifier.

$$V_{out} = \left(1 + \frac{R_F}{R_1}\right)$$

$$= \left(1 + \frac{1\ K\Omega}{1\ K\Omega}\right)$$

$$= 2$$

It has a maximum gain of 2. and if this capacitor is connected to V_{in} so it will only pass the signal of high frequency. So it is α H.P.F. with α a gain of 2.

100. Ans (d)

Solution:

The given circuit is Schmitt trigger

Since initially

$V_0 = V_{sat} = 10V$

Therefore, voltage at positive terminal of ops-amplifier

$$V_+ = \frac{1}{1+9} \times 10 = 1\ V$$

When input increases from 1 V then $V_- > V_+$

Then output changes to $-V_{sat}$ i.e. $-10\ V$

101. Ans (b)

Solution:

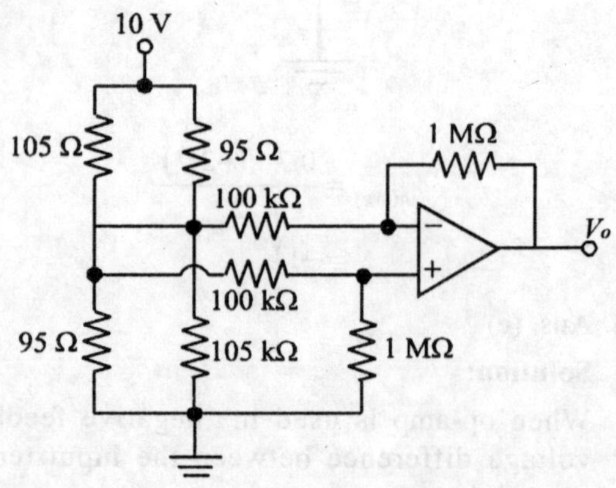

Breaking figure in two parts as to obtain input voltage applied to two terminal

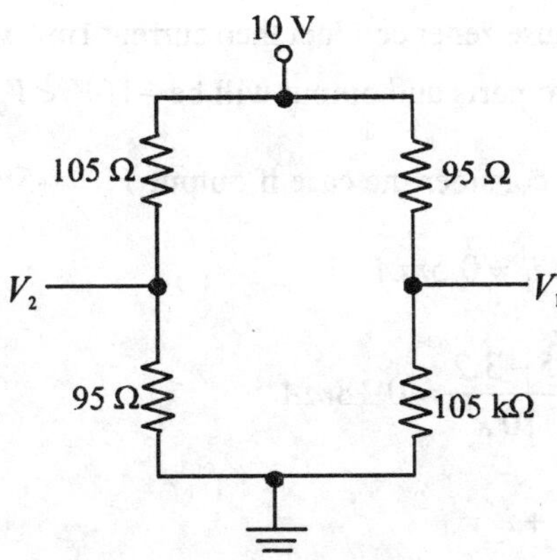

Figure: (a) Potential divider section

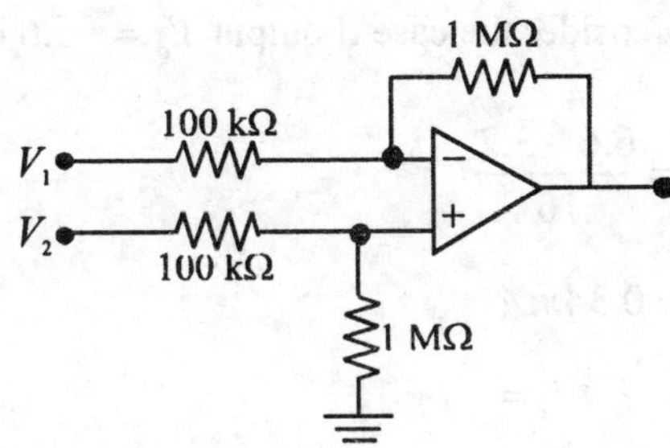

Figure: (b) Diference amplifier section

From figure (a) apply voltage divider rule

$$V_2 = \frac{95}{200} \times 10$$

$$= 4.75V$$

and $V_1 = \frac{105}{200} \times 10$

$$= 5.25V$$

$\therefore$ Since $R_1 = R_2$ and $R_f = R_3$

Due to summetry,

$$V_0 = A_V(V_1 - V_2)$$

$$A_V = \frac{R_f}{R_1} = -\frac{10^6}{10^5} = -10$$

therefore $V_0 = -10(5.25 - 4.75) = -5V$

102. Ans (a)

Solution:

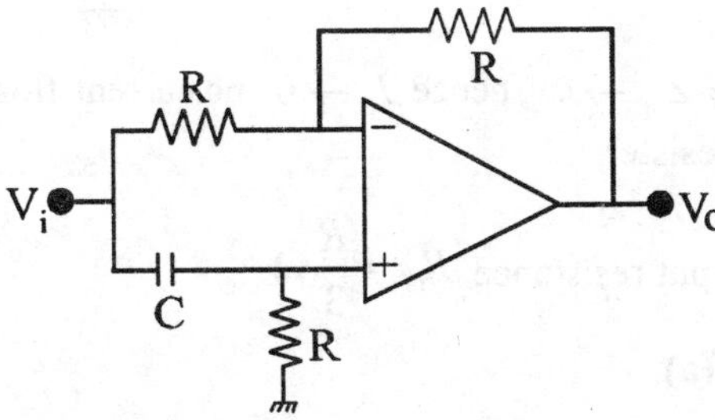

This can be solved as

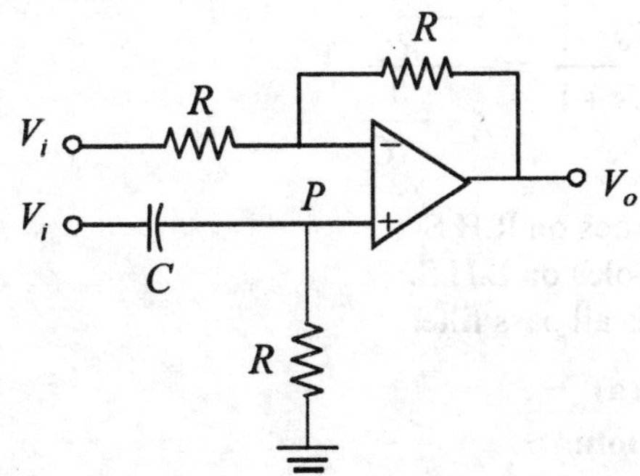

According to superposition theorem. if input voltage of inverting amplifier is zero,

Then voltage at non-inverting amplifier

$$V_P = \frac{V_i \times R}{R + \frac{1}{Cs}}$$

$$V_o = \left[1 + \frac{R_F}{R_1}\right] V_P$$

$$= \left[1 + \frac{R}{R}\right]\left[\frac{V_i \times R}{R + \frac{1}{Cs}}\right]$$

$$= \frac{2V_i \times R}{R + \frac{1}{CS}}$$

Now put $V_i = 0$ for non-inverting amplifier

Now gain of inverting terminal

$$\frac{V_o}{V_i} = \frac{-R}{R} = -1$$

$$V_o = -V_i$$

$$V_o = \left(\frac{2R}{R + \frac{1}{sC}}\right) V_i - V_i$$

$$\frac{V_o}{V_i} = \left[\frac{2R}{R + \frac{1}{Cs}} - 1\right]$$

$$\frac{V_o}{V_i} = \frac{2RCs}{RCs + 1} - 1$$

$$\frac{V_o}{V_i} = \frac{2RCs - (RCs + 1)}{RCs + 1}$$

$$=\frac{RCs-1}{RCs+1}=\frac{s-\frac{1}{RC}}{s+\frac{1}{RC}}$$

So zeroes on R.H.S
and poles on L.H.S.
So it is all pass filter

103. Ans. (a)

Solution:

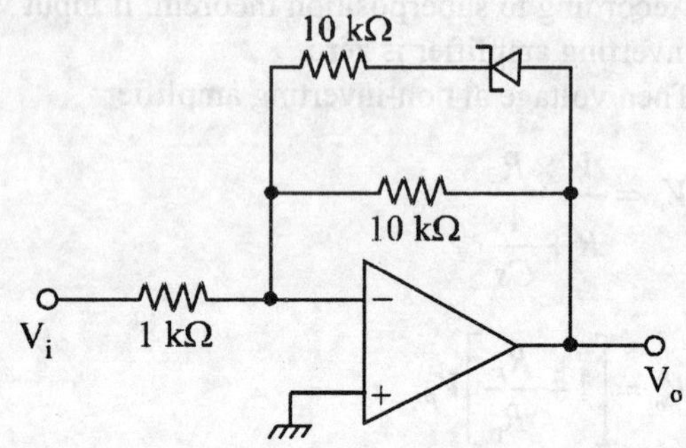

For $V_i \leq 3.2$ V, zener diode does not conduct, i.e. open circuit . The equivalent circuit will be as shown.

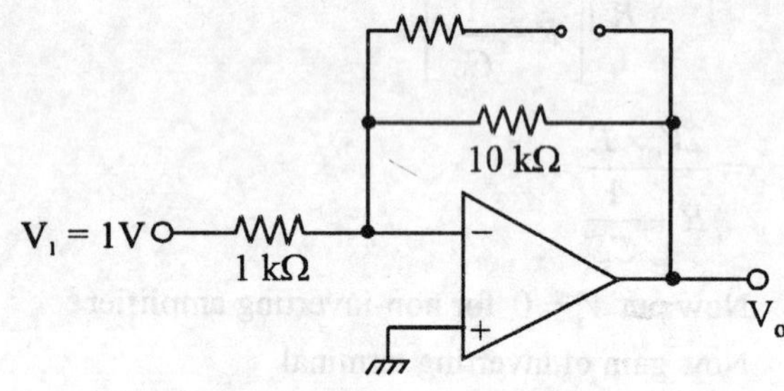

$$\frac{V_i-0}{1\,k\Omega}=\frac{0-V_o}{10\,k\Omega}$$

or $$\frac{V_o}{V_i}=-10\text{ V}$$

or $$V_o=-10\,V$$

and $$I=1mA$$

Now since $V_{OC}=10V>V_z$

Hence zener diode will conduct and circuit will become as

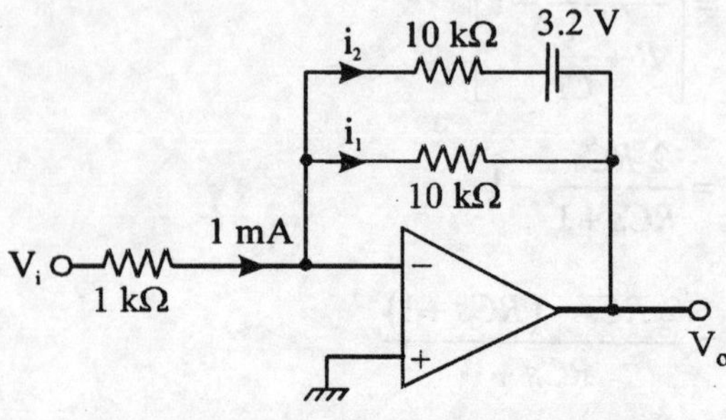

Because zener conduct then current 1mA will be divided in two parts and output will be $-10V<V_0<-3.2V$

Now consider the case if output $V_0=-5V$

then, $i_1=0.5mA$

$$i_2=\frac{5-3.2}{10\,k}=0.18mA$$

so, $i_1+i_2\neq i$

Hence, this is not possible

Now cosider the case if output $V_o=-6.6\,V$

$$i_2=\frac{6.6-3.2}{10}$$

$$=0.34mA$$

here, $i_1+i_2=i$

Hence this option is correct,

104. Ans (a)

Solution:

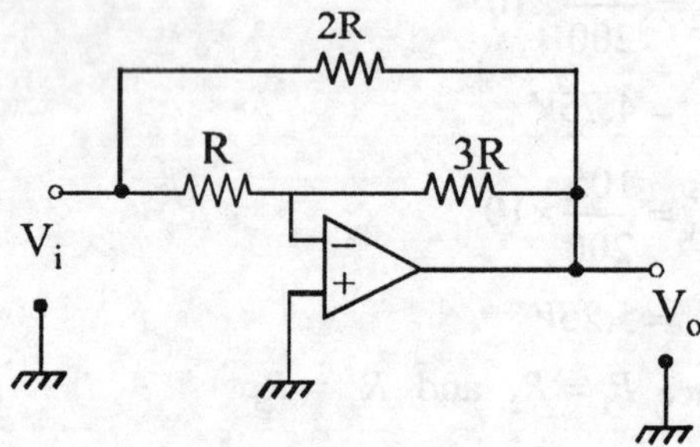

Now convert Dalta network to start network and circuit becomes as given below.

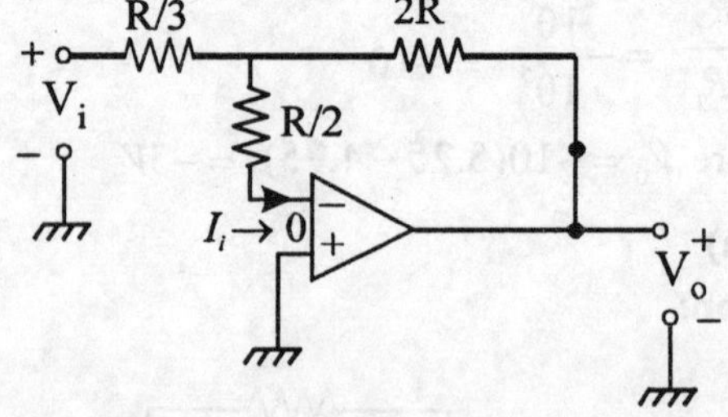

Since $Z_{in}\to\infty$, hence $I_i\to 0$, no current flow through R/2 resister,

$\therefore$ input resistance $R_{in}=\frac{R}{3}\Omega$

105. Ans (a)

Solution:

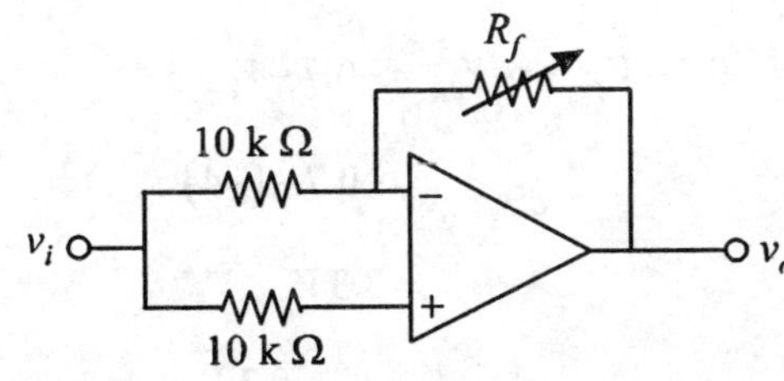

$$G = \frac{V_o}{V_i} = \left[1+\frac{R_f}{R_1}\right] + \left[\frac{-R_f}{R_1}\right]$$

$$\frac{v_o}{v_i} = 1$$

So gain remain always equal to 1

106. Ans. (d)

Solution:

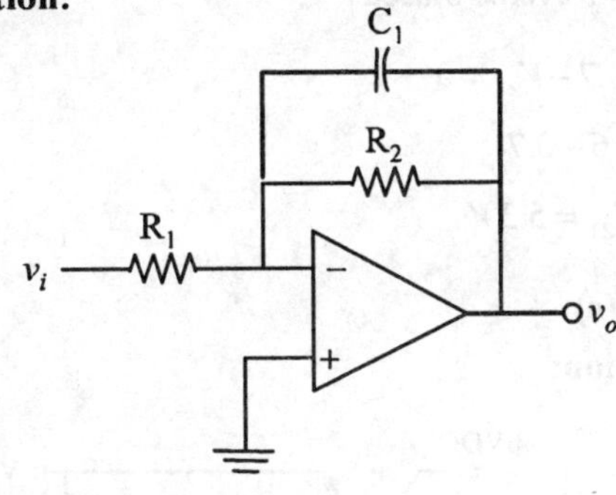

$R_1 = 15.9\ k\Omega$

$R_2 = 159\ k\Omega$

$C_1 - 1nF$

For D.C. supply, capacitor will be open cut

$$\frac{V_o}{V_i} = \left|\frac{-R_2}{R_1}\right| = \frac{159}{15.9} \Rightarrow 10$$

Gain in dB = 20 log10

$= 20 \times 1 = 20dB$

The 3 dB cut-off frequency

$$f = \frac{1}{2 \times \pi \times R_2 \times C_1}$$

$$\frac{1}{2 \times 3.14 \times 159\ k\Omega \times 1 \times 10^{-9}}$$

$\Rightarrow 1\ kHz$

107. Ans (c)

Solution:

Input impedance seen looking into the terminal V_1 and V_2 w.r.t. ground

Apply superposition theorem

Put $V_2 = 0$

So voltage at $P = 0$

By virtual ground concept

Voltage at $Q = 0$

So input imedance seen from V_1 w.r.t. ground

$R_1 = 47\ k\Omega$

When $V_1 = 0$

Then input impedance from V_2 w.r.t ground

$= R_1 + R_2$

$= 47 + 470 = 517\ k\Omega$

108. Ans (a)

Solution:

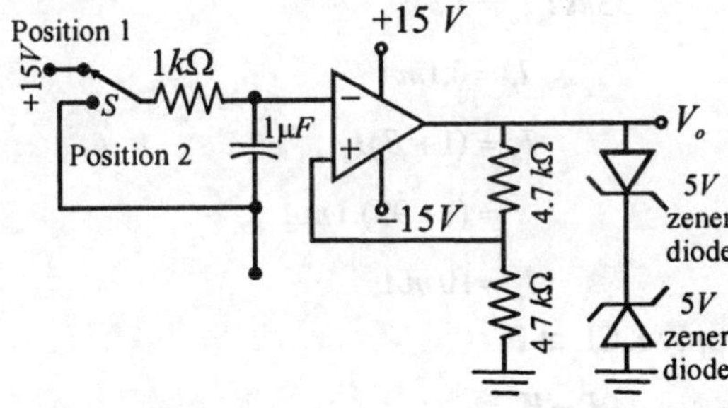

Switch at position 1

Hence, initially, $V_c = 15V$

One diode is forward biased and other is reverse biased

and $V_o = +5V$

therefore, $V_+ = +\frac{5}{2} = +2.5V$

so transition occurs when V_c reaches 2.5V after changing the position of switch to position 2.

$$I = \frac{V_+}{1k\Omega} = V_+ mA = 2.5mA$$

$$C\frac{dV_c}{dt} = I$$

$$\text{or } 10^{-6}\left[\frac{15-2.5}{dt}\right] = 2.5 \times 10^{-3}$$

$$\text{or} \quad dt = 10^{-3} \times \frac{12.5}{2.5}$$

$$= 5 \times 10^{-3} = 5m\sec$$

109. Ans. (c)

Solution:

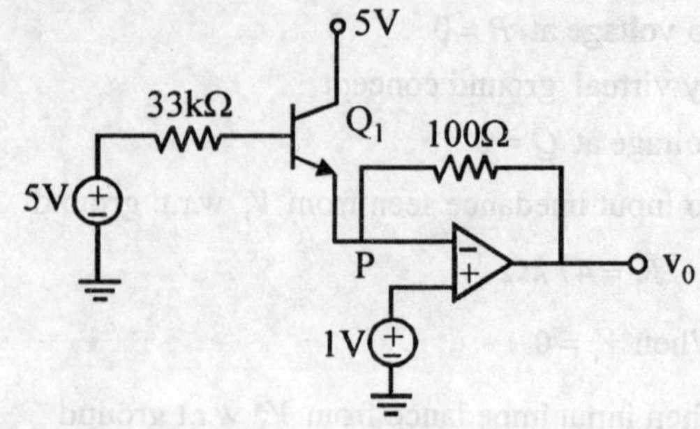

$\beta = 99$

$V_{\beta E} = .7$

According to virtual ground concept voltage at P = 1 volt

Apply KVL in 5V source to 1V source.

$5 - 33k\Omega I_b - 0.7 = 1V$

$33k\Omega\ I_b = 3.3V$

$I_b = 0.1mA$

$I_e = (1+\beta)I_b$

$= (1+99).1mA$

$I_e = 10\,mA$

Apply KCL at P

$\frac{P - V_0}{100} \Rightarrow 10\ mA$

$1 - V_0 = 10^3 m = 1$

$V_0 = 0V$

110. Ans. (d)

Solution:

When input voltage $V_i > 0$ (During +ve cycle)

$Z_1 \rightarrow$ Forward biased

$Z_2 \rightarrow$ Reverse biased

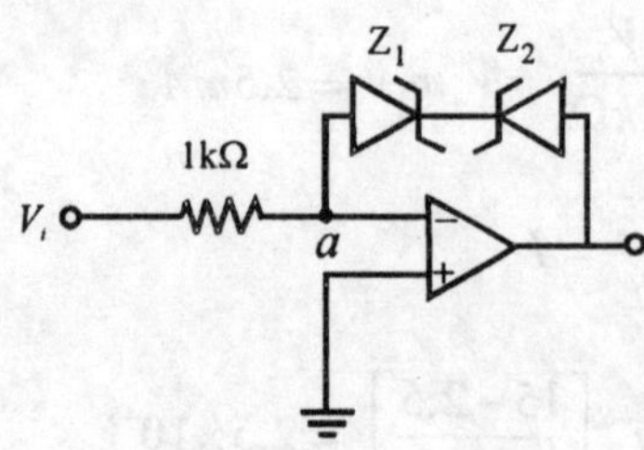

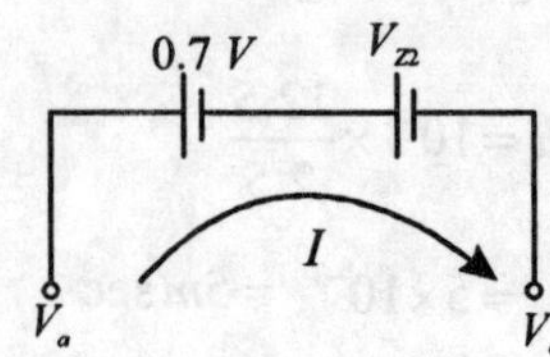

$V_a - 0.7 - V_{Z2} - V_o = 0$

$V_o = -0.7 - V_{Z2}$ or $V_{Z2} = -0.7 - V_o$

$= -0.7 - (-4)$

$= 3.3\,V$

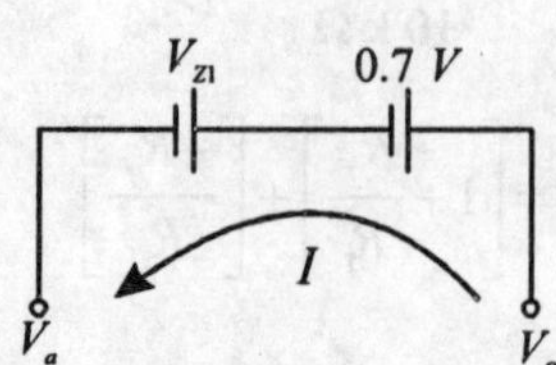

When input voltage $V_i = 0$ (During −ve cycle)

$\left.\begin{array}{l} Z_1 \rightarrow \text{Reverse biased} \\ Z_2 \rightarrow \text{Reverse biased} \end{array}\right\}$ then $V_o = 6\,V$

$V_o - 0.7 - V_{Z1} = 0$

$V_{Z1} = 6 - 0.7$

or, $V_{Z1} = 5.3\,V$

111. Ans. (c)

Solution:

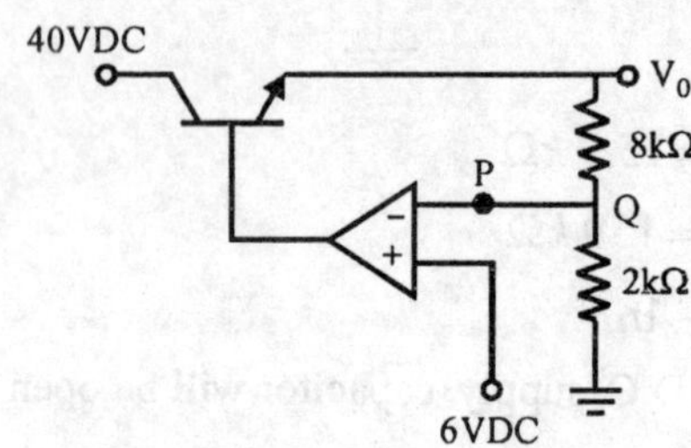

$V_P = 6$ V by using virtual ground concept

$V_Q = 6$ V

$$V_Q = \frac{2k\Omega \times V_o}{2k\Omega + 8k\Omega}$$

$$6 = \frac{2K\Omega \times V_o}{10\,k\Omega}$$

$V_0 = 30\,V$

112. Ans. (c)

Solution:

When $f = 0$, $X_c = \frac{1}{2\pi f_c} = \infty$

All the capacitors are replaced by open circuit then equivalent circuit is

No signal transmitted to output terminals

2. When $f = \infty$ $X_c = 0$

All capacitors are replaced by short circuit. Then the equivalent circuit is

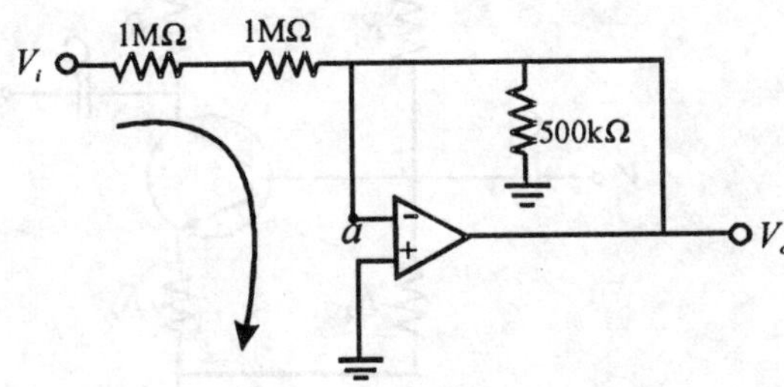

or Band pass filfter

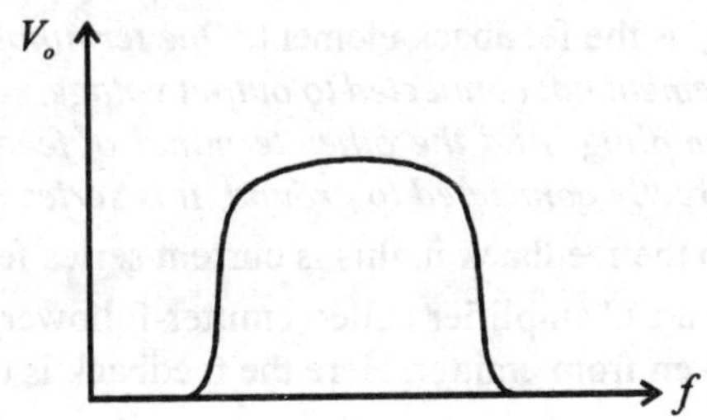

113. Ans. (b)

Solution:

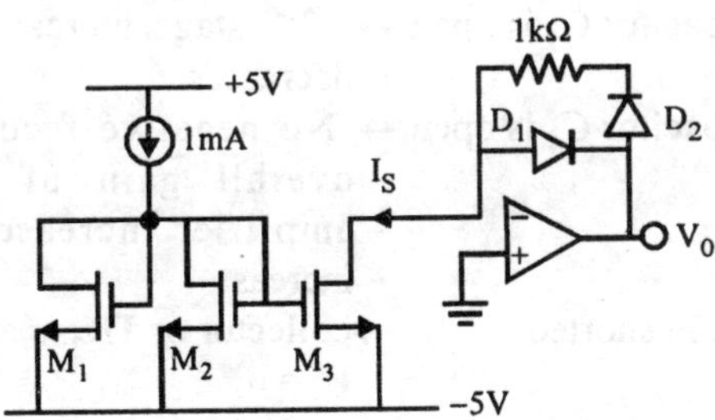

In the given cirucit diagram, the N-channel enhancement mode MOSFETs M_1 and M_2 are matched.

Hence, due to symmetry

$I_{D_1} = I_{D_2} = 0.5\ mA$

MOSFETs M_2 and M_3 are the working as current mirror to each other

$\therefore\ I_{D_2} = I_{D_3} = 0.5\ mA$

Hence, $I_{D_3} = I_s = 0.5\ mA$

114. Ans. (a)

Solution:

Now as observed from the direction of I_s, D_1 is OFF and D_2 is ON. Hence for present case the circuit becomes

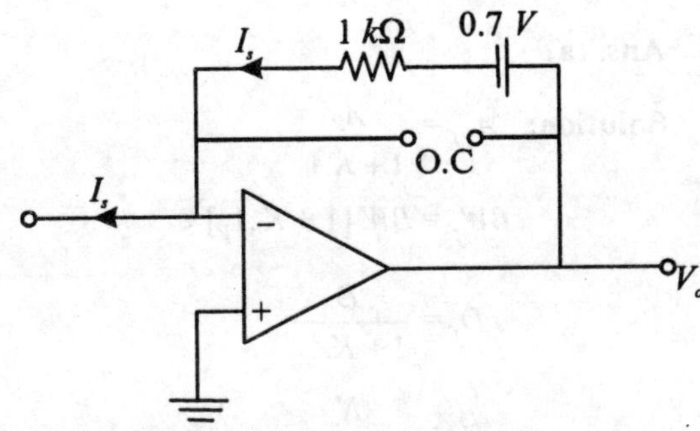

Apply KVL in output to input circuit of op-amplifier

$V_o = 0.7\ V + I_s \times 10^3$

$= 0.7 + 0.5 \times 10^{-3} \times 10^3$

$= 1.2\ V$

115. Ans.(b)

Solution:

Gain of simple non-inverting amplifier

$$A_{vf} = \frac{V_o}{V_{in}} = \frac{Z_f}{Z} = \frac{R_2}{R_1 + \frac{1}{sC}}$$

$$= -\frac{sCR_2}{sCR_1 + 1} = \frac{sR_2 / R_1}{s + \frac{1}{R_1 C}}$$

Thus, it is HPF transfer function with cut-off frequency

$$= \frac{1}{R_1 C}$$

7.6 Feedback Amplifier

1. Ans. (c)

Solution: Non-inverting amplifier

Gain 1	Gain 20
$\beta = 1$	$\beta = \frac{1}{20}$
	less feedback

($G \times BW =$ Constant)

$1 \times BW_1 = 20\ BW_2 = K$

$BW_2 = \frac{BW_1}{20} =$ Less the BW_1

2. **Ans. (b)**

3. **Ans. (a)**

Solution: In a shunt-shunt negative feedback amplifier, i.e. voltage shunt amplifier input impedance and output impedance decrease.

4. **Ans. (a)**

Solution: $A_{Vf} = \dfrac{A_V}{1 + KA_V}$

$BW_f = BW[1 + KA_V]$

$D_f = \dfrac{D}{1 + KA_V}$

$N_f = \dfrac{N}{1 + KA_V}$

Negative feedback in an amplifier reduces gain stabilizing it.

5. **Ans. (c)**

Solution: For voltage series input impedance increases and output impedance decreases by

$$Z_{inf} = Z_{in}(1 + KA_V)$$
$$= 1\text{ K}[1 + 100 \times 0.99] = 1\text{ K} \times 100$$
$$= 100\text{ k}\Omega$$

$$Z_{out\,f} = \frac{Z_{out}}{(1 + KA_v)}$$
$$= \frac{100\,\Omega}{(1 + 100 \times 0.99)} = \frac{100}{1 + 99}$$
$$= \frac{100}{100} = 1\,\Omega$$

6. **Ans. (c)**

Solution: In voltage series amplifier

$R_{if} = R_i(1 + KA) \rightarrow$ increases

$R_{of} = \dfrac{R_o}{(1 + KA)} \rightarrow$ decreases

7. **Ans. (b)**

Solution: The input resistance R_i is effective value of parallel combination of 10 k and input impedance of amplifier, i.e.

$$R_i = 10\text{ k} \parallel \infty = 10\text{ k}\Omega$$

8. **Ans. (c)**

Solution: For voltage series feedback, the input impedance increase by factor of $(1 + AB)$ and output impedance decrease by a factor of $\dfrac{1}{(1 + AB)}$

Where, A = Gain of amplifier

B = feedback factor

9. **Ans. (a)**

10. **Ans. (b)**

Solution: There are two circuits with emitter resistance unbypassed, i.e. CE and CC in *CE* amplifier, the output is taken from collector.

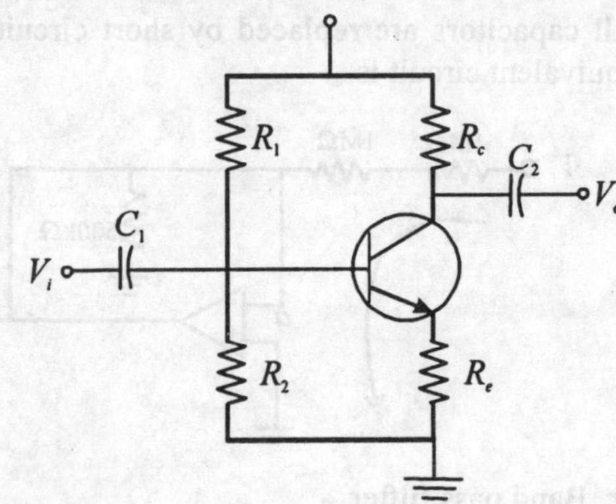

R_e is the feedback element. *One terminal of feedback element not connected to output votage, so this is curent sampling. And the other terminal of feedback element directly connected to ground, it is series mixing.*

So the feedback in this is current series feedback.

In a *CC* amplifier called emitter-follower, the output is taken from emitter. Here the feedback is current-series.

11. **Ans. (c)**

Solution:

(a) Capacitor C_1 is open → No input connected to circuit $V_o = 0$

(b) Capacitor C_3 is open → 2nd stage increases, V_o decreases

(c) Capacitor C_4 is open → No negative feedback, overall gain of the amplifier increases, V_o increase

(d) Rc_2 in shorted →collector or TR_2 is at V_{CC}, $V_o = 0$

12. **Ans. (c)**

Solution:

Closed loop gain, $A_{CL} = \dfrac{A_{oL}}{1 + KA_o}$

Given, $A_{CL} = 100$, and $A_{oL} = 10^5$

$\therefore \quad 1 + KA_o = 10^3$

So, upper cut-off frequency $f_{HF} = f_H(1 + KA_o)$

$= 10 \times 10^3 = 10^4$ Hz.

13. Ans (a)

Solution:

Gain without error = 10

Gain with +10% error = 10.1

Gain with −10% error = 9.9

Hence over all system gain it is 10 ±1%

Second Method

$$G_f = \frac{G}{1+GH} = \frac{100}{1+100\times\frac{9}{100}}$$

$$= \frac{100}{1+9} = \frac{100}{10} = 10$$

14. Ans. (d)

Solution: To realize a phase shift oscillator, collector of the common emitter amplifier connected the R-C phase shift network.

Second

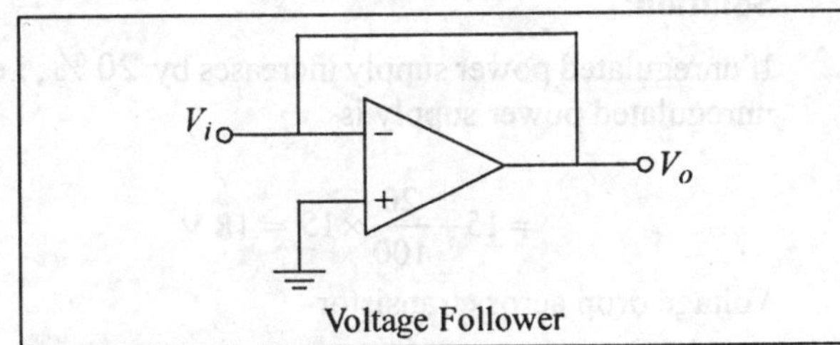

Voltage Follower

$V_o = -V_i$, hence 180° phase shift

Phase shift network gives 180° then total become 360°

15. Ans. (d)

Solution:

It is negative voltage feedback

7.7 Oscillators

1. Ans. (a)

Solution: $\omega = \frac{1}{\sqrt{(LC)}} = \frac{1}{\sqrt{(10\times10^{-3}\times0.01\times10^{-6})}}$

$= 10^5$ rad/sec

For oscillation, $\beta = 1$

or $\left(\frac{1+100\text{ k}}{5\text{k}}\right)\left(\frac{1000}{1000+R}\right) = 1$

$\therefore$ R = 20000 Ω = 2 × 10⁴ Ω

2. Ans. (a)

Solution: There are two types of feedback applied to input. Both are to be considered.

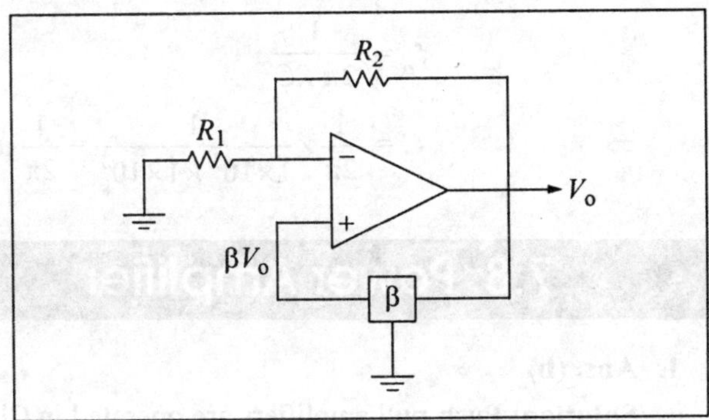

KCL at Node 1

$$\frac{\beta V_o - 0}{R_1} + \left(\frac{\beta V_o - V_o}{R_2}\right) = 0$$

$$\beta\left(\frac{1}{R_1}+\frac{1}{R_2}\right) = \frac{1}{R_2}$$

or $\beta\left(\frac{R_1+R_2}{R_1}\right) = 1$

or $1+\frac{R_1}{R_1} = \frac{1}{\beta} = 6$

or $\frac{R_2}{R_1} = 5$

$\Rightarrow$ $R_2 = 5\,R_1$

3. Ans. (a)

Solution: $F = -\frac{1}{2\pi RC\sqrt{6}}$

This is *RC* phase shift *AF* oscillator.

4. Ans. (a)

Solution: The given figure represent Wein Bridge oscillator.

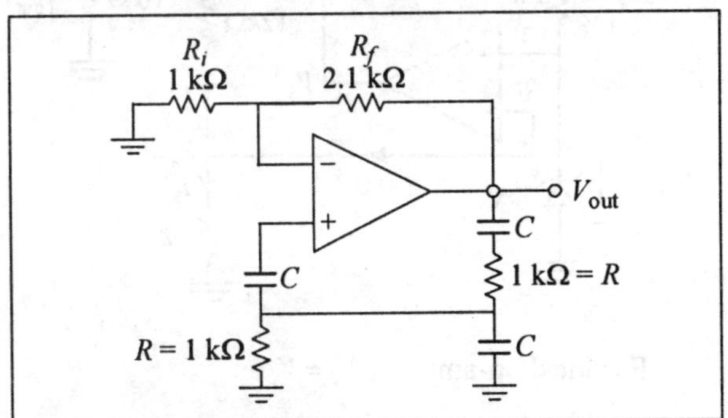

For Wein Bridge oscillator

Frequency of oscillation $f_o = \frac{1}{2\pi\,RC}$

The ratio of $\frac{R_f}{R_i} \simeq 2$

Gain of amplifier $A_V \geq 3$

$$f_o = \frac{1}{2\pi RC}$$

$$\Rightarrow \quad C = \frac{1}{2\pi} \times \frac{1}{1\times10^3 \times 1\times10^3} = \frac{1}{2\pi}\,\mu F$$

7.8 Power Amplifier

1. Ans. (b)

Solution: Push-pull amplifiers are operated in Class *B*. In Class *B*, there is no quiescent (dc) current as biasing is at cut-off and the output current does not contain even harmonic.

$$i_L = 2(A_1 \cos \omega t + A_3 \cos 3\omega t + \ldots)$$

2. Ans. (c)

Solution: In transformer coupled amplifier, efficiency is 50 %

$P_{ac} = 10$ watts $\quad \therefore P_{dc} = 20$ watts.

For Class *A*, $\eta = 50$

$$\therefore P_{D\max} \geq \left[\frac{10\times100}{50}\right] = 20 \text{ watts.}$$

3. Ans. (b)

7.9 Power Supplies

1. Ans. (c)

Solution:

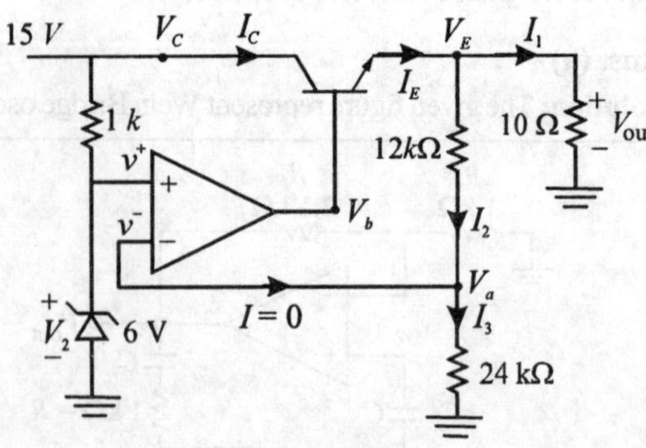

For ideal op-amp $\quad V_- = V_+$

Here $\quad V_+ = V_z = 6V$

$\therefore \quad V_- = 6V = V_a$

Applying voltage divider rule, we have

$$V_a = \frac{24}{24+12} \times V_{out}$$

or $$V_{out} = \frac{24+12}{24} \times V_a$$

$$= \frac{36}{24} \times 6$$

$$= 9\,V$$

Collector current of BJT $I_C \cong I_E = I_1 + I_2$

$$= \frac{9.0}{10} + \frac{9.0}{36\,k} \approx 0.9\,A$$

Voltage across BJT, $V_{CE} = 15 - V_{out}$

$$= 15 - 9 = 6\,V$$

Power loss in BJT, $= V_{CE} I_C \approx V_{CE} I_E$

$$= 6.0 \times 0.90 = 5.4\,W$$

2. Ans. (b)

Solution:

If unregulated power supply increases by 20 %, i.e. new unregulated power supply is

$$= 15 + \frac{20}{100} \times 15 = 18 \text{ V}$$

Voltage drop across transistor

$$V_{CE} = V_C - V_E$$

$$= 18 - 9 = 9V$$

Power loss

$$P_D = V_{CE} \times I_C$$

Power loss $\quad = 9 \times 0.9 = 8.1$ W

$$\% \text{ increase in power loss} = \frac{8.1 - 5.4}{5.4} = 50\%$$

3. Ans. (c)

Solution:

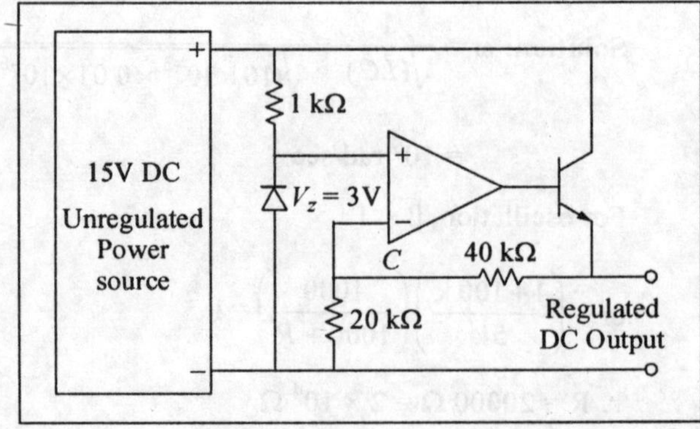

In an amp series regulator the o/p is

$$V_o = \left(1+\frac{R_1}{R_2}\right)V_Z$$

$$= \left(1+\frac{40}{20}\right)\times 3 = 9\text{ V}$$

Other way, the voltage at op-amp input

$$V_+ = V_- = V_z = 3\text{ Volt}$$

Then, by voltage division rule

$$\left(\frac{20}{20+40}\right)V_o = 3,\ V_o = 9\text{V}$$

Another Approach

As voltage at non-inverting terminal is 3 V due to Zener diode, voltage at inverting terminal will be same as 3 V because of vertual ground concept.

∴ Current through 20 kΩ resistor is

$$\frac{3}{20\text{ k}} = \frac{2}{20}\text{ mA}$$

$$\therefore \quad V_o = \frac{3}{20\text{ k}}\times 60\text{ k} = 9\text{ V}$$

4. **Ans. (d)**

Solution:

$$\text{Load regulation} = \frac{V_{(\text{no load})} - V_{(\text{Full load})}\times 100}{V_{(\text{Full load})}}$$

$$= \frac{30-25}{25}\times 100 = 20\%$$

$$\text{Output resistance} = \frac{V_{(\text{Full load})}}{I_{(\text{Full load})}} = \frac{25}{1} = 25\ \Omega$$

5. **Ans. (c)**

Solution: The pass transistor in a series regulator is in *CC* configuration having gain unity.

6. **Ans. (b)**

Solution:

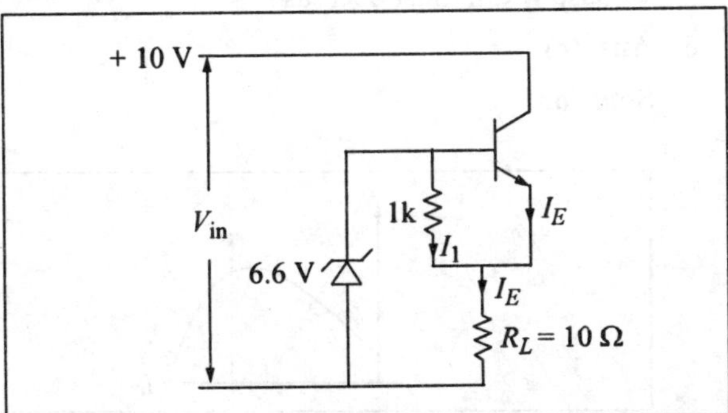

Power dissipated in transistor

$$= V_{CE}\times I_C$$

$$V_{BE} = 0.6\text{ V}$$

$$I_1 = \frac{0.6}{1k}$$

$$= 0.5\ m\text{A}$$

∴ Voltage across $R_L = 6.6 - 0.6 = 6$ V

$$\therefore \quad I_E = \frac{6}{10} = 0.6\text{ A}$$

$I_E \cong I_C$ as I_B is nearly 0.

$$\therefore \quad V_{CE} = 10 - 6 = 4\text{ V}$$

$$\therefore \quad P = 4\times 0.6 = 2.4\text{ W}$$

7. **Ans. (c)**

Solution:

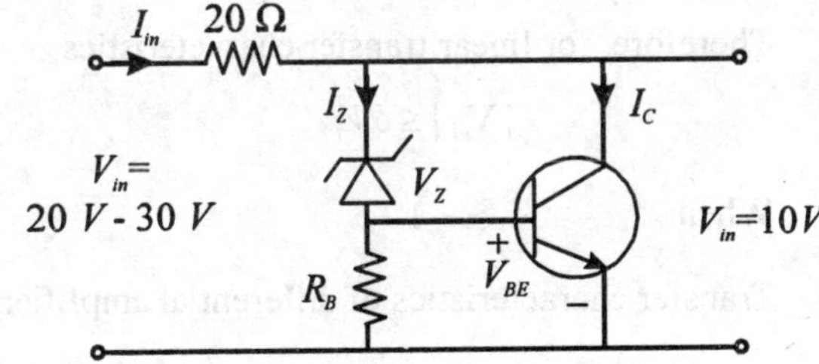

Maximum input current,

$$I_{in,\max} = \frac{V_{\max}-V_o}{20} = \frac{30-10}{20} = 1A \qquad \text{...(1)}$$

When current through R_B is neglected,

Then base current of transistor

$$I_B = I_Z$$

Current gain of BJT,

$$\beta = \frac{I_C}{I_B} = \frac{I_C}{I_Z} = 99$$

or, $\quad I_C = 99I_Z$

Input current I_{in} given by

$$I_{in} = I_C + I_Z = 99I_Z + I_Z = 100\ I_Z$$

$$I_Z = \frac{I_{in}}{100}$$

Zener current is maximum when input current is maximum. So, the maximum zener current can be given by

$$I_{Z,\max} = \frac{I_{in,\max}}{100} \qquad \text{...(2)}$$

From equation (1) and (2) we have

$$I_{Z,\max} = \frac{1}{100} = 0.001\ A$$

Then maixmum collector current

$$I_{C,\max} = 99 I_{Z,\max} = 99 \times 0.01 = 09.9 A$$

Then maximum power dissipation in zener diode

$$P_Z = V_Z I_{Z,\max} = 9.5 \times 0.01 = 95\ mW$$

The maixmum power dissipation in BJT,

$$P_T = V_C I_{C,\max} = 10 \times 0.99 = 9.9\ W$$

7.10 Miscellaneous

1. **Ans. (d)**

Solution:

Transfer characteristics of emitter coupled differential amplifier is linear when differential input voltage is less than or equal to thermal voltage.

Therefore, for linear transfer characteristics

$$|V_{id}| \leq \alpha V_T$$

When $\alpha = 1$

Transfer characteristics of differential amplifier,

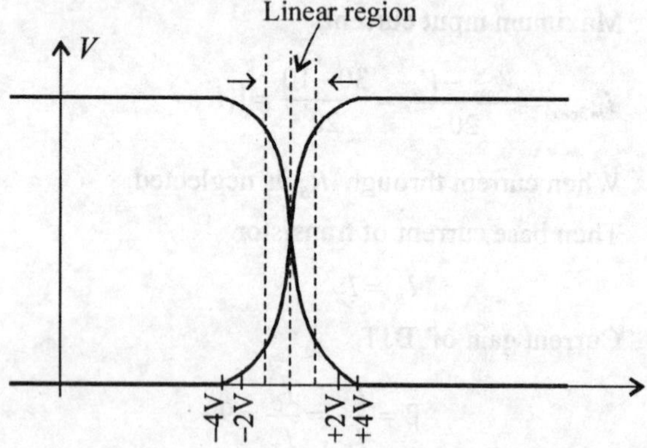

The circuit behaves as good limitter for $V_{in} > 4\ V$

2. **Ans. (a)**

Solution:

In cascading amplifiers gain is increased and bandwidth is decreased which results in higher low 3dB frequency and lower high 3 dB frequency.

3. **Ans. (c)**

Solution: The pass transistor in a series regulator is in *CC* configuration having voltage gain unity.

4. **Ans. (d)**

Solution:

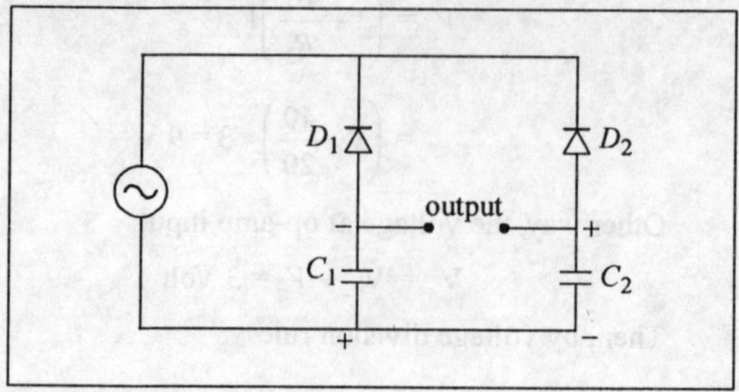

During the positive half-cycle D_2 conducts due to C_2 charged to positive V volt. During the negative half-cycle C_1 gets charged to negative V volt. So that output voltage across AB is V + V = 2V Volt, i.e. voltage doubler.

5. **Ans. (d)**

Solution:

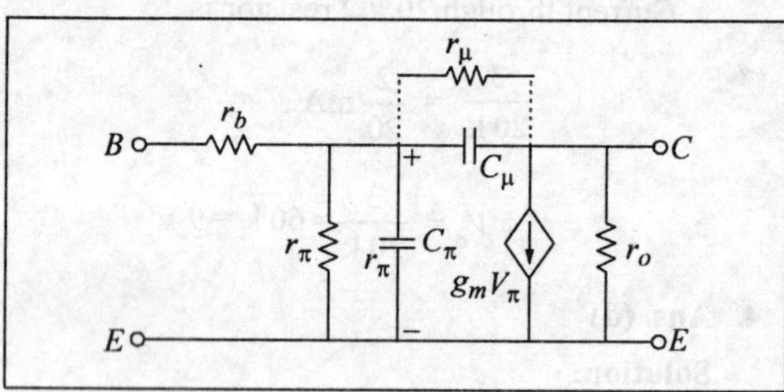

$$f_\beta = \frac{1}{h_{fe}} = \frac{g_m}{2\pi(C_\mu + C_\pi)} ; C_\mu = C_{be} ; C_\pi = C_{be}$$

$$f_\beta = f_\alpha (1 - \alpha)$$

If α is fixed for a transistor, $f_\beta = \alpha f_a$

For f_B to increase $C_\mu + C_\pi$ should decrease.

For capacitance to decrease, separation of width must increase

$$C = \frac{\varepsilon A}{d} ; C \propto \frac{1}{d}, \text{ i.e.}$$

Depletion layer of E_{jn} and C_{jn} must increase

⇒ base width must decrease.

6. **Ans. (c)**

Solution:

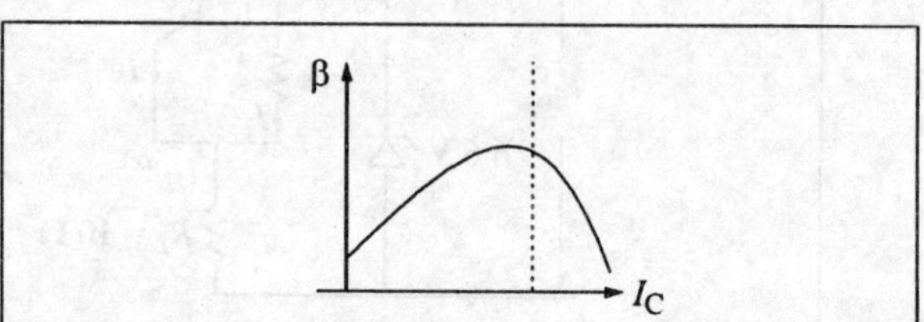

$$\beta_{\text{force}} \rightarrow \beta_{\text{reverse}}$$

$$\beta_F \rightarrow \beta_{FB}$$

$\beta_{\text{forced}} = \beta_F$ in the active region

In the saturation region, β_{forced} is less than β_F

7. Ans. (b)

Solution: Half-wave rectifier circuit diagram is given below

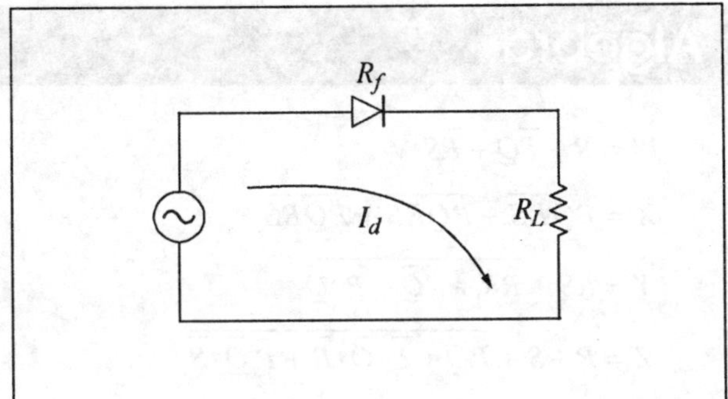

$$V_{dc} = \frac{V_m}{\pi}$$

$$I_{dc} = \frac{V_m}{\pi(R_f + R_L)}$$

8. Ans. (a)

9. Ans. (b)

Solution: Since poles are present on left half-plane, the amplifier will always be stable as it has negative feed back.

10. Ans. (c)

Solution:

$$V_o = -\frac{R_f}{R_1} V_{in} = -\left(\frac{\omega_c}{\omega_c}\right) V_1 \sin \omega t - \left(\frac{\omega_c}{\omega_c}\right) V_2 \sin \omega t$$

$$V_o = -V_1 \sin \omega t - V_2 \sin \omega t$$

$$= -(V_1 + V_2) \sin \omega t$$

11. Ans. (a)

Solution: Ramp input is applied

$$R(s) = \frac{1}{s^2}$$

$$e_{ss} = \lim_{s \to 0} sE(s)$$

$$sE(s) = \frac{sR(s)}{1+G(s)} = \frac{s \cdot [1/s^2]}{1+G(s)} = \frac{1}{s + s\,G(s)}$$

$$e_{ss} = \lim_{s \to 0} sE(s) = \lim_{s \to 0} \frac{1}{s + G(s)}$$

The above expression represent type 1 system, therefore, for type 1 system is given by

$$e_{ss} = \frac{1}{K} = 5\,\% = \frac{1}{20}$$

$$\therefore \quad K = 20$$

12. Ans. (c)

Solution: 3 dB BW is given by determinate of the characteristic equation for G_1 and G_2

$$BW = \sqrt{a^2 - 4b}$$

13. Ans. (a)

Solution: The frequency of a stable multi-vibrator is given by

$$f = \frac{1}{T}$$

$$\because \quad T = 1.386 \text{ RC}$$

$$\because \quad f = \frac{0.72}{RC}$$

Thus, the frequency of a stable multi-vibrator depends on R_2 and C.

14. Ans. (d)

Solution:

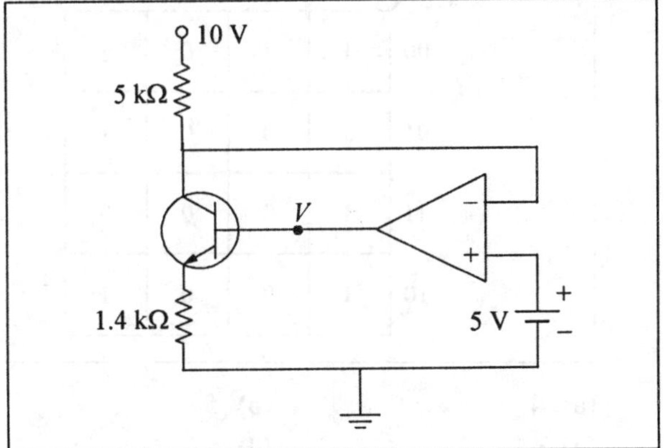

As applied voltage at inverting terminal is 5 V op-amplifier is an ideal. Hence voltage at positive terminal is 5 V. Current through the op-amplifier is zero

Hence $\quad I_C = \frac{10-5}{5\text{ k}} = 1 \text{ mA}$

$$V_E = I_e R_e = 1\, m \times 1.4\text{ k} = 1.4 \text{ V}$$

$$\text{V} = V_{be} + V_e = 0.6 + 1.4 = 2 \text{ V}$$

8 Digital Electronics

8.1 Boolean Algebra

1. The number of **Boolean function** that can be generated by n variable, is equal to
   (a) 2^n (b) 2^{2^n}
   (c) 2^{n-1} (d) 2^{n-1}

2. **Two 2's complement numbers having sign bits x and y are added and the sign bit of the result is z. Then the occurrence of over-flow is indicated by the Boolean function:**
   (a) xyz (b) $\bar{x}\,\bar{y}\,\bar{z}$
   (c) $\bar{x}\,\bar{y}\,z + xy\bar{z}$ (d) $xy + yz + zx$

3. The K-map for a Boolean function is shown in Figure. The number of essential prime implicants for this function is:

CD \ AB	00	01	11	10
00	1	1	0	1
01	0	0	0	1
11	1	0	0	0
10	1	0	0	1

   (a) 4 (b) 5
   (c) 6 (d) 8

4. The logical expression $y = A + \bar{A}B$ is equivalent to
   (a) $y = AB$ (b) $y = \bar{A}B$
   (c) $y = \bar{A} + B$ (d) $y = A + B$

5. The number of distinct Boolean expressions of 4 variables is:
   (a) 16 (b) 256
   (c) 1024 (d) 65536

6. The functions W, X, Y and Z are as follows:

$$W = R + \overline{P}Q + \overline{R}S$$

$$X = PQ\overline{RS} + \overline{PQRS} + P\overline{QRS}$$

$$Y = RS + \overline{PR + \overline{PQ} + \overline{P}\cdot\overline{Q}}$$

$$Z = R + S + \overline{PQ + \overline{P}\cdot\overline{Q}\cdot\overline{R} + P\,\overline{Q}\cdot\overline{S}}$$

   Then
   (a) $W = Z, X = \overline{Z}$ (b) $W = Z, X = Y$
   (c) $W = Y$ (d) $W = Y = \overline{Z}$

7. **The Boolean expression $AC + B\hat{C}$ is equivalent to**
   (a) $\overline{A}C + B\overline{C} + AC$
   (b) $\overline{B}C + A\overline{C} + BC + \overline{\overline{A}}C\overline{B}$
   (c) $AC + B\overline{C} + \overline{B}C + ABC$
   (d) $ABC + \overline{A}B\overline{C} + AB\overline{C} + A\overline{B}C$

8. **A Boolean function f of two variable x and y is defined as follows:**

$$f(0,0) = f(0,1) = f(1,1) = 1; f(1,0) = 0$$

   Assuming complements of x and y are not available, a minimum cost solution for realizing using only 2-input NOR gates and 2-input OR gate (each having unit cost) would have a total cost of
   (a) 1 unit (b) 4 unit
   (c) 3 unit (d) 2 unit

9. **The Boolean expression for the truth table shown in**

A	B	C	F
0	0	0	0
0	0	1	0
0	1	0	0
0	1	1	1
1	0	0	0
1	0	1	0
1	1	0	1
1	1	1	0

(a) $B(A+C)(\overline{A}+\overline{C})$ (b) $B(A+\overline{C})(\overline{A}+C)$
(c) $\overline{B}(A+\overline{C})(\overline{A}+C)$ (d) $\overline{B}(A+C)(\overline{A}+\overline{C})$

10. The number of product terms in the minimized sum of product expression obtained through the following.

K – map is (where, "d" denotes do not care states)

1	0	0	1
0	d	0	0
0	0	d	1
1	0	0	1

(a) 2 (b) 3
(c) 4 (d) 5

11. The point P in the following figure is struck at –1. The output f will be:

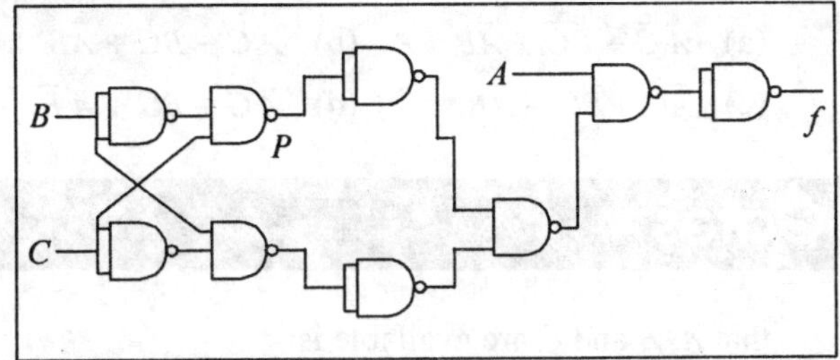

(a) $\overline{A\overline{BC}}$ (b) $\overline{A}$
(c) $AB\overline{C}$ (d) A

12. The Boolean function $Y = AB + CD$ is to be realized using only 2-input NAND gates. The minimum number of gates required is:
(a) 2 (b) 3
(c) 4 (d) 5

13. The Boolean expression $Y=\overline{A}\,\overline{B}\,\overline{C}\,D+\overline{A}\,B\,C\,\overline{D}+A\,\overline{B}\,\overline{C}\,D+A\,B\,\overline{C}\,\overline{D}$ can be minimized to
(a) $Y=\overline{A}\,\overline{B}\,\overline{C}\,D+\overline{A}\,B\,\overline{C}+A\,\overline{C}\,D$
(b) $Y=\overline{A}\,\overline{B}\,\overline{C}\,D+B\,C\,\overline{D}+A\,\overline{B}\,\overline{C}\,D$
(c) $Y=\overline{A}\,B\,C\,\overline{D}+\overline{B}\,\overline{C}\,D+A\,\overline{B}\,\overline{C}\,D$
(d) $Y=\overline{A}\,B\,C\,\overline{D}+\overline{B}\,\overline{C}\,D+A\,B\,\overline{C}\,\overline{D}$

14. If $X = 1$ in the logic equation
$$[X+Z\{\overline{Y}+(\overline{Z}+X\overline{Y})\}]\{\overline{X}+\overline{Z}(X+Y)\}=1,$$
then
(a) $Y = Z$ (b) $Y=\overline{Z}$
(c) $Z = 1$ (d) $Z = 0$

15. The Bollean function $A + BC$ is a reduced form of
(a) $AB + BC$ (b) $(A + B).(A + C)$
(c) $AB + A\overline{B}C$ (d) $(A + C).B$

16. For the identity $AB + \overline{A}C + BC = AB + \overline{A}C$, the **dual form is**
(a) $(A+B)(\overline{A}+C)(B+C)=(A+B)(\overline{A}+C)$
(b) $(\overline{A}+\overline{B})(A+\overline{C})(\overline{B}+\overline{C})=(\overline{A}+\overline{B})(A+\overline{C})$
(c) $(A+B)(\overline{A}+C)(B+C)=(\overline{A}+B)(A+\overline{C})$
(d) $\overline{A}\,\overline{B}+A\overline{C}+\overline{B}\,\overline{C}=\overline{A}\,\overline{B}+A\overline{C}$

17. The simplified form of the Boolean expression $Y=\left(\overline{A}.BC+D\right)\left(\overline{A}.D+\overline{B}.\overline{C}\right)$ can be written as
(a) $\overline{A}.D+\overline{B}.\overline{C}.D$
(b) $AD+B.\overline{C}.D$
(c) $(\overline{A}.+D)(\overline{B}.C+\overline{D})$
(d) $A.\overline{D}+BC.\overline{D}$

18. A minimized form of the function F is
(a) $F=\overline{X}Y+YZ$
(b) $F=\overline{X}\,\overline{Y}+YZ$
(c) $F=\overline{X}\,\overline{Y}+Y\overline{Z}$
(d) $F=\overline{X}\,\overline{Y}+\overline{Y}Z$

19. In the sum of products function $f(X,Y,Z)=\sum(2,3,4,5)$, the prime implicates are
(a) $\overline{X}\,Y,\ X\,\overline{Y}$
(b) $\overline{X}\,Y,\ X\,\overline{Y}\,\overline{Z},\ X\,\overline{Y}\,Z$
(c) $\overline{X}\,Y\,\overline{Z},\ \overline{X}\,Y\,Z,\ X\,\overline{Y}$
(d) $\overline{X}\,Y\,\overline{Z},\ \overline{X}\,Y\,Z,\ X\,\overline{Y}\,\overline{Z},\ X\,\overline{Y}\,Z$

20. The logic function $f=\overline{(x.\overline{y})+(\overline{x}.y)}$ is the same as
(a) $f=(x+y)(\overline{x}+\overline{y})$
(b) $f=\overline{(\overline{x}+\overline{y})+(x+y)}$
(c) $f=\overline{(x.y)}.(\overline{x}.\overline{y})$
(d) None of (A), (B), (C)

21. The Boolean expression $\overline{A+B+C}$ is equal to
(a) $\overline{A}+\overline{B}+\overline{C}$ (b) $\overline{A}.\overline{B}.\overline{C}$
(c) $\overline{A}+\overline{B+C}$ (d) $A.(B+C)$.

22. The expression $Y=\overline{AB}$ is equivalent to
(a) $\overline{A}+\overline{B}$ (b) $AB + A$
(c) $A + B$ (d) AB

23. The simplest form of the Boolean expression $AB\overline{C}\overline{D}+ABC\overline{D}+AB\overline{C}D+ABCD$ is

(a) AD (b) BC
(c) $\overline{A}B$ (d) AB

24. For the Boolean expression

$$f = +\overline{a}\,\overline{b}\,\overline{c} + \overline{a}\,b\,\overline{c} + a\,\overline{b}\,\overline{c} + a\,bc + ab\overline{c}$$

the minimized Product of Sum (PoS) expression is
(a) $f = (b + \overline{c}).(a + \overline{c})$
(b) $f = (\overline{b} + c).(\overline{a} + c)$
(c) $f = (\overline{b} + c).(a + \overline{c})$
(d) $f = \overline{c} + abc$

25. The minimal sum of products form of

$$f = A\overline{B}CD + \overline{A}BC + \overline{A}\,\overline{B}\,C + BCD \text{ is}$$

(a) $\overline{A}C + BD$ (b) $\overline{A}\,C + CD$
(c) $AC + \overline{B}D$ (d) $A\overline{B} + C\overline{D}$

26. Min-term (Sum of Products) expression for a Boolean function is follows.

$$f(A, B, C) = \Sigma m\,(0, 1, 2, 3, 5, 6)$$

where A is the MSB and C is the LSB. The minimized expression for the function is
(a) $A + (B \oplus C)$ (b) $(A \oplus B) + C$
(c) $\overline{A} + (B \oplus C)$ (d) $\overline{ABC}$

27. The minimum sum of products form of the Boolean expression

$$Y = \begin{bmatrix} \overline{P}\,\overline{Q}\,\overline{R}\,\overline{S} + P\,\overline{Q}\,\overline{R}\,\overline{S} + P\,\overline{Q}\,\overline{R}\,S \\ +P\,\overline{Q}\,R\,S + P\,\overline{Q}\,R\,\overline{S} + \overline{P}\,\overline{Q}\,R\,\overline{S} \end{bmatrix} \text{ is}$$

(a) $Y = P\,\overline{Q} + \overline{Q}\,\overline{S}$ (b) $Y = P\,\overline{Q} + \overline{Q}\,R\,\overline{S}$
(c) $Y = P\,\overline{Q} + \overline{Q}\,\overline{R}\,\overline{S}$ (d) $Y = \overline{Q}\,\overline{S} + P\,\overline{Q}\,R$

28. The minimized form of the logical expression

$$(\overline{A}\,\overline{B}\,\overline{C} + \overline{A}\,B\,\overline{C} + \overline{A}\,B\,C + AB\,\overline{C}) \text{ is}$$

(a) $A\,\overline{C} + B\overline{C} + \overline{A}B$ (b) $A\,\overline{C} + BC + \overline{A}B$
(c) $\overline{A}C + \overline{B}C + \overline{A}B$ (d) $A\,\overline{C} + BC + A\,\overline{B}$

8.2 Logic Gates

1. Boolean expression for the output of X-NOR (equivalent) logic gate with input A and B is
(a) $A\,\overline{B} + \overline{AB}$ (b) $A\,\overline{B} + A\,\overline{B}$
(c) $(\overline{A} + B)(A + \overline{B})$ (d) $(A + B)(A + B)$

2. The output of a logic gate is '1' when all its inputs are at logic '0'. Then gate is either
(a) A NAND or an EX-OR gate
(b) A NOR or an EX-NOR gate
(c) An OR or an EX-NOR gate
(d) An AND or an EX-OR gate

3. The minimum number of NAND gates required to implement the Boolean function $A + A\overline{B} + A\overline{B}C$, is equal to:
(a) Zero (b) 1
(c) 4 (d) 7

4. The output of the logic gate in this Figure is:

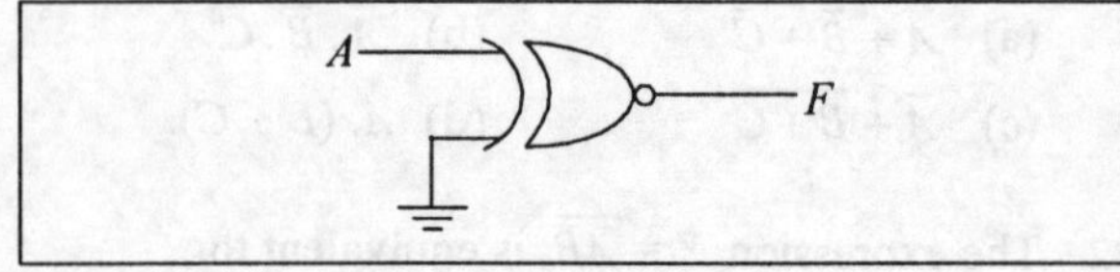

(a) 0 (b) 1
(c) $\overline{A}$ (d) A

5. The minimum number of 2-input NAND gates required to implement the Boolean function $Z = A\,\overline{B}\,C$, assuming that A, B and C are available is:
(a) Two (b) Three
(c) Five (d) Six

6. In Figure $A = 1$ and $B = 1$, the input B is now replaced by a sequence 101010..., the outputs x and y will be:

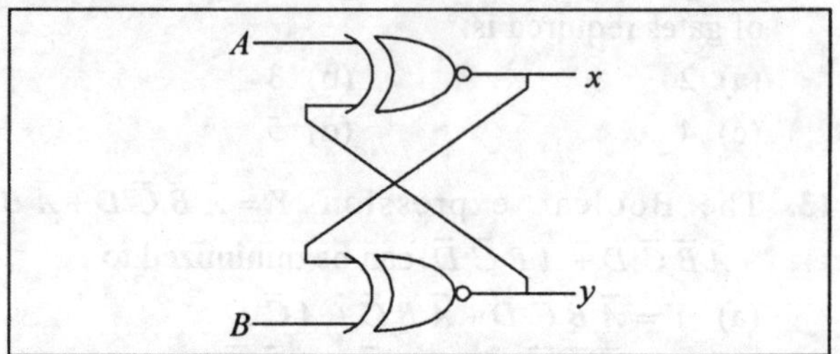

(a) Fixed at 0 and 1, respectively
(b) $x = 1010$ ...... while $y = 0101$
(c) $x = 1010$ ...... and $y = 1010$
(d) Fixed at 1 and 0, respectively

7. For the logic circuit shown in Figure, the required input condition (A, B, C) to make the output (X) = 1 is:

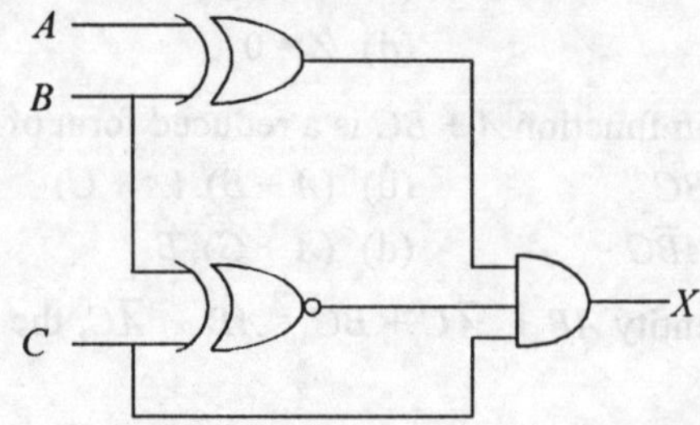

(a) 1, 0, 1 (b) 0, 0, 1
(c) 1, 1, 1 (d) 0, 1, 1

8. For the logic circuit shown in figure, the simplified Boolean expression for the output Y is

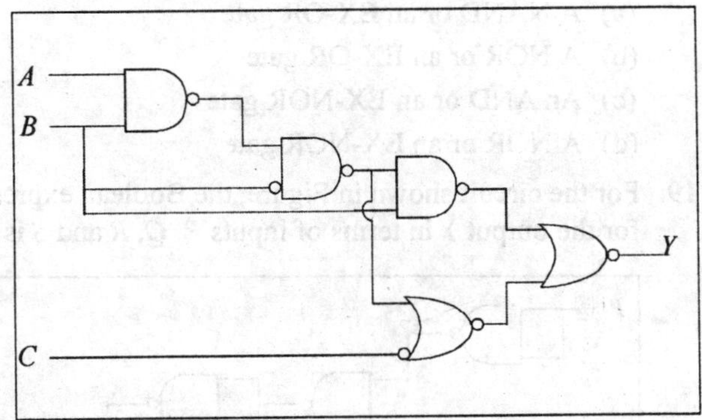

(a) $A+B+C$ (b) A
(c) B (d) C

9. If the input to the digital circuit shown in figure consisting of a cascade of 20 *XOR*-gates is X, then the output Y is equal to:

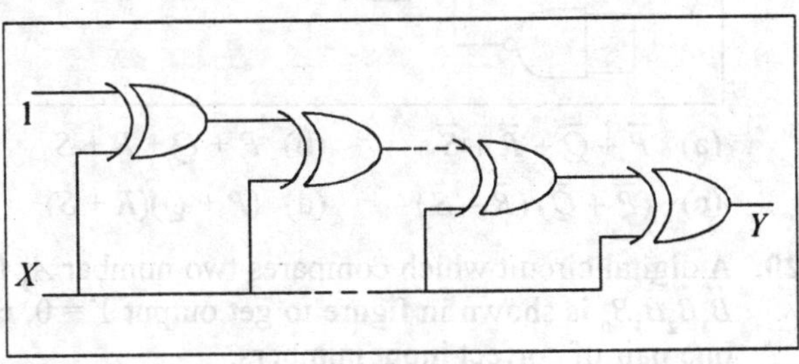

(a) 0 (b) 1
(c) X (c) $\bar{X}$

10. The gate G_1 and G_2 in figure have propagation delays of 10 n sec and 20 n sec respectively. If the input V_i makes an abrupt change form logic 0 to 1 at time $t = t_0$, then the output waveform V_0 is:

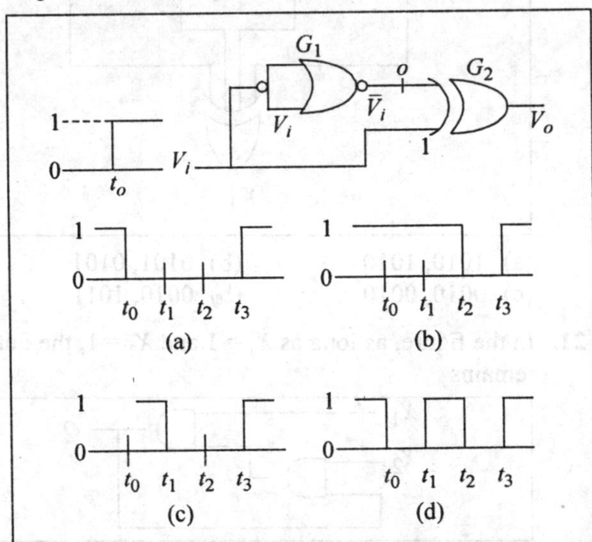

$[t_1 = t_0 + 10\ n\ \text{sec},\ t_2 = t_0 + 20\ n\ \text{sec},\ t_3 = t_0 + 30\ n\ \text{sec}]$

11. The following binary values were applied to the X and Y inputs of the NAND latch shown in the figure in the sequence indicated below:

$$X = 0, Y = 1; \quad X = 0, Y = 0; \quad X = 1, Y = 1$$

The corresponding stable P, Q outputs will be:

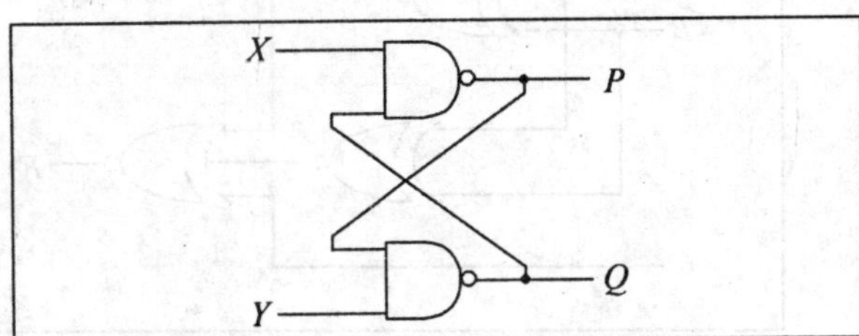

(a) $P = 1, Q = 0$; $P = 1, Q = 0$ $P = 1, Q = 0$ or $P = 0, Q = 1$
(b) $P = 1, Q = 0$; $P = 0, Q = 1$ or $P = 0, Q = 1$ $P = 0, Q = 1$
(c) $P = 1, Q = 0$; $P = 1, Q = 1$ $P = 1, Q = 0$ or $P = 0, Q = 1$
(d) $P = 1, Q = 0$; $P = 1, Q = 1$ $P = 1, Q = 1$

12. Which of the following Boolean Expressions correctly represents the relation between P, Q, R and M_1?

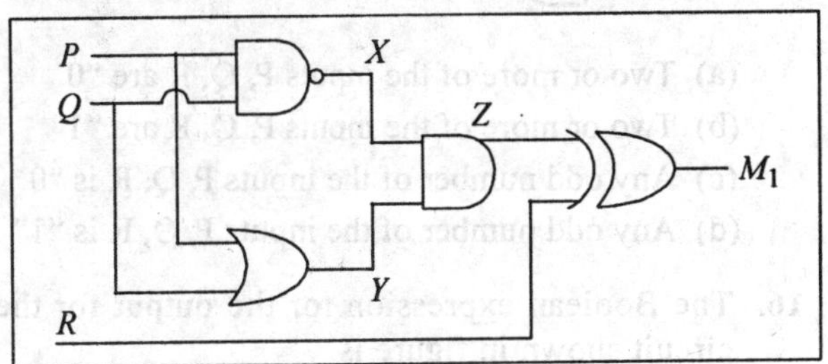

(a) $M_1 = (P \text{ OR } Q)\ XOR\ R$
(b) $M_1 = (P \text{ AND } Q)\ XOR\ R$
(c) $M_1 = (P \text{ NOR } Q)\ XOR\ R$
(d) $M_1 = (P \text{ XOR } Q)\ XOR\ R$

13. Match the logic gates in **Column A** with their equivalents in **Column B.**

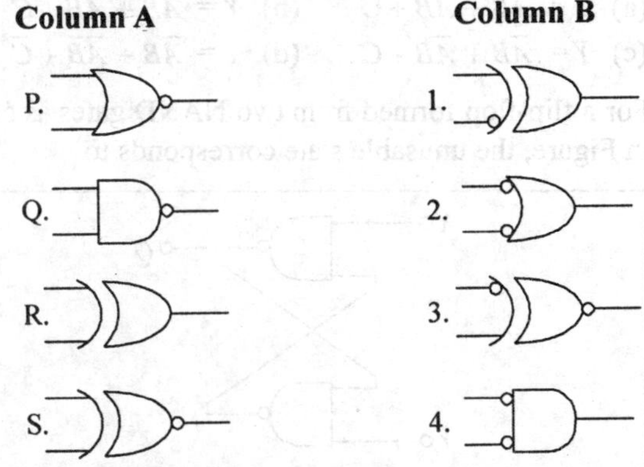

(a) P-2, Q-4, R-1, S-3 (b) P-4, Q-2, R-1, S-3
(c) P-2, Q-4, R-3, S-1 (d) P-4, Q-2, R-3, S-1

14. For the output F to be 1 in the logic circuit shown, the input combination should be

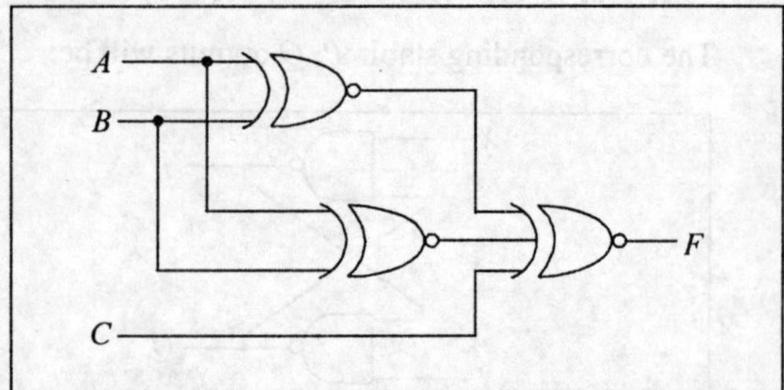

(a) $A = 1, B = 1, C = 0$ (b) $A = 1, B = 0, C = 0$
(c) $A = 0, B = 1, C = 0$ (d) $A = 0, B = 0, C = 1$

15. The output Y in the circuit belw is always "1' when

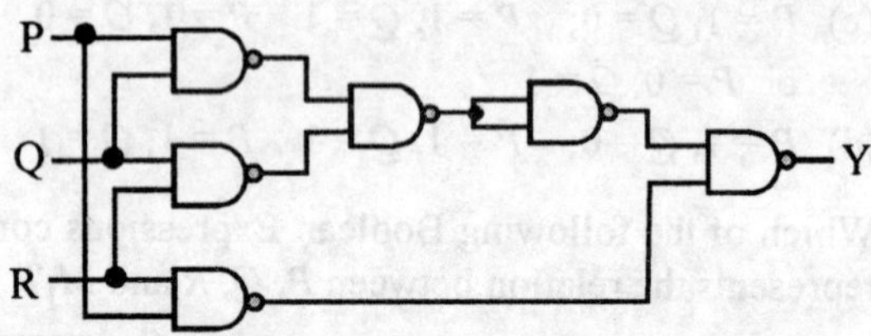

(a) Two or more of the inputs P, Q, R are "0'
(b) Two or more of the inputs P, Q, R are "1"
(c) Any odd number of the inputs P, Q, R is "0'
(d) Any odd number of the inputs P, Q, R is "1"

16. The Boolean expression for the output for the logic circuit shown in figure is

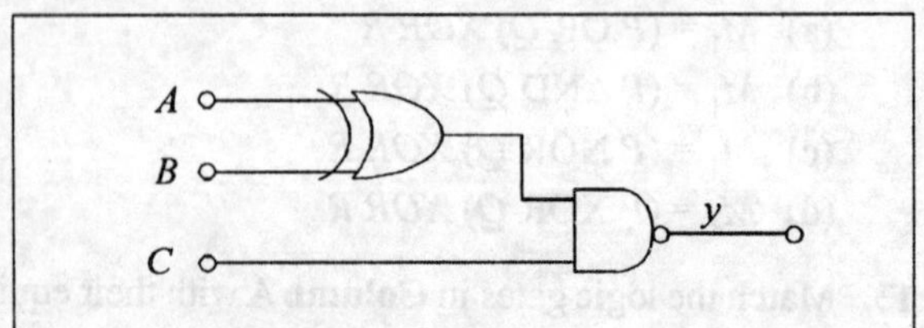

(a) $Y = \overline{AB} + AB + \overline{C}$ (b) $Y = A\overline{B} + AB + \overline{C}$
(c) $Y = \overline{A}B + \overline{A}B + C$ (d) $Y = \overline{AB} + \overline{AB} + \overline{C}$

17. For a flip-flop formed from two NAND gates as shown in Figure, the unusable state corresponds to

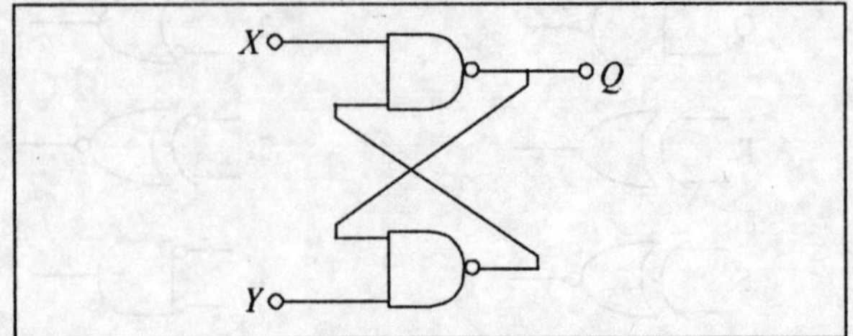

(a) $X = 0, Y = 0$ (b) $X = 0, Y = 1$
(c) $X = 1, Y = 0$ (d) $X = 1, Y = 0$

18. The output of a logic gate is "1" when all its inputs are at logic "0". The gate is either.

(a) A NAND or an EX-OR gate
(b) A NOR or an EX-OR gate
(c) An AND or an EX-NOR gate
(d) A NOR or an EX-NOR gate

19. For the circuit shown in Figure, the Boolean expression for the output Y in terms of inputs P, Q, R and S is

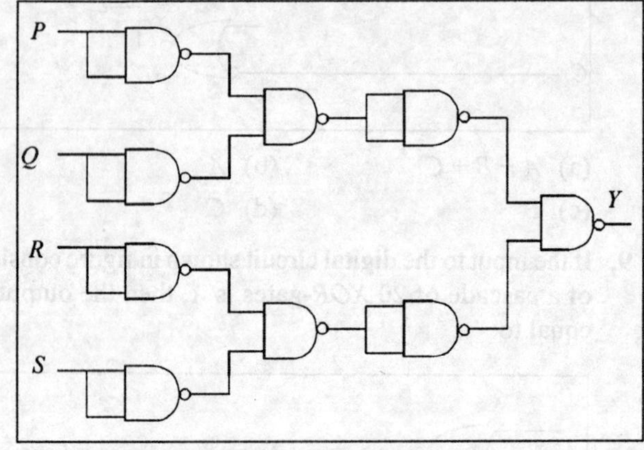

(a) $\overline{P} + \overline{Q} + \overline{R} + \overline{S}$ (b) $P + Q + R + S$
(c) $(\overline{P} + \overline{Q})(\overline{R} + \overline{S})$ (d) $(P + Q)(R + S)$

20. A digital circuit which compares two number $A_3A_2A_1A_0$, $B_3B_2B_1B_0$ is shown in figure to get output $Y = 0$, choose one pair of correct input numbers.

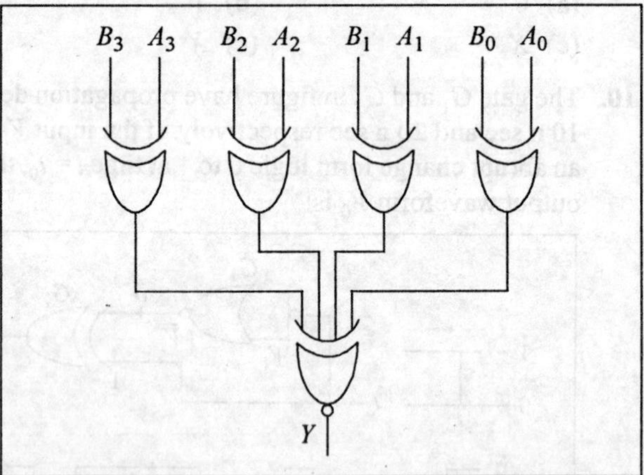

(a) 1010, 1010 (b) 0101, 0101
(c) 0010, 0010 (b) 0010, 1011

21. In the figure, as long as $X_1 = 1$ and $X_2 = 1$, the output Q remains

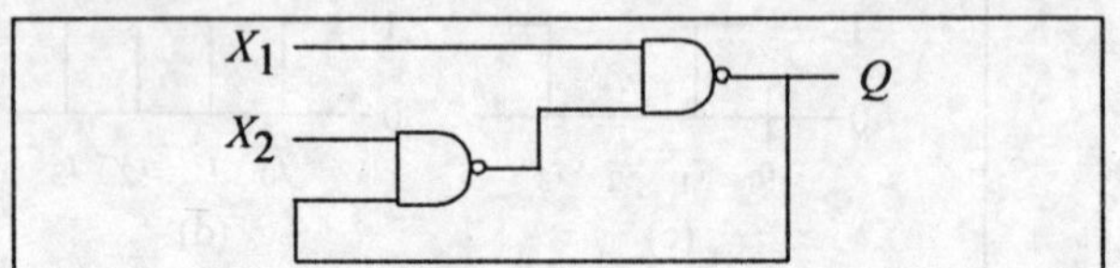

(a) At 1

(b) At 0

(c) At its initial value

(d) Unstable

22. *A*, *B*, *C* and *D* are input and *Y* is the output bit in the XOR gate circuit of the figure below. Which of the following statements about the sum *S* of *A*, *B*, *C*, *D* and *Y* is correct?

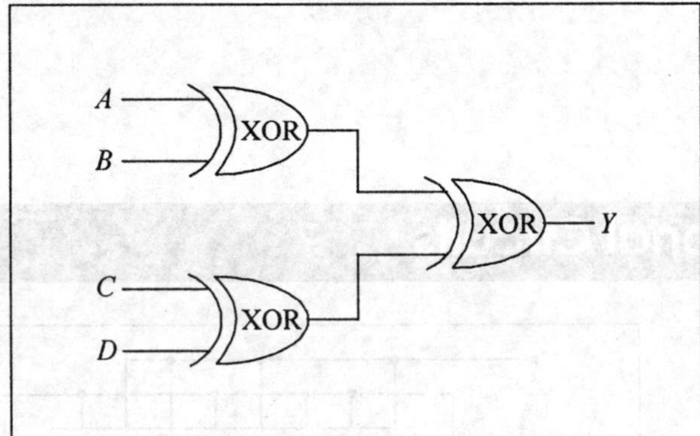

(a) *S* is always either zero or odd

(b) *S* is always either zero or even

(c) *S* = 1 only if the sum of *A*, *B*, *C* and *D* is even

(d) *S* = 1 only if the sum of A, *B*, *C* and *D* is odd

23. The complete set of only those Logic Gates desig nated as Universal Gates is

(a) NOT, OR and AND Gates

(b) XNOR, NOR and NAND Gates

(c) NOR and NAND Gates

(d) XOR, NOR and NAND Gates

24. The digital circuit using two inverters shown in figure will act as

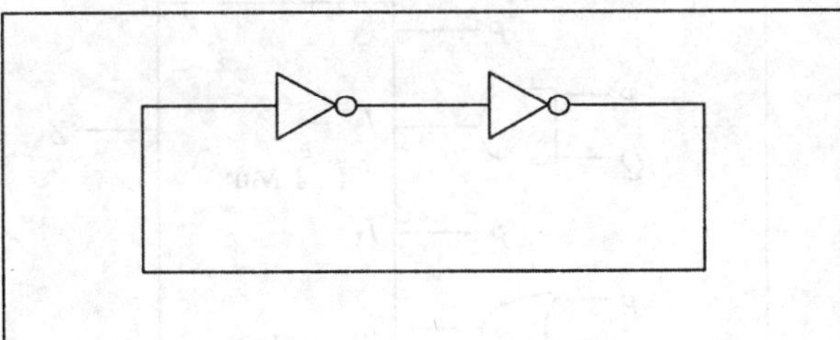

(a) A bistable multi-vibrator

(b) An astable muti-vibrator

(c) A monostable multi-vibrator

(d) An oscillator

25. In the digital circuit shown in figure, the output *f* is found to be logic 1 when *A* is logic '0'. The values of *B* and *C* are:

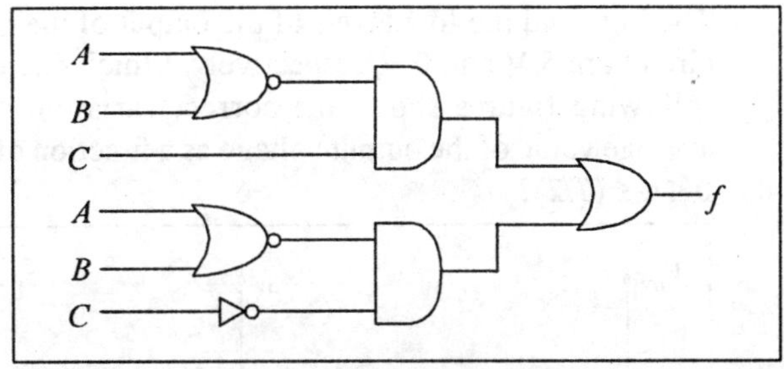

(a) $B = 1, C = 0$ (b) $B = 0, C = 0$ or 1

(c) $B = 1, C = 1$ (d) Indeterminate

26. If *A* and *B* are inputs of a logic gate, then match the logic gate with its output and select the correct answer using the codes:

List-I	*List-II*
A. NAND	1. $\overline{\bar{A}+\bar{B}}$
B. NOR	2. $\bar{A}+\bar{B}$
C. XNOR	3. $\bar{A}\,\bar{B}+AB$
D. AND	4. $\overline{AB}$

Codes:

	A	*B*	*C*	*D*
(a)	2	4	3	1
(b)	4	1	3	2
(c)	3	1	4	2
(d)	4	2	1	3

27. Two square waves of equal period *T*, but with a time delay τ are applied to a digital circuit whose truth table is shown in the following figure.

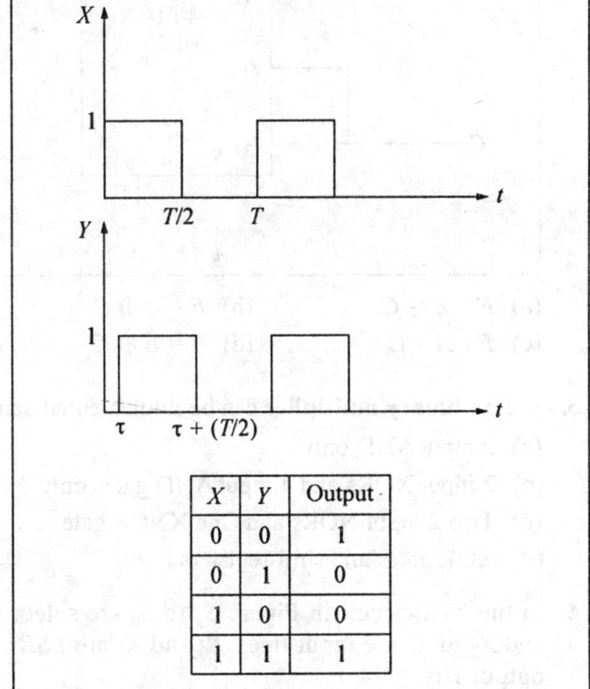

X	*Y*	Output
0	0	1
0	1	0
1	0	0
1	1	1

The high and the low levels of the output of the digital circuit are 5 V and 0 V, respectively. Which one of the following figures shows the correct variation of the average value of the output voltage as a function of τ for $0 \le \tau \le (T/2)$?

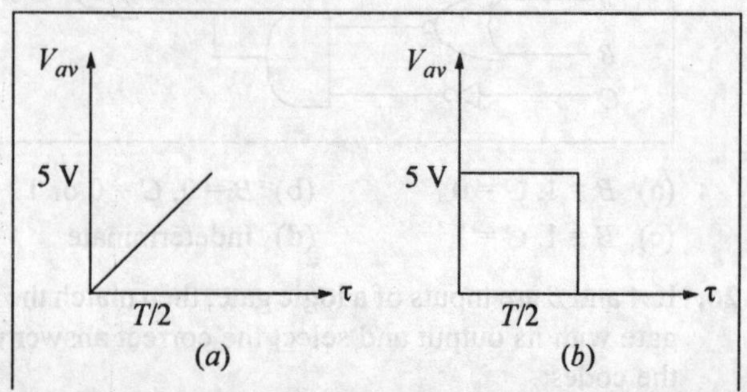

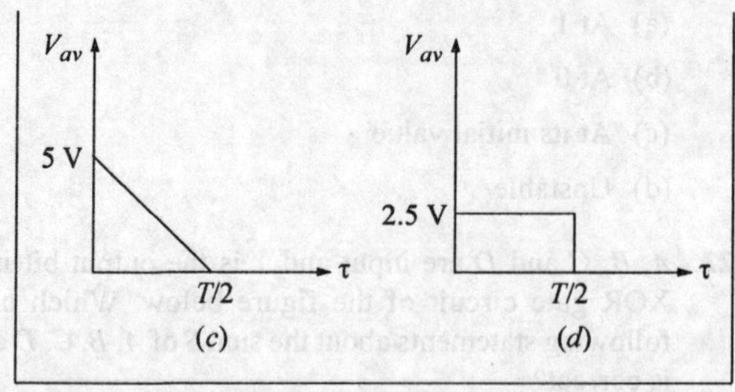

8.3 Combinational Circuits

1. The minimal function that can detect a "divisible by 3" 8421 BCD code digit (representation is $D_8\, D_4\, D_2\, D_1$) is given by:
   (a) $D_8D_1 + D_4D_2 + D_8D_2D_1$
   (b) $D_8D_1 + \bar{D}_4D_2D_1 + D_4D_2\bar{D}_1 + \bar{D}_8\bar{D}_4\bar{D}_2\bar{D}_1$
   (c) $D_4D_1 + D_4D_2 + D_8\bar{D}_1\bar{D}_2D_1$
   (d) $D_4 + D_2\bar{D}_1 + \bar{D}_4D_2D_1 + D_8D_4D_2D_1$

2. The logic realized by the circuit shown in Figure is

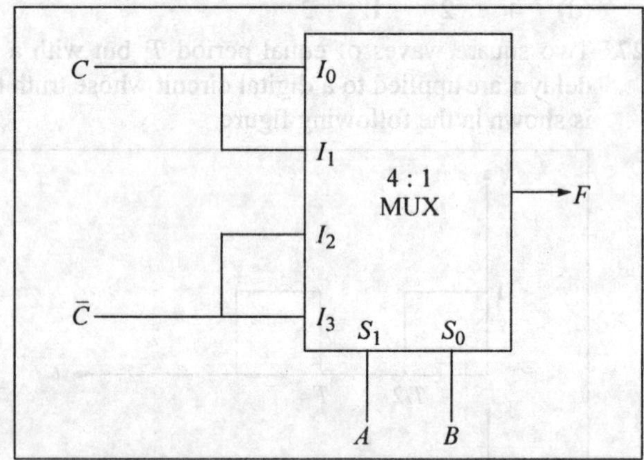

   (a) $F = A \odot C$ (b) $F = A \oplus C$
   (c) $F = A + C$ (d) $F = B \oplus C$

3. A 2 bit **binary multiplier** can be implemented using
   (a) 2 input AND only
   (b) 2 input XORs and 4 input AND gates only
   (c) Two 2 input NORs and One XNOR gate
   (d) XOR gates and shift registers.

4. In the *TTL* circuit in Figure S_2 to S_0 are select lines and X_7 to X_0 are input lines. S_0 and X_0 are *LSBs*. The output *Y* is:

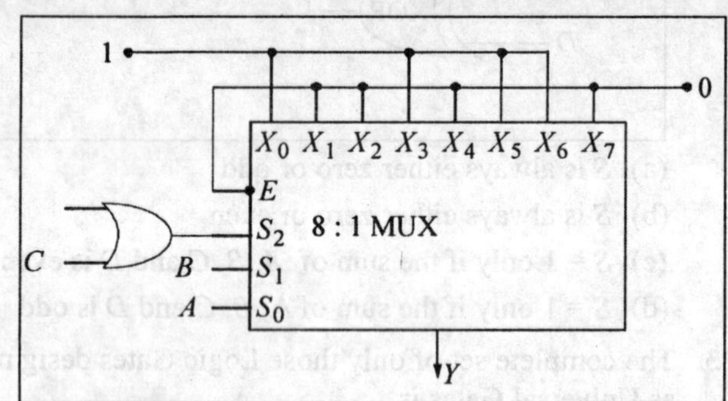

   (a) Indeterminate
   (b) $A \oplus B$
   (c) $\overline{A \oplus B}$
   (d) $\overline{C}\,\overline{(A \oplus B)} + C \cdot (A \oplus B)$

5. For the circuit shown in the following figure, I_0–I_3 are inputs to the 4 : 1 multiplexer. *R* (MSB) and *S* are control bits.

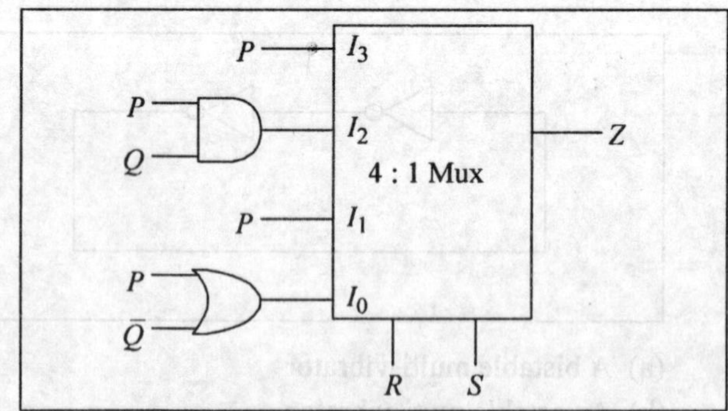

The output *Z* can be represented by
   (a) $PQ + P\bar{Q}S + \bar{Q}\bar{R}\bar{S}$
   (b) $P\bar{Q} + PQ\bar{R} + \bar{P}\bar{Q}\bar{S}$
   (c) $P\bar{Q}\bar{R} + \bar{P}QR + PQRS + \bar{Q}\bar{R}\bar{S}$

(d) $PQ\overline{R} + PQR\overline{S} + P\overline{Q}\,\overline{R}S + \overline{Q}\,\overline{R}\,\overline{S}$

6. What are the minimum number of 2-to-1 multiplexers required to generate a 2-input AND gate and a 2-input Ex-OR gate ?

(a) 1 and 2

(b) 1 and 3

(c) 1 and 1

(d) 2 and 2

7. The Boolean function realized by the logic circuit shown is

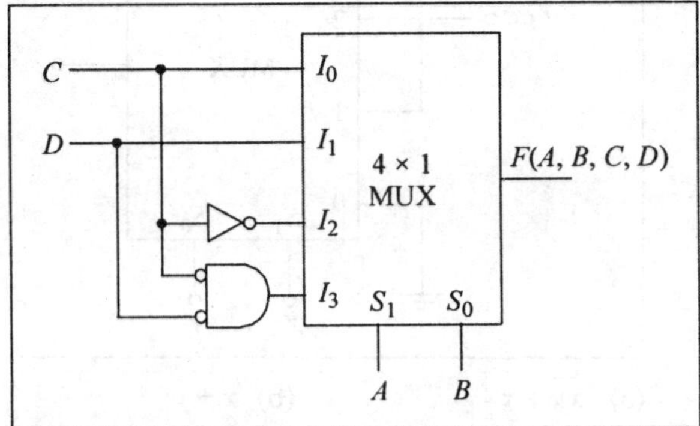

(a) $F = \Sigma\, m\,(0, 1, 3, 5, 9, 10, 14)$

(b) $F = \Sigma\, m\,(2, 3, 5, 7, 8, 12, 13)$

(c) $F = \Sigma\, m\,(1, 2, 4, 5, 11, 14, 15)$

(d) $F = \Sigma\, m\,(2, 3, 5, 7, 8, 9, 12)$

8. The logic function implemented by the circuit below is (ground implies a logic "0")

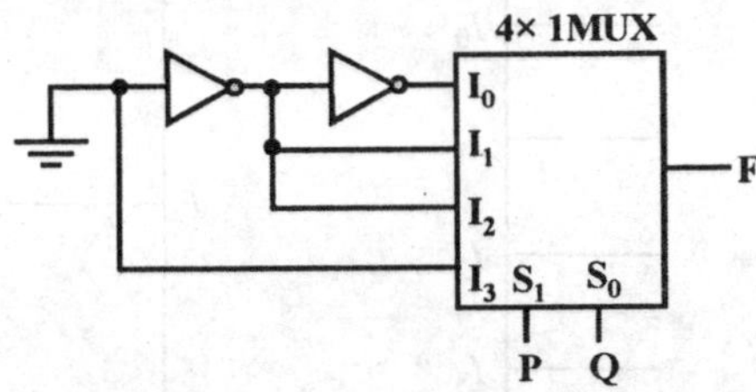

(a) F = AND (P, Q) (b) F = OR (P, Q)

(c) F = XNOR(P, Q) (d) F = XOR(P, Q)

9. The output Y of 2-bit comparator is logic 1 whenever the 2-bit input A is greater than the 2-bit input B. The number of combinations for which the output is logic 1, is

(a) 4 (b) 6

(c) 8 (d) 10

10. Without any additional circuitry, an 8 : 1 MUX can be used to obtain

(a) Some but not all Boolean functions of 3 variables

(b) All function of 3 variables but none of 4 variables

(c) All function of 3 variables and some but not all of 4 variables

(d) All function of 4 variable.

11. The circuit shown in Figure has 4 boxes each described by inputs P, Q, R and outputs Y, Z with

$Y = P \oplus Q \oplus R,\qquad Z = RQ + \overline{P}R + Q\overline{P}$

The circuit acts as a

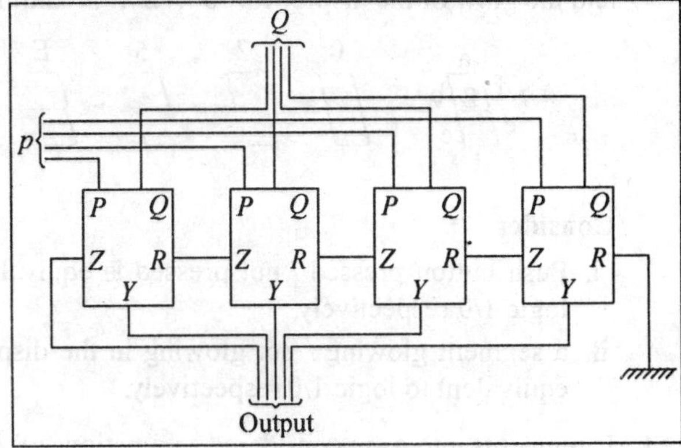

(a) 4 bit adder giving $P + Q$

(b) 4 bit subtractor giving $P - Q$

(c) 4 bit subtractor giving $Q - P$

(d) 4 bit adder giving $P + Q + R$

12. The minimum number of 2-to-1 multiplexers required to realize a 4-to-1 multiplexers is

(a) 1 (b) 2

(c) 3 (d) 4

13. In the following circuit, X is given by:

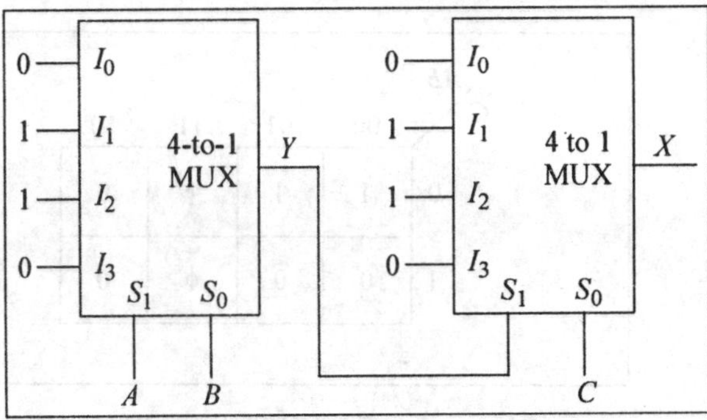

(a) $X = A\,\overline{B}\,\overline{C} + \overline{A}\,B\,\overline{C} + \overline{A}\,\overline{B}\,C + A\,B\,C$

(b) $X = \overline{A}\,B\,C + A\,\overline{B}\,C + A\,B\,\overline{C} + \overline{A}\,\overline{B}\,\overline{C}$

(c) $X = A\,B + B\,C + A\,C$

(d) $X = \overline{A}\,\overline{B} + \overline{B}\,\overline{C} + \overline{A}\,\overline{C}$

Statement for Linked Answers Questions 14 and 15

Two products are sold from a vending machine, which has two push buttons P_1 and P_2. When a button is pressed, the price of the corresponding product is displayed in a 7 - segment display

If no buttons are pressed, '0' is displayed, signifying 'Rs . 0'.

If only P_1 is pressed '2' is displayed, signifying 'Rs. 2 '.

If only P_2 is pressed, '5' is displayed, signifying 'Rs. 5'.

If both P_1 and P_2 are pressed, 'E' is displayed, signifying 'Error'.

The names of the segment in the 7 - segment display, and the glow of the display for '0' , '2' , '5' and 'E' are

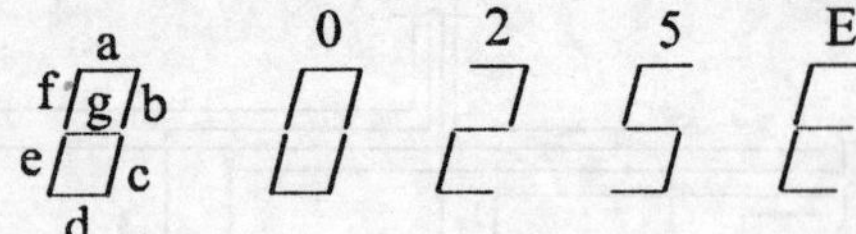

Consider

i. Push button pressed / not pressed is equivalent to logic 1/0 respectively,

ii. a segment glowing / not glowing in the display is equivalent to logic 1/0 respectively.

14. If segments ***a*** to ***g*** are considered as functions of P_1 and P_2 , then which of the following is correct ?

(a) $g = \overline{P_1} + P_2,\ d = c + e$ (b) $g = P_1 + P_2,\ d = c + e$

(c) $g = \overline{P_1} + P_2,\ e = b + c$ (d) $g = P_1 + P_2,\ e = b + c$

15. What are the minimum numbers of NOT gates and 2-input OR gates required to design the logic of the driver for this 7 - segment display ?

(a) 3 NOT and 4 OR (b) 2 NOT and 4 OR

(c) 1 NOT and 3 OR (d) 2 NOT and 3 OR

16. The minimal product-of-sums function described by the K-map given in Figure

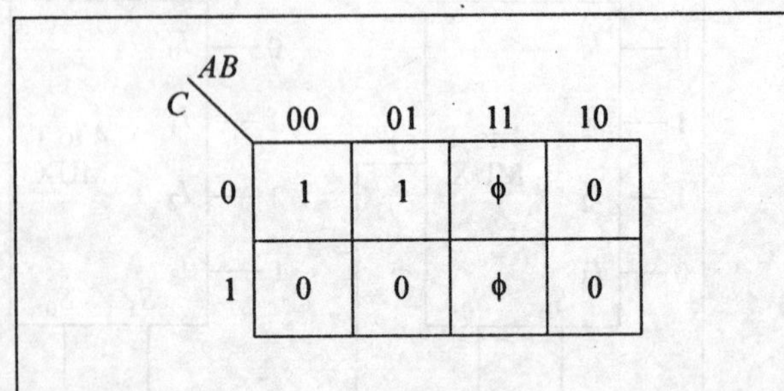

(a) $A'C'$ (b) $A' + C'$

(c) $A + C$ (d) AC

17. Consider the given circuit

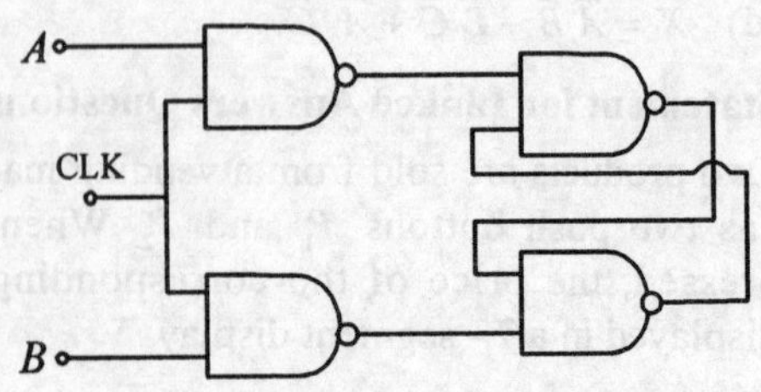

In this circuit, the race around

(a) Does not occur

(b) Occurs when CLK = 0

(c) Occur when CLK = 1 and A = B = 1

(d) Occurs when CLK = 1 and A = B = 0

18. The output f of the 4-to-1 MUX shown in Figure is

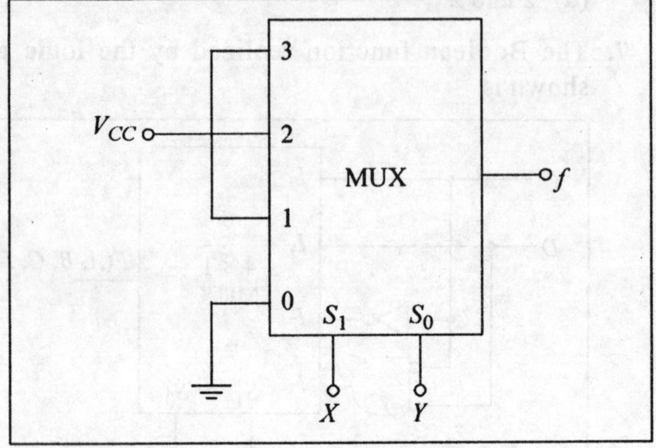

(a) $\overline{xy} + x$ (b) $x + y$

(c) $\overline{x} + \overline{y}$ (d) $x\overline{y} + x$

19. Figure shown a 4-to-1 MUX to be used to implement the sum S of a 1-bit full adder with input bits P and Q the carry input C_{in}. Which of the following combination of inputs to I_0, I_1, I_2 and I_3 of the MUX will realize the sum S?

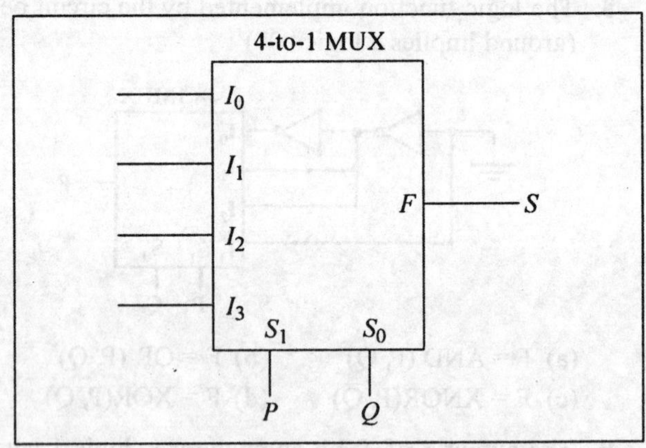

(a) $I_0 = I_1 = C_{in}$; $I_2 = I_3 = \overline{C}_{in}$

(b) $I_0 = I_1 = \overline{C}_{in}$; $I_2 = I_3 = C_{in}$

(c) $I_0 = I_3 = C_{in}$; $I_1 = I_2 = \overline{C}_{in}$

(d) $I_0 = I_3 = \overline{C}_{in}$; $I_1 = -I_2 = C_{in}$

20. A 4 × 1 MUX is used to implement a 3-input Boolean function as shown in figure. The Boolean function $F(A, B, C)$ implemented is

(a) $F(A, B, C) = \Sigma(1, 2, 4, 6)$
(b) $F(A, B, C) = \Sigma(1, 2, 6)$
(c) $F(A, B, C) = \Sigma(2, 4, 5, 6)$
(d) $F(A, B, C) = \Sigma(1, 5, 6)$

21. **A 3 line to 8 line decoder, with active low outputs, is used to implement a 3-variable Boolean function as shown in the figure.**

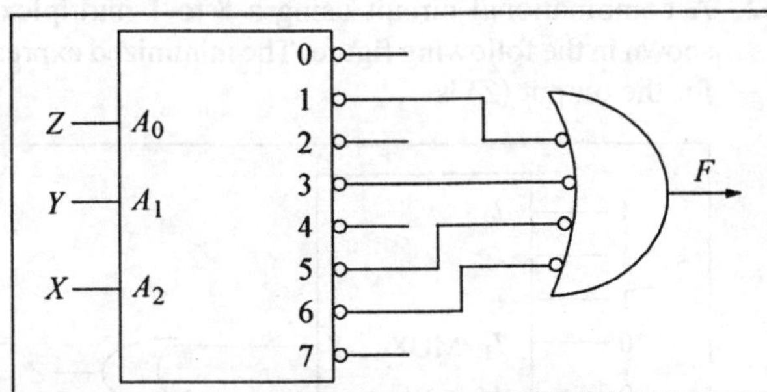

The simplified form of Boolean function $F(A, B, C)$ implemented in 'Product of Sum' form will be

(a) $(X+Z)\cdot(\bar{X}+\bar{Y}+\bar{Z})\cdot(Y+Z)$
(b) $(\bar{X}+\bar{Z})\cdot(X+Y+Z)\cdot(\bar{Y}+\bar{Z})$
(c) $\left[\begin{array}{l}(\bar{X}+\bar{Y}+\bar{Z})\cdot(\bar{X}+Y+\bar{Z})\cdot(X+Y+\bar{Z})\\ \cdot(X+\bar{Y}+\bar{Z})\end{array}\right]$
(d) $\left[\begin{array}{l}(\bar{X}+\bar{Y}+Z)\cdot(\bar{X}+Y+Z)\cdot(X+Y+Z)\\ \cdot(X+\bar{Y}+\bar{Z})\end{array}\right]$

22. **The output Y of the logic circuit given below is**

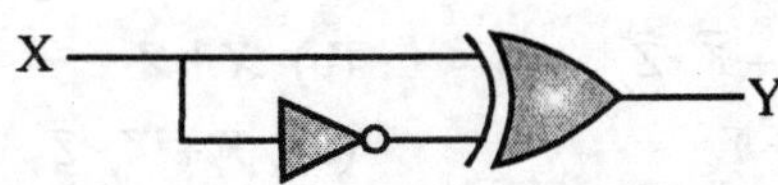

(a) 1 (b) 0
(c) x (d) $\bar{x}$

23. **The output Y of 2-bit comparator is logic 1 whenever the 2-bit input A is greater than the 2-bit input B. The number of combinations for which the output is logic 1, is**

(a) 4 (b) 6
(c) 8 (d) 10

24. **A combinational circuit has inputs A, B and C and its Karnaugh map is given in figure. The output of the circuit is:**

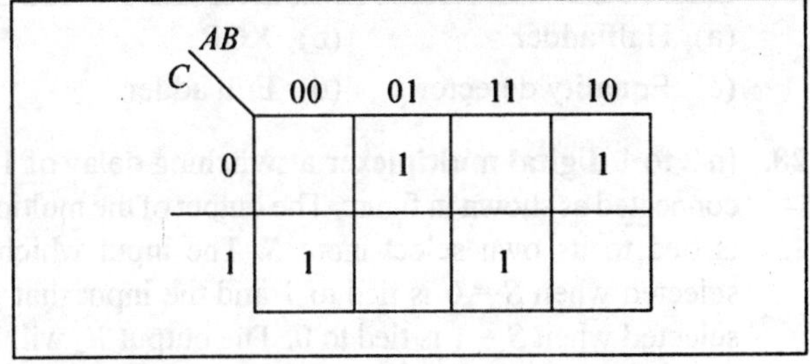

C \ AB	00	01	11	10
0		1		1
1	1		1	

(a) $(\overline{AB} + AB)\bar{C}$ (b) $(\bar{A}B + A\bar{B})C$
(c) $A \oplus B \oplus C$ (d) $\bar{A}\,\bar{B}\,\bar{C}$

25. **The combinatorial circuit shown in figure employs a 4-to-1 multiplexer. The output Q of the circuit is:**

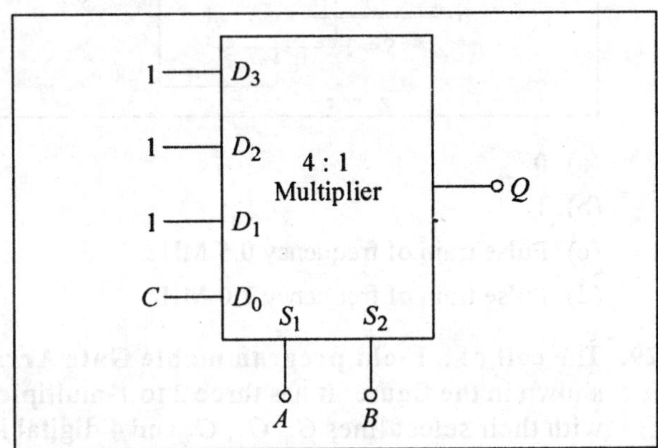

(a) $\bar{A}BC$ (b) $A + B + C$
(c) $A \oplus B \oplus C$ (d) $\overline{ABC}$

26. **In the logic circuit shown in figure the output x is**

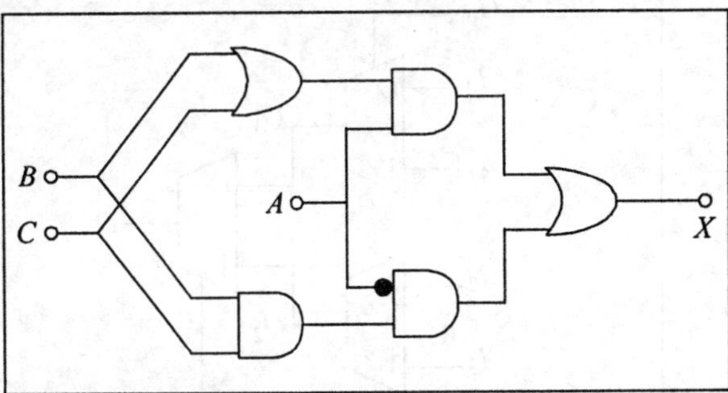

(a) $A\bar{B} + B\bar{C} + C\bar{A}$
(b) $A + B + C$
(c) $AB + BC + CA$
(d) $\overline{AB} + \overline{BC} + \overline{CA}$

27. The logic circuit of figure is a

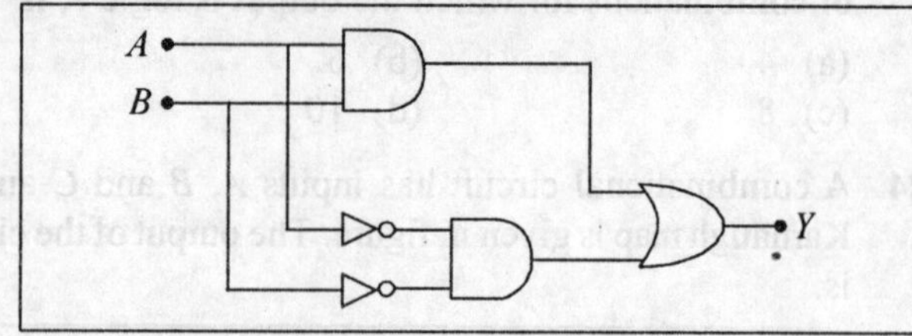

(a) Half adder (b) XOR
(c) Equality detector (d) Full adder

28. In 2 to 1 **digital multiplexer** a switching delay of 1 μs is connected as shown in figure. The output of the multiplexer is tied to its own select input *S*. The input which gets selected when $S = 0$ is tied to 1 and the input that get is selected when $S = 1$ is tied to 0. The output V_o will be

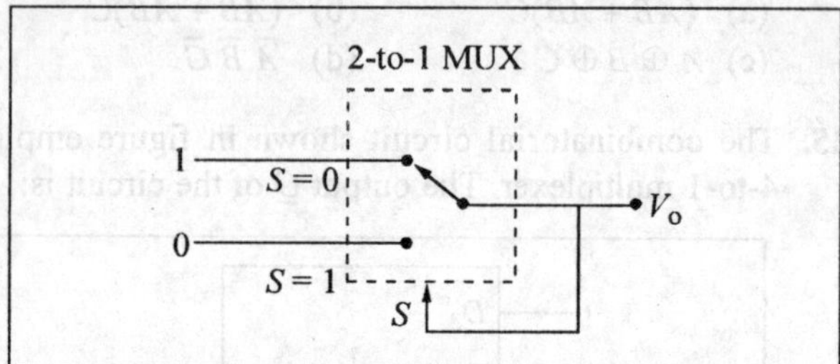

(a) 0
(b) 1
(c) Pulse train of frequency 0.5 MHz.
(d) Pulse train of frequency 1.0 MHz.

29. The cell of a **Field programmable Gate Array** is shown in the figure. It has three 2 to 1-multiplexers with their select lines G_0, G_1, G_2 and 4 digital input lines I_0, I_1, I_2 and I_3. The logical function that relates the output *O* to the select and signal input lines is

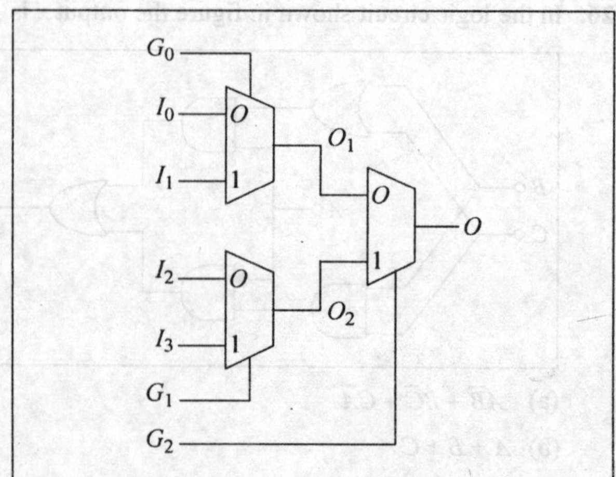

(a) $\overline{G_0}\,\overline{G_1}I_2 + \overline{G_0}\,G_1I_3 + \overline{G_2}\,\overline{G_1}I_0 + \overline{G_2}\,G_1I_1$

(b) $\overline{G_0}\,I_2 + \overline{G_0}\,G_1 + \overline{G_2}\,I_0 + G_2G_1I_1 + G_0$

(c) $\overline{G_0}\,\overline{G_2}\,I_0 + G_0\,\overline{G_2}\,I_1 + G_2\overline{G_1}I_2 + G_2G_1I_3$

(d) $G_2G_1\overline{I_2} + \overline{G_2}\,\overline{G_1}\,\overline{I_3} + G_2\,\overline{G_0}\,I_0 + G_0\,\overline{G_2}\,I_1$

30. A combinational logic circuit has three input *A*, *B* and *C* one output *Y*. The output $Y = 1$ when at least two input are 1. Otherwise, $Y = 0$. In its minimized *SOP* realization, the maximum number of two-input terms is

(a) 1 (b) 2
(c) 3 (d) 4

31. Notwithstanding overflow, the addition and subtraction of K-bit signed binary numbers can be realized using

(a) *K* 2-input AND Gates
(b) *K* 2-input NOR Gates
(c) *K* 2-input OR Gates
(d) *K* 2-input XOR Gates

32. A combinational circuit using a 8-to-1 multiplexer is shown in the following figure. The minimized expression for the output (*Z*) is

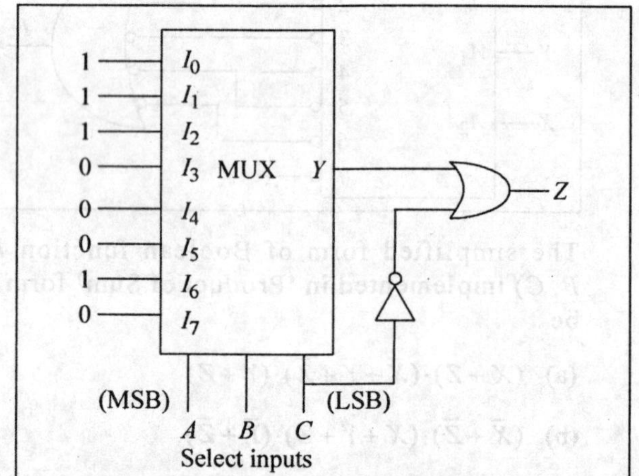

(a) $C(\overline{A}+\overline{B})$ (b) $C(A+B)$
(c) $\overline{C}+\overline{A}\,\overline{B}$ (d) $\overline{C}+AB$

33. A logic circuit implements the Boolean function $F = \overline{X}\cdot Y + X\cdot\overline{Y}\cdot\overline{Z}$. It is found that the input combination $X = Y = 1$ can never occur. Taking this into account, a simplified expression for F is given by

(a) $\overline{X}+\overline{Y}\cdot\overline{Z}$ (b) $X+Z$
(c) $X+Y$ (d) $Y+X\cdot\overline{Z}$

34. Let $X = X_1\,X_0$ and $Y = Y_1\,Y_0$ be unsigned 2-bit numbers. The function $F = 1$ if $X > Y$ and otherwise. The minimized sum of products expression for *F* is

(a) $Y_1\cdot Y_0 + X_0\cdot Y_0 + \overline{X}_1\cdot\overline{X}_0\cdot\overline{Y}_1$

(b) $X_0 \cdot \overline{Y_1} + Y_1 \cdot \overline{Y_0} + X_1 \cdot \overline{X_0}$

(c) $Y_1 \cdot \overline{X_1} + Y_0 \cdot \overline{X_1} + Y_1 \cdot Y_0 \cdot \overline{X_0}$

(d) $X_1\overline{Y_1} + X_0 \cdot \overline{Y_0} \cdot \overline{Y_1} + X_0 \cdot X_1 \cdot \overline{Y_0}$

35. A MUX circuit shown in the figure below implements a logic function F_1.

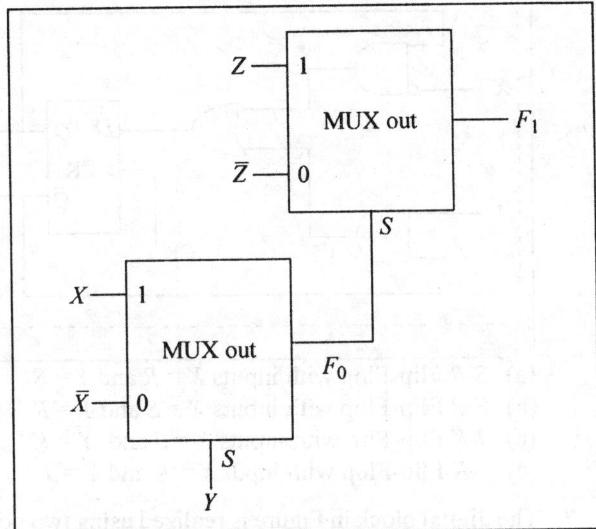

The correct expression for F_i is

(a) $(\overline{X \oplus Y}) \oplus Z$ (b) $\overline{(\overline{X \oplus Y}) \oplus Z}$

(c) $(X \oplus Y) \oplus \overline{Z}$ (d) $(X \oplus Y) + Z$

36. The minimal sum-of-products expression for the logic function f represented by the given Karnaugh map is

(a) $QS + P\overline{R}S + PQR + \overline{P}RS + \overline{P}Q\overline{R}$

(b) $\overline{QS} + \overline{P}R\overline{S} + \overline{PQ}R + P\overline{R}\,\overline{S} + P\overline{Q}R$

(c) $\overline{P}R\overline{S} + \overline{P}\,\overline{Q}\,\overline{R} + P\overline{R}\,\overline{S} + P\overline{Q}R$

(d) $P\overline{R}S + PQR + \overline{P}RS + \overline{P}QR + \overline{P}Q\overline{R}$

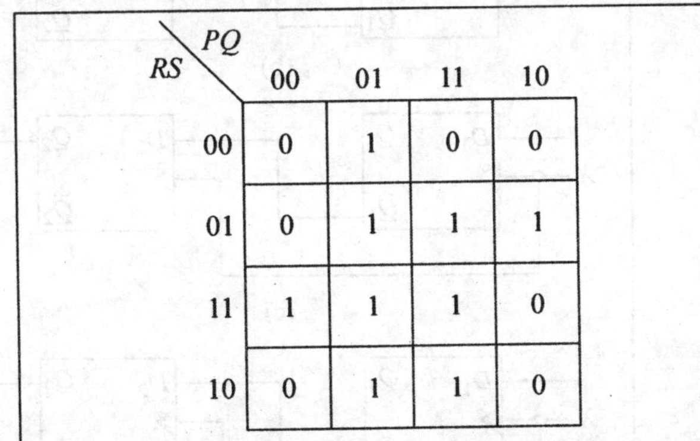

RS \ PQ	00	01	11	10
00	0	1	0	0
01	0	1	1	1
11	1	1	1	0
10	0	1	1	0

37. The output F of the multiplexer circuit shown below expressed in terms of the inputs P, Q and R is

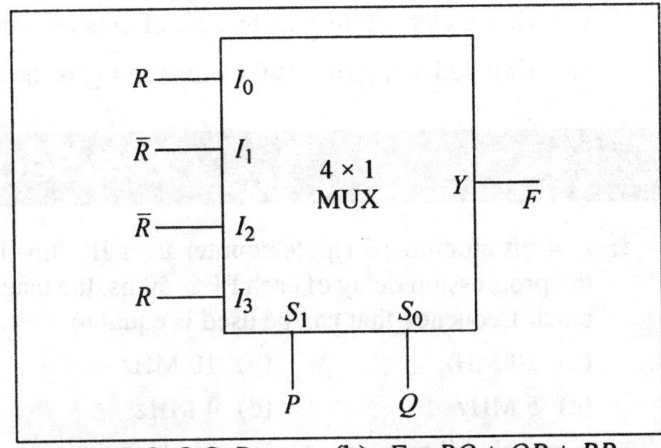

(a) $F = P \oplus Q \oplus R$ (b) $F = PQ + QR + RP$

(c) $F = (P \oplus Q)R$ (d) $F = (P \oplus Q)\overline{R}$

38. The logic gate circuit shown in the adjoining figure realizes the function

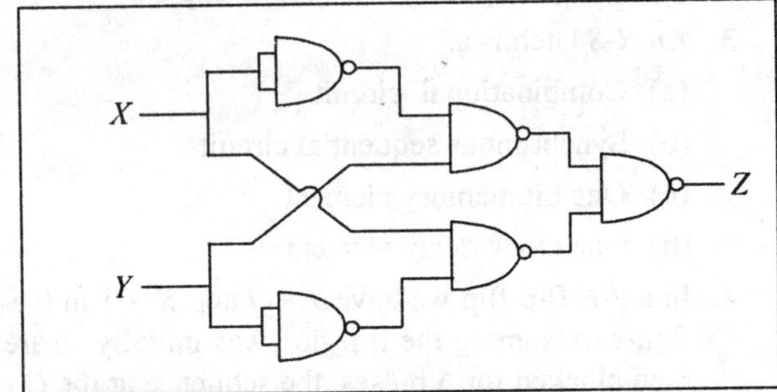

(a) XOR (b) XNOR

(c) Half adder (d) Full adder

39. For a binary half-subtractor having two inputs A and B, the correct set of logical expression for the outputs D (= A minus B) and X (= borrow are)

(a) $D = \overline{A}B + A\overline{B} + AB, \overline{X} = A\overline{B}$

(b) $D = \overline{A}B + A\overline{B} + A\overline{B}, X = A\overline{B}$

(c) $D = \overline{A}B + A\overline{B}, X = \overline{A}B$

(d) $D = AB + \overline{A}\,\overline{B}, X = A\overline{B}$

40. The minimum number of 2-to-1 multiplexers required to realize a 4-to-1 multiplexers is

(a) 1 (b) 2

(c) 3 (d) 4

41. In the circuit of figure A, B, C are the inputs and P, Q are the two outputs. The circuit is a

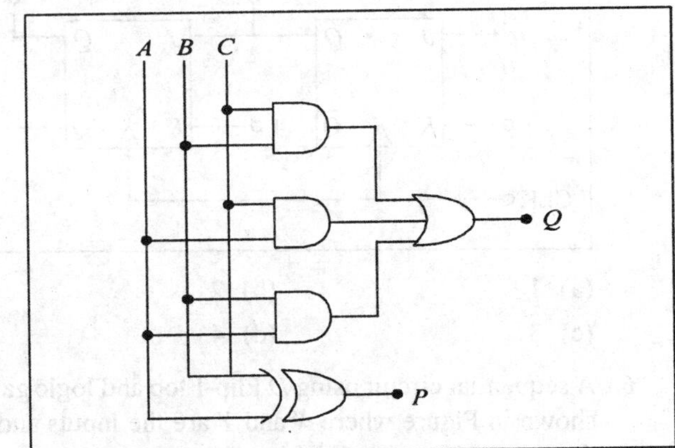

(a) Half adder where P is the sum of Q is the carry
(b) Half adder where P is the carry and Q is the sum
(c) Full adder where P is the sum and Q is the carry
(d) Full adder where P is the carry and Q is the sum

8.4 Sequential Circuits

1. A 4-bit modulo-16 ripple counter uses JK flip-flop. If the progression delay of each FF is 50 ns, the maximum clock frequency that can be used is equal to
(a) 20 MHz (b) 10 MHz
(c) 5 MHz (d) 4 MHz

2. A switch tail ring counter is made by using a single D flip-flop. The resulting circuit is a
(a) SR flip-flop (b) JK flip-flop
(c) D flip-flop (d) T flip-flop

3. An R-S latch is a
(a) Combinational circuit
(b) Synchronous sequential circuit
(c) One bit memory element
(d) One clock delay element

4. In a J-K flip-flip we have $J = \bar{Q}$ and $K = 1$ in the given figure. Assuming the flip flop was initially cleared and then clocked for 6 pulses, the sequence at the Q output will be

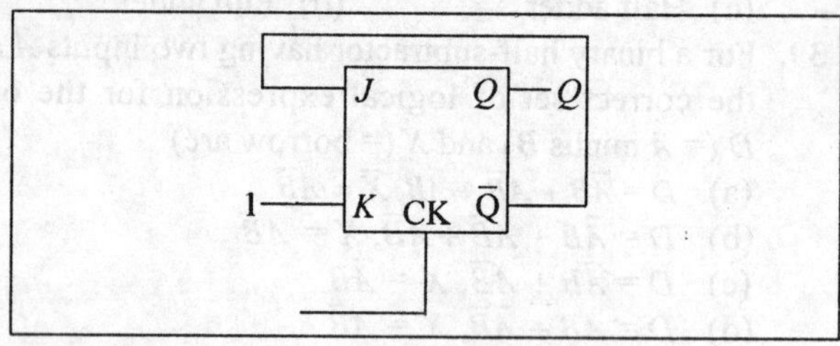

(a) 010000 (b) 011001
(c) 010010 (d) 010101

5. Figure shows a mod-K-counter, here K is equal to

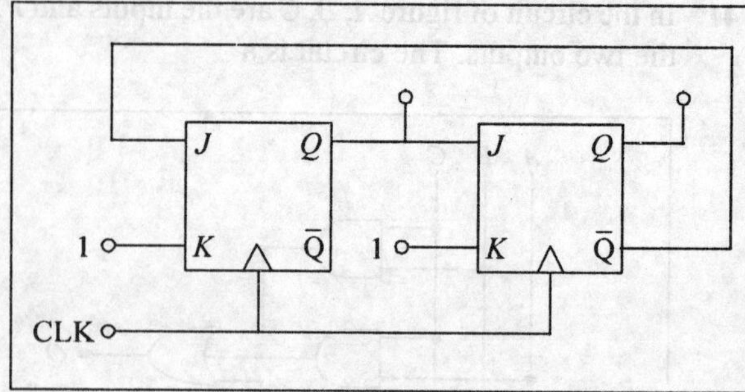

(a) 1 (b) 2
(c) 3 (d) 4

6. A sequential circuit using D Flip-Flop and logic gates is shown in Figure where X and Y are the inputs and Z is the output. The circuit is:

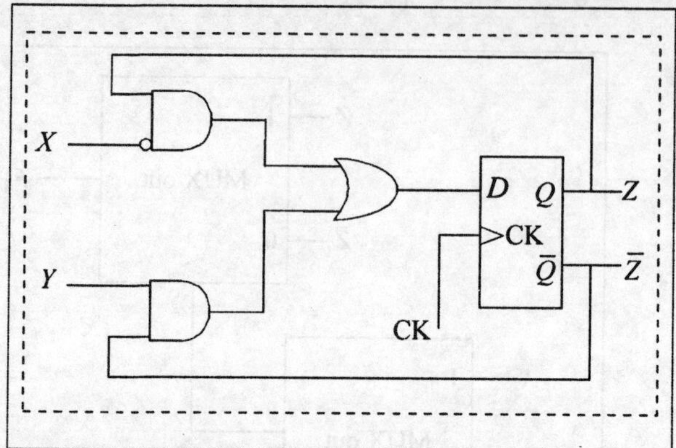

(a) S-R Flip-Flop with inputs $X = R$ and $Y = S$
(b) S-R Flip-Flop with inputs $X = S$ and $Y = R$
(c) J-K Flip-Flip with inputs $X = J$ and $Y = K$
(d) J-K Flip-Flop with input $X = K$ and $Y = J$

7. The digital block in Figure is realized using two positive edge triggered D-flip-flops. Assume that for $t < t_0$, $Q_1 = Q_2 = 0$. The circuit in the digital block is given by:

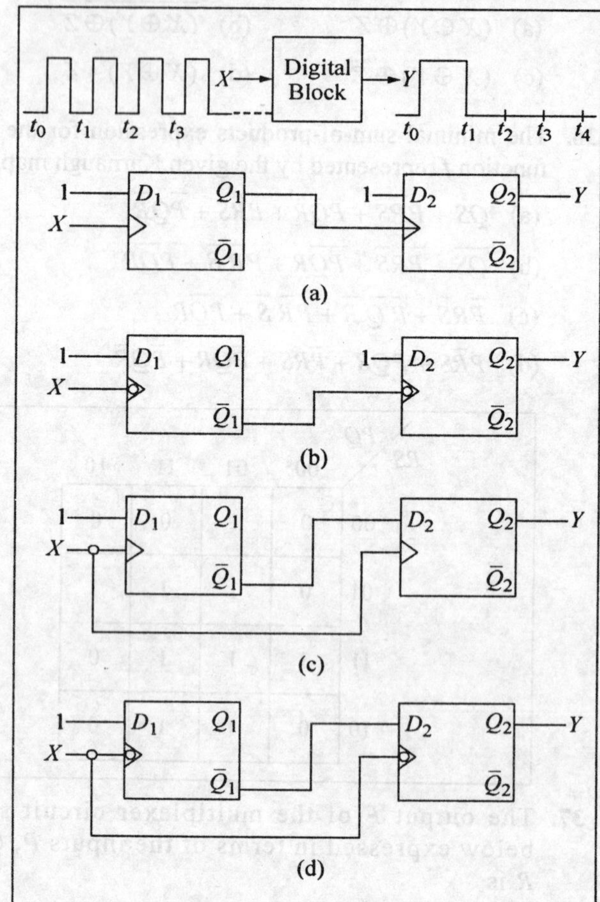

(a) Figure (a) (b) Figure (b)
(c) Figure (c) (d) Figure (d)

8. A 0 to 6 counter consists of 3 flip flops and a combination circuit of 2 input gate(s). The combination circuit consists of:
(a) One AND gate
(b) One OR gate
(c) One AND gate and one OR gate
(d) Two AND gates

9. A 4 bit ripple counter and a 4 bit synchronous counter are made using flip flops having a propagation delay of 10 ns each. If the worst case delay in the ripple counter and the synchronous counter be R and S respectively then:
(a) $R = 10$ ns, $S = 40$ ns (b) $R = 40$ ns, $S = 10$ ns
(c) $R = 10$ ns, $S = 30$ ns (d) $R = 30$ ns, $S = 10$ ns

10. A master-slave flip-flop has the characteristic that
(a) Change in the input immediately reflected in the output
(b) Change in the output occurs when the state of master is affected
(c) Change in the output occurs when the state of the slave is affected
(d) Both the master and the slave states are affected at the same time.

11. Choose the correct one from among the alternatives *a*, *b*, *c*, *d* after matching an item from Group 1 with the most appropriate item in Group 2.

Group-1	Group-2
P : Shift register	1 : Frequency division
Q : Counter	2 : Addressing in memory chips
R : Decoder	3 : Serial to parallel data conversion

	P	Q	R
(a)	3	2	1
(b)	3	1	2
(c)	2	1	3
(d)	1	2	2

12. In the modulo-6 ripple counter shown in figure, the output of the 2-input gate is used to clear the J-K flip-flops:

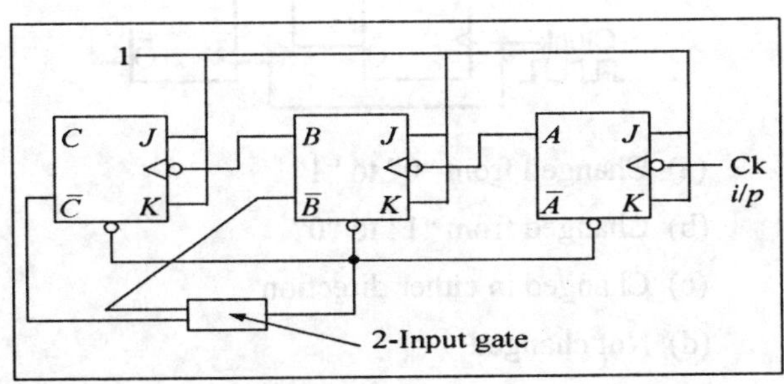

The 2-input gate is
(a) a NAND gate (b) a NOR gate
(c) an OR gate (d) an AND gate

13. In figure, the J and K inputs of all the four Flip-Flops are made high. The frequency of the signal at output Y is:

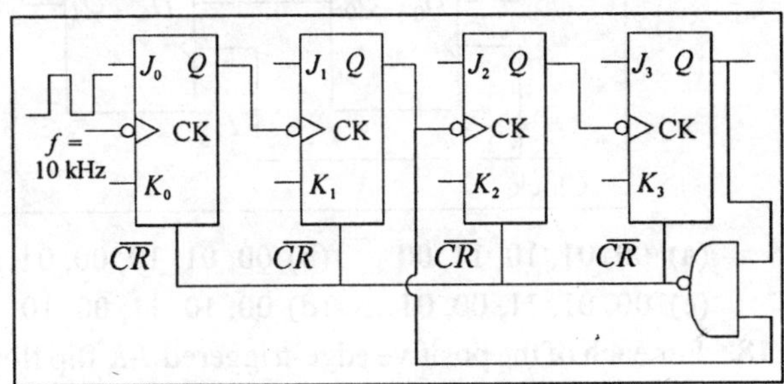

(a) 0.833 kHz (b) 1.0 kHz
(c) 0.91 kHz (d) 0.77 kHz

14. The present output Q_n of an edge triggered JK flip-flop is logic 0. If $J = 1$, then Q_{n+1}
(a) Cannot be determined (b) Will be logic 0
(c) Will be logic 1 (d) Will race around.

15. The given figure shows a ripple counter using positive edge triggered flip-flops. If the present state of the counter is $Q_2\,Q_1\,Q_0 = 011$, then its next state $(Q_2\,Q_1\,Q_0)$ will be:

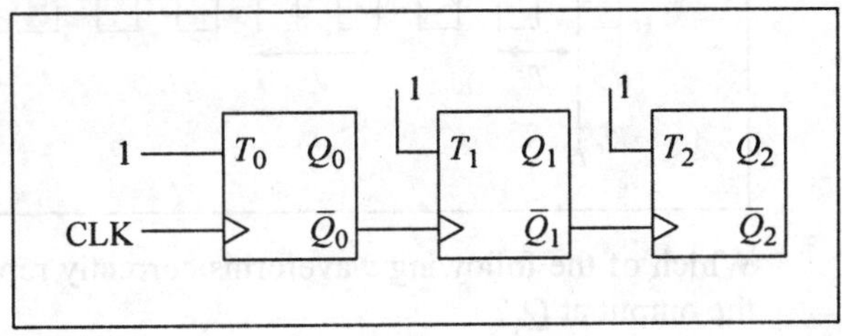

(a) 010 (b) 100
(c) 111 (d) 101

16. Two D-flip-flops as shown below, are to be connected as a synchronous counter that goes through the following $Q_1\,Q_0$ sequence.

$00 \rightarrow 01 \rightarrow 11 \rightarrow 10 \rightarrow 00$ ....

The inputs D_0 and D_1, respectively should be connected as:

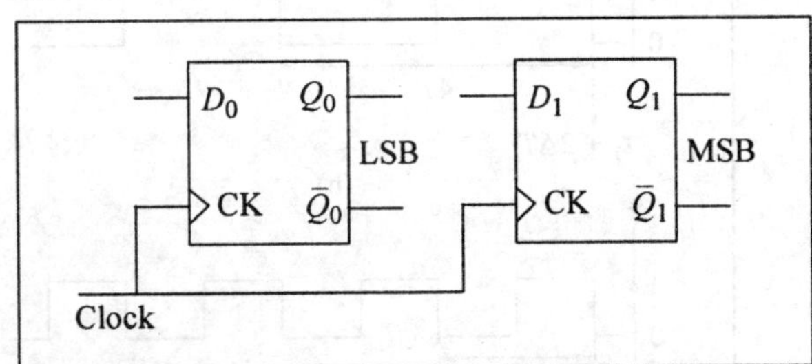

(a) $\bar{Q}_1$ and Q_0 (b) $\bar{Q}_0$ and Q_1
(c) $\bar{Q}_1\,Q_0$ and $\bar{Q}Q_0$ (d) $\bar{Q}_1\,\bar{Q}$ and $Q_1\,Q_0$

17. For the circuit shown, the counter state (Q_1, Q_0) follows the sequence

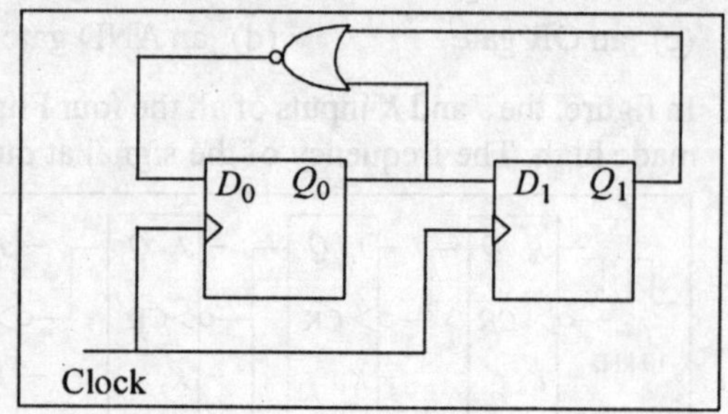

(a) 00, 01, 10, 11, 00… (b) 00, 01, 10, 00, 01…
(c) 00, 01, 11, 00, 01… (d) 00, 10, 11, 00, 10…

18. For each of the positive edge-triggered *J-K* flip flop used in the following figure, the propagation delay is ΔT.

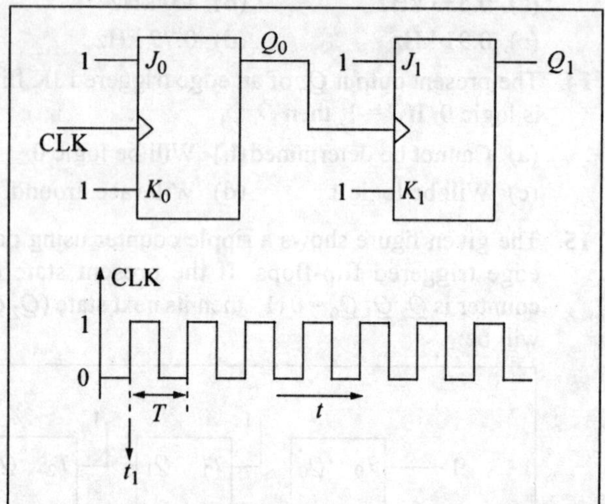

Which of the following waveforms correctly represent the output at Q_1?

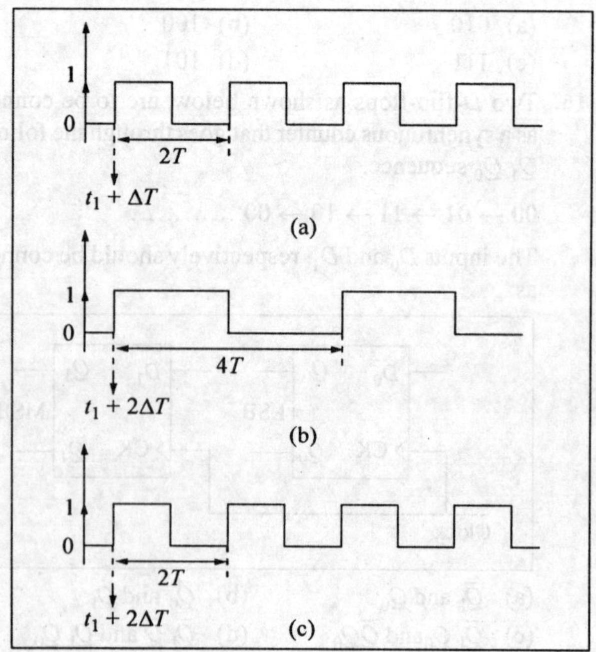

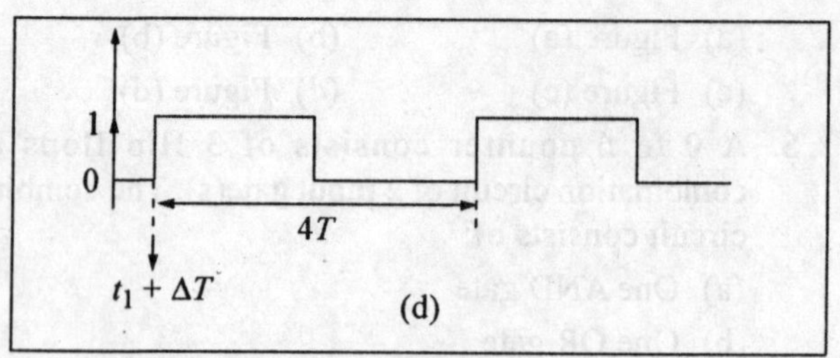

19. What are the counting states (Q_1, Q_2) for the counter shown in the figure below ?

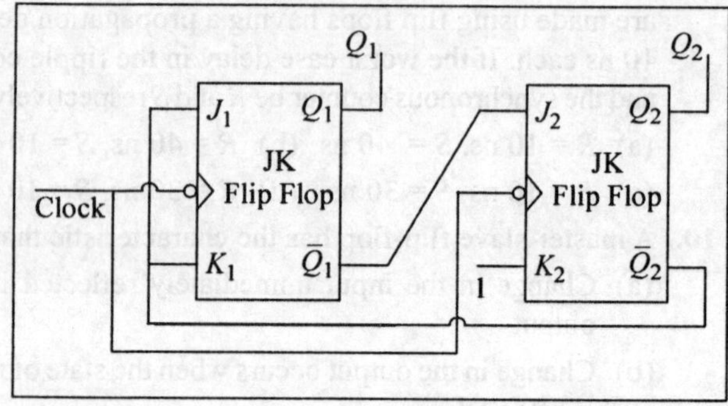

(a) 11, 10, 00, 11, 10, ... (b) 01, 10, 11, 00, 01, ...
(c) 00, 11, 01, 10, 00, ... (d) 01, 10, 00, 01, 10, ...

20. Assuming that all flip-flops are in reset condition initially, the count sequence observed at Q_A in the circuit sown is

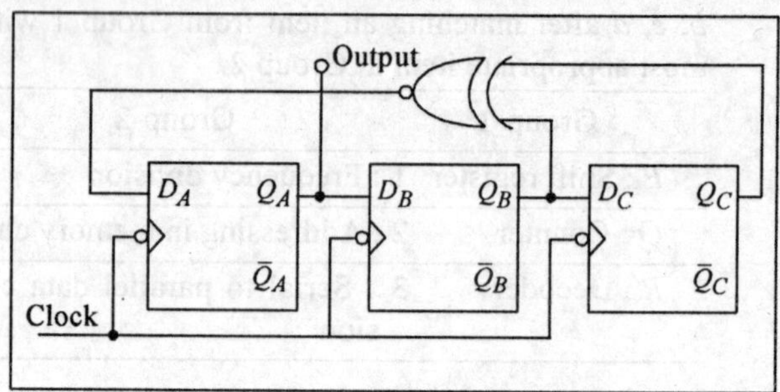

(a) 0010111 (b) 0001011
(c) 0101111 (d) 0110100

21. When the output Y in the circuit below is "1", it implies that data has

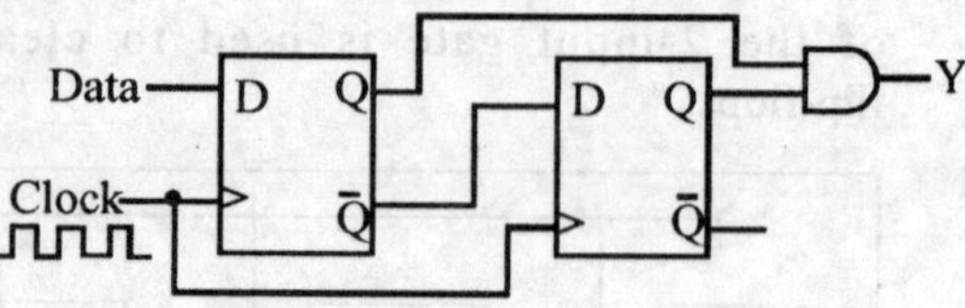

(a) Changed from "0" to "1"
(b) Changed from "1" to "0"
(c) Changed in either direction
(d) Not changed

22. Consider the given circuit

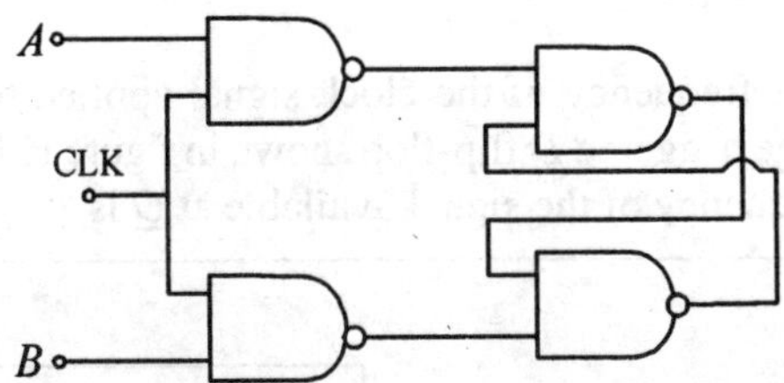

In this circuit, the race around

(a) Does not occur

(b) Occurs when CLK = 0

(c) Occur when CLK = 1 and A = B = 1

(d) Occurs when CLK = 1 and A = B = 0

23. The output of a 3-stage Johnson (twisted - ring) counter is fed to a digital-to-analog (D/A) converter as shown in the figure below. Assume all states of the counter to be unset initially. The waveform which represents the D/A converter output V_0 is

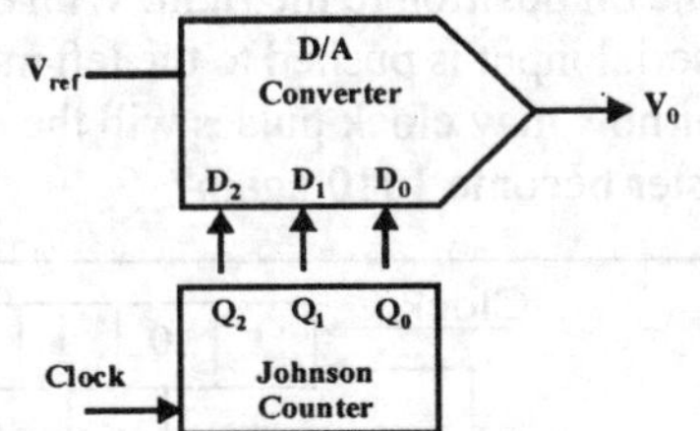

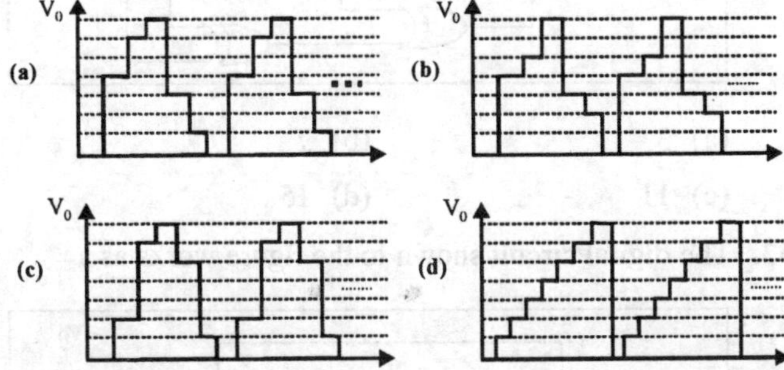

24. Two D flip-flops are connected as a synchronous counter that goes through the following $Q_B\ Q_A$ sequence 00→ 11→ 01→ 10 →100 → ...

The connections to the inputs D_A and D_B are

(a) $D_A = Q_B,\ D_B = Q_A$

(b) $D_A = \bar{Q}_A,\ D_B = \bar{Q}_B$

(c) $D_A = (Q_A\bar{Q}_B + \bar{Q}_A Q_B),\ D_B = Q_A$

(d) $D_A = (Q_A Q_B + \bar{Q}_A\bar{Q}_B),\ D_B = \bar{Q}_B$

25. The state transition diagram for the logic circuit shown is

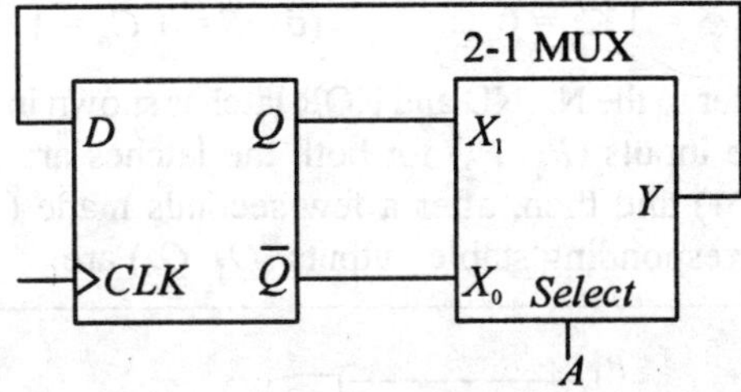

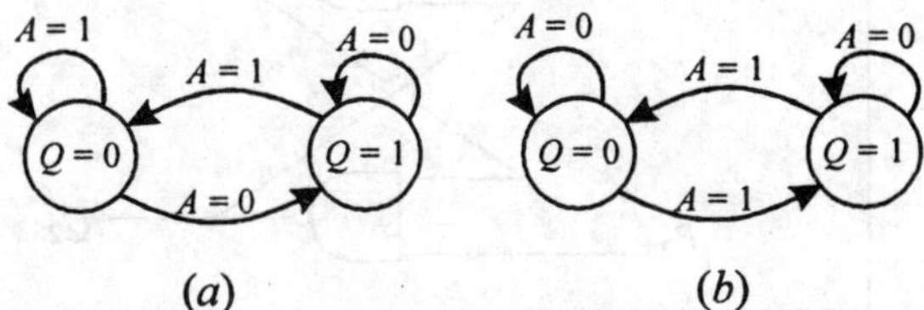

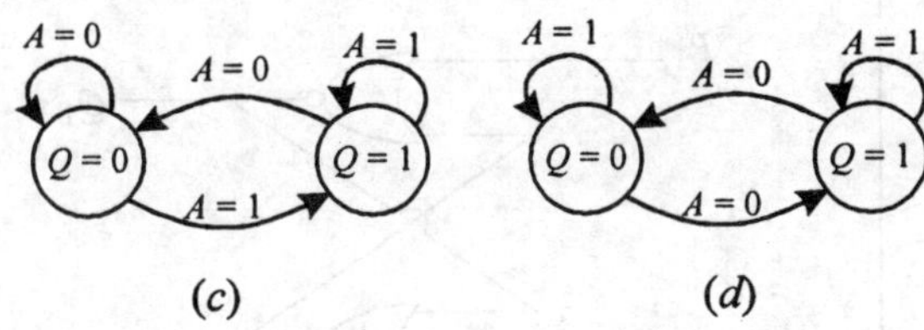

26. The initial contents of the 4-bit serial-in- parallel-out, right shift, shift Register shown in Figure is 0110. After three clock pulses applied, the contents of the Shift Register will be

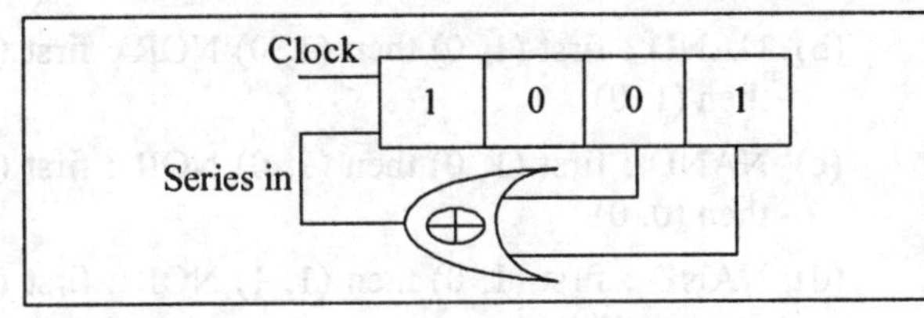

(a) 0 0 0 0 (b) 0 1 0 1

(c) 1 0 1 0 (d) 1 1 1 1

27. For the circuit shown in figure below, two 4-bit parallel-in serial-out shift registers loaded with the data shown are used to feed the data to a full adder. Initially, all the flip-flops are in clear state. After applying two clock pulses, the outputs of the full-adder should be

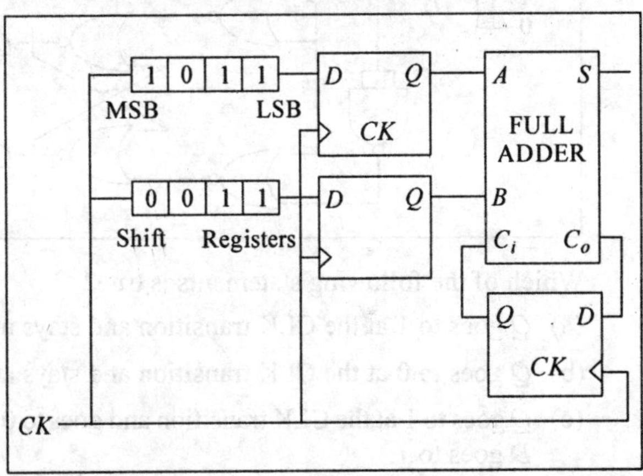

(a) $S = 0$ $C_o = 0$ (b) $S = 0$ $C_o = 1$
(c) $S = 1$ $C_o = 0$ (d) $S = 1$ $C_o = 1$

28. Refer to the NAND and NOR latches shown in the figure. The inputs (P_1, P_2) for both the latches are first made (0, 1) and then, after a few seconds made (1, 1). The corresponding stable outputs (Q_1, Q_2) are

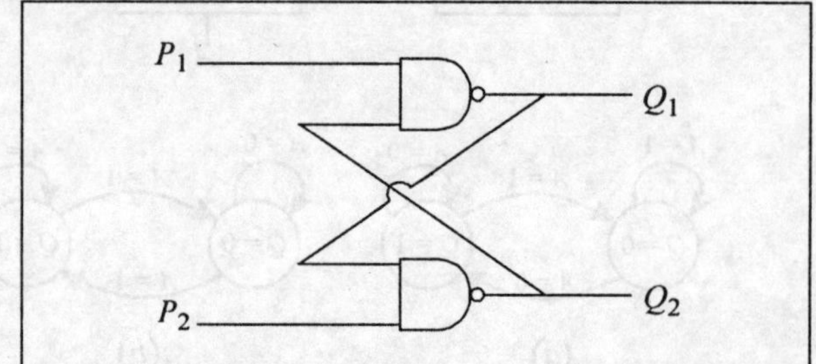

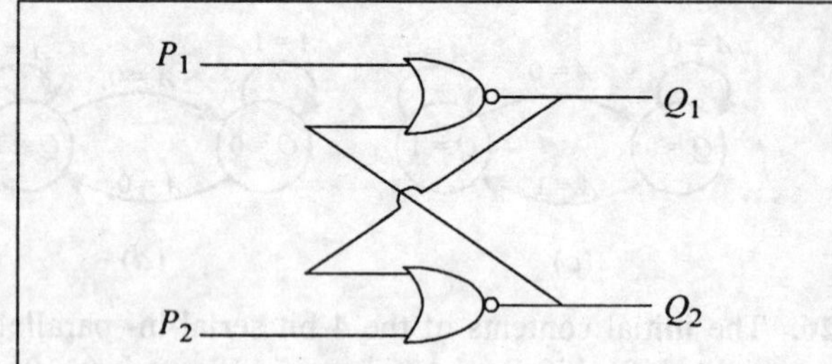

(a) NAND : first (0, 1) then (0, 1) NOR : first (1, 0) then (0, 0)

(b) NAND : first (1, 0) then (1, 0) NOR : first (1, 0) then (1, 0)

(c) NAND : first (1, 0) then (1, 0) NOR : first (1, 0) then (0, 0)

(d) NAND : first (1, 0) then (1, 1) NOR : first (0, 1) then (0, 1)

29. For the circuit shown in the figure, D has a transition from 0 to 1 after CLK changes from 1 to 0. Assume gate delays to be negligible.

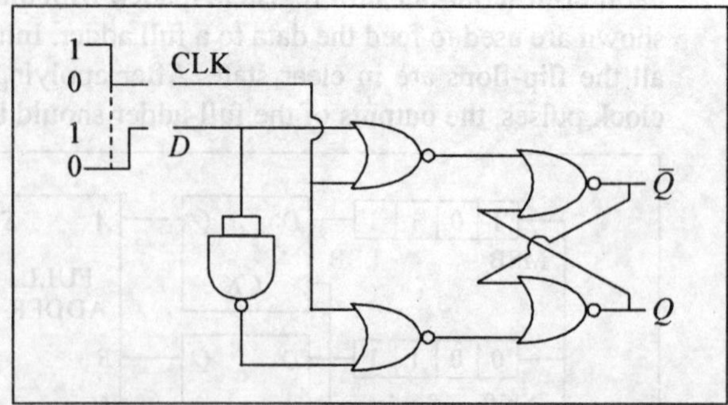

Which of the following statements is true?

(a) Q goes to 1 at the CLK transition and stays at 1.

(b) Q goes to 0 at the CLK transition and stays at 0.

(c) Q goes to 1 at the CLK transition and goes to 0 when D goes to 1.

(d) Q goes to 0 at the CLK transition and goes to 1 when D goes to 1.

30. The frequency of the clock signal applied to the rising edge triggered D flip-flop shown in figure is 10 kHz. The frequency of the signal available at Q is

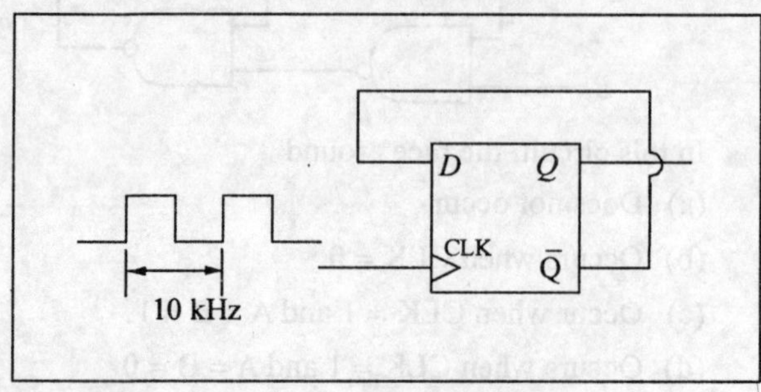

(a) 10 kHz (b) 2.5 kHz
(c) 20 kHz (d) 5 kHz

31. The shift register shown in figure is initially loaded with the pattern 1010. Subsequently, the shift register is clocked and with each clock pulse the pattern gets shifted by one bit position to the right. With each shift, the bit at the serial input is pushed to the left most position (msb). After how may clock pulses will the content of the shift register become 1010 again?

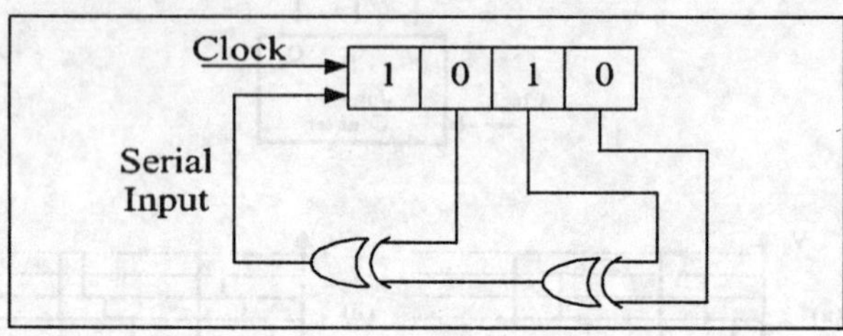

(a) 3 (b) 7
(c) 11 (d) 15

32. The digital circuit shown in the figure works as a

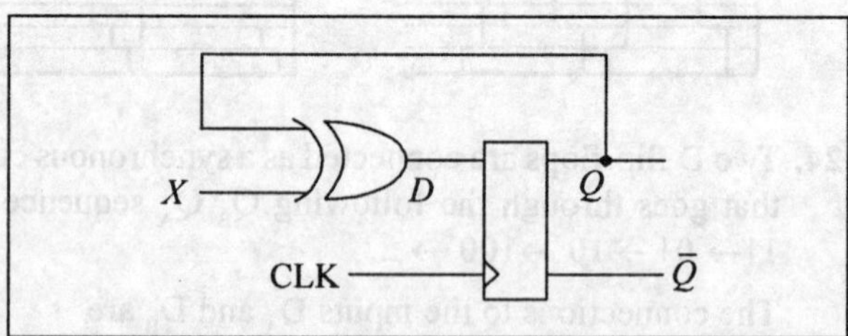

(a) JK flip-flop

(b) Clocked RS flip-flop

(c) T flip-flop

(d) Ring counter

33. Select the circuit which will produce the given output Q for the input signals X_1 and X_2 given in the figure

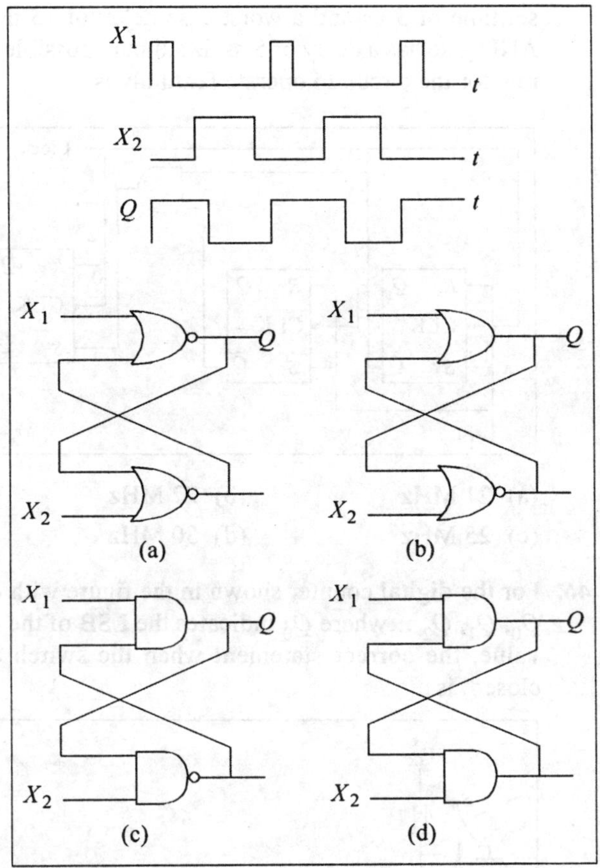

34. A two-bit counter circuit is shown below.

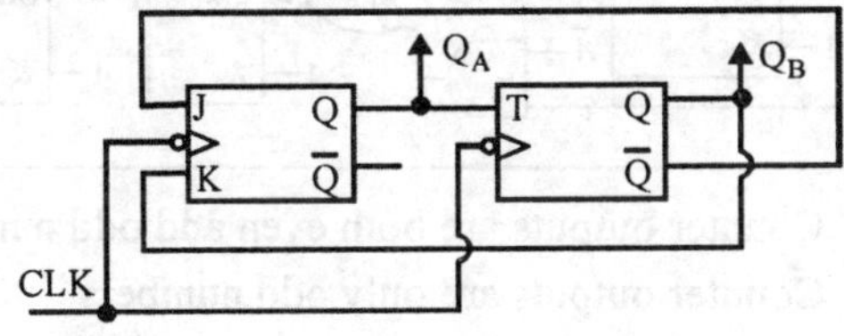

If the state Q_AQ_B of the counter at the clock time t_n is "10" then the state Q_AQ_B of the counter at $t_n + 3$ (after three clock cycles) will be

(a) 00 (b) 01
(c) 10 (d) 11

35. A dual-slope analog-to-digital converter uses an N-bit counter. When the input signal V_a is being integrated, the counter is allowed to count upto a value

(a) Equal to $2^N - 2$
(b) Equal to $2^N - 1$
(c) Proportional to V_a
(d) Inversely proportional to V_a

36. The output Q_{n+1} of a *J-K* flip-flop for the input $J = 1$, $K = 1$ is

(a) *O* (b) 1
(c) Q_n (d) $\bar{Q}_n$

37. The minimum number of flip-flops required to design a MOD-10 counter is

(a) 3 (b) 10
(c) 4 (d) 5

38. A 4-bit synchronous counter with a series carry, uses flip- flops and gates, having a propagation delay of a 30 ns and 10 ns, respectively. The maximum time-interval required between two successive clock pulses for reliable operation of the counter is:

(a) 10 ns (b) 30 ns
(c) 40 ns (d) 50 ns

39. The minimum number *f* flip-flops needed to make mod-2 counter is

(a) 1 (b) 2
(c) 3 (d) 10

40. In the circuit shown in figure when inputs $A = B = 0$, the possible logic states of *C* and *D* are.

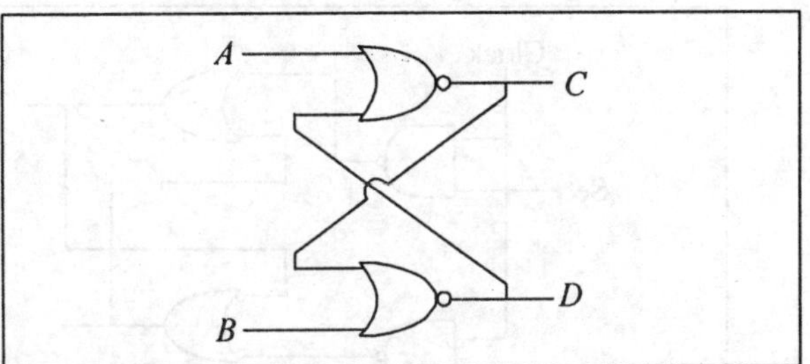

(a) $C = 0, D = 1$ or $C = 1, D = 0$
(b) $C = 1, D = 1$ or $C = 0, D = 0$
(c) $C = 1, D = 0$
(d) $C = 0, D = 1$

41. The correct cyclic sequence of the outputs (Q_0Q_1) for the JK master slave flip flop circuit shown in figure when the input clock is applied is

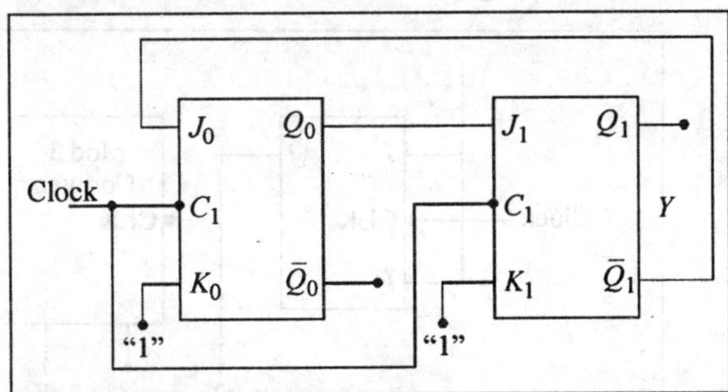

(a) 00, 01, 10, 11, 00, 01 ...
(b) 00, 10, 01, 00, 10, 01 ...
(c) 00, 01, 10, 00, 01, 10 ...
(d) 00, 11, 00, 11, 00, 11...

42. The square wave C_1 shown in figure is given to the clock input of a 4-bit binary up/down counter whose UP/DN input is fed with pulse train P_u. The counter is a negative edge triggered one. This counter starts with 0000 and will reach 0000 again at the

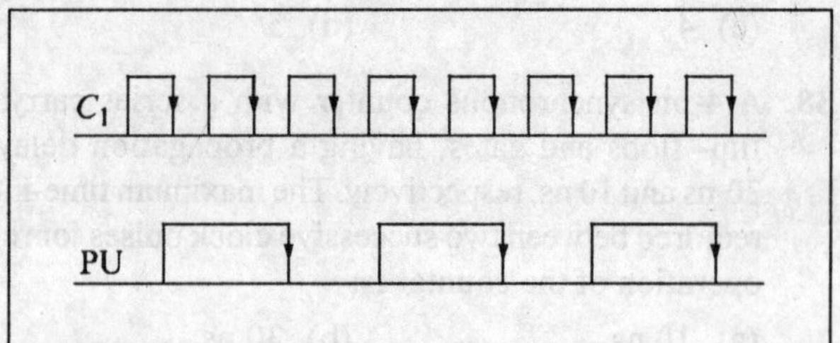

(a) 15th clock pulse
(b) 16th clock pulse
(c) 44th clock pulse
(d) 48th clock pulse

43. The two NAND gates before the latch circuit shown in figure are used to

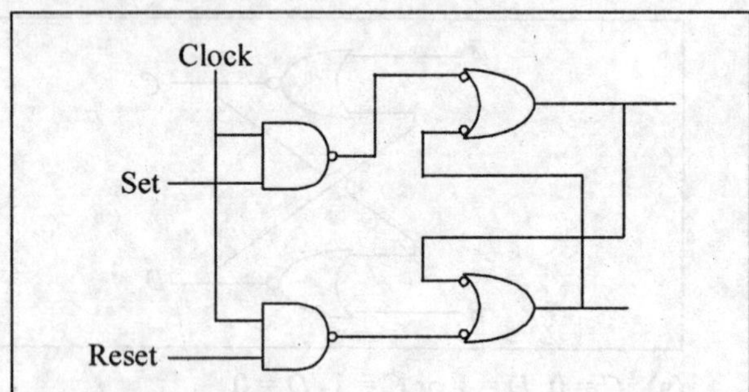

(a) Act as buffers
(b) Operate the latch faster
(c) Avoid racing problem
(d) Invert the latching action

44. In figure initially $Q = A = B = 0$. After three clock triggers, the states of Q, A and B will be respectively

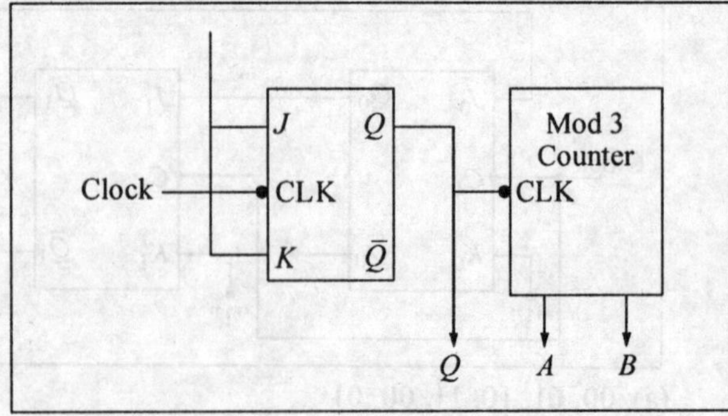

(a) 110 (b) 011
(c) 001 (d) 101

45. In the digital circuit shown in figure the flip-flop have set time of 5 ns and a worst case delay of 15 ns. The AND gate has a delay of 5 ns. Maximum possible clock rate for the circuit to operate faithfully is

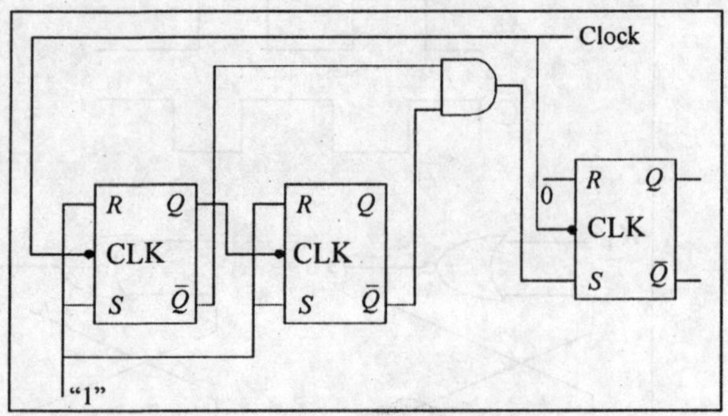

(a) 21 MHz (b) 22 MHz
(c) 25 MHz (d) 30 MHz

46. For the digital counter shown in the figure with ouput $Q_0, Q_1, Q_2, \ldots$ where Q_0 indicates the LSB of the count value, the correct statement when the switch *Sw* is closed, is

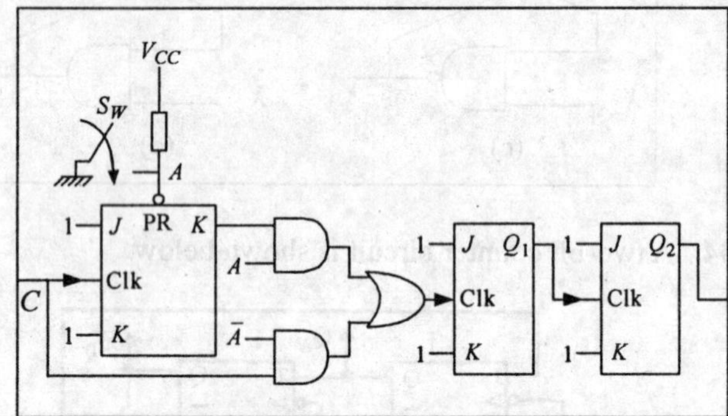

(a) Counter outputs are both even add odd numbers
(b) Counter outputs are only odd numbers
(c) Counter outputs are only even numbers
(d) Counter stops counting

47. Given that the initial state (Q_1 Q_0) is 00, the counting sequence of the counter shown in the following figure is, Q_1 Q_0 =

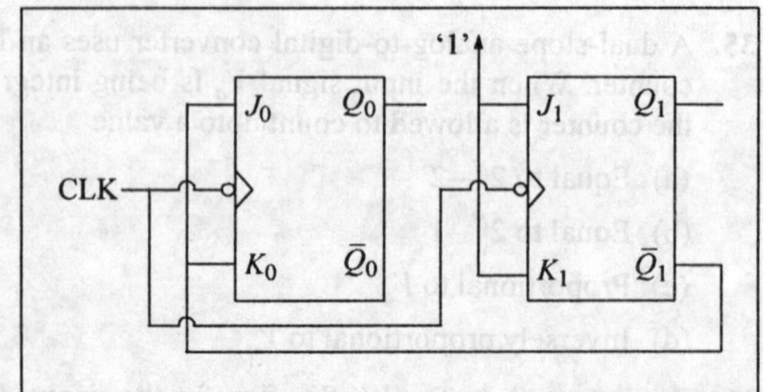

(a) 00-11-01-10-00 (b) 00-01-11-10-00
(c) 00-11-10-01-00 (d) 00-10-01-11-00

48. All the logic gates in the circuit shown below have finite propagation delay. The circuit can be used as a clock generator, if

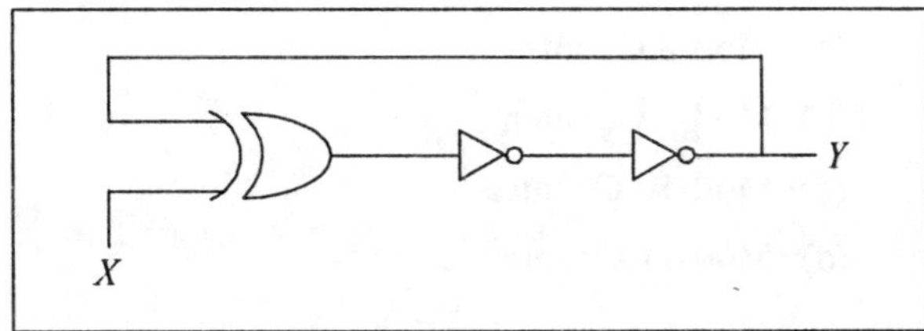

(a) $X = 0$ (b) $X = 1$
(c) $X = 0$ or 1 (d) $X = Y$

49. A sequential circuit is shown in the figure below. Let the state of the circuit be encoded as $Q_A Q_B$. The notation $X ® Y$ implies that state Y is reachable form state X in a finite number of clock transitions t.

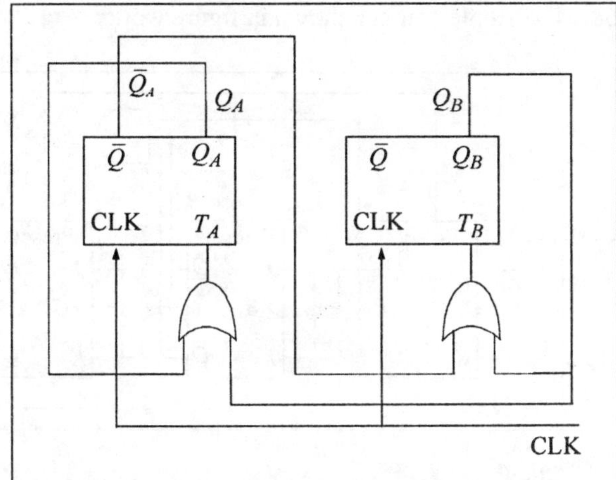

Identify the INCORRECT statement

(a) 01 → 00 (b) 11 → 01
(c) 01 → 11 (d) 01 → 10

50. The pulse width T of an asynchronous pulse is measured by a counter with an edge triggered clock of known frequency f_c as shown in the figure below:

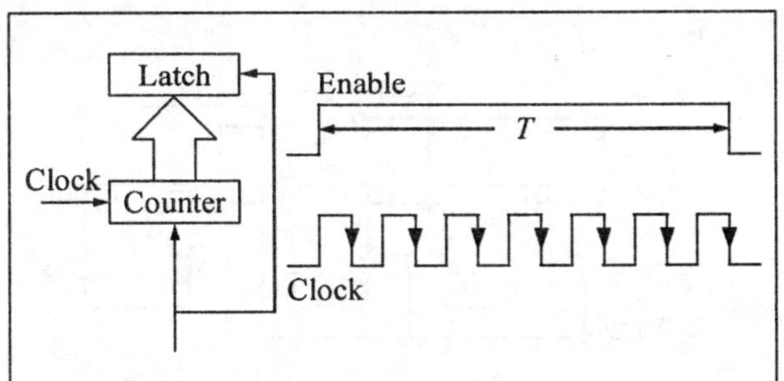

The pulse, whose width is to be measured, is applied to the Enable pin of the counter. The counter counts while the Enable is high and is held reset to zero otherwise. The counter output is latched by the negative edge of the Enable signal. The measured pulse width is taken to be N times the clock period, where N is the count reached at the end of a count cycle

Assuming no overflow, the measurement error will be limited to x % of T if

(a) $T > \frac{100}{x f_c}$

(b) $T < \frac{100}{x f_c}$

(c) $T > \frac{200}{x f_c}$

(d) $T < \frac{200}{x f_c}$

51. The inverters in the ring oscillator circuit shown be low are identical. If the output waveform has a frequency of 10 MHz, the propagation delay of each inverter is

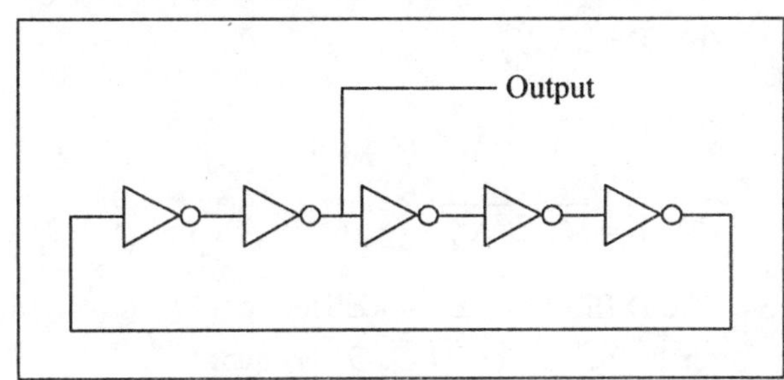

(a) 5 ns

(b) 10 ns

(c) 20 ns

(d) 50 ns

52. In the figure shown, the initial state of Q is 0. The output is observed after the application of each clock pulse. The output sequence at Q is

(a) 0 0 0 0 ...

(b) 1 0 1 0 ...

(c) 1 1 1 1 ...

(d) 1 0 0 0 ...

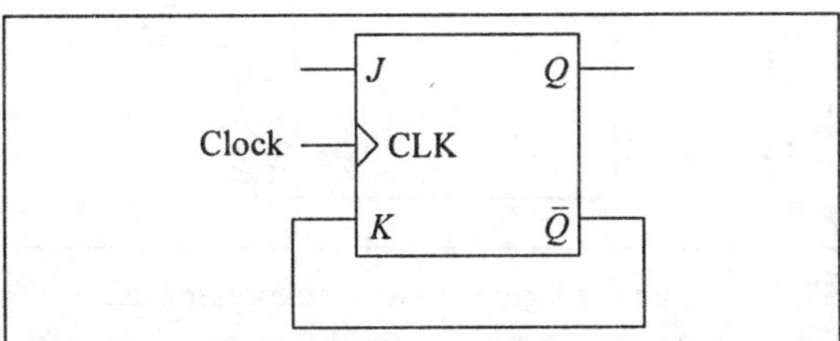

For Linked Answer Questions 53 and 54

Consider the circuit shown in the following figure.

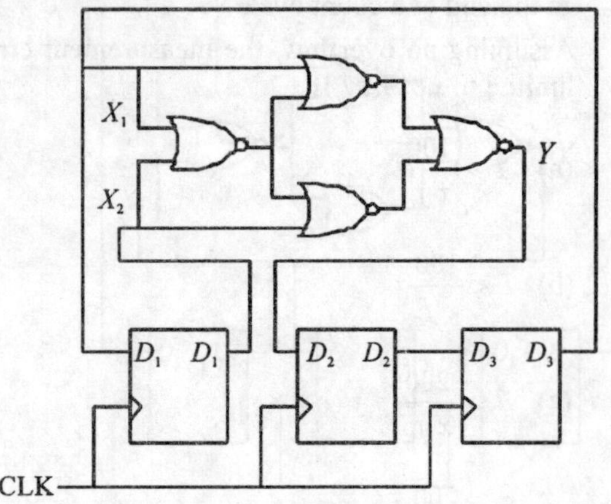

53. The correct input-output relationship between Y and (X_1, X_2) is

(a) $Y = X_1 + X_2$

(b) $Y = X_1X_2$

(c) $Y = X_1 \oplus X_2$

(d) $Y = \overline{X_1 \oplus X_2}$

54. The D flip-flops are initialized to $Q_1Q_2Q_3 = 100$. After 1 clock cycle, $Q_1Q_2Q_3$ is equal to

(a) 011 (b) 010

(c) 100 (d) 101

Statement for linked Answer Questions 55 and 56

Consider the counter circuit shown below:

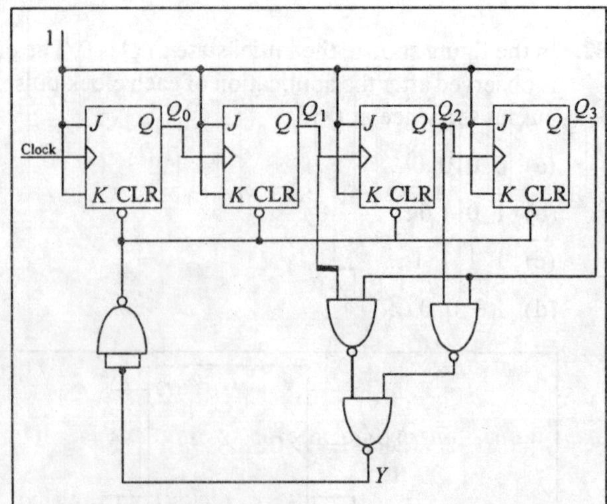

55. In the above figure, Y can be expressed as

(a) $Q_3(Q_2 + Q_1)$ (b) $Q_3 + Q_2\,Q_1$

(c) $Q_3(Q_2 + Q_1)$ (d) $Q_3 + Q_2Q_1$

56. The above circuit is a

(a) Mod-8 Counter

(b) Mod-9 Counter

(c) Mod-10 Counter

(d) Mod-11 Counter

57. A pulse train can be delayed by a finite number of clock periods using

(a) A serial-in serial-out shift register

(b) A serial-in parallel-out shift register

(c) A parallel-in serial -out shift register

(d) A parallel -in parallel -out shift register

58. The ripple counter shown in figure works as a:

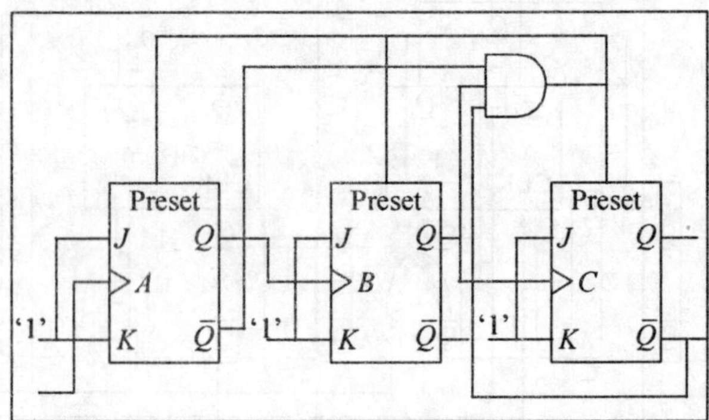

Clock

(a) Mod-3 up counter

(b) Mod-5 up counter

(c) Mod-3 down counter

(d) Mod-5 down counter

59. Assuming that all flip-flops are in reset condition initially, the count sequence observed at Q_A in the circuit shown is

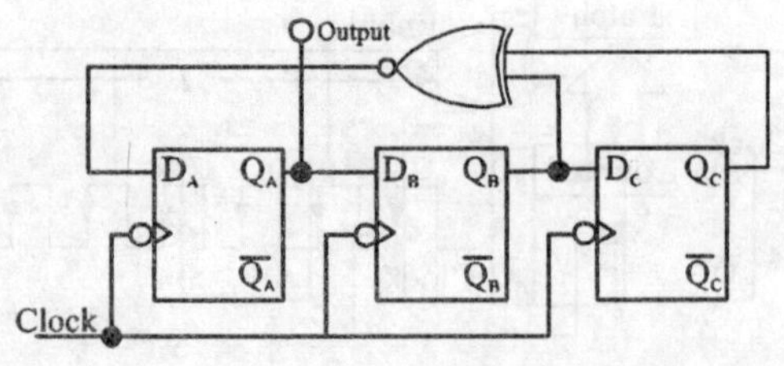

(a) 0010111 (b) 0001011

(c) 0101111 (d) 0110100

8.5 Semiconductor Memories

1. Choose the correct statement (s) from the following:
   1. PROM contains a programmable AND array and a fixed OR array
   2. PLA contains a fixed AND array and a programmable OR array
   3. PROM contains a fixed AND array and a programmable OR array
   4. PLA contains a programmable AND array and a programmable OR array.

   Of these
   (a) Only 1 is correct
   (b) 2 and 3 are correct
   (c) 3 and 4 are correct
   (d) 1 and 4 are correct

2. A PLA can be used
   (a) As a microprocessor
   (b) As a dynamic memory
   (c) To realize a sequential logic
   (d) To realize a combinational logic

3. A dynamic RAM consists of
   (a) 6 transistors
   (b) 2 transistors and 2 capacitors
   (c) 1 transistors and 1 capacitors
   (d) 2 capacitors only

4. The maximum number of MOS transistors required to make a dynamic RAM cell is
   (a) 1
   (b) 2
   (c) 3
   (d) 4

5. Each cell of a static Random Access Memory contains
   (a) 6 MOS transistors
   (b) 4 MOS transistors and 2 capacitors
   (c) 2 MOS transistors and 4 capacitors
   (d) 1 MOS transistors and 1 capacitors

6. If the inputs $X_3X_2X_1X_0$ to the ROM in Figure are 8-4-2-1 BCD numbers, then the outputs $Y_3Y_2Y_1Y_0$ are:
   (a) Gray code numbers
   (b) 2-4-2-1 BCD numbers
   (c) Excess-3 code numbers
   (d) None of the above.

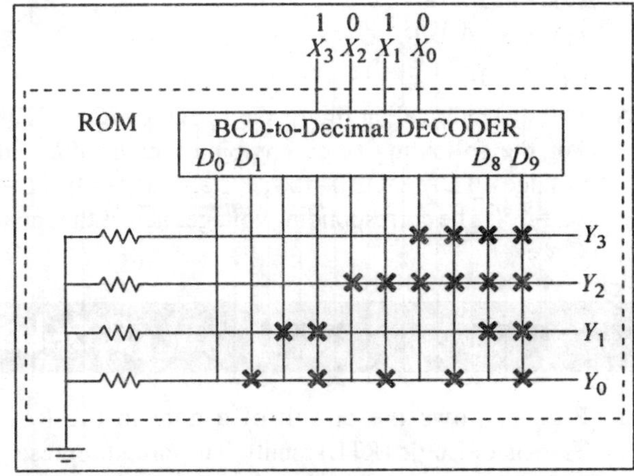

7. In the circuit shown in figure, A is a parallel-in-parallel-out 4-bit register, which loads at the rising edge of the clock C. The input lines are connected to a 4 bit bus, W. Its output acts as the input to a 16 × 4 ROM whose output is floating when the enable E is 0. A partial table of the contents of the ROM is as follows

Add	0	2	4	6	8	10	11	14
Data	0011	1111	0100	1010	1011	1000	0010	1000

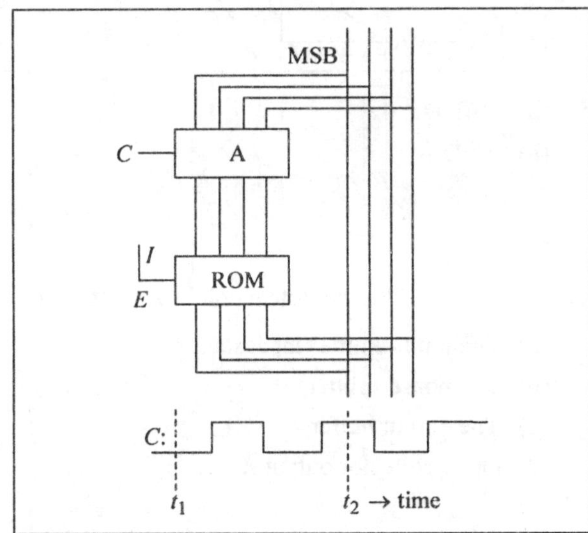

The clock to the register is shown, and the data on the W bus at time t_1 is 0110. The data on the bus at time t_2 is
(a) 1111 (b) 1011
(c) 1000 (d) 0010

8. A dynamic RAM cell which hold 5 V has to be refreshed every 20 m secs, so that the stored voltage does not fall by more than 0.5 V. If the cell has a constant discharge current of 0.1 pA, the storage

capacitance of the cell is:

(a) 4×10^{-6} F
(b) 4×10^{-9} F
(c) 4×10^{-12} F
(d) 4×10^{-15} F

9. In the *DRAM* cell in figure, the V_t of the MOSFET is 1V. For the following three combinations of *WL* and *BL* voltages $WL = 5$V, $BL = 5$V; $WL = 5$, $BL = 3$; $WL = 5$V, $BL = 7$V; The corresponding voltages across the capacitor *C* are:

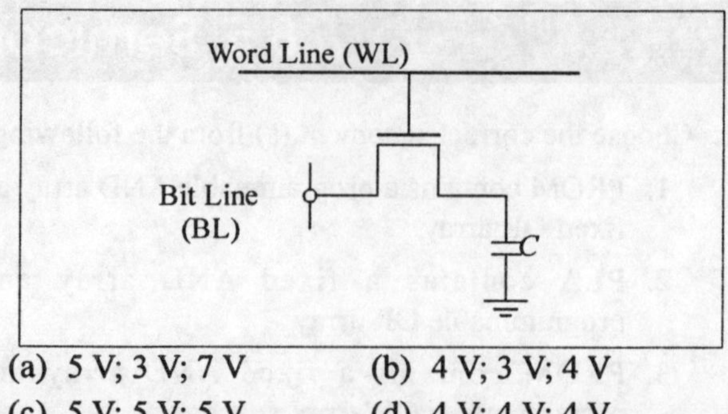

(a) 5 V; 3 V; 7 V (b) 4 V; 3 V; 4 V
(c) 5 V; 5 V; 5 V (d) 4 V; 4 V; 4 V

8.6 Logic Families

1. Figure shows the circuit of a gate in the Resistor Transistor Logic (RTL) family. The circuit represent an:

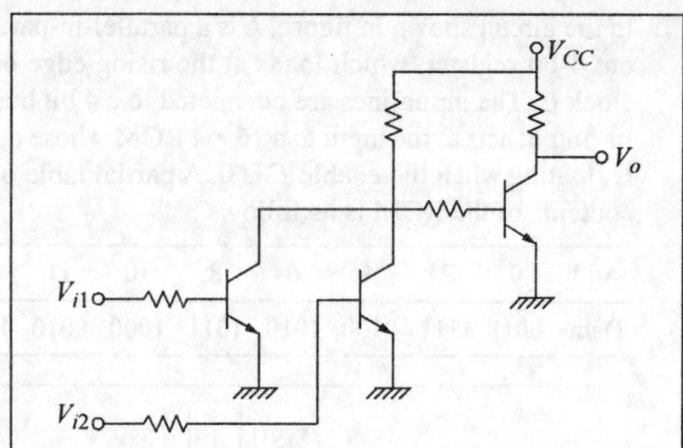

(a) NAND
(b) AND
(c) NOR
(d) OR

2. In standard TTL, the 'totem pole' stage refers to

(a) The multi-emitter input stage
(b) The phase splitter
(c) The output buffer
(d) Open collector output stage.

3. The inverter 74 ALS O4 has the following specifications:

$$I_{OH} \max = -0.4 \text{ mA}, I_{OL} \max = 8 \text{ mA}$$

$$I_{iH} \max = 20 \ \mu\text{A}, I_{iL} \max = -0.1 \text{ mA}$$

The fanout based on the above will be.

(a) 10 (b) 20
(c) 80 (d) 100

4. For the NMOS logic gate shown in figure, the logic function implemented is

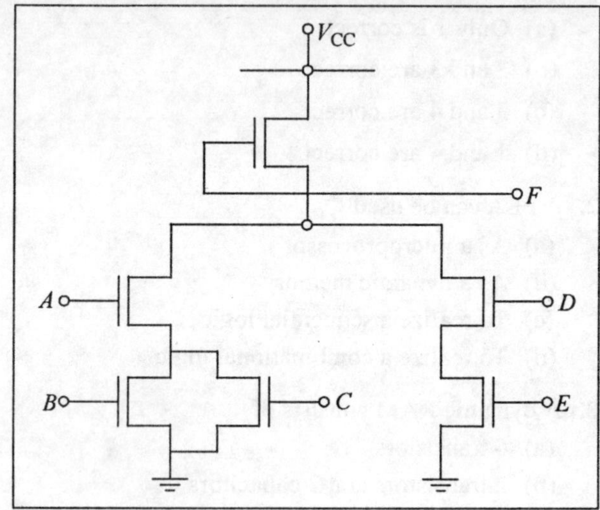

(a) $\overline{ABCD}$

(b) $(AB + C)\overline{(D + E)}$

(c) $\overline{A \cdot (B + C) + D \cdot E}$

(d) $\overline{(A + B) \cdot C + \overline{DE}}$

5. The gate delay of an NMOS inverter is dominated by charge time rather discharge time because:

(a) The driver transistor has larger threshold voltage than the load transistor
(b) The driver transistor has larger leakage current compared to the load transistor
(c) The load transistor has smaller *W*/*L* ratio compared to the driver transistor
(d) Non of the above.

6. The noise margin of a *TTL* gate is about

(a) 0.2 V (b) 0.4 V
(c) 0.6 V (d) 0.8 V

7. The threshold voltage for each transistor in figure is 2 V. For this circuit to work as an inverter, V_i must take the values:

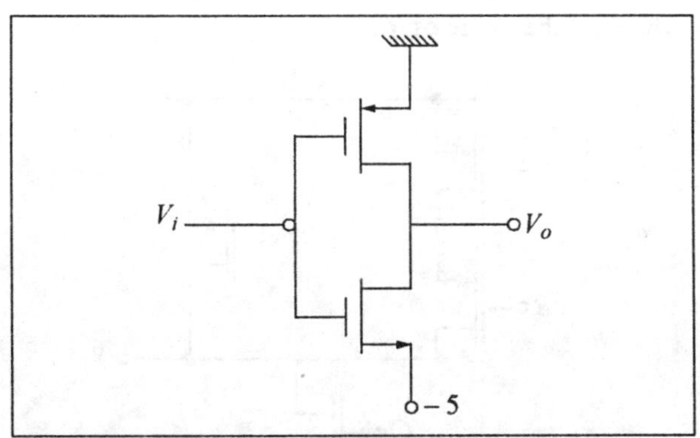

(a) –5 V and 0 V (b) –5 V and 5 V
(c) 0 V and 5 V (d) –3 V and 3 V

8. A Darlington emitter-follower circuit is sometimes used in the output stage of *TTL* gate in order to
(a) Increase its I_{OL}
(b) Reduces its I_{OH}
(c) Increase its speed of operation
(d) Reduce power dissipation.

9. Commercially available ECL gate use two ground lines and one negative supply in order to
(a) Reduce power dissipation
(b) Increases fan-out
(c) Reduces loading effect
(d) Eliminate the effect of power line glitches or the biasing circuit.

10. The output of the 74 series of GATE of TTL gates is taken from a BJT in:
(a) Totem pole and common collector configuration
(b) Either totem pole or open collector configuration
(c) Common base configuration
(d) Common collector configuration.

11. Figure show the internal schematic of s **TTL AND-OR-invert (AOI)** gate. For the inputs shown in Figure, the output *Y* is:

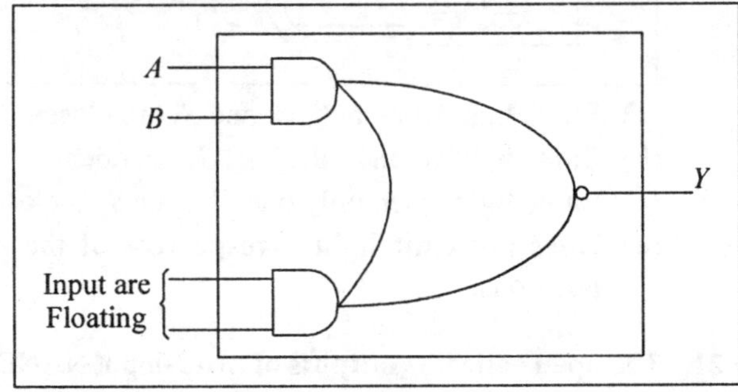

(a) 0 (b) 1
(c) AB (d) $\overline{AB}$

12. The transistors used in a portion of th TTL gate shown in the figure have a β = 100. The base-emitter voltage of is 0.7 V for a transistor in active region and 0.75 V for a transistor in saturation. If the sink current *I* = 1 mA and the output is at logic 0, then the current I_R will be equal to:

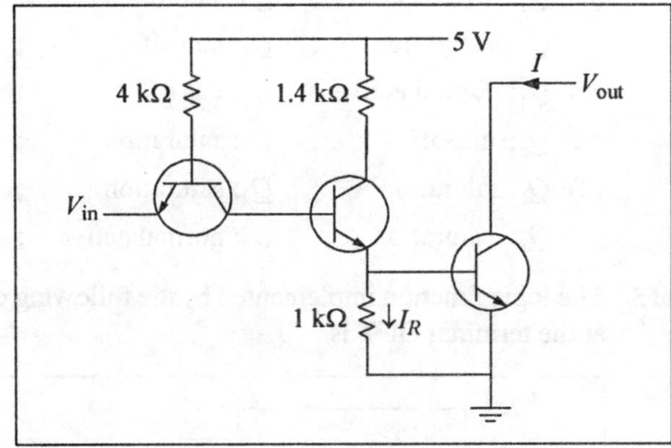

(a) 0.65 mA (b) 0.70 mA
(c) 0.75 mA (d) 1.00 mA

13. Both transistors T_1 and T_2 shown in the figure, have a threshold voltage of 1 volts. The device parameters K_1 and K_2 of T_1 and T_2 are, respectively, 36 μA/V^2 and 9 μA/V^2. The output V_o is:

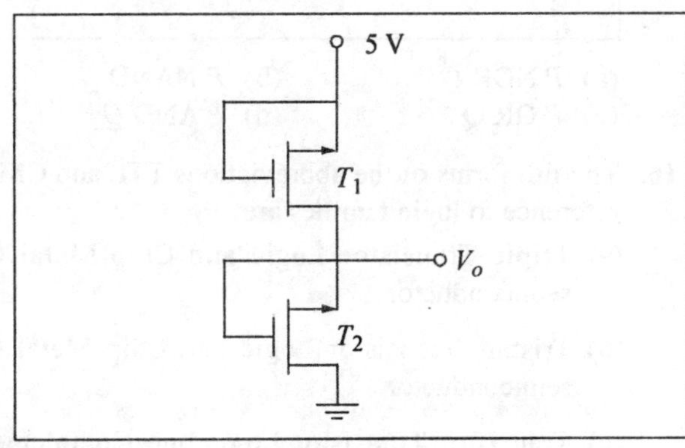

(a) 1 V (b) 2 V
(c) 3 V (d) 4 V

14. The circuit diagram of a standard TTL NOT gate is shown in the figure. When V_i = 2.5 *V*, modes of operation of the transistors will be:

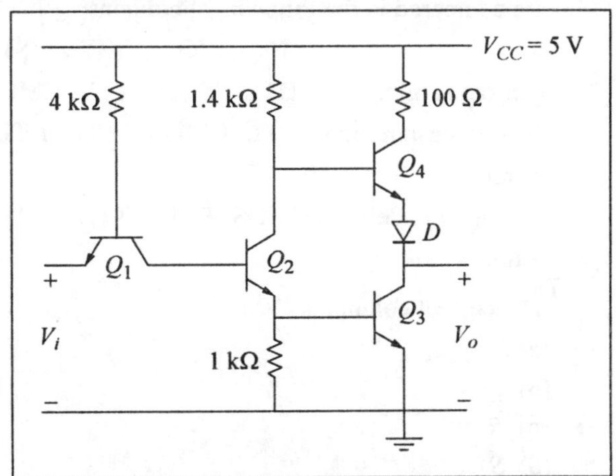

(a) Q_1: reverse active ; Q_2: normal active ;
Q_3: saturation ; Q_4: cut-off ;

(b) Q_1: reverse active ; Q_2: saturation ;
Q_3: saturation ; Q_4: cut-off ;

(c) Q_1: normal active ; Q_2: cut-off ;
Q_3: cut-off ; Q_4: saturation ;

(d) Q_1: saturation ; Q_2: saturation ;
Q_3: saturation ; Q_4: normal active ;

15. The logic function implemented by the following circuit at the terminal OUT is

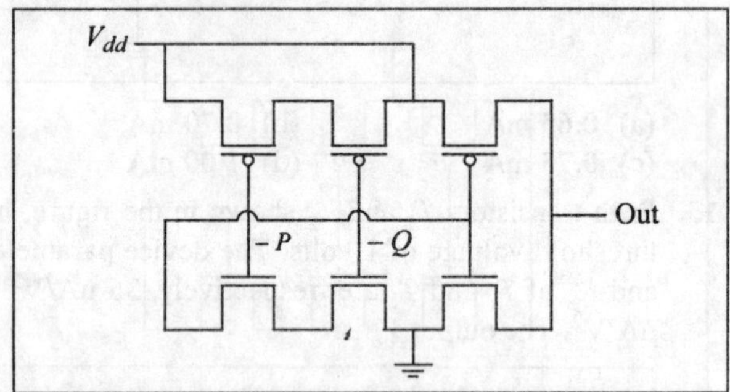

(a) P NOR Q (b) P NAND Q
(c) P OR Q (d) P AND Q

16. The full forms of the abbreviations TTL and CMOS in reference to logic families are

(a) Triple Transistor Logic and Chip Metal Oxide semiconductor

(b) Tristate Transistor Logic and Chip Metal Oxide semiconductor

(c) Transistor Transistor Logic and Complementary Metal Oxide Semiconductor

(d) Tristate Transistor Logic and Complementary Metal Oxide Silicon

17. The DTL, TTL, ECL and CMOS families of digital ICs are compared in the following 4 columns

	(P)	(Q)	(R)	(S)
Fan-out is min.	DTL	DTL	TTL	CMOS
Power consumption is min.	TTL	CMOS	ECL	DTL
Propagation delay is min.	CMOS	ECL	TTL	TTL

The correct column is

(a) P
(b) Q
(c) R
(d) S

18. Find the value of Y?

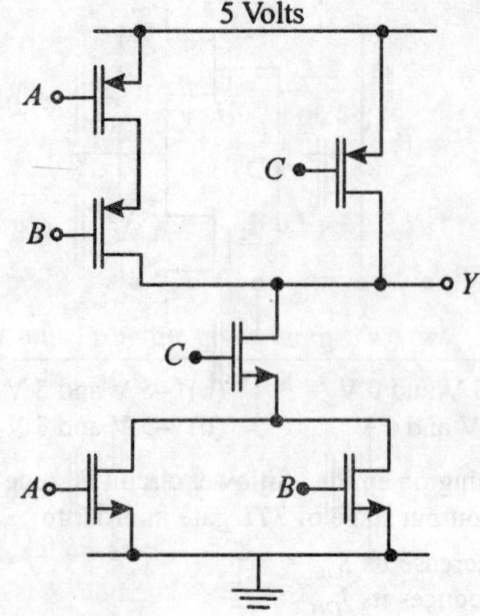

(a) $Y=\overline{A}\,\overline{B}+\overline{C}$ (b) $Y=(A+B)C$
(c) $Y=(\overline{A}+\overline{B})\overline{C}$ (d) $Y=AB+C$

19. Schottky clamping is resorted to in TTL gates

(a) To reduce propagation delay
(b) To increase noise margins
(c) To increase pecking density
(d) To increase fan-out

20. In figure the LED

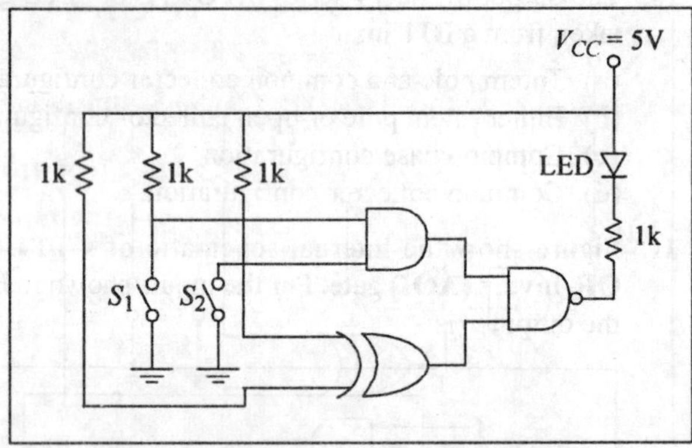

(a) Emits light when both S_1 and S_2 are closed
(b) Emits light when both S_1 ad S_2 are open
(c) Emits light when only one of S_1 or S_2 is closed
(d) Does not emit light, irrespective of the switch positions.

21. The **open collector outputs** of two 2-input NAND gates are connected to a common pull up resistor. If the input to the gates are P, Q and R, S respectively, the output is equal to

(a) $\overline{P.Q.R.S.}$ (b) $\overline{PQ}+\overline{RS}$
(c) $\overline{P.Q}+\overline{R.S}$ (d) $\overline{P.Q}.\overline{R.S}$

22. In standard TTL gates, the **totem pole** stage is primarily used to

(a) Increase the noise margin of the gate

(b) Decrease the o----utput switching delay

(c) Facilitate a wired OR logic connection

(d) Increases the output impedance of the circuit

23. If X_1 and X_2 are the inputs to the circuit shown in the figure, the output Q is.

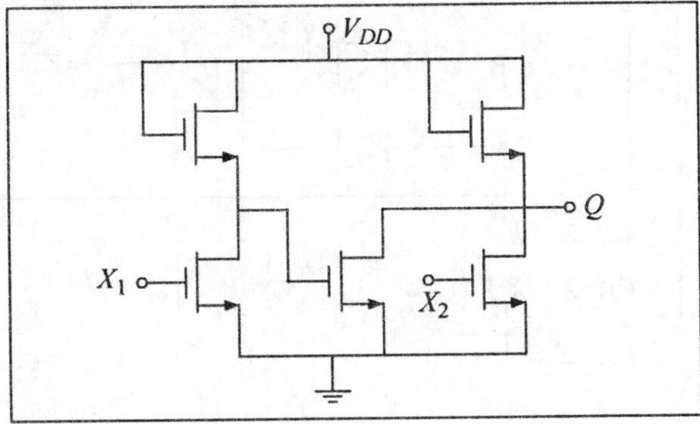

(a) $\overline{X_1 + X_2}$

(b) $\overline{X_1 \cdot X_2}$

(c) $\overline{X}_1 \cdot X_2$

(d) $X_1 \cdot \overline{X}_2$

24. A TTL NOT gate circuit is shown in figure. Assuming $V_{BE} = 0.7$ V of both the transistors, if $V_i = 3.0$ V then the states of the two transistors will be

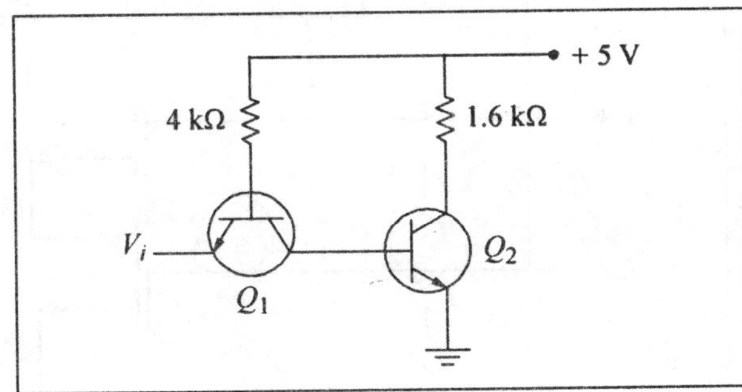

(a) Q_1 ON and Q_2 OFF

(b) Q_1 reverse ON and Q_2 OFF

(c) Q_1 reverse ON and Q_2 ON

(d) Q_1 OFF and Q_2 reverse ON

25. The TTL circuit shown in the figure is fed with the waveform X (also shown). All gates have equal propagation delay of 10 ns. The output Y of the circuit is

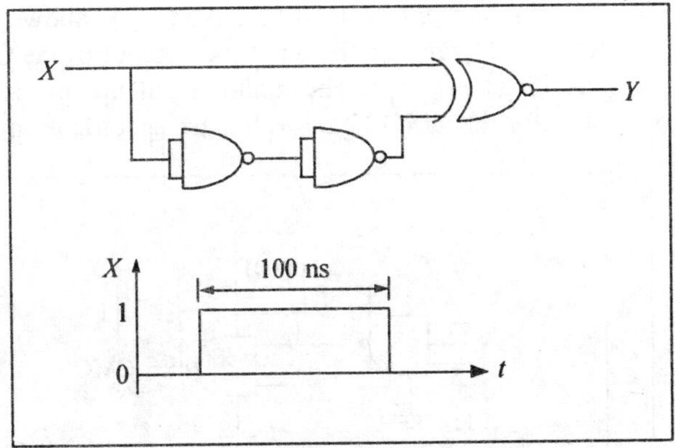

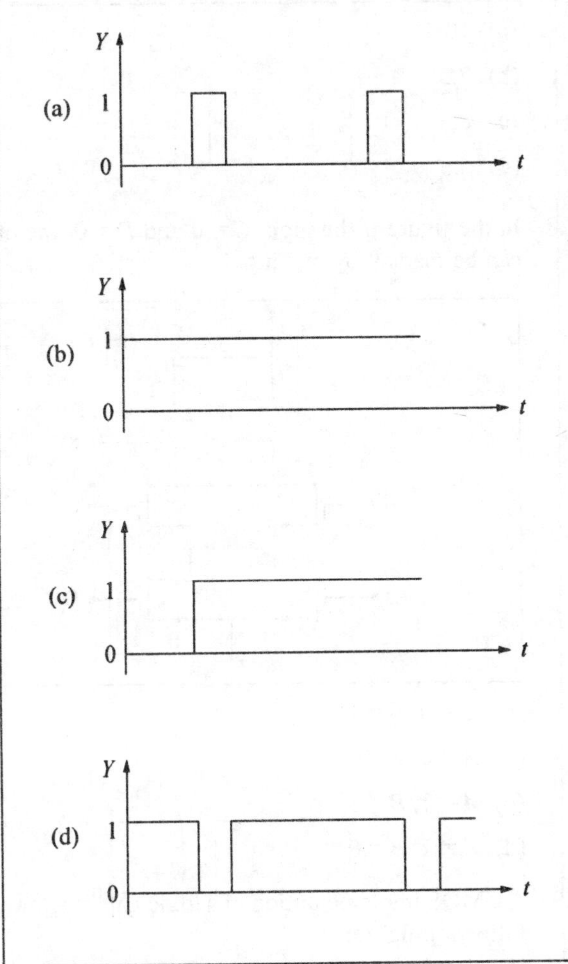

26. Introducing a **schottky diode** between the base and collector of the output transistor in a TTL circuit.

(a) Increases the speed of operation by inhibiting saturation

(b) Decreases the speed of operation by inhibiting saturation

(c) Increases the FAN OUT by enabling saturation

(d) Increases the speed of operation by enabling saturation.

27. The sink current of the TTL NAND gate shown in the figure is 16 mA and the input capacitance of the CMOS NAND gate is 10 pF. The minimum pull up time constant in ns for the CMOS gate with these specification is

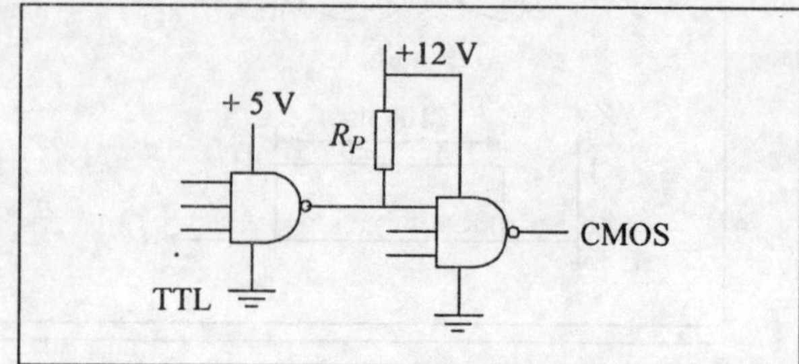

(a) 10.2
(b) 7.5
(c) 8.5
(d) 6.5

28. In the figure if the input $C = 0$ and $D = 0$, the output Y can be made 0 by making

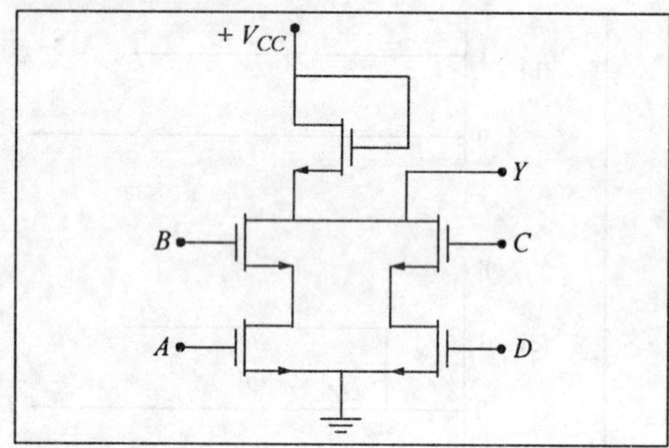

(a) $A = 0, B = 1$
(b) $A = 0, B = 0$
(c) $A = 1, B = 1$
(d) $A = 1, B = 0$

29. A CMOS implementation of a logic gate is shown in the following figure:

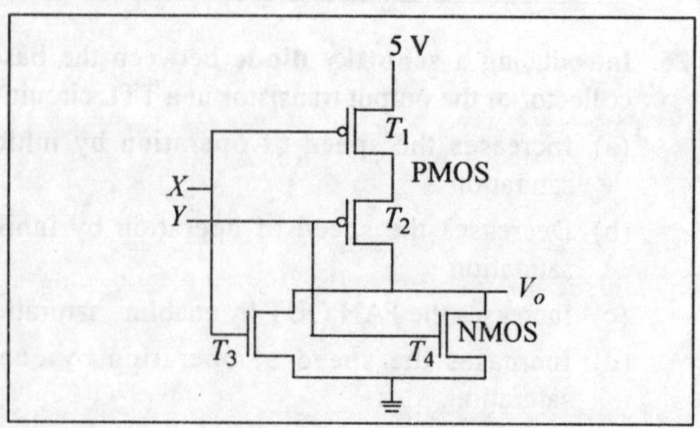

The Boolean logic function realized by the circuit is

(a) AND (b) NAND
(c) NOR (d) OR

30. The diodes in the circuit shown are ideal. A voltage of 0 V represents logic 0 and + 5 V represents logic 1. The logic function Z realized by the circuit for logic inputs X and Y is

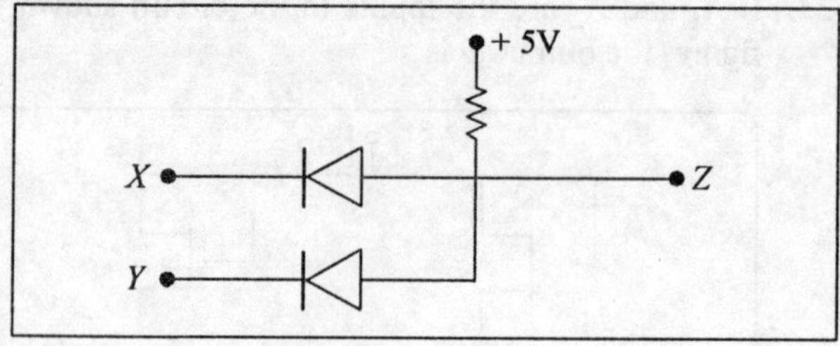

(a) $Z = X + Y$
(b) $Z = X\,Y$
(c) $Z = \overline{X + Y}$
(d) $Z = \overline{XY}$

31. Identify the logic gate given in the figuer.

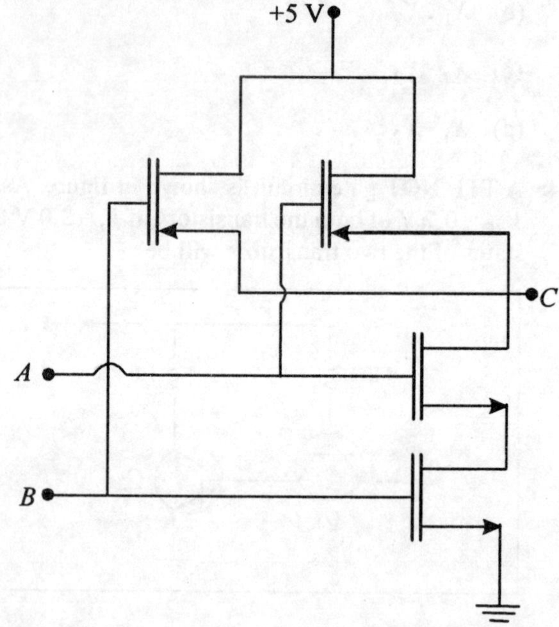

(a) NOR
(b) NAND
(c) AND
(d) OR

32. Consider the following statements in connection with the CMOS inverter in Figure where both the MOSFETs are of enhancement type and both have a threshold voltage of 2 V.

Statement 1: T_1 conducts with $V_i \geq 2$ V.
Statement 2: T_1 is always in saturation when $V_o = 0$ V.
Which of the following is correct?

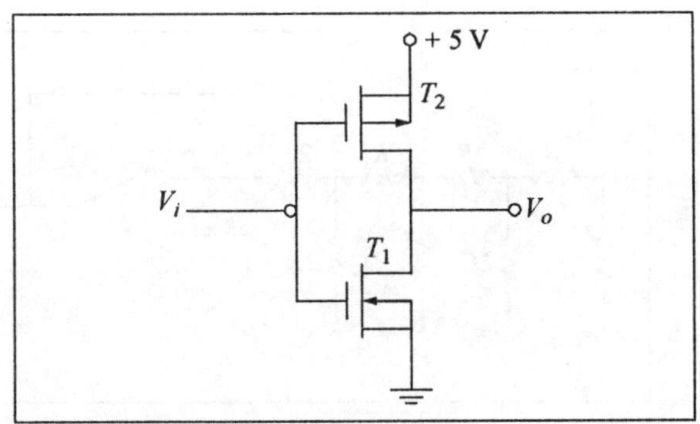

(a) Only Statement 1 is TRUE
(b) Only Statement 2 is TRUE
(c) Both the Statements are TRUE
(d) Both the Statements are FLASE

33. Figure is the **voltage transfer characteristic** of

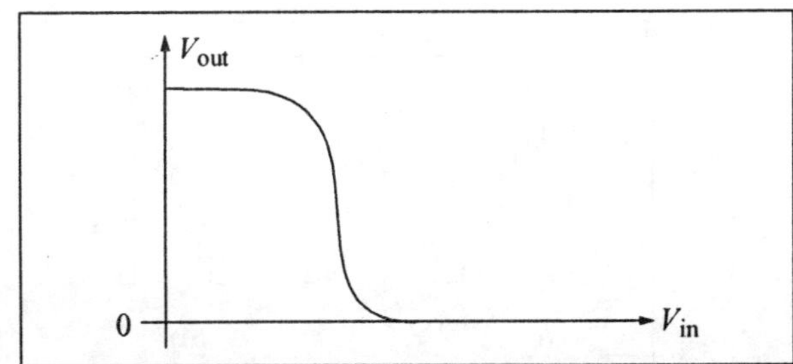

(a) An NMOS inverter with enhancement mode transistor as load
(b) An NMOS inverter wiith depletion mode transistor as load
(c) A CMOS inverter (d) A BJT inverter

34. The desirable characteristics of a digital logic family are:
(1) Large fan-out (2) Large threshold voltage
(3) Large noise (4) Margin small threshold voltage

Choose correct option
(a) 1, 2 are correct (b) 2, 3 are correct
(c) 1, 3, 4 are correct (d) 1, 3 are correct

8.7 A/D and D/A Converters

1. Which of the resistance network of figure can be used as 3 bit *R*-2*R* ladder DAC. Assume V_0 corresponds to LSB:

(a)

(b)

(c)

(a) Both (a) and (b)
(b) Both (a) and (c)
(c) Only (c)
(d) Only (b).

2. The resolution of a 4-bit counting *ADC* is 0.5 volts. For an analog input of 6.6 volts, the digital output of the *ADC* will be
(a) 1011
(b) 1101
(c) 1100
(d) 1110

3. A 4-bit D/A converter is connected to a free-running 3-bit UP counter, as shown in the following figure. Which of the following waveforms will be observed at V_o?

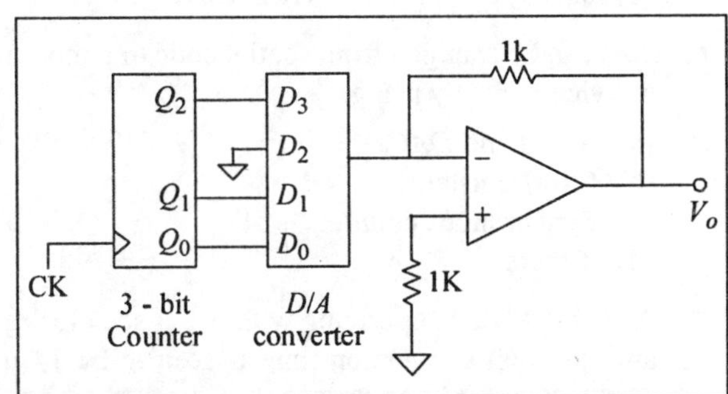

In the figure shown above, the ground has been shown by the symbol Δ.

Figure:

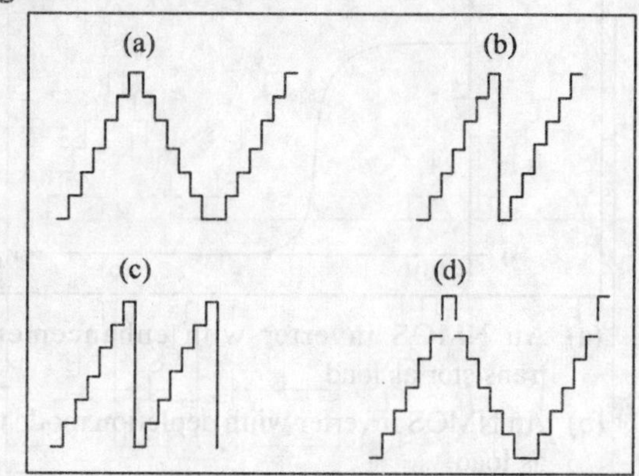

Statement for linked Questions 4 and 5

In the follwing, circuit the comparator output is logic "1" if $V_1 > V_2$ and is logic "0" otherwise. The D/A conversion is done as per realtion $V_{DAC} = \sum_{n=0}^{3} 2^{n-1} b_n$ Volts, where b_3(MSB), b_2, b_1 and b_0 (LSB) are the counter outputs. The counter starts from the clear state.

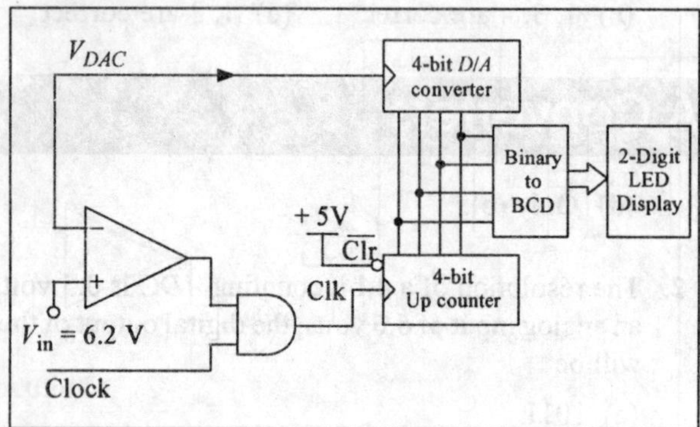

4. The stable reading of the LED displays is
   (a) 06 (b) 07
   (c) 12 (d) 13

5. The magnitude of the error between V_{DAC} and V_{in} at steady state in volt is
   (a) 0.2 (b) 0.3
   (c) 0.5 (d) 1.0

6. Data can be changed from spatial code to temporal code and vice-versa by using
   (a) *ADCs* and *DACs*
   (b) Shift-registers
   (c) Synchronous counters
   (d) Timers

7. A 12-bit ADC is operating with a 1 μ sec clock period and the total conversion time is seen to be 14 μ secs. The ADC must be of the
   (a) Flash type
   (b) Counting type
   (c) Integrating type
   (d) Successive approximation type

8. The current *I* though resistance *r* in the circuit shown in Figure:

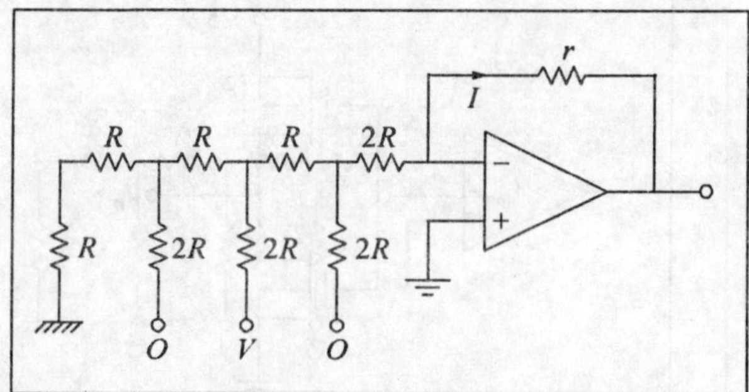

   (a) $\frac{-V}{12R}$
   (b) $\frac{V}{12R}$
   (c) $\frac{V}{6R}$
   (d) $\frac{V}{3T}$

9. An 8 bit successive approximation analog to digital converter has a full scale reading of 2.55 V and its conversion time for an analog input of 1 V is 20 μs. The conversion time for a 2 V input will be:
   (a) 10 μs
   (b) 20 μs
   (c) 40 μs
   (d) 50 μs

10. For the 4 bit *DAC* shown in Figure, the output voltage V_o is:

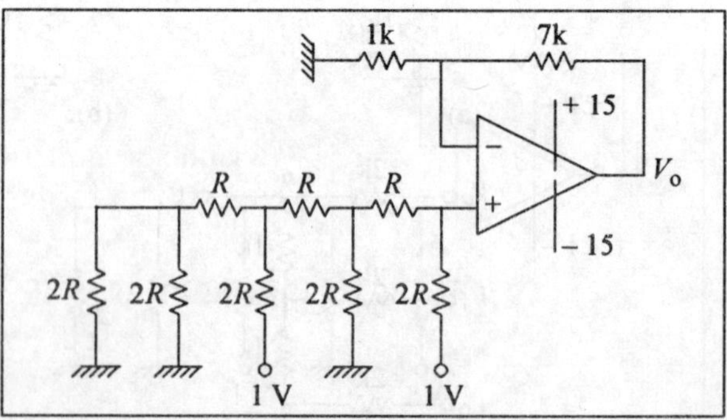

   (a) 10 V (b) 5 V
   (c) 4 V (d) 8 V

11. The number of comparators required in a 3-bit comparator type *ADC* is:
   (a) 2 (b) 3
   (c) 7 (c) 8

12. The circuit shown in Figure is a 4 bit DAC. The input bits 0 and 1 are represented by 0 and 5V respectively. The OP AMP is ideal, but all the resistances and the 5V inputs have a tolerance of 10%. The specification (rounded to the nearest multiple of 5 %) for the tolerance of the DAC is:

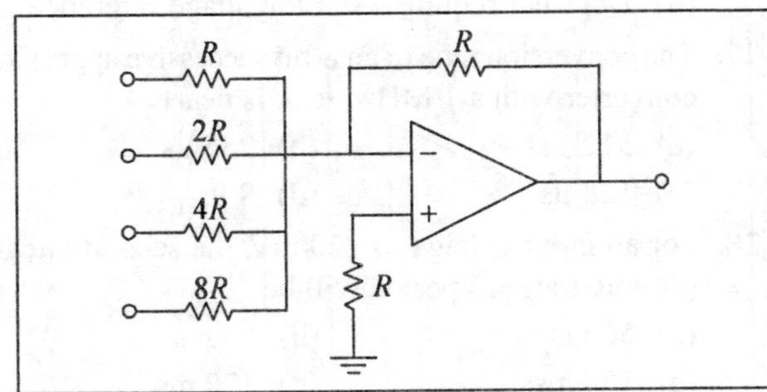

(a) ± 35 % (b) ± 20 %
(c) ± 10 % (d) ± 5 %

13. The number of comparators in a 4 bit flash ADC is

(a) 4 (b) 5
(c) 15 (d) 16

14. Dual-slope integration type Analog-to-digital converters provide

1. Higher speeds compared to all other types of *A*/*D* converters
2. Very good accuracy without putting extreme requirements on component stability
3. Good rejection of power supply hum
4. Better resolution compared to all other types of *A*/*D* converters for the same number of bits.

Codes:

(a) Only 1 is correct
(b) 2 and 3 correct
(c) 1, 2 , 3 and 4 correct
(d) 2 and 4 correct

Statement for linked Answer Question 15 and 16

In the Digital-to-Analog converter circuit shown in the figure below, V_R = 10 V and R = 10 kW

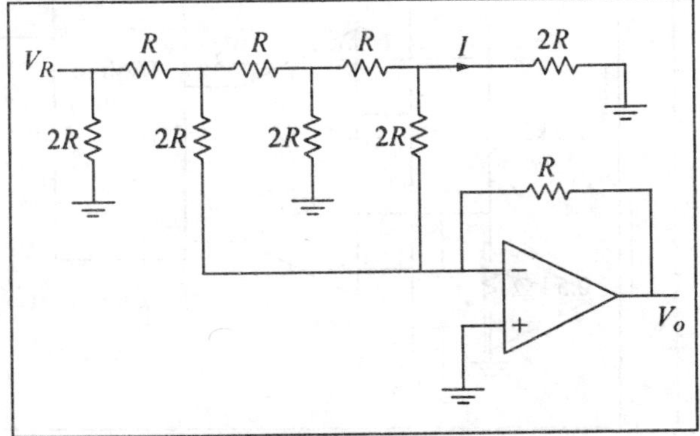

15. The current I is

(a) 31.25 μA (b) 62.5 μA
(c) 125 μA (d) 250 μA

16. The voltage V_0 is

(a) –0.781 V (b) –1.562 V
(c) –3.125 V (d) –6.250 V

17. The number of comparisons carried out in a 4-bit flash-type *A*/*D* converter is

(a) 16
(b) 15
(c) 4
(d) 3

18. Among the following four, the slowest ADC (analog-to-digital converter) is

(a) Parallel-comparator (i.e., flash) type
(b) Successive approximation type
(c) Integrating type
(d) Counting type slow

19. The voltage comparator shown in figure can be used in the analog-to-digital conversion as

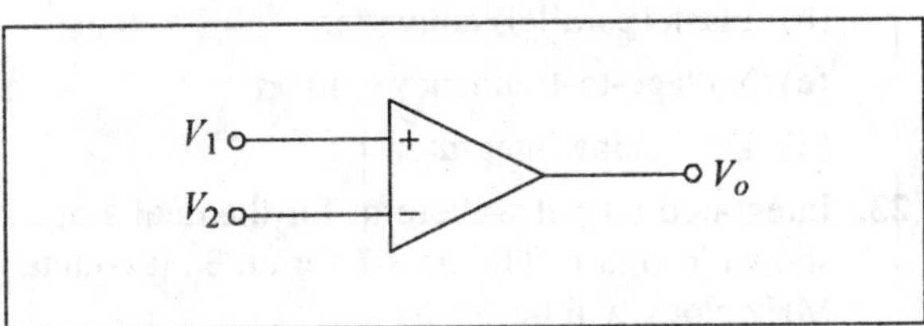

(a) a 1-bit quantizer
(b) a 2-bit quantizer
(c) a 4-bit quantizer
(d) a 8-bit quantizer

20. A student has made a 3-bit binary down counter and connected to the *R*-2*R* ladder type DAC [Gain = (–1 kΩ /2*R*)] as shown in figure to generate a staircase wave form. The output achieved is different as shown in figure. What could be the possible cause of this error?

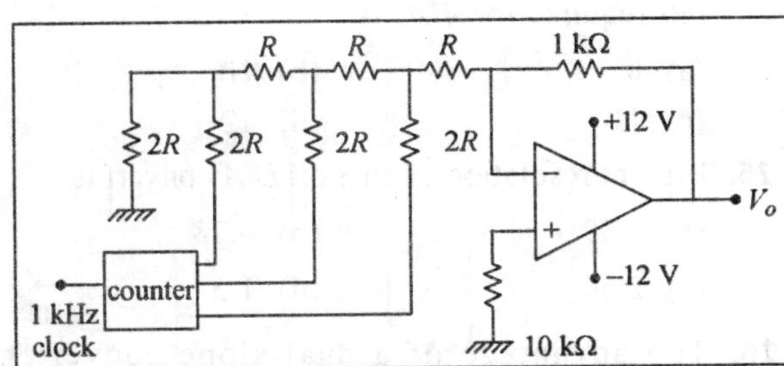

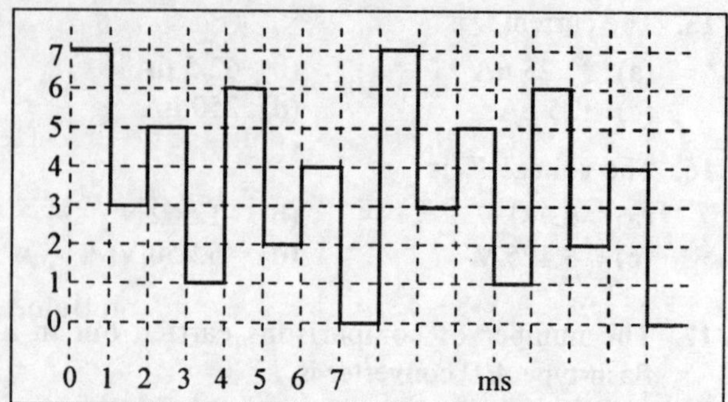

(a) The resistance values are incorrect

(b) The counter is not working properly

(c) The connection from the counter to DAC is not proper

(d) The R and $2R$ resistances are interchanged

21. A 3-bit resistance ladder *D*/*A* (*R*-2*R* network) has resistor of $R = 10$ Wk and 20 Wk V_{REF} equals 8 volts. What is I_{out} for a digital input of 111?

(a) 0.7 mA (b) 0.2 mA
(c) 0.1mA (d) 1 mA

22. A **sample and hold circuit** is normally required before the following type of *A*/*D* converter.

(a) Successive approximation

(b) Flash (parallel) converter

(c) Voltage-to-frequency converter

(d) Dual slope integrator.

23. Integrated output waveform for the dual slope *ADC* is shown in figure. The time *T* for an 8 bit counter with 4 MHz clock will be:

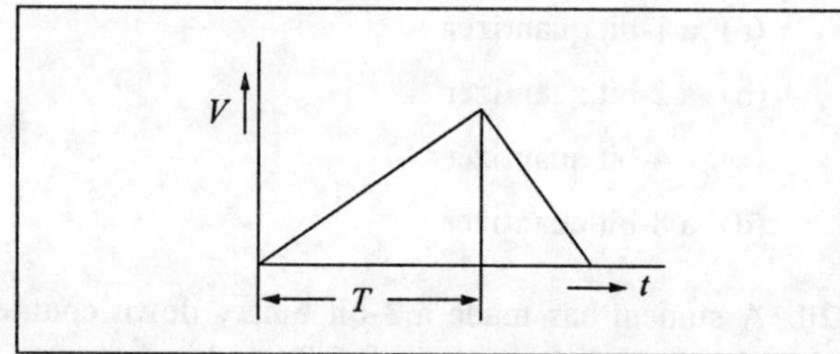

(a) 0.032 ms (b) 0.064 ms
(c) 0. 64 ms (d) 0.024 ms

24. The full scale input voltage to an *ADC* is 10 V. The resolution required is 0.5 mV. The minimum number of bits required for *ADC* is.

(a) 8 (b) 10
(c) 11 (d) 12

25. Percent resolution of an 8 bit *D*/*A* convert is

(a) 0.39 (b) 0.78
(c) 2.56 (d) 1.5

26. The advantage of a dual slope converter over successive approximation converter is that the dual slope converter.

(a) Is faster

(b) Eliminates error due to drift

(c) Can reduce the errors due to power supply

(d) Does not require a stable voltage reference

27. The conversion time of an 8-bit successive approximation converter with a 1 MHz clock is nearly

(a) 512 μs (b) 256 μs
(c) 128 μs (d) 8 μs

28. For an input voltage of 120 mV, the second integration (de-integration) period will be.

(a) 50 ms (b) 60 ms
(c) 100 ms (d) 120 ms

29. The number of comparators required in an 8-bit flash-type *A*/*D* converter is

(a) 256 (b) 255
(c) (8 + 2) (d) 8

30. For N-bit **Successive Approximation** ADCs, other parameters such as clock frequency remaining constant, the conversion time is proportional to

(a) N^2 (b) $\sqrt{N}$
(c) $\log N$ (d) N

31. A single channel signal acquisition system with 0-10 V range consists of a sample and hold circuit with worst case **drop rate of 100 μ V/ms** and 10-bit ADC. The maximum conversion time for the ADC is

(a) 49 μs (b) 0.49 ms
(c) 4.9 ms (d) 49 ms

32. The circuit shown in the figure works as a 2-bit analog-to-digital converter for $0 \le V_{in} \le 3$ V

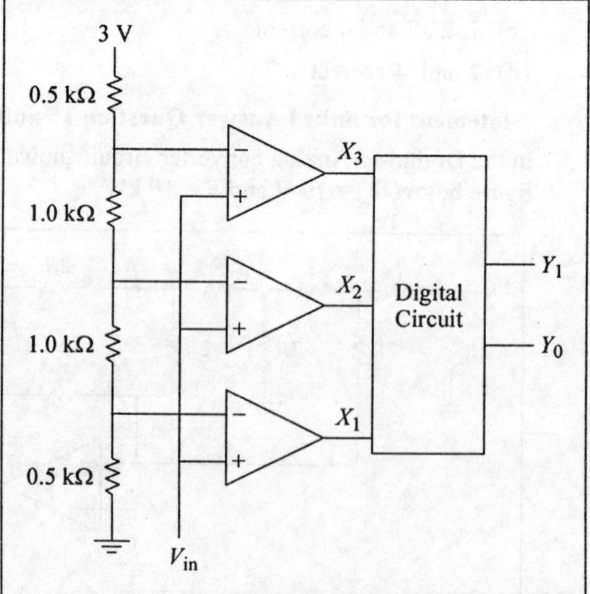

The MSB of the output Y_1, expressed as a Boolean function of the inputs X_1, X_2, X_3 is given by

(a) X_1

(b) X_2

(c) X_3

(d) $X_1 + X_2$

33. An 8-bit ADC, with 2's complement output, has a nominal input range of –2 V to + 2 V. It generates a digital code of 00H for an analog input in the range –7.8125 mV to + 7.815 mV. An input of – 1.5 V will produces a digital output of

(a) 90 H

(b) 96 H

(c) 9 BH

(d) A 0 H

34. A 4-bit successive approximation type ADC has a full scale value of 15 V. The sequence of the states, the SAR will traverse, for the converion of an inut of 8.15 V is.

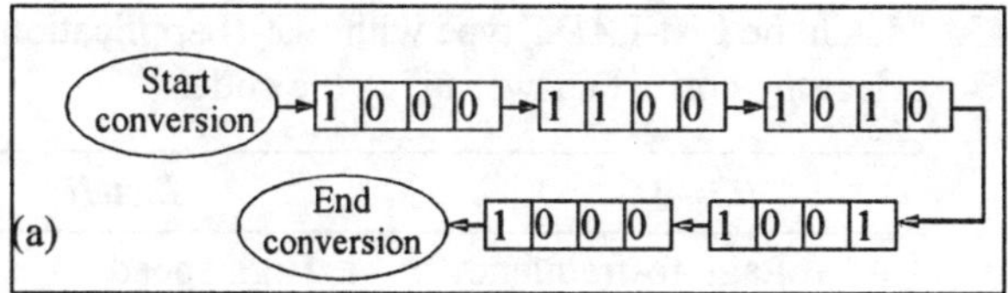

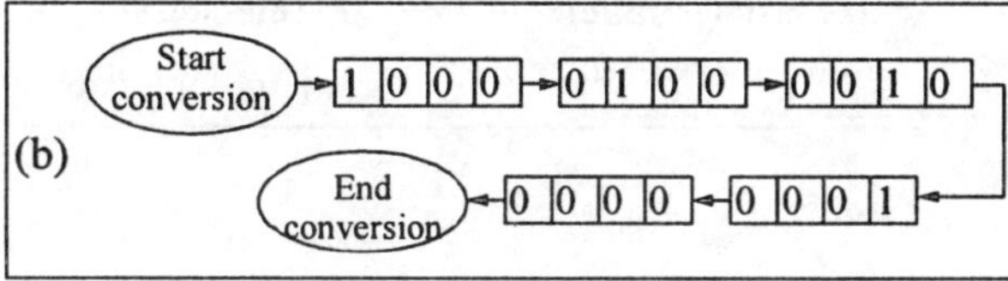

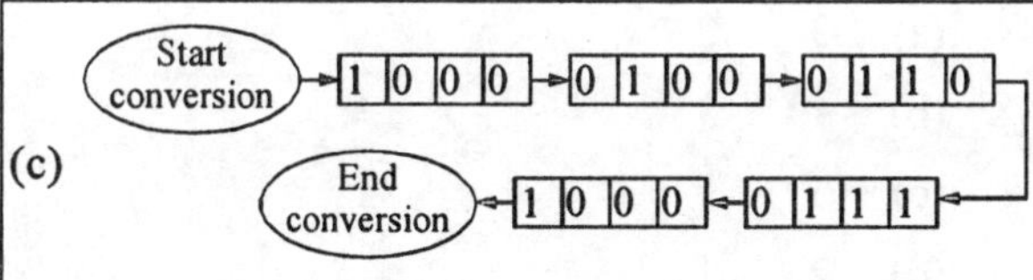

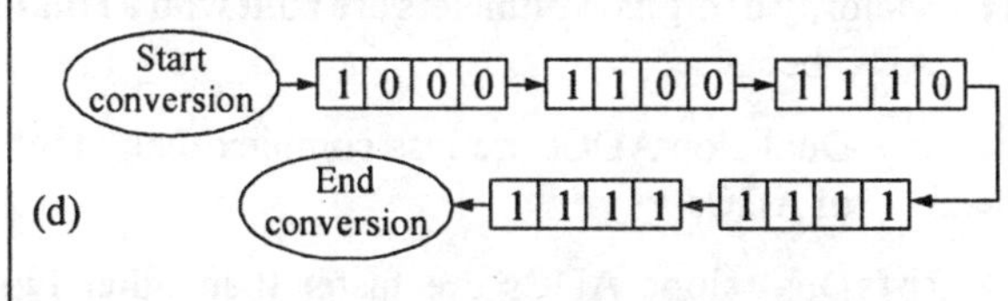

35. A 4 - bit successive approximation type of A/D converter has an input rang of 0 to 15 volts. The output bit b_1 next to the LSB has a stuck - at - zero fault. The pair of input voltages that produces the same output code word is

(a) 2 V and 4 V

(b) 4 V and 6 V

(c) 1 V and 2 V

(d) 8 V and 9 V

36. Figure beow shows a circuit for implementing an 8-bit digirtal - to - Analog converter (DAC) using two identical 4 - bit DACs with equal reference voltages. Assume that b_0 represents LSB, b_7 MSB and the opamp is ideal. To obtain correct analog values corresponding to an 8 -bit DAC at the output V_0, the value of resistor R is:

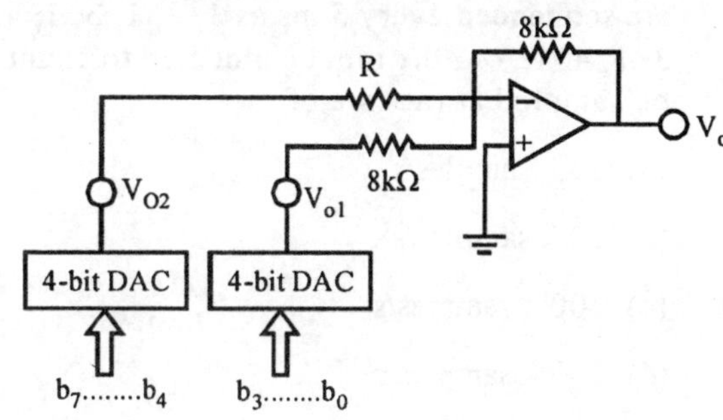

(a) 0.25 kΩ

(b) 0.5 kΩ

(c) 1 kΩ

(d) 8 kΩ

Common Data Questions

Common Data for Questions 37, 38 and 39:

A data acquisition system (DAS) shown below employs a successive approximation type 12-bit ADC having a conversion time of 5 ms.

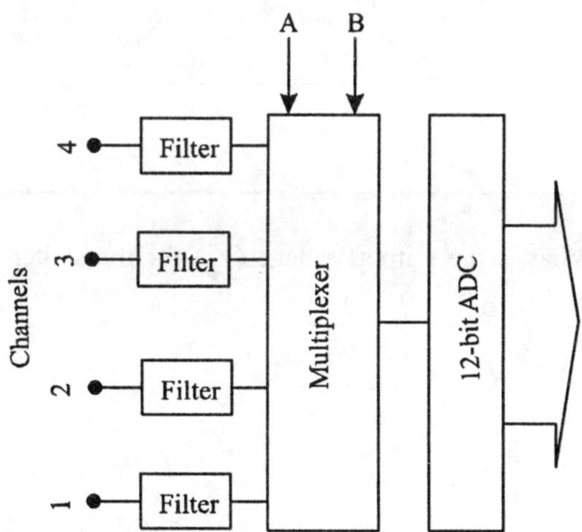

37. The quantization error of the ADC is

(a) 0 %

(b) $\pm 0.012\%$

(c) $\pm 0.024\%$

(d) $\pm 0.048\%$

38. The system is used as a single channel DAS with channel 1 selected as input to the ADC which is set in the continuous conversion mode. For avoiding aliasing error, the cutoff frequency f_c of the filter in channel 1 should be

(a) $f_c < 100\ kHz$

(b) $f_c = 100\ kHz$

(c) $100\ kHz < f_c < 200\ kHz$ 1

(d) $f_c = 200\ kHz$

39. If the multiplexer is controlled such that the channels are sequenced every 5 ms as 1, 2, 1, 3, 1, 4, 1, 2, 1, 3, 1, 4, 1, ..., the input connected to channel 1 will be sampled at the rate of

(a) 25 k samples/s

(b) 50 k samples/s

(c) 100 k samples/s

(d) 200 k samples/s

Common data for questions 40 and 41

An R-$2R$ ladder type *DAC* is shown below. If a switch status is '0' 0 V is applied and if a switch status is '1', 5 V is applied to corresponding terminals of the DAC.

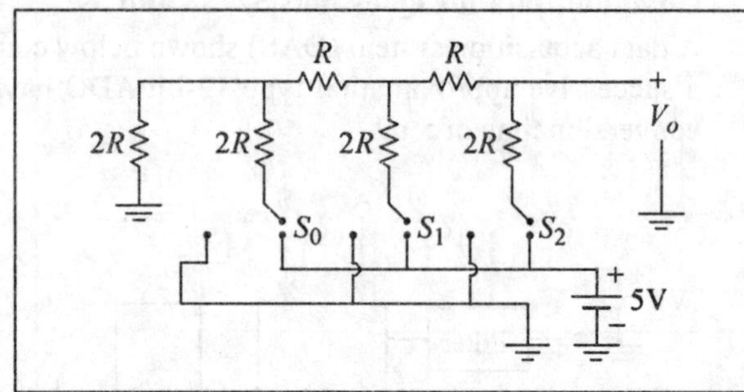

40. What is the output voltage (V_o) for the switch status $S_0 = 0, S_1 = 1, S_2 = 1$?

(a) $\frac{5}{4}$ V

(b) $\frac{15}{4}$ V

(c) $\frac{17.5}{4}$ V

(d) $\frac{22.5}{4}$ V

41. What is the step size of the DAC ?

(a) 0.125 V

(b) 0.525 V

(c) 0.625 V

(d) 0.75 V

42. Two 8-bit ADCs, one of single slope integrating type and other of successive approximation type, take T_A and T_B times to convert 5 V analog input signal to equivalent digital output.If the input analog signal is reduced to 2.5 V, the approximate time take by the two ADCs will respectively. be

(a) T_A, T_B

(b) $\frac{T_A}{2}, T_B$

(c) $T_A, \frac{T_B}{2}$

(d) $\frac{T_A}{2}, \frac{T_B}{2}$

43. Match the List-I *ADC* type with List-II applications and select the correct answer using the codes:

List-I	*List-II*
A. Voltage-to-frequency converter	1. High speed
B. Flash (parallel)	2. Telemetry
	3. High precision

Codes:

	A	B
(a)	3	2
(b)	1	3
(c)	3	1
(d)	2	1

44. Majority of digital voltmeters are built with a **Dual-slope ADC** because.

(a) Dual slop ADCs are less complex than other types of ADCs.

(b) Dual slope ADCs are faster than other types of ADCs.

(c) Dual slope ADCs can be designed to be insensitive to noise and interference.

(d) Dual slop ADCs provide BCD outputs.

8.8 Miscellaneous

1. The digital circuit shown in figure generates a modified clock pulse at the output. Choose the correct output waveform from the option given below:

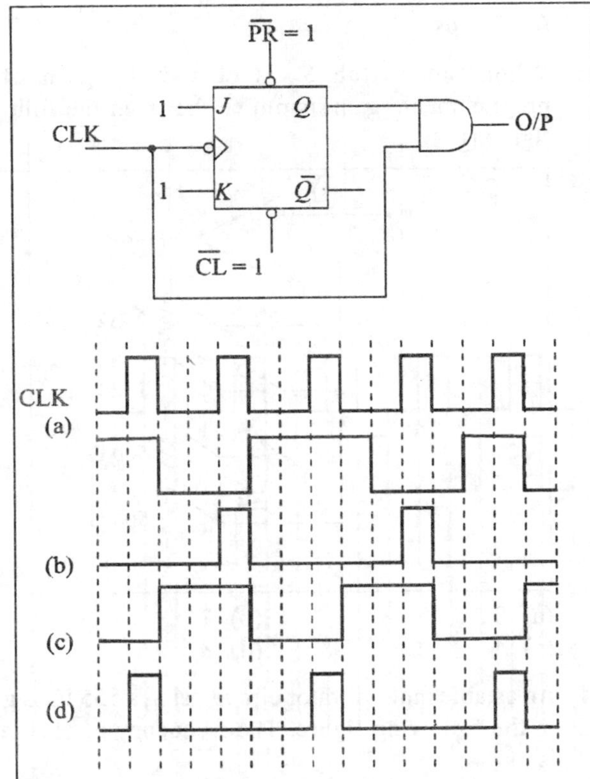

2. For the two square inputs in figure the PMMC meter will read maximum when the phase difference between them is:

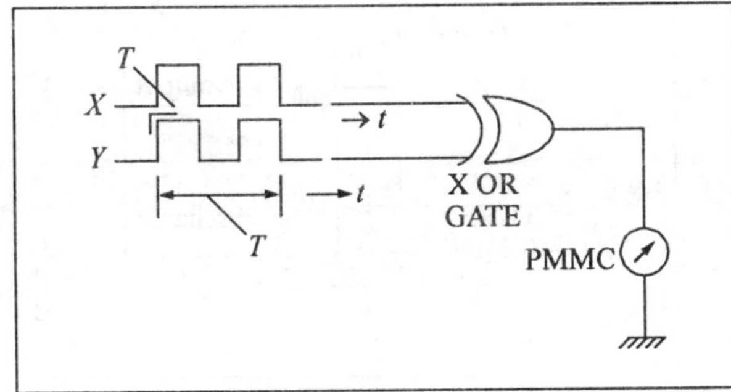

(a) 0 (b) $\frac{T}{4}$

(c) $\frac{T}{2}$ (d) T.

3. $A\ 3\left(\frac{1}{2}\right)$ digit digital multimeter can display

(a) 999.5 (b) 5.999
(c) 5555 (d) 1.999

4. A voltmeter has been connected between the input of a TTL inverting gate and ground. The gate is powered by 5 V. The voltmeter reading will be approximately

(a) 0 V (b) 2 V
(c) 4 V (d) 5 V

5. **A sample and hold circuit** has two buffers, one at the input and the other at the output. The primary requirements for the buffers are.
   (a) The input buffer should have high slew rate and the output buffer should have low bias current.
   (b) The input buffer should have low bias current and the output buffer should have high slew rate
   (c) Both the buffers should have low bias currents
   (d) Both the buffers should have high slew rate.

6. In a **dual slope type** digital voltmeter, an unknown signal voltage is integrated over 100 cycles of the clock. if the signal has a 50 Hz pick up. The maximum clock frequency can be.

(a) 50 Hz (b) 5 kHz.
(c) 10 kHz (d) 50 kHz.

7. An 8-bit **DAC** with a reference voltage V, when tested gave the following results

Input	20H	21H	22H	23H	24H
Output V	0.320	0.330	0.325	0.350	0.360

The DAC is
(a) Monotonic (b) Non monotonic
(c) Bi-polar (d) Quadratic

8. The karnaugh map for a four variable Boolean function is given in. The correct Boolean sum of product is

RS \ PQ	00	01	11	10
00	0	0	0	0
01	1	0	0	1
11	1	0	0	1
10	0	1	0	0

(a) $PQRS + \bar{Q}S$

(b) $\bar{P}QR\bar{S} + \bar{Q}S$

(c) $PQR + Q\bar{S}$

(d) $PQRS + Q$

9. For the CMOS analog switch shown in figure the positive supply is $+V_{DD}$ and the negative supply is $-V_{EE}$. The

input V_i is bipolar. The switch will be ON and V_o will be equal to V_i provided.

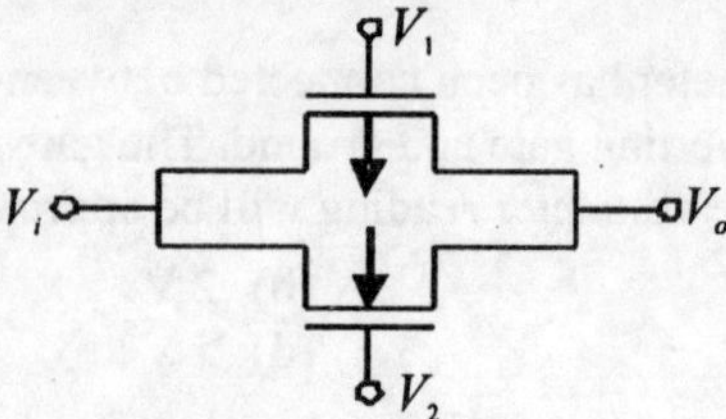

(a) $V_1 = V_2 = +V_{DD}$
(b) $V_1 = V_2 = -V_{EE}$
(c) $V_1 = +V_{DD}$ and $V_2 = -V_{EE}$
(d) $V_1 = -V_{EE}$ and $V_2 = +V_{DD}$

10. Figure (a) shows a **Schmitt trigger** circuit and figure (b) the corresponding hysteresis characteristics. The values of V_{TL} and V_{TH} are

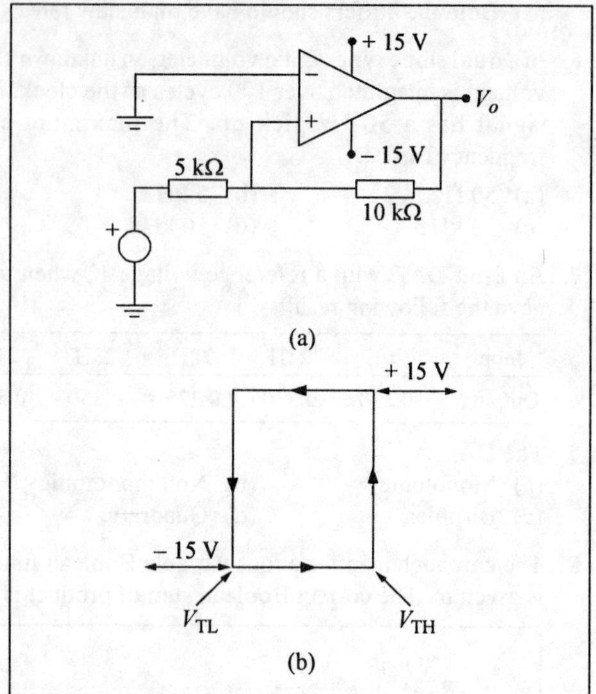

(a) $V_{TL} = -3.75$, $V_{TH} = +3.75$ V
(b) $V_{TL} = -1$ V, $V_{TH} = +5$ V
(c) $V_{TL} = -5$ V, $V_{TH} = +1$ V
(d) $V_{TL} = -5$ V, $V_{TH} = +5$ V

11. $7E_H$ and $5F_H$ are *XO* Red. The output is multiplied by 10_H. The result is

(a) 0210_H
(b) $7E\,5F_H$
(c) $5F\,7E_H$
(d) 2100_H

12. The input characteristic of a Schmidt trigger has a Hystersis output band of ± 0.1 V. If the input voltage is 5 sin (1000 πt), the delay between the corresponding zero cross-over points of the output and input signals is

(a) 6.37 μs
(b) 0.02 μs
(c) 63.7 μs
(d) 2.0 μs

13. When the switch S_2 is closed the gain of the programmable gain amplifier shown in the following figure is

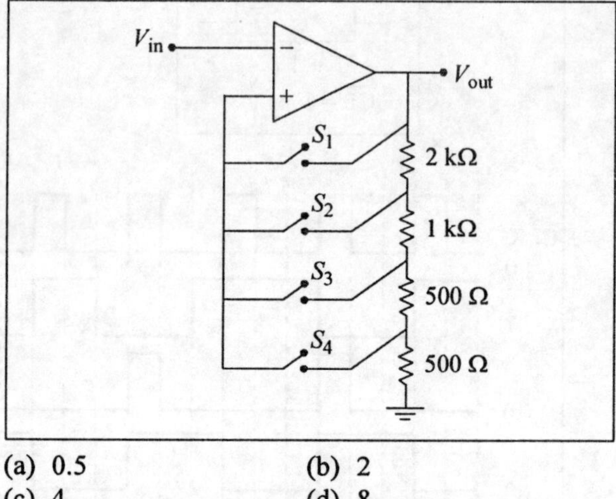

(a) 0.5
(b) 2
(c) 4
(d) 8

14. An astable multivibrator circuit using a 555 IC is given in the following figure. The frequency of oscillation is

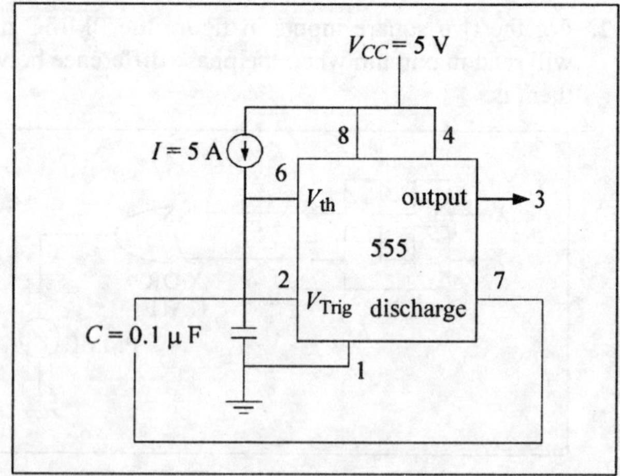

(a) 20 k Hz
(b) 30 kHz
(c) 40 k Hz
(d) 45 kHz

15. In the Schmitt trigger circuit shown below, the zener diodes have V_z (reverse saturation voltage) = 6 V and V_D (forward voltage drop) = 0.7 V.

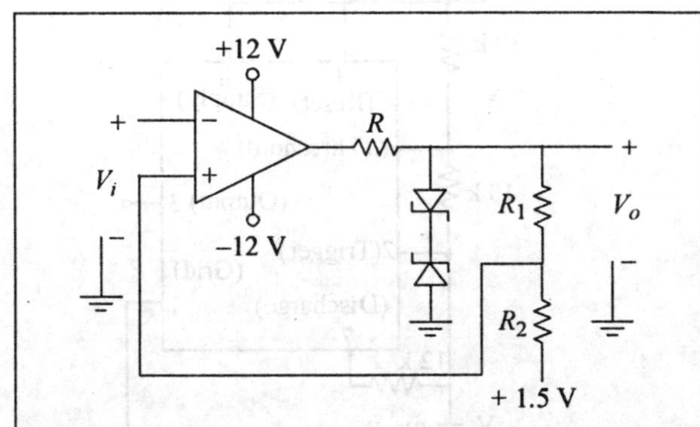

If the circuit has the input lower trip point (LTP) = 0V, then value of (R_1/R_2) is given as

(a) 0.223
(b) 2.67
(c) 4.67
(d) ∞

16. In the Schmitt trigger circuit shown below, the zener diodes have Vz (reverse saturation voltage) = 6 V and V_D (forward voltage drop) = 0.7 V.

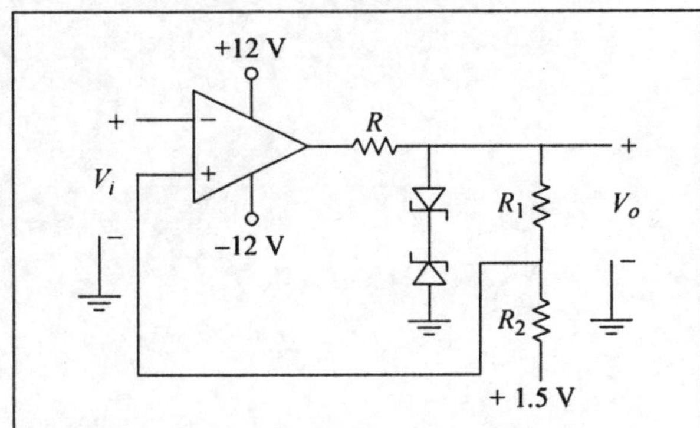

In input upper trip point (UTP) of the Schmitt trigger is

(a) 1.5 V
(b) 2.1 V
(c) 2.42 V
(d) 6.7 V

17. A 555 astable mulitivibrator circuit is shown in the figure below:

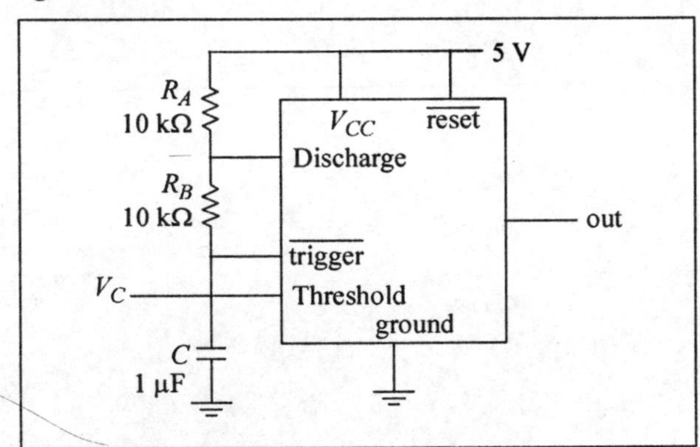

If R_B is shorted, the waveform at V_C is

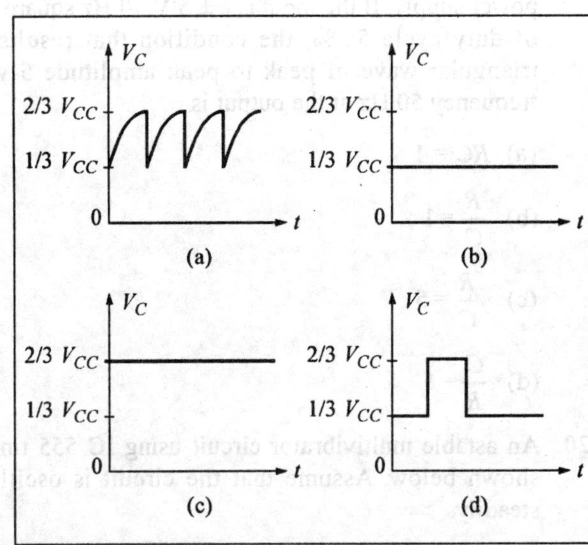

18. The input signal shown in the figure below is fed to Schmitt trigger. The signal has a square wave amplitude of 6 V p-p. It is corrupted by an additive high frequency noise of amplitude 8 V p-p

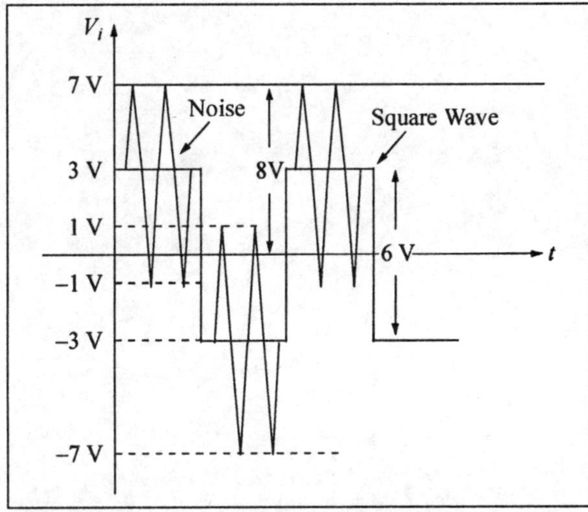

Which one of the following is an appropriate choice for the upper and lower trip points of the Schmitt trigger to recover a square wave of the same frequency from the corrupted input signal V_i?

(a) ± 8.0 V
(b) ± 2.0 V
(c) ± 0.5 V
(d) 0 V

19. Consider the triangular wave generator shown below:

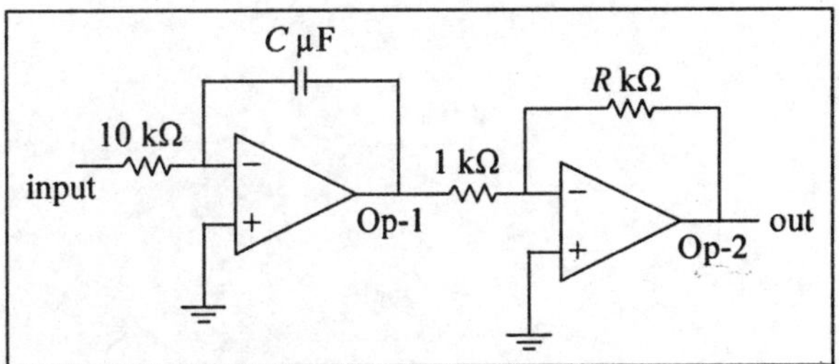

Assume that the Op-amps are ideal and have ± 12 V power supply. If the input is a ± 5 V 50 Hz square wave of duty cycle 50 %, the condition that results in a triangular wave of peak to peak amplitude 5 V and frequency 50 Hz at the output is

(a) $RC = 1$

(b) $\frac{R}{C} = 1$

(c) $\frac{R}{C} = 5$

(d) $\frac{C}{R} = 5$

20. An astable multivibrator circuit using IC 555 timer is shown below. Assume that the circuit is oscillating steadily.

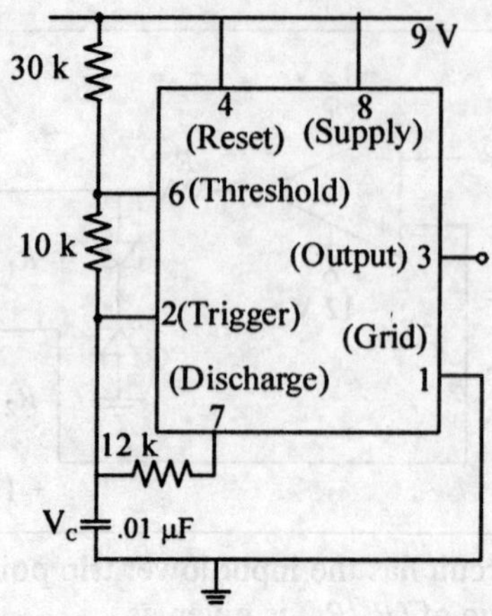

The voltage V_c across the capacitor varies between

(a) 3 V to 5 V (b) 3 V to 6 V

(c) 3.6 V to 6 V (d) 3.6 V to 5 V

Answers

8.1 Boolean Algebra

1. **Ans. (b)**

Solution:

If the number of Boolean variables are 'n' then Number of Booken function possible are

$$2^nC_o + 2^nC_1 + 2^nC_2 \ldots + 2^nC_{2n} = 2^{2^n}$$

Hence n variable will give 2^{2^n} combination

2. **Ans. (c)**

Soluton:

* The addition, subtraction of signed binary numbers can most conveniently be perofrmed using 2's complement representatin of both the operands
* If the two operands are of the opposite sign, the result is to be obtained by rule of subtraction using 2's complement. There is no overflow in this case.
* If the operands are of the same sign, the sign bit of the result (MSB) is to be compared with the sign bit of the operand. In case the sign bit are same, the result is correct and is the 2's complement form. If the sign bits are not same there is problem of overflow, i.e. the result cannot be accomodated using eitht bits.
* Thus, overflow occurs when all the three sign bits are not equal, i.e. $x = y \neq z$. The truth table for overflow can be drawn as under

x	y	z	overflow	
0	0	0	0	No
0	0	1	1	Yes ←
0	1	0	0	No
0	1	1	0	No
1	0	0	0	No
1	0	1	0	No
1	1	0	1	Yes ←
1	1	1	0	No

z \ xy	00	01	11	10
0	0	0	1	0
1	1	0	0	0

The Boolean expression for overflow can be written as under $x\,y\,\overline{z} + \overline{x}\,\overline{y}\,z$

3. **Ans. (a)**

Solution: Essential prime implicates for the function whose Boolean function is shown:

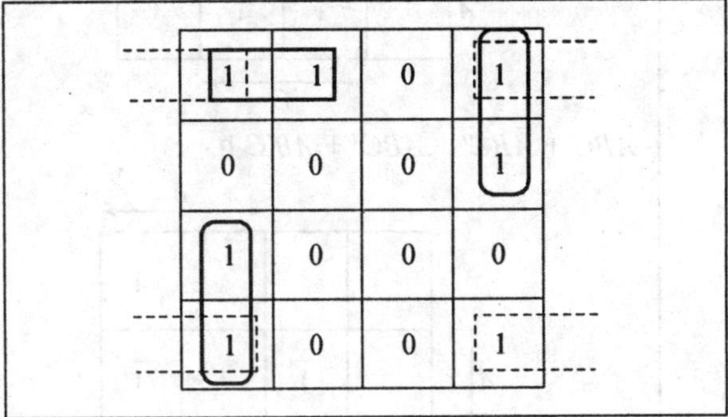

4. **Ans. (d)**

Solution:

A	B	$\overline{A}$	$\overline{A}B$	$A+\overline{A}B$	$A+B$
0	0	1	0	0	0
0	1	1	1	1	1
1	0	0	0	1	1
1	1	0	1	1	1

$\therefore \quad A + \overline{A}B = A + B$

5. **Ans. (d)**

Solution: Let ABCD be 4 variables

⇒ Total number of variables

$= A\overline{A}\ B\overline{B}\ C\overline{C}\ D\overline{D}$ i.e. 8

The total elements in 4 variable K – map $= 2^4 = 16$

$\therefore$ Number of distinct Boolean expression $= 2^{16} = 65536$

6. **Ans. (a)**

Solution: $W = R + \overline{P}Q + \overline{R}S = R + S\,\overline{P}Q$

$$Z = R + S(\overline{P}+\overline{Q})(P+Q+R)(\overline{P}+Q+S)$$

$$= R + S + \overline{P}Q + \overline{P}QS + \overline{P}R + \overline{P}RQ + \overline{P}RS + \overline{Q}PS + \overline{P}QS + \overline{Q}RS$$

$$= R + S + \overline{P}Q$$

$\therefore \quad W = Z$

$$\overline{Z} = \overline{W} = \overline{R + \overline{R}S + \overline{P}Q}$$

$$\overline{W} = \overline{R + \overline{R}S + \overline{P}Q}$$

$$= \overline{R}\,(R+\overline{S})(P+\overline{Q}) = \overline{R}\,\overline{S}\,(P+\overline{Q})$$

$$X = PQ\,\overline{RS} + \overline{P}Q\overline{RS} + \overline{P}\overline{Q}\overline{RS}$$

$$= P\overline{R}\overline{S} + \overline{Q}\overline{R}\overline{S} = \overline{R}\overline{S}\,(P + \overline{Q})$$

$\therefore \quad X = \overline{Z}$

7. Ans. (d)

Solution: $AC + BC' \Rightarrow$

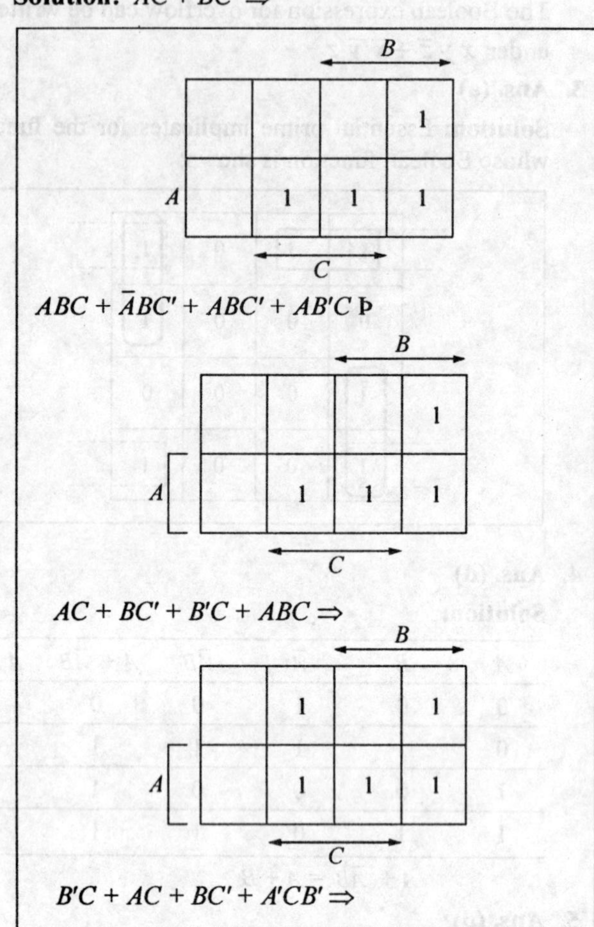

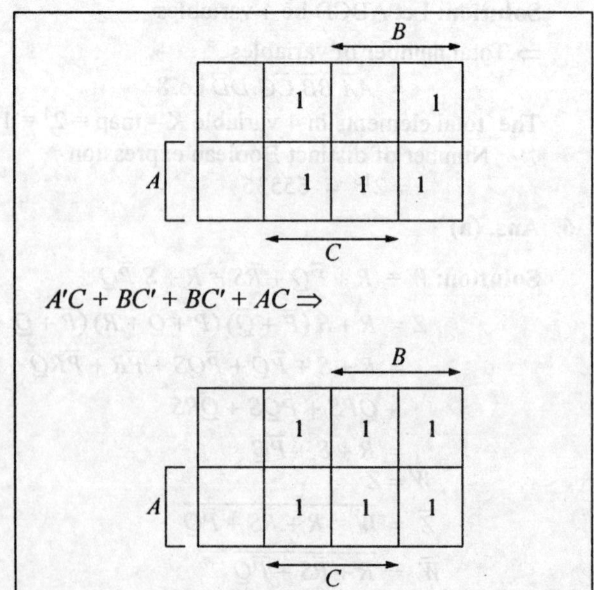

On comparing all, we find that ∞ is correct.

8. Ans. (d)

Solution:

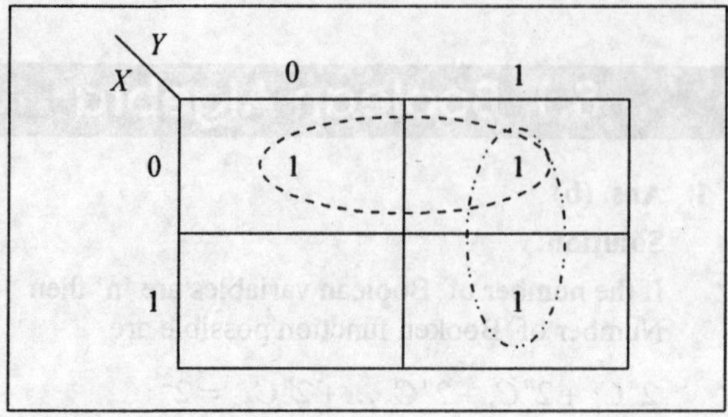

$f = \overline{X}Y$

x	y	F
0	0	1
0	1	1
1	0	0
1	1	1

Since complement are not available

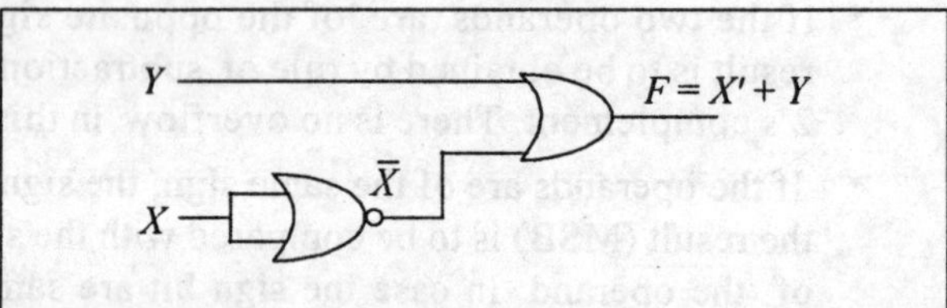

Hence 2 unit are required.

9. Ans. (a)

Solution: $F = \overline{A}\,BC + AB\overline{C}$

$= B\,[\overline{A}C + A\overline{C}]$

$F = B[A + C]\,[\overline{A} + \overline{C}]$

10. Ans. (a)

Solution:

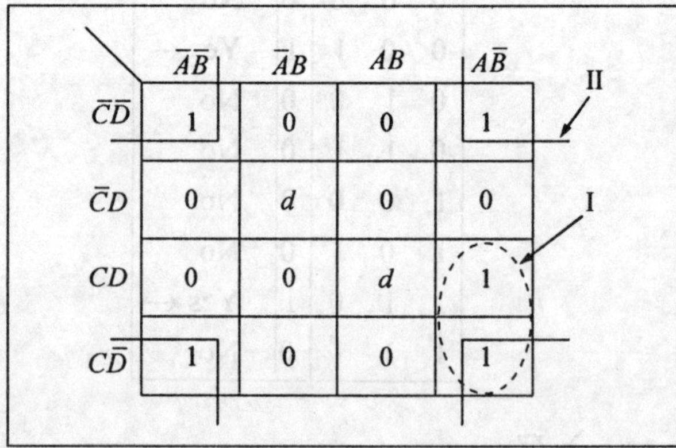

$F = \overline{B}\,\overline{D} + A\,\overline{B}C$

11. Ans. (d)

Solution:

For anlysing the gate for boolen expresion consider the following points

For NAND Gate

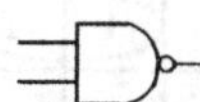

* If both inputs are tied together, the NAND gate acts as an inverter.
* If one of the inputs is 1, the output will be complement of other input.
* If one of the inputs is 0, the output will be 1 irrespective of other input.

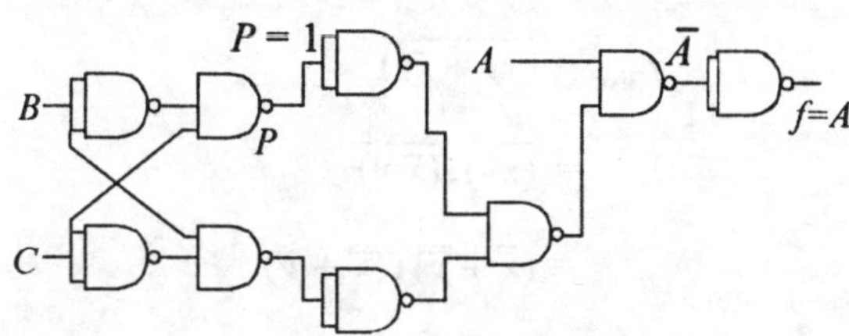

12. **Ans. (b)**

Solution:

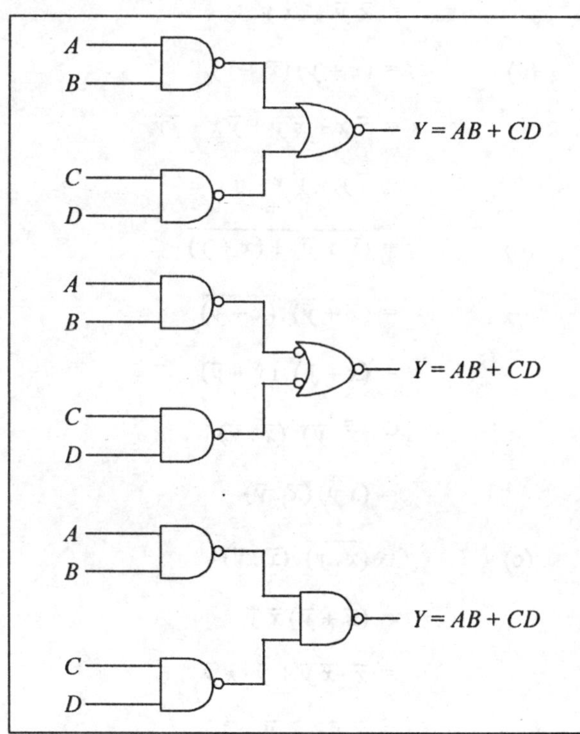

For realization using NAND the minimum number of gates required is three.

13. **Ans. (d)**

Solution: $Y = \bar{A}\bar{B}\bar{C}D + \bar{A}BC\bar{D} + A\bar{B}\bar{C}D + AB\bar{C}\bar{D}$

$Y = \bar{B}\bar{C}\,D + AB\bar{C}\bar{D} + \bar{A}BC\bar{D}$

1 2 3

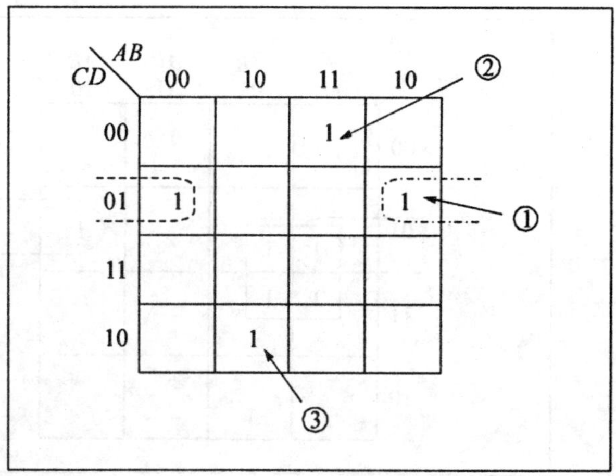

14. **Ans. (d)**

Solution: $\left[\begin{array}{c}[X + Z\{\bar{Y} + (\bar{Z} + X\bar{Y})\}] \\ \{\bar{X} + \bar{Z}(X+Y)\} = 1\end{array}\right]$

$\because \quad X = 1$

$\therefore \quad [1 + Z\{\bar{Y} + (\bar{Z} + \bar{Y})\}]\{0 + \bar{Z}(1+Y)\} = 1$

$[1 + Z(\bar{Y} + \bar{Z})][\bar{Z}] = 1$

$[1 + \bar{Y}Z]\bar{Z} = 1$

$\bar{Z} = 1$

$\therefore \quad Z = 0$

15. **Ans. (b)**

Solution:

$F = A + BC = (A + B)(A + C)$

By the use of distributive propery

16. **Ans. (a)**

Solution:

Given $AB + \bar{A}C + BC = AB + \bar{A}C$

The dual of an expression is find in circuit by replacing each OR gate by and AND gaten and vice versa.

Hence$(A + B)(\bar{A} + C)(B + C) = (A + B)(\bar{A} + C)$

17. **Ans. (a)**

Solution: $4 = (\bar{A}.BC + D)(\bar{A}D + \bar{B}\bar{C})$

$= \bar{A}BCD + \bar{A}B.\bar{B}C\bar{C} + \bar{A}D.D + \bar{B}\bar{C}D$

$= \bar{A}BCD + 0 + \bar{A}D + \bar{B}\bar{C}D$

$= \bar{A}D + \bar{B}\bar{C}D$

Drawing k-map, we have

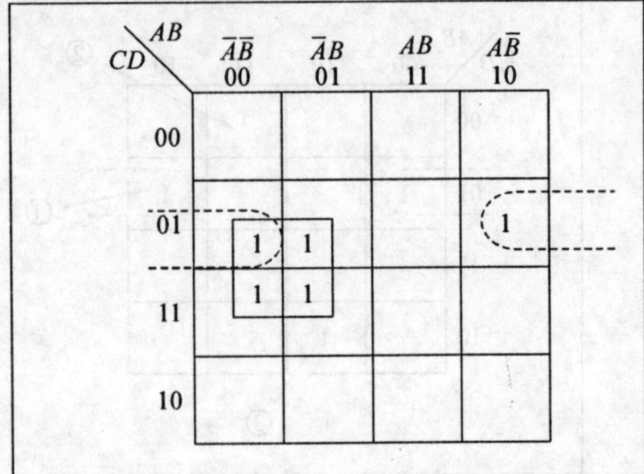

From the k-map, it can be inferred that, simplified expression is $Y = \bar{A}D + \bar{B}\,\bar{C}D$.

18. Ans. (b)

Solution:

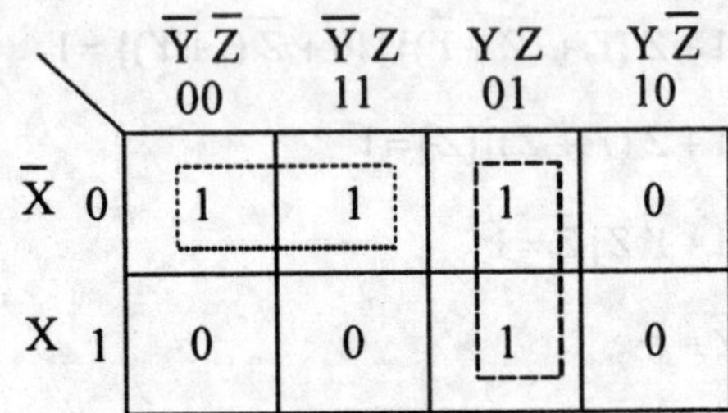

$$F = \bar{X}\,\bar{Y} + YZ$$

19. Ans. (a)

Solution:

	X Y Z	
0	0 0 0	$\bar{X}\ \bar{Y}\ \bar{Z}$
1	0 0 1	$\bar{X}\ \bar{Y}\ Z$
2	0 1 0	$\bar{X}\ Y\ \bar{Z}$
3	0 1 1	$\bar{X}\ Y\ Z$
4	1 0 0	$X\ \bar{Y}\ \bar{Z}$
5	1 0 1	$X\ \bar{Y}\ Z$
6	0 1 1	$\bar{X}\ Y\ Z$
7	1 1 1	$X\ Y\ Z$

$$f(X\,Y\,Z) = \Sigma(2,\ 3,\ 4,\ 5)$$

$$= \Sigma\,\bar{X}\ Y\ \bar{Z},\ \bar{X}\ Y\ Z,\ X\ \bar{Y}\ \bar{Z},\ X\ \bar{Y}\ Z$$

X \ YZ	$\bar{X}\bar{Y}$ 00	$\bar{X}Y$ 01	XY 11	$X\bar{Y}$ 10
$\bar{X}$ 0	0	1	3	2
X 1	4	5	7	6

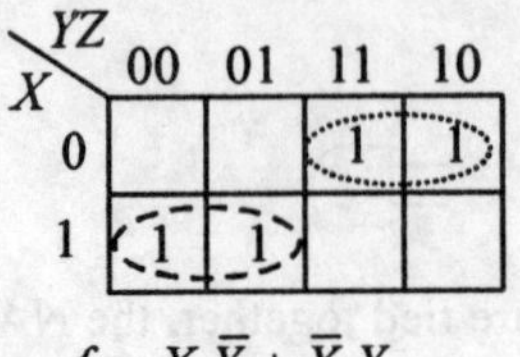

$$f = X\,\bar{Y} + \bar{X}\,Y$$

Implicants are $\bar{X}\,Y\,\bar{Z},\ \bar{X}\,Y\,Z,\ X\,\bar{Y}\,\bar{Z},\ X\,\bar{Y}\,Z$

The prime implicants are $\bar{X}Y$ and $X\bar{Y}$

20. Ans. (d)

Solution: The given function is EX-OR Gate

$$f_{\text{Give}} = \overline{(x\bar{y} + \bar{x}y)}$$

$$= \overline{(x\bar{y})}\,.\,\overline{(\bar{x}\,y)}$$

$$= (\bar{\bar{x}} + \bar{\bar{y}})\,.\,(\bar{x} + \bar{y})$$

$$= (x + \bar{y})\,.\,(x + \bar{y})$$

$$= \bar{x}\,x + \bar{x}\,\bar{y} + x\,y + y\,\bar{y}$$

$$= \bar{x}\,\bar{y} + .\,x\,y$$

(a) $f = (x + y)\,(\bar{x} + \bar{y})$

$$= \bar{x}\,x + \bar{x}\,y + \bar{y}\,x + \bar{y}y$$

$$= \bar{x}\,y + \bar{y}\,x$$

(b) $f = \overline{(\bar{x} + \bar{y}) + (x + y)}$

$$= (\bar{\bar{x}} + \bar{\bar{y}})\,.\,\overline{(x + y)}$$

$$= (x + \bar{y})\,.\,(x + \bar{y})$$

$$= (\bar{\bar{x}}\,.\,\bar{y})\,.\,(\bar{x}\,.\,\bar{y})$$

$$= (x\,y)\,(\bar{x}\,.\,\bar{y})$$

(c) $f = \overline{(x\,.\,y)}\,.\,(\bar{x}\,.\,\bar{y})$

$$= (\bar{x} + \bar{y})\,\bar{x}\,\bar{y}$$

$$= \bar{x}\cdot\bar{x}\,\bar{y} + \bar{y}\cdot\bar{x}\,\bar{y}$$

$$= \bar{x}\,\bar{y} + \bar{x}\,\bar{y}$$

$$= \bar{x}\,\bar{y}$$

Therefore, none of the above (a) (b) (c) match

21. Ans. (b)

Solution: Demorgan's theorem

$$\overline{A + B + C} = \bar{A}\cdot\bar{B}\cdot\bar{C}$$

22. Ans. (a)

Solution:

According to Demorgan's rule

$\overline{AB} = \overline{A} + \overline{B}$

23. Ans. (d)

Solution:

$AB\overline{C}\,\overline{D} + ABCD + AB\overline{C}D + ABC\overline{D}$

$= AB\overline{C}\left(\overline{D} + D\right) + ABC\left(D + \overline{D}\right)$

$\left[\because \overline{D} + D = 1\right]$

$= AB\overline{C} + ABC = AB\left[\overline{C} + C\right] = AB$

24. Ans. (a)

Solution:

$F = \overline{a}\,\overline{b}\,\overline{c} + \overline{a}\,b\,\overline{c} + a\,\overline{b}\,\overline{c} + abc + ab\overline{c}$

SO P form

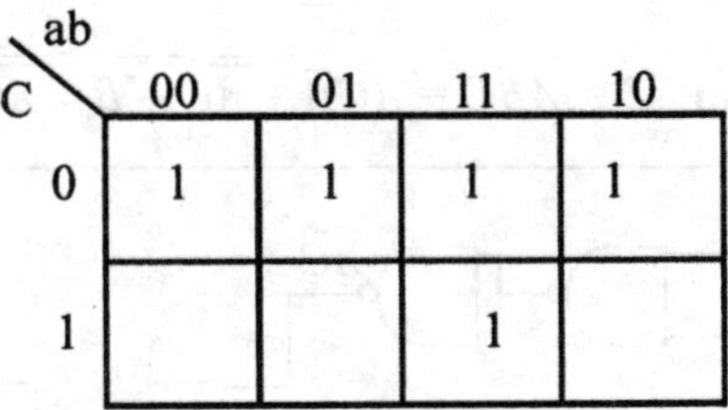

POS form

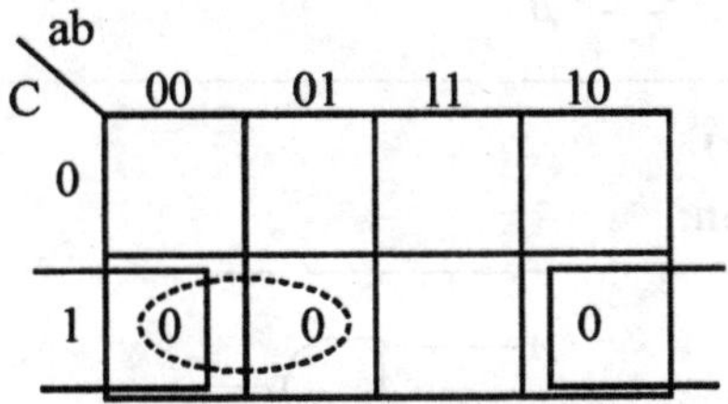

$\Rightarrow (a + \overline{c})(b + \overline{c})$

$\Rightarrow (b + \overline{c})(a + \overline{c})$

25. Ans. (b)

Solution: Apply the Karnaugh's map for given function.

$f = A\overline{B}CD + \overline{A}BC + \overline{A}\overline{B}C + BCD$

$= A\overline{B}CD + \overline{A}BCD + \overline{A}BC\overline{D} + \overline{A}\overline{B}CD + \overline{A}BC\overline{D}$

$+ ABCD + \overline{A}BCD$

$f = \overline{A}C + CD$

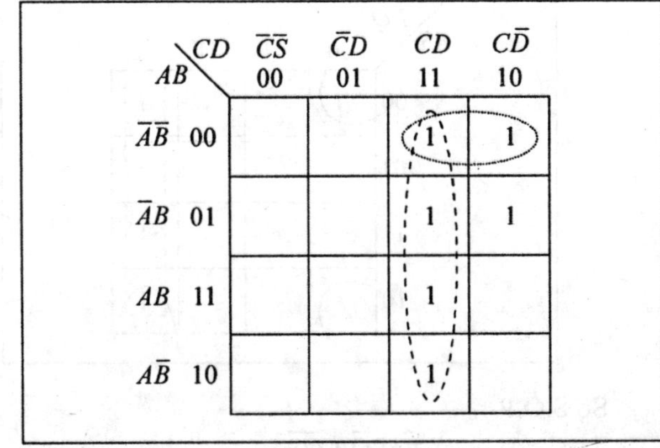

26. Ans. (c)

Solution:

Decimal No.	A	B	C	Min-term
0	0	0	0	$\overline{A}\,\overline{B}\,\overline{C}$
1	0	0	1.	$\overline{A}\,\overline{B}\,C$
2	0	1	0	$\overline{A}\,B\,\overline{C}$
3	0	1	1	$\overline{A}\,B\,C$
4	1	0	0	$A\,\overline{B}\,\overline{C}$
5	1	0	1	$A\,\overline{B}\,C$
6	1	1	0	$A\,B\,\overline{C}$
7	1	1	1	$A\,B\,C$

$$\bar{f}(A\,B\,C) = \sum m\,(0, 1, 2, 3, 5, 6)$$

$f = \overline{A}\overline{B}\overline{C} + \overline{A}\overline{B}C + \overline{A}B\overline{C} + \overline{A}BC$

$+ A\overline{B}C + AB\overline{C}$

$= A + \overline{B}C + B\overline{C}$

$f = \overline{A} + B \oplus C$

27. Ans. (a)

Solution: $Y = \overline{P}\overline{Q}\overline{R}\overline{S} + P\overline{Q}\overline{R}\overline{S} + P\overline{Q}\overline{R}S + P\overline{Q}RS$

$+ P\overline{Q}R\overline{S} + \overline{P}\overline{Q}R\overline{S}$

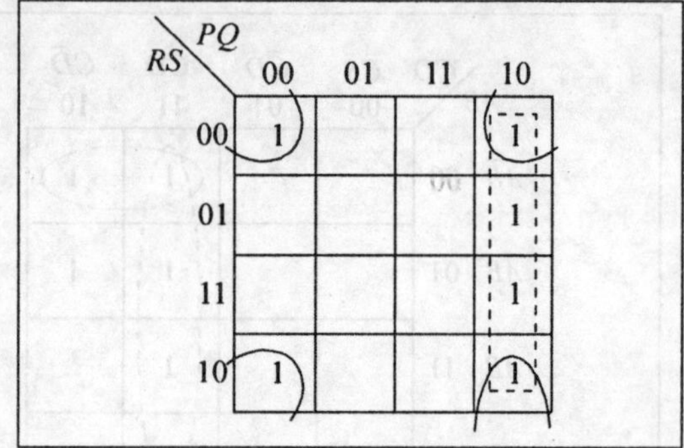

So S.O.P

$$Y = P\overline{Q} + \overline{Q}\overline{S}$$

So Ans = A

28. Ans. (a)

Solution:

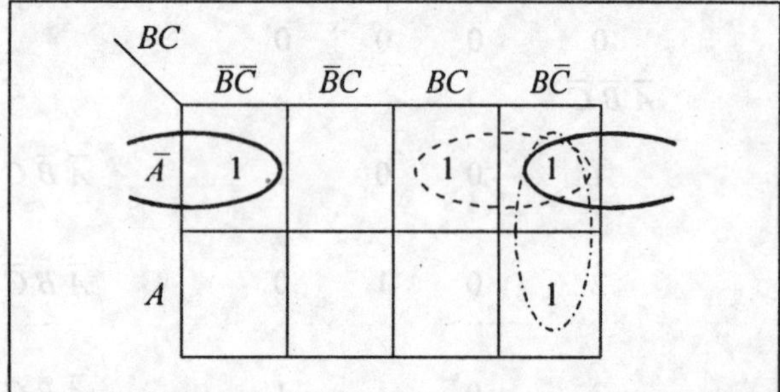

$$y = \overline{A}\overline{C} + \overline{A}B + B\overline{C}$$

8.2 Logic Gates

1. Ans. (c)

Solution: $X\text{-}OR = A\overline{B} + \overline{A}B$

$$X\text{-}NOR = \overline{A\overline{B} + \overline{A}B} = (\overline{A\overline{B}}) \cdot (\overline{\overline{A}B})$$

$$= \left(\overline{A} + \overline{\overline{B}}\right) \cdot \left(\overline{\overline{A}} + \overline{B}\right) = \left(\overline{A} + B\right) \cdot \left(A + \overline{B}\right)$$

2. Ans. (b)

Solution: Truth Table

NOR = $\overline{A+B}$

Input		$A + B$	$\overline{A+B}$
0	0	0	1
0	1	1	0
1	0	1	0
1	1	1	0

E-NOR = EX-OR → NOR

$$Y = \overline{\overline{A}B + A\overline{B}}$$

Input		$A + B$	$\overline{A + B}$
0	0	0	1
0	1	1	0
1	0	1	0
1	1	0	1

3. Ans. (a)

Solution: $A + A\overline{B} + A\overline{B}C$

$$= A + A\overline{B}\,(1 + C) = A(1 + A\overline{B}) = A$$

The output is directly taken from A, Hence we do not required any gate i.e. zero GATE.

4. Ans. (b)

Solution:

Given diagram is XNOR with output connected to ground i.e. B = 0

$$A \odot B = AB + \overline{A}\,\overline{B}$$

$$A \odot 0 = A0 + \overline{A}\,1$$

$$= \overline{A}$$

5. Ans. (c)

Solution: $Z = A\overline{B}C = AC\overline{B} = \overline{\overline{AC} + B} = \overline{\overline{\overline{AC}} + \overline{B}}$

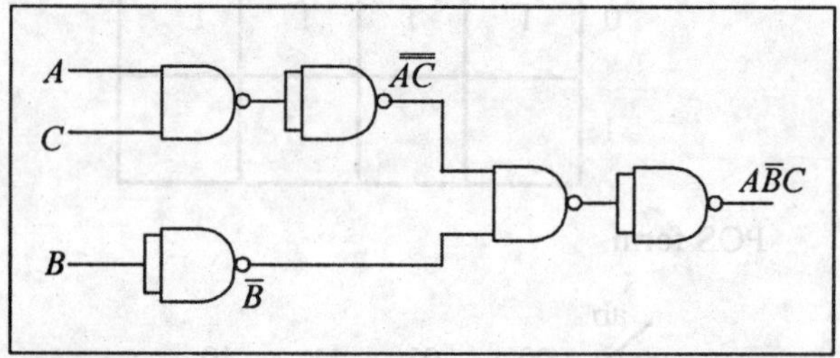

6. Ans. (a)

Solution:

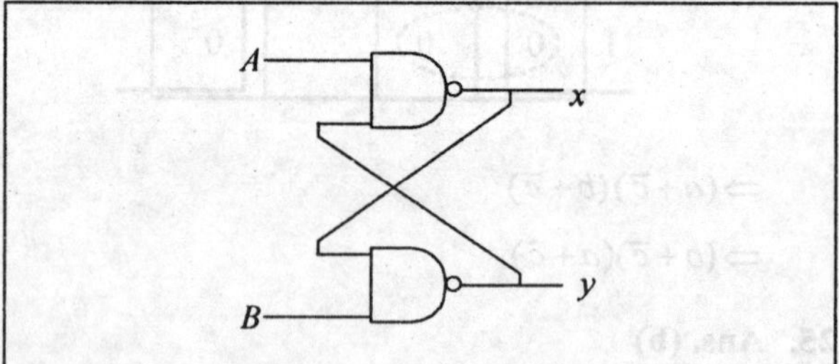

The given circuit is shown below:

If $A = 1$ $X = 0$

$B = 1$ $Y = 1$

$A = 1$ $X = 0$

$B = 0$ $Y = 1$

$A = 1$ $X = 0$

$B = 1$ $Y = 1$

$A = 1$ $X = 0$

$B = 1$ $Y = 1$

7. **Ans. (d)**

Solution:

$$X = (A\overline{B} + \overline{A}B)(\overline{B}\,\overline{C} + BC)C$$
$$= (A\overline{B} + \overline{A}B)\,BC = \overline{A}BC = \overline{0}11 = 1$$
$$A = 0,\ B = 1,\ C = 1$$

8. **Ans. (c)**

Solution:

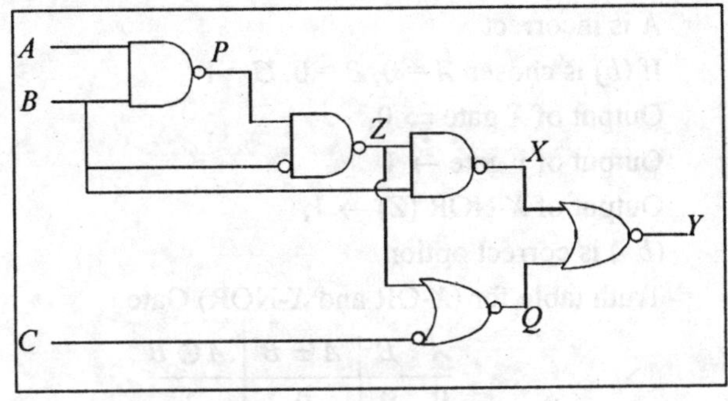

$$P = \overline{AB}$$
$$Z = \overline{P \cdot \overline{B}} = \overline{P} + B$$
$$= \overline{\overline{AB}} + B = AB + B$$
$$= B(A+1) = B$$
$$X = \overline{B.B} = \overline{B}$$
$$Q = \left(\overline{Z+\overline{C}}\right) = \overline{Z} \cdot C = \overline{B}\,C$$
$$F = \overline{X+Q} = \overline{\overline{B} + \overline{B}C}$$
$$= \overline{\overline{B}(1+C)} = \overline{\overline{B}} = B$$

9. **Ans. (b)**

Solution: Output of 1st EX-OR $= 1 \oplus X = \overline{X}$

Output of 2nd EX-OR $= \overline{X} \oplus X = 1$

Output for even (i.e. 20) EX-OR = 1

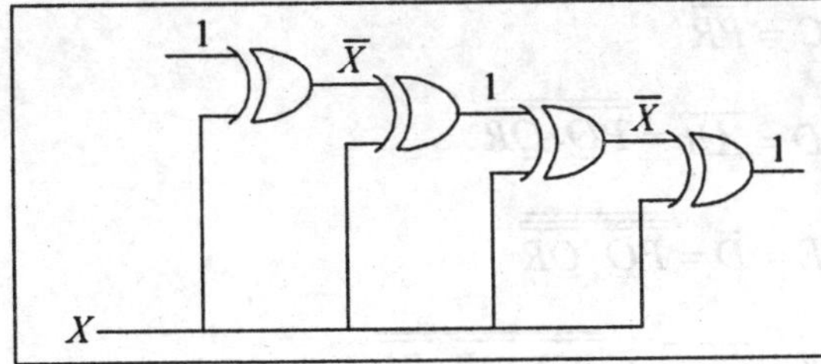

10. **Ans. (b)**

Solution:

	$t < t_0$	t_0	t_1	t_2
Input to G_1	0	1	1	1
V_i	1	0	0	0
Input to G_2	0	1	1	1
V_0	1	1	1	0

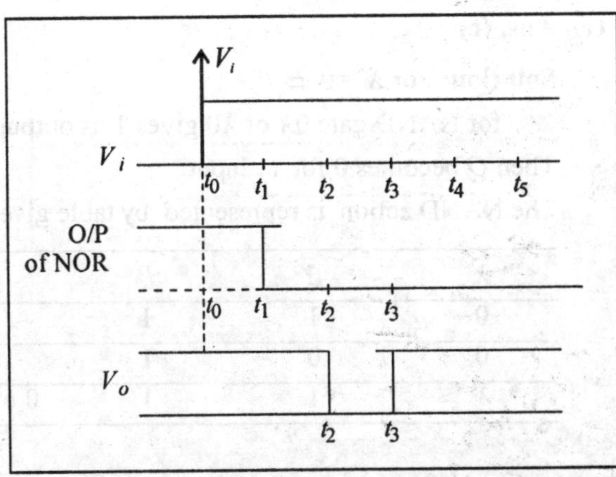

Second Method

Since both the inputs of NOR gates are tied together, the NOR gate behave like an inverter.

Let analyze it in two steps.

Let us consider that the propgation delay for EXOR gate (G_2) is zero. Let the output in this case be V_1.

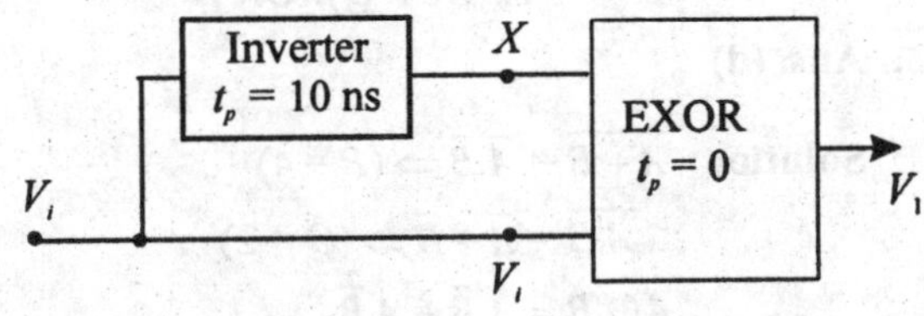

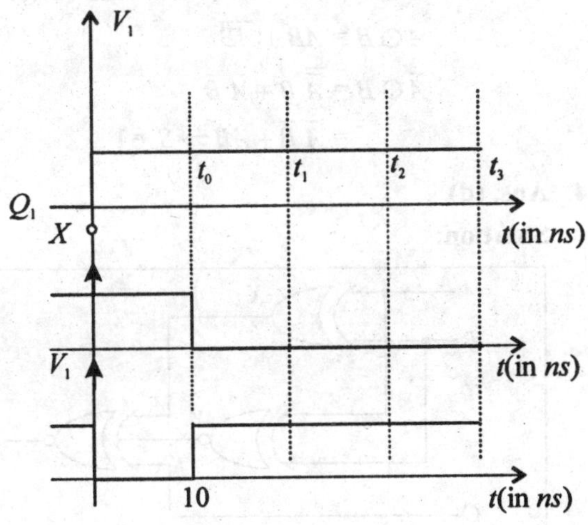

Now, if the propgation delay of G_2 is 20 ns. The output - V_0 will be delayed by 20 ns.

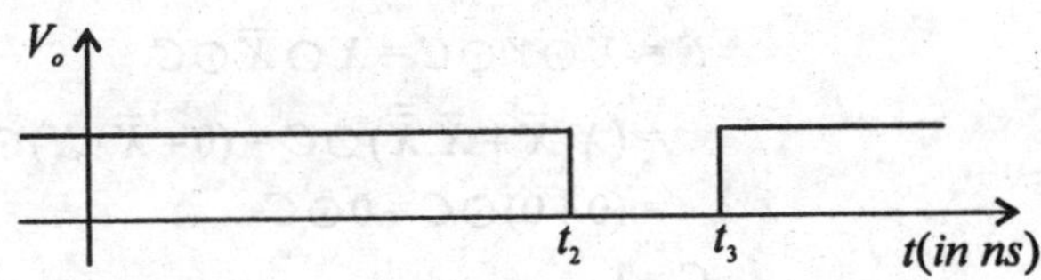

11. Ans. (c)

Solution: For $X = 0 \Rightarrow P = 1$

$\therefore$ for NAND gate $0X$ or $X0$ gives 1 as output

Then Q becomes 0 for 11 input

The NAND action is represented by table given below

X	Y	P	Q
0	1	1	0
0	0	1	1
1	1	1	0 or 0/1

12. Ans. (d)

Solution: $M_1 = Z\ XOR\ R$

$M_1 = (X \cdot Y)\ XOR\ R$

$M_1 = [\overline{P.Q}.(P+Q)]XOR\ R$

$= [(\bar{P}+\bar{Q})(P+Q)XOR\ R]$

$= (\bar{P}P + \bar{P}Q + P\bar{Q} + Q\bar{Q})\ XOR\ R$

$= (\bar{P}Q + P\bar{Q})XOR\ R$

$= (P\ XOR\ Q)XOR\ R$

13. Ans. (d)

Solution: $\overline{A+B} = \bar{A}.\bar{B} \Rightarrow (P=4)$

$\overline{A.B} = \bar{A}+\bar{B} \Rightarrow (Q=2)$

$A \oplus B = A\,\bar{B} + \bar{A}\,B$

$\bar{A} \odot B = \bar{\bar{A}}\bar{B} + \bar{A}B = A\bar{B} + \bar{A}B \Rightarrow (R=3)$

$A \odot B = AB + \bar{A}\bar{B}$

$\bar{A} \oplus B = \bar{\bar{A}}\,B + \bar{A}\,\bar{B}$

$= \bar{A}\,\bar{B} + AB \Rightarrow S = 1$

14. Ans. (d)

Solution:

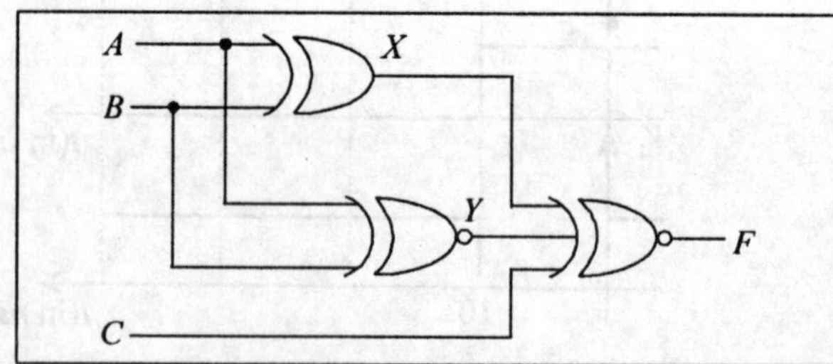

$X = \bar{A}\,B + A\,\bar{B}$

$Y = A\,B + \bar{A}\,\bar{B} = \bar{X}$

$C = C$

$F = X \odot Y \odot C = X \odot \bar{X} \odot C$

$= (X\,\bar{X} + \bar{X}\,\bar{\bar{X}}) \odot C = (0 + \bar{X} \cdot X) \odot C$

$= (0+0) \odot C = 0 \odot C$

$C = 1$

Second Method

For F to be 1

Number of 1's or 0's at the X-NOR Gate should be even output of Y-gate is 1 when A and B are either 0 or 1 both

From ans option (a)

$A = 1, B = 1, C = 0$

Output of X Gate $\rightarrow$ 0, Output of Y Gate $\rightarrow$ 1

C input is 0, Output of Z is

A is incorrect

If (b) is chosen $A = 0, B = 0, C = 1$

Output of X gate $\Rightarrow$ 0,

Output of y gate $\rightarrow$ 1

Output of X-NOR (Z) $\rightarrow$ 1,

(b) is correct option

Truth table for (X-OR and X-NOR) Gate

A	B	$A \oplus B$	$A \odot B$
0	0	0	1
0	1	1	0
1	0	1	0
1	1	0	1

15. Ans. (b)

Solution:

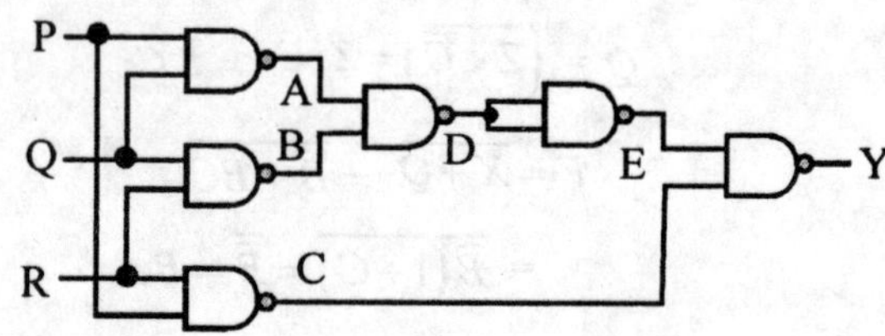

$A = \overline{PQ}$

$B = \overline{QR}$

$C = \overline{PR}$

$D = \overline{AB} = \overline{\overline{PQ} \cdot \overline{QR}}$

$E = \bar{D} = \overline{\overline{\overline{PQ} \cdot \overline{QR}}}$

$Y = \overline{C \cdot E} = \overline{\overline{PQ} \cdot \overline{QR} \cdot \overline{PR}}$

$= \overline{\overline{PQ}} + \overline{\overline{QR}} + \overline{\overline{PR}}$

$= PQ + QR + PQ$

The output Y expression in the circuit

$Y = PQ + PR + RQ$

So that if two or more inputs are '1', Y is always '1'.

16. Ans. (a)

Solution: $\overline{(A\oplus B).C} = \overline{(\bar{A}B + A\bar{B}).C}$

$= \overline{(\bar{A}B + A\bar{B})} + \bar{C}$

$= (\overline{\bar{A}B})(\overline{A\bar{B}}) + \bar{C}$

$= (\bar{\bar{A}} + \bar{B})(\bar{A} + \bar{\bar{B}}) + \bar{C}$

$= (A + \bar{B})(\bar{A} + B) + \bar{C}$

$= \bar{A}A + AB + \bar{A}\bar{B} + \bar{B}B + \bar{C}$

$= 0 + \bar{A}\bar{B} + AB + 0 + \bar{C}$

$= AB + \bar{A}\bar{B} + \bar{C}$

Second Method

$\overline{A\oplus B\cdot C} = \overline{A\oplus B} + \bar{C}$

$= A\odot B + \bar{C}$

$= \bar{A}\,\bar{B} + AB + C$

17. Ans. (a)

Solution:

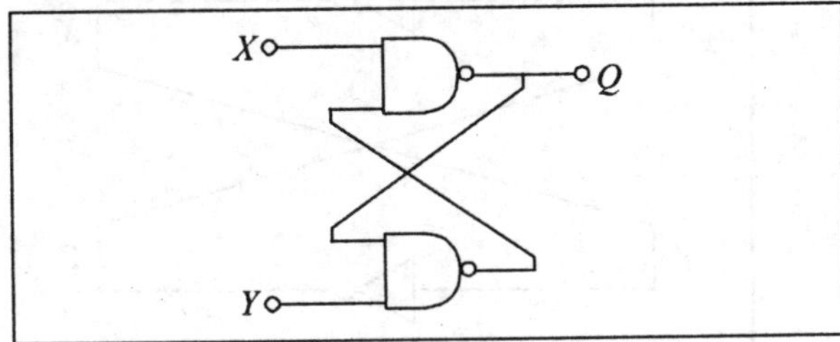

Truth table of the circuit can be drawn as,

X	Y	$Q_{(n+1)}$
0	0	?
0	1	0
1	0	1
1	1	Q_n

Thus the flio-flop formed with NAND gate cirucit will be unstable for $X = 0$ and $Y = 0$

18. Ans. (d)

Solution:

Analyzing all the gate output for given condition

Gate	A	B	Output
NAND	0	0	$\overline{A.B}=\overline{0.0}=1\leftarrow$
Ex-OR	0	0	$A\bar{B}=\bar{A}\,B$ $=0.1+1.0=0$
Ex-NOR	0	0	$A.B=\bar{A}\,\bar{B}$ $=0.0+\bar{0}.\bar{0}=1\leftarrow$
OR	0	0	$A+B=0$
AND	0	0	$A.B=0$
NOR	0	0	$\overline{A+B}=\bar{0}=1\leftarrow$

The ouptut of NAND Ex-NOR or NOR gaes is 1 when all its input are at logic '0' Hence correct option is (d)

19. Ans. (b)

Solution:

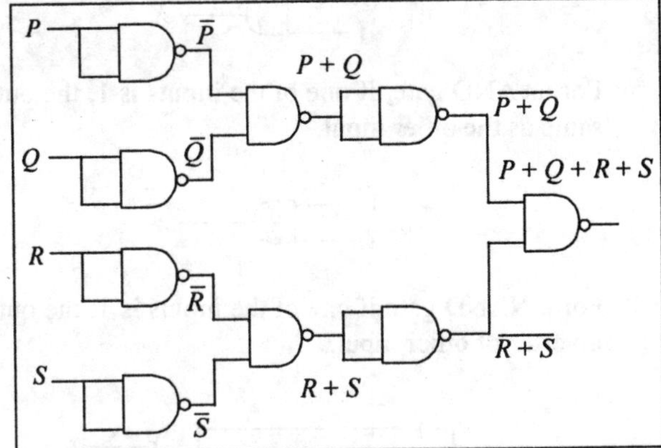

Second Method

NAND–NAND Logic = AND – OR Logic

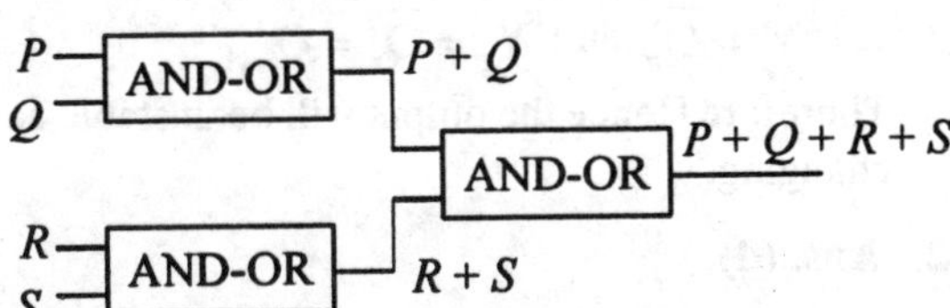

20. Ans. (d)

Solution: Exclusive OR gate gives output 1 when the input are different.

* For an EXOR gate, output = 1 only if total number of 1's at the input is odd, and for a EXOR gate, output = 1 only if total number of 1's at the input is even.

or For Y = 0, total number of 1's at the input of EXNOR gate (5) should be odd.

or Among four digits of the two input binary numbers, either '1' or '3' bits should be different.

The above critatin is satisfied only by option (d) where the two numbers are 1010 and 1011.

21. Ans. (d)

Solution:

Case 1:

$$\left.\begin{array}{l} X_1 = 1 \to 1 \\ \left.\begin{array}{l} X_2 = 1 \\ Q_2 = 0 \end{array}\right\} \xrightarrow{\text{NAND}} 1 \end{array}\right\} \to 1$$

Case 2:

$$\left.\begin{array}{l} X_1 = 1 \to 1 \\ \left.\begin{array}{l} X_2 = 1 \\ Q_2 = 1 \end{array}\right\} \xrightarrow{\text{NAND}} 0 \end{array}\right\} \to 0$$

Therefore output is unstable

Second Method

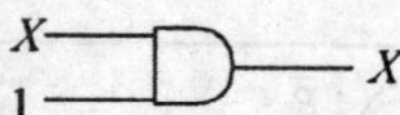

For an AND gate, if one of the inputs is 1, the output is same as the other input.

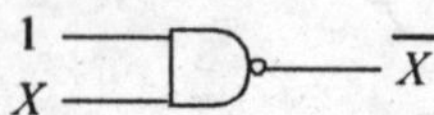

For a NAND gate if one of the inputs is 1, the output is inverter of other input.

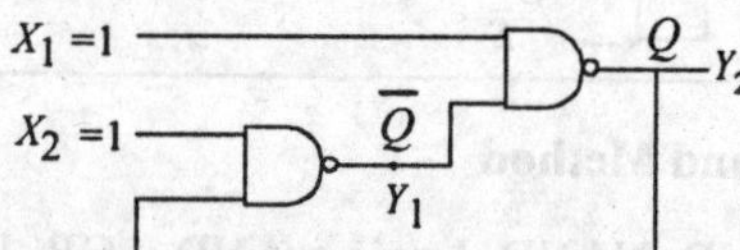

$\Rightarrow Y_1 = \overline{Q}_n$ and $Y_2 = \overline{Q}_n = Q_{n+1}$

Therefore Hence the output will be unstable and keep on changing.

22. Ans. (d)

Solution:

A	B	C	D	Y	$S = A + B + C + D + Y$
0	0	0	0	0	0
0	0	0	1	1	0
0	0	1	0	1	0
0	0	1	1	0	0
0	1	0	0	1	0
0	1	0	1	0	0
0	1	1	0	0	0
0	1	1	1	1	0
1	0	0	0	1	0
1	0	0	1	0	0
1	0	1	0	0	0
1	0	1	1	1	0
1	1	0	0	0	0
1	1	0	1	1	0
1	1	1	0	1	0
1	1	1	1	0	0

Sum is always zero and zero is a even term therefore, option (b) is correct.

Second Methoed

The circuit in the question can be replaced by a single 4 - input XOR gate as given below

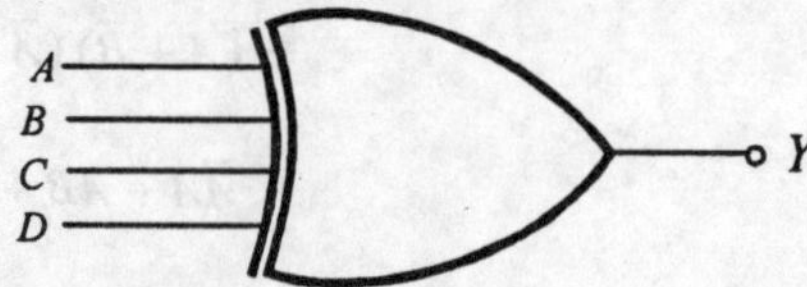

* $Y = 1$ only when there are odd no. of 1 s at the input

or

* $Y = 1$ only if the sum of A, B, C, D is odd

23. Ans (c)

Solution:

All gates can be derived from NOR, or, NAND gates,

24. Ans. (a)

Solution:

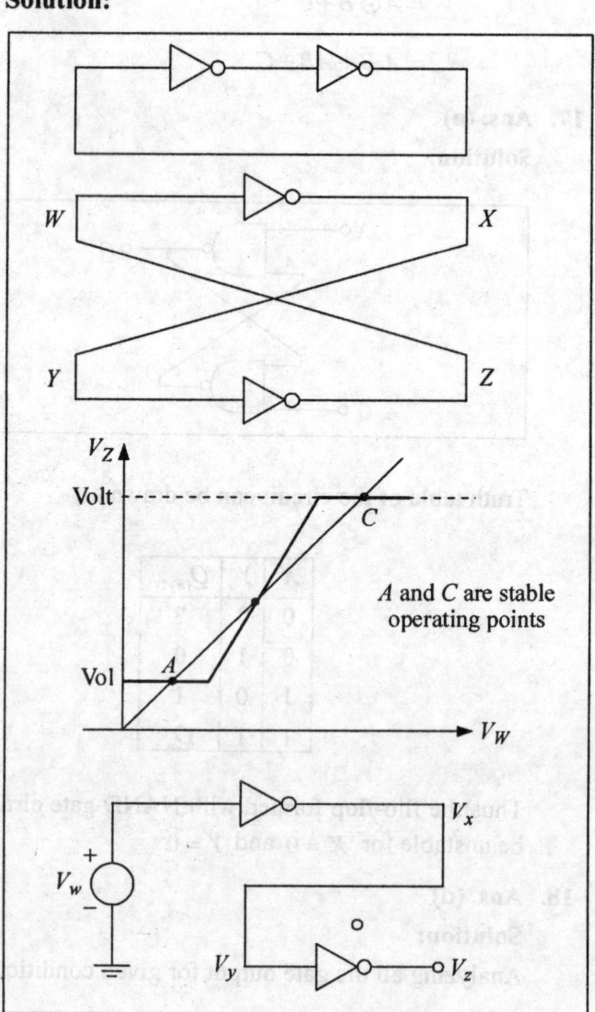

The latch has two stable operating point A and C. Then the latch is a bistable circuit having two complementary output.

25. Ans. (b)

Solution: From the diagram backward analysis. The result is as output will be 1 when $A=0, B=0, C=X=0$ or 1

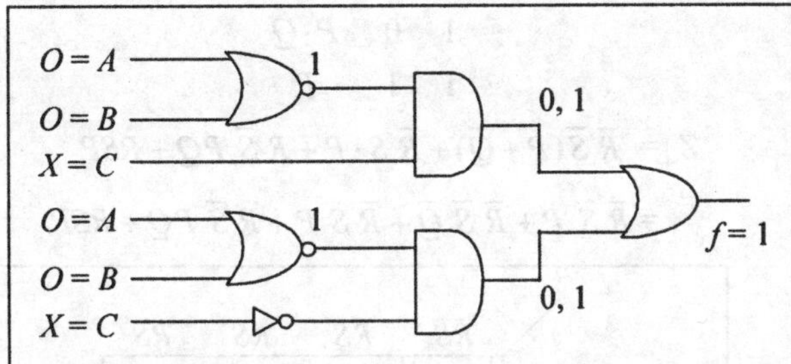

Second Method

Given condition $f=1$ and $A=0$

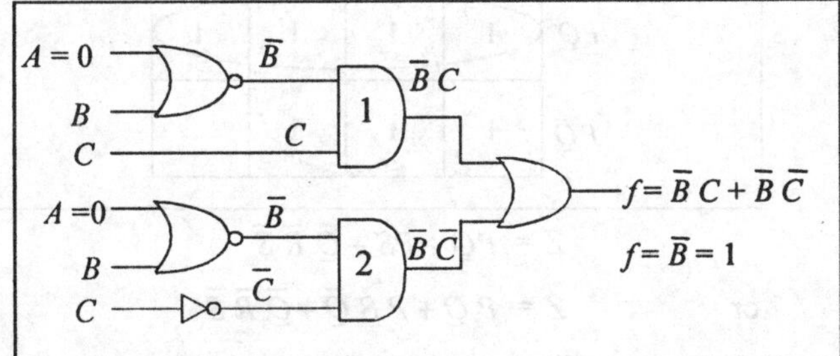

Hence $B=0$ and C can be 0 or 1

26. Ans. (a)

Solution: The correct match is given below:

A. NAND	1. $\bar{A}+\bar{B}=\overline{A\cdot B}$
B. NOR	2. $\bar{A}\cdot\bar{B}=\overline{A+B}$
C. XNOR	3. $\bar{A}\bar{B}+AB$
D. AND	4. $\overline{\bar{A}\cdot\bar{B}}=\overline{\overline{A\cdot B}}=A\cdot B$

27. Ans. (d)

Solution:

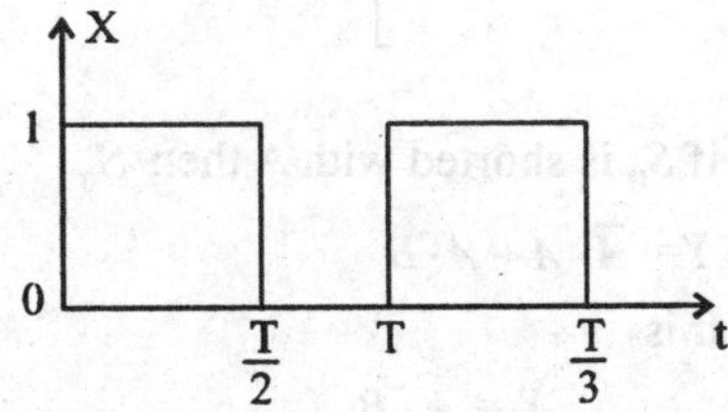

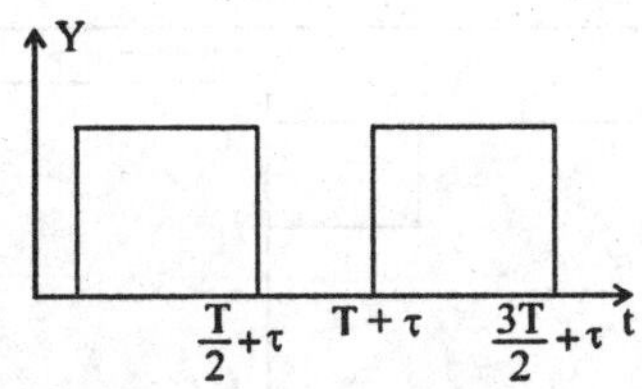

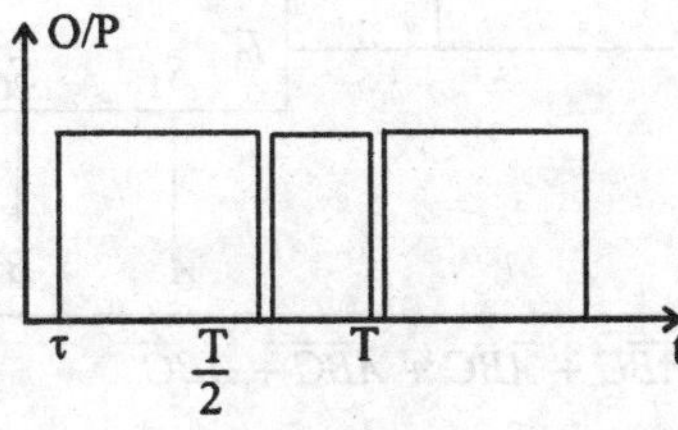

$$V_{av}=\int_{t}^{T/2}\frac{at}{T/2}=1-\frac{2t}{T}$$

8.3 Combinational Circuits

1. Ans. (b)

Solution: 8421 code represent 0 to 9 only

Divisible by the i.e. 0. 3, 6, 9, which need 4 variable, then from k-map do not care conditions are 10, 11, 12, 13, 14, 15

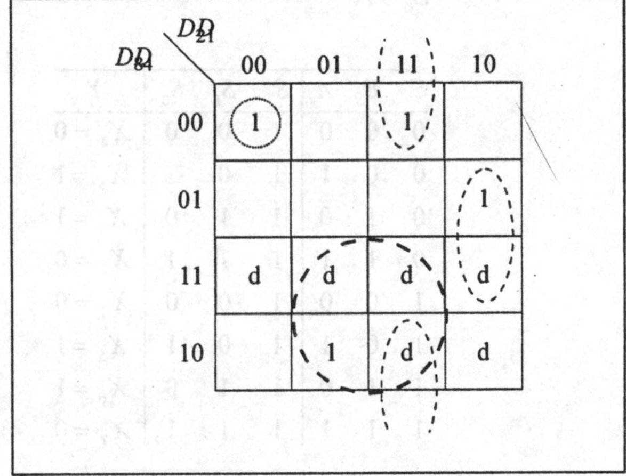

$$M=\bar{D}_8\bar{D}_4\bar{D}_2\bar{D}_1+\bar{D}_4D_2D_1+D_4D_2\bar{D}_1+D_1D_8$$

2. Ans. (b)

Solution:

	S_1	S_0	F
	A	B	
I_0	0	0	C
I_1	0	1	C
I_2	1	0	C
I_3	1	1	C

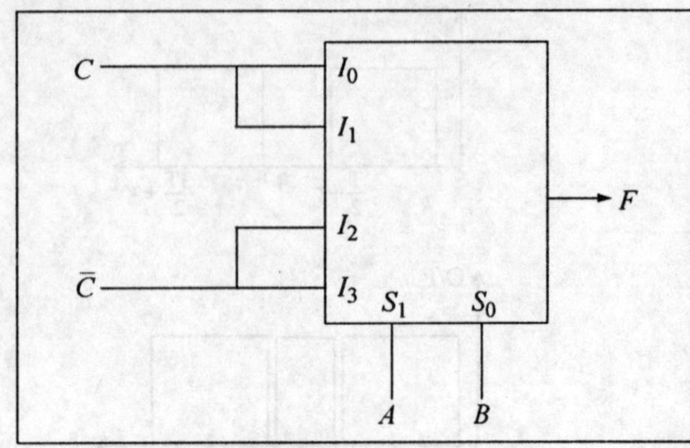

$$F = \bar{A}\bar{B}C + \bar{A}BC + A\bar{B}\bar{C} + AB\bar{C}$$

$$= \bar{A}C[\bar{B} + B] + A\bar{C}[\bar{B} + B]$$

$$= \bar{A}C + A\bar{C} = A \oplus C$$

3. **Ans. (d)**

4. **Ans. (d)**

Solution:

In TTL circuit, if any of the inputs is floating, i.e. not connecte to any input that input is considered as 1.

$\Rightarrow$ $S_2 = 1 + C = 1$

$S_1 = B$

$S_0 = A$

C	B	A	S_2	S_1	S_0	Y
0	0	0	1	0	0	$X_4 = 0$
0	0	1	1	0	1	$X_5 = 1$
0	1	0	1	1	0	$X_6 = 1$
0	1	1	1	1	1	$X_7 = 0$
1	0	0	1	0	0	$X_4 = 0$
1	0	1	1	0	1	$X_5 = 1$
1	1	0	1	1	0	$X_6 = 1$
1	1	1	1	1	1	$X_7 = 0$

$\Rightarrow$ $Y = \Sigma(1, 2, 5, 6)$

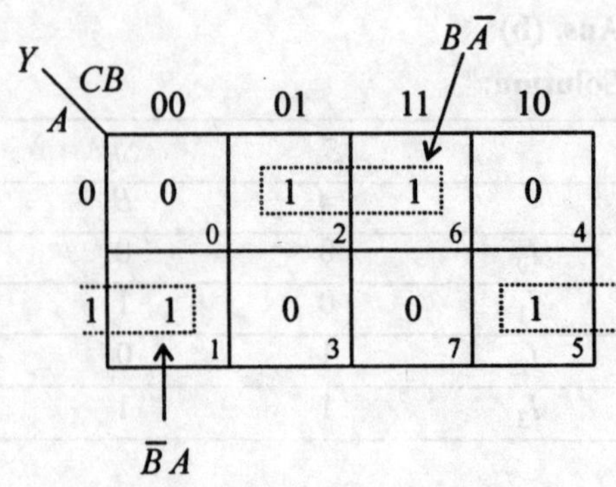

$\Rightarrow$ $Y = B\bar{A} + \bar{B}A$

$\Rightarrow$ $Y = B \oplus A$

5. **Ans. (a)**

Solution: Use input , output relation for solution

R	S	Z
0	0	$P+\bar{Q}$
0	1	P
1	0	$P \cdot Q$
1	1	P

$$Z = \bar{R}\bar{S}(P+\bar{Q}) + \bar{R}S \cdot P + R\bar{S}PQ + RSP$$

$$= \bar{R}\bar{S}P + \bar{R}\bar{S}\bar{Q} + \bar{R}SP + R\bar{S}PQ + RSP$$

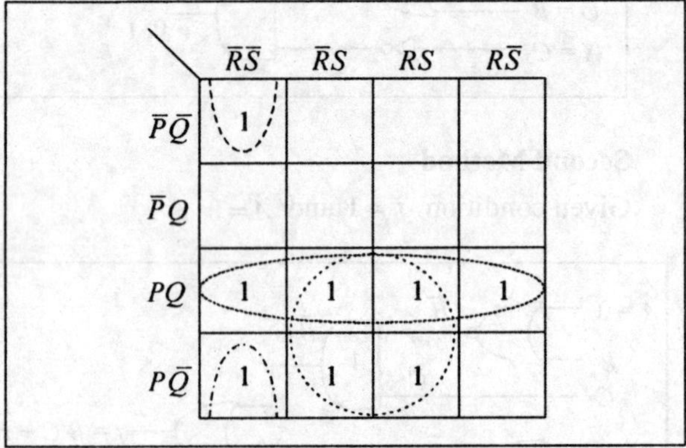

$$Z = PQ + PS + \bar{Q}\bar{R}\bar{S}$$

or $$Z = PQ + PS\bar{Q} + \bar{Q}\bar{R}\bar{S}$$

6. **Ans. (a)**

Solution: Minimum number of 2:1 MUX required to generate 2 input AND gate.

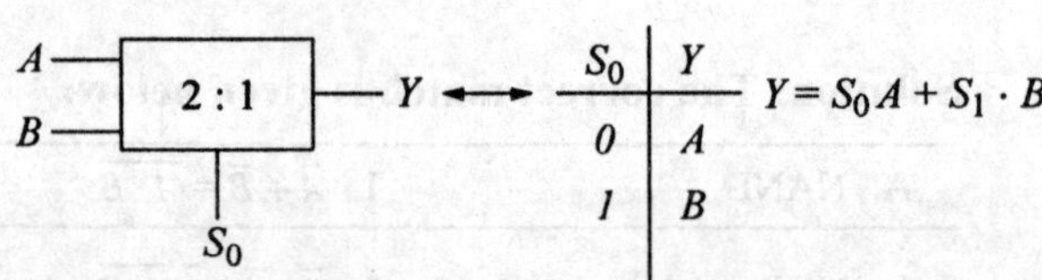

AND Gate Table

A	B	$Y = A \cdot B$	
0	0	0	$\rightarrow$ same as A
0	0	0	
1	0	0	$\rightarrow$ same as B
1	1	1	

Now if S_0 is shorted with A then S_0

Then $Y = \bar{A} \cdot A + A \cdot B$

Now this

$$Y = A \cdot B$$

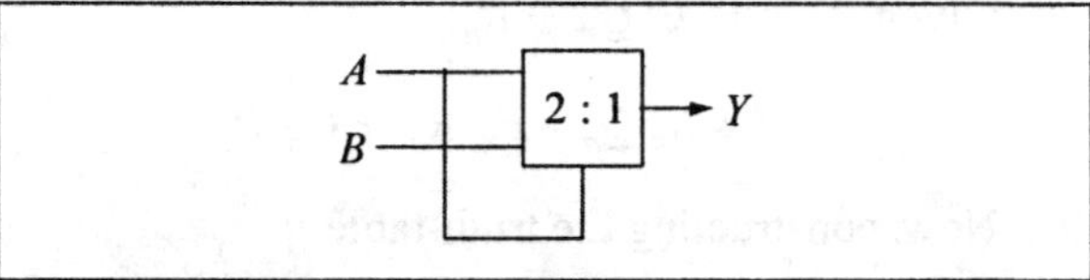

Is an AND Gate

Mux required to get EX-OR gate

EX-OR Table

A	B	Y
0	0	0
0	1	1←
1	0	1←
1	1	0

$$Y = \bar{A}B + A\bar{B}$$

Concept is to get NOT gate using 2 : 1 MUX

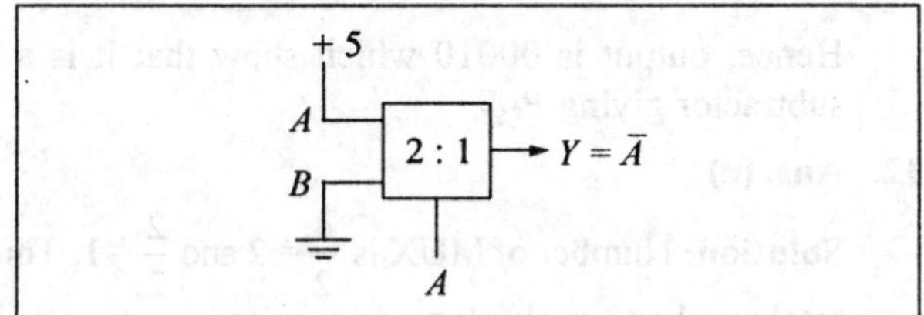

Now

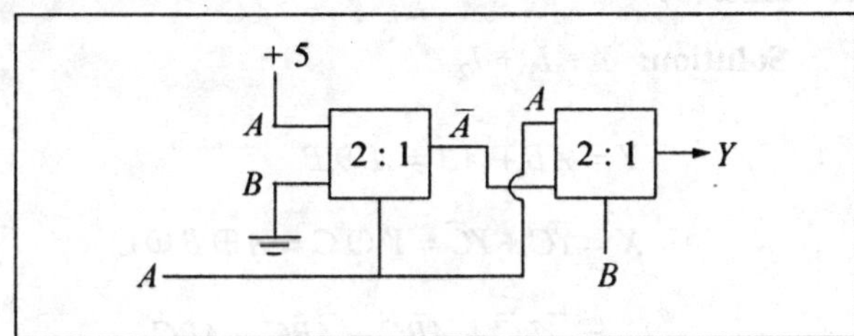

$Y = \bar{B}A + B\cdot\bar{A}$ = EX-OR output

Hence 2 (2 : 1) MUX are required to realise EX-OR gate.

7. **Ans. (d)**

Solution:

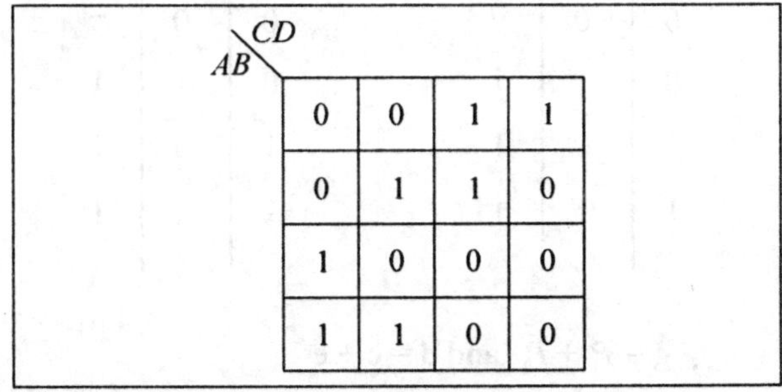

$$F = \Sigma\, m\,(2, 3, 5, 7, 8, 9, 12)$$

Second Method

MUX Table

S_1	S_0		
A	B	In	Additional term
0	0	C	$(D+\bar{D})$
0	1	D	$(C+\bar{C})$
1	0	$\bar{C}$	$(D+\bar{D})$
1	1	$\overline{CD}$	

$$f = \bar{A}\,\bar{B}\,C\,D + \bar{A}\,\bar{B}\,C\,\bar{D} + \bar{A}\,B\,C\,D + \bar{A}\,B\,\bar{C}\,D + A\,\bar{B}\,\bar{C}\,D + A\,\bar{B}\,\bar{C}\,\bar{D} + A\,B\,\bar{C}\,\bar{D}$$

$$f = \Sigma m(3, 2, 7, 5, 9, 8, 12)$$

$$= \Sigma m(2, 3, 5, 7, 8, 9, 12)$$

8. **Ans. (d)**

Solution:

P	Q	F
0	0	I_0
0	1	I_1
1	0	I_2
1	1	I_3

From the given circuit diagram
O is connected to 'I_0' and 'I_3'
And '1' is connected to 'I_0' and 'I_2'

$\therefore F = P\bar{Q} + P\bar{Q} = XOR(P, Q)$

Hence $F = I_1\bar{P}Q + P\bar{Q}I_2$

$$= \bar{P}Q + P\bar{Q}$$

$$= X\text{-}OR(PQ)$$

Second Method

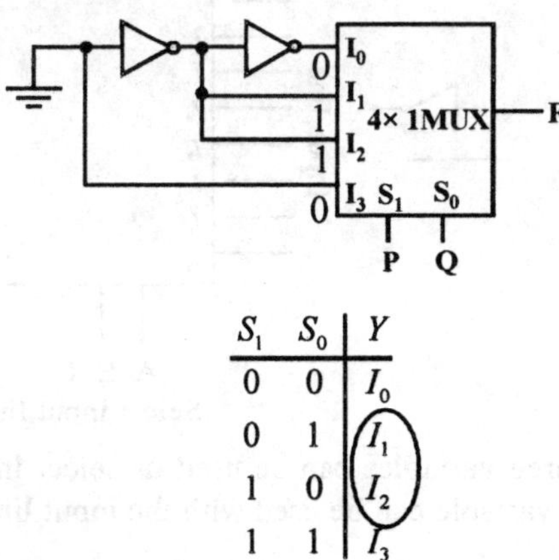

S_1	S_0	Y
0	0	I_0
0	1	I_1
1	0	I_2
1	1	I_3

$F = \bar{P}\,Q + P\,\bar{Q} = P \oplus Q$

$= \text{X-OR } (P\,Q)$

9. Ans. (b)

Solution:

Input A		Input B		
A_2	A_1	B_2	B_1	
0	0	1	0	0
0	0	0	1	0
0	0	1	0	0
0	0	1	1	0
0	1	0	0	1
0	1	0	1	0
0	1	1	0	0
0	1	1	1	0
1	0	0	0	1
1	0	0	1	1
1	0	1	0	0
1	0	1	1	0
1	1	0	0	1
1	1	0	1	1
1	1	1	0	0
1	1	1	1	0

Thus for 6 combinations output in logic 1

So that the inputs of AND gate is '1' ⇒ y = '1'

10. Ans. (b)

Solution: Without any addition circuit i.e. without any input variable only selector expression will pass i.e. expression of 3 variable.

$8 = 2^n$, $n = 3$ select lines and a function with $n + 1 = 4$ can be implemented with addition circuit.

Second Method

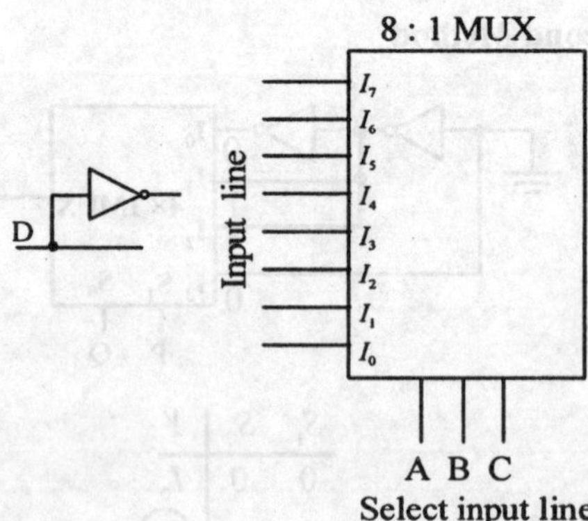

The three variables can be used as select inputs. The fourth variable can be used with the input lines.

11. Ans. (b)

Solution: Let $P = 1011$ and $Q = 1001$

Given, $Y_n = P_n \oplus Q_n \oplus R_n$

$$Z = \overline{R_n}\, Q_n + P_n\, \overline{R_n} + Q_n\, \overline{P_n}$$

Now, constructing the truth-table

	P_n	Q_n	R_n	Z_n	Y_n
$n = 1$	1	1	0	0	0
$n = 2$	1	0	0	0	1
$n = 3$	0	0	0	0	0
$n = 4$	1	1	0	0	0
					0

From the circuited given, we have

$$R_{n+1} = Z_n,\ 1 \le n \le 3$$

And $\quad Z_4 = Y_5$ (MSB)

Hence, output is 00010 which show that it is a 4-bit subtractor giving P-Q.

12. Ans. (c)

Solution: Number of MUX is $\frac{4}{2} = 2$ and $\frac{2}{2} = 1$. Thus the total number 3 multiplexer is required.

13. Ans. (a)

Solution: $Y = I_1 + I_2$

$$Y = \bar{A}B + A\bar{B} = A \oplus B$$

$$X = \bar{Y}C + Y\bar{C} = Y \oplus C = A \oplus B \oplus C$$

$$= \bar{A}B\bar{C} + A\bar{B}\bar{C} + \bar{A}\bar{B}C + ABC$$

14. Ans. (b)

Solution:

According to given conditions , we have

P_1	P_2	g
0	0	0
0	1	1
1	0	1
1	1	1

c	e	d
0	0	-
0	1	1
1	0	1
1	1	1

∴ $g = P_1 + P_2$ and $d = c + e$

15. Ans. (b)

Solution:

The given conditions can be represented as follows:

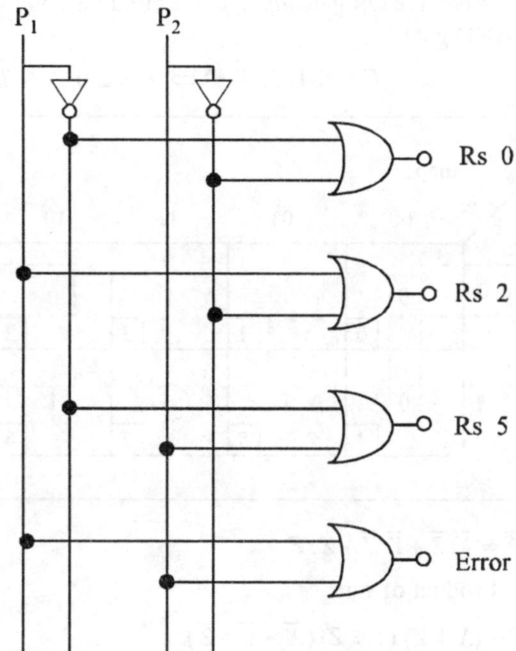

16. Ans. (c)

Solution: $\overline{F} = \overline{A}\overline{C}$

$$\overline{\overline{F}} = \overline{\overline{A}\,\overline{C}}$$

$$F = \overline{\overline{A}} + \overline{\overline{C}} = A + C$$

Note: Minimal product of sums $F = A + C$

17. Ans. (a)

Solution:

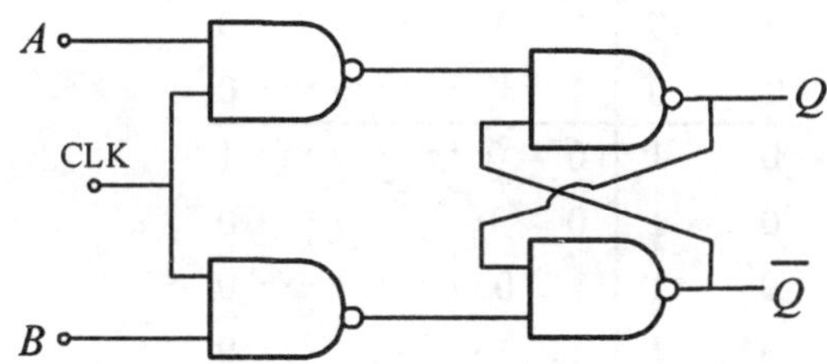

Race arround condition occurs when the output at gate toggle for steady input

From figure obtain the output

$$Q_{n+1} = \overline{\overline{A.CLK.\overline{Q}_n}}$$

$$= A.CLK + Q_n$$

$$\overline{Q}_{n+1} = B.CLK + \overline{Q}_n$$

if $CLK = 1$ and A and $B = 1$

$$\left.\begin{array}{l} Q_{n+1} = 1 \\ \overline{Q}_{n+1} = 1 \end{array}\right\} \begin{array}{l} \Rightarrow \text{Stable} \\ \text{No race around} \end{array}$$

if $CLK = 1$ and A and $B = 0$

Then $\left.\begin{array}{l} Q_{n+1} = 1 \\ \overline{Q}_{n+1} = 1 \end{array}\right\} \begin{array}{l} \Rightarrow \text{Stable} \\ \text{No race around} \end{array}$

Thus race around does not occur in the circuit

18. Ans. (b)

Solution:

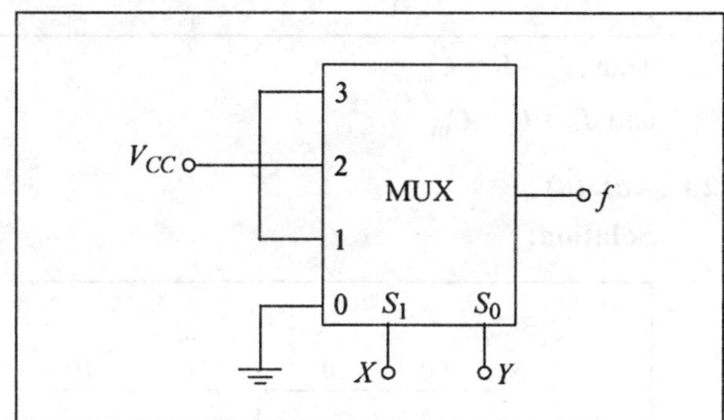

x	y	f
0	0	1
0	1	1
1	0	1
1	1	0

$$f = \overline{X}\,\overline{Y} + \overline{X}\,Y + X\,\overline{Y}$$

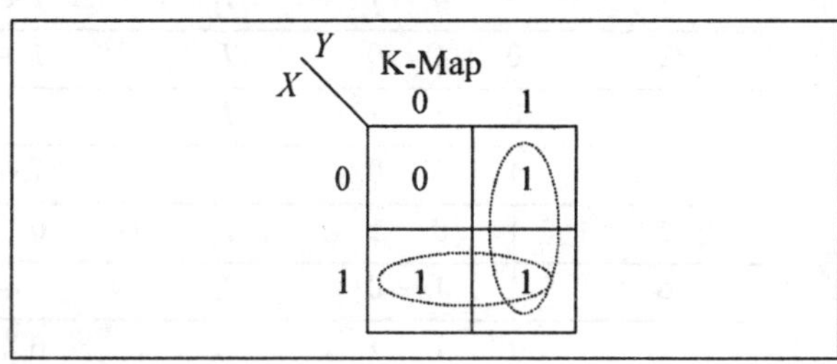

$$f = X + Y$$

19. Ans. (c)

Solution: We draw the Truth-table for full adder is as shown below

S_1 (P)	S_o (Q)	C_{in}	SUM = S
0	0	0	0
0	0	1	1
0	1	0	1
0	1	1	0
1	0	0	1

1	0	1	0
1	1	0	0
1	1	1	1

Implementing the truth-table as for conditions

S_1	S_o	S	
0	0	C_{in}	I_o
0	1	$\overline{C}_{in}$	I_1
1	0	$\overline{C}_{in}$	I_2
1	1	C_{in}	I_3

Thus, $I_o = I_3 = C_{in}$

and $I_1 = I_2 = \overline{C}_{in}$

20. Ans. (a)

Solution:

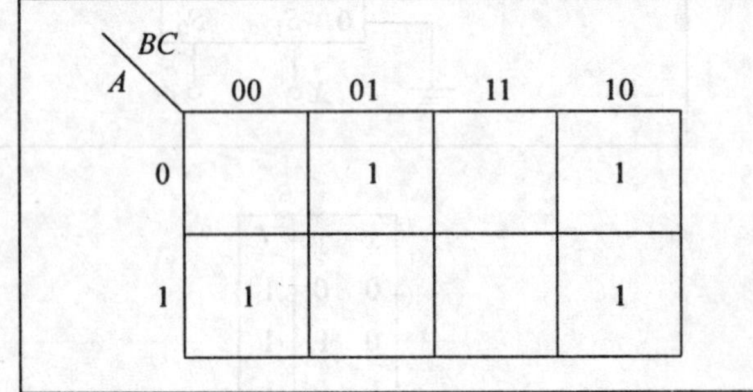

Decimal	*A*	*B*	*C*	*Uni selected*	*Output*
0	0	0	0	I_0	0
1	0	0	1	I_1	1 ←
2	0	1	0	I_2	1 ←
3	0	1	1	I_3	0
4	1	0	0	I_4	1 ←
5	1	0	1	I_5	0
6	1	1	0	I_6	1 ←
7	1	1	1	I_7	0

$F(A, B, C) = \Sigma\,(1, 2, 4, 6)$

Second Method

From the MUX Boolen expresion

$$F = \overline{B}\overline{C}I_0 + \overline{B}CI_1 + B\overline{C}I_2 + BCI_3$$

$$= \overline{B}\overline{C}A + \overline{B}C\overline{A} + B\overline{C}$$

$$= A\overline{B}\overline{C} + \overline{A}\overline{B}C + (A + \overline{A})B\overline{C}$$

$$= A\overline{B}\overline{C} + \overline{A}\overline{B}C + \overline{A}B\overline{C} + AB\overline{C}$$

$$\quad 100 \qquad 001 \qquad 010 \qquad 110$$

$$\Rightarrow F = \Sigma(1, 2, 4, 6)$$

21. Ans. (a)

Solution:

The negative OR gate given in the circuit is equivalent to NAND gate

$$F = \Sigma(1, 3, 5, 6) \Rightarrow \overline{F} = \Sigma\, 0, 2, 4, 7$$

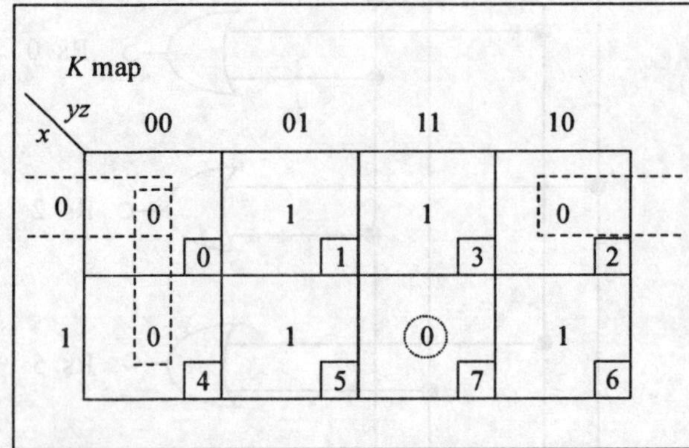

$$\overline{F} = \overline{X}\,\overline{Z} + \overline{Y}\,\overline{Z} + ZYZ$$

∴ Product of sum

$$F = (X + Z)(Y + Z)(\overline{X} + \overline{Y} + \overline{Z})$$

22. Ans. (a)

Solution: $Y = \overline{\overline{x}.x} + \overline{x}.\overline{x} = x + \overline{x} = 1$

23. Ans.(b)

Solution:

Input A		Input B		
A_2	A_1	B_2	B_1	
0	0	1	0	0
0	0	0	1	0
0	0	1	0	0
0	0	1	1	0
0	1	0	0	1
0	1	0	1	0
0	1	1	0	0
0	1	1	1	0
1	0	0	0	1
1	0	0	1	1
1	0	1	0	0
1	0	1	1	0
1	1	0	0	1
1	1	0	1	1
1	1	1	0	0
1	1	1	1	0

Thus for 6 combinations output in logic 1

24. Ans. (c)

Solution:

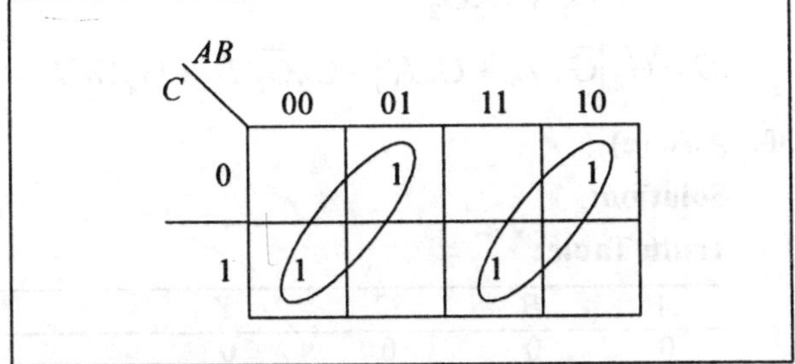

1st pair (010 to 001) = $\overline{A}\,(B \oplus C)$

2nd pair (100 to 100) = $A[B \odot C]$

$$Y = \overline{A}\,(B \oplus C) + A\,[B \odot C]$$
$$= \overline{A}\,(B \oplus C) + A\overline{(A \oplus C)}$$
$$= A \oplus B \oplus C$$

Second Method

$$F(A, B, C) = \Sigma m(1, 2, 4, 7) \Rightarrow F$$
$$= \overline{A}\,\overline{B}\,C + \overline{A}\,B\,\overline{C} + A\,\overline{B}\,\overline{C} + ABC$$
$$= \overline{A}(B \oplus C) + A(B \odot C)$$
$$F = A\overline{X} + A\overline{X}$$
$$F = A \oplus X$$

Hence $F = A \oplus B \oplus C$

25. Ans. (b)

Solution:

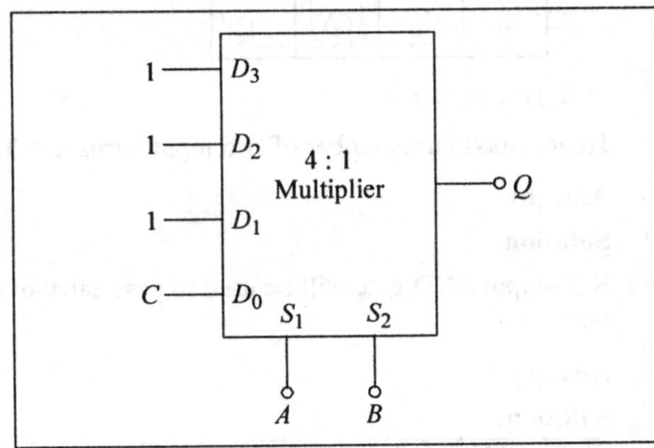

A	B	D_n	
0	0	C	D_0
0	1	1	D_1
1	0	1	D_2
1	1	1	D_3

$A = 0,\ B = 0,\ D_0$ is selected

Therefore, $A = 0, B = 0, Q = C$

For the rest of A and B. $Q = 1$

Therefore, $Q = A + B + C$

Second Method

Given, $S_1 = A,\ S_0 = B,\ D_0 = C,\ D_1 = D_2 = D_3 = 1$

$$Q = \overline{S}_1\overline{S}_0D_0 + \overline{S}_1S_0D_2 + S_1\overline{S}_0D_2 + S_1S_0D_3$$
$$= \overline{A}\,\overline{B}\,C + \overline{A}B\cdot 1 + A\overline{B}\cdot 1 + AB\cdot 1$$
$$= \overline{A}B + A\overline{B} + AB + \overline{A}\,\overline{B}\,C$$
$$= B\left(A + \overline{A}\right) + A\overline{B} + \overline{A}\,\overline{B}\,C$$
$$= B + \overline{B}\left(A + \overline{A}C\right)$$
$$= B + \overline{B}\left(A + C\right)$$
$$= B + \overline{B}A + \overline{B}\,C$$
$$= \left(B + \overline{B}\right)\left(A + B\right) + \overline{B}\,C$$
$$= A + B + \overline{B}\,C$$
$$= A + \left(B + \overline{B}\right)\left(B + C\right)$$
$$= A + B + C$$

26. Ans. (c)

Solution:

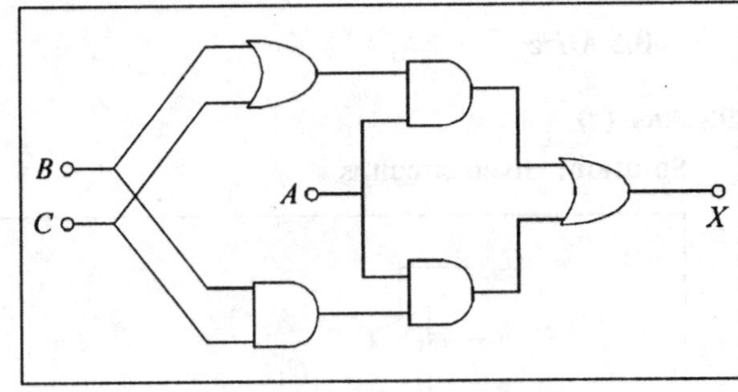

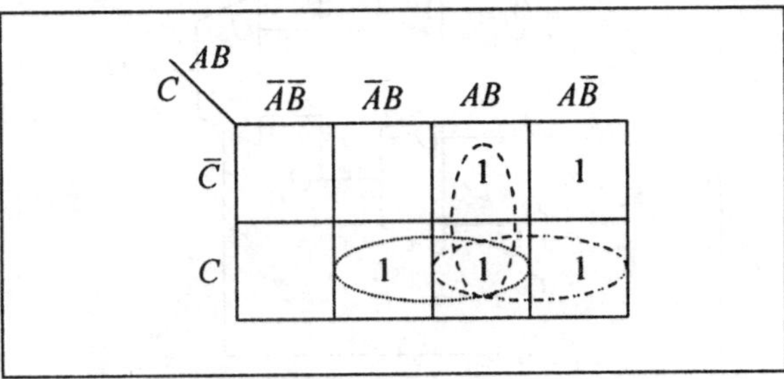

$$x = AB + AC + BC$$

27. Ans. (c)

Solution:

A	B	Y
0	0	1
0	1	0←
1	0	0
1	1	1←

$$Y = AB + \overline{A}\,\overline{B}$$

This is basically truth table of EX-NOR gate which is known as **equality detector**

28. Ans. (c)

Solution:

As can be seen from the figure switch is connected to output which provide output 1, then S will be also 1 and S will go to 1 after 1 μsec and output become zero, again S becomes 0 and switch will go to after 1 μ sec and output goes to 1 this will continue which give the waveform at output as

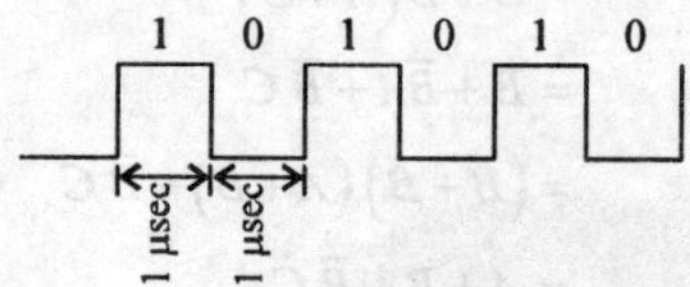

So, $T = 2\,\mu\text{sec}$

Then, $f = \frac{1}{T} = \frac{1}{2\times10^{-6}}$

$= 0.5\times10^{6}$

$= 0.5\ MHz$

29. Ans. (c)

Solution: Given circuit is

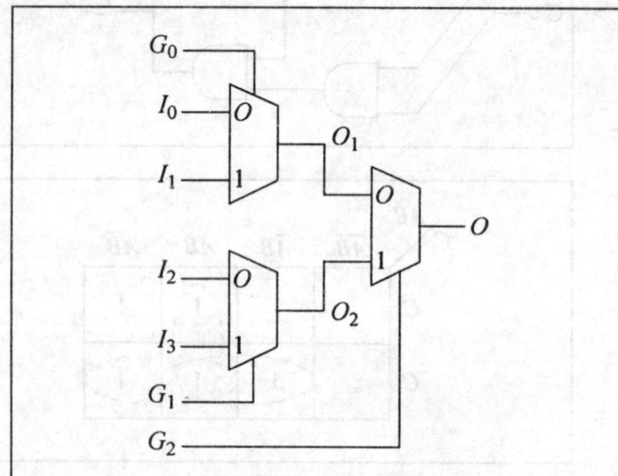

Function of the field programmable array

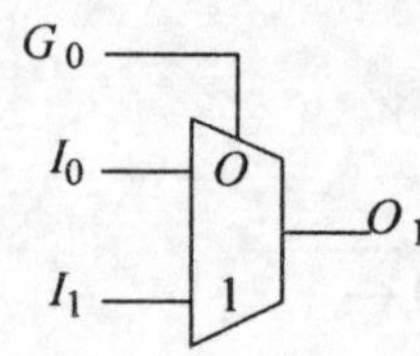

$$O_1 = \overline{G}_0\, I_0 + G_0 I_1$$

$$O_2 = \overline{G}_1\, I_2 + G_1 I_2$$

$$O = \overline{G}_2\, O_1 + G_2 O_2$$

$$O = \overline{G}_2\left[\overline{G}_0\, I_0 + G_0 I_1\right] + \overline{G}_2 \overline{G}_1\, I_2 + G_2 G_1 `I_3$$

30. Ans. (c)

Solution:

Truth Table

A	B	C	Y
0	0	0	0
0	0	1	0
0	1	0	0
0	1	1	1←
1	0	0	0
1	0	1	1←
1	1	0	1←
1	1	1	1←

$$Y = \overline{A}BC + AB\overline{C} + A\overline{B}C + ABC$$

Reduction by K-map

$A \backslash BC$	00	01	11	10
0	0	0	1	0
1	0	1	1	1

$A \backslash BC$	00	01	11	10
0	0	0	1	0
1	0	1	1	1

$$Y = AB + BC + CA$$

Hence maximum number of two input terms is = 3

31. Ans. (a)

Solution:

K 2 –input AND gate will be used to pass earry of each bit next bit

32. Ans. (c)

Solution:

Decimal No.	A	B	C	Y
0	0	0	0	1
1	0	0	1	1
2	0	1	0	1
3	0	1	1	0
4	1	0	0	0
5	1	0	1	0
6	1	1	0	1
7	1	1	1	0

Constructing k –map,

A \ BC	00	01	11	10
0	1	1		1
1				1

From K- map

$y = \overline{A}\ \overline{B} + B\,\overline{C}$

Output, $Z = Y + \overline{C} = \overline{A}\,\overline{B} + B\,\overline{C} + \overline{C} = \overline{A}\,\overline{B} + \overline{C}$

33. Ans. (d)

Solution:

$$F = \overline{X}Y + X\,\overline{Y}\,\overline{Z} = \overline{X}Y\left[Z + \overline{Z}\right] + X\,\overline{Y}\,\overline{Z}$$
$$= \overline{X}\,Y\,Z + \overline{X}\,Y\,\overline{Z} + X\,\overline{Y}\,\overline{Z}$$

Z	Y	Z	F	
0	0	0	0	
0	0	1	0	
0	1	0	1	$\leftarrow \overline{X}\,Y\,\overline{Z}$
0	1	1	1	$\leftarrow \overline{X}\,Y\,Z$
1	0	0	1	$\leftarrow X\,Y\,\overline{Z}$
1	0	1	0	
1	1	0	X	$\leftarrow$ *do'nt care*
1	1	1	X	$\leftarrow$ *do'nt care*

Reduction by K-map

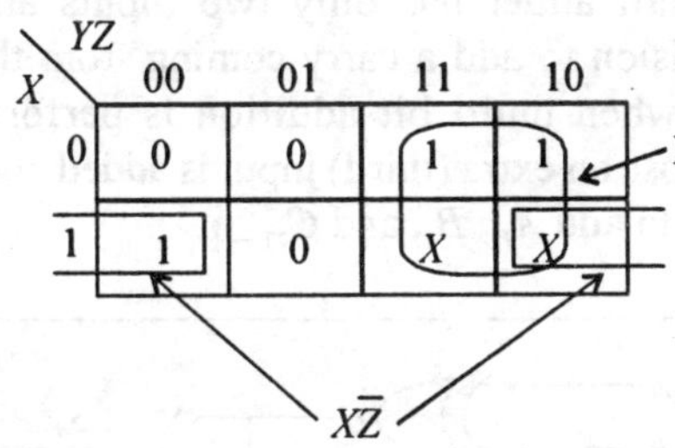

$F = Y + X\overline{Z}$

34. Ans. (d)

Solution:

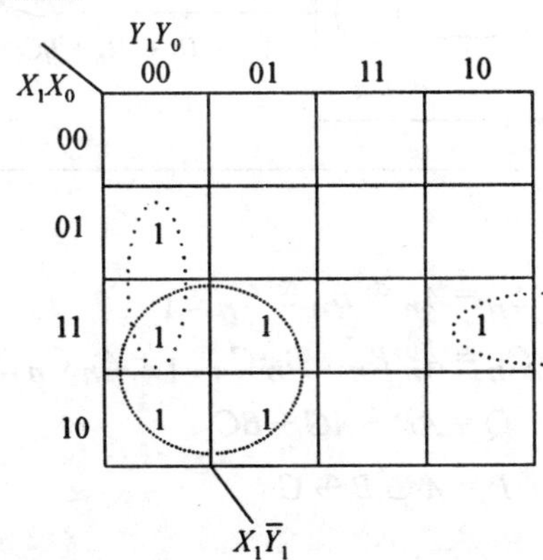

From the K-map,

$$F = X_1X_0Y_1\overline{Y_0} + X_1X_0Y_1\overline{Y_0}$$
$$+X_0\overline{Y_0}\,\overline{Y_1}X_1 + X_0\overline{Y_0}\,\overline{Y_1}\,\overline{X_1} + X_1\overline{Y_1}$$
$$= X_1\overline{Y_1} + X_0\overline{Y_0}\,\overline{Y_1} + X_0X_1\overline{Y_0}$$

35. Ans. (b)

Solution:

$$F = ZS + \overline{ZS}$$
$$= Z\left(XY + \overline{X}\overline{Y}\right) + \overline{Z}\left(XY + \overline{X}\overline{Y}\right)$$
$$F = \left(\overline{X} \oplus \overline{Y}\right) \oplus Z$$

Second Method

Y	F_o
0	X

$\Rightarrow \overline{Y}\ X + Y\ \overline{X} = F_o$

F_o	F_1
0	z
0	$\overline{z}$

$$F_1 = \left(\overline{X \oplus Y}\right)Z + \left(X \oplus Y\right)\overline{Z}$$

Let $X \oplus Y = A$

$$F_1 = \overline{A}Z + A\overline{Z}$$
$$= X \oplus Y \oplus Z$$

36. Ans. (d)

Solution: $f = P\overline{R}S + PQR + \overline{P}RS + \overline{P}\,Q\,\overline{R}$

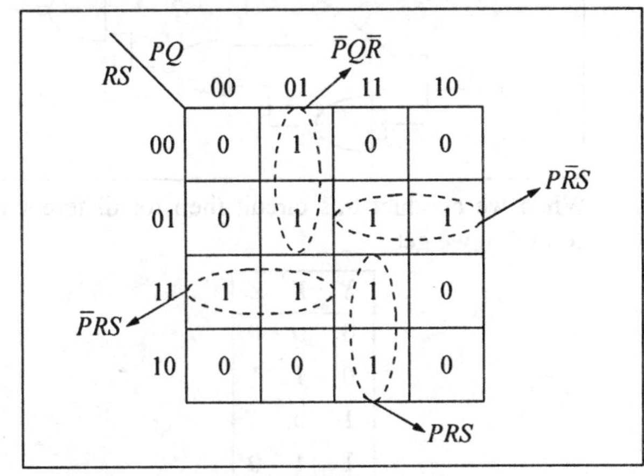

37. Ans. (a)

Solution:

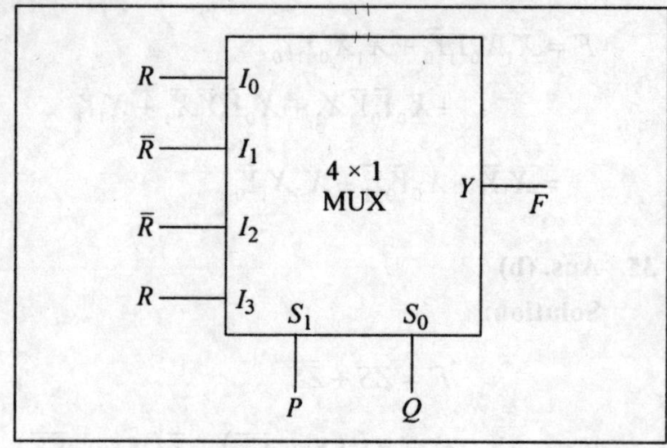

P	Q	F
0	0	R
0	1	$\overline{R}$
1	0	$\overline{R}$
1	1	R

$F = \overline{P}\,\overline{Q}R + \overline{P}Q\overline{R} + P\overline{Q}\,\overline{R} + PQR$ ...(1)

Equation (1) is derived because it is a AND-OR Logic

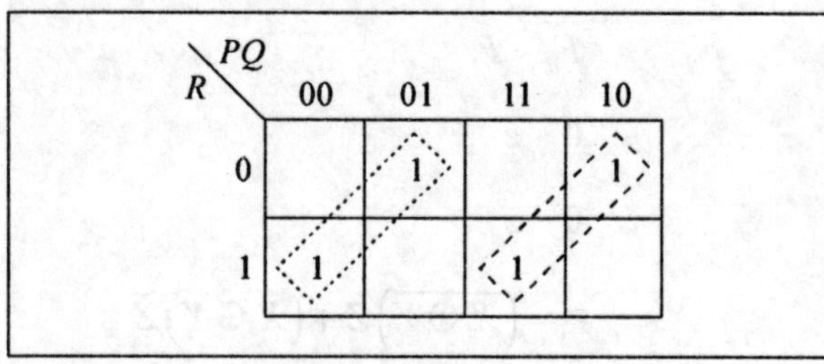

$F = P \oplus Q \oplus R$

38. Ans. (a)

Solution: Find the gate circuit function

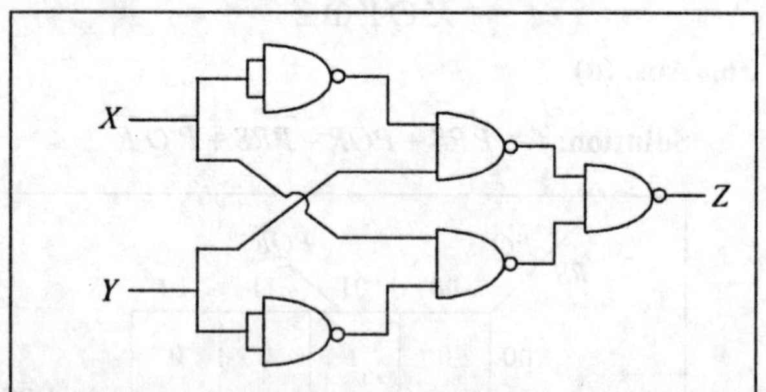

When we conduct this circuit then for different input condition we get

X	Y	Z
0	0	0
0	1	1
1	0	1
1	1	0

The given logic circuit is a XOR gate.

39. Ans. (d)

Solution: 3 flip flips are used and it has –ve edge triggering. Hence it is mod 5 down counter

Apply the Trick as below:

- When ever preset is activate it indicate down counter
- When ever clear is activated it indicate up counter

up/down $= CP \oplus Q$, 1 for up and 0 for down

CP = (clock pulse)

Q = (O/P)

0 = –ve edge; $Q = 1$

1 = +ve; $\overline{Q} = 1$

$= 1 \oplus 1 = 0$ (dowm counter)

Counting sequence

C	B	A	
1	1	1	
1	1	0	
1	0	1	
1	0	0	
0	1	1	(present state) so Mod 5
0	1	0	$\rightarrow \overline{C}\,B\,\overline{A} = 1$ preset
1	1	1	

40. Ans. (c)

Solution: Number of MUX is $\frac{4}{2} = 2$ and $\frac{2}{2} = 1$. Thus the total number 3 multiplexer is required.

41. Ans. (c)

Solution: Full adder

An half adder has only two inputs and there is no provision to add a carry coming from the lower order bits when multi bit addition is performed. For this purpose an extra (third) input is added and this circuit is used to add A_n, B_n and C_{n-1}.

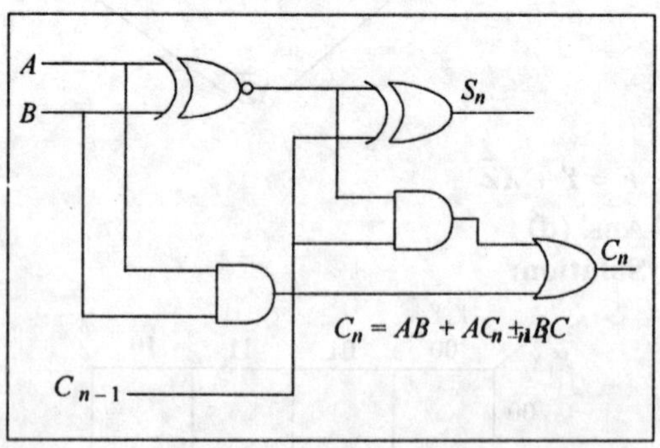

We have

$$S_n = A_n \oplus B_n \oplus C_{n-1}$$

$$C_n = A_n B_n + B_n.C_{n-1} + A_n.C_{n-1}$$

Carry $Q = AB + AC + BC$

and Sum, $P = A \oplus B \oplus C$

8.4 Sequential Circuits

1. **Ans. (c)**

Solution: 4-Bit/modulo 16 ripple counter

T_d Total delay = $N \times t_d = 4 \times 50 \times 10^{-9}$

$$f = \frac{1}{T_d} = \frac{1}{4 \times 50 \times 10^{-9}} = \frac{1}{200 \times 10^{-9}}$$

$$= \frac{1000}{200} \times 10^6 = 5 \times 10^6 = 5 \text{ MHz}$$

2. **Ans. (d)**

Solution:

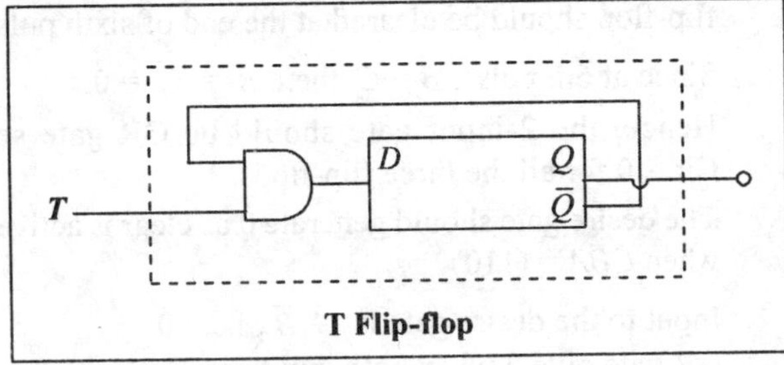

T Flip-flop

In a switch ring counter using *D-FF*, the complementary output *Q* is connected to *D* input. For single input, it becomes *T - FF*.

3. **Ans. (c)**

Solution:

A 1-Bit Memory Cell

The basic digital memory circuit is known as *flip-flop*. It has two stable states, 1 state and 0 state.

It consists of two inverters G_1 and G_2 (NAND gates based).

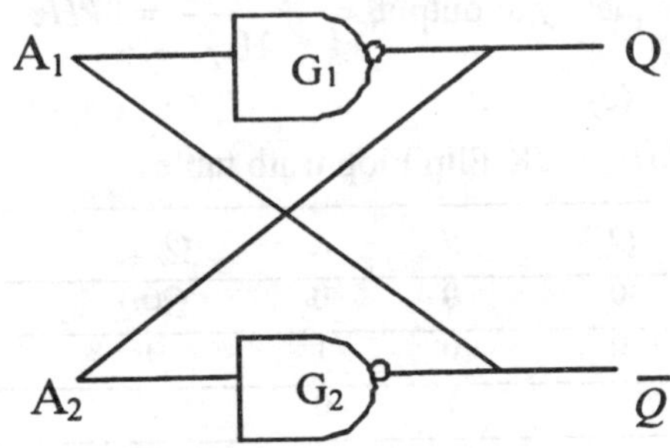

- The output Q and $\overline{Q}$ are complementary.
- The circuits has two states, in one of the stable state Q = 1 which is known as (Set state), where as on the other hand, other stable Q = 0 known as (reset state).
- If the circuit is in state 1, it continues to remain in this state and similarly if it is in 0 state. This property is known as memory, it can store 1-bit of digital information.
- This circuit is also known as latch.

4. **Ans. (d)**

Solution:

Q_n	J	Q_{n+1}
1	1	1
0	0	0
1	1	1
0	0	0
1	1	1

Sequence is 010101

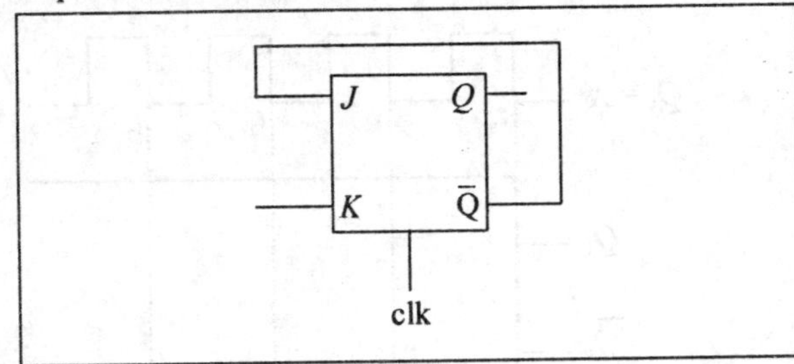

5. **Ans. (c)**

Solution:

	$\bar{Q}_{n-1}$ J_1	K_1	Q_{n-1} J_2	K_2	Q_1	Q_2	$\bar{Q}_2$
1st clock					0	0	1
2nd clock	1	1	0	1	1	0	1
3rd clock	1	1	1	1	0	1	0
4th clock	0	1	0	1	0	0	1

6. **Ans. (d)**

Solution:

X	Y	Q_n	Q_{n+1}	
0	0	1	1	Hold
0	0	0	0	
0	1	1	1	Set
0	1	0	1	$X = K, Y = J$
1	0	1	0	Reset
1	0	0	0	
1	1	1	0	Toggle
1	1	0	1	

In short analysis

$$D = \bar{X}Z + Y\bar{Z}$$

For JK FF $Q_{n+1} = \bar{K}Q + J\bar{Q}$

In present case, $D = Q$ for D FF

Hence $Y = J$ and $X = K$

7. Ans. (c)

Solution:

This question may be solved in a reverse way, i.e. verifying each option to get the correct output.

For option (c)

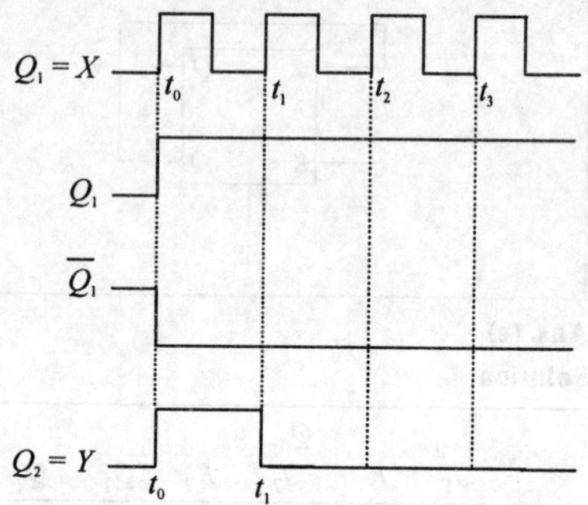

8. Ans. (b)

Solution: 0 to 6 counter:

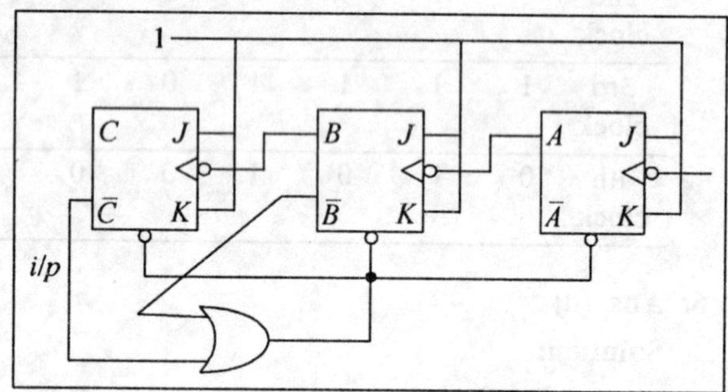

9. Ans. (b)

Solution: Total propagation delay in 4-bit ripple counter $R = 4\,t_d = 40$ ns

In synchronous counter all flip flops are given clock simultaneously

$\therefore \quad R = 10$ ns

10. Ans. (c)

Solution: Change in output occurs when the state of the slave is affected.

11. Ans. (b)

Solution:

Shift register→ Serial to parallel data conversion
Counter→ frequency division
Decoder→ Addressing in memory chips

12. Ans. (c)

Solution:

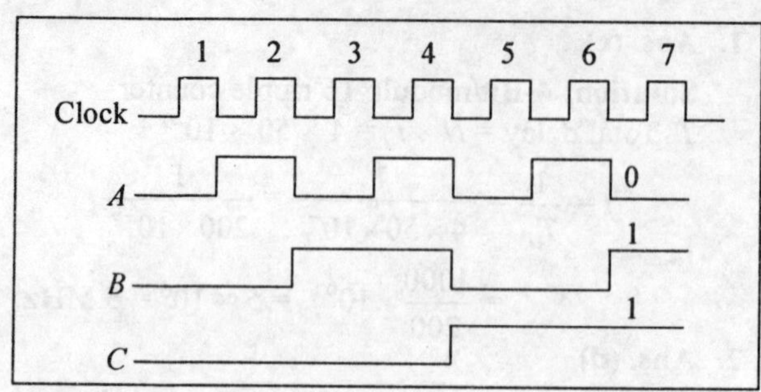

To perform the given ripple counter as module-6, the flip-flop should be cleared at the end of sixth pulse.

Since at 6th pulse, $B = C$, then $\bar{B} = \bar{C} = 0$.

Hence, the 2-input gate should be OR gate so that Clr = 0 for all the three flip-flops.

The desire gate should generate 0 as clear is active Low when $CBA = (110)_2$

Input to the desire gate is $\bar{C}\,\bar{B}$, i.e. 00
OR gate give 0 when both input are 0.

13. Ans. (b)

Solution:

The given circuit is an asynchronous or ripple counter whose clear is activated for

$$\overline{CLR} = \overline{Q_3 Q_1}$$

The counter is cleared every time when $Q_3 = 1$ and $Q_1 = 1$

This first situation occurs when counter's state is 1010

Hence, counter counts from 0000 to 1001, (i.e. 0 to 9) and it is MOD-10 counter

$$\text{Frequency at output} = \frac{10 \times 10^3}{10} = 1\,kHz$$

14. Ans. (c)

Solution: JK Flip Flop truth table

Q_n	J	K	Q_{n+1}	Q_{n+1}
0	0	0	$Q(n)$	0
0	0	1	0	0
0	1	0	1	1
0	1	1	$[\overline{Q(n)}]$	1

15. Ans. (b)

Solution: T Flip-flop will toggle with each trigger pulse clock if $T = 1$

Q_n	$Q_{n+1} = \bar{Q}_n$
$Q_0 = 1$	0
$Q_1 = 1$	0
$Q_2 = 0$	1

Output $Q_2\,Q_1\,Q_0 = 100$

16. Ans. (a)

Solution:

Present state		*Next state*	
Q_1	Q_0	D_1	D_0
0	0	0	1
0	1	1	1
1	1	1	0
1	0	0	0

$D_0 = \bar{Q}_1$; $D_1 = Q_0$

17. Ans. (b)

Solution:

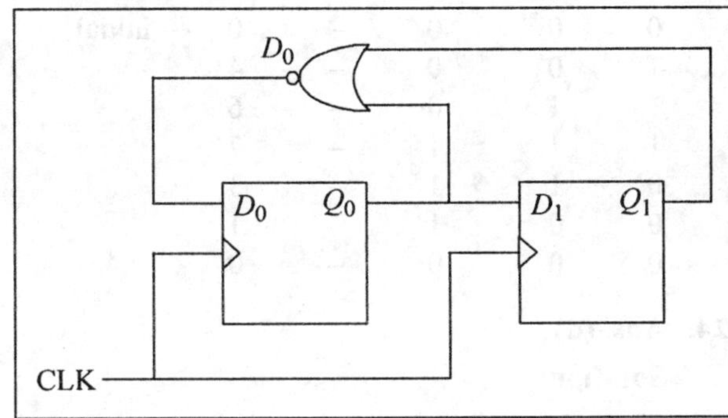

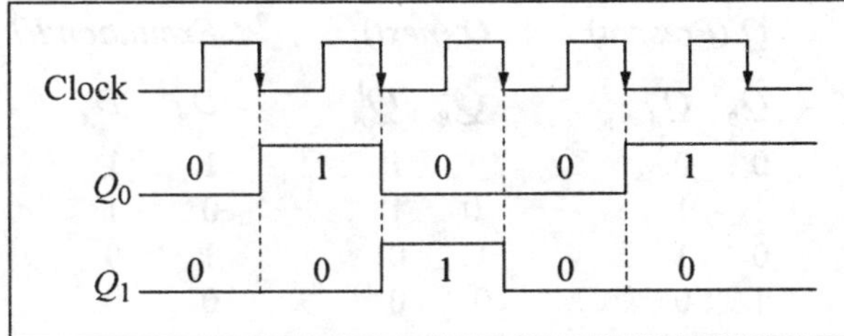

NOR GATE		
Q_0	Q_1	D_0
0	0	1
1	0	0
0	1	0
0	0	1
1	0	0

Sequence is

$\therefore Q_1Q_2$ 00, 01, 10, 00, 01...

18. Ans. (b)

Solution: Two JK flip flops in series. Thus, propagation delay will be $2\Delta T$, frequency is getting reduced by a factor of 4.

19. Ans. (a)

Solution:

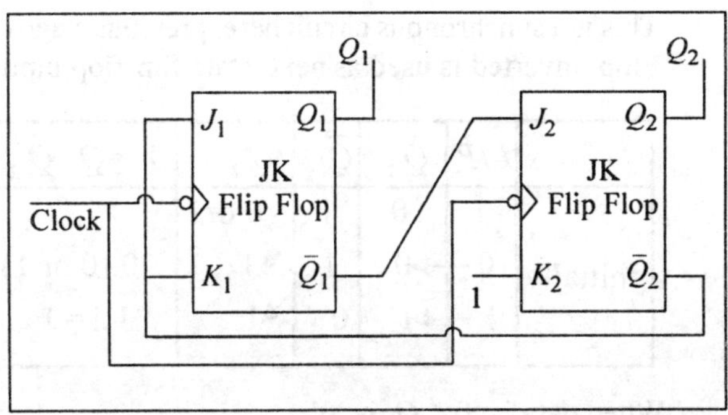

It is a –ve edge of clock trigger synchronous counter

Let initially $\quad Q_2\,Q_1 = 0\ 0$

Falling edge of clock

CLK	Q_1	Q_2	$\bar{Q}_1$	$\bar{Q}_2$	Result in	J_1 ($\bar{Q}_2$)	K_1 ($\bar{Q}_2$)	J_2 ($\bar{Q}_1$)	K_2 (1)
–	0	0	1	1	→	1	1	1	1
↓	1	1	0	0	→	0	0	0	1
↓	1	0	0	1	→	1	1	0	1
↓	0	0	1	1	→	1	1	1	1

Hence, sequence moves in the way

$Q_1\,Q_2 \Rightarrow 0\,0 \rightarrow 1\,1 \rightarrow 1\,0 \rightarrow 0\,0 \rightarrow 11$

and so on.

20. Ans. (d)

Solution:

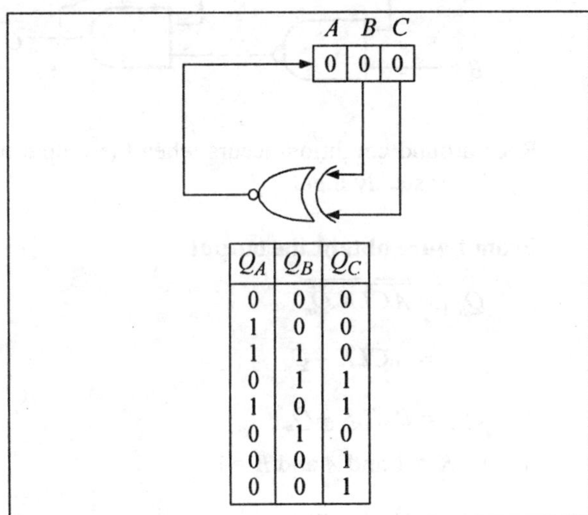

Q_A	Q_B	Q_C
0	0	0
1	0	0
1	1	0
0	1	1
1	0	1
0	1	0
0	0	1
0	0	1

21. Ans. (a)

Solution:

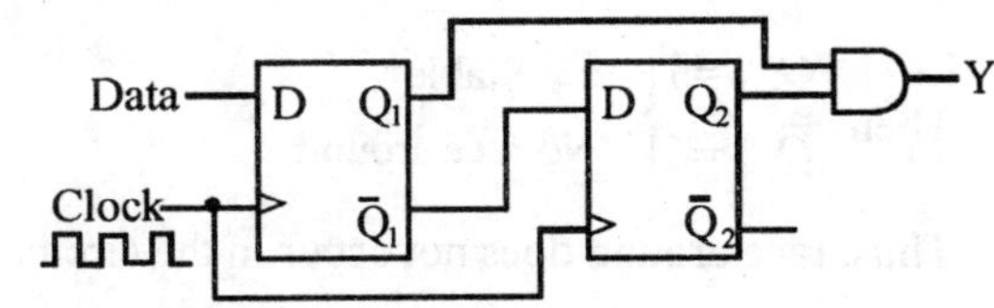

This is a synchronous circuit here, previous stage of flip-Flop inverted is used as next state flip-flop input.

	I/P	Q_1	$\overline{Q}_1$	Q_2	$Y = Q_1 Q_2$
		0	1	0 or 1	
initially	0 →	0	1	1	0.(0 or 1)=0
	1 →	1	0	1	1.1=1

When data is '0', Q_1 is '0'

And $\overline{Q}_1$ is '1' for first flip flop

If $Q_2 = 1$

Then $y = 0$

Data is changed to 1

Q_1 is 1 → first 'D'

$\overline{Q}_1$ zero is connected to 2nd flip flop

So that $Q_2 = 1$

So that the inputs of AND gate is '1' ⇒ Y = '1'

22. Ans. (a)

Solution:

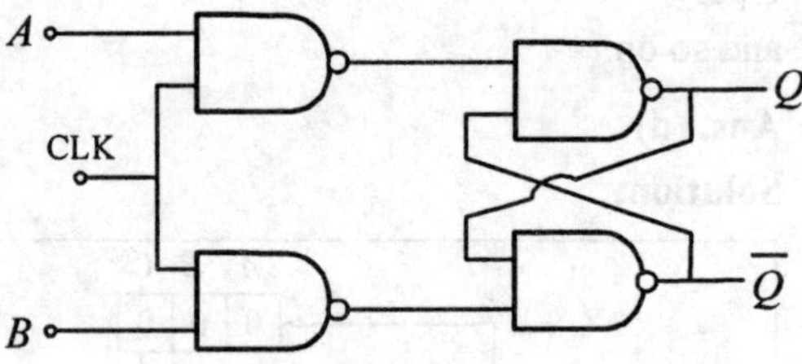

Race around condition occurs when the output at gate toggle for steady input

From figure obtain the output

$$Q_{n+1} = \overline{A.CLK.\overline{Q}_n}$$

$$= A.CLK + Q_n$$

$$\overline{Q}_{n+1} = B.CLK + \overline{Q}_n$$

if $CLK = 1$ and A and $B = 1$

$$\left.\begin{aligned} Q_{n+1} &= 1 \\ \overline{Q}_{n+1} &= 1 \end{aligned}\right\} \Rightarrow \text{Stable, No race around}$$

if $CLK = 1$ and A and $B = 0$

Then $$\left.\begin{aligned} Q_{n+1} &= 1 \\ \overline{Q}_{n+1} &= 1 \end{aligned}\right\} \Rightarrow \text{Stable, No race around}$$

Thus, race around does not occur in the circuit.

23. Ans. (a)

Solution:

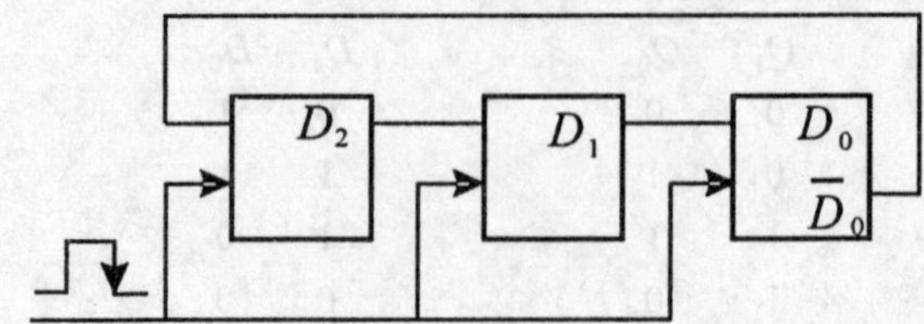

Figure: Johnson counter

For the Johnson counter sequence, table and its corresponding analog output is given below

D_2	D_1	D_0		V_0	
0	0	0	–	0	← initial
1	0	0	–	4	
1	1	0	–	6	
1	1	1	–	7	
0	1	1	–	3	
0	0	1	–	1	
0	0	0	–	0	

24. Ans. (d)

Solution:

Excitation table

Q (Present)		Q (next)		Excitation I/P	
Q_B	Q_A	Q_B^1	Q_A^1	D_B	D_A
0	0	1	1	1	1
1	1	0	1	0	1
0	1	1	0	1	0
1	0	0	0	0	0

K-map for D_A input

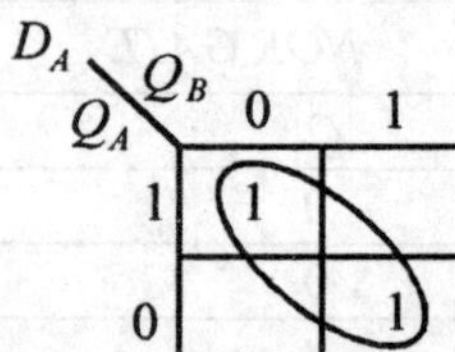

K-map for D_B input

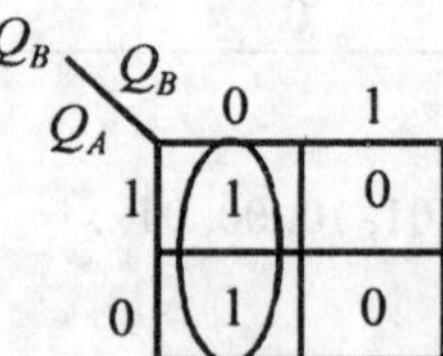

25. Ans.(d)

Solution:

If $A = 0,\ Y = X_0 = \overline{Q}$

$A = 1,\ Y = X_1 = Q$

* Whenever $A = 1$, output gets into same state
* Whenever $A = 0$, output gets toggle.

Present state (Q)	A	Next state
0	0	1
1	0	0
0	1	0
1	1	1

State machine

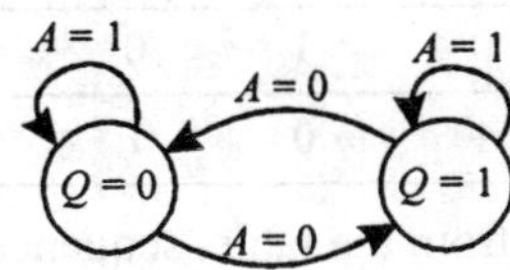

26. Ans. (c)

Solution:

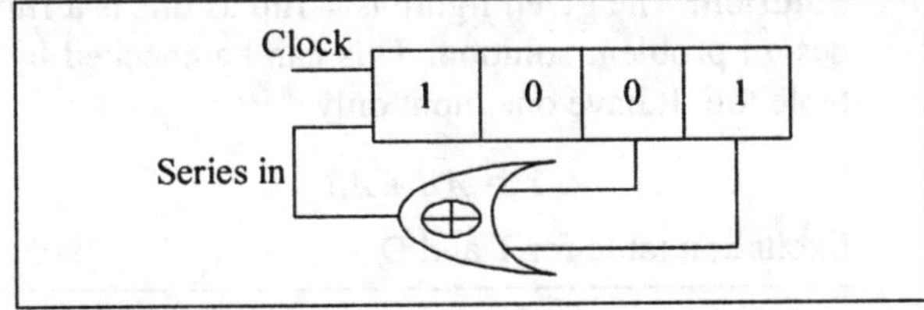

4-bit SIPO

Pulse		
	1 X OR (1,0) = 1	1011
	2 X OR (1,1) = 0	0101
	3 X OR (0,1) = 1	1011

Contents will be 1010.

27. Ans. (c)

Solution:

CLK	A	B	C_i	S	C_o
0	0	0	0	0	0
1	1	1	0	0	1
2	1	1	1	1	1

Afer 2nd CLK : $S = 1$ and $C_o = 1$

28. Ans. (c)

Solution: For NAND latch

1. When $P_1 = 0, P_2 = 1$

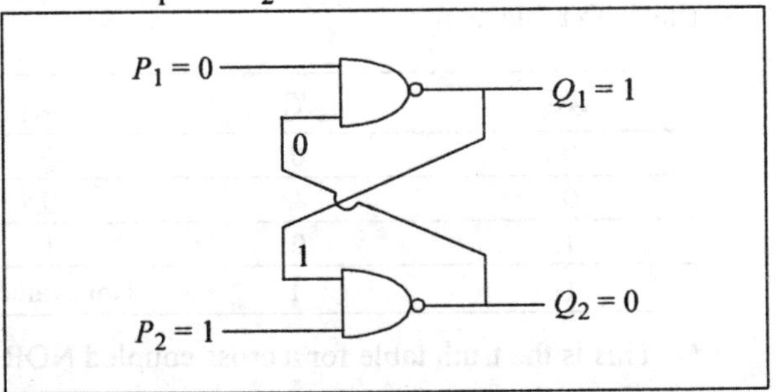

According to NAND operation

A	B	Y
0	0	1
0	1	1
1	0	1
1	1	0

If any input is low , the output is high

$$Q_1 = 1 \text{ and } Q_2 = 1$$

2. When after a few seconds

$$P_1 = 1, P_2 = 1$$

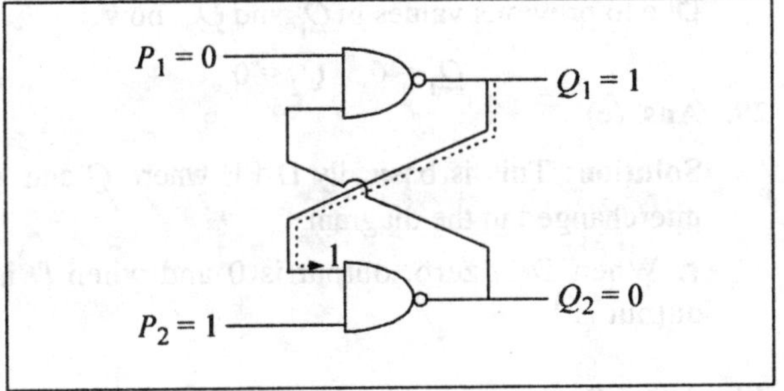

	Previous stage	Present stage
Q_1	1	1
Q_2	0	0

Due to previous values of Q_1 and Q_2, now

$$Q_1 = 1 \quad Q_2 = 0$$

For NOR latch:

1. When $P_1 = 0 \quad P_2 = 1$

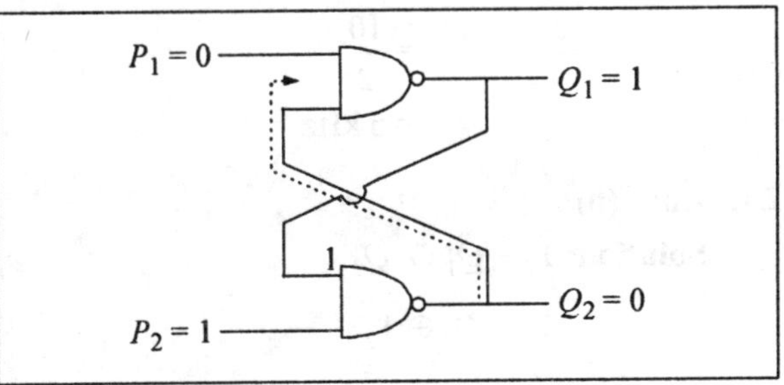

According to NOR operation

A	B	Y
0	0	1
0	1	0
1	0	0
1	1	0

If any input is high , then output is low

$$\therefore \quad Q_2 = 0 \quad Q_1 = 1$$

2. When after a few seconds

$$P_1 = 1, P_2 = 1$$

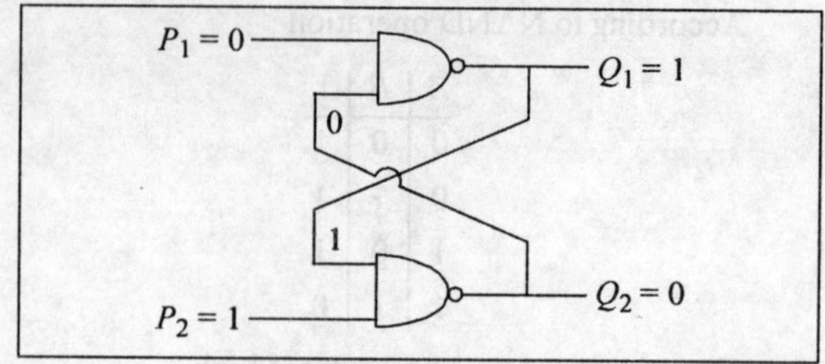

	Previous stage	Present stage
Q_1	1	0
Q_2	0	0

Due to previous values of Q_1 and Q_2, now

$$Q_1 = 0, \quad Q_2 = 0$$

29. Ans. (c)

Solution: This is basically D-FF where Q and $\overline{Q}$ are interchanged in the diagram.

$\therefore$ When D is zero, output is 0 and when D is one, output is 1.

30. Ans. (d)

Solution:

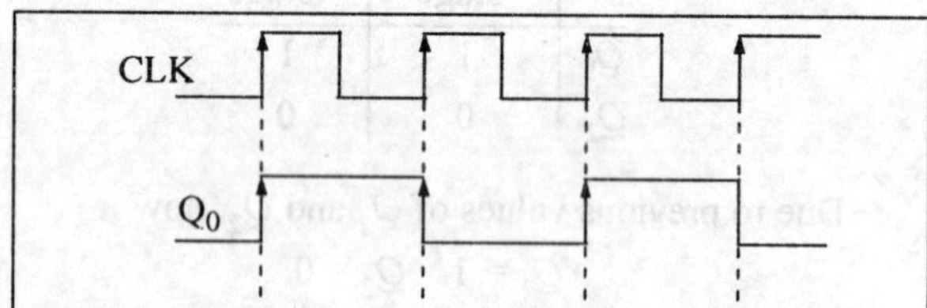

The frequency of $Q = \dfrac{f_o}{2}$

$$= \frac{10}{2}$$

$$= 5 \text{ kHz}$$

31. Ans. (b)

Solution: $Y_2 = Q_1 \oplus Q_0$

$$Y_1 = Q_2 \oplus Y_2$$

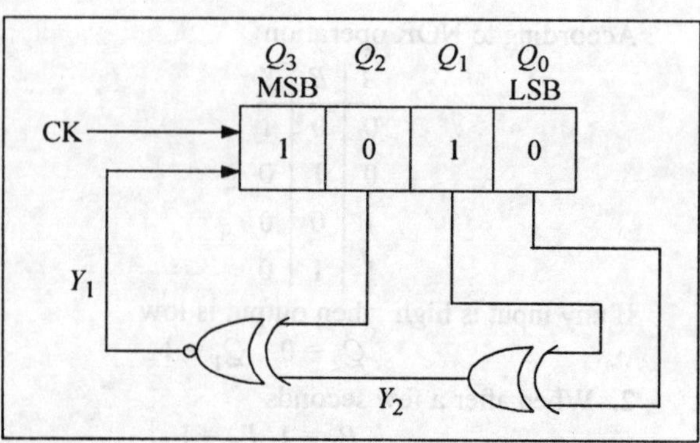

The sequences are as follows:

	$Q_3\ (Y_1)$	Q_2	Q_1	Q_0	Y_2	Y_1
1	1	0	1	0	1	1
2	1	1	0	1	1	0
3	0	1	1	0	1	0
4	0	0	1	1	0	0
5	0	0	0	1	1	1
6	1	0	0	0	0	0
7	0	1	0	0	0	1
	1	0	1	0		

As we see from the table, sequence 1010 repeats after 7 cycles.

32. Ans. (c)

Solution: The given figure is T flip as this is a flip-flop design problem solution. This can be checked by truth table too. It have one input only.

$$Y = A\overline{B} + \overline{A}B$$

Excitation table for T and D

Q_n	Q_{n+1}	D	T
0	0	0	0
0	1	1←	1
1	0	0	1
1	1	1←	0

Expression for D in terms of T

$$D = \overline{Q_n}T + Q_n\overline{T}$$

33. Ans. (a)

Solution:

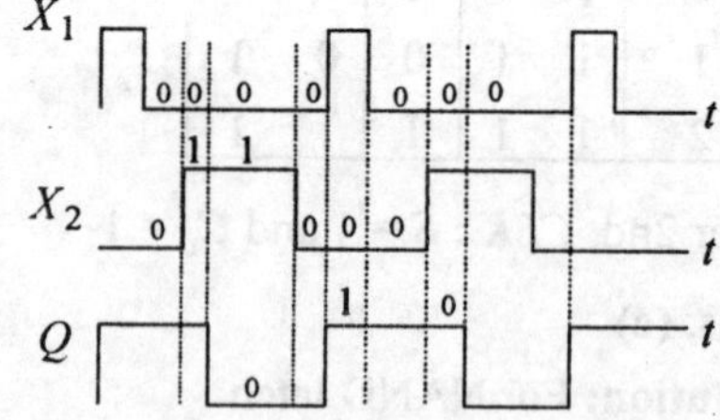

Looking at the waveform of X_1, X_2 and Q we can write the truth table as :

X_1	X_2	Q_{n+1}
0	0	Q
0	1	0
1	0	1
1	1	Not available

* This is the truth table for a cross coupled NOR latch.

34. Ans. (c)

Solution:

Clock $J_A = \overline{Q_B}$		*Input* $K_A = Q_B$	$T_B = Q_B$	*Output* Q_A	Q_B
Initial state				1	0
1	1	0	1	1	1
2	0	1	1	0	0
3	1	0	0	1	0

35. Ans. (d)

36. Ans. (d)

Solution: For J-K flip-flop when $J = 1, K = 1$, then output toggle, i.e. $Q_{n+1} = \bar{Q}_n$

Second Method

J	K	Q_{n+1}
0	0	Q_n
0	1	0
1	0	1
1	1	$\bar{Q}_n$

37. Ans. (c)

Solution: $2^n \geq$ Mode of counter

Given MOD No is 10

therefore, $2^n = 10$ or $n \approx 4$

By using 4 FFs, we can design MOD -10 counter

38. Ans. (c)

Solution: The minimum interval required = total sum of propagation delay = 30 ns + 10 ns = 40 ns

Second Method

Frequency of operation of synchronous counter is

$$f \leq \frac{1}{(30+10)\,ns}$$

Hence, time interval between two successive clock pulses $= \frac{1}{f} = 40\,ns$

39. Ans. (a)

Solution: $2^n = 2 \Rightarrow n = 1$

The minimum number of FFs needed to make a mod-2 counter is 1. The states may be 0 and 1 by use of 'one' FF.

40. Ans. (a)

41. Ans. (b)

Solution:

$J_0 = (\bar{Q}_1)$	$K_0(1)$	$J_0 = (\bar{Q}_1)$	$K_0(1)$	Q_0 Q_1
				1 1←Initially
1	1	0	1	1 0
1	1	1	1	0 1
0	1	0	1	0 0

So, the correct sequence is
00, 10, 01, 00, 10, 01,...

42. Ans. (c)

Solution:

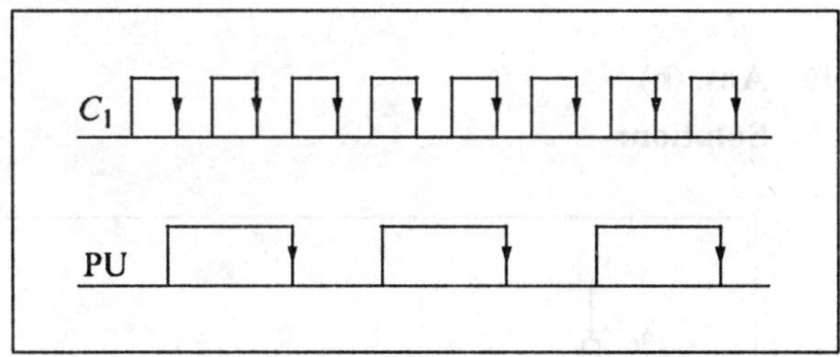

From the given figure, it is seen that every P_u pulse contain 3 clock pulses. Initially two clock pulses counter work as up counter as P_u is high and for 3rd clock pulse it work as down counter so for every three clock pulse it is incremented by only 1 (2 increments and one decrement). Hence after $14 \times 3 = 42$ clock pulses counter goes to 14. After that counter goes high for two clock pulses and counter will be incremented by 2, i.e. $42 + 2 = 44$, i.e. equivalent to 0000. Therefore, after clock pulses counter will be to 0000 again.

43. Ans. (d)

Solution:

When clock is high then NAND gate used invert the S and R so invert the latching action.

44. Ans. (a)

Solution:

JK Flip-Flop toggle for each clock if input is 1, i.e. same. Therefore, the waveforms at different state are shown below.

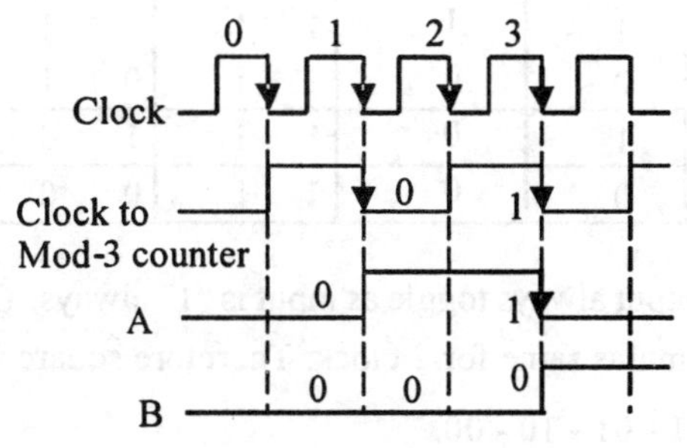

45. Ans. (d)

Solution: Since,

Where, t_g = AND gate delay time = 5 ns

t_d = Flip-flop set time = 5 ns

t_s = Worst delay time = 15 ns

$$\therefore f_{max} = \frac{1}{N t_d + T_s + t_g}$$

$$= \frac{1}{5\times10^{-9}\times3+15\times10^{-9}+5\times10^{-9}} \text{ Hz}$$

= 30 MHz

46. Ans. (b)

Solution:

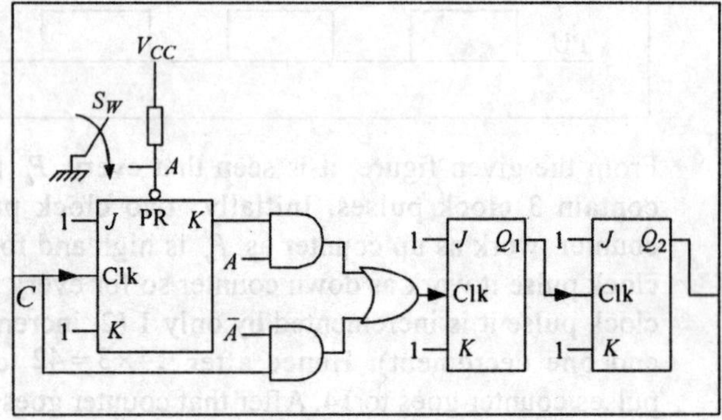

Presets gets priority under all condition

When switch is closed A becomes zero and low active present pin PR gets active and Q_o become equal to1.

And if LSB of any number is 1 then it will be definitly an odd number

47. Ans. (a)

Solution:

To represent the state, draw the table

$j_0(\overline{Q_1})$	$K_0(\overline{Q_1})$	J_1	K_1	Q_1	Q_0
				0	0
1	1	1	1	1	1
0	0	1	1	0	1
1	0	1	1	1	0
0	0	1	1	0	0

Q_1 output always toggle as input is "1" always. Q_0 output remains same for 2 clock. Therefore square is

00 - 11 - 01 - 10 - 00.

48. Ans. (b)

Solution:

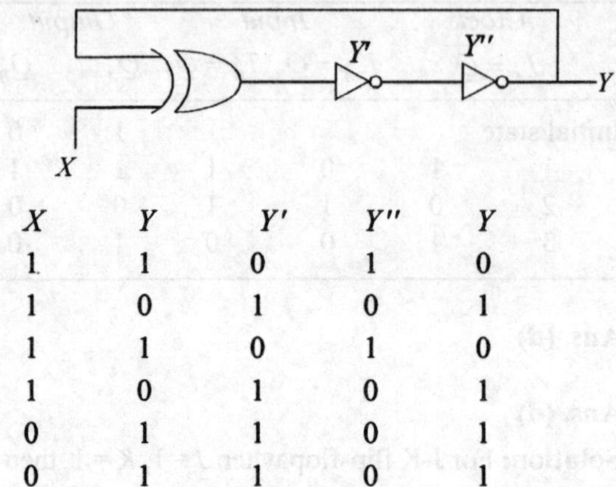

X	Y	Y'	Y''	Y
1	1	0	1	0
1	0	1	0	1
1	1	0	1	0
1	0	1	0	1
0	1	1	0	1
0	1	1	0	1

Hence if $X = 1$ it generate clock pulse, i.e. 0 and 1 with a fixed time-period as shown below

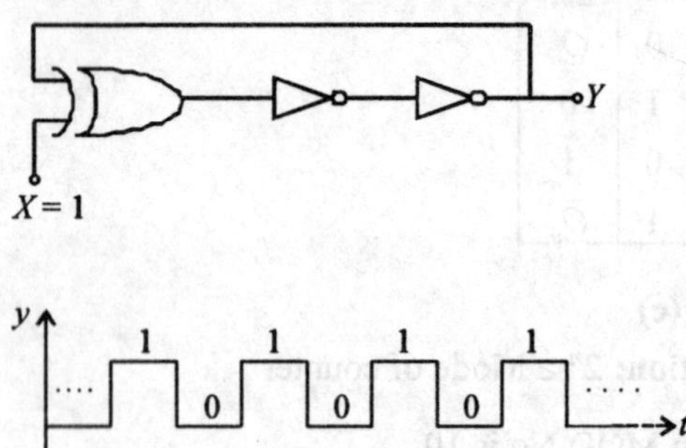

Square wave output can be obtained if odd number of NOT get converted in series with some time delay. Hence XOR gate must work as inverter for Y. Only if $Y \oplus 1 = \overline{Y}$

49. Ans. (c)

Solution:

From the given figure

$$T_B = \overline{Q}_A + Q_B$$

$$T_A = Q_A + Q_B$$

output of T flip flop $= T \oplus Q$

i.e. it toggle for input 1

Start with $Q_A Q_B = 01$

$$T_B = \overline{Q}_A + Q_B = 1+1 = 1$$

$$T_A = Q_A + Q_B = 1+1 = 1$$

Then $Q_A Q_B = 10$

Now $T_B = \overline{Q}_A + Q_B = 0+0 = 0$

$$T_A = Q_A + Q_B = 1+0 = 1$$

Then $Q_A Q_B = 00$

$$T_B = 1$$
$$T_A = 0$$
$$Q_AQ_B = 01$$

So sequence is repeated as 01, 10, 00, 01. So it will never reach to 11.

$$\text{And } Q_AQ_B = 11$$
$$T_B = 1$$
$$T_A = 1$$
$$Q_AQ_B = 00$$

Hence correct option is (c)

50. Ans. (a)

Solution:

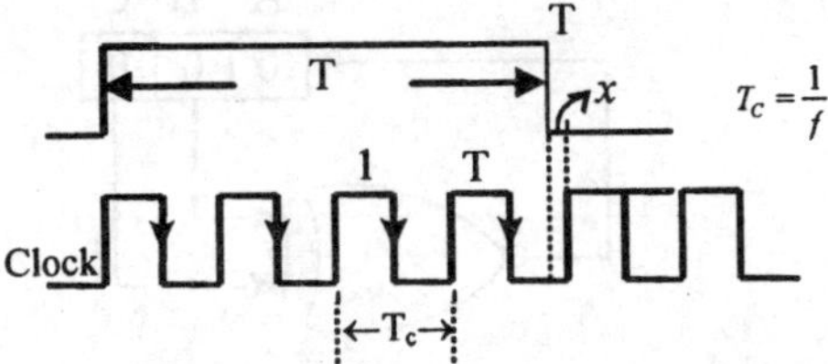

Maximum error possible is one clock period, if this occurs when negative edge of enable occurs just before negative edge of clock so that count is 1 less than actual count.
Since error should be limited to $x\%$ of T,
So maximum error should not be greater than $x\%$ of T.

$$T_c < x\%\,T$$
$$T_c < \frac{x \times T}{100}$$
$$\frac{1}{f_c} < \frac{xT}{100}$$
$$T > \frac{100}{xf_c}$$

51. Ans. (b)

Solution:

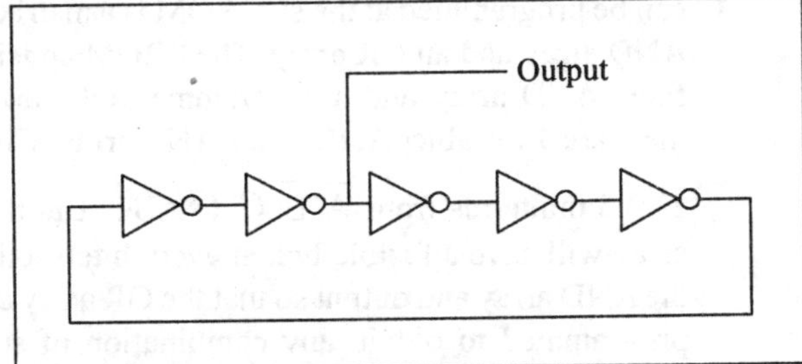

In closed loop ring counter $T_d = 2 \times n \times P_d$
Where P_d = Propagation delay of each inverter
n = number of inverters

$$T_d = 2 \times 5 \times P_d$$
$$T_d = 10\, P_d \quad \text{...(1)}$$
$$T_d = \frac{1}{\text{frequency}} = \frac{1}{10 \times 10^6 \text{ Hz}} \quad \text{...(2)}$$

From equations 1 and 2

$$10\, P_d = \frac{1}{10 \times 10^6 \text{ Hz}}$$
$$P_d = 10 \text{ n sec}$$

52. Ans. (c)

Solution:

J	$K(\bar{Q})s$	Q_{n+1}
1	1	0
1	0	1
1	0	1
		1

J	K	Q_{n+1}
		0 (Initial)
1	1	1
1	0	1
1	0	1
1	0	1

53. Ans. (d)

Solution:

The output of NOR gate start from left side to right side.
The output of first NOR gate

$$= \overline{X_1 + X_2}$$

The output of upper NOR gate in second stage

$$= \overline{X_1 + \overline{X_1 + X_2}} = \bar{X}_1 \cdot (X_1 + X_2) = \bar{X}_1 X_2$$

The output of lower NOR gate in second stage

$$= \overline{X_2 + \overline{X_1 + X_2}}$$
$$= \bar{X}_2 \cdot (X_1 + X_2)$$
$$= X_1 \bar{X}_2$$

The output of right most NOR gate is

$$Y = \overline{\bar{X}_1 X_2 + X_1 \bar{X}_2}$$
$$Y = \overline{X_1 \oplus X_2}$$

54. Ans. (b)

Solution

Given,

$Q_1 = X_2$

$Q_2 = Y$

$Q_3 = D_1$

$Y = D_2$

D-flip truth table

Q_n	Q_{n+1}
0	0
1	1

NOR gate output is 1 when both inputs are same

Q_1	Q_2	Q_3	
1	0	0	Q_n
0	1	0	Q_{n+1}

55. Ans. (a)

Solution:

From the figure, two NAND output is used as input to obtain Y from a NAND gate. Hence

$$Y = \overline{\overline{(Q_1Q_3)} \cdot \overline{(Q_2Q_3)}}$$

$$= \overline{\overline{(Q_1Q_3)}} + \overline{\overline{(Q_2Q_3)}}$$

$$= Q_1Q_3 + Q_2Q_3$$

$$= Q_3(Q_1 + Q_2)$$

56. Ans. (c)

Solution:

As it can be seen in figure clear is inverted actives Hence to reset the counter output Y must be one.

$Y = Q_3(Q_1 + Q_2)$

For this requirement Q_3 must be 1 and $Q_1 = 1$ or $Q_2 = 0$ or Q_1 and Q_2 both are 1.

Here, 4FF are used, Hence it must be counting above 8. Therefore,

Q_3	Q_2	Q_1	Q_0
1	0	0	0
1	0	0	1
1	0	1	0 ←

When the counter output $Q_3\ Q_2\ Q_1\ Q_0 = 1010$ then the output Y will give output 1 and it will given to inverter so it will reset the counter and again counting will start and hence there is no need to go for further combinations.

57. Ans. (b)

Solution: Since a pulse train is applied to shift register it should be serial input shift register. For integral number of clock delay parallel output so that any number of clock period can be introduced.

58. Ans. (b)

Solution: Dual slope A/D is

1. Very slow
2. Highly accurate
3. Capable of eliminating hum noise

59. Ans. (d)

Solution:

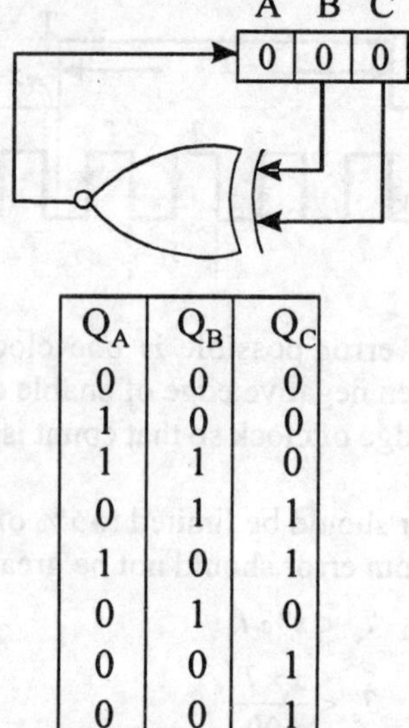

Q_A	Q_B	Q_C
0	0	0
1	0	0
1	1	0
0	1	1
1	0	1
0	1	0
0	0	1
0	0	1

8.5 Semiconductor Memories

1. Ans. (c)

Solution:

* ***PROM:*** Programmable Read Only Memory which can be programmed at the site. ROM is a matrix of an AND array and an OR array. The PROM consists of fixed AND array and a programmed OR array. If there are 3 variables A, B, C, the AND array will have $2^3 = 8$ minterms from $\overline{A}\,\overline{B}\,\overline{C}$ to ABC. But the OR array will have a fusible link at every intersection of the AND array and output so that the OR array can be programmed to obtain any combination of sum of products (SOP). For instance, if there link is broken (fused) for a 0 and left intact for a 1.
* ***PLA:*** Programmable Logic Array has both the AND and OR arrays which are programmable. There will

be a fusible link at each intersection. If are 4 inputs A, B, C, D, the AND array will have as before $2^4 = 16 \times 8 = 128$ fusible links. The OR array will have 4 ′ 16 = 64 fusible links.

2. **Ans. (d)**

Solution:

PLA: The Programmable Logic Array has both the AND and OR arrays which are programmable. There will be a fusible link at each intersection. If are 4 inputs A, B, C, D, the AND array will have as before $2^4 = 16 \times 8 = 128$ fusible links. The OR array will have 4 × 16 = 64 fusible links.

3. **Ans. (c)**

Solution:

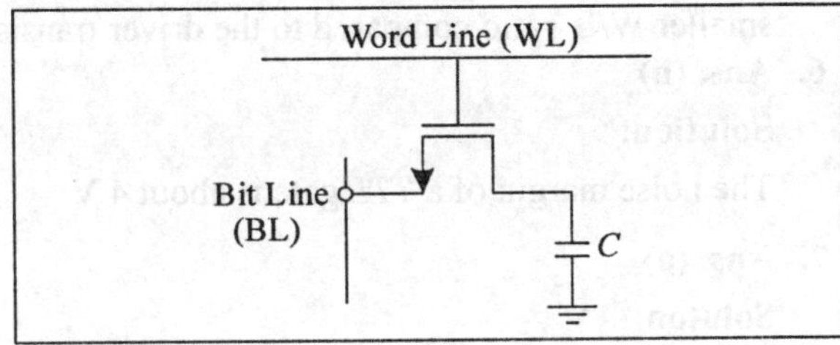

4. **Ans. (a)**

Solution:

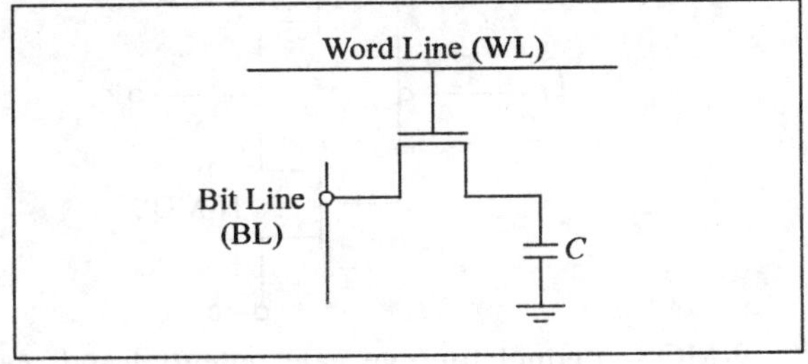

5. **Ans. (a)**

Solution:

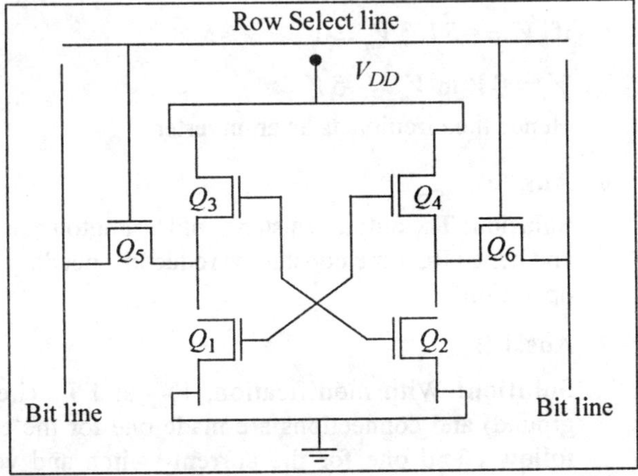

Each cell of a static Random Access Memory contains six MOS transistors In N-MOS RAM or CMOS RAM.

In CMOS RAM each bit on static RAM is stored on four transistors out of which two are PMOS and two are NMOS, that form cross-coupled inverters. Remaining two transistors are used to control reading from or writing into cell.

6. **Ans. (b)**

Solution:

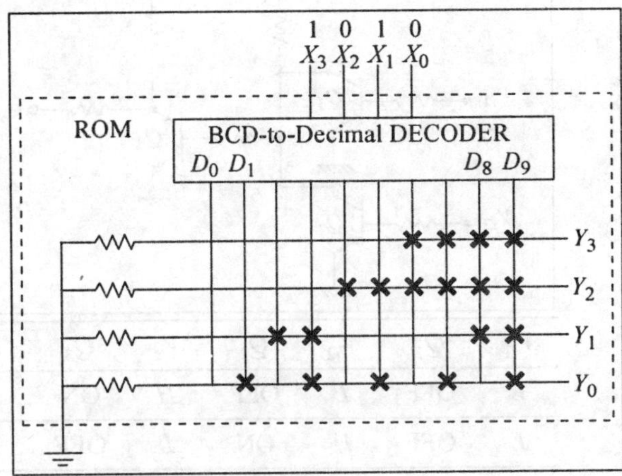

Y_3, Y_2, Y_1, Y_0 can have different value which depends upon the input X_3, X_2, X_1, X_0 value.

Y_3	Y_2	Y_1	Y_0	F	
0	0	0	0	0	
0	0	0	1	1	
0	0	1	0	2	
0	0	1	1	3	
0	1	0	0	4	
0	1	0	1	5	for 2-4-2-1
1	1	0	0	6	for 2-4-2-1
1	1	0	1	7	for 2-4-2-1
1	1	1	0	8	for 2-4-2-1

7. **Ans. (c)**

Solution: When W has data 0110, i.e. 6 in decimal its data value at address is 1010

New 1010, i.e. 10 is acting as address at time t_2 and data at that moment is 1000.

8. **Ans. (d)**

Solution: $C = \frac{Q}{V} = \frac{It}{V} = \frac{0.1 \times 10^{-12} \times 20 \times 10^{-3}}{0.5}$

$= 4 \times 10^{-15}$ F

9. **Ans. (a)**

Solution: The cell is written into by applying the bit line voltage through the transistor to the capacitor, Reading is done by connecting bit line through the gate and sensing the capacitor voltage level.

The read out is destructive. Hence the voltage across the capacitor and bit line is same all the time.

8.6 Logic Families

1. **Ans. (d)**

Solution:

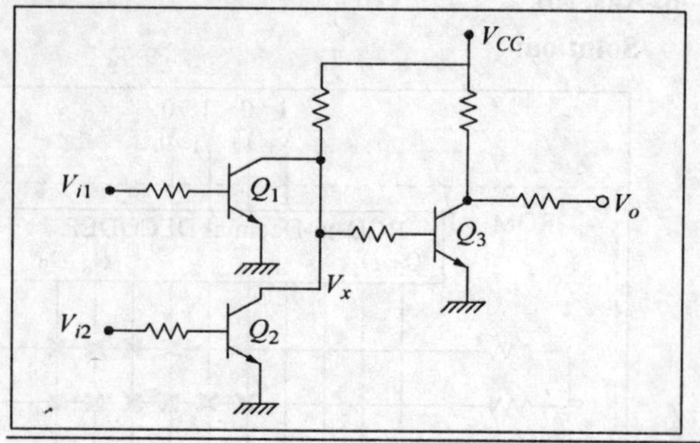

V_{i1}	Q_1	V_{i2}	Q_2	V_x	Q_3	V_0
L	OFF	L	OFF	H	ON	L
L	OFF	H	ON	L	OFF	H
H	ON	L	OFF	L	OFF	H
H	ON	H	ON	L	OFF	H

The circuit represents an OR.

2. **Ans. (c)**

Solution: Totem-pole transistor Q_4 is used because they produce low output impedance which act as power driver or power buffer.

3. **Ans. (b)**

Solution:

Given,

$I_{OL}(\max) = 8\text{ mA},\ I_{IL} = -0.1\text{ mA}$

$I_{OL}(\max) = -0.4\ mA, \qquad I_{IH}(\max) = 20\ \mu A$

The number of inputs that can be driven in the Low state is

$$\text{Fan out (Low)} = \frac{I_{OL}(\max)}{I_{IL}(\max)}$$

$$= \frac{8}{0.1} = 80$$

The number of inputs that can be driven in High state is

$$\text{fan out (High)} = \frac{I_{OH}(\max)}{I_{IH}(\max)}$$

$$= \frac{400\ \mu A}{20\ \mu A}$$

$$= 20$$

* If fan out (Low) and fan out (High) are not the same, the fan out is chosen as the smaller of the two.

Thus, fan out based on given data is 20.

4. **Ans. (c)**

Solution:

Given circuit consist of NMOS logic gate

In NOMS logic gate

Parallel combination = OR

Series combination = AND

The output F is low

* When A in ON and B is ON, i.e. $A(B+C)$
* When OR D is ON and E is ON

Hence $f = \overline{A(B+C)+DE}$

5. **Ans. (c)**

Solution:

Delay of an NMOS inverter is dominated by charge time rather than discharge time because load transistor has a smaller W/L ratio compared to the driver transistor.

6. **Ans. (b)**

Solution:

The noise margin of a *TTL* gate is about 4 V

7. **Ans. (a)**

Soluton:

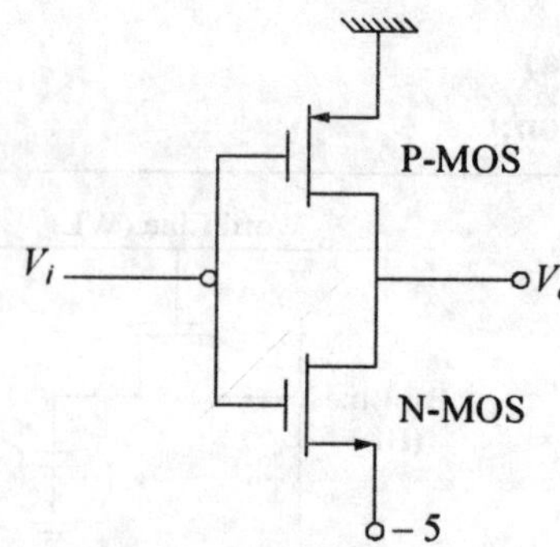

P-MOS conduct for –ve gate potential and earth is extended to output, hence output $V_o = 0$

Given $V_T = 2\ V$

if $V_i = -5\ V$; $V_o = 0$

$V_o = 0\ V$ m $V_o = -5\ V$

Hence the circuit acts as an inverter.

8. **Ans. (c)**

Solution: The output resistance of Darlington pair is low and hence RC time constant is reduced, speeding up the operation.

9. **Ans. (d)**

Solution: With modification, V_{CC} and V_R (i.e. two ground) and connections are made one for the emitter follower and one for the current switch and voltage reference.

The reason for the separate connections is to isolate the

switching transients (large current and voltage spikes caused by charging and discharging the external load and parasitic output capacitance) that appear at the output of the emitter follower from affecting the differential pair and reference circuits (the internal stages)
Since ECL gates have low noise margin (0.3V) this configuration helps to suppress the noise levels, i.e. eliminate the effect of power line glitches on the biasing circuit.

Second Method

Glitches are nothing but spikes produced at some of the output of the decoding circuits. This can be eliminated by connecting ECL gates.

10. Ans. (b)

Solution:

74 series refer to TTL logic using NAND gates.
There are two methods of taking output from a BJT in 74 series logic gates

1. Either totem pole and
2. Open collector configuration.

11. Ans. (a)

Solution: For TTL logic floating input represent logic 1

$$Y = \overline{(AB+1)} = \overline{AB} \cdot 0 = 0$$

12. Ans. (c)

Solution: V_o is at logic 0. So transistor is in saturation

$$V_{BE} = 0.75 \text{ V}$$

$$I_R = \frac{0.75}{1\text{k}}$$

$$= 0.75 \text{ mA}$$

$$I_R = 0.75 \text{ mA}$$

13. Ans. (c)

Solution:

Since two transistor are connected in sereis.

$$V_{GS_1} = 5 - V_o$$

$$V_{GS_2} = V_o$$

$$I_{DS}(T_1) = I_{DS}(T_2)$$

$$\left(\frac{1}{2}\right) K_1 \ (V_{GS1} - V_t)^2 = \left(\frac{1}{2}\right) K_2 \ (V_{GS2} - V_t)^2$$

$$36\ (5 - V_o - 1)^2 = 9\ (V_o - 1)^2$$

$$6\ (4 - V_o) = 3\ (V_o - 1) \Rightarrow 2\ (4 - V_o) = V_o - 1$$

$$8 + 1 = 3\ V_o = 9$$

$$V_o = 3$$

14. Ans. (b)

Solution:

As given TTL circuit is NOT gate $V_i = 2.5$ V high is applied. Output is low only when the earth is extended to output. It means Q_3 is in saturation mode and Q_4 is cut off as Q_3 and Q_4 always behave negatively. Q_2 work in saturaton region as TTL is saturated logic family but Q_1 in reverse active as $V_i = 2.5$ V applied to emitter of Q_1.

15. Ans. (d)

Solution:

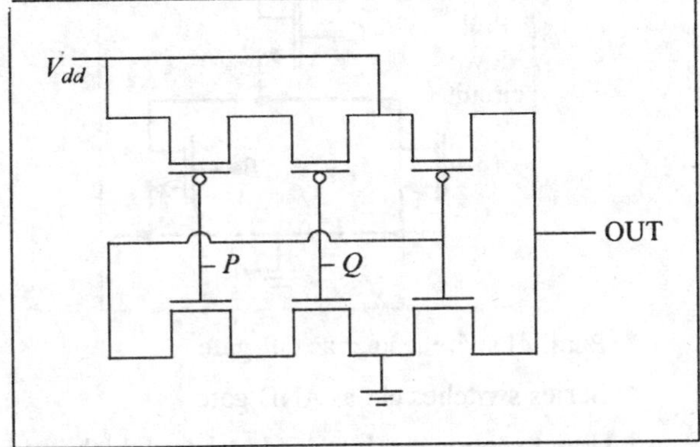

From the given figure, it may be seen that upper MOSFET are shorted and connected to V_{DD}. Hence, OUT is 1 only when node S is 0.
Since lower MOSFET are shorted to ground, node S is 0 only when input P are 1. Thus the circuit behave as AND gate.
Both input and output are in series that result AND gate

16. Ans. (c)

Solution:

TTL stands for Transistor Transistor Logic.
CMOS stands for Complementary Metal Oxide Semiconductor.

17. Ans. (b)

Solution:

Logic Families Comparison

Logic	RTL	DTL	HTL	TTL	ECL	MOS	CMOS
Gate	N1	N2	N2	N2	N1/N2	N2	N1/N2
Fan-out	5	8	10	10	25	20	50
Power loss mW	12	8	50	12	40	0.2	0.01
Noise	N	G	E	VG	G	N	VG
Propagation nsec	12	30	90	12	4	300	70

* N → Normal, * S → Satisfactory

* G → Good * VG → Very good

* E → Excellent * N1 → NOR

* N2 → NAND * N3 → OR.

18. Ans. (a)

Solution:

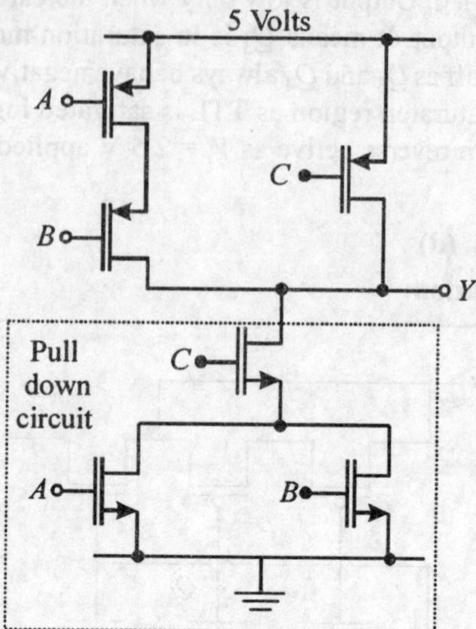

* Parallel switches act as OR gate
* Series switches act as AND gate
* Upper section work as load resistor which invert the input
* The p-channel circuit is taken as the complementary network where again parallel n-channel sub-circuit transforms into series p-channel sub-circuit.

$$Y = \overline{(A+B).C}$$

$$= \overline{A+B}+\overline{C}$$

$$= \overline{A}\cdot\overline{B}+\overline{C}$$

19. Ans. (a)

Solution:

Schottky TTL

Speed limitation of TTL is mainly due to the turn-off time delay involved in transistor while making transition from saturation to cut-off (storage time delay). This can be eliminated by replacing the transistors of TTL gate by Schottky transistor.

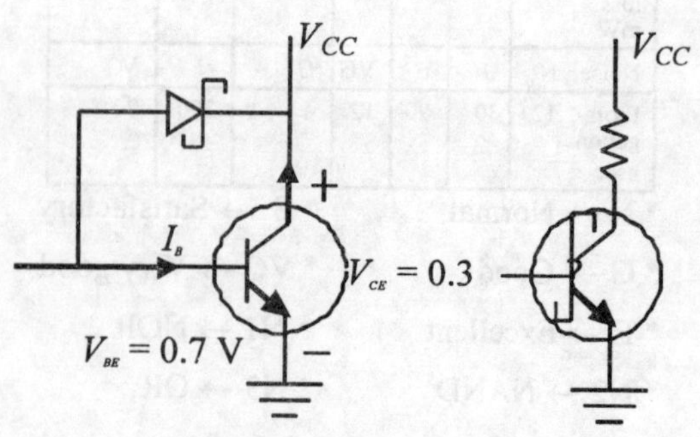

This Schottky transistor prevent transistor from entering saturation and hence, there is saving in turn-off time (Schottky diode is clamped between base and collector).

A Schottky diode works faster as compared to commercial diode because of fact the electron which have crossed the junction and entered the metal, when the current flowing are similar to the conduction electrons of the metal.

Reverse recovery time for Schottky diode is few p sec rang.

20. Ans. (d)

Solution: For LED to emit light, NAND output should be low (0) and input 1, for which AND input to be 1 and EX-OR gate input 0.1 or 1, 0. When S_1 and or S_2 are closed, input to the AND gate is 0. When S_1, S_2 are open input to the AND gate is 1 as also to X OR gate.

Second Method

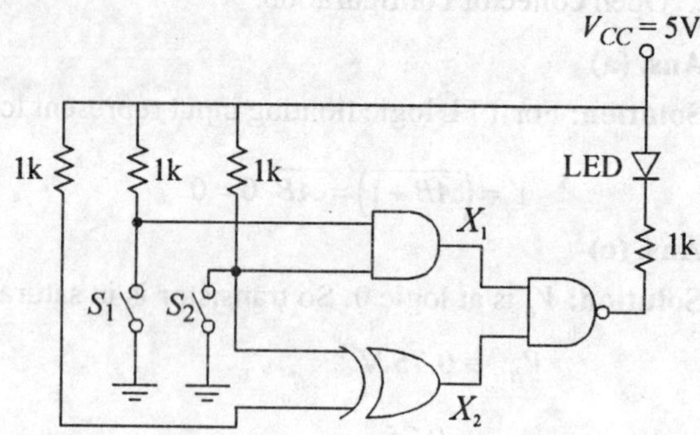

LED will be ON only when Y = 0

But $Y = \overline{X_1 X_2}$

Y = 0 only when both X_1 and X_2 are 1.

$X_1 = 1$ when both switches S_1 and S_2 are off i.e. both inputs to AND gate are 1.
But this also make both the inputs at EXOR gate equal to 1.

$\therefore\ X_2 = 0$

Hence, both X_1 and X_2 cannot be 1 simultaneously.

21. Ans. (c)

Solution:

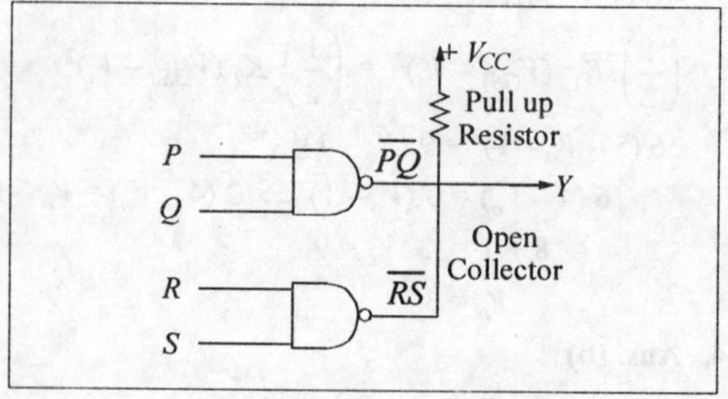

Open collector is AND gate

Wired AND

i.e. $\overline{(PQ)}.\overline{(RS)} \Rightarrow \overline{PQ+RS}$

Output of NAND gate is $\overline{PQ}$ and $\overline{RS}$.

Open collector output is $\overline{PQ}$, $\overline{RS}$.

This will get inverted by pull up resistor

$$\overline{\overline{PQ}.\overline{RS}} = \overline{\overline{PQ}} + \overline{\overline{RS}} = PQ + RS$$

22. Ans. (b)

Solution: Totem pole output stage is used in standard TTL gates to decrease the output switching delay by decreasing the output impedance and provide sufficient power level for a high fan out. It also provide low output impedance. But a totem pole has problem of internally generated current spikes and its output cannot be wired AND unlike open collector configuration of TTL

23. Ans. (d)

Solution:

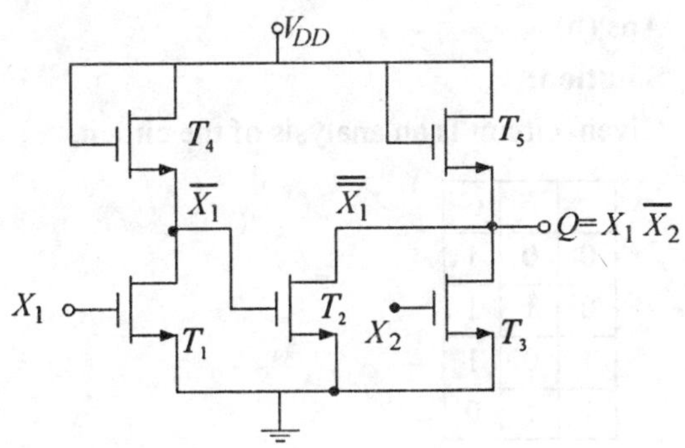

Transistors T_4 and T_5 work as load; and T_1, T_2, T_3 act as inverters/ driver switch

or Output of T_1 is $\overline{X}_1$

Output of T_2 is $\overline{\overline{X}}_1 = X_1$

Output of T_3 is $\overline{X}_2$

$\therefore$ $Q = X_1 . \overline{X}_2$

24. Ans. (c)

Solution: Base of Q_2 always at 0.7 V

So Q_2 is ON.

Q_1 will conduct in reverse active region as collector potential should always base potential.

$V_{CB} \geq 0$, $V_{CE} > 0$

So, Q_1 is reverse ON.

25. Ans. (a)

Solution:

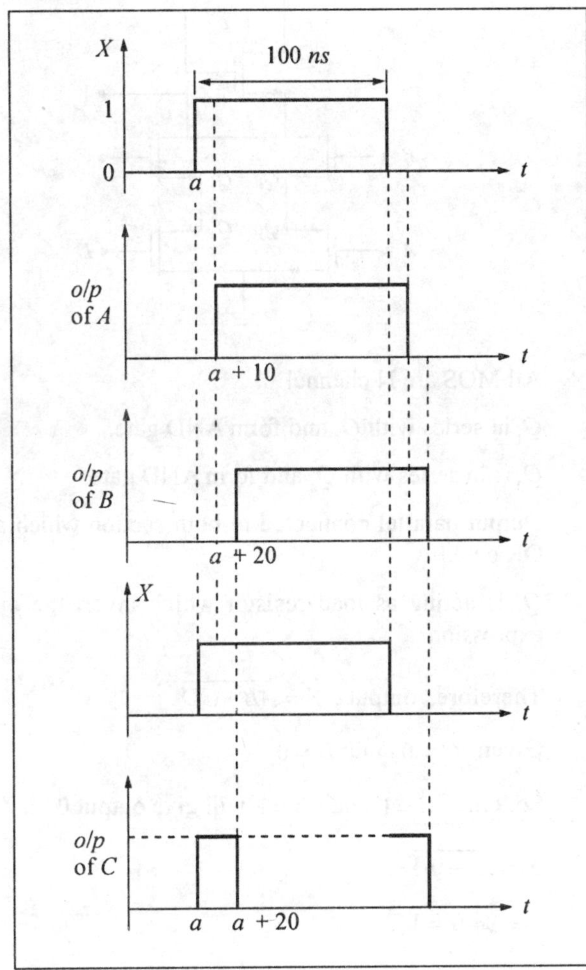

26. Ans. (a)

Solution:

Schottky diode is used to speed up the operation by inhibiting the saturation and brings the transistor into active region.

27. Ans. (b)

Solution:

Given , $V_{CC} = 12\ V$ sink current I = 16 mA

Input capacitance = 10 pF

Pull-up time constant ,

$$T = \frac{CV}{I} = \frac{10^{-12} \times 12}{16 \times 10^{-3}} = 7.5$$

28. Ans. (c)

Solution:

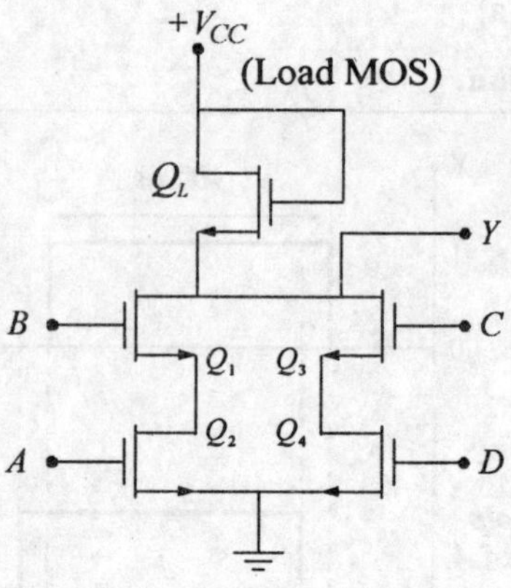

All MOS are N channel

Q_1 in series with Q_2 and form AND gate.

Q_3 is in series with Q_4 and form AND gate.

Output parallel connected to both section which act as OR gate.

Q_L is acting as load resistor which invert the overall expression

Therefore, output, $Y = \overline{AB + CD}$

Given $C = 0$ and $D = 0$

So, only $A = 1$ and $B = 1$ will give output 0

$Y = \overline{1.1 + 0.1}$

$= \overline{1 + 0} = \overline{1}$

$= 0$

29. Ans. (c)

Solution:

X	Y	T_1	T_2	T_3	T_4	V_o
0	0	*ON*	*ON*	*OFF*	*OFF*	*high*
0	1	*ON*	*OFF*	*OFF*	*ON*	*low*
1	0	*OFF*	*ON*	*ON*	*OFF*	*low*
1	1	*OFF*	*OFF*	*ON*	*ON*	*low*

$Y = \overline{A + B} \Rightarrow$ NOR

30. Ans. (b)

Solution:

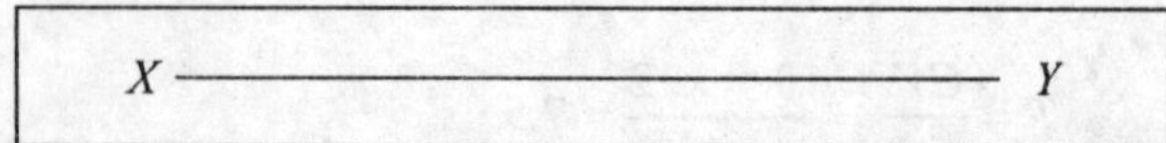

When inputs to the diode are zero, the diode conducts and when input is high, the diodes become non-conducting

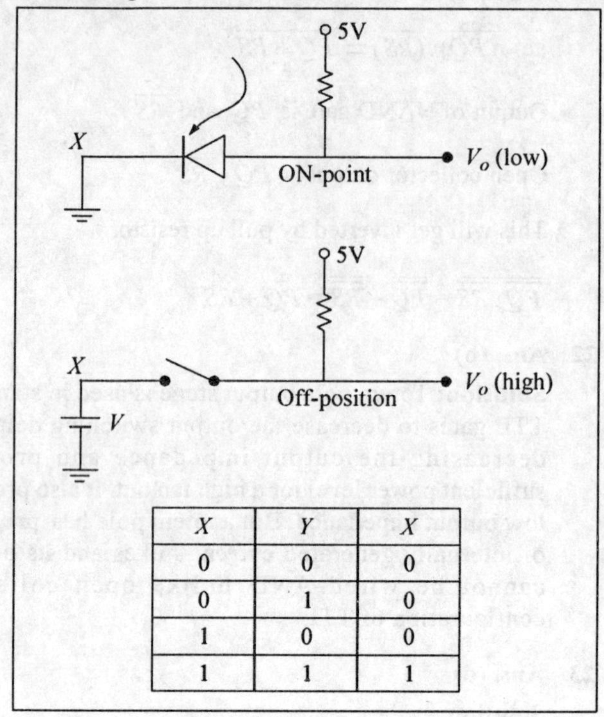

X	Y	X
0	0	0
0	1	0
1	0	0
1	1	1

$Z = XY$

31. Ans (b)

Solution:

Given circuit is an analysis of the cirucit,

A	B	C
0	0	1
0	1	1
1	0	1
1	1	0

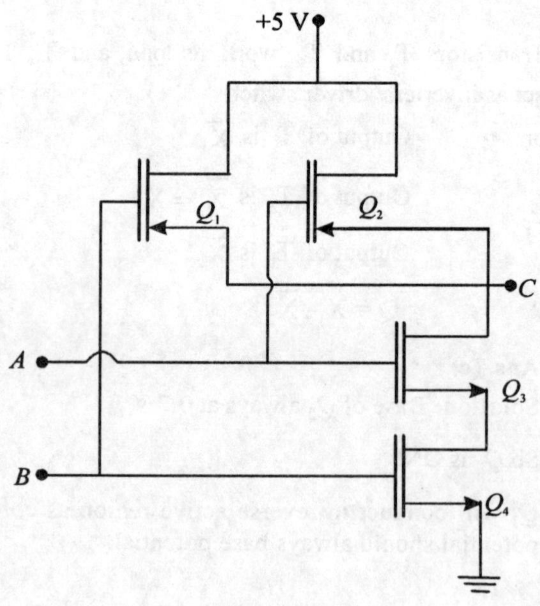

Only when $A = B = 1$; Q_3 and Q_4 are ON and output $C = 0$, otherwise one of MOSFET Q_3 and Q_4 is OFF making output $C = +5\,V$

32. Ans. (a)

Solution:

T_1 is N-MOSFET

T_2 is P-MOSFET

T_1 conducts when $V_i > V_{th}$

$V_o = 0V$, when $V_i = 5\,V$ for COMS inverter

T_1 is always inlinear when $V_o = 0V$

And T_2 is in cut-off

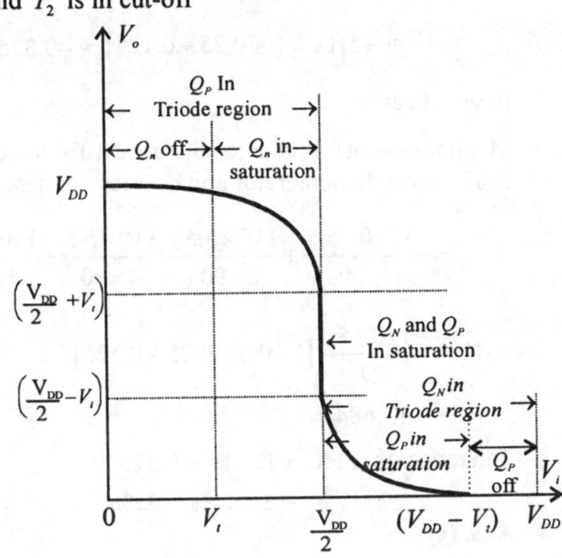

33. Ans. (c)

Solution:

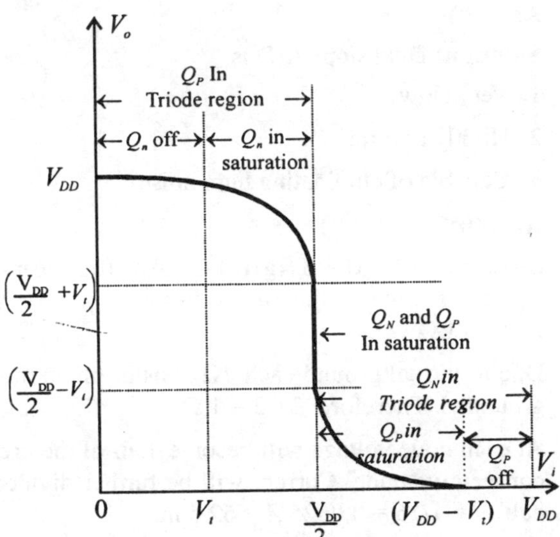

The given transfer characteristic is very close to an ideal inverter. Hence, it must be a CMOS inverter.

34. Ans. (a)

Solution: Digital families should have noise immunity

8.7 A/D and D/A Converters

1. Ans. (c)

2. Ans. (b)

Solution: Given resolution = 0.5 V

$$\text{No. of step} = \frac{V_{in}}{\text{Resolution}} = \frac{6.6}{0.5} = 13.2 = 13$$

Binary output = 1101

3. Ans. (b)

Solution:

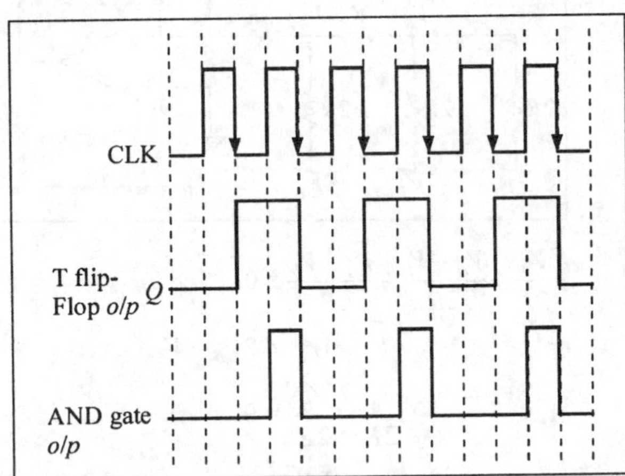

Since J = K = 1; Q will toggle with every CLK. Ouptut = 1 only when both Q and CLK are high.

4. Ans. (d)

Solution: $V_{DC} = \sum_{n=0}^{3} 2^{n-1} b_n$

b_3	b_2	b_1	b_0	V_{DAC}
0	0	0	0	0
0	0	0	1	0.5
0	0	1	1	1.5
1	1	0	0	⋮
1	1	0	1	6.5
1	1	1	0	7
1	1	1	1	7.5

Using the given V_{DAC} relation we can get t_c

Thus, stable reading of LED is 6.5 × 2 = 13

5. Ans. (b)

Solution: Magnitude error = 6.5 – 6.2 = 0.3

6. Ans. (b)

Solution:

Shift-registers can be used to change spatial code to temporal code and vice-versa

7. Ans. (d)

Solution:

In successive approximation type ADC required for conversion = nt

$$= 12 \times 1\,\mu s$$

$$= 12\,\mu s$$

8. Ans. (b)

Solution:

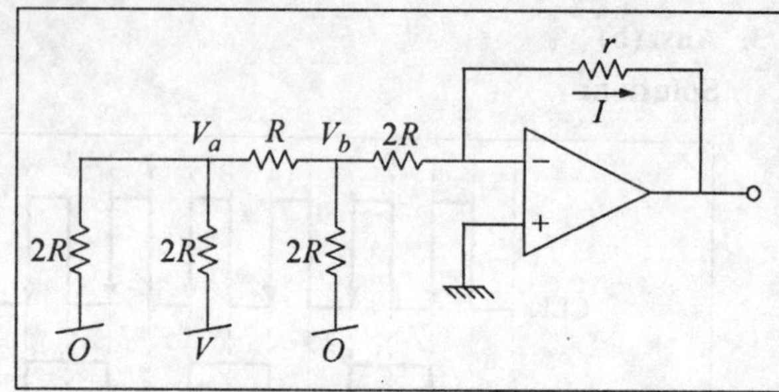

$$\frac{V_a}{2R} + \frac{V_a - V}{2R} + \frac{V_a - V_b}{R} = 0$$

or $\quad 4V_a = 2V_b + V$

or, $\quad \frac{V_b - V_a}{R} + \frac{V_b}{2R} + \frac{V_b}{2R} = 0$

or, $\quad 2V_b = V_a$

Since $\quad 3V_a = V_a$

$\therefore \quad V_b = \frac{V}{6}$

$\therefore \quad I = \frac{V}{12R}$

9. Ans. (b)

Solution:

Conversion time for SAR is independent of input voltage. It depends on bit only.

10. Ans. (b)

Solution: V_{in} at non-inverting terminal

$$\frac{V_R}{2^4}\left(2^3 + 2^1\right) = \frac{1}{16} \times 10$$

Gain of non-inverting amplifier

$$= \left(1 + \frac{R_f}{R}\right)$$

$$V_o = \left(1 + \frac{R_f}{R}\right)V_{in}$$

$$= \frac{10}{16}\left(1 + \frac{7}{1}\right) = 5\,\text{V}$$

11. Ans. (c)

Solution: Number of comparator = $2^n - 1$

$$= 2^3 - 1 = 8 - 1 = 7$$

12. Ans. (a)

Solution: $V_o = -V_R\left[\frac{R}{R}d_3 + \frac{R}{2R}d_2 + \frac{R}{4R}d_1 + \frac{R}{8R}d_0\right]$

$$= -V_R\left[\frac{d_3}{1} + \frac{d_2}{2} + \frac{d_1}{4} + \frac{d_0}{8}\right]$$

When $V_R = 5$V **and** $d_3 = d_2 = d_1 = d_0$

$$V_o = -5[1 + 0.5 + 0.25 + 0.125] = -9.375$$

In worst case,

Maximum value V_o due to tolerance 10% increase and 5% increase in numerator and decrease in denominator

$$V_o = -\left[\frac{110 \times 5.5}{90} + \frac{110 \times 5.5}{2 \times 90} + \frac{110 \times 5.5}{4 \times 90} + \frac{110 \times 5.5}{8 \times 90}\right]$$

$$V_o = \frac{-110 \times 5.5}{90}[1 + 0.5 + 0.25 + 0.125]$$

$$= -12.604$$

Tolerance of $DAC = 12.604 - 9.375$

$$= 34.44\% = 35\%$$

13. Ans. (c)

Solution: Number of comparators = $2^n - 1$

$$= 2^4 - 1 = 15$$

14. Ans. (b)

Solution: Dual slope A/D is

1. Very slow
2. Highly accurate
3. Capable of eliminating hum noise

15. Ans. (b)

Solution: In the Digital-to-Analog converter circuit shown in the figure below, $V_R = 10$ V and $R = 10$ kΩ

Due to virtual ground each 2Ω resistor one terminal is grounded. Therefore $2 \parallel 2 = 1\,\Omega$

At each node voltage will become half of the previous consecutive node. Current will be further divided into half. $I = 1/16 = 0.0625\,A = 62.5$ mA.

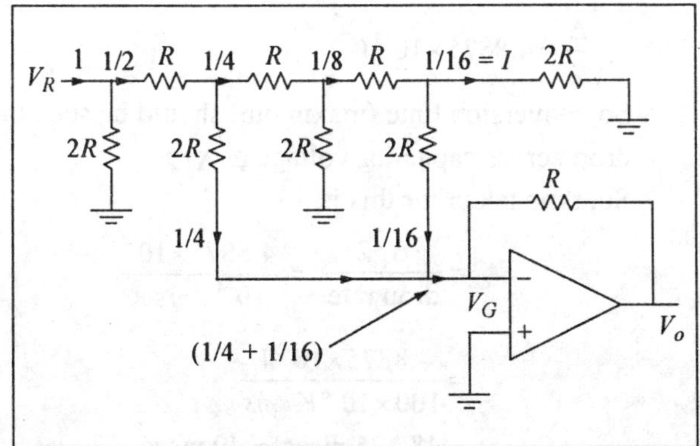

16. Ans. (c)

Solution: The current through the Ops-Amp resistance

R is $\left(\frac{1}{4}+\frac{1}{16}\right) \times 10\text{ k} = -3.125\text{ V}$

17. Ans. (b)

Solution: For flash type *A*/*D* converter, the number of comparators = $2^n - 1 = 2^4 - 1 = 15$

18. Ans. (c)

Solution:

The correct sequence of decreasing conversion times of ADCs is

Dual slope integrating type >Counting> Succesive Approximatin type> Simulatneous or Flash type ADC.

* The Flash type or simultaneous ADC is fastest
* Counting type ADC is faster then counter type
* In successive approximation type A/D converter the conversion time is constant and proportional to the number of bits inthe digital output, unlike the counter and continous tpe A/D converters
* Conversion time of single slope and dual slope type ADC is maximum.

19. Ans. (a)

Solution: Even when $V_1 > V_2$, the output V_o is high, i.e. $+V_{CC}$ and for the next case when $V_1 < V_2$, output is low, i.e. $-V_{CC}$.

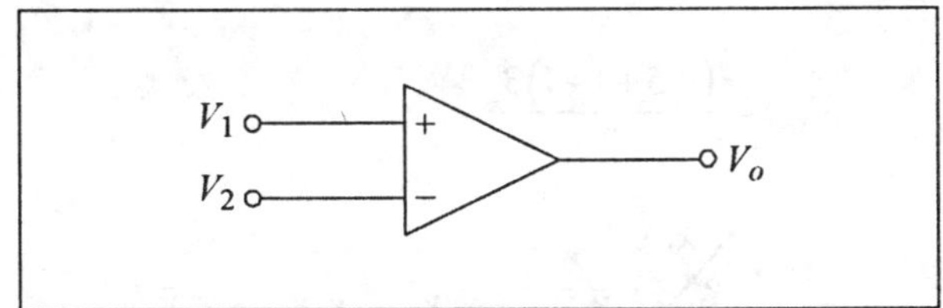

For representing two levels, only one bit is required, and hence the voltage comparator shown in the figure is a one - bit quantizer.

20. Ans. (c)

Solution: The general circuit diagram of the DAC is:

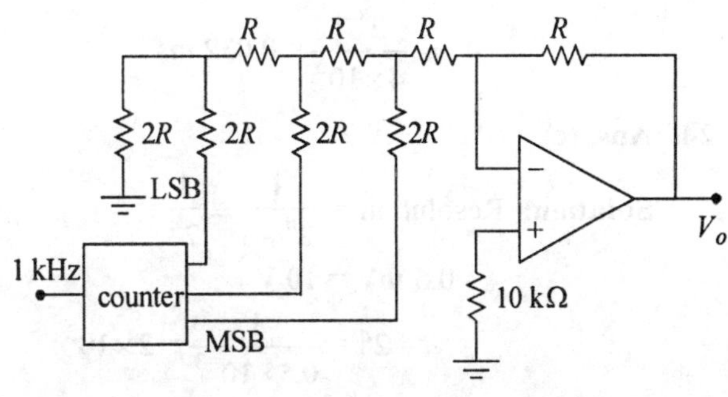

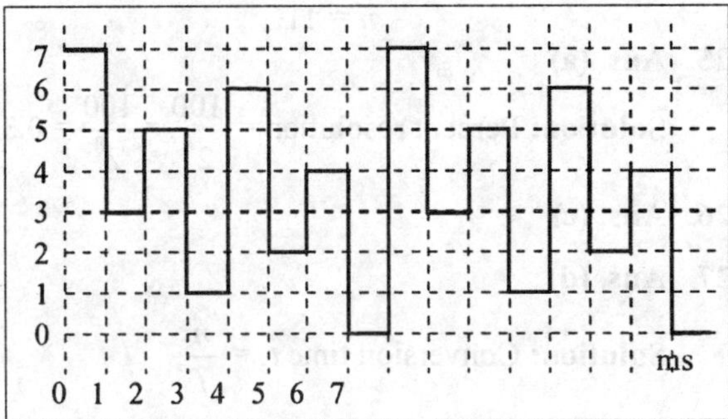

Here LSB and MSB are not correct.

The interconnection from the counter to DAC is not proper. The lines corresponding to bits D_2 and D_0 are interchanged.

Counter output	111	110	101	100	011	010	001	000
ADD output	7	3	5	1	6	2	4	0
Correct Sequence	7	6	5	4	3	2	1	0

21. Ans. (d)

Solution:

$$I_{out} = \frac{4}{3R} + \frac{2}{3R} + \frac{1}{3R}$$

$$= \frac{7}{3R} = \frac{7}{30\times10^{-3}}$$

$$= 0.233\ mA$$

22. Ans. (a)

Solution: A sample and hold is normally required before successive approximation analog to digital converter as it is most useful if multiplexing is required.

23. Ans. (a)

$$T_1\text{ (Fixed time)} = 2^{n-1} \times T_C \; \frac{2^{n-1} - 1}{\text{Clock frequency}}$$

$$= \frac{2^{8-1}}{4\times10^6} = 0.032 \text{ ms}$$

24. Ans. (c)

Solution: Resolution $= \frac{1}{2^n - 1} \cong \frac{1}{2^n}$

$$0.5 \text{ mV} = 10 \text{ V}$$

$$2^n = \frac{1}{0.5\times10^{-3}} = 2\times10^3$$

$$2^n = 2000$$

$$n = 11$$

25. Ans. (a)

Solution: Percent resolution $= \frac{100}{2^n} = \frac{100}{2^8} = 0.39$

26. Ans. (c)

27. Ans. (d)

Solution: Conversion time $t_c = \frac{n}{f}$,

Where n is number of bits

f is Clock frequency, $\frac{8}{1\times10^6} = 8\ \mu s$

28. Ans. (b)

Solution:

De-integration period, $T_2 - T_1 = \frac{V_a}{V_R}\cdot T_1$

$$= \frac{120}{100}\cdot 50 = 60\ ms$$

29. Ans. (b)

Solution: Number of comparator

$$= 2^N - 1$$

$$= 2^8 - 1 = 255$$

30. Ans. (d)

Solution:

Conversion time $= N\tau$

Where

$N =$ number of bits

$\tau =$ time period of clock

31. Ans. (d)

Solution:

In a ADC with sample and hold circuit.

For avoiding error at output, voltage across capacitor

$$\geq \pm\frac{\Delta}{2}$$

where Δ is step size

From given data

Here, $\Delta = \frac{10-0}{(2^{10}-1)} = 9.775\times10^{-3} V$

$$\therefore\ \frac{\Delta}{2} = 4.8875\times10^{-3} V$$

So conversion time (maximum) should be such that the drop across capacitor voltage $\geq \Delta/2$.

So, time taken for this is

$$t_C = \frac{\Delta/2}{\text{drop rate}} = \frac{4.8875\times10^{-3}}{10^{-4}\ /m\sec}$$

$$= \frac{4.8875\times10^{-3} V}{100\times10^{-6} V/ms}$$

$$= 48.875\ m\sec \simeq 49\ m\sec$$

32. Ans. (b)

Solution:

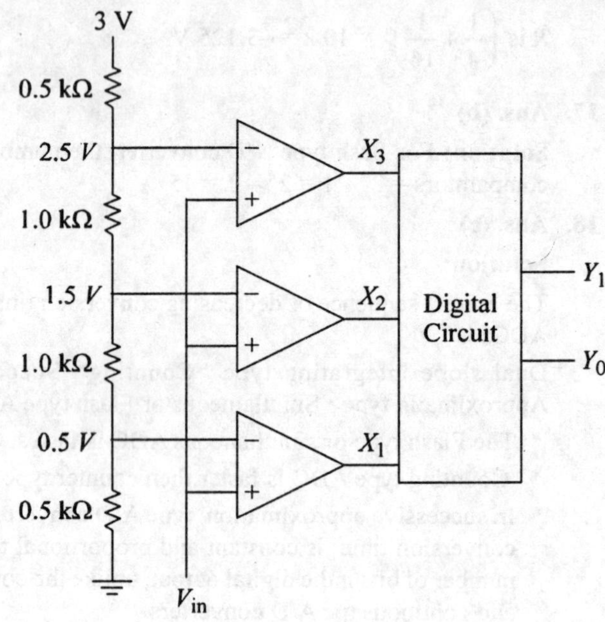

$$X_{1ref} = \frac{0.5\times3}{0.5+1+1+0.5}$$

$$= \frac{1.5}{3} = 0.5V,$$

$$X_{2ref} = \frac{(0.5+1)3}{3} = 1.5V$$

$$X_{3ref} = \frac{(0.5+1+1)3}{3}$$

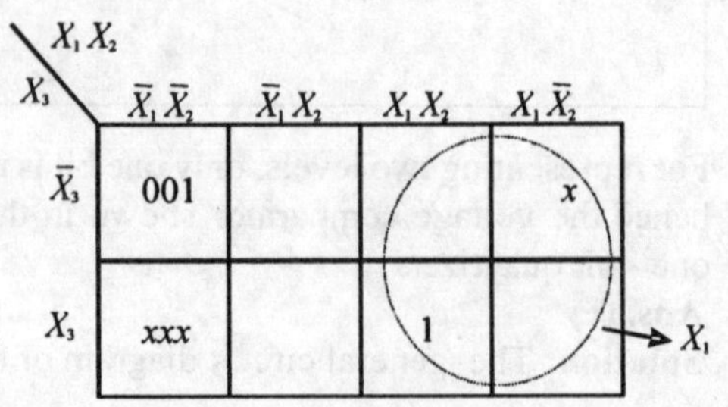

V_{in}	X_3	X_2	X_1	X_1	X_0
$2.5\,V < V_{in} \le 3$	1	1	1	1	1
$1.5\,V < V_{in} \le 2.5$	0	1	1	1	0
$0.5\,V < V_{in} \le 1.5$	0	0	0	0	1
$0\,V < V_{in} \le 0.5$	0	0	0	0	0

Hence $Y_1 = X_1$

33. Ans. (d)

Solution:

8-bit ADC, circuit is in 2's complement form.

i.e. it represents $+127_{10}$ to -128_{10}

Input voltage range $= -2V$ to $+2V$

i.e. $2V \to 128$

$$1.5\,V \to \frac{1.5}{2} \times 128 = 96_{10}$$

For -1.5 analog input, its equivalent is given as

$-96_{10} = 2's$ complement of -96

$= 2's$ complement of $0110\ 0000$

$= 1001\ 1111 + 1$

$= 1010\ 0000$

$= A0H$

34. Ans. (a)

Solution: In this 4-bit successive approximation type ADC is having full value = 15 V we have to measure 8.15 V, we know that

[At start of conversion - MSB is set, i.e. 1000

$(1000)_2 = (8)_{10}$

Now $8.15 > 8$

So 4th bit remain 1 and 3rd will be incremented by 1.

Then $1100 \Rightarrow 12$

Now $12 > 8.15$

Then 4th will be remain same 3rd will change 0 and 2nd will change to 1

$1010 \Rightarrow 10$

Now $10 > 8.15$

Then 3rd bit will change to 0 but 1st bit change to 1

$\Rightarrow 1001 = 9$

$9 > 8.15$

So 4th bit will change to 0. Hence result $\Rightarrow 1000$

35. Ans. (b)

Solution: New Type ACE

For 4 V, 6V the output expected is

$b_3\ b_2\ b_1\ b_0$

0 1 0 0

0 1 0 0

and but as b_1 is struck at 0 both 4 V, 6V shoe the same value

36. Ans. (c)

Solution:

For $4V, 6V$ the output expected is

	MSB b_3	b_2	b_1	LSB b_0
$4 \to$	0	1	0	0
$6 \to$	0	1	1	0

and but as b_1 is struck at 0

i.e.	b_3	b_2	b_1	b_0
$4 \to$	0	1	0	0
$6 \to$	0	1	0	0

Hence, both $4V, 6V$ shown the same value.

37. Ans. (c)

Solution:

$$< q_e >= \frac{1}{2^N - 1} = \frac{1}{2^{12} - 1} = \pm 0.024\%$$

38. Ans. (d)

Solution:

$$f_c = \frac{1}{T_c} = \frac{1}{5 \times 10^{-6}} = 200\text{ kHz}$$

39. Ans. (c)

Solution:

Sampling period, $T_s = 2 \times 5\mu s = 10\mu s$

Sampling rate $f_s = \frac{1}{T_s} = 100$ k samples/s.

40. Ans. (b)

Solution:

At switch status $S_0 = 0, S_1 = 1\ S_2 = 1$

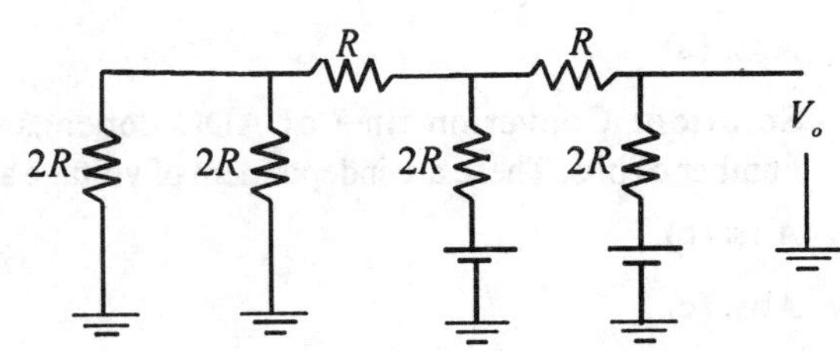

Thevenin equivalent

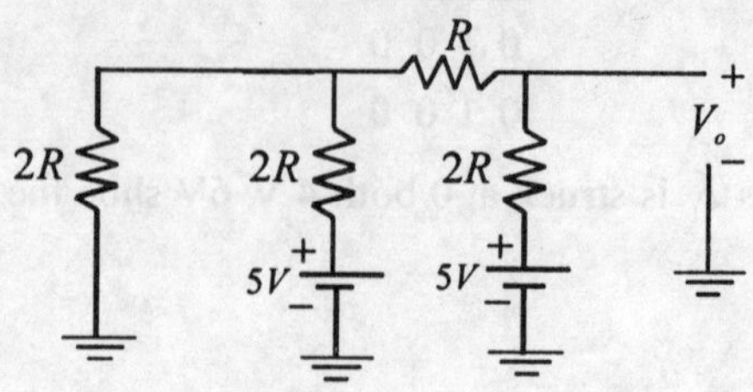

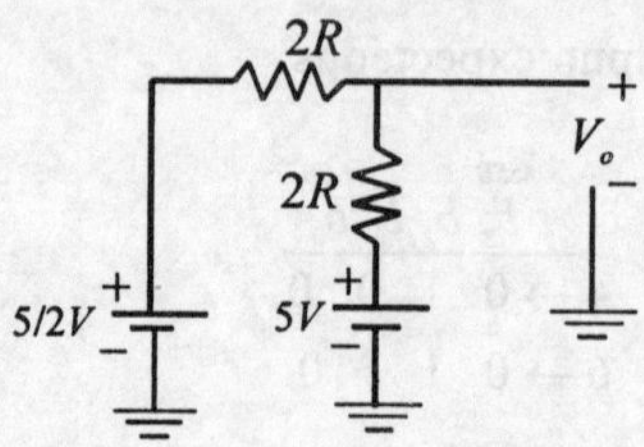

Second Method

$$= \frac{V_{ref}}{2r}\left[D_o \times 2^0 + D_1 \times 2^1 + D_2 \times 2^2\right]$$

$$= \frac{5}{8}[2+4]$$

$$= \frac{5\times 6}{8} = \frac{15}{4}$$

41. Ans. (c)

Solution:

$V_{ref} = 5V$; $n = 3 =$ number of bits

Step size, $S = \frac{V_{ref}}{2^n} = \frac{5}{2^3}$

$$= \frac{5}{8} = 0.625$$

42. Ans. (a)

Solution: Converion time of ADC depends on the number of bits. These are independent of voltage applied.

43. Ans. (c)

44. Ans. (c)

8.8 Miscellaneous

1. **Ans. (b)**

Solution:

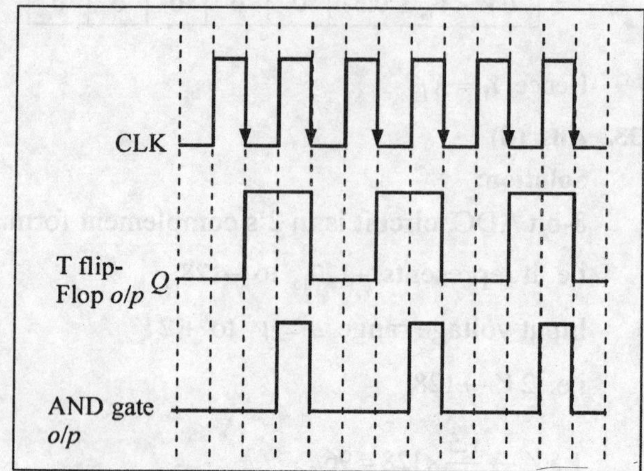

Since J = K = 1 ; Q will toggle with every CLK. Ouptut = 1 only when both Q and CLK are high.

2. **Ans. (c)**

Solution: Output of EX-OR gate if both input are different, i.e. in present case both expect waveforms have a phase difference of $T/2$ for mean reading of PMMC.

3. **Ans. (d)**

Solution: For (1/2) digit MSB can be maximum and other full digit gives 9, respectively.

$$3 \text{ and } \frac{1}{2} \text{ digit} = 1.999$$

4. **Ans. (a)**
5. **Ans. (c)**
6. **Ans. (b)**
7. **Ans. (a)**
8. **Ans. (b)**

Solution:

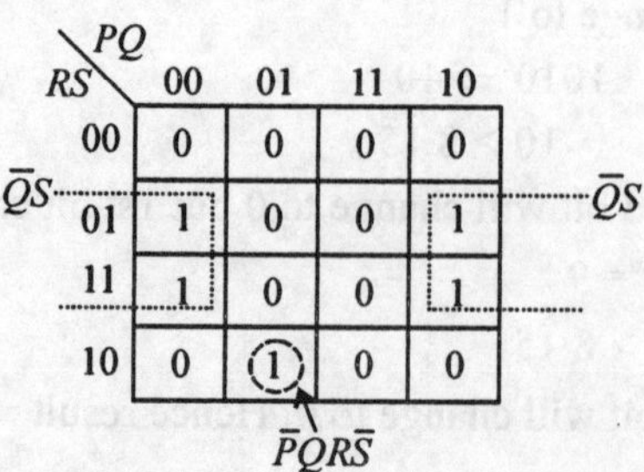

The given expression have a group of four and pair from the k-map,

Output(Y) $= \overline{Q}\,S + \overline{P}\,Q\,R\,\overline{S}$

9. **Ans. (d)**

Solution:

From the given figure it is indicated that V_1 is connected to PMOS and V_2 is connected to NMOS as can be seen by arrow direction.

V_i is bipolar and it is equal to V_o provided any one of the MOS is ON. NMOS conduct for +ve gate voltage and PMOS conduct for –ve gate voltage. Hence for $V_o = V_i$ the requirement is $V_1 = -V_{EE}$ and $V_2 = +V_{DD}$

10. **Ans. (d)**

Solution:

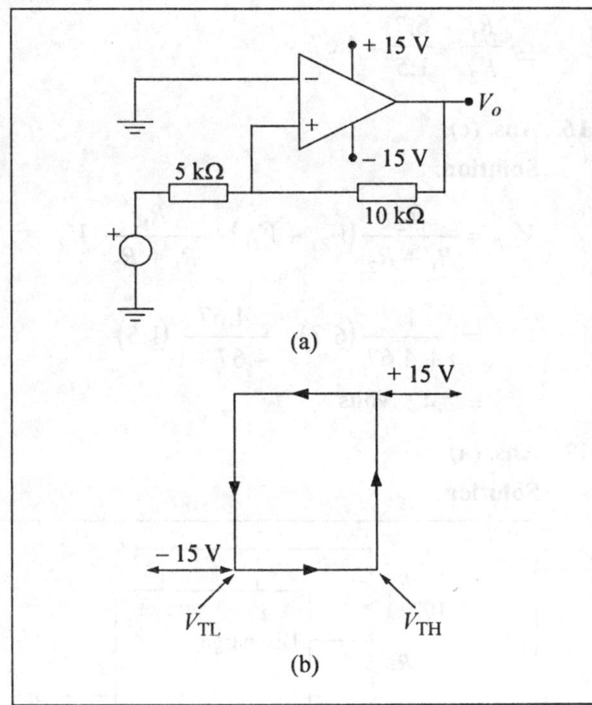

Given, $R_1 = 10\ k\Omega$; $R_2 = 5\ k\Omega$

$$V_{TL} = \frac{-R_2}{R_1 + R_2}(V_{Z1} + V_D)$$

$$= -\frac{-5}{10+5} \times 15$$

$$= -5\ V$$

$$V_{TH} = \frac{R_2}{R_1 + R_2}(V_{Z2} + V_D)$$

$$= \frac{5}{10+5} \times 15$$

$$= +5\ V$$

11. **Ans. (a)**

Solution:

$$(7E)_H \rightarrow 0111\ \ 1110$$

$$(5F)_H \rightarrow 0101\ \ 1111$$

$$0010\ \ 0001 \rightarrow (21)_H$$

$$(21)_H \times (10)_H = (0210)_H$$

12. **Ans. (a)**

Solution:

Hysteresis loop of Schmitt trigger follow the characteristic shown in following figure.

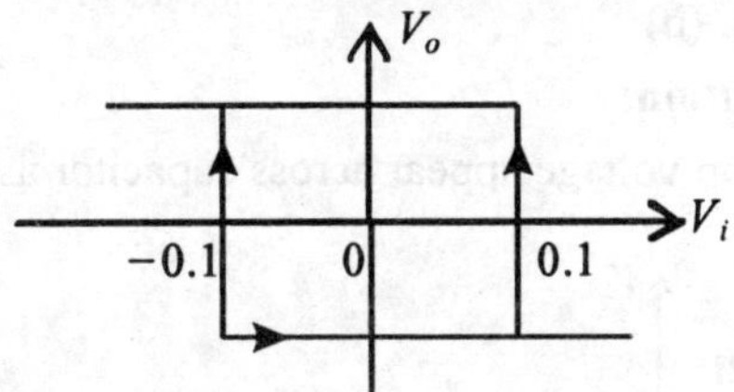

$$0.1 = 5 \times \sin(1000\,\pi\, t)$$

$$\Rightarrow 1000\,\pi t = \sin^{-1}\frac{0.1}{5} = 0.02 \text{ rad}$$

$$\Rightarrow \quad t = \frac{0.02}{1000\,\pi}$$

$$= 6.37 \times 10^{-6}$$

$$= 6.37\ \mu s$$

13. **Ans. (b)**

Solution:

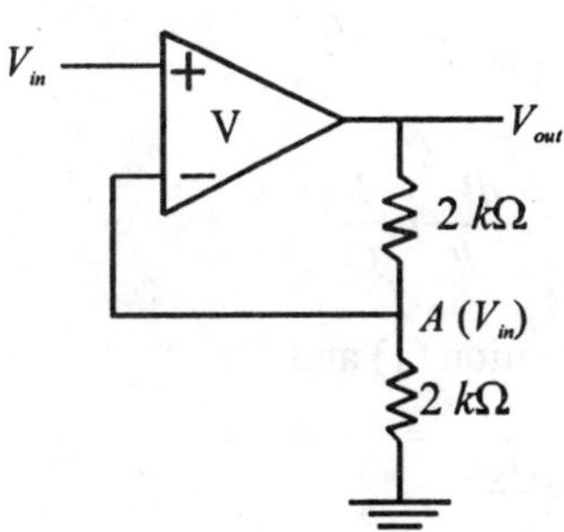

KCL At node A

$$\frac{V_{out} - V_{in}}{2\,k} = \frac{V_{in}}{1\,k + 500\,\Omega + 5000\,\Omega} = \frac{V_{in}}{2\,k}$$

$$V_{out} = 2\,V_{in}$$

$$\Rightarrow \quad \frac{V_{out}}{V_{in}} = 2$$

Another Method

The given amplifier is non-inverting amplifier whose gain is given

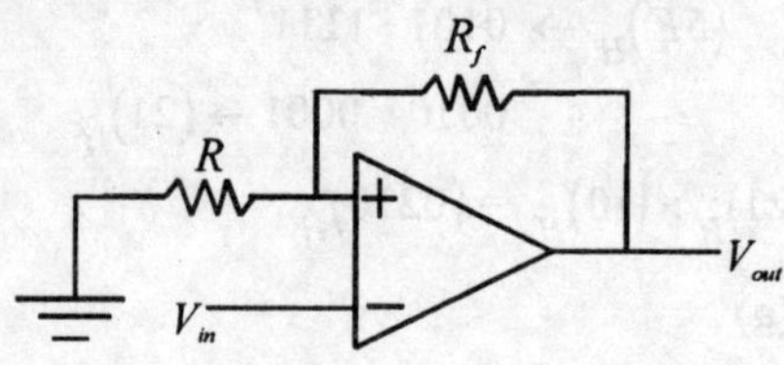

$$A_{vf}=1+\frac{R_f}{R}=\left(1+\frac{2}{2}\right)=2$$

14. Ans. (b)

Solution:

Ramp voltage appear across capacitor as shown below

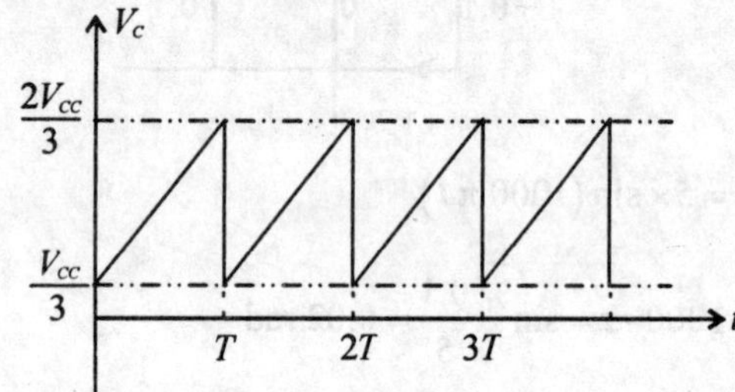

From the figure

$$\frac{dV_c}{dt}=\frac{\frac{2V_{CC}}{3}-\frac{V_{CC}}{3}}{T}$$

$$=\frac{V_{CC}}{3T} \quad ...(1)$$

It is know as $I=C\frac{dV_c}{dt}$

$$\frac{dV_c}{dt}=\frac{I}{C} \quad ...(2)$$

From equation (1) and (2)

$$\frac{V_{CC}}{3T}=\frac{I}{C}$$

$$=\frac{5\times10^{-3}}{0.1\times10^{-4}}$$

or $\frac{V_{CC}}{3T}=5\times10^4$

or $f=\frac{1}{T}=\frac{3\times5\times10^4}{5}$

$=30\ kHz$

15. Ans. (c)

Solution:

$$V_{UT}=\frac{R_2}{R_1+R_2}(V_{Z1}+V_D)+\frac{R_1}{R_1+R_2}V_R$$

$$V_{LT}=\frac{R_2}{R_1+R_2}(V_{Z2}+V_D)+\frac{R_1}{R_1+R_2}\cdot V_R$$

By gain

$$\Rightarrow 0=\frac{-R_2}{R_1+R_2}(6+0.7)+\frac{R_1}{R1+R_2}(1.5)$$

$$\Rightarrow \frac{R_1}{R_2}=\frac{6.7}{1.5}=4.67$$

16. Ans. (c)

Solution:

$$V_{UT}=\frac{R_2}{R_1+R_2}(V_{Z1}+V_D)+\frac{R_1}{R_1+R_2}\cdot V_R$$

$$=\frac{1}{1+4.67}(6.7)+\frac{4.67}{4.67+1}(1.5)$$

$=2.42$ Volts

17. Ans. (a)

Solution:

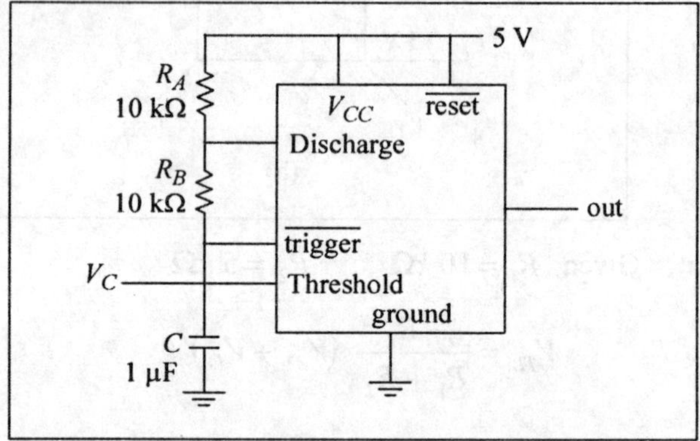

Capacitor C charges toward V_{CC} through R_A and R_B until in raises upto $2V_{CC}/B$.

V C
2/3 V CC
1/3 V CC
0
t

This voltage is the threshold voltage of pin 6, which drives comparator 1 to trigger the flip-flop so that the output at pin 3 goes low. In addition, the discharge transistor is driven on . Causing the output at pin 7 to discharge the capacitor through resistor R_B. But R_B is shorted out.

Capacitor voltage fall sharply as discharge time is zero. This process continue and option is (a)

18. Ans. (b)

19. Ans. (b)

Solution:

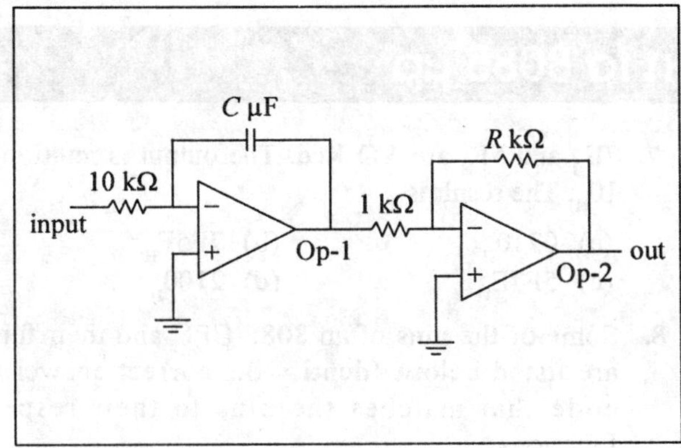

Output should be

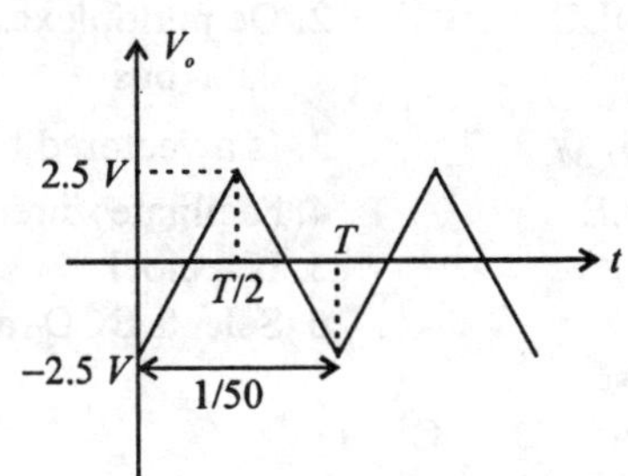

Slope in +ve half is

$$\frac{dV_o}{dt} = \frac{5}{(T/2)}$$

$$= \frac{10}{T}$$

Or, $10f = 10 \times 50$

$$= 500\ V/\text{sec} \qquad \dots(1)$$

Reverse action of integrator

$$-RC\frac{dV_1}{dt} = 5\ V$$

$$\frac{dV_1}{dt} = \frac{-5}{RC}$$

$$\frac{dV_1}{dt} = \frac{-5 \times 10^{-4}}{C} \qquad \dots(2)$$

From second op-amplifiers action

$$V_o = -RV_1$$

$$\frac{dV_o}{dt} = -R\frac{dV_1}{dt} \qquad \dots(3)$$

$$= \frac{R}{C} \times 5 \times 10^{-4} \qquad \dots(4)$$

From equations (1), (2), (3) and (4)

$$\frac{R}{C} \times 5 \times 10^{-4} = 500$$

or $\dfrac{R}{C} = 10^6$

or $\dfrac{R}{C \times 10^{-6}} = 1$

or, $\dfrac{R}{C(\mu F)} = 1$

20. Ans. (b)

Solution:

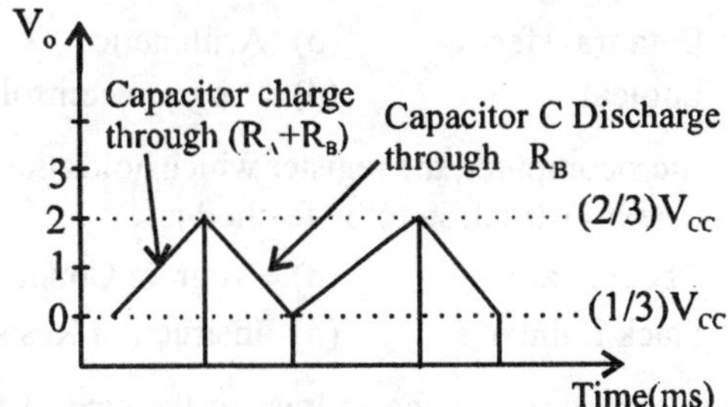

Charging and discharging level of capacitor will be voltage across it

This is eqal to $\frac{1}{3}V_{CC}$ and $\frac{2}{3}V_{CC}$. Thus 3V to 6V is the voltage V_C across the capacitor.

9 Microprocessor

9.1 Introduction to 8085 μp

1. In a microcomputer, wait states are used to
   (a) Make the processor wait during a DMA operation
   (b) Make the processor wait during an interrupt processing
   (c) Make the processor wait during a power shunt-down
   (d) Interface slow peripherals to the processor

2. An 'Assembler' for a microprocessor is used for
   (a) Assembly of processors in production line
   (b) Creation of new programs using different modules
   (c) Translation of a program from assembly to machine language
   (d) Translation of a higher level language into English text.

3. An instruction used to reset the carry Flag in a computer can be classified as
   (a) Data transfer (b) Arithmetic
   (c) Logical (d) Program control

4. In a microcomputer, the register which holds the address of the next instruction to be fetched is
   (a) Accumulator (b) Program Counter
   (c) Stack Pointer (d) Instruction Resistor

5. In a microprocessor, the address of the next instruction to be executed is stored in
   (a) Stack pointer
   (b) Address latch
   (c) Program counter
   (d) General purpose register

6. When a program is being executed in an 8085 microprocessor, its program counter contains
   (a) The number of instuctions in the current that have already been executed
   (b) The total number of instructions in the program being executed
   (c) The memory address of the instructions that is being currently executed
   (d) The memory address of the instructions that is to be executed next

7. $7E_H$ and $5F_H$ are XO Red. The output is multiplied by 10_H. The result is
   (a) 0210_H (b) $7E5F_H$
   (c) $5F7E_H$ (d) 2100_H

8. Some of the pins of an 8085 CPU and their function are listed below. Identify the correct answer using code that matches the pins to their respective functions.

A. RST 7.5 — 1. Selects IO or memory
B. HOLD — 2. De multiplexes the address and data bus
C. $IO/\overline{M}$ — 3. Is a vectored interrupt
D. ALE — 4. Facilitates direct memory access
5. Is a clock
6. Selects BCD mode of operation

Codes:

	A	B	C	D
(a)	3	2	1	4
(b)	4	1	5	3
(c)	3	4	1	2
(d)	2	3	6	1

9. In an 8085 microprocessor, which one of the following is the correct sequence of the machine cycles for the execution of the DCR M instruction?
   (a) Op-code fetch
   (b) Op-code fetch, memory read, memory write
   (c) Op-code fetch, memory read
   (d) OP-code fetch, memory write, memory read

10. The stack pointer in a microprocessor is a register containing
    (a) The address of the next operand
    (b) The current size of the stack
    (c) The address of the top of the stack
    (d) The address for storing the result of arithmetic operations.

11. In a microprocessor with 16 address and 12 data lines, the maximum number of opcodes is
    (a) 2^6 (b) 2^8
    (c) 2^{12} (d) 2^{16}

12. In 8085 microprocessor, CY flag may be set by the instruction
    (a) SUB (b) INX
    (c) CMA (d) ANA

13. Microprocessor 8085 regains control of the bus
    (a) Immediately after HOLD goes low
    (b) Immediately after HOLD goes high
    (c) After half-clock cycle after HLDA goes low
    (d) After half-clock cycle after HLDA goes high

14. In an INTEL 8085 microprocessor the ADDRESS – DATA bus and the DATA bus are
    (a) Non multiplexed
    (b) Multiplexed
    (c) Duplicated
    (d) Same as CONTROL bus

9.2 Instruction Set

1. When a CPU is interrupted, it
   (a) Stops execution of instructions
   (b) Acknowledges interrupt and branches to a subroutine
   (c) Acknowledges interrupt and continues
   (d) Acknowledges interrupt and wait for the next instruction from the interrupting device

2. The total number of memory accesses involved (inclusive of the op-code fetch) when an 8085 processor executes the instruction LDA 2003 is
   (a) 1 (b) 2
   (c) 3 (d) 4

3. The contents of Register (*B*) and Accumulator (*A*) of 8085 microprocessor are 49 H and 3 A H, respectively. The contents of *A* and the status of carry flag (*CY*) and sing flag (*S*) after executing SUB *B* instruction are
   (a) $A = F1, CY = 1, S = 1$
   (b) $A = 0F, CY = 1, S = 1$
   (c) $A = F0, CY = 0, S = 0$
   (d) $A = 1F, CY = 1, S = 1$

4. In an 8085 microprocessor, the instruction CMP *B* has been executed while the content of the accumulator is less than that of register *B*. As a result
   (a) Carry flag will be set but zero flag will be reset
   (b) Carry flag will be reset but zero flag will be set
   (c) Both Carry flag and Zero flag will be reset
   (d) Both Carry flag and Zero flag will be set

5. The number of memory cycles required to execute the following 8085 instruction
   (I) LDA 3000 H
   (II) LXI D, F0F 1 H
   would be
   (a) 2 for (I) and 2 for (II)
   (b) 4 for (I) and 3 for (II)
   (c) 3 for (I) and 3 for (II)
   (d) 3 for (I) and 4 for (II)

6. In a microcomputer, the register which holds the address of the next instruction to be fetched is:
   (a) Accumulator (b) Program Counter
   (c) Stack Pointer (d) Instruction Resistor

7. If the HLT instruction of a 8085 microprocessor is executed
   (a) The microprocessor is disconnected from the system bus till Reset is pressed.
   (b) The microprocessor enters into a Halt state and the buses are tri-stated
   (c) The microprocessor halts execution of the program and returns to monitor
   (d) The microprocessor reloads the program from the locations 0024 and 0025 H.

8. In an 8085 microprocessor, after the execution of XRA A instruction
   (a) The carry flag is set
   (b) The accumulator contains FFH
   (c) The zero flag is set
   (d) The accumulator contents are shifted left by one bit

9. The 8085 assemble language instruction that stores the content of *H* and *L* registers into the memory location 2050_H and 2051_H, respectively, is
   (a) SPHL 2050_H (b) SPHL 2051_H
   (c) SHLD 2050_H (d) STAX 2050_H

10. In an 8085 A microprocessor based system it is desired to increment the contents of memory location whose address is available in (*D*, *E*) register pair and store the result in same location. The sequence of instruction is
    (a) XCH G
INR M
    (b) XCH G
INX H
    (b) INX D
XCH G
    (d) INR M
XCH G

11. When a "CALL Addr" instruction is executed, the CPU carries out the following sequential operations internally

Note:

(*R*) means content of register *R*

(*R*) means content of memory location pointed to by *R*

PC means Program counter

SP means Stack Pointer

(a) (*SP*) incremented
(*PC*) ← Addr
((*SP*)) ← (*PC*)

(b) (*PC*) ← Addr
((*SP*)) ← (*PC*)
(*SP*) incremented

(c) (*PC*) ← Addr
(*SP*) incremented
((*SP*)) ← (*PC*)

(d) ((*SP*)) ← (*PC*)
(*SP*) incremented
(*PC*) ← Addr

12. Three devices *A*, *B* and *C* have to be connected to a 8085 microprocessor. Device A has highest priority and device *C* has the lowest priority. In this context which of the following is correct assignment of interrupt inputs?

(a) A uses TRAP, B uses RST 5.5 and C uses RST 6.5
(b) A uses RST 7.5, B uses RST 6.5 and C uses RST 5.5
(c) A uses RST 5.5, B uses RST 6.5 and C uses RST 7.5
(d) A uses RST5.5, B uses RST 6.5 and C uses TRAP

13. The contents of the accumulator in an 8085 microprocessor is altered after the execution of the instruction

(a) CMPC (b) CPI 3 A
(c) ANI 5 C (d) ORA A

14. Find the correct match among the following pairs in the context of a 8085 microprocessor:

List I	List II
A. DAA	1. Program control instruction
B. LXI	2. Data movement instruction
C. RST	3. Interrupt instruction
D. JMP	4. Arithmetic instruction

Codes:

	A	B	C	D
(a)	1	2	3	4
(b)	4	2	3	1
(c)	4	3	2	1
(d)	2	4	3	1

15. For an 8085 microprocessor, the Stack Pointer (SP) and Program Counter (PC) registers contain the number F000 and 2400 in Hex, respectively. The contents of the register after execution of the instruction CALL E000 would be

a. PC : F003 SP : 2400
b. PC : E000 SP. : 2400
c. PC : E000 SP. : 2401
d. PC : E000 SP. : 23 FE

16. A memory mapped I/O device has an address of 00F0H. Which of the following 8085 instruction outputs the content of the accumulator to the I/O device?

(a) LXI H, 00F0H
MOV M, A

(b) LXI H, 00F0H
OUT M

(c) LXI H, 00F0H
OUT F0H

(d) LXI H, 00F0H
MOV A, M

9.3 Memory Mapping

1. In an 8085 microprocessor system with mapped *I/O*,

1. *I/O* devices have 16-bit addresses
2. *I/O* devices are addresses using IN and OUT instructions
3. There can be a maximum of 256 input devices and 256 output devices
4. Arithmetic and logic operations can be directly performed with the *I/O* data.

Of these

(a) 1, 2 and 3 are correct (b) 2 and 3 are correct
(c) only 1 is correct (d) 1 and 4 are correct

2. In the circuit shown, the device connected to Y5 can have address in the range

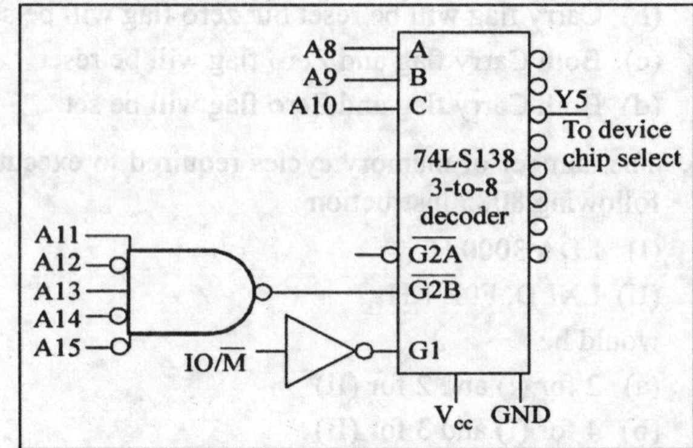

(a) 2000 – 20FF (b) 2DD00 – 2DFF
(c) 2E00 – 2EFF (d) FD00 – FDFF

3. The decoding circuit shown in figure has been used to generate the active low chip select signal for a microprocessor peripheral. (The address lines are designated as *A*0 to *A*7 for *I-O* addresses).
The peripheral will correspond to IO addresses in the range

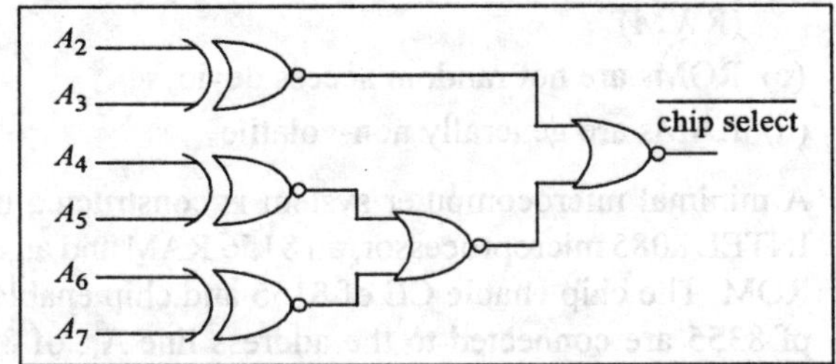

(a) 60H to 63H (b) *A*4H to *A*7H
(c) 30H to 33H (d) 70H to 73H

4. What memory address range is NOT represented by chip # 1 and chip # 2 in the figure? A_0 to A_{15} in this figure are the address lines and CS means Chip Select

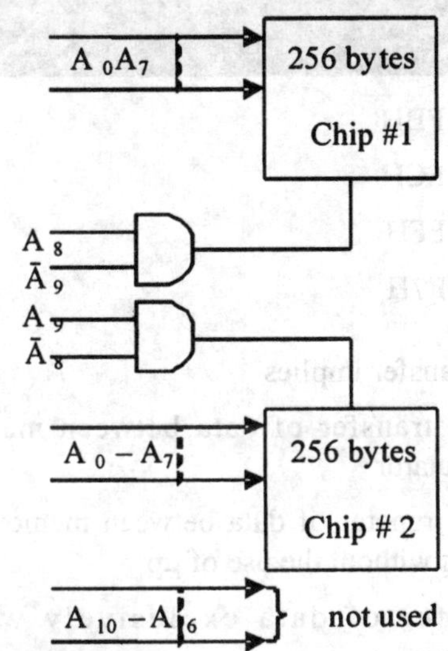

(a) 0100-02FF
(b) 1500-16FF
(c) F900-FAFF
(d) F800-F9FF

5. A memory system of size 16 K bytes is required to be designed using memory chips which have 12 address lines and 4 data lines each. The number of such chips required to design the memory system is
(a) 2 (b) 4
(c) 8 (d) 16

6. If $CS = \bar{A}_{15} A_{14} A_{13}$ is used as the chip select logic of a 4 K RAM and 8085 in a system, then it memory range will be
(a) 3000 H – 3 FFF H
(b) 7000 H – 7 FFF H
(c) 5000 H – 5 FFF H and 6000 H – 6 FFFH
(d) 6000 H – 6 FFF H and 7000 H – 7 FFFH

7. An 8085 microprocessor based system uses a 4K × 8 bit RAM whose starting address is AA00 H. The address of the last byte in this RAM is
(a) 0FFFH (b) 1000 H
(c) B9FFH (d) BA00H

8. The logic circuit used to generate the active low chip select (CS) by an 8085 microprocessor to address a peripheral is shown in figure. The peripheral will respond to address in the range

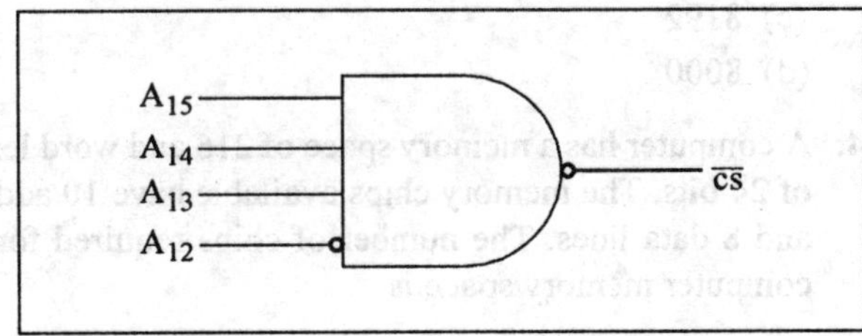

(a) E000-EFFF (b) 000E-FFFE
(c) 1000-FFFF (d) 0001-FFF1

9. A memory system has a total memory of 8 chips, each with 12 address lines and 4 data lines. The total size of the memory system is
(a) 6 kbytes (b) 32 kbytes
(c) 48 kbytes (d) 64 kbytes

10. A ROM is interfaced to an 8085 CPU as indicated in figure. The address range occupied by the ROM is

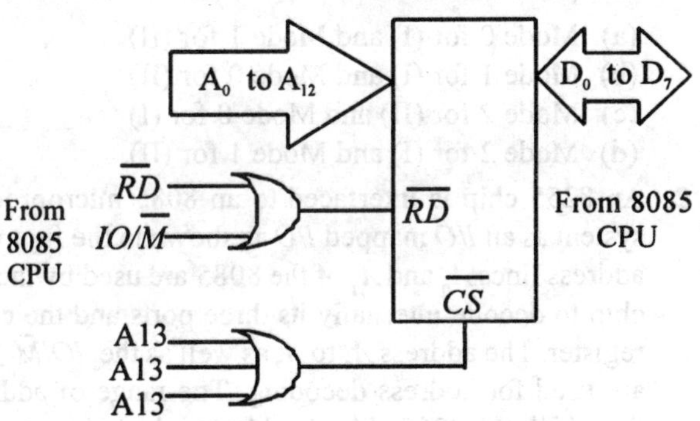

(a) 0000 - 0FFF (b) 0000 - 1FFF
(c) 0000 - 2 FFF (d) 8000 - 9FFF

11. A 2*K* × 8 bit RAM is interfaced to an 8-bit microprocessor. If the address of the first memory location in the RAM is 0800H, the address of the last memory location will be.
(a) 1000 H (b) 0FFFH
(c) 4800H (d) 47FFH

12. An 8K × 8 bit RAM is interfaced to an 8085 microprocessor. In a fully decoded scheme if the address of the last memory location of this RAM is 4FFFH, the address of the first memory location of the RAM will be
 (a) 1000 H
 (b) 2000 H
 (c) 3000 H
 (d) 4000 H

13. An 8-bit microcontroller has an external RAM in the memory map from 8000H to 9FFFH. Number of bytes this RAM can store is
 (a) 8193
 (b) 8191
 (c) 8192
 (d) 8000

14. A computer has a memory space of 216 and word length of 24 bits. The memory chips available have 10 address and 8 data lines. The number of chips required for the computer memory space is
 (a) 192
 (b) 256
 (c) 512
 (d) 1024

15. Indicate which one of the following statements is correct:
 (a) Static RAMs are faster that dynamic RAMs
 (b) RAMs connot be used to realize Read Write Memory (RWM)
 (c) ROMs are not random access devices
 (d) RAMs are generally non-volatile

16. A minimal microcomputer system is constructed using INTEL 8085 microprocessor, an 8156 RAM and an 8355 ROM. The chip enable CE of 8156 and chip enable CE pf 8355 are connected to the address line A_{12} of 8085. The address of port A of the 8156 chip is.
 (a) 21 H
 (b) 12 H
 (c) 11 H
 (d) 20 H

9.4 Interfacing

1. The 8255 Programmable Peripheral Interface is used as described below
 I. An A/D converter is interfaced to a microprocessor through an 8255. The Port C causes data to be stored into Port A.
 II. Two computers exchange data using a pair of 8255s. Port A works as bi-directional data port supported by appropriate handshaking signals.

 The appropriate modes of operation of the 8255 for (I) and (II) would be
 (a) Mode 0 for (I) and Mode 1 for (II)
 (b) Mode 1 for (I) and Mode 0 for (II)
 (c) Mode 2 for (II) and Mode 0 for (I)
 (d) Mode 2 for (I) and Mode 1 for (II)

2. An 8255 chip is interfaced to an 8085 microprocessor system as an *I/O* mapped *I/O* as shown in the figure. The address lines A_0 and A_1 of the 8085 are used by the 8255 chip to decode internally its three ports and the control register. The address A_3 to A_7 as well as the $IO/\bar{M}$ signal are used for address decoding. The range of addresses for which the 8255 chip would get selected is

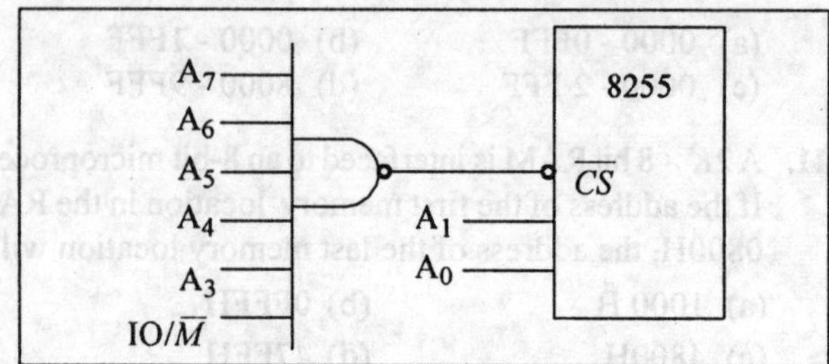

 (a) F8H – FBH
 (b) F8H – FCH
 (c) F8H – FFH
 (d) F0H – F7H

3. A 'DMA' transfer implies
 (a) Direct transfer of data between memory and accumulator
 (b) Direct transfer of data between memory and *I/O* devices without the use of μp
 (c) Transfer of data exclusively within μp registers
 (d) A fast transfer of data between μp registers

4. An I/O processor control the flow of information between
 (a) Cache memory and *I/O* devices
 (b) Main memory and *I/O* devices
 (c) Two *I/O* devices
 (d) Cache and main memories

5. An *I/O* peripheral device shown in Fig. (b) below is to be interfaced to an 8085 microprocessor. To select the *I/O* address range D4 H – D7 H, its chip – select ($\overline{CS}$) should be connected to the output of the decoder shown in Fig. (a) below.

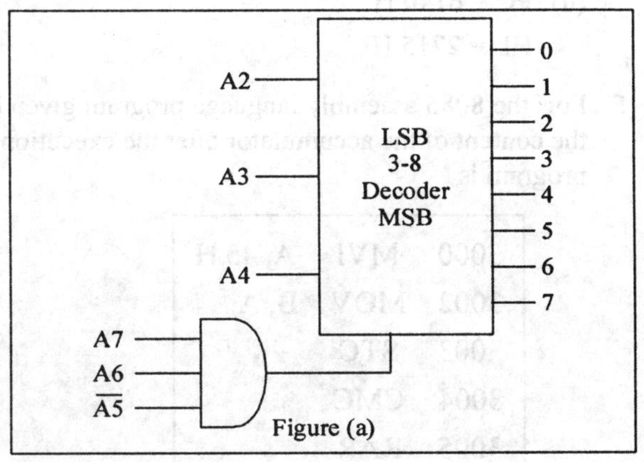

Figure (a)

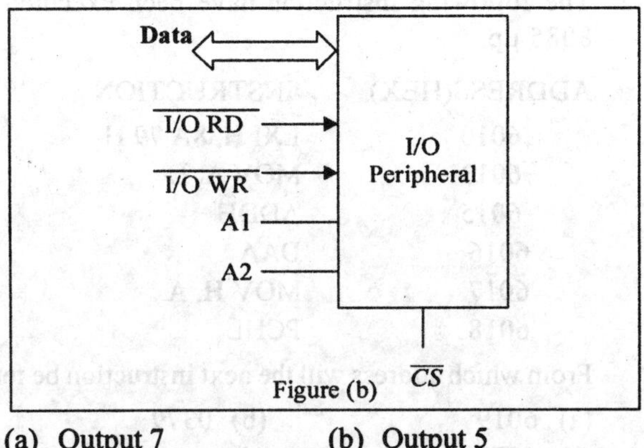

Figure (b)

(a) Output 7 (b) Output 5
(c) Output 2 (d) Output 0

9.5 Interrupts

1. In an 8085 μP system, the RST instruction will cause an interrupt
(a) Only in an interrupt service routine is not being executed
(b) Only if a bit in the interrupt mask is made 0.
(c) Only if interrupts have been enabled by an EI instruction
(d) None of above

2. In an 8085 microprocessor, the RST6 instruction transfers the program execution to the following location.
(a) 30 H (b) 24 H
(c) 48 H (d) 60 H

3. The number of hardware interrupts (which require an external signal to interrupt) present in an 8085 microprocessor are:
(a) 1 (b) 4
(c) 5 (d) 13

4. In a microprocessor, the service routine for a certain interrupt starts from a fixed location of memory which cannot be externally set, but the interrupt can be delayed or rejected. Such an interrupt is
(a) Non-maskable and non-vectored
(b) Maskable and non-vectored
(c) Non-maskable and vectored
(d) Maskable and vectored

5. Which of the following in not a vectored interrupt?
(a) TRAP (b) INTR
(c) RST7.5 (d) RST3

6. The vectored address corresponding to the software interrupt common RST7 in 8085 microprocessor is
(a) 0017H (b) 0027H
(c) 0038H (d) 0700H

7. The time-period of a square wave in the audio frequency range is measured using an 8085 microprocessor by feeding the square wave to one of the four interrupts, namely, RST 7.5, RST 6.5, RST 5.5, or INTR. The algorithm used starts a timer at the beginning of a time period, stops the timer at the beginning of the next time period and reads the timer values for time measurement. Which of the following interrupts should be selected for this application?
(a) INTR (b) RST 5.5
(c) RST 6.5 (d) RST 7.5

8. Identify the software interrupt instruction in 8085 microprocessor .
(a) INT (b) RST 5
(c) RST 7.5 (d) RST 6.5

9.6 Assembly Language Programming

1. The following sequence of instructions are executed by an 8085 microprocessor:

1000	LXI SP, 27 FF
1000	CALL
1006	POP H

The contents of the stack pointer (SP) and the HL, register pair on completion of execution of these instruction are
(a) $SP = 27\ FF, HL = 1003$
(b) $SP = 27\ FD, HL = 1003$
(c) $SP = 27\ FF, HL = 1006$
(d) $SP = 27\ FD, HL = 1006$

2. The following instruction have been executed by an 8085 μp

ADDRESS(HEX)	INSTRUCTION
6010	LXI H, 8A 79 H
6013	MOV A, L
6015	ADDH
6016	DAA
6017	MOV H, A
6018	PCHL

From which address will the next instruction be fetched?

(a) 6019 (b) 0379
(c) 6979 (d) one of the above

3. Consider the following assembly language program:

```
        MVI    B, 87H
        MOV    A, B
START:  JMP    NEXT
        MVI    B, 00H
        OUT    PORT1
        HLT
NEXT:   XRA    B
        JP     START
        OUT    PORT 2
        HLT
```

The execution of the above program in an 8085 microprocessor will result in

(a) An output of 87H at PORT1
(b) An output of 87H at PORT2
(c) Infinite looping of the program execution with accumulator data remaining 00H
(d) Infinite looping of the program execution with accumulator data alternating between 00H and 87 H

4. An 8085 executes the following instructions

2710 LXI H, 30A0H
2713 DAD H
2714 PCHL

All addresses and constants are in Hex. Let PC be the contents of the program counter and HL be the contents of the HL register pair just after excuting PCHL.

Which of the following statement is correct

(a) PC = 2715 H
HL = 30A0 H
(b) PC = 30A0 H
HL = 2715 H
(c) PC = 6140 H
HL = 6140 H
(d) PC = 6140 H
HL = 2715 H

5. For the 8085 assembly language program given below, the content of the accumulator after the execution of the program is

```
3000  MVI  A, 45 H
3002  MOV  B, A
3003  STC
3004  CMC
3005  RAR
3006  XRA  B
```

(a) 00 H (b) 45 H
(c) 67 H (d) E 7 H

6. An 8085 assembly language program is given below. Assume that the carry flag is initially unset. The content of the accumulator after the execution of the program is

```
MVI A, 07 H
RLC
MOV B, A
RLC
RLC
ADD B
RRC
```

(a) 8CH (b) 64H
(c) 23 H (d) 15 H

Common data for Questions 7 and 8

Consider an 8085 microprocessor

The following program starts at location 0100H.

```
LXI  SP, 00FF
LXI  H, 0701
MVI A, 20H
SUB  M
```

7. The content of an accumulator when the program counter reaches 0109 H is

(a) 20 H (b) 02 H
(c) 00 H (d) FF H

8. If in addition following code exists from 019 onwards,

```
ORI  40 h
ADD  M
```

What will be the result in the accumulator after the last instruction executed ?

(a) 40 H (b) 20 H
(c) 60 H (d) 42 H

9. Following is the segment of an 8085 assembly language program.

LXI SP, EFFFH
CALL 3000H
:
:
3000H : LXI, 3CF4 H
PUSH PSW
SPHL
POP PSW
RET

On completion of RET execution, the contents of SP is

(a) 3 CFO H (b) 3CF 8 H
(c) EFFD H (d) EFFF H

Statement for Linked Answer Questions 10 and 11

An 8085 assembly language program is given below.

Line 1: MVI A,B5H
2: MVI B,0EH
3: XRI 69H
4: ADD B
5: ANI 9BH
6: CPI 9FH
7: STA 3010H
8: HLT

10. The contents of the accumulator just after the execution of the ADD instruction in line 4 will be

(a) C3H (b) EAH
(c) DCH (d) 69H

11. After execution of line 7 of the program, the status of the CY and Z flags will be

(a) CY = 0, Z = 0 (b) CY = 0, Z =1
(c) CY =1, Z = 0 (d) CY = 1, Z = 1

12. Consider the sequence of 8085 instruction given below.

LXI H, 9258,
MOV A, M,
CMA,
MOV M, A

Which one of the following is performed by this sequence

(a) Contents of location 9258 are moved to the accumulator
(b) Contents of location 9258 are compared with the contents of the accumulator
(c) Contents of location 8529 are complemented and stored in location 8529.
(d) Contents of location 5892 are complement and stored in location 5892

13. It is desired to multiply the numbers $0AH$ by $0BH$ and store the result in the accumulator. The numbers are available in registers B and C respectively. A part of the 8085 program for this purpose is give below

MVI A, 00H
Loop ; --------------------
....
....
HLT END

The sequence of instruction to complete the program would be

(a) JNZ LOOP, ADD B, DCR C
(b) ADD B, JNZ LOOP, DCR C
(c) DCR C, JNZ LOOP, ADD B
(d) ADD B, DCR C, JNZ LOOP

14. The number of memory cycles required to execute the following 8085 instruction

(I) LDA 3000 H
(II) LXI D, F0F 1 H

would be

(a) 2 for (I) and 2 for (II)
(b) 4 for (I) and 3 for (II)
(c) 3 for (I) and 3 for (II)
(d) 3 for (I) and 4 for (II)

15. An Intel 8085 processor is executing the program given below.

MVI A, 10 H
MVI B, 10 H
BACK: NOP
ADD B
RLC
JNC BACK
HLT

The number of times that the operation NOP will be executed is equal to

(a) 1 (b) 2
(c) 3 (d) 4

16. A software delay subroutine is written as given below:

DELAY: MVI H, 255 D
MVI L, 255 D
LOOP: DCR L
JNZ LOOP
DCR H
JNZ LOOP

How many times DCRL instruction will be executed?

(a) 255 (b) 510
(c) 65025 (d) 65279

17. An input device is interfaced with Intel 8085 microprocessor as memory mapped *I/O*. The address of the device is 2500 H. In order to input data from the device to accumulator, the sequence of instruction will be

(a) LXI H, 2500 H
MOV A, M

(b) LXI H, 2500 H
MOV M, A

(c) LHLD 2500 H
MOV A, M

(d) LHLD 2500 H
MOV M, A

18. The contents (in Hexadecimal) of some of the memory locations in an 8085A based system are given below:

Address	Contents
..	..
26 FE	00
26 FF	01
2700	02
2701	03
2702	04
..	..

The contents of stack pointer (SP), program counter (PC) and (HL) are 2700H, 2100H and 0000H, respectively. When the following sequence of instructions are executed,

```
2100 H:    DAD SP
2101 H:    PCHL
```

The contents of (SP) and (PC) at the end of execution will be

(a) $(PC) = 2102H, (SP) = 2700H$

(b) $(PC) = 2700H, (SP) = 2700H$

(c) $(PC) = 2800H, (SP) = 26FEH$

(d) $(PC) = 2A02H, (SP) = 2702H$

19. In an 8085 microprocessor, the contents of the Ac cumulator, after the following instructions are executed will become.

```
XRA A
MVIB F0H
SUB B
```

(a) 01 H (b) 0F H
(c) F0 H (d) 10 H

20. The following program is written for an 8085 microprocessor to add two bytes located at memory addresses 1FFF and 1FFF

```
LXI    H, 1FFE
MOV    B, M
INR    L
MOV    A, M
ADD    B
INR    L
MOV    M, A
XOR    A
```

On completion of the execution of the program, the result of addition is found

(a) In the register A

(b) At the memory address 1000

(c) At the memory address 1F00

(d) At the memory address 2000

21. A portion of the main program to call a subroutine SUB in an 8085 environment is given below.

```
    :
    :
LP: LXI   D, DISP
    CALL  SUB
    :
    :
```

It is desired that control be returned to LP + DISP + 3 when the RET instruction is executed in the subroutine. The set of instructions that precede the RET instruction in the subroutine are

(a) POP D
DAD H
PUSH D

(b) POP H
DAD D
INX H
INX H
INX H
PUSH H

(c) POP H
DAD D
PUSH H

(d) XTHL
INX D
INX D
INX D
XTHL

22. If the following program is executed in a microprocessor, the number of instruction cycles it will take from START to HALT is

```
START   MVIA, 14H ;  Move 14H to register A
SHIFT   RLC       ;  Rotate left without carry
JNZ     SHIFT     ;  Jump on non-zero to SHIFT
        HALT
```

(a) 4 (b) 8
(c) 13 (d) 16

23. In an 8085 system containing 8 KB of ROM and 8 KB of RAM, the ROM is selected when A_{15} is 0 and the

RAM is selected when A_{15} is 1.A_{13} and A_{14} are unused. The CPU executed the following program

```
Prog 1  MVI A,00H
        STA   8080H
        DCR   A
        STA   C080H
        RET
```

The content of memory location 8080H after the execution of the RETURN instruction is

(a) FFH (b) FEH
(c) 01H (d) 00H

24. The following 8085 instructions are executed sequentially.

```
PROG:   XRA    A
        MOV    L, A
        MOV    H, L
        INX    H
        DAD    H
```

After execution, the content of HL register pair is

(a) 0000H (b) 0101 H
(c) 0001H (d) 0002H

25. A microprocessor has an instruction XOR (r_1, r_2), which performs an Exclusive OR operation of register r_1, r_2 and stores the result in r_1. After the following instruction are executed,

XOR (r_2, r_1)
XOR (r_1, r_2)
XOR (r_2, r_1)

Which one of the following is true?

(a) Content of register r_1 is half sum of r_1 and r_2
(b) Content of register r_2 is half sum of r_1 and r_2
(c) Content of register r_1 and r_2 remain unaltered
(d) Content of register r_1 and r_2 are swapped

26. In a 8085 microprocessor the value of the stack pointer (SP) is 2010H and that of DE register pair is 1234H before the following code is executed. The value of the DE register pair after the following code is executed is

```
LXI H, 0000H
PUSH  H
PUSH  H
POP   B
DAD   SP
XCHG
```

(a) 200EH (b) 200CH
(c) 2010H (d) 1232H

27. An 8085 assembly language program is given as follows. The execution time of each instruction is given against the instruction in terms of T–state.

Instruction	T-states
MVI A, 0AH	7T
LOOP : MVIC, 05H	7T
DCR C	4T
DCR A	4T
JNZ LOOP	10T/7Y

The execution time of the program in terms of T–states is

(a) 247 T (b) 250 T
(c) 254 T (d) 257 T

28. A snapshot of the address, data an control buses of an 8085 microprocessor executing is given below:

Address	**2020 H**
Data	24 H
IO/$\overline{M}$	Logic high
$\overline{RD}$	Logic high
$\overline{WR}$	Logic low

The assembly language instructin begin execute is

(a) IN 24 H (b) IN 20 H
(c) OUT 24 H (d) OUT 20 H

29. 8-bit signed integers in 2's complement form are read into the accumulator of an 8085 microprocessor from an input port using the following assembly language program segment with symbolic addresses

```
BEGIN:  IN PORT
RAL
JNC     BEGIN
RAR
END:    HLT
```

The program

(a) Halts upon reading a negative number
(b) Halts upon reading a positive number
(c) Halts upon reading a zero
(d) Never halts

30. A part of a porgram written for an 8085 microprocessor is shown below. When the program execution reaches LOOP2, the value of register C will be

```
        SUB A
        MOV C, A
LOOPI:  INR A
        DAA
        JC LOOP2
        INR C
        JNC LOOP1
LOOP2:  NOP
```

(a) 63 H (b) 64 H
(c) 99 H (d) 100 H

31. The following is an assembly language program for 8085 microprocessors:

Address	Instruction Code	Mnemonic
1000H	3E 06	MVI A, 06H
1002H	C6 70	ADI 70II
1004H	32 07 10	STA 1007H
1007H	AF	XRA
1008H	76	HLT

When this program halts, the accumulator contains
(a) 00 H
(b) 06 H
(c) 70 H
(d) 96 H

32. In an 8085 processor, the main program calls the subroutine SUBI given below. When the program returns to the main program after executing SUBI, the value in the accumulator is

Address	Opcode	Mnemonic
2000	3E 00	SUB1: MVI A, 00h
2002	CD 05 20	CALL SUB2
2005	3C	SUB2: INR A
2006	C9	RET

(a) 00 (b) 01
(c) 02 (d) 03

33. The subroutine SBX given below is executed by an 8085 processor. The value in the accumulator immediately after the execution of the subroutine will be:

```
SBX : MVI A, 99 H
      ADI 11 H
      MOV C,A
      RET
```

(a) 00 H (b) 11 H
(c) 99 H (d) AA H

34. In an 8085 based system the subroutine TEST given below is called by another program. When the processor returns from the subroutine TEST , the value in the accumulator will be.

```
TEST   MOV A, # 00H
       CALL TESK
TESK   INR A
       RET
```

(a) 02 (b) 00
(c) FF (d) 20

9.7 Miscellaneous

1. Which one of the following statements regarding the INT (interrupt) and the BRQ (bus request) pins in a CPU is true?
(a) The BRQ pin is sampled after every instruction cycle, but the INT is sampled after every machine cycle
(b) Both INT and BRQ are sampled after every machine cycle
(c) The INT pin in sampled after every instruction cycle, but the BRQ is sampled after every machine cycle
(d) Both INT and BRQ are sampled after every instruction cycle.

2. An 8-bit DAC is interfaced with a microprocessor having 16 address lines $(A_0.....A_{15})$ as shown in the adjoining figure. A possible valid address for this DAC is

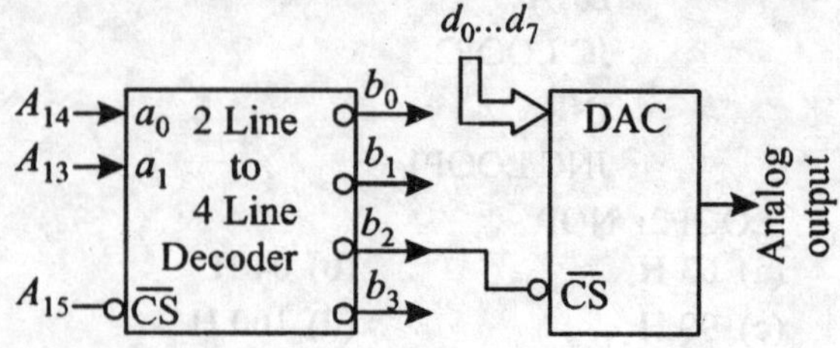

(a) 3000h
(b) 4FFFh
(c) AFFFh
(d) C000h

3. The number of objects crossing a window sequentially at variable speed is to be counted using an interrupt in the 8085 microprocessor. The objects are sensed by an optical source and a detector with associated signal conditioning circuit. The circuit produces a logic high output as long as the object is in front of the window, and the output is used to interrupt the processor. The duration of an object being in front of the window is in the range of 100 ms to 2 s. The processor takes 1 ms to process the input after an interrupt. The best choice of interrupt for an error free counting is
(a) RST 5.5
(b) RST 6.5
(c) RST 7.5
(d) INTR

4. Consider a system consisting of a microprocessor,

memory and peripheral devices connected by a common bus. During DMA data transfer, the microprocessor

(a) Only reads from the bus
(b) Only writes to the bus
(c) Both reads from and writes to the bus
(d) Neither reads from nor writes to the bus

5. The important feature of a microcontroller is that it has an on-chip

(a) Math co-processor
(b) Program memory
(c) Interface for I/O devices
(d) Hardware multiplier

6. An m-bit microprocessor has an m-bit

(a) Flag register
(b) Instruction register
(c) Data counter
(d) Program counter

7. The 14-bit timer of 8156 is loaded with the counter value of 07DOH. The timer input is connected to a clock with a frequency of 800 kHz. The timer is programmed to produce a continuous signature wave output. The frequency of the square wave output is

(a) 400 kHz
(b) 800 kHz
(c) 400 Hz
(d) 2000 kHz

Answers

9.1 Introduction to 8085 µp

1. Ans. (d)

Solution:

In a mp, wait states are used to interface slow peripherals to the processor. When READY signal is low the microprocessor waits for integral number of clock cycles. It is used to provide proper WAIT state when the microprocessor is communicating with a slow peripheral device.

2. Ans. (c)

Solution:

The assembler is a program that translates the mnemonics entered by the ASCII keyboard into the corresponding binary machine codes of the microprocessor. Each microprocessor has its own assembler because the mnemonics and machine codes are specific to the microprocessor being used, and each assembler has rules that must be followed by the programmer, i.e. the instructions written in assembly language are converted to machine language by assemblers.

3. Ans. (b)

Solution:

CY-Carry: If an *arithmetic operation* results in a carry, the CY flag is set: otherwise it is reset.

4. Ans. (b)

Solution: The program counter is an operational register which always holds the address of either the next instructions to be executed or the address of the next word of a multiword instruction that has not been completely fetched.

5. Ans. (c)

Solution:

The program counter is an operational register which always holds the address of either the next instructions to be executed or the address of the next word of a multiword instruction that has not been completely fetched.

6. Ans. (d)

Solution:

The program counter is an operational register which always holds the address of either the next instructions to be executed or the address of the next word of a multiword instruction that has not been completely fetched.

7. Ans. (a)

Solution:

$(7E)_H \rightarrow 0111 \quad 1110$

$(5F)_H \rightarrow 0101 \quad 1111$

$0111\ 1110 \oplus 0101\ 1111 = \underset{2}{\underset{\downarrow}{0010}}\ \ \underset{1}{\underset{\downarrow}{0001}}$

$0010 \quad 0001 \rightarrow (21)_H$

$(21)_H \times (10)_H = (0210)_H$

8. Ans. (c)

Solution: the correct match is given below

RST 7.5	Is a vectored interrupt Facilitates
HOLD	direct memory access DMA
IO/$\overline{M}$	Selects IO or memory i.e. if IO/$\overline{M}$ = 1 then it selects I/O and if IO/ $\overline{M}$= 0it selects memory
ALE	De multiplexes the address and data bus, i.e. ALE is used to demultiplex lower order address bus and data bus.

9. Ans. (b)

Solution:

DCR M→ This instruction reduces the content of memory by one, whose address is pointed by the content of HL register pair.

By default the first operation is op-code fetching, so the first operation is op-code fetch, then second the content of memory is read by the microprocessor, so second machine cycle is memory read and again it is written back to the same memory location after decrementing the content of memory by one, so third machine cycle is memory write.

10. Ans. (c)

Solution:

The stack pointer in a microprocessor is a register containing the address of the top of the stack

11. Ans. (c)

12. Ans. (a)

13. Ans. (b)

14. Ans. (b)

9.2 Instruction Set

1. Ans. (b)

Solution:

When CPU is interrupted, it acknowledges interrupt through $\overline{INTA}$ and branches to a subroutine.

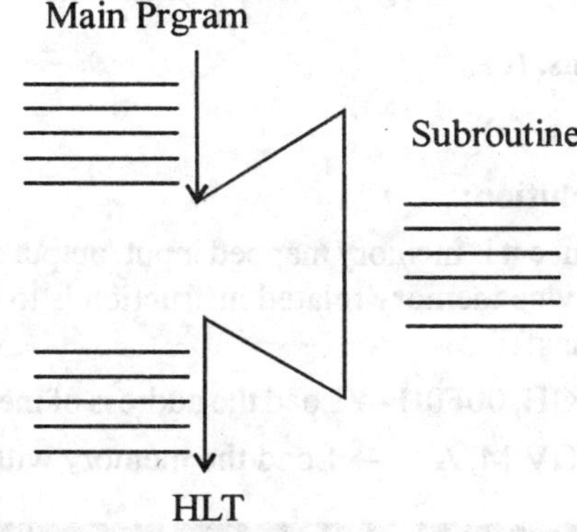

2. Ans. (d)

Solution:

Op-code fetch → 1 machine cycle,
Memory Read → 2 machine cycle,
Memory Read → 3 machine cycle,
Memory Read → 4 machine cycle.

Therefore, four memory accesses are involved in the execution of LDA 2003.

3. Ans. (a)

Solution:

A	0011	1010
2's complement of B	1011	0111
	01111	0001
	F	1

CY = 1 ; S = 1, ⇒ A = F1 H

4. Ans. (a)

Solution: CMP B: Contents of B and A are compared and result is stared in A that is indicated by flag $A < B$

$\therefore \quad CY = 1, Z = 0$

5. Ans. (b)

Solution:

LDA 3000 H → Fetch, (Address) read, read, read ⇒ $4M/C$

LXI F0F1H → Fetch data, read, read ⇒ 3M/C

6. Ans. (b)

Solution: The program counter is an operational register which always holds the address of either the next instructions to be executed or the address of the next word of a multiword instruction that has not been completely fetched

7. Ans. (b)

Solution:

If the HLT instruction of a 8085 microprocessor is executed the microprocessor enters into a Halt state and the buses are tri-stated

8. Ans. (c)

Soluton:

Let contents of A are 62 H

XRA A: Performs XOR operation the contents of A and Store result in A

$A \oplus A$: 01100010
01100010
00000000

or, $A = 00_H$

Contents of A are 00H after execution of XRA" X R A A" Instruction exectution clears the Accumulator and set zero flag

* XRA A instruction can be used to reset CY and AC flags and set Z flag.

9. Ans. (c)

Solution:

SHLD 2050 H

This instruction stands for 'Store HL register pari direct', i.e. the contents of H and L are stored at memory locations 2050 H and 2051 H, respectively

10. Ans. (a)

Solution:

XCHG; [DE] ↔ [HL], i.e. the address is now stored in HL

INR M; The contents of the memory location is incremented by 1

Exchange the content of HL - pair with content of DE pair.

11. Ans (d)

Solution:

'CALL addr' instruction is used to call a subroutine or the program sequence is transfered to the address specified by the operand. Before the control is transfered to the subroutine, the address of the next instruction of the main program is saved in the stack. The content of the stack pointer is deremented by two to indicate the new stack top. Then the program counter jumps to subroutine staring at address specified by the label.

Add r_1
CALL addr

Add r_2

PC – Add r_1

$((SP)) \leftarrow (PC)$

$(SP) \leftarrow (SP + 1)$ Add 2. incremented

$(PC) \leftarrow$ Add r.

12. Ans. (b)

Solution:

There are five interrupt inputs having a fixed multilevel structure. From highest to lowest they are TRAP, RST 7.5, RST 5.5, INTR

TRAP is usually reserved to handle catastrophic events such as power supply failure

I/O devices are connected to the other four interrupt inputs in such a way that the highest priority device is connected to RST 7.5, the next to RST 6.5, and so on.

13. Ans. (c)

Solution:

8085 has
- 1 accumulator A register
- 3 Secondary accumulators B, D, H
- 2 Secondary data counter C, E
- 1 Primary data counter L
- 1 Stack pointer SP
- 1 Program counter PC
- 1 Program status word for flags PSW

CMP C

Compare action taken; only flags are modified but the register/accumulator contents are not altered.

CPI 3 A

Compare accumulator A contents with immediate data 3A16. Accumulator contents are not altered only flags will be modified

ANI 5 C

Logical AND operation of immediate data with Accumulator A Accumulator contents are the results of logical AND operation of contents of Accumulator A and immediate data 5C16.

ORA A

Logical OR operation of the contents of accumulator with the contents of Register specified (in this case accumulator A) and the results are stored in accumulator A. Even though the results modify the contents of accumulator A, because of logical 'OR' ing of A with A results in no change in the contents of accumulator. Hence this is not the answer.

14. Ans. (b)

Solution: The correct match is given below

DAA → Arithmetic instruction

LXI → Date movement instruction

RST → Interrupt instruction

JMP → Program control instruction

15. Ans. (c)

16. Ans. (a)

Solution:

Since it is memory mapped input /output device therefore, device memory related instruction is to be used for data transfer

LXIH, 00F0H → Load the address of memory in HL pair

MOV M, A → Load the memory with content of A.

9.3 Memory Mapping

1. Ans. (d)

Solution:

Comparison of memory

Mapped input and peripherial input and output

Characteristic	Memory-Mapped I/O	Peripheral I/O
1. Device address	16 bit	8-bit = 2^8 = 256
2. Control signals for I/O	$\overline{MEMR}$ / $\overline{MWMW}$	$\overline{IOR}$ / $\overline{IOW}$
3. Instructions available	Memory-related instrucions such as STA ; LDA LDAX; STAX MOV M,R: ADD M; SUB M; ANA M: etc.	IN and OUT (Few)
4. Data transfer	Between any resistors and I/O	Only between I/O accumulator
5. Maximum No. Of I/Os possible	The memory map (64K) is shared between I/Os and system memory	The I/O map is independent of the memory map; 256 output devices can be connected
6. Execution Speed	13 T-staes (STA, LDA) 7 T-states (MOV, M,R)	10 T-states
7. Hardware requirements	More hardware is needed to decode 16-bit address	Less hardware is needed to decode 8-bit address
8. Other features	Arithmetic or logical operations can be directly performed With I/O data	Not availabl

Memory pointer (H, L rp) can be used to access memory mapped I/O addresses.

In memory mapped I/O

* The I/O devices are also treated as memory locations, under assumption they will be given 16-bit address,
* Microprocessor uses memory related instructions to communicate with I/O devices.
* Arithmetic and Logic operations can be directly performed wih the I/O data

2. Ans (b)

Solution:

$$\underbrace{A_{15}A_{14}A_{13}A_{12}}\ \underbrace{A_{11}A_{10}A_9A_8}\ \underbrace{A_7A_6A_5A_4}\ \underbrace{A_3A_2A_1A_0}$$

$$\underbrace{0\ 0\ 1\ 0}_{2}\ \underbrace{1\ 1\ 0\ 1}_{D}\ \underbrace{0\ 0\ 0\ 0}_{0}\ \underbrace{0\ 0\ 0\ 0}_{0}$$

$$\Downarrow \qquad \Downarrow \qquad \Downarrow \qquad \Downarrow$$

$$\underbrace{0\ 0\ 1\ 0}_{2}\ \underbrace{1\ 1\ 0\ 1}_{D}\ \underbrace{1\ 1\ 1\ 1}_{F}\ \underbrace{1\ 1\ 1\ 1}_{F}$$

* Here Y_5 output is selected, i.e. only posible when $ABC = 101$ or $A_{10} = 1, A_9 = 0, A_8 = 1$ are fixed

Only variables are $A_7 \ldots A_0$

3. Ans. (a)

Solution: A_0 and A_1 can take any value from 00 to 11

Chip select =

$$A_2\overline{A_3} + \overline{A_2}A_3 + \overline{A_4}\,\overline{A_5} + A_4A_5 + \overline{A_6}\,\overline{A_7} + A_6A_7$$

To get active low, chip select term should be zero

$$A_2 = A_3$$
$$A_4 \neq A_5$$
$$A_6 \neq A_7$$

This conditions are satisfied by address 60 H to 63 H.

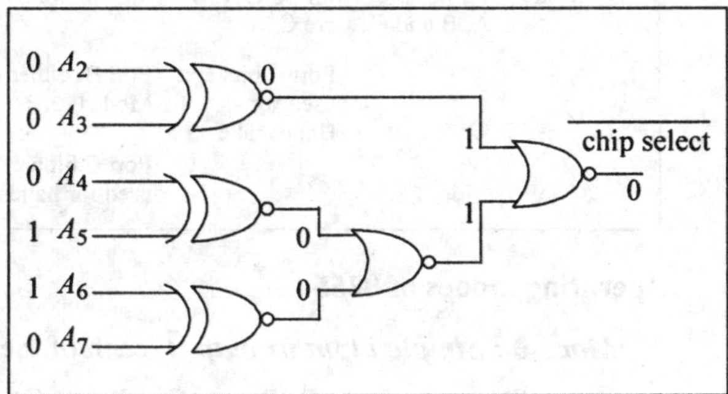

4. Ans. (d)

Solution: Address range of chip 1 and chip 2 →

$$\begin{array}{l} A_0 - A_7 \\ 00 - FF \end{array} \begin{cases} A_9A_8A_7 - A_0 \\ 00FF - 00 \\ \left.\begin{array}{l} 01FF - 00 \\ 10FF - 00 \end{array}\right\} \begin{array}{l} \overline{A}_9\ A_8 \\ A_9\ \overline{A}_8 \end{array} \\ 11FF - 00 \end{cases}$$

$A_8\bar{A}_9$ → selecting chip # 1 [Address range is 100 – 1FF]

$\bar{A}_8A_9$ → selecting chip # 2 [Address range is 200 – 2FF]

Address range is 100 – 2 FF

Since A $_{10}$ – A $_{15}$ are not used, the address range can be 100 – 2 FF (or) 500 – 6FF (or) 900 – AFF.

For the address space F800 – F9FF chip # 1 and chip # 2 are not selected.

5. Ans. (c)

Solution:

Word length

Bits = $2^{12} \times 4 = 16\ k$

Memory = $16\ k = 16 \times 8$ Bits

Chip required = $\dfrac{16k \times 8}{16k} = 8$

6. Ans. (d)

Solution:

Chip selection is done by $A_{14}\,A_{13}$ i.e. from 00 to 11., A_{15} $A_{14}\,A_{13}$ stands for 0.11.

If A_{12} is 0, the memory address starts from 6000 to 6 FFF.

If $A_{12} = 1$, the memory address will be from 7000 to 7FFF.

7. Ans. (d)

Solution: Initial Address

A	A	0	0	
1010	1010	0000	000 ----	(A A 00)H
0001	0000	0000	000 ----	(1000) H
1011	1011	0000	000	
B	A	0	0	H

8. Ans. (a)

Solution:

A_{15}	A_{14}	A_{13}	A_{12}	A_{11}	A_{10}	...	A_0
1	1	1	0	0	0		0
1	1	1	1	1	1		1

Address range is (E 000 - EFFF)

9. Ans. (b)

Solution:

Each clip has 12 address lines which give address location 2^{12} = 4096 bytes = 4 k bytes. There are so many chips

Then the memory system has 8 × 4 k bytes memory space

10. Ans. (b)

Solution:

Chip select lines ↓			Register select lines ↓													Ranges ↓
A_{15}	A_{14}	A_{13}	A_{12}	A_{11}	A_{10}	A_9	A_8	A_7	A_6	A_5	A_4	A_3	A_2	A_1	A_0	
0	0	0	0	0	0	0	0	0	0	0	0	0	0	0	0	← 0000H
0	0	0	1	1	1	1	1	1	1	1	1	1	1	1	1	← 1FFFH

To select the chip address line A_{13} to A_{15} has to be low under all conditions. Address lines A_0 to A_{12} only chargeable overall, address range is decided by A_0 to A_{15}

11. Ans (b)

Solution:

$2\,K \times 8\,RAM$

Address line $= 2K$

i.e. from 0 to $2\times 2^{10} - 1$

$\Rightarrow 0$ to 2047

i.e. in Hexadecimal form

$\underbrace{0000}_{0}\ \underbrace{0111}_{7}\ \underbrace{1111}_{F}\ \underbrace{1111}_{F}$

Initial Address

0800

0000 1000 0000 0000

So Address of last memory location

1st location :0000 1000 0000 0000

No. of location: 0000 0111 1111 1111

$\underbrace{0000}_{0}\ \underbrace{1111}_{F}\ \underbrace{1111}_{F}\ \underbrace{1111}_{F}$

In Hexadecimal Form OFFF

12. Ans. (c)

Solution:

Number of memory locations required

$= 8$ k

$= 8 \times 1024$

$= (8192)_{10}$

In hexagonal system $= (2000)_H$

Then, the address of the first memory location

$= (4FFF - 2000) + 1$

$= 2\,FFF + 1$

$= (3000)_H$

13. Ans. (c)

Solution:

Memory mapped is in range 8000 H – 9 FFFH.

$$\begin{array}{r} 9FFF \\ -\ 7FFF \\ \hline \end{array}$$

Number of byte = 2000

Hence, in decimal $= 2 \times 16^3 = 8192\ bytes$

14. Ans. (a)

Solution: Memory bits = $2^{10} \times 24 = 65536 \times 24$

Chips size = $2^{10} \times 8 = 1024 \times 8$

Number of chips required = $\dfrac{65536 \times 24}{1024} = 192$

15. Ans. (a)

16. Ans. (d)

9.4 Interfacing

1. Ans. (c)

Solution:

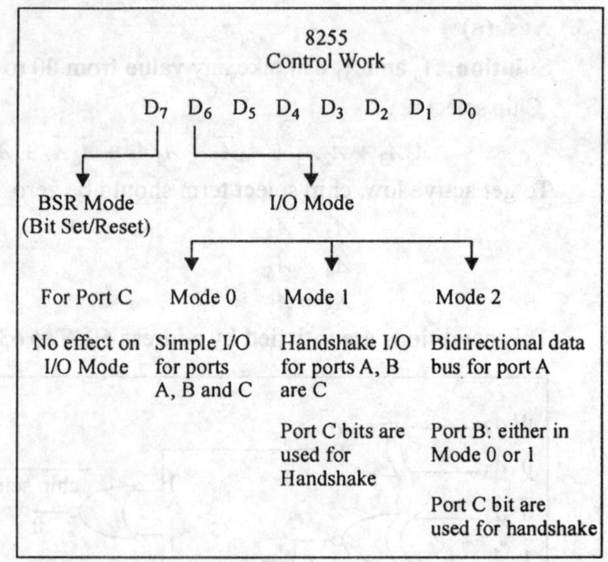

Operating modes of 8255:

* ***Mode 0 : Simple input or output*** (each of the four parts - Port - A, Port -B, Port - C_u, Port - C_L) can be used as either input or output.

* ***Mode 1 : Stored input/ output***

  (Ports A and B can be designed to work in this mode: PC_0, PC_1, and PC_2 are used to control port B).

  PC_3, PC_4, and PC_5 are used for the control of Port - A if Port - A is used as an input port. PC_6 and PC_7 can be used as either input or output lines. PC_3, PC_6 and PC_7 are used for the control of port - A, if port - A is operated as output part. Then PC_4 and PC_5 can be either input or output.

* ***Mode 2 : Bidirectional Data Transfer***

  This mode is used primarily in applications such as data between two computers or floppy disk controller interface. In this mode, port A can be configured as the bidirectional port and port B either in Mode 0 or Mode 1. Port A uses five signal from port C as handshake signals for data transfer. The remaining three signals from port C can be used either as simple I/O or as handshake for Port B.

2. **Ans. (c)**

Solution: For $\overline{CS}$ should be 0, i.e. all input should be high for GATE

A_7	A_6	A_5	A_4	A_3	A_2	A_1	A_0
1	1	1	1	1	0	X	X

← F →← 8 →

A_1	A_0
0	0
0	1
1	0

⇕

A_7	A_6	A_5	A_4	A_3	A_2	A_1	A_0
1	1	1	1	1	1	1	1

← F →← F →

Thus, the address range is F8H to FFH

3. **Ans. (b)**

Soluton:

The DMA is a process of direct data transfer between memory and external peripherals (I/O device) without programming intervention and use of microprocessor. The DMA controller takes control of the buses and transfer data directly between memory and external peripherals.

4. **Ans. (b)**

Solution:

An I/O processor control the flow of information between main memory and I/O devices.

5. **Ans. (b)**

Solution: Address range (D4)H – (D7)H

	A_7	A_6	A_5	A_4	A_3	A_2	A_1	A_0
$(D4)_H \rightarrow$	1	1	0	1	0	1	0	0
$(D7)_H \rightarrow$	1	1	0	1	0	1	1	1

($A_4 A_3 A_2$) = 5

$A_1 A_0$ vary from 00 to 11

$A_4 A_3 A_2$ remains as it is = 101 = 5 will be connected to CS

$A_7 A_6 A_5$ remains as it is 110

Second method

	A_7	A_6	A_5	A_4	A_3	A_2	A_1	A_0
$D4_H \Rightarrow$	1	1	0	1	0	1	0	0
$D7_H \Rightarrow$	1	1	0	1	0	1	1	1

($A_4 A_3 A_2$) = 5

Hence, the address lines connected to the input of the decoder (A_4 A_3 A_2) represents 5.

⇒ Output 5 will be connected to the chip select input of input output peripheral.

9.5 Interrupts

1. **Ans. (c)**

Solution:

When RST instruction is given via external hardware it requires INTR. RST will cause an interrupt only if INTR is enabled by EI instruction.

The RST instruction calls the subroutine stored at loction corresponding to the instruction.

2. **Ans. (a)**

Solution: $6 \times 8 = 48 = 30$ H

3. **Ans. (c)**

Solution:

Hardware interrupt with their memory location are given below

TRAP	0024 H
RST 7.5	003C H
RST 6.5	0034 H
RST 5.5	002C H
INTR	Dependent on RST code and it follow RST 0 to RST7

4. **Ans (d)**

Solution:

- The microprocessor can ignore or delay a maskable interrupt request if it is performing some critical task, however, it has to respond to a non maskable request immediately.
- Vectored interrupts are automatically transferred to specific locations on memory pages without any external hardware. They do not require the $\overline{\text{INTA}}$ signal or an input points the necessary hardware is already implemented inside the processor.

5. **Ans. (b)**

Solution

Interrupts of 8085 microprocessor,

* RST 7.5 vectored, maskable and positive edge triggered

* RST 6.5 vectored, maskable and level triggered
* RST 5.5 vectored, maskable and level triggered
* TRAP is vectored, non-maskable and level as well as edge Triggered
* INTR is non-vectored, maskable and level triggered inrerrupt. It is used to incresases number of interrupts

Hardware and software interrupts of 8085 and their respective memory locations

Hardware Interrupts

Instruction	*Location*
TRAP	0024_H
*RST*5.5	$002C_H$
*RST*6.5	0034_H
*RST*7.5	$003C_H$
INTR	–
	Software Interrupts
*RST*0	0000_H
*RST*1	0008_H
*RST*2	0010_H
*RST*3	0018_H
*RST*4	0020_H
*RST*5	0028_H
*RST*6	0030_H
*RST*7	0038_H

6. Ans. (c)

Solution:

Vectored address corresponding to RST 7 is Hexadecimal equivalent of,

$8 \times 7 = (56)_{10} = 0038H$

7. Ans. (a)

Solution:

INTR is best suitable for measurement of frequency of square wave

8. Ans. (a)

Solution:

INT is software interrupt

9.6 Assembly Language Programming

1. Ans. (c)

Solution:

1000 LXI SP, 27FF ⇒ SP ← 27FF

: Initializes the stack pointer with 27FFH

1003 CALL 1006 ⇒ HL ←1006

Saves the contents of program counter (i.e. address of next instruction 1006 H) on top of stack and call the subroutine stored at address 1006 H. During that period the stack pointer is decremented by one and 10 H is saved at 27FEH and stock pointer is again decremented by one and 06H is saved at 27FDH.

1006 POP H ⇒ SP ←27FF

Retrieves the contents of top of stack in HL pair. During poping up of stack the contents of 27FDH (i.e. 06H) are stored in L and stack pointer is incremented by one to 27FDH.

Then the contents of next memory location 27FEH (i.e. 10 H) are stoted in H and stack pointer is again incremented by one to 27FFH.

∴ The contents of the stack pointer (SP) and the LH register pair on completion of execution of these instruction are SP = 27FFH and HL = 1006H

SP = 27FF and HL = 1006

2. Ans. (c)

Solution:

6010: LXI H, 8A79 H HL←8A79

: Load HL pair with 8A79 H. With H = 8A H and L = 79 H

6013: MOV A, L A←79

: Moves the contents of L(i.e. 79 H) to A.

So, A = 79 H

6014: ADD H A←A + (H)

: Adds the content of H (i.e. 8 AH) to contents of A and store the result in A.

$$79H = 0111\ 10001$$
$$8AH = 1000\ 1010$$

1111

1 0000 0011

↓ ↓ ↓

CY 0 3 *H*

6015: DDA A +

: As both AC and CY flags are set during execution of ADDH, So, DAA, adds 0110 H to lower order bit and 0110 H to higher order bits to A to adjust binary result to B inary coded decimal

$$03\,H = 0000\ \ 0011$$

0110 0110

0110 1001 $= 69\,H$

So, A = 69 H

Since CY = 1 and AC = 1

DAA instruction adds 66 to content of A

A← 03 +66

A← 69

6061 MOV H, A

: Moves the contents of A (i.e. 69 H) to H. So, H - 69 H

6017 PCHL

: Copies the contents of HL pair (i.e. H = 69 H and L = 79 H) to program counter so, PC = 6979 H

After the execution of PCHL the program counter has 6979 H. So the address of next instruction to be fetched will be 6979 H.

3. Ans. (b)

Solution:

The sequence of execution of above porgram will be as under,

MVI B, 87 H → B → 87H

: Moves 87 H to B

MOV A, B A → A → 87H

: Moves the contents of B to A.

So , A = 87 H

STATE JMP NEXT

: Jumps to NEXT unconditionlly

XRA → a 00H Hence s = 0

: Contents of B are exored with contents of (i.e. 00 H) and result is stored in A.

$A \oplus B$: 00000000

10000111

10000111

⇒ A = 87 H

As D_7 bit of A is high so sign (S) flag is set.

JP START → jump to start as s = 0

: Jumps to START if sign flag is reset.

As XRA B in previous step sets the sign flag so the program comes out of loop

JMP NEXT → jump unconditionally to label Next

XRA B → A → 87 H

JP START → not jump as D_7 = 1].

Out Port 2 → 87H

: Send contents of A (i.e. 87 H) to I/O device of PORT 2

HLT : Halts the execution of program.

Thus, program displays 87 H at port 2.

4. Ans. (c)

Solution:

LXI H, 30A0 H → H L [30|40]

DAD H → [61|40]

Adds the contents of H-L pair to itself

PCHL → PC → 61 40 H

Loads the contents of H-L pair to PC.

5. Ans. (c)

Solution:

A → 0100 0101 ⇒ 45 H

B → 0100 0101 ⇒ 45 H

Carry flag → 0

RAR will A → 0010 0010

XRA B 0110 0111 (↓ 6, ↓ 7)

6. Ans. (c)

Solution:

MVI A, 07 H ⇒ 0 0 0 0 0 1 1 1

→ The content in 'A'

RLC ⇒ 0 0 0 0 1 1 1 0

→ The content of 'A'

MOV B, A ⇒ 0 0 0 0 1 1 1 0

→ The content in 'B'

RLC ⇒ 0 0 0 1 1 1 0 0

→ The content in 'B'

RLC ⇒ 0 0 1 1 1 0 0 0

→ The content in 'B'

ADD B

$A =$ 0000 1110

$+B =$ 0011 1000

0100 0110

$RRC \Rightarrow 0010\ \ 0011 = 23\ H$

7. Ans (a)

Solution:

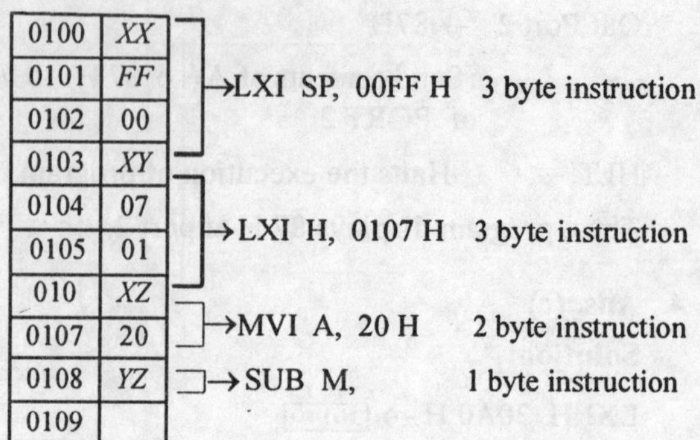

SUB M subtract to content of memory location "M" whose address is shown by HL pair from accumulator content. HL pair is pointing to 0107 the content of location 0107 is 20 H

8. Ans (c)

Solution:

ORI 40 H // A = 00 logically 0 Read with 40 H

or A = 40 H

ADD M H // (A) + (HL) = 40 H +20 = 60 H

9. Ans. (b)

Solution:

	SP
After SPHL	3 C F 4
After POP PSW	3 C F 6
After RET	3 C F 8

POP and RET are 2 Byte instruction each

10. Ans. (b)

Solution

1. A → B5 H
2. B → 0EH
3. AXOR 69H ⇒

$$\begin{array}{r} 1101\ 1100 \\ +1101\ 1100 \\ \hline 1110\ 1010 \\ \underset{E}{\downarrow}\quad \underset{A}{\downarrow} \end{array}$$

4. ADD B

$$A + B \Rightarrow DC + 0E = EAH \leftarrow A$$

5. ANI $99B_H$; $A \leftarrow 8A_H$
6. CPI $9F_H$; $A \leftarrow 8A_H$
7. STA 3010_H
8. HLT

$$\begin{array}{lll} B5_H & \Rightarrow & 1011\ 0101 \\ 69_H & \Rightarrow & 0110\ 1001 \\ \hline \text{Logical XOR} & \Rightarrow & \underbrace{1101}_{D}\ \underbrace{1100}_{C} \end{array}$$

11. Ans. (c)

Solution

$$\begin{array}{lll} EAH & \Rightarrow & 1110\ 1010 \\ 9BH & \Rightarrow & 1001\ 1011 \\ \hline \text{Logical AND} & \Rightarrow & \underbrace{1000}_{8}\ \underbrace{1010}_{A} \end{array}$$

Execution of line 6 will compare 8 AH with 9 BH

Since $|A| < 9BH$

⇒ After the execution :

CY = 1 and $Z = 0$

12. Ans. (a)

Solution:

LXI H 9258	⇒ HL ← 9258
MOV A M	⇒ A ← contents of address 9258
CMA	⇒ Content of accumulator complemented
MOV M, A	⇒ Complemented content of A stored in M(9258)

* The above program moves the contents of memory location 9258 H to accumulator and complement it and stores the result back to 9258 H.

13. Ans. (d)

Soltuion:

```
        MVI,  A,  00 H
Loop:
        ADD   B
        DCR   C
        JNZ   Loop
        HLT
```

* In 8085 μp, multiplication is performed through successive addition.

14. Ans. (b)

Solution:

LDA 3000 H → Fetch, (Address) Read, Read, Read ⇒ 4*M/C*

LXI F0F1H → Fetch data, read, read ⇒ 3M/C

15. Ans. (c)

Solution:

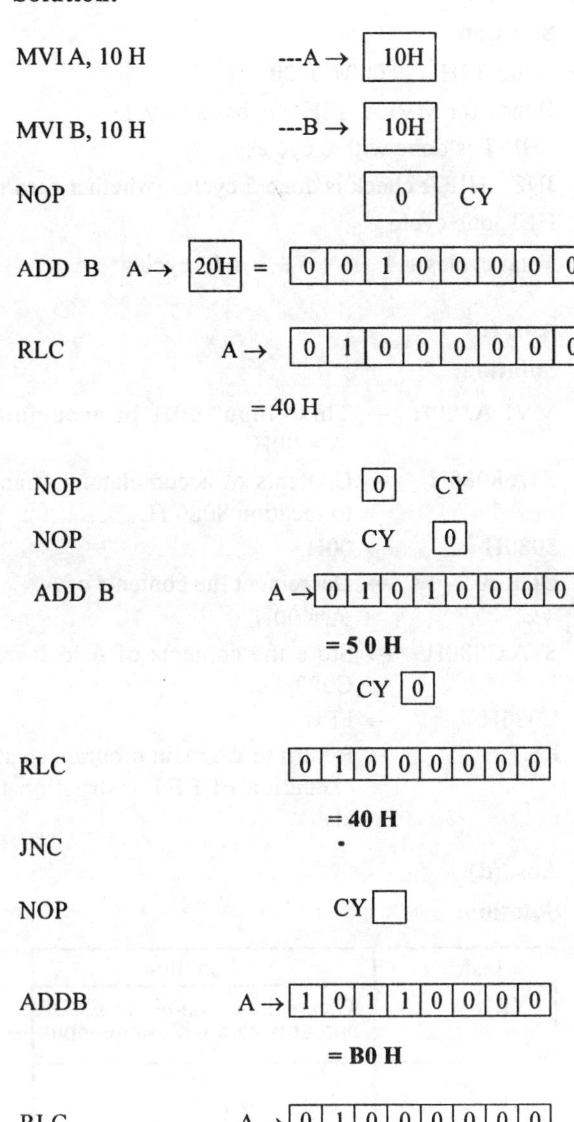

Second Method

			1st iteration	2nd iteration	3rd iteration
MVI	*A,10 H*	;	$A \leftarrow 10\,H$		
MVI	*B,10 H*	;	$B \leftarrow 10\,H$		
BACK :	*NOP*	;			
	ADD B	;	$A \leftarrow 20\,H$	$A \leftarrow 50\,H$	$A \leftarrow BO_H$
	RLC	;	$A \leftarrow 40\,H$	$A \leftarrow AO_H$	This time
	JNC BACK	;			*RLC* will set
	HLT	;			*CY* flay

Hence, NOP will be executed 3 times

16. Ans. (d)

Solution:

```
DELAY : MVI   H,    225D
        MVI   L,    255D
LOOP  : DCR   L
        JNR   LOOP
        DCR   H
        JNZ   LOOP
```

At first, 255_D is stored in H as well as L. ($255_D = FF_H$)

When we give the command DCR L this will execution for 255 time unit zero will come in 'L' register. Now executed will come to DCR H and the content of H will be decreased by '1' the execution will again come to DCR L, and this time the conent of 'L' will be modified to FF_H. This time the execution of DCRL will go for 256 times until we get 'zero' in register 'L'.

The process will go on till we get 'zero' in register H.

Hence, total number of times, DCR L will be executed is

$256 \times 254 + 255 + 1 = 65279$

17. Ans. (a)

Solution:

LXI H, 2500 H ⇒ Load the memory location 2500 H to HL pair.

MOV A M ⇒ Here M stand for memory location where data is kept. Move the data form M, i.e. HL pair location to accumulator

18. Ans. (b)

Solution:

Given data is (SP) = 2700 H

(PC) = 2100 H

(HL) = 0000 H

DAD SP: It adds the contents of strack pointer (SP) to the content of HL pair and store the result in HL pair

∴ 2700 H + 0000 H stored in HL pair

PCHL: The content of HL pair are transfered to program counter. So now PC has 2700 H and contents of SP remains unchanged

∴ (PC) = 2700 H

(SP) = 2700 H

19. Ans. (d)

Solution:

XRA A → Accumulator is cleared, $A \leftarrow 00H$

MVI BF0H $\leftarrow F0H$

SUB B $\leftarrow (A) \leftarrow (A) - (B)$

$(A) \leftarrow 10H$

20. Ans. (c)

Solution:

LXI H, 1FFE
MOV B, M } B ← First byte

INR L
MOV A, M } A ← Second byte

ADD B ⟶ Sum of byte first and second

INR L
MOV A, M } Memory location 1F00H
XOR A

21. Ans. (c)

Solution:

The portion of program to call subroutine is as below

⋮

$LXI\ \ D,\ DISP\ ;\ \ [DE] \leftarrow DISP$

$LP\ :\ CALL\ \ SUB\ \ ;\ \ SP \leftarrow SP - 2$

⋮

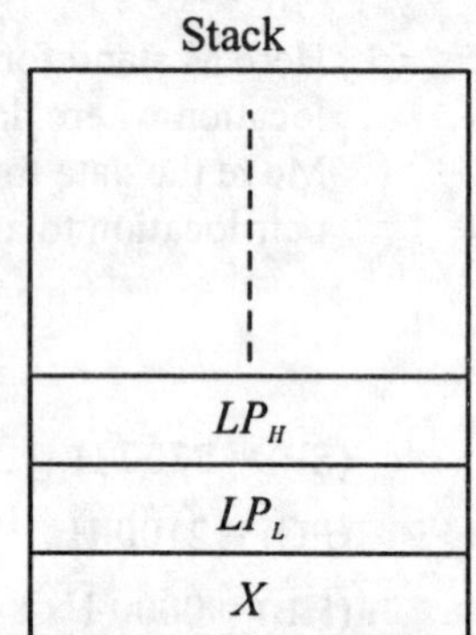

To return the control to LP-DISP-3 when RET instruction is executed; it is required that the contents of the stack should be $(LP + DISP + 3)$.

$POP\ \ \ H\ ;\ [HL] \leftarrow LP$

$DAD\ \ \ D\ ;\ [HL] \leftarrow LP + DISP$

$INX\ \ \ H\ ;\ [HL] \leftarrow [HL] + 1$

$INX\ \ \ H\ ;\ [HL] \leftarrow [HL] + 1$

$INX\ \ \ H\ ;\ [HL] \leftarrow [HL] + 1$

That result as $[HL] \leftarrow LP + DISP + 3$

$PUSH\ \ \ H\ ;\ [Stack] \leftarrow [HL]$

22. Ans. (c)

Solution:

A has 14H, i.e. 0001 0100

Hence for MVI A, 14H; we have 1 cycle

SHIFT is done with 6 cycles,

JNZ SHIFT; check is done 5 cycles (whether zero/not)

HLT; one cycle.

Total cycles = 1 + 6 + 5 + 1 = 13 cycles

23. Ans. (d)

Solution:

MVI A, 00H → This input 00H in accumulator A = 00H

STA 8080H → Contents of accumulator is transfer to location 8080 H.

8080H → 00H

DCR A → Decrement the contents of A. A = 00H

STA C080H → Store the contents of A to location C080H

C080H → FFH

RET → Return to the main program so after execution of RET instruction then also

24. Ans. (d)

Solution:

Instruction	Function
XRA A	Clears the accumulator EXOR output is zero with same input
MOV L, A	L → 00 Load L with accumulator data i.e. 00.
MOV H, L	H → 00 Move register L data to register H i.e. 00 to H register
INX H	HL → 00 01 Increment the HL pair
DAD H	HL → 00 02 This instruction increases the contents of HL pair by 1 in Hexadecimal

25. Ans. (d)

Solution:

r_2	r_1	$\dfrac{X(r_2 r_1)}{(New\ r_2)}$	$\dfrac{X(New\ r_1\ r_2)}{(New\ r_1)}$	r_2 *final*
0	0	0	0	0
1	0	1	1	0
0	1	1	0	1
1	1	0	1	1

The above table can be clearly understood by mathematical expression

1st instruction $XOR(r_2,r_1) \rightarrow r_2 \leftarrow r_2 \oplus r_1$

2nd instruction $XOR(r_1,r_2) \rightarrow r_1 \leftarrow r_1 \oplus (r_2 \oplus r_1)$

or $r_1 \leftarrow r_2$

3rd instruction $XOR(r_2,r_1) \rightarrow r_2 \leftarrow (r_2 \oplus r_1) \oplus r_2$

or $r_2 \leftarrow r_1$

It is clear that content of r_1 and r_2 are swapped

26. Ans. (a)

Solution:

LXI H, 0000H → Load HL register pair with value 0000H.

PUSH H → Store the content of HL register pair to stack memory, i.e. SP = 0000H

POP B → Retrieve the content of BC register pair from stack memory

DAD SP → Add content of HL register pair with the content of SP register and store HL

XCHG → Exchange content of DE register pair with HL register pair.

PUSH and POP are two byte instruction with each PUSH instruction SP is decremented by two byte and with POP incremented by the two result as explained below

PUSH H : SP – 2

2010 – 2 = 2008

PUSH H : SP –2

2008 – 2 = 2006

POP B : SP +2

2006 + 2 = 2008

DAD SP : SP + HL

2008 – 0000 = 2008 H

XCHG HL ⟷ DE

2008 H

27. Ans. (c)

Solution: The loop will be executed until value in the register equals to zero, i.e. 10 times, then, execution time

$= 10 \times (7T + 4T + 4T + 10T) + 7T$

$= 257T - 3[7T \text{ in case of yes}]$

$\therefore$ 254

28. Ans (c)

Solution:

Address 2020H: The contents of lower and higher address lines are same, i.e. 20*H*. So, this indicates that the technique is memory mapped input/output

Data 24*H*: This indicate that the content of accumulator is 24*H*, which have to sent to input/output port.

$10/\bar{M}$ **(Logic high):** Indicate that I/O operation active

$\overline{RD}$ **(Logic High):** Logic high indicates that write operation is active.

(Logic low): Logic, low indicates that write operation is active.

Therefore, by observation of all the operations the appropriate instruction is OUT 20H

29. Ans (a)

Solution:

Let 01H positive number → $(0000\ 0001)_b$

After RAL

CY	Accumulator
0	0000 0010 or 00000011

JNC is TRUE and again IN instruction will load accumulator with same port address and microprocessor will be in infinite loop.

Negative number $(1000\ 0000)_b$ after RAL

CY	Accumulator
1	00000000 or 0000 0001

JNC is false and microprocessor will come out from the loop

Therefore this program HALT after upon reading a negative number.

30. Ans (c)

Solution:

SUB A → A = 00H and CY = 0

MOV C,A → C = 00H and A = 00H and CY = 0

Loop1: INR A → increament A by 1.

DAA → decimal adjust accumulator if any of the bytes are greater than 9.

JC LOOP 2 → go to LOOP 2 if carry is set.

INR C → increament C but there is no effect on carry flag.

The only way to go to LOOP 2 is when carry get set. And it get when carry is generated due to DAA instruction, when A and C become 99H and control goes to LOOP 1 then INRA will make.

A = 9AH.

Now DAA will adjust the lower order byte of accumulator since it is greater than 9.

Now, 9 AH
+ 06 H
A 0H

Now agin if we will adjust the higher order byte by adding 6 to higher data byte, that is

A0H
+60H
Carry generated ← 1 00H

therefore CY = 1 and control goes to LOOP 2.
So, content of C = 99H and content to A= 00H.

31. Ans (a)

Solution:

MVI A, 06H $A \rightarrow 06$ H

ADI 70H $A \rightarrow 06 + 70 = 76H$

STA 1007H 1007H → 70 H

XRA $A \rightarrow 00$ H

HLT

32. Ans (b)

Solution:

SUB 1: MVI A OO_H → Move immediately data OOH to accumulator A.

CALL SUB2 → Transfer the sequence to subroutine 2 address

SUB 2: INR A → At subroutine 2 increase is to increment the accumulator that mean as accumulator had 00_H incremented to 01_H as SUB-2 is part of SUB-1

33. Ans (d)

Solution:

SBX:

$MVI\ A\ 99H \Rightarrow A \leftarrow 99\ H$

$ADI\quad 11\,H \Rightarrow A \leftarrow 99\,H + 11\,H = AA\,H$

$MOV\quad C, A \Rightarrow C \leftarrow 11\,H$

RET

$ADI \Rightarrow$ It will add the content immediately

So

99 H
+11 H
AA H

34. Ans. (a)

Solution:

MOV A, 00H → A← H

CALL TESK → program will go to TESK location

TESK INR A → $\overset{A}{\boxed{01H}}$

RET

The contents of accumulator after the execution of subroutine TEST is 01 H

INR A A← 02 H

RET

9.7 Miscellaneous

1. Ans. (c)

Solution:

In a microprocessor the INT pin is sampled after every instruction cycle, but the BRQ is sampled after every machine cycle.

2. Ans (d)

Solution:

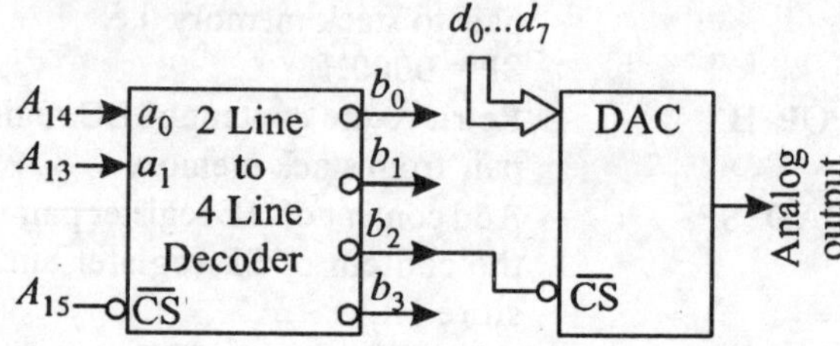

For 2 : 4 decoder

a_1	a_0	→	Output
0	0	→	b_0
0	1	→	b_1
1	0	→	b_2
1	1	→	b_3

Chip select $\overline{CS}$ must be low means A_{15} high

$A_{15} = 1$

$a_1 = 1$ $a_0 = 0$ $a_{15} = 1$

$A_{13} = 1$ $A_{14} = 0$ $a_{15} = 1$

$A_{15}\ A_{14}\ A_{13}\ A_{12}$ 0 0 0 0 0 0 0 0 0 0 0 0

$\underbrace{1\ 0\ 1\ 0}_{C}\ \underbrace{0\ 0\ 0\ 0}_{0}\ \underbrace{0\ 0\ 0\ 0}_{0}\ \underbrace{0\ 0\ 0\ 0}_{0}$

$(A000)\,H$

3. **Ans (d)**
4. **Ans (c)**

Solution:

In DMA, transfer of data between memory and peripheral takes place without any involvement of microprocessor, both in read/write mode.

5. **Ans. (c)**

Solution:

A typical microcontroller (also called single chip microcomputer) such as Motorola 6801 contains 2 K bytes of factor y programmed ROM, 128 bytes of RAM and for general purpose IO ports which are user programmable so that they can be configured in various ways to match the characteristic of the particular IO devices being controlled.

6. **Ans. (c)**
7. **Ans. (c)**

Solution:

Given, Timer count = 07DOH = $(2000)_d$

$$\text{And Timer count} = \frac{Pulse\ Period}{Clock\ Period}$$

$$\text{Or Pulse frequency} = \frac{Clock\ frequency}{linear\ count}$$

$$= \frac{800\ kHz}{2000} = 400\ Hz$$

10 Control Systems and Process Control

10.1 Transfer Function, Block Diagram and SFG

1. In the signal flow graph of figure, the gain c/r will be:

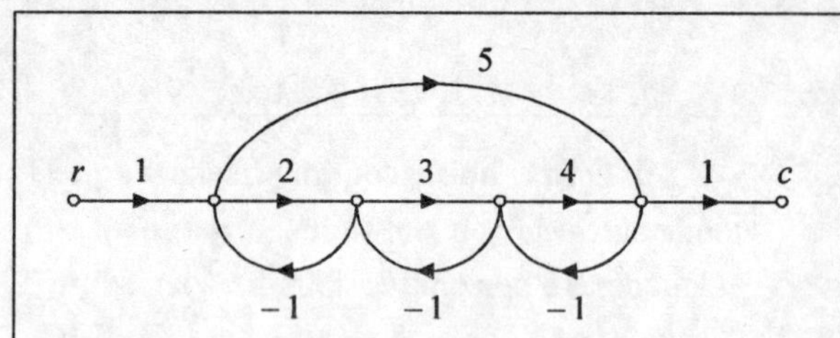

(a) 11/9 (b) 22/15
(c) 24/23 (d) 44/23

2. Signal flow graph is used to find
(a) Stability of the system
(b) Controllability of the system
(c) Transfer function of the system
(d) Poles of the system

3. The transfer function of a linear system is the
(a) Ratio of the output, $V_o(t)$, and input, $V_i(t)$
(b) Ratio of the derivative of the output and the input
(c) Ratio of the Laplace transform of the output and that of the input with all initial conditions zeros
(d) None of these

4. In the signal flow graph of figure y/x equals

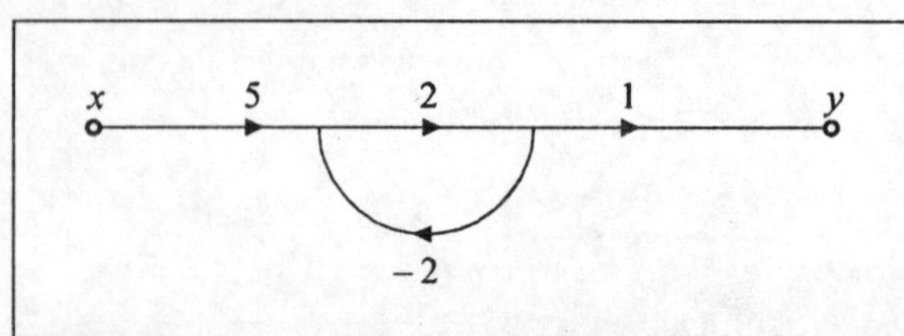

(a) 3 (b) $\frac{5}{2}$
(c) 2 (d) None of the above

5. The transfer function of a system is

$$\frac{2s^2 + 6s + 5}{(s+1)^2\,(s+2)}$$

The characteristic equation of the system is
(a) $2s^2 + 6s + 5 = 0$
(b) $(s + 1)^2 (s + 2) = 0$
(c) $2s^2 + 6s + 5 + (s + 1)^2 (s + 2) = 0$
(d) $2s^2 + 6s + 5 - (s + 1)^2 (s + 2) = 0$

6. An electrical system and its signal-flow graph representation are shown in figure (A) and (B) respectively. The values of G_2 and H, respectively, are:

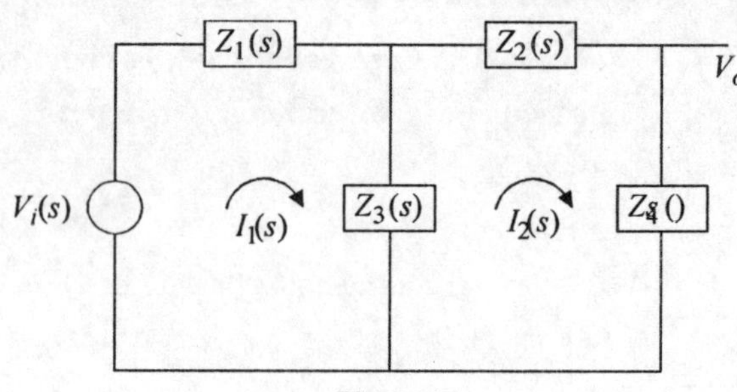

Figure A

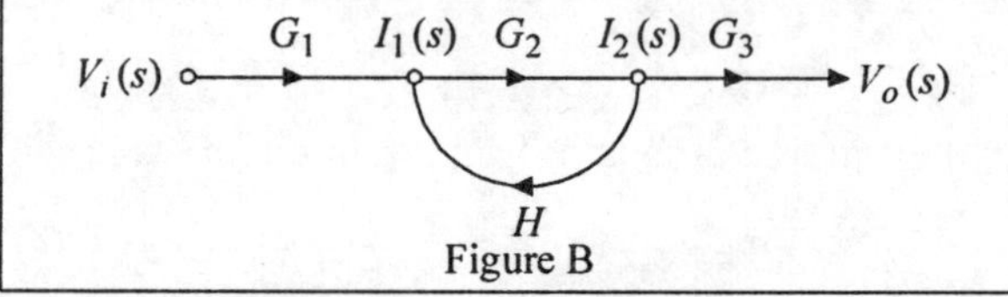

Figure B

(a) $\dfrac{Z_3(s)}{Z_2(s)+Z_3(s)+Z_4(s)}, \dfrac{-Z_3(s)}{Z_1(s)+Z_3(s)}$

(b) $\dfrac{-Z_3(s)}{Z_2(s)-Z_3(s)+Z_4(s)}, \dfrac{-Z_3(s)}{Z_1(s)+Z_3(s)}$

(c) $\dfrac{Z_3(s)}{Z_2(s)+Z_3(s)+Z_4(s)}, \dfrac{Z_3(s)}{Z_1(s)+Z_3(s)}$

(d) $\dfrac{-Z_3(s)}{Z_2(s)-Z_3(s)+Z_4(s)}, \dfrac{Z_3(s)}{Z_1(s)+Z_3(s)}$

7. The signal flow graph of a system is shown in figure. The transfer function $\dfrac{C(s)}{R(s)}$ of the system is:

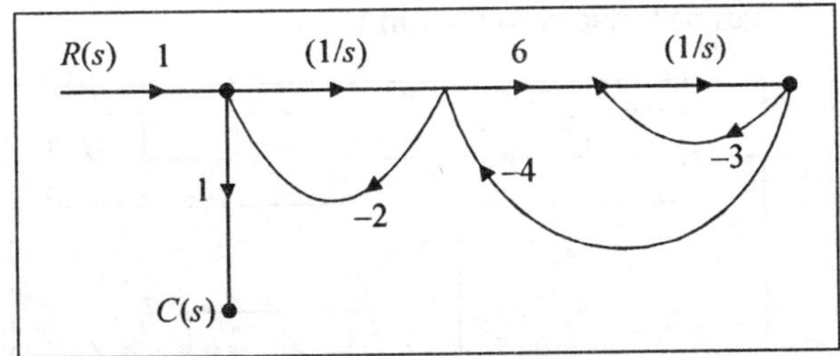

(a) $\dfrac{6}{s^2+29s+6}$ (b) $\dfrac{6s}{s^2+29s+6}$

(c) $\dfrac{s(s+2)}{s^2+29s+6}$ (d) $\dfrac{s(s+27)}{s^2+29s+6}$

8. Consider the signal flow graph shown in figure. The gain $[x_5/x_1]$ is:

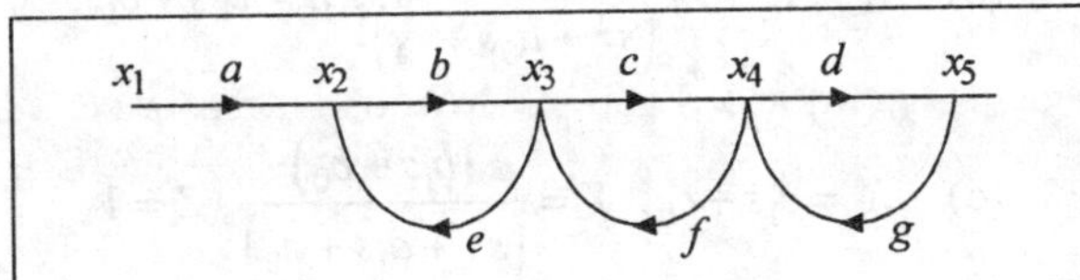

(a) $\dfrac{1-(be+cf+dg)}{abcd}$

(b) $\dfrac{bedg}{1-(be+cf+dg)}$

(c) $\dfrac{abcd}{1-(be+cf+dg)+bedg}$

(d) $\dfrac{1-(be+cf+dg)+bedg}{bedg}$

9. The transfer function $\dfrac{Y(s)}{R(s)}$ of the system shown is

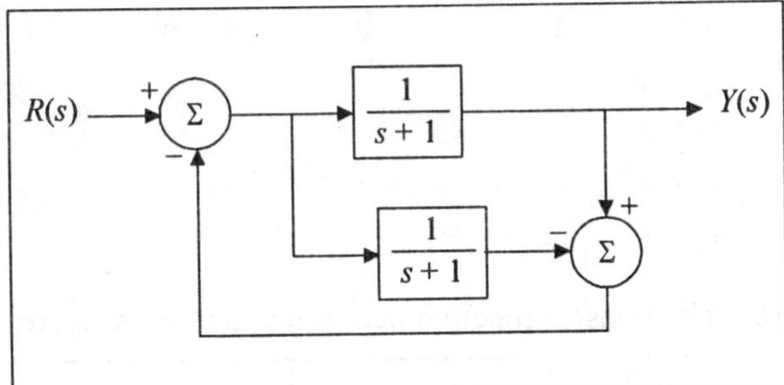

(a) 0 (b) $\dfrac{1}{s+1}$

(c) $\dfrac{2}{s+1}$ (c) $\dfrac{2}{s+3}$

10. Despite the presence of negative feedback, control systems still have problems of instability because the

(a) Components used have non-lineararities.

(b) Dynamic equations of the subsystem are not known exactly

(c) Mathematical analysis involves approximations

(d) System has large negative phase angle at high frequencies.

11. The overall transfer function of the system in figure is

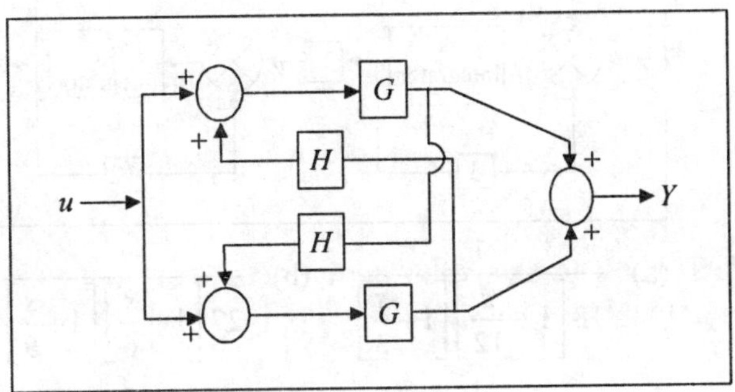

(a) $\dfrac{G}{1-GH}$ (b) $\dfrac{2G}{1-GH}$

(c) $\dfrac{GH}{1-GH}$ (d) $\dfrac{2G}{1-H}$

12. Signal flow graph is used to obtain the

(a) Stability of a system

(b) Transfer function of a system

(c) Controllability of a system

(d) Observability of a system

13. For block diagram shown in figure, $\dfrac{C(s)}{R(s)}$ is given by

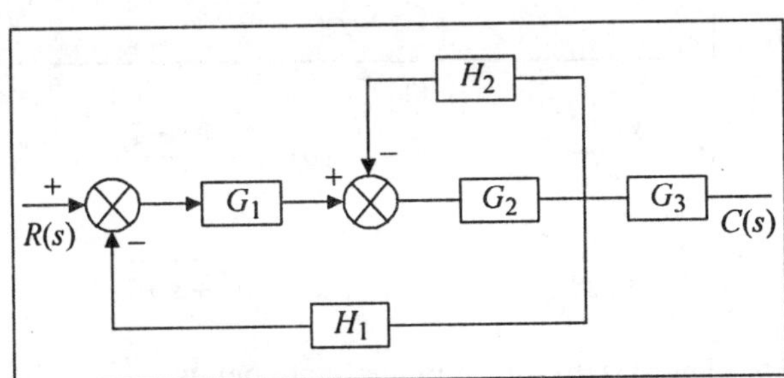

(a) $\dfrac{G_1G_2G_3}{1+H_2G_2+H_1G_1G_2}$

(b) $\dfrac{G_1G_2G_3}{1+G_1G_2G_3\ H_1G_2}$

(c) $\dfrac{G_1G_2G_3}{1+G_1G_2G_3H_1+G_1G_2G_3H_2}$

(d) $\dfrac{G_1G_2G_3}{1+G_1G_2G_3H_1}$

14. The block diagram of a control system is shown in figure.

The transfer function $G(s) = \dfrac{Y(s)}{U(s)}$ of the system is

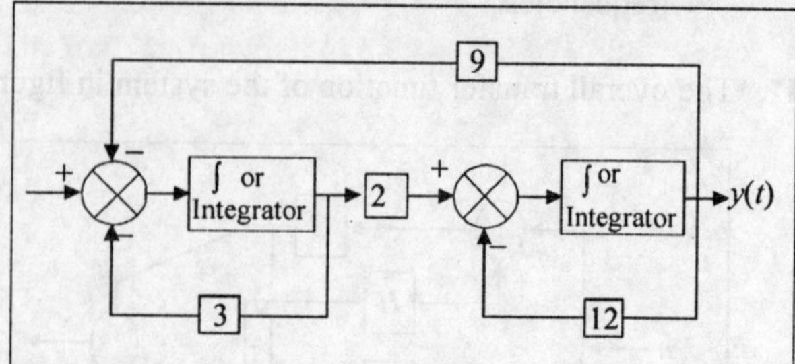

(a) $\dfrac{1}{18\left[1+\dfrac{s}{12}\right]\left[1+\dfrac{s}{3}\right]}$ (b) $\dfrac{1}{27\left[1+\dfrac{s}{6}\right]\left[1+\dfrac{s}{9}\right]}$

(c) $\dfrac{1}{27\left[1+\dfrac{s}{12}\right]\left[1+\dfrac{s}{9}\right]}$ (d) $\dfrac{1}{27\left[1+\dfrac{s}{9}\right]\left[1+\dfrac{s}{3}\right]}$

15. For block diagram shown in figure, the transfer function $\dfrac{C(s)}{R(s)}$ is equal to

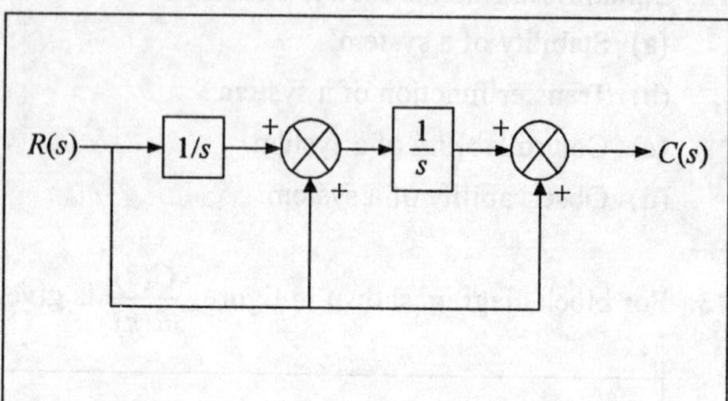

(a) $\dfrac{s^2+1}{s^2}$ (b) $\dfrac{s^2+s+1}{s^2}$

(c) $\dfrac{s^2+s+1}{s^3}$ (d) $\dfrac{1}{s^2+s+1}$

16. The system shown in the figure below

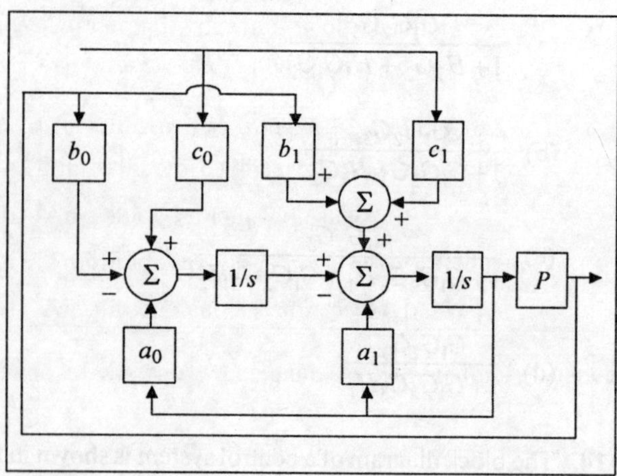

can be reduced to the form

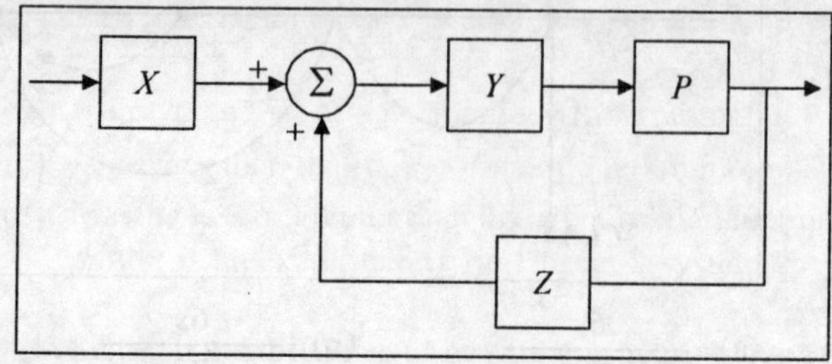

With

(a) $X = c_0 s + c_1,\ Y = \dfrac{1}{(s^2 + a_0 s + a_1)},\ Z = b_0 s + b_1$

(b) $X = 1,\ Y = \dfrac{(c_0 s + c_1)}{(s^2 + a_0 s + a_1)},\ Z = b_0 s + b_1$

(c) $X = c_1 s + c_0,\ Y = \dfrac{(b_1 s + b_0)}{(s^2 + a_1 s + a_0)},\ Z = 1$

(d) $X = c_1 s + c_0,\ Y = \dfrac{1}{(s^2 + a_1 s + a_0)},\ Z = b_1 s + b_0$

(d) $\dfrac{G_1G_2G_3}{1+G_1G_2G_3H_1}$

17. If the open loop transfer function of negative feedback control system is:

$$G(s)\,H(s) = \left(\frac{K}{(s+1)^3}\right)$$

then the gain K for a closed loop pole to be at $\left(-\dfrac{1}{2}+j\dfrac{\sqrt{3}}{2}\right)$ is:

(a) 1
(b) 2
(c) 3
(d) 4

18. The transfer function between y_2 and y_1 in figure,

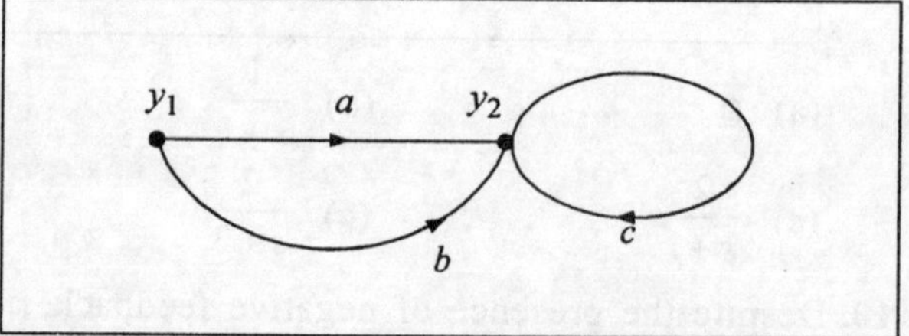

(a) $a + b$ (b) $(a + b)\,c$

(c) $\dfrac{(a+b)}{(1-c)}$ (d) $\dfrac{(a+b)}{(1+c)}$

19. The signal flow graph of a system is given below.

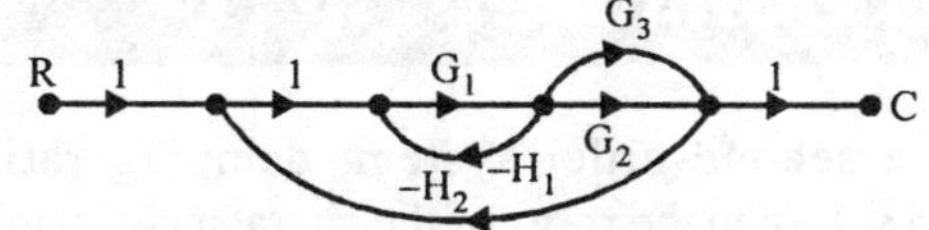

The transfer function (C/R) of the system is

(a) $(G_1G_2 + G_1G_3) / (1 + G_1G_2H_2)$

(b) $(G_1G_2 + G_1G_3) / (1 - G_1H_1 + G_1G_2H_2)$

(c) $(G_1G_2 + G_1G_3) / (1 - G_1H_1 + G_1G_2H_2 + G_1G_3H_2)$

(d) $\dfrac{G_1G_2 + G_1G_3}{1 - G_1H_1 + G_1G_2H_2 + G_1G_2H_2 + G_1G_2G_3H_2}$

20. A filter is represented by the signal flow graph shown in the figure. Its input is $x(t)$ and output is $y(t)$. The transfer function of the filter is

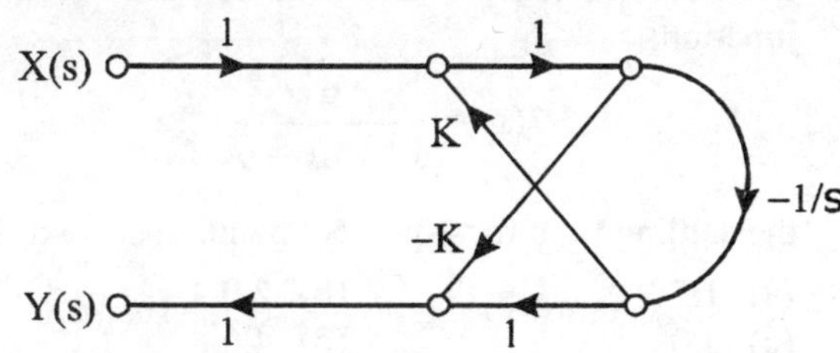

(a) $\dfrac{-(1+ks)}{s+k}$

(b) $\dfrac{(1-ks)}{s-k}$

(c) $\dfrac{-(1-ks)}{s+k}$

(d) $\dfrac{(1-ks)}{s+k}$

21. In the feedback scheme shown in figure the time constant of the closed loop system will be

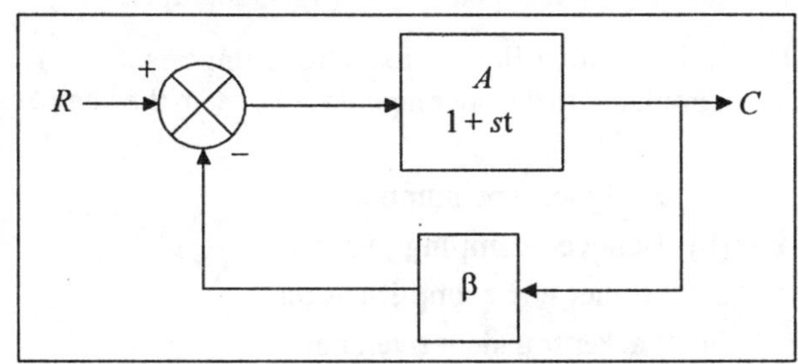

(a) $A\beta\tau$ (b) $[(1 + A(b)\tau]$

(c) τ (d) $\dfrac{\tau}{(1 + A\beta)}$

22. The transfer function of the open loop system $G(s)$ which is represented by the signal flow graph shown in the figure below is

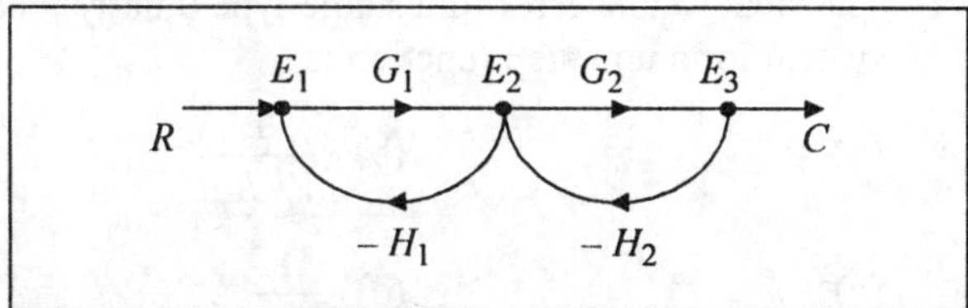

(a) $\dfrac{G_1G_2}{1+G_1H_1+G_2H_2}$

(b) $\dfrac{G_1G_2}{1+G_1H_1G_2H_2}$

(c) $\dfrac{G_1G_2}{(1+G_1H_1)(1+G_2H_2)}$

(d) $\dfrac{G_1G_2}{1+(G_1+H_1)(G_2+H_2)}$

23. The transfer function of a **position servo system** is given as $G(s) = \dfrac{1}{s(s+1)}$. **A first order compensator** is designed in a unity feedback configuration so that the poles of the compensated system are placed at $-1 \pm j1$ and -4. The transfer function of compensated system is

(a) $\dfrac{s+3}{2(s+5)}$ (b) $\dfrac{2s+3}{s+2}$

(c) $\dfrac{5(s+1.6)}{s+5}$ (d) $\dfrac{3(2s+3)}{s+4}$

24. A feedback control system with high gain K is shown in the figure below:

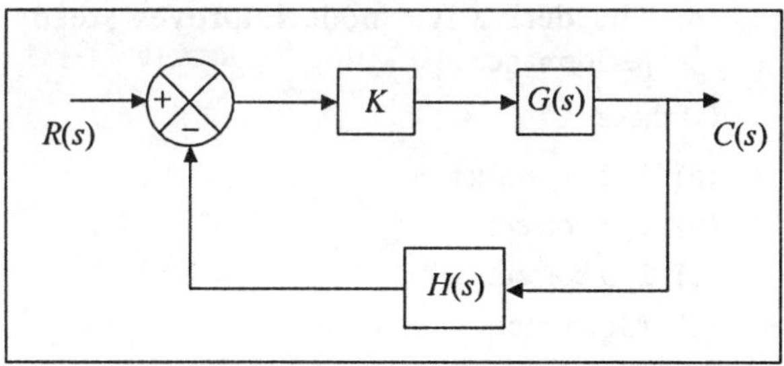

Then the closed loop transfer function is

(a) Sensitive to perturbations in $G(s)$ and $H(s)$

(b) Sensitive to perturbations in $G(s)$ but not to perturbations in $H(s)$

(c) Sensitive to perturbations in $H(s)$ but not to perturbations in $G(s)$

(d) Insensitive to perturbations in $G(s)$ and $H(s)$

10.2 Time Response and Controller

1. The steady state error of a stable type 0 unity feedback system for a unit step function is

(a) 0 (b) $\frac{1}{1+K_p}$

(c) ∞ (d) $\frac{1}{K_p}$

Where K_p is the position error constant.

2. A second-order system has a transfer function given by

$$G(s) = \frac{25}{s^2+8s+25}$$

If the system, initially at rest is subject to unit step input at $t = 0$, the second peak in the response will occur at:

(a) π sec (b) $\frac{\pi}{3}$ sec

(c) $\frac{2\pi}{3}$ sec (d) $\frac{\pi}{2}$ sec

3. A unity feedback control system has the open-loop transfer function:

$$G(s) = \frac{4(1+2s)}{s^2(s+2)}$$

If the input to the system is a unit ramp, the steady- state error will be:

(a) 0 (b) 0.5

(c) 2 (d) Infinity

4. A process with open-loop model $G(s) = (ke^{-sT_y}/\tau s + 1)$ is controlled by a PID controller, For this process:

1. The integral mode improves transient performance
2. The integral mode improves steady-state performance
3. The derivative mode improves transient performance
4. The derivative mode improves steady-state performance

Of these

(a) 1, 2, 3 correct

(b) 1, 2 correct

(c) 2, 3 correct

(d) All correct

5. The step error coefficient of a system

$$G(s) = \frac{1}{(s+6)(s+1)}$$

with unity feedback is

(a) $\frac{1}{6}$ (b) ∞

(c) 0 (d) 1

6. For a second-order system, damping ratio (δ) is $0 < \delta < 1$, then the roots of the characteristic polynomial are

(a) Real but not equal (b) Real and equal

(c) Complex conjugates (d) Imaginary

7. Consider a unity feedback control system with open-loop transfer function $G(s) = \frac{K}{s(s+1)}$. The steady state error of the system due to a unit step input is:

(a) Zero (b) K

(c) $\frac{1}{K}$ (d) Infinite

8. For a second-order system with the closed-loop transfer junction:

$$T(s) = \frac{9}{s^2+4s+9}$$

the settling time for 2-percent band, in second, is

(a) 1.5 (b) 2.0

(c) 3.0 (d) 4.0

9. If the closed-loop transfer function $T(s)$ of a unity negative feedback system is given by

$$T(s) = \frac{a_{n-1}s + a_n}{s^n + a_1 s^{n-1} + \ldots + a_{n-1}s + a_n}$$

Then the steady state error for a unit ramp input is

(a) $\frac{a_n}{a_{n-1}}$ (b) $\frac{a_n}{a_{n-2}}$

(c) $\frac{a_{n-1}}{a_{n-2}}$ (d) zero

10. If the characteristic equation of a closed-loop system is $s^2 + 2s + 2 = 0$, then the system is:

(a) Over-damped (b) Critically damped

(c) Under-damped (d) Undamped

11. A PD controller is used to compensate a system. Compared to the uncompensated system the compensated system has

(a) A higher type number

(b) Reduced damping

(c) Higher noise amplification

(d) Larger transient overshoot

12. A second-order system has the transfer function

$$\frac{C(s)}{R(s)} = \frac{4}{s^2+4s+4}$$

with $r(t)$ as the unit-step function, the response $c(t)$ of the system is represented by:

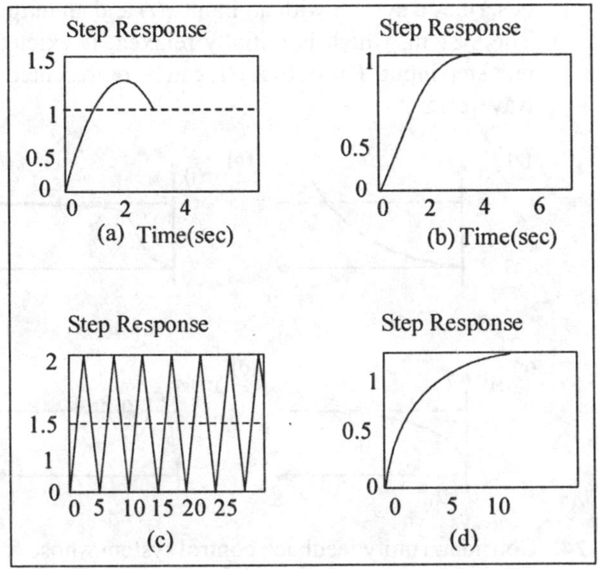

(a) Figure (a) (b) Figure (b)
(c) Figure (c) (d) Figure (d)

13. In the derivation of expression for peak percent overshoot, $M_P = \exp\dfrac{-\pi\delta}{\sqrt{(1-\delta^2)}} \times 100\ \%$ which one of the following condition is NOT required?

(a) System is linear and time invariant.
(b) The system transfer function has a pair of complex conjugate poles and no zeros.
(c) There is no transportation delay in the system.
(d) The system has zero initial conditions

14. A ramp input applied to an unity feedback system results in 5 % steady state error. The type number and zero frequency gain of the system are respectively.

(a) 1 and 20 (b) 0 and 20
(c) 0 and $\dfrac{1}{20}$ (d) 1 and $\dfrac{1}{20}$

15. The unit impulse response of a system is:

$$h(t) = e^{-t},\ t \geq 0$$

For this system, the steady-state value of the output for unit step input is equal to:

(a) –1 (b) 0
(c) 1 (d) ∞

16. If the closed-loop transfer function of a control system is given as:

$$T(s) = \frac{s-5}{(s+2)(s+3)},\ \text{then it is}$$

(a) An unstable system
(b) An uncontrollable system
(c) A minimum phase system
(d) A non-minimum phase system

17. A control system with a *PD* controller is shown in the figure. If the velocity error constant $K_v = 1000$ and the damping ratio $\delta = 0.5$, then the values of K_p and K_D are:

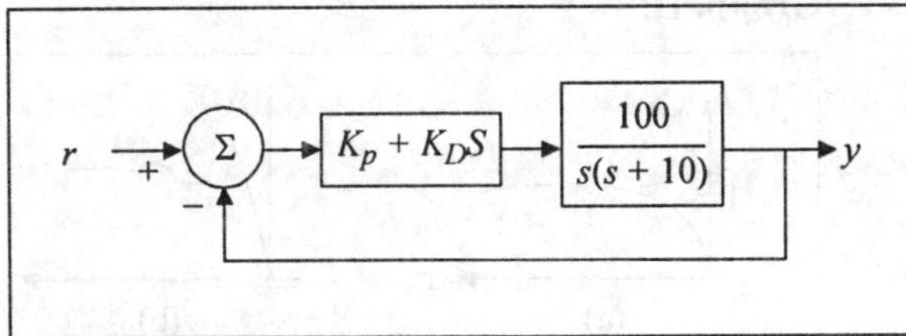

(a) $K_P = 100, K_D = 0.09$
(b) $K_P = 100, K_D = 0.9$
(c) $K_P = 10, K_D = 0.09$
(d) $K_P = 10, K_D = 0.9$

18. The transfer function of a plant is

$$T(s) = \frac{5}{(s+5)(s^2+s+1)}.$$

The second order approx-imation of $T(s)$ using dominant pole concept is

(a) $\dfrac{1}{(s+5)(s+1)}$ (b) $\dfrac{5}{(s+5)(s+1)}$
(c) $\dfrac{5}{s^2+s+1}$ (d) $\dfrac{1}{s^2+s+1}$

19. Step responses of a set of three second-order underdamped systems all have the same percentage overshoot. Which of the following diagrams represents the poles of the three system?

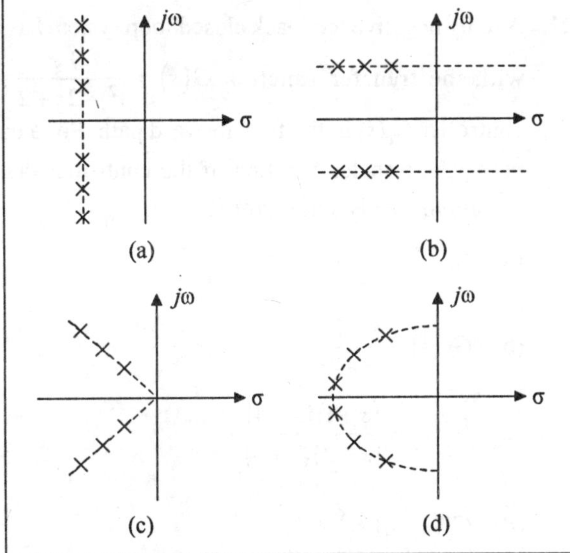

20. Group I lists set of four transfer functions. Group II gives a list of possible step responses $y(t)$. Match the step responses with the corresponding transfer functions.

Group I: $P = \frac{25}{s^2 + 25}$, $Q = \frac{36}{s^2 + 20s + 36}$

$R = \frac{36}{s^2 + 12s + 36}$, $S = \frac{49}{s^2 + 7s + 49}$

Group. II:

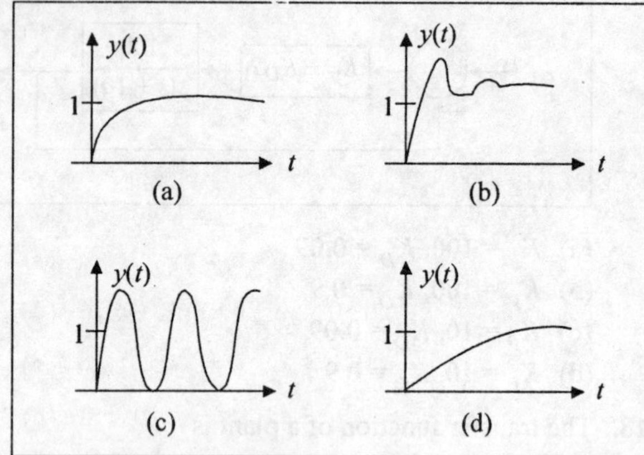

(a) P-3, Q-1, R-4, S-2
(b) P-3, Q-2, R-4, S-1
(c) P-2, Q-1, R-4, S-3
(d) P-3, Q-4, R-1, S-2

21. The unit step response of an under-damped second order system has steady state value of –2. Which one of the following transfer functions has these properties?

(a) $\frac{-2.24}{s^2 + 2.59s + 1.12}$ (b) $\frac{-3.82}{s^2 + 1.91s + 1.91}$

(c) $\frac{-2.24}{s^2 - 2.59s + 1.12}$ (d) $\frac{-3.82}{s^2 - 1.91s + 1.91}$

22. A unity negative feedback closed loop system has a plant with the transfer function $G(s) = \frac{1}{s^2 + 2s + 2}$ and a controller $G_c(s)$ in the feed forward path. For a unit step input, the transfer function of the controller that gives minimum steady state error is

(a) $G_c(s) = \frac{s+1}{s+2}$

(b) $G_c(s) = \frac{s+2}{s+1}$

(c) $G_c(s) = \frac{(s+1)(s+4)}{(s+2)(s+3)}$

(d) $G_c(s) = 1 + \frac{2}{s} + 3s$

23. The differential equation $10\frac{d^2y}{dt^2} - 20\frac{dy}{dt} + y = x(t)$ describes a system with an input $x(t)$ and an output $y(t)$. The system, which is initially relaxed, is exicted by a unit step input. The output $y(t)$ can be represented by the waveform.

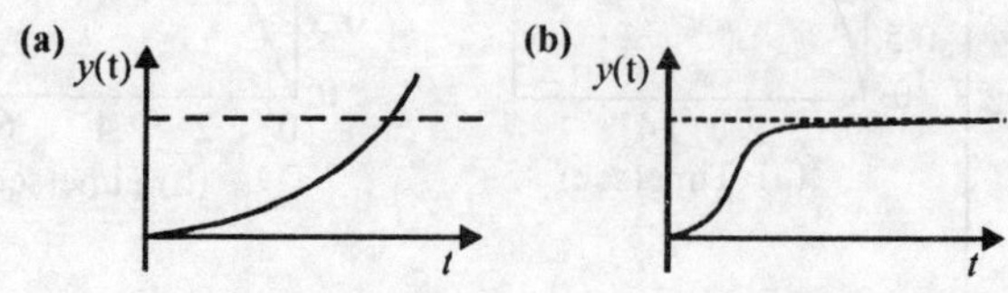

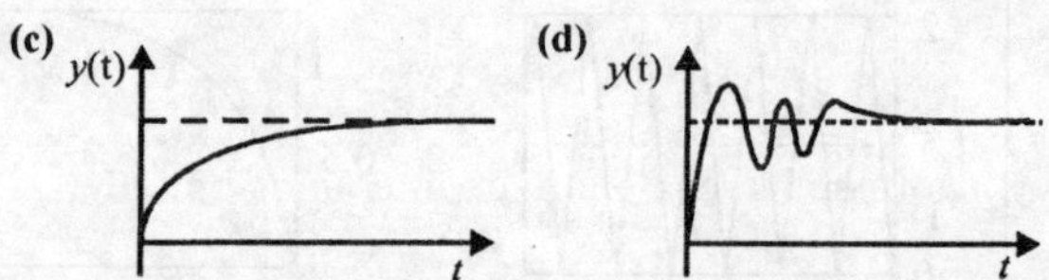

24. Consider a unity feedback control system whose forward path transfer function $G(s) = K/s^2$. The steady state error, for a step input is

(a) 1.0 (b) ∞
(c) 0.0 (d) Does not exist

25. For what values of 'a' does the system shown in figure have a zero **steady state error** [i.e. $\lim_{t\to\infty} e(t)$] for a step input?

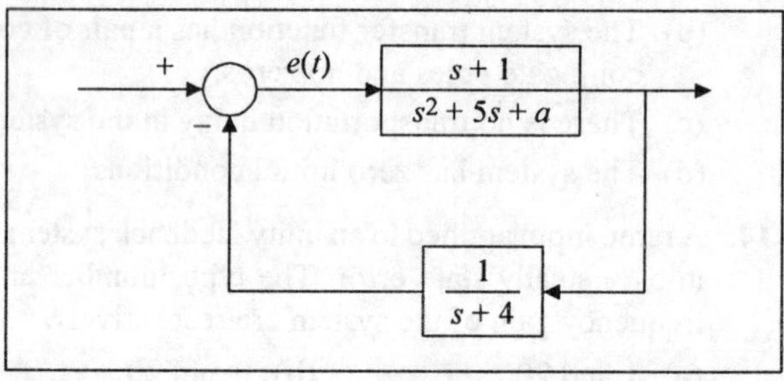

(a) $a = 0$ (b) $a = 1$
(c) $a \geq 4$ (d) For no value of 'a'

26. The **closed-loop transfer** function of a control system is given by

$$\frac{C(s)}{R(s)} = \frac{2(s-1)}{(s+2)(s+1)}$$

For a unit step input, the output is

(a) $-3\,e^{-2t} + 4\,e^{-t} - 1$ (b) $-3\,e^{-2t} - 4\,e^{-t} + 1$

(c) Zero (d) Infinity

27. The **unit-impulse response** of a **unit-feedback control system** is given by

$$c(t) = -t\,e^{-t} + 2\,e^{-t},\ (t \geq 0)$$

the open loop transfer function is equal to

(a) $\dfrac{s+1}{(s+2)^2}$ (b) $\dfrac{2s+1}{s^2}$

(c) $\dfrac{s+1}{(s+1)^2}$ (d) $\dfrac{s+1}{s^2}$

28. Consider the unit-step response of a unity-feedback control system whose open-loop transfer function is $G(s) = \dfrac{1}{s(s+1)}$. The **maximum overshoot** is equal to

(a) 0.143 (b) 0.153
(c) 0.163 (d) 0.173

29. For a feedback control system of type 2, the **steady error for a ramp** input is

(a) Infinite (b) Constant
(c) Zero (d) Indeterminate

30. For the system shown in figure, with a **damping ratio** of 0.7 and an undamped natural frequency ω_n of 4 rad/sec, the value of K and a are

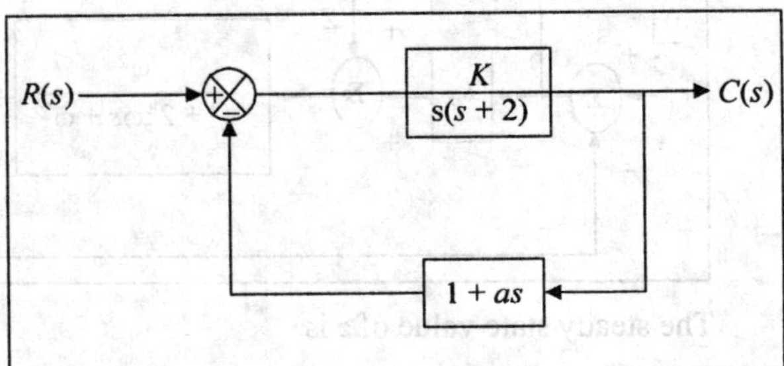

(a) $K = 4$, $a = 0.35$
(b) $K = 8$, $a = 0.455$
(c) $K = 16$, $a = 0.225$
(d) $K = 64$, $a = 0.9$

31. Introduction of **integral action** in the forward path of a unity feedback system results in a

(a) Marginally stable system
(b) System with no steady state rerror
(c) System with increased stability margin
(d) System with better speed o response

32. The output of a **linear time invariant control system** is $c(t)$ for a certain input $r(t)$. If $r(t)$ is modified by passing it through a block whose transfer function is e^{-s} and then applied to the system, the modified output of the system would be

(a) $\dfrac{c(t)}{1+e^t}$ (b) $\dfrac{c(t)}{1+e^{-t}}$

(c) $c(t-1)\,u(t-1)$ (d) $c(t)\,u(t-1)$

33. A **unity feedback system** has open-loop transfer function $G(s)$. The **steady-state** error is zero for

(a) Step input and type 1 $G(s)$
(b) Ramp input and type 1 $G(s)$
(c) Step input and type 2 $G(s)$
(d) Ramp input and type 0 $G(s)$

34. A **unity feedback system** has **open-loop transfer** function $G(s) = \dfrac{25}{s(s+6)}$. The **peak over-shoot** in the step-input response of the system is approximately equal to

(a) 5 % (b) 10 %
(c) 15 % (d) 20 %

35. A control system is defined by the following mathematical relationship

$$\frac{d^2x}{dt^2} + 6\frac{dx}{dt} + 5x = 12(1 - e^{-2t})$$

The response of the system as $t \to \infty$ is

(a) $x = 6$ (b) $x = 2$
(c) $x = 2.4$ (d) $x = -2$

36. A control system with certain excitation is governed by the following mathematical equation

$$\frac{d^2x}{dt^2} + \frac{1}{2}\frac{dx}{dt} + \frac{1}{18}x = 10 + 5e^{-4t} + 2e^{-5t}$$

The natural time constants of the response of the system are

(a) 2s and 5s
(b) 3s and 6s
(c) 4s and 5s
(d) $\dfrac{1}{3}$ s and $\dfrac{1}{6}$ s

37. The block diagram shown in figure gives a unity feedback closed loop control system. The steady state error in the response of the above system to unit step input is

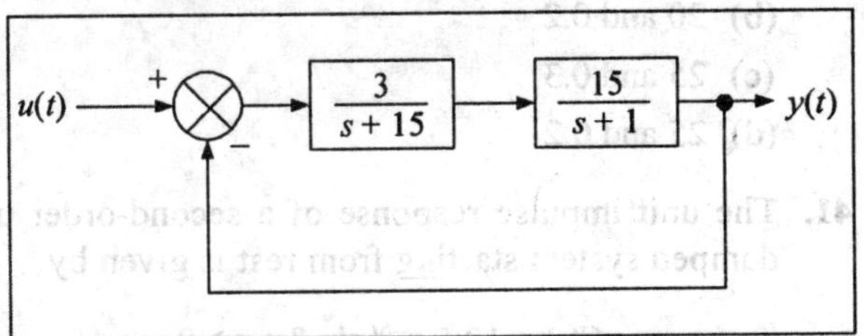

(a) 25 %
(b) 0.75 %
(c) 6 %
(d) 33 %

38. The roots of the closed loop characteristic equation of the system shown in figure are

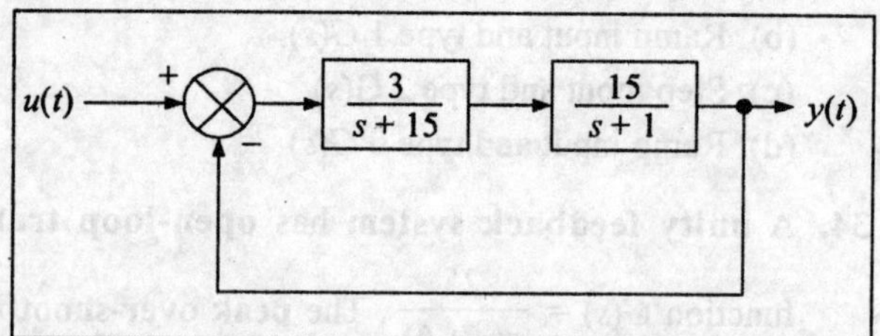

(a) -1 and -15 (b) 6 and 10

(c) -4 and -15 (d) -6 and -10

39. Consider the function, $F(s) = \dfrac{5}{s\left(s^2 + 3s + 2\right)}$

where $F(s)$ is the Laplace transform of the function $f(t)$. The initial value of $f(t)$ is equal to

(a) 5 (b) $\dfrac{5}{2}$

(c) $\dfrac{5}{3}$ (d) 0

40. The block diagram of a closed loop control system is given by figure. The values of K and P, such that the system has a damping ratio of 0.7 and an undamped natural frequency $f\omega_n$ of 5 rad/sec, are respectively equal to

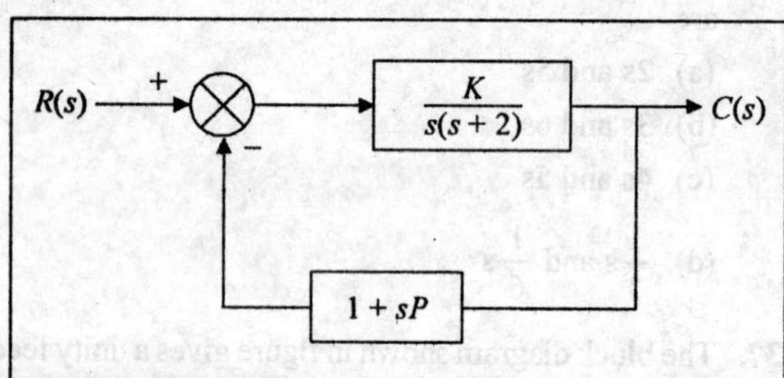

(a) 20 and 0.3

(b) 20 and 0.2

(c) 25 and 0.3

(d) 25 and 0.2

41. The unit impulse response of a second-order under-damped system starting from rest is given by

$$C(t) = 12.5\, e^{-6t} \sin 8\, t,\ t \geq 0$$

The steady-state value of the unit step response of the system is equal to

(a) 0 (b) 0.25

(c) 0.5 (d) 1.0

42. When subjected to a unit step input, the closed loop control system shown in the figure will have a steady state error of

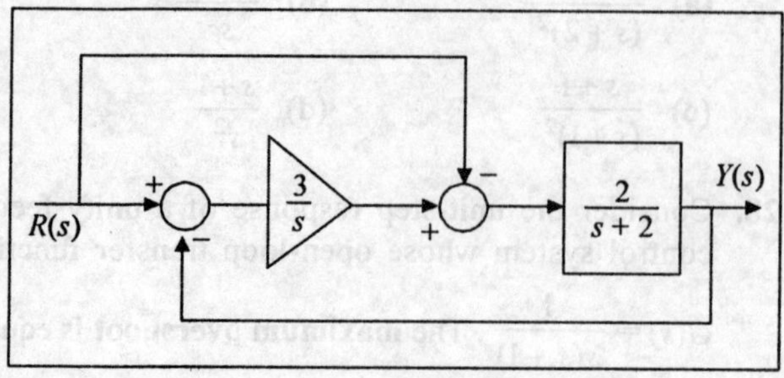

(a) -1.0 (b) -0.5

(c) 0 (d) 0.5

43. Consider the feedback control system shown below which is subjected to a unit step input. The system is stable and has the following parameters $k_p = 4$, $k_1 = 10$, $\omega = 500$ and $\delta = 0.7$

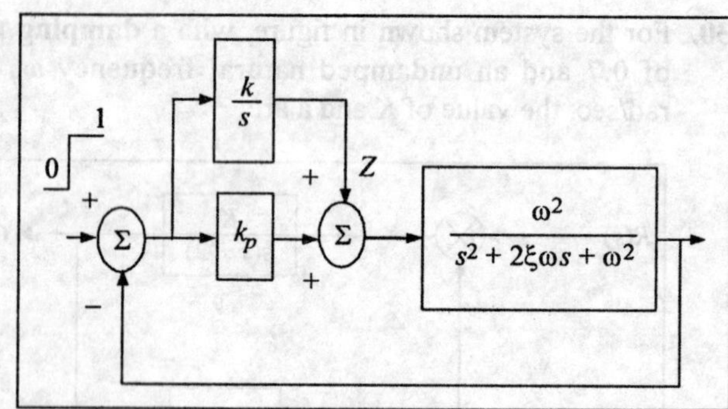

The steady state value of z is

(a) 1 (b) 0.25

(c) 0.1 (d) 0

44. Consider the R-L-C circuit shown in figure

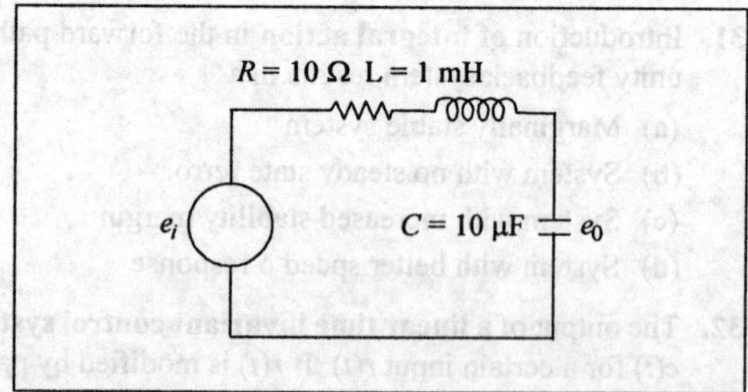

For a step-input e_i, the overshoot in the output e_o will be

(a) 0, since the system is not under-damped

(b) 5 %

(c) 16 %

(d) 48 %

45. Consider the R-L-C circuit shown in figure

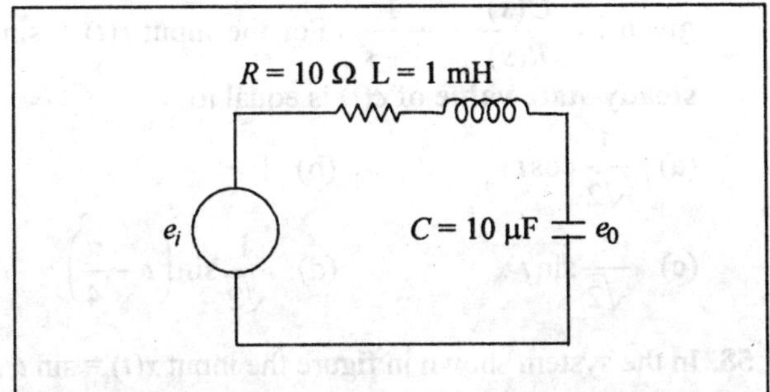

If the above step response is to be observed on a non-storage CRO, then it would be best to have the e_i as a

(a) Step function
(b) Square wave of frequency 50 Hz
(c) Square wave of frequency 300 Hz
(d) Square wave of frequency 2.0 kHz

46. A function $y(t)$ satisfies the following differential equation: $\frac{dy(t)}{dt}+y(t)=\delta(t)$

where $\delta(t)$ is the delta function. Assuming zero initial condition and denoting the unit step function by $u(t) \cdot y(t)$ can be of the from

(a) e^t (b) e^{-t}
(c) $e^t u(t)$ (d) $e^{-t}u(t)$

47. The transfer function of a linear time invariant system is given as

$$G(s)=\frac{1}{s^2+3s+2}$$

The steady state value of the output of this system for a unit impulse input applied at time instant $t=1$ will be

(a) 0 (b) 0.5
(c) 1 (d) 2

48. The transfer function of a system is given as

$$\frac{100}{s^2+20s+100}$$

This system is

(a) An overdamped system
(b) An underdamped system
(c) A cirtically damped system
(d) An unstable system

Statement for Linked Answer Questions 49 and 50

The state equation of a system is described by

$$x' = Ax + Bu$$
$$y = Cx$$

where x is state vector, u is input, y is output and

$$A=\begin{bmatrix}0 & 1\\ 0 & -2\end{bmatrix}, B=\begin{bmatrix}0\\ 1\end{bmatrix}, C=\begin{bmatrix}1 & 0\end{bmatrix}$$

49. The transfer fucntion $G(s)$ of this system will be

(a) $\frac{s}{(s+2)}$
(b) $\frac{s+1}{s(s-2)}$
(c) $\frac{s}{(s-2)}$
(d) $\frac{1}{s(s+2)}$

50. A unity feedback is provided to the above system $G(s)$ to make it a closed loop system as shown in figure

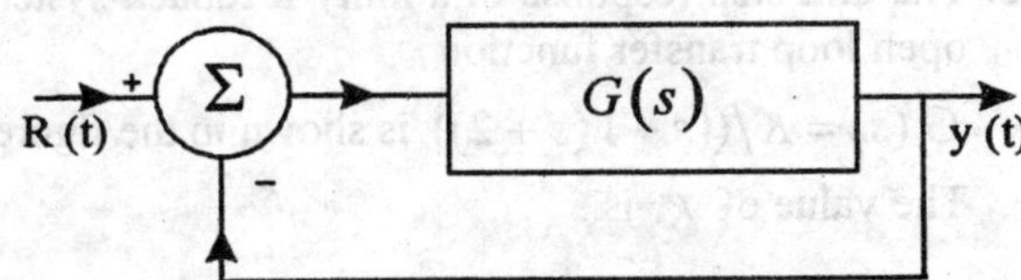

For a unit step input $r(t)$, the steady state error in the output will be

(a) 0 (b) 1
(c) 2 (d) ∞

51. For the system $2/(s+1)$, the approximate time taken for a step response to reach 98% of its final value is

(a) 1 s (b) 2 s
(c) 4 s (d) 8 s

52. The steady state error of a unity feedback linear system for a unit step input is 0.1. The steady state error of the same system, for a pulse input $r(t)$ having a magnitude of 10 and a duration of one second, as shown in the figure is

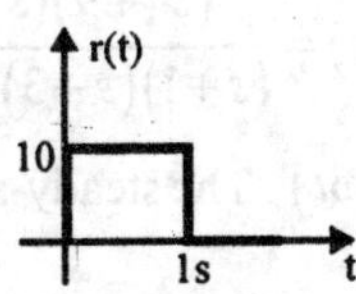

(a) 0 (b) 0.1
(c) 1 (d) 10

53. The response $h(t)$ of a linear time invariant system to an impulse $\delta(t)$ under initially relaxed condition is $h(t) = e^{-t} + e^{-2t}$. The response of this system for a unit step input $u(t)$ is

(a) $u(t) + e^{-t} + e^{-2t}$ (b) $(e^{-t} + e^{-2t})u(t)$

(c) $(1.5 - e^{-t} - 0.5e^{-2t})u(t)$ (d) $e^{-t}\delta(t) + e^{-2t}u(t)$

54. A two-loop position control system is shown below.

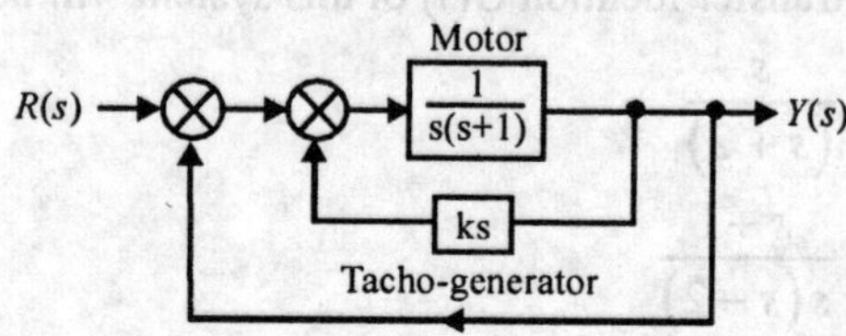

The gain k of the Tacho-generator influences mainly the

(a) Peak overshoot

(b) Natural frequency of oscillation

(c) Phase shift of the closed loop transfer function at very low frequencies $(\omega \to 0)$

(d) Phase shift of the closed loop transfer function at very high frequencies $(\omega \to \infty)$

55. The unit step response of a unity feedback system with open loop transfer function $G(s) = K/((s+1(s+2))$ is shown in the figure. The value of K is.

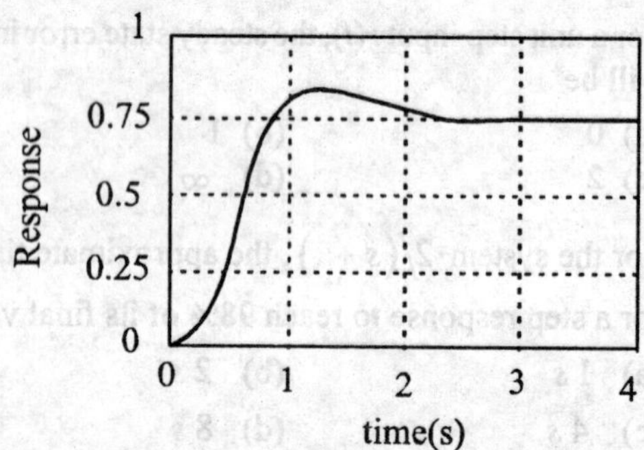

(a) 0.5 (b) 2

(c) 4 (d) 6

56. A system with transfer function

$$G(s) = \frac{(s^2+9)(s+2)}{(s+1)(s+3)(s+4)}$$

is excited by $\sin(\omega t)$. The steady-state output of the system is zero at

(a) $\omega = 1$ rad/s (b) $\omega = 2$ rad/s

(c) $\omega = 3$ rad/s (d) $\omega = 4$ rad/s

57. The **closed-loop transfer** function of a control system is given by $\frac{C(s)}{R(s)} = \frac{1}{1+s}$. For the input $r(t) = \sin t$, the **steady state value** of $c(t)$ is equal to

(a) $\frac{1}{\sqrt{2}}\cos t$ (b) 1

(c) $\frac{1}{\sqrt{2}}\sin t$ (d) $\frac{1}{\sqrt{2}}\sin\left(t - \frac{\pi}{4}\right)$

58. In the system shown in figure the input $x(t) = \sin t$. In the steady-state, the response $y(t)$ will be

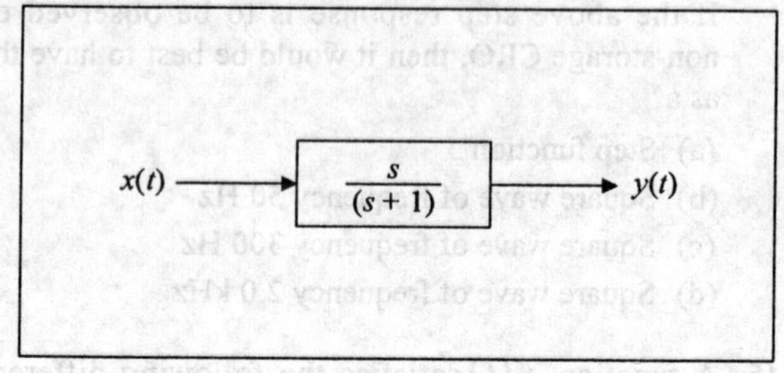

(a) $\frac{1}{\sqrt{2}}\sin(t - 45°)$ (b) $\frac{1}{\sqrt{2}}\sin(t + 45°)$

(c) $\sin(t - 45°)$ (d) $\sin(t - 45°)$

59. A system with zero initial conditions has the closed loop transfer function. $T(s) = \frac{s^2+4}{(s+1)(s+4)}$. The system output is zero at the frequency

(a) 0.5 rad/sec (b) 1 rad/sec

(b) 2 rad/sec (d) 4 rad/se

60. A unity feedback closed loop second order system has a transfer function $\frac{81}{s^2 + 0.6s + 9}$ and is excited by a step input of 10 units. The steady state error of the output is:

(a) 10 (b) 0.0

(c) 1.0 (d) 0.1

61. For a unit step input, a system with a closed-loop transfer function of $G(s) = \frac{20}{s^2 + 2s + 5}$ has a steady-state output of

(a) 10 (b) 5

(c) 2 (d) 4

62. For a unity feedback system with open loop gain $G(s) = \frac{9}{s(s+3)}$, the closed loop system will have a damping ratio of

(a) 0.33 (b) 0.5

(c) 1.0 (d) 3.

63. The transfer function of a control system is given by

$$\frac{C(s)}{R(s)} = \frac{25}{s^2+6s+25}$$

The first maximum value of the response occurs at a time t_{max} given by

(a) $\frac{\pi}{8}$ (b) $\frac{\pi}{4}$

(c) $\frac{\pi}{2}$ (d) π

64. For a unity feedback system with open loop transfer function $G(s) = \frac{9}{s(s+2)}$, the damping ratio is

(a) $\frac{1}{3}$ (b) $\frac{1}{2}$

(c) 1 (d) 2

65. A unity feed back system has a forward path transfer function $G(s) = \frac{10}{s(s+5)}$. It is subjected to a unit ramp input. The **integral of the squared** error (ISE) is.

(a) Zero (b) Infinity

(c) Always negative (d) 0.5

66. The loop transfer function of a system is given by

$$G(s)\,H(s) = \frac{10e^{-0.1s}}{s(s+5)(s^2+4s+16)}$$

When the system is subjected to a **parabolic input** of $t^2/2$, the **steady state error** is.

(a) Zero (b) 0.1

(c) Infinity (d) $e^{-0.1}/0.1$

67. The forward path transfer function of an unity feedback system is given by.

$$G(s) = \frac{(1+5s)(1+10s)(1+2s)}{(1+s)(1+8s)(1+20s)}$$

If $e(t)$ is the error to a unit impulse input the value of the performance index $J = \int_0^\infty e(t)\,dt$ is equal to

(a) Zero

(b) Infinity

(c) −12

(d) 0.5

68. A plane with a transfer function $\frac{2}{s(s+3)}$ is controlled by a PI controller with $K_p = 1$ and $K_i \geq 0$ in a unity feedback configuration. The lowest value of K_i that ensure zero steady state error for a step change in the reference input is

(a) 0 (b) $\frac{1}{3}$

(c) $\frac{1}{2}$ (d) 1

69. $u(t)$ represents the unit step function. The Laplace transform of $u(t-\tau)$ is

(a) $\frac{1}{s\tau}$ (b) $\frac{1}{s-\tau}$

(c) $\frac{e^{-s\tau}}{s}$ (d) $e^{-s\tau}$

70. A unit ramp input applied to the system shown in the adjoining figure. The steady state error in its output is

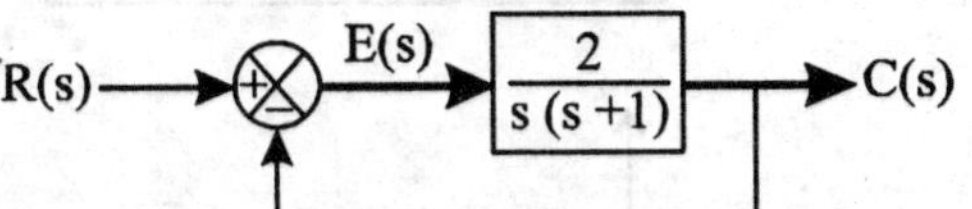

(a) 0 (b) 0.5

(c) 1 (d) 2

71. A unity feedback system has an open loop transfer function $G(s) = \frac{K}{s(s+3)}$. The value of K that yields a damping ratio of 0.5 for the closed loop system is

(a) 1 (b) 3

(c) 5 (d) 9

72. The characteristics of a thermometer measuring ambient temperature is $2\frac{dTi}{dt} + T_i - T_a = 0$, where T_i and T_a are the indicated and ambient temperatures, respectivley, both in °C, and time is in seconds. The –3dB cut - off frequency response of the thermometer is

(a) $\frac{1}{4\pi}$ Hz (b) $\frac{1}{2}$ Hz

(c) 1 Hz (d) 2π Hz

73. The unit - step response of a negative feedback system with the open - loop transfer function $G(s) = \frac{6}{s+5}$ is

(a) $1-e^{-5t}$ (b) $6-6e^{-5t}$

(c) $\frac{6}{5}-\frac{6}{5}e^{-5t}$ (d) $\frac{6}{11}-\frac{6}{11}e^{-11t}$

74. The step response of a linear time imvariant system is $y(t) = 5e^{-10t}u(t)$, where $u(t)$ is the unit step function. If the output of the system corresponding to an impulse input $\delta(t)$ is $h(t)$, then $h(t)$ is.

(a) $-50e^{-10t}\,u(t)$

(b) $5e^{-10t}\,\delta(t)$

(c) $5u(t)-50e^{-10t}\,\delta(t)$

(d) $5\delta(t)-50e^{-10t}\,u(t)$

75. If a first order system and its time response to a unit step input are as shown below, the gain K is.

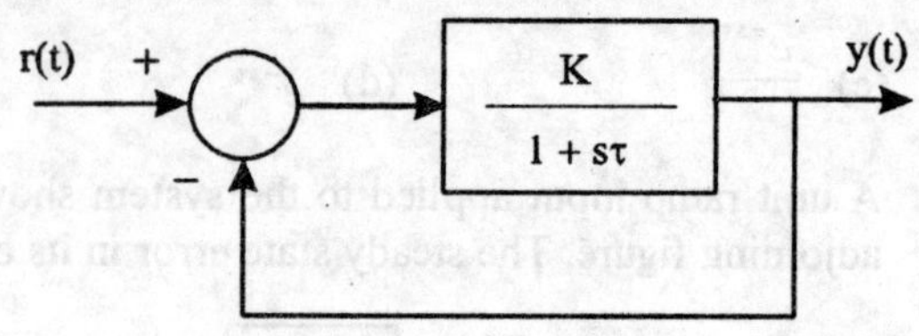

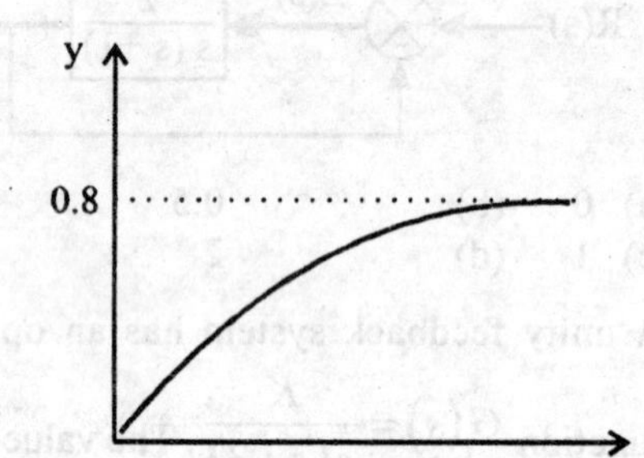

(a) 0.25 (b) 0.8
(c) 1 (d) 4

Statements for Linked Answer Questions 76 and 77:
A unit feedback system has open loop transfer function $G(s)=\dfrac{100}{s(s+p)}$. The time at which the response to a unit step input reaches its peak is [π/8] seconds.

76. The damping coefficient for the closed loop system is.

(a) 0.4 (b) 0.6
(c) 0.8 (d) 1

77. The value of p is

(a) 6 (b) 12
(c) 14 (d) 16

78. A system with transfer function

$$G(s)=\frac{(s^2+9)(s+2)}{(s+1)(s+3)(s+4)}$$

is excited by $\sin(\omega t)$. The steady-state output of the system is zero at

(a) ω = 1 rad/s (b) ω = 2 rad/s
(c) ω = 3 rad/s (d) ω = 4 rad/s

79. The feedback system shown below oscillates at 2 rad/s when

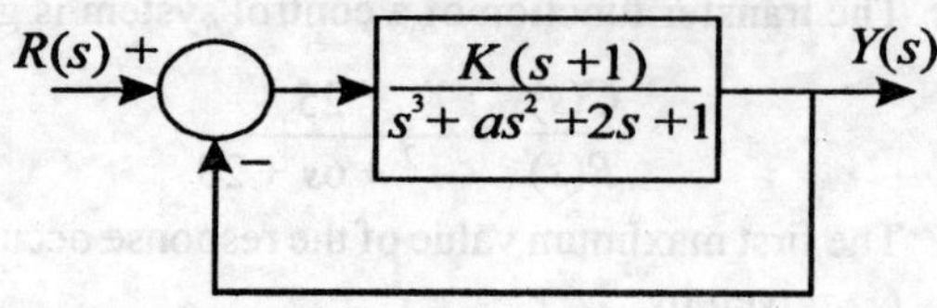

(a) $K=2$ and $a=0.75$

(b) $K=3$ and $a=0.75$

(c) $K=4$ and $a=0.5$

(d) $K=2$ and $a=0.5$

80. The open loop transfer function of a unity gain negative feedback control system is given by $G(s)=\dfrac{s^2+4s+8}{s(s+2)(s+8)}$. The angle θ, at which the root locus approaches the zeros of the system, satisfies

(a) $|\theta|=\pi-\tan^{-1}\left(\frac{1}{4}\right)$ (b) $|\theta|=\frac{3\pi}{4}-\tan^{-1}\left(\frac{1}{3}\right)$

(c) $|\theta|=\frac{\pi}{2}-\tan^{-1}\left(\frac{1}{4}\right)$ (d) $|\theta|=\frac{\pi}{4}-\tan^{-1}\left(\frac{1}{3}\right)$

81. The steady state error for the feedback control system shown in figure when subjected to a unit step input is.

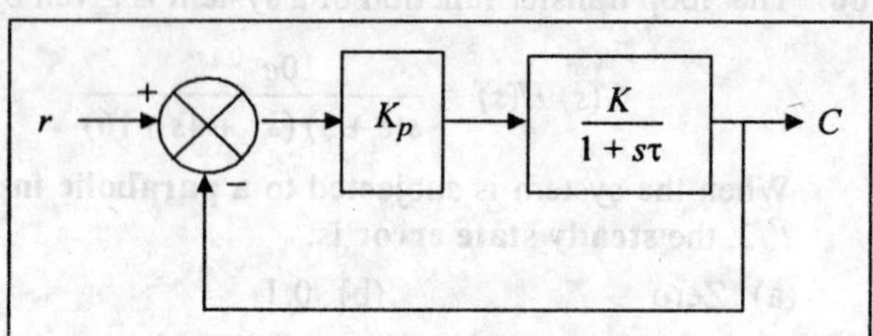

(a) Zero (b) $\dfrac{1}{1+KK_p}$

(c) $\dfrac{KK_p}{1+KK_p}$ (d) Infinite

82. The transfer function of a system is given by $\dfrac{Y(s)}{X(s)}=\dfrac{e^{-0.1s}}{1+s}$. If $x(t)$ is 0.5 sin t, then the phase angle between the output and the input will be

(a) – 39.27° (b) – 45°
(c) – 50.73° (d) – 90°

83. The dominant poles of a servo system are located at $s=(-2\pm j2)$. The damping ratio of the system is

(a) 1
(b) 0.8
(c) 0.707
(d) 0.6

84. The **steady state error** due to a unit-step disturbance input $D(s)$ in figure is

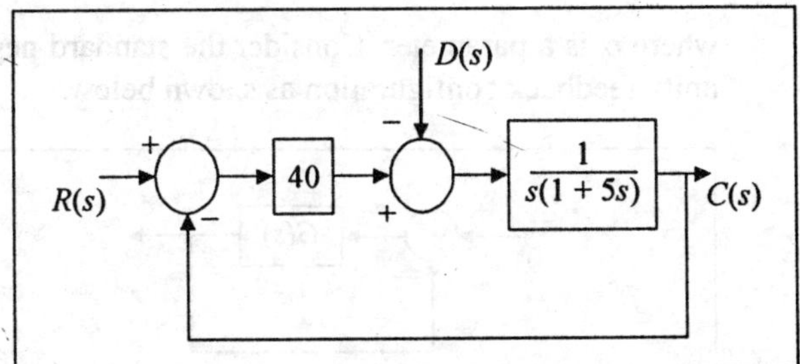

(a) 0 (b) 0.012
(c) 0.021 (d) 0.025

85. Consider the following systems:

System 1: $G(s)=\dfrac{1}{(2s+1)}$

System 2: $G(s)=\dfrac{1}{(5s+1)}$

The true statement regarding the system is

(a) Bandwidth of system 1 is greater than the bandwidth of system 2
(b) Bandwidth of system 1 is lower than the bandwidth of system 2
(c) Bandwidth of both the systems are the same
(d) Bandwidth of both the systems are infinite

86. Consider a unity feedback control system whose forward path transfer function $G(s)=K/s^2$. The **steady state error,** for a step input is

(a) 1.0 (b) ∞
(c) 0.0 (d) does not exist

87. The transfer function of a system is $\dfrac{A}{\left(s^2+\omega^2\right)}$. The **steady state gain** of the system to a unit step input is.

(a) A/ω^2
(b) 0
(c) ∞
(d) Not possible to be determined

88. For a first order instrument a 5% settling time is equal to.

(a) Three times the time constant.
(b) Two times the time constant.
(c) The time constant.
(d) Time required for the output signal to reaches 5% of the final value.

Statement for Linked Answer Questions 89 and 90

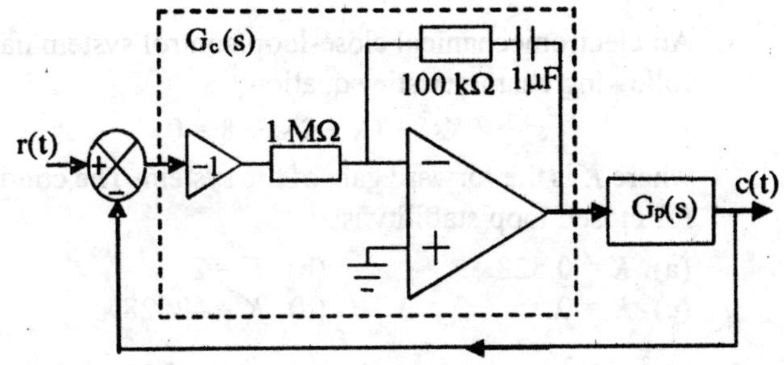

$$G_p(s)=\frac{1}{(s+1)(1+0.1s)}$$

Given figure shows a closed loop unity feedback system. The controller block has transfer function denoted by $G_c(s)$. The controller is cascaded to the plant, which is denoted by $G_p(s)$

89. The transfer function $G_C(s)$ is

(a) $\dfrac{1-0.1s}{s}$ (b) $\dfrac{1+0.1s}{s}$
(c) $-\dfrac{s}{s+1}$ (d) $\dfrac{s}{s+1}$

90. The peak percentage overshoot of the closed loop system is

(a) 5.0 % (b) 10.0 %
(c) 16.3 % (d) 1.63 %

Common Data for Questions 91 and 92

The following figure represents a proportional control scheme of a first order system with transportation lag

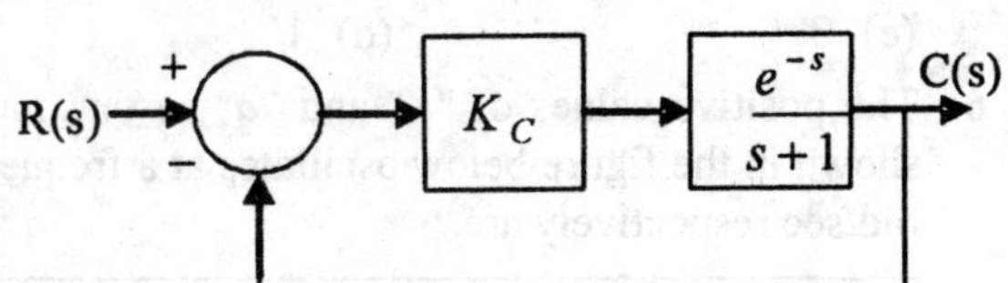

91. The angular frequency in radian/s at which the loop phase lag becomes 180° is

(a) 0.408 (b) 0.818
(c) 1.56 (d) 2.03

92. The steady state error for a unit step input when the gain $K_c=1$ is

(a) 1/4 (b) 1/2
(c) 1 (d) 2

10.3 Stability: Routh-Hurwitz Criterion

1. An electromechanical close-loop control system has the following characteristic equation

$$s^3 + 6Ks^2 + (K+2)s + 8 = 0$$

where K is the forward gain of the system. The condition for closed loop stability is:

(a) $K = 0.528$ (b) $K = 2$
(c) $K = 0$ (d) $K = -2528$

2. If $s^3 + 3s^2 + 4s + A = 0$, then all roots of this equation are in the left half plane provided that

(a) $A > 12$ (b) $-3 < A < 4$
(c) $0 < A < 12$ (d) $5 < A < 12$

3. The number or roots of $s^3 + 5s^2 + 7s + 3 = 0$ in the left half of the s-plane is

(a) Zero (b) One
(c) Two (d) Three

4. The open-loop transfer function of a unity feedback system is $G(s) = \dfrac{K}{s(s^2+s+2)(s+3)}$

The range of K for which the system is stable is

(a) $\dfrac{21}{4} > K > 0$
(b) $13 > K > 0$
(c) $\dfrac{21}{4} < K < \infty$
(d) $-6 < K < \infty$

5. For the polynomial $P(s) = s^5 + s^4 + 2s^3 + 2s^2 + 3s + 15$, the number of roots which lie in the right half of the s-plane is:

(a) 4 (b) 2
(c) 3 (d) 1

6. The positive values of "K" and "a" so that the system shown in the figure below oscillates at a frequency of 2 rad/sec respectively are.

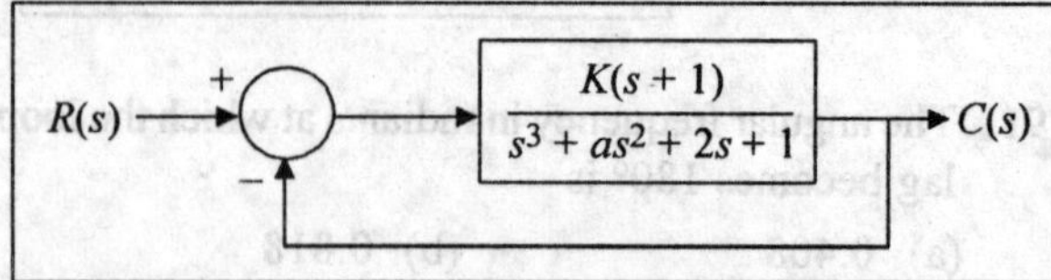

(a) 1, 0.75 (b) 2, 0.75
(c) 1, 1 (d) 2, 2

7. A certain system has transfer function

$$G(s) = \frac{s+8}{s^2 + \alpha s - 4},$$

where α is a parameter. Consider the standard negative unity feedback configuration as shown below.

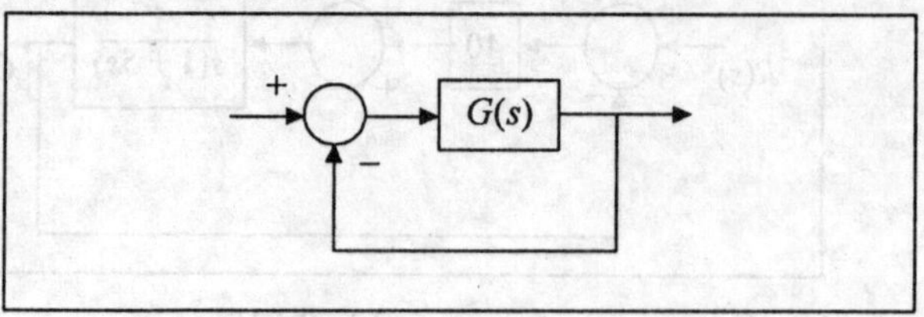

Which of the following statements is true?

(a) The closed loop system is never stable for any value of α
(b) For some positive values of α, the closed loop system is stable, but not for all positive values.
(c) For all positive values of α, the closed loop system is stable.
(d) The closed loop system is stable for all values the of α both positive and negative.

8. The feedback system shown below oscillates at 2 rad/s when

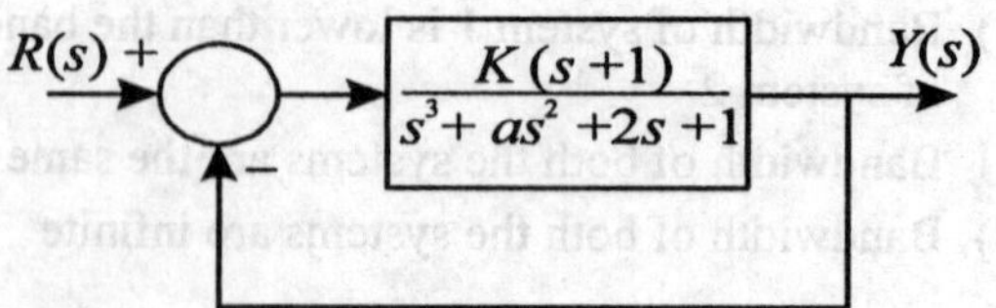

(a) $K = 2$ and $a = 0.75$
(b) $K = 3$ and $a = 0.75$
(c) $K = 4$ and $a = 0.5$
(d) $K = 2$ and $a = 0.5$

9. The number of open right half plane poles of $G(s) = \dfrac{10}{s^5 + 2s^4 + 3s^3 + 6s^3 + 5s + 3}$ is

(a) 0
(b) 1
(c) 2
(d) 3

10. None of the poles of **linear control system** lie in the right half of s-plane. For a bounded input, the output of this system

(a) Is always bounded
(b) Could be unbounded
(c) Always tends to zero
(d) None of the above

11. The **number of roots** of the equation

$$2s^4 + s^3 + 3s^2 + 5s + 7 = 0$$

that lie in the right half of S plane is;

(a) Zero (b) One
(c) Two (d) Three

12. The loop gain GH of a closed loop system is given by the following expression $\frac{K}{s(s+2)(s+4)}$

The value of K for which the system just becomes unstable is

(a) $K = 6$
(b) $K = 8$
(c) $K = 48$
(d) $K = 96$

13. For the equation, $s^3 - 4s^2 + s + 6 = 0$. The number of roots in the left half of s-plane will be

(a) Zero (b) One
(c) Two (d) Three

14. A unity feedback system, having an open loop gain $G(s)H(s) = \frac{K(1-s)}{(1+s)}$, becomes stable when

(a) $|K| > 1$ (b) $K > 1$
(c) $|K| < 1$ (d) $K < -1$

15. The system shown in the figure is

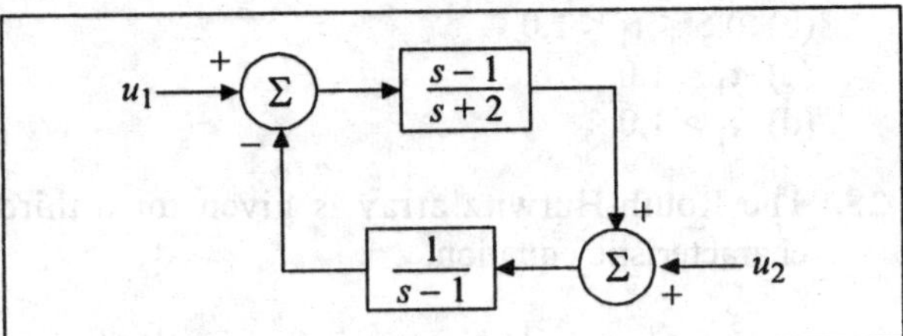

(a) Stable
(b) Unstable
(c) Conditionally stable
(d) Stable for input u_1, but unstable for input u_2.

16. If the loop gain K of a negative feedback system having a loop transfer function $\frac{K(s+3)}{(s+8)^2}$ is to be adjusted to induce a sustained oscillation then

(a) The frequency of this oscillation must be $\frac{4}{\sqrt{3}}$ rad/s
(b) The frequency of this oscillation must be 4 rad/s
(c) The frequency of this oscillation must be 4 or $\frac{4}{\sqrt{3}}$ rad/s
(d) Such a K does not exist

17. The first two rows of Routh's tabulation of a third order equation are as follows.

s^3	2	2
s^2	4	4

This means there are

(a) Two roots at $s = \pm j$ and one root in right half s-plane
(b) Two roots at s= ± j2 and one root in left half s-plane
(c) Two roots at s= ± j2 and one root in right half s-plane
(d) Two roots at $s = \pm j$ and one root in left half s-plane

18. Figure shows a feedback system where K > 0.

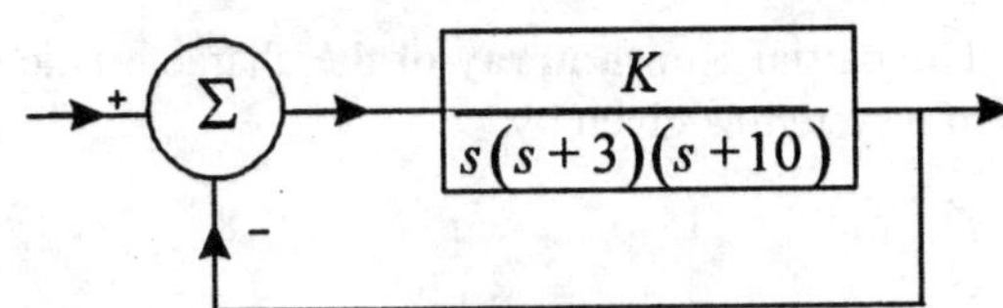

The range of K for which the system is stable will be zgiven by

(a) $0 < K < 30$
(b) $0 < K < 39$
(c) $0 < K < 390$
(d) $K < 390$

19. The feedback system shown below oscillates at 2 rad/s when

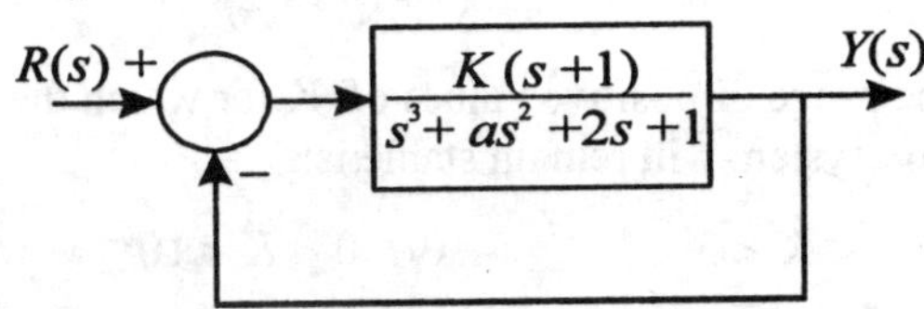

(a) $K = 2$ and $a = 0.75$
(b) $K = 3$ and $a = 0.75$
(c) $K = 4$ and $a = 0.5$
(d) $K = 2$ and $a = 0.5$

20. A continuous-time system is described by $y(t) = e^{-|x(t)|}$, where y(t) is the output and $x(t)$ is the input. $y(t)$ is bounded

(a) Only when $x(t)$ is bounded
(b) Only when $x(t)$ is non-negative
(c) Only for $t \leq 0$ if $x(t)$ is bounded $t \geq 0$
(d) Even when x(t) is not bounded

21. The algebraic equation

$F(s) = s^5 - 3s^4 + 5s^3 - 7s^2 + 4s + 20$ is given.

F(s) = 0 has

(a) A single complex root with the remaining roots being real

(b) One positive real root and four complex roots, all with positive real parts

(c) One negative real root, two imaginary roots, and two roots with positive real parts

(d) One positive real root, two imaginary roots, and two roots with negative real parts

22. The characteristic equation of a **feedback control system** is

$$2s^4 + s^3 + 3s^2 + 5s + 10 = 0$$

The number of roots in the right half of s-plane are

(a) Zero (b) 1

(c) 2 (d) 3

23. The partial **Routh array** of the characteristic equation of a system is given by.

s^4	1	a	8
s^3	3	12	

The system oscillates with a frequency of 2 rad/s. The value of the parameter 'a' of the system is.

(a) 6 (b) 2

(c) 8 (d) 12

24. The open loop transfer function of a unity gain fedback system is given by:

$$G(s) = \frac{K(s+3)}{(s+1)(s+2)}$$

The range of positive values of K for which the closed loop system will remain stable is:

(a) $1 < K < 3$ (b) $0 < K < 10$

(c) $5 < K < \infty$ (d) $0 < K < \infty$

25. The first two rows of Routh's table of a third - order characteristic equation are

s^3	3	3
s^2	4	4

It can be inferred that the system has

(a) One real pole in the right - half of s -plane

(b) A pair of complex conjugate poles in the right - half of s-plane

(c) A pair of real poles symmetrically placed around s = 0

(d) A pair of complex conjugate poles on the imaginary axis of the s-plane.

26. A closed loop control system is shown below. The range of the controller gain K_C which will make the real parts of all the closed loop poles more negative than –1 is.

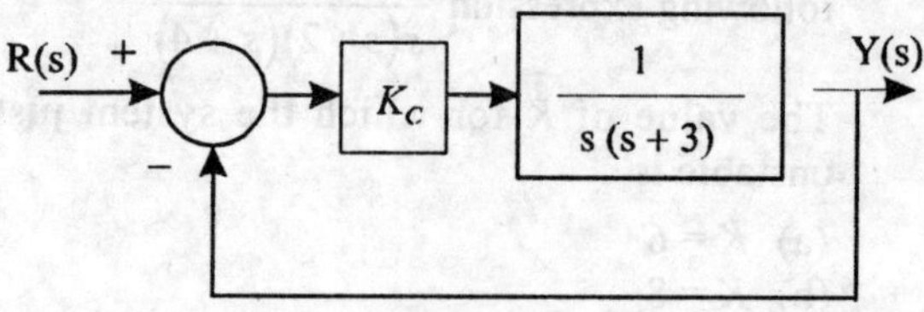

(a) $K_C > -4$ (b) $K_C > 0$

(c) $K_C > 2$ (d) $K_C < 2$

27. The closed loop control system shown in figure has $\tau_1 > 0$. The system will remain stable for all τ_1 in the range

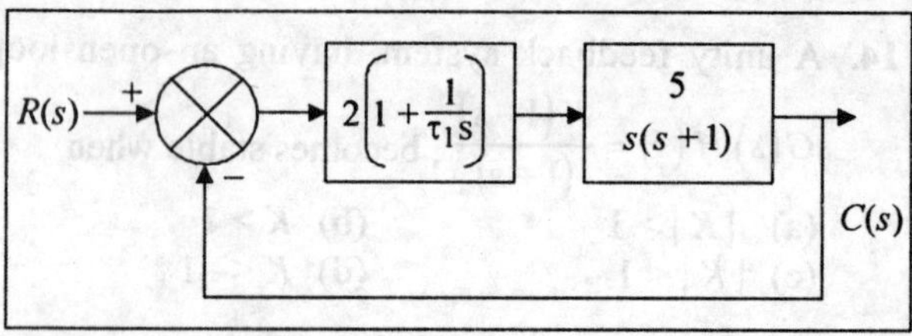

(a) $\tau_1 < 0.5$

(b) $0.5 < \tau_1 < 1.0$

(c) $\tau_1 < 1.0$

(d) $\tau_1 > 1.0$

28. The Routh-Hurwitz array is given for a third order characteristic equation.

s^3	1	b	0	0
s^2	a	c	0	0
s^1	$\frac{(6-K)}{3}$	0		
s^0	K			

The coefficients a, b, c and K are such that $a = 3$, $b > 0$, $c > 0$ and c is a function of K. The root locus for the corresponding equations is as shown in the given figure. The values of K and c for stability respectively are

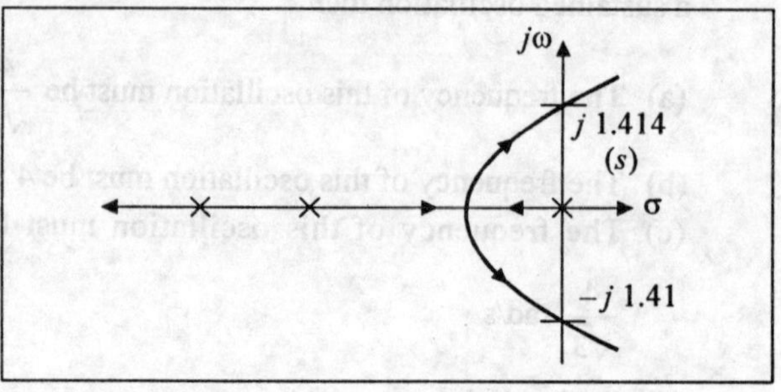

(a) 6 and 6
(b) 6 and 2
(c) 3 and 2
(d) 3 and 3

29. The range of the controller gain $\left(K_p, K_1\right)$ that makes the closed loop control system (shown in the following figure) stable is given as

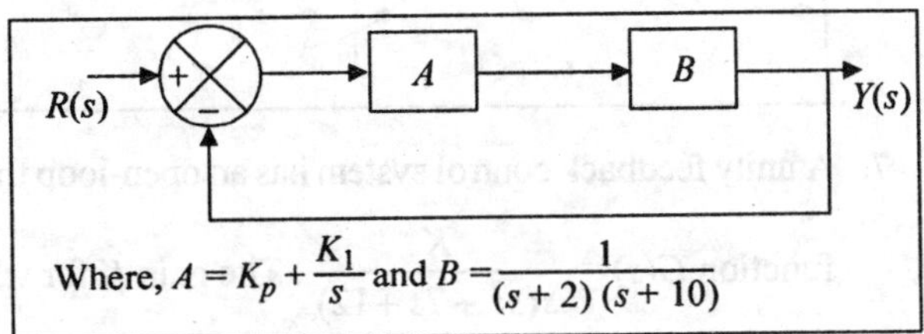

(a) $K_I < 0$ and $K_P < \frac{K_I}{12} - 20$

(b) $K_I < 0$ and $K_P > 0$

(c) $K_I < 0$ and $K_P > \frac{K_I}{12} - 20$

(d) $K_I > 0$ and $K_P > \frac{K_I}{12} - 20$

30. A certain closed loop system with unity feedback has the forward path transfer function given by $G(s) = \frac{K}{s(s+1)(s+5)}$. With the gain set at the ultimate value, the system will oscillate at an angular frequency of

(a) $\sqrt{6}$ rad/s

(b) $\sqrt{5}$ rad/s

(c) $\sqrt{2}$ rad/s

(d) $\sqrt{3}$ rad/s

10.4 Root Locus

1. The transfer function of a closed loop system is:

$$T(s) = \frac{K}{s^2 + (3-K)s + 1}$$

Where K is forward path gain. The root locus plot of the system is:

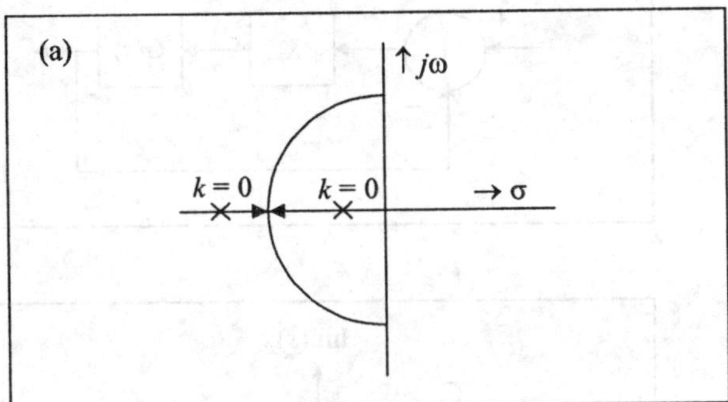

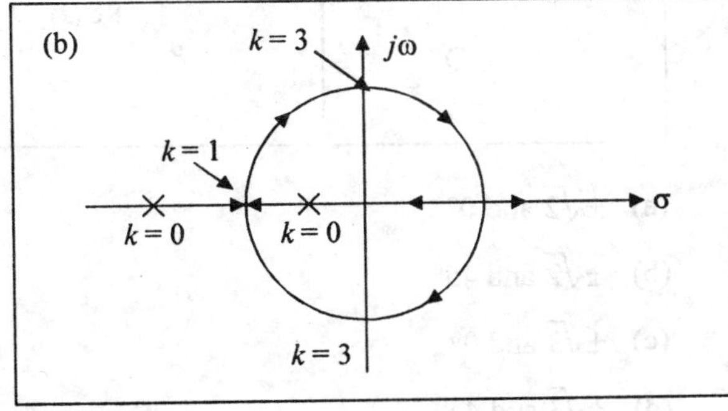

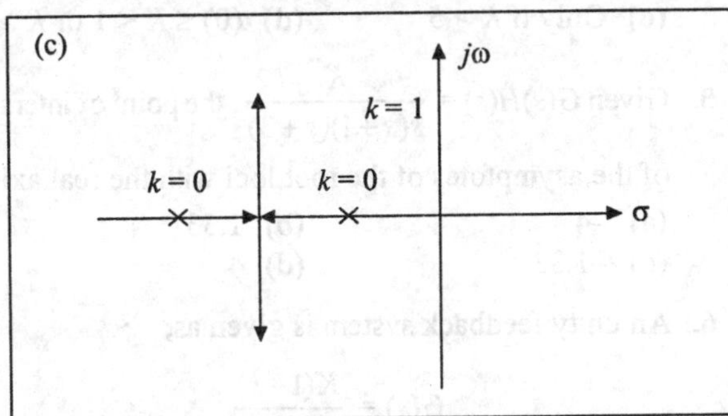

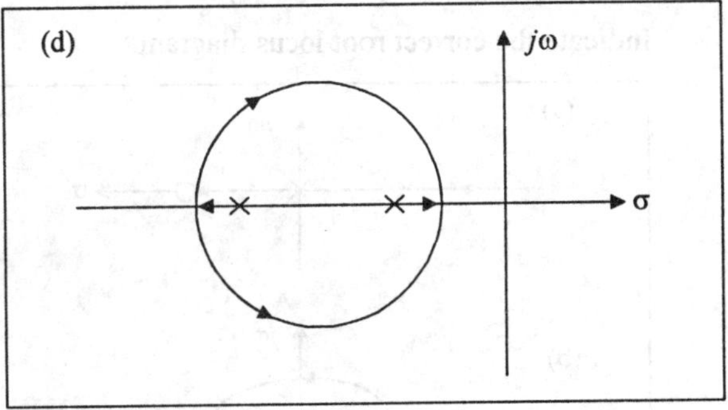

2. The characteristic equation of a feedback control system is given by

$$s^3 + 5s^2 + (K+6)\,s + K = 0$$

Where $K > 0$ is scalar variable parameter In the root-loci diagram of the system the asymptodes of the root-locus for large values of K meet at a point in the s-plane, whose coordinates are:

(a) (−3, 0)
(b) (−2, 0)
(c) (−1, 0)
(d) (2, 0)

3. Consider the points $s_1 = -3 + j4$ and $s_2 = -3 - j2$ in the s-plane. Then, for a system with the open-loop transfer function:

$$G(s)H(s) = \frac{K}{(s+1)^4}$$

(a) s_1 is on the root locus, but not s_2
(b) s_2 is on the root locus, but not s_1
(c) both s_1 and s_2 are on the root locus
(d) neither s_1 nor s_2 is on the root locus.

4. The root-locus diagram for a closed-loop feedback system is shown in Figure. The system is over-damped:

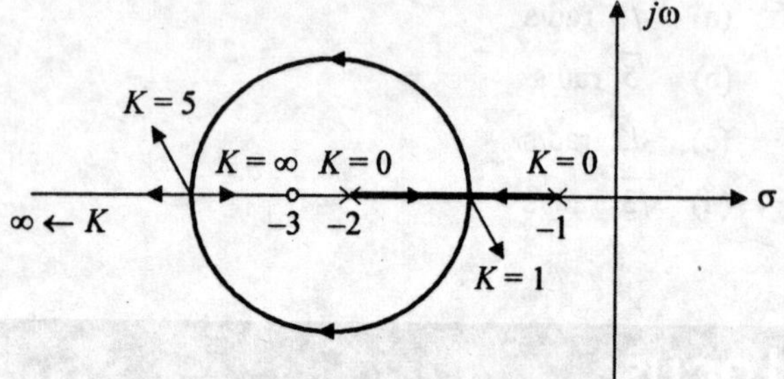

(a) Only if $0 \le K \le 1$ (b) Only if $1 < K < 5$
(c) Only if $K > 5$ (d) $(0) \le K < 1$ or $K > 5$

5. Given $G(s)H(s) = \frac{K}{s(s+1)(s+3)}$, the point of intersection of the asymptotes of the root loci with the real axis is
(a) −4 (b) 1.33
(c) −1.33 (d) 4

6. An unity feedback system is given as,

$$G(s) = \frac{K(1-s)}{s(s+3)}$$

Indicate the correct root locus diagram

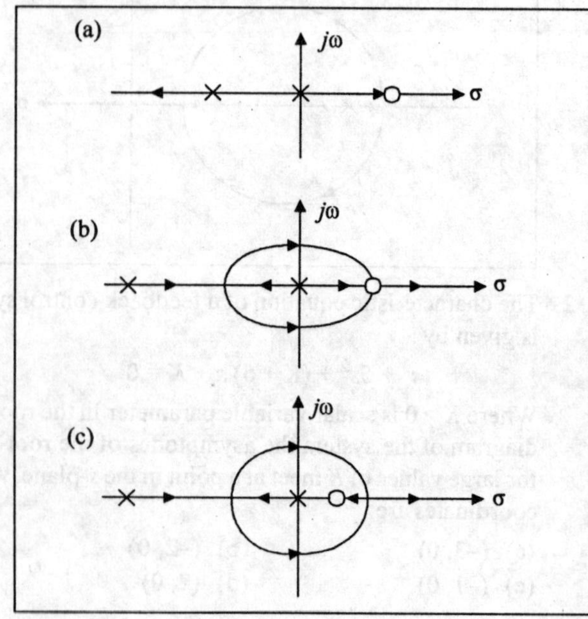

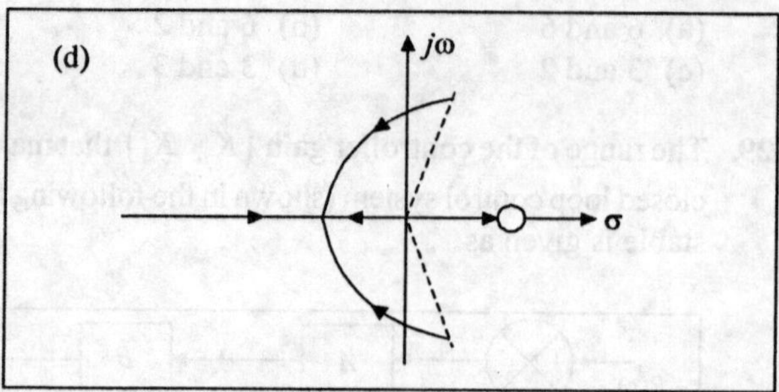

7. A unity feedback control system has an open-loop transfer function $G(s) = \frac{K}{s(s^2+7s+12)}$. The gain K for which $s = -1 + j1$ will lie on the root locus of this system is:
(a) 4
(b) 5.5
(c) 6.5
(d) 10

8. The feedback configuration and the pole-zero locations of $G(s) = \frac{s^2-2s+2}{s^2+2s+2}$ are shown below.
The root locus for **negative values** of K, i.e. for $-\infty < K < 0$, has break away/break-in points and angle of departure at pole P (with respect to the positive real axis) equal to

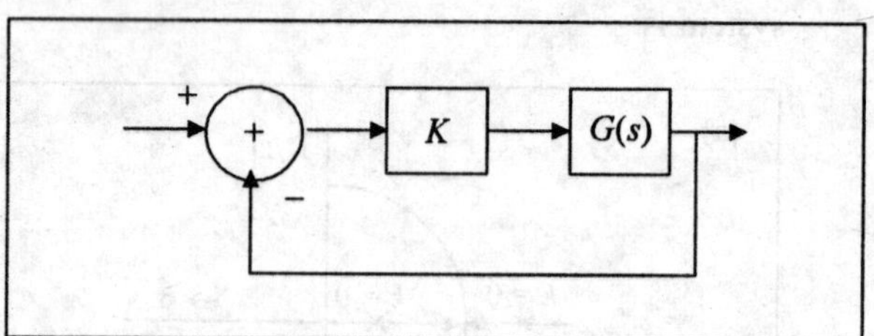

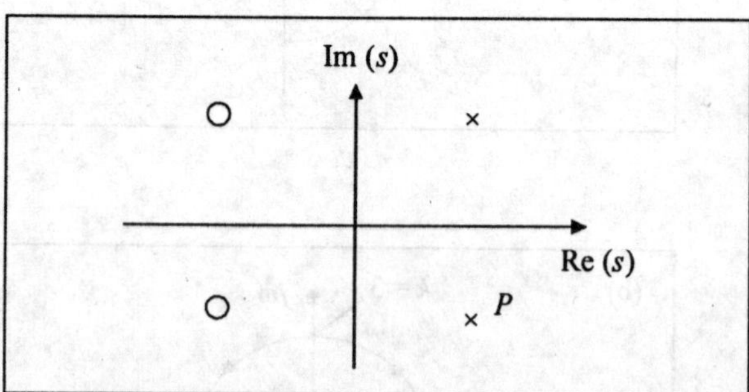

(a) $\pm\sqrt{2}$ and 0°
(b) $\pm\sqrt{2}$ and 45°
(c) $\pm\sqrt{3}$ and 0°
(d) $\pm\sqrt{3}$ and 45°

9. The root locus plot for a system is given below. The open loop transfer function corresponding to this plot is given by

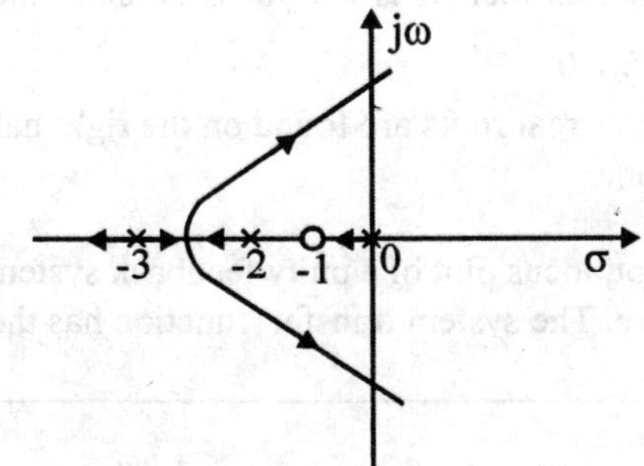

(a) $G(s)H(s) = k\dfrac{s(s+1)}{(s+2)(s+3)}$

(b) $G(s)H(S) = k\dfrac{(s+1)}{s(s+2)(s+3)^2}$

(c) $G(s)H(s) = k\dfrac{1}{s(s-1)(s+2)(s+3)}$

(d) $G(s)H(s) = k\dfrac{(s+1)}{s(s+2)(s+3)}$.

10. A unity feedback system has an open-loop transfer function of the form

$$KG(s) = \frac{K(s+a)}{s^2(s+b)};\ b > a$$

Which of the loci shown in figure can be valid root-loci for the system?

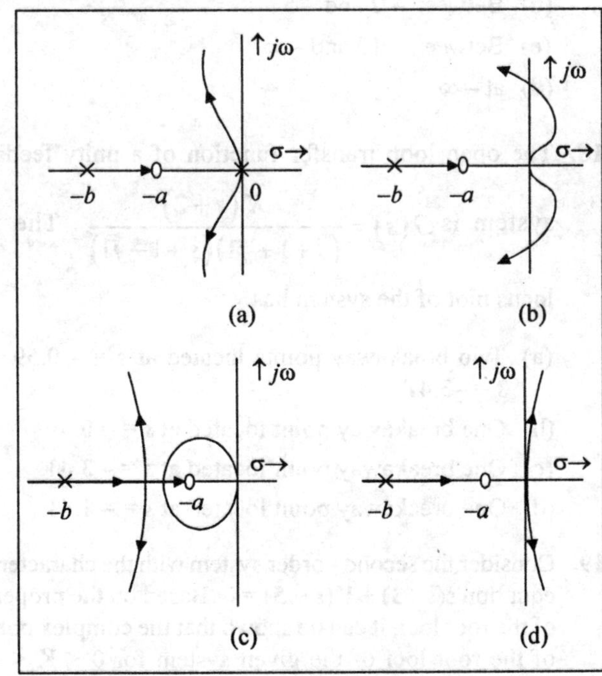

11. Which of the following figure (s) represent valid root loci in the *s*-plane for positive *K*? Assume that the system has a transfer function with real co-efficient.

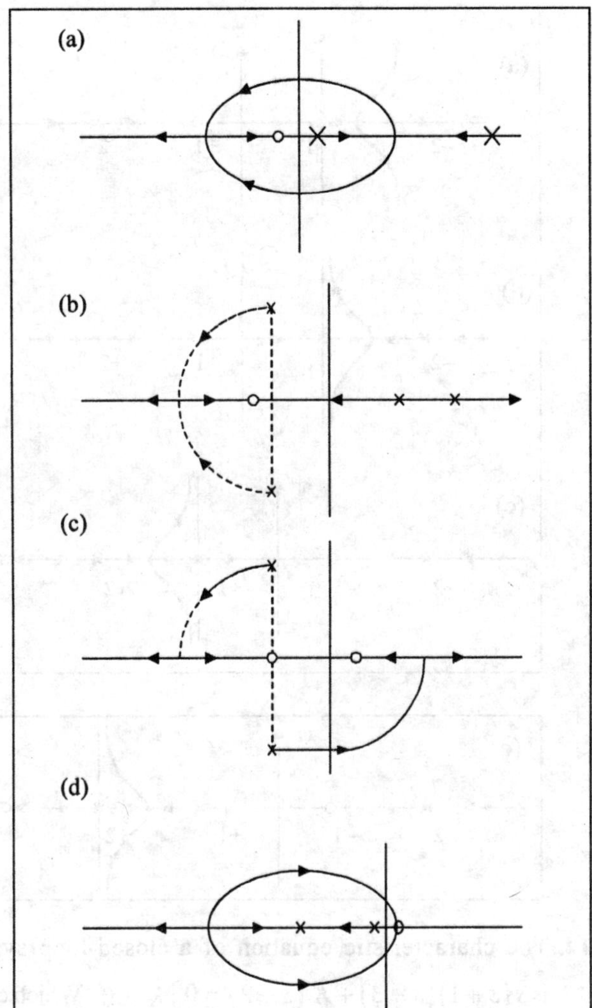

12. A unity feedback system as an open loop transfer function, $G(s) = [K/s^2]$. The root locus plot is:

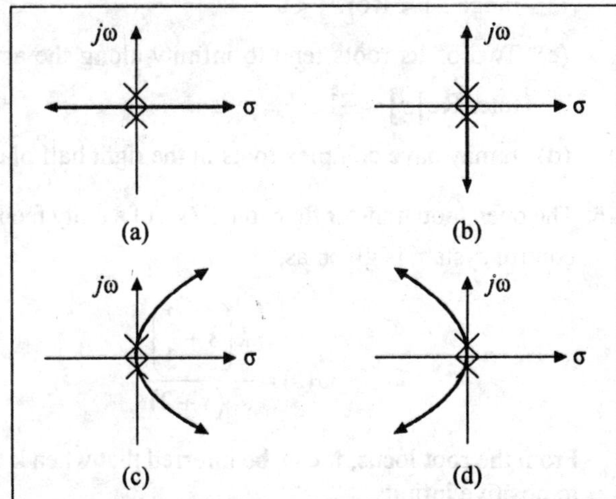

13. A closed-loop system has the characteristic function. $(s^2 - 4)(s + 1) + K(s - 1) = 0$ Its root locus plot against K is

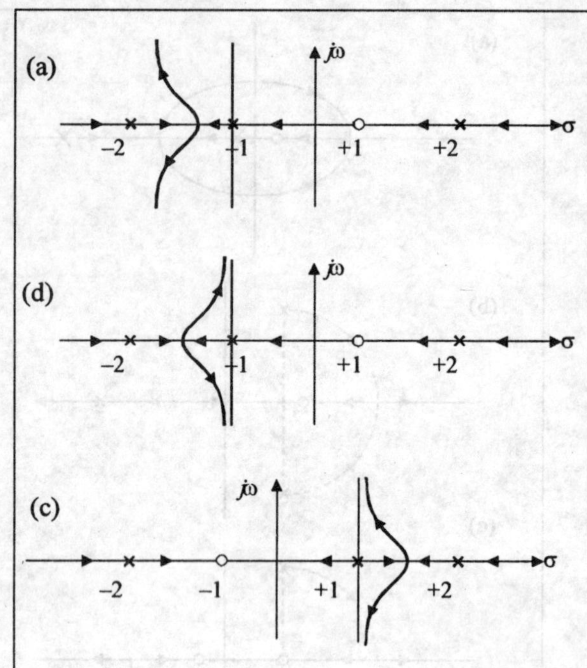

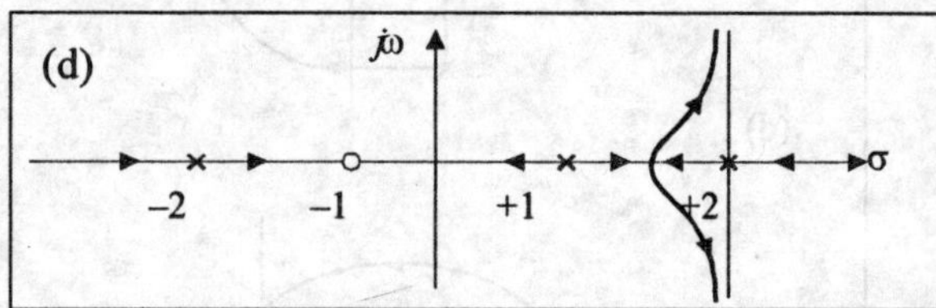

14. The characteristic equation of a closed-loop system is $s(s+1)(s+3)+K(s+2)=0$, $K>0$. Which of the following statements is true ?

(a) Its roots are always real

(b) It cannot have a breakaway point in the range $-1<\text{Re}[s]<0$

(c) Two of its roots tend to infinity along the asymptotes $\text{Re}[s]=-1$

(d) It may have complex roots in the right half plane.

15. The open loop transfer function $G(s)$ of a unity feedback control system is given as,

$$G(s)=\frac{k\left(s+\frac{2}{3}\right)}{s^2(s+2)}$$

From the root locus, it can be inferred that when k tends to positive infinity,

(a) Three roots with nearly equal real parts exist on the left half of the s-plane

(b) One real root is found on the right half of the s-plane

(c) The root loci cross the $j\omega$ axis for a finite value of $k; k \neq 0$

(d) Three real roots are found on the right half of the s-plane

16. The root, locus plot of a unity feedback system is shown in figure. The system transfer function has the form:

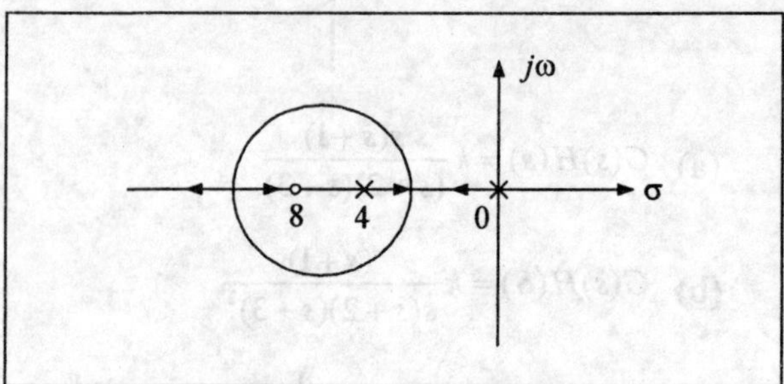

(a) $\dfrac{K}{s(s+4)(s+8)}$ (b) $\dfrac{K(s+8)}{s(s+4)}$

(c) $\dfrac{K(s+4)}{s(s+8)}$ (d) $\dfrac{Ks}{s(s+4)(s+8)}$

17. The break way point of the root locus from the real axis for a closed-loop system with a loop gain

$$G(s)\,H(s)=\frac{K(s+10)}{(s+2)\,(s+5)} \text{ lies}$$

(a) Between – 2 and origin

(b) Between – 2 and –5

(c) Between – 10 and – ∞

(d) at – ∞

18. The open loop transfer function of a unity feedback system is $G(s)=\dfrac{K(s+2)}{(s+1+j1)(s+1-j1)}$. The root locus plot of the system has

(a) Two breakaway points located at $s = -0.59$ and $s=-3.41$

(b) One breakaway point located at $s = -0.59$

(c) One breakaway point located at $s = -3.41$

(d) One breakaway point located at $s = -1.41$

19. Consider the second - order system with the characteristic equation $s(s+3) + K(s+5) = 0$. Based on the properties of the root loci, it can be shown that the complex portion of the root loci of the given system for $0 < K < \infty$ is described by a circle, and the two breakaway points on the real axis are

(a) $-5 \pm \sqrt{5}/2$

(b) $-5 \pm \sqrt{5}$

(c) $-5 \pm \sqrt{10}$

(d) $-5 + 2\sqrt{5}$

Common Dats for Questions 20 and 21

The root locus plot of a system with transfer function $\frac{2}{s(s+1)(s+2)}$, is shown in figure. A unity feedback proportional system is built using this system.

20. The maximum possible controller gain, for which the unity feedback system is stable, is approximately.

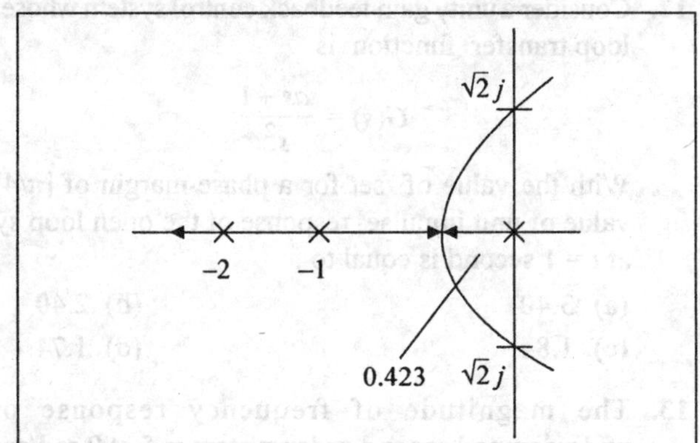

(a) 6.0 (b) 3.0
(c) 0.4 (d) 0.2

21. The maximum possible controller gain, for which the unity feedback system exhibits a non-oscillatory response to a unit step input, is

(a) 6.4

(b) 3.0

(c) 0.4

(d) 0.2

22. The root locus of a plant is given in following figure. The root locus crosses imaginary axis at $\omega = 4\sqrt{2}$ rad/s with gain $K = 384$. It is observed that the point $= -1.5 + j1.5$ lies in the root locus. The gain K at $s = -1.5 + j1.5$ is computed as

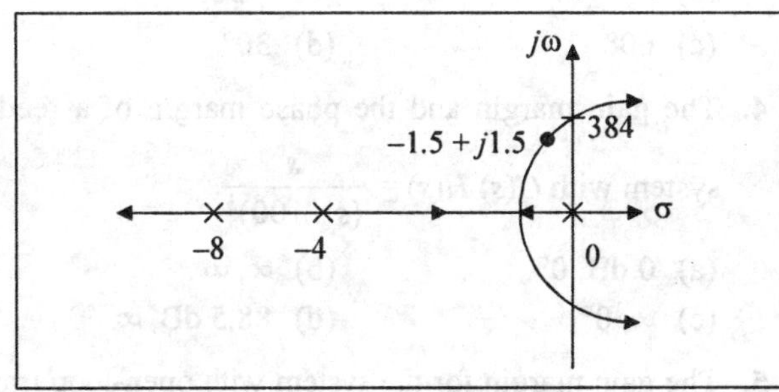

(a) 11.3 (b) 21.2
(c) 41.25 (d) 61.2

Statement for linked Answers Questions 23 and 24

A transfer function with unity DC gain has three-poles at −1, −2 and −3 and no finite zeros. A plant with this transfer function is connected with a proportional controller of gain K in the forward path, in a unity feedback configuration

23. The transfer function is

(a) $\frac{s}{(s-1)(s-2)(s-3)}$

(b) $\frac{6}{(s+1)(s+2)(s+3)}$

(c) $\frac{s}{(s+1)(s+1)(s+3)}$

(d) $\frac{6}{(s-1)(s-2)(s-3)}$

24. If the root locus plot of the closed loop system passes through the points $\pm j\sqrt{11}$, the maximum value of K for stability of the unity feedback closed loop system is

(a) $\sqrt{11}$ (b) 6
(c) 10 (d) $6\sqrt{11}$

10.5 Frequency Response

1. The open-loop transfer function of a feedback control system is:

$$G(s)\,H(s) = \frac{1}{(s+1)^3}$$

The gain margin of the system is:

(a) 2 (b) 4
(c) 8 (d) 16

2. The gain margin (in dB) of a system having the loop transfer function

$$G(s)\,H(s) = \frac{\sqrt{2}}{s(s+1)} \text{ is}$$

(a) 0 (b) 3
(c) 6 (d) ∞

3. The phase margin (in degrees) of a systems having the loop transfer function

$$G(s)\,H(s) = \frac{2\sqrt{3}}{s(s+1)} \text{ is}$$

(a) 45° (b) – 30°
(c) 60° (d) 30°

4. The gain margin and the phase margin of a feedback system with $G(s)\,H(s) = \dfrac{s}{(s+100)^3}$

(a) 0 dB, 0° (b) ∞, ∞
(c) ∞, 0° (d) 88.5 dB, ∞

5. The gain margin for the system with open-loop transfer function $G(s)\,H(s) = \dfrac{2(1+s)}{s^2}$ is

(a) ∞ (b) 0
(c) 1 (d) –∞

6. A system has poles at 0.01 Hz, 1 Hz and 80 Hz; zeros at 5Hz. 100 Hz and 200 Hz. The approximate phase of the system response at 20 Hz is:

(a) –90° (b) 0°
(c) 90° (d) –180°

7. The gain and phase crossover frequencies in rad/sec are, respectively

(a) 0.632 and 1.26
(b) 0.632 and 0.485
(c) 0.485 and 0.632
(d) 1.26 and 0.632

8. Based on the above results, the gain and phase margins of the system will be

(a) – 7.09 and 87.5°
(b) 7.09 and 87.5°
(c) 7.09 dB and – 87.5°
(d) – 7.09 dB and – 87.5°

9. The open-loop transfer function of a unity-gain feedback control system is given by

$$G(s) = \frac{K}{(s+1)\,(s+2)}$$

The gain margin of the system in dB is given by.

(a) 0 (b) 1
(c) 20 (d) ∞

10. Consider two transfer function

$$G_1(s) = \frac{1}{s^2 + as + b} \text{ and } G_2(s) = \frac{s}{s^2 + as + b}$$

The 3 dB bandwidths of their frequency responses are, respectively.

(a) $\sqrt{(a^2-4b)}, \sqrt{(a^2+4b)}$

(b) $\sqrt{(a^2+4b)}, \sqrt{(a^2-4b)}$

(c) $\sqrt{(a^2-4b)}, \sqrt{(a^2-4b)}$

(d) $\sqrt{(a^2+4b)}, \sqrt{(a^2+4b)}$

11. Consider a unity gain feedback control system whose open loop transfer function is

$$G(s) = \frac{as+1}{s^2}$$

The value of "a" so that the system has a phase margin equal to $\pi/4$ is approximately equal to

(a) 2.40 (b) 1.40
(c) 0.84 (c) 0.74

12. Consider a unity gain feedback control system whose open loop transfer function is

$$G(s) = \frac{as+1}{s^2}$$

With the value of set for a phase margin of [$\pi/4$], the value of unit impulse response of the open loop system at $t = 1$ second is equal to .

(a) 3.40 (b) 2.40
(c) 1.84 (d) 1.74

13. The magnitude of frequency response of an underdamped second order system is 5 at 0 rad/sec and peak to $\dfrac{10}{\sqrt{3}}$ at $5\sqrt{2}$ rad/sec. The transfer function of the system is

(a) $\dfrac{500}{s^2+10s+100}$ (b) $\dfrac{375}{s^2+5s+75}$

(c) $\dfrac{720}{s^2+12s+144}$ (d) $\dfrac{1125}{s^2+25s+225}$

Common Data for Questions 14 and 15

The input - output transfer function of a plant $H(s) = \dfrac{100}{s(s+10)^2}$. The plant is placed in a unity negative feedback configuration as shown in the figure below.

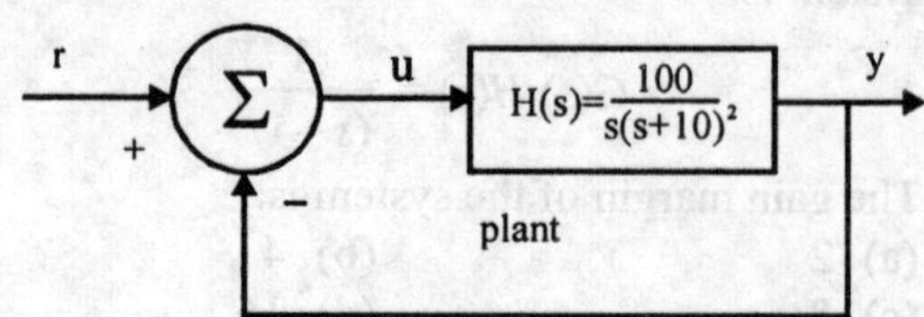

14. The signal flow graph that DOES NOT model th planet transfer function H(s) is

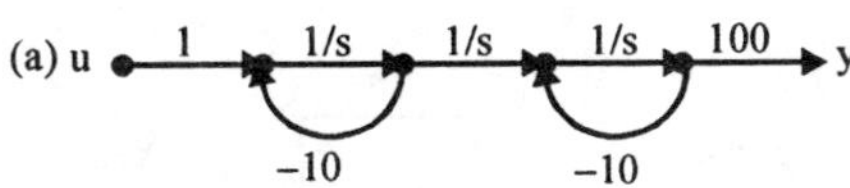

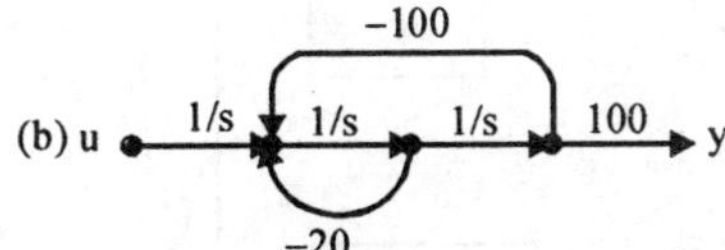

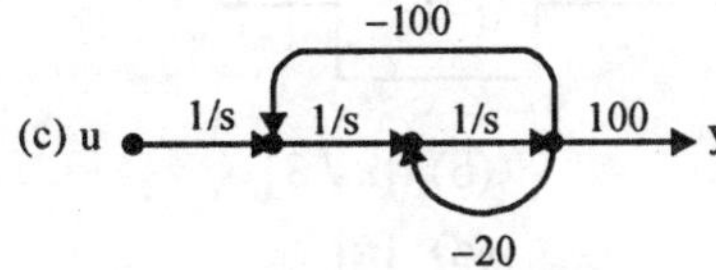

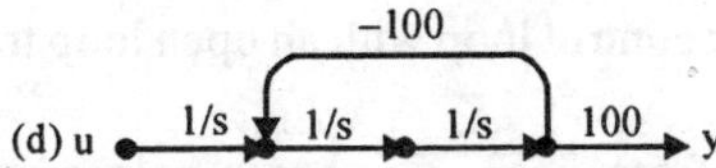

15. The gain margin of the system under closed loop unity negative feedback is

(a) 0 dB (b) 20 dB
(c) 26 dB (d) 46 dB

16. A system with the transfer function $\frac{Y(s)}{X(s)} = \frac{s}{s+p}$ has an output $y(t) = \cos\left(2t - \frac{\pi}{2}\right)$ for the input signal $x(t) = p\cos\left(2t - \frac{\pi}{2}\right)$. Then, the system parameter 'p' is

(a) $\sqrt{3}$

(b) $\frac{2}{\sqrt{3}}$

(c) 1

(d) $\frac{\sqrt{3}}{2}$

17. A system with transfer function

$$G(s) = \frac{(s^2+9)(s+2)}{(s+1)(s+3)(s+4)}$$

is excited by $\sin(\omega t)$. The steady-state output of the system is zero at

(a) $\omega = 1$ rad/s
(b) $\omega = 2$ rad/s
(c) $\omega = 3$ rad/s
(d) $\omega = 4$ rad/s

18. The open loop transfer function of a unity feedback control system is given as $G(s) = \frac{as+1}{s^2}$. The value of 'a' to give a phase margin of 45° is equal to

(a) 0.141
(b) 0.441
(c) 0.841
(d) 1.141

19. If the compensated system shown in the figure has a phase margin of 60° at the crossover frequency of 1 rad/sec, then value of the gain K is

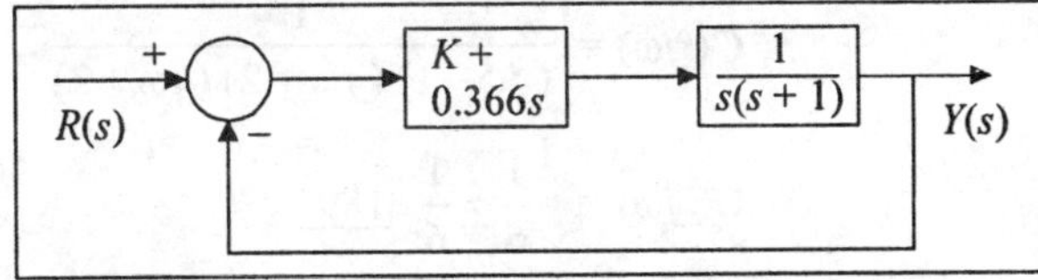

(a) 0.366 (b) 0.732
(c) 1.366 (d) 2.738

20. The open loop transfer function of a unity feedback system is given by $G(s) = (e^{-0.1s})/s$. The gain margin of this system is .

(a) 11·95 dB

(b) 17·67 dB

(c) 21·33 dB

(d) 23·9 dB

21. An open loop system represented by the transfer function $G(s) = \frac{(s-1)}{(s+2)(s+3)}$ is

(a) Stable and of the minimum phase type
(b) Stable and of the non-minimum phase type
(c) Unstable and of the minimum phase type
(d) Unstable and of the non-minimum phase type

22. The frequency response of a linear system $G(j\omega)$ is provided in the tabular form below

|G(jω)|	1.3	1.2	1.0	0.8	0.5	0.3
∠G(jω)	-130°	-140°	-150°	-160°	-180°	-200°

The gain margin and phase margin of the system are

(a) 6 dB and 30°
(b) 6 dB and –30°
(c) –6 dB and 30°
(d) –6 dB and –30°

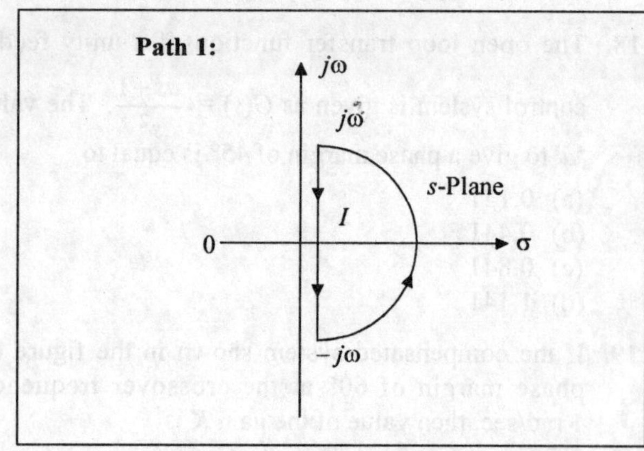

$$G(j\omega) = \frac{1}{(j\omega - 1)(j\omega + 2)(j\omega + 3)}$$

$$G(j\omega) = -\frac{1}{6} = \frac{1}{6}\underline{|189°}$$

Path II:

$$G(j\omega) = \frac{1}{(j\omega)^3} = \frac{1}{(R3^{j0})^3} \quad R \to \infty$$

$$= \delta e^{-j30} \quad \delta \to 0$$

Hence as ϕ goes from –90° to + 90° in path II, $G(j\omega)$ goes from 270° to –270° as shown below.

23. The open-loop transfer function of a feedback Control system is $\frac{1}{(s+1)^3}$. The gain margin of the system is:

(a) 16
(b) 8
(c) 4
(d) 2

24. The phase margin of the system for which loop gain $GH(j\omega) = \frac{1}{(j\omega + 1)^3}$ is

(a) $-\pi$
(b) π
(c) 0
(d) $\frac{\pi}{2}$

25. The loop transfer function of a system is given by $G(s)H(s) = \frac{10e^{-Ls}}{s}$ the phase cross-over frequency is 5 rad/s. The value of the dead time L is

(a) $\frac{\pi}{20}$ (b) $\frac{\pi}{10}$
(c) $-\frac{\pi}{20}$ (d) Zero

26. For the closed loop system shown below to be stable, the value of time delay T_D (in seconds) should be less than.

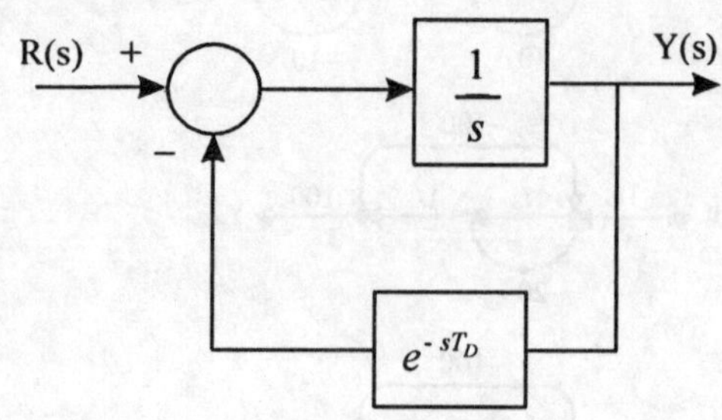

(a) $[\pi / 4]$ (b) $[\pi / 3]$
(c) $[\pi / 2]$ (d) $[\pi]$

27. A unity feedback control loop with an open loop transfer function of the from $\frac{K}{s(s+a)}$ has a gain crossover frequency of 1 rad/s and a phase margin of 60°. If an element having a transfer function $\frac{s-\sqrt{3}}{s+\sqrt{3}}$ is inserted into the loop, the phase margin will become.

(a) 0° (b) 30°
(c) 45° (d) 60°

Common Data for Questions 28 and 29:

The open-loop transfer function of a unity negative-feedback control system is given by

$$G(s) = \frac{K}{(s+5)^3}$$

28. The value of K for the phase margin of the system to be 45° is

(a) $250\sqrt{5}$ (b) $250\sqrt{2}$
(c) $125\sqrt{5}$ (d) $125\sqrt{2}$

29. The value of K for the damping ratio ζ to be 0.5, corresponding to the dominant closed - loop complex conjugate pole pair is

(a) 250
(b) 125
(c) 75
(d) 50

30. In control system design, gain and phase margins are usually provided to.

(a) Account for the uncertainties in the system.
(b) Make the system respond fast.
(c) Reduce the overshoot in step response
(d) Reduce the steady state offset.

10.6 Bode Plot

1. Non-minimum phase transfer function is defined as the transfer function:
   (a) Which has zeros in the right-half S-plane
   (b) Which has zeros only in the left-half S-plane
   (c) Which has poles in the right-half S-plane
   (d) Which has poles in the left-half S-plane.

2. In the Bode-plot of unity feedback control system, the value of phase of $G(\omega)$ at the gain cross over frequency is $-125°$. The phase margin of the system is
   (a) $-125°$ (b) $-55°$
   (c) $55°$ (d) $125°$

3. The approximate Bode magnitude plot of minimum-phase system is shown in Figure. The transfer function of the system is:

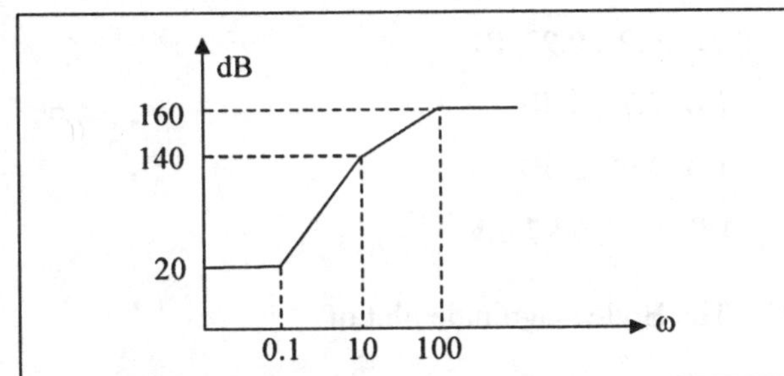

   (a) $\frac{10^8(s+0.1)^3}{(s+10)^2(s+100)}$ (b) $\frac{10^7(s+0.1)^3}{(s+10)(s+100)}$
   (c) $\frac{10^8(s+0.1)^2}{(s+10)^2(s+100)}$ (d) $\frac{10^9(s+0.1)^3}{(s+10)(s+100)^2}$

4. Consider the Bode magnitude plot shown in figure. The transfer function $H(s)$ is:

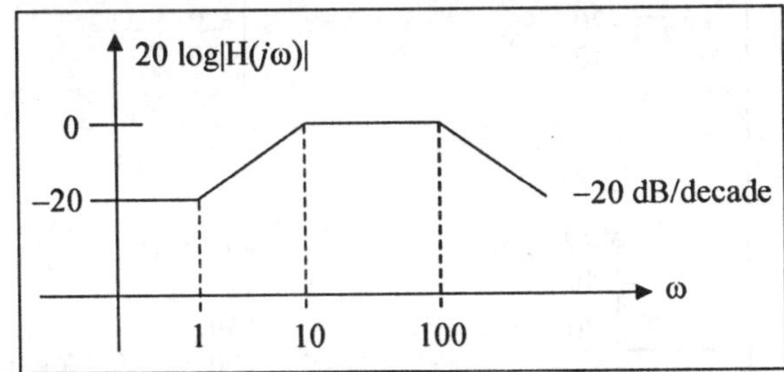

   (a) $\frac{(s+10)}{(s+1)(s+100)}$ (b) $\frac{10(s+1)}{(s+1)(s+100)}$
   (c) $\frac{10^2(s+1)}{(s+1)(s+100)}$ (d) $\frac{10^3(s+100)}{(s+1)(s+10)}$

5. The asymptotic Bode plot of a transfer function is as shown in the figure. The transfer function $G(s)$ corresponding to this Bode plot is:

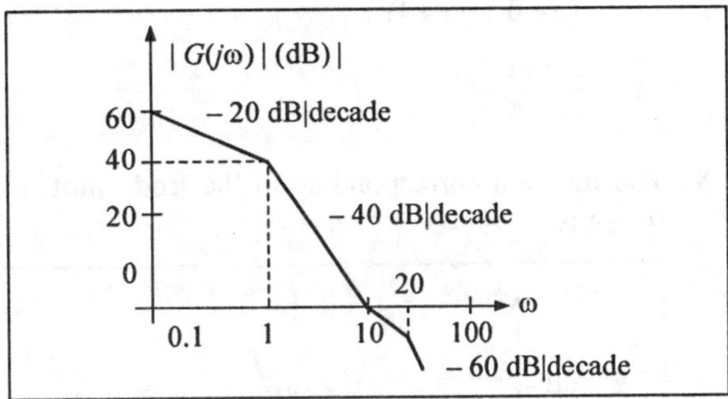

   (a) $\frac{1}{(s+1)(s+20)}$ (b) $\frac{1}{s(s+1)(s+20)}$
   (c) $\frac{100}{(s+1)(s+20)}$ (d) $\frac{100}{s(s+1)(1+0.05s)}$

6. For the asymptotic Bode magnitude plot shown below, the system transfer function can be

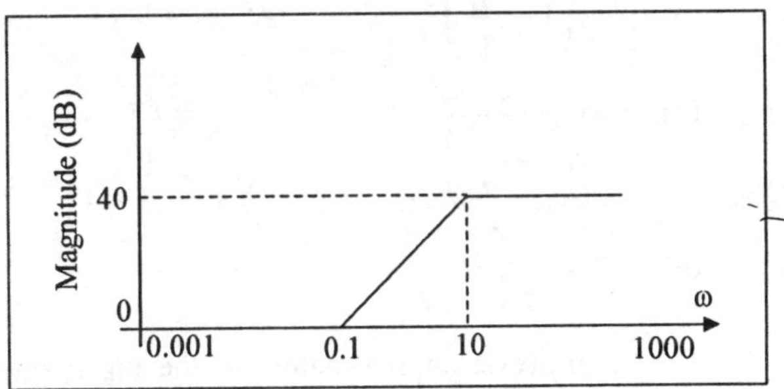

   (a) $\frac{10s+1}{0.1s+1}$ (b) $\frac{100s+1}{0.1s+1}$
   (c) $\frac{100s}{10s+1}$ (d) $\frac{0.1s+1}{10s+1}$

7. The system having the **Bode magnitude plot** shown in figure has the transfer function.

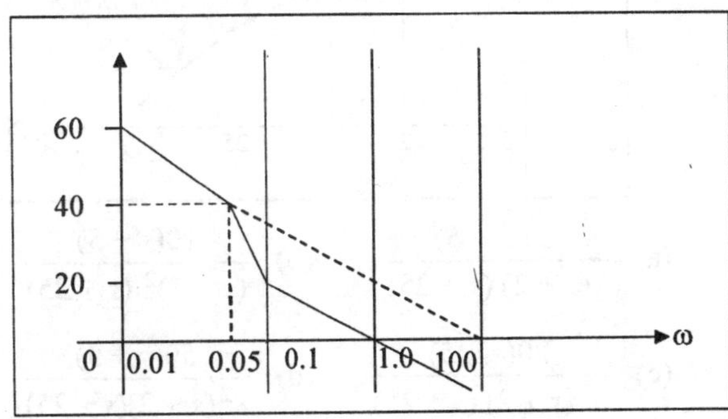

(a) $\frac{60(s+0.01)(s+0.1)}{s^2(s+0.05)^2}$

(b) $\frac{10(1+10s)}{s(1+20s)}$

(c) $\frac{3(s+0.05)}{s(s+0.1)(s+1)}$

(d) $\frac{5(s+0.1)}{s(s+0.05)}$

8. The function corresponding to the **Bode plot** of given figure is

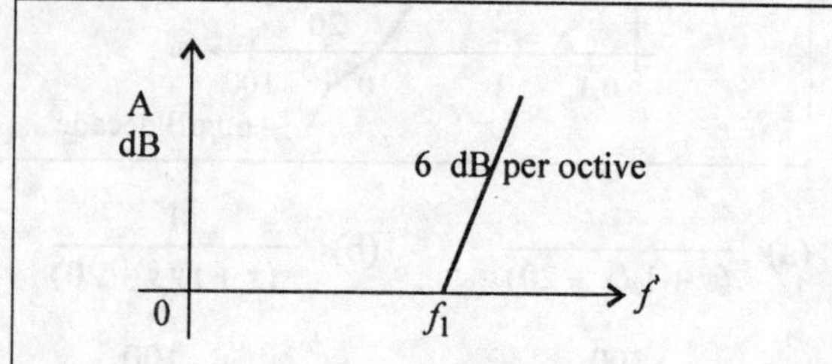

(a) $A = \frac{jf}{f_1}$

(b) $A = \frac{1}{\left(1 - \frac{jf_1}{1}\right)}$

(c) $A = \frac{1}{\left(1 + \frac{jf_1}{f}\right)}$

(d) $A = 1 + \frac{jf}{f_1}$

9. The **asymptotic** approximately of the **log-magnitude versus frequency plot** of a minimum phase system with real poles and one zero is shown in Figure. Its transfer function is

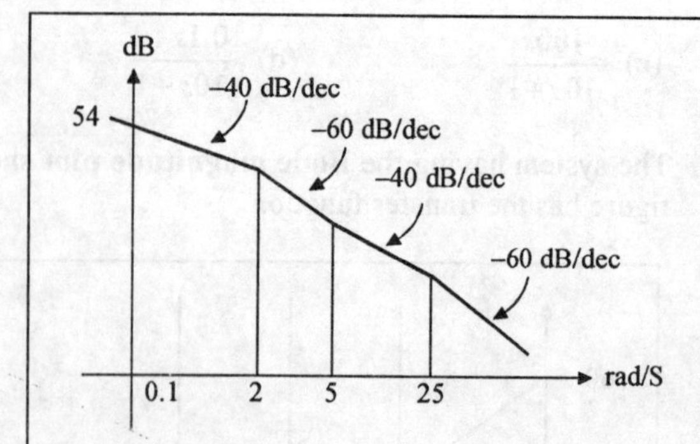

(a) $\frac{20(s+5)}{s(s+2)(s+25)}$ (b) $\frac{10(s+5)}{(s+2)^2(s+25)}$

(c) $\frac{20(s+5)}{s^2(s+2)(s+25)}$ (d) $\frac{50(s+5)}{s^2(s+2)(s+25)}$

10. The asymptotic Bode plot of the transfer function $\frac{K}{1+(s/a)}$ is given in figure. The error in phase angle and dB gain at a frequency of $\omega = 0.5$ a are respectively

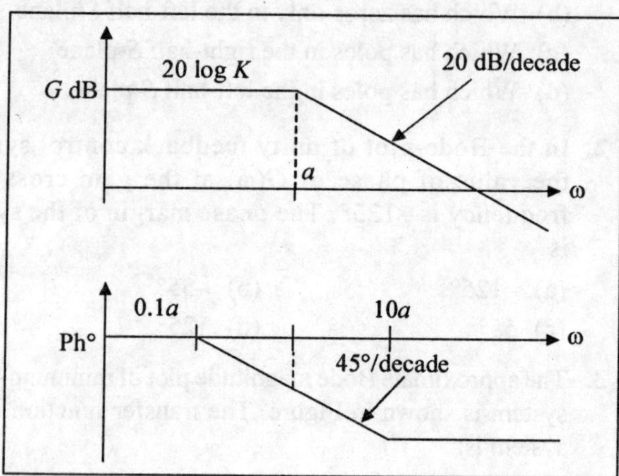

(a) 4.9°, 0.97 dB

(b) 5.7°. 3 dB

(c) 4.9°, 3 dB

(d) 5.7°, 0.97 dB

11. The Bode magnitude plot of

$$H(j\omega) = \frac{10^4(1+j\omega)}{(10+j\omega)(100+j\omega)^2} \text{ is}$$

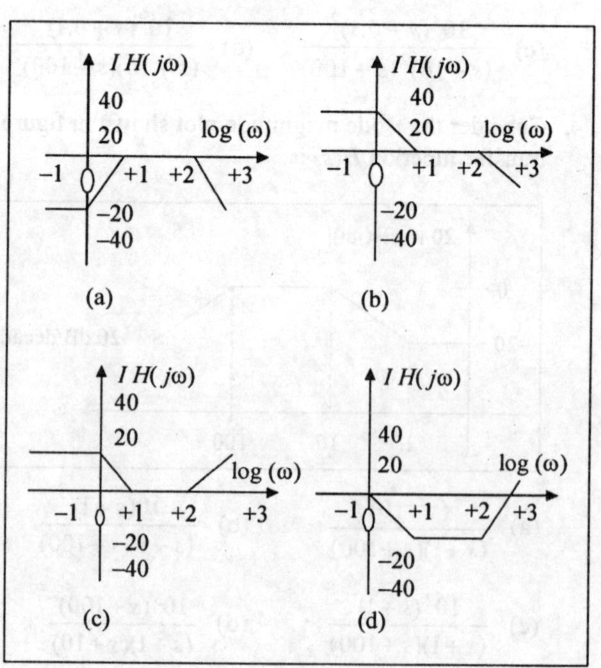

12. The asymptotic approximation of the log-magnitude vs frequency plot of a system containing only real poles and zeros is shown. Its transfer function is

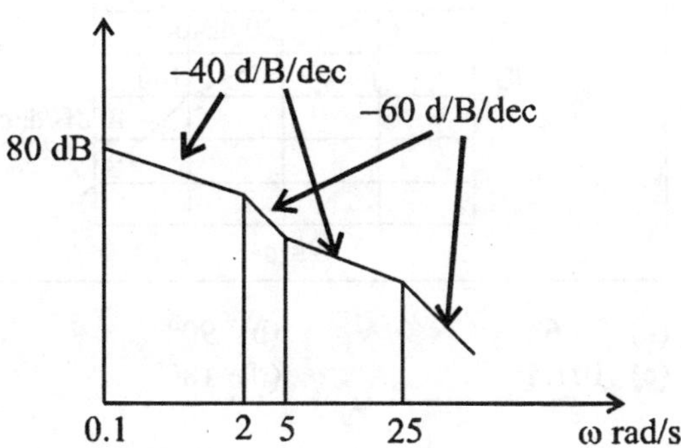

(a) $\frac{10(s+5)}{s(s+2)(s+25)}$

(b) $\frac{1000(s+5)}{s^2(s+2)(s+25)}$

(c) $\frac{100(s+5)}{s^2(s+2)(s+25)}$

(d) $\frac{80(s+5)}{s^2(s+2)(s+25)}$

13. The asymptotic Bode plot for the gain magnitude of a minimum phase system $G(s)$ is shown in figure:

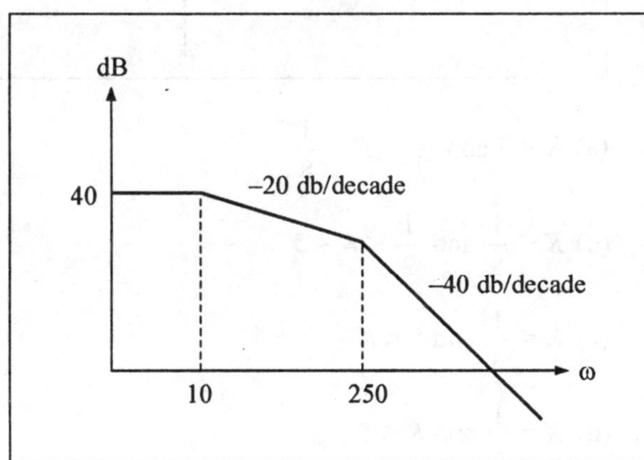

The transfer function is $G(s) =$

(a) $\frac{100}{\left(1+\frac{s}{10}\right)\left(1+\frac{s}{250}\right)}$ (b) $\frac{40}{s(s+250)}$

(c) $\frac{100}{(s+10)(s+250)}$ (d) $\frac{100s}{(s+1)(s+250)}$

14. The Bode asymptotic plot of a transfer function is given below. In the frequency range shown, the transfer function has.

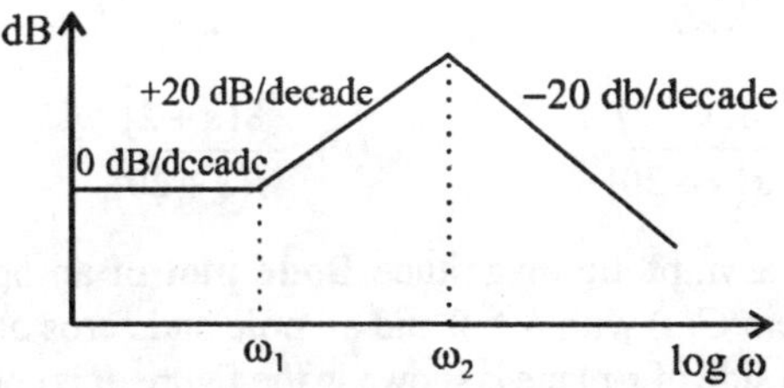

(a) 3 poles and 1 zero

(b) 1 pole and 2 zeros

(c) 2 poles and 1 zero

(d) 2 poles and 2 zeros

15. The asymptotic Bode magnitude plot of a lead network with its pole and zero on the left half of the s-plane is shown in the ajoining figure. The frequency at which the phase angle of the network is maximum (in rad/s) is

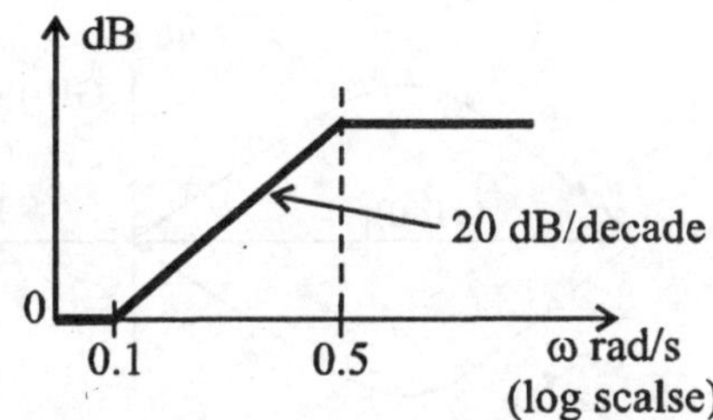

(a) $\frac{3}{\sqrt{10}}$

(b) $\frac{1}{\sqrt{20}}$

(c) $\frac{1}{20}$

(d) $\frac{1}{30}$

16. Figure shows the Bode magnitude plot of a system. The minimum phase transfer function of the system is given by

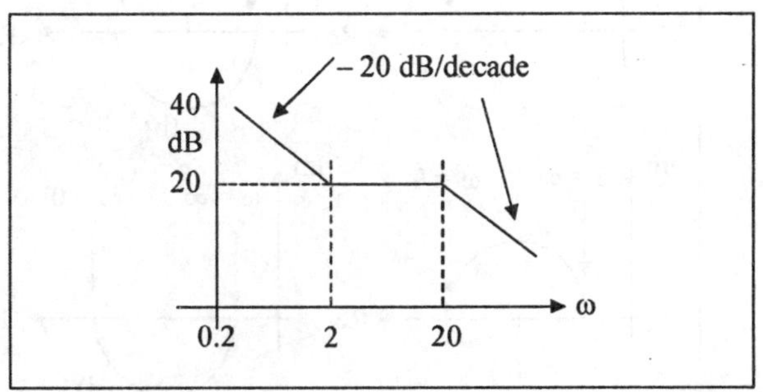

(a) $\dfrac{80\left(\dfrac{s}{2}+1\right)}{s\left(\dfrac{s}{20}+1\right)}$ (b) $\dfrac{20\left(\dfrac{s}{2}+1\right)}{s\left(\dfrac{s}{20}+1\right)}$

(c) $\dfrac{8(s+2)}{s(s+20)}$ (d) $\dfrac{8(s+2)}{s(s+20)}$

17. The asymptotic magnitude **Bode plot** of an open loop system $G(s)$ with $K > 0$ and all pole and zeros on the left hand side of s-plane is shown in the figure. It is completely symmetric about ω_c. The minimum absolute phase angle contribution by $G(s)$ is given by

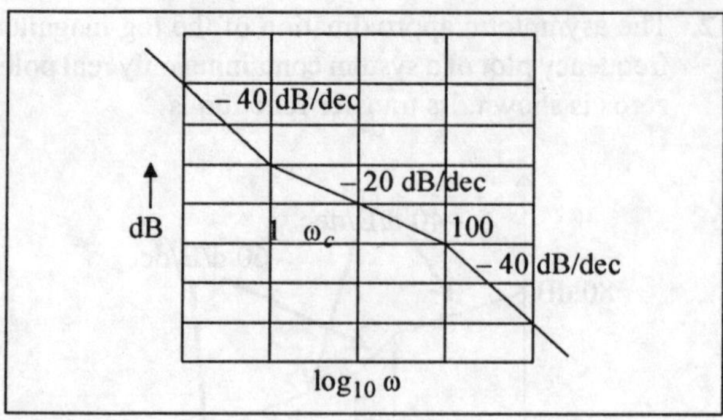

(a) 78.6° (b) 90°
(c) 101.4° (d) 180°

10.7 Nyquist Plot

1. Figure shows the Nyquist plot of the open-loop transfer function $G(s)\ H(s)$ of a system. If $G(s)\ H(s)$ has one right-hand pole, the closed-loop system is:

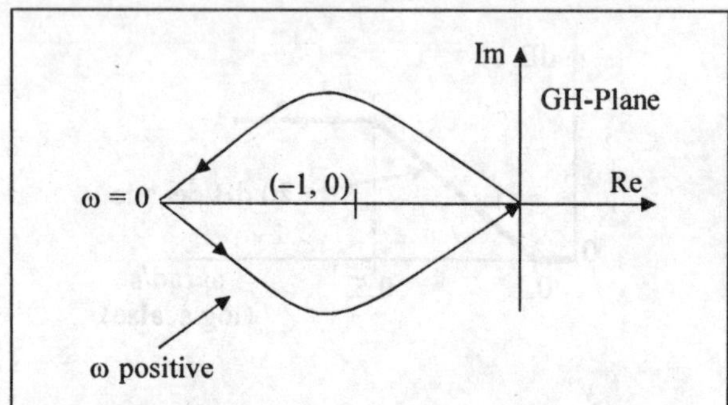

(a) Always stable
(b) Unstable with one closed-loop right hand pole
(c) Unstable with two closed-loop right poles
(d) Unstable with three closed-loop right hand poles

2. Which one of the following polar diagrams corresponds to a lag network?

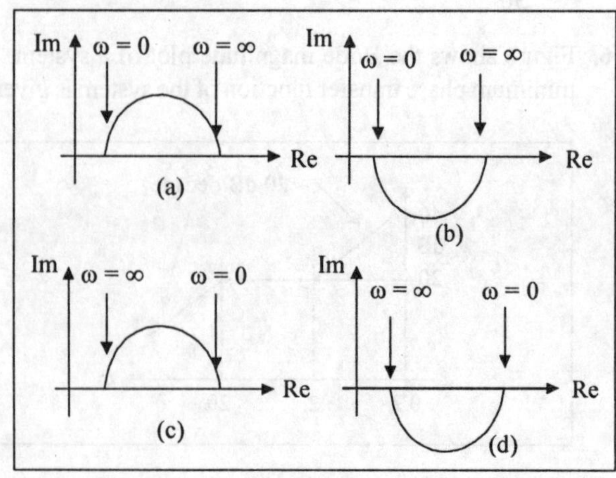

3. The polar diagram of a conditionally stable system for open gain $K = 1$ is shown in the figure. The open loop transfer function of the system is known to be stable. The closed loop system is stable for

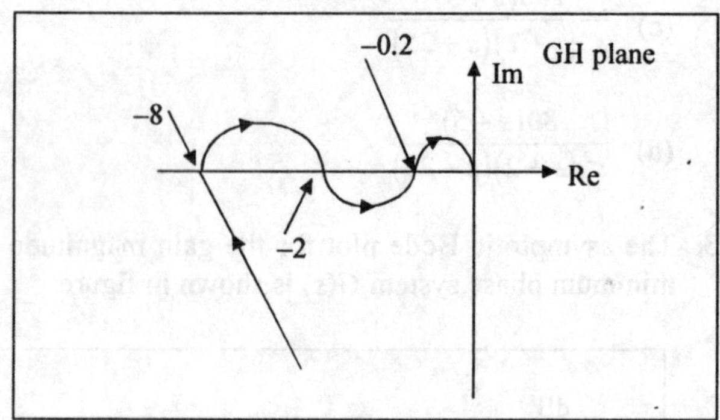

(a) $K < 5$ and $\dfrac{1}{2} < K < \dfrac{1}{8}$.

(b) $K < \dfrac{1}{8}$ and $\dfrac{1}{2} < K < 5$.

(c) $K < \dfrac{1}{8}$ and $5 < K$.

(d) $K > \dfrac{1}{8}$ and $K < 5$.

4. The Nyquist plot of $G(j\omega)\ H(j\omega)$ for a closed loop control system, passes through $(-1, j0)$ point in the GH plane. The gain margin of the system is dB is equal to:

(a) Infinite
(c) Greater than zero
(c) Less than zero
(d) Zero

Common Data for Questions 5 and 6

The Nyquist plot of a stable transfer function $G(s)$ is shown in the figure. We are interested in the stability of the closed loop system in the feedback configuration shown.

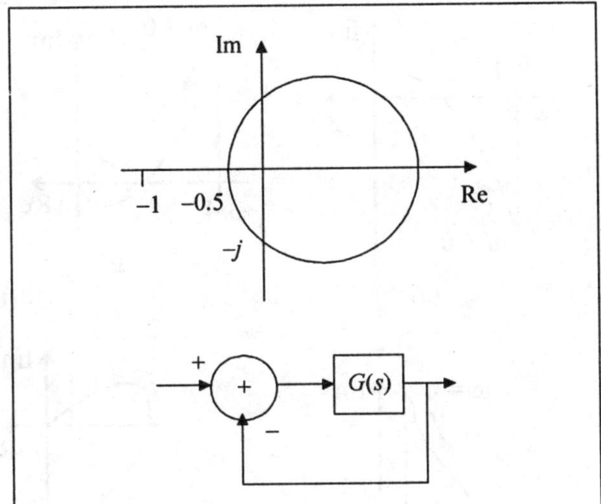

5. Which of the following statements is true ?
   (a) $G(s)$ is an all pass filter
   (b) $G(s)$ has a zero in the right half plane
   (c) $G(s)$ is the impedance of a passive network
   (d) $G(s)$ is marginally stable
6. The gain and phase margins of $G(s)$ for closed loop stability are
   (a) 6 dB and 180°
   (b) 3 dB and 180°
   (c) 6 dB and 90°
   (d) 3 dB and 90°
7. For the tranfer function $G(j\omega) = 5 + j\omega$, the corresponding Nyquist plot for positive freqeuncy has the form

(a) 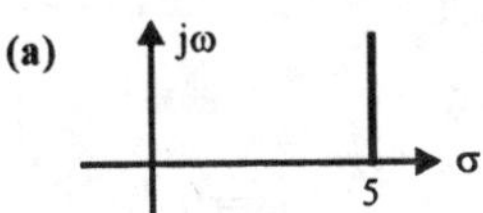

(b) 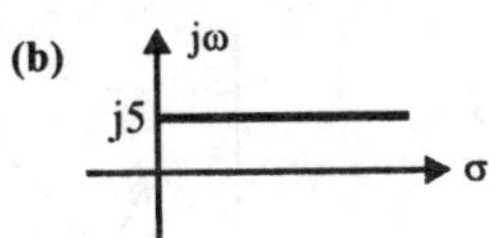

(c) 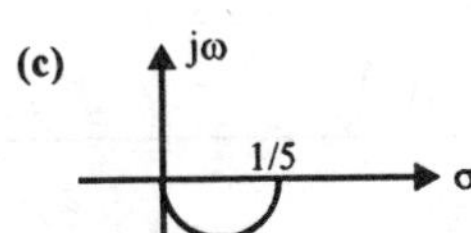

(d) 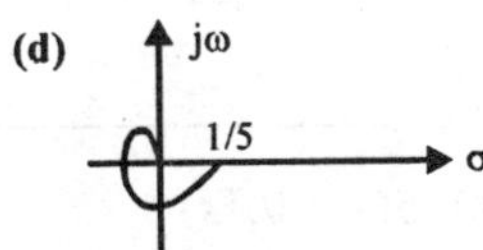

8. Which of the following is the transfer function of a system having the **Nyquist plot** in figure.

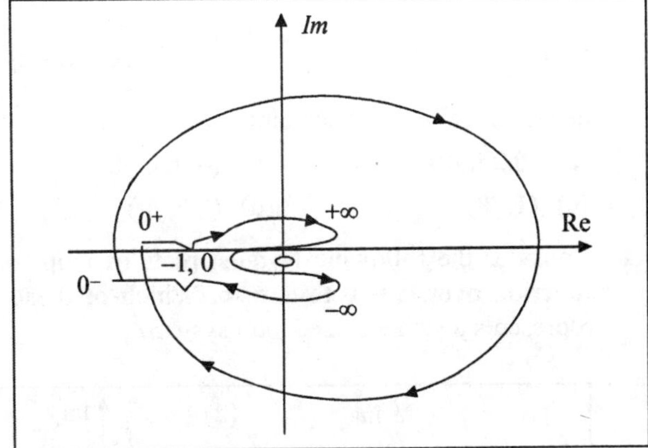

(a) $\dfrac{K}{s(s+2)^2(s+5)}$

(b) $\dfrac{K}{s^2(s+2)(s+5)}$

(c) $\dfrac{K(s+1)}{s^2(s+2)(s+5)}$

(d) $\dfrac{K(s+1)(s+3)}{s^2(s+2)(s+5)}$

9. A unity feedback system has the open loop transfer function

$$G(s) = \frac{1}{(s-1)(s+2)(s+3)}$$

The **Nyquist plot** of G encircles the origin
   (a) Never
   (b) Once
   (c) Twice
   (d) Thrice
10. The **polar plot** of a type-1, 3-pole, open-loop system is shown in Figure. The closed-loop system is

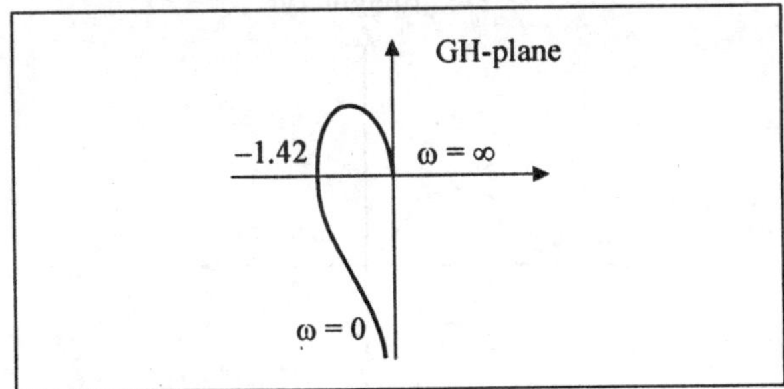

   (a) Always stable
   (b) Marginally stable
   (c) Unstable with one pole on the right half s-plane
   (d) Unstable with two poles on the right s-plane

11. In the $GH(s)$ plane, the Nyquist plot of the loop transfer function $G(s)H(s) = \frac{\pi e^{-0.25s}}{s}$ passes through the negative real axis at the point

(a) $(0.25, j0)$ (b) $(-0.5, j0)$
(c) $(1, j0)$ (d) $(-2, j0)$

12. Consider the following Nyquist plots of loop transfer function over $\omega = 0$ to $\omega = \infty$. Which of these plots represents a stable closed loop system?

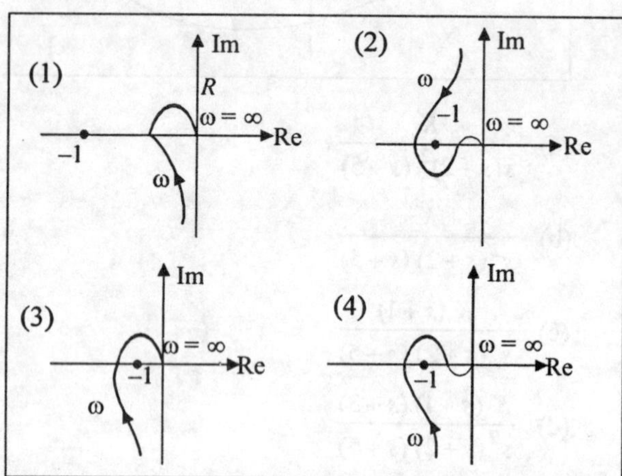

(a) (1) only (b) All, except (1)
(c) All, except (3) (d) (1) and (2) only

13. If $x = \text{Re } G(j\omega)$, and $y = \text{Im } G(j\omega)$ then for $\omega \to 0^+$, the Nyquist for $G(s) = \frac{1}{s(s+1)(s+2)}$

(a) $x = 0$ (b) $x = -3/4$
(c) $x = y - 1/6$ (d) $x = y/\sqrt{3}$

14. The polar plot of an open loop stable system is shown below. The closed loop system is

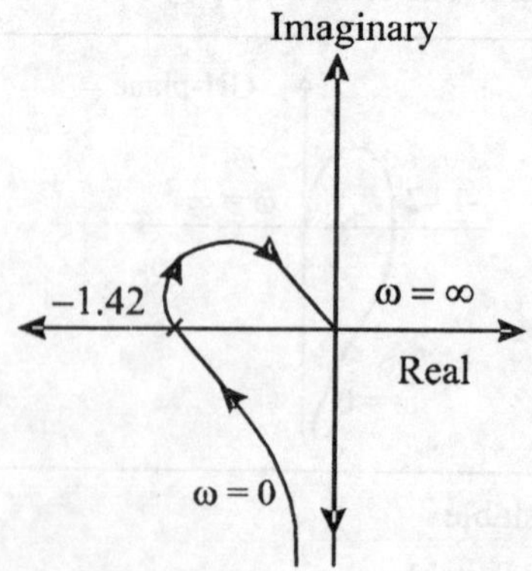

(a) Always stable
(b) Marginally stable
(c) Unstable with one pole on the RH s-plane
(d) Unstable with two poles on the RH-s-plane

15. The frequency response of $G(s) = 1/[s(s+1)(s+2)]$ plotted in the complex $G(j\omega)$ plane (for $0 < \omega < \infty$) is

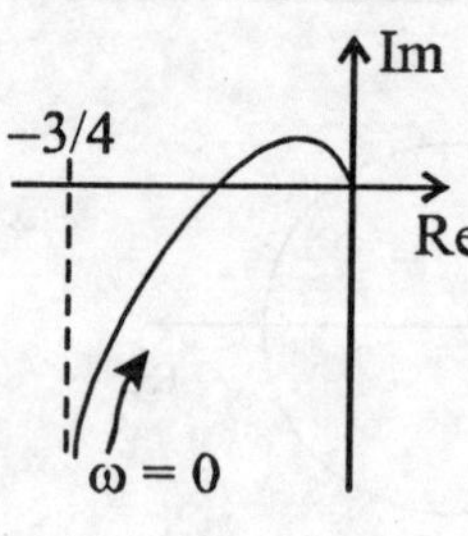

(a)

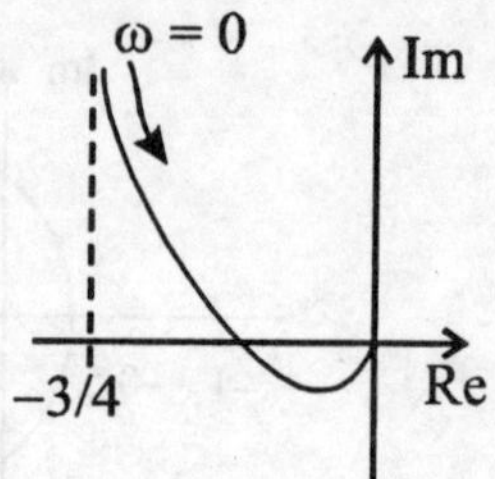

(b)

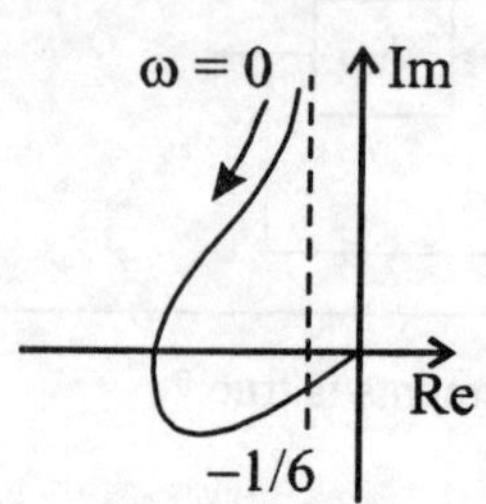

(c)

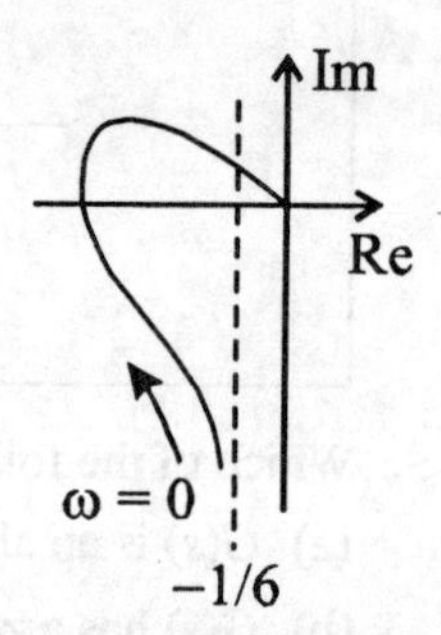

(d)

16. The Nyquist plot of loop transfer function $G(s) H(s)$ of a closed loop control system passes through the point $(-1, j0)$ in the $G(s) H(s)$ plane. The phase margin of the system is

(a) 0° (b) 45°
(c) 90° (d) 180°

17. Figure shows the polar plot of a system. The transfer function of the system is

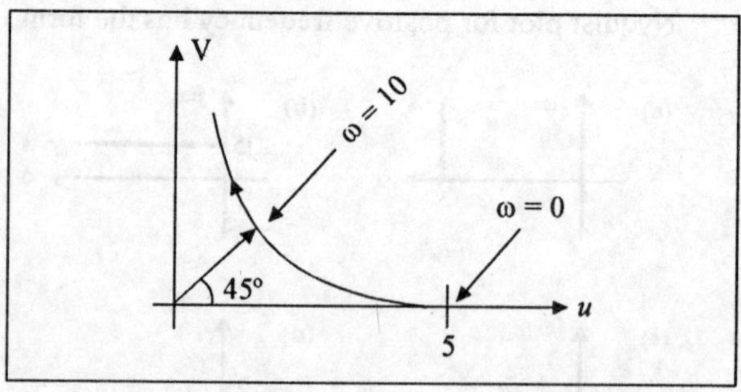

(a) $5(1 + 0.1s)$
(b) $(1 + 0.5s)$
(c) $5(1 + 10s)$
(d) $5(1 + s)$

18. The loop transfer function of a system has a pole in the right half of the s-plane. The **Nyquist plot** makes one clock wise encirclement of (–1, 0) point. The closed loop system is.

(a) Unstable
(b) Stable
(c) Marginally stable
(d) Asymptotically stable

10.8 Controller or Compensator

1. The transfer function of a simple RC network function as a controller is:

$$G_c(s) = \frac{s+z_1}{s+P_1}$$

The condition for the RC network to act as a **phase lead** controller is:

(a) $p_1 < z_1$
(b) $p_1 = 0$
(c) $p_1 = z_1$
(d) $p_1 > z_1$

2. The transfer function of a phase lead controller is $\frac{1+3Ts}{1+Ts}$. The maximum value of phase provided by this controller is

(a) 90°
(b) 45°
(c) 45°
(d) 30°

3. A double integrator plant, $G(s) = \frac{K}{s^2}$, $H(s) = 1$ is to be compensated to achieve the damping ratio $\delta = 0.5$, and an undamped natural frequency, $\omega_n = 5$ rad/s. Which one of the following compensator $G_e(s)$ will be suitable?

(a) $\frac{s+3}{s+9.9}$ (b) $\frac{s+9.9}{s+3}$

(c) $\frac{s-6}{s+8.33}$ (d) $\frac{s-6}{s}$

4. The open-loop transfer function of a plant is given as $G(s) = \frac{1}{s^2-1}$. If the plant is operated in a unity feedback configuration, then the lead compensator that can stabilize this control system is:

(a) $\frac{10(s-1)}{s+2}$ (b) $10\frac{(s+4)}{s+2}$

(c) $\frac{10(s+2)}{s+10}$ (d) $\frac{2(s+2)}{s+10}$

5. The magnitude plot of a rational transfer function $G(s)$ with real coefficients is shown below. Which of the following compensators has such a magnitude plot?

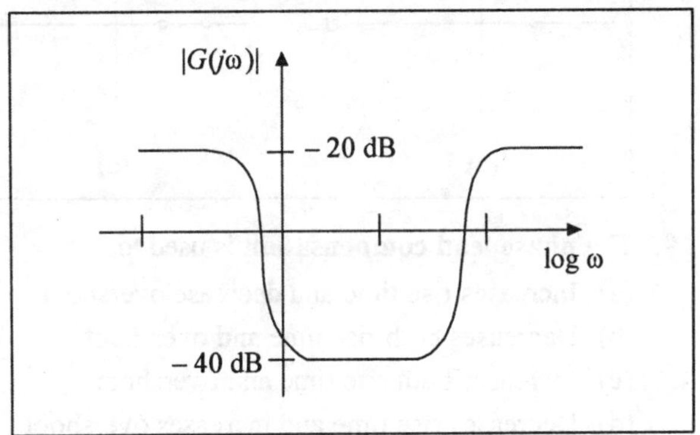

(a) Lead compensator
(b) Lag compensator
(c) PID compensator
(d) Lead-lag compensator

Linked Answer Questions

Statement for linked Answers Questions 6 and 7

The transfer function of a compensator is given as

$$G_C(s) = \frac{s+a}{s+b}$$

6. $G_C(s)$ is a lead compensator if

(a) $a = 1$, $b = 2$ (b) $a = 3$, $b = 2$
(c) $a = -3$, $b = -1$ (d) $a = 3$, $b = 1$

7. The phase of the above lead compensator is maximum at

(a) $\sqrt{2}\ rad/s$
(b) $\sqrt{3}\ rad/s$
(c) $\sqrt{6}\ rad/s$
(d) $\frac{1}{\sqrt{3}}\ rad/s$

8. The pole-zero configuration of a **phase-lead compensator** is given by

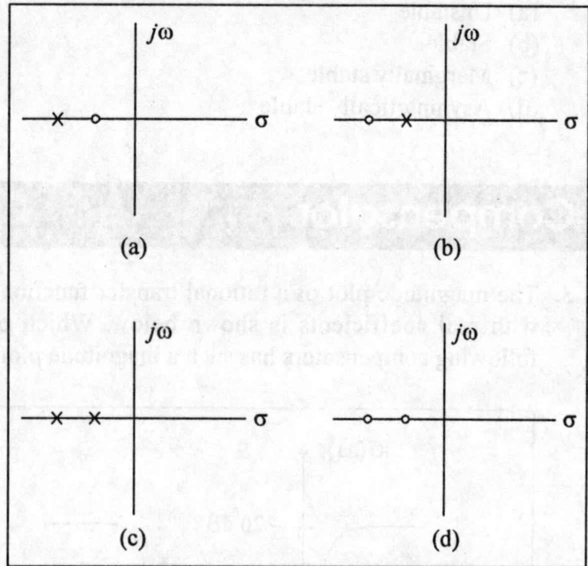

9. The **phase lead compensation** is used to
(a) Increases rise time and decrease overshoot
(b) Decreases both rise time and overshoot
(c) Increases both rise time and overshoot
(d) Decreases rise time and increases overshoot

10. Maximum **phase-lead** of the compensator

$$D(s) = \frac{(0.5s + 1)}{(0.05s + 1)}, \text{ is}$$

(a) 52 deg at 4 rad/sec
(b) 52 deg at 10 rad/sec
(c) 55 deg at 12 rad/sec
(d) None of the answers is correct

11. A lead compensator used for a closed loop controller has the following transfer function

$$\frac{K\left(1+\frac{s}{a}\right)}{\left(1+\frac{s}{b}\right)}$$

For such a lead compensator

(a) $a < b$ (b) $b < a$
(c) $a > Kb$ (d) $a < Kb$

12. The system $\frac{900}{s(s+1)(s+9)}$ is to be such that its gain-crossover frequency becomes same as its uncompensated phase-crossover frequency and provides a 45° phase margin. To achieve this, one may use
(a) A lag compensator that provides an attenuation of 20 dB and a phase lag of 45° at the frequency of $3\sqrt{3}$ rad/s
(b) A lead compensator that provides an amplification of 20 dB and a phase laed of 45° at the frequency of 3 rad/s
(c) A lag-lead compensator that provides an amplification of 20 dB and a phase lag of 45° at the frequency of $\sqrt{3}$ rad/s
(d) A lag-lead compensator that provides an attenuation of 20 dB and a phase lead of 45° at the frequency of 3 rad/s

13. The transfer fucntion of two compensators are given below:

$$C_1 = \frac{10(s+1)}{(s+10)}, \quad C_2 = \frac{s+10}{10(s+1)}$$

Which one of the following statement is correct?
(a) C_1 is a lead compensator and C_2 is a lag compensator
(b) C_1 is a lag compensator and C_2 are lead compensators
(c) Both C_1 amd C_2 are lead compenstors
(d) Both Q_1 amd C_2 are lag compensators

Statement for Linked Answer Questions 14 and 15

The transfer function of a compensator is given as

$$G_C(s) = \frac{s+a}{s+b}$$

14. $G_C(s)$ is a lead compensator if
(a) $a=1,\ b=2$
(b) $a=3,\ b=2$
(c) $a=-3,\ b=-1$
(d) $a=3,\ b=1$

15. The phase of the above lead compensator is maximum at
(a) $\sqrt{2}\ rad/s$ (b) $\sqrt{3}\ rad/s$
(c) $\sqrt{6}\ rad/s$ (d) $\frac{1}{\sqrt{3}} rad/s$

16. A **temperature control** system is usually very sluggish to improve its dynamics if:
(a) A PI controller can be used
(b) An I controller can be used
(c) A PID controller with large I and a negligible D action can be used
(d) A PD controller can be used

17. **A First order system with a proportional controller** exhibits an offset to step input. The offset can be reduced by:
(a) Increasing the gain

(b) Adding integral mode

(c) Adding derivative mode

(d) Decreasing the gain.

18. A plant is controlled by **a proportional controller.** If a **time delay element** is introduced in the loop then

1. Gain margin remains the same
2. Phase margin remains the same
3. Gain margin decreases
4. Decreasing the gain.

Chose the correct option

(a) 1, 2 are correct

(b) 2, 3 are correct

(c) 2, 3 are correct

(c) 3, 4 are correct

19. The transfer function of a basic PD controller is given by (K's are real constants)

(a) $K_0 + \dfrac{K_1}{s} + K_2 s$ (b) $K_0 + K_2 s$

(c) $K_1 s + K_2 s$ (d) $K_0 + \dfrac{K_1}{s}$

20. First order system with a proportional controller in the negative feedback loop has an offset to a step/input. This offset can be eliminated by

(a) Adding a derivative mode to the controller

(b) Adding an integral mode to the controller

(c) Decreasing the magnitude of the gain of the proportional controller

(d) Adding a delay in the controller loop

21. A process in a feedback loop with a proportional controller with ultimate gain $k_u = 10$ is oscillating at a frequency of P_u = 8Hz. The **Ziegler-Nichols** setting for the proportional controller is

(a) 8

(b) 5

(c) 1.25

(d) 10

22. The transfer function of a PID Controller is given by

$G(s) = 4\left(1 + \dfrac{1}{2s} + 0.5s\right)$ as tends to infinity

(a) Magnitude of $G(j\omega)$ tends to zero and phase angle of $G(j\omega)$ tends to infinity

(b) Magnitude of $G(j\omega)$ tends to infinity and phase angle of $G(j\omega)$ tends to zero

(c) Magnitude tends of $G(j\omega)$ to infinity and phase angle of $G(j\omega)$ tends to –90°.

(d) Magnitude tends of $G(j\omega)$ to zero and phase angle of $G(j\omega)$ tends to + 90°.

23. In a **synchro pair,** the control transmitter excites the three stator windings of the control transformer. The stator winding voltages will have

(a) Equal magnitudes but different phases

(b) Different magnitudes and different phases

(c) Equal magnitudes and phase

(d) Different magnitudes but equal phases

24. A three stack **variable reluctance stepper motor** has 12 numbers of rotor and stator teeth on each stack. The resolution of angular rotation for the stepper motor, assuming number two stacks are excited simultaneously, is

(a) 10°

(b) 15°

(c) 20°

(d) 30°

25. The **proportional band** of PI controller is 50% and the reset time is 0.2s. The transfer function of the controller is given by

(a) $50\left(1 + \dfrac{1}{0.2s}\right)$ (b) $50\left(1 + \dfrac{1}{5s}\right)$

(c) $2\left(1 + \dfrac{1}{5s}\right)$ (d) $2\left(1 + \dfrac{1}{s}\right)$

26. In figure K_1 and K_2 are proportional controllers. For the control, system to remain stable, the upper limits of K_1 and K_2 are.

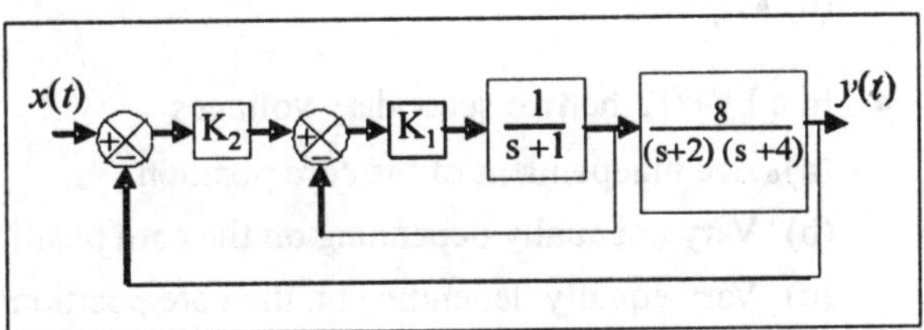

(a) ∞, ∞ (b) 1, ∞

(c) ∞, 10 (d) 10, 1

27. A PID controller has the transfer function $2 + \dfrac{0.4}{s}$ with the unit of time expressed in minutes. The parameters **proportional band and reset time** for the above controller are respectively

(a) 200% and 0.4 min (b) 50% and 0.4 min.

(c) 200% and 5 min (d) 50% and 5 min.

10.9 Control Components

1. The transfer function of a tachometer is of the form

(a) Ks

(b) $\frac{K}{s}$

(c) $\frac{K}{s+1}$

(d) $\frac{K}{s(s+1)}$

2. In a synchro error detector, the output voltage is proportional to $[\omega(t)]^n$, where $\omega(t)$ is the rotor velocity and n equals:

(a) –2

(b) –1

(c) 1

(d) 2

3. For a tachometer, if $\theta(t)$ is the rotor displacement in radians, $e(t)$ is the output voltage and K_t is the tachometer constant V/rad/sec, then the transform function, $\frac{E(s)}{Q(s)}$ will be

(a) $K_t s^2$

(b) $\frac{K_t}{s}$

(c) $K_t s$

(d) K_t

4. In a LVDT, the two secondary voltages

(a) Are independent of the core position

(b) Vary unequally depending on the core position

(c) Vary equally depending on the core position

(d) Are always in the phase quadrature

5. A **differentiator** has transfer function whose

(a) Phase increases linearly with frequency

(b) Amplitude remains constant

(c) Amplitude increases linearly with frequency

(d) Amplitude decreases linearly with frequency

6. Ratio of the rotor reactance X to the rotor resistance R for a **two-phase servomotor**

(a) Is equal to that of a normal induction motor

(b) Is less than that of a normal induction motor

(c) Is greater than that of a normal induction motor.

(d) May be less or greater than that of normal induction motor is an example of type 1 order 2 system.

7. A **synchro transmitter** is used with a synchro receiver for the purpose of

(a) Amplification

(b) Error detection

(c) Remote sensing

(d) Addition

8. Match the List I elements and the List II transfer function and select the correct answer using codes

List-I	List-II
A. Pneumatic actuator	1. e^{-sT}
B. Transport delay	2. K
C. A servo-moter	3. $\left(\frac{K}{as^2+bs+c}\right)$
D. A nozzle-flapper	4. $\left(\frac{K}{s(1+as)}\right)$

Codes:

	A	B	C	D
(a)	1	2	3	4
(b)	3	1	4	2
(c)	1	2	4	3
(d)	4	1	2	3

9. The rotor of the control transformer of a synchro pair gives a maximum voltage of 1.0 V at a particular position of the rotor of the control transmitter. The transmitter rotor is now rotated by 30° anticlockwise keeping the transformer rotor stationary. The transformer rotor voltage for this position is

(a) 1.0 V

(b) 0.866 V

(c) 0.5 V

(d) 0 V

10. The two secondary coils of an LVDT have wrongly been connected as shown in figure. Then the input-output relationship would be:

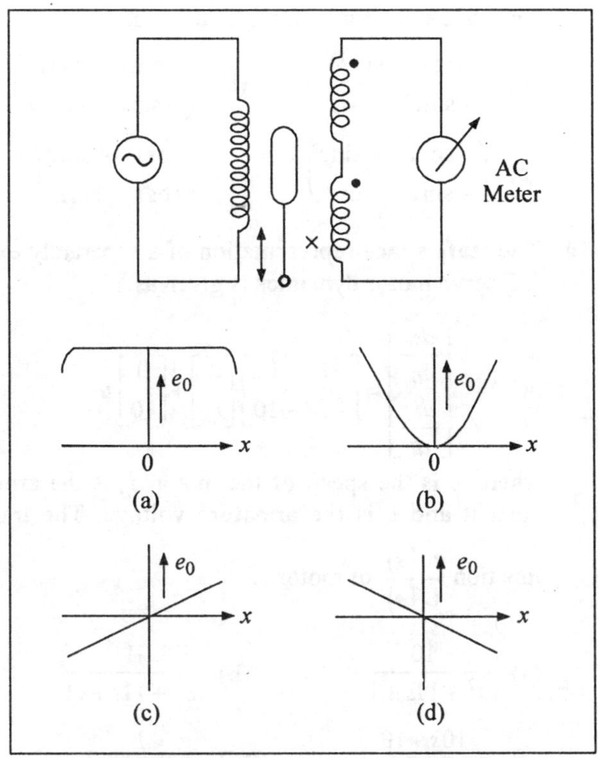

11. Linear variable differential transformer has
(a) Two primary coils connected in phase and a secondary coil.
(b) Two primary coils connected in opposition and a secondary coil.
(c) One primary coil and two secondary coils connected in phase.
(d) One primary coil and two secondary coils connected in opposition.

12. In a **synchro pair,** the control transmitter excites the three stator windings of the control transformer. The stator winding voltages will have:
(a) Equal magnitudes but different phases
(b) Different magnitudes and different phases
(c) Equal magnitudes and phase
(d) Different magnitudes but equal phases

13. A three stack **variable reluctance stepper motor** has 12 number of rotor and stator teeth on each stack. The resolution of angular rotation for the stepper motor, assuming number two stacks are excited simultaneously, is:
(a) 10° (b) 15°
(c) 20° (d) 30°

14. In a second order instrument, 'K' is the spring constant, 'M' is the mass of the instrument, and 'B' is the damping constant. The damping constant necessary to the just prevent overshoot in the step response is
(a) Equal to $\sqrt{(KM)}$
(b) Less than $\sqrt{(KM)}$
(c) Greater than $\sqrt{(KM)}$
(d) Equal to 2 $\sqrt{(KM)}$

10.10 State Space Analysis

1. A linear second-order single input continuous-time system is described by the following set of differential equations:

$$X'_1(t) = -2x_1(t) + 4x_2(t)$$
$$X'_2(t) = 2x_1(t) - x_2(t) + u(t)$$

Where $x_1(t)$ and $x_2(t)$ are the state variables and $u(t)$ is the control variable. The system is
(a) Controllable and stable
(b) Controllable but unstable
(c) Uncontrollable and unstable
(d) Uncontrollable but stable

2. e^{At} can be expanded as

(a) $\sum_{k=0}^{\infty} \frac{A^k t^k}{(k+1)!}$ (b) $\sum_{k=0}^{\infty} \frac{A^k t^k}{k!}$

(c) $\sum_{k=0}^{\infty} \frac{A^k t^{k+1}}{(k+1)!}$ (d) $\sum_{k=0}^{\infty} \frac{A^k t^{k+1}}{k!}$

3. The solution of $X' = A(t)\, X(t)$ is

(a) $e^{At}.\, X_0$ (b) $e\left(\int_{te}^{t} A(Z)\, dz\, X_0\right)$

(c) $[I + \int_{t_0}^{t} A(\tau)\, d\tau].X_0$ (d) None of these

4. For the system described by the state equation:

$$x' = \begin{pmatrix} 0 & 1 & 0 \\ 0 & 0 & 1 \\ 0.5 & 1 & 2 \end{pmatrix} x + \begin{pmatrix} 0 \\ 0 \\ 1 \end{pmatrix} u$$

if the control signal u is given by

$$u = [-0.5\ \ -3\ \ -5]\ \ x + v,$$

then the eigen values of the closed system will be

(a) 0, –1, –2
(b) 0, –1, –3
(c) –1, –1, –2
(d) 0, –1, –1

5. The zero-input response of a system given by the state-space equation:

$$\begin{pmatrix} x_1' \\ x_2' \end{pmatrix} = \begin{pmatrix} 1 & 0 \\ 1 & 1 \end{pmatrix}\begin{pmatrix} x_1 \\ x_2 \end{pmatrix} \text{ and } \begin{pmatrix} x_1(0) \\ x_2(0) \end{pmatrix} = \begin{pmatrix} 1 \\ 0 \end{pmatrix} \text{ is}$$

(a) $\begin{pmatrix} te^t \\ t \end{pmatrix}$ (b) $\begin{pmatrix} e^t \\ t \end{pmatrix}$

(c) $\begin{pmatrix} e^t \\ te^t \end{pmatrix}$ (d) $\begin{pmatrix} t \\ te^t \end{pmatrix}$

6. Given $A = \begin{bmatrix} 1 & 0 \\ 0 & 1 \end{bmatrix}$, the state transition matrix e^{At} is given by

(a) $\begin{pmatrix} 0 & e^{-t} \\ e^{-t} & 0 \end{pmatrix}$ (b) $\begin{pmatrix} e^{-t} & 0 \\ 0 & e^t \end{pmatrix}$

(c) $\begin{pmatrix} e^{-t} & 0 \\ 0 & e^{-t} \end{pmatrix}$ (d) $\begin{pmatrix} 0 & e^t \\ e^t & 0 \end{pmatrix}$

7. The state variable equations of a system are:

$$x_1' = -3x_1 + x_2 + u;$$
$$x_2' = 2x_1 \text{ and } y = x_1 + u$$

The system is:

(a) Controllable but not observable
(b) Observable but not controllable
(c) Neither controllable nor observable
(d) Controllable and observable

8. A linear system is equivalently represented by two sets of state equations.

$$X' = AX + BU \text{ and } W = CW + DU$$

The eigen values of the representations, are also computed as (λ) and (μ). Which one of the following statements is true?

(a) $[\lambda] = [\mu]$ and $X = W$
(b) $[\lambda] = [\mu]$ and $X \neq W$
(c) $[\lambda] \neq [\mu]$ and $X = W$
(d) $[\lambda] \neq [\mu]$ and $X \neq W$

9. A linear system is described by the following state equation.

$$X'(t) = AX(t) + BU(t);\ A = \begin{pmatrix} 0 & 1 \\ -1 & 0 \end{pmatrix}$$

The state-transition matrix of the system is

(a) $\begin{pmatrix} \cos t & \sin t \\ -\sin t & \cos t \end{pmatrix}$ (b) $\begin{pmatrix} -\cos t & \sin t \\ -\sin t & -\cos t \end{pmatrix}$

(c) $\begin{pmatrix} -\cos t & -\sin t \\ -\sin t & \cos t \end{pmatrix}$ (d) $\begin{pmatrix} \cos t & -\sin t \\ \cos t & \sin t \end{pmatrix}$

10. The state space representation of a separately excited DC servo motor dynamics is given as:

$$\begin{bmatrix} \dfrac{d\omega}{dt} \\ \dfrac{di_o}{dt} \end{bmatrix} = \begin{bmatrix} -1 & 1 \\ -1 & -10 \end{bmatrix}\begin{bmatrix} \omega \\ i_a \end{bmatrix} + \begin{bmatrix} 0 \\ 10 \end{bmatrix} u$$

where ω is the speed of the motor, i_a is the armature current and u is the armature voltage. The transfer function $\dfrac{\omega(s)}{U(s)}$ of motor is

(a) $\dfrac{10}{s^2 + 11s + 11}$ (b) $\dfrac{1}{s^2 + 11s + 11}$

(c) $\dfrac{10s + 10}{s^2 + 11s + 11}$ (d) $\dfrac{1}{s^2 + s + 1}$

11. Consider a linear system whose space representation is $x(t) = Ax(t)$. If the initial state vector of the system is $x(0) = \begin{bmatrix} 1 \\ -2 \end{bmatrix}$, then the system response is $x(t) = \begin{bmatrix} e^{-2t} \\ -2e^{-2t} \end{bmatrix}$. If the initial state vector of the system changes to $x(0) = \begin{bmatrix} 1 \\ -1 \end{bmatrix}$, then the system response becomes $x(t) = \begin{bmatrix} e^{-t} \\ -e^{-t} \end{bmatrix}$.

The eigen value and eigen vector pairs (λ_i, v_i) for the system are

(a) $\left(-1, \begin{bmatrix} 1 \\ -1 \end{bmatrix}\right)$ and $\left(-2, \begin{bmatrix} 1 \\ -2 \end{bmatrix}\right)$

(b) $\left(-1, \begin{bmatrix} 1 \\ -1 \end{bmatrix}\right)$ and $\left(-1, \begin{bmatrix} 1 \\ -2 \end{bmatrix}\right)$

(c) $\left(-1, \begin{bmatrix} 1 \\ -1 \end{bmatrix}\right)$ and $\left(2, \begin{bmatrix} 1 \\ -2 \end{bmatrix}\right)$

(d) $\left(-2, \begin{bmatrix} 1 \\ -1 \end{bmatrix}\right)$ and $\left(1, \begin{bmatrix} 1 \\ -2 \end{bmatrix}\right)$

12. Consider a linear system whose space representation is $x(t) = Ax(t)$. If the initial state vector of the system is $x(0)=\begin{bmatrix}1\\-2\end{bmatrix}$, then the system response is $x(t)=\begin{bmatrix}e^{-2t}\\-2e^{-2t}\end{bmatrix}$. If the initial state vector of the system changes to $x(0)=\begin{bmatrix}1\\-1\end{bmatrix}$, then the system response becomes $x(t)=\begin{bmatrix}e^{-t}\\-e^{-t}\end{bmatrix}$.

The system matrix A is

(a) $\begin{bmatrix}0 & 1\\-1 & 1\end{bmatrix}$ (b) $\begin{bmatrix}1 & 1\\-1 & -2\end{bmatrix}$

(c) $\begin{bmatrix}2 & 1\\-1 & -1\end{bmatrix}$ (d) $\begin{bmatrix}0 & 1\\-2 & -3\end{bmatrix}$

13. A signal flow graph of a system is given below:

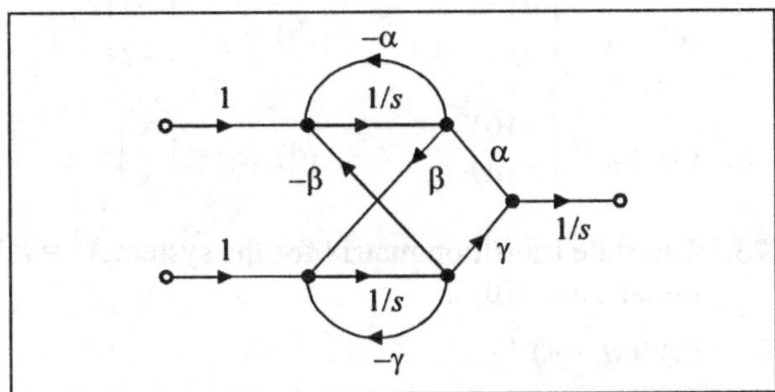

The set of equations that correspond to this signal flow graph is

(a) $\frac{d}{dt}\begin{pmatrix}x_1\\x_2\\x_3\end{pmatrix}=\begin{bmatrix}\beta & -\gamma & 0\\\gamma & \alpha & 0\\-\alpha & -\beta & 0\end{bmatrix}\begin{pmatrix}x_1\\x_2\\x_3\end{pmatrix}+\begin{bmatrix}1 & 0\\0 & 0\\0 & 1\end{bmatrix}\begin{pmatrix}u_1\\u_2\end{pmatrix}$

(b) $\frac{d}{dt}\begin{pmatrix}x_1\\x_2\\x_3\end{pmatrix}=\begin{bmatrix}0 & \alpha & \gamma\\0 & -\alpha & -\gamma\\0 & \beta & -\beta\end{bmatrix}\begin{pmatrix}x_1\\x_2\\x_3\end{pmatrix}+\begin{bmatrix}0 & 0\\0 & 1\\1 & 0\end{bmatrix}\begin{pmatrix}u_1\\u_2\end{pmatrix}$

(c) $\frac{d}{dt}\begin{pmatrix}x_1\\x_2\\x_3\end{pmatrix}=\begin{bmatrix}-\alpha & \beta & 0\\-\beta & -\gamma & 0\\\alpha & \gamma & 0\end{bmatrix}\begin{pmatrix}x_1\\x_2\\x_3\end{pmatrix}+\begin{bmatrix}1 & 0\\0 & 1\\0 & 0\end{bmatrix}\begin{pmatrix}u_1\\u_2\end{pmatrix}$

(d) $\frac{d}{dt}\begin{pmatrix}x_1\\x_2\\x_3\end{pmatrix}=\begin{bmatrix}-\gamma & 0 & \beta\\\gamma & 0 & \alpha\\-\beta & 0 & -\alpha\end{bmatrix}\begin{pmatrix}x_1\\x_2\\x_3\end{pmatrix}+\begin{bmatrix}0 & 1\\0 & 0\\1 & 0\end{bmatrix}\begin{pmatrix}u_1\\u_2\end{pmatrix}$

14. Consider the system $\frac{dx}{dt} = Ax + Bu$ with

$$A=\begin{bmatrix}1 & 0\\0 & 1\end{bmatrix} \text{ and } B=\begin{bmatrix}p\\q\end{bmatrix}$$

where p and q are arbitrary real numbers. Which of the following statements about the controllability of the system is true ?

(a) The system is completely state controllable for any non-zero values of p and q

(b) Only $p = 0$ and $q = 0$ result in controllability

(c) The system is uncontrollable for all values of p and q

(d) We cannot conclude about controllability from the given data

15. The signal flow graph of a system is shown below:

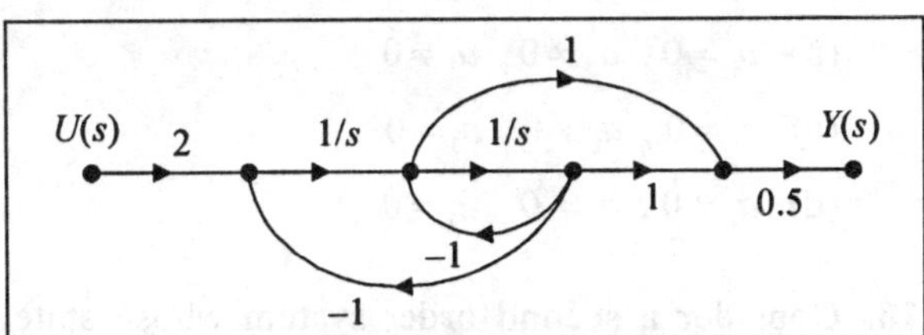

The state variable representation of the system can be

(a) $x' = \begin{bmatrix}1 & 1\\-1 & 0\end{bmatrix}x+\begin{bmatrix}0\\2\end{bmatrix}u$, $y=\begin{bmatrix}0 & 0.5\end{bmatrix}x$ (b) $x' = \begin{bmatrix}-1 & 1\\-1 & 0\end{bmatrix}x+\begin{bmatrix}0\\2\end{bmatrix}u$, $y=\begin{bmatrix}0 & 0.5\end{bmatrix}x$

(c) $x' = \begin{bmatrix}1 & 1\\-1 & 0\end{bmatrix}x+\begin{bmatrix}0\\2\end{bmatrix}u$, $y=\begin{bmatrix}0.5 & 0.5\end{bmatrix}x$ (d) $x' = \begin{bmatrix}-1 & 1\\-1 & 0\end{bmatrix}x+\begin{bmatrix}0\\2\end{bmatrix}u$, $y=\begin{bmatrix}0.5 & 0.5\end{bmatrix}x$

16. The block diagram of a system with one input u and two outputs y_1 and y_2 is given below.

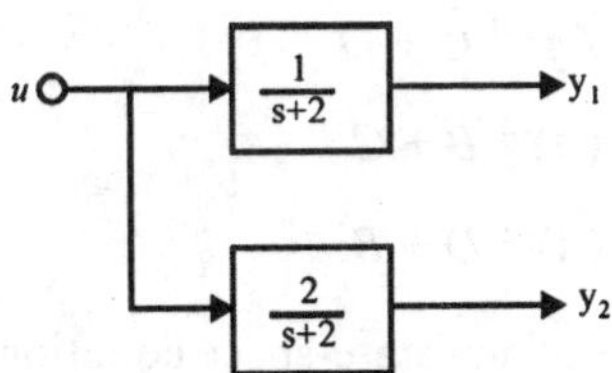

A state space model of the above system in terms of the state vector x and the output vector y $[y_1\ y_2]^T$ is

(a) $\dot{\underline{x}}=[2]\underline{x}+[1]u;$ $\underline{y}=[1\ \ 2]\underline{x}$

(b) $\dot{\underline{x}}=[-2]\underline{x}+[1]u;$ $\underline{y}=[1\ \ 2]\underline{x}$

(c) $\dot{\underline{x}}=\begin{bmatrix}-2 & 0\\0 & -2\end{bmatrix}\underline{x}+\begin{bmatrix}1\\1\end{bmatrix}u;$ $\underline{y}=[1\ \ 2]\underline{x}$

(d) $\dot{\underline{x}} = \begin{bmatrix} 2 & 0 \\ 0 & 2 \end{bmatrix} \underline{x} + \begin{bmatrix} 1 \\ 1 \end{bmatrix} u; \quad y = \begin{bmatrix} 1 \\ 2 \end{bmatrix} \underline{x}$

17. The state variable description of an LTI system is given by

$$\begin{pmatrix} \dot{x}_1 \\ \dot{x}_2 \\ \dot{x}_3 \end{pmatrix} = \begin{pmatrix} 0 & a_1 & 0 \\ 0 & 0 & a_2 \\ a_3 & 0 & 0 \end{pmatrix} \begin{pmatrix} x_1 \\ x_2 \\ x_3 \end{pmatrix} + \begin{pmatrix} 0 \\ 0 \\ 1 \end{pmatrix} u$$

$$y = \begin{pmatrix} 1 & 0 & 0 \end{pmatrix} \begin{pmatrix} x_1 \\ x_2 \\ x_3 \end{pmatrix}$$

Where y is the output and u is the input. The system is controllable for

(a) $a_1 \neq 0,\ a_2 = 0,\ a_3 \neq 0$

(b) $a_1 = 0,\ a_2 \neq 0,\ a_3 \neq 0$

(c) $a_1 = 0,\ a_2 \neq 0,\ a_3 = 0$

(d) $a_1 \neq 0,\ a_2 \neq 0,\ a_3 = 0$

18. Consider a second order system whose state space representation is of the form

$$X' = AX + Bu$$

If $x_1(t) = x_2(t)$, the system is

(a) Controllable
(b) Un-controllable
(c) Observable
(d) Unstable

19. The transfer function for the state variable representation $X' = AX + Bu$, $Y = CX + Du$, is given by

(a) $D + CsI - (A)^{-1} B$

(b) $B\, sI - (A)^{-1} C + D$

(c) $DsI - (A)^{-1} B + C$

(d) $CsI - (A)^{-1} D + B$

20. The matrix of any state-space equations for the transfer function $[C(s)/R(s)]$ of the system shown below in Figure is

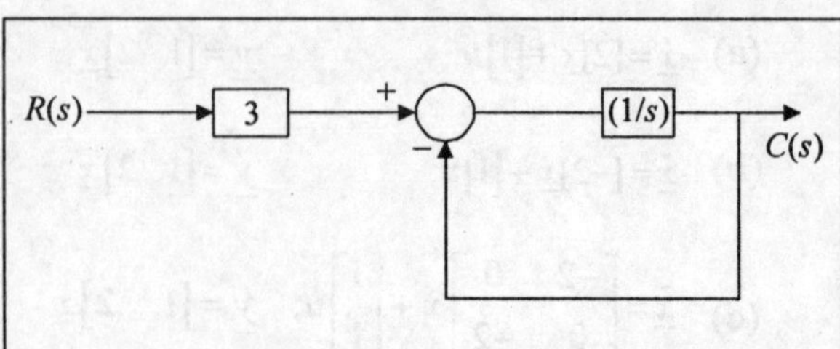

(a) $\begin{pmatrix} -1 & 0 \\ 0 & -1 \end{pmatrix}$ (b) $\begin{pmatrix} 0 & 1 \\ 0 & -1 \end{pmatrix}$

(c) [–1] (d) [3]

21. A system is described by the state equation $X' = AX + BU$. The output is given by $Y = CX$

Where $A = \begin{pmatrix} -4 & -1 \\ 3 & -1 \end{pmatrix}$, $B = \begin{pmatrix} 1 \\ 1 \end{pmatrix}$ $C = [1, 0]$.

Transfer function $G(s)$ of the system is

(a) $\dfrac{s}{s^2 + 5s + 7}$ (b) $\dfrac{s}{s^2 + 5s + 7}$

(c) $\dfrac{s}{s^2 + 5s + 7}$ (d) $\dfrac{s}{s^2 + 3s + 2}$

22. Given the homogeneous state-space equation

$$x' = \begin{pmatrix} -3 & 1 \\ 0 & -2 \end{pmatrix} x$$

the steady state value of $x_{ss} = \lim_{t \to \infty} x(t)$, given the initial state value of $x(0) = [10 - 10]^T$, is

(a) $x_{ss} = \begin{pmatrix} 0 \\ 0 \end{pmatrix}$ (b) $x_{ss} = \begin{pmatrix} -3 \\ -2 \end{pmatrix}$

(c) $x_{ss} = \begin{pmatrix} -10 \\ 10 \end{pmatrix}$ (d) $x_{ss} = \begin{pmatrix} \infty \\ \infty \end{pmatrix}$

23. The state transition matrix for the system $X' = AX$ with initial state $X(0)$ is

(a) $(sI - A)^{-1}$

(b) $e^{At} X(0)$

(c) Laplace inverse of $[(sI - A)^{-1}]$

(d) Laplace inverse of $[(sI - A)^{-1} X(0)]$

24. For the system

$$X' = \begin{pmatrix} 2 & 0 \\ 0 & 4 \end{pmatrix} X + \begin{pmatrix} 1 \\ 1 \end{pmatrix} u;$$

$y = [4\ 0]\, X$, with u as unit impulse and with zero initial state, the output, y, becomes

(a) $2e^{2t}$ (b) $4e^{2t}$
(c) $2e^{4t}$ (d) $4e^{4t}$

25. A second order system starts with an initial condition of $\begin{bmatrix} 2 \\ 3 \end{bmatrix}$ without any external input. The state transition matrix for the system is given by $\begin{bmatrix} e^{-2t} & 0 \\ 0 & e^{-t} \end{bmatrix}$. The state of the system at the end of 1 second is given by

(a) $\begin{bmatrix} 0.271 \\ 1.100 \end{bmatrix}$ (b) $\begin{bmatrix} 0.135 \\ 0.368 \end{bmatrix}$

(c) $\begin{bmatrix} 0.271 \\ 0.736 \end{bmatrix}$ (b) $\begin{bmatrix} 0.135 \\ 1.100 \end{bmatrix}$

26. The following equation defines a separately excited dc motor in the form of a differential equation

$$\frac{d^2\omega}{dt^2} + \frac{B}{J}\frac{d\omega}{dt} + \frac{K^2}{LJ}\omega = \frac{K}{LJ}V_a$$

The above equation may be organized in the state-space form as follows

$$\begin{bmatrix} \frac{d^2\omega}{dt^2} \\ \frac{d\omega}{dt} \end{bmatrix} = P\begin{bmatrix} \frac{d\omega}{dt} \\ \omega \end{bmatrix} + QV_a$$

Where the matrix P is given by

(a) $\begin{bmatrix} -\frac{B}{J} & -\frac{K^2}{LJ} \\ 1 & 0 \end{bmatrix}$ (b) $\begin{bmatrix} -\frac{K^2}{LJ} & -\frac{B}{J} \\ 0 & 1 \end{bmatrix}$

(c) $\begin{bmatrix} 0 & 1 \\ -\frac{K^2}{LJ} & -\frac{B}{J} \end{bmatrix}$ (d) $\begin{bmatrix} 0 & 1 \\ -\frac{B}{J} & -\frac{K^2}{LJ} \end{bmatrix}$

27. The state variable description of a linear autonomous system is $X' = AX$, where X is the two dimensional state vector and A is the system matrix given by $A = \begin{bmatrix} 0 & 2 \\ 2 & 0 \end{bmatrix}$.

The roots of the characteristic equation are

(a) –2 and +2 (b) –j2 and +j2

(c) –2 and –2 (d) +2 and +2

28. A state variable system $X'(t) = \begin{bmatrix} 0 & 1 \\ 0 & -3 \end{bmatrix} X(t) + \begin{bmatrix} 1 \\ 0 \end{bmatrix} u(t)$,

with the initial condition $X(0) = [-1 \ 3]^T$ and the unit step input $u(t)$ has the state transition matrix

(a) $\begin{bmatrix} 1 & \frac{1}{3}\left(1-e^{-3t}\right) \\ 0 & e^{-3t} \end{bmatrix}$ (b) $\begin{bmatrix} 1 & \frac{1}{3}\left(e^{-t}-e^{-3t}\right) \\ 0 & e^{-t} \end{bmatrix}$

(c) $\begin{bmatrix} 1 & \frac{1}{3}\left(e^{-t}-e^{-3t}\right) \\ 0 & e^{-3t} \end{bmatrix}$ (d) $\begin{bmatrix} 1 & \left(1-e^{-t}\right) \\ 0 & e^{-t} \end{bmatrix}$

29. A state variable system $X'(t) = \begin{bmatrix} 0 & 1 \\ 0 & -3 \end{bmatrix} X(t) + \begin{bmatrix} 1 \\ 0 \end{bmatrix} u(t)$,

with the initial condition $X(0) = [-1 \ 3]^T$ and the unit step input $u(t)$ has the state transition equation

(a) $X(t) = \begin{bmatrix} t-e^{-t} \\ e^{-t} \end{bmatrix}$ (b) $X(t) = \begin{bmatrix} t-e^{-t} \\ 3e^{-3t} \end{bmatrix}$

(c) $X(t) = \begin{bmatrix} t-e^{-3t} \\ 3e^{-3t} \end{bmatrix}$ (d) $X(t) = \begin{bmatrix} t-e^{-3t} \\ e^{-t} \end{bmatrix}$

30. For a system with the transfer function $H(s) = \frac{3(s-2)}{4s^2-2s+1}$, the matrix A in the state space form $x' = Ax + Bu$ is equal to

(a) $\begin{bmatrix} 1 & 0 & 0 \\ 0 & 1 & 0 \\ -1 & 2 & -4 \end{bmatrix}$

(b) $\begin{bmatrix} 0 & 1 & 0 \\ 0 & 1 & 0 \\ -1 & 2 & -4 \end{bmatrix}$

(c) $\begin{bmatrix} 0 & 1 & 0 \\ 3 & -2 & 1 \\ 1 & -2 & 4 \end{bmatrix}$

(d) $\begin{bmatrix} 1 & 0 & 0 \\ 0 & 0 & 1 \\ -1 & 2 & -4 \end{bmatrix}$

Statement for Linked Answer Questions 31 and 32

The state equation of a system is described by

$$x' = Ax + Bu$$

$$y = Cx$$

Where x is state vector, u is input, y is output and

$$A = \begin{bmatrix} 0 & 1 \\ 0 & -2 \end{bmatrix}, B = \begin{bmatrix} 0 \\ 1 \end{bmatrix}, C = \begin{bmatrix} 1 & 0 \end{bmatrix}$$

31. The transfer function $G(s)$ of this system will be

(a) $\frac{s}{(s+2)}$ (b) $\frac{s+1}{s(s-2)}$

(c) $\frac{s}{(s-2)}$ (d) $\frac{1}{s(s+2)}$

32. A unity feedback is proivded to the above system $G(s)$ to make it a closed loop system as shown in figure

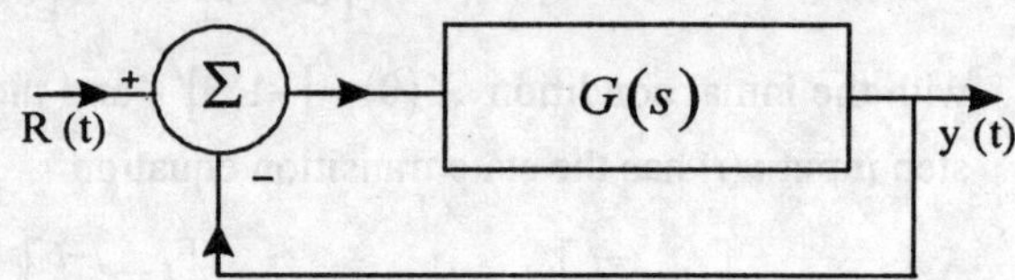

For a unit step input $r(t)$, the steady state error in the output will be

(a) 0

(b) 1

(c) 2

(d) ∞

Common Data for Questions 33 and 34 :

A system is described by the following state and output equations .

$$\frac{dx_1(t)}{dt} = -3x_1(t) + x_2(t) + 2u(t)$$

$$\frac{dx_2(t)}{dt} = -2x_2(t) + u(t)$$

$$y(t) = x_1(t)$$

where $u(t)$ is the input and $y(t)$ is the output

33. The system transfer function is

(a) $\frac{s+2}{s^2+5s-6}$ (b) $\frac{s+3}{s^2+5s+6}$

(c) $\frac{2s+5}{s^2+5s+6}$ (d) $\frac{2s-5}{s^2+5s-6}$

34. The state transition matrix of the above system is

(a) $\begin{bmatrix} e^{-3t} & 0 \\ e^{-2t}+e^{-3t} & e^{-2t} \end{bmatrix}$

(b) $\begin{bmatrix} e^{-3t} & e^{-2t}-e^{-3t} \\ 0 & e^{-2t} \end{bmatrix}$

(c) $\begin{bmatrix} e^{-3t} & e^{-2t}+e^{-3t} \\ 0 & e^{-2t} \end{bmatrix}$

(d) $\begin{bmatrix} e^{3t} & e^{-2t}-e^{-3t} \\ 0 & e^{-2t} \end{bmatrix}$

35. The eigen values of the system represented by

$$X' = \begin{pmatrix} 0 & 1 & 0 & 0 \\ 0 & 0 & 1 & 0 \\ 0 & 0 & 0 & 1 \\ 0 & 0 & 0 & 1 \end{pmatrix} X \text{ are}$$

(a) 0, 0, 0, 0 (b) 1, 1, 1, 1

(c) 0, 0, 0, –1 (d) 1, 0, 0, 0

36. The minimum number of states required to describe the network shown in Figure.

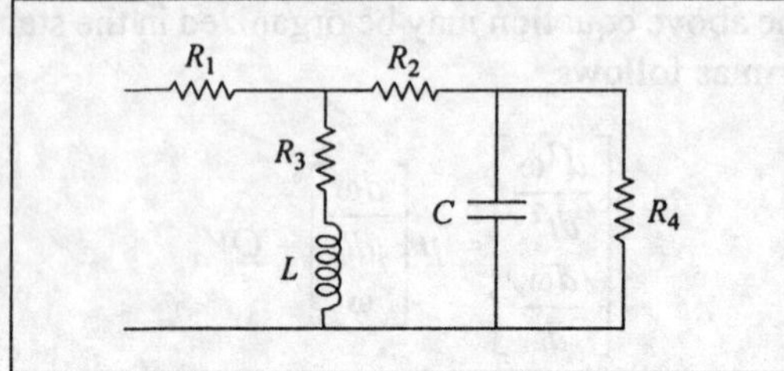

(a) 1 (b) 2

(c) 3 (d) 4

37. The state space representation of a system is given by

$$x' = \begin{bmatrix} 0 & 1 \\ 0 & -3 \end{bmatrix} x + \begin{bmatrix} 1 \\ 0 \end{bmatrix} u$$

$$y = \begin{bmatrix} 1 & 0 \end{bmatrix} x$$

The transfer function $\frac{y(s)}{u(s)}$ of the system will be

(a) $\frac{1}{s}$ (b) $\frac{1}{s(s+3)}$

(c) $\frac{1}{s+3}$ (d) $\frac{1}{s^2}$

38. A linear time-invariant single-input single-output system has a state space model given by

$$\frac{dx}{dt} = Fx + Gu;$$

$$y = Hz$$

where $F = \begin{bmatrix} 0 & 1 \\ -4 & -2 \end{bmatrix}$; $G = \begin{bmatrix} 0 \\ 1 \end{bmatrix}$; $H = \begin{bmatrix} 1 & 0 \end{bmatrix}$

Here x is the state vector,

u is the input,

and y is the output.

The damping ratio of the system is

(a) 0.25 (b) 0.5

(c) 1 (d) 2

39. The transfer function of the system described by the state space equations

$$\begin{bmatrix}\dot{x}_1\\ \dot{x}_2\end{bmatrix}=\begin{bmatrix}-4 & -1\\ -3 & -1\end{bmatrix}\begin{bmatrix}1\\ 1\end{bmatrix}u,\quad y=\begin{bmatrix}1 & 0\end{bmatrix}\begin{bmatrix}x_1\\ x_2\end{bmatrix}$$

(a) $\dfrac{s}{s^2+5s+1}$ (b) $\dfrac{2s}{s^2+5s+1}$

(c) $\dfrac{3s}{s^2+5s+1}$ (d) $\dfrac{4s}{s^2+5s+1}$

40. For a dynamic system, various matrices for a state space model are given by $A=\begin{pmatrix}0 & 1\\ -2 & -3\end{pmatrix}$ $B=\begin{pmatrix}0\\ 1\end{pmatrix}$ and $C = [1\ \ 0]$. The system poles are located at

(a) −2, −1 (b) −2, −3

(c) −3, −1 (d) −1, 0

41. A state space representation for the transfer function

$\dfrac{Y(s)}{U(s)}=\dfrac{s+6}{s^2+5s+6}$ is $x' = Ax + Bu, y = Cx$

where $A=\begin{bmatrix}0 & 1\\ -6 & -5\end{bmatrix}$, $B=\begin{bmatrix}0\\ 1\end{bmatrix}$ and the value of C is

(a) $\begin{bmatrix}1\\ 2\end{bmatrix}$ (b) $\begin{bmatrix}6 & 1\end{bmatrix}$

(c) $\begin{bmatrix}1 & 2\\ -1 & 0\end{bmatrix}$ (d) $\begin{bmatrix}1 & -5\end{bmatrix}$

42. A single input single output linear system as described by the equations $\dfrac{dx}{dt}=Ax+bu$ and $y=cx+du$, where A is an $[n\times n]$ matrix and x and b are $[n\times 1]$ vectors. The transfer function is given by

(a) $c\,[sI-A]^{-1}\,b$

(b) $c\,[sI-A]^{-1}\,b+d$

(c) $c\,[sI-A]\,b$

(d) $c[sI-A]\,b+d$

10.11 Miscellaneous

1. Match the List I characteristic with the List II system transfer function and select the correct answer using the codes:

List-I	List-II
A. All pass	1. $e^{-0.5s}$
B. Transport delay	2. $\dfrac{1-s}{1+s}$
C. Lead compensator	3. $\dfrac{1}{s^2+2s-3}$
D. Unstable	4. $\dfrac{1}{s^2+s+100}$
	5. $\dfrac{10(1+0.3s)}{1+0.1s}$

Codes:

	A	B	C	D
(a)	1	2	5	3
(b)	5	2	4	3
(c)	2	1	5	3
(d)	5	1	2	3

Common Data for Questions 2 and 3:

A disturbance input $d(t)$ is injected into the unity feedback control loop shown in the figure. Take the reference input $r(t)$ to be a unit step.

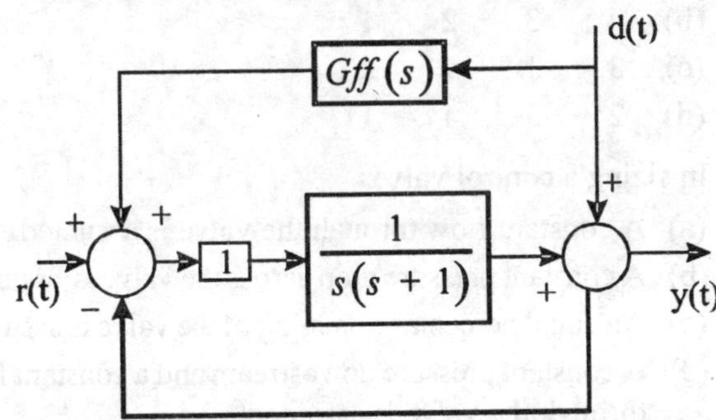

2. If the disturbance is measurable, its effect on the output can be minimized significantly using a feed forward controller $G_{ff}(s)$. To eliminate the component of the output due to $d(t)=\sin t$, $G_{ff}(j\omega)\big|_{\omega=1}$ should be

(a) $\dfrac{1}{\sqrt{2}}\angle-\dfrac{3\pi}{4}$

(b) $\dfrac{1}{2}\angle\dfrac{\pi}{4}$

(c) $\sqrt{2}\angle\pi$

(d) $\sqrt{2}\angle-\dfrac{\pi}{4}$

3. Let $G_{ff}(s)$ be a PD controller. If $d(t) = \sin 2t$, the amplitude of the frequency component of $y(t)$ due to $d(t)$ is

(a) $\sqrt{\frac{5}{13}}$ (b) $\sqrt{\frac{9}{13}}$

(c) $\sqrt{\frac{17}{13}}$ (d) $\sqrt{\frac{20}{13}}$

4. Match the *List-I* transfer functions with the *List-II* systems and select the correct answer using the codes:

List-I	*List-II*
A. All pass filter	1. e^{-sT}
B. Transport delay	2. $\frac{1-s}{1+s}$
C. Lag network	3. $\frac{1+as}{1+bs}$, $a < b$
D. Servomotor	4. $\frac{K}{s(1+as)}$

Codes:

	A	B	C	D
(a)	2	1	3	4
(b)	4	3	2	1
(c)	3	1	4	2
(d)	2	3	4	1

5. In sizing a control valve

(a) A constant flow through the valve is assumed
(b) A constant pressure drop across the valve is assumed
(c) An equal percentage opening of the valve is assumed
(d) A constant pressure downstream and a constant flow through the valve is assumed

6. In a seismic pick-up for getting an output proportional to acceleration, it is desirable to have

(a) A natural frequency very small in comparison to frequency of input and a damping ratio around 0.7
(b) A natural frequency very large in comparison to frequency of input and a damping ratio around 0.7
(c) A natural frequency equal to the frequency of input signal and a damping ratio around 0.7
(d) A natural frequency very small in comparison to frequency of input and a damping ratio higher that unity.

7. A current of 5 mA in a standard process controller indicate while zero current output indicates

(a) Live zero
(b) Dead zero
(c) Full scale
(d) Threshold

8. A cascade controller is used when the process

(a) Gain is too small
(b) Gain is too large
(c) Has widely different time constant
(d) Oscillation at the output is not permitted

9. Of the following transfer functions, the function that has a minimum phase characteristic is

(a) $\frac{1+s}{1-s}$ (b) $\frac{1-s}{1+s}$

(c) $\frac{e^{-3}}{s}$ (d) $\frac{s}{(s+1)(s+2)}$

10. The number of roots, lying outside the unit circle, of the polynomial

$$F(z) = 4z^3 - 8z^2 - z + 2 \text{ is}$$

(a) 3 (b) 2
(c) 1 (d) 0

11. The input error voltage to a PD-controller, e_i is shown in figure. The output voltage of the controller e_o is:

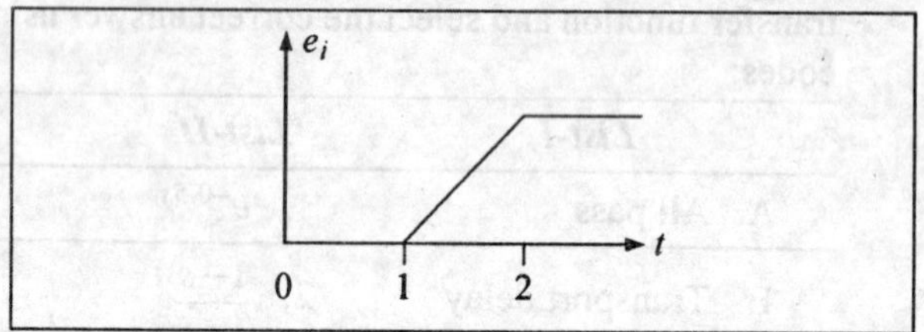

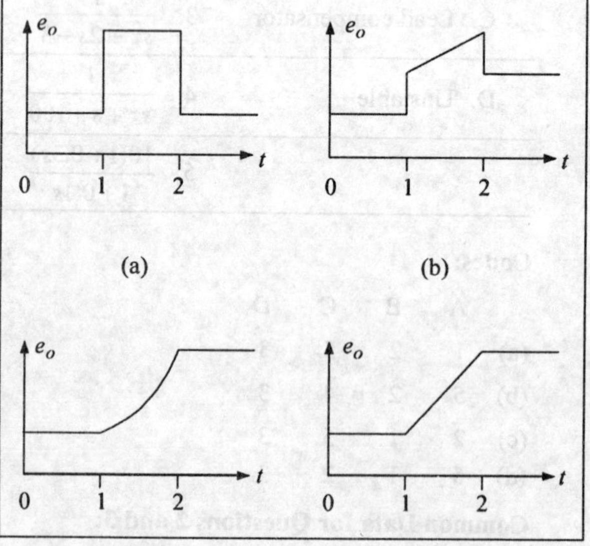

12. The proportional gain of a PID controller can be expressed in terms of its proportional band (PB) as

(a) PB (b) $\frac{100}{PB}$

(c) $\frac{PB}{100}$ (d) $\frac{10}{PB}$

13. A second order feedback system is found to be oscillating with a high frequency. The oscillations

(a) Can be reduced by increasing the proportional action
(b) Can be reduced by increasing the integral action
(c) Can be reduced by increasing the derivative action
(d) Cannot be reduced.

14. The closed loop control system shown in figure has $\tau_1 > 0$. The system will remain stable for all τ_1 in the range

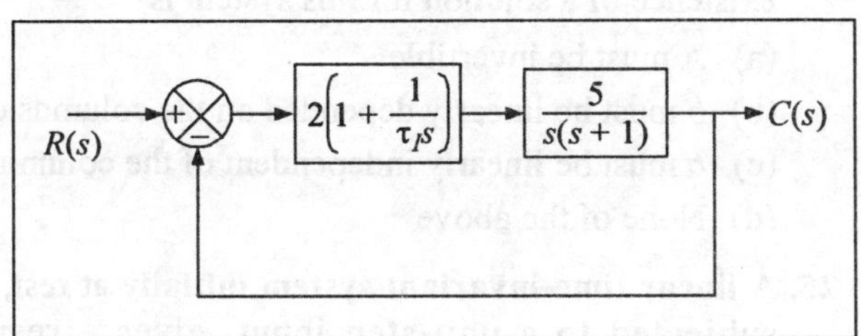

(a) $\tau_1 < 0.5$ (b) $0.5 < \tau_1 < 1.0$
(c) $\tau_1 < 1.0$ (d) $\tau_1 > 1.0$

15. In figure, K_1 and K_2 are proportional controllers. For the control system to remain stable, the upper limits of K_1 and K_2 are

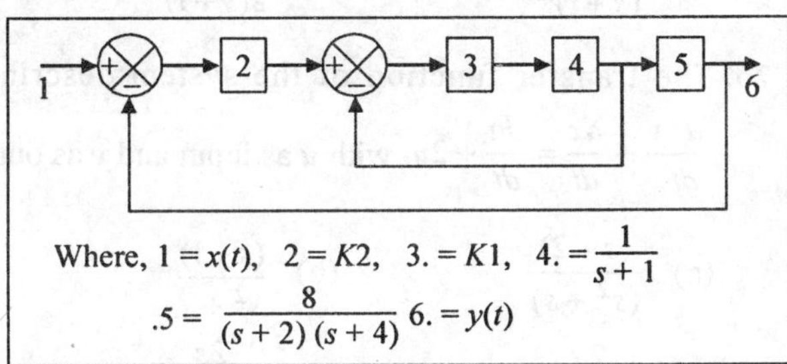

(a) ∞, ∞ (b) 1, ∞
(c) ∞, 10 (d) 10, 1

16. The response of a system to a unit impulse is $y(t) = e^{-5(t-1)}$. Which one of the following is the correct statement about the system?

(a) The system is non-linear
(b) The system is unstable in open loop.
(c) The steady state gain of the system for a unit step input is 0.2.
(d) The steady state gain of the system for a unit step input is 1.0

17. A PID controller has the transfer function $2 + \frac{0.4}{s}$ with the unit of time expressed in minutes. The parameters proportional band and reset time for the above controller are respectively

(a) 200% and 0.4 min (b) 50% and 0.4 min.
(c) 200% and 5 min (d) 50% and 5 min.

18. The transfer function of second order band pas filter, having a centre frequency of 1000 rad/s, selectivity of 100 and a gain of 0 dB at the center frequency, is

(a) $\frac{10s}{s^2+10s+10^6}$ (b) $\frac{s}{s^2+s+10^6}$

(c) $\frac{100s}{s^2+100s+10^7}$ (d) $\frac{100s}{s^2+100s+10^6}$

Statement for Linked Answers Questions 19 and 20

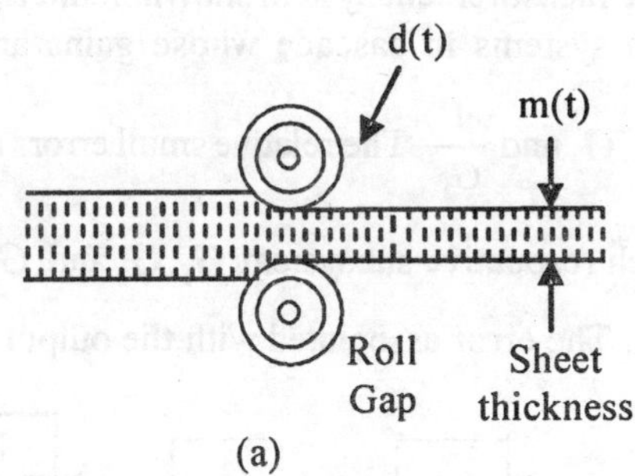

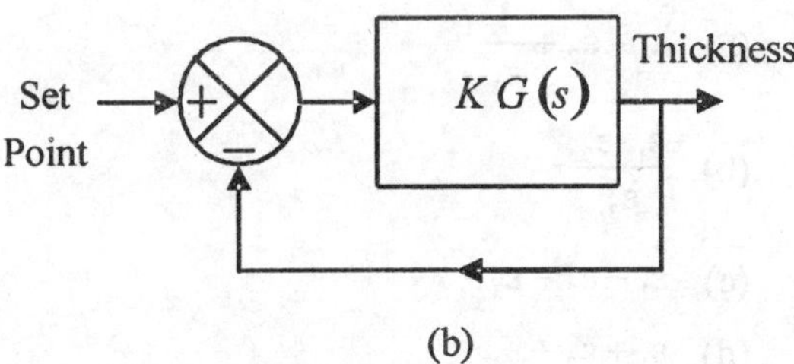

(b)

The thickness of a metal sheet from a rolling mill is manipulated by controlling the roll gap d(t) and measuring the sheet thickness m(t) at a certain distance from the rolls as shown in the figure. The thickness of the sheet at the roll out point is equal to the roll gap d(t) as shown and remains invariant during transit

19. Given the distance at which the thickness is measured is 0.5 m from the roll gap and the velocity of the sheet 0.2 m/s, the transfer function system G(s) = M(s) / D(s), is

(a) $e^{-2.5s}$ (b) $e^{+2.5s}$
(c) $e^{+0.4s}$ (d) $e^{-0.4s}$

20. The gap d(t) is controlled by a proportional controller along with a hydraulic actuator of combined gain K and a unity feedback loop exists around K G(s) figure (b). Increasing K beyond a certain value causes the system to become unstable. The frequency of oscillations at the critical value of K is

(a) $\omega = \pi$ (b) $\omega = 25\,\pi$

(c) $\omega = 0.4\,\pi$ (d) $\omega = \dfrac{\pi}{2.5}$

21. Feedback control systems are

(a) Insensitive to both forward and feedback-path parameter changes

(b) Less sensitive to feedback-path parameter changes than to forward-path parameter changes.

(c) Less sensitive to forward-path parameter changes than to feedback-path parameter changes.

(d) Equally sensitive to forward and feedback-path parameter changes.

22. The measurement system shown in the figure uses three sub systems in cascade whose gains are specified as G_1, G_2 and $\dfrac{1}{G_3}$. The relative small errors associated with each respective subsystem G_1, G_2 and G_3 are $\varepsilon_1, \varepsilon_2$ and ε_3. The error associated with the output is :

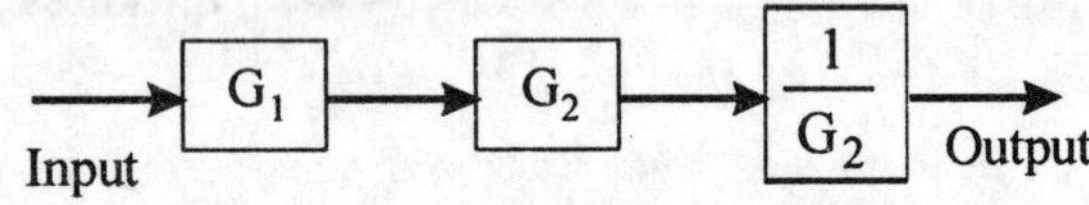

(a) $\varepsilon_1 + \varepsilon_2 + \dfrac{1}{\varepsilon_3}$

(b) $\dfrac{\varepsilon_1 \cdot \varepsilon_2}{\varepsilon_3}$

(c) $\varepsilon_1 + \varepsilon_2 - \varepsilon_3$

(d) $\varepsilon_1 + \varepsilon_2 + \varepsilon_3$

23. As shown in the figure, a negative feedback system has an amplifier of gain 100 with $\pm 10\%$ tolerance in the forward path and an attenuator of value 9/100 in the feedback path. The overall system gain is approximately

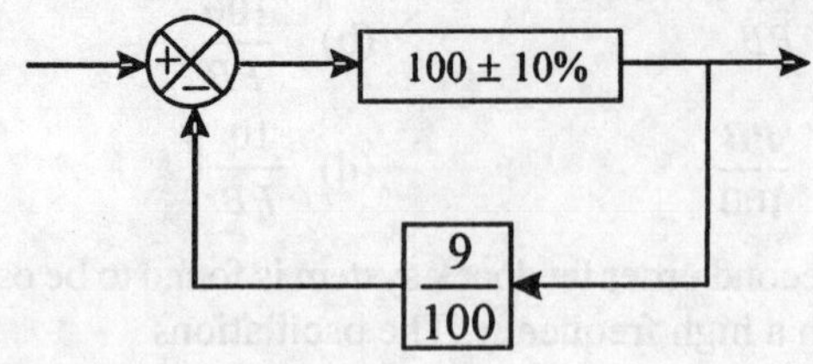

(a) $10 \pm 1\%$

(b) $10 \pm 2\%$

(c) $10 \pm 5\%$

(d) $10 \pm 10\%$

24. A set of linear equations is represented by the matrix equation. $Ax = b$. The necessary condition for the existence of a solution for this system is

(a) A must be invertible

(b) b must be linearly depended on the columns of A

(c) b must be linearly independent of the columns of A

(d) None of the above

25. A linear time-invariant system initially at rest, when subjected to a unit-step input, gives a response $y(t) = te^{-t}, t > 0$. The transfer function of the system is

(a) $\dfrac{1}{(s+1)^2}$ (b) $\dfrac{1}{s(s+1)^2}$

(c) $\dfrac{s}{(s+1)^2}$ (d) $\dfrac{1}{s(s+1)}$

26. The transfer function of the system described by $\dfrac{d^2y}{dt^2} + \dfrac{dx}{dt} = \dfrac{du}{dt} + 2u$ with u as input and y as output is

(a) $\dfrac{(s+2)}{(s^2+s)}$ (b) $\dfrac{(s+1)}{s^2+s}$

(c) $\dfrac{2}{s^2+s}$ (d) $\dfrac{2s}{(s^2+s)}$

Answers

10.1 Transfer Function, Block Diagram and SFG

1. Ans. (d)

Solution: Masson's gain formula, $[C/R] = [\Sigma P_R \Delta_R/\Delta]$

Forward path 1 Gain $P_1 = 1.2.3.4 = 24$

Forward path 3 Gain $P_2 = 1.5 = 5$

$\Delta_1 = 1$

$\Delta_2 = 1 - (3 \times -1)$

$= 1 + 3 = 4$

Gains of Loop 1 -2

Gains of Loop 2 -3

Gains of Loop 3 -4

Gains of Loop 4 -5

Non touching loop gain $= (-2)(-4) = 8$

$$\Delta = 1 - (-2 -3 -5 - 4) + 8$$
$$= 1 + 14 + 8 = 23$$

$$\Sigma P_R \Delta_R = P_1\Delta_1 + P_2\Delta_2 = 24 \times 1 + 5 \times 4 = 24 + 20 = 44$$

$$\frac{C}{R} = \frac{\Sigma P_R \Delta_R}{\Delta} = \frac{44}{23}.$$

2. Ans. (c)

Solution: Transfer function of system

$$\frac{C}{R} = \sum_k \frac{M_k \Delta_k}{\Delta}$$

3. Ans. (c)

Solution: $H(s) = \dfrac{F_o(s)}{F_i(s)}$

4. Ans. (c)

Solution: F.P. Gain $M_1 = 5 \cdot 2 \cdot 1 = 10$

Loop gain $= -4$ $\quad \Delta = 1 - [-4] = 5$

$$\frac{C}{R} = \frac{10 \cdot 1}{5} = \sum \frac{M_1\Delta_1}{\Delta}$$
$$= 2$$

5. Ans. (b)

Solution: The transfer function

$$T(s) = \frac{C(s)}{R(s)} = \frac{G}{1+GH}$$

$$= \frac{2s^2 + 6s + 5}{(s+1)^2(s+2)}$$

Characteristic equation $1 + GH = 0$

$\Rightarrow \quad (s + 1)^2 (s + 2)$

6. Ans. (c)

Solution:

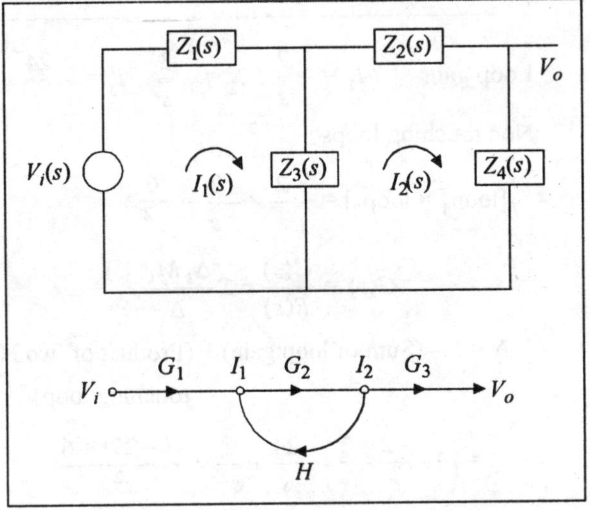

KVL in loop 1

$$V_i(s) = Z_1(s)\, I_1(s) + Z_3(s)\,[I_1(s) - I_2(s)]$$

$$I_1(s) = V_i(s) = [Z_1(s) + Z_3(s) - I_1(s) Z_2(s)]$$

$$V_i(s) + Z_3(s)\, I_2(s) = I_1(s)\,[Z_1 + Z_3]$$

$$\frac{V_i(s)}{Z_1 + Z_2} + \frac{I_2 Z_3}{Z_1 + Z_3} \quad \text{...(1)}$$

From signal Flow graph

$$I_1 = V_1 G_1 + I_2 H \quad \text{...(2)}$$

Comparer equation (1) and (2)

$$H = \frac{Z_3}{Z_1 + Z_3}$$

KVL in loop 2

$$[I_2(s) - I_1(s)]\, Z_3(s) + I_2(s)\, Z_2(s) + I_2(s)\, Z_4(s) = 0$$

$$I_2(s)\,[Z_2 + Z_3 + Z_4] = I_1 Z_3$$

$$G_2 = \frac{I_2}{I_1} = \frac{Z_3}{Z_2 + Z_3 + Z_4}$$

7. Ans. (d)

Solution:

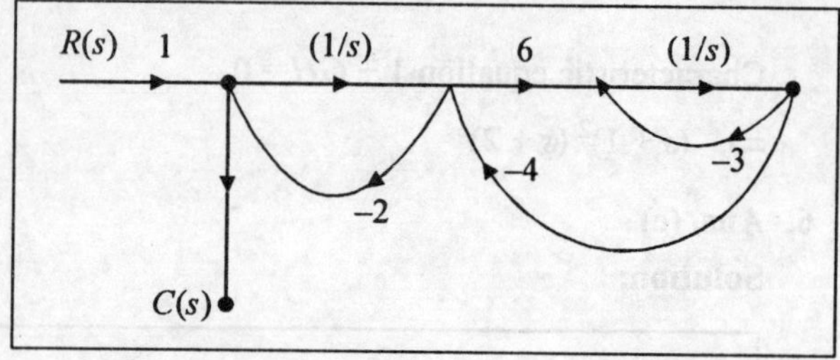

Loop gain $\quad l_1 = -\dfrac{2}{s} \quad l_2 = -\dfrac{3}{s} \quad l_3 = -\dfrac{24}{s}$

Non touching loops

$$[\text{loop}_1 \times \text{loop}_2] = -\frac{2}{s} \times -\frac{3}{s} = \frac{6}{s^2}$$

$$G(s) = \frac{C(s)}{R(s)} = \Sigma \frac{\Delta_k M_k}{\Delta}$$

$\Delta = 1 -$ (Sum of loop gain) + (Product of two Non touching loop)

$$= \left(1 - \frac{-2}{s} - \frac{3}{s}\right) - \frac{24}{s} + \frac{6}{s^2} = \frac{s^2 + 29s + 6}{s^2}$$

$M_1 = 1$

For path No touching $= 1 + \dfrac{3}{s} + \dfrac{24}{s} = \dfrac{s+27}{s}$

$$\frac{C(s)}{R(s)} = \frac{M_1 \Delta_1}{\Delta} = 1 \times \frac{(s+27)/s}{\Delta}$$

$$= \frac{(s+27)/s}{\dfrac{s^2+29+6}{s^2}} = \frac{s(S+27)}{s^2+29s+6}$$

8. Ans. (c)

Solution: Forward path gain, $P_1 = abcd$

And, as it touches all the three loops, $\Delta_1 = 1$.

Now, Three individual loops with loop gains

$$P_{11} = be, P_{21} = cf, P_{31} = dg$$

And combination of two non-touching loops with loop gains products,

$$P_{12} = bedg$$

Then, $\dfrac{C}{R} = \dfrac{x_5}{x_1} = \dfrac{P_1 \Delta_1}{1 - (P_{11} + P_{21} + P_{31}) + P_{12}}$

$$= \frac{abcd}{1 - (be + c.f + dg) + be.Dg}$$

9. Ans. (b)

Solution: Construct a signal flow graph

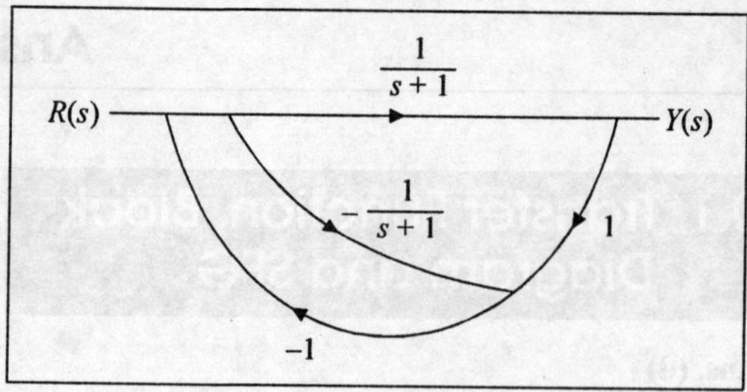

$$\frac{Y(s)}{R(s)} = \frac{\dfrac{1}{s+1}[1]}{1 - \left[\dfrac{1}{s+1} - \dfrac{1}{s+1}\right]} = \frac{1}{s+1}$$

10. Ans. (a)

Solution: Despite the presence of negative feedback, control system still have problems of instability because components used have nonlinearities. There are always some variation as compared to ideal characteristics.

11. Ans. (b)

Solution:

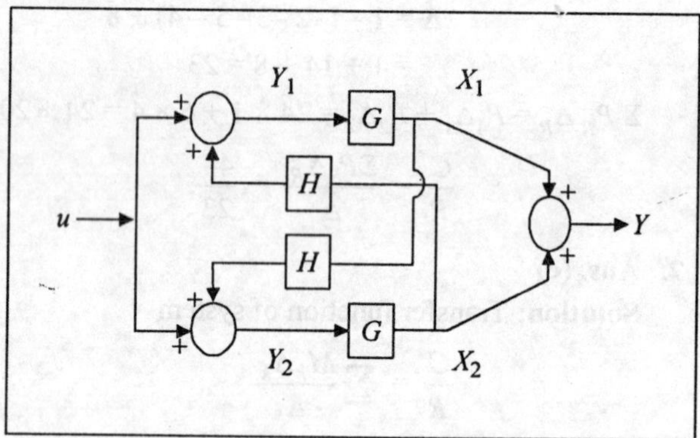

This problem can be solved by letting the point $Y_1 Y_2$, and $X_1 X_2$

$$Y_1 = U + X_2 H$$

$$Y_2 = U + X_1 H$$

And $\quad X_1 = GY_1 = GU + GX_2 H$

$$X_1 = GY_2 = GU + GX_1 H$$

And $\quad Y = X_1 + X_2$

$$= 2GU + GH(X_1 + X_2)$$

$$= 2GU + GHY$$

$$Y - GHY = 2GU$$

$$Y[1 - GH] = 2GU$$

or $\quad Y = [2GU(1 - GH)]$

or $\quad \dfrac{Y}{U} = \dfrac{2G}{(1 - GH)}$

Second Method

Just see the block diagram as signal flow graph and

analyze the result, the number of forward paths = 4

Loops = 1

$$\frac{Y}{U} = \frac{G + GHG + G + GHG}{1 - [GHGH]}$$

$$= \frac{2G[1+GH]}{1-G^2H^2} = \frac{2G(HGH)}{(1+GH)(1-GH)} = \frac{2G}{1-GH}$$

12. Ans. (b)

Solution: SFG is used to find the transfer function of the system.

13. Ans. (a)

Solution:

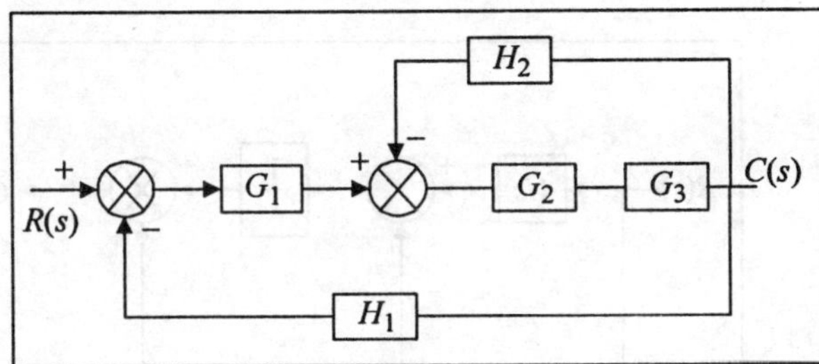

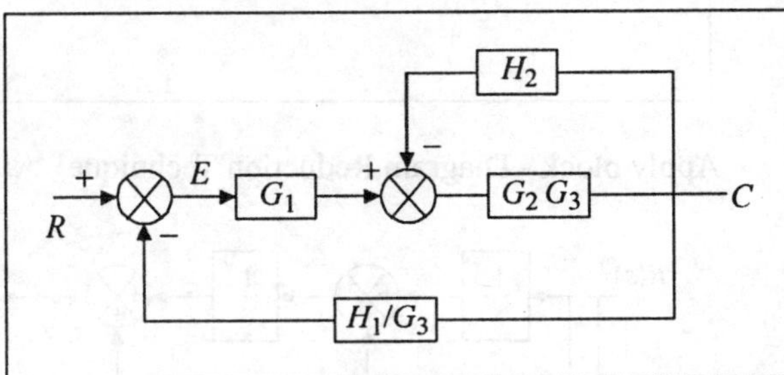

$$\frac{C}{E} = \frac{G_1G_2G_3}{1+G_2G_3H_2}$$

$$\frac{C}{R} = \frac{\dfrac{G_1G_2G_3}{1+G_2G_3H_2}}{1+\left(\dfrac{G_1G_2G_3}{1+G_2G_3H_2}\right)\times\dfrac{H_1}{G_3}}$$

$$= \frac{G_1G_2G_3}{1+G_2G_3H_2+G_1G_2H_1}$$

14. Ans. (b)

Solution:

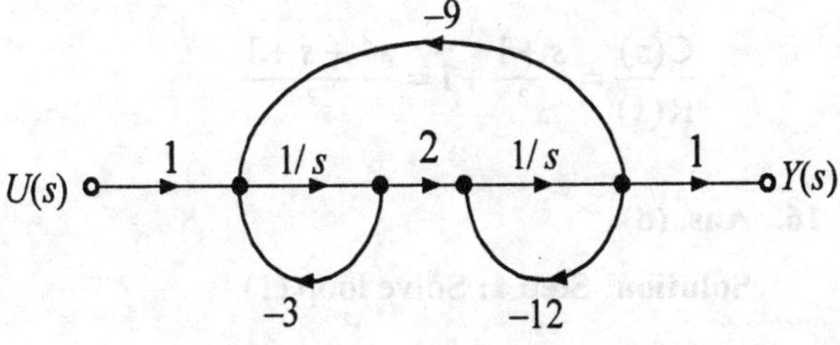

Forward path gain, $P_1 = \frac{2}{s^2}$

Individual loop gains = $\left(\frac{-3}{s}\right), \left(\frac{-12}{s}\right), \left(\frac{-18}{s^2}\right)$

P_{12} (Non-touching loop gain) = $\frac{36}{s^2}$

$$\frac{Y(s)}{U(s)} = T(s) = \frac{P_1\Delta_1}{1-P_{11}-P_{21}-P_{23}+P_{12}}$$

$$= \frac{\dfrac{2}{s^2}}{1+\dfrac{15}{s}+\dfrac{18}{s^2}+\dfrac{36}{s^2}} = \frac{2}{s^2+15s+54}$$

$$= \frac{2}{(s+6)(s+9)} = \frac{\dfrac{2}{(6\times 9)}}{\dfrac{\{(s+6)(s+9)\}}{(6\times 9)}}$$

$$= \frac{1}{27\left(1+\dfrac{s}{6}\right)\left(1+\dfrac{s}{9}\right)}$$

Second Method

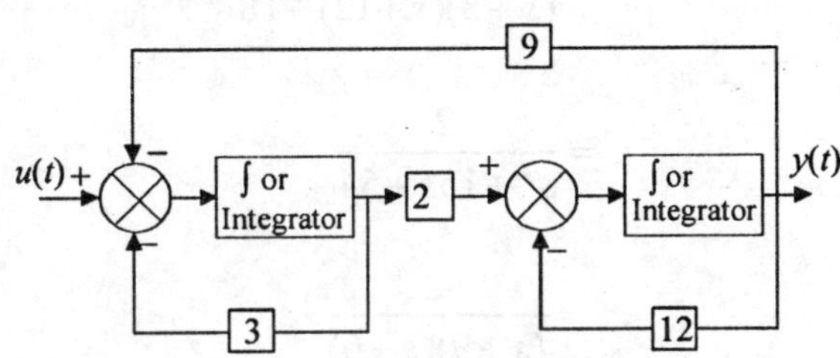

Apply block diagram reduction technique

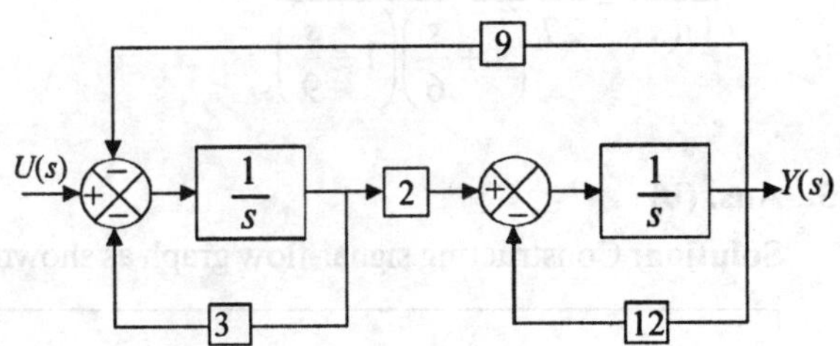

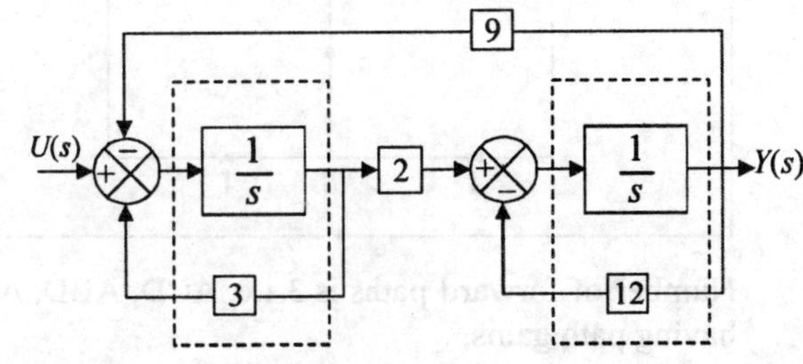

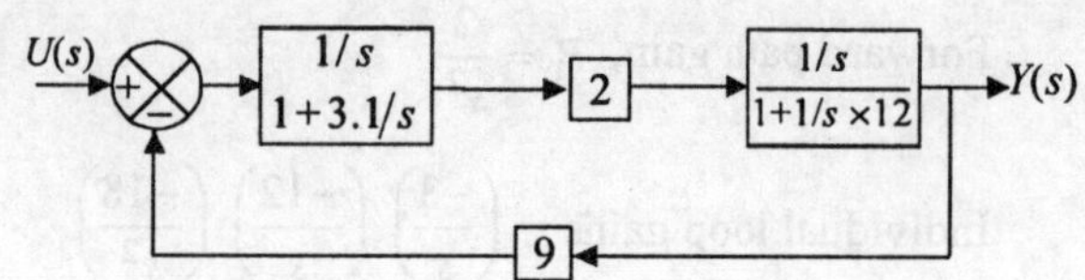

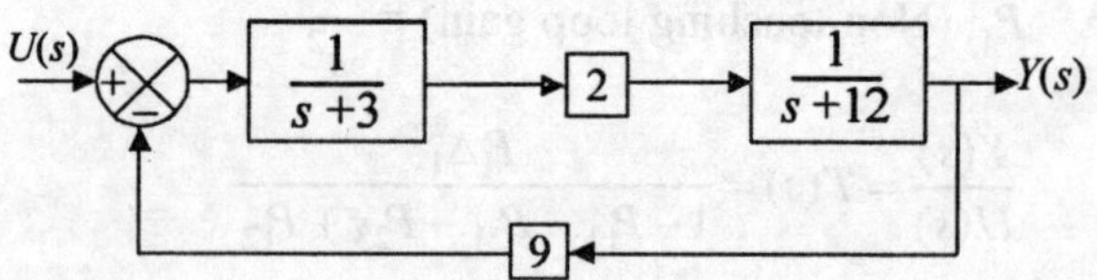

$$\frac{Y(s)}{U(s)} = \frac{\dfrac{2}{(s+3)(s+12)}}{1+9\dfrac{2}{(s+3)(s+12)}}$$

$$= \frac{2}{(s+3)(s+12)+18}$$

$$= \frac{2}{s^2+15s+54}$$

$$= \frac{2}{(s+9)(s+6)}$$

$$\frac{Y(s)}{U(s)} = \frac{1}{27}\frac{1}{\left(1+\dfrac{s}{6}\right)\left(1+\dfrac{s}{9}\right)}$$

15. Ans. (b)

Solution: Constructing signal-flow graph as shown here,

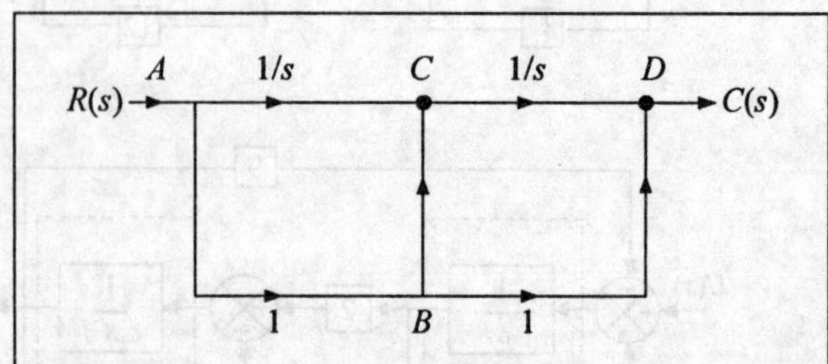

Number of forward paths is 3 i.e. ACD, ABD, ABCD, having path gains,

$$P_1 = \frac{1}{s^2},\ P_2 = 1,\ P_3 = \frac{1}{s}$$

There are no loops, then, transfer function is given a by Masson's gain formula

$$T = \frac{P_1\Delta_1 + P_2\Delta_2 + P_3\Delta_3}{\Delta}$$

Where, $[\Delta_1 = \Delta_2 = \Delta_3 = \Delta = 1]$

$$= \frac{1}{s^2} + 1 + \frac{1}{s} = \frac{s^2+s+1}{s^2}$$

Second Method

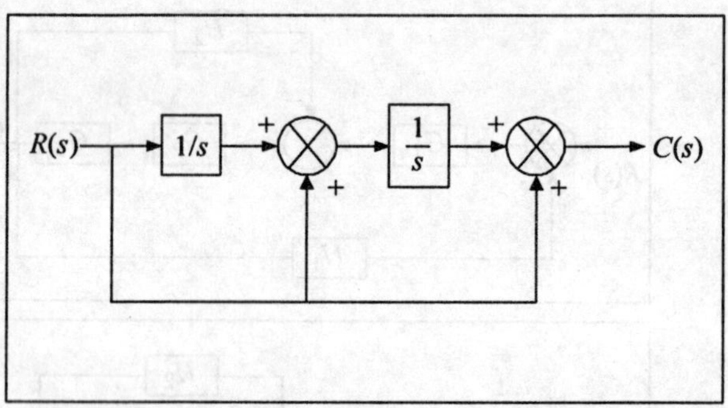

Apply block - Diagram Reduction Technique

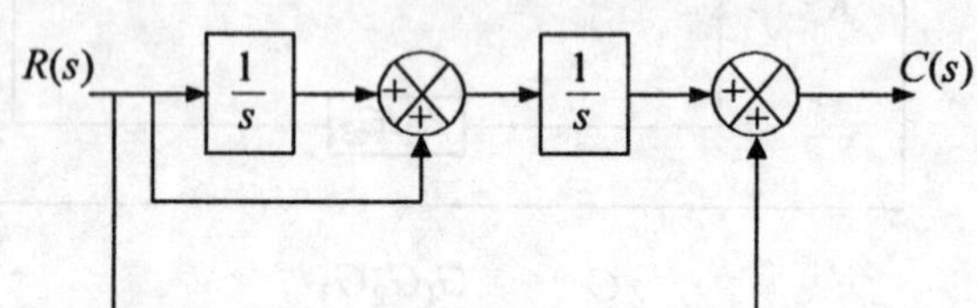

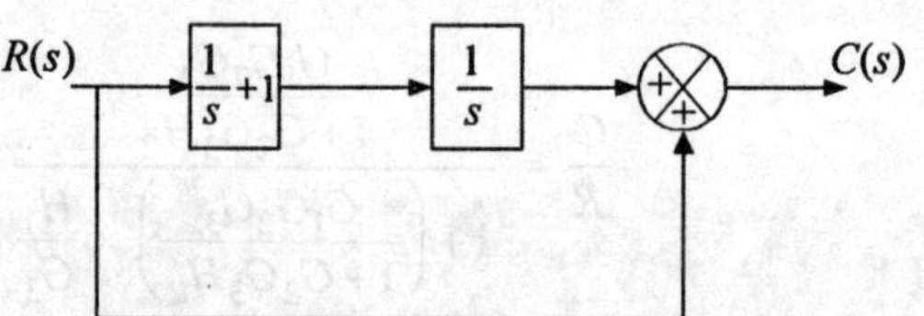

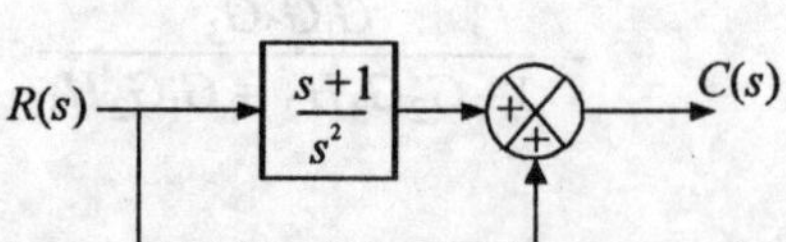

$$\frac{C(s)}{R(s)} = \frac{s+1}{s^2} + 1 = \frac{s^2+s+1}{s^2}$$

16. Ans. (d)

Solution: Step 1: Solve loop (1)

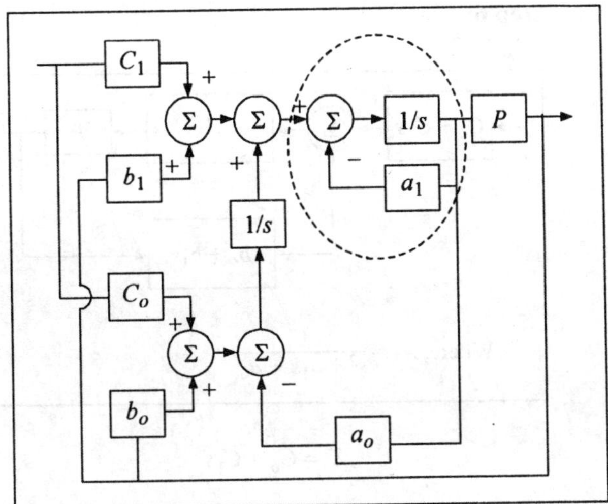

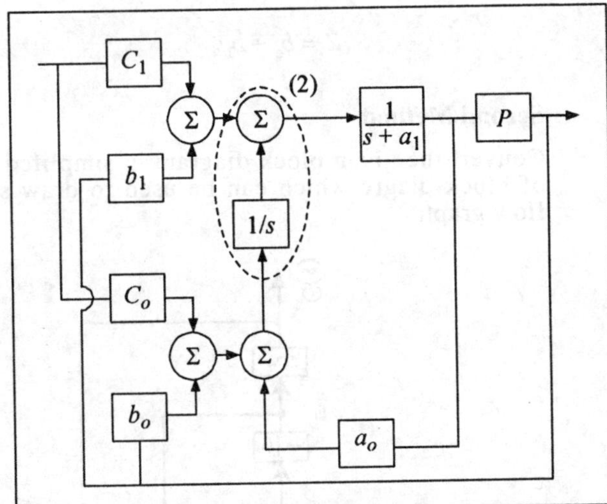

Step 2: Solve loop (2) bring summing point before (1/s)

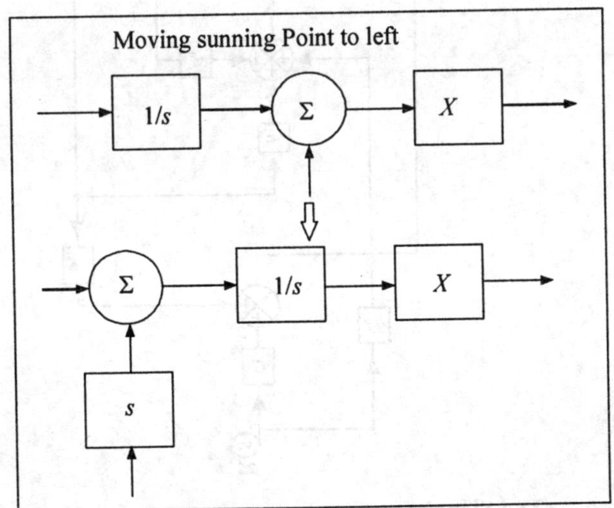

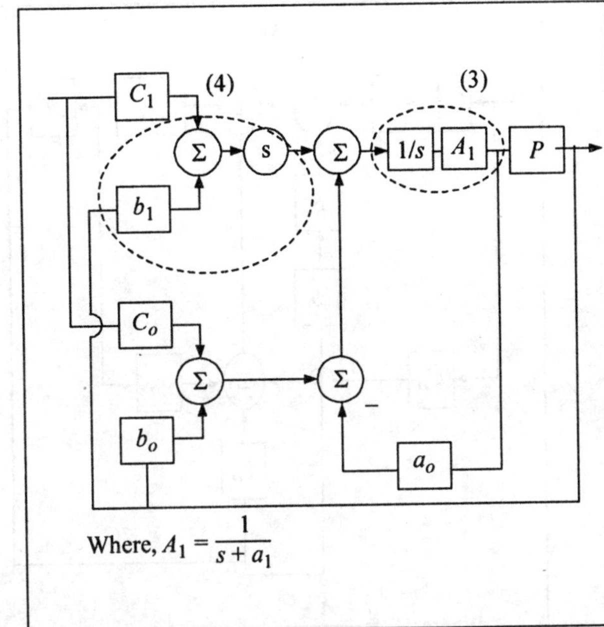

Step 3: Solve loop (3) and (4)

In (4) move (Σ) point to right of (S)

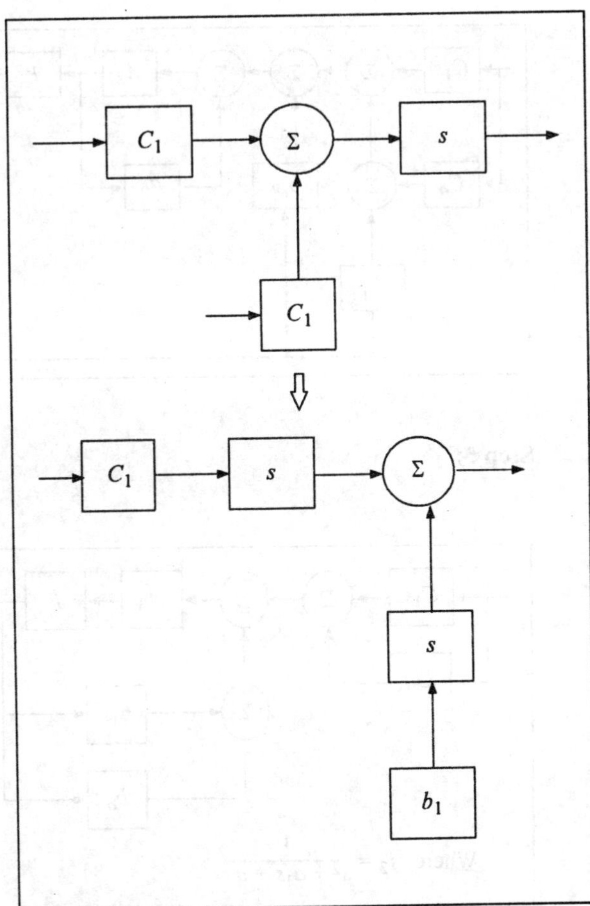

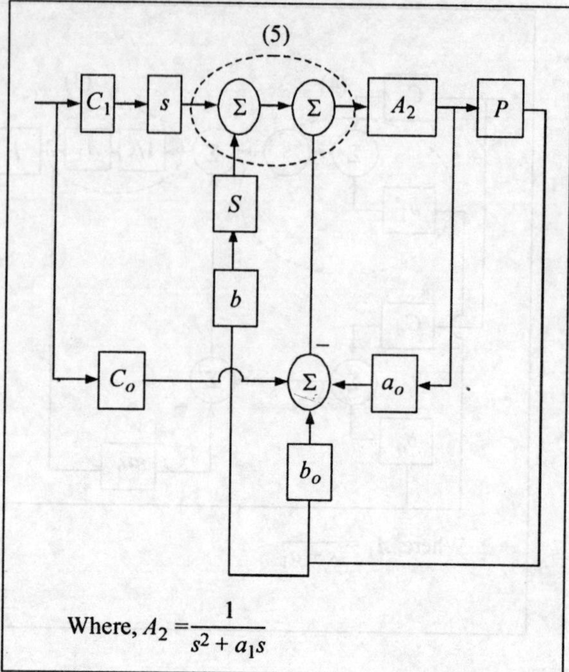

Step 4: Interchange summing Point in loop 5.

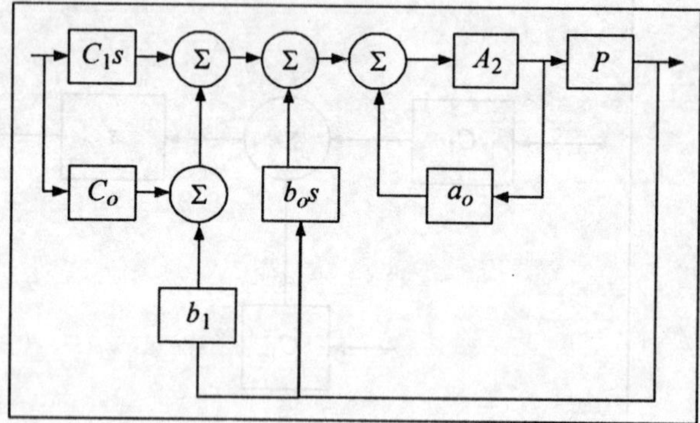

Step 5:

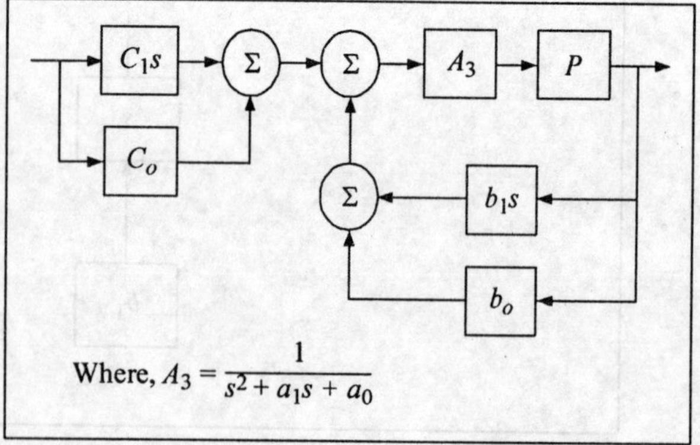

Step 6:

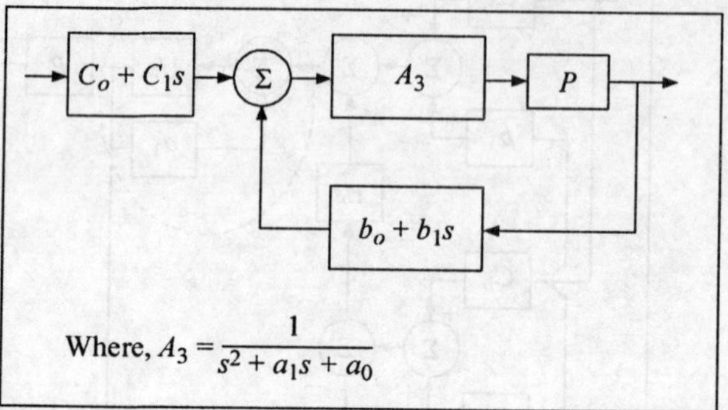

$$X = C_o + C_1 s$$

$$Y = \frac{1}{s^2 + a_1 s + a_o}$$

$$Z = b_o + b_1 s$$

Second Method

Convert the given block diagram in simplifed form of block diagre which can be used to draw signal flow graph

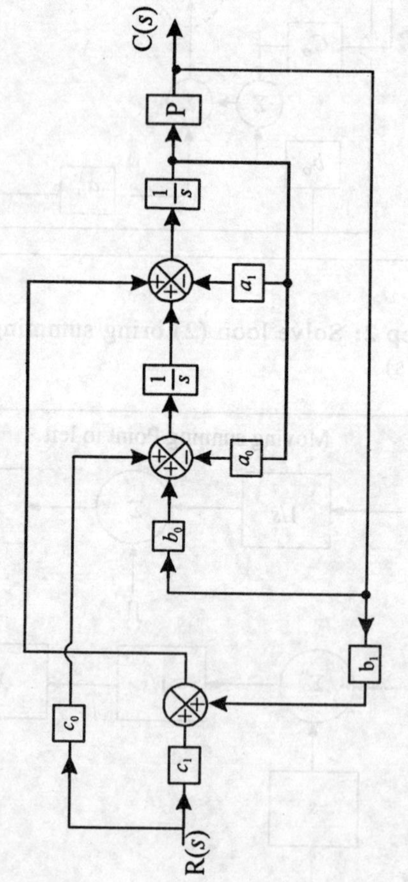

The Signal flow graph

Number of forwards paths n = 2

Gain of forward paths

$$M_1 = 1 \times c_0 \times \frac{1}{s} \times \frac{1}{s} \times P = \frac{c_0 P}{s^2}$$

$$M_2 = 1 \times c_1 \times 1 \times \frac{1}{s} \times P = \frac{c_1 P}{s^2}$$

There are four loops.

1st Loop DED $= \frac{a_1}{s}$,

2nd Loop CDEC $= -\frac{a_0}{s^2}$

3rd Loop CDEFC $= -\frac{b_0 P}{s^2}$

4th Loop BDEFB $= \frac{b_1 P}{s}$

There is no. non-touching loop

Overall T.F using Mason is Gain Formula

$$\frac{C(s)}{R(s)} = \frac{\frac{c_0 P}{s^2} \times 1 + \frac{c_1 P}{s} \times 1}{1 + \frac{a_1}{s} + \frac{a_0}{s^2} - \frac{b_0 P}{s^2} - \frac{b_1 P}{s}}$$

$$= P \times \frac{c_1 s + c_0}{s^2 + s(a_1 - b_1 P) + (a_0 - b_0)P} \quad \text{... (1)}$$

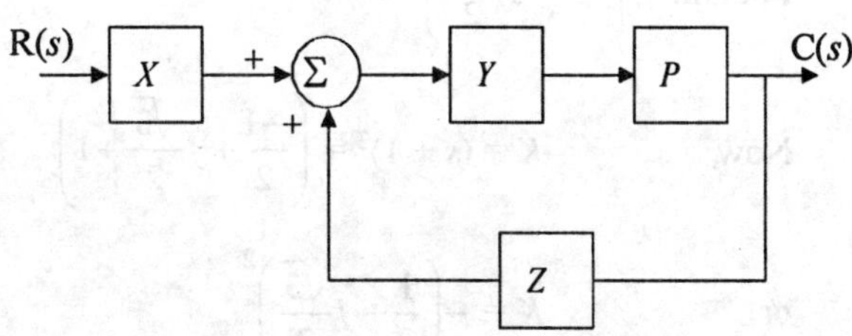

Overall Transfer fucntion

$$\frac{C(s)}{R(s)} = X \times \frac{PY}{1PYX}$$

$$= \frac{PY}{1 - PYX} \quad \text{... (2)}$$

Comparing equation (1) and (2)

$$\frac{C(s)}{R(s)} = P \times \frac{c_1 s + c_0}{s^2 + a_1 s + a_0 - (sb_1 + b_0)P}$$

$$= P \times \frac{\frac{c_1 s + c_0}{s^2 + a_1 s + a_0}}{1 - \frac{b_1 s + b_0 P}{s^2 + a_1 s + a_0}}$$

$$XY = \frac{c_1 s + c_0}{s^2 + a_1 s + a_0}$$

$$YZ = \frac{b_1 s + b_0}{s^2 + a_1 s + a_0}$$

$$X = c_1 s + c_0$$

$$Y = \frac{1}{s^2 + a_1 s + a_0}$$

$$Z = b_1 s + b_0$$

17. Ans. (a)

Solution: $1 + GH = 0$

Substitute $\quad s = -\frac{1}{2} + j\frac{\sqrt{3}}{2}$

Second Method

Characteristic equation

$$C(s)\,H(s) = \frac{K}{(s+1)^3}$$

$$1 + G(S)\,H(S) = 0$$

$$= \frac{(s+1)^3 + K}{(s+1)^3} = 0$$

$$(s+1)^3 + K = 0$$

or $\quad -K = (s+1)^3$

Put and $= \left(\frac{-1}{2} + j\frac{\sqrt{3}}{2}\right)$

Now, $\quad -K = (s+1)^3 = \left(\frac{-1}{2} + j\frac{\sqrt{3}}{2} + 1\right)^3$

or $\quad K = -\left(\frac{1}{2} + j\frac{\sqrt{3}}{2}\right)^2$

$$= \frac{1}{8} - j\frac{3\sqrt{3}}{8} + \frac{3\times 1}{2} \times j\frac{\sqrt{3}}{2}\left(\frac{1}{2} + j\frac{\sqrt{3}}{2}\right)$$

$$= \frac{1}{8} - j\frac{3\sqrt{3}}{8} + j\frac{3\sqrt{3}}{8}\left(\frac{1+j}{2} + \frac{\sqrt{3}}{2}\right)$$

$$= \frac{1}{8} - j\frac{3\sqrt{3}}{8} + j\frac{3\sqrt{3}}{8} - \frac{9}{8} = \frac{1}{8} - \frac{9}{8} = \frac{-8}{8} = -1$$

$-K = -1; K = 1$

18. Ans. (c)

Solution:

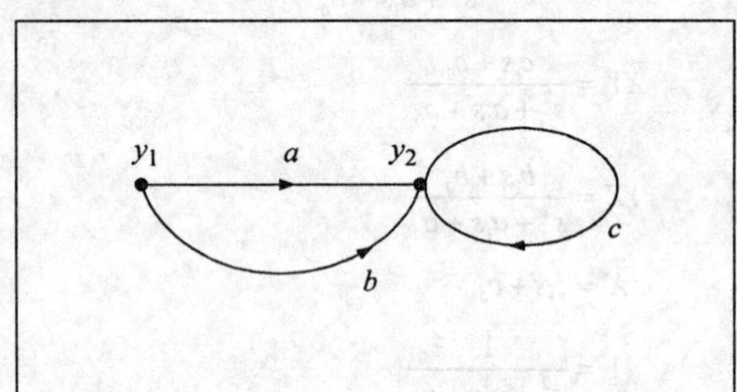

$$y_2 = (a+b)\,y_1 + cy_2$$

i.e. $\quad y_2(1-c) = (a+b)\,y_1$

i.e. $\quad \frac{y_2}{y_1} = \frac{a+b}{1-c}$

19. Ans. (c)

Solution:

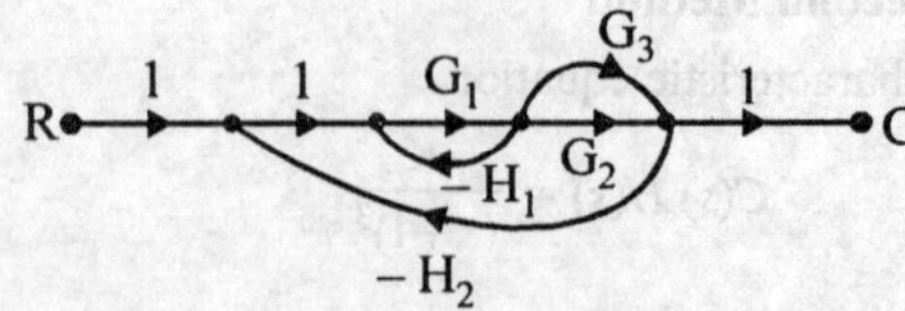

No of forward path = 2

(1) $M_1 = G_1G_2$

(2) $M_2 = G_1G_3$

No. of individual Loop	Gain Product of Two individual Loop $\Rightarrow 0$
(1) $+G_1H_1$	
(2) $-G_1G_3H_2$	
(3) $-G_1G_2H_2$	

$\Delta_1 = 1$

$\Delta_2 = 1$

$$\frac{C}{R} = \frac{M_1\Delta_1 + M_2\Delta_2}{\Delta}$$

$$\frac{C}{R} \Rightarrow \frac{G_1G_2 + G_1G_3}{1 - G_1H_1 + G_1G_3H_2 + G_1G_2H_2}$$

20. Ans (a)

Solution:

Forward path gain

$$P_1 = -\frac{1}{s}, \quad P_2 = -k$$

Here $\quad \Delta_1 = \Delta_2 = 1$

loop gain $P_{11} = -\frac{k}{s}$

Transfer function

$$T(s) = \frac{P_1\Delta_1 + P_2\Delta_2}{1 - P_{11}} = \frac{-(1+ks)}{s+k}$$

21. Ans. (d)

Solution:

$$T(s) = \frac{C(s)}{R(s)} = \frac{G(s)}{1 + G(s)H(s)} = \frac{A}{1 + s\tau + A\beta}$$

$$= \frac{A}{s\tau + A\beta + 1} = \frac{A/(A\beta+1)}{\frac{s\tau}{(A\beta+1)} + 1}$$

or $\quad T(s) = \frac{a}{\tau' s + 1}$

$$\alpha = \frac{A}{A\beta + 1} \text{ and } \tau' = \frac{\tau}{(1 + A\beta)}$$

Thus, time constant of $T(s), \tau' = \frac{\tau}{(1 + A\beta)}$

22. Ans. (c)

Solution:

This not a cascaded system but linked or connected system whose transform function are found by masson's gain formula.

Forward path gain, $P_1 = G_1G_2$

Loop $\quad L_1 = -G_1H_1$ and $\quad L_2 = -G_2H_2$

There are No non touching loop

$$\Delta = 1 + G_1H_1 + G_2H_2$$

$$\Delta_1 = 1$$

By maisson's gain formula

$$T(s) = \frac{C(s)}{R(s)} = \frac{P_1\Delta_1}{\Delta}$$

$$= \frac{G_1G_2}{1 + G_1H_1 + G_2H_2}$$

23. Ans. (c)

Solution:

$$G(s) = \frac{1}{s(s+1)}$$

$$H(s) = \frac{1}{(s+4)(s+1+j)(s+1-j)}$$

$$= \frac{1}{(s+4)(s^2 + s - js + s + 1 - j + js + j - j^2)}$$

$$= \frac{1}{(s+4)(s^2 + 2s + 2)}$$

$$\frac{C(s)}{R(s)} = \frac{G(s)}{1 + G(s)H(s)}$$

$$= \frac{\frac{1}{s(s+1)}}{1 + \frac{1}{s(s+1)(s+4)(s^2+2s+1)}}$$

$$= \frac{(s+4)(s^2+2s+2)}{s(s+1)(s+4)(s^2+2s+2)+1}$$

24. Ans. (c)

Solution:

$$\frac{C}{R} = \frac{KG}{1 + KGH}$$

When $K >> 1$

$$\frac{C}{R} \approx \frac{1}{H}$$

10.2 Time Response and Controller

1. Ans. (b)

Solution:

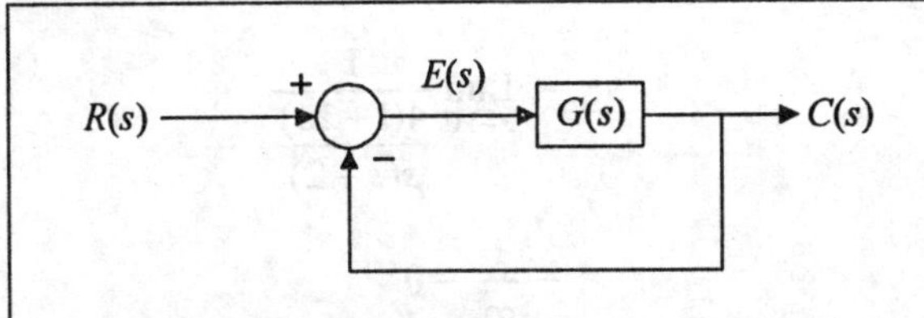

$r(t)$ = unit steps

$$R(s) = \frac{1}{S}$$

$$C(s) = G(s)\,[R(s) - C(s)]$$

$$C(s) = G(s)\frac{R(s)}{1 + G(s)}$$

$$E(s) = R(s) - C(s) = \frac{R(s)}{1 + G(s)}$$

Steady state error

$$= e(\infty) = \lim_{s \to 0} sE(s) = \lim_{s \to 0} \frac{s \times \frac{1}{s}}{1 + G(s)}$$

$$= \lim_{s \to 0} \frac{1}{1 + G(s)}$$

$$K_p = G(0)$$

$$SSE = e(\infty) = \frac{1}{1 + K_p}$$

2. Ans. (a)

Solution: $G(s) = \dfrac{25}{s^2 + 8s + 25}$

Standard second order equation

$$\frac{\omega_n^2}{s^2 + 2\delta\omega_n s + \omega_n^2}$$

Comparer the above two equations

$$\omega_n = 5$$

$$\delta\omega_n = 4$$

$$\delta . 5 = 4 \quad ; \quad \delta = \frac{4}{5} = 0.8$$

$$t_{(\text{max or min})} = \frac{n\pi}{\omega n\sqrt{1 - \delta^2}},$$

for second peak, $n = 3$

$$t_{2\text{nd max}} = \frac{3\pi}{5\sqrt{1 - 0.8^2}} = \frac{3\pi}{5 \times .6} = \frac{3\pi}{3} = \pi$$

3. Ans. (a)

Solution: $E_{ss} = \lim_{s \to 0} \dfrac{sR(s)}{1 + GH}$

$$= \lim_{s\to 0} \frac{[s \times (1/s^2)]}{s + sGH} = \lim_{s\to 0} \frac{1}{1+GH}$$

$$= \lim_{s\to 0} \frac{1}{s(G)} = \lim_{s\to 0} \frac{1}{\frac{s \times 4(1+2s)}{s^2(s+2)}}$$

$$= \lim_{s\to 0} \frac{1}{\frac{4(1+2s)}{s(s+2)}}$$

$$= \frac{1}{\infty} = 0$$

4. Ans. (c)

Solution: Integral mode improves steady state response. Derivative mode improves transient response.

5. Ans. (a)

Solution: $K_p = \lim_{S\to 0} G(s)\,H(s) = \lim_{S\to 0} \frac{1}{(S+6)(S+1)}$

$$= \frac{1}{6}$$

6. Ans. (c)

Solution: 2nd order system $G(s) = \frac{\omega_n^2}{s^2 + 2\delta\,\omega_{ns}\,s + \omega_n^2}$

Characteristic equation of the system

$$s^2 + 2\delta\,\omega_n\,s + \omega_n^2 = 0$$

$$s = -\delta\,\omega_n \pm j\,\omega_n\sqrt{1-\delta^2}$$

For $0 < \delta < 1$ roots are complex conjugate. i.e. system is under damped:

7. Ans. (a)

Solution:

Type	K_p	K_v	K_a	$e_{ss} = \frac{R}{1+K_p}$	$e_{ss} = \frac{R}{K_v}$	$e_{ss} = \frac{R}{K_a}$
0	K	0	0	$R/(1+K_p)$	∞	∞
1	∞	K	0	0	R/K_v	∞
2	∞	∞	K	0	0	R/K
3	∞	∞	∞	0	0	0

8. Ans. (b)

Solution: Setting times $t_s = \frac{4}{\delta\,\omega_n}$ (for 2% band)

$$T(s) = \frac{9}{s^2 + 4s + 9}$$

Standard second order equation

$$= \frac{\omega_n^2}{S^2 + 2\delta\,\omega_n S + \omega_n^2}$$

Compare the above two equation

$$\delta\,\omega_n = 2$$

Setting times $t_s = \frac{4}{\delta\omega_n}$ (for 2% band)

$$= \frac{4}{2} = 2 \text{ second}$$

9. Ans. (d)

Solution:

$$\frac{C(s)}{R(s)} = T(s) = \frac{a_{n-1}\,s + a_n}{s^n + a_1\,s^{n-1} + \ldots + a_{n-1}\,s + a_n}$$

For unit ramp input $R(s) = \frac{1}{s^2}$

$$= \frac{G(s)}{1 + G(s)\,H(s)}$$

$$\text{GH} = \frac{a_{n-1}\,s + a_n}{s^n + a_1\,s^{n-1} + \ldots + a_{n-2}\,s^2}$$

$$1 + \text{GH} = \left(1 + \frac{a_{n-1}\,s + a_n}{s^n + a_1\,s^{n-1} + \ldots + a_{n-1}s^2}\right)$$

$$e_{ss} = \lim_{s\to 0} \frac{sR(s)}{1+G(s)}; \qquad (\because H(s) = 1)$$

$$= \lim_{s\to 0} \frac{s \times \frac{1}{s^2}}{1 + \left(\frac{a_{n-1}s + a_n}{s^n + a_1\,s^{n-1} + \ldots\, a_{n-1}\,s^2}\right)}$$

$$= \lim_{s\to 0} \frac{1}{s}\left[\frac{s^n + a_1 s^{n-1} + \ldots\, a_{n-1}\,s^2}{s^n + a_1\,s^{n-1} + \ldots\, a_{n-1}\,s^2 + a_{n-1}\,s + a_n}\right]$$

$$= \lim_{s\to 0} \frac{s^{n-1} + a_1\,s^{n-2} + \ldots + a_{n-1}\,s}{\text{Dr}} = \frac{0}{a_n} = 0$$

Second Method

Close loop transfer fucntion of unity feedback system

$$\text{T}(s) = \frac{\text{C}(s)}{\text{R}(s)} = \frac{a_{n-1}s + a_n}{s^n + a_1 s^{n-1} + \ldots\ldots + a_{n-1}s + a_n}$$

For unit ramp input $\text{R}(s) = \frac{1}{s^2}$

Error signal

$$\text{E}(s) = \text{R}(s) - \text{C}(s)$$

$$= \text{R}(s)[1 - \text{T}(s)]$$

$$= \frac{1}{s^2}\left[1 - \frac{a_{n-1}s + a_n}{s^n + a_1 s^{n-1} + ... + a_{n-1}s + a_n}\right]$$

$$= \frac{1}{s^2}\left[\frac{s^n + a_1 s^{n-1} + ... + a_{n-2}s^2}{s^n + a_1 s^{n-1} + ... + a_n}\right]$$

S.S error

$$e_{ss} = \lim_{s \to 0} sE(s)$$

$$= \lim_{s \to 0} \frac{s^{n-1} + a_1 s^{n-1} + ... + a_{n-2}s}{s^n + ... + a_n} = 0$$

10. Ans. (c)

Solution:

Characteristic equation of close loop system is

$$s^2 + 2s + 2 = 0$$

$$s^2 + 2\,\delta\omega_n\,s + 2 = 0$$

Compare the above two equation

$$\omega_n = \sqrt{2}$$

Than, $\delta\,\omega_n = 1$

$$\delta = \frac{1}{\omega_n} = \frac{1}{\sqrt{2}} < 1$$

Hence under damped.

Second Method

Given characteristic equation

$$s^2 + 2s + 2 = 0$$

its roots are $s = -1 \pm j$

There are complex conjugate roots in left side of s-plane so the system is under damped.

11. Ans. (c)

Solution: Differentiator amplify the noise.

12. Ans. (b)

Solution: $2\,\omega_n = 4$

$$\delta\omega_n = 2$$

and $\omega_n = 2$

then, $\delta = 1$ (critically-damped system)

and t_{ss} (Setting time) $= \dfrac{4}{\delta\omega_n} = 2$ sec

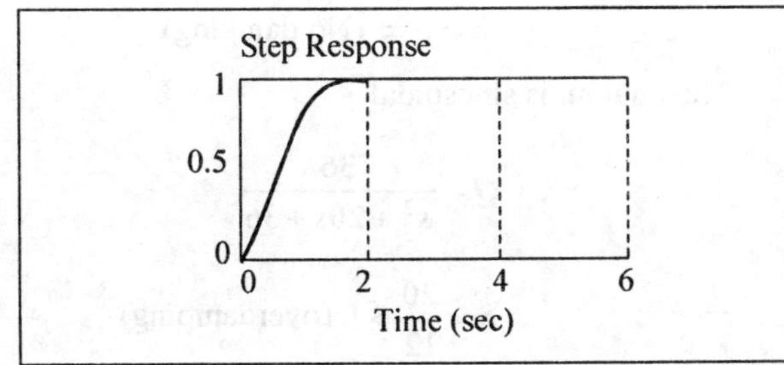

Thus, only response is as shown in Figure (b).

Figure (d) is also critically-damped system, but its setting time is about 5 sec. Hence, the only choice is (b).

13. Ans. (c)

Solution: The peak percent overshoot is determined for LTI second order closed loop system with zero initial condition. It's transfer function is

$$T(s) = \frac{\omega_n^2}{s^2 + 2\delta\omega_n s + \omega_n^2}$$

Transfer function has a pair of complex conjugate poles and zeros.

14. Ans. (a)

Solution: Ramp input is applied

$$R(s) = \frac{1}{s^2}\ ;\ e_{ss} = \lim_{s \to 0} sE(s)$$

$$sE(s) = \frac{sE(s)}{1+G(s)} = \frac{s.[1/s^2]}{1+G(s)} = \frac{1}{s + sG(s)}$$

$$e_{ss} = \lim_{s \to 0} sE(s) = \lim_{s \to 0} \frac{1}{sG(s)}$$

The above expression represent type 1 system therefore, type = 1. e_{ss} for type 1 system is given by

$$e_{ss} = \frac{1}{K} = 5\% = \frac{1}{20}$$

$\therefore$ K = 20

15. Ans. (c)

Solution: $h(t) = e^{-t}\ ;\ H(s) = \dfrac{1}{s+1}$

$$Y(s) = H(s)\,X(s) = \frac{1}{s+1} \times \frac{1}{s}$$

$$y(t) = 1 - e^{-t}$$

In steady state $t \to \infty\ ;\ y(\infty) = 1$

16. Ans. (c)

Solution:

If O.L.T.F. $= G(s)$, Feedback $= H(s)$

The closed loop transfer function

$$\text{C.L.T.F}\ \ T(s) = \frac{G(s)}{1 + G(s)\,H(s)}$$

Given

$$T(s) = \frac{s-5}{(s+2)\,(s+3)}$$

$$= \frac{\dfrac{s-5}{s^2+5s}}{1 + \dfrac{6}{s^2+5s}}$$

$$= \frac{\frac{s-5}{s(s+5)}}{1+\frac{6}{s(s+5)}}$$

$$T(s) = \frac{\frac{(s-5)}{s(s+5)}}{1+\frac{s-5}{s(s+5)} \times \frac{6}{s-5}}$$

$$G(s) = \frac{s-5}{s(s+5)}, \quad H(s) = \frac{6}{s-5}$$

Loop Transfer Function

$$G(s)H(s) = \frac{6}{s(s+5)}$$

$G(s)\,H(s)$ has two poles in Let Half so s-plane and no zero.

So it is minimum phase system.

17. Ans. (b)

Solution:

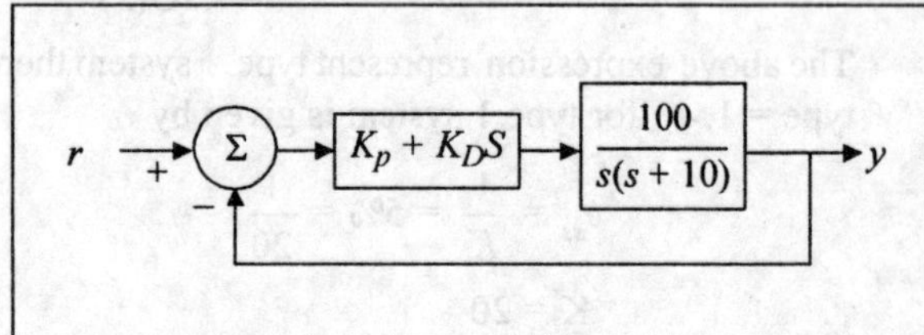

Given that,

$$K_V = 1000, \quad \delta = 0.5$$

$$T(s) = \frac{100\left(K_P + K_D s\right)}{s^2 + \left(100 K_D + 10\right)s + K_P 100}$$

$$1 + G(s)H(s) = s^2 + \left(100\ K_D + 10\right)s + 100\ K_P$$

$$\therefore\ \omega_n^2 = 100\ K_P$$

$$2\delta\,\omega_n = 2 \times 0.5\,\omega_n = \omega_n = 100\ K_D + 10$$

$$\therefore\ \sqrt{K_P 100} = 100 K_D + 10$$

$$\sqrt{K_P} = 10 K_D + 1 \qquad \ldots(1)$$

Also $\quad K_V = 1000$

$$K_V = \underset{s\to 0}{Lt}\ sG(s)H(s)$$

$$1000 = \underset{s\to 0}{Lt}\ \frac{s\,100\left(K_P + sK_D\right)}{s(s+10)}$$

$$1000 = 10\ K_P$$

$$K_P = 100$$

From (1)

$$\sqrt{K_P} = 10\ K_D + 1 \Rightarrow 10\ K_D = \sqrt{100} - 1 = 9$$

$$K_D = 0.9$$

18. Ans. (d)

Solution: $T(s) = \dfrac{5}{(s+5).\left(s^2+s+1\right)}$

Dominant poles are poles near origin. or imaginary axis in LHS of s-plane which give rise to transient response that decay relatively slow.

Thus $s + 5$ nearly equal to 5 .

Which result as $T(s) = \dfrac{1}{\left(s^2+s+1\right)}$

Another way

Dominant poles are poles near origin. So eliminate $(s + 5)$ but in dominant pole concept, the factor that has to be eliminated should be in time domain form

$$= \frac{5}{(s+5)\left(s^2+s+1\right)} = \frac{5}{5\left(1+\frac{s}{5}\right)\left(s^2+s+1\right)}$$

$$= \frac{1}{s^2+s+1}$$

19. Ans. (c)

Solution: Since pole are trying to touch the negative real axis and system is stable under all condition.

Another way

For same overshoot dambing ratio δ must be constant

$$\because \qquad \cos\theta = \delta$$

$$\theta = \text{constant}$$

20. Ans. (d)

Solution: For $\quad P = \dfrac{25}{s^2+25}$

$$\delta = 0 \Rightarrow \text{(No damping)}$$

Then output is sinusoidal

$$Q = \frac{36}{s^2+20s+36}$$

$$\Rightarrow \qquad \delta = \frac{20}{12} > 1 \text{ (overdamping)}$$

⇒ output is slow

$$R = \frac{36}{s^2 + 12s + 36}$$

⇒ $\delta = 1$ (critical damping)

⇒ output is optimum

$$S = \frac{49}{s^2 + 7s + 49}$$

⇒ $\delta < 1$

⇒ Fast but gives overshoot.

21. Ans (b)

Solution: Given that

$$X(s) = \frac{1}{s}$$

$$H(s) = \frac{-3.82}{s^2 + 1.91s + 1.91}$$

$$Y(s) = H(s).X(s)$$

$$Y(s) = \frac{-3.82}{s(s^2 + 1.91s + 1.91)}$$

Steady state output, $y(t) = \underset{s \to 0}{Lt}\,[sY(s)]$

$$= \underset{s \to 0}{Lt} \frac{-3.82}{s^2 + 1.91s + 1.91}$$

$$y(t) = \frac{-3.82}{1.91}$$

$$y(t) = -2$$

Second Method

For under damped case δ must be less than 1.

$$\delta_a = \frac{2.59}{2 \times \sqrt{1.2}} > 1$$

$$\delta_b = \frac{1.91}{2 \times \sqrt{1.91}} < 1$$

Option (c) and (d) does not be consider as denomi nator contian Negative term.

There for option (b) is correct as it give $e_{ss} = -2$.

22. Ans. (d)

Solution:

$$E(s) = \frac{sR(s)}{1 + [G(s) + G_C(s)]H(s)}$$

$$E(s) = \left.\frac{sR(s)}{1 + [G(s) + G_C(s)]}\right|_{s \to 0}$$

so, $G_C(s)|_{s \to 0}$ should be maximum for low error

$$\Rightarrow G_C(s)|_{s \to 0} = \infty$$

Second Method

Steady state error,

$$e_{ss} = \lim_{s \to 0} \frac{sR(s)}{1 + G(s)G_c(s)}$$

For step input $r(t) = u(t)$

$$R(s) = \frac{1}{s}$$

$$e_{ss} = \lim_{s \to 0} \frac{s.\frac{1}{s}}{1 + G(s)G_c(s)}$$

$$e_{ss} = \lim_{s \to 0} \frac{1}{1 + G(s)G_c(s)}$$

For $G_c(s) = \frac{s+1}{s+2}$, ⇒ $e_{ss} = \frac{2}{3}$

For $G_c(s) = \frac{s+2}{s+1}$, ⇒ $e_{ss} = \frac{1}{3}$

For $G_c(s) = \frac{(s+1)(s+4)}{(s+2)(s+3)}$, ⇒ $e_{ss} = \frac{3}{5}$

For $G_c(s) = 1 + \frac{2}{5} + 3s$, ⇒ $e_{ss} = 0$

23. Ans. (a)

Solution:

$$100\frac{d^2Y}{dt^2} - 20\frac{dy}{dt} + y = x(t)$$

$$100s^2Y(s) - 20\,sY(s) + Y(s) = X(s)$$

$$(100s^2 - 20\,sY + 1)\,Y(s) = X(s)$$

$$\frac{Y(s)}{X(s)} = \frac{1}{100\,s^2 - 20\,s + 1}$$

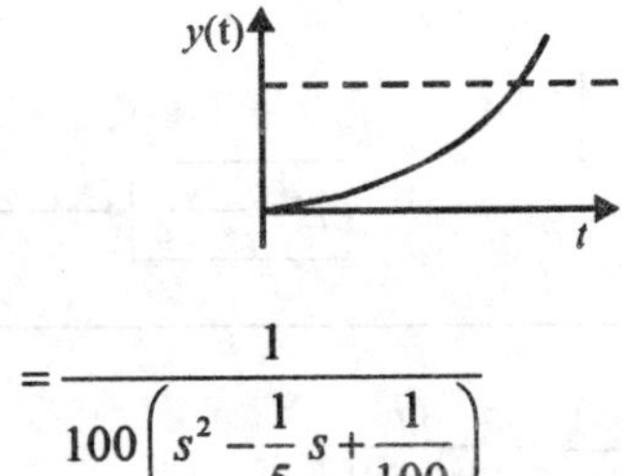

$$= \frac{1}{100\left(s^2 - \frac{1}{5}s + \frac{1}{100}\right)}$$

$$\omega_n = \frac{1}{10}$$

$$\delta_1 = -1\,;\; 2\delta_1\omega_n = \frac{-1}{5}$$

Initially relaxed

$$Y'(0) = 0$$

$$Y''(0) = 0$$

Second Method

$$\frac{100d^2y}{dt^2} - \frac{20dy}{dt} + y = x(t)$$

Take Laplace transform both sides

$$(100s^2 - 20s + 1)Y(s) = \frac{1}{s}$$

$$\left[\because x(t) = u(t) \Rightarrow X(s) = \frac{1}{s}\right]$$

$$Y(s) = \frac{1}{s(100s^2 - 20s + 1)}$$

So we have poles with positive real part $\Rightarrow$ system is unstable

24. Ans. (c)

Solution:

$$E(s) = \frac{R(s)}{1+G(s)}$$

$$\underset{t\to\infty}{Lt}\, e(t) = \underset{s\to 0}{Lt}\, \frac{sR(s)}{1+G(s)}$$

Here, $G(s) = \frac{K}{s^2}$ and $R(s) = \frac{1}{s}$

$$\underset{t\to\infty}{Lt}\, \frac{s\cdot\frac{1}{s}}{1+\frac{K}{s^2}} = \frac{1}{\infty} = 0$$

So, $\underset{t\to\infty}{Lt}\, e(t) = 0$

25. Ans. (a)

Solution:

X(s) → ⊗ → E(s) → $G = \frac{s+1}{s^2+5s+a}$ → Y(s); feedback $H = \frac{1}{s+4}$

$$\frac{Y(s)}{X(s)} = \frac{G}{1+GH}$$

$$Y(s) = \frac{G}{1+GH}.X(s)$$

$$E(s) = X(s) - Y(s)\,H$$

$$= X(s) - \left\{\frac{G}{1+GH}\right\}H.X(s)$$

$$= \frac{X(s)\,[1+GH-GH]}{1+GH}$$

$$= X(s)\frac{1}{1+GH}$$

For step input, $X(s) = \frac{1}{s}$

Hence,

$$E(s) = \frac{1}{s(1+GH)} = \frac{1}{s}\left(\frac{1}{1+\frac{s+1}{(s+4)\,(s^2+5s+a)}}\right)$$

the steady state error,

$$e_{ss} = \lim_{t\to\infty} e(t)$$

$$= \lim_{s\to 0} sE(s)$$

$$= \lim_{s\to 0} \frac{1}{1+\frac{s+1}{(s+4)\,(s^2+5s+a)}}$$

$$= \frac{1}{1+[1/4a]} = \frac{4a}{4a+1} = 0$$

For $e_{ss} = 0$ or $a = 0$

26. Ans. (a)

Solution: $\frac{C(s)}{R(s)} = \frac{2(s-1)}{(s+2)\,(s+1)}$; $R(s) = \frac{1}{s}$

Hence, $C(s) = \frac{2(s-1)}{s(s+2)\,(s+1)}$

$$= \frac{K_1}{s} + \frac{K_2}{s+1} + \frac{K_3}{s+2}$$

$$K_1 = \left.\frac{2(s-1)}{(s+2)\,(s+1)}\right|_{s=0} = \frac{-2}{2} = -1$$

$$K_2 = \left.\frac{2(s-1)}{s\,(s+2)}\right|_{s=-1} = \frac{2(-2)}{-1(1)} = 4$$

$$K_3 = \left.\frac{2(s-1)}{s\,(s+1)}\right|_{s=-2} = \frac{2(-3)}{-2(-1)} = -3$$

Hence $C(s) = \dfrac{-1}{s} + \dfrac{4}{s+1} - \dfrac{3}{s+2}$

and the output $c(t) = [-1 + 4e^{-t} - 3e^{-2t}]\, u(t)$.

27. Ans. (b)

Solution: input is $\delta(t) \to 1$

$$C(t) = -te^{-t} + 2e^{-t}\,;\ (t \geq 0)$$

$$G(s) \frac{1}{(s+1)^2} + \frac{2}{(s+1)} = \frac{-1+2(s+1)}{(s+1)^2}$$

$$C(s) = \frac{2s+1}{(s+1)^2}$$

$$C(s) = \frac{G(s)}{1+G(s)},$$

$$C(s)\,[1+G(s)] = G(s)$$

$$C(s) = G(s) - C(s)\,G(s)$$

$$= G(s)\,[1 - C(s)]$$

$$G(s) = \frac{G(s)}{1-C(s)}$$

$$G(s) = \frac{\dfrac{2s+1}{(s+1)^2}}{1-\left(\dfrac{2s+1}{(s+1)^2}\right)} = \frac{2s+1}{s^2}$$

28. Ans. (c)

Solution: $G(s) = \dfrac{1}{s(s+1)}$

Close loop transfer function

$$\frac{C(s)}{R(s)} = \frac{C(s)}{1+G(s)}$$

$$\frac{C(s)}{1+G(s)} = \frac{\dfrac{1}{s\,(s+1)}}{1+\dfrac{1}{s\,(s+1)}} = \frac{1}{s^2+s+1}$$

This is similar to second order transfer function

$$= \frac{\omega_n^2}{s^2 + 2\delta\omega_n s + \omega_n^2}$$

$$\omega_n = 1; \quad 2\delta\omega_n = 1;\ \delta = \frac{1}{2}$$

$$M_p = \exp - \frac{\delta\pi}{\sqrt{(1-\delta^2)}}$$

$$= \exp - \frac{\dfrac{-\pi}{2}}{\sqrt{1-\dfrac{1}{4}}}$$

$$= \exp \frac{-\pi}{\sqrt{3}} = 0.163$$

29. Ans. (c)

Solution:

* For a type 0 sysem

  ⇒ Error to a ramp input is '∞'

* For a type 1 system

  ⇒ Error to a ramp input is finite

* For a type 2 and higher type

  ⇒ Errror to a ramp input is 0

Type	K_p	K_v	K_a	$e_{ss} = R/1+K_p$	$e_{ss} = R/K_V$	$e_{ss} = R/K_A$
0	K	0	0	$R/1+K_p$	∞	∞
1	∞	K	0	0	R/K_v	∞
2	∞	∞	K	0	0	R/K_A
3	∞	∞	∞	0	0	0

30. Ans. (c)

Solution: For feedback control system transfer function

$$T(s) = \frac{G(s)}{1+G(s)\,H(s)}$$

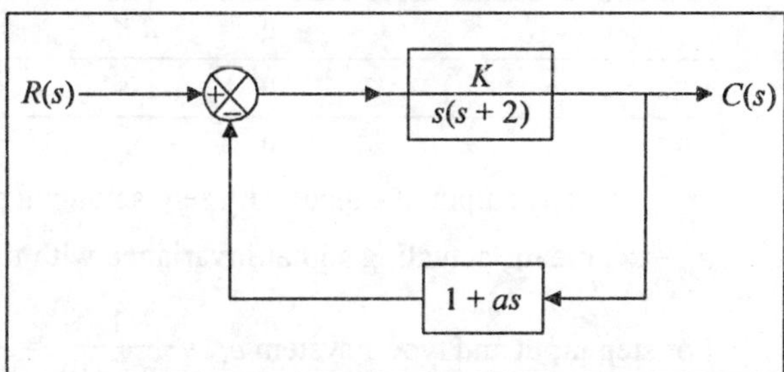

$$\frac{C(s)}{R(s)} = \frac{\dfrac{K}{s(s+2)}}{1+\dfrac{K}{s(s+2)}(1+as)}$$

$$= \frac{K}{s(s+2)+(1+as)+K}$$

$$K = \omega_n^2 = 16$$

$$2\delta\omega_n = 2 + Ka = 2 \times 0.7 \times 4 = 5.6$$

$$a = \frac{5.6-2}{16} = 0.225$$

31. Ans. (b)

Solution: Introduction of **integral action** in the forward path of a unity feedback system results in a System with no steady state error.

32. Ans. (c)

Solution: Linear time invariant control system in shown as below:

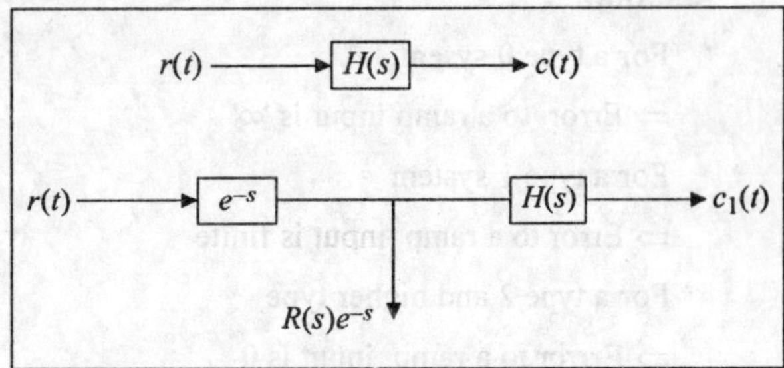

We know, $C(s) = H(s)\, R(s)$

$\therefore$ output is $H(s)\, R(s)\, e^{-s} = C_1(s)$

or $\quad C_1(s) = e^{-S}\, C(s)$

$\Rightarrow \quad c_1(t) = c(t-1)\, u(t-1)$

33. Ans. (a)

Solution:

Type	K_p	K_v	K_a	$e_{ss}=$ $R/1+K_p$	$e_{ss}=$ R/K_V	$e_{ss}=$ R/K_A
0	K	0	0	$R/1+K_p$	∞	∞
1	∞	K	0	0	R/K_v	∞
2	∞	∞	K	0	0	R/K_A
3	∞	∞	∞	0	0	0

$e_{ss} = 0$ means output like input with zero actuating signal.

$e_{ss} = \infty$, means actuating signal invariance with time

For step input and type 1 system $e_{ss} = \dfrac{1}{1+G(0)} = \dfrac{1}{\infty} = 0$

34. Ans. (b)

Solution: $\quad \text{T.F} = \dfrac{2s}{(s^2+6s+25)}$

$$\omega_n = 5$$

$$2\delta\omega_n = 6$$

$$2\,\delta\,5 = 6$$

$$\delta = 0.6$$

$$M_P = \exp - \frac{\pi.\delta}{\sqrt{(1-\delta)^2}}$$

$$= \exp - \frac{\pi \times 0.6}{\sqrt{(1-\delta)^2}}$$

$$= e^{-2.355} = 10\%$$

35. Ans. (c)

Solution: By applying Laplace-transform on both sides, we get

$$\left[s^2 X(s) - sx(0) - x'(0)\right] + 6\left[s\,X(s) - x(0)\right] + 5X(s) = 12\left(\frac{1}{s} - \frac{1}{s+2}\right)$$

Its ignition condition not given, assume zero

$$[s^2 + 6s + 5]\, X(s) = 12\left(\frac{1}{s} - \frac{1}{s+2}\right) = \frac{24}{s(s+2)}$$

or $\quad X(s) = \dfrac{24}{s(s^2+6s+5)\,(s+2)}$

then, $\lim\limits_{t\to\infty} x(t) = \lim\limits_{s\to 0}\left[sX(s)\right] = \lim\limits_{s\to 0} \dfrac{24}{(s^2+6s+5)\,(s+2)}$

$$= \frac{24}{5\times 2} = 2.4$$

36. Ans. (b)

Solution: Natural response is when differential equation is zero i.e. forcing function is zero

$$\left(s^2 + \frac{1}{2}s + \frac{1}{18}\right) = 0.$$

$$s_{1,2} = \frac{-\frac{1}{2} \pm \sqrt{\frac{1}{4} - \frac{4}{18}}}{12} = -\frac{1}{4} \pm \frac{1}{12} = \frac{1}{3}, \frac{1}{6}$$

The time constant $T_1 = 3$ sec

$T_2 = 6$ sec

Second Method

$$T(s) = \frac{C(s)}{R(s)} = \frac{1}{s^2 + \frac{s}{2} + \frac{1}{18}} = \frac{18}{18s^2 + 9s + 1}$$

$$= \frac{18}{(3s+1)(6s+1)}$$

As, $T(s)$ have the form, $\dfrac{A}{(T_1 s+1)(T_2 s+1)}$

The time – constant are $T_1 = 3$ sec,

$T_2 = 6$ sec.

Third Method

For natural response

$$\frac{dx^2}{dt^2}+\frac{1}{2}\frac{dx}{dt}+\frac{x}{18}=\delta(t)$$

$$X(s)=\frac{1}{s^2+\frac{s}{2}+\frac{1}{18}}=\frac{1}{\left(s+\frac{1}{3}\right)\left(s+\frac{1}{6}\right)}$$

Thus $s=-\frac{1}{3}$ and $s=-\frac{1}{6}$

Natural time constant is 3 second and 6 second .

37. Ans. (a)

Solution: $e_{ss}=\lim_{s\to 0}\frac{s.R(s)}{1+G(s)H(s)}$ *as* $R(s)=\frac{1}{s}$

$$=\lim_{s\to 0}\frac{s\times\frac{1}{s}}{1+\left(\frac{3}{s+15}\right)\left(\frac{15}{s+1}\right)}$$

$$=\lim_{s\to 0}\frac{s^2+16s+15}{s^2+16s+60}=\frac{15}{60}$$

$$=\frac{1}{4}\times 100=25\%$$

Second Method

From given figure

$$G(s)=\frac{3}{s+15}\times\frac{15}{s+1}=\frac{45}{(s+1)(s+15)}$$

In time constnt form $G(s)=\frac{3}{(1+s)\left(1+\frac{s}{15}\right)}$

Comparing with standared

$$G(s)=\frac{K}{s^n(1+sT_1)(1+sT_2)\cdots(1+sT_m)}$$

Here $n=0$ so it is Type-0 system

Position error consnat $K_p=3$

Steady-state error to the unit-step input

$$e_{ss}=\frac{1}{1+K_p}=\frac{1}{1+3}=0.25=25\%$$

38. Ans. (d)

Solution: Characteristic equation $=1+G(s)H(s)=0$

$$=s^2+16s+60=0$$

$$s=\frac{-16\pm\sqrt{16^2-240}}{2}=\frac{-16\pm 4}{2}=-6,-10$$

39. Ans. (d)

Solution: Apply initial value theorem $=f(0)=\lim_{t\to 0}f(t)$

$$=\lim_{s\to\infty}sF(s)=\lim_{s\to\infty}=\frac{s\,5}{s(s^2+3s+2)}=0$$

40. Ans. (d)

Solution: $T(s)=\frac{C(s)}{R(s)}=\frac{K/s(s+2)}{1+\frac{K(1+sP)}{s(s+2)}}$

$$=\frac{K}{s^2+(2+Pk)s+K} \quad \text{...(1)}$$

Standard 2nd order equation

$$=\frac{\omega_{n^2}}{s^2+2\delta\omega_n s+\omega_{n^2}} \quad \text{...(2)}$$

Compare equation (1) and (2)

Now, $K=\omega_n^2$, and $(PK+2)=2\delta\omega_n$

Then, $K=25$ and $25P+2=2\times.7\times 5=7$

$$25\,P=5$$

$$P=\frac{5}{25}=\frac{1}{5}=0.2$$

41. Ans. (d)

Solution: $H(s)=C(s)=\frac{12.5\times 8}{(s+6)^2+8}$

$$=\frac{100}{s^2+12s+100}$$

Then, output for unit step response

$$C_u(s)=\frac{1}{s}\cdot\frac{100}{s^2+12s+100}$$

and steady-state value $=\lim_{s\to 0}sC_u(s)$

$$=\lim_{s\to 0}\frac{100}{s^2+12s+100}=1$$

42. Ans. (c)

Solution:

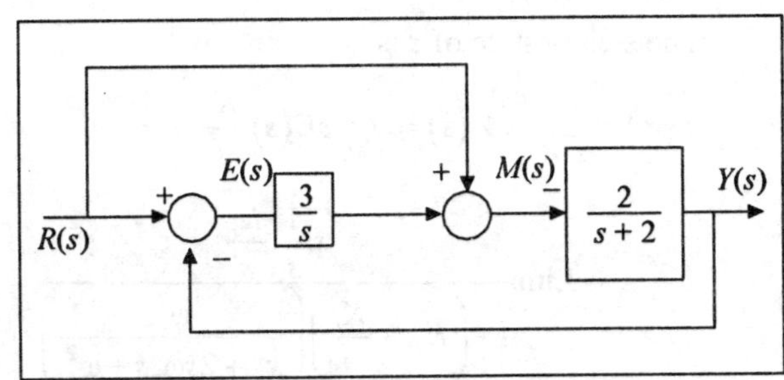

$$R(s) = \frac{1}{s}$$

$$M(s) = R(s) + [R(s) - Y(s)]\ \frac{3}{s}$$

And $$Y(s) = \frac{2}{s+2}\left[R(s)\left(1+\frac{3}{s}\right)-\frac{3}{s}Y(s)\right]$$

or $$Y(s)+\frac{6}{s(s+2)}Y(s) = R(s)\frac{(s+3)}{s}\frac{2}{(s+2)}$$

or $$Y(s)\left[\frac{s(s+2)+6}{s(s+2)}\right] = R(s)\frac{2(s+3)}{s(s+2)}$$

$$\frac{Y(s)}{R(s)} = \frac{2(s+3)}{s(s+2)}\times\frac{s(s+2)}{s^2+2s+6}$$

$$\therefore \quad \frac{Y(s)}{R(s)} = \frac{2(s+3)}{s^2+2s+6}$$

$$E(s) = R(s)-Y(s) = R(s)\left[1-\frac{2(s+3)}{s^2+2s}+6\right]$$

$$E(s) = R(s)\frac{s^2}{s^2+2s+6} = \frac{s}{s^2+25+6}$$

$$e_{ss} = \underset{s\to 0}{Lim}\ sE(s) = 0$$

43. Ans. (a)

Solution:

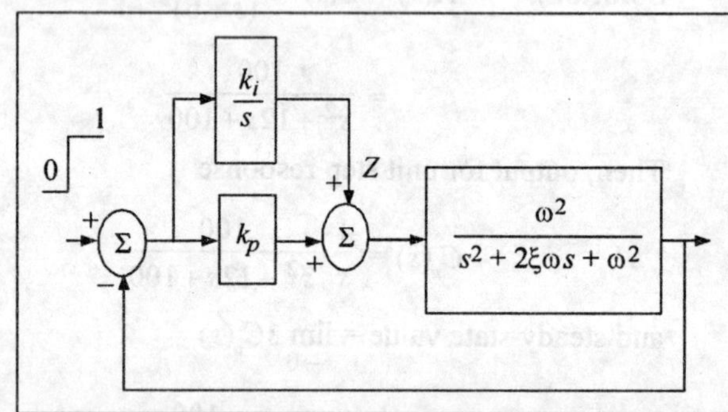

$$E(s) = \frac{R(s)}{1+G(s)\,H(s)}$$

$r(t) \rightarrow$ Unit step

$R(s) \rightarrow (1/s)$

The steady state of z is

$$z(\infty) = \underset{s\to 0}{Lt}\ s\,Z(s) = \underset{s\to 0}{Lt}\ sE(s)\frac{K_i}{s}$$

$$= \underset{s\to 0}{\text{Lim}}\frac{s.\frac{1}{s}.\frac{K_i}{s}}{1+\left(K_P+\frac{K_i}{s}\right)\left(\frac{\omega^2}{s^2+2s\omega_r s+\omega^2}\right)}$$

$$= \underset{s\to 0}{\text{Lim}}\frac{\frac{K_i}{s}}{1+\left(K_P+\frac{K_i}{s}\right)\left(\frac{\omega^2}{s^2+2s\omega_r s+\omega^2}\right)}$$

$$\underset{s\to 0}{\text{Lim}}\frac{K_i}{s\left(1+\left(K_P+\frac{K_i}{s}\right)\left(\frac{\omega^2}{s^2+2s\omega_r s+\omega^2}\right)\right)} = 1$$

44. Ans. (c)

Solution: Characteristic equation

$$LC\ s^2 + RCs + 1 = 0$$

$$s^2 + \frac{R}{L}s + \frac{1}{LC} = 0$$

$$s^2 + 2\delta\omega_n s + \omega_n^2 = 0$$

$$\omega_n = \frac{1}{\sqrt{LC}} = \frac{1}{\sqrt{10^{-3}\times 10\times 10^{-6}}} = \frac{1}{\sqrt{10^{-8}}}$$

$$= \sqrt{10^8} = 10^4$$

$$2\delta\omega_n = \frac{R}{L}$$

$$2\times\delta\times 10^4 = \frac{10}{10^{-3}} = 10^4$$

$$\delta = \frac{1}{2} = 0.5$$

$$\text{overshoot} = e^{-\frac{n\pi}{\sqrt{1-\delta^2}}} = 0.16 = 0.16 \Rightarrow 16\%$$

45. Ans. (c)

Solution:

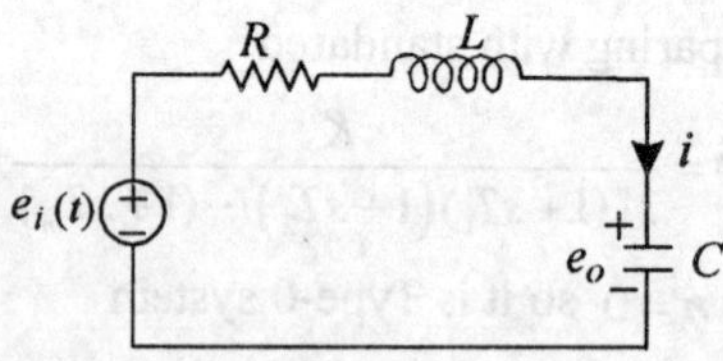

From given figure

$$e_i = Ri + L\frac{di}{dt} + e_o$$

Capacitor voltage $e_o = \frac{1}{C}\int idt$

or $i = C\frac{de_o}{dt}$

$$LC\frac{d^2e_o}{dt^2} + RC\frac{de_o}{dt} + e_o = e_i$$

$$\frac{d^2e_o}{dt^2}+\frac{R}{L}\frac{de_o}{dt}+\frac{1}{LC}e_o=\frac{1}{LC}e_i$$

In Laplace domain

$$\left[s^2+\frac{R}{L}+\frac{1}{LC}\right]E_o(s)=\frac{1}{LC}E_i(s)$$

$$\frac{E_o}{E_i}=\frac{1}{LC}\times\frac{1}{s^2+\frac{R}{L}s+\frac{1}{LC}}$$

Characterisitc equation $s^2+\frac{R}{L}s+\frac{1}{LC}=0$

Compare $s^2+2\delta\omega_n s+\omega_n^2=0$

$$\omega_n=\frac{1}{\sqrt{LC}}$$

$$2\delta\omega_n=\frac{R}{L}$$

$$\delta=\frac{R}{2L}\times\frac{1}{\omega_n}=\frac{R}{2}\sqrt{\frac{C}{L}}$$

$R=10\Omega$

$L=1\,mH=1\times10^{-3}\,H$

$C=10\,\mu F=10^{-5}\,F$

$$\delta=\frac{10}{2}\sqrt{\frac{10^{-5}}{10^{-3}}}=\frac{1}{2}=0.5$$

$\delta<1\Rightarrow$ **underdamped response**

Peak overshoot $M_p=e^{\frac{-\pi\rho}{\sqrt{1-\rho^2}}}$

$=0.163=16.3\%$

$$\omega_n=\frac{1}{\sqrt{LC}}10^{+4}$$

Setting time $t_s=\frac{4}{\delta\omega_n}=\frac{4}{0.5\times10^{+4}}=0.8\,ms$

Peak time $t_p=\frac{\pi}{\omega_n\sqrt{1-\delta^2}}=0.36\,ms$

To observe peak overshoot, the input must be applied at least upto $t_p=0.36\,ms$.

To observe the transient response input must be applied atleast upto $t_s=0.8\,ms$

If step input is applied

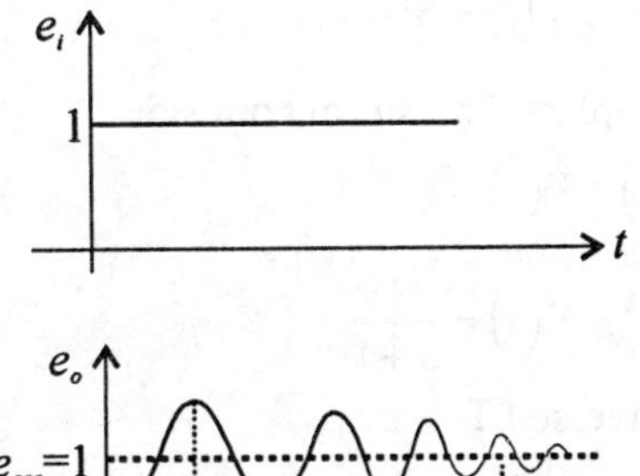

The given CRO is non - storage means it can't record the response. So it is not possible to read the transient response parctically as it will appear only for one t_s, i.e. $0.8\,ms$.

So the constant input $e_i=1$ should be applied again and again (i.e. square wave) so that transient response appears always on CRO screen.

The square wave input

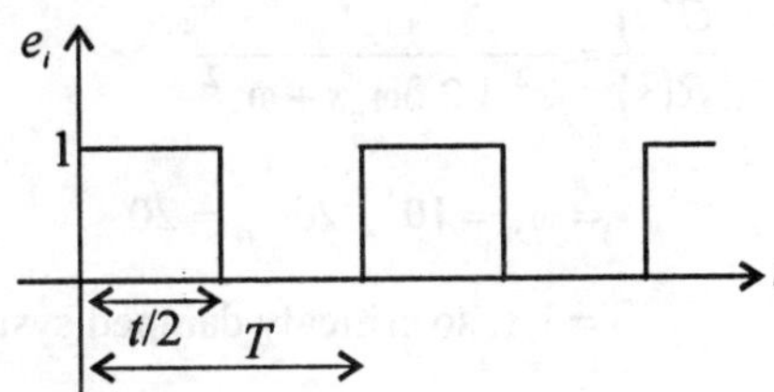

For 50 Hz frequency	Time period $T=\frac{1}{f}=20\,ms$
For 300 Hz frequency	$T=\frac{10}{3}=3.33\,ms$
For 2 kHz frequency	$T=0.5\,ms$

For obtaining transient response $\frac{T}{2}>(t_p \text{and } t_s)$ hence 300 Hz frequency square wave is the most suitable as $\frac{T}{2}=1.67\,ms$ because for

$2\text{ kHz}, \frac{T}{2}=0.2\,ms<t_p$

That's why incase of $\frac{T}{2}=1.67\,ms$, only transient response will always appear on CRO screen.

46. Ans. (d)

Solution: Given differetial equation

$$\frac{dy(t)}{dt}+y(t)=\delta(t)$$

Take Laplace Transform both side

$$sY(s)+Y(s)=1$$

$$Y(s)=\frac{1}{s+1}$$

Take inverse LT

$$y(t)=e^{-t}u(t)$$

47. Ans. (a)

Solution: $\frac{C(s)}{R(s)}=G(s)$; $R(s)=1e^{-s}$

$$\underset{x\to\infty}{Lt}\ C(t)=\underset{s\to 0}{Lt}\ sC(s)$$

$$=\underset{s\to 0}{Lt}\frac{s\cdot e^{-s}}{s^2+3s+2}=0$$

48. Ans. (c)

Solution:

Compare the given equation with standard equation

$$\frac{C(s)}{R(s)}=\frac{\omega_n^{\ 2}}{s^2+2\delta\omega_n s+\omega_n^{\ 2}}$$

$$=\omega_n=10\ ;\ 2\delta\omega_n=20$$

$\delta=1$,. so criticaly damped system

49. Ans. (d)

Solution: $T.F=C(sI-A)^{-1}B+D$

$$(sI-A)=\begin{bmatrix} s & 0\\ 0 & s\end{bmatrix}-\begin{bmatrix} 0 & 1\\ 0 & -2\end{bmatrix}$$

$$=\begin{bmatrix} s & -1\\ 0 & s+2\end{bmatrix}$$

$$s(sI-A)^{-1}=\frac{1}{s(s+2)}\begin{bmatrix} s+2 & 1\\ 0 & s\end{bmatrix}$$

$$=[1\ \ 0]\frac{1}{s(s+2)}\begin{bmatrix} s+2 & 1\\ 0 & s\end{bmatrix}\begin{bmatrix} 0\\ 1\end{bmatrix}$$

$$=\frac{1}{s(s+2)}[1\ \ 0]\begin{bmatrix}(0+1)\\(0+s)\end{bmatrix}$$

$$=\frac{1}{s+2}[+1\ \ +0]=\frac{1}{s(s+2)}$$

50. Ans.(a)

Solution:

For type 1 system, i.e. $G(s)=\frac{1}{s(s+2)}$ and unit step input

$$e_{ss}=0$$

51. Ans (c)

Solution:

$$H(s)=\frac{2}{s+1}$$

Input step $X(s)=\frac{1}{s}$

Step response, $Y(s)=\frac{2}{s(s+1)}$

$$\Rightarrow y(t)=2(1-e^{t})u(t)$$

$$\Rightarrow 2(1-e^{t})=0.98\times 2$$

$$\Rightarrow t=3.9\ see$$

52. Ans. (a)

Solution: For step input $e_{ss}=0.1=\frac{1}{1+K}$

or, $K=9$

$$G(s)=\frac{9}{s+1}$$

Now the input is pulse $r(t)=10[u(t)-u(t-1)]$

$$R(s)=10\left[\frac{1-e^{-s}}{s}\right]$$

$$e_{ss}=\underset{s\to 0}{Lt}\frac{S.R(s)}{1+G(s)H(s)}$$

$$=\underset{s\to 0}{Lt}\frac{\frac{s\ 10[1-e^{-s}]}{s}}{\frac{s+10}{s+1}}$$

$$=\frac{0}{10}=0$$

53. Ans. (c)

Solution:

LT (Impulse response) = Transfer Function

$$=\frac{1}{(s+1)}+\frac{1}{(s+2)}$$

$$\text{Step response}=LT^{-1}\left[\frac{1}{s(s+1)}+\frac{1}{s(s+2)}\right]$$

$$=LT^{-1}\left[\left(\frac{1}{s}-\frac{1}{s+1}\right)+\left(\frac{0.5}{s}-\frac{0.5}{(s+2)}\right)\right]$$

$$= (1 - e^{-t} + 0.5 - 0.5e^{-2t})u(t)$$
$$= (1.5 - e^{-t} - 0.5e^{-2t})u(t)$$

54. Ans. (a)

Solution:

$$\frac{Y(s)}{R(s)} = \frac{1}{s^2 + (K+1)s + 1}$$

$$= \frac{\omega_n^2}{s^2 + 2\delta\omega_n s + \omega_n^2}$$

$$2\delta\omega_n = K + 1$$

$$\Rightarrow \quad \delta = \left(\frac{K+1}{2}\right)$$

Peak over shoot $= e^{-\pi\delta/(1-\delta^2)}$

Second method

$$\frac{Y(s)}{R(s)} = \frac{1}{s^2 + (k+1)s + 1}$$

$\omega_n = 1$ No change with k

$$\delta\,\omega_n = \frac{k+1}{2}$$

$$\delta = \frac{k+1}{2}$$

$$\delta \propto k$$

$M_p = e^{-\delta\pi/\sqrt{1-\delta^2}}$ depends upon δ

Phase shift $= -\tan^{-1}\dfrac{\omega(k+1)}{(\omega^2+1)}$

$$= -\tan^{-1}\frac{(k+1)}{\left(\omega + \frac{1}{\omega}\right)}$$

influences But not mainly

55. Ans. (d)

Solution:

From the given graph

$$e_{ss} = 1 - 0.75 = 0.25$$

$$e_{ss} = \lim_{s\to 0} \cdot s \cdot E(s)$$

$$= \lim_{s\to 0} s \cdot \frac{R(s)}{1 + G(s)H(s)}; \text{ Here } R(s) = \frac{1}{s}$$

$$= \lim_{s\to 0} \frac{s \cdot \frac{1}{s}}{1 + \frac{K}{s^2 + 3s + 2}}$$

$$= \lim_{s\to 0} \frac{s^2 + 3s + 2}{s^2 + 3s + 2 + K}$$

$$= \frac{2}{2+K} - 0.25$$

or, $2 = 0.5 + 0.25K$

or, $1.5 = 0.25K$

or, $K = \dfrac{1.5}{0.25} = 6$

56. Ans. (c)

Solution:

The magnitude of sinsoidal steady state response depends on magnitude of transfer fucntion only.

output

$$y(t) = |G(j\omega)| \sin(\omega t + \angle G(j\omega))$$

Puting $s = j\omega$ in given equation

$$G(j\omega) = \frac{\left[(j\omega)^2 + 9\right](j\omega + 2)}{(j\omega+1)(j\omega+3)(j\omega+4)}$$

$$|G(j\omega)| = \frac{(9-\omega^2)\sqrt{4+\omega^2}}{\sqrt{\omega^2+1}\sqrt{\omega^2+6}\sqrt{16+\omega^2}} = 0$$

For $y(t) = 0$

$|G(j\omega)|$ can be zero

Hence $G(j\omega) = 0$

at $\omega^2 = 9$

or, $\omega = 3rad/s$

57. Ans. (d)

Solution: $\dfrac{C(s)}{R(s)} = \dfrac{1}{1+s}$; $\dfrac{C(j\omega)}{R(j\omega)} = \dfrac{1}{1+j\omega}$

From $r(t) = \sin t$; $\omega = 1$,

$$\left|\frac{C(j\omega)}{R(j\omega)}\right| = \frac{1}{\sqrt{1+1}} = \frac{1}{\sqrt{2}}$$

$$\angle \frac{C(j\omega)}{R(j\omega)} = -\tan^{-1} = -\frac{\pi}{4}$$

$\therefore$ SS value of $c(t) = \dfrac{1}{\sqrt{2}}\sin\left(t - \dfrac{\pi}{4}\right)$

For sinusoidal function steady state value will not defined for $t \to \infty$.

58. Ans. (b)

Solution: $H(s) = \frac{s}{(s+1)}$

$$\phi = 90 - 45 = 45°$$

$$|H(s)| = \frac{1}{\sqrt{1+1}} = \frac{1}{\sqrt{2}}$$

Therefore this transfer function circuit will give amplitude change of $\frac{1}{\sqrt{2}}$ and phase 45°

Hence, in steady state output is $\frac{1}{\sqrt{2}}$ sin $(t + 45°)$

Second Method

$$x(t) = \sin t \,;\, x(s) = \frac{1}{s^2+1} \,;\, Y(s) = X(s)\frac{s}{s+1}$$

$$Y(s) = \frac{1}{(s^2+1)} \cdot \frac{s}{s+1} = \frac{A}{(s+j)} + \frac{B}{(s-j)} + \frac{C}{(s+1)}$$

$$= \frac{1}{2(+j)}\left(\frac{1}{s-j}\right) + \frac{1}{2(1-j)(s+j)} - \frac{1}{2(s+1)}$$

Taking inverse Laplace transformation, we have

$$y(t) = \frac{1}{2(1+j)}e^{jt} + \frac{1}{2(1-j)}e^{-jt} - \frac{1}{2}e^{-t}$$

$$= \frac{1}{2}(\sin t + \cos t) - \frac{1}{2}e^{-t}$$

$$= \frac{1}{\sqrt{2}}\sin(t+45^o) - \frac{1}{2}e^{-t}$$

Then, steady state value = $\lim_{t\to\infty} y(t)$

$$\frac{1}{\sqrt{2}}\sin(t+45°)$$

59. Ans. (c)

Solution: $T(s) = \frac{s^2+4}{(s+1)(s+4)}$;

At the frequency ω, gain is zero i.e.

$$T(j\omega) = \frac{|(j\omega)^2 + 4^1|}{|(j\omega+1)(j\omega+4)|} = 0$$

$$-\omega^2 + 4 = 0$$

$$\omega = 2 \text{ rad/sec.}$$

60. Ans. (d)

Solution: Steady state error for second order system is given by $\frac{2\delta}{\omega_n}$ =

From given function $\omega_n = 3$;

$$2\delta\,\omega_n = 0.6 \to \delta = \frac{0.6}{2\times 3} = 0.1$$

$$\text{Steady state error} = \frac{2\times 0.1}{3} = 0.15$$

61. Ans. (d)

Solution:

$$F_o(s) = \frac{1}{s}\left(\frac{20}{s^2+2s+5}\right)$$

$$f(\infty) = \underset{s\to 0}{Lt}\ sF_o(s) = \underset{s\to 0}{Lt}\ s\times\frac{1}{s}\times\frac{20}{(s^2+2s+5)}$$

$$= \underset{s\to 0}{Lt}\frac{20}{s^2+2s+5} = \frac{20}{5} = 4$$

62. Ans. (b)

Solution: Close loop gain = $\frac{G(s)}{1+G(s)} = T(s)$

$1 + G(s)$ gives the stability criteria

$$G(s) = \frac{9}{s(s+3)}$$

$$\frac{C(s)}{R(s)} = \frac{9}{9+s(s+3)}$$

or $s^2 + 3s + 9 = 0$

Therefore $\omega_n^2 = 9$

i.e. $\omega_n = \sqrt{9} = 3$

Now $2\delta\,.\,\omega_n = 3$

Now $2\delta\,.\,\omega_n = 3$

i.e. $\delta = \frac{3}{2\times 3} = \frac{1}{2} = 0.5$

63. Ans. (b)

Solution:

$$\frac{C(s)}{R(s)} = \frac{25}{s^2+6s+25}$$

$$\cong \frac{\omega_n^2}{s^2+2\delta\omega_n s+\omega_n^2}$$

Here, $\omega_n = 5$; $2\delta\omega_n = 6$

$$\delta = \frac{6}{2\times 5} = 0.6$$

$$t_{peak} = \frac{\pi}{\omega_n\sqrt{1-\delta^2}} = \frac{\pi}{5\sqrt{1-(0.6)^2}}$$

$$= \frac{\pi}{5\sqrt{1-0.36}} = \frac{\pi}{5\sqrt{0.64}}$$

$$= \frac{\pi}{5\times 0.8} = \frac{\pi}{4}$$

64. Ans. (a)

Solution: $\frac{C(s)}{R(s)} = \frac{9}{s^2+2s+9}$

$$\cong \frac{\omega_n^2}{s^2+2\delta_n+\omega_n^2}$$

$$\omega_n = \sqrt{9} = 3$$

$$2\delta\omega_n = 2$$

$$\Rightarrow \quad 2\times 3\delta = 2$$

$$\delta = \frac{1}{3}$$

65. Ans. (b)

Solution: $G(s) = \frac{10}{s(s+5)}$

$H(s = 1)$

$$G(s)\,H(s) = \frac{10}{s(s+5)}$$

This is type 1 system

Ramp error constant $k_v = \underset{s\to 0}{Lt}\,(s\cdot G(s)\,H(s))$

$$= \underset{s\to 0}{Lt}\left(s\times\frac{10}{s(s+5)}\right) = \frac{10}{5} = 2$$

$$\Rightarrow \quad \frac{R}{k_v} = \frac{1}{2} = 0.5$$

66. Ans. (c)

Solution: $e_{ss} = \underset{s\to 0}{Lt}\,sE(s) = \underset{s\to 0}{Lt}\,\frac{sR(s)}{1+G(s)\,H(s)}$

For $r(t) = \frac{t^2}{2},\ R(s) = \frac{1}{s^3},\ H(s) = 1$

Then, $e_{ss} = \underset{s\to 0}{Lt}\,\frac{1}{s^2+\frac{10se^{-0.1s}}{(s+5)(s2+45+16)}} = \infty$

67. Ans. (a)

Solution: $e(t) = \lim_{s\to 0}$

$$s\,.\,E(s) = \underset{s\to 0}{Lt}\,s\cdot\frac{R(s)}{1+GH}$$

$$= \underset{s\to 0}{Lt}\,s\cdot\frac{1}{1+\frac{(1+5s)(1+10s)(1+2s)}{(1+s)(1+8s)(1+20s)}}$$

$$= \underset{s\to 0}{Lt}\,s\cdot\frac{(1+s)(1+8s)(1+20s)}{(1+s)(1+8s)(1+20s)+(1+5s)(1+10s)(1+2s)} = 0$$

$$J = \int_0^\infty e(t)\,dy = \int_0^\infty 0\,dt = 0$$

68. Ans (b)

Solution:

Forward path gain

$$G'(s) = \left[k_p+\frac{k_i}{s}\right]\left[\frac{2}{s(s+3)}\right]$$

$$e_{ss} = \underset{s\to 0}{Lt}\,\frac{sR(s)}{1+G'(s)}$$

$$\left[\text{for step input } R(s) = \frac{1}{s}\right]$$

$$= \underset{s\to 0}{Lt}\,\frac{1}{1+\left[k_p+\frac{k_i}{s}\right]\left[\frac{2}{s(s+3)}\right]}$$

$$= \underset{s\to 0}{Lt}\,\frac{s^2(s+3)}{s^2(s+3)+2(k_ps+k_i)^2}$$

Lowest value of $k_i = \frac{1}{3}$ for e_{ss} to be zero.

69. Ans (c)

Solution:

$$f(t) = u(t-\tau)$$

$$F(s) = ?$$

If $f(t-\tau) = F(s)e^{-\tau s}$

If $f(t) = u(t)$ Then $F(s) = \frac{1}{s}$

$$FT[u(t-\tau)] = \frac{1}{s}e^{-\tau s}$$

70. Ans (b)

Solution:

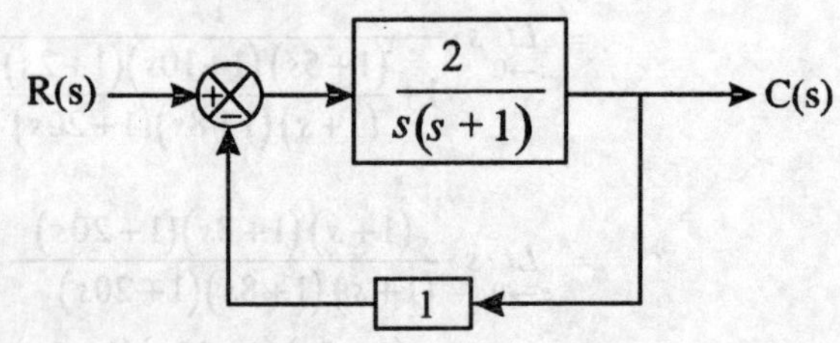

$$\frac{C(s)}{R(s)} = \frac{\frac{2}{s(s+1)}}{1+\frac{2}{s(s+1)}} \Rightarrow \frac{2}{s^2+s+2}$$

Given as ramp input is applied

Hence, $R(s) = \frac{1}{s^2}$

$$e_{ss} = \lim_{s\to 0} \frac{sR(s)}{1+GH}$$

$$= \lim_{s\to 0} \frac{(s)\left[\frac{1}{s^2}\right]}{\frac{2+s+s^2}{s(s+1)}}$$

$$= \lim_{s\to 0} \frac{1+(s+1)s}{s(s^2+s+2)}$$

$$e_{ss} \Rightarrow \frac{1}{2}$$

$$= 0.5$$

71. Ans (d)

Solution:

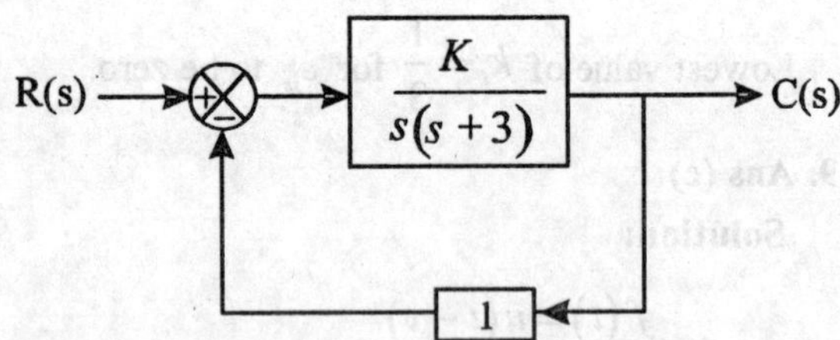

$$\frac{C(s)}{R(s)} \Rightarrow \frac{\frac{K}{s(s+3)}}{1+\frac{K}{s(s+3)}} \Rightarrow \frac{K}{s^2+3s+K}$$

$$\frac{C(s)}{R(s)} \Rightarrow \frac{K}{s^2+3s+K}$$

Compare it with standard 2nd order equation

$$\frac{\omega_n^2}{s^2+2\delta\omega_n s+\omega_n^2}$$

Given $\delta = 0.5$

$$2\delta\omega_n = 3$$

$$2[0.5]\omega_n = 3$$

$$\omega_n = 3$$

$$K = \omega_n^2$$

$$K = (3)^2 \Rightarrow 9$$

72. Ans. (a)

Solution:

Characteristic of thermometer

$$\frac{2dT_i}{dt} = T_a - T_i$$

Take the Laplace transform

or, $\quad 2sT_i(s) = T_a(s) - T_i(s)$

$$2sT_i(s) + T_i(s) = T_a(s)$$

or, $\quad \frac{T_i(s)}{T_o(s)} = \frac{1}{2s+1} = \frac{0.5}{s+0.5}$

$\Rightarrow \quad 0.5\,e^{-0.5t}$

Hence, $\quad \omega = 0.5$

$$f = \frac{\omega}{2\pi} = \frac{1}{4\pi}\,Hz$$

73. Ans. (d)

Solution:

$$u(t) = \frac{1}{s} \longrightarrow \boxed{H.S.} \longrightarrow Y(s)$$

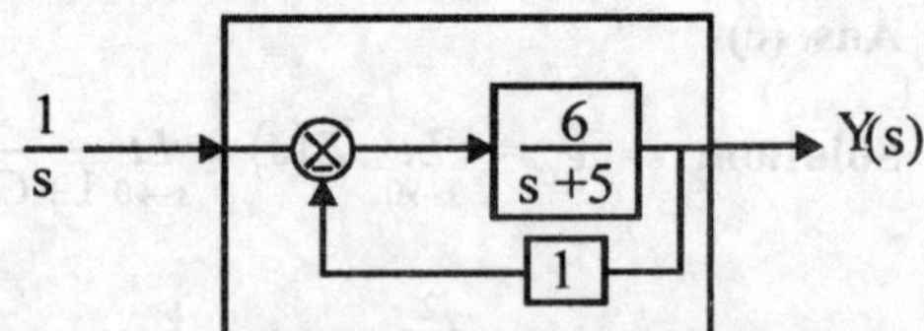

$$Y(s) = \frac{1}{s}\left[\frac{(6/s+5)}{1+\left(\frac{6}{s+5}\right)}\right]$$

$$Y(s) = \frac{1}{s}\left[\frac{6}{s+5+6}\right]$$

$$Y(s)=\frac{6}{(s)\,(s+11)}$$

$$\frac{6}{(s)(s+11)} \Rightarrow \frac{A}{s}+\frac{B}{(s+11)}$$

6 = A(s + 11) + B(s)

Put s = 0 Put s = −11

$$A=\frac{6}{11} \qquad B=-\frac{6}{11}$$

$$Y(s)=\frac{6}{11s}-\frac{6}{11(s+11)}$$

$$Y(t)\Rightarrow\frac{6}{11}-\frac{6}{11}e^{-11t}$$

74. Ans. (d)

Solution:

$$y(t)=5e^{-10t}u(t)$$

we know that

$$y(t)=x(t)*h(t)$$

Here $x(t)=\delta(t)$

$$y(t)=\delta(t)*h(t)$$

Convolution in time domain is equal to multiplication in frequency domain

$$Y(s)=H(s)$$

$$H(s)=Y(s)\Rightarrow\frac{5s}{s+10}$$

$h(t)=?$

Because degree of Numerator & Denominator is same so will firstly divide *N* by *D*

$$H(s)=s+10)\,5s\;(5$$

$$\underline{5s+50}$$

$$\underline{-50}$$

$$H(s)\Rightarrow 5-\frac{50}{s+10}$$

$$h(t)=5\delta(t)-50e^{-10t}u(t)$$

75. Ans. (d)

Solution:

$$\frac{y(s)}{U(s)}=\frac{\dfrac{K}{(1+s\tau)}}{1+\left(\dfrac{K}{1+s\tau}\right)}=\frac{K}{1+s\tau+K}$$

$$\therefore\ y(s)=\frac{K}{s+\dfrac{K+1}{\tau}}\cdot\frac{1}{s}=\left[\frac{1}{s}-\frac{1}{s+\dfrac{K+1}{\tau}}\right]\cdot\frac{K}{K+1}$$

$$y(t)=\frac{K}{K+1}\left[1-e^{\frac{-K+1}{\tau}t}\right]$$

At steady-state $y(\infty)=0.8=\dfrac{K}{K+1}$

or $K=4$

76. Ans. (b)

Solution:

$$G(s)=\frac{100}{s(s+p)}$$

Closed loop transfer function is

$$\frac{C(s)}{R(s)}=\frac{100}{s^2+sp+100}$$

Compare with standard second order function i.e.

$$\frac{C}{R}=\frac{\omega_n^2}{s^2+2\delta\omega_n s+\omega_n^2}$$

$$\omega_n^{\,2}=100$$

or $\omega_n=10$ rad/sec

It is know that

$$t_p\ \text{(peak time)}=\frac{\pi}{\omega_n\sqrt{1-\delta^2}}=\frac{\pi}{8}$$

$$\text{or,}\quad\frac{\pi}{10\sqrt{1-\delta^2}}=\frac{\pi}{8}$$

After solving $\delta=0.6$

77. Ans. (b)

Solution:

$$P=3\delta\omega_n=2\times0.6\times10=12$$

78. Ans. (c)

Solution:

The magnitude of sinusoidal steady state response depends on magnitude of transfer function only.

output

$$y(t)=|G(j\omega)|\sin\left(\omega t+\angle G(j\omega)\right)$$

Putting $s = j\omega$ in given equation

$$G(j\omega) = \frac{\left[(j\omega)^2 + 9\right](j\omega+2)}{(j\omega+1)(j\omega+3)(j\omega+4)}$$

$$|G(j\omega)| = \frac{(9-\omega^2)\sqrt{4+\omega^2}}{\sqrt{\omega^2+1}\sqrt{\omega^2+6}\sqrt{16+\omega^2}} = 0$$

For $y(t) = 0$

$|G(j\omega)|$ can be zero

Hence $G(j\omega) = 0$

at $\quad \omega^2 = 9$

or, $\quad \omega = 3 rad/s$

79. Ans. (a)

Solution:

From the given block diagram

Characteristic equation is $1 + G(s)H(s) = 0$

$$1 + \frac{K(s+1)}{s^3 + as^2 + 2s + 1} = 0$$

or, $s^3 + s^2 a + 2(K+2) + (K+1) = 0$

Apply RH criterion

s^3	1	$K+2$
s^2	a	$K+1$
s^1	$\frac{(K+2)a-(K+1)}{a}$	
s^0	$K+1$	

For system to oscillate at ω = 2 rad/s get auxilary equation

$$as^2 + (K+1) = 0 \qquad ...(1)$$

Put, $\quad s = j\omega$

$$s^2 = -\omega^2 \qquad ...(2)$$

From equation (1) and (2)

$-a\omega^2 + (k+1) = 0$

$a\omega^2 = K+1$

As given $\omega = 2$

$4a = K+1 \qquad ...(3)$

From option $K = 2,\ a = 0.75$ satisfy equation (3) for sysem to oscillate

$a(2+K) - (1+K) = 0$

$$a = \left(\frac{1+K}{2+K}\right)$$

Derive the auxialry equation

A.E $\Rightarrow as^2 + (1+K) = 0$

or, $\quad \omega = \sqrt{\frac{1+K}{a}} = 2$

or, $\quad \left(\frac{1+K}{a}\right) = a$

or, $\quad 2+K = 4$

or, $\quad K = 2$

Thus $\quad a = 0.75$

80. Ans. (d)

Solution:

$$G(s) = \frac{s^2+4s+8}{s(s+2)(s+8)}$$

Zeros: $s^2 + 4s + 8$

$$s = \frac{-4 \pm \sqrt{4^2 - 4\times 8}}{2}$$

$$= \frac{-4 \pm \sqrt{16-32}}{2}$$

$$= \frac{-4 \pm 4j}{2}$$

$$= -2 \pm 2j$$

Poles: $s = 0, -2, -8$

Pole zero polt:

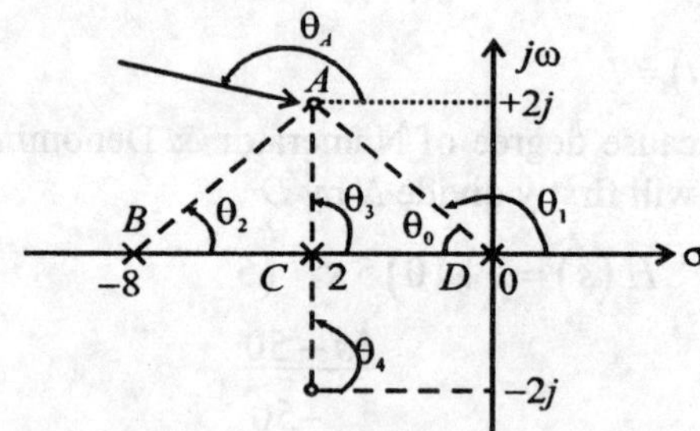

$\Sigma\theta(z) - \Sigma\theta(p) = 180$

$\theta_A(z) = \pi$ – sum (angle to zero) + sum (angle to poles)

$$= \pi + [\theta_1 + \theta_3 + \theta_2] - \theta_4$$

$$\theta_1 = \pi - \theta_0$$

$$= \pi - \tan^{-1}\left(\frac{2}{2}\right)$$

$$= \pi - \frac{\pi}{4} = \frac{3\pi}{4}$$

$$\theta_2 = \tan^{-1}\frac{AC}{BC}$$

$$= \tan^{-1}\left(\frac{2}{6}\right) = \tan^{-1}\left(\frac{1}{3}\right)$$

$$\theta_3 = \frac{\pi}{2}$$

$$\theta_4 = \frac{\pi}{2}$$

$$\theta_A = \pi + \left[\frac{3\pi}{4} + \frac{\pi}{2} + \tan^{-1}\left(\frac{1}{3}\right)\right] - \frac{\pi}{2}$$

$$= \pi + \frac{3\pi}{4} + \tan^{-1}\left(\frac{1}{3}\right)$$

$$= \frac{7\pi}{4} + \tan^{-1}\left(\frac{1}{3}\right)$$

To make angle w.r.t. to 0°

$$= \frac{7\pi}{4} - 2\pi + \tan^{-1}\left(\frac{1}{3}\right)$$

$$= \frac{7\pi - 8\pi}{4} + \tan^{-1}\left(\frac{1}{3}\right)$$

$$= -\frac{\pi}{4} + \tan^{-1}\left(\frac{1}{3}\right)$$

$$|\theta_A| = \frac{\pi}{4} - \tan^{-1}\left(\frac{1}{3}\right)$$

$$\theta_{\text{arrival}} = \frac{\pi}{4} - \tan^{-1}\left(\frac{1}{3}\right)$$

81. Ans. (b)

Solution:

$$G(s)H(s) = \frac{K K_p}{1 + s\tau}$$

Positional error constant

$$K_{pec} = \underset{s\to 0}{Lt}\left(\frac{K K_p}{1 + s\tau}\right)$$

$$= K K_p$$

Steady state error e_{ss}

$$= \frac{1}{1 + K_{pec}} = \frac{1}{1 + K K_p}$$

82. Ans. (b)

Solution:

Phase angle $= -\tan^{-1}\omega - \dfrac{0.1\omega \times 180^o}{\pi}$

For $\omega = 1$

$$\angle Y(j\omega) - \angle X(j\omega) = -\tan^{-1} 1 - \frac{0.1\omega \times \pi}{180^o}$$

$$= -45^o$$

83. Ans. (c)

Solution:

Poles at $s = (-2 \pm j2)$

This is 2nd order system with

$$T(s) = \frac{1}{(s+2)^2 + 4} = \frac{1}{s^3 + 4s + 8}$$

We have $\sqrt{8} = \omega_n$

And $2\delta\omega_n = 4$

$$\delta = \frac{4}{2\sqrt{8}} = \sqrt{12} = 0.707$$

84. Ans. (a)

Solution:

Here $D(s)$ is considered as input

$$E(s) = \frac{D(s)}{1 + G(s)H(s)}$$

Then, steady-state error $e_{SS} = \underset{s\to t}{Lt}\, sE(0)$

$$= \underset{s\to 0}{Lt}\frac{s \times 1/s}{1 + \dfrac{40}{s(1+5s)}} = 0$$

85. Ans. (a)

Solution:

$$G_1(s) = \frac{1}{2s+1}$$

Time-constant $\tau_1 = 2\,\text{sec}$

$$G_2(s) = \frac{1}{5s+1}$$

Time constant, $\tau_2 = 5\,\text{sec}$

Bandwidth is inversely proportional to time constant.

B. $W_1 \propto \dfrac{1}{2}$

B. $W_2 \propto \dfrac{1}{5}$

Hence, system 1 is having more bandwidth than system 2.

86. Ans. (c)

Solution: $E(s) = \dfrac{R(s)}{1+G(s)}$

$$\underset{t\to\infty}{Lt}\, e(t) = \underset{s\to 0}{Lt}\, \frac{sR(s)}{1+G(s)}$$

Here, $G(s) = \dfrac{K}{s^2}$ and $R(s) = \dfrac{1}{s}$

$$\underset{t\to\infty}{LT} \frac{s\cdot\frac{1}{s}}{1+\frac{K}{s^2}} = \frac{1}{\infty} = 0$$

So, $\underset{t\to\infty}{Lt}\, e(t) = 0$

87. Ans. (d)

Solution:

$$T.F. = \frac{A}{s^2+\omega^2}$$

Poles located at

$s^2+\omega^2 = 0$

$s^2 = -\omega^2$

$s = \pm j\omega$ (Complex conjugate, i.e. marginal sable)

The steady state gain of the marginally sable system is not possible to determined.

88. Ans. (a)

Solution:

Settling time $t_s\,(5\%) = \dfrac{3}{\delta\omega_n} = 3\tau$

For 5%, $t_s = 3RC$, and for 2%, $t_s = 4RC$

89. Ans. (b)

Solution:

The controller block is

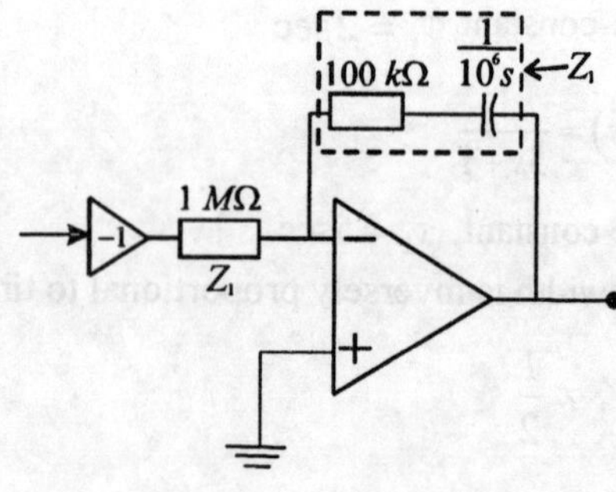

$$Z_f = 10^5 + \frac{1}{10^{-6}s}$$

$$G_C(s) = \frac{V_o}{V_{in}}$$

$$= -\frac{Z_f}{Z_1}$$

$$= -\left(\frac{10^5+\frac{1}{10^{-6}s}}{-10^6}\right)$$

$$= \frac{1}{10}+\frac{1}{s} = \frac{s+10}{10s} = \frac{0.1\,s+1}{s}$$

$$G_C(s) = \frac{0.1s+1}{s}$$

90. Ans. (c)

Solution:

Using the concept of above answers the simplified block diagram is

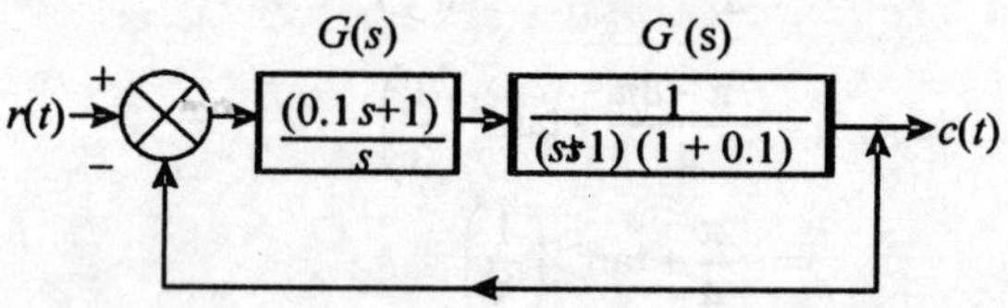

$$\frac{C(s)}{R(s)} = \frac{\frac{1}{s(s+1)}}{1+\frac{1}{s(s+1)}} = \frac{1}{s^2+s+1} \qquad \text{... (1)}$$

The standard second order equation is given as

$$\frac{C(s)}{R(s)} = \frac{\omega_n^2}{s^2+2\delta\omega_n+\omega_n^2} \qquad \text{... (2)}$$

On comparing equation (1) and (2)

$\omega_n = 1\, rad/\sec$

$2\delta\omega_n = 1$

$\delta = 0.5$

Percentage of peak overshoot

$$M_P = 100\times e^{\frac{-\pi\delta}{\sqrt{1-\delta^2}}}$$

$$= 100\times e^{\frac{-\pi 0.5}{\sqrt{1-(0.5)^2}}}$$

$$= 100\times 0.163$$

$$= 16.3\%$$

91. Ans. (d)

Solution:

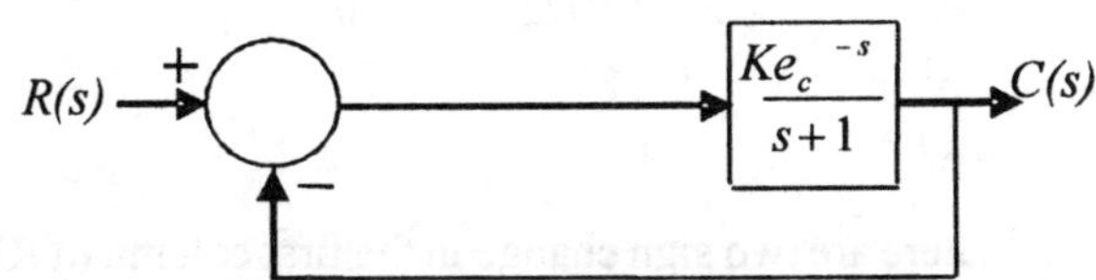

$$G(s) = \frac{K_c e^{-s}}{s+1}$$

$$G(j\omega) = \frac{K_c e^{-j\omega}}{1+j\omega}$$

$$= \frac{K_c(\cos\omega - j\sin\omega)}{1+j\omega}$$

Phase angle ϕ

$$\phi = -\tan^{-1}(\tan\omega) - \tan^{-1}\tan\omega$$

$$= -\tan^{-1}\omega - \omega$$

or, $-\tan^{-1}\omega - \dfrac{180^\circ\omega}{\pi} = -180^\circ$

or, $\tan^{-1}\omega + \dfrac{180^\circ\omega}{\pi} = 180^\circ$

After solving

$\omega = 2.03\ rad/\sec$

92. Ans. (b)

Solution:

$$\frac{C(s)}{R(s)} = \frac{G(s)}{1+G(s)} = \frac{K_c\, e^{-s}}{(s+1)+K_c\, e^{-s}}$$

$$E(s) = R(s) - C(s)$$

$$= R(s) - \frac{R(s)K_c\, e^{-s}}{(s+1)+K_c\, e^{-s}}$$

$$= R(s)\left[1 - \frac{K_c\, e^{-s}}{(s+1)+K_c\, e^{-s}}\right]$$

$$= R(s)\left[\frac{(s+1)}{(s+1)+K_c\, e^{-s}}\right]$$

$$e_{ss} = \lim_{s\to 0} s\cdot E(s)$$

$$= \lim_{s\to 0}\frac{s\cdot R(s)\cdot(s+1)}{(s+1)+K_c e^{-s}}$$

Since $R(s) = \dfrac{1}{s}$ and $K_c = 1$

$$e_{ss} = \lim_{s\to 0}\frac{s\cdot\frac{1}{s}(s+1)}{s+1+e^{-s}}$$

$$= \lim_{s\to 0}\frac{(s+1)}{s+1+e^{-s}}$$

$$= \frac{0+1}{0+1+1} = \frac{1}{1+1}$$

$$= \frac{1}{2}$$

10.3 Stability: Routh-Hurwitz Criterion

1. Ans. (a)

Solution: Characteristic equation given by

$$s^3 + 6\,Ks^2 + (K+2)\,s + 8 = 0$$

Apply RH table

S^3			$K+2$
S^2			8
S^1		$\dfrac{6K(K+2)-8.1}{6K}$	0
S^0	8		

For stability,

$$\frac{6K(K+2)-8}{6K} \geq 0 \text{ or } K > 0$$

$$6K^2 + 12K - 8 = 0$$

$$3K^2 + 6K - 4 = 0$$

$$K = \frac{-6 \pm \sqrt{36+48}}{2.3} = \frac{-6\pm\sqrt{84}}{6}$$

$$= 0.528 \text{ or } -2.58$$

2. Ans. (c)

Solution: Characteristic equation $= s^3 + 3s^2 + 4s + A = 0$

Routh-Hurwitz table

s^3	1	4	0
s^2	3	A	0
s^1	$[12-A]$	0	
s^0	A		

$$12 - A \geq 0$$

$$12 \geq A$$

or, $\quad 0 \leq A \leq 12$

3. Ans. (d)

Solution: The number of roots in LHS of *S*-plane can be easily found by RH table

Characteristic equation: $s^3 + 5s^2 + 7s + 3 = 0$

Routh-Hurwitz table

s^3	1	7
s^2	5	3
s^1	[35 – 3/5]	0
s^0	3	

There is no sign change in the first column of all the row. Hence all roots lie in the L.H.S of *S*-plane, i.e $n = 3$.

4. Ans. (a)

Solution: The characteristic equation is, $1 + G(s) = 0$

or, $s^4 + 4s^2 + 5^2 + 6s + K = 0$

Apply Routh-Hurwitz array, we have

s^4	1	5	K
s^3	4	6	0
s^2	(7 / 2)	K	
s^1	(21-4K) (2/7)	0	
s^0	K		

For the system to be stable

$21 - 4K > 0$ and $K > 0$; $[21/4] > K$ and $K > 0$

On combining we get $\frac{21}{4} > K > 0$

5. Ans. (b)

Solution: $P(s) = s^5 + s^4 + 2s^3 + 2s^2 + 3s + 15 = 0$

Apply RH Criterion

S^5	1	2	3
S^4	1	2	15
S^3	0	–15	

This test method fails then apply another method

Put $S = \frac{1}{p}$, and test it again with RH table

$$\left(\frac{1}{p}\right)^5 + \left(\frac{1}{p}\right)^4 + 2\left(\frac{1}{p}\right)^3 + 2\left(\frac{1}{p}\right)^2 + 3\frac{1}{p} + 15 = 0$$

$15p^4 + 3p^4 + 2p^3 + 2p^2 + p^1 + 1 = 0$

R.H table

p^5	15	2	1	0
p^4	3	2	1	0
p^3	([6 – 30)/3]	[(3 – 15)/3]		
	= [–24/3]			
p^3	= –8	= –4	0	
p^2	[(–16 + 12)/ –8]	1	0	
p^2	= [1 /2]			
p^1	–(2 + 8)/1/2	0		
p^1	= 12			
p^0	1			

There are two sign change in the first column of RH table which indicate two roots in the RHS of *s* plan.

6. Ans. (b)

Solution: Characteristic equation $1 + GH = 0$

or $s^3 + as^2 + 2s + 1 + K(s + 1) = 0$

or $s^3 + as^2 + (2 + K)s + (1 + K) = 0$

Routh-Hurwitz array

s^3	1	$2 + k$
s^2	a	$1 + k$
s^1	$\frac{(1 + k) - a(2 + k)}{a}$	0
s^0	$1 + k$	

For oscillating at 2 rad/sec.

$$\frac{2+k}{1} = \frac{1+k}{a} = 2^2$$

$\Rightarrow k = 2$ and $a = 0.75$

The system is oscillatory when all element of s^1 row becomes zero.

$$2+k-1\frac{(1+k)}{a} = 0$$

$$a = \frac{1+k}{2+k} \qquad ...(1)$$

$$as^2 + (1+k) = 0 \qquad ...(2)$$

$s = j\omega$

at $\omega = 2$ (given)

Hence $k = 2$; $a = 0.75$

7. Ans. (c)

Solution: $G(s) = \frac{s+8}{s^2 + \alpha s - 4}$

$$T(s) = \frac{C(s)}{R(s)} = \frac{G(s)}{1+G(s)\,H(s)}$$

$$= \frac{s+8}{(s+8)+(s^2+\alpha s-4)}$$

$$= \frac{s+8}{s^2+s(\alpha+1)+4}$$

Characteristic equation $= s^2 + s(\alpha+1) + 4 = 0$

Apply Routh-Hurwitz criteria

s^2	1	4
s	$(\alpha+1)$	0
s^o	4	

For the close loop to be stable

$$\alpha + 1 \geq 0 \Rightarrow \alpha \geq -1$$

i.e. positive value of α is requried for stable system.

8. **Ans. (a)**

Solution:

From the given block diagram

Characteristic equation is $1 + G(s)H(s) = 0$

$$1 + \frac{K(s+1)}{s^3 + as^2 + 2s + 1} = 0$$

or, $s^3 + s^2 a + 2(K+2) + (K+1) = 0$

Apply RH criterion

s^3	1	$K+2$
s^2	a	$K+1$
s^1	$\frac{(K+2)a-(K+1)}{a}$	
s^0	$K+1$	

For system to oscillate at $\omega = 2$ rad/s get auxiliary equation

$$as^2 + (K+1) = 0 \quad \text{...(1)}$$

Put, $s = j\omega$

$$s^2 = -\omega^2 \quad \text{...(2)}$$

From equation (1) and (2)

$-a\omega^2 + (k+1) = 0$

$a\omega^2 = K+1$

As given $\omega = 2$

$4a = K+1 \quad$ ...(3)

From option $K = 2,\ a = 0.75$ satisfy equation (3)

For system to oscillate all elements s^1 row becomes zero

$a(2+K) - (1+K) = 0$

$$a = \left(\frac{1+K}{2+K}\right)$$

Derive the auxiliary equation

A.E $\Rightarrow as^2 + (1+K) = 0$

or, $\quad \omega = \sqrt{\frac{1+K}{a}} = 2$

or, $\quad \left(\frac{1+K}{a}\right) = a$

or, $\quad 2+K = 4$

or, $\quad K = 2$

Thus $\quad a = 0.75$

9. **Ans. (c)**

Solution: For open loop system

$$s^5 + 2s^4 + 3s^3 + 6s^3 + 5s + 3 = 0$$

Apply RH criterion

s^5	1	3	5
s^4	2	6	3
s^3	0	7/2	

Due to first element in a row is zero, the Analysis fails

Now, Modify the equation by $s \to \frac{1}{p}$

Then $3p^5 + 5p^4 + 6p^3 + 3p^2 + 2p + 1$

p^5	3	6	2
p^4	5	3	1
p^4	21/5	7/5	0
p^3	(63 − 21/5)	21/5	0
p^3	42/5	21/5	0
p^2	294/25	−441/25	
p^2	−147/25	0	
p^1	21/5		

Ther are two sign changes in first column of RH table two roots in RH of s-plane.

10. **Ans. (b)**

Solution:

Stability depends on the poles on the $j\omega$ axis and at the origin

Therefore, Output may be bounded (or) unbounded to a bounded input

11. **Ans. (c)**

Solution: Routh-Hurwitz criterion

s^4	2	3	7
s^3	1	5	
	→		
s^2	− 7	7	
	→		
s^1	6		
s^0	7		

Since sign changes twice, therefore 2 roots in RHS of s-plane.

12. Ans. (c)

Solution: The characteristic equation is, $1 + GH = 0$

$$\text{or} \quad 1+\frac{K}{s(s+2)(s+4)}=0$$

$$\text{or} \quad 1+\frac{K}{s(s^2+6s+8)}=0$$

$$\text{or} \quad s^3 + 6s^2 + 8s + K = 0$$

Apply Routh-Hurwitz array

s^3	1	8 0
s^2	6	K
s^1	$K-48-0$	
s	K	

The system will be just if first column of all Row have same sign, i.e. $K-48 \geq 0$ or $K \geq 48$

13. Ans. (b)

Solution: Characteristic equation

$$s^3 - 4s^2 + s + 6$$

Constructing Routh-Hurwitz array

s^3	1	1	0
	→		
s^2	−4	6	0
	→		
s^1	2	0	
s^0	6		

Number of sign changes in the first column is two, therefore the number of roots in the right half s-pane is 2, total roots = 3

Roots in left half = 3 − 2 = 1

14. Ans. (c)

Solution: Given equation $G(s)\,H(s) = \dfrac{K(1-s)}{(1+s)}$

Characteristic equation = $1 + G(s)\,H(s) = 0$

$$1+\frac{K(1-s)}{(1+s)}=0$$

$$\text{or} \quad (1+s)+K(1-s)=0$$

$$\text{or} \quad s-Ks+K+1=0$$

$$\text{or} \quad (1-k)s+(K+1)=0$$

RH Table

s^1	$1-K$
s^o	$1+K$

For stability $1-K>0$

$1+K>0$ Thus $|K|<|$

15. Ans. (d)

Solution: For U_1 only

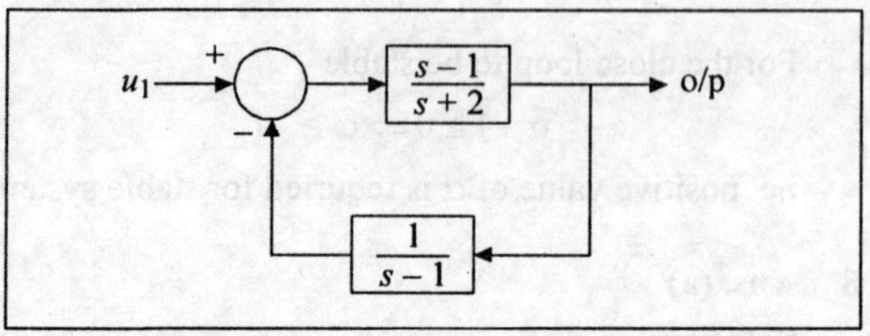

For U_1 transfer function

$$TF=\frac{\dfrac{(s-1)}{(s+2)}}{1+\dfrac{(s-1)}{(s+2)}\cdot\dfrac{1}{(s-1)}}=\frac{(s-1)}{s+2+1}$$

$$=\frac{(s-1)}{(s+3)} \Rightarrow \text{Stable}$$

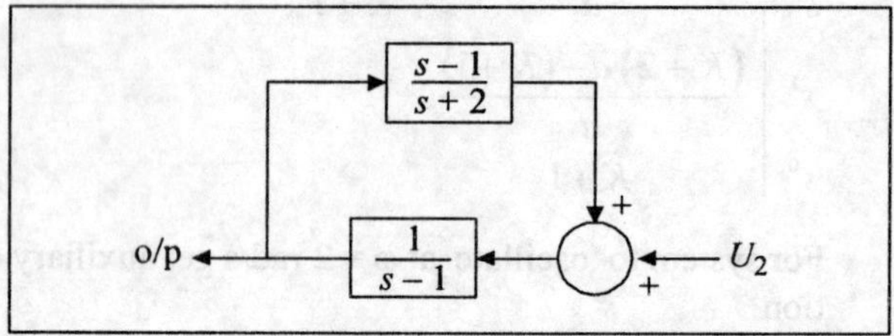

For U_2 transfer function

$$TF=\frac{\dfrac{1}{s-1}}{1-\dfrac{1}{(s-1)}\times\dfrac{(s-1)}{(s+2)}}=\frac{\dfrac{1}{(s-1)}}{\dfrac{s+2-1}{s+2}}$$

$$=\frac{\dfrac{1}{(s-1)}}{\dfrac{(s+1)}{(s+2)}}=\frac{\dfrac{s+2}{(s+1)}}{(s+2)}=\frac{s+2}{(s-1)(s+1)}$$

$\Rightarrow$ Unstable.

16. Ans. (b)

Solution: Characteristic equation

$$1+\frac{K(s+3)}{(s+8)^2}=0$$

$$\text{or} \quad s^2+(16+K)s+3K+64=0$$

Routh array

s^2	1	$(3K+64)$
s^1	$(16+K)$	
s^0	$(3K+64)$	

$16+K=0$; $K=-16$

$$3K+64=0\ ;\ K=\frac{-64}{3}=-21$$

$$K>-16$$

Put the value $K=-16$ which both the condition in auxilary equation

$$s^2+3(-16)+64$$
$$s^2+16=0$$
$$s=\pm j4=\pm j\omega$$
$$s=\pm j4=\pm j\omega$$
$$\omega=4 \text{ rad/sec}$$

17. Ans. (d)

Solution:

From Routh table , the equation is

$$2s^3+4s^2+2s+4=0$$
$$s^2[4+2s]+1[2s+4]=0$$
$$(s^2+1)(2s+4)=0$$
$$s=\pm j$$

and -0.5

It means two roots on $j\omega$ axis and one root at -0.5, i.e. left half of plane.

From RH table

s^3	2	2
s^2	4	4
s^1	0	0
s^0		

As the terms in the S^1 row are all zero, the Routh's test break down .

Now, obtain the Auxillary Polynomial ,

$$A(s)=4s^2+4$$
$$\frac{d}{ds}A(s)=8s$$

Now , rearrange the Routh Hurwitz Array with

S^1 8 0

i.e.

s^3	2	2
s^2	4	4
s^1	8	0
s^0	4	0

Now from the auxilliary polynomial ,

$$4s^2+4=0$$
$$4s^2=-4$$
$$s^2=-1$$
$$s=\pm j$$

The mumber of roots an $j\omega$ axis $=2$

Since all the elements in 1st column of Routh Hurutiz array are positive, so no root lie in right hand side of s plane.

There are total 3 roots of this equation. Two roots at $s=\pm j$ and one root in left half of s plane.

18. Ans. (c)

Solution: $G(s)=\dfrac{K}{s(s+3)(s+10)}$

Characteristic equation

$$1+GH=1+\frac{K}{s(s+3)(s+10)}=0$$

Characteristic equation

$s^3+13s^2+30s+k=0$

Apply RH criteria

s^3	1	30
s^2	13	K
s^1	$\dfrac{-13\times 30-K}{13}$	0
s^3	K	

For stability $390-K\geq 0$

$$390\geq K$$
$$0<K<390$$

19. Ans. (a)

Solution:

From the given block diagram

Charcteristic equation is $1+G(s)H(s)=0$

$$1+\frac{K(s+1)}{s^3+as^2+2s+1}=0$$

or, $s^3+s^2a+2(K+2)+(K+1)=0$

Apply RH criterion

s^3	1	$K+2$
s^2	a	$K+1$
s^1	$\dfrac{(K+2)a-(K+1)}{a}$	
s^0	$K+1$	

For system to oscillate at $\omega = 2$ rad/s get auxilary equation

$$as^2 + (K+1) = 0 \quad \text{... (1)}$$

Put, $s = j\omega$

$$s^2 = -\omega^2 \quad \text{... (2)}$$

From equation (1) and (2)

$$-a\omega^2 + (k+1) = 0$$

$$a\omega^2 = K+1$$

As given $\omega = 2$

$$4a = K+1 \quad \text{... (3)}$$

From option $K=2, a=0.75$ satisfy equation (3)

for sysem to oscillate

$$a(2+K)-(1+K)=0$$

$$a = \left(\frac{1+K}{2+K}\right)$$

Derive the auxialry equation

A.E $\Rightarrow as^2 + (1+K) = 0$

or, $\omega = \sqrt{\dfrac{1+K}{a}} = 2$

or, $\left(\dfrac{1+K}{a}\right) = a$

or, $2+K=4$

or, $K=2$

Thus $a = 0.75$

20. Ans. (d)

Solution:

$y(t) = e^{-|x(t)|}$

When $|x(t)| \to \infty$; then $y(t) \to 0$

when $|x(t)| \to 0$; then $y(t) \to 1$

i.e. $y(t)$ is bounded irrespective of $x(t)$

* Therefore, e^{-x} is always convergent even when x is not bounded

Second Method

For any input, i.e. bounded or unbounded $y(t) = e^{-|x(t)|}$ is always bounded

21. Ans. (d)

Solution:

$F(s) = s^5 - 3s^4 + 5s^3 - 7s^2 + 4s + 20$

Apply Routh-Hurwitz table

$s^5 \to$	1	5	4
$s^4 \to$	–3	–7	20
s^3	(8 / 3)	(20 / 3)	0
s^2	5	20	0
s^1	0	0	0

Auxilary equation $A(s) = 5s^2 + 20$

$$\frac{dA(s)}{ds} = 10s + 0$$

s^1	10	0 0
s^0	20	0 0

If we observe the 1st column, sign is changing two times So we have two poles on right half side of imaginary axis and $5s^2 + 20 = 0$

So, $s = \pm 2j$ and 1 pole on left side of imaginary axis

$s^5 - 3s^4 + 5s^3 - 7s^2 + 4s + 20 = 0$

For, $s = -1$ the equation becomes

$-1 - 3 - 5 - 7 - 4 + 20 = 0$

It indicate factor of s + 1 that means one root on negative real axis

Second Method

Apply RH tabulation

s^5	1	5	4
s^4	–3	–7	20
s^3	(8/3)	(32/2)	0
s^2	5	20	0
s^1	0(10)	0	0
s^0	20	0	0

$AE = 5s^2 + 20$

$$\frac{dAE}{ds} = 10s + 0$$

AE signal $= s = \pm j2$

Two sign changes in the RH tabulation.

$\therefore$ Number of right hand roots = 2

Number of $j\omega$ axis roots = 2

Number of left hand root = 1(real)

22. Ans. (c)

Solution: There are two sign change in R-H table.

23. Ans. (a)

Solution: Constructing Routh-array

s^4	1	a	8
s^3	3	12	0
s^2	$\frac{3a-12}{3}$	8	
s^1			
s^0			

24. Ans (d)

Solution:

$$G(s) \Rightarrow \frac{K(s+3)}{(s+1)(s+2)}$$

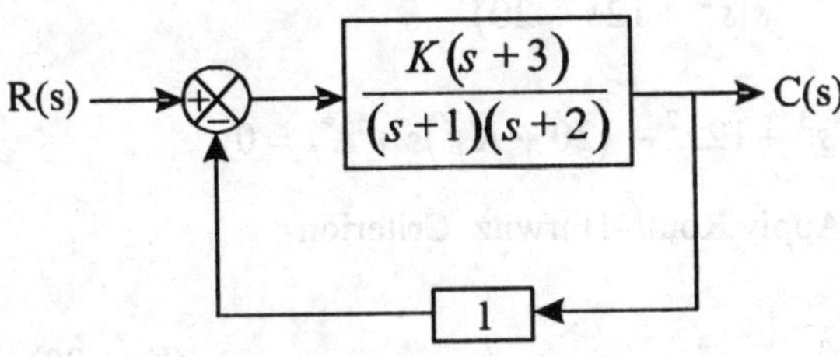

$$\frac{C(s)}{R(s)} = \frac{\frac{K(s+3)}{(s+1)(s+2)}}{1+\frac{K(s+3)}{(s+1)(s+2)}}$$

$$= \frac{K(s+3)}{(s+1)(s+2)+K(s+3)}$$

Characteristics Equation is $1+GH=0$

$$(s+1)(s+2)+K(s+3)=0$$

$$s^2+s+2s+2+Ks+3K=0$$

$$s^2+3s+Ks+2+3K=0$$

$$s^2+(s+K)s+(2+3K)=0$$

Apply RH criterion

s^2	1	$3K+2$
s^1	$(3+K)$	0
s^0	$\frac{(3+K)(3K+2)}{(3+K)}$	$=3K+2$

1st row must be positive. For system to be stable

$$3+K \geq 0$$

$$K \geq -3$$

$$3K+2 \geq 0$$

$$3K \geq -2$$

$$K \geq \frac{-2}{3}$$

so $0 < K < \infty$

25. Ans. (d)

Solution:

Apply RH criterion

s^3	3	3
s^2	4	4
s^1	0	0

so $4s^2 + 4 = 0$

Take derivtive

8s

s^1	8	0
s^0	4	

$4s^2 + 4 = 0$

$4s^2 = -4$

$s^2 = -4$

$s^2 = -1$

$s = +i$

So pair of complex conjucate poles on imaginare axis

26. Ans (c)

Solution:

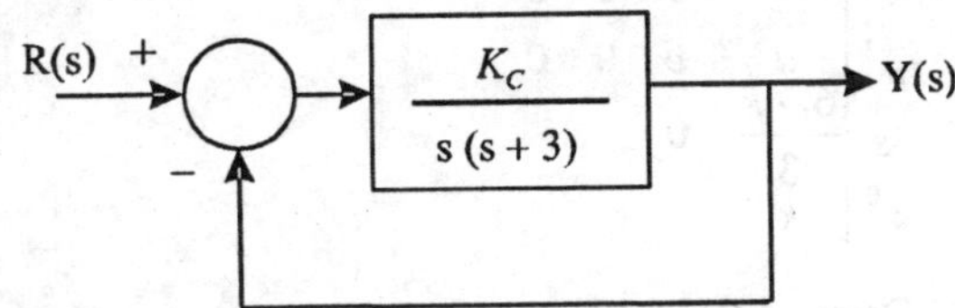

$$\frac{Y(s)}{R(s)} = \frac{K_C}{s^2+3s+K_C}$$

Closed-loop poles are given dy,

$$s_1, s_2 = \frac{-3 \pm \sqrt{9-K_C.4}}{2}$$

For real parts more negatine then -1

$$\frac{-3 \pm \sqrt{9-4K_C}}{2} < -1$$

$$-3 \pm \sqrt{9-4K_c} < -2$$

$$\pm\sqrt{9-4K_c} < 1$$

$4K_c > 9$

or $K_c > \frac{9}{4} = 2.25$

or $K_C > 2$

27. Ans. (d)

Solution:

The characteristic equation is,

$1 + G(s) = 0$

or $1 + \frac{10(\tau_1 s + 1)}{\tau_1 s^2 (s+1)} = 0$

or $\tau, s^3 + \tau_1 s^2 + 10\tau_1 s + 10 = 0$

Constructing routh-array,

s^3	τ_1	$10\tau_1$
s^2	τ_1	10
s^1		0
s^0	10	

For stability, $\tau_1 - 1 > 0$

or $\tau_1 > 1$

Thus, the stability range will be $\tau_1 > 1.0$

28. Ans. (a)

Solution:

Given, Routh Hurwitz Array

s^3	1	b 0	0
s^3	a	c 0	0
s	$\frac{6-k}{3}$	0	
s^0	k		

Derived RH table

s^3	1	b
s^2	a	c
s^1	$\frac{ab-c}{a}$	0
s^0	c	

Compare the given RH table with derived RH table, we get

$c = K$

$\frac{ab-c}{a} = \frac{6-k}{3}$

For critical stability $6 - K \geq 0$

$K = 6$

Hence, $c = K = 6$

$a = 3$ and $ab = 6$

$3.b = 6$

$b = \frac{6}{3} = 2$

For critical stability, K = 6

A = 3, c = 6, so the answer will be 6, 6.

29. Ans. (d)

Solution:

Characteristic equation 1 + GH = 0

$1 + \left(\frac{K_p s + K_I}{s}\right)\frac{1}{(s+2)(s+10)} = 0$

$1 + \frac{s K_P + K_I}{s(s^2 + 12s + 20)} = 0$

$s^3 + 12s^2 + (20 + K_P)s + K_I = 0$

Apply Routh-Hurwitz Criterion

s^3	1	$(K_p + 20)$
s^2	12	K_I
s^1	$(12 K_p + 12 \times 20 - K_I) / 12$	0
s^0	K_I	

For stability, RH criterion to be satisfied,

$\frac{12 K_p + 12 \times 20 - K_I}{12} > 0$ and $K_I > 0$

or, $K_p > \frac{K_I}{12} - 20$

30. Ans. (b)

Solution:

Given

$G(s)H(s) = \frac{K}{s(s+1)(s+5)}$

In the root locus plot the when either the roots are on the $j\omega$ axis or near to origin, then system will oscillate. Here ultimate value of gain means the value of K when root locus intersects imaginary axis at the frequency. Characteristic equation,

$1 + G(s)H(s) = 0$

$1 + \frac{K}{s(s+1)(s+5)} = 0$

$s(s+1)(s+5)+K=0$

$s^3+6s^2+5s+K=0$

Apply the RH criterion array

s^3	1	5
s^2	6	K
s^1	$\frac{(30-K)}{K}$	0
s^0	6	0

For the oscillation

$$\frac{30-K}{6}=0$$

or, $K=30$

The crossing point of RL on the imaginary axis are found by solving the auxiliary equation obtained from the s^2 row.

$6s^2+30=0$

or $6s^2=-30$

or $s^2=-5$

or, $s=\pm j\sqrt{5}$

or, Angular frequency $\omega=\sqrt{5}$ rad/sec

10.4 Root Locus

1. **Ans. (b)**

Solution: The characteristic equation is

$$s^2+(3-K)s+1=0$$

At $K=0$,

$$s^2+3s+1=0$$

$$s=\frac{-3\pm\sqrt{9-4}}{2}$$

$$=\frac{-3\pm\sqrt{5}}{2}$$

$$=-2.62,\,-0.38.$$

These are the poles of $KG(s)$ since the root locus stands at the poles of $KG(s)$.

At $K=1$,

$$s^2+2s+1=0 \text{ or } (s+1)^2=0$$

$$s=-1$$

i.e. meeting of 2 branches or root locus is at $s=-1$

At $K=3$, we have $s^2+1=0$, $\quad s=\pm j$

At $K=\infty$,

$$\lim_{K\to\infty} s^2+(3-K)s+1=0$$

$$\lim_{K\to\infty}\frac{s^2}{K}+\frac{3s}{K}-s+\frac{1}{K}=0 \text{ or } s=0$$

i.e. a zero exist at $s=0$

Hence, satisfying all these conditions, choice (b) is correct.

2. **Ans. (b)**

Solution: Characteristic equation

$$=s^3+5s^2+(K+6)s+K$$

$$=s^3+5s^2+6s+K(s+1)=0$$

$$=1+GH=1+\frac{1+K(s+1)}{s^3+5s^2+6s}=1+\frac{K(s+1)}{s(s+2)(s+3)}$$

Point of intersection of asymptotes with real axis is

$$\sigma=\frac{\Sigma\text{ finite of GH pole }-\Sigma\text{ finite of GH zero}}{P-Z}$$

$$=\frac{(0-2-3)-(-1)}{3-1}=\frac{-4}{2}=-2$$

Coordinate are $(-2, 0)$.

3. **Ans. (d)**

Solution: Characteristic equation $1+G(s)H(s)=0$

Apply RH table

$(s+1)^4+K=0 \Rightarrow s^4+4s^3+6s^2+4s+1+K=0$

s^4	1	6	$1+K$
s^3	4	4	0
s^2	5	$1+K$	0
s^1	$\frac{20-4(1+K)}{5}$	0	
s^0	$1+K$		

For stability $K>-1$.

$$20>4(1+K),\ 4K<16$$

$$-1<K<4,\quad K<4$$

The root locus crosses the imaginary axis is found by solution of auxiliary equation $5s^2+(1+K)=5s^2+1+4=0$, $s=\pm j$. There are 4 poles at -1 and the root locus has to be the right of the poles. So both s_1 and s_2 are not on the root locus.

4. **Ans. (d)**

Solution: The value of k which lie left of $j\omega$ axis makes system stable if these value lie with in the close circle then system becomes over damped, the value of over damped is $0\le k<1$ or $k>5$

Second Method

Characteristic equation:

$$1 + GH = 0$$

$$\Rightarrow \quad 1 + \frac{K(s+3)}{(s+1)(s+2)} = 0$$

$$\Rightarrow \quad s^2 + 3s + 2 + Ks + 3K = 0$$

$$\Rightarrow \quad s^2 + (3 + K)s + (2 + 3K) = 0$$

Standard second order characteristic equation

$$s^2 + 2\,\delta\omega_n s + \omega_n^2 = 0$$

Comparing the above two equation, we get

$$\omega_n = \sqrt{2+3K}$$

$$2\,\delta\omega_n = 3 + K$$

$$\delta = \frac{3+K}{2\sqrt{2+3K}}$$

For stability, $\delta > 1$

$$\Rightarrow \quad \frac{3+K}{2\sqrt{2+3K}} > 1$$

$$\Rightarrow \quad 9 + K^2 + 6K > 4(2 + 3K)$$

$$9 + K^2 + 6K - 12K - 8 > 0$$

$$K^2 - 6K + 1 > 0$$

$$K = \frac{6 \pm \sqrt{36-4}}{2}$$

$$K = 5.82 \text{ and } 0.17$$

Which satisfy the condition (d).

5. Ans. (c)

Solution: Point of interaction of Asymptotes with real axis

$$\text{Centroid, } (-\sigma_A) = \frac{\sum_{i=1}^{m} P_i - \sum_{j=1}^{n} Z_j}{m-n}$$

Where m and n are the number of poles and zero respectively.

$$= \frac{-4-0}{3} = -1.33$$

6. Ans. (c)

Solution:

The Rough sketch of the root locus is a circle.

The breakaway point is found by $\frac{dK}{ds} = 0$ roots

Characteristic equation, $1 + GH = 0$

$$K = \frac{s(s+3)}{(s-1)}$$

$$\frac{dK}{ds} = 0$$

$$\text{or,} \quad \frac{s(s+3)1-(s-1)(2s+3)}{(s-1)^2} = 0$$

$$\text{or,} \quad s^2 + 3s - 2s^2 - s + 3 = 0$$

$$\text{or,} \quad -s^2 + 2s + 3 = 0$$

$$\text{or,} \quad -(s^2 - 2s + 3) = 0$$

$$\text{or,} \quad (s-3)(s-1) = 0$$

$$s = 3 \text{ or } -1$$

7. Ans. (d)

Solution: We known that characteristic equation is

$$1 + G(s)\,H(s) = 0$$

$$\text{or} \quad G(s)\,H(s) = -1$$

$$\text{or} \quad |G(s)\,H(s)| = 1$$

unity feed back system

Hence put $H(s) = 1$ in above equation then

$$\therefore \quad \left|\frac{K}{s(s^2+7s+12)}\right|_{s=-1+j} = 1$$

$$\Rightarrow \quad \left|\frac{K}{(-1+j)(5+5j)}\right| = 1$$

$$\Rightarrow \quad \left|\frac{K}{-5+5j-5j-5}\right| = 1$$

$$\because (-1+j)\left[(-1+j)^2 + 12 + 7(-1+j)\right]$$

$$= (-1+j)(5+5j)$$

$$\left|\frac{K}{-10}\right| = 1$$

$$K = 10$$

8. Ans. (b)

Solution: Using Characteristic equation $1 + GH = 0$

$$1 + \frac{K(s^2 - 2s + 2)}{s^2 + 2s + 2} = 0$$

$$\frac{K(s^2 - 2s + 2)}{s^2 + 2s + 2} = -1$$

$$K = -\frac{(s^2 + 2s + 2)}{s^2 - 2s + 2}$$

Take $\frac{dK}{ds} = 0$

$$\begin{bmatrix} (s^2 - 2s + 2)(2s + 2) \\ -(s^2 + 2s + 2)(2s - 2) = 0 \end{bmatrix}$$

$$-4s^2 + 8 = 0$$
$$4s^2 - 8 = 0$$
$$s^2 - 2 = 0$$
$$s^2 = 2$$
$$s = \pm\sqrt{2}$$

Angle of departure of complete root locus at a pole of $G(a)\,H(s)$ the behavior of the root loci near that pole. For the RL these angle can be determined by the equation.

$$\angle G_1(s)\,H_1(s) = \sum_{i=1}^{m} s + Z_1 - \sum_{j=1}^{m} s + P_1$$
$$= (2K+1)\pi \quad \text{for} \quad 0 \le K < \infty \quad (1)$$

$$\angle G_1(s)\,H_1(s) = \sum_{i=1}^{m} s + Z_1 - \sum_{j=1}^{m} s + P_1$$
$$= 2K\pi \quad \text{for} \quad -\infty < K \le 0 \quad (2)$$

Where $K = 0, \pm1, \pm2...$

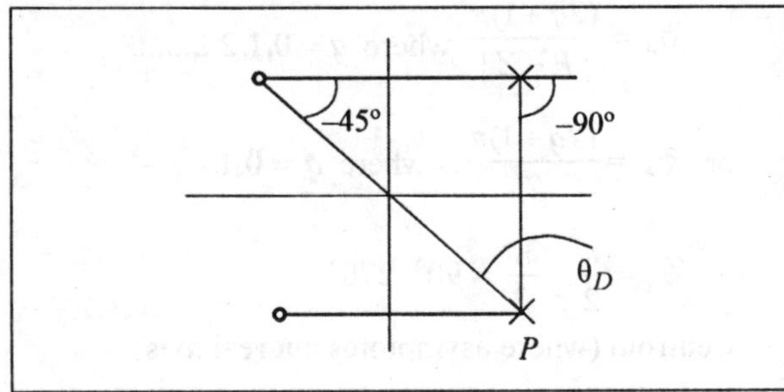

Using equation (2)

$$(-45^\circ - 0^\circ) - (-90^\circ + \theta_D) = 0$$
$$\theta_D = -45^\circ + 90^\circ = 45^\circ$$

9. Ans. (b)

Solution:

'X' $\rightarrow$ indicates pole

'O' $\rightarrow$ indicates zero

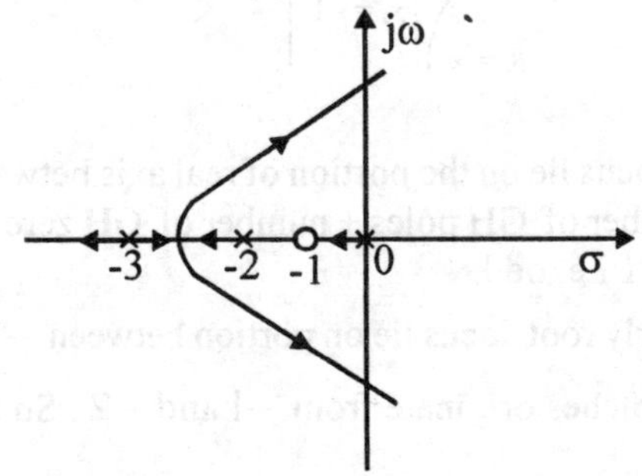

From the root loces plote, At $s = -3$, two root locus start one towards origin and other towards infinity on the real axis itself which indicate a multiple pole or two poles at $s = -3$. Hence the correct option is (b)

10. Ans. (a)

Solution: $\frac{K(s+a)}{s^2(s+b)}$; $b > a$

Poles are at $s = 0, 0, -b \Rightarrow P = 3$

Zeros are at $s = -a \quad \Rightarrow Z = 1$

Number of asymptote $|P - Z| = 3 - 1 = 2$

The asymptote's angles are,

$$\theta_K = \frac{[2K+1]}{P-Z}\pi; \quad K = 0, 1$$
$$= \frac{\pi}{2}, \frac{3\pi}{2} \text{ radians}$$
$$= 90^\circ, 270^\circ$$

The centroid is at, $\quad \sigma = \frac{-b+a}{2}$

The characteristic equation is

$$1 + KG(s) = 0$$

or $\quad \frac{1 + K(s+a)}{s^2(s+b)} = 0$

i.e. $\quad s^2(s+b) + K(s+a) = 0$

$$s^3 + bs^2 + Ks + Ka = 0$$

The Routh array is

s^3	1	K
s^2	b	Ka
s^1	$\frac{Kb - Ka}{b}$	0
s^0	Ka	0

The auxiliary equation is for $K = 0$

$$bs^2 = 0\,;\; s = 0$$

Hence the root locus meets the imaginary axis at $s = 0$

Also $\frac{a-b}{2} > \frac{-b-a}{2}$

Hence the centroid lies to the right of $a = -a$

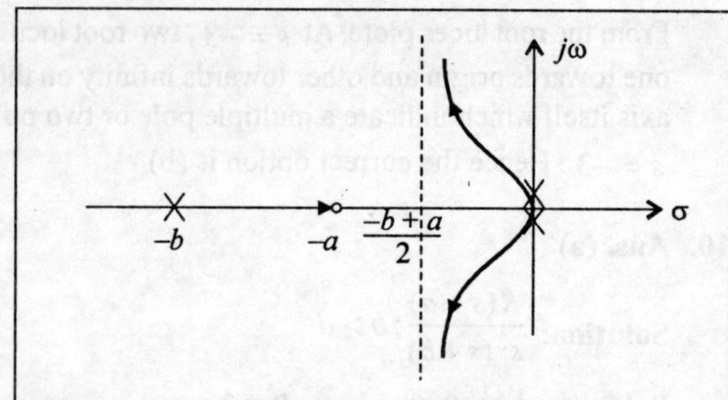

11. Ans. (a)

Solution: On the even section of real axis the root loci lie if the total number of poles and zeros of $G(s)\,H(s)$ to the right of section is odd. This is only applicable in the case of (a) therefore only (a) is a valid root loci since in (b), (c) and (d) the root loci lies to the left of an even number of zeros and poles.

12. Ans. (b)

Solution:

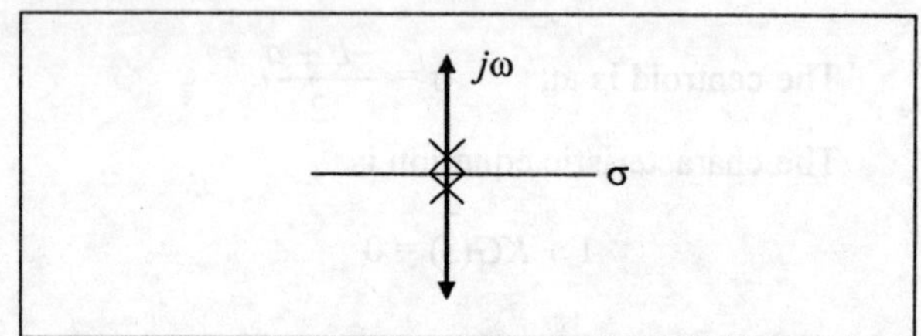

OLTF $\quad G(s) = \dfrac{K}{s^2}$

Number of asymptodes = |P–Z|

= 2

Centriod = 0

Angle of asymptote, $\phi_A = \dfrac{\pm(2K+1)\,180}{2-0}$

$= \pm 90$

Asxymptotes lie on imaginary axis

Hence root locus can be option (b).

13. Ans. (b)

Solution: Characteristics equation

$(s^2-4)\,(s+1) + K\,(s-1) = 0 \quad$ i.e. $\; 1 + GH = 0$

i.e. $1 + \dfrac{(s-1)}{(s^2-4)\,(s+1)} = 0$

zero of GH = 1

Poles of GH = –1; 2

Therefore the Root locus plot is as

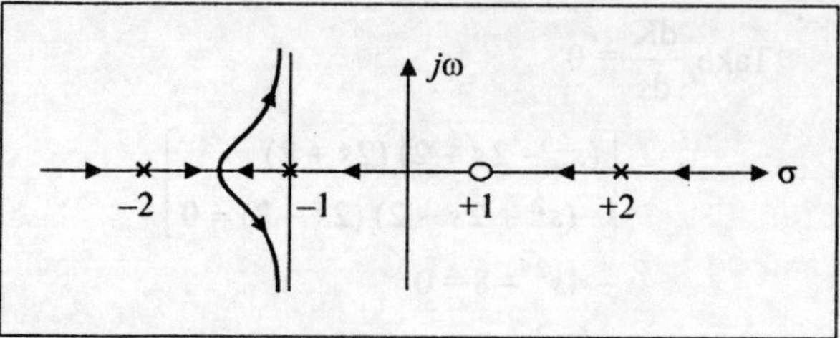

Second Method

Given characteristic equation

$$(s^2-4)(s+1) + K(s-1) = 0$$

or $\; 1 + \dfrac{K(s-1)}{(s^2-4)(s+1)} = 0$

Compare with Standared $1 + G(s)\,H(s) = 0$

$$GH(s) = \frac{K(s-1)}{(s^2-4)(s+1)}$$

Zero at $s = 1, \qquad Z = 1$

Poles at $s = \pm 2, -1 \quad P = 3$

Number of braches terminating to infinity

$= |P - Z| = 3 - 1 = 2$

$\phi_A = \dfrac{(2q+1)\pi}{|P-Z|}$ **where** $q = 0, 1, 2, \ldots\ldots,$

or $\phi_A = \dfrac{(2q+1)\pi}{2}$ **where** $q = 0, 1$

$\phi_A = \dfrac{\pi}{2}, \; \dfrac{3\pi}{2} = 90°, 270°$

Centroid (where asymptotes cut real axis)

$$\sigma_A = \frac{\Sigma(\text{Poles of GH}) - \Sigma(\text{Zeros of GH})}{|P-Z|}$$

$$= \frac{(2-2-1)-1}{2} = -1$$

Root locus lie on the portion of real axis between 1 and 2 as number of GH poles + number of GH zero in its R.H. side is 1 i.e. odd.

Similarly root locus lie on portion between −1 and −2.

Two braches originate from −1 and −2. So there must

be a breakaway point between -1 and -2. Breakaway point is given by $\frac{dK}{ds}=0$.

14. Ans. (c)

Solution:

Characteristic equation

$$\Rightarrow s(s+1)(s+3)+K(s+2)=0$$

$$1+GH=0$$

$$GH=\frac{K(s+2)}{s(s+1)(s+3)}$$

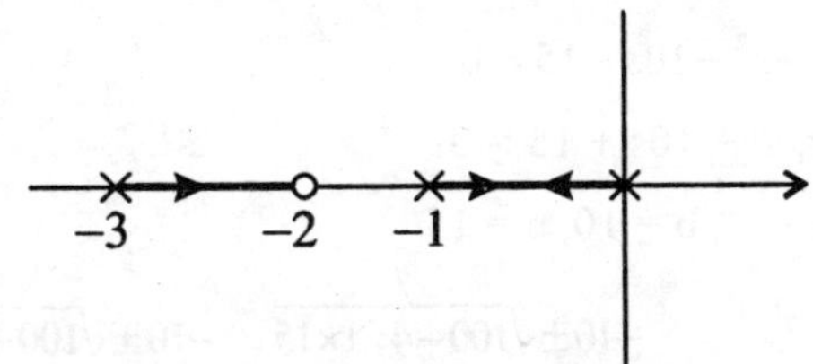

Intersection of assymptodes with real axis

$$\sigma=\frac{\Sigma P-\Sigma Z}{|P-Z|}$$

$$=\frac{(0-1,-3)-(-2)}{|3-1|}$$

$$=\frac{-4+2}{2}$$

$$=\frac{-2}{2}=-1$$

15. Ans. (a)

Solution:

Number of asyptotes

$=|P-Z|$

$=|3-1|=2$

Centroid, $\sigma=\frac{\Sigma P-\Sigma Z}{P-Z}$

$$\sigma=\frac{-2-\left(-\frac{2}{3}\right)}{3-1}=\frac{-6+2}{6}=\frac{-4}{6}=-\frac{2}{3}$$

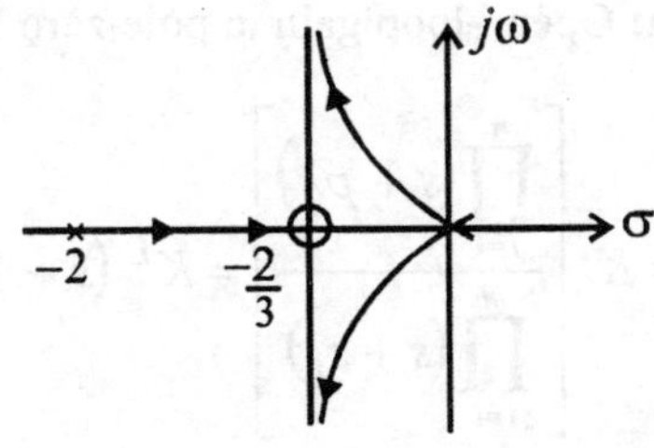

Angle of Asmptotes $=\frac{(2n\pm1)180}{P-Z}$

$$\theta_1=\frac{180}{2}=90$$

$$\theta_2=\frac{18\times3}{2}=270^\circ$$

If $K\to\infty$, roots are $\frac{-2}{3}$, $\frac{-2}{3}$, $\pm j\infty$

It is clear from the root locus diagram that when $K\to\infty$ al three roots are in left side of s-plane and have equal real parts.

16. Ans. (b)

Solution: Root locus start from poles and terminate at zero of GH. In present case pole are at $s=0$ and $s=-4$ and zero at $s=-8$.

17. Ans. (b)

Solution: $G(s)\,X(s)=\frac{K(s+10)}{(s+2)(s+5)}$

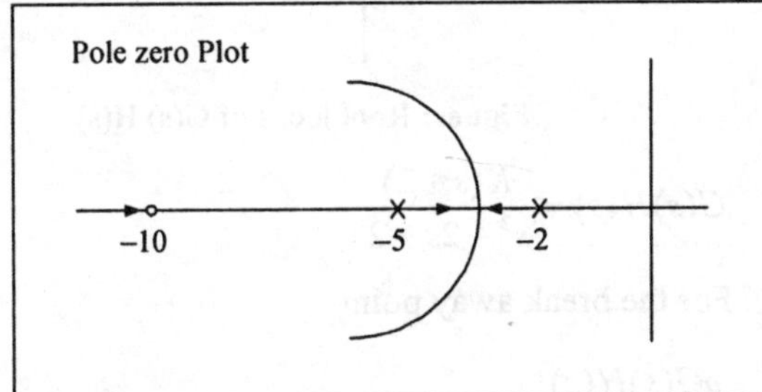

The characteristic equation:

$$1+G(s)\,H(s)=0$$

$$\frac{1+K(s+10)}{(s+2)(s+5)}=0$$

or $(s+2)(s+5)+K(s+10)=0$

or $s^2+7s+10+K(s+10)=0$

$$\Rightarrow K=\frac{-(s^2+7s+10)}{(s+10)}$$

For breakaway point, put $\frac{dK}{ds}=0$

$$\frac{dK}{ds}=-\left[\frac{(s+10)(2s+7)-(s^2+7s+10)(1)}{(s+10)^2}\right]$$

$$=-\left[\frac{2s^2+7s+20s+70-s^2-7s+0}{(s+10)^2}\right]$$

$$= -\left[\frac{s^2+20s+60}{(s+10)^2}\right]$$

$$\Rightarrow \quad -\frac{(s^2+20s+60)}{(s+10)^2} = 0$$

$s^2 + 20s + 60 = 0$

$$= s + -20 \pm \sqrt{\frac{(20)^2 - 4(1)(60)}{2}}$$

$$= \frac{-20 \pm \sqrt{160}}{2}$$

$$= -3.6, 16.4$$

18. Ans (c)

Solution:

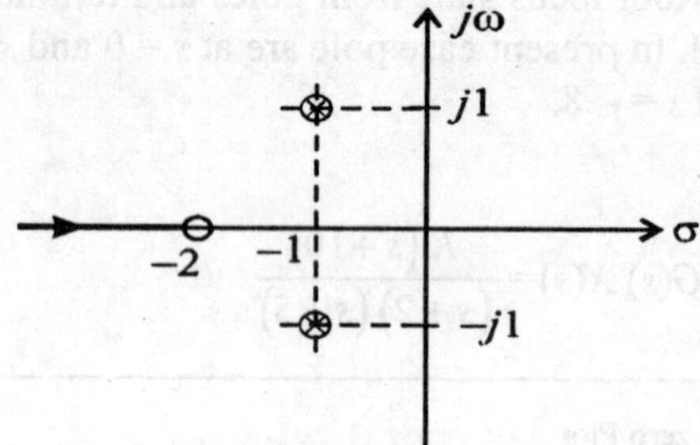

Figuer: Root locus of G(s) H(s)

$$G(s)H(s) = \frac{K(s+2)}{s^2+2s+2}$$

For the break away point

$$\frac{dG(s)H(s)}{ds} = 0$$

$$\frac{K(s^2+2s+2)1 - K(s+2)(2s+2)}{(S^2+2s+2)^2} = 0$$

or $s^2 + 4s + 2 = 0$

$$s = \frac{-4 \pm 2\sqrt{2}}{2} = -2 \pm \sqrt{2}$$

$s = -3.414$ and -0.59

From the root locus plot it indicate that the vaild break - away point is $= -3.414$

19. Ans. (c)

Solution: $s(s+3) + k(s+5) = 0$

$k(s+5) = -s(s+3) = -s^2 - 3s$

$$k = \frac{-s^2 - 3s}{(s+5)}$$

$$\frac{dk}{ds} = \frac{(-2s-3)(s+5) - (-s^2-3s)(1)}{(s+5)^2}$$

$$\frac{dk}{ds} = \frac{-2s^2 - 3s - 10s - 15 + s^2 + 3s}{(s+5)^2}$$

$$\frac{dK}{ds} = \frac{-s^2 - 10s - 15}{(s+5)^2}$$

put $\frac{dk}{ds} = 0$

$-s^2 - 10s - 15 = 0$

or, $s^2 + 10s + 15 = 0$

$a = 1$, $b = 10$, $c = 15$

$$s = \frac{-10 \pm \sqrt{100 - 4\times1\times15}}{2} = \frac{-10 \pm \sqrt{100-60}}{2}$$

$$= \frac{-10 \pm 2\sqrt{10}}{2} = -5 + \sqrt{10}$$

20. Ans. (b)

Solution:

At $s = 0 \pm \sqrt{2}\,j$, the system will have maximum controller gain and is,

$$k = \left|\frac{s(s+1)(s+2)}{2}\right|_{s=\sqrt{2}j}$$

$$= \frac{\left|\left(\sqrt{2}j\right)\right|\left(1+\sqrt{2}j\right)\left(2+\sqrt{2}j\right)\Big|}{2} = 3$$

21. Ans. (d)

Solution:

At $s = -0.425$, the system have maximum non-oscillatory response, because after the system will become oscillatory since roots will be become complex.

$$\text{Then } k = \left|\frac{s(s+1)(s+2)}{2}\right|_{s=-0.425} = 0.2$$

22. Ans. (c)

Solution: Open –loop gain in pole-zero form

$$K = K'\left[\frac{\prod_{j=1}^{n}(s+p_j)}{\prod_{i=1}^{m}(s+z_i)}\right] = K's(s+4)(s+8)$$

At $\omega = 4\sqrt{2}$ rad/s and $K = 384$

$$\therefore\ 384 = K'\left(4\sqrt{2}\right)\left(4+4\sqrt{2}j\right)\left(8+4\sqrt{2}\,j\right)$$

$$384 = K'\left(4\sqrt{2}\right)\sqrt{4^2+\left(4\sqrt{2}\right)^2}\sqrt{8^2+\left(4\sqrt{2}\right)^2}$$

$$= K'\left(4\sqrt{2}\right)\sqrt{4\sqrt{2}}\sqrt{16+32}\sqrt{64+32}$$

$$= K'\left(4\sqrt{2}\right)\sqrt{48}\sqrt{96}$$

$$= K'\left(4\sqrt{2}\right)4\sqrt{3}\,4\sqrt{6}$$

$$= K'\left(4\sqrt{2}\right)16\sqrt{3\times 6}$$

$$= K'\left(4\times\sqrt{2}\right)16\sqrt{18}$$

$$= K'4\sqrt{2}\times 48\sqrt{2}$$

$$= K'4\times 48\times 2$$

$$\Rightarrow\ K' = 1$$

At, $s = -15 + j\,1.5$

$$K = \left|(1)(-1.5+j1.5)(2.5+j1.5)(6.5+j1.5)\right|$$

$$= 41.25$$

23. Ans. (b)

Solution:

From the given statement

$$T(s) = \frac{k}{(s+1)(s+2)(s+3)}$$

Dc gain is unity

For DC gain, put $s = 0$

$$\therefore\ 1 = \frac{K}{6}$$

or $k = 6$

Therefore $T(s) = \dfrac{6}{(s+1)(s+2)(s+3)}$

24. Ans. (c)

Solution:

Characteristic equation

$$1 + G(s)H(s) = 0$$

$$1 + \frac{6K}{(s+1)(s+2)(s+3)} = 0$$

$$s^3 + 6s^2 + 11s + 6 + 6K = 0$$

Apply RH criterion

s^3	1	11
s^2	6	$6+6K$
s^1		
s^0		

Root locus intersection with imaginary axis is obtained from 2nd order simple auxiliary equation, i.e.

According to question

$$6s^2 + 6 + 6K = 0$$

$$s = \left(\pm j\sqrt{11}\right)$$

$$\Rightarrow\ j\sqrt{11}$$

$$6\left(j\sqrt{11}\right)^2 + 6 + 6K = 0$$

$$-66^2 + 6 + 6K = 0$$

$$\therefore\ \mathrm{K} = 10$$

10.5 Frequency Response

1. Ans. (c)

Solution:

$$G(s)\,H(s) = \frac{1}{(s+1)^3} = \frac{1}{1-3\omega^2+3j\omega-j\omega^3}$$

$$\angle GH(j\omega) = \tan^{-1}\frac{3\omega-\omega^3}{1-3\omega^2} = -3\tan^{-1}\omega$$

$$\phi_m = 180° + \angle GH\,(j\omega)$$

At phase cross over frequency the phase margin is zero

Therefore, $\phi_m = 0°$

$$0 = 180° + \angle GH\,(j\omega)$$

$$-180 = \angle GH\,(j\omega),$$

therefore,

$$-3\tan^{-1}\omega = -180°$$

$$\mathrm{Tan}^{-1}\,\omega = 60°$$

$$\omega = \tan 60° = \sqrt{3}$$

Let $\omega = \tan x$

$$\text{Gain Margin} = \frac{1}{|G(s)\,H(s)|}$$

$$= |\,1 - 3\omega^2 + 3\,j\omega - j\omega^3\,|$$

$$= |\,1 - 3\,(\sqrt{3}\,)^2 + 3\times\sqrt{3} - (\sqrt{3}\,)^3\,|$$

$$= |\,1 - 3\times 3 + 3\sqrt{3} - 3\sqrt{3}\,|$$

$$= |\,1 - 9\,| = 8$$

2. Ans. (d)

Solution: $G(j\omega) = \dfrac{\sqrt{2}}{j\omega\,(j\omega+1)} = \dfrac{\sqrt{2}}{\omega(j-\omega)} \times \dfrac{(j+\omega)}{(j+\omega)}$

Equating imaginary part = 0

$$= \frac{-\sqrt{2}\,(j+\omega)}{\omega\,(1+\omega^2)}$$

$$= \frac{\sqrt{2}(j+\omega)}{\omega\,(1+\omega^2)} = \frac{\sqrt{2}\omega}{\omega\,(1+\omega^2)} - \frac{\sqrt{2}j}{\omega\,(1+\omega^2)}$$

$$\frac{-\sqrt{2}\,j}{\omega\,(1+\omega^2)} \to 0$$

Only when $\omega \to \infty$

$$\text{GM} = -20\log\left|\frac{-\sqrt{(2\omega)}}{\omega\,(1+\omega^2)}\right|_{\omega=\infty}$$

$$= -20\log 0 = \infty$$

Second Method

Given open loop transfer functin

$$G(s)H(s) = \frac{\sqrt{2}}{s(s+1)}$$

The polar plot of function $\dfrac{K}{s(1+sT)}$

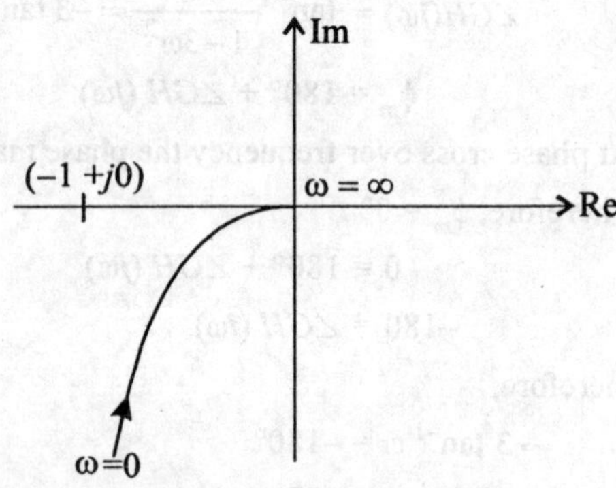

The phase cross over frequency $\omega_p = \infty$

At $\omega_p = \infty$ with phase $\angle 180°$

The gain of the system at $\omega_p = 0$

So gain margin $= \dfrac{1}{|\text{gain}|_{\omega=\omega_p}}$

$$\Rightarrow \text{G.M.} = \frac{1}{0} = \infty$$

3. Ans. (d)

Solution: $G(s)\,H(s) = \dfrac{2\sqrt{3}}{s(s+1)}$

For frequency response put $s = j\omega$

At gain cross over frequency $\omega = 1$

or, $\left|\dfrac{2\sqrt{3}}{j\omega(1+j\omega)}\right| = 1$

or, $\left|\dfrac{2\sqrt{3}}{\omega\sqrt{1+\omega^2}}\right| = 1$

or, $2\sqrt{3} = \omega\sqrt{1+\omega^2}$

$$\omega_g^2\left(1+\omega_g^2\right) = 12$$

$$\omega_g^4 + \omega_g^2 = 12$$

$$\omega_g^2 = \frac{-1 \pm \sqrt{1+48}}{2}$$

$$= \frac{-1 \pm 7}{2} = -4, 3$$

$$\omega_g = \sqrt{3}\text{ rad}$$

$$\angle G(j\omega)\,H(j\omega) = -90 - \tan^{-1}\omega$$

$$= -90° - \tan^{-1}\sqrt{3}$$

$$= -90 - 60° = -150°$$

$$PM = 180 + \angle G(j\omega)\,H(j\omega)$$

$$= 180 - 150 = 30°$$

4. Ans. (b)

Solution:

Given

$$G(s)H(s) = \frac{s}{(s+10)^3}$$

or $$G(j\omega)H(j\omega) = \frac{j\omega}{(j\omega+100)^3}$$

or $$|G(j\omega)H(j\omega)| = \frac{\omega}{\left(j\sqrt{\omega+10^4}\right)^3}$$

and $$\angle G(j\omega)H(j\omega) = 90° - 3\left[\tan^{-1}\frac{\omega}{100}\right]$$

At phase crossover

$$\angle G\left(j\omega_{pc}\right)H\left(j\omega_{pc}\right) = -180°$$

$$90° - 3\tan^{-1}\frac{\omega_{pc}}{100} = -180°$$

$$3\tan^{-1}\frac{\omega_{pc}}{100} = 270°$$

$$\tan^{-1}\frac{\omega_{pc}}{100} = 90°$$

or $$\frac{\omega_{pc}}{100} = \tan 90° = \infty$$

or $$\omega_{pc} = \infty$$

Gain margin of system

$$GM = \frac{1}{|G(\omega_{pc})\,H(\omega_{pc})|}$$

or $$GM = \frac{\left(\sqrt{\omega_{pc}^2 + (10^4)}\right)^3}{\omega_{pc}}$$

$$GM = \infty$$

At gain crossover frequency

$$G(j\omega_{gc})\,H(j\omega_{gc}) = 1$$

$$|G(j\omega_{gc})\,H(j\omega_{gc})| = \frac{1}{\omega_{gc}^2\left[1+\frac{10^4}{\omega_{gc}}\right]^{3/2}}$$

Here magnitude of G(jω) H(jω) can never be more than or equal to unity when frequency is varied from 0 to ∞. So , Phase margin PM = ∞

So gain over frequency does not exist so gain is never unity or 0 dB.

5. **Ans. (b)**

Solution: $$GH(s) = \frac{2(1+s)}{s^2}$$

$$GH(j\omega) = \frac{2(1+j\omega)}{(j\omega)^2} = \frac{2(1+j\omega)}{-\omega^2}$$

The interaction with real axis gives the phase cross over frequency.

$$Im(GH(j\omega)) = 0 \Rightarrow \omega = 0$$

$$\text{Real } G\,H(j0) = \frac{2}{-0} = -\infty$$

$$GM = \frac{1}{|\text{Real } GH(j\omega)|} = \frac{1}{\infty} = 0$$

Another way

$$\angle G(s)\,H(s) = -180 + \tan^{-1}\omega$$

For phase cross over frequency

$$-180 + \tan^{-1}\omega = -180$$

$$\omega = 0$$

$$|G(s)\,H(s)|_{\omega=0} = \frac{2\sqrt{1+\omega^2}}{\omega^2} = \infty = a$$

$$GM = \frac{1}{|a|} = \frac{1}{\infty} = 0$$

In present case GM is not asked into db. Hence correct option is (b) only.

6. **Ans. (a)**

Solution: Transfer function

$$= G(s) = \frac{\left(\frac{s}{5}+1\right)\left(\frac{s}{100}+1\right)\left(\frac{s}{200}+1\right)}{\left(\frac{s}{.01}+1\right)\left(\frac{s}{1}+1\right)\left(\frac{s}{80}+1\right)}$$

$$\angle G(j\omega) = \tan^{-1}\frac{\omega}{5} + \tan^{-1}\frac{\omega}{100} + \tan^{-1}\frac{\omega}{200} - \tan^{-1}\frac{\omega}{.01} - \tan^{-1} - \tan^{-1}\frac{\omega}{80}$$

At ω = 20, transfer function. $\angle G(j\omega)$

$$= \tan^{-1}(4) + \tan^{-1}\frac{1}{5} + \tan^{-1}\frac{1}{10} - \tan^{-1}(2000) - \tan^{-1}20 - \tan^{-1}\frac{1}{4} = -90°$$

7. **Ans (d)**

Solution: Gain crossover frequency gain of

$$\frac{3e^{-2s}}{s(s+2)} = \left|\frac{3e^{-2s}}{s(s+2)}\right| = 1$$

$$e^{-2s} = \cos(2\omega) - j\,\text{Sin}\,2\omega\ ;\ |e^{-2s}| = 1$$

$$\left|\frac{3}{j\omega\,(2+j\omega 2)}\right| = 1\ ;\ \left|\frac{3}{-\omega^2 + j2\omega}\right| = 1$$

$$\omega = 1.26$$

Phase crossover frequency

$$\text{Phase angle} = -90 - \tan^{-1}\frac{\omega}{2} - 2\omega\frac{180}{\pi}$$

$$\text{Phase} = -180 \Rightarrow -90 - \tan^{-1}\frac{\omega}{2} - 2\omega\frac{180}{\pi} = -180$$

$$\omega = 0.632$$

8. **Ans. (d)**

Solution: Gain margin

$$A = \text{Gain} = \left|\frac{3e^{-2s}}{S(s+2)}\right|_{\omega=0.632} = 2.2631$$

$$\text{GM} = 20\log\frac{1}{a} = -7.094\text{ dB}$$

$$\text{Phase margin} = 180 + [-90 - \tan^{-1}\frac{\omega}{2} - 2\omega\frac{180}{\pi}]$$

$$\omega = 1.26$$

Phase margin = – 86.59°

9. Ans. (d)

Solution: $\phi = -\tan^{-1}\frac{\omega}{1} - \tan^{-1}\frac{\omega}{2}$

So either $\omega = 0$ or $\omega = \infty$ for ϕ to be $-180°$

But $\omega = \infty$ will given G.M zero and in dB it will ∞ db

10. Ans. (b)

Solution:

For $G_1(s) = \frac{1}{s^2 + as + b}$

The characteristic equation

$s^2 + as + b = 0$

$$s = \frac{-b \pm \sqrt{a^2 - 4b}}{2}$$

Roots $s_1 = \frac{-b \pm \sqrt{a^2 - 4b}}{2} = -\alpha$ (say)

and $s_2 = \frac{-b - \sqrt{a^2 - 4b}}{2} = -\beta$

Where $\alpha = \frac{b - \sqrt{a^2 - 4b}}{2}$

and $\beta = \frac{b + \sqrt{a^2 - 4b}}{2}$

$$G_1(s) = \frac{1}{(s - s_1)(s - s_2)}$$

$$= \frac{1}{\alpha\beta} \frac{1}{\left(1 + j\frac{\omega}{\alpha}\right)\left(1 + j\frac{\omega}{\beta}\right)}$$

At frequencies $\omega = \alpha, \beta$ f gain is reduced –3dB

So β = upper 3-dB frequency

α = Lower 3-dB frequency

3-dB Bandwidth

$$BW = \beta - \alpha = \sqrt{a^2 - 4b}$$

Similarly for $G_2(s) = \frac{s}{s^2 + as + b}$

$$= \frac{s}{(s + \alpha)(s + \beta)}$$

Due to zero at origin, i.e. s = 0 gain increase with frequency. While due to poles α and β gain decrease

So α, β are 3-dB frequencies

Hence $BW = \beta - \alpha = \sqrt{a^2 - 4b}$

11. Ans. (c)

Solution: We have $G(s) = \frac{as + 1}{s^2}$

$$\angle G(j\omega) = \tan^{-1}(\omega a) - \pi$$

or $\frac{\pi}{4} = \pi + \tan^{-1}(\omega_g a) - \pi$

or $\frac{\pi}{4} = \tan^{-1}(\omega_g a)$

or $a\omega_g = 1$

At gain crossover frequency $|G(j\omega_g)| = 1$

Thus $\frac{\sqrt{1 + a^2\omega_g^2}}{\omega_g^2} = 1$

or $\sqrt{1+1} = \omega_g^2$

or $\omega_g = (2)^{1/4}$

Now $a = \frac{1}{2^{1/4}} = 0.84$

12. Ans. (c)

Solution: $Y(s) = G(s)\ U(s) = \frac{as + 1}{s^2} U(s)$

$LT\ (t) = 1$

Output $LT \rightarrow \frac{as + 1}{s^2} = \frac{a}{s} + \frac{1}{s^2}$

$\Rightarrow$ $au(t) + r(t) = a\,u(t) + t\,.\,u(t)$

$\Rightarrow$ $a + 1 \Rightarrow 1.84$

13. Ans. (a)

Solution: Given, at $\omega = 0$

Magnitude of frequency response $K = 5$

Resonant frequency $\rightarrow \omega_r = \omega_n\sqrt{1 - 2\delta^2}$

Resonant Peak, $M_r = \frac{K}{2\delta\sqrt{1 - \delta^2}}$

Given, $\omega_r = 5\sqrt{2}$, and $M_r = \frac{10}{\sqrt{3}}$

Checking the option

$$T(s) = \frac{500}{s^2 + 10s + 100} \; ; \; \omega_n = 10, \; \delta = \frac{1}{2}$$

At $\omega = 0$

T (0) = 5

$K = 5$

$$\therefore \quad \omega_r = 10\sqrt{1 - 2\left(\frac{1}{4}\right)} = 5\sqrt{2}$$

$$\text{and } M_r = \frac{5}{2 \cdot \frac{1}{2}\sqrt{1 - \frac{1}{4}}} = \frac{10}{\sqrt{3}}$$

Hence this option satisfie the answer.

14. Ans. (c)

Solution: $H(s) = \dfrac{100}{s(s+10)^2}$

Phase cross over frequency

$$= -90 - 2\tan^{-1}\left(\frac{\omega}{10}\right) = -180^o$$

$$\text{or, } -2\tan^{-1}\left(\frac{\omega}{10}\right) = -90^o$$

$$\text{or } \tan^{-1}\left(\frac{\omega}{10}\right) = 45^o \text{ rad/sec}$$

or, $\omega = 10$ rad/sec

$$|H(j\omega)|_{\omega=\omega_c} = \left|\frac{100}{j10(j10+10)^2}\right| = \left|\frac{100}{10\sqrt{(100)^2 + (10)^2}}\right|$$

$$= \frac{100}{10\left(\sqrt{200}\right)^2} = \frac{100}{10 \times 200}$$

$$= \frac{1}{10 \times 2} = \frac{1}{20}$$

$$GM\, 20\log\frac{1}{|H(j\omega_c)|} = 20\log\frac{1}{1/20}$$

$$= 20\log 20 = 26dB$$

15. Ans. (d)

Solution: Option (D) does not fix for the given transfer function.

16. Ans. (b)

Solution:

$$|y(t)| = |M(\omega)||x(t)|$$

$$M(\omega) = \frac{y(\omega)}{X(s)} = \frac{1}{P}$$

$$|M(\omega)| = \frac{\omega}{\sqrt{\omega^2 + P^2}}$$

$$1 = \frac{\omega p}{\sqrt{\omega^2 + p^2}}$$

$$\omega^2 + p^2 = \omega^2 p^2$$

$$4 + p^2 = 4p^2$$

$$3p^2 = 4, \; p = \frac{2}{\sqrt{3}}$$

Second Method

Given $\omega = 2\,\text{rad/s}$

Phase difference between input and output,

$$\phi = -\frac{\pi}{3} - \left(-\frac{\pi}{2}\right) = \frac{\pi}{6} = 30^\circ$$

Phase angle from transfer function,

$$\phi = 90^\circ - \tan^{-1}\frac{\omega}{p}$$

$$90^\circ - \tan^{-1}\frac{2}{p} = 30^\circ$$

$$\tan^{-1}\frac{2}{p} = 60^\circ$$

$$\frac{2}{p} = \tan 60^\circ = \sqrt{3}$$

$$p = \frac{2}{\sqrt{3}}$$

17. Ans. (c)

Solution:

The magnitude of sinusoidal steady state response depends on magnitude of transfer function only.

output

$$y(t) = |G(j\omega)|\sin\left(\omega t + \angle G(j\omega)\right)$$

Putting $s = j\omega$ in given equation

$$G(j\omega) = \frac{\left[(j\omega)^2 + 9\right](j\omega + 2)}{(j\omega + 1)(j\omega + 3)(j\omega + 4)}$$

$$|G(j\omega)| = \frac{(9 - \omega^2)\sqrt{4 + \omega^2}}{\sqrt{\omega^2 + 1}\sqrt{\omega^2 + 6}\sqrt{16 + \omega^2}} = 0$$

For $y(t) = 0$

$|G(j\omega)|$ can be zero

Hence $G(j\omega)=0$

at $\omega^2=9$

or, $\omega=3 rad/s$

18. Ans. (c)

Solution: For phase margin, find the cut off frequency for gain = 1

The value of ω at which $|G(j\omega)|=1$ is given by,

$$1=\frac{\sqrt{(\omega_c a)^2+1}}{\omega_c^2} \quad \text{... (1)}$$

$$PM=\angle G(j\omega)|_{\omega=\omega_c}+180°=45°$$

or $\tan^{-1}(\omega_c a)-180°+180°=45°$

or $\omega_c a=1\Rightarrow a=\frac{1}{\omega_C}$

From equation (i), we have,

$$\omega_c^2=\sqrt{(\omega_c a)^2+1}$$

$$\omega_c^4=(\omega_c a)^2+1$$

$$\omega_c^4=1+1=2$$

$$\omega_c=(2)^{1/4}$$

$$\omega_c=2^{1/4}=1.189 \text{ rad/sec.}$$

Then, $a=\frac{1}{\omega_C}$

$$=\frac{1}{1.189}$$

$$=0.841.$$

19. Ans. (c)

Solution: Here $G(s)=\frac{K+0.366s}{s(s+1)}$, in present case, unity feedback is applied

Since, phase margin = 180° + ∠G(jω) H(jω)

$$60°=180°+\tan^{-1}\left(\frac{0.366\omega}{K}\right)-90°-\tan^{-1}\omega$$

As given gain crossover frequency $\omega_g=1$

or $60=180-135+\tan^{-1}\left(\frac{0.366}{K}\right)$

∴ $K=1.366$

20. Ans. (d)

Solution:

At Phase crossover frequency

$$-90-0.1\omega\times\frac{180}{\pi}=-180$$

$$\therefore \omega=15.7$$

$$a=|G(j\omega)|_{\omega=15.7}$$

$$GM=20\log a=23.9dB$$

Second Method

$$G(s)=\frac{e^{-0.1s}}{s}$$

put $s=j\omega$

$$G(j\omega)=\frac{e^{-0.1j\omega}}{j\omega}$$

For gain margin we need to find phase crossover frequency,

Since, there is a pole at origin the initial slope of magnitude plot will – 20 db/decade. $e^{-0.1\omega}$ has no effect an magnitude pole.

Phase Plot:

$$\angle e^{-0.1j\omega}=-57.3\times 0.1\omega \quad \left[\begin{matrix}1\,radian\\=57.3°\end{matrix}\right]$$

$$=-5.7\omega$$

$$\angle G(j\omega)H(j\omega)=-5.7\omega-\tan^{-1}\frac{\omega}{0}$$

$$-180°=-5.7\omega-\tan^{-1}\frac{\omega}{0}$$

$$-180°=-5.7\omega-90°$$

$$+90°=+5.7\omega$$

$$\omega=\frac{90}{5.7}$$

$$\omega=15.78\, radian/\sec$$

phase crossover frequency = 15.78 radian/sec

Now, calculate magnitude at this frequency

$$G(j\omega)H(j\omega)=\frac{e^{-0.1j\times15.78}}{j\times15.78}$$

$$=\frac{e^{-1.578j}}{15.78j}$$

$$=\frac{\cos(-1.578)-j\sin(-1.578)}{15.78j}$$

$$= \frac{\cos 1.578 + j \sin 1.578}{j15.78}$$

$$G(j\omega)H(j\omega) = \frac{0.99 + j0.027}{j15.78}$$

$$|G(j\omega)H(j\omega)| = \sqrt{\frac{(0.99)^2 + (0.027)^2}{\sqrt{(15.78)^2}}}$$

$$= \sqrt{\frac{0.9801 + 0.000729}{15.78}}$$

$$= \frac{\sqrt{0.98}}{15.78}$$

$$|G(j\omega)H(j\omega)|_{\omega=\omega_p} = 0.0627$$

$$\text{Gain margin} = 20 \log_{10} \frac{1}{|G(j\omega)H(j\omega)|_{\omega=\omega_p}}$$

$$= 20 \log_{10} \frac{1}{0.0627}$$

$$= 20 \log_{10} 15.94$$

$$= 20 \times 1.202$$

$$= 24\,dB \cong 23.9\,dB$$

21. Ans. (b)

Solution: Open loop system stability is depends only on pole locations ⇒ system is stable

There is one zero on right half of s-plane so system is non-minimum phase type

22. Ans. (a)

Solution:

Gain Margin are find at phase crossover frequency at which phase plote cross the $-180°$ phase line.

Hence $M = G(j\omega)|_{\omega_p} = 0.5$

$At\ \angle G(j\omega) = -180$

⇒ magnitude M = 0.5

So G.M $= 20 \log\left(\frac{1}{0.5}\right) = 6\,dB$

At $|G(j\omega)| = 1$

⇒ Phase angle $\angle G(j\omega) = -150$

So P.M $= 180 + \phi_m$

$= 180 + (-150)$

$= 30°$

23. Ans. (b)

Solution: The inter section with the real axis gives the phase cross over frequency, at this

$$I_m[G(j\omega)H(j\omega)] = 0$$

OLTP, $$G(s)H(s) = \frac{1}{(s+1)^3}$$

Putting $s = j\omega$, we have

$$G(j\omega)H(j\omega) = \frac{1}{(j\omega+1)^3}$$

$$= \frac{(1-j\omega)^3}{(j\omega+1)^3(1-j\omega)^3} = \frac{(1-j\omega)^3}{(1+\omega^2)^3}$$

$$= \frac{(1-j\omega)(1-\omega^2-2j\omega)}{(1+\omega^2)^3}$$

$$= \frac{(1-j\omega-\omega^2+j\omega^3-2j\omega-2\omega^2)}{(1+\omega^2)^3}$$

$$= \frac{1-3\omega^2+j(\omega^2-3\omega)}{(1+\omega^2)^3}$$

Equation imaginary part to zero, we have

$$(\omega^2 - 3\omega) = 0$$

or $$\omega^2 = 3$$

or $$\omega_c = \pm\sqrt{3}$$

Gain margin, $$GM = \left|\frac{1}{G(j\omega_c)H(j\omega_c)}\right|_{\omega=\sqrt{3}}$$

$$= \left|\frac{(1+\omega_c^2)^3}{1-3\omega_c^2}\right|_{\omega_c=\sqrt{3}}$$

$$= \left|\frac{(1+3)^3}{-8}\right| = \frac{4^3}{8} = 8$$

24. Ans. (b)

Solution: $$GH = \frac{1}{(j\omega+1)^3}$$

$$\phi = -3\tan^{-1}\omega$$

$$|G(j\omega)| = 1$$

$$[1+\omega^2]^3 = 1$$

Put $$\omega^2 = x$$

$$(1+x)^3 = 1$$
$$x = 0$$
$$\omega = 0$$
$$\phi = -3\tan^{-1} 0 = 0$$
$$PM = 180 + \phi = 180 + 0 = 180^\circ$$

25. Ans. (b)

Solution: $GH = \dfrac{10e^{-Ls}}{s}$

$$e^{jx} = \cos x + j\sin x$$
$$\phi = \tan^{-1}\frac{\sin x}{\cos x} = x$$
$$\angle G(s)\,H(s)\,\Big|_{s=j\omega} = -90^o - \frac{(\omega L)180^o}{\pi}$$

At phase cross-over frequency,

$$\angle GH = -180^\circ$$

Then, $-90^o - \dfrac{(\omega L)180^o}{\pi} = -180^\circ$

At $\omega = 5$

Or $90^\circ = \dfrac{5(L)\cdot 180^o}{\pi}$

$\pi = 10\,L$

Or $L = \dfrac{\pi}{10}$

26. Ans. (c)

Solution:

$$\angle G(j\omega)\big|_{\omega=\omega_1} - \frac{(\omega T)180^0}{\pi} = -180^0$$
$$-90^0 + \frac{\omega T 180^0}{\pi} = 180^0$$

or $\omega T = \dfrac{\pi}{2}$

27. Ans (a)

Solution:

Given: $T(s) = \dfrac{k}{s(s+a)}$

$$\text{PM} = 60^\circ = 180^\circ + \angle T(jw)\big|_{\omega=1}$$

$\therefore$ $60^\circ = 180^\circ + \left(-90^\circ - \tan^{-1}\dfrac{\omega}{a}\right)$

$\Rightarrow$ $\tan^{-1}\dfrac{\omega}{a} = 30^\circ$

$\Rightarrow$ $\dfrac{\omega}{a} = \tan 30^\circ$

$\Rightarrow$ $a = \sqrt{3}\omega$

$\therefore$ $T'(s) = \dfrac{s-\sqrt{3}}{s+\sqrt{3}}\cdot\dfrac{k}{s(s+\sigma)}$

$$\angle T'(j\omega) = -2\tan^{-1}\frac{\omega}{\sqrt{3}} - 90^\circ - \tan^{-1}\frac{\omega}{a}$$

$\therefore$ $\text{PM} = 180^\circ + \angle T'(j\omega)\big|_{\omega=1}$

$$= 180^\circ - 2\tan^{-1}\frac{1}{\sqrt{3}} - 90^\circ - \tan^{-1}\frac{1}{\sqrt{3}} = 0^\circ$$

28. Ans. (b)

Solution:

$$G(s) = \frac{k}{(s+5)3}$$

At w_{gc}

$$\angle G(j\omega) = \phi$$
$$P.M. = 180 + \phi$$

Given $P.M. = 45^0$

$$45^0 = 180 + \phi$$
$$\phi = -135^0$$
$$\angle G(j\omega) = -3\tan^{-1}\left(\frac{\omega_{gC}}{5}\right) = -135$$
$$\tan^{-1}\left(\frac{\omega_{gc}}{5}\right) = 45$$
$$\frac{\omega_{gc}}{5} = 1$$
$$w_{gc} = 5$$

At w_{gc}

$$|G(j\omega) \Rightarrow 1|$$
$$G(j\omega) = \frac{K}{\left(\sqrt{\omega^2+25}\right)^3} = 1$$
$$K = \left(\sqrt{50}\right)^3$$
$$K = 50\times\sqrt{50} = 50\times 5\sqrt{2}$$
$$K = 250\sqrt{2}$$

29. Ans. (b)

30. Ans. (a)

10.6 Bode Plot

1. **Ans. (a)**

Solution: A transfer function which has one or more zero in the right half *S*-plane is known as non minimum-phase transfer function.

2. **Ans. (c)**

Solution: Given that in Bode plot $\angle G(j\omega) = -125°$

Phase margin = $180 + \angle G(j\omega)$

$= 180 - 125° = 55°$

3. **Ans. (a)**

Solution:

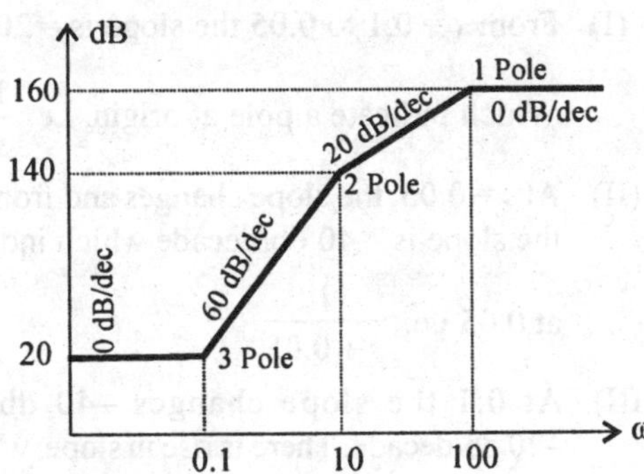

There are three corner frequencies i.e. at $\omega = 0.1, 10, 100\ rad/\sec$

The lowest corner frequency $= 0.1$

The initial slope i.e. slope at frequency less than lowest corner frequency $= 0\ \text{dB/decade}$

$-20n = 0$

$\Rightarrow n = 0$

So it is type –0 system.

The gain changes form +20 dB to +140 dB from 0.1 to 10 i.e. in two decades.

$(-0.1 \rightarrow 10 \Rightarrow 100)$

So there is charge in slope $= \dfrac{140-20}{2} = 60$ dB/dec.

Hence at $\omega = 0.1$, there are 3 - zeros as slope changes by +60 dB/decade.

The gain changes from + 140 dB to +160 dB from 10 to 100 i.e. in one decade.

So slope change = +20 dB/dec

At $\omega = 10$, slope changes from +60 to +20 i.e. changes by –40 dB dec. Hence 2 - poles are added.

At $\omega = 100$ slope changes from 20 dB to 0 dB i.e. –20 dB/dec Hence one pole is added.

Finally open lop transfer function can be considered as

$$G(s) = \frac{K\left(1+\frac{s}{0.1}\right)^3}{\left(1+\frac{s}{10}\right)^2\left(1+\frac{s}{100}\right)}$$

At a frequency $\omega \le 0.1$ there is gain contribution only by K in asymptotic or Bode plot

$G(j\omega) = K$ if $\omega \le 0.1$

Magnitude

$M = |G(j\omega)| = K$

Gain = 20 dB

$20 \log M = 20 \log K = 20$

$K = 10$

$$G(s) = \frac{10\left(1+\frac{s}{0.1}\right)^3}{\left(1+\frac{s}{10}\right)^2\left(1+\frac{s}{100}\right)}$$

$$= \frac{10(s+0.1)^3}{(0.1)^3\left(\frac{1}{10}\right)^2\left(\frac{1}{100}\right)(s+10)^2(s+100)}$$

$$= \frac{10(s+0.1)^3}{(0.1)^3\,(0.1)^2(0.01)\,(s+10)^2(s+100)}$$

$$= 10^8\frac{(s+0.1)^2}{(s+10)^2(s+100)}$$

4. **Ans. (c)**

Solution: At ω = 1, slope changes form 0dB/decade to 20 dB/decade, and, at ω = 10, slope changes from 20 dB/decade to 0dB/decade and at ω = 100, slope changes form odB/decade to – 20 dB/ decade.

Hence, there is a zero at ω = 1 and poles at ω = 10,100.

Now, $$G(s) = \frac{K(s+1)}{(s+10)(s+100)}$$

$$G(j\omega) = \frac{K(j\omega+1)}{(j\omega+10)(j\omega+100)}$$

$$= \frac{K(j\omega+1)\cdot 10^{-3}}{\left(\frac{j\omega}{10}+1\right)\left(\frac{j\omega}{100}+1\right)}$$

Then, $20 \log (10^{-3} K) = -20$

or, $K = 10^2$

Hence, $G(s) = \dfrac{10^2(s+1)}{(s+10)(s+100)}$

Another way

$$20\log_{10}\left|G(j\omega)\,H(j\omega)\right|_{\omega=0} = -20$$

$$20\log_{10} 10^{-1} = -20$$

$$\therefore\ \left|G(j\omega)\,H(j\omega)\right|_{\omega=0} = (10)^{-1} = 0.1$$

$$T(0) = \frac{K(1)}{(10)(100)} = 0.1$$

$$\Rightarrow K = 10^2$$

5. Ans. (d)

Solution: At $\omega = 0.1$ gain is 60 db that is due to K and $\dfrac{1}{s}$

$\dfrac{1}{s}$ give gain= 20 db at 0.1 and 40 db due to K

Therefore, $20 \log K = 40 \Rightarrow K = 100$

$$T_1 = \frac{1}{1}$$

$$T_2 = \frac{1}{20} = 0.05$$

$$G(s) = \frac{K}{s(1+sT_1)(1+sT_2)}$$

$$= \frac{100}{s(1+s)\,(1+0.05s)}$$

Another way

$$G(s) = \frac{K}{s(s+1)(s+20)} = \frac{K}{s(s+1)20(s+0.05\,s)}$$

as there is slop in the begining of plot

$$G(s) = \frac{K/20}{s(s+1)\left(1+\dfrac{s}{20}\right)} = \frac{K^1}{s(s+1)\,(1+0.5\,s)}$$

As the term are in the $(1 + sT)$ form

$$20\log\left.\frac{K'}{\omega}\right|_{\omega=0.1} = 60\ db$$

$$= \left.\frac{K'}{\omega}\right|_{\omega=0.1} = (10)^3$$

or $\dfrac{K'}{0.1} = (10)^3$

or $K' = 100$

$$G(s) = \frac{100}{s(s+1)\,(1+0.5\,s)}$$

6. Ans. (a)

Solution: Standard transfer function from bode plot given

$$GH = \frac{K(1+j\omega T_1)}{(1+j\omega T_2)}$$

$$\omega_z = \frac{1}{T_1} = 0.1 \Rightarrow T_1 = 10$$

$$\omega_p = \frac{1}{T_2} = 0.1 \Rightarrow T_2 = 0.1$$

At $\omega \to 0,\ G_{dB} = 0$

Hence $K = 1$

$$G(s)\,H(s) = \frac{1+s\times10}{1+s\times0.1} = \frac{10\,s+1}{0.1s+1}$$

7. Ans. (b)

Solution:

(I) From $s = 0.1$ to 0.05 the slope is – 20 db/decade

Which indicate a pole at origin, i.e. $\dfrac{1}{s}$

(II) At $s = 0.05$, the slope changes and from 0.05 to 0.1 the slope is – 40 db/decade which indicate at pole at 0.05 i.e. $\dfrac{1}{s+0.05}$

(III) At 0.1 the slope changes –40 db/decade to –20 db/decade. There is rise in slope, which indicate a zero at 0.1, i.e. $(s + 0.1)$

Finally the transform function is

$$\frac{K(s+0.01)}{s(s+0.05)} = \frac{K_1(s+0.1)\,[10/10]}{s(s+0.05)\,[20/20]}$$

$$= \frac{2K_1(10s+1)}{s(20s+1)}$$

where $2K_1 = K$.

at 0.01 gain is 60 db = $20 \log K + 40$ db

$20 \log K = 20$ db; $K = 10$

$\therefore$ the correct answer is (b)

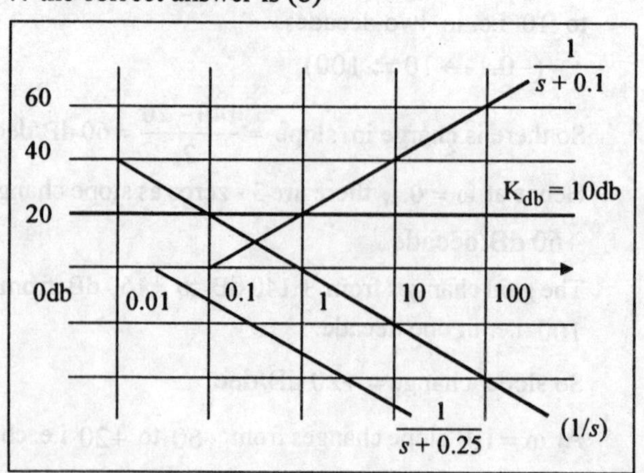

Second Method

From the given Bode Plot

$$G(s)\,H(s)=\frac{K\left(1+\frac{s}{0.1}\right)}{s\left(1+\frac{s}{0.05}\right)}$$

From the figure,

$$20\log K-20\log\omega\big|_{\omega=0.01}=60\ dB$$

$$K=10$$

$$G(s)\,H(s)=\frac{10(1+10s)}{s(1+20s)}$$

8. **Ans. (d)**
Solution: 6 db/octave = 20 db/decad
The Slop is + 20 db/dec is equal to placing a zero at f_1

$$A=1+\frac{jf}{f_1}$$

Second Method
Zero is present whose corner frequency is f_1

$$A=K(1+sT_1)$$

$$=K(1+j\omega T_1)$$

$$=K\left(1+\frac{j2\pi f}{2\pi f_1}\right)$$

$$20\log K=0\ dB$$

$\Rightarrow$ K = 1

$$A=\left(1+\frac{if}{f_1}\right)$$

9. **Ans. (d)**
Solution: Since at $\omega=2$ and $\omega=25$, slope changes from 40 dB to – 60 dB/dec at both the value. So there are poles at ω = 2 and ω = 25. Also at ω = 0.1 slope is – 40 dB/sec. It means, there is two poles at origin. Hence the transfer function, should be of the form

$$T(s)=\frac{K(s+5)}{s^2(s+2)(s+25)}$$

Now, at $\omega=0.1$

$$54=20\log\frac{5K}{(0.1)^2\times 50}$$

or $K=50$

thus $$T(s)=\frac{50\,(s+5)}{s^2(s+2)\,(s+25)}$$

Second Method
From given bode plot

$$G(s)H(s)=\frac{K\left(1+\frac{s}{5}\right)}{s^2\left(1+\frac{s}{5}\right)\left(1+\frac{s}{25}\right)} \qquad (1)$$

$$|G(s)H(s)|=20\log K-2(20\log\omega)$$

$$=54dB\big|_{\omega=0.1}$$

$$20\log K=54\ dB+2(20\log 0.1)$$

$$20\log K=14\ dB$$

$$K=5$$

Puting the value K and solving equaion (1)

$$G(s)H(s)=\frac{5(s+5)\,(2)\,(25)}{(5)s^2(s+2)(s+25)}$$

$$=\frac{50(s+5)}{s^2(s+2)(s+25)}$$

10. **Ans. (a)**

Solution: $\angle GH\big|_{\omega=0.5a}=-\tan^{-1}\frac{\omega}{a}$

$$=-\tan^{-1}0.5=26.56^\circ$$

From the plot,

At $\omega=a$, $\angle GH=22.5^\circ$

Then error $=-22.50^\circ-(-26.56^\circ)=4.06^\circ$

At $\omega=0.5\,a$, $|GH|_{dB}=20\log K-20\log|1+j0.5|$

$$=20\log K-0.97$$

From the plot, at $\omega=0.5\,a\;|GH|_{dB}$

$$=20\log|GH|$$

$$=20\log K$$

Then, error in dB gain = 0.97 dB

Second method
Given

For $G(s)=\dfrac{K}{1+\left(\frac{s}{a}\right)}$

Frequency response put $s=j\omega$

$$G(j\omega)=\frac{K}{1+j\left(\frac{\omega}{a}\right)}$$

At $\omega=0.5\,a$

Magnitude $M = |G(j\omega)| = \dfrac{K}{\sqrt{1+\left(\dfrac{\omega}{a}\right)^2}}$

$$= \frac{K}{\sqrt{1+(0.5)^2}}$$

$$= \frac{K}{\sqrt{1.25}}$$

$$M_{dB} = 20 \log M = 20 \log K - 20 \log (1.25)^{\frac{1}{2}}$$

$$= 20 \log K - 0.97$$

The approximate value from asymptotic plot

$$= 20 \log K$$

Error in magnitude $= 0.97\ dB$

Phase $\phi = G(j\omega)$

$$= -\tan^{-1}\left(\frac{\omega}{a}\right)$$

at $\omega = 0.5\,a$

$$\phi = -\tan^{-1}(0.5)$$

$$= -26.56^\circ$$

* Phase angle at corner frequency for pole not a orign is –45°.

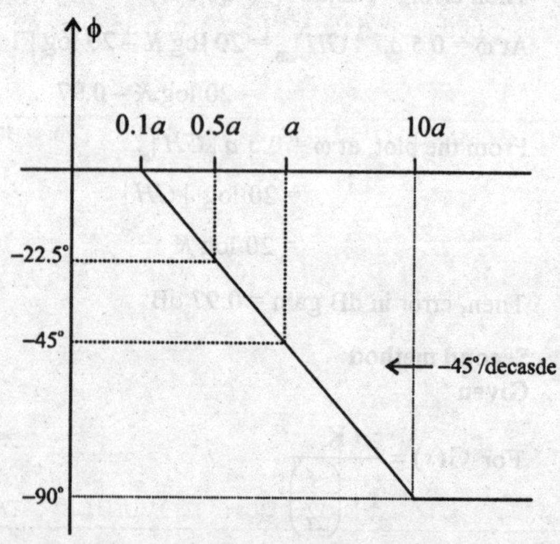

The aproximate phase at $\omega = 0.5\,a$ is

$$\phi = -22.5^\circ$$

Error $= -22.5^\circ + 26.56^\circ$

$$= 4.06^\circ$$

11. Ans. (a)

Solution: Given $H(j\omega) = \dfrac{10^4\,(1+j\omega)}{(10+j\omega)\,(100+j\omega)^2}$

Convert the given equation is standard form

i.e. $\dfrac{K\left(1+j\omega T_1\right)}{(j\omega)^n\,(1+j\omega T_a)\,(1+j\omega T_b)^2}$

$$= \frac{10^4\,(1+j\omega)}{10\left(1+\dfrac{j\omega}{10}\right)(100)^2\left(1+\dfrac{j\omega}{100}\right)^2}$$

$$= \frac{0\cdot 1\,(1+j\omega)}{\left(1+\dfrac{j\omega}{10}\right)\left(1+\dfrac{j\omega}{100}\right)^2}$$

Value of $K_{db} = +20\,\log_{10}(0.1) = -20$ db.

So plot will start from – 20 db.

In other way this problem can be seen as

When $\omega \to 0$

$$\text{Gain} \to H(0) = \frac{10^4}{10\times 100}$$

$$= 10$$

$$20 \log \frac{1}{10} = -20 \text{ dB}$$

Second Method

Given

$$H(j\omega) = \frac{10^4(1+j\omega)}{(10+j\omega)(100+j\omega)^2}$$

$$H(s) = \frac{10^4(1+s)}{(10+s)(100+s)^2}$$

Transform into standared formate

$$H(s) = \frac{(1+s)}{10\left(1+\dfrac{s}{10}\right)\left(1+\dfrac{s}{100}\right)^2}$$

Constant $K = \dfrac{1}{10} = 0.1$

Gain $= 20 \log K$

$= -20$ dB

i.e. intial slope $= 0$

Corner frequencies 1, 10, 100

Lowerst corner frequency $= 1$

Slope at frequency less than $1\ rad/\sec$

At $\omega = 1$, one zero present

Hence slope $=$ +20 dB/decade

At $\omega = 10$, pole exist

Hence slope $=$ −20 dB/decade

At $\omega = 100$, double pole exist slope $= -40$ dB/decade

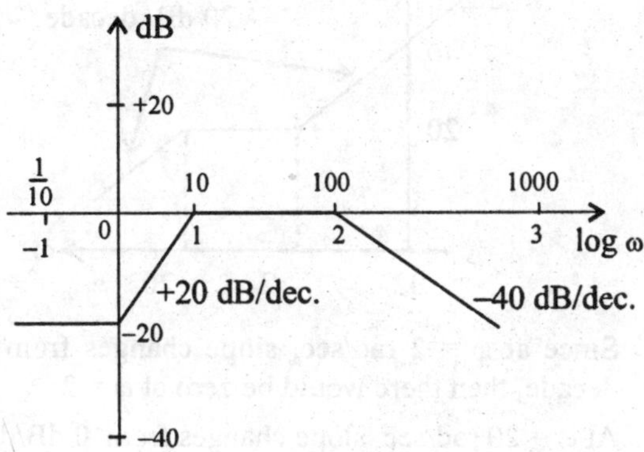

12. Ans (b)

Solution:

Since at $\omega = 2$ and $\omega = 25$, slope changes from $-40\ dB/dec$ to $-60\ dB/dec$ at both the value. So there are poles at $\omega = 2$ and $\omega = 25$.

Also at $\omega = 0.1$ slope is $-40 dB/dec$.

$\therefore$ It means there are two poles at origin.

Hence the transfer function of the form,

$$T(s) = \frac{K(s+5)}{s^2(s+2)(s+25)}$$

At lower value of frequency lowest order's term is consider of

$$80 = 20\log\frac{5K}{(0.1)^2 \times 50} \Rightarrow K = 1000$$

Second Method

Since there is −40 db/decade slope initially, so there is s^2 term in denominator.

There are 3 corner frequencies,

$\omega_1 = 2$ rad/sec

$\omega_2 = 5$ rad/sec

$\omega_3 = 25$ rad/sec

At $\omega_1 = 2$ rad/sec, slope changes from −40 db/ decade to −60 db/decade. So there is increase in −20 db/decade which indicate a term $\left(1+\frac{s}{2}\right)$ in denominator.

At $\omega_2 = 5$ rad/sec, slop changes from −60 db/decade to −40 db/decade, so there is in crease in +20 db/decade which indicate a term $\left(1+\frac{s}{5}\right)$ in numerator.

At $\omega_3 = 25$ rad/sec, slope changes from −40 db/decade to −60 db/decade.

So there is increase in −20 db/decade which indicatea term $\left(1+\frac{s}{25}\right)$ in denominator.

So, Transfer function becomes

$$G(s) = \frac{K\left(1+\frac{s}{5}\right)}{s^2\left(1+\frac{s}{2}\right)\left(1+\frac{s}{25}\right)}$$

$$= \frac{K\left(\frac{s+5}{5}\right)}{s^2\left(\frac{s+2}{2}\right)\left(\frac{s+25}{25}\right)}$$

$$= K\left(\frac{s+5}{5}\right) \times \frac{50}{s^2(s+2)(s+25)}$$

$$= 10\,K\frac{(s+5)}{s^2(s+2)(s+25)}$$

Now further for K

$$80 = 20\log_{10}\frac{10\,K\,(s+5)}{s^2\,(s+2)\,(s+25)}$$

Put $\omega = 0.1$ at lower frequency

$\omega = 0$ at higher frequency

i e. $80 = 20\log_{10}\dfrac{10\,K\,(0+5)}{(0.1)^2\,(0+2)\,(0+25)}$

$$4 = \log_{10}\frac{10K \times 5}{(0.1)^2 \times 50}$$

$$4 = \log_{10}\frac{100\,K}{1}$$

$$100\,K = 10^4$$

$$K = 10^2$$

$$G(s) = \frac{1000\,(s+5)}{s^2\,(s+2)\,(s+25)}$$

13. Ans. (a)

Solution: $20 \log_{10} K = 40$

$$\log_{10} K = 2$$

$$\therefore \quad K = 100$$

Corner frequencies = 10,250

$$\therefore \quad = \frac{100}{(1+j\omega T_1)(1+j\omega T_2)}$$

$$G(s) = \frac{100}{\left(1+\frac{s}{10}\right)\left(1+\frac{5}{250}\right)}$$

14. Ans (c)

Solution:

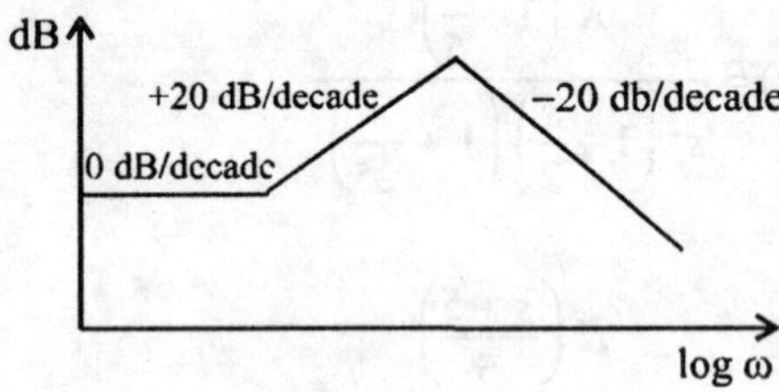

0 dB /decade is due to Gain K

+20 dB /decade is due to zero

From 20 dB /decade to 0 dB /decade is due to pole

From 0 dB /decade to –20 dB / decade is due to pole

So it has 1 zero and 2 pole.

15. Ans (b)

Solution:

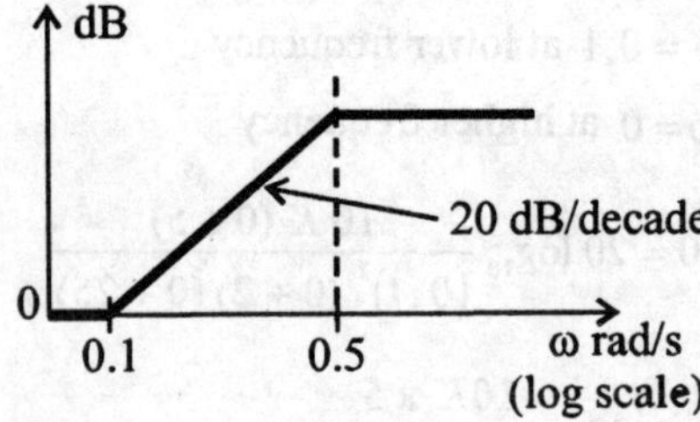

$$\omega_m = \sqrt{T_1 T_2}$$

T_1 = Lower cut off frequency

T_2 = upper cut off frequency

$$\omega_m = \sqrt{T_1 T_2}$$

$$\Rightarrow \sqrt{\frac{1}{10} \times 0.5} = \sqrt{\frac{1}{10} \times \frac{1}{2}}$$

$$= \frac{1}{\sqrt{20}}$$

16. Ans. (b)

Solution:

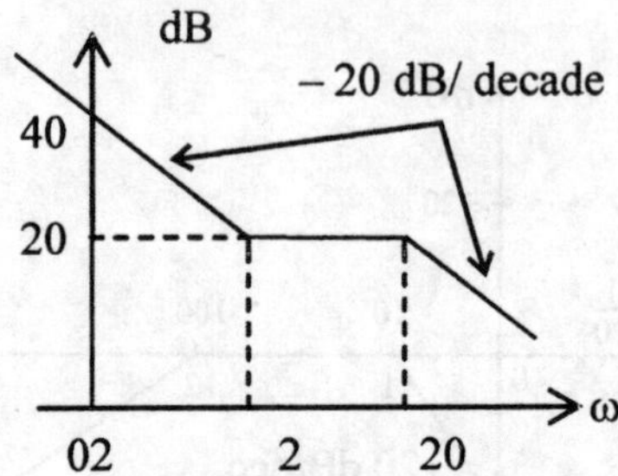

Since at ω = 2 rad/sec, slope changes from –20 dB/ decade, then there would be zero at ω = 2

At ω = 20 rad/sec, slope changes from 0 dB/decade, to –20 dB/decade then there would be zero at ω = 20

The slope cut vertical axis at 40 dB

$$\text{The, } 20 \log \frac{k}{0.2} = 40 \Rightarrow \log \frac{k}{0.2} = 2$$

$$\frac{k}{0.2} = 10^2 = 100$$

$$k = 0.2 \times 100 = 20$$

$$\text{Hence} \quad T(s) = \frac{20\left(\frac{s}{2}+1\right)}{s\left(\frac{s}{20}+1\right)}$$

17. Ans. (c)

Solution:

Transfer function of the system

$$= \frac{K(s+1)}{s^2\left(\frac{s}{100}+1\right)}$$

$$\angle\phi = -180° + \tan^{-1}\omega - \tan^{-1}\frac{\omega}{100}$$

On calculating, we find that minimum absolute phase being 180°

$$\theta(\omega) = \angle G(j\omega) = -180° + \tan^{-1}(\omega) - \tan^{-1}\left(\frac{\omega}{100}\right)$$

θ(ω) is a lagging angle and is minimum at

$\omega = \sqrt{\omega_1\omega_2} = 10r/s$

$-\tan^{-1}(\omega) > \tan^{-1}\left(\frac{\omega}{100}\right)$

or, Minimum phase angle = 180° + 84.3° – 5.7°
= –101.4°

10.7 Nyquist Plot

1. **Ans. (a)**

Solution: The encirclement of critical point (–1, *j*0) is once in anticlockwise direction

$\therefore \quad N = 1$

G(*s*) *H*(*s*) has one right side pole

$\therefore \quad P = 1$

$Z = P - N = 1 - 1 = 0$

No zero in the right hand side of *s*-plane

So system is stable.

2. **Ans. (d)**

Solution: Lag network is an RC network

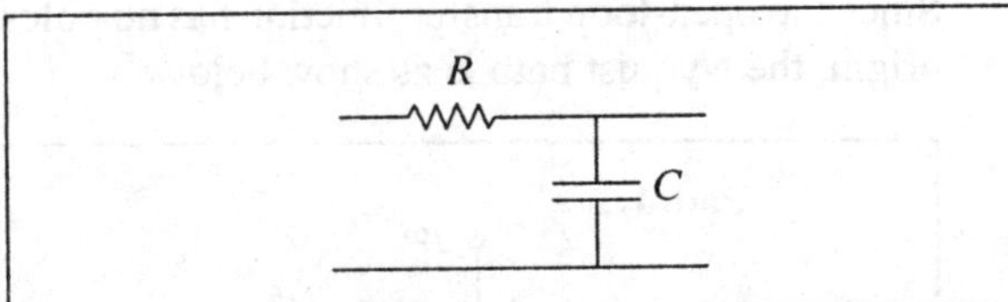

$$G(s) = \frac{1}{(1+sT)}$$

$$G(j\omega) = \frac{1}{1+j\omega T}$$

$$= \frac{1}{\sqrt{1+\omega^2T^2}} \angle -\tan^{-1}\omega T$$

$$= M\angle\phi$$

When $\omega = 0$, $M = 1$ and $\phi = 0$

The phasor at $\omega = 0$ has unit length and lies along the positive real axis. As ω increases *M* decreases ε[1] phase angle increases negatively.

When $\omega = \frac{1}{T}$, $M = \frac{1}{\sqrt{2}}$, $\phi = -45$

$\omega \to \infty$, $M \to 0$, $\phi = -90$

3. **Ans. (b)**

Solution: The closed loop system is stable for

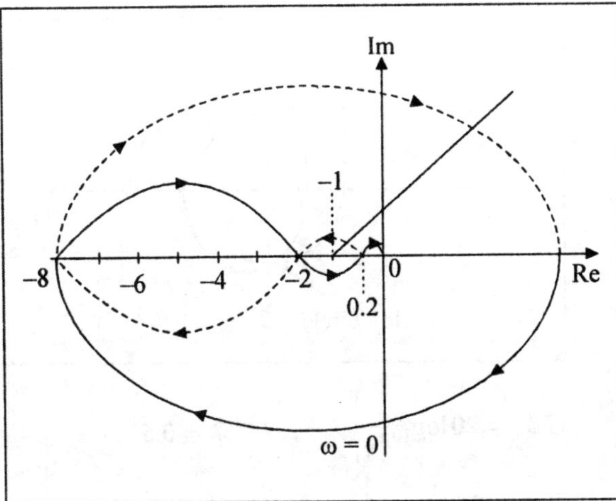

System is stable in region –0.2 to –2 and on the left side of –8 as the number of encirclement is zero

$$0.2\,K < 1$$

$$\Rightarrow \quad K < 5$$

$$2K > 1$$

$$\Rightarrow \quad K > 0.5$$

$$\therefore \quad 0.5 < K < 5$$

and $8K < 1$

$$K < \frac{1}{8}$$

4. **Ans. (d)**

Solution: As the system Nyquest plot passes through $-1+j0$, $|G(s)\,H(s)| = 1$

So, G.M = 1, but in dB this will be zero

5. **Ans (b)**

Solution:

For all pass filter, i.e. same again at all frequencies all poles (in L.H of s-plane) and all zeos (in R.H. of s-plane) are mirro image about jω-axis, e.g.

$$G(s) = K\frac{1-sT}{1+sT}$$

So $M = |G(j\omega)|$ = K for all ω

i.e. $0 < \omega < \infty$

The given Nyquist plot does not represent same again (the distance from origin) at different frequency $M_1 \neq M_2$ as $\omega_1 \neq \omega_2$

So it in not all pass filter

The probable choise is one zero in RHS of s-plane 6.

6. **Ans. (c)**

Solution: Nyquist plot of *G*(*s*) is gives.

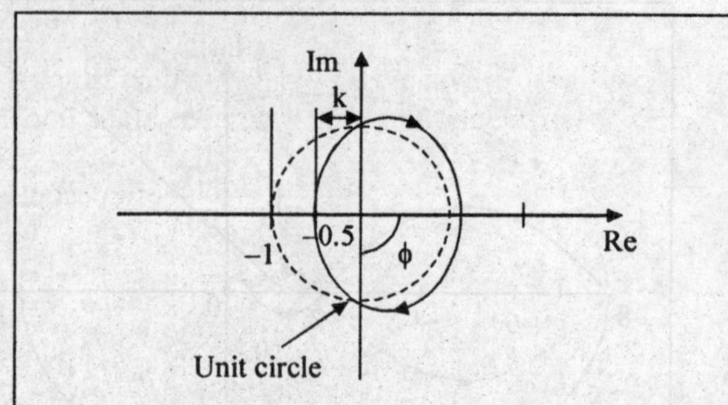

$G.M = 20\log_{10}\left|\frac{1}{k}\right| \qquad k = 0.5$

$= 20\log_{10}\frac{1}{0.5} = 20\log_{10}2$

$= 20 \times 0.3010 = 6.02$ dB

It an be clearly seen that Nyquist plot and unit circle interest on imaginary axis so $\phi = -90°$

Phase margin = 180 + ϕ = 180 − 90 = 90

7. **Ans. (a)**

Solution:

$G(j\omega) = 5 + j\omega$

$|G(j\omega)| = \sqrt{25+\omega^2}$

$\phi = \tan^{-1}\frac{\omega}{5}$

at $\omega = 0 \quad G(j\omega) = 5 \quad \phi = 0°$

$\omega = \infty \quad |G(j\omega)| = \infty \quad \phi = 90°$

$j\omega$, σ, 5

Second Method

Real part '5' is fixed only imaginary part increases. Which indicate that graph (a) is correct

8. **Ans. (b)**

Solution: Nyquist path $s = -j\infty \ldots +j\infty$ anti-clockwise

Total angle covered = 720° clock wise

Hence $n = \frac{720}{180} = 4$

i.e., $H(s) \ldots \frac{K_1}{s^4}$, $s = \text{Re}^{j\theta}$ as $R \ldots \infty$

Nyquist path $s = j0 \ldots -j\,0$ clockwise

Total angle covered = 360° anti-clockwise

Hence $m = \frac{360}{180} = 2$

i.e., $H(s) \ldots \frac{K_2}{s^4}$, $s = \text{Re}^{j\theta}$ as $R \ldots 0$

Hence choice (b) is correct

$$H(s) = \frac{K}{s^2(s+2)(s+5)}$$

Second Method

From the given Nyquist plote

At $\omega = 0 \qquad \infty\angle -180°$

At $\omega = \infty \qquad 0\angle -360°$

Initial angle (at ω = 0) −180°

Therefore, it is a type '2' system

At $\omega \to \infty \quad \angle -360°$

Indicate that the difference between poles and zeros must be '4'

$|P - Z|(-90°) = -360° \Rightarrow P = 4$ and $Z = 0$

Hence correction options is (b) only

9. **Ans. (b)**

Solution: Given: $G(s) = \frac{1}{(s-1)(s+2)(s+3)}$

Since the open-loop transfer function has no poles at the origin, the Nyquist path is as show below:

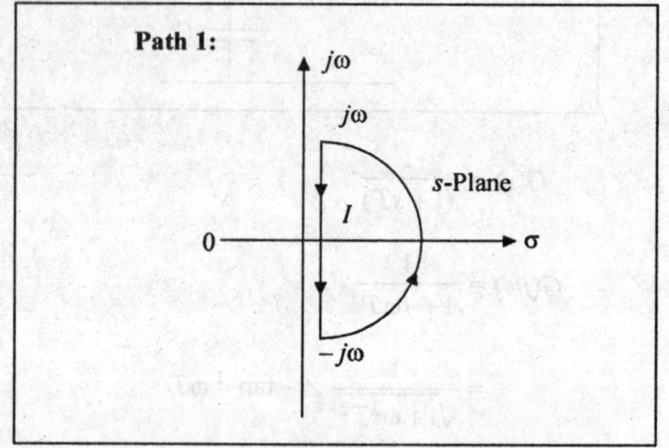

$$G(j\omega) = \frac{1}{(j\omega - 1)(j\omega + 2)(j\omega + 3)}$$

$$G(j\omega) = -\frac{1}{6}$$

$$= \frac{1}{6}\underline{|189°}$$

Path II:

$$G(j\omega) = \frac{1}{(j\omega)^3} = \frac{1}{(R3^{j0})^3} \quad R \to \infty$$

$$= \delta e^{-j30} \qquad \delta \to 0$$

Hence as ϕ goes from – 90° to + 90° in path II, $G(j\omega)$ goes from 270° to – 270° as shown below.

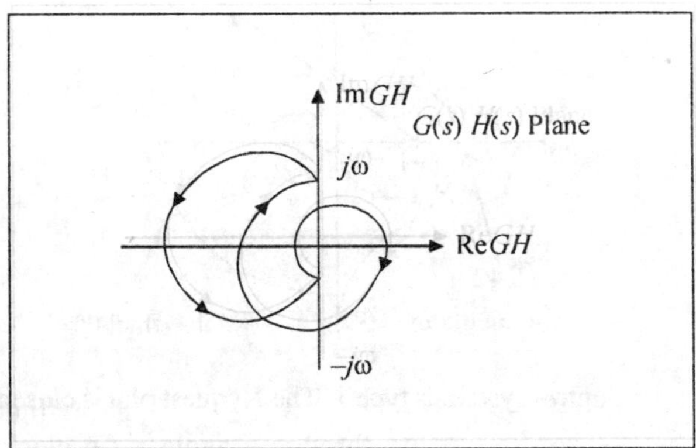

Second Method

$$G(s) = \frac{1}{(s-1)(s+2)(s+3)} \text{ and } H(s) = 1$$

The Nyquist plot of $G(s)H(s)$ will encircle the origin as may number of times as the differenc beween the number of right side poles and zeros of $G(s)H(s)$.

$N_{(0,0)} = P - Z$

$N_{(0,0)}$ = Numbner of encirclements of the origin by the Nyquist plot of $G(s)H(s)$.

P = Number of right side poles of $G(s)H(s)$.

Z = Number of right side zeros of OLTF $G(s)H(s)$.

From the given Transfer fucntion

$P = 1$

$Z = 0$

$N_{(0,0)} = 1 - 0 = 1$

It encircle once only

10. Ans. (d)

Solution:

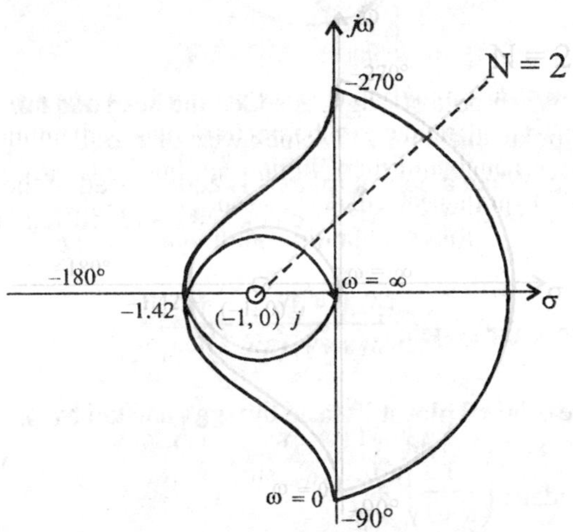

$$|a| = +1.42$$

$$\text{GM (gain margin)} = 20\log\frac{1}{|a|}$$

Since $a > 1$

So GM will be – ve and system unstable

Second Method

From the given, Nyquist plot

Encirclement of (–1, j0) $N = -2$

Number of right side poles of open loop system P = 0

$N = P - Z$

$Z = P - N$

$= 0 - (-2) = 2$

$Z = 2$ implies, closed loop system is unstable with two right hand poles.

11. Ans. (b)

Solution: $G(s)\,H(s) = \dfrac{\pi\, e^{-0.25s}}{s}$

$$G(j\omega)H(j\omega) = \frac{\pi[\cos(0.25\omega) - j\sin(0.25\omega)]}{j\omega}$$

$$= \frac{-\pi}{\omega}\sin(0.25\omega) - j\frac{\pi}{\omega}\cos(0.25\omega)$$

To get the interception on real axis put imaginary part = 0

$$\frac{\pi}{\omega}\cos(0.25\omega) = 0;\ \cos\frac{\omega}{4} = \cos\frac{\pi}{2}$$

$$\frac{\omega}{4} = \frac{\pi}{2};\ \omega = 2\pi$$

intercept on Negative real axis.

$$\therefore\ |G(j\omega)H(j\omega)|_{\omega=2\pi} = \left|-\frac{\pi}{2\pi}\sin\left(\frac{2\pi}{4}\right)\right| = \left|-\frac{1}{2}\right| = 0.5$$

12. Ans. (d)

Solution:

* If the GH plot in enclosed (–1, j0) that is the critical point, then the system becomes unstable. So in option (1) and (2) that does not enclose the (–1, j0)
* **Enclosed:** If a path is traversed in the clockwise direction and the point is found to lie to the right of the path, the point is said to be enclosed by the path.
* A system is stable when point (–1, j0) lies towards the left while moving along the polar plot in the direction of increasing frequency. This satisfy option (d)
* Final and correct analysis is nyquist criterion.

13. Ans. (b)

Solution: $G(s) = \dfrac{1}{s(s+1)(s+2)}$

$$G(j\omega) = \frac{1}{j\omega\,(1+j\omega)(2+j\omega)}$$

$$= \frac{-j\,(1-j\omega)(2-j\omega)}{\omega\,(1+\omega^2)(4+\omega^2)}$$

$$= \frac{-j\,(2-3j\omega-\omega^2)}{\omega\,(1+\omega^2)(4+\omega^2)}$$

$$= \frac{-3\omega - j\,(2-\omega^2)}{\omega\,(1+\omega^2)(4+\omega^2)}$$

$$= \frac{-3}{(1+\omega^2)(4+\omega^2)} - j\frac{(2-\omega^2)}{\omega\,(1+\omega^2)(4+\omega^2)}$$

$$\therefore \quad X = -\frac{3}{(1+\omega^2)(4+\omega^2)},$$

$$y = -\frac{(2-\omega^2)}{\omega\,(1+\omega^2)(4+\omega^2)}$$

at $\omega \to 0$

$x \to [-3/4]$ and $y \to \infty$

14. Ans. (d)

Solution:

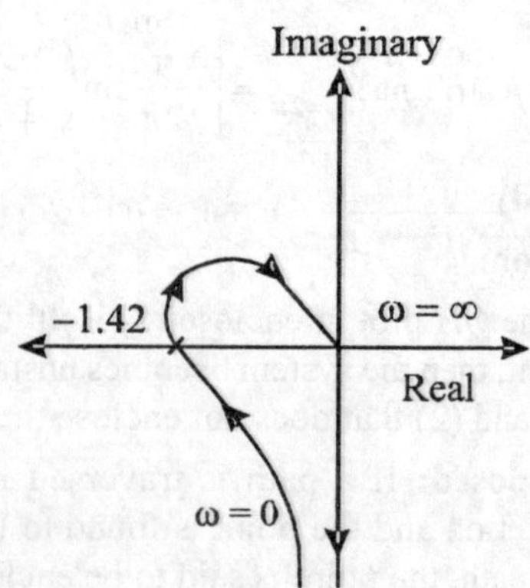

$a = 1.42$;

$GM = -20 \log a$

Since $a > 1$, GM will be negative and unstable

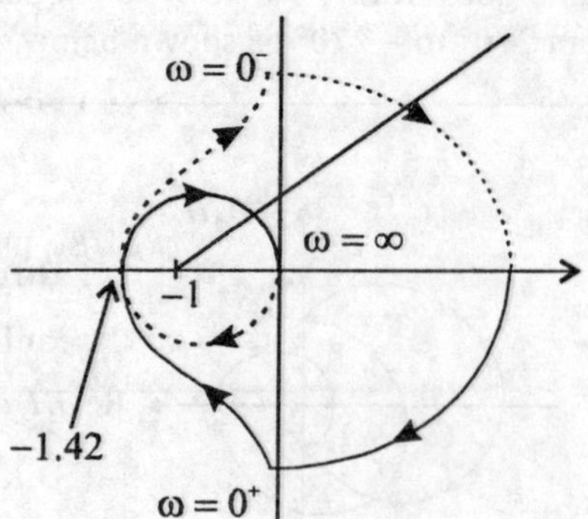

Control system is type 1. The Nyquest plot is closed from $\omega = 0^-$ to $\omega = 0^+$ through a angle π (in clock wise direction) with ∞ redius

The point $(-1 + j0)$ is encircled 2 times in clock wise direction. Thus N = − 2

From open loop fucntion

$$P_+ = 0$$

as $\quad N = P_+ - Z_+$

$$Z_+ = 2$$

The close loop characteristive equation has 2 roots with positive real part

15. Ans. (a)

Solution:

Standard plot for $\left(\dfrac{1}{s+1}\right)$ is

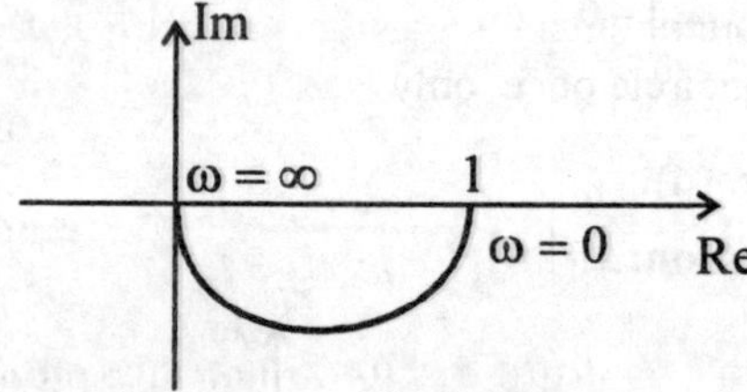

* When pole at origin is added the head and tail of the polar shift by 90° in clock wise direction

* 2. When a pole at ral axis is added, head of the polar plote shift by 90° in clock wise and tail remian the same.

Hence for system $\dfrac{1}{s(s+1)(s+2)}$

The original plot shift head by 180° and tail by 90° with standard $\left(\dfrac{1}{s+1}\right)$ for

Convert the given equation to standard form

i.e. $\frac{1/2}{s(s+1)(1+0.5s)} = \frac{1}{2}\left(\frac{1}{s(s+1)(1+0.5s)}\right)$

$T_1 + T_2 = 1 + 0.5 = 1.5 = \frac{15}{10} = \frac{3}{2}$

for standard are $= \frac{-3}{2} \times \frac{1}{2} = \frac{-3}{4}$

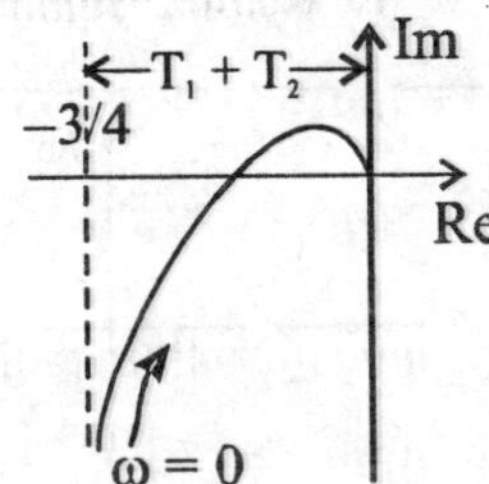

16. Ans. (a)

Solution: At point (–1, *j*0), $\angle G(j\omega).H(j\omega)|_{\omega=\omega_1} = -180°$

Phase margin, $\phi = \angle G(j\omega)\,H(j\omega)|_{\omega=\omega_1} + 180°$

$= -180° + 180°$

$= 0°$

17. Ans. (a)

Solution:

$T(s) = 5(1 + 0.1\,s)$

$T(j\omega) = 5(1 + 0.1\,j\omega)$

$\angle T(j\omega) = \tan^{-1}\frac{\text{Im}}{\text{Reel}}$

$= \tan^{-1}\frac{0.1\,\omega}{1}$

$= \tan^{-1} 0.1\,\omega$

At $\omega = 10$, $\angle T(j\omega) = 45°$

At $\omega = 0$, $\angle T(j\omega) = 0°$

18. Ans. (a)

Solution: For the closed loop to be stable, contour τ_{GH} of *G*(*s*) *H*(*s*) does not encircle (–1 + *j*0) point, i.e. the net encirclement is zero.

10.8 Controller or Compensator

1. Ans. (d)

Solution: $G_e(s) = \frac{S + Z}{S + p_1}$

$\tan^{-1} A - \tan^{-1} B = \tan^{-1}\left[\frac{A - B}{1 + A \cdot B}\right]$

$\angle G_e(s) = \tan^{-1}\left[\frac{\omega}{Z_1}\right] - \tan^{-1}\left[\frac{\omega}{P_1}\right]$

$= \tan^{-1}\frac{\left[\frac{\omega}{Z_1}\right] - \left[\frac{\omega}{P_1}\right]}{1 + \left(\frac{\omega}{Z_1}\right)\cdot\left(\frac{\frac{\omega}{P_1}}{Z_1 P_1}\right)}$

$= \tan^{-1}\left(\frac{\frac{\omega(P_1 - Z_1)}{Z_1 P_1}}{\frac{(Z_1 P_1 + \omega^2)}{Z_1 P_1}}\right)$

$\Rightarrow P_1 > Z_1$

Phase lead network

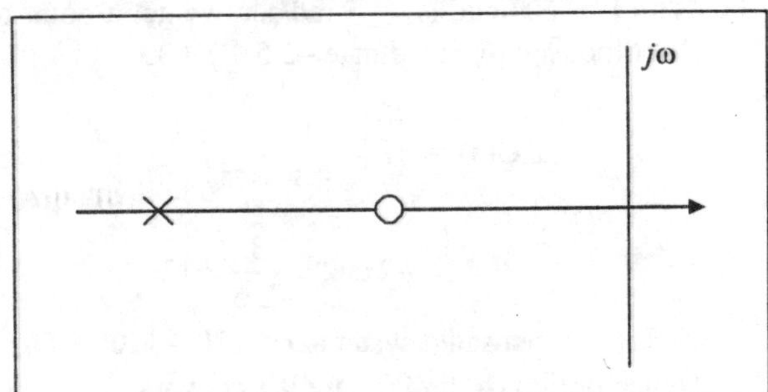

$H(s) = \left(\frac{1}{a}\right)\cdot\left(\frac{1 + aTs}{1 + Ts}\right)$

Where $a > 1$.

2. Ans. (d)

Solution:

$F(s) = \frac{1 + 3Ts}{1 + Ts}$

$\phi = \tan^{-1} 3\,T\omega - \tan^{-1} T\omega$

$$\tan^{-1} A - \tan^{-1} B = \tan^{-1}\left(\frac{A-B}{1+A\cdot B}\right)$$

$$= \tan^{-1}\left(\frac{2T\omega}{1+3T^2\omega^2}\right) = \tan^{-1} x$$

For max and min $\frac{dx}{d\omega} = 0$

$2T(1 + 3T^2\omega^2) - (2T\omega)(3T)^2(2\omega) = 0$

or, $2T + 6T^3\omega^2 - 12T^3\omega^2 = 0$

or, $2T - 6T^3\omega^2 = 0$

or $2T(1 - 3T^2\omega^2) = 0$

$3T^2\omega^2 = 1$ or $T\omega = \frac{1}{\sqrt{3}}$

$$\phi_{max} = \tan^{-1}\frac{2\times\frac{1}{\sqrt{3}}}{1+3\times\frac{1}{3}} = \tan^{-1}\frac{2\times\frac{1}{\sqrt{3}}}{2}$$

$$= \tan^{-1}\frac{1}{\sqrt{3}} = 30°$$

3. Ans. (a)

Solution: $\delta = 0.5$

$\omega_n = 5$ rad/sec

$G(s)\,H(s) = \frac{K}{s^2}$

$\cos^{-1}\delta = 60°$

$\cos^{-1} 0.5 = 60°$

$\theta = 60°$

With $\theta = 60°$ and $\omega_n = 5$ rad/sec we get a point on the Root locus with coordinate $-2.5 + j\,4.33$

$$\angle G(s) = \left.\frac{K}{s^2}\right|_{s=-2.5+j4.33}$$

$$= -2\tan^{-1}\frac{4.33}{-2.5} = 120°$$

$\therefore$ For compensated system $D = 180 - 120° = 60°$

Hence option (b) and (c) are lag network

For compensating lag network $\frac{K}{s^2}$, we need a lead network

Putting $s = -2.5 + j4.33$ in option (a) network gives

$$\left.\frac{K(s+3)}{s^2(s+9.9)}\right|_{s=-2.5+j4.33}$$

$$= \frac{0.5+j4.33}{7.4+j4.33} = 53° \doteq 60°$$

4. Ans. (a)

Solution:

$$G(s) = \frac{1}{s^2-1} = \frac{1}{(s+1)(s-1)}$$

Uncompensated system have pole in the Right side of *s*-plane, i.e. unstable system. The right side pole is to be neutralized by compensator of $\frac{10(s-1)}{s+1}$

$$G_C(s) = \frac{1}{(s+1)(s-1)} \times \frac{10(s-1)}{(s+2)}$$

$$= \frac{10}{(s+1)(s+2)}$$ **is stable system**

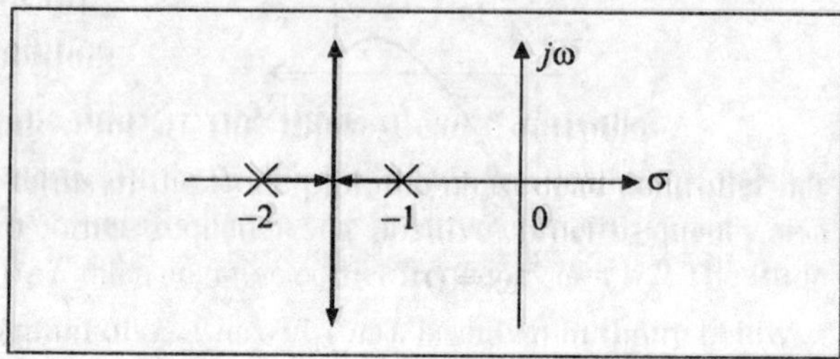

5. Ans. (d)

Solution: Lead lag compensator is as shown below:

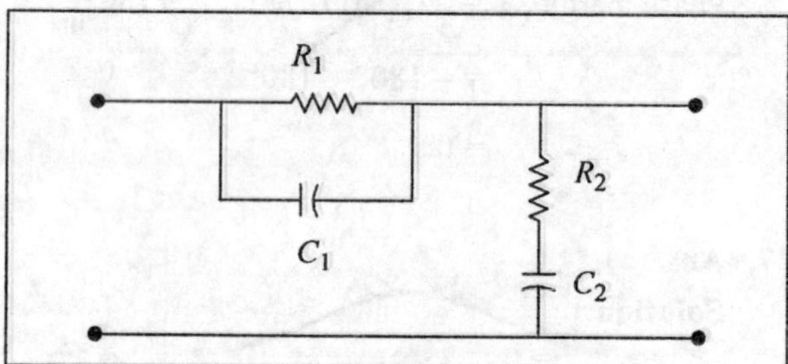

The transfer function is given by

$$G(s) = \left(\frac{1+aT_1 s}{1+T_1 s}\right)\left(\frac{1+bT_2 s}{1+T_2 s}\right)$$

It has zeroes at $-\frac{1}{aT_1}, -\frac{1}{bT_2}$

It has poles at $-\frac{1}{T_1}, -\frac{1}{T_2}$

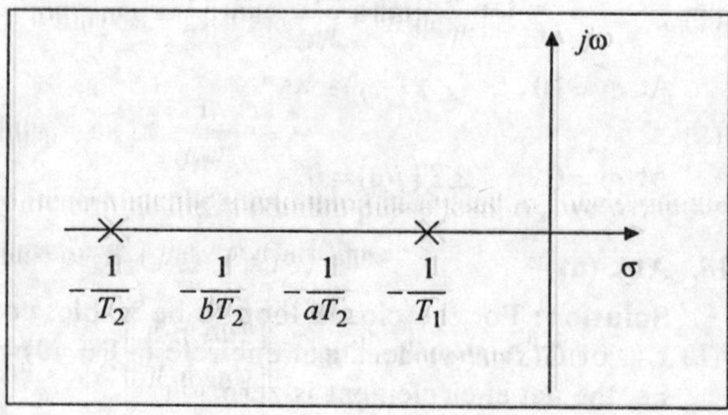

Its magnitude plot is given below:

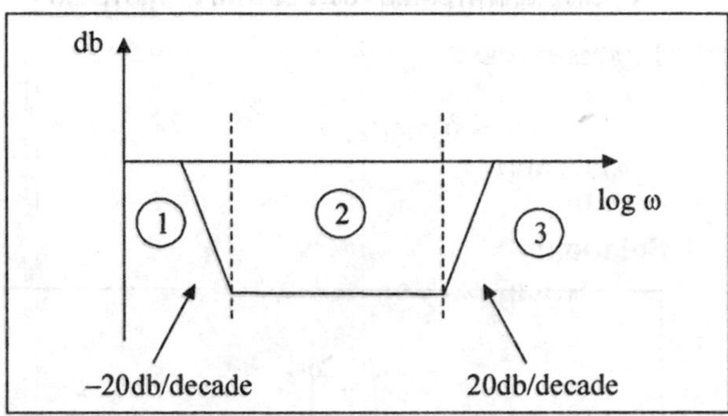

So if

1. $b < 1 \rightarrow$ lag network
2. $a > 1 \rightarrow$ Lead network
3. $ab = 1 \rightarrow$ Lead-lag network

In Nute cell we say that

Since magnitude plot shows both increasing as well as decreasing plot, it is lead-lag compensator.

6. **Ans. (a)**

Solution:

PHASE-LEAD CONTROLLER

The phase-lead controller is used for the compensation of control systems, and the design is carried out in the time domain or the *s*-plans via the root loci. In this section the phase-lead controller is to be designed using frequency domain techniques.

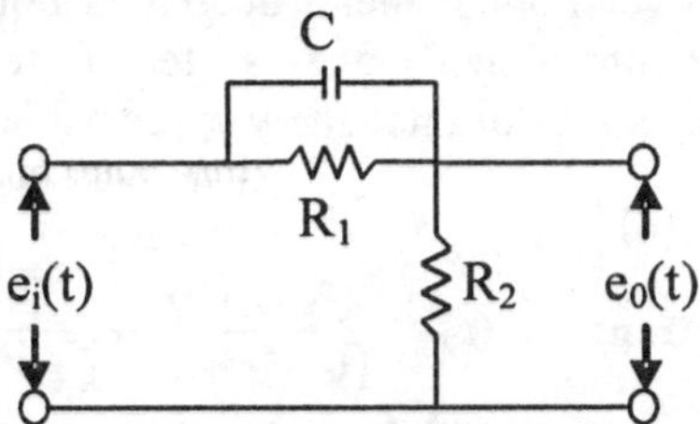

Fig. Phase lead network

The transfer of the network is derived as follows by assuming that the source impedance which the lead network sees is zero, and the output load impedance is infinite. This assumption is necessary in the derivation of the transfer function of any four terminal network.

$$\frac{E_2(s)}{E_1(s)} = \frac{R_2 + R_1R_2C_s}{R_1 + R_2 + R_1R_2C_s} \quad \text{...(1)}$$

or

$$\frac{E_2(s)}{E_1(s)} = \frac{R_2}{R_1 + R_2} \frac{1 + R_1C_s}{\frac{R_1R_2}{R_1 + R_2}C_s} \quad \text{...(2)}$$

Let $$a = \frac{R_2}{R_1 + R_2} \quad a > 1$$

and $$T = \frac{R_1R_2}{R_1 + R_2}C$$

Then equation (3) becomes

$$\frac{E_2(s)}{E_1(s)} = \frac{s + 1/aT}{s + 1/T} \quad a > 1 \quad \text{...(3)}$$

As seen from equation (3), the transfer function of the phase-lead network has a real zero at $s = -1/aT$ and a real pole at $s = -1/T$. These are represented in the s-plane as shown in figure below. By varying the value of a and T, the pole and zero may be located at any point on the negative real axis inthe *s*-plane. Since $a > 1$, the zero is always located to the right of the pole, and the distance between them is determined by the constant a.

"The zero being to the right of the pole is the reason that the phase-lead controller can improve the relative and absolute stability of a closed-loop control system".

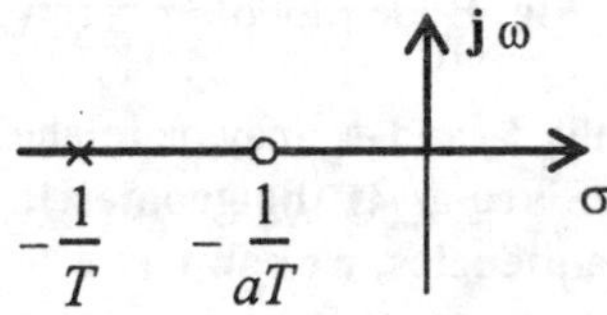

Fig. Pole zero configuration of phase-lead (HPF) network

$$\frac{E_2(s)}{E_2(s)} = \frac{\left(s + \frac{1}{aT}\right)}{\left(s + \frac{1}{T}\right)}$$

In Present Case

Phase Lead Angle

$$\phi = \tan^{-1}\frac{\omega}{a} - \tan^{-1}\frac{\omega}{b}$$

For phase lead ϕ should be positive

or, $\tan^{-1}\frac{\omega}{a} > \tan^{-1}\frac{\omega}{b}$

or, $a < b$

both option (a) and (c) satisfies but option (c) will plot pole and zero in the RHS of s-plane thus not possible Hence option (a) correct choice.

7. **Ans. (a)**

Solution:

Bode Plot of the Phase-Lead Controller

In terms of the Bode plot, the phase-lead controller has two corner frequencies: a positive corner frequency at ω

$= 1/aT$ and a negative corner frequency $\omega = 1/T$. The Bode diagram of $E_2(j\omega)/E_1(j\omega)$ is shown in figure below.

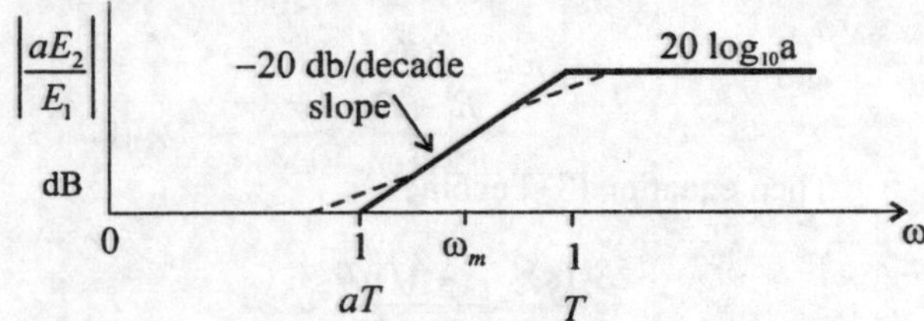

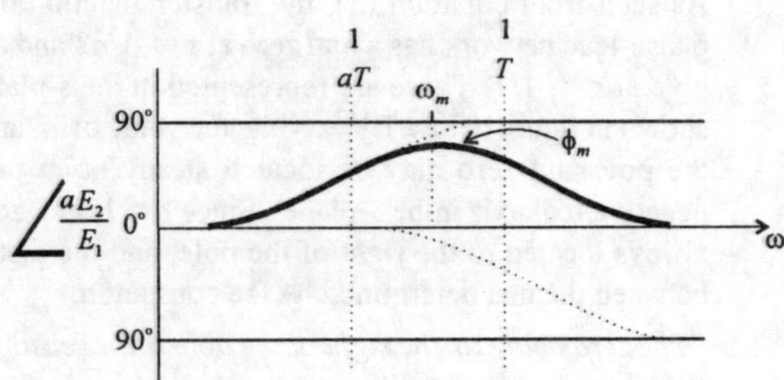

Fig. Bode plot of $\frac{1+aTs}{1+Ts}$: $a > 1$

Analytically, ϕ_m and ω_m may be related to the parameters a and T. **Since ω_m is the geometric mean of the two corner frequencies, we can write**

$$\log_{10}\omega_m = \frac{1}{2}\left(\log_{10}\frac{1}{aT} + \log_{10}\frac{1}{T}\right) \qquad \dots(1)$$

Thus, $\omega_m = \frac{1}{\sqrt{a}T}$ ...(2)

To determine the maximum phase lead, ϕ_m, we write the phase of $E_2(j\omega)/E_1(j\omega)$ sas

$$\phi = Arg\left[\frac{E_2(j\omega)}{E_1(j\omega)}\right] = \tan^{-1} aT\omega - \tan^{-1} T\omega \qquad \dots(3)$$

from which we have $\tan = \frac{aT\omega - T\omega}{1+(aT\omega)(T\omega)}$ ...(4)

When $\phi = \phi_m$

$$\omega = \omega_m = \frac{1}{\sqrt{a}\,T} \qquad \dots(5)$$

Therefore, equation (4) gives

$$\tan\phi_m = \frac{(a-1)\left(1/\sqrt{a}\right)}{1+1} = \frac{a-1}{2\sqrt{a}} \qquad \dots(6)$$

$$\text{or } \sin\phi_m = \frac{a-1}{a+1} \qquad \dots(7)$$

This last expression is a very useful relationship in the proper selection of the value of a for compensation design.

In present case

$$a = 1,\ b = 2\ \omega_{max} = \sqrt{ab} = \sqrt{2}\ rad/s$$

8. Ans. (a)

Solution:

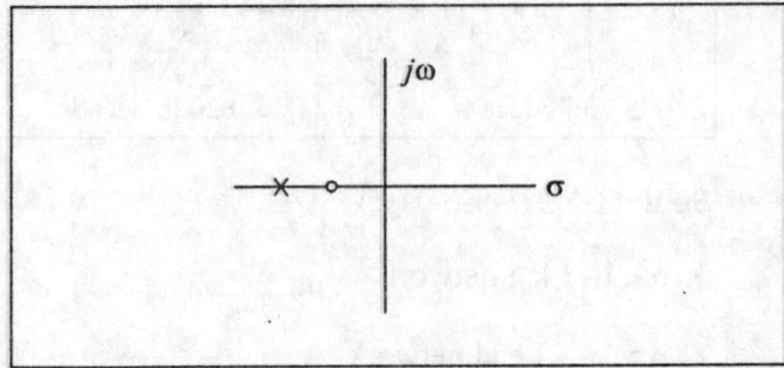

The transfer function of a phase-lead compensator is

$$\frac{E_2(s)}{E_1(s)} = \frac{1}{a}\frac{(1+aTs)}{1+Ts},\ a > 1$$

Zero is at $s = -\frac{1}{aT}$ Pole is at $s = -\frac{1}{T}$

Hence choice (a) is correct, since $a > 1$.

9. Ans. (b)

Solution: In phase lead compensation the zero is near to origin as compared to pole. Hence the effect of zero is dominant. This network allows to pass high frequencies and alternates the low frequencies. This network increase the band width. It also improves the speed of response and overshoot it mean decreases both rise time and overshoot which makes system faster. However, the steady state does not show appreciable improvement.

10. Ans. (d)

Solution: $D(s) = \frac{(0.5s+1)}{(0.05s+1)} = \left(\frac{1+\tau s}{1+\alpha\tau s}\right)$

Comparing equation $\tau = 0.5$, $\alpha = 0.1$

$$\varphi_m = \sin^{-1}\frac{\alpha-1}{\alpha+1} = \sin^{-1}\frac{0.1-1}{0.1+1} = \sin\frac{0.9}{1.1} = 55°$$

$$\omega_m = \frac{1}{\sqrt{\tau.(\tau\alpha)}} = \frac{1}{\tau\sqrt{\alpha}} = \frac{1}{0.5\sqrt{0.1}}$$

$$= \frac{1}{0.5\times 0.316} = \frac{1}{0.158} = 6.32$$

Max. phase lead is 55 deg. At 6.32 rad/sec

11. Ans. (a)

Solution: Phase of circuit

$$\theta = \tan^{-1}\frac{\text{Im}}{\text{Re}}$$

$$= \frac{\tan^{-1}\frac{\omega}{a}}{\tan^{-1}\frac{\omega}{b}}$$

$$\theta = \tan^{-1}\frac{\omega}{a} - \tan^{-1}\frac{\omega}{b}$$

For lead compensator, $\theta > 0$

Then, $b > a$

12. **Ans. (d)**

Solution:

Let, response of the un-compensated system is

$$H_{UC}(s) = \frac{900}{s(s+1)(s+9)}$$

Response of compensated system.

$$H_C(s) = \frac{900}{s(s+1)(s+9)}G_C(s)$$

Where $G_C(s) \rightarrow$ Response of compensator

Given that gain-crossover frequency of compensated system is same as phase crossover frequency of un-compensated system

So, $(\omega_g)_{\text{compensated}} = (\omega_p)_{\text{uncompensated}}$

$$-180^\circ = \angle H_{UC}(j\omega_p)$$

$$-180^\circ = -90^\circ - \tan^{-1}(\omega_p) - \tan^{-1}\left(\frac{\omega_p}{9}\right)$$

$$90^\circ = \tan^{-1}\left(\frac{\omega_p + \frac{\omega_p}{9}}{1 - \frac{\omega_p^2}{9}}\right)$$

$$1 - \frac{\omega_p^2}{9} = 0$$

$$\omega_p = 3 \text{ rad/sec.}$$

Therefore, $(\omega_g)_{\text{compensated}} = 3$ rad/sec.

At this frequency phase margin of compensated system is

$$\text{PM} = 180^\circ + \angle H_C(j\omega_g)$$

$$45^\circ = 180^\circ - 90^\circ - \tan^{-1}(\omega_g) - \tan^{-1}(\omega_g/9) + \angle G_C(j\omega_g)$$

$$45^\circ = 180^\circ - 90^\circ - \tan^{-1}(\omega_g) - \tan^{-1}(1/3) + \angle G_C(j\omega_g)$$

$$45^\circ = 90^\circ - \tan^{-1}\left[\frac{3+\frac{1}{3}}{1-3\left(\frac{1}{3}\right)}\right] + \angle G_C(j\omega_g)$$

$$45^\circ = 90^\circ - 90^\circ + \angle G_C(j\omega_g) \quad \angle G_C(j\omega_g) = 45^\circ$$

The gain cross over frequency of compensated system is lower than un-compensated system, Hence , one can use phase lag-lead compensator.

At gain cross over frequency gain of compensated system is unity so.

$$|H_C(j\omega_g)| = 1$$

$$\frac{900|G_C(j\omega_g)|}{\omega_g\sqrt{\omega_g^2+1}\sqrt{\omega_g^2+81}} = 1$$

$$|G_C(j\omega_g)| = \frac{3\sqrt{9+1}\sqrt{9+81}}{900}$$

$$= \frac{3\times 30}{900} = \frac{1}{10}$$

In db, $$G_C(\omega_g) = 20\log\frac{1}{10}$$

$$= -20\,dB \text{ (attenuation)}$$

13. **Ans.(a)**

Solution: $C_1 = \dfrac{10(s+1)}{(s+10)}$; $\quad C_2 = \dfrac{s+10}{10(s+1)}$

s-plane plot of C_1

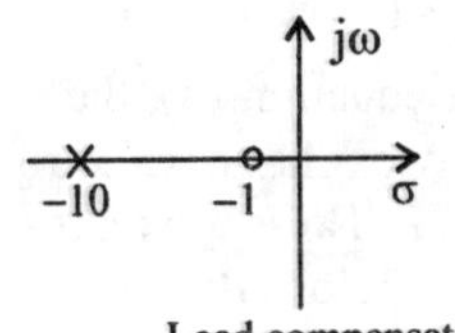

Lead compensator

s-plane plot of C_2

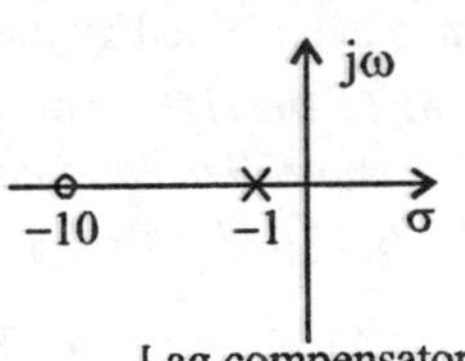

Lag compensator

14. **Ans. (a)**

Solution:

PHASE-LEAD CONTROLLER

The phase-lead controller is used for the compensation of control systems, and the design is carried out in the time domain or the *s*-plans via the root loci. In this section the phase-lead controller is to be designed using frequency domain techniques.

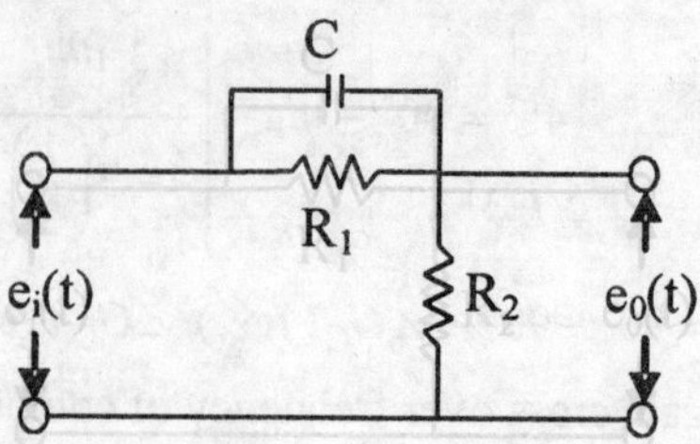

Fig. Phase lead network

The transfer of the network is derived as follows by assuming that the source impedance which the lead network sees is zero, and the output load impedance is infinite. This assumption is necessary in the derivation of the transfer function of any four terminal network.

$$\frac{E_2(s)}{E_1(s)} = \frac{R_2 + R_1R_2C_s}{R_1 + R_2 + R_1R_2C_s} \quad ...(1)$$

or

$$\frac{E_2(s)}{E_1(s)} = \frac{R_2}{R_1 + R_2}\frac{1 + R_1C_s}{\frac{R_1R_2}{R_1 + R_2}C_s} \quad ...(2)$$

Let $a = \frac{R_2}{R_1 + R_2} \quad a > 1$

and $T = \frac{R_1R_2}{R_1 + R_2}C$

Then equation (3) becomes

$$\frac{E_2(s)}{E_1(s)} = \frac{s + 1/aT}{s + 1/T} \quad a > 1 \quad ...(3)$$

As seen from equation (3), the transfer function of the phase-lead network has a real zero at $s = -1/aT$ and a real pole at $s = -1/T$. These are represented in the s-plane as shown in figure below. By varying the value of a and T, the pole and zero may be located at any point on the negative real axis inthe s-plane. Since $a > 1$, the zero is always located to the right of the pole, and the distance between them is determined by the constant a.

"The zero being to the right of the pole is the reason that the phase-lead controller can improve the relative and absolute stability of a closed-loop control system".

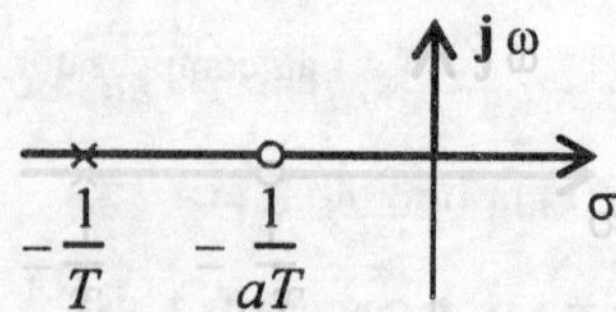

Fig. Pole zero configuration of phase-lead (HPF) network

$$\frac{E_2(s)}{E_2(s)} = \frac{\left(s + \frac{1}{aT}\right)}{\left(s + \frac{1}{T}\right)}$$

In Present Case

Phase Lead Angle

$$\phi = \tan^{-1}\frac{\omega}{a} - \tan^{-1}\frac{\omega}{b}$$

For phase lead ϕ should be positive

or, $\tan^{-1}\frac{\omega}{a} > \tan^{-1}\frac{\omega}{b}$

or, $a < b$

both option (a) and (c) satisfies but option (c) will plot pole and zero in the RHS of s-plane thus not possible Hence option (a) correct choise.

15. Ans.(a)

Solution:

Bode Plot of the Phase-Lead Controller

In terms of the Bode plot, the phase-lead controller has two corner frequencies: a positive corner frequency at $\omega = 1/aT$ and a negative corner frequency $\omega = 1/T$. The Bode diagram of $E_2(j\omega)/E_1(j\omega)$ is shown in figure below.

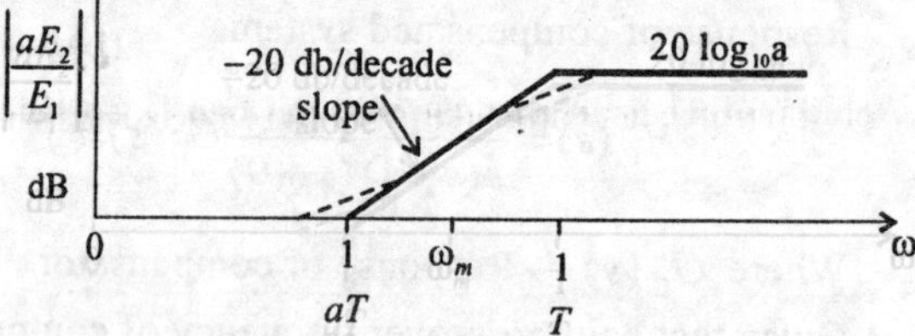

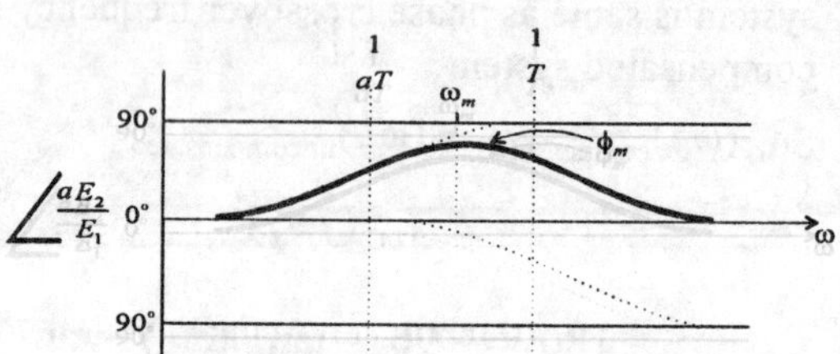

Fig. Bode plot of $\frac{1 + aTs}{1 + Ts}$: $a > 1$

Analytically, ϕ_m and ω_m may be related to the parameters a and T. **Since ω_m is the geometric mean of the two corner frequencies, we can write**

$$\log_{10}\omega_m = \frac{1}{2}\left(\log_{10}\frac{1}{aT} + \log_{10}\frac{1}{T}\right) \quad ...(1)$$

Thus, $\omega_m = \frac{1}{\sqrt{aT}} \quad ...(2)$

To determine the maximum phase lead, ϕ_m, we write the phase of $E_2(j\omega)/E_1(j\omega)$ sas

$$\phi = Arg\left[\frac{E_2(j\omega)}{E_1(j\omega)}\right] = \tan^{-1} aT\omega - \tan^{-1} T\omega \quad ...(3)$$

from which we have $\tan = \dfrac{aT\omega - T\omega}{1+(aT\omega)(T\omega)}$ ... (4)

When $\phi = \phi_m$

$$\omega = \omega_m = \frac{1}{\sqrt{a}\,T} \quad \text{... (5)}$$

Therefore, equation (4) gives

$$\tan\phi_m = \frac{(a-1)\left(1/\sqrt{a}\right)}{1+1} = \frac{a-1}{2\sqrt{a}} \quad \text{... (6)}$$

or $\sin\phi_m = \dfrac{a-1}{a+1}$ ... (7)

This last expression is a very useful relationship in the proper selection of the value of a for compensation design.

In present case

$a = 1,\ b = 2\ \omega_{max} = \sqrt{ab} = \sqrt{2}\ rad/s$

16. Ans. (d)

Solution: A temperature control system is usually Very sluggish, To improve its dynamics a PD controller can be used.

An integral system always makes a system sluggish (slow). The response can be made faster by increasing the gain but oscillation will increase which will tend the system towards instability. This is not desirable. Hence to improve the dynamics, an derivative controller is always introduced.

17. Ans. (b)

Solution: Integration reduces the offset caused by proportional controller action. It integrates area under curve thus minimizing the offset.

18. Ans. (a)

Solution: When proportional controller action is used and time delay is introduce then gain margin is not affected and phase margin also remain the saem.

19. Ans. (b)

Solution: $\dfrac{E_o(s)}{E_i(s)} = K_1 + K_2 s$

20. Ans. (a)

21. Ans. (b)

Solution: For proportional controller

$K_p = 0.5\ K_u$

$K_u = 0.5 \times 10$

$= 5$

22. Ans. (b)

Solution: For $\omega \to \infty$ take higher order term of s and for $\omega \to \infty$ take the lower order term of s.

In present case one s is in denominator and other in numerator which given phase angle of zero.

23. Ans. (a)

Solution:

Synchro Transmitter

It has a Y-connected stator winding which resembles the stator of a 3-phase induction motor. The rotor is a *salient pole, dumbbell shaped magnet* with a single winding. A single phase ac voltage is applied to the rotor through two slip rings. Let the ac voltage applied to the rotor of synchro transmitter be

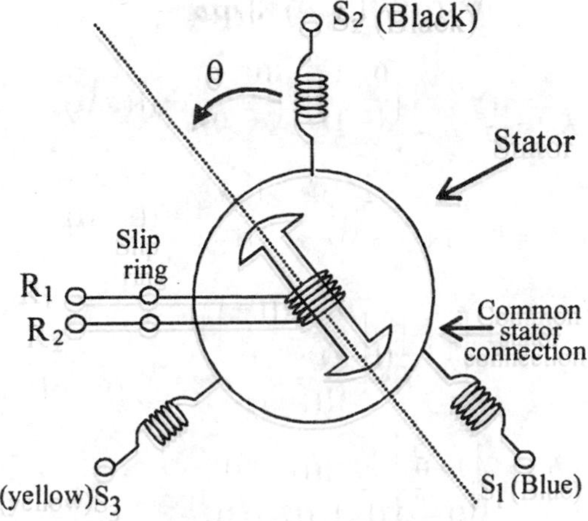

$e_r(t) = E_r \sin\omega_c t$

$Es_2n(t) = K\ E\ (\sin\omega t\ \ \cos 0°)$

$Es_1n\ (t) = K\ E\ (\sin\omega t\ \ \cos 120°)$

$Es_3n\ (t) = K\ E\ (\sin\omega t\ \ \cos 240°)$

where, n '! neutral if rotor with S_2 at 0°.

The voltage in each stator winding will vary as function of the cosine of rotor displacement. This head to the use of synchros Transmitter to identify angular position by measuring and identify the set of voltage at three stator terminal.

Synchro control Transformer

Basically, the principle of operation of synchro control Transformer is identical to that of synchros Transmitter except that the rotor is cylindrically shaped so that the air gap flux is uniformly distributed around the rotor. Rotor terminals are usually connected to an amplifier.

There are basically four type synchros i.e. Transmitter, Receiver, Transformers and differentials.

* Stator winding voltage are equal in magnitude but 120° of successive phase difference.

24. Ans. (a)

Solution:

Resolution, $\alpha = \frac{360^o}{nT}$,

Where, n = number of stacks

T = number of teeths

$$\therefore \quad \alpha = \frac{360^o}{12 \times 3} = 10^o$$

25. Ans. (c)

Solution:

Proportional Band (PB) = 50%

$$K_P = \frac{100}{\text{P.B}} = \frac{100}{50} = 2$$

$$T_i = 0.2$$

$$K_J = \frac{1}{T_i} = \frac{1}{0.2} = 5$$

$$T.F = K_P\left[1 + \frac{K_I}{s}\right]$$

$$= K_P\left[1 + \frac{1}{T_i s}\right]$$

$$T.F \Rightarrow 2\left[1 + \frac{5}{s}\right]$$

So, $T.F = 2\left[1 + \frac{5}{s}\right]$

26. Ans. (d)

Solution:

For system to be stable

1. The inner loop gain > outer loop gain
2. Both gain should be finite

For Block diagram reduction

$$\text{Inner loop} = \frac{K_1/(s+1)}{1 + \frac{K_1}{(s+1)}} = \frac{K_1}{1 + K_1 + s}$$

$$\text{Outer loop } GH = \left[\frac{K_1 K_2}{1 + K_1 + s}\right]\left[\frac{8}{s^2 + 6s + 8}\right]$$

Characteristic equation $1 + GH = 0$

$$\text{or} \quad \begin{bmatrix} s^3 + 6s^2 + 8s + (1+K_1)s^2 + 6(1+K_1)s \\ +8(1+K_1) + 8K_1K_2 \end{bmatrix} = 0$$

$$\text{or} \quad \begin{bmatrix} s^3 + (6+1+K_1)s^2 + (8+6+6K_1)s \\ +8(1+K_1+K_1K_2) \end{bmatrix} = 0$$

$$\text{or} \quad \begin{bmatrix} s^3 + (7+K_1)s^2 + (14+6K_1)s \\ +8(1+K_1+K_1K_2) \end{bmatrix} = 0$$

Apply RH criterion

$$\begin{array}{c|ll} s^3 & 1 & 14+6K_1 \\ s^2 & 7+K_1 & 8(1+K_1+K_1K_2) \\ s^1 & \frac{(7+K_1)(14+6K_1) - 8(1+K_1+K_1K_2)}{7+K_1} & \\ s^0 & 8(1+K_1+K_1K_2) & \end{array}$$

For stability

$$\frac{(7+K_1)(14+6K_1) - 8(1+K_1+K_1K_2)}{7+K_1} > 0 \quad \text{...(1)}$$

$$\text{And} \quad 8(1+K_1+K_1K_2) > 0 \quad \text{...(2)}$$

Equation (1) is true only for $K_1 = 10$ and $K_2 = 1$

Equation (2) is true for all the given value options

27. Ans. (d)

Solution:

$$T.F. = 2 + \frac{0.4}{s}$$

$$T.F. = 2\left[1 + \frac{0.2}{s}\right]$$

Compare it with $K_P\left[1 + \frac{K_I}{5}\right]$

$$K_P = 2, \quad K_I = -2$$

We know that $K_I = \frac{1}{T_i}$

Where T_i = Reset time

$$0.2 = \frac{1}{T_i}$$

$$T_i = 5 \text{ minutes}$$

$$P.B. = \frac{100}{K_P}$$

$$= \frac{100}{2} = 50\%$$

10.9 Control Components

1. Ans. (a)

Solution: Tachometer output is proportional to the angular velocity, i.e. $e_t \propto \omega_n$

Tachometer is basically a differentiator

Or, $e_t = K_t\,\omega$

$$= K_t \frac{d\theta}{dt}$$

$$E_t(s) = K_t\, s\, \theta(s)$$

$$\frac{E_t(s)}{\theta(s)} = s\, K_t$$

2. **Ans. (c)**

Solution: In synchro the output voltage is proportional to sin $(\theta_r - \theta_L)$

3. **Ans. (c)**

Solution: Transfer function is derived as, $e(t) = K_t \frac{d\theta}{dt}$

$$E(s) = K_t\, sQ(s)$$

$$\frac{F(s)}{Q(s)} = s\, K_t$$

4. **Ans. (b)**

Solution: In a LVDT, the two secondary voltages Vary unequally depending on the core position.

5. **Ans. (c)**

Solution: Let, $f(t) = A \sin \omega t$

$$\frac{df(t)}{dt} = A\,\omega \cos \omega t$$

A **differentiator** has transfer function whose amplitude increases linearly with frequency.

6. **Ans. (b)**

Solution:

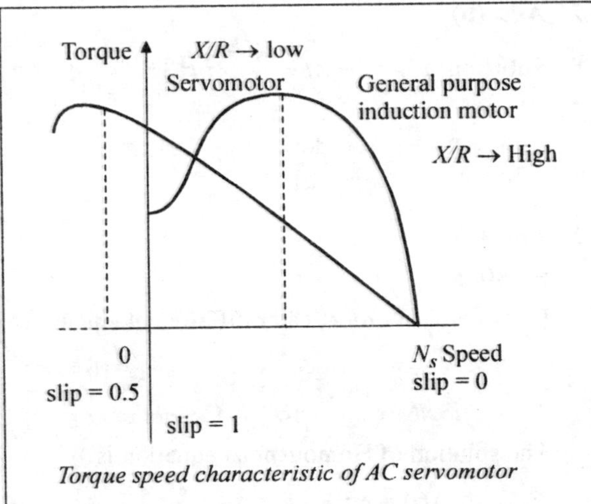

Torque speed characteristic of AC servomotor

$R \rightarrow$ resistance of rotor

$X \rightarrow$ Inductance of rotor

Difference between general purpose Induction motor and AC servomotor:

1. $\left(\frac{X}{R}\right)_{servo\ motor} < \left(\frac{X}{R}\right)_{General\ purpose\ induction\ motor}$
2. Torque speed characteristics is linear for A.C. servomotor whereas it is non-linear for general purpose induction motor.
3. Rotor is generally drag cup type or squirrel cage type.
4. AC ser

7. **Ans. (b)**

Solution: Selsyn used for error detection

8. **Ans. (b)**

9. **Ans (b)**

Solution:

Synchros output $\propto \cos(\theta_1 - \theta_2)$

$\therefore \quad v = K \cos\theta$

For $\theta = 0^o$

$\therefore v = K = 1.0$ volt

$\therefore \quad v = 1 \cos\theta$

For $\Rightarrow \theta = 30^o$

$\therefore v = \cos 30$

$$= \frac{\sqrt{3}}{2} = 0.866\,V$$

10. **Ans. (a)**

11. **Ans. (d)**

Solution: Linear variable differential transformer has one primary coil and two secondary coils connected in opposition.

12. **Ans. (a)**

13. **Ans. (a)**

Solution:

Resolution, $\alpha = \frac{360^o}{nT}$,

Where, n = number of stacks

T = number of teeths

$$\therefore \quad \alpha = \frac{360^o}{12 \times 3} = 10^o$$

14. **Ans. (d)**

Solution:

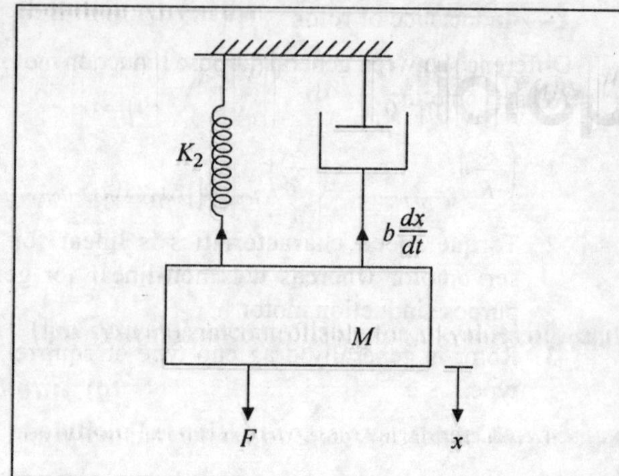

$$F - \frac{b\,dx}{dt} - K_x = \frac{Md^2x}{dt^2}$$

$$F = \frac{M\,d^2x}{dt^2} + b\frac{dx}{dt} + K_x$$

$$F(s) = Ms^2X(s) + bs\,X(s) + K\,X(s)$$

$$I.F = \frac{X(s)}{F(s)} = \frac{1}{Ms^2 + bs + K}$$

$$Ms^2 + bs + K = 0$$

$$s^2 + 2\delta\omega_n s + \omega_n^2$$

$$s^2 + \frac{b}{M}s + \frac{K}{M} = 0$$

$$\omega_n = \sqrt{\frac{K}{M}}\text{rad/sec}$$

$$2\delta\omega_n = \frac{b}{M}$$

$$\delta = \frac{b}{2M\cdot\omega_n}$$

$$\delta = \frac{b}{2M\cdot\sqrt{\frac{K}{M}}} = \frac{b}{2\sqrt{KM}}$$

$$\delta = \frac{b}{2\sqrt{KM}}$$

If $\quad b = 2\sqrt{KM}$

Then $\delta = 1$ critically damped system.

Overshoot is observed if $\delta < 1$, i.e. under damped system.

This is n overshoot in $\delta = 1$ (critically damped system) a for $\delta > 1$ (over-damped system.

$\therefore b = 2\sqrt{KM}$ to just prevent overshoot in step response.

10.10 State Space Analysis

1. Ans. (b)

Solution: In Matrix form

$$\begin{pmatrix} x_1'(t) \\ x_2'(t) \end{pmatrix} = \underbrace{\begin{pmatrix} -2 & 4 \\ 2 & -1 \end{pmatrix}}_{A}\begin{pmatrix} x_1(t) \\ x_2(t) \end{pmatrix} + \underbrace{\begin{pmatrix} 0 \\ 1 \end{pmatrix}}_{B}u(t)$$

Test for Controllability $[B : AB]$ = Finite

$$\left|\begin{pmatrix} 0 \\ 1 \end{pmatrix} : \begin{pmatrix} -2 & 4 \\ 2 & -1 \end{pmatrix}\begin{pmatrix} 0 \\ 1 \end{pmatrix}\right|$$

$$= \begin{vmatrix} 0 : (-2)(0) + (4)\ (1) \\ 1 : (2)\ (0) + (-1)(1) \end{vmatrix}$$

$$= \begin{vmatrix} 0 & 4 \\ 1 & -1 \end{vmatrix}$$ finite or Rank is 2,

Hence system is Controllable

Test for stability $|(sI - (A)| = 0$ i.e. characteristic equation

$$\begin{pmatrix} s & 0 \\ 0 & s \end{pmatrix} - \begin{pmatrix} -2 & 4 \\ 2 & -1 \end{pmatrix} = \begin{pmatrix} s+2 & -4 \\ -2 & s+1 \end{pmatrix}$$

$$s^2 + 3s - 6 = 0$$

Apply the RH criterion to check the stability

Routh Hurwitz table

s^2	1	−6
s^1	3	0
s^0	−6	

There is one sign change in first column, Hence system is unstable.

2. Ans. (b)

Solution: $e^{At} = I + At + \frac{1}{2!}A^2t^2 + \ldots \frac{1}{k!}A^kt^k$

$$= \sum_{k=0}^{\infty}\frac{A^kt^k}{k!}$$

3. Ans. (a)

Solution:

Consider first order scalar differential equation

$$\frac{dx}{dt} = ax\ ;\ X(o) = X_0$$

The solution of Homogenous equation is

$$X(t) = e^{at}X_0$$

For same way

$$X'(t) = A(t)\,X(t) \text{ solution is}$$

$$X(t) = e^{At}\,X_0$$

If the function is **Non Homogenous,** i.e.

$$X'(t) = A\,X(t) + B\,u(t)\,;\, X(0) = X_0$$

The solution of this equation is given by

$$X(t) = e^{At}\,X(o) + \int_0^t e^{-A\tau} Bu(d\tau)$$

$$X(t) = \phi(t)\,X(0) + \int_0^t \phi(t-\tau)\,B\,r(\tau)\,d\tau$$

Where, $\phi(t) = LT^{-1}\,[sI - A]^{-1}$

Where $\phi(t)$ is *STM.*

4. **Ans. (a)**

Solution: Consider first order scalar differential equation

$$\frac{dx}{dt} = ax\,;\, X(o) = X_0$$

The solution of Homogenous equation is

$$X(t) = e^{at}\,X_0$$

For same way

$$X'(t) = A(t)\,X(t) \text{ solution is}$$

$$X(t) = e^{At}\,X_0$$

If the function is **Non Homogenous** i.e.

$$X'(t) = A\,X(t) + B\,u(t)\,;\, X(0) = X_0$$

The solution of this equation is given by

$$X(t) = e^{At}\,X(o) + \int_0^t e^{-A\tau} Bu(d\tau)$$

$$X(t) = \phi(t)\,X(0) + \int_0^t \phi(t-\tau)\,B\,r(\tau)\,d\tau$$

Where, $\phi(t) = LT^{-1}\,[sI - A]^{-1}$

Where $\phi(t)$ is *STM.*

$$C(n)\text{ (final)} = \lim_{z\to 1}(z-1) \times \frac{1}{4}\frac{(z^2+1)(z-1)(z+1)}{z^3(z-1)^2}$$

$$= \lim_{Z\to 1}\frac{1}{4}\frac{(z^2+1)(z+1)}{z^3} = \frac{1}{4} \times 4 = 1$$

5. **Ans. (c)**

Solution: If the function is Non Homogenous i.e.

$$X(t) = A\,X(t) + B\,u(t)\,;\, X(o) = X_o$$

The solution of this equation is given by

$$X(t) = e^{At}X(o) + \int_0^t e^{-A\tau} B\,u\,(d\tau)$$

$$X(t) = \phi(t)\,X(0) + \int_0^t \phi\,(t-\tau)\,Br\,(\tau)\,d\tau$$

Where, $\phi(t) = LT^{-1}\,[sI - A]^{-1}$

Where $\phi(t)$ is *STM.*

In present case

$$A = \begin{pmatrix} 1 & 0 \\ 1 & 1 \end{pmatrix} \text{ and } X(t_o) = \begin{pmatrix} 1 \\ 0 \end{pmatrix} \quad B = \begin{pmatrix} 0 \\ 0 \end{pmatrix}$$

$$x(t) = \exp\,[A(t-t_o)]\;X(t_o) + \int_{to}^{t} \exp\,[A(t-\tau)]\,B\,u\,(\tau)\,d\tau$$

$$= \exp\,[A\,(t-t_o]\,X(t_o) + 0$$

$$(sI - A) = \begin{pmatrix} s & 0 \\ 0 & s \end{pmatrix} - \begin{pmatrix} 1 & 0 \\ 1 & 1 \end{pmatrix}$$

$$(sI - A) = \begin{pmatrix} s-1 & 0 \\ -1 & s-1 \end{pmatrix}$$

$$(sI - A)^{-1} = \frac{\begin{pmatrix} (s-1) & 0 \\ 1 & (s-1) \end{pmatrix}}{(s-1)^2}$$

$$= \begin{pmatrix} [1/(s-1)] & 0 \\ [1/(s-1)^2] & [1/(s-1)] \end{pmatrix}$$

$$LT^{-1}\,[(sI - A)^{-1}] = e^{At} = \begin{pmatrix} e^t & 0 \\ te^t & e^t \end{pmatrix}$$

$$x(t) = e^{At} \times [0]$$

$$= \begin{pmatrix} e^t & 0 \\ te^t & e^t \end{pmatrix}\begin{pmatrix} 1 \\ 0 \end{pmatrix} = \begin{pmatrix} e^t \\ te^t \end{pmatrix}$$

6. **Ans. (b)**

Solution:

$$[sI - A]^{-1} = \begin{pmatrix} s-1 & 0 \\ 0 & s-1 \end{pmatrix}^{-1}$$

$$\text{STM, } \phi(t) = LT^{-1}\,[sI - A]^{-1}$$

$$= LT^{-1}\begin{Bmatrix} [1/(s-1)] & 0 \\ 0 & [1/(s-1)] \end{Bmatrix}$$

$$= \begin{pmatrix} e^t & 0 \\ 0 & e^t \end{pmatrix}$$

7. Ans. (d)

Solution: Rewriting state equations in matrix form, we get

$$\begin{bmatrix} X_1' \\ X_2' \end{bmatrix} = \begin{bmatrix} -3 & -1 \\ 2 & 0 \end{bmatrix}\begin{bmatrix} X_1 \\ X_2 \end{bmatrix} + \begin{bmatrix} 1 \\ 0 \end{bmatrix} u(t)$$

$$Y = [1 \ \ 0]\begin{bmatrix} X_1 \\ X_2 \end{bmatrix} + \begin{bmatrix} 1 \\ 0 \end{bmatrix} u(t)$$

Then,

$$A = \begin{bmatrix} -3 & -1 \\ 2 & 0 \end{bmatrix}; B = \begin{bmatrix} 1 \\ 0 \end{bmatrix}; D = [1 \ \ 0]$$

Now, $Q_c = [B : BA]$

$$= \begin{bmatrix} -3 & -1 \\ 0 & 20 \end{bmatrix}$$

$$\det Q_c = \begin{bmatrix} -3 & -1 \\ 0 & 2 \end{bmatrix} \neq \text{ i.e. controllable}$$

$$Q_c = \left[D^T \ \ D^T \ \ A^T\right]$$

$$\begin{bmatrix} D \\ DA \end{bmatrix} = \begin{bmatrix} -3 & -1 \\ 0 & 20 \end{bmatrix}$$

$$\det Q_c = \begin{bmatrix} 1 & -3 \\ 0 & -1 \end{bmatrix} \neq \text{ i.e. observable}$$

i.e. controllable

Hence the system is controllable and observable.

8. Ans. (b)

Solution: We have

$\dot{X} = AX + BU$ Where λ is set of Eigen values

$\dot{W} = CW + DU$ Where μ set of Eigen values

If a linear system is equivalently represented by two set of state equation, then for both sets, states will be same but their sets of Eigen values will not be same, i.e.

$$X = W \text{ but } \lambda \neq \mu$$

9. Ans. (a)

Solution: State transition matrix = $[SI - A]^{-1}$

$$\phi(s) = \begin{pmatrix} s & -1 \\ 1 & -2 \end{pmatrix}^{-1} = \frac{1}{s^2+1}\begin{pmatrix} s & 1 \\ -1 & s \end{pmatrix}$$

Another way

$$\phi(t) = LT^{-1}\ [SI - A]^{-1}$$

$$= LT^{-1}\left[\begin{pmatrix} s & 0 \\ 0 & s \end{pmatrix} - \begin{pmatrix} 0 & 1 \\ -1 & 0 \end{pmatrix}\right]$$

$$= LT^{-1}\left[\begin{pmatrix} s & -1 \\ 1 & s \end{pmatrix}\right]^{-1}$$

$$= \frac{LT^{-1}}{s^2+1}\begin{pmatrix} s & 1 \\ -1 & s \end{pmatrix} = \begin{pmatrix} \cos t & \sin t \\ -\sin t & \cos t \end{pmatrix}$$

10. Ans. (a)

Solution: Transfer function

$$\frac{\omega(s)}{U(s)} = C(sI - A)^{-1}\ B + D$$

Where $x' = Ax + Br\ (t)$

$$y = [1 \ \ 0]\begin{bmatrix} \omega \\ i_a \end{bmatrix} = CX$$

$$\frac{\omega(s)}{U(s)} = [1 \ \ 0]\begin{bmatrix} \frac{s+10}{s^2+11s+11} & \frac{+1}{s^2+11s+11} \\ \frac{-1}{s^2+11s+11} & \frac{s+1}{s^2+11s+11} \end{bmatrix}\begin{bmatrix} 0 \\ 10 \end{bmatrix}$$

$$= \begin{bmatrix} \frac{s+10}{s^2+11s+11} & \frac{1}{s^2+11s+11} \end{bmatrix}\begin{bmatrix} 0 \\ 10 \end{bmatrix}$$

$$= \frac{10}{s^2+11s+11}$$

Where, STM $\phi(x) = [sI - A]^{-1}$

$$= \frac{\begin{bmatrix} s+10 & -1 \\ 1 & s+1 \end{bmatrix}}{s^2+11s+11}$$

11. Ans. (c)

Solution:

If $\quad x(0) = \begin{bmatrix} 1 \\ -2 \end{bmatrix}$

Then response $x(t) = \begin{bmatrix} e^{-2t} \\ -2e^{-2t} \end{bmatrix} \rightarrow \begin{bmatrix} 1/(s+2) \\ -2/(s+2) \end{bmatrix}$

If $\quad x(0) = \begin{bmatrix} 1 \\ -1 \end{bmatrix}$

Then response $x(t) = \begin{bmatrix} e^{-2t} \\ -e^{-2t} \end{bmatrix} \rightarrow \begin{bmatrix} 1/(s+2) \\ -1/(s+2) \end{bmatrix}$

Let, $\quad A = \begin{bmatrix} A_0 & A_1 \\ A_2 & A_3 \end{bmatrix}$

Then, $\quad x(t) = [sI - A]^{-1}\ x(0)$

Multiply both sides by $(sI - A)$ to above equation

$$[sI - A]\ x(t) = x(0)$$

Case 1:

$$\begin{bmatrix} s-A_0 & -A_1 \\ -A_2 & s-A_3 \end{bmatrix}\begin{bmatrix} 1/(s+2) \\ -2/(s+2) \end{bmatrix}=\begin{bmatrix} 1 \\ -2 \end{bmatrix}$$

$$\frac{s-A_o}{s+2}+\frac{2A_1}{s+2}=1$$

or, $s-A_o+2A_1=s+2$

or, $-A_o+2A_1=2$

or, $2A_1-A_o=2$

Finally,

$$2A_1-A_0=2 \quad \text{...(1)}$$

$$A_2-2A_3=4 \quad \text{...(2)}$$

Case 2

$$\begin{bmatrix} s-A_0 & -A_1 \\ -A_2 & s-A_3 \end{bmatrix}\begin{bmatrix} 1/(s+1) \\ -1/(s+1) \end{bmatrix}=\begin{bmatrix} 1 \\ -1 \end{bmatrix}$$

$$A_1-A_0=1 \quad \text{...(3)}$$

$$A_2-A_3=1 \quad \text{...(4)}$$

From (1), (2), (3) and (4)

$A_o=0$; $A_1=1$

$A_2=-2$; $A_3=-3$

Then, $A=\begin{bmatrix} 0 & 1 \\ -2 & -3 \end{bmatrix}$

STM, $\phi(\tau)=[sI-A]^{-1}$

$$[sI-A]=\begin{bmatrix} s & -1 \\ +2 & s+3 \end{bmatrix}$$

For eigen values

$$|(sI-A)|=0$$

$$s(s+3)+2=0$$

$$s^2+3s+2=0$$

$$(s+2)(s+1)=0$$

$$s=-1,-2$$

Eigen vectors can be find out from

$$[sI-A]x=0$$

If for eigen value $s=-1$ then eigen vector

$$\begin{bmatrix} -1 & -1 \\ 2 & 2 \end{bmatrix}\begin{bmatrix} x_1 \\ x_2 \end{bmatrix}=0$$

$$-x_1-x_2=0$$

$$2x_1+2x_2=0$$

$$\frac{x_1}{1}=\frac{-x_2}{1}$$

$$\text{Vector}=\begin{bmatrix} 1 \\ -1 \end{bmatrix}$$

If Eigen value then eigen vector $s=-2$

$$\begin{bmatrix} -2 & -1 \\ +2 & 2 \end{bmatrix}\begin{bmatrix} x_1 \\ x_2 \end{bmatrix}=0$$

$$-2x_1-x_2=0$$

$$\frac{x_1}{1}=\frac{-x_2}{1}$$

$$\text{Vector}=\begin{bmatrix} 1 \\ -2 \end{bmatrix}$$

12. Ans. (b)

Solution:

See above Solution No. (45)

Shich result as

$$A=\begin{bmatrix} 0 & 1 \\ -2 & -3 \end{bmatrix}$$

13. Ans. (d)

Solution:

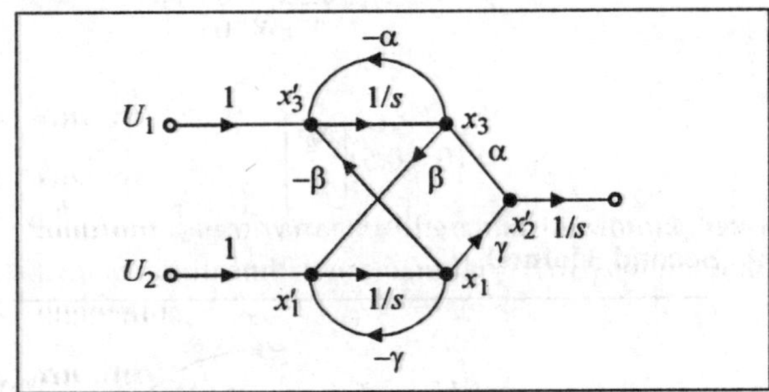

$$x_1'=-\gamma\, x_1+0\,x_2+\beta x_3+U_2$$

$$x_2'=\gamma\, x_1+0\,x_2+\alpha\, x_3+0$$

$$x_2'=\gamma\, x_1+0\,x_2-\alpha\, x_3+U_1$$

$$x_3'=-\beta x_1+0-\alpha x_3+U_1$$

$$\frac{d}{dt}\begin{pmatrix} x_1 \\ x_2 \\ x_3 \end{pmatrix}=\begin{bmatrix} -\gamma & 0 & \beta \\ \gamma & 0 & \alpha \\ -\beta & 0 & -\alpha \end{bmatrix}\begin{pmatrix} x_1 \\ x_2 \\ x_3 \end{pmatrix}+\begin{bmatrix} 0 & 1 \\ 0 & 0 \\ 1 & 0 \end{bmatrix}\begin{pmatrix} u_1 \\ u_2 \end{pmatrix}$$

Thus option (d) satisfies the solution.

14. Ans. (c)

Solution: Given that

$$A=\begin{bmatrix}1&0\\0&1\end{bmatrix},\ B=\begin{bmatrix}p\\q\end{bmatrix},\ AB=\begin{bmatrix}1&0\\0&1\end{bmatrix}\begin{bmatrix}p\\q\end{bmatrix},\ AB=\begin{bmatrix}p\\q\end{bmatrix}$$

$$\left[B:AB\right]=\begin{bmatrix}p&p\\q&q\end{bmatrix}$$

$$\left|B:AB\right| = pq-pq=0$$

Thus system is uncontrollable for all values of p and q.

15. Ans. (b)

Solution: Let us take two state variable as shown x_1 and x_2

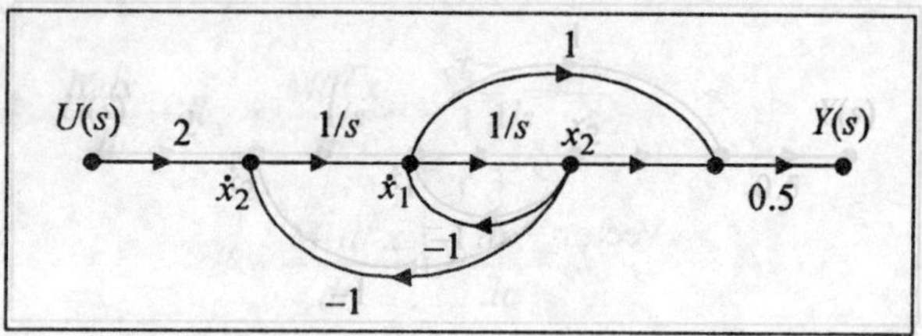

$$\dot{x}_2=-x_2+x_1$$

$$\dot{x}_1=-x_2+2\,U(s)$$

$$\Rightarrow\qquad \begin{bmatrix}\dot{x}_2\\\dot{x}_1\end{bmatrix}=\begin{bmatrix}-1&1\\-1&0\end{bmatrix}\begin{bmatrix}x_2\\x_1\end{bmatrix}+\begin{bmatrix}0\\2\end{bmatrix}U(s)$$

$$y(s)=0.5x_2'+0.5x_2$$

$$=0.5(-x_2+x_1)+0.5x_2$$

$$=-0.5x_2+0.5x_1+0.5x_2$$

$$=0+0.5x_2$$

$$=\begin{bmatrix}0&0.5\end{bmatrix}\begin{bmatrix}x_2\\x_1\end{bmatrix}$$

Second Method

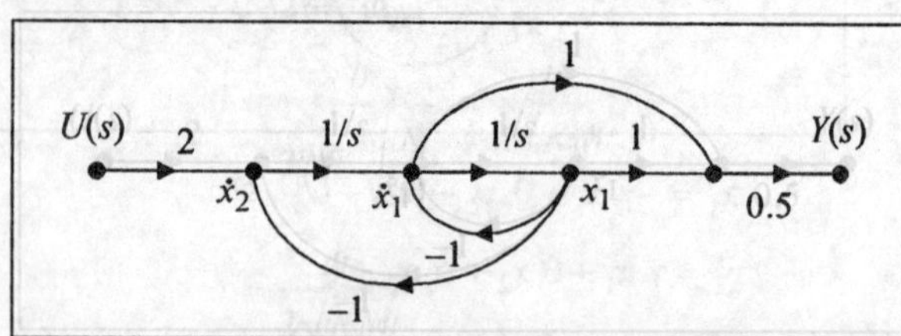

$$x_1'=-x_1+x_2$$

$$x_2'=-x_1+2U(s)$$

$$\begin{bmatrix}x_1'\\x_2'\end{bmatrix}=\begin{bmatrix}-1&1\\-1&0\end{bmatrix}\begin{bmatrix}x_1\\x_2\end{bmatrix}+\begin{bmatrix}0\\2\end{bmatrix}U(s)$$

$$Y(s)=0.5\,x_1+0.5\,x_1'$$

$$=0.5\,x_1+0.5\left(-x_1+x_2\right)$$

$$=0\,x_1+0.5\,x_2$$

$$=\begin{bmatrix}0&0.5\end{bmatrix}\begin{bmatrix}x_1\\x_2\end{bmatrix}$$

16. Ans. (b)

Solution: Draw the signal flow graph

From the graph

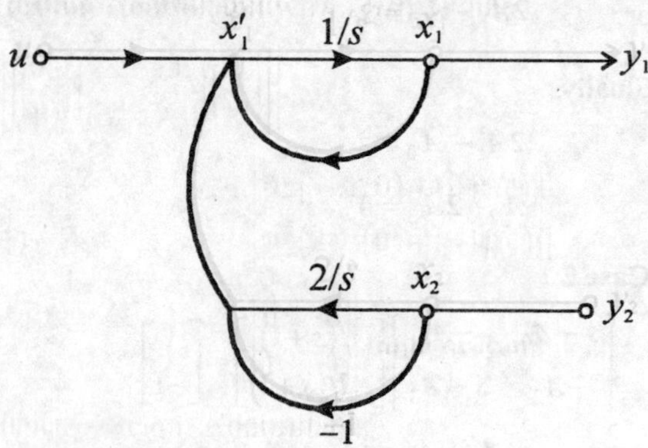

$$x'=-2x+4 \text{ and } y_1=x_1=x$$

$$y_2=2\int x_1'dx=2x_1$$

$$x'=[-2]x+[1]u$$

$$\begin{bmatrix}y_1\\y_2\end{bmatrix}=\begin{bmatrix}1\\2\end{bmatrix}x$$

17. Ans. (d)

Solution:

From the given state Dynamic equation

$$A=\begin{bmatrix}0&a_1&0\\0&0&a_2\\a_3&0&0\end{bmatrix};\quad B=\begin{bmatrix}0\\0\\1\end{bmatrix}$$

$$AB=\begin{bmatrix}0\\a_2\\0\end{bmatrix};\quad A^2B=\begin{bmatrix}a_1a_2\\0\\0\end{bmatrix}$$

System is controllable if

$$Q_C=\left[B\ \ AB\ \ A^2B\right]\neq 0$$

$$Q_C=\begin{bmatrix}0&0&a_1a_2\\0&a_2&0\\1&0&0\end{bmatrix}$$

Here rank of $Q_C=3$ order of matrix then Q_C is controllable

For system to be controllable determinant of controllability matrix should not be zero
Hence

$$\left.\begin{aligned} a_1 &\neq 0 \\ a_2 &\neq 0 \\ a_3 &= 0 \end{aligned}\right\} \text{then } |Q_C| \neq 0$$

18. Ans. (b)

Solution: The state equation can be written as

$$\begin{pmatrix} x_1'(t) \\ x_2'(t) \end{pmatrix} = \begin{pmatrix} a & b \\ c & d \end{pmatrix}\begin{pmatrix} x_1(t) \\ x_2(t) \end{pmatrix} + \begin{pmatrix} e \\ e \end{pmatrix} u(t)$$

Since $x_1(t) = x_2(t)$, therefore $a + b = c + d$

The controllability matrix is,

$$\phi = [B \quad AB]$$

$$B = \begin{pmatrix} e \\ e \end{pmatrix}$$

and

$$AB = \begin{pmatrix} a & b \\ c & d \end{pmatrix}\begin{pmatrix} e \\ e \end{pmatrix} = \begin{pmatrix} (a+b)e \\ (c+d)e \end{pmatrix}$$

$$\phi = \begin{pmatrix} e & (a+b)e \\ e & (c+d)e \end{pmatrix}$$

Let $\quad \phi = e^2 (c + d) - e^2 (a + b)$

Since $\quad a + b = c + d$

$$\phi = 0$$

Hence the system is uncontrollable.

19. Ans. (a)

Solution: $\dfrac{Y(s)}{U(s)} = C\,[sI - A]^{-1} B + D$

20. Ans. (c)

Solution:

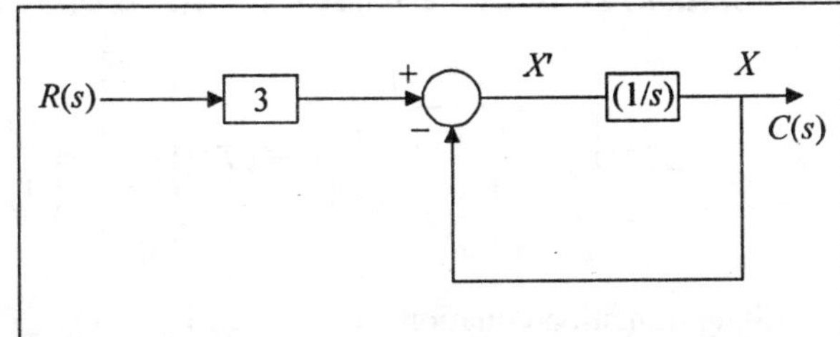

Form the figure $x_1' = -x_1 + 3r(t)$

The A matrix is (–1) and hence choice (c) is correct.

Second Method

$$X' = AX(s) + BU(s)$$

$$= \frac{C(s)}{R(s)} = \frac{3}{s+1}$$

Pole $\Rightarrow s = -1$

Therefore, $a = [-1]_{1\times 1}$

Eigen value of A = poles of the system

21. Ans. (a)

Solution: The transfer function $G(s)$ is:

$$G(s) = C(sI - A)^{-1} B + D$$

$$(sI - A)\begin{pmatrix} s & 0 \\ 0 & s \end{pmatrix} - \begin{pmatrix} -4 & -1 \\ 3 & -1 \end{pmatrix} = \begin{pmatrix} s+4 & 1 \\ -3 & s+1 \end{pmatrix}$$

$$(sI - A)^{-1} = \frac{\begin{pmatrix} s+1 & -1 \\ 3 & s+4 \end{pmatrix}}{s^2 + 5s + 4 + 3} = \frac{\begin{pmatrix} s+1 & -1 \\ 3 & s+4 \end{pmatrix}}{s^2 + 5s + 7}$$

$$G(s) = C(sI - A)^{-1} B + D$$

Hence

$$G(s) = \frac{[1 \;\; 0]\begin{pmatrix} s+1 & 3 \\ -1 & s+4 \end{pmatrix}}{s^2 + 5s + 7}\begin{pmatrix} 1 \\ 1 \end{pmatrix}$$

$$= \frac{[(s + 1 + 0) - (-1)]}{s^2 + 5s + 7}\begin{pmatrix} 1 \\ 1 \end{pmatrix}$$

$$= \frac{[(s + 1 + 1]}{s^2 + 5s + 7}\begin{pmatrix} 1 \\ 1 \end{pmatrix} = \frac{s}{s^2 + 5s + 7}$$

22. Ans. (a)

Solution: $X' = \begin{pmatrix} -3 & 1 \\ 0 & -2 \end{pmatrix} X = AX$

$$x(t) = [LT^{-1}\,\{sI - A)^{-1}]\, x(0)$$

$$= LT^{-1}\begin{pmatrix} s+3 & -1 \\ 0 & s+2 \end{pmatrix}^{-1} x(0)$$

$$= LT^{-1}\left(\frac{\begin{pmatrix} s+2 & 1 \\ 0 & s+3 \end{pmatrix}}{(s+2)(s+3)}\right) x(0)$$

$$= \begin{pmatrix} LT^{-1}\dfrac{1}{s+3} & LT^{-1}\left(\dfrac{1}{s+2}\cdot\dfrac{1}{s+3}\right) \\ 0 & LT^{-1}\dfrac{1}{s+2} \end{pmatrix} x(0)$$

$$= \begin{pmatrix} e^{-3t} & e^{-2t} - e^{-3t} \\ 0 & e^{-2t} \end{pmatrix} \begin{pmatrix} 10 \\ -10 \end{pmatrix}$$

$$X_{ss} = \lim_{t \to \infty} x(t)$$

$$= \lim_{t \to \infty} \begin{pmatrix} e^{-3t} & e^{-2t} - e^{-3t} \\ 0 & e^{-2t} \end{pmatrix} \begin{pmatrix} 10 \\ -10 \end{pmatrix} = \begin{pmatrix} 0 \\ 0 \end{pmatrix}$$

23. Ans. (c)

Solution: $STM = e^{at} = LT^{-1}(sI - A)^{-1}$

irrespective of $x(0)$

24. Ans. (b)

Solution: $X' = \begin{pmatrix} 2 & 0 \\ 0 & 4 \end{pmatrix} X + \begin{pmatrix} 1 \\ 1 \end{pmatrix} u;\ Y = [4\ 0]\, X$

$$A = \begin{pmatrix} 2 & 0 \\ 0 & 4 \end{pmatrix}; C = [4\ 0]$$

$$[sI - A] = \begin{pmatrix} s-2 & 0 \\ 0 & s-4 \end{pmatrix}$$

$$[sI - A]^{-1} = \phi(s) = \begin{pmatrix} \frac{1}{s-2} & 0 \\ 0 & \frac{1}{s-2} \end{pmatrix}$$

$$Y(s) = \phi(s)\, C \,.\, X(s)$$

$$\frac{Y(s)}{X(s)} = \begin{pmatrix} \frac{1}{s-2} & 0 \\ 0 & \frac{1}{s-2} \end{pmatrix} [4\ \ 0] = \frac{4}{s-2}$$

With input as $\delta(t)$, $L[\delta(t)] = 1$

$$Y(s) = \frac{4}{s-2} \,.\, [1]$$

$$Y(s) = \frac{4}{s-2}$$

$$y(t) = 4\, e^{2t}$$

25. Ans. (a)

Solution: The state of the system at time t is given by formula

$$x(t) = \phi(t)\, X(0)$$

$$= LT^{-1}\, [sI - A]^{-1}\, x(0)$$

$$= \begin{bmatrix} e^{-2t} & 0 \\ 0 & e^{-t} \end{bmatrix} \begin{bmatrix} 3 \\ 2 \end{bmatrix} = \begin{bmatrix} 2e^{-2t} \\ 2e^{-t} \end{bmatrix}$$

At, $t = 1$, $\quad x(1) = \begin{bmatrix} 2e^{-2} \\ 3e^{-1} \end{bmatrix} = \begin{bmatrix} 0.271 \\ 1.100 \end{bmatrix}$

26. Ans. (a)

Solution: $x_1 = \omega$,

$$x_1' = x_2 = \frac{d\omega}{dt},$$

$$x_2' = x_1'' = \frac{d^2\omega}{dt^2}$$

$$\frac{d^2\omega}{dt^2} = -\frac{B}{J}\frac{d\omega}{dt} - \frac{K^2}{LJ}\omega + \frac{K}{LJ}V_a$$

$$x_2' = -\frac{B}{J}x_2 - \frac{K^2}{LJ}x_1 + \frac{K}{LJ}V_a$$

$$\begin{bmatrix} x_2' \\ x_1' \end{bmatrix} = \begin{bmatrix} -\frac{B}{J} & \frac{K^2}{LJ} \\ 1 & 0 \end{bmatrix} \begin{bmatrix} x_2 \\ x_1 \end{bmatrix} + \begin{bmatrix} \frac{K}{LJ} \\ 0 \end{bmatrix} V_a$$

27. Ans. (a)

Solution: Characteristic equation for the system is, given by

$$| sI - A | = 0$$

or $\quad \begin{vmatrix} s & -2 \\ -2 & s \end{vmatrix} = 0$

or $\quad s^2 - 4 = 0$ or $s = \pm\, 2$.

28. Ans. (a)

Solution: $(sI - A)^{-1} = \begin{bmatrix} s & +1 \\ 0 & s+3 \end{bmatrix} = \dfrac{Adj(sI - A)}{|sI - A|}$

$$= \frac{\begin{bmatrix} s+3 & -1 \\ 0 & s \end{bmatrix}}{\begin{bmatrix} s & 1 \\ 0 & s+3 \end{bmatrix}} = \frac{\begin{bmatrix} s+3 & -1 \\ 0 & s \end{bmatrix}}{s(s+3)} = \begin{bmatrix} \frac{1}{s} & \frac{-1}{s(s+3)} \\ 0 & \frac{1}{s+3} \end{bmatrix}$$

$$\therefore\ \varphi(t) = LT^{-1}(sI - A)^{-1} = \begin{bmatrix} 1 & \frac{1}{3}(1 - e^{-3t}) \\ 0 & e^{-3t} \end{bmatrix}$$

29. Ans. (c)

Solution: Zero state response = $LT^{-1}\phi(s)\, B\, U(s)$

$$= LT^{-1} \begin{bmatrix} \frac{1}{s} & \frac{-1}{s(s+3)} \\ 0 & \frac{1}{s+3} \end{bmatrix} \begin{bmatrix} 1 \\ 0 \end{bmatrix} \frac{1}{s} = LT^{-1} \begin{bmatrix} \frac{1}{s^2} \\ 0 \end{bmatrix} = \begin{bmatrix} t \\ 0 \end{bmatrix}$$

State transition equation

= Zero input response + zero state response

$$\therefore \quad x(t) = \phi(t)\, x(0) + t$$

$$\begin{bmatrix} 1 & \frac{1}{3}(1-e^{-3t}) \\ 0 & e^{-3t} \end{bmatrix}\begin{bmatrix} -1 \\ 3 \end{bmatrix}+\begin{bmatrix} t \\ 0 \end{bmatrix}$$

$$=\begin{bmatrix} -1+1-e^{-3t} \\ 0+3e^{-3t} \end{bmatrix}+\begin{bmatrix} t \\ 0 \end{bmatrix}=\begin{bmatrix} t-e^{-3t} \\ 3e^{-3t} \end{bmatrix}$$

30. Ans. (b)

Solution: Given transfer function

$$\mathrm{H}(s)=\frac{\mathrm{Y}(s)}{\mathrm{U}(s)}=\frac{3(s-2)}{s^3+4s^2-2s+1}$$

Let $\mathrm{Y}_1(s)=\dfrac{1}{s^3+4s^2-2s+1}\mathrm{U}(s)$

or $s^3\mathrm{Y}_1(s)+4s^2\mathrm{Y}_1(s)-2s\mathrm{Y}_1(s)+\mathrm{Y}_1(s)=\mathrm{U}(s)$

$$y_1'''(t)+4y_1''(t)-2y_1'(t)+y_1(t)=u(t) \qquad \ldots(1)$$

Let $y_1=x_1$ ...(2)

$y_1'=x_2=x_1'$ ...(3)

$y_1''=x_3=x_2'$ ...(4)

$y_1'''=x_3'$ ...(5)

From equation (1) (2) (3) (4) and (5)

$$x_3'+4x_3-2x^2+x_1=u(t)$$

or $x_3'=-x_1+2x_2-4x_3+u$

In state space model

$x_1=x_2$

$x_2=x_3$

$x_3=-x_1+2x_2-4x_2+u$

or $\begin{bmatrix} x_1' \\ x_2' \\ x_3' \end{bmatrix}=\begin{bmatrix} 0 & 1 & 0 \\ 0 & 0 & 1 \\ -1 & 2 & -4 \end{bmatrix}\begin{bmatrix} x_1 \\ x_2 \\ x_3 \end{bmatrix}+\begin{bmatrix} 0 \\ 0 \\ 1 \end{bmatrix}u$

matrix $\mathrm{A}=\begin{bmatrix} 0 & 1 & 0 \\ 0 & 0 & 1 \\ -1 & 2 & -4 \end{bmatrix}$

and $\mathrm{Y}(s)=-6\mathrm{Y}_1(s)+3s\mathrm{Y}_1(s)$

or $y(t)=-6_{y1}(t)+3_{y1}(t)$

$=-6x_1+3x_2$

and $y=\begin{bmatrix} -6 & 3 & 0 \end{bmatrix}\begin{bmatrix} x_1 \\ x_2 \\ x_3 \end{bmatrix}$

31. Ans. (d)

Solution: $T.F=C(sI-A)^{-1}B+D$

$$(sI-A)=\begin{bmatrix} s & 0 \\ 0 & s \end{bmatrix}-\begin{bmatrix} 0 & 1 \\ 0 & -2 \end{bmatrix}$$

$$=\begin{bmatrix} s & -1 \\ 0 & s+2 \end{bmatrix}$$

$$s(sI-A)^{-1}=\frac{1}{s(s+2)}\begin{bmatrix} s+2 & 1 \\ 0 & s \end{bmatrix}$$

$$=\begin{bmatrix} 1 & 0 \end{bmatrix}\frac{1}{s(s+2)}\begin{bmatrix} s+2 & 1 \\ 0 & s \end{bmatrix}\begin{bmatrix} 0 \\ 1 \end{bmatrix}$$

$$=\frac{1}{s(s+2)}\begin{bmatrix} 1 & 0 \end{bmatrix}\begin{bmatrix} (0+1) \\ (0+s) \end{bmatrix}$$

$$=\frac{1}{s+2}\begin{bmatrix} +1 & +0 \end{bmatrix}=\frac{1}{s(s+2)}$$

32. Ans. (a)

Solution:

For type 1 system, i.e. $G(s)=\dfrac{1}{s(s+2)}$ and unit step input

$e_{ss}=0$

33. Ans. (c)

Solution:

$$\frac{dx_1(t)}{dt}=-3X_1(t)+x_2(t)+2u(t)$$

$$\Rightarrow sX_1(s)=-3X_1(s)+X_2(s)+2 \qquad \ldots(1)$$

$$\frac{dx_2(t)}{dt}=-2x_2(t)+\mu(t)$$

$$\Rightarrow sX_2(s)=-2x_2(s)+1$$

$$(s+2)X_2=1 \quad x_2=\frac{1}{(s+2)}$$

From [1], $(s+3)x_1$

$$=2+\frac{1}{s(s+2)} \quad x_1=\frac{2s+5}{(s+2)(s+3)}$$

$$y(t)=x_1(t)\Rightarrow H(s)=\frac{Y(s)}{X(s)}$$

$$=\frac{2s+5}{(s+2)(s+3)}=\frac{2s+5}{s^2+5s+6}$$

Second Method

$$x_i'(t) = -3x_1(t) + x_2(t) + 2u(t)$$

$$x_2'(t) = -2x_2(t) + u(t)$$

$$y(t) = x_1(t)$$

$$\begin{bmatrix} x_1' \\ x_2' \end{bmatrix} = \underbrace{\begin{bmatrix} -3 & 1 \\ 0 & -2 \end{bmatrix}}_{A}\begin{bmatrix} x_1 \\ x_2 \end{bmatrix} + \underbrace{\begin{bmatrix} 2 \\ 0 \end{bmatrix}}_{B} u(t)$$

$$y(t) = \begin{bmatrix} \underbrace{1 \quad 0}_{C} \end{bmatrix}\begin{bmatrix} x_1 \\ x_2 \end{bmatrix}$$

$$I = \begin{bmatrix} 1 & 0 \\ 0 & 1 \end{bmatrix}$$

$$[sI - A] = \begin{bmatrix} s & 0 \\ 0 & s \end{bmatrix} - \begin{bmatrix} -3 & 1 \\ 0 & -2 \end{bmatrix}$$

$$= \begin{bmatrix} s+3 & -1 \\ 0 & s+2 \end{bmatrix}$$

$$[sI - A]^{-1} = \frac{1}{(s+3)(s+2)}\begin{bmatrix} s+2 & 1 \\ 0 & s+3 \end{bmatrix}$$

$$= \begin{bmatrix} \frac{1}{s+3} & \frac{1}{(s+3)(s+2)} \\ 0 & \frac{1}{s+2} \end{bmatrix}$$

$$C = [1 \quad 0]$$

$$H(s) = C[sI - A]^{-1} B$$

$$= [1 \quad 0]\begin{bmatrix} \frac{1}{s+3} & \frac{1}{(s+3)(s+2)} \\ 0 & \frac{1}{(s+2)} \end{bmatrix}\begin{bmatrix} 2 \\ 0 \end{bmatrix}$$

$$= \begin{bmatrix} \frac{1}{s+3} & \frac{1}{(s+3)(s+2)} \end{bmatrix}\begin{bmatrix} 2 \\ 0 \end{bmatrix}$$

$$= \begin{bmatrix} \frac{2}{s+3} + 0 + \frac{2}{(s+3)(s+2)} + 0 \end{bmatrix}$$

$$= \frac{2(s+2)+2}{(s+3)(s+2)}$$

$$= \frac{2s+4+2}{s^2+5s+6}$$

$$= \frac{2s+6}{s^2+5s+6}$$

∴ The required transfer function

$$= \frac{2s+6}{s^2+5s+6}$$

34. Ans (b)

Solution:

$$e^{At} = L^{-1}[\phi(s)]$$

$$\phi(s) = [[sI - A]^{-1}]$$

From given states equations

$$x_1(t) = -3x_1(t) + x_2(t) + 2\mu(t)$$

$$x_2(t) = -2x_2(t) + \mu(t)$$

$$\phi(s) = \left[\begin{bmatrix} s & 0 \\ 0 & s \end{bmatrix} - \begin{bmatrix} -3 & 1 \\ 0 & -2 \end{bmatrix}\right]^{-1}$$

$$= \begin{bmatrix} s+3 & -1 \\ 0 & s+2 \end{bmatrix}^{-1}$$

$$\phi(s) = \begin{bmatrix} \frac{1}{s+3} & \frac{1}{(s+2)(s+3)} \\ 0 & \frac{1}{(s+2)} \end{bmatrix}$$

$$= \begin{bmatrix} e^{-3t} & e^{-2t} - e^{-3t} \\ 0 & e^{-2t} \end{bmatrix}$$

Second Method

Solution:

State Transition matrix

$$\phi(t) = \lambda^{-1}[SI - A]^{-1}$$

$$= \lambda^{-1}\begin{bmatrix} \frac{1}{s+3} & \frac{1}{(s+3)(s+2)} \\ 0 & \frac{1}{s+2} \end{bmatrix}$$

$$\lambda^{-1}\left[\frac{1}{(s+3)(s+2)}\right]$$

Solve it by partial fraction

$$\frac{1}{(s+3)(s+2)}$$

$$= \frac{A}{s+3} + \frac{B}{s+2}$$

$$I = A(s+2) + B(s+3)$$

$$I = 2A + 3B + (A+B)s$$

Equating coefficients of similar power of 's'

$$2A + 3B = 1 \text{.......}1$$

$$A + B = 0$$

$$2A + 2B = 0 \text{......}2$$

$$2A + 3B = 1$$

$$\underline{2A + 2B = 0}$$

$$B = 1$$

Hence from 1

$$2A + 3 = 1$$

$$2A = -2$$

$$A = -1$$

Put value of A and B in 0 equation

$$\frac{1}{(s+3)(s+2)} = \frac{-1}{s+3} + \frac{1}{s+2}$$

$\therefore$ $STM, \phi(t) =$

$$\lambda^{-1}\begin{bmatrix} \frac{1}{s+3} & \frac{-1}{s+3} + \frac{1}{s+2} \\ 0 & \frac{1}{s+2} \end{bmatrix}$$

$$\text{STM}, \phi(t) = \begin{bmatrix} e^{-3t} & -e^{-3t} + e^{-2t} \\ 0 & e^{-2t} \end{bmatrix}$$

$$\phi(t) = \begin{bmatrix} e^{-3t} & e^{-2t} - e^{-3t} \\ 0 & e^{-2t} \end{bmatrix}$$

35. Ans. (d)

Solution: Roots of $(sI - A)$ are given value.

$$\begin{bmatrix} s & 0 & 0 & 0 \\ 0 & s & 0 & 0 \\ 0 & 0 & s & 0 \\ 0 & 0 & 0 & s \end{bmatrix} - \begin{bmatrix} 0 & 1 & 0 & 0 \\ 0 & 0 & 1 & 0 \\ 0 & 0 & 0 & 1 \\ 0 & 0 & 0 & 1 \end{bmatrix}$$

$$= \begin{bmatrix} s & -1 & 0 & 0 \\ 0 & s & -1 & 0 \\ 0 & 0 & s & -1 \\ 0 & 0 & 0 & s-1 \end{bmatrix}$$

$$s^3(s-1) = 0$$

$$s = 1, 0, 0, 0$$

36. Ans. (b)

Solution: Number of state variable depends on the number of storage element, i.e. Total number of L and C is 2.

37. Ans. (a)

Solution:

$$T(s) = C(sI - A)^{-1} B = \begin{bmatrix} 1 & 0 \end{bmatrix} \begin{bmatrix} s & -1 \\ 0 & s+3 \end{bmatrix}^{-1} \begin{bmatrix} 1 \\ 0 \end{bmatrix}$$

$$= \frac{\begin{bmatrix} 1 & 0 \end{bmatrix} \begin{bmatrix} s+3 & 1 \\ 0 & s \end{bmatrix} \begin{bmatrix} 1 \\ 0 \end{bmatrix}}{s(s+3)} = \frac{1}{s}$$

38. Ans. (b)

Solution:

$$T(s) = C[sI - A]^{-1} B$$

As given, $A = F = \begin{bmatrix} 0 & 1 \\ -4 & -2 \end{bmatrix};$

$$B = G = \begin{bmatrix} 0 \\ 1 \end{bmatrix} \text{ and}$$

$$C = H = \begin{bmatrix} 1 & 0 \end{bmatrix};$$

$$[sI - A] = \begin{bmatrix} s & 0 \\ 0 & s \end{bmatrix} - \begin{bmatrix} 0 & 1 \\ -4 & -2 \end{bmatrix}$$

$$T(s) = \begin{bmatrix} 1 & 0 \end{bmatrix} \begin{bmatrix} s & -1 \\ 4 & (s+2) \end{bmatrix}^{-1} \begin{bmatrix} 0 \\ 1 \end{bmatrix}$$

$$= \frac{\begin{bmatrix} 1 & 0 \end{bmatrix} \begin{bmatrix} (s+2) & -1 \\ -4 & s \end{bmatrix} \begin{bmatrix} 0 \\ 1 \end{bmatrix}}{s^2 + 2s + 4}$$

$$\frac{\begin{bmatrix} 1 & 0 \end{bmatrix} \begin{bmatrix} 1 \\ s \end{bmatrix}}{s^2 + 2s + 4}$$

$$= \frac{1}{s^2 + 2s + 4}$$

$$\frac{1}{s^2 + 2\delta\omega_{ns} + \omega_n^2}$$

Hence $\omega_n = 2$

$$2\delta\omega_n = 2$$

$$\delta = \frac{1}{2} = 0.5$$

39. Ans. (a)

Solution:

$$\frac{s}{s^2+5s+1}$$

$$\begin{bmatrix} x_1' \\ x_2' \end{bmatrix} = \begin{bmatrix} -4 & -1 \\ -3 & -1 \end{bmatrix}\begin{bmatrix} x_1 \\ x_2 \end{bmatrix} + \begin{bmatrix} 1 \\ 1 \end{bmatrix} u$$

$$y = [1 \quad 0]\begin{bmatrix} x_1 \\ x_2 \end{bmatrix}$$

$$T.F. = C(sI - A)^{-1} B$$

$$C = [1 \quad 0],\ A = \begin{bmatrix} -4 & -1 \\ -3 & -1 \end{bmatrix}$$

$$I = \begin{bmatrix} 1 & 0 \\ 0 & 1 \end{bmatrix}, B = \begin{bmatrix} 1 \\ 1 \end{bmatrix}$$

$$T.F. = [1 \quad 0], \left[\begin{bmatrix} S & 0 \\ 0 & S \end{bmatrix} - \begin{bmatrix} -4 & -1 \\ -3 & -1 \end{bmatrix}\right]^{-1}\begin{bmatrix} 1 \\ 1 \end{bmatrix}$$

$$= [1 \quad 0]\begin{bmatrix} s+4 & 1 \\ 3 & s+1 \end{bmatrix}^{-1}\begin{bmatrix} 1 \\ 1 \end{bmatrix}$$

$$\Rightarrow [1 \quad 0]\left[\frac{Adj\begin{bmatrix} s+4 & 1 \\ 3 & s+1 \end{bmatrix}}{\begin{vmatrix} s+4 & 1 \\ 3 & s+1 \end{vmatrix}}\right]\begin{bmatrix} 1 \\ 1 \end{bmatrix}$$

$a_{11} = s + 1$

$a_{12} = -3$

$a_{21} = -1$

$a_{22} = s + 4$

$$= \left(\frac{1}{s^2+5s+1}\right)[1 \quad 0]\begin{bmatrix} s+1 & -1 \\ -3 & s+4 \end{bmatrix}\begin{bmatrix} 1 \\ 1 \end{bmatrix}$$

$$= \left(\frac{1}{s^2+5s+1}\right)[s+1+0 \quad -1]\begin{bmatrix} 1 \\ 1 \end{bmatrix}$$

$$= \left[\frac{1}{s^2+5s+1}\right] \Rightarrow \frac{s}{s^2+5s+1}$$

40. Ans. (a)

Solution:

$$T(s) = C|sI - A|^{-1} B$$

$$= [1 \quad 0]\begin{bmatrix} s & -1 \\ 2 & s+3 \end{bmatrix}^{-1}\begin{bmatrix} 0 \\ 1 \end{bmatrix}$$

$$= [1 \quad 0]\begin{bmatrix} s+3 & 1 \\ -2 & s \end{bmatrix}\begin{bmatrix} 0 \\ 1 \end{bmatrix}$$

$$s^2 + 3s + 2$$

$$= \frac{1}{s^2+3s+2}$$

The system poles are given by

$$s^2 + 3s + 2 = 0$$

or $$(s+1)(s+2) = 0$$

or $$s = -1,\ s = -2$$

41. Ans. (b)

Solution:

$$\frac{Y(s)}{U(s)} = \frac{s+6}{s^2+5s+6}$$

or, $$\frac{X_1(s)}{U(s)} \times \frac{Y(s)}{X_1(s)} = \frac{s+6}{s^2+5s+6}$$

Here, $$\frac{X_1(s)}{U(s)} = \frac{1}{s^2+5s+6}$$

Let the state variable be defined as

$$x_1 = y(t)$$

$$x_1' = x_2$$

$$x_2' = x_1'' \text{ then } \ddot{x}_1 + 5\dot{x}_1 + 6x_1 = u(t) \qquad \text{...(1)}$$

or, $$\dot{x}_2 = -5x_2 - 6x_1 + u(t)$$

and, $$\frac{Y(s)}{X_1(s)} = s + 6$$

or, $$y(t) = \dot{x}_1 + 6x_1 = x_2 + 6x_1 \qquad \text{...(2)}$$

Writing equation (1) and equation (2) in matrix from, we have

$$y = [6 \quad 1]\begin{bmatrix} x_1 \\ x_2 \end{bmatrix}$$

and $$\begin{bmatrix} \dot{x}_1 \\ \dot{x}_2 \end{bmatrix} = \begin{bmatrix} 0 & 1 \\ -6 & -5 \end{bmatrix}\begin{bmatrix} x_1 \\ x_2 \end{bmatrix} + \begin{bmatrix} 0 \\ 1 \end{bmatrix} u(t)$$

42. Ans. (b)

10.11 Miscellaneous

1. **Ans. (c)**
2. **Ans (d)**

   Solution:

   Output due to

$$d(t) = d(t) + d(t).G_{ff}(s)\frac{1}{s(s+1)}$$

$$= d(t)\left[1 + \frac{G_{ff}(s)}{s(s+1)}\right]$$

$$0 = \sin t\left[1 + \frac{G_{ff}(j\omega)}{j\omega(j\omega+1)}\right]$$

$$\therefore \quad G_{ff}|j\omega|_{\omega=1} = -1(1+j) = 1-j$$

$$= \sqrt{2}\angle -\frac{\pi}{4}$$

Resistance (R_g) and (R_1)

therefore error = drop across $(R_g + R_1)$

$$= I_g(R_g + R_1)$$

$$= 10\times 10^{-6}(100)$$

$$= 10^{-3} volts = 1mV$$

% age error $= \frac{1mV}{200mV}\times 100 = 05\%$

3. **Ans (a)**

   Solution:

   For the figuer ouput due to $d(t)$ signal only.

$$Y(s) = \left[D(s)G_{ff}(s) - Y(s)\right]\left[\frac{1}{s(s+1)}\right] + D(s)$$

$$D(s)G_{ff}(s) + (s^2+s)D(s) = (s^2+s+1)Y(s)$$

$$\frac{Y(s)}{D(s)} = H(s) = \frac{G_{ff}(s)+s^2+s}{(s^2+s+1)}$$

As given let PD controller, $G_{ff}(s) = (1+s)$

therefore $H(s) = \frac{(s^2+2s+1)}{(s^2+s+1)}$

$$\left[H(j\omega)\right]_{\omega=2} = \frac{5}{\sqrt{9+4}} = \frac{5}{\sqrt{13}}$$

4. **Ans (a)**
5. **Ans. (c)**
6. **Ans. (b)**

   Solution: In a **seismic pick-up for getting** an output proportional to acceleration it is desirable to have a natural frequency very large in comparison to frequency of input and a damping ratio around 0.7.
7. **Ans. (a)**
8. **Ans. (d)**
9. **Ans. (c)**
10. **Ans. (c)**
11. **Ans. (d)**

    Solution: Proportional will give the output as the error signal and derivative will output as a step for a the ramp, so result is figure:

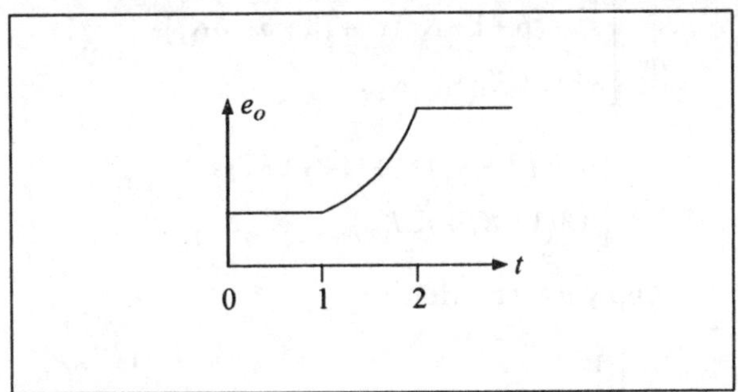

12. **Ans. (b)**
13. **Ans. (b)**
14. **Ans. (d)**

    Solution:

    The characteristic equation is,

$$1 + G(s) = 0$$

or $1 + \frac{10(\tau_1 s+1)}{\tau_1 s^2(s+1)} = 0$

or $\tau_1 s^3 + \tau_1 s^2 + 10\tau_1 s + 10 = 0$

Constructing routh-array,

s^3	τ_1	$10\tau_1$
s^2	τ_1	10
s^1	$10(\tau_1 - 1)$	0
s^0	10	

For stability, $t_1 - 1 > 0$

or $\tau_1 > 1$

Thus, the stability range will be $\tau_1 > 1.0$

15. Ans. (d)

Solution:

For system to be stable

1. The inner loop gain > outer loop gain
2. Both gain should be finite

For Block diagram reduction

$$\text{Inner loop} = \frac{K_1/(s+1)}{1+\dfrac{K_1}{(s+1)}} = \frac{K_1}{1+K_1+s}$$

$$\text{Outer loop } GH = \left[\frac{K_1 K_2}{1+K_1+s}\right]\left[\frac{8}{s^2+6s+8}\right]$$

Characteristic equation $1+GH=0$

$$\text{or, } \begin{bmatrix} s^3+6s^2+8s+(1+K_1)s^2+6(1+K_1)s \\ +8(1+K_1)+8K_1K_2 \end{bmatrix}=0$$

$$\text{or, } \begin{bmatrix} s^3+(6+1+K_1)s^2+(8+6+6K_1)s \\ +8(1+K_1+K_1K_2) \end{bmatrix}=0$$

$$\text{or, } \begin{bmatrix} s^3+(7+K_1)s^2+(14+6K_1)s \\ +8(1+K_1+K_1K_2) \end{bmatrix}=0$$

Apply RH criterion

s^3	1	$14+6K_1$
s^2	$7+K_1$	$8(1+K_1+K_1K_2)$
s^1	$\dfrac{(7+K_1)(14+6K_1)-8(1+K_1+K_1K_2)}{7+K_1}$	
s^0	$8(1+K_1+K_1K_2)$	

For stability

$$\frac{(7+K_1)(14+6K_1)-8(1+K_1+K_1K_2)}{7+K_1}>0 \quad \text{...(1)}$$

$$\text{And} \quad 8(1+K_1+K_1K_2)>0 \quad \text{...(2)}$$

Equation (1) is true only for $K_1=10$ and $K_2=1$

Equation (2) is true for all the given value options

16. Ans. (a)

17. Ans. (d)

Solution:

$$T.F.=2+\frac{0.4}{s}$$

$$T.F.=2\left[1+\frac{0.2}{s}\right]$$

$$\text{Compare it with } K_P\left[1+\frac{K_I}{s}\right]$$

$$K_P=2,\ K_I=-2$$

$$\text{We know that } K_I=\frac{1}{T_i}$$

Where T_i = Reset time

$$0.2=\frac{1}{T_i}$$

$$T_i = 5 \text{ minutes}$$

$$P.B.=\frac{100}{K_P}=\frac{100}{2}=50\%$$

18. Ans. (a)

Solution:

Transfer function second order band-pass filter

$$A_v(s)=\frac{(\omega_0/Q)A_0 s}{s^2+(\omega_0/Q)s+\omega_0^2}$$

Given, Q = 100, A_0 = 1, ω0 = 1000, we get

$$A_v(s)=\frac{10s}{s^2+10s+10^6}$$

19. Ans. (a)

Solution:

Thickness measured = 0.5 *m* and

Speed of sheet = 0.2 m/sec

So time taken to move distance

$$=L/v=\frac{0.5}{0.2}=2.5 \text{ sec}$$

Therefore, *m(t)* is delayed signal of $d(t)$ 2.5 sec

So, $m(t)=d(t-2.5)$

$$\text{So } G(s)=\frac{M(s)}{D(s)}=\frac{e^{-2.5s}\cdot D(s)}{D(s)}=e^{-2.5s}$$

20. Ans. (a)

Solution:

Characteristic equation of system is

$$1+KG(s)=0$$

For the oscillatory output,

$$s=j\omega$$

$$1+Ke^{-2.5j\omega}=0$$

It will be imaginary at $\omega=\omega_c$

$$1+K\,e^{-2.5j\omega_c}=0$$

Neglecting 1, as does not playoll in phase

$$\text{Phase angle for imaginary axis}=\pm\frac{\pi}{2}$$

So, when $\omega_c = \pi$

Then, $e^{-2.5j\omega_c} = e^{-2.5j\pi}$

$$e^{-0.5j\pi} = e^{-\frac{\pi}{2}j} = e^{-j\theta}$$

When $\theta = \frac{\pi}{2}$

21. Ans. (c)

Solution: Feed back path sensitivity i.e. sensitivity with change in G.

$$S_H^T = \frac{-GH}{1+GH} = -1$$

Forward path sensitivity, i.e. sensitivity with change in G.

$$S_G^T = \frac{1}{1+GH} = 0$$

Feed back path is more sensitive then forward path.

22. Ans. (c)

Solution:

On reducting this block diagram, we get

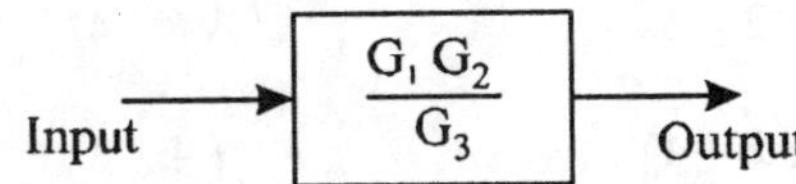

Let $y = \frac{u}{v}$

% error in y = % error in u − % error in v

here $u = G_1 + G_2, V = G_3$

23. Ans. (a)

Solution:

Gain without error = 10

Gain with +10% error = 10.1

Gain with − 10% error = 9.9

Hence over all system gain it is 10 ±1%

Second Method

$$G_f = \frac{G}{1+GH} = \frac{100}{1+100\times\frac{9}{100}}$$

$$= \frac{100}{1+9}$$

$$= \frac{100}{10}$$

$$= 10$$

24. Ans. (b)

Solution: Exists $Ax = b$, b should be in column space of A.

25. Ans. (c)

Solution: $x(t) = u(t) \Rightarrow X(s) = \frac{1}{s}$

$$y(t) = t\,e^{-t} \Rightarrow Y(s) = \frac{1}{(s+1)^2}$$

$$G(s) = \frac{Y(s)}{X(s)} = \frac{s}{(s+1)^2}$$

26. Ans. (a)

Solution: $\frac{d^2y}{dt^2} + \frac{dy}{dt} = \frac{du}{dt} + 2u$

Taking Laplace Transform and initial value zero.

$$s^2\,Y(s) + sY(s) = sU(s) + 2U(s)$$

$$(s^2 + s)\,Y(s) = U(s)\,(s+2)$$

$$\frac{Y(s)}{U(s)} = \frac{(s+2)}{s^2+s}$$

11 Engineering Mathematics

11.1 Linear Algebra

1. Given the matrix $\begin{pmatrix}-4\\4\end{pmatrix}\begin{pmatrix}2\\3\end{pmatrix}$, the eigenvector is

(a) $\begin{pmatrix}3\\2\end{pmatrix}$ (b) $\begin{pmatrix}4\\3\end{pmatrix}$

(c) $\begin{pmatrix}2\\-1\end{pmatrix}$ (d) $\begin{pmatrix}-1\\2\end{pmatrix}$

2. Given an orthogonal matrix

$$A=\begin{bmatrix}1 & 1 & 1 & 1\\ 1 & 1 & -1 & -1\\ 1 & -1 & 0 & 0\\ 0 & 0 & 1 & -1\end{bmatrix}$$

$[AA^T]^{-1}$ is

(a) $\begin{bmatrix}[1/4] & 0 & 0 & 0\\ 0 & [1/4] & 0 & 0\\ 0 & 0 & [1/2] & 0\\ 0 & 0 & 0 & [1/2]\end{bmatrix}$

(b) $\begin{bmatrix}[1/2] & 0 & 0 & 0\\ 0 & [1/21] & 0 & 0\\ 0 & 0 & [1/2] & 0\\ 0 & 0 & 0 & [1/2]\end{bmatrix}$

(c) $\begin{bmatrix}1 & 0 & 0 & 0\\ 0 & 1 & 0 & 0\\ 0 & 0 & 1 & 0\\ 0 & 0 & 0 & 1\end{bmatrix}$

(d) $\begin{bmatrix}[1/4] & 0 & 0 & 0\\ 0 & [1/4] & 0 & 0\\ 0 & 0 & [1/4] & 0\\ 0 & 0 & 0 & [1/4]\end{bmatrix}$

3. The eigen values and the corresponding eigen vectors of a 2 × 2 matrix are given by

Eigen value	Eigen vector
$\lambda_1 = 8$	$v_1 = \begin{pmatrix}1\\1\end{pmatrix}$
$\lambda_2 = 4$	$v_2 = \begin{pmatrix}1\\-1\end{pmatrix}$

The matrix is

(a) $\begin{pmatrix}6 & 2\\2 & 6\end{pmatrix}$ (b) $\begin{pmatrix}4 & 6\\6 & 4\end{pmatrix}$

(c) $\begin{pmatrix}2 & 4\\4 & 2\end{pmatrix}$ (d) $\begin{pmatrix}4 & 8\\8 & 4\end{pmatrix}$

4. For the matrix $\begin{pmatrix}4 & 2\\2 & 4\end{pmatrix}$, the eigen value corresponding to the eigen vector $\begin{pmatrix}101\\101\end{pmatrix}$ is

(a) 2 (b) 4

(c) 6 (d) 8

5. In the matrix equation Px = q, which of the following is a necessary condition for the existence of at least one solution for the unknown vector x :

(a) Augmented matrix [Pq] must have the same rank as matrix P

(b) Vector q must have only non-zero elements

(c) Matrix P must be singular

(d) Matrix P must be square

6. For the matrix $p=\begin{bmatrix}3 & -2 & 2\\ 0 & -2 & 1\\ 0 & 0 & 1\end{bmatrix}$, one of the eigen values is equal to –2. Which of the following is an eigen vector?

(a) $\begin{bmatrix} 3 \\ -2 \\ 1 \end{bmatrix}$ (b) $\begin{bmatrix} -3 \\ 2 \\ -1 \end{bmatrix}$

(c) $\begin{bmatrix} 1 \\ -2 \\ 3 \end{bmatrix}$ (d) $\begin{bmatrix} 2 \\ 5 \\ 0 \end{bmatrix}$

7. If $R = \begin{bmatrix} 1 & 0 & -1 \\ 2 & 1 & -1 \\ 2 & 3 & 2 \end{bmatrix}$, then top row of R^{-1} is

(a) [5 6 4] (b) [5 –3 1]
(c) [2 0 –1] (d) [2 –1 1/2]

8. Let x and y be two vectors in a 3 dimensional space and $<x, y>$ denote their dot product. Then the determinant

$$\det\begin{vmatrix} <x,x> & <x,y> \\ <y,x> & <y,x> \end{vmatrix}$$

(a) Is zero when x and y are linearly independent
(b) Is positive when x and y are linearly independent
(c) Is non-zero for all non-zero x and y
(d) Is zero only when either x or y is zero

9. Let A be a 3 × 3 matrix with rank 2. Then AX = 0 has
(a) Only the trivial solution X = 0
(b) One independent solution
(c) Two independent solutions
(d) Three independent solutions

10. Identify which one of the following is an eigen vector of the matrix $A = \begin{bmatrix} 1 & 0 \\ -1 & -2 \end{bmatrix}$

(a) $[-1 \quad 1]^T$
(b) $[3 \quad -1]^T$
(c) $[1 \quad -1]^T$
(d) $[-2 \quad 1]^T$

11. For a given 2 × 2 matrix A, it is observed that

$$A\begin{bmatrix} 1 \\ -1 \end{bmatrix} = -\begin{bmatrix} 1 \\ -1 \end{bmatrix} \text{ and } A\begin{bmatrix} 1 \\ -2 \end{bmatrix} = -2\begin{bmatrix} 1 \\ -2 \end{bmatrix}$$

Then, matrix A is

(a) $A = \begin{bmatrix} 2 & 1 \\ -1 & -1 \end{bmatrix}\begin{bmatrix} -1 & 0 \\ 0 & -2 \end{bmatrix}\begin{bmatrix} 1 & 1 \\ -1 & -2 \end{bmatrix}$

(b) $A = \begin{bmatrix} 1 & 1 \\ -1 & -2 \end{bmatrix}\begin{bmatrix} 1 & 0 \\ 0 & 2 \end{bmatrix}\begin{bmatrix} 2 & 1 \\ -1 & -1 \end{bmatrix}$

(c) $A = \begin{bmatrix} 1 & 1 \\ -1 & -2 \end{bmatrix}\begin{bmatrix} -1 & 0 \\ 0 & -2 \end{bmatrix}\begin{bmatrix} 2 & 1 \\ -1 & -1 \end{bmatrix}$

(d) $A = \begin{bmatrix} 0 & -2 \\ 1 & -3 \end{bmatrix}$

12. Let A be an $n \times n$ real matrix such that $A^2 = 1$ and y be an n-dimensional vector . Then, the linear system of equations $Ax = y$ has
(a) No solution
(b) A unique solution
(c) More than one but finitely many independent solutions
(d) Infinitely many independent solutions

13. Let $A = [a_{ij}]$, $1 \le i,\ j \le n$, with $n \ge 3$ and $a_{ij} = i \cdot j$. Then, rank of A is
(a) 0 (b) 1
(c) n – 1 (d) n

14. Consider the following system of linear equations

$$\begin{bmatrix} 2 & 1 & -4 \\ 4 & 3 & -12 \\ 1 & 2 & -8 \end{bmatrix}\begin{bmatrix} x \\ y \\ z \end{bmatrix} = \begin{bmatrix} \alpha \\ 5 \\ 7 \end{bmatrix}$$

Notice that the second and the third columns of the coefficient matrix are linearly dependent. For how many values of α, does this system of equations have infinitely many solutions?
(a) 0 (b) 1
(c) 2 (d) Infinitely many

15. The number of different n × n symmetric matrices with each element being either 0 or 1 is: (Note: power (2, x) is same as 2^x)
(a) Power (2, n) (b) Power (2, n^2)
(c) Power (2, $(n^2 + n)/2$)
(d) Power (2, $(n^2 - n)/2$)

16. Let A, B, C, D be n × n matrices, each with non-zero determinant. If ABCD = I, then B^{-1} is
(a) $D^{-1} C^{-1} A^{-1}$ (b) CDA
(c) ADC (d) Does not necessarily exist

17. What are the eigen values of the following 2 × 2 matrix

$$\begin{bmatrix} 2 & -1 \\ -4 & 5 \end{bmatrix}$$

(a) −1 and 1 (b) 1 and 6
(c) 2 and 5 (d) 4 and −1

18. It is given that $X_1, X_2, \ldots, X_M$ are M non-zero, orthogonal vectors. The dimension of the vector space spanned by the 2M vectors $X_1, X_2, \ldots, X_M$, $-X_1, -X_2, \ldots, -X_M$ is

(a) 2M
(b) M + 1
(c) M
(d) Dependent on the choice of $X_1, X_2, \ldots, X_M$

19. The following vector is linearly dependent upon the solution to the previous problem

(a) $[8 \ \ 9 \ \ 3]$ (b) $[-2 \ \ 17 \ \ 30]$
(c) $[4 \ \ 4 \ \ 5]$ (d) $[13 \ \ 2 \ \ -3]$

20. $x = [x_1 \ x_2 \ \ldots \ x_n]^T$ is an n-tuple non-zero vector. The n × n matrix $V = xx^T$

(a) Has rank zero (b) Has rank 1
(c) Is orthogonal (d) Has rank n

21. The linear operator $L(x)$ is defined by the cross product $L(x) = bX_x$ where $b = [0 \ 1 \ 0]^T$ and $x = [\dot{x}_1 \ x_2 \ x_3]^T$ are three dimensional vectors. The 3 × 3 matrix M of this operation satisfies

$$L(x) = M \begin{bmatrix} x_1 \\ x_2 \\ x_3 \end{bmatrix}$$

Then the eigen values of M are

(a) 0, +1, −1 (b) 1, −1, 1
(c) $i, -i$, 1 (d) $i, -i$, 0

22. Cayley –Hamilton theorem states that a square matrix satisfies its own characteristic equation.

Consider a matrix

$$A = \begin{bmatrix} -3 & 2 \\ -1 & 0 \end{bmatrix}$$

A satisfies the relation

(a) $A + 3I + 2A^{-1} = 0$
(b) $A^2 + 2A + 2I = 0$
(c) $(A+I)(A+2I) = 0$
(d) $e^{(A)} = 0$

23. Cayley –Hamilton theorem states that a square matrix satisfies its own characteristic equation.

Consider a matrix

$$A = \begin{bmatrix} -3 & 2 \\ -1 & 0 \end{bmatrix}$$

A^9 equals

(a) $511A + 510I$ (b) $309A + 104I$
(c) $154A + 155I$ (d) $e^{(9A)}$

24. How many solutions does the following system of linear equations have?

$$-x + 5y = -1$$
$$x - y = 2$$
$$x + 3y = 3$$

(a) Infinitely many solution
(b) Two distinct solutions
(c) Unique solution
(d) None

25. Consider the following system of equations in three real variables x_1, x_2 and x_3

$$2x_1 - x_2 + 3x_3 = 1$$
$$3x_1 - 2x_2 + 5x_3 = 2$$
$$-x_1 - 4x_2 + x_3 = 3$$

This system of equations has

(a) No solution
(b) A unique solution
(c) More than one but a finite number of solutions
(d) An infinite number of solutions

26. Consider the set of (column) vectors defined by $X = \{x \in R^3 \mid x_1 + x_2 + x_3 = 0$

where, $x^T = [x_1, x_2, x_3]^T\}$

Which of the following is TRUE?

(a) $\{[1, -1, 0]^T, [1, 0, -1]^T\}$ is basis for the sub-space X

(b) $\{[1, -1, 0]^T, [1, 0, -1]^T\}$ is linearly independent set, but it does not span X and therefore is not a basis of X

(c) X is not a subspace of R^3

(d) None of the above

27. All the four entries of the 2 × 2 matrix $P = \begin{bmatrix} p_{11} & p_{12} \\ p_{21} & p_{22} \end{bmatrix}$ are non-zero, and one of its eigen values is zero. Which of the following statements is true?

(a) $p_{11}p_{22} - p_{12}p_{21} = 1$ (b) $p_{11}p_{22} - p_{12}p_{21} = -1$
(c) $p_{11}p_{22} - p_{12}p_{21} = 0$ (d) $p_{11}p_{22} + p_{12}p_{21} = 0$

28. The system of linear equations
$4x + 2y = 7$
$2x + y = 6$ has
(a) A unique solution
(b) No solution
(c) An infinite number of solutions
(d) Exactly two distinct solutions

29. The eigen values of the following matrix are

$$\begin{bmatrix} -1 & 3 & 5 \\ -3 & -1 & 6 \\ 0 & 0 & 3 \end{bmatrix}$$

(a) $3, 3+5j, 6-j$ (b) $-6+5j, 3+j, 3-j$
(c) $3+j, 3-j, 5+j$ (d) $3, -1+3j, -1-3j$

30. The eigen values of a skew-symmetric matrix are
(a) always zero
(b) always pure imaginary
(c) either zero or pure imaginary
(d) always real

31. The system of equations
$x + y + z = 6$
$x + 4y + 6z = 20$
$x + 4y + \lambda z = \mu$
has no solution for values of λ and μ given by
(a) $\lambda = 6, \mu = 20$ (b) $\lambda = 6, \mu \neq 20$
(c) $\lambda \neq 6, \mu = 20$ (d) $\lambda \neq 6, \mu \neq 20$

32. The eigen value of a (2 × 2) matrix X are –2 and –3. The eigen value of the matrix $(X+1)^{-1}(X+5I)$ are
(a) – 3, – 4 (b) – 1, – 2
(c) –1, – 3 (d) –2, – 4

33. A real $n \times n$ matrix $A = [a_{ij}]$ is defined as follows:
$a_{ij} = i$, if $i = j$
$= 0$, otherwise
The summation of all n eigen values of A is
(a) $n(n+1)/2$ (b) $n(n-1)/2$
(c) $\frac{n(n+1)(2n+1)}{6}$ (d) n^2

34. X and Y are non-zero square matrices of size $n \times n$. If $\mathbf{XY} = 0_{n \times n}$ then
(a) $|\mathbf{X}| = 0$ and $|\mathbf{Y}| \neq 0$ (b) $|\mathbf{X}| \neq 0$ and $|\mathbf{Y}| = 0$
(c) $|\mathbf{X}| = 0$ and $|\mathbf{Y}| = 0$ (d) $|\mathbf{X}| \neq 0$ and $|\mathbf{Y}| \neq 0$

35. The matrix $M = \begin{bmatrix} -2 & 2 & -3 \\ 2 & 1 & -6 \\ -1 & -2 & 0 \end{bmatrix}$ has eigen values –3, –3, 5. An eigen vector corresponding to the eigen value 5 is $[1\ \ 2\ \ -1]^T$. One of the eigen vectors of the matrix M^3 is:

(a) $[1\ \ 8\ \ -1]^T$ (b) $[1\ \ 2\ \ -1]^T$
(c) $[1\ \ \sqrt[3]{2}\ \ -1]^T$ (d) $[1\ \ 1\ \ -1]^T$

36. An eigen vector of $P = \begin{pmatrix} 1 & 1 & 0 \\ 0 & 2 & 2 \\ 0 & 0 & 3 \end{pmatrix}$ is

(a) $[-1\ \ 1\ \ 1]^T$ (b) $[1\ \ 2\ \ 1]^T$
(c) $[1\ \ -1\ \ 2]^T$ (d) $[2\ \ 1\ \ -1]^T$

37. The trace and determinant of a 2×2 matrix are known to be –2 and –35 respectively. Its eigen values are
(a) –30 and –5 (b) –37 and –1
(c) –7 and 5 (d) 17.5 and –2

38. The characteristic equation of a (3×3) matrix P is defined as
$\alpha(\lambda) = |\lambda I - P| = \lambda^3 + \lambda^2 + 2\lambda + 1 = 0$. If I denote identity matrix, then the inverse of matrix P will be

(a) $(P^2 + P + 2I)$ (b) $(P^2 + P + I)$
(c) $-(P^2 + P + I)$ (d) $-(P^2 + P + 2I)$

39. If the rank of a (5×6) matrix Q is 4, then which one of the following statements is correct?

(a) Q will have four linearly independent rows and four linearly independent columns

(b) Q will have four linearly independent rows and five linearly independent columns

(c) QQ^T will be invertible

(d) Q^TQ will be invertible

40. Consider the matrices $X_{(4\times3)}$, $Y_{(4\times3)}$ and $P_{(2\times3)}$. The order of $[P(X^TY)^{-1}P^T]^T$ will be

(a) (2 × 2) (b) (3 × 3)
(c) (4 × 3) (d) (3 × 4)

41. Real matrices $[A]_{3\times1}$, $[B]_{3\times3}$, $[C]_{3\times5}$, $[D]_{5\times3}$, $[E]_{5\times5}$ and $[F]_{5\times1}$ are given. Matrices [B] and [E] are symmetric. Following statements are made with respect to these matrices.

1. Matrix product $[F]^T [C]^T [B] [C] [F]$ is a scalar.
2. Matrix product $[D]^T [F] [D]$ is always symmetric.

With reference to above statements, which of the following applies?

(a) Statement 1 is true but 2 is false
(b) Statement 1 is false but 2 is true
(c) Both the statements are true
(d) Both the statements are false

42. A square matrix B is skew-symmetric if

(a) $B^T = -B$ (b) $B^T = B$
(c) $B^{-1} = B$ (d) $B^{-1} = B^T$

43. The product of matrices $(PQ)^{-1} P$ is

(a) P^{-1} (b) Q^{-1}
(c) $P^{-1} Q^{-1} P$ (d) $PQ P^{-1}$

44. For a matrix $[M] = \begin{bmatrix} \frac{3}{5} & \frac{4}{5} \\ x & \frac{3}{5} \end{bmatrix}$, the transpose of the matrix is equal to the inverse of the matrix, $[M]^T = [M]^{-1}$. The value of x is given by

(a) $-\frac{4}{5}$ (b) $-\frac{3}{5}$
(b) $\frac{3}{5}$ (d) $\frac{4}{5}$

45. Let A, B, C, D be n × n matrices, each with non-zero determinant, if ABCD = I, then B^{-1} is

(a) $D^{-1} C^{-1} A^{-1}$
(b) CDA
(c) ADC
(d) does not necessarily exist

46. Given matrix $[A] = \begin{bmatrix} 4 & 2 & 1 & 3 \\ 6 & 3 & 4 & 7 \\ 2 & 1 & 0 & 1 \end{bmatrix}$, the rank of the matrix is

(a) 4 (b) 3
(c) 2 (d) 1

47. Solution for the system defined by the set of equations
$4y + 3z = 8$;
$2x - z = 2$ and
$3x + 2y = 5$ is

(a) x = 0; y = 1; z = 4/3 (b) x = 0; y = 1/2; z = 2
(c) x = 1; y = 1/2; z = 2 (d) non-existent

48. A is a 3 × 4 matrix and A x = b is inconsistent system of equation. The highest possible rank of A is

(a) 1 (b) 2
(3) 3 (d) 4

49. For the matrix $\begin{bmatrix} 4 & 1 \\ 1 & 4 \end{bmatrix}$ the eigen value are

(a) 3 and – 3 (b) – 3 and – 5
(b) 3 and 5 (d) 5 and 0

50. The minimum and the maximum eigen values of the matrix $\begin{bmatrix} 1 & 1 & 3 \\ 1 & 5 & 1 \\ 3 & 1 & 1 \end{bmatrix}$ are –2 and 6, respectively.

What is the other eigen value?

(a) 5 (b) 3
(c) 1 (d) – 1

51. The sum of the eigen values of the matrix given below is $\begin{bmatrix} 1 & 2 & 3 \\ 1 & 5 & 1 \\ 3 & 1 & 1 \end{bmatrix}$

(a) 5 (b) 7
(c) 9 (d) 18

52. If a square matrix A is real and symmetric, then the eigen values.

(a) Are always real
(b) Are always real and positive
(c) Are always real and non-negative
(d) Occur in complex conjugate pairs

53. The matrix $\begin{bmatrix} 1 & 2 & 4 \\ 3 & 0 & 6 \\ 1 & 1 & p \end{bmatrix}$ has one eigen value equal to 3.

The sum of the other two eigen values is

(a) p (b) p – 1
(c) p – 2 (d) p – 3

54. Multiplication of matrices E and F is G. Matrices E and G are

$$E = \begin{bmatrix} \cos\theta & -\sin\theta & 0 \\ \sin\theta & \cos\theta & 0 \\ 0 & 0 & 1 \end{bmatrix} \text{and } G = \begin{bmatrix} 1 & 0 & 0 \\ 0 & 1 & 0 \\ 0 & 0 & 1 \end{bmatrix}$$

What is the matrix F?

(a) $\begin{bmatrix} \cos\theta & -\sin\theta & 0 \\ \sin\theta & \cos\theta & 0 \\ 0 & 0 & 1 \end{bmatrix}$ (b) $\begin{bmatrix} \cos\theta & \cos\theta & 0 \\ -\cos\theta & \sin\theta & 0 \\ 0 & 0 & 1 \end{bmatrix}$

(c) $\begin{bmatrix} \cos\theta & \sin\theta & 0 \\ -\sin\theta & \cos\theta & 0 \\ 0 & 0 & 1 \end{bmatrix}$ (d) $\begin{bmatrix} \sin\theta & -\cos\theta & 0 \\ \cos\theta & \sin\theta & 0 \\ 0 & 0 & 1 \end{bmatrix}$

55. Match **List-I** with **List-II** and select the correct answer using the codes given below the lists:

List -I	List-II
A. Singular matrix	1. Determinant is not defined
B. Non-square matrix	2. Determinant is always one
C. Real symmetric	3. Determinant is zero
D. Orthogonal matrix	4. Eigen values are always real
	5. Eigen values are not defined

Codes:

	A	B	C	D
(a)	3	1	4	2
(b)	2	3	4	1
(c)	3	2	5	4
(d)	3	4	2	1

56. The inverse of the 2 × 2 matrix $\begin{bmatrix} 1 & 2 \\ 5 & 7 \end{bmatrix}$ is

(a) $\frac{1}{3}\begin{bmatrix} -7 & 2 \\ -5 & 1 \end{bmatrix}$ (b) $\frac{1}{3}\begin{bmatrix} -7 & 2 \\ 5 & 1 \end{bmatrix}$

(c) $\frac{1}{3}\begin{bmatrix} 7 & -2 \\ -5 & 1 \end{bmatrix}$ (d) $\frac{1}{3}\begin{bmatrix} -7 & -2 \\ -5 & -1 \end{bmatrix}$

57. The inverse of the matrix $\begin{bmatrix} 3+2i & -i \\ i & 3-2i \end{bmatrix}$ is

(a) $\frac{1}{12}\begin{bmatrix} 3+2i & -i \\ i & 3-2i \end{bmatrix}$ (b) $\frac{1}{12}\begin{bmatrix} 3-2i & -i \\ i & 3+2i \end{bmatrix}$

(c) $\frac{1}{14}\begin{bmatrix} 3+2i & -i \\ i & 3-2i \end{bmatrix}$ (d) $\frac{1}{14}\begin{bmatrix} 3-2i & -i \\ i & 3+2i \end{bmatrix}$

58. For what value of x will the matrix given below become singular?

$$\begin{bmatrix} 8 & x & 20 \\ 4 & 0 & 2 \\ 12 & 6 & 0 \end{bmatrix}$$

(a) 4 (b) 6
(c) 8 (d) 12

59. For what values of a and b, the following simultaneous equations have an infinite number of solutions?

$x + y + z = 5;$
$x + 3y + 3z = 9;$
$x + 2y + az = b$

(a) 2, 7 (b) 3, 8
(c) 8, 3 (d) 7, 2

60. The following simultaneous equations

$x + y + z = 3$
$x + 2y + 3z = 4$
$x + 4y + kz = 6$

will not have a unique solution for k equal to

(a) 0 (b) 5
(c) 6 (d) 7

61. Consider the system of simultaneous equations

$x + 2y + z = 6$
$2x + y + 2z = 6$
$x + y + z = 5$

This system has

(a) unique solution
(b) infinite number of solutions
(c) no solution
(d) exactly two solutions

62. For what value of a, if any, will the following system of equations in x, y and z have a solution?

$2x + 3y = 4$
$x + y + z = 4$
$x + 2y - z = a$

(a) Any real number (b) 0
(c) 1 (d) There is no such value

63. The eigen values of the matrix $\begin{bmatrix} 4 & -2 \\ -2 & 1 \end{bmatrix}$ are

(a) 1 and 4 (b) –1 and 2
(c) 0 and 5 (d) Cannot be determined

64. The eigen values of the matrix $[P] = \begin{bmatrix} 4 & 5 \\ 2 & -5 \end{bmatrix}$ are

(a) –7 and 8 (b) –6 and 5
(c) 3 and 4 (d) 1 and 2

65. Which one of the following is an eigen vector of the matrix $\begin{bmatrix} 5 & 0 & 0 & 0 \\ 0 & 5 & 5 & 0 \\ 0 & 0 & 2 & 1 \\ 0 & 0 & 3 & 1 \end{bmatrix}$?

(a) $\begin{bmatrix} 1 \\ -2 \\ 0 \\ 0 \end{bmatrix}$ (b) $\begin{bmatrix} 0 \\ 0 \\ 1 \\ 0 \end{bmatrix}$

(c) $\begin{bmatrix} 1 \\ 0 \\ 0 \\ -2 \end{bmatrix}$ (d) $\begin{bmatrix} 1 \\ -1 \\ 2 \\ 1 \end{bmatrix}$

66. The number of lineraly independent eigen vectors of $\begin{bmatrix} 2 & 1 \\ 0 & 2 \end{bmatrix}$ is

(a) 0 (b) 1
(c) 2 (d) Infinite

67. The eigen vectors of the matrix $\begin{bmatrix} 1 & 2 \\ 0 & 2 \end{bmatrix}$ are written in the form $\begin{bmatrix} 1 \\ a \end{bmatrix}$ and $\begin{bmatrix} 1 \\ b \end{bmatrix}$. What is the value of a + b?

(a) 0 (b) 1/2
(c) 1 (d) 2

68. One of the eigen vectors of the matrix $A = \begin{bmatrix} 2 & 2 \\ 1 & 3 \end{bmatrix}$ is

(a) $\begin{bmatrix} 2 \\ -1 \end{bmatrix}$

(b) $\begin{bmatrix} 2 \\ 1 \end{bmatrix}$

(c) $\begin{bmatrix} 4 \\ 1 \end{bmatrix}$

(d) $\begin{bmatrix} 1 \\ -1 \end{bmatrix}$

69. Consider the system of equations $A_{(n\times n)} X_{(n\times l)} = \lambda_{(n\times l)}$ where λ is scalar. Let (λ_i, x_i) be an eigen-pair of an eigen value and its corresponding eigen vector for real matrix A. Let I be a (n × n) unit matrix. Which one of the following statement is NOT correct?

(a) For a homogeneous n × n system of linear equations, $(A - \lambda I)\, x = 0$ having a non-trivial solution, the rank of $(A - \lambda I)$ is less than n

(b) For matrix A^m, m being a positive integer, (λ_i^m, x_i^m) will be the eigen-pair for all i

(c) If $A^T = A^{-1}$, then $|\lambda_i| = 1$ for all i

(d) If $A^T = A$, then λ_i is real for all i

70. For a given matrix

$$A = \begin{bmatrix} 2 & -2 & 3 \\ -2 & -1 & 6 \\ 1 & 2 & 0 \end{bmatrix}$$

One of the eigen value is 3. The other two eigen values are

(a) 2, –5 (b) 3, –5
(c) 2, 5 (d) 3, 5

71. Eigenvalues of a matrix $S = \begin{bmatrix} 3 & 2 \\ 2 & 3 \end{bmatrix}$ are 5 and 1. What are the eigenvalues of the matrix.

$S_2 = SS$?

(a) 1 and 25
(b) 6 and 4
(c) 5 and 1
(d) 2 and 10

11.2 Calculus

1. The value of the integral

$I = [1/(2\pi)] \int_0^\infty \exp[-(x)^2/8]\, dx$ is

(a) 1 (b) π
(c) 2 (d) 2π

2. The following plot shows a function y which varies linearly with x. The value of the integral $I = \int_1^2 y\, dx$ is

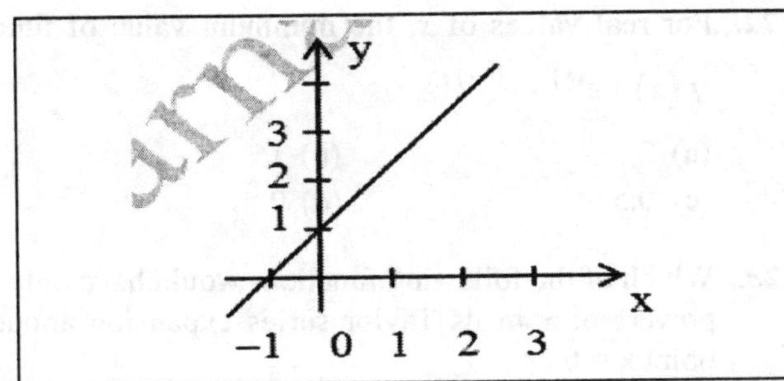

(a) 1.0 (b) 2.4
(c) 4.0 (d) 5.0

3. $\lim_{\theta \to 0} \sin \frac{(\theta/2)}{\theta}$ is

(a) 0.5 (b) 1
(c) 2 (d) Not defined

4. For the function e^{-x}, the linear approximation around $x = 2$ is

(a) $(3-x)e^{-2}$
(b) $1-x$
(c) $[3+2\sqrt{2}-(1+\sqrt{2})x]\,e^{-2}$
(d) e^{-2}

5. Consider the function $f(x) = x^2 - x - 2$. The maximum of $f(x)$ in the closed interval [–4, 4] is

(a) 18 (b) 10
(c) –2.25 (d) Indeterminate

6. If $S = \int_1^\infty x^{-3} dx$, then S has the value

(a) $-1/3$
(b) $1/4$
(c) $1/2$
(d) 1

7. For the function $f(x)=x^2e^{-x}$ the maximum occurs when x is equal to

(a) 2 (b) 1
(c) 0 (d) –1

8. The integral $\frac{1}{2\pi}\int_0^{2\pi} \sin(t-\tau)\cos\tau\, d\tau$ equal

(a) $\sin t \cos t$ (b) 0
(c) $\left(\frac{1}{2}\right)\cos t$ (d) $\left(\frac{1}{2}\right)\sin t$

9. The value of the integral $\int_{-1}^{1} \frac{1}{x^2}\, dx$ is

(a) 2 (b) Does not exist
(c) –2 (d) ∞

10. For real x, the maximum value of $\frac{e^{\sin x}}{e^{\cos x}}$ is

(a) 1 (b) e
(c) $e^{\sqrt{2}}$ (d) ∞

11. Consider the function $f(x) = |x|^3$, where x is real. Then the function $f(x)$ at $x = 0$ is

(a) Continuous but not differentiable
(b) Once differentiable but not twice
(c) Twice differentiable but not thrice
(d) Thrice differentiable

12. The value of the integral $\int_0^\infty \int_0^\infty e^{-x^2} e^{-y^2}\, dx\, dy$ is

(a) $\sqrt{\pi}/2$ (b) $\sqrt{\pi}$
(c) π (d) $\pi/4$

13. $\nabla \times \nabla \times P$, where P is a vector, is equal to

(a) $P \; \nabla \times P - \nabla^3 P$
(b) $\nabla^2 P + \nabla(\nabla . P)$
(c) $\nabla^2 P + \nabla \times P$
(d) $\nabla(\nabla . P) - \nabla^2 P$

14. The integral $\int_0^{\pi} \sin^3\theta \, d\theta$ is given by

(a) [1 / 2] (b) [2 / 3]
(c) [4 / 3] (d) [8 / 3]

15. As x is increased from $-\infty$ to ∞, the function. (x) = [$e^x / (1 + e^x)$]

(a) Monotonically increases
(b) Monotonically decreases
(c) Increases to a maximum value and then decreases
(d) Decreases to a minimum value and then increases.

16. Which one of the following function is strictly bounded ?

(a) $1/x^2$ (b) e^x
(c) x^2 (d) e^{-x^2}

17. The expression $V = \int_0^H \pi R^2[(1 - h/H)]^2 \, dh$ dh for the volume of a cone is equal to

(a) $\int_0^R \pi R^2(1 - h/H) \, dr$ (b) $\int_0^R \pi R^2(1 - h/H)^2 \, dh$
(c) $\int_0^R 2\pi rH(1 - r/R) \, dh$ (d) $\int_0^R 2\pi rH\left(1 - \frac{r}{R}\right)^2 dr$

18. A surface S (x, y) = 2x+5y–3 is integrated once over a path consisting of the points that satisfy $(x+1)^2 + (y-1)^2 = \sqrt{2}$. The integral evaluates to

(a) $17\sqrt{2}$
(b) $17/\sqrt{2}$
(c) $\sqrt{2}/17$
(d) 0

19. For the scalar field $u = \frac{x^2}{2} + \frac{y^2}{3}$, magnitude of the gradient at the point (1, 3) is

(a) $\sqrt{13/9}$ (b) $\sqrt{9/2}$
(c) $\sqrt{5}$ (d) $\frac{9}{2}$

20. If a vector $\bar{R}(t)$ has a constant magnitude, then

(a) $\bar{R} \cdot \frac{d\bar{R}}{dt} = 0$ (b) $\bar{R} \times \frac{d\bar{R}}{dt} = 0$
(c) $\bar{R} \cdot \bar{R} = \frac{d\bar{R}}{dt}$ (d) $\bar{R} \times \bar{R} = \frac{d\bar{R}}{dt}$

21. A scalar field is given by $f = x^{2/3} + y^{2/3}$, where x and y are the Cartesian coordinates. The derivative of f along the line $y = x$ directed away from the origin, at the point (8,8) is

(a) $\sqrt{2/3}$ (b) $\sqrt{3/2}$
(c) $2/\sqrt{3}$ (d) $3/\sqrt{2}$

22. For real values of x, the minimum value of function $f(x) = e^{(x)} + e^{(-x)}$ is

(a) 2 (b) 1
(c) 0.5 (d) 0

23. Which of the following functions would have only odd powers of x in its Taylor series expansion about the point x = 0 ?

(a) sin (x^3) (b) sin (x^2)
(c) cos (x^3) (d) cos (x^2)

24. In the Taylor series expansion of $\exp(x) + \sin(x)$ about the point $x = \pi$, the coefficient of $(x - \pi)^2$ is

(a) exp (π) (b) 0.5 exp (π)
(c) exp(π) + 1 (d) exp (π) –1

25. $P_x(x) = M\exp(-2|x|) + N\exp(-3|x|)$ is the probability density function for the real random variable X, over the entire x axis. M and N are both positive real numbers. The equation relating M and N is

(a) $M + \frac{2}{3}N = 1$
(b) $2M + \frac{1}{3}N = 1$
(c) $M + N = 1$
(d) $M + N = 3$

26. The Taylor series expansion of $\frac{\sin x}{x - \pi}$ at $x = \pi$ is given by

(a) $1 + \frac{(x-\pi)^2}{3!} + \ldots\ldots$ (b) $-1 - \frac{(x-\pi)^2}{3!} + \ldots.$
(c) $1 - \frac{(x-\pi)^2}{3!} + \ldots.$ (d) $-1 + \frac{(x-\pi)^2}{3!} + \ldots.$

27. If $e^y = x^{\frac{1}{x}}$, then y has a

(a) Maximum at $x = e$ (b) Minimum at $x = e$
(c) Maximum at $x = e^{-1}$ (d) Minimum at $x = e^{-1}$

28. The maximum value of $f(x) = x^3 - 9x^2 + 24x + 5$ in the interval [1, 6] is

(a) 21 (b) 25
(c) 41 (d) 46

29. Given $y = x^2 + 2x + 10$, the value of $\left.\frac{dy}{dx}\right|_{x=1}$ is equal to

(a) 0 (b) 4
(c) 12 (d) 13

30. $\lim_{x \to 0} \frac{\sin x}{x}$ is

(a) Indeterminate (b) 0
(c) 1 (d) ∞

31. Consider the function $y = x^2 - 6x + 9$. The maximum value of y obtained when x varies over the interval 2 to 5 is.

(a) 1 (b) 3
(c) 4 (d) 9

32. The integral $\int_{-\infty}^{\infty} \delta(t - \pi/6) 6 \sin(t)\, dt$ evaluates to

(a) 6 (b) 3
(c) 1.5 (d) 0

33. The series $\sum_{m=0}^{\infty} \frac{1}{4^m}(x-1)^{2m}$ converges for

(a) $-2 < x < 2$ (b) $-1 < x < 3$
(c) $-3 < x < 1$ (d) $x < 3$

34. The value of the quantity P, where $P = \int_0^1 xe^x dx$, is equal to

(a) 0 (b) 1
(c) e (d) $1/e$

35. At $t = 0$, the function $f(t) = \frac{\sin t}{t}$ has

(a) A minimum
(b) A discontinuity
(c) A point of inflection
(d) A maximum

36. Value of the integral $\oint_c (xydy - y^2dx)$, where, c is the square cut from the first quadrant by the lines x = 1 and y = 1 will be.

(a) $\frac{1}{2}$ (b) 1
(c) $\frac{3}{2}$ (d) $\frac{5}{3}$

37. $f(x, y)$ is a continuous function defined over $(x, y) \in [0\ \ 1] \times [0\ \ 1]$. Given the two constraints, $x > y^2$ and $y > x^2$, the volume under $f(x\ y)$ is

(a) $\int_{y=0}^{y=1} \int_{x=y^2}^{x=\sqrt{y}} f(x, y)\, dxdy$

(b) $\int_{y=x^2}^{y=1} \int_{x=y^2}^{x=1} f(x, y)\, dxdy$

(c) $\int_{y=0}^{y=1} \int_{x=0}^{x=1} f(x, y)\, dxdy$

(d) $\int_{y=0}^{y=\sqrt{x}} \int_{x=0}^{x=\sqrt{y}} f(x, y)\, dxdy$

38. The value of the function f(x) = $\lim_{x \to 0} \frac{x^3 + x^2}{2x^3 - 7x^2}$ is

(a) 0 (b) $-\frac{1}{7}$
(b) $\frac{1}{7}$ (d) ∞

39. The $\lim_{x \to 0} \frac{\sin\left[\frac{2}{3}x\right]}{x}$ is

(a) 2/3 (b) 1
(c) 3/2 (d) ∞

40. $\mathrm{Lt}_{x \to 0} \frac{\sin^2 x}{x}$ is equal to

(a) 0
(b) ∞
(c) 1
(d) -1

41. The value of $\lim_{x \to 8} \frac{x^{1/3} - 2}{(x-8)}$

(a) $\frac{1}{16}$ (b) $\frac{1}{12}$

(c) $\frac{1}{8}$ (d) $\frac{1}{4}$

42. If x = a (θ + sin θ) and y = a (1 – cos θ), then dy/dx will be equal to

(a) $\sin\left(\frac{\theta}{2}\right)$ (b) $\cos\left(\frac{\theta}{2}\right)$

(c) $\tan\left(\frac{\theta}{2}\right)$ (d) $\cot\left(\frac{\theta}{2}\right)$

43. The minimum value of function $y = x^2$ in the interval [1, 5] is

(a) 0 (b) 1

(c) 25 (d) Indefined

44. In the Taylor series expansion of e^x about x = 2, the coefficient of $(x-2)^4$ is

(a) 1/4! (b) $2^4/4!$

(c) $e^2/4!$ (d) $e^4/4!$

45. The value of the integral $\int_{-\infty}^{\infty} \frac{dx}{1-x^2} dx$ is

(a) – π (b) –π/2

(b) π/2 (d) π

46. The parabolic arc $y = \sqrt{x}$, $1 \le x \le 2$ is revolved around the x-axis. The volume of the solid of the revolution is

(a) π/4 (b) π/2

(c) 3π/4 (d) 3π/2

47. Changing the order of the integration in the double integral $I = \int_0^8 \int_{x/4}^2 f(x, y)\, dy dx$ leads to

(a) 4y (b) $16y^2$

(c) x (d) 8

48. For a scalar function $f(x, y, z) = x^2 + 3y^2 + 2z^2$, the gradient at the point P(1, 2, –1) is

(a) $2\vec{i} + 6\vec{j} + 4\vec{k}$

(b) $2\vec{i} + 12\vec{j} - 4\vec{k}$

(c) $2\vec{i} + 12\vec{j} + 4\vec{k}$

(d) $\sqrt{56}$

49. The divergence of the vector field $3xz\hat{i} + 2zy\hat{j} - yz^2\hat{k}$ at a point (1, 1, 1) is equal to

(a) 7 (b) 4

(c) 3 (d) 0

50. Stokes theorem connects

(a) A line integral and a surface integral

(b) A surface integral and a volume integral

(c) A line integral and a volume integral

(d) Gradient of a function and its surface integral

51. If $f(x) = \frac{2x^2 - 7x + 3}{5x^2 - 12x - 9}$, then $\lim_{x \to 3} f(x)$ will be

(a) –1/3 (b) 5/18

(c) 0 (d) 2/5

52. $\underset{x \to 0}{\text{Lt}} \frac{e^x - \left(1 + x + \frac{x^2}{2}\right)}{x^3} =$

(a) 0 (b) 1/6

(c) 1/3 (d) 1

53. A rail engine accelerates from its stationary position for 8 seconds and travels a distance of 280m. According to the Mean Value Theorem, the speedometer at a certain time during acceleration must read exactly.

(a) 0 (b) 8 kmph

(c) 75 kmph (d) 126 kmph

54. Given a function $f(x, y) = 4x^2 - 6y^2 - 8x - 4y + 8$. The optimal value of f(x, y).

(a) Is a minimum equal to 10/3

(b) Is a maximum equal to 10/3

(c) Is a minimum equal to 8/3

(d) Is a maximum equal to 8/3

55. The distance between the origin and the point nearest to it on the surface $z^2 = 1 + xy$ is

(a) 1 (b) $\frac{\sqrt{3}}{2}$

(c) $\sqrt{3}$ (d) 2

56. The function $f(x) = 2x^3 - 3x^2 - 36x + 2$ has its maxima at

(a) x = – 2 only (b) x = 0 only

(c) x = –3 only (d) both x = – 2 and x = –3

57. The right circular cone of largest volume that can be enclosed by a sphere of 1 m radius has a height of

(a) 1/3 m (b) 2/3 m

(c) $\frac{2\sqrt{2}}{3}$ m (d) 4/3 m

58. Let $f = y^x$. What is $\frac{\partial^2 f}{\partial x \partial y}$ at x = 2, y = 1?

(a) 0 (b) ln 2

(c) 1 (d) $\frac{1}{\ln 2}$

59. Assuming $i = \sqrt{-1}$ and t is a real number, $\int_0^{\pi/3} e^{it} dt$ is

(a) $\frac{\sqrt{3}}{2} + i\frac{1}{2}$ (b) $\frac{\sqrt{3}}{2} - i\frac{1}{2}$

(c) $\frac{1}{2} + i\left(\frac{\sqrt{3}}{2}\right)$ (d) $\frac{1}{2} + i\left(1 - \frac{\sqrt{3}}{2}\right)$

60. Evaluate $\int_0^{\infty} \frac{\sin t}{t} dt$

(a) π (b) $\frac{\pi}{2}$

(c) $\frac{\pi}{4}$ (d) $\frac{\pi}{8}$

61. Which of the following integrals is unbounded?

(a) $\int_0^{\pi/4} \tan x\, dx$ (b) $\int_0^{\infty} \frac{1}{x^2 + 1} dx$

(c) $\int_0^{\infty} xe^{-x} dx$ (d) $\int_0^{1} \frac{1}{1 - x} dx$

62. What is the area common to the circles r = a and r = 2a cos θ?

(a) 0.524 a^2 (b) 0.614 a^2

(c) 1.047 a^2 (d) 1.228 a^2

63. The length of the curve $y = \frac{2}{3}x^{2/3}$ between x = 0 and x = 1 is

(a) 0.27 (b) 0.67

(c) 1 (d) 1.22

64. A path AB in the form of one quarter of a circle of unit radius is shown in the figure. Integration of $(x + y)^2$ on path AB traversed in a counter-clockwise sense is

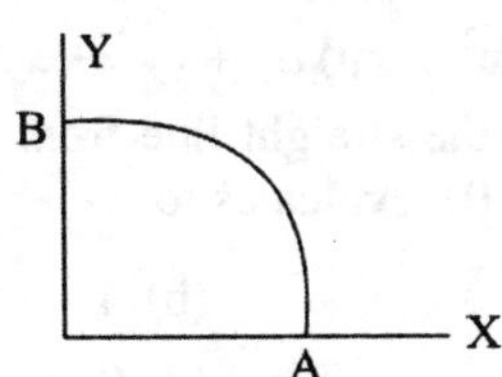

(a) $\frac{\pi}{2}$ (b) $\frac{\pi}{2} + 1$

(c) $\frac{\pi}{2}$ (d) 1

65. The area enclosed between the curves $y^2 = 4x$ and $x^2 = 4y$ is

(a) $\frac{16}{3}$ (b) 8

(c) $\frac{32}{3}$ (d) 16

66. The value of $\int_0^3 \int_0^x (6 - x - y)\, dx\, dy$ is

(a) 13.5 (b) 27.0

(c) 40.5 (d) 54.0

67. The volume of an object expressed in spherical co-ordinates is given by $V = \int_0^{2\pi} \int_0^{\pi/3} \int_0^1 r^2 \sin\phi\, dr\, d\phi\, d\theta$

The value of the intergal is

(a) $\frac{\pi}{3}$ (b) $\frac{\pi}{6}$

(c) $\frac{2\pi}{6}$ (d) $\frac{\pi}{4}$

68. By a change of variable x (u, v) = uv, y (u, v) = v/u is double integral, the integrand f(x, y) changes to ϕ (uv, v/u) ϕ (u, v). Then, ϕ (u, v) is

(a) 2 u/v (b) 2 uv

(c) v^2 (d) 1

69. Consider the shaded triangular region P shown in the figure. What is $\iint_P xy\,dx\,dy$?

(a) $\frac{1}{6}$ (b) $\frac{2}{9}$

(c) $\frac{7}{16}$ (d) 1

70. The area of a triangle formed by the tips of vectors $\bar{a}$, $\bar{b}$ and $\bar{c}$ is

(a) $\frac{1}{2}(\bar{a} - \bar{b}).(\bar{a} - \bar{c})$

(b) $\frac{1}{2}\left|(\bar{a} - \bar{b}).(\bar{a} - \bar{c})\right|$

(c) $\frac{1}{2}\left|a \times b \times c\right|$

(d) $\frac{1}{2}(a \times b).\bar{c}$

71. Let x and y be two vectors in 3 dimensional space and <x, y> denoe their dot product. Then the determinant det $\begin{vmatrix} \langle x, x\rangle & \langle x, y\rangle \\ \langle y, x\rangle & \langle y, y\rangle \end{vmatrix}$.

(a) Is zero when x and y are linearly independent
(b) Is positive when x and y are linearly independent
(c) Is non-zero for all non-zero x and y
(d) Is zero only when either x or y is zero

72. The directional derivative of $f(x, y, z) = 2x^2 + 3y^2 + z^2$ at the point P (2, 1, 3) in the direction of the vector $a = i - 2k$ is

(a) –2.785 (b) –2.145
(c) –1.789 (d) 1.000

73. For a scalar function $f(x, y, z) = x^2 + 3y^2 + 2z^2$, the directional derivative at the point P (1, 2, –1) in the direction of a vector $\vec{i} - \vec{j} + 2\vec{k}$ is

(a) –18 (b) $-3\sqrt{6}$
(c) $3\sqrt{6}$ (d) 18

74. Equation of the line normal to function $f(x) = (x-8)^{2/3} + 1$ at P (0, 5) is

(a) y = 3x – 5 (b) y = 3x + 5
(c) 3y = x + 15 (d) 3y = x – 15

75. Velocity vector of a flow field is given as $\vec{V} = 2xy\hat{i} - x^2z\hat{j}$. The vorticity vector at (1, 1, 1) is

(a) $4\hat{i} - \hat{j}$ (b) $4\hat{i} - \hat{k}$
(c) $\hat{i} - 4\hat{j}$ (d) $\hat{i} - 4\hat{k}$

76. The line integral $\int \vec{V}.d\vec{r}$ of the vector $\vec{V}.(\vec{r}) = 2xyz\hat{i} + x^2z\hat{j} + x^2y\hat{k}$ from the origin to the point P(1, 1, 1).

(a) Is 1
(b) Is –1
(c) Is zero
(d) Cannot be determined without specifying the path

77. Consider point P and Q in the x-y plane, with $p = (1.0)$. The line integral $2\int_P^Q (xdx + ydx)$ along the semicircle with the line segment P Q as its diameter

(a) Is –1 (b) Is 0
(c) Is 1
(d) Depends on the direction (clockwise or anti-clockwise) of the semicircle

78. If $\vec{A} = xy\,\hat{a}_x + x^2\hat{a}_y$ then $\oint_c \vec{A}\bullet\vec{dl}$ over the path shown in the figure is

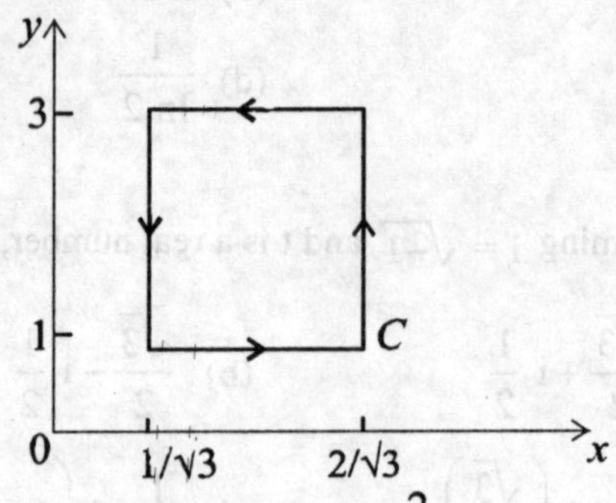

(a) 0 (b) $\frac{2}{\sqrt{3}}$
(c) 1 (d) $2\sqrt{3}$

79. The direction of vector *A* is radially outward from the origin, with $|\mathbf{A}| = k\,r^n$ where $r^2 = x^2 + y^2 + z^2$ and k is a constant. The value of *n* for which $\nabla \cdot \mathbf{A} = 0$ is

(a) –2 (b) 2
(c) 1 (d) 0

80. A sphere of unit radius is centered at the origin. The unit normal at a point (x, y, z) on the surface of the sphere is the vector.

(a) (x, y, z)
(b) $\left(\frac{1}{\sqrt{3}}, \frac{1}{\sqrt{3}}, \frac{1}{\sqrt{3}}\right)$
(c) $\left(\frac{x}{\sqrt{3}}, \frac{y}{\sqrt{3}}, \frac{z}{\sqrt{3}}\right)$
(d) $\left(\frac{x}{\sqrt{2}}, \frac{y}{\sqrt{2}}, \frac{z}{\sqrt{2}}\right)$

81. Divergence of the three-dimensional radial vector field $\vec{r}$ is

(a) 3
(b) $1/r$
(c) $\hat{i} + \hat{j} + \hat{k}$
(d) $3(\hat{i} + \hat{j} + \hat{k})$

82. $F(x, y) = (x^2 + xy)\,\hat{a}_x + (y^2 + xy)\,\hat{a}_y$. It's line integral over the straight line from $(x, y) = (0, 2)$ to $(x, y) = (2, 0)$ evaluates to

(a) – 8 (b) 4
(c) 8 (d) 0

11.3 Differential Equation

1. The following differential equation has

$$3\left(\frac{d^2y}{dt}\right)+4\left(\frac{dy}{dt}\right)^3+y^2+2=x$$

(a) Degree = 2, order = 1
(b) Degree = 1, order = 2
(c) Degree = 4, order = 3
(d) Degree = 2, order = 3

2. A solution of the following differential equation is given by

$$\frac{d^2y}{dx^2}-5\frac{dy}{dx}+6y=0$$

(a) $y = e^{2x} + e^{-3x}$ (b) $y = e^{2x} + e^{3x}$
(c) $y = e^{-2x} + e^{3x}$ (d) $y = e^{-2x} + e^{-3x}$

3. For the differential equation $[d^2 y / dx^2] + k^2 y = 0$ the boundary conditions are
1. y = 0 for x = 0 and
2. y = 0 for x = a

The form of non- zero solutions of y (where m varies over all integers) are
(a) $y = \sum_m A_m \sin(m\pi x / a)$
(b) $y = \sum_m A_m \cos(m\pi x / a)$
(c) $y = \sum_m A_m x^{(m\pi / a)}$
(d) $y = \sum_m A_m \exp(-m\pi x / a)$

4. The solution of the differential equation $k^2\frac{d^2y}{dx^2}=y-y_2$, under the boundary conditions (1) $y=y_1$ at $x=0$ and (2) $y=y_2$ at $x=\infty$, where k, y_1 and y_2 are constant, is

(a) $y=(y_1-y_2)e^{(-x/k^2)}+y_2$
(b) $y=(y_2-y_1)e^{(-x/k)}+y_1$
(c) $y=(y_1-y_2)\sinh(x/k)+y_1$
(d) $y=(y_1-y_2)e^{(-x/k)}+y_2$

5. The solution of the first order differential equation $x'(t) = -3x(t)$, $x(0) = x_0$ is

(a) $x(t)=x_0e^{-3t}$
(b) $x(t)=x_0e^{-3}$
(c) $x(t)=x_0e^{-1/3}$
(d) $x(t)=x_0e^{-1}$

6. For the equation $\ddot{x}(t)+3\dot{x}(t)+2x(t)=5$ the solution x (t) approaches which of the following values as $t\to\infty$?

(a) 0 (b) $\frac{5}{2}$
(c) 5 (d) 10

7. The general solution of the differential equation $(D^2-4D+4)y=0$, is of the form (given D = d/dx), and C_1 and C_2 are constants

(a) C_1e^{2x}
(b) $C_1e^{2x}+C_2e^{-2x}$
(c) $C_1e^{2x}+C_2xe^{2x}$
(d) $C_1e^{2x}+C_2xe^{-2x}$

8. For an initial value problem $\ddot{y}+2\dot{y}+101y=10.4e^x$ $y(0)=1.1$ and $\dot{y}(0)=-0.9$. Various solution are written in the following group. Match the type of solution with the correct expression

Group 1	Group 2
A. General solution of homogeneous equations	1. $0.1e^x$
B. Particular integral	2. $e^{-x}(A\cos 10x+B\sin 10x)$
C. Total solution satisfying boundary conditions	3. $e^{-x}\cos 10x+0.1e^x$

Codes:

	A	B	C
(a)	2	1	3
(b)	1	3	2
(c)	1	2	3
(d)	3	2	1

9. Which of the following is a solution to the differential equation $\frac{dx(t)}{dt}+3x(t)=0$?

(a) $x(t)=3e^{-t}$ (b) $x(t)=2e^{-3t}$
(c) $x(t)=-\frac{3}{2}t^2$ (d) $x(t)=3t^2$

10. The order of the differential equation $\frac{d^2y}{dt^2}+\left(\frac{dy}{dt}\right)^3+y^4=e^{-t}$ is

(a) 1 (b) 2
(c) 3 (d) 4

11. Match each differential equation in Group I to its family of solution curves from Group II.

Group I	Group II
P. $\frac{dy}{dx}=\frac{y}{x}$	1. Circles
Q. $\frac{dy}{dx}=-\frac{y}{x}$	2. Straight lines
R. $\frac{dy}{dx}=\frac{x}{y}$	3. Hyperbolas
S. $\frac{dy}{dx}=-\frac{x}{y}$	

(a) $P-2, Q-3, R-3, S-1$
(b) $P-1, Q-3, R-2, S-1$
(c) $P-2, Q-1, R-3, S-3$
(d) $P-3, Q-2, R-1, S-2$

12. A function $n(x)$ satisfies the differential equation $\frac{d^2n(x)}{dx^2}-\frac{n(x)}{L^2}=0$ where L is a constant. The boundary conditions are: $n(0)=K$ and $n(\infty)=0$. The solution to this equation is

(a) $n(x)=K\exp(x/L)$
(b) $n(x)=K\exp(-x/\sqrt{L})$
(c) $n(x)=K^2\exp(-x/\sqrt{L})$
(d) $n(x)=K\exp(-x/L)$

13. The solution of the differential equation $\frac{dy}{dx}=ky, y(0)=c$ is

(a) $x=ce^{-ky}$ (b) $x=ke^{cy}$
(c) $y=ce^{kx}$ (d) $y=ce^{-kx}$

14. With initial condition $x(1)=0.5$, the solution of the differential equation, $t\frac{dx}{dt}+x=t$ is

(a) $x=t-\frac{1}{2}$ (b) $x=t^2-\frac{1}{2}$
(c) $x=\frac{t^2}{2}$ (d) $x=\frac{t}{2}$

15. Consider the differential equation $\frac{dy}{dx}=1+y^2$. Which one of the following can be a particular solution of this differential equation?

(a) $y=\tan(x+3)$ (b) $y=\tan x+3$
(c) $x=\tan(y+3)$ (d) $x=\tan y+3$

16. Consider the differential equation

$\frac{dy}{dx}+y=e^x$ with $y(0)=1$. The value of $y(1)$ is

(a) $e+e^{-1}$ (b) $\frac{1}{2}\left(e-e^{-1}\right)$
(c) $\frac{1}{2}\left(e+e^{-1}\right)$ (d) $2\left(e-e^{-1}\right)$

17. For the differential equation $\frac{d^2x}{dt^2}+6\frac{dx}{dt}+8x=0$ with initial conditions $x(0)=1$ and $\left.\frac{dx}{dt}\right|_{t=0}=0$, the solution is

(a) $x(t)=2e^{-6t}-e^{-2t}$ (b) $x(t)=2e^{-2t}-e^{-4t}$
(c) $x(t)=-e^{-6t}+2e^{-4t}$ (d) $x(t)=e^{-2t}+2e^{-4t}$

18. The degree of the differential equation $\frac{d^2x}{dt^2}+2x^3=0$ is

(a) 0 (b) 1
(c) 2 (d) 3

19. The order and degree of the differential equation $\frac{d^3y}{dx^3}+4\sqrt{\left(\frac{dy}{dx}\right)^3}+y^2=0$ are respectively.

(a) 3 and 2 (b) 2 and 3
(c) 3 and 3 (d) 3 and 1

20. The Blasius equation, $\frac{d^3f}{d\eta^3}+\frac{f}{2}\frac{d^2f}{d\eta^2}=0$ is a

(a) Second order nonlinear ordinary differential equation
(b) Third order nonlinear ordinary differential equation
(c) Third order linear ordinary differential equation
(d) Mixed order nonlinear ordinary differential equation

21. The solution for the differential equation $\frac{dy}{dx} = x^2y$ with the condition that y = 1 at x = 0 is
(a) $y = e^{\frac{1}{2x}}$ (b) $in(y) = \frac{x^3}{3} + 4$
(c) $\ln(y) = \frac{x^2}{2}$ (d) $y = e^{\frac{x^3}{3}}$

22. The solution of the differential equation $\frac{dy}{dt} + 2xy = e^{-x^2}$ with y (0) = 1 is
(a) $(1 + x)\, e^{+x^2}$ (b) $(1 + x)\, e^{-x^2}$
(c) $(1 - x)\, e^{+x^2}$ (d) $(1 - x)\, e^{-x^2}$

23. The general solution of $\frac{d^2y}{dx^2} + y = 0$ is
(a) y = p cos x + Q sin x
(b) y = P cos x
(c) y = P sin x
(d) $y = P \sin^2 x$

24. Given that $\ddot{x} + 3x = 0$, and x (0) = 1, $\dot{x}$ (0) = 0, what is x (1)?
(a) – 0.99 (b) – 0.16
(c) 0.16 (d) 0.99

25. Biotransformation of an organic compound having concentration (x) can be modeled using an ordinary differential equation $\frac{dx}{dt} + kx^2 = 0$, where k is the reaction rate constant. If x = a at t = 0, the solution of the equation is
(a) $x = ae^{-kt}$ (b) $\frac{1}{x} = \frac{1}{a} + kt$
(c) $a = a(1 - e^{-kt})$ (d) x = a + kt

26. A spherical naphthalene ball exposed to the atmosphere loses volume at a rate proportional to its instantaneous surface area due to evaporation. If the initial diameter of the ball is 2 cm and the diameter reduces to 1 cm after 3 months, the ball completely evaporates in
(a) 6 months (b) 9 months
(c) 12 months (d) infinite time

27. Solution of $\frac{dy}{dx} = -\frac{x}{y}$ at x = 1 and y = $\sqrt{3}$ is
(a) $x - y^2 = -2$ (b) $x + y^2 = 4$
(c) $x^2 - y^2 = -2$ (d) $x^2 + y^2 = 4$

28. Solution of the differential equation $3y\frac{dy}{dx} + 2x = 0$ represents a family of
(a) Ellipses (b) Circles
(c) Parabolas (d) Hyperbolas

29. The solution of the differential equation $\frac{dy}{dx} + y^2 = 0$ is
(a) $y = \frac{1}{x + c}$
(b) $y = \frac{-x^3}{3} + c$
(c) ce^x
(d) unsolvable as equation is non-linear

30. The solution of dy/dx = y^2 with initial value y (0) = 1 bounded in the interval
(a) $-\infty \le x \le \infty$ (b) $-\infty \le x \le 1$
(c) $x < 1, x > 1$ (d) $-2 \le x \le 2$

31. Transformation to linear form by substituting $v = y^{1-n}$ of the equation $\frac{dy}{dt} + p(t)y = q(t)y^n$; n > 0 will be
(a) $\frac{dv}{dt} + (1 - n)\, pv = (1 - n)q$
(b) $\frac{dv}{dt} + (1 - n)\, pv = (1 + n)q$
(c) $\frac{dv}{dt} + (1 + n)\, pv = (1 - n)q$
(d) $\frac{dv}{dt} + (1 + n)\, pv = (1 + n)q$

32. The solution of $x\frac{dy}{dx} + y = x^4$ with the condition $y(1) = \frac{6}{5}$ is
(a) $y = \frac{x^4}{5} + \frac{1}{x}$ (b) $y = \frac{4x^4}{5} + \frac{4}{5x}$
(c) $y = \frac{x^4}{5} + 1$ (d) $y = \frac{x^5}{5} + 1$

33. The solution of $\frac{d^2y}{dx^2} + 2\frac{dy}{dx} + 17y = 0$; y(0) = 1, $\frac{dy}{dx}\left(\frac{\pi}{4}\right) = 0$ in the range $0 < x < \frac{\pi}{4}$ is given by
(a) $e^{-x}\left(\cos 4x + \frac{1}{4}\sin 4x\right)$

(d) $e^{x}\left(\cos 4x - \frac{1}{4}\sin 4x\right)$

(c) $e^{-4x}\left(\cos 4x - \frac{1}{4}\sin x\right)$

(d) $e^{-4x}\left(\cos 4x - \frac{1}{4}\sin 4x\right)$

34. It is given that $y'' + 2y' + y = 0$, $y(0) = 0$, $y(1) = 0$. What is $y(0.5)$?

(a) 0 (b) 0.37
(c) 0.62 (d) 1.13

35. For $\frac{d^2y}{dx^2} + 4\frac{dy}{dx} + 3y = 3e^{2x}$, the particular integral is

(a) $\frac{1}{15}e^{2x}$ (b) $\frac{1}{5}e^{2x}$
(c) $3e^{2x}$ (d) $C_1e^{-x} + C_2e^{-3x}$

36. An analytic function of a complex variable $z = x + iy$ is expressed as $f(z) = u(x, y) + i\,v(x, y)$ where $i = \sqrt{-1}$. If $u = xy$, the expression for v should be

(a) $\frac{(x+y)^2}{2} + k$ (b) $\frac{x^2 - y^2}{2} + k$
(d) $\frac{y^2 - y^2}{2} + k$ (d) $\frac{(x-y)^2}{2} + k$

11.4 Complex Variables

1. For the function of a complex variable $W = \ln Z$ (where, $W = u + jv$ and $Z = x + jy$).
The u = constant lines get mapped in Z – plane as
(a) Set of radial straight lines
(b) Set of concentric circles
(c) Set of confocal hyperbolas
(d) Set confocal ellipses

2. The value of the contour integral

$\oint_{|z-j|=2} [1/(z^2+4)]\,dz$ in positive sense is

(a) $[j\pi/2]$ (b) $[-\pi/2]$
(c) $[-j\pi/2]$ (d) $[\pi/2]$

3. If the semi-circular contour D of radius 2 is as shown in the figure, then the value of the integral $\oint_D \frac{1}{(s^2-1)}ds$ is

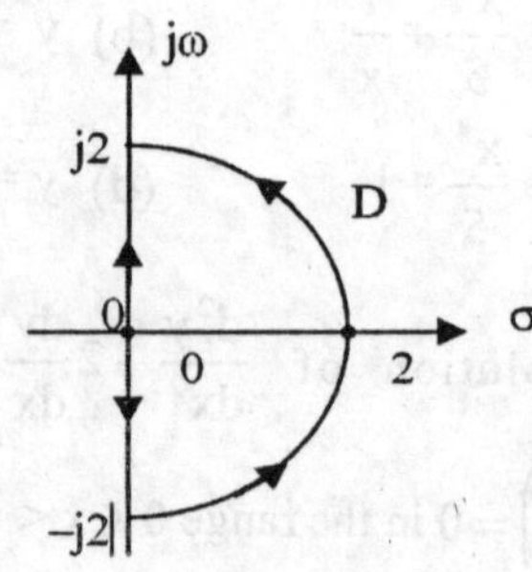

(a) $j\pi$ (b) $-j\pi$
(c) $-\pi$ (d) π

4. Consider the circle $|z - 5 - 5i| = 2$ in the complex plane (x, y) with $z = x + iy$. The minimum distance from the origin to the circle is

(a) $5\sqrt{2} - 2$ (b) $\sqrt{54}$
(c) $\sqrt{34}$ (d) $5\sqrt{2}$

5. Let $z^3 = \bar{z}$, where z is a complex number not equal to zero. Then z is a solution of

(a) $z^2 = 1$ (b) $z^3 = 1$
(c) $z^4 = 1$ (d) $z^9 = 1$

6. Let $j = \sqrt{-1}$. Then one value of j^j is

(a) $\sqrt{j}$ (b) -1
(c) $\pi/2$ (d) $e^{-\pi/2}$

7. The value of $\oint_D \frac{dz}{(1-Z^2)}$ where C is the contour $\left|z - \frac{i}{2}\right| = 1$ is

(a) $2\pi i$ (b) π
(c) $\tan^{-1} z$ (d) $\pi i \tan^{-1} z$

8. The equation $\sin(z) = 10$ has
(a) No real or complex solution
(b) Exactly two distinct complex solutions
(c) A unique solution
(d) An infinite number of complex solutions

9. If $f(z) = c_0 + c_1 z^{-1}$, then $\oint_{\text{unit circle}} \frac{1+f(z)}{z} dz$ is given by

(a) $2\pi c_1$ (b) $2\pi(1+c_0)$

(c) $2\pi jc_1$ (d) $2\pi j(1+c_0)$

10. If a vector field $\vec{V}$ is related to another vector field $\vec{A}$ through $\vec{V} = \nabla \times \vec{A}$, which of the following is true? Note: C and S_C refer to any closed contour and any surface whose boundary is C.

(a) $\oint_C \vec{V}.\vec{dl} = \iint_{S_C} \vec{A}.\vec{dS}$

(b) $\oint_C \vec{A}.\vec{dl} = \iint_{S_C} \vec{V}.\vec{dS}$

(c) $\oint_C \nabla\times\vec{V}.\vec{dl} = \iint_{S_C} \nabla\times\vec{A}.\vec{dS}$

(d) $\oint_C \nabla\times\vec{A}.\vec{dl} = \iint_{S_C} \vec{V}.\vec{dS}$

11. The residues of a complex function $X(z) = \frac{1-2z}{z(z-1)(z-2)}$ at its poles are

(a) $\frac{1}{2}, -\frac{1}{2}$ and 1 (b) $\frac{1}{2}, \frac{1}{2}$ and -1

(c) $\frac{1}{2}$, 1 and $-\frac{3}{2}$ (d) $\frac{1}{2}$, -1 and $\frac{3}{2}$

12. The value of the integral $\oint_C \frac{-3z+4}{(z^2+4z+5)} dz$ where c is the circle $|z| = 1$ is given by

(a) 0 (b) 1/10

(c) 4/5 (d) 1

13. If $x = \sqrt{-1}$, then the value of x^x is

(a) $e^{-\pi/2}$ (b) $e^{\pi/2}$

(c) x (d) 1

14. Given $f(z) = \frac{1}{z+1} - \frac{2}{z+3}$. If C is a counter clock wise path in the z-plane such that $|z+1| = 1$, then value of $\frac{1}{2\pi j}\oint_C f(z)\, dz$ is

(a) -2 (b) -1

(c) 1 (d) 2

15. A complex variable $Z = x + j0.1$ has its real part x varying in the range $-\infty$ *to* $+\infty$. Which one of the following is the locus (shown in thick lines) of $\frac{1}{Z}$ in the complex plane?

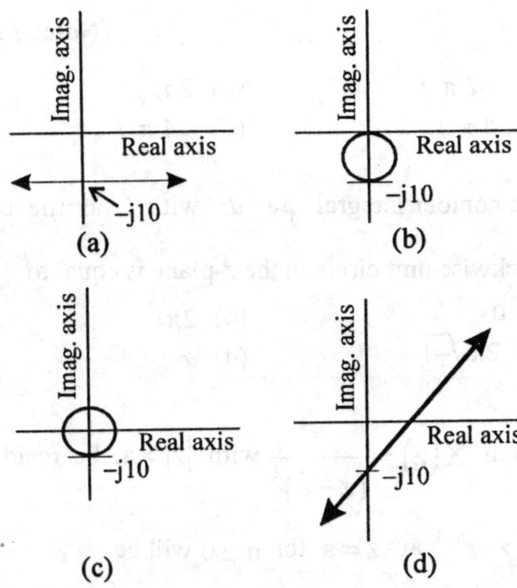

16. If $z = x + jy$, where x and y are real, the value of $|e^{jz}|$ is

(a) 1 (b) $e^{\sqrt{x^2+y^2}}$

(c) e^y (d) e^{-y}

17. The value of $\oint \frac{\sin z}{z} dz$, where the contour of integration is a simple closed curve around the origin, is

(a) 0 (b) $2\pi j$

(c) ∞ (d) $1/(2\pi j)$

18. One of the roots of the equation $x^3 = j$, where j is the positive square root of -1, is

(a) j (b) $\frac{\sqrt{3}}{2} + j\frac{1}{2}$

(c) $\frac{\sqrt{3}}{2} - j\frac{1}{2}$ (d) $-\frac{\sqrt{3}}{2} - j\frac{1}{2}$

19. The contour C in the adjoining figure is described by $x^2 + y^2 = 16$

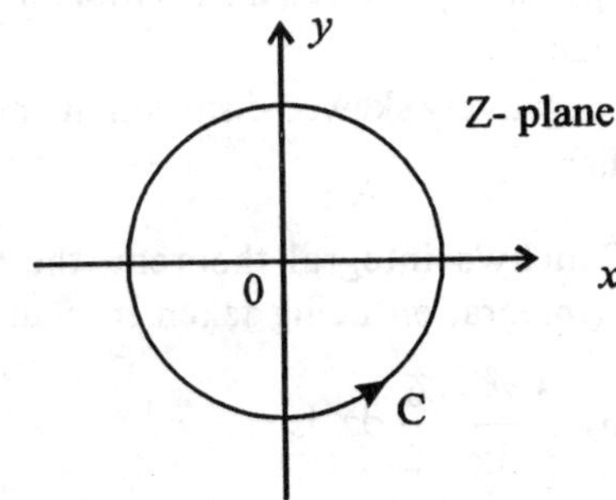

The value of $\oint_C \frac{z^2+8}{0.5z-1.5j} dz$ is

(**Note:** $j=\sqrt{-1}$)

(a) $-2\pi j$ (b) $2\pi j$
(c) $4\pi j$ (d) $-4\pi j$

20. The contour integral $\oint_c e^{1/z} dz$ with C as the counter-clockwise unit circle in the z-plane is equal to

(a) 0 (b) 2π
(c) $2\pi\sqrt{-1}$ (d) ∞

21. Given $X(z)=\frac{z}{(z-a)^2}$ with $|z|>a$. the residue of $X(z)z^{n-1}$ at $z=a$ for $n\geq 0$ will be

(a) a^{n-1} (b) a^n
(c) na^n (d) na^{n-1}

22. The modulus of the complex number $\left(\frac{3+4i}{1-2i}\right)$ is

(a) 5 (b) $\sqrt{5}$
(c) $1/\sqrt{5}$ (d) 1/5

23. The analytic function $f(z)=\frac{z-1}{z^2+1}$ has singularities at

(a) 1 and – 1
(b) 1 and i
(c) 1 and – i
(d) i and – i

24. Which one of the following statements is NOT true?

(a) The measure of skewness is dependent upon the amount of dispersion
(b) In a symmetric distribution, the values of mean, mode and median are the same
(c) In a positively skewed distribution: mean > median > mode
(d) In a negatively skewed distribution: mode > mean > median

25. Using Cauchy's integral theorem, the value of the integral (integration being taken in counterclockwise direction) $\oint_C \frac{z^3-6}{3z-i} dz$ is

(a) $\frac{2\pi}{81}-4\pi i$ (b) $\frac{\pi}{8}-6\pi i$
(c) $\frac{4\pi}{81}-6\pi i$ (d) 1

26. The value of the integral $\int_C \frac{\cos 2\pi z}{(2z-1)(z-3)} dz$ (where C is a closed curve given by $|z|=1$) is

(a) $-\pi i$ (b) $\frac{\pi}{5}$
(c) $\frac{2\pi i}{5}$ (d) πi

27. Consider likely applicability of Cauchy's Integral Theorem to evaluate the following integral counter clockwise around the unit circle c.

$$I=\oint_C \sec z\,dz,$$

z being a complex variable. The value of I will be

(a) $I=0$: singularities set = ϕ
(b) $I=0$: singularities set = $\left\{\pm\frac{2n+1}{2}\pi;\ n=0,1,2...\right\}$
(c) $I=\pi/2$: singularities set = $\{\pm n\pi;\ n=0,1,2...\}$
(d) None of above

28. The integral $\oint f(z)\,dz$ evaluated around the unit circle on the complex plane for $f(z)=\frac{\cos z}{z}$ is

(a) $2\pi i$
(b) $4\pi i$
(c) $-2\pi i$
(d) 0

29. The value of the contour integral $\oint_{|z-i|=2} \frac{1}{z^2+4} dz$ in positive sense is

(a) $i\pi/2$ (d) $-\pi/2$
(c) $-i\pi/2$ (d) $\pi/2$

30. The residue of the function $f(z)=\frac{1}{(z+2)^2(z-2)^2}$ at $z=2$ is

(a) $-\frac{1}{32}$

(b) $-\frac{1}{16}$

(c) $\frac{1}{16}$ (d) $\frac{1}{32}$

31. A point Z has been plotted in the complex plane, as shown in figure below.

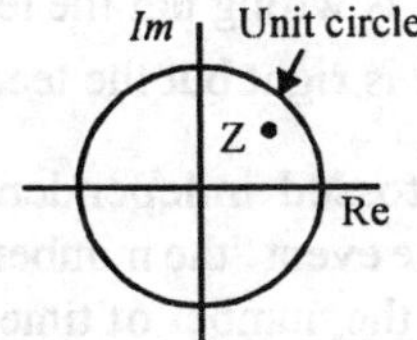

The plot of the complex number $y = \frac{1}{z}$ is

(a)
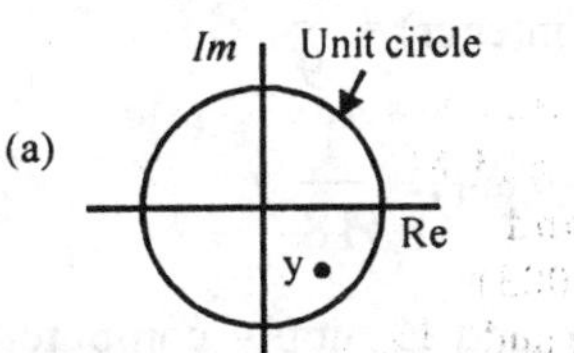

(b)
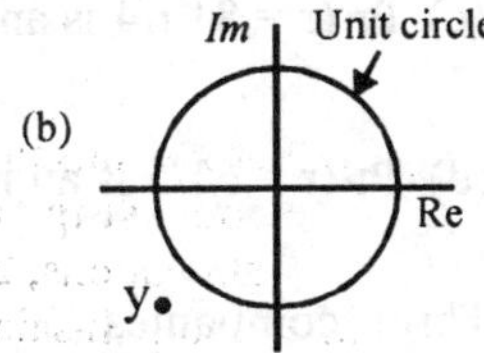

(c)
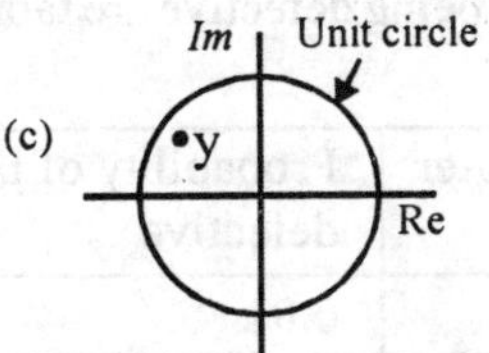

(d)
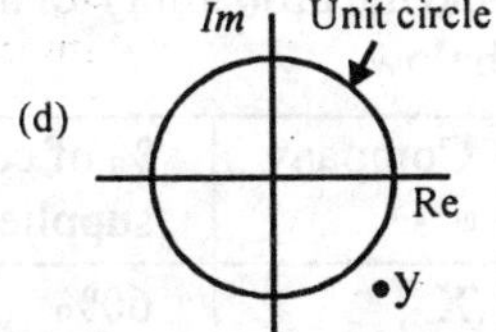

11.5 Probability and Statistics

1. A fair dice is rolled twice. The probability that an odd number will follow an even number is

(a) [1 / 2] (b) [1 / 6]
(c) [1 / 3] (d) [1 / 4]

2. If P and Q are two random events, then the following is true

(a) Independence of P and Q implies that probability $(P \cap Q) = 0$
(b) Probability $(P \cup Q) \geq$ Probability (P) + Probability (Q)
(c) If P and Q are mutually exclusive, then they must be independent
(d) Probability $(P \cap Q) \leq$ Probability (P)

3. A fair coin is tossed three times in succession. If the first toss produces a head, then the probability of getting exactly two hands in three tosses is

(a) $\frac{1}{8}$ (b) $\frac{1}{2}$
(c) $\frac{3}{8}$ (d) $\frac{3}{4}$

4. If a fair coin is tossed four times. What is the probability that two heads and two tails will result?

(a) 3/8
(b) 1/2
(c) 5/8
(d) 3/4

5. An examination paper has 150 multiple-choice questions of one mark each, with each question having four choices. Each incorrect answer fetches – 0.25 mark. Suppose 1000 students choose all their answers randomly with uniform probability. The sum total of the expected marks obtained by all these students is

(a) 0
(b) 2550
(c) 7525
(d) 9375

6. An examinations consists of two papers, Paper 1 and Paper 2. The probability of failing in Paper 1 is 0.3 and that in Paper 2 is 0.2. Given that student has failed in Paper 2, the probability failing in Paper 1 is 0.6. The probability of student failing in both the paper is

(a) 0.5 (b) 0.18
(c) 0.12 d) 0.06

7. A loaded dice has following probability distribution of occurrences

DiceValue	1	2	3	4	6	7
Probability	1/4	1/8	1/8	1/8	1/8	1/4

If three identical dice as the above are thrown, the probability of occurrence of values 1, 5 and 6 on the re dice is

(a) Same as that of occurrence of 3, 4, 5
(b) Same as that of occurrence of 1,2,5
(c) 1/128
(d) 5/8

8. Two fair dice are rolled and the sum r of the numbers turned up is considered

(a) $\Pr(r > 6) = \frac{1}{6}$

(b) $\Pr(r / 3 \text{ is an integer}) = \frac{5}{6}$

(c) Pr (r = 81 r/4 is an integer) = $\frac{5}{9}$

(d) Pr (r = 61/5 is an integer) = $\frac{1}{18}$

9. Three companies X, Y and Z supply computers to a university. The percentage of computers supplied by them and the probability of those being defective are tabulated below.

Company	% of computer supplied	Probability of being defective
X	60%	0.01
Y	30%	0.02
Z	10%	0.03

Given that a computer is defective, the probability it was supplied by Y is.

(a) 0.1 (b) 0.2
(c) 0.3 (d) 0.4

10. A probability density function is of the form

$P(x) = Ke^{-\alpha|x|}$, $x \in (-\infty, \infty)$

The value of K is .

(a) 0.5
(b) 1
(c) 0.5 α
(d) α

11. A fair coin is tossed 10 times. What is the probability that only the first two tosses will yield heads?

(a) $\left(\frac{1}{2}\right)^2$ (b) ${}^{10}C_2\left(\frac{1}{2}\right)^2$

(c) $\left(\frac{1}{2}\right)^{10}$ (d) ${}^{10}C_2\left(\frac{1}{2}\right)^{10}$

12. Consider two independent random variables X and Y with identical distributions. The variables X and Y take values 0, 1 and 2 with probabilities $\frac{1}{2}, \frac{1}{4}$ and $\frac{1}{4}$ respectively. What is the conditional probability $P(X+Y=2 | X-Y=0)$?

(a) 0 (b) $\frac{1}{16}$

(c) $\frac{1}{6}$ (d) 1

13. A discrete random variable X takes values from 1 to 5 with probabilities as shown in the table.

k	1	2	3	4	5
P(X = k)	0.1	0.2	0.4	0.2	0.1

A student calculates the mean of X as 3.5 and her teacher calculates the variance of X as 1.5. Which of the following statements is true?

(a) Both the student and the teacher are right
(b) Both the student and the teacher are wrong
(c) The student is wrong but the teacher is right
(d) The student is right but the teacher is wrong

14. A fair coin is tossed independently four times. The probability of the event "the number of times heads show up is more than the number of times tails show up" is

(a) $\frac{1}{16}$ (b) $\frac{1}{8}$

(c) $\frac{1}{4}$ (d) $\frac{5}{16}$

15. A fair dice is tossed two times. The probability that the second toss results in a value that is higher than the first toss is

(a) 2/36 (b) 2/6
(c) 5/12 (d) 1/2

16. Two independent random variable X and Y are uniformly distributed in the interval [–1, 1]. The probability that max [X, Y] is less than 1/2 is

(a) 3/4 (b) 9/16
(c) 1/4 (d) 2/3

17. A fair coin is tossed till a head appears for the first time. The probability that the number of required tosses is odd, is

(a) 1/3 (b) 1/2
(c) 2/3 (d) 3/4

18. Consider a Gaussian distributed random variable with zero mean and standard deviation s. The value of its cumulative distribution function at the origin will be

(a) 0 (b) 0.5
(c) 1 (d) 10 σ

19. A random variable is uniformly distributed over the interval 2 to 10. Its variance will be

(a) $\frac{16}{3}$ (b) 6

(c) $\frac{256}{9}$ (d) 36

20. A screening test is carried out to detect a certain disease. It is found that 12% of the positive reports and 15% of the negative reports are incorrect. Assuming that the probability of a person getting a positive report is 0.01, the probability that a person tested gets an incorrect report is

(a) 0.0027 (b) 0.0173
(c) 0.1497 (d) 0.2100

21. The box 1 contains chips numbered 3, 6, 9, 12 and 15. The box 2 contains chips numbered 6, 11, 16, 21 and 26. Two chips, one from each box, are drawn at random. The numbers written on these chips are multiplied. The probability for the product to be an even number is

(a) 6/25 (b) 2/5
(c) 3/5 (d) 19/25

22. A box contains 4 white balls and 3 red balls. In succession, two balls are randomly selected and removed from the box.Given that the first removed ball is white, the probability that the second removed ball is red is

(a) 1/3 (b) 3/7
(c) 1/2 (d) 4/7

23. Assume for simplicity that N people, all born in April (a month of 30 days), are collected in a room. Consider the event of at least two people in the room being born on the same date of the month, even if in different years, e.g. 1980 and 1985. What is the smallest N so that the probability of this event exceeds $0 \cdot 5$?

(a) 20 (b) 7
(c) 15 (d) 16

24. X is a uniformly distributed random variable that takes values between 0 and 1. The value of $E\{X^3\}$ will be

(a) 0 (b) 1/8
(c) 1/4 (d) 1/2

25. A box contains 10 screws, 3 of which are defective. Two screws are drawn at random with replacement. The probability that none of the two screws is defective will be

(a) 100% (b) 50%
(c) 49% (d) None of these

26. A lot has 10% defective items. Ten item are chosen randomly from this lot. The probability that exactly 2 of the chosen items are defective is

(a) 0.0036
(b) 0.1937
(c) 0.2234
(d) 0.3874

27. A coin is tossed 4 times. What is the probability of getting heads exactly 3 times?

(a) 1/4 (b) 3/8
(c) 1/2 (d) 3/4

28. If three coins are tossed simultaneously, the probability of getting at least one head is

(a) 1/8 (b) 3/8
(c) 1/2 (d) 7/8

29. A box contains 20 defective items and 80 non-defective items. If two items are selected at random without replacement, what will be the probability that both items are defective?

(a) $\frac{1}{5}$ (b) $\frac{1}{25}$
(c) $\frac{20}{99}$ (d) $\frac{19}{495}$

30. A single die is thrown twice. What is the probability that the sum is neither 8 nor 9?

(a) 1/3 (b) 5/36
(c) 1/4 (d) 3/4

31. A hydraulic structure has four gates which operate independently. The probability of failure of each gate is 0.2. Given that gate 1 has failed, the probability that both gates 2 and 3 will fail is

(a) 0.240 (b) 0.200
(c) 0.040 (d) 0.008

32. A person on a trip has a choice between private car and public transport. The probability of using a private car is 0.45. While using the public transport, further choices available are bus and metro, out of which the probability of commuting by a bus is 0.55. In such a situation, the probability (rounded up to two decimals) of using a car, bus and metro, respectively would be

(a) 0.45, 0.30 and 0.25
(b) 0.45, 0.25 and 0.30
(c) 0.45, 0.55 and 0.00
(d) 0.45, 0.35 and 0.20

33. A box contains 2 washers, 3 nuts and 4 bolts. Items are drawn from the box at random one at a time without replacement. The probability of drawing 2 washers first followed by 3 nuts and subsequently the 5 bolts is

(a) 2/315 (b) 1/630
(c) 1/1260 (d) 1/250

34. If the standard deviation of the spot speed of vehicles in a highway is 8.8 kmph and the mean speed of the vehicles is 33 kmph, the coefficient of variation in speed is

(a) 0.1517 (b) 0.1867
(c) 0.2666 (d) 0.3646

35. There are 25 calculators in a box. Two of them are defective. Suppose 5 calculators are randomly picked for inspection (i.e. each has the same chance of being selected), what is the probability that only one of the defective calculators will be included in the inspection?

(a) $\frac{1}{2}$ (b) $\frac{1}{3}$
(c) $\frac{1}{4}$ (d) $\frac{1}{5}$

36. A box contains 5 black and 5 red balls. Two balls are randomly picked one after another from the box, without replacement. The probability for both balls being red is

(a) $\frac{1}{90}$ (b) $\frac{1}{2}$

(c) $\frac{19}{90}$ (d) $\frac{2}{9}$

37. From a pack of regular playing cards, two cards are drawn at random. What is the probability that both cards will be Kings, if first card is NOT replaced?

(a) $\frac{1}{26}$ (b) $\frac{1}{52}$

(c) $\frac{1}{169}$ (d) $\frac{1}{122}$

38. If probability density function of a random variable X is

$$f(x) = x^2 \text{ for } -1 \le x \le 1, \text{ and}$$
$$= 0 \text{ for any other value of x}$$

then, the percentage probability $P\left(-\frac{1}{3} \le x \le \frac{1}{3}\right)$ is

(a) 0.247 (b) 2.47

(c) 24.7 (d) 247

39. Consider the continuous random variable with probability density function

$$f(t) = 1 + 1 \text{ for } -1 \le t \le 0$$
$$= 1 - t \text{ for } 0 \le t \le 1$$

The standard deviation of the random variable is

(a) $\frac{1}{\sqrt{3}}$ (b) $\frac{1}{\sqrt{6}}$

(c) $\frac{1}{3}$ (d) $\frac{1}{6}$

40. Let X and Y be two independent random variables. Which one of the relations between expectation (E), variance (Var) and covariance (Cov) given below is FALSE?

(a) E (XY) = E (X) E (Y)

(b) Cov (X, Y) = 0

(c) Var (X + Y) = Var (X) + Var (Y)

(d) $E(X^2 Y^2) = (E(X))^2 (E(Y))^2$

41. The standard deviation of a uniformly distributed random variable between 0 and 1 is

(a) $\frac{1}{\sqrt{12}}$ (b) $\frac{1}{\sqrt{3}}$

(c) $\frac{5}{\sqrt{12}}$ (d) $\frac{7}{\sqrt{12}}$

42. A class of first year B. Tech. students is composed of four batches A, B, C and D, each consisting of 30 students. It is found that the sessional marks of students in Engineering Drawing in batch C have a mean of 6.6 and standard deviation of 2.3. It is decided by the course instructor to normalize the marks of the students of all batches to have the same mean and standard deviation as that of the entire class. Due to this, the marks of a student in batch C are changed from 8.5 to

(a) 6.0 (b) 7.0

(c) 8.0 (d) 9.0

11.6 Numerical Methods

1. Match the following and choose the correct combination:

Group (1)

A. Newton-Raphson method
B. Runge - Kutta method
C. Simpson's Rule
D. Gauss elimination

Group (2)

1. Solving nonlinear equations
2. Solving linear simultaneous equations
3. Solving ordinary differential equations
4. Numerical integration
5. Interpolation
6. Calculation of Eigen values

	A	B	C	D
(a)	6	1	5	3
(b)	1	6	4	3
(c)	1	3	4	2
(d)	5	3	4	1

2. The equation $x^3 - x^2 + 4x - 4 = 0$ is to be solved using the **Newton-Raphson method.** If $x = 2$ is taken as the initial approximation of the solution, then the next approximation using this method will be

(a) 2/3

(b) 4/3

(c) 1

(d) 3/2

3. The polynomial $P(x)=x^5+x+2$ has

(a) All real roots
(b) 2 real and 2 complex roots
(c) 1 real and 4 complex roots
(d) All complex roots

4. The differential equation $\frac{dx}{dt}=\frac{1-x}{\tau}$ is discretised using **Euler's numerical integration** method with a time step $\Delta T>0$. What is the maximum permissible value of ΔT to ensure stability of the solution of the corresponding discrete time equation?

(a) 1 (b) $\tau/2$
(c) τ (d) 2τ

5. The recursion relation to solve $x=e^{-x}$ using Newton-Raphson method is

(a) $x_{n+1}=e^{-x_n}$

(b) $x_{n+1}=x_n-e^{-x_n}$

(c) $x_{n+1}=(1+x_n)\frac{e^{-x_n}}{1+e^{-x_n}}$

(d) $x_{n+1}=\frac{x_n^2-e^{-x_n}(1+x_n)-1}{x_n-e^{-x_n}}$

6. Consider a differential equation $\frac{dy(x)}{dx}-y(x)=x$ with the initial condition $y(0)=0$. Using Euler's first order method with a step size of 0.1, the value of $y(0.3)$ is

(a) 0.01 (b) 0.031
(c) 0.0631 (d) 0.1

7. A numerical solution of the equation

$$f(x)=x+\sqrt{x}-3=0$$

can be obtained using Newton-Raphson method. If the starting value is $x=2$ for the iteration, the value of x that is to be used in the next step is

(a) 0.306
(b) 0.739
(c) 0.694
(d) 2.306

8. It is known that two roots of the nonlinear equation $x^3-6x^2+11x-6=0$ are 1 and 3. The third root will be

(a) j (b) $-j$
(c) 2 (d) 4

9. Let $x^2-117=0$. The iterative steps for the solution using Newton-Raphson's method is given by

(a) $x_{k+1}=\frac{1}{2}\left(x_k+\frac{117}{x_k}\right)$

(b) $x_{k+1}=x_k-\frac{117}{x_k}$

(c) $x_{k+1}=x_k-\frac{x_k}{117}$

(d) $x_{k+1}=x_k-\frac{1}{2}\left(x_k+\frac{117}{x_k}\right)$

10. Equation $e^x-1=0$ is required to be solved using Newton's method with an initial gauges $x_0=-1$. Then, after one step of Newton's method estimate x_1 of the solution will be given by

(a) 0.71828 (b) 0.36784
(c) 0.20587 (d) 0.00000

11. The accuracy of Simpson's rule quadrature for a step size h is

(a) $O(h^2)$ (b) $O(h^3)$
(c) $O(h^4)$ (d) $O(h^5)$

12. The differential equation $(dy/dx)=0.25y^2$ is to be solved using the backward (implicit) Euler's method with the boundary condition y = 1 at x = 0 and with a step size of 1. What would be the value of y at x = 1?

(a) 1.33 (b) 1.67
(c) 2.00 (d) 2.33

13. In the solution of the following set of linear equations by Gauss elimination using partial pivoting $5x+y+2z=34$; $4y-3z=12$; and $10x-2y+z=-4$; the pivots for elimination of x and y are

(a) 10 and 4 (b) 10 and 2
(c) 5 and 4 (d) 5 and – 4

14. Given that one root of the equation $x^3-10x^2+31x-30=0$ is 5 the other two roots are

(a) 2 and 3 (b) 2 and 4
(c) 3 and 4 (d) –2 and –3

15. Starting from $x_0=1$, one step of Newton-Raphson method in solving the equation $x^3+3x-7=0$ gives the next value (x_1) as

(a) $x_1=0.5$
(b) $x_1=1.406$
(c) $x_1=1.5$
(d) $x_1=2$

16. The table below gives values of a function F(x) obtained for values of x at intervals of 0.25.

x	0	0.25	0.5	0.75	1.0
F(x)	1	0.9412	0.8	0.64	0.50

The value of the integral of the function between the limits 0 to 1 using Simpson's rule is

(a) 0.7854 (b) 2.3562
(c) 3.1416 (d) 7.5000

17. A 2nd degree polynomial, f(x) has values of 1, 4 and 15 at x = 0, 1 and 2, respectively. The integral $\int_0^2 f(x)\,dx$ is to be estimated by applying the trapezoidal rule to this data. What is the error (defined as "true value-approximate value") in the estimate?

(a) $-\frac{4}{3}$ (b) $-\frac{2}{3}$
(c) 0 (d) $\frac{2}{3}$

18. A calculator has accuracy up to 8 digits after decimal place. The value of $\int_0^{2\pi} \sin x\, dx$ when evaluated using this calculator by trapezoidal method with 8 equal intervals, to 5 significant digits is

(a) 0.00000
(b) 1.0000
(c) 0.00500
(d) 0.00025

19. Torque exerted on a flywheel over a cycle is listed in the table. Flywheel energy (in J per unit cycle) using Simpson's rule is

Angle (degree)	0	60	120	180	240	300	360
Torque (N m)	0	1066	−323	0	323	−355	0

(a) 542 (b) 993
(c) 1444 (d) 1986

11.7 Laplace Transforms

1. The Laplace transform of $i(t)$ is given by $I(s) = \frac{2}{s(1+s)}$. As $t \to \infty$, the value of tends to

(a) 0 (b) 1
(c) 2 (d) ∞

2. In what range should Re(s) remain so that the Laplace transform of the function $e^{(a+2)t+5}$ exits ?

(a) Re (s) > a + 2 (b) Re (s) > a + 7
(c) Re (s) < 2 (d) Re (s) > a + 5

3. A solution for the differential equation

$$x'(t) + 2x(t) = \delta(t)$$

With initial condition x (0 -) = 0 is

(a) $e^{-2t}u(t)$
(b) $e^{2t}u(t)$
(c) $e^{-1t}u(t)$
(d) $e^{t}u(t)$

4. Given that $F(s)$ is the one-sided Laplace transform of $f(t)$, the Laplace transform of $\int_0^t f(\tau)\,d\tau$ is

(a) $sF(s) - f(0)$ (b) $\frac{1}{s}F(s)$
(c) $\int_0^s F(\tau)\,d\tau$ (d) $\frac{1}{s}[F(s) - f(0)]$

5. The unilateral Laplace transform of $f(t)$ is $\frac{1}{s^2+s+1}$. The unilateral Laplace transform of $tf(t)$ is

(a) $-\frac{s}{(s^2+s+1)^2}$ (b) $-\frac{2s+1}{(s^2+s+1)^2}$
(c) $\frac{s}{(s^2+s+1)^2}$ (d) $\frac{2s+1}{(s^2+s+1)^2}$

6. If $x[n] = \left(\frac{1}{3}\right)^{|n|} - \left(\frac{1}{2}\right)^n u[n]$, then the region of convergence (ROC) of its Z-transform in the Z-plane will be

(a) $\frac{1}{3} < |z| < 3$ (b) $\frac{1}{3} < |z| < \frac{1}{2}$
(c) $\frac{1}{2} < |z| < 3$ (d) $\frac{1}{3} < |z|$

7. $u(t)$ represents the unit step function. The Laplace transform of $u(t-\tau)$ is

(a) $\frac{1}{s\tau}$ (b) $\frac{1}{s-\tau}$
(c) $\frac{e^{-s\tau}}{s}$ (d) $e^{-s\tau}$

8. The inverse Laplace transform of $\dfrac{1}{(s^2+s)}$ is

(a) $1+e^{t}$ (b) $1-e^{t}$
(c) $1-e^{-t}$ (d) $1+e^{-t}$

9. If F(s) is the Laplace transform of function f (t), then Laplace transform of $\int_0^{t} f(\tau)\,d\tau$ is

(a) $\dfrac{1}{s}F(s)$ (b) $\dfrac{1}{s}F(s)-f(0)$
(c) $sF(s)-f(0)$ (d) $\int F(s)\,ds$

10. Evaluate $\int_0^{\infty}\dfrac{\sin t}{t}\,dt$

(a) π (b) $\dfrac{\pi}{2}$
(c) $\dfrac{\pi}{4}$ (d) $\dfrac{\pi}{8}$

11. The Laplace transform of a function f(t) is $\dfrac{1}{s^2(s+1)}$. The function f(t) is

(a) $t-1+e^{-1}$ (b) $t+1+e^{-1}$
(c) $-1+e^{-1}$ (d) $2t+e^{t}$

12. A delayed unit step function is defined as

$$u(t-a)=\begin{cases}0, & \text{for } t<a\\ 1, & \text{for } t\ge a\end{cases}$$

Its Laplace transform is

(a) a.e–as (b) $\dfrac{e^{-as}}{s}$
(c) $\dfrac{e^{as}}{s}$ (d) $\dfrac{e^{as}}{a}$

13. Given $f(t)=LT^{-1}\left[\dfrac{3s+1}{s^3+4s^2+(K-3)s}\right]$. If $\lim_{t\to\infty} f(t)=1$, then the value of K is

(a) 1 (b) 2
(c) 3 (d) 4

14. Given $f(t)$ and $g(t)$ as shown below:

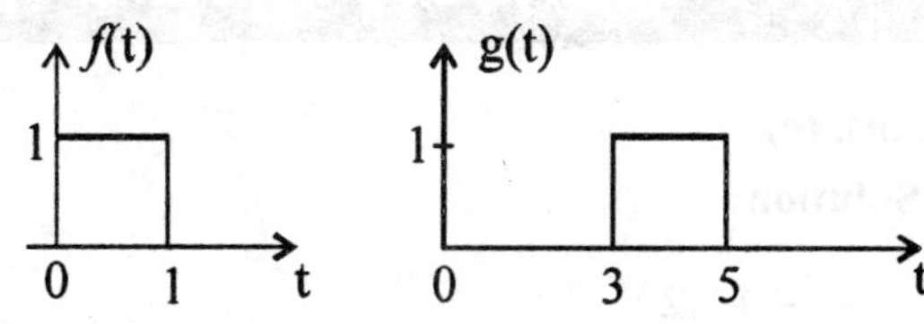

$g(t)$ can be expressed as

(a) $g(t)=f(2t-3)$ (b) $g(t)=f\left(\dfrac{t}{2}-3\right)$
(c) $g(t)=f\left(2t-\dfrac{3}{2}\right)$ (d) $g(t)=f\left(\dfrac{t}{2}-\dfrac{3}{2}\right)$

15. Given $f(t)$ and $g(t)$ as shown below:

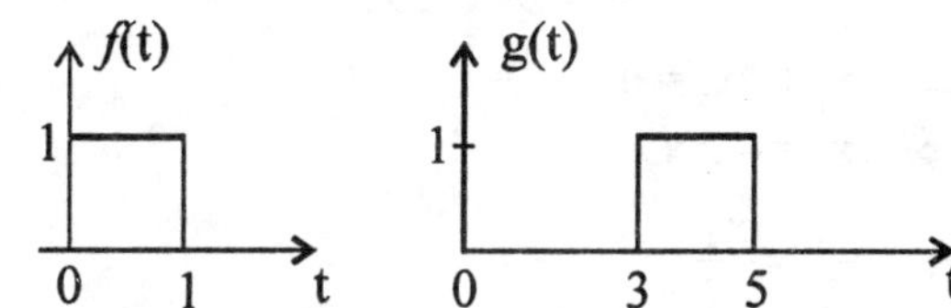

The Laplace transform of $g(t)$ is

(a) $\dfrac{1}{s}\left(e^{3s}-e^{5s}\right)$ (b) $\dfrac{1}{s}\left(e^{-5s}-e^{-3s}\right)$
(c) $\dfrac{e^{-3s}}{s}\left(1-e^{-2s}\right)$ (d) $\dfrac{1}{s}\left(e^{5s}-e^{3s}\right)$

16. Consider the differential equation

$$\frac{d^2y(t)}{dt^2}+2\frac{dy(t)}{dt}+y(t)=\delta(t) \text{ with } y(t)\big|_{t=0^-}=-2 \text{ and } \left.\frac{dy}{dt}\right|_{t=0^-}=0.$$

The numerical value of $\left.\dfrac{dy}{dt}\right|_{t=0^+}$ is

(a) –2 (b) –1
(c) 0 (d) 1

Answers

11.1 Linear Algebra

1. Ans. (c)

Solution:

$$A=\begin{pmatrix}-4 & 2\\ 4 & 3\end{pmatrix}$$

$\therefore \quad (A-\lambda I)=0$

$\Rightarrow \quad (4-\lambda)(3-\lambda)-2\times 4=0$

$\Rightarrow \quad -12+l^2-3l+4l-2\times 4=0$

$\Rightarrow \quad l^2+l-20=0$

$\therefore \quad l=-5, 4$

Putting $l=-5$ $\begin{pmatrix}1 & 2\\ 4 & 8\end{pmatrix}\begin{pmatrix}x_1\\ x_2\end{pmatrix}=0$

$\Rightarrow \quad x_1+2x_2=0$

$\Rightarrow \quad x_1=-2x_2$

$\Rightarrow \quad \frac{x_1}{2}=\frac{x_2}{-2}$

Hence $\begin{pmatrix} \\ -1\end{pmatrix}$ is eigen vector

2. Ans. (d)

Solution: Given Matrix

$$A=\begin{bmatrix}1 & 1 & 1 & 1\\ 1 & 1 & -1 & -1\\ 1 & -1 & 0 & 0\\ 0 & 0 & 1 & -1\end{bmatrix}$$

$$A^T=\begin{bmatrix}1 & 1 & 1 & 0\\ 1 & 1 & -1 & 0\\ 1 & -1 & 0 & 1\\ 1 & -1 & 0 & -1\end{bmatrix}$$

$$AA^T=\begin{bmatrix}4 & 0 & 0 & 0\\ 0 & 4 & 0 & 0\\ 0 & 0 & 4 & 0\\ 0 & 0 & 0 & 4\end{bmatrix}$$

$$\therefore \left(AA^T\right)^{-1}=\begin{bmatrix}1/4 & 0 & 0 & 0\\ 0 & 1/4 & 0 & 0\\ 0 & 0 & 1/4 & 0\\ 0 & 0 & 0 & 1/4\end{bmatrix}$$

3. Ans. (b)

Solution:

Given $\lambda_1=8$, $\lambda_2=4$ and

$$V_1=\begin{bmatrix}1\\1\end{bmatrix},\ V_2=\begin{bmatrix}1\\-1\end{bmatrix}$$

If $P=[V_1 \ \ V_2]$ and $D=\begin{bmatrix}8 & 0\\ 0 & 4\end{bmatrix}$

Then Matrix $A=PDP^{-1}$

$$=\begin{bmatrix}1 & 1\\ 1 & -1\end{bmatrix}\begin{bmatrix}8 & 0\\ 0 & 4\end{bmatrix}\begin{bmatrix}1 & 1\\ 1 & -1\end{bmatrix}^{-1}$$

$$=\begin{bmatrix}8 & 4\\ 8 & -4\end{bmatrix}\begin{bmatrix}1 & 1\\ 1 & -1\end{bmatrix}\frac{1}{2}$$

$$=\frac{1}{2}\begin{bmatrix}12 & 4\\ 4 & 12\end{bmatrix}$$

$$=\begin{bmatrix}6 & 2\\ 2 & 6\end{bmatrix}$$

4. Ans. (c)

Solution:

Given $X=\begin{bmatrix}101\\101\end{bmatrix}$ is an eigen vector of $A=\begin{bmatrix}4 & 2\\ 2 & 4\end{bmatrix}$ corresponding to some an eigen value λ.

By definition of an eigen vector of A, $AX=\lambda X$.

$$AX=\begin{bmatrix}4 & 2\\ 2 & 4\end{bmatrix}\begin{bmatrix}101\\101\end{bmatrix}$$

$$=\begin{bmatrix}606\\606\end{bmatrix}$$

$$=6\begin{bmatrix}101\\101\end{bmatrix}=\lambda X$$

$\therefore X=\begin{bmatrix}101\\101\end{bmatrix}$ **is an eigen vector of A corresponding to an eigen value $\lambda=6$**

5. Ans. (a)

Solution:

The system is consistent if and only if coefficient matrix and the argumented matrix are of the same rank, otherwise the system is in consistent.

6. Ans. (d)

Solution:

$$P = \begin{bmatrix} 3 & -2 & 2 \\ 0 & -2 & 1 \\ 0 & 0 & 1 \end{bmatrix}$$

$$|P - vI| = \begin{bmatrix} 3-v & -2 & 2 \\ 0 & -2v & 1 \\ 0 & 0 & 1-v \end{bmatrix}$$

$$= (3-v)[(-2-v)(1-v)] - 0; \quad (p - vI) = 0$$

$$= (3-v)(-2+2v-v+v^2) = 1; \quad v = 3$$

$v^2 + v - 2 = 0 \quad ; \quad v^2 + 2v - v - 2 = 0$

$v(v+2) + (v+1) \; ; \quad v+2 \quad vv-1$

$v = -2$

$v = 1$

$v = 1, 2, 3$

$\therefore$ eigen values of P are 1, 2, 3

for eigen vector corresponding to v = 1

$$[A - vI]X = \begin{bmatrix} 3-v & -2 & 2 \\ 0 & -2-v & 1 \\ 0 & 0 & 1-v \end{bmatrix} \begin{bmatrix} x \\ y \\ z \end{bmatrix} = 0$$

Put v = 1

$$= \begin{bmatrix} 3-1 & -2 & 2 \\ 0 & -2-1 & 1 \\ 0 & 0 & 0 \end{bmatrix} \begin{bmatrix} x \\ y \\ z \end{bmatrix} = 0$$

or, $$= \begin{bmatrix} 2 & -2 & 2 \\ 0 & -3 & 1 \\ 0 & 0 & 0 \end{bmatrix} \begin{bmatrix} x \\ y \\ z \end{bmatrix} = 0$$

$2x - 2y + 3v = 0$ ...(1)

$-3y + z = 0$ ...(2)

Form equation (1) and (2)

–3y = – z

$\frac{y}{1} = \frac{z}{3}$ ...(3)

Put,

z = 3y equation (1) from (3) (2) and (4)

$2x - 2y + 6y = 0 \qquad \frac{x}{-2} = \frac{y}{1} = \frac{z}{3}$

$2x + 4y = 0$ eigen vector (–2, 1, 3)

$2x = 4y$

$\frac{x}{-2} = \frac{y}{1}$ ...(4)

Case II

When 1 = –2

$$\begin{bmatrix} 3+2 & -2 & 2 \\ 0 & -2+2 & 1 \\ 0 & 0 & 1+2 \end{bmatrix} \begin{bmatrix} x \\ y \\ z \end{bmatrix} = 0$$

$$\begin{bmatrix} 5 & -2 & 2 \\ 0 & 0 & 1 \\ 0 & 0 & 3 \end{bmatrix} \begin{bmatrix} x \\ y \\ z \end{bmatrix} = 0$$

5 x – 2y = 0

5x = 2y

$\frac{x}{2} = \frac{y}{5} = \frac{z}{0}$

$5x - 2y + 2z = 0$ ...(1)

Z = 0 ...(2)

3z = 0 ...(3)

$\therefore$ Eigen vector (2, 5, 0)

Second Method

Let $x = \begin{pmatrix} x_1 \\ x_2 \\ x_3 \end{pmatrix}$ be an eigen vector corresponding to – 2

$Ax = \lambda x$ gives $3x_1 - 2x_2 + 2x_3 = -2x_1$

$-2x_2 + x_3 = -2x_2$

$x_3 = -2x_3$

$\therefore \; x_3 = 0$

Let $x_2 = t$ then $x_1 = \frac{2}{5}t$

$\therefore$ An eigen vector $\begin{bmatrix} 2 \\ 5 \\ 0 \end{bmatrix}$ corresponding to eigen value (–2)

7. Ans. (b)

Solution:

$$\text{If } R = \begin{bmatrix} 1 & 0 & -1 \\ 2 & 1 & -1 \\ 2 & 3 & 2 \end{bmatrix}$$

Co-factors are as follows

$= 1(-1+3) + 0(1+(-1))(6-2)2 + 0 - 4 = -2$

$C_{11} = +(2+3) = +5$ $C_{12} = -(4+2) = -6$ $C_{13} = +(6-2) = 4$

$C_{21} = -(0+3) = -3$ $C_{22} = +(2+2) = +4$ $C_{23} = -(3-0) = -3$

$C_{31} = +(0+13)$ $C_{32} = -(-1+2) = -1$ $C_{33} = +(1-0) = 1$

$$C = \begin{bmatrix} 5 & -6 & 4 \\ -3 & 4 & -3 \\ 1 & -1 & 1 \end{bmatrix}$$

$$\therefore \ A+ = \frac{1}{|A|} Adj\ A$$

$$C' = \begin{bmatrix} 5 & -3 & 1 \\ -6 & 4 & -1 \\ 4 & -3 & 1 \end{bmatrix}$$

$$= -\frac{1}{2}\begin{bmatrix} 5 & -3 & 1 \\ -6 & 4 & -1 \\ 4 & -3 & +1 \end{bmatrix}$$

$$= \begin{bmatrix} -5/2 & 3/2 & -1/2 \\ 3 & -2 & 1/2 \\ -2 & 3/22 & -1/2 \end{bmatrix}$$

Tor row of $C' = [-5/2 \ \ 3/2 \ \ -1/2]$

$\Rightarrow \ [5 \ \ -3 \ \ 1]$

8. Ans. (b)

Solution:

Given that $< x.y >$ denote their dot product of vector x and y.

$$\det\begin{bmatrix} <x\cdot x> & <x\cdot y> \\ <y\cdot x> & <y\cdot y> \end{bmatrix} = (x\cdot x)(y\cdot y) - (x\cdot y)(y\cdot x)$$

$$= x^2y^2 - (xy)^2 \quad (1)$$

But dot product is scalar

$\therefore$ Equation (1) becomes positive when x and y are linearly independent

Second Method

Let $\overline{x}$ and $\overline{y}$ are 3-dimensinal vectors

$\overline{x}\cdot\overline{x} = |\overline{x}||\overline{x}| \cos 0 = |x|^2 = x^2$

$\overline{y}\cdot\overline{y} = y^2$

$\overline{x}\cdot\overline{y} = \overline{y}\cdot\overline{x} = |x||y|\cos\alpha = xy\cos\alpha$

Where a is the angle between vectors $\overline{x}$ and $\overline{y}$

Determinant

$$|A| = \begin{vmatrix} \overline{x}\cdot\overline{x} & \overline{x}\cdot\overline{y} \\ \overline{y}\cdot\overline{x} & \overline{x}\cdot\overline{x} \end{vmatrix} = \begin{vmatrix} x^2 & xy\cos\alpha \\ xy\cos\alpha & y^2 \end{vmatrix}$$

$$= xy\begin{vmatrix} x & y\cos\alpha \\ x\cos\alpha & y \end{vmatrix}$$

$$= xy\left[xy - xy\cos^2\alpha\right]$$

$$= x^2y^2\sin^2\alpha$$

$$|A| = x^2y^2\sin^2\alpha = 0$$

Case 1: If either $x = 0$ or $y = 0$ or $\alpha = 0$

$|A| = 0$ even if $x \neq 0$, $y \neq 0$ but $\alpha = 0$

Hence option (d) is incorrect

Case 2: If $x \neq 0$, $y \neq 0$ but $\alpha = 0$

$|A| = 0$; Hence option (c) is incorrect

Case 3: If vector $\overline{x}$ and $\overline{y}$ are linearly independent means they are not parallel i.e. $\alpha = 0$

$|A| = (xy\sin\alpha)^2 > 0$ i.e. +ve; Hence option (b) is correct

Case 4: If $\overline{x}$ and $\overline{y}$ are linearly dependent, i.e. parallel vectors $\overline{y} = c\overline{x}$ and c is a constant

$\overline{y} = c\overline{x}$, i.e. $\alpha = 0$

$|A| = 0$; Hence option (c) is also incorrect

9. Ans. (c)

Solution:

Since. A is a 3 × 3 matrix having rank 2, so, A can be written as

$$A = \begin{bmatrix} a_1 & a_2 & a_0 \\ b_1 & b_2 & b_3 \\ 0 & 0 & 0 \end{bmatrix}$$

Where,

$$a_1 \neq k\ b_1 \neq 0 \ \ a_2 \neq k\ b_2 \neq 0 \ \ a_3 \neq k\ b_3 \neq 0$$

and $X = \begin{bmatrix} x_1 \\ x_2 \\ x_3 \end{bmatrix}$

Equation formed by AX = 0 are

$$a_1x_1 + a_2x_2 + a_3x_3 = 0 \quad \ldots(1)$$

$$b_1x_1 + b_2x_2 + b_3x_3 = 0 \quad \ldots(2)$$

On analysising equation (1) and (2),we get 2 independent solution.

10. Ans. (b)

Solution:

Given: $A = \begin{bmatrix} 1 & 0 \\ -1 & -2 \end{bmatrix}$

Characteristic equations is ,

$$[A-\lambda I]=\begin{bmatrix}1-\lambda & 0\\ -1 & -2-\lambda\end{bmatrix}$$

i.e. $(1-\lambda)(-2-\lambda)=0$

$$\lambda=1,-2$$

Thus, the eigen value are 1, – 2

If x, y, be component of eigen vector corresponding to the eigen values λ, we have

$$[A-\lambda I]\ X=\begin{bmatrix}1-\lambda & 0\\ -1 & -2-\lambda\end{bmatrix}\begin{bmatrix}x\\ y\end{bmatrix}=0$$

For λ = 1, we get the eigen vector as $\begin{bmatrix}3\\ -1\end{bmatrix}$

Hence, the answer will be $[3-1]^T$

11. Ans. (c)

Solution:

Let $A=\begin{bmatrix}a & b\\ c & d\end{bmatrix}$

given $A\begin{bmatrix}1\\ -1\end{bmatrix}=\begin{bmatrix}a & b\\ c & d\end{bmatrix}\begin{bmatrix}1\\ -1\end{bmatrix}=\begin{bmatrix}a-b\\ c-d\end{bmatrix}=\begin{bmatrix}1\\ -1\end{bmatrix}$

$$A\begin{bmatrix}1\\ -2\end{bmatrix}=\begin{bmatrix}a & b\\ c & d\end{bmatrix}\begin{bmatrix}1\\ -2\end{bmatrix}=\begin{bmatrix}a-2b\\ c-2d\end{bmatrix}=\begin{bmatrix}-2\\ +4\end{bmatrix}$$

$$a-b=-1 \qquad c-d=1$$

$$\underline{a-2b=-2}\ ; \qquad \underline{c-2d=4}$$

$$b=1 \qquad d=3$$

$$a=0 \qquad c=-2$$

$$a=0 \qquad b=1$$

$$d=-3 \qquad c=-2$$

$$\begin{bmatrix}0 & 1\\ -2 & -3\end{bmatrix}$$

12. Ans. (b)

Solution:

$AX=y$

$X=A^{-1}y$

U se De-movies

Thus $\left(\sqrt{-1}\right)^{(-1)1/2}$

$\Rightarrow e^{-\pi/2}$

13. Ans. (b)

Solution:

Frequency $3\times 3\begin{bmatrix}1 & 2 & 3\\ 2 & 4 & 6\\ 3 & 6 & 9\end{bmatrix}_{3\times 3}$

All rows are L.I., so rank is 1

14. Ans (b)

Solution:

$$\begin{bmatrix}2 & 1 & -4\\ 4 & 3 & -12\\ 1 & 2 & -8\end{bmatrix}\begin{bmatrix}X\\ Y\\ Z\end{bmatrix}=\begin{bmatrix}\alpha\\ 5\\ 7\end{bmatrix}$$

Augmented matrix $=\begin{bmatrix}2 & 1 & -4\\ 4 & 3 & -12\\ 1 & 2 & -8\end{bmatrix}\begin{bmatrix}\alpha\\ 5\\ 7\end{bmatrix}$

Applying Gauss Elimination

R_2-2R_1

$$\begin{bmatrix}2 & 1 & -4\\ 0 & 1 & -4\\ 1 & 1 & -8\end{bmatrix}\begin{bmatrix}\alpha\\ 5-2\alpha\\ 7\end{bmatrix}$$

$$\xrightarrow{R_3-\frac{1}{2}R_1}\begin{bmatrix}2 & 1 & -4\\ 0 & 1 & -4\\ 0 & 3/2 & -6\end{bmatrix}\begin{bmatrix}\alpha\\ 5-2\alpha\\ 7-2\alpha\end{bmatrix}$$

$$\xrightarrow{R_3-\frac{3}{2}R_1}\begin{bmatrix}2 & 1 & -4\\ 0 & 1 & -4\\ 0 & 0 & 0\end{bmatrix}\begin{bmatrix}\alpha\\ 5-2\alpha\\ \frac{5\alpha-1}{2}\end{bmatrix}$$

If one row is completely zero, then we have infinitely many solution. So Here R_3 is 0

$$\frac{5\alpha-1}{2}=0$$

$$\alpha=\frac{1}{5}$$

Thus, we have only one value by a for infinite solution

15. Ans. (d)

Solution:

Symmetric matrix is a square matrix whose element in lower triangle is equal to the element in upper triangle.

Say, we filled upper triangle. So first row is filled by n element, 2nd is filled by n–1 element ...nth row is filled by 1 element

Total number $= n+(n-1)+(n-2)+\ldots+1$

$$= \frac{n(n-1)}{2}$$

As there are 0's and 1's thus total ways is $= 2^{\frac{n(n+1)}{2}}$

16. Ans. (b)

Solution:

We have $ABCD = I$

$$A^{-1}\cdot ABCD = A^{-1}\cdot I$$

$$BCD = A^{-1}\cdot I \; \{A^{-1}\cdot A = I\}$$

$$B.C.DD^{-1} = A^{-1}.D^{-1}$$

$$BC\cdot C^{-1} = A^{-1}\, D^{-1}\, C^{-1}$$

$$B = A^{-1}\, D^{-1}\, C^{-1}$$

$$B^{-1} = \left(C^{-1}\, D^{-1}\, A^{-1}\right)^{-1}$$

$$B^{-1} = \left(C^{-1}\right)^{-1}\left(D^{-1}\right)^{-1}\left(A^{-1}\right)^{-1}$$

$$B^{-1} = CDA$$

17. Ans. (b)

Solution:

$$A = \begin{bmatrix} 2 & -1 \\ -4 & 5 \end{bmatrix}$$

For eigen values

$$|A-\lambda I| = \begin{vmatrix} 2-\lambda & -1 \\ -4 & 5-\lambda \end{vmatrix}$$

$$= (2-\lambda)(5-\lambda)-4$$

$$= 10-7\lambda+\lambda^2-4$$

$$\Rightarrow \lambda^2-6\lambda-\lambda+6=0$$

$$(\lambda-6)(\lambda-1)=0$$

$$\lambda = 1, 6$$

Eigenvlaues are 1 and 6.

18. Ans. (c)

Solution:

$x_1, -x_1$ for one axis:

$x_2, -x_2$ for another axis:

$\vdots \quad \vdots$

$x_m, -x_m$ for m^{th} axis

Since $(-X_1, -X_2, \ldots X_M)$ are linearly dependent of, $X_1, X_2, \ldots X_M$ the set $(X_1, X_2, X_3 \ldots X_M, -X_1, -X_2, \ldots X_M)$ one also span a vector space of dimension *M* only

Since $(X_1, X_2, \ldots X_M)$ are orthogonal, they span a vector space of dimension *M*.

19. Ans. (b)

Solution:

The set of n-vectors

$X_1 = (x_{11}, x_{12}, \ldots\ldots x_{1m})$, $X_2 = (x_{21}, x_{22}, \ldots\ldots x_{2m})$ and so on.

Is linearly independent if the matrix formed by $X_1, \ldots, X_n$ (rows are $X_1, X_2, \ldots X_{nm}$)

$$A = \begin{bmatrix} x_{11} & x_{12}\ldots x_{1m} \\ \vdots & \\ x_{n1} & x_{n2}\ldots x_{nm} \end{bmatrix}$$

is of rank *n*

If rank of $A < n$

The set of vectors $X_1, \ldots X_n$ is linearly dependent.

For given option (8, 9, 3)

The set (–6, –3, 6), (4, –2, 3) and (8, 9, 3)

$$\text{Matrix } A = \begin{bmatrix} -6 & -3 & 6 \\ 4 & -2 & 3 \\ 8 & 9 & 3 \end{bmatrix}$$

Rank $A = 3$

Hence the set is linearly independent.

So (8, 9, 3) is linearly independent of (–6, –3, 6) and (–4, –2, 3)

For given option (–2, 17, 30)

$$A = \begin{bmatrix} -6 & -3 & 6 \\ 4 & -2 & 3 \\ -2 & -17 & 30 \end{bmatrix}$$

Rank of $A = 2$

Hence the set of (–6, –3, 6), (4, –2, 3) and (–2, –17, 30) is linearly dependent.

So (–2, –17, 30) is linearly dependent on (–6, –3, 6) and (4, –2, 3)

20. Ans. (b)

Solution:

Given $X = [x_1 \quad x_2 \quad \ldots \quad x_n]^T$

$$\text{Then } X = \begin{bmatrix} x_1 \\ x_2 \\ \vdots \\ x_n \end{bmatrix}$$

$$V = XX^T = \begin{bmatrix} x_1 \\ x_2 \\ \vdots \\ x_n \end{bmatrix} \begin{bmatrix} x_1 & x_2 & \cdots & x_n \end{bmatrix}$$

$$= \begin{bmatrix} x_1^2 & x_1x_2 & \cdots & x_1x_n \\ x_1x_2 & x_2^2 & \cdots\cdots\cdots & \\ \vdots & \vdots & x_{n-1}^2 & x_{n-1}\,x_n \\ x_1x_n & \vdots & x_{n-1}x_n & x_n^2 \end{bmatrix}_{n\times n}$$

For example let n = 3

$$V = \begin{bmatrix} x_1^2 & x_1x_2 & x_1x_3 \\ x_1x_2 & x_2^2 & x_2x_3 \\ x_1x_3 & x_2x_3 & x_3^2 \end{bmatrix}$$

Rank = 1 become all (3 × 3) and (2 × 2) minors are zero therefore $V_{n\times n}$ has Rank = 1

21. Ans. (a)

Solution:

Three dimensinal vectors given are

$\bar{b} = [0 \quad 1 \quad 0]^T$ and $\bar{x} = [x_1 \quad x_2 \quad x_3]^T$

or $\bar{b} = 0.\hat{i} + 1.\hat{j} + 0.\hat{k}$

$\bar{x} = x_1\hat{i} + x_2\hat{j} + x_3\hat{k}$

$L(x) = \bar{b} \times \bar{x}$

i.e. Cross-Product

$$= \begin{vmatrix} \hat{i} & \hat{j} & \hat{k} \\ 0 & 1 & 0 \\ x_1 & x_2 & x_3 \end{vmatrix}$$

$$= \hat{i}[1\cdot x_3 - 0\cdot x_2] - \hat{j}[0\cdot x_3 - 0\cdot x_1] + \hat{k}[0\cdot x_2 - 1\cdot x_1]$$

$$= x_3\hat{i} + 0\hat{j} - x_1\hat{k}$$

$$L(x) = [x_3 \quad 0 \quad x_1]^T = \begin{bmatrix} x_3 \\ 0 \\ -x_1 \end{bmatrix}$$

$$L(x) = \begin{bmatrix} 0 & 0 & 1 \\ 0 & 0 & 0 \\ -1 & 0 & 0 \end{bmatrix} \begin{bmatrix} x_1 \\ x_2 \\ x_3 \end{bmatrix}$$

Compare $L(x) = M \begin{bmatrix} x_1 \\ x_2 \\ x_3 \end{bmatrix}$

Then, Matrix $M = \begin{bmatrix} 0 & 0 & 1 \\ 0 & 0 & 0 \\ -1 & 0 & 0 \end{bmatrix}$

Characteristic Equation

$|M - \lambda I| = 0$

or, $|M - \lambda I| = \begin{bmatrix} 0 & 0 & 1 \\ 0 & 0 & 0 \\ -1 & 0 & 0 \end{bmatrix} - \lambda \begin{bmatrix} 1 & 0 & 0 \\ 0 & 1 & 0 \\ 0 & 0 & 1 \end{bmatrix}$

$$= \begin{bmatrix} -\lambda & 0 & 1 \\ 0 & -\lambda & 0 \\ -1 & 0 & -\lambda \end{bmatrix}$$

$$= \begin{vmatrix} -\lambda & 0 & 1 \\ 0 & -\lambda & 0 \\ -1 & 0 & -\lambda \end{vmatrix} = 0$$

or $|-\lambda| = \begin{vmatrix} -\lambda & 0 \\ 0 & -\lambda \end{vmatrix} + \begin{vmatrix} 0 & -\lambda \\ -1 & 0 \end{vmatrix} = 0$

or $-\lambda[-\lambda^2] + [0 - \lambda] = 0$

or $\lambda^3 + \lambda = 0$

or $\lambda(\lambda^2 + 1) = 0$

∴ eigen values are

$\lambda = 0, \pm 1$

22. Ans. (c)

Solution: Given $A = \begin{bmatrix} -3 & 2 \\ -1 & 0 \end{bmatrix}$

Then characteristic equation is

$$|A - \lambda I| = 0$$

or, $\begin{vmatrix} -3-\lambda & 2 \\ -1 & -\lambda \end{vmatrix} = 0$

or, $(-3-\lambda)\ (-\lambda) + 2 = 0$

or, $3\lambda + \lambda^2 + 2 = 0$

or, $\lambda^2 + 3\lambda + 2 = 0$ ...(1)

or, $(\lambda+1)\ (\lambda+2) = 0$

or, $\lambda = -1, -2$ [Eigen values]

23. Ans. (a)

Solution:

$$A = \begin{bmatrix} -3 & 2 \\ -1 & 0 \end{bmatrix}$$

$$A-\lambda I=\begin{bmatrix}-3 & 2\\-1 & 0\end{bmatrix}-\lambda\begin{bmatrix}1 & 0\\0 & 1\end{bmatrix}$$

$$=\begin{bmatrix}-3-\lambda & 2\\-1 & \lambda\end{bmatrix}$$

$$|A-\lambda I|=\begin{bmatrix}-3-\lambda & 2\\-1 & \lambda\end{bmatrix}=0$$

or, $\begin{bmatrix}\lambda+3 & -2\\1 & \lambda\end{bmatrix}=0$

$$\lambda(\lambda+3)+2=0$$

or, $(\lambda+1)(\lambda+2)=0$

$\lambda=-1$ and -2

Using Caylay Hamilton Theorem λ can be replaced A

or $(A+I)(A+2I)=0$

or $A^2+3A+2I=0$

or $A^2=-3A-2I$

$$A\cdot A^2=-A(3A+2I)$$

$$A^3=-3A^2-2A$$

$$=-3(-3A-2I)-2A$$

$$=7A+6I$$

$$A^3\cdot A^3=(7A+6I)(7A+6I)$$

$$=49A^2+84A+36I$$

$$=-49(3A+2I)+84A+36I$$

$$A^6=-63A-62I$$

$$A^3\cdot A^6=-(7A+6I)(63A-62I)$$

$$A^9=-\left[441A^2+812A+372I\right]$$

$$=-441(-3A-2I)-(812A-372I)$$

$$=511A+510I$$

24. Ans. (c)

Solution:

$$-x+5y=-1$$
$$x-y=2$$
$$x+3y=3$$

Augmented matrix $=\begin{bmatrix}-1 & 5\\1 & -1\\1 & 3\end{bmatrix}\begin{bmatrix}-1\\2\\3\end{bmatrix}$

Applying Gauss Elimination

R_2+R_1

$$=\begin{bmatrix}-1 & 5\\0 & 4\\1 & 8\end{bmatrix}\begin{bmatrix}-1\\2\\3\end{bmatrix}$$

$$=\xrightarrow{R_3+R_1}\begin{bmatrix}-1 & 5\\0 & 4\\0 & 8\end{bmatrix}\begin{bmatrix}-1\\1\\3\end{bmatrix}$$

$$=\xrightarrow{R_3-2R_2}\begin{bmatrix}-1 & 5\\0 & 4\\0 & 0\end{bmatrix}\begin{bmatrix}-1\\1\\0\end{bmatrix}$$

Rank of augmented matrix is 2 because it has 2 non-zero rows

Rank of matrix A is also 2.

Thus $\rho(A|B)=\rho(A)$ means unique solution

25. Ans. (b)

Solution:

$$2x_1-x_2+3x_3=1$$
$$3x_1-2x_2+5x_3=2$$
$$-x_1-4x_2+x_3=3$$

Augmented matrix $\left[\begin{array}{ccc|c}2 & -1 & 3 & 1\\3 & -2 & 5 & 2\\-1 & -4 & 1 & 3\end{array}\right]$

Apply Gauss elimination method

$$\xrightarrow{R_2-\frac{3}{2}R_1}\left[\begin{array}{ccc|c}2 & -1 & 3 & 1\\0 & -1/2 & 1/2 & 1/2\\-1 & -4 & 1 & 3\end{array}\right]$$

$$\xrightarrow{R_3+\frac{1}{2}R_1}\left[\begin{array}{ccc|c}2 & -1 & 3 & 1\\0 & -1/2 & 1/2 & 1/2\\0 & -9/2 & 5/2 & 7/2\end{array}\right]$$

$$\xrightarrow{R_3-9R_2}\left[\begin{array}{ccc|c}2 & -1 & 3 & 1\\0 & -1/2 & 1/2 & 1/2\\0 & 0 & -2 & -1\end{array}\right]$$

Rank of augmented matrix is 3 as there are 3 non-zero rows.

Rank of matrix A is 3

Rank $\rho(A|B)=\rho(A)$

Thus, it has a unique solution.

26. Ans. (b)

Solution:

For the basis for subspace X, two conditions are to be satisfied i.e.

1. The vector have to be linearly independent.
2. The vector must have span X.

Given

$$X = \left[X \,\varepsilon\, R^3 \mid X_1 + X_2 + X_3 = 0\right]$$

$$X^T = \left[\, X_1, X_2, X_3\right]^T$$

- Step 1: Now $\left[(1, -1, 0)^T, (1, 0, -1)^T\right]^T$ is linearly independent set because one can be obtained from another by scalar multiplication.
- The fact independent can also be established by seeing that rank of $\begin{bmatrix} 1 & -1 & 0 \\ 1 & 0 & -1 \end{bmatrix}$ is 2
- Next, one need to check if the set spans X.

Hence $X = \left\{X \,\varepsilon\, R^2 \mid X_1 + X_2 + X_3 = 0\right\}$

The general infinite solution of $X = \begin{bmatrix} -k_1 - k_2 \\ k_1 \\ k_2 \end{bmatrix}$

Choose k_1, k_2 as $\begin{bmatrix} k_1 \\ k_2 \end{bmatrix} = \begin{bmatrix} 0 \\ k \end{bmatrix}$ and $\begin{bmatrix} k_1 \\ k_2 \end{bmatrix} = \begin{bmatrix} k \\ 0 \end{bmatrix}$

One gets 2 linearly independent solutions, for X,

$$X = \begin{bmatrix} -k \\ 0 \\ k \end{bmatrix} \text{ or } \begin{bmatrix} -k \\ k \\ 0 \end{bmatrix}$$

- Now gets since both of these can be generated by linearly combinations of $\begin{bmatrix} 1 & -1 & 0 \end{bmatrix}^T$ and $\begin{bmatrix} 1 & 0 & -1 \end{bmatrix}^T$, the set spans X.
- One can show that the set is not only linearly independent but also spans X, therefore by definition it is a basis for the subspace X.

27. Ans. (c)

Solution: Use $\left[\lambda I - P\right] = 0$

$$\begin{bmatrix} \lambda & 0 \\ 0 & \lambda \end{bmatrix} - \begin{pmatrix} P_{11} & P_{12} \\ P_{21} & P_{22} \end{pmatrix} = 0$$

$$\begin{bmatrix} \lambda - P_{11} & -P_{12} \\ -P_{21} & \lambda - P_{22} \end{bmatrix} = 0$$

$$\Rightarrow (\lambda - P_{11})(\lambda - P_{22}) - P_{21}P_{12} = 0$$

Putting $\lambda = 0$, we get $P_{11}P_{22} - P_{21} = 0$

28. Ans. (b)

Solution: Given $4x + 2y = 7$ ...(1)

$2x + y = 6$ ...(2)

Equation (1) Can be written as

$2x + y = 3.5$ ...(3)

Now LHS of equation (2) and (3) are same but RHS is different which is not possible. Hence above equations have no solution.

29. Ans. (d)

Solution:

Given that, $A = \begin{bmatrix} -1 & 3 & 5 \\ -3 & -1 & 6 \\ 0 & 0 & 3 \end{bmatrix}$

The characteristic equation of A is

$|A - \lambda I| = o$

Where I is an identity matrix

$$\lambda I = \lambda \begin{bmatrix} 1 & 0 & 0 \\ 0 & 1 & 0 \\ 0 & 0 & 1 \end{bmatrix}$$

$$= \begin{bmatrix} \lambda & 0 & 0 \\ 0 & \lambda & 0 \\ 0 & 0 & \lambda \end{bmatrix}$$

$$|A - \lambda I| = \begin{bmatrix} -1 & 3 & 5 \\ -3 & -1 & 6 \\ 0 & 0 & 3 \end{bmatrix} - \begin{bmatrix} \lambda & 0 & 0 \\ 0 & \lambda & 0 \\ 0 & 0 & \lambda \end{bmatrix}$$

$$= \begin{bmatrix} -1-\lambda & 3 & 5 \\ -3 & -1-\lambda & 6 \\ 0 & 0 & 3-\lambda \end{bmatrix}$$

$$|A - \lambda I| = \lambda^3 - \lambda^2 + 4\lambda - 30$$

From $|A - \lambda I| = 0$

$$\lambda^3 - \lambda^2 + 4\lambda - 30 = 0$$

$$(\lambda - 3)(\lambda^2 + 2\lambda + 10) = 0$$

$$(\lambda - 3)(\lambda^2 + 2\lambda + 10) = 0$$

$$(\lambda - 3)(\lambda + 1 - 3j)(\lambda + 1 + 3j) = 0$$

$\lambda = 3,\ -1 + 3j,\ -1 - 3j$

The eigen value are 3, $-1 + 3j$, $-1 - 3j$

30. Ans. (c)

Solution:

Let, $A_{n\times n}$ is real and skew-symmetric matrix.

As A is real

$\bar{A} = A$ (complex conjugate)

As A is skew-symmetric also

$A^T = -A$ (Transpose)

Transpose conjugate

$A^\theta = (\bar{A})^T = A^T = -A$

So A is Skewalso.

Hence eigen values of A matrix are either zero or pure imaginary.

31. Ans (b)

Solution:

$$\text{Given} \begin{bmatrix} 1 & 1 & 1 \\ 1 & 4 & 6 \\ 1 & 4 & \lambda \end{bmatrix}\begin{bmatrix} x \\ y \\ z \end{bmatrix} = \begin{bmatrix} 6 \\ 20 \\ \mu \end{bmatrix}$$

i.e. $AX = B$

$$[A|B] = \left[\begin{array}{ccc|c} 1 & 1 & 1 & 6 \\ 1 & 4 & 6 & 20 \\ 1 & 4 & \lambda & \mu \end{array}\right]$$

$$\Rightarrow \left[\begin{array}{ccc|c} 1 & 1 & 1 & 6 \\ 0 & 3 & 5 & 14 \\ 0 & 3 & \lambda-1 & \mu-6 \end{array}\right]$$

$$\Rightarrow \left[\begin{array}{ccc|c} 1 & 1 & 1 & 6 \\ 0 & 3 & 5 & 14 \\ 0 & 0 & \lambda-6 & \mu-20 \end{array}\right]$$

For no solution $\lambda = 6$ and $\mu \neq 20$

32. Ans (c)

Solution:

Here, $X = \begin{bmatrix} -2 & 0 \\ 0 & -3 \end{bmatrix}$, $I = \begin{bmatrix} 1 & 0 \\ 0 & 1 \end{bmatrix}$

$$\therefore \quad A = (X+1)^{-1}(X+5I)$$

$$\Rightarrow \quad A = \begin{bmatrix} -1 & 0 \\ 0 & -2 \end{bmatrix}^{-1}\begin{bmatrix} 3 & 0 \\ 0 & 2 \end{bmatrix}$$

$$= \frac{1}{2}\begin{bmatrix} -2 & 0 \\ 0 & -1 \end{bmatrix}\begin{bmatrix} -3 & 0 \\ 0 & -1 \end{bmatrix} = \begin{bmatrix} -3 & 0 \\ 0 & -1 \end{bmatrix}$$

Now $\quad |\lambda I - A| = 0$

$$\Rightarrow \begin{bmatrix} \lambda+3 & 0 \\ 0 & \lambda+1 \end{bmatrix} = 0$$

$$\Rightarrow (\lambda+3)(\lambda+1) = 0$$

$$\Rightarrow \lambda = -1, -3$$

33. Ans (a)

Solution:

$$A = [a_{ij}]$$

where A = Real $n \times n$ matrix

i.e. square matrix

$a_{ij} = i$ if $i = j$

$a_{ij} = 0$ if $i \neq j$

$$a_{nn} = \begin{bmatrix} i & 0 & 0 & 0 & \cdots & n \\ 0 & i & 0 & 0 & \cdots & n \\ 0 & 0 & i & 0 & \cdots & n \\ 0 & 0 & 0 & & i & 0 \\ n & n & n & & 0 & i \end{bmatrix}$$

We know that the value of a_{ij} when $i = j = i$

Then for $n \times n$ matrix the sum of n eigen value is equal to

$i + i \ldots$ up to n

The sum of first n eigen value

is $\Rightarrow \dfrac{n(n+1)}{2}$

34. Ans (c)

Solution:

$$xy = (0)_{n\times n}$$

Let x and y is 2×2 square matrix

$$x = \begin{bmatrix} a_{11} & a_{12} \\ a_{21} & a_{22} \end{bmatrix}$$

$$y = \begin{bmatrix} b_{11} & b_{12} \\ b_{21} & b_{22} \end{bmatrix}$$

$$xy = 0$$

$$y = \begin{bmatrix} a_{11} & a_{12} \\ a_{21} & a_{22} \end{bmatrix}\begin{bmatrix} b_{11} & b_{12} \\ b_{21} & b_{22} \end{bmatrix} = 0$$

$$\begin{bmatrix} a_{11}b_{11} + a_{12}b_{21} & a_{11}b_{12} + a_{12}b_{22} \\ a_{21}b_{11} + a_{22}b_{21} & a_{21}b_{12} + a_{22}b_{22} \end{bmatrix} = \begin{bmatrix} 0 & 0 \\ 0 & 0 \end{bmatrix}$$

Comparing the matrix elements

$a_{11}b_{11}+a_{12}b_{21}=0$

$a_{11}b_{11}=-a_{12}b_{21}$

$$\frac{a_{11}}{a_{12}}=\frac{-b_{21}}{b_{11}} \quad ...(1)$$

$a_{11}b_{12}=-a_{12}b_{22}$

$$\frac{a_{11}}{a_{12}}=\frac{-b_{22}}{b_{12}} \quad ...(2)$$

$a_{21}b_{11}=-a_{22}b_{21}$

$$\frac{a_{21}}{a_{22}}=\frac{-b_{21}}{b_{11}} \quad ...(3)$$

$a_{21}b_{12}=-a_{22}b_{22}$

$$\frac{a_{21}}{a_{22}}=\frac{-b_{22}}{b_{12}} \quad ...(4)$$

Divide equation (1) by (2)

$$1=-\left(\frac{b_{21}}{b_{11}}\right)\left(\frac{-b_{12}}{b_{22}}\right)$$

$b_{11}b_{22}-b_{21}b_{12}=0$

$|y|=0$

Divide equation (2) by equation (4)

$$\frac{a_{11}a_{22}}{a_{12}a_{21}}=1$$

$a_{11}a_{22}-a_{12}a_{21}=0$

$|x|=0$

So $|x|=0$ and $|y|=0$

35. Ans. (a)

Solution: $M=\begin{bmatrix}-2 & 2 & -3\\ 2 & 1 & 6\\ -1 & -1 & 0\end{bmatrix}$

Eigen value = [–3, –3, 5]

Eigen vector corresponding to 5.

$[1 \;\; 2 \;\; -1]^T$

Eigen vector correspopnding to M^3

$= [(1)^3 \;\; (2)^3 \;\; (-1)^3]^T$

$[1 \;\; 8 \;\; -1]^T$

36. Ans (b)

Solution:

given matrix

$$P=\begin{pmatrix}1 & 1 & 0\\ 0 & 2 & 2\\ 0 & 0 & 3\end{pmatrix}$$

$|P-\lambda I|=0$

$$\begin{vmatrix}1-\lambda & 1 & 0\\ 0 & 2-\lambda & 2\\ 0 & 0 & 3-\lambda\end{vmatrix}$$

$(3-\lambda)[(1-\lambda)(2-\lambda)-0]=0$

$\lambda=3$, $1,2\rightarrow$ eigen values

Now eigen vector for $\lambda=3$

$|A-3I|X=0$

$$\text{or} \quad \begin{vmatrix}-2 & 1 & 0\\ 0 & -1 & 2\\ 0 & 0 & 0\end{vmatrix}\begin{vmatrix}x_1\\ x_2\\ x_3\end{vmatrix}=0$$

or $-2x_1+x_2+0x_3$ ...(1)

$0x_1-x_2+2x_3=0$ ...(2)

Using equations (1) and (2)

$$\frac{x_1}{2-0}=\frac{x_2}{0+4}=\frac{x_3}{2-0}$$

$$=\frac{x_1}{2}=\frac{x_2}{4}=\frac{x_3}{2}$$

$$\text{or} \quad \frac{x_1}{1}=\frac{x_2}{2}=\frac{x_3}{1}$$

So eigne vectors are $X=\begin{vmatrix}1\\ 2\\ 1\end{vmatrix}$

or $[1 \;\; 2 \;\; 1]^T$

37. Ans (c)

Solution:

Letd λ_1 and λ_2 are eigen given values of matrix

$\lambda_1+\lambda_2=$ Trace

$\lambda_1\times\lambda_2=$ Determinant

$$\lambda_1 + \lambda_2 = -2 \quad ...(1)$$

$$\lambda_1 \times \lambda_2 = -35 \quad ...(2)$$

$$\lambda_2 = -2 - \lambda_1$$

Put this value in (2) equation

$$\lambda_1 \times (-2 - \lambda_1) = -35$$

$$-2\lambda_1 - \lambda_1^2 = -35$$

$$\lambda_1^2 + 2\lambda_1 - 35 = 0$$

$$\lambda_1^2 + 7\lambda_1 - 5\lambda_1 - 35 = 0$$

$$\lambda_1(\lambda_1 + 7) - 5(\lambda_1 + 7) = 0$$

$$(\lambda_1 - 5)(\lambda_1 + 7) = 0$$

$$\lambda_1 = 5 \text{ or } -7$$

Put value $\lambda_1 = 5$ in equation 1

$$5 + \lambda_2 = -2$$

$$\lambda_2 = -7$$

∴ Eigen values are $5, -7$

38. Ans. (d)

Solution: The characteric equation

$\lambda^3 + \lambda^2 + 2\lambda + I = 0$, must satisfy the P-matrix

$$\Rightarrow P^3 + P^2 + 2P + 1 = 0$$

$$\Rightarrow P\left(P^2 + P + 2I + P^{-1}\right) = 0$$

$$P \neq 0 \; and \; P^2 + P + 2I + P^{-1} = 0$$

$$P^{-1} = -\left(P^2 + P + 2I\right)$$

39. Ans. (a)

Solution:

Given $r(Q_{5\times6}) = 4$

Therefore Q has four linearly independent row and four linearly independent columns.

$$r(Q) = 4$$

or $r(Q^T) = 4$

As $r(AB) \leq r(A) \times r(B)$

$$r(QQ^T) < 4$$

or $|QQ^T| = 0$

or QQ^T is Singular Matrix

Hence it is not invertible

40. Ans. (a)

Solution:

From the given order of matrices

$X^T \rightarrow 3 \times 4$

$Y \rightarrow 4 \times 3$

$X^TY \rightarrow 3 \times 3$

$(X^TY)^{-1} \rightarrow 3 \times 3$

$P \rightarrow 2 \times 3$

$P^T \rightarrow 3 \times 2$

$P(X^TY)^{-1}P^T \rightarrow (2 \times 3)(3 \times 3)(3 \times 2) \rightarrow 2 \times 2$

$(P(X^TY)^{-1}P^T)^T \rightarrow 2 \times 2$

41. Ans. (d)

Solution:

Since the matrix product when it exists is always another matrix and not a scalar. Hence Statement 1 is false, 2: D^T F D is always symmetric.

Now D F D does not exist since $D_{3\times5}$, $F_{5\times1}$ and $D_{5\times3}$ are not compatible for multiplication since, $D_{3\times5}\, F_{5\times1} = X_{3\times1}$ and $X_{3\times1}\, D_{5\times3}$ does not exist.

Hence, statement 2 is false.

42. Ans. (a)

Solution:

A square matrix B is defined as skew-symmetric if and only if $B^T = -B$.

43. Ans. (b)

Solution:

$$(PQ)^{-1} P = (Q^{-1} P^{-1}) P$$
$$= (Q^{-1})(P^{-1} P)$$
$$= Q^{-1}$$

44. Ans. (a)

Solution:

$$A^{-1} = \frac{\text{adj } A}{|A|}$$

$A^T = A^{-1}$, then A is orthogonal matrix.

Therefore $A.A^{-1} = A^{-1} A = I$ and $A^T A = A.A^T = I$

Since M is orthognal matrix.

$M^T M = I$

$$\begin{bmatrix} \frac{3}{5} & x \\ \frac{4}{5} & \frac{3}{5} \end{bmatrix} \begin{bmatrix} \frac{3}{5} & \frac{4}{5} \\ x & \frac{3}{5} \end{bmatrix} = \begin{bmatrix} 1 & 0 \\ 0 & 1 \end{bmatrix}$$

or $\begin{bmatrix} \left(\frac{3}{5}\right)^2 + x^2 & \left(\frac{3}{5}\cdot\frac{4}{5}\right)+\frac{3}{5}x \\ \left(\frac{4}{5}\cdot\frac{3}{5}\right)+\frac{3}{5}\cdot x & \left(\frac{4}{5}\right)^2+\left(\frac{3}{5}\right)2 \end{bmatrix} = \begin{bmatrix} 1 & 0 \\ 0 & 1 \end{bmatrix}$

or Compare both sides a_{12}

$$a_{12} = \left(\frac{3}{5}\right)\left(\frac{4}{5}\right) + \left(\frac{3}{5}\right) x = 0$$

or $\frac{3}{5}x = -\frac{3}{5}\cdot\frac{4}{5}$

or $x = -\frac{4}{5}$

45. Ans. (b)

Solution:

Given

A, B, C, D is n × n matrix.

and	ABCD	= I
or	$ABCDD^{-1}C^{-1}$	= $D^{-1}C^{-1}$
or	AB	= D–1 C–1
or	A–1 AB	= A–1 D–1 C–1
or	B	= A–1 D–1 C–1
		= $(C^{-1})^{-1}.(D^{-1})^{-1}.(A^{-1})^{-1}$
		= CDA

46. Ans. (c)

Solution:

Since maximum possible rank is 3 Consider first 3 × 3 minors.

$$\begin{vmatrix} 4 & 2 & 1 \\ 6 & 3 & 4 \\ 2 & 1 & 0 \end{vmatrix} = 0$$

$$\begin{vmatrix} 2 & 1 & 3 \\ 3 & 4 & 7 \\ 1 & 0 & 1 \end{vmatrix} = 0$$

$$\begin{vmatrix} 4 & 1 & 3 \\ 6 & 4 & 7 \\ 2 & 0 & 1 \end{vmatrix} = 0$$

$$\begin{vmatrix} 4 & 2 & 3 \\ 6 & 3 & 7 \\ 2 & 1 & 1 \end{vmatrix} = 0$$

Since all 3 × 3 minors are zero, now consider 2 × 2 minors.

$$\begin{vmatrix} 4 & 2 \\ 6 & 3 \end{vmatrix} = 0$$

$$\begin{vmatrix} 2 & 1 \\ 3 & 4 \end{vmatrix} = 8 - 3 = 5 \neq 0$$

rank = 2

47. Ans. (d)

Solution:

On augmented matrix apply gausss elimination method

$$\left[\begin{array}{ccc|c} 0 & 4 & 3 & 8 \\ 2 & 0 & -1 & 2 \\ 3 & 2 & 0 & 5 \end{array}\right] \xrightarrow{R_1 \leftrightarrow R_2} \left[\begin{array}{ccc|c} 2 & 0 & -1 & 2 \\ 0 & 4 & 3 & 8 \\ 3 & 2 & 0 & 5 \end{array}\right]$$

$$\left[\begin{array}{ccc|c} 2 & 0 & -1 & 2 \\ 0 & 4 & 3 & 8 \\ 3 & 2 & 0 & 5 \end{array}\right] \xrightarrow{R_3 - \frac{3}{2}R_1} \left[\begin{array}{ccc|c} 2 & 0 & -1 & 2 \\ 0 & 4 & 3 & 8 \\ 0 & 2 & 3/2 & 2 \end{array}\right]$$

$$\xrightarrow{R_3 - \frac{2}{4}R_2} \left[\begin{array}{ccc|c} 2 & 0 & -1 & 8 \\ 0 & 4 & 3 & 8 \\ 0 & 0 & 0 & -2 \end{array}\right]$$

For last row 0 = – 2 which is inconsistent. Also notice that r(A) = 2, while r(A) | B) = 3, (r(A) ≠ r(A | B) means inconsistent).

48. Ans. (c)

Solution:

$$r(A_{m\times n}) \leq \min(m, n)$$

Highest possible rank = Least value of 3 and 4.

49. Ans. (c)

Solution:

Given $A = \begin{bmatrix} 4 & 1 \\ 1 & 4 \end{bmatrix}$

$$A - \lambda I = 0$$

Therefore $\begin{bmatrix} 4-\lambda & 1 \\ 1 & 4-\lambda \end{bmatrix} = 0$

$$(4-\lambda)^2 - 1 = 0$$

$$(4-\lambda)^2 - (1)^2 = 0$$

or, $(4-\lambda+1)(4-\lambda-1) = 0$

or, $(5-\lambda)(3-\lambda) = 0$

Therefore $\lambda = 3, \lambda = 5$

50. Ans. (b)

Solution:

Sum of eigen values of given matrix = sum of diagonal element of given matrix

$$\Sigma\lambda_i = \text{Trace (A)}$$
$$\lambda_i = 1 + 5 + 1 = 7$$

Given $\lambda_1 = -2, \lambda_2 = 6$

Therefore $-2 + 6 + \lambda_3 = 7$

$$\lambda_3 = 3$$

51. Ans. (b)

Solution:

Sum of eigen values of given matrix = sum of diagonal element of given matrix = 1 + 5 + 1 = 7.

52. Ans. (a)

Solution:

The eigen values of any real matrix is always real.

53. Ans. (c)

Solution:

Sum of the eigen values of matrix = sum of diagonal elements of given matrix

Therefore $1 + 0 + p = 3 + \lambda_2 + \lambda_3$

or $p + 1 = p + 1 - 3$

$= p - 2$

54. Ans. (c)

Solution:

According to the given problem

$$E \times F = G$$

$$\therefore \begin{bmatrix} \cos\theta & \sin\theta & 0 \\ -\sin\theta & \cos\theta & 0 \\ 0 & 0 & 1 \end{bmatrix} \times F = \begin{bmatrix} 1 & 0 & 0 \\ 0 & 1 & 0 \\ 0 & 0 & 1 \end{bmatrix}$$

Hence we see that product of (E × F) is unit matrix so F will be the inverse of E.

$$F = E^{-1} = \frac{\text{Adj(E)}}{|E|} = \begin{bmatrix} \cos\theta & \sin\theta & 0 \\ -\sin\theta & \cos\theta & 0 \\ 0 & 0 & 1 \end{bmatrix}$$

55. Ans. (a)

Solution:

The correct match is given below:

Singular matrix → Determinant is zero

Non-square matrix → Determinant is not defined

Real symmetric → Eigen values are always real

Orthogonal matrix → Determinant is always one

56. Ans. (b)

Solution:

The inverse of $\begin{bmatrix} a & b \\ c & d \end{bmatrix}$ is

$$\begin{bmatrix} a & b \\ c & d \end{bmatrix}^{-1} = \frac{1}{(ad - bc)}\begin{bmatrix} d & -b \\ -c & a \end{bmatrix}$$

$$\therefore \begin{bmatrix} 1 & 2 \\ 5 & 7 \end{bmatrix} = \frac{1}{(7-10)}\begin{bmatrix} 7 & -2 \\ -5 & 1 \end{bmatrix}$$

$$= \frac{1}{3}\begin{bmatrix} -7 & 2 \\ 5 & -1 \end{bmatrix}$$

57. Ans. (b)

Solution:

$$\begin{bmatrix} a & b \\ c & d \end{bmatrix}^{-1} = \frac{1}{(ad - bc)}\begin{bmatrix} d & -b \\ -c & a \end{bmatrix}$$

$$\therefore \begin{bmatrix} 3+2i & i \\ -i & 3-2i \end{bmatrix} = \frac{1}{[(3+2i)(3-2i)+i^2]}\begin{bmatrix} 3-2i & -i \\ i & 3+2i \end{bmatrix}$$

$$= \frac{1}{12}\begin{bmatrix} 3-2i & -i \\ i & 3+2i \end{bmatrix}$$

58. Ans. (a)

Solution:

$$\text{For singularity matrix} = \begin{bmatrix} 8 & x & 20 \\ 4 & 0 & 2 \\ 12 & 6 & 0 \end{bmatrix} = 0$$

or $8(0 - 12) - x(0 - 2 \times 12) = 0$

Therefore $x = 4$

59. Ans. (a)

Solution:

The augmented matrix for this system is as below:

$$\begin{bmatrix} 1 & 1 & 1 \\ 1 & 3 & 3 \\ 1 & 2 & \alpha \end{bmatrix}\begin{bmatrix} 5 \\ 9 \\ \beta \end{bmatrix}$$

Apply Gauss-elimination method

$$\begin{bmatrix} 1 & 1 & 1 & | & 5 \\ 1 & 3 & 3 & | & 9 \\ 1 & 2 & \alpha & | & \beta \end{bmatrix} \xrightarrow[R_3 - R_1]{R_2 - R_1} \begin{bmatrix} 1 & 1 & 1 & | & 5 \\ 0 & 2 & 2 & | & 4 \\ 0 & 1 & \alpha - 1 & | & \beta - 5 \end{bmatrix}$$

$$\xrightarrow{R_3 - \frac{1}{2}R_2} \begin{bmatrix} 1 & 1 & 1 & | & 5 \\ 0 & 2 & 2 & | & 4 \\ 0 & 1 & \alpha - 1 & | & \beta - 7 \end{bmatrix}$$

For infinite solution last row must be zero

i.e. $\alpha - 2$ and $\beta - 7 = 0$

or $\quad \alpha = 2$ and $\beta = 7$

60. Ans. (d)

Solution:

The augmented matrix for given equations is

$$\begin{bmatrix} 1 & 1 & 1 & | & 3 \\ 1 & 2 & 3 & | & 4 \\ 1 & 4 & k & | & 6 \end{bmatrix}.$$

Apply Gauss elimination method.

$$\begin{bmatrix} 1 & 1 & 1 & | & 3 \\ 1 & 2 & 3 & | & 4 \\ 1 & 4 & k & | & 6 \end{bmatrix} \xrightarrow[R_3 - R_1]{R_2 - R_1} \begin{bmatrix} 1 & 1 & 1 & | & 3 \\ 0 & 1 & 2 & | & 1 \\ 0 & 3 & k - 1 & | & 3 \end{bmatrix}$$

$$\xrightarrow[R_3 - 3R_3]{} \begin{bmatrix} 1 & 1 & 1 & | & 3 \\ 0 & 1 & 2 & | & 1 \\ 0 & 0 & k - 7 & | & 0 \end{bmatrix}$$

if $k \neq 7$

rank (A) = rank (A | B) = 3

$\Rightarrow$ unique solution

If $k = 7$, rank (A) = rank (A | B) = 2

which is less than number or variables

$\Rightarrow$ unique solution is not possible and only infinite solution is possible.

61. Ans. (c)

Solution:

Given equations can be written as

$$\begin{bmatrix} 1 & 2 & 1 \\ 2 & 1 & 2 \\ 1 & 1 & 1 \end{bmatrix} \begin{bmatrix} x \\ y \\ z \end{bmatrix} = \begin{bmatrix} 6 \\ 6 \\ 5 \end{bmatrix}$$

Augmented matrix is $\begin{bmatrix} 1 & 2 & 1 & | & 6 \\ 2 & 1 & 2 & | & 6 \\ 1 & 1 & 1 & | & 5 \end{bmatrix}$

By gauss elimination method

$$\begin{bmatrix} 1 & 2 & 1 & | & 6 \\ 2 & 1 & 2 & | & 6 \\ 1 & 1 & 1 & | & 5 \end{bmatrix} \xrightarrow[R_3 - R_1]{R_2 - 2R_1} \begin{bmatrix} 1 & 2 & 1 & | & 6 \\ 0 & -3 & 0 & | & -6 \\ 0 & -1 & 0 & | & -1 \end{bmatrix}$$

$$\xrightarrow{R_3 - \frac{1}{3}R_2} \begin{bmatrix} 1 & 2 & 1 & | & 6 \\ 0 & -3 & 0 & | & -6 \\ 0 & 0 & 0 & | & 1 \end{bmatrix}$$

rank r(A) = 2

rank r(A | B) = 3

Since the rank of coefficient matrix is 2 and rank of argument matrix is 3, which is not equal. Hence system has no solution, i.e. system is inconsitent.

62. Ans. (b)

Solution:

Augmented matrix is $\begin{bmatrix} 2 & 3 & 0 & | & 4 \\ 1 & 1 & 1 & | & 4 \\ 1 & 2 & -1 & | & a \end{bmatrix}$.

Apply guess-elimination method,

$$\begin{bmatrix} 2 & 3 & 0 & | & 4 \\ 1 & 1 & 1 & | & 4 \\ 1 & 2 & -1 & | & a \end{bmatrix} \xrightarrow[R_3 - \frac{1}{2}R_1]{R2 - \frac{1}{2}R_1} \begin{bmatrix} 2 & 3 & 0 & | & 4 \\ 0 & -1/2 & 1 & | & 2 \\ 0 & 1/2 & -1 & | & a - 2 \end{bmatrix}$$

$$\xrightarrow{R_3 + R_2} \begin{bmatrix} 2 & 3 & 0 & | & 4 \\ 0 & -1/2 & 1 & | & 2 \\ 0 & 0 & 0 & | & a \end{bmatrix}$$

If $a \neq 0$, rank (A) = 2 and rank (A | B) = 3, hence equations have no solutions.

If $a = 0$, rank (A) = rank (A | B) = 2, then equations infinite solution.

63. Ans. (c)

Solution:

Characteristic equation is $|A - II| = 0$

$$\therefore \quad \begin{vmatrix} 4-\lambda & -2 \\ -2 & 1-\lambda \end{vmatrix} = 0$$

$$(4-\lambda)\times(1-\lambda) - [(-2)\times(-2)] = 0$$

$$\lambda^2 - 5\lambda = 0$$

or $\lambda(\lambda - 5) = 0$

Hence, $\lambda = 0$, and 5 are the eigen values.

64. Ans. (b)

Solution:

$$A = \begin{bmatrix} 4 & 5 \\ 2 & -5 \end{bmatrix}$$

Characteristic equation is $|A - I\,I| = 0$

$$\begin{vmatrix} 4\ \lambda & -5 \\ -2 & -5-\lambda \end{vmatrix} = 0$$

or $(4-\lambda)\times(-5-\lambda) - 2\times 5 = 0$

or $\lambda^2 - \lambda - 30 = 0$

$$\lambda = 5, -6$$

These are the eigen values.

65. Ans. (a)

Solution:

Charactreristi equation is $|A - \lambda I| = 0$

$$= \begin{bmatrix} 5-\lambda & 0 & 0 & 0 \\ 0 & 5-\lambda & 5 & 0 \\ 0 & 0 & 2-\lambda & 1 \\ 0 & 0 & 3 & 1-\lambda \end{bmatrix} = 0$$

$$= (5-\lambda)(5-\lambda)[(2-\lambda)(1-\lambda) - 3] = 0$$

$$= (5-\lambda)(5-\lambda)(\lambda^2 - 3\lambda - 1) = 0$$

$$\lambda = 5, 5, \frac{3 \pm \sqrt{13}}{2}$$

Use these eigen value to get eign vector

put $l = 5$ in $[A - lI]\ \hat{X} = 0$

$$\begin{bmatrix} 5-\lambda & 0 & 0 & 0 \\ 0 & 5-5 & 5 & 0 \\ 0 & 0 & 2-5 & 1 \\ 0 & 0 & 3 & 1-5 \end{bmatrix} \begin{bmatrix} x_1 \\ x_2 \\ x_3 \\ x_4 \end{bmatrix} = \begin{bmatrix} 0 \\ 0 \\ 0 \\ 0 \end{bmatrix}$$

or
$$\begin{bmatrix} 0 & 0 & 0 & 0 \\ 0 & 0 & 5 & 0 \\ 0 & 0 & -3 & 1 \\ 0 & 0 & 3 & -4 \end{bmatrix} \begin{bmatrix} x_1 \\ x_2 \\ x_3 \\ x_4 \end{bmatrix} = \begin{bmatrix} 0 \\ 0 \\ 0 \\ 0 \end{bmatrix}$$

$$5x_3 = 0$$
$$-3x_3 + x_4 = 0$$
$$3x_3 - 4x_4 = 0$$

On solving $x_3 = 0$, x_1 and x_2 may be anything. The eigen vector corresponding to $\lambda = 5$, may be written as

$$\hat{X}_1 = \begin{bmatrix} x_1 \\ x_2 \\ x_3 \\ x_4 \end{bmatrix} = \begin{bmatrix} k_1 \\ k_2 \\ 0 \\ 0 \end{bmatrix}$$

where k_1, k_2 may be any real number. Since correct choice (a) as the matrix in this form with both x_3 and $x_4 = 0$,.

66. Ans. (b)

Solution:

$$A = \begin{bmatrix} 2 & 1 \\ 0 & 2 \end{bmatrix}$$

$$[A - \lambda I] = 0$$

$$\begin{bmatrix} 2-2 & 1 \\ 0 & 2-2 \end{bmatrix} = 0$$

or $(2-\lambda)^2 = 0$

or $\lambda = 2$

Use this eign value to set eigen vector.

put $\lambda = 2$, in $[A - \lambda I]\ \hat{X} = 0$

$$\begin{bmatrix} 2-\lambda & 1 \\ 0 & 2-\lambda \end{bmatrix} \begin{bmatrix} x_1 \\ x_2 \end{bmatrix} = \begin{bmatrix} 0 \\ 0 \end{bmatrix}$$

$$\begin{bmatrix} 0 & 1 \\ 0 & 0 \end{bmatrix} \begin{bmatrix} x_1 \\ x_2 \end{bmatrix} = \begin{bmatrix} 0 \\ 0 \end{bmatrix}$$

$$x_2 = 0$$
$$0 = 0$$

The solution is therefore $x_2 = 0$, and x_1 = anything

$$\begin{bmatrix} x_1 \\ x_2 \end{bmatrix} = \begin{bmatrix} k \\ 0 \end{bmatrix}$$

$\therefore$ There is only one linearly independent eigen vector for this problem which may be written as $\begin{bmatrix} k \\ 0 \end{bmatrix}$.

67. Ans. (b)

Solution: Characteristic equation is

$$|A - \lambda I| = 0$$

$$\begin{vmatrix} (1-\lambda) & 2 \\ 0 & (2-\lambda) \end{vmatrix} = 0$$

or $(1-\lambda)(2-\lambda) = 0$

therefore $\lambda = 1, 2$

putting the value of $\lambda = 1$

or $$\begin{bmatrix} 0 & 2 \\ 0 & 1 \end{bmatrix}\begin{bmatrix} 1 \\ a \end{bmatrix} = 0$$

or $a = 0$

putting the value of $\lambda = 2$

$$\begin{bmatrix} -1 & 2 \\ 0 & 0 \end{bmatrix}\begin{bmatrix} 1 \\ b \end{bmatrix} = 0$$

or $-1 + 2b = 0$

and $0 = 0$

or $b = \frac{1}{2}$

therefore $a + b = \frac{1}{2}$

68. Ans. (a)

Solution:

$$A = \begin{bmatrix} 2 & 2 \\ 1 & 3 \end{bmatrix}$$

Characteristic equation of A is $|A - \lambda I| = 0$

$$\begin{bmatrix} 2-\lambda & 2 \\ 1 & 3-\lambda \end{bmatrix} = 0$$

$$(2-\lambda)(3-\lambda) - 2 = 0$$

$$\lambda^2 - 5\lambda = 0$$

$$\lambda = 1, 4$$

The eigen vector problem is $[A - \lambda I]\,\hat{X} = 0$

$$\begin{bmatrix} 2-\lambda & 2 \\ 1 & 3-\lambda \end{bmatrix}\begin{bmatrix} x_1 \\ x_2 \end{bmatrix} = \begin{bmatrix} 0 \\ 0 \end{bmatrix}$$

Putting $\lambda = 1$

$$\begin{bmatrix} 1 & 2 \\ 1 & 2 \end{bmatrix}\begin{bmatrix} x_1 \\ x_2 \end{bmatrix} = \begin{bmatrix} 0 \\ 0 \end{bmatrix}$$

$$x_1 + 2x_2 = 0 \quad \ldots(1)$$

$$x_1 + 2x_2 = 0 \quad \ldots(2)$$

For solution if $x_2 = k$, Then $2y = -2$

$$\hat{X} = \begin{bmatrix} -2k \\ k \end{bmatrix} \text{ i.e. } x_1 : x_2 = -2:1$$

Therefore solution is

$$= \begin{bmatrix} 2y \\ x^2 \end{bmatrix} = \begin{bmatrix} -2 \\ 1 \end{bmatrix}$$

69. Ans. (b)

Solution:

λ_i^m may be the correcsponding eigen values of A^m, x_i^m may not be corresponding eigen vectors.

70. Ans. (b)

Solution:

$$\Sigma\lambda_i = \text{Trace (A)}$$

$$\lambda_1 + \lambda_2 + \lambda_3 = \text{Trace (A)} = 2 + (-1) + 0 = 1$$

given, $\lambda_1 = 3$

therefore $3 + \lambda_2 + \lambda_3 = 1$

or $\lambda_2 + \lambda_3 = -2$

Here choice (b) only satisfies this condition.

71. Ans. (a)

Solution:

If $\lambda_1, \lambda_2, \lambda_3 \ldots \lambda_n$ are the eigen values of A. Then the eigen values of

A^m are $\lambda_1^m, \lambda_2^m, \lambda_3^m, \ldots$

$\therefore$ S^2 matrix has eigen values 1^2 and 5^2 i.e. 1 and 25

11.2 Calculus

1. Ans. (a)

Solution: $\int_0^\infty e^{[-x^2/8]}\,dx = \int_0^\infty e^{[-bx^2]}\,dt$

$b = [1/8]$,

Let, $bx^2 = t$; $2bx\,dx = dt$

$dx = [dt/(2\sqrt{bt})]$

$$\frac{1}{2\sqrt{b}}\int_0^\infty e^{-t}t^{(-1/2)}\,dt = \frac{\pi^{(1/2)}}{2\sqrt{b}}$$

$\pi(1/2)$ Gamma function

$$I = \frac{1}{\sqrt{(2)}} \times \frac{\sqrt{\sqrt{8}}}{2} = \frac{(2\sqrt{2}\sqrt{\pi})}{(2\sqrt{2}\sqrt{\pi})} = 1$$

$I = 1$

Another method

Given integral, $I = \frac{1}{\sqrt{2\pi}}\int_0^\infty \exp - x^2\,dx$

Let, $\frac{x}{2\sqrt{2}} = Z$,

then, $\frac{dx}{2\sqrt{2}} = dZ \Rightarrow dx = 2\sqrt{2}dZ$

$$\therefore \quad I = \frac{2\sqrt{2}}{\sqrt{2\pi}} \int_0^{\infty} \exp(-Z^2)\, dZ$$

$$= \frac{2\sqrt{2}}{\sqrt{\pi}} \int_0^{\infty} \exp(-Z^2)\, dZ$$

$$= \frac{2\sqrt{2}}{\sqrt{\pi}} \int_0^{\infty} \exp(-Z^2)\, dZ$$

$= \text{erf}(\infty) = 1$

2. Ans. (b)

Solution:

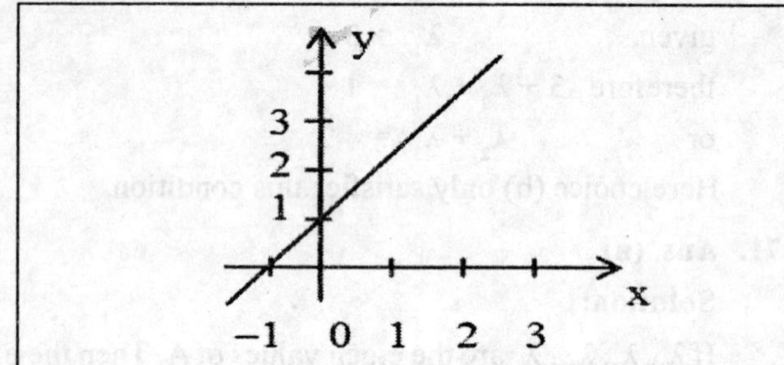

$$\int_1^2 y\,dx = ?$$

$\Rightarrow$ Equation of line $\Rightarrow y = mx + C$

Slope, $m = \frac{y_2 - y_1}{x_2 - x_1} = \frac{1-0}{0-(-1)} = 1$

Intercept on y at x =0 is 1 i..e C = 1

$\therefore \; y = x + 1$

$$\int_1^2 y\,dx = \int_1^2 (x+1)\,dx = \left|\frac{x^2}{2} + x\right|_1^2 = \left(\frac{4}{2} + 2\right) - \left(\frac{1}{2} + 1\right)$$

$$= 4 - \frac{3}{2} = 2.5$$

3. Ans. (a)

Solution: $\underset{\theta \to 0}{Lt} \frac{\sin \theta/2}{\theta}$

Apply L' Hospital rule

$$\underset{\theta \to 0}{Lt} \frac{\frac{d}{d\theta}\sin\frac{\theta}{2}}{\frac{d}{d\theta}\theta}$$

$$= \underset{\theta \to 0}{Lt} \frac{1}{2}\cos\frac{\theta}{2}$$

$$= 0.5 = \frac{1}{2}$$

$$[\because \cos 0 = 1]$$

4. Ans. (a)

Solution:

$$f(x) = e^{-x}$$

$$= e^{-(x-2)-2} = e^{-(x-2)}e^{-2}$$

$$= \left|1 - (x-2) + \frac{(x-2)^2}{2!} + \ldots\right| e^{-2}$$

$$= \left|1 - (x-2)\right| e^{-2}$$

$$= \left|(3-x)\right| e^{-2}$$

5. Ans. (a)

Solution:

Given $f(x) = x^2 - x - 2$

Maximum value of $y = f(x)$ in the interval $[-4, 4]$

i.e. $-4 \le x \le 4$

$$\frac{dy}{dx} 2x - 1 = 0$$

$\Rightarrow x = \frac{1}{2}$ is a stationary point

At $x = \frac{1}{2} \Rightarrow \frac{d^2y}{dx} = 2 = +ve$ minima

$\therefore$ The greatest value lie at extreme points.

$$\therefore f(-4) = (-4)^2 - (-4) = 2$$

$$= 16 + 4 - 2 = 18$$

And $f(4) = 4^2 - (4) - 2$

$$= 16 - 4 - 2 = 2$$

6. Ans. : (c)

Solution:

Given integral is

$$s = \int_1^{\infty} x^{-3}\,dx$$

$$= \left[\frac{x^{-2}}{-2}\right]_1^{\infty} = \left[\frac{\infty^{-2} - 1^{-2}}{-2}\right]$$

$$= \frac{\left[\frac{1}{\infty} - \frac{1}{1}\right]}{-2} = \frac{0-1}{-2} = \frac{1}{2}$$

7. **Ans. : (a)**

Solution:

Let given Function $y = f(x) = x^2 e^{-x}$

To obtain Maxima or Minima

$$\frac{dy}{dx} = 0$$

or, $2xe^{-x} - x^2 e^{-x} = 0$

or, $(2x - x^2)e^{-x} = 0$

or, $xe^{-x}(2-x) = 0$

Then $x = 0, 2$

$$\frac{dy}{dx} = (2x - x^2)e^{-x}$$

$$\frac{d^2y}{dx^2} = -(2x - x^2)e^{-x} + (2-2x)e^{-x}$$

$$= (x^2 - 4x + 2)e^{-x}$$

Now at $x = 0$

$$\frac{d^2y}{dx^2} = 2 = +ve\text{, i.e. minima}$$

And at $x = 2$

$$\frac{d^2y}{dx^2} = -2e^{-2} = -ve\text{, i.e. maxima}$$

$\therefore$ y is maximum at $x = 2$

8. **Ans. (d)**

Solution:

$$\frac{1}{2\pi}\int_0^{2\pi} \sin(t-\tau)\cos\tau\, d\tau$$

$$= \frac{1}{2\pi}\left(\sin t\cos^2\tau - \sin\tau\cos\tau\cos t\right)d\tau$$

$$= \frac{1}{2\pi}\int \frac{1}{2}\sin t(\cos 2\tau + 1)\,d\tau - \frac{1}{2\pi}\int \frac{1}{2}\sin 2\tau\cos t\, d\tau$$

$$= \frac{1}{2\pi}\frac{\sin t}{2}\int(\cos 2\tau + 1)\,d\tau - \frac{1}{2\pi}\frac{\cos t}{2}\int \sin 2\tau\, d\tau$$

$$= \begin{bmatrix} \frac{1}{4\pi}\sin t\left|\frac{\sin 2\tau}{2}\right|_0^{2\pi} + \frac{\sin t}{4\pi}\cdot t\Big|_0^{2\pi} \\ -\frac{1}{4\pi}\cos t\left|-\frac{\cos 2\tau}{2}\right|_0^{2\pi} \end{bmatrix}$$

$$= \frac{1}{4\pi}\sin t\,(0) - \frac{1}{4\pi}\cos t\,(0) + \frac{2\pi}{4}\sin t$$

$$= \frac{1}{2}\sin t$$

Second Method

Let $I = \frac{1}{2\pi}\int_0^{2\pi}\sin(t-\tau)\cos\tau\, d\tau$

$$I = \frac{1}{2\pi}\int_0^{2\pi}\sin(t - 2\pi + \tau)\cos\tau\, d\tau$$

$$I = \frac{1}{2\pi}\int_0^{2\pi} -\sin(t+\tau)\cos\tau\, d\tau$$

$$2I = \frac{1}{2\pi}\int_0^{2\pi}[\sin(t-\tau) - \sin(t+\tau)]\cos\tau\, d\tau$$

$$2I = \frac{1}{2\pi}\int_0^{2\pi}[2\cos\tau\sin t]\cos\tau\, d\tau$$

$$2I = \frac{1}{2\pi}\sin t\left(\int_0^{2\pi}(1+\cos 2\tau)\,d\tau\right)$$

$$I = \frac{1}{2}\sin t$$

9. **Ans. (d)**

Solution: Given integral is,

$$I = \int_{-1}^{1}\frac{1}{x^2}dx$$

Let $f(x) = \frac{1}{x^2}$

So curve of $\frac{1}{x^2}$ will be

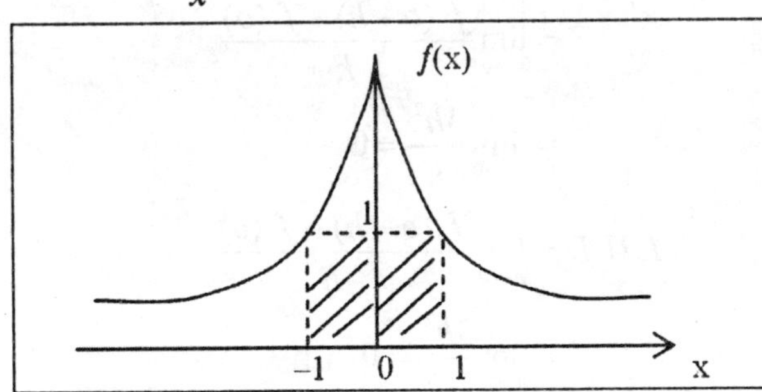

Even if it is discontinuous at $x = 0$ but the integral is the sum of area.

Second Method

$\frac{1}{x^2}$ is discontinuous at $x = 0$, but limit exist $x = 0$

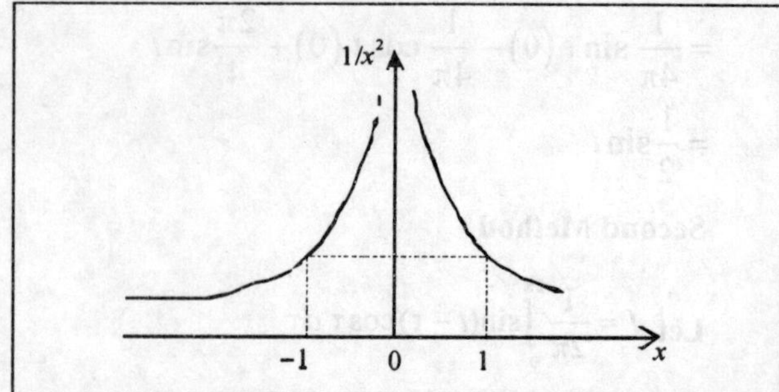

So integral $\int_{-1}^{1} \frac{1}{x^2} = \infty$

As clear from the curve area under curve is ∞

10. Ans. (c)

Solution:

Given $y = e^{\sin x - \cos x}$

For maximum and minima

$$\therefore \quad \frac{dy}{dx} = \left(e^{\sin x - \cos x}\right).(\cos x + \sin x) = 0$$

$\Rightarrow \tan x = -1$

or $x = \frac{3\pi}{4}$

y will be maximum at $x = \frac{3\pi}{4}$

$$\therefore \quad y = e^{\sin x - \cos x}$$

$$\therefore \quad = y = e^{\sin 3\pi/4 - \cos 3\pi/4} = e^{\sqrt{2}}$$

11. Ans (a)

Solution:

It is differentiable once.

For checking 2nd order differentiation $R.H.D$

$$= \lim_{h \to 0} \frac{f'(o+h) - f'(o)}{h}$$

$$= \lim_{h \to 0} \frac{3h^2}{h} = 0$$

$$L.H.D = \lim_{h \to 0} \frac{f'(o-h) - f'(o)}{-h}$$

$$= \lim_{h \to 0} \frac{3h^2}{-h} = 0$$

So second order differentiation exist and

$$\frac{d^2 f}{dx^2} = 6|x| \, at \; x = 0$$

For checking 3rd order differentiation

$$R.H.D = \lim_{h \to 0} \frac{f''(o+h) - f''(o)}{h}$$

$$= \lim_{h \to 0} \frac{6h}{h} = 6$$

$$L.H.D = \lim_{h \to 0} \frac{f''(o-h) - f''(o)}{-h}$$

$$= \lim_{h \to 0} \frac{6h}{-h} = -6$$

$\because R.H.D. \neq L.H.D.$, so not differentiable thrice.

Second Method

Given $f(x) = |x|^3$

$$= \begin{cases} x^3 & \text{if } x > 0 \\ -x^3 & \text{if } x < 0 \\ 0 & \text{if } x = 0 \end{cases}$$

$f(x)$ is continuous at $x = 0$

$L.H.D = 0$ and $R.H.D = 0$

- Therefore, $f(x)$ is differentiable at $x = 0$

$$f'(x) = \begin{cases} 3x^2 & \text{if } x > 0 \\ -3x^2 & \text{if } x < 0 \\ 0 & \text{if } x = 0 \end{cases}$$

Similarly $f''(x) = \begin{cases} 6x & \text{if } x > 0 \\ -6x & \text{if } x < 0 \\ 0 & \text{if } x = 0 \end{cases}$

$$f'''(x) = \begin{cases} 6 & \text{if } x > 0 \\ -6 & \text{if } x < 0 \\ \text{Does not exist} & \text{if } x = 0 \end{cases}$$

12. Ans. (d)

Solution:

$$\int_0^\infty \int_0^\infty e^{-x^2} e^{-y^2} dx\, dy$$

$$\int_0^\infty e^{-x^2} dx \int_0^\infty e^{-y^2} dy \qquad \text{...(1)}$$

Take $\int_0^\infty e^{-x^2} dx$

Put $x^2 = t$

$$x = \sqrt{t}$$

$$dx = \frac{1}{2\sqrt{t}} dt$$

So $\int_0^\infty -e^t \cdot \frac{1}{2}(t)^{-1/2} dt$

$$\boxed{\Gamma(n) = \int_0^{\infty} e^{-x} x^{n-1} dx}$$

$$= \frac{1}{2}\int_0^{\infty} e^{-t}(t)^{(1/2)-1}\,dt$$

$$= \frac{1}{2}\,\Gamma\left(\frac{1}{2}\right) = \frac{1}{2}\sqrt{\pi} \qquad \ldots(2)$$

Similarly $\int_0^{\infty} e^{-y^2} dy = \frac{1}{2}\sqrt{\pi}$

From (1)

$$\int_0^{\infty} e^{-x^2} dx \int_0^{\infty} e^{-y^2} dy = \frac{\sqrt{\pi}}{2} \times \frac{\sqrt{\pi}}{2}$$

$$= \frac{\pi}{4}$$

13. Ans. (d)

Solution: $\nabla \times \nabla \times P = \nabla(\nabla \cdot P) - \nabla^2 P$

It is vector identity.

14. Ans. (c)

Solution:

$$\int_0^{\pi} \sin^3\theta\, d\theta \qquad \ldots(1)$$

As we know that

$$\sin 3\theta = \sin 3\theta - \frac{1}{4}\sin^3\theta$$

So $\sin^3\theta = \frac{3}{4}\sin\theta - 4\sin 3\theta$

$$\int_0^{\pi} \sin^3 d\theta = \int_0^{\pi}\left(\frac{3}{4}\sin\theta - \frac{1}{4}\sin 3\theta\right)d\theta$$

$$= \frac{3}{4}[-\cos\theta]_0^{\pi} - \frac{1}{4\times 3}[-\cos 3\theta]_0^{\pi}$$

$$= \frac{3}{4}[1+1] - \frac{1}{12}[1+1]$$

$$= \frac{3}{2} - \frac{1}{6} = \frac{9-1}{6} = \frac{4}{3}$$

15. Ans. (a)

Solution: F (- ∞) = 0

F(0) = [1 / 2]

F(∞) = 1

Which represent $f(x) = [e^x / (1 + e^x)]$ is Monotonically increases

16. Ans. (d)

Solution: Strictly bounded function is e^{-x^2}

In given options only 2 function are bounded

i.e. $\frac{1}{x^2}$ and e^{-x^2}

at $x = 0$; $\frac{1}{x^2} = \infty$; $e^{-x^2} = 1$

at $x = 10$; $\frac{1}{x^2} = .01$

$$e^{-x^2} = 3.7 \times 10^{-44}$$

Therefore e^{-x^2} is bounded

17. Ans. (d)

Solution:

We consider only option (a) and (d), because these contains variable r, as variable of integration.

By integrating (d), we get

$\frac{1}{3}\pi a^2 H$, which is volume of a cone

$$= \int_0^{R} 2\pi R^2\left(1 - \frac{r}{R}\right)$$

$$dr = \frac{\pi R^2 H}{3}$$

18. Ans. (d)

Solution:

Path equation given as

$$(x+1)^2 + (y-1)^2 = \sqrt{2}$$

But $X = x + 1$, $Y = y - 1$

So it can be written as $X^2 + Y^2 = \sqrt{2}$ ...(1)

Which is the equation of circle and path of circle is closed and $s(x, y)$ is integrated once over this path.

So $\oint s(x, y) = 0$

19. Ans. (c)

Solution:

$$\nabla.u = \text{Grad } u = \hat{i}\frac{\partial u}{\partial x} + \hat{j}\frac{\partial u}{\partial y} = x\hat{i} + \frac{2}{3}y\hat{j}$$

At (1, 3), Grad $u = \sqrt{1 + \left(\frac{2}{3}\times 3\right)^2} = \sqrt{5}$

20. Ans. (a)

Solution:

Let $\vec{R} = f(t)$

$\vec{R}$ is constant in magnitude

$\therefore\ |\vec{R}| =$ constant

Now $\vec{R}\cdot\vec{R} = |\vec{R}|^2$ = constant

Differentiate both side w.r.t. t

$$\frac{d}{dt}(\vec{R}\cdot\vec{R}) = \frac{d}{dt}(\text{constant}) = 0$$

$$= \vec{R}\cdot\frac{d\vec{R}}{dt} + \frac{d\vec{R}}{dt}\cdot\vec{R} = 0$$

$$= 2\vec{R}\cdot\frac{d\vec{R}}{dt} = 0$$

$$= \vec{R}\cdot\frac{d\vec{R}}{dt} = 0$$

If $\vec{R}$ is constant in direction then $\vec{R}\times\frac{d\vec{R}}{dt} = 0$

21. Ans. (a)

Solution: $f = x^{2/3} + y^{2/3}$

Derivative of f directed along the line, $y = x$ at the point (8, 8) will be the Δf (gradient of f)

$$\Delta f = \left.\frac{\partial f}{\partial x}\hat{i} + \frac{\partial f}{\partial y}\hat{j}\right|_{(8,8)} = \frac{1}{3}\hat{i} + \frac{1}{3}\hat{j}$$

$$= \frac{\sqrt{2}}{3} = \left[\left(\frac{1}{3}\right)^2 + \left(\frac{1}{3}\right)^2\right]^{1/2}$$

22. Ans. (a)

Solution: Given function

$f(x) = e^x + e^{-x}$

$f'(x) = e^x - e^{-x}$

For maximum or minimum

$f'(x) = 0 \Rightarrow e^x - e^{-x} = 0$

$f''(x) = e^x + e^{-x}$

$f''(0) \Rightarrow$ Positive

Thus minimum of

$f(x) = e^0 + e^0 = 2$

23. Ans. (a)

Solution: $\sin x = x - \frac{x^3}{3!} + \frac{x^5}{5!} + \cdots$

$$\cos(x) = 1 - \frac{x^2}{2!} + \frac{x^2}{4!} + \cdots$$

Thus $\sin(x^3)$ will have odd powers

$\sin(x^2)$ will have even powers

$\cos(x^3)$ will have even powers

$\cos(x^2)$ will have even powers

24. Ans. (b)

Solution: $f(x) = e^x + \sin x$

Taylor's series,

$$f(x) = f(a) + (x-a)f'(a) + \frac{(x-a)^2}{2!}f''(a)$$

Where a = π

$$f(x) = f(\pi) + (x-\pi)f'(\pi) + \frac{(x-\pi)^2}{2!}f''(\pi)$$

Coefficient of $(x-\pi)^2 = \frac{1}{2!}f''(x)$

$f'(x) = e^x + \cos x$

$f''(x) = e^x - \sin x$

$f''(x)\big|_{x=\pi} = e^{\pi}$

Thus coefficient of $(x-\pi)^2 = 0.5$

25. Ans. (a)

Solution: Total sum of porbability density of function is one

i.e. $\int_{-\infty}^{\infty} p_x\,dx = 1$

or $\int_{-\infty}^{\infty}\left[M\,exp(-2|x|) + N\exp(-3|x|)\right]dx = 1$

or $\frac{M\,e^{-2|x|}\big|_{-\infty}^{\infty}}{-2} + \frac{N\,e^{-3|x|}\big|_{-\infty}^{\infty}}{-3} = 1$

$\Rightarrow \frac{M}{-2}e^{-2(\infty-\infty)} + \frac{N\,e^{-3(\infty-\infty)}}{-3} = 1$

$\Rightarrow M + \frac{2}{3}N = 1$

26. Ans (d)

Solution: $\sin x = x - \frac{x^3}{3!} + \frac{x^5}{5!}$

$\sin(x-\pi) = (x-\pi) - \frac{(x-\pi)^3}{3!} + \frac{(x-\pi)^5}{5!}$

$$\Rightarrow -\frac{\sin x}{(x-\pi)} = 1-\frac{(x-\pi)^2}{3!}+\frac{(x-\pi)^4}{5!}$$

$$\Rightarrow \frac{\sin x}{(x-\pi)} = -1+\frac{(x-\pi)^2}{3!}+----$$

27. Ans. (a)

Solution:

As given $e^y = x^{\frac{1}{x}}$

To find the maxima and minima find it differentiation and equate to zero

i.e. $e^y = x^{\frac{1}{x}}$

take log both side

$$\log e^y = \log x^{\frac{1}{x}}$$

$$y = \frac{1}{x}\log x$$

Now $\frac{dy}{dx} = \frac{1}{x^2} - \frac{1}{x^2}\log x$

put $\frac{dy}{dt} = 0$ for maximum or minima

$$\frac{1}{x^2} - \frac{1}{x^2}\log x = 0$$

$$\frac{1}{x^2} = 0 \text{ or } 1-\log x = 0$$

$\log x = 1$

or maxima at $x = e$

28. Ans. (c)

Solution:

Given, $f(x) = x^3 - 9x^2 + 24x + 5$

$f'(x) = 0$ for stationary values

or, $3x^2 - 18x + 24 = 0$

or, $3x^2 - 6x - 12x + 24 = 0$

or, $3x(x-2) - 12(x-2) = 0$

or, $(3x-12)-(x-2) = 0$

or, $x = 2, 4$

$f''(x) = 6x - 18$

$f''(2) = 12 - 18 < 0$

$f''(4) = 24 - 18 > 0$

Hence $f(x)$ has maximum value at $x = 2$

The maximum value is $2^3 - 9\times 2^2 + 24\times 2 + 5 = 25$

But we have to find the maximum value in the interval [1, 6]

$f(6) = 6^3 - 9\times 6^2 + 24\times 6 + 5 = 41$

Second Method

Given

$f(x) = x^3 - 9x^2 + 24x + 5$ in range of [1, 6]

$f(1) = 21$, $f(6) = 41$

$f(x) = 3x^2 - 18x + 24 = 0$

or, $x = 2, 4$ are the stationary point of function

$f''(x) = 6x - 18$

$f''(2) = -6 < 0$

$f''(4) = 6 > 0$

$f(2) = 25$ and

$f(4) = 21$

Maximum value of $f(x)$ in range of [1, 6] is $f(6) = 41$

29. Ans (b)

Solution:

$$y = x^2 + 2x + 10$$

$$\frac{dy}{dx} = 2x + 2$$

$$\left|\frac{dy}{dx}\right|_{x=1} = 2(1) + 2 \Rightarrow 2 + 2 = 4$$

30. Ans (c)

Solution:

$$\lim_{x\to 0}\frac{\sin x}{x}$$

Put the Limit

* $\lim_{x\to 0}\frac{\sin 0}{0} = \%$ form

Because it is a zero/zero form

So use L' Hospital Rule

$$\lim_{x\to 0}\frac{d/_{d}x(\sin x)}{\frac{d}{dx}(x)} = \lim_{x\to 0}\frac{\cos x}{1} \Rightarrow \cos(0)$$

$\Rightarrow 1$

31. Ans (c)

Solution:

$$y = x^2 - 6x + 9$$

$$\frac{dy}{dx} = 2x - 6$$

$$\frac{d^2y}{dx^2} = 2[+ve]$$

So $\frac{dy}{dx} = 2x - 6$

Put $\frac{dy}{dx} = 0$

$$x = 3$$

Because secound derivative is positive. So it is will give point. of minima. But we have to Find point. of maxima Form 2 to 5 range.

$$y[2] = [2]^2 - 6[2] + 9$$
$$= 4 - 12 + 9 = 1$$
$$y[3] = [3]^2 - 6[3] + 9 = 0$$
$$y[4] = [4]^2 - 6[4] + 9 = 1$$
$$y[5] = [5]^2 - 6[5] + 9 = 4$$

So value of y at $x = 5$ is 4

32. Ans (b)

Solution:

$$\int_{-\infty}^{\infty} \delta\left(t \frac{-\pi}{6}\right) 6 \sin(t)\, dt$$

$$= 6 \int_{-\infty}^{\infty} \delta\left(t \frac{-\pi}{6}\right) 6 \sin t\, dt$$

[Using Property $6 \int_{-\infty}^{\infty} x(t)\ \delta(t - t_0) dt = x(t_0)$]

$$= 6 \sin\left(\frac{\pi}{6}\right) = 6\left(\frac{1}{2}\right) \Rightarrow 3$$

33. Ans. (b)

Solution:

Given $\sum_{m=0}^{\infty} \frac{1}{4m}(x-1)^{2m}$

Then m^{th} term, $u_m = \frac{1}{4^m}(x-1)^{2m}$

or $u_{m+1} = \frac{(x-1)^{2m+2}}{4^{m+1}}$

$$\underset{m\to\alpha}{Lim} \frac{u_{m+1}}{u_m} = \underset{m\to\infty}{Lim} \frac{(x-1)^2}{4} = \frac{(x-1)^2}{4}$$

The series converges if $= \frac{(x-1)^2}{4} < 1$

or, $-1 < x < 3$

34. Ans (b)

Solution:

As given $P = \int_0^1 x\ e^x\ dx$

Let, $x = I$

$e^x = II$

So using product rule of integration.

$$\int I \cdot II\ dx = I \int II\ dx - \int\left[\left(\frac{dI}{dx}\right)\int II\ dx\right] dx$$

$$\int_0^1 xe^x\ dx = xe^x\Big|_0^1 - \int_0^1 e^x\ dx$$

$$= [e - 0] - e^x\Big|_0^1$$
$$= e - [e - 0]$$
$$= e - e + 1 = 1$$

35. Ans (a)

Solution:

The function

$$f(t) = \frac{\sin t}{t} = \text{sinc } t$$

It is know as sinc function

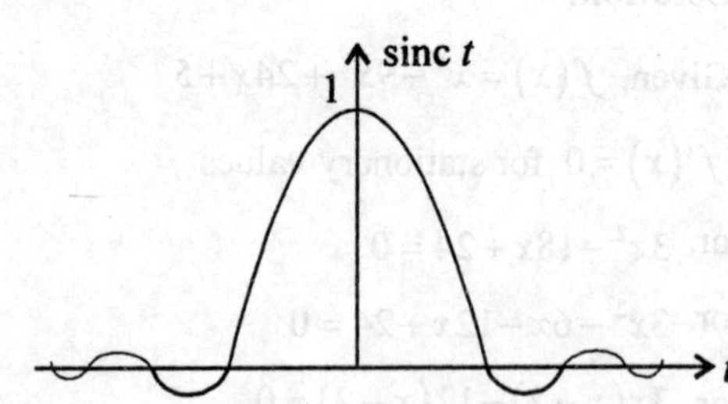

At $t = 0$

$$f(0) = \lim_{t\to 0} \frac{\sin t}{t} = 1 \text{ i.e. Maxima}$$

36. Ans. (c)

Solution:

By Green's Theorem

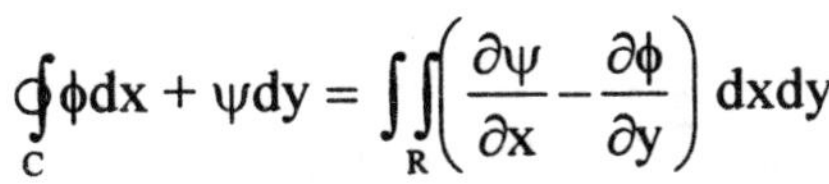

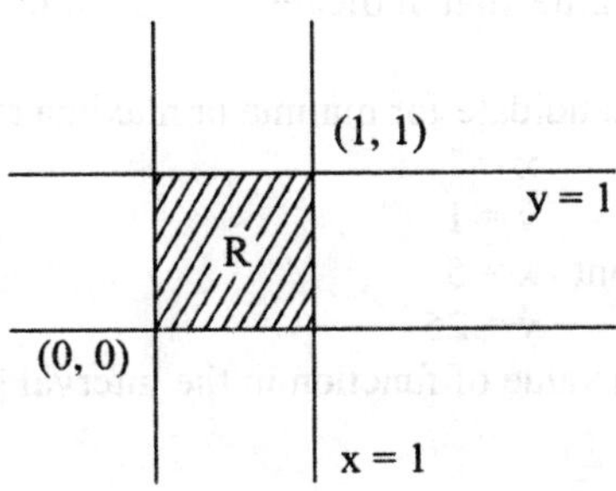

Here

$$I = \oint_C (xydy - y^2dx = \oint_C (-y^2)\,dx + (xy)\,dy$$

Therefore $\phi = -y^2, \psi = xy$

$$\frac{\partial \psi}{\partial x} = y, \frac{\partial \phi}{\partial y} = -2y$$

Using Green's theorem,

$$I = \int_{y=0}^{1}\int_{x=0}^{1} [y - (2y)]\,dx\,dy = \int_{y=0}^{1}\int_{x=0}^{1} 3y\,dx\,dy$$

$$= \int_{y=0}^{1} [3xy]_{x=0}^{1}\,dy = \int_{y=0}^{1} 3ydy = \frac{3}{2}$$

37. Ans. (a)

Solution:

$f(x, y)$ is defined over $0 \le x, y \le 1$

The region of integratin is between the parabols $y^2 = x$ and $x^2 = y$

$x > y^2 \rightarrow$ The region on Right hand side of parabolas $y^2 = x$

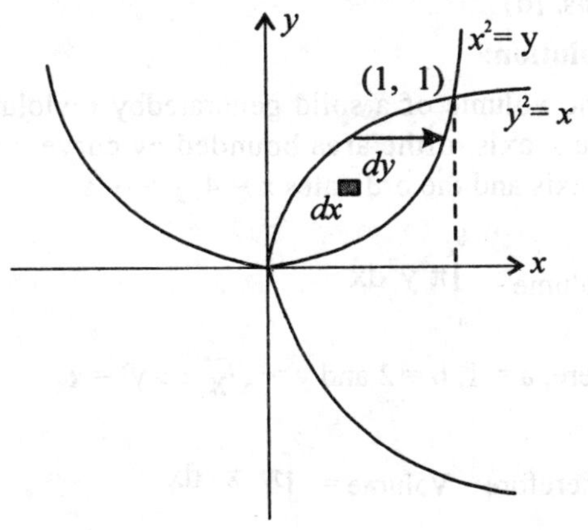

$y > x^2 \rightarrow$ The region above the parabolas $x^2 = y$

$\therefore$ the volume under $f(x, y)$

$I = \iint f(x, y)\,dx\,dy$

x varies from parabola $y^2 = x$ to parabolas $x^2 = y$

i.e. $x = y^2$ to $x = \sqrt{y}$

and y varies from $y = 0$ to $y = 1$

$$I = \int_{y=0}^{1}\int_{x=y^2}^{\sqrt{y}} f(x, y)\,dx\,dy$$

38. Ans. (b)

Solution:

Given $f(x) = \lim_{x \to 0}\left[\frac{x^3 + x^2}{2x^3 - 7x^2}\right]$

Has $\frac{0}{0}$ form,

Apply L' Hospitals rule.

$$f(x) = \lim_{x \to 0}\left[\frac{3x^2 + 2x}{6x^2 - 14x}\right]$$

$$= \lim_{x \to 0}\left[\frac{6x + 2}{12x - 14}\right]$$

$$= \left[\frac{6 \times 0 + 2}{12 \times 0 - 14}\right] = -\frac{1}{7}$$

39. Ans. (a)

Solution:

Modify the seven limit as

$$\lim_{x \to 0} \frac{\sin\left[\frac{2}{3}x\right]}{x} = \lim_{\frac{2}{3}x \to 0} \frac{\frac{2}{3}\sin\left[\frac{2}{3}x\right]}{\frac{3}{2}x}$$

Let $\theta \frac{2}{3}x$

$$\lim_{\frac{2}{3}x \to 0} \frac{\frac{2}{3}\sin\left[\frac{2}{3}x\right]}{\frac{3}{2}x} = \frac{2}{3}\lim_{\theta \to 0}\frac{\sin\theta}{\theta}$$

$$= \frac{2}{3} \times 1 = \frac{2}{3}$$

40. Ans. (a)

Solution:

$$\underset{x\to 0}{\text{Lt}} \frac{\sin^2 x}{x} = \underset{x\to 0}{\text{Lt}} \left(\frac{\sin x}{x^2}\right)^2 x$$

$$= 1 \times 0 = 0$$

41. Ans. (b)

Solution:

Let $h = (x - 8)$

or $x = 8 + h$

therefore $\underset{h\to 0}{\text{Lt}} \frac{(8+h)^{1/3} - 2}{h}$

Since this becomes $\left(\frac{0}{0}\right)$ form

Apply L' hospital rule

$$\lim_{h\to 0} \frac{\frac{1}{3}(8+h)^{\left(\frac{1}{3}-1\right)} - 2}{1} = \frac{1}{3}(8)^{-\frac{2}{3}}$$

$$= \frac{1}{12}$$

42. Ans. (c)

Solution:

Given, $x = a(\theta + \sin\theta)$,

$y = a(1 - \cos\theta)$

$$\frac{dx}{d\theta} = a(1 + \cos\theta),$$

$$\frac{dy}{d\theta} = a\sin\theta$$

$$\frac{dy}{dx} = \frac{a\sin\theta}{a(1+\cos\theta)}$$

$$= \frac{2a\sin\left(\frac{\theta}{2}\right).\cos\left(\frac{\theta}{2}\right)}{a \times 2\cos^2\left(\frac{\theta}{2}\right)}$$

$$= \tan\theta/2$$

43. Ans. (b)

Solution:

Given, $y = x^2$

or $\frac{dy}{dx} = 2x$

$= 0$ at $x = 0$

or $\left.\frac{d^2y}{dx^2}\right]_{x=0} = 2$

which is + ve value that indicate a local minima at $x = 0$

but it is not a in candidate for minima or maxima range

At the end point $x = 1$

$y = 1$

at second end point $x = 5$

$y = 25$

Hence, minimum value of function in the interval [1, 5] is 1.

44. Ans. (c)

Solution:

f(x) in the neighbourhood of a is,

$$f(x) = \sum_{n=0}^{\infty} b_n (x-a)^n$$

$$b_n = \frac{f^n(a)}{n!}$$

$f^4(x) = e^x$;

$f^4(2) = e^2$

Therefore Coefficient of

$$(x-2)^4 = b_4 = \frac{f^4(2)}{4!} = \frac{e^2}{4!}$$

45. Ans. (d)

Solution:

Given, $$\int_{-\infty}^{\infty} \frac{dx}{1-x^2} = \left[\tan^{-1} x\right]_{-\infty}^{\infty}$$

$$= \tan^{-1}(\infty) - \tan^{-1}(\infty)$$

$$= \frac{\pi}{2} - \left[\frac{-\pi}{2}\right]$$

$$= \pi$$

46. Ans. (d)

Solution:

The volume of a solid generatedby revlolution about the x-axis, ofthe area bounded by curve y = f(x), the x-axis and the ordinates x = 4, y = b is

$$\text{Volume} = \int_a^b \pi y^2 dx$$

Here, a = 1, b = 2 and $y = \sqrt{x} \Rightarrow y^2 = x$

Therefore $$\text{Volume} = \int_1^2 \pi \cdot x \cdot dx$$

$$= \pi \cdot \left[\frac{x^2}{2}\right]_1^2$$

$$= \frac{\pi}{2}\left[x^2\right]_1^2 = \frac{\pi}{2}\left[2^2 - 1^2\right]$$

$$= \frac{3}{2}\pi$$

47. Ans. (a)

Solution:

When $I = \int_0^8 \int_{x/4}^2 f(x \cdot y)\, dydx$

i.e.

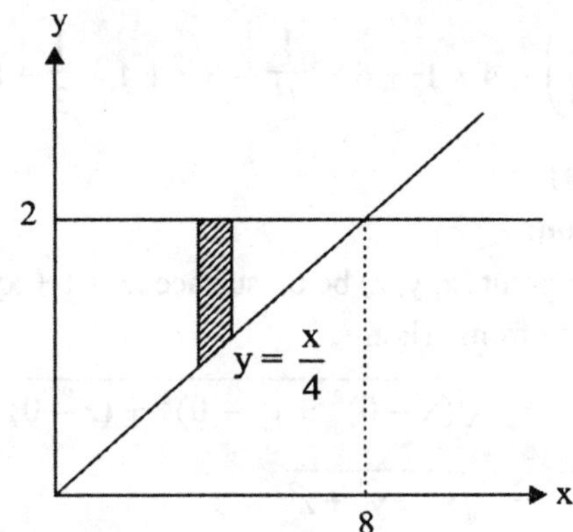

Now,

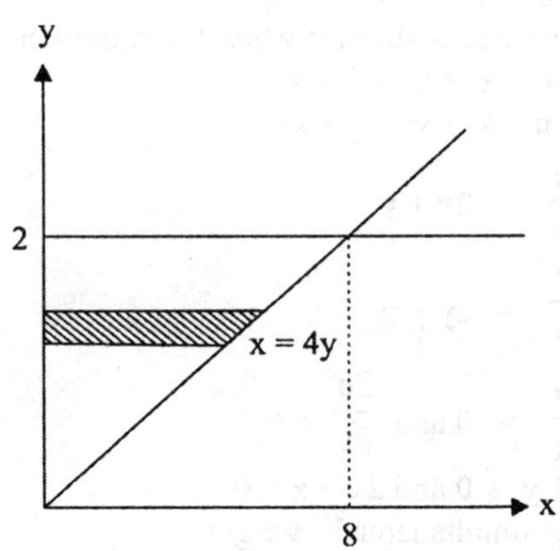

$$I = \int_0^2 \int_0^{4y} f(x, y)\, dy\, dx$$

therefore $q = 4y$

48. Ans. (b)

Solution:

Given, $f = x^2 + 3y^2 + 2z^2$

$$\Delta f = \text{grad } f = i\frac{\partial f}{\partial x} + j\frac{\partial f}{\partial y} + k\frac{\partial f}{\partial z}$$

$$= i\,2x + j6y + k4z$$

The gradient at P(1, 2, –1) is

$$= i\,(2 \times 1) + j\,(6 \times 2) + k\,(4\,x\,-1)$$

$$= 2i + 12j - 4k$$

49. Ans. (c)

Solution:

Given, Vector field,

$$\vec{f} = 3xz\hat{i} + 2zy\hat{j} - yz2\hat{k}$$

$$= v_x i + v_y j + v_z k$$

Divergence of vector field

$$\text{Div (f)} = \Delta.f$$

$$= \frac{\partial v_x}{\partial x} + \frac{\partial v_y}{\partial y} + \frac{\partial v_z}{\partial z}$$

$$= \frac{\partial}{\partial x}[3xz] + \frac{\partial}{\partial y}[2xy] + \frac{\partial}{\partial z}[-2yz^2]$$

$$= 3z + 2x - 2zy$$

$$\text{Div (f)}\,|_{(1,1,1)} = 3(1) + 2(1) - 2(1)(1) = 3$$

50. Ans. (a)

Solution:

A line integral and a surface integral is related by stoke's theorem, i.e. $\oint A \cdot \overline{dl} = \int\int_S (\nabla \times A)\, ds$

51. Ans. (b)

Solution:

$$\underset{x \to 3}{\text{Lt}}\, f(x) = \underset{x \to 3}{\text{Lt}} \left(\frac{2x^2 - 7x + 3}{5x^2 - 12x - 9}\right)$$

This is of the form of $\left(\frac{0}{0}\right)$

Applying L Hospitals's rule

$$\underset{x \to 3}{\text{Lt}} \left(\frac{4x - 7}{10x - 12}\right) = \frac{5}{18}$$

52. Ans. (b)

Solution: The given limit is of the form of $\left(\frac{0}{0}\right)$

Applying L' Hospital rule

$$\lim_{x \to 0} \frac{e^x - \left(1 + x + \frac{x^2}{2}\right)}{x^3}$$

$$\lim_{x \to 0} \frac{e^x - 1}{6x}$$

$$\lim_{x \to 0} \frac{e^x}{6} = \frac{1}{6}$$

53. Ans. (d)

Solution:

Since the position of rail engine x(t) is continuous and differentiable function, according to lagranges mean value theorem.

t where $0 \le t \le 8$ such that

$$x'(t) = v(t) = \frac{x(8) - x(0)}{8 - 0} = \frac{(280 - 0)}{(8 - 0)} \text{ m/sec}$$

$$= \frac{280}{8} \text{ m/sec}$$

$$= \frac{280}{8} \times \frac{3600}{1000} \text{ kmph}$$

$$= 126 \text{ kmph}$$

54. Ans. (a)

Solution:

The sixth function is

$$f(x, y) = 4x^2 + 6y^2 - 8x - 4y + 8$$

$$\frac{\partial f}{\partial x} = 8x - 8$$

$$\frac{\partial f}{\partial y} = 12y - 4y$$

Put, $\frac{\partial f}{\partial x} = 0$ and $\frac{\partial f}{\partial y} = 0$

$$8x - 8 = 0 \text{ and } 12y - 4y = 0$$

$$\therefore \quad x = 1 \text{ and } y = \frac{1}{3}$$

$\left(1, \frac{1}{3}\right)$ is the stationary point.

$$r = \left[\frac{\partial^2 f}{\partial x^2}\right]_{\left(1, \frac{1}{3}\right)} = 8$$

$$s = \left[\frac{\partial^2 f}{\partial x dy}\right]_{\left(1, \frac{1}{3}\right)} = 0$$

$$t = \left[\frac{\partial^2 f}{\partial y^2}\right]_{\left(1, \frac{1}{3}\right)} = 12$$

Since, $rt = 8 \times 12 = 48$

$s^2 = 0$

Since, $rt > s^2$,

Hence either a maxima or minima occure at $\left(1, \frac{1}{3}\right)$.

also since, $r = \left[\frac{\partial f}{\partial x^2}\right]_{\left(1, \frac{1}{3}\right)} = 8 > 0$, the point $\left(1, \frac{1}{3}\right)$ is a point of minima.

The minimum value is

$$f\left(1, \frac{1}{3}\right) = 4 \times 1^2 + 6 \times \frac{1}{3^2} - 8 \times 1\ 4 \times \frac{1}{3} + 8 = \frac{10}{3}$$

55. Ans. (a)

Solution:

Let the point (x, y, z) be on surface $z^2 = 1 + xy$

Distance from origin = l

$$= \sqrt{(x - 0)^2 + (y - 0)^2 + (z - 0)^2}$$

$$= \sqrt{x^2 + y^2 + z^2}$$

$$l = \sqrt{x^2 + y^2 + 1 + xy}$$

[since $z^2 = 1 + xy$ is given]

This distance is shortest when l is minimum. Hence find minima of $x^2 + y^2 + 1 + xy$

Let, $u = x^2 + y^2 + 1 + xy$

$$\frac{\partial u}{\partial x} = 2x + y$$

$$\frac{\partial u}{\partial y} = 2y + z$$

$$\frac{\partial u}{\partial x} = 0 \text{ and } \frac{\partial u}{\partial y} = 0$$

or $2x + y = 0$ and $2y + x = 0$

Solving simultaneously, we get

$x = 0$ and $y = 0$

so (0, 0) is the only stationary point.

$$r = \frac{\partial^2 u}{\partial x^2} = 2$$

$$s = \frac{\partial^2 u}{\partial x \partial y^2} = 1$$

$$t = \frac{\partial^2 u}{\partial y^2} = 2$$

Since $rt = 2 \times = 4 > S^2 = 1$

We have case 1, i.e. either a maximum or minimum exists at (0, 0).

Now, since $r = 2 > 0$, so it is a minima at (0, 0).

Now at $x = 0, y = 0, \; z = \sqrt{1+xy} = \sqrt{1+0} = 1$

The distance $l = \sqrt{0^2 + 0^2 + 1^2} = 1$

56. Ans. (a)

Solution:

Putting

$$f'(x) = 6x^2 - 6x - 36 = 0$$

or $x^2 - x - 6 = 0$

or $x = 3 \text{ or} - 2$

$$f''(x) = 12x - 6$$

$$f''(3) = 30 > 0 \text{ (minima)}$$

$$f''(-2) = -30 < 0 \text{ (minima)}$$

Hence maxima is at $x = -2$ only.

57. Ans. (c)

58. Ans. (c)

Solution:

$$f = y^x$$

Treating x as constant, we get

$$\frac{\partial f}{\partial y} = xy^{x-1}$$

Now we treat y as a constant and get,

$$\frac{\partial^2 f}{\partial x \partial y} = \frac{\partial}{\partial x}\left(y^{x-1}\right) = y^{x-1} + x\,y^{x-1}\ \ln y$$

at x = 2 and y = 1

$$= \frac{\partial^2 f}{\partial x \partial y} = 1^{(2-1)}(1 + 2.\,n1) = 1$$

59. Ans. (a)

Solution:

$$I = \int_0^{\pi/3} e^{it}\,dt = \left[\frac{e^{it}}{i}\right]_0^{\pi/3}$$

$$= \left[\frac{\cos t + i \sin t}{i}\right]_0^{\pi/3} = \frac{1}{i}\left[\frac{1}{2} + i\,\frac{\sqrt{3}}{2} - 1\right]$$

$$= \left[-\frac{1}{2i} + \frac{\sqrt{3}}{2}\right] = \left[\frac{\sqrt{3}}{2} + i\,\frac{1}{2}\right]$$

60. Ans. (b)

Solution:

Let $$I(\alpha) = \int_0^{\infty} e^{-\alpha x}\,\frac{\sin x}{x}\,dx \qquad ...(1)$$

$$\frac{\partial I}{\partial \alpha} = \int_0^{\infty}(-x)\,e^{-\alpha x}\,\frac{\sin x}{x}\,dx = -\int_0^{\infty} e^{-\alpha x}\ \sin x\,dx$$

Then integrating by parts we get,

$$\frac{\partial I}{\partial \alpha} = -\left[\frac{e^{-\alpha x}}{1+\alpha^2}(-\alpha \sin x - 1.\cos x)\right]_0^{\infty} = -\frac{1}{1+\alpha^2}$$

Therefore $$\frac{\partial I}{\partial \alpha} = -\frac{1}{1+\alpha^2}$$

$$\partial I = -\frac{\partial \alpha}{1+\alpha^2}$$

Integrating, $I = -\tan^{-1}\alpha + C$

$$I(\infty) = -\frac{\pi}{2} + C \text{ but from (1), } I(\infty) = 0$$

Therefore $$-\frac{\pi}{2} + C = 0$$

or $$C = \frac{\pi}{2}$$

Therefore $$I(a) = -\tan^{-1} a + \frac{\pi}{2}$$

or $$I(0) = \frac{\pi}{2} - \tan^{-1}\ 0$$

Therefore $$I(0) = \frac{\pi}{2}$$

But from equation (1),

$$I(0) = \int_0^{\infty}\frac{\sin}{x}\,dx$$

Therefore $$\int_0^{\infty}\frac{\sin}{x}\,dx = \frac{\pi}{2}$$

61. Ans. (d)

Solution:

$$\int_0^{\pi 4} \tan x\,dx = \log\sqrt{2}$$

$$\int_0^{\infty}\frac{dx}{x^2+1} = \frac{\pi}{2}$$

$$\int_0^{\infty} x\,e^{-x}\,dx$$

Integrating by part, taking u = x and dv = e^{-x} dx

$$\int_0^{\infty} x\, e^{-x} dx = x(-e^{-x}) - \int -e^{-x}\, dx$$

$$= -x\, e x^{-x} - e^{-x}$$

$$= -e^{-x}(x+1)$$

Now $= \left[-e - x\,(x+1)\right]_0^{\infty} = 1$

$$\int_0^1 \frac{1}{1-x} dx = \ln 0 - \ln 1 = -\infty - 0 = -\infty$$

62. Ans. (d)

Solution:

Area common to circles, r = a and r = 2a cosθ is 1.228a².

63. Ans. (d)

Solution:

$$y = \frac{2}{3}x^{2/3}$$

$$\frac{dy}{dx} = x^{1/2}$$

length of the curve is given by

$$\int_0^1 \sqrt{1 + \left(\frac{dy}{dx}\right)^2}\, dx = \int_0^1 \sqrt{1+x}\, dx$$

$$= \left[\frac{2}{3}(1+x)^{3/2}\right]_{x=0}^{x=1} = 1.22$$

64. Ans. (b)

Solution:

Path AB: $x^2 + y^2 = 1$

x = cos θ

y = sin θ

Along path ABθ varies from [0 to π/2]

$$\int_{\text{Path AB}} (x+y)^2\,(rd\theta) = \int_0^{\pi/2} (\cos\theta + \sin\theta)^2 1.d\theta$$

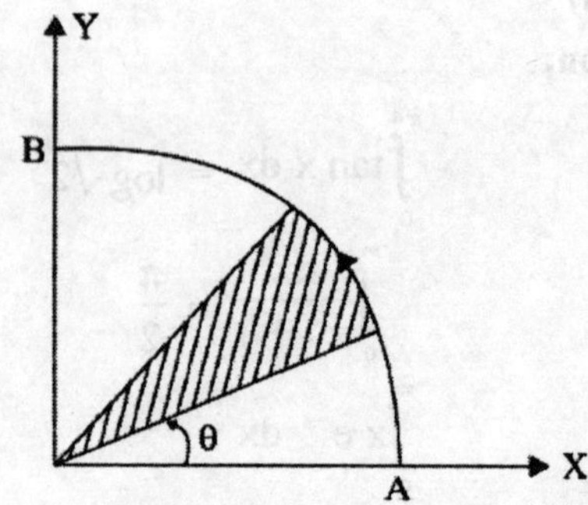

$$= \int_0^{\pi/2} (\sin^2\theta + \cos^2\theta + 2\sin\theta\cos\theta)d\theta$$

$$= \int_0^{\pi/2} (1 + \sin 2\theta)\, d\theta$$

$$= \theta + \left.\frac{(-\cos 2\theta)}{2}\right|_0^{\pi/2}$$

$$= \frac{\pi}{2} - \frac{1}{2}\left[\cos 2\frac{\pi}{2} - \cos 0\right]$$

$$= \frac{\pi}{2} - \frac{1}{2}[-1-1] = \frac{\pi}{2} + 1$$

65. Ans. (a)

Solution:

Curve 1: $y^2 = 4x$

Curve 2: $x^2 = 4y$

Intersection points of curve 1 and 2

$y^2 = 4x = 4\sqrt{4} = 8\sqrt{y}$

$y^4 = 8 \times 8y$

or $y(y^3 - 64) = 0$

y = 4 and y = 0

then x = 4 x = 0

Therefore intersection point are A(4, 4) and 0 (0, 0)

The area enclosed between curves 1 and 2 are given by

$$\text{Area} = \int_{x_1}^{x_2} y_1\, dx - \int_{x_1}^{x_2} y_2\, dx$$

$$= -\int_0^4 \frac{x^2}{4} dx$$

$$= 2\left.\frac{x^{3/2}}{3/2}\right|_0^4 - \left.\frac{x^3}{3\times 4}\right|_0^4$$

$$= \frac{4}{3}(4)^{3/2} - \frac{(4)^3}{3\times 4} = \frac{16}{3}$$

66. Ans. (a)

Solution:

$$\int_0^3\int_0^x (6-x-y)\, dxdy$$

$$= \int_0^3 \left[(6-x)y - \frac{y^2}{2}\right]_0^x dx$$

$$= \int_0^3 \left[(6-x)x - \frac{x^2}{2}\right] dx = 13.5$$

67. Ans. (a)

Solution:

$$V = \int_0^{2\pi}\int_0^{\pi/3}\int_0^1 r^2 \sin\phi \cdot dr \cdot d\phi \cdot d\theta$$

$$\int_0^{2\pi}\int_0^{\pi/3}\left[\frac{r^3}{3}\right]_0^1 \sin\phi d\phi\, d\theta$$

$$= \frac{1}{3}\int_0^{2\pi}[-\cos\phi]_0^{\pi/3} d\theta$$

$$= \frac{1}{3}\times\frac{1}{2}\times\int_0^{2\pi} d\theta$$

$$= \frac{1}{3}\times\frac{1}{2}\times 2\pi = \frac{\pi}{3}$$

68. Ans. (a)

Solution:

$$\frac{\partial x}{\partial u} = v \qquad \frac{\partial x}{\partial v} = u$$

and

$$\frac{\partial y}{\partial u} = -\frac{v}{u^2} \qquad \frac{\partial y}{\partial v} = \frac{1}{u}$$

and

$$\phi(u, v) = \begin{vmatrix} \frac{\partial x}{\partial u} & \frac{\partial x}{\partial v} \\ \frac{\partial y}{\partial u} & \frac{\partial y}{\partial v} \end{vmatrix} = \begin{vmatrix} v & u \\ -\frac{v}{u^2} & \frac{1}{u} \end{vmatrix} = \frac{v}{u} + \frac{v}{u} = \frac{2v}{u}$$

69. Ans. (a)

Solution:

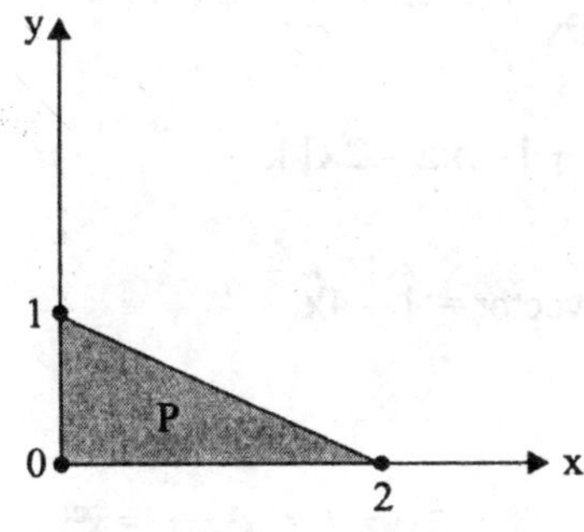

The equation of the straight line with x-intercept = 2 and y-intercept = 1 is

$$\frac{x}{2} + \frac{y}{1} = 1$$

or $$y = 1 - \frac{x}{2}$$

or $$x = 2 - 2y$$

or $$\int_0^1\int_0^{(2-2y)} (xy\, dx)\, dy = \int_0^1\left[\left(\frac{yx^2}{2}\right)\right]_0^{2-2y} dy$$

$$= \int_0^1 \frac{y}{2}(2-2y)^2\, dy = \int_0^1 2y(1-y)^2\, dy = \frac{1}{6}$$

70. Ans. (b)

Solution:

Form C, draw CN ⊥ AB. From right-angled ΔCAN.

$$\sin A = \frac{|CN|}{|AC|} \Rightarrow |CN| = |AC| \sin A.$$

$$\text{Area of } \Delta ABC = \frac{1}{2}|AB| \times |CN|$$

$$= \frac{1}{2}|AB|.|CN| \sin A = \frac{1}{2}\left|\overrightarrow{AB} \times \overrightarrow{AC}\right|$$

From above figure, $\overrightarrow{AB} = \vec{b} - \vec{a}$ and $\overrightarrow{AC} = \vec{c} - \vec{a}$.

$$\text{Area of } \Delta ABC = \frac{1}{2}\left|(\vec{b} - \vec{a}) \times (\vec{c} - \vec{a})\right|$$

$$= \frac{1}{2}\left|(\bar{a} - \bar{b})\,(\bar{a} - c)\right|$$

71. Ans. (b)

Solution:

Let $$D = \begin{vmatrix} x.x & x.y \\ y.x & y.y \end{vmatrix}$$

Let
$$x = x_1 i + x_2 j$$
$$y = y_1 i + y_2 j$$
$$x \,.\, y = x_1^2 + x_2^2$$
$$y \,.\, y = y_1^2 + y_2^2$$
$$x \,.\, y = x_1 x_2 + y_1 y_2$$

Therefore $$D = \begin{vmatrix} x_1^2 + x_2^2 & x_1 x_2 + y_1 y_2 \\ x_1 x_2 + y_1 y_2 & y_1^2 + y_2^2 \end{vmatrix}$$

$$= (x_1^2 + x_2^2)(y_1^2 + y_2^2) - (x_1 x_2 + y_1 y_2)^2$$
$$= x_2^2 y_1^2 + x_1^2 y_2^2 - 2x_1 y_1 x_2 y_2$$
$$= (x_2 y_1 - x_1 y_2)^2$$

Now, $$D = 0$$
$$x_2 y_1 - x_1 y_2 = 0$$

Or $$\frac{x_1}{x_2} = \frac{y_1}{y_2}$$

Or Vector $x_1 i + x_2 j$ and $y_1 i + y_2 j$ are linearly dependent.

Therefore Linear dependence ⇒ D = 0

So, $D \neq 0$, i.e. is negative or positive.

[notice that since $D = (x_2 y_1 - x_1 y_2)^2$, it cannot be negative].

So, Linear independence $\Rightarrow$ D is positive

72. Ans. (c)

Solution:

Given

$$f = 2x^2 + 3y^2 + z^2$$

$$\Delta f = i\frac{\partial f}{\partial x} + j\frac{\partial f}{\partial y} + k\frac{\partial f}{\partial z}$$

$$= 4xi + 6yj + 2zk$$

at P (2, 1, 3)

$$\Delta f = 4 \times 2 \times i + 6 \times 1 \times j + 2 \times 3 \times k = 8i + 6j + 6k$$

directional derivative of f in direction of vector

$a = i \times 2k$ is

nothing but the component of grad f in the direction of vector a and is given by $\frac{a}{|a|} \cdot$ grad f

$$= \left(\frac{i - 2k}{\sqrt{1^2 + (-2)^2}}\right) \cdot (8i + 6j + 6k)$$

$$= \frac{1}{\sqrt{5}}(1.8 + 0.6 + (-2)6)$$

$$= \frac{-4}{\sqrt{5}} = -1.789$$

73. Ans. (b)

Solution:

$$\Delta f = i\frac{\partial f}{\partial x} + j\frac{\partial f}{\partial y} + k\frac{\partial f}{\partial z}$$

here $f = x^2 + 3y^2 + 2z^2$

therefore $\Delta f = i(2x) + j(6y) + k(4 \times -1)$

$$= 2i + 12j - 4k$$

The directional derivative in direction of vector $a = i - j + 2k$ is given by

$$\frac{a}{|a|} \cdot \text{grad } f = \frac{i + j + 2k}{\sqrt{1^2 + (-1)^2 + 2^2}} \cdot (2i + 12j - 4k)$$

$$= \frac{1}{\sqrt{6}}(1.2 + (-1).\ 12 + 2(-4))$$

$$= -\frac{18}{\sqrt{6}} = -\sqrt{6}$$

74. Ans. (b)

Solution:

Given $f(x) = (x - 8)^{2/3} + 1$

$$f'(x) = \frac{2}{3}(x - 8)^{-1/3}$$

Slope of tangent at point (0, 5)

$$m = \frac{2}{3}(0 - 8)^{-1/3} = -\frac{1}{3}$$

Slope of normal at point (0, 5)

$$m_1 = \frac{-1}{m} = 3$$

$$y - 5 = 3(x - 0)$$

or $y = 3x + 5$

75. Ans. (d)

Solution:

Velocity vector $\vec{V} = 2xy\hat{i} - x^2z\hat{j}$

The vorticity vector = curl (velocity vector)

= curl ($\vec{V}$)

$$= \Delta \times \vec{V}$$

$$= \begin{vmatrix} \hat{i} & \hat{i} & \hat{k} \\ \frac{\partial}{\partial x} & \frac{\partial}{\partial y} & \frac{\partial}{\partial z} \\ 2xy & -x^2z & 0 \end{vmatrix}$$

$$= \left[\frac{\partial}{\partial y}(0) - \frac{\partial}{\partial z}(-x^2z)\right] i - \left[\frac{\partial}{\partial x}(0) - \frac{\partial}{\partial z}(2xy)\right]\hat{j}$$

$$+ \left[\frac{\partial}{\partial x}(-x^2z) - \frac{\partial}{\partial y}(2xy)\right]\hat{k}$$

$$= x^2\,\hat{i} + [-2xz - 2x]\,\hat{k}$$

at (1, 1, 1),

vorticity vector = $\hat{i} - 4\hat{k}$

76. Ans. (a)

Solution:

$$f_x = 2xyz,\ f_y = x^2z,\ f_z = x^2y$$

By using line integration

f = Potential function of V = x^2yz

Therefore, Line integral of the vector function from point A (0, 0, 0) to the point B (1, 1, 1) is

$$= f(B) - f(A)$$

$$= (x^2yz)_{(1,1,1)} - (x^2yz)_{(0,0,0)}$$

$$= 1 - 0 = 1$$

77. Ans. (b)

Solution: $Y = mx + c$

$C = 1$

$m = -1$

$Y = -1 \cdot x + c$

or $Y + x = 1$

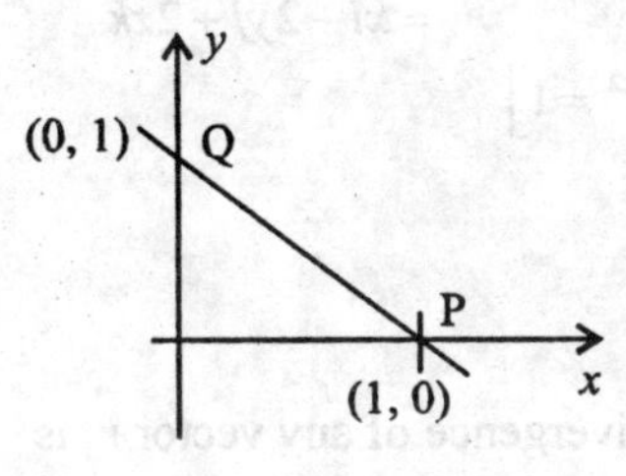

$$y = mx + C$$

$$C = 1$$

$$m = -1$$

$\therefore \quad y = -1 \cdot x + C = -x + 1 = 1 - x$

or, $\quad y + x = 1$

The straight line equation is $x + y = 1$

$$I = \int_P^Q (x\,dx + y\,dy)$$

Then, $I = 2\int_0^1 (1-y)\cdot(-dy) + 2\int_0^1 y\,dy$

$$= 2\left[\frac{y^2}{2} - y\right]_0^1 + 2\left[\frac{y^2}{2}\right]_0^1 = 0$$

78. Ans. (c)

Solution:

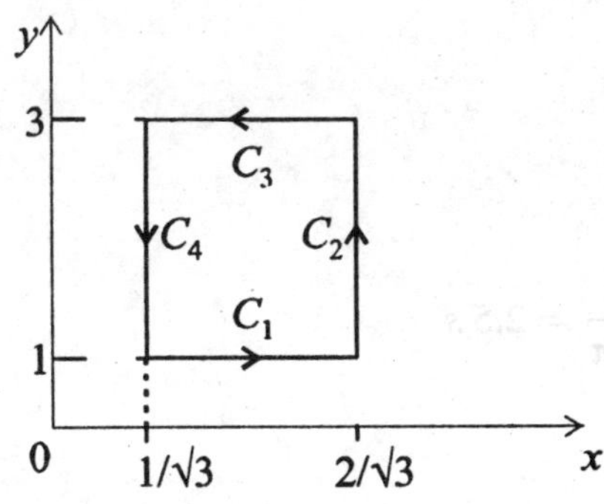

As $\quad \overline{A} = xy\,\hat{a}_x + x^2\hat{a}_y$

$$\oint_C \vec{A}.\vec{dl} = \oint_C \left(xy\hat{a}_x + x^2\hat{a}_y\right).\left(\hat{a}_x dx + \hat{a}_y dy\right)$$

$$= \oint_C xy\,dx + x^2 dy.$$

Now along this closed loop total integral will be.

$$= \left[\begin{array}{l} \int_{C_1} xydx + x^2dy + \int_{C_2} xy\,dx + x^2dy \\ + \int_{C_3} xy\,dx + x^2dx + \int_{C_4} xy\,dx + x^2dy \end{array}\right]$$

Along $\quad C_1 : y = 1$ and $\quad x = \frac{1}{\sqrt{3}}$ to $\frac{2}{\sqrt{3}}$

$$dy = 0$$

$$\int xydx + x^2dy = \int_{\frac{1}{\sqrt{3}}}^{\frac{2}{\sqrt{3}}} x\,dx + 0$$

$$= \frac{1}{2}x^2\Big|_{\frac{1}{\sqrt{3}}}^{\frac{2}{\sqrt{3}}} = \frac{1}{2}\left(\frac{4}{3} - \frac{1}{3}\right) = \frac{1}{2} \qquad \text{...(1)}$$

Along $\quad C_2 : x = \frac{2}{\sqrt{3}}$ and $y = 1$ to 3

$$dx = 0$$

so, $\int_{C_2} xydx + x^2dy = \int_1^3 0 + \left(\frac{2}{\sqrt{3}}\right)^2 dy$

$$= \frac{4}{3}[y]_1^3 = \frac{4}{3}[3-1] = \frac{8}{3} \qquad \text{...(2)}$$

Along $C_3 : y = 3$ and $x = \frac{2}{\sqrt{3}}$ to $\frac{1}{\sqrt{3}}$

$$dy = 0$$

$$\int_{C_3} xydx + x^2dy = \int_{\frac{2}{\sqrt{3}}}^{\frac{1}{\sqrt{3}}} 3x\,dx \qquad \text{...(3)}$$

$$= \frac{3}{2}x^2\Big|_{\frac{2}{\sqrt{3}}}^{\frac{1}{\sqrt{3}}} = \frac{3}{2}\left[\frac{1}{3} - \frac{4}{3}\right] = \frac{-3}{2}$$

Along $\quad C_4 : x = \frac{1}{\sqrt{3}}$ and $y = 3$ to 1

$$\int_{C_4} xydx + x^2dy = \int_3^1 \left(\frac{1}{\sqrt{3}}\right)^3 dy$$

$$= \frac{1}{3}[y]_3^1 = \frac{1}{3}[1-3] = \frac{-2}{3} \qquad \text{...(4)}$$

Put all the. value from equation 1, 2, 3, and 4

we get $\oint_C A.dl = \frac{1}{2} + \frac{8}{3} - \frac{3}{2} - \frac{2}{3}$

$$= \frac{3+16-9-4}{6}$$

$$= \frac{6}{6}$$

$$\oint_C A.dl = 1$$

79. Ans.(a)

Solution:

$$\underset{DiV}{\nabla \cdot \bar{A}} = \frac{1}{h_1 h_2 h_3}\left(\frac{\delta}{\delta u}(h_2 h_3 Au) + \frac{\delta}{\delta v}(h_3 h_1 Av) + \frac{\delta}{\delta w}(h_1 h_2 Aw)\right)$$

Where, $\vec{A} = A_u\ \hat{a}_u + A_v\ \hat{a}_v + A_w\ \hat{a}_w$.

	u	v	w	h_1	h_2	h_3
Cartesian	x	y	z	1	1	1
Cylindrical	r	ϕ	z	1	r	1
Spiracle	r	θ	ϕ	1	r	r.sin θ

Given

$A = k\, r^n$

It is known $\nabla \cdot \bar{A} = \frac{1}{r^2}\frac{\partial}{\partial r}(r^2 A_r)$

Now $\nabla \cdot \bar{A} = \frac{1}{r^2}\frac{\partial}{\partial r}(r^2 A_r)$

$$= \frac{1}{r^2}\frac{\partial}{\partial r}(r^2\, k\, r^n)$$

$$= \frac{1}{r^2}\frac{\partial}{\partial r}(kr^{n+2})$$

$$= \frac{k}{r^2}(n+2)r^{n+1}$$

$$= k(n+2)r^{n-1}$$

$$\nabla \cdot \bar{A} = 0$$

$$(n+2) = 0$$

$$n = -2$$

80. Ans. (b)

Solution:

The Equation of sphere is represented as $x^3 + y^2 + z^2 = 1$

Let $\phi(x, y, z) = x^2 + y^2 + z^2$

Normal to the surface is $\nabla\phi - 2xi + 2yi + 2z\bar{k}$

Unit normal vector is $\frac{\nabla\phi}{|\nabla\phi|} = \frac{2(xi + 2yi + 2z\bar{k})}{\sqrt{4x^2 + 4y^2 + 4z^2}}$

$$= xi + 2yj + 2z\bar{k}$$

$\left[\text{since } x^2 + y^2 + z^2 = 1\right]$

81. Ans (a)

Solution:

As we know that divergence of any vector $\vec{r}$ is

$\nabla \cdot \vec{r}$

$\vec{r}$ is given three dimensional

so suppose vector $\vec{r}$ as

$\vec{r} = x\hat{i} + y\hat{j} + z\hat{k}$

Where

$\hat{i}, \hat{j}, \hat{k}$ are unit vector in the direction of (x, y, z) respectively, so

$$\nabla \cdot \vec{r} = \left(\frac{d}{dx}\hat{i} + \frac{d}{\partial y}\hat{j} + \frac{d}{\partial z}\hat{k}\right) \cdot \left(x\hat{i} + y\hat{j} + z\hat{k}\right)$$

$$= \frac{d}{dx}x + \frac{d}{dy}y + \frac{d}{dz}z = 1 + 1 + 1 = 3$$

$$x(t) = \sin(2\pi + \omega t)$$

$$x(t) = 8\sin\left(0.8\,\pi t + \frac{\pi}{4}\right)$$

$$\omega t = 0.8\,\pi t$$

$$\omega = \frac{2\,\pi}{T}$$

$$0.8\,\pi = \frac{2\,\pi}{T}$$

$$T = \frac{2\,\pi}{0.8\,\pi} = 2.5\ s$$

82. Ans (d)

Solution:

The straight line passes through these points.

$(0,2) \rightarrow (2,0)$

$$m = \frac{0-2}{2-0} = -1$$

$$y - 2 = -1[x - 0] = -x$$

$$x+y=2$$

$$\int_{(x,y)} F(x,y)\,dxdy$$

$$=\int_0^2 x^2+x(2-x)dx+\int_2^0 y^2+y(2-y)dy=0$$

11.3 Differential Equation

1. **Ans. (b)**

Solution:

$$3\left(\frac{d^2y}{dt^2}\right)+4\left(\frac{dy}{dt}\right)^3+y^2+2=x$$

Order is the highest derivative term present in the given equation and degree is the power of highest derivative term in the given equation

Hence order = 2 and

degree = 1

2. **Ans. (b)**

Solution:

For the given equation, auxiliary equation is

$m^2-5m+6=0$

$(m-3)(m-2)=0$

$m=3, 2$

$y=e^{2x}+e^{3x}$

3. **Ans. (a)**

Solution:

The given differential equation is

$$\frac{\partial^2 y}{\partial x^2}+K^2y=0$$

$$(D^2+K^2)y=0$$

$$D=\pm iK$$

So $y=C_1\cos Kx+C_2\sin Kx$ ...(1)

1st boundary condition

$y=0$ for $x=0$

From equation (1) $0=C_1+0$

or $C_1=0$ ...(2)

2nd boundary condition

$y=0$ for $x=a$

From equation (1)

$y=C_1\cos Ka+C_2\sin Ka$

using equation (2)

$0=0+C_2\sin Ka$

or $C_2=0$

So Over all solution becomes zero for the second boundary condition

When we put in all the option only option (a) get vanish so correct option is (a)

4. **Ans. (d)**

Solution:

The Given differential is

$$K^2\frac{d^2y}{dx^2}=y-y_2$$

or $$\frac{\partial^2 y}{dx^2}=\frac{y}{K^2}-\frac{y_2}{K^2}$$

$$=\frac{d^2y}{dx^2}-\frac{y}{K^2}=\frac{-y}{K^2}$$

$$=\left(D^2-\frac{1}{K^2}\right)y=\frac{-y_2}{K^2}$$

A.E equation

$$D^2-\frac{1}{K^2}=0$$

$$D=\pm\frac{1}{K}$$

$$\text{C.F}=C_1e^{+\frac{1}{K}x}+C_2e^{-\frac{1}{K}x}$$

$$\text{P.I}=\frac{1}{\left(D^2-\frac{1}{K^2}\right)}\cdot\frac{-y^2e^{0x}}{K^2}$$

$$=\frac{-y_2}{K^2}\cdot\frac{1}{D^2-\frac{1}{K^2}}\cdot e^{0x}$$

$$=\frac{-y_2}{K^2}\times\frac{1}{\frac{-1}{K^2}}$$

$$\text{P.I}=y_2$$

Complete solution $= CF + P.I$

$$y=C_1e^{\frac{1}{K}x}+C_2e^{\frac{-1}{K}x}+y_2 \quad ...(1)$$

Boundary condition

$y=y_1$ at $x=0$

From equation (1)

$$y_1=C_1+C_2+y_2 \quad ...(2)$$

Also $y=y_2$ at $x=\infty$

$$y_2=C_1e^{\infty}+0+y_2$$

or $$\frac{y_2}{e^{\infty}}=C_1+\frac{y}{e^{\infty}}$$

or $C_1 = 0$ ...(3)

Put in equation (2)

$$y_1 = C_2 + y_2$$

$$C_2 = y_1 - y_2 \quad ...(4)$$

Use (3) and (4) in equation (1)

$$y_2 = (y_1 - y_2)\, e^{\frac{-1}{K}x} + y^2$$

5. Ans. (a)

Solution:

Given $x'(t) = -3x(t)$ or

$x'(t) + 3x(t) = 0$

A.E. $D + 3 = 0$

Thus solution is

$$x(t) = C_1 e^{-3t}$$

Fromwe $x(0) = x_0$ get

$$C_1 = x_0$$

Thus $x(t) = x_0 e^{-3t}$

6. Ans. (b)

Solution:

$$(s^2 + 3s + 2)\,X(s) = \frac{5}{s}$$

$$\therefore X(s) = \frac{5/s}{s^2 + 3s + 2}$$

Apply final value theorem.

$$x(\infty) = \lim_{t\to\infty} x(t)$$

$$= \lim_{s\to 0} sX(s) = \frac{5}{2}$$

7. Ans. (c)

Solution:

Given differential equation is

$$(D^2 - 4D + 4)\,y = 0$$

$$A.E = f(D) = D^2 - 4D + 4 = 0$$

$$D^2 - 2D - 2D + 4 = 0$$

$$D(D-2) - 2(D-2) = 0$$

$$= (D-2)(D-2) = 0$$

Roots are real and equal

So complementary function

$$C.F = (C_1 x + C_2)\,e^{mx}$$

$$= (C_1 x + C_2)\,e^{2x}$$

or $C_1 x e^{2x} + C_2\, e^{2x}$

or $C_1 x e^{2x} + C_2 e^{2x}$

or $C_1 e^{2x} + C_2 x e^{2x}$

8. Ans. (a)

Solution:

Given differential equation

$$y'' + 2y' + 101\, y = 10.4\, e^x$$

$$= (D^2 + 2D + 101)\, y = 10.4\, e^x$$

Auxiliary equation $f(D) = 0$

$$D^2 + 2D + 101 = 0$$

$$D = \frac{-2 \pm \sqrt{4 - 404}}{2}$$

$$D = -2 \pm 101^0$$

Compare with $D = \alpha + j\,\beta$

$$\alpha = -1,$$

General solution for complex root is

$$e^{\alpha x}\left[A\cos\beta x + B\sin\beta x\right]$$

So $e^{-x}\left[A\cos 10x + B\sin 10x\right]$ ...(1)

Particular integral

When $X = e^{\alpha x}$

$$P.I = \frac{1}{f(D)}e^{\alpha x} = \frac{1}{f(\alpha)}e^{\alpha x}.$$

So $\frac{1}{D^2 + 2D + 101}\cdot 10.4\, e^x = 10.4\frac{1}{(D^2 + 2D + 101)}\cdot e^x$

$$= 10.4\frac{1}{(1 + 2 + 101)}\cdot e^x = \frac{10.4\cdot e^x}{104}$$

$$= \frac{1}{10}e^x = 0.1e^x$$

Total solution

$$y(x) = e^{-x}(A\cos 10x + B\sin 10x) + 0.1e^x \quad ...(2)$$

$$y(0) = e^0(A\cos 0 + B\sin 0) + 0.1e^0 = 1.1$$

$$= A + 0.1 = 1.1$$

$$\therefore A = 1$$

Put in (2)

$$y(x)\cdot e^{-x}[\cos 10x + B\sin 10x] + 0.1e^x \quad ...(3)$$

$$y'(x) = \begin{bmatrix} e^{-x}[-\sin 10x + \beta\sin 10x] \\ -e^{-x}[\cos 10x + \beta\sin 10x] + 0.01e^x \end{bmatrix}$$

$y'(0) = B - 1 + 0.1 = -0.9$

$B = 0$

From equation (3) Total solution

$y(x) = e^{-x}\cos 10x + 0.01e^{x}$

Second Method

The correct is given below

General solution of homogeneous equations	$e^{-x}\left(A\cos 10x + B\sin 10x\right)$
Particular integral	$0.1\, e^{x}$
Total solution satisfying boundary conditions	$e^{-x}\cos 10x + 0.1e^{x}$

9. Ans. (b)

Solution: $\dfrac{dx(t)}{dt} + 3x(t) = 0$

$$\frac{dx(t)}{dt} = -3x(t)$$

$$\frac{dx(t)}{x(t)} = -3dt$$

or $$\int \frac{dx(t)}{x(t)} = \int 3dt$$

$$\log x(t) = 3t$$

$$x(t) = k.e^{-3t}$$

10. Ans (b)

Solution:

The order of a differential equation is the order of the highest ordered derivative present in the equation. The given equation is .

$$\frac{d^2y}{dt^2} + \left(\frac{dy}{dx}\right)^3 + y^4 = e^{-t}$$

Highest ordered derivative = 2

Therefore, the order of differential equation is 2.

11. Ans (a)

Solution:

Using $\dfrac{dy}{dx} = \dfrac{y}{x}$

$$\frac{1}{y}\,dy = \frac{1}{x}\,dx$$

Integrate both side

$\log y = \log x + \log c$

$\log y - \log x = \log c$

$$\log\frac{y}{x} = \log c$$

$$\frac{y}{x} = c$$

or $y = cx$

Which is equation of straight line .

Using $\dfrac{dy}{dx} = \dfrac{-y}{x}$

$$\frac{1}{y}\,dy = -\frac{1}{x}\,dx$$

$$\log y = -\log x + \log c$$

$$\log x + \log y = \log c$$

$$xy = c$$

Using $\dfrac{dy}{dx} = \dfrac{x}{y}$ $\quad y\,dy = x\,dx$

Integrating on both side, we get

$$\frac{y^2}{2} = \frac{x^2}{2} + c$$

$$\frac{y^2}{2} - \frac{x^3}{2} = c$$

$$y^2 - x^2 = 2c$$

$y^2 - x^2 =$ constant

Which is an equation of hyperbola

Using $\dfrac{dy}{dx} = \dfrac{-x}{y}$

$$y\,dy = -x\,dx$$

Integrating on both side, we get

$$\frac{y^2}{2} = \frac{-x^2}{2} + c$$

$$x^2 + y^2 = 2c$$

$x^2 + y^2 =$ constant

Which is an equation of circle.

12. Ans. (d)

Solution:

Given differential equation.

$$\frac{d^2n(x)}{dx^2} - \frac{n(x)}{L^2} = 0$$

Put $\dfrac{\partial}{\partial x} = D$ or $\dfrac{\partial^2}{\partial x^2} = D^2$

Then $D^2\, n(x) - \dfrac{1}{L^2} n(x) = 0$

$$\left(D^2-\frac{1}{L^2}\right)n(x)=0$$

A.E $D^2-\frac{1}{L^2}=0$

$$D^2=\frac{1}{L^2}$$

$$D=\pm\frac{1}{L}$$

So Complementary function

$$CF=n(x)=C_1e^{\frac{1}{L}x}+C_2e^{\frac{-1}{L}x} \quad ...(1)$$

given data $n(0)=K$

from equation (1) $C_1e^0+C_2e^0=K$

or $C_1+C_2=K$ ...(2)

also $n(\infty)=0$

from equation (1) $C_1e^{\infty}+C_2e^{-\infty}=0$

$C_1e^{\infty}=0$

$C_1=0$ ...(3)

Put in equation (2) $C_2=K$ ...(4)

using equation (3) and (4) in equation (1)

$$n(x)=0+Ke^{\frac{-1}{L}x}$$

$$=Ke^{\frac{-1}{L}x}$$

13. Ans. (c)

Solution:

Given $y(0)=C$

and $\frac{dy}{dx}=ky \Rightarrow \frac{dy}{y}=kdx$

In $y=kx+c$

$\Rightarrow y=e^{kx}e^{c}$

When $(0)=C, Y=k^1e^0$

$\therefore y=ce^{kx}\ (\because\ k_1=C)$

14. Ans. (d)

Solution:

Given,

$$\frac{dx}{dt}+x=t$$

or $\frac{dx}{dt}+\frac{x}{t}=1$

$IF=e^{\int\frac{1}{t}dt}=e^{\log t}=t$

solution is $x(IF)=\int(IF)\,tdt$

$xt=\int t\cdot tdt$

or, $xt=\frac{t^2}{2}+C$

Given that $x(1)=0.5$

or, $0.5=\frac{1}{2}+C$

or $C=0$

The required solution is $xt=\frac{t^2}{2}$

or $x=\frac{t}{2}$

15. Ans (a)

Solution:

$$\frac{dy}{dx}=1+y^2$$

$$\int\frac{dy}{1+y^2}=\int 1\,dx$$

$\tan^{-1}y=x+C$

$y=\tan(x+C)$

But C is constant in every option

$y=\tan(x+3)$

16. Ans (c)

Solution:

The General Equation is given as

$$\frac{dy}{dx}+py=Q$$

Given

$y(x=0)\Rightarrow 1$

$y(1)=?$

I.F. $=e^{\int Pdx}$

Given

$$\frac{dy}{dx}+y=e^x$$

Compare it with General equation

we get $P=1$ & $Q=e^x$

$$e^{\int 1\,dx} \Rightarrow e^x$$

$$y.(I.F.)=\int P.\ I.F.$$

$$y.e^x=\int e^x.e^x$$

$$y.e^x=\int e^{2x}$$

$$y.e^x=\frac{e^{2x}}{2}+c$$

Given

$$y(x=0)\Rightarrow 1$$

$$y(e)^0=\frac{e^0}{2}+c$$

Put $y=1$

$$1=\frac{1}{2}+c$$

$$c=\frac{1}{2}$$

$$y.e^x=\int e^x.e^x$$

$$y.e^x=\frac{e^{2x}}{2}+c$$

$$y.e^x=\frac{e^{2x}}{2}+\frac{1}{2}$$

$$y=\frac{e^x}{2}+\frac{1}{2}e^{-x}$$

$$y=\frac{1}{2}\left[e^x+e^{-x}\right]$$

Put $x=1$

$$y=\frac{1}{2}\left[e+e^{-1}\right]$$

17. Ans (b)

Solution:

Given differential equation is

$$\frac{d^2x}{dt^2}+\frac{6dx}{dt}+8x=0 \qquad \text{... (1)}$$

This differential equation is homogeneous so it will only have complementary function. No particular integral will be true

Put $\frac{d}{dE}=D$

from equation (1) $\left(D^2x+6Dx+8x\right)=0$

$$\left(D^2+6D+8\right)x=0$$

Auxiliary equation $D^2+6D+8=0$

$$D^2+4D+2D+8=0$$

$$D(D+4)+2(D+4)=0$$

$$(D+2)(D+4)=0$$

$$D=-2,-4$$

As roots are real and distinct so solution will be

$$x(t)=Ae^{-2t}+A_2e^{-4t} \qquad (2)$$

Now $x(0)=A_1e^0+A_2e^0$ $\qquad A_1+A_2=1$

Equation (2) of (3) satisfies the option (b) only

Second Method

$$\frac{d^2x}{dt^2}+6\frac{dx}{dt}+8x=0 \qquad \text{...(1)}$$

Initial condition $x(0)=1$ and $x'(0)=0$

Take Laplace Transform both side to equation (1)

$$\left[s^2X(s)-sX(0)-x'(0)\right]+6\left[sX(s)-x(0)\right]+8X(s)=0$$

or, $\left[s^2+6s+8\right]X(s)=s$

or, $X(s)=\frac{s}{s^2+6s+8}=\frac{s}{(s+2)(s+4)}$

$$=\frac{2}{s+2}-\frac{1}{s+4}$$

$$x(t)=2e^{-2t}-e^{-4t}$$

18. Ans. (b)

Solution:

Degree of a differential equation is the power of its highest order derivative after the differential equation is made free of radicals and fractions if any, in derivative power.

Here the degree is 1, which is power of $\frac{d^2x}{dt^2}$.

19. Ans. (a)

Solution:

Degree of a differential equation is the power of its highest order derivative after the differential equation is made free of radeicals and fractions if any, in derivative power.

$$\frac{d^3y}{dx^3}+4\sqrt{\left(\frac{dy}{dx}\right)^3+y^2}=0$$

Remove radicals

$$\left(\frac{d^3y}{dx^3}\right)^2 = 16\left[\left(\frac{dy}{dx}\right)^3 + y^2\right]$$

The order is 3 since highest differential is $\frac{d^3y}{dx^3}$

The degree is 2 since power of highest differential is 2.

20. Ans. (b)

Solution:

Given equation

$$\frac{d^3f}{dn^3} + \frac{f}{2}\frac{d^2f}{dn^2} = 0 \text{ it is third order } \left(\frac{d^3f}{dn^3}\right)$$

It is non linear, the product $f \times \frac{d^2f}{dn^2}$

21. Ans. (d)

Solution:

Given

$$\frac{dy}{dx} = x^2y$$

This is solved by variable separable form

$$\frac{dy}{y} = x^2dx$$

$$\int\frac{dy}{y} = \int x^2dx$$

or $\quad \log_e y = \frac{x^3}{3} + C_1$

or $\quad y = e^{\frac{x^3}{3}+C_1}$

$$= e^{C1} \times e^{\frac{x^3}{3}}$$

at $x = 0$, $y = 1$

$$1 = Ce^{\frac{0}{3}}$$

or $\quad C = 1$

Therefore $\quad y = e^{\frac{x^3}{3}}$

22. Ans. (b)

Solution:

Given

$$\frac{dv}{dt} + 2xy = e^{-x^2}$$

This is a first order linear differential equation Integrating factor I.F. $= e^{\int 2xdx} = e^{x^2}$

Solution is $y(\text{I.F.}) = \int Q(\text{IF})dx + C$

$$ye^{x^2} = \int e^{-x^2}\, e^{x^2}\, dx + C$$

$$ye^{x^2} = x + C$$

at $\quad x = 0, y = 1$ (given)

therefore $\quad 1e^0 = 0 + C$

or $\quad C = 1$

So, $\quad ye^{-x^2} = x + 1$

or $\quad y = e^{x^2}(x + 1)$

23. Ans. (a)

Solution:

Given

$$\frac{d^2y}{dx^2} + y = 0$$

$$D^2 + 1 = 0$$

$$D = \pm i$$

Therefore, General solution is

$$y = e^{0x}[C_1 \cos(1\,x) + C_2 \sin(1\,x)]$$

$$= C_1 \cos x + C_2 \sin x$$

$$= P \cos x + Q \sin x$$

where P and Q are constants.

24. Ans. (d)

Solution:

Given

$$\ddot{x} + 3x = 0$$

A.E. is

$$D^2 + 3 = 0$$

i.e. $\quad D = \pm\sqrt{3}\, i$

Therefore $\quad x = A\cos\sqrt{3}t + B\sin\sqrt{3}t$

at $\quad t = 0, x = 1$

or $\quad A = 1$

Now, $\quad \dot{x} = \sqrt{3}\,(B\cos\sqrt{3}t - A\sin\sqrt{3}t)$

At $\quad t = 0, \dot{x} = 0$

or $\quad B = 0$

So, $\quad x = \cos\sqrt{3}t$

$$x(1) = \cos\sqrt{3}t = 0.99$$

25. Ans. (b)

Solution:

$$\frac{dx}{dt} = -kx^2$$

or $\frac{dx}{x^2} = -kdt$

Integrate both sides, $\int \frac{dx}{x^2} = -\int kdt$

$-\frac{1}{x} = -kt + C$

or $\frac{1}{x} = kt + C'$

at $t = 0, x = a$

or $\frac{1}{a} = k \times 0 + C'$

or $C' = \frac{1}{a}$

therefore $\frac{1}{x} = kt + \frac{1}{a}$

26. Ans. (a)

Solution:

From question given as

$$\frac{dV}{dt} = -kA \quad \text{...(1)}$$

If volume of spherical

$$V = \frac{4}{3}\pi r^3$$

Outer surface Area of sphere $A = 4\pi r^2$

or $\frac{dV}{dt} = \frac{4}{3}\pi \times 3r^2 \frac{dr}{dt} = 4\pi r^2 \frac{dr}{dt}$

Substituting these in (1),

$$4\pi r^2 \frac{dr}{dt} = -k(4\pi r^2)\frac{dr}{dt}$$

$$\frac{dr}{dt} = -k$$

or $dr = -kdt$

integrating both side

$r = -kt + C$

at $t = 0, r = 1$

or $1 = -k \times 0 + C$

or $C = 1$

therefore $r = -kt + 1$ ... (2)

Now at $t = 3$ months

$r = 0.5$ cm

or $k = \frac{0.5}{3}$

Now substituting this value of r in equation (2)

$$r = -\frac{0.5}{3}t + 1$$

putting $r = 0$ (ball completely evaporates)

in above and solving for t gives $0 = -\frac{0.5}{3}t + 1$

or $t = 6$ months

27. Ans. (d)

Solution:

$$\frac{dy}{dx} = \frac{-x}{y}$$

or $y\,dy = -x\,dx$

or $\int y\,dy = \int -x\,dx$

$$\frac{y^2}{2} = -\frac{x^2}{2} + C$$

at $x = 1, y = \sqrt{3}$

therefore $\frac{(\sqrt{3})^2}{2} = \frac{-1^2}{2} + C$

or $C = 2$

$\therefore$ Solution is $\frac{y^2}{2} = \frac{x^2}{2} + 2$

or $x^2 + y^2 = 4$

28. Ans. (a)

Solution:

Given differential equation.

$$3y\frac{dy}{dx} + 2x = 0$$

or $\frac{dy}{dx} = \frac{-2}{3y}$

or $3ydy = -2xdx$

or $\int 3ydy = \int -2xdx$

or $\frac{3}{2}y^2 = -2 \times \frac{x^2}{2} + C$

or $3y^2 + 2x^2 = C$

or $\frac{x^2}{\left(\frac{1}{2}\right)} + \frac{y^2}{\left(\frac{1}{3}C\right)} = C$

which is the equation of a family of ellipses.

29. Ans. (a)

Solution:

Given differential equation

$$\frac{dy}{dx} + y^2 = 0$$

or $-\frac{dy}{y^2} = dx$

or $-\int\frac{dy}{y^2} = \int dx$

$\frac{1}{y} = x + C$

$y = \frac{1}{x+C}$

30. Ans. (c)

Solution:

Given $\frac{dy}{dx} = y^2$

or $\int\frac{dy}{y^2} = \int dx$

or $-\frac{1}{y} = x + C$

Therefore $y = -\frac{1}{x+C}$

When $x = 0$
$y = 1$

Therefore $C = -1$

Therefore $y = -\frac{1}{x-1}$

y is bounded when $x - 1 \neq 0$

i.e. $x \neq 1$

i.e. $x < 1$ or $x > 1$

31. Ans. (a)

Solution:

Given equation is $\frac{dy}{dt} + p(t)\,y = q(t)\,y^n\ ;\ n > 0$ ... (1)

put $v = y^{1-n}$

$\frac{dv}{dt} = (1-n)\,y^{-n}\,\frac{dy}{dt}$

$\frac{dy}{dt} = \frac{1}{(1-n)y^{-n}}\frac{dv}{dt}$

Substituting in equation (1)

$\frac{1}{(1-n)y^{-n}}\frac{dv}{dt} + p(t)y = q(t)y^n$

Multiplying by $(1-n)y^{-n}$,

$\frac{dv}{dt} + p(t)\,(1-n)y^{1-n} = q(t)\,(1-n)$

Since $y^{1-n} = v$,

Therefore $\frac{dv}{dt} + (1-n)\,pv = (1-n)q$

(which is linear with v as dependent variable and t as independent variable).

32. Ans. (a)

Solution:

Given equation is

$x\frac{dy}{dx} + y = x^4$

or $\frac{dy}{dx} + \left(\frac{y}{x}\right) = x^3$ ... (1)

Standard form of linear equation is

$\frac{dy}{dx} + Py = Q$ ...(2)

where P and Q function of x only and solution is given by

$y(\text{I.F.}) = \int Q.\,(\text{I.F.})\,dx + C$

where, $(\text{I.F.}) = e^{\int P dx}$

Here in equation (1)

$P = \frac{1}{x}$ and $Q = x^3$

$\text{I.F.} = e^{\int\frac{1}{x}dx} = e^{\ln x} = x$

Solution $y(x) = \int x^3\,x\,dx + C$

$yx = \frac{x^5}{5} + C$

given condition $y(1) = \frac{6}{5}$

means at $x = 1; y = \frac{6}{5}$

or $\frac{6}{5}\times 1 = \frac{1}{5} + C$

or $C = \frac{6}{5} - \frac{1}{5} = 1$

Therefore $y = \frac{x^4}{5} + \frac{1}{x}$

or $y = \frac{x^4}{5} + \frac{1}{x}$

33. Ans. (a)

Solution:

Given equation is

$$\frac{d^2y}{dx^2} + 2\frac{dy}{dx} + 17 = 0$$

$$y(0) = 1$$

$$\frac{dy}{dx}\left(\frac{\pi}{4}\right) = 0$$

This is a linear differential equation

$$D^2 + 2D + 17 = 0$$

$$D = -1 \pm 4i$$

$$Y = C_1e^{(-1+4i)} + C_2e^{(-1-4i)x}$$

$$= e^{-x}C_1e^{4xi} + C_2e^{e-4xi}$$

$$= e^{-x}[C_1(\cos 4x + i \sin 4x)]$$

$$+ C_2[\cos(-4x) + i \sin(-4x)]$$

$$= e^{-x}[(C_1+C_2)\cos 4x$$

$$+ (C_1-C_2)\, i \sin 4x]$$

Let $C_1 + C_2 = C_3$ and $(C_1-C_2)\, i = C_4$

$$y = e^{-x}(C_3 \cos 4x + C_4 \sin 4x$$

since $y(0) = 1$

or $1 = e^{-0}(C_3 \cos 0 + C_4 \sin 0)$

or $C_3 = 1$

$$\frac{dy}{dx} = e^{-x}(-4C_3 \sin 4x + 4C_4 \cos 4x)$$

$$-e^{-x}[C_3 \cos 4x + C_4 \sin 4x]$$

$$= e^{-x}[(-4C_3 - C_4)\sin 4x + (4C_4 - C_3)\cos 4x]$$

$\frac{dy}{dx}$ at $x = \frac{\pi}{4}$ is 0

therefore

$$-(-4C_4 - C_3)\, e^{-\pi/4} = 0$$

$$4C_4 = C_3$$

$$C_4 = \frac{C_3}{4} = \frac{1}{4}$$

therefore $C_3 = 1$ and $C_4 = \frac{1}{4}$

$$y = e^{-x}(\cos 4x + \frac{1}{4}\sin 4x)$$

34. Ans. (a)

Solution:

Given equation is

$$y'' + 2y' + y = 0$$

$$(D^2 + 2D + 1)y = 0$$

or $D^2 + 2D + 1 = 0$

or $(D + 1)^2 = 0$

or $D = -1, -1$

Therefore $y = (C_1 + C_2x)e^{-x}$

$y(0) = 0 \Rightarrow$ $0 = (C1 + C2\,(0)e^{-0}$

or $C_1 = 0$

$y(1) = 0$ $0 = (C_1 + C_2)e^{-1}$

or $C_1 + C_2 = 0$

or $C_2 = 0$

Therefore $y = (0 + 0x)e^{-x} = 0$ is the solution

Therefore $y(0.5) = 0$

35. Ans. (b)

Solution:

Given equations is

$$\frac{d^2y}{dx^2} + 4\frac{dy}{dx} + 3y = 3e^{2x}$$

or $(D^2 + 4D + 3)y = 3e^{2x}$

Particular integral $\quad P.I. = \frac{1}{D^2 + 4D + 3}3e^{2x}$

Now since, $\quad \frac{1}{f(D)}e^{ax} = \frac{1}{f(a)}e^{ax}$

$$P.I. = 3\frac{e^{2x}}{(2)2 + 4(2) + 3} = \frac{3e^{2x}}{15} = \frac{e^{2x}}{5}$$

36. Ans. (c)

Solution:

Given equation is

$$f(z) = u + iv$$

Therefore, It must satisfy the Cauchy-Reimann equation

$u_x = v_y$ ...(1)

and $v_x = -u_y$ ...(2)

Since, given as $u = xy$

Substitute u_x and u_y in equation (1) and (2)

$v_y = y$ ...(3)

and $v_x = -x$ ...(4)

Integrating (3) and (4) get v as follow:

$$v_y = y$$

or $\frac{\partial v}{\partial y} = y$

or $\int \partial v = \int y\, \partial y$

or $v = \frac{y^2}{2} + f(x)$ ...(5)

from (5), $v_x = f'(x)$ ...(6)

Since from (4), $v_x = -x$

Substituting this in (6),

$$f'(x) = -x$$

or $$\frac{df}{dx} = -x$$

or $$\int df = \int -x\,dx$$

or $$f = \frac{-x^2}{2} + k$$

Substitute this in (5),

$$v = \frac{y^2}{2} - \frac{x^2}{2} + k$$

or $$v = \frac{y^2 - x^2}{2} + k$$

11.4 Complex Variables

1. **Ans. (b)**

Solution: $W = \log_e z$

$\Rightarrow u + jv = \log(x + jy)$

$= [1/2] \log(x^2 + y^2) + j\tan^{-1}[y/x]$

u = constant

$\Rightarrow [1/2] \log(x^2 + y^2) = C$

$\Rightarrow x^2 + y^2 = C$

2. **Ans. (d)**

Solution: $|Z - j| = 2$ is circle with center (0, 1) and radius 2.

$$\frac{1}{z^2 + 4} = \frac{1}{(z + 2j)(z - 2j)}$$

z = (0, 2) lies inside circle.

So, by Cauchy integral formula,

$$\text{Integral} = \frac{2\pi j}{(2j + 2j)} = \frac{\pi}{2}$$

3. **Ans. (a)**

Solution:

$$\oint_D \frac{1}{(S^2 - 1)} dS$$

$$\oint_0 \frac{1}{(S-1)(S-1)} dS$$

It has two poles as $S = -1$ and $+1$ only pole $S = +1$ in the given contour then according to Cauchy's integral theorem

$$\oint_D \frac{1}{(S-1)(S-1)} dS = 2\pi j \left.\frac{1}{(S+1)}\right|_{S=1} + 0$$

$$= \frac{2\pi j}{2} = \pi j$$

4. **Ans. (a)**

Solution:

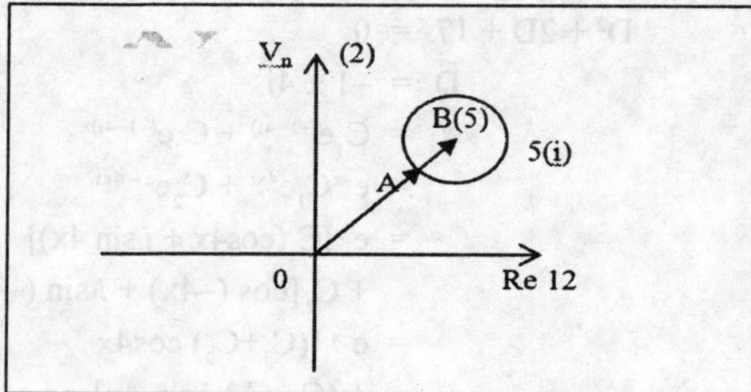

Given complex equation is

$|z - 5 - 5i| = 2$

Which is equation of circle having Center at (5, 5) and radius 2.

$\therefore$ Minimum distance $= 5\sqrt{2} - 2$

5. **Ans. (a)**

Solution: Given, $z^3 = \bar{z}$

$\Rightarrow z^4 = |z|^2$ (on multiplying z both side)

$z^4 - z^2 = 0$

$|z|^2 \left(|z|^2 - 1\right) = 0$ (possible is there $z = 0$ given)

or $z = \pm 1$

$\therefore$ z is a

or $z^2 = 1$

Now by hit and trial method we see the solution being

$z^2 = 1$

6. **Ans (d)**

Solution:

Let $X = j^j = \left(e^{j\pi/2}\right)^{c/\pi/2}$

$\Rightarrow \log {}^{X} = c^{r/\pi/2} = \log c^{r/\pi/2} = j^j \frac{\pi}{2}$

$\Rightarrow \log {}^{X}_{c} = -\frac{\pi}{2}$

$\Rightarrow X = c^{r/\pi/2}$

7. Ans. (b)

Solution: $\oint_C \frac{dz}{1+z^2}$

Where C is the contour $|z - i/2| = 1$

$|z - i/2| = 1$; $z = x + iy$

$|x + iy - i/2| = 1$; $|x + i(y - 1/2)| = 1$

$x^2 + (y - 1/2)^2 = 1$; This represent a circle

Whose centre = $(0, i/2)$ and radius = 1

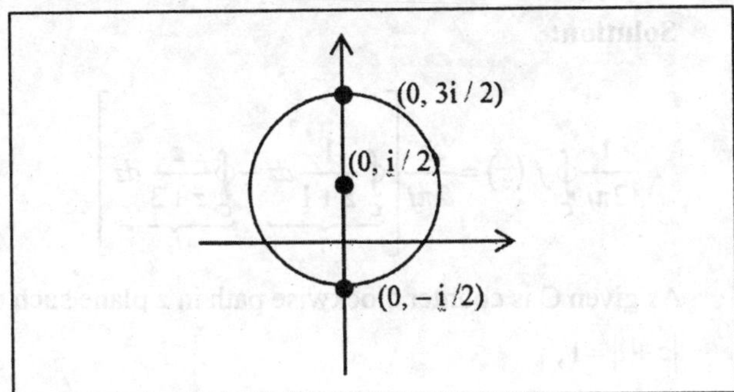

8. Ans. (d)

Solution:

Let, $z = x + jy$

$\sin z = 10$

or, $\sin(x + jy) = 10$

or, $\sin x \cos(jy) + \sin(jy) \cos x = 10$

or, $\sin x \cosh y + j \cos x \sin y = 10$

Compare real and imaginary parts

$\sin x \cosh y = 0$ ...(1)

$\sin x \cosh y = 10$ ...(2)

By equion (1)

either $\cos x = 0$

or, $x = \frac{(2n+1)\pi}{2}$ Where, $n = 0, \pm 1, \pm 2, \ldots$

or, $\sinh y = 0$

or $y = 0$

For $y = 0$ from eqution (2)

$\sin x \cosh 0 = 10$

or, $\sin x = 10 \Rightarrow$ no solution possible

$\therefore$ $y \neq 0$ as no x exist for this

For $x = \pm\frac{\pi}{2}$, $= \pm\frac{3\pi}{2}, \ldots$

If x is +ve

$x = \frac{3\pi}{2}, \frac{7\pi}{2}, \ldots$

$\sin x = -1$

From equation (2)

$\sin x \cosh y = 10$

or, $\cosh y = -10$

No y exist.

For, $x = \frac{\pi}{2}, \frac{5\pi}{2}, \frac{9\pi}{2}, \ldots$

$\sin x = 1$

$\cosh y = 10$

or, $y = 2.993$

Hence infinite number of complex solutions exist

$$z = \frac{\pi}{2} + j2.99$$

$$= \frac{5\pi}{2} + j2.99$$

$$= \frac{9\pi}{2} + j2.99, \ldots$$

Similarly if x –ve

$$x = \frac{-3\pi}{2}, = \frac{-7\pi}{2}, \ldots$$

$\sin x = 1$

$\cosh y = 10$

or, $y = 2.993$

Infinite number of complex solutions exist

$$x = \frac{-3\pi}{2} + j2.993,$$

$$= \frac{-7\pi}{2} + j2.99, \ldots$$

9. Ans (d)

Solution:

$$f'(z) = \left(\frac{1 + c_o + z^{-1}}{z}\right)$$

$$f'(z) = \left(\frac{(1 + c_o) + c_1 z^{-1}}{z}\right)$$

$$\int f'(z) = \frac{d}{d^2} z^2 \left(\frac{(1 + c_o) z + c_1}{z^2}\right)$$

$$= 2\pi i(1 + c_o)$$

10. Ans (b)

Solution: According to stoke's Theorem,

$$\oint_C \vec{A}.\vec{dl} = \int_s \left(\nabla \times \vec{A}\right).\vec{ds}$$

Given that, $\vec{V} = \nabla \times \vec{A}$

$$\oint_C \vec{A}.\vec{dl} = \int_s \vec{V}.\vec{ds}$$

11. Ans. (c)

Solution:

$$\text{Residue at pole } = \left.\frac{X(z)}{(\text{Pole})}\right|_{z=\text{pole}}$$

i.e. $\left. z\,X(z)\right|_{z=0} = \left.\dfrac{1-2z}{(z-1)(z-2)}\right|_{z=0}$

$$= \frac{1}{-1\times-2} = \frac{1}{2}$$

$$\left.(z-1)X(z)\right|_{z=1} = \left.\frac{1-2z}{z(z-2)}\right|_{z=1} = \frac{1-2\times(1)}{1(1-2)}$$

$$= \frac{1-2}{1\times-1} = \frac{-1}{-1} = 1$$

$$\left.(z-2)X(z)\right|_{z=2} = \left.\frac{1-2z}{z(z-1)}\right|_{z=2} = \frac{1-2\times(+2)}{+2(+2-1)}$$

$$= \frac{1-4}{2\times1} = \frac{-3}{2}$$

12. Ans. (a)

Solution:

When factor the Numerator and denominator getting the value $z=-2\pm j$ for denominator which lie out side the unity circle. Hence integration value is zero.

$$\oint_C \frac{-3z+4}{z^2+4z+5}dz = 0$$

$(\because z^2+4z+5 = (z+2)^2+1 = 0)$

$z=-2\pm j$ will be outside the unit circle

So that integration value is 'zero

13. Ans.(a)

Solution:

Given, $x=\sqrt{-1}$ that is equal to j

$$x^x = \left(\sqrt{-1}\right)^{\sqrt{-1}} = (j)^j$$

It is known that $e^{i\theta} + i\sin\theta = \cos\theta + j\sin\theta$

$$\therefore \quad e^{j\frac{\pi}{2}} = \cos\frac{\pi}{2} + j\sin\frac{\pi}{2} = 0 + j.1$$

$$= j$$

Therefore, $(j)^j = e^{(j\pi/2)^j}$

$$= e^{(j\pi/2)\times j}$$

$$= e^{(j^2\pi/2)} = e^{-\pi/2}$$

14. Ans. (c)

Solution:

$$\frac{1}{2\pi j}\oint_C f(z) = \frac{1}{2\pi j}\left[\underbrace{\oint_C \frac{1}{z+1}dz}_{I_1} - \underbrace{\oint_C \frac{z}{z+3}dz}_{I_2}\right]$$

As given C is counter clockwise path in z plane such that $|z+1| = 1$,

Hence $z=-1$ is singularity in C and $z=-3$ is not in C

By cauchy's integral formula

$$I_2 = \frac{1}{2\pi j}\oint_C \frac{z}{z+3}dz = 0$$

and

$$I_1 = \frac{1}{2\pi j}\oint_C \frac{1}{z+1}dz = 1$$

$$\therefore \quad I_1 - I_2 = 1$$

15. Ans (b)

Solution:

$$Z = x + j$$

or $\dfrac{1}{Z} = \dfrac{1}{x+j}$

$$\therefore \quad \frac{1}{Z} = \frac{x}{x^2+1} - \frac{j}{x^2+1}$$

16. Ans (b)

Solution:

$$z = x + iy$$

$$p = \left|e^{jz}\right|$$

Taking log,

$$\log_e p = \log\left|e^{jz}\right|$$

$$\Rightarrow \log_e p = |jz|.\log_e e$$

$$= \sqrt{x^2+Y^2}$$

$$p = e^{\sqrt{x^2+y^2}}$$

17. Ans (a)

Solution:

Cauchy's integral formula is

$$f(a)=\frac{1}{2\pi i}\int_c \frac{f(z).dz}{z-a}$$

Here $a=0,$

then $f(0)=\sin 0=0$

Second Method

$$\oint \frac{\sin z}{z}\ dz$$

$$F(a)=\frac{1}{2\pi i}\int_c \frac{F(z)dz}{z-a}$$

Here $a=0$

$$\oint \frac{F(z)dz}{z-a}=2\pi i\ F(a)$$

$$F(z)=\sin z$$

$$F(a)=\sin(a)$$

$$F(0)=\sin(0)\Rightarrow 0$$

$$\oint \frac{F(z)dz}{(z-a)}\Rightarrow 2\pi i\ F(a)$$

$$\Rightarrow 2\pi i\ (0)\Rightarrow 0$$

So Ans $=0$

18. Ans (b)

Solution:

Given $x^3=j=e^{+j\pi/2}$

$\Rightarrow$ $x=e^{j\pi/6}$

$\Rightarrow$ $x=\cos\frac{\pi}{6}+j\sin\frac{\pi}{6}=\frac{\sqrt{3}}{2}+j.\frac{1}{2}$

19. Ans (d)

Solution:

$$\oint \frac{z^2+8}{0.5z-1.5j}\ dz$$

Take 0.5 common from denominator

$$\oint \frac{\left(\frac{z^2+8}{0.5}\right)dz}{z-3j}$$

Compare it with

$$\oint \frac{F(z)\ dz}{z-a}$$

we get $a=3j$

$$F(z)\Rightarrow\left(\frac{z^2+8}{0.5}\right)$$

$$\oint \frac{F(z)}{(z-a)}=2\pi j\ F(a)$$

$$F(z)\Rightarrow\left(\frac{z^2+8}{0.5}\right)$$

$$F(a)\text{ i.e. }F(3j)=\left(\frac{(3j)^2+8}{0.5}\right)$$

$$\Rightarrow\left(\frac{-9+8}{0.5}\right)=\left(\frac{-1}{0.5}\right)=-2$$

so result $\Rightarrow 2\pi j(-2)$

$\Rightarrow -4\pi j$

20. Ans. (c)

Solution:

Let $f(z)=e^{\frac{1}{z}}$

$$=1+\frac{1}{z}+\frac{1}{2!z^2}+\frac{1}{3!z^3}+\ldots\ldots$$

$$=1+\frac{1}{z-0}+\frac{1}{2!(z-0)}+\frac{1}{3!(z-0^3)}+\ldots\ldots$$

$$R_1=R_{es}\left(f(z):\ z=0\right)=1$$

$$\therefore\ \int_C e^{\frac{1}{z}}dz=2\pi i\ R_1=2\pi i(1)$$

$$=2\pi\sqrt{-1}$$

21. Ans. (c)

Solution: Reside of

$$z^P\ X(z)=\frac{1}{(n-1)!}\frac{d^n}{dz^{n-1}}z^P X(z)(z-a)^n\Big|_{z=a}$$

where $(z-a)^n$ at $(z=a)$ $\quad n=$ order of pole

given problem $n=2$

$$\text{Resides}\left(z^{n-1}X(z)\right)=\frac{1}{(2-1)!}\frac{d}{dz}\left(z^{n-1}X(z)\right)\Big|_{z=a}$$

$$=\frac{1}{1!}\frac{d}{dz}\left(z^{n-1}\right)\frac{z}{(z-a)^2}(2-a)^2$$

$$=\frac{d}{dz}\left(z^{n}\right)\Big|_{z=a}=nz^{n-1}\Big|_{z=a}=na^{n-1}$$

22. Ans. (b)

Solution:

Let, $Z=\frac{3+4i}{1-2i}$

$$=\frac{(3+4i)(1+2i)}{(1-2i)(1+2i)}$$

$$=\frac{-5+10i}{5}=1+2i$$

$$|Z|=\sqrt{(-1)^2+(2)^2}=\sqrt{5}$$

23. Ans. (d)

Solution:

$$f(z)=\frac{z-1}{z^2+1}=\frac{z-1}{z^2-i^2}=\frac{z-1}{(z-i)(z+i)}$$

Therefore, The gingularities arc at z = i and –i

24. Ans. (d)

Solution:

In a negatively skewed distribution, mode > median > mean.

25. Ans. (a)

Solution:

Cauchy's integral formula is

$$f(a)=\frac{1}{2\pi i}\oint_C\frac{f(z)}{z-a}\,dz$$

$$=\oint_C\frac{f(2)}{2-a}\,dz=2\pi i\,f(a)$$

i.e. $\oint_C\frac{z^3-6}{3z-i}\,dz=\frac{1}{3}\oint_C\frac{z^3-6}{\left(z-\frac{i}{3}\right)}\,dz$

Apply Cauchy's integral theorem, using $f(z)=z^3-6$.

$$=\frac{1}{3}\left(2\pi i\,f\left(\frac{i}{3}\right)\right)=\frac{1}{3}\left(2\pi i\left[\left(\frac{i}{3}\right)^3-6\right]\right)$$

$$=\frac{1}{3}\left[2\pi i\left(\frac{i^3}{27}-6\right)\right]$$

$$=\frac{2\pi}{81}i^4-4\pi i$$

$$=\frac{2\pi}{81}-4\pi i$$

26. Ans. (c)

Solution:

Given integral.

$$I=\int_C\frac{\cos(2\pi z)}{(2z-1)(z-3)}\,dz$$

$$=\frac{1}{2}\int_C\frac{\left[\frac{\cos(2\pi z)}{(z-3)}\right]}{\left[z-\frac{1}{2}\right]}$$

Since, z = 1/2 is a point within | z | = 1 (the closed curve C) Apply Cauchy's integral theorem

Using $f(z)=\frac{\cos(2\pi z)}{(z-3)}$

$$I=\frac{1}{2}f\left(\frac{1}{2}\right)$$

Therefore $I=\frac{1}{2}\frac{\cos\left(2\pi\frac{1}{2}\right)}{\left(\frac{1}{2}-3\right)}$

$$=\frac{2\pi i}{5}$$

27. Ans. (a)

Solution:

$$\int\sec z\,dz=\int\frac{1}{\cos z}\,dz$$

The poles are at $z_0=(n+1/2)\pi=\ldots-3\pi/2,-\pi/2,\pi/2,+3\pi/2\ldots$

None of these poles lie inside the unit circle | z | = 1

Hence, sum of residues at poles = 0

Therefore, singularities set = ϕ and

I = 2πi [sum of residues of f(z) at the poles]

$= 2\pi i\times 0=0$

28. Ans. (a)

Solution:

Given function

$$f(z) = \frac{\cos z}{z}$$

has simple pole at z = 0 and z = 0 is inside the unit circle on complex plane

Therefore,

Residue of f(z) at z = 0

$$\underset{z\to 0}{\text{Lt}}\, f(z)\cdot z\cdot = \underset{z\to 0}{\text{Lt}} \cos z = 1$$

$$\int_C f(z)\, dz = 2\pi i \text{ (Residue at } z = 0)$$

$$= 2\,\pi\, i.\, 1 = 2\pi i$$

29. Ans. (d)

Solution:

$$\frac{1}{z^2+4} = \frac{1}{(z+2i)(z-2i)}$$

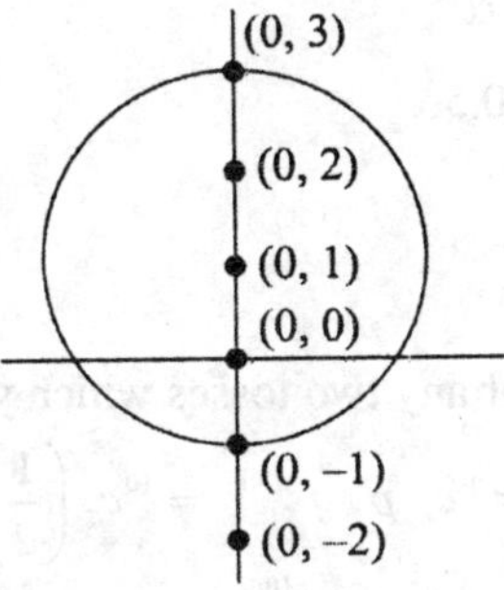

Pole (0, 2) lies inside the circle | z – i | = 2 while pole (0, –2) is out side the circle | 2 – i | = 2 as can be seen from figure below:

$$\int_c f(z)\, dz = 2\pi i \text{ [Residue at those poles which are inside]}$$

$$= 2\pi i \text{ Res } f(2i)$$

$$= 2\pi i \frac{f}{(2i+2i)} = \frac{\pi}{2}$$

30. Ans. (a)

Solution: Residue at z = 2

$$= \lim_{z-2} \frac{1}{1!} \frac{d}{d^2}\left[(z-2)^2 f(z)\right]$$

$$\lim_{z\to 2} \frac{1}{1!} \frac{d}{dz}(z-2)^2 \frac{1}{(z+2)^2 (z-2)^2}$$

$$= \lim_{z\to 2} \frac{-Z}{(z+2)^3} \Rightarrow \text{Residue } = \frac{-1}{32}$$

31. Ans. (d)

Solution:

$|Z| < 1,\ so\ |Y| > 1$

Z is having positive real part and positive imaginary part (∴ from the characteristics)

So Y should have positive real part and negative imaginary part.

11.5 Probability and Statistics

1. Ans. (d)

Solution: n (E) = (1 , 2) (3 , 4) (5 , 6)
(1 , 4) (3 , 2) (5 , 2)
(1 , 6) (3 , 6) (5 , 4)

n (s) = 6 × 6 = 36

P(E) = [n(E) / n(s)]
= [9 / 36] = [1 / 4]

2. Ans. (d)

Solution:

For two random events conditional probability is given by probability $(P \cap Q)$ = probability (P) probability (Q)

$$\text{Probability (Q)} = \frac{\text{probability}(P \cap Q)}{\text{probability (P)}} \le 1$$

So probability

$(P \cap Q) \le$ probability (P)

3. Ans. (d)

Solution:

The coin is tossed thrice. The first toss products Head.

The 2nd and 3rd toss give.

HT, TH, HH, TT

Sample space = (HHT, HTH, HHH, HTT)

Two heads in 3-toss, i.e. only one Head in The and toss either HT or TH

$$\text{Required Probability } = \frac{2}{4} = \frac{1}{2}$$

4. Ans. (a)

Solution:

Probability of getting a head, $P(H) = \frac{1}{2}$

Probability of getting a tail, $P(T) = \frac{1}{2}$

For getting exactly 2 heads in 4 tosses we have

$$= {}^4C_2 \left(\frac{1}{2}\right)^2 \left(\frac{1}{2}\right)^2$$

$$= \frac{4!}{2!\,2!} \times \frac{1}{4} \times \frac{1}{4}$$

$$= \frac{4 \times 3}{2} \times \frac{1}{4} \times \frac{1}{4} = \frac{3}{8}$$

5. Ans. (d)

Solution:

Let the marks obtained per question be a random variable X.

As, there are 4 choices and only one in correct.

So, probability of correct answer is $\frac{1}{4}$

Probability of incorrect answer $= 1 - \frac{1}{4} = \frac{3}{4}$

So, for correct answer $X = 1$

For incorrect answers $X = 0.25$

So, expected marks per question $= \Sigma \times p(x)$

$$= 1 \times \frac{1}{4} + (-0.25) \times \frac{3}{4} = \frac{1}{4} - \frac{3}{16} = \frac{1}{16}$$

Total marks for 150 question per student $= \frac{1}{16} \times 150$

Total expected marks obtained by 1000 student

$$= \frac{75}{8} \times 100$$

= 9375 marks

6. Ans. (c)

Solution:

Probability of failing in Paper 1 = $\overline{P_1} = 0.3 = P(A')$

Probability of failing in Paper 2 = $\overline{P_2} = 0.2 = P(B)$

Probability of failing in Paper 1

When failed in 2 = $P\left(\frac{\overline{P_1}}{\overline{P_2}}\right) = 0.6 = P(A/B)$

$$P(A \cap B) = P(A/B) \cdot P(B)$$

$$P(\overline{P_1} \cap \overline{P_2}) = P\left(\frac{\overline{P_1}}{\overline{P_2}}\right) \times P(\overline{P_2}) = 0.6 \times 0.2 = 0.12$$

7. Ans. (c)

Solution:

The probability of occurrence of value 1, 5 and 6 on three dice = Product of probability of individual

occurrence = $\frac{1}{4} \times \frac{1}{8} \times \frac{1}{4} = \frac{1}{128}$

8. Ans. (c)

9. Ans. (d)

Solution:

Probability that defective part is supplied by

$$y' = \frac{0.3 \times 0.02}{0.6 \times 0.01 + 0.3 \times 0.02 + 0.1 \times 0.03}$$

$$y = \frac{6}{6+6+3} = \frac{6}{15} = \frac{2}{5} = 0.4$$

10. Ans. (c)

Solution: For a pdf the total probability is given by

$$\int_{\infty}^{\infty} p(x)\, dx = 1$$

$$\Rightarrow \int_{\infty}^{\infty} k\, e^{-\alpha|x|}\, dx$$

$$\Rightarrow \frac{-k}{\alpha}\left(e^{-\infty} - e^{0}\right)$$

$$\Rightarrow \frac{k}{\alpha} + \frac{k}{\alpha} \Rightarrow \frac{2k}{\alpha} = 1$$

$$\Rightarrow k = \frac{1}{2}\alpha = 0.5\alpha$$

11. Ans (c)

Solution:

Probability for any two tosses which yield head

$$= {}^{n}c_r\, p^r\, q^{n-r} = {}^{10}c_2 \left(\frac{1}{2}\right)^2 \left(\frac{1}{2}\right)^8$$

$$= {}^{10}C_2 \left(\frac{1}{2}\right)^{10}$$

But here it is required only for first two tosses. Thus in this case it is equal to $\frac{1}{2} \cdot \frac{1}{2}$ up to 10 times. $= \left(\frac{1}{2}\right)^{10}$

12. Ans (c)

Solution:

$$P\left(\frac{x - y = 2}{x - y = 0}\right) = ?$$

Probability of

$X = 0$	$Y = 0$	(1/2)
$X = 1$	$Y = 1$	(1/4)
$X = 2$	$Y = 2$	(1/4)

Let $P(X + Y = 2) = P(A)$

$P(X - Y = 0) = P(B)$

and we are suppose to find $P(A/B)$

So, $P(X+Y\ \ 2)=P(A)$

$$=P(X=0)P(Y=2)+P(X=1)P(Y=1)+P(X=2)P(Y=0)$$

Because X and Y are independent variable

So, $P(A)=\frac{1}{2}\cdot\frac{1}{4}+\frac{1}{4}\cdot\frac{1}{4}+\frac{1}{2}\cdot\frac{1}{4}=\frac{5}{16}$

$P(X-Y=0)=P(B)$

$$=P(X=0)\ P(Y=0)+P(X=1)\ P(Y=1)+P(X=2)\ P(Y=2)$$

$$=\frac{1}{2}\cdot\frac{1}{2}+\frac{1}{4}\cdot\frac{1}{4}+\frac{1}{4}\cdot\frac{1}{4}=\frac{6}{16}$$

Now $P\left(\frac{A}{B}\right)=\frac{P(A\cap B)}{P(B)}$

$P(A\cap B)=P(X+Y=2)\cap(X-Y=0)$

$$=P(X=1)\ P(Y=1)$$

$$=\frac{1}{4}\cdot\frac{1}{4}=\frac{1}{16}$$

As only one case is possible when both $X\neq Y=1$

Hence $P\left(\frac{A}{B}\right)=\frac{(1/16)}{(6/16)}=\frac{1}{6}$

$P\left(\frac{X+Y=2}{X-Y=0}\right)=\frac{1}{6}$

13. Ans (b)

Solution:

k(x)	p(f)	kp	$x=X-\overline{X}$	x^2	px^2
1	0.1	0.1	−2	4	0.4
2	0.2	0.4	−1	1	0.2
3	0.4	1.2	0	0	0
4	0.2	0.8	1	1	0.2
5	0.1	0.5	2	4	0.4

$\Sigma p(f)=N=1$

$\Sigma kp=3$

$\Sigma px^2=1.2$

Mean $=\frac{\sum k\,p}{N}=\frac{3}{1}=3$

Strandard deviation $\sigma=\sqrt{\frac{\sum P\,x^2}{N}}$

Variance $\sigma^2=\frac{\sum p\,x^2}{N}=\frac{1.2}{1}$

Variance = 1.2

14. Ans. (d)

Solution:

4 heads and 0 tail or 3 head and 1 tail

$$P={}^4C_0\left(\frac{1}{2}\right)^2\left(\frac{1}{2}\right)^0+{}^4C_3\left(\frac{1}{2}\right)^3\left(\frac{1}{2}\right)^1=\frac{5}{16}$$

Second Method

In this question we can use the binomial probability distribution in which P be the probability of success and q that failure

then $P(X=r)={}^nC_rP^rq^{n-r}$

we required either 3 Head or all four head in outcome for sauces

Probaliestoy of $P(H)=P(T)=\frac{1}{2}$

Required probability

$$={}^4C_4\left(\frac{1}{2}\right)^4\cdot\left(\frac{1}{2}\right)^0+{}^4C_3\left(\frac{1}{2}\right)^3\cdot\left(\frac{1}{2}\right)^1$$

$$=\frac{1}{16}+\frac{4}{16}=\frac{5}{16}$$

15. Ans. (c)

Solution: Total number of cases = 36

Total number of favorable cases

$=5+4+3+2+1=15$

Then probability $=\frac{15}{36}=\frac{5}{12}$

(1, 1)	(2, 1)	(3, 1)	(4, 1)	(5, 1)	(6, 1)
(1, 2)	(2, 2)	(3, 2)	(4, 2)	(5, 2)	(6, 2)
(1, 3)	(2, 3)	(3, 3)	(4, 3)	(5, 3)	(6, 3)
(1, 4)	(2, 4)	(3, 4)	(4, 4)	(5, 4)	(6, 4)
(1, 5)	(2, 5)	(3, 5)	(4, 5)	(5, 5)	(6, 5)
(1, 6)	(2, 6)	(3, 6)	(4, 6)	(5, 6)	(6, 6)

16. Ans. (b)

Solution:

Distribution function

$$F_X(X)=[X\leq x]=\int_{-\infty}^{x}f_X(\alpha)d\alpha$$

$f_X(x)\rightarrow$ Density function

Joint distribution function is given as

$$F_{X,Y}(x, y) = P[X \le x,\ Y \le y]$$

$$\int_{-\infty}^{y} \int_{-\infty}^{x} f_{X,Y}(\alpha, \beta)\, d\alpha\ d\beta$$

Uniform random variable since both X and Y are uniform random variables

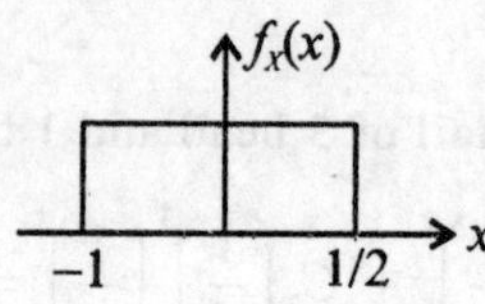

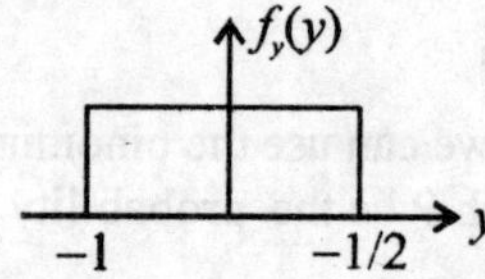

Uniform distution of x, y $[-1, 1]$;

$$P\left(\max(X,Y) \le \frac{1}{2}\right) = \begin{bmatrix} P\left(X = \frac{1}{2}, -1 \le Y \le \frac{1}{2}\right) \\ \cdot P\left(-1 \le X \le \frac{1}{2}, Y = \frac{1}{2}\right) \end{bmatrix}$$

$$= \int_{-1}^{1/2} \frac{1}{2} dx \int_{-1}^{1/2} \frac{1}{2} dy$$

$$= \frac{3}{4} \times \frac{3}{4} = \frac{9}{16}$$

[Since X, Y are independent random variable

$$\int_{-\infty}^{\infty} \int_{-\infty}^{\infty} f_{XY}(x, y)\, dx\, dy = P(-\infty < X < \infty,\ -\infty < Y < \infty)$$

$$= \int_{-\infty}^{\infty} f_X(x)\, dx \int_{-\infty}^{\infty} f_Y(y)\, dy$$

$$P\left(|X| \le \frac{1}{2}\right) P\left(|Y| \le \frac{1}{2}\right) = P\left(|X\ \ Y| \le \frac{1}{2}\right)$$

$$P(X\ \ Y) = P(X)P(Y)$$

X and Y being independent random variable]

Second Method

The range of $-1 \le x \le 1$ and $-1 \le y \le 1$ is entire rectangle.

The region in which maximum of $\{x, y\}$ is less than 1/2 is shown below as shaded region inside this rectangle.

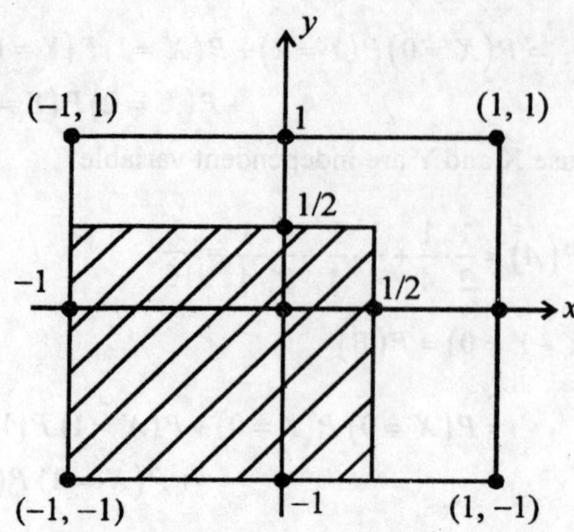

$$p\left(\max(x, y) < \frac{1}{2}\right) = \frac{\text{Avea of shaded region}}{\text{Avea of entire rectangle}}$$

$$= \frac{\frac{3}{2} \times \frac{3}{2}}{2 \times 2} = \frac{9}{16}$$

17. Ans.(c)

Solution:

P(odd tosses) = $P(H) + P(TTH) + P(TTTTH) + \ldots$

$$= \frac{1}{2} + \left(\frac{1}{2}\right)^3 + \left(\frac{1}{2}\right)^5 + \ldots = \frac{1}{2} + \left[1 + \left(\frac{1}{2}\right)^2 + \left(\frac{1}{2}\right)^4 + \ldots\right]$$

$$= \frac{1}{2} + \left[1 + \left(\frac{1}{4}\right) + \left(\frac{1}{4}\right)^2 + \ldots\right] = \frac{1}{2} + \left(\frac{1}{1 - \frac{1}{4}}\right)$$

$$= \frac{1}{2} \times \frac{4}{3} = \frac{2}{3}$$

18. Ans. (b)

Solution:

As the given function curve is symmetric

19. Ans. (a)

Solution:

Variance $(x) = \dfrac{(b-a)^2}{12}$ for $a < x < b$

$$= \frac{(10-2)^2}{12} = \frac{8^2}{12} = \frac{64}{12}$$

$$= \frac{16}{3}$$

20. Ans. (c)

Solution:

Given

Probability of getting positive report $= 0.01$

Probability of getting negative report $(1-0.01) = 0.99$

Probability of getting incorrect report when it is test positive or negative

$$= (0.01)(0.12) + (0.99)(0.15)$$

$$= \frac{12}{10000} + \frac{99 \times 15}{10000}$$

$$= 0.1497$$

21. Ans. (d)

Solution: We know that

Odd × odd = odd

even × even = even

odd × even = even

even × even = even

3, 6, 9, 12, 15	6, 11, 16, 21, 26
Box - 1	Box - 2
P_1 (Even) 2/5	P_2(Even) = 3/5
P_1 (odd) = 3/5	P_2 I(odd) = 2/5

P (Even) = P_1(E) P_2 (E) + P_1 (E) P_2 (0) + P_2(E) P_1(0)

$$\Rightarrow \frac{2}{5} \times \frac{3}{5} + \frac{2}{5} \times \frac{2}{5} + \frac{3}{5} \times \frac{3}{5}$$

$$= \frac{19}{25}$$

Second Method:

$$P_{even} = 1 - P(odd)$$

$$= 1 - \frac{2}{5} \times \frac{3}{5}$$

$$1 - \frac{6}{25} = \frac{19}{25}$$

22. Ans (c)

Solution:

Given that there are 4 white and 3 red balls in a box

Two balls are re-moved in succession, the first ball is white.

After revmoving first ball there are 3 white and 3 red balls.

Number of ways to remove red ball $= {}^3C_1$

Total number of ways to remove one ball $= {}^6C_1$

The probability that 2nd ball is red $= \frac{{}^3C_1}{{}^6C_1} = \frac{3}{6} = \frac{1}{2}$

23. Ans (b)

Solution:

Given there are N-people

$M_1, M_2, \ldots M_N$

All one born in the month of April.

The probability that at least two people born on the same date.

Let the two people born on 1st April.

Either [M_1 and M_2 OR M_2 and M_3 … OR M_1 and M_N] and on similar cases

The probability that a person born on a particular date

$$= \frac{1}{360}$$

If $N \rightarrow$ Natural number

So the probability that two people born on 1st April

$$= \left[\frac{1}{30} \times \frac{1}{30} + \frac{1}{30} \times \frac{1}{30} + \ldots + \frac{1}{30} \times \frac{1}{30} (N-1) \text{times}\right]$$

$$+ \left[\frac{1}{30} \times \frac{1}{30} + \ldots + \frac{1}{30} \times \frac{1}{30} (N-2) \text{times}\right]$$

$$\vdots$$

$$+ \left[\frac{1}{30} \times \frac{1}{30}\right] \text{once}$$

$$= \frac{1}{(30)^2}\left[(N-1) + (N-2) + \ldots + 3 + 2 + 1\right]$$

$$= \frac{1}{900}\left[\text{Sum of } (N-1)\right]$$

$$= \frac{1}{900} \frac{N(N-1)}{2}$$

$$= \frac{N(N-1)}{1800}$$

Similarly the probability that 2-peoples born on 2nd April

$$= \frac{N(N-1)}{1800}$$

and so on a for each date

The probability that 2-people born on any data of April

$$= 30 \times \frac{N(N-1)}{1800}$$

$$= \frac{N(N-1)}{60}$$

For $\frac{N(N-1)}{60} > 0.5$

or $N(N-1) > 30$

or $N^2 - N - 30 > 0$

$(N-6)(N+5) > 0$

or $N > 6$

or $N = 7$ at least.

Second Method

Let, $N = 2$

Probability that none of them born on the same day

$= 1 \times \frac{29}{30}$

Required probability $= 1 - \frac{29}{30}$

$= \frac{1}{30}$ is not greater than 0.5

Let, $N = 3$

Required probability $= 1 - \frac{29}{30} \times \frac{28}{30} = 0.098$

Similary $N = 7$

Required probability $= 1 - \frac{29}{30} \times \frac{28}{30} \times \frac{27}{30} \times \frac{26}{30} \times \frac{25}{30} \times \frac{24}{30}$

$= 0.53 > 0.5$

24. Ans. (c)

Solution:

Give $f(x) = \begin{matrix} 1 & 0 < x < 1 \\ 0 & \text{otherwise} \end{matrix}$

$E(X) = \int x dx$

$E(X)^3 = \int x^3 dx = \left.\frac{x^4}{4}\right|_0^1 = \frac{1}{4}$

25. Ans. (c)

Solution:

Since population is finite, sampling is done with replacement and so probability does not change from trial to trial. This problem can is to be solved by binormal distribution, as

Here, $n = 2$

$x = 0$ (no defective)

$p = p(\text{defective}) = \frac{3}{10}$

So, $p(x = 0) = \left(1 - \frac{3}{10}\right)^2$

$= 0.49 = 49\%$

Therefore, p (none of 2 screws is defective)

$= \frac{3C_0 \times 7C_2}{10C_2} \times 100\%$

$= \frac{7}{15} \times 100\%$

$= 46.6 \simeq 47\%$

26. Ans. (b)

Solution:

Since population is infinite. This problem can be done using binomial distrubution.

Probability of defective item,

$p = 0.1$

Probability of non-defective item,

$q = 1 - p$

$= 1 - 0.1 = 0.9$

Probability that exactly 2 of the chosen items are defective

$= {}^{10}C_2 (p)^2 (q)^8$

$= {}^{10}C_2 (0.1)^2 (0.9)^8 = 0.1937$

27. Ans. (a)

Solution:

Since this problem involves coins. Binomial distribution is used.

$p = p(H) = 0.5$

Probability of getting head exactly 3 times is

$p(X = 3) = {}^4C_3 (0.5)^3 (0.5)^1 = 1/4$

28. Ans. (d)

Solution:

Since this problem involves coins. Binomial distribution is used.

Here, $n = 3$

$p = p(H) = 1/2$

Now, $p(x \geq 1) = 1 - p(x = 0)$

$= 1 - 3C_0 \left(\frac{1}{2}\right)^0 \left(1 - \frac{1}{2}\right)^3$

$= 1 - \frac{1}{8} = \frac{7}{8}$

29. Ans. (d)

Solution:

Use hypergeometric distribution

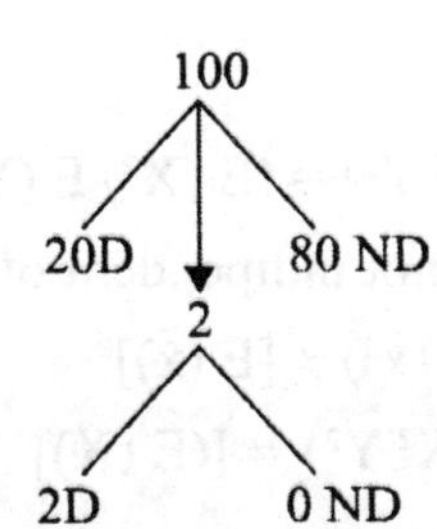

$$p(X=2)=\frac{20C_2\times 80C_0}{100C_2}=\frac{19}{495}$$

30. Ans. (d)

Solution:

For die sample space = $(6)^2 = 36$

Total ways in which sum is either 8 or 9 is

(2, 6), (3, 5), (3, 6) (4, 4) (4, 5) (5, 3), (5, 4) (6, 2) (6, 3) = 9 ways

Therefore, Probability of coming 8 or 9 $=\frac{9}{36}=\frac{1}{4}$

So probability of not coming 8 or 9 $=1-\frac{1}{4}=\frac{3}{4}$

31. Ans. (c)

Solution:

As given all three gates are independent

p (gate 2 and gate 3 fail when gate 1 already failed)

= p (gate 2 and gate 3 fail)

[As gate 2 and 3 fail independently]

= p (gate 2) × p (gate 3)

= 0.2 × 0.2 = 0.04

32. Ans. (a)

Solution:

Given data are,

p(private car) = 0.45

p(bus public transport) = 0.55

A person has a choice between private car and public transport

p(public transport) = 1 – p(private car) = 1 – 0.45

= 0.55

p(bus) = p(bus ∩ public transport) × p(public transport)

= 0.55 × 0.55 = 0.3025 ≈ 0.30

p(metro) = 0.45

Therefore p(bus) = 0.30 and

p(metro) = 0.25

33. Ans. (c)

Solution:

Given data are,

Box contains 2 washers, 3 nuts and 4 bolts

Total no. of i item = 9

p(2 washers, then 3 nuts, then 4 bolts)

$$=\left(\frac{2}{9}\times\frac{1}{8}\right)\times\left(\frac{3}{7}\times\frac{2}{6}\times\frac{1}{5}\right)\times\left(\frac{4}{4}\times\frac{3}{3}\times\frac{2}{2}\times\frac{1}{1}\right)=\frac{1}{1260}$$

34. Ans. (c)

Solution:

The coefficient of variation in speed

$$=\frac{\sigma}{\mu}=\frac{8.8}{33}=0.2666$$

35. Ans. (b)

Solution:

Population is finite, hypergeometric distribution can be applied as.

$$p\,(1 \text{ defective in 5 calculators})=\frac{2C_1\times 23C_4}{25C_5}=\frac{1}{3}$$

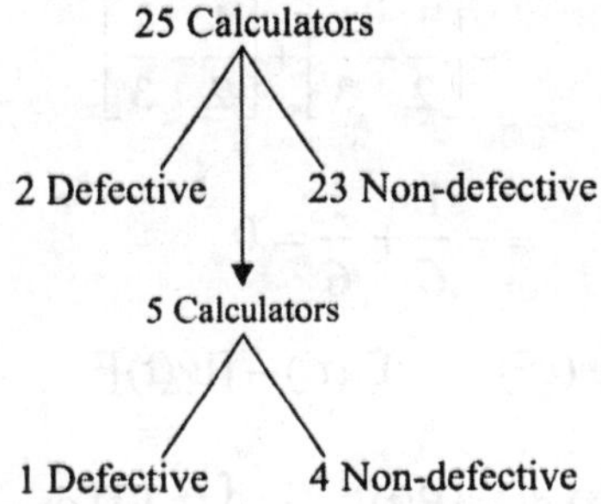

36. Ans. (d)

Solution:

Probability of drawing two red balls

= p (first is red) × p (second is red)

$$=\frac{5}{10}\times\frac{4}{9}=\frac{2}{9}$$

37. Ans. (d)

Solution:

Hypergeometric distribution can be applied as:

$$p(X=2)=\frac{4C_2\times 48C_0}{52C_2}=\frac{1}{221}$$

38. Ans. (b)

Solution:

$$P\left(-\frac{1}{3} \le x \le \frac{1}{3}\right) = \int_{-\frac{1}{3}}^{\frac{1}{3}} f(x)\,dx = \int_{-\frac{1}{3}}^{\frac{1}{3}} x^2\,dx$$

$$= \left[\frac{x^3}{3}\right]_{-\frac{1}{3}}^{\frac{1}{3}} = \frac{2}{81}$$

The probability in percentage,

$$= \frac{2}{81} \times 100 = 2.469\% = 2.47\%$$

39. Ans. (b)

Solution:

$$\text{Mean } E(t) = \int_{-\infty}^{\infty} t \cdot f(t) \cdot dt = \int_{-1}^{1} t \cdot f(t) \cdot dt$$

$$= \int_{-1}^{0} t(1+t)\,dt + \int_{0}^{1} t(1-t)\,dt$$

$$= \left[\frac{t^2}{2} + \frac{t^3}{3}\right]_{-1}^{0} + \left[\frac{t^2}{2} - \frac{t^3}{3}\right]_{0}^{1}$$

$$= -\left[\frac{1}{2} - \frac{1}{3}\right] + \left[\frac{1}{2} - \frac{1}{3}\right]$$

$$= -\frac{1}{6} + \frac{1}{6} = 0$$

$$\text{Variance } (\sigma^2) = E(t^2) - [E(t)]^2$$

$$= \int_{-\infty}^{\infty} t^2\, f(t)\,dt - [E(t)]^2 = \int_{-\infty}^{\infty} t^2\, f(t)\,dt - (0)^2$$

$$= \int_{-\infty}^{\infty} t^2\, f(t)\,dt = \int_{-1}^{0} t^2\,(1+t)\,dt + \int_{0}^{1} t^2\,(t-1)\,dt$$

$$= \int_{-1}^{0} (t^2 + t^3) \cdot dt + \int_{0}^{1} t^2\,(-t^3)\,dt$$

$$= \left[\frac{t^3}{3} + \frac{t^4}{4}\right]_{-1}^{0} + \left[\frac{t^3}{3} - \frac{t^4}{4}\right]_{0}^{1} = -\left[\frac{-1}{3} + \frac{1}{4}\right] + \left[\frac{1}{3} - \frac{1}{4}\right]$$

$$= \frac{1}{12} + \frac{1}{12} = \frac{1}{6}$$

$$\text{Standard deviation } (\sigma) = \sqrt{\text{variance}} = \frac{1}{\sqrt{6}}$$

40. Ans. (d)

Solution:

$$E(X^2 Y^2) = (E(X^2)\, E(Y^2))$$

But since x is not indipendent of x,

$$E(x^2) \ne [E(x)]^2$$

Therefore, $E(X^2 Y^2) \ne [(E(X)]^2\, [(E(Y)]^2$

41. Ans. (a)

Solution:

$$\sigma = \sqrt{\frac{(\beta-\alpha)^2}{12}} = \sqrt{\frac{(1-0)^2}{12}} = \frac{1}{\sqrt{12}}$$

42. Ans. (d)

Solution:

Let μ_c and σ_c are mean and standard deviation of the students of batch C respectively, μ and σ mean and standarrd deviation of entire class of first year students respectively.

Given, $\mu_c = 6.6$

$\sigma_c = 2.3$

$\mu = 5.5$

$\sigma = 4.2$

In order to normalise batch C to entire class, the normalised score (Z scores) must be equated.

since $$Z = \frac{x-\mu}{\sigma} = \frac{x-5.5}{4.2} \quad (1)$$

$$Z_c = \frac{x_c - \mu_c}{\sigma_c} = \frac{8.8-6.6}{2.3} \quad (2)$$

Equate equation (1) and (2) and solve it

$$\frac{8.5-6.6}{2.3} = \frac{x-5.5}{4.2}$$

Or $x = 8.969 \simeq 9.0$

11.6 Numerical Methods

1. Ans. (c)

Solution: The correct match is given below:

Newton-Raphson method	Solving nonlinear equations
Ruge – kutta Method	Solving ordinarydifferential equations
Simpson's Rule Gauss elimination	Numerical integration Solving linearsimultaneous equations

2. Ans. (b)

Solution: $f(x) = x^3 - x^2 + 4x + 4$

$$f'(x) = 3x^2 - 2x + 4$$
$$f(2) = 8 - 4 + 8 - 4 = 8$$
$$f'(2) = 12 - 4 + 4 = 12$$
$$x_{n+1} = x_n - \frac{f(x_n)}{f'(x_n)}$$
$$x_{n+1} = 2 - \frac{8}{12}$$
$$= 2 - (2/3) = 4/3$$

3. Ans (c)

Solution:

$$P(x) = x^5 + x + 2$$

$x = -1$ is one and the root.

So $$\begin{bmatrix} (x+1)(x^2 - 1.947x + 1.567) \\ (x^2 + 0.947x + 1.276) \end{bmatrix} = 0$$

Real root

$$x = -1$$

Complex roots

$$x = 0.973 + 0.7871j$$
$$x = 0.973 - 0.7871j$$
$$x = -0.473 + 1.0261j$$
$$x = -0.473 - 1.0261j$$

1 Real , 4 complex

4. Ans. (d)

Solution:

Given differential equation is

$$\frac{dx}{dt} = \frac{1-x}{\tau}$$

Using Euler's numerical integration with step $\Delta T > 0$.

For stability analysis using Euler's methods i.e.

$$\delta t \le \left| \frac{2}{\partial f / \partial x} \right|$$

Let Equation $f(x,t) = \dfrac{dx}{dt} = \dfrac{1-x}{\tau}$

As $$f = \frac{1-x}{\tau}$$

or, $$\frac{\partial f}{\partial t} = \left| \frac{-1}{\tau} \right|$$

so, $$\Delta T \le \frac{2}{-1/\tau}$$

or, $\Delta T \le (-2\tau)$

On Magnitude basis $\Delta T \le 2\tau$

$\therefore$ maximum possible step size for stability is

5. Ans.(c)

Solution: $f(x_n) = x_n e^{-x_n}$

$$f'(x_n) = 1 + e^{-x_n}$$
$$x_{n+1} = x_n - \frac{x_n - e^{-x_n}}{1 + e^{-x_n}}$$
$$= \frac{x_n + x_n e^{-x_n} - x_n + e^{-x_n}}{1 + e^{-x_n}} = \frac{(x_n + 1)e^{-x_n}}{1 + e^{-x_n}}$$

6. Ans. (b)

Solution:

Given that

$$\frac{dy(x)}{dx} - y(x) = x$$

$y(0) = 0$ and $h = 0.1$

$$x_1 = 0$$
$$y_1 = 0$$
$$\therefore \left(\frac{dy}{dx}\right) = 0 - 0 = 0$$
$$y_2 = y_1 + h\left(\frac{dy}{dx}\right)_1$$
$$= 0 + 0.1 \times 0 = 0$$
$$x_2 = x_1 + h = 0.1$$
$$y_2 = 0$$
$$= x_2 + y_2(x)$$
$$\therefore \left(\frac{dy}{dx}\right)_2 = x_2 + y_2(x)$$
$$= 0.1 + 0 = 0.1$$
$$y_3 = y_2 + h\left(\frac{dy}{dx}\right)_2 = 0 + 0.1 \times 0.1 = 0.01$$
$$x_3 = x_2 + h = 0.2$$
$$y_3 = 0.01$$
$$\therefore \left(\frac{dy}{dx}\right)_3 = x_3 + y_3(x)$$
$$= 0.2 + 0.01$$
$$= 0.21$$

$$\therefore \quad y_4 = y_3 + h\left(\frac{dy}{dy}\right)$$

$$= 0.01 + 0.1 \times 0.21$$
$$= 0.01 + 0.021$$
$$= 0.031$$

$$x_4 = x_3 + h = 0.3$$

$$y_4 = 0.31$$

$$\therefore \quad y(0.3) = 0.031$$

7. Ans. (c)

Solution: $X_{n+1} = X_n - \dfrac{f(X_n)}{f'(X_n)}$

$$f(2) = (2+\sqrt{2}-3) = \sqrt{2}-1$$

and $$f'(2) = 1 + \frac{1}{2\sqrt{2}} = \frac{2\sqrt{2}+1}{2\sqrt{2}}$$

$$X_{n+1} = 2 - \frac{(\sqrt{2}-1)}{\dfrac{x\sqrt{2}+1}{2\sqrt{2}}} = 1.694$$

8. Ans (c)

Solution:

$x^3 - 6x^2 + 11x - 6 = 0$

Two roots are 1,3

means $(x-1)(x-3)$ are two factor

$x^2 - x - 3x + 3$

$x^2 - 4x + 3$

So third root will be

$$x^2 - 4x + 3)\,x^3 - 6x^2 + 11x - 6\,(x-2$$
$$\underline{x^3 - 4x^2 + 3x}$$
$$-2x^3 + 8x - 6$$
$$\underline{-2x^3 + 8x - 6}$$
$$X$$

means

$(x-2) = 0$

$x = 2$

9. Ans. (a)

Solution:

Given $f(x) = x^2 - 117 = 0$

or, $f'(x) = 2x$

Newton-Raphson's formula is given as

$$x_{k+1} = x_k - \frac{f(x_k)}{f'(x_k)}$$

$$x_{k+1} = x_k - \frac{(x_k^2 - 117)}{2x_k}$$

$$= \frac{2x_k^2 - x_k^2 + 117}{2x_k}$$

$$= \frac{x_k^2 + 117}{2x_k}$$

$$x_{k+1} = \frac{1}{2}\left(x_k + \frac{117}{x_k}\right)$$

10. Ans. (a)

Solution:

Given $f(x) = e^x - 1 = 0$ and Initial solution $x_o = -1$

or $e^x = -1$

$f'(x) = e^x$

Using Newtron-Raphson formula, the $(n+1)^{th}$ iteration is

$$x_{n+1} = x_n - \frac{f(x_n)}{f'(x_n)}$$

$$= x_n - \frac{e^{x_n} - 1}{e^{x_n}}$$

$$= \frac{x_n e^{x_n} - e^{x_n} + 1}{e^{x_n}}$$

$$\therefore \quad x_1 = \frac{x_o e^{x^0} - e^{x^0} + 1}{e^{x^0}}$$

$$= x_o - 1 + e^{-x_o}$$

$$= -1 - 1 + e^{+1}$$

$$= e - 2$$

$$= 0.71828$$

11. Ans. (d)

Solution:

Here quadrature mean fifth order, Hance Accuracy = O(h^5).

12. Ans. (c)

Solution:

Given

$$\frac{dy}{dx} = 0.25y^2 \qquad (y = 1 \text{ at } x = 0)$$

$h = 1$

Iterative equation for backward (implicit) Euler's methods is

$$y_{k+1} = y_k + h\, f(x_{k+1}, y_{k+1})$$

$$y_{k+1} = y_k + h \times 0.25\, y^2_{k+1}$$

or

$$0.25h\, y^2_{k+1} - y_{k+1} + y_k = 0$$

put $k = 0$ in above equation

$$2.25\, y_1^2 - y_1 + y_0 = 0$$

or $$y_1 = \frac{1 \pm \sqrt{1-1}}{2 \times 0.25} = 2$$

or $$y_1 = 2$$

13. Ans. (a)

Solution:

The given equations are

$$5x + y + 2z = 34$$
$$0x + 4y - 3z = 12$$
$$10x - 2y + z = -4$$

The augmented maxtrix is $\left[\begin{array}{ccc|c} 5 & 1 & 2 & 34 \\ 0 & 4 & -3 & 12 \\ 10 & -2 & 1 & -4 \end{array}\right]$

Since in the first column maximum element in absolute value is 10, exchange row 1 with row 3.

$$\left[\begin{array}{ccc|c} 5 & 1 & 2 & 34 \\ 0 & 4 & -3 & 12 \\ 10 & -2 & 1 & -4 \end{array}\right] \xrightarrow{R(1,3)} \left[\begin{array}{ccc|c} 10 & -2 & 1 & -4 \\ 0 & 4 & -3 & 12 \\ 5 & 1 & 2 & 34 \end{array}\right]$$

So the pivot for eliminating x is $a_{11} = 10$.

Now to eliminate y, compare the elements in second column at and below the diagonal element.

Since, $a_{22} = 4$ is already larger in absolute value compares to $a_{32} = 1$

$\therefore$ The pivot element for eliminating y is $a_{22} = 4$

14. Ans. (a)

Solution:

Since 5 is a root, f(x) is divisible by $x - 5$.

$$x - 5 \overline{)\, x^2 - 10x^2 + 31x - 30\,} \left(x^2 - 5x + 6 \right.$$

$$\begin{array}{r} x^2 - 5x \\ \hline -5x^2 + 31x - 30 \\ -5x^2 + 25x \\ \hline 6x - 30 \\ 6x - 30 \\ \hline 0 \end{array}$$

Therefore $(x - 5)(x^2 - 5x + 6) = 0$

Roots of $x^2 - 5x + 6$ are 2 and 3.

15. Ans. (c)

Solution:

Newton–Rephson formula is

$$x_1 = x_0 - \frac{f(x_0)}{f'(x_0)} \qquad \ldots (1)$$

Given $$f(x) = x^3 + 3x - 7$$
$$f'(x) = 3x^2 + 3$$

Put $$x_0 = 1$$
$$f(x_0) = f(1) = (1)^3 + 3 \times (1) - 7 = -3$$
$$f'(x_0) = f'(1) = 3 \times (1)^2 + 3 = 6$$

Subsite x_0, $f(x_0)$ and $f'(x_0)$ values into (1)

$$x_1 = 1 - \left(\frac{-3}{6}\right) \times 1 = 1.5$$

16. Ans. (a)

Solution:

Simpson's formula is

$$I = \frac{h}{3}\,[f_0 + 4(f_1 + f_3 + f_5 + \ldots + y_{n-1}) + 2(f_2 + f_4 + f_6 \ldots + y_{n-2})] + y_n$$

$$I = \frac{1}{3} h\,(f_0 + 4f_1 + 2f_2 + 4f_3 + f_4)$$

$$= \frac{1}{3} \times 0.25\,(1 + 4 \times 0.9412 + 2 \times 0.8 + 4 \times 0.64 + 0.5)$$

$$= 0.7854$$

17. Ans. (a)

Solution:

Given

f(x) = 1, 4 15 at x = 0, 1 and 2 respectively

Trapezoidal rule

$$\int_{x_0}^{2_y} y\,dx = \frac{h}{2}[y_0 + 2(y_1 + y_2 + \ldots y_{n-1}) + y_n]$$

$$\int_0^2 f(x)\,dx = \frac{h}{2}(f_1 + 2f_2 + f_3)$$

(3 point Trapezoidal Rule)

Here $h = 1$

Therefore $\int_0^2 f(x)\,dx = \frac{1}{2}(1 + 2 \times 4 + 15) = 12$

Since f(x) is second degree polynomial, let

$$f(x) = a_0 + a_1x + a_2x^2$$

$$f(0) = 1$$

or $a_0 + 0 + 0 = 1$

or $a_0 = 1$

$$f(1) = 4$$

or $a_0 + a_1 + a_2 = 4$

or $1 + a_1 + a_2 = 4$

or $a_1 + a_2 = 3$ ...(1)

$$f(2) = 15$$

or $a_0 + 2a_1 + 4a_2 = 15$

or $1 + 2a_1 + 4a_2 = 15$

or $2a_1 + 4a_2 = 14$ ...(2)

Solving (1) and (2) $a_1 = -1$ and $a_2 = 4$

$$f(x) = 1 - x + 4x^2$$

Now, exact value of $\int_0^2 f(x)\,dx = \int_0^2 (1 - x + 4x^2)\,dx$

$$= \left[x - \frac{x^2}{2} + \frac{4x^3}{3}\right]_0^2 = \frac{32}{3}$$

Error = Exact – Approximate vaule

$$= \frac{32}{3} - 12 = -\frac{4}{3}$$

18. Ans. (a)

Solution:

$$h = \frac{2\pi - 0}{8} = \frac{\pi}{4}$$

$$y_0 = \sin(0) = 0$$

$$y_1 = \sin\left(\frac{\pi}{4}\right) = 0.70710$$

$$y_2 = \sin\left(\frac{\pi}{2}\right) = 1$$

$$y_3 = \sin\left(\frac{3\pi}{2}\right) = 0.70710$$

$$y_4 = \sin(\pi) = 0$$

$$y_5 = \sin\left(\frac{5\pi}{4}\right) = -0.70710$$

$$y_6 = \sin\left(\frac{6\pi}{4}\right) = -1$$

$$y_7 = \sin\left(\frac{7\pi}{4}\right) = -0.70710$$

$$y_8 = \sin\left(\frac{8\pi}{4}\right) = -0$$

By Trapezoidal rule

$$\int_{x_0}^{x_0+nh} f(x)\cdot dx = \frac{h}{2}[y_0 + y_n + 2(y_1 + y_2 + \ldots + y_{n-1})]$$

$$\int_0^{2\pi} \sin x \cdot dx = \frac{\pi}{8} \times [(0+0) + 2(0.70710 + 1 + 0.70710$$

$$+ 0 - 0.70710 - 0.70710)]$$

$$= 0.00000$$

19. Ans. (b)

Solution:

$$\text{Flywheel energy} = \int_0^{2\pi} T(\theta)\,d\theta$$

where T(θ) is torque exerted.

The integral by using simpson's rule is

$$I = \frac{h}{3}(f_0 + 4f_1 + 2f_2 + 4f_3 + 2f_4 + 4f_5 + f_6)$$

$$h = 60 \text{ degrees} = \frac{\pi}{3} \text{ radious}$$

$$I = \frac{1}{3} \times \frac{\pi}{3} \times [0 + 4 \times 1066 + 2(-323) + 4(0)$$

$$+ 2(323) + 4(-355) + 0]$$

$$= 993$$

11.7 Laplace Transforms

1. Ans. (c)

Solution:

$$I(s) = \frac{2}{s(1+s)}, \text{ By final value theorem}$$

$$\lim_{t\to\infty} i(t) = \lim_{s\to 0} sI(s) = 2$$

2. Ans. (a)

Solution: $e^{(a+2t)+5} = e^5 \cdot e^{(a+2)t}$

$\therefore$ Laplace transform exists if Rs (s) > a + 2

Condition for LT to exist

$$\int_{-\infty}^{\infty} |f(t)\, e^{-st}|\, dt < \infty$$

3. Ans. (a)

Solution: $\frac{dx}{dt} + 2x = \delta(t)$

Taking Laplace Transform both side

$s X(s) - x(0) + 2X(s) = 1$

assume $x(0) = 0$

$$X(s)\frac{1}{s+2}$$

Taking inverse Laplace $x(t) = e^{-2t}\, u(t)$

4. Ans (b)

Solution:

For both sided,

The laplace transform of $\int_{-\infty}^{t} f(\tau)\, dz$

$$= \frac{1}{s}\left[F(s) - f^{-1}(0)\right]$$

Since it is one sided, take initial conditions equal to zero

$$\int_0^t f(\tau)\, d\tau = \frac{1}{s} F(s)$$

5. Ans.(d)

Solution:

If $f(t) \leftrightarrow F(s)$,

$$LT[f(t)] = F(s)$$

Then $tf(t) \leftrightarrow -\frac{d}{ds} F(s)$

$$LT[tf(t)] = (-1)\frac{dF(s)}{ds}$$

Thus if $F(s) = \frac{1}{s^2+s+1}$

$$tf(t) \to -\frac{d}{ds}\left(\frac{1}{s^2+s+1}\right)$$

$$= (-1)\left[\frac{-(2s+1)}{(s^2+s+1)^2}\right] = \frac{2s+1}{(s^2+s+1)^2}$$

6. Ans. (c)

Solution:

$$x(n) = \left[\left(\frac{1}{3}\right)^{|n|} - \left(\frac{1}{2}\right)^n\right] u(n)$$

$$= \left(\frac{1}{3}\right)^n u(n) + \left(\frac{1}{3}\right)^{-n} u(-n-1) - \left(\frac{1}{2}\right)^n u(n)$$

$$x(n) = \left(\frac{1}{3}\right)^n u(n) + (3)^n u(-n-1) - \left(\frac{1}{2}\right)^n u(n)$$

As known $a^{|n|}\, u[n]$

$$= a^n\, u[n] + a^{-n} u[-n-1]$$

$$= a^n\, u[n] \to \frac{1}{1 - a z^{-1}},\ ROC|z| > |a|$$

and $a^n u[-n-1] \to \frac{a^{-1} z}{1 - a^{-1} z}$

$$\to \frac{z}{a-z} ROC,\ |z| < a$$

$$\left(\frac{1}{3}\right)^n u(n) \to ROC: |z| > \left|\frac{1}{3}\right|$$

$$3^n u(-n-1)\ ROC: |z| < |3|$$

$$\left(\frac{1}{2}\right)^n u(n)\ ROC: |z| > \left|\frac{1}{2}\right|$$

Common ROC: $\frac{1}{2} < |z| < 3$

7. Ans (c)

Solution:

$$f(t) = u(t-\tau)$$

$$F(s) = ?$$

If $\quad f(t-\tau) = F(s) e^{-\tau s}$

If $\quad f(t) = u(t)$ Then $F(s) = \frac{1}{s}$

$$FT[u(t-\tau)] = \frac{1}{s} e^{-\tau s}$$

8. Ans. (c)

Solution:

$$LT^{-1}\left(\frac{1}{s^2+s}\right) = ?$$

$$\frac{1}{s^2+s} = \frac{1}{s(s^2+1)} = \frac{1}{s} - \frac{1}{s+1}$$

$$LT^{-1}\left(\frac{1}{s^2+s}\right) = LT^{-1}\left(\frac{1}{s}\right) - L^{-1}\left(\frac{1}{s+1}\right)$$

$$= 1 - e^{-t}$$

9. Ans. (a)

Solution:

$$LT\left[\int_0^t\int_0^t \ldots \int_0^t f(t)\, dt^n\right] = \frac{1}{s^n}F(s)$$

Given n = 1

$$LT\left[\int_0^t f(\tau)\, d\tau\right] = \frac{1}{s}F(s)$$

10. Ans. (b)

Solution:

$$LT(\sin mt) = \frac{m}{(s^2+m^2)} = F(s),$$

$$LT\left(\frac{\sin mt}{t}\right) = \int_s^\infty F(s)\, ds = \int_s^\infty \frac{mds}{s_2+m_2} = \left|\tan^{-1}\frac{s}{m}\right|_s^\infty$$

or by Definition,

$$\int_0^\infty e^{-st}\frac{\sin mt}{t}\, dt = \frac{\pi}{2}\tan^{-1}\frac{s}{m}$$

Now $\underset{s\to 0}{Lt}\tan^{-1}\left(\frac{s}{m}\right) = 0$ if $m > 0$ or π if $m < 0$.

Thus taking limits as $s \to 0$, we get

$$\int_0^\infty \frac{\sin mt}{t}\, dt = \frac{\pi}{2} \text{ if } m > 0 \text{ or} -\frac{\pi}{2} \text{ if } m < 0.$$

Here = 1 which is > 0 therefore the answer is $\frac{\pi}{2}$.

11. Ans. (a)

Solution:

$$f(t) = LT^{-1}\left[\frac{1}{s^2(s+1)}\right]$$

$$\frac{1}{s^2(s+1)} = \frac{A}{s} + \frac{B}{s^2} + \frac{C}{s+1}$$

$$\frac{1}{s^2(s+1)} = \frac{As(s+1)+B(s+1)+C(s^2)}{s^2(s+1)}$$

Matching coefficient of s^2, s and constant in numerator,

$A + C = 0$ ...(1)

$A + B = 0$ ...(2)

$B = 1$ ...(3)

On solving $A = -1, B = 1, C = 1$

So, $$f(t) = LT^{-1}\left[\frac{-1}{s} + \frac{1}{s^2} + \frac{1}{s+1}\right]$$

$$= -1 + t + e^{-1}$$

$$= t - 1 + e^{-t}$$

12. Ans. (d)

Solution:

$$L[u(t-a)] = \int_0^\infty e^{-st}\, u(t-a)\, dt$$

$$= \int_0^a e^{-st}\cdot 0\cdot dt + \int_0^\infty e^{-st}\cdot 1\cdot dt$$

$$= 0 + \int_a^\infty e^{-st} dt$$

$$= \left[\frac{e^{-st}}{-s}\right]_a^\infty = \frac{e^{-as}}{s}$$

13. Ans. (d)

Solution:

Apply final value therom $\lim_{t\to\infty} f(t) = \lim_{s\to 0} s\cdot F(s)$

$$1 = \left.\frac{s(3s+1)}{s^3+4s^2+(k-3)s}\right|_{s\to 0} = \left.\frac{3s+1}{s^2+4s+(k-3)}\right|_{s\to 0}$$

or $$1 = \frac{1}{K-3}$$

or $$K - 3 = 1$$

$$K = 4$$

14. Ans (d)

Solution:

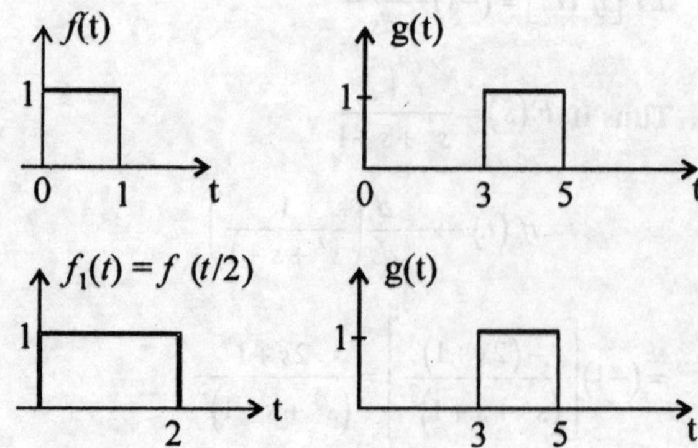

$$g(t) = f_1(t-3)$$

$$= f\left(\frac{t}{2} - \frac{3}{2}\right)$$

15. Ans. (c)

Solution:

$$g(t) = f\left(\frac{t}{2} - \frac{3}{2}\right)$$

$$= f_1(t-3)$$

$$= u(t-3) - u(t-5)$$

$$= \frac{e^{-3s}}{s} - \frac{e^{-5s}}{s}$$

$$= \frac{1}{s}\left(e^{-3s} - e^{-5s}\right)$$

$$= \frac{e^{-3s}}{s}\left(1 - e^{-2s}\right)$$

16. Ans. (d)

Solution:

Given differential equation

$$\frac{d^2y(t)}{dt^2} + \frac{2dy(t)}{dt} + y(t) = \delta(t)$$

and $y(0^-) = -2$

$$y'(0^-) = 0$$

Taking Laplace transform

$$s^2Y(s) - sy(0^-) - y'(0) + 2\left[sY(s) - y(0^-)\right] + Y(s) = 1$$

$$\left(s^2 + 2 + 1\right)Y(s) + 2s + 4 = 1$$

$$Y(s) = \frac{-3-2s}{\left(s^2 + 2s + 1\right)}$$

$$= \frac{-2s-3}{(s+1)^2} = \frac{-1}{(s+1)^2} + \frac{-2}{(s+1)}$$

Taking inverse Laplace transform

$$y(t) = \left[-2e^{-t} - te^{-t}\right]u(t)$$

$$\frac{dy(t)}{dt} = -2e^{-t} \times -1 - \left[te^{-t} \times -1 + e^{-t}\right]$$

$$= -2e^{-t} + te^{-t} - e^{t}$$

$$\left.\frac{dy(t)}{dt}\right|_{t=0^+} = 2 - 1 = 1$$

Bibliography

Communication Systems

1. Simon Haykin, Communication System, (New York, John Wiley & Sons, INC, 2001)
2. Simon Haykin, Digital Communications (New York, John Wiley& Sons, INC, 1999)
3. Herbert Taub & Donald L. Schilling, Principles of Communication System (New Delhi Tata McGraw-Hill Publishing Company Limited, 199)
4. James Martin, Telecommunication and the Computer, (New Delhi Prentice-Hall of India Private Limited, 2000)
5. Thiagarajan Viswanathan , Telecommunication Switching Systems and Networks, (New Delhi Prentice-Hall of India Private Limited, 2000)
6. George Kennedy, Electronic Communication Systems,(New Delhi Tata McGraw-Hill Publishing Company Limited, 1993)
7. George Kennedy, Electronic Communication Systems,(New Delhi Tata McGraw-Hill Publishing Company Limited, 1999)
8. Wayne Tomasi, Electronic Communication Systems, Fundamental Through Advanced. (New Delhi Prentice-Hall of India Private Limited, 2005)
9. B.P. Lathi, Communication Systems, (New Delhi Wiley Eastern Limited, 1968)
10. John C. Hancock The Principles of Communication Theory, (New Delhi Tata McGraw-Hill Publishing Company Limited, 1992)
11. R P Singh & S D Sapre, Communication systems (Analog and digital) (New Delhi Tata McGraw-Hill Publishing Company Limited, 1995)
12. Anokh Singh, Principles of communication Engineering, (New Delhi S. Chand & Company LTD, 1984)
13. R.R. Gulati, Monochrome and Colour Television, (New Delhi New age International (P) Limited, Publishers, 2002)
14. Willian C.Y.Lee Mobile Cellular Telecommunications Analog and Digital Systems, (McGraw-Hill,Inc, 1995)
15. Voice and Data Communications Handbook (New Delhi Tata McGraw-Hill Publishing Company Limited, 2000)
16. Sanjeeva Gupta, Dinesh Arora , Satya Bhushan Sarna, Prashant Lonkar, Microwave Engineering, (Delhi, Khanna Publishers, 2003)
17. Dr. D.C. Agarwal Satellite communications, (Delhi Khanna Publishers, 2002)
18. A. Bruce Carlson, Paul-B.Crilly, Ganted C. Ruat Rutledge Communication Systems (MC Graw International Edition 2002)
19. Athanasois Paopulis, S. Unnikrishna Pillai. Probability, Random Variables, and Stochastic Processes (New Delhi, Tata McGraw-Hill Publishing Company Limited, Edition 2006)
20. G.K Mithal , Radio Engineering Khanna Pulisher, new Delhi 1996
21. John G. Proakis, Digital Communications (New Delhi McGraw Hall International Edition, 2001
22. Dr. Dharma Raj Cheruku Satellite Communication (New Delhi Galgotia Publication Private Limited)
23. Bernard Sklar, 'Digital Communications fundamentals and Appliations' (New Delhi Pearson Education in South Asia)
24. Ian A. Glover & Peter M. Grant 'Digital Communications' (New Delhi, Pearson Education in South Asia)
25. William Stallings 'Wireless Communications and Networks' (New Delhi, Pearson Education in South Asia)
26. Simon Haykin & Michael Moher 'Modern Wireless Communications' (New Delhi, Pearson Education in South Asia)
27. K. Sam Shanmugam 'Digital and Analog communication Systems' (New Delhi, John Wiley & Sons (Asia) Private Limited)
28. R. Gopal 'Problems and Solution in Communication Systems' (New Delhi, CBS Publishers & Distributions)
29. Forrest Barker 'Communication Electronics: Systems, Circuits, and Devices' (New Delhi, Prentice-Hall of India Private Limited)
30. Alberto Leon-Garcia & Indra Widjaja 'Communication Networks Fundamental Concepts and Key Architectures' (New Delhi, Tata McGraw-Hill Publishing Company Limited)
31. J.S. Chitode 'Communications Systems-1' (PUNE Technical Publications)
32. Richard A. Williams 'Communication Systems Analysis and Design: A System approach' (New Delhi, Prentice-Hall of India Private Limited)
33. Behroag A. FORUZAN "Data commnication & Networkding" (New Delhi The Mc-Graw Hill Company)
34. J. Das, Sk Mullick and PK chahagee "Digital communication" (New Delhi, New Age international publication)

Electronic Devices and Circuits

1. Mathur Kulshreshtha Chadha Electronic Devices Applications and integrated Circuits (New Delhi Umesh Publications)
2. Allen Mottershead, electronic Devices and circuits (New Delhi Prentice-Hall of Indial Private Limited, 2004)
3. Jacob Millman and Herbert Taub, Pulse, Digital and Switching Waveforms, (New Delhi McGraw-Hall Book Company, 1965)
4. Jimmie J. Cathey, Schaum's out line series Electronic Devices and Circuits (New Delhi Tata McGraw-Hill Publishing Company Limited)
5. Adel S. Sedra and Kenneth C. Smith, Microelectronic Circuits (New York Oxford University Press, 2004)
6. Ben G. Streetman and Sanjay Banerjee, Solid State Electronic Devices(New Delhi Pearson Aducation, PTE Limited Indian Branch, New Delhi)
7. Jacob Millman, and Arvin Grabel, Microelectronics, (New Delhi Tata McGraw-Hill Publishing Company Limited)

8. Jacob Millman, Ph.D., Christos C.HalKias. Ph.D. Integrated Electronics Analog and Digital Circuit and Systems, (New Delhi Tata McGraw-Hill Publishing Company Limited, 1991)
9. S.M. Sze, Physics of Semiconductor Devices, (New YorkA wiley - Interscience Publication John Wiley & Sons, 2004)
10. Robert L. Boylestad Louis Nashelsky, Electronic Devices and Circuit Theory (New Delhi, Published by Pearson Eduaction (Singapore) Private Limited, Indian Branch, 482 F.I.E. Patparganj, 2003)
11. J. B Gupta, Electronic Devices and Circuits (New Delhi S.K. Kataria & Sons, 2004)
12. B.S. Sonde, Introduction to System design using Integrated circuits (New Delhi, 1999)
13. R. Gopal, Problem and Solution in Integrated electronics (New Delhi, CBS Publisher and Distributor)
14. Ramakant A. Gayakwad, OP-AMPS and Linear Integrated Circuits (New Delhi Dhanpat Rai & Sons, 2006)
15. Samuely Y.Liao microwave Devices and Circuits(New Delhi Prentice-Hall, 2000)
16. P. John Paul Micro Electronics (New Delhi Galgotia Publication Private Limited)
17. K. "Gopal" Gopalan 'Introduction To Digital Microelectronic Circuuts' (New Delhi, Pearson Education in South Asia)
18. Theodore F. Bogart Jr., Jeffrey S. Beasley & Guillermo Rico 'Electronic Devices and Circuits' (New Delhi, Pearson Education in South Asia)
19. Pallab Bhattacharya 'Semiconductor Optoelectronic Devices' (New Delhi, Pearson Education in South Asia)
20. Sanjay Sharma 'Electronics Engineering' (New Delhi, S.K. Kataria & Sons)
21. U.A Bakshi and A.P. Godse 'Analog Integrated Circuits' (Techincal Publications)
22. Donald A Neamen 'Electronic Circuits Analysis and Design' Jacob Millman, and Arvin Grabel, Microelectronics, (New Delhi Tata McGraw-Hill Education Private Limited)
23. Jacob Millman, Christos C. Halkias 'Integrated Electronics Analog and Digital Circuits and Systems' (New Delhi Tata McGraw-Hill Publishing Company Limited)
24. Randall L. Geiger Phillip E. Allen and Noel R strader "VLSI Design technique for analog and Digital circuits" (New Delhi, Tata Mc Graw- Hill)
25. Philips E. Allen Douglas R Holbarg "CMOS analog circuit design" (New Delhi, Oxford university press)

Network/Circuit Theory

1. Joseph A. Edminister, M.S.E, Schaum's out line series Theory and Problems of Electric circuit (New Delhi Tata McGraw-Hill Publishing Company Limited, 1991)
2. M.E. Van Valkenburg, Network Analysis (New Delhi Prentice-Hall of India Private Limited, 1989)
3. Franklin F. Kuo, Network Analysis and Synthesis, (John Wiley & Sons (ASIA) Private Limited)
4. T S K V Iyer, Theory and Problems in Circuit Analysis, (New Delhi Tata McGraw-Hill Publishing Company Limited, 2002)
5. M.L. Soni, M.E. and J.C Gupta, M.S., A course in Electrical Circuit Analysis, (New Delhi Dhanpat Rai & Sons, 1990)
6. Willian H. Hayt, Jr. and Jack E. Kemmerly Engineering Circuit Analysis (New Delhi Tata McGraw-Hill Publishing Company Limited, 2001)
7. D. Roy Choudhary, Network and System (New Delhi, Wiley Eastern Ltd.)
8. A Chakrabarti, Circuit theory (Analysis and Synthesis) Dhanpat Rai 2005
9. A sudhakar, Shyammohan S. Palli, Circuit and Networks. (New Delhi Tata McGraw-Hill Publishing Company Limited, 2007)
10. K.S. Suresh Kumar 'Electric Circuits and Networks' (New Delhi, Pearson Education in South Asia)
11. Vincent Del Toro 'Electrical Engineering Fundaments' (New Delhi, Pearson Education in South Asia)
12. D.P Kothari & I J Nagrath 'Basic Electrical Engineering' (New Delhi, Tata McGraw-Hill Publishing Company Limited)
13. Vincent Del Toro 'Electrical Engineering Fundaments' (New Delhi, Pearson Education in South Asia)
14. Ashfaq Husain 'Network and Systems' (New delhi, Khanna Book Publishing Company Private Limited)
15. Charles A. Desoer and Ernest S. Kuh 'Basic Circuit theory'(New Delhi Tata McGraw-Hill Publishing Company Limited, 2001)
16. CL Wadhwa 'Network Analysis' (New Delhi New Age International (P) Limited, Publishers)
17. B.Subramanyam 'Electric Circuit Analysis' (New Delhi I. K. International Publishing House Private Limited)
18. Gopal G Bhise, Prem R. Chandha and Durgesh C. Kulshreshtaha 'Engineering Network Analysis and Filter design' (New Delhi Umesh Publications)
19. Russell M. Mersereau and Joel R. Jackson 'Circuit Analysis A Systems Approach' (New Delhi Pearson Education Private Limited)
20. K. Channa Venkatesh and D. Ganesh Rao "Network Theory" '(New Delhi Pearson Education Pvt. Ltd.)
21. K. Rajewaran "Electrical Circuit Theory" (New Delhi Pearson Education Private Limited)
22. C.P. Kuria Kose "Circuit Theory" (New Delhi Pearson Education Private Limited)
23. Samarajit Ghosh "Network Theory" (New Delhi Pearson Education Private Limited)
24. Charles K Alexander and Mathew Saidikue "Electrical Circuits" (New Delhi Tata McGraw-Hill Publishing Company Limited)
25. James W. Nilsson and Susan A Riedel "Electrical Circuits" (New Delhi Pearson Education Private Limited)

Control Systems

1. Joseph J. Distefano, III PH.D, Allen R. Stubberud, PH.D and Ivan J. Williams, PH.D. Schaum's Outline of Theory and Problems of Feedback and Control System (New Delhi Tata McGraw-Hill Publishing Company Limited)
2. Benjamin C. Kuo, Automatic control System,(New Delhi Prentice-Hall of India Private Limited, 1990)
3. I.J. Nagrath and G.Gopal, Control System Engineering, (New Delhi New Age International Private Limited, Publishers, 1999)
4. Katsuhiko Ogata, Modern Control Engineering, (New Delhi Prentice-Hall of India, 2000)
5. Dr. S. Das Gupta, Control System Theory (New Delhi Khanna Publishers, 1997
6. M Gopal Digital Control and State Variable Methods (New Delhi Tata McGraw-Hill Publishing Company Limited-2005)
7. Richard C. Dorf & Robert H. Bishop 'Modern Control Systems' (New Delhi, Pearson Education in South Asia)

8. A.K. Jairath 'Problems & Solution of Control Systems' (New Delhi, CBS Publishers & Distributions)
9. Graham C. Goodwin, Stefan F. Graebe & Mario E. Salgado 'Control Systems Design' (New Delhi, Pearson Education in South Asia)
10. Smarajit Ghosh 'Control Systems Theory and Applications' (New Delhi, Pearson Education in South Asia)
11. S.P. Eugene Xavier & Joseph Cyril Babu, J. 'Principles of Control Systems' (New Delhi, S. Chand & Company Limited)
12. Norman S. Nise 'Control Systems Engineering Systems' (New Delhi, John Wiley & Sons (Asia) Private Limited)
13. Dr. Yaduvir Singh & Amit Agrawal 'Control Systems Theory, Problems & Solutions' (New Delhi Galgotia Publication Private Limited)
14. Dr. A. K. Tandan , Dr. A. Subba rao, Parag R. Desai & Dr. S.K. Kulkarni 'Control Engineering' (New Delhi, Dhanpat Rai & Sons)
15. Francis H. Raven 'Automatic Control Engineering' (New Delhi, McGraw-Hill Publishing Company Limited)
16. Dr. H.M. Rai & Dr. S.N. Sivanandam 'Control System Engineering' (New Delhi, Satay prakashan)
17. A.K. Sawhnecy Instrumentation and Process control (New Delhi, Dhanpet Rai and Co.)
18. Santanice Chakraborty "A Text Book of Process control and Contorl Systems" (New Delhi, Dhanpat Rai and Co.)
19. D. Patranbis "Principles of Process Control" (New Delhi, Tata Mc-Graw-Hill)
20. Krishna Kant "Computer Based Industrial control" (New Delhi, PHI Learmy Private Limited)

Digital Electronics

1. Albert Paul Malvino, Ph.D. and Jerald A. Brown, Digital Computer Electronics, (New Delhi Tate McGraw-Hill Publishing Company Limited, 1995)
2. Willian I. Fletcher, An Engineering Approach To Digital Design, (New Delhi Prentice-Hall of India Private Liminted, 2003)
3. M. Morris Mano, Digital Design(New Delhi Tate McGraw-Hill Publishing Company Limited, 1995)
4. V. Rajaraman, Ph. D. and T. Radhakrishnan, Ph. D. An introduction to Digital computer Design, (New Delhi Prentice-Hall of India Private Limited, 1988)
5. Albert Paul Malvind, PH.D. and Donald P. Leach, PH.D, Digital Principal and Applications, (New Delhi Tata McGraw-Hill International Editions, 1987)
6. C.K. Nagpal. Elements of Computer Science, (New Delhi Dhanpat Rai & Sons)
7. Herbert Taub, Donald Schilling Digital Integrated Electronics. (New Delhi Tate McGraw-Hill Internation Editions Singapore, 1977)
8. A. Anand Kumar 'Pulse and Digital Circuits' (New Delhi, Prentice-Hall of India Private Limited, 2008)
9. A. Anand Kumar 'Fundaments of Digital Circuits' (New Delhi, Pearson Education in South Asia)
10. M. Morris Mano & Michael D. Ciletti 'Digital Design' (New Delhi, Pearson Education in South Asia)
11. R P Jain 'Modern Digital Electronics' (New Delhi Tata McGraw-Hill Education Private Limited)
12. Albert Paul Malvino and Donald P. Leach 'Digital Principles' (New Delhi Tata McGraw-Hill Editions)
13. Gopalan 'Introduction to Digital Microelectronic Circuits' (New Delhi Tata McGraw-Hill Editions)
14. I.J Nagrath 'Electronics, Analog and Digital' (New Delhi PHI Learning Private Limited)

Microprocessor

1. Ramesh S. Gaonkar, Microprocessor Architecture, Programming, and Applications with the 8085, (Penram International Publishing (India Private Limited)
2. M. Morris Mano, Computer System Architecture, (New Delhi Prentice-Hall of India Private Limited, 2001)
3. B.Ram, Fundaments of Microprocessor and Microcomputers, (New Delhi Dhanpat Rai Publications Private Limited)
4. Douglas V. Hall, Microprocessors and Interfacing Programming and Hardware, (New Delhi Tata McGraw-Hill Publishing Company Limited, 2000)
5. V. Rajaraman, Analog computation and simulation, (New Delhi Prentice-Hall of India Private Limited, 1989)
6. Muhammad Ali Mazidi, Janice Gillispie Mazidi and Rolin D. McKinlay The 8051 Microcontroller and Embedded Systems (New Delhi Prentice-Hall of India Private Limited, 2006)
7. Kenneth J. Ayala The 8051Microcontroller (Thomson Delmar Learning- Penram International Publishing (I) PVT. LTD. Mumbai 1997)
8. K. A Parthasarathy, A. Ramachandran and R. Purushothaman Computer Architecture (Chennai, THOMSON -Vijay Nicole imprints Private Limited, 2004).
9. Andrew S. Tanenbaum 'Computer Networks' (New Delhi, S. Chand & Company Limited)
10. Rafiquzzaman Mohamed & Chandra Rajan 'Modern Computer Architecture' (New Delhi, Galgotia Publications Private Limited)
11. Badri Ram 'Advanced Microprocessors and Interfacing' (New Delhi Tata McGraw-Hill Education Private Limited)
12. Y. Rajasree "Advance Micro Processors" (New Delhi New age International Private Limited)
13. A.K. Mukhopadhyay "Micro processors Micro Computer and their applications" (New Delhi, Narosa publishing House)

Electrical and Electronic Measurements and Instrumentation

1. Umesh Sinha, Electrical & Electronics Measurement and Instrumentation, (New Delhi Satya Prakashan, 1990)
2. A.K. Sawhney, A course In Electrical and Electronic Measurements and Instrumentation, (New Delhi Dhanapat Rai & Sons, 1991)
3. J.B. Gupta, Electronic and Electrical Measurements and Instrumentation (New Delhi S.K. Kataria and Sons, 2007).
4. Albert D. Helfrick & William D. Cooper 'Modern Electronics Instrumentation and Measurement Techniques' (New Delhi, Pearson Education in South Asia)
5. Ernest O. Doebelin 'Measurement Systems Application and Design' (New Delhi, McGraw-Hill Publishing Company Limited)
6. D. Patranabis "Principles of Industrial Instrumentation" (New Delhi, MC Graw Hill Education)

Mathematics

1. Dr. B.S. Grewal Higher Engineering Mathematics (Khanna Publishers)
2. M.L. Aggarwal, Mathematics (Arya Publications, 2006)
3. K.K. Gupta, Ravinder Guru and M.L. Sharma, Prime Mathematics, (Sharma Publications)
4. Manjeet Singh, Pioneer Mathematics (Dhanpat Rai & Company Private Limited, 2002)

5. Parmanand Gupta Comprehensive Mathematics (Laxmi Publications (P) LTD New Delhi, 1999)

Signals and Systems

1. Farooq husain and Mrs. Shahnaz husain, Digital Signal Processing and Applications,(New Delhi Umesh Publications, 2003)
2. Farooq Husain and Mrs. Shahnaz Husain A Text Book of Signals and Systems (New Delhi Umesh Publications, 2002)
3. Smarajit Ghosh 'Signal and Systems'
4. Nagoor Kani 'Signal and System' (New Delhi Tata McGraw-Hill Education Private Limited)
5. S Salivahanan, A Vallaveraj and C Gnanapriya, Digital Signal Processing (New Delhi Tata McGraw-Hill Publishing Company Limited, 2001)
6. John G.Proakis and Dimitris G.Manolakis Digital Signal Processing Principle, Algorithms, and Applications (New Delhi Pearson Education Private Limited, 2003)
7. Alan V. Oppenheim & Alan S. Willsky, Signals and systems, (New Delhi Prentice-Hall of Indial Private Limited, 2000)
8. Alan V. Oppenheim, Ronald W. Schaffer and John R. Buck Discrete-Time Signal Processing (New Delhi PEARSON Prentice-Hall of Indial Private Limited, 2005)
9. Simon Haykin, Barry Van Veen, Signals and Systems (Singapore, John Wiley and Sons (ASIA) Private Limited, 2005)
10. Johnny R. Johnson 'Introduction to Digital Signal Processing' (New Delhi, Prentice-Hall of India Private Limited, 1997)
11. Emmanuel C. Ifeachor & Barrie W. Jervis 'Digital Signal Processing' A Practical Approach (New Delhi, Pearson Education in South Asia)
12. Johnny R. Johnson 'Introduction to Digital Signal Processing' (New Delhi, Prentice-Hall of India Private Limited, 1997)
13. D.P Kothari & I J Nagrath 'Basic Electrical Engineering' (New Delhi, Tata McGraw-Hill Publishing Company Limited)
14. Vincent Del Toro 'Electrical Engineering Fundaments' (New Delhi, Pearson Education in South Asia)
15. Rolf Schaumannm M.S. Ghausi & Kenineth R. Laker 'Design of Analog filters Passive, Active RC, and Switched Capacitor' (New Delhi, Prentice-Hall of India Private Limited)

Analytical, Optical and Biomedical Instrumentation

1. John Gowar, Optical Communication Systems, (New Delhi Prentice-Hall of India Private Limited, 2001)
2. John M Senior, Optical Fiber Communications, Principle and practice(New Delhi Prentice-Hall of India Private Limited, 1992)
3. Govind P. Agrawal "Fiber-optical communication system" (New Delhi, Wiley India)
4. Gerd Keiser "Optical Fiber Communication" (New Delhi Tata McGraw-Hill Publishing Company Limited)
5. Dr. Subir Kumar Sarkar "Optical Fibers and Fiber Optic Communication System" (New Delhi S.Chand and Company)
6. Dr. N. Subrahmavnayam, Brij Lal and DR M.N. Avadhnulu "A textbook of opticics" (New Delhi S. Chand and Company)
7. K.N. Scott and A.K. Mathur "Text book of Biomedical Instrumentation (New Delhi, CBS Publishers & Distributors"
8. Leslie Cromwell, Fred J. Weibell & Erich A. Pfeiffer "Biomedical Instrumentation and Measurements" (New Delhi, Pearson Education in South Asia)
9. Prof. S.K. Venkata Ram "Biomedical Electronics & Instrumentation" (New Delhi Galgotia Publications Private Limited)
10. R.S. Khandpur Analytical Instruments (New Delhi Tate McGraw-Hill Publishing Company Limited, 2004)
11. R.S. Khandpur Biomedical Instrumentation (New Delhi Tate McGraw-Hill Publishing Company Limited, 2002).
12. A.K. Srivastava and P.C. Jain "Instrumental Approach to Chemical Analysis" (New Delhi, S. Chand & Company Limited)
13. Ajay Ghatak and K.Thyagrajan "Introduction to Fiber Optics" (New Delhi, Cambridge university press)
14. J.H. From 2, Vk. Jain "Optical communication Components and System" (New Delhi, Narosa publishing House)
15. John G Webster "Medical Instrumentation Application and Design" (New delhi, Wiley India)

Mechanical Measurement

1. A.K. Tayal & Akash Tayal "Insturmentation, Mechanical Measurmentss & Contorl" (New Delhi Galgotia Publications Private Limited)
2. Dr. S. Kumar "Mechanical Measurements and Contorl" (New Delhi Metropolita Book Company Private Limited)
3. A.K. Sawhney, Puneet Sawhney Mechanical Measurements and Instrumentation (New Delhi Dhanapat Rai & Sons, 2007)
4. D. Patranatirs "Sensors and Transducer" (New Delhi PHI)
5. S.K. Singh "Industrial Instrumentation and Control" (New Delhi, Tata McGraw Hill Education Private Limited)

General Ability

1. Abul Hashem "General English" (New Delhi, Ramesh Publishing House)
2. R.S. Agrawal "Objective Arithmetic" (New Delhi, S. Chand & company Limited)
3. Dr. Ravi Chopra "Reasoning and Reasoning" (New Delhi, Galgotia publication)
4. R.S Agrawal "Verbal and Non Verbal reasonin" (New Delhi, S.Chand & Private Limited)
5. Navratan singh and Ms Manpreet Cheema, "Numerical Ability" (New Delhi, Dhillon Group of Publication)
6. R.S. Agrawal "Quantitative Aptitude" (New Delhi, S.Chand & Company Limited)
7. Sujit Kumar "Reading comprehension for the CAT" (New Delhi Pearson Education Private Limited)
8. Dr. R.S Aggrawal and Vikas Aggrawal "Objective General English" (New Delhi, S.Chand & Company Limited)
9. Dr. R.S Aggrawal, Deepak Aggrawal and Vikas Aggrawal "Quantitative Aptitude" (New Delhi, S. Chand & Company Limited)
10. Bhagwant Swerap sharma, Sumira Aggrawal, Neeta sharma, Monik Ahuja and suchiya Dahall "National Talent Search Examination" (New Delhi, S. Chand & Company Limited)
11. Dr. N.K. Parwal "General Intelligent and Test of Reasoning" (Agra, pritiyogita sahitya)
12. R.D. Sharma "Mathematics for Class IX " (New Delhi, Dhanpat Rai and Company)
13. R.D. Sharma "Mathematics for Class X" (New Delhi, Dhanpat Rai and Company)
14. R.D. Sharma "Mathematics for Class XI" (New Delhi, Dhanpat Rai and Company)
15. R.D. Sharma "Mathematics for Class XII" (New Delhi, Dhanpat Rai and Company).

Reader's Notes

Reader's Notes